NEUROSCIENCE
SEVENTH EDITION

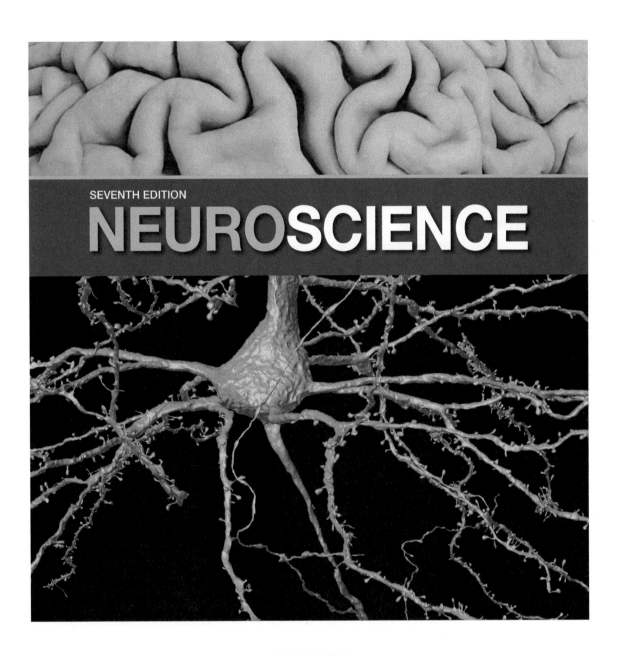

SEVENTH EDITION

NEUROSCIENCE

EDITORS

George J. Augustine

Jennifer M. Groh | Scott A. Huettel

Anthony-Samuel LaMantia | Leonard E. White

OXFORD

UNIVERSITY PRESS

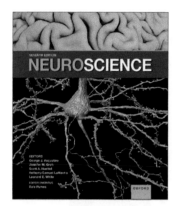

About the Cover
The upper image shows a photograph of the dorsal-medial surface of the right hemisphere of a human brain, courtesy of Leonard E. White and S. Mark Williams. The lower image is a 3D reconstruction of a pyramidal neuron from mouse visual cortex, obtained by transmission electron microscope imaging. Data from microns-explorer.org, courtesy of the Allen Institute for Brain Science, Princeton University, and Baylor College of Medicine.

Oxford University Press is a department of the University of Oxford. It furthers the University's objective of excellence in research, scholarship, and education by publishing worldwide. Oxford is a registered trademark of Oxford University Press in the UK and certain other countries.

Published in the United States of America by Oxford University Press
198 Madison Avenue, New York, NY 10016, United States of America.

© 2024, 2018 Oxford University Press
© 2012, 2008, 2004, 2001, 1997 Sinauer Associates

Library of Congress Cataloging-in-Publication Data

Names: Augustine, George J., editor. | Groh, Jennifer M., editor. | Huettel, Scott A., editor. | LaMantia, Anthony-Samuel, editor. | White, Leonard E., editor.

Title: Neuroscience / editors George J. Augustine, Jennifer M. Groh, Scott A. Huettel, Anthony-Samuel LaMantia, Leonard E. White.

Description: Seventh edition. | New York, NY: Oxford University Press, [2024] | Revised edition of: Neuroscience / editors, Dale Purves, George J. Augustine, David Fitzpatrick, William C. Hall, Anthony-Samuel LaMantia, Richard D. Mooney, Michael L. Platt, Leonard E. White. Sixth edition. [2018]. | Includes bibliographical references and index. | Summary: "For over 25 years, Neuroscience has been the most comprehensive and clearly written neuroscience textbook on the market. This level of excellence continues in the Seventh Edition, with a balance of animal, human, and clinical studies that discuss the dynamic field of neuroscience from cellular signaling to cognitive function. New learning objectives, and more concise sections make the content even more accessible than before. Neuroscience, Seventh Edition is intended primarily for medical, premedical, and undergraduate students. The book's length and accessible writing style make it suitable for both medical students and undergraduate neuroscience courses."-- Provided by publisher.

Identifiers: LCCN 2022040688 (print) | LCCN 2022040689 (ebook) | ISBN 9780197616246 (hardcover) | ISBN 9780197616253 (print other) | ISBN 9780197616680 (epub)
Subjects: LCSH: Neurosciences.
Classification: LCC QP355.2 .N487 2024 (print) | LCC QP355.2 (ebook) | DDC 612.8--dc23/eng/20221128

LC record available at https://lccn.loc.gov/2022040688
LC ebook record available at https://lccn.loc.gov/2022040689

Contents in Brief

1 Studying the Nervous System   1

Unit I Neural Signaling   35
2 Electrical Signals of Nerve Cells   37
3 Voltage-Dependent Membrane Permeability   54
4 Ion Channels and Transporters   70
5 Synaptic Transmission   93
6 Neurotransmitters and Their Receptors   122
7 Molecular Signaling within Neurons   155
8 Synaptic Plasticity   179

Unit II Sensation and Sensory Processing   203
9 Vision   205
10 Hearing   236
11 The Vestibular System   261
12 Touch and Proprioception   278
13 Pain and Temperature   303
14 Olfaction   324
15 Taste   356

Unit III Movement and Its Central Control   377
16 Lower Motor Neuron Circuits and Motor Control   379
17 Upper Motor Neuron Control of the Brainstem and Spinal Cord   403
18 Modulation of Movement by the Basal Ganglia   430
19 Modulation of Movement by the Cerebellum   451
20 Eye Movements and Sensorimotor Integration   471
21 The Visceral Motor System   490

Unit IV The Changing Brain   515
22 Early Brain Development   517
23 Construction of Neural Circuits   555
24 Experience-Dependent Plasticity in the Developing Brain   598
25 Sex Differences and Neural Circuit Development   629
26 Repair and Regeneration in the Nervous System   658

Unit V Complex Brain Functions and Cognitive Neuroscience   695
27 Cognitive Functions and the Organization of the Cerebral Cortex   697
28 Cortical States   714
29 Attention   738
30 Memory   754
31 Speech and Language   775
32 Emotion   797
33 Thinking, Planning, and Deciding   819

Appendix   A-1
Atlas: The Human Central Nervous System   AT-1
Glossary   G-1
Box References   BR-1
Illustration Credits   IC-1
Index   I-1

Contributors

George J. Augustine

Nirupa Chaudhari

Jennifer M. Groh

Scott A. Huettel

Alfredo Fontanini

Michael A. Fox

Anthony-Samuel LaMantia

Lauren L. Orefice

Leonard E. White

Unit Editors

Unit I: George J. Augustine

Unit II: Jennifer M. Groh

Unit III: Leonard E. White

Unit IV: Anthony-Samuel LaMantia

Unit V: Scott A. Huettel

Contents

CHAPTER 1

Studying the Nervous System 1

Overview 1

Neurons and Glia Are the Primary Cell Types of All Nervous Systems 2

Neurons Are Interconnected in Ensembles Called Neural Circuits 9

Circuits That Process Related Information Constitute a Neural System 15

The Genome Controls Brain Organization and Function 20

■ BOX 1A | Model Organisms in Neuroscience 24

Neural Circuits and Systems Can Be Analyzed in the Human Brain 26

Summary 32

Additional Reading 33

Unit I Neural Signaling 35

CHAPTER 2

Electrical Signals of Nerve Cells 37

Overview 37

Nerve Cells Generate Electrical Signals to Encode Information 37

Neuronal Electrical Signals Can Be Transmitted Over Long Distances 39

■ CLINICAL APPLICATIONS
Anesthesia and Neuronal Electrical Signaling 41

Ion Movements Produce Electrical Signals 44

Electrical and Chemical Forces Create Membrane Potentials 45

Potassium Ions Generate the Resting Membrane Potential 47

■ BOX 2A | The Remarkable Giant Nerve Cells of Squid 48

More Than One Type of Permeant Ion Can Generate Electrical Signals 49

Action Potentials Arise from Sequential Changes in Sodium and Potassium Permeability 51

■ BOX 2B | Action Potential Form and Nomenclature 52

Summary 53

Additional Reading 53

CHAPTER 3

Voltage-Dependent Membrane Permeability 54

Overview 54

Ion Currents Flow across Nerve Cell Membranes 54

■ BOX 3A | The Voltage Clamp Method 55

Depolarization Activates Two Types of Voltage-Dependent Ion Currents 56

Voltage-Dependent Ion Currents Arise from Two Voltage-Dependent Membrane Conductances 59

Action Potentials Can Be Reconstructed Based on the Properties of Voltage-Dependent Na^+ and K^+ Conductances 60

Action Potentials Enable Long-Distance Electrical Signaling 62

Myelination Increases Action Potential Conduction Velocity 64

■ CLINICAL APPLICATIONS
Multiple Sclerosis 67

Summary 68

Additional Reading 69

CHAPTER 4
Ion Channels and Transporters 70

Overview 70

Ion Channels Generate Electrical Currents 70

■ BOX 4A | The Patch Clamp Method 72

■ BOX 4B | Toxins That Poison Ion Channels 75

There Are Many Types of Ion Channels 76

Ions Permeate through Channels via Ion-Selective Pores 78

Molecular Specializations Permit Gating of Ion Channels by Different Types of Stimuli 80

■ CLINICAL APPLICATIONS
Neurological Diseases Caused by Altered Ion Channels 82

Active Transporters Create and Maintain Ion Gradients 86

Summary 90

Additional Reading 91

CHAPTER 5
Synaptic Transmission 93

Overview 93

There Are Two Mechanisms of Synaptic Signaling 94

Calcium Ions Regulate the Release of Discrete Packets of Neurotransmitters 98

A Cycle of Membrane Trafficking Is Responsible for Neurotransmitter Release 101

Many Proteins Are Required for Synaptic Vesicle Trafficking 104

■ CLINICAL APPLICATIONS
Disorders That Affect the Presynaptic Terminal 107

There Are Two Families of Neurotransmitter Receptors 110

Postsynaptic Membrane Permeability Changes during Synaptic Transmission 111

Postsynaptic Ion Fluxes Determine Whether Synapses Are Excitatory or Inhibitory 116

■ BOX 5A | The Tripartite Synapse 119

Summary 120

Additional Reading 121

CHAPTER 6
Neurotransmitters and Their Receptors 122

Overview 122

There Are Many Neurotransmitters 122

Acetylcholine Serves as the Prototype Neurotransmitter 124

■ BOX 6A | Neurotoxins That Act on Neurotransmitter Receptors 126

■ CLINICAL APPLICATIONS
Myasthenia Gravis: An Autoimmune Disease of Neuromuscular Synapses 129

Glutamate Is the Main Excitatory Neurotransmitter 131

GABA and Glycine Are Inhibitory Transmitters 135

■ BOX 6B | Excitatory Actions of GABA in the Developing Brain 138

Biogenic Amines Are Neuromodulatory Transmitters 140

ATP and Other Purines Can Serve as Co-Transmitters 145

Neuropeptides Are Exceptionally Diverse Neurotransmitters 146

Interneuronal Signaling Can Also Occur via Unconventional Neurotransmitters 149

■ BOX 6C | Marijuana and the Brain 150

Summary 153

Additional Reading 153

CHAPTER 7
Molecular Signaling within Neurons 155

Overview 155

There Are Several Modes of Chemical Signaling 155

Cellular Responses Are Determined by Several Families of Receptors 158

G-Proteins Connect Receptor Activation to Intracellular Signaling Pathways 159

Second Messengers Are Involved in Many Intracellular Signaling Pathways 161

■ BOX 7A | Dynamic Imaging of Intracellular Signaling 163

Intracellular Signaling Often Involves Adding or Removing Phosphates from Proteins 165

■ **CLINICAL APPLICATIONS**
Molecular Basis of Psychiatric Disorders   169

■ **BOX 7B** | Dendritic Spines   171

Long-Lasting Responses Involve Changes in Nuclear Signaling   172

There Are Many Ways to Signal Changes in Neuronal Structure and Function   175

Summary   177

Additional Reading   178

CHAPTER 8
Synaptic Plasticity   179

Overview   179

Some Forms of Synaptic Plasticity Last for a Few Minutes or Less   179

Synaptic Plasticity Can Produce Long-Lasting Changes in Behavior   182

■ **BOX 8A** | Genetics of Learning and Memory in the Fruit Fly   186

Long-Term Synaptic Plasticity Is Found in the Mammalian Hippocampus   186

NMDA-Type Glutamate Receptors Serve as Coincidence Detectors for Hippocampal Long-Term Potentiation   189

■ **BOX 8B** | Silent Synapses   193

Long-Term Depression Weakens Synapses via Multiple Mechanisms   194

Some Forms of Plasticity Depend on the Timing of Synaptic Activity   197

■ **CLINICAL APPLICATIONS**
Epilepsy: The Effect of Pathological Activity on Neural Circuitry   198

Summary   200

Additional Reading   200

Unit II Sensation and Sensory Processing   203

CHAPTER 9
Vision   205

Overview   205

The Eye Collects Light and Focuses It onto Specialized Photoreceptive Cells   205

Photoreceptors Convert Light into Electrical Signals in the Retina   209

■ **BOX 9A** | The Importance of Context in Color Perception   215

Retinal Circuitry Extracts Information about Features of the Visual Scene   216

■ **BOX 9B** | The Perception of Light Intensity   220

Retinal Ganglion Cells Convey Feature Information to the Brain   221

■ **CLINICAL APPLICATIONS**
Visual Field Deficits   223

Visual Centers in the Cerebral Cortex Detect Increasingly Complex Visual Features   227

Optimizing Vision and Visual Behaviors Requires Autonomic, Motor, and Cognitive Control   231

Summary   233

Additional Reading   234

CHAPTER 10
Hearing   236

Overview   236

Sound Is a Pressure Wave Composed of Different Frequencies Important for Speech, Music, and Other Natural Sounds   236

■ **CLINICAL APPLICATIONS**
Hearing Loss: Causes and Treatments   238

The Structures of the Ear Filter the Frequencies of Sound and Transmit Vibrations from Air to Fluid   240

Sound Waves Are Transduced into Neural Signals by Cochlear Hair Cells   245

■ **BOX 10A** | The Sweet Sound of Distortion   247

Transduction Is Controlled by Active Mechanisms Involving Middle Ear Muscles and Cochlear Outer Hair Cells   248

Auditory Pathways Involve Signals Traveling Bilaterally and in Both Feedforward and Feedback Directions   250

Auditory Perception Involves the Synthesis of Multiple Aspects of Sound   252

Neural Codes for Sound Frequency Are Based on Resonance and Synchrony　252

Sounds Are Localized Based on Frequency, Relative Loudness, and Relative Time of Arrival at the Ears　255

Hearing Coordinates with Vision via Numerous Signal Transformations and Robust Intersensory Cross-talk　257

Summary　258

Additional Reading　259

CHAPTER 11
The Vestibular System　261

Overview　261

The Vestibular System Helps Us Sense Our Position and Movement in Space　261

The Utricle and the Saccule Sense Static Tilt and Dynamic Linear Movements　263

The Semicircular Canals Sense Rotations of the Head in Three Dimensions　266

The Vestibular System Works with Visual Cues to Evaluate Self Motion and to Stabilize Eye Gaze　269

■ CLINICAL APPLICATIONS CLINICAL
Evaluation of the Vestibular System　271

The Brain Synthesizes Information to Support Perception of Translational and Rotational Movements and Maintain Equilibrium　273

■ BOX 11A | Mauthner Cells in Fish　274

Summary　276

Additional Reading　277

CHAPTER 12
Touch and Proprioception　278

Overview　278

Mechanical Forces on the Skin Are Conveyed to the CNS via an Array of Somatosensory Afferent Neurons　278

■ CLINICAL APPLICATIONS
Dermatomes　280

Somatosensory Afferent Neurons Form Specialized Endings and Interact with Other Mechanosensitive Cells in the Skin　284

Sensory Transduction from the Skin Involves Converting Mechanical Forces into Electrical Signals　287

Proprioception Involves Sensing Forces in Muscles, Joints, and Connective Tissue　288

A Variety of Neural Pathways Convey Different Aspects of Somatosensory Information to the Brain　289

■ BOX 12A | Specialized Mechanosensation in Animals　296

Central Representations of the Body Are Plastic and Modified by Experience　298

Somatosensory Dysfunction Is an Important Factor in Multiple Diseases　299

Summary　300

Additional Reading　301

CHAPTER 13
Pain and Temperature　303

Overview　303

Potentially Damaging Stimuli Are Detected by Nociceptors　303

Noxious Stimuli Are Transduced into Neural Signals via Various Ion Channel Receptors　305

■ BOX 13A | Capsaicin　306

Central Pain Pathways Are Distinct from Mechanosensory Pathways and Transmit Different Aspects of Pain in Parallel　307

■ BOX 13B | Referred Pain　309

■ BOX 13C | A Dorsal Column Pathway for Visceral Pain　310

Signals from a Variety of Innocuous Stimuli, as Well as From Internal Organs, Are Also Carried by Pain Pathways　314

Sensitivity to Pain Is Subjective and Can Be Modified by a Variety of Factors　315

■ CLINICAL APPLICATIONS
Phantom Limbs and Phantom Pain　318

Summary　322

Additional Reading　322

CHAPTER 14
Olfaction　324

Overview　324

The Olfactory and Vomeronasal Systems Process Airborne Molecules That Influence a Wide Range of Behaviors　325

Olfactory and Vomeronasal Transduction Occurs via G-Protein-Coupled Receptors　327

■ CLINICAL APPLICATIONS
Only One Nose　329

■ BOX 14A | The "Dogtor" Is In　335

Olfactory and Vomeronasal Information Is Relayed Directly to the Main and Accessory Olfactory Bulbs, and from There to Multiple Forebrain Targets 341

In Many Species, Including Humans, Olfactory Capacity Reflects the Size and Complexity of the Olfactory System 348

Summary 354

Additional Reading 355

CHAPTER 15
Taste 356

Overview 356

Detection of Taste Qualities Is Mediated by Specialized Cells in the Taste Buds and Conveyed to the Brain by Three Cranial Nerves 356

■ CLINICAL APPLICATIONS
Ageusia and Dysgeusia: Taste Loss and Taste Alterations from COVID-19 361

■ BOX 15A | Extraoral Taste Receptors and the Microbiome 364

Gustatory Information Flows to the Brain through Diverging and Converging Pathways and is Encoded via the Spatio-Temporal Firing Patterns of Neurons 365

Perception of Food and Beverages Also Involves Olfaction, Somatosensation, Audition, and Vision 369

Palatability or Aversiveness of a Food Is Mediated by the Gustatory and Limbic Systems and Can Be Modified by Experience 372

Summary 374

Additional Reading 375

Unit III Movement and Its Central Control 377

CHAPTER 16
Lower Motor Neuron Circuits and Motor Control 379

Overview 379

Interacting Subsystems in the CNS Make Essential and Distinct Contributions to Motor Control 379

Lower Motor Neurons in the Spinal Cord and Brainstem Map the Body's Musculature 381

Motor Units of Varying Size Produce Appropriate Movements 383

■ BOX 16A | Motor Unit Plasticity 385

Local Circuitry Mediates Reflexes That Rapidly Adjust Muscle Tension in Response to Sensory Input 389

Local Circuitry Coordinates the Output of Lower Motor Neurons for Rhythmic, Stereotyped Behavior 394

■ BOX 16B | Locomotion in the Leech and the Lamprey 395

Damage to Lower Motor Neurons Results in "Lower Motor Neuron Syndrome" 399

■ CLINICAL APPLICATIONS
Amyotrophic Lateral Sclerosis 399

Summary 401

Additional Reading 401

CHAPTER 17
Upper Motor Neuron Control of the Brainstem and Spinal Cord 403

Overview 403

Upper Motor Neurons Give Rise to Lateral Tracts in the Spinal Cord that Govern Skilled Movements and Medial Tracts that Influence Posture, Balance, and Locomotion 404

■ CLINICAL APPLICATIONS
Patterns of Facial Weakness and Their Importance for Localizing Neurological Injury 408

Neurons in the Primary Motor Cortex Encode Intentions for Body Movements in Central Personal Space 409

■ BOX 17A | What Do Motor Maps Represent? 411

Neurons in the Premotor Cortex Encode Intentions for Body Movements that Are Oriented toward Extrapersonal Space 414

■ BOX 17B | Minds and Machines 415

Upper Motor Neurons in the Brainstem Help Maintain Balance, Govern Posture, Initiate Locomotion, and Orient Visual Gaze 419

■ BOX 17C | The Reticular Formation 422

Damage to Upper Motor Neurons Produces "Upper Motor Neuron Syndrome"   425

■ BOX 17D | Muscle Tone   427

Summary   428

Additional Reading   428

CHAPTER 18

Modulation of Movement by the Basal Ganglia   430

Overview   430

The Basal Ganglia Comprise a Set of Nuclei Deep in the Cerebral Hemispheres   431

■ BOX 18A | Basal Ganglia Loops and Non-Motor Brain Functions   433

■ BOX 18B | Making and Breaking Habits   435

The Basal Ganglia Influence Movement by Regulating the Activity of Upper Motor Neuronal Circuits   437

Direct and Indirect Pathways Regulate the Initiation of Voluntary Movement and the Suppression of Unwanted Movement   440

Dopamine Modulates Basal Ganglia Circuits by Increasing or Decreasing the Excitability of Medium Spiny Neurons   442

Hypokinetic Movement Disorders Are Associated with Excessive Inhibition of Motor Nuclei in the Thalamus and Brainstem   443

■ CLINICAL APPLICATIONS
Deep Brain Stimulation   445

Hyperkinetic Movement Disorders Are Associated with Insufficient Inhibition of Motor Nuclei in the Thalamus and Brainstem   447

Summary   449

Additional Reading   449

CHAPTER 19

Modulation of Movement by the Cerebellum   451

Overview   451

The Cerebellum Comprises Three Major Subdivisions: The Cerebrocerebellum, Spinocerebellum, and Vestibulocerebellum   451

The Cerebellar Hemispheres Coordinate Movements of the Ipsilateral Body   453

Efferent Output from the Cerebellum to the Brainstem and Thalamus Originates in the Deep Cerebellar Nuclei and the Vestibulocerebellum   455

Purkinje Neurons Integrate Afferent Input and Modulate the Output of Deep Cerebellar Nuclei   458

The Cerebellum Coordinates Ongoing Movement by Reducing Motor Error   462

■ CLINICAL APPLICATIONS
Prion Diseases   463

Cerebellar Injury Compromises the Coordination of Movement, with or without Impacts on Cognitive or Affective Regulation   465

■ BOX 19A | Genetic Analysis of Cerebellar Function   466

Summary   469

Additional Reading   470

CHAPTER 20

Eye Movements and Sensorimotor Integration   471

Overview   471

Eye Movements Are Necessary to Acquire and Foveate a New Visual Target and to Maintain Foveal Fixation   471

■ BOX 20A | The Perception of Stabilized Retinal Images   472

Eye Movements Are Generated around Three Axes of Rotation by Three Pairs of Striated Muscles   473

Conjugate Eye Movements Rotate the Eyes in the Same Direction, and Disconjugate Eye Movements Rotate the Eyes in Opposite Directions   475

■ CLINICAL APPLICATIONS
Eye Movements and Neurological Injury, Disease, and Disorder   477

Neural Circuits in the Cerebral Cortex and Brainstem Govern the Amplitude, Direction, and Velocity of Eye Movements   479

■ BOX 20B | Sensorimotor Integration in the Superior Colliculus   482

■ BOX 20C | From Place Codes to Rate Codes   485

Summary   488

Additional Reading   489

CHAPTER 21

The Visceral Motor System   490

Overview   490

The Visceral (Autonomic) Motor System Controls Involuntary Bodily Functions   490

The Sympathetic Division Prepares the Body to Mobilize Resources in Challenging Situations   495

■ BOX 21A | The Hypothalamus   496

The Parasympathetic Division Serves to Increase Metabolic Resources and Conserve Energy   498

The Enteric Division Is a Semi-Autonomous Network of Gastrointestinal Neurons That Promotes Digestion   501

Visceral Sensory Signals Serve Local Visceral Motor Reflexes and a Central Autonomic Network   502

■ CLINICAL APPLICATIONS
Horner's Syndrome   504

■ BOX 21B | Obesity and the Brain   505

Visceral Motor Neurons Use Small-Molecule and Neuropeptide Neurotransmitters to Mediate a Variety of Effects   507

Summary   513

Additional Reading   514

Unit IV The Changing Brain   515

CHAPTER 22

Early Brain Development   517

Overview   517

Neural Stem Cells, Derived from Pluripotent Stem Cells, Generate the Entire Nervous System   517

■ BOX 22A | Stem Cells: Promise and Peril   519

Neural Stem Cells Generate the Central and Peripheral Nervous Systems   523

Transcription Factor Patterning Regulated by Cell-Cell Signaling Establishes Distinct Brain Regions   529

■ CLINICAL APPLICATIONS
Inductive Signals and Neurodevelopmental Disorders   537

Neurogenesis Is an Irreversible Termination of the Cell Cycle That Constrains Neuron Identity   540

Nerve Cells Often Migrate from Their Site of Neurogenesis to Their Final Position   547

Summary   552

Additional Reading   554

CHAPTER 23

Construction of Neural Circuits   555

Overview   555

Neural Circuit Construction Relies on Basic Mechanisms of Cell Polarity   556

Neuronal Growth Cones Are Critical for Establishing Connections   559

Neuronal Growth and Synapse Formation Depend on Signaling Molecules   564

■ BOX 23A | Choosing Sides: Axon Guidance at the Optic Chiasm   569

■ CLINICAL APPLICATIONS
Axon Guidance Disorders   571

Axon, Dendrite, and Synapse Development and Numbers Are Regulated by Trophic Interactions   580

BOX 23B | Why Do Neurons Have Dendrites?   586

Axon, Dendrite, and Synaptic Growth Results in Orderly Patterns of Connections, Including Topographic Maps   592

Summary   596

Additional Reading   596

CHAPTER 24

Experience-Dependent Plasticity in the Developing Brain   598

Overview   598

Electrical Activity in New Neural Circuits Determines Final Numbers and Patterns of Functional Connections   599

Electrical Activity Reflects Initial Experience and Defines Connections during Critical Periods   602

■ BOX 24A | Built-In Behaviors   603

Relative Levels of Electrical Activity across
  Inputs Determine Final Connections in Neural
  Circuits   611

■ CLINICAL APPLICATIONS
  Dancing in the Dark   614

Ion Channels, Neurotransmitters and Their
  Receptors, and Neurotrophins Regulate
  Activity-Dependent Circuit Development   621

Summary   627

Additional Reading   628

CHAPTER 25
Sex Differences and Neural Circuit Development   629

Overview   629

Systemic, Secreted Signals Influence Neural
  Circuit Development and Maintenance   629

■ BOX 25A | The Science of Love (or, Love As
              a Drug)   632

Sexual Dimorphisms Reflect Systemic Signaling
  in Peripheral Organs and Related Neural
  Circuits   636

Systemic Signals Target Neurons and Circuits for
  Reproductive and Parenting Behaviors   642

■ CLINICAL APPLICATIONS
  The Good Mother   647

Complex Human Behaviors Are Difficult to
  Associate with Sex, Gender, or Early Systemic
  Signaling   649

Summary   656

Additional Reading   656

CHAPTER 26
Repair and Regeneration in the Nervous System   658

Overview   658

Neural Tissue Has a Distinct Response to Injury
  and Limited Potential for Regeneration   659

The Peripheral Nervous System Retains the
  Capacity for Axon Regrowth and Synaptic
  Reinnervation   665

■ BOX 26A | Specific Regeneration of Synaptic
Connections in Autonomic Ganglia   674

CNS Axons and Dendrites in Most Adult
  Mammals Lack the Capacity for Extensive
  Regrowth   676

■ CLINICAL APPLICATIONS
  Casualties of War and Sports   676

Adult Vertebrate Nervous Systems Retain Some
  Neural Stem Cells for Limited Replacement of
  Neurons   684

■ BOX 26B | Nuclear Weapons and
              Neurogenesis   691

Summary   693

Additional Reading   694

 Unit V Complex Brain Functions and
              Cognitive Neuroscience   695

CHAPTER 27
Cognitive Functions and the Organization of the Cerebral Cortex   697

Overview   697

The Cerebral Cortices Are Organized into
  Subregions   697

■ BOX 27A | Cortical Lamination   699

■ BOX 27B | Large-Scale Neuroscience:
              Meta-Analyses and Consortium
              Studies   701

The Parietal Cortex Has Many Functions   703

The Temporal Cortex Plays a Critical Role
  in Object Processing   705

The Prefrontal Cortex Supports Executive Control,
  Planning, and Goal-Directed Action   707

■ BOX 27C | Neuropsychological Testing   708

■ CLINICAL APPLICATIONS
  Psychosurgery   710

Summary   712

Additional Reading   713

CHAPTER 28

Cortical States   714

Overview   714

Circadian Cycles of Function Are Regulated by Neural Circuits   714

■ BOX 28A | Electroencephalography   718

Sleep Supports Physiological Functions Critical for Health and Behavior   718

■ CLINICAL APPLICATIONS
Sleep Disorders and Their Treatment   720

Sleep Progresses through Stages of Brain Activity   723

■ BOX 28B | Dreaming   726

Transitions between Sleep and Wakefulness Rely on Brain Circuits   726

Selective Impairments in Cortical Function Can Alter Conscious Experiences   731

A Distributed Set of Brain Regions Becomes Active When People Disengage from Active Tasks   735

Summary   736

Additional Reading   736

CHAPTER 29

Attention   738

Overview   738

Attention Prioritizes Some Stimuli over Others   738

Attention Alters Activity in Brain Regions Associated with Perception   742

Damage to Key Brain Regions Can Disrupt Attentional Processes   746

■ CLINICAL APPLICATIONS
Balint's Syndrome   747

■ BOX 29A | Attention and the Frontal Eye Fields   749

A Frontal-Parietal Network Supports the Allocation of Attention   750

Summary   752

Additional Reading   752

CHAPTER 30

Memory   754

Memory Processes Can Be Categorized by Function   754

Memory Encoding Involves Creating Associations That Support Later Recall   758

■ BOX 30A | Savant Syndrome   760

The Medial Temporal Lobe Supports Declarative Memory   762

■ CLINICAL APPLICATIONS
Clinical Cases That Illustrate the Neural Basis of Memory   764

Memories Are Stored in a Distributed Manner throughout the Cerebral Cortex   766

■ BOX 30B | Alzheimer's Disease   768

■ BOX 30C | Place Cells and Grid Cells   770

Nondeclarative Memory Relies on Brain Systems Distinct from Those Supporting Declarative Memory   771

As Humans Age, Changes in the Brain Alter Memory Processes   772

Summary   773

Additional Reading   774

CHAPTER 31

Speech and Language   775

Overview   775

Language Production Relies on Both the Vocal Apparatus and Cortical Regions   776

■ BOX 31A | Sign Language   778

■ CLINICAL APPLICATIONS
Clinical Presentations of Aphasia   779

Language Comprehension Relies on a Distributed Brain Network   781

■ BOX 31B | Semantics: Extracting Meaning from Language   782

The Right Hemisphere Makes Important Contributions to Language   784

■ BOX 31C | Language and Handedness   788

Language Development Includes a Critical Period During Childhood   790

Nonhuman Animals Exhibit Complex Communicative Abilities   792

Summary   795

Additional Reading   795

CHAPTER 32
Emotion   797

   Overview   797

   Emotions Integrate Feelings, Physiology, and Behavior   797

   ■ BOX 32A | Determination of Facial Expressions   802

   The Amygdala Plays a Central Role in Emotional Processing   803

   ■ BOX 32B | Anatomy of the Amygdala   805

   ■ BOX 32C | Fear and the Human Amygdala   807

   The Cerebral Cortices Support Emotional Processing   811

   ■ CLINICAL APPLICATIONS Affective Disorders   812

   Emotions Interact with Other Cognitive Processes   815

   Summary   817

   Additional Reading   818

CHAPTER 33
Thinking, Planning, and Deciding   819

   Overview   819

   The Prefrontal Cortex Supports Processes Related to Cognitive Control   819

   Lateral Prefrontal Cortex Supports Cognitive Control   822

   Orbitofrontal Cortex Supports the Evaluation of the Outcomes of Behavior   825

   ■ CLINICAL APPLICATIONS Addiction   826

   ■ BOX 33A | Dopamine and Reward Prediction Errors   828

   Anterior Cingulate Cortex Supports Regulation of Activity in Other Brain Regions   832

   The Anterior Insula Incorporates Information about Body States into Decision Processes   834

   Posterior Cingulate Cortex Supports Internally Directed Processes   834

   ■ BOX 33B | What Does Neuroscience Have to Say about Free Will?   835

   Summary   838

   Additional Reading   839

Preface

The field of neuroscience, like the brain itself, is constantly changing. The same is true of our textbook, *Neuroscience*. The First Edition, published in 1997, was based upon a collection of chapters written by multiple instructors while developing a course that presented an integrated view of the function, structure, and plasticity of the human brain. Subsequent editions chronicled significant new understanding of how the brain might work, as well as substantial advances in the technologies used to analyze neural functions and behaviors. The completely reimagined format and content of the Seventh Edition signals our sustained commitment to ensure that our ever-accelerating understanding of the brain continues to be reflected in an up-to-date, authoritative textbook.

The roster of *Neuroscience* editors and authors has also changed. From the time the book began through the Sixth Edition published in 2018, our now-emeritus editor, Dale Purves, provided keen insight and indefatigable editorial review of the entire text. His commitment to this project at its inception insured its initial completion. His continued guidance over the subsequent two decades established a standard of scientific precision and narrative clarity that we have strived to maintain. For the Seventh Edition, three veteran editors—George Augustine, Anthony-Samuel LaMantia and Leonard White—have been joined by two new editors, Jennifer Groh and Scott Huettel. In addition, the contents of several chapters have been refreshed and enriched by new authors whose expertise deepens and expands the content developed over previous editions. Although some names have changed, as we assembled this new edition we have continued to embrace the rigor and excitement of neuroscience. Our primary goal remains to provide students and instructors with an accessible and engaging account of what is known about the structure, function, and dysfunction of neural systems.

To facilitate the use of *Neuroscience* in the classroom, as well as its use as a reference text, we changed our presentation style. For each chapter, we now identify a small number of key concepts, each with concrete learning objectives to support better understanding. The glossary of key terms has been expanded, where students can easily search for terms and discover where they appear in the text. Our coverage of sensory systems is also reconfigured. Vision, previously covered in two chapters, is now presented as an integrated narrative in a single chapter. Olfaction and taste, previously covered in one chapter, are now presented in two. We have also extensively revised the online study area, with the assessment questions, and have created new learning activities for *Sylvius* that extend the exploration of relevant human neuroanatomy.

Through these student-focused changes, *Neuroscience* provides a renewed, yet familiar, resource for learning, teaching, and understanding the ever-changing narrative of how brains are put together, how they work, and how they sometimes fail to work. We are indebted to the many medical, graduate, and undergraduate students who have given us classroom-tested feedback to improve *Neuroscience*. We are also grateful to the many colleagues in every branch of neuroscience for their rigorous review of the content of the book. *Neuroscience*, like its namesake discipline, has always been a collaboration: a dialogue among students, teachers, investigators, authors and editors, past and present. We look forward to continuing that dialogue as neuroscientists, present and future, read the Seventh Edition of *Neuroscience*.

Acknowledgments

We are grateful to the many colleagues who provided helpful contributions, criticisms, and suggestions to this and previous editions. We particularly wish to thank Paul Adams, Ralph Adolphs, David Amaral, Dora Angelaki, Eva Anton, Gary Banker, Marlene Behrmann, Ursula Bellugi, Carlos Belmonte, Staci Bilbo, Dan Blazer, Alain Burette, Bob Burke, Roberto Cabeza, Jean-Pierre Changeux, John Chapin, Milt Charlton, Michael Davis, Rob Deaner, Bob Desimone, Allison Doupe, Sasha du Lac, Jens Eilers, Chagla Eroglu, Anne Fausto-Sterling, Howard Fields, Elizabeth Finch, Nancy Forger, Jannon Fuchs, David Gadsby, Michela Gallagher, Dana Garcia, Steve George, Josh Gooley, Henry Greenside, William Guido, Zach Hall, Kristen Harris, Christopher Harvey, Ben Hayden, Bill Henson, John Heuser, Bertil Hille, Miguel Holmgren, Jonathan Horton, Ron Hoy, Alan Humphrey, Jon Kaas, Kai Kaila, Jagmeet Kanwal, Herb Killackey, Len Kitzes, Marc Klein, Chieko Koike, Kevin LaBar, Arthur Lander, Story Landis, Simon LeVay, Jeff Lichtman, Alan Light, Steve Lisberger, Arthur Loewy, Ron Mangun, Eve Marder, Carol Mason, Robert McCarley, Greg McCarthy, Jim McIlwain, Daniel Merfeld, Steve Mitroff, Sulochana Naidoo, Ron Oppenheim, Larysa Pevny, Franck Polleux, Scott Pomeroy, Louis Reichardt, Sidarta Ribiero, Marnie Riddle, Jamie Roitman, Steve Roper, John Rubenstein, David Rubin, Josh Sanes, Cliff Saper, Lynn Selemon, Paul Selvin, Carla Shatz, Sid Simon, Rich Simerly, Bill Snider, Larry Squire, Peter Strick, Joe Takahashi, Leng Zoo Tan, Stephen Traynelis, Christopher Walsh, Fan Wang, Xiaoqin Wang, Richard Weinberg, Christina Williams, S. Mark Williams, Joel Winston, Marty Woldorff, Rachel Wong, and Ryohei Yasuda. Of course, these individuals are in no way responsible for any errors that may be present.

We are also grateful for the support our ongoing and ever-evolving project has received from the staff and managers at Oxford University Press. For editorial guidance, we thank Jessica Fiorillo and Chelsea Noack. For shepherding the production of the revision, we thank Senior Production Editor Martha Lorantos, Production Managers Meg Clark and Joan Gemme, and Production Specialist Donna DiCarlo, who guided this revision's fresh design and cover. We thank our Copy Editor Elizabeth Pierson. We would also like to thank Digital Resource Development Editor Peter Lacey, Photo Researcher Cailen Swain, and the team at Dragonfly Media Group for rendering the illustrations.

ACCESSIBLE COLOR CONTENT   Every opportunity has been taken to ensure that the content herein is fully accessible to those who have difficulty perceiving color. Exceptions are cases where the colors provided are expressly required because of the purpose of the illustration.

Digital Resources

to Accompany *Neuroscience*, Seventh Edition

Optimize Student Learning with Oxford Insight

Neuroscience, Seventh Edition, is available in **Oxford Insight. Oxford Insight** delivers the trusted content of *Neuroscience* within a powerful, data-driven learning experience designed to increase student success. A guided and curated learning environment—delivered either via LMS/VLE integration or standalone—**Oxford Insight** provides access to the e-book, multimedia resources, assignable/gradable activities and exercises, and analytics on student achievement and progress. As students work through the course material, **Oxford Insight** automatically sets personalized learning paths for them, based on their specific performance.

Developed with applied social, motivational, and personalized learning research, **Oxford Insight** enables instructors to deliver an immersive experience that empowers students by actively engaging them with assigned reading. This approach, paired with real-time actionable data about student performance, helps instructors ensure that all students are best supported along their unique learning paths.

With Oxford Insight, Instructors Can:

- Assign auto-scored multiple-choice, fill-in, and other machine-gradable questions
- Score specific items (including open-ended questions) with feedback
- Export grades and change grading points
- Establish a course roster and add/drop students
- Share courses and resources with students and faculty
- Sync real-time assignments with LMS/VLE gradebooks
- Author new content and/or customize the publisher-provided content

For more information on how *Neuroscience*, Seventh Edition, powered by **Oxford Insight**, can enrich the teaching and learning experience in your course, please visit oxfordinsight.oup.com or contact your Oxford University Press representative.

For the Instructor
(Available at www.oup.com/he/purves7e)

Instructors using *Neuroscience*, Seventh Edition, have access to a wide variety of resources to aid in course planning, lecture development, and student assessment.

Content includes:

- **Figure PowerPoint Presentations** enable instructors to enhance lectures with images and walk students through illustrations from the text. Each presentation includes all figures from each chapter, with figure numbers and titles on each slide. Artwork has been optimized for exceptional image quality when projected in class.
- **Atlas Images** include all of the images from the book's Atlas of the Human Central Nervous System (from *Sylvius*) are included in PowerPoint Format, for use in lecture.

- *Sylvius* provides students an opportunity to explore and to apply the rich content of *Sylvius 4 Online: An Interactive Atlas and Visual Glossary of Human Neuroanatomy.*

- **Test Banks** provide a large collection of multiple-choice and short-answer questions—all aligned to Learning Objectives—for use in assessing student mastery of key facts and concepts. Available in multiple formats, including MS Word and Common Cartridge (for import into learning management systems).

- **Chapter Quizzes** enable a lower-stakes assessment of student grasp of content and concepts, assignable through your learning management system.

- **Animation Quizzes** offer an assignable tool for checking student comprehension of the animations from the e-book.

Enhanced E-Book for the Student

(ISBN 9780197616680)

Ideal for self-study, the *Neuroscience*, Seventh Edition, enhanced e-book delivers the full suite of digital resources in a format independent from any courseware or learning management system platform, making *Neuroscience*'s online resources more accessible for students.

The enhanced e-book is available via RedShelf, VitalSource, and other leading higher education e-book vendors and includes the following student resources:

- **Learning Objectives** outline the important takeaways of every major section.

- **Animations** give students detailed, narrated depictions of some of the complex processes described in the textbook.

Studying the Nervous System

Overview

The major challenge facing students of neuroscience is to integrate knowledge derived from various levels of analysis into a coherent sense of how the brain is organized and how it works. A key aspect of understanding brain organization and function is determining how the principal cells of all nervous systems—neurons and glia—perform their functions. Cellular and molecular specializations of neurons facilitate their capacity to transmit and process information, and parallel specializations of glia facilitate the stability and reliability of information processing. Variation in these features provides cellular diversity necessary to process a broad spectrum of information and generate a range of behaviors. Diverse subsets of neurons and glia form ensembles called neural circuits that process distinct types of information and generate behavioral responses. Neural circuits that analyze similar classes of information are aggregated into neural systems. Neural systems serve three general purposes: Sensory systems report the state of the organism and its environment; motor systems organize and generate actions; and associational systems integrate information. Neurons, circuits, and systems change their electrical activity over time in response to external stimuli, to represent or store information and to direct behavioral responses. Ultimately, brain organization and function reflect the same genomic information that instructs genesis and differentiation of all cells, tissues, and organs that constitute an individual animal, including neurons and glia in the brain and peripheral nervous system. In parallel with cellular, molecular, physiological, and genetic analyses of the brain, neuroscientists strive to understand "higher-order" brain functions, especially in ourselves and fellow humans. These complex abilities—perception, attention, memory, emotions, language, and thinking—constitute the broadly defined capacity of cognition. Insight into cognition has grown by using minimally invasive approaches to image the human brain in living individuals, including while these individuals perform demanding behavioral tasks. Clearly, for neuroscientists—and students of neuroscience—the most essential complex ability is "multitasking" to unite diverse information into a singular view of how brains work.

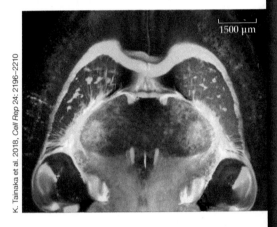

1500 μm

K. Tainaka et al. 2018, *Cell Rep* 24: 2196–2210

KEY CONCEPTS

- **1.1** Neurons and glia are the primary cell types of all nervous systems

- **1.2** Neurons are interconnected in ensembles called neural circuits

- **1.3** Circuits that process related information constitute a neural system

- **1.4** The genome controls brain organization and function

- **1.5** Neural circuits and systems can be analyzed in the human brain

Neurons and Glia Are the Primary Cell Types of All Nervous Systems

LEARNING OBJECTIVES

1.1.1   Explain what the neuron doctrine is.

1.1.2   Identify the key features that distinguish neurons from all other cell classes.

1.1.3   Identify dendrites and axons as input and output domains of neurons.

1.1.4   Discuss how glial cells interact with neurons to maintain the integrity of neural tissue and the efficiency of neuronal signaling.

1.1.5   Discuss the role of neuronal and glial diversity for brain function.

Cellular components of the nervous system

The story of modern neuroscience began with the recognition that the brain, like every other organ, is made up of individual cells. Early in the nineteenth century, the cell was recognized as the fundamental unit of all living organisms. It took until the middle of the twentieth century, however, to reach an agreement that the brain, like all other organs in the bodies of animals, including humans, is also made up of these fundamental, discrete units. The major reason for this late realization was that the first generation of "modern" neuroscientists in the nineteenth century had difficulty resolving the unitary nature of nerve cells with the microscopes and cell-staining techniques then available. The complex shapes and extensive branches of individual nerve cells—all of which are packed together and thus difficult to distinguish from one another—obscured their resemblance to the geometrically simpler cells of other tissues (Figure 1.1). Some biologists of that era even concluded that each apparent nerve cell was connected to

FIGURE 1.1   Nerve cells are remarkably diverse in vertebrate nervous systems, including in humans These drawings are tracings of actual nerve cells (neurons) from the brains of humans or other mammals that were stained by impregnation with silver salts (the so-called Golgi technique, used in the classic studies of Golgi and Cajal, and thereafter to visualize the details of individual neuronal morphology). The variety of dendritic arbors (green) extending from the cell body (purple) of different classes of neurons is striking, and is a key aspect of how neuronal diversity accommodates the functional demands of complex information processing in the brain. Neurons typically have only one axon (orange) that is usually less extensively branched than the dendritic arbor. Asterisks indicate that the axon of a particular class of neuron extends much farther than shown. Note, however, that some cells, such as the retinal bipolar cell, have very short axons, while others, such as the retinal amacrine cell, have no axon at all. The drawings are not all at the same scale.

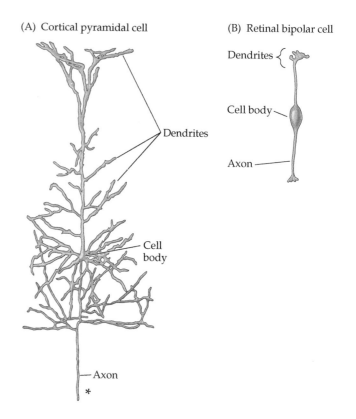

(A) Cortical pyramidal cell

(B) Retinal bipolar cell

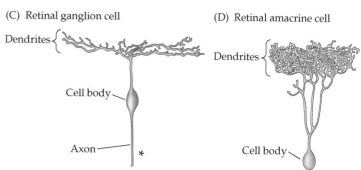

(C) Retinal ganglion cell

(D) Retinal amacrine cell

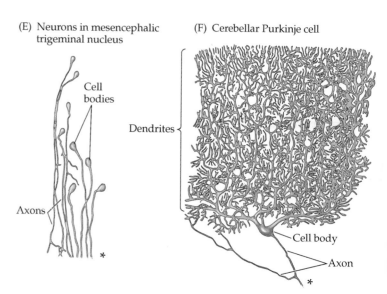

(E) Neurons in mesencephalic trigeminal nucleus

(F) Cerebellar Purkinje cell

its neighbors by protoplasmic links, forming a continuous, directly interconnected network, or *reticulum* (Latin, "net"), to relay information in the brain rather than a tissue composed of individual cells like every other organ.

The Italian pathologist Camillo Golgi articulated and championed this "reticular theory" of nerve cell communication in the late 1800s, based on his assessment of sections (very thin slices) of brain tissue stained with silver solutions (similar to those used in the photographic processes of the era) and viewed with the microscopes available at the time. The alternative, the **neuron doctrine**, was championed by Spanish neuroanatomist Santiago Ramón y Cajal (sometimes called the "father of modern neuroscience") and British physiologist Charles Sherrington. Based on light microscopic examination of sections of nervous tissue stained with silver solutions according to Golgi's pioneering

method, Cajal argued that nerve cells are discrete entities, that they are "polarized" to receive and transmit information, and that they communicate with one another by means of specialized contacts that are not sites of continuity between cells. Indeed, in Cajal's precise drawings of the sections (very thin slices) of brain tissue that he saw through the microscope, he included arrows showing what he thought was the direction of information flow across what he intuited (correctly) were nerve cells that used specialized junctions to signal from one cell to the next, rather than cytoplasmic continuity (Figure 1.2A). Sherrington, who had been working on the apparent transfer of electrical signals via reflex pathways in the nervous system, called these specialized sites **synapses** and recognized that the electrical activity at them was different than that necessary to conduct electrical impulses over the distance covered by an axon.

(A)

(B)

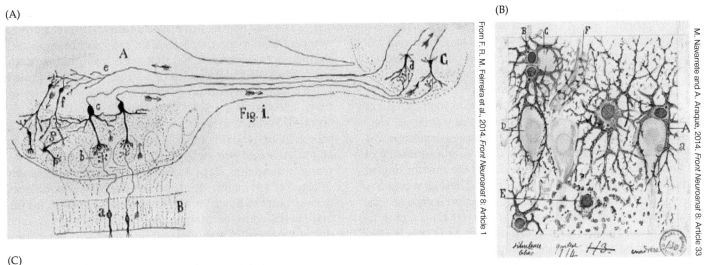

From F. R. M. Ferreira et al., 2014. *Front Neuroanat* 8: Article 1

M. Navarrete and A. Araque, 2014. *Front Neuroanat* 8: Article 33

(C)

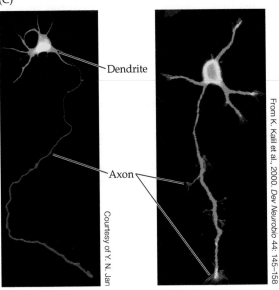

Dendrite

Axon

Courtesy of Y. N. Jan

From K. Kalil et al., 2000. *Dev Neurobio* 44: 145–158

FIGURE 1.2   Neurons and glial cells are the primary cell classes of all nervous systems (A) A drawing by Santiago Ramón y Cajal showing the complexity of individual neurons in the periphery and in the brain that relay information via synaptic connections in a singular direction—in this case from the peripheral neurons in the nose ("B", bottom left), through the olfactory bulb ("A", top left) to a distinct region of the cerebral cortex ("C", top right). (B) Cajal recognized that glial cells, the other major cell class in neural tissue, are intimately related to neurons. Glial cells ensheath the cell bodies of neighboring neurons and extend highly branched processes into the surrounding regions where dendrites and synapses are located. (C) Cajal's sense that neurons are polarized—with receiving and transmitting ends to establish directional information flow—is reflected by molecular specializations, shown here in two neurons grown in cell culture. At left, specific molecules are localized to dendrites (green) versus the single axon (red). At right, the distal domains of dendrites and axons that are actively growing to establish synaptic contacts with other neurons are molecularly distinct (red), reinforcing the conclusion that dendrites and axons make cell-cell contacts without direct continuity to form a nervous system composed of discrete but functionally interconnected neurons.

The histological studies of Cajal and a host of successors also led to the consensus that the cells of the nervous system can be divided into two broad categories: nerve cells, or **neurons**, and supporting **glial cells** (also called **neuroglia**, or simply **glia**) (Figure 1.2B). Cajal recognized that glial cells and their processes were integral elements of the specialized junctions he inferred were sites of communication between neurons. More than a century of experimental analysis has shown that neurons are indeed polarized cells and that their processes are distinguished by cellular specializations and molecules limited to distinct regions (Figure 1.2C). These cellular features reflect the capacity of neurons for electrical signaling, primarily with other neurons, either locally or over long distances. Glial cells, which in the brains of most species are more numerous than neurons, had sometimes been considered "support cells." This notion has since been challenged, and it is now clear that the contributions of multiple types of glial cells are essential for brain function as well as brain development. Finally, Cajal and his successors recognized that variation of the basic properties of just these two cell classes—neurons and glia—accounted for the full range of cellular diversity seen in the nervous system.

As the capacity to measure small electrical signals in the tissues of living animals improved during the first half of the twentieth century, it became clear that transfer of electrical signals at synaptic junctions between nerve cells was a distinct physiological process involving a sequence of changes in current and voltage across the membranes of the **presynaptic** and **postsynaptic** cells. These observations provided strong—but still indirect—support for the neuron doctrine. Indeed, challenges to the discrete nature of individual neurons and their communication via specialized junctions rather than cytoplasmic continuity were raised during most of the first half of the twentieth century. The advent of electron microscopy in the 1950s finally resolved any lingering doubts about the discrete identities of neurons. The high-magnification, high-resolution images obtained with electron microscopes (Figure 1.3) clearly established that nerve cells are functionally independent units, even though they are embedded in a dense matrix of additional cellular processes. Electron micrographs also established that the junctions Cajal had shown in his drawings, and that Sherrington had inferred from his physiological observations as synapses, were indeed specialized sites of communication between two discrete nerve cells, one presynaptic, the other postsynaptic. As belated consolation for Golgi, however, electron microscopic studies also demonstrated relatively rare intercellular continuities between some neurons. These continuities, called **gap junctions**, are similar to those found between cells in epithelia such as the lung, heart, and intestine. In addition to neuronal gap junctions (sometimes called **electrical synapses**, reflecting the direct transfer of current from one cell to the other), glial cells in some regions of the brain are also interconnected with one another by gap junctions. In the nervous system, the cells between which a gap junction forms remain as separate entities—their membranes are not fused, and only the proteins in each cell membrane interact to allow direct cell-cell contact. Nevertheless, gap junctions do allow for cytoplasmic continuity and transfer of electrical and chemical signals between cells in the nervous system.

Neurons

Even though neurons come in many different shapes and sizes, they all share several basic cell biological features. Neuronal cell bodies (see Figure 1.3A,B) are defined by the cell's nucleus and by the high frequency of mitochondria as well as organelles for protein synthesis (endoplasmic reticulum) and protein trafficking (Golgi apparatuses). Neurons are further specialized for long-distance electrical signaling and intercellular communication, reflecting their cellular and molecular polarization (see Figure 1.2C). The key features that distinguish neurons from all other cell classes are the local, often extensive branching of **dendrites** (see Figure 1.3C) that arise from the neuronal cell body, and the presence of a single **axon** (see Figure 1.3D) that can extend far beyond the location of the cell body. The variation in local branching of dendrites is substantial. Some neurons lack dendrites altogether, while others have dendritic branches that rival the complexity of a mature tree (see Figure 1.1). Indeed, the branched network of dendrites from an individual neuron is usually referred to as a dendritic **arbor** because of its resemblance to the branches of a tree. Most neuronal axons extend unbranched toward their targets, branching only within their target location. In addition, many axons that extend for long distances become myelinated by a class of glial cells called oligodendroglia (Figure 1.3E) in the **central nervous system** (**CNS**, the brain and spinal cord), and Schwann cells in the **peripheral nervous system** (**PNS**, the network of neurons and axon pathways distributed throughout an animal's body). Finally, the local branches and the presynaptic terminals made by a neuron's single axon define that axon's **terminal field**. The terminals of axons are secretory specializations that establish presynaptic domains for synapses between interconnected neurons (Figure 1.3F). Numerous **synaptic vesicles** are found in the presynaptic domain. These vesicles fuse with the regions in the presynaptic membrane called active zones and release their contents, **neurotransmitters**, to act on the postsynaptic specialization. Postsynaptic specializations are most frequently found in dendritic regions adjacent to presynaptic terminals in the CNS, or on several target tissues in the body in the PNS, including muscles and glands. The postsynaptic specialization is characterized by a concentration of **neurotransmitter receptors** that detect the neurotransmitters released by the presynaptic cell.

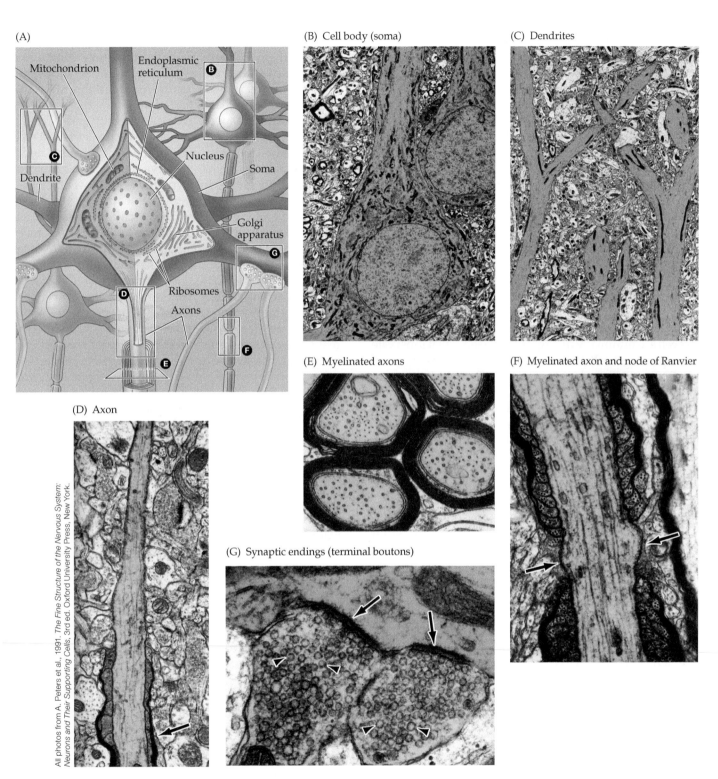

(A)

Mitochondrion

Endoplasmic reticulum

B

Nucleus

Soma

C

Dendrite

Golgi apparatus

G

Ribosomes

D

Axons

E

F

(D) Axon

All photos from A. Peters et al., 1991. *The Fine Structure of the Nervous System: Neurons and Their Supporting Cells*, 3rd ed. Oxford University Press, New York.

(B) Cell body (soma)

(C) Dendrites

(E) Myelinated axons

(F) Myelinated axon and node of Ranvier

(G) Synaptic endings (terminal boutons)

FIGURE 1.3   The major features of neurons visualized with electron microscopy (A) Diagram of nerve cells and their component parts. The circled letters correspond to the micrographs in the figure. (B) Nerve cell bodies (purple) occupied by large round nuclei, dense networks of endoplasmic reticulum in the cytoplasm, and numerous mitochondria (darker tubular structures). (C) Apical dendrites (purple) of cortical pyramidal cells. (D) Axon initial segment (blue), the region of the axon that emerges from the neuronal cell body, entering a myelin sheath (gold). (E) Transverse section of axons (blue) ensheathed by the processes of oligodendrocytes (gold); the surrounding myelin is black. (F) Portion of a myelinated axon (blue) illustrating the intervals that occur between adjacent segments of myelin (gold and black) referred to as nodes of Ranvier (arrows). (G) Terminal boutons (blue) loaded with synaptic vesicles (arrowheads) forming synapses (arrows) with a dendrite (purple).

The variation of dendritic arbors and differences in the extent of axonal terminal fields in specific neuron classes provides the cell biological foundation for connecting neurons with one another to establish neural circuits (see the next section) with flexible capacities for processing information. Increased dendritic arbors provide additional postsynaptic space that can be innervated by either one or a few inputs, increasing the ability of the input to elicit a robust and reliable postsynaptic response in the target neuron. Alternatively, elaborate dendritic arbors can also accommodate large numbers of different inputs, allowing for greater integration of information from a variety of sources. The degree to which a dendrite is innervated by a singular input or multiple inputs is referred to as the level of **convergence** for the target neuron. In parallel, axons can make connections with a variable number of target neurons, from one to many (the actual limits are unknown). This feature of a single neuron and its axonal connections is referred to as its **divergence**. Neurons whose axons innervate only one target cell are minimally divergent, while those that innervate multiple target cells, either in the same or diverse areas of the CNS or PNS, are highly divergent. Thus, the fundamental cellular properties of neurons, and the extent to which they vary, reflect both the molecular and cellular specializations of neurons and underlie the flexibility of neurons to become connected to one another to accommodate a broad range of information processing.

Glial cells

Glial cells—usually referred to more simply as glia—are quite different from neurons (Figure 1.4). No reliable absolute numbers of glia versus neurons are available, but it is likely that glia outnumber neurons in the mammalian, including human, brain. Glia do not participate directly in synaptic transmission. Instead, they interact with the pre- and postsynaptic domains of neurons to define synaptic contacts and with their axons to maintain signaling capacity. Like nerve cells, many glial cells have complex processes extending from their cell bodies, but these processes do not function like dendrites or axons. Instead, glial processes envelope synapses, intercalate between different classes of neurons, form "capsules" (a continuous boundary) around subsets of neurons in different brain regions, and surround axons and dendrites, including via myelination. Glial cells also are highly responsive to brain injury. Finally, cells with glial characteristics appear to be the only stem cells retained in the mature brain, and are capable of giving rise both to new glia and, in a few instances, new neurons. Thus, glial cells serve multiple essential functions for maintaining the integrity of neural tissue and the efficiency of neuronal signaling throughout life.

The word *glia* is Greek for "glue" and reflects the nineteenth-century presumption that these cells "held the nervous system together." The term has survived despite the lack of any evidence that glial cells alone bind nerve cells together. The validity of considering glia to be the glue of a nervous system has been further diminished by a still incomplete characterization of how these cells influence multiple aspects of tissue integrity and information processing in the brain. Glial functions that *are* well established include maintaining the ionic milieu of nerve cells; modulating the rate of nerve signal propagation via myelination; modulating synaptic action by controlling the uptake and metabolism of neurotransmitters at or near the synaptic cleft; providing a scaffold for several aspects of neural development; aiding (or in some instances impeding) recovery from neural injury; providing an interface between the brain and the immune system; and facilitating the convective flow of interstitial fluid through the brain during sleep, a process that is thought to ensure efficient removal of metabolic waste throughout the brain.

There are three types of differentiated glial cells in the mature nervous system: astrocytes, oligodendrocytes, and microglial cells. **Astrocytes**, which are restricted to the CNS (i.e., the brain and spinal cord), have elaborate local processes that give these cells a starlike ("astral") appearance (see Figure 1.4A,F). Astrocytes or their processes can be seen at many CNS synapses. A major function of astrocytes is to maintain, in a variety of ways, an appropriate chemical environment for neuronal signaling, including formation of the **blood-brain barrier**, which prevents circulating immune cells, molecules, or pathogens that might interfere with neural function from entering the brain. (see the Appendix). In addition, recent observations suggest that astrocytes secrete substances that influence the construction of new synaptic connections, and that a subset of astrocytes in the adult brain are apparently stem cells (see Figure 1.4D). In certain regions of the mammalian brain, these cells, which have the key characteristics of astrocytes, retain the ability to generate neurons in the developing or mature brain, as well as when they are isolated and grown in cell culture.

Oligodendrocytes, which are also restricted to the CNS, establish a laminated, lipid-rich wrapping called **myelin** around some, but not all, axons (see Figure 1.4B,E,G,H). Myelin has important effects on the speed of transmission of electrical signals (see Chapter 3). In the PNS, the cells that provide myelin are called **Schwann cells** (see Figure 1.4H). In the mature nervous system, subsets of oligodendrocytes and Schwann cells retain neural stem cell properties and can generate new oligodendrocytes and Schwann cells in response to injury or disease. These oligodendrocyte or Schwann cell precursors (see Figure 1.4E), if grown in cell culture, can also generate neurons.

Microglial cells (see Figure 1.4C,J) are derived primarily from hematopoietic precursor cells; nevertheless, they become resident in the brain during early development and

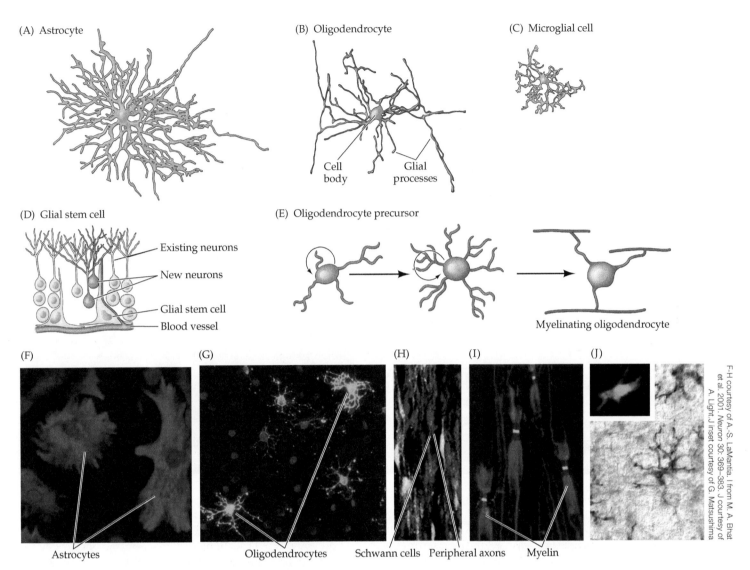

(A) Astrocyte

(B) Oligodendrocyte

Cell body   Glial processes

(C) Microglial cell

(D) Glial stem cell

Existing neurons
New neurons
Glial stem cell
Blood vessel

(E) Oligodendrocyte precursor

Myelinating oligodendrocyte

(F) (G) (H) (I) (J)

Astrocytes   Oligodendrocytes   Schwann cells   Peripheral axons   Myelin

F–H courtesy of A.-S. LaMantia, I from M. A. Bhat et al. 2001. *Neuron* 30: 369–383. J courtesy of A. Light. J inset courtesy of G. Matsushima

FIGURE 1.4   Glial cell types (A–C) Tracings of differentiated glial cells in the mature nervous system visualized using the Golgi method include an astrocyte (A), an oligodendrocyte (B), and a microglial cell (C). The three tracings are at approximately the same scale. (D) Glial stem cells in the mature nervous system include stem cells with properties of astrocytes that can give rise to neurons, astrocytes, and oligodendrocytes. (E) Another class of glial stem cell, the oligodendrocyte precursor, has a more restricted potential, giving rise primarily to differentiated oligodendrocytes. (F) Astrocytes (red) in tissue culture are labeled with an antibody against an astrocyte-specific protein. (G) Oligodendrocytes (green) in tissue culture labeled with an antibody against an oligodendrocyte-specific protein. (H) Schwann cells (green) are seen immediately adjacent to fascicles of peripheral axons (red). (I) Peripheral axons are ensheathed by myelin (red) except at nodes of Ranvier (see Figure 1.3F). The green label indicates ion channels (see Chapter 4) concentrated in the node; the blue label indicates a molecularly distinct region called the paranode. (J) Microglial cells from the spinal cord labeled with a cell type–specific antibody. Inset: Higher-magnification image of a single microglial cell labeled with a macrophage-selective marker. (A–C after E. G. Jones and M. W. Cowan, 1983. The nervous tissue. In *The Structural Basis of Neurobiology*, E. G. Jones [Ed.]. New York: Elsevier; D,E, after A. Nishiyama et al., 2009. *Nat Rev Neurosci* 10: 9–22.)

integrate into neural tissue. Microglia share many properties with macrophages found in other tissues: They are primarily scavenger cells that remove cellular debris from sites of injury or normal turnover of cellular constituents. In addition, microglia, like their macrophage counterparts, secrete signaling molecules—particularly a wide range of cytokines that are also produced by cells of the immune system—that can modulate local inflammation and influence whether other cells survive or die. Indeed, some neurobiologists prefer to categorize microglia as a type of macrophage. Following brain damage, the number of microglia at the site of injury increases dramatically. Some of these cells proliferate from microglia resident in the brain, while others come from macrophages that migrate to the injured area and enter the brain via local disruptions in the cerebral vasculature (the blood-brain barrier).

Visualizing neurons in the CNS and PNS

Although the cellular constituents of the human nervous system are in many ways similar to those of other organs, they are unusual in their extraordinary diversity as well as differences in their frequency and distribution in distinct regions of the CNS and PNS. The nervous system has a greater range of distinct cell types—whether categorized by morphology (shape and size), molecular identity, or physiological role—than any other organ system. For much of the twentieth century, neuroscientists relied on

the techniques developed by Cajal, Golgi, and other pioneers of histology (the microscopic analysis of cells and tissues) and pathology (the study of the cellular basis of disease) to describe and categorize the cell types—particularly neurons—in the nervous system. Labeling individual nerve cells and their processes sparsely and randomly with silver salts using the original method developed by Golgi allowed these cells' dendrites and axons, as well as the processes of glial cells, to be analyzed (Figure 1.5A,B). Subsequently, techniques were developed for injecting fluorescent dyes and other soluble molecules into individual central or peripheral neurons—often after physiological recording—to visualize their features in the context of information about their function (Figure 1.5C,D). As a complement to these methods (which provide a sample of

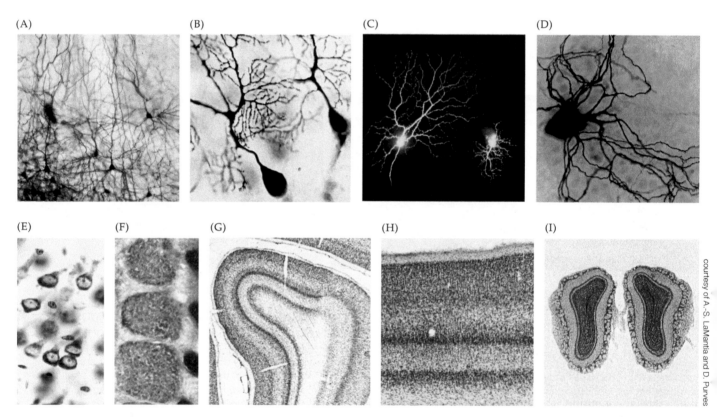

courtesy of A.-S. LaMantia and D. Purves

FIGURE 1.5   Visualizing neurons in the CNS and PNS
(A) Cortical neurons stained using the Golgi method (impregnation with silver salts). (B) Golgi-stained Purkinje cells in the cerebellum. Purkinje cells have a single, highly branched apical dendrite (as diagrammed in Figure 1.1F). (C) Intracellular injection of fluorescent dye labels two retinal neurons that vary dramatically in the size and extent of their dendritic arborizations. (D) Intracellular injection of an enzyme labels a neuron in a ganglion of the autonomic (involuntary control of internal organs) nervous system. (E) The dye cresyl violet stains RNA in all cells in a tissue, labeling the nucleolus (but not the nucleus) as well as the ribosome-rich endoplasmic reticulum. Dendrites and axons are not labeled, which explains the "blank" spaces between these neurons.

(F) A series of Nissl-stained dorsal root ganglia intercalated between vertebrae. (G) Nissl-stained section of the cerebral cortex reveals lamination—cell bodies arranged in layers of differing densities. The different laminar densities define boundaries between cortical areas with distinct functions. (H) Higher magnification of the primary visual cortex, seen on the left side of panel G. Differences in cell density define the laminae of the primary visual cortex and differentiate this region from other cerebral cortical areas. (I) Nissl stain of the olfactory bulbs reveals a distinctive distribution of cell bodies, particularly those cells arranged in rings on each bulb's outer surface. These structures, including the cell-sparse tissue contained within each ring, are called glomeruli.

specific subsets of neurons), other stains reveal the distribution of all cell bodies—but not details of their processes or connections—in neural tissue. The widely used Nissl method is one example; this technique stains the nucleolus and other organelles (e.g., ribosomes) where DNA or RNA is found so that much of a neuron's cytoplasm and even large proximal dendrites can be visualized (Figure 1.5E). Such stains demonstrate that the size, density, and distribution of the total population of nerve cells are not uniform within the PNS or CNS. In the periphery, accumulations of neurons into dorsal root ganglia can be seen adjacent to vertebrae (Figure 1.5F). In several regions of the brain, such as the cerebral cortex, neurons are arranged in layers (Figure 1.5G,H), each of which is defined by differences in cell density. Structures such as the olfactory bulbs display even more complicated arrangements of cell bodies (Figure 1.5I). These differences in distribution, frequency, and density of neurons from brain region to brain region are referred to as the **cytoarchitecture** of that region.

CONCEPT 1.2 | Neurons Are Interconnected in Ensembles Called Neural Circuits

LEARNING OBJECTIVES

1.2.1 Describe the key features of neural circuits.

1.2.2 Identify the three functional types of neurons that constitute all neural circuits.

1.2.3 Define the roles of excitatory, inhibitory, and modulatory neurons in neural circuits.

1.2.4 Describe the approaches used to define neural circuit connections and functions.

Neural circuits

Neurons never function in isolation; they are organized into ensembles called **neural circuits** that process specific kinds of information. The synaptic connections that underlie neural circuits are typically made in a dense tangle of dendrites, axon terminals, and glial cell processes that together constitute what is called **neuropil** (Greek *pilos*, "felt") (see Figure 1.3G). The neuropil constitutes the regions between nerve cell bodies where most synaptic connectivity occurs.

Although the arrangement of neural circuits varies greatly according to the function of the circuit, some features are characteristic of all such ensembles. Preeminent is the direction of information flow in any particular circuit (see Figure 1.2A), which is obviously essential to understanding its purpose. Nerve cells that carry information from the periphery *toward* the brain or spinal cord (or deeper centrally within the spinal cord and brain) are called **afferent neurons**; nerve cells that carry information

away from the brain or spinal cord (or away from the circuit in question) are **efferent neurons**. Both afferent and efferent neurons are often referred to as **projection neurons**, because their axons project (extend) for a significant distance beyond their cell body and connect with distal targets. **Interneurons** (also called **local circuit neurons**) participate only in local aspects of circuit function, based on their relatively short axons and the restricted targets with which they connect. These three functional classes—afferent neurons, efferent neurons, and interneurons—are the basic constituents of all neural circuits.

The other dimension of cellular identity in neural circuits is the type of signal that afferent, efferent, and interneurons transmit to the neurons with which they connect. Signals in neural circuits can be **excitatory**, **inhibitory**, or **modulatory**, based on the chemical neurotransmitter released at presynaptic terminals and the identity and functional characteristics of neurotransmitter receptors in the postsynaptic domain. Excitatory neurotransmitters, when they activate excitatory neurotransmitter receptors, generate signals that enhance electrical activity in the target neuron and make it more likely that the target neuron will relay signals to additional neurons in the circuit. Inhibitory neurotransmitters, when they activate inhibitory neurotransmitter receptors, diminish electrical activity in the target neuron far below the threshold necessary for it to transmit electrical signals to additional neurons in the circuit. Modulatory neurotransmitters do not enhance target neuron activity to a level where the target neuron can transmit signals, nor do they diminish activity below a threshold where signaling is possible. Instead, they modify the thresholds in target neurons, thus modulating the effectiveness of either excitatory or inhibitory signals.

Defining the organization and function of neural circuits

Neural circuits have two key features: connections between an identified group of neurons, and the capacity to process a distinct type of information. In addition, neural circuits are defined by the information that comes into the circuit, the *input*, and that which emerges, the *output*. Thus, to define a neural circuit, one must demonstrate the connections of a group of neurons as well as show that those connections result in processing of signals that convert the inputs of the circuit into a novel output.

A simple example of a neural circuit is one that mediates the **myotatic reflex**, commonly known as the knee-jerk reflex (Figure 1.6). The afferent neurons—carrying the input to the circuit that elicits the reflex movement—are sensory neurons whose cell bodies lie in the **dorsal root ganglia** and send axons peripherally that terminate in sensory endings in skeletal muscles. (The ganglia that

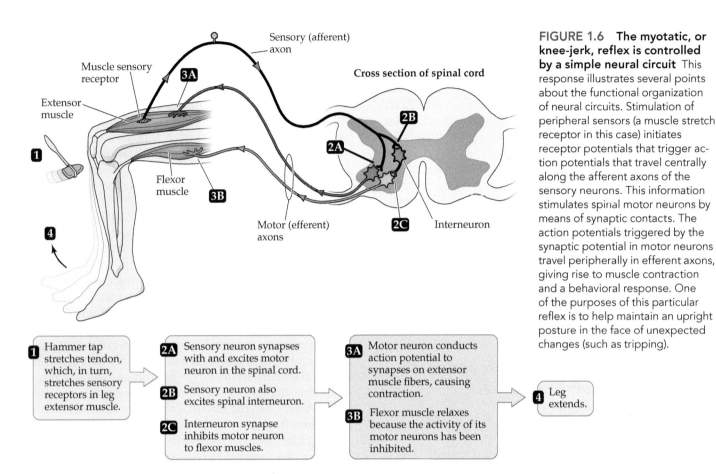

FIGURE 1.6 The myotatic, or knee-jerk, reflex is controlled by a simple neural circuit This response illustrates several points about the functional organization of neural circuits. Stimulation of peripheral sensors (a muscle stretch receptor in this case) initiates receptor potentials that trigger action potentials that travel centrally along the afferent axons of the sensory neurons. This information stimulates spinal motor neurons by means of synaptic contacts. The action potentials triggered by the synaptic potential in motor neurons travel peripherally in efferent axons, giving rise to muscle contraction and a behavioral response. One of the purposes of this particular reflex is to help maintain an upright posture in the face of unexpected changes (such as tripping).

1 Hammer tap stretches tendon, which, in turn, stretches sensory receptors in leg extensor muscle.

2A Sensory neuron synapses with and excites motor neuron in the spinal cord.

2B Sensory neuron also excites spinal interneuron.

2C Interneuron synapse inhibits motor neuron to flexor muscles.

3A Motor neuron conducts action potential to synapses on extensor muscle fibers, causing contraction.

3B Flexor muscle relaxes because the activity of its motor neurons has been inhibited.

4 Leg extends.

serve this same function for much of the head and neck are called **cranial nerve ganglia**; see the Appendix.) The central axons of these sensory neurons enter the spinal cord, where they terminate on a variety of central neurons concerned with the regulation of muscle tone. Peripheral sensory neurons are exceptions to the general rule of polarized neurons having multiple dendrites for inputs and one axon for outputs—they have two axons that have two distinct targets. The ultimate targets of these afferents, and the input information they carry, are the **motor neurons** that relay the output activity of the circuit and control the muscles that make the knee jerk. There are distinct types of efferent motor neurons in the circuit: One type of motor neuron projects to the flexor muscles in the limb, and the other to extensor muscles. Thus, there is divergent output from this reflex circuit.

This divergent output has functional implications for the myotatic reflex circuit: Flexors and extensors can't be activated together; instead, one must be contracted while the other relaxes. Thus, the divergent output can deliver different signals to each of these muscle targets to coordinate their opposing responses. This functional demand establishes the need for a third type of element in the circuit: spinal cord interneurons. The interneurons receive synaptic

contacts from sensory afferent neurons and make synapses on the efferent motor neurons that project to the flexor muscles; thus, they are capable of modulating the input–output linkage. The interneurons "convert" the excitatory input from the sensory afferent into an inhibitory signal. The excitatory synaptic connections between the sensory afferents and the extensor efferent motor neurons cause the extensor muscles to contract. At the same time, interneurons activated by the same afferents are inhibitory, and their activation diminishes electrical activity in flexor efferent motor neurons and causes the flexor muscles to relax (Figure 1.7). The result is a complementary activation (contraction) and inactivation (relaxation) of the synergistic and antagonistic muscles that control the position of the leg.

A distinct, dynamic perspective on the connections and function of the myotatic or any other neural circuit can be obtained by **electrophysiological recording**, which measures the electrical activity of a neuron. There are two approaches to this method: **extracellular recording**, whereby an electrode is placed *near* a nerve cell of interest to detect its activity; and **intracellular recording**, whereby the electrode is placed *inside* a cell of interest. Extracellular recording is particularly useful for detecting temporal patterns of action potential activity and relating

FIGURE 1.7 Physiological recording shows the pattern and timing of electrical signals that are relayed via the neural circuit for the myotatic reflex (A) In this drawing of extracellular recording from the neurons in the myotatic reflex circuit, action potentials are indicated by individual vertical lines. As a result of the stimulus, the sensory neuron is triggered to fire at higher frequency (i.e., more action potentials per unit of time). This increase triggers a higher frequency of action potentials in both the extensor motor neurons and the interneurons. Concurrently, the inhibitory synapses made by the interneurons onto the flexor motor neurons cause the frequency of action potentials in these cells to decline. (B) Intracellular recordings from each of the neurons in the myotatic reflex circuit (left) show the temporal relationships between electrical signals in the circuit (panels at right): (i) action potential measured in a sensory neuron; (ii) postsynaptic potential recorded in an extensor motor neuron; (iii) postsynaptic potential recorded in an interneuron; (iv) postsynaptic potential recorded in a flexor motor neuron. Such intracellular recordings are the basis for understanding the cellular mechanisms that coordinate the sequence and timing of action potential generation in a circuit, and the sensory receptor and synaptic potentials that trigger these conducted signals.

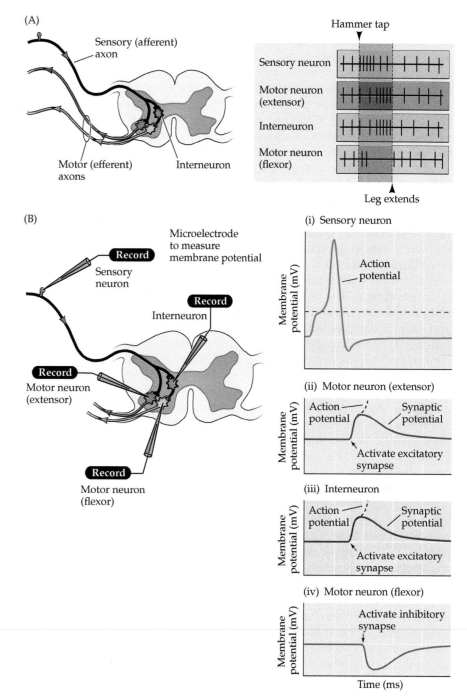

those patterns to stimulation by other inputs or in response to specific behavioral events. Intracellular recording can detect the smaller, graded changes in electrical potential that trigger action potentials, and thus allows a more detailed analysis of communication among neurons within a circuit. These graded triggering potentials can arise at either sensory receptors or synapses and are called **receptor potentials** or **synaptic potentials**, respectively.

For the myotatic circuit, electrical activity can be measured both extracellularly and intracellularly, thus defining the functional relationships among the neurons in the circuit. With electrodes placed near—but still outside—individual cells, the pattern of action potential activity can be recorded extracellularly for each element of the circuit (afferents, efferents, and interneurons) before, during, and after a stimulus (see Figure 1.7A). By comparing the onset, duration, and frequency of action potential activity in each cell, a functional picture of the circuit emerges. Using intracellular recording, it is possible to observe directly the

changes in membrane potential underlying the synaptic connections of each element of the myotatic reflex (or any other) circuit, and the temporal relationships of their onset and termination (see Figure 1.7B).

Visualizing and analyzing neural circuits

The organization and function of the neural circuit for the myotatic reflex were first inferred from behavior (the knee jerk in response to a hammer tap), and then using electrophysiological as well as basic histological analysis (see Figure 1.5). Today, several additional approaches are available to better define the connections and functions of neural circuits. The neurons that constitute a circuit are defined not only by proximity but by demonstrable synaptic connections. Cajal and other pioneer neuroscientists inferred connections based on the proximity of individual presynaptic endings (referred to as boutons) of an afferent neuron to a target cell whose electrical activity was presumed to be responsive to signals from that afferent neuron. These inferences remained the standard for visualization of neural circuits for much of the last century. Improved methods for labeling individual neurons, including injecting dyes to label afferent neurons while recording their electrical activity and that of their targets, allowed for the assessment of individual neurons and their connections. The recognition that neurons, like other cells, could not only secrete but also internalize proteins and other molecules led to development of a variety of transsynaptic tracing approaches that allow a presynaptic neuron and its postsynaptic partner to be identified. The most recent example of this sort of transsynaptic labeling to define interconnected neurons in a neural circuit takes advantage of genetically engineered viruses (see Concept 1.4) to label specific presynaptic afferent neurons and their postsynaptic targets via high-fidelity viral transfer (no widespread leakage of the virus into extracellular spaces) from the afferent neuron or neurons to target neurons within a particular neural circuit (Figure 1.8A).

Parallel advances in physiological analysis allow the activity of entire populations of neurons that define neural circuits to be monitored. One approach, known as **calcium imaging**, records the transient changes in intracellular concentration of calcium ions (see Chapter 7) that are associated with action potential firing in arrays of individual neurons (Figure 1.8B–D). Because calcium channels establish currents that lead to voltage changes in neurons, and because calcium is an important second messenger, calcium imaging can visualize neuronal activity in large numbers of individual cells by detecting calcium transients in the cytoplasm of multiple cells with relatively high temporal resolution. A related approach uses voltage-sensitive fluorescent dyes that insert into the neuronal cell membrane and report on the transmembrane potential, thereby imaging the consequences of action potentials and other electrical signals in many neurons at once. Calcium indicators or voltage-sensitive dyes can be introduced directly into neurons in living slices or into primary cultured neurons based on their osmotic properties in solution. In addition, viral vectors can be used to locally transfect subpopulations of cells by limited injection in the intact brain in a living animal.

After a substantial period to allow for expression of the genetically encoded sensor introduced into the cell by the virus, labelled cells can be visualized and their activity monitored in the brain of the intact animal or in live brain slices. Finally, genes that encode calcium- or voltage-sensitive proteins can be introduced into transgenic animals for more precise, non-invasive control of where and when the proteins are available for measuring activity in the living animal (see Concept 1.4).

The most specific and arguably the most effective way to manipulate the function of neuronal connections that define circuits, however, is to use molecular genetic tools that control the activity of specific inputs, an approach called **optogenetics**. Optogenetic methods emerged as a consequence of the identification and cloning of bacterial channels referred to as opsins, similar to the opsins in animal retinas. Bacterial opsins use the same chromophore found in retinal opsins to transduce light energy into a chemical signal that activates channel proteins. Since opsins modulate membrane currents when they absorb photons, light can be used to control nerve cell activity when bacterial chromophores are incorporated into the membrane of any neuron. Three bacterial opsins have been used to modify neuronal excitability: **bacteriorhodopsin**, **halorhodopsin**, and **channelrhodopsin** (Figure 1.9A). Both bacteriorhodopsin and halorhodopsin have a net hyperpolarizing effect on cells: Bacteriorhodopsin conducts H^+ ions from inside the cell to outside, and halorhodopsin conducts Cl^- ions from outside to inside the cell. In contrast, channelrhodopsin conducts cations (Na^+, K^+, Ca^{2+}, H^+) as well as anions (Cl^-), providing for either depolarizing or hyperpolarizing modulation, depending on the channelrhodopsin variant and the wavelengths of light used.

The genes for opsins can be introduced into neurons either in living brain slices or intact animals. In brain slices, a variety of viral transduction methods, similar to those discussed earlier, are used. In whole animals, transgenic methods are used (see Concept 1.4). Once the opsins are expressed in living neurons, these neurons can be illuminated by specific wavelengths of light, and neural activity can be manipulated with a high degree of spatial and temporal resolution, due to microscopic illumination of one or more opsin-labeled nerve cells or axon terminals. In awake and behaving animals, this approach can be used during the performance of specific tasks to evaluate the role of the optogenetically modified neurons in neural circuits that mediate task performance (Figure 1.9B). When optogenetic methods are applied in brain slices, synaptic activity in axon terminals and dendrites can be modified locally by illuminating only those regions of the opsin-expressing nerve cell; the resulting change in local circuit activity can then be recorded electrophysiologically or optically

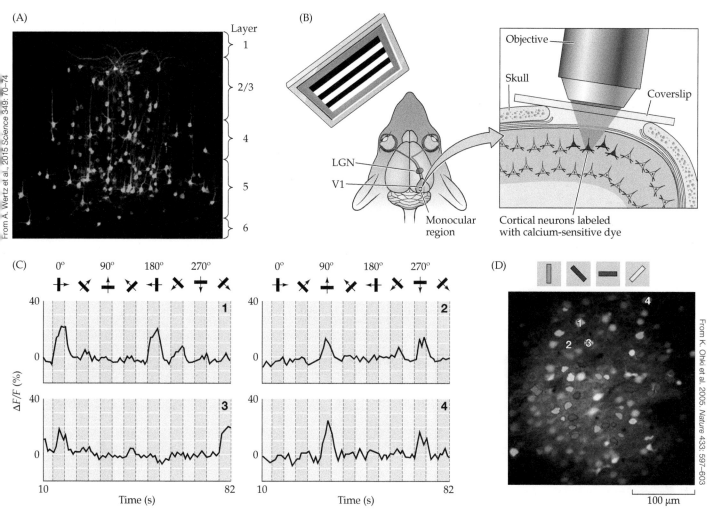

FIGURE 1.8 Visualizing and monitoring the activity of individual neurons in neural circuits (A) The apparent full set of neurons (green) synaptically interconnected with a single neuron (red) in a circuit in the cerebral cortex, visualized using a virus that can be introduced into a single cell (the red neuron) and transferred to all of its postsynaptic targets via release of the virus at the presynaptic terminal and its reuptake by the postsynaptic domain. The virus then activates a green reporter in all of the target neurons. In this example, a single cortical projection neuron (red) in the upper part of layer 2 of the 6 layered cerebral cortex is synaptically connected to multiple presumed projection neurons as well as interneurons (green) in all of the adjacent layers, except for layer 1 which has very few neurons. (B–D) Imaging cortical neurons responding to visual stimuli using calcium-sensitive dyes. (B) Ca²⁺ imaging was done in a live mouse presented with visual stimuli at different orientations (a horizontally oriented series of high-contrast stripes is shown here). A small "window" of bone was removed over the visual cortex for application of the dyes and subsequent imaging; changes in fluorescence intensity were detected using a microscope with the objective over the exposed cortical surface. (LGN = lateral geniculate nucleus; V1 = primary visual cortex.) (C) The change in fluorescence intensity ($\Delta F/F$ [%]) of four cells imaged this way while the mouse viewed stripes in the orientations shown at the top of the graphs, moving in directions indicated by the arrows (numbers 1–4 on the graphs identify the cells, which can be localized in panel D). Each separate graph shows the response over time of one cortical cell. In each graph, the peaks in fluorescence signal indicate robust responses when the cell's preferred stimulus orientation was presented; little response was elicited by nonpreferred stimuli. (D) The distribution of cells with preferred responses to stripes oriented at different angles (colors indicate preferred orientation). Activated cells preferring different orientations were interspersed, with each orientation represented by several cells in different positions within this small cortical area. (B after M. Mank, et al., 2008. *Nat Meth* 5: 805–811; C after K. Ohki et al. 2005. *Nature* 433: 597–603.)

(Figure 1.9C,D). Thus, optogenetic approaches can modify neuronal activity at a variety of scales—from single neurons to local neural circuits and even to more widely distributed neural networks that influence specific behaviors.

The ability to modify specific synaptic connections using optogenetic methods thus allows for circuit connections to be mapped functionally, and the specific influence of those connections on behavior to be assessed.

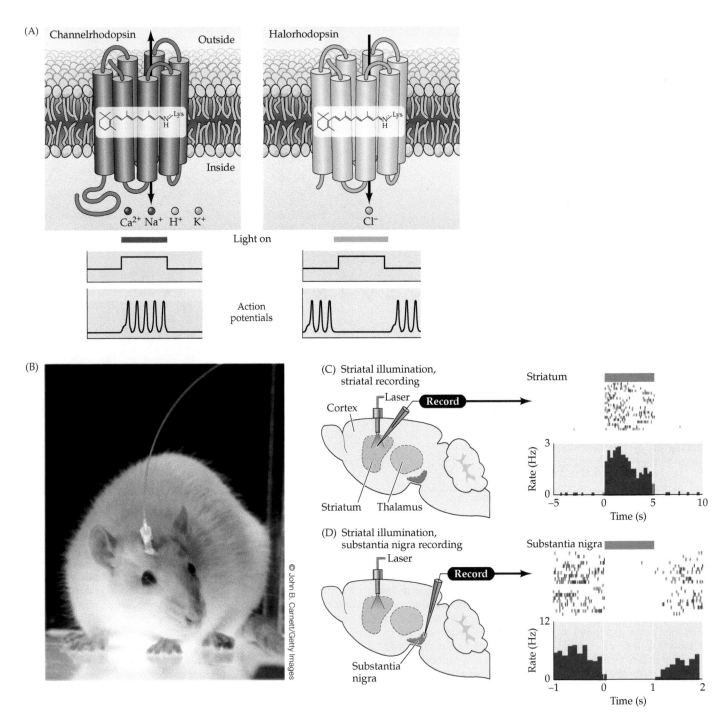

FIGURE 1.9 Optogenetic methods used to control electrical activity in nerve cells (A) Two bacterial opsins, showing their 7-transmembrane domains. The light-sensing all-*trans* retinal transduces a change in illumination that transiently opens the channels. (B) A fiber-optic probe, stabilized with a permanent head mount, can use a laser to deliver a narrow bandwidth of light to specific opsin-expressing neurons. (C) Illumination of bacterial opsins expressed in striatal neurons that regulate movement. Neurons expressing channelrhodopsin in the striatum, where neurons have little or no spontaneous action potential activity (regions on the graph with very few marks), fire robustly when illuminated (the histogram indicates action potential frequency when light is on and channelrhodopsin is activated). (D) Neurons in the substantia nigra pars reticulata, where neurons have a high frequency of spontaneous action potential activity (rasters and histograms in the left- and right-flanking regions), can be "silenced" transiently by illumination in the striatum. The striatal axons release the inhibitory neurotransmitter GABA. Thus, when the striatal neurons are stimulated optogenetically, the result of the "activation" is increased inhibition in the substantia nigra. Thus, optogenetic mechanisms can assess the physiology of neural circuits based on the activation of neuronal populations. (A after F. Zhang et al., 2011. *Cell* 147: 1446–1457; C,D after A. V. Kravitz et al., 2010. *Nature* 466: 622–626.)

CONCEPT 1.3 | Circuits That Process Related Information Constitute a Neural System

LEARNING OBJECTIVES

1.3.1 Define neural systems as interconnected circuits across multiple brain regions that process specific types of information.

1.3.2 Explain how single neurons process information in neural systems.

1.3.3 Assess how neural systems represent information.

1.3.4 Describe the basic anatomical features of neural systems in the human brain.

Neural systems

Individual neural circuits that process similar types information are integrated into broader networks that analyze larger volumes of the same general type of information and generate more complex behaviors. These networks are referred to as **neural systems**. Neural systems are defined both by anatomy—the location of constituent neurons and circuits; by function—the class of information processed by a system, and how the system processes that information. Thus, a neural system comprises a specific set of interconnected brain regions with a unitary function. Neural systems serve broader purposes than the individual neural circuits they comprise: Neural systems are concerned with multiple aspects of discrete classes of information from the environment, integrating and interpreting that information with other types of information, and generating commands for observable behaviors: when an animal moves to perform tasks for survival (e.g., eating, drinking, evading predators), manipulates objects in its environment, or interacts with conspecifics (social interactions, including mating and parenting). The most general functional distinctions divide neural systems into **sensory systems** that acquire and process information from the internal and external environments (e.g., the visual system or the auditory system, both described in Unit II); and **motor systems** that respond to such information by generating movements (described in Unit III). There are, however, large numbers of neurons, circuits, and brain regions that lie between these relatively well-defined input and output systems. These are collectively referred to as **associational systems**, and they mediate the most complex and least well characterized brain functions (see Unit V).

The essential defining characteristic of all neural systems is unity of function. For example, the visual system is defined by all the neurons and connections primarily dedicated to detecting and analyzing light, the auditory system by those dedicated to sound, the somatosensory system by those dedicated to touch, the pyramidal motor system by those dedicated to voluntary movement. In many instances, a system's components are distributed throughout the body and brain. Thus, sensory systems include peripheral sensory specializations in eye, ear, skin, nose, and tongue, while motor systems include the peripheral motor nerves and synapses made on target muscles whose concerted contraction and relaxation execute various actions, and the autonomic nerves that innervate organs that will secrete signals in response to neural stimulation to maintain internal homeostasis. Accordingly, sensory and motor systems are composed of pathways that connect the entire body with neurons distributed across the entire PNS (sensory organs, sensory neurons, autonomic neurons, glands, and muscles) as well as the CNS (the spinal cord, brainstem, thalamus, relevant areas of the cerebral cortex, and other forebrain structures such as the basal ganglia). In contrast, associational systems are defined by highly interconnected neural circuits within the brain that have no direct inputs or outputs to or from the periphery.

Two other features of many neural systems are orderly *representation* of information at various levels, and a division of the function of the system into submodalities that are typically relayed and processed in *parallel pathways*. For neural systems that distinguish differences between neighboring points in external space—locations in visual space or different places on the skin surface—the representation of information is *topographic*. Thus, individual neurons in these systems respond selectively to particular points in the sensory fields. Such representations form **topographic maps** that reflect a point-to-point correspondence between the sensory periphery (the visual field projected on the retina or the body surface) and neighboring neurons within the central components of the system (in the spinal cord, thalamus, and multiple regions of the cerebral cortex). Motor systems also entail topographic representations of movements, although the direction of topographic information flow is from the CNS to the periphery.

Parallel pathways arise because virtually all neural systems have identifiable subsystems. The human visual system, for instance, has subsystems that emphasize stimulus characteristics such as color, form, or motion, with each class of information relayed and processed separately to some degree. Similar segregation of information subtypes into parallel pathways is apparent in other sensory and motor systems. For somatosensation, different subsystems distinguish between regular touch (e.g., detection of a texture at the fingertip, the laying of a hand on a shoulder) and pain (excessive force, heating, cooling, or rupture on the skin surface). For motor systems, the most dramatic division into parallel pathways is the segregation of control of external movements using skeletal muscles for the pyramidal, or corticospinal, pathway versus the movements that regulate the internal state of the organism (breathing, cardiovascular function, digestion) controlled by the autonomic pathway.

For neural systems where the representation and processing of information do not depend on discriminating neighboring points in a field—for example, hearing, smell, and taste—multiple stimulus attributes are still analyzed in an orderly way that facilitates the extraction and processing of essential information. These representations, many of which remain only partially understood, are collectively referred to as **computational maps**. In a computational map, the timing and pattern of inputs, the responses they elicit in neighboring cells, and the coordination of outputs are integrated to extract key information. Thus, computational maps in the auditory system represent sequences of sounds that form a word, or indicate the direction from which a siren is approaching. The organization of even more complex information such as perception, attention, emotions, multiple aspects of language, and memories is also unclear, and presumably involves processing that engages additional dimensions of information representation (see Unit V) beyond the relatively rudimentary level of topographic or computational maps in sensory and motor cortices.

Structural analysis of neural systems

Early neuroscientists made inferences of functional localization in the human brain (i.e., which region of the nervous system serves which function) by correlating behavioral deficits (e.g., selective sensory loss, paralysis of a specific extremity, difficulty speaking or understanding language) to damaged brain structures observed post mortem. This sort of information suggested the existence of what are now referred to as neural systems. Using structure to infer function based upon focal damage to different brain regions was soon adapted to experimental animals to more precisely define the localization of apparent systems dedicated to particular functions. Indeed, the foundations of the neuroscience of neural systems rest on observations made by purposefully damaging a brain region, nerve, or tract and observing a subsequent loss of discrete functions. Such **lesion studies** provided a framework for understanding the anatomy and likely flow of information in multiple neural systems.

Subsequent neuroanatomical studies defined neural systems with higher resolution based upon techniques that can trace connections of individual neurons or defined groups of neurons from their source to their termination (**anterograde**), as well as from terminus to source (**retrograde**), without first making damaging lesions in any brain region. Initially these techniques relied on injecting single neurons or a brain region with visible dyes or other molecules that could be taken up by local cell bodies and transported to axon terminals, or taken up by local axon terminals and transported back to the parent cell body (Figure 1.10A,B). Such tracers can also demonstrate an entire network of axonal projections in different regions of a neural system from a single group of nerve cells

(Figure 1.10C). Additional approaches use viral proteins or entire viruses, made innocuous via genetic engineering, that are readily taken up and transported to target cells (Figure 1.10D). When replication-competent (active but nontoxic) viruses are used, the viruses use the host cell to replicate and be released and can then be taken up by afferent terminals or target cells, thereby tracing circuitry beyond the direct target of any particular set of nerve cells. Together these approaches permit assessment of the extent of connections from a single population of nerve cells to their targets throughout the nervous system.

Analyses of connectivity have been augmented by molecular and histochemical techniques that demonstrate biochemical and genetic distinctions among nerve cells and their processes. Whereas conventional cell staining methods (see Figure 1.5) show differences in cell size and distribution, molecular labeling by *transgenic reporters* (see Concept 1.4) for specific genes (Figure 1.10E) and *antibody stains* to recognize specific proteins (Figure 1.10F) can further define the identity of subsets of neurons in a neural system and can be used in combination with the tracing methods described earlier. Probes for specific messenger RNA (mRNA) transcripts that detect the presence of that message and thus the expression of a particular gene, a technique called *in situ hybridization*, can also be used to define transcriptional states of neurons with distinct connections or additional molecular properties (Figure 1.10G,H). Thus, a variety of molecular and genetic engineering approaches now allow researchers to trace connections between molecularly defined populations of neurons and their targets.

Functional analysis of neural systems

Functional analyses—what neurons within a system do from moment to moment in a living animal—provide a complementary approach to understanding the organization of neural systems. A variety of physiological methods evaluate the electrical and metabolic activity of the neuronal circuits that make up a neural system. The most widely used electrophysiological method is single-cell, or single-unit, electrophysiological recording with microelectrodes (see Figure 1.7). Using a microelectrode, changes of electrical activity in individual neurons in the intact brain can be monitored over time and correlated with a wide range of environmental stimuli, internal states, or behavioral actions. The use of microelectrodes to record action potential activity allows cell-by-cell analysis of the organization of topographic maps (see earlier in this concept) and can give specific insight into the type of stimulus to which the neuron is "tuned" (i.e., the stimulus that elicits a maximal change in neuronal activity from a baseline state over a specific time period). Such tuning defines a neuron's **receptive field**—the region in sensory space (e.g., the body surface or a specialized structure such as the retina)

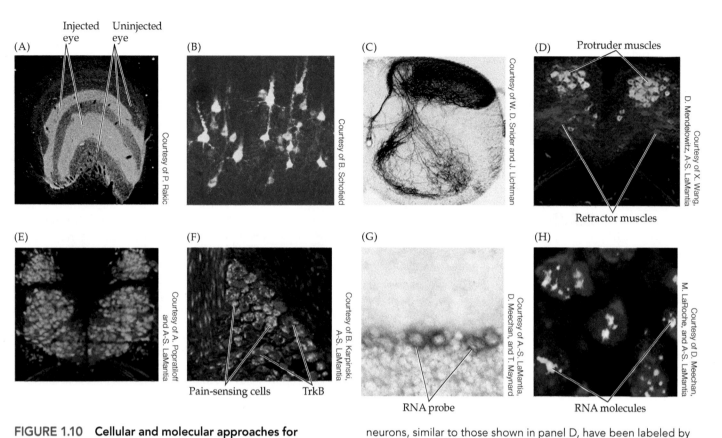

Injected eye   Uninjected eye

(A) Courtesy of P. Rakic

(B) Courtesy of B. Schofield

(C) Courtesy of W. D. Snider and J. Lichtman

(D) Protruder muscles   Retractor muscles   Courtesy of X. Wang, D. Mendelowitz, A.-S. LaMantia

(E) Courtesy of A. Popratiloff and A.-S. LaMantia

(F) Pain-sensing cells   TrkB   Courtesy of B. Karpinski, A.-S. LaMantia

(G) RNA probe   Courtesy of A.-S. LaMantia, D. Meechan, and T. Maynard

(H) RNA molecules   Courtesy of D. Meechan, M. LaRoche, and A.-S. LaMantia

FIGURE 1.10   Cellular and molecular approaches for studying connectivity and molecular identity of nerve cells (A–C) Tracing connections and pathways in the brain. (A) Radioactive amino acids can be taken up by one population of nerve cells (in this case, injection of a radioactively labeled amino acid into one eye) and transported to the axon terminals of those cells in the target region in the brain. (B) Fluorescent molecules injected into nerve tissue are taken up by the axon terminals at the site of the injection (the dark layers evident in the thalamus). The molecules are then transported, labeling cell bodies and dendrites of the nerve cells that project to the injection site. (C) Tracers that label axons can reveal complex pathways in the nervous system. In this case, a dorsal root ganglion has been injected with tracer that labels multiple axon pathways from the ganglion into the spinal cord. (D) Topographic projections can be resolved using fluorescent tracers of different colors (reflecting their excitation and emission wavelengths). Here, the hypoglossal motor neurons that innervate muscles at the front of the tongue that mediate protrusion, and the hypoglossal neurons that innervate muscles at the base that mediate retraction, have been labeled with a fragment of a cholera toxin protein that also is labeled by a green (protruder muscles) or red (retractor muscles) fluorescent molecule. (E–H) Molecular differences among nerve cells. (E) Hypoglossal motor neurons, similar to those shown in panel D, have been labeled by a transgenic reporter (green) that identifies neurons that express an enzyme involved in the metabolism of acetylcholine, the neurotransmitter used by all motor neurons. Thus, the hypoglossal motor neurons, arranged into the bilaterally symmetrical hypoglossal nuclei, are the larger group of reporter-labelled neurons toward the bottom of the panel. The dorsal motor nuclei of the vagus nerve, the smaller group of reporter-labelled cranial motor neurons toward the top of the panel is separated from the hypoglossal motor neurons by a broad unlabeled region. (F) Mechanosensory neurons in the trigeminal ganglion (a cranial sensory ganglion similar to a dorsal root ganglion) have been labeled by an antibody against TrkB (red), a receptor for a neurotropic molecule that acts selectively on these cells. In addition, a transgenic reporter has been used to label the pain-sensing (nociceptive) cells in the trigeminal ganglion green. (G) Purkinje cells in the cerebellum have been labeled with an RNA probe (blue) for a specific gene transcript that is expressed only by Purkinje cells. (H) Pyramidal neurons in layer 2 of the cerebral cortex have been labeled with an antibody that recognizes a molecule expressed only by these cells, and individual mRNA molecules of a transcript found in these neurons have been labeled as well (green dots primarily in the blue-labeled nucleus).

within which a specific stimulus elicits action potential response (Figure 1.11). The preferences of individual neurons to respond optimally to specific information can be identified for multiple aspects of a stimulus. Thus, a neuron in the visual system that responds to light in a distinct point of visual space, for example, can be further tested for preferences to color, motion, and so, on presented in

that spot. These physiological techniques can assess the dynamic properties of individual neurons across multiple brain regions that define a circuit. This view of receptive field organization throughout a system provides granular insight into how information is processed and movements are matched to assess environmental stimuli and respond to behavioral demands.

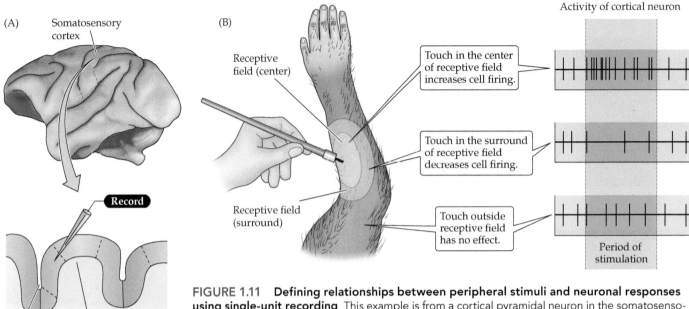

FIGURE 1.11 Defining relationships between peripheral stimuli and neuronal responses using single-unit recording This example is from a cortical pyramidal neuron in the somatosensory cortex of a rhesus monkey, showing the firing pattern in response to a specific peripheral stimulus—pressure applied to a discrete point on the skin surface—in the anesthetized animal. (A) Typical experiment setup, in which a recording electrode is inserted into the somatosensory cortex in the postcentral gyrus. (B) Defining neuronal receptive fields based on the location of the peripheral stimulus in sensory space (in this case, a particular region of the skin of the arm) that elicits the maximal response from the CNS neuron whose action potential activity is being recorded, referred to as the receptive field center. The flanking region, called the surround, is the region of the skin surface that when stimulated will actually diminish the response of the same neuron.

This way of understanding neural systems was pioneered by Stephen Kuffler and Vernon Mountcastle in the early 1950s and has been used by successive generations of neuroscientists to evaluate the relationship between peripheral stimuli (primarily sensory information) or actions (single movements or movement sequences) and neuronal responses. For neurons that are not concerned with topographic, point-by-point information, stimulus-selective responses can still be used to define receptive field–like properties that are used to categorize individual neurons. Thus, the frequency of tones categorizes "receptive field properties" of neurons in the auditory system, and the response to different odorant molecules can be used to define the preferences of individual neurons in the olfactory system. Electrical recording techniques at the single-cell level have been extended and refined to include simultaneous analysis of the individual responses of multiple neurons in animals performing sensory and motor as well as more complex tasks. It is also possible to make recordings of neuronal activity in intact animals performing multiple types of behaviors. Finally, intracellular recordings can be made to link receptive field properties to synaptic inputs and outputs, and patch electrodes can be used to detect and monitor the activity of the individual membrane molecules and the ions they let in or out of the neuron that ultimately underlie key aspects of neural signaling (see Chapters 2 and 3).

Organization of the human nervous system

The nervous system of humans, and all vertebrates, is divided into two anatomically defined subdivisions (Figure 1.12). The CNS comprises the **brain** (cerebral hemispheres, diencephalon, cerebellum, and brainstem) and the **spinal cord**. The PNS includes the sensory neurons that link sensory receptors on the body surface or deeper within it, as well as those for the special senses (vision, hearing, olfaction, taste), with relevant processing circuits in the CNS. The motor portion of the PNS in turn consists of two components. The motor axons that connect the brain and spinal cord to skeletal muscles make up the **somatic motor division** of the PNS, whereas the cells and axons that innervate smooth muscle, cardiac muscle, and glands make up the **visceral**, or **autonomic**, **motor division**. Neurons in the PNS are located in **ganglia**, which are local accumulations of nerve cell bodies and supporting cells. Peripheral axons are gathered into bundles called **nerves**, many of which are enveloped by Schwann cells that either myelinate these axons or provide a single glial covering that protects otherwise unmyelinated axons within peripheral nerves. A final distinction between the CNS and PNS reflects their development. The CNS is derived from neural stem cells in an embryonic region called the neural tube, while the PNS is derived primarily from a population of neural stem cells that migrate away from

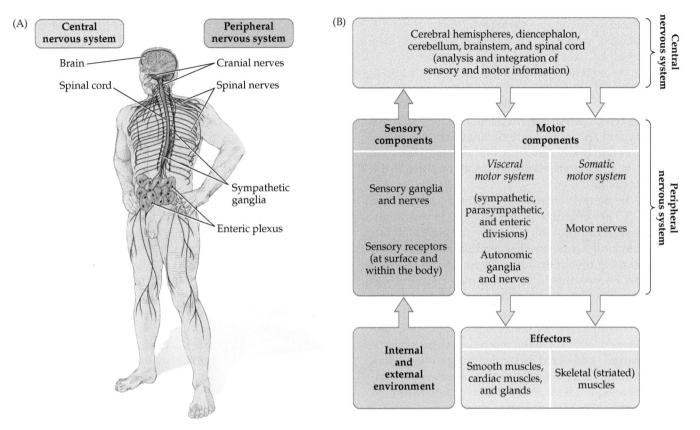

FIGURE 1.12   Major anatomical components of the nervous system and their functional relationships (A) The CNS (brain and spinal cord) and PNS (spinal and cranial nerves, autonomic ganglia, and enteric nervous system). (B) Diagram of the major components of the CNS and PNS and their functional relationships. Stimuli from the environment convey information to processing circuits in the brain and spinal cord, which in turn interpret their significance and send signals to peripheral effectors that move the body and adjust the workings of its internal organs.

the neural tube into the developing body called the neural crest (see Chapter 22).

Two anatomical terms, derived from early observations of fixed brain specimens without using microscopes, distinguish brain regions rich in neuronal cell bodies versus regions rich in axons. **Gray matter** refers to any accumulation of cell bodies and neuropil in the brain and spinal cord. **White matter** (named for its relatively light appearance, the result of the lipid content of myelin) refers to axon tracts and commissures. Within gray matter, nerve cells are arranged in two different ways. A local accumulation with neurons that have roughly similar connections and functions is called a **nucleus** (plural *nuclei*, not to be confused with the nucleus of a cell); such collections are found throughout the cerebrum, diencephalon, brainstem, and spinal cord. The nuclei of the thalamus and hypothalamus are clear examples of this organizational principle. In contrast, **cortex** (plural *cortices*) describes sheetlike arrays of nerve cells. The cortices of the cerebral hemispheres, the hippocampus, and the cerebellum provide the clearest examples of this organizational principle. Within the white matter of the CNS, axons are gathered into **tracts** that are more or less analogous to nerves in the periphery. Each tract contains axons that typically originate in the same gray matter structure, are organized in parallel, and often terminate in the same division of gray matter at some distance from their origin. Tracts that cross the midline of the brain, such as the corpus callosum that interconnects the cerebral hemispheres, are referred to as **commissures**. The sensory tracts of the dorsal spinal cord are referred to as **columns**.

The organization of the visceral motor division of the PNS (the nerve cells that control the functions of the visceral organs, including the heart, lungs, gastrointestinal tract, and genitalia) is a bit more complicated (see Chapter 21). Visceral motor neurons in the brainstem and spinal cord—the so-called preganglionic neurons—form synapses with peripheral motor neurons that lie in the **autonomic ganglia**. The peripheral motor neurons in autonomic ganglia innervate smooth muscle, glands, and cardiac muscle, thus controlling most involuntary (visceral) behavior. In the **sympathetic** division of the autonomic motor system, the ganglia lie along or in front of the vertebral column and send their axons to a variety of peripheral targets. In the **parasympathetic** division, the ganglia are found in or adjacent to the organs they innervate. Another component of the visceral motor system, called the **enteric system**, comprises small ganglia as well as individual neurons scattered throughout the wall of the gut. These neurons and their intrinsic axonal connections comprise vast neural networks that influence gut motility and secretion. More details about the physical structures and overall anatomy of the human nervous system can be found in the Appendix and the Atlas in the back of this book.

Finally, the neural systems that comprise the human PNS and CNS are organized hierarchically to accommodate the flow of information from environmental detection to behavioral action (see Figure 1.12B). Information is gathered from the environment through the sensory systems that encode this information into electrical signals and then relay it to so-called higher-order processing regions, particularly the cerebral cortex, for analysis and generation of commands for behavioral responses. These regions then send signals to the motor systems to modulate the internal state of the individual (organ functions, hormonal secretion) or to elicit coordinate action of an individual via a sequence of movements targeted toward the external world—what we recognize as behavior.

CONCEPT
1.4 | **The Genome Controls Brain Organization and Function**

LEARNING OBJECTIVES

1.4.1 Establish that the genome has all information necessary to build the nervous system.

1.4.2 Assess the percentage of the genome used to build the nervous system.

1.4.3 Analyze how genetic disruptions result in nervous system dysfunction and disease.

1.4.4 Define how the genome can be manipulated experimentally to analyze nervous system organization and function.

Genetics and genomics

The nervous system, like all other organs, is the product of expression of the genes encoded in the DNA within the nucleus of every neuron and glial cell. **Genetic analysis** is thus fundamental to understanding the structure, function, and development of organs and organ systems—including, and perhaps especially, the brain, because of its remarkable cellular diversity and the requirement for coordinating connections between identified neurons that form neural circuits and systems. The expression of these genes begins at the outset of embryogenesis (see Unit IV), and differences in expression—which genes are turned off, which are turned on, and the level of expression of those

genes that are turned on—underlie the genesis of all cell classes, tissues, and organs, including the brain. A **gene** comprises both *coding* DNA sequences (exons) that are the templates for mRNA that will ultimately be translated into a protein, and *regulatory* DNA sequences (promoters and introns) that control whether and in what quantities a gene is expressed in a given cell type (i.e., transcribed into mRNA). These components of genes that encode proteins make up a surprisingly small portion of the entire complement of DNA arranged into chromosomes within the nucleus of any cell. Indeed, they constitute only about 5% of the nuclear DNA sequence. The remaining 95% is presumed to serve regulatory functions that are both essential and at present are not fully understood. The advent of **genomics**, which focuses on the analysis of complete DNA sequences (both coding and regulatory) for a species or an individual, has thus provided insight into how nuclear DNA helps determine the assembly and operation of the brain and the rest of the nervous system.

Based on current estimates, the human genome comprises about 20,000 protein-coding genes, of which nearly 17,000 (approximately 84%) are expressed in the developing or mature human nervous system (Figure 1.13A). While subsets of these genes may also be expressed in non-neural cells and tissues in additional organs, there is nevertheless a substantial portion of the genome that must be engaged to build and maintain any individual's brain. Most "nervous system–specific" genetic information (where a gene is expressed and how it contributes to development or ongoing neural function) resides in the regulatory sequences and introns that control location, timing, quantity, variability, and cellular specificity of gene expression in the nervous system. Thus, despite the number of genes shared by the nervous system and other tissues, individual genes are regulated differentially throughout the nervous system, as measured by the amount of mRNA expressed from region to region and from one cell type to another (Figure 1.13B). Moreover, variable messages transcribed from the same gene, called **splice variants**, add diversity by allowing a single gene to encode information for a variety of related protein products. All of these differences play a part in the diversity and complexity of brain structure and function.

A dividend of sequencing the human genome, and the genomic analysis it has facilitated, is the realization that altered (mutated) genes, sometimes even one, can underlie neurological and psychiatric disorders. For example, mutation of a single gene that regulates mitosis can result in microcephaly, a condition in which the brain and head fail to grow and brain function is dramatically diminished (Figure 1.13C; also see Chapter 22). In addition to genes that disrupt brain development, mutant genes can either cause (or are risk factors for) degenerative disorders of the adult brain such as Huntington's and Parkinson's diseases.

(A)

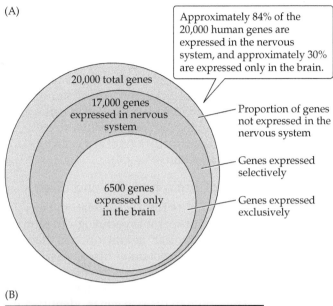

Approximately 84% of the 20,000 human genes are expressed in the nervous system, and approximately 30% are expressed only in the brain.

20,000 total genes

17,000 genes expressed in nervous system

6500 genes expressed only in the brain

— Proportion of genes not expressed in the nervous system

— Genes expressed selectively

— Genes expressed exclusively

(B)

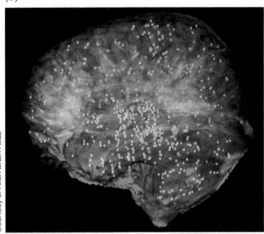

(C)

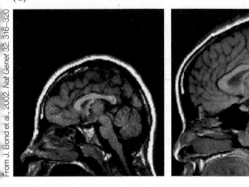

FIGURE 1.13 The genome and the nervous system (A) In this diagram of the human genome, nested circles indicate the comparative size of the human genome (approximately 20,000 genes that encode proteins, outer pink circle), the subset of those genes that are expressed in the developing or adult brain as well as other tissues (17,000 genes, middle purple circle) and the subset of those genes that are expressed only in the brain (6,000 genes; inner blue circle). (B) The locations and levels of expression of a single gene in the human brain. Dots indicate brain regions where mRNA for this particular gene is found, while their color (blue to orange) indicates the relative level (lower to higher) of mRNA detected at each location. (C) The consequence of a single-gene mutation for brain development. The gene, *ASPM* (*Abnormal SPindle-like Microcephaly-associated*), affects the function of a protein associated with mitotic spindles and results in microcephaly. In an individual carrying the *ASPM* mutation (left), the size of the brain is dramatically reduced and its anatomical organization is distorted compared with the brain of a typical control individual (right) of similar age and sex. (A data from S. K. Negi and C. Guda, 2017. *Sci Rep* 7: 897.)

The relationship between genotype and phenotype, however, is clearly not just the result of following genetic instructions, and genomic information on its own will not fully explain how the brain develops and operates, or how disease processes disrupt normal brain functions. To understand how the brain and the rest of the nervous system grow as an individual matures and works in health and disease, neuroscientists and clinicians must also understand the cell biology, anatomy, and physiology of the constituent cells, the neural circuits they form, how the structure and function of such circuits change with use across the life span, and the details of the behaviors that neural systems mediate. Whereas understanding the operating principles of most other organ systems has long been clear—thus allowing for more rapid integration of genomic, cellular, anatomical, and physiological analysis—this challenge has yet to be met for the nervous system, and in particular the human brain.

Genetic analysis of neural systems

The challenge of integrating genomics and the study of the brain is focused on how individual genes and gene networks map onto the cellular features of neurons and glia, their differentiation, and ongoing function in neural circuits and systems. Insights into this relationship between genes and the brain they build and maintain rely primarily on genetic analyses that associate single genes with single features of neurons, circuits, or systems. One approach examines families in which a brain malformation, behavioral distinction or disease appears to be genetic because it is inherited by offspring from their parents in a Mendelian fashion. **Mendelian analysis** (named for the pioneer geneticist Gregor Mendel, who intuited the existence of genes and mutations and their influence on specific biological traits in the late 1850s, long before DNA was identified)

Using genetics and genomics to understand diseases of the developing and adult nervous system permits deeper insight into the pathology, and raises the hope for gene-based therapies. It also indicates the extent to which networks of genes, influenced by key genes that define single nodes in those networks, ultimately control every aspect of brain development and function.

defines **mutations** in single genes (changes in their DNA sequences) as **recessive** or **dominant** based on their capacity to cause variation from a typical or frequent **phenotype** (a trait such as eye color or brain size) when two copies (recessive, one from each parent) or one copy (dominant, transferred from one parent to an offspring) of the mutant gene is present. When phenotypic variation caused by a mutation threatens viability, that mutation is considered to be *lethal*; mutations that disrupt homeostasis (health) are considered to be *disease-associated*.

In families in which both parents carry one copy of a disease-associated recessive gene, the presence of a non-mutant gene on the complementary chromosome of each parent prevents that mutation from causing disease. In 25% of the offspring of such parents, however, homozygosity—two copies in one individual—of the recessive gene will lead to disease. A gene-mapping strategy referred to as **homozygosity mapping** has been used to identify multiple genes associated with rare brain disorders, including the single gene, *ASPM*, that causes microcephaly because of the failure of neural stem cells to divide efficiently during brain development (see Figure 1.13C). Dominant genes, which act heterozygously, can also be identified using analysis of families. This has been the case for several dominant genes associated with brain diseases. These diseases are not lethal in utero or during early life, but nevertheless result in pathology at some point in the life span. If an individual who has a disease-associated dominant mutation reaches an age where reproduction is possible, the mutation can be passed to offspring, and there is a 50% chance those offspring will also have one copy of the disease-associated gene. Several well-known brain diseases: Huntington's disease, Parkinson's disease, and some forms of dementia can be explained, in part, by dominant mutations passed to offspring from carriers who will eventually develop the disease later in life.

An alternative approach to this single-gene analysis based upon transmission in individual families is statistical correlation of likely disease genes drawn from analyses of large cohorts of individuals with the same clinical diagnoses. The idea with **genome-wide association studies**, or **GWAS**, is that if a genetic variant occurs with a greater than random frequency in individuals with a clinically diagnosed condition such as Alzheimer's disease, schizophrenia, or autism, it probably contributes to that pathology. GWAS analyses also can detect a broad range of variations in the numbers of copies of individual genes or small chromosomal segments due to deletions, duplications, or translocations. These **copy number variants** are thought to influence pathogenesis based on altered amounts of gene products rather than anomalies in individual proteins encoded by the relevant genes. GWAS has been effective for identifying rare mutations and several copy number variants that are statistically highly predictive of brain diseases; however, most of the genetic associations identified by GWAS are not absolutely predictive: Some individuals with the mutation do not have the disease. Despite its limitations, GWAS assessment in large populations has identified multiple gene candidates that may contribute to brain pathology in several clinically defined neurological and psychiatric diseases. The identification of genes associated with diseases of the nervous system also provides insight into the genomic foundations of typical or optimal brain structure and function.

Genetic models of brain function and disease

To extend genetic insights from a correlation between a mutant gene and an observable pathology in humans (and therefore also the "typical" neural function that gene regulates), the biological function of disease-associated genes must be studied. Such studies are difficult, if not impossible, in developing or mature humans, especially for the brain. Thus, human disease genes, identified using Mendelian or GWAS methods, are modelled in experimental animals that have **orthologous genes** (identical or similar genes based on sequence and chromosomal location). Most analyses of the biological function of human disease genes have been done in mice, although some have been done in fruit flies (*Drosophila melanogaster*), a nematode worm (*Caenorhabditis elegans*), zebrafish (*Danio rerio*) (Box 1A). In all of these model organisms, complete genome sequences are available, and additional species-specific advantages can be exploited to generate clear explanations of gene function in healthy and diseased cells and tissues. In the fly and worm, "forward" genetic analysis can be used, in which flies or worms are randomly mutagenized using chemicals, ultraviolet light, or X-ray irradiation and the resulting phenotypes are evaluated. Mutants with genes orthologous to human genes are then identified based on phenotypes that parallel those in humans.

An alternative to mutagenesis is **genetic engineering**, or "reverse" genetics. This approach is used in several model species; however, for neuroscience, genetically engineered mice have become particularly important, as mutations that parallel those found in humans can be made in orthologous mouse genes (Figure 1.14A). The most commonly used techniques of genetic engineering rely on introducing a novel gene into the genome either in newly fertilized mouse zygotes (**transgenic mice** are made this way) or in mouse embryonic stem (ES) cells ("**knock-out**" and "**knock-in**" mice are made this way). In each instance, the murine genome is altered so that all cells in the mouse will carry the engineered gene and transmit it to its offspring. For transgenic mice, the engineered gene inserts randomly into the genome, while for knock-in or knock-out mice, the engineered gene is targeted to a specific site in the genome.

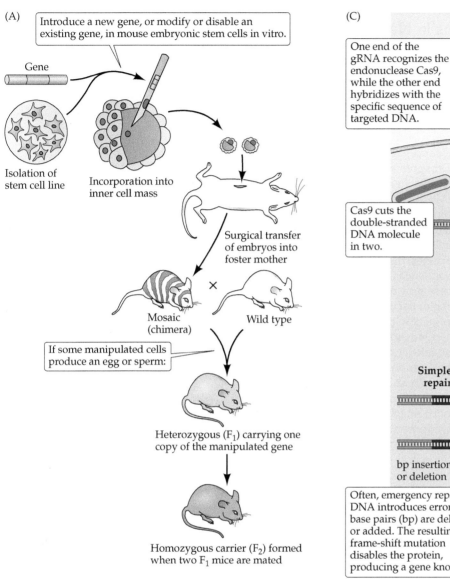

(A)

Introduce a new gene, or modify or disable an existing gene, in mouse embryonic stem cells in vitro.

Gene

Isolation of stem cell line

Incorporation into inner cell mass

Surgical transfer of embryos into foster mother

Mosaic (chimera) × Wild type

If some manipulated cells produce an egg or sperm:

Heterozygous (F_1) carrying one copy of the manipulated gene

Homozygous carrier (F_2) formed when two F_1 mice are mated

(C)

5′

gRNA

3′

Cas9

One end of the gRNA recognizes the endonuclease Cas9, while the other end hybridizes with the specific sequence of targeted DNA.

Into cell/zygote

Cas9

5′

3′

gRNA

Cas9 cuts the double-stranded DNA molecule in two.

Double-stranded break

Repair

Simple repair

Homology-directed repair

Re-combination

bp insertion or deletion → Frame shift → Premature stop codon

Homology

DNA fragment for insertion

Often, emergency repair of DNA introduces error; a few base pairs (bp) are deleted or added. The resulting frame-shift mutation disables the protein, producing a gene knockout.

If synthesized DNA is also introduced, it can hybridize with the two ends of the break and introduce a new sequence in between. In this way, a disease allele can be introduced into a model animal, or a dysfunctional gene in a human can be replaced with a functional version.

FIGURE 1.14 Genetic engineering to model the function of genes that influence nervous system development and function
(A) Creation of a homozygous carrier mouse for embryonic stem cells that have had a specific gene deleted, mutated, or engineered for conditional expression or deletion. (B) Conditional mutagenesis using the Cre/lox system, which in this case has been used to excise a genetic sequence of interest flanked by loxP elements. Upon cell class specific expression of the Cre recombinase, the loxP sites will be excised, a key exon deleted, and the flanking ends of the DNA joined. This approach is often used to completely eliminate a gene by preventing its mRNA from being transcribed, or translated into protein. (C) DNA editing using CRISPR-Cas9 can introduce specific mutations into a target gene. (A after T. A. Stewart and B. Mintz, 1981. *Proc Nat Acad Sci USA* 78: 6314–6318.)

(B)

In most cells: No recombination

Promoter for nestin

Cre recombinase

"Floxed" allele of androgen receptor gene

Exon 1   Exon 2   Exon 3

loxP binding sites for Cre recombinase

In cells that do not express Cre recombinase, the floxed gene is left intact.

In nervous system only (expressing nestin)

Promoter for nestin

Cre recombinase

Exon 2

Cre recombinase

Exon 1   Exon 3

Disrupted androgen receptor gene

Exon 1   Exon 3

The targeted gene is disrupted only in those cell types that express the Cre transgene.

■ BOX 1A | Model Organisms in Neuroscience

Much of modern neuroscience focuses on understanding the organization and function of the human nervous system, as well as the pathological bases of neurological and psychiatric diseases. These issues, however, and often difficult to address by studying the human brain alone; therefore, neuroscientists have relied on the nervous systems of other animals as a guide. A wealth of information about the anatomy, biochemistry, physiology, cell biology, and genetics of neural systems has been gleaned by studying the brains of a variety of species.

Often the choice of model species studied reflects the assumptions about enhanced functional capacity in that species; for example, from the 1950s through the 1970s, cats were the subjects of pioneering studies on visual function because they are highly "visual" animals, and therefore could be expected to have well-developed brain regions devoted to vision—regions similar to those found in primates, including humans. Much of what is currently known about human vision is based on studies carried out in cats. Studies on invertebrates such as the squid and the sea slug *Aplysia californica* yielded similarly critical insights into the basic cell biology of neurons, synaptic transmission, and synaptic plasticity (the basis of learning and memory). Both the squid and the sea slug were chosen because of certain exceptionally large nerve cells with stereotypic identity and connections that were well suited to physiological measurements. In each case, the advantages offered by these cells made it possible to perform experiments that helped answer key questions.

Biochemical, cellular, anatomical, physiological, and behavioral studies continue to be conducted on a wide range of animals. However, the complete sequencing of the genomes of invertebrate and vertebrate species, including mammals, has led to the informal adoption by many neuroscientists of four "model" organisms based on the ability to do genetic analysis and manipulation in each of these species. A majority of the genes in the human genome (approximately 84%) are expressed in the developing and adult nervous system (see Figure 1.13A). The same is true in the nematode worm *Caenorhabditis elegans*; the fruit fly *Drosophila melanogaster*; the zebrafish (*Danio rerio*); and the house mouse (*Mus musculus*)—the four species commonly used in modern genetics, and used increasingly in neuroscience. Despite certain limitations in each of these species, the availability of the complete sequences of their genomes facilitates research on a range of questions at the molecular, cellular, anatomical, and physiological levels.

One advantage of these model species is that the wealth of genetic and genomic information for each one permits sophisticated manipulation of gene expression and function. Thus, once an important gene for brain development or later function is identified, it can be specifically manipulated in the worm, fly, fish, or mouse. Large-scale screens of mutant animals whose genomes have been modified randomly by chemical mutagens allow investigators to search for changes from typical development structure and function (the phenotype) and to identify genes related to specific aspects of brain architecture or behavior. Similar efforts, although more limited in scope, have identified spontaneous or induced mutations in the mouse that disrupt brain development or function. In addition, manipulations that result in so-called *transgenic animals* permit genes to be introduced into the genome ("knocked in"), or to be deleted or mutated ("knocked out"), using the remarkable capacity of genomes to splice in new sequences that are similar to endogenous genes (see Figure 1.14). This capacity, referred to as *homologous recombination*, allows DNA constructs that disrupt or alter the expression of specific genes to be inserted into the location of the normal gene in the host species. These approaches

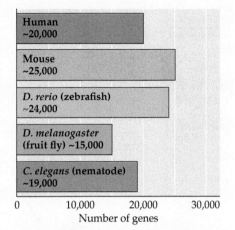

Estimated genome sizes of humans and several model species. Note that the number of genes in an organism's genome does not correlate with cellular or organismal complexity; the simple nematode Caenorhabditis elegans, for example, has almost the same number of genes as a human. Much genetic activity is dependent on transcription factors that regulate when and to what degree a given gene is expressed.

allow assessment of the consequences of eliminating or altering gene function.

Neuroscientists study the nervous systems and behaviors of other species as well, but with somewhat different aims. Crustaceans such as crayfish and lobsters and insects such as grasshoppers and cockroaches have been useful for discerning basic rules that govern neural circuit function. Avians and amphibians (chickens and frogs) continue to be useful for studying early neural development, and mammals such as rats are used extensively for neuropharmacological and behavioral studies of adult brain function. Finally, non-human primates (the rhesus monkey in particular) provide opportunities to study complex functions that closely approximate those carried out in the human brain. None of these species, however, is as amenable to genetic and genomic manipulations as are the four species mentioned above, each of which has made significant contributions to understanding the human brain.

Transgenic mice can express a wild type or mutant gene at higher levels under the control of a promoter sequence that is either broadly activated or activated in a specific cell class. Thus, one can assess the function of a particular gene when it is expressed at higher levels or in cells beyond those in which it is typically expressed. Transgenic mice can also express a **reporter protein**, usually a protein encoded by a gene that is not found in the mouse genome. Transgenic reporter proteins are expressed under the control of a promoter sequence that is selectively activated in a subset of cells. In such reporter transgenic mice (see Figure 1.10D), specific populations of neurons or glia can be identified based upon detection of the novel reporter protein. These reporter proteins include bacterial enzymes that can be visualized using histochemical techniques or endogenously fluorescent proteins from marine animals.

Knock-in and knock-out mice are made by **homologous recombination**. This approach relies on endogenous cellular mechanisms for DNA replication and repair. DNA polymerases and ligases can integrate a synthetic DNA sequence targeted by homologous 3′ and 5′ sequences into a specific region of the genome, thus substituting ("recombining") the exogenous DNA sequence (at a low frequency) for the sequence normally found at that location. ES cells that have undergone homologous recombination for the targeted gene sequence are then injected into the blastocyst of a newly fertilized in vitro mouse embryo, where these cells carrying the homologously recombined gene integrate into the developing embryo, including germ cells (embryonic cells that give rise to egg or sperm precursors). A knock-out gene can be a "null" mutation that eliminates the function of that gene. Knock-in genes can be targeted mutations that approximate disease-causing gene variants, or novel genes expressed in specific cell classes that do not usually express that particular gene.

Knock-in and knock-out mice can be engineered for **conditional mutations** using the **Cre/lox** system, in which an exogenous recombinase recognizes unique DNA excision sequences (loxP sequences) that are introduced at the 5′ and 3′ ends of an endogenous gene and eliminates the intervening sequence (Figure 1.14B). The loxP sequences are not found in mammalian genomes but occur in genomes of bacteria targeted by certain viruses. The viruses use a unique DNA cutting enzyme, called **Cre recombinase** (Cre stands for *Causes recombination*), to cut pieces of DNA out of the bacterial genome, and then recombine the cut ends. Cre recombinase is also not found in any vertebrate genome, so in applying the Cre/lox system to murine models, the gene encoding Cre recombinase must first be introduced into the mouse genome. The Cre insert is engineered so that it has DNA sequences on the 5′ and 3′ ends that are homologous to an endogenous mouse gene. During mitotic DNA replication, the Cre DNA gets recombined into the genome at that locus and is then expressed under the control of the promoter and other regulatory sequences for that gene. With expression of the Cre DNA, the resulting Cre recombinase engages the loxP binding sites, and the intervening endogenous exon targeted for elimination (the so-called floxed sequence) is excised. In a further refinement of this technique, Cre recombinase has been reengineered with a genetically modified estrogen receptor that cannot bind endogenous estrogen (the gonadal steroid) and can only be activated by an exogenous chemical (tamoxifen), a synthetic estrogen analogue. This approach, referred to as the Cre:ERT method (ERT stands for *estrogen receptor reengineered for tamoxifen* activation), allows for temporal control of recombination. Tamoxifen is given at the time during development or in the adult that one wishes to assess gene function, and the target gene is excised or activated by the Cre recombinase only at that time.

An even newer approach to genetic engineering uses CRISPR-Cas9 DNA editing, which allows specific mutations to be inserted into targeted genes (Figure 1.14C). CRISPR-Cas9 relies on a specific RNA guide sequence (gRNA) that combines with **tracrRNA** recognized by the bacterial Cas9 DNA excision and repair enzyme. This RNA–enzyme complex cleaves the DNA at the genomic location recognized by the guide sequence. Following Cas9 excision, the DNA is repaired by non-homologous end joining, yielding a microdeletion mutation. Alternatively, a donor DNA sequence can be inserted following Cas9 cleavage via a mechanism similar to homologous recombination. CRISPR-Cas9 gene editing provides opportunities to efficiently introduce a broad range of mutations into any gene, including multiple disease associated variants of the sort identified in GWAS analyses. The consequences of these modifications can then be studied to gain insight into how the mutation influences specific phenotypes in these genetically engineered mice.

Despite these remarkable techniques, identifying mutant genes responsible for diseases, and the dysfunction that specific mutations cause in neurons, neural circuits, and neural systems, has been much harder than anticipated. Indeed, it remains unclear whether targeted null mutations or knock-in mutations of disease variants in mice provide valid models of human diseases. Newer methods use human somatic cells (from adult tissues) that have been "reprogrammed" to become stem cells (see Unit IV). These induced pleuripotent stem cells can then be instructed to make neurons or brainlike aggregates (referred to as **organoids**) in cell culture systems, offer alternative approaches to assessing some aspects of disease-associated gene functions in cells and tissues that are derived from humans.

CONCEPT
1.5 Neural Circuits and Systems Can Be Analyzed in the Human Brain

LEARNING OBJECTIVES

1.5.1 Evaluate the utility of observing anatomy and physiology in the living human brain.

1.5.2 Identify key methods for visualizing brain structure and function in humans.

1.5.3 Discuss how the use of human brain imaging contributes to the study of typical and pathological behavior.

Imaging the living human brain

Until the late 1970s, most understanding of the structure and function of the human brain was derived from analyzing postmortem human specimens or inferred from animal studies. While these approaches were informative, the information they provided failed to convey the structural complexity of the human brain, let alone its physiology. Furthermore, most anatomical and functional correlations were based on the consequences of brain damage, raising additional uncertainties: When one part of a complex system is broken, it's hard to be sure that the deficits seen reflect what that one part does, versus a novel, if suboptimal, capacity of the altered system that remains. The advent of techniques for imaging the anatomical and functional details of the human brain in healthy living individuals as well as those with brain lesions caused by traumatic injury, surgical resection for tumors, strokes, and neurological or psychiatric diseases has produced observations that confirm many of the conclusions drawn from postmortem lesion analysis in humans as well as animal studies. In addition to confirming older conclusions in a more rigorous way, these approaches have opened new avenues for exploring how the brain carries out complex functions such as language, reading, math, music, and more. In parallel, human brain imaging has become indispensable for studying brain disease pathology as well as for clinical diagnostic purposes to localize lesions, plan surgical interventions, and monitor neurodegenerative changes. Thus, human brain imaging is now a key approach in the study of the human brain.

Minimally invasive analysis of human brain function

Approaches for studying the human brain in live, behaving humans vary based on technology as well as the type of information—physiological, anatomical, or both—and the resolution that can be realized. One common, minimally invasive approach provides information about broad changes in the overall electrical activity of the brain by recording from electrodes that are placed painlessly on the surface of the scalp to generate signals referred to

as an **electroencephalogram**, or **EEG**. The EEG is used clinically to diagnose seizure disorders (epilepsy) and to assess issues with sleep and wakefulness. It also provides an easy-to-administer, although low-resolution, approach to mapping brain activity in behaving humans of all ages. The advantage of the EEG is that it can be recorded using fairly small portable equipment and can be done without the supervision of a physician or health care professional. Methods have been developed using a "cap" with scalp EEG electrodes placed in an ordered spatial array across the head to obtain a map of changes in electrical activity across the brain (Figure 1.15A). This approach, referred to as **event-related potential** analysis, or **ERP**, uses neither radioactivity (used for CT and PET scans; see the following sections) or strong magnetic fields (used for MRI and fMRI, which require dedicated, heavy equipment and significant computational resources). Instead, the net electrical activity from each point in the scalp electrode array is detected, amplified, and mapped digitally with reference to each electrode's position on the head (Figure 1.15B). Individuals can be presented with sensory stimuli or directed to execute a motor task, and time-locked EEG signals from each collecting electrode can then be averaged with respect to stimulus duration or task onset. ERPs can be recorded from adults as well as children, facilitating activity-based analysis of developmental and behavioral changes. These analyses allow, indirectly, for general localization during the performance of different tasks by examining anterior–posterior or medial–lateral differences in EEG activity across the inferred brain surface (see Figure 1.15B). While ERP lacks the ability to define specific cortical areas, let alone structures that are deeper in the brain, its relative ease of use makes it possible to perform experiments in standard laboratory settings on a broad range of individuals.

A noninvasive approach for transiently (and harmlessly) *modifying*, rather than mapping, local human brain activity has been introduced into both clinical practice and physiological and behavioral research. **Transcranial magnetic stimulation (TMS)** (Figure 1.15C) uses magnetic pulses delivered by a paddlelike device held at different positions near the scalp. When the magnetic pulses are delivered locally in this way, activity in the underlying cortical tissue (and potentially deeper structures) is briefly disrupted, leading to a transient change in behavioral performance. This transient "lesion" of activity causes no detectable harm to patients or to research volunteers. TMS can correlate the likely role of broad regions of the brain (adjacent to the location of the stimulator above the head) with specific ongoing behaviors. It is also potentially useful for "resetting" disrupted patterns of activity that lead to behavioral deficits associated with brain diseases—indeed, it has been used successfully to treat some forms of major depression as well as migraine headaches for nearly a decade. Additional

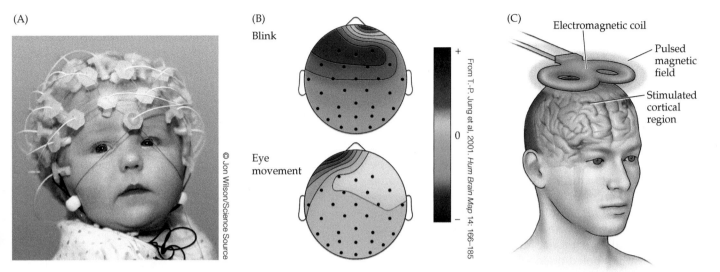

(A)

© Jon Wilson/Science Source

(B)

Blink

Eye
movement

+

0

−

From T.-P. Jung et al. 2001. *Hum Brain Map* 14: 166–185

(C)

Electromagnetic coil

Pulsed
magnetic
field

Stimulated
cortical
region

FIGURE 1.15   Analyzing brain function using noninvasive neurophysiological approaches (A) Event-related potential (ERP) recording in an awake, alert child. The array of scalp electrodes in the "cap" indicates the recording locations of individual electroencephalographic (EEG) traces. (B) The response intensity of each of the EEG traces recorded at each individual electrode in the scalp cap can be compared computationally to generate a low-resolution map of localized cortical activity. The blue regions indicate areas of lower intensity responses, while the yellow and red indicate regions of higher intensity responses to a single eye blink (top) or a more sustained eye movement (bottom). (C) Transcranial magnetic stimulation (TMS) relies on a handheld device that generates a magnetic field based on current flow through a magnetic coil. The device can deliver a pulse of current to the surface of the cerebral cortex, causing a brief disruption of electrical activity in that area. TMS has gained some acceptance as a clinical treatment for depression and other mood disorders, as well as being used to assess normal function.

therapeutic applications of TMS are under active investigation for a variety of neurological and neuropsychiatric conditions. TMS has also been adapted for use in standard nonclinical laboratory settings, often in combination with ERP analysis, to assess and then modify typical activity patterns during complex behavioral tasks. Thus, TMS provides an approach to manipulate local function of the human brain to gain further insight into relationships between structure, physiology, and behavior in typical individuals and those with a variety of brain disorders.

Radiographic methods for brain imaging

With the advent of radiologic imaging—beginning in the late 1800s (!)—efforts to use X-rays to visualize the human brain were at best only modestly successful. The soft tissue of the brain provided little contrast, and the most useful images were of the vasculature after injection of X-ray opaque dyes, a fairly invasive procedure with significant risk to the patient. These limitations changed dramatically in the 1970s when **computerized tomography,** or **CT,** ushered in a new era in noninvasive imaging using X-irradiation, sensitive detector arrays, and computer processing technology to visualize the living brain in three dimensions. CT uses a narrow X-ray beam and a row of very sensitive detectors placed on opposite sides of the head to probe small portions of tissue with limited radiation exposure over a series of brief pulses (Figure 1.16A). In order to form an image, the X-ray tube and detectors rotate around

the head, collecting radiodensity information from every orientation around a narrow slice. Computer processing techniques then calculate the radiodensity of each point within the slice plane, producing a tomographic image (Greek *tomo*, "cut" or "slice"). If the individual is moved through the scanner slowly while the X-ray tube rotates in this manner, a three-dimensional radiodensity matrix can be created, allowing images corresponding to serial sections to be computed for any plane through the brain. CT scans can readily distinguish gray matter and white matter, differentiate the ventricles quite well, and show many other brain structures with a spatial resolution of several millimeters. Thus, major anatomical structures can be identified with relative confidence (Figure 1.16B), and lesions can be recognized if they are within the limits of the CT resolution (a few millimeters with newer techniques). CT has been particularly valuable for clinical diagnoses: Brain lesions that are not visible using standard X-ray methods that produce low-resolution, one-dimensional images can be resolved using CT. For example, metastatic lesions in the brain (tumor cells that migrate from their origin to distal target tissues) related to a malignancy in another tissue can be localized fairly precisely, and can be correlated more definitively with functional loss as well as with subsequent response to chemotherapeutic, radiological, or surgical treatments (Figure 1.16C). CT remains a valued diagnostic tool; however, its use in fundamental brain research on healthy individuals is limited because of the risks of unnecessary radiation exposure and the relatively low resolution of brain structure.

Radioactive signals have been used with a different method to image the dynamic physiology of the human brain in individuals performing behavioral tasks. **Positron emission tomography**, or **PET**, is based on the capacity of highly active cells in any organ to use specific metabolites to support demands of transiently increased function. In PET scanning, an individual is injected intravenously with a radiolabeled metabolite that will be taken up into active cells and emit a radioactive signal that can be monitored by specialized detectors. The transient regional increase of uptake and emission during the decay of the radiolabeled compound can be associated with locally increased function. Individuals injected with the radiolabeled metabolite probe and placed in a PET scanner can be presented with behavioral tasks. The regional change in uptake of the probe (often a radiolabeled glucose analogue that can detect locally increased cellular metabolism) during performance of a task can be measured and mapped onto a brain image obtained using either CT or MRI (see the next section) to infer functional localization. PET can also be used to detect differences in brain function in several neurodegenerative diseases, including Parkinson's disease, where the uptake of a radiolabeled metabolic analogue of dopamine (whose availability is reduced in the brains of individuals with Parkinson's disease) can be compared with that of individuals without the disease (Figure 1.16D). The utility of the dynamic physiological mapping and molecular specificity possible with PET is balanced by its technical demands and risks. Radiolabeled ligands for PET must be synthesized immediately before scanning (because of their rapid radioactive decay), requiring an on-site cyclotron. In addition, the detection instrumentation, and the risks of injection of radioligands limits the feasibility of PET in typical individuals as well as those with known diseases. Nevertheless, prior to the development of less invasive functional imaging approaches, PET imaging

provided some of the most compelling evidence for regional differences in brain activation that confirmed or extended inferences from postmortem lesion analyses in individuals with brain damage.

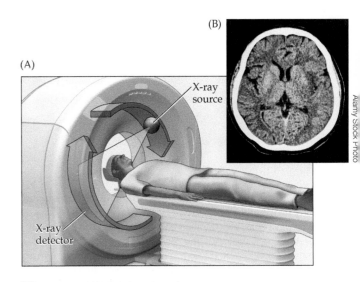

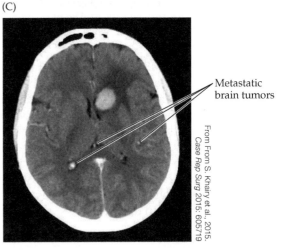

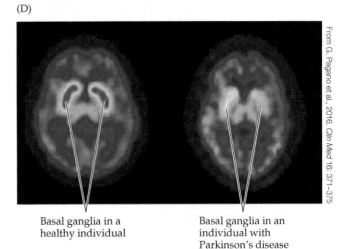

FIGURE 1.16 Radiological approaches for imaging the living human brain (A) In computerized tomography (CT), the X-ray source and detector are moved around the individual's head. (B) Horizontal CT section of a typical adult brain. (C) CT scan of an individual with multiple sites of a metastatic brain tumor (white spots throughout the cortical gray and white matter). CT scans are very useful in detecting brain lesions where the damage has a different tissue density than the normal brain tissue. (D) Positron emission tomography (PET) scan of a typical individual (left) who has been injected with a radiolabeled isotope of a metabolic analogue of the neurotransmitter dopamine, which is taken up by cells in the basal ganglia (focal red regions). An individual with Parkinson's disease who has been injected with the same radiolabeled isotope (right) shows diminished dopamine localization as a result of the disease.

Magnetic resonance imaging

Understanding human brain structure and function in typical individuals and those with brain disease took a huge step forward in the 1980s with the development of **magnetic resonance imaging (MRI)** for biomedical applications. Unlike CT, MRI is based on the physics of relatively benign atomic motion intrinsic to brain tissue rather than exposure to X-rays or injection of radioactive compounds to visualize the brain. The nuclei of some atoms act as spinning magnets. If placed in a strong magnetic field, these atoms will line up with the field and spin at a frequency that is dependent on the field strength. If a brief radiofrequency pulse tuned to the atoms' spinning frequency is applied, the atoms are knocked out of alignment with the field and subsequently emit energy in an oscillatory fashion as they gradually realign themselves with the field. The strength of the emitted signal depends on how many atomic nuclei are affected by this process. These oscillations can be recorded by sensitive detectors and the records computationally analyzed to generate high-resolution three-dimensional images of living tissue.

In an MRI scanner (Figure 1.17A), the magnetic field is distorted slightly by imposing magnetic gradients along three different spatial axes so that only nuclei at certain locations are tuned to the detector's frequency at any given time. Almost all MRI scanners use detectors tuned to the radio frequencies of spinning hydrogen nuclei in water molecules, creating images based on the distribution of water in different tissues (Figure 1.17B). Safety (there is no high-energy radiation), noninvasiveness (no dyes are injected), and versatility (applicable to individuals in a variety of conditions) have made MRI the technique of choice

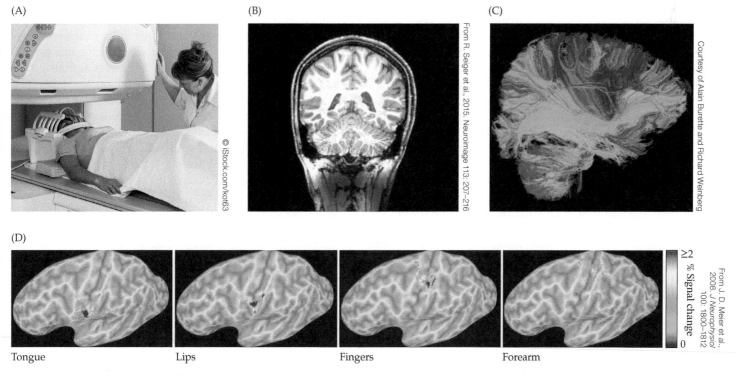

(A) © iStock.com/kot63

(B) From R. Seiger et al., 2015. NeuroImage 113: 207–216

(C) Courtesy of Alain Burette and Richard Weinberg

(D) ≥2 ... % Signal change ... 0   From J. D. Meier et al., 2008. J Neurophysiol 100: 1800–1812

Tongue   Lips   Fingers   Forearm

FIGURE 1.17   Anatomical and functional magnetic resonance imaging (A) The MRI scanner has a portal for the individual's head (or other region of the body to be imaged). A magnetic coil is placed around the head to activate and record magnetic resonance signal. Digital screens, virtual reality goggles, or earphones can be used to present visual or auditory stimuli. (B) MRI image showing the white matter of the cerebral cortex as white and the gray matter as gray, generated using an imaging method that detects changes in the distribution of water molecules in different tissues based on their magnetic properties. (C) Diffusion tensor imaging (DTI), a variant of MRI, identifies major axon pathways in the brain by detecting similarities in the direction of water molecules within specific regions of the magnetic field. This information is then assembled into an image that indicates the degree of alignment of water molecules in the tissue. Maximally aligned molecules (shown in different colors, depending upon shared directions) define regions where major axon tracts are found in the cortical white matter and other brain regions. (D) fMRI mapping of physiological activation in the human motor and sensory cortices during movements of different body parts. Focal activity, expressed as percentage change from rest (scale at far right), is indicated in colors from blue (no change) to red (maximal change) on an MRI image of the lateral surface of the brain of a single individual. The motor cortex is to the left of the large central sulcus (the deep groove in the middle of each image) and the sensory cortex is to the right. Motor and sensory cortical activation is topographically mapped, and the focal activation for each body part (red; showing maximal percentage change from the resting state) changes from ventral, where the face (tongue and lips) is represented, toward the dorsal, where the fingers and then the forearm are represented when this individual is asked to move each body part.

for imaging brain structure, especially in typical individuals without known brain disease, in whom the risks versus benefits of more invasive methods remain uncertain. The magnetic field and radiofrequency pulses used in scanning are harmless (although ferromagnetic objects in or near the scanner are a safety concern).

MRI can be used in a variety of ways to generate different anatomical views of a living brain. By changing the scanning parameters, images based on a wide variety of different contrast mechanisms can be generated. This results in remarkably detailed images that show structural features of the human brain (see Figure 1.17B) and, for several important clinical conditions, the presence of pathology. Another advantage is that, like CT scans but with far better resolution, MRI data from each individual represent the equivalent of a "whole brain" image in three dimensions. Software has been developed so that from one detailed scan, it is possible to create detailed views in all the cardinal planes of section for the brain (see Appendix Figure A1), as well as three-dimensional renderings of volumes, such as those of the cortical surface or the intracerebral ventricles.

The alignment of the magnetic fields of water molecules in axon tracts also makes it possible to visualize axon pathways using a variant of MRI referred to as **diffusion tensor imaging (DTI)** (Figure 1.17C). DTI can establish differences in major axon pathway connectivity, making it possible to study individuals with genetic disorders that result in major alterations of axon projections (see Clinical Applications, Chapter 23). Additional settings can generate images in which gray matter and white matter are relatively invisible but the brain vasculature stands out in sharp detail. Thus, variations of MRI can clearly and safely visualize human neuroanatomy, including cerebral cortical regions, some subcortical structures, axon tracts, and cerebral vasculature.

Functional brain imaging

Imaging specific functions in the brain has become possible with the development of techniques for detecting local changes in cerebral metabolism or blood flow that can be measured by scanners that detect changes in magnetic fields. To conserve energy, the brain regulates its blood flow such that active neurons with relatively high metabolic demands receive more oxygen and nutrients than do relatively inactive neurons. Two noninvasive approaches (versus far more invasive options such as PET) have emerged to assess regional differences in activity in the brain, either during specific behavioral tasks in typical individuals as well as those with brain diseases. The first (and most commonly used) is **functional magnetic resonance imaging (fMRI)**. fMRI has all of the advantages of safety enumerated for structural MRI imaging and can be done using the same scanners. Currently considered the safest approach for imaging activity correlated with behavior in the living brain, fMRI can be done with sufficient spatial resolution so that distinct brain regions can be associated with specific behaviors. Because fMRI signals rely on averaging over time, however, this approach lacks temporal resolution.

fMRI relies on the fact that hemoglobin in blood slightly distorts the magnetic resonance properties of hydrogen nuclei in its vicinity. The amount of magnetic distortion changes depending on whether the hemoglobin has oxygen bound to it, thus providing a dynamic signal in brain regions where blood flow and oxygen use increase. When a brain area is activated by a specific task, it begins to use more oxygen, and within seconds the brain microvasculature responds by increasing the flow of oxygen-rich blood to the active area. These changes in the concentration of oxygen and blood flow lead to localized **blood oxygenation level–dependent (BOLD)** changes in the magnetic resonance signal. Such fluctuations are analyzed using image-processing techniques to produce maps of active brain regions during performance of specific movements or tasks (Figure 1.17D). Because fMRI uses signals intrinsic in the brain, tracer injections are not necessary, and repeated observations can be made on the same individual in the same scanning session or over multiple sessions—a major advantage over imaging methods such as PET. The spatial resolution (2 to 3 mm) and temporal resolution (a few seconds) of fMRI are superior to those of PET, and can be combined with structural images generated by MRI using computational and statistical methods. fMRI has thus emerged as the technology of choice for functional imaging of the human brain. Its uses extend from clinical characterization of functional change in disease states like depression, to the networks of brain regions active in healthy individuals performing complex, distinctively human behaviors like language. The necessity of averaging fMRI signals over seconds, plus the statistical assumptions necessary to fit highly variable individual brains into standardized neuroanatomical templates, imposes some significant limitations, but these have not reduced the use of fMRI for studies of functional activation of brain regions while performing a broad range of behaviors as well as changes in brain activity associated with multiple brain diseases.

The second noninvasive approach that has emerged for assessing regional differences in brain activity is **magnetoencephalography (MEG)**. MEG has better temporal resolution than fMRI and thus is sometimes used to map changes in brain function with millisecond resolution in tasks requiring highly dynamic processing rather than the lower temporal resolution of seconds afforded by fMRI. MEG's comparatively lower spatial resolution, however, limits its utility in many assessments of structure and function. As its name suggests, MEG records the magnetic consequences of brain electrical activity rather than the electrical signals themselves. For MEG recording, an array of individual detector devices called SQUIDs (*superconducting quantum interference devices*) is arranged as a helmet and fitted onto the individual. The individual is then placed in a biomagnetometer scanner that amplifies the signals detected by each SQUID to reconstruct an image (Figure 1.18). Thus, unlike EEG, MEG detects independent sources of current flow, without reference to other currents. The magnetic signals that MEG recordings detect are local, and the resolution can be as accurate as a few millimeters around the signal origin—sufficient for basic localization but not as precise as fMRI. Thus, one can detect dynamic electrical activity in the brain with temporal resolution that approximates the key events in neuronal electrical signaling (i.e., action potentials and synaptic potentials) and identify the general location of the source of that activity in the brain.

These signals, like the far lower resolution signals collected from EEG using ERP approaches, provide a map of current sources across the brain, using reference points (usually the ears and nose) to create a three-dimensional space. Given its millisecond (or even faster) temporal resolution combined with its spatial resolution, MEG can be used to evaluate local activity changes over time in individuals performing a variety of discrete, rapid tasks. This allows for comparison of baseline activity before task initiation, changes during task performance, and activity after task completion (see Figure 1.18). In addition, MEG can be used to map the temporal characteristics as well as the brain localization of epileptic foci, reducing the need for intrasurgical brain surface electrode mapping. Even though MEG has reasonable spatial resolution, MEG maps alone often lack sufficient anatomical detail for some applications. Thus, MEG is often combined with structural MRI (see Figure 1.17B), a combination referred to as **magnetic source imaging**, or **MSI**.

In sum, the use of modern structural and functional imaging methods has revolutionized human neuroscience. It is now possible to obtain images of the developing brain as it grows and changes, and of the living brain in action, assessing brain activity both in typical individuals and in individuals with neurological disorders.

Analyzing complex behavior

Many advances in modern neuroscience have involved reducing the complexity of the brain to more readily analyzed components—cells, circuits, genes, and molecules. Nevertheless, more complex brain functions, especially those that seem particularly human, such as perception, language, emotions, memory, and consciousness, remain a challenge, and a topic of fundamental curiosity, for

(A)

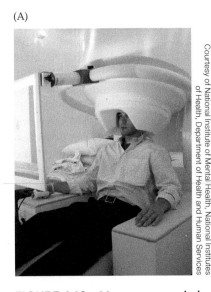

Courtesy of National Institute of Mental Health, National Institutes of Health, Department of Health and Human Services

(B)

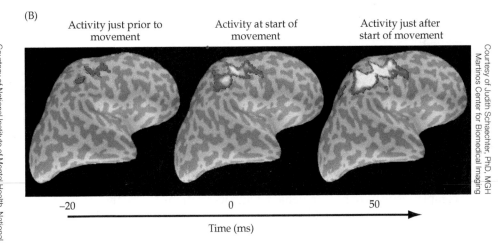

Courtesy of Judith Schaechter, PhD, MGH Martinos Center for Biomedical Imaging

FIGURE 1.18 Magnetoencephalography (MEG) provides greater temporal resolution than fMRI (A) The individual is fitted with a helmet that includes several magnetic detectors (a SQUID array) and then is placed in a biomagnetometer (the large cylindrical structure) that can amplify the small local changes in magnetic field orientation or strength that indicate fast temporal current flow changes in ensembles of neurons. MEG, like MRI and fMRI, is noninvasive. (B) The temporal resolution of MEG permits millisecond resolution of electrical activity in the human brain before, during, and after performance of a variety of tasks (displayed in color over a structural MRI of the same brain that has been "inflated" to reveal cortex folded into sulci and fissures). In this case, a digit was moved in response to a specific sensory cue. There was some "anticipatory" activity in the region of the somatosensory as well as motor cortices where the digit was represented. As the movement began, this baseline activity increased so that a wider region of both the somatosensory and motor cortices was activated. As the movement proceeded, the activity increased over the time of performance.

contemporary neuroscientists. Over the last 40 years or so, a new field called **cognitive neuroscience** has emerged that is specifically devoted to understanding these issues (see Unit V). Cognitive neuroscience focuses on human behaviors and their underlying structural and functional representation in the human brain. Much of cognitive neuroscience focuses on human behaviors analyzed in a variety of typical individuals as well as those with brain disorders. Nevertheless, animal models—especially non-human primates—can also be used to gain insight into complex behaviors and their underlying neural circuits and systems. These analyses provide complimentary assessments of key questions in cognitive neuroscience. One approach to cognitive neuroscience is to design and validate specific behavioral tasks that assess quantitatively complex information processing and behavior in humans or model animals. These tasks can be used in humans and analyzed based on correct versus incorrect responses, numbers of trials needed to learn the task, or the reaction time between the presentation of a stimulus and the individual's response. The development of these tasks can also engage a range of statistical models that capture more subtle aspects of behavior.

Cognitive neuroscience has grown in its scope by combining these sophisticated behavioral approaches with functional imaging of the human brain. Many of the complex behavioral tasks that probe human cognitive functions can be adapted for presentation when the individual is wearing an EEG cap, or placed in an MRI scanner. Using these combined approaches, investigators can evaluate brain regions that are active when individuals are engaging in tasks that involve language, mathematics, music, aesthetics, and even abstract thinking and social appraisals. Carefully constructed behavioral tasks can also be used to study the functional pathology of complex neurological disorders that compromise cognition, such as Alzheimer's disease, schizophrenia, and depression. Although there is clearly a long way to go, these increasingly powerful approaches are beginning to unravel even the most complex aspects of human behavior and its relationship to neural circuits and systems.

Summary

The structure and function of the brain can be understood at multiple levels based on the questions that need to be answered and the technical approaches used. The foundation of all neuroscience rests on the recognition that neurons and glial cells are the essential structural and functional units of a nervous system. The organization of neurons and glia into neural circuits that relay information from neuron to neuron via synaptic connections provides the substrate for processing specific aspects of sensory, motor, and cognitive information. The aggregation of these circuits into neural systems, whose components are often distributed across multiple brain regions as well as in distinct regions of the body, allows for more complex analysis of information. The physiological characterization of the neurons and glia in neural circuits and systems, particularly the detailed analysis of the electrical properties of individual neurons, can be recorded with electrodes placed near or inserted inside the cell. These recordings provide insight into the dynamic nature of sensory maps, motor commands, and the integration of complex information to guide behavior. Genetic analysis, made possible by the availability of the full human genome sequence as well as similar sequences in animal models, has begun to provide an outline of the genetic instructions necessary to generate neurons and glia, assemble them into a nervous system, and maintain their function. To gain more insight into complex capacities in the human brain, minimally invasive and noninvasive imaging methods can be used in living humans. N-invasive imaging can used in to individuals with brain damage or disease to address questions of how disrupted brain function underlies altered behavior and to diagnose specific brain illnesses. These methods can help integrate observations from animal models and human postmortem analysis with structure and function in the living human brain. Among the goals that remain are understanding how basic molecular genetic phenomena are linked to cellular, circuit, and system functions; understanding how these processes go awry in neurological and psychiatric diseases; and understanding the especially complex cognitive functions of the brain that make us human.

■ Additional Reading

Reviews

Baker, M. (2014) Gene editing at CRISPR speed. *Nat. Biotechnol.* 32: 347–355.

Dennis, C., R. Gallagher and P. Campbell (Eds.) (2001) Special issue on the human genome. *Nature* 409: 745–964.

Ferriera, F.R.M., Noguiera, M.I., and J. DeFelipe (2014) The influende of James and Darwin on Cajal and his research into the neuron theory and evolution of the nervous system. *Front. Neuroanat.* 8:1.

Jasny, B. R. and D. Kennedy (eds.) (2001) Special issue on the human genome. *Science* 291: 1153.

Kim, H. and J.-S. Kim (2014) A guide to genome engineering with programmable nucleases. *Nat. Rev. Genet.* 5: 321–334.

Kravitz, A. V. and 6 others (2010) Regulation of parkinsonian motor behaviours by optogenetic control of basal ganglia circuitry. *Nature* 466: 622–626.

Navarrete, M., and A. Araque (2014) The Cajal school and the physiological role of astrocytes: a way of thinking. *Front. Neuroanat.* 8: 33.

Negi, S.K. and C. Guda (2017) Global gene expression profiling of healthy human brain and its application in studying neurological disorders. *Sci. Rep.* 7: 897.

Ohki, K. and 4 others (2005) Functional imaging with cellular resolution reveals precise micro-architecture in visual cortex. *Nature* 433: 597–603.

Raichle, M. E. (1994) Images of the mind: Studies with modern imaging techniques. *Annu. Rev. Psychol.* 45: 333–356.

Shepherd, G.M. and S.D. Erulkar (1997) Centenary of the synapse: from Sherrington to the molecular biology of the synapse and beyond. *Trends Neurosci.* 20: 385 – 392.

Wheless, J. W. and 6 others (2004) Magnetoencephalography (MEG) and magnetic source imaging (MSI). *Neurologist* 10: 138–153.

Zhang, F. and 12 others (2011) The microbial opsin family of optogenetic tools. *Cell* 147: 1446–1457.

Books

Brodal, P. (2010) *The Central Nervous System: Structure and Function*, 4th Edition. New York: Oxford University Press.

Gibson, G. and S. Muse (2009) *A Primer of Genome Science*, 3rd Edition. Sunderland, MA: Sinauer Associates/Oxford University Press.

Huettel, S. A., A. W. Song and G. McCarthy (2009) *Functional Magnetic Resonance Imaging*, 2nd Edition. Sunderland, MA: Sinauer Associates/Oxford University Press.

Oldendorf, W. and W. Oldendorf, Jr. (1988) *Basics of Magnetic Resonance Imaging*. Boston: Kluwer Academic Publishers.

Peters, A., S. L. Palay and H. de F. Webster (1991) *The Fine Structure of the Nervous System: Neurons and Their Supporting Cells*, 3rd Edition. New York: Oxford University Press.

Posner, M. I. and M. E. Raichle (1997) *Images of Mind*, 2nd Edition. New York: W. H. Freeman.

Purves, D. and 6 others (2013) *Principles of Cognitive Neuroscience*, 2nd Edition. Sunderland, MA: Sinauer Associates/Oxford University Press.

Ramón y Cajal, S. (1990) *New Ideas on the Structure of the Nervous System in Man and Vertebrates.* (Transl. by N. Swanson and L. W. Swanson.) Cambridge, MA: MIT Press.

Ropper, A. H. and N. Samuels (2009) *Adams and Victor's Principles of Neurology*, 9th Edition. New York: McGraw-Hill Medical.

Schild, H. (1990) *MRI Made Easy (...Well, Almost).* Berlin: H. Heineman.

Schoonover, C. (2010) *Portraits of the Mind: Visualizing the Brain from Antiquity to the 21st Century.* New York: Abrams.

Shepherd, G. M. (1991) *Foundations of the Neuron Doctrine.* History of Neuroscience Series, no. 6. Oxford, UK: Oxford University Press.

UNIT I
Neural Signaling

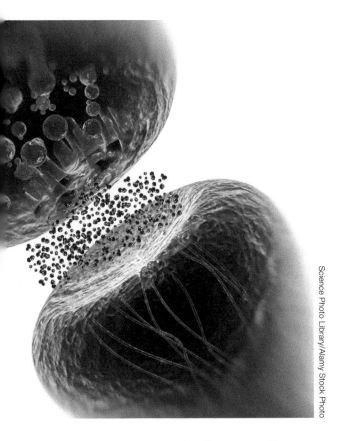

Science Photo Library/Alamy Stock Photo

The brain is remarkably adept at acquiring, coordinating, and disseminating information about the body and its environment. Such information must be processed within milliseconds, yet it also can be stored as memories that endure for years. Neurons perform these functions by generating sophisticated electrical and chemical signals. This unit describes these signals and how they are produced. It explains how one type of electrical signal, the action potential, allows information to travel along the length of a nerve cell. It also explains how other types of signals—both electrical and chemical—are generated at synaptic connections between nerve cells. Synapses permit information transfer by interconnecting neurons to form the circuitry on which brain information processing depends. Finally, this unit describes the intricate biochemical signaling events that take place within neurons and how such signaling can produce activity-dependent changes in synaptic communication and brain circuit function. Understanding these fundamental forms of neural signaling provides a foundation for appreciating the higher-level processes considered in the rest of the book.

The cellular and molecular mechanisms that give neurons their unique signaling abilities are targets for disease processes that compromise the function of the nervous system, as well as targets for anesthetics and many other clinically important drugs. A working knowledge of the cellular and molecular biology of neurons is therefore fundamental to understanding a variety of brain pathologies, and for developing novel approaches to diagnosing and treating these all too prevalent problems.

CHAPTER 2   **Electrical Signals of Nerve Cells**

CHAPTER 3   **Voltage-Dependent Membrane Permeability**

CHAPTER 4   **Ion Channels and Transporters**

CHAPTER 5   **Synaptic Transmission**

CHAPTER 6   **Neurotransmitters and Their Receptors**

CHAPTER 7   **Molecular Signaling within Neurons**

CHAPTER 8   **Synaptic Plasticity**

On the previous page:

Structure of a chemical synapse within the cerebral cortex. A presynaptic terminal (top) forms a synapse with a dendritic spine of the postsynaptic neuron (bottom). Colors indicate different organelles found within these structures.

Electrical Signals of Nerve Cells

Overview

Nerve cells generate a variety of electrical signals that transmit and store information. Although neurons are not intrinsically good conductors of electricity, they have elaborate mechanisms that generate electrical signals based on the flow of ions across their plasma membranes. Ordinarily, neurons generate a negative potential, called the *resting membrane potential*, that can be measured by recording the voltage between the inside and outside of nerve cells. The action potential is a fundamental electrical signal that transiently abolishes the negative resting potential and makes the transmembrane potential positive. Action potentials are propagated along the length of axons and carry information from one place to another within the nervous system. Still other types of electrical signals are produced by the activation of synaptic contacts between neurons or by the actions of external forms of energy, such as light and sound, on sensory neurons. All of these electrical signals arise from ion fluxes brought about by the selective ion permeability of nerve cell membranes, produced by ion channels, and the nonuniform distribution of these ions across the membrane, created by active transporters.

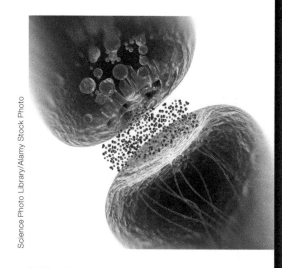

Science Photo Library/Alamy Stock Photo

KEY CONCEPTS

2.1 Nerve cells generate electrical signals to encode information

2.2 Neuronal electrical signals can be transmitted over long distances

2.3 Ion movements produce electrical signals

2.4 Electrical and chemical forces create membrane potentials

2.5 Potassium ions generate the resting membrane potential

2.6 More than one type of permeant ion can generate electrical signals

2.7 Action potentials arise from sequential changes in sodium and potassium permeability

CONCEPT
2.1 | **Nerve Cells Generate Electrical Signals to Encode Information**

LEARNING OBJECTIVES

2.1.1 List the basic types of electrical signals used to process information within the brain.

2.1.2 Describe the properties that differentiate these electrical signals from each other.

Types of neural electrical signals

Neurons employ several different types of electrical signals to encode and transfer information. The best way to observe these signals is to use an intracellular microelectrode to measure the electrical potential across the neuronal plasma membrane. A typical microelectrode is a piece of glass tubing pulled to a very fine point (with an opening less than 1 μm in diameter) and filled with a good electrical conductor, such as a concentrated salt solution. This conductive core can then be connected to a voltmeter, typically a computer, that records the transmembrane voltage of the nerve cell.

The first type of electrical phenomenon can be observed as soon as a microelectrode is inserted through the membrane of the neuron. Upon entering the cell, the microelectrode reports a negative potential, indicating that neurons have a means of generating a constant voltage across their membranes when at rest. This voltage, called the **resting membrane potential**, depends on the type of neuron being examined, but it is always a fraction of a volt (typically –40 to –90 mV).

Neurons encode information via electrical signals that result from transient changes in the resting membrane potential. **Receptor potentials** are due to the activation of sensory neurons by external stimuli, such as light, sound, or heat. For example, touching the skin activates nerve endings in Pacinian corpuscles, receptor neurons that sense mechanical disturbances of the skin. These neurons respond to touch with a receptor potential that changes the resting potential for a fraction of a second (Figure 2.1A). These transient changes in potential are the first step in generating the sensation of vibrations of the skin in the somatosensory system (see Chapter 12). Similar sorts of receptor potentials are observed in all other sensory neurons during transduction of sensory stimuli (see Unit II).

Another type of electrical signal is associated with communication between neurons at synaptic contacts. Activation of these synapses generates **synaptic potentials**, which allow transmission of information from one neuron to another. An example of such a signal is shown in Figure 2.1B. In this case, activation of a synaptic terminal innervating a hippocampal pyramidal neuron causes a very brief change in the resting membrane potential in the pyramidal neuron. Synaptic potentials serve as the means of exchanging information in the complex neural circuits found in both the central and peripheral nervous systems (see Chapter 5).

Finally, many neurons generate a special type of electrical signal that travels along their long axons. Such signals are called **action potentials** and are also referred to as *spikes* or *impulses*. An example of an action potential recorded from the axon of a spinal motor neuron is shown in Figure 2.1C. Action potentials are responsible for long-range transmission of information within the nervous system and allow the nervous system to transmit information to its target organs, such as muscle.

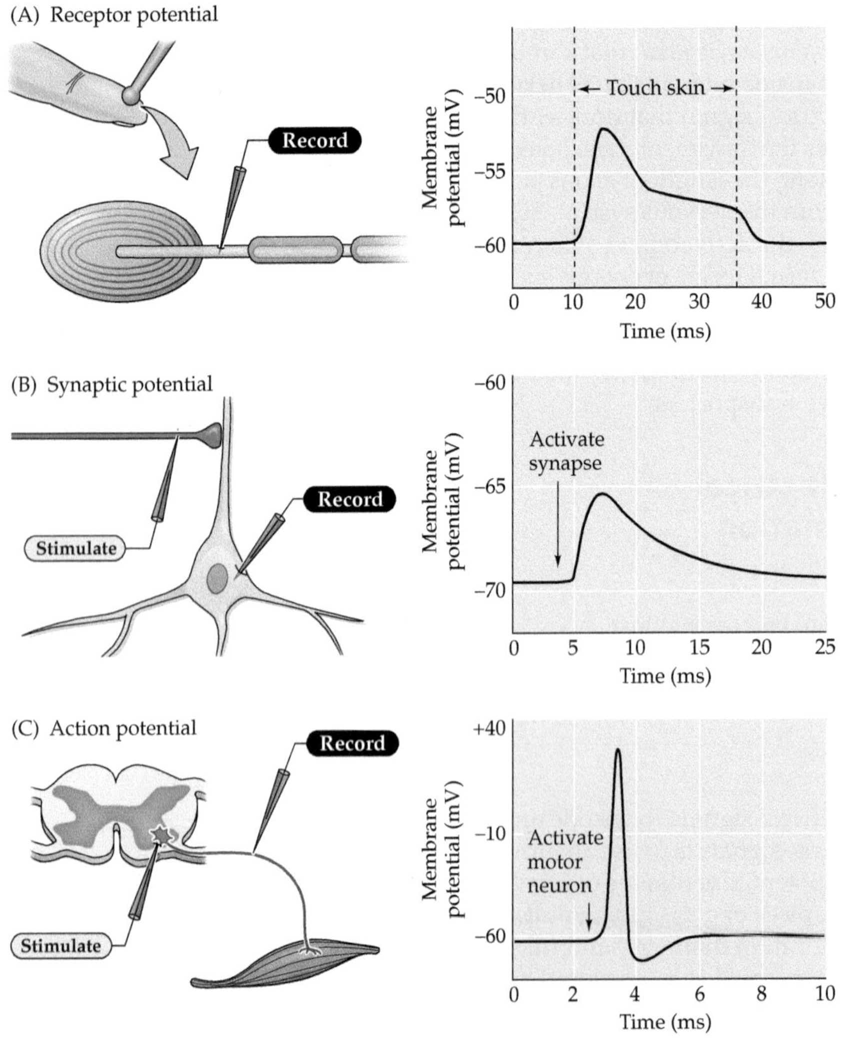

Passive and active electrical signals

One way to elicit an action potential is to pass electrical current across the membrane of the neuron. In normal circumstances, this current would be generated by receptor potentials or by synaptic potentials. In the laboratory, however, electrical current suitable for initiating an action potential can be readily produced by inserting a microelectrode into a neuron and then connecting the electrode to a battery (Figure 2.2A). A second microelectrode can be inserted to measure the membrane potential changes produced by the applied current. If the current delivered in this way makes the membrane potential more negative (**hyperpolarization**), nothing very dramatic happens. The membrane potential simply changes in proportion to the magnitude of the injected current (Figure 2.2B, central part). Such hyperpolarizing responses do not require any unique property of neurons and are therefore called **passive electrical responses**. A much more interesting phenomenon is seen if current of the opposite polarity is delivered, so that the membrane potential of the

FIGURE 2.1 Types of neuronal electrical signals In all cases, microelectrodes are used to measure changes in the resting membrane potential during the indicated signals. (A) A brief touch causes a receptor potential in a Pacinian corpuscle in the skin. (B) Activation of a synaptic contact onto a hippocampal pyramidal neuron elicits a synaptic potential. (C) Stimulation of a spinal reflex produces an action potential in a spinal motor neuron.

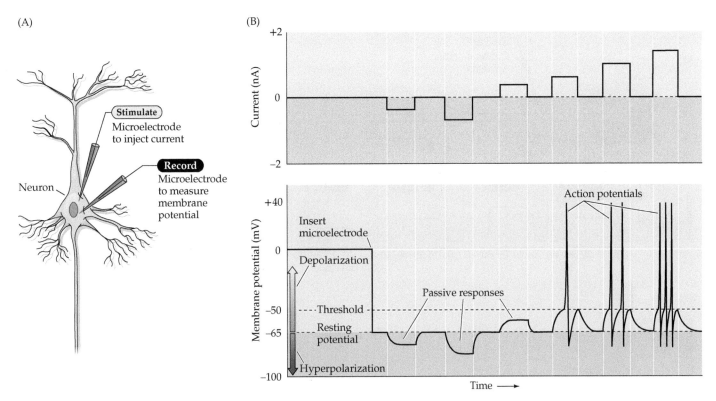

FIGURE 2.2   Recording passive and active electrical signals in a nerve cell (A) Two microelectrodes are inserted into a neuron; one of these measures membrane potential while the other injects current into the neuron. (B) Inserting the voltage-measuring microelectrode into the neuron (bottom) reveals a negative potential, the resting membrane potential.

Injecting current through the other microelectrode (top) alters the neuronal membrane potential. Hyperpolarizing current pulses produce only passive changes in the membrane potential. While small depolarizing currents also elicit only passive responses, depolarizations that cause the membrane potential to meet or exceed threshold additionally evoke action potentials.

nerve cell becomes more positive than the resting potential (**depolarization**). In this case, at a certain level of membrane potential, called the **threshold potential**, action potentials occur (see Figure 2.2B, right side).

The action potential is an active response generated by the neuron and typically is a brief (about 1 ms) change in the transmembrane potential, from negative to positive. Action potentials are considered active responses because they are generated by selective changes in the permeability of the neuronal membrane. Importantly, the amplitude of the action potential is independent of the magnitude of the current used to evoke it; that is, larger currents do not elicit larger action potentials. The action potentials of a given neuron are therefore said to be *all-or-none*—that is, they occur fully or not at all. If the amplitude or duration of the stimulus current is increased sufficiently, multiple action potentials occur, as can be seen in the responses to the three depolarizing current intensities shown in Figure 2.2B (right side). It follows, therefore, that the intensity of a stimulus is encoded in the *frequency* of action potentials rather than in their amplitude. This arrangement differs dramatically from that of receptor potentials, whose amplitudes are graded in proportion to the magnitude

of the sensory stimulus; and from that of synaptic potentials, whose amplitudes vary according to the number of synapses activated, the strength of each synapse, and the previous amount of synaptic activity.

CONCEPT
2.2

Neuronal Electrical Signals Can Be Transmitted Over Long Distances

LEARNING OBJECTIVES

2.2.1 Explain the difference between passive and active current flow in a neuron.

2.2.2 Explain the significance of active current flow for information spread within the long axons of neurons.

Poor spread of passive electrical signals

The use of electrical signals—as in sending electricity over wires to provide power or information—presents a series of challenges in electrical engineering. A fundamental problem for neurons is that their axons, which can be quite long (remember that a spinal motor neuron can extend for a meter or more), are not good electrical conductors.

PASSIVE CONDUCTION DECAYS OVER DISTANCE

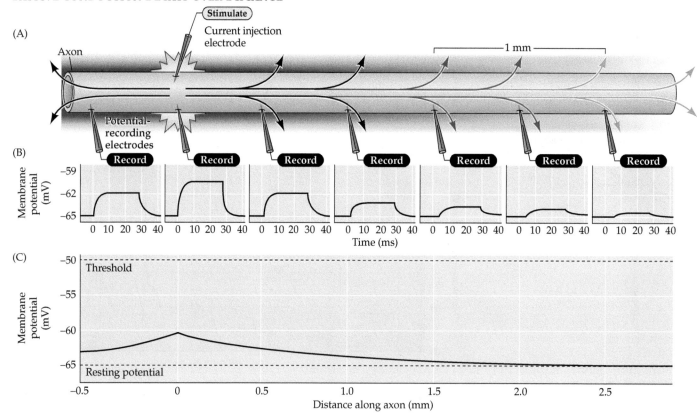

ACTIVE CONDUCTION IS CONSTANT OVER DISTANCE

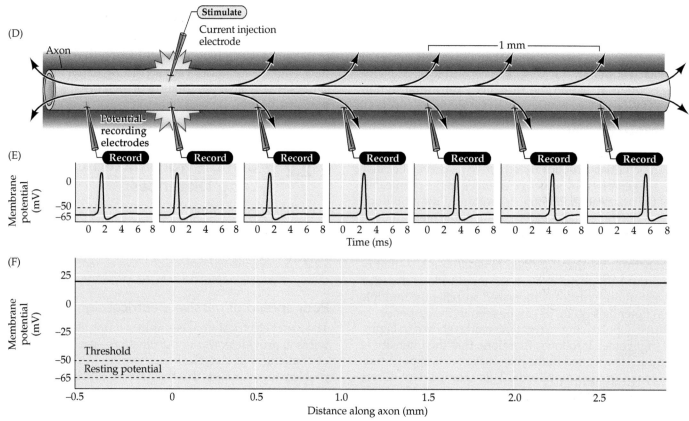

FIGURE 2.3   Passive and active current flow in an axon
(A) Experimental arrangement for examining passive flow of electrical current in an axon. A current-passing electrode produces a current that yields a subthreshold change in membrane potential, which spreads passively along the axon. (B) Potential responses recorded by microelectrodes at the positions indicated. With increasing distance from the site of current injection, the amplitude of the potential change is attenuated as current leaks out of the axon. (C) Relationship between the amplitude of potential responses and distance. (D) If the experiment shown in (A) is repeated with a suprathreshold current, an active response, the action potential, is evoked. (E) Action potentials recorded at the positions indicated by microelectrodes. The amplitude of the action potential is constant along the length of the axon, although the time of appearance of the action potential is delayed with increasing distance. (F) The constant amplitude of an action potential (solid black line) measured at different distances. (After A. L. Hodgkin and W. A. H. Rushton. 1946. *Proc R Soc Lond B* 133: 444–478.)

Although neurons and wires are both capable of passively conducting electricity, the electrical properties of neurons compare poorly with that of an ordinary wire. This can be seen by measuring the passive electrical properties of a nerve cell axon by determining the voltage change resulting from a current pulse passed across the axonal membrane (Figure 2.3A). If this current pulse is below the threshold for generating an action potential, then the magnitude of the resulting potential change will decay with increasing distance from the site of current injection (Figure 2.3B). Typically, the potential falls to a small fraction of its initial value at a distance of no more than a few millimeters away from the site of injection (Figure 2.3C).

For comparison, a wire would typically allow passive current flow over distances many thousands of times longer. The progressive decrease in the amplitude of the induced potential change occurs because the injected current leaks out across the axonal membrane; accordingly, farther along the axon less current is available to change the membrane potential. This leakiness of the axonal membrane prevents effective passive conduction of electrical signals along the length of all but the shortest axons (those 1 mm or less in length). To compensate for this deficiency, action potentials serve as a "booster system" that allows neurons to conduct electrical signals over great distances despite the poor passive electrical properties of axons.

Long-distance propagation via action potentials

The ability of action potentials to boost the spatial spread of electrical signals can be seen if the experiment shown in Figure 2.3A is repeated with a depolarizing current pulse that is large enough to produce an action potential (Figure 2.3D). In this case, the result is dramatically different. Now an action potential of constant amplitude is observed along the entire length of the axon (Figure 2.3E). The fact that electrical signaling now occurs without any decrement (Figure 2.3F) indicates that active conduction via action potentials is a very effective way to circumvent the inherent leakiness of neurons.

Action potentials are the basis of information transfer in the nervous system and are targets of many clinical treatments, including anesthesia (Clinical Applications). For these reasons, it is essential to understand how these and other neuronal electrical signals arise. Remarkably, all

---

■ Clinical Applications

Anesthesia and Neuronal Electrical Signaling

Anesthesia is an essential part of clinical practice and refers to procedures that reduce sensation during surgical procedures, most often to alleviate pain (see Chapter 13) or to create a state of unconsciousness. The drugs that produce anesthesia are called **anesthetics**, and these agents usually work by interfering with the electrical signaling mechanisms of neurons. There are three broad categories of anesthesia.

The mildest form of anesthesia, *local anesthesia*, is used to prevent pain sensation in localized parts of the body. Local anesthetics, such as lidocaine, ropivacaine, and bupivacaine,

block action potential propagation along peripheral nerves (Figure A) by blocking the Na^+ channels involved in action potential generation (see Chapter 4). This causes a loss of sensory perception commonly called numbing. Local anesthesia can be highly restricted in its range of action, due to topical application of local anesthetics onto (or injection into) the target tissue. Perhaps the most widespread example of local anesthesia is the injection of lidocaine or other local anesthetics into the mandible to block action potential conduction in the inferior alveolar nerve, thereby preventing pain

sensation in a portion of the mouth during dental procedures.

Regional anesthesia desensitizes a larger region of the body and is typically produced by injecting local anesthetics to prevent pain sensation in the region where a surgical procedure will be performed. These anesthetics are injected near the spinal cord, nerve plexuses, and major nerves. For example, more than 50% of women giving birth in U.S. hospitals have local anesthetics injected into the epidural space of their spinal canal to prevent pelvic pain associated with labor. Although patients receiving regional anesthe-

(Continued)

■ Clinical Applications (continued)

sia remain conscious, they sometimes additionally receive a **sedative** agent (also called a tranquilizer) to reduce anxiety or induce sleep. Many sedatives enhance the activity of the postsynaptic GABA receptors found at most inhibitory synapses (see Chapter 6); the resulting strengthening of synaptic inhibition reduces neuronal activity (see Chapter 5) and presumably causes sedation (Figure B). One class of sedative is the barbiturates, which include pentobarbital and thiopentone. Because of their addictive properties, as well as their lethality at high doses, barbiturates have largely been replaced by benzodiazepines. The best-known benzodiazepine is diazepam, more commonly known as Valium; other clinically usefully, shorter-acting benzodiapines include midaz-

olam and lorezepam. A related compound is zolpidem (Ambien), an imidazopyridine sedative that is also used as a **hypnotic** agent to induce sleep. Propofol is yet another type of sedative that works primarily by enhancing GABA receptor activity, though it also blocks Na⁺ channels in a way similar to that of local anesthetics. An overdose of propofol and benzodiazepines was responsible for the death of the popular musician Michael Jackson.

General anesthesia causes unconsciousness, an absence of sensation and muscular relaxation. It is used for major surgical procedures, such as those of long duration or involving substantial loss of blood. General anesthetic agents can be injected into the circulatory system (*intravenous anesthetics*; Figure C) or inhaled through

the respiratory system (*inhalation anesthetics*; Figure D).

General anesthesia can be induced by injection of higher doses of some of the sedative agents described above that work on synaptic GABA receptors. Because these agents do not prevent pain sensation, they are often administered along with analgesic agents, such as fentanyl, that alleviate pain by acting on the receptors for opiate peptides (see Chapter 6) of peripheral and central pain pathways (see Chapter 13). Fentanyl abuse is part of an "opioid epidemic" that killed more than 50,000 people in the U.S. in 2020. Longer-acting opiate receptor drugs are used for postoperative analgesia; the best-known example of such drugs is morphine. Another intravenous anesthetic is ketamine (see Figure C), a drug

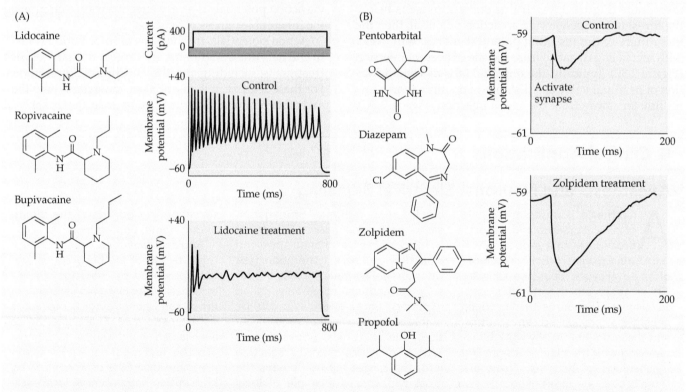

(A) Left: Chemical structures of the local anesthetics lidocaine, ropivacaine, and bupivacaine. Right: A depolarizing current pulse (top) ordinarily evokes a train of action potentials (control, center) in a sensory neuron of the dorsal root ganglion. Treatment with lidocaine (bottom) greatly reduces the ability of the neuron to generate action potentials. (B) Left: Sedatives that

act on brain GABA receptors. Right: Postsynaptic potentials at an inhibitory synapse between an interneuron and a pyramidal neuron in the hippocampus are enhanced by treatment with the sedative zolpidem. (A after A. Scholz, 2002. *Br J Anaesth* 89: 52–61; B after A. M. Thomson et al. 2000. *Eur J Neurosci* 12: 425–436.)

■ Clinical Applications　(*continued*)

that blocks the NMDA-type glutamate receptors found at many excitatory synapses (see Chapter 6) but that also works on GABA and opiate receptors. The resulting reduction in excitatory synaptic transmission presumably underlies the anesthetic actions of ketamine. Use of ketamine is limited by its diverse side effects, including production of hallucinations. Subanesthetic doses of ketamine are increasingly being used to treat depression; these antidepressant effects may be mediated by metabolites of ketamine.

During general anesthesia, intravenous paralytic agents such as rocuronium and vecuronium are often used to facilitate surgery by relaxing skeletal muscles. These agents work by imparing cholinergic neuromuscular trans-mission (see Chapter 6). One attractive feature of these agents is that their actions can be reversed by postoperative administration of reversal agents, such as neostigmine and sugammadex, that neutralize their neuromuscular blocking effects and promote rapid recovery of muscle function.

Inhalation anesthetics are volatile liquids that vaporize at room temperature and can then be inhaled. The first and best-known example of an inhalant anesthetic is diethyl ether, which is no longer used clinically because it is highly flammable. Second-generation inhalant anesthetics, such as halothane and isoflurane, have been replaced for clinical use by desflurane and sevoflurane, among others. Although the mechanism of action of inhalation anesthetics is not estab-lished, one leading hypothesis is that they hyperpolarize the resting membrane potential of neurons and thereby make it more difficult to fire action potentials (see Figure D). This action is thought to be caused by the anesthetics opening the 2-P K^+ channels that create the resting membrane potential (see Chapter 4). Inhalation anesthetics also enhance the activity of synaptic GABA receptors.

In conclusion, understanding the mechanisms underlying anesthetic action both illuminates how these clinically valuable agents work and emphasizes the fundamental importance of neuronal electrical signaling for the function of the nervous system.

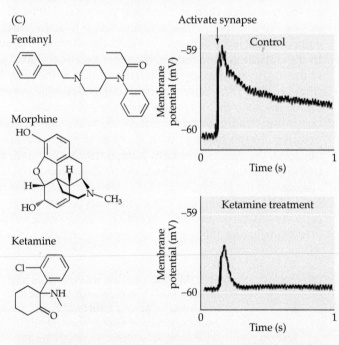

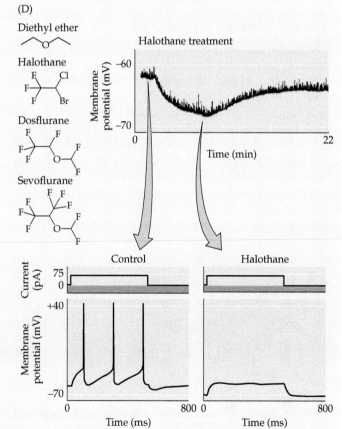

(C) Left: Examples of analgesics and intravenous anesthetics. Right: Postsynaptic potentials at an excitatory synapse of a spinal cord motor neuron are reduced by treatment with ketamine. (D) Left: Four inhalation anesthetics. Right: Halothane treatment hyperpolarizes the resting membrane potential of a spinal cord motor neuron (top). This moves the resting membrane potential away from the threshold for firing action potentials, thereby reducing the ability of a depolarizing current pulse (center) to evoke action potentials (bottom). (C after I. Lizarraga et al. 2008. *Br J Pharmacol* 153: 1030–1042; D after J. E. Sirois et al. 1998. *J Physiol* 512: 851–862.)

types of neuronal electrical signals are produced by similar mechanisms that rely on the movement of ions across the neuronal membrane. The remainder of this chapter addresses the question of how nerve cells use ions to generate electrical potentials. Chapter 3 explores more specifically the means by which action potentials are produced and how these signals solve the problem of long-distance electrical conduction within nerve cells. Chapter 4 examines the properties of membrane molecules responsible for electrical signaling. Finally, Chapters 5–8 consider how electrical signals are transmitted from one nerve cell to another at synaptic contacts.

CONCEPT 2.3 | Ion Movements Produce Electrical Signals

LEARNING OBJECTIVES

2.3.1 Describe the types of membrane proteins that control ion fluxes across a cellular membrane.

2.3.2 Explain how ion fluxes across membranes generate electrical signals.

2.3.3 Define electrochemical equilibrium and describe how it determines ion flux across a membrane.

Two requirements for generating cellular electrical signals

Electrical potentials are generated across the membranes of neurons—and, indeed, across the membranes of all cells—because (1) there are *differences in the concentrations of specific ions* across nerve cell membranes, and (2) these *membranes are selectively permeable* to some of these ions. These two conditions depend in turn on two different kinds of proteins in the plasma membrane (Figure 2.4). The ion concentration gradients are established by proteins known as **active transporters**, which, as their name suggests, actively move ions into or out of cells against their concentration gradients. The selective permeability of membranes is due largely to **ion**

channels, proteins that allow only certain kinds of ions to cross the membrane in the direction of their concentration gradients. Thus, channels and transporters basically work against each other, and in so doing they generate the resting membrane potential, action potentials, and the synaptic potentials and receptor potentials that trigger action potentials. Chapter 4 describes the structure and function of these channels and transporters.

To appreciate the roles of ion gradients and selective permeability in generating a membrane potential, consider a simple system in which an artificial membrane separates two compartments containing solutions of ions. For comparison with the situation in neurons, we will refer to the left compartment as the inside and call the right compartment the outside. In such a system, it is possible to control the composition of the two solutions and thereby control the ion gradients across the membrane. It is also possible to control the ion permeability of the membrane.

As a first example, consider the case of a membrane that is permeable only to potassium ions (K^+). If the concentration of K^+ on each side of this membrane is equal, then no electrical potential will be measured across it (Figure 2.5A). However, if the concentration of K^+ is not the same on the two sides, then an electrical potential will be generated. For instance, if the concentration of K^+ in the inside compartment is ten times higher than the K^+ concentration in the outside compartment, then the electrical potential of the inside will be negative relative to that of the outside (Figure 2.5B). This difference in electrical potential is generated because the potassium ions flow down their concentration gradient and take their electrical charge (one positive charge per ion) with them as they go. Because neuronal membranes contain pumps that accumulate K^+ in the cell cytoplasm, and because potassium-permeable channels in the plasma membrane allow a transmembrane flow of K^+, an analogous situation exists in living nerve cells. As will be proved in Concept 2.5, such an efflux of K^+ is responsible for the resting membrane potential.

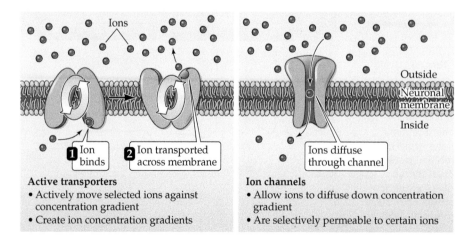

Active transporters
- Actively move selected ions against concentration gradient
- Create ion concentration gradients

Ion channels
- Allow ions to diffuse down concentration gradient
- Are selectively permeable to certain ions

FIGURE 2.4 Active transporters and ion channels are responsible for ion movements across neuronal membranes Transporters create ion concentration differences by actively transporting ions against their chemical gradients. Channels take advantage of these concentration gradients, allowing selected ions to move, via diffusion, down their chemical gradients.

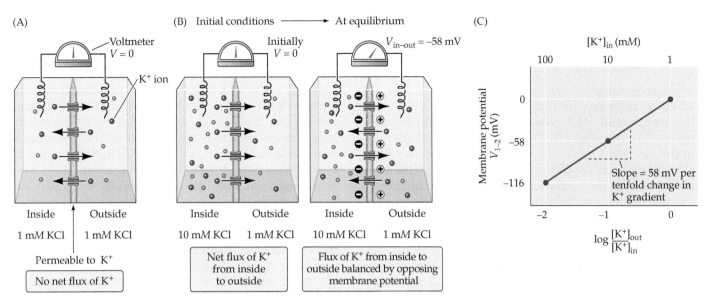

FIGURE 2.5 Electrochemical equilibrium (A) A membrane permeable only to K$^+$ (gold spheres) separates the inside and outside compartments, which contain the indicated concentrations of KCl. (B) Increasing the KCl concentration of the inside compartment to 10 mM initially causes a small movement of K$^+$ into the outside compartment (initial conditions) until the electromotive force acting on K$^+$ balances the concentration gradient, and there is no further net movement of K$^+$ (at equilibrium). (C) The relationship between the transmembrane concentration gradient ([K$^+$]$_{out}$/[K$^+$]$_{in}$) and the membrane potential. As predicted by the Nernst equation, this relationship is linear when plotted on semilogarithmic coordinates, with a slope of 58 mV per tenfold difference in the concentration gradient.

Electrochemical equilibrium

In the hypothetical case just described, an equilibrium will quickly be reached. As K$^+$ moves from the inside compartment to the outside (the initial conditions; shown on the left in Figure 2.5B), a potential is generated that tends to impede further flow of K$^+$. This impediment results from the fact that the potential gradient across the membrane tends to repel the positive potassium ions that would otherwise move across the membrane. Thus, as the outside becomes positive relative to the inside, the increasing positivity makes the outside less attractive to the positively charged K$^+$. The net movement (or flux) of K$^+$ will stop at the point ("At equilibrium" in the right panel of Figure 2.5B) where the potential change across the membrane (the relative positivity of the outside compartment) exactly offsets the concentration gradient (the tenfold excess of K$^+$ in the inside compartment). At this **electrochemical equilibrium**, there is an exact balance between two opposing forces (see Figure 2.5B): (1) the concentration gradient that causes K$^+$ to move from inside to outside, taking along positive charge, and (2) an opposing electrical gradient that increasingly tends to stop K$^+$ from moving across the membrane. The number of ions that needs to flow to generate this electrical potential is very small (approximately 10^{-12} moles of K$^+$ per cm^2 of membrane, or less than one millionth of the K$^+$ ions present on each side). This last fact is significant in two ways. First, it

means that the concentrations of permeant ions on each side of the membrane remain essentially constant, even after the flow of ions has generated the potential. Second, the tiny fluxes of ions required to establish the membrane potential do not disrupt chemical electroneutrality because each ion has an oppositely charged counter-ion (chloride ions in Figure 2.5) to maintain the neutrality of the solutions on each side of the membrane. The concentration of K$^+$ remains equal to the concentration of Cl$^-$ in the solutions in both compartments, meaning that the separation of charge that creates the potential difference is restricted to the immediate vicinity of the membrane.

CONCEPT
2.4

Electrical and Chemical Forces Create Membrane Potentials

LEARNING OBJECTIVES

2.4.1 Explain how the Nernst equation uses ion concentration gradients to predict equilibrium potentials.

2.4.2 Apply the Nernst equation to predict electrochemical equilibrium conditions for any situation where a membrane is permeable to only a single type of ion.

The Nernst equation and electrochemical equilibrium

The electrical potential generated across the membrane at electrochemical equilibrium—the **equilibrium potential**

—can be predicted by a simple formula called the **Nernst equation**. This relationship is generally expressed as

$$E_X = \frac{RT}{zF} \ln \frac{[X]_{out}}{[X]_{in}}$$

where E_X is the equilibrium potential for any ion X, R is the gas constant, T is the absolute temperature (in degrees on the Kelvin scale), z is the valence (electrical charge) of the permeant ion, and F is the Faraday constant (the amount of electrical charge contained in one mole of a univalent ion). The brackets indicate the concentrations of ion X on each side of the membrane, with "in" referring to the inside compartment and "out" referring to the outside, and the symbol ln indicates the natural logarithm of the concentration gradient. Because it is easier to perform calculations using base 10 logarithms and to perform experiments at room temperature, this relationship is usually simplified to

$$E_X = \frac{58}{z} \log \frac{[X]_{out}}{[X]_{in}}$$

where log indicates the base 10 logarithm of the concentration ratio. (The constant of 58 becomes 61 mV at mammalian body temperatures.) Thus, for the example in Figure 2.5B, the potential across the membrane at electrochemical equilibrium is

$$E_K = \frac{58}{z} \log \frac{[K]_{out}}{[K]_{in}} = 58 \log \frac{1}{10} = -58 \text{ mV}$$

The equilibrium potential is conventionally defined in terms of the potential difference between the outside and inside compartments. Thus, when the concentration of K^+ is higher inside than out, an inside-negative potential is measured across the K^+-permeable neuronal membrane.

For a simple hypothetical system with only one permeant ion species, the Nernst equation allows the electrical potential across the membrane at equilibrium to be predicted exactly. For example, if the concentration of K^+ on the inside is increased to 100 mM, the membrane potential will be –116 mV. More generally, if the membrane potential is plotted against the logarithm of the K^+ concentration gradient ($[K^+]_{out}/[K^+]_{in}$), the Nernst equation predicts a linear relationship with a slope of 58 mV (actually 58/z) per tenfold change in the K^+ gradient (Figure 2.5C).

Roles of permeant ion gradients and electrical potentials in electrochemical equilibrium

To reinforce and extend the concept of electrochemical equilibrium, consider some additional experiments on the influence of ion species and ion permeability that could be performed on the simple model system in Figure 2.5. What would happen to the electrical potential across the membrane (i.e., the potential of the inside

relative to the outside) if the potassium on the outside were replaced with 10 mM sodium (Na^+) and the K^+ in the inside compartment were replaced with 1 mM Na^+? No potential would be generated because no Na^+ could flow across the membrane (which was defined as being permeable only to K^+). However, if under these ionic conditions (ten times more Na^+ outside) the K^+-permeable membrane were to be magically replaced by a membrane permeable only to Na^+, a potential of +58 mV would be measured at equilibrium. If 10 mM calcium (Ca^{2+}) were present outside and 1 mM Ca^{2+} inside, and a Ca^{2+}-selective membrane separated the two sides, what would happen to the membrane potential? A potential of +29 mV would develop—half that observed for Na^+, because the valence of calcium is +2. Finally, what would happen to the membrane potential if 10 mM Cl^- were present inside and 1 mM Cl^- were present outside, with the two sides separated by a Cl^--permeable membrane? Because the valence of this anion is –1, the potential would again be +58 mV.

The balance of chemical and electrical forces at equilibrium means that the electrical potential can determine ion fluxes across the membrane, just as the ion gradient can determine the membrane potential. To examine the influence of membrane potential on ion flux, imagine connecting a battery across the two sides of the membrane to control the electrical potential across the membrane without changing the distribution of ions on the two sides (Figure 2.6). As long as the battery is off, things will be just as in Figure 2.5B, with the flow of K^+ from inside to outside causing a negative membrane potential (see Figure 2.6A, left). However, if the battery is used to make the inside compartment initially more negative relative to the outside, there will be less K^+ flux, because the negative potential will tend to keep K^+ in the inside compartment. How negative will the inside need to be before there is no net flux of K^+? The answer is –58 mV, the voltage needed to counter the tenfold difference in K^+ concentrations on the two sides of the membrane (see Figure 2.6A, center). If the inside is initially made more negative than –58 mV, then K^+ will actually flow from the outside into the inside because the positive ions will be attracted to the more negative potential of the inside (see Figure 2.6A, right). This example demonstrates that both the direction and magnitude of ion flux depend on the membrane potential. Thus, in some circumstances the electrical potential can overcome an ion concentration gradient.

The ability to alter ion flux experimentally by changing either the potential imposed on the membrane (see Figure 2.6B) or the transmembrane concentration gradient for an ion (see Figure 2.5C) provides convenient tools for studying ion fluxes across the plasma membranes of neurons, as will be evident in many of the experiments described in the chapters that follow.

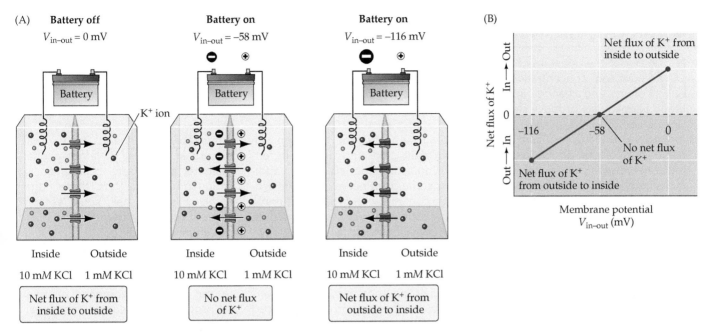

FIGURE 2.6   Membrane potential influences ion fluxes
(A) Connecting a battery across the K⁺-permeable membrane allows direct control of membrane potential. When the battery is turned off (left), K⁺ ions (gold) flow simply according to their concentration gradient. Setting the initial membrane potential (V_{in-out}) at the equilibrium potential for K⁺ (center) yields no net flux of K⁺, whereas making the membrane potential more negative than the K⁺ equilibrium potential (right) causes K⁺ to flow against its concentration gradient. (B) Relationship between membrane potential and direction of K⁺ flux.

CONCEPT 2.5	Potassium Ions Generate the Resting Membrane Potential

LEARNING OBJECTIVES

2.5.1 List the experimental evidence for the conclusion that the negative resting potential of neurons arises from a high resting permeability to potassium ions.

2.5.2 Explain how manipulation of transmembrane ion gradients can be used to determine membrane ion permeability.

Ion concentration gradients across neural membranes

The hypothetical scenarios considered in Concepts 2.3 and 2.4 are highly relevant to the case of living neurons. In these cells—indeed, in all cells in the body—ion transporters create substantial transmembrane gradients for most ions. Table 2.1 summarizes the ion concentrations measured directly in an exceptionally large neuron found in the nervous system of the squid (Box 2A). These measurements indicate that there is much more K⁺ inside a neuron than out, and much more Na⁺ outside than in. Similar concentration gradients occur in the neurons of most animals, including humans. However, because the ionic strength of mammalian blood is lower than that of sea-dwelling animals such as squid, in mammals the concentrations of each ion are several times lower (see Table 2.1). These transporter-dependent

concentration gradients enable the resting membrane potential and other electrical signals of neurons.

Once the ion concentration gradients across various neuronal membranes are known, the Nernst equation can be used to calculate the equilibrium potential for K⁺ and other major ions (see the right column in Table 2.1). Since

TABLE 2.1   Extracellular and Intracellular Ion Concentrations and Resultant Equilibrium Potentials

Ion	Concentration (mM)		Equilibrium Potential (mV)
	Intracellular	Extracellular	
Squid neuron			
Potassium (K⁺)	400	20	−75
Sodium (Na⁺)	50	440	+55
Chloride (Cl⁻)	110	560	−41
Calcium (Ca²⁺)	0.0001	10	+145
Mammalian neuron			
Potassium (K⁺)	140	5	−88
Sodium (Na⁺)	12	145	+66
Chloride (Cl⁻)	8	110	−69
Calcium (Ca²⁺)	0.0001	1.2	+124

■ BOX 2A | The Remarkable Giant Nerve Cells of Squid

Many of the initial insights into how ion concentration gradients and changes in membrane permeability produce electrical signals came from experiments performed on the extraordinarily large nerve cells of squid. The axons of these nerve cells can be up to 1 mm in diameter—100 to 1000 times larger than mammalian axons. Thus, squid axons are large enough to allow experiments that would be impossible on most other nerve cells. For example, it is not difficult to insert simple wire electrodes inside these giant axons and make reliable electrical measurements. The relative ease of this approach yielded the first direct recordings of action potentials from nerve cells and, as we will discuss in Chapter 3, the first experimental measurements of the ion currents that produce action potentials. It also is practical to extrude the cytoplasm from giant axons and measure its ionic composition (see Table 2.1). In addition, some giant nerve cells form synaptic contacts with other giant nerve cells, producing very large synapses that have been extraordinarily valuable in understanding the fundamental mechanisms of synaptic transmission (see Chapter 5).

Giant neurons evidently evolved in squid because they enhanced the animal's survival. These neurons participate in a simple neural circuit that activates the contraction of the mantle muscle, producing a jet propulsion effect that allows the squid to move away from predators at a remarkably fast speed. As we will discuss in Chapter 3, larger axonal diameter allows faster conduction of action potentials. Thus, these huge nerve cells must help squid escape more successfully from their numerous enemies.

Today—nearly 90 years after their discovery by John Z. Young at Oxford University—the giant nerve cells of squid remain useful experimental systems for probing basic neuronal functions.

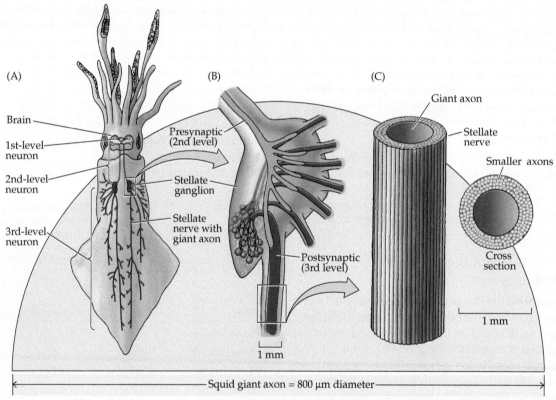

(A) Diagram of a squid, showing the location of its giant nerve cells. Different colors indicate the neuronal components of the escape circuitry. The first- and second-level neurons originate in the brain, while the third-level neurons are in the stellate ganglion and innervate muscle cells of the mantle. (B) Giant synapses within the stellate ganglion. The second-level neuron forms a series of fingerlike processes, each of which makes an extraordinarily large synapse with a single third-level neuron. (C) Structure of a giant axon of a third-level neuron lying within its nerve. The enormous difference in the diameters of a squid giant axon and a mammalian axon are shown below.

(A)

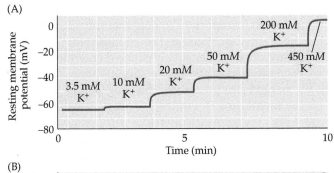

(B)

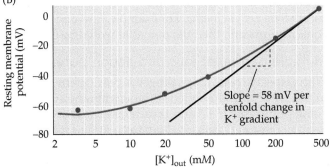

FIGURE 2.7 **The resting membrane potential of a squid giant axon is determined by the K⁺ concentration gradient across the membrane** (A) Increasing the external K⁺ concentration depolarizes the resting membrane potential. (B) Relationship between resting membrane potential and external K⁺ concentration, plotted on a semilogarithmic scale. The straight line represents a slope of 58 mV per tenfold change in concentration, as given by the Nernst equation. (Data from A. L. Hodgkin and B. Katz. 1949. *J Physiol* 108: 37–77.)

the resting membrane potential of the squid neuron is approximately –65 mV, K⁺ is the ion that is closest to being in electrochemical equilibrium when the cell is at rest. This fact implies that the resting membrane is more permeable to K⁺ than to the other ions listed in Table 2.1, and that this permeability is the source of resting potentials.

Role of K⁺ in neuronal resting membrane potentials

It is possible to test this hypothesis, as Alan Hodgkin and Bernard Katz did in 1949, by asking what happens to the resting membrane potential if the concentration of K⁺ outside the neuron is altered. If the resting membrane were permeable only to K⁺, then the Nernst equation predicts that the membrane potential will vary in proportion to the logarithm of the K⁺ concentration gradient across the membrane. Assuming that the internal K⁺ concentration is unchanged during the experiment, a plot of membrane potential against the logarithm of the external K⁺ concentration should yield a straight line with a slope of 58 mV per tenfold change in external K⁺ concentration at room temperature (see Figure 2.5C).

When Hodgkin and Katz analyzed experiments done on a living squid neuron by Howard Curtis and Kenneth Cole, they found that the resting membrane potential did indeed change when the external K⁺ concentration was modified, becoming less negative as external K⁺ concentration was raised (Figure 2.7A). When the external K⁺ concentration was raised high enough to equal the concentration of K⁺ inside the neuron, thus making the K⁺ equilibrium potential 0 mV, the resting membrane potential was also approximately 0 mV. In short, the resting membrane potential varied as predicted with the logarithm of the K⁺ concentration, with a slope that approached 58 mV per tenfold change in K⁺ concentration (Figure 2.7B). The value obtained was not exactly 58 mV because other ions, such as Cl⁻ and Na⁺, are also slightly permeant and thus influence the resting potential to a small degree. The contribution of these other ions is particularly evident at low external K⁺ levels. In general, however, manipulation of the external concentrations of these other ions has only a small effect (see Figure 2.9E), emphasizing that permeability to K⁺ is indeed the primary source of the resting membrane potential.

In summary, Hodgkin and Katz showed that the inside-negative resting potential arises because (1) the membrane of the resting neuron is more permeable to K⁺ than to any of the other ions present, and (2) there is more K⁺ inside the neuron than outside. The selective permeability to K⁺ is caused by K⁺-permeable membrane channels that are open in resting neurons, while the large K⁺ concentration gradient is produced by membrane transporters that selectively accumulate K⁺ within neurons. Many subsequent studies have confirmed the general validity of these principles.

CONCEPT **2.6** | # More Than One Type of Permeant Ion Can Generate Electrical Signals

LEARNING OBJECTIVES

2.6.1 Understand how the Goldman equation uses ion concentration gradients and relative permeability to predict membrane potential.

2.6.2 Be able to apply the Goldman equation to predict the potential generated across a membrane that is permeable to multiple types of ions.

2.6.3 Understand how changes in ion permeability can change membrane potential.

The Goldman equation for multiple permeant ions

While the resting potential is caused by the neuronal membrane being permeable to K⁺, many other neuronal electrical signals rely on this membrane being permeable to other ions. For example, consider a situation in which both Na^+ and K^+ are unequally distributed across the membrane, as in Figure 2.8A. What would happen if 10 mM K^+ and 1 mM Na^+ were present inside, and 1 mM K^+ and 10 mM Na^+ were present outside? If the membrane were permeable only to K^+, the membrane potential would be −58 mV; if the membrane were permeable only to Na^+, the potential would be +58 mV. But what would the potential be if the membrane were permeable to both K^+ and Na^+? In this case, the potential would depend on the relative permeability of the membrane to K^+ and Na^+. If it were more permeable to K^+, the potential would approach −58 mV, and if it were more permeable to Na^+, the potential would be closer to +58 mV. Because there is no permeability term in the Nernst equation, which considers only the simple case of a single permeant ion species, a more elaborate equation is needed. This equation must take into account both the concentration gradients of the permeant ions and the relative permeability of the membrane to each permeant species.

Such an equation was developed by David Goldman in 1943. For the case most relevant to neurons, in which K^+, Na^+, and Cl^- are the primary permeant ions at room temperature, the **Goldman equation** is written

$$V_m = 58 \log \frac{P_K[K]_{out} + P_{Na}[Na]_{out} + P_{Cl}[Cl]_{in}}{P_K[K]_{in} + P_{Na}[Na]_{in} + P_{Cl}[Cl]_{out}}$$

where V is the voltage across the membrane (again, the inside compartment, relative to the reference outside compartment) and P_x indicates the permeability of the membrane to each ion of interest. The Goldman equation is thus an extended version of the Nernst equation that takes into account the relative permeabilities of each of the ions involved. The relationship between the two equations becomes obvious in the situation where the membrane is permeable only to one ion, such as K^+; in this case, the Goldman expression collapses back to the simpler Nernst equation. In this context, it is important to note that the valence factor (z) in the Nernst equation has been eliminated; this is why the concentrations of negatively charged chloride ions, Cl^-, have been inverted relative to the concentrations of the positively charged ions [remember that −log (A/B) = log (B/A)].

If the membrane in Figure 2.8A is permeable only to K^+ and Na^+, the terms involving Cl^- drop out because P_{Cl} is 0. In this case, solution of the Goldman equation yields a potential of −58 mV when only K^+ is permeant, +58 mV when only Na^+ is permeant, and some intermediate value if both ions are permeant. For example, if K^+ and Na^+ were equally permeant, then the potential would be 0 mV.

Multiple permeant ions and action potentials

With respect to neural signaling, it is particularly pertinent to ask what would happen if the membrane started out being permeable to K^+ and then temporarily switched to become most permeable to Na^+. In this circumstance, the membrane potential would start out at a negative level, become positive while the Na^+ permeability remained high, and then fall back to a negative level as the Na^+ permeability decreased again. As it turns out, this case essentially describes what goes on in a neuron during the generation of an action potential. In the resting state, P_K of the neuronal plasma membrane is much higher than P_{Na}; since, as a result of the action of ion transporters, there is always more K^+ inside the cell than outside, the resting potential is negative (Figure 2.8B). As the membrane potential is depolarized (by synaptic action, for example), P_{Na} increases. The transient increase in Na^+ permeability causes the membrane potential to become even more positive (red region in Figure 2.8B), because Na^+ rushes in (there is much more Na^+ outside a neuron than inside, again as a result of ion pumps). Because of this positive feedback relationship,

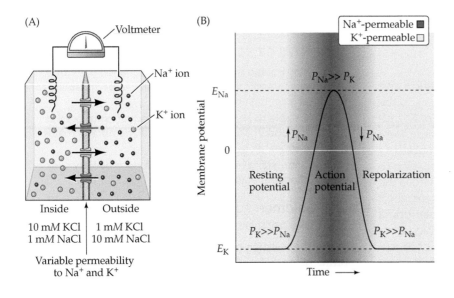

FIGURE 2.8   **Resting and action potentials rely on permeabilities to different ions**
(A) Hypothetical situation in which a membrane variably permeable to Na^+ (red) and K^+ (gold) separates two compartments that contain both ions. For simplicity, Cl^- ions are not shown in the diagram. (B) Schematic representation of the membrane ion permeabilities associated with resting and action potentials. At rest, neuronal membranes are more permeable to K^+ (gold) than to Na^+ (red); accordingly, the resting membrane potential is negative and approaches the equilibrium potential for K^+, E_K. During an action potential, the membrane becomes very permeable to Na^+ (red); thus, the membrane potential becomes positive and approaches the equilibrium potential for Na^+, E_{Na}. The rise in Na^+ permeability is transient, however, so the membrane again becomes primarily permeable to K^+, causing the potential to return to its negative resting value.

an action potential occurs. The rise in Na$^+$ permeability during the action potential is transient, however; as the membrane permeability to K$^+$ is restored, the membrane potential quickly returns to its resting level.

Appreciation of the influence of ion permeability on membrane potential will make it easy to understand the key experiment in Concept 2.7 that proved how neurons generate action potentials.

<table>
<tr><td>CONCEPT
2.7</td><td>## Action Potentials Arise from Sequential Changes in Sodium and Potassium Permeability</td></tr>
</table>

LEARNING OBJECTIVES

2.7.1 Explain the experimental evidence for the conclusion that overshooting action potentials of neurons can arise from a transient increase in membrane permeability to sodium ions.

2.7.2 Describe the different phases of an action potential.

Role of Na$^+$ in action potential generation

What causes the membrane potential of a neuron to depolarize during an action potential? Although a general answer to this question has been given (i.e., increased permeability to Na$^+$), it is well worth examining the most persuasive experimental support for this concept. The data in Table 2.1 indicate that the equilibrium potential for Na$^+$ (E_{Na}) in neurons, and indeed in most cells, is positive. Thus, if the membrane were to become highly permeable to Na$^+$, the membrane potential would become positive. Based on these considerations, Hodgkin and Katz hypothesized that the action potential arises because the neuronal membrane becomes temporarily permeable to Na$^+$.

Taking advantage of the same style of ion substitution experiment they used to assess the ionic basis of the resting potential, Hodgkin and Katz tested the role of Na$^+$ in generating the action potential by asking what happens to the action potential when Na$^+$ is removed from the external medium. They found that lowering the external Na$^+$ concentration reduces both the rate of rise of the action potential and its peak amplitude (Figure 2.9A–C). Indeed, when they examined this Na$^+$ dependence quantitatively, they found a more-or-less linear relationship between the amplitude of the action potential and the logarithm of the external Na$^+$ concentration (Figure 2.9D). The slope of this relationship approached a value of 58 mV per tenfold change in Na$^+$ concentration, as expected for a membrane selectively permeable to Na$^+$. In contrast, lowering Na$^+$ concentration had very little effect on the resting membrane potential (Figure 2.9E). Thus, while the resting neuronal membrane is only slightly permeable to Na$^+$, the membrane

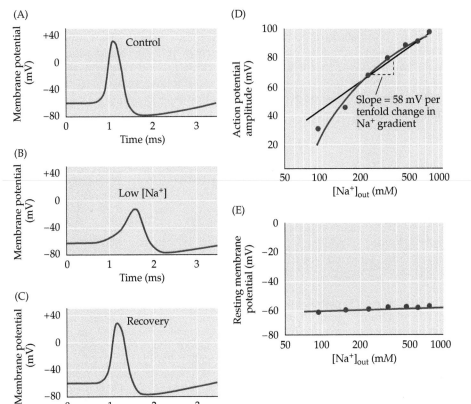

FIGURE 2.9 The role of Na$^+$ in generating an action potential in a squid giant axon (A) An action potential evoked with the normal ion concentrations inside and outside the cell. (B,C) The amplitude and rate of rise of the action potential (B) diminish when external Na$^+$ concentration is reduced to one-third of normal, but recover (C) when the Na$^+$ is replaced. (D,E) Although the amplitude of the action potential (D) is quite sensitive to the external concentration of Na$^+$, the resting membrane potential (E) is little affected by changing the concentration of this ion. (After A. L. Hodgkin and B. Katz. 1949. *J Physiol* 108: 37–77.)

becomes extraordinarily permeable to Na⁺ during the **rising phase** and the **overshoot phase** of an action potential. (Box 2B further explains action potential nomenclature.) This temporary increase in Na⁺ permeability results from the opening of Na⁺-selective channels that are closed in the resting state. Membrane pumps maintain a large electrochemical gradient for Na⁺, which is more concentrated outside the neuron than inside. This causes Na⁺ to flow into the neuron when the Na⁺ channels open, making the membrane potential depolarize and approach E_{Na}.

■ BOX 2B | Action Potential Form and Nomenclature

The action potential of the squid giant axon has a characteristic shape, or waveform, with several different phases (Figure A). During the rising phase, the membrane potential rapidly depolarizes. In fact, action potentials cause the membrane potential to depolarize so much that the membrane potential transiently becomes positive with respect to the external medium, producing an overshoot. The overshoot of the action potential gives way to a falling phase, in which the membrane potential rapidly repolarizes. Repolarization takes the membrane potential to levels even more negative than the resting membrane potential for a short time; this brief period of hyperpolarization is called the undershoot.

Although the waveform of the squid action potential is typical, the detailed form of the action potential varies widely from neuron to neuron in different animals. In myelinated axons of vertebrate motor neurons (Figure B), the action potential is virtually indistinguishable from that of the squid axon. However, the action potential recorded in the cell body of this same motor neuron (Figure C) looks rather different. Thus, action potential waveform can vary even within the same neuron. More complex action potentials are seen in other central neurons. For example, action potentials recorded from the cell bodies of neurons in the mammalian inferior olive (a region of the brainstem involved in motor control) last tens of milliseconds (Figure D). These action potentials exhibit a pronounced plateau during their falling phase, and their undershoot lasts even longer than that of the motor neuron. One of the most dramatic types of action potentials occurs in cerebellar Purkinje neurons (Figure E). These potentials, termed *complex spikes*, are well named because they have several phases that result from the summation of multiple, discrete action potentials generated in different regions of the neuron.

The variety of action potential waveforms could mean that each type of neuron has a different mechanism of action potential production. Fortunately, however, these diverse waveforms all result from relatively minor variations in the scheme used by the squid giant axon. For example, plateaus in the repolarization phase result from the presence of ion channels that are permeable to Ca²⁺, and long-lasting undershoots result from the presence of additional types of membrane K⁺ channels. The complex action potential of the Purkinje cell results from these extra features plus the fact that different types of action potentials are generated in various parts of the Purkinje neuron—cell body, dendrites, and axons—and are summed together in recordings from the cell body. Thus, the lessons learned from the squid axon are applicable to, and indeed essential for, understanding action potential generation in all neurons.

(A) 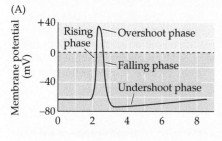 (B) (C) 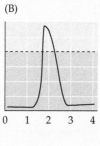 (D) 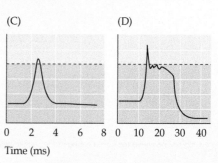 (E)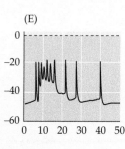

Time (ms)

(A) The phases of an action potential of the squid giant axon. (B) Action potential recorded from a myelinated axon of a frog motor neuron. (C) Action potential recorded from the cell body of a frog motor neuron. The action potential is smaller and the undershoot prolonged in comparison with the action potential recorded from the axon of this same neuron (B). (D) Action potential recorded from the cell body of a neuron from the inferior olive of a guinea pig. This action potential has a pronounced plateau during its falling phase. (E) Action potential recorded from the dendrite of a Purkinje neuron in the cerebellum of a mouse. (A after A. L. Hodgkin and A. F. Huxley. 1939. *Nature* 144: 710–711; B after F. A. Dodge and B. Frankenhaeuser. 1958. *J Physiol* 143: 76–90; C after E. F. Barrett and J. N. Barrett. 1976. *J Physiol* 255: 737–774; D after R. Llinás and Y. Yarom. 1981. *J Physiol* 315: 549–567; E after S. Chen et al. 2016. *eLife* 5: e10509.)

The length of time the membrane potential lingers near E_{Na} (about +50 mV) during the overshoot phase of an action potential is brief because the increased membrane permeability to Na^+ itself is short-lived. The membrane potential rapidly repolarizes to resting levels and is followed by a transient **undershoot**. As we will describe in Chapter 3, these latter phases of the action potential are due to an inactivation of the Na^+ permeability and an increase in the K^+ permeability of the membrane. During the undershoot, the membrane potential is transiently hyperpolarized because K^+ permeability becomes even greater than it is at rest. The action potential ends when this phase of enhanced K^+ permeability subsides, and the membrane potential thus returns to its normal resting level.

The ion substitution experiments carried out by Hodgkin and Katz provided convincing evidence that (1) the resting membrane potential results from a high resting membrane permeability to K^+, and (2) depolarization during an action potential results from a transient rise in membrane Na^+ permeability. Although these experiments identified the ions that flow during an action potential, they did not establish *how* the neuronal membrane is able to change its ion permeability to generate the action potential, or what mechanisms trigger this critical change.

Chapter 3 addresses these issues, documenting the surprising conclusion that the neuronal membrane potential itself affects membrane permeability.

Summary

Nerve cells generate electrical signals to convey information over substantial distances and to transmit it to other cells by means of synaptic connections. These signals ultimately depend on changes in the resting electrical potential across the neuronal membrane. A negative membrane potential at rest results from a net efflux of K^+ across neuronal membranes that are predominantly permeable to K^+. In contrast, an action potential occurs when a transient rise in Na^+ permeability allows a net influx of Na^+. The brief rise in membrane Na^+ permeability is followed by a secondary, transient rise in membrane K^+ permeability that repolarizes the neuronal membrane and produces a brief undershoot of the action potential. As a result of these processes, the membrane is depolarized in an all-or-none fashion during an action potential. When these active permeability changes subside, the membrane potential returns to its resting level because of the high resting membrane permeability to K^+.

■ Additional Reading

Reviews

Barnett, M. W. and P. M. Larkman (2007) The action potential. *Pract. Neurol.* 7: 192–197.

Cardozo, D. (2016) An intuitive approach to understanding the resting membrane potential. *Adv. Physiol. Educ.* 40: 543–547.

Hodgkin, A. L. (1958) The Croonian Lecture: Ionic movements and electrical activity in giant nerve fibres. *Proc. R. Soc. Lond. (B)* 148: 1–37.

Important Original Papers

Baker, P. F., A. L. Hodgkin and T. I. Shaw (1962) Replacement of the axoplasm of giant nerve fibres with artificial solutions. *J. Physiol.* 164: 330–354.

Cole, K. S. and H. J. Curtis (1939) Electric impedence of the squid giant axon during activity. *J. Gen. Physiol.* 22: 649–670.

Curtis, H. J. and K. S. Cole (1942) Membrane resting and action potentials from the squid giant axon. *J. Cell. Comp. Physiol.* 19: 135–144.

Goldman, D. E. (1943) Potential, impedence, and rectification in membranes. *J. Gen. Physiol.* 27: 37–60.

Hodgkin, A. L. and P. Horowicz (1959) The influence of potassium and chloride ions on the membrane potential of single muscle fibres. *J. Physiol.* 148: 127–160.

Hodgkin, A. L. and B. Katz (1949) The effect of sodium ions on the electrical activity of the giant axon of the squid. *J. Physiol.* 108: 37–77.

Hodgkin, A. L. and R. D. Keynes (1953) The mobility and diffusion coefficient of potassium in giant axons from *Sepia*. *J. Physiol.* 119: 513–528.

Hodgkin, A. L. and W. A. H. Rushton (1946) The electrical constants of a crustacean nerve fibre. *Proc. R. Soc. Lond. B* 133: 444–479.

Keynes, R. D. (1951) The ionic movements during nervous activity. *J. Physiol.* 114: 119–150.

Nernst, W. (1888). Zur Kinetik der Lösung befindlichen Körper: Theorie der Diffusion. *Z. Phys. Chem.* 3: 613–637.

Books

Campenot, R. B. (2017) *Animal Electricity*. Cambridge, MA: Harvard University Press.

Hodgkin, A. L. (1992) *Chance and Design*. Cambridge, UK: Cambridge University Press.

Voltage-Dependent Membrane Permeability

Science Photo Library/Alamy Stock Photo

KEY CONCEPTS

3.1 Ion currents flow across nerve cell membranes

3.2 Depolarization activates two types of voltage-dependent ion currents

3.3 Voltage-dependent ion currents arise from two voltage-dependent membrane conductances

3.4 Action potentials can be reconstructed based on the properties of voltage-dependent Na^+ and K^+ conductances

3.5 Action potentials enable long-distance electrical signaling

3.6 Myelination increases action potential conduction velocity

Overview

The action potential is a fundamental electrical signal generated by nerve cells and arises from changes in membrane permeability to specific ions. Present understanding of these changes in ion permeability is based on evidence obtained by the voltage clamp technique, which permits detailed characterization of permeability changes as a function of membrane potential and time. For most types of neurons, these changes consist of a rapid and transient rise in sodium (Na^+) permeability, followed by a slower but more prolonged rise in potassium (K^+) permeability. Both permeabilities are voltage-dependent, increasing as the membrane potential depolarizes. The measured kinetics and voltage dependence of Na^+ and K^+ permeabilities are sufficient to explain action potential generation. Depolarizing the membrane potential to the threshold level causes a rapid, self-sustaining increase in Na^+ permeability that produces the rising phase of the action potential; however, the Na^+ permeability increase is short-lived and is followed by a slower increase in K^+ permeability that restores the membrane potential to its usual negative resting level. A mathematical model that describes the behavior of these ion permeabilities accurately predicts the observed properties of action potentials. Importantly, voltage-dependent Na^+ and K^+ permeabilities also permit action potentials to be propagated along the length of axons, explaining how electrical signals are conveyed within neurons throughout the nervous system.

CONCEPT
3.1 | Ion Currents Flow across Nerve Cell Membranes

LEARNING OBJECTIVES

3.1.1 Understand how the voltage clamp technique allows measurement of ion currents flowing across a membrane.

3.1.2 Explain how depolarizing and hyperpolarizing a neuronal membrane elicit different types of current flow.

Voltage clamp reveals the relationship between membrane potential and permeability

Chapter 2 introduced the idea that nerve cells generate electrical signals by virtue of a membrane that is differentially permeable to various ion species. In particular, a transient increase in the permeability of the neuronal membrane to Na^+ initiates the action potential. This chapter considers exactly how this increase in Na^+ permeability occurs. A key to understanding this phenomenon is the observation that action potentials are initiated *only* when the neuronal membrane potential becomes more positive than a threshold level. This observation suggests that the mechanism responsible for the increase in Na$^+$ permeability is sensitive to the membrane potential. Therefore, if one could understand how a change in membrane potential activates Na$^+$ permeability, it should be possible to explain how action potentials are generated.

The fact that the Na$^+$ permeability that generates the membrane potential change is itself sensitive to the membrane potential presents both conceptual and practical obstacles to studying the mechanisms underlying the action potential. A practical problem is the difficulty of systematically varying the membrane potential to study the permeability change, because such changes in membrane potential will produce an action potential, which causes further, uncontrolled changes in the membrane potential. Historically, then, it was not possible to understand action potentials until a technique was developed that allowed experimenters to control membrane potential *and* simultaneously measure the underlying permeability changes. This technique, the **voltage clamp method** (Box 3A), provides the information needed to define the ion permeability of the membrane at any level of membrane potential.

■ BOX 3A │ The Voltage Clamp Method

Breakthroughs in scientific research often rely on the development of new technologies. In the case of the action potential, detailed understanding came only after the invention of the voltage clamp technique by Kenneth Cole in the 1940s. This device is called a voltage clamp because it controls, or clamps, membrane potential (or voltage) at any level desired by the experimenter. The method measures the membrane potential with an electrode placed inside the cell (1) and electronically compares this voltage with the voltage to be maintained (called the *command voltage*) (2). The clamp circuitry then passes a current back into the cell though another intracellular electrode (3). This electronic feedback circuit holds the membrane potential at the desired level, even in the face of permeability changes that would normally alter the membrane potential (such as those generated during the action potential). Most importantly, the device permits the simultaneous measurement of the current needed to keep the cell at a given voltage (4). This current is exactly equal to the amount of current flowing across the neuronal membrane, allowing direct measurement of these membrane currents. Therefore, the voltage clamp technique can indicate how membrane potential influences ion current flow across the membrane. This information gave Hodgkin and Huxley the key insights that led to their model for action potential generation.

Today, the voltage clamp method remains widely used to study ion currents in neurons and other cells. The most popular contemporary version of this approach is the patch clamp technique, a method that can be applied to virtually any cell and has a resolution high enough to measure the minute electrical currents flowing through single ion channels (see Box 4A).

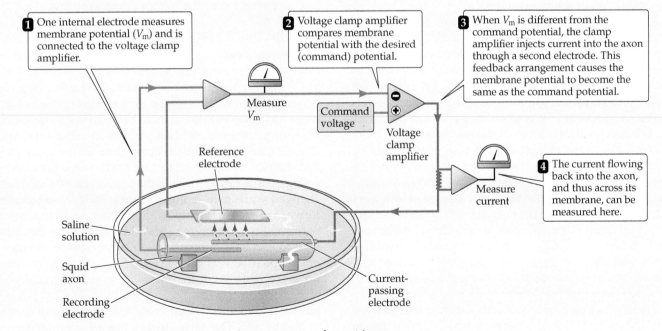

1 One internal electrode measures membrane potential (V_m) and is connected to the voltage clamp amplifier.

2 Voltage clamp amplifier compares membrane potential with the desired (command) potential.

3 When V_m is different from the command potential, the clamp amplifier injects current into the axon through a second electrode. This feedback arrangement causes the membrane potential to become the same as the command potential.

4 The current flowing back into the axon, and thus across its membrane, can be measured here.

Measure V_m

Command voltage

Voltage clamp amplifier

Reference electrode

Measure current

Saline solution

Squid axon

Recording electrode

Current-passing electrode

Voltage clamp technique for studying membrane currents of a squid axon.

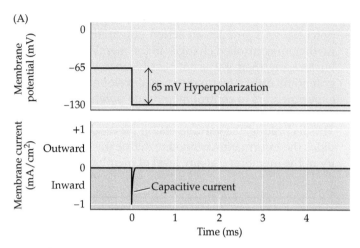

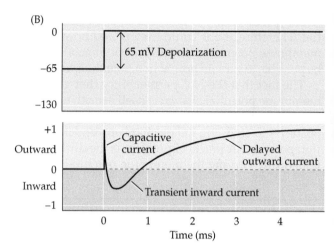

FIGURE 3.1   Current flow across a squid axon membrane during a voltage clamp experiment (A) A 65-mV hyperpolarization of the membrane potential produces only a very brief capacitive current. (B) A 65-mV depolarization of the membrane potential also produces a brief capacitive current, which is followed by a longer lasting but transient phase of inward current and a delayed but sustained outward current. (After A. Hodgkin et al. 1952. *J Physiol* 116: 424–448.)

Voltage-sensitive ion currents

In the late 1940s, Alan Hodgkin and Andrew Huxley, from the University of Cambridge, used the voltage clamp technique to work out the permeability changes underlying the action potential. They again chose to use the giant axon of a squid because its large size (up to 1 mm in diameter; see Box 2A) allowed insertion of the electrodes necessary for voltage clamping. They were the first investigators to test directly the hypothesis that potential-sensitive Na^+ and K^+ permeability changes are both necessary and sufficient for the production of action potentials.

Hodgkin and Huxley's first goal was to determine whether neuronal membranes do, in fact, have voltage-dependent permeabilities. To address this issue, they asked whether ion currents flow across the membrane when its potential is changed. The result of one such experiment is shown in Figure 3.1. Figure 3.1A illustrates the currents produced by a squid axon when its membrane potential, V_m, is hyperpolarized from the resting level (–65 mV) to –130 mV. The initial response of the axon results from the redistribution of charge across the axonal membrane. This capacitive current is nearly instantaneous, ending within a fraction of a millisecond. Aside from this brief event, very little current flows when the membrane is hyperpolarized. However, when the membrane potential is depolarized from –65 mV to 0 mV, the response is quite different (see Figure 3.1B). Following the capacitive current, the axon produces a rapidly rising inward ion current (inward refers to a positive charge entering the cell—that is, cations in or anions out), which gives way to a more slowly rising, delayed outward current. The fact that membrane depolarization elicits these ion currents establishes that the membrane permeability of axons is indeed voltage-dependent.

<table>
<tr><td>CONCEPT
3.2</td><td>## Depolarization Activates Two Types of Voltage-Dependent Ion Currents</td></tr>
</table>

LEARNING OBJECTIVES

3.2.1 Recall that membrane depolarization activates Na^+ and K^+ currents.

3.2.2 Explain the properties of voltage-dependent Na^+ and K^+ currents.

3.2.3 Describe how Na^+ and K^+ currents can be distinguished in a voltage clamp experiment.

Voltage dependence of ion currents

The results shown in Figure 3.1 demonstrate that the ion permeability of neuronal membranes is voltage-sensitive, but the experiments do not identify how many types of permeability exist, or which ions are involved. As discussed in Chapter 2 (see Figure 2.6), varying the potential across a membrane makes it possible to deduce the equilibrium potential for the ion fluxes through the membrane, and thus to identify the ions that are flowing. Because the voltage clamp method allows the membrane potential to be changed while measuring ion currents, it was a straightforward matter for Hodgkin and Huxley to determine ion permeability by examining how the properties of the initial inward and later outward currents changed as the membrane potential was varied (Figure 3.2). As already noted, no appreciable ion currents flow at membrane potentials more negative than the resting potential. At more positive potentials, however, the currents not only flow but change in magnitude. The early current has a U-shaped dependence on membrane potential, increasing over a range of depolarizations up to approximately 0 mV but decreasing

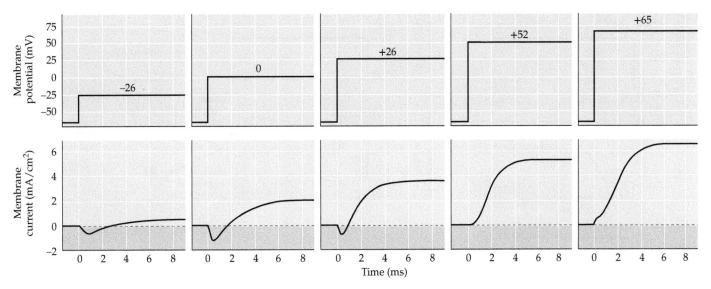

FIGURE 3.2   Currents produced by membrane depolarizations to several different potentials The early current first increases, then decreases in magnitude as depolarization increases; note that this current reverses its polarity at potentials more positive than about +55 mV. The later outward current increases monotonically with increasing depolarization. (After A. Hodgkin et al. 1952. *J Physiol* 116: 424–448.)

as the potential is depolarized further. In contrast, the late current increases monotonically with increasingly positive membrane potentials. These different responses to membrane potential can be seen more clearly when the magnitudes of the two current components are plotted as a function of membrane potential, as in Figure 3.3.

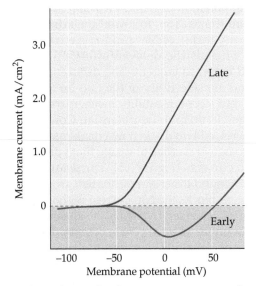

FIGURE 3.3   Relationship between current amplitude and membrane potential Experiments such as the one shown in Figure 3.2 indicate that the late outward current increases steeply with increasing depolarization, whereas the early inward current first increases in magnitude but then decreases and reverses to outward current at about +55 mV (the sodium equilibrium potential). (After A. Hodgkin et al. 1952. *J Physiol* 116: 424–448.)

Role of Na⁺ in the early current

The voltage sensitivity of the early current gives an important clue about the nature of the ions carrying the current—namely, that no current flows when the membrane potential is clamped at +52 mV. For the squid neurons studied by Hodgkin and Huxley, the external Na⁺ concentration is 440 mM, and the internal Na⁺ concentration is 50 mM (see Table 2.1). For this concentration gradient, the Nernst equation predicts that the equilibrium potential for Na⁺ should be +55 mV. Recall further from Chapter 2 that at the Na⁺ equilibrium potential there is no net flux of Na⁺ across the membrane, even if the membrane is highly permeable to Na⁺. Thus, the experimental observation that no early current flows at the membrane potential where Na⁺ cannot flow is a strong indication that the current is carried by entry of Na⁺ into the axon.

An even more demanding way to test whether Na⁺ carries the early current is to examine the behavior of this current after *removing* external Na⁺. Removing Na⁺ outside the axon makes E_{Na} negative; if the permeability to Na⁺ is increased under these conditions, current should flow outward as Na⁺ leaves the neuron, due to the reversed electrochemical gradient. Hodgkin and Huxley performed this experiment, and found that removing external Na⁺ caused the early current to reverse its polarity and become an outward current at a membrane potential that gave rise to an inward current when external Na⁺ was present (Figure 3.4). This result demonstrates convincingly that the early inward current measured when Na⁺ is present in the external medium must be due to Na⁺ entering the neuron.

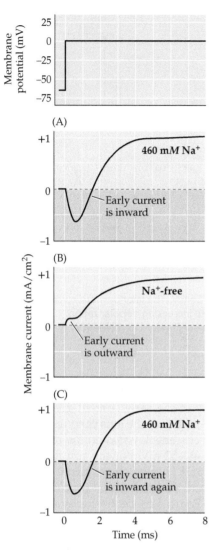

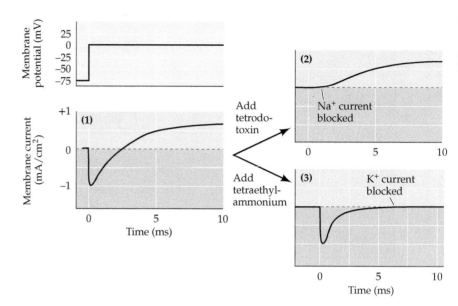

FIGURE 3.5 Pharmacological separation of Na⁺ and K⁺ currents into sodium and potassium components Panel (1) shows the current that flows when the membrane potential of a squid axon is depolarized to 0 mV in control conditions. (2) Treatment with tetrodotoxin causes the early Na⁺ current to disappear but spares the late K⁺ current. (3) Addition of tetraethylammonium blocks the K⁺ current without affecting the Na⁺ current. (After J. W. Moore et al. 1967. *J Gen Physiol* 50: 1401–1411 and C. M. Armstrong and L. Binstock. 1965. *J Gen Physiol* 48: 859–872.)

FIGURE 3.4 Dependence of the early current on sodium (A) In the presence of normal external concentrations of Na⁺, depolarization of a squid axon to 0 mV (top) produces an inward initial current. (B) Removal of external Na⁺ causes this initial inward current to become outward, an effect that is reversed (C) by restoration of external Na⁺. (After A. L. Hodgkin and A. F. Huxley. 1952. *J Physiol* 116: 449–472.)

Role of K⁺ in the late current

In the experiment shown in Figure 3.4, removal of external Na⁺ has little effect on the outward current that flows after the neuron has been kept at a depolarized membrane voltage for several milliseconds. This shows that the late outward current must be due to the flow of an ion other than Na⁺. Several lines of evidence presented by Hodgkin, Huxley, and others showed that this outward current is caused by K⁺ exiting the neuron. Perhaps the most compelling demonstration of K⁺ involvement is that the amount of K⁺ efflux from the neuron (measured by loading the

neuron with radioactive K⁺) is closely correlated with the magnitude of the late outward current.

Taken together, these experiments show that changing the membrane potential to a level more positive than the resting potential produces two effects: an early influx of Na⁺ into the neuron, followed by a delayed efflux of K⁺. The early influx of Na⁺ produces a transient inward current, whereas the delayed efflux of K⁺ produces a sustained outward current. The differences in the time course and ion selectivity of the two fluxes suggest that two different ion permeability mechanisms are activated by changes in membrane potential. Confirmation that there are indeed two distinct mechanisms has come from pharmacological studies of drugs that specifically affect these two currents (Figure 3.5). *Tetrodotoxin*, an alkaloid neurotoxin found in certain puffer fish, tropical frogs, and salamanders, blocks the Na⁺ current without affecting the K⁺ current. Conversely, *tetraethylammonium* ions block K⁺ currents without affecting Na⁺ currents. The differential sensitivity of Na⁺ and K⁺ currents to these drugs provides strong additional evidence that Na⁺ and K⁺ flow through independent permeability pathways. As we will discuss in Chapter 4, it is now known that these pathways are ion channels that are selectively permeable to either Na⁺ or K⁺. In fact, tetrodotoxin, tetraethylammonium, and other drugs that interact with specific types of ion channels have been extraordinarily useful tools in characterizing these channel proteins (see Box 4B).

CONCEPT
3.3
Voltage-Dependent Ion Currents Arise from Two Voltage-Dependent Membrane Conductances

LEARNING OBJECTIVES

3.3.1 Describe the properties of voltage-dependent Na$^+$ and K$^+$ conductances.

3.3.2 Summarize how these voltage-dependent conductances produce the voltage-dependent Na$^+$ and K$^+$ currents.

Membrane conductance

The next goal Hodgkin and Huxley set for themselves was to describe Na$^+$ and K$^+$ permeability changes mathematically. To do this, they assumed that the ion currents are due to a change in **membrane conductance**, defined as the reciprocal of the membrane resistance. Membrane conductance is thus closely related, although not identical, to membrane permeability. When evaluating ion movements from an electrical standpoint, it is convenient to describe them in terms of ion conductances rather than ion permeabilities. For present purposes, permeability and conductance can be considered synonymous.

If membrane conductance (g) obeys Ohm's Law (which states that voltage is equal to the product of current and resistance), then the ion current that flows during an increase in membrane conductance is given by

$$I_{ion} = g_{ion} (V_m - E_{ion})$$

where I_{ion} is the ion current, V_m is the membrane potential, and E_{ion} is the equilibrium potential for the ion flowing through the conductance, g_{ion}. The difference between V_m and E_{ion} is the electrochemical driving force acting on the ion.

Properties of Na$^+$ and K$^+$ conductances

Hodgkin and Huxley used this simple relationship to calculate the dependence of Na$^+$ and K$^+$ conductances on time and membrane potential. They knew V_m, which was set by their voltage clamp device (Figure 3.6A), and they could

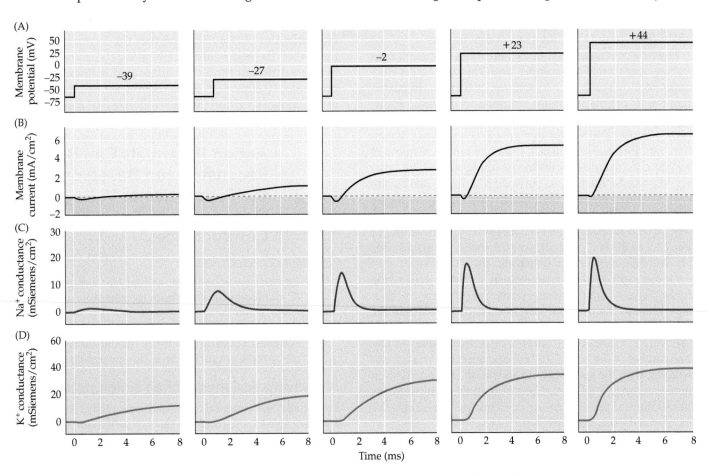

FIGURE 3.6 Membrane conductance changes underlying the action potential are time- and voltage-dependent Depolarizations to various membrane potentials (A) elicit different membrane currents (B). Shown below are the conductances of Na$^+$ (C) and K$^+$ (D) calculated from these currents. Both peak Na$^+$ conductance and steady-state K$^+$ conductance increase as the membrane potential becomes more positive. In addition, the activation of both conductances, as well as the rate of inactivation of the Na$^+$ conductance, occurs more rapidly with larger depolarizations. (After A. L. Hodgkin and A. F. Huxley. 1952. *J Physiol* 116: 473–493.)

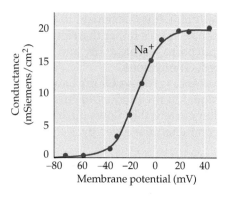

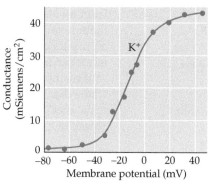

FIGURE 3.7 **Depolarization increases Na⁺ and K⁺ conductances of the squid giant axon** The peak magnitude of Na⁺ conductance and steady-state value of K⁺ conductance both increase steeply as the membrane potential is depolarized. (After A. L. Hodgkin and A. F. Huxley. 1952. *J Physiol* 116: 473–493.)

determine E_{Na} and E_K from the ion concentrations on the two sides of the axonal membrane (see Table 2.1). The currents carried by Na⁺ and K⁺—that is, I_{Na} and I_K—could be determined separately from recordings of the membrane currents resulting from depolarization (Figure 3.6B) by measuring the difference between currents recorded in the presence and absence of external Na⁺ (as shown in Figure 3.4). From these measurements, Hodgkin and Huxley were able to calculate g_{Na} and g_K (Figure 3.6C,D), from which they drew two fundamental conclusions.

The first conclusion is that the Na⁺ and K⁺ conductances change over time. For example, both Na⁺ and K⁺ conductances require some time to activate, or turn on. In particular, the K⁺ conductance has a pronounced delay, requiring several milliseconds to reach its maximum (see Figure 3.6D), whereas the Na⁺ conductance reaches its maximum more rapidly (see Figure 3.6C). The more rapid activation of the Na⁺ conductance allows the resulting inward Na⁺ current to precede the delayed outward K⁺ current (see Figure 3.6B). Although the Na⁺ conductance rises rapidly, it quickly declines, even though the membrane potential is kept at a depolarized level. This fact shows that depolarization not only causes the Na⁺ conductance to activate, but also causes it to decrease over time, or inactivate. The K⁺ conductance of the squid axon does not inactivate in this way; thus, while the Na⁺ and K⁺ conductances share the property of time-dependent **activation**, only the Na⁺ conductance exhibits **inactivation**. (Inactivating K⁺ conductances have since been discovered in other types of nerve cells; see Chapter 4.) The time courses of the Na⁺ and K⁺ conductances are voltage-dependent, with the speed of both activation and inactivation increasing at more depolarized potentials. This accounts for the more rapid membrane currents measured at more depolarized potentials (see Figure 3.6B).

Voltage dependence of conductances

The second conclusion derived from Hodgkin and Huxley's calculations is that both the Na⁺ and K⁺ conductances are voltage-dependent—that is, both conductances increase progressively as the neuron is depolarized. Figure 3.7 illustrates this by plotting the relationship between peak value of the conductances (from Figure 3.6C,D) against the membrane potential. Note the similar voltage dependence for each conductance; both conductances are quite small at negative potentials, maximal at very positive potentials, and exquisitely dependent on membrane voltage at intermediate potentials. The observation that these conductances are sensitive to changes in membrane potential shows that the mechanism underlying the conductances somehow "senses" the voltage across the membrane.

All told, the voltage clamp experiments carried out by Hodgkin and Huxley showed that the ion currents that flow when the neuronal membrane is depolarized are due to three different time-dependent and voltage-sensitive processes: (1) activation of Na⁺ conductance, (2) activation of K⁺ conductance, and (3) inactivation of Na⁺ conductance.

CONCEPT **3.4**

Action Potentials Can Be Reconstructed Based on the Properties of Voltage-Dependent Na⁺ and K⁺ Conductances

LEARNING OBJECTIVES

3.4.1 Explain how voltage-dependent Na⁺ and K⁺ conductances contribute to generation of action potentials.

3.4.2 Describe the feedback relationships between these conductances and how these relationships confer all-or-none properties on action potentials.

A mathematical model of an action potential

From their experimental measurements, Hodgkin and Huxley were able to construct a detailed mathematical model of the Na⁺ and K⁺ conductance changes. The goal of these modeling efforts was to determine whether the Na⁺ and K⁺ conductances alone are sufficient to produce an action potential. Using this information, they could in fact generate the form and time course of the action potential with remarkable accuracy (Figure 3.8A). The Hodgkin–Huxley model could simulate many other features of action potential behavior in the squid axon. For example, it was well known that,

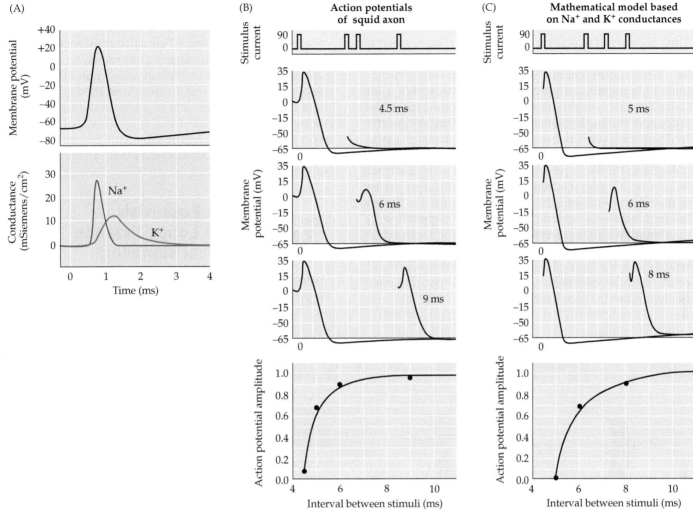

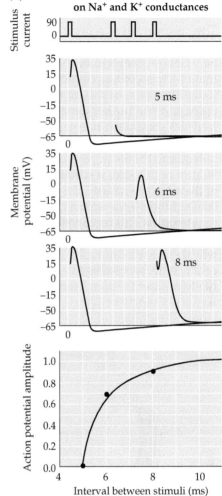

FIGURE 3.8 Mathematical simulation of the action potential (A) Simulation of an action potential (black curve) together with the underlying changes in Na⁺ (red curve) and K⁺ (blue curve) conductance. The size and time course of the action potential were calculated using only the properties of g_{Na} and g_K measured in voltage clamp experiments. (B) The refractory period can be observed by stimulating an axon with two current pulses that are separated by variable intervals. Whereas the first stimulus reliably evokes an action potential, during the refractory period the second stimulus will generate only a small action potential or no response at all. (C) The mathematical model accurately predicts responses of the axon during the refractory period. (After A. L. Hodgkin and A. F. Huxley. 1952. *J Physiol* 116: 507–544.)

following an action potential, the axon becomes refractory to further excitation for a brief period of time, termed the **refractory period** (Figure 3.8B). The model was capable of closely mimicking such behavior (Figure 3.8C).

The Hodgkin–Huxley model also provided many insights into how action potentials are generated. Figure 3.8A compares the time courses of a reconstructed action potential and the underlying Na⁺ and K⁺ conductances. The coincidence of the initial increase in Na⁺ conductance with the rapid rising phase of the action potential demonstrates that a selective increase in Na⁺ conductance is responsible for action potential initiation. The increase in Na⁺ conductance causes Na⁺ to enter the neuron, thus depolarizing the membrane potential, which approaches E_{Na}. The rate of depolarization subsequently falls both because the electrochemical

driving force on Na⁺ decreases and because the Na⁺ conductance inactivates. At the same time, depolarization slowly activates the voltage-dependent K⁺ conductance, causing K⁺ to leave the cell and repolarizing the membrane potential toward E_K. Because the K⁺ conductance becomes temporarily higher than it is in the resting condition, the membrane potential briefly becomes more negative than the normal resting potential, to yield the undershoot. The hyperpolarization of the membrane potential causes the voltage-dependent K⁺ conductance (and any Na⁺ conductance not inactivated) to turn off, allowing the membrane potential to return to its resting level. The relatively slow time course of turning off the K⁺ conductance, as well as the persistence of Na⁺ conductance inactivation, is responsible for the refractory period (see also Figure 3.10).

Feedback loops underlying action potential generation

This mechanism of action potential generation represents a positive feedback loop: Activating the voltage-dependent Na⁺ conductance increases Na⁺ entry into the neuron, which makes the membrane potential depolarize, which leads to the activation of still more Na⁺ conductance, more Na⁺ entry, and still further depolarization (Figure 3.9). Positive feedback continues unabated until Na⁺ conductance inactivation and K⁺ conductance activation restore the membrane potential to the resting level. Because this positive feedback loop, once initiated, is sustained by the intrinsic properties of the neuron—namely, the voltage dependence of the ion conductances—the action potential is self-supporting, or **regenerative**. This regenerative quality explains why action potentials exhibit all-or-none behavior (see Figure 2.2) and why they have a threshold. The delayed activation of the K⁺ conductance represents a negative feedback loop that eventually restores the membrane to its resting state.

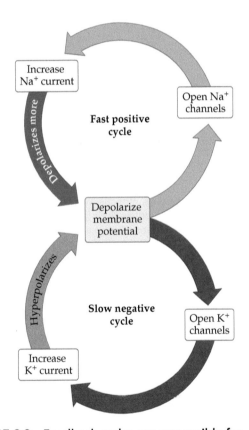

FIGURE 3.9   Feedback cycles are responsible for membrane potential changes during an action potential Membrane depolarization rapidly activates a positive feedback cycle fueled by the voltage-dependent activation of Na⁺ conductance. This phenomenon is followed by the slower activation of a negative feedback loop as depolarization activates a K⁺ conductance, which helps repolarize the membrane potential and terminate the action potential.

Hodgkin and Huxley's reconstruction of the action potential and all its features shows that the properties of the voltage-sensitive Na⁺ and K⁺ conductances, together with the electrochemical driving forces created by ion transporters, are sufficient to explain action potentials. Their use of both empirical and theoretical methods brought an unprecedented level of rigor to a long-standing problem, setting a standard of proof that is achieved only rarely in biological research. As will be shown in Chapter 4, we now know that these conductances arise from voltage-sensitive Na⁺ and K⁺ channels.

CONCEPT 3.5 Action Potentials Enable Long-Distance Electrical Signaling

LEARNING OBJECTIVE

3.5.1 Describe how the mechanisms involved in action potential generation permit propagation of an action potential.

How voltage-dependent conductances allow action potentials to propagate

The voltage-dependent mechanisms of action potential generation also explain the long-distance transmission of these electrical signals. Recall from Chapter 2 that neurons are relatively poor passive conductors of electricity, at least compared with a wire. Nonetheless, action potentials can traverse great distances along axons despite these poor passive properties. How does this occur?

The mechanism of action potential propagation is easy to grasp once one understands how action potentials are generated and how current passively flows along an axon. A depolarizing stimulus—a synaptic potential or a receptor potential in an intact neuron, or an injected current pulse in an experiment such as the one depicted in Figure 3.10—locally depolarizes the axon, thus opening the voltage-sensitive Na⁺ channels in that region. The opening of Na⁺ channels causes inward movement of Na⁺, and the resultant depolarization of the membrane potential generates an action potential at that site. Some of the local current generated by the action potential will then flow passively down the axon, in the same way that subthreshold currents spread along an axon (see Figure 2.3). Note that this passive current flow does not require the movement of Na⁺ along the axon but instead occurs by a shuttling of charge, somewhat similar to what happens when wires passively conduct electricity by transmission of electron charge. This passive current flow depolarizes the membrane potential in the adjacent region of the axon, thus opening the Na⁺ channels in the neighboring membrane. The local depolarization triggers an action potential in this region, which then spreads again in a continuing cycle until the action

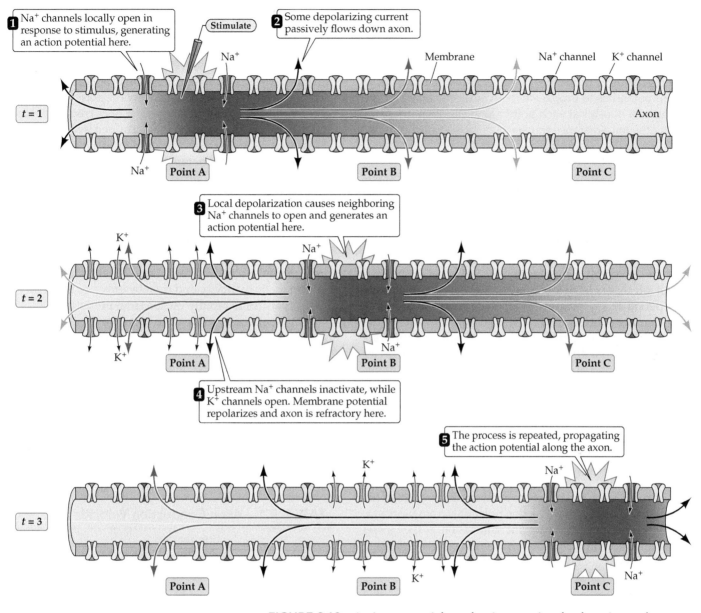

1 Na⁺ channels locally open in response to stimulus, generating an action potential here.

Stimulate

2 Some depolarizing current passively flows down axon.

Membrane   Na⁺ channel   K⁺ channel

Na⁺

t = 1

Axon

Na⁺   Point A   Point B   Point C

3 Local depolarization causes neighboring Na⁺ channels to open and generates an action potential here.

K⁺

Na⁺

t = 2

K⁺   Point A   Na⁺   Point B   Point C

4 Upstream Na⁺ channels inactivate, while K⁺ channels open. Membrane potential repolarizes and axon is refractory here.

5 The process is repeated, propagating the action potential along the axon.

K⁺   Na⁺

t = 3

Point A   Point B   K⁺   Point C   Na⁺

FIGURE 3.10 Action potential conduction requires both active and passive current flow Depolarization opens Na⁺ channels locally and produces an action potential at point A of the axon (time *t* = 1). The resulting inward current flows passively along the axon, depolarizing the adjacent region (point B) of the axon. At a later time (*t* = 2), the depolarization of the adjacent membrane has opened Na⁺ channels at point B, resulting in the initiation of the action potential at this site and additional inward current that again spreads passively to an adjacent point (point C) farther along the axon. At a still later time (*t* = 3), the action potential has propagated even farther. This cycle continues along the full length of the axon. Note that as the action potential spreads, the membrane potential repolarizes due to K⁺ channel opening and Na⁺ channel inactivation, leaving a "wake" of refractoriness behind the action potential that prevents its backward propagation. The lower panel shows the time course of membrane potential changes at the points indicated.

t = 1   *t* = 2   *t* = 3

0 mV

Point A   Threshold

−65

Point B   0

−65

Point C   0

−65

potential reaches the end of the axon. Thus, action potential propagation requires the coordinated action of two forms of current flow: the passive flow of current as well as active currents flowing through voltage-dependent ion channels. The regenerative properties of Na⁺ channel opening allow action potentials to propagate in an all-or-none fashion by acting as a booster at each point along the axon, thus ensuring the long-distance transmission of electrical signals.

Role of refractoriness

Recall that the axons are refractory following an action potential: Generation of an action potential briefly makes it harder for the axon to produce subsequent action potentials (see Figure 3.8B). Refractoriness limits the number of action potentials that a neuron can produce per unit of time, with different types of neurons having different maximum rates of action potential firing due to different types and densities of ion channels. As described in Concept 3.4, the refractory period arises because the depolarization that produces Na⁺ channel opening also causes delayed activation of K⁺ channels and Na⁺ channel inactivation, which temporarily makes it more difficult for the axon to produce another action potential. This refractoriness also has important implications for action potential conduction along axons. As the action potential sweeps along the length of an axon, in its wake the action potential leaves the Na⁺ channels inactivated and K⁺ channels activated for a brief time. The resulting refractoriness of the membrane region where an action potential has been generated prevents subsequent re-excitation of this membrane as action potentials are generated in adjacent regions of the axon (see Figure 3.10). This important feature prevents action potentials from propagating backward, toward their point of initiation, as they travel along an axon. Thus, refractory behavior ensures polarized propagation of action potentials from their usual point of initiation near the neuronal cell body, toward the synaptic terminals at the distal end of the axon.

CONCEPT **3.6** | ## Myelination Increases Action Potential Conduction Velocity

LEARNING OBJECTIVES

3.6.1 Identify what limits the speed of action potential propagation.

3.6.2 Explain how myelin greatly accelerates action potential conduction velocity.

Conduction velocity and how to speed it up

As a consequence of their mechanism of propagation, action potentials occur later and later at greater distances along the axon (see Figure 3.10, bottom left). Thus, the action potential has a measurable rate of propagation, called the **conduction velocity**. Conduction velocity is an important parameter because it defines the time required for electrical information to travel from one end of a neuron to another, and thus limits the flow of information within neural circuits. It is not surprising, then, that various mechanisms have evolved to optimize the propagation of action potentials along axons. Because action potential conduction requires passive and active flow of current, the rate of action potential propagation is determined by both of these phenomena. One way of improving passive current flow is to increase the diameter of an axon, which effectively decreases the internal resistance to passive current flow. For example, comparison of the conduction velocities of the axons of Aα and Aγ types of human motor neurons (Table 3.1) illustrates this point: Increasing axon diameter only 2.5-fold yields a 20-fold increase in conduction velocity. The even larger giant axons of invertebrates such as squid presumably evolved because they increase action potential conduction velocity and thereby enhance the ability of these creatures to rapidly escape from predators.

A more efficient strategy to improve the passive flow of electrical current is to insulate the axonal membrane, reducing the ability of current to leak out of the axon and thus increasing the distance along the axon that a given local current can flow passively. This strategy is evident in the **myelination** of axons, a process by which oligodendrocytes in the central nervous system (and Schwann cells in the peripheral nervous system) wrap the axon in **myelin**, which consists of multiple layers of closely opposed glial membranes (Figure 3.11A; see also Chapter 1). By acting

TABLE 3.1 Axon Conduction Velocities

Axon	Conduction velocity (m/s)	Diameter (µm)	Myelination
Squid giant axon	25	500	No
Human			
Motor axons			
Aα type	80–120	13–20	Yes
Aγ type	4–24	5–8	Yes
Sensory axons			
Aα type	80–120	13–20	Yes
Aβ type	35–75	6–12	Yes
Aδ type	3–35	1–5	Thin
C type	0.5–2.0	0.2–1.5	No
Autonomic			
Preganglionic B type	3–15	1–5	Yes
Postganglionic C type	0.5–2.0	0.2–1.5	No

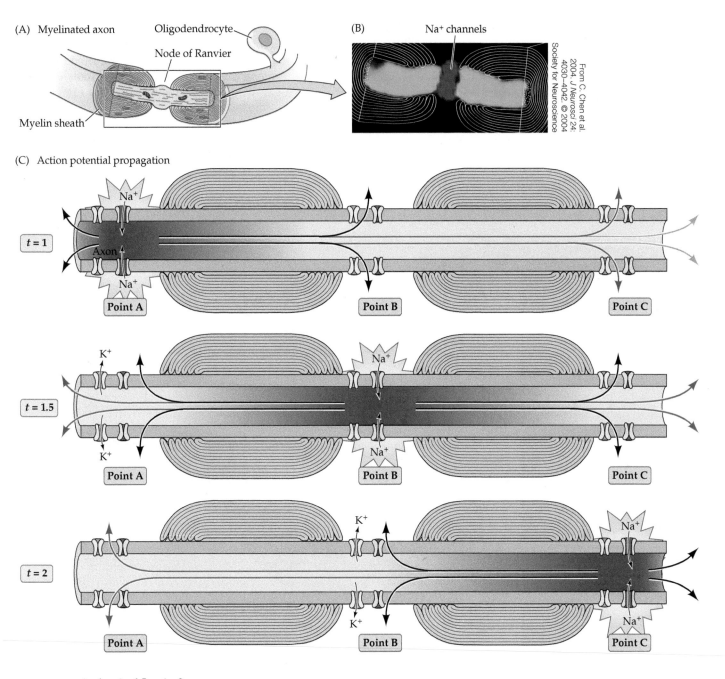

(A) Myelinated axon
Oligodendrocyte
Node of Ranvier
Myelin sheath

(B) Na⁺ channels

From C. Chen et al. 2004. *J Neurosci* 24: 4030–4042. © 2004 Society for Neuroscience

(C) Action potential propagation

t = 1   Na⁺   Axon   Na⁺   Point A   Point B   Point C

t = 1.5   K⁺   Na⁺   K⁺   Na⁺   Point A   Point B   Point C

t = 2   K⁺   Na⁺   K⁺   Na⁺   Point A   Point B   Point C

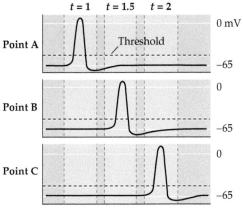

FIGURE 3.11 **Saltatory action potential conduction along a myelinated axon** (A) Diagram of a myelinated axon. (B) Localization of voltage-gated Na⁺ channels (red) at a node of Ranvier in a myelinated axon of the optic nerve. Green indicates the protein Caspr, which is located adjacent to the node of Ranvier. (C) Local current in response to action potential initiation at a particular site flows locally, as described in Figure 3.10. However, the presence of myelin prevents the local current from leaking across the internodal membrane; it therefore flows farther along the axon than it would in the absence of myelin, reaching several adjacent nodes (for clarity, only one node is shown). Moreover, voltage-gated Na⁺ channels are present only at the nodes of Ranvier (voltage-gated K⁺ channels are present at the nodes of some axons but not others). This arrangement means that the generation of active, voltage-gated Na⁺ currents need occur only at these unmyelinated regions. Bottom panel: The more rapid conduction of action potentials is illustrated by the more rapid timing of action potentials at the points indicated.

as an electrical insulator, myelin greatly speeds up action potential conduction (Figure 3.12). For example, Table 3.1 shows that whereas unmyelinated axon conduction velocities range from about 0.5 to 2 m/s, myelinated axons can conduct action potentials at velocities of up to 120 m/s (faster than a Formula 1 racing car). The major cause of this marked increase in speed is that the time-consuming process of action potential generation occurs only at specific points along the axon, called **nodes of Ranvier**, where there is a gap in the myelin wrapping (see Figure 1.3F).

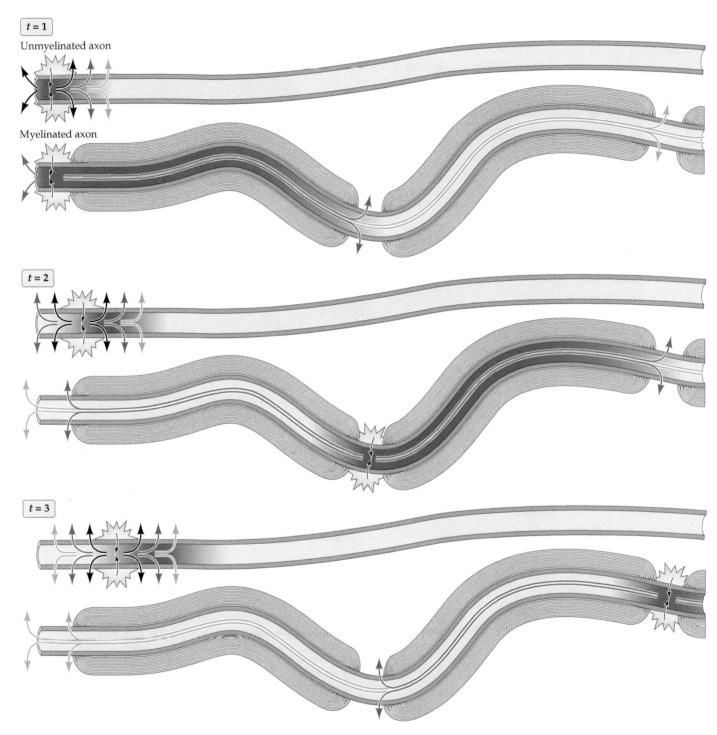

t = 1

Unmyelinated axon

Myelinated axon

t = 2

t = 3

FIGURE 3.12 Myelin increases action potential conduction speed The diagram compares action potential conduction at three different time points in unmyelinated (upper panel in each pair) and myelinated (lower panels) axons. Passive conduction of current is shown by arrows.

If the entire surface of an axon were insulated, there would be no place for current to flow out of the axon, and action potentials could not be generated. The voltage-gated Na^+ channels required for action potentials are found only at these nodes of Ranvier (Figure 3.11B). An action potential generated at one node of Ranvier elicits current that flows passively within the axon until the next few nodes are reached. This local current flow then generates an action potential in the neighboring nodes, and the cycle is repeated along the length of the axon. Because current flows across the neuronal membrane only at the nodes (Figure 3.11C), this type of propagation is called **saltatory**, meaning that the action potential jumps from node to node. Loss of myelin—as occurs in diseases such as multiple sclerosis, Guillain–Barré syndrome, and others—causes a variety of serious neurological problems because of the resulting defects in axonal conduction of action potentials (Clinical Applications).

■ Clinical Applications

Multiple Sclerosis

Multiple sclerosis (MS) is a disease of the central nervous system characterized by a variety of clinical problems that arise from demyelination and inflammation of axonal pathways. The disorder commonly appears in individuals between the ages of 20 and 40, with an abrupt onset of neurological deficits that persist for days or weeks and then subside. The clinical course ranges from patients with no persistent neurological loss, some of whom experience only occasional later symptoms, to others who gradually deteriorate as a result of extensive and relentless progression of the disease into the central nervous system.

The signs and symptoms of MS are determined by the location of the affected regions. Particularly common are monocular blindness (due to lesions of the optic nerve), motor weakness or paralysis (due to lesions of the corticospinal tracts), abnormal somatic sensations (due to lesions of somatosensory pathways, often in the posterior columns), double vision (due to lesions of medial longitudinal fasciculus), and dizziness (due to lesions of vestibular pathways). Abnormalities are often apparent in the cerebrospinal fluid, which usually contains an abnormal number of cells associated with inflammation and an increased content of antibodies (a sign of an altered immune response). MS is difficult to diagnose, generally relying on the presence of a neurological problem that subsides, only to return at an unrelated site. Confirmation can sometimes be obtained from magnetic resonance imaging (MRI) or from functional evidence of lesions in a particular pathway. The postmortem histological hallmark of MS is multiple lesions at different sites showing loss of myelin associated with infiltration of inflammatory cells and, in some instances, loss of axons themselves.

The concept of MS as a demyelinating disease is deeply embedded in the clinical literature, although precisely how the demyelination translates into functional deficits is poorly understood. The loss of the myelin sheath surrounding many axons (see figure) clearly compromises action potential conduction, and the abnormal patterns of nerve conduction that result presumably produce most of the disease symptoms. However, MS may have effects that extend beyond loss of the myelin sheath. It is clear that some axons are actually destroyed, probably as a result of inflammatory processes in the overlying myelin and/or loss of trophic support of the axon by oligodendrocytes. Thus, axon loss contributes to functional deficits in MS, especially in the chronic, progressive forms of the disease.

The ultimate cause of MS remains unclear. The immune system undoubtedly contributes to the damage, and new immunoregulatory therapies provide substantial benefits to some patients. Precisely how the immune system is activated to cause the injury is not known. The most popular hypothesis is that MS is an autoimmune disease (i.e., the immune system attacks the body's proper constituents). The fact that immunization of experimental animals with any one of several molecular constituents of the myelin sheath can induce a demyelinating disease (called experimental allergic encephalomyelitis)

(Continued)

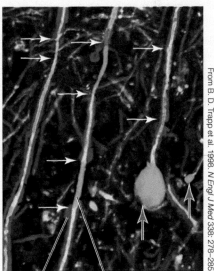

Myelin   Axon

Demyelination of axons in multiple sclerosis. Myelin (red) ordinarily surrounds axons (green). In MS, axons lose myelin (horizontal arrows) and can break as a result (vertical arrows).

From B. D. Trapp et al. 1998. *N Engl J Med* 338: 278–285

■ **Clinical Applications** (*continued*)

shows that an autoimmune attack on the myelin membrane is sufficient to produce a picture similar to MS. Very recent evidence suggests that one important target of autoimmune attack is contactin-2, a protein found on glia and axons at nodes of Ranvier. A possible cause is that a genetically susceptible individual becomes transiently infected (by a minor viral illness, for example) with a microorganism that expresses a molecule structurally similar to contactin-2 or some other component of myelin. An immune response to this antigen is mounted to attack the invader, but the failure of the immune system to discriminate between the foreign protein and self results in destruction of otherwise normal myelin.

An alternative hypothesis is that MS is caused by a persistent infection by a virus or other microorganism. In this scenario, the ongoing efforts of the immune system to get rid of the pathogen damage myelin. Tropical spastic paraparesis (TSP) provides a precedent for this idea. TSP is characterized by the gradual progression of weakness of the legs and impaired control of bladder function associated with increased deep tendon reflexes and a positive Babinski sign (see Figure 17.17). This clinical picture is similar to that of rapidly advancing MS, and TSP is known to be caused by persistent infection with a retrovirus (human T lymphotropic virus-1). Proving the persistent viral infection hypothesis for MS, however, requires unambiguous demonstration of the presence of a virus. Despite periodic reports of a virus associated with MS, convincing evidence has not been forthcoming. In sum, despite the benefits to some patients of immunomodulatory therapies, both the diagnosis and treatment of multiple sclerosis remain daunting clinical challenges.

Summary

The action potential and all its complex properties can be explained by time- and voltage-dependent changes in the Na^+ and K^+ permeabilities of neuronal membranes. This conclusion derives primarily from evidence obtained by a device called the voltage clamp. The voltage clamp technique is an electronic feedback method that allows control of neuronal membrane potential and, simultaneously, direct measurement of the voltage-dependent fluxes of Na^+ and K^+ that produce the action potential. Voltage clamp experiments show that a transient rise in Na^+ conductance activates rapidly and then inactivates during a sustained depolarization of the membrane potential. Such experiments also demonstrate a rise in K^+ conductance that activates in a delayed fashion and, in contrast to the Na^+ conductance, does not inactivate. Mathematical modeling of the properties of these conductances indicates that they, and they alone, are responsible for the production of all-or-none action potentials in the squid axon. Action potentials propagate along the nerve cell axons, initiated by the voltage gradient between the active and inactive regions of the axon by virtue of the local current flow. In this way, action potentials compensate for the relatively poor passive electrical properties of nerve cells and enable neural signaling over long distances. The molecular underpinnings of these signaling mechanisms will be revealed in Chapter 4, which describes the properties of ion channels and transporters.

■ Additional Reading

Reviews

Armstrong, C. M. and B. Hille (1998) Voltage-gated ion channels and electrical excitability. *Neuron* 20: 371–380.

Bean, B. P. (2007) The action potential in mammalian central neurons. *Nat. Rev. Neurosci.* 8: 451–465.

Brown, A. M. (2019) The classics updated, or an act of electrophysiological sacrilege? *J. Physiol.* 597: 2821–2825.

Salzer, J. L. (2003) Polarized domains of myelinated axons. *Neuron* 40: 297–318.

Important Original Papers

Armstrong, C. M. and L. Binstock (1965) Anomalous rectification in the squid giant axon injected with tetraethylammonium chloride. *J. Gen. Physiol.* 48: 859–872.

Chen, C. and 17 others (2004) Mice lacking sodium channel beta1 subunits display defects in neuronal excitability, sodium channel expression, and nodal architecture. *J. Neurosci.* 24: 4030–4042.

Hodgkin, A. L. and A. F. Huxley (1952a) Currents carried by sodium and potassium ions through the membrane of the giant axon of *Loligo*. *J. Physiol.* 116: 449–472.

Hodgkin, A. L. and A. F. Huxley (1952b) The components of membrane conductance in the giant axon of *Loligo*. *J. Physiol.* 116: 473–496.

Hodgkin, A. L. and A. F. Huxley (1952c) The dual effect of membrane potential on sodium conductance in the giant axon of *Loligo*. *J. Physiol.* 116: 497–506.

Hodgkin, A. L. and A. F. Huxley (1952d) A quantitative description of membrane current and its application to conduction and excitation in nerve. *J. Physiol.* 116: 507–544.

Hodgkin, A. L., A. F. Huxley and B. Katz (1952) Measurements of current–voltage relations in the membrane of the giant axon of *Loligo*. *J. Physiol.* 116: 424–448.

Hodgkin, A. L. and W. A. H. Rushton (1938) The electrical constants of a crustacean nerve fibre. *Proc. R. Soc. Lond.* 133: 444–479.

Huxley, A. F. and R. Stämpfli (1949) Evidence for saltatory conduction in peripheral myelinated nerve fibres. *J. Physiol.* 108: 315–339.

Moore, J. W., M. P. Blaustein, N. C. Anderson and T. Narahashi (1967) Basis of tetrodotoxin's selectivity in blockage of squid axons. *J. Gen. Physiol.* 50: 1401–1411.

Tasaki, I. and T. Takeuchi (1941) Der am Ranvierschen Knoten entstehende Aktionsstrom und seine Bedeutung für die Erregungsleitung. *Pflügers Arch.* 244: 696–711.

Books

Aidley, D. J. and P. R. Stanfield (1996) *Ion Channels: Molecules in Action*. Cambridge: Cambridge University Press.

Brown, A. M. (2019) *A Companion Guide to the Hodgkin-Huxley Papers*. London: The Physiological Society.

Campenot, R. B. (2017) *Animal Electricity*. Cambridge, MA: Harvard University Press.

Hille, B. (2001) *Ion Channels of Excitable Membranes*, 3rd Edition. Sunderland, MA: Sinauer/Oxford University Press.

Hodgkin, A. L. (1967) *The Conduction of the Nervous Impulse*. Springfield, IL: Charles C. Thomas.

Johnston, D. and S. M.-S. Wu (1995) *Foundations of Cellular Neurophysiology*. Cambridge, MA: MIT Press.

Matthews, G. G. (2003) *Cellular Physiology of Nerve and Muscle*, 4th Edition. Malden, MA: Blackwell Publishing.

Ion Channels and Transporters

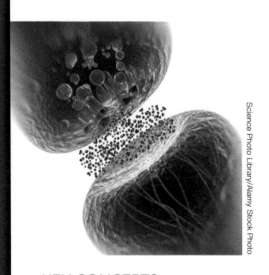

Science Photo Library/Alamy Stock Photo

KEY CONCEPTS

4.1 Ion channels generate electrical currents

4.2 There are many types of ion channels

4.3 Ions permeate through channels via ion-selective pores

4.4 Molecular specializations permit gating of ion channels by different types of stimuli

4.5 Active transporters create and maintain ion gradients

Overview

The generation of electrical signals in neurons requires that plasma membranes establish concentration gradients for specific ions and that these membranes undergo rapid and selective changes in their permeability to these ions. The membrane proteins that create and maintain ion gradients are called active transporters; other proteins, called ion channels, give rise to selective ion permeability changes. Ion channels are transmembrane proteins that contain a narrow pore that selectively permits particular ions to permeate the membrane. Some ion channels also contain voltage sensors that are able to detect the electrical potential across the membrane. Such voltage-gated channels open or close in response to the magnitude of the membrane potential, allowing the membrane's permeability to be regulated by changes in this potential. Other types of ion channels are gated by chemical signals, either by extracellular signals such as neurotransmitters or by intracellular signals such as second messengers. Still other channels respond to mechanical stimuli, temperature changes, or a combination of signals. Different combinations of ion channels are found in different cell types, yielding a wide spectrum of electrical characteristics. In contrast to ion channels, active transporters are membrane proteins that produce and maintain ion concentration gradients. The most important of these is the Na^+ pump, which hydrolyzes ATP to regulate the intracellular concentrations of both Na^+ and K^+. Other active transporters produce concentration gradients for the full range of physiologically important ions, including Cl^-, Ca^{2+}, and H^+. From the perspective of electrical signaling, active transporters and ion channels are complementary: Transporters create the concentration gradients that help drive ion fluxes through open ion channels, thereby generating electrical signals.

CONCEPT
4.1 | Ion Channels Generate Electrical Currents

LEARNING OBJECTIVES

4.1.1 Explain how the patch clamp method helped prove the existence of ion channels.

4.1.2 Discuss the relationship between microscopic and macroscopic ion currents.

4.1.3 Recognize that ion permeability and gating are two different properties that are common to ion channels.

Measurement of currents flowing through single ion channels

Although Hodgkin and Huxley had no knowledge of the physical nature of the conductance mechanisms underlying action potentials, they nonetheless proposed that nerve cell membranes have channels that allow ions to pass selectively from one side of the membrane to the other (see Chapter 3). The properties of the ion conductances and currents measured in voltage clamp experiments indicated that the postulated channels should have several properties. First, because the ion currents are quite large, the channels had to be capable of allowing ions to move across the membrane at high rates. Second, because the currents depend on the electrochemical gradient across the membrane, the channels had to make use of these gradients. Third, because Na^+ and K^+ flow across the membrane independently of each other, different channel types had to be capable of discriminating between Na^+ and K^+, allowing only one of these ions to flow across the membrane under the relevant conditions. Finally, given that the ion conductances are voltage-dependent, the channels had to be able to sense the membrane potential, opening only when the voltage reached appropriate levels. While this concept of channels was highly speculative at the time, later experimental work established beyond any doubt that transmembrane proteins called voltage-sensitive ion channels indeed exist and are responsible for action potentials and other types of electrical signals.

The first direct evidence for the presence of voltage-sensitive, ion-selective channels in nerve cell membranes came from measurements of the ion currents flowing through individual ion channels. The voltage clamp apparatus used by Hodgkin and Huxley could only resolve the *aggregate* current resulting from the flow of ions through many thousands of channels. A technique capable of measuring the currents flowing through single channels was devised in 1976 by Erwin Neher and Bert Sakmann at the Max Planck Institutes in Germany. This remarkable approach, called **patch clamp recording**, revolutionized the study of membrane currents (Box 4A). In particular, the patch clamp method provided the means to test directly Hodgkin and Huxley's deductions about the characteristics of ion channels.

Currents flowing through Na^+ channels are best examined in experimental circumstances that prevent the flow of current through other types of channels that are present in the membrane (e.g., K^+ channels). Under such conditions, depolarizing a patch of membrane from a squid giant axon causes tiny inward currents to flow, but only occasionally (Figure 4.1). The size of these currents is

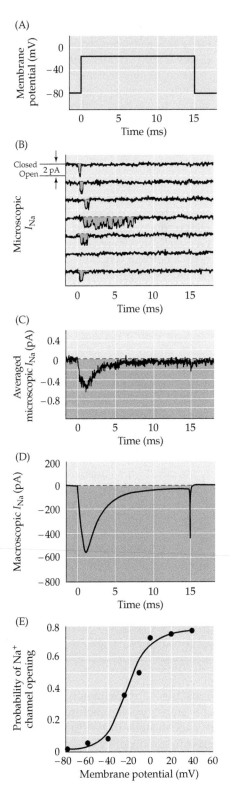

FIGURE 4.1 Patch clamp measurements of ion currents flowing through single Na^+ channels in a squid giant axon In these experiments, Cs^+ was applied to the axon to block voltage-gated K^+ channels. Depolarizing voltage pulses (A) applied to a patch of membrane containing a single Na^+ channel result in brief currents (B, downward deflections) in the seven successive recordings of membrane current (I_{Na}). (C) The average of many such current records shows that most channels open in the initial 1 to 2 ms following depolarization of the membrane, after which the probability of channel openings diminishes because of channel inactivation. (D) A macroscopic current measured from another axon shows the close correlation between the time courses of microscopic and macroscopic Na^+ currents. (E) The probability of a Na^+ channel opening depends on the membrane potential, increasing as the membrane is depolarized. (B,C after F. Bezanilla and A. M. Correa, 1995. In *Cephalopod Neurobiology*, N. J. Abbott, R. Williamson and L. Maddock [Eds.]. New York: Oxford University Press, pp. 131–151. © 1995 Oxford University Press; D after C. A. Vanderberg and F. Bezanilla, 1991. *Biophys J* 60: 1511–1533; E after A. M. Correa and F. Bezanilla, 1994. *Biophys J* 66: 1864–1878.)

■ BOX 4A | The Patch Clamp Method

A wealth of information about the function of ion channels has resulted from the invention of the patch clamp method. This technique is based on a very simple idea. A glass pipette with a very small opening is used to make tight contact with a tiny area, or patch, of neuronal membrane. After the application of a small amount of suction to the back of the pipette, the seal between the pipette and membrane becomes so tight that no ions can flow between the pipette and the membrane. Thus, all the current that flows when a single ion channel opens must flow into the pipette. This minute electrical current can be measured with an ultrasensitive electronic amplifier connected to the pipette. This arrangement is the *cell-attached patch clamp recording method*. As with the conventional voltage clamp method, the patch clamp method allows experimental control of the membrane potential to characterize the voltage dependence of membrane currents.

Minor manipulations allow other recording configurations. For example, if the membrane patch within the pipette is disrupted by briefly applying strong suction, the interior of the pipette becomes continuous with the cell cytoplasm. This arrangement allows measurements of electrical potentials and currents from the entire cell and is therefore called *whole-cell recording*. The whole-cell configuration also allows diffusional exchange

between the solution in the pipette and the cytoplasm, producing a convenient way to inject substances into the interior of a "patched" cell.

Two other variants of the patch clamp method originate from the finding that once a tight seal has formed between the membrane and the glass pipette, small pieces of membrane can be pulled away from the cell without disrupting the seal; this yields a preparation that is free of the complications

imposed by the rest of the cell. Simply retracting a pipette that is in the cell-attached configuration causes a small vesicle of membrane to remain attached to the pipette. By briefly exposing the tip of the pipette to air, the vesicle opens to yield a small patch of membrane with its (former) intracellular surface exposed. This arrangement, called the *inside-out patch recording configuration*, makes it possible to change the medium to which the intracellular surface of the membrane is exposed. Thus, the inside-out configuration is particularly valuable when studying the influence of intracellular molecules on ion channel function.

Alternatively, if the pipette is retracted while it is in the whole-cell configuration, the membrane patch produced has its extracellular surface exposed. This arrangement, called the *outside-out recording configuration*, is optimal for studying how channel activity is influenced by extracellular chemical signals such as neurotransmitters (see Chapter 5). This range of possible configurations makes the patch clamp method an unusually versatile technique for studies of ion channel function. Robotic versions of the patch clamp technique have made their way into industry, serving as a very sensitive and rapid means of screening therapeutic drugs that act on ion channels.

Cell-attached recording

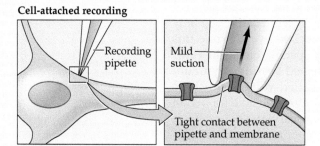

Recording pipette

Mild suction

Tight contact between pipette and membrane

Inside-out recording

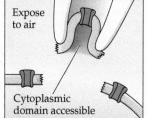

Expose to air

Cytoplasmic domain accessible

Whole-cell recording

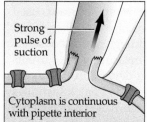

Strong pulse of suction

Cytoplasm is continuous with pipette interior

Outside-out recording

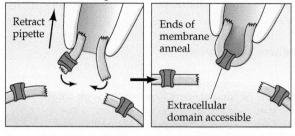

Retract pipette

Ends of membrane anneal

Extracellular domain accessible

Four configurations in patch clamp measurements of ion currents.

minuscule—approximately 1 to 2 pA (i.e., 10^{-12} ampere)—but is stereotyped, reaching discrete values that suggest ion flux though individual, open ion channels. These currents are orders of magnitude smaller than the Na^+ currents measured by voltage clamping the entire axon. The currents flowing through single channels are called **microscopic currents** to distinguish them from the **macroscopic currents** flowing through a large number of channels distributed over a much more extensive region of surface membrane. Although microscopic currents are certainly small, a current of 1 pA nonetheless reflects the flow of thousands of ions per millisecond. Thus, as predicted, a single channel can let many ions pass through the membrane in a very short time.

Several observations further proved that the microscopic currents in Figure 4.1B are due to the opening of single, voltage-activated Na^+ channels. First, the currents are carried by Na^+; thus, they are directed inward when the membrane potential is more negative than E_{Na}, reverse their polarity at E_{Na}, are outward at more positive potentials, and are reduced in size when the Na^+ concentration of the external medium is decreased. This behavior exactly parallels that of the macroscopic Na^+ currents described in Chapter 3 (see Figure 3.4). Second, the channels have a time course of opening, closing, and inactivating that matches the kinetics of macroscopic Na^+ currents. This correspondence is difficult to appreciate in the measurement of microscopic currents flowing through a single open channel, because individual channels open and close in a stochastic (random) manner, as can be seen by examining the individual traces in Figure 4.1B. However, repeated depolarization of the membrane potential causes each Na^+ channel to open and close many times. When the current responses to a large number of such stimuli are averaged together, the collective response has a time course that looks much like that of the macroscopic Na^+ current (see Figure 4.1C). In particular, the channels open mostly at the beginning of a prolonged depolarization, showing that they activate and subsequently inactivate, as predicted from the macroscopic Na^+ current (compare Figures 4.1C,D). Third, both the opening and closing of the channels are voltage-dependent; thus, the channels are closed at −80 mV but open when the membrane potential is depolarized. In fact, the probability that any given channel will be open varies with membrane potential (see Figure 4.1E), again as predicted from the macroscopic Na^+ conductance (see Figure 3.7). Finally, tetrodotoxin, which blocks the macroscopic Na^+ current (see Figure 3.5), also blocks microscopic Na^+ currents. Taken together, these results show that the macroscopic Na^+ current measured by Hodgkin and Huxley does indeed arise from the aggregate effect of many millions of microscopic Na^+ currents, each representing the opening of a single voltage-sensitive Na^+ channel.

Patch clamp experiments also revealed the properties of the channels responsible for the macroscopic K^+ currents associated with action potentials. When the membrane potential is depolarized (Figure 4.2A), microscopic outward

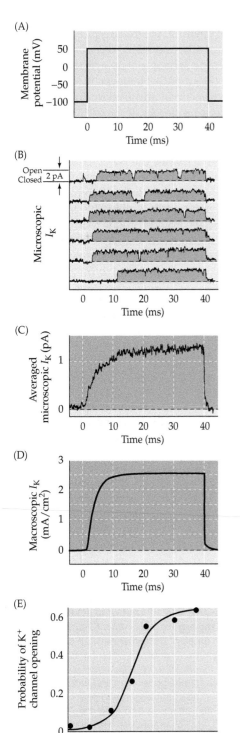

FIGURE 4.2 Patch clamp measurements of ion currents flowing through single K^+ channels in a squid giant axon In these experiments, tetrodotoxin was applied to the axon to block voltage-gated Na^+ channels. Depolarizing voltage pulses (A) applied to a patch of membrane containing a single K^+ channel result in brief currents (B, upward deflections) whenever the channel opens. (C) The average of such current records shows that most channels open with a delay, but remain open for the duration of the depolarization. (D) A macroscopic current measured from another axon shows the correlation between the time courses of microscopic and macroscopic K^+ currents. (E) Membrane potential controls K^+ channel opening, with the probability of a channel opening increasing as the membrane is depolarized. (A–C after B. Hille, 2001. *Ion Channels of Excitable Membranes*, pp. 61–93. Sinauer/Oxford University Press: Sunderland, MA. Courtesy of C.K. Augustine and F. Bezanilla; D after C. K. Augustine and F. Bezanilla, 1990. *J Gen Physiol* 95: 245–271; E after E. Perzo et al., 1991. *J Gen Physiol* 98: 19–34.)

currents (Figure 4.2B) can be observed under conditions that block Na$^+$ channels. The microscopic outward currents exhibit all the features expected for currents flowing through action potential–related K$^+$ channels. Thus, the microscopic currents (Figure 4.2C), like their macroscopic counterparts (Figure 4.2D), fail to inactivate during brief depolarizations. Moreover, these single-channel currents are sensitive to ionic manipulations and drugs that affect the macroscopic K$^+$ currents and, like the macroscopic K$^+$ currents, are voltage-dependent (Figure 4.2E). This and other evidence show that macroscopic K$^+$ currents associated with action potential repolarization arise from the opening of many voltage-sensitive K$^+$ channels.

In summary, patch clamping has allowed direct observation of microscopic currents flowing through single ion channels. Such observations confirmed that voltage-sensitive Na$^+$ and K$^+$ channels are responsible for the macroscopic conductances and currents that underlie the action potential.

Ion permeability and gating of channels

Measurements of the behavior of single ion channels have also provided insight into the molecular attributes of these channels. Such studies have shown that the squid axon membrane contains at least two distinct types of

channels—one selectively permeable to Na$^+$ and a second selectively permeable to K$^+$. This **ion selectivity** means that these channels are able to discriminate between Na$^+$ and K$^+$ and allow only one ion or the other to permeate through the membrane.

Furthermore, because their opening is influenced by membrane potential, both Na$^+$ and K$^+$ channels are **voltage gated**. For each channel, depolarization increases the probability of channel opening, whereas hyperpolarization closes them (see Figures 4.1E and 4.2E). Thus, both channel types must have a **voltage sensor** that detects the potential across the membrane (Figure 4.3). However, these channels differ in important respects. In addition to differences in ion selectivity, the kinetic properties of the gating of the two channels differ as expected from the macroscopic behavior of the Na$^+$ and K$^+$ currents described in Chapter 3. Furthermore, depolarization inactivates the Na$^+$ channel but not the K$^+$ channel, causing Na$^+$ channels to pass into a non-conducting state. The Na$^+$ channel must therefore have an additional molecular mechanism responsible for *inactivation*. Finally, these channels provide unique binding sites for drugs and for various neurotoxins known to block specific subclasses of ion channels (Box 4B). This information about the physiology of ion channels set the stage for subsequent studies of their workings at the molecular level.

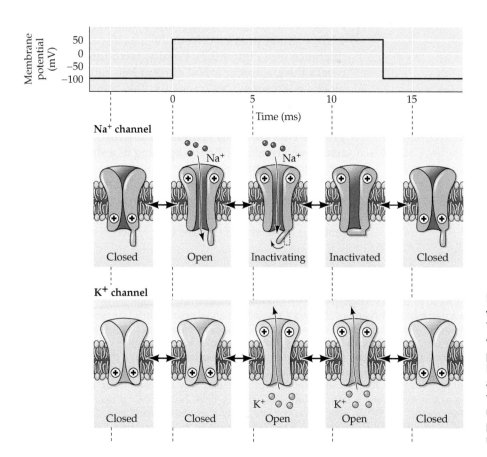

FIGURE 4.3 **Functional states of voltage-gated Na$^+$ and K$^+$ channels** The gates of both channels are closed when the membrane potential is hyperpolarized. When the potential is depolarized, voltage sensors (indicated by +) allow the channel gates to open—first the Na$^+$ channels and then the K$^+$ channels. Na$^+$ channels also inactivate during prolonged depolarization, whereas many types of K$^+$ channels do not.

■ BOX 4B | Toxins That Poison Ion Channels

Given the importance of Na⁺ and K⁺ channels for neuronal excitation, it is not surprising that channel-specific toxins have evolved in several organisms as mechanisms for self-defense or for capturing prey. A rich collection of natural toxins selectively target the ion channels of neurons and other cells. These toxins are valuable not only for survival, but also as tools for studying the function of cellular ion channels. The best-known channel toxin is *tetrodotoxin*, produced by certain puffer fish and other animals. Tetrodotoxin produces a potent and specific blockade of the pore of the Na⁺ channels responsible for action potential generation (see Figure 3.5), thereby paralyzing the animals unfortunate enough to ingest it. *Saxitoxin*, a chemical homologue of tetrodotoxin produced by "red tide" dinoflagellates, has a similar action on Na⁺ channels. The potentially lethal effects of eating shellfish that have ingested these dinoflagellates are due to the potent neuronal actions of saxitoxin.

Fish-eating cone snails (see Box 6A) paralyze their prey by producing a potent venom consisting of tens or hundreds of peptide neurotoxins. One group of these toxins, called μ-conotoxins, produces paralysis by blocking the pore of voltage-gated Na⁺ channels. Scorpions similarly paralyze their prey by injecting a potent mix of peptide toxins that also affect ion channels. Among these are the α-toxins, which slow the inactivation of Na⁺ channels (Figure A1); exposure of neurons to α-toxins prolongs the action potential (Figure A2), thereby scrambling information flow within the nervous system of the soon-to-be-devoured victim. Other peptides in scorpion venom, called β-toxins, shift the voltage dependence of Na⁺ channel activation (Figure

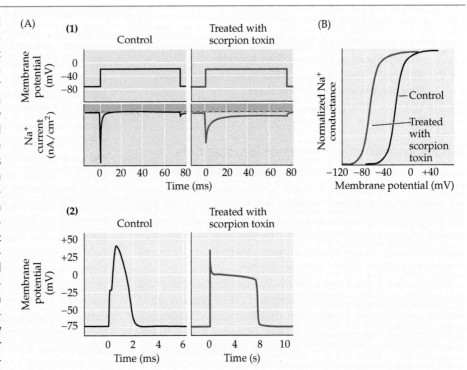

(A) Effects of toxin treatment on Na⁺ channels of frog axons. (1) α-Toxin from the scorpion *Leiurus quinquestriatus* prolongs Na⁺ currents recorded with the voltage clamp method. (2) As a result of the increased Na⁺ current, α-toxin greatly prolongs the duration of the axonal action potential. Note the change in timescale after treating with toxin. (B) Treatment with β-toxin from another scorpion, *Centuroides sculpturatus*, shifts the activation of Na⁺ channels, so that Na⁺ conductance begins to increase at potentials much more negative than usual. (A after O. Schmitt and H. Schmidt, 1972. *Pflügers Arch* 333: 51–61; B after M. Cahalan, 1975. *J Physiol* 244: 511–534.)

B). These toxins cause Na⁺ channels to open at potentials much more negative than normal, inducing uncontrolled action potential firing. *Batrachotoxin* is an alkaloid toxin, produced by a species of frog, that is used by some tribes of indigenous peoples of South American to poison their arrow tips. This toxin works both by removing inactivation and shifting activation of Na⁺ channels. Several plants produce similar toxins, including *aconitine*, from buttercups; *veratridine*, from lilies; and several insecticidal toxins (pyrethrins) produced by plants such as chrysanthemums and rhododendrons.

Potassium channels have also been targeted by toxin-producing organisms. Peptide toxins affecting K⁺ channels include *dendrotoxin*, from wasps; *apamin*, from bees; and *charybdotoxin*, yet another toxin produced by scorpions. All of these toxins block K⁺ channels as their primary action; no toxin is known to affect the activation or inactivation of these channels, although such agents may simply be awaiting discovery.

CONCEPT
4.2 # There Are Many Types of Ion Channels

LEARNING OBJECTIVES

4.2.1 Name four different groups of ion channel genes and the types of channels they encode.

4.2.2 Recognize that channels differ in the types of ions that can permeate through them.

4.2.3 List four different types of stimuli that can gate ion channels.

Genes encode many types of ion channels

Many subsequent insights into ion channels have come from molecular genetic studies: More than 200 ion channel genes have been discovered, a remarkable number that could not have been anticipated from the work of Hodgkin and Huxley. To understand the functional significance of this multitude of ion channel genes, individual channel types can be selectively expressed in well-defined experimental systems, such as cultured cells or frog oocytes, and then studied with patch clamping and other physiological techniques. Channel genes can also be mutated in genetically tractable organisms, such as mice or fruit flies, to determine the roles these channels play in the intact organism. Such studies have established that these channels are proteins that are selectively permeable to a single type of ion, such as Ca^{2+} or Cl^-, though some permit flow of a wider range of cations or anions. While some ion channels are always open, most are gated by one or more types of stimuli.

More than 100 genes encode voltage-gated channels that respond to membrane potential in much the same way as the Na^+ and K^+ channels that underlie the action potential. There are ten human Na^+ channel genes (called **SCN genes**) that produce proteins that differ in their structure, function, and distribution in specific tissues. In addition to the rapidly inactivating Na^+ channels that underlie the action potential in many types of neurons, including the squid neurons studied by Hodgkin and Huxley, a voltage-sensitive Na^+ channel that does not completely inactivate has been identified in mammalian neurons. This channel gives rise to a "persistent" Na^+ current that helps regulate action potential threshold and repetitive firing. Ten different genes encoding voltage-gated Ca^{2+} channels (**CACNA genes**) have been identified (Figure 4.4A). Different types of Ca^{2+} channels vary in their activation and inactivation properties, allowing subtle variations in both electrical and chemical signaling processes mediated by Ca^{2+}. In some neurons, Ca^{2+} channels contribute to action potential generation. In all cases, open Ca^{2+} channels increase the intracellular concentration of Ca^{2+}, which regulates an enormous range of biochemical signaling processes within cells (see Chapter 7). K^+ channels are by far the largest and most diverse class of voltage-gated ion channels: There are 78 human K^+ channel genes (**KCN genes**). These fall into several distinct groups that differ substantially in their activation, gating, and inactivation properties, as well as their roles in neuronal electrical signaling. Finally, several types of Cl^- channel genes (**CLCN genes**), encoding different voltage-gated Cl^- channels, have been identified. These channels are present in every type of neuron, where they control excitability, contribute to the resting membrane potential, and help regulate cell volume.

Other types of channels are insensitive to membrane potential, instead being gated by the binding of chemical signals to the channels. Among these **ligand-gated ion channels**, the most important for nervous system function are neurotransmitter receptors, which are activated by binding of neurotransmitters to their extracellular surface (Figure 4.4B). These ligand-gated channels are essential for synaptic transmission and other forms of signaling discussed in Chapters 5–8. Neurotransmitter receptors and other channels activated by extracellular ligands often allow multiple types of ions to flow. For example, the neurotransmitter receptors involved in excitatory synaptic transmission typically are permeable to both Na^+ and K^+, as well as to other cations. Another important class of channels activated by extracellular chemical signals is the acid-sensing ion channels (ASICs). These Na^+ channels are gated by external H^+, rather than by voltage, and are important for a wide range of functions, including taste and pain sensation.

Some ligand-gated ion channels are distinguished by ligand-binding domains on their *intracellular* surfaces that interact with second messengers such as Ca^{2+}, the cyclic nucleotides cAMP and cGMP, or protons (see Chapter 7). The main function of these channels is to convert intracellular chemical signals into electrical information. This process is particularly important in sensory transduction, where intracellular second-messenger signals associated with sensory stimuli—such as odors and light—are transduced into electrical signals by cyclic nucleotide-gated channels. For example, closure of a cyclic nucleotide-gated cation channel is involved in detection of light in retinal photoreceptor cells (Figure 4.4C; see also Chapter 9). Although many ligand-gated ion channels are located in the plasma membrane, others are found in the membranes of intracellular organelles such as mitochondria or the endoplasmic reticulum, where they regulate intracellular signaling and metabolism.

Still other channels are gated by other types of stimuli. **Thermosensitive** ion channels (Figure 4.4D) respond to changes in temperature and contribute to the sensations of pain and body temperature (see Chapters 12 and 13). David Julius and colleagues at the University of California, San Francisco have found that some thermosensitive channels are members of the **transient receptor potential (TRP)** gene family, a group of cation channels (28 in mammals) that represent the second-largest ion channel family (after K^+ channels). Thermosensitive TRP channels are gated

(A) Voltage-gated Ca²⁺ channel

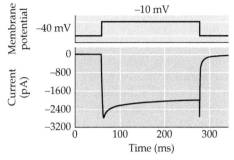

(B) Neurotransmitter-gated channel

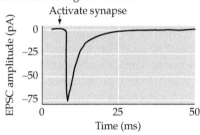

(C) Cyclic nucleotide-gated channel

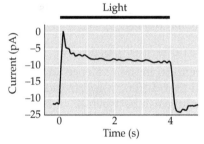

(D) Thermosensitive channel

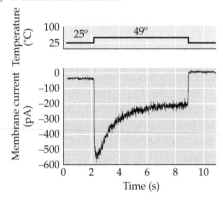

(E) Mechanosensitive channel

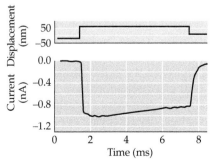

FIGURE 4.4 Ion channels can be gated by a variety of stimuli (A) Depolarization of the membrane potential of a dorsal root ganglion neuron, under conditions where voltage-gated Na⁺ and K⁺ channels are blocked, reveals the presence of a slowly inactivating inward current that is carried by Ca²⁺. (B) Excitatory postsynaptic current (EPSC) in a postsynaptic interneuron mediated by AMPA-type glutamate receptors at a synapse in the visual cortex. The presynaptic pyramidal neuron was activated at the time indicated by the arrow. (C) Outward current generated in a retinal cone cell by closure of cyclic nucleotide-gated cation channels in response to light (at time indicated by bar). (D) Inward cation current produced by a nociceptive neuron from the rat dorsal root ganglion in response to a temperature change from 25°C to 49°C (top). (E) Inward cation current generated by minute displacement (top) of the hair bundle of a hair cell from the turtle auditory system. (A after Y. Kitano et al., 2019. *Pharmazie* 74: 147–149; B after J. Watanabe, 2005. *J Neurosci* 25: 1024–1033; C after N. T. Ingram et al., 2020. *Proc Natl Acad Sci USA* 117: 19599–19603; D after P. Cesare et al., 1999. *Neuron* 23: 617–624; E after A. J. Ricci et al., 2005. *J Neurosci* 25: 7831–7839.)

over specific temperature ranges (Figure 4.5), with some activated by cold temperatures (down to 0°C in the case of TRPA1) and others by heat (hotter than 50°C for TRPV2). Many thermosensitive TRP channels also are activated by ligands and are used to detect chemical signals. For example, the TRPV1 channel responds to temperatures above 40°C and is also sensitive to capsaicin, the ingredient that makes chili peppers spicy (see Box 13A). Similarly, the TRPM8 channel responds to cool temperatures below 30°C and also responds to menthol, an organic compound in mints that produces a sensation of coolness. Thus, these TRP channels transduce both thermal and chemical stimuli into the sensation of "hot" or "cool." More generally, many TRP channels serve as multimodal integrators of different types of stimuli.

Mechanosensitive channels respond to mechanical displacement and are the critical transducers in stretch receptors and neuromuscular stretch reflexes (see Chapters 12,

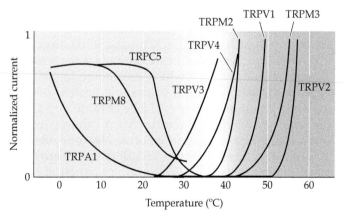

FIGURE 4.5 Temperature sensitivity of TRP channels Effects of temperature on current flow through members of the TRP channel family. While some TRP channels, such as TRPA1, TRPM8 and TRPC5, are activated by cold temperatures, many other TRP channels are activated by warm or hot temperatures. (After H. Wang and J. Siemens, 2015. *Temperature* 2: 178–187.)

16, and 17). A specialized form of these channels, termed the mechanoelectrical transduction channel, enables hearing by allowing auditory hair cells to respond to sound waves (Figure 4.4E; see also Chapter 10). These sound-induced responses are probably mediated by TMC cation channels. Many other types of ion channels, including certain TRP, TREK, Piezo, and TMEM channels, also respond to mechanical distortion of the plasma membrane.

Thus, although the basic electrical signals of the nervous system are relatively stereotyped, the ion channels responsible for generating these signals are remarkably diverse, conferring specialized signaling properties to the many different types of neurons that populate the nervous system.

CONCEPT 4.3 Ions Permeate through Channels via Ion-Selective Pores

LEARNING OBJECTIVES

4.3.1 Describe how the structure of pores allows ions to diffuse across a membrane.

4.3.2 Describe how selectivity filters determine the ion permeability of a channel.

Studies of the atomic-level structure of ion channels have answered many fundamental questions about how channels work. These studies reveal that all channels are integral membrane proteins that span the plasma membrane repeatedly and share a common transmembrane architecture.

Ion channel pores

The interior of plasma membranes is extremely hydrophobic and represents a hostile environment for charged and water-associated ions, such as Na^+ and K^+. How do channels conduct ions across a membrane? The ion permeation mechanism of channels was first revealed by studies done by Rod MacKinnon and colleagues at Rockefeller University. They examined a bacterial K^+ channel (Figure 4.6A–C), which was chosen for analysis because the large quantity of channel protein needed for X-ray crystallography could be obtained by growing large numbers of bacteria. MacKinnon's results showed that this K^+ channel is formed by four subunits (colored structures in Figure 4.6A,B) that each cross the plasma membrane twice. Between the two helical membrane-spanning structures (pore helices) of each subunit is a **pore loop** that inserts into the plasma membrane (thin structures in Figure 4.6B). In the center of the assembled channel is a **pore** (see Figure 4.6A) that serves as a narrow tunnel to allow K^+ to flow through the protein and thus cross the membrane (see Figure 4.6C). This pore is formed by the assembled pore loops and adjacent pore helices of all four subunits. Subsequent work has established that many other types of K^+ channels,

including those of humans, have remarkably similar pore-forming structures.

Na^+ channels consist of motifs of six membrane-spanning regions that are repeated four times, yielding a total of 24 transmembrane regions (Figure 4.6D,E). Thus, one Na^+ channel protein forms a structure very similar to that produced by four K^+ channel subunits. In the center of the Na^+ channel is a pore that connects the extracellular and intracellular sides of the membrane and allows Na^+ to permeate (Figure 4.6F). This voltage-gated Na^+ channel is much larger than the bacterial K^+ channel shown in Figure 4.6A,B because it includes additional domains that are involved in voltage sensing (discussed in Concept 4.4).

Although the structure of Cl^- channels is somewhat different from that of K^+ and Na^+ channels, these anion channels also have pores. The Cl^- channel shown in Figure 4.6G,H comprises two different protein monomers (colors), with each monomer containing a pore. These pores consist of two funnel-shaped structures that open to the extracellular and intracellular sides of the membrane to permit transmembrane Cl^- flux (Figure 4.6I).

In summary, *all* ion channels have transmembrane pores that mediate ion fluxes.

Ion selectivity in channel pores

K^+ and Na^+ are identical in their valence and very similar in size, with radii that differ by approximately 0.04 nm. How do pores distinguish between K^+ or Na^+ and allow only one to selectively permeate through a channel? How do pores of cation channels exclude anions, such as Cl^-? MacKinnon's structural studies of the bacterial K^+ channel provided a clear answer to these questions too. The pore of this channel is well suited for conducting K^+; its narrowest part is near the outside mouth of the channel and is so constricted that only a nonhydrated K^+ ion can fit through the bottleneck (see Figure 4.6C). Larger cations, such as Cs^+, are too large to traverse this region of the pore, while smaller cations such as Na^+ cannot enter the pore because the pore "walls" are too far apart to stabilize a dehydrated Na^+ ion. This part of the channel complex is responsible for the selective permeability to K^+ and is therefore called the **selectivity filter**. Deeper within the channel is a water-filled cavity that connects to the interior of the cell. Because of negative charges on a nearby pore helix, which attracts K^+ ions and repels anions, this cavity collects K^+ from the cytoplasm and these ions then dehydrate and enter the selectivity filter. These "naked" ions are able to move through four K^+ binding sites within the selectivity filter to eventually reach the extracellular space (recall that the physiological electrochemical gradient normally drives K^+ out of cells). The presence of multiple (up to four) K^+ ions within the selectivity filter causes electrostatic repulsion between the ions that helps speed their transit through the selectivity filter, thereby permitting rapid ion flux through the channel.

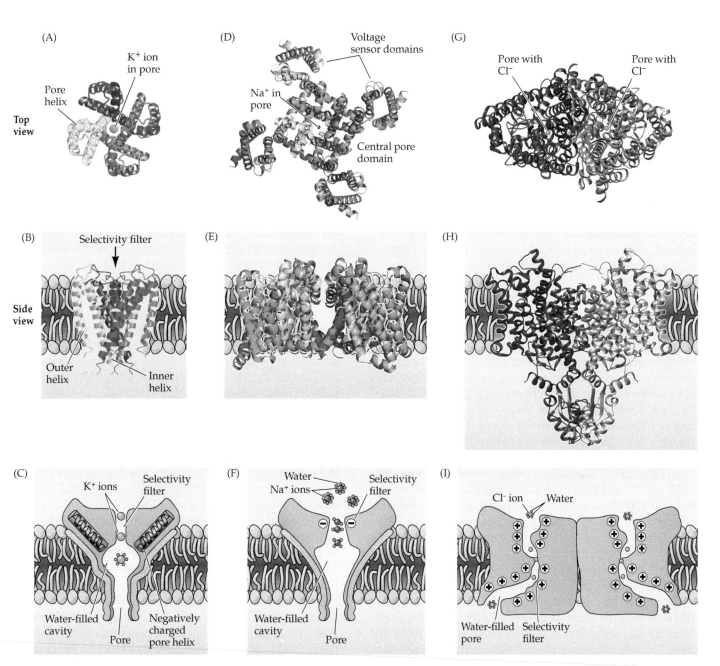

FIGURE 4.6 Comparison of structures and ion-selectivity mechanisms of three ion channels (A–C) Structure of a simple bacterial K^+ channel. Four channel subunits (different colors), each consisting of two membrane-spanning domains and a pore loop that inserts into the membrane, are assembled to form a K^+ channel. (A) View from the top of the channel, showing a K^+ ion (yellow) in the channel pore. (B) Side view, showing the channel's eight membrane-spanning helices. (C) Cut-away side view of the K^+ permeation pathway, which consists of a large aqueous cavity connected to a narrow selectivity filter. Helical domains of the channel direct negative charges (green) toward this cavity, allowing K^+ ions (yellow) to become dehydrated and then move through the selectivity filter. (D–F) Structure of a human Nav1.7 voltage-gated Na^+ channel. (D) View from the top of the channel, showing a Na^+ ion (red) in the channel pore. (E) Side view, showing the channel's 24 membrane-spanning helices. (F) Cut-away side view of the Na^+ permeation pathway, which consists of a large

aqueous cavity connected to a narrow selectivity filter. Negative charges near the external surface attract Na^+ ions (red), which lose some of their associated water (blue) and then move through the selectivity filter. For clarity, the voltage-sensor domains of the channel are not shown. (G–I) Structure of a bovine CLC-K Cl^- channel. (G) View from the top of the channel, showing Cl^- ions (green) in the pores of the two channel monomers (each a different shade of blue). (H) Side view, showing that each channel monomer has 18 membrane-spanning helices. (I) The Cl^- permeation pathway of each channel monomer consists of two hourglass-shaped pore structures that are filled with water; these meet at the narrow selectivity filter. Positive charges along the surface of the pore attract Cl^- ions (green), which lose their associated water (blue) to move through the selectivity filter. (A,B from D. A. Doyle et al., 1998. *Science* 280: 69–77; D,E after S. Ahuja et al., 2015. *Science* 350: aac5464; G after R. Dutzler et al., 2002. *Nature* 415: 287–294; H from E. Park et al., 2017. *Nature* 541: 500–505.)

We now know that very similar principles apply to ion permeation through Na^+ channels. For example, the pore of the Na^+ channel shown in Figure 4.6D–F also has a central water-filled cavity and negative charges—in this case on the external end of the channel pore—that attract Na^+ to enter the pore. The selectivity filter of the Na^+ channel pore, like that of K^+ channels, is formed by four pore loops. However, the selectivity filter of Na^+ channels is much wider than that of K^+ channels, with dimensions that are optimal for the selective entry and permeation of partially hydrated Na^+ (see Figure 4.6F). Multiple Na^+ can reside within the selectivity filter, again speeding ion flow though the channel pore. Although Ca^{2+} channels have selectively filters that are permeable to Ca^{2+}, these channels are otherwise remarkably similar in structure to Na^+ and K^+ channels.

Permeation of Cl^- through Cl^- channels is based on similar structural principles. The pore of these channels is lined with positive charges that attract Cl^- (see Figure 4.6I). Although this pore does not include a central water-filled cavity, the pore is wide enough to contain water along nearly its entire length. Near the center of the pore is a narrow constriction that serves as a selectivity filter. This filter is small enough to fit dehydrated or partially hydrated Cl^- ions, which interact with nearby amino acids to enable flux of Cl^- though the filter. Similar to the other channels we have described, the presence of multiple Cl^- in the pore promotes the rapid flux of these ions through the channel.

In conclusion, atomic-level structural characterization of K^+ and other types of ion channels has yielded considerable insight into how channels use pores equipped with selectivity filters to enable a specific type of ion to rapidly diffuse from one side of the plasma membrane to the other. The selective permeability of channels to specific ions is based on several structural features: a water-filled pore that allows charged ions to make their way into the hydrophobic interior of the membrane, charges that attract the ions into the pore, and a selectivity filter that is optimal to accommodate dehydrated or partially hydrated ions. The simultaneous presence of multiple ions, which push each other through the channel pore, allows the prodigious rates of ion flux associated with channels.

CONCEPT 4.4 | Molecular Specializations Permit Gating of Ion Channels by Different Types of Stimuli

LEARNING OBJECTIVES

4.4.1 Describe the various types of sensors used to control the gating of ion channels.

4.4.2 Explain how detection of stimuli by sensors is coupled to the opening and closing of ion channel pores.

Given that different types of channels are opened or closed by different types of stimuli, these channels must have structural specializations that confer sensivity to these stimuli. Here we consider how voltage, chemical signals, heat, and mechanical displacement gate ion channels.

Gating of channels by voltage

Studies of the structure of voltage-gated channels have provided insights into how these channels are gated by the transmembrane potential. MacKinnon and colleagues initiated such crystallographic studies by examining a mammalian voltage-gated K^+ channel. As is the case for the bacterial K^+ channel shown in Figure 6A–C, four subunits assemble to form the voltage-gated K^+ channel (Figure 4.7A). The central pore region of this channel is very similar to that of the bacterial K^+ channel, confirming the generality of the ion permeation mechanism established from studies of the bacterial K^+ channel (compare Figures 4.7A and 4.6A). The voltage-gated K^+ channel is larger because it has additional structures on its cytoplasmic side, such as a regulatory β subunit and a T1 domain that links the β subunit to the channel (Figure 4.7B). Most important, each channel subunit has four *additional* transmembrane structures that form the voltage sensors of this channel. These voltage sensors can be observed as separate domains that each consist of four transmembrane helices that extend into the plasma membrane and are linked to the central pore of the channel (see Figure 4.7A). These sensors are connected to the central pore of the channel via helical linkers (see Figure 4.7B).

The voltage sensors contain positive charges that enable movement within the membrane in response to changes in membrane potential. The nature of these movements has been the subject of considerable debate. It is likely that the voltage sensors slide through the membrane, with depolarization pushing the sensors outward, while hyperpolarization pulls them inward (Figure 4.7C, black arrows). Such movements of the sensors then exert force on the helical linkers connecting the sensors to the channel pore (violet structures in Figure 4.7C), pulling the channel pore open or pushing it closed. Very similar voltage sensor structures are found in Na^+ channels (see Figure 4.6D,E) and Ca^{2+} channels; in all cases, they are responsible for the voltage-dependent gating of these channels. Voltage-gated Cl^- channels employ a different structure that nonetheless relies on movement of a charged voltage sensor for channel gating.

In conclusion, voltage-gated ion channels have charged voltage sensors that move in response to the transmembrane potential and allow the channel pore to open or close. Defects in the function of these channels, associated with gene mutations, lead to a variety of neurological disorders (Clinical Applications).

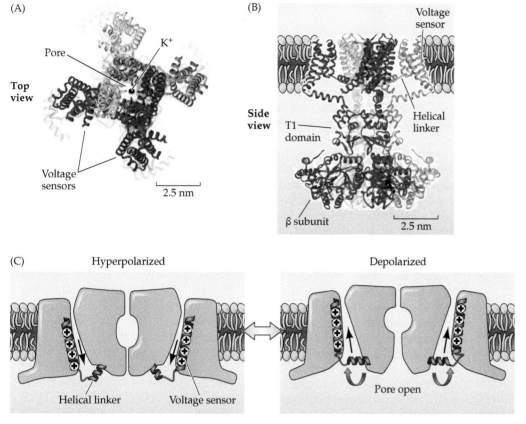

FIGURE 4.7   Structure of a mammalian voltage-gated K⁺ channel (A) The Kv1.2 channel includes four subunits (in different colors), each possessing two transmembrane domains (pore helices). When viewed from above, the four transmembrane domains can be seen to have separate domains for voltage sensing and for forming the K⁺-conducting pore (K⁺ indicated by black sphere in the middle of the pore). (B) Side view of the Kv1.2 channel, illustrating its transmembrane domains as well as cytoplasmic T1 domains and β subunits. (C) Model for voltage-dependent gating of the K⁺ channel. Depolarization causes the positively charged voltage sensor domains to move toward the extracellular surface of the membrane (straight black arrows), pulling on the helical linker (curved red) and thereby opening the channel pore. Hyperpolarization causes the voltage sensor domains to move toward the intracellular surface, pushing down on the helical linker and shutting the channel pore. (A,B after S. B. Long et al. 2005. *Science* 309: 903–908.)

Gating of channels by chemical signals

Ligand-gated ion channels respond to chemical signals present on either their extracellular or intracellular surfaces. The best-studied ligand-gated channels are neurotransmitter receptors, which are activated by extracellular binding of neurotransmitters. A good example is the AMPA receptor, a channel gated by the neurotransmitter glutamate which mediates the response shown in Figure 4.4B. Similar to many K⁺ channels, the AMPA receptor is composed of four subunits. Each AMPA receptor subunit contains three transmembrane helices that form a transmembrane domain. This domain contains both the channel pore and a gate that occludes the pore when glutamate is not bound to the receptor (Figure 4.8A). Most of each subunit is extracellular and includes a "clamshell"-shaped ligand-binding domain where glutamate and other ligands bind (see Figure 4.8A). The

assembled AMPA receptor is Y-shaped (Figure 4.8B), with the large extracellular domains of the subunits narrowing down as the receptor passes through the plasma membrane (Figure 4.8C,D). Binding of glutamate causes the clamshell structure to "shut"; this movement then causes the gate helices in the transmembrane domain to move and thereby opens the channel pore (Figure 4.8E). Thus, in the same way that voltage causes conformational changes that allow ions to flow through the pore of voltage-gated channels, neurotransmitter binding changes the conformation of the pore of AMPA receptors. Very similar mechanisms apply to activation of other ligand-gated ion channels that serve as neurotransmitter receptors (see Chapter 6).

Similar structural principles also apply to ion channels gated by intracellular ligands. A good example is the cyclic nucleotide-gated channel responsible for the light

■ **Clinical Applications**

Neurological Diseases Caused by Altered Ion Channels

Numerous genetic diseases, collectively called *channelopathies*, result from mutations in ion channel genes. For example, more than 20 different inherited diseases are associated with mutations in Na$^+$ channel genes. Numerous other diseases result from mutations in other voltage-gated ion channels, as well as mutations in ligand-gated ion channels, such as receptors for the neurotransmitters acetylcholine, glutamate, and GABA. Here we focus on neurological disorders caused by defects in voltage-gated ion channels.

Epilepsy

Epilepsy is a class of neurological disorders associated with recurrent seizures, which are spontaneous bouts of rhythmic firing of large groups of neurons caused by hyperexcitability of brain networks (see Clinical Applications, Chapter 8). While many types of epilepsy are sporadic, meaning they have no known genetic origin, other forms of epilepsy are inherited. Mutations in no fewer than five different Na$^+$ channel genes (SCNAs), seven K$^+$ channel genes (KCNs), and two Ca^{2+} channel genes (CACNs) have been implicated in epilepsies. *Severe myoclonic epilepsy of infancy* causes intractable seizures that begin in the first postnatal year. Most cases of this disorder are associated with missense or nonsense mutations in the *SCNA1* or *SCNA2* genes that reduce Na$^+$ channel function. This causes a preferential impairment in the ability of inhibitory interneurons to fire action potentials, leading to the hyperexcitability that causes the severe seizures. Defects in any of three Na$^+$ channel genes are known to cause *generalized epilepsy with febrile seizures* (GEFS), which begins in infancy and usually continues through early puberty. These mutations cause a slowing of Na$^+$ channel inactivation, which may explain the neuronal hyperexcitability underlying GEFS. Other

types of epilepsies associated with Na$^+$ channel defects include *migrating partial seizures of infancy*, *benign familial neonatal infantile seizures*, and *infantile epileptic encephalopathy*.

Another type of seizure, benign familial neonatal convulsion (BFNC), is due to K$^+$ channel mutations. This disease is characterized by frequent brief seizures commencing within the first week of life and disappearing spontaneously within a few months. BFNC has been mapped to at least two voltage-gated K$^+$ channel genes, *KCNQ2* and *KCNQ3*, with a reduction in K$^+$ current flow through the mutated channels probably causing the neuronal hyperexcitability associated with BFNC. Mutations in other KCN genes cause numerous other forms of epilepsy, including generalized epilepsy with paroxysmal movement disorder and myoclonic epilepsy.

Mutations in at least two Ca^{2+} channel genes also cause epilepsy: Mutations in either *CACNA1H* or *CACNA1A* cause childhood absence epilepsy, which presents with loss of awareness or responsiveness without overt movements.

Ataxia

Ataxia refers to loss of voluntary motor movement, often associated with impairment of cerebellar function (see Chapter 19). *Episodic ataxia type 1* (EA1) is characterized by brief episodes of ataxia and has been linked to defects in the gene for a voltage-gated K$^+$ channel, *KCNA1*. These are typically missense mutations that may produce clinical symptoms by impairing action potential repolarization. In *episodic ataxia type 2* (EA2), affected individuals have recurrent attacks of abnormal limb movements and severe ataxia. These problems are sometimes accompanied by vertigo, nausea, and headache. EA2 is caused by a variety of types of mutations in the *CACNA1A* gene, which may cause the clinical manifestations of the disease

by reducing current flow through Ca^{2+} channels. *Spinocerebellar ataxia type 6* (SCA6) also is caused by *CACNA1A* mutations; in this case, the mutations encode additional glutamine residues on Ca^{2+} channels, leading to a progressive degeneration of cerebellar Purkinje cells that is the cause of ataxia. SCA6 is thus an example of *polyglutamine expansion* that underlies many types of neurodegenerative disorders.

Migraine headaches

Migraines are recurrent headaches that typically last for hours and affect half of the head. *Familial hemiplegic migraine type 1* (FHM1) is characterized by migraine attacks that can last 1 to 3 days. During such episodes, individuals experience severe headaches and vomiting. Several mutations in a voltage-gated Ca^{2+} channel gene (*CACNA1A*) have been identified in families with FHM1, each producing different clinical symptoms. For example, a mutation in the pore-forming region of the channel produces hemiplegic migraine along with progressive cerebellar ataxia, whereas other mutations cause only the usual FHM1 symptoms. These are gain-of-function mutations that increase the amount of current flowing through Ca^{2+} channels. A mutation in the *SCNA1* gene causes another form of FHM, *familial hemiplegic migraine type 3*. It is unknown how these changes in Ca^{2+} or Na$^+$ channel properties lead to migraine attacks.

Pain

Numerous channelopathies are associated with either increases or decreases in pain perception. These are typically diseases of peripheral nerve, particularly in the nociceptive neurons of the dorsal root ganglia (see Chapter 13). These neurons express a unique type of Na$^+$ channel, encoded by the *SCN9A* gene, that regulates their excitability. Mutations in *SCN9A* underlie two syndromes associated with increased pain

■ Clinical Applications *(continued)*

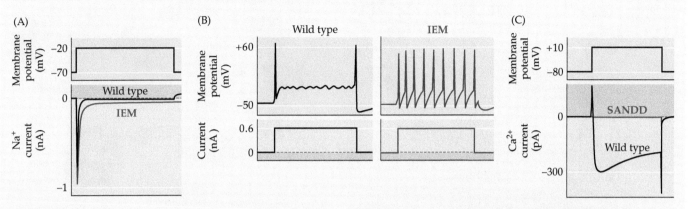

(A) Enhancement in voltage-gated Na⁺ current caused by IEM-associated mutation in the *SCN9A* gene. Wild type refers to the normal, unmutated form of channel. (B) Increased action potential firing caused by IEM-associated mutation in the *SCN9A* gene. (C) Loss of voltage-gated Ca²⁺ current caused by SANDD-associated mutation in the *CACNA1D* gene. (A,B after S. G. Waxman and G. W. Zamponi, 2014. *Nat Neurosci* 17: 153–163 and J. G. Hoeijmakers et al., 2012. *Brain* 135: 345-358; C after S. M. Baig et al., 2011. *Nat Neurosci* 14: 77–84.)

perception: *inherited erythromelalgia* (IEM) and *paroxysmal extreme pain disorder* (PEPD). Individuals with IEM experience intense burning pain. The *SCN9A* mutation associated with this syndrome is a gain-of-function mutation that shifts the voltage dependence of Na⁺ channel gating to more hyperpolarized potentials, increasing the amount of Na⁺ current produced by depolarization (Figure A). This enhances repetitive firing of action potentials (Figure B), which presumably causes the painful burning sensation. Individuals with PEPD experience severe visceral pain and have a different mutation in *SCN9A* that impairs Na⁺ channel inactivation, leading to longer-lasting Na⁺ current. How this persistent Na⁺ current leads to PEPD symptoms is not yet known. Loss-of-function mutations in *SCN9A* are associated with *congenital insensitivity to pain*, which causes individuals to lack the ability to sense painful stimuli that indicate bodily harm. Likewise, mutations in another Na⁺ channel gene, *SCN11A*, can also produce congenital insensitivity to pain.

Deafness

Hearing loss is the most common sensory disorder in humans. Congenital hearing impairment is a genetically diverse spectrum of disorders; more than half of these cases of deafness are linked to mutations in more than 50 genes, including the genes for ion channels and active transporters. *Sinoatrial node dysfunction and deafness* (SANDD) is caused by a mutation in the *CACNA1D* gene that encodes a voltage-gated Ca²⁺ channel expressed in both heart muscle cells and cochlear hair cells (see Chapter 10). As indicated by the name, people with SANDD exhibit both cardiac dysfunction as well as congenital deafness. The SANDD-associated mutation in the *CACNA1D* gene disrupts Ca²⁺ permeation through the Ca²⁺ channel, eliminating Ca²⁺ influx (Figure C). This causes deafness via loss of Ca²⁺-dependent neurotransmitter release from the hair cells, while cardiac dysfunction results from disrupted action potential generation. At least two other hearing disorders also are caused by ion channel mutations. The progressive loss of hearing associated with *nonsyndromic sensorineural deafness type 2* is due to mutations in *KCNQ4*, a voltage-gated K⁺ channel found in hair cells of the auditory and vestibular systems. Deafness in *Bartter syndrome type IV* is caused by mutations in barttin, a β subunit of the ClC Cl⁻ channels. Forty percent of newborns with nonsyn-dromic hearing loss have mutations in connexin-26, a gap junction channel found in the cochlea (see Chapter 5).

Blindness

X-linked *congenital stationary night blindness* (CSNB) is a recessive retinal disorder that causes night blindness, decreased visual acuity, myopia, nystagmus, and strabismus. Complete CSNB causes retinal rod photoreceptors to be nonfunctional. Incomplete CSNB causes subnormal (but measurable) functioning of both rod and cone photoreceptors. Like EA2, the incomplete type of CSNB is caused by mutations producing truncated Ca²⁺ channels, in this case the CACNA1F channel. Abnormal retinal function may arise from decreased Ca²⁺ currents and neurotransmitter release from photoreceptors (see Chapter 11).

In summary, ion channels are frequent targets of neurological disorders, emphasizing the value of understanding ion channel function for elucidating the etiology of these disorders. In turn, channelopathies are a valuable window for further understanding the roles of ion channels in the function of the brain and peripheral nervous system.

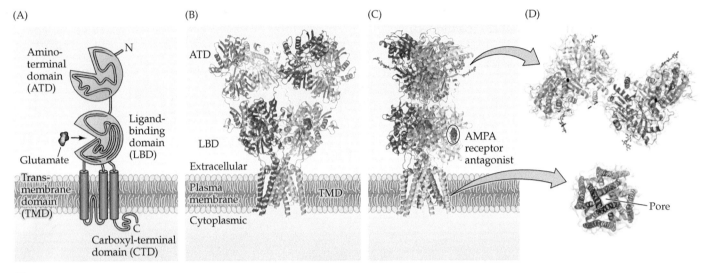

(A)

Amino-terminal domain (ATD)

N

Ligand-binding domain (LBD)

Glutamate

Trans-membrane domain (TMD)

C

Carboxyl-terminal domain (CTD)

(B)

ATD

LBD

Extracellular

Plasma membrane

Cytoplasmic

TMD

(C)

AMPA receptor antagonist

TMD

(D)

Pore

(E)

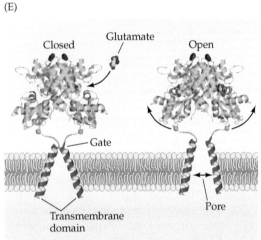

Closed

Glutamate

Open

Gate

Transmembrane domain

Pore

FIGURE 4.8 **Structure and gating of the AMPA receptor** (A) Domain structure of an AMPA receptor subunit. The largest part of each subunit is extracellular and consists of two domains, the amino-terminal domain (ATD) and the ligand-binding domain (LBD). In addition, a transmembrane domain (TMD) forms part of the ion channel pore, and an intracellular carboxyl-terminal domain (CTD) connects the receptor to intracellular proteins. (B–D) Crystallographic structure of the AMPA receptor. Each of the four subunits is indicated in a different color. (B) From this perspective, the Y shape of the AMPA receptor is evident. (C) After the receptor is rotated 90 degrees, the asymmetrical dimensions of the receptor are evident. One ligand-binding domain is visible and is occupied by an antagonist drug (circled). (D) Cross-section views of the AMPA receptor at two different positions (gray arrows) reveal the spatial relationships between subunits and also illustrate the changes in shape that occur along the length of the receptor. (E) Model for gating of the AMPA receptor by glutamate. The transmembrane domain (blue helices) and part of the extracellular ligand-binding domain are shown. Binding of glutamate closes the clamshell structure of the ligand-binding domain (side arrows), leading to movement of the gate helices that opens the channel pore. (A,E after S. F. Traynelis et al., 2010. *Pharmacol Rev* 62: 405–496; B,C after A. I. Sobolevsky et al., 2009. *Nature* 462: 745–756; D,E after K. B. Hansen et al., 2021. *Pharmacol Rev* 73: 298–487. CC BY 4.0.)

response shown in Figure 4.4C. The structure of cyclic nucleotide-gated channels is reminiscent of that of voltage-gated K^+ channels, being an assembly of four subunits, each possessing six transmembrane helices (Figure 4.9A,B). Furthermore, these four subunits include pore loops that form the channel pore, as well as a structure similar to the voltage sensor domain of voltage-gated channels (Figure 4.9C). Notably, each subunit also includes an extensive cytoplasmic domain that binds cyclic nucleotides; this is connected to the channel domain via a linker. Binding of cyclic nucleotides alters the structure of the cyclic nucleotide-binding domain, similar to the shutting of the ligand-binding domain of the AMPA receptor. This conformational change is then propagated, via the linker, to the channel domain and opens the pore gate (Figure 4.9D). Once the gate is opened, cations flow through the channel pore.

In conclusion, ligand-gated ion channels have ligand-binding domains that move in response to chemical signals, thereby opening gates that permit ions to flow through the channel pore.

Gating of channels by heat

Temperature-sensitive TRP channels have attracted considerable attention because of their roles in thermosensation (see Figures 4.4D and 4.5) and many other physiological responses. The structure of TRP channels is remarkably similar to that of voltage-gated K^+ channels (see Figure 4.6) and cyclic nucleotide-gated channels (see Figure 4.9), consisting of four subunits that each possess six transmembrane helices. Each subunit contributes pore loops that form a central cation-selective pore and also has a peripheral structure reminiscent of the voltage sensors of K^+ channels (Figure 4.10A).

Gating of the TRPV1 channel by heat, or capsaicin, is apparently mediated by displacement of membrane lipids. Ordinarily, phosphatidylinositol lipids occupy a binding site for capsaicin that is located in the vicinity of the helical

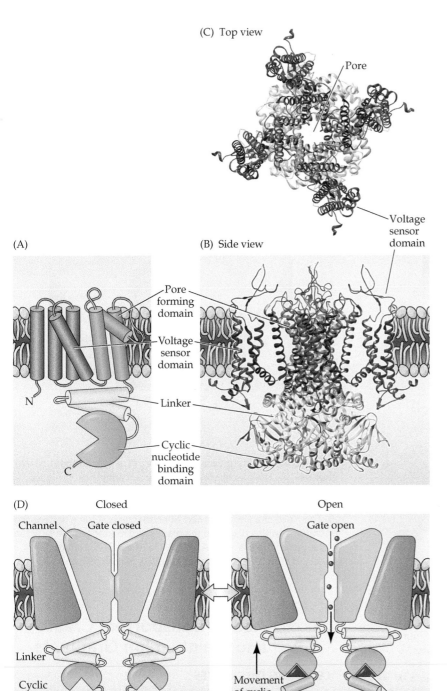

(C) Top view

Pore

Voltage sensor domain

(A)

(B) Side view

Pore forming domain

Voltage sensor domain

Linker

Cyclic nucleotide binding domain

N

C

(D) Closed

Channel   Gate closed

Linker

Cyclic nucleotide binding domain

Open

Gate open

Movement of cyclic nucleotide binding domain and linker

Cyclic nucleotide

FIGURE 4.9 Structure and gating of cyclic nucleotide-gated channels (A) Diagram of a subunit of a cyclic nucleotide-gated channel. Each subunit consists of several domains: a transmembrane pore-forming domain (blue), a transmembrane domain similar to the voltage sensor domain of voltage-gated channels (red), a cyclic nucleotide-binding domain (green), and a linker (yellow) that connects the binding domain to the pore-forming domain. (B,C) Side (B) and top (C) views of a cyclic nucleotide-gated channel, which is formed by four subunits. (D) Model for gating of cyclic nucleotide-gated channels. Left: In the absence of cyclic nucleotides, the channel gate is closed. Right: Upon binding of cyclic nucleotides to the binding domain, the binding domain undergoes a conformational change that moves the linkers, thereby opening the channel gate and allowing cations to flow through the channel pore. (A–C after L. M. R. Napolitano et al., 2021. *Pflugers Arch—Eur J Physiol* 473: 1423–1435; D after J. Xue et al. 2022. *Neuron* 110: 86–95.)

the nature of the stimuli that move this linker in the two types of channels.

Gating of channels by mechanical displacement

Among mechanosensitive channels (see Figure 4.4E), members of the Piezo family are of particular interest because of their roles in sensing light touch and allodynia (painful sensitivity to mechanical stimuli). The structure of Piezo channels is unique compared with that of most other ion channels. These channels are extremely large proteins, with 38 transmembrane helices that are embedded within curved regions of the plasma membrane (Figure 4.11A,B). In addition to forming a central pore that is permeable to cations, the transmembrane domains of Piezo channels form three blades that serve as levers for sensing membrane curvature. Mechanical forces that flatten the membrane change its curvature, thereby creating tension that pushes the levers to open the channel pore (Figure 4.11B). While other types of mechanosensitive channels are gated via a variety of other mechanisms, many of these mechanisms similarly rely on mechanical forces modifying the interaction between membrane lipids and the channel protein.

In summary, a tremendous variety of ion channels allows neurons to generate electrical signals in response to a broad range of stimuli, including changes in membrane potential, synaptic input, intracellular second

linkers that connect the channel pore to the sensorlike peripheral domains (Figure 4.10B, left). These lipids keep the channel pore closed. However, both heat (see Figure 4.10B, right) and capsaicin displace the lipids from this binding site, causing the linker to move and open the channel pore. Thus, heat gates the TRPV1 channel via a mechanism analogous to the voltage-dependent movement of the linker of voltage-gated K^+ channels, with the key difference being

(A)

Top view

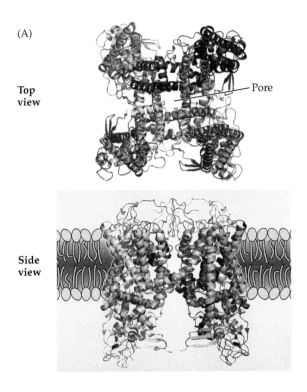

Pore

Side view

(B)

Channel closed   Channel open

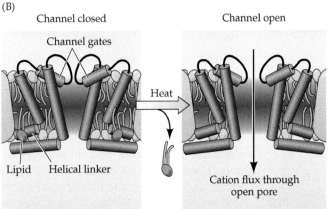

Channel gates

Heat

Lipid   Helical linker

Cation flux through open pore

FIGURE 4.10   Structure and gating of TRPV1, a thermo-sensitive channel (A) The TRPV1 channel consists of four sub-units (different colors), which form a central cation-selective pore, as well as structures similar to the voltage sensors of voltage-gated channels. (B) Heat opens the channel pore by displacing membrane lipids closely associated with the helical linker that connects the sensorlike domains to the channel pore, leading to a conformational change that opens channel gates (red). Capsaicin also activates this channel by displacing lipids from their binding sites. (After Y. Gao et al., 2016. *Nature* 534: 347–351.)

messengers, light, odors, heat, sound, touch, pressure, pH, and many other stimuli. Stimuli activate relevant sensors on the channel—be they charged voltage sensors, ligand-binding sites, or sites of interaction with membrane lipids—which then generate an internal movement that leads to channel opening and resultant ion diffusion through the channel pore.

(A)

Top view

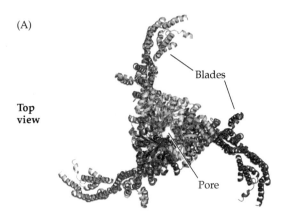

Blades

Pore

(B)

Side view

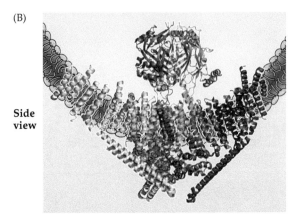

FIGURE 4.11   Structure and gating of a mechanosensitive channel (A) The mechanosensitive Piezo channel has a unique structure that includes three subunits that form a pore, as well as large blades that are embedded within curved membranes. (B) The Piezo channel is opened when mechanical force flattens the membrane, which displaces the blades and rearranges the structure of the rest of the channel. (A after J. Ge et al., 2015. Nature 527: 64–69.; B after J. M. Kefauver et al., 2020. *Nature* 587: 567–576.)

CONCEPT **4.5** | **Active Transporters Create and Maintain Ion Gradients**

LEARNING OBJECTIVES

4.5.1 Describe the different mechanisms used to transport ions by ATPase pumps and ion exchangers.

4.5.2 Explain the role of the Na+ pump in generating transmembrane gradients for Na+ and K+.

Up to this point, the discussion of the molecular basis of electrical signaling has taken for granted the remarkable fact that nerve cells maintain ion concentration gradients across their surface membranes: None of the ions of physiological importance (Na^+, K^+, Cl^-, H^+, and Ca^{2+}) are in electrochemical equilibrium. Furthermore, because channels produce electrical signals by allowing one or more of these ions to diffuse down their electrochemical gradients, there would be a gradual dissipation of these concentration gradients unless nerve cells could restore ions displaced during the current flow that occurs as a result of both neural signaling and the continual leakage of ions that occurs even at rest. The work of generating and maintaining concentration gradients for particular ions is carried out by plasma membrane proteins known as active transporters. They are called active transporters because they must consume energy as they transport ions uphill against their electrochemical gradients.

Active transporters carry out their task by forming complexes with the ions they are translocating. The process of ion binding and unbinding during transport typically requires several milliseconds. As a result, ion translocation by active transporters is orders of magnitude slower than the diffusion of ions through channel pores (recall that a single ion channel can conduct thousands of ions across a membrane each millisecond). In short, active transporters gradually store energy in the form of ion concentration gradients, whereas the opening of ion channels rapidly dissipates this stored energy during relatively brief electrical signaling events. Although the specific jobs of active transporters are highly diverse,

active transporters can be sorted into two main types on the basis of the source of energy used for ion movement: **ATPase pumps** and **ion exchangers**.

ATPase pumps

ATPase pumps acquire energy for ion translocation directly from the hydrolysis of ATP. The most prominent example of an ATPase pump is the Na^+ pump (or more properly, the Na^+/K^+ ATPase pump), which is responsible for maintaining transmembrane concentration gradients for both Na^+ and K^+ (Figure 4.12A). The Na^+ pump is a large integral membrane protein made up of at least two subunits, α (encoded by the **ATP1A gene**) and β (**ATP1B gene**). The α subunit is responsible for ion translocation and spans the membrane ten times, with most of the molecule found on the cytoplasmic side, whereas the β subunit spans the membrane only once and is predominantly extracellular.

Ca^{2+} pumps are another class of ATPase pump (Figure 4.12B). Ca^{2+} pumps are an important mechanism for removing Ca^{2+} from cells (see Chapter 7). Two different types of Ca^{2+} pumps have been identified. One, called **PMCA**, is found on the plasma membrane; the other, termed **SERCA**, is used to store Ca^{2+} in the endoplasmic reticulum. The structure of SERCA (see Figure 4.12B) is remarkably similar to that of the Na^+ pump, aside from having a binding site for Ca^{2+}.

The crucial importance of the Na^+ pump for brain function is evident from the fact that this transporter accounts for up to two-thirds of the brain's total energy consumption. The Na^+ pump of neurons was first discovered in the 1950s, when Richard Keynes at Cambridge University used radioactive Na^+ to demonstrate the energy-dependent efflux of Na^+ from squid giant axons. Keynes and his collaborators found that this efflux depended on extracellular K^+ (Figure 4.13, point 2) and ceased when the axon's supply of ATP was interrupted by treatment with metabolic poisons (see Figure 4.13, point 4), indicating that removal of intracellular Na^+ requires K^+ and cellular metabolism. Further studies with radioactive K^+ demonstrated that Na^+ efflux is associated with the simultaneous ATP-dependent influx of K^+. These energy-dependent movements of Na^+ and K^+ generate transmembrane gradients for both ions. Subsequent work by Jens Christian Skou in Denmark established that these fluxes of Na^+ and K^+ are due to an ATP-hydrolyzing Na^+/K^+

FIGURE 4.12   Examples of ATPase pumps (A) Structure of the Na^+ pump. Domains responsible for nucleotide binding (NB), phosphorylation (P), and an actuator domain (AD) are evident. In this conformation, ADP occupies the NB domain of the pump, and two K^+ (inside square) can be observed in the transmembrane domain. Activity of the pump leads to transfer of Na^+ from inside to outside, and of K^+ in the opposite direction. (B) Structure of the SERCA Ca^{2+} pump. Domains responsible for nucleotide binding (NB), phosphorylation (P), and ion translocation activity (TA) are indicated. Shown is the structure of the pump when bound to ADP; in this state, two Ca^{2+} (purple spheres within red circle) are sequestered within the membrane-spanning regions of the pump. Note the similarity between this structure and that of the Na^+/K^+ pump shown in (A). (A after T. Shinoda et al., 2009. *Nature* 459: 446–450; B after C. Toyoshima et al., 2004. *Nature* 432: 361–368.)

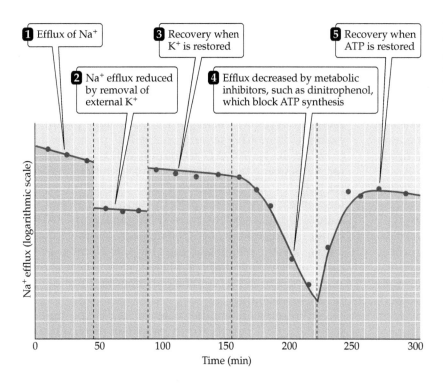

1 Efflux of Na⁺

2 Na⁺ efflux reduced by removal of external K⁺

3 Recovery when K⁺ is restored

4 Efflux decreased by metabolic inhibitors, such as dinitrophenol, which block ATP synthesis

5 Recovery when ATP is restored

FIGURE 4.13 **Ion movements produced by the Na⁺ pump** Measurement of radioactive Na⁺ efflux from a squid giant axon. This efflux depends on external K⁺ and intracellular ATP. (After A. L. Hodgkin and R. D. Keynes, 1955. *J Physiol* 128: 28–60.)

pump. Quantitative studies indicate that Na⁺ and K⁺ are not pumped at identical rates: The rate of K⁺ influx is only about two-thirds that of Na⁺ efflux, indicating that the pump transports two K⁺ into the cell for every three Na⁺ that are removed.

The Na⁺ pump is thought to alternately shuttle Na⁺ and K⁺ across the membranes in a cycle fueled by binding of ATP and transfer of a phosphate group from ATP to the pump (Figure 4.14A). ATP binding promotes binding of intracellular Na⁺ and release of K⁺, while pump phosphorylation leads to extracellular release of Na⁺ and binding of K⁺. In between these two states of ion translocation are occluded states that prevent leakage of ions in

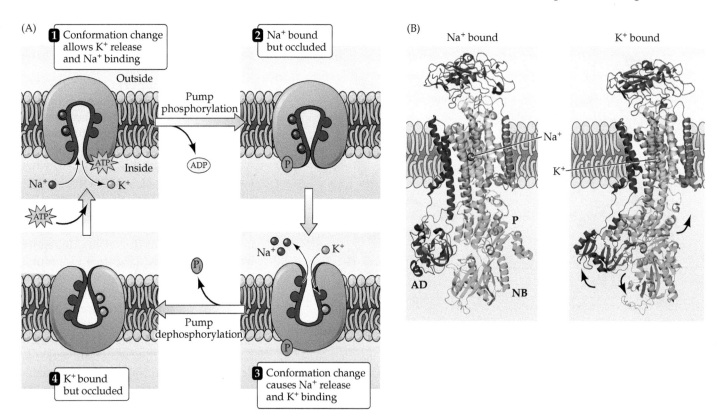

FIGURE 4.14 **Translocation of Na⁺ and K⁺ by the Na⁺ pump** (A) A model for the movement of ions by the Na⁺ pump. Uphill movements of Na⁺ and K⁺ are driven by binding and hydrolysis of ATP, which phosphorylates the pump (indicated by P). These ion fluxes are asymmetrical, with three Na⁺ carried out for every two K⁺ brought in. (B) Comparison of structure of the Na⁺ pump in Na⁺-bound state (left), corresponding to step 2 in (A), and K⁺-bound state (right), corresponding to step 4. Changes in location of phosphorylation (P), nucleotide binding (NB), and activator (AD) domains (arrows) are associated with switch between Na⁺-bound and K⁺-bound conformations. (A after J. B. Lingrel et al., 1994. *Kidney Internat Suppl* 44: S32–S39; B after M. Nyblom et al., 2013. *Science* 342: 123–127.)

the reverse direction, with subsequent hydrolysis of ATP leading to dissociation of ADP, which toggles the pump between accumulating intracellular K^+ and removing intracellular Na^+. While the precise mechanism of ion translocation is still being elucidated, many of the structures involved have been identified (Figure 4.14B). Na^+ and K^+ alternately bind to sites in the interior of the pump, within the transmembrane domain of the α subunit. A nucleotide-binding domain is responsible for binding ATP, with hydrolysis of ATP transferring a phosphate onto a phosphorylation domain that subsequently changes the location of an activator domain thought to regulate ion binding to the pump (see Figure 4.14B). This ATP-dependent structural rearrangement explains how the pump is able to move Na^+ and K^+ uphill, against their steep electrochemical gradients. Very similar structural changes are also thought to be involved in removal of cytoplasmic Ca^{2+} by the SERCA Ca^{2+} pump.

Ion exchangers

The second class of active transporter does not use ATP directly but instead uses the electrochemical gradients of other ions as an energy source. Such ion exchangers carry one or more ions *up* their electrochemical gradient while simultaneously taking another ion (most often Na^+) *down* its gradient. These transporters can be further subdivided into two types according to the direction of ion movement. As their name implies, **antiporters** exchange intracellular and extracellular ions. An example of an antiporter is the Na^+/Ca^{2+} exchanger, which shares with the Ca^{2+} pump the important job of keeping intracellular Ca^{2+} concentration low (Figure 4.15A). Another antiporter, the Na^+/H^+ exchanger, regulates intracellular pH (Figure 4.15B). Ion exchangers of the second type, the **co-transporters**, work by carrying multiple ions in the same direction. Two such co-transporters regulate intracellular Cl^- concentration by translocating Cl^- along with extracellular Na^+ and/or K^+; these are the $Na^+/K^+/Cl^-$ co-transporter, which transports Cl^- along with Na^+ and K^+ into cells (Figure 4.15C), and the K^+/Cl^- co-transporter, which removes

intracellular Cl^- (Figure 4.15D). As you will see in Chapter 6, neurotransmitters are transported into synaptic terminals and glial cells via other co-transporters (Figure 4.15E). Although the electrochemical gradient of Na^+ (or other counter-ions) is the proximate source of energy for both ion exchangers and co-transporters, these gradients ultimately depend on the hydrolysis of ATP by ATPase pumps such as the Na^+ pump.

The structure of the Na^+/Ca^{2+} exchanger (**SLC8A genes**) illustrates a mechanism that ion exchangers use to translocate ions across membranes. This antiporter has ten transmembrane helices, eight of which comprise a core domain responsible for ion binding and translocation, and two of which serve as a gating bundle (Figure 4.16A). The core domain has three negatively charged ion-binding sites that are alternately occupied by three Na^+ or one Ca^{2+}. Like the Na^+ pump, the Na^+/Ca^{2+} exchanger assumes different structures; unlike with the Na^+ pump, the conformation of the Na^+/Ca^{2+} exchanger depends on its ion-binding status rather than phosphorylation state. When bound to intracellular Ca^{2+}, the exchanger undergoes a remarkable conformational change that causes its gating bundle to slide across the inner core helices, thereby causing the inner core to face outward (Figure 4.16B, left). This allows the ion-binding sites to be exposed to the extracellular side of the membrane and bound Ca^{2+} to be released into the extracellular medium. This frees up the unoccupied ion-binding sites and permits extracellular Na^+ to bind to them. This reverses the position of the gating bundle and causes the binding sites to face inward, toward the cytoplasm (see Figure 4.16B right). This allows Na^+ to be released into the cytoplasm and allows intracellular Ca^{2+} to once again bind to the ion-binding sites, triggering a switch back to the outward-facing conformation. Thus, ion exchange by the Na^+/Ca^{2+} exchanger represents an ion binding/unbinding cycle similar to that of the Na^+ pump (see Figure 4.14A). The fundamental difference is that the conformational changes produced in the Na^+/Ca^{2+} exchanger during ion translocation are driven by ion binding, rather than ATP hydrolysis.

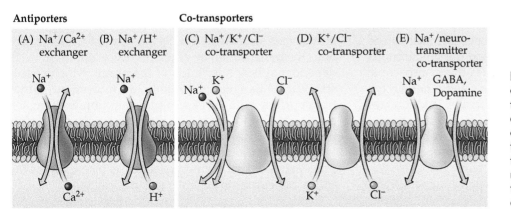

Antiporters

(A) Na^+/Ca^{2+} exchanger **(B)** Na^+/H^+ exchanger

Na^+ Na^+

Ca^{2+} H^+

Co-transporters

(C) $Na^+/K^+/Cl^-$ co-transporter **(D)** K^+/Cl^- co-transporter **(E)** $Na^+/$neuro-transmitter co-transporter

Na^+ K^+ Cl^- Na^+ GABA, Dopamine

K^+ Cl^-

FIGURE 4.15 Examples of ion exchangers Ion exchangers use the electrochemical gradients of co-transported ions as a source of energy. These exchangers can be further subdivided into antiporters that swap ions on the two sides of the membrane (A,B) and co-transporters that carry multiple ions in the same direction (C–E).

(A)

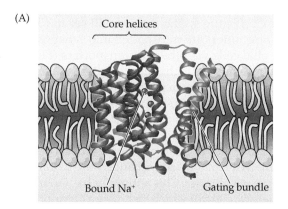

FIGURE 4.16   Structure and function of the Na⁺/Ca²⁺ exchanger (A) The Na⁺/Ca²⁺ exchanger includes ten transmembrane helices, eight associated with a central ion-binding core and two that form a gating bundle that controls the conformation of the exchanger. Bound Na⁺ are shown as red spheres. (B) Model for the transmembrane movement of ions by the Na⁺/Ca²⁺ exchanger. Binding of Ca²⁺ and Na⁺ toggles the exchanger between outward-facing and inward-facing conformations, which in turn controls binding and translocation of these ions. Ion movement is asymmetrical, with three Na⁺ brought in for every one Ca²⁺ carried out of the cell. (A after J. Liao et al., 2016. *Nat Struct Mol Biol* 23: 590–599; B after M. Iwaki et al., 2020. *FEBS J* 287: 4678–4695.)

(B)

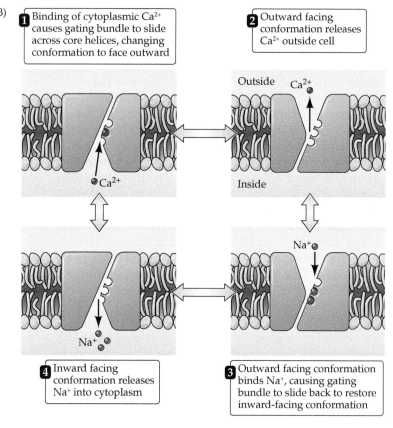

1 Binding of cytoplasmic Ca²⁺ causes gating bundle to slide across core helices, changing conformation to face outward

2 Outward facing conformation releases Ca²⁺ outside cell

4 Inward facing conformation releases Na⁺ into cytoplasm

3 Outward facing conformation binds Na⁺, causing gating bundle to slide back to restore inward-facing conformation

In summary, both ATPase pumps and ion exchangers transfer ions across membranes by having ion-binding sites that are deep within the core of the transporter protein. Unlike the selectivity filters of ion channel pores, these binding sites can be occluded to prevent free diffusion between the intracellular and extracellular compartments. This arrangement enables active, uphill transport of ions to create transmembrane ion concentration gradients.

Summary

Ion channels and active transporters have complementary functions. The primary purpose of transporters is to generate transmembrane concentration gradients, which are then exploited by ion channels to generate electrical signals. The flow of ions through single open channels can be detected as tiny electrical currents, and the synchronous opening of many channels generates the macroscopic currents that produce action potentials and other electrical signals. A large number of ion channel genes creates channels with a correspondingly wide range of functional characteristics, thus allowing different types of neurons to have a remarkable spectrum of electrical properties. All channels are integral membrane proteins that share certain structural characteristics, such as pores and selectivity filters, responsible for permeation of specific types of ions across the membrane. Voltage-gated ion channels are responsible for the voltage-dependent conductances that underlie the action potential and open or close their pores in response to the membrane potential. Voltage-gated channels have voltage sensors that are responsible for voltage sensing. Other types of channels have specialized structures that enable their gating to be controlled by chemical signals, such as neurotransmitters or second messengers, or other types of signals such as heat or mechanical distortion of the plasma membrane. Active transporter proteins are quite different from ion channels because they move ions against a concentration gradient. The energy required for ion translocation is provided either by the hydrolysis of ATP or by the electrochemical gradient of other ions, such as Na⁺. The Na⁺ pump produces and maintains the transmembrane gradients of Na⁺ and K⁺ by ATP-dependent phosphorylation of the pump, which causes structural changes that enable ion movement. Other transporters, both ATPase pumps and ion exchangers, are responsible for the electrochemical gradients for other physiologically important ions, including Cl⁻, Ca²⁺, and H⁺. Together, transporters and channels provide a comprehensive and satisfying molecular explanation for the ability of neurons to generate electrical signals.

■ Additional Reading

Reviews

Bezanilla, F. and A. M. Correa (1995) Single-channel properties and gating of Na[+] and K[+] channels in the squid giant axon. In *Cephalopod Neurobiology*, N. J. Abbott, R. Williamson and L. Maddock (Eds.). New York: Oxford University Press, pp. 131–151.

Cesare P., A. Moriondo, V. Vellani and P. A. McNaughton (1999) Ion channels gated by heat. *Proc. Natl. Acad. Sci. U.S.A.* 96: 7658–7663.

Hansen, K. B. and 18 others (2021) Structure, function, and pharmacology of glutamate receptor ion channels. *Pharmacol Rev.* 73: 298–487.

Jentsch, T. J. and M. Pusch (2018) CLC chloride channels and transporters: Structure, function, physiology, and disease. *Physiol. Rev.* 98: 1493–1590.

Kefauver, J. M., A. B. Ward and A. Patapoutian (2020) Discoveries in structure and physiology of mechanically activated ion channels. *Nature* 587: 567–576.

Napolitano, L. M. R., V. Torre and A. Marchesi (2021) CNG channel structure, function, and gating: A tale of conformational flexibility. *Pflugers Arch.* 473: 1423–1435.

Traynelis, S. F. and 9 others (2010) Glutamate receptor ion channels: structure, regulation, and function. *Pharmacol. Rev.* 62: 405–496.

Wang, H. and J. Siemens (2015) TRP ion channels in thermosensation, thermoregulation and metabolism. *Temperature* 2: 178–187.

Important Original Papers

Ahuja, S. and 34 others (2015) Structural basis of Nav1.7 inhibition by an isoform-selective small-molecule antagonist. *Science* 350: aac5464.

Caterina, M. J. and 5 others (1997) The capsaicin receptor: A heat-activated ion channel in the pain pathway. *Nature* 389: 816–824.

Doyle, D. A. and 7 others (1998) The structure of the potassium channel: Molecular basis of K[+] conduction and selectivity. *Science* 280: 69–77.

Gao, Y., E. Cao, D. Julius and Y. Cheng (2016) TRPV1 structures in nanodiscs reveal mechanisms of ligand and lipid action. *Nature* 534: 347–351.

Ge, J. and 9 others (2015) Architecture of the mammalian mechanosensitive Piezo1 channel. *Nature* 527: 64–69.

Gonzalez, E. B., T. Kawate and E. Gouaux (2009) Pore architecture and ion sites in acid sensing ion channels and P2X receptors. *Nature* 460: 599–604.

Hite, R. K., X. Tao and R. MacKinnon (2017) Structural basis for gating the high-conductance Ca[2+]-activated K[+] channel. *Nature* 541: 52–57.

Hodgkin, A. L. and R. D. Keynes (1955) Active transport of cations in giant axons from *Sepia* and *Loligo*. *J. Physiol.* 128: 28–60.

Ingram, N. T., G. L. Fain and A. P. Sampath (2020) Elevated energy requirement of cone photoreceptors. *Proc. Natl. Acad. Sci. U.S.A.* 117: 19599–19603.

Iwaki, M. and 6 others (2020) Structure-affinity insights into the Na[+] and Ca[2+] interactions with multiple sites of a sodium-calcium exchanger. *FEBS J.* 287: 4678–4695.

Kitano, Y. and 8 others (2019) Effects of mirogabalin, a novel ligand for the $\alpha2\delta$ subunit of voltage-gated calcium channels, on N-type calcium channel currents of rat dorsal root ganglion culture neurons. *Pharmazie* 74: 147–149.

Liao, J. and 5 others (2016) Mechanism of extracellular ion exchange and binding-site occlusion in a sodium/calcium exchanger. *Nat. Struct. Mol. Biol.* 23: 590–599.

Llano, I., C. K. Webb and F. Bezanilla (1988) Potassium conductance of squid giant axon. Single-channel studies. *J. Gen. Physiol.* 92: 179–196.

Long, S. B., E. B. Campbell and R. MacKinnon (2005) Crystal structure of a mammalian voltage-dependent *Shaker* family K[+] channel. *Science* 309: 897–903.

Nyblom, M. and 7 others (2013) Crystal structure of Na[+], K[+]-ATPase in the Na[+]-bound state. *Science* 342: 123–127.

Park, E., E. B. Campbell and R. MacKinnon (2017) Structure of a CLC chloride ion channel by cryo-electron microscopy. *Nature* 541: 500–505.

Ricci, A. J., H. J. Kennedy, A. C. Crawford and R. Fettiplace (2005) The transduction channel filter in auditory hair cells. *J. Neurosci.* 25: 7831–7839.

Shinoda, T., H. Ogawa, F. Cornelius and C. Toyoshima (2009) Crystal structure of the sodium-potassium pump at 2.4 Å resolution. *Nature* 459: 446–450.

Sigworth, F. J. and E. Neher (1980) Single Na[+] channel currents observed in cultured rat muscle cells. *Nature* 287: 447–449.

Sobolevsky, A. I., M. P. Rosconi and E. Gouaux (2009) X-ray structure, symmetry and mechanism of an AMPA-subtype glutamate receptor. *Nature* 462: 745–756.

Toyoshima, C., H. Nomura and T. Tsuda (2004) Luminal gating mechanism revealed in calcium pump crystal structures with phosphate analogues. *Nature* 432: 361–368.

Vanderberg, C. A. and F. Bezanilla (1991) A sodium channel model based on single channel, macroscopic ionic, and gating currents in the squid giant axon. *Biophys. J.* 60: 1511–1533.

Waldmann, R. and 4 others (1997) A proton-gated cation channel involved in acid-sensing. *Nature* 386: 173–177.

Watanabe, J., A. Rozov and L. P. Wollmuth (2005) Target-specific regulation of synaptic amplitudes in the neocortex. *J. Neurosci.* 25: 1024–1033.

Wisedchaisri, G. and 6 others (2019) Resting-state structure and gating mechanism of a voltage-gated sodium channel. *Cell* 178: 993–1003.

Wu, J. and 6 others (2015) Structure of the voltage-gated calcium channel CaV1.1 complex. *Science* 350: aad2395–1.

Xue, J., Y. Han, W. Zeng and Y. Jiang (2021) Structural mechanisms of assembly, permeation, gating, and pharmacology of native human rod CNG channel. *Neuron* 109: 1302–1313.

Books

Gribkoff, V. K. and L. K. Kaczmarek (2009) *Structure, Function, and Modulation of Neuronal Voltage-Gated Ion Channels*. New York: Wiley.

Hille, B. (2001) *Ion Channels of Excitable Membranes*, 3rd Edition. Sunderland, MA: Sinauer/Oxford University Press.

Zheng, J. and M. C. Trudeau (2015) *Handbook of Ion Channels*. Boca Raton, FL: CRC Press.

Synaptic Transmission

Overview

The human brain contains 86 billion neurons, each with the ability to influence many other cells. Clearly, sophisticated and highly efficient mechanisms are needed to enable communication among this astronomical number of elements. Such communication is made possible by synapses, the functional contacts between neurons. Synapses allow neurons to form circuits that are responsible for processing and storing information within the brain. Two different types of synapses—electrical and chemical—can be distinguished on the basis of their mechanism of transmission. At electrical synapses, current flows through connexons, which are specialized membrane channels that connect two cells at gap junctions. In contrast, chemical synapses enable cell-to-cell communication via the secretion of neurotransmitters; these chemical agents released by the presynaptic neurons produce secondary current flow in postsynaptic neurons by activating specific neurotransmitter receptors. Virtually all neurotransmitters undergo a similar cycle of use: synthesis and packaging into synaptic vesicles; release from the presynaptic cell; binding to postsynaptic receptors; and finally, rapid removal or degradation. The influx of Ca^{2+} through voltage-gated channels triggers the secretion of neurotransmitters; this, in turn, gives rise to a transient increase in Ca^{2+} concentration in the presynaptic terminal. The rise in Ca^{2+} concentration causes synaptic vesicles to fuse with the presynaptic plasma membrane and release their contents into the space between the pre- and postsynaptic cells. Proteins on the surface of the synaptic vesicle and the presynaptic plasma membrane mediate the triggering of exocytosis by Ca^{2+}, as well as the subsequent retrieval of vesicle components from the plasma membrane via endocytosis. Neurotransmitters evoke postsynaptic electrical responses by binding to members of a diverse group of neurotransmitter receptors. There are two major classes of receptors: those in which the receptor molecule is a ligand-gated ion channel, and those in which the receptor and ion channel are separate entities. These receptors give rise to electrical signals by transmitter-induced opening or closing of the ion channels. Whether the postsynaptic actions of a particular neurotransmitter are excitatory or inhibitory is determined by the ion permeability of the ion channel affected by the transmitter, and by the electrochemical gradient for the permeant ions.

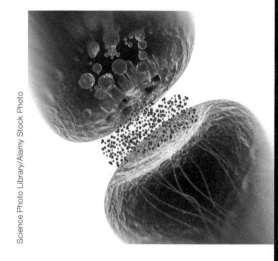

Science Photo Library/Alamy Stock Photo

KEY CONCEPTS

5.1 There are two mechanisms of synaptic signaling

5.2 Calcium ions regulate the release of discrete packets of neurotransmitters

5.3 A cycle of membrane trafficking is responsible for neurotransmitter release

5.4 Many proteins are required for synaptic vesicle trafficking

5.5 There are two families of neurotransmitter receptors

5.6 Postsynaptic membrane permeability changes during synaptic transmission

5.7 Postsynaptic ion fluxes determine whether synapses are excitatory or inhibitory

| **There Are Two Mechanisms of Synaptic Signaling**

LEARNING OBJECTIVES

5.1.1 Distinguish between electrical and chemical synapses.

5.1.2 Describe the cellular mechanisms involved in electrical synaptic transmission.

5.1.3 Describe the cellular mechanisms involved in chemical synaptic transmission.

Two classes of synapses

The many kinds of synapses in the human brain fall into two general classes: electrical synapses and chemical synapses. These two classes of synapses can be distinguished based on their structures and the mechanisms they use to transmit signals from the "upstream" neuron, called the **presynaptic** element, and the "downstream" neuron, termed **postsynaptic**.

The structure of an electrical synapse is shown schematically in Figure 5.1A. Electrical synapses permit direct, passive flow of electrical current from one neuron to another. The usual source of this current is the potential difference generated locally by the presynaptic action potential (see Chapter 3). Current flow at electrical synapses arises at an intercellular specialization called a **gap junction**, where membranes of the two communicating neurons are extremely close to one another and are linked together (Figure 5.1B). Gap junctions contain a unique type of channel, termed a **connexon**, which provides the path for electrical current to flow from one neuron to another (see Figure 5.2).

The general structure of a chemical synapse is shown schematically in Figure 5.1C. The space between the pre- and postsynaptic neurons is substantially greater at chemical synapses than at electrical synapses and is called the **synaptic cleft**. However, the most important structural feature of all chemical synapses is the presence of small, membrane-bounded organelles called **synaptic vesicles**

(A) Electrical synapse

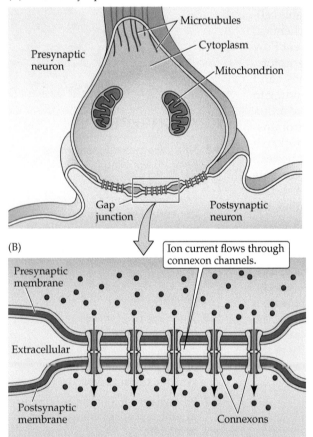

(C) Chemical synapse

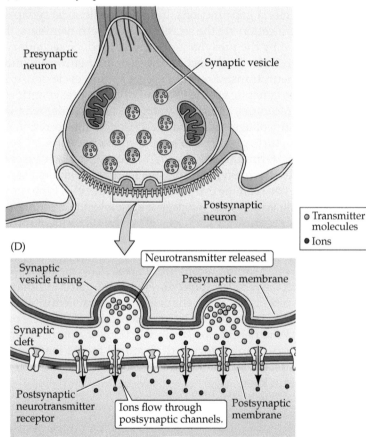

FIGURE 5.1 Electrical and chemical synapses differ fundamentally in their transmission mechanisms
(A) At electrical synapses, gap junctions occur between pre- and postsynaptic membranes. (B) Gap junctions contain connexon channels that permit current to flow passively from the presynaptic cell to the postsynaptic cell. (C) At chemical synapses, there is no intercellular continuity, and thus no direct flow of current from pre- to postsynaptic cell. (D) Synaptic current flows across the postsynaptic membrane only in response to the secretion of neurotransmitters, which open or close postsynaptic ion channels after binding to receptor molecules on the postsynaptic membrane.

within the presynaptic terminal. These spherical organelles are filled with one or more **neurotransmitters**, chemical signals that are secreted from the presynaptic neuron and detected by specialized receptors on the postsynaptic cell (Figure 5.1D). These chemical agents act as messengers between the communicating neurons and give this type of synapse its name.

Signaling at electrical synapses

Figure 5.2A shows an electron micrograph of an electrical synapse from a mammalian brain. As in the diagrams in Figure 5.1A and B, it can be seen that the processes of the presynaptic and postsynaptic neurons are connected via a gap junction (Figure 5.2B). The connexons contained within gap junctions are key to understanding how electrical synapses work (Figure 5.2C). Connexons are composed of a unique family of ion channel proteins, the **connexins**, which serve as subunits to form connexon channels. There are 21 different types of human connexin genes (*GJA–GJE*)

that are expressed in different cell types and yield connexons with diverse physiological properties. All connexins have four transmembrane domains, and all connexons consist of six connexins that come together to form a hemi-channel in both the pre- and postsynaptic neurons (Figure 5.2D). These hemi-channels are precisely aligned to form a pore that connects the two cells and permits electrical current to flow. The pore of a connexon channel is more than 1 nm in diameter, which is much larger than the ion channels described in Chapter 4. As a result, a variety of substances can simply diffuse between the cytoplasm of the pre- and postsynaptic neurons. In addition to ions, substances that diffuse through connexon pores include molecules with molecular weights as great as several hundred daltons. This permits important intracellular metabolites, such as ATP and second messengers (see Chapter 7), to be transferred between neurons.

Although they are a distinct minority, electrical synapses have several functional advantages. One is that

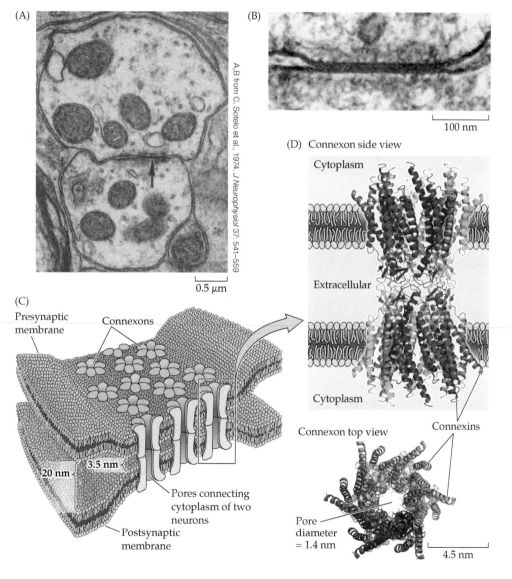

(A) *A,B from C. Sotelo et al., 1974. J Neurophysiol 37: 541–559*

0.5 μm

(B) 100 nm

(C)
Presynaptic membrane
Connexons
20 nm
3.5 nm
Pores connecting cytoplasm of two neurons
Postsynaptic membrane

(D) Connexon side view
Cytoplasm
Extracellular
Cytoplasm

Connexon top view
Connexins
Pore diameter = 1.4 nm
4.5 nm

FIGURE 5.2 Structure of electrical synapses (A) Electron micrograph of an electrical synapse (arrow) connecting two neurons within the inferior olive of a mammalian brain. (B) Higher-magnification electron micrograph of another electrical synapse, showing the gap junction structure characteristic of electrical synapses. (C) Gap junctions consist of connexons, hexameric complexes present in both the pre- and postsynaptic membranes. Channels assembled from connexons in these two membranes form pores that create electrical continuity between the two cells. (D) Crystallographic structure of connexons. Colors indicate individual connexins, integral membrane proteins that form the subunits of connexons. Side view shows the channels spanning the pre- and postsynaptic membranes; top view illustrates how six connexin subunits assemble in each membrane to form a channel with an exceptionally large pore. (D from S. Maeda et al., 2009. *Nature* 458: 597–602.)

transmission is extraordinarily fast: Because passive current flow across connexons is virtually instantaneous, communication can occur without the delay that is characteristic of chemical synapses. The high speed of electrical synaptic transmission is apparent in the operation of the first electrical synapse to be discovered, which resides in the crayfish nervous system. A postsynaptic electrical signal is observed at this synapse within a fraction of a millisecond after the generation of a presynaptic action potential (Figure 5.3A). In fact, at least part of this brief synaptic delay is caused by propagation of the action potential into the presynaptic terminal, so there may be essentially no delay at all in the transmission of electrical signals across the synapse. Such synapses interconnect many of the neurons within the circuit that allows the crayfish to escape from its predators, thus minimizing the time between the presence of a threatening stimulus and a potentially lifesaving motor response.

Another unique advantage of electrical synapses is that transmission can be bidirectional; although some connexons have special features for unidirectional transmission, in most cases current can flow in either direction, depending on which member of the coupled pair is invaded by an action potential. This allows electrical synapses to synchronize electrical activity among populations of neurons. For example, the brainstem neurons that generate rhythmic electrical activity underlying breathing are synchronized by electrical synapses, as are populations of interneurons

in the cerebral cortex, thalamus, cerebellum, and other brain regions (Figure 5.3B). Electrical transmission between vasopressin- and oxytocin-secreting neurons in the hypothalamus ensures that all cells fire action potentials at about the same time, thus facilitating a synchronized burst of secretion of these hormones into the circulation (see Box 21A). The fact that connexon pores are large enough to allow second messengers to diffuse between cells also permits electrical synapses to synchronize the intracellular signaling of coupled cells. This feature may be particularly important for glial cells, which form large intracellular signaling networks via their gap junctions.

Signaling at chemical synapses

Figure 5.4A shows an electron micrograph of a chemical synapse in the cerebral cortex. This image illustrates the presynaptic terminal, with its abundance of synaptic vesicles, as well as the postsynaptic cell separated by a synaptic cleft. A three-dimensional rendering of this chemical synapse, constructed from many images including the one in Figure 5.4A, reveals these features as well as many more structures, including filamentous elements in both pre- and postsynaptic processes, and structures in the synaptic cleft (Figure 5.4B). In the presynaptic terminal, dense projections (dark blue) are associated with the **active zone**, the place where synaptic vesicles discharge their neurotransmitters into the synaptic cleft, while the

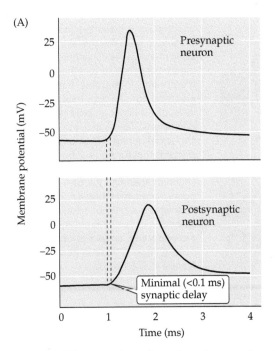
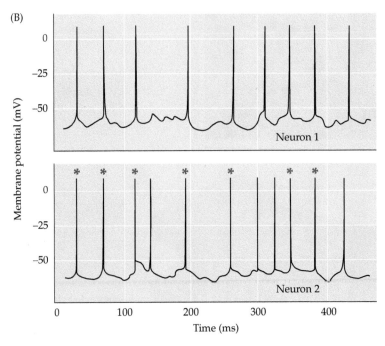

FIGURE 5.3 Function of gap junctions at electrical synapses
(A) Rapid transmission of signals at an electrical synapse in the crayfish. An action potential in the presynaptic neuron causes the postsynaptic neuron to be depolarized within a fraction of a millisecond. (B) Electrical synapses allow synchronization of electrical activity in hippocampal interneurons. In a pair of interneurons connected by electrical synapses, generation of an action potential in one neuron often results in the synchronized firing of an action potential in another neuron (asterisks). (A after E. J. Furshpan and D. D. Potter, 1959. *J Physiol* 145: 289–325; B after M. Beierlein et al., 2000. *Nat Neurosci* 3: 904–910.)

(A)

— Presynaptic

— Synaptic cleft

Postsynaptic density

Filaments

Filament branches

Postsynaptic

250 nm

(B)

Presynaptic

Postsynaptic

FIGURE 5.4 **Structure and function of chemical synapses** (A) Structure of a chemical synapse in the cerebral cortex. A presynaptic terminal (pink) forms a synapse with a postsynaptic dendrite (green). (B) Three-dimensional reconstruction of the synapse shown in (A). Inside the presynaptic terminal, spheres indicate synaptic vesicles at various stages of their trafficking cycle, linear elements indicate intracellular filaments, and dark blue indicates dense projections associated with the active zone. Inside the postsynaptic neuron, the blue structure is the postsynaptic density, green structures represent filaments, and red spheres indicate points where the filaments branch. Green material within the synaptic cleft indicates structures of unknown function. (C) Sequence of events involved in transmission at a typical chemical synapse.

(C)

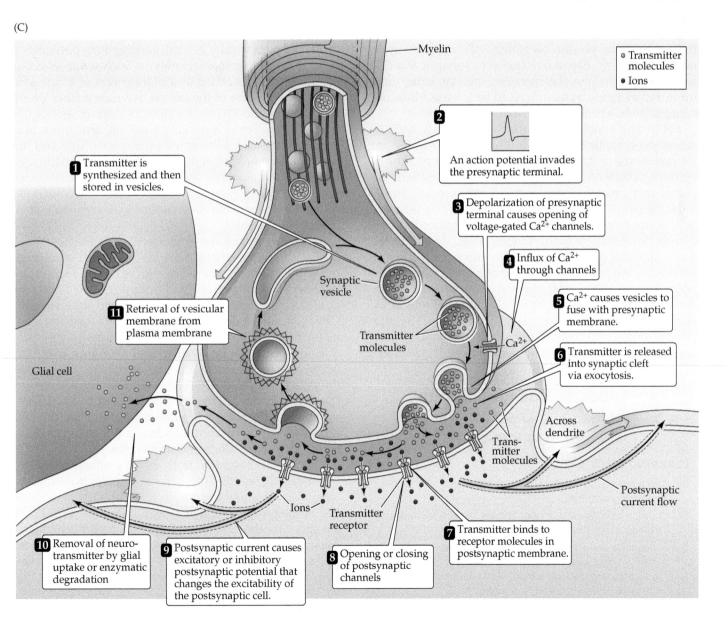

— Myelin

• Transmitter molecules
• Ions

2 An action potential invades the presynaptic terminal.

1 Transmitter is synthesized and then stored in vesicles.

3 Depolarization of presynaptic terminal causes opening of voltage-gated Ca²⁺ channels.

4 Influx of Ca²⁺ through channels

5 Ca²⁺ causes vesicles to fuse with presynaptic membrane.

6 Transmitter is released into synaptic cleft via exocytosis.

11 Retrieval of vesicular membrane from plasma membrane

Synaptic vesicle

Transmitter molecules

Ca²⁺

Across dendrite

Transmitter molecules

Glial cell

Postsynaptic current flow

Ions

Transmitter receptor

7 Transmitter binds to receptor molecules in postsynaptic membrane.

10 Removal of neurotransmitter by glial uptake or enzymatic degradation

9 Postsynaptic current causes excitatory or inhibitory postsynaptic potential that changes the excitability of the postsynaptic cell.

8 Opening or closing of postsynaptic channels

blue structure on the postsynaptic side represents the **postsynaptic density**, a structure important for postsynaptic signaling at excitatory synapses (see Box 7B).

Transmission at chemical synapses is based on the elaborate sequence of events depicted in Figure 5.4C. Prior to transmission, synaptic vesicles are formed and filled with neurotransmitter. Synaptic transmission is initiated when an action potential invades the terminal of the presynaptic neuron. The change in membrane potential caused by the arrival of the action potential leads to the opening of voltage-gated calcium channels in the presynaptic membrane. Because of the steep gradient of Ca^{2+} across the presynaptic membrane (the external Ca^{2+} concentration is approximately $10^{-3}\,M$, whereas the internal Ca^{2+} concentration is approximately $10^{-7}\,M$), the opening of these channels causes a rapid influx of Ca^{2+} into the presynaptic terminal, with the result that the Ca^{2+} concentration of the cytoplasm in the terminal transiently rises to a much higher value. Elevation of the presynaptic Ca^{2+} concentration, in turn, allows synaptic vesicles to fuse with the plasma membrane of the presynaptic neuron. The Ca^{2+}-dependent fusion of synaptic vesicles with the terminal membrane causes their contents, most importantly neurotransmitters, to be released into the synaptic cleft, a process called **exocytosis**.

Following exocytosis, transmitters rapidly diffuse across the synaptic cleft and bind to specific receptors on the membrane of the postsynaptic neuron. The binding of neurotransmitter to the receptors causes channels in the postsynaptic membrane to open (or sometimes to close), thus changing the ability of ions to flow across the postsynaptic membrane. The resulting neurotransmitter-induced current flow alters the conductance and (usually) the membrane potential of the postsynaptic neuron, increasing or decreasing the probability that the neuron will fire an action potential. Subsequent removal of the neurotransmitter from the synaptic cleft, by uptake into glial cells or by enzymatic degradation, terminates the action of the neurotransmitter. In this way, information is transmitted transiently from one neuron to another at chemical synapses.

CONCEPT **5.2**	**Calcium Ions Regulate the Release of Discrete Packets of Neurotransmitters**

LEARNING OBJECTIVES

5.2.1 Explain the concept of quantal release of neurotransmitters.

5.2.2 Describe how calcium ions trigger neurotransmitter release from presynaptic terminals.

5.2.3 Describe the criteria to establish that presynaptic calcium ions are necessary and sufficient for neurotransmitter release.

Quantal release of neurotransmitters

The notion that information can be transferred from one neuron to the next by means of chemical signaling was the subject of intense debate through the first half of the twentieth century. In 1926, the German physiologist Otto Loewi performed a key experiment that supported this idea. Acting on an idea that allegedly came to him in the middle of the night, Loewi proved that electrical stimulation of the vagus nerve slows the heartbeat by releasing a chemical signal that was later shown to be **acetylcholine (ACh)**. ACh is now known to be a neurotransmitter that acts not only in the heart but also at a variety of postsynaptic targets in the central and peripheral nervous systems, preeminently at the neuromuscular junction of striated muscles and in the visceral motor system (see Chapters 6 and 21).

Much of the evidence leading to the present understanding of chemical synaptic transmission was obtained from experiments examining the release of ACh at neuromuscular junctions. These synapses between spinal motor neurons and striated muscle cells are simple, large, and peripherally located, making them particularly amenable to experimental analysis. Such synapses occur at specializations called **end plates** because of the saucerlike appearance of the site on the muscle fiber where the presynaptic axon elaborates its terminals (Figure 5.5A). Most of the pioneering work on neuromuscular transmission was performed by Bernard Katz and his collaborators at University College London. Although Katz worked primarily on the frog neuromuscular junction, numerous subsequent experiments have confirmed the applicability of his observations to transmission at all chemical synapses.

When an intracellular microelectrode is used to record the membrane potential of a muscle cell, an action potential in the presynaptic motor neuron can be seen to elicit a transient depolarization of the postsynaptic muscle fiber. This change in membrane potential, called an **end plate potential (EPP)**, is normally large enough to bring the membrane potential of the muscle cell well above the threshold for producing a postsynaptic action potential (Figure 5.5B). The postsynaptic action potential triggered by the EPP causes the muscle fiber to contract. Unlike at electrical synapses, there is a pronounced delay between the time that the presynaptic motor neuron is stimulated and when the EPP occurs in the postsynaptic muscle cell. This **synaptic delay** is characteristic of all chemical synapses.

One of Katz's seminal findings, in studies carried out with Paul Fatt, was that spontaneous changes in muscle cell membrane potential occur even in the absence of stimulation of the presynaptic motor neuron (Figure 5.5C). These changes have the same shape as EPPs but are much smaller (typically less than 1 mV in amplitude, compared with an EPP of more than 50 mV). Both EPPs and these small, spontaneous events are sensitive to pharmacological agents that

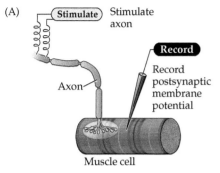

(A)

Stimulate
Stimulate axon

Record

Axon

Record postsynaptic membrane potential

Muscle cell

FIGURE 5.5 Synaptic transmission at the neuromuscular junction
(A) Experimental arrangement: The axon of the motor neuron innervating the muscle fiber is stimulated with an extracellular electrode, while an intracellular microelectrode is inserted into the postsynaptic muscle cell to record its electrical responses. (B) End plate potentials (shaded area) evoked by stimulation of a motor neuron are normally above threshold and therefore produce an action potential in the postsynaptic muscle cell. (C) Spontaneous miniature EPPs (MEPPs) occur in the absence of presynaptic stimulation. (D) When the neuromuscular junction is bathed in a solution that has a low concentration of Ca^{2+}, stimulating the motor neuron evokes EPPs whose amplitudes are reduced to about the size of MEPPs. (After P. Fatt and B. Katz, 1952. *J Physiol* 117: 109–128.)

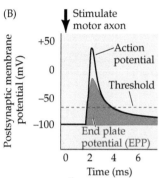

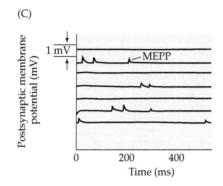

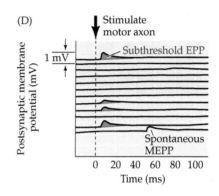

block postsynaptic acetylcholine receptors, such as curare (see Box 6A). These and other parallels between EPPs and the spontaneously occurring depolarizations led Katz and his colleagues to call these spontaneous events **miniature end plate potentials,** or **MEPPs.**

The relationship between the full-blown end plate potential and MEPPs was clarified by careful analysis of the EPPs. The magnitude of the EPP provides a convenient electrical assay of neurotransmitter secretion from a motor neuron terminal; however, measuring it is complicated by the need to prevent muscle contraction from dislodging the microelectrode. The usual means of eliminating muscle contractions is either to lower Ca^{2+} concentration in the extracellular medium or to partially block the postsynaptic ACh receptors with the drug curare. As expected from the scheme illustrated in Figure 5.4C, lowering the Ca^{2+} concentration reduces neurotransmitter secretion, thus reducing the magnitude of the EPP below the threshold for postsynaptic action potential production and allowing it to be measured more precisely. Under such conditions, stimulation of the motor neuron produces very small EPPs that fluctuate in amplitude from trial to trial (Figure 5.5D). These fluctuations give considerable insight into the mechanisms responsible for neurotransmitter release. In particular, the variable response evoked in low Ca^{2+} is now known to result from the release of unit amounts of ACh by the presynaptic nerve terminal. Indeed, the amplitude of the smallest evoked EPP response is strikingly similar to the size of single MEPPs (compare Figure 5.5C and D). Further supporting this similarity, increments in the EPP response (Figure 5.6A) occur in units about the

size of single MEPPs (Figure 5.6B). These "quantal" fluctuations in the amplitude of EPPs indicated to Katz and his colleague Jose del Castillo that EPPs are made up of individual units, each equivalent to a MEPP.

The idea that EPPs represent the simultaneous release of many MEPP-like units can be tested statistically. A method of statistical analysis based on the independent occurrence of unitary events (called Poisson statistics) predicts what the distribution of EPP amplitudes would look like during a large number of trials of motor neuron stimulation, under the assumption that EPPs are built up from unitary events represented by MEPPs (see Figure 5.6B). The distribution of EPP amplitudes determined experimentally was found to be just that expected if transmitter release from the motor neuron is indeed quantal (the red curve in Figure 5.6A). Such analyses confirmed the idea that release of acetylcholine does indeed occur in discrete packets, each equivalent to a MEPP. In short, a presynaptic action potential causes a postsynaptic EPP because it synchronizes the release of many transmitter quanta.

The role of calcium in transmitter secretion

How is it that presynaptic action potentials lead to release of quantal packets of neurotransmitters? An important clue came from the observation that lowering the concentration of Ca^{2+} outside a presynaptic motor nerve terminal reduces the size of the EPP (compare Figure 5.5B and D). Moreover, measurement of the number of transmitter quanta released under such conditions shows that the EPP gets smaller because lowering Ca^{2+} concentration decreases the number of quanta that are released from the presynaptic terminal. We

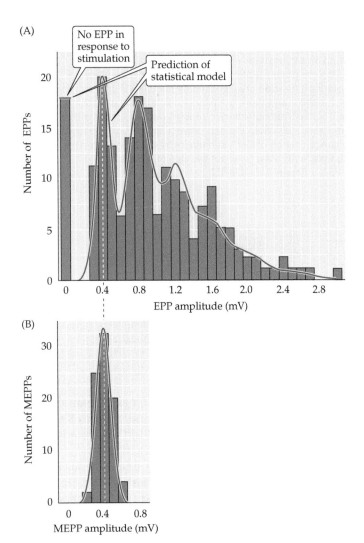

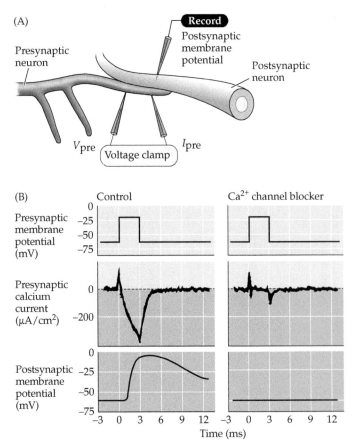

FIGURE 5.6 Quantized distribution of EPP amplitudes evoked in a low-Ca²⁺ solution Peaks of EPP amplitudes (A) tend to occur in integer multiples of the mean amplitude of MEPPs, whose amplitude distribution is shown in (B). The leftmost bar in the EPP amplitude distribution shows trials in which presynaptic stimulation failed to elicit an EPP in the muscle cell. The red curve indicates the prediction of a statistical model based on the assumption that the EPPs result from the independent release of multiple MEPP-like quanta. The observed match, including the predicted number of failures, supports this interpretation. (After I. A. Boyd and A. R. Martin, 1956. *J Physiol* 132: 74–91.)

FIGURE 5.7 Entry of Ca²⁺ through presynaptic voltage-gated calcium channels causes transmitter release (A) Experimental setup using an extraordinarily large synapse in the squid. The voltage clamp method detects currents flowing across the presynaptic membrane when the membrane potential is depolarized. (B) Pharmacological agents that block currents flowing through Na⁺ and K⁺ channels reveal a remaining inward current flowing through Ca²⁺ channels. This influx of calcium triggers transmitter secretion, as indicated by a change in the postsynaptic membrane potential. Treatment of the same presynaptic terminal with cadmium, a calcium channel blocker, eliminates both the presynaptic calcium current and the postsynaptic response. (B after G. J. Augustine and R. Eckert, 1984. *J Physiol* 346: 257–271.)

now know that Ca²⁺ regulates neurotransmitter release because the presynaptic action potential opens voltage-gated Ca²⁺ channels (see Chapter 4).

The first indication of presynaptic Ca²⁺ channels was provided by Bernard Katz and Ricardo Miledi. They observed that presynaptic terminals treated with tetrodotoxin (which blocks voltage-gated Na⁺ channels; see Chapter 3) could still produce a peculiarly prolonged type of action potential. The explanation for this surprising finding was that inward current was flowing through Ca²⁺

channels, substituting for the current ordinarily carried by the blocked Na⁺ channels. Subsequent voltage clamp experiments, performed by Rodolfo Llinás and others at a giant presynaptic terminal of the squid (Figure 5.7A), confirmed the presence of voltage-gated Ca²⁺ channels in the presynaptic terminal (Figure 5.7B). Such experiments showed that the amount of neurotransmitter released is very sensitive to the exact amount of Ca²⁺ that enters. Furthermore, blockade of these Ca²⁺ channels with drugs also inhibits transmitter release (see Figure 5.7B, right). These observations establish that the voltage-gated Ca²⁺ channels are directly involved in synaptic transmission: Opening of these Ca²⁺ channels by

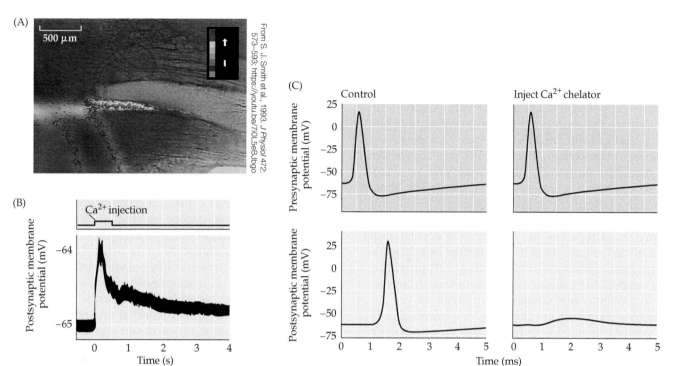

From S. J. Smith et al., 1993. *J Physiol* 472:
573–593; https://youtu.be/7i0L5eBJbgo

FIGURE 5.8   Evidence that a rise in presynaptic Ca²⁺ concentration triggers transmitter release from presynaptic terminals (A) Fluorescence microscopy measurements of presynaptic Ca²⁺ concentration at the squid giant synapse (see Figure 5.7A). A train of presynaptic action potentials causes a rise in Ca²⁺ concentration, as revealed by a dye (called fura-2) that fluoresces more strongly when the Ca²⁺ concentration increases (colors). For video of this experiment, see https://youtu.be/7i0L5eBJbgo. (B) Microinjection of Ca²⁺ into a squid giant presynaptic terminal triggers transmitter release, measured as a depolarization of the postsynaptic membrane potential. (C) Microinjection of BAPTA, a Ca²⁺ chelator, into a squid giant presynaptic terminal prevents transmitter release. (B after R. Miledi, 1973. *Proc R Soc Lond B* 183: 421–425; C after E. M. Adler et al., 1991. *J Neurosci* 11: 1496–1507.)

presynaptic action potentials causes an influx of Ca²⁺ into the presynaptic terminal.

As is the case for many other forms of neuronal signaling (see Chapter 7), Ca²⁺ serves as a second messenger during transmitter release. Ca²⁺ entering into presynaptic terminals accumulates within the terminal, as can be seen with microscopic imaging of terminals filled with Ca²⁺-sensitive dyes (Figure 5.8A). The presynaptic second messenger function of Ca²⁺ has been established in two complementary ways. First, microinjection of Ca²⁺ into presynaptic terminals triggers transmitter release even in the absence of presynaptic action potentials (Figure 5.8B). Second, presynaptic microinjection of calcium chelators (chemicals that bind Ca²⁺ and keep its concentration buffered at low levels) prevents presynaptic action potentials from causing transmitter secretion (Figure 5.8C). These results prove beyond any doubt that a rise in presynaptic Ca²⁺ concentration is both necessary and sufficient for neurotransmitter release. While Ca²⁺ is a universal trigger for transmitter release, not all transmitters are released with the same speed. For example, while secretion of ACh from motor neurons requires only a millisecond (see Figure 5.5), release of neuropeptides requires high-frequency bursts of action potentials for many

seconds. These differences in the rate of release probably arise from differences in the spatial arrangement of synaptic vesicles relative to presynaptic Ca²⁺ channels, yielding differences in the time course of local Ca²⁺ signaling.

CONCEPT
5.3

A Cycle of Membrane Trafficking Is Responsible for Neurotransmitter Release

LEARNING OBJECTIVES

5.3.1 Understand the experimental evidence that neurotransmitters are released via exocytosis from synaptic vesicles.

5.3.2 Know that synaptic vesicles are locally recycled within presynaptic terminals.

5.3.3 List the sequence of membrane trafficking events involved in synaptic vesicle recycling.

Release of transmitters from synaptic vesicles

The discovery of the quantal release of packets of neurotransmitter immediately raised the question of how such quanta are formed and discharged into the synaptic cleft. At about the time Katz and his colleagues were using physiological methods to discover quantal release of neurotransmitter, electron microscopy revealed, for the first time, the presence of synaptic vesicles in presynaptic terminals. Putting these two discoveries together, Katz and others proposed that synaptic vesicles loaded with transmitter are the source of the quanta. Subsequent biochemical studies confirmed that synaptic vesicles are the repositories of transmitters. These studies have shown that ACh is highly concentrated in the synaptic vesicles of motor neurons, where it is present at a concentration of about 100 mM. Given the diameter of a synaptic vesicle (~50 nm), approximately 10,000 molecules of neurotransmitter are contained in a single vesicle. This number corresponds quite nicely to the amount of ACh that must be applied to a neuromuscular junction to mimic an MEPP, providing further support for the idea that quanta arise from discharge of the contents of single synaptic vesicles.

To prove that quanta are caused by the fusion of individual synaptic vesicles with the plasma membrane, it is necessary to show that each fused vesicle produces a single quantal event in the postsynaptic cell. This challenge was met when John Heuser, Tom Reese, and colleagues

correlated measurements of vesicle fusion with the quantal content of EPPs at the neuromuscular junction (Figure 5.9A). They used electron microscopy to determine the number of vesicles that fused with the presynaptic plasma membrane at the active zones of presynaptic terminals (Figure 5.9B). By treating terminals with different concentrations of a drug (4-aminopyridine, or 4-AP) that enhances the number of quanta released by single action potentials, it was possible to vary the amount of quantal release, determined from parallel electrical measurements of the quantal content of the EPPs. A comparison of the number of synaptic vesicle fusions observed with the electron microscope and the number of quanta released at the synapse showed a good correlation between these two measures (Figure 5.9C). These results remain one of the strongest lines of support for the idea that a quantum of transmitter release is due to fusion of a single synaptic vesicle with the presynaptic membrane. Subsequent evidence, based on other means of measuring vesicle fusion, has left no doubt about the validity of this interpretation, thereby establishing that chemical synaptic transmission results from the discharge of neurotransmitters from synaptic vesicles.

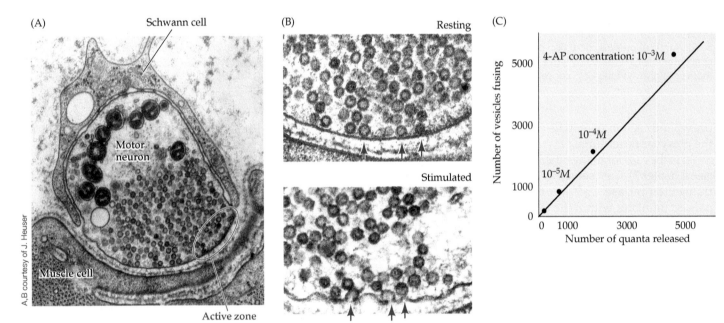

FIGURE 5.9   Relationship between synaptic vesicle exocytosis and quantal transmitter release (A) Electron micrograph of a frog neuromuscular synapse. This synapse includes a presynaptic motor neuron that innervates a postsynaptic muscle cell and is covered by a type of glial cell called a Schwann cell. The active zone of the presynaptic terminal is the site of synaptic vesicle exocytosis. (B) Top: Active zone of an unstimulated presynaptic terminal. While many synaptic vesicles are present, including several that are docked at the active zone (arrows), none are fusing with the presynaptic plasma membrane. Bottom: Active zone of a terminal stimulated by an action potential; stimulation causes fusion (arrows) of synaptic vesicles with the presynaptic membrane. (C) Comparison of the number of observed vesicle fusion events with the number of quanta released by a presynaptic action potential. Transmitter release was varied by using different concentrations of a drug (4-AP) that affects the duration of the presynaptic action potential, thus changing the amount of calcium that enters during the action potential. The diagonal line is the 1:1 relationship expected if opening of each vesicle released a single quantum of transmitter. (C after J. E. Heuser et al., 1979. *J Cell Biol* 81: 275–300.)

Local recycling of synaptic vesicles

The fusion of synaptic vesicles causes new membrane to be added to the plasma membrane of the presynaptic terminal, but the addition is not permanent. Although a bout of exocytosis can dramatically increase the surface area of presynaptic terminals, this extra membrane is removed within a few minutes or less. Heuser and Reese performed another important set of experiments showing that the fused vesicle membrane is retrieved and taken back into the cytoplasm of the nerve terminal via a process called **endocytosis**. The experiments, again carried out at the frog neuromuscular junction, were based on filling the synaptic cleft with horseradish peroxidase (HRP), an enzyme that produces a dense reaction product that is visible in an electron microscope. Under appropriate experimental conditions, endocytosis could then be visualized by the uptake of HRP into the nerve terminal (Figure 5.10). To activate endocytosis, exocytosis was first triggered by stimulating the presynaptic terminal with a train of action potentials, and the subsequent fate of the HRP was followed by electron microscopy. Immediately following stimulation, the HRP was found in special endocytotic organelles called coated

vesicles, which form from membrane budded off via coated pits (see Figure 5.10B). A few minutes later, however, the coated vesicles had disappeared, and the HRP was found in a different organelle, the endosome (see Figure 5.10C). Finally, within an hour after the terminal had been stimulated, the HRP reaction product appeared inside synaptic vesicles (see Figure 5.10D).

These observations indicate that synaptic vesicle membrane is recycled within the presynaptic terminal via the sequence summarized in Figure 5.10E. In this process, called the **synaptic vesicle cycle**, the retrieved vesicular membrane passes through several intracellular compartments—such as coated vesicles and endosomes—and is eventually used to make new synaptic vesicles. After synaptic vesicles are re-formed, most are stored in a reserve pool within the cytoplasm until they need to participate again in neurotransmitter release. These vesicles are mobilized from the reserve pool, docked at the presynaptic plasma membrane, and primed to participate in exocytosis once again. More recent experiments, employing a fluorescent label rather than HRP, have determined the time course of synaptic vesicle recycling. These studies

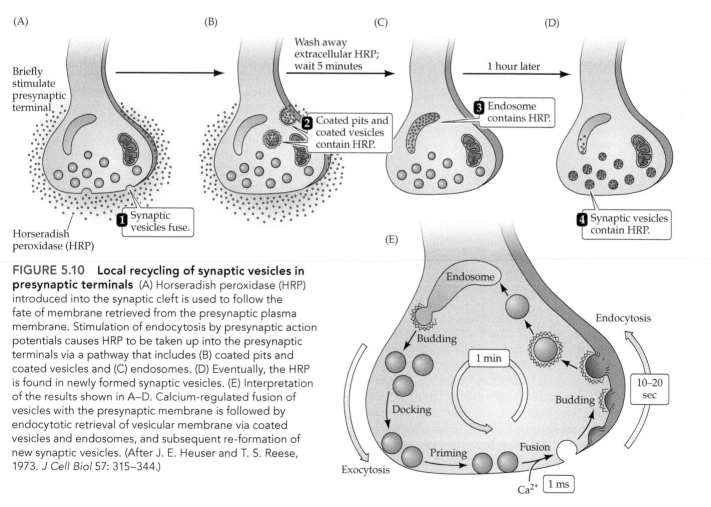

FIGURE 5.10   Local recycling of synaptic vesicles in presynaptic terminals (A) Horseradish peroxidase (HRP) introduced into the synaptic cleft is used to follow the fate of membrane retrieved from the presynaptic plasma membrane. Stimulation of endocytosis by presynaptic action potentials causes HRP to be taken up into the presynaptic terminals via a pathway that includes (B) coated pits and coated vesicles and (C) endosomes. (D) Eventually, the HRP is found in newly formed synaptic vesicles. (E) Interpretation of the results shown in A–D. Calcium-regulated fusion of vesicles with the presynaptic membrane is followed by endocytotic retrieval of vesicular membrane via coated vesicles and endosomes, and subsequent re-formation of new synaptic vesicles. (After J. E. Heuser and T. S. Reese, 1973. *J Cell Biol* 57: 315–344.)

indicate that the entire vesicle cycle requires approximately 1 minute, with membrane budding during endocytosis requiring 10 to 20 seconds of this time. As can be seen from the 1-millisecond delay in transmission following excitation of the presynaptic terminal (see Figure 5.5B), membrane fusion during exocytosis is much more rapid than budding during endocytosis. Thus, all of the recycling steps interspersed between membrane fusion and subsequent regeneration of a new vesicle are completed in less than a minute.

The precursors to synaptic vesicles *originally* are produced in the endoplasmic reticulum and Golgi apparatus in the neuronal cell body. Because of the long distance between the cell body and the presynaptic terminal in most neurons, transport of vesicles from the soma would not permit rapid replenishment of synaptic vesicles during continuous neural activity. Thus, local recycling is well suited to the peculiar anatomy of neurons, giving nerve terminals the means to provide a continual supply of synaptic vesicles.

CONCEPT 5.4 | Many Proteins Are Required for Synaptic Vesicle Trafficking

LEARNING OBJECTIVES

5.4.1 Recall the proteins involved in synaptic vesicle trafficking.

5.4.2 Explain the molecular mechanisms of synaptic vesicle exocytosis.

5.4.3 Explain the molecular mechanisms of synaptic vesicle endocytosis.

Much is now understood about the molecular basis of synaptic vesicle trafficking. Molecular studies have identified and characterized the proteins found on synaptic vesicles (Figure 5.11A) and their binding partners on the presynaptic plasma membrane and cytoplasm. Most, if not all, of these proteins act at one or more steps in the synaptic vesicle cycle (Figure 5.11B).

Molecular mechanisms of synaptic vesicle exocytosis

Several lines of evidence indicate that the protein **synapsin**, which reversibly binds to synaptic vesicles, may keep these vesicles tethered within the reserve pool by cross-linking vesicles to each other. Mobilization of these reserve pool vesicles is caused by phosphorylation of synapsin by protein kinases, most notably the **Ca^{2+}/calmodulin-dependent protein kinase, type II** (CaMKII; see Chapter 7), which allows synapsin to dissociate from the vesicles. Once vesicles are free from their reserve pool tethers, they make their way to the plasma membrane and are then attached to

this membrane by docking reactions that involve SNARE proteins (see below). A series of priming reactions then prepares the vesicular and plasma membranes for fusion. A large number of proteins are involved in priming, including some proteins that are also involved in other types of membrane fusion events common to all cells (see Figure 5.11B). For example, two proteins originally found to be important for the fusion of vesicles with membranes of the Golgi apparatus, the ATPase **NSF** (*NEM-sensitive fusion protein*) and **SNAPs** (*soluble NSF-attachment proteins*), are also involved in priming synaptic vesicles for fusion. These two proteins work by regulating the assembly of other proteins that are called **SNAREs** (*SNAP receptors*). Many of the other proteins involved in priming—such as munc13, munc18, complexin, snapin, syntaphilin, and tomosyn—also interact with the SNAREs.

One of the main purposes of priming is to organize SNARE proteins into the correct conformation for membrane fusion. One of the SNARE proteins, **synaptobrevin**, is in the membrane of synaptic vesicles, while two other SNARE proteins called **syntaxin** and **SNAP-25** are found primarily on the plasma membrane. These SNARE proteins can form a macromolecular complex that spans the two membranes, thus bringing them into close apposition (Figure 5.12A). Recent evidence indicates that a symmetrical array of six SNARE complexes links synaptic vesicles and the presynaptic plasma membrane (Figure 5.12B). Such an arrangement is well suited to promote the fusion of the two membranes, and several lines of evidence suggest that this is what actually occurs. One important observation is that toxins that cleave the SNARE proteins block neurotransmitter release (Clinical Applications). In addition, putting SNARE proteins into artificial lipid membranes and allowing these proteins to form complexes with each other causes the membranes to fuse.

Because the SNARE proteins do not bind Ca^{2+}, still other molecules must be responsible for Ca^{2+} regulation of neurotransmitter release. Numerous presynaptic proteins, including calmodulin, CAPS, and munc13, are capable of binding Ca^{2+}. However, Ca^{2+} regulation of neurotransmitter release usually is conferred by **synaptotagmins**, a family of proteins found in the membrane of synaptic vesicles (see Figure 5.12A). Synaptotagmin binds Ca^{2+} at concentrations similar to those required to trigger vesicle fusion within the presynaptic terminal, and this property allows synaptotagmin to act as a Ca^{2+} sensor that triggers vesicle fusion by signaling the elevation of Ca^{2+} within the terminal. In support of this idea, disruption of synaptotagmin in the presynaptic terminals of mice, fruit flies, squid, and other experimental animals impairs Ca^{2+}-dependent neurotransmitter release. In fact, deletion of only 1 of the 17 synaptotagmin (*SYT*) genes of mice is a lethal mutation, causing the mice to die soon after birth.

FIGURE 5.11   Presynaptic proteins and their roles in synaptic vesicle cycling
(A) Model of the molecular organization of a synaptic vesicle. The cytoplasmic surface of the vesicle membrane is densely covered by proteins, only 70% of which are shown here. (B) The vesicle trafficking cycle shown in Figure 5.10E is now known to be mediated by numerous presynaptic proteins, including some of those shown in (A), with different proteins participating in different reactions. (A from S. Takamori et al., 2006. *Cell* 127: 831–846.)

(A)

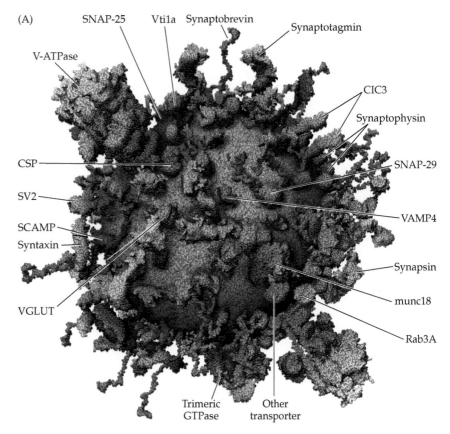

(B)

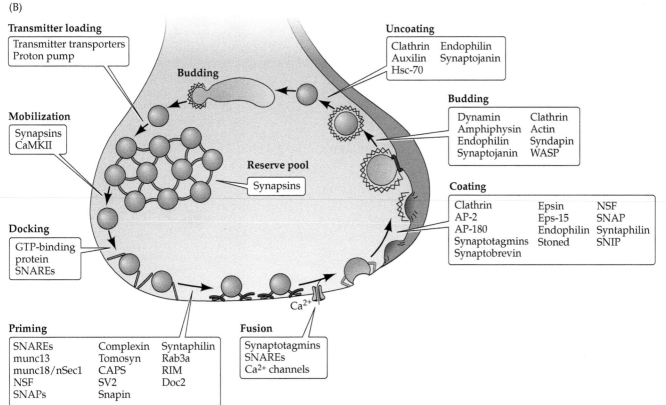

(A)

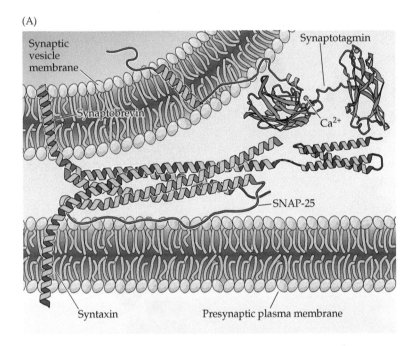

Synaptic vesicle membrane

Synaptobrevin

Synaptotagmin

Ca²⁺

SNAP-25

Syntaxin

Presynaptic plasma membrane

(B)

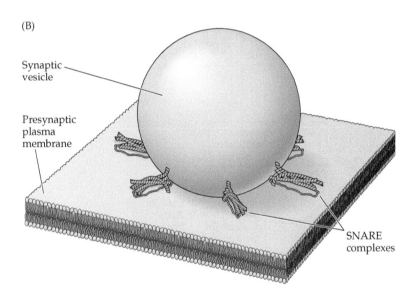

Synaptic vesicle

Presynaptic plasma membrane

SNARE complexes

FIGURE 5.12 Molecular mechanisms of exocytosis during neurotransmitter release (A) Structure of the SNARE complex. The vesicular SNARE, synaptobrevin (blue), forms a helical complex with the plasma membrane SNAREs syntaxin (red) and SNAP-25 (green). Also shown is the structure of synaptotagmin, a vesicular Ca²⁺-binding protein, with bound Ca²⁺ indicated by spheres. (B) Three-dimensional organization of SNARE complexes of a docked synaptic vesicle. Six of these complexes are symmetrically arranged to form a ring around the interface between the synaptic vesicle and the presynaptic plasma membrane. (C) A model for Ca²⁺-triggered vesicle fusion. During docking of the synaptic vesicle, SNARE proteins on the vesicle and plasma membranes form a complex (as in A) that brings together the two membranes. Synaptotagmin binds to this SNARE complex. Subsequent binding of Ca²⁺ to synaptotagmin causes the cytoplasmic region of this protein to insert into the plasma membrane to produce the membrane curvature that catalyzes membrane fusion. (A from R. B. Sutton et al., 1998. *Nature* 395: 347–353; B after A. Radhakrishnan et al., 2021. *Proc Natl Acad Sci USA* 118: e2024029118; C after Q. Zhou et al., 2015. *Nature* 525: 62–67.)

(C) (1) Free SNARES on vesicle and plasma membranes

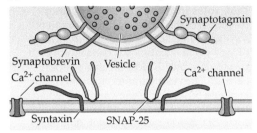

Synaptobrevin

Synaptotagmin

Ca²⁺ channel

Vesicle

Ca²⁺ channel

Syntaxin

SNAP-25

(2) SNARE complexes form as vesicle docks

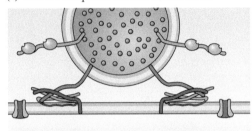

(3) Synaptotagmin binds to SNARE complex

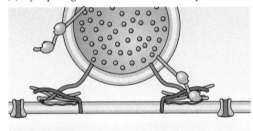

(4) Entering Ca²⁺ binds to synaptotagmin, leading to curvature of plasma membrane, which brings membranes together

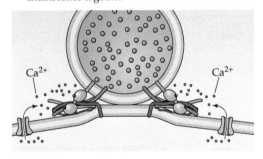

Ca²⁺   Ca²⁺

(5) Fusion of membranes leads to exocytotic release of neurotransmitter

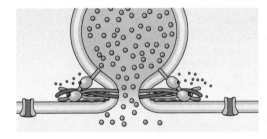

■ Clinical Applications

Disorders That Affect the Presynaptic Terminal

Defects in various steps in the exocytosis and endocytosis of synaptic vesicles have been shown to be at the root of several rare but debilitating neurological diseases.

Myasthenic syndromes

In myasthenic syndromes, abnormal transmission at neuromuscular synapses leads to weakness and fatigability of skeletal muscles. Some of these syndromes affect the acetylcholinesterase that degrades acetylcholine in the synaptic cleft; others arise from autoimmune attack of acetylcholine receptors (see Clinical Applications, Chapter 6); and yet others affect acetylcholine release from motor neurons. One of the best understood myasthenic syndromes is *Lambert– Eaton myasthenic syndrome*, or *LEMS*, an occasional complication in patients with certain kinds of cancers. Biopsies of muscle tissue removed from LEMS patients allow intracellular recordings identical to those in Figure 5.5. These recordings from affected tissues show that when a motor neuron is stimulated, the number of quanta contained in EPPs is greatly reduced while the size of individual quanta is unaffected. Several lines of evidence indicate that this reduction in neurotransmitter release is due to a loss of voltage-gated Ca^{2+} channels in the presynaptic terminal of motor neurons (Figure A); perhaps most compelling are anatomical studies that reveal a lowered density of Ca^{2+} channels in the presynaptic plasma membrane.

The loss of presynaptic Ca^{2+} channels in LEMS patients apparently arises from an autoimmune disorder. Their blood has a very high concentration of antibodies that bind to Ca^{2+} channels, and it seems likely that these antibodies are the primary cause of LEMS. For example, removal of Ca^{2+} channel antibodies from the blood of LEMS patients by plasma exchange reduces muscle weakness.

Similarly, immunosuppressant drugs also can alleviate LEMS symptoms. Perhaps most telling, injecting these antibodies into experimental animals elicits muscle weakness and abnormal neuromuscular transmission.

Why the immune system generates antibodies against Ca^{2+} channels is not clear. Most LEMS patients have small-cell carcinoma, a form of lung cancer that may somehow initiate the immune response to Ca^{2+} channels. Whatever the origin, the binding of antibodies to Ca^{2+} channels reduces the ability of these channels to carry Ca^{2+} current. It is this antibody-induced defect in presynaptic Ca^{2+} entry that accounts for the muscle weakness associated with LEMS.

Congenital myasthenic syndromes are genetic disorders that, like LEMS, cause muscle weakness by affecting neuromuscular transmission. Several congenital myasthenic syndromes also arise from defects in acetylcholine release from the motor neuron terminal. Neuromuscular synapses in some of these patients have EPPs with reduced quantal content, a deficit that is especially prominent when the synapse is activated repeatedly. These symptoms are consistent with genetic analyses showing that mutations in genes encoding several presynaptic proteins—including SNAP-25, synaptotagmin and choline acetyltransferase, an enzyme responsible for synthesis of ACh (see Chapter 6)—are found in patients with congenital myasthenic syndromes. Electron microscopy shows that presynaptic motor nerve terminals also have a greatly reduced number of synaptic vesicles. The origins of this shortage of synaptic vesicles are not clear, but could result either from an impairment in endocytosis in the nerve terminal (see Figure A) or from a reduced supply of vesicle components from the motor neuron cell body.

Still other patients suffering from *familial infantile myasthenia* appear to have neuromuscular weakness that arises from reductions in the size of

(Continued)

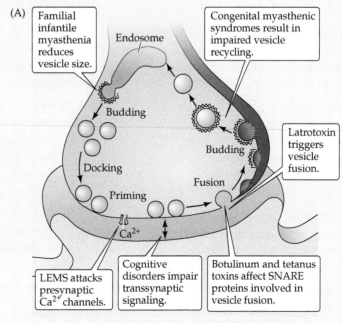

(A) Presynaptic targets of several neurological pathologies.

■ Clinical Applications (continued)

individual quanta rather than the number of quanta released. Motor nerve terminals from these patients have synaptic vesicles that are typical in number but smaller than usual in diameter. This finding suggests a different type of genetic lesion that somehow alters formation of new synaptic vesicles following endocytosis, leading to less acetylcholine in each vesicle.

Botulism and tetanus

Impairment of synaptic transmitter release also results from poisoning by anaerobic *Clostridium* bacteria. This genus of microorganisms produces some of the most potent toxins known, including several botulinum toxins and tetanus toxin. Both botulism and tetanus are potentially deadly disorders.

Botulism can occur by consuming food containing *Clostridium* bacteria, or by infection of wounds with the spores of these ubiquitous organisms. In either case, the presence of the toxin can cause paralysis of peripheral neuromuscular synapses due to impaired neurotransmitter release. This interference with neuromuscular transmission causes skeletal muscle weakness; in extreme cases respiratory failure results from paralysis of the diaphragm and other muscles required for breathing. Botulinum toxins also block synapses innervating the smooth muscles of several organs, giving rise to visceral motor dysfunction. This paralysis of neuromuscular transmission also serves as the basis for clinical use of botulinum toxin in cosmetic surgery and other applications where highly local relaxation of muscle contraction is of therapeutic benefit to the patient.

Tetanus typically results from the contamination of puncture wounds by *Clostridium* bacteria that produce tetanus toxin. In contrast to botulism, tetanus poisoning blocks the release of inhibitory transmitters from interneurons in the spinal cord. This effect causes a loss of synaptic inhibition on spinal motor neurons (see Chapter 16),

producing hyperexcitation of skeletal muscle and tetanic contractions in affected muscles (hence the name of the disease).

Although the clinical consequences of tetanus toxin are dramatically different from those of botulinum toxins, clever and patient biochemical work has shown that these toxins have a common mechanism of action: They are highly specific proteases that inhibit neurotransmitter release by cleaving the SNARE proteins involved in fusion of synaptic vesicles with the presynaptic plasma membrane (Figure B). Tetanus toxin and botulinum toxin types B, D, F, and G specifically cleave the vesicle SNARE protein synaptobrevin. Other botulinum toxins cleave syntaxin (type C) and SNAP-25 (types A and E), SNARE proteins found on the presynaptic plasma membrane. Destruction of these presynaptic proteins is the basis for the inhibitory actions of clostridial toxins on neurotransmitter release.

The different actions of these toxins on synaptic transmission at excitatory

motor versus inhibitory synapses apparently result from the fact that these toxins are taken up by different types of neurons: Whereas the botulinum toxins are taken up by motor neurons, tetanus toxin is preferentially targeted to interneurons. The differential uptake of toxins presumably arises from the presence of different types of toxin receptors on the two types of neurons. Recent work shows that botulinum toxins enter presynaptic terminals by binding to synaptic vesicle proteins, such as synaptotagmin and SV2, as well as gangliosides in the presynaptic membrane. Tetanus toxin receptors include other types of presynaptic gangliosides.

Insights from α-latrotoxin

Latrodectism refers to the severe muscle pain and cramping experienced by individuals bitten by the black widow spider, *Latrodectus*. These symptoms arise from the presynaptic neurotoxin α-latrotoxin, present in black widow spider venom; this toxin causes a massive discharge of neurotransmitter

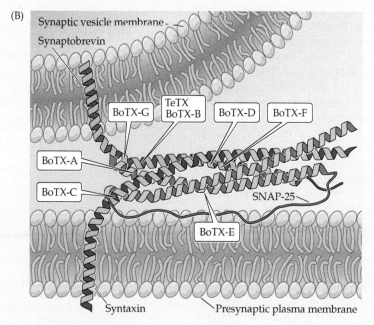

(B) Cleavage of SNARE proteins by clostridial toxins. Indicated are the sites of proteolysis by tetanus toxin (TeTX) and various types of botulinum toxin (BoTX). (From R. B. Sutton et al., 1998. *Nature* 395: 347–353.)

from presynaptic terminals. Although neurotransmitter release ordinarily requires Ca^{2+}, α-latrotoxin is capable of releasing neurotransmitter even when Ca^{2+} is absent from the extracellular medium. Although it is not yet clear how the toxin triggers Ca^{2+}-independent exocytosis, we know that α-latrotoxin binds to two different types of presynaptic proteins that may mediate its actions. One group of binding partners for α-latrotoxin is the neurexins, a group of integral membrane proteins found in presynaptic terminals. Neurexins bind to synaptotagmin, the presynaptic Ca^{2+} sensor, and this interaction may allow α-latrotoxin to bypass the usual Ca^{2+} requirement for triggering vesicle fusion. Another type of presynaptic protein that can bind to α-latrotoxin is called CL1 (based on its previous names, Ca^{2+}-independent receptor for latrotoxin and *latrophilin-1*). CL1 is a relative of the G-protein-coupled

receptors that mediate the actions of neurotransmitters and other extracellular chemical signals (see Figure 5.14B). Thus, the binding of α-latrotoxin to CL1 could activate an intracellular signal transduction cascade involved in the Ca^{2+}-independent actions of α-latrotoxin. While more work is needed to establish the definitive roles of neurexins and CL1 in α-latrotoxin action, these two proteins are probably the basis for the toxin's potent presynaptic activity.

In addition to their possible role in latrodectism, neurexins have been linked to a variety of cognitive disorders. Mutations in the neurexin gene have been identified in several individuals with schizophrenia, a psychiatric disease causing delusions, hallucinations, and loss of emotional expression (see Box 18B). Neurexins are known to play an important role in signaling across the synaptic cleft

by binding to a family of postsynaptic membrane proteins called neuroligins. Remarkably, mutations in neuroligins also have been associated with autism, a spectrum of psychiatric disorders characterized by impaired social interaction, communication problems, and other behavioral disorders. While a direct connection between neurexins or neuroligins and these psychiatric disorders has not yet been established, it seems likely that defects in these transsynaptic signaling partners may serve as a central mechanism underlying several psychiatric disorders.

In summary, not only does research into synaptic vesicle trafficking illuminate the causes of numerous neurological and psychiatric disorders, but research into these diseases in turn has provided tools—such as clostridial toxins and α-latrotoxin—that have proven valuable in elucidating the basic mechanisms of synaptic vesicle trafficking.

It is thought that Ca^{2+} binding to synaptotagmin leads to exocytosis by changing the chemical properties of synaptotagmin, thereby allowing it to insert into the plasma membrane. This causes the plasma membrane to locally curve and leads to fusion of the two membranes. Thus, SNARE proteins bring the two membranes close together, while Ca^{2+}-induced changes in synaptotagmin then produce the final curvature that enables rapid fusion of these membranes (Figure 5.12C).

Molecular mechanisms of synaptic vesicle endocytosis

Still other proteins appear to be involved at the endocytosis steps of the synaptic vesicle cycle (Figure 5.13). The most important protein involved in endocytotic budding of vesicles from the plasma membrane is **clathrin** (*CLTA, CLTC*). Clathrin has a unique structure that is called a triskelion because of its three-legged appearance; these triskelia can assemble to form a cagelike coating around the vesicle membrane (see Figure 5.13A). Several adaptor proteins, such as AP-2 and AP-180, connect clathrin to the proteins and lipids of this membrane. These adaptor proteins, as well as other proteins such as amphiphysin, epsin, and Eps-15, help assemble individual triskelia into structures that resemble geodesic domes (see Figure 5.13A, bottom). Such domelike structures form coated pits that initiate membrane budding, working

together with amphiphysin to increase the curvature of the budding membrane until it forms a coated vesicle–like structure that remains connected to the plasma membrane via a narrow lipid stalk (see Figure 5.13C). Another protein, called **dynamin**, forms a ringlike coil that surrounds the lipid stalk (see Figure 5.13B). This coil causes the final pinching-off of membrane that severs the stalk and completes the production of coated vesicles. Coated vesicles then are transported away from the plasma membrane by the cytoskeletal protein **actin**. This allows the clathrin coats to be removed by an ATPase, **Hsc70**, with another protein, **auxilin**, serving as a co-factor that recruits Hsc70 to the coated vesicle. Other proteins, such as **synaptojanin**, are also important for vesicle uncoating. Uncoated vesicles can then continue their journey through the recycling process, eventually becoming refilled with neurotransmitter due to the actions of neurotransmitter transporters in the vesicle membrane. These transporters exchange protons within the vesicle for neurotransmitter; the acidic interior of the vesicle is produced by a proton pump that also is located in the vesicle membrane.

In summary, a complex cascade of proteins, acting in a defined temporal and spatial order, allows neurons to secrete transmitters. This molecular cascade underlies the powerful ability of the brain to use its synapses to process and store information.

(A)

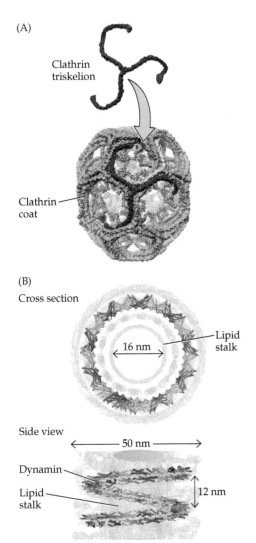

Clathrin triskelion

Clathrin coat

(B)

Cross section

16 nm

Lipid stalk

Side view

50 nm

Dynamin

Lipid stalk

12 nm

(C)

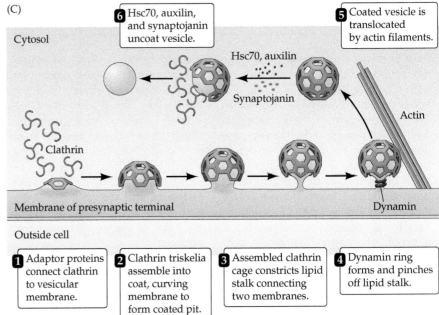

Cytosol

6 Hsc70, auxilin, and synaptojanin uncoat vesicle.

5 Coated vesicle is translocated by actin filaments.

Hsc70, auxilin

Synaptojanin

Actin

Clathrin

Membrane of presynaptic terminal

Dynamin

Outside cell

1 Adaptor proteins connect clathrin to vesicular membrane.

2 Clathrin triskelia assemble into coat, curving membrane to form coated pit.

3 Assembled clathrin cage constricts lipid stalk connecting two membranes.

4 Dynamin ring forms and pinches off lipid stalk.

FIGURE 5.13 Molecular mechanisms of endocytosis following neurotransmitter release (A) Individual clathrin triskelia assemble together to form membrane coats involved in membrane budding during endocytosis. (B) Dynamin forms ringlike coils around the lipid stalks of budding membranes; these rings disconnect vesicle membrane from plasma membrane during endocytosis. (C) A model for membrane budding during endocytosis. Following addition of synaptic vesicle membrane during exocytosis, clathrin triskelia attach to the vesicular membrane. Adaptor proteins, such as AP-2 and AP-180, aid their attachment. Polymerization of clathrin, acting in concert with amphiphysin, causes the membrane to curve and constrict, allowing dynamin to pinch off the coated vesicle. Subsequent uncoating of the vesicle, by Hsc70, auxilin, and synaptojanin, yields a recycled synaptic vesicle. (A from A. Fotin et al., 2004. *Nature* 432: 573–579; B from T. F. Ruebold et al., 2015. *Nature* 525: 404–408; C after O. Shupliakov et al., 2010. *Exp Cell Res* 316: 1344–1350.)

CONCEPT 5.5

There Are Two Families of Neurotransmitter Receptors

LEARNING OBJECTIVES

5.5.1 Recall that there are two types of neurotransmitter receptors.

5.5.2 Distinguish between the properties of ionotropic and metabotropic receptors.

Neurotransmitter receptors

The generation of postsynaptic electrical signals is also understood in considerable depth. Such studies began in 1907, when the British physiologist John N. Langley introduced the concept of **receptor molecules** to explain the specific and potent actions of certain chemicals on muscle and nerve cells. We now know that neurotransmitter receptors are proteins that are embedded in the plasma membrane of postsynaptic cells and have extracellular neurotransmitter binding sites that detects the presence of neurotransmitters in the synaptic cleft.

There are two broad families of receptor proteins that differ in their mechanism of transducing transmitter binding into postsynaptic responses. The receptors in one family are ligand-gated ion channels (see Chapter 4) that contain a membrane-spanning domain that forms an ion channel (Figure 5.14A). These receptors combine transmitter-binding and channel functions into a single molecular entity and thus are called **ionotropic receptors** (Greek *tropos*, "to move in response to a stimulus").

The second family of neurotransmitter receptors is **metabotropic receptors**, so called because the eventual movement of ions through a channel depends on intervening metabolic steps. These receptors do not have ion channels as part of their structure; instead, they have an intracellular domain that indirectly affects channels though the activation of intermediate molecules called

(A) Ligand-gated ion channels

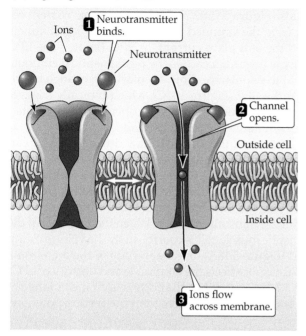

(B) G-protein-coupled receptors

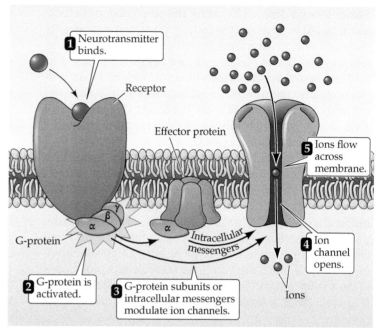

FIGURE 5.14   Two different types of neurotransmitter receptors (A) Ligand-gated ion channels combine receptor and channel functions in a single protein complex. (B) Metabotropic receptors usually activate G-proteins, which modulate ion channels directly or indirectly through intracellular effector enzymes and second messengers.

G-proteins (Figure 5.14B). Neurotransmitter binding to these receptors activates G-proteins, which then dissociate from the receptor and interact directly with ion channels or bind to other effector proteins, such as enzymes, that make intracellular messengers that open or close ion channels. Thus, G-proteins can be thought of as transducers that couple neurotransmitter binding to a receptor with regulation of postsynaptic ion channels. For this reason, metabotropic receptors are also called **G-protein-coupled receptors**. The postsynaptic signaling events initiated by metabotropic receptors are described in Chapter 7. Metabotropic receptors are tremendously important in the pharmaceutical industry because they are the targets of approximately one-third of all therapeutic drugs.

These two families of postsynaptic receptors give rise to postsynaptic actions that range from less than a millisecond to minutes, hours, or even days. Ionotropic receptors generally mediate rapid postsynaptic effects. Examples are the EPP produced at neuromuscular synapses by ACh (see Figure 5.5A,B), as well as the postsynaptic responses produced at many glutamatergic synapses and GABAergic synapses (see Figure 5.19A,B). In these cases, the postsynaptic potentials arise within a millisecond or two of an action potential invading the presynaptic terminal and last for only a few tens of milliseconds or less. In contrast, the activation of metabotropic receptors typically produces much slower responses, ranging from

hundreds of milliseconds to minutes or even longer. The comparative slowness of metabotropic receptor actions reflects the fact that multiple proteins need to bind to each other sequentially in order to produce the final physiological response. Importantly, many transmitters can activate both ionotropic and metabotropic receptors to produce both fast and slow postsynaptic potentials (sometimes even at the same synapse).

CONCEPT
5.6

Postsynaptic Membrane Permeability Changes during Synaptic Transmission

LEARNING OBJECTIVES

5.6.1 Describe the relationship between postsynaptic currents and postsynaptic potentials.

5.6.2 Explain the ionic basis of postsynaptic currents and potentials.

Postsynaptic ion currents

Just as studies of the neuromuscular synapse paved the way for understanding neurotransmitter release mechanisms, this peripheral synapse has been equally valuable for understanding the mechanisms that allow neurotransmitter receptors to generate postsynaptic signals. The binding of ACh to postsynaptic receptors opens ion channels in the muscle fiber membrane. This effect can be demonstrated directly by using the patch clamp method (see Box 4A) to measure the minute postsynaptic currents that flow when two molecules of individual ACh bind to receptors, as Erwin Neher and Bert Sakmann first did in 1976. Exposure of the extracellular surface of a patch of postsynaptic membrane to ACh causes single-channel currents to flow for a few milliseconds (Figure 5.15A). This shows that ACh binding to its receptors opens ligand-gated ion channels (see Chapter 4).

The electrical actions of ACh are greatly multiplied when an action potential in a presynaptic motor neuron causes the release of millions of molecules of ACh into the synaptic cleft. In this more physiological case, the transmitter molecules bind to many thousands of ACh receptors packed in a dense array on the postsynaptic membrane, transiently opening a very large number of postsynaptic ion channels. Although individual ACh receptors generate a microscopic current of only a few picoamperes (Figure 5.15B1), a large number of channels are opened synchronously when ACh is secreted from presynaptic terminals (Figure 5.15B2,3). The macroscopic current resulting from the summed opening of many ion channels is called the **end plate current**, or **EPC**. Because the EPC normally is inward, it causes the postsynaptic membrane potential to depolarize. This depolarizing change in potential is the EPP (Figure 5.15C), which typically triggers a postsynaptic action potential by opening voltage-gated Na$^+$ and K$^+$ channels (see Figure 5.5B).

The identity of the ions that flow during the EPC can be determined via the same approaches used to identify the roles of Na$^+$ and K$^+$ fluxes in the currents underlying action potentials (see Chapter 3). Key to such an analysis is identifying the membrane potential at which no current flows during transmitter action. When the potential of the postsynaptic muscle cell is controlled by the voltage clamp method (Figure 5.16A), the magnitude of the membrane potential clearly affects the amplitude and polarity of EPCs (Figure 5.16B). Thus, when the postsynaptic membrane potential is made more negative than the resting potential,

FIGURE 5.15 Activation of ACh receptors at neuromuscular synapses (A) Patch clamp measurement of current flowing through a single ACh receptor from a patch of membrane removed from the postsynaptic muscle cell. When ACh is applied to the extracellular surface of the membrane, the repeated brief opening of a single channel can be observed as downward deflections corresponding to inward current (i.e., positive ions flowing into the cell). (B) Synchronized opening of many ACh-activated channels at a synapse being voltage clamped. (1) If a single channel is examined during the release of ACh from the presynaptic terminal, the channel can be seen to open transiently. (2) If several channels are examined together, ACh release opens the channels almost synchronously. (3) The opening of a very large number of postsynaptic channels produces a macroscopic EPC. (C) In a muscle cell not being voltage clamped, the inward EPC depolarizes the postsynaptic muscle cell, giving rise to an EPP. Typically, this depolarization generates an action potential (not shown; see Figure 5.5B).

(A) Patch clamp measurement of current flowing through a single ACh receptor

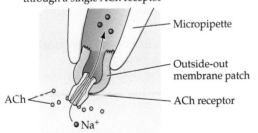

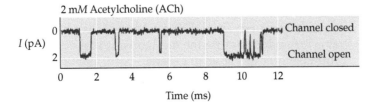

(B) Synaptic currents produced by:

(1) SINGLE OPEN CHANNEL

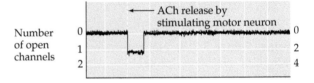

(2) FEW OPEN CHANNELS

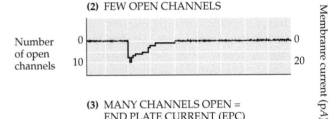

(3) MANY CHANNELS OPEN = END PLATE CURRENT (EPC)

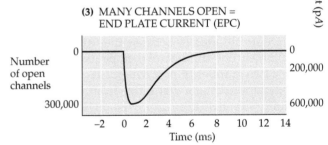

(C) Postsynaptic potential change (EPP) produced by EPC

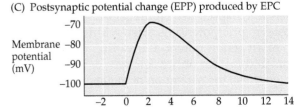

the amplitude of the EPC becomes larger, whereas this current is reduced when the membrane potential is made more positive. At approximately 0 mV, no EPC is detected, and at even more positive potentials, the current reverses its polarity, becoming outward rather than inward (Figure 5.16C). The potential where the EPC reverses, about 0 mV in the case of the neuromuscular junction, is called the **reversal potential**.

As was the case for currents flowing through voltage-gated ion channels (see Chapter 3), the magnitude of the EPC at any membrane potential is given by the product of the ion conductance activated by ACh (g_{ACh}) and the electrochemical driving force on the ions flowing through ligand-gated channels. Thus, the value of the EPC is given by the relationship

$$EPC = g_{ACh}(V_m - E_{rev})$$

where E_{rev} is the reversal potential for the EPC. This relationship predicts that the EPC will be an inward current at potentials more negative than E_{rev} because the electrochemical driving force, $V_m - E_{rev}$, is a negative number.

(A) Scheme for voltage clamping postsynaptic muscle fiber

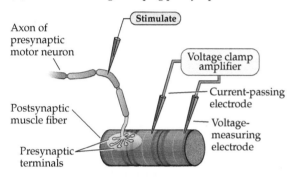

FIGURE 5.16   Influence of the postsynaptic membrane potential on end plate currents (A) A postsynaptic muscle fiber is voltage clamped using two electrodes, while the presynaptic neuron is electrically stimulated to cause the release of ACh from presynaptic terminals. This experimental arrangement allows the recording of EPCs produced by ACh. (B) Amplitude and time course of EPCs generated by stimulating the presynaptic motor neuron while the postsynaptic cell is voltage clamped at four different membrane potentials. (C) The relationship between the peak amplitude of EPCs and postsynaptic membrane potential is nearly linear, with a reversal potential (the voltage at which the direction of the current changes from inward to outward) close to 0 mV. Also indicated on this graph are the equilibrium potentials of Na^+, K^+, and Cl^- ions. (D) The identity of the ions permeating postsynaptic receptors is revealed by the reversal potential (E_{rev}). Activation of postsynaptic channels permeable only to K^+ (brown) results in currents reversing at E_K, near −100 mV, while activation of postsynaptic Na^+ channels results in currents that reverse at E_{Na}, near +70 mV (red). Cl^--selective currents reverse at E_{Cl}, near −50 mV (green). (A–C after A. Takeuchi and N. Takeuchi, 1960. *J Physiol* 154: 52–67.)

(B) Effect of membrane voltage on postsynaptic end plate currents (EPCs)

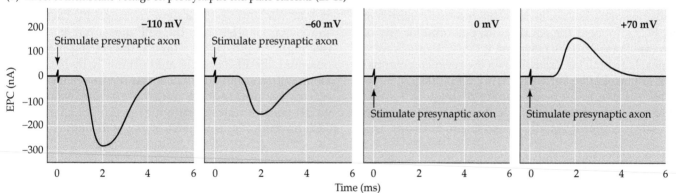

(C)

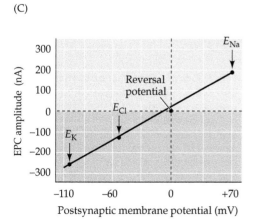

(D)

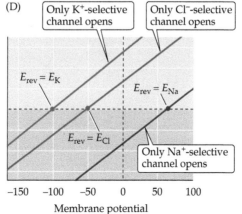

Furthermore, the EPC will become smaller at potentials approaching E_{rev} because the driving force is reduced. At potentials more positive than E_{rev}, the EPC is outward because the driving force is reversed in direction (that is, positive). Because the channels opened by ACh are largely insensitive to membrane voltage, g_{ACh} will depend only on the number of channels opened by ACh, which depends in turn on the concentration of ACh in the synaptic cleft. Thus, the magnitude and polarity of the postsynaptic membrane potential determine the direction and amplitude of the EPC solely by altering the driving force on ions flowing through the receptor channels opened by ACh.

When V_m is at the reversal potential, $V_m - E_{rev}$ is equal to 0 and there is no net driving force on the ions that can permeate the receptor-activated channel. The identity of the ions that flow during the EPC can be deduced by observing how the reversal potential of the EPC compares with the equilibrium potential for various ion species (Figure 5.16D). For example, if ACh were to open an ion channel permeable only to K^+, then the reversal potential of the EPC would be at the equilibrium potential for K^+, which for a muscle cell is close to –100 mV. If the ACh-activated channels were permeable only to Na^+, then the reversal potential of the current would be approximately +70 mV, the Na^+ equilibrium potential of muscle cells; if these channels were permeable only to Cl^-, then the reversal potential would be approximately –50 mV. By this reasoning, ACh-activated channels cannot be permeable to only one of these ions, because the reversal potential of the EPC is not near the equilibrium potential for any of them (see Figure 5.16C). However, if these channels were permeable to both Na^+ and K^+, then the reversal potential of the EPC would be between +70 mV and –100 mV.

The fact that EPCs reverse at approximately 0 mV is therefore consistent with the idea that ACh-activated ion channels are almost equally permeable to both Na^+ and K^+. This hypothesis was tested in 1960, by the Japanese husband and wife team of Akira and Noriko Takeuchi, by altering the extracellular concentration of these two ions. As predicted, the magnitude and reversal potential of the EPC were changed by altering the concentration gradient of each ion. Lowering the external Na^+ concentration, which makes E_{Na} more negative, produces a negative shift in E_{rev} (Figure 5.17A), whereas elevating external K^+ concentration, which makes E_K more positive, causes E_{rev} to shift to a more positive potential (Figure 5.17B). Such experiments establish that the ACh-activated ion channels are in fact permeable to both Na^+ and K^+.

Relationship between ion fluxes and postsynaptic potential changes

Defining the ion fluxes occurring during the EPC permits understanding of how these ion fluxes generate the EPP. If the membrane potential of the muscle fiber is kept at E_K (approximately –100 mV), the EPC will arise entirely from an influx of Na^+ because at this potential there is no driving force on K^+ (Figure 5.18A, left). In the absence of a voltage clamp to prevent postsynaptic membrane potential changes, such an influx of Na^+ would yield a large depolarizing EPP (Figure 5.18A, right). At the usual resting membrane potential of –90 mV, there is a small driving force on K^+, but a much greater one on Na^+. This means that much more Na^+ flows into the muscle cell than K^+ flows out (Figure 5.18B, left); the net influx of cations causes an EPC somewhat smaller than that measured at –100 mV and yields a depolarizing EPP that is also somewhat smaller than the EPP measured at –100 mV (see Figure 5.18B, right). Thus, at the resting membrane potential, the EPP is generated primarily by Na^+ influx, along with a small efflux of K^+. At the reversal potential of 0 mV, Na^+ influx and K^+ efflux are exactly balanced, so no net current flows during the opening of channels by ACh binding (Figure 5.18C). This yields neither an EPC

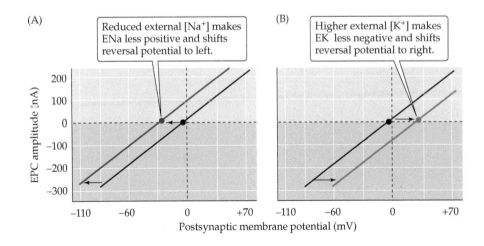

(A) Reduced external [Na⁺] makes ENa less positive and shifts reversal potential to left.

(B) Higher external [K⁺] makes EK less negative and shifts reversal potential to right.

Postsynaptic membrane potential (mV)

EPC amplitude (nA)

FIGURE 5.17 Reversal potential of the end plate current changes when ion gradients change (A) Lowering the external Na^+ concentration causes EPCs to reverse at more negative potentials. (B) Raising the external K^+ concentration makes the reversal potential more positive. (After A. Takeuchi and N. Takeuchi, 1960. *J Physiol* 154: 52–67.)

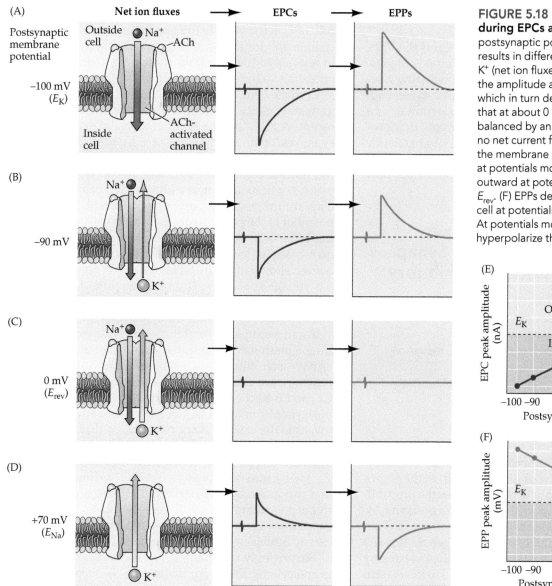

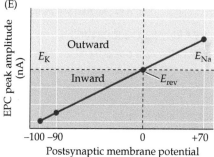

FIGURE 5.18   Na⁺ and K⁺ movements during EPCs and EPPs (A–D) Each of the postsynaptic potentials indicated at the left results in different relative fluxes of Na⁺ and K⁺ (net ion fluxes). These ion fluxes determine the amplitude and polarity of the EPCs, which in turn determine the EPPs. Note that at about 0 mV the Na⁺ flux is exactly balanced by an opposite K⁺ flux, resulting in no net current flow, and hence no change in the membrane potential. (E) EPCs are inward at potentials more negative than E_{rev} and outward at potentials more positive than E_{rev}. (F) EPPs depolarize the postsynaptic cell at potentials more negative than E_{rev}. At potentials more positive than E_{rev}, EPPs hyperpolarize the cell.

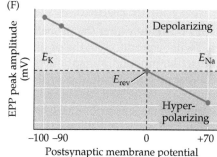

nor an EPP. At potentials more positive than E_{rev} the balance reverses; for example, at E_{Na} there is no influx of Na⁺ and a large efflux of K⁺ because of the large driving force on K⁺ (Figure 5.18D). This produces an outward EPC and a hyperpolarizing EPP. In summary, the polarity and magnitude of the EPC (Figure 5.18E) depend on the electrochemical driving force on the permeant ions, which in turn determines the polarity and magnitude of the EPP (Figure 5.18F). EPPs will depolarize when the membrane potential is more negative than E_{rev}, and hyperpolarize when the membrane potential is more positive than E_{rev}. The general rule, then, is that *the action of a transmitter drives the postsynaptic membrane potential toward E_{rev} for the particular ion channels being activated.*

Although this discussion has focused on the neuromuscular junction, similar mechanisms generate postsynaptic responses at all chemical synapses: Transmitter binding to postsynaptic receptors produces a postsynaptic conductance change as ion channels are opened (or sometimes closed). The postsynaptic conductance is increased if—as at the neuromuscular junction—channels are opened, and is decreased if channels are closed. This conductance change typically generates an electrical current, the **postsynaptic current (PSC)**, which in turn changes the postsynaptic membrane potential to produce a **postsynaptic potential (PSP)**. As in the specific case of the EPP at the neuromuscular junction, PSPs are depolarizing if their reversal potential is more positive than the resting membrane potential and hyperpolarizing if their reversal potential is more negative.

The postsynaptic conductance changes and the PSPs that typically accompany them are the ultimate outcome of most chemical synaptic transmission, concluding a sequence of electrical and chemical events that begins with the invasion of an action potential into the terminals of a presynaptic neuron. In many ways, the events that produce PSPs at synapses are similar to those that generate action potentials in axons; in both cases, conductance changes produced by ion channels lead to ion current flow that changes the membrane potential.

CONCEPT **5.7**
Postsynaptic Ion Fluxes Determine Whether Synapses Are Excitatory or Inhibitory

LEARNING OBJECTIVES

5.7.1 Explain the difference between excitatory and inhibitory synaptic transmission.

5.7.2 Explain the ionic basis for postsynaptic excitation and inhibition.

5.7.3 Outline the role of temporal and spatial summation in integrating the synaptic activity of neurons.

Excitatory and inhibitory postsynaptic potentials

PSPs ultimately alter the probability that an action potential will be produced in the postsynaptic cell. At the neuromuscular junction, synaptic action increases the probability that an action potential will occur in the postsynaptic muscle cell; indeed, the large amplitude of the EPP ensures that an action potential always is triggered. At many other synapses, PSPs similarly increase the probability of firing a postsynaptic action potential. However, still other synapses actually *decrease* the probability that the postsynaptic cell will generate an action potential. PSPs are called **excitatory postsynaptic potentials** (or **EPSPs**) if they increase the likelihood of a postsynaptic action potential occurring, and **inhibitory postsynaptic potentials** (or **IPSPs**) if they decrease this likelihood. Given that most neurons receive inputs from both excitatory and inhibitory synapses, it is important to understand more precisely the mechanisms that determine whether a particular synapse excites or inhibits its postsynaptic partner.

The principles of excitation just described for the neuromuscular junction are pertinent to all excitatory synapses. The principles of postsynaptic inhibition are much the same as for excitation, and they are also quite general. In both cases, neurotransmitters binding to receptors open or close ion channels in the postsynaptic cell. Whether a postsynaptic response is an EPSP or an IPSP depends on the type of channel that is coupled to the receptor, and on

the concentration of permeant ions inside and outside the cell. In fact, the only distinction between postsynaptic excitation and inhibition is the reversal potential of the PSP in relation to the threshold voltage for generating action potentials in the postsynaptic cell.

Consider, for example, a neuronal synapse that uses glutamate as the transmitter. Many such synapses have receptors that, like the ACh receptors at neuromuscular synapses, open ion channels that are nonselectively permeable to cations (see Chapter 6). When these glutamate receptors are activated, both Na^+ and K^+ flow across the postsynaptic membrane, yielding an E_{rev} of approximately 0 mV for the resulting postsynaptic current. If the resting potential of the postsynaptic neuron is −60 mV, the resulting EPSP will depolarize by bringing the postsynaptic membrane potential toward 0 mV. For the hypothetical neuron shown in Figure 5.19A, the action potential threshold voltage is −40 mV. Thus, a glutamate-induced EPSP will increase the probability that this neuron produces an action potential, defining the synapse as excitatory.

As an example of inhibitory postsynaptic action, consider a neuronal synapse that uses the amino acid GABA as its transmitter. At such synapses, the GABA receptors typically are channels that are selectively permeable to Cl^-, and the action of GABA causes Cl^- to flow across the postsynaptic membrane. Consider a case where E_{Cl} is −70 mV, as is the case for some neurons, so that the postsynaptic resting potential of −60 mV is less negative than E_{Cl}. The resulting positive electrochemical driving force ($V_m − E_{rev}$) will cause negatively charged Cl^- to flow into the cell and produce a hyperpolarizing IPSP (Figure 5.19B). This hyperpolarizing IPSP will take the postsynaptic membrane away from the action potential threshold of −40 mV, clearly inhibiting the postsynaptic cell.

Surprisingly, inhibitory synapses need not produce hyperpolarizing IPSPs. For instance, if E_{Cl} were −50 mV instead of −70 mV, then the negative electrochemical driving force would cause Cl^- to flow out of the cell and produce a depolarizing IPSP (Figure 5.19C). However, the synapse would still be inhibitory: Given that the reversal potential of the IPSP still is more negative than the action potential threshold (−40 mV), the depolarizing IPSP would inhibit because the postsynaptic membrane potential would be kept more negative than the threshold for action potential initiation. Another way to think about this peculiar situation is that if another excitatory input onto this neuron brought the cell's membrane potential to −41 mV—just below the threshold for firing an action potential—the IPSP would then hyperpolarize the membrane potential toward −50 mV, bringing the potential away from the action potential threshold. Thus, while EPSPs depolarize the postsynaptic cell, IPSPs can either hyperpolarize or depolarize; indeed, an inhibitory conductance change may produce no potential change at all and still exert an inhibitory

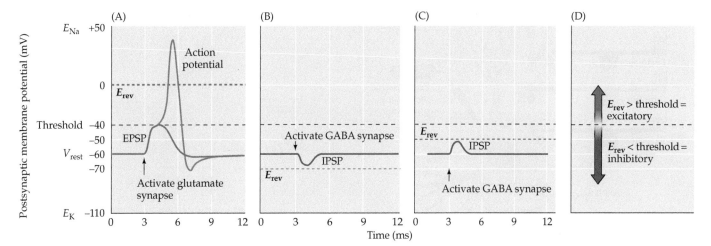

FIGURE 5.19 Reversal potentials and threshold potentials determine postsynaptic excitation and inhibition (A) If the reversal potential for a PSP (0 mV) is more positive than the action potential threshold (–40 mV), the effect of a transmitter is excitatory, and it generates EPSPs. (B) If the reversal potential for a PSP is more negative than the action potential threshold, the transmitter is inhibitory and generates IPSPs. (C) IPSPs can nonetheless depolarize the postsynaptic cell if their reversal potential is between the resting potential and the action potential threshold. (D) The general rule of postsynaptic action is: If the reversal potential is more positive than threshold, excitation results; inhibition occurs if the reversal potential is more negative than threshold.

effect by making it more difficult for an EPSP to evoke an action potential in the postsynaptic cell.

Although the particulars of postsynaptic action can be complex, a simple rule distinguishes postsynaptic excitation from inhibition: An EPSP has a reversal potential more positive than the action potential threshold, whereas an IPSP has a reversal potential more negative than threshold (Figure 5.19D). Intuitively, this rule can be understood by realizing that an EPSP will tend to depolarize the membrane potential so that it exceeds threshold, whereas an IPSP will always act to keep the membrane potential more negative than the threshold potential.

Summation of synaptic potentials

The PSPs produced at most synapses in the brain are much smaller than those at the neuromuscular junction; indeed, EPSPs produced by individual excitatory synapses may be only a fraction of a millivolt and are usually well below the threshold for generating postsynaptic action potentials. How can such synapses transmit information if their PSPs are subthreshold? The answer is that neurons in the central nervous system are typically innervated by thousands of synapses, and the PSPs produced by each active synapse can *sum together*—in space and in time—to determine the behavior of the postsynaptic neuron.

Consider the highly simplified case of a neuron that is innervated by two excitatory synapses, each generating a subthreshold EPSP, and an inhibitory synapse that produces an IPSP (Figure 5.20A). While activation of either one of the excitatory synapses alone (E1 or E2 in Figure 5.20B) produces a subthreshold EPSP, activation of both

excitatory synapses at about the same time causes the two EPSPs to sum together. If the sum of the two EPSPs (E1 + E2) depolarizes the postsynaptic neuron sufficiently to reach the threshold potential, a postsynaptic action potential will result. **Summation** thus allows subthreshold EPSPs to influence action potential production. Likewise, an IPSP generated by an inhibitory synapse (I) can sum (algebraically speaking) with a subthreshold EPSP to reduce its amplitude (E1 + I) or can sum with suprathreshold EPSPs to prevent the postsynaptic neuron from reaching threshold (E1 + I + E2).

In short, the summation of EPSPs and IPSPs by a postsynaptic neuron permits a neuron to integrate the electrical information provided by all the inhibitory and excitatory synapses acting on it at any moment. Whether the sum of active synaptic inputs results in the production of an action potential depends on the balance between excitation and inhibition. If the sum of all EPSPs and IPSPs results in a depolarization of sufficient amplitude to raise the membrane potential above threshold, then the postsynaptic cell will produce an action potential. Conversely, if inhibition prevails, then the postsynaptic cell will remain silent. Typically, the balance between EPSPs and IPSPs changes continually over time, depending on the number of excitatory and inhibitory synapses active at a given moment and the magnitude of the current at each active synapse. Summation is therefore a tug-of-war between all excitatory and inhibitory postsynaptic currents; the outcome of the contest determines whether or not a postsynaptic neuron fires an action potential and thereby becomes an active element in the neural circuits to which it belongs.

(A)

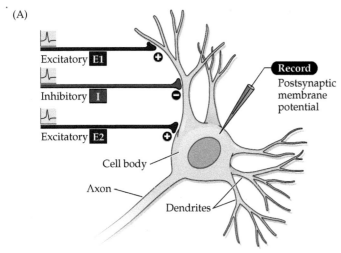

(B)

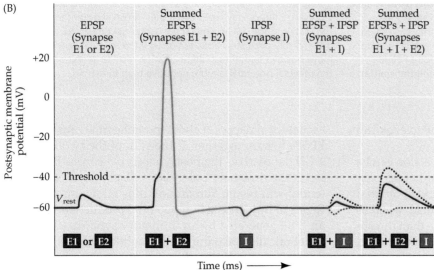

FIGURE 5.20 Summation of postsynaptic potentials (A) A microelectrode records the postsynaptic potentials produced by the activity of two excitatory synapses (E1 and E2) and an inhibitory synapse (I). (B) Electrical responses to synaptic activation. Stimulating either excitatory synapse (E1 or E2) produces a subthreshold EPSP, whereas stimulating both synapses at the same time (E1 + E2) produces a suprathreshold EPSP that evokes a postsynaptic action potential (shown in blue). Activation of the inhibitory synapse alone (I) results in a hyperpolarizing IPSP. Summing this IPSP (dashed red line) with the EPSP (dashed purple line) produced by one excitatory synapse (E1 + I) reduces the amplitude of the EPSP (solid purple line), while summing it with the suprathreshold EPSP produced by activating synapses E1 and E2 keeps the postsynaptic neuron below threshold, so that no action potential is evoked.

In conclusion, at chemical synapses neurotransmitter release from presynaptic terminals initiates a series of postsynaptic events that culminate in a transient change in the probability of a postsynaptic action potential occurring (Figure 5.21). Such synaptic signaling allows neurons to form the intricate synaptic circuits that play fundamental roles in brain information processing. Recent work further suggests that glial cells also contribute to synaptic signaling, adding another dimension to information processing in the brain (Box 5A).

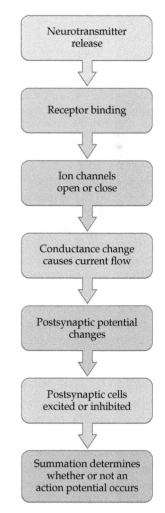

FIGURE 5.21 Overview of postsynaptic signaling Neurotransmitter released from presynaptic terminals binds to its cognate postsynaptic receptor, which causes the opening or closing of specific postsynaptic ion channels. The resulting conductance change causes current to flow, which may change the membrane potential. The postsynaptic cell sums (or integrates) all of the EPSPs and IPSPs, resulting in moment-to-moment control of action potential generation.

■ BOX 5A | The Tripartite Synapse

Up to this point, the discussion has considered signaling between presynaptic neurons and their postsynaptic targets to be a private dialogue between these two cells. However, much recent work suggests that this synaptic conversation may involve glial cells as well.

As mentioned in Chapter 1, glial cells support neurons in several ways. For example, it is well established that glia regulate the extracellular environment by removing K$^+$ that accumulates during action potential generation and by removing neurotransmitters at the conclusion of synaptic transmission. Consistent with such roles, glia seem to occupy virtually all of the non-neuronal volume of the brain. This means that glia are found in very close association with neurons (Figure A) and synapses (Figure B). Indeed, a given synapse typically is no more than a few hundred nanometers away from a glial cell. Glia form exceedingly fine processes that completely envelop synapses (Figure C), an intimate association that raises the possibility of a signaling role for glia at synapses.

The first support for such a role came from the discovery that glia respond to application of neurotransmitters. The list of neurotransmitters that elicit responses in glia now includes acetylcholine, glutamate, GABA, and many others. These responses are mediated by the same sorts of neurotransmitter receptors that are employed in synaptic signaling, most

(Continued)

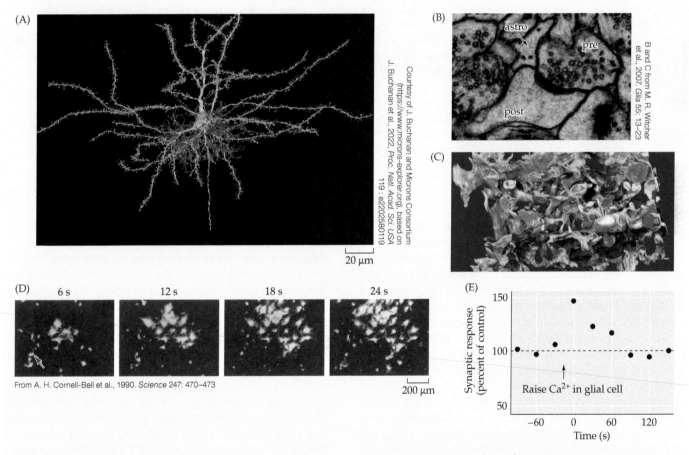

Courtesy of J. Buchanan and Microns Consortium (https://www.microns-explorer.org), based on J. Buchanan et al., 2022, *Proc. Natl. Acad. Sci. USA* 119 : e2202580119

B and C from M. R. Witcher et al., 2007. *Glia* 55: 13–23

From A. H. Cornell-Bell et al., 1990. *Science* 247: 470–473

(A) Three-dimensional structure of the contact between a neuron (white) and two glial cells (pink and purple). The glia are in close contact with the neuronal cell body and dendrites. (B) Electron microscopy image of a synapse between a presynaptic terminal (pre) and postsynaptic neuron (post), with a glial cell (astrocyte; astro) immediately adjacent to the synapse. cell (blue) surrounding dendrites of four postsynaptic neurons (different colors). (D) Application of glutamate (at arrow) increases intracellular Ca^{2+} (white) in cultured glia. Examination of these cells at different times indicates that the calcium signals spread as a wave through neighboring glia. (E) Transient elevation of Ca^{2+} within a single glial cell (arrow) enhances transmission at an excitatory synapse in the hippocampus. (E after G. Perea and A. Araque, 2007. *Science* 317: 1083–1086.)

■ BOX 5A | The Tripartite Synapse *(continued)*

often the metabotropic receptors that are coupled to intracellular signaling cascades (see Figure 5.14B). In some cases, neurotransmitters produce changes in the membrane potential of glia. More often, these neurotransmitters cause transient changes in intracellular calcium concentration within the glial cell. These intracellular calcium signals often are observed to trigger calcium waves that both spread within a single glial cell and also propagate between glia (Figure D).

These transient rises in intracellular calcium serve as second messenger signals (see Chapter 7) that trigger several physiological responses in glia. The most remarkable response is the release of several molecules—such as glutamate, GABA, and ATP—that are traditionally considered to be neurotransmitters. Release of such "gliotransmitters" occurs both via the calcium-triggered exocytotic

mechanisms employed in presynaptic terminals of neurons (see Figure 5.4C), as well as via unconventional release mechanisms such as permeation through certain ion channels.

The ability to both respond to and release neurotransmitters potentially makes glia participants in synaptic signaling. Indeed, release of neurotransmitters from a variety of presynaptic terminals has been found to elicit responses in glia. Furthermore, release of gliotransmitters has been found to regulate transmission at numerous synapses (Figure E). In some cases, glia regulate release of transmitters from presynaptic terminals, while in other cases they alter postsynaptic responsiveness. Glia can also alter the ability of synapses to undergo activity-dependent, plastic changes in synaptic transmission (see Chapter 8). Furthermore, there is accumulating evidence that astrocytes, microglia,

and other glial cells can also remove synapses by engulfing them. Such pruning of synapses may play a role in developmental synapse elimination (see Chapter 23), as well as in experience-dependent changes in adult brain circuits and even in forgetting stored memories (see Chapter 30).

This ability of glia to participate in synaptic signaling has led to the concept of the *tripartite synapse*, a three-way junction involving the presynaptic terminal, the postsynaptic process, and neighboring glia. While there is still debate about the physiological significance of such neuron–glia interactions, the ability of glia to release neurotransmitters, similar to presynaptic terminals, and to respond to neurotransmitters, similar to postsynaptic neurons, is dramatically changing our view of brain signaling mechanisms.

Summary

Synapses communicate the information carried by action potentials from one neuron to the next in neural circuits. The mechanisms underlying postsynaptic potentials generated during synaptic transmission are closely related to the mechanisms that generate other types of neuronal electrical signals, namely, ionic flow through membrane channels. In the case of electrical synapses, these channels are connexons; direct but passive flow of current through connexons is the basis for transmission. In the case of chemical synapses, channels with smaller and more selective pores are activated by the binding of neurotransmitters to postsynaptic receptors following release of the neurotransmitters from the presynaptic terminal. Transmitters are released in units, or quanta, reflecting their storage within synaptic vesicles. Vesicles discharge their contents into the synaptic cleft when the presynaptic depolarization generated by the invasion of an action potential opens voltage-gated calcium channels, allowing Ca^{2+} to enter the presynaptic terminal. Calcium triggers neurotransmitter release by binding to the Ca^{2+} sensor

protein, synaptotagmin, working in concert with SNARE proteins found on both vesicle and plasma membranes. Postsynaptic receptors are a diverse group of proteins that transduce binding of neurotransmitters into electrical signals by opening or closing postsynaptic ion channels. Two broadly different families of neurotransmitter receptors have evolved to carry out the postsynaptic signaling actions of neurotransmitters. The postsynaptic currents produced by the synchronous opening or closing of ion channels change the conductance of the postsynaptic cell, thus increasing or decreasing its excitability. Conductance changes that increase the probability of firing an action potential are excitatory, whereas those that decrease the probability of generating an action potential are inhibitory. Because postsynaptic neurons are usually innervated by many different inputs, the integrated effect of the conductance changes underlying all EPSPs and IPSPs produced in a postsynaptic cell at any moment determines whether or not the cell fires an action potential. The response elicited at a given synapse depends on the type of neurotransmitter released and the postsynaptic complement of receptors and associated channels.

■ Additional Reading

Reviews

Chanaday, N. L. and 4 others (2019) The synaptic vesicle cycle revisited: New insights into the modes and mechanisms. *J. Neurosci.* 39: 8209–8216.

Dolphin, A. C. and A. Lee (2020) Presynaptic calcium channels: Specialized control of synaptic neurotransmitter release. *Nat. Rev. Neurosci.* 21: 213–229.

Lefkowitz, R. J. (2007) Seven transmembrane receptors: something old, something new. *Acta Physiol. (Oxf.)* 190: 9–19.

McEnery, M. W. and R. E. Siegel (2014) Neurotransmitter receptors. Chapter 1 in *Encyclopedia of the Neurological Sciences*, 2nd Edition, M. J. Aminoff and R. B. Daroff (Eds.). Cambridge: Academic Press.

Mochida, S. (2021) Stable and flexible synaptic transmission controlled by the active zone protein interactions. *Int. J. Mol. Sci.* 22: 11775.

Important Original Papers

Adler, E. and 4 others (1991) Alien intracellular calcium chelators attenuate neurotransmitter release at the squid giant synapse. *J. Neurosci.* 11: 1496–1507.

Augustine, G. J. and R. Eckert (1984) Divalent cations differentially support transmitter release at the squid giant synapse. *J. Physiol.* 346: 257–271.

Beierlein, M., J. R. Gibson and B. W. Connors (2000) A network of electrically coupled interneurons drives synchronized inhibition in neocortex. *Nat. Neurosci.* 3: 904–910.

Boyd, I. A. and A. R. Martin (1955) The end-plate potential in mammalian muscle. *J. Physiol.* 132: 74–91.

Burette, A. C. and 6 others (2012) Electron tomographic analysis of synaptic ultrastructure. *J. Comp. Neurol.* 520: 2697–2711.

del Castillo, J. and B. Katz (1954) Quantal components of the end plate potential. *J. Physiol.* 124: 560–573.

Fatt, P. and B. Katz (1951) An analysis of the end plate potential recorded with an intracellular electrode. *J. Physiol.* 115: 320–370.

Fatt, P. and B. Katz (1952) Spontaneous subthreshold activity at motor nerve endings. *J. Physiol.* 117: 109–128.

Fotin, A. and 6 others (2004) Molecular model for a complete clathrin lattice from electron cryomicroscopy. *Nature* 432: 573–579.

Furshpan, E. J. and D. D. Potter (1959) Transmission at the giant motor synapses of the crayfish. *J. Physiol.* 145: 289–325.

Heuser, J. E. and T. S. Reese (1973) Evidence for recycling of synaptic vesicle membrane during transmitter release at the frog neuromuscular junction. *J. Cell Biol.* 57: 315–344.

Heuser, J. E. and 5 others (1979) Synaptic vesicle exocytosis captured by quick freezing and correlated with quantal transmitter release. *J. Cell Biol.* 81: 275–300.

Imig, C. and 8 others (2014) The morphological and molecular nature of synaptic vesicle priming at presynaptic active zones. *Neuron* 84: 416–431.

Loewi, O. (1921) Über humorale Übertragbarkeit der Herznerven-wirkung. *Pflügers Arch.* 189: 239–242.

Maeda, S. and 6 others (2009) Structure of the connexin 26 gap junction channel at 3.5 Å resolution. *Nature* 458: 597–602.

Miledi, R. (1973) Transmitter release induced by injection of calcium ions into nerve terminals. *Proc. R. Soc. Lond. B* 183: 421–425.

Neher, E. and B. Sakmann (1976) Single-channel currents recorded from membrane of denervated frog muscle fibres. *Nature* 260: 799–802.

Radhakrishnan, A. and 5 others (2021) Symmetrical arrangement of proteins under release-ready vesicles in presynaptic terminals. *Proc. Natl. Acad. Sci. U.S.A.* 118: e2024029118.

Reubold, T. F. and 12 others (2015) Crystal structure of the dynamin tetramer. *Nature* 525: 404–408.

Smith, S. J., and 4 others (1993) The spatial distribution of calcium signals in squid presynaptic terminals. *J. Physiol.* 472: 573–593.

Sotelo, C., R. Llinas and R. Baker (1974) Structural study of inferior olivary nucleus of the cat: morphological correlates of electrotonic coupling. *J. Neurophysiol.* 37: 541–559.

Sutton, R. B., D. Fasshauer, R. Jahn and A. T. Brünger (1998) Crystal structure of a SNARE complex involved in synaptic exocytosis at 2.4 Å resolution. *Nature* 395: 347–353.

Takamori, S. and 21 others (2006) Molecular anatomy of a trafficking organelle. *Cell* 127: 831–846.

Takeuchi, A. and N. Takeuchi (1960) On the permeability of end-plate membrane during the action of transmitter. *J. Physiol.* 154: 52–67.

Zhou, Q. and 19 others (2015) Architecture of the synaptotagmin-SNARE machinery for neuronal exocytosis. *Nature* 525: 62–67.

Books

Katz, B. (1969) *The Release of Neural Transmitter Substances*. Liverpool: Liverpool University Press.

Nestler, E., S. Hyman, D. M. Holtzman and R. Malenka (2015) *Molecular Neuropharmacology: A Foundation for Clinical Neuroscience*, 3rd Edition. New York: McGraw Hill.

Pickel, V. and M. Segal (2014) *The Synapse: Structure and Function*. Cambridge: Academic Press.

6

Neurotransmitters and Their Receptors

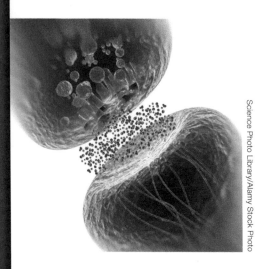

KEY CONCEPTS

6.1 There are many neurotransmitters

6.2 Acetylcholine serves as the prototype neurotransmitter

6.3 Glutamate is the main excitatory neurotransmitter

6.4 GABA and glycine are inhibitory transmitters

6.5 Biogenic amines are neuromodulatory transmitters

6.6 ATP and other purines can serve as co-transmitters

6.7 Neuropeptides are exceptionally diverse neurotransmitters

6.8 Interneuronal signaling can also occur via unconventional neurotransmitters

Overview

Neurons in the human brain communicate with one another, for the most part, by releasing chemical messengers called neurotransmitters. A large number of neurotransmitters are known. The main excitatory neurotransmitter in the brain is the amino acid glutamate, while the main inhibitory neurotransmitter is γ-aminobutyric acid (GABA). These and all other neurotransmitters evoke postsynaptic responses by binding to and activating neurotransmitter receptors. Most neurotransmitters are capable of activating several different receptors, yielding many possible modes of synaptic signaling. After activating their postsynaptic receptors, neurotransmitters are removed from the synaptic cleft by neurotransmitter transporters or by degradative enzymes. Abnormalities in the function of neurotransmitter systems contribute to a wide range of neurological and psychiatric disorders; thus, many neuropharmacological therapies are based on drugs that affect neurotransmitters, their receptors, or their removal from the synaptic cleft.

CONCEPT
6.1 | **There Are Many Neurotransmitters**

LEARNING OBJECTIVES

6.1.1 Understand what defines a neurotransmitter.

6.1.2 Know the different types of neurotransmitters.

What defines a neurotransmitter?

As introduced in Chapter 5, a neurotransmitter is a molecule that is used as a signal at chemical synapses. Formal criteria have been established to prove that a substance is used as a neurotransmitter at a synapse. These criteria include the presence of the neurotransmitter in the presynaptic neuron, the release of the neurotransmitter during synaptic activity, and the presence of postsynaptic receptors to detect the transmitter.

Applying such criteria has led to the identification of more than 100 different neurotransmitters. This rich diversity of neurotransmitters broadens the physiological repertoire of synapses. For example, multiple neurotransmitters can produce different types of responses on individual postsynaptic cells: A neuron can be excited by one type of neurotransmitter and inhibited by another type of neurotransmitter. The speed of postsynaptic responses produced by different transmitters also differs, allowing for control of electrical signaling over different timescales. In some cases, neurons synthesize and release two or more different neurotransmitters; in this case, the molecules are called **co-transmitters**. Co-transmitters can be differentially released according to the pattern of synaptic activity, so that the signaling properties of such synapses change dynamically according to the rate of activity.

SMALL-MOLECULE NEUROTRANSMITTERS

Acetylcholine $(CH_3)_3\overset{+}{N}$—CH_2—CH_2—O—$\overset{\displaystyle O}{\overset{\|}{C}}$—$CH_3$

AMINO ACIDS

Glutamate $H_3\overset{+}{N}$—$\overset{\displaystyle H}{\underset{\displaystyle |}{C}}$—$COO^-$
$|$
CH_2
$|$
CH_2
$|$
$COOH$

Aspartate $H_3\overset{+}{N}$—$\overset{\displaystyle H}{\underset{\displaystyle |}{C}}$—$COO^-$
$|$
CH_2
$|$
$COOH$

GABA $H_3\overset{+}{N}$—CH_2—CH_2—CH_2—COO^-

Glycine $H_3\overset{+}{N}$—$\overset{\displaystyle H}{\underset{\displaystyle |}{\underset{\displaystyle H}{C}}}$—$COO^-$

PURINES

ATP

$$O^-\!-\!\overset{\displaystyle O}{\overset{\|}{\underset{\underset{\displaystyle O^-}{|}}{P}}}\!-\!O\!-\!\overset{\displaystyle O}{\overset{\|}{\underset{\underset{\displaystyle O^-}{|}}{P}}}\!-\!O\!-\!\overset{\displaystyle O}{\overset{\|}{\underset{\underset{\displaystyle O^-}{|}}{P}}}\!-\!O\!-\!CH_2$$

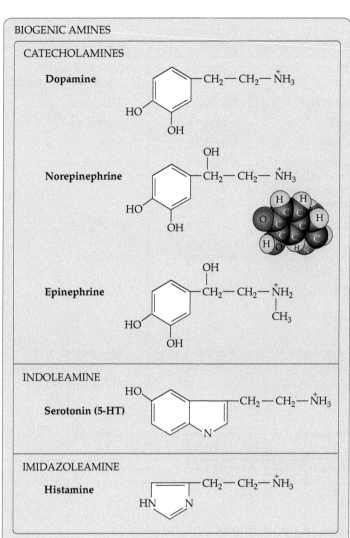

BIOGENIC AMINES

CATECHOLAMINES

Dopamine   —CH_2—CH_2—$\overset{+}{N}H_3$

Norepinephrine   —$\overset{\displaystyle OH}{\underset{\displaystyle |}{CH}}$—$CH_2$—$\overset{+}{N}H_3$

Epinephrine   —$\overset{\displaystyle OH}{\underset{\displaystyle |}{CH}}$—$CH_2$—$\overset{+}{N}H_2$
$|$
CH_3

INDOLEAMINE

Serotonin (5-HT)   —CH_2—CH_2—$\overset{+}{N}H_3$

IMIDAZOLEAMINE

Histamine   —CH_2—CH_2—$\overset{+}{N}H_3$

PEPTIDE NEUROTRANSMITTERS (more than 100 peptides, usually 3–36 amino acids long)

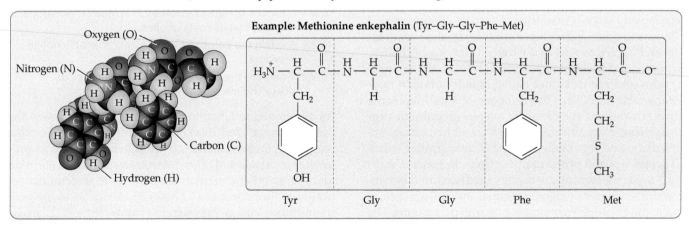

Example: Methionine enkephalin (Tyr–Gly–Gly–Phe–Met)

Oxygen (O)
Nitrogen (N)
Carbon (C)
Hydrogen (H)

Tyr   Gly   Gly   Phe   Met

FIGURE 6.1   Examples of small-molecule and peptide neurotransmitters
Small-molecule transmitters can be subdivided into acetylcholine, amino acids, purines, and biogenic amines. Size differences between the small-molecule neurotransmitters and the peptide neurotransmitters are indicated by the space-filling models for glycine, norepinephrine, and methionine enkephalin.

TABLE 6.1   Functional Features of the Major Neurotransmitters

Neurotransmitter	Postsynaptic effect[a]	Precursor(s)	Rate-limiting step in synthesis	Removal mechanism	Type of vesicle
ACh	Excitatory	Choline + acetyl CoA	ChAT	AChE	Small, clear
Glutamate	Excitatory	Glutamine	Glutaminase	Transporters	Small, clear
GABA	Inhibitory	Glutamate	GAD	Transporters	Small, clear
Glycine	Inhibitory	Serine	Phosphoserine	Transporters	Small, clear
Catecholamines (epinephrine, norepinephrine, dopamine)	Excitatory	Tyrosine	Tyrosine hydroxylase	Transporters, MAO, COMT	Small, dense-core or large, irregular dense-core
Serotonin (5-HT)	Inhibitory	Tryptophan	Tryptophan hydroxylase	Transporters, MAO	Large, dense-core
Histamine	Excitatory	Histidine	Histidine decarboxylase	Transporters	Large, dense-core
ATP	Excitatory	ADP	Mitochondrial oxidative phosphorylation; glycolysis	Hydrolysis to AMP and adenosine	Small, clear
Neuropeptides	Excitatory and inhibitory	Amino acids (protein synthesis)	Synthesis and transport	Proteases	Large, dense-core
Endocannabinoids	Inhibits inhibition	Membrane lipids	Enzymatic modification of lipids	Hydrolysis by FAAH	None
Nitric oxide	Excitatory and inhibitory	Arginine	Nitric oxide synthase	Spontaneous oxidation	None

[a]The most common postsynaptic effect is indicated; the same transmitter can elicit postsynaptic excitation or inhibition, depending on the nature of the receptors and ion channels activated by transmitter.

Types of neurotransmitter molecules

It is useful to separate the panoply of transmitters into two broad categories based simply on size (Figure 6.1). **Neuropeptides**, also called **peptide neurotransmitters**, are relatively large transmitter molecules composed of 3 to 36 amino acids. Individual amino acids, such as glutamate and GABA, as well as the transmitters acetylcholine, serotonin, and histamine, are much smaller than neuropeptides and are therefore called **small-molecule neurotransmitters**. Within the category of small-molecule neurotransmitters, the **biogenic amines** (dopamine, norepinephrine, epinephrine, serotonin, and histamine) are often discussed separately because of their similar chemical properties and postsynaptic actions. In general, most small-molecule neurotransmitters mediate rapid synaptic actions, whereas biogenic amines and neuropeptides tend to modulate slower, ongoing neuronal functions. The particulars of synthesis, packaging, release, and removal differ for each neurotransmitter (Table 6.1). This chapter describes the main features of these transmitters and their postsynaptic receptors.

CONCEPT **6.2** | # Acetylcholine Serves as the Prototype Neurotransmitter

LEARNING OBJECTIVES

6.2.1 Describe the steps involved in synthesis, release, and degradation of acetylcholine.

6.2.2 Explain the composition of nicotinic acetylcholine receptors.

6.2.3 Compare the properties of the two types of acetylcholine receptors.

As mentioned in Chapter 5, acetylcholine (ACh) was the first substance identified as a neurotransmitter. For this reason, experimental analyses of ACh have served as prototypes for studies of other neurotransmitters. In addition to its function as the neurotransmitter at skeletal neuromuscular junctions (see Chapter 5), as well as at the neuromuscular synapse between the vagus nerve and cardiac muscle fibers, ACh serves as a transmitter at synapses in the ganglia of the visceral motor system and at a variety of sites in the CNS. A great deal is known about the function of cholinergic transmission at neuromuscular junctions and ganglionic synapses. Although the actions of ACh in the CNS are not as well understood, the basal forebrain sends diffuse cholinergic input to many brain regions. ACh is likely to fine-tune the function of neurons throughout these regions, thereby contributing to attention and arousal.

Acetylcholine synthesis and metabolism

Acetylcholine is synthesized in nerve terminals from the precursors acetyl coenzyme A (acetyl CoA, which is synthesized from glucose) and choline, in a reaction catalyzed by choline acetyltransferase (ChAT) (Figure 6.2). Choline is present in plasma at a high concentration (about 10 mM) and is taken up into cholinergic neurons by a high-affinity, Na$^+$-dependent choline co-transporter (ChT). After synthesis in the cytoplasm of the neuron, a vesicular ACh transporter (VAChT) loads approximately 10,000 molecules of ACh into each cholinergic vesicle. The energy required to concentrate ACh within the vesicle is provided by the acidic pH of the vesicle lumen, which allows the VAChT to exchange H$^+$ for ACh.

In contrast to most other small-molecule neurotransmitters, the postsynaptic actions of ACh at many cholinergic synapses (the neuromuscular junction in particular) are not terminated by reuptake but by a powerful hydrolytic enzyme, **acetylcholinesterase** (**AChE**). This enzyme is concentrated in the synaptic cleft, ensuring a rapid decrease in ACh concentration after its release from the presynaptic terminal. AChE has a very high catalytic activity (about 5000 molecules of ACh per AChE molecule per second) and rapidly hydrolyzes ACh into acetate and choline. The choline produced by ACh hydrolysis is recycled by being transported back into nerve terminals, where it is used to resynthesize ACh.

Among the many interesting drugs that interact with cholinergic enzymes are the organophosphates. This group includes some potent chemical warfare agents. One such compound is the nerve gas sarin, made notorious in 1995 when a group of terrorists released it in Tokyo's underground rail system. Organophosphates can be lethal because they inhibit AChE, allowing ACh to accumulate at cholinergic synapses. This buildup of ACh depolarizes the postsynaptic muscle cell and renders it refractory to subsequent ACh release, causing neuromuscular paralysis and other effects. The high sensitivity of insects to AChE inhibitors has made organophosphates popular insecticides.

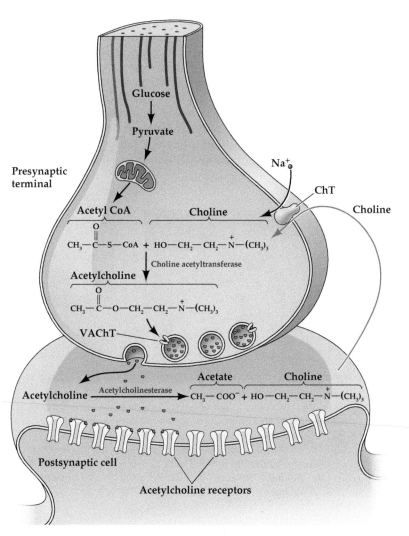

FIGURE 6.2   Acetylcholine metabolism in cholinergic nerve terminals The synthesis of acetylcholine from choline and acetyl CoA requires choline acetyltransferase. Acetyl CoA is derived from pyruvate generated by glycolysis, while choline is transported into the terminals via an Na$^+$-dependent co-transporter (ChT). Acetylcholine is loaded into synaptic vesicles via a vesicular transporter (VAChT). After release, acetylcholine is rapidly metabolized by acetylcholinesterase, and choline is transported back into the terminal via the ChT.

Nicotinic acetylcholine receptors

Many of the postsynaptic actions of ACh are mediated by the **nicotinic ACh receptor** (**nAChR**), so named because the CNS stimulant nicotine also binds to these receptors. Nicotine consumption produces some degree of euphoria, relaxation, and eventually addiction, effects believed to be mediated by nAChRs. nAChRs are nonselective cation channels that generate excitatory postsynaptic responses, such as those illustrated in Figures 5.15–5.18. Several toxins, with remarkably diverse chemical structures, specifically bind to and block nicotinic receptors (Box 6A). The availability of these highly specific ligands—particularly a component of snake venom called α-bungarotoxin—has provided a valuable way to isolate and purify nAChRs. As a result, nAChRs are the best-studied type of ionotropic

■ BOX 6A │ Neurotoxins That Act on Neurotransmitter Receptors

Poisonous plants and venomous animals are widespread in nature. The toxins they produce have been used for a variety of purposes, including hunting, healing, mind altering and, more recently, research. Many of these toxins have potent actions on the nervous system, often interfering with synaptic transmission by targeting neurotransmitter receptors. The poisons found in some organisms contain a single type of toxin, whereas others contain a mixture of tens or even hundreds of toxins.

Consistent with the essential role of ACh receptors in mediating muscle contraction at neuromuscular junctions in numerous species, a large number of natural toxins interfere with signaling mediated by these receptors. In fact, the classification of nicotinic and muscarinic ACh receptors is based on the sensitivity of these receptors to the toxic plant alkaloids nicotine and muscarine, which activate nicotinic and muscarinic ACh receptors, respectively. Nicotine is derived from the dried leaves of the tobacco plant *Nicotinia tabacum*, and muscarine is from the poisonous red mushroom *Amanita muscaria*. Both toxins are stimulants that produce nausea, vomiting, mental confusion, and convulsions. Muscarine poisoning can also lead to circulatory collapse, coma, and death.

The poison α-bungarotoxin, one of many peptides that together make up the venom of the banded krait (*Bungarus multicinctus*) (Figure A), blocks transmission at neuromuscular junctions and is used by the snake to paralyze its prey. This 74-amino–acid toxin (Figure B) blocks neuromuscular transmission by irreversibly binding to nicotinic ACh receptors, thus preventing ACh from opening postsynaptic ion channels. Paralysis ensues because skeletal muscles can no longer be activated by motor neurons. As a result of its specificity and its high affinity for nACh receptors, α-bungarotoxin has contributed greatly to understanding the ACh receptor. Other snake toxins that block nicotinic ACh receptors are cobra α-neurotoxin and the sea snake peptide erabutoxin. The same strategy used by these snakes to paralyze prey was adopted by South American natives who used curare, a mixture of plant toxins from *Chondrodendron tomentosum*, as an arrowhead poison to immobilize their quarry. Curare also blocks nACh receptors; the active agent is an alkaloid, δ-tubocurarine.

Another interesting class of animal toxins that selectively block nACh and

(A)

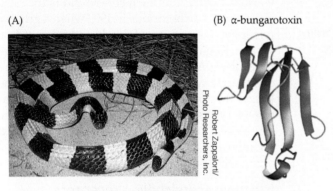

Robert Zappalorti/
Photo Researchers, Inc.

(B) α-bungarotoxin

(C)

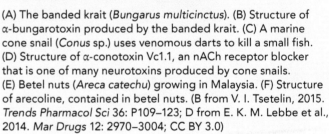

(D) α-conotoxin Vc1.1

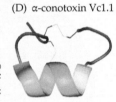

© Alex Kerstitch/Getty Images

(E)

Fletcher & Baylis/Photo Researchers, Inc.

(F) Arecoline

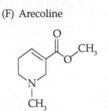

(A) The banded krait (*Bungarus multicinctus*). (B) Structure of α-bungarotoxin produced by the banded krait. (C) A marine cone snail (*Conus* sp.) uses venomous darts to kill a small fish. (D) Structure of α-conotoxin Vc1.1, an nACh receptor blocker that is one of many neurotoxins produced by cone snails. (E) Betel nuts (*Areca catechu*) growing in Malaysia. (F) Structure of arecoline, contained in betel nuts. (B from V. I. Tsetelin, 2015. *Trends Pharmacol Sci* 36: P109–123; D from E. K. M. Lebbe et al., 2014. *Mar Drugs* 12: 2970–3004; CC BY 3.0)

■ BOX 6A | Neurotoxins That Act on Neurotransmitter Receptors (continued)

other receptors includes the peptides produced by fish-hunting marine cone snails (Figure C). These colorful snails kill small fish by "shooting" venomous darts into them. The venom contains hundreds of peptides, known as conotoxins, many of which target proteins that are important in synaptic transmission. Figure D illustrates α-conotoxin Vc1.1, a 14-amino–acid peptide that blocks nACh receptors. Other conotoxin peptides block numerous other types of channels and receptors, including voltage-gated Ca^{2+} and Na^{+} channels, as well as glutamate receptors. The various physiological responses produced by these peptides all serve to immobilize any prey unfortunate enough to encounter a cone snail. Many other organisms, including other mollusks, corals, worms, and frogs, also use toxins containing specific blockers of ACh receptors.

Other natural toxins have mind- or behavior-altering effects and in some cases have been used for thousands of years by shamans and, more recently, physicians. Two examples are plant alkaloid toxins that block muscarinic ACh receptors: atropine from deadly nightshade (belladonna), and scopolamine from henbane. Because these plants grow wild in many parts of the world, exposure is not unusual, and poisoning by either toxin can be fatal.

Another postsynaptic neurotoxin that, like nicotine, is used as a social drug is found in the betel nut, the seed of the areca palm (*Areca catechu*) (Figure E). Betel nut chewing, although virtually unknown in the United States, is practiced by up to 25% of the population in India, Bangladesh, Sri Lanka, Malaysia, and the Philippines. Chewing these nuts produces a euphoria caused by arecoline, an alkaloid agonist of nACh receptors (Figure F). Like nicotine, arecoline is an addictive CNS stimulant.

Many other neurotoxins alter transmission at noncholinergic synapses. For example, amino acids found in certain mushrooms, algae, and seeds are potent glutamate receptor agonists. The excitotoxic amino acids kainate, from the red alga *Digenea simplex*, and quisqualate, from the seed of *Quisqualis indica*, are used to distinguish two families of glutamate receptors (see Concept 6.3). Other neurotoxic amino acid activators of glutamate receptors include ibotenic acid and acromelic acid, both found in mushrooms, and domoate, which occurs in algae, seaweed, and mussels. Another large group of peptide neurotoxins blocks glutamate receptors. These include the α-agatoxins from the funnel web spider, NSTX-3 from the orb weaver spider, jorotoxin from the Joro spider, and β-philanthotoxin from wasp venom, as well as many cone snail toxins.

All the toxins discussed so far target excitatory synapses. The inhibitory GABA and glycine receptors, however, have not been overlooked by the exigencies of survival. Strychnine, an alkaloid extracted from the seeds of *Strychnos nux-vomica*, is the only drug known to have specific actions on transmission at glycinergic synapses. Because the toxin blocks glycine receptors, strychnine poisoning causes overactivity in the spinal cord and brainstem, leading to seizures. Strychnine has long been used commercially as a poison for rodents, although alternatives such as the anticoagulant warfarin (Coumadin) are now more popular because they are safer for humans. Neurotoxins that block $GABA_A$ receptors include plant alkaloids such as bicuculline from Dutchman's breeches and picrotoxin from *Anamirta cocculus*. Dieldrin, a commercial insecticide, also blocks these receptors. Like strychnine, these agents are powerful CNS stimulants. Muscimol, a mushroom toxin that is a powerful depressant as well as a hallucinogen, activates $GABA_A$ receptors. A synthetic analogue of GABA, baclofen, is a $GABA_B$ receptor agonist that is used clinically to reduce the frequency and severity of muscle spasms.

Chemical warfare between species has thus given rise to a staggering array of molecules that target synapses throughout the nervous system. Although these toxins are designed to defeat normal synaptic transmission, they have also provided a set of powerful tools to understand postsynaptic mechanisms.

---

neurotransmitter receptor, and unraveling their molecular organization has provided deep insights into the workings of ionotropic receptors.

Nicotinic receptors are large protein complexes consisting of five subunits. At the neuromuscular junction, the nAChR contains two α subunits (encoded by *CHRNA* genes), each of which has a binding site that binds a single molecule of ACh. Both ACh binding sites must be occupied for the receptor to be activated, so only relatively high concentrations of ACh activate these receptors. These subunits also bind other ligands, such as nicotine and α-bungarotoxin. The two α subunits are combined with three other subunits from among the four other types of nAChR subunits—

β (*CHRNB* genes), δ (*CHRND* gene), and either γ (*CHRNG* gene) or ε (*CHRNE* gene)—in the ratio 2α:1β:1δ:1γ/ε. Neuronal nAChRs differ from those of muscle in that they (1) lack sensitivity to α-bungarotoxin and (2) comprise only two receptor subunit types (α and β), in a ratio of 3α:2β, with the $α_4$ and $β_2$ subunits being most common in the brain.

Each subunit of the receptor contains a large extracellular region (which in α subunits contains the ACh binding site) as well as four membrane-spanning domains (Figure 6.3A). The transmembrane domains of the five individual subunits together form a channel with a central membrane-spanning pore (Figure 6.3B,C). The width of this pore (Figure 6.3D) is substantially larger than that of

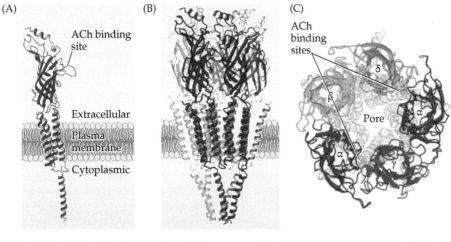

(A) ACh binding site / Extracellular / Plasma membrane / Cytoplasmic

(B)

(C) ACh binding sites / Pore / δ / β / α / γ

(D) Extracellular / Pore / Gate / Cytoplasmic

(E) ACh

FIGURE 6.3 Structure of the nicotinic ACh receptor (A) Structure of the α subunit of the receptor. Each subunit crosses the membrane four times; the α subunit additionally contains a binding site for ACh in its extracellular domain. (B) Five subunits come together to form a complete nAChR. (C) View of the nAChR from the perspective of the synaptic cleft. The arrangement of the five subunits is evident, with each subunit contributing one transmembrane helix that forms the channel pore. (D) Cross-section view of the transmembrane domain of the nAChR. The openings at either end of the channel pore are very large, and the pore narrows at the channel gate. The turquoise sphere indicates the dimension of a sodium ion (0.3 nm diameter). (E) Model for gating of the nAChR. Binding of ACh to its binding sites on the two α subunits causes a conformational change in part of the extracellular domain, which causes the pore-forming helices to move and open the pore gate. (F) A diversity of subunits come together to form ionotropic neurotransmitter receptors. (A–C from J. Unwin, 2005. *J Mol Biol* 346: 967–989; D,E from A. Miyazawa et al., 2003. *Nature* 423: 949–955.)

(F)

Receptor	nACh	AMPA	NMDA	Kainate	GABA	Glycine	Serotonin	Purines
Subunits (combination of 3–5 required for each receptor type)	α_{1-10}	GluA1	GluN1	GluK1	α_{1-6}	α_{1-6}	5-HT$_{3A}$	P2X$_1$
	β_{1-4}	GluA2	GluN2A	GluK2	β_{1-3}	β	5-HT$_{3B}$	P2X$_2$
	γ	GluA3	GluN2B	GluK3	γ_{1-3}		5-HT$_{3C}$	P2X$_3$
	δ	GluA4	GluN2C	GluK4	δ		5-HT$_{3D}$	P2X$_4$
	ε		GluN2D	GluK5	ε		5-HT$_{3E}$	P2X$_5$
			GluN3A		θ			P2X$_6$
			GluN3B		η			P2X$_7$
					ρ_{1-3}			

the pores of voltage-gated ion channels (see Figure 4.6), consistent with the relatively poor ability of nACh receptors to discriminate between different cations. Within this pore is a constriction that may represent the gate of the receptor. Binding of ACh to the α subunits is thought to cause a twisting of the extracellular domains of the receptor, which causes some of the receptor transmembrane domains to tilt to open the channel gate and permit ions to diffuse through the channel pore (Figure 6.3E).

In summary, the nACh receptor is a ligand-gated ion channel. The intimate association of the ACh binding sites

of this receptor with the pore of the channel permits the rapid response to ACh that is characteristic of nACh receptors. This general arrangement—several receptor subunits coming together to form a ligand-gated ion channel—is characteristic of *all* the ionotropic receptors at fast-acting synapses employing glutamate, GABA, serotonin, and other neurotransmitters. Thus, the nicotinic receptor has served as a paradigm for studies of other ionotropic receptors, at the same time leading to a much deeper appreciation of several neuromuscular diseases (Clinical Applications). The subunits used to make nAChRs and

■ Clinical Applications

Myasthenia Gravis: An Autoimmune Disease of Neuromuscular Synapses

Myasthenia gravis is a disease that interferes with transmission between motor neurons and skeletal muscle fibers and afflicts approximately 1 of every 10,000 people. The hallmark of the disorder, which was originally described by the British physician Thomas Willis in 1685, is muscle weakness, particularly during sustained activity (Figure A). Although the course is variable, myasthenia commonly affects muscles controlling the eyelids (resulting in drooping of the eyelids, or ptosis) and eye movements (resulting in double vision, or diplopia). Muscles controlling facial expression, chewing, swallowing, and speaking are other common targets.

An important indication of the cause of myasthenia gravis came from the clinical observation that the muscle weakness improves following treatment with neostigmine and other inhibitors of AChE, the enzyme that normally degrades ACh at the neuromuscular junction (see Figure A). Studies of muscle obtained by biopsy

showed that both end plate potentials (EPPs) and miniature end plate potentials (MEPPs) are much smaller than normal (Figure B).

Because both the frequency of MEPPs and the quantal content of EPPs are normal, it seemed likely that myasthenia gravis affects the postsynaptic muscle cells. Indeed, electron microscopy shows that the structure of neuromuscular junctions is altered, with obvious changes being a widening of the synaptic cleft and an apparent reduction in the number of ACh receptors in the postsynaptic membrane.

A chance observation in the early 1970s led to the discovery of the underlying cause of these changes. Jim Patrick and Jon Lindstrom, then working at the Salk Institute, were attempting to raise antibodies to nicotinic ACh receptors by immunizing rabbits with the receptors. Unexpectedly, the immunized rabbits developed muscle weakness that improved after treatment with AChE inhibitors. Subsequent work showed that the blood of

people with myasthenia gravis contains antibodies directed against the ACh receptor, and that these antibodies are present at neuromuscular synapses. Removal of antibodies by plasma exchange improves the weakness. Finally, injecting the serum of people with myasthenia gravis into mice produces myasthenic effects, because the serum carries circulating antibodies.

These findings indicate that myasthenia gravis is an autoimmune disease that targets nicotinic ACh receptors. The immune response reduces the number of functional receptors at the neuromuscular junction and eventually destroys them altogether, diminishing the efficiency of synaptic transmission; muscle weakness occurs because motor neurons are less capable of exciting the postsynaptic muscle cells. This causal sequence also explains why cholinesterase inhibitors alleviate the symptoms: The inhibitors increase the concentration of ACh in the synaptic cleft, allowing more effective activation of those postsynaptic receptors not yet destroyed by the immune system. However, it is still not clear what triggers this autoimmune response to ACh receptors. Some other myasthenic syndromes weaken neuromuscular transmission by affecting ACh release from presynaptic terminals (see Clinical Applications, Chapter 5).

(A)

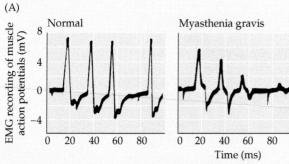

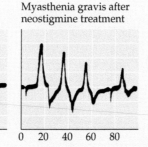

(B)

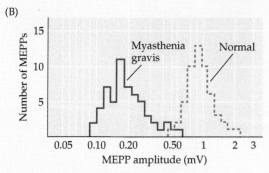

Myasthenia gravis reduces the efficiency of neuromuscular transmission. (A) Electromyographs (EMGs) show muscle responses elicited by stimulating motor nerves. In typical individuals, each stimulus in a train evokes the same contractile response. In people with myasthenia gravis, transmission rapidly fatigues, although it can be partially restored by administration of AChE inhibitors such as neostigmine. (B) Distribution of MEPP amplitudes in muscle fibers from people with myasthenia gravis and from healthy controls. The smaller size of MEPPs in people with myasthenia gravis is due to a diminished number of postsynaptic receptors. (A after A. M. Harvey et al., 1942. *J Clin Invest* 21: 579–588; B after D. Elmqvist et al., 1964. *J Physiol* 174: 417–434.)

other types of ionotropic neurotransmitter receptors are summarized in Figure 6.3F.

Muscarinic acetylcholine receptors

A second class of ACh receptors is activated by muscarine, a poisonous alkaloid found in some mushrooms (see Box 6A), and thus they are referred to as **muscarinic ACh receptors (mAChRs)**. mAChRs are metabotropic and mediate most of the effects of ACh in the brain. Like other metabotropic receptors, mAChRs have seven helical membrane-spanning domains (Figure 6.4A). ACh binds to a single binding site on the extracellular surface of the mAChR; this binding site is within a deep pocket that is formed by several of the transmembrane helices (Figure 6.4B). Binding of ACh to this site causes a conformational change that permits G-proteins to bind to the cytoplasmic domain of the mAChR, which is only partially shown in Figure 6.4A.

Five subtypes of mAChRs are known (Figure 6.4C) and are coupled to different types of G-proteins, thereby causing a variety of slow postsynaptic responses. Muscarinic ACh receptors are highly expressed in the corpus striatum and various other forebrain regions, where they activate K^+ channels to exert an inhibitory influence on dopamine-mediated motor effects. In other parts of the brain, such as the hippocampus, mAChRs are excitatory and act by closing other K^+ channels. These receptors are also found in the ganglia of the peripheral nervous system. Finally, mAChRs mediate peripheral cholinergic responses of autonomic effector organs such as heart, smooth muscle, and exocrine glands and are responsible

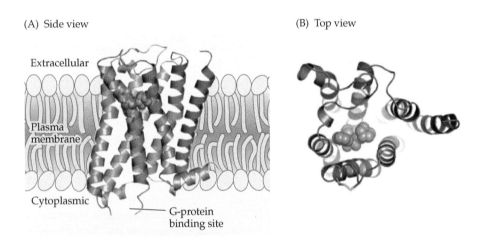

(A) Side view

(B) Top view

FIGURE 6.4 Muscarinic and other metabotropic receptors (A,B) Structure of the human M_2 mAChR. (A) This receptor spans the plasma membrane seven times and has a cytoplasmic domain (only partially shown here) that binds to and activates G-proteins, as well as an extracellular domain that binds ACh. In this view, the ACh binding site is occupied by 3-quinuclidinyl-benzilate (QNB, colored spheres), a muscarinic receptor antagonist. (B) View of the extracellular surface of the mAChR showing QNB bound to the ACh binding site. (C) Muscarinic and other metabotropic neurotransmitter receptors. (A,B after Haga et al., 2012. *Nature* 482: 547–551.)

(C)

Receptor class	Muscarinic	Glutamate	GABA$_B$	Dopamine	Adrenergic	Histamine	Serotonin	Purines
Receptor subtype	M_1	Class I	GABA$_{B1}$	D1	Alpha	H_1	5-HT$_{1A}$	Adenosine
	M_2	mGlu$_1$	GABA$_{B2}$	D2	α_{1A}	H_2	5-HT$_{1B}$	A_1
	M_3	mGlu$_5$		D3	α_{1B}	H_3	5-HT$_{1D}$	A_{2A}
	M_4	Class II		D4	α_{1D}	H_4	5-HT$_{1E}$	A_{2B}
	M_5	mGlu$_2$		D5	α_{2A}		5-HT$_{1F}$	A_3
		mGlu$_3$			α_{2B}		5-HT$_{2A}$	P2Y
		Class III			α_{2C}		5-HT$_{2B}$	P2Y$_1$
		mGlu$_4$			Beta		5-HT$_{2C}$	P2Y$_2$
		mGlu$_6$			β_1		5-HT$_4$	P2Y$_4$
		mGlu$_7$			β_2		5-HT$_{5A}$	P2Y$_6$
		mGlu$_8$			β_3		5-HT$_6$	P2Y$_{11}$
							5-HT$_7$	P2Y$_{12}$
								P2Y$_{13}$
								P2Y$_{14}$

for the inhibition of heart rate by the vagus nerve. Numerous drugs act as mAChR agonists or antagonists; mAChR blockers that are therapeutically useful include atropine (used to dilate the pupil), scopolamine (effective in preventing motion sickness), and ipratropium (useful in the treatment of asthma).

CONCEPT 6.3 | Glutamate Is the Main Excitatory Neurotransmitter

LEARNING OBJECTIVES

6.3.1 Know how glutamate is synthesized and degraded.

6.3.2 Understand the roles that different types of glutamate receptors play in synaptic signaling.

6.3.3 Know the unique physiological properties of NMDA-type glutamate receptors and understand how these properties allow NMDA receptors to play unique roles in brain information storage.

Glutamate is the most important transmitter for normal brain function. Nearly all excitatory neurons in the CNS are glutamatergic, and it is estimated that more than half of all brain synapses release this neurotransmitter. During brain trauma, there is excessive release of glutamate that can produce *excitotoxic* brain damage.

Glutamate synthesis and metabolism

Glutamate is a nonessential amino acid that does not cross the blood-brain barrier and therefore must be synthesized in neurons from local precursors. The most prevalent precursor for glutamate synthesis is glutamine, which is taken up into presynaptic terminals by the system A transporter 2 (SAT2) and is then metabolized to glutamate by the mitochondrial enzyme glutaminase (Figure 6.5). Glucose metabolized by neurons also can be used to synthesize glutamate by transamination of 2-oxoglutarate, an intermediate of the tricarboxylic acid (Krebs) cycle. Glutamate synthesized in the presynaptic cytoplasm is packaged into synaptic vesicles by vesicular glutamate transporters (VGLUTs). At least three different *VGLUT* genes have been identified, with different VGLUTs involved in packaging glutamate into vesicles at different types of glutamatergic presynaptic terminals.

Once released, glutamate is removed from the synaptic cleft by the excitatory amino acid transporters (EAATs). EAATs are a family of five different Na^+-dependent glutamate co-transporters. Some EAATs are present in glial cells and others in presynaptic terminals. Glutamate transported into glial cells via EAATs is converted into glutamine by the enzyme glutamine synthetase. Glutamine is then transported out of the glial cells by a different transporter, the system N transporter 1 (SN1), and transported into nerve terminals via SAT2. This overall sequence of events is referred to as the **glutamate–glutamine cycle** (see Figure 6.5). This cycle allows glial cells and presynaptic terminals to cooperate both to maintain an adequate supply of glutamate for synaptic transmission and to rapidly terminate postsynaptic glutamate action.

Ionotropic glutamate receptors

There are several types of **ionotropic glutamate receptors** (see Figure 6.3F). **AMPA receptors**, **NMDA receptors**, and **kainate receptors** are named after the agonists that activate them: AMPA (α-amino-3-hydroxyl-5-methyl-4-isoxazole-propionate), NMDA (*N*-methyl-d-aspartate), and kainic acid. All of these receptors are glutamate-gated

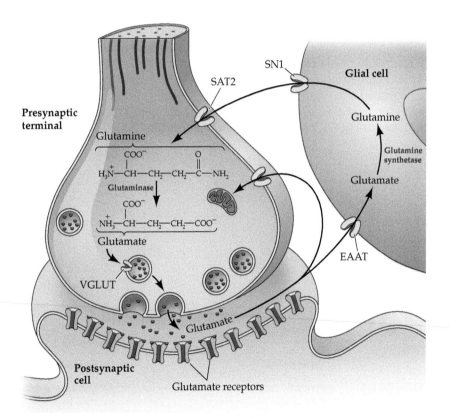

FIGURE 6.5 Glutamate synthesis and cycling between neurons and glia The action of glutamate released into the synaptic cleft is terminated by uptake into surrounding glial cells (and neurons) via excitatory amino acid transporters (EAATs). Within glial cells, glutamate is converted to glutamine by glutamine synthetase and released by glial cells through the SN1 transporter. Glutamine is taken up into nerve terminals via SAT2 transporters and converted back to glutamate by glutaminase. Glutamate is then loaded into synaptic vesicles via vesicular glutamate transporters (VGLUTs) to complete the cycle.

cation channels that allow the passage of Na^+ and K^+, similar to the nAChR. Hence AMPA, kainate, and NMDA receptor activation always produces excitatory postsynaptic responses.

Most central excitatory synapses possess both AMPA and NMDA receptors. Antagonist drugs that selectively block either AMPA or NMDA receptors are often used to identify synaptic responses mediated by each receptor type. Such experiments reveal that excitatory postsynaptic currents (EPSCs) produced by NMDA receptors are slower and last longer than those produced by AMPA receptors (Figure 6.6A). EPSCs generated by AMPA receptors usually are much larger than those produced by other types of ionotropic glutamate receptors, so that AMPA receptors are the primary mediators of excitatory transmission in the brain. The physiological roles of kainate receptors are less well defined; in some cases, these receptors are found on presynaptic terminals and serve as a feedback mechanism to regulate glutamate release. When found on postsynaptic cells, kainate receptors generate EPSCs that rise quickly but decay more slowly than those mediated by AMPA receptors (Figure 6.6B).

Unlike nAChRs, AMPA receptors are tetramers composed of four subunits. The structure of AMPA receptors has already been described in Chapter 4 and is reviewed here. There are four different AMPA receptor subunits, designated GluA1 through GluA4 (see Figure 6.3F) and encoded by *GRIA* genes, with each subunit conferring unique functional properties to AMPA receptors. AMPA

receptor subunits have several different domains, including an extracellular ligand-binding domain that is responsible for binding glutamate, and a transmembrane domain that forms part of the ion channel (Figure 6.7A). There is also an amino-terminal domain that is involved in trafficking of these receptors to the plasma membrane. AMPA receptors are Y-shaped (Figure 6.7B), with most of the protein being extracellular. Unlike nAChRs, AMPA receptors are asymmetrical (Figure 6.7C) and look different when viewed from their front and side surfaces (compare Figure 6.7B and C). The transmembrane domain consists of helices that form both the channel pore and a gate that occludes the pore when glutamate is not bound to the receptor. A characteristic feature of AMPA receptors and other glutamate receptors is the "clamshell" shape of the ligand-binding domains that bind glutamate and other ligands (circle in Figure 6.7C). Binding of glutamate causes the clamshell structure to "shut"; this movement pulls on the gate helices within the transmembrane domain, causing the gates to move and thereby open the channel pore (Figure 6.7D). Kainate receptors are remarkably similar in structure and are activated by kainite or other ligands binding to their ligand-binding domains (Figure 6.7E).

NMDA receptors have physiological properties that set them apart from the other ionotropic glutamate receptors. Perhaps most significant is that the pore of the NMDA receptor channel allows the entry of Ca^{2+} in addition to Na^+ and K^+. As a result, excitatory postsynaptic potentials (EPSPs) produced by NMDA receptors increase the concentration of Ca^{2+} in the postsynaptic neuron, with Ca^{2+} then acting as a second messenger to activate intracellular signaling processes (see Chapter 7). Another key property is that Mg^{2+} blocks the pore of this channel at hyperpolarized membrane potentials, while depolarization pushes Mg^{2+} out of the pore (Figure 6.8A). This imparts a peculiar voltage dependence to current flow through the receptor (Figure 6.8B, red line); removing extracellular Mg^{2+} eliminates this behavior (blue line), which demonstrates that Mg^{2+} confers the voltage dependence. Because of this property, NMDA receptors pass cations (most notably Ca^{2+}) only when the postsynaptic membrane potential is depolarized, such as during activation of strong excitatory inputs or during action potential firing in the postsynaptic cell. This requirement for the coincident presence of

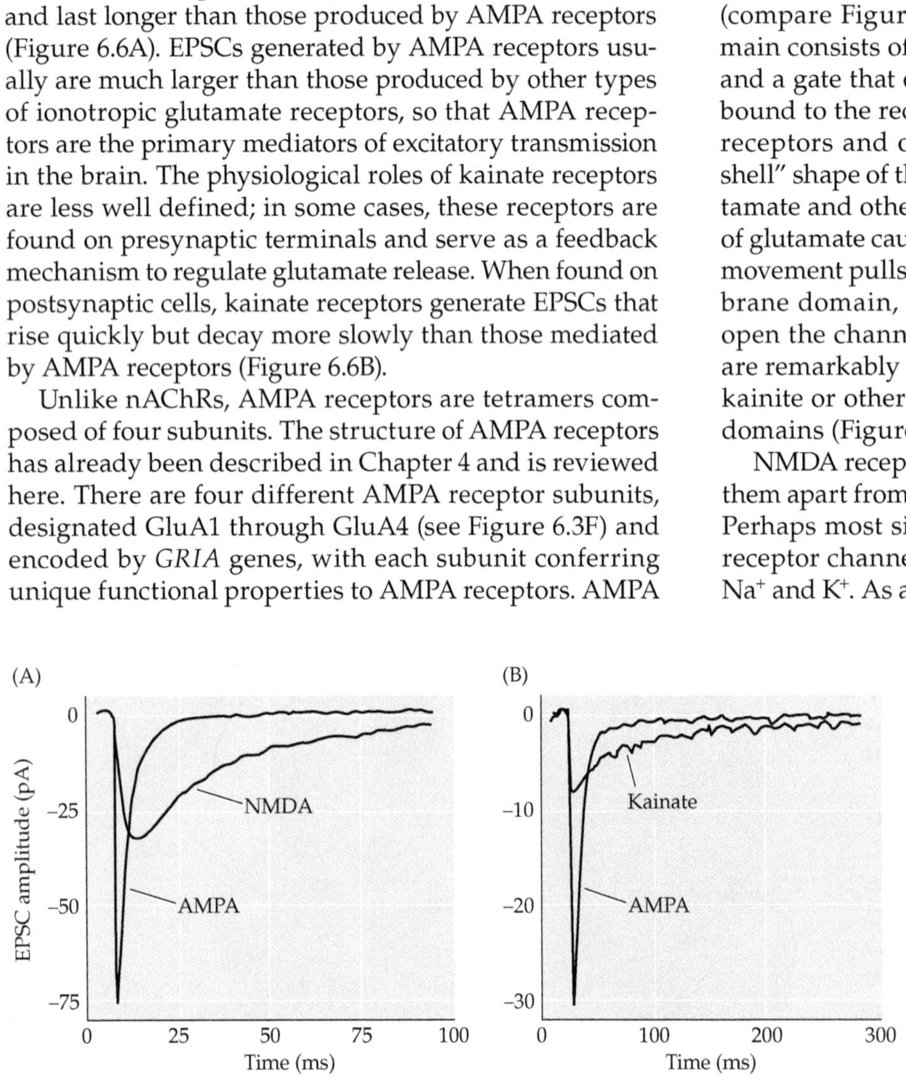

FIGURE 6.6 Postsynaptic responses mediated by ionotropic glutamate receptors (A) Contributions of AMPA and NMDA receptors to EPSCs at a synapse between a presynaptic pyramidal cell and a postsynaptic interneuron in the visual cortex. Blocking NMDA receptors reveals a large and fast EPSC mediated by AMPA receptors, while blocking AMPA receptors reveals a slower EPSC component mediated by NMDA receptors. (B) Contributions of AMPA and kainate receptors to miniature EPSCs at the excitatory synapse formed between mossy fibers and CA3 pyramidal cells in the hippocampus. Pharmacological antagonists reveal that the component of EPSCs mediated by AMPA receptors is larger and decays faster than that mediated by kainate receptors. (A after J. Watanabe et al., 2005. *J Neurosci* 25: 1024–1033; B from D. Mott et al., 2008. *J Neurosci* 28: 1659–1671. © 2005 Society for Neuroscience.)

(A)

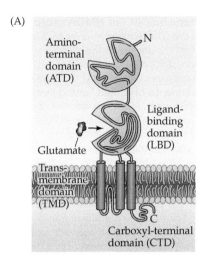

(B)

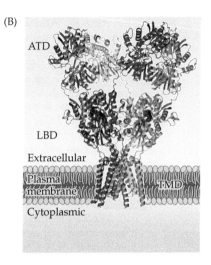

(C)

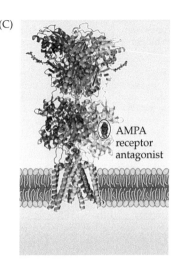

(D)

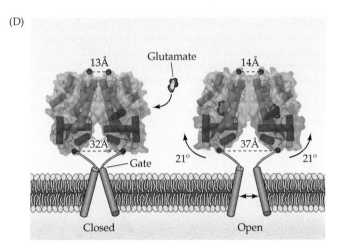

(E)

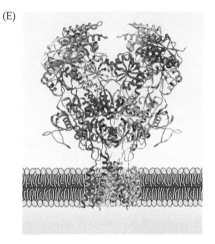

FIGURE 6.7 Structure of the AMPA and kainate receptors
(A) Domain structure of an AMPA receptor subunit. The largest part of each subunit is extracellular and consists of two domains, the amino-terminal domain (ATD) and the ligand-binding domain (LBD). In addition, a transmembrane domain (TMD) forms part of the ion channel pore, and an intracellular carboxyl-terminal domain (CTD) connects the receptor to intracellular proteins. (B–C) Crystallographic structure of the AMPA receptor. Each of the four subunits is indicated in a different color. (B) From this perspective, the Y shape of the AMPA receptor is evident. (C) After rotating the receptor 90 degrees, the asymmetrical dimensions of the receptor are evident. One ligand-binding domain is visible and is occupied by an antagonist drug (circled). (D) Model for gating of the AMPA receptor by glutamate. The transmembrane domain (blue cylinders) and extracellular domains are shown. Binding of glutamate closes the clamshell structure of the ligand-binding domain (curved red arrows), leading to movement of the gate helices (double-headed arrow) that opens the channel pore. (E) Crystallographic structure of the kainate receptor. Each of the four subunits is indicated in a different color and black arrow indicates a bound agonist molecule. (A after S. F. Traynelis et al., 2010. *Pharmacol Rev* 62: 405–496; B,C from A. I. Sobolevsky et al., 2009. *Nature* 462: 745–756; D,E from K. B. Hansen et al., 2021. *Pharmacol Rev* 73: 298–487. CC BY 4.0)

both glutamate and postsynaptic depolarization to open NMDA receptors is widely thought to underlie some forms of synaptic information storage, such as long-term synaptic plasticity (see Chapter 8). Another unusual feature of NMDA receptors is that their gating requires a co-agonist—the amino acid glycine, which is present in the ambient extracellular environment of the brain.

NMDA receptors are tetrameric assemblies of subunits with many similarities to AMPA receptors. There are three groups of NMDA receptor subunits (GluN1, GluN2, and GluN3, encoded by *GRIN1*, *GRIN2*, and *GRIN3* genes, respectively), with a total of seven different types of subunits (see Figure 6.3F). While GluN2 subunits bind glutamate, GluN1 and GluN3 subunits bind glycine. NMDA receptor tetramers typically comprise two glutamate-binding subunits (GluN2) and two glycine-binding subunits (GluN1). In some cases, GluN3 replaces one of the two GluN2 subunits. This mix of subunits ensures that the receptor binds both to glutamate released from presynaptic terminals and to the ambient glycine co-agonist.

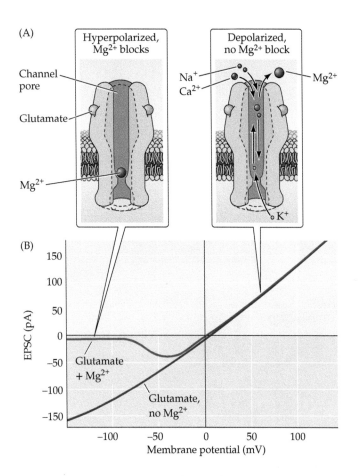

(A)

(B)

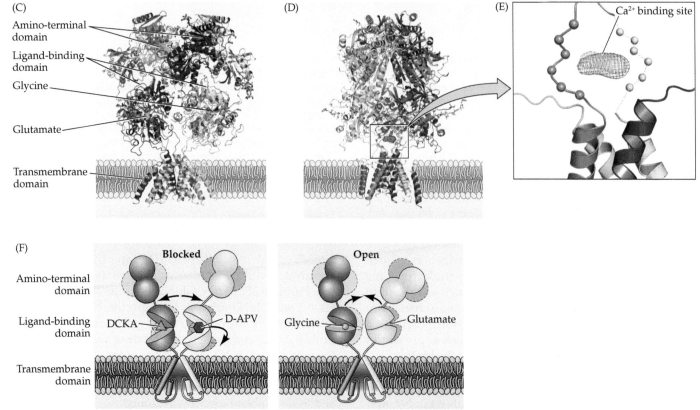

FIGURE 6.8 **Function and structure of the NMDA receptor**
(A) Voltage-dependent block of the NMDA receptor pore by Mg^{2+}. At hyperpolarized potentials, Mg^{2+} resides within the channel pore and blocks it (left). Depolarization of the membrane potential pushes Mg^{2+} out of the pore, so that current can flow through the NMDA receptor (right). (B) Voltage dependence of current flowing through NMDA receptors activated by glutamate. In the presence of Mg^{2+} (red line), Mg^{2+} block of the channel pore prevents current flow at hyperpolarized membrane potentials. If extracellular Mg^{2+} is removed (blue line), there is no block of the channel pore. (C–E) Crystallographic structure of the NMDA receptor. Each of the four subunits is indicated in a different color: GluN1 subunits are orange and yellow, GluN2 subunits cyan and purple. (C) The structure of the NMDA receptor is similar to that of the AMPA receptor, with an amino-terminal domain, ligand-binding domain, transmembrane domain, and carboxyl-terminal domain. The ligand-binding domain of GluN2A binds to glutamate (green spheres), while the ligand-binding domain of GluN1 binds to the co-agonist, glycine (green spheres). (D) Rotating the receptor 90 degrees reveals the location of the putative binding site for Ca^{2+}. (E) Close-up view of the putative Ca^{2+} binding site (red and green mesh) in the extracellular vestibule of the receptor. (F) Model for gating of NMDA receptors. Proposed movements (arrows) of the amino-terminal domain and ligand-binding domain regions of the receptor when bound to antagonists, such as DCKA and D-APV (left), or agonists (glycine and glutamate, right). (C–E from E. Karakas and H. Furukawa, 2014. *Science* 344: 992–997; F from S. Zhu et al., 2016. *Cell* 165: P704–714.)

The structure of an NMDA receptor resembles a hot-air balloon (Figure 6.8C,D). Similar to AMPA receptor subunits, NMDA receptor subunits possess clamshell-shaped ligand-binding domains that bind to glutamate and to glycine, as well as transmembrane domains that form the channel pore and gate. One unique feature of NMDA receptors is a structure in the extracellular vestibule, adjacent to the transmembrane domain, that is postulated to bind Ca^{2+} and may help confer Ca^{2+} permeability to NMDA receptors (Figure 6.8E). Gating of NMDA receptors is proposed to arise from closure of the ligand-binding domains upon binding of glutamate and glycine, leading to a conformational change that opens the channel pore; in contrast, binding of antagonists to the ligand-binding domains displaces them and prevents channel opening (Figure 6.8F). The site at which Mg^{2+} binds to block the pore of the NMDA receptor has not yet been identified.

Metabotropic glutamate receptors

In addition to the ionotropic glutamate receptors, there are three classes of metabotropic glutamate receptors (mGluRs; see Figure 6.4C), encoded by *GRM* genes. These receptors differ in their coupling to intracellular signal transduction pathways (see Chapter 7) and in their sensitivity to pharmacological agents. Unlike the excitatory ionotropic glutamate receptors, mGluRs cause slower postsynaptic responses that can either excite or inhibit postsynaptic cells. As a result, the physiological roles of mGluRs are quite varied. Activation of many mGluRs leads to inhibition of postsynaptic Ca^{2+} and Na^+ channels.

Although they possess a transmembrane domain that spans the membrane seven times, characteristic of all G-protein-coupled receptors, mGluRs are structurally unique because they are dimers of two identical subunits (Figure 6.9). Each subunit possesses a *venus flytrap domain*, a glutamate-binding domain similar to the clamshell-shaped ligand-binding domains of ionotropic glutamate receptors (see Figures 6.7 and 6.8). This venus flytrap domain is connected to the transmembrane domain via a linker domain rich in the amino acid cysteine (see Figure 6.9, left). Binding of glutamate causes the venus flytrap domains to close, with the resulting movement causing the transmembrane domains to rotate and thereby activate the receptor (see Figure 6.9, right). Binding of G-proteins to the activated receptor then initiates intracellular signaling.

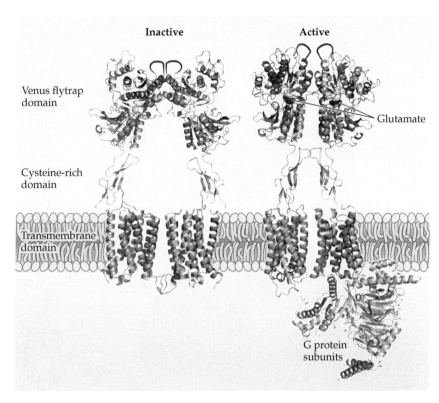

FIGURE 6.9 Structural model of metabotropic glutamate receptors (Left) Metabotropic glutamate receptors consist of a pair of identical subunits, each containing a venus flytrap domain, a cysteine-rich linker domain, and a transmembrane domain consisting of the canonical seven membrane-spanning helices. (Right) Binding of glutamate (red spheres) to the venus flytrap domains causes the transmembrane domains to rotate and bind to G-proteins, thereby activating intracellular signaling processes. (From J.-P. Pin and B. Bettler, 2016. *Nature* 540: 60–68.)

CONCEPT 6.4 | GABA and Glycine Are Inhibitory Transmitters

LEARNING OBJECTIVES

6.4.1 Know how GABA and glycine are synthesized and degraded.

6.4.2 Understand the roles that GABA and glycine play in the nervous system.

6.4.3 Know that GABA receptors are the targets of many therapeutic drugs.

Most inhibitory synapses in the brain and spinal cord use either γ-aminobutyric acid (GABA) or glycine as neurotransmitters (Figure 6.10). GABA was identified in brain tissue during the 1950s (as was glutamate). It is now known that as many as one-third of the synapses in the brain use GABA as their inhibitory neurotransmitter. GABA is most commonly found in local circuit interneurons, although medium spiny neurons of the striatum (see Chapter 18) and cerebellar Purkinje cells (see Chapter 19) are examples of GABAergic projection neurons.

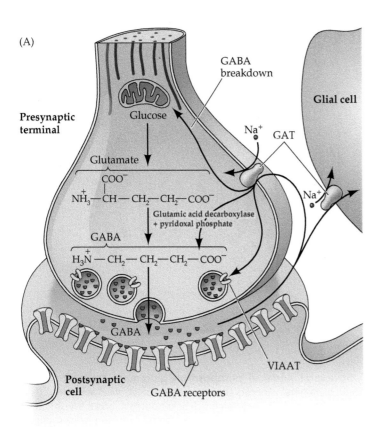

(A)

Presynaptic terminal

GABA breakdown

Glial cell

Glucose

Na⁺ GAT

Glutamate

COO^-

$NH_3^+ - CH - CH_2 - CH_2 - COO^-$

Glutamic acid decarboxylase + pyridoxal phosphate

Na⁺

GABA

$H_3N^+ - CH_2 - CH_2 - CH_2 - COO^-$

GABA

Postsynaptic cell   GABA receptors

VIAAT

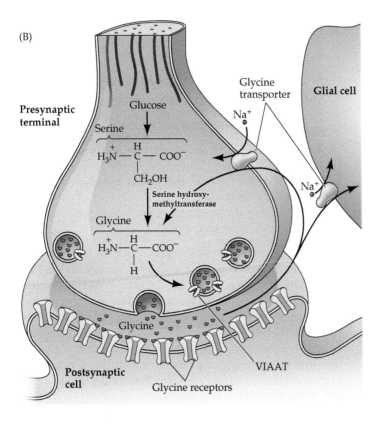

(B)

Presynaptic terminal

Glycine transporter   Glial cell

Glucose

Na⁺

Serine

$H_3N^+ - \overset{H}{\underset{CH_2OH}{C}} - COO^-$

Serine hydroxy-methyltransferase

Na⁺

Glycine

$H_3N^+ - \overset{H}{\underset{H}{C}} - COO^-$

Glycine

Postsynaptic cell

Glycine receptors

VIAAT

GABA synthesis and metabolism

The predominant precursor for GABA synthesis is glucose, which is metabolized to glutamate by the tricarboxylic acid cycle enzymes. (Pyruvate and glutamine can also act as GABA precursors.) The enzyme glutamic acid decarboxylase (GAD, encoded by *GAD1* and *GAD2* genes), which is found almost exclusively in GABAergic neurons, catalyzes the conversion of glutamate to GABA (see Figure 6.10A). GAD requires a co-factor, pyridoxal phosphate, for activity. Because pyridoxal phosphate is derived from vitamin B₆, a deficiency of this vitamin can lead to diminished GABA synthesis. The significance of this fact became clear after a disastrous series of infant deaths was linked to the omission of vitamin B₆ from infant formula. The absence of vitamin B₆ greatly reduced the GABA content of the brain, and the subsequent loss of synaptic inhibition caused seizures that in some cases were fatal. Once GABA is synthesized, it is transported into synaptic vesicles via a vesicular inhibitory amino acid transporter (VIAAT), also called the vesicular GABA transporter (VGAT; *SLC36A1* gene).

The mechanism of GABA removal is similar to that for glutamate: Both neurons and glia contain high-affinity Na⁺-dependent co-transporters for GABA. These co-transporters are termed GATs, and several forms of GAT have been identified. Most GABA is eventually converted to succinate, which is metabolized further in the tricarboxylic acid cycle that mediates cellular ATP synthesis. Two mitochondrial enzymes are required for this degradation: GABA transaminase and succinic semialdehyde dehydrogenase. There are also other pathways for degradation of GABA, the most noteworthy of which results in the production of γ-hydroxybutyrate, a GABA derivative that has been abused as a "date rape" drug. Oral administration of γ-hydroxybutyrate can cause euphoria, memory deficits, and unconsciousness. Presumably these effects arise from actions on GABAergic synapses in the CNS: Inhibition of GABA breakdown causes a rise in tissue GABA content and an increase in inhibitory synaptic activity.

Ionotropic GABA receptors

GABAergic synapses employ two types of postsynaptic receptors, called GABA_A and GABA_B. GABA_A are ionotropic receptors, while GABA_B are metabotropic

FIGURE 6.10 Synthesis, release, and reuptake of the inhibitory neurotransmitters GABA and glycine (A) GABA is synthesized from glutamate by the enzyme glutamic acid decarboxylase, which requires pyridoxal phosphate. (B) Glycine can be synthesized by several metabolic pathways; in the brain, the major precursor is serine. High-affinity transporters terminate the actions of these transmitters and return GABA or glycine to the synaptic terminals for reuse, with both transmitters being loaded into synaptic vesicles via the vesicular inhibitory amino acid transporter (VIAAT).

receptors. The ionotropic GABA$_A$ receptors are GABA-gated anion channels, with Cl$^-$ being the main permeant ion under physiological conditions. The reversal potential for Cl$^-$ usually is more negative than the threshold for action potential firing (see Figure 5.19) due to the action of the K$^+$/Cl$^-$ co-transporter (see Figure 4.15D), which keeps intracellular Cl$^-$ concentration low. Thus, activation of these GABA receptors causes an influx of negatively charged Cl$^-$ that inhibits postsynaptic cells (Figure 6.11A). In cases where postsynaptic Cl$^-$ concentration is high—in developing neurons, for example—GABA$_A$ receptors can excite their postsynaptic targets (Box 6B).

Like nACh receptors, GABA$_A$ receptors are pentamers (Figure 6.11B). There are 19 types of GABA$_A$ subunits (see Figure 6.3F); this diversity of subunits causes the composition and function of GABA$_A$ receptors to differ widely among neuronal types. Typically, GABA$_A$ receptors consist of two α subunits (encoded by *GABRA* genes), two β subunits (*GABRB* genes), and one other subunit, most often a γ subunit (*GABRG* genes). A specialized type of GABA$_A$

receptor, found exclusively in the retina, consists entirely of ρ subunits (*GABRR* genes) and is called the GABA$_{Aρ}$ receptor (formerly the GABA$_C$ receptor). The five GABA$_A$ receptor subunits are assembled into a structure quite similar to that of the nAChR (see Figure 6.11B). The transmembrane domains of the subunits form a central pore that includes a ring of positive charges that presumably serve as the binding site for Cl$^-$ (Figure 6.11C). GABA binds in pockets found at the interface between the extracellular domains of the subunits; many other types of ligands also bind to these sites (Figure 6.11D). Benzodiazepines such as diazepam (Valium) and chlordiazepoxide (Librium) are anxiety-reducing drugs that enhance GABAergic transmission by binding to the extracellular domains of α and δ subunits of GABA$_A$ receptors. The same site binds the hypnotic zolpidem (Ambien), which is widely used to induce sleep. Barbiturates such as phenobarbital and pentobarbital are other hypnotics that also bind to the extracellular domains of the α and β subunits of some GABA receptors and potentiate GABAergic transmission; these drugs are used

(A)

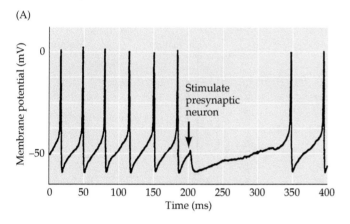

FIGURE 6.11 Ionotropic GABA receptors (A) Stimulation of a presynaptic GABAergic interneuron, at the time indicated by the arrow, causes a transient inhibition of action potential firing in the postsynaptic target. This inhibitory response is caused by activation of postsynaptic GABA$_A$ receptors. (B–D) Crystallographic structure of a GABA$_A$ receptor. (B) The receptor is formed from five subunits, each containing an extracellular domain and a transmembrane domain. One subunit is highlighted in darker blue. (C) This extracellular perspective shows the pore formed by the transmembrane domains of the receptor subunits. (D) View of two receptor subunits, indicating the binding sites for numerous ligands. Here the GABA binding site is occupied by benzamidine, a GABA$_A$ receptor agonist. (A after J. Chavas and A. Marty, 2003. *J Neurosci* 23: 2019–2031; © 2003 Society for Neuroscience; B,C from P. S. Miller and A. R. Aricescu, 2014. *Nature* 512: 270–275; D from R. Puthenkalam et al., 2016. *Front Mol Neurosci* 9: 44, CC BY 4.0.)

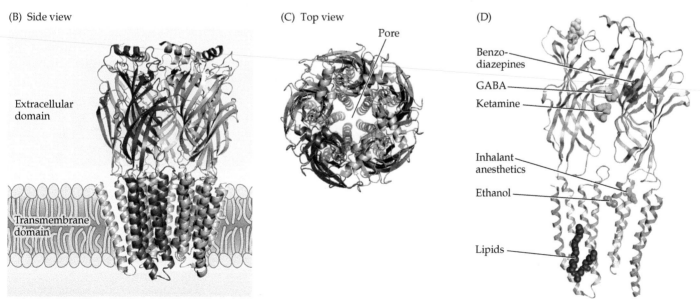

■ BOX 6B | Excitatory Actions of GABA in the Developing Brain

Although GABA normally functions as an inhibitory neurotransmitter in the mature brain, in the developing brain GABA excites its target cells. This remarkable reversal of action arises from developmental changes in intracellular Cl⁻ homeostasis. The mechanisms involved in this switch have been studied most extensively in cortical neurons. In young neurons, intracellular Cl⁻ concentration is controlled mainly by the Na⁺/K⁺/Cl⁻ co-transporter, which pumps Cl⁻ into the neurons and yields a high $[Cl^-]_i$ (Figure A, left). As the neurons continue to develop, they begin to express a K⁺/Cl⁻ co-transporter that pumps Cl⁻ out of the neurons and lowers $[Cl^-]_i$ (see Figure A, right). Such shifts in Cl⁻ homeostasis can cause $[Cl^-]_i$ to drop several-fold over the first 1 to 2 postnatal weeks of development (Figure B).

Because ionotropic GABA receptors are Cl⁻-permeable channels, ion flux through these receptors varies according to the electrochemical driving force on Cl⁻. In the young neurons, where $[Cl^-]_i$ is high, E_{Cl} is more positive than the resting potential. As a result, GABA depolarizes these neurons. In addition, E_{Cl} often is more positive than threshold, so GABA is able to excite these neurons to fire action potentials (Figure C). As described in Concept 6.4, the lower $[Cl^-]_i$ of mature neurons causes E_{Cl} to be more negative than the action potential threshold (and often more negative than the resting potential), resulting in inhibitory responses to GABA.

Why does GABA undergo such a switch in its postsynaptic actions? While the logic of this phenomenon is not yet completely clear, it appears that depolarizing GABA responses produce electrical activity that controls neuronal proliferation, migration, growth, and maturation, as well as determining synaptic connectivity. Once these developmental processes are completed, the resulting neural circuitry requires inhibitory transmission that can then also be provided by GABA. Further work will be needed to fully appreciate the significance of the excitatory actions of GABA, as well as to understand the mechanisms underlying the expression of the K⁺/Cl⁻ co-transporter that ends the brief career of GABA as an excitatory neurotransmitter.

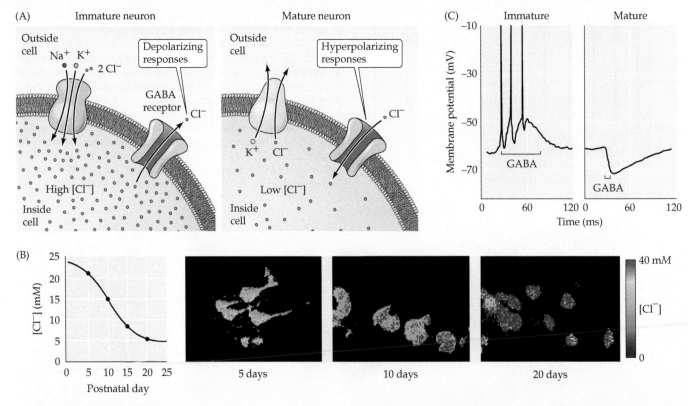

(A) The developmental switch in expression of Cl⁻ transporters lowers $[Cl^-]_i$, thereby reversing direction of Cl⁻ flux through GABA receptors. (B) Imaging $[Cl^-]_i$ between postnatal days 5 and 20 (right) demonstrates a progressive reduction in $[Cl^-]_i$ (left). (C) Developmental changes in $[Cl^-]_i$ cause GABA responses to shift from depolarizing in young (6-day-old) neurons (left) to hyperpolarizing in older (10-day-old) neurons (right) cultured from the chick spinal cord. (B from K. Berglund et al., 2006. *Brain Cell Biol* 35: 207–228; C after K. Obata et al., 1978. *Brain Res* 144: 179–184.)

therapeutically for anesthesia and to control epilepsy. The injection anesthetic ketamine also binds to the extracellular domain of GABA receptors (see Figure 6.11D). The transmembrane domains of GABA$_A$ receptors also serve as the targets for numerous ligands, such as inhalant anesthetics and steroids. Another drug that binds to the transmembrane domain of GABA receptors is ethanol; at least some aspects of drunken behavior are caused by ethanol-mediated alterations in ionotropic GABA receptors.

Metabotropic GABA receptors

The metabotropic GABA$_B$ receptors (encoded by *GABBR* genes) are also widely distributed in the brain. Like the ionotropic GABA receptors, GABA$_B$ receptors are inhibitory. Rather than relying on Cl$^-$-selective channels, however, GABA$_B$-mediated inhibition is often due to the activation of K$^+$ channels. A second action of GABA$_B$ receptors is to block Ca^{2+} channels, which also inhibits postsynaptic cells. The structure of GABA$_B$ receptors is similar to that of other metabotropic receptors, although GABA$_B$ receptors assemble as heterodimers of B1 and B2 subunits (Figure 6.12). Like the mGluRs, GABA$_B$ receptors possess venus flytrap domains (see Figure 6.12, left), but these bind GABA rather than glutamate. Binding of GABA to the venus flytrap domain of the B1 subunit causes this domain to close, leading to conformational changes in the transmembrane domains of both subunits that permit binding of G-proteins (see Figure 6.12, right).

Glycine synthesis and metabolism

The distribution of the neutral amino acid glycine in the CNS is more restricted than that of GABA. About half of the inhibitory synapses in the spinal cord use glycine; most other inhibitory synapses use GABA. Glycine is synthesized from serine by the mitochondrial isoform of serine hydroxymethyltransferase (see Figure 6.10B) and is transported into synaptic vesicles via the same vesicular inhibitory amino acid transporter that loads GABA into vesicles. Once released from the presynaptic cell, glycine is rapidly removed from the synaptic cleft by glycine transporters in the plasma membrane. Mutations in the genes coding for some of these transporters result in hyperglycinemia, a devastating neonatal disease characterized by lethargy, seizures, and mental retardation.

Ionotropic glycine receptors

Glycine receptors are exclusively ionotropic and are pentamers consisting of mixtures of four types of α subunits along with an accessory β subunit (encoded by *GLRA* and *GLRB* genes, respectively) (see Figure 6.3F). These receptors are potently blocked by strychnine, which may account for the toxic properties of this plant alkaloid (see Box 6A). Glycine receptors are ligand-gated Cl$^-$ channels whose general structure closely mirrors that of the GABA$_A$ receptors (Figure 6.13). Gating of glycine receptors by ligands is well understood. Binding of glycine to a ligand-binding site on the extracellular domains causes a conformational change that opens the pore, increasing the pore radius from 1.4 Å (smaller than a Cl$^-$ ion, which has a radius of 1.8 Å) to 4.4 Å, thereby enabling Cl$^-$ and other permeant anions to flow through the pore formed by the transmembrane domains of the five subunits (see Figure 6.13, right). This Cl$^-$ flux inhibits the postsynaptic neuron. Blocking these receptors, by binding of strychnine to the same ligand-binding site, closes the pore (see Figure 6.13, left).

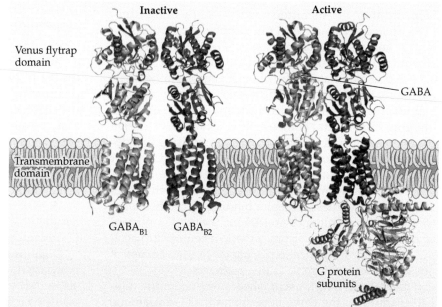

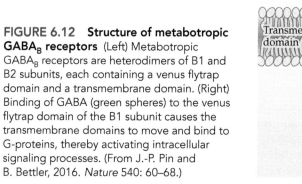

FIGURE 6.12 **Structure of metabotropic GABA$_B$ receptors** (Left) Metabotropic GABA$_B$ receptors are heterodimers of B1 and B2 subunits, each containing a venus flytrap domain and a transmembrane domain. (Right) Binding of GABA (green spheres) to the venus flytrap domain of the B1 subunit causes the transmembrane domains to move and bind to G-proteins, thereby activating intracellular signaling processes. (From J.-P. Pin and B. Bettler, 2016. *Nature* 540: 60–68.)

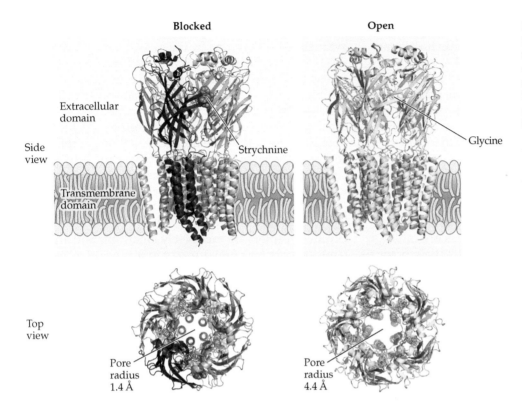

Blocked   Open

Side view

Extracellular domain

Strychnine

Glycine

Transmembrane domain

Top view

Pore radius 1.4 Å

Pore radius 4.4 Å

FIGURE 6.13 Gating of glycine receptors Similar to GABA$_A$ receptors, glycine receptors are pentamers consisting of five subunits. Each subunit (one of which is highlighted in darker blue) consists of an extracellular domain and a pore-forming transmembrane domain. (Left) Binding of strychnine (orange) to a ligand-binding site on the extracellular domain closes the channel pore. (Right) Binding of glycine to the same ligand-binding site causes a conformational change that opens the pore. (From J. Du et al., 2015. *Nature* 526: 224–229.)

CONCEPT **6.5**

Biogenic Amines Are Neuromodulatory Transmitters

LEARNING OBJECTIVES

6.5.1 Be able to identify biogenic amine neurotransmitters.

6.5.2 Know how biogenic amine neurotransmitters are synthesized and degraded.

6.5.3 Understand the roles that biogenic amine transmitters play in the nervous system.

Biogenic amine transmitters regulate many brain functions and are also active in the peripheral nervous system. In contrast to ACh, glutamate, GABA, and glycine, biogenic amine transmitters usually have slow, diffuse actions that modulate the function of neurons rather than mediating rapid synaptic transmission. Because biogenic amines are implicated in such a wide variety of behaviors (ranging from central homeostatic functions to cognitive phenomena such as attention), it is not surprising that defects in biogenic amine function are implicated in most psychiatric disorders. The pharmacology of amine synapses is critically important in psychotherapy, with drugs affecting the synthesis, receptor binding, or catabolism of these neurotransmitters being among the most important agents in the armamentarium of modern neuropharmacology (see Clinical Applications, Chapter 7). Many drugs of abuse also act on biogenic amine pathways.

There are five well-established biogenic amine neurotransmitters: the three **catecholamines—dopamine,**

norepinephrine (noradrenaline), and **epinephrine (adrenaline)** —and **histamine** and **serotonin** (see Figure 6.1). All the catecholamines (so named because they share the catechol moiety) are derived from a common precursor, the amino acid tyrosine (Figure 6.14). The first step in catecholamine synthesis is catalyzed by tyrosine hydroxylase in a reaction requiring oxygen as a co-substrate and tetrahydrobiopterin as a co-factor to synthesize dihydroxyphenylalanine (DOPA). Histamine and serotonin are synthesized via other routes, as described later in this concept.

Dopamine

Dopamine is present in several brain regions (Figure 6.15A), although the major dopamine-containing area of the brain is the corpus striatum, which receives major input from the substantia nigra and plays an essential role in the coordination of body movements. In Parkinson's disease, the dopaminergic neurons of the substantia nigra degenerate, leading to a characteristic motor dysfunction (see Chapter 18). Dopamine produced in the ventral tegmental area is also believed to be involved in motivation, reward, and reinforcement (see Chapter 32); many drugs of abuse work by affecting dopaminergic circuitry in the CNS. In addition to these roles in the CNS, dopamine also plays a poorly understood role in some sympathetic ganglia.

Dopamine is produced by the action of DOPA decarboxylase on DOPA (see Figure 6.14). Following its synthesis in the cytoplasm of presynaptic terminals, dopamine is loaded into synaptic vesicles via a vesicular monoamine

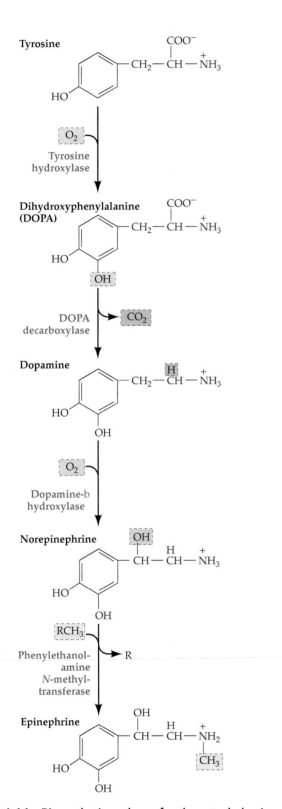

FIGURE 6.14 Biosynthetic pathway for the catecholamine neurotransmitters The amino acid tyrosine is the precursor for all three catecholamines. The first step in this reaction pathway, catalyzed by tyrosine hydroxylase, is rate-limiting.

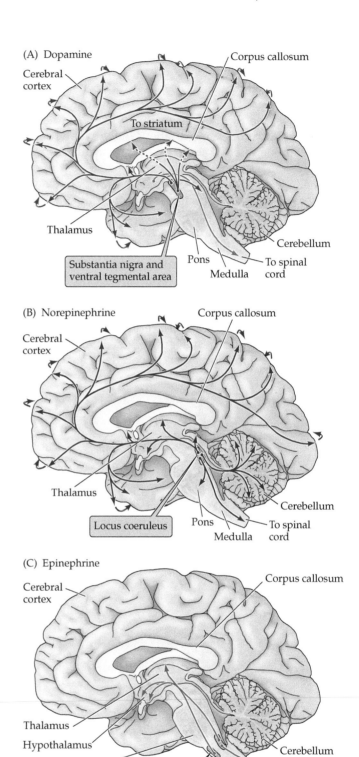

FIGURE 6.15   Distribution of catecholamine neurotransmitters in the human brain Shown are neurons and their projections (arrows) that contain catecholamine neurotransmitters. Curved arrows along the perimeter of the cortex indicate the innervation of lateral cortical regions not shown in this midsagittal plane of section.

transporter (VMAT). Dopamine action in the synaptic cleft is terminated by reuptake of dopamine into nerve terminals or surrounding glial cells by a Na⁺-dependent dopamine co-transporter, termed DAT. Cocaine apparently produces its psychotropic effects by inhibiting DAT, thereby increasing dopamine concentrations in the synaptic cleft. Amphetamine, another addictive drug, also inhibits DAT, as well as the transporter for norepinephrine (see the next section in this concept). The two major enzymes involved in the catabolism of dopamine are monoamine oxidase (MAO) and catechol O-methyltransferase (COMT). Both neurons and glia contain mitochondrial MAO and cytoplasmic COMT. Inhibitors of these enzymes, such as phenelzine and tranylcypromine, are used clinically as antidepressants (see Clinical Applications, Chapter 7).

Once released, dopamine acts exclusively by activating metabotropic receptors encoded by five dopamine receptor genes (*DRDs*). One of these receptors, the D3 dopamine receptor, is shown in Figure 6.16A. The monomeric structure of this receptor closely parallels that of other metabotropic receptors, such as the mACh receptor (see Figure 6.4A), except that its ligand-binding site is optimized for binding to dopamine. Most dopamine receptor subtypes (see Figure 6.4C) act by either activating or inhibiting adenylyl cyclase (see Chapter 7). Activation of these receptors generally contributes to complex behaviors; for example, administration of dopamine receptor agonists causes hyperactivity and repetitive, stereotyped behavior in laboratory animals. Activation of another type of dopamine receptor in the medulla inhibits vomiting. Thus, antagonists of these receptors are used as emetics to induce vomiting after poisoning or a drug overdose. Dopamine receptor antagonists can also elicit catalepsy, a state in which it is difficult to initiate voluntary motor movement, suggesting a basis for this aspect of some psychoses.

Norepinephrine

Norepinephrine (also called noradrenaline) is synthesized by neurons within the locus coeruleus, a brainstem nucleus that projects diffusely to a variety of forebrain targets (Figure 6.15B). These projections allow norepinephrine to modulate the activity of neurons throughout the brain, influencing sleep and wakefulness, arousal, attention, and feeding behavior. Perhaps the most prominent noradrenergic neurons are sympathetic ganglion cells, which employ norepinephrine as the major peripheral transmitter in this division of the visceral motor system (see Chapter 21).

Norepinephrine synthesis requires dopamine β-hydroxylase (encoded by the *DBH* gene), which catalyzes the production of norepinephrine from dopamine (see Figure 6.14). Norepinephrine is then loaded into synaptic vesicles via the same VMAT involved in vesicular dopamine transport. Norepinephrine is cleared from the synaptic cleft by the norepinephrine transporter (NET), a Na⁺-dependent co-transporter that also is capable of taking up dopamine. As mentioned earlier, NET is a molecular target of amphetamine, which acts as a stimulant by producing a net increase in the release of norepinephrine and dopamine. A mutation in the *NET* gene is a cause of orthostatic intolerance, a disorder that produces lightheadedness while standing up. Like dopamine, norepinephrine is degraded by MAO and COMT.

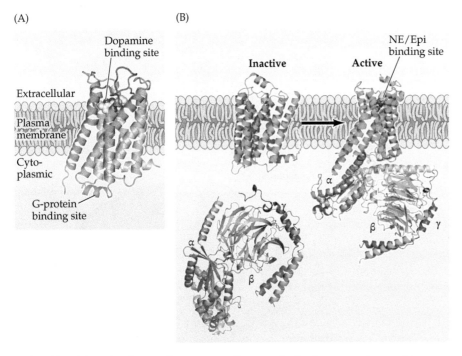

FIGURE 6.16 Metabotropic receptors for catecholamine neurotransmitters (A) Structure of the D3 dopamine receptor. Like all metabotropic receptors, the D3 receptor spans the plasma membrane seven times and has a cytoplasmic domain that binds to and activates G-proteins, as well as an extracellular domain that binds dopamine. (B) Structure of the β₂-adrenergic receptor and its associated G-protein. (Left) In the absence of ligand, the cytoplasmic domain of the β₂ receptor is not bound to the G-protein (α, β, and γ subunits). (Right) Binding of ligand (β agonist, indicated by colored spheres) to the extracellular binding site for norepinephrine (NE) and epinephrine (Epi) causes the β₂ receptor to bind to the α subunit of the G-protein, which in turn induces a dramatic change in the structure of this subunit. (A from E. Y. T. Chien et al., 2010. *Science* 330: 1091–1095; B after K. M. Betke et al., 2012. *Prog Neurobiol* 96: 304–321 and S. G. F. Rasmussen et al., 2011. *Nature* 477: 549–555.)

Both norepinephrine and epinephrine act on α- and β-adrenergic receptors (see Figure 6.4C). Both types of receptors are G-protein-coupled; in fact, work done by Robert Lefkowitz at Duke University established the β-adrenergic receptor (encoded by *ADRB* genes) as the first metabotropic neurotransmitter receptor. As shown in Figure 6.16B, the structure of this receptor is very similar to that of other metabotropic receptors (such as the dopamine receptor in Figure 6.16A). Binding of norepinephrine or epinephrine causes small changes in the structure of this receptor, which permits the G-protein to bind (see Figure 6.16B, right). This, in turn, causes larger changes in the shape of the α subunit of the G-protein, the first step in a series of reactions that allow the G-protein to regulate intracellular signaling cascades (see Chapter 7).

There are three subtypes of β-adrenergic receptors, two of which are expressed in many types of neurons. Agonists and antagonists of adrenergic receptors, such as the β-blocker propranolol (Inderol), are used clinically for a variety of conditions ranging from cardiac arrhythmias to migraine headaches. However, most of the actions of these drugs are on β-adrenergic receptors on muscle, particularly in the cardiovascular and respiratory systems (see Chapter 21).

Two subclasses of α-adrenergic receptors have been identified and are encoded by *ADRA* genes. Activation of α_1 receptors usually results in a slow depolarization linked to the inhibition of K^+ channels, while activation of α_2 receptors produces a slow hyperpolarization due to the activation of a different type of K^+ channel.

Epinephrine

Epinephrine (also called adrenaline) is found in the brain at lower levels than the other catecholamines and also is present in fewer brain neurons than other catecholamines. Epinephrine-containing neurons in the CNS are primarily in the lateral tegmental system and in the medulla and project to the hypothalamus and thalamus (Figure 6.15C). These epinephrine-secreting neurons regulate respiration and cardiac function.

The enzyme that synthesizes epinephrine, phenylethanolamine-*N*-methyltransferase (encoded by *PNMT* gene) (see Figure 6.14), is present only in epinephrine-secreting neurons. Otherwise, the metabolism of epinephrine is very similar to that of norepinephrine. Epinephrine is loaded into vesicles via the VMAT. No plasma membrane transporter specific for epinephrine has been identified, although the NET is capable of transporting epinephrine. As already noted, epinephrine acts on both α- and β-adrenergic receptors.

Histamine

Histamine is found in neurons in the hypothalamus that send sparse but widespread projections to almost all regions of the brain and spinal cord (Figure 6.17A). The central

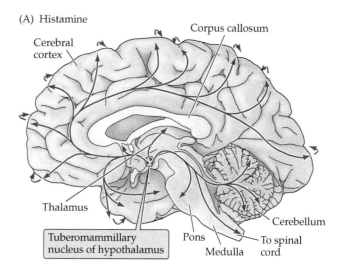

(A) Histamine

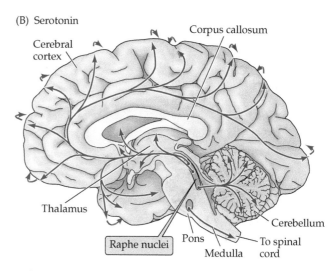

(B) Serotonin

FIGURE 6.17   Distribution of histamine and serotonin in the human brain Diagrams show the distribution of neurons and their projections (arrows) containing histamine (A) or serotonin (B). Curved arrows along the perimeter of the cortex indicate the innervation of lateral cortical regions not shown in this midsagittal plane of section.

histamine projections mediate arousal and attention, similar to central ACh and norepinephrine projections. Histamine also controls the reactivity of the vestibular system. Allergic reactions or tissue damage cause release of histamine from mast cells in the bloodstream. The close proximity of mast cells to blood vessels, together with the potent actions of histamine on blood vessels, raises the possibility that histamine may influence brain blood flow.

Histamine is produced from the amino acid histidine by a histidine decarboxylase (encoded by the *HDC* gene) (Figure 6.18A) and is transported into vesicles via the same VMAT as the catecholamines. Released histamine may be taken up by a plasma membrane monoamine transporter.

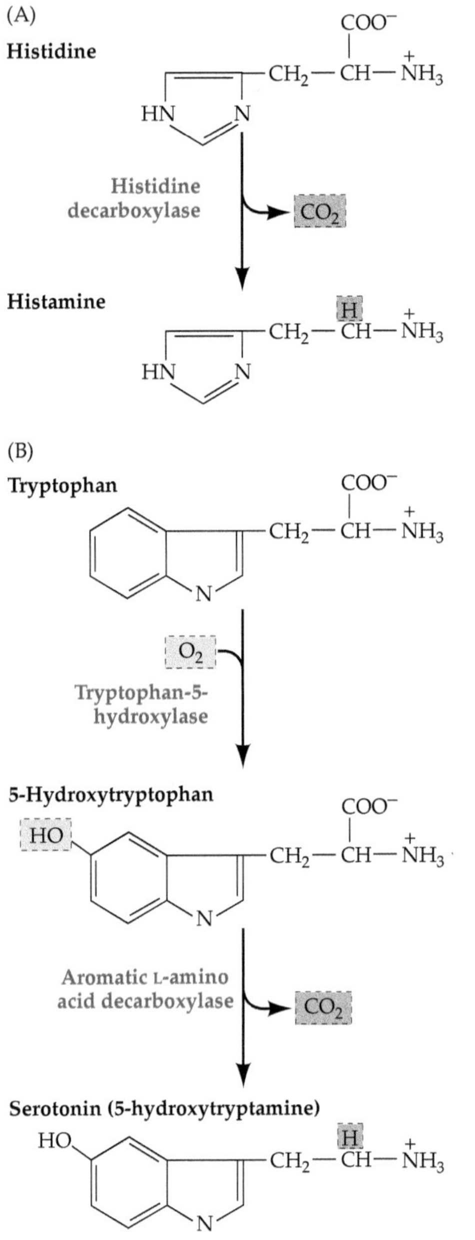

(A)

Histidine

Histidine
decarboxylase

CO_2

Histamine

(B)

Tryptophan

O_2

Tryptophan-5-
hydroxylase

5-Hydroxytryptophan

Aromatic L-amino
acid decarboxylase

CO_2

Serotonin (5-hydroxytryptamine)

FIGURE 6.18 Synthesis of histamine and serotonin
(A) Histamine is synthesized from the amino acid histidine.
(B) Serotonin is derived from the amino acid tryptophan by a two-step process that requires the enzymes tryptophan-5-hydroxylase and a decarboxylase.

Histamine is degraded by the combined actions of histamine methyltransferase and MAO.

The four known types of histamine receptors (encoded by *HRH* genes) are all metabotropic receptors (see Figure 6.4C). Because of the role of histamine receptors in mediating allergic responses, many histamine receptor antagonists have been developed as antihistamine agents. Antihistamines that cross the blood-brain barrier, such as diphenhydramine (Benadryl), act as sedatives by interfering with the roles of histamine in CNS arousal. Antagonists of the H_1 receptor also are used to prevent motion sickness, perhaps because of the role of histamine in controlling vestibular function. H_2 receptors control the secretion of gastric acid in the digestive system, allowing H_2 receptor antagonists to be used in treating a variety of upper gastrointestinal disorders (e.g., peptic ulcers).

Serotonin

Serotonin, or 5-hydroxytryptamine (5-HT), was initially thought to increase vascular tone by virtue of its presence in blood serum (hence the name serotonin). Serotonin is found primarily in groups of neurons in the raphe region of the pons and upper brainstem, which have widespread projections to the forebrain (Figure 6.17B) and regulate sleep and wakefulness (see Chapter 28). 5-HT occupies a place of prominence in neuropharmacology because a large number of antipsychotic drugs that are valuable in the treatment of depression and anxiety act on serotonergic pathways (see Clinical Applications, Chapter 7).

5-HT is synthesized from the amino acid tryptophan, which is an essential dietary requirement. Tryptophan is taken up into neurons by a plasma membrane transporter and hydroxylated in a reaction catalyzed by the enzyme tryptophan-5-hydroxylase (encoded by *TPH* genes) (Figure 6.18B), the rate-limiting step for 5-HT synthesis. Loading of 5-HT into synaptic vesicles is done by the VMAT that is also responsible for loading other monoamines into synaptic vesicles. The synaptic effects of serotonin are terminated by transport back into nerve terminals via a specific serotonin transporter (SERT) that is present in the presynaptic plasma membrane and is encoded by the *SLC6A4* (or *5HTT*) gene. Many antidepressant drugs are **selective serotonin reuptake inhibitors (SSRIs)** that inhibit transport of 5-HT by SERT. Perhaps the best-known example of an SSRI is the antidepressant drug Prozac (see Clinical Applications, Chapter 7). The primary catabolic pathway for 5-HT is mediated by MAO.

A large number of 5-HT receptors (encoded by *HTR* genes) have been identified. Most 5-HT receptors are metabotropic (see Figure 6.4C), with a monomeric structure typical of G-protein-coupled receptors (Figure 6.19A). Metabotropic 5-HT receptors have been implicated in a wide range of behaviors, including circadian rhythms, motor behaviors, emotional states, and state of mental arousal. Impairments in the function of these receptors have been implicated in numerous psychiatric disorders, such as depression, anxiety disorders, and schizophrenia (see Chapter 31), and drugs acting on serotonin receptors are effective treatments for several of these conditions. The psychedelic drug LSD (lysergic acid diethylamide) presumably causes hallucinations by activating multiple types of metabotropic 5-HT receptors (see Figure 6.19A). Activation of 5-HT receptors also mediates satiety and

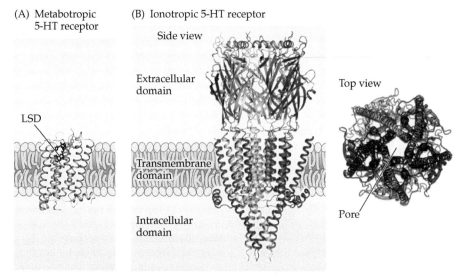

(A) Metabotropic 5-HT receptor

(B) Ionotropic 5-HT receptor

Side view

Extracellular domain

LSD

Transmembrane domain

Intracellular domain

Top view

Pore

FIGURE 6.19 Serotonin receptors (A) Structure of the human 5-HT$_{2B}$ receptor, a metabotropic 5-HT receptor. The pink structure indicates LSD bound to the 5-HT binding site of the receptor. (B) Structure of the human 5-HT$_3$ receptor, an ionotropic 5-HT receptor consisting of five subunits (each in a different color), each of which has an extracellular domain, transmembrane domain, and intracellular domain. An ion channel is formed by the transmembrane domains of the five subunits (right). (A from D. Wacker et al., 2017. *Cell* 168: P377–389; B from G. Hassaine et al., 2014. *Nature* 512: 276–281.)

decreased food consumption, which is why serotonergic drugs are sometimes useful in treating eating disorders.

One group of serotonin receptors, the 5-HT$_3$ receptors, are ligand-gated ion channels formed from combinations of the five 5-HT$_3$ subunits (see Figure 6.3F). Their pentameric structure is very similar to that of other ionotropic receptors, with functional channels formed by the transmembrane domains of the five subunits (Figure 6.19B). 5-HT$_3$ receptors are nonselective cation channels and therefore mediate excitatory postsynaptic responses. Ligand-binding sites reside within the extracellular domains of these receptors and serve as targets for a wide variety of therapeutic drugs, including ondansetron (Zofran) and granisetron (Kytril), which are used to prevent postoperative nausea and chemotherapy-induced emesis.

<table>
<tr><td>CONCEPT
6.6</td><td>### ATP and Other Purines Can Serve as Co-Transmitters</td></tr>
</table>

LEARNING OBJECTIVES

6.6.1 Be able to identify purinergic neurotransmitters.

6.6.2 Know how purinergic neurotransmitters are synthesized, released, and degraded.

6.6.3 Understand the roles that purinergic transmitters play in the nervous system.

All synaptic vesicles contain ATP, which is co-released with one or more "classic" neurotransmitters. This observation

raises the possibility that ATP acts as a co-transmitter. It has been known since the 1920s that the extracellular application of ATP (or its breakdown products AMP and adenosine) can elicit electrical responses in neurons. The idea that some purines (so named because all these compounds contain a purine ring; see Figure 6.1) are also neurotransmitters is now well established. ATP acts as an excitatory neurotransmitter in motor neurons of the spinal cord, as well as in sensory and autonomic ganglia. Postsynaptic actions of ATP have also been demonstrated in the CNS, specifically for dorsal horn neurons and in a subset of hippocampal neurons. Extracellular enzymes degrade released ATP to adenosine, which has its own set of signaling actions. Thus, adenosine cannot be considered a classic neurotransmitter because it is not stored in synaptic vesicles or released in a Ca^{2+}-dependent manner. Several enzymes, including apyrase, ecto-5′ nucleotidase, and nucleoside transporters, are involved in the rapid catabolism and removal of purines from extracellular locations.

Ionotropic purinergic receptors

Receptors for both ATP and adenosine are widely distributed in the nervous system as well as in many other tissues. Three classes of these purinergic receptors are known. One class consists of ionotropic receptors called **P2X receptors** (encoded by *P2RX* genes) (see Figure 6.3F). The structure of these receptors is unique among ionotropic receptors because each subunit has a transmembrane domain that crosses the membrane only twice (Figure 6.20A). Furthermore, only three of these subunits are required to form a trimeric receptor (Figure 6.20B). As in all ionotropic receptors, a pore is located in the center of the P2X receptor (Figure 6.20C) and forms a nonselective cation channel. Thus, P2X receptors mediate excitatory postsynaptic responses. Ionotropic purinergic receptors are widely distributed in central and peripheral neurons. In sensory nerves, they evidently play a role in mechanosensation and pain; their function in most other cells, however, is not known.

Metabotropic purinergic receptors

The other two classes of purinergic receptors (encoded by *P2Y* genes) are G-protein-coupled metabotropic receptors (see Figure 6.4C). The two classes differ in their sensitivity to agonists—one type is preferentially stimulated by adenosine, whereas the other is preferentially activated by ATP. An example of the former, the A$_{2A}$ adenosine

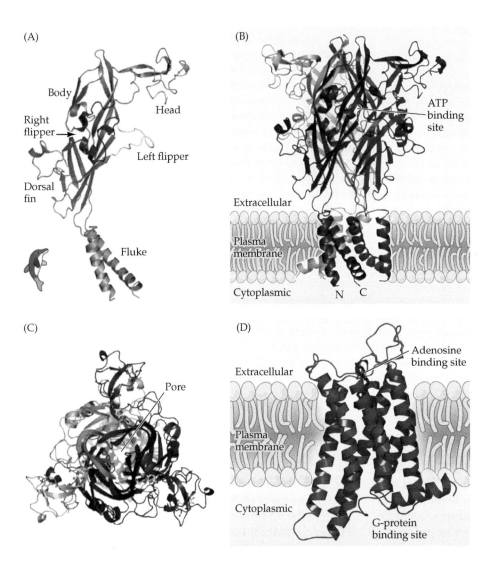

(A)

Body

Head

Right flipper

Left flipper

Dorsal fin

Fluke

(B)

ATP binding site

Extracellular

Plasma membrane

Cytoplasmic   N   C

(C)

Pore

(D)

Adenosine binding site

Extracellular

Plasma membrane

Cytoplasmic

G-protein binding site

FIGURE 6.20 Purinergic receptors
(A) Subunit of anionotropic P2X$_4$ receptor. Each subunit has a transmembrane domain consisting of two helical structures that form part of a channel, as well as a large extracellular domain that includes the ATP binding site. The shape of the subunit is reminiscent of a dolphin, with the structures color-coded as indicated in the inset. (B) Side view of a P2X$_4$ receptor; this receptor is a trimer of three subunits, with each subunit shown in a different color. The ATP binding site is proposed to be in the center of the extracellular domain. (C) Top view of the P2X$_4$ receptor, indicating the centrally located channel pore. (D) Structure of a metabotropic A$_{2A}$ adenosine receptor. This receptor has the seven-membrane-spanning domain structure characteristic of metabotropic receptors and is shown with an antagonist drug (purple structure) occupying the adenosine binding site. (A–C from T. Kawate et al., 2009. *Nature* 460: 592–598; D from V.-P. Jaakola and A. P. Ijzerman, 2010. *Curr Opin Struct Biol* 20: 401–414.)

receptor, is shown in Figure 6.20D. Both receptor types are found throughout the brain, as well as in peripheral tissues such as the heart, adipose tissue, and the kidney. Xanthines such as caffeine and theophylline block adenosine receptors, and this activity is thought to be responsible for the stimulant effects of these agents.

CONCEPT 6.7

Neuropeptides Are Exceptionally Diverse Neurotransmitters

LEARNING OBJECTIVES

6.7.1 Understand how neuropeptides are synthesized, stored, released, and degraded.

6.7.2 Know the different categories of neuropeptides.

6.7.3 Understand the different types of opioid peptides.

Many peptides known to be hormones also act as neurotransmitters and thus are also called neuropeptides.

Some neuropeptides have been implicated in modulating emotions. Others, such as substance P and the opioid peptides, are involved in the perception of pain (see Chapter 13). Still other neuropeptides, such as melanocyte-stimulating hormones, adrenocorticotropin, and β-endorphin, regulate complex responses to stress.

Neuropeptide synthesis and packaging

The mechanisms responsible for the synthesis and packaging of peptide transmitters are fundamentally different from those used for the small-molecule neurotransmitters and are much like those used for the synthesis of proteins that are secreted from non-neuronal cells (pancreatic enzymes, for instance). Peptide-secreting neurons generally synthesize polypeptides that are much larger than the final, "mature" peptide. Processing these polypeptides, which are called **pre-propeptides** (or pre-proproteins), takes place within the neuron's cell body by a sequence of reactions that occur in several intracellular organelles. Pre-propeptides are synthesized in the rough endoplasmic reticulum, where the signal sequence—that is, the sequence of amino acids indicating that the peptide is to be secreted—is removed. The remaining polypeptide, called a **propeptide** (or pro-protein), then traverses the Golgi apparatus and is packaged into vesicles in the *trans*-Golgi network. The final stages of peptide neurotransmitter processing occur after packaging into vesicles and involve proteolytic cleavage, modification of the ends of the peptide, glycosylation, phosphorylation, and disulfide bond formation.

Propeptide precursors are typically larger than their active peptide products and can give rise to more than one species of neuropeptide (Figure 6.21), which means that multiple neuroactive peptides can be released from a single vesicle. In addition, neuropeptides often are co-released with small-molecule neurotransmitters. Thus, peptidergic synapses often elicit complex postsynaptic responses. Peptides are catabolized into inactive amino acid fragments by enzymes called peptidases, usually located on the extracellular surface of the plasma membrane.

The biological activity of neuropeptides depends on their amino acid sequence (Figure 6.22). Based on their sequences, neuropeptides have been loosely grouped into five categories: the brain–gut peptides; opioid peptides; pituitary peptides; hypothalamic releasing hormones; and a catch-all category containing other, not easily classified, peptides.

Substance P, a brain–gut peptide

The study of neuropeptides began more than 90 years ago with the accidental discovery of **substance P** (see Figure 6.22A), a powerful hypotensive agent and an example of a brain–gut peptide. (The peculiar name derives from the fact that this molecule was an unidentified component of *p*owder extracts from brain and intestine.) Substance P is an 11-amino–acid peptide present in high concentrations in the human hippocampus and neocortex and also in the gastrointestinal tract—hence its classification as a brain–gut peptide. It is also released from C fibers (see Tables 3.1 and 12.1), the small-diameter afferents in peripheral nerves that convey information about pain and temperature (as well as postganglionic autonomic signals). Substance P is a sensory neurotransmitter in the spinal cord, where its release can be inhibited by opioid peptides released from spinal cord interneurons, resulting in the suppression of pain (see Chapter 13). The diversity of neuropeptides is highlighted by the finding that the gene coding for substance P (*TAC1*) also encodes several other neuroactive peptides, including neurokinin A, neuropeptide K, and neuropeptide γ.

Opioid peptides

An especially important category of peptide neurotransmitters is the family of opioids (see Figure 6.22B), so named because they bind to the same postsynaptic receptors that are activated by opium. The opium poppy has been cultivated for some 5000 years, and its derivatives have been used as an analgesic since at least the Renaissance. The active ingredients in opium are a variety of plant alkaloids, predominantly morphine. Morphine, named for Morpheus, the Greek god of dreams, is still in use today and is one of the most effective analgesics, despite its addictive potential. Synthetic opiates such as meperidine, methadone, and fentanyl are also powerful analgesics.

The opioid peptides were discovered in the 1970s during a search for **endorphins**—*endo*genous compounds that mimicked the actions of *mo*rphine. It was hoped that such compounds would be analgesics, and that understanding them would shed light on drug addiction. The endogenous ligands of the opioid receptors have now been identified as a

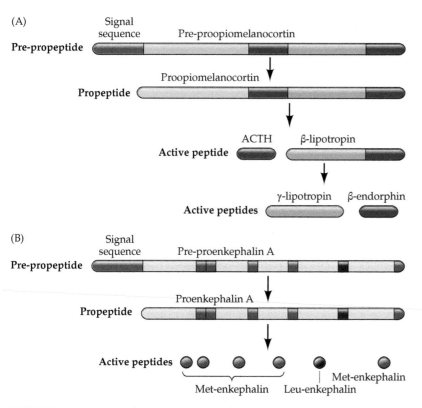

FIGURE 6.21   Proteolytic processing of pre-propeptides Shown here are pre-proopiomelanocortin (A) and pre-proenkephalin A (B). For each pre-propeptide, the signal sequence is indicated at the left; the locations of active peptide products are indicated by darker colors. The maturation of the pre-propeptides involves cleaving the signal sequence and other proteolytic processing. Such processing can result in several different neuroactive peptides such as ACTH, γ-lipotropin, and β-endorphin (A), or multiple copies of the same peptide, such as methionine enkephalin (B).

(A) Brain–gut peptides

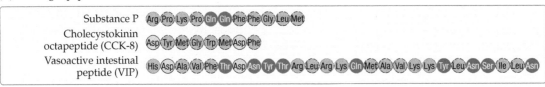

(B) Opioid peptides

Amino acid properties

- Hydrophobic
- Polar, uncharged
- Acidic
- Basic

(C) Pituitary peptides

(D) Hypothalamic–releasing peptides

(E) Miscellaneous peptides

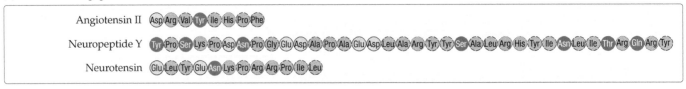

FIGURE 6.22 Amino acid sequences of neuropeptides These neuropeptides vary in length, usually containing between 3 and 36 amino acids. The sequence of amino acids determines the biological activity of each peptide.

family of more than 20 opioid peptides that fall into three classes: endorphins, enkephalins, and dynorphins (Table 6.2). Each class is liberated from an inactive pre-propeptide (pre-proopiomelanocortin, pre-proenkephalin A, or pre-prodynorphin) derived from distinct genes (see Figure 6.21). Opioid precursor processing is carried out by tissue-specific processing enzymes that are packaged into vesicles, along with the precursor peptide, in the Golgi apparatus.

Opioid peptides are widely distributed throughout the brain and are often co-localized with small-molecule neurotransmitters, such as GABA and 5-HT. In general, the opioids tend to be depressants. When injected intracerebrally in experimental animals, they act as analgesics; on the basis of this and other evidence, opioids are

likely to be involved in the mechanisms underlying acupuncture-induced analgesia. Opioids are also involved in complex behaviors such as sexual attraction, as well as aggressive and submissive behaviors. They have also been implicated in psychiatric disorders such as schizophrenia and autism, although the evidence for this is debated. Unfortunately, repeated administration of opioids leads to tolerance and addiction.

Neuropeptide receptors

Virtually all neuropeptides initiate their effects by activating G-protein-coupled receptors. Studying these metabotropic peptide receptors in the brain has been difficult because few specific agonists and antagonists are known.

TABLE 6.2 Endogenous Opioid Peptides

Name	Amino acid sequence[a]
Endorphins	
α-Endorphin	Tyr-Gly-Gly-Phe-Met-Thr-Ser-Glu-Lys-Ser-Gln-Thr-Pro-Leu-Val-Thr
α-Neoendorphin	Tyr-Gly-Gly-Phe-Leu-Arg-Lys-Tyr-Pro-Lys
β-Endorphin	Tyr-Gly-Gly-Phe-Met-Thr-Ser-Glu-Lys-Ser-Gln-Thr-Pro-Leu-Val-Thr-Leu-Phe-Lys-Asn-Ala-Ile-Val-Lys-Asn-Ala-His-Lys-Gly-Gln
γ-Endorphin	Tyr-Gly-Gly-Phe-Met-Thr-Ser-Glu-Lys-Ser-Gln-Thr-Pro-Leu-Val-Thr-Leu
Enkephalins	
Leu-enkephalin	Tyr-Gly-Gly-Phe-Leu
Met-enkephalin	Tyr-Gly-Gly-Phe-Met
Dynorphins	
Dynorphin A	Tyr-Gly-Gly-Phe-Leu-Arg-Arg-Ile-Arg-Pro-Lys-Leu-Lys-Trp-Asp-Asn-Gln
Dynorphin B	Tyr-Gly-Gly-Phe-Leu-Arg-Arg-Gln-Phe-Lys-Val-Val-Thr

[a]Note the initial homology, indicated by italics.

Peptides activate their receptors at low (nM to μM) concentrations compared with the concentrations required to activate receptors for small-molecule neurotransmitters. These properties allow the postsynaptic targets of peptides to be quite far removed from presynaptic terminals and to modulate the electrical properties of neurons that are simply in the vicinity of the site of peptide release. Neuropeptide receptor activation is especially important in regulating the postganglionic output from sympathetic ganglia and the activity of the gut (see Chapter 21). Peptide receptors, particularly the neuropeptide Y receptor, are also implicated in the initiation and maintenance of feeding behavior leading to satiety or obesity.

Other behaviors ascribed to peptide receptor activation include anxiety and panic attacks; antagonists of cholecystokinin receptors are used to treat these afflictions. Other useful drugs have been developed by targeting the opioid receptors. Three well-defined opioid receptor subtypes (μ, δ, and κ)—encoded by the *OPRM1*, *OPRD1*, and *OPRK1* genes, respectively—play a role in reward mechanisms as well as addiction. The μ-opioid receptor has been specifically identified as the primary site for drug reward mediated by opiate drugs. Fentanyl, a selective agonist of μ-opioid receptors, has 80 times the analgesic potency of morphine. This synthetic opiate is widely used as a clinical analgesic agent to alleviate pain (see Clinical Applications, Chapter 2) and is an increasingly popular recreational drug, often with fatal consequences.

Interneuronal Signaling Can Also Occur via Unconventional Neurotransmitters

LEARNING OBJECTIVES

6.8.1 Understand the unique features of signaling via unconventional neurotransmitters.

6.8.2 Be able to identify NO and endocannabinoids as unconventional neurotransmitters.

6.8.3 Understand the mechanisms involved in NO and endocannabinoid signaling.

In addition to the conventional neurotransmitters already described, some unusual molecules are used for signaling between neurons and their targets. These chemical signals can be considered as neurotransmitters because of their roles in interneuronal signaling and because their release from neurons is regulated by Ca^{2+}. However, they are unconventional in comparison with other neurotransmitters because they are not stored in synaptic vesicles and are not released from presynaptic terminals via exocytotic mechanisms. In fact, these unconventional neurotransmitters need not be released from presynaptic terminals at all and are often associated with retrograde signaling (that is, from postsynaptic cells back to presynaptic terminals).

Endocannabinoids

Endocannabinoids are a family of related endogenous signals that interact with cannabinoid receptors. These receptors are the molecular targets of Δ^9-tetrahydrocannabinol, the psychoactive component of the marijuana plant, *Cannabis* (Box 6C). While some members of this emerging group of chemical signals remain to be determined, anandamide and 2-arachidonoylglycerol (2-AG) have been established as endocannabinoids. These signals are unsaturated fatty acids with polar head groups and are produced by enzymatic degradation of membrane lipids (Figure 6.23A,B). Production of endocannabinoids is stimulated by a second messenger within postsynaptic neurons, typically a rise in postsynaptic Ca^{2+} concentration, allowing these hydrophobic signals to diffuse through the postsynaptic membrane to reach cannabinoid receptors on other nearby cells. Endocannabinoid action is terminated by carrier-mediated transport of these signals back into the postsynaptic neuron, where they are hydrolyzed by the enzyme fatty acid hydrolase (FAAH).

At least two types of cannabinoid receptors have been identified, with most actions of endocannabinoids in the CNS mediated by the CB_1 type (encoded by the *CNR1* gene) (see Box 6C). The CB_1 receptor is a G-protein-coupled receptor related to the metabotropic receptors for ACh, glutamate, and other conventional neurotransmitters. Several compounds that are structurally related to

■ BOX 6C | Marijuana and the Brain

Medicinal use of the marijuana plant (*Cannabis sativa*) (Figure A) dates back thousands of years. Ancient civilizations—including both Greek and Roman societies in Europe, as well as Indian and Chinese cultures in Asia—appreciated that this plant was capable of producing relaxation, euphoria, and several other psychopharmacological actions. In more recent times, medicinal use of marijuana has revived and the recreational use of marijuana has become so popular that some societies have decriminalized its use.

Understanding the brain mechanisms underlying the actions of marijuana was advanced by the discovery that a cannabinoid, Δ^9-tetrahydrocannabinol (THC; Figure B), is the active component of marijuana. This finding led to the development of synthetic derivatives, such as WIN 55,212-2 and rimonabant (see Figure 6.23C), that have served as valuable tools for probing the brain actions of THC. Of

particular interest is that receptors for these cannabinoids exist in the brain. The best studied of these receptors, called CB_1, is a metabotropic receptor that activates G-protein signaling pathways (Figure C). CB_1 exhibits marked regional variations in distribution, being especially enriched in brain areas—such as the substantia nigra and caudate putamen—that have been implicated in drug abuse (Figure D). The presence of these brain receptors for cannabinoids led in turn to a search for endogenous cannabinoid compounds in the brain, culminating in the discovery of endocannabinoids such as 2-AG and anandamide (see Figure 6.23A,B). This path of discovery closely parallels the identification of endogenous opioid peptides, which resulted from the search for endogenous morphine-like compounds in the brain (see Concept 6.7 and Table 6.2).

Given that THC interacts with brain endocannabinoid receptors, particularly the CB_1 receptor, it is likely that

such actions are responsible for the behavioral consequences of marijuana use. Indeed, many of the well-documented effects of marijuana are consistent with the distribution and actions of brain CB_1 receptors. For example, marijuana's effects on perception could be due to CB_1 receptors in the neocortex, effects on psychomotor control due to endocannabinoid receptors in the basal ganglia and cerebellum, effects on short-term memory due to cannabinoid receptors in the hippocampus, and the well-known effects on stimulating appetite due to hypothalamic actions. While formal links between these behavioral consequences of marijuana and the underlying brain mechanisms are still being forged, studies of the actions of this drug have shed substantial light on basic synaptic mechanisms, which promise to further elucidate the mode of action of one of the world's most popular drugs.

(A)

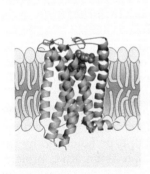

Cannabis sativa

(B)

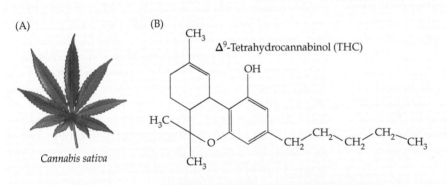

Δ^9-Tetrahydrocannabinol (THC)

(C)

(D)

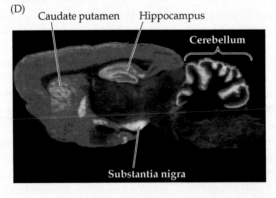

Caudate putamen Hippocampus Cerebellum Substantia nigra

(A) Leaf of *Cannabis sativa*, the marijuana plant. (B) Structure of Δ^9-tetrahydrocannabinol (THC), the active ingredient of marijuana. (C) Structure of the human CB_1 receptor, bound to the ligand taranabant (colored spheres). (D) Distribution of brain CB_1 receptors, visualized by examining the binding of CP-55,940, a CB_1 receptor ligand. (B after L. Iversen, 2003. *Brain* 126: 1252–1270; C from Z. Shao et al., 2016. *Nature* 540: 602–606; D courtesy of M. Herkenham, NIMH.)

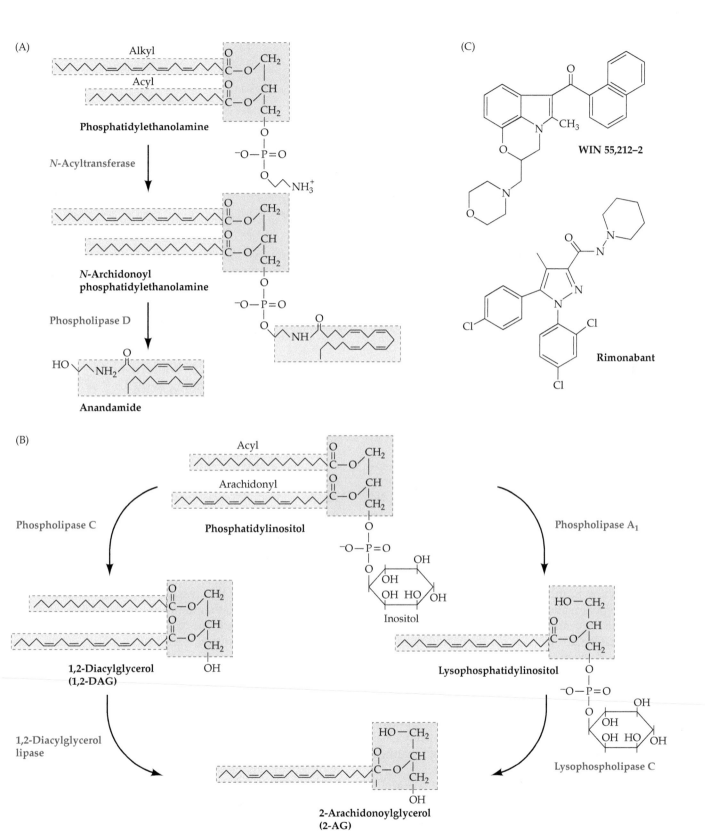

FIGURE 6.23 Endocannabinoid signals involved in synaptic transmission Possible mechanism of production of the endocannabinoids (A) anandamide and (B) 2-AG. (C) Structures of the endocannabinoid receptor agonist WIN 55,212-2 and the antagonist rimonabant. (A,B after T. F. Freund et al., 2003. *Physiol Rev* 83: 1017–1066; C after L. Iversen, 2003. *Brain* 126: 1252–1270.)

(A)

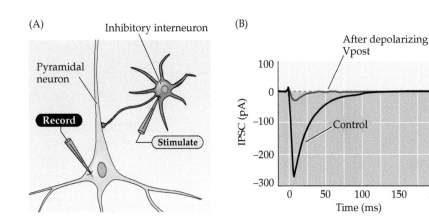

(B)

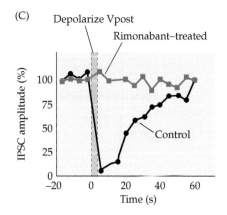

FIGURE 6.24 Endocannabinoid-mediated retrograde control of GABA release (A) Experimental arrangement. Stimulation of a presynaptic interneuron causes release of GABA onto a postsynaptic pyramidal neuron. (B) Inhibitory postsynaptic currents (IPSCs) elicited by the inhibitory synapse (control) are reduced in amplitude following a brief depolarization of the postsynaptic neuron (Vpost). This reduction in the IPSC is due to less GABA being released from the presynaptic interneuron. (C) The reduction in IPSC amplitude produced by postsynaptic depolarization lasts a few seconds and is mediated by endocannabinoids, because it is prevented by the endocannabinoid receptor antagonist rimonabant. (B,C after T. Ohno-Shosaku et al., 2001. *Neuron* 29: P729–738.)

endocannabinoids and that bind to the CB_1 receptor have been synthesized (Figure 6.23C). These compounds act as agonists or antagonists of the CB_1 receptor and serve both as tools for elucidating the physiological functions of endocannabinoids and as targets for developing therapeutically useful drugs.

Endocannabinoids participate in several forms of synaptic regulation. The best-documented action of these agents is the inhibition of communication between presynaptic inputs and their postsynaptic target cells. In both the hippocampus and the cerebellum (among other brain regions), endocannabinoids serve as retrograde signals that regulate GABA release at certain inhibitory synapses. At such synapses, depolarization of the postsynaptic neuron causes a transient reduction in inhibitory postsynaptic responses (Figure 6.24). Depolarization reduces synaptic transmission by elevating the concentration of Ca^{2+} in the postsynaptic neuron; this rise in Ca^{2+} triggers synthesis and release of endocannabinoids from the postsynaptic cells. The endocannabinoids then bind to CB_1 receptors on presynaptic terminals, inhibiting the amount of GABA released in response to presynaptic action potentials, and thereby reducing inhibitory transmission. The mechanisms responsible for the reduction in GABA release are not entirely clear but probably involve effects on voltage-gated Ca^{2+} channels and/or K^+ channels in the presynaptic neurons.

Nitric oxide

Nitric oxide (NO) is an unusual and especially interesting chemical signal. It is a gas produced by the action of nitric oxide synthase, an enzyme that converts the amino acid arginine into a metabolite (citrulline) and simultaneously generates NO (Figure 6.25). Within neurons, NO synthase is regulated by Ca^{2+} binding to the Ca^{2+} sensor protein calmodulin (see Chapter 7). Once produced, NO can permeate the plasma membrane, meaning that NO generated inside one cell can travel through the extracellular medium and act inside nearby cells. Thus, this gaseous signal has a range of influence that extends well beyond the cell of origin, diffusing a few tens of micrometers from its site of production before it is degraded. This property makes NO a potentially useful agent for coordinating the activities of multiple cells in a localized region and may mediate certain forms of synaptic plasticity that spread within small networks of neurons.

All of the known actions of NO are mediated within its cellular targets; for this reason, NO often is considered a second messenger rather than a neurotransmitter. Some of the actions of NO are due to the activation of the enzyme guanylyl cyclase, which then produces the second messenger cGMP within target cells (see Chapter 7). Other actions of NO are the result of covalent modification of target proteins via nitrosylation, the addition of a nitryl group to selected amino acids within the proteins. NO decays spontaneously by reacting with oxygen to produce inactive nitrogen oxides; thus, its signals last for only a short time (seconds or less). NO signaling evidently regulates a variety of synapses that also employ conventional neurotransmitters; so far, presynaptic terminals that release glutamate are the best-studied NO targets in the CNS. NO may also be involved in some neurological diseases. For example, it has been proposed that an imbalance between nitric oxide and superoxide generation underlies some neurodegenerative diseases.

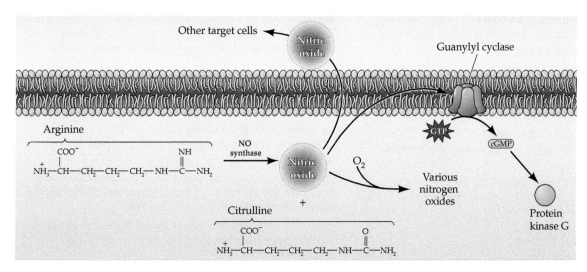

FIGURE 6.25   Synthesis, release, and termination of nitric oxide (NO)

Summary

The complex synaptic computations occurring at neural circuits throughout the brain arise from the actions of a large number of neurotransmitters, which act on an even larger number of postsynaptic neurotransmitter receptors. Glutamate is the major excitatory neurotransmitter in the brain, whereas GABA and glycine are the major inhibitory neurotransmitters. The actions of these small-molecule neurotransmitters are typically faster than those of the neuropeptides. Thus, most small-molecule transmitters mediate synaptic transmission when a rapid response is essential, whereas the neuropeptide transmitters, as well as the biogenic amines and some small-molecule neurotransmitters, tend to modulate ongoing activity in the brain or in peripheral target tissues in a more gradual and ongoing way. The postsynaptic signaling actions of neurotransmitters are mediated by ionotropic and metabotropic neurotransmitter receptors. Ionotropic or ligand-gated ion channels combine the neurotransmitter receptor and ion channel in one molecular entity, and therefore give rise to rapid postsynaptic electrical responses. Metabotropic receptors regulate the activity of postsynaptic ion channels indirectly, usually via G-proteins, and induce slower and longer-lasting electrical responses. Metabotropic receptors are especially important in regulating behavior, and drugs targeting these receptors have been clinically valuable in treating a wide range of behavioral disorders. The postsynaptic response at a given synapse is determined by the combination of receptor subtypes, G-protein subtypes, and ion channels that are expressed in the postsynaptic cell. Because each of these features can vary both within and among neurons, a tremendous diversity of transmitter-mediated effects is possible. Drugs that influence transmitter actions have enormous importance in the treatment of neurological and psychiatric disorders, as well as in a broad spectrum of other medical problems.

■ Additional Reading

Reviews

Beaulieu, J. M., S. Espinoza and R. R. Gainetdinov (2015) Dopamine receptors – IUPHAR Review 13. *British J. Pharmacol.* 172: 1–23.

Betke, K. M., C. A. Wells and H. E. Hamm (2012) GPCR mediated regulation of synaptic transmission. *Prog. Neurobiol.* 96: 304–321.

Carlsson, A. (1987) Perspectives on the discovery of central monoaminergic neurotransmission. *Annu. Rev. Neurosci.* 10: 19–40.

Cristino, L., T. Bisogno and V. Di Marzo (2020) Cannabinoids and the expanded endocannabinoid system in neurological disorders. *Nat. Rev. Neurol.* 16: 9–29.

Hansen, K. B. and 18 others (2021) Structure, function, and pharmacology of glutamate receptor ion channels. *Pharmacol. Rev.* 73: 298–487.

Iversen, L. (2003) *Cannabis* and the brain. *Brain* 126: 1252–1270.

Jaakola, V. P. and A. P. Ijzerman (2010) The crystallographic structure of the human adenosine A_{2A} receptor in a high-affinity antagonist-bound state: Implications for GPCR drug screening and design. *Curr. Opin. Struct. Biol.* 20: 401–414.

Pierce, K. L., R. T. Premont and R. J. Lefkowitz (2002) Seven-transmembrane receptors. *Nat. Rev. Mol. Cell Biol.* 3: 639–650.

Pin, J.-P. and B. Bettler (2016) Organization and functions of mGlu and GABA_B receptor complexes. *Nature* 540: 60–68.

Puthenkalam, R. and 6 others (2016) Structural studies of GABA_A receptor binding sites: Which experimental structure tells us what? *Front. Mol. Neurosci.* 9: 44.

Rosenbaum, D. M., S. G. Rasmussen and B. K. Kobilka (2009) The structure and function of G-protein-coupled receptors. *Nature* 459: 356–363.

Wong, K. L. L., A. Nair and G. J. Augustine. (2021) Changing the cortical conductor's tempo: Neuromodulation of the claustrum. *Front. Neural Circ.* 15: 658228.

Important Original Papers

Chavas, J. and A. Marty (2003) Coexistence of excitatory and inhibitory GABA synapses in the cerebellar interneuron network. *J. Neurosci.* 23: 2019–2031.

Chien, E. Y. and 10 others (2010) Structure of the human dopamine D3 receptor in complex with a D2/D3 selective antagonist. *Science* 330: 1091–1095.

Curtis, D. R., J. W. Phillis and J. C. Watkins (1959) Chemical excitation of spinal neurons. *Nature* 183: 611–612.

Dale, H. H., W. Feldberg and M. Vogt (1936) Release of acetylcholine at voluntary motor nerve endings. *J. Physiol.* 86: 353–380.

Du, J. and 4 others (2015) Glycine receptor mechanism elucidated by electron cryo-microscopy. *Nature* 526: 224–229.

Gupta, S., S. Chakraborty, R. Vij and A. Auerbach (2017) A mechanism for acetylcholine receptor gating based on structure, coupling, phi, and flip. *J. Gen. Physiol.* 149: 85–103.

Haga, K. and 10 others (2012) Structure of the human M2 muscarinic acetylcholine receptor bound to an antagonist. *Nature* 482: 547–551.

Hassaine, G. and 14 others (2014) X-ray structure of the mouse serotonin 5-HT3 receptor. *Nature* 512: 276–281.

Karakas, E. and H. Furukawa (2014) Crystal structure of a heterotetrameric NMDA receptor ion channel. *Science* 344: 992–997.

Kawate, T., J. C. Michel, W. T. Birdsong and E. Gouaux (2009) Crystal structure of the ATP-gated P2X_4 ion channel in the closed state. *Nature* 460: 592–598.

Miller, P. S. and A. R. Aricescu (2014) Crystal structure of a human GABA_A receptor. *Nature* 512: 270–275.

Miyazawa, A., Y. Fujiyoshi and N. Unwin (2003) Structure and gating mechanism of the acetylcholine receptor pore. *Nature* 423: 949–955.

Ohno-Shosaku, T., T. Maejima and M. Kano (2001) Endogenous cannabinoids mediate retrograde signals from depolarized postsynaptic neurons to presynaptic terminals. *Neuron* 29: 729–738.

Rasmussen, S. G. and 12 others (2007) Crystal structure of the human β2 adrenergic G-protein-coupled receptor. *Nature* 450: 383–387.

Rasmussen, S. G. and 19 others (2011) Crystal structure of the β_2 adrenergic receptor–Gs protein complex. *Nature* 477: 549–555.

Sobolevsky, A. I., M. P. Rosconi and E. Gouaux (2009) X-ray structure, symmetry and mechanism of an AMPA-subtype glutamate receptor. *Nature* 462: 745–756.

Thal, D. M. and 13 others (2016) Crystal structures of the M1 and M4 muscarinic acetylcholine receptors. *Nature* 531: 335–340.

Unwin, N. (2005) Refined structure of the nicotinic acetylcholine receptor at 4 Å resolution. *J. Mol. Biol.* 346: 967–989.

Wacker D. and 12 others (2017) Crystal structure of an LSD-bound human serotonin receptor. *Cell* 168: 377–389.

Watanabe, J., A. Rozov and L. P. Wollmuth (2005) Target-specific regulation of synaptic amplitudes in the neocortex. *J. Neurosci.* 25: 1024–1033.

Zhu, S. and 6 others (2016) Mechanism of NMDA receptor inhibition and activation. *Cell* 165: 704–714.

Books

Cooper, J. R., F. E. Bloom and R. H. Roth (2003) *The Biochemical Basis of Neuropharmacology*, 8th Edition. New York: Oxford University Press.

Iversen, L., S. Iversen, F. E. Bloom and R. H. Roth (2009) *Introduction to Neuropsychopharmacology*. New York: Oxford University Press.

Nestler, E., P. J. Kenny, S. J. Russo and A. Schaefer (2020) *Nestler, Hyman & Malenka's Molecular Neuropharmacology: A Foundation for Clinical Neuroscience*, 4th Edition. New York: McGraw Hill.

Siegel, G. J., R. W. Albers, S. Brady and D. Price (2012) *Basic Neurochemistry: Principles of Molecular, Cellular, and Medical Neurobiology*, 8th Edition. Burlington, MA: Elsevier Academic Press.

Van Dongen, A. M. (2009) *Biology of the NMDA Receptor*. Boca Raton, FL: CRC Press.

Molecular Signaling within Neurons

Overview

Electrical and chemical signaling mechanisms allow one nerve cell to receive and transmit information to another. This chapter focuses on the related events within neurons and other cells that are triggered by the interaction of a chemical signal with its receptor. This intracellular processing typically begins when extracellular chemical signals (such as neurotransmitters, hormones, and trophic factors) bind to specific receptors located either on the surface or in the cytoplasm or nucleus of the target cells. Such binding activates the receptors and in so doing stimulates cascades of intracellular reactions involving GTP-binding proteins, second-messenger molecules, protein kinases, ion channels, and many other effector proteins whose modulation temporarily changes the physiological state of the target cell. These same intracellular signal transduction pathways can also cause longer-lasting changes by altering the transcription of genes, thus affecting the protein composition of the target cells on a more permanent basis. The large number of components involved in intracellular signaling pathways allows precise temporal and spatial control over the function of individual neurons, thereby enabling control and coordination of the activity of neurons that comprise neural circuits and systems.

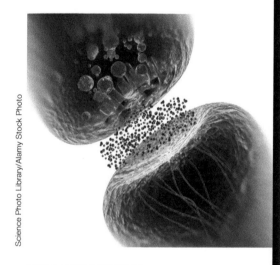

Science Photo Library/Alamy Stock Photo

KEY CONCEPTS

7.1 There are several modes of chemical signaling

7.2 Cellular responses are determined by several families of receptors

7.3 G-proteins connect receptor activation to intracellular signaling pathways

7.4 Second messengers are involved in many intracellular signaling pathways

7.5 Intracellular signaling often involves adding or removing phosphates from proteins

7.6 Long-lasting responses involve changes in nuclear signaling

7.7 There are many ways to signal changes in neuronal structure and function

CONCEPT
7.1

There Are Several Modes of Chemical Signaling

LEARNING OBJECTIVES

7.1.1 Distinguish between synaptic, paracrine, and endocrine forms of chemical signaling.

7.1.2 Explain how amplification occurs during intracellular signal transduction.

7.1.3 Compare the properties of the three different classes of chemical signaling molecules.

Strategies of molecular signaling

Chemical communication coordinates the behavior of individual nerve and glial cells in physiological processes that range from neural differentiation to learning and memory. Indeed, molecular signaling ultimately mediates and modulates all brain functions. To carry out such communication, a series of extraordinarily diverse and complex chemical signaling pathways has evolved. The preceding chapters have described in some detail the electrical signaling mechanisms that allow neurons to generate action potentials for conduction of information. Those chapters also described synaptic transmission, a special form of chemical signaling that transfers information from one neuron to another. But chemical signaling is not limited to synapses. Other

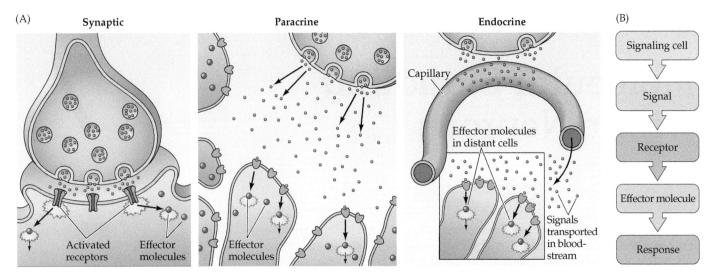

FIGURE 7.1 Chemical signaling (A) Forms of chemical communication include synaptic transmission, paracrine signaling, and endocrine signaling. (B) The essential components of chemical signaling are: cells that initiate the process by synthesizing and releasing signaling molecules; specific receptors on target cells; intracellular effector molecules; and subsequent cellular responses.

well-characterized forms of chemical communication include **paracrine** signaling, which acts over a longer range than synaptic transmission and involves the secretion of chemical signals onto a group of nearby target cells, and **endocrine** signaling, which refers to the secretion of hormones into the bloodstream, where they can affect targets throughout the body (Figure 7.1).

Chemical signaling of any sort requires three components: a molecular *signal* that transmits information from one cell to another; a *receptor* molecule that transduces the information provided by the signal; and an *effector* molecule that mediates the cellular response (see Figure 7.1B). The part of this process that takes place within the confines of the target cell is called **intracellular signal transduction**. A good example of transduction in the context of *intercellular* communication is the sequence of events triggered by chemical synaptic transmission (see Figure 7.1A and Chapter 5): Neurotransmitters serve as the signal, neurotransmitter receptors serve as the transducing receptor, and the effector molecule is an ion channel that is opened or closed to produce the electrical response of the postsynaptic cell. In many cases, however, synaptic transmission activates additional *intracellular* pathways that have a variety of functional consequences. For example, the binding of the neurotransmitter norepinephrine to its receptor activates GTP-binding proteins (see Figure 6.16B), which produces second messengers within the postsynaptic target, activates enzyme cascades, and eventually changes the chemical properties of numerous effector molecules within the affected cell.

A general advantage of chemical signaling in both intercellular and intracellular contexts is **signal amplification**. Amplification occurs because individual signaling

reactions can generate a much larger number of molecular products than the number of molecules that initiate the reaction. In the case of norepinephrine signaling, for example, a single norepinephrine molecule binding to its receptor can generate many thousands of second-messenger molecules (such as cyclic AMP), yielding an amplification of tens of thousands of phosphates transferred to effector proteins (Figure 7.2). Similar amplification occurs in all signal transduction pathways. Because the transduction processes often are mediated by a sequential set of enzymatic reactions, each with its own amplification factor, a small number of signal molecules ultimately can activate a very large number of effector molecules. Such amplification guarantees that a physiological response is evoked in the face of other, potentially countervailing, influences.

Another rationale for these complex signal transduction schemes is that they permit precise control of cell behavior over a wide range of times. Some molecular interactions allow information to be transferred rapidly, while others are slower and longer lasting. For example, the signaling cascades associated with synaptic transmission at neuromuscular junctions allow a person to respond to rapidly changing cues, such as the trajectory of a kicked ball, while the slower responses triggered by adrenal medullary hormones (epinephrine and norepinephrine) secreted during a challenging game produce slower and longer-lasting effects on muscle metabolism (see Chapter 21) and emotional state (see Chapter 32). To encode information that varies so widely over time, the concentration of the relevant signaling molecules must be carefully controlled. On one hand, the concentration of every signaling molecule within the signaling cascade must return to subthreshold values before the arrival of another stimulus. On the

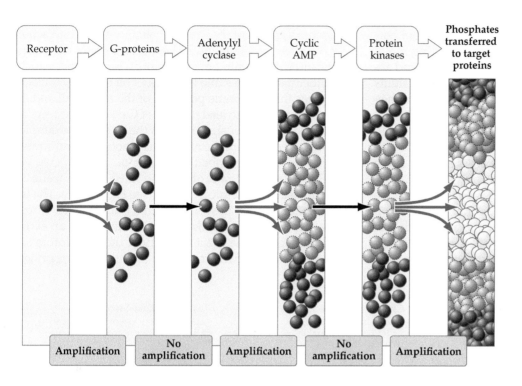

FIGURE 7.2 Amplification in signal transduction pathways The activation of a single receptor by a signaling molecule, such as the neurotransmitter norepinephrine, can lead to the activation of numerous G-proteins inside cells. These activated proteins can bind to other signaling molecules, such as the enzyme adenylyl cyclase. Each activated enzyme molecule generates a large number of cAMP molecules. cAMP binds to and activates another family of enzymes —the protein kinases—that can phosphorylate many target proteins. Although not every step in this signaling pathway involves amplification, overall the cascade results in a tremendous increase in the potency of the initial signal.

other hand, prolonged activation of the intermediates in a signaling pathway is critical for a sustained response. Having multiple levels of molecular interactions facilitates the intricate timing of these signaling events.

Activation of signaling pathways

The molecular components of intracellular signal transduction pathways are always activated by a chemical signaling molecule. Such signaling molecules can be grouped into three classes: **cell-impermeant**, **cell-permeant**, and **cell-associated signaling molecules** (Figure 7.3). The first two classes are secreted molecules and thus can act on target cells removed from the site of signal synthesis or release. Cell-impermeant signaling molecules typically bind to receptors associated with the plasma membrane. Hundreds of secreted molecules have now been identified, including the neurotransmitters discussed in Chapter 6; proteins such as neurotrophic factors (see Chapter 23); and peptide hormones such as glucagon, insulin, and various reproductive hormones. These signaling molecules are typically short-lived, either because they are rapidly metabolized or because they are internalized by endocytosis once bound to their receptors.

Cell-permeant signaling molecules can cross the plasma membrane to act directly on receptors that are inside the cell. Examples include numerous steroid hormones

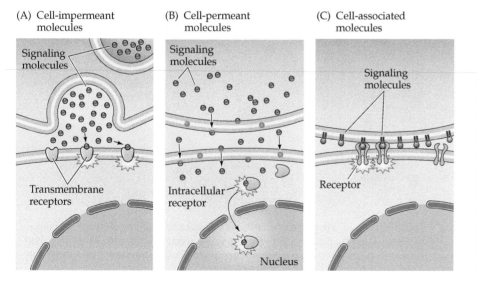

(A) Cell-impermeant molecules

Signaling molecules

Transmembrane receptors

(B) Cell-permeant molecules

Signaling molecules

Intracellular receptor

Nucleus

(C) Cell-associated molecules

Signaling molecules

Receptor

FIGURE 7.3 Three classes of cell signaling molecules (A) Cell-impermeant molecules, such as neurotransmitters, cannot readily traverse the plasma membrane of the target cell and must bind to the extracellular portion of transmembrane receptor proteins. (B) Cell-permeant molecules are able to cross the plasma membrane and bind to receptors in the cytoplasm or nucleus of target cells. (C) Cell-associated molecules are presented on the extracellular surface of the plasma membrane. These signals activate receptors on target cells only if they are directly adjacent to the signaling cell.

(glucocorticoids, estradiol, and testosterone), thyroid hormones (thyroxin), and retinoids. These signaling molecules are relatively insoluble in aqueous solutions and are often transported in blood and other extracellular fluids by binding to specific carrier proteins. In this form, they may persist in the bloodstream for hours or even days.

The third group of chemical signaling molecules, cell-associated signaling molecules, is arrayed on the extracellular surface of the plasma membrane. As a result, these molecules act only on other cells that are physically in contact with the cell that carries such signals. Examples include proteins such as the integrins and neural cell adhesion molecules (NCAMs) that influence axonal growth (see Chapter 23). Membrane-bound signaling molecules are more difficult to study, but are clearly important in neuronal development and other circumstances where physical contact between cells provides information about cellular identities.

CONCEPT 7.2 | Cellular Responses Are Determined by Several Families of Receptors

LEARNING OBJECTIVES

7.2.1 List the four main types of cellular receptors.

7.2.2 Compare the properties of different types of receptors.

Regardless of the nature of the initiating signal, cellular responses are determined by the presence of receptors that specifically bind to the signaling molecules. Binding of signal molecules causes a conformational change in the receptor, which then triggers a subsequent signaling cascade within the affected cell. The receptors for impermeant signal molecules are proteins that span the plasma membrane. The extracellular domain of such receptors includes the binding site for the signal, while the intracellular domain activates intracellular signaling cascades after the signal binds. A large number of these receptors have been identified and are grouped into families defined by the mechanism used to transduce signal binding into a cellular response (Figure 7.4).

Receptor types

Channel-linked receptors (see Figure 7.4A), also called ligand-gated ion channels (see Figure 4.8), have the receptor and transducing

functions as part of the same protein molecule. Interaction of the chemical signal with the binding site of the receptor causes the opening or closing of an ion channel pore in another part of the same molecule. The resulting ion flux changes the membrane potential of the target cell and, in some cases, can also lead to entry of Ca^{2+} ions that serve as a second-messenger signal within the cell. Good examples of such receptors are the numerous ionotropic neurotransmitter receptors described in Chapter 6.

Enzyme-linked receptors also have an extracellular binding site for chemical signals (see Figure 7.4B). The intracellular domain of such receptors is an enzyme whose catalytic activity is regulated by the binding of an extracellular signal. The great majority of these receptors are **protein kinases**, often tyrosine kinases, that phosphorylate

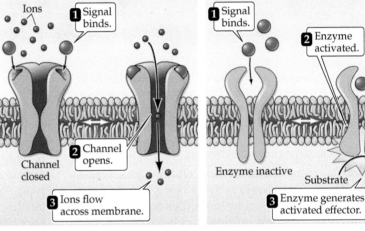

(A) Channel-linked receptors

Ions
1 Signal binds.
Channel closed
2 Channel opens.
3 Ions flow across membrane.

(B) Enzyme-linked receptors

1 Signal binds.
2 Enzyme activated.
Enzyme inactive
Substrate / Effector
3 Enzyme generates activated effector.

(C) G-protein-coupled receptors

1 Signal binds.
Receptor
G-protein
2 G-protein binds.
3 G-protein activated.

(D) Intracellular receptors

Signaling molecule
2 Activated receptor regulates transcription.
1 Signal binds.
Receptor

FIGURE 7.4 Categories of cellular receptors Cell-impermeant signaling molecules can bind to and activate channel-linked receptors (A), enzyme-linked receptors (B), or G-protein-coupled receptors (C). Cell-permeant signaling molecules activate intracellular receptors (D).

intracellular target proteins, thereby changing the physiological function of the target cells. Noteworthy members of this group of receptors are the Trk family of neurotrophin receptors (see Chapter 23) and other receptors for growth factors.

G-protein-coupled receptors (see Figure 7.4C), also called metabotropic receptors (see Chapter 5), regulate intracellular reactions by an indirect mechanism involving an intermediate transducing molecule, called a **GTP-binding protein** (or **G-protein**). Hundreds of different G-protein-linked receptors have been identified. Well-known examples include the β-adrenergic receptor, the muscarinic type of acetylcholine receptor, and metabotropic glutamate receptors, all discussed in Chapter 6, as well as the receptors for odorants in the olfactory system, and many types of receptors for peptide hormones. Rhodopsins, the light-sensitive proteins of retinal photoreceptors, are another type of G-protein-linked receptor whose activating signal is photons of light, rather than a chemical signal (see Figure 9.7).

Intracellular receptors are activated by cell-permeant or lipophilic signaling molecules (see Figure 7.4D). Many of these receptors lead to the activation of signaling cascades that produce new mRNA and protein within the target cell. Often such receptors comprise a receptor protein bound to an inhibitory protein complex. When the signaling molecule binds to the receptor, the inhibitory complex dissociates to expose a DNA-binding domain on the receptor. This activated form of the receptor can then move into the nucleus and directly interact with nuclear DNA, resulting in altered transcription. Some intracellular receptors are located primarily in the cytoplasm, while others are in the nucleus. In either case, once these receptors are activated they affect gene expression by altering DNA transcription.

<table>
<tr><td>CONCEPT
7.3</td><td>**G-Proteins Connect Receptor Activation to Intracellular Signaling Pathways**</td></tr>
</table>

LEARNING OBJECTIVES

7.3.1 Identify the types of GTP-binding proteins involved in intracellular signal transduction.

7.3.2 Compare the molecular mechanisms of regulation of the activity of heterotrimeric and monomeric G-proteins.

7.3.3 List three downstream effectors of activated G-proteins.

7.3.4 Describe the molecular components of one G-protein-coupled effector pathway.

Both G-protein-coupled receptors and enzyme-linked receptors can activate biochemical reaction cascades that ultimately modify the function of target proteins. For both of these receptor types, the coupling between receptor activation and their subsequent effects is provided by GTP-binding proteins.

Two classes of GTP-binding proteins

There are two general classes of GTP-binding proteins (Figure 7.5). **Heterotrimeric G-proteins** consist of three distinct subunits (α, β, and γ). There are many different α, β, and

(A) Heterotrimeric G-proteins

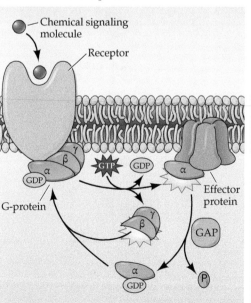

(B) Monomeric G-proteins

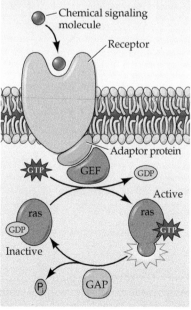

FIGURE 7.5 Types of GTP-binding proteins (A) Heterotrimeric G-proteins consist of three distinct subunits (α, β, and γ). Receptor activation causes the binding of the G-protein and the α subunit to exchange GDP for GTP, leading to a dissociation of the α and βγ subunits. The biological actions of these G-proteins are terminated by hydrolysis of GTP, which is enhanced by GTPase-activating (GAP) proteins. (B) Monomeric G-proteins use similar mechanisms to relay signals from activated cell surface receptors to intracellular targets. The biological actions of these G-proteins depend on binding of GTP, which is activated by guanine nucleotide exchange factors (GEFs) that bind to the receptor in association with adaptor proteins, and their activity is terminated by hydrolysis of GTP, which is also regulated by GAP proteins.

γ subunits, allowing a bewildering number of G-protein permutations. Regardless of the specific composition of the heterotrimeric G-protein, its α subunit binds to guanine nucleotides, either GTP or GDP. Binding of GDP then allows the α subunit to bind to the β and γ subunits to form an inactive trimer. Binding of an extracellular signal to a G-protein-coupled receptor in turn allows the G-protein to bind to the receptor and causes GDP to be replaced with GTP (see Figure 7.5A). When GTP is bound to the G-protein, the α subunit dissociates from the βγ complex and activates the G-protein. Following activation, both the GTP-bound α subunit and the free βγ subunit complex can bind to downstream effector molecules that mediate a variety of responses in the target cell.

The second class of GTP-binding proteins is the **monomeric G-proteins** (also called **small G-proteins**). These monomeric GTPases also relay signals from activated cell surface receptors to intracellular targets such as the cytoskeleton and the vesicle trafficking apparatus of the cell. The first small G-protein was discovered in a virus that causes *rat* sarcoma tumors and was therefore called **ras**. Ras helps regulate cell differentiation and proliferation by relaying signals from receptor kinases to the nucleus; the viral form of ras is defective, which accounts for the ability of the virus to cause the uncontrolled cell proliferation that leads to tumors. Ras is known to be involved in many forms of neuronal signaling, including long-term synaptic potentiation (see Chapter 8). Since the discovery of ras, a large number of small GTPases have been identified and can be sorted into five different subfamilies with different functions. For instance, some are involved in vesicle trafficking in the presynaptic terminal or elsewhere in the neuron, while others play a central role in protein and RNA trafficking in and out of the nucleus.

Similar to heterotrimeric G-proteins, monomeric G-proteins function as molecular timers that are active in their GTP-bound state, becoming inactive when they have hydrolyzed the bound GTP to GDP (see Figure 7.5B). Monomeric G-proteins are activated by replacement of bound GDP with GTP; this reaction is controlled by a group of proteins called *guanine nucleotide exchange factors* (GEFs). There are more than 100 GEFs, some of which specifically activate single types of monomeric G-proteins, while others can activate multiple monomeric G-proteins. GEFs, in turn, are activated by binding to activated receptors. This binding is promoted by yet another protein, termed an adaptor protein.

Termination of signaling by both heterotrimeric and monomeric G-proteins is determined by hydrolysis of GTP to GDP. The rate of GTP hydrolysis is an important property of a particular G-protein and can be regulated by other proteins, termed GTPase-activating proteins (GAPs). By replacing GTP with GDP, GAPs return G-proteins to their inactive form. GAPs were first recognized as regulators of small G-proteins (see Figure 7.5B), but similar proteins are now known to regulate the α subunits of heterotrimeric G-proteins (see Figure 7.5A).

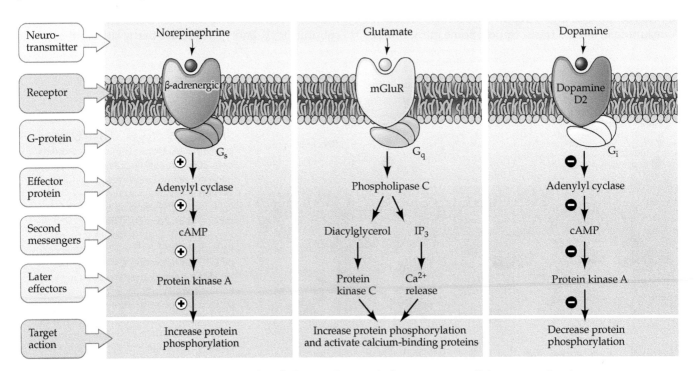

FIGURE 7.6 Effector pathways associated with G-protein-coupled receptors In all three examples shown here, binding of a neurotransmitter to such a receptor leads to activation of a G-protein and subsequent recruitment of second-messenger pathways. G_s, G_q, and G_i refer to three different types of heterotrimeric G-proteins.

Effectors activated by G-proteins

Activated G-proteins alter the function of many downstream effectors. Most of these effectors are enzymes that produce intracellular second messengers. Effector enzymes include adenylyl cyclase, guanylyl cyclase, phospholipase C, and others (Figure 7.6). The second messengers produced by these enzymes trigger the complex biochemical signaling cascades discussed in Concept 7.4. Because each of these cascades is activated by specific G-protein subunits, the pathways activated by a particular receptor are determined by the specific identity of the G-protein subunits associated with it.

G-proteins can also directly bind to and activate ion channels. For example, activation of muscarinic receptors by acetylcholine can open K^+ channels, thereby reducing the rate at which the neurons fire action potentials. Such inhibitory responses are believed to be the result of $\beta\gamma$ subunits of G-proteins binding to the K^+ channels. The activation of α subunits can also lead to the rapid closing of voltage-gated Ca^{2+} and Na^+ channels. Because these channels carry inward currents involved in generating action potentials, closing them also makes it more difficult for target cells to fire (see Chapters 3 and 4). Thus, by directly regulating the gating of ion channels, G-proteins can influence the electrical signaling of target cells.

In summary, the binding of chemical signals to their receptors activates cascades of signal transduction events in the cytosol of target cells. Within such cascades, G-proteins serve a pivotal function as the molecular transducing elements that couple membrane receptors to their molecular effectors within the cell. The diversity of G-proteins and their downstream targets leads to many types of physiological responses.

CONCEPT
7.4 | ## Second Messengers Are Involved in Many Intracellular Signaling Pathways

LEARNING OBJECTIVES

7.4.1 Describe the sources of cytoplasmic Ca^{2+} signals, the molecular targets of these signals, and the mechanisms for removing these signals.

7.4.2 Describe the sources of cyclic nucleotide signals, the molecular targets of these signals, and the mechanisms for removing these signals.

7.4.3 Describe the sources of diacylglycerol and IP_3 signals, the molecular targets of these signals, and the mechanisms for removing these signals.

Neurons use many different second messengers as intracellular signals. These messengers differ in the mechanism by which they are produced and removed, as well as in their downstream targets and effects (Figure 7.7A). This

concept summarizes the attributes of some of the principal second messengers.

Calcium

The calcium ion (Ca^{2+}) is perhaps the most common intracellular messenger in neurons. Indeed, few neuronal functions are immune to the influence—direct or indirect—of Ca^{2+}. In all cases, information is transmitted by a transient rise in the cytoplasmic calcium concentration, which allows Ca^{2+} to bind to and activate a large number of Ca^{2+}-binding proteins that serve as molecular targets. One of the most thoroughly studied targets of Ca^{2+} is **calmodulin**, a Ca^{2+}-binding protein abundant in the cytosol of all cells. Binding of Ca^{2+} to calmodulin activates this protein, which then initiates its effects by binding to still other downstream targets, such as protein kinases. Another important family of intracellular Ca^{2+}-binding proteins is the synaptotagmins, which serve as Ca^{2+} sensors during neurotransmitter release and other forms of intracellular membrane fusion (see Chapter 5).

Ordinarily the concentration of Ca^{2+} ions in the cytosol is extremely low, typically 50 to 100 nanomolar (10^{-9} M). The concentration of Ca^{2+} ions outside neurons—in the bloodstream or cerebrospinal fluid, for instance—is several orders of magnitude higher, typically several millimolar ($10^{-3}M$). This steep Ca^{2+} gradient is maintained by several mechanisms (Figure 7.7B). Most important in this maintenance are two proteins that translocate Ca^{2+} from the cytosol to the extracellular medium: an ATPase called the **calcium pump**; and an **Na^+/Ca^{2+} exchanger**, which is a protein that replaces intracellular Ca^{2+} with extracellular sodium ions (see Chapter 4). In addition to these plasma membrane mechanisms, Ca^{2+} is also pumped into the endoplasmic reticulum and mitochondria. These organelles can thus serve as storage depots of Ca^{2+} ions that are later released to participate in signaling events. Finally, nerve cells contain other Ca^{2+}-binding proteins—such as **calbindin**—that serve as Ca^{2+} buffers. Such buffers reversibly bind Ca^{2+} and thus blunt the magnitude and slow the kinetics of Ca^{2+} signals within neurons.

The Ca^{2+} ions that act as intracellular signals enter the cytosol by means of one or more types of Ca^{2+}-permeable ion channels (see Chapter 4). These can be voltage-gated Ca^{2+} channels or ligand-gated channels in the plasma membrane, both of which allow Ca^{2+} to flow down its concentration gradient and into the cell from the extracellular medium. In addition, other channels allow Ca^{2+} to be released from the interior of the endoplasmic reticulum into the cytosol. These intracellular Ca^{2+}-releasing channels are also gated, so they can be opened or closed in response to various intracellular signals. One such channel is the **inositol trisphosphate (IP_3) receptor**. As the name implies, this channel is regulated by IP_3, a second messenger described in more detail later in this concept. A second type of intracellular Ca^{2+}-releasing

(A)

Second messenger	Sources	Intracellular targets	Removal mechanisms
Ca²⁺	Plasma membrane: Voltage-gated Ca²⁺ channels, Various ligand-gated channels; Endoplasmic reticulum: IP₃ receptors, Ryanodine receptors	Calmodulin, Protein kinases, Protein phosphatases, Ion channels, Synaptotagmins, Many other Ca²⁺-binding proteins	Plasma membrane: Na⁺/Ca²⁺ exchanger, Ca²⁺ pump; Endoplasmic reticulum: Ca²⁺ pump; Mitochondria
Cyclic AMP	Adenylyl cyclase acts on ATP	Protein kinase A, Cyclic nucleotide-gated channels	cAMP phosphodiesterase
Cyclic GMP	Guanylyl cyclase acts on GTP	Protein kinase G, Cyclic nucleotide-gated channels	cGMP phosphodiesterase
IP₃	Phospholipase C acts on PIP₂	IP₃ receptors on endoplasmic reticulum	Phosphatases
Diacylglycerol	Phospholipase C acts on PIP₂	Protein kinase C	Various enzymes

FIGURE 7.7 Neuronal second messengers (A) Mechanisms responsible for producing and removing second messengers, and the downstream targets of these messengers. (B) Proteins involved in delivering calcium to the cytoplasm and in removing calcium from the cytoplasm. (C) Mechanisms for producing and degrading cyclic nucleotides. (D) Pathways involved in producing and removing diacylglycerol and IP₃.

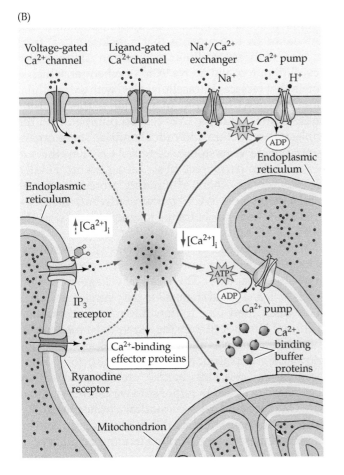

(B)

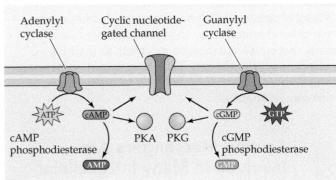

(C)

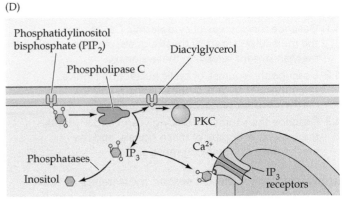

(D)

channel is the **ryanodine receptor**, named after a drug that binds to and partially opens these receptors. Among the biological signals that activate ryanodine receptors are cytoplasmic Ca^{2+} and, at least in muscle cells, depolarization of the plasma membrane.

These various mechanisms for elevating and removing Ca^{2+} ions allow precise control of both the timing and location of Ca^{2+} signaling within neurons, which in turn permits Ca^{2+} to control many different signaling events. For example, voltage-gated Ca^{2+} channels allow Ca^{2+} concentrations to rise very rapidly and locally within presynaptic terminals to trigger neurotransmitter release, as described in Chapter 5. Slower and more widespread rises in Ca^{2+} concentration regulate a wide variety of other responses, including gene expression in the cell nucleus.

Cyclic nucleotides

Another important group of second messengers is the cyclic nucleotides, specifically cyclic adenosine monophosphate (cAMP) and cyclic guanosine monophosphate (cGMP) (Figure 7.7C). Cyclic AMP is a derivative of the abundant cellular energy storage molecule ATP, and is produced when G-proteins activate adenylyl cyclase in the plasma membrane. Adenylyl cyclase converts ATP into cAMP by removing two phosphate groups from the ATP. Cyclic GMP is similarly produced from GTP by the action of guanylyl cyclase. Once the intracellular concentration of cAMP or cGMP is elevated, these nucleotides can bind to two different classes of targets. The most common targets of cyclic nucleotide action are protein kinases, either the cAMP-dependent protein kinase (PKA) or the cGMP-dependent protein kinase (PKG). These enzymes mediate many physiological responses by phosphorylating target proteins, as described in Concept 7.5. In addition, cAMP and cGMP can influence neuronal signaling by binding to certain ligand-gated ion channels (see Figure 4.9D). These cyclic nucleotide-gated channels are particularly important in phototransduction and other sensory transduction processes, such as olfaction (see Chapters 9 and 14). Cyclic nucleotide signals are degraded by phosphodiesterases, enzymes that cleave phosphodiester bonds and convert cAMP into AMP or cGMP into GMP.

Diacylglycerol and IP$_3$

Remarkably, membrane lipids can also be converted into intracellular second messengers (Figure 7.7D). The two most important messengers of this type are produced from phosphatidylinositol bisphosphate (PIP$_2$). This lipid component is cleaved by phospholipase C, an enzyme activated by certain G-proteins and by calcium ions. Phospholipase C splits the PIP$_2$ into two smaller molecules, each of which acts as a second messenger. One of these messengers is diacylglycerol (DAG), a molecule that remains within the membrane and activates protein kinase C, which phosphorylates substrate proteins in both the plasma membrane and elsewhere. The other messenger is IP$_3$, which leaves the plasma membrane and diffuses within the cytosol. IP$_3$ binds to IP$_3$ receptors, channels that release calcium from the endoplasmic reticulum. Thus, the action of IP$_3$ is to produce yet another second messenger (perhaps a third messenger, in this case!) that triggers an entire spectrum of reactions in the cytosol. The actions of DAG and IP$_3$ are terminated by enzymes that convert these two molecules into inert forms that can be recycled to produce new molecules of PIP$_2$.

The intracellular concentration of these second messengers changes dynamically over time, allowing precise control over their downstream targets. These signals can also be localized to small compartments within single cells or can spread over great distances, some even spreading between cells via gap junctions (see Chapter 5). Understanding of the complex temporal and spatial dynamics of these second-messenger signals has been greatly aided by the development of imaging techniques that visualize second messengers and other molecular signals within cells (Box 7A).

■ BOX 7A | Dynamic Imaging of Intracellular Signaling

Dramatic breakthroughs in our understanding of the brain often rely on development of new experimental techniques. This certainly has been true for our understanding of intracellular signaling in neurons, which has benefited enormously from the invention of imaging techniques that allow direct visualization of signaling processes within living cells. The first advance—and arguably the most significant—came from the development, by Roger Tsien and his colleagues, of the fluorescent dye fura-2 (Figure A). Calcium ions bind to fura-2 and cause the dye's fluorescence properties to change. When fura-2 is introduced inside cells and the cells are then imaged with a fluorescence microscope, this dye serves as a reporter of intracellular Ca^{2+} concentration. Fura-2 imaging has allowed investigators to detect the spatial and temporal dynamics of the Ca^{2+} signals that trigger innumerable processes within neurons and glial cells; for example, fura-2 was used to obtain the image of Ca^{2+} signaling during neurotransmitter release shown in Figure 5.8A.

Subsequent refinement of the chemical structure of fura-2 has yielded many other fluorescent Ca^{2+} indicator dyes with different fluorescence properties and different sensitivities to Ca^{2+}. One of these dyes is Calcium Green, which

(Continued)

■ BOX 7A | Dynamic Imaging of Intracellular Signaling (*continued*)

was used to image the dynamic changes in Ca^{2+} concentration produced within the dendrites of cerebellar Purkinje cells by the intracellular messenger IP$_3$ (Figure B). Further developments have led to indicators for visualizing the spatial and temporal dynamics of other second-messenger signals, such as cAMP.

Another tremendous advance in the dynamic imaging of signaling processes came from the discovery of a green fluorescent protein that was first isolated from the jellyfish *Aequorea victoria* by Osamu Shimomura. Green fluorescent protein, or GFP, is (as its name indicates) a protein that is brightly fluorescent (Figure C). Molecular cloning of the *GFP* gene allows imaging techniques to visualize expression of gene products labeled with GFP fluorescence. The first such use of GFP was in experiments with the worm *Caenorhabditis elegans*, in which Martin Chalfie and his colleagues rendered neurons fluorescent by inducing GFP expression in these cells. Many

subsequent experiments have used expression of GFP to image the structure of individual neurons in the mammalian brain (Figure D).

Molecular genetic strategies make it possible to attach GFP to almost any protein, thereby allowing fluorescence microscopy to image the spatial distribution of labeled proteins. In this way, it has been possible to visualize dynamic changes in the location of neuronal proteins during signaling events. Related techniques also allow visualization of the location of second-messenger signals or the biochemical activity of signaling proteins. For example, Figure 8.12 illustrates the use of this approach to monitor the activation of CaMKII during long-term synaptic potentiation.

As was the case with fura-2, subsequent refinement of GFP has led to numerous improvements. One significant improvement, also pioneered by Roger Tsien, was the production of proteins that fluoresce in colors other

than green, thus permitting the simultaneous imaging of multiple types of proteins and/or neurons. A particularly vivid demonstration of the power of multicolor imaging of fluorescent proteins can be seen in Brainbow, a technique that uses differential expression of combinations of several fluorescent proteins to label neurons with one of nearly 100 different colors (Figure E). This permits the axons of individual neurons to be identified and followed, even through the complex tangle of neuronal processes typically found in the CNS, thereby defining the circuits formed by the labeled neurons.

Just as development of the Golgi staining technique opened our eyes to the cellular composition of the brain (see Chapter 1), study of intracellular signaling in the brain has been revolutionized by fura-2, GFP, and other fluorescent tools. There is no end in sight for the potential of such imaging methods to illuminate new and important aspects of brain signaling dynamics.

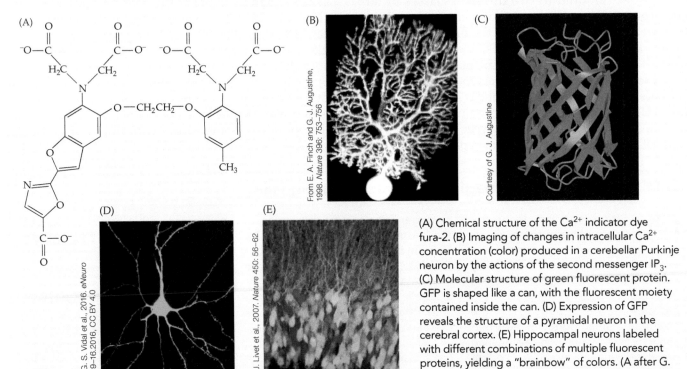

(A) Chemical structure of the Ca^{2+} indicator dye fura-2. (B) Imaging of changes in intracellular Ca^{2+} concentration (color) produced in a cerebellar Purkinje neuron by the actions of the second messenger IP$_3$. (C) Molecular structure of green fluorescent protein. GFP is shaped like a can, with the fluorescent moiety contained inside the can. (D) Expression of GFP reveals the structure of a pyramidal neuron in the cerebral cortex. (E) Hippocampal neurons labeled with different combinations of multiple fluorescent proteins, yielding a "brainbow" of colors. (A after G. Grynkiewicz et al., 1985. *J Biol Chem* 260: 3440–3450.)

From E. A. Finch and G. J. Augustine, 1998. *Nature* 396: 753–756

Courtesy of G. J. Augustine

From G. S. Vidal et al. 2016. *eNeuro* 3: 0089–16.2016, CC BY 4.0

From J. Livet et al., 2007. *Nature* 450: 56–62

<table>
<tr><td>CONCEPT
7.5</td><td># Intracellular Signaling Often Involves Adding or Removing Phosphates from Proteins</td></tr>
</table>

LEARNING OBJECTIVES

7.5.1 Explain the difference between protein kinases and phosphatases.

7.5.2 List three different protein kinases and the second-messenger signals that activate them.

7.5.3 List two different protein phosphatases and the second-messenger signals that activate them.

Phosphorylation and dephosphorylation of proteins

As already mentioned, second messengers typically regulate neuronal functions by modulating the phosphorylation state of intracellular proteins (Figure 7.8). Phosphorylation (the addition of phosphate groups) is a post-translational modification that rapidly and reversibly changes protein function. Proteins are phosphorylated by a wide variety of protein kinases; phosphate groups are removed by other enzymes called **protein phosphatases**. The importance of protein phosphorylation as a regulatory mechanism is emphasized by the fact that the human genome contains more than 500 protein kinase genes and approximately 200 protein phosphatase genes. This means that nearly 3% of our genome is directly dedicated to control of the phosphorylation state of proteins.

The degree of phosphorylation of a target protein reflects a balance between the competing actions of protein kinases and phosphatases, thereby integrating a host of cellular signaling pathways. The substrates of protein kinases and phosphatases include enzymes, neurotransmitter receptors, ion channels, and structural proteins. Protein kinases and phosphatases typically act either on the serine and threonine residues (Ser/Thr kinases or phosphatases) or on the tyrosine residues (Tyr kinases or phosphatases) of their substrates. Some of these enzymes act specifically on only one or a handful of protein targets, while others are multifunctional and have a broad range of substrate proteins.

Protein kinases

The activity of most protein kinases is regulated either by second messengers or by extracellular chemical signals such as growth factors (see Chapter 23). Typically, second messengers activate Ser or Thr kinases, whereas extracellular signals activate Tyr kinases. Each protein kinase has catalytic domains responsible for transferring phosphate groups to the relevant amino acids of their target proteins. Kinases that are regulated by second messengers typically have an additional regulatory domain that inhibits the catalytic site. Binding of second messengers (such as cAMP, DAG, or Ca^{2+}) to the regulatory domain removes the inhibition, allowing the catalytic domain to phosphorylate the substrate protein. Other kinases are activated by phosphorylation by another protein kinase; these kinases typically do not have inhibitory regulatory domains. Among the hundreds of protein kinases expressed in the brain, a relatively small number are the main regulators of neuronal signaling.

- *cAMP-dependent protein kinase (PKA).* PKA is the primary effector of cAMP action in neurons and consists of two catalytic subunits and two regulatory subunits (Figure 7.9A). cAMP activates PKA by binding to the regulatory subunits, releasing active catalytic subunits that can phosphorylate many different target proteins. Although the catalytic subunits are similar to the catalytic domains of other protein kinases, distinct amino acids allow PKA to bind to specific target proteins, thus allowing only those targets to be phosphorylated in response to intracellular cAMP signals. Signaling specificity is also achieved by using specific anchoring proteins, called A kinase anchoring proteins (AKAPs), to localize PKA activity to specific compartments within cells.

- *Ca^{2+}/calmodulin-dependent protein kinase, type II (CaMKII).* Ca^{2+} ions binding to calmodulin can regulate numerous protein kinases. In neurons, CaMKII is the predominant Ca^{2+}/calmodulin-dependent protein kinase; this kinase is the most abundant component of the postsynaptic density, a structure important for postsynaptic signaling (see Box 7B). CaMKII comprises 12 subunits that are connected by a central association domain to form a wheel-like structure (Figure 7.9B). Each subunit contains a catalytic domain and a regulatory domain. When intracellular Ca^{2+} concentration is low, binding of the regulatory domain to the catalytic domain inhibits kinase activity. Elevated Ca^{2+} allows Ca^{2+}/calmodulin to bind to the regulatory

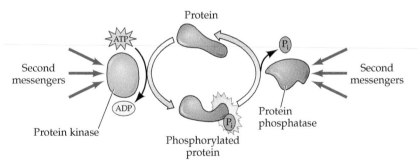

FIGURE 7.8 Regulation of cellular proteins by phosphorylation Protein kinases transfer phosphate groups (P_i) from ATP to serine, threonine, or tyrosine residues on substrate proteins. This phosphorylation reversibly alters the structure and function of cellular proteins. Removal of the phosphate groups is catalyzed by protein phosphatases. Both kinases and phosphatases are regulated by a variety of intracellular second messengers.

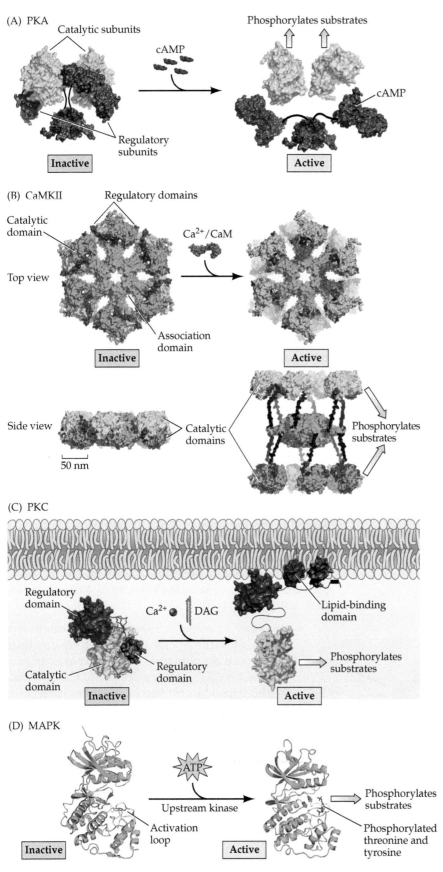

(A) PKA

Catalytic subunits

cAMP

Phosphorylates substrates

cAMP

Regulatory subunits

Inactive

Active

(B) CaMKII

Regulatory domains

Catalytic domain

Top view

Ca²⁺/CaM

Association domain

Inactive

Active

Side view

50 nm

Catalytic domains

Phosphorylates substrates

(C) PKC

Regulatory domain

Ca²⁺ • ‖ DAG

Lipid-binding domain

Catalytic domain

Regulatory domain

Phosphorylates substrates

Inactive

Active

(D) MAPK

ATP

Upstream kinase

Activation loop

Inactive

Active

Phosphorylates substrates

Phosphorylated threonine and tyrosine

FIGURE 7.9 Activation of protein kinases
(A) In the inactive state (left), the catalytic subunits of PKA are inhibited by the regulatory subunits. Binding of cAMP to the regulatory subunits relieves the inhibition and frees the catalytic subunits to phosphorylate their targets. Black lines indicate flexible structures that connect parts of the regulatory subunits. (B) CaMKII is a large wheel-shaped structure comprising 12 subunits, each with a catalytic (tan/yellow) and a regulatory (blue) domain, held together by a central association domain (green). Binding of Ca²⁺/calmodulin to the regulatory domain allows the catalytic domain to extend and phosphorylate its substrates. (C) Binding of Ca²⁺ allows lipid-binding domains of PKC (small blue structures) to insert into the plasma membrane and bind to DAG and other membrane lipids. This change in structure and location displaces the regulatory (blue) and allows the catalytic domain of PKC (yellow) to phosphorylate its substrates. Black lines indicate flexible structures that connect the various domains of PKC. (D) Activation of MAPK is caused by phosphorylation of an activation loop (yellow) by upstream kinases. Phosphorylation of the activation loop allows the catalytic domain of MAPK to assume its active conformation and phosphorylate downstream targets. (A after H. M. Berman et al., 2000. *Nucleic Acids Research* 28: 235–242. CC BY 4.0. https://pdb101.rcsb.org/motm/152; B after T. J. A. Craddock et al., 2012. *PLOS Comput Biol* 8: e1002421, CC BY; C after T. A. Leonard et al., 2011. *Cell* 144: 55–66; D from B. E. Turk, 2007. *Biochem J* 402: 405–417.)

domain, relieving its inhibition of the catalytic domain. This allows the catalytic domain to extend to form a barrel-shaped structure that can phosphorylate substrate proteins. CaMKII substrates include ion channels and numerous proteins involved in intracellular signal transduction. The multimeric structure of CaMKII also allows neighboring subunits to phosphorylate each other, leading to sustained activation of CaMKII even after intracellular Ca^{2+} concentration returns to basal levels. This process of autophosphorylation is thought to serve as a cellular memory mechanism.

- *Protein kinase C (PKC).* Another important group of protein kinases is PKC, a diverse family of monomeric kinases activated by the second messengers DAG and Ca^{2+}. Ca^{2+} causes PKC to move from the cytosol to the plasma membrane, where the regulatory domains of PKC then bind to DAG and to membrane phospholipids (Figure 7.9C). These events separate the regulatory and catalytic domains, allowing the catalytic domain to phosphorylate various protein substrates. PKC also diffuses to sites other than the plasma membrane—such as the cytoskeleton, perinuclear sites, and the nucleus—where it phosphorylates still other substrate proteins. Tumor-promoting compounds called phorbol esters mimic DAG and cause a prolonged activation of PKC that is thought to trigger tumor formation.

- *Protein tyrosine kinases.* Two classes of protein kinases transfer phosphate groups to tyrosine residues on substrate proteins. Receptor tyrosine kinases are transmembrane proteins with an extracellular domain that binds to protein ligands (growth factors, neurotrophic factors, or cytokines) and an intracellular catalytic domain that phosphorylates the relevant substrate proteins. Non-receptor tyrosine kinases are cytoplasmic or membrane-associated enzymes that are indirectly activated by extracellular signals. Tyrosine phosphorylation is less common than Ser or Thr phosphorylation, and it often serves to recruit signaling molecules to the phosphorylated protein. Tyrosine kinases are particularly important for cell growth and differentiation (see Chapters 22 and 23).

- *Mitogen-activated protein kinase (MAPK).* MAPKs, also called extracellular signal-regulated kinases (ERKs), are important examples of protein kinases that are activated via phosphorylation by another protein kinase. MAPKs were first identified as participants in the control of cell growth but are now known to have many other signaling functions. MAPKs are inactive at rest but become activated when they are phosphorylated. In fact, MAPKs are part of a kinase cascade in which one protein kinase phosphorylates and activates the next protein kinase in the cascade. The signals that trigger these kinase cascades are often extracellular growth factors that bind to receptor tyrosine

kinases that, in turn, activate monomeric G-proteins such as ras. However, MAPKs are also activated by other types of signals, such as osmotic stress and heat shock. Activation of MAPKs is caused by phosphorylation of one or more amino acids within a structure called the activation loop; phosphorylation changes the structure of the activation loop, enabling the catalytic domain of MAPK to phosphorylate downstream targets (Figure 7.9D). These targets include transcription factors—proteins that regulate gene thanscription and protein expression (see Figure 7.12)—as well as various enzymes, including other protein kinases, and cytoskeletal proteins.

Protein phosphatases

Among the several different families of protein phosphatases, the best characterized are the Ser/Thr phosphatases 1, 2A, and 2B. Like many protein kinases, these protein phosphatases consist of both catalytic and regulatory subunits (Figure 7.10). The catalytic subunits remove phosphates from proteins. Because of the remarkable similarity of their catalytic subunits, protein phosphatases display less substrate specificity than protein kinases. Furthermore, most regulatory subunits of phosphatases do not bind to second messengers, making most protein phosphatases constitutively active. Instead, expression of different regulatory subunits determines which substrates are dephosphorylated and where the phosphatase is located within cells.

- *Protein phosphatase 1 (PP1).* PP1 is a dimer consisting of one catalytic subunit and one regulatory subunit (see Figure 7.10A). PP1 activity is determined by which of the more than 200 different regulatory subunits binds to its catalytic subunit. PP1 also is regulated by phosphorylation of its regulatory subunits by PKA. PP1 is one of the most prevalent Ser/Thr protein phosphatases in mammalian cells and dephosphorylates a wide array of substrate proteins. In addition to its well-studied regulation of metabolic enzymes, PP1 can also influence neuronal electrical signaling by dephosphorylating K^+ and Ca^{2+} channels, as well as neurotransmitter receptors such as AMPA-type and NMDA-type glutamate receptors.

- *Protein phosphatase 2A (PP2A).* PP2A is one of the most abundant enzymes in the brain and accounts for approximately 1% of the total amount of protein found within cells. PP2A is a multisubunit enzyme consisting of both catalytic and regulatory subunits, as well as an additional scaffold subunit that brings together the catalytic and regulatory subunits (see Figure 7.10B). There are two different versions of the catalytic and scaffold subunits and approximately 25 different regulatory subunits. Different combinations of these three subunits yield more than 80 different versions of PP2A. PP2A has a broad range of substrates

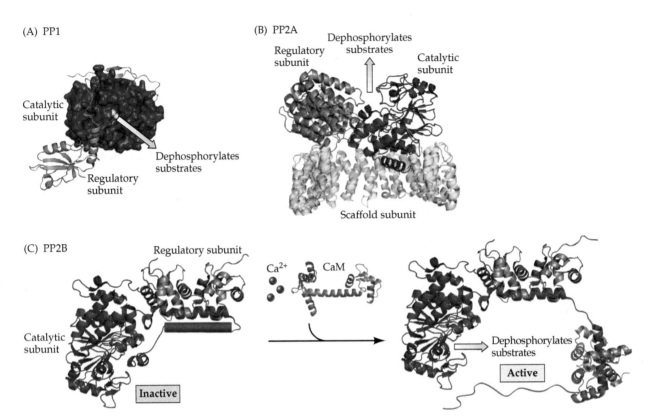

(A) PP1

Catalytic subunit

Dephosphorylates substrates

Regulatory subunit

(B) PP2A

Dephosphorylates substrates

Regulatory subunit

Catalytic subunit

Scaffold subunit

(C) PP2B

Regulatory subunit

Catalytic subunit

Ca^{2+} CaM

Dephosphorylates substrates

Active

Inactive

FIGURE 7.10 Types of protein phosphatases (A) Structure of PP1, consisting of a catalytic subunit (red) bound to spinophilin (blue), a regulatory subunit found in dendritic spines. (B) PP2A is a trimeric enzyme that has a scaffold subunit (yellow), in addition to the catalytic (red) and regulatory (blue) subunits common to other phosphatases. (C) The activity of PP2B, or calcineurin, is regulated by Ca^{2+}/calmodulin. In the absence of Ca^{2+}/calmodulin (left), part of the regulatory subunit blocks the active site of the catalytic subunit, preventing phosphatase activity. Binding of Ca^{2+}/calmodulin to the catalytic subunit relieves this blockade, allowing PP2B to dephosphorylate its substrate proteins. (A from M. Bollen et al., 2010. *Trends Biochem Sci* 35: 450–458; B from U. S. Cho and W. Xu, 2007. *Nature* 445: 53–57; C after H. Li et al., 2011. *Trends Cell Biol* 21: 91–103.)

that overlap with those of PP1. One of its best-studied substrates is tau, a protein associated with microtubules in the cytoskeleton. Alzheimer's disease is associated with excessive phosphorylation of tau, perhaps due to defects in PP2A. Alterations in PP2A activity also have been implicated in other neurodegenerative diseases, as well as in cancer and diabetes. Although PP2A is constitutively active, its activity can be regulated by phosphorylation and other post-translational modifications of both the catalytic and regulatory subunits.

• *Protein phosphatase 2B (PP2B)*. PP2B, or calcineurin, is present at high levels in neurons and comprises a catalytic subunit and a regulatory subunit (see Figure 7.10C). Unlike the activity of PP1 and PP2A, PP2B activity is acutely controlled by intracellular Ca^{2+} signaling: Ca^{2+}/calmodulin activates PP2B by binding to the catalytic subunit and displacing the inhibitory regulatory domain, thereby activating PP2B. Even though both PP2B and CaMKII are activated by Ca^{2+}/calmodulin, they generally have different molecular targets. Substrates of PP2B include

a transcriptional regulator, NFAT, and ion channels. Dephosphorylation of AMPA-type glutamate receptors by PP2B is thought to play a central role in signal transduction during long-term depression of hippocampal synapses (see Chapter 8).

In summary, activation of membrane receptors can elicit complex cascades of enzyme activation, resulting in second-messenger production and protein phosphorylation or dephosphorylation. These cytoplasmic signals produce a variety of physiological responses by transiently regulating enzyme activity, ion channels, cytoskeletal proteins, and many other cellular processes. Intracellular signaling mechanisms also serve as targets for numerous psychiatric disorders (Clinical Applications). At excitatory synapses, these signaling components are often contained within dendritic spines, which appear to serve as specialized signaling compartments within neurons (Box 7B). In addition, such signals can propagate to the nucleus and cause long-lasting changes in gene expression.

■ Clinical Applications

Molecular Basis of Psychiatric Disorders

Defective molecular signaling at synapses has been strongly implicated in several psychiatric diseases, including psychoses, mood disorders, and anxiety disorders. In fact, our knowledge of synaptic signaling mechanisms has played a fundamental role in advancing our understanding of these disorders and in developing therapeutic strategies to treat them.

Psychoses

Individuals suffering from psychoses have a loss of contact with reality; this can include both positive symptoms, such as hallucinations and delusions, as well as negative symptoms such as apathy. The most prevalent form of psychosis is *schizophrenia*, which affects approximately 1% of the world's population. Antagonists of dopamine receptors serve as *antipsychotic drugs* that reduce the positive symptoms of many people with schizophrenia. First-generation antipsychotic drugs include chlorpromazine, haloperidol, and fluphenazine. These drugs produce several side effects, most notably motor control problems—including tremors and immobility—that are also associated with dopamine-deficiency disorders such as Parkinson's disease (see Chapter 18). Much effort has gone into developing more effective antipsychotic drugs that have fewer side effects. The resulting second-generation antipsychotic drugs, such as clozapine, melperone, and olanzapine, reportedly cause fewer motor control problems but can produce other side effects such as weight gain and diabetes.

While the fact that dopamine receptor blockers ameliorate some symptoms implicates dopamine signaling defects in schizophrenia, the pathogenesis of schizophrenia remains unknown. Studies of twins indicate that the heritability of schizophrenia is 80%. This strong genetic linkage has motivated studies that have identified mutations in more than 100 schizophrenia susceptibility genes. Several schizophrenia-associated genes encode proteins involved in dopaminergic signaling. These include genes encoding the type 2 dopamine receptor (*DRD2*); catechol O-methyltransferase (*COMT*), an enzyme that degrades dopamine; intracellular signal transduction molecules such as a regulatory subunit of PP1 (*DARPP-32*); another protein phosphatase, calcineurin (*PPP3CC*); and a GAP for G-protein signaling (*RGS4*). Still other schizophrenia-associated genes encode other types of neurotransmitter receptors, such as the NMDA-type (*GRIN2A*) and Abb-type (*GRIA3*) glutamate receptors, the 5-HT$_{2A}$ receptor (*HTR2A*), nicotinic acetylcholine receptors (*CHRNA7*), a receptor for vasoactive intestinal peptide (*VIPR2*), and a receptor for dopamine-like trace amino acids (*TAAR6*). Another schizophrenia susceptibility gene encodes the D-amino acid oxidase activator (*DAOA*). This activates D-amino acid oxidase (*DAO*), an enzyme that metabolizes D-serine, an activator of NMDA receptors. Finally, numerous other schizophrenia-associated genes encode proteins associated with brain or synapse development. These include Disrupted in Schizophrenia 1 (*DISC1*), dysbindin (*DTNBP1*), neuregulin (*NRG1*), the C4 complement factors (*C4A* and *C4B*), and the AKT1 protein kinase (*AKT1*). The *Armadillo* repeat gene deleted in Velo-Cardio-Facial syndrome (*ARVCF*) is also associated with brain or synapse development.

These studies indicate that schizophrenia is a complex disorder that results from the interaction of many genes, as well as other factors such as traumatic brain injury. Collectively, these factors produce defects in synapse formation and regulation that yield the alterations in brain dopamine signaling that underlie at least some of the symptoms of schizophrenia.

Mood disorders

As their name indicates, mood (or affective) disorders are characterized by prolonged elevation or lowering of an individual's mood. The two most prevalent types of mood disorders are depression and bipolar disorder.

Depression, also called major depressive disorder, is associated with a low mood that persists for 2 weeks or longer. Approximately 17% of the world's population experiences depression sometime in their lives, with women being afflicted at approximately twice the rate of men. The three major classes of *antidepressant drugs* all influence aminergic synaptic transmission. Inhibitors of monoamine oxidase (MAO), such as phenelzine, block the breakdown of amines, whereas tricyclic antidepressants, such as desipramine, block the reuptake of norepinephrine and other amines. The extraordinarily popular antidepressant fluoxetine (Prozac) selectively blocks the reuptake of serotonin without affecting the reuptake of catecholamines, such as dopamine. Amphetamine is a stimulant that also is used to treat some depressive disorders; this drug has multiple actions, including inhibition of dopamine and norepinepherine uptake into presynaptic terminals and stimulation of norepinephrine release from nerve terminals.

Although the genetic heritability of depression is significant—approximately 30% to 40%—there have been relatively few genetic studies of depression. Among the depression-associated gene mutations identified thus far, several are associated with synaptic molecular signaling: the beta subunit of CaMKII (*CAMK2B*), synapse-associated protein 102 (*DLG3*), a membrane-associated guanylate kinase found at glutamatergic synapses, and MAPK phosphatase-1 (*DUSP1*), a negative regulator of MAPK signaling. The relationship between these genetic

(Continued)

■ **Clinical Applications** *(continued)*

defects and the actions of antidepressant drugs is not yet clear.

Individuals with *bipolar disorder* have moods that alternate between depression and mania, an abnormally elevated mood state; this led to the previous designation of manic depression. Less than 1% of the population suffers from bipolar disorder. The drugs most frequently used to treat bipolar disorder are the antipsychotic compounds that alter dopaminergic or serotonergic synaptic transmission.

Genetically, bipolar disorder has many parallels to schizophrenia. First, as with schizophrenia, the genetic heritability of bipolar disorder is very high (80% to 90%). Also as with schizophrenia, bipolar disorder seems to be a complex disease that is associated with numerous gene defects, some of which are also found in people with schizophrenia. Several of these genes encode proteins associated with dopaminergic or serotonergic signaling: MAO (*MAOA*) and catechol *O*-methyltransferase (*COMT*), the two major enzymes involved in dopamine degradation (see Chapter 6), as well as the G-protein receptor kinase (*GRK3*), which is also involved in dopamine metabolism, and the plasma membrane transporter responsible for clearing 5-HT from the synaptic cleft (*5HTT*). Several other genes encode proteins associated with glutamatergic synaptic transmission, such as NMDA receptors (*GRIN2B*), the synapse organizing proteins synapse-associated protein 102 (*DLG3*) and *PSD-95* (*DLG4*; see Box 7B), D-amino acid oxidase activator (*DAOA*), and CaMKII (*CAMK2B*). Also implicated are genes encoding proteins involved in synapse development, such as the brain-derived neurophic

factor (*BDNF*; see Chapter 23), Disrupted in Schizophrenia 1 (*DISC1*), and neuregulin (*NRG1*). Finally, bipolar disorder is also associated with mutations in voltage-gated chloride channels (*VDAC1* and *VDAC2*) and calcium channels (*CACNG2*).

In summary, although the pathogenesis of bipolar disorder remains unclear, the genetic and psychotherapeutic links between biopolar disorder and schizophrenia suggest that both disorders share some common origins.

Anxiety disorders

Anxiety disorders are estimated to afflict 10% to 35% of the population, making them the most common psychiatric problem. As in depression, the prevalence of anxiety disorders is approximately twice as high in females as in males. Two of the major forms of pathological anxiety—panic disorder and generalized anxiety disorder—differ in their durations.

Panic disorder is associated with sudden, and often unprovoked, attacks of fear that are associated with changes in heart rate and other responses associated with panic. Agents used to treat panic disorder include drugs that affect aminergic transmission, such as MAO inhibitors and blockers of serotonin receptors. Although a heritability of approximately 40% suggests a significant genetic component to panic disorder, there have been few genetic studies of panic disorder. Two initial genetic associations indicate mutations in a membrane-associated regulatory subunit of PP1 (*TMEM132D*) and catechol *O*-methyltransferase (*COMT*), consistent with the efficacy of drugs that affect aminergic transmission.

Generalized anxiety disorder causes bouts of extreme anxiety that can persist for weeks or longer and are associated with other symptoms, such as restlessness or difficulties in concentrating. The most effective drugs for treatment of generalized anxiety disorder are agents that increase the efficacy of inhibitory transmission at synapses employing $GABA_A$ receptors. Examples of these *anxiolytic drugs* include benzodiazepines, such as chlordiazepoxide (Librium) and diazepam (Valium; see Chapter 5 and Clinical Applications, Chapter 2). Generalized anxiety disorder has a heritability of 30% to 40%, again indicating a genetic contribution to this disease. Among the sparse genetic studies to date, the most promising links to generalized anxiety disorder are mutations in the gene encoding glutamatic acid decarboxylase (*GAD2*), the enzyme responsible for synthesis of GABA. This fits well with the clinical effectiveness of the benzodiazepines in treating generalized anxiety disorder.

In conclusion, psychiatric disorders are complex, both in terms of their genetic origins as well as their symptoms. While some therapies are effective in treating some of the symptoms of these disorders, in no case do we currently have a comprehensive molecular explanation for any psychiatric disorder. The common theme linking these disorders is defects in synaptic molecular signaling, most prominently involving aminergic transmission, but also glutamatergic and GABAergic transmission as well. Eventually, these molecular clues will lead to elucidation of the causes of psychiatric disorders.

■ BOX 7B | Dendritic Spines

Many excitatory synapses in the brain involve microscopic dendritic protrusions known as spines (Figure A). Spines are distinguished by the presence of globular tips called spine heads, which serve as the postsynaptic site of innervation by presynaptic terminals. Spine heads are connected to the main shafts of dendrites by narrow links called spine necks (Figure B). Just beneath the site of contact between presynaptic terminals and spine heads are intracellular structures called postsynaptic densities. The number, size, and shape of spines are quite variable along some dendrites (Figure C). At least in some cases, spine shape can change dynamically over time (see Figures 8.15 and 25.9), and is altered in several neurodegenerative disorders (such as Alzheimer's disease) and psychiatric disorders (such as autism and schizophrenia).

Since the earliest description of these structures by Santiago Ramón y Cajal in the late 1800s, dendritic spines have fascinated generations of neuroscientists and have inspired much speculation about their function. One of the earliest conjectures was that the narrow spine neck electrically isolates synapses from the rest of the neuron. While this has been the subject of a protracted debate, the most recent experimental measurements indicate that the high electrical resistance of the spine neck is sufficient to attenuate EPSPs as they spread from spine heads to dendrites.

Another theory—currently the most popular functional concept—postulates that spines create biochemical compartments. This idea is based on the idea that the spine neck could impede diffusion of biochemical signals from the spine head to the rest of the dendrite. Several observations are consistent with this notion. First, measurements show that the spine neck does indeed serve as a barrier to diffusion, in some cases slowing the rate of molecular movement by a factor of 100 or more. Second, spines are found at excitatory synapses, where it is known that synaptic transmission generates many diffusible signals, most notably the second messenger Ca^{2+}. Finally, fluorescence imaging shows that postsynaptic Ca^{2+} signals can indeed be restricted to dendritic spines in some circumstances (Figure D).

Nevertheless, there are counterarguments to the hypothesis that spines provide relatively isolated biochemical compartments. For example, it is known that other second messengers, such as IP_3, as well as other signaling molecules, can readily diffuse out of the spine head and into the dendritic shaft. This difference in diffusion presumably is due to the fact that such signals last longer than Ca^{2+} signals, allowing them sufficient time to overcome the diffusion barrier of the spine neck. Another relevant point is that postsynaptic Ca^{2+} signals are highly localized, even at excitatory synapses that do not have spines. Thus, in at least some instances, spines are neither necessary nor sufficient for localization of synaptic second-messenger signaling.

A final and less controversial idea is that the purpose of spines is to serve as reservoirs where signaling proteins—such as the downstream molecular targets of second-messenger signals—

(Continued)

(A)   (B)   (C)

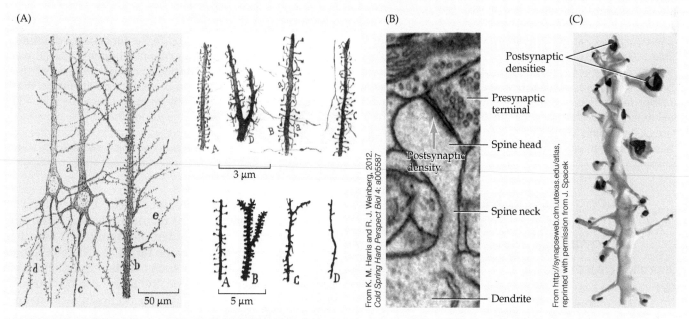

From K. M. Harris and R. J. Weinberg, 2012.
Cold Spring Harb Perspect Biol 4: a005587

From http://synapseweb.clm.utexas.edu/atlas, reprinted with permission from J. Spacek

(A) Cajal's classic drawings of dendritic spines. Left, dendrites of cortical pyramidal neurons. Right, higher-magnification images of several different types of dendritic spines. (B) Electron micrograph of an excitatory synapse in the hippocampus. Green arrow indicates postsynaptic density. (C) Reconstruction of a small region of the dendrite of a hippocampal pyramidal neuron, revealing a remarkable diversity in spine structure. Red structures indicate postsynaptic densities within each spine.

■ BOX 7B │ Dendritic Spines *(continued)*

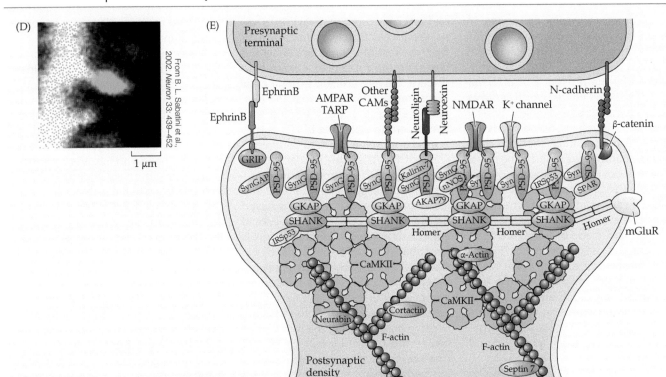

(D)

From B. L. Sabatini et al., 2002. *Neuron* 33: 439–452

1 μm

(D) Localized Ca²⁺ signal (green) produced in the spine of a hippocampal pyramidal neuron following activation of a glutamatergic synapse. (E) Postsynaptic densities include dozens of signal transduction molecules, including glutamate receptors (AMPA-type [AMPAR], NMDA-type [NMDAR], mGluR), other ion channels such as K⁺ channels, and many intracellular signal transduction molecules, most notably the protein kinase CaMKII. Cytoskeletal elements, such as actin and its numerous binding partners, also are prominent and help create the structure of dendritic spines. (E after M. Sheng and E. Kim, 2011. *Cold Spring Harb Perspect Biol* 3: a005678.)

can be concentrated. Consistent with this possibility, glutamate receptors are highly concentrated on spine heads, and the postsynaptic density comprises dozens of proteins involved in intracellular signal transduction (Figure E). According to this view, the spine head is the destination for these signaling molecules during the assembly of synapses, as well as the target of the second messengers that are produced by the local activation of glutamate receptors. Spines also can trap molecules that are diffusing along the dendrite, which could be a means of concentrating these molecules within spines.

Although the function of dendritic spines remains enigmatic, Cajal undoubtedly would be pleased at the enormous amount of attention that these tiny synaptic structures continue to command, and the real progress that has been made in understanding the variety of tasks of which they are capable.

CONCEPT
7.6

Long-Lasting Responses Involve Changes in Nuclear Signaling

LEARNING OBJECTIVES

7.6.1 Describe the sequence of molecular events that regulate gene transcription.

7.6.2 Compare the mechanisms of gene regulation produced by CREB, nuclear receptors, and c-fos.

Second messengers elicit prolonged changes in neuronal function by regulating gene expression, which promotes the synthesis of new RNA and protein. The resulting accumulation of new proteins requires at least 30 to 60 minutes, a time frame that is orders of magnitude slower than the responses mediated by ion fluxes or phosphorylation. Likewise, the reversal of such events requires hours to days. In some cases, genetic "switches" can be "thrown" to permanently alter a neuron, as occurs in neuronal differentiation (see Chapter 22).

The amount of protein present in cells is determined primarily by the rate of transcription of DNA into RNA (Figure 7.11). The first step in RNA synthesis is the decondensation of the structure of chromatin to provide binding

FIGURE 7.11 **Steps in the transcription of DNA into RNA** Condensed chromatin (left) is decondensed into a beads-on-a-DNA-string array (right) in which an upstream activator site (UAS) is free of proteins and is bound by a sequence-specific transcriptional activator protein (a transcription factor). The transcriptional activator protein then binds co-activator complexes that enable the RNA polymerase with its associated factors to bind at the start site of transcription and initiate RNA synthesis.

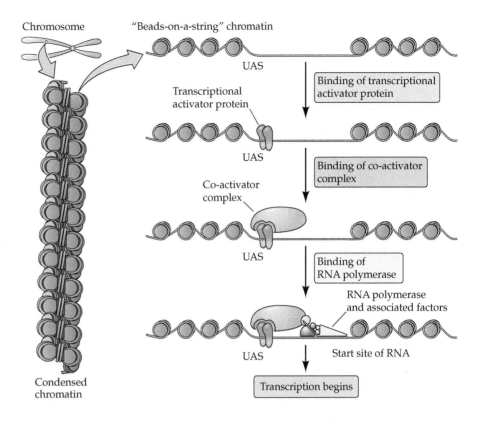

sites for the RNA polymerase complex and for **transcriptional activator proteins**, also called **transcription factors**. Transcriptional activator proteins attach to binding sites that are present on the DNA molecule near the start of the target gene sequence; they also bind to other proteins that promote unwrapping of DNA. The net result of these actions is to allow RNA polymerase, an enzyme complex, to assemble on the **promoter** region of the DNA and begin transcription. In addition to clearing the promoter for RNA polymerase, activator proteins can stimulate transcription by interacting with the RNA polymerase complex or by interacting with other activator proteins that influence the polymerase.

Regulation of gene expression by three key activator proteins

Intracellular signal transduction cascades regulate gene expression by converting transcriptional activator proteins from an inactive state to an active state in which they are able to bind to DNA. This conversion comes about in several ways. The remainder of this concept briefly summarizes three key activator proteins and the mechanisms that allow them to regulate gene expression in response to signaling events.

- *CREB.* The **cAMP response element binding protein**, usually abbreviated **CREB**, is a ubiquitous transcriptional activator (Figure 7.12). CREB is typically bound to its binding site on DNA (called the cAMP response element, or

CRE), either as a homodimer or bound to another, closely related transcription factor. In unstimulated cells, CREB is not phosphorylated and has little or no transcriptional activity. However, phosphorylation of CREB greatly potentiates transcription. Several signaling pathways are capable of causing CREB to be phosphorylated. Both PKA and the ras pathway, for example, can phosphorylate CREB. CREB can also be phosphorylated in response to increased intracellular calcium, in which case the CRE site is also called the CaRE (*calcium response element*) site. The calcium-dependent phosphorylation of CREB is primarily caused by Ca^{2+}/calmodulin kinase IV (a relative of CaMKII) and by MAPK, which leads to prolonged CREB phosphorylation. CREB phosphorylation must be maintained long enough for transcription to ensue, even though neuronal electrical activity only transiently raises intracellular calcium concentration. Such signaling cascades can potentiate CREB-mediated transcription by inhibiting a protein phosphatase that dephosphorylates CREB. CREB is thus an example of the convergence of multiple signaling pathways onto a single transcriptional activator.

Many genes whose transcription is regulated by CREB have been identified. CREB-sensitive genes include the immediate early gene *c-fos* (discussed shortly), the neurotrophin BDNF (see Chapter 23), the enzyme tyrosine hydroxylase (which is important for synthesis of catecholamine neurotransmitters; see Chapter 6), and many neuropeptides (including somatostatin, enkephalin, and corticotropin-releasing hormone). CREB also is thought

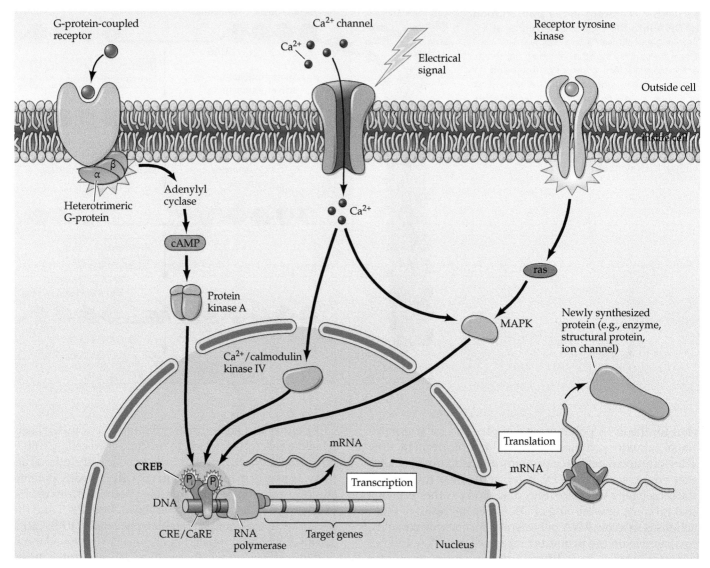

FIGURE 7.12 **Transcriptional regulation by CREB** Multiple signaling pathways converge by activating kinases that phosphorylate CREB. These include PKA, Ca^{2+}/calmodulin kinase IV, and MAPK. Phosphorylation of CREB allows it to bind co-activators (not shown here), which then stimulate RNA polymerase to begin synthesis of RNA. RNA is then processed and exported to the cytoplasm, where it serves as mRNA for translation into protein.

to mediate long-lasting changes in brain function. For example, CREB has been implicated in spatial learning, behavioral sensitization, long-term memory of odorant-conditioned behavior, and long-term synaptic plasticity (see Chapters 8, 25, and 26).

• *Nuclear receptors.* Nuclear receptors for membrane-permeant ligands also are transcriptional activators. The receptor for glucocorticoid hormones illustrates one mode of action of such receptors. In the absence of glucocorticoid

hormones, the receptors are located in the cytoplasm. Binding of glucocorticoids causes the receptor to unfold and move to the nucleus, where it binds a specific recognition site on the DNA. This DNA binding activates the relevant RNA polymerase complex to initiate transcription and subsequent gene expression. Thus, a critical regulatory event for steroid receptors is their translocation to the nucleus to allow DNA binding.

The receptor for thyroid hormone (TH) illustrates a second mode of regulation. In the absence of TH, the receptor is bound to DNA and serves as a potent repressor of transcription. Upon binding TH, the receptor undergoes a conformational change that ultimately opens the promoter for polymerase binding. Hence, TH binding switches the receptor from being a repressor to being an activator of transcription. Several hormones regulate gene expression via such nuclear receptors.

- *c-fos*. A different strategy of gene regulation is apparent in the function of the transcriptional activator protein **c-fos**. In resting cells, c-fos is present at a very low concentration. Stimulation of the target cell causes c-fos to be synthesized, and the amount of this protein rises dramatically over 30 to 60 minutes. Therefore, *c-fos* is considered to be an **immediate early gene** because its synthesis is directly triggered by the stimulus. Once synthesized, c-fos protein can act as a transcriptional activator to induce synthesis of second-order genes. These are termed **delayed response genes** because their activity is delayed by the fact that an immediate early gene—*c-fos* in this case—must be activated first.

Multiple signals converge on *c-fos*, activating different transcription factors that bind to at least three distinct sites in the promoter region of the gene. The regulatory region of the *c-fos* gene contains a binding site that mediates transcriptional induction by cytokines and ciliary neurotropic factor. Another site is targeted by growth factors such as neurotrophins through ras and protein kinase C, and a CRE/CaRE that can bind to CREB and thereby respond to cAMP or calcium entry resulting from electrical activity. In addition to synergistic interactions among these *c-fos* sites, transcriptional signals can be integrated by converging on the same activator, such as CREB.

Nuclear signaling events typically result in the generation of a large and relatively stable complex composed of a functional transcriptional activator protein, additional proteins that bind to the activator protein, and the RNA polymerase and associated proteins bound at the start site of transcription. Most of the relevant signaling events act to "seed" this complex by generating an active transcriptional activator protein by phosphorylation, by inducing a conformational change in the activator upon ligand binding, by fostering nuclear localization, by removing an inhibitor, or simply by making more activator protein.

CONCEPT
7.7

There Are Many Ways to Signal Changes in Neuronal Structure and Function

LEARNING OBJECTIVES

7.7.1 Explain how the neurotrophin NGF regulates neuron growth via changes in gene expression.

7.7.2 Describe the signal transduction cascade involved in long-term synaptic depression in the cerebellum.

7.7.3 Explain how phosphorylation of tyrosine hydroxylase regulates catecholamine neurotransmitter synthesis.

Understanding the general properties of signal transduction processes at the plasma membrane, in the cytosol, and in the nucleus makes it possible to consider how these processes work in concert to mediate specific functions in

the brain. Three important signal transduction pathways illustrate some of the roles of intracellular signal transduction processes in the nervous system.

- *NGF/TrkA*. The first of these is signaling by the **nerve growth factor** (**NGF**). This protein is a member of the neurotrophin growth factor family and is required for the differentiation, survival, and synaptic connectivity of sympathetic and sensory neurons (see Chapter 23). NGF works by binding to a high-affinity tyrosine kinase receptor, TrkA, found on the plasma membrane of these target cells (Figure 7.13). NGF binding causes TrkA receptors to dimerize, and the intrinsic tyrosine kinase activity of each receptor then phosphorylates its partner receptor. Phosphorylated TrkA receptors trigger the ras

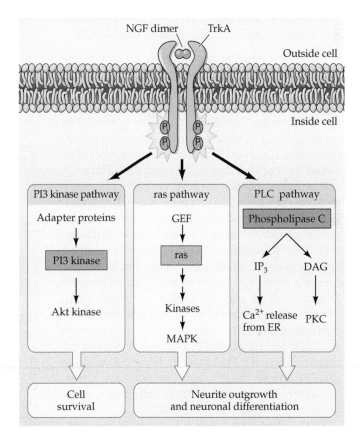

FIGURE 7.13   Mechanism of action of NGF NGF binds to a high-affinity tyrosine kinase receptor, TrkA, on the plasma membrane to induce phosphorylation of TrkA at two different tyrosine residues. These phosphorylated tyrosines serve to tether various adapter proteins or phospholipase C (PLC), which in turn activate three major signaling pathways: the PI3 kinase pathway leading to activation of Akt kinase, the ras pathway leading to MAPK, and the PLC pathway leading to release of intracellular Ca^{2+} from the endoplasmic reticulum and activation of PKC. The ras and PLC pathways primarily stimulate processes responsible for neuronal differentiation, whereas the PI3 kinase pathway is primarily involved in cell survival.

cascade, resulting in the activation of multiple protein kinases, including MAPK. Some of these kinases translocate to the nucleus to activate transcriptional activators, such as CREB. This ras-based component of the NGF pathway is primarily responsible for inducing and maintaining differentiation of NGF-sensitive neurons. Phosphorylation of TrkA also causes this receptor to stimulate the activity of phospholipase C, which increases production of IP_3 and DAG. IP_3 induces release of Ca^{2+} from the endoplasmic reticulum, and DAG activates PKC. These two second messengers appear to target many of the same downstream effectors as ras. Finally, activation of TrkA receptors also causes activation of other protein kinases (such as Akt kinase) that inhibit cell death. This pathway, therefore, primarily mediates the NGF-dependent survival of sympathetic and sensory neurons described in Chapter 23.

• *Long-term synaptic depression (LTD)*. The interplay between several intracellular signals can be observed at the excitatory synapses that innervate Purkinje cells in the cerebellum. These synapses are central to information flow through the cerebellar cortex, which coordinates motor movements (see Chapter 19). One of the synapses is between the parallel fibers (PFs) and their Purkinje cell targets: LTD is a form of synaptic plasticity that weakens these synapses (see Chapter 8). When PFs are active, they release the neurotransmitter glutamate onto the dendrites of Purkinje cells. This activates AMPA-type receptors, which are ligand-gated ion channels (see Chapter 6), and causes a small EPSP that briefly depolarizes the Purkinje cell. In addition to this electrical signal, the glutamate released by PFs also activates metabotropic glutamate receptors. These receptors stimulate phospholipase C to generate two second messengers in the Purkinje cell, IP_3 and DAG (Figure 7.14). When the PF synapses alone are active, these intracellular signals are insufficient to open IP_3 receptors or to stimulate PKC.

LTD is induced when PF synapses are activated at the same time as the glutamatergic climbing fiber synapses that also innervate Purkinje cells. The climbing fiber synapses produce large EPSPs that strongly depolarize the membrane potential of the Purkinje cell. This depolarization allows Ca^{2+} to enter the Purkinje cell via voltage-gated Ca^{2+} channels. When both synapses are simultaneously activated, the rise in intracellular Ca^{2+} concentration caused by the climbing fiber synapse enhances the sensitivity of IP_3 receptors to the IP_3 produced by PF synapses and allows the IP_3 receptors in the Purkinje cell to open. This releases Ca^{2+}

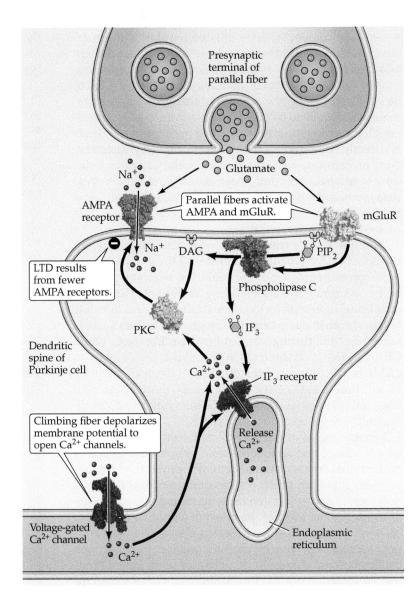

FIGURE 7.14 Signaling at cerebellar parallel fiber synapses during long-term synaptic depression Glutamate released by parallel fibers activates both AMPA-type and metabotropic receptors. The latter produce IP_3 and DAG in the Purkinje cell. When paired with a rise in Ca^{2+} associated with activity of climbing fiber synapses, the IP_3 causes Ca^{2+} to be released from the endoplasmic reticulum, while Ca^{2+} and DAG together activate protein kinase C. These signals together change the properties of AMPA receptors to produce long-term depression.

from the endoplasmic reticulum and further elevates Ca^{2+} concentration locally near the PF synapses. This larger rise in Ca^{2+}, in conjunction with the DAG produced by the PF synapses, activates PKC. PKC in turn phosphorylates several substrate proteins, including the AMPA-type receptors, ultimately altering trafficking of these receptors and leading to fewer AMPA-type receptors being located at the PF synapse (see Figure 8.17D). As a result, glutamate released from the PFs produces smaller EPSPs, causing LTD.

In short, transmission at Purkinje cell synapses produces brief electrical signals and chemical signals that last much longer. The temporal interplay between these signals allows LTD to occur only when both PF and climbing fiber synapses are active. The actions of IP$_3$, DAG, and Ca^{2+} also are restricted to small parts of the Purkinje cell dendrite, which is a more limited spatial range than the EPSPs, which spread throughout the entire dendrite and cell body of the Purkinje cell. Thus, in contrast to the electrical signals, the second-messenger signals can impart information about the location of active synapses and allow LTD to occur only in the vicinity of active PFs.

- *Phosphorylation of tyrosine hydroxylase.* A third example of intracellular signaling in the nervous system is the regulation of the enzyme tyrosine hydroxylase. Tyrosine hydroxylase governs the synthesis of the catecholamine neurotransmitters: dopamine, norepinephrine, and epinephrine (see Chapter 6). Several signals, including electrical activity, other neurotransmitters, and NGF, increase the rate of catecholamine synthesis by increasing the catalytic activity of tyrosine hydroxylase (Figure 7.15). The rapid increase of tyrosine hydroxylase activity is largely due to phosphorylation of this enzyme.

Tyrosine hydroxylase is a substrate for several protein kinases, including PKA, CaMKII, MAPK, and PKC. Phosphorylation causes conformational changes that increase the catalytic activity of tyrosine hydroxylase. Stimuli that elevate cAMP, Ca^{2+}, or DAG can all increase tyrosine hydroxylase activity and thus increase the rate of catecholamine biosynthesis. This regulation by several different signals allows for close control of tyrosine hydroxylase activity and illustrates how several different pathways can converge to influence a key enzyme involved in synaptic transmission.

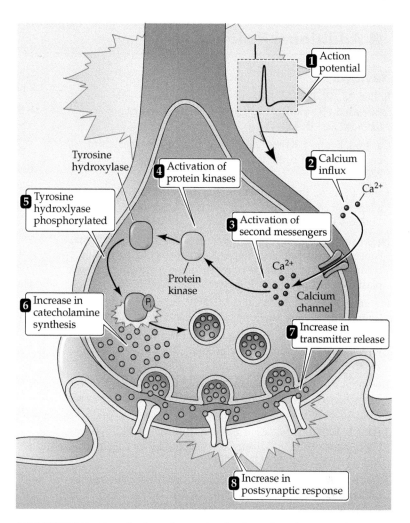

FIGURE 7.15 Regulation of tyrosine hydroxylase by protein phosphorylation Tyrosine hydroxylase governs the synthesis of the catecholamine neurotransmitters and is stimulated by several intracellular signals. In the example shown here, neuronal electrical activity (1) causes influx of Ca^{2+} (2). The resultant rise in intracellular Ca^{2+} concentration (3) activates protein kinases (4), which phosphorylates tyrosine hydroxylase (5), stimulating catecholamine synthesis (6). This increased synthesis in turn increases the release of catecholamines (7) and enhances the postsynaptic response produced by the synapse (8).

Summary

Diverse signal transduction pathways exist within all neurons. Activation of these pathways typically is initiated by chemical signals such as neurotransmitters and hormones, which bind to receptors that include ligand-gated ion channels, G-protein-coupled receptors, and tyrosine kinase receptors. Many of these receptors activate either heterotrimeric or monomeric G-proteins that regulate intracellular enzyme cascades or ion channels. A common outcome of the activation of these receptors is the production of second messengers, such as cAMP, Ca^{2+}, and IP$_3$, that bind to effector proteins. Particularly important effectors are protein kinases and phosphatases that regulate the phosphorylation state of their substrates, and thus their function. These substrates can be metabolic enzymes or other signal transduction molecules, such as ion channels, protein kinases, or transcription factors that regulate gene expression. Examples of such transcription factors include CREB, steroid hormone receptors, and c-fos. This plethora of molecular components allows intracellular signal transduction pathways to generate responses over a wide range of times and distances, greatly augmenting and refining the information-processing ability of neuronal circuits and, ultimately, brain systems.

■ Additional Reading

Reviews

Bollen, M., W. Peti, M. J. Ragusa and M. Beullens (2010) The extended PP1 toolkit: Designed to create specificity. *Trends Biochem. Sci.* 35: 450–458.

Carafoli, E. and J. Krebs (2016) Why calcium? How calcium became the best communicator. *J. Biol. Chem.* 291: 20849–20857.

Finch, E. A., K. Tanaka and G. J. Augustine (2012) Calcium as a trigger for cerebellar long-term synaptic depression. *Cerebellum* 11: 706–717.

Hagenston, A. M., H. Bading and C. Bas-Orth (2020) Functional consequences of calcium-dependent synapse-to-nucleus communication: Focus on transcription-dependent metabolic plasticity. *Cold Spring Harb. Perspect. Biol.* 12: a035287.

Hamada, K. and K. Mikoshiba (2020) IP_3 receptor plasticity underlying diverse functions. *Annu. Rev. Physiol.* 82: 151–176.

Li, H., A. Rao and P. G. Hogan (2011) Interaction of calcineurin with substrates and targeting proteins. *Trends Cell Biol.* 21: 91–103.

Rosenbaum, D. M., S. G. Rasmussen and B. K. Kobilka (2009) The structure and function of G-protein-coupled receptors. *Nature* 459: 356–363.

Taylor, S. S., R. Ilouz, P. Zhang and A. P. Kornev (2012) Assembly of allosteric macromolecular switches: lessons from PKA. *Nat. Rev. Mol. Cell. Biol.* 13: 646–658.

Turk, B. E. (2007) Manipulation of host signalling pathways by anthrax toxins. *Biochem. J.* 402: 405–417.

Woll, K. A. and F. Van Petegem (2022) Calcium-release channels: structure and function of IP_3 receptors and ryanodine receptors. *Physiol. Rev.* 102: 209–268.

Important Original Papers

Bueno-Carrasco, M. T. and 12 others. (2022) Structural mechanism for tyrosine hydroxylase inhibition by dopamine and reactivation by Ser40 phosphorylation. *Nat. Commun.* 13: 74.

Burgess, G. M. and 5 others (1984) The second messenger linking receptor activation to internal calcium release in liver. *Nature* 309: 63–66.

Cho, U. S. and W. Xu (2007) Crystal structure of a protein phosphatase 2A heterotrimeric holoenzyme. *Nature* 445: 53–57.

Craddock, T. J. A., J. A. Tuszynski and S. Hameroff (2012) Cytoskeletal signaling: Is memory encoded in microtubule lattices by CaMKII phosphorylation? *PLOS Comput. Biol.* 8: e1002421.

Crerar, H. and 9 others (2019) Regulation of NGF signaling by an axonal untranslated mRNA. *Neuron* 102: 553–563.

Finch, E. A. and G. J. Augustine (1998) Local calcium signaling by IP_3 in Purkinje cell dendrites. *Nature* 396: 753–756.

Garcia-Marcos, M. (2021) Complementary biosensors reveal different G-protein signaling modes triggered by GPCRs and non-receptor activators. *eLife* 10: e65620.

Harward, S. C. and 9 others (2016) Autocrine BDNF-TrkB signalling within a single dendritic spine. *Nature* 538: 99–103.

Hilger, D. and 12 others (2020) Structural insights into differences in G protein activation by family A and family B GPCRs. *Science* 369: eaba3373.

Jones-Tabah, J. and 4 others (2021) High-content single-cell Förster resonance energy transfer imaging of cultured striatal neurons reveals novel cross-talk in the regulation of nuclear signaling by protein kinase A and extracellular signal-regulated kinase 1/2. *Mol. Pharmacol.* 100: 526–539.

Lee, H. K., A. Cording, J. Vielmetter and K. Zinn (2013) Interactions between a receptor tyrosine phosphatase and a cell surface ligand regulate axon guidance and glial-neuronal communication. *Neuron* 78: 813–826.

Leonard, T. A. and 4 others (2011) Crystal structure and allosteric activation of protein kinase C βII. *Cell* 144: 55–66.

Books

Alberts, B. and 8 others (2022) *Molecular Biology of the Cell*, 7th Edition. New York: Garland Science.

Carafoli, E. and C. Klee (1999) *Calcium as a Cellular Regulator*. New York: Oxford University Press.

Synaptic Plasticity

Overview

Synaptic connections between neurons provide the basic "wiring" of the brain's circuitry. However, unlike the wiring of an electronic device such as a computer, the strength of synaptic connections between neurons is a dynamic entity that is constantly changing in response to neural activity and other influences. Such changes in synaptic transmission arise from several forms of plasticity that vary in timescale from milliseconds to years. Most short-term forms of synaptic plasticity affect the amount of neurotransmitter released from presynaptic terminals in response to a presynaptic action potential. Several forms of short-term synaptic plasticity—including facilitation, augmentation, and potentiation—enhance neurotransmitter release and are caused by persistent actions of calcium ions within the presynaptic terminal. Another form of short-term plasticity, synaptic depression, decreases the amount of neurotransmitter released and appears to be due to an activity-dependent depletion of synaptic vesicles that are ready to undergo exocytosis. Long-term forms of synaptic plasticity alter synaptic transmission over timescales of 30 minutes or longer. Examples of such long-lasting plasticity include long-term potentiation and long-term depression. These long-lasting forms of synaptic plasticity arise from molecular mechanisms that vary over time: The initial changes in synaptic transmission arise from post-translational modifications of existing proteins, most notably changes in the trafficking of glutamate receptors, while later phases of synaptic modification result from changes in gene expression and synthesis of new proteins. These changes produce enduring changes in synaptic transmission, including synapse growth, that can yield essentially permanent modifications of brain function.

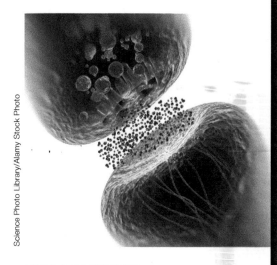

Science Photo Library/Alamy Stock Photo

KEY CONCEPTS

8.1 Some forms of synaptic plasticity last for a few minutes or less

8.2 Synaptic plasticity can produce long-lasting changes in behavior

8.3 Long-term synaptic plasticity is found in the mammalian hippocampus

8.4 NMDA-type glutamate receptors serve as coincidence detectors for hippocampal long-term potentiation

8.5 Long-term depression weakens synapses via multiple mechanisms

8.6 Some forms of plasticity depend on the timing of synaptic activity

---

CONCEPT
8.1 | **Some Forms of Synaptic Plasticity Last for a Few Minutes or Less**

LEARNING OBJECTIVES

8.1.1 Compare the time course of different forms of short-term synaptic plasticity.

8.1.2 Distinguish the mechanisms of synaptic facilitation, depression, augmentation, and post-tetanic potentiation.

Types of short-term synaptic plasticity

Chemical synapses are capable of undergoing plastic changes that either strengthen or weaken synaptic transmission. Synaptic plasticity mechanisms occur on timescales ranging from milliseconds to days, weeks, or longer. The short-term forms of plasticity—those lasting for a few minutes or less—are readily observed during repeated activation of any chemical synapse. There are several forms of short-term synaptic plasticity that differ in their time courses and their underlying mechanisms.

Synaptic facilitation is a rapid increase in synaptic strength that occurs when two or more action potentials invade the presynaptic terminal within a few milliseconds of each other (Figure 8.1A). By varying the time interval between presynaptic action potentials, it can be seen that

facilitation produced by the first action potential lasts for tens of milliseconds (Figure 8.1B). Many lines of evidence indicate that facilitation is the result of prolonged elevation of presynaptic calcium levels following synaptic activity. Although the entry of Ca^{2+} into the presynaptic terminal occurs within 1 to 2 ms after an action potential invades (see Figure 5.7B), the mechanisms that return Ca^{2+} to

(A)

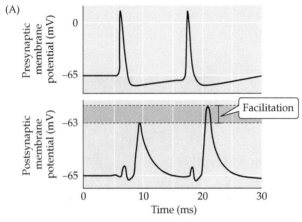

(B)

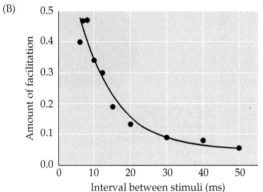

(C)

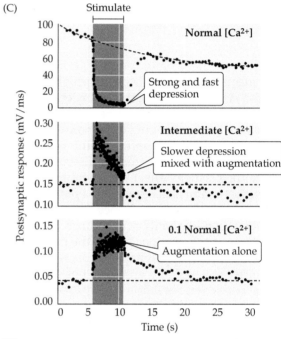

(D)

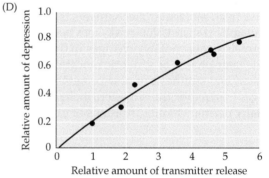

(E)

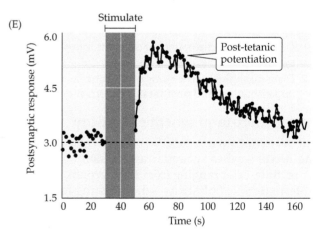

FIGURE 8.1 Forms of short-term synaptic plasticity
(A) Facilitation at the squid giant synapse. A pair of presynaptic action potentials elicits two excitatory postsynaptic potentials (EPSPs). Because of facilitation, the second EPSP is larger than the first. (B) By varying the time interval between pairs of presynaptic action potentials, it can be seen that facilitation decays over a time course of tens of milliseconds. (C) Under typical physiological conditions, a high-frequency tetanus (shading) causes pronounced depression of EPSPs at the squid giant synapse (top). Lowering the external Ca^{2+} concentration to an intermediate level reduces transmitter release and causes a mixture of depression and augmentation (middle). Further reduction of the external Ca^{2+} eliminates depression, leaving only augmentation (bottom). (D) Synaptic depression at the frog neuromuscular synapse increases in proportion to the amount of transmitter released from the presynaptic terminal. (E) Application of a high-frequency tetanus (shading) to presynaptic axons innervating a spinal motor neuron causes a post-tetanic potentiation that persists for a few minutes after the tetanus ends. (A,B after M. P. Charlton and G. D. Bittner, 1978. *J Gen Physiol* 72: 487–511; C after D. Swandulla et al., 1991. *Neuron* 7: P915–926; D from W. J. Betz, 1970. *J Physiol* 206: 629–644; E after A. Lev-Tov et al., 1983. *J Neurophysiol* 50: 379–398.)

resting levels are much slower. Thus, when action potentials arrive close together in time, calcium builds up in the terminal and allows more neurotransmitter to be released by a subsequent presynaptic action potential. Recent evidence indicates that the target of this residual Ca^{2+} signal is synaptotagmin 7, a Ca^{2+}-binding protein that is found on the plasma membrane and is related to the synaptotagmins that are present on synaptic vesicles and serve as Ca^{2+} sensors for triggering neurotransmitter release (see Chapter 5).

Opposing facilitation is **synaptic depression**, which causes neurotransmitter release to decline during sustained synaptic activity (Figure 8.1C). An important clue to the cause of synaptic depression comes from observations that depression depends on the amount of neurotransmitter that has been released. For example, lowering the external Ca^{2+} concentration, to reduce the number of quanta released by each presynaptic action potential, causes the rate of depression to be slowed (see Figure 8.1C). Likewise, the total amount of depression is proportional to the amount of transmitter released from the presynaptic terminal (Figure 8.1D). These results have led to the idea that depression is caused by progressive depletion of a pool of synaptic vesicles that are available for release: When rates of release are high, these vesicles deplete rapidly and cause a lot of depression; depletion slows as the rate of release is reduced, yielding less depression. According to this vesicle depletion hypothesis, depression causes the strength of transmission to decline until this pool is replenished by mobilization of vesicles from a reserve pool. Consistent with this explanation are observations that more depression is observed after the size of the reserve pool is reduced by impairing synapsin, a protein that maintains vesicles in the reserve pool (see Chapter 5).

Still other forms of synaptic plasticity, such as synaptic **potentiation** and **augmentation**, also are elicited by repeated synaptic activity and serve to increase the amount of transmitter released from presynaptic terminals. Both augmentation and potentiation enhance the ability of incoming Ca^{2+} to trigger fusion of synaptic vesicles with the plasma membrane, but the two processes work over different timescales. While augmentation rises and falls over a few seconds (see Figure 8.1C, bottom), potentiation acts over a timescale of tens of seconds to minutes (Figure 8.1E). As a result of its slow time course, potentiation can greatly outlast the tetanic stimulus that induces it, and is often called **post-tetanic potentiation (PTP)**. Although both augmentation and potentiation are thought to arise from prolonged elevation of presynaptic calcium levels during synaptic activity, the mechanisms responsible for these forms of plasticity are poorly understood. It has been proposed that augmentation results from Ca^{2+} enhancing the actions of the presynaptic SNARE-regulatory protein munc13 (see Figure 5.11), while potentiation may arise when Ca^{2+} activates presynaptic protein kinases that go on to phosphorylate substrates (such as synapsin) that regulate transmitter release.

During repetitive synaptic activity, the various forms of short-term plasticity can interact to cause synaptic transmission to change in complex ways. For example, at the peripheral neuromuscular synapse, repeated activity first causes an accumulation of Ca^{2+} in the presynaptic terminal that allows facilitation and then augmentation to enhance synaptic transmission (Figure 8.2). The ensuing

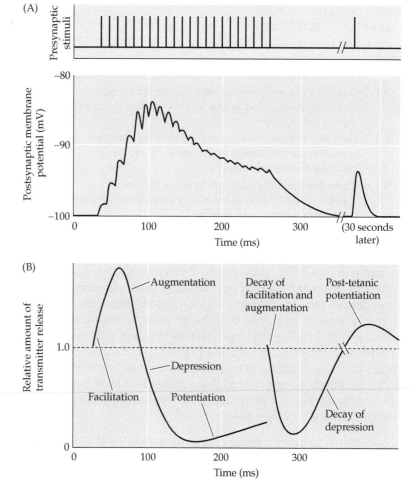

FIGURE 8.2 Short-term plasticity at the neuromuscular synapse (A) A train of electrical stimuli (top) applied to the presynaptic motor nerve produces changes in the end plate potential (EPP) amplitude (bottom). (B) Dynamic changes in transmitter release caused by the interplay of several forms of short-term plasticity. Facilitation and augmentation of the EPP occur at the beginning of the stimulus train and are followed by a pronounced depression of the EPP. Potentiation begins late in the stimulus train and persists for many seconds after the end of the stimulus, leading to post-tetanic potentiation. (A after B. Katz, 1966. *Nerve, Muscle, and Synapse*. New York: McGraw-Hill; B after R. C. Malenka and S. A. Siegelbaum, 2001. In W. M. Cowan et al. [eds.], pp. 393–413. Baltimore: Johns Hopkins University Press.)

depletion of synaptic vesicles then causes depression to dominate and weaken the synapse. Presynaptic action potentials that occur within 1 to 2 minutes after the end of the tetanus release more neurotransmitter because of the persistence of post-tetanic potentiation. Although their relative contributions vary from synapse to synapse, these forms of short-term synaptic plasticity collectively cause transmission at all chemical synapses to change dynamically as a consequence of the recent history of synaptic activity.

CONCEPT 8.2 | Synaptic Plasticity Can Produce Long-Lasting Changes in Behavior

LEARNING OBJECTIVES

8.2.1 Identify the forms of synaptic plasticity involved in habituation and sensitization of *Aplysia* gill withdrawal behavior.

8.2.2 Describe the mechanisms responsible for synaptic depression occurring at the sensory–motor synapse during habituation.

8.2.3 Describe the mechanisms responsible for synaptic enhancement occurring at the sensory–motor synapse during sensitization.

8.2.4 Compare the molecular mechanisms involved in synaptic modifications underlying short-term versus long-term changes in behavior.

Facilitation, depression, augmentation, and potentiation modify synaptic transmission over timescales of a few minutes or less. While these mechanisms probably are responsible for many short-lived changes in brain circuitry, they cannot provide the basis for changes in brain function that persist for weeks, months, or years. Many synapses exhibit long-lasting forms of synaptic plasticity that are plausible substrates for more permanent changes in brain function. Because of their duration, these forms of synaptic plasticity may be cellular correlates of learning and memory. Thus, a great deal of effort has gone into understanding how they are generated.

Long-term synaptic plasticity and behavioral modification in *Aplysia*

An obvious obstacle to exploring synaptic plasticity in the brains of humans and other mammals is the enormous number of neurons and the complexity of synaptic connections. One way to circumvent this dilemma is to examine plasticity in far simpler nervous systems. Eric Kandel and his colleagues at Columbia University have used the marine mollusk *Aplysia californica* (Figure 8.3A) to identify several forms of long-term synaptic plasticity and demonstrate that such forms of synaptic plasticity underlie simple forms of learning. This sea slug has only a few tens of thousands

of neurons, many of which are quite large (up to 1 mm in diameter) and in stereotyped locations within the ganglia that make up the animal's nervous system (Figure 8.3B). These attributes make it practical to monitor the electrical activity and synaptic connectivity of specific, identifiable nerve cells, and thus to define the synaptic circuits involved in mediating the limited behavioral repertoire of *Aplysia*.

Aplysia exhibit several elementary forms of behavioral plasticity. One form is **habituation**, a process that causes the animal to become less responsive to repeated occurrences of a stimulus. Habituation is found in many other species, including humans. For example, when dressing we initially experience tactile sensations due to clothes stimulating our skin, but habituation quickly causes these sensations to fade. Similarly, a light touch to the siphon of an *Aplysia* results in withdrawal of the animal's gill, but habituation causes the gill withdrawal to become weaker during repeated stimulation of the siphon (Figure 8.3C). The gill withdrawal response of *Aplysia* also exhibits another form of plasticity called **sensitization**. Sensitization is a process that allows an animal to generalize an aversive response—elicited by a noxious stimulus—to a variety of other, non-noxious stimuli. In *Aplysia* that have habituated to siphon touching, sensitization of gill withdrawal is elicited by pairing a strong electrical stimulus to the animal's tail with another light touch to the siphon. This pairing causes the siphon stimulus to again elicit a strong withdrawal of the gill (see Figure 8.3C, right) because the noxious stimulus to the tail sensitizes the gill withdrawal reflex to light touch. Even after a single stimulus to the tail, the gill withdrawal reflex remains enhanced for at least an hour (Figure 8.3D). This can be viewed as a simple form of short-term memory. With repeated pairing of tail and siphon stimuli, this behavior can be altered for days or weeks (Figure 8.3E), thus demonstrating a simple form of long-term memory.

The small number of neurons in the *Aplysia* nervous system makes it possible to define the synaptic circuits involved in gill withdrawal and to monitor the activity of individual neurons in these circuits. Although hundreds of neurons are ultimately involved in producing this simple behavior, the activities of only a few different types of neurons can account for gill withdrawal and its plasticity during habituation and sensitization. These critical neurons include mechanosensory neurons that innervate the siphon, motor neurons that innervate muscles in the gill, and interneurons that receive inputs from a variety of sensory neurons (Figure 8.4A). Touching the siphon activates the mechanosensory neurons, which form excitatory synapses that release glutamate onto both the interneurons and the motor neurons; thus, touching the siphon increases the probability that both of these postsynaptic targets will produce action potentials. The interneurons form excitatory synapses on motor neurons, further increasing the

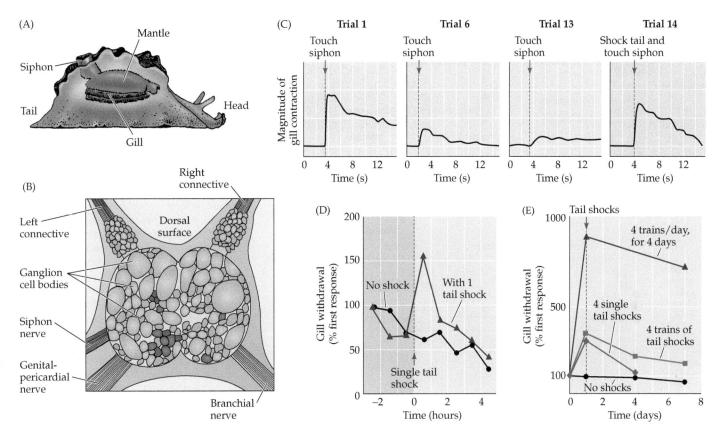

FIGURE 8.3 **Short-term sensitization of the *Aplysia* gill withdrawal reflex** (A) Drawing of an *Aplysia* (commonly known as a sea slug). (B) The abdominal ganglion of *Aplysia*. The cell bodies of many of the neurons involved in gill withdrawal can be recognized by their size, shape, and position within this ganglion. (C) Changes in the gill withdrawal behavior due to habituation and sensitization. The first time the siphon is touched, the gill contracts vigorously. Repeated touches elicit smaller gill contractions because of habituation. Subsequently pairing a siphon touch with an electrical shock to the tail restores a large and rapid gill contraction, the result of short-term sensitization. (D) Time course of short-term sensitization of the gill withdrawal response following the pairing of a single tail shock with a siphon touch. (E) Repeated applications of tail shocks cause prolonged sensitization of the gill withdrawal response. (After L. R. Squire and E. R. Kandel, 1999. *Memory: From Mind to Molecules.* New York: Scientific American Library.)

likelihood of the motor neurons firing action potentials in response to mechanical stimulation of the siphon. When the motor neurons are activated by the summed synaptic excitation of the sensory neurons and interneurons, they release acetylcholine that excites the muscle cells of the gill, producing gill withdrawal.

Both habituation and sensitization appear to arise from plastic changes in synaptic transmission in this circuit. During habituation, transmission at the glutamatergic synapse between the sensory and motor neurons is depressed (see Figure 8.4B, left). This synaptic depression persists for many minutes and is thought to be responsible for the decreasing ability of siphon stimuli to evoke gill contractions during habituation. Much like the short-term form of synaptic depression described in Concept 8.1, this depression is presynaptic and is due to a reduction in the number of synaptic vesicles available for release. Sensitization modifies the function of this circuit by recruiting additional sensory neurons that innervate the tail. These

sensory neurons in turn excite modulatory interneurons that release serotonin onto the presynaptic terminals of the sensory neurons of the siphon (see Figure 8.4A). Serotonin enhances transmitter release from the siphon sensory neuron terminals, leading to increased synaptic excitation of the motor neurons (Figure 8.4B). This modulation of the sensory neuron–motor neuron synapse lasts approximately 1 hour (Figure 8.4C), which is similar to the duration of the short-term sensitization of gill withdrawal produced by applying a single stimulus to the tail (see Figure 8.3D).

The mechanism responsible for the enhancement of glutamatergic transmission during short-term sensitization is shown in Figure 8.5A. Serotonin released by the modulatory interneurons binds to G-protein-coupled receptors on the presynaptic terminals of the siphon sensory neurons (step 1), which stimulates production of the second messenger, cAMP (step 2). Cyclic AMP binds to the regulatory subunits of protein kinase A (PKA; step 3),

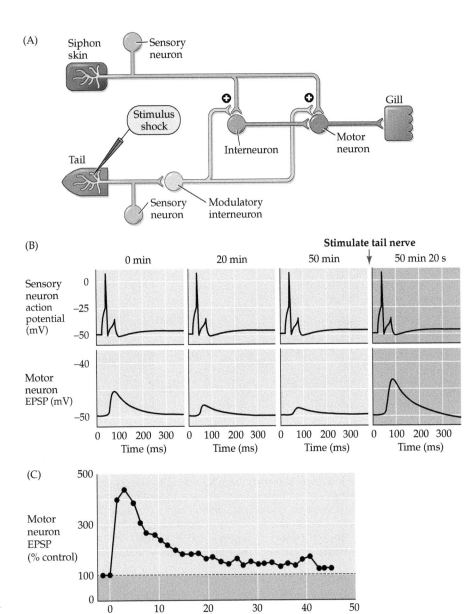

FIGURE 8.4 Synaptic mechanisms underlying short-term sensitization. (A) Neural circuitry involved in sensitization. Touching the siphon skin activates sensory neurons that excite interneurons and gill motor neurons, yielding a contraction of the gill muscle. A shock to the animal's tail stimulates modulatory interneurons that alter synaptic transmission between the siphon sensory neurons and gill motor neurons, resulting in sensitization. (B) Changes in synaptic efficacy at the sensory neuron–motor neuron synapse during short-term sensitization. Prior to sensitization, activating the siphon sensory neurons causes an EPSP to occur in the gill motor neurons. Repetitive activation of this synapse causes synaptic depression, indicated by a reduction in EPSPs in motor neurons. Activation of the serotonergic modulatory interneurons enhances release of transmitter from the sensory neurons onto the motor neurons, increasing the EPSP in the motor neurons and causing the motor neurons to more strongly excite the gill muscle. (C) Time course of the serotonin-induced facilitation of transmission at the sensory–motor synapse. (After L. R. Squire and E. R. Kandel, 1999. *Memory: From Mind to Molecules.* New York: Scientific American Library.)

liberating catalytic subunits of PKA that are then able to phosphorylate several proteins, probably including K⁺ channels (step 4). The net effect of PKA action is to reduce the probability that the K⁺ channels open during a presynaptic action potential. This prolongs the presynaptic action potential, thereby opening more presynaptic Ca^{2+} channels (step 5). There is evidence that the opening of presynaptic Ca^{2+} channels is also directly enhanced by serotonin. Finally, the enhanced influx of Ca^{2+} into the presynaptic terminals increases the amount of transmitter released onto motor neurons during a sensory neuron action potential (step 6). In summary, a signal transduction cascade that involves a modulatory neurotransmitter, second messengers, protein kinases, and ion channels mediates short-term sensitization of gill withdrawal. This cascade ultimately causes a short-term enhancement of

synaptic transmission between the sensory and motor neurons in the gill withdrawal circuit.

Gene expression and long-lasting memory in *Aplysia*

The same serotonin-induced enhancement of glutamate release that mediates short-term sensitization is also thought to underlie long-term sensitization. However, during long-term sensitization this circuitry is affected for up to several weeks. The prolonged duration of this form of plasticity is evidently due to changes in gene expression and thus protein synthesis (Figure 8.5B). With repeated training (i.e., additional tail shocks), the serotonin-activated PKA involved in short-term sensitization also phosphorylates—and thereby activates—the transcriptional activator CREB. As described in Chapter 7, CREB

(A)

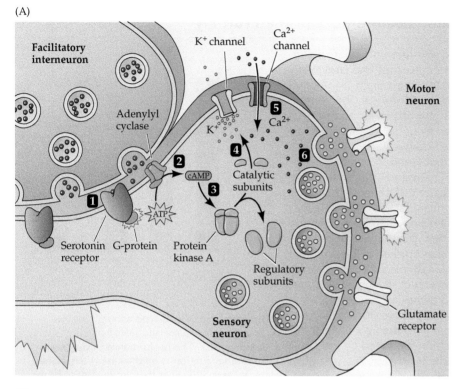

(B)

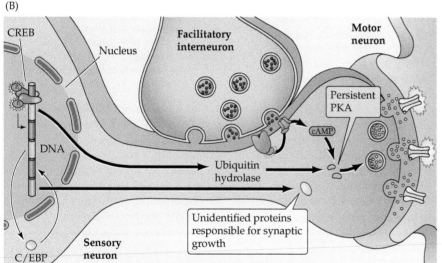

FIGURE 8.5 Mechanisms of presynaptic enhancement underlying behavioral sensitization (A) Short-term sensitization is due to an acute, PKA-dependent enhancement of glutamate release from the presynaptic terminals of sensory neurons. See text for explanation. (B) Long-term sensitization is due to changes in gene expression, resulting in the synthesis of proteins that change PKA activity and lead to changes in synapse growth. (After L. R. Squire and E. R. Kandel, 1999. *Memory: From Mind to Molecules*. New York: Scientific American Library.)

genes that cause addition of synaptic terminals, yielding a long-term increase in the number of synapses between the sensory and the motor neurons. Such structural increases are not seen following short-term sensitization and therefore may represent the ultimate cause of the long-lasting enhancement of the gill withdrawal response. Another protein involved in the long-term synaptic facilitation is a cytoplasmic polyadenylation element binding protein, somewhat confusingly called CPEB. CPEB activates mRNAs and may be important for local control of protein synthesis. Most intriguing, CPEB has self-sustaining properties like those of prion proteins (see Clinical Applications, Chapter 19), which could allow CPEB to remain active in perpetuity and thereby mediate permanent changes in synaptic structure to generate long-term sensitization.

These studies of *Aplysia* and related work on other invertebrates, such as the fruit fly (Box 8A), have led to several generalizations about synaptic plasticity. First, synaptic plasticity clearly can lead to changes in circuit function and, ultimately, to behavioral plasticity. This conclusion has triggered intense interest in synaptic plasticity mechanisms. Second, these plastic changes in synaptic function can be either short-term effects that rely on post-translational modification of existing synaptic proteins, or they can be long-term changes that require changes in gene expression, new protein synthesis, and growth of new synapses (as well as enlarging or eliminating existing synapses). Thus, it appears that short- and long-term changes in synaptic function have different mechanistic underpinnings. As the following Concepts will show, these generalizations apply to synaptic plasticity in the mammalian brain and have helped guide our understanding of these forms of synaptic plasticity.

binding to the cAMP response elements (CREs) in regulatory regions of nuclear DNA increases the rate of transcription of downstream genes. Although the changes in genes and gene products that follow CRE activation have been difficult to sort out, several consequences of gene activation have been identified. First, CREB stimulates the synthesis of an enzyme, ubiquitin hydrolase, that stimulates degradation of the regulatory subunit of PKA. This causes a long-lasting increase in the amount of free catalytic subunit, meaning that some PKA is persistently active and no longer requires serotonin to be activated. CREB also stimulates another transcriptional activator protein called C/EBP. C/EBP stimulates transcription of other, unknown

■ BOX 8A | Genetics of Learning and Memory in the Fruit Fly

As part of a renaissance in the genetic analysis of simple organisms in the mid-1970s, several investigators recognized that the genetic basis of learning and memory might be effectively studied in the fruit fly *Drosophila melanogaster*. In the intervening 50 years, this approach has yielded some fundamental insights. Although the mechanisms of learning and memory have certainly been among the more difficult problems tackled by *Drosophila* geneticists, their efforts have been surprisingly successful. Several genetic mutations that alter learning and memory have been discovered, and the identification of these genes has provided a valuable framework for studying the cellular mechanisms of these processes.

The initial problem in this work was to develop behavioral tests that could identify atypical learning or memory defects in large populations of flies. This challenge was met by Seymour Benzer and his colleagues Chip Quinn and Bill Harris at the California Institute of Technology, who developed the olfactory and visual learning tests that have become the basis for most subsequent analyses of learning and memory in the fruit fly (see figure). Behavioral paradigms pairing odors or light with an aversive stimulus allowed Benzer and his colleagues to assess associative learning in flies. The design of an ingenious testing apparatus controlled for non-learning-related sensory cues that had previously complicated such behavioral testing. Moreover, the apparatus allowed large numbers of flies to be screened relatively easily, expediting the analysis of mutagenized populations.

These studies led to the identification of an ever-increasing number of single gene mutations that disrupt learning or memory in flies. The behavioral and molecular studies of the mutants (given whimsical but descriptive names such as *dunce*, *rutabaga*, and *amnesiac*) suggested that a central pathway for learning and memory in the fly is signal transduction mediated by the cyclic nucleotide cAMP. Thus, the gene products of the *dunce*, *rutabaga*, and *amnesiac* loci are, respectively, a phosphodiesterase (which degrades cAMP), adenylyl cyclase (which converts ATP to cAMP), and a peptide transmitter that stimulates adenylyl cyclase. This conclusion about the importance of cAMP has been confirmed by the finding that genetic manipulation of the CREB transcription factor also interferes with learning and memory in typical flies.

These observations in *Drosophila* accord with conclusions reached in studies of *Aplysia* and mammals (see Concept

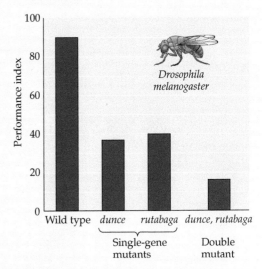

Performance of wild-type and mutant fruit flies (*Drosophila melanogaster*) on an olfactory learning task. The performance of *dunce* and of *rutabaga* mutants on this task is diminished by at least 50%. Flies that are mutant at both the *dunce* and *rutabaga* loci exhibit an even larger decrease in performance, suggesting that the two genes disrupt different but related aspects of learning. (After T. Tully, 1996. *Proc Natl Acad Sci USA* 93: 13460–13467. © 1996 National Academy of Sciences.)

8.2) and have emphasized the importance of cAMP-mediated learning and memory in a wide range of additional species. More generally, the genetic accessibility of *Drosophila* continues to make it a powerful experimental system for understanding the genetic underpinnings of learning and memory.

CONCEPT
8.3

Long-Term Synaptic Plasticity Is Found in the Mammalian Hippocampus

LEARNING OBJECTIVES

8.3.1 Distinguish long-term synaptic potentiation from other forms of synaptic plasticity.

8.3.2 Explain the type of electrical activity that is required to evoke long-term synaptic potentiation in the hippocampus.

8.3.3 List three properties of long-term potentiation that make this form of long-lasting synaptic plasticity an attractive neural mechanism for information storage.

Long-term plasticity at a hippocampal synapse

Long-term synaptic plasticity has also been identified in the mammalian brain. Here, some patterns of synaptic activity produce a long-lasting increase in synaptic strength known as **long-term potentiation** (LTP), whereas other patterns of activity produce a long-lasting decrease in synaptic strength, known as **long-term depression** (LTD). LTP and LTD are broad terms that describe only the direction of change in synaptic efficacy; in fact, different cellular and molecular mechanisms can be involved in producing LTP or LTD at different synapses throughout the brain. In general, LTP and LTD are produced by different histories of activity and are mediated by different intracellular signal transduction pathways in the nerve cells involved.

Long-term synaptic plasticity has been most thoroughly studied at excitatory synapses in the mammalian hippocampus. The hippocampus is especially important in the formation and retrieval of some forms of memory (see Chapter 30). In humans, functional imaging shows that the hippocampus is activated during certain kinds of memory tasks and that damage to this brain region results in an inability to form certain types of new memories. Although many other brain areas are involved in the complex process of memory formation, storage, and retrieval, these observations from imaging and lesion (damage) studies have led many investigators to study long-term synaptic plasticity of hippocampal synapses.

Work on LTP began in the late 1960s, when Terje Lomo and Timothy Bliss, working in the laboratory of Per Andersen in Oslo, Norway, discovered that a few seconds of high-frequency electrical stimulation can enhance synaptic transmission in the rabbit hippocampus for hours or longer. More recently, however, progress in understanding the mechanism of LTP has relied heavily on in vitro studies of slices of living hippocampus. The arrangement of neurons allows the hippocampus to be sectioned such that most of the relevant circuitry is left intact. In such preparations, the cell bodies of the pyramidal neurons lie in a single densely packed layer that is readily apparent (Figure 8.6). This layer

is divided into several distinct regions, the major ones being CA1 and CA3. "CA" refers to *cornu Ammonis*, Latin for Ammon's horn—the ram's horn that resembles the shape of the hippocampus. The dendrites of pyramidal cells in the CA1 region form a thick band (the stratum radiatum), where they receive synapses from Schaffer collaterals, the axons of pyramidal cells in the CA3 region.

Much of the work on LTP has focused on the synaptic connections between the Schaffer collaterals and CA1 pyramidal cells. Electrical stimulation of Schaffer collaterals generates EPSPs in the postsynaptic CA1 cells (Figure 8.7A,B). If the Schaffer collaterals are stimulated only two or three times per minute, the size of the EPSP elicited in the CA1 neurons remains constant. However, a brief, high-frequency train of stimuli to the same axons causes LTP, which is evident as a long-lasting increase in EPSP amplitude (Figure 8.7B,C). While the maximum duration of LTP is not known, in some cases LTP can last for more than a year (Figure 8.7D). The long duration of LTP shows that this form of synaptic plasticity is capable of serving as a mechanism for long-lasting storage of information. LTP occurs at each of the three excitatory synapses of the hippocampus shown in Figure 8.6. LTP also is found at excitatory synapses in a variety of brain regions—including the cortex, amygdala, and cerebellum—and at some inhibitory synapses as well.

Properties of long-term potentiation

LTP of the Schaffer collateral synapse exhibits several properties that make it an attractive neural mechanism for information storage. First, LTP requires strong activity in both presynaptic and postsynaptic neurons. If action potentials in a small number of presynaptic Schaffer collaterals—which evoke transmitter release that produces subthreshold EPSPs that would not normally yield LTP—are paired with strong depolarization of the postsynaptic CA1 cell, the activated Schaffer collateral synapses undergo LTP (Figure 8.8). This increase in synaptic transmission occurs only if the paired activities of the presynaptic and postsynaptic cells are tightly linked in time, such that the strong postsynaptic depolarization occurs within about 100 ms of transmitter release from the Schaffer collaterals. Such a requirement for coincident presynaptic and postsynaptic activity is the central postulate of a theory of learning devised by Donald Hebb in 1949. Hebb proposed that coordinated activity of a presynaptic terminal and a postsynaptic neuron would strengthen the synaptic connection between them, precisely as is observed for LTP. This indicates the involvement of a **coincidence detector** that allows LTP to occur only when both presynaptic and postsynaptic neurons are active. Hebb's postulate has also been useful in thinking about the role of neuronal activity in other brain functions, most notably development of neural circuits (see Chapter 24).

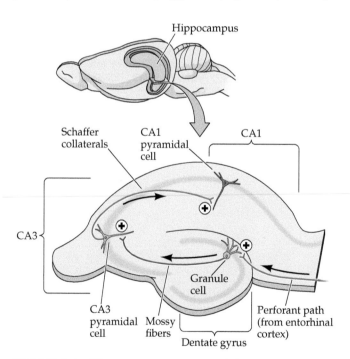

FIGURE 8.6 The trisynaptic circuit of the hippocampus A section through the rodent hippocampus diagrams excitatory pathways and synaptic connections. Long-term potentiation (plus signs) has been observed at each of the three synaptic connections shown here.

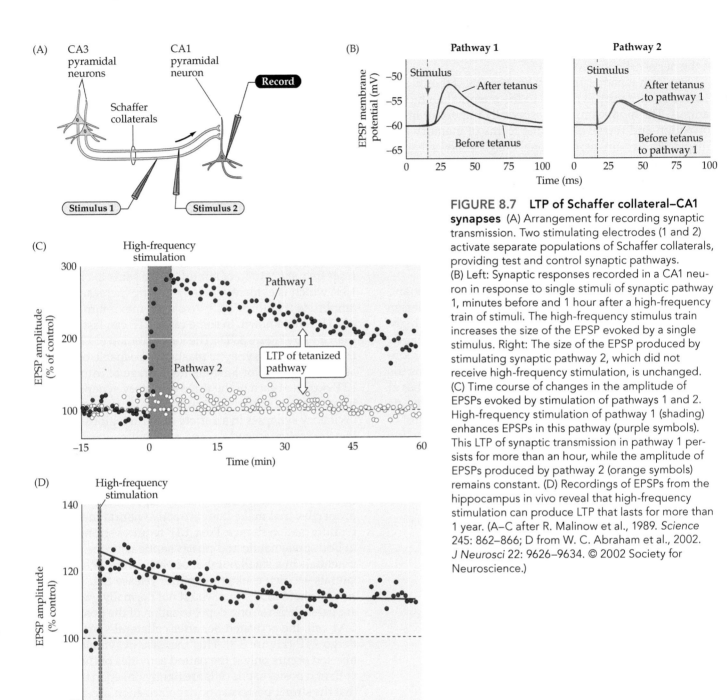

FIGURE 8.7 LTP of Schaffer collateral–CA1 synapses (A) Arrangement for recording synaptic transmission. Two stimulating electrodes (1 and 2) activate separate populations of Schaffer collaterals, providing test and control synaptic pathways. (B) Left: Synaptic responses recorded in a CA1 neuron in response to single stimuli of synaptic pathway 1, minutes before and 1 hour after a high-frequency train of stimuli. The high-frequency stimulus train increases the size of the EPSP evoked by a single stimulus. Right: The size of the EPSP produced by stimulating synaptic pathway 2, which did not receive high-frequency stimulation, is unchanged. (C) Time course of changes in the amplitude of EPSPs evoked by stimulation of pathways 1 and 2. High-frequency stimulation of pathway 1 (shading) enhances EPSPs in this pathway (purple symbols). This LTP of synaptic transmission in pathway 1 persists for more than an hour, while the amplitude of EPSPs produced by pathway 2 (orange symbols) remains constant. (D) Recordings of EPSPs from the hippocampus in vivo reveal that high-frequency stimulation can produce LTP that lasts for more than 1 year. (A–C after R. Malinow et al., 1989. *Science* 245: 862–866; D from W. C. Abraham et al., 2002. *J Neurosci* 22: 9626–9634. © 2002 Society for Neuroscience.)

A second property that makes LTP a particularly attractive neural mechanism for information storage is that LTP is input specific: When LTP is induced by activation of one synapse, it does not occur in other, inactive synapses that contact the same neuron (see Figure 8.7C). Thus, LTP is restricted to activated synapses rather than to all of the synapses on a given cell (Figure 8.9A). Such specificity of LTP is consistent with its involvement in memory formation (or at least the selective storage of information

at synapses). If activation of one set of synapses led to all other synapses—even inactive ones—being potentiated, it would be difficult to selectively enhance particular sets of inputs, as is presumably required to store specific information. The synapse specificity of LTP means that each of the tens of thousands of synapses on a hippocampal neuron can store information, thereby making it possible for the millions of neurons in the hippocampus to store a vast amount of information.

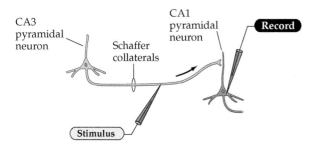

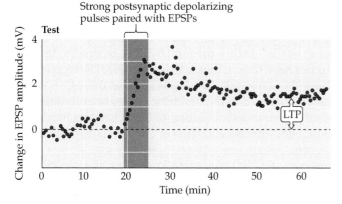

FIGURE 8.8 **Pairing presynaptic and postsynaptic activity causes LTP** Single stimuli applied to a Schaffer collateral synaptic input evoke EPSPs in the postsynaptic CA1 neuron. These stimuli alone do not elicit any change in synaptic strength. However, brief polarization of the CA1 neuron's membrane potential (by applying current pulses through the recording electrode), in conjunction with the Schaffer collateral stimuli, results in a persistent increase in the EPSPs. (After B. Gustafsson et al., 1987. *J Neurosci* 7: 774–780. © 1987 Society for Neuroscience.)

Another important property of LTP is **associativity** (Figure 8.9B). As noted, weak stimulation of a pathway will not by itself trigger LTP. However, if one pathway is weakly activated at the same time that a neighboring pathway onto the same cell is strongly activated, both synaptic pathways undergo LTP. Associativity is another consequence of the coincidence detection feature of LTP, specifically pairing of the activity of the weak synapse with the coincident generation of action potentials by the strong synapse. The selective enhancement of conjointly activated sets of synaptic inputs is often considered a cellular analogue of associative learning, whereby two stimuli are required for learning to take place. The best-known type of associative learning is classical (Pavlovian) conditioning. More generally, associativity is expected in any network of neurons that links one set of information with another.

Although there is clearly a gap between understanding LTP of hippocampal synapses and understanding learning, memory, or other aspects of behavioral plasticity in mammals, this form of long-term synaptic plasticity provides a plausible mechanism for producing enduring neural changes within a part of the brain that is known

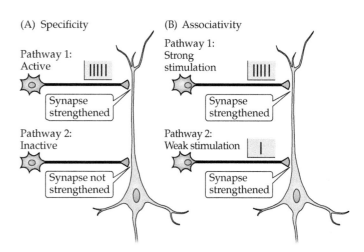

FIGURE 8.9 **Properties of LTP at a CA1 pyramidal neuron receiving synaptic inputs from two independent sets of Schaffer collateral axons** (A) Strong activity initiates LTP at active synapses (pathway 1) without initiating LTP at nearby inactive synapses (pathway 2). (B) Weak stimulation of pathway 2 alone does not trigger LTP. However, when the same weak stimulus to pathway 2 is activated together with strong stimulation of pathway 1, both sets of synapses are strengthened.

to be involved in certain kinds of memories. Indeed, genetic manipulations that affect LTP often produce parallel changes in memory in mice, indicating that this form of long-term synaptic plasticity is likely to serve as a memory storage mechanism.

CONCEPT
8.4

NMDA-Type Glutamate Receptors Serve as Coincidence Detectors for Hippocampal Long-Term Potentiation

LEARNING OBJECTIVES

8.4.1 Describe how Mg^{2+} blockade of NMDA receptors enables these receptors to serve as molecular coincidence detectors.

8.4.2 List the evidence that Ca^{2+} serves as a postsynaptic second messenger to induce LTP.

8.4.3 Describe the molecular mechanisms that convert a postsynaptic Ca^{2+} signal into a strengthening of synaptic transmission during LTP.

Induction of long-term potentiation

The underpinnings of LTP now are understood at the molecular level. A key advance was the discovery that antagonists of the NMDA-type glutamate receptor prevent LTP during high-frequency stimulation of the Schaffer collaterals, but have no effect on the synaptic response evoked by low-frequency stimulation. At about the same time, the unique biophysical properties of the NMDA receptor were first appreciated and provided a critical insight into how LTP is selectively induced by high-frequency activity. As described in Chapter 6, the NMDA receptor channel is permeable to Ca^{2+} but is blocked by Mg^{2+} at the normal resting membrane potential. Thus, the NMDA receptor is a molecular coincidence detector: The channel of this receptor opens (to induce LTP) only when glutamate is bound to the receptor and the postsynaptic cell is depolarized to relieve the Mg^{2+} block of the channel pore (Figure 8.10). During low-frequency synaptic transmission, glutamate released by the Schaffer collaterals binds to both NMDA-type and AMPA-type glutamate receptors; however, Mg^{2+} blockade prevents current flow through the NMDA receptors, so that the EPSP is mediated entirely by the AMPA receptors (see Figure 8.10, left). Because blockade of the NMDA receptor by Mg^{2+} is voltage-dependent, summation of EPSPs during high-frequency stimulation (as in Figure 8.7) leads to a prolonged depolarization that expels Mg^{2+} from the NMDA channel pore (see Figure 8.10, right). Removal of Mg^{2+} then allows Ca^{2+} to enter the dendritic spines of postsynaptic CA1 neurons.

These properties of the NMDA receptor can account for many of the characteristics of LTP. First, the requirement for strong coincident presynaptic and postsynaptic activity to induce LTP (see Figure 8.8) arises because presynaptic activity releases glutamate, while the coincident postsynaptic depolarization relieves the Mg^{2+} block of the NMDA receptor. The specificity of LTP (see Figure 8.9A) can be explained by the fact that NMDA channels will be opened only at synaptic inputs that are active and releasing glutamate, thereby confining LTP to these sites even though EPSPs generated at active synapses depolarize the postsynaptic neuron. With respect to associativity (see Figure 8.9B), a weakly stimulated input releases glutamate but cannot sufficiently depolarize the postsynaptic cell to relieve the Mg^{2+} block. If a large number of other inputs are strongly stimulated, however, they provide the depolarization necessary to relieve the block. Thus, associative induction of LTP relies on activation of NMDA receptors.

Numerous observations have established that induction of LTP is due to accumulation of postsynaptic Ca^{2+} as a result of Ca^{2+} influx through NMDA receptors. Imaging studies have shown that activation of NMDA receptors increases postsynaptic Ca^{2+} levels and that these Ca^{2+} signals can be restricted to the dendritic spines of individual synapses. Furthermore, injection of Ca^{2+} chelators blocks LTP induction, whereas elevation of Ca^{2+} levels in postsynaptic neurons potentiates synaptic transmission. Thus, a rise in postsynaptic Ca^{2+} concentration serves as a second-messenger signal that induces LTP. The fact that these postsynaptic Ca^{2+} signals are highly localized (see Box 7B) can account for the input specificity of LTP (see Figure 8.9A).

Expression of long-term potentiation

After a postsynaptic Ca^{2+} signal induces LTP, the subsequent *expression* of LTP relies on dynamic changes in AMPA receptors. Excitatory synapses can dynamically regulate their postsynaptic AMPA receptors via the same sorts of membrane trafficking processes that occur in presynaptic neurons during neurotransmitter release (see Chapter 5). LTP apparently is due to synaptotagmin-mediated insertion of AMPA receptors into the postsynaptic membrane. These AMPA receptors come from an intracellular organelle known as the recycling endosome. The resulting increase in the number of AMPA receptors increases the response of the postsynaptic cell to released glutamate (Figure 8.11A), yielding a strengthening of synaptic transmission that can last for as long as LTP is maintained (Figure 8.11B). LTP does not affect the number of postsynaptic NMDA receptors; thus, while these receptors are crucial for induction of LTP, they do not play a major role in LTP expression. Stimulus-induced changes in AMPA receptor trafficking can even add new AMPA receptors to "silent" synapses that did not previously have postsynaptic AMPA receptors (Box 8B). At silent synapses, where synaptic activity generates no postsynaptic response at

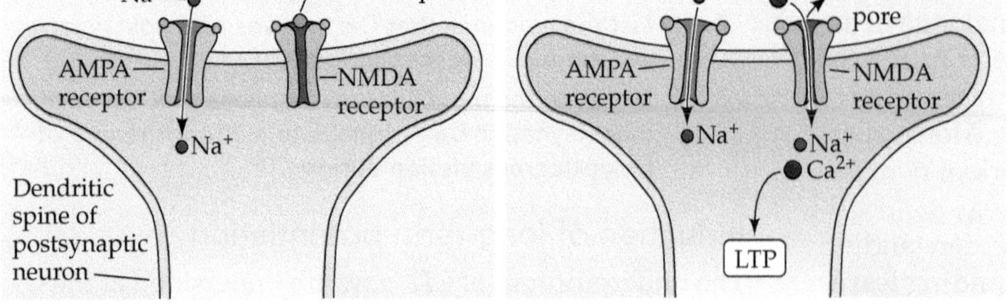

FIGURE 8.10   The NMDA receptor channel can open only during depolarization of the postsynaptic neuron from its normal resting potential Depolarization expels Mg^{2+} from the NMDA channel, allowing current to flow into the postsynaptic cell. This leads to Ca^{2+} entry, which in turn triggers LTP. (After R. A. Nicoll et al., 1988. *Neuron* 1: 97–103.)

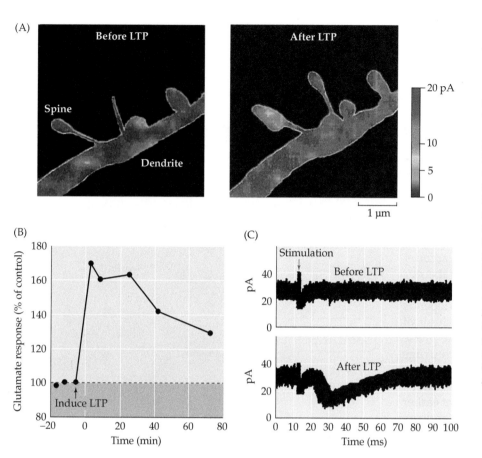

FIGURE 8.11 Addition of postsynaptic AMPA receptors during LTP (A) Spatial maps of the glutamate sensitivity of a hippocampal neuron dendrite before and 120 minutes after induction of LTP. The color scale indicates the amplitude of responses to highly localized glutamate application. LTP causes an increase in the glutamate response of a dendritic spine due to an increase in the number of AMPA receptors on the spine membrane. (B) Time course of changes in glutamate sensitivity of dendritic spines during LTP. Induction of LTP at time = 0 causes glutamate sensitivity to increase for more than 60 minutes. (C) LTP induces AMPA receptor responses at silent synapses in the hippocampus. Prior to inducing LTP, no excitatory postsynaptic currents (EPSCs) are elicited at −65 mV at this silent synapse (upper trace). After LTP induction, the same stimulus produces EPSCs that are mediated by AMPA receptors (lower trace). (A,B from M. Matsuzaki et al., 2004. *Nature* 429: 761–766; C after D. Liao et al., 1995. *Nature* 375: 400–404.)

the normal resting potential, LTP adds AMPA receptors so that the synapse can produce postsynaptic responses (Figure 8.11C). Thus, the strengthening of synaptic transmission during LTP arises from an increase in the sensitivity of the postsynaptic cell to glutamate. Under some circumstances, LTP also can increase the ability of presynaptic terminals to release glutamate.

Ca²⁺ also activates complex postsynaptic signal transduction cascades that include at least two Ca²⁺-activated protein kinases: Ca²⁺/calmodulin-dependent protein kinase, type II (CaMKII) and protein kinase C (PKC) (see Chapter 7). CaMKII, which is the most abundant postsynaptic protein at Schaffer collateral synapses, seems to play an especially important role. This enzyme is activated by stimuli that induce LTP (Figure 8.12), and pharmacological inhibition or genetic deletion of CaMKII prevents LTP. The autophosphorylation property of CaMKII (see Chapter 7) may also serve to prolong the duration of LTP. It is thought that CaMKII and PKC phosphorylate downstream targets, including both AMPA receptors and other targets,

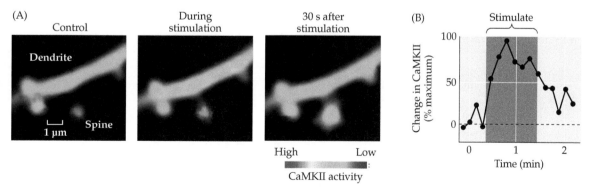

FIGURE 8.12   CaMKII activity in the dendrite of a CA1 pyramidal neuron during LTP (A) The degree of CaMKII activation (indicated by the pseudocolor scale below) in a dendritic spine dramatically increases during stimulation that induces LTP. (B) Time course of transient changes in CaMKII activity associated with LTP. (From S. J. Lee et al., 2009. *Nature* 458: 299–304.)

that collectively facilitate delivery of extrasynaptic AMPA receptors to the synapse.

In summary, the molecular signaling pathways involved in LTP at the Schaffer collateral–CA1 synapse are well understood. Ca^{2+} entering through postsynaptic NMDA receptors leads to activation of synaptotagmins and protein kinases that regulate trafficking of AMPA receptors, thereby enhancing the postsynaptic response to glutamate released from the presynaptic terminal (Figure 8.13). Still other forms of LTP are observed at other synapses and, in some cases, rely on signaling mechanisms different from those involved in LTP at the Schaffer collateral–CA1 synapse.

The scheme depicted in Figure 8.13 can account for the changes in synaptic transmission that occur over the first 1 to 2 hours after LTP is induced. However, there is also a later phase of LTP that depends on changes in gene expression and the synthesis of new proteins. The contributions of this late phase can be observed by treating synapses with drugs that inhibit protein synthesis: Blocking protein synthesis prevents LTP measured several hours after a stimulus but does not affect LTP measured at earlier times (Figure 8.14). This late phase of LTP appears to be initiated by protein kinase A, which goes on to activate transcription factors such as CREB, which stimulate the expression of other proteins. Although most of these newly synthesized proteins have not yet been identified, they include other transcriptional regulators, protein kinases, and AMPA receptors (Figure 8.15A). How these proteins contribute to the late phase of LTP is not yet known. There is evidence that the number and size of synaptic contacts increase during LTP (Figure 8.15B,C). Thus, it is likely that some of the proteins newly synthesized during the late phase of LTP are involved in construction of new synaptic contacts that serve to make LTP essentially permanent (as in Figure 8.7D).

In conclusion, it appears that LTP in the mammalian hippocampus has many parallels with the long-term changes in synaptic transmission underlying behavioral sensitization

in *Aplysia*. Both consist of an early, transient phase that relies on protein kinases to produce post-translational changes in membrane ion channels, and both have later, long-lasting phases that require changes in gene expression mediated by CREB. Both forms of long-term synaptic plasticity are likely to be involved in long-term storage of information.

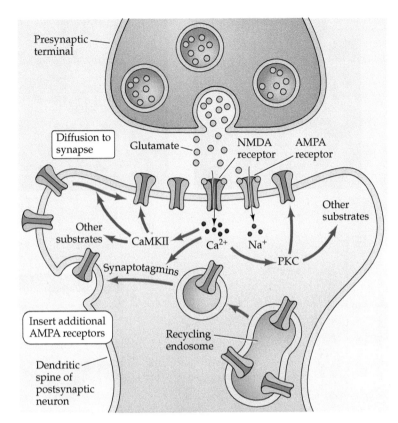

FIGURE 8.13   Signaling mechanisms underlying LTP During glutamate release, the NMDA receptor channel opens only if the postsynaptic cell is sufficiently depolarized. The Ca^{2+} ions that enter the cell through the channel activate postsynaptic protein kinases, such as CaMKII and PKC, that trigger a series of phosphorylation reactions. These reactions regulate trafficking of postsynaptic AMPA receptors through recycling endosomes, leading to insertion of new AMPA receptors into the postsynaptic spine. Subsequent diffusion of AMPA receptors to the subsynaptic region yields an increase in the spine's sensitivity to glutamate, which causes LTP.

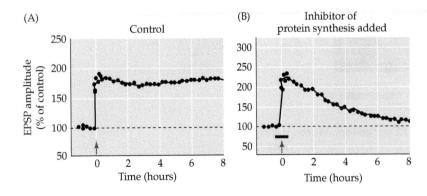

FIGURE 8.14   Role of protein synthesis in maintaining LTP (A) Repetitive high-frequency stimulation (arrow) induces LTP that persists for many hours. (B) Treatment with anisomycin (at bar), an inhibitor of protein synthesis, causes LTP to decay within a few hours after the high-frequency stimulation (arrow). (After U. Frey and R. G. Morris, 1997. *Nature* 385: 533–536.)

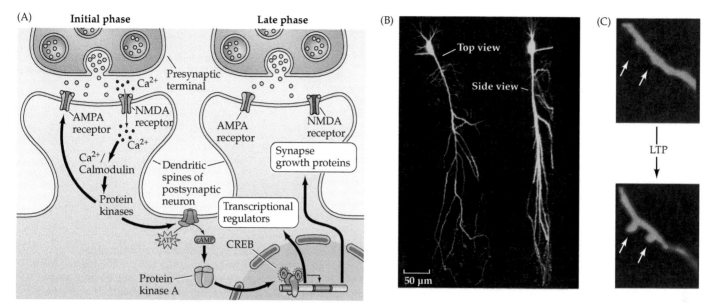

FIGURE 8.15 Mechanisms responsible for long-lasting changes in synaptic transmission during LTP (A) The late component of LTP is the result of PKA activating the transcriptional regulator CREB, which turns on expression of several genes that produce long-lasting changes in PKA activity and synapse structure. (B,C) Structural changes associated with LTP in the hippocampus. (B) The dendrites of a CA1 pyramidal neuron were visualized by filling the cell with a fluorescent dye. (C) New dendritic spines (white arrows) can be observed approximately 1 hour after a stimulus that induces LTP. The presence of novel spines raises the possibility that LTP may arise, in part, from formation of new synapses. (A after L. R. Squire and E. R. Kandel, 1999. *Memory: From Mind to Molecules.* New York: Scientific American Library; B and C after F. Engert and T. Bonhoeffer, 1999. *Nature* 399: 66–70.)

■ BOX 8B | Silent Synapses

Several observations indicate that postsynaptic glutamate receptors are dynamically regulated at excitatory synapses. Early insight into this process came from the finding that stimulation of some glutamatergic synapses generates no postsynaptic electrical signal when the postsynaptic cell is at its normal resting membrane potential (Figure A). However, once the postsynaptic cell is depolarized, these "silent synapses" can transmit robust postsynaptic electrical responses. The fact that transmission at such synapses can be turned on or off according to the postsynaptic membrane potential suggests an interesting and simple means of modifying neural circuitry.

Silent synapses are especially prevalent in development and have been found in many brain regions, including the hippocampus, cerebral cortex, and spinal cord. The silence of these synapses is evidently due to the voltage-dependent blockade of NMDA

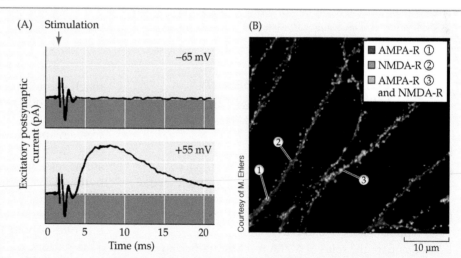

(A) Electrophysiological evidence for silent synapses. Stimulation of some axons fails to activate synapses when the postsynaptic cell is held at a negative potential (–65 mV, upper trace). However, when the postsynaptic cell is depolarized (+55 mV, lower trace), stimulation produces a robust response. (B) Immuno-fluorescent localization of NMDA receptors (green) and AMPA receptors (red) in a cultured hippocampal neuron. Many dendritic spines are positive for NMDA receptors but not AMPA receptors, indicating NMDA receptor–only synapses. (A after D. Liao et al., 1995. *Nature* 375: 400–404.)

receptors by Mg^{2+} (see Concept 8.4 and Chapter 6). At the normal resting membrane potential, presynaptic release of glutamate evokes no postsynaptic response at such synapses because their

(Continued)

■ **BOX 8B** | **Silent Synapses** *(continued)*

NMDA receptors are blocked by Mg^{2+}. However, depolarization of the postsynaptic neuron displaces the Mg^{2+}, allowing glutamate release to induce postsynaptic responses mediated by NMDA receptors.

Glutamate released at silent synapses evidently binds only to NMDA receptors. How, then, does glutamate release avoid activating AMPA receptors? The most likely explanation is that these synapses may have only NMDA receptors. Immunocytochemical experiments demonstrate that some excitatory synapses have only NMDA receptors (green spots in Figure B). Such NMDA receptor–only synapses are particularly abundant early in postnatal development and decrease in adults (Figure C). Silencing of glutamatergic synapses has also been reported to result from the use of the recreational drug cocaine. Thus, at least some silent synapses are not a separate class of excitatory synapses that lack AMPA receptors, but rather an early stage in the ongoing maturation of the glutamatergic synapse (Figure D). Evidently, AMPA and NMDA receptors are not

(C) Electron microscopy of excitatory synapses in CA1 stratum radiatum of the hippocampus from 10-day-old (juvenile) or 5-week-old (adult) rats double-labeled for AMPA receptors and NMDA receptors. The presynaptic terminal (pre), synaptic cleft, and postsynaptic spine (post) are indicated. AMPA receptors are abundant at the adult synapse but are absent from the younger synapse. (From R. S. Petralia et al., 1999. *Nat Neurosci* 2: 31–36.) (D) Diagram of glutamatergic synapse maturation. Early in postnatal development, many excitatory synapses contain only NMDA receptors. As synapses mature, AMPA receptors are recruited.

inextricably linked at excitatory synapses, but are targeted via independent cellular mechanisms. Such synapse-specific glutamate receptor composition implies sophisticated mechanisms for regulating the localization of each type of receptor. Dynamic changes in the trafficking of AMPA and NMDA

receptors can strengthen or weaken synaptic transmission and are important in LTP and LTD, as well as in the maturation of glutamatergic synapses.

In summary, although silent synapses have begun to whisper their secrets, much remains to be learned about their physiological importance.

CONCEPT
8.5

Long-Term Depression Weakens Synapses via Multiple Mechanisms

LEARNING OBJECTIVES

8.5.1 List the steps that couple glutamate release from presynaptic terminals to LTD of synaptic transmission at hippocampal synapses.

8.5.2 Contrast the roles of NMDA receptors in hippocampal LTP and LTD.

8.5.3 Compare the roles of AMPA receptor trafficking in hippocampal LTP and LTD.

8.5.4 List the steps that couple glutamate release from presynaptic terminals to LTD of synaptic transmission at cerebellar synapses.

8.5.5 Compare the roles of AMPA receptor trafficking in hippocampal and cerebellar LTD.

Long-term depression in the hippocampus

If synapses simply continued to increase in strength as a result of long-term potentiation, eventually they would

reach some level of maximum efficacy, making it difficult to encode new information. Thus, to make synaptic strengthening useful, other processes must selectively weaken specific sets of synapses, and LTD is such a process. In the late 1970s, LTD was found to occur at the synapses between the Schaffer collaterals and the CA1 pyramidal cells in the hippocampus. Whereas LTP at these synapses requires brief, high-frequency stimulation, LTD occurs when the Schaffer collaterals are stimulated at a low rate—about 1 Hz—for long periods (10–15 minutes). This pattern of activity depresses the EPSP for several hours and, like LTP, is specific to the activated synapses (Figure 8.16A,B). Moreover, LTD can erase the increase in EPSP size that is due to LTP, and conversely, LTP can erase the decrease in EPSP size that is due to LTD. This complementarity suggests that LTD and LTP reversibly affect synaptic efficiency by acting at a common site.

LTP and LTD at the Schaffer collateral–CA1 synapses actually share several key elements. Both require activation of NMDA-type glutamate receptors and the resulting entry of Ca^{2+} into the postsynaptic cell. The major

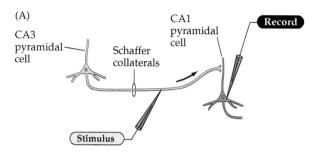

(A)

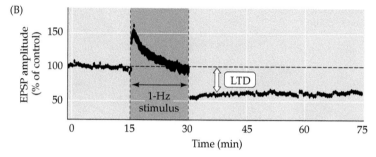

(B)

(C)

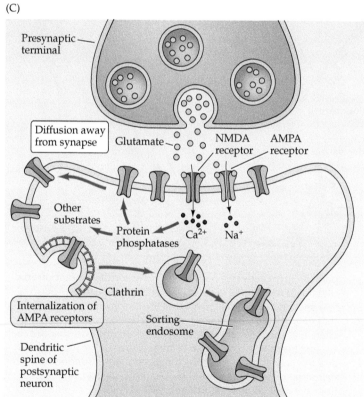

FIGURE 8.16 **Long-term synaptic depression in the hippocampus** (A) Electrophysiological procedures used to monitor transmission at the Schaffer collateral synapses on CA1 pyramidal neurons. (B) Low-frequency stimulation (one per second) of the Schaffer collateral axons causes a long-lasting depression of synaptic transmission. (C) Mechanisms underlying LTD. A low-amplitude rise in Ca²⁺ concentration in the postsynaptic CA1 neuron activates postsynaptic protein phosphatases, which cause internalization of postsynaptic AMPA receptors, thereby decreasing the sensitivity to glutamate released from the Schaffer collateral terminals. (B after R. M. Mulkey et al., 1993. *Science* 261: 1051–1055.)

determinant of whether LTP or LTD arises appears to be the nature of the Ca^{2+} signal in the postsynaptic cell: Small and slow rises in Ca^{2+} lead to depression, whereas large and fast increases in Ca^{2+} trigger potentiation. As noted in Concept 8.4, LTP is at least partially due to activation of protein kinases, which phosphorylate their target proteins. LTD, in contrast, appears to result from activation of phosphatases, specifically PP1 and PP2B (calcineurin), the latter a Ca^{2+}-dependent phosphatase (see Chapter 7). Evidence in support of this idea is that inhibitors of these phosphatases prevent LTD but do not block LTP. The different effects of Ca^{2+} during LTD and LTP may arise from the selective activation of protein phosphatases and kinases by the different types of postsynaptic Ca^{2+} signals occurring during these two forms of synaptic plasticity. Although the phosphatase substrates important for LTD have not yet been identified, it is possible that LTP and LTD phosphorylate and dephosphorylate the same set of regulatory proteins to control the efficacy of transmission at the Schaffer collateral–CA1 synapse. Just as LTP at this synapse is associated with insertion of AMPA receptors, LTD is often associated with a loss of synaptic AMPA receptors. This loss probably arises from internalization of AMPA receptors into sorting endosomes in the postsynaptic cell (Figure 8.16C), due to the same clathrin-dependent endocytosis mechanisms important for synaptic vesicle recycling in the presynaptic terminal (see Chapter 5). As is also the case for LTP, there is a late phase of LTD that requires synthesis of new proteins.

Long-term depression in the cerebellum

A quite different form of LTD is observed in the cerebellum. LTD of synaptic inputs onto cerebellar Purkinje cells was first described by Masao Ito and Masanobu Kano in Japan in the early 1980s. Purkinje neurons in the cerebellum receive two distinct types of excitatory input: climbing fibers and parallel fibers (Figure 8.17A) (see Chapter 19). LTD reduces the strength of transmission at the parallel fiber synapse (Figure 8.17B) and subsequently was found to depress transmission at the climbing fiber synapse as well. This form of LTD has been implicated in the motor learning that mediates the coordination, acquisition, and storage of complex movements in the cerebellum. Although the role of LTD in cerebellar motor learning remains controversial, it has nonetheless been a useful model system for understanding the cellular mechanisms of long-term synaptic plasticity.

Like many forms of motor learning, cerebellar LTD is associative because it occurs only when climbing fibers and parallel fibers are activated at the same time (Figure 8.17C). In this case, associativity arises from

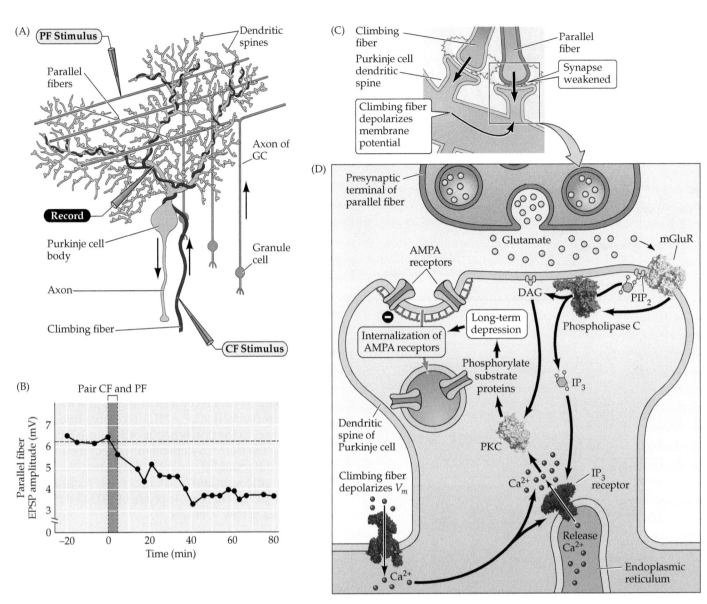

FIGURE 8.17 Long-term synaptic depression in the cerebellum (A) Experimental arrangement. Synaptic responses were recorded from Purkinje cells following stimulation of parallel fibers and climbing fibers. (B) Pairing stimulation of climbing fibers (CF) and parallel fibers (PF) causes LTD that reduces the parallel fiber EPSP. (C) LTD requires depolarization of the Purkinje cell membrane potential (V_m), produced by climbing fiber activation, as well as signals generated by active parallel fiber synapses. (D) Mechanism underlying cerebellar LTD. Glutamate released by parallel fibers activates both AMPA receptors and metabotropic glutamate receptors (mGluRs). The activated mGluRs produce two second messengers, DAG and IP_3, which interact with Ca^{2+} that enters when climbing fiber activity opens voltage-gated Ca^{2+} channels. The resultant release of Ca^{2+} from the endoplasmic reticulum leads to a further rise in intracellular Ca^{2+} concentration and the activation of PKC, which triggers clathrin-dependent internalization of postsynaptic AMPA receptors, weakening the parallel fiber synapse. (B after M. Sakurai, 1987. *J Physiol* 394: 463–480.)

the combined actions of two distinct intracellular signal transduction pathways that are activated in the postsynaptic Purkinje cell in response to activity of the climbing fiber and parallel fiber synapses. In the first pathway, glutamate released from the parallel fiber terminals activates two types of receptors, the AMPA-type and metabotropic glutamate receptors (see Chapter 6). Glutamate binding to the AMPA receptor results in mild membrane depolarization, whereas binding to the metabotropic receptor produces the second messengers inositol triphosphate (IP_3) and diacylglycerol (DAG) (see Chapter 7). Glutamate released by climbing fibers also activates AMPA receptors, which strongly depolarizes the membrane potential of the Purkinje cell and initiates a second signal transduction pathway: an influx of Ca^{2+} through voltage-gated channels and a subsequent increase in intracellular Ca^{2+} concentration.

These two second messengers—IP$_3$ and Ca^{2+}—cause an amplified rise in intracellular Ca^{2+} concentration by acting together on IP$_3$ receptors. This triggers release of Ca^{2+} from intracellular stores and leads to synergistic activation of PKC by Ca^{2+} and DAG (Figure 8.17D). Thus, the associative property of cerebellar LTD appears to arise from both IP$_3$ receptors and PKC serving as coincidence detectors. Other protein kinases also act to sustain the activation of PKC beyond the time that IP$_3$ and Ca^{2+} concentrations are elevated.

PKC phosphorylates several downstream substrate proteins, including AMPA receptors. The main consequence of PKC-dependent phosphorylation is to cause an internalization of AMPA receptors via clathrin-dependent endocytosis (see Figure 8.17D). This loss of AMPA receptors decreases the response of the postsynaptic Purkinje cell to glutamate released from the presynaptic terminals of the parallel fibers. Thus, in contrast to LTD in the hippocampus, cerebellar LTD requires the activity of protein kinases, rather than phosphatases, and does not involve Ca^{2+} entry through NMDA receptors. However, the net effect is the same in both cases: Internalization of postsynaptic AMPA receptors is a common mechanism for decreased efficacy of both hippocampal and cerebellar synapses during LTD. As is the case for LTP at the hippocampal Schaffer collateral synapse, as well as for long-term synaptic plasticity in *Aplysia*, CREB appears to be required for a late phase of cerebellar LTD. It is not yet known which proteins are synthesized as a consequence of CRE activation.

CONCEPT 8.6 | Some Forms of Plasticity Depend on the Timing of Synaptic Activity

LEARNING OBJECTIVES

8.6.1 Define spike timing-dependent synaptic plasticity.

8.6.2 Compare the timing requirements for spike timing-dependent long-term potentiation and long-term depression.

8.6.3 Explain the computational value of spike timing-dependent synaptic plasticity.

Spike timing-dependent plasticity

The preceding concepts have shown that LTP and LTD are preferentially initiated by different rates of repetitive synaptic activity, with LTP requiring high-frequency activity and LTD being induced by low-frequency activity. However, the precise temporal relationship between activity in the pre- and postsynaptic cells can also be an important determinant of the amount and direction of long-term synaptic plasticity. At a given (low) frequency of synaptic activity, LTD will occur if presynaptic activity is preceded by a postsynaptic action potential, while LTP will occur if the postsynaptic action potential follows presynaptic

activity (Figure 8.18A,B). The relationship between the time interval and the magnitude of the synaptic change is a very sensitive function of the time interval, with no changes observed if the presynaptic and postsynaptic activities are separated by 100 ms or longer (Figure 8.18C).

Because precise timing of presynaptic and postsynaptic activity determines the polarity of these forms of long-lasting synaptic plasticity, they are called **spike timing-dependent plasticity (STDP)**. Although the mechanisms involved are not yet well understood, it appears that the properties of STDP arise from timing-dependent differences in postsynaptic Ca^{2+} signals. Specifically, if a postsynaptic action potential occurs *after* presynaptic activity, the resulting depolarization will relieve the block of NMDA receptors by Mg^{2+} and cause a relatively large amount of Ca^{2+} influx through postsynaptic NMDA receptors, yielding LTP. In contrast, if the postsynaptic action potential occurs *before* the presynaptic action potential, then the depolarization associated with the postsynaptic action potential will subside by the time an EPSP occurs. This sequence of events will reduce the amount of Ca^{2+} entry through the NMDA receptors, leading to LTD. It has been postulated that other signals, such as endocannabinoids (see Chapter 6), may also be required for LTD induction during STDP.

The requirement for a precise temporal relationship between presynaptic and postsynaptic activity means that STDP can perform several novel types of neuronal computation. STDP can provide a means of encoding information about causality. For example, if a synapse generates a suprathreshold EPSP, the resulting postsynaptic action potential would rapidly follow presynaptic activity, and the resulting LTP would encode the fact that the postsynaptic action potential resulted from the activity of that synapse. STDP could also serve as a mechanism for competition between synaptic inputs: Stronger inputs would be more likely to produce suprathreshold EPSPs and be reinforced by the resulting LTP, whereas weaker inputs would not generate postsynaptic action potentials that were correlated with presynaptic activity. There is evidence that STDP is important for neural circuit function, such as determining orientation preference in the visual system (see Chapter 9).

In summary, activity-dependent forms of synaptic plasticity cause changes in synaptic transmission that modify the functional connections within and among neural circuits. These changes in the efficacy and local geometry of synaptic connectivity can provide a basis for learning, memory, and other forms of brain plasticity. Activity-dependent changes in synaptic transmission may also be involved in some pathologies. Abnormal patterns of neuronal activity, such as those that occur in epilepsy, can stimulate abnormal changes in synaptic connections that may further increase the frequency and severity of seizures (Clinical Applications). Despite the substantial

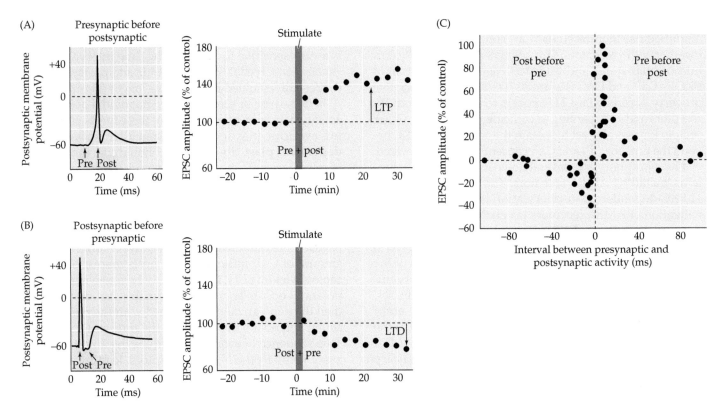

FIGURE 8.18   Spike timing-dependent synaptic plasticity in cultured hippocampal neurons (A) Left: Stimulating a presynaptic neuron (Pre) causes an EPSP in the postsynaptic neuron; applying a subsequent stimulus to the postsynaptic neuron (Post) causes an action potential that is superimposed on the EPSP. Right: Repetitive application of this stimulus paradigm causes LTP of the EPSP. (B) Reversing the order of stimulation, so that the postsynaptic neuron is excited before the presynaptic neuron, causes LTD of the EPSP. (C) Complex dependence of STDP on the interval between presynaptic activity and postsynaptic activity. If the presynaptic neuron is activated 40 ms or less before the postsynaptic neuron, then LTP occurs. Conversely, if the postsynaptic neuron is activated 40 ms or less before the presynaptic neuron, LTD occurs. If the interval between the two events is longer than 40 ms, no STDP is observed. (After G. Q. Bi and M. M. Poo, 1998. *J Neurosci* 18: 10464–10472. © 1998 Society for Neuroscience.)

■ Clinical Applications

Epilepsy: The Effect of Pathological Activity on Neural Circuitry

Epilepsy is a brain disorder characterized by periodic and unpredictable seizures. The behavioral manifestations of epileptic seizures range from mild twitching of an extremity to loss of consciousness and uncontrollable convulsions. Although many highly accomplished people have had epilepsy (Alexander the Great, Julius Caesar, Napoleon, van Gogh and musicians Prince and Lil Wayne, to name a few), seizures of sufficient intensity and frequency obviously can interfere with many aspects of daily life. Moreover, uncontrolled convulsions can lead to excitotoxicity. As much as 1% of the population is afflicted, making epilepsy one of the most common neurological problems.

Epilepsy is mediated by the rhythmic firing of large groups of neurons. It seems likely that this atypical neuronal activity generates plastic changes in cortical circuitry that are critical to the pathogenesis of the disease. The importance of neuronal plasticity in epilepsy is indicated most clearly by an animal model of seizure production called *kindling*. To induce kindling, a stimulating electrode is implanted in the brain, often in the amygdala (a component of the limbic system that makes and receives connections with the cortex, thalamus, and other limbic structures, including the hippocampus; see Chapter 32). At the beginning of such an experiment, weak electrical stimulation, in the form of a low-amplitude train of electrical pulses, has no discernible effect on behavior or on the pattern of electrical activity in the brain. As this weak stimulation is repeated once a day for several weeks, it begins to produce behavioral and electrical indications of seizures. By the end of the experiment, the same weak stimulus that initially had no effect now causes full-blown seizures. This phenomenon is essentially permanent;

■ Clinical Applications (*continued*)

even after an interval of a year, the same weak stimulus will again trigger a seizure. Thus, repetitive weak activation produces long-lasting changes in the excitability of the brain that time cannot reverse. The word *kindling* is therefore quite appropriate—a single match can start a devastating fire.

The changes in the electrical patterns of brain activity detected in kindled animals resemble those in human epilepsy. Modern thinking about the causes (and possible cures) of epilepsy has focused on where seizures originate and the mechanisms that make the affected region hyperexcitable. Most of the evidence suggests that atypical activity in small areas of the cerebral cortex (called foci) provide the triggers for a seizure that then spreads to other synaptically connected regions. For example, a seizure originating in the thumb area of the right motor cortex will first be evident as uncontrolled movement of the left thumb that subsequently extends to other more proximal limb muscles, whereas a seizure originating in the visual association cortex of the right hemisphere may be heralded by complex hallucinations in the left visual field. The behavioral manifestations of seizures therefore provide important clues for the neurologist seeking to pinpoint the atypical region of cerebral cortex.

Epileptic seizures can be caused by a variety of acquired or congenital factors, including cortical damage from trauma, stroke, tumors, congenital cortical dysgenesis (failure of the cortex to grow properly), and congenital vascular malformations. One rare form of epilepsy, Rasmussen's encephalitis, is an autoimmune disease that arises when the immune system attacks the brain, using both humoral agents (i.e., antibodies) and cellular agents (lymphocytes and macrophages) that can destroy neurons. Some forms of epilepsy are heritable, and more than a dozen distinct genes have been demonstrated to underlie unusual types of epilepsy (see Clinical Applications, Chapter 4). Most forms of familial epilepsy (such

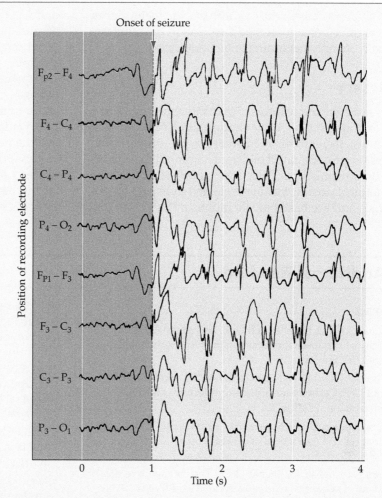

Electroencephalogram (EEG) recorded during a seizure (at arrow). The traces show atypical rhythmic activity that reflects the synchronous firing of large numbers of cortical neurons. This pattern of activity persisted for much longer than the 3 seconds shown. Designations at left are various positions of electrodes on the head. (After F. M. Dyro, 1989. *The EEG Handbook*. Boston: Little, Brown.)

as juvenile myoclonic epilepsy and petit mal epilepsy) are caused by the simultaneous inheritance of more than one mutant gene.

No effective prevention or cure exists for epilepsy. Pharmacological therapies that successfully inhibit seizures are based on two general strategies. One approach is to enhance the function of inhibitory synapses that use the neurotransmitter GABA; the other is to limit action potential firing by acting on voltage-gated Na^+ channels. Commonly used antiseizure medications include carbamazepine, phenobarbital, phenytoin (Dilantin),

and valproic acid. These agents, which must be taken daily, successfully inhibit seizures in 60% to 70% of people who have epilepsy. In a small fraction of people, the epileptogenic region can be surgically excised. In extreme cases, physicians resort to cutting the corpus callosum to prevent the spread of seizures (most of the "split-brain" individuals described in Chapter 32 had intractable epilepsy). One of the major reasons for controlling epileptic activity is to prevent the more permanent plastic changes that would ensue as a consequence of atypical and excessive neural activity.

advances in understanding the cellular and molecular bases of some forms of plasticity, the means by which selective changes of synaptic strength encode memories or other complex behavioral modifications in the mammalian brain are only beginning to be understood.

Summary

Synapses exhibit many forms of plasticity that occur over a broad temporal range. At the shortest times (milliseconds to minutes), facilitation, augmentation, potentiation, and depression provide rapid but transient modifications in synaptic transmission. These forms of plasticity change the amount of neurotransmitter released from presynaptic terminals and are based on alterations in Ca^{2+} signaling and synaptic vesicle pools at recently active terminals. Longer-lasting forms of synaptic plasticity such as LTP and LTD are also based on Ca^{2+} and other intracellular second messengers. At least some of the synaptic changes produced by these long-lasting forms of plasticity are postsynaptic, caused by changes in neurotransmitter receptor trafficking, although alterations in neurotransmitter release from the presynaptic terminal can also occur. In these more enduring forms of plasticity, protein phosphorylation and changes in gene expression greatly outlast the period of synaptic activity and can yield changes in synaptic strength that persist for hours, days, or even longer. Long-lasting synaptic plasticity can serve as a neural mechanism for many forms of brain plasticity, such as learning new behaviors or acquiring new memories.

■ Additional Reading

Reviews

Abraham, W. C., O. D. Jones and D. L. Glanzman (2019) Is plasticity of synapses the mechanism of long-term memory storage? *NPJ Sci. Learn.* 4: 9.

Díaz-Alonso, J. and R. A. Nicoll (2021) AMPA receptor trafficking and LTP: Carboxy-termini, amino-termini and TARPs. *Neuropharmacol.* 197: 108710.

Ito, M., K. Yamaguchi, S. Nagao and T. Yamazaki (2014) Long-term depression as a model of cerebellar plasticity. *Prog. Brain Res.* 210: 1–30.

Jackman, S. L. and W. G. Regehr (2017) The mechanisms and functions of synaptic facilitation. *Neuron* 94: 447–464.

Kandel, E. R., Y. Dudai and M. R. Mayford (2014) The molecular and systems biology of memory. *Cell* 157: 163–186.

Malenka, R. C. and S. A. Siegelbaum (2001) Synaptic plasticity: Diverse targets and mechanisms for regulating synaptic efficacy. In *Synapses*, W. M. Cowan, T. C. Sudhof and C. F. Stevens (eds.). Baltimore, MD: Johns Hopkins University Press, pp. 393–413.

Pereyra, M. and J. H. Medina (2021) AMPA receptors: A key piece in the puzzle of memory retrieval. *Front. Hum. Neurosci.* 15: 729051.

Important Original Papers

Abraham, W. C., B. Logan, J. M. Greenwood and M. Dragunow (2002) Induction and experience-dependent consolidation of stable long-term potentiation lasting months in the hippocampus. *J. Neurosci.* 22: 9626–9634.

Ahn, S., D. D. Ginty and D. J. Linden (1999) A late phase of cerebellar long-term depression requires activation of CaMKIV and CREB. *Neuron* 23: 559–568.

Betz, W. J. (1970) Depression of transmitter release at the neuromuscular junction of the frog. *J. Physiol.* 206: 629–644.

Bi, G. Q. and M. M. Poo (1998) Synaptic modifications in cultured hippocampal neurons: Dependence on spike timing, synaptic strength, and postsynaptic cell type. *J. Neurosci.* 18: 10464–10472.

Bliss, T. V. P. and T. Lomo (1973) Long-lasting potentiation of synaptic transmission in the dentate area of the anaesthetized rabbit following stimulation of the perforant path. *J. Physiol.* 232: 331–356.

Charlton, M. P. and G. D. Bittner (1978) Presynaptic potentials and facilitation of transmitter release in the squid giant synapse. *J. Gen. Physiol.* 72: 487–511.

Cheng, Q., S. H. Song and G. J. Augustine (2018) Molecular mechanisms of short-term plasticity: Role of synapsin phosphorylation in augmentation and potentiation of spontaneous glutamate release. *Front. Synaptic Neurosci.* 10: 33.

Chung, H. J., J. P. Steinberg, R. L. Huganir and D. J. Linden (2003) Requirement of AMPA receptor GluR2 phosphorylation for cerebellar long-term depression. *Science* 300: 1751–1755.

Collingridge, G. L., S. J. Kehl and H. McLennan (1983) Excitatory amino acids in synaptic transmission in the Schaffer collateral-commissural pathway of the rat hippocampus. *J. Physiol.* 334: 33–46.

Engert, F. and T. Bonhoeffer (1999) Dendritic spine changes associated with hippocampal long-term synaptic plasticity. *Nature* 399: 66–70.

Frey, U. and R. G. Morris (1997) Synaptic tagging and long-term potentiation. *Nature* 385: 533–536.

Gustafsson, B., H. Wigstrom, W. C. Abraham and Y. Y. Huang (1987) Long-term potentiation in the hippocampus using depolarizing current pulses as the conditioning stimulus to single volley synaptic potentials. *J. Neurosci.* 7: 774–780.

Junge, H. J. and 7 others (2004) Calmodulin and Munc13 form a Ca^{2+} sensor/effector complex that controls short-term synaptic plasticity. *Cell* 118: 389–401.

Katz, B. and R. Miledi (1968) The role of calcium in neuromuscular facilitation. *J. Physiol.* 195: 481–492.

Konnerth, A., J. Dreessen and G. J. Augustine (1992) Brief dendritic calcium signals initiate long-lasting synaptic depression in cerebellar Purkinje cells. *Proc. Natl. Acad. Sci. U.S.A.* 89: 7051–7055.

Lee, S. J., Y. Escobedo-Lozoya, E. M. Szatmari and R. Yasuda (2009) Activation of CaMKII in single dendritic spines during long-term potentiation. *Nature* 458: 299–304.

Lev-Tov, A., M. J. Pinter and R. E. Burke (1983) Posttetanic potentiation of group Ia EPSPs: Possible mechanisms for differential distribution among medial gastrocnemius motoneurons. *J. Neurophysiol.* 50: 379–398.

Liao, D., N. A. Hessler and R. Malinow (1995) Activation of postsynaptically silent synapses during pairing-induced LTP in CA1 region of hippocampal slice. *Nature* 375: 400–404.

Malinow, R., H. Schulman and R. W. Tsien (1989) Inhibition of postsynaptic PKC or CaMKII blocks induction but not expression of LTP. *Science* 245: 862–866.

Matsuzaki, M., N. Honkura, G. C. Ellis-Davies and H. Kasai (2004) Structural basis of long-term potentiation in single dendritic spines. *Nature* 429: 761–766.

Mulkey, R. M., C. E. Herron and R. C. Malenka (1993) An essential role for protein phosphatases in hippocampal long-term depression. *Science* 261: 1051–1055.

Murakoshi, H. and 5 others (2017) Kinetics of endogenous CaMKII required for synaptic plasticity revealed by optogenetic kinase inhibitor. *Neuron* 94: 37–47.

Rossetti, T. and 8 others (2017) Memory erasure experiments indicate a critical role of CaMKII in memory storage. *Neuron* 96: 207–216.

Sakurai, M. (1987) Synaptic modification of parallel fibre-Purkinje cell transmission in *in vitro* guinea-pig cerebellar slices. *J. Physiol.* 394: 463–480.

Tanaka, K. and G. J. Augustine (2008) A positive feedback signal transduction loop determines timing of cerebellar long-term depression. *Neuron* 59: 608–620.

Books

Bliss, T., G. Collingridge and R. Morris (eds.) (2004) *Long-term Potentiation: Enhancing Neuroscience for 30 Years.* New York: Oxford University Press.

Kandel, E. R. (2007) *In Search of Memory: The Emergence of a New Science of Mind.* New York: W. W. Norton.

Katz, B. (1966) *Nerve, Muscle, and Synapse.* New York: McGraw-Hill.

Squire, L. R. and E. R. Kandel (1999) *Memory: From Mind to Molecules.* New York: Scientific American Library.

UNIT II
Sensation and Sensory Processing

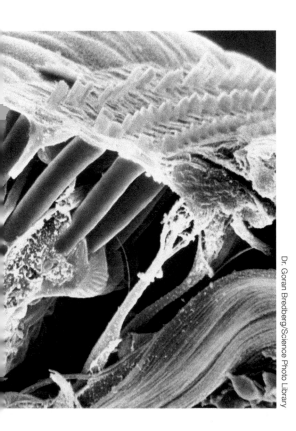

Dr. Goran Bredberg/Science Photo Library

CHAPTER 9 **Vision**

CHAPTER 10 **Hearing**

CHAPTER 11 **The Vestibular System**

CHAPTER 12 **Touch and Proprioception**

CHAPTER 13 **Pain and Temperature**

CHAPTER 14 **Olfaction**

CHAPTER 15 **Taste**

Much of the information our brains process is sensory. Our bodies contain sensory receptors that sense light, sound, mechanical forces, and chemical composition, allowing us to see, hear, feel, taste, smell, and balance ourselves as we move our bodies through space. Sensory receptors raise the alarm regarding potentially damaging events—producing our sense of pain—and also help our brains keep track of our physiological needs such as to stay warm and well fed.

Different types of sensory receptors are needed depending on the nature of the physical energy involved. This physical energy is first transduced into electrical signals in these specialized neurons, then sent via nerves to the brain. Sensory areas in the brain must in turn represent what the stimuli are, where and when they occur, and how strong they are. The nature of these representations varies depending on the type of information and involves a combination of two general properties: which neurons are active (different stimuli can activate different neural populations) and how active they are (different stimuli can activate the same population of neurons to varying degrees). Sensory representations can produce either perceptual awareness of events (such as your ability to consciously perceive and think about the words on this page) or be more subconscious. Indeed, for some types of sensory information, you may hardly be aware of the operation of your sensory system unless something goes wrong, such as if a dysfunction in your sense of balance impairs your ability to stand or walk. Finally, sensation is not a purely feedforward process but a circular one, guided by the brain via a variety of mechanisms that affect what input actually reaches our sensory receptors and how corresponding signals are ultimately represented in the brain.

While the chapters in this unit focus on individual sensory systems, all sensory systems interact with other sensory systems, particularly when they are devoted to evaluating the same general kind of information. Vision and hearing, for example, are both concerned with detecting events that occur in external space. Smell and taste are both concerned with evaluating the chemical composition of stimuli. The vestibular, visual, and proprioceptive systems all play a role in monitoring our position and movement in space. Thus, sensation is not a series of parallel monologues in individual senses but rather a rich conversation among them.

Vision

Overview

Much of our ability to navigate and respond to the environment depends on our visual system and its capacity to detect visible light. The human visual system is extraordinary in the quantity and quality of information it detects and processes from the world around us, and the speed at which it does so. Not only must it be capable of collecting and conveying information about object location, size, color, texture, and motion, but it operates over a wide range of luminance values and stimulus intensities. How can it accomplish such a challenging task? This chapter aims to answer this question by addressing: (1) how the eye, the sensory organ of the visual system, collects and focuses visible light onto a light-sensitive layer of CNS tissue; (2) how photoreceptive cells transduce light energy into electrical signals that can be processed and transmitted by neurons; (3) how specialized circuits among neurons in the retina extract information about features in the visual scene and transmit it to the brain; and (4) how visual centers in the brain integrate feature-specific information to facilitate object recognition and visually guided behaviors. It is important to note that while a quick glance is sufficient to capture information about a visual scene, the true power of the visual system requires active control to regulate how much light enters our eye and what areas of the visual field we are collecting information from. This last fact should be apparent as you read this chapter—without coordinated eye movements, the information you can gather from the words and images on this page is severely limited.

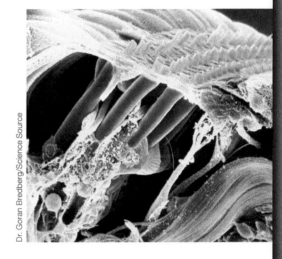

Dr. Goran Bredberg/Science Source

KEY CONCEPTS

9.1 The eye collects light and focuses it onto specialized photoreceptive cells

9.2 Photoreceptors convert light into electrical signals in the retina

9.3 Retinal circuitry extracts information about features of the visual scene

9.4 Retinal ganglion cells convey feature information to the brain

9.5 Visual centers in the cerebral cortex detect increasingly complex visual features

9.6 Optimizing vision and visual behaviors requires autonomic, motor, and cognitive control

CONCEPT **9.1** | **The Eye Collects Light and Focuses It onto Specialized Photoreceptive Cells**

LEARNING OBJECTIVES

9.1.1 Illustrate the anatomy of the eye.

9.1.2 Describe the structures that refract and focus light onto the retinal surface.

9.1.3 Describe the projection of the visual field onto the retinal surface.

Visible light

The visual system is built to detect electromagnetic radiation, a form of electromagnetic energy that propagates in waves and carries radiant energy. While the full spectrum of electromagnetic radiation waves is broad, the human visual system can only detect electromagnetic radiation with wavelengths falling between 380 and 750 nanometers, which is termed **visible light** (referred to here simply as light). As this light propagates from its source (which can include the sun, fires, and more recently, electric lights and electronic devices), it is reflected off objects in our environment. The first task of the visual system is to collect this light and focus it onto specialized photoreceptive cells in the retina. This task is accomplished by the eye.

Anatomy of the eye

The eye is a fluid-filled sphere enclosed by three layers of tissue (Figure 9.1). The innermost layer of the eye, the **retina**, is part of the CNS and contains more than 100 million specialized photoreceptive cells, termed **photoreceptors**. Projection neurons in the retina send their axons through the optic disk and along the optic nerve to connect with central targets in the brain. Adjacent to the retina is a layer of tissue referred to as the uveal tract. The largest component of the uveal tract is the choroid, which consists of a dense capillary bed that nourishes the outer retina and a pigmented epithelial layer rich in the light-absorbing pigment melanin. Extending from the choroid near the front of the eye is the **ciliary body**, which contains a muscular component that adjusts the shape of the **lens** and a vascular component that produces the fluid that fills the front of the eye. The most anterior component of the uveal tract is the iris, the colored portion of the eye that can be seen through the cornea. The iris contains two sets of muscles with opposing actions, which regulate **pupil** diameter and thereby control the quantity of light entering the eye. The **sclera** forms the outermost tissue layer of the eye and is composed of a tough, white, fibrous tissue. At the front of the eye, however, this opaque outer layer is transformed into the **cornea**, a transparent tissue that serves as the main refractive element of the eye.

Light enters the eye through the transparent cornea and must pass through the pupil, lens, and two distinct fluid environments before striking the retina. In the anterior chamber, just behind the cornea and in front of the lens, is the **aqueous humor**, a clear, watery fluid that supplies nutrients to both the lens and cornea. Aqueous humor is produced by the ciliary processes within the ciliary body, and the rate of production must be balanced by a comparable rate of drainage in order to ensure a constant intraocular pressure. A meshwork of cells lying at the junction of the iris and the cornea is responsible for aqueous drainage; dysfunction of this drainage causes **glaucoma**, a group of disorders associated with elevated intraocular pressure and loss of projection neurons that connect the retina to the brain. The space between the back of the lens and the surface of the retina is filled with a thick, gelatinous fluid called the **vitreous humor**, which accounts for about 80% of the volume of the eye. In addition to maintaining the shape of the eye, the vitreous humor contains phagocytic cells that remove blood and other debris that might otherwise interfere with light transmission.

Optics and the eye

Normal vision requires that the optical media of the eye be transparent, and both the cornea and lens are remarkable examples of such transparency. A dominant feature of both elements is their lack of blood vessels (i.e., avascular nature), the presence of which could scatter light as it enters the eye. Alterations in the composition of the cornea or lens can significantly reduce their transparency and affect the path of light onto the retina. Indeed, clouding of the lens as proteins break down because of aging or disease accounts for roughly half the cases of blindness in the world. This condition is known as **cataracts**. Fortunately, cataracts are treatable, and successful surgical intervention can restore vision in most individuals.

In addition to efficiently transmitting light energy, the primary function of the optical components of the eye is to generate a focused image on the surface of the retina. The cornea contributes most of the necessary refraction of the eye. This can be appreciated by considering the hazy, out-of-focus images experienced when swimming underwater. Water, unlike air, has a refractive index close to that of the cornea; as a result, immersion in water virtually eliminates the refraction that normally occurs at the air–cornea interface; thus, the image is no longer focused on the retina. The majority of the human population has some form of refractive error, or **ametropia**, usually resulting from discrepancies in the shape of the corneal surface or the shape of the eye. In nearsightedness (**myopia**), too much

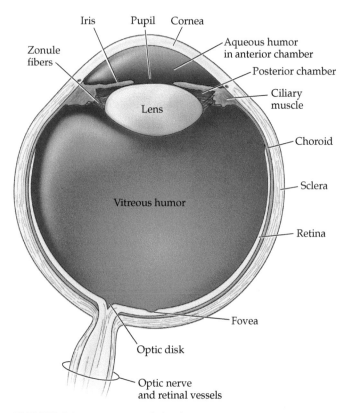

Iris Pupil Cornea

Zonule
fibers

Aqueous humor
in anterior chamber

Posterior chamber

Ciliary
muscle

Lens

Choroid

Sclera

Vitreous humor

Retina

Fovea

Optic disk

Optic nerve
and retinal vessels

FIGURE 9.1 Anatomy of the human eye

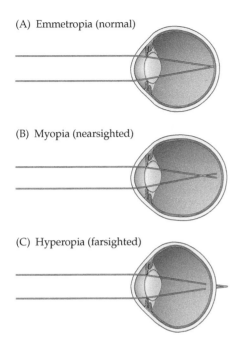

FIGURE 9.2   **Refractive errors caused by discrepancies in the shape of the human eye** (A) In the normal eye, with ciliary muscles relaxed, an image of a distant object is focused on the retina. (B) In myopia, light rays from a distant object are focused in front of the retina. (C) In hyperopia, light rays from a distant object are focused at a point beyond the retina. (After G. Westheimer, 1974. In *Medical Physiology*, 13th ed. V. B. Mountcastle [Ed.] St. Louis: Mosby.)

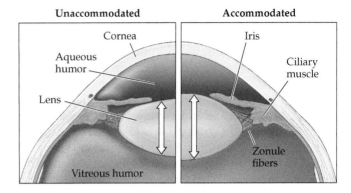

FIGURE 9.3   **Accommodation in the human eye** The diagram on the left shows the anterior part of the eye in the unaccommodated state for viewing a distant object. The diagram on the right shows the anterior part of the eye in the accommodated state for viewing a near object. Note that accommodation involves contraction of the ciliary muscle, which reduces tension in the zonule fibers and allows the lens to increase its curvature (white arrows).

refractive power of the eye or an eyeball that is too long causes light to be focused in front of the retina, and such individuals cannot focus on distant objects. In farsightedness (**hyperopia**), a weak refracting system or an eyeball that is too short causes light to be focused behind the retina, and these individuals cannot focus on near objects (Figure 9.2). **Astigmatism** is a condition that distorts or blurs vision because of defects in the curvature of the cornea (or lens). Fortunately, myopia, hyperopia, and astigmatism are all correctable with appropriate lenses or with laser-assisted in situ keratomileusis (LASIK) corneal surgery.

The lens has considerably less refractive power than the cornea; however, its refractive capabilities are adjustable, allowing the observer to bring objects at various distances into sharp focus. Dynamic changes in the shape of the lens are referred to as **accommodation**. When viewing distant objects, the lens is made relatively thin and flat and has the least refractive power. For near vision, the lens becomes thicker and rounder and has the most refractive power (Figure 9.3). The lens begins to lose its elasticity with age, and at some point, usually during middle age, the accommodative ability of the eye is impaired and near-vision activities, such as reading, become difficult. This condition is referred to as **presbyopia**. Presbyopia can be corrected by convex lenses for near-vision tasks.

Pupil diameter can also affect both the quantity and quality of light that falls on the retina. The pupil acts as a pinhole aperture, an optical concept called the **pinhole effect**, whereby reducing the size of the opening through which light can pass prevents unfocused light from entering the eye, falling on the retina, and blurring vision. This is why vision improves when we squint, making the opening between the eyelids (i.e., the eyelid fissure) smaller than the opening of the pupil.

Projections of the visual field onto the retina

While the cornea and lens focus light onto the retina, it is important to note that not all regions of the retina are equal in their ability to resolve fine details of the visual field. The **macula lutea**, a circular region containing yellow pigment (xanthophyll) and roughly 3 mm in diameter near the center of the retina, supports high **visual acuity** (the ability to resolve fine details). Acuity is greatest at the center of the macula, at a small depression called the **fovea** (see Figure 9.1). Several features of the fovea facilitate its high resolving power, including a high density of certain types of photoreceptive cells, specialized circuits for generating small receptive fields for neurons in these regions, and an avascular zone so that the path of light to the photoreceptors is unimpeded. These specializations are discussed in more detail in Concept 9.2. Damage to this small region of the retina, such as in **age-related macular degeneration** where there is a progressive loss of central vision, has a devastating impact on visual perception.

Moving away from the macula lutea toward the peripheral retina, the resolving power of the retina diminishes significantly, a fact that can be easily observed by trying to identify and describe objects in your peripheral vision. The differences in the resolving power of the central and peripheral retina result from differences in the distribution, function, and connectivity of different types of photoreceptors, a topic discussed in detail in Concept 9.2. It is also important to point out that there is a region of peripheral retina that contains no photoreceptors and thus is a natural "blind spot," or **scotoma**. This region occurs at the optic disk where retinal axons leave the eye and enter the optic nerve (see Figure 9.1). It is logical to suppose that you might be able to notice a region where you receive no visual information, but scotomas (even pathological ones) often go unnoticed. In the case of the natural blind spot,

one reason it goes unnoticed is that with both eyes open, you receive information about this portion of the visual field from the other eye. However, the blind spot is still difficult to notice even when one eye is closed. The blind spot falls on a portion of peripheral retina where visual information is sparsely coded, allowing the visual system to "fill in" the missing details.

It is also important to highlight that refraction of light by the cornea and lens inverts and left–right reverses images as they fall on the retina. Each eye sees a part of visual space that defines its **visual field** (Figure 9.4). For descriptive

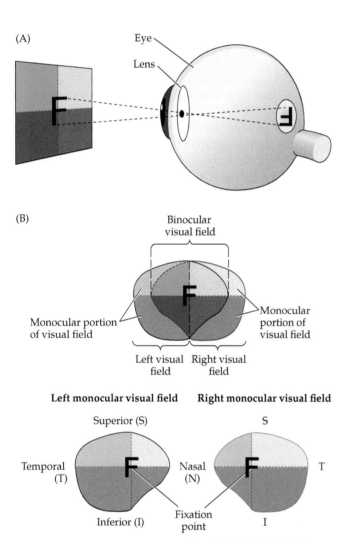

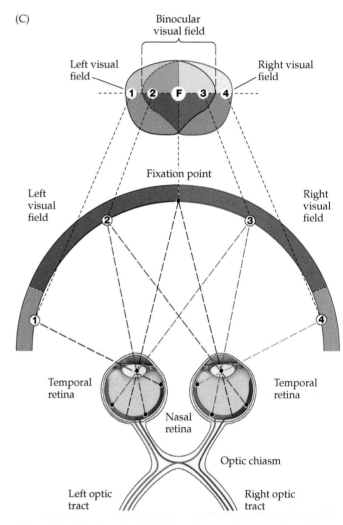

FIGURE 9.4 Projection of the visual fields onto the left and right retinas (A) Projection of an image onto the surface of the retina results in images that are inverted and left–right reversed. (B) The visual field can be divided into regions viewed by one or both eyes (monocular vs. binocular visual fields) or into either superior, inferior, left and right quandrants (for binocular fields) or superior, inferior, nasal and temporal quandrants (for monocular fields). (F = point of fixation.) (C) Projection of the binocular visual field onto the two retinas and the axonal projections of ganglion cells into the ipsi- and contralateral optic tracts. Points in the binocular

portion of the left visual field (2) fall on the nasal retina of the left eye and the temporal retina of the right eye. Points in the binocular portion of the right visual field (3) fall on the nasal retina of the right eye and the temporal retina of the left eye. Points that lie in the monocular portions of the left and right visual fields (1 and 4) fall on the left and right nasal retinas, respectively. Axons of ganglion cells in the nasal retina cross in the optic chiasm, whereas those from the temporal retina remain ipsilateral. As a result, the right optic tract carries information from the left visual field, and the left optic tract carries information from the right visual field.

purposes, each retina and its corresponding visual field are divided into quadrants. The vertical line divides the retina into **nasal division of the retina** and the **temporal division of the retina**, and the horizontal line divides it into **superior division of the retina** and **inferior division of the retina**. The foveae of both eyes normally align on a single target in visual space, causing the visual fields of the two eyes to overlap extensively. This **binocular field** consists of two symmetrical visual hemifields (left and right). The left binocular hemifield includes the nasal visual field of the right eye and the temporal visual field of the left eye; the right hemifield includes the temporal visual field of the right eye and the nasal visual field of the left eye. Note that information from the peripheral visual field is strictly monocular, mediated by the most medial portion of the ipsilateral nasal retina. You can easily demonstrate this by closing one eye; you will immediately notice the loss of your ipsilateral peripheral vision.

| CONCEPT **9.2** | **Photoreceptors Convert Light into Electrical Signals in the Retina** |

LEARNING OBJECTIVES

9.2.1 Define and describe the principal cell types in the retina.

9.2.2 Define phototransduction and describe graded potentials and hyperpolarization of rods and cones.

9.2.3 Describe the differences in the structure, distribution, and function of rods and cones.

9.2.4 Summarize the mechanisms of color vision.

Cell types in the retina

Although light travels as a wave, it also acts as a particle and thus the energy of light can be absorbed at a singular location. This absorbed energy is known as a photon and represents the minimal amount of radiant energy capable of interacting with atoms and molecules. In the visual system, a photon is the minimal amount of energy required for **phototransduction**, the process by which this radiant energy is converted into electrical signals by specialized cells in the retina.

The retina is an intricately layered tissue consisting of five primary classes of neurons whose cell bodies and processes are stacked in alternating nuclear and synaptic layers (Figure 9.5). The most abundant neurons in the retina are the photoreceptors, whose cell bodies reside in the **outer nuclear layer** and are divided into two functionally and morphologically distinct types, **rods** and **cones** (see Figure 9.5B,C). The outer segments of rods and cones contain membranous disks with light-sensitive photopigment and lie directly adjacent to the pigment epithelium. The spatial arrangement of photoreceptors within the retina

may seem counterintuitive: Light rays must pass through various non-light-sensitive elements of the retina before reaching the outer segments of the photoreceptors where photons are absorbed (see Figure 9.5A,B). One reason for this curious anatomy is the special relationship that exists between photoreceptors and the adjacent pigment epithelium. The cells of the pigment epithelium have long processes that extend into the photoreceptor layer and surround the tips of the outer segments of each photoreceptor. Together with the rich capillary bed in the choroid, the pigment epithelium clears degenerating fragments of outer segments, regenerates photopigment molecules, and provides nourishment for retinal photoreceptors, all of which are crucial for photoreceptor function.

The inner segments of photoreceptors give rise to synaptic terminals that contact interneuron dendrites in the **outer plexiform layer**. These interneurons include both **bipolar cells** and **horizontal cells** (see Figure 9.5A,B), both of which reside in the outer portion of the **inner nuclear layer**. Bipolar cells are excitatory neurons whose short axonal processes make synaptic contacts on the dendritic processes of the projection neurons of the retina, **retinal ganglion cells**, in the **inner plexiform layer**. This three-neuron chain—photoreceptor cell to bipolar cell to ganglion cell—represents the most direct path of information flow from photoreceptors to the optic nerve.

The final class of neurons in the retina is the **amacrine cells**, which have their cell bodies mainly in the inner nuclear layer and have processes that are limited to the inner plexiform layer (see Figure 9.5B). The processes of amacrine cells are postsynaptic to bipolar cell terminals and presynaptic to the dendrites of retinal ganglion cells.

Although the retina has only five primary classes of neurons, many of these classes can be divided into a multitude of morphologically and functionally distinct types, generating considerable cellular diversity within the retina. This diversity is instrumental for the generation of specific circuits that allow the retina to extract information about features in the visual scene and transmit it to the brain. Beyond neurons, the retina also contains a diversity of glial cells, including microglia, astrocytes, and Müller glial cells. Myelin-forming oligodendrocytes are largely absent from the mammalian retina, although they populate the optic nerve.

Phototransduction

In the retina, the conversion of photons into electrical signals (i.e., phototransduction) occurs by two distinct mechanisms. The first (and dominant) phototransduction mechanism is achieved by rods and cones. These canonical photoreceptors differ from most sensory receptors in that they do not depolarize in response to an appropriate stimulus, nor do they generate action potentials. Instead, light activation causes a graded *hyperpolarization* of

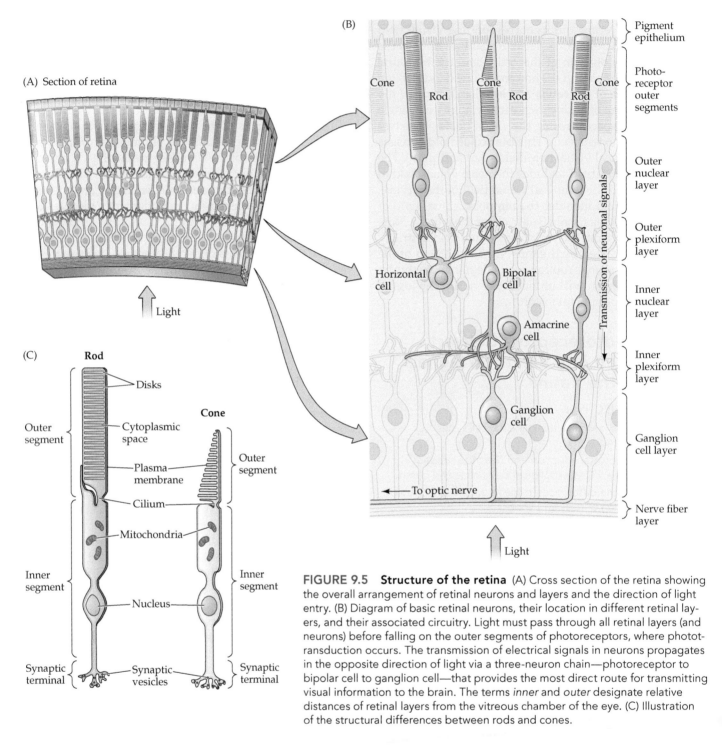

(A) Section of retina

Light

(C) **Rod**

Outer segment
Disks
Cytoplasmic space
Plasma membrane
Cone
Outer segment
Cilium
Mitochondria
Inner segment
Inner segment
Nucleus
Synaptic terminal
Synaptic vesicles
Synaptic terminal

(B)
Pigment epithelium
Cone
Rod
Cone
Rod
Cone
Rod
Photo-receptor outer segments
Outer nuclear layer
Outer plexiform layer
Horizontal cell
Bipolar cell
Transmission of neuronal signals
Inner nuclear layer
Amacrine cell
Inner plexiform layer
Ganglion cell
Ganglion cell layer
To optic nerve
Nerve fiber layer
Light

FIGURE 9.5 **Structure of the retina** (A) Cross section of the retina showing the overall arrangement of retinal neurons and layers and the direction of light entry. (B) Diagram of basic retinal neurons, their location in different retinal layers, and their associated circuitry. Light must pass through all retinal layers (and neurons) before falling on the outer segments of photoreceptors, where phototransduction occurs. The transmission of electrical signals in neurons propagates in the opposite direction of light via a three-neuron chain—photoreceptor to bipolar cell to ganglion cell—that provides the most direct route for transmitting visual information to the brain. The terms *inner* and *outer* designate relative distances of retinal layers from the vitreous chamber of the eye. (C) Illustration of the structural differences between rods and cones.

the membrane potential (Figure 9.6) and a corresponding change in the rate of transmitter release onto postsynaptic neurons (recall from Chapter 5 that typical neuronal excitation involves depolarization, not hyperpolarization). The second phototransduction mechanism occurs in a small subset of specialized ganglion cells that not only receive signals from rods and cones (through retinal interneurons) but also express a light-sensitive photopigment

(**melanopsin**) that endows them with the ability to be directly *depolarized* by light. These **intrinsically photosensitive ganglion cells**, which account for approximately 5% of all ganglion cells, project axons to target regions in the hypothalamus (among other brain regions) that regulate the ability to photoentrain circadian rhythms to the environmental day–night cycle. The presence of intrinsically photosensitive ganglion cells helps explain why some

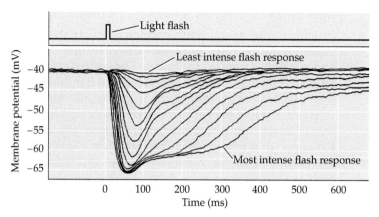

FIGURE 9.6 **Hyperpolarization of a photoreceptor** Intracellular recording from a single cone stimulated with different amounts of light. Each trace represents the response to a brief flash that was varied in intensity. At the highest light levels, the response amplitude saturates (at about −65 mV). (After D. A. Baylor et al., 1974. *J Physiol* 242: 685–727.)

people (and animals) who have lost rod- and cone-dependent phototransduction can maintain normal circadian rhythms and pupillary light responses.

Our discussion of phototransduction focuses solely on rods and cones, which outnumber intrinsically photosensitive ganglion cells 20,000 to 1. Phototransduction occurs when photons are absorbed by specialized photopigments in the outer segments of rods and cones. These photopigments contain an organic, light-absorbing chromophore, **11-*cis*-retinal**, coupled to one of several 7-pass transmembrane proteins called opsins. The different opsins—**rhodopsin** in rods and one of three different **cone opsins** in cones—tune the molecule's absorption of light to a particular region of the light spectrum; indeed, it is the differing protein components of the photopigments in rods and cones that allow the functional specialization of these two receptor types.

When retinal absorbs a photon of light, one of the double bonds between its carbon atoms breaks, and its configuration changes from the 11-*cis* isomer to all-*trans* isomer (Figure 9.7); this change triggers a series of alterations in the opsin component of the molecule. The changes in opsin lead, in turn, to the activation of an intracellular messenger called **transducin**. Transducin activates a phosphodiesterase (PDE) that hydrolyzes the nucleotide cyclic guanosine monophosphate (cGMP), which binds and opens cation-permeable channels in the outer segment. Thus, absorption of light by the photoreceptor reduces the concentration of cGMP, leading to a closure of the cGMP-gated channels and,

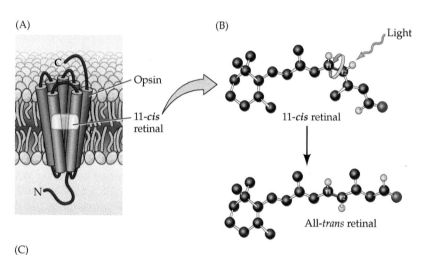

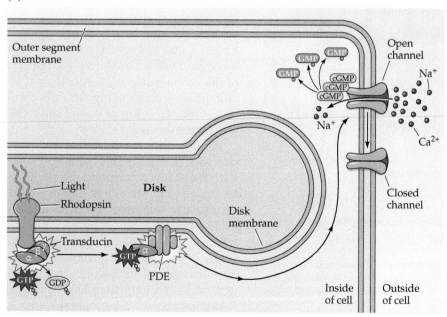

FIGURE 9.7 **Phototransduction in rod photoreceptors** (A) The 7-pass transmembrane protein rhodopsin resides in the disk membrane of the rod outer segment and encloses the light-sensitive molecule retinal. (B) Absorption of a photon of light by retinal leads to a change in configuration from the 11-*cis* to the all-*trans* isomer. (C) The second-messenger cascade of phototransduction. The change in the retinal isomer activates transducin, which in turn activates a phosphodiesterase (PDE). The PDE then hydrolyzes cGMP, reducing its concentration in the outer segment and leading to the closure of channels in the outer segment membrane. (A after E. A. Dratz and P. A. Hargrave, 1983. *Trends Biochem Sci* 8: 128–131; B after L. Stryer, 1987. *Sci Am* 257: 42–50.)

consequently, a reduction in the inward flow of Na⁺ and Ca²⁺. As a result, positive charge (carried by K⁺) flows out of the cell more rapidly than positive charge (carried by Na⁺ and Ca²⁺) flows in, and the cell becomes hyperpolarized (see Figure 9.7C). One of the important features of this complex biochemical cascade is that it provides enormous signal amplification. Competing mechanisms exist to limit the duration of this amplifying cascade and ultimately restore the various molecules to their inactivated states.

As in other neurons, neurotransmitter release from the synaptic terminals of the photoreceptor is dependent on voltage-sensitive Ca²⁺ channels in the terminal membrane. Thus, in the dark, when photoreceptors are relatively depolarized because of the high concentration of cGMP and open cation-permeable channels in the outer segment (Figure 9.8), the number of open Ca²⁺ channels in the synaptic terminal is high. In this state, the rate of neurotransmitter release is correspondingly high. In the presence of increasing light, photoreceptor cGMP levels drop, cation-permeable channels in the outer segment close, and the cell becomes hyperpolarized (see Figure 9.8). This hyperpolarization reduces the number of open voltage-gated Ca²⁺ channels in the photoreceptor's nerve terminal, thereby reducing the rate of neurotransmitter release.

Specialization of rod and cone photoreception

In addition to being distinguished by their shape and photopigments, rods and cones differ in their number and distribution across the retina. The human retina contains approximately 120 million rods and 6 million cones (together accounting for more than 80% of all retinal neurons). As a result, the density of rods is much greater than that of cones throughout most of the retina (Figure 9.9A). However, this relationship changes dramatically in the fovea, where cone density increases almost 200-fold. This high density is achieved by decreasing the diameter of the cone outer segments, allowing as many cones as possible to squeeze into this small region of retina. The increased density of cones in the fovea is accompanied by an equally dramatic decline in the density of rods. In fact, the central region of the fovea, called the foveola, is rod-free (Figure 9.9B). The inner layers of the retina are also displaced in the fovea so that photons are subjected to minimum scattering before they strike the photoreceptors.

In addition to these dramatic differences in number and spatial distribution, rods and cones differ significantly in their connectivity. The connectivity of a rod bipolar cell (the specific type of bipolar cell innervated by rods) exhibits a high degree of **convergence**, meaning a single rod bipolar cell receives input from as many as 15 to 30 rods (Figure 9.10A). In contrast, the cone system exhibits far less convergence, meaning bipolar cells in the fovea are driven by the activity of single cones. The benefit of convergence in the rod system is that it improves the ability to detect light (especially small amounts of light), because small signals from many rods are pooled to generate a response in the bipolar cell. At the same time, convergence reduces the spatial resolution of the rod system, because the source of a signal in a rod bipolar cell could have come from anywhere within a relatively large area of the retinal surface. The one-to-one relationship of cones to bipolar cells maximizes the resolving power of the retina and increases visual acuity.

Finally, differences in their transduction mechanisms endow rods and cones with the ability to respond differentially to ranges of light intensity. Rods are exceptionally sensitive and produce reliable responses to single photons of light, whereas current changes in response to a single photon of light in cones are small and difficult to distinguish from background. In fact, more than 100 photons are required to produce a comparable current

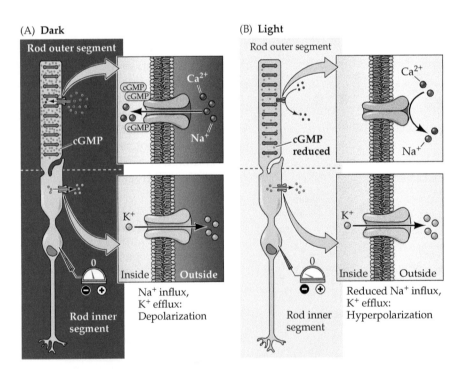

FIGURE 9.8 Cyclic GMP-gated channels and light-induced changes in the electrical activity of photoreceptors This diagram shows a rod, but the same scheme applies to cones. (A) In the dark, cGMP levels in the outer segment membrane are high; cGMP binds to the Na⁺-permeable channels in the membrane, keeping them open and allowing sodium and other cations to enter, thus depolarizing the cell. (B) Absorption of photons leads to a decrease in cGMP levels, closing the cation channels and resulting in receptor hyperpolarization.

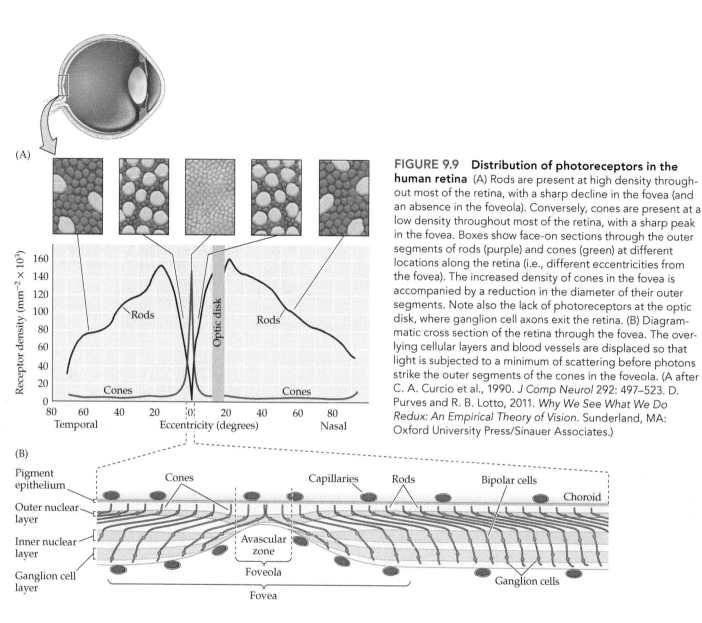

FIGURE 9.9   Distribution of photoreceptors in the human retina (A) Rods are present at high density throughout most of the retina, with a sharp decline in the fovea (and an absence in the foveola). Conversely, cones are present at a low density throughout most of the retina, with a sharp peak in the fovea. Boxes show face-on sections through the outer segments of rods (purple) and cones (green) at different locations along the retina (i.e., different eccentricities from the fovea). The increased density of cones in the fovea is accompanied by a reduction in the diameter of their outer segments. Note also the lack of photoreceptors at the optic disk, where ganglion cell axons exit the retina. (B) Diagrammatic cross section of the retina through the fovea. The overlying cellular layers and blood vessels are displaced so that light is subjected to a minimum of scattering before photons strike the outer segments of the cones in the foveola. (A after C. A. Curcio et al., 1990. *J Comp Neurol* 292: 497–523. D. Purves and R. B. Lotto, 2011. *Why We See What We Do Redux: An Empirical Theory of Vision*. Sunderland, MA: Oxford University Press/Sinauer Associates.)

response in a cone. While cones are less sensitive than rods, the adaptation mechanisms of cones are more effective: Cones recover four times faster than rods (Figure 9.10B).

All of these properties reflect the fact that the rod and cone systems (i.e., the receptor cells and their connections within the retina) are specialized for different aspects of vision. The rod system has very low spatial resolution but is extremely sensitive to light; it is therefore specialized for sensitivity. Conversely, the cone system has very high spatial resolution (and allows us to see color; see the next section) but is less sensitive to low levels of illumination. Thus, these two phototransduction systems operate at different ranges of illumination (Figure 9.11). Rod-mediated vision (**scotopic vision**) occurs at the lowest levels of illumination. At this level of illumination, the ability to make fine-detailed visual discrimination is poor based on the properties of the rod phototransduction system. As illumination

increases, cones (and the central retina) become more dominant in determining what is seen. Cones drive visual perception in normal indoor lighting or sunlight (**photopic vision**), where the contributions of rods nearly drop out because their response to light saturates. Thus, what we think of as normal "seeing" is mediated by the cone system, despite our retinas containing 20 times more rods than cones. The loss of cone function (and central vision) results in an individual being legally blind. In contrast, the loss of rod function is less devastating to vision, leading to the loss of peripheral vision and night blindness.

Color vision

Detecting differences in the spectral quality, or color, of light endows us with additional feature-specific information to distinguish objects in our environment. This is the job of cones. The human retina contains three types of cones

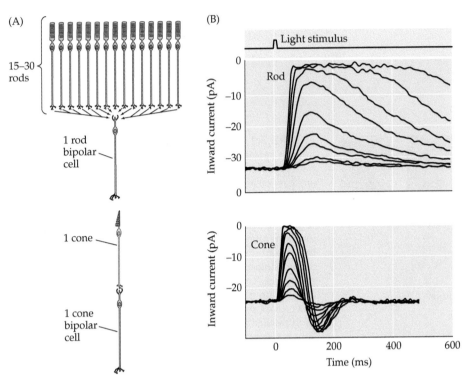

FIGURE 9.10 Circuitry and differential responses of rods and cones (A) Difference in the amount of convergence in the rod and cone pathways. Each rod bipolar cell receives synapses from 15 to 30 rods. In contrast, in the center of the fovea, each bipolar cell receives its input from a single cone. (B) Suction electrode recordings of the reduction in inward current produced by flashes of successively higher light intensity on primate rods and cones. For moderate to long flashes, the rod response continues for more than 600 ms, whereas even for the brightest flashes tested, the cone response returns to baseline (with an overshoot) in roughly 200 ms. (B after D. A. Baylor, 1987. *Invest Ophthalmol Vis Sci* 28: 34–49)

with each expressing a unique photopigment that is sensitive to select wavelengths of light. Human vision is thus **trichromatic.** Although human cones are often described as being sensitive to *blue, green,* or *red* light, it is important to note that each of these cones is sensitive to a broad range of wavelengths of light and that their peak sensitivities do not actually coincide with these three colors. A more appropriate nomenclature for cones is short- (S-), medium- (M-), and long- (L-) wavelength cones (Figure 9.12A).

Viewing the absorption spectra for cones in Figure 9.12A, it is obvious that many wavelengths of visible light can excite more than one type of cone. Green light, for example, activates both M- and L-wavelength cones. How then is green detected? Such ambiguity can be resolved by *comparing* the activity in different classes of cones. In the case of detecting green, the M-wavelength cones are more strongly excited than the L-wavelength cones. Downstream neurons in the retina (such as ganglion

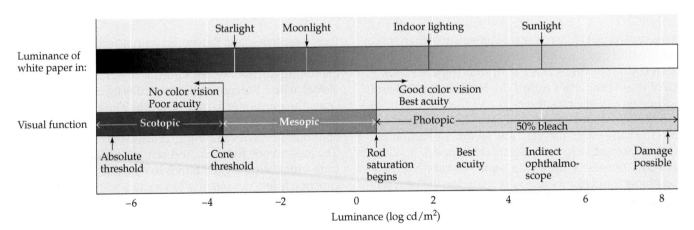

FIGURE 9.11 Range of luminance values over which the visual system operates At the lowest levels of illumination, only rods are activated (scotopic vision). Cones begin to contribute to perception at about the level of starlight and are the only receptors that function under relatively bright conditions (photopic vision). Mesoscopic vision, a transition stage between scotopic and photopic vision, occurs in a range of low-light conditions in which both cones and rods can be activated.

(A) Normal (trichromat)

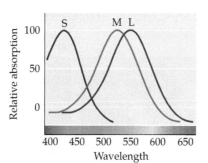

(B) Protanopia

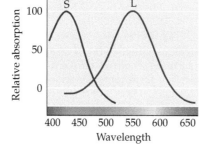

(C) Deuteranopia

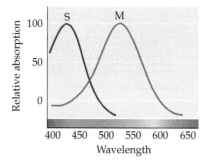

Photos by M. H. Siddall

FIGURE 9.12   Normal and abnormal color vision
Simulation of the image of a flower as it would appear to an observer with normal color vision (A); an observer with protanopia (loss of long-wavelength-sensitive cones) (B); and an observer with deuteranopia (loss of medium-wavelength-sensitive cones) (C). The graphs show the corresponding absorption spectra of retinal cones in typical individuals and in those with defective color vision.

cells) and in central brain regions (such as the lateral geniculate nucleus; see Concept 9.4) are sensitive to differences in activity elicited by these two cone types, or are color opponent, and serve critical roles in how the visual system extracts color information from spectral stimuli. A full understanding of the neural mechanisms that underlie color perception remains an important question that is being addressed by visual scientists and that has puzzled the public (e.g., see the optical illusion of #*theDress* that became a viral internet phenomenon in 2015). Box 9A highlights the complexity of color perception and the importance of context, illumination, and prior experiences in this process. While cones play a critical role in the first steps of color vision, signal processing in the retina and brain is critical for how color percepts are built.

While normal human color vision is fundamentally trichromatic, 8% of the male population in the United States and less than 1% of the female population have a deficiency in color vision—commonly referred to as **color blindness**. Color vision deficiencies

---

■ BOX 9A | The Importance of Context in Color Perception

Seeing color logically demands that retinal responses to different wavelengths in some way be compared. The three human cone types and their different absorption spectra are therefore correctly regarded as the basis for human color vision. Nevertheless, how these human cone types and the higher-order neurons they contact produce the sensations of color is still unclear.

A fundamental problem has been that, although the relative activities of three cone types can more or less explain the colors perceived in color-matching experiments performed

in the laboratory, the perception of color is strongly influenced by context. For example, a patch returning the exact same spectrum of wavelengths to the eye can appear quite different depending on its surround, a phenomenon called *color contrast* (Figure A). Moreover, test patches returning different spectra to the eye can appear to be the same color, an effect called *color constancy* (Figure B). Color constancy allows us to distinguish familiar objects regardless of the amount or type of illumination on them and the subsequent amount or wavelength of

light reflected off them. For example, a tennis ball appears the same to us at midday and at dusk, despite differences in the illumination at those times and the subsequent differences in wavelength of light reflected off the tennis ball.

The phenomena of color contrast and color constancy have led to a scientific debate about how color percepts are generated. Multiple mechanisms have been proposed, some of which include low-level, subconscious

(Continued)

■ BOX 9A | The Importance of Context in Color Perception *(continued)*

adaptations (such as cone adaptation, in which cones adjust their activity in response to light levels in the visual environment) and some of which require higher-level processing whereby color perception is generated empirically according to what spectral stimuli or the context that they appear in have typically signified in past experiences.

This is a debate that captivates not only the scientific experts. The phenomena of color contrast and constancy have recently given rise to debate among non-experts on social media. In 2015, #theDress created heated online debate over an image of a dress that was purchased for a wedding and shared for feedback (initially with family and friends, and later with the world). Some people perceived the dress as being "blue and black," while others vehemently disagreed, describing it as "white and gold." As just described, context is important for color perception.

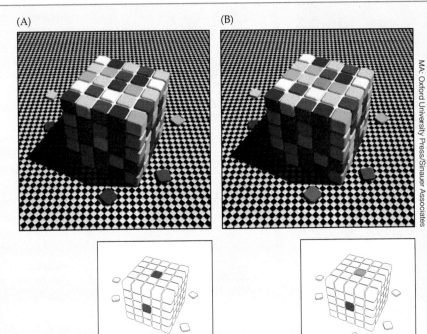

From D. Purves and R. B. Lotto, 2011, *Why We See What We Do Redux: An Empirical Theory of Vision*, Sunderland MA: Oxford University Press/Sinauer Associates

The genesis of contrast and constancy effects by exactly the same context. The two panels demonstrate the effects on apparent color when two *similarly* reflective target surfaces (A) or two *differently* reflective target surfaces (B) are presented in the *same* context in which all the information provided is consistent with illumination that differs only in intensity. The appearances of the relevant target surfaces in a neutral context are shown in the insets below.

---

most often result from either the inherited loss of one of the genes that encode cone photopigments or from an alteration in the absorption spectra of one of these cone pigments (rather than from lesions in central brain regions that process color information). The genes that encode the L- and M-wavelength pigments lie adjacent to each other on the X chromosome, thus explaining the prevalence of color deficiency in males. The S-wavelength-sensitive pigment gene is located on chromosome 7. For individuals with mutations in one of these genes, color vision is **dichromatic**: Only two cone types remain able to detect colors and contribute to a color percept. The most prevalent forms of dichromacy are **protanopia** (Figure 9.12B), characterized by impairment in perception of long wavelengths, and **deuteranopia** (Figure 9.12C), impairment in the perception of medium wavelengths. Although there are differences in the color discrimination capabilities of people with protanopia and those with deuteranopia, both have difficulty discriminating red and green, and for this reason dichromacy is commonly called red–green color blindness (see Figure 9.12).

CONCEPT
9.3

Retinal Circuitry Extracts Information about Features of the Visual Scene

LEARNING OBJECTIVES

9.3.1 Summarize how neuronal diversity and retinal circuits give rise to increasingly complex receptive fields.

9.3.2 Explain how types of bipolar cells are differentially depolarized or hyperpolarized when photoreceptors are stimulated by light.

9.3.3 Describe how lateral interactions by GABAergic horizontal cells and amacrine cells shape receptive fields.

Neurons, circuits, and visual receptive fields

Every neuron in the retina or in visual centers of the brain has a region of visual space where both the presence and properties of light will alter its activity or firing pattern. This is the neuron's **visual receptive field**. For a canonical photoreceptor, the receptive field is exceptionally simple, a mere point of light in the visual field that falls on the outer segment of the photoreceptor. Neurons downstream of this photoreceptor have increasingly complex receptive fields that can include additional elements of the stimulus, such as spatial structure. Therefore, these downstream neurons encode higher-level features of the visual scene. Generating higher-order, complex receptive fields is accomplished by neural circuits that facilitate both the integration of information from earlier levels of visual processing and the exclusion of information that is not necessary for the receptive field of a particular neuron. Great diversity within each class of retinal neuron (e.g., 40 types of ganglion cells and 60 types of amacrine cells have recently been identified in the rodent retina) provides the cellular substrate to generate an enormous number of specific neural circuits to extract information about simple visual features, such as contrast, color, and motion. Here we do not provide a comprehensive description of all the retinal circuits that are currently understood or hypothesized, but rather highlight just a few to introduce a perspective on how the retina can detect spatial and temporal changes in light intensity.

ON and OFF pathways in the retina

One of the most fundamental functions of the retina (or any other sensory system) is to detect changes in the environment. In the visual system, such change includes both spatial and temporal changes in light intensity. Early insight into how the retina detects increments and decrements of light stimuli came from studies performed more than half a century ago. Two broad categories of ganglion cells were identified that respond in opposite ways to increments and decrements of light striking a small circular patch of the retinal surface. Turning on a spot of bright light in the receptive field of an **ON ganglion cell** produces a burst of action potentials (Figure 9.13A). The same stimulus applied to the receptive field of an **OFF ganglion cell** reduces the rate of firing; when the spot of light is turned off, the cell responds with a burst of action potentials (see Figure 9.13A). When a stimulus darker than the background illumination is placed on the same patch of retina, inverse patterns of activity are found for each cell type (Figure 9.13B). Thus, types of ganglion cells can respond differently to changes in light intensity. ON ganglion cells intensify their discharge rate to luminance increases within the receptive field, whereas OFF ganglion cells intensify their discharge rate to luminance decreases in the receptive field. Although this text will not detail how ON and OFF ganglion cells can be classified into many distinct cell types based on features such as anatomy, morphology, molecular identity, connectivity, and function, two specific features of these cells do warrant mention here. First, functional specializations of some ON and OFF ganglion cells are based on an

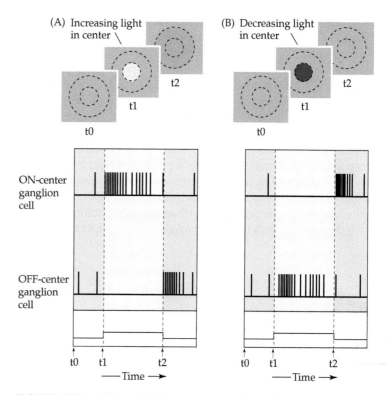

FIGURE 9.13 ON- and OFF-center retinal ganglion cell responses to stimulation of different regions of their receptive fields Diagrams depicting the effects of increasing (A) and decreasing (B) the intensity of light in the receptive field center. Upper panels indicate the time sequence of stimulus changes (with increments or decrements of light in the receptive field center occurring at t1). Lower panels depict activity (i.e., spikes) of ON-center and OFF-center ganglion cells in response to the stimuli represented in the upper panels.

antagonistic center-surround arrangement of their receptive fields. Stimulation of the center of the receptive field of an **ON-center ganglion cell** or an **OFF-center ganglion cell** (with increases or decreases in light, respectively) evoke responses. However, if the same stimulus falls in regions surrounding the center of the receptive field of one of these cells it will have the opposite effect. Second, some types of ganglion cells can respond transiently to *both* increases and decreases in light; these cells are termed **ON/OFF ganglion cells.**

Understanding the mechanisms that underlie the different ON and OFF ganglion cell responses to changes in light intensity requires looking closely at the neurons that innervate these ganglion cells. As described in Concept 9.2, the principal excitatory input to ganglion cells is from bipolar cells. Like most other cells in the retina, bipolar cells have graded potentials rather than action potentials. While more than a dozen types of bipolar cells have been characterized, they can be generally divided into two broad classes: those that are depolarized in response to photoreceptor activation by light, termed **ON bipolar cells,** and those that are hyperpolarized in response to photoreceptor activation by light, termed **OFF bipolar cells**. The selective response of ON and OFF bipolar cells to light increments and decrements is explained by the fact that they express different types of glutamate receptors (Figure 9.14A). OFF bipolar cells have ionotropic AMPA and kainate receptors that cause the cells to depolarize in response to glutamate released from photoreceptor terminals. ON bipolar cells express mGluR6, a G-protein-coupled metabotropic glutamate receptor that causes the cells to hyperpolarize in response to glutamate. Recall that photoreceptors hyperpolarize in response to light increments, decreasing their release of neurotransmitter. This in turn frees ON bipolar cells from the hyperpolarizing influence of the photoreceptor's transmitter, and they depolarize, releasing glutamate

to stimulate ganglion cells. In contrast, for OFF bipolar cells, the reduction in glutamate represents the withdrawal of a depolarizing influence, leading the cells to hyperpolarize (Figure 9.14B). As one might expect, decrements in light intensity naturally have the opposite effect on these two classes of bipolar cells, hyperpolarizing ON bipolar cells and depolarizing OFF bipolar cells (Figure 9.14C).

In addition to expressing different types of glutamate receptors, ON and OFF bipolar cells generate distinct axonal projections into the inner plexiform layer that allow them to specifically innervate ON or OFF ganglion cells, respectively. This precise connectivity endows ON and OFF ganglion cells with the ability to respond to luminance increases and decreases.

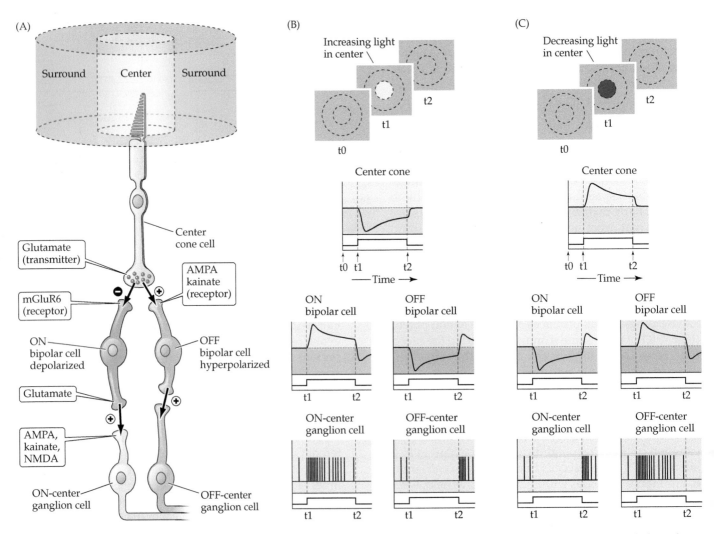

FIGURE 9.14   Circuitry responsible for responses of ON and OFF retinal ganglion cells (A) Functional anatomy of cone inputs to the center of a ganglion cell receptive field. A plus indicates a sign-conserving synapse; a minus represents a sign-inverting synapse. (B) Responses of the various cell types to increasing light in the center of the ganglion cell receptive field.

(C) Responses of the various cell types to decreasing light in the center of the ganglion cell receptive field. Responses in (B,C) are color-coded to match the retinal neuron with which they are associated. Note the graded responses in cones and bipolar cells and the action potentials generated in ganglion cells.

Lateral interactions in the retina

Horizontal cells, which are GABAergic, and amacrine cells, which use GABA, glycine, dopamine, acetylcholine, and even glutamate as neurotransmitters, provide the main sources of lateral interactions within the retina. The dendrites of these cells extend laterally across the retina (perpendicular to the pathway of light), and all processing by these cells occurs locally in their dendrites. Dendrites of horizontal cells arborize in the outer plexiform layer, whereas amacrine cell dendrites arborize in the inner plexiform layer. This arrangement places horizontal and amacrine cells in prime position to shape the spatial structure of receptive fields within the retina. In the case of some amacrine cells, it also allows them to influence the responses of retinal ganglion cells to moving stimuli.

One well understood example of lateral interactions that influence signaling in the retina relates to how retinal ganglion cells are sensitive to **contrast**, that is differences in the level of illumination that falls on the receptive field center versus on the surrounding area. When stimulated, regions that surround the receptive field of a retinal ganglion cell antagonize the response to stimulation of the receptive field center (Figure 9.15A). Because of their antagonistic surrounds, most ganglion cells respond much more vigorously

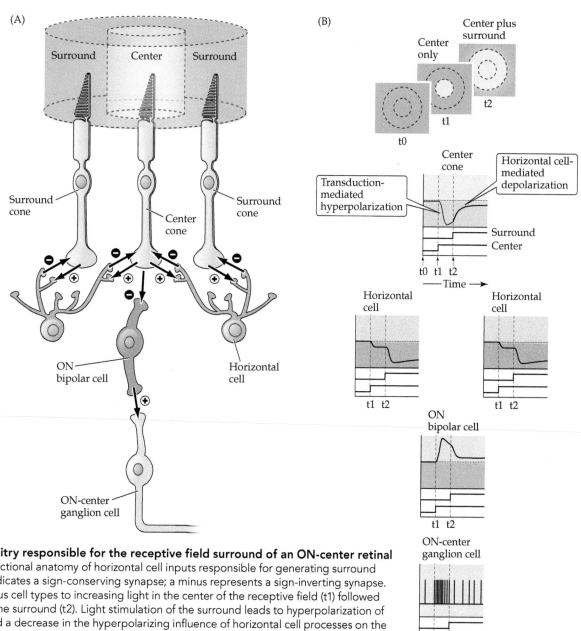

FIGURE 9.15 Circuitry responsible for the receptive field surround of an ON-center retinal ganglion cell (A) Functional anatomy of horizontal cell inputs responsible for generating surround antagonism. A plus indicates a sign-conserving synapse; a minus represents a sign-inverting synapse. (B) Responses of various cell types to increasing light in the center of the receptive field (t1) followed by increasing light in the surround (t2). Light stimulation of the surround leads to hyperpolarization of the horizontal cells and a decrease in the hyperpolarizing influence of horizontal cell processes on the photoreceptor terminals. The net effect is to depolarize the center cone terminal, offsetting much of the hyperpolarization induced by the transduction cascade in the center cone's outer segment.

to small spots of light confined to their receptive field centers than to either large spots or to uniform illumination of the visual field. Thus, the human retina is more sensitive to detecting regions of the visual scene where there are contrasts in light intensities, such as object boundaries, than it is to detecting absolute luminance values.

This surround antagonism of the receptive field center arises from lateral connections established by horizontal cells (Figure 9.15B). Horizontal cells both receive and provide synaptic input from photoreceptor terminals and are linked via gap junctions with a vast network of other horizontal cells distributed over a wide area of the retinal surface. Horizontal cell signaling can therefore have wide-ranging impacts on retinal signaling. Glutamate release from photoreceptor terminals has a depolarizing effect on horizontal cells, while GABA release from horizontal cells has a hyperpolarizing influence on photoreceptor terminals (see

Figure 9.15B). As a result, the net effect of inputs from the horizontal cell network is to oppose changes in the membrane potential of the photoreceptor that are induced by phototransduction events in the outer segment. The more activation there is of these wide horizontal cell networks from illumination falling on both the receptive field centers and surrounds, the more membrane potential changes associated with phototransduction in the receptive field center are counteracted. This ultimately reduces the firing rate of the downstream bipolar and ganglion cells associated with the receptive field center, and therefore provides a mechanism underlying luminance contrast.

This example demonstrates that even at the earliest stages in visual processing, neural signals may not represent simply the absolute numbers of photons that are captured by a receptor, but rather the relative spatial and temporal changes in light intensity (Box 9B).

■ BOX 9B | The Perception of Light Intensity

Understanding the link between retinal stimulation and what we perceive is arguably the central problem in vision. The relation between luminance (a physical measurement of light intensity) and brightness (the sensation elicited by light intensity) is probably the simplest place to consider this challenge.

As indicated in the text, how we see the brightness differences (i.e.,

contrast) between adjacent territories having distinct luminances depends, in part, on the relative firing rate of retinal ganglion cells, modified by lateral interactions. However, there is a problem with the assumption that the central nervous system simply "reads out" these relative rates of ganglion cell activity to sense brightness. The difficulty is that the brightness of a given target is markedly affected by its context in

ways that are difficult or impossible to explain in terms of the retinal output (just as was discussed for color perception in Box 9A). The accompanying figures in this box illustrate this point. In Figure A, two photometrically identical (*equiluminant*) gray squares appear differently bright as a function of the background in which they are presented. A conventional interpretation of this phenomenon is that the receptive

(A)

(B)

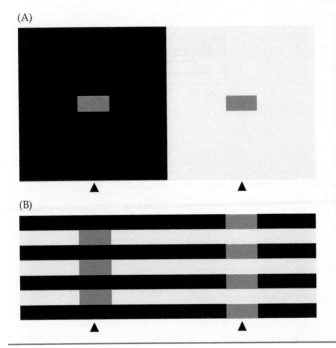

(C)

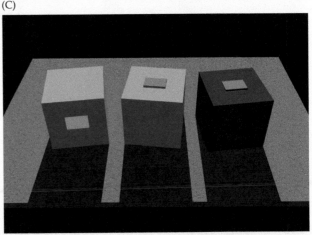

(A) Standard illusion of simultaneous brightness contrast. (B) Another illusion of simultaneous brightness contrast that is difficult to explain in conventional terms. (C) Cartoons of some possible sources of the standard simultaneous brightness contrast illusion depicted in (A). (Courtesy of R. B. Lotto and D. Purves.)

field properties illustrated in Figures 9.13–9.15 cause ganglion cells to fire differently depending on whether the surround of the equiluminant target is dark or light. The demonstration in Figure B, however, undermines this explanation, because in this case the target surrounded by more dark area actually looks darker than the same target surrounded by more light area.

An alternative interpretation of luminance perception that can account for these puzzling phenomena is that brightness percepts are generated on the basis of experience as a means of contending with the fact that biological vision does not have the ability to measure the physical parameters of objects and conditions in the world (in this case, surface reflectance and illumination). Because an observer has

to respond to the real-world sources of luminance and not to light intensity alone, this ambiguity of the retinal stimulus presents a quandary. A plausible solution to the inherent uncertainty of the relationship between luminance values and their actual sources would be to generate the perception of brightness using both an object's given luminance and an observer's prior experiences of viewing similar objects in similar settings. To understand this sort of strategy, consider Figure C, which shows the same equiluminant target patches from Figure A on different surfaces of three adjacent cubes. The surfaces of these cubes are either painted differently or are differentially illuminated. These conditions all affect one's perception of brightness of the three identical target patches. An

expedient way for the visual system to cope with this profound uncertainty is to generate the perception of the stimuli in Figures A and B empirically—that is, based on what the target patches typically turned out to signify in the past. Because the equiluminant targets will have arisen from a variety of possible sources, it makes sense to have the lightness elicited by the patches determined by the relative frequency of occurrence of that luminance in the particular context in which it is presented. The advantage of seeing luminance according to the relative probabilities of the possible sources of the stimulus is that percepts generated in this way give the observer the best chance of making appropriate behavioral responses to real-world sources that cannot be measured.

CONCEPT
9.4

Retinal Ganglion Cells Convey Feature Information to the Brain

LEARNING OBJECTIVES

9.4.1 Discuss how morphologically and functionally distinct types of retinal ganglion cells convey feature-specific information to the hypothalamus, thalamus, and midbrain.

9.4.2 Define the retino-geniculo-cortical pathway for image formation and describe retinotopic projections to the brain.

9.4.3 Recognize that retinal ganglion cells in each eye project to both hemispheres of the brain.

9.4.4 Explain how binocular disparity underlies stereopsis.

Types of retinal ganglion cells

Concept 9.3 illustrated how the retina extracts feature-specific information about the visual scene to send to the brain and highlighted how two broad types of ganglion cells, the ON and OFF cells, convey such information. But this is just the tip of the iceberg in terms of ganglion cell diversity. Distinct ganglion cell types exist that convey information about color, object movement, luminance, and other features (Figure 9.16A). More than 40 types of ganglion cells have been identified in the rodent retina, each with unique morphologies, molecular compositions, functions, or projections into the brain. Interestingly, ganglion

cell diversity in the rodent retina appears to be at least twofold higher than that in primates (where the number of ganglion cell types is between 12 and 20, depending on the species). This suggests that the rodent retina may rely more on a broad array of feature detectors for image formation, whereas primates may rely more on complex higher-order processing in the brain for image formation. Higher-order processing may allow for more flexibility in building a visual percept of the world, although the cost of higher-order processing may be reduced speed of object or motion detection. While the diversity of ganglion cells may differ between rodents and primates, a concept that is shared is that the different types of ganglion cells generate parallel pathways into the brain. This ensures that details about features in the visual world remain separate, in parallel channels, until they are processed and integrated in higher visual centers of the brain.

Where do these parallel pathways go in the brain? Retinal ganglion cells innervate dozens of different brain regions (termed **retino-recipient brain regions**) scattered across the hypothalamus, thalamus, and midbrain (Figure 9.16B). It is important to note that visual, or more precisely light-derived, information is used both to generate visual percepts of our environment (i.e., image formation) and to regulate critical physiological functions not associated with image formation. Thus, retino-recipient brain regions can be classified as those that process image-forming information and those that process non-image-forming visual

(A)

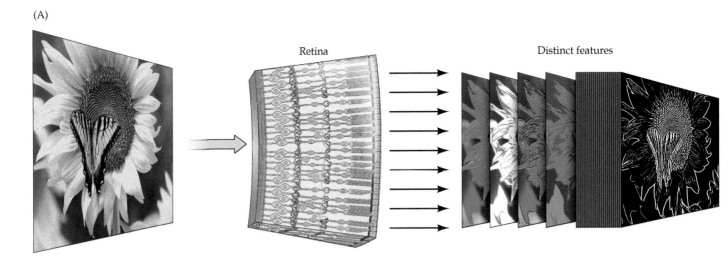

Retina

Distinct features

(B)

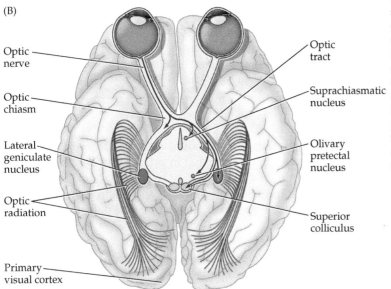

Optic nerve

Optic chiasm

Lateral geniculate nucleus

Optic radiation

Primary visual cortex

Optic tract

Suprachiasmatic nucleus

Olivary pretectal nucleus

Superior colliculus

FIGURE 9.16 Retinal ganglion cells transmit feature-specific information to the brain in parallel channels (A) Schematic depicting how the retina decodes a visual scene, such as this image of a sunflower and butterfly, into information about different features such as object color, contrast, luminance, and motion. Retinal ganglion cells then transmit feature-specific information in parallel channels to the brain. (B) Ganglion cell axons terminate in more than 40 regions of the brain, including the lateral geniculate nucleus of the thalamus, the superior colliculus, the olivary pretectal nucleus, and the suprachiasmatic nucleus. For clarity, only the crossing axons of the right eye are shown.

information. In humans, the **lateral geniculate nucleus** in the dorsal thalamus is the largest target of ganglion cell axons and is important for processing and relaying image-forming visual information to **primary visual cortex** (also called *striate cortex*; see Concept 19.5) in the **occipital lobe** of the cerebrum. The second largest target of ganglion cell axons in humans (and the largest in rodents) is the **superior colliculus**, a dorsal region of the midbrain that plays critical roles in coordinating head and eye movements to visual targets (see Chapter 20). Two of the retino-recipient brain regions that process non-image-forming visual information are the **suprachiasmatic nucleus** of the hypothalamus and the **olivary pretectal nucleus** of the midbrain. Both regions are innervated by axons of intrinsically photosensitive retinal ganglion cells that express melanopsin and are directly depolarized by light. The

suprachiasmatic nucleus is the master regulator of circadian rhythms and uses light-derived signals to photoentrain these rhythms to daily fluctuations in the day–night cycle. The olivary pretectal nucleus is important for regulating the pupillary light reflex, an important mechanism for controlling the amount of light permitted into the eye (see Concept 9.6). Axons from retinal ganglion cells also target several brain regions associated with other sensory systems, allowing the sharing of information among these systems. For example, retinal axons innervate the inferior colliculus, a brain region essential for auditory processing. Similarly, many brain regions innervated by retinal ganglion cells are also associated with the vestibular system, ensuring that image-stabilizing eye movements can be performed during head and neck movement, to prevent blurring of the visual scene during motion.

Each of these different retino-recipient brain regions, whether it processes image-forming or non-image-forming visual information, receives parallel channels of information from specific types of ganglion cells. In some regions, such as the suprachiasmatic nucleus, a single type of ganglion cell provides all of the retinal input. In other regions, such as the lateral geniculate nucleus, many types of ganglion cells (conveying distinct visual information) provide input. Conversely, some types of ganglion cells project to just one or two retino-recipient brain regions, while others project to a dozen or more.

Parallel and retinotopic projections in the retino-geniculo-cortical pathway

In humans, parallel visual pathways connect specific types of ganglion cells to **thalamocortical relays cells** in the lateral geniculate nucleus. These thalamocortical relays cells then project axons to layer 4 of primary visual cortex. Together these pathways make the **retino-geniculo-cortical pathway**, which is essential for image-forming vision. Clinical Applications provides examples of visual deficits that result from various injuries to the retino-geniculo-cortical pathway.

■ Clinical Applications

Visual Field Deficits

A variety of retinal or central pathologies that affect the retino-geniculo-cortical pathway can cause visual impairments that lead to the loss of detection or perception of particular regions of the visual field. Because spatial relationships in the retinas are maintained in central brain regions, a careful analysis of the visual fields can often indicate the site of neurological damage. Relatively large visual field deficits are called **anopsias**; smaller ones are called *scotomas*. The former term is combined with various prefixes to indicate the specific region of the visual field from which sight has been lost.

(Continued)

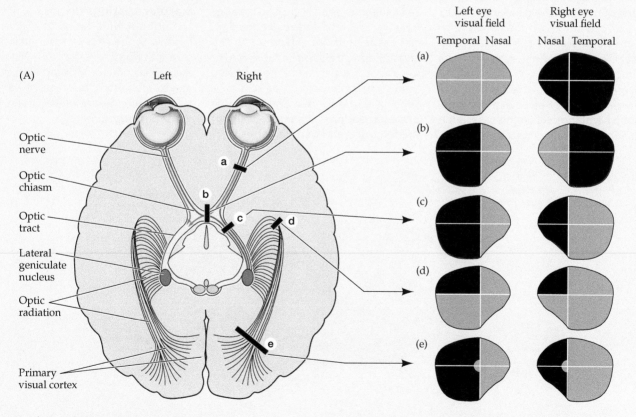

(A) Visual field deficits resulting from damage along the retino-geniculo-cortical pathway . The diagram on the left illustrates the basic organization of the retino-geniculo-cortical pathway and indicates the location of various lesions. The right panels illustrate the visual field deficits associated with each lesion. (a) Loss of vision in right eye. (b) Bitemporal (heteronomous) hemianopsia. (c) Left homonymous hemianopsia. (d) Left superior quadrantanopsia. (e) Left homonymous hemianopsia with macular sparing.

■ Clinical Applications (continued)

Damage to the retina or optic nerve results in a loss of vision that is limited to the eye of origin (Figure A; part a). Damage at the center of the optic chiam, leads to a loss of the contralateral projecting retinal ganglion cells from both eyes and results in a loss of peripheral vision (Figure A; part b). Damage beyond the optic nerve and chiasm results in several distinct deficits. While these deficits may differ depending on the site of damage, they share two similarities: First, they involve the visual fields observed by both eyes; and second, they result in deficits that are limited to the contralateral visual hemifield (see Figure A, parts c–e). For example, interruption of the optic tract in the right hemisphere of the brain (see Figure A, part c) results in a loss of sight in the left visual field. Because such damage affects corresponding parts of the visual field in each eye (i.e., in part c, blindness in the temporal visual field of the left eye and the nasal visual field of the right eye), there is a complete loss of vision in the affected region of the binocular visual field, and the deficit is referred to as a **homonymous hemianopsia** (in this case, a *left* homonymous hemianopsia).

In contrast, damage to the optic chiasm results in visual field deficits that involve non-corresponding parts of the visual field of each eye. For example, damage to the middle portion of the optic chiasm (often the result of pituitary tumors) can affect the fibers that cross from the nasal retina of each eye, leaving the uncrossed fibers from the temporal retinas intact. The resulting loss of vision, confined to the temporal visual field of each eye, is known as **bitemporal hemianopsia** (see Figure A, part b). It is also called **heteronomous hemianopsia** to emphasize that the parts of the visual field that are lost in each eye do not overlap. Individuals with this condition are able to see the central, binocular region of the visual field, but all information from the monocular, peripheral parts of visual field is lost.

Damage to central visual structures is rarely complete. As a result, the deficits associated with damage to the chiasm, optic tract, optic radiation, or visual cortex are typically more limited than those shown in Figure A. This is especially true for damage along the optic radiation, which fans out under the temporal and parietal lobes in its

course from the lateral geniculate nucleus to primary visual cortex cortex. Some of the optic radiation axons run out into the temporal lobe on their route to primary visual cortex cortex, a branch called **Meyer's loop** (Figure B). Meyer's loop carries information from the superior portion of the contralateral visual field. More medial parts of the optic radiation, which pass under the cortex of the parietal lobe, carry information from the inferior portion of the contralateral visual field. Damage to parts of the temporal lobe with involvement of Meyer's loop can thus result in a superior **homonymous quadrantanopsia** (see Figure A, part d); damage to the optic radiation underlying the parietal cortex results in an inferior homonymous quadrantanopsia.

Injury to central visual structures can also lead to a phenomenon called **macular sparing**: the loss of vision throughout wide areas of the visual field, with the exception of foveal vision (see Figure A, part e). Macular sparing is commonly found with damage to the cortex, but it can be a feature of damage anywhere along the visual pathway.

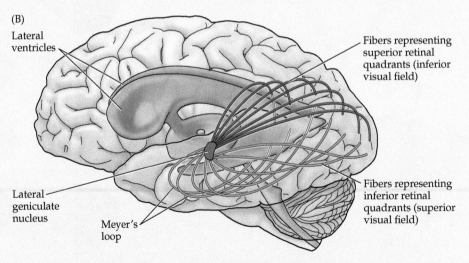

(B)
Lateral ventricles

Fibers representing superior retinal quadrants (inferior visual field)

Lateral geniculate nucleus

Meyer's loop

Fibers representing inferior retinal quadrants (superior visual field)

(B) Course of the optic radiation to primary visual cortex. Axons carrying information about the superior portion of the visual field sweep around the lateral horn of the ventricle in the temporal lobe (Meyer's loop) before reaching the occipital lobe. Axons carrying information about the inferior portion of the visual field travel in the parietal lobe.

Due to its importance for human vision, it is worth exploring in more detail the cells and circuits associated with the retino-geniculo-cortical pathway. The most abundant types of ganglion cells present in humans are midget cells, termed P cells, and parasol cells, termed M cells. This nomenclature appears confusing, or perhaps even looks like a typo, but it is not. Midget ganglion cells are termed

P cells because they project to the four **parvocellular layers** of the lateral geniculate nucleus; parasol cells are termed M cells because they project to the two **magnocellular layers** of this nucleus (Figure 9.17). Thus, even at this gross anatomical level, the separation of parallel channels to transmit visual information to the human brain is apparent. While both M and P ganglion cells exhibit

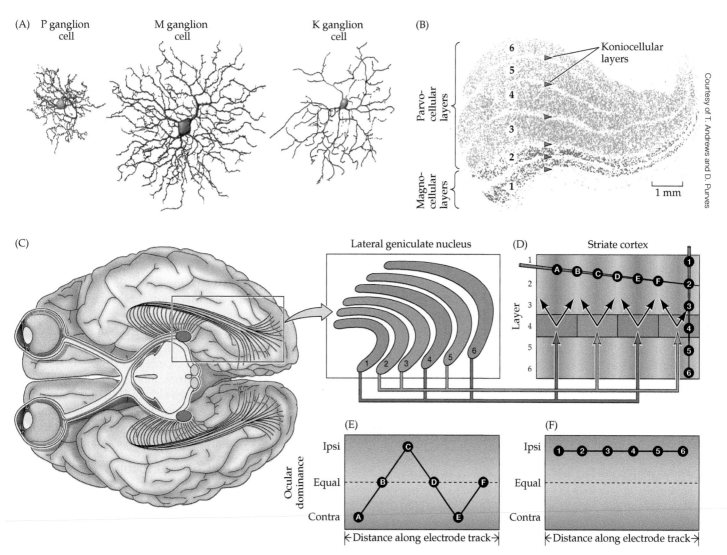

Courtesy of T. Andrews and D. Purves

FIGURE 9.17   Retino-geniculo-cortical pathways (A) Tracings of M, P, and K ganglion cells as seen in flat mounts of the retina. (B) The human lateral geniculate nucleus showing the magnocellular (blue; layers 1,2), parvocellular (green; layers 3-6), and koniocellular layers (orange arrowheads). (C) Although the lateral geniculate nucleus receives inputs from both eyes (labelled blue from the contralateral retina and green from the ipsilateral retina), the inputs are segregated into distinct magnocellular (layers 1,2) and parvocellular layers (layers 3-6). (D) In many species, inputs from the two eyes remain segregated in ocular dominance columns of layer 4, where neurons vary in the strength of their response to the inputs from the two eyes, from complete domination by one eye to equal influence of the two eyes. Layer 4 neurons send their axons to other cortical layers; it is at this stage that the information from the two eyes converges onto individual neurons. Schematic shows two electrode tracks in ocular dominance columns of primary (striate) visual cortex: one tangential electrode track (with points labeled A–E) and one electrode track perpendicular to the cortical surface (or vertical) (with points labeled 1–5). (E,F) depict representations of the physiological responses along the two electrode tracks from (D): the tangential electrode penetration (E) reveals a gradual shift in the strength of response to the inputs from the two eyes, from complete domination by one eye to equal influence of the two eyes, whereas the vertical electrode penetration (F) encounters neurons that tend to have similar ocular dominance. (A after M. Watanabe and R. W. Rodieck, 1989. *J Comp Neurol* 289: 434–454.)

the ON and OFF organization described in Concept 9.3, there are several important differences in their response properties which provide clues about the contributions of these pathways to visual perception, including: (1) M cells have larger receptive fields than P cells; (2) M cells respond only for a brief time (i.e., *transiently*) during the presentation of a visual stimulus, while P cells respond in a sustained fashion; (3) P cells can transmit information about color, whereas M cells cannot; and (4) P cells respond poorly in detecting low-contrast stimuli compared with M cells. These differences suggest that the visual information conveyed by the parvocellular pathway is particularly important for color discrimination and detecting the finest details of an object. The magnocellular pathway, by contrast, appears critical for tasks that require movement detection or high temporal resolution, such as evaluating the location, speed, and direction of a rapidly moving object, and for detecting low-contrast stimuli.

The magno- and parvocellular pathways are not the only parallel channels contributing to the retino-geniculo-cortical pathway. One additional, but not as well understood, pathway is the **koniocellular pathway** (see Figure 9.17A,B). Neurons contributing to the koniocellular pathway are directly innervated by retinal ganglion cells, exhibit distinct receptive field properties, reside largely in the interlaminar zones of the lateral geniculate layers, and project in a patchy fashion to the superficial layers of primary visual cortex. Although the contribution of the koniocellular pathway to visual perception is not as well understood as that of the magnocellular and parvocellular pathways, it appears that some aspects of color vision, especially information derived from short-wavelength-sensitive cones, and direction selectivity for moving stimuli may be transmitted via the koniocellular pathway.

Another important aspect of conveying light-derived information from the retina to the brain is that the information must also convey spatial details, informing central brain regions about the location within the visual field where a stimulus occurred. This is critically important for image-processing centers in the brain, such as the lateral geniculate nucleus (and the superior colliculus). This is achieved by transferring the spatial relationships that exist among ganglion cells onto the target cells they innervate in the brain. Thus, ganglion cells adjacent to each other in the retina innervate thalamic relay cells that are adjacent to each other in the lateral geniculate nucleus. Conversely, ganglion cells that reside in distant regions of the retina innervate thalamic relay cells that reside in distant regions of the lateral geniculate nucleus. This arrangement ensures that orderly representations, called **retinotopic maps**, of visual space are transferred from the retina to visual centers in the brain (Figure 9.18).

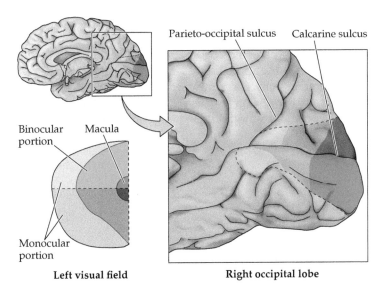

FIGURE 9.18 Visuotopic organization of primary visual cortex Seen in midsagittal view, the primary visual cortex (also termed *striate cortex*) occupies a large part of the occipital lobe. The area of central vision (the fovea) is represented over a disproportionately large part of the caudal portion of the lobe, whereas peripheral vision is represented more anteriorly. The upper visual field is represented below the calcarine sulcus, the lower field above the calcarine sulcus.

Last, as one views the retino-geniculo-cortical pathway, another important organizational feature emerges: Although maps convey an orderly representation of sensory stimuli to the brain, they are distorted so that space is allocated based on use (something you will see for other sensory systems in subsequent chapters of this text). In the visual system, although the fovea makes up only about 1% of the retinal surface, it supplies almost 50% of the visual information to primary visual cortex. This reveals how much cortical processing power we dedicate to understanding features in a very small portion of our visual field.

Eye-specific projections, binocular disparity, and stereopsis

Recall that the central region of the visual field is viewed by both eyes. However, each lateral geniculate nucleus receives information that relates only to the contralateral visual hemifield. This is because ganglion cells that reside in the nasal retina project axons to the contralateral hemisphere of the brain, and those in the temporal retina project axons to the ipsilateral hemisphere (see Figure 9.4C). The midline crossing, or **decussation**, of axons from nasal ganglion cells occurs in the **optic chiasm**, a region that lies directly under the hypothalamus and demarcates the end of the **optic nerve** and the beginning of the **optic tract**. In humans, about 60% of ganglion cell axons decussate in the chiasm; the other 40% continue toward the ipsilateral thalamic and midbrain targets. As ipsilateral- and contralateral-projecting ganglion cell axons extend into the lateral geniculate nucleus, they not only are segregated by "type" into the four parvocellular and two magnocellular layers described in the previous section, they also are segregated based on eye-of-origin. Two parvocellular layers receive input from the contra-lateral retina and two receive input from the ipsilateral retina. Likewise, one magnocellular layer receives input from the contralateral retina and one receives input from the ipsilateral retina. Thus, even information from each eye is segregated into distinct parallel channels as it is conveyed into the brain.

The segregation of eye-specific information persists as geniculate neurons relay visual information to primary visual cortex. The axons of geniculate neurons terminate in alternating eye-specific **ocular dominance columns** in layer 4 of primary visual cortex (see Figure 9.17D). Beyond this point, however, signals from the two eyes begin to converge, and therefore most of the neurons outside of layer 4 are binocular. Bringing together the inputs from the two eyes at the level of visual cortex provides a basis for **stereopsis**, the perception of depth that arises from viewing objects with two eyes instead of one. Because the two eyes look at the world from slightly different angles, objects that lie in front of or behind the plane of fixation project to non-corresponding points on the two retinas. To convince yourself of this fact, hold your hand at arm's length and fixate on the tip of one finger. Maintain fixation on the finger as you hold a pencil in your other hand about half an arm's length away from you. At this distance, the image of the pencil falls on non-corresponding points on the two retinas and will therefore be perceived as two separate pencils (a phenomenon called double vision, or **diplopia**). If you move the pencil toward the finger (the point of fixation), the two images of the pencil fuse into one. Thus, for a small distance on either side of the plane of fixation, where the disparity between the two views of the world remains modest, a single image is perceived; the disparity between the two eye views (**binocular disparity**) of objects nearer or farther than the point of fixation is perceived as *depth*.

CONCEPT
9.5

Visual Centers in the Cerebral Cortex Detect Increasingly Complex Visual Features

LEARNING OBJECTIVES

9.5.1 Describe the structural composition of primary visual cortex and cortical columns.

9.5.2 Explain what columnar response properties, orientation selectivity, and ocular dominance columns are.

9.5.3 Define the role of extrastriate cortical regions in object recognition and visually guided behaviors.

Cerebral cortex and visual percepts

While the retina plays a role in decoding elements of the visual scene to convey to the brain, the role of the cerebral cortex is to integrate this information and generate usable percepts of the visual scene. This process begins in the primary visual cortex and then radiates out to several other cortical regions that process higher-level visual information. These regions extend well outside the occipital lobe of the cerebrum, illustrating just how much of our brain power is designated to decoding and processing visual information.

Cellular and columnar organization of primary visual cortex

Like all neocortex, primary visual cortex is a sheet approximately 2 mm thick and divided into six cellular layers (layers 1–6; Figure 9.19). The composition of these layers is similar to that in other regions of neocortex and includes several broad classes of neurons. Pyramidal neurons, which are present in layers 2, 3, 5, and 6, employ the excitatory neurotransmitter glutamate and are the principal source of axonal projections that leave the cortex to target subcortical and other cortical areas. Pyramidal neurons in superficial layers of visual cortex project to other cortical areas that process visual information (termed *extrastriate visual areas*; see later in this concept), while those in the deeper cortical layers send their axons to subcortical targets, including providing modulatory feedback to the lateral geniculate nucleus. **Spiny stellate cells** are excitatory neurons that reside in layer 4. These cells are the primary target of axons from principal relay cells in the lateral geniculate nucleus and convey this information to other neurons in the cortex. The density of thalamocortical relay cell axons in layer 4 is so high that it serves as a defining morphological feature of primary visual cortex compared with other neocortical regions. The fibers of thalamocortical relay cells stain densely and appear as a stripe throughout primary visual cortex, leading early anatomists to name this region **striate cortex**. Twenty percent of neurons in visual cortex are interneurons and have local axonal arbors that are the principal source of cortical inhibition, employing the neurotransmitter GABA.

What cannot be discerned from a cursory examination of individual cells in primary visual cortex is that this brain region (like other cortical regions) has a columnar organization. In a **cortical column**, neurons within a domain that extends down perpendicularly from the cortical surface share similar receptive field properties or response preferences regardless of which layer they reside in. Cortical columns can be observed by recording neuronal activity with microelectrode penetrations. Electrodes that are implanted perpendicular to the cortical surface are confined to a single cortical column and therefore encounter neurons with similar receptive field properties or response

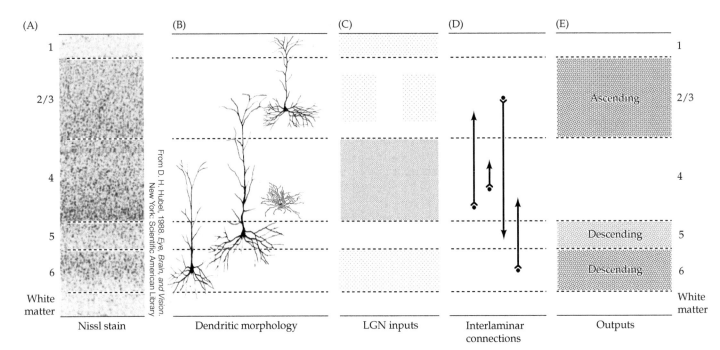

From D. H. Hubel, 1988. *Eye, Brain, and Vision.* New York: Scientific American Library.

FIGURE 9.19 **Organization of primary visual cortex** Visual cortex is divided into six principal cellular layers that differ in cell density, neuronal morphology, and connectivity. (A) Primary visual cortex visualized using a histological stain that reveals neuronal cell bodies. (B) Excitatory pyramidal neurons with prominent apical and basilar dendrites are the most numerous cell type in neocortex; they are located in most layers except layers 1 and 4. Layer 4 is populated by spiny stellate neurons. (C) Laminar organization of inputs from the lateral geniculate nucleus (LGN). Lateral geniculate axons terminate most heavily in layer 4, with less dense projections to layers 1, 2/3, and 6. (D) Laminar organization of major intracortical connections. Neurons in layer 4 give rise to axons that terminate in more superficial layers. Axons of layer 2/3 neurons terminate heavily in layer 5. Axons of layer 6 neurons terminate in layer 4. (E) Laminar organization of neurons projecting to different targets. Connections with extrastriate cortex arise primarily from neurons in layers 2/3 (red). Descending projections to the lateral geniculate nucleus arise from layer 6 neurons (blue), while those projecting to the superior colliculus reside in layer 5 (green).

preferences. Electrodes that are implanted at oblique angles to the cortical surface penetrate multiple cortical columns and therefore encounter neurons with different receptive field properties or response preferences. Examples of this can be seen in the ocular dominance columns described in Concept 9.4 (see Figure 9.17D–F) and in the orientation-selective neurons described in the next section. From the description in Concept 9.4 of retinotopic mapping in brain centers that process visual information, it should come as no surprise that adjacent cortical columns in primary visual cortex have similar but slightly shifted receptive field locations, consistent with the global mapping of visual space.

Edge detection and orientation preference in primary visual cortex

While responses of neurons in the lateral geniculate nucleus are quite similar to those in the retina, with a center–surround receptive field organization and selectivity for luminance increases or decreases, presentations of small spots of light are largely ineffective in activating neurons in primary visual cortex. Instead, most cortical neurons respond vigorously to light–dark bars or edges, and only if the bars are presented at a particular range of orientations within the cell's receptive field. The responses of cortical neurons are thus tuned to the orientation of edges, much as cone photoreceptors are tuned to the wavelength of light; the peak in the **tuning curve** (i.e., the orientation to which a cell is most responsive) is referred to as the neuron's preferred orientation (Figure 9.20). As a result, a given orientation in a visual scene appears to be "encoded" in the activity of a distinct population of orientation-selective neurons.

To appreciate how the properties of an image might be represented by populations of neurons that are tuned to different orientations, an image can be decomposed into its frequency components and then filtered to create a set of images whose spectral composition simulates the information that would be conveyed by neurons tuned to different orientations (Figure 9.21). Each class of orientation-selective neuron transmits only a small fraction of the information in the scene—the part that matches its filter properties—but the information from these different filters contains all the spatial information necessary to generate a faithful representation of the original image.

(A) Experimental setup

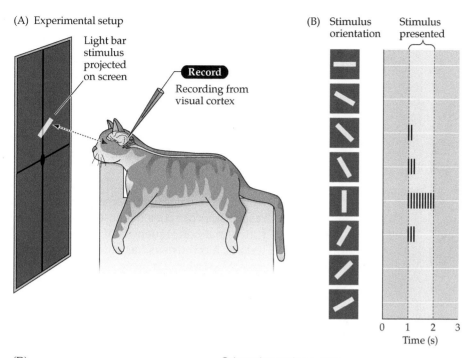

(B) Stimulus orientation   Stimulus presented

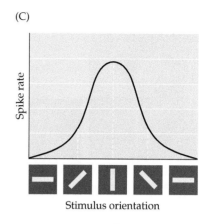

(C)

(D)

(E)

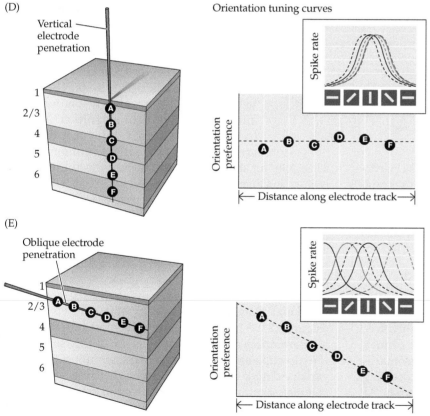

FIGURE 9.20 Neurons in the primary visual cortex respond selectively to oriented edges (A) An anesthetized animal is fitted with contact lenses to focus the eyes on a screen, where images can be projected; an extracellular electrode records the neuronal responses. (B) Neurons in the primary visual cortex typically respond vigorously to a bar of light oriented at a particular angle and less strongly—or not at all—to other orientations. (C) Orientation tuning curve for a neuron in primary visual cortex. In this example, the highest rate of action potential discharge occurs for vertical edges—the neuron's "preferred" orientation. (D) Neurons within cortical columns exhibit similar orientation preferences. At left is a depiction of a microelectrode penetration perpendicular to the surface of primary visual cortex. The orientation tuning curves (right panel, top) and preferred orientation (right panel, bottom) for neurons encountered along the electrode track show that there is little variation in the orientation preference of the neurons. (E) Neurons displaced along the tangential axis of the cortex (i.e., in different cortical columns) exhibit orientation preferences (right panel) that shift in an orderly progressive fashion.

Orientation preference is only one of the qualities that define the filter properties of neurons in primary visual cortex. A substantial fraction of cortical neurons are also tuned to the direction of stimulus motion, for example, responding much more vigorously when a stimulus moves to the right than when it moves to the left. Neurons can also be characterized by their preference for spatial frequency (the coarseness or fineness of the variations in contrast that fall within their receptive fields) as well as temporal frequency (rate of change in contrast). Why

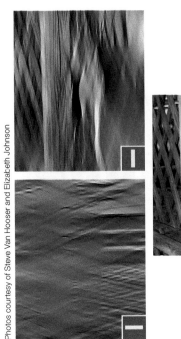

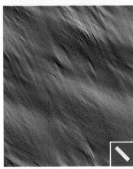

FIGURE 9.21 Representation of a visual image by neurons selective for different stimulus orientations This simulation uses image mathematics and selective filtering to illustrate the attributes of a visual image (greyhound and fence) that would be represented in the responses of populations of cortical neurons tuned to different preferred orientations. The panels surrounding the image illustrate the components of the image that would be detected by neurons tuned to vertical, horizontal, and oblique orientations (blue boxes). In ways that are still not understood, the activity in these different populations of neurons is integrated to yield a coherent representation of the image features.

should cortical neurons show selectivity for these particular stimulus dimensions? Computational analyses suggest that receptive fields with properties such as these are well matched to the statistical structure of natural scenes and would therefore maximize the amount of information transferred with a minimum of redundancy.

From detection to object recognition and visually guided behaviors

Beyond primary visual cortex, a multitude of other areas in the occipital, parietal, and temporal lobes are involved in processing visual information. In fact, in some primate species up to 50% of the entire cerebral cortex is primarily (and sometimes exclusively) dedicated to visual processing. These areas are classified as **extrastriate visual areas**. All of these areas depend on the primary visual cortex for their activation and retain retinotopic maps of the visual world. The theme of parallel processing of visual information partially continues into these areas (i.e., the parvocellular and magnocellular pathways described in Concept 9.4 remain partially segregated); however, **hierarchical processing** also becomes important for signal processing in extrastriate areas. Progressive integration and bidirectional signaling in

these areas shape responses in subsequent regions. Ultimately, what emerges is that extrastriate visual areas integrate information from centers involved in earlier stages of visual processing and are specialized to respond to different, and in many cases quite specific, objects in the visual scene. As an extreme example, cells in the **fusiform gyrus** (a region in the inferior temporal cortex) are precisely tuned to detect faces.

Based on the anatomical connections between extrastriate visual areas, differences in response properties of these areas, and the functional consequences of certain cortical lesions, a consensus has emerged that these cortical areas are organized into two largely separate systems that feed information into cortical association areas in the temporal and parietal cerebral lobes. These systems can be summarized as the "what" and "where" visual pathways.

In the "what" pathway, called the **ventral stream**, projections from primary visual cortex are sent sequentially through several visual areas in the occipital lobe (V1 → V2 → V3 → V4) before reaching the inferior temporal (IT) cortex. This system is thought to be responsible for high-resolution vision and high-level object recognition. Neurons in the ventral stream exhibit properties that are important for object recognition, such as selectivity for shape, color, and texture. As already noted, regions in IT can be precisely tuned to very specific objects in the visual world, including objects as specific as faces, animals, plants, or trees.

In the "where" pathway, called the **dorsal stream**, projections from primary visual cortex are sent through V2 to the **middle temporal area** (MT, or V5) and then into the parietal lobe. This system is thought to be responsible for spatial aspects of vision, such as the analysis of motion, and positional relationships between objects in the visual scene (Figure 9.22). The MT is perhaps the most well studied of all extrastriate areas, and its cells are highly selective to the direction and speed of a moving stimulus. Just as the earlier example of the face cell highlighted the increasingly complex stimuli that activate neurons in extrastriate areas, cells in the MT respond to more complex types of movements than cells in earlier stages of visual processing.

The functional dichotomy between these two visual streams is supported by the effects of lesions to specific extrastriate areas. Three examples highlight this point. First, a woman who suffered a stroke that damaged the MT was unable to appreciate the motion of objects, a rare disorder called **cerebral akinetopsia**. The neurologist who

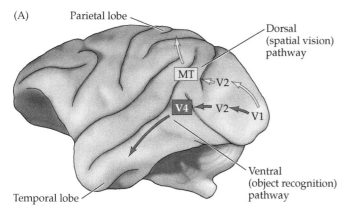

(A)
Parietal lobe
Dorsal (spatial vision) pathway
MT
V2
V4
V2
V1
Temporal lobe
Ventral (object recognition) pathway

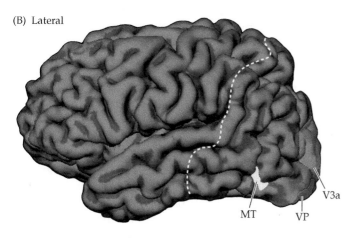

(B) Lateral
V3a
MT   VP

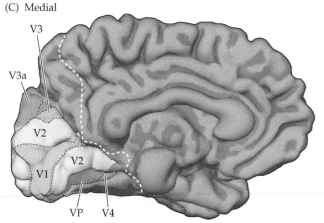

(C) Medial
V3
V3a
V2
V2
V1
VP   V4

FIGURE 9.22   Visual areas beyond primary visual cortex (A) Outside the occipital lobe, visual areas are broadly organized into two pathways: a ventral pathway that leads to the temporal lobe, and a dorsal pathway that leads to the parietal lobe. The ventral pathway plays an important role in object recognition, the dorsal pathway in spatial vision. (B,C) Functional MRI yields lateral and medial views (respectively) of the human brain, illustrating the location of primary visual cortex (V1) and additional visual areas V2, V3, V3a, VP (ventral posterior area), V4, and MT (middle temporal area). (After M. I. Sereno et al., 1995. *Science* 268: 889–893.)

to individuals with hereditary loss or mutation in cone opsins, individuals with achromatopsia know the normal colors of objects but can no longer see them. Finally, lesions in the fusiform gyrus, mentioned earlier in this section, cause **prosopagnosia**, or face-blindness, a cognitive disorder that affects face perception.

As already mentioned, some aspects of parallel processing persist in extrastriate cortical areas. The dorsal stream appears to be largely dominated by inputs from the magnocellular pathway, while the ventral stream appears to be largely dominated by inputs from the parvocellular pathway. However, in contrast to earlier stages of visual processing, the separation of these channels is blurred in extrastriate areas. Thus, as one might expect, to build a perceptual image of the world, the functions of higher visual areas involve the integration of information derived from these magnocellular and parvocellular pathways, as well as from other cortical regions.

CONCEPT **9.6**

Optimizing Vision and Visual Behaviors Requires Autonomic, Motor, and Cognitive Control

LEARNING OBJECTIVES

9.6.1 Discuss the significance of active control of pupil size, lens accommodation, and the "near response" for optimizing vision.

9.6.2 Explain the importance of eye movements for redirecting the region of visual scene that falls on the fovea.

Active control of the visual system

Proper function of the visual system requires active motor control from centers in the brain. For example, autonomic circuits help govern changes in the eye that regulate both the light entering the eye and the shape of the lens, two processes that have significant effects on the quality and quantity of light that land on the retina. In addition to autonomic functions, active control of the direction of gaze is critical for vision. In the chapter Overview we alluded to the fact that a quick glance is sufficient to capture information about a specific location in the visual scene. However,

treated the woman noted that she had difficulty pouring tea into a cup because the fluid seemed to be "frozen." In addition, she could not stop pouring at the right time because she was unable to perceive when the fluid level had risen to the brim. Second, in **cerebral achromatopsia**, individuals lose the ability to see the world in color, although other aspects of vision remain in good working order. The normal colors of a visual scene are described as being replaced by "dirty" shades of gray, much like looking at a poor-quality black-and-white movie. As opposed

such a glance it is not sufficient to build a full percept of the world around us. To build a comprehensive percept of the world around us, we must execute rapid, coordinated eye movements to change the location of our focus in the visual field. Thus, despite the extraordinary cells and circuitry of the visual system described in detail in this chapter, optimizing vision and visual behaviors requires autonomic, motor, and cognitive control. We highlight just three of these processes here, but additional examples are described in later chapters.

Autonomic control of pupil size

Active control of the visual system begins with autonomic control of the quantity of light entering the eye. A reflex arc, called the **pupillary light reflex**, is present in an area of the midbrain that receives direct afferent information from the retina and returns efferent signals that regulate the diameter of the pupil (Figure 9.23). Specifically, neurons in the olivary pretectal nucleus are innervated by both intrinsically photosensitive and conventional retinal ganglion cells. These pretectal neurons, in turn, project bilaterally to the **Edinger–Westphal nucleus**, a small group of nerve cells that lie close to the nucleus of the oculomotor nerve (cranial nerve III) in the midbrain. The Edinger–Westphal nucleus contains the **preganglionic parasympathetic neurons** that send their axons via the oculomotor nerve to terminate on neurons in the **ciliary ganglion** (not to be confused with the ciliary body, discussed in Concept

9.1). Neurons in the ciliary ganglion innervate the circumferential constrictor muscle in the iris, which decreases the diameter of the pupil when activated (an action called **miosis**). Thus, shining light into the eye leads to an increase in the activity of parasympathetic neurons in the midbrain that govern pupil constriction and limit the quantity of light entering the eye.

While parasympathetic neurons govern pupil constriction, pupil dilation is controlled by the sympathetic nervous system and a set of radially arranged dilator muscles in the iris. The sympathetic neurons regulating dilation, or widening, of the pupil (**mydriasis**) are activated by diminishing light conditions, attention, arousal, or activation of the flight-or-fight response. Recall that Concept 9.1 mentioned that changing pupil size has the potential of improving vision by the pinhole effect. Thus, active control of pupil diameter by these parasympathetic and sympathetic inputs affects both the quantity and quality of light falling on the retina. The balance between constriction and dilation of the pupil is an active trade-off between enhanced acuity and sensitivity that depends on the task and situation.

It is noteworthy that in addition to its role in regulating the quantity and quality of light that enters the eye, the pupillary reflex provides an important diagnostic tool for physicians. Because of the bilateral projections of the olivary pretectal neurons, light stimulus in one eye produces constriction of both the stimulated eye (the direct response) and the unstimulated eye (the consensual

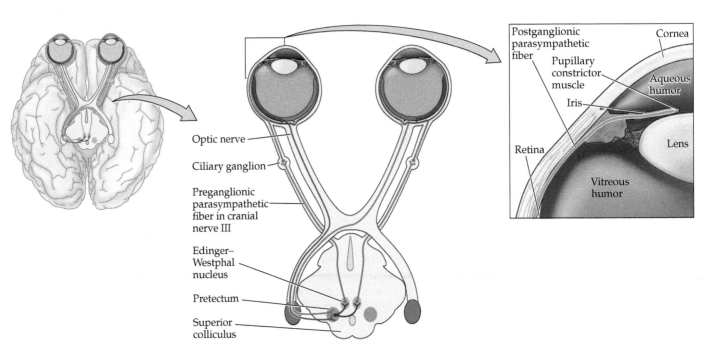

FIGURE 9.23 Circuitry responsible for the pupillary light reflex This pathway includes bilateral projections from the retina to the pretectum and projections from the pretectum to the Edinger–Westphal nucleus. Neurons in the Edinger–Westphal nucleus terminate in the ciliary ganglion, and neurons in the ciliary ganglion innervate the pupillary constrictor muscle. Notice that the afferent axons activate both Edinger–Westphal nuclei via the neurons in the pretectum.

response). Comparing the responses in the two eyes is often helpful in localizing a lesion following trauma. For example, a direct response in the left eye without a consensual response in the right eye suggests a problem with the visceral motor outflow to the right eye, possibly as a result of damage to the oculomotor nerve or Edinger–Westphal nucleus in the brainstem. Failure to elicit a response (either direct or indirect) to stimulation of the left eye if both eyes respond normally to stimulation of the right eye suggests damage to the sensory input from the left eye, possibly to the left retina or optic nerve.

Accommodation and the "near response"

Just as intraocular muscles are required for pupillary constriction, a different set of intraocular muscles, the ciliary muscles, regulate the thickening and rounding up of the lens for near vision, so that the lens exhibits increased refractive power (see Figure 9.3). These changes arise from the tension of the ciliary muscle that surrounds the lens and the radially arranged zonule fibers that attach the two. The shape of the lens is determined by two opposing forces: the elasticity of the lens, which tends to keep it rounded up, and the tension exerted by the zonule fibers, which tends to flatten it.

The ciliary muscles that regulate accommodation are under the control of the Edinger–Westphal nucleus, just as is the pupillary light reflex. While these are separate processes, they are tightly linked. In fact, when we accommodate our lens to view a near object, we also converge our eyes (a type of **disconjugate eye movement**) and constrict our pupils. Together these three events—accommodation, convergence, and constriction—are known as the **near response** (or the accommodation reflex) and are important for viewing close objects with high acuity.

Maintaining the image of a visual target on the fovea

Hold out one of your arms so it is fully extended, and now extend your thumb straight up in the air. Focus your gaze on the nail of your thumb. This is approximately the size of the visual field where you have high-acuity vision—at arm's length it is the size of your thumbnail. Objects everywhere else in your visual field (even those close by, like your wrist) fall onto surfaces of the retina that are outside the fovea, where spatial acuity drops off precipitously. Thus, active control of redirecting our gaze is essential to viewing the world around us.

Eye movements for redirecting focus are accomplished by a set of six **extraocular muscles** that attach to the sclera and rotate the globe of the eye within the orbit of the skull. The contractions of muscles associated with each eye are highly controlled; different muscles must be activated simultaneously in each of the eyes to facilitate **conjugate eye movements**—movements that allow for retaining

binocular fixation on an object in the visual field. Eye movements for redirecting gaze are accomplished by two types of movements: **smooth pursuit eye movements** and **saccadic eye movements**. In smooth pursuit movements, the two eyes shift their positions in a fluid, slow, continuous motion to track a moving object. Smooth pursuit movements are not only triggered by a slowly moving object, but *cannot* be executed in the absence of such a stimulus. Saccadic eye movements are rapid, ballistic changes in eye position to redirect visual focus from one region to another. This form of eye movement is how we explore the visual scene (and rapidly shift our region of focus to sample many regions of the visual scene). In fact, we are constantly using saccadic eye movements to shift our gaze while our eyes are open. Even as you read this text you are using saccadic eye movements intermingled with very brief **fixations** (or pauses). A deeper dive into the types and mechanisms of eye movements can be found in Chapter 20.

While saccadic eye movements are essential for us to scan the visual scene, they do create a problem for the visual system. Rapid eye movements cause images to sweep across the surface of retina, which should be detected as object movement or cause a blurring of the visual percept. Yet this does not happen. This is because of **saccadic suppression**. Mechanisms underlying saccadic suppression are still being actively investigated and probably include both motor and visual mechanisms. For the former, movement-related signals from motor or premotor cortical areas may act to suppress visual neurons during saccadic eye movements; for the latter, retinal and cortical mechanisms may directly modulate image processing during saccadic eye movements. Regardless of the exact mechanism, what is clear is that we are unable to detect a visual percept during saccadic eye movements. You can observe this phenomenon by standing in front of a mirror and focusing your attention on your right eye. Now quickly shift your gaze to your left eye. You will not see your eyes move during the redirection of your gaze. This experiment will not work if you try it with a camera phone in selfie mode—the time delay in capturing the video is greater than the time of the saccadic suppression during eye movements.

Summary

The visual system is perhaps the most investigated sensory system. This is largely due to its importance for human behavior (and survival) and our inherent interest in understanding how it detects and processes information from the world around us so quickly; but it is also because of its accessibility for manipulation and scientific investigation. This chapter introduced how

- light is focused onto the neural retina by the refractive power of the cornea and lens;

- photons of light are converted into electrical signals by specialized photopigments in the outer segments of rods and cones;

- rod and cone pathways are specialized for distinct aspects of vision, with rod pathways being extremely sensitive to low levels of light and cone pathways providing high spatial resolution and color vision;

- neuronal diversity and intraretinal circuits provide the cellular substrate for extracting information about features of the visual scene;

- distinct types of glutamate receptors endow types of bipolars cells with the ability to respond differentially to increases or decreases of light;

- parallel circuits transmit light-derived signals to retino-recipient regions of the hypothalamus, thalamus and midbrain in spatially ordered representations (i.e., topographical maps);

- bringing together information about the visual world from both eyes, at the level of primary visual cortex, provides a basis for binocular vision;

- extrastriate visual areas integrate visual information and generate usable percepts of the visual scene; and

- autonomic and motor control of the pupil, lens, and eye are all active mechanisms to optimize vision.

In addition to understanding specific details about the functioning of the visual system, many of the general concepts introduced in this chapter will be applicable to understanding sensory systems that will be introduced in subsequent chapters; this includes general concepts of receptive fields and how common motifs in neural circuits (such as lateral inhibition through local interneurons) shape these receptive fields, the orderly representation and parallel processing of sensory information, and the role of active motor control in optimizing sensory systems. However, there are also unique features of the visual system that distinguish it from how other systems process sensory information. The most obvious of these differences is in how canonical vertebrate photoreceptors (i.e., rods and cones) are hyperpolarized by light and signal through graded potentials rather than through the generation of action potentials.

■ Additional Reading

Reviews

Arshavsky, V. Y., T. D. Lamb and E. N. Pugh Jr. (2002) G proteins and phototransduction. *Annu. Rev. Physiol.* 64: 153–187.

Berson, D. M. (2003) Strange vision: Ganglion cells as circadian photoreceptors. *Trends Neurosci.* 26: 314–320.

Burns, M. E. and D. A. Baylor (2001) Activation, deactivation, and adaptation in vertebrate photoreceptor cells. *Annu. Rev. Neurosci.* 24: 779–805.

Courtney, S. M. and L. G. Ungerleider (1997) What fMRI has taught us about human vision. *Curr. Opin. Neurobiol.* 7: 554–561.

Demb, J. S. and J. H. Singer (2015) Functional circuitry of the retina. *Annu. Rev. Vis. Sci.* 1: 236–289.

Dhande, O. S., B. K. Stafford, J.-H. A. Lim and A. D. Huberman (2015) Contributions of retinal ganglion cells to subcortical visual processing and behaviors. *Annu. Rev. Vis. Sci.* 1: 291–328.

Diamond, J. S. (2017) Inhibitory interneurons in the retina: Types, circuitry, and function. *Annu. Rev. Vis. Sci.* 3: 1–24.

Duchaine, B. and G. Yovel (2015) A revised neural framework for face processing. *Annu. Rev. Vis. Sci.* 1: 393–416.

Euler, T., S. Haverkamp, T. Schubert and T. Baden (2014) Retinal bipolar cells: Elementary building blocks of vision. *Nat. Rev. Neurosci.* 15: 507–519.

Felleman, D. J. and D. C. Van Essen (1991) Distributed hierarchical processing in primate cerebral cortex. *Cereb. Cortex* 1: 1–47.

Grill-Spector, K. and R. Malach (2004) The human visual cortex. *Annu. Rev Neurosci.* 27: 649–677.

Lamb, T. D. and E. N. Pugh Jr. (2004) Dark adaptation and the retinoid cycle of vision. *Prog. Retin. Eye Res.* 23: 307–380.

Masland, R. H. (2012) The neuronal organization of the retina. *Neuron.* 76: 266–280.

Maunsell, J. H. R. (1992) Functional visual streams. *Curr. Opin. Neurobiol.* 2: 506–510.

Nassi, J. J. and E. M. Callaway (2009) Parallel processing strategies of the primate visual system. *Nat. Rev. Neurosci.* 10: 360–372.

Nathans, J. (1987) Molecular biology of visual pigments. *Annu. Rev. Neurosci.* 10: 163–194.

Rieke, F. and M. E. Rudd (2009) The challenges natural images pose for visual adaptation. *Neuron* 64: 605–616.

Sanes, J. R. and R. H. Masland (2015) The types of retinal ganglion cells: Current status and implications for neuronal classification. *Annu. Rev. Neurosci.* 38: 221–246.

Thoreson, W. B. and S. C. Mangel (2012) Lateral interactions in the outer retina. *Prog. Retin. Eye Res.* 31: 407–441.

Wassle, H. (2004) Parallel processing in the mammalian retina. *Nat. Rev. Neurosci.* 5: 747–757.

Important Original Papers

Baden, T., and 5 others (2016) The functional diversity of retinal ganglion cells in the mouse. *Nature* 529: 345–350.

Bao, P., L. She, M. McGill and D. Y. Tsao (2020) A map of object space in primate inferotemporal cortex. *Nature* 583: 103–108.

Basole, A., L. E. White and D. Fitzpatrick (2003) Mapping multiple features in the population response of visual cortex. *Nature* 423: 986–990.

Enroth-Cugell, C. and R. M. Shapley (1973) Adaptation and dynamics of cat retinal ganglion cells. *J. Physiol.* 233: 271–309.

Fasenko, E. E., S. S. Kolesnikov and A. L. Lyubarsky (1985) Induction by cyclic GMP of cationic conductance in plasma membrane of retinal rod outer segment. *Nature* 313: 310–313.

Glasser, M. F. and 11 others (2016) A multi-modal parcellation of human cerebral cortex. *Nature* 536: 171–178.

Hattar, S. and 4 others (2002) Melanopsin-containing retinal ganglion cells: Architecture, projections, and intrinsic photosensitivity. *Science* 295: 1065–1070.

Hubel, D. H. and T. N. Wiesel (1962) Receptive fields, binocular interaction and functional architecture in the cat's visual cortex. *J. Physiol.* 160: 106–154.

Hubel, D. H. and T. N. Wiesel (1968) Receptive fields and functional architecture of monkey striate cortex. *J. Physiol.* 195: 215–243.

Hung, C. P., G. Kreiman, T. Poggio and J. J. DiCarlo (2005) Fast readout of object identity from macaque inferior temporal cortex. *Science* 310: 863–866.

Kuffler, S. W. (1953) Discharge patterns and functional organization of mammalian retina. *J. Neurophysiol.* 16: 37–68.

Mancuso, K. and 7 others (2009) Gene therapy for red-green colorblindness in adult primates. *Nature* 461: 784–787.

Nathans, J. and 4 others (1986) Molecular genetics of inherited variation in human color vision. *Science* 232: 203–211.

Nathans, J., D. Thomas and D. S. Hogness (1986) Molecular genetics of human color vision: The genes encoding blue, green, and red pigments. *Science* 232: 193–202.

Peng, Y. R. and 9 others (2019) Molecular classification and comparative taxonomics of foveal and peripheral cells in primate retina. *Cell* 176(5):1222–1237.

Sakmann, B. and O. D. Creutzfeldt (1969) Scotopic and mesopic light adaptation in the cat's retina. *Pflügers Arch.* 313: 168–185.

Schiller, P. H., J. H. Sandell and J. H. R. Maunsell (1986) Functions of the "on" and "off" channels of the visual system. *Nature* 322: 824–825.

Sereno, M. I. and 7 others (1995) Borders of multiple visual areas in humans revealed by functional magnetic resonance imaging. *Science* 268: 889–893.

Tsao, D. Y., W. A. Freiwald, R. B. H. Tootell and M. S. Livingstone (2006) A cortical region consisting entirely of face-selective cells. *Science* 311: 670–674.

Books

Chalupa, L. M. and J. S. Werner (Eds.) (2013) *The New Visual Neurosciences.* Cambridge, MA: MIT Press.

Chalupa, L. M. and R. W. Williams (Eds.) (2008) *Eye, Retina, and Visual System of the Mouse.* Cambridge, MA: MIT Press.

Dowling, J. E. (1987) *The Retina: An Approachable Part of the Brain.* Cambridge, MA: Belknap Press.

Fain, G. L. (2003) *Sensory Transduction.* Sunderland, MA: Sinauer/Oxford University Press.

Hubel, D. H. (1988) *Eye, Brain, and Vision.* Scientific American Library Series. New York: W. H. Freeman.

Rodieck, R. W. (1998) *First Steps in Seeing.* Sunderland, MA: Sinauer/Oxford University Press.

Wandell, B. A. (1995) *Foundations of Vision.* Sunderland, MA: Sinauer/Oxford University Press.

CHAPTER

10

Dr. Goran Bredberg/Science Photo Library

KEY CONCEPTS

10.1 Sound is a pressure wave composed of different frequencies important for speech, music, and other natural sounds

10.2 The structures of the ear filter the frequencies of sound and transmit vibrations from air to fluid

10.3 Sound waves are transduced into neural signals by cochlear hair cells

10.4 Transduction is controlled by active mechanisms involving middle ear muscles and cochlear outer hair cells

10.5 Auditory pathways involve signals traveling bilaterally and in both feedforward and feedback directions

10.6 Auditory perception involves the synthesis of multiple aspects of sound

10.7 Neural codes for sound frequency are based on resonance and synchrony

10.8 Sounds are localized based on frequency, relative loudness, and relative time of arrival at the ears

10.9 Hearing coordinates with vision via numerous signal transformations and robust intersensory cross-talk

Hearing

Overview

The auditory system is one of the engineering masterpieces of the human body. The ear contains an array of miniature acoustical detectors packed into a space no larger than a pea. These detectors can transduce vibrations as small as the diameter of an atom, and they respond 1000 times faster than visual photoreceptors. Such rapid responses to acoustical cues, which are paralleled by rapid signaling in the auditory brainstem, are critical for both analysis of rapidly varying sounds such as music and speech and for computing sound location based in part on the difference in sound arrival time across the two ears. Human social communication is largely mediated by the auditory system, making the auditory system at least as critical for well being as the visual system; indeed, loss of hearing can be more socially debilitating than blindness. For these and other reasons, audition represents a fascinating and especially important mode of sensation.

CONCEPT 10.1	**Sound Is a Pressure Wave Composed of Different Frequencies Important for Speech, Music, and Other Natural Sounds**

LEARNING OBJECTIVES

10.1.1 Explain what sound is and how its frequency content is analyzed scientifically.

10.1.2 Define pitch, loudness, and timbre.

10.1.3 Discuss the range of intensities and frequencies of sound that humans can hear, and how it compares with that of other species.

Sound and its properties

Sound refers to pressure waves generated by vibrating air molecules. The waveform of a sound stimulus is its amplitude plotted against time. The simplest type of sound is a pure tone, or single sine wave, such as the sound generated by a tuning fork, as shown in Figure 10.1. The vibrating tines of the tuning fork produce local displacements of the surrounding molecules, such that when the tine moves in one direction, the air molecules are compressed; when the tine moves in the other direction, the air molecules are spread out, creating a region of rarefaction. These changes in density of the air molecules are equivalent to local changes in air pressure across time and distance.

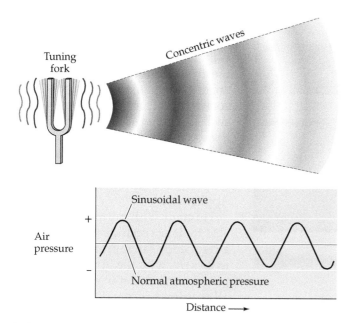

FIGURE 10.1 **A sine wave** Diagram of the periodic condensation and rarefaction of air molecules produced by the vibrating tines of a tuning fork. The molecular disturbance of the air is pictured as if frozen at the instant the constituent molecules responded to the resultant pressure wave. Shown below is a plot of the air pressure versus distance from the fork. Note its sinusoidal quality.

When a sound consists of a single sine wave, its frequency corresponds to the perceptual attribute known as **pitch**, which can be expressed in cycles per second (or Hertz, abbreviated Hz). The overall amplitude of the sine wave corresponds to the perceptual property of **loudness**, which is expressed on the logarithmic decibel scale (abbreviated dB).

But single sine waves are rarely found in nature. Instead, most natural sounds such as speech and music involve waveforms far more complicated than a single sine wave. Such complex waveforms can nevertheless be described as "containing" multiple sound frequencies via a computational analysis known as a Fourier decomposition (Figure 10.2A). The Fourier decomposition expresses any complex waveform as the sum of sine waves of different frequencies. How much of each frequency is present is illustrated in a graph called a **power spectrum** (Figure 10.2B), and how the frequency content varies across time is plotted in a graph called a **spectrogram** (Figure 10.2C).

The waveforms and spectrograms of a variety of naturally occurring sounds are illustrated in Figure 10.3. Speech, music, and environmental stimuli contain energy distributed across a broad frequency spectrum. When one frequency in a complex sound is particularly dominant (usually the lowest or fundamental frequency), as occurs in musical notes, it will have a distinct perceptual pitch corresponding to that frequency. The full panoply of frequencies present in a sound nevertheless contributes in important ways to sound identification. Speech phonemes are distinguishable from one another based on the full pattern of frequencies present. Different musical instruments can play notes having the same pitch but a different **timbre**, a perceptual quality that is based again on the overall pattern of frequencies.

The final major perceptual attribute of hearing is the ability to tell *where* a sound is coming from. Sound location is usually considered not to be contained in the stimulus itself, but must be *computed* by the brain—something that happens with the aid of vision. We cover this fascinating process in Concept 10.8 and 10.9.

(A) Fourier decomposition

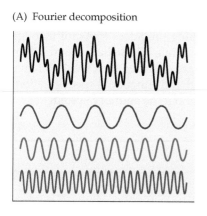

(B) Power spectrum

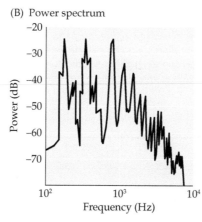

(C) Spectrogram

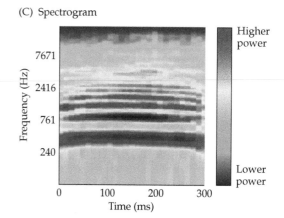

FIGURE 10.2 **Frequencies of natural sounds** (A) Most natural sounds have complex waveforms, which can be expressed as the sum of simple sine waves in a process known as Fourier decomposition. In this example, the complex wave shown in black can be broken down into the three sine waves in the bottom panel, which, if added together, would reproduce the complex wave. (B) This power spectrum of a "coo" call of a rhesus monkey shows the amount of energy at different frequencies. (C) How the power spectrum varies across time is shown in spectrogram plots. "Coo" calls have a stable pattern of frequencies across time. (A based on an image from AnaesthesiaUK. https://www.frca.co.uk/article.aspx?articleid=100500; B,C after Y. Kikuchi et al., 2014. *Front Neurosci* 8: 204. CC BY 4.0.)

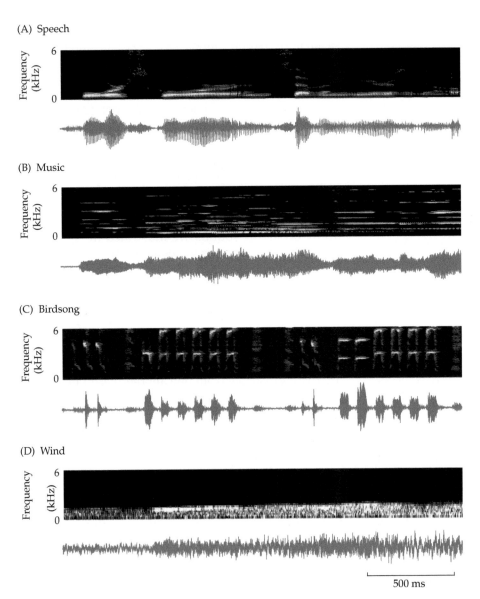

(A) Speech

(B) Music

(C) Birdsong

(D) Wind

500 ms

FIGURE 10.3 Examples of different natural sounds In each case, the top panel is a spectrogram and the bottom panel shows the corresponding sound amplitude versus time. Note that animal vocalizations, speech, and music can contain highly periodic (tonal and harmonic) elements, whereas environmental sounds such as wind lack such periodic structure. (Courtesy of Timothy Warren.)

The audible spectrum

The human ear is extraordinarily sensitive to changes in sound pressure. At the threshold of hearing, air molecules are displaced an average of only 10 picometers (10^{-11} m), and the intensity of such a sound is about one-trillionth of a watt per square meter! Even dangerously high sound pressure levels (>100 dB) have power at the eardrum that is only in the milliwatt range (Clinical Applications).

Humans with normal hearing are able to detect sounds that fall within a frequency range from about 20 Hz to 20 kHz, with the upper limit dropping off somewhat in adulthood. Not all mammalian species are sensitive to the same range of frequencies, and most small mammals are sensitive to very high frequencies but not to low frequencies. For instance, some species of bats are sensitive to tones as high as 200 kHz, but their lower limit is around 20 kHz—the upper limit for young people with normal hearing. Different animal species also tend to emphasize certain frequency bandwidths in both their vocalizations and their range of hearing. Animals that echolocate, such as bats and dolphins, rely on very high frequency vocal sounds to maximally resolve spatial features of the target, while animals intent on avoiding predation have auditory systems "tuned" to the low-frequency vibrations that approaching predators transmit through the substrate. These behavioral differences are mirrored by a wealth of anatomical and functional specializations throughout the auditory system.

■ Clinical Applications

Hearing Loss: Causes and Treatments

Acquired hearing loss is an increasingly common sensory deficit that currently affects more than 30 million people in the United States, a number that is anticipated to grow dramatically over the next several decades. Major causes of acquired hearing loss are acute acoustical trauma, such as that caused by proximity to gunfire or explosions; chronic exposure to high intensity noise, as occurs in industrial and certain musical settings; use of ototoxic drugs; and presbyacusis (literally, "the hearing of the old"), which may stem in part from atherosclerotic

■ Clinical Applications (continued)

damage to the especially fine micro-vasculature of the inner ear, as well as from genetic predispositions to hair cell damage. Increasing public awareness about these risk factors and developing therapies for restoring normal hearing are of great importance, as hearing loss can impair communication and lead to social isolation, which in turn has been linked to premature cognitive decline.

By far the most common forms of hearing loss involve the peripheral auditory system—namely, those structures that transmit and transduce sounds into neural impulses. Monaural hearing deficits are the defining symptom of a peripheral hearing loss, because unilateral damage at or above the auditory brainstem results in a binaural deficit (because of the extensive bilateral organization of the central auditory system). Peripheral hearing insults can be further divided into conductive hearing losses, which involve damage to the outer or middle ear, and sensorineural hearing losses, which stem from damage to the inner ear, typically the cochlear hair cells or the auditory nerve itself. Although both forms of peripheral hearing loss manifest themselves as a raised threshold for hearing on the affected side, their diagnoses and treatments differ.

Conductive hearing loss can be due to occlusion of the ear canal by wax or foreign objects, rupture of the tympanic membrane itself, or arthritic ossification of the middle ear bones. In contrast, sensorineural hearing loss usually is due to congenital or environmental insults that lead to hair cell death or damage to the auditory nerve. As hair cells are relatively few in number and do not regenerate in humans, their depletion leads to a diminished ability to detect sounds. The Weber test, a simple test involving a tuning fork, can be used to distinguish between these two forms of hearing loss. If a resonating tuning fork (~256 Hz) is placed on the vertex, an individual with conductive

hearing loss will report that the sound is louder in the affected ear. In the "plugged" state, sounds propagating through the skull do not dissipate as freely back out through the auditory meatus, and thus a greater amount of sound energy is transmitted to the cochlea on the blocked side. In contrast, an individual with a monaural sensorineural hearing loss will report that a Weber test sounds louder on the intact side, because even though the inner ear may vibrate equally on the two sides, the damaged side cannot transduce this vibration into a neural signal.

Treatment also differs for these two types of deafness. In conductive hearing losses, an external hearing aid is used to boost sounds to compensate for the reduced efficiency of the conductive apparatus. These miniature devices, which are inserted into the ear canal, contain a microphone, a speaker, and an amplifier (Figure A). Although often helpful in quiet environments, external hearing aids can be less effective in noisy environments; moreover, they do not achieve a high degree of directionality, interfering with sound localization, which is an important aid in distinguishing different sound sources. The use of digital signal-processing strategies partly overcomes these problems, and hearing aids obviously provide significant benefits to many people.

The treatment of profound sensorineural hearing loss is more complicated and invasive; conventional hearing aids are useless, because no amount of mechanical amplification can compensate for the inability to generate or convey a neural impulse from the cochlea.

However, if the auditory nerve is intact, cochlear implants can partially restore hearing. The implant consists of a peripherally mounted microphone and digital signal processor that decomposes a sound into its spectral components. Additional electronics use this information to activate different combinations of contacts on a

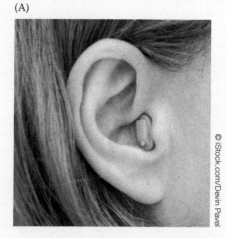

(A) An external hearing aid nestled in the ear canal. These compact devices amplify sound and are used to treat conductive hearing loss.

multisite stimulating electrode array. The electrode is inserted into the cochlea through the round window (Figure B) and positioned along the length of the tonotopically organized basilar membrane and auditory nerve endings. This placement enables electrical stimulation of the nerve in a manner that mimics some aspects of the spectral decomposition naturally performed by the cochlea.

Cochlear implants can be remarkably effective in restoring hearing to people with hair cell damage, permitting them to engage in spoken communication. Despite such success in treating those who have lost their hearing *after* having learned to speak, children with severe to profound hearing loss who receive cochlear implants exhibit highly variable spoken language development outcomes. Although cochlear implants cannot help people who have auditory nerve damage, brainstem implants currently in development use a conceptually similar approach to stimulate the cochlear nuclei directly, bypassing the auditory periphery altogether.

(Continued)

■ **Clinical Applications** *(continued)*

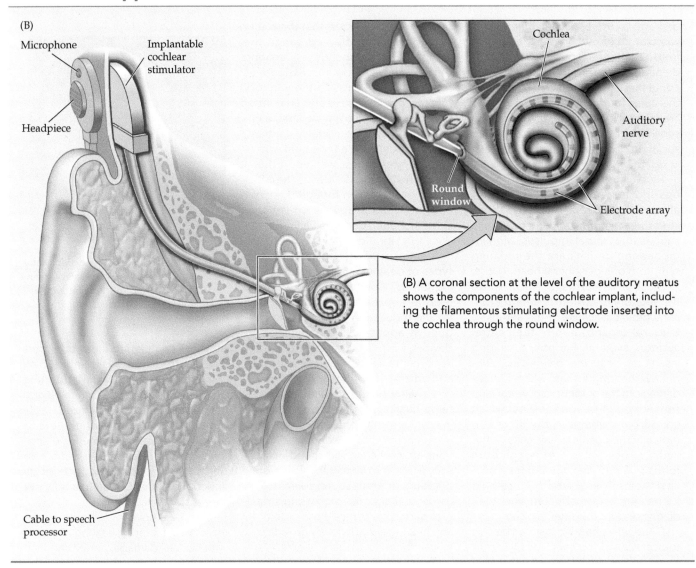

(B) A coronal section at the level of the auditory meatus shows the components of the cochlear implant, including the filamentous stimulating electrode inserted into the cochlea through the round window.

CONCEPT
10.2

The Structures of the Ear Filter the Frequencies of Sound and Transmit Vibrations from Air to Fluid

LEARNING OBJECTIVES

10.2.1 Identify the structures of the ear and their respective roles with regard to sound frequency, sound amplitude, and transmission from air to a fluid medium.

10.2.2 Explain how sounds of different frequencies activate different regions within the cochlea.

The external ear and filtration by sound frequency

The external ear, which consists of the **pinna**, **concha**, and **auditory meatus**, gathers sound energy and focuses it on the eardrum, or **tympanic membrane** (Figure 10.4). One consequence of the configuration of the human auditory meatus is that it selectively boosts the sound pressure 30- to 100-fold for frequencies around 3 kHz via passive resonance effects. This amplification makes humans especially sensitive to frequencies in the range of 2 to 5 kHz—and also explains why we are particularly prone to hearing loss near this frequency following exposure to high-intensity broadband noise, such as that generated by heavy machinery or explosives (see Clinical Applications). In humans, the sensitivity to this frequency range appears to be directly related to speech perception. Although human speech is a broadband signal (meaning it contains many frequencies), important spectral cues used for discriminating different speech sounds, including plosive consonants

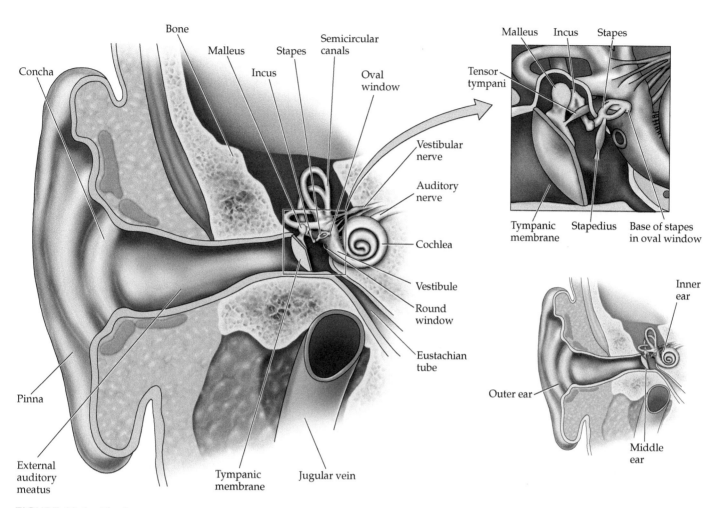

FIGURE 10.4 **The human ear** Note the large surface area of the tympanic membrane (eardrum) relative to the oval window. This feature, along with the lever action of the malleus, incus, and stapes, facilitates transmission of airborne sounds to the fluid-filled cochlea.

(e.g., *ba* and *pa*), are concentrated around 3 kHz (see Figure 31.2). Therefore, selective hearing loss in the range of 2 to 5 kHz disproportionately degrades speech recognition.

A second important function of the pinna and concha is to selectively filter different sound frequencies in order to provide cues about the elevation of the sound source. The vertically asymmetrical convolutions of the pinna are shaped so that the external ear transmits more high-frequency components from an elevated source than from the same source at ear level. This effect can be demonstrated by recording identical sounds from different elevations after they have passed through an "artificial" external ear; when the recorded sounds are played back via earphones, so that the whole series is presented from a source at the same elevation relative to the listener, the recordings from higher elevations are perceived as coming from positions higher in space than the recordings from lower elevations.

The middle ear and regulation of sound amplitude

Sounds impinging on the external ear are airborne; however, the environment within the inner ear, where the sound-induced vibrations are converted to neural impulses, is aqueous. The major function of the middle ear is to match relatively low-impedance airborne sounds to the higher-impedance fluid of the inner ear. Normally, when sound waves travel from a low-impedance medium such as air to a much higher-impedance medium such as water, almost all (>99.9%) of the acoustical energy is reflected. The middle ear (see Figure 10.4) overcomes this problem and ensures transmission of the sound energy across the air–fluid boundary by boosting the pressure measured at the tympanic membrane almost 200-fold by the time it reaches the inner ear.

Two mechanical processes occur within the middle ear to achieve this large pressure gain. The first and major boost is achieved by focusing the force impinging on the relatively large-diameter tympanic membrane onto the much smaller-diameter **oval window**, the site where the bones of the middle ear contact the inner ear. A second and related process relies on the mechanical advantage gained by the lever action of the three small, interconnected middle ear bones, or **ossicles** (i.e., the malleus, incus, and stapes; see Figure 10.4), which connect the tympanic membrane to

the oval window. **Conductive hearing loss**, which involves damage to the external or middle ear, lowers the efficiency at which sound energy is transferred to the inner ear and can be partially overcome by artificially boosting sound pressure levels with an external hearing aid (see Clinical Applications). In normal hearing, the efficiency of sound transmission to the inner ear also is regulated by two small muscles in the middle ear, the tensor tympani, innervated

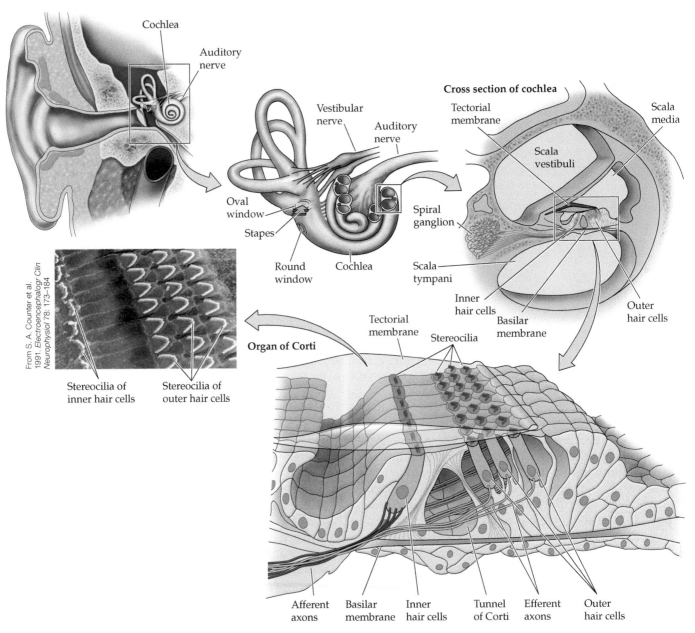

FIGURE 10.5 **The cochlea** The cochlea is here viewed face-on (upper left) and in cross section (subsequent panels). The stapes (shown as an orange arrow), along with other ossicles of the middle ear, transfers force from the tympanic membrane to the oval window. The cross section of the cochlea shows the scala media between the scalae vestibuli and tympani. Blowup of the organ of Corti shows that the hair cells are located between the basilar and tectorial membranes; the latter is rendered transparent in the line drawing and removed in the scanning electron micrograph (above). The hair cells are named for their tufts of stereocilia; inner hair cells receive afferents from cranial nerve VIII, whereas outer hair cells receive mostly efferent innervation.

by cranial nerve V, and the stapedius, innervated by cranial nerve VII (see the Appendix). Contraction of these muscles, which is triggered automatically by loud noises or during self-generated vocalization, counteracts the movement of the ossicles and reduces the amount of sound energy transmitted to the cochlea, serving to protect the inner ear. Conversely, conditions that lead to flaccid paralysis of either of these muscles, such as Bell's palsy (nerve VII), can trigger a painful sensitivity to moderate or even low-intensity sounds known as **hyperacusis**.

Bony tissues and soft tissues, including those surrounding the inner ear, have impedance values close to that of water. Therefore, even without an intact tympanic membrane or middle ear ossicles, acoustical vibrations of sufficient energy, such as those arising from a tuning fork directly touching the head, can still be transferred directly through the bones and tissues of the head to the inner ear. In the clinic, the Weber test uses a tuning fork placed against the scalp to determine whether hearing loss is due to conductive problems or to damage either to the hair cells of the inner ear or to the auditory nerve itself (**sensorineural hearing loss**; see Clinical Applications).

The inner ear and frequency analysis

The **cochlea** of the inner ear is the site where the energy from vibrating pressure waves is transformed into neural impulses. The cochlea not only amplifies sound waves and converts them into neural signals, but it also acts as a mechanical frequency analyzer, decomposing complex acoustical waveforms into simpler elements in a manner resembling Fourier decomposition. Many features of auditory perception accord with aspects of the physical properties of the cochlea; hence, it is important to consider this structure in some detail.

The cochlea (from the Latin for "snail") is a small (~10 mm wide) coiled structure, which, were it uncoiled, would form a tube about 35 mm long (Figures 10.5 and 10.6). Both the oval window and the **round window**, another region where the bone is absent surrounding the cochlea, are at the basal end of this tube. The cochlea is bisected from its basal end almost to its apical end by the cochlear partition, a flexible structure that supports the **basilar membrane** and the **tectorial membrane**. There are fluid-filled chambers on each side of the cochlear partition, called the **scala vestibuli** and the **scala tympani**. A distinct channel, the **scala media**, runs within the cochlear partition. The cochlear partition does not extend all the way to the apical end of the cochlea; instead, an opening known as the **helicotrema** joins the scala vestibuli to the scala tympani, allowing their fluid, known as **perilymph**, to mix. One consequence of this structural arrangement is that inward movement of the oval window displaces the fluid of the inner ear, causing the round window to bulge out slightly and deforming the cochlear partition.

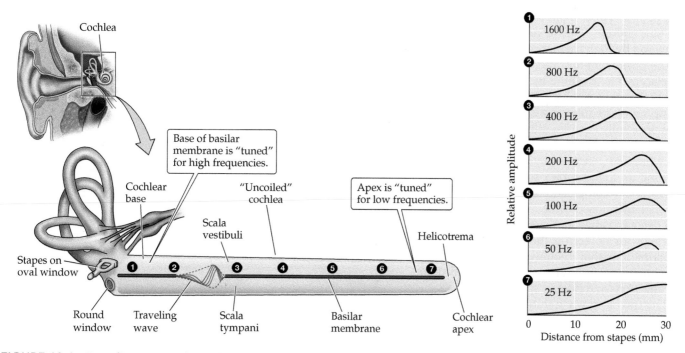

FIGURE 10.6 Traveling waves along the cochlea A traveling wave is shown at a given instant along the cochlea, which has been uncoiled for clarity. The graphs on the right profile the amplitude of the traveling wave along the basilar membrane for different frequencies. The position (labeled 1–7 in the figure) at which the traveling wave reaches its maximum amplitude varies directly with the frequency of stimulation: Higher frequencies map to the base, and lower frequencies map to the apex. (Drawing after P. Dallos, 1992. *J Neurosci* 12: 4575–4585; graphs after G. Von Békésy, 1960. *Experiments in Hearing*. New York: Mcgraw Hill.)

The manner in which the basilar membrane vibrates in response to sound is the key to understanding how hearing is initiated. Measurements of the vibration of different parts of the basilar membrane, as well as the discharge rates of individual auditory nerve fibers that terminate along its length, show that both of these features are tuned; that is, although they respond to a broad range of frequencies, they respond most intensely to a specific frequency. Frequency tuning within the inner ear is attributable in part to the geometry of the basilar membrane, which is wider and more flexible at the apical end and narrower and stiffer at the basal end. Georg von Békésy, a Hungarian biophysicist working at Harvard University, showed that a membrane that varies systematically in its width and flexibility vibrates maximally at different positions as a function of the stimulus frequency (see Figure 10.6). Using models and human cochleas taken from cadavers, von Békésy found that an acoustical stimulus such as a sine wave initiates a **traveling wave** in the cochlea that propagates from the base toward the apex of the basilar membrane, growing in amplitude and slowing in velocity until a point of maximum displacement is reached. The point of maximum displacement is determined by the frequency of the stimulus and persists vibrating in that pattern as long as the tone endures. The maximum displacements triggered by high frequencies occur at the base of the basilar membrane, and the maximum displacements triggered by low frequencies occur at the apex, giving rise to a topographical mapping of frequency, also known as **tonotopy**. Spectrally complex stimuli cause a pattern of vibration equivalent to the superposition of the vibrations generated by the individual tones making up that complex sound, thus accounting for the decompositional aspects of cochlear function mentioned earlier. This process of spectral decomposition appears to be an important strategy for detecting the various harmonic combinations that distinguish natural sounds that have a periodic character, such as animal vocalizations and human speech sounds.

The traveling wave initiates sensory transduction by displacing the sensory hair cells that sit atop the basilar membrane. Because the basilar membrane and the overlying tectorial membrane are anchored at different positions, the vertical component of the traveling wave is translated into a shearing motion between these two membranes (Figure 10.7). This

(A) Resting position

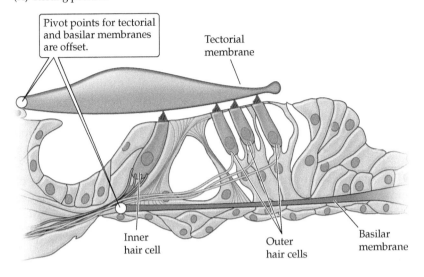

(B) Sound-induced vibration

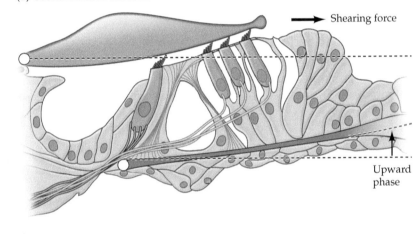

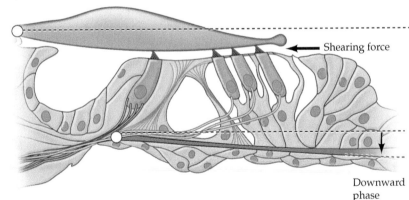

FIGURE 10.7 Traveling waves initiate auditory transduction Vertical movement of the basilar membrane is translated into a shearing force that bends the stereocilia of the hair cells. The pivot point of the basilar membrane is offset from the pivot point of the tectorial membrane so that when the basilar membrane is displaced, the tectorial membrane moves across the tops of the hair cells, bending the stereocilia.

motion bends the tiny processes, called **stereocilia**, that protrude from the apical ends of the hair cells, leading to voltage changes across the hair cell membrane. How the bending of stereocilia leads to receptor potentials in hair cells is considered in the next concept.

| CONCEPT **10.3** | Sound Waves Are Transduced into Neural Signals by Cochlear Hair Cells |

LEARNING OBJECTIVES

10.3.1 Recognize inner hair cells as the source of auditory sensory input to the brain.

10.3.2 Explain the role of stereocilia and tip links in the transduction of sound waves into electrical signals.

10.3.3 Describe the patterns of ionic flow that occur during transduction of sound waves into electrical signals in inner hair cells.

Hair cells and the mechanotransduction of sound waves

The cochlear hair cells in humans consist of one row of inner hair cells and three rows of outer hair cells (see Figures 10.5 and 10.7). The inner hair cells are the sensory receptors, and 95% of the fibers of the auditory nerve that project to the brain arise from this subpopulation. In contrast, the outer hair cells receive *descending* input from the brain, playing an important role in modulating basilar membrane motions. We have more to say about the outer hair cells in Concept 10.4.

Hair cells are named for the bundles of hairlike processes that protrude from their apical ends into the scala media. Each hair bundle contains anywhere from 30 to a few hundred stereocilia (Figure 10.8A). Each stereocilium tapers where it inserts into the apical membrane, forming a hinge about which each stereocilium pivots. The stereocilia are graded in height and are arranged in a bilaterally symmetrical fashion. Fine filamentous structures, known as tip links, run in parallel to the plane of bilateral symmetry, connecting the tips of adjacent stereocilia (Figure 10.8B).

The tip links provide the means for rapidly translating hair bundle movement into a receptor potential. Displacement of the hair bundle parallel to the plane of bilateral symmetry in the direction of the tallest stereocilia stretches the tip links, directly opening cation-selective channels referred to as hair cell mechanoelectrical transduction channels (hair cell MET or hcMET). These channels are located at the end of the link and depolarizing the hair cell (Figure 10.9). Movement in the opposite direction compresses the tip links, closing the hcMET channels and hyperpolarizing the hair cell. As the linked stereocilia pivot back and forth, the tension

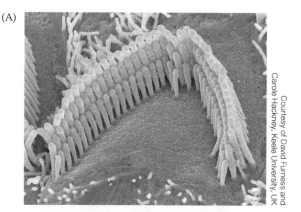

(A)

Courtesy of David Furness and Carole Hackney, Keele University, UK

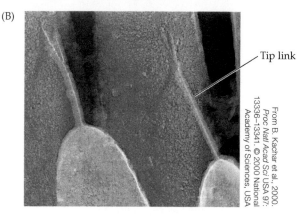

(B)

Tip link

From B. Kachar et al. 2000. *Proc Natl Acad Sci USA* 97: 13336–13341. © 2000 National Academy of Sciences, USA

FIGURE 10.8 Structure of the hair bundle in cochlear hair cells (A) Scanning electron micrograph of a cochlear outer hair cell bundle viewed along the plane of mirror symmetry. (B) Tip links that connect adjacent stereocilia are believed to be mechanical linkages that open and close transduction channels.

on the tip link varies, modulating the ionic flow and resulting in a graded receptor potential that follows the movements of the stereocilia. This receptor potential in turn leads to transmitter release from the basal end of the inner hair cell, which triggers action potentials in cranial nerve VIII fibers that follow the up-and-down vibration of the basilar membrane at relatively low frequencies (the role of the outer hair cells in this process is discussed in Concept 10.4).

The transduction of mechanical forces by inner hair cells is both fast and remarkably sensitive. The hair bundle movements at the threshold of hearing are approximately 0.3 nm—about the diameter of an atom of gold. Hair cells can convert the displacement of the stereociliary bundle into an electrical potential in as little as 10 µs; such speed is required for the accurate localization of the source of the sound. The need for microsecond resolution requires direct mechanical gating of the transduction channel, rather than the relatively slow second-messenger pathways used in visual and olfactory transduction (see Chapters 9 and 14). Although mechanotransduction in inner hair cells is extremely fast, springiness of the tip link introduces

(A)

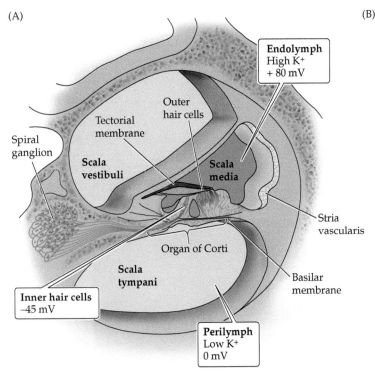

Endolymph
High K⁺
+ 80 mV

Outer
hair cells

Tectorial
membrane

Spiral
ganglion

**Scala
vestibuli**

**Scala
media**

Stria
vascularis

Organ of Corti

**Scala
tympani**

Basilar
membrane

Inner hair cells
−45 mV

Perilymph
Low K⁺
0 mV

(B)

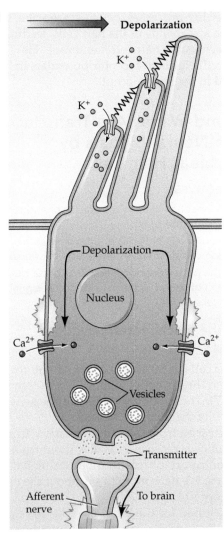

Depolarization

K⁺

K⁺

Depolarization

Nucleus

Ca^{2+}

Vesicles

Ca^{2+}

Transmitter

Afferent
nerve

To brain

FIGURE 10.9 Cochlear structures and mechanism of auditory transduction (A) Cross section of the cochlea showing the organ of corti, which contains the hair cells, and three fluid filled compartments, the scala tympani, scala vestibuli, and scala media. The fluid in the scala tympani and scala vestibuli is called perilymph whereas the fluid in the scala media is known as endolymph. Endolymph is high in K⁺ and has an electrical potential of +80 mV relative to the perilymph. This voltage difference is known as the endochlear potential. (B) The stereocilia of the hair cells protrude into the endolymph of the scala media. When the hair bundle is deflected toward the tallest stereocilium, cation-selective hcMET channels open near the tips of the stereocilia, allowing K⁺ to flow into the hair cell down their electrochemical gradient (see text for the explanation of this peculiar situation). The resulting depolarization of the hair cell opens voltage-gated Ca^{2+} channels in the cell soma, allowing calcium entry and release of neurotransmitter onto the nerve endings of the auditory nerve. (B after R. Lewis and A. Hudspeth, 1983. *Nature* 304: 538–541.)

distortion effects that, in some cases, are audible (Box 10A; also see Concept 10.4). Moreover, the exquisite mechanical sensitivity of the stereocilia also presents substantial risks. High-intensity sounds can break the tip links and destroy the hair bundle, resulting in profound hearing deficits. Because human stereocilia, unlike those in fish and birds, do not regenerate, such damage is irreversible. The small number of hair cells (a total of ~30,000 in a human, or 15,000 per ear) further compounds the sensitivity of the inner ear to environmental and genetic insults. An important goal of current research is to identify stem cells and factors that could contribute to the regeneration of human hair cells, thus affording a possible therapy for some forms of sensorineural hearing loss.

The ionic basis of mechanotransduction in hair cells

Intracellular recordings reveal that the hair cell has a resting potential between −45 and −60 mV relative to the fluid that bathes the basal end of the cell. At the resting potential, only a small fraction of the transduction channels are open. When the hair bundle is displaced in the direction of the tallest stereocilium, more transduction channels open, which causes depolarization as K⁺ and Ca^{2+} enter the cell (see Figure 10.9). Depolarization in turn opens voltage-gated calcium channels in the hair-cell membrane, and the resultant Ca^{2+} influx causes transmitter release from the basal end of the cell onto the auditory nerve endings (see Figure 10.9B), similar to chemical

■ BOX 10A | The Sweet Sound of Distortion

As early as the first half of the eighteenth century, musical composers such as G. Tartini and W. A. Sorge discovered that, upon playing pairs of tones, other tones not present in the original stimulus are also heard. These combination tones, fc, are mathematically related to the played tones, f_1 and f_2 ($f_2 > f_1$), by the formula

$$fc = mf_1 + nf_2$$

where m and n are positive integers. Combination tones have been used for a variety of compositional effects, as they can strengthen the harmonic texture of a chord. Furthermore, organ builders sometimes use the difference tone ($f_2 - f_1$) created by two smaller organ pipes to produce the extremely low tones that would otherwise require building one especially large pipe.

Modern experiments suggest that this distortion product is due at least in part to the nonlinear properties of the inner ear. M. Ruggero and his colleagues placed small glass beads (10–30 nm in diameter) on the basilar membrane of an anesthetized animal and then determined the movement of the basilar membrane in response to different combinations of tones by measuring the Doppler shift of laser light reflected from the beads. When two tones were played into the ear, the basilar membrane vibrated not only at those two frequencies but also at other frequencies predicted by the above formula. Related experiments on hair cells studied in vitro suggest that these nonlinearities result from the properties of the mechanical linkage of the transduction apparatus. By moving the hair bundle sinusoidally with a metal-coated glass fiber, A. J. Hudspeth and his coworkers found that the hair bundle exerts a force at the same frequency. However, when two sinusoids were applied simultaneously, the forces exerted by the hair bundle occurred not only at the primary frequencies but at several combination frequencies as well. These distortion products are due to the transduction apparatus, because blocking the transduction channels causes the forces exerted at the combination frequencies to disappear, even though the forces at the primary frequencies remain unaffected. It seems that the tip links add a certain extra springiness to the hair bundle in the small range of motions over which the transduction channels are changing between closed and open states. If nonlinear distortions of basilar membrane vibrations arise from the properties of the hair bundle, then it is likely that hair cells can indeed influence basilar membrane motion, thereby accounting for the cochlea's extreme sensitivity. When we hear difference tones, we may be paying the price in distortion for an exquisitely fast and sensitive transduction mechanism.

neurotransmission elsewhere in the central and peripheral nervous systems (see Chapters 5 and 6). Because some transduction channels are open at rest, the receptor potential is biphasic: Movement toward the tallest stereocilia depolarizes the cell, whereas movement in the opposite direction leads to hyperpolarization. This situation allows the hair cell to generate a sinusoidal receptor potential in response to a sinusoidal stimulus, thus preserving the temporal information present in the original signal, up to frequencies of less than about 3 kH (Figure 10.10). Hair cells still can signal at frequencies above this zone, but they do not preserve the exact temporal structure of the stimulus: The cell's membrane time-constant filters the asymmetrical displacement-receptor current function of the hair cell bundle to produce a tonic depolarization of the soma, augmenting transmitter release and thus exciting auditory nerve terminals.

The high-speed demands of mechanotransduction have resulted in some impressive specializations of the ion fluxes within the inner ear. An unusual adaptation of the hair cell in this regard is that K^+ serves both to depolarize and repolarize the cell, enabling the hair cell's K^+ gradient to be largely maintained by passive ion movement alone. As with other epithelial cells, the basal and apical surfaces of the hair cell are separated by tight junctions, allowing separate extracellular ionic environments at these two surfaces. The apical end (including the stereocilia) protrudes into the scala media and is exposed to the K^+-rich, Na^+-poor **endolymph** produced by dedicated ion-pumping cells in the **stria vascularis** (see Figure 10.9A). The basal end of the hair cell body is bathed in perilymph—the same fluid that fills the scala tympani. Perilymph resembles other extracellular fluids in that it is K^+-poor and Na^+-rich. However, the compartment containing endolymph is about 80 mV more positive than the perilymph compartment (this difference is known as the *endocochlear potential*), while the inside of the hair cell is about 45 mV more negative than the perilymph and about 125 mV more negative than the endolymph. The resulting electrical gradient across the membrane of the stereocilia (~125 mV) drives K^+ through open transduction channels into the hair cell, which depolarizes the hair cell, opening voltage-gated K^+ and Ca^{2+} channels located in the membrane of the hair cell soma. The opening of *somatic* K^+ channels favors K^+ efflux, and thus repolarization, whereas Ca^{2+} entry triggers transmitter release and opens Ca^{2+}-dependent K^+ channels, which provide another avenue for K^+ to enter the perilymph. Indeed, the interaction of Ca^{2+} influx and Ca^{2+}-dependent K^+ efflux can lead to electrical resonances that enhance the tuning of response properties of hair cells.

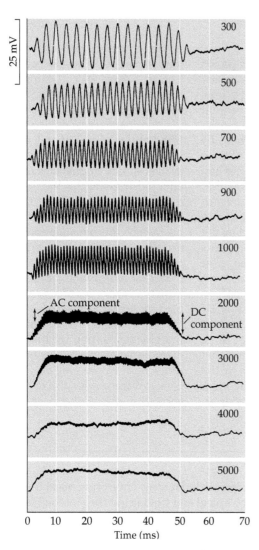

FIGURE 10.10 Frequency-dependent characteristics of receptor potentials in hair cells Receptor potentials generated by an individual hair cell in the cochlea in response to pure tones (indicated in Hz, right). Note that the hair cell potential faithfully follows the waveform of the stimulating sinusoids for lower frequencies (< 3 kHz) but still responds with a direct current offset to higher frequencies because of the asymmetrical stimulus-response function and the electrical filtering properties of the hair cells. (After Palmer and Russell, 1986. *Hearing Res* 24: 1–15)

In essence, the hair cell operates as two distinct compartments, each dominated by its own Nernst equilibrium potential for K^+; this arrangement ensures that the hair cell's ion gradient does not run down, even during prolonged stimulation. The rupture of Reissner's membrane (which normally separates the scalae media and vestibuli) or the presence of compounds such as ethacrynic acid that selectively poison the ion-pumping cells of the stria vascularis can cause the endocochlear potential to dissipate, resulting in a sensorineural hearing deficit (see Clinical

Applications). In short, the hair cell exploits the different ionic milieus of its apical and basal surfaces to provide extremely fast and energy-efficient repolarization.

The hcMET channel that underlies these ion conductances has yet to be isolated and the genes that encode the channel have yet to be identified, despite extensive knowledge of the hcMET's physiological properties and intensive genetic analyses that have led to the promotion of several potential candidates. Sheer paucity of material is one of the major challenges to isolating the hcMET protein: A single hair bundle may possess as few as 200 functional channels, which represent a tiny fraction (<0.001%) of all hair bundle proteins, factors that have rendered biochemical purification impractical. A further challenge is the complexity of the transduction apparatus, with current evidence indicating that the pore-forming molecule must interact with a variety of other accessory proteins to enable mechanotransduction. Despite these challenges, the genetic analysis of heritable forms of deafness has identified numerous genes that are important to normal hearing, including candidate hcMET channels. Currently, four especially promising candidates are *TMC1, TMC2, TMIE,* and *LHFPL5*, all of which localize to the apical end of stereocilia and mutations in which reduce or abolish mechanotransduction currents in auditory hair cells. However, none of these genes have been shown to sustain mechanotransduction currents when expressed in heterologous systems, which may reflect that mechanotransduction in hair cells is the product of a multi-molecular machine comprising these and other molecules, including associated tip links. Despite the difficulty in isolating the hcMET channel and identifying the genes that encode it, these topics are of intense interest, as fully understanding how we hear will depend on a full molecular and genetic characterization of this channel.

CONCEPT 10.4 | Transduction Is Controlled by Active Mechanisms Involving Middle Ear Muscles and Cochlear Outer Hair Cells

LEARNING OBJECTIVES

10.4.1 Explain why active forces are needed to account for the responsiveness of the living cochlea to sound.

10.4.2 Describe the role of outer hair cells in generating such active forces.

How active processes contribute to hearing

Von Békésy's model of cochlear mechanics was a passive one, resting on the premise that the basilar membrane acts like a series of linked resonators, much as a concatenated set of tuning forks. More recent studies made from the intact, living cochlea, however, indicate that normal hearing depends on some means of amplification within the cochlea. That something active must be at work can be deduced from two observations. First, the tuning of the auditory periphery, whether measured at the basilar membrane or recorded as the electrical activity of auditory nerve fibers, is too sharp to be explained by passive mechanics alone. Second, at very low sound intensities, the basilar membrane vibrates 100-fold more than would be predicted by extrapolation from the motion measured at high intensities.

A direct indication that active forces are at play is that the ear can actually *generate* sounds (Figure 10.11). These self-generated sounds, or **otoacoustic emissions**, are thought to arise at least in part from the *outer* hair cells,

which constitute about three-fourths of the hair cells in the cochlea. Outer hair cells contract and expand in response to small electrical currents, thus inducing or altering the vibrations of the basilar membrane. The outer hair cells receive descending signals via the auditory nerve from the brain, providing a route by which the brain can control these contractions and expansions and thus modulate the response of the cochlea to incoming sound. The actions of the outer hair cells and their supervision by the brain probably contribute to sharpening frequency sensitivity, enhancing responsiveness to low-intensity sounds, and modulating input based on the focus of attention.

Detecting the collective actions of the outer hair cells is surprisingly simple. Because the internal structures of the ear are mechanically coupled to one another, vibrations induced or modified in the cochlea by the expansion and contraction of outer hair cells propagate backward through the middle ear to produce vibrations of the eardrum, and can be recorded with small microphones placed in the ear canal. Otoacoustic emissions testing provides a useful means to assess cochlear function in newborns, and is now done routinely to rule out congenital deafness. Such emissions can also occur spontaneously and are thus one possible source of **tinnitus** (ringing in the ears; most cases of tinnitus, however, are thought to be due to anomalies within the brain's auditory pathway).

The outer hair cells and middle ear muscles together provide the known means of active modification of the ear's sensitivity to sound. Conventionally, the middle ear muscles have been seen as providing a dampening of responsiveness to higher-intensity or anticipated sounds

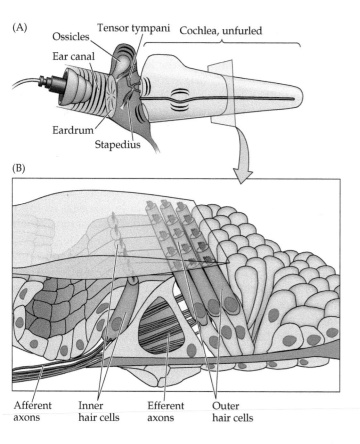

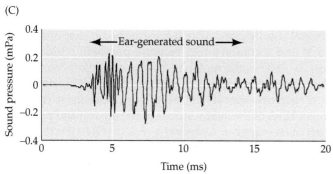

Figure 10.11   The brain controls how the ear responds to sound, and can even cause it to make sound (A) The ear contains actuators that exert mechanical control over auditory transduction. The middle ear muscles, the tensor tympani and stapedius, are attached to the ossicles (bones) that connect the eardrum to the cochlea. The brain controls the action of these muscles via the trigeminal and facial nerves, respectively (not shown). It is thought that these muscles play a role in protecting the ear from loud sounds and limiting the responsiveness of the cochlea to self-generated sounds such as our own speech. (B) Within the cochlea, the outer hair cells act like tiny muscles—they are capable of expanding and contracting and are stimulated both by incoming sound and by descending signals from the brain. (C) When a brief auditory click is delivered, the outer hair cells continue to respond after the click is over. Their motion causes the basilar membrane to continue oscillating, and these cochlear-generated vibrations are propagated backward along the ossicular chain. This causes the eardrum to continue vibrating, creating a sound that can be detected with a microphone in the ear canal. In general, the outer hair cells are thought to amplify the responsiveness of the cochlea to faint sounds. (A after J. M. Groh, 2014. *Making Space: How the Brain Knows Where Things Are.* Cambridge, MA: Harvard University Press. © 2014 Jennifer M. Groh; B,C after L. Zheng et al., 1999. *IEEE Trans Biomed Eng* 46: 1098–1106.)

(such as our own speech), and the outer hair cells have been seen as enhancing sensitivity to lower-intensity sounds. However, it should be noted that both systems may also play more complex roles, and the dividing line between the two may ultimately prove less than distinct. Ultimately, active processes under top-down brain control are needed to explain the remarkable dynamic range of the auditory system to both faint and louder sounds.

For more on the role of active mechanisms in frequency perception, see Box 10A.

CONCEPT 10.5 Auditory Pathways Involve Signals Traveling Bilaterally and in Both Feedforward and Feedback Directions

LEARNING OBJECTIVE

10.5.1 Identify the major areas of the ascending and descending auditory pathways.

Anatomical pathways in the central auditory system

As in the visual system, the anatomical connections of the auditory system involve multiple branches and pathways that both *ascend* and *descend*, ultimately both conveying information about sound to the brain and allowing the brain in turn to control how those signals are processed via the actions of the middle ear muscles, outer hair cells, and processing within brain regions at "lower" stages of the pathway.

The ascending portion of the auditory system begins with the auditory nerve conveying signals from the cochlea to the cochlear nucleus, which is located in the brain stem (Figure 10.12). Each ascending auditory nerve fiber receives input from one or a few inner hair cells, although each inner hair cell forms synapses with multiple auditory nerve fibers, thus exhibiting a few-to-many mapping. Within the brain stem, the auditory nerve branches to innervate three divisions of the cochlear nucleus: the anteroventral cochlear nucleus, the posteroventral cochlear nucleus, and the dorsal cochlear nucleus (see Figure 10.12). The frequency map, or **tonotopic organization**, of the cochlea (see Figure 10.6) is maintained in the three parts of the cochlear nucleus, each of which contains different populations of cells with quite different properties. In addition, the patterns of termination of the auditory nerve axons differ in density and type; thus, there are several opportunities at this level for transformation of the information from the hair cells.

Just as the auditory nerve branches to innervate several different targets in the cochlear nuclei, the neurons in these nuclei give rise to several different pathways (see Figures 10.12, 10.17, 10.18), notably the medial superior olive (MSO), the lateral superior olive (LSO), which together from the superior olivary complex, and the medial nucleus of the trapezoid body (MNTB). These connections arise from the cochlear nuclei on both sides of the brain and therefore play a key role in the localization of sound, a topic covered in greater detail in Concept 10.8. From these structures, signals continue ascending to the inferior colliculus in the midbrain and thence to the medial geniculate complex (MGC) of the thalamus and finally to primary or core auditory cortex and other auditory cortical regions. In short, there are many more pre-cortical stops on the auditory processing stream than is the case for other sensory systems.

The descending portion of the auditory pathway has received less attention but undoubtedly contributes substantially to sound processing (see Figure 10.12). The auditory cortex projects back to the inferior colliculus, which in turn projects back to the superior olivary complex. The superior olivary complex is the locus from which signals to the outer hair cells originate, via the medial olivocochlear bundle, which is a branch of the auditory nerve. Descending control over the middle ear muscles is achieved via the facial nerve, which innervates the stapedius muscle, and the trigeminal nerve, which innervates the tensor tympani muscle. Many of these descending circuits have crossed components, as do their ascending counterparts. Thus, there is no place in the auditory system that is purely monaural (receiving sound from only one ear), as input from one ear can enter the brain and affect responsiveness of the other ear through these crossed and descending connections. Clinically, this means that damage to central auditory structures is almost never manifested as a monaural hearing loss. Indeed, a monaural hearing loss implicates peripheral damage on one side, either to the middle or inner ear or to the auditory nerve itself (see Clinical Applications). Overall, these descending pathways are positioned to exert considerable control over the processing of sound by the brain, although many aspects of their function remain to be discovered.

The description just outlined captures the broad strokes of the auditory system's ascending and descending pathways. Many of the individual stages in these pathways can themselves be divided into subregions (as discussed for the cochlear nucleus). Notably, the auditory cortex has several subdivisions, which are grouped into primary (i.e., core) and secondary (i.e., belt and parabelt) regions (Figure 10.13). The core region in macaque monkeys comprises three divisions, including **auditory area 1 (A1)**, **rostral (R)**, and **rostrotemporal (RT)**, all of which are located on the lower bank of the lateral sulcus in the medial and posterior part of the superior temporal gyrus (STG) in the temporal lobe. Imaging studies in humans indicate that the core region is located in the transverse temporal gyri (Heschl's gyri, or Brodmann's areas 41 and 42), buried in the lateral

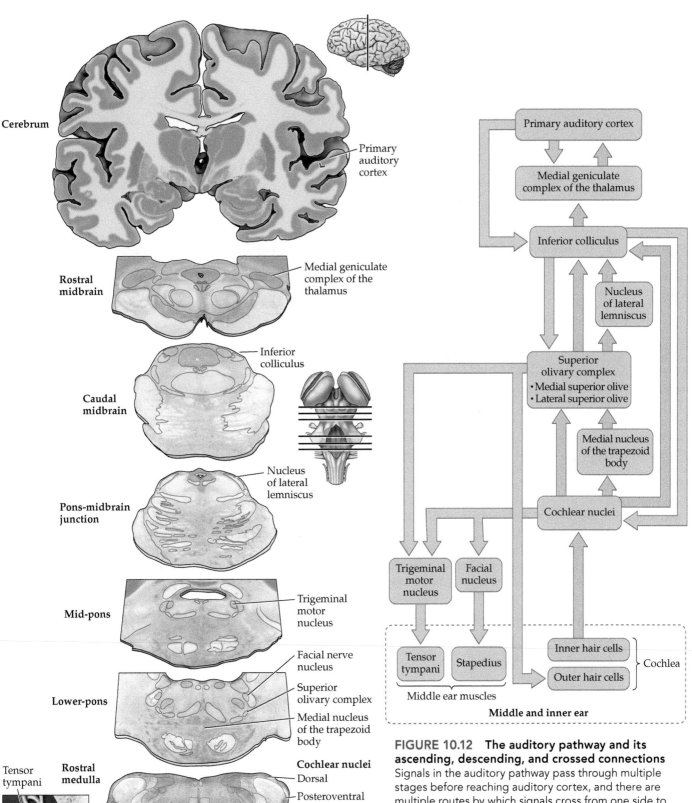

FIGURE 10.12   The auditory pathway and its ascending, descending, and crossed connections
Signals in the auditory pathway pass through multiple stages before reaching auditory cortex, and there are multiple routes by which signals cross from one side to the other or are transmitted back to "lower" stages from "higher" stages. The recurrent nature of the connection patterns affords opportunities for numerous interactions among brain regions. These patterns also make identifying a unique role for any one brain region challenging.

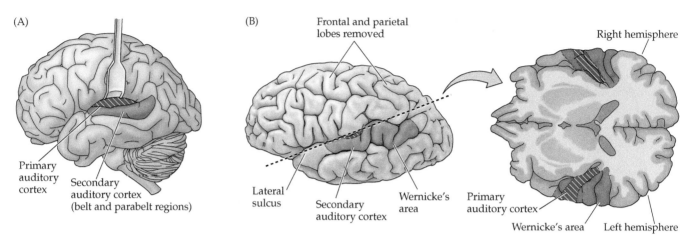

FIGURE 10.13 **The human auditory cortex** (A) Diagram showing the brain in left lateral view, including the depths of the lateral sulcus, where part of the auditory cortex occupying the superior temporal gyrus normally lies hidden. The core region is shown in blue; the surrounding belt regions of the auditory cortex are in red. (B) Diagram of the brain in left lateral view, showing locations of human auditory cortical regions related to processing speech sounds in the intact hemisphere. Right: An oblique section (plane of dashed line) shows the cortical areas on the superior surface of the temporal lobe. Note that Wernicke's area, a region important in comprehending speech, is just posterior to the primary auditory cortex.

sulcus. Input to the core region derives from a particular division of the MGC, whereas the **belt and parabelt regions** of the auditory cortex receive input from the belt division of the MGC, as well as input from the primary auditory cortex, and are less precise in their tonotopic organization compared with the primary area. Additionally, these various auditory cortical areas are strongly interconnected, with reciprocal connections between the core and belt regions, between the belt and parabelt regions, and between these latter two regions and auditory-related cortical areas in the superior temporal gyrus (STG) and the superior temporal sulcus (STS), suggestive of a processing hierarchy. Subdivisions also occur at the level of the inferior colliculus, which is thought to contain a central nucleus and at least one surrounding beltlike region as well.

example, is related to the intensity of sound. The neural basis of this perceptual attribute likely involves attributes such as the vigor of neural responses in the cochlea and later stages of the pathway.

Other aspects of auditory perception involve more complex computational processes in the brain. In particular, the ability to perceive sound frequency and location underlie a rich array of higher-order perceptual capacities such as speech and music, and involve sophisticated neural computations that interact with one another. Frequency and loudness both contribute to perception of location, and location helps us segregate the multitude of different sounds in the environment, thus helping us perceive the frequencies of sounds from different sources. Concepts 10.7 and 10.8 describe the neural computations that support our abilities to parse sound frequency and location.

CONCEPT 10.6 Auditory Perception Involves the Synthesis of Multiple Aspects of Sound

LEARNING OBJECTIVE

10.6.1 Discuss different aspects of sound perception and how they interact with one another.

Auditory perception

The mechanisms from ear to brain described thus far in the chapter support our perceptual ability to hear. Auditory perception relies on the physical qualities of sound and how the brain detects and encodes these physical attributes. Some aspects of auditory perception have a relatively straightforward relationship to the physical qualities of sound – our perception of sound loudness, for

CONCEPT 10.7 Neural Codes for Sound Frequency Are Based on Resonance and Synchrony

LEARNING OBJECTIVES

10.7.1 Describe the two types of neural representation of sound frequency.

10.7.2 Discuss some of the key puzzles in how these representations account for perception.

Two methods of encoding sound frequency: Place and time (and a perceptual mystery)

Information about what frequencies are present in a sound is represented in the activity patterns of the auditory nerve in two ways: *which* nerve fibers are active, and *when* they are active. These two properties are referred to as tonotopic organization (introduced in Concept 10.5) and phase locking, respectively.

Phase locking occurs when the sound frequency is low enough that the rapid response time of the transduction apparatus allows the membrane potential of the hair cell to follow deflections of the hair bundle. As a result, such frequencies can be encoded by the temporal patterns of activity of hair cells and their associated auditory nerve fibers: Their activity fluctuates in phase with the sound.

At higher frequencies, phase locking is not possible, despite the extraordinary rapidity of the mechanotransduction process (see Concept 10.3 and Figure 10.10). For such frequencies, the resonance properties of the basilar membrane become particularly important: Different frequencies produce maximum displacements at different locations along the basilar membrane and drive different hair cells and their associated auditory nerve fibers to different degrees, with the apical end of the cochlea most responsive to low frequencies and the basal end most responsive to high frequencies (see Figure 10.6). This sorting by frequency along a physical dimension, or tonotopy, is maintained at many stages of the auditory pathway, as noted in Concept 10.5. Tonotopy is the auditory system's parallel to the visual systems's retinotopy or the somatosensory system's somatotopy, and falls under the more general heading of *labeled-line* coding.

How these two forms of coding account for our extraordinary ability to perceive the frequencies of sound remains puzzling. Humans can distinguish between sounds that differ by only a few hertz (Figure 10.14A), and this ability

(A)

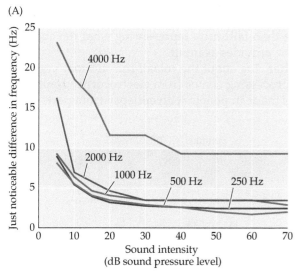

(B)

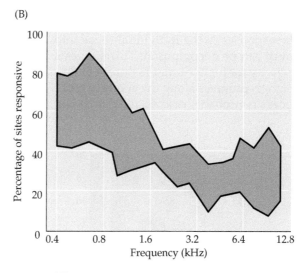

(C)

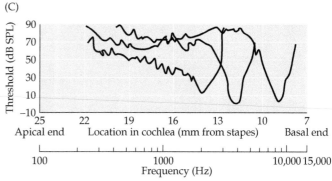

FIGURE 10.14 Neural response properties cannot account for perceptual abilities in an obvious way (A) The human ability to distinguish sounds of different frequencies is considerably sharper than the tuning curve width observed in cats and monkeys, and actually improves with increasing sound loudness; "just noticeable differences" of a few hertz have been observed for sound levels above 30 dB SPL with little indication of a decline in frequency acuity with increasing loudness. (B) Frequency tuning curves of various fibers in the auditory nerve of anesthetized cats. Each curve depicts the minimum sound level required to increase the fiber's firing rate above its spontaneous firing level, across all frequencies to which the fiber responds. The lowest point in the plot is the weakest sound intensity to which the neuron will respond. The frequency at this point is called the neuron's characteristic frequency. Note that the breadth of the tuning curves at the level of quiet conversation, about 50 dB SPL, is very broad. (C) Percentage of recording sites in the inferior colliculus of awake monkeys that respond to tones of different frequencies at 50 dB SPL (shading indicates the range across different individual animals). At 0.5 kHz, between 40% and 80% of the sites are responsive. (A based on data from J. R. Pierce, 1983. *The Science of Musical Sound*. New York: Scientific American Library, distributed by W.H. Freeman; B fter D. A. Bulkin and J. M. Groh, 2011. *J Neurophysiol* 105: 1785-1797; C after E. Javel, 1994. *Hear Res* 81: 167-188.)

improves for louder compared with softer sounds. And yet it is only at the very softest sound intensities that the tuning of auditory nerve fibers (as measured in animals) appears to approach this perceptual threshold (Figure 10.14B). At the sound intensities where perceptual acuity for frequency is greatest, the basilar mesplmbrane's traveling wave activates a wide range of auditory nerve fibers even for a stimulus containing only a single frequency. This broad activity probably carries forward along the auditory pathway: In the monkey inferior colliculus, a 500-Hz tone at 50 decibels of sound pressure level (dB SPL) activates more than 40% of the neurons (Figure 10.14C).

As signals ascend in the hierarchy of the auditory pathway, the tuning curve as assessed with pure tones appears to become an imperfect predictor of how neurons will respond to more complicated stimuli such as communication calls (Figure 10.15): Neural responses to these complex stimuli can often be quite different from the sum of their responses to the individual tones into which such stimuli can be decomposed. This suggests burgeoning selectivity for higher-order features of auditory stimuli as signals progress along the auditory pathway. Such a process is probably needed for perceiving the combinations of different frequencies across time, as in speech and music. Indeed, in marmosets, which are small New World monkeys with a complex vocal repertoire, the core and belt cortical regions contain neurons that are strongly responsive to spectral combinations that characterize certain vocalizations. Recent studies in marmosets and humans also implicate secondary regions of the auditory cortex in the perception of pitch. This percept is especially important to our musical sense and to vocal communication, because it enables us to hear two speech sounds as distinct even when they have overlapping spectral content and arise from the same location. A curious feature of pitch perception is that, for the harmonically complex sounds that typify speech and music, pitch corresponds to the fundamental frequency, even when it is absent from the actual stimulus. This ability of pitch processing to "fill in" a missing frequency further underscores the idea that the auditory cortex is doing far more than faithfully representing what the auditory periphery provides as input.

Another clue about the role of the auditory cortex in speech processing comes from electrocorticographic recordings made in people with epilepsy from parts of the

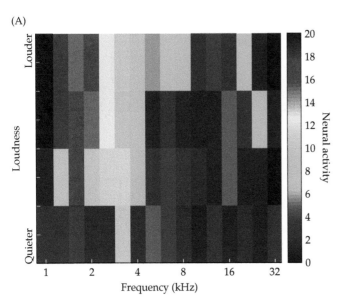

(A)

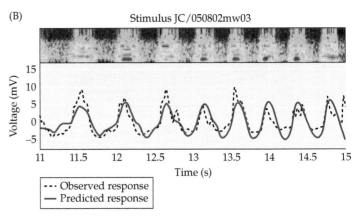

(B) Stimulus JC/050802mw03

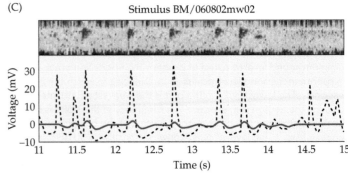

(C) Stimulus BM/060802mw02

FIGURE 10.15 Frequency tuning curves measured with individual pure tones may or may not successfully predict responses to more complex stimuli (A) Example of a frequency response area plot from rat auditory cortex. Activity is plotted as a function of sound frequency and loudness; brighter colors indicate more activity. (B,C) When tested with more complex natural stimuli such as communication calls (spectrograms), the responses of auditory cortical neurons (black lines) may be well predicted from the frequency response area (red lines), as in (B), or poorly predicted, as in (C). The poor match indicates that by this stage of auditory processing, neural response patterns are incompletely described by assessment with simple stimuli such as tones. (After C. K. Machens, 2004. *J Neurosci* 24: 1089–1100.)

STG that most likely correspond to parabelt regions in non-human primates (i.e., Brodmann's area 22). These recordings reveal that STG population activity strongly correlates with syllable onsets and offsets, which are especially important to speech intelligibility, and is sensitive to acoustical cues, such as voice onset times, that are important to the perceptual categorization of different speech sounds (e.g., distinguishing between *ba* and *pa*). Neural activity in the STG also depends strongly on context and attention: When individuals are instructed to attend to only one of two people speaking simultaneously, their STG neurons robustly encode fine spectrotemporal features of the attended voice but display little or no responsiveness to the other voice. Thus, auditory cortical activity is strongly influenced both by linguistic features and cognitive context, consistent with an influence of experience and task demands on auditory cortical processing of speech.

CONCEPT 10.8 Sounds Are Localized Based on Frequency, Relative Loudness, and Relative Time of Arrival at the Ears

LEARNING OBJECTIVE

10.8.1 Name the three cues used to localize sound and explain what they entail.

Localizing sound: A computational process

Identifying where sounds are coming from is also a critical perceptual ability. Sound localization is fundamentally a *computational* process. Information about the location of origin of a sound is not reflected in the sound wave itself, but has to be inferred from differences in the waveforms detected at the two ears. There are three types of location cues: **interaural timing differences, interaural level differences,** and **spectral cues** (Figure 10.16).

(A) Interaural timing difference

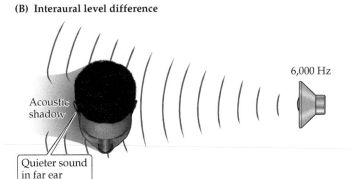

Greater distance to far ear leads to delay in sound arrival time

Sound source

(B) Interaural level difference

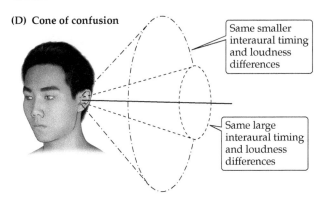

Acoustic shadow

6,000 Hz

Quieter sound in far ear

(D) Cone of confusion

Same smaller interaural timing and loudness differences

Same large interaural timing and loudness differences

(C) Spectral cues

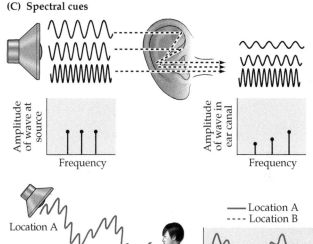

Amplitude of wave at source

Frequency

Amplitude of wave in ear canal

Frequency

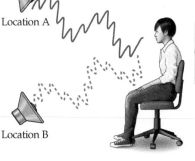

Location A

Location B

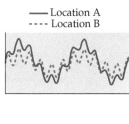

—— Location A
---- Location B

FIGURE 10.16 Computing a sound's location The locations of sounds are computed by the brain from three types of cues: (A) the delay it takes sounds to reach the more distant ear, which is a function of the combination of sound angle and head size (ear separation); (B) the loudness differences that occur due to the acoustic shadow cast by the head; and (C) the filtering of sounds of different frequencies in a direction-dependent manner by the external ear. (D) Interaural timing and level differences alone are ambiguous: The same loudness and timing differences can occur for sounds that are above or below as well as in front or in back. This cone of confusion can be resolved by incorporating spectral cues or by assessing how the timing and level differences change when the head turns. (After J. Groh, 2014. *Making Space: How the Brain Knows Where Things Are.* Harvard University Press. © 2014 Jennifer M. Groh.)

Interaural timing differences occur because sound is not instantaneous and our ears are separated by the width of our head. A sound located directly to the left will arrive in the left ear about 0.5 ms sooner than it will arrive in the right ear. Comparing sound arrival time across the two ears can thus provide information about where the sound is coming from, but only in the horizontal dimension—that is, relative to the axis connecting our two ears.

Interaural level differences occur because our head casts an acoustic shadow, so that leftward sound is louder in the left ear than in the right ear. Again, this cue provides information only about where the sound is located with respect to the horizontal axes connecting the two ears. Interaural loudness differences are greater for higher-frequency than for lower-frequency sounds, because shorter wavelength stimuli diffract around the head less than do longer ones.

Spectral cues operate differently, and involve resonance properties of the external ear, or pinna. Folds in the pinna filter the frequency content of sound, and they do so in a fashion that depends on the direction of incidence of the sound. Deducing a sound's location based on its frequency content is a complex computation that may involve comparison across the two ears or may be accomplished using one ear, provided the sound is familiar and its typical frequency content has been learned previously. Nevertheless, spectral cues can provide both horizontal and vertical information about sound location, and provide an effective means of localizing even in individuals who are deaf in one ear.

All of these cues are indirect, and the relationship between a given cue value, such as a 0.3-ms timing difference or a 10-dB sound level difference, and an external spatial location, such as 20 degrees to the left, has to be learned from experience, and relearned continually during development as the head grows and the ears move farther apart. Thus, the neural computational processes involved in sound localization concern (a) detecting and representing these cue values, (b) relating them to external space, and (c) keeping them calibrated across the life span.

Circuits that are optimized for assessing the two types of binaural difference cues are found in the superior olivary complex. In particular, neurons in the medial superior olive (MSO) receive input from both ears and are exquisitely sensitive to the relative delay between the two sides (Figure 10.17). Binaural neurons in the lateral

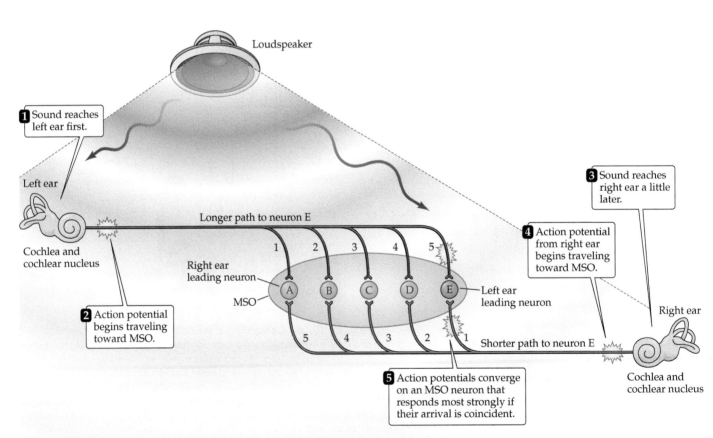

FIGURE 10.17 A model of how the MSO computes the location of a sound by interaural timing differences A given MSO neuron responds most strongly when the two inputs arrive simultaneously, as occurs when the contralateral and ipsilateral inputs precisely compensate (via their different lengths) for differences in the time of arrival of a sound at the two ears. The systematic (and inverse) variation in the delay lengths of the two inputs creates a map of sound location. In this model, neuron E in the MSO would be most sensitive to sounds located to the left, and neuron A to sounds from the right; neuron C would respond best to sounds coming from directly in front of the listener. (After L. A. Jeffress, 1948. *J Comp Physiol Psychol* 41: 35–39.)

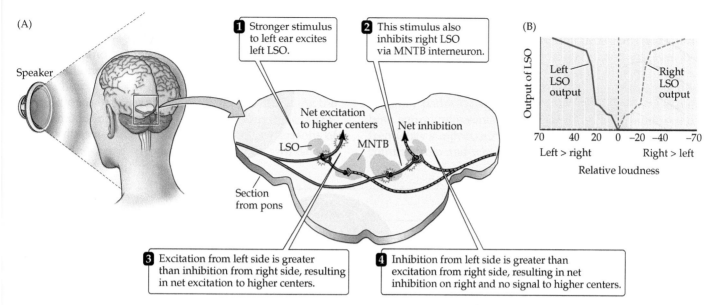

FIGURE 10.18 **LSO neurons encode sound location through interaural intensity differences** (A) LSO neurons receive direct excitation from the ipsilateral cochlear nucleus; input from the contralateral cochlear nucleus is relayed via inhibitory interneurons in the medial nucleus of the trapezoid body (MNTB). (B) This arrangement of excitation–inhibition makes LSO neurons fire most strongly in response to sounds arising directly lateral to the listener on the same side as the LSO, because excitation from the ipsilateral input will be great and inhibition from the contralateral input will be small. In contrast, sounds arising from in front of the listener, or from the opposite side, will silence the LSO output, because excitation from the ipsilateral input will be minimal, but inhibition driven by the contralateral input will be great. Note that LSOs are paired and bilaterally symmetrical; each LSO encodes only the location of sounds arising from the ipsilateral hemifield.

superior olive (LSO), in contrast, are less concerned with relative timing but exhibit sensitivity to the relative intensities of the input they receive from the two ears (Figure 10.18). These timing and level difference sensitivity patterns are thought to be recombined in the inferior colliculus, producing a representation of space.

CONCEPT
10.9

Hearing Coordinates with Vision Via Numerous Signal Transformations and Robust Intersensory Cross-talk

LEARNING OBJECTIVE

10.9.1 Explain what is meant by the differences in reference frame and coding format between the auditory and visual systems.

Codes for sound location in relation to vision

A final level of computational processing concerns how the brain represents sound location and how it reconciles visual and auditory spatial information (see Figure 10.19). The visual system plays an important role in sound localization because, as noted in Concept 10.8, the relationship between a particular time delay or level difference and location in the external world has to be learned from experience—and that experience often involves vision. Plausible visual stimuli—such as the performance of a skilled ventriloquist—can capture the perceived locations of sounds, making it seem as if a puppet is actually speaking. Under normal circumstances, in which a sound truly does come from the location of a visible source, such visual capture of perceived sound locations would be advantageous and could contribute to successfully learning how to interpret sound cues on their own.

Several computational problems underlie the process of linking visual and auditory spatial representations. One key problem is that sound localization cues are informative about the location of a sound with respect to the head and ears, whereas the retina receives information about the location of light sources with respect to the eyes. In some species, such as Barn Owls, the eyes are largely fixed in orientation with respect to the head, and these two reference frames are in alignment with each other. Indeed, the Barn Owl optic tectum is thought to contain aligned maps of both visual and auditory space. However, in primates and many other mammals, the eyes can move about (Figure 10.19A; also see Chapter 9). Saccadic eye movements can span a range of plus or minus 40 degrees—meaning that the eyes can move 40 degrees to the left, right, up, or down—and occur at a frequency of three or four times per second. This eye mobility disrupts the possibility of any fixed relationship between visual and auditory spatial signals, and requires the involvement of signals from the oculomotor system to permit a computation of the locations of sounds in relation to the visual scene. Indeed, sensitivity

Groh J. M., A. S Trause, A. M. Underhill, K. R. Clark, S. Inati S. (2001) Eye position influences auditory responses in primate inferior colliculus. Neuron 29:509-518.

Gruters, K. G. and 5 others (2018) The eardrums move when the eyes move: A multisensory effect on the mechanics of hearing. *Proc. Natl. Acad. Sci. U.S.A.* 115: E1309–E1318.

Jay, M. F. and D. L. Sparks (1984) Auditory receptive fields in primate superior colliculus shift with changes in eye position. *Nature* 309: 345–347.

Jeffress, L. A. (1948) A place theory of sound localization. *J. Comp. Physiol. Psychol.* 41: 35–39.

Knudsen, E. I. and M. Konishi (1978) A neural map of auditory space in the owl. *Science* 200: 795–797.

Lee, J. and J. M. Groh (2014) Different stimuli, different spatial codes: A visual map and an auditory rate code for oculomotor space in the primate superior colliculus. *PLoS One* 9: e85017.

Litovsky, R. Y., H. S. Colburn, W. A. Yost and S. J. Guzman (1999) The precedence effect. *J. Acoust. Soc. Am.* 106: 1633–1654.

Machens, C. K., M. S. Wehr and A. M. Zador (2004) Linearity of cortical receptive fields measured with natural sounds. *J. Neurosci.* 24(5): 1089–1100.

Mesgarani, N. and E. F. Chang (2012) Selective cortical representation of attended speaker in multi-talker speech perception. *Nature* 485: 233–236.

Mesgarani, N., C. Cheung, K. Johnson and E. F. Chang (2014) Phonetic feature encoding in human superior temporal gyrus. *Science* 343: 1006–1010.

Middlebrooks, J. C., A. E. Clock, L. Xu and D. M. Green (1994) A panoramic code for sound location by cortical neurons. *Science* 264: 842–844.

Oertel, D. and 4 others (2000) Detection of synchrony in the activity of auditory nerve fibers by octopus cells of the mammalian cochlear nucleus. *Proc. Natl. Acad. Sci. U.S.A.* 97: 11773–11779.

Overath, T., J. H. McDermott, J. M. Zarate and D. Poeppel (2015) The cortical analysis of speech-specific temporal structure revealed by responses to sound quilts. *Nat. Neurosci.* 18: 903–911.

Patterson, R. D., S. Uppenkamp, I. S. Johnsrude and T. D. Griffiths (2002) The processing of temporal pitch and melody information in auditory cortex. *Neuron* 36: 767–776.

Romanski, L. M. and 5 others (1999) Dual streams of auditory afferents target multiple domains in the primate prefrontal cortex. *Nat. Neurosci.* 2: 1131–1136.

Rothschild, G., I. Nelken and A. Mizrahi (2010) Functional organization and population dynamics in the mouse primary auditory cortex. *Nat. Neurosci.* 13: 353–360.

Ruggles, D., H. Bharadwaj and B. G. Shinn-Cunningham (2012) Why middle-aged listeners have trouble hearing in everyday settings. *Curr. Biol.* 22: 1417–1422.

Salminen, N. H., P. J. May, P. Alku and H. Tiitinen (2009) A population rate code of auditory space in the human cortex. *PLoS One* 4: e7600.

Schneider, D. M., A. Nelson and R. Mooney (2014) A synaptic and circuit basis for corollary discharge in the auditory cortex. *Nature* 513: 189–194.

Suga, N., W. E. O'Neill and T. Manabe (1978) Cortical neurons sensitive to combinations of information-bearing elements of biosonar signals in the mustache bat. *Science* 200: 778–781.

Teki, S. and 4 others (2013) Segregation of complex acoustic scenes based on temporal coherence. *ELife* 2: e00699.

Books

Blauert, J. (1997) *Spatial Hearing.* Cambridge, MA: MIT Press.

Groh, J. M. (2014) *Making Space: How the Brain Knows Where Things Are.* Cambridge, MA: Harvard University Press.

Moore, B. C. J. (2003) *An Introduction to the Psychology of Hearing.* London: Academic Press.

Pickles, J. O. (2013) *An Introduction to the Physiology of Hearing,* 4th Edition. Leiden: Brill.

Schnupp, J., I. Nelken and A. King (2011) *Auditory Neuroscience.* Cambridge, MA: MIT Press.

von Békésy, G. (1960) *Experiments in Hearing.* New York: McGraw-Hill. (A collection of von Békésy's original papers.)

The Vestibular System

Overview

The vestibular system processes sensory information related to self-motion, head position, and spatial orientation relative to gravity. Vestibular signals help stabilize gaze, head, and posture. The peripheral portion of the vestibular system is located in the inner ear and contains sensors that function as an inertial guidance system sensitive to linear acceleration and angular velocity. These sensors report information about the motions and position of the head to integrative centers in the brainstem, cerebellum, and cerebral cortices. The central portion of the system includes the vestibular nuclei, which make extensive connections with brainstem and cerebellar structures. The vestibular system is closely connected to motor systems, specifically innervating motor neurons that control extraocular, cervical, and postural muscles. This sensory–motor loop helps stabilize gaze, head orientation, and posture during movement. The vestibular system works in tandem with vision for their mutual benefit. The vestibular system aids vision by stabilizing the visual scene on the retina through precise counter-rotations of the eye that compensate for head movements, and is aided by vision via sensing of optic flow patterns that assist with computing self-motion. The vestibular system is plastic, undergoing recalibration, such as in response to a change in an eyeglasses prescription, and can be improved through practice. Although we are normally unaware of its functioning, the vestibular system and its multimodal partners are critical to our perception of self-motion, spatial orientation, and body representation. The vestibular system gives rise to a "sixth sense" that is critical both to automatic behaviors and to perception, with the consequence that balance, gaze stabilization during head movement, and sense of orientation in space are all adversely affected if the system is damaged.

Dr. Goran Bredberg/Science Photo Library

KEY CONCEPTS

11.1 The vestibular system helps us sense our position and movement in space

11.2 The utricle and the saccule sense static tilt and dynamic linear movements

11.3 The semicircular canals sense rotations of the head in three dimensions

11.4 The vestibular system works with visual cues to evaluate self motion and to stabilize eye gaze

11.5 The brain synthesizes information to support perception of translational and rotational movements and maintain equilibrium

CONCEPT
11.1 | **The Vestibular System Helps Us Sense Our Position and Movement in Space**

LEARNING OBJECTIVES

11.1.1 Describe the types of motions to which the vestibular system is sensitive.

11.1.2 Identify the physical structures of the vestibular labyrinth and which types of motion sensing they contribute to.

11.1.3 Explain how vestibular hair cells transduce physical movement into electrical signals.

Anatomy of the vestibular labyrinth

The vestibular system is responsible for sensing self-motion involving multiple degrees of freedom: **translational movements**, or linear motion forward, backward, and sideways, such as occurs when walking through space, and rotational movements, meaning head turning such as nodding, shaking, or cocking the head to one side. In addition to these two types of motions, the vestibular system is also sensitive to the time course of movements, ranging from static posture to active changes in position.

These motions are sensed through structures located in the inner ear. The main peripheral component of the vestibular system is a set of interconnected chambers—the **labyrinth**—that has much in common, and is in fact continuous, with the cochlea (see Chapter 10). Like the cochlea, the labyrinth uses **hair cells** to transduce physical motion into neural impulses (see next section). In the cochlea, the motion arises from airborne sound stimuli; in the labyrinth, the pertinent motions arise from the effects of gravity and from translational and rotational movements of the head.

The labyrinth is buried deep in the temporal bone and consists of the two **otolith organs**—the **utricle** and **saccule** —and three **semicircular canals** (Figure 11.1A). The utricle and saccule are specialized primarily to respond to translational movements of the head and static head position relative to the gravitational axis (i.e., head tilts), whereas the semicircular canals, as their shapes suggest, are specialized for responding to rotations of the head.

The labyrinth involves specialized ionic environments much like those found in the cochlea. The outer portion of the labyrinth, the bony labyrinth, is filled with perilymph similar in composition to cerebrospinal fluid (Figure 11.1B; see also Chapter 10). Inside the bony labyrinth are membranous sacs collectively called the membranous labyrinth. The membranous labyrinth is filled with a different fluid, endolymph. The vestibular hair cells are located in the utricle and saccule and in three jug-like swellings called **ampullae**, located at the base of the semicircular canals next to the utricle. As in the cochlea, tight junctions seal the apical surfaces of the vestibular hair cells, ensuring that endolymph selectively bathes the hair cell bundle while remaining separate from the perilymph surrounding the basal portion of the hair cell.

Vestibular hair cells

The vestibular hair cells, like cochlear hair cells, transduce minute physical displacements of their hair bundles into electrical receptor potentials. Vestibular and auditory hair

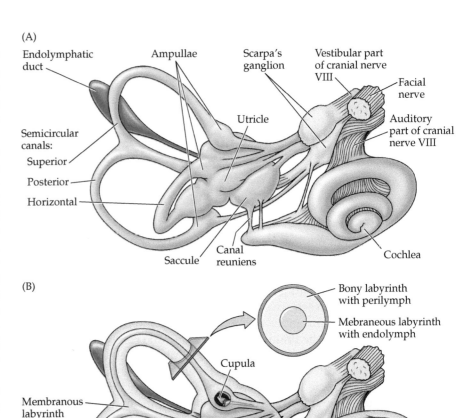

FIGURE 11.1 The labyrinth and its innervation (A) The vestibular and auditory portions of cranial nerve VIII are shown; the small connection from the vestibular nerve to the cochlea contains auditory efferent fibers. General orientation within the head is shown in Figure 10.4; see also Figure 11.8. (B) The labyrinth consists of two fluid filled compartments, the interior membranous labyrinth which is filled with endolymph, and the surrounding bony labyrinth which is filled with perilymph.

cells are quite similar; Chapter 10 gave a detailed description of hair cell structure and function. As in the case of auditory hair cells, movement of the stereocilia in one direction opens mechanically gated mechanoelectrical transduction channels located at the tips of the stereocilia, depolarizing the hair cell and causing neurotransmitter release onto (and excitation of) the vestibular nerve fibers (Figure 11.2). Movement of the stereocilia in the opposite direction closes the channels, hyperpolarizing the hair cell and thus reducing vestibular nerve activity. The biphasic nature of the receptor potential means that some transduction channels are open in the absence of stimulation, with the result that hair cells tonically release transmitter, thereby generating considerable spontaneous activity in vestibular nerve fibers (see Figure 11.6). One consequence of these spontaneous action potentials is that the firing

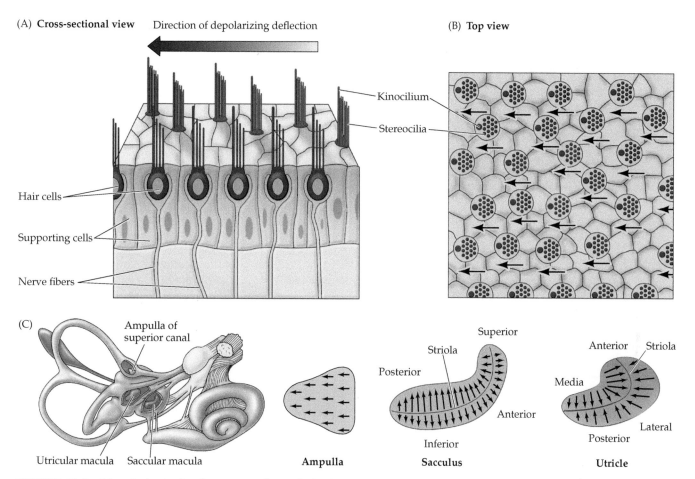

(A) Cross-sectional view   Direction of depolarizing deflection

Kinocilium

Stereocilia

Hair cells

Supporting cells

Nerve fibers

(B) Top view

(C)   Ampulla of superior canal

Utricular macula   Saccular macula   **Ampulla**

Superior

Striola

Posterior

Anterior

Inferior

Sacculus

Anterior   Striola

Media

Lateral

Posterior

Utricle

FIGURE 11.2 Morphological polarization of vestibular hair cells and polarization maps of the vestibular organs (A) A cross section of hair cells shows that the kinocilia of the bundles of hairs are all located on the same sides of the hair cells. The arrow indicates the direction of deflection that depolarizes the hair cell. (B) View looking down on the hair bundles. (C) In the ampulla located at the base of each semicircular canal, the hair bundles are oriented in the same direction. In the sacculus and utricle, the striola divides the hair cells into populations with opposing hair bundle polarities.

rates of vestibular fibers can increase or decrease in a manner that faithfully mimics the receptor potentials produced by the hair cells. Adaptation in vestibular hair cells, mediated by calcium entering through mechanoelectrical transduction and voltage-gated calcium channels, is especially important to vestibular function, as it allows hair cells to continue to signal small changes in head position despite much larger tonic forces of gravity.

CONCEPT
11.2

The Utricle and the Saccule Sense Static Tilt and Dynamic Linear Movements

LEARNING OBJECTIVES

11.2.1 Explain how otoconia deflect hair bundles in the otolith organs.

11.2.2 Describe how the utricle and the saccule work together to evaluate the direction of head motion, and the contribution of the striola to this process.

11.2.3 Explain the difference between head tilt and linear motion.

11.2.4 Discuss the similarities and difference between sensing static position and dynamic motion and their consequences on neural activity patterns.

Otolith organs: The utricle and saccule

The two otolith organs, the utricle and the saccule, detect static tilt and dynamic translational (i.e, linear, as opposed to rotational) movements of the head. Both of these organs contain a sensory epithelium, the **macula**, which consists of hair cells and associated supporting cells. Overlying the hair cells and their hair bundles is a gelatinous layer; above this layer is a fibrous structure, the **otolithic membrane**, in which are embedded crystals of calcium carbonate called **otoconia** (Figure 11.3). The crystals give the otolith organs their name (*otolith* is Greek for "ear stones"). The otoconia make the otolithic membrane heavier than the structures and fluids surrounding it; thus, when the head tilts, gravity causes the membrane to shift relative to

From J. D. Dickman et al., 2004. Hear Res 188: 89–103

FIGURE 11.3 **Calcium carbonate crystals (otoconia) in the utricular macula of a quail** Each otoconium in this scanning electron micrograph is about 50 μm long.

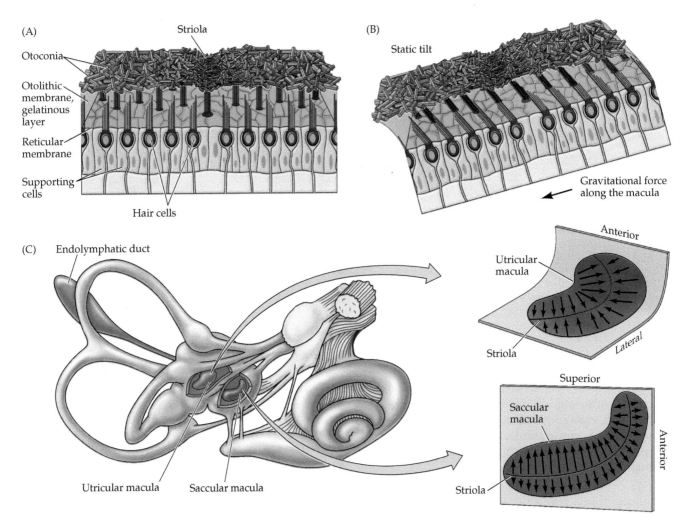

FIGURE 11.4 **Morphological polarization of hair cells in the utricular and saccular maculae** (A) Cross section of the utricular macula showing hair bundles projecting into the gelatinous layer when the head is level. (B) Cross section of the utricular macula when the head is tilted. The hair bundles are deflected by the otoconia in the direction of the gravitational force along the macular plane. An equivalent linear acceleration opposite to this force would induce the same deflection of the otoconia and is referred to as the *equivalent acceleration*. (C) Orientation of the utricular and saccular maculae in the head; black arrows show orientation of the kinocilia, as in Figure 11.2. The saccules on either side are oriented more or less vertically, and the utricles more or less horizontally. The striola is a structural landmark consisting of small otoconia arranged in a narrow trench that divides each otolith organ. In the utricular macula, the kinocilia are directed toward the striola. In the saccular macula, the kinocilia point away from the striola. Note that, given the utricle and sacculus on both sides of the body, there is a continuous representation of all directions of head movement.

Upright

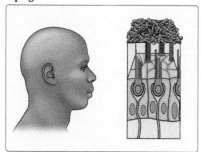

Head tilt; sustained

Backward

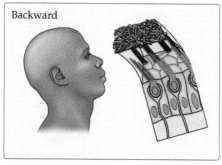

Forward

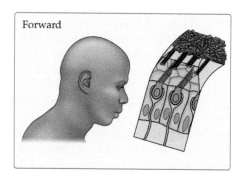

No head tilt; transient

Forward acceleration

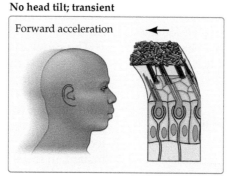

Deceleration

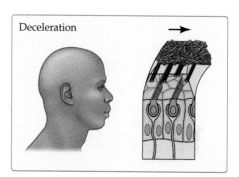

FIGURE 11.5   Forces acting on the head displace the otolithic membrane of the utricular macula For each of the positions and accelerations caused by translational movements, some set of hair cells will be maximally excited, while another set will be maximally inhibited. Note that head tilts produce displacements similar to certain accelerations.

the macula (Figure 11.4). The resulting shearing motion between the otolithic membrane and the macula displaces the hair bundles, which are embedded in the lower, gelatinous surface of the membrane. This displacement of the hair bundles generates a receptor potential in the hair cells. A shearing motion between the macula and the otolithic membrane also occurs when the head undergoes translational movements (Figure 11.5); the greater relative mass of the otolithic membrane causes it to lag behind the macula temporarily, leading to transient displacement of the hair bundle.

One consequence of the similar effects exerted on otolithic hair cells by certain head tilts and translational movements is that otolith afferents cannot convey information that distinguishes between these two types of stimuli. Consequently, one might expect that these different stimuli would be rendered perceptually equivalent when visual feedback is absent, as occurs in the dark or when the eyes are closed. Nevertheless, blindfolded individuals can discriminate between these two stimulus categories, a feat that depends on the integration of information from the otolith organs and the semicircular canals in the central vestibular system, as described in more detail in Concept 11.6.

Importantly, the hair cell bundles in the utricle and saccule have specific orientations (see Figure 11.2C). As a result, the organs as a whole are responsive to displacements in all directions. A specialized area called the **striola** divides the hair cells into two populations having opposing polarities (see Figures 11.2C and 11.4C). The striola forms an axis of mirror symmetry such that hair cells on opposite sides of the striola have opposing morphological polarizations. Thus, a head tilt along the axis of the striola will excite the hair cells on one side while inhibiting the hair cells on the other side. The saccular macula is oriented vertically and the utricular macula horizontally, with continuous variation in the morphological polarization of the hair cells located in each macula (as shown in Figure 11.4C, where the black arrows indicate the direction of movement that produces excitation). Inspection of the excitatory orientations in the maculae indicates that the utricle responds to translational movements of the head in the horizontal plane and to sideways head tilts, whereas the saccule responds to vertical translational movements of the head and to upward or downward head tilts.

Note that the saccular and utricular maculae on one side of the head are mirror images of those on the other side. Thus, a tilt of the head to one side has opposite effects on corresponding hair cells of the two utricular maculae. This concept is important in understanding how the central connections of the vestibular periphery mediate the interaction of inputs from the two sides of the head.

How otolith neurons sense static head tilts and dynamic translational head movements

The structure of the otolith organs enables them to sense both sustained and transient aspects of head position and movement, and both rotational and linear (translational) directions, albeit with different temporal response profiles.

Figure 11.5 illustrates some of the forces produced by head tilts and translational movements on the utricular macula.

The mass of the otolithic membrane relative to the surrounding endolymph, as well as the otolithic membrane's physical uncoupling from the underlying macula, means that hair bundle displacement will occur transiently in response to translational head acceleration, and tonically in response to steady tilting of the head. The resulting hair bundle displacements are reflected in the responses of the vestibular nerve fibers that innervate the otolith organs. These nerve fibers have a steady and relatively high firing rate when the head is upright. Figure 11.6 shows these responses recorded from an otolith afferent fiber, or axon, in a monkey seated in a chair that could be tilted for several seconds to produce a steady force on the head. Prior to the tilt, the axon has a high firing rate, which increases

or decreases depending on the direction of the tilt. Notice also that the response remains at a high level as long as the tilting force remains constant; thus, such neurons faithfully encode the static force being applied to the head (see Figure 11.6A). When the head is returned to the original position, the firing level of the neurons returns to baseline value. Conversely, when the tilt is in the opposite direction, the neurons respond by decreasing their firing rate below the resting level (see Figure 11.6B) and remain depressed as long as the static force continues. In a similar fashion, transient increases or decreases in firing rate from spontaneous levels signal the direction of translational accelerations of the head, as occurs when you are riding in a car that is speeding up or slowing down.

The range of orientations of hair bundles within the otolith organs enables them to transmit information about linear forces in every direction the head might move (see Figure 11.4C). The utricle, which is primarily concerned with motion in the horizontal plane, and the saccule, which is concerned with vertical motion, combine to effectively gauge the linear forces acting on the head at any instant, in three dimensions. Tilts of the head off the horizontal plane and translational movements of the head in any direction stimulate a distinct subset of hair cells in the saccular and utricular maculae, while simultaneously suppressing the responses of other hair cells in these organs. Ultimately, variations in hair cell polarity in the otolith organs produce patterns of vestibular nerve fiber activity that, at a population level, encode head position and the forces that influence it.

(A)

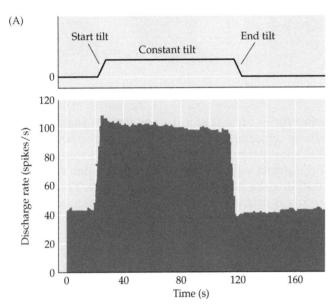

(B)

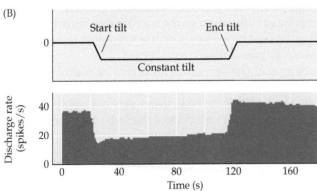

FIGURE 11.6   Response of a vestibular nerve axon from an otolith organ in a monkey The utricle is the example shown here. (A) The stimulus (top) is a change in head tilt. The spike histogram shows the neuron's response to tilting in a particular direction. (B) A response of the same fiber to tilting in the opposite direction. (After C. Fernández and J. M. Goldberg, 1976. *J Neurophys* 39: 970–1008.)

<table>
<tr><td>CONCEPT
11.3</td><td>**The Semicircular Canals Sense Rotations of the Head in Three Dimensions**</td></tr>
</table>

LEARNING OBJECTIVES

11.3.1 Explain how the cupula stimulates semicircular canal hair cells during rotation of the head.

11.3.2 Compare how the otoliths and semicircular canals decompose direction of motion.

11.3.3 Discuss the temporal dynamics of neural signals originating in the semicircular canals and how they depend on the velocity and acceleration of rotational movements.

11.3.4 Summarize the neural pathways by which vestibular signals travel to the brain.

How the semicircular canals sense dynamic head rotations

The semicircular canals are optimized for sensing rotational movements of the head. Each of the three semicircular canals has at its base a bulbous expansion—the ampulla—that houses the sensory epithelium, or **crista**, that contains the hair cells (Figure 11.7). The structure of

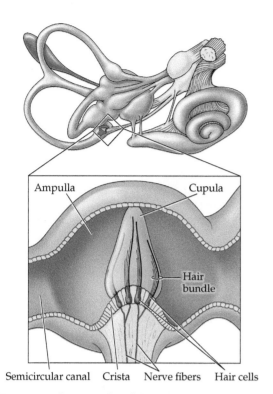

FIGURE 11.7   The ampulla of the posterior semicircular canal The crista, hair bundles, and cupula are diagrammed. When the head rotates, fluid in the membranous canal distorts the cupula.

the canals suggests how they detect the angular accelerations that arise through rotation of the head. The hair bundles extend out of the crista into a gelatinous mass, the **cupula**, that bridges the width of the ampulla, forming a viscous barrier through which endolymph cannot circulate. As a result, movements of the endolymphatic fluid distort the relatively compliant cupula. When the head turns in the plane of one of the semicircular canals, the inertia of the endolymph produces a force across the cupula, distending it away from the direction of head movement and causing a displacement of the hair bundles within the crista (Figure 11.8A,B). Note that semircircular canals can be excited by rotations that occur during the initiation of a head tilt, overlapping in function with the otoliths. In contrast, translational movements of the head produce equal forces on the two sides of the cupula, so the hair bundles within the ampulla are not displaced.

Unlike the saccular and utricular maculae, all of the hair cells in the crista within each semicircular canal are organized with their hair bundles oriented in the same direction (see Figure 11.2). Thus, when the cupula moves in the appropriate direction, the entire population of hair cells is depolarized and activity in all of the innervating axons increases. When the cupula moves in the opposite direction, the population is hyperpolarized and neuronal activity decreases. Deflections orthogonal to the excitatory–inhibitory direction produce little or no response.

Each semicircular canal works in concert with the partner located on the other side of the head that has its hair cells aligned oppositely. There are three such pairs: the two (right and left) horizontal canals, and the anterior canal on each side working with the posterior canal on the other side (Figure 11.8C). Head rotation deforms the cupula in opposing directions for the two partners, resulting in opposite changes in their firing rates. Thus, the orientation of the horizontal canals makes them selectively sensitive to rotation in the horizontal plane. More specifically, the hair cells in the canal toward which the head is turning are depolarized, while those on the other side are hyperpolarized.

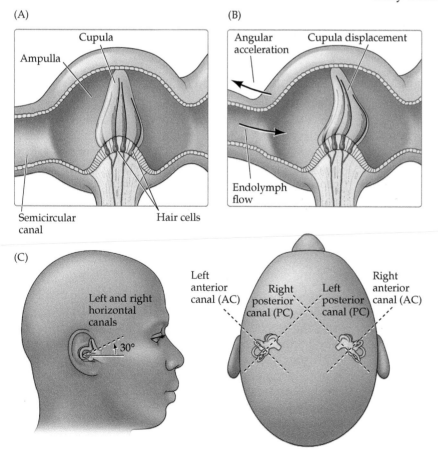

FIGURE 11.8   Functional organization of the semicircular canals (A) The position of the cupula without angular acceleration. (B) Distortion of the cupula during angular acceleration. When the head is rotated in the plane of the canal (arrow outside canal), the inertia of the endolymph creates a force (arrow inside canal) that displaces the cupula. (C) Arrangement of the canals in pairs. The two horizontal canals form a pair; the right anterior canal (AC) and the left posterior canal (PC) form a pair; and the left AC and the right PC form a pair.

For example, when the head rotates to the left, the cupula is pushed toward the kinocilium (tallest cilium) in the left horizontal canal, and the firing rate of the relevant axons in the left vestibular nerve increases. In contrast, the cupula in the right horizontal canal is pushed away from the kinocilium, with a concomitant decrease in the firing rate of the related neurons. If the head rotation is to the right, the result is just the opposite. This push–pull arrangement operates for all three pairs of canals; the pair whose activity is modulated is in the plane of the rotation, and the member of the pair whose activity is increased is on the side toward which the head is turning. The net result is a system that provides information about the rotation of the head in any direction.

Signaling of head rotation by semicircular canal neurons

Like axons that innervate the otolith organs, the vestibular fibers that innervate the semicircular canals exhibit a high level of spontaneous activity. As a result, they can transmit information by either increasing or decreasing their firing rate, thus more effectively encoding rotational head movements (see the previous section). The bidirectional responses of fibers innervating the hair cells of the semicircular canal have been studied by recording the axonal firing rates in a monkey's vestibular nerve. Seated in a chair, the monkey was rotated continuously in one direction during three phases: an initial period of acceleration, then a period of several seconds at constant angular velocity, and finally a period of sudden deceleration to a stop (Figure 11.9). The maximum firing rate observed corresponds to the period of acceleration, when the cupula is deflected; the minimum firing rate corresponds to the period of deceleration, when the cupula is deflected in the opposite direction. During the constant angular velocity phase, firing rates return to a baseline level as the cupula returns to its undeflected state over a time course that is related to the cupular elasticity and the viscosity of the endolymph (about 15 seconds). Note that the time it takes the cupula to return to its undistorted state (and for the hair bundles to return to their undeflected position) can occur while the head is still turning, as long as a constant angular velocity is maintained. Such constant forces are rarely found in nature, although they are encountered aboard ships, airplanes, space vehicles, and amusement park rides, where prolonged acceleratory arcs can occur.

An interesting aspect of the cupula–endolymph system dynamic is that it "smooths" the transduction of head accelerations into neural signals. For example, when the head is angularly accelerated to a constant velocity rather rapidly (corresponding to high-frequency rotational movements of the head), vestibular units associated with the affected canal generate a velocity signal; note that the axon firing rate in Figure 11.9 rises linearly during the

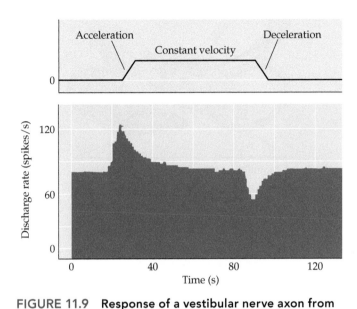

FIGURE 11.9   Response of a vestibular nerve axon from the semicircular canal in a monkey to angular acceleration The stimulus (top) is a rotation that first accelerates, then maintains constant velocity, and then decelerates the head. The stimulus-evoked change in firing rate of this vestibular unit (bottom) reflects the fact that the endolymph has viscosity and inertia and that the cupola has elasticity. Thus, during the initial acceleration, the deflection of the cupula causes the unit activity to rapidly increase. During constant angular velocity, the cupula returns to its undeflected state over a time course related to its elasticity and the viscosity of the fluid, and the unit activity returns to the baseline rate. During deceleration, the cupula is deflected in the opposite direction, causing a transient decrease in the unit firing rate. This behavior can be thought of as the cupula–endolymph system dynamic; the inertia of the fluid plays a minor role in this dynamic, coming into play only at very high frequencies of head movement. (After Fernández and Goldberg, 1976.)

acceleration phase. However, when the head is moving at a constant angular velocity (i.e., low-frequency rotational movements), the rate decays to the spontaneous level (corresponding to an acceleration of zero). This transduction process results in a velocity signal at high frequencies and an acceleration signal at low frequencies, a behavior that can be seen clearly in response to sinusoidal stimuli applied over a wide frequency range.

Pathways for transmission of vestibular signals to the brain

The vestibular end organs communicate, via the vestibular branch of cranial nerve VIII, with targets in the brainstem and the cerebellum that process much of the information necessary to compute head position and motion. As with the cochlear nerve, the vestibular nerves arise from a population of bipolar neurons, the cell bodies of which in this instance reside in the **vestibular nerve ganglion** (also called **Scarpa's ganglion**; see Figure 11.1). The distal processes of these cells innervate the semicircular canals and the otolith organs, while the central processes project via the vestibular portion of the **vestibulocochlear nerve** (cranial nerve VIII) to the **vestibular nuclei** (and also directly to the cerebellum). Although the canal and otolith afferents are largely segregated in the periphery, a large amount of canal–otolith convergence is found in the vestibular nuclei, a feature that ultimately enables the unambiguous encoding of head orientation and motion through the environment. Indeed, although head tilts and translational movements of the head can similarly excite otolith organs, the semicircular canals are excited only by rotations that accompany head tilts and not by purely translational movements. Therefore, integration of information from the otolith organs and semicircular canals in the vestibular nuclei and cerebellum can be used to distinguish head tilts from translational head movements. The vestibular nuclei also integrate a broad range of vestibular and non-vestibular information, receiving input from the vestibular nuclei of the opposite side as well as from the cerebellum and the visual and somatosensory systems, topics we discuss further in the next two concepts.

CONCEPT
11.4

The Vestibular System Works with Visual Cues to Evaluate Self Motion and to Stabilize Eye Gaze

LEARNING OBJECTIVES

11.4.1 Explain what optic flow is and how it contributes to our sense of self-motion.

11.4.2 Explain why and how vestibular signals contribute to eye movement, notably the vestibulo-ocular reflex (VOR).

11.4.3 Describe the circumstances in which the gain of the VOR must be adjusted to ensure stability of visual inputs despite head movements.

Interactions between vestibular signals, vision, and eye movements

The vestibular system operates in tandem with the visual system in several key ways (Figure 11.10). First, vision contributes to our sense of self-motion. As we walk along a forest trail, for example, translational movements activate the vestibular system and also generate a characteristic visual motion pattern across the retina known as **optic flow** (see Figure 11.10A). If such flow patterns occur without self-motion, such as when a moviegoer watches a film in wide-screen format or when a train passenger sees an adjacent train start to leave the station, the resultant visual flow generates a strong sense of self-motion, a perceptual process known as **vection**. Optic flow normally helps the brain maintain a sense of self-motion during sustained translational movement, such as when a car is travelling

at a steady speed—a situation in which the otolith signals themselves would have returned to baseline firing. Anatomically, visual and vestibular signals first converge at the vestibular nuclei—the earliest point in central vestibular processing, reflecting the critical synergy between these two sensory systems.

A second key vision-related function of the vestibular system is to help stabilize images on the retina by triggering eye movements that precisely compensate for head movements. The **vestibulo-ocular reflex** (**VOR**) (see Figure 11.10B) in particular refers to eye movements that counter head movements, thus permitting the gaze to remain fixed on a particular point (Clinical Applications; see also Chapter 20). For example, activity in the left horizontal canal induced by leftward rotary acceleration of the head excites neurons in the left vestibular nucleus and results in compensatory eye movements to the right.

Figure 11.10D illustrates the circuits mediating this reflex. Vestibular nerve fibers originating in the left horizontal semicircular canal project to the medial and superior vestibular nuclei. Excitatory fibers from the medial vestibular nucleus cross to the contralateral abducens nucleus, which has two outputs. One of these is a motor pathway that causes the lateral rectus of the right eye to contract; the other is an excitatory projection that crosses the midline and ascends via the **medial longitudinal fasciculus** to the left oculomotor nucleus, where it activates neurons that cause the medial rectus of the left eye to contract. Finally, inhibitory neurons project from the medial vestibular nucleus to the left abducens nucleus, directly causing the motor drive on the lateral rectus of the left eye to decrease and also indirectly causing the right medial rectus to relax. The consequence of these several connections is that excitatory input from the horizontal canal on one side produces eye movements toward the opposite side. Therefore, turning the head to the left causes eye movements to the right. In a similar fashion, head turns in other planes activate other semicircular canals, causing other appropriate compensatory eye movements. Thus, the VOR also plays an important role in vertical gaze stabilization in response to the linear vertical head oscillations that accompany locomotion and in response to vertical angular accelerations of the head, as can occur when riding on a swing. Notably, voluntary movements to redirect gaze transiently diminish the VOR, preventing vestibular reflexes from interfering with goal-directed movements. In the clinic, caloric testing provides a useful way to activate the VOR without moving the head, and it is a valuable tool for diagnosing peripheral and central lesions to the vestibular system (see Clinical Applications).

Loss of the VOR can have severe consequences. An individual with vestibular damage finds it difficult or impossible to fixate on visual targets while the head is moving, a condition called **oscillopsia** ("bouncing vision"). If the damage is unilateral, the individual usually recovers the ability to fixate

(A) Optic flow

(B) Vestibulo-ocular reflex (VOR) stabilizes visual input

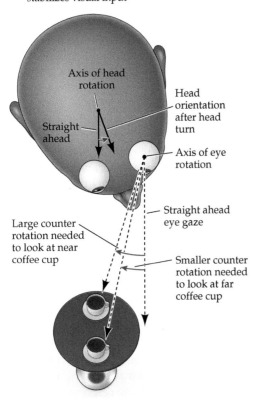

Axis of head rotation

Straight ahead

Head orientation after head turn

Axis of eye rotation

Straight ahead eye gaze

Large counter rotation needed to look at near coffee cup

Smaller counter rotation needed to look at far coffee cup

(C) Corrective lenses affect the VOR

(D) Neural pathways

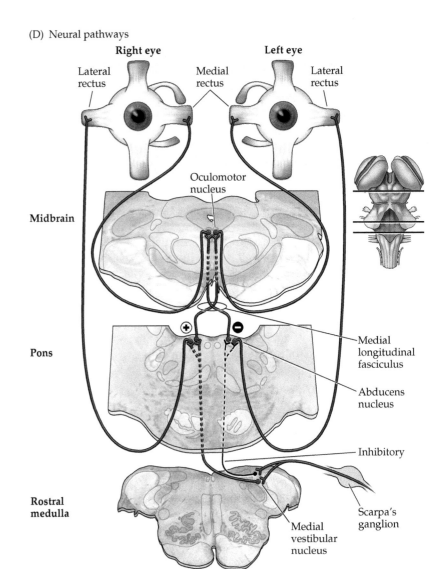

Right eye   **Left eye**

Lateral rectus   Medial rectus   Lateral rectus

Oculomotor nucleus

Midbrain

Pons

Rostral medulla

Medial longitudinal fasciculus

Abducens nucleus

Inhibitory

Scarpa's ganglion

Medial vestibular nucleus

FIGURE 11.10   Vestibular processes interact with vision and eye movements (A) Self-motion is detected in part by vision, based on optic flow, the large-scale patterns of motion of the visual scene that occur when one moves through space. (B) Eye movements help minimize visual motion on the retina as one moves. In particular, the vestibulo-ocular reflex (VOR) involves a counter-rotation of the eyes to compensate for head movement and stabilize visual input. The gain of the VOR has to change dynamically with viewing distance, because the eyes and head rotate around different axes. (C) The gain of the VOR also changes when one wears eyeglasses, which change the size of the visual scene. Shown here is a view of a newspaper through the lens of a pair of glasses for a person who is nearsighted. (D) The VOR is mediated by projections of the vestibular nucleus to the nuclei of cranial nerves III (oculomotor) and VI (abducens). The connections to the oculomotor nucleus and to the contralateral abducens nucleus are excitatory (red), whereas the connections to the ipsilateral abducens nucleus are inhibitory (black). There are connections from the oculomotor nucleus to the medial rectus of the left eye and from the abducens nucleus to the lateral rectus of the right eye. This circuit moves the eyes to the right—that is, in the direction away from the left horizontal canal—when the head rotates to the left. Turning to the right, which causes increased activity in the right horizontal canal, has the opposite effect on eye movements. The projections from the right vestibular nucleus are omitted for clarity. (A after J. J. Gibson, 1947. Motion picture testing and research, Report No 7, Army Air Force Aviation Psychology Program Research Reports. Washington, DC: US Government Printing Office; B after J. M. Groh, 2014. *Making Space: How the Brain Knows Where Things Are*. Cambridge, MA: Harvard University Press. © 2014 Jennifer M. Groh.)

■ Clinical Applications

Clinical Evaluation of the Vestibular System

The vestibulo-ocular reflex provides an important means of assessing the function of the vestibular, abducens, and oculomotor nerves and connections between their associated cell bodies in the brainstem. When the head is rotated in the horizontal plane, the vestibular afferents on the side toward the turning motion increase their firing rate, while the afferents on the opposite side decrease their firing rate (Figures A and B). The net difference in firing rates then leads to slow movements of the eyes counter to the turning motion; in a conscious person with normal vestibular function, a fast saccade (see Chapter 20) resets the eye position when the eye reaches its far excursion. This process is referred to as *physiological nystagmus*, which means

"nodding" or oscillatory movements of the eyes (see Figure B1). Physiological nystagmus is an adaptive process that enables the individual to fixate on a visual target despite ongoing rotational movements of the head and body.

Pathologic *spontaneous nystagmus* can occur if there is unilateral damage to the vestibular system. In this case, the silencing of output from the damaged side results in an abnormal difference in firing rate between the two sides (see Figure B2). This difference causes nystagmus even though no head movements are being made, often resulting in vertigo sufficiently severe to trigger falls and vomiting. Meniere's disease, named after the nineteenth-century French physician Prosper Meniere, is one cause of acute

unilateral vestibular dysfunction and can be particularly disabling for the individual; another is vestibular nerve section, which is often an unavoidable consequence of surgical removal of a vestibular Schwannoma (also known as an acoustical neuroma, a benign tumor of the nerve sheath around the vestibular and acoustical nerves). Interestingly, spontaneous nystagmus slowly diminishes following unilateral vestibular nerve section, presumably because of central compensation and plasticity. In contrast, Meniere's disease is typically progressive, initially affecting only one ear but slowly encompassing both ears, and is also usually accompanied by tinnitus and a sensorineural hearing deficit in the affected ear(s).

(*Continued*)

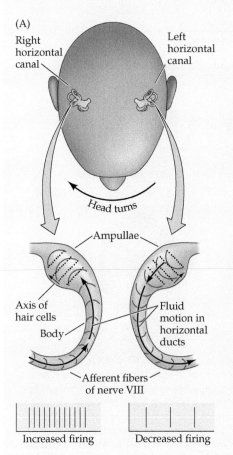

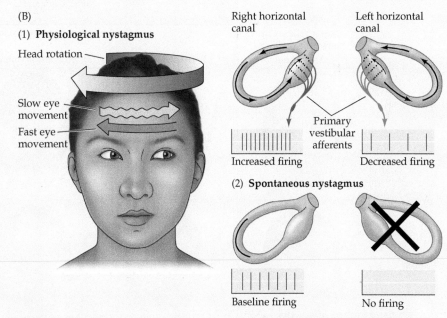

(A) View looking down on the top of a person's head illustrates the fluid motion generated in the left and right horizontal canals and the changes in vestibular nerve firing rates when the head turns to the right. (B) In typical individuals, rotating the head elicits physiological nystagmus (1), which consists of a slow eye movement counter to the direction of head turning. The slow component of the eye movements is due to the net differences in left and right vestibular nerve firing rates acting via the central circuit diagrammed in Figure 11.10D. Spontaneous nystagmus (2), where the eyes move rhythmically from side to side in the absence of any head movements, occurs when one of the canals or nerves is damaged. In this situation, net differences in vestibular nerve firing rates exist even when the head is stationary because the vestibular nerve innervating the intact canal fires steadily at rest, in contrast to a lack of activity on the damaged side.

■ Clinical Applications (continued)

Responses to vestibular stimulation are also clinically important because they can be useful in assessing the integrity of the brainstem in unconscious patients. If the individual is placed on his or her back and the head is elevated to about 30° above horizontal, the horizontal canals lie in an almost vertical orientation. Irrigating one ear with cold water will then lead to spontaneous eye movements because convection currents in the canal and direct cooling of the nerve mimic rotational head movements away from the irrigated ear (Figure C). In typical individuals, these eye movements consist of a slow movement toward the irrigated ear and a fast movement away from it. In patients who are comatose because of dysfunction of both cerebral hemispheres but whose brainstem is intact, saccadic movements are no longer made, and the response to cold water consists of only the slow

movement component of the eyes toward the side of the irrigated ear (Figure D). In the presence of brainstem lesions involving either the vestibular nuclei themselves, the connections from the

vestibular nuclei to oculomotor nuclei (cranial nerves III, IV, or VI), or the peripheral nerves exiting these nuclei, vestibular responses are abolished or altered, depending on the severity of the lesion.

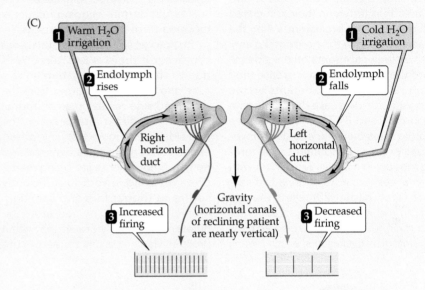

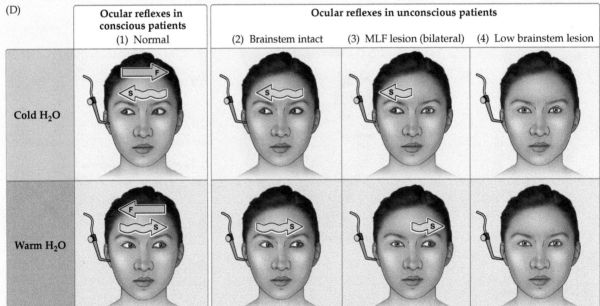

(C) Caloric testing of vestibular function is possible because irrigating an ear with water slightly warmer than body temperature generates convection currents in the canal that mimic the endolymph movement induced by turning the head to the irrigated side. Irrigation with cold water induces the opposite effect. These currents result in changes in the firing rate of the associated vestibular nerve, with an increased rate on the warmed side and a decreased rate on the chilled side. As in head rotation and spontaneous nystagmus, net differences

in firing rates generate eye movements. (D) Caloric testing can be used to test the function of the brainstem in an unconscious patient. The figures show eye movements resulting from cold or warm water irrigation in one ear for (1) a typical individual, and in three different conditions in an unconscious patient: (2) with the brainstem intact; (3) with a lesion of the medial longitudinal fasciculus (MLF; note that irrigation in this case results in lateral movement of the eye only on the less active side); and (4) with a low brainstem lesion (see Figure 11.10D).

objects during head movements. However, an individual with bilateral loss of vestibular function has the persistent and disturbing sense that the world is moving when the head moves. The underlying problem in such cases is that information about head movements normally generated by the vestibular organs is not available to the oculomotor centers, so that compensatory eye movements cannot be made.

In people who wear corrective lenses, the ideal gain of the VOR depends on the lens prescription (see Figure 11.10C). Corrective lenses can either magnify or minimize the apparent size of the visual scene, in turn increasing or decreasing the size of the eye rotation that is needed to compensate for a given amount of head rotation. Fortunately, the gain of the VOR is plastic, and individuals learn to compensate for these effects over a period of time. VOR plasticity also comes into play during development, when the rapid growth of the head changes the transform between head and eye movements.

Because the eyes and head rotate around different axes, the gain of the VOR is also adjusted on a moment-by-moment basis depending on whether one is looking at a nearby or more distant object (see Figure 11.10B). Turning one's head while looking at one's coffee cup requires a larger counter-rotation of the eyes than when gazing at a distant mountain range.

| CONCEPT **11.5** | The Brain Synthesizes Information to Support Perception of Translational and Rotational Movements and Maintain Equilibrium |

LEARNING OBJECTIVES

11.5.1 Describe the neural pathways involved in postural reflexes for balance.

11.5.2 Discuss the close connection between vestibular and cerebellar pathways and their role in synthesizing the different aspects of vestibular sensing into a coherent whole.

11.5.3 Describe the pathways by which vestibular signals reach the thalamus and cortex and the role of these regions in perception of self-motion.

11.5.4 Recognize that perception of self-motion requires synthesis and interpretation of the physical cues detected by the peripheral organs of the vestibular system.

Contributions to posture and equilibrium

The vestibular system also participates in reflexes responsible for maintaining posture and equilibrium during body movement. Descending projections from the vestibular nuclei are essential for postural adjustments of the head, mediated by the vestibulocervical reflex (VCR), and body, mediated by the vestibulospinal reflex (VSR). As with the VOR,

these postural reflexes are extremely fast, in part because of the small number of synapses interposed between the vestibular organ and the relevant motor neurons (Box 11A). Like the VOR, the VCR and the VSR are both compromised in people with bilateral damage to the vestibular periphery. Such individuals exhibit diminished head and postural stability, resulting in gait deviations; they also have difficulty balancing. These balance defects become more pronounced in low light or while walking on uneven surfaces, providing confirmation that balance normally is the product of vestibular, visual, and proprioceptive inputs.

The anatomical substrate for the VCR involves the medial vestibular nucleus; axons from this nucleus descend in the medial longitudinal fasciculus to reach the upper cervical levels of the spinal cord (Figure 11.11). This pathway regulates head position by reflex activity of neck muscles in response to stimulation of the semicircular canals caused by rotations of the head. For example, during a downward pitch of the body (e.g., tripping), the superior canals are activated and the head muscles reflexively pull the head up. The dorsal flexion of the head initiates other reflexes, such as forelimb extension and hindlimb flexion, to stabilize the body and protect against a fall (see Chapter 17).

The VSR is mediated by a combination of pathways, including the **lateral** and **medial vestibulospinal tracts** and the reticulospinal tract. The inputs from the otolith organs project mainly to the lateral vestibular nucleus, which in turn sends axons in the lateral vestibulospinal tract to the ipsilateral ventral horn of the spinal cord (see Figure 11.11). These axons terminate monosynaptically on extensor motor neurons, and they disynaptically inhibit flexor motor neurons; the net result is a powerful excitatory influence on the extensor (antigravity) muscles. When hair cells in the otolith organs are activated, signals reach the medial part of the ventral horn. By activating the ipsilateral pool of motor neurons innervating extensor muscles in the trunk and limbs, this pathway mediates balance and the maintenance of upright posture.

Decerebrate rigidity, characterized by rigid extension of the limbs, arises when the brainstem is transected above the level of the vestibular nucleus. Decerebrate rigidity in experimental animals is relieved when the vestibular nuclei are lesioned, underscoring the importance of the vestibular system to the maintenance of muscle tone. The tonic activation of extensor muscles in decerebrate rigidity suggests further that descending projections from higher levels of the brain, especially the cerebral cortex, normally suppress the vestibulospinal pathway (see also Chapter 17).

Vestibular–cerebellar pathways

The cerebellum is a major target of ascending vestibular pathways and also provides descending input to the vestibular nuclei, resulting in a recurrent circuit architecture that plays an important role in modulating vestibular activity. These vestibular–cerebellar circuits play a critical role in integrating and modulating vestibular signals to enable adaptive changes to the VOR, distinguish head tilts from translational movements, and distinguish passive movements of the head and body from those that are self-generated. Major vestibular targets in the cerebellum include the flocculus, paraflocculus, nodulus, uvula, and rostral fastigial nucleus, all of which play distinct roles in vestibular plasticity and multimodal integration.

A previously mentioned, integration of signals from the otolith organs and semicircular canals is needed to disambiguate head tilts from purely translational movements. Recordings made in the nodulus and uvula reveal that individual Purkinje cells integrate signals from these two vestibular sources to unambiguously encode either head tilts or translational movements, suggesting that the nodulus and uvula are critical sites of computation for making this distinction. Another major function of vestibular–cerebellar circuitry is helping distinguish vestibular signals that arise from self-generated motion from those that are triggered by external forces. An interesting feature of neurons in the rostral fastigial nucleus is that they do not respond to self-generated head or body movements, despite receiving both vestibular and proprioceptive signals from the head and body. An influential idea to account for this specialized responsiveness is that predictive signals generated in the cerebellum cancel out vestibular and proprioceptive signals in rostral fastigial neurons generated from self-motion, helping distinguish active from passive movements of the head.

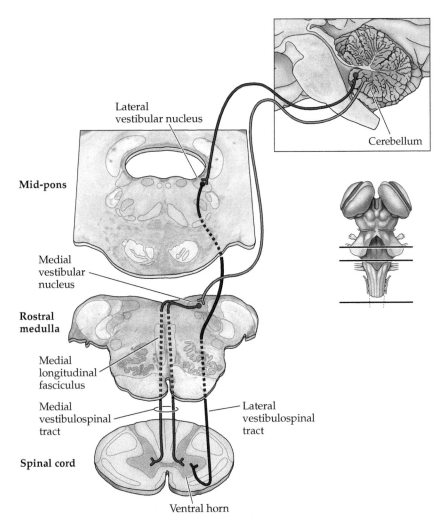

FIGURE 11.11   Descending projections from the medial and lateral vestibular nuclei to the spinal cord underlie the VCR and VSR The medial vestibular nuclei project bilaterally in the medial longitudinal fasciculus to reach the medial part of the ventral horns and mediate head reflexes in response to activation of semicircular canals. The lateral vestibular nucleus sends axons via the lateral vestibular tract to contact ventral horn cells innervating the axial and proximal limb muscles. Neurons in the lateral vestibular nucleus receive input from the cerebellum, allowing the cerebellum to influence posture and equilibrium.

■ BOX 11A | Mauthner Cells in Fish

A primary function of the vestibular system is to provide information about the direction and speed of ongoing movements, ultimately enabling rapid, coordinated reflexes to compensate for both self-induced and externally generated forces. One of the most impressive and speediest vestibular-mediated reflexes is the tail-flip escape behavior of fish (and larval amphibians), a stereotyped response that allows a potential prey to elude its predators (Figure A) (tap on the side of a fish tank if you want to observe the reflex). In response to a perceived risk, fish flick their tail and are thus propelled laterally away from the approaching threat.

The circuitry underlying the tail-flip escape reflex includes a pair of giant medullary neurons called Mauthner cells, their vestibular inputs, and the spinal cord motor neurons to which the Mauthner cells project. (Most fish possess one pair of Mauthner cells in a stereotypical location. Thus, these cells can be consistently visualized and studied from animal to animal.) Movements in the water, such as might be caused by an approaching predator,

■ BOX 11A | Mauthner Cells in Fish (continued)

excite saccular hair cells in the vestibular labyrinth. These receptor potentials are transmitted via the central processes of vestibular ganglion cells in cranial nerve VIII to the two Mauthner cells in the brainstem. As in the vestibulospinal pathway in humans, the Mauthner cells project directly to spinal motor neurons. The small number of synapses intervening between the receptor cells and the motor neurons is one of the ways that this circuit has been optimized for speed by natural selection, an arrangement evident in humans as well. The large size of the Mauthner axons is another; the axons from these cells in a goldfish are about 50 μm in diameter.

The optimization for speed and direction in the escape reflex also is reflected in the synapses that vestibular nerve afferents make on each Mauthner cell (Figure B). These connections are electrical synapses that allow rapid and faithful transmission of the vestibular signal.

An appropriate direction for escape is promoted by two features: (1) each Mauthner cell projects only to contralateral motor neurons; and (2) a local network of bilaterally projecting interneurons inhibits activity in the Mauthner cell away from the side on which the vestibular activity originates. In this way, the Mauthner cell on one side

faithfully generates action potentials that command contractions of contralateral tail musculature, thus moving the fish out of the path of the oncoming predator. Conversely, the local inhibitory network silences the Mauthner cell on the opposite side during the response (Figure C).

The Mauthner cells in fish are analogous to the reticulospinal and vestibulospinal pathways that control balance, posture, and orienting movements in mammals. The equivalent behavioral responses in humans are evident in a friendly game of tag, or more serious escape endeavors.

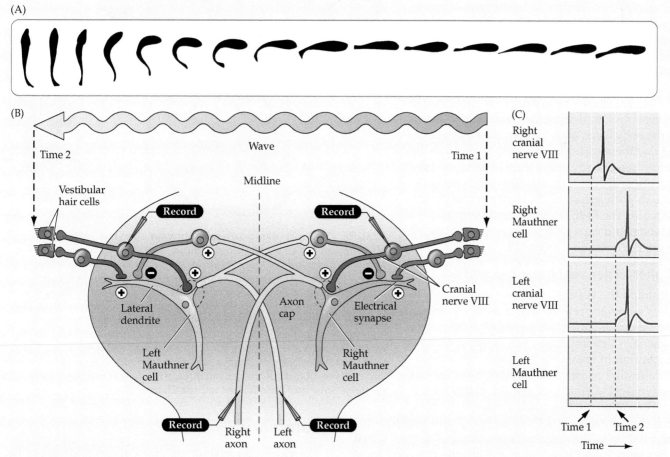

(A) Bird's-eye view of the sequential body orientations of a fish engaging in a tail-flip escape behavior, with time progressing from left to right. This behavior is mediated largely by vestibular inputs to Mauthner cells. (B) Diagram of synaptic events in the Mauthner cells of a fish in response to a disturbance in the water coming from the right. (C) Complementary responses of the right and left Mauthner cells mediating the escape response. Times 1 and 2 correspond to those indicated in Figure B. (A after R. C. Eaton et al., 1977. *J Exp Biol* 66: 65–81; B after E. J. Furshpan and T. Furukawa, 1962. *J Neurophysiol* 25: 732–771.)

Vestibular pathways to the thalamus and cortex

In addition to the several projections of the vestibular nuclei already mentioned, the superior and lateral vestibular nuclei send axons to the ventral posterior nuclear complex of the thalamus. This, in turn, projects to a variety of cortical areas relevant to the perceptions arising from the processing of vestibular information, including Brodmann's area 2v just posterior of the face representation in the somatosensory cortex, two regions in Brodmann's area 3a in the fundus of the central sulcus, and the parietoinsular vestibular cortex (PIVC), which may be especially important to our sense of self-motion and orientation in space (Figure 11.12). Indeed, in the 1950s Wilder Penfield found that electrically stimulating the PIVC could elicit strong vestibular sensations, and more modern imaging studies indicate that this region is activated by vestibular stimulation. Electrophysiological studies of individual neurons in these various cortical areas show that the relevant cells respond to proprioceptive and visual stimuli as well as to vestibular stimuli, reflecting the multisensory nature of central vestibular processing. Many of these cortical neurons are activated by moving visual stimuli as well as by rotation of the body (even with the eyes closed), suggesting that these cortical regions are involved in the perception of body orientation in extrapersonal space. Consistent with this interpretation, individuals with lesions of the right parietal cortex, including the PIVC, suffer altered perception of personal and extrapersonal space, as discussed in greater detail in Chapter 26.

Higher-order vestibular perception

Although the vestibular system contributes to many automatic reflexes and most of us remain unaware of its functioning unless it is damaged, it also plays an important role in our conscious perception of spatial orientation and self-motion. Vestibular percepts do not simply mirror the physical attributes of the associated stimulus. For example, a blindfolded individual riding on a chair rotating at a constant speed will perceive that the rotation is slowing and, after about 30 seconds, that it has stopped entirely. Interestingly, the time course of this perceptual decay is similar to, but more prolonged than, the decrement of the signal transmitted from the semicircular canal to the brain, suggesting that the brain is somehow compensating for the diminishing signal to generate a percept that more closely approximates the actual rotation.

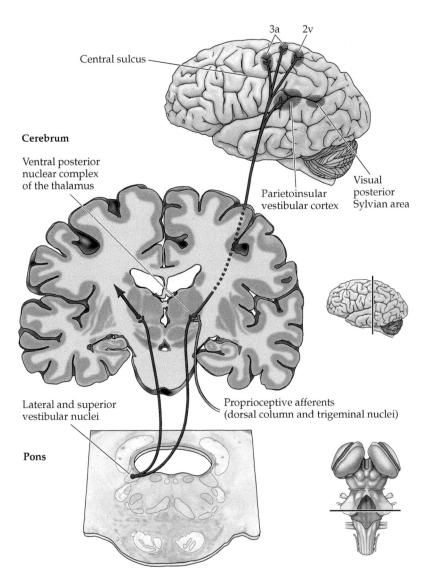

FIGURE 11.12 Thalamocortical pathways carrying vestibular information Unlike somatosensory, visual, and auditory systems, there is no single, canonical "primary vestibular cortex." Rather, there is a "vestibular cortical system" involving a distributed set of cortical areas in the parietal and posterior insular regions (purple ovals). Each of these areas contains neurons that are modulated by vestibular signals, interconnected across areas, and give rise to subcortical connections with the vestibular nuclear complex of the brainstem. Of particular importance is the parietoinsular vestibular cortex, which integrates multimodal proprioceptive signals and generates a "head-in-space" frame for body orientation and motor control. Other areas contributing to this vestibular cortical system include the ventral premotor cortex and the cingulate motor area (not shown; see Chapter 17).

Summary

The vestibular system provides information about head motion and the orientation of the head with respect to gravity. The sensory receptor cells are located in the otolith organs and the semicircular canals of the inner ear. The otolith organs provide information for ocular reflexes and postural adjustments when the head tilts in various directions or undergoes translational movements, and for the perception of these tilts and translations. The semicircular canals, in contrast, provide information about head rotations; these stimuli initiate reflex movements that adjust the eyes, head, and body during motor activities. Among the best studied of these reflexes are eye movements that compensate for head movements, thereby stabilizing the visual scene when the head moves. Information from the vestibular system also plays a central role in our perception of spatial orientation and our ability to navigate through the environment. Vestibular processing is inherently multisensory: Input from all the vestibular organs is integrated with input from the visual and somatosensory systems to provide perceptions of body position and orientation in space.

■ Additional Reading

Reviews

Angelaki, D. E. and K. E. Cullen (2008) Vestibular system: The many facets of a multimodal sense. *Annu. Rev. Neurosci.* 31: 125–150.

Benson, A. (1982) The vestibular sensory system. In *The Senses,* H. B. Barlow and J. D. Mollon (Eds.). New York: Cambridge University Press, pp. 333–368.

Brandt, T. (1991) Man in motion: Historical and clinical aspects of vestibular function. A review. *Brain* 114: 2159–2174.

Cullen, K. E. (2011) The neural encoding of self-motion. *Curr. Opin. Neurobiol.* 21: 587–595.

Cullen, K. E. (2012) The vestibular system: Multimodal integration and encoding of self-motion for motor control. *Trends Neurosci.* 35: 185–196.

Cullen K. E. (2019) Vestibular processing during natural self-motion: Implications for perception and action. *Nat. Rev. Neurosci.* 20: 346–363.

Cullen, K. E. and J. X. Brooks (2015) Neural correlates of sensory prediction errors in monkeys: Evidence for internal models of voluntary self-motion in the cerebellum. *Cerebellum* 14: 31–34.

Eatock, R. A. and J. E. Songer (2011) Vestibular hair cells and afferents: Two channels for head motion signals. *Annu. Rev. Neurosci.* 34: 501–534.

Furman, J. M. and R. W. Baloh (1992) Otolith-ocular testing in human subjects. *Ann. N. Y. Acad. Sci.* 656: 431–451.

Goldberg, J. M. (2000) Afferent diversity and the organization of the central vestibular pathways. *Exp. Brain Res.* 130: 277–297.

Goldberg, J. M. and C. Fernandez (1984) The vestibular system. In *Handbook of Physiology. Section 1: The Nervous System, Volume III: Sensory Processes, Part II,* J. M. Brookhart, V. B. Mountcastle, I. Darian-Smith and S. R. Geiger (Eds.). Bethesda, MD: American Physiological Society, pp. 977–1022.

Green, A. M. and D. E. Angelaki (2010) Multisensory integration: Resolving sensory ambiguities to build novel representations. *Curr. Opin. Neurobiol.* 20: 353–360.

Hess, B. J. (2001) Vestibular signals in self-orientation and eye movement control. *News Physiol. Sci.* 16: 234–238.

Raphan, T. and B. Cohen (2002) The vestibulo-ocular reflex in three dimensions. *Exp. Brain Res.* 145: 1–27.

Important Original Papers

Angelaki, D. E., A. G. Shaikh, A. M. Green and J. D. Dickman (2004) Neurons compute internal models of the physical laws of motion. *Nature* 430: 560–564.

Brooks, J. X., J. Carriot and K. E. Cullen (2015) Learning to expect the unexpected: Rapid updating in primate cerebellum during voluntary self-motion. *Nat. Neurosci.* 18: 1310–1317.

Chen, A. and 4 others (2016) Evidence for a causal contribution of macaque vestibular, but not intraparietal, cortex to heading perception. *J. Neurosci.* 36: 3789–3798.

Cullen, K. E. and R.-H. Wei (2021) Differences in the structure and function of the vestibular efferent system among vertebrates. *Front. Neurosci.* 15: 679.

Fetsch, C. R., G. C. DeAngelis and D. Angelaki (2010) Visual–vestibular cue integration for heading perception: Applications of optimal cue integration theory. *Eur. J. Neurosci.* 31: 1721–1729.

Fetsch, C. R., A. H. Turner, G. C. DeAngelis and D. E. Angelaki (2009) Dynamic reweighting of visual and vestibular cues during self-motion perception. *J. Neurosci.* 29: 15601–15612.

Goldberg, J. M. and C. Fernandez (1971) Physiology of peripheral neurons innervating semicircular canals of the squirrel monkey, Parts 1, 2, 3. *J. Neurophysiol.* 34: 635–684.

Goldberg, J. M. and C. Fernandez (1976) Physiology of peripheral neurons innervating otolith organs of the squirrel monkey, Parts 1, 2, 3. *J. Neurophysiol.* 39: 970–1008.

Laurens, J., H. Meng and D. E. Angelaki (2013) Neural representation of orientation relative to gravity in the macaque cerebellum. *Neuron* 80: 1508–1518.

Lindeman, H. H. (1973) Anatomy of the otolith organs. *Adv. Otorhinolaryngol.* 20: 405–433.

Merfeld, D. M. (1995) Modeling the vestibular-ocular reflex of the squirrel monkey during eccentric rotation and roll tilt. *Exp. Brain. Res.* 106: 123–134.

Books

Baloh, R. W. (1998) *Dizziness, Hearing Loss, and Tinnitus.* Philadelphia, PA: F. A. Davis Company.

Baloh, R. W. and V. Honrubia (2001) *Clinical Neurophysiology of the Vestibular System,* 3rd Edition. New York: Oxford University Press.

Bronstein, A. 2013. *Oxford Textbook of Vertigo and Imbalance.* New York: Oxford University Press.

12

Touch and Proprioception

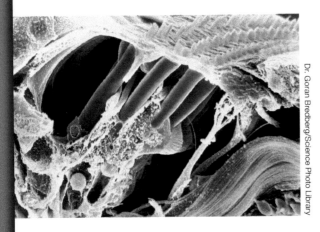

KEY CONCEPTS

12.1 Mechanical Forces on the Skin Are Conveyed to the CNS via an Array of Somatosensory Afferent Neurons

12.2 Somatosensory Afferent Neurons Form Specialized Endings and Interact with Other Mechanosensitive Cells in the Skin

12.3 Sensory Transduction from the Skin Involves Converting Mechanical Forces into Electrical Signals

12.4 Proprioception Involves Sensing Forces in Muscles, Joints, and Connective Tissue

12.5 A Variety of Neural Pathways Convey Different Aspects of Somatosensory Information to the Brain

12.6 Central Representations of the Body Are Plastic and Modified by Experience

12.7 Somatosensory Dysfunction Is an Important Factor in Multiple Diseases

Overview

The somatosensory system is arguably the most diverse of the sensory systems, mediating a range of sensations—touch, pressure, vibration, limb position, heat, cold, itch, and pain—that are transduced by receptors within the skin, muscles, or joints and conveyed to a variety of CNS targets. Not surprisingly, this complex neurobiological machinery can be divided into functionally distinct subsystems with different sets of peripheral neurons that possess complex end organs and transmit somatosensory information through multiple central pathways. One subsystem transmits information from cutaneous mechanoreceptors and mediates the sensations of fine touch, vibration, and pressure. Another originates in specialized receptors that are associated with muscles, tendons, and joints and is responsible for proprioception—our ability to sense the position of our own limbs and other body parts in space. A third subsystem arises from receptors that supply information about painful stimuli and changes in temperature as well as non-discriminative affective touch. This chapter focuses on the tactile and proprioceptive subsystems. The mechanisms responsible for sensations of pain and temperature are considered in Chapter 13.

CONCEPT **12.1**	**Mechanical Forces on the Skin Are Conveyed to the CNS via an Array of Somatosensory Afferent Neurons**

LEARNING OBJECTIVES

12.1.1 Describe the type of physical energy that skin detects.

12.1.2 Identify the two types of mammalian skin.

12.1.3 Describe the anatomical and morphological features of different classes of somatosensory afferent neurons.

12.1.4 Discuss the concept of a receptive field in the somatosensory system: how it varies with somatosensory afferent type and location on the body surface, and how its size relates to the spatial acuity of touch.

12.1.5 Describe the different temporal response properties of somatosensory afferent neurons.

How the skin detects mechanical forces

Sensory systems detect energy in the physical world. Our eyes detect light, our ears detect sound, and our skin detects mechanical forces that impinge on it. The detection of mechanical forces on the skin relies on the activation of sensory neurons that innervate the skin, muscles, and joints and then transmit this tactile information to the CNS. Sensory neuron endings are embedded within tissues where they form highly specialized endings and interact with other cells, including epithelial and glial cells. There are several different types of epithelial and glial cell types that, together with the sensory neurons that innervate them, are uniquely adapted for detecting different types of mechanical forces—such as strong pressure or the lightest of breezes—whether they are brief, sustained, or even rapidly fluctuating. Through various pathways, these cells report to the CNS what is happening on the body surface, where on the body it is occurring, and when.

To understand this process, we begin by describing two key components of it: the different types of skin, and the sensory afferent neural fibers that carry touch information to the brain.

Two types of skin

Mammalian skin is divided into two types: glabrous (hairless) skin, which is found predominantly on the palms of the hands and feet and on the lips, and hairy skin, which is found on the majority of the body. In humans, the term *hairy skin* refers not only to parts of the body with thick hair, such as the scalp, but also to areas that may appear to

have very little hair, such as the back of the hands. Each of these hairy skin regions is covered with hair follicles that help detect different aspects of tactile stimuli.

Each type of skin contains unique combinations of somatosensory afferent neurons with specialized mechanosensory end organs that permit distinctive functions for each body region. Glabrous skin is highly specialized for discriminative touch, to accurately assess the shape and texture of objects and provide feedback to the CNS regarding sensorimotor behaviors such as gripping and reaching. Hairy skin covers more than 90% of the body and transmits both discriminative touch and non-discriminative affective touch (touch that evokes an emotional response).

Somatosensory afferent neurons: Mechanoreceptors

Innocuous mechanical stimuli acting on the skin are detected by sensory neurons known as low-threshold mechanoreceptors (LTMRs). These somatosensory afferents (**mechanoreceptors**) are a heterogeneous group of neurons that play two roles: (1) They transduce the forces impinging on the skin into neural signals, and (2) they serve as the route by which such signals reach the CNS (Figure 12.1). Here we review the anatomy and morphology of these

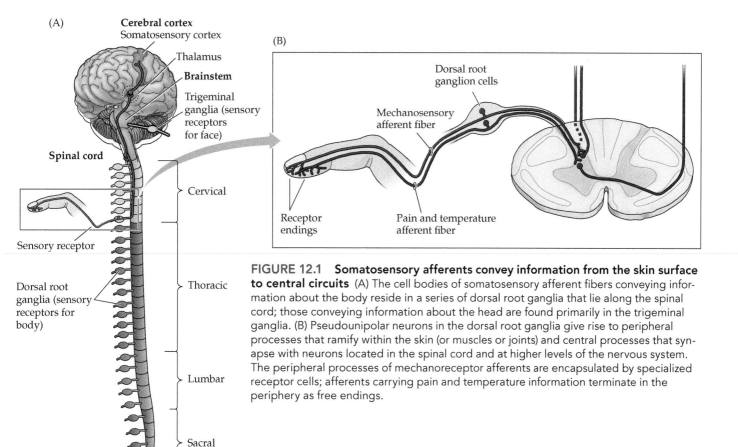

FIGURE 12.1 Somatosensory afferents convey information from the skin surface to central circuits (A) The cell bodies of somatosensory afferent fibers conveying information about the body reside in a series of dorsal root ganglia that lie along the spinal cord; those conveying information about the head are found primarily in the trigeminal ganglia. (B) Pseudounipolar neurons in the dorsal root ganglia give rise to peripheral processes that ramify within the skin (or muscles or joints) and central processes that synapse with neurons located in the spinal cord and at higher levels of the nervous system. The peripheral processes of mechanoreceptor afferents are encapsulated by specialized receptor cells; afferents carrying pain and temperature information terminate in the periphery as free endings.

neurons. An important aspect of neurological assessment may involve testing the functions of these different classes of mechanoreceptive afferents and noting geographically constrained zones, called **dermatomes**, that may present sensory loss in patients with nerve or spinal cord injury (Clinical Applications). The unique properties of the specialized epithelial and glial cell types that contribute to mechanosensation are described in Concept 12.2.

The cell bodies of afferent somatosensory neurons reside in a series of ganglia that lie alongside the spinal cord and

■ Clinical Applications

Dermatomes

Each dorsal root (sensory) ganglion and its associated spinal nerve arises from an iterated series of embryonic tissue masses called *somites* (see Chapter 22). This fact of development explains the overall segmental arrangement of somatic nerves and the targets they innervate in the adult. The territory innervated by each spinal nerve is called a *dermatome*. In humans, the cutaneous area of each dermatome has been defined in individuals in whom specific dorsal roots were affected (as in herpes zoster, or shingles) or after surgical interruption (for relief of pain or for other reasons). Such studies show that dermatomal maps vary among individuals. Moreover, dermatomes overlap substantially, so that injury to an individual dorsal root does not lead to complete loss of sensation in the relevant skin region. The overlap is more extensive for sensations of touch, pressure, and vibration than for pain and temperature. Thus, testing for pain sensation provides a more precise assessment of a segmental nerve injury than does testing for responses to touch, pressure, or vibration. The segmental distribution of proprioceptors, however, does not follow the dermatomal map but is more closely aligned with the pattern of muscle innervation. Despite these limitations, knowledge of dermatomes is essential in the clinical evaluation of neurological patients, particularly in determining the level of a spinal lesion.

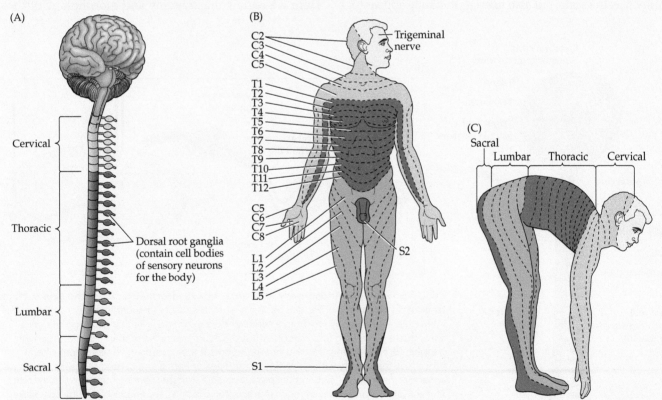

The innervation arising from a single dorsal root ganglion and its spinal nerve is called a dermatome. The full set of sensory dermatomes is shown here for a typical adult. Knowledge of this arrangement is particularly important in defining the location of suspected spinal (or other) lesions. The numbers refer to the spinal segments by which each nerve is named. (A after M. R. Rosenzweig et al., 2005. *Biological Psychology*, 3rd ed. Sunderland, MA: Sinauer Associates; B,C after W. Haymaker and B. Woodhall, 1967. *Peripheral Nerve Injuries: Principles of Diagnosis*. New York: American Association of Neurological Surgeons.)

the brainstem (the dorsal root ganglia and trigeminal ganglia, respectively; see Figure 12.1A). These neurons possess an unusual pseudounipolar structure, meaning they possess a single axonal branch that bifurcates and extends one fiber or process into peripheral tissues, while the other fiber branch innervates the CNS. Action potentials generated in afferent fibers by events that occur in the skin, muscles, or joints propagate along the peripheral fibers and past the locations of the cell bodies in the ganglia until they reach a variety of targets in the CNS (see Figure 12.1B).

Somatosensory afferents vary in their cell body size, conduction velocity, types of stimuli to which they respond, and response properties (adaptation) to stimulation. These differences, taken together, define distinct classes of afferents, each of which makes unique contributions to somatic sensation. Cell body size, axon diameter, and degree of myelination are among the factors that help differentiate classes of somatosensory afferents (Table 12.1). The largest-diameter (and most heavily myelinated) sensory afferents (designated Ia) are those that supply the sensory receptors in the muscles, discussed further in Concept 12.4. Most of the information subserving touch from the skin is conveyed by slightly smaller-diameter fibers (Aβ-LTMRs) that are also heavily myelinated and have large cell bodies. Information about pain, temperature, and some forms of light touch is conveyed by even smaller-diameter fibers (Aδ- and C-LTMRs, which are lightly myelinated or unmyelinated, respectively). The axon diameter and degree of myelination determine the action potential conduction speed and are well matched to the properties of the central circuits and the various behavioral demands for which each type of sensory afferent is employed (see Chapter 3).

Another distinguishing feature of cutaneous sensory afferents is the size of their **receptive field**—the area of the skin surface over which stimulation results in the generation of action potentials (Figure 12.2A). A given region of the body surface is served by sensory afferents that vary significantly in the size of their receptive fields. The size of the receptive field is, in part, governed by the branching characteristics of the afferent axon within the skin; smaller arborizations result in smaller receptive fields. Moreover, there are systematic regional variations in the average size of afferent receptive fields that largely reflect the density of afferent fibers supplying the area. The receptive fields in regions with dense innervation (fingers, lips, toes) are relatively small compared with those in the forearm or back that are innervated by a smaller number of afferent fibers (Figure 12.2B). Receptive fields of mechanosensory neurons often overlap, thus improving the accuracy of stimulus localization and making the somatosensory system less vulnerable to damage.

Regional differences in receptive field size and innervation density are among the major factors that limit the spatial accuracy with which tactile stimuli can be sensed. Thus, measures of **two-point discrimination**—the minimum interstimulus distance required to perceive two simultaneously applied stimuli as distinct—vary dramatically across the skin surface (Figure 12.2C). On the fingertips, stimuli (the indentation points produced by the tips of a caliper, for example) are perceived as distinct if they are separated by roughly 2 mm, but the same stimuli applied to the upper arm are not perceived as distinct until they are at least 40 mm apart.

Sensory afferents are further differentiated by the temporal dynamics of their response to sensory stimulation (Figure 12.3). **Slowly adapting afferents** fire continuously during a sustained stimulus. These neurons are well suited to provide information about the spatial attributes of the stimulus, such as size and shape. In contrast, **rapidly adapting afferents** respond at the onset (and sometimes the offset) of a stimulus. These afferents are thought to be particularly effective in conveying information about changes in ongoing stimulation such as those produced by stimulus movement. At least for some classes of afferent fibers, the adaptation characteristics are attributable to the properties of the receptor cells that encapsulate them. For example, rapidly adapting afferents that are associated with Pacinian corpuscles (see Concept 12.2) become slowly adapting when the corpuscle is removed.

Finally, sensory afferent types have unique sensitivities to somatosensory stimuli, because of differences in the properties of the channels they express as well as the properties of the specialized receptor cells that encapsulate their endings in peripheral tissues. For example, the afferents encapsulated within specialized receptor cells (see Concept 12.2) in the skin respond vigorously to mechanical deformation of the skin surface, but not to changes in temperature or to the presence of mechanical forces or chemicals that are known to elicit painful sensations. The latter stimuli are especially effective in driving the responses of sensory afferents known as *nociceptors* (see Chapter 13) that terminate in the skin as free nerve endings. Other subtypes of mechanoreceptors and nociceptors are identified on the basis of their distinct responses to somatic stimulation.

While a single sensory afferent can give rise to multiple peripheral branches, the transduction properties of all the branches of a single fiber are identical. As a result, classes of somatosensory afferents constitute **parallel pathways** that differ in conduction velocity, receptive field size, dynamics, and sensitivity to stimulus features. As will become apparent, multiple pathways transmit different aspects of somatosensory information through several stages of central processing, which is necessary for perceiving the complex features of somatosensory stimuli as well as for the appropriate control of both goal-oriented and reflexive movements.

(A)

(B)

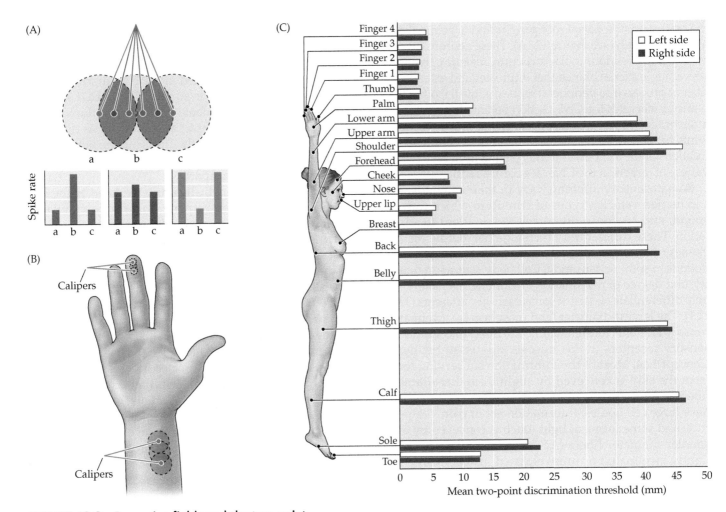

(C)

FIGURE 12.2 **Receptive fields and the two-point discrimination threshold** (A) Patterns of activity in three mechanosensory afferent fibers with overlapping receptive fields a, b, and c on the skin surface. When two-point discrimination stimuli are closely spaced (green dots and histogram), there is a single focus of neural activity, with afferent b firing most actively. As the stimuli are moved farther apart (red dots and histogram), the activity in afferents a and c increases and the activity in b decreases. At some separation distance (blue dots and histogram), the activity in a and c exceeds that in b to such an extent that two discrete foci of stimulation can be identified. This differential pattern of activity forms the basis for the two-point discrimination threshold. Stimulation applied to the center of the receptive field tends to evoke stronger responses than stimuli applied at more eccentric locations within the receptive field (see Figure 1.14). (B) The two-point discrimination threshold in the fingers is much finer than that in the wrist because of differences in the sizes of afferent receptive fields—that is, the separation distance necessary to produce two distinct foci of neural activity in the population of afferents innervating the lower arm is much greater than that for the afferents innervating the fingertips. (C) Differences in the two-point discrimination threshold across the surface of the body. Somatic acuity is much higher in the fingers, toes, and face than in the arms, legs, and torso. (C after S. Weinstein, 1968. Intensive and Extensive Aspects of Tactile Sensitivity as a Function of Body Part, Sex, and Laterality. In D.R. Kenshalo [Ed.], *The Skin Senses*. Springfield, IL: Charles C. Thomas, pp.195–222.)

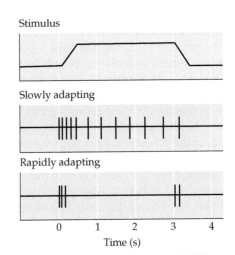

FIGURE 12.3 **Slowly and rapidly adapting mechanoreceptors provide different information** Slowly adapting receptors continue responding to a stimulus, whereas rapidly adapting receptors respond only at the onset (and sometimes the offset) of stimulation. These functional differences allow mechanoreceptors to provide information about both the static (via slowly adapting receptors) and dynamic (via rapidly adapting receptors) qualities of a stimulus.

Table 12.1   Properties of cutaneous mechanoreceptor afferents and their end organ structures

Physiological subtype[a]	Axon diameter	Conduction velocity	Skin type[b,c]	End organ/ ending type[b]	Location[b]	Optimal Stimulus[b]	Response properties[b]
Aβ SAI-LTMR	7–11 μm	16–96 m/s	Glabrous	Merkel cells	Basal layer of epidermis	Indentation	
			Hairy	Merkel cells (touch dome)	Around guard hair follicles		
Aβ SAII-LTMR[b]	6–12 μm	20–100 m/s	Glabrous	Ruffini?	Dermis?	Stretch	
			Hairy				
Aβ RAI-LTMR	6–12 μm	26–91 m/s	Glabrous	Meissner corpuscle	Dermal papillae	Hair deflection, skin stroke and indentation	
			Hairy	Longitudinal lanceolate ending	Guard and awl/ auchene hair follicles		
Aβ RAII-LTMR	6–12 μm	26–91 m/s	Glabrous	Pacinian corpuscle	Deep dermis and periosteum of certain bones	Indentation and vibration (high frequency)	
			Hairy				
Aβ Field-LTMR	6–12 μm	15–25 m/s	Hairy	Circumferential ending	Guard, awl/auchene and zigzag hair follicles	Skin stroke	
Aδ LTMR	1–5 μm	5–30 m/s	Hairy	Longitudinal lanceolate ending	Awl/auchene and zigzag hair follicles	Direction-selective hair deflection, skin stroke and indentation	
C-LTMR	0.2–1.5 μm	0.2–2 m/s	Hairy	Longitudinal lanceolate ending	Awl/auchene and zigzag hair follicles	Direction-selective hair deflection, skin stroke and indentation	
Aβ-HTMR/ Aδ-HTMR/ C-HTMR	0.2–1.5 μm	0.5–100 m/s	Glabrous	Free nerve ending	Epidermis	Noxious mechanical stimuli (high-force indentation and hair pull)	
			Hairy	Free nerve ending and circumferential ending	Epidermis and guard, awl/ auchene and zigzag hair follicles		

Source: Based on data in V. Abraira and D. D. Ginty, 2013. *Neuron* 79: 618–639 and A. Handler and D. D. Ginty, 2021. *Nat Rev Neurosci* 22: 521–537.

[a]The conduction velocities of LTMR and HTMR subtypes vary across species. Please see the following references: Cain et al. (2001), Schmidt et al. (1995), Leem et al. (1993), Brown and Iggo (1967), Burgess et al. (1968), Perl (1968), and Knibestol (1973).

[b]For reference on skin and end organ type for each mechanoreceptor subtype please see: Iggo and Muir (1969), Moll, Moll and Franke (1984), Pare, Smith and Rice (2002), Pare, Behets and Cornu (2003), Cuana and Ross (1960), Lishi and Ginty (2014), Bai, Lehnert et al. (2015), Pease and Quilliam (1957), Li et al. (2011), Halata (1993), and Ghitani et al. (2017).

[c]In addition to indentation and hair deflection, in the case of LTMRs innervating hairy skin, many LTMR subtypes are sensitive to vibratory stimuli in a frequency-dependent (Aβ RAI- and RAII-LTMR) and frequency-independent (Ab SAI-LTMR) manner. The frequency-dependence in force sensitivity of lanceolate endings is unknown. Please see the following references for further description of response properties: Zotterman (1939), Loewenstein and Mendelson (1965), Mountcastle et al. (1967), Bessou and Perl (1969), Iggo and Muir (1969), Chambers et al. (1972), Iggo and Ogawa (1976), Gottschaldt and Vahle-Hinz (1981), Li, Rutlin et al. (2011), Rutlin et al. (2014), Bai, Lehnert et al. (2015), and Ghitani et al. (2017).

CONCEPT
12.2
Somatosensory Afferent Neurons Form Specialized Endings and Interact with Other Mechanosensitive Cells in the Skin

LEARNING OBJECTIVES

12.2.1 Describe the four different types of mechanosensory cells found in glabrous skin, the types of tactile stimuli they detect, and the types of mechanoreceptor neurons with which they interact.

12.2.2 Describe the properties of mechanoreceptors in hairy skin.

Specialized cell types in the skin transduce unique aspects of touch information

Afferent fiber terminals that detect and transmit touch sensory stimuli possess complex endings that are often encapsulated by specialized receptor cells to help tune the afferent fiber to particular features of somatic stimulation. Afferent fibers that lack specialized receptor cells are referred to as **free nerve endings** and are especially important in the sensation of pain (see Chapter 13). Afferents that have encapsulated endings generally have lower thresholds for action potential generation and are thus more sensitive to sensory stimulation than are free nerve endings. In this concept we consider four distinct classes of mechanosensory end organs associated with specific somatosensory afferent types in the glabrous skin—Merkel cells, Meissner corpuscles, Pacinian corpuscles, and Ruffini corpuscles. Merkel cells, Pacinian corpuscles, and Ruffini corpuscles are also found in hairy skin, along with hair follicle receptors that are found only in hairy skin (Figure 12.4; also see Table 12.1).

Merkel cells are groups of specialized, oval-shaped cells. These end organs are especially enriched in the fingertips, where spatial acuity is highest, and are sparse in skin regions where spatial acuity is low, such as the calf. Merkel cells are anchored to other cells in the epidermis via cytoplasmic protrusions and adhesion proteins. These physical connections with other skin cells link compression or movement of the skin with mechanical changes in the Merkel cells. Both Merkel cells and the afferent neurons that innervate them contain ion channels in their membranes that are sensitive to mechanical forces and are collectively responsible for transducing forces into electrical signals for transmission to the brain (detailed in Concept 12.3).

Merkel cell afferents are slowly adapting (Aβ-SAI LTMRs) and thus tend to respond steadily to sustained indentation. Their fibers are heavily myelinated, and their cell bodies are comparatively large. Merkel cell afferents report tactile events with the highest spatial resolution of all the sensory

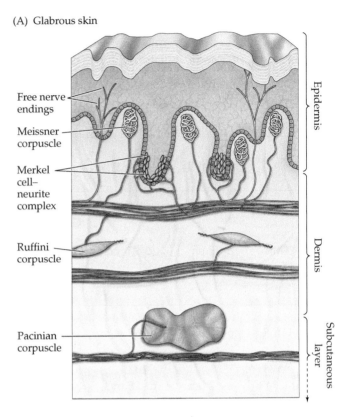

(A) Glabrous skin

Free nerve endings
Meissner corpuscle
Merkel cell–neurite complex
Ruffini corpuscle
Pacinian corpuscle

Epidermis
Dermis
Subcutaneous layer

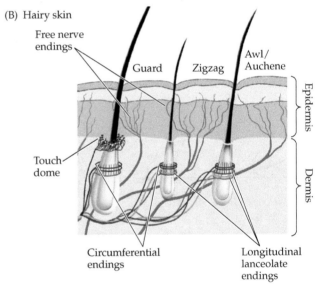

(B) Hairy skin

Free nerve endings
Guard
Zigzag
Awl/Auchene
Touch dome
Circumferential endings
Longitudinal lanceolate endings

Epidermis
Dermis

FIGURE 12.4 The skin harbors a variety of morphologically distinct mechanoreceptors (A) This diagram represents the smooth, hairless (glabrous) skin of the fingertip. (B) In hairy skin, tactile stimuli are transduced through a variety of mechanosensory afferents innervating different types of hair follicles. These arrangements are best known in mouse skin (illustrated here); see text for details. Similar mechanosensory afferents are believed to innervate hair follicles in human skin. Table 12.1 summarizes the major characteristics of the various mechanoreceptor types found in glabrous and hairy skin. (A after R. S. Johansson and A. B. Vallbo, 1983. *Trends Neurosci* 6: 27–32; B after V. E. Abraira and D. D. Ginty, 2013. *Neuron* 79: 618–639.)

afferents—individual Merkel afferents are densely packed within certain skin regions (e.g., the fingertips) and can therefore resolve spatial details of 0.5 mm. Overall, they represent about 25% of the mechanosensory innervation to the human hand. Merkel cell–Aβ-SAI LTMR complexes are also highly sensitive to points, edges, and curvature, which makes them ideally suited for processing information about shape and texture.

Meissner corpuscles are located within the tips of the dermal papillae close to the skin surface (see Figure 12.4A). Meissner corpuscles and the complexes they form with sensory afferents are even more densely packed than Merkel afferents, accounting for about 40% of the mechanosensory innervation of the human hand. Meissner corpuscles are formed by a connective tissue capsule that contains a set of flattened lamellar cells derived from Schwann cells, with the capsule and the lamellar cells suspended from the basal epidermis by collagen fibers. The center of the capsule contains two to six sensory afferents that terminate between and around the lamellar cells. These Aβ sensory afferents are similar in size and myelination to those associated with the Merkel cells, but they are rapidly adapting (Aβ-RAI LTMRs), such that they respond preferentially to *changes* in tactile stimulation. With indentation of the skin, the dynamic tension transduced by the collagen fibers provides the transient mechanical force that deforms the corpuscle and triggers generator potentials that may induce a volley of action potentials in the afferent fibers. When the stimulus is removed, the indented skin relaxes and the corpuscle returns to its resting configuration, generating another burst of action potentials. Thus, Meissner afferents display characteristic rapidly adapting, on–off responses (see Figure 12.3). At least in part because of their close proximity to the skin surface, Meissner afferents are more than four times as sensitive to skin deformation as Merkel afferents. However, because of the size of their peripheral arbors and differences in end-organ innervation patterns, the receptive fields of Meissner afferents are larger than those of Merkel afferents, and thus they transmit signals with reduced spatial resolution.

Meissner corpuscles are particularly efficient at transducing information about the relatively low-frequency vibrations (< 40 Hz) that occur when textured objects are moved across the skin. Several lines of evidence suggest that the information conveyed by Meissner afferents is responsible for detecting slippage between the skin and an object held in the hand, essential feedback information for the efficient control of grip.

Pacinian corpuscles are located deep in the dermis or in the subcutaneous tissue; their appearance resembles that of a small onion, with concentric layers of membranes surrounding a single afferent fiber (see Figure 12.4A). This laminar capsule acts as a high-pass filter that dampens low-frequency mechanical stimuli, allowing only transient disturbances at high frequencies (optimal activation ~200 Hz). Signals arising in Pacinian corpuscles are transmitted to the CNS via a class of rapidly adapting sensory afferent fibers that make up 10% to 15% of the mechanosensory innervation in the human hand (Aβ-RAII LTMRs). Pacinian corpuscles adapt more rapidly than Meissner corpuscles and have a lower response threshold for activation. The most sensitive Pacinian afferents generate action potentials for displacements of the skin as small as 10 nanometers. Because Pacinian afferents are so sensitive, their receptive fields are often large and their boundaries are difficult to define. The properties of Pacinian afferents make them well suited to detect vibrations transmitted through objects that contact the hand or are being grasped in the hand, especially when making or breaking contact. These properties are important for the skilled use of tools (e.g., using a wrench, cutting bread with a knife, writing).

Ruffini corpuscles are hypothesized to be innervated by slowly adapting fibers and are the least understood of the cutaneous mechanoreceptors. Ruffini endings are elongated, spindle-shaped, capsular specializations located deep in the skin, as well as in ligaments and tendons (see Figure 12.4A). The long axis of the corpuscle is usually oriented parallel to the stretch lines in skin. Although there is still some question as to their function, Ruffini corpuscles are thought to be particularly sensitive to skin stretches, such as those that occur during digit or limb movements. They account for about 20% of the mechanoreceptors in the human hand. Ruffini corpuscles are presumed to be innervated by a second class of slowly adapting Aβ-LTMRs (Aβ-SAII LTMRs). Information supplied by Ruffini afferents contributes, along with muscle receptors, to providing an accurate representation of finger position and the conformation of the hand (see Concept 12.4).

The different kinds of information that sensory afferents convey to central structures were first illustrated in experiments conducted by K. O. Johnson and colleagues, who compared the responses of different afferents as a fingertip was moved across a row of raised Braille letters (Figure 12.5). Clearly, all of the afferent types are activated by this stimulation, but the information supplied by each type varies enormously. The pattern of activity in the Merkel afferents is sufficient to recognize the details of the Braille pattern, and the Meissner afferents supply a slightly coarser version of this pattern. However, these details are lost in the responses of the Pacinian and Ruffini afferents, and presumably these responses have more to do with tracking the movement and position of the finger than with the specific identity of the Braille characters. The dominance of Merkel afferents in transducing textural information is probably because Braille letters are coarse. Human fingers are also exquisitely sensitive to fine textures. For example, we can easily distinguish silk from satin. The microgeometries of different fine textures

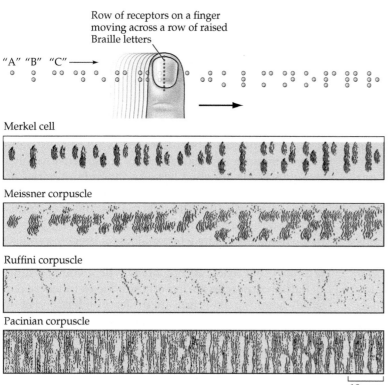

Row of receptors on a finger moving across a row of raised Braille letters

"A" "B" "C" →

Merkel cell

Meissner corpuscle

Ruffini corpuscle

Pacinian corpuscle

10 mm

FIGURE 12.5 Simulation of activity patterns in different mechanosensory afferents in the fingertip Each dot in the response records represents an action potential recorded from a single mechanosensory afferent fiber innervating the human finger as it moves across a row of Braille type. A horizontal line of dots in the raster plot represents the pattern of activity in the afferent as a result of moving the pattern from left to right across the finger. The position of the pattern (relative to the tip of the finger) was then displaced by a small distance, and the pattern was once again moved across the finger. Repeating this pattern multiple times produces a record that simulates the pattern of activity that would arise in a population of afferents whose receptive fields lie along a line in the fingertip (red dots). Only slowly adapting Merkel cell afferents (top panel) provide a high-fidelity representation of the Braille pattern—that is, the individual Braille dots can be distinguished only in the pattern of Merkel afferent neural activity. (After J. R. Phillips et al., 1990. *Exp Brain Res* 81: 589–592.)

produce different patterns of vibrations on the skin while the finger is scanning across the textured surface, which are best detected by the rapidly adapting afferents.

Mechanoreceptor endings in hairy skin

Ruffini corpuscles, Pacinian corpuscles, and their associated sensory afferents are also present in hairy skin. There are also several types of mechanoreceptive afferents that are found only in hairy skin and that innervate hair follicles found only in such skin (see Figure 12.4B). These include Merkel cell afferents (Aβ-SAI LTMRs) innervating **touch domes** (Merkel cell-neurite complexes) that contain dozens of Merkel cells and are associated with the apical collars of specific hair follicles (these structures are anatomically different than the Merkel cell–Aβ-SAI LTMRs

that innervate the glabrous skin; see Figure 12.4A). Other hair follicle receptors include the **circumferential endings** and **longitudinal lanceolate endings** that surround the base of different hair follicle types. The longitudinal lanceolate endings form a palisade around the follicle that is exquisitely sensitive to the deflection of the hair by stroking the skin or simply the movement of air over the skin surface. These longitudinal lanceolate endings are derived from Aβ-, Aδ-, or C-LTMRs, all of which form rapidly adapting LTMRs associated with the hairs. Circumferential endings around hair follicles are also sensitive to skin stroking (Aβ-field LTMRs). These responses of longitudinal lanceolate and circumferential endings should be distinguished from the responses of free nerve endings in the epidermis, which have different physiological properties and respond to painful stimuli at much higher activation thresholds than touch-sensitive receptors associated with hair follicles (see Chapter 13).

Interestingly, lanceolate endings appear to be important for mediating forms of non-discriminative touch, such as a gentle caress. A unique class of these fibers, C-LTMRs, are found exclusively in hairy skin and respond preferentially to gentle stroking of the skin. These neurons are proposed to mediate non-discriminative affective touch. Using microneurography techniques to record from single somatosensory afferents in humans, researchers identified that C-LTMRs preferentially respond to gentle stroking of the skin at very slow speeds (0.1–2 m/s), which was perceived by individuals as the most pleasant stroking speed. Individuals with reduced C-LTMR innervation of the hairy skin because of a genetic mutation reported that slow arm stroking, optimal for activating C-LTMRs, was less pleasant than reported by controls. These individuals also showed lower activation in insular cortex, a region of the brain implicated in social perception and social cognition, during skin stroking compared with controls. Thus, increasing evidence suggests that C-LTMRs promote non-discriminative affective touch in mammals. As C-LTMRs terminate in lamina II of the dorsal horn, there is also evidence that these neurons may be involved in the modulation of spinal responses to nociceptive stimuli (see Chapter 13). In addition to C-LTMRs, another class of C fibers that express Mas-related G-protein-coupled receptor B4 (MrgprB4) are also sensitive to stroking of hairy skin in rodents. Activation of MrgprB4-expressing neurons promotes conditioned place preference in mice, which suggests that activation of these neurons is positively reinforcing or reduces anxiety in rodents.

CONCEPT
12.3

Sensory Transduction from the Skin Involves Converting Mechanical Forces into Electrical Signals

LEARNING OBJECTIVES

12.3.1 Describe the process of sensory transduction in mechanoreceptors.

12.3.2 Describe the role of Piezo channels in sensory transduction.

Sensory transduction

The fundamental mechanism of **sensory transduction**—the process of converting the energy of a stimulus into an electrical signal—is similar in all somatosensory afferents: A stimulus alters the permeability of cation channels in the afferent nerve endings, generating a depolarizing current known as a **receptor** (or **generator**) **potential** (Figure 12.6). If sufficient in magnitude, the receptor potential reaches threshold for the generation of action potential(s) in the afferent fiber. In most cases, the resulting rate of action potential firing is roughly proportional to the magnitude of the depolarization, as described in Chapters 2 and 3. The sense of touch requires an applied force to be turned into an electrical signal in the endings of sensory

neurons throughout the body. The first family of mammalian mechanosensitive channels identified consists of two members: Piezo1 and Piezo2 (Greek *piesi*, "pressure"; see Chapter 4). Piezo channels exhibit a unique propeller-shaped architecture with dozens (24–38) of transmembrane domains. Piezo channels allow positive ions to flow across the surface membrane of cells in response to force applied to the plama membrane. Piezo ion channels are critical for several biological functions, including light touch, proprioception, and vascular blood flow.

Piezo channels can be expressed in both the specialized receptor cells and the sensory afferents that innervate them. For example, Merkel cells and their afferents (Aβ-SAI LTMRs) both express the mechanosensitive channel Piezo2. As a result, Merkel cells and their afferent axons can both directly sense mechanical stimuli. The expression and function of Piezo2 channels in the two cell types shape the temporal pattern of the afferent neuron responses to tactile stimuli. Removing Piezo2 selectively in Merkel cells significantly reduces the sustained or static firing of the innervating afferents. Thus, Merkel cells signal the static aspect of a touch stimulus, such as pressure, whereas the terminal portions of the Merkel afferents in these complexes transduce the dynamic aspects of stimuli (including stimuli with ~5 Hz or less vibration).

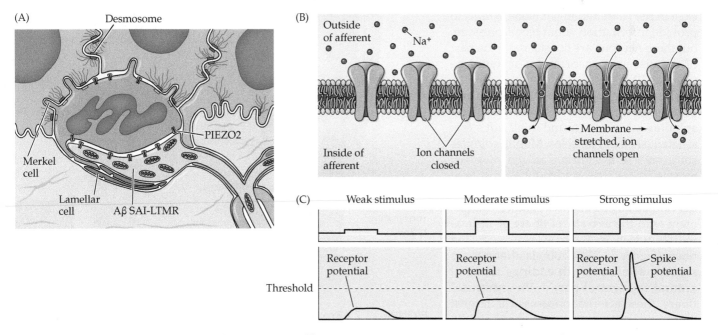

FIGURE 12.6 Transduction in a mechanosensory afferent The process is illustrated here for an Aβ-SAI neuron innervating a Merkel cell. (A) Deformation of the skin activates Piezo2 mechanosensitive ion channels in both Merkel cells and Aβ-SAI afferents. Desmosomes are intercellular junctions that provide strong adhesion between cells. (B) Opening of these cation channels leads to depolarization of the afferent fiber (receptor potential). (C) If the afferent is sufficiently depolarized, an action potential is generated and propagates to central targets. (A after A. Handler and D. D. Ginty, 2021. *Nat Rev Neurosci* 22: 521–537.)

| Proprioception Involves Sensing Forces in Muscles, Joints, and Connective Tissue

LEARNING OBJECTIVES

12.4.1 Explain what proprioception is and what it is used for.

12.4.2 Explain what muscle spindle receptors, Golgi tendon organs, and joint receptors are, and describe the types of stimuli they detect.

Mechanoreceptors for proprioception

While cutaneous mechanoreceptors provide information derived from external stimuli (exteroception), another major class of receptors provides information about mechanical forces arising within the body itself (interoception; see Chapter 13), particularly from the musculoskeletal system. The purpose of these proprioceptors ("receptors for self") is primarily to give detailed and continuous information about the position of the limbs and other body parts in space. Proprioceptors are mechanosensory neurons located within the muscles, tendons, and joints. Proprioception is essential for accurate performance of complex movements and postural control. Information about the position and motion of the head is integrated with the highly specialized vestibular system, which we considered in Chapter 11. Specialized proprioceptors also exist in the heart and major blood vessels to provide information about blood pressure, but these neurons are considered to be part of the visceral sensorimotor system (see Chapter 21).

The most detailed knowledge about proprioception derives from studies of **muscle spindles**, which are found in all but a few striated (skeletal) muscles. Muscle spindles consist of four to eight specialized **intrafusal muscle fibers** surrounded by a capsule of connective tissue. The intrafusal fibers are distributed among and in a parallel arrangement with the **extrafusal fibers** of skeletal muscle, which are the true force-producing fibers (Figure 12.7A). Group Ia afferents are sensory afferents with endings that coil around the central part of the intrafusal fibers; these afferents possess the largest myelinated sensory axons and have rapidly adapting responses to changes in muscle length as well as to the rate of change (velocity) of muscle length. When the muscle is stretched, the tension on the intrafusal fibers activates mechanically gated ion channels in

the Ia-fiber endings, triggering bursts of action potentials. These properties of group Ia afferents enable rapid detection of postural changes, even before large changes in position occur. At the edge of the spindles, and to the sides of the Ia fibers, are the sensory endings of a secondary group of proprioceptors called group II afferents. Group II afferents produce graded and sustained responses to constant muscle lengths. Thus, primary endings (Ia) are thought to transmit information about limb dynamics—the velocity and direction of movement—whereas secondary endings (II) provide information about the static position of limbs. Piezo2 is the major mechanosensitive ion channel expressed by group Ia and group II proprioceptors and is required for functional proprioception. Loss of Piezo2 in proprioceptive neurons causes abnormal limb positions and severely uncoordinated body movements in mice.

(A) Muscle spindle

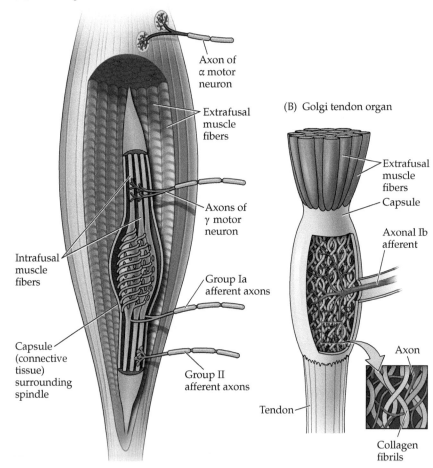

(B) Golgi tendon organ

FIGURE 12.7 Proprioceptors in the musculoskeletal system These "self-receptors" provide information about the position of the limbs and other body parts in space. (A) A muscle spindle and several extrafusal muscle fibers. The specialized intrafusal muscle fibers of the spindle are surrounded by a capsule of connective tissue. (B) Golgi tendon organs are LTMRs found in tendons; they provide information about changes in muscle tension. (A after P. B. C. Matthews, 1964. *Physiol Rev* 44: 219–289.)

Muscle spindles also receive innervation from efferent **gamma (γ) motor neurons**, whose cell bodies reside in the ventral horn of the spinal cord. Gamma motor neurons regulate the tension on the spindles, and in this way are able to modify the gain of the proprioceptor afferent responses. For a more detailed explanation of the interaction of γ motor neurons and the activity of spindle afferents, see Chapter 16.

The density of spindles in human muscles varies. Large muscles that generate coarse movements have relatively few spindles. In contrast, extraocular muscles and the muscles of the hand and neck are richly supplied with spindles, reflecting the importance of accurate eye movements, the need to manipulate objects with great finesse, and the continuous demand for precise positioning of the head. This relationship between receptor density and muscle size is consistent with the generalization that the sensorimotor apparatus at all levels of the nervous system is much richer for the hands, head, speech organs, and other parts of the body that are used to perform especially important and demanding tasks. Spindles are lacking altogether in a few muscles, such as those of the middle ear, that do not require the kind of feedback that these receptors provide.

Whereas muscle spindles are specialized to signal changes in muscle length, LTMRs that innervate tendons inform the CNS about changes in muscle tension. Each encapsulated tendon organ contains the sensory endings of a single mechanosensory neuron, called a **Golgi tendon organ**. These afferents are **group Ib afferents**, and they wrap around strands of collagen, which are attached to individual muscle fibers (Figure 12.7B). Golgi tendon organs are arranged in series with a small number (10–20) of extrafusal muscle fibers. Group Ib afferents encode muscle force: At rest, these neurons are silent, but they increase their firing frequency as tension in the muscle increases.

In addition to the mechanosensory neurons that innervate muscle spindles and tendons, there are proprioceptive neurons that innervate each of the joints. These **joint receptors** resemble LTMRs that innervate the skin, including those that innervate Ruffini endings and Pacinian corpuscles. These mechanoreceptors embedded in the joint are generally considered to be of three major types: type I, slowly adapting neurons in the outer layers of the joint capsule; type II, rapidly adapting neurons in the deeper layers of the joint capsule; and type III, slowly adapting neurons in the ligaments and terminal regions of the tendons close to the joint capsule. A fourth type (type IV) of nociceptive sensory neuron possesses free nerve endings distributed throughout the joint capsule. Type IV neurons have a higher mechanical threshold for firing and contribute to pain sensations. The firing responses of joint receptors are often highest at the extremes of joint position, suggesting that they play a protective role in signaling positions that lie near the limits of normal joint range of motion. How each of these proprioceptive afferents contributes to the perception of limb position, movement, and force remains an area of active investigation.

CONCEPT 12.5 | A Variety of Neural Pathways Convey Different Aspects of Somatosensory Information to the Brain

LEARNING OBJECTIVES

12.5.1 Identify the various neural pathways that convey touch information from the body and face to the brain.

12.5.2 Describe the somatotopic organization of touch stimuli present at various stages in the somatosensory circuits, and the over-representation of especially sensitive areas of the body surface in the CNS.

12.5.3 Describe the differences between the direct and indirect dorsal column pathways for processing touch stimuli.

12.5.4 Explain what first, second, and third order neurons are in the somatosensory pathways.

12.5.5 Explain how circuits in the spinal cord and descending input from somatosensory cortex modulate tactile information.

Spinal cord pathways conveying tactile information

As noted in Concept 12.1, sensory information from the trunk and limbs enters the spinal cord via dorsal root ganglia neuron axons. All LTMRs (*first order neurons*) thus converge onto the dorsal horn, where their inputs are arranged in a columnar manner that maintains a topographic, or *somatotopic*, organization. In the spinal cord there are multiple pathways through which various aspects of tactile information are transmitted to the brain, but there is also a high degree of integration and processing of LTMR inputs in the spinal cord.

The spinal cord is composed of a central core region of gray matter surrounded by white matter. The gray matter is subdivided into dorsal and ventral horns that include ten Rexed laminae (I–X) based on variations in cell density. (Rexed's laminae are the descriptive divisions of the spinal cord gray matter in cross section, named after the neuroanatomist who described these details in the 1950s; see Table A1 and Figure A7 in the Appendix.) The dorsal horn functions as an intermediary processing center for somatosensory information, comprising a complex network of excitatory and inhibitory interneurons as well as projection neurons that transmit the processed somatosensory information from the spinal cord to the brain (Figure 12.8A). In contrast, the ventral horn receives

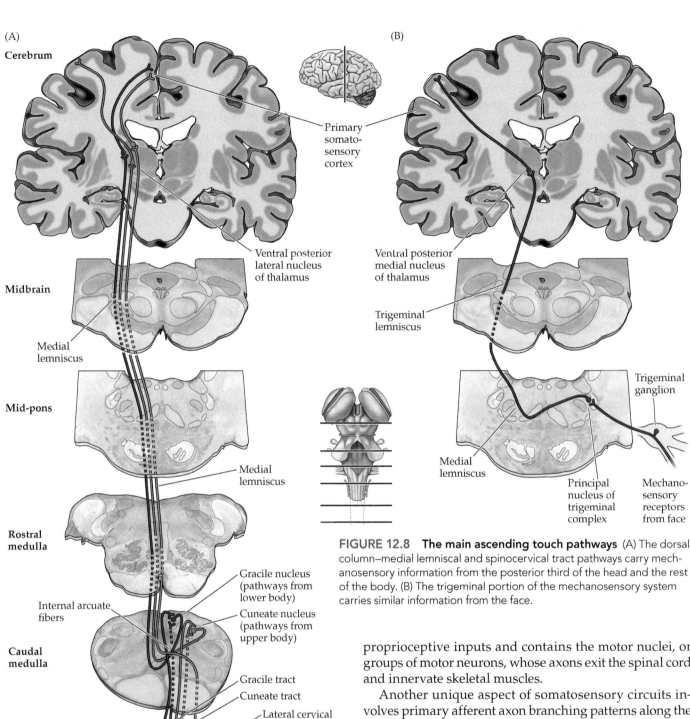

(A)

Cerebrum

Midbrain

Mid-pons

Rostral medulla

Caudal medulla

Cervical spinal cord

Lumbar spinal cord

Medial lemniscus

Internal arcuate fibers

Gracile nucleus (pathways from lower body)

Cuneate nucleus (pathways from upper body)

Gracile tract

Cuneate tract

Lateral cervical nucleus (LCN)

Mechanosensory receptors from upper body

Spinocervical tract (SCT)

Mechanosensory receptors from lower body

(B)

Primary somato-sensory cortex

Ventral posterior lateral nucleus of thalamus

Ventral posterior medial nucleus of thalamus

Trigeminal lemniscus

Trigeminal ganglion

Medial lemniscus

Principal nucleus of trigeminal complex

Mechano-sensory receptors from face

Medial lemniscus

FIGURE 12.8   The main ascending touch pathways (A) The dorsal column–medial lemniscal and spinocervical tract pathways carry mechanosensory information from the posterior third of the head and the rest of the body. (B) The trigeminal portion of the mechanosensory system carries similar information from the face.

proprioceptive inputs and contains the motor nuclei, or groups of motor neurons, whose axons exit the spinal cord and innervate skeletal muscles.

Another unique aspect of somatosensory circuits involves primary afferent axon branching patterns along the rostral–caudal axis of the spinal cord that differ according to the afferent fiber type. C and Aδ fibers enter the spinal cord through the dorsal roots and turn rostrally before they arborize in the dorsal horn. In contrast, Aβ fibers extend branches in both the rostral and caudal directions and can extend for many spinal cord segments (see Figure 12.9A. The role of these complex primary afferent arborizations is an area of active investigation and is hypothesized to contribute to precise localization of sensory stimuli. The input to the dorsal horn at a given level of the spinal cord arises from a particular region of the body, maintaining a somatotopic organization. However, the particular projection patterns vary among the different mechanosensory afferents. Many (but not all) Aβ-RA and Aβ-SA mechanosensory afferents also possess a "direct pathway" projection to the brainstem: These axons bifurcate upon exiting the dorsal root into the spinal cord, with a rostral branch that extends ipsilaterally through the dorsal columns (also called the **posterior funiculi**), while a second branch arborizes in the spinal cord dorsal horn, as described earlier. These fibers are topographically organized such that the fibers conveying information from lower limbs lie most medial and travel in a circumscribed bundle known as the **gracile tract**. Those fibers that convey information from the upper limbs, trunk, and neck lie in a more lateral bundle known as the **cuneate tract**. In turn, the fibers in these two tracts end in different subdivisions of the dorsal column nuclei: a medial subdivision, the nucleus gracilis or **gracile nucleus**; and a lateral subdivision, the nucleus cuneatus or **cuneate nucleus** (see Figure 12.8A).

Historically, much emphasis has been placed on the direct pathway for the propagation and processing of somatosensory information, while the spinal cord has previously been described as a "relay" in somatosensory circuits. However, recent studies reveal that the dorsal horn is a key step for integration and processing of mechanosensory information through a complex, modular architecture in which innocuous, nociceptive, and pruritic (itch) stimuli are processed by distinct, molecularly defined spinal cord cell types. In support of this new model, only a subset of mechanosensory afferents possess axonal branches that extend directly via the dorsal columns to the dorsal column nuclei. In contrast, all mechanosensory neurons (and nociceptors) possess axonal branches that make synaptic connections with interneurons and projection neurons in the dorsal horn. These spinal cord neuron types include excitatory neuron populations that transmit information about both innocuous and painful touch, as well as inhibitory neuron populations that serve as a gate to prevent innocuous stimuli from activating the nociceptive and pruritic transmission pathways. Cutaneous afferents project to and synapse with interneurons and projection neurons in the upper layers of the dorsal horn (laminae II$_{iv}$, III, IV, and V; Figure 12.9), whereas proprioceptive afferents synapse onto interneurons and motor neurons throughout the intermediate spinal cord and ventral horn.

(A) LTMR-processing units

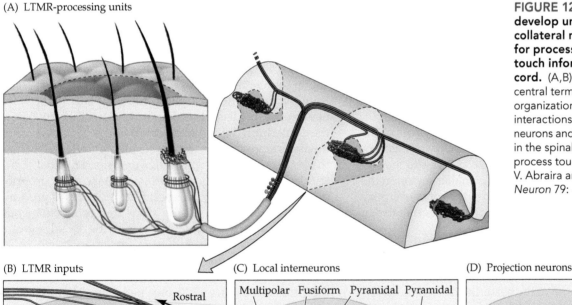

(B) LTMR inputs

(C) Local interneurons

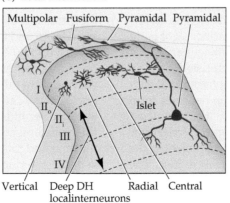

(D) Projection neurons

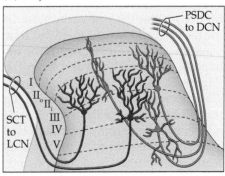

FIGURE 12.9 LTMR subtypes develop unique branching and collateral morphologies critical for processing of innocuous touch information by the spinal cord. (A,B) LTMRs have unique central terminations and columnar organization, as well as (C) distinct interactions with different interneurons and (D) projection neurons in the spinal cord dorsal horn that process touch information. (After V. Abraira and D. D. Ginty, 2013. *Neuron* 79: 618–639.)

Mechanosensory information in the dorsal horn is processed and integrated by two principal outputs: the postsynaptic dorsal column (PSDC) neurons and spinocervical tract (SCT) neurons, whose cell bodies are located in lamina III through V. Electrophysiological recordings and lesion studies demonstrate that the direct dorsal column pathway, together with PSDC and SCT neurons, transmits innocuous touch information to the brain. PSDC neurons project to the gracile and cuneate nuclei (dorsal column nuclei, located in the medulla), where sensory information converges with that of the direct dorsal column pathway.

The *second-order neurons* in the dorsal column nuclei send their axons to the somatosensory portion of the thalamus. The axons exiting from dorsal column nuclei are identified as the **internal arcuate fibers**. The internal arcuate fibers subsequently cross the midline and then form a dorsoventrally elongated tract known as the **medial lemniscus**. In contrast to the dorsal column/PSDC neuron pathway, SCT neurons receive information almost exclusively from hairy skin. SCT projections travel through the dorsolateral white matter and terminate in the lateral cervical nucleus (LCN) at cervical levels 1 through 3. From there, LCN second-order neurons decussate through the dorsal commissure and join the dorsal column/PSDC neuron pathway in the medial lemniscus. In addition to the SCT and PSDC pathways, the anterolateral tract is also involved in processing some aspects of non-discriminative touch (see Chapter 13).

The axons of the medial lemniscus synapse with thalamic neurons located in the ventral posterior lateral nucleus (VPL). Thus, the VPL receives input from contralateral dorsal column nuclei. In turn, *third-order neurons* in the VPL send their axons to the ipsilateral postcentral gyrus of the cerebral cortex, a region known as the primary somatosensory cortex, or SI. VPL neurons also send axons to the secondary somatosensory cortex (SII), a smaller region in the upper bank of the lateral sulcus. Thus, the somatosensory cortex represents mechanosensory signals first generated in the cutaneous surfaces of the contralateral body.

Somatosensation of the face

Thus far, the pathways we have described mostly concern the body below the level of the head. Tactile signals involving the head and face are mediated by a distinct set of pathways involving a cranial nerve, the **trigeminal nerve**, that bypasses the spinal cord and enters the brainstem directly (Figure 12.8B). The peripheral processes of trigeminal ganglion neurons (akin to dorsal root ganglion neurons) form three subdivisions of the trigeminal nerve (the *ophthalmic*, *maxillary*, and *mandibular* branches). Each branch innervates a well-defined territory on the face and head, including the teeth and the mucosa of the oral and nasal cavities. The central processes of trigeminal ganglion neurons form the sensory roots of the trigeminal nerve; they enter the

brainstem at the level of the pons to terminate on neurons in the **trigeminal brainstem complex**.

The trigeminal complex has two major components: the **principal nucleus** and the **spinal nucleus**. (A third component, the mesencephalic trigeminal nucleus, is considered in the next section.) Most of the afferents conveying information from cutaneous LTMRs terminate in the principal nucleus. In effect, this nucleus corresponds to the dorsal column nuclei that transmit mechanosensory information from the rest of the body. The spinal nucleus contains several subnuclei, and all of them receive inputs from collaterals of mechanoreceptors. Trigeminal neurons that are sensitive to pain, temperature, and non-discriminative touch do not project to the principal nucleus; instead, they project to the spinal nucleus of the trigeminal complex (discussed more fully in Chapter 13). The second-order neurons of the trigeminal brainstem nuclei give off axons that cross the midline and ascend to the **ventral posterior medial (VPM) nucleus** of the thalamus by way of the **trigeminal lemniscus**. Neurons in the VPM send their axons to ipsilateral cortical areas SI and SII.

Central pathways conveying proprioceptive information

Like their counterparts for cutaneous sensation, the axons of proprioceptive afferents enter the spinal cord through the dorsal roots, and many of the fibers from proprioceptive afferents also bifurcate into ascending and descending branches, which in turn send collateral branches to several spinal segments (Figure 12.10). Collateral branches penetrate the dorsal horn of the spinal cord and synapse on neurons in the deep dorsal, intermediate (directly below the dorsal horn region), and ventral horns. These synapses mediate, among other things, segmental reflexes such as the knee-jerk, or myotatic, reflex described in Chapters 1 and 16. There are two main pathways that convey proprioceptive information: Some ascending branches of proprioceptive axons travel with the axons conveying cutaneous mechanosensory information through the dorsal column, while others travel along the spinocerebellar tracts.

First-order proprioceptive afferents from the upper limbs have a course that is similar to that of cutaneous mechanoreceptors (see Figure 12.10, blue pathway). They enter the spinal cord and travel via the dorsal column (fasciculus cuneatus) up to the level of the medulla, where they synapse on proprioceptive neurons in the dorsal column nuclei, including a lateral nucleus among the tier of dorsal column nuclei in the caudal medulla called the **external cuneate nucleus**. Second-order neurons then send their axons into the ipsilateral cerebellum, while other branches cross the midline and join the medial lemniscus, ascending to the ventral posterior lateral nucleus (VPL) of the thalamus.

The information supplied by proprioceptive afferents is important not only for the ability to sense limb position;

FIGURE 12.10 Proprioceptive pathways for the upper and lower body Proprioceptive afferents for the lower part of the body synapse on neurons in the dorsal and ventral horn of the spinal cord and on neurons in Clarke's nucleus. Neurons in Clarke's nucleus send their axons via the dorsal spinocerebellar tract to the cerebellum, with a collateral to the dorsal column nuclei. Proprioceptive afferents for the upper body also have synapses in the dorsal and ventral horns, but then ascend via the dorsal column to the dorsal column nuclei; the external cuneate nucleus, in turn, transmits signals to the cerebellum. Proprioceptive target neurons in the dorsal column nuclei send their axons across the midline and ascend through the medial lemniscus to the ventral posterior nucleus (see Figure 12.8).

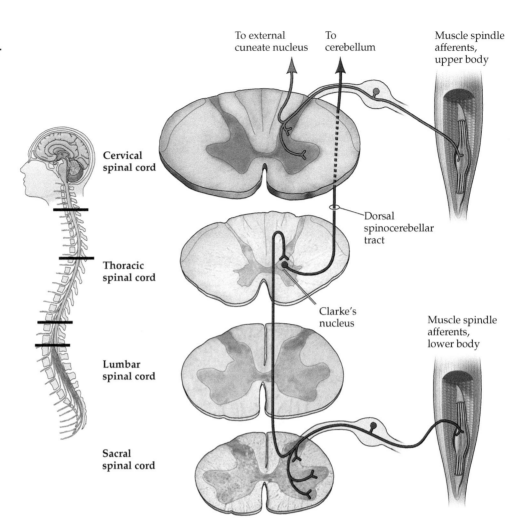

it is also essential for the functions of the cerebellum, a structure that regulates the timing of muscle contractions necessary for the performance of voluntary movements. As a consequence, proprioceptive information reaches higher cortical circuits as branches of pathways that are also targeting the cerebellum.

The association with cerebellar pathways is especially clear for the route that conveys proprioceptive information for the lower part of the body to the dorsal column nuclei. First-order proprioceptive afferents that enter the spinal cord between the mid-lumbar and thoracic levels (L2–T1) synapse on neurons in **Clarke's nucleus**, located in the medial aspect of the dorsal horn (see Figure 12.10, red pathway). Afferents that enter below this level ascend through the dorsal column and then synapse with neurons in Clarke's nucleus. Second-order neurons in Clarke's nucleus then send their axons into the ipsilateral posterior lateral column of the spinal cord, where they travel up to the level of the medulla in the **dorsal spinocerebellar tract**. These axons continue into the cerebellum, but in their course they give off collaterals that synapse with neurons lying just outside the nucleus gracilis (for the present purpose,

proprioceptive neurons of the dorsal column nuclei). Axons of these third-order neurons decussate and join the medial lemniscus, accompanying the fibers from cutaneous mechanoreceptors in their course to the VPL of the thalamus.

Like the information from cutaneous mechanoreceptors, proprioceptive information from the face is conveyed through the trigeminal nerve. However, the cell bodies of the first-order proprioceptive neurons for the face have an unusual location. Instead of residing in the trigeminal ganglia, they are found in the CNS, in the **mesencephalic trigeminal nucleus**, a well-defined array of neurons lying at the lateral extent of the periaqueductal gray matter of the dorsal midbrain. Like their counterparts in the trigeminal and dorsal root ganglia, these pseudounipolar neurons have peripheral processes that innervate muscle spindles and Golgi tendon organs associated with facial musculature (especially the jaw muscles) and central processes that include projections to brainstem nuclei responsible for reflex control of facial muscles. Although the exact route is not clear, information from proprioceptive afferents in the mesencephalic trigeminal nucleus also reaches the thalamus and is represented in somatosensory cortex.

The role of thalamic neurons in somatosensation

The somatosensory pathways described thus far make synaptic connections in various regions of the ventral posterior complex of the thalamus (Figure 12.11). These thalamic neurons then project to primary somatosensory cortex to enable conscious perception of tactile stimuli. While the thalamus has previously been classified solely as a "relay station" for sensory information to cortex, recent studies suggest that reciprocal inter-actions between cortical and thalamic neurons are essential for the proper de-velopment of functional thalamocortical circuits, sensory processing, and sen-sorimotor behaviors. Other neurons in the thalamus receive descending projec-tions from primary somatosensory cor-tex (mainly layer-5 neurons), as well as some inputs from ascending pathways. This thalamocortical circuit architecture suggests that sensory signals are trans-mitted by thalamic neurons and further processed in the cortex, with sensory information then conveyed via distinct thalamic neurons to different sensory-related cortical areas.

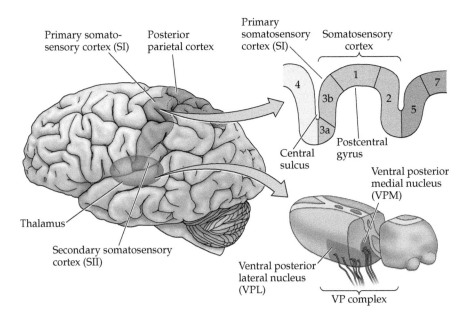

FIGURE 12.11   **Somatosensory portions of the thalamus and their cortical targets in the postcentral gyrus**   The ventral posterior nuclear complex comprises the VPM, which transmits somatosensory information carried by the trigeminal sys-tem from the face, and the VPL, which transmits somatosensory information from the rest of the body. The diagram at the upper right shows the organization of the soma-tosensory cortex in the postcentral gyrus, shown here in a section cutting across the gyrus from anterior to posterior. (After P. Brodal, 1992. *The Central Nervous System: Structure and Function.* New York: Oxford University Press, p. 151; E. G. Jones and D. P. Friedman, 1982. *J Neurophys* 48: 521–544.)

Primary somatosensory cortex

The majority of the axons arising from neurons in the ven-tral posterior complex of the thalamus project to neurons located in layer 4 of the primary somatosensory cortex (see Box 27A for a description of cortical lamination). In humans, the SI is located in the postcentral gyrus of the parietal lobe and is known as Brodmann's area 3b (Figure 12.12A). Al-though only area 3b is considered SI, nearby Brodmann's areas 1, 2, and 3a are also involved in tactile perception.

Mapping studies in humans and other primates show that each of these four cortical areas contains a separate and complete representation of the body. In these **soma-totopic maps**, the foot, leg, trunk, forelimbs, and face are represented in a medial to lateral arrangement, as shown in Figure 12.12B. As noted earlier in this concept, somato-topy refers to the correspondence between a particular area on the body and a specific functional area in the CNS. At each processing stage in the somatosensory system, the arrangement of the inputs preserves the spatial relations of the peripheral neuron endings on the body surface. This topographic consistency creates a neural map of the body surface at each processing point (e.g., spinal cord, brain-stem, thalamus, and somatosensory cortex) so that body map associations are preserved.

A salient feature of somatotopic maps, recognized soon after their discovery, is their failure to represent the human body in its actual proportions. When neurosurgeons deter-mined the representation of the human body in the primary somatosensory (and motor) cortex, the homunculus ("little man") defined by such mapping procedures had a grossly enlarged face and hands compared with the torso and prox-imal limbs (Figure 12.12C). These anomalies arise because manipulation, facial expression, and speech are extraordi-narily important for humans and require a great deal of circuitry, both central and peripheral, to govern them. Thus, in humans the cervical spinal cord is enlarged to accom-modate the extra circuitry related to the hand and upper limb, and as stated earlier, the density of sensory receptors is greater in regions such as the hands and lips.

Such distortions are also apparent when topographical maps are compared across species. In the rat and mouse brains, for example, an inordinate amount of somato-sensory cortex is devoted to representing the large facial whiskers that are key components of the somatosensory input for rats and mice. By contrast, raccoon SI possesses an overrepresentation of the paws, the platypus SI its bill, and the star-nosed mole SI its specialized star sensory or-gan (Box 12A). In short, the sensory input (or motor out-put) that is particularly significant to a given species gets relatively more cortical representation.

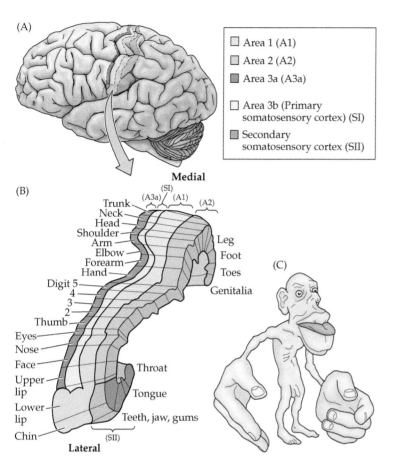

FIGURE 12.12 Somatotopic order in the human primary somatosensory cortex (A) Diagram showing the region of the human cortex from which electrical activity is recorded following mechanosensory stimulation of different parts of the body. (The individuals in the study were undergoing neurosurgical procedures for which such mapping was required.) Although modern imaging methods are now refining these classic data, the human somatotopic map defined in the 1930s has remained generally valid. (B) Diagram showing the somatotopic representation of body parts from medial to lateral. (C) Cartoon of the homunculus constructed on the basis of such mapping. Note that the amount of somatosensory cortex devoted to the hands and face is much larger than the relative amount of body surface in these regions. A similar disproportion is apparent in the primary motor cortex, for much the same reasons (see Chapter 17). (After W. Penfield. and T. Rasmussen, 1950. *The Cerebral Cortex of Man: A Clinical Study of Localization of Function.* New York: Macmillan; P. Corsi, P. 1991. *The Enchanted Loom: Chapters in the History of Neuroscience,* P. Corsi [Ed.]. New York: Oxford University Press.)

Although the topographic organization of the several somatosensory areas is similar, the functional properties of the neurons in each region are distinct. Experiments carried out in non-human primates indicate that neurons in areas 3b and 1 respond primarily to cutaneous stimuli, whereas neurons in 3a respond mainly to stimulation of proprioceptors; area 2 neurons process both tactile and proprioceptive stimuli. These differences in response properties reflect, at least in part, parallel sets of inputs from functionally distinct classes of neurons in the ventral posterior complex. In addition, a rich pattern of corticocortical connections between other somatosensory cortex areas, including areas 3a, 1, and 2, contribute significantly to the elaboration of SI response properties. Area 3b receives the bulk of the input from the ventral posterior complex and provides a particularly dense projection to areas 1 and 2. This arrangement of connections establishes a functional hierarchy in which area 3b serves as an obligatory first step in cortical processing of somatosensory information (Figure 12.13). Consistent with this view, lesions of area 3b in non-human primates result in profound deficits in all forms of tactile sensations mediated by cutaneous mechanoreceptors, while lesions limited to areas 1 or 2 result in partial deficits and an inability to use tactile information to discriminate either the texture of objects (area 1 deficit) or the size and shape of objects (area 2 deficit).

Pioneering work by Vernon Mountcastle using vertical microelectrode penetrations in SI of animals provided the first evidence that neurons in somatosensory cortex are organized into functional units, known as cortical columns, that traverse the depths of the cortical layers. Each column contains neurons that are responsive to particular types of tactile stimuli and have similar receptive field locations

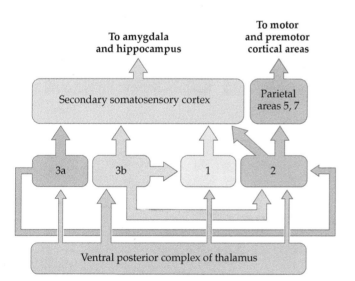

FIGURE 12.13 Connections within the somatosensory cortex establish functional hierarchies Inputs from the ventral posterior complex of the thalamus terminate in Brodmann's areas 3a, 3b, 1, and 2, with the greatest density of projections in area 3b. Area 3b in turn projects heavily to areas 1 and 2, and the functions of these areas are dependent on the activity of area 3b. Somatosensory cortex areas 3a, 3b, 1, and 2 all project to SII; the functions of SII are dependent on the activity of SI.

■ BOX 12A | Specialized Mechanosensation in Animals

Evolution has endowed vertebrates with remarkable and specialized abilities to explore the world through the sense of touch. The star-nosed mole lives in moist, muddy wetlands and forages using its "star" organ, which is formed by two sets of 11 fleshy mechanosensory appendages surrounding the nostrils, containing Eimer's organs (Figures A–C). Each organ is associated with a Merkel cell–Aβ-SAI LTMR complex, a lamellated corpuscle, and a series of five to ten free nerve endings that form a circle of terminal swellings, with an average of 6110 fibers per ray. Together, these structures constitute the densest population of mechanoreceptive end organs found in mammals, which enables the mole to recognize small shapes and textures with high precision and accuracy. Upon encountering an object, appendages 1 through 10 perform saccade-like movements to

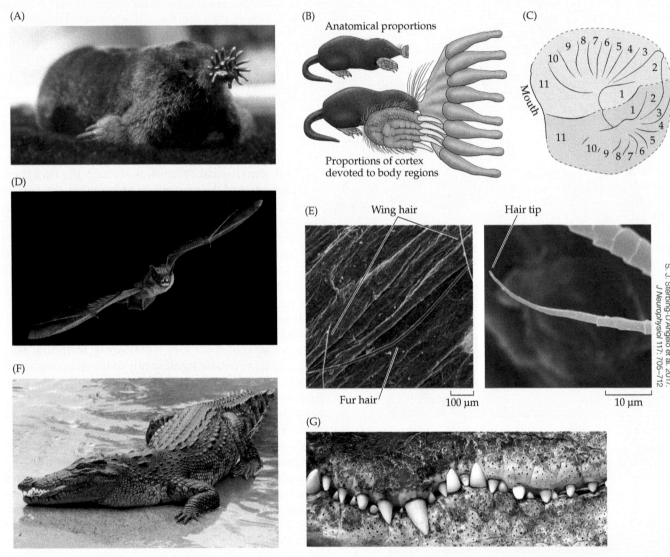

S. J. Sterbing-D'Angelo et al. 2017. *J Neurophysiol* 117: 705–712

(A) Photograph of a star-nosed mole (*Condylura cristata*). (B) Drawings showing proportions of the body, and proportions of the cortex devoted to processing information from various parts of the body. (C) A region of cerebral cortex, showing where sensory information arrives from the numbered appendages of the star organ on the mole's nose. (D) Photograph of a Big Brown Bat (*Eptesicus fuscus*). (E) Scanning electron microscope photomicrographs of hairs on the wing of the Big Brown Bat. Left: Two short, tactile hairs and one cut, long fur hair. Right: Difference in diameter between the tip and the base of wing hairs. (F) Photograph of a crocodile. (G) Photograph of a Nile crocodile (*Crocodylus niloticus*) jaw. The small, melanin-pigmented ISOs seen here on the gums are present on virtually every scale of the head and body. (B,C after K. C. Catania andJ/ H. Kaas, 1996. *Bioscience* 46: 578-586. Reprinted by permission of Oxford University Press on behalf of the American Institute of Biological Sciences.)

■ BOX 12A | Specialized Mechanosensation in Animals (*continued*)

direct the object to appendage 11, referred to as the fovea of the star organ, which performs detailed investigation of the object (which can be < 1 mm in diameter). As may be expected from its specialized functions, appendage 11 has the highest LTMR innervation density and the largest representation in primary somatosensory cortex (SI) of all the mole's appendages. The elaborate star organ is, in general, overrepresented in SI, because of its critical roles in enabling the fastest known mammalian predator to identify and eat its prey in times as short as 100 to 300 ms.

Bats are the only mammals capable of sustained flight. The remarkable agility of bat flight requires precisely timed coordination of limb and body movements in response to changes in air flow and body position. In addition to relying on hearing and vision, the muscle movements controlling air maneuvers rely on the abundant, rapidly changing tactile information from the mechanosensory system in the wing.

Unlike the wings of birds and insects, bats' thin, flexible wings stretch and reshape during flight. Histological analyses of the wing membrane of the Big Brown Bat indicate that approximately half of the hair follicles are dually innervated by lanceolate nerve endings and Merkel cell–Aβ-SAI LTMR complexes (Figures D and E). This is in contrast to rodent hairy skin, in which Merkel cells are restricted to guard hairs, which comprise only about 2% of all hairs. The high density of slowly adapting and rapidly adapting mechanosensory neurons on the bat wing are thought to be critical for monitoring precise changes in the speed and direction of airflow, as removal of hair follicles (and hence loss of mechanosensory neuron sensitivity) negatively affects flight behavior. The expansion of guard hairs and the elimination of the drag-inducing coat hairs on the bat wing could represent a unique evolutionary adaptation for flight in the bat somatosensory system.

Crocodiles are highly efficient hunters that rely heavily on the sense of touch. Behavioral studies show that crocodiles can detect vibration from a water drop in the absence of auditory or visual cues. All crocodiles have specialized miniature domelike pigmented structures, known as integumentary sensory organs (ISOs), distributed on the skin throughout the body and head (Figures F and G). Each ISO is covered with a 5-μm-thick stratum corneum, which allows for highly sensitive detection of mechanical stimuli. Within the ISOs are unmyelinated free nerve endings that penetrate the upper level of epidermis, as well as mechanosensory end organs such as Merkel cell complexes and Pacinian-like corpuscles that are innervated by their respective myelinated afferents. ISOs are thought to play multiple critical functions during prey capture, including detection of water vibration and physical contact, as well as prey analysis after the capture.

in the periphery. These cortical neurons exhibit complex receptive field and response properties; for example, there are SI neurons that respond to a variety of tactile features, including specific stimulus textures (e.g., silk vs. burlap) and orientation (e.g., the edges of a textbook) as well as stimulus direction, velocity, and speed. In addition, there are multiple populations of SI neurons that respond to proprioceptive signals. It was previously assumed that signals from sensory afferent types are segregated into distinct ascending pathways, with cortical neurons receiving segregated inputs from rapidly adapting (Aβ-RAI) or slowly adapting (Aβ-SAI) mechanosensory neurons. Instead, recent evidence indicates that there is a convergence of submodality-specific inputs onto individual neurons in SI. These convergent inputs onto SI neurons result from substantial processing and nonlinear transformation of tactile-evoked signals in subcortical areas, which are likely to include the spinal cord, thalamus, and brainstem.

Beyond SI: Corticocortical and descending pathways

Somatosensory information is distributed from the primary somatosensory cortex as well as areas 1, 2 and 3a, to "higher-order" cortical fields, including SII (see Figures 12.11 and 12.12). SII receives convergent projections from somatosensory cortex areas 3a, 3b, 1, and 2, and these inputs are necessary for normal SII function, as lesions of these areas eliminate the somatosensory responses of SII neurons. SII sends projections in turn to limbic structures such as the amygdala and hippocampus (see Chapters 30 and 31). This latter pathway is believed to play an important role in tactile learning and memory.

Neurons in SI, as well as somatosensory cortex areas 3a, 1, and 2, also project to parietal areas posterior to area 2, especially areas 5a and 7b. These areas receive direct projections from area 2 and, in turn, supply inputs to neurons in motor and premotor areas of the frontal lobe. This is a major route by which information derived from proprioceptive afferents signaling the current state of muscle contraction gains access to circuits that initiate voluntary movements. More generally, the projections from parietal cortex to motor cortex are critical for the integration of sensory and motor information (see Chapters 17, 27, and 29 for discussion of sensorimotor integration in the parietal and frontal lobes).

Finally, a fundamental but often neglected feature of the somatosensory system is the presence of massive descending projections. These pathways originate in somatosensory cortices and project to the thalamus, brainstem, and spinal cord. Indeed, descending projections from the somatosensory cortex *outnumber* ascending somatosensory pathways! While their physiological role is an area of active investigation, it is generally thought that descending projections modulate the ascending flow of sensory information at the level of the thalamus, brainstem, and spinal cord. For example, corticospinal tract fibers have been shown to modulate sensory signals in the spinal cord through a mechanism called primary afferent depolarization. Primary afferent depolarization allows for presynaptic inhibition of primary afferent inputs to the spinal cord, which is thought to enhance gain (e.g., signal to noise) and acuity of somatosensory signals.

CONCEPT 12.6 | Central Representations of the Body Are Plastic and Modified by Experience

LEARNING OBJECTIVES

12.6.1 Discuss how the brain's representation of the body surface is plastic and changes in response to altered input and experience.

12.6.2 Describe some of the ways in which the somatosensory system interacts with other sensory and motor systems.

Plasticity in the adult cerebral cortex

The analysis of maps of the body surface in primary somatosensory cortex and of the responses to altered patterns of activity in peripheral afferents has been instrumental in understanding the potential for the reorganization of cortical circuits in adults. Jon Kaas and Michael Merzenich were the first to explore this issue, by examining the impact of peripheral lesions (e.g., cutting a nerve that innervates the hand, or amputating a digit) on the topographic maps in somatosensory cortex. Immediately after the lesion, the corresponding region of the cortex was found to be unresponsive. After a few weeks, however, the unresponsive area became responsive to stimulation of neighboring regions of the skin (Figure 12.14). For example, if digit 3 was amputated, cortical neurons that formerly responded to stimulation of digit 3 now responded to stimulation of digits 2 and 4. Thus, the central representation of the remaining digits had expanded to take over the cortical territory that had lost its main input. Such "functional remapping" also occurs in the somatosensory nuclei in the thalamus and brainstem. Indeed, some of the reorganization of cortical circuits probably depends, at least in part, on concurrent subcortical plasticity. This sort of adjustment in the somatosensory system may contribute to the altered sensation of phantom limbs after amputation (see Clinical Applications, Chapter 13). Similar plastic changes

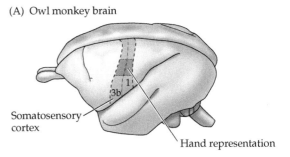

(A) Owl monkey brain

Somatosensory cortex

Hand representation

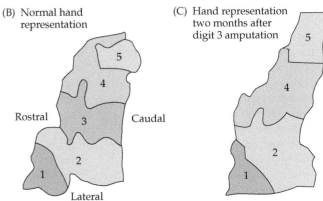

(B) Normal hand representation

Rostral — Caudal

Lateral

(C) Hand representation two months after digit 3 amputation

FIGURE 12.14 Functional changes in the somatosensory cortex following amputation of a digit (A) Diagram of the somatosensory cortex in the owl monkey, showing the approximate location of the hand representation. (B) The hand representation in the animal before amputation; the numbers correspond to different digits. (C) The cortical map determined in the same animal 2 months after amputation of digit 3. The map has changed substantially; neurons in the area formerly responding to stimulation of digit 3 now respond to stimulation of digits 2 and 4. (After M. M. Merzenich et al., 1984. *J Comp Neurol* 224: 591–605.)

have been demonstrated in the visual, auditory, and motor cortices, suggesting that some ability to reorganize after peripheral deprivation or injury is a general property of the mature neocortex.

Appreciable changes in cortical representation also can occur in response to physiological changes in sensory or motor experience. For instance, if a monkey is trained to use a specific digit for a particular task that is repeated many times, the functional representation of that digit, determined by electrophysiological mapping, can expand at the expense of the other digits (Figure 12.15). In fact, significant changes in receptive fields of somatosensory neurons can be detected when a peripheral nerve is blocked temporarily by a local anesthetic. The transient loss of sensory input from a small area of skin induces a reversible reorganization of the receptive fields of both cortical and subcortical neurons. During this period, the neurons assume new receptive fields that respond to tactile stimulation of the skin surrounding the anesthetized region. Once the effects of the local anesthetic subside, the receptive fields of cortical and subcortical neurons return

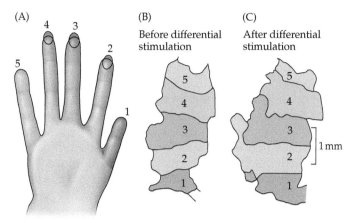

FIGURE 12.15   Functional expansion of a cortical representation by a repetitive behavioral task (A) An owl monkey was trained in a task that required heavy usage of digits 2, 3, and occasionally 4. (B) The map of the digits in the primary somatosensory cortex prior to training. (C) After several months of "practice," a larger region of the cortex contained neurons activated by the digits used in the task. Note that the specific arrangements of the digit representations are somewhat different from those for the owl monkey shown in Figure 12.14, indicating the variability of the cortical representation in individual animals. (After J. Jenkins et al., 1990. *J Neurophysiol* 63: 82–104.)

to their usual size. The common experience of an anesthetized area of skin feeling disproportionately large—as experienced, for example, following dental anesthesia—may be a consequence of this temporary change.

Despite these intriguing observations, the mechanism, purpose, and significance of the reorganization of sensory and motor maps that occurs in adult cortex are not known. Clearly, changes in cortical circuitry occur in the adult brain. Centuries of clinical observations, however, indicate that these changes may be of limited value for recovery of function following brain injury, and they may well lead to symptoms that detract from rather than enhance the quality of life following neural damage. Given their rapid and reversible character, most of these changes in cortical function probably reflect alterations in the strength of synapses already present. Indeed, finding ways to prevent or redirect the synaptic events that underlie injury-induced plasticity could reduce the long-term impact of acute brain damage.

Interactions between the somatosensory system and other sensory and motor systems

The somatosensory system works in tandem with other sensory and motor systems to control a range of complex behaviors. As we saw in Concept 12.4, proprioceptive feedback helps us move accurately through the world. Indeed, the interactions between somatosensory and motor systems are extensive: Humans and other primates actively explore objects with their hands, as rodents and felines do with their whiskers. The combination of such exploratory

movement and resulting tactile feedback is known as **haptic touch**. In addition, the somatosensory system can augment lost function in other systems, as evidenced by the use of Braille by individuals who are blind, or by feeling rather than hearing musical vibrations by individuals with hearing loss. Somatosensory information can also help us interpret information acquired through other sensory systems. For example, auditory cortex receives information from the somatosensory system, a connection that may play a role in helping distinguish self-generated sounds from external sounds, such as the sound of one's own versus someone else's footsteps.

Integrating tactile information with other sensory modalities poses a computational challenge for the nervous system. For example, given that the body and eyes can move with respect to each other, computations must be performed by the brain to factor in the relative positions of the body and eyes to permit temporally precise integration of visual and tactile information regarding oneself and physical objects in the environment. The neural pathways involved in such computations may originate in somatosensory cortex, and probably also involve additional parietal and oculomotor regions of the brain, where modulation of tactile responses by eye movements has been observed. This spatial, multisensory information is important for enacting appropriate motor programs in response to environmental cues and interactions with the physical environment.

CONCEPT 12.7 | Somatosensory Dysfunction Is an Important Factor in Multiple Diseases

LEARNING OBJECTIVE

12.7.1 Discuss how somatosensory dysfunction contributes to various diseases and disorders.

Dysfunction of the somatosensory system

The sense of touch is critical for our interactions with the external world, and disruptions to mechanosensory processing can significantly affect a person's quality of life and behavior. Friedreich's ataxia is a neurodegenerative disorder caused by mutations in the gene that encodes frataxin (FXN), a protein that is necessary for normal mitochondrial function. This progressive disorder leads to a significant loss of large, heavily myelinated neurons of the dorsal root ganglia, resulting in an impaired ability to coordinate movements such as posture, walking, and manipulating objects with the hands.

Parkinson's disease is a progressive disease of the nervous system marked by tremor, muscular rigidity, and slow, imprecise movement that mostly affects middle-aged and elderly people. Accumulation and aggregation of the protein α-synuclein in the form of Lewy bodies and Lewy neurites leads to degeneration of the basal ganglia of the brain and a deficiency of the neurotransmitter dopamine. Although Parkinson's disease is often associated with motor symptoms, poor balance, and trembling, the first symptoms of the disease are often sensory issues, including a reduced sense of touch and increased pain, as well as a loss of smell. These sensory symptoms are thought to arise from disruptions in basal ganglia function, but increasing evidence indicates that changes in peripheral sensory neurons are also common in individuals with Parkinson's disease. Alpha-synuclein deposits have been observed in skin nerve fibers, raising the possibility of disease-related peripheral neurodegeneration in Parkinson's disease. Indeed, researchers have observed a loss of both myelinated and unmyelinated sensory nerve axons, as well as Meissner corpuscles, in skin samples taken from people with Parkinson's disease. The loss of skin sensory nerve fibers and Meissner corpuscles appears to occur early in the stages of disease pathology and is proposed to account for somatosensory issues in patients.

Autism spectrum disorder (ASD) is a highly prevalent class of neurodevelopmental disorders characterized by impairments in social interaction, alterations in verbal and nonverbal communication, and restrictive, repetitive behaviors. ASD is highly varied in terms of etiology and severity; individuals with autism also exhibit a complicated array of comorbid symptoms, among them abnormal reactions to sensory stimuli, including tactile and pain stimuli. In addition, several studies have identified that individuals exhibiting the greatest expression of archetypal ASD traits have the highest degree of impairment in neural processing of non-discriminative affective touch, as well as heightened responses to light touch stimuli. Thus, the severity of somatosensory processing alterations correlates with ASD symptom severity. While ASD is commonly attributed to alterations in the brain, and somatosensory alterations may be due to changes in brain function (e.g., thalamus or primary somatosensory cortex), increasing evidence also suggests that alterations in peripheral sensory neurons are present in individuals with autism. For example, research in rodents indicates that genetic or environmental disruptions to peripheral sensory neurons can cause hyperresponsiveness to sensory stimuli, and these alterations in sensory processing can disrupt social interactions in animals. How disruptions in tactile processing at the earliest stages of somatosensory processing may lead to alterations in social behaviors is an open research question.

Summary

The components of the somatosensory system process information conveyed by mechanical stimuli that either impinge on the body surface (cutaneous mechanoreception) or are generated within the body itself (proprioception). Somatosensory processing is performed by neurons distributed across the spinal cord as well as several brain structures that are connected by both ascending and descending pathways. Transmission of afferent mechanosensory information from the periphery to the brain begins with a variety of specialized receptor types that initiate action potentials. This activity is then processed and propagated centrally via diverse neuron types that are organized into multiple gray matter structures and white matter tracts. First-order neurons in these circuits are the primary sensory neurons located in the dorsal root and cranial nerve ganglia. The next set of neurons conveying ascending mechanosensory signals are located in spinal cord and brainstem nuclei. The final link in the pathway from periphery to cerebral cortex consists of neurons found in the thalamus, which in turn project to the postcentral gyrus. These pathways are topographically arranged throughout the system, with the amount of cortical and subcortical space allocated to various body parts being largely proportional to the density of peripheral receptors. Studies of non-human primates show that specific cortical regions correspond to each functional submodality; area 3b, for example, processes information from cutaneous LTMRs, while area 3a processes inputs from proprioceptors. Thus, at least two broad criteria operate in the organization of the somatosensory system: modality and somatotopy. The end result of this complex interaction is the unified perceptual representation of the body and its ongoing interaction with the environment. Major research efforts are now aimed at identifying how touch stimulus information is detected, processed, and transmitted along somatosensory pathways in a nonredundant manner, from the primary afferents to the spinal cord, brainstem, thalamus, and somatosensory cortices. Discovering how somatosensory stimuli are processed at each stage of this hierarchy will allow us to understand how touch pathways ultimately give rise to accurate perceptions of the rich and complex somatosensory world.

■ Additional Reading

Reviews

Abraira, V. E. and D. D. Ginty (2013) The sensory neurons of touch. *Neuron* 79: 618–639.

Barnes, S. J. and G. T. Finnerty (2010) Sensory experience and cortical rewiring. *Neuroscientist* 16: 186–198.

Chapin, J. K. (1987) Modulation of cutaneous sensory transmission during movement: Possible mechanisms and biological significance. In *Higher Brain Function: Recent Explorations of the Brain's Emergent Properties*, S. P. Wise (Ed.). New York: John Wiley and Sons, pp. 181–209.

Darian-Smith, I. (1982) Touch in primates. *Annu. Rev. Psychol.* 33: 155–194.

Delhaye, B. P., K. H. Long and S. Bensmaia (2018) Neural basis of touch and proprioception in primate cortex. *Compr. Physiol.* 8: 1575–1602.

Elias, L. J. and I. Abdus-Saboor (2022) Bridging skin, brain, and behavior to understand pleasurable social touch. *Curr. Opin. Neurobiol.* 73: 102527.

Johansson, R. S. and J. R. Flanagan (2009) Coding and use of tactile signals from the fingertips in object manipulation tasks. *Nat. Rev. Neurosci.* 10: 345–359.

Johnson, K. O. (2002) Neural basis of haptic perception. In *Steven's Handbook of Experimental Psychology*, 3rd Edition, H. Pashler and S. Yantis (Eds.). Vol 1: *Sensation and Perception*. New York: Wiley, pp. 537–583.

Kaas, J. H. (2004) Somatosensory system. In *The Human Nervous System*, G Paxinos (Ed.). San Diego: Academic Press, pp. 1059–1092.

Lumpkin, E. A. and M. J. Caterina (2007) Mechanisms of sensory transduction in the skin. *Nature* 445 (7130): 858–865.

McGlone, F., J. Wessberg and H. Olausson (2014) Discriminative and affective touch: Sensing and feeling. *Neuron* 82: 737–755.

Moehring, F., P. Halder, R. P. Seal and C. L. Stucky (2018) Uncovering the cells and circuits of touch in normal and pathological settings. *Neuron* 100 (2): 349–360.

Mountcastle, V. B. (1975) The view from within: Pathways to the study of perception. *Johns Hopkins Med. J.* 136: 109–131.

O'Connor, D. H., L. Krubitzer and S. Bensmaia (2021) Of mice and monkeys: Somatosensory processing in two prominent animal models. *Prog. Neurobiol.* 201: 102008.

Petersen, R. S., S. Panzeri and M. E. Diamond (2002) Population coding in somatosensory cortex. *Curr. Opin. Neurobiol.* 12: 441–447.

Ranade, S. S., R. Syeda and A. Patapoutian (2015) Mechanically activated ion channels. *Neuron* 87: 1162–1179.

Saal, H. P. and S. J. Bensmaia (2014) Touch is a team effort: Interplay of submodalities in cutaneous sensibility. *Trends Neurosci.* 37: 689–697.

Wolff, M. and 4 others (2021) A thalamic bridge from sensory perception to cognition. *Neurosci. Biobehav. Rev.* 120: 222–235.

Woolsey, C. (1958) Organization of somatosensory and motor areas of the cerebral cortex. In *Biological and Biochemical Bases of Behavior*, H. F. Harlow and C. N. Woolsey (Eds.). Madison: University of Wisconsin Press, pp. 63–82.

Important Original Papers

Adrian, E. D. and Y. Zotterman (1926) The impulses produced by sensory nerve endings. II. The response of a single end organ. *J. Physiol.* 61: 151–171.

Chen, J. L. and 6 others (2015) Pathway-specific reorganization of projection neurons in somatosensory cortex during learning. *Nat. Neurosci.* 18: 1101–1108.

Dooley, J. C. and L. A. Krubitzer (2019) Alterations in cortical and thalamic connections of somatosensory cortex following early loss of vision. *J. Comp. Neurol.* 527(10): 1675–1688.

Emanuel, A. J. and 4 others (2021) Cortical responses to touch reflect subcortical integration of LTMR signals. *Nature* 600: 680–685.

Friedman, R. M., L. M. Chen and A. W. Roe (2004) Modality maps within primate somatosensory cortex. *Proc. Natl. Acad. Sci. U.S.A.* 101: 12724–12729.

Groh, J. M. and D. L. Sparks (1996) Saccades to somatosensory targets. III. Eye-position-dependent somatosensory activity in primate superior colliculus. *J. Neurophysiol.* 75(1): 439–453.

Johansson, R. S. (1978) Tactile sensibility of the human hand: Receptive field characteristics of mechanoreceptive units in the glabrous skin. *J. Physiol.* 281: 101–123.

Johnson, K. O. and G. D. Lamb (1981) Neural mechanisms of spatial tactile discrimination: Neural patterns evoked by Braille-like dot patterns in the monkey. *J. Physiol.* 310: 117–144.

Jones, E. G. and D. P. Friedman (1982) Projection pattern of functional components of thalamic ventrobasal complex on monkey somatosensory cortex. *J. Neurophysiol.* 48: 521–544.

Jones, E. G. and T. P. S. Powell (1969) Connexions of the somatosensory cortex of the rhesus monkey. I. Ipsilateral connexions. *Brain* 92: 477–502.

Lamotte, R. H. and M. A. Srinivasan (1987) Tactile discrimination of shape: Responses of rapidly adapting mechanoreceptive afferents to a step stroked across the monkey fingerpad. *J. Neurosci.* 7: 1672–1681.

Li, L. and 11 others (2011) The functional organization of cutaneous low-threshold mechanosensory neurons. *Cell* 7: 1615–1627.

Maksimovic, S. and 11 others (2014) Epidermal Merkel cells are mechanosensory cells that tune mammalian touch receptors. *Nature* 509: 617–621.

Maricich, S. M. and 6 others (2009) Merkel cells are essential for light-touch responses. *Science* 324(5934): 1580–1582.

Moore, C. I. and S. B. Nelson (1998) Spatiotemporal subthreshold receptive fields in the vibrissa representation of rat primary somatosensory cortex. *J. Neurophysiol.* 80: 2882–2892.

Nicolelis, M. A. L., L. A. Baccala, R. C. S. Lin and J. K. Chapin (1995) Sensorimotor encoding by synchronous neural ensemble activity at multiple levels of the somatosensory system. *Science* 268: 1353–1359.

Orefice, L. L. and 5 others (2016) Peripheral mechanosensory neuron dysfunction underlies tactile and behavioral deficits in mouse models of ASDs. *Cell* 166(2): 299–313.

Padberg, J. and 6 others (2009) Thalamocortical connections of parietal somatosensory cortical fields in macaque monkeys are highly divergent and convergent. *Cereb. Cortex* 19(9): 2038–2064.

Ranade, S. S. and 16 others (2014) Piezo2 is the major transducer of mechanical forces for touch sensation in mice. *Nature* 516: 121–125.

Schroeder, C. E. and 5 others (2001) Somatosensory input to auditory association cortex in the macaque monkey. *J. Neurophysiol.* 85(3): 1322–1327.

Wall, P. D. and W. Noordenhos (1977) Sensory functions which remain in man after complete transection of dorsal columns. *Brain* 100: 641–653.

Wang, X., M. Zhang, I. S. Cohen and M. E. Goldberg (2007) The proprioceptive representation of eye position in monkey primary somatosensory cortex. *Nat. Neurosci.* 10(5): 640–646.

Weber, A. I. and 6 others (2013) Spatial and temporal codes mediate the tactile perception of natural textures. *Proc. Natl. Acad. Sci. U.S.A.* 110: 17107–17112.

Woo, S. H. and 11 others (2014) Piezo2 is required for Merkel-cell mechanotransduction. *Nature* 509: 622–626.

Zhu, J. J. and B. Connors (1999) Intrinsic firing patterns and whisker-evoked synaptic responses of neurons in the rat barrel cortex. *J. Neurophysiol.* 81: 1171–1183.

Books

Hertenstein, M. J. and S. J. Weiss (eds.) (2011) *The Handbook of Touch: Neuroscience, Behavioral, and Health Perspectives.* New York: Springer.

Linden, D. J. (2015) *Touch: The Science of Hand, Heart, and Mind.* New York: Viking Penguin.

Mountcastle, V. B. (1998) *Perceptual Neuroscience: The Cerebral Cortex.* Cambridge, MA: Harvard University Press.

Pain and Temperature

Overview

Pain is the unpleasant sensory and emotional experience associated with stimuli that cause actual or potential tissue damage. Although it may be natural to assume that the sensations associated with injurious stimuli arise from excessive stimulation of the same receptors that generate other somatic sensations (i.e., those discussed in Chapter 12), this is not the case. The perception of injurious stimuli, called nociception, depends on an array of specifically dedicated peripheral neurons that detect nociceptive stimuli produced in internal tissues or by the external world. Pain information is then transmitted to a series of neural circuits in the spinal cord dorsal horn or brainstem, and then to the brain. Pain can manifest with a variety of qualities (e.g., sharp, dull, burning), and the response of an organism to noxious stimuli is multidimensional, involving discriminative, non-discriminative (affective), and motivational components. The central distribution of nociceptive information is correspondingly complex, involving multiple areas in the brainstem, thalamus, and forebrain. Chronic pain is a highly prevalent condition that can be debilitating, and treatment options are often inadequate. The overriding importance of pain in clinical practice (both as a diagnostic tool and as a focus of treatment), as well as the many aspects of pain physiology and pharmacology that remain imperfectly understood, continues to make nociception an extremely active area of research.

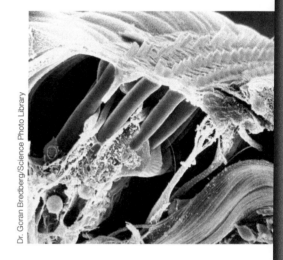

Dr. Goran Bredberg/Science Photo Library

KEY CONCEPTS

13.1 Potentially damaging stimuli are detected by nociceptors

13.2 Noxious stimuli are transduced into neural signals via various ion channel receptors

13.3 Central pain pathways are distinct from mechanosensory pathways and transmit different aspects of pain in parallel

13.4 Signals from a variety of innocuous stimuli, as well as from internal organs, are also carried by pain pathways

13.5 Sensitivity to pain is subjective and can be modified by a variety of factors

---

CONCEPT
13.1 | **Potentially Damaging Stimuli Are Detected by Nociceptors**

LEARNING OBJECTIVES

13.1.1 Explain what a nociceptor is and what kind of information it can transmit.

13.1.2 Recognize that nociception and touch involve distinct neural substrates.

13.1.3 Explain the difference between first and second pain.

13.1.4 Discuss the different types of injurious stimuli detected by different sensory afferent fiber types.

Nociceptors

The relatively unspecialized nerve cell endings that initiate the sensation of pain are called **nociceptors** (Latin *nocere*, "to hurt"). Like other somatosensory afferents, nociceptors innervate the skin, deep tissues, and internal organs and transduce a variety of stimuli into receptor potentials, which in turn trigger afferent action potentials. Moreover, nociceptors, like other somatosensory afferents, are pseudounipolar neurons that arise from cell bodies in dorsal root ganglia or the trigeminal ganglion. Nociceptor neurons send one axonal branch to the periphery and the other into the spinal cord (dorsal root ganglia neurons) or the brainstem (trigeminal ganglia neurons; see Figure 12.1).

Because peripheral nociceptive axons terminate in morphologically unspecialized "free nerve endings," it is conventional to categorize nociceptors according to the properties of the axons associated with them (see Table 12.1). As described in Chapter 12, most of the somatosensory afferents responsible for the perception of innocuous mechanical stimuli possess myelinated axons that have relatively rapid conduction velocities. In contrast, the axons associated with nociceptors conduct relatively slowly, being only lightly myelinated or, more commonly, unmyelinated. For the most part, axons conveying information about pain fall into either the **Aδ fiber group** of lightly myelinated axons, which conduct at 5 to 30 m/s, or into the **C fiber group** of unmyelinated axons, which conduct at velocities generally less than 2 m/s, although a small number of Aβ neurons that conduct at 16 to 100 m/s are reported to transmit nociceptive signals. Thus, even though the conduction of nociceptive information is relatively slow, pain transmission can be generally classified as either fast or slow. At their peripheral endings, A fibers lose their myelin sheath and the unmyelinated branches

cluster in separated, small spots within a restricted area, which is the anatomical substrate for their receptive field (see Chapter 12). In contrast, C-fiber branches are more broadly distributed, which impedes precise localization of a stimulus.

Studies carried out in both humans and experimental animals demonstrated some time ago that the rapidly conducting axons that subserve somatic sensation are not involved in the transmission of pain. Figure 13.1 illustrates a typical experiment of this sort. The peripheral axons responsive to nonpainful mechanical or thermal stimuli do not discharge at a greater rate when painful stimuli are delivered to the same region of the skin surface. The nociceptive axons, by contrast, begin to discharge only when the strength of the stimulus (a thermal stimulus in the example in Figure 13.1) reaches high levels. At this same stimulus intensity, other thermoreceptors discharge at a rate no different from the maximum rate already achieved within the nonpainful temperature range, indicating the presence of both nociceptive and non-nociceptive thermoreceptors. When more slowly conducting Aδ and C fibers are stimulated electrically in humans, these fibers produce sensations of pain. There are also C fibers that mediate non-discriminative touch (discussed in Chapter 12), as

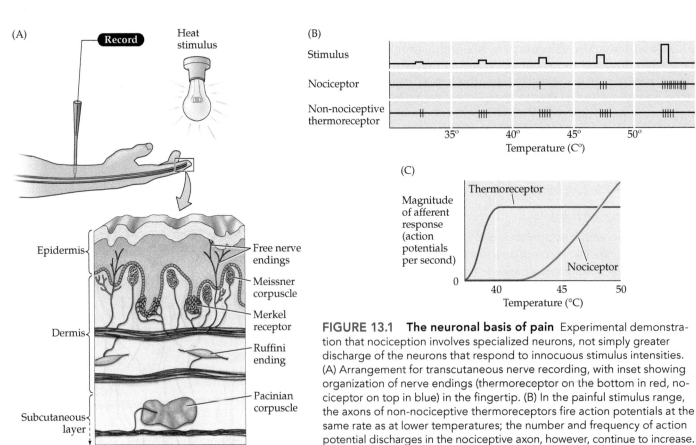

FIGURE 13.1 The neuronal basis of pain Experimental demonstration that nociception involves specialized neurons, not simply greater discharge of the neurons that respond to innocuous stimulus intensities. (A) Arrangement for transcutaneous nerve recording, with inset showing organization of nerve endings (thermoreceptor on the bottom in red, nociceptor on top in blue) in the fingertip. (B) In the painful stimulus range, the axons of non-nociceptive thermoreceptors fire action potentials at the same rate as at lower temperatures; the number and frequency of action potential discharges in the nociceptive axon, however, continue to increase. (Note that 43°C is the approximate threshold for pain.) (C) Summary of results. (After H. L. Fields, 1987. *Pain*, p. 19. New York: McGraw-Hill.)

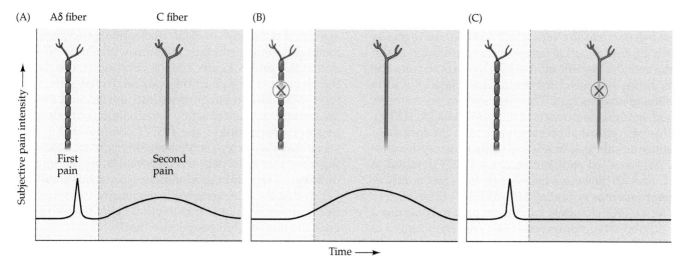

FIGURE 13.2   First and second pain Pain can be separated into an early perception of sharp pain and a later sensation that is described as having a duller, burning quality. (A) First and second pain, as these sensations are called, are carried by different axons, as can be shown by (B) the selective blockade of the more rapidly conducting myelinated axons that carry the sensation of first pain, or (C) blockade of the more slowly conducting C fibers that carry the sensation of second pain. (After H. L. Fields, 1990. *Pain Syndromes in Neurology*. London: Butterworths.)

well as the non-painful sensations of warmth, coolness, and itch that are discussed in Concept 13.4.

How, then, do different classes of nociceptors lead to the perception of pain? As mentioned, one way of determining the answer has been to stimulate different nociceptors in human volunteers while noting the sensations reported. In general, two categories of pain perception have been described: a sharp **first pain** and a more delayed, diffuse, and longer-lasting sensation that is generally called **second pain** (Figure 13.2A). Stimulation of the large, rapidly conducting Aα and Aβ axons in peripheral nerves does not elicit the sensation of pain. When investigators raise the stimulus intensity to a level that also activates a subset of Aδ fibers, however, a tingling sensation or, if the stimulation is intense enough, a feeling of sharp pain is reported. If the stimulus intensity is increased still further, so that the small-diameter, slowly conducting C-fiber axons are activated, then individuals report a duller, longer-lasting sensation of pain. By selectively blocking C fibers or Aδ fibers, researchers have demonstrated that, in general, Aδ fibers are responsible for first pain and C fibers are responsible for the duller, longer-lasting second pain (Figure 13.2B,C).

The faster-conducting Aδ nociceptors are now known to fall into two main classes. Type I Aδ fibers respond to dangerously intense mechanical (e.g., pinch or pin-prick) and chemical stimulation but have relatively high heat thresholds, while type II Aδ fibers have complementary sensitivities—that is, much lower thresholds for heat but very high thresholds for mechanical stimulation. Thus, the Aδ system has specialized pathways for the transmission of heat or mechanical nociceptive stimuli. Many of the slower-conducting, unmyelinated C-fiber nociceptors respond

to all forms of nociceptive stimuli—thermal, mechanical, and chemical—and are therefore said to be polymodal. However, C-fiber nociceptors are also heterogeneous, with subsets that respond preferentially to heat or chemical stimulation rather than mechanical stimulation. Other subtypes of C-fiber nociceptors are especially responsive to chemical irritants, acidic substances, or cold. These specialized response properties are due to unique expression patterns of receptors and ion channels that detect specific nociceptive stimuli (see Concept 13.2). In short, each of the major classes of nociceptive afferents is composed of multiple subtypes with distinct sensitivity profiles.

CONCEPT
13.2

Noxious Stimuli Are Transduced into Neural Signals via Various Ion Channel Receptors

LEARNING OBJECTIVES

13.2.1 Name three different types of ion channels that are involved in nociception and discuss what types of stimuli they transduce.

13.2.2 Explain how nociceptive stimuli are transmitted to the CNS.

Transduction and transmission of nociceptive signals

Nociceptors are tuned to detect a wide range of noxious mechanical, thermal (hot or cold temperatures), and chemical stimuli through the activation of modality-specific sensory transduction molecules. While many factors involved in nociception have been identified, the precise mechanisms through which pain signaling occurs are still an area of active investigation. Although many puzzles remain, significant insights have come from the identification of a specific receptor associated with the sensation of noxious heat. The threshold for perceiving a thermal stimulus as noxious heat is around 43°C (110°F), and this pain threshold corresponds with the threshold for activating subtypes of Aδ- and C-fiber nociceptive endings. The so-called vanilloid receptor (TRPV1), found in both C and Aδ fibers, is a member of the larger family of **transient receptor potential (TRP)** channels (discussed in Chapter 4). TRP channels are now known to comprise a large number of receptors sensitive to different ranges of heat and cold. Structurally, TRP channels resemble voltage-gated potassium or cyclic nucleotide-gated channels, having six transmembrane domains with a pore between domains 5 and 6. Under resting conditions, the pore of the channel is closed. In the open, activated state, these receptors allow an influx of sodium and calcium that initiates the generation of action potentials in the nociceptive neurons. Interestingly, the TRPV1 channel that confers sensitivity to heat also confers sensitivity to capsaicin, the ingredient in chili peppers responsible for the tingling or burning sensation produced by spicy foods (Box 13A). Given that the same receptor is responsive to heat as well as capsaicin, it is not surprising that many people experience the taste of chili peppers as "hot."

■ BOX 13A | Capsaicin

Capsaicin, the principle ingredient responsible for the pungency of hot peppers, is eaten daily by more than a third of the world's population. Capsaicin activates responses in a subset of nociceptive neurons by opening ligand-gated ion channels that permit the entry of Na+ and Ca2+. One of these channels, TRPV1, has been cloned and has been found to be activated by capsaicin, acid, and anandamide (an endogenous compound that also activates cannabinoid receptors), or by heating the tissue to about 43°C. It follows that anandamide and temperature are probably the endogenous activators of these channels. Mice whose TRPV1 receptors have been knocked out drink capsaicin solutions as if they were water. Receptors for capsaicin have been found in polymodal nociceptors of all mammals, but they are not present in birds (leading to the production of squirrel-proof birdseed laced with capsaicin).

When applied to the mucus membranes of the oral cavity, capsaicin acts as an irritant, producing protective reactions. When injected into skin, it produces a burning pain and elicits hyperalgesia to thermal and mechanical stimuli. Repeated applications of capsaicin also desensitize pain fibers and prevent neuromodulators such as substance P, vasoactive intestinal peptide (VIP), and somatostatin from being released by peripheral and central nerve terminals. Consequently, capsaicin is used clinically as an analgesic and anti-inflammatory agent; it is usually applied topically in a cream (0.075%) to relieve the pain associated with arthritis, postherpetic neuralgia, mastectomy, and trigeminal neuralgia. Thus, this remarkable chemical irritant not only gives gustatory pleasure on an enormous scale, but it is also a useful pain reliever.

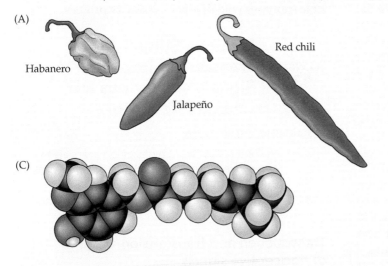

(A)

Habanero

Jalapeño

Red chili

(B) Capsaicin

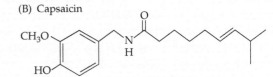

(C)

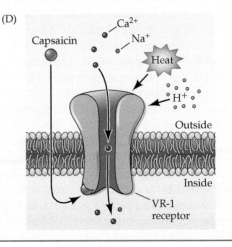

(D)

Capsaicin

Ca2+

Na+

Heat

H+

Outside

Inside

VR-1 receptor

(A) Some widely used peppers that contain capsaicin. (B) The chemical structure of capsaicin. (C) The capsaicin molecule. (D) Schematic of the TRPV1 (VR-1)/capsaicin receptor channel. This channel can be activated by capsaicin intracellularly, or by heat or protons (H+) at the cell surface.

Innocuous cool temperatures can be perceived when the skin is cooled as little as 1°C from normal body temperature. As temperatures descend below 15°C (59°F), the perception of cold pain can be felt, which is often described as including pricking, burning, and aching sensations in the affected region. Cold-responsive neurons (which account for about 10% to 20% of all dorsal root and trigeminal ganglia neurons) include both Aδ- and C-fiber types, and the temperature at which these neurons are activated varies. Most neurons that are sensitive to cold are also sensitive to the cooling compound menthol. Menthol is a cyclic terpene alcohol, which in small amounts can produce a pleasant cooling effect, as is the case with candy and gum. However, high concentrations of menthol can lead to pain sensations. Studies have identified that both menthol and cold stimuli evoke rapid opening of a nonselective cation channel, TRPM8, which leads to depolarization and the generation of an action potential in these cold-sensitive neurons.

Ion channels responsible for the transduction of mechanical and chemical forms of nociceptive stimulation have also been identified. These include other members of the TRP family (TRPV4), a rapidly adapting ion channel called Piezo2 (discussed in Chapter 12), and some members of the ASIC (acid-sensing ion channels) family. TRP channels also appear to be responsible for the detection of chemical irritants in the environment. TRPA1 in particular has been shown to be sensitive to a diverse group of chemical irritants, including the pungent ingredients in mustard and garlic plants, as well as volatile irritants present in tear gas, vehicle exhaust, and cigarettes. The ASIC3 channel subtype is specifically expressed in nociceptors, including neurons that innervate skeletal and cardiac muscle. ASIC3 channels are thought to be responsible for the muscle or cardiac pain that results from changes in pH associated with ischemia.

The graded potentials arising from receptors in the distal branches of nociceptive neurons must be transformed into action potentials in order to be conveyed to synapses in the CNS. Voltage-gated sodium and potassium channels are critical in this process (see Chapter 4), and several subtypes of sodium channels appear to be especially important for the transmission of nociceptive information. Altered activity of Nav1.7 is responsible for a variety of human pain disorders. Mutations of the *NAV1.7* gene (known as *SCN9A* in humans) that lead to a loss of this channel's function result in an inability to detect noxious stimulation, while mutations leading to hyperexcitability of the channel are associated with pain disorders that cause intense burning sensations. The *NAV1.8* gene is highly expressed by most C-fiber nociceptors and some Aδ neurons. Based on studies of mice, the NAV1.8 protein has been associated with the transmission of noxious mechanical and thermal information. Indeed, recent evidence indicates that genetic variations in the *NAV1.7*, *NAV1.8*, and *NAV1.9* genes are linked to a range of human pain disorders such as peripheral neuropathy. The development of local anesthetics specific to these subtypes of sodium channels may help treat a variety of intractable pain syndromes.

CONCEPT 13.3 | Central Pain Pathways Are Distinct from Mechanosensory Pathways and Transmit Different Aspects of Pain in Parallel

LEARNING OBJECTIVES

13.3.1 Explain how touch and pain signals are organized differently in the spinal cord dorsal horn.

13.3.2 Identify which pathways transmit pain information to the brain.

13.3.3 Describe the sensory–discriminative aspects of pain.

13.3.4 Describe the affective–motivational aspects of pain.

Central pain pathways

Pathways responsible for pain from the body originate with a subset of sensory neurons in dorsal root ganglia, and as with other sensory nerve cells, the central axons of nociceptive nerve cells enter the spinal cord via the dorsal roots (Figure 13.3A). When these centrally projecting axons reach the dorsal horn of the spinal cord, they branch into ascending and descending collaterals, forming the **dorsolateral tract of Lissauer** (named after the German neurologist who first described this pathway in the late nineteenth century). Axons in Lissauer's tract typically run up and down for one or more spinal cord segments before they penetrate the gray matter of the dorsal horn. Once within the dorsal horn, the axons give off branches that contact second-order neurons in multiple laminae. These afferent terminations are organized in a lamina-specific fashion: Most C and Aδ nociceptive neurons form synaptic connections in the superficial laminae. C-fiber nociceptors primarily terminate in laminae I and II, with some terminations in lamina V, while Aδ nociceptors terminate in laminae I and V. In contrast, low-threshold Aδ and Aβ afferents generally project to deeper laminae (III, IV, and V). As discussed in Chapter 12, the dorsal horn comprises a large number of excitatory and inhibitory interneuron subtypes throughout the Rexed's laminae, as well as a smaller population of projection neurons that are located in laminae I, III, IV, and V. Projection neurons have axons that travel to brainstem and thalamic targets. While there are interneurons in all laminae of the spinal cord, they are especially abundant and diverse morphologically and histochemically in lamina II. A subset of the lamina V neurons called

(A)

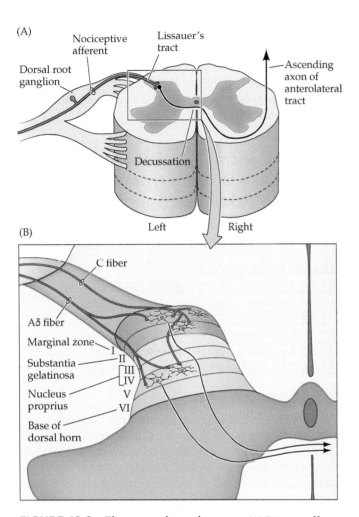

(B)

FIGURE 13.3 **The anterolateral system** (A) Primary afferents in the dorsal root ganglia send their axons via the dorsal roots to terminate in the dorsal horn of the spinal cord. Afferents branch and course for one or more segments up and down the spinal cord in Lissauer's tract, giving rise to collateral branches that terminate in the dorsal horn. Second-order neurons in the dorsal horn send their axons (black) across the midline to ascend to higher levels in the anterolateral column of the spinal cord. (B) C-fiber nociceptive afferents primarily terminate in Rexed's laminae I and II of the dorsal horn (with some terminations in lamina V), while Aδ nociceptive fibers terminate in laminae I and V. The axons of second-order neurons in laminae I and V cross the midline and ascend to higher centers.

wide-dynamic-range neurons are multimodal and receive converging inputs from nociceptive and non-nociceptive afferents. Some of them receive visceral sensory input as well, making them a likely substrate for referred pain (i.e., pain that arises from damage to visceral organs but is misperceived as coming from a somatic location). The most common clinical example is angina, in which poor perfusion of the heart muscle is misperceived as pain in the chest wall, the shoulder, and the left arm and hand (Box 13B). Dorsal horn interneurons and descending pathways

originating from the brain modulate the transmission of nociceptive signals, which can contribute to pain perception and the prioritization of pain relative to homeostatic demands or competing behavioral needs.

First-order neurons that convey pain, temperature, and crude (less well defined) touch terminate in the dorsal horn, and second-order neurons in the dorsal horn send their axons across the midline and ascend on the contralateral side of the cord (in the anterolateral [also called ventrolateral] column) to their targets in the thalamus and brainstem (Figure 13.3B). For this reason, this neural pathway that conveys pain, temperature, and crude touch information to higher centers is often referred to as the **anterolateral system**, to distinguish it from the dorsal column–medial lemniscal system (and also the spinocervical tract) that convey mechanosensory information (see Chapter 12).

Axons conveying information for the anterolateral system and the dorsal column–medial lemniscal system travel in different parts of the spinal cord white matter. This difference provides a clinically relevant sign that is useful for defining the locus of a spinal cord lesion. The first-order and projection neurons for the dorsal column–medial lemniscal system possess axons that ascend in the ipsilateral dorsal column all the way to the medulla, where they synapse on neurons in the dorsal column nuclei (Figure 13.4, left panel). The axons of neurons in the dorsal column nuclei then cross the midline and ascend to the contralateral thalamus. In contrast, the crossing point for information conveyed by the anterolateral system lies within the spinal cord.

Because of this anatomical difference in the site of decussation for touch versus pain pathways, a unilateral spinal cord lesion results in dorsal column–medial lemniscal symptoms (loss of sensation of touch, pressure, vibration, and proprioception) on the side of the body *ipsilateral* to the lesion, and anterolateral symptoms (deficits of pain and temperature perception) on the *contralateral* side of the body (see Figure 13.4, right panel). The deficits are due to the interruption of fibers ascending from lower levels of the cord; for this reason, they include all regions of the body (on either the contralateral or ipsilateral side) that are innervated by spinal cord segments that lie below the level of the lesion. This pattern of **dissociated sensory loss** (contralateral pain and temperature, ipsilateral touch and pressure) is a signature of spinal cord lesions and, together with local dermatomal signs (see Clinical Applications, Chapter 12), can be used to define the level of the lesion. (Box 13C discusses an important exception to the functional dissociation of the dorsal column–medial lemniscal and anterolateral systems for visceral pain.) It is worth noting, however, that while distinct nociceptor and low-threshold mechanosensory afferents convey different aspects of the tactile world through unique pathways, recent studies

■ BOX 13B | Referred Pain

Surprisingly, few if any neurons in the dorsal horn of the spinal cord are specialized solely for the transmission of *visceral* (internal) pain. Obviously, we recognize such pain, but it is conveyed centrally via dorsal horn neurons that may also convey *cutaneous* pain. As a result of this economical arrangement, the disorder of an internal organ is sometimes perceived as cutaneous pain. An individual may therefore present to the physician with the complaint of pain at a site other than its actual source, a potentially confusing phenomenon called *referred pain*. The most common clinical example is anginal pain (pain arising from heart muscle that is not being adequately perfused with blood) referred to the upper chest wall, with radiation into the left arm and hand. Other important examples are gallbladder pain referred to the scapular region, esophageal pain referred to the chest wall, ureteral pain (e.g., from passing a kidney stone) referred to the lower abdominal wall, bladder pain

referred to the perineum, and the pain from an inflamed appendix referred to the anterior abdominal wall around the umbilicus. Understanding referred pain can lead to an astute diagnosis that might otherwise be missed.

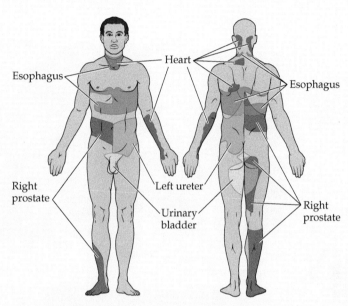

Examples of pain arising from a visceral disorder referred to a cutaneous region (color and label).

FIGURE 13.4 Nociceptive and mechanosensory pathways
As diagrammed here, the anterolateral system (blue) crosses and ascends in the contralateral anterolateral column of the spinal cord, while the dorsal column–medial lemniscal system (red) ascends in the ipsilateral dorsal column. A lesion restricted to the left half of the spinal cord results in dissociated sensory loss and mechanosensory deficits on the left half of the body, with pain and temperature deficits experienced on the right.

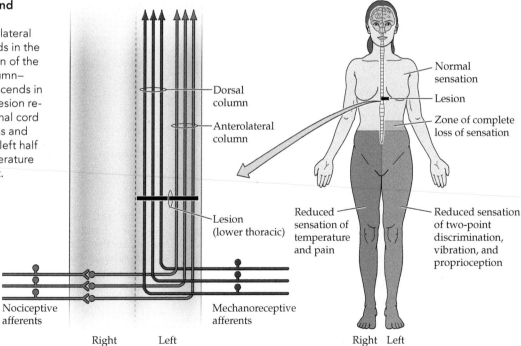

■ BOX 13C │ A Dorsal Column Pathway for Visceral Pain

Chapters 12 and 13 present a framework for considering the central neural pathways that convey innocuous mechanosensory signals and painful signals from cutaneous and deep somatic sources. Considering just the signals derived from the body below the head, discriminative mechanosensory and proprioceptive information travels to the ventral posterior thalamus via the dorsal column–medial lemniscal system (see Figure

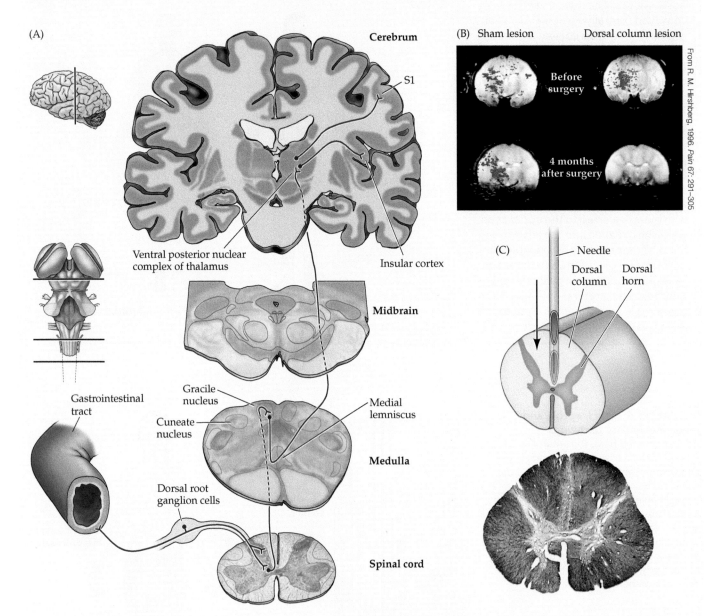

(A) A visceral pain pathway in the dorsal column–medial lemniscal system. For simplicity, only the pathways that mediate visceral pain from the pelvis and lower abdomen are illustrated. (B) Empirical evidence supporting the existence of the visceral pain pathway shown in (A). Increased neural activity was observed with fMRI techniques in the thalamus of monkeys that were subjected to noxious distention of the colon and rectum, indicating the processing of visceral pain. This activity was abolished by lesion of the dorsal columns at T10, but not by "sham" surgery. (C) Top: One method of punctate midline myelotomy for the relief of severe visceral pain. Bottom: Myelin-stained section of the thoracic spinal cord (T10) from a patient who underwent midline myelotomy for the treatment of colon cancer pain that was not controlled by analgesics. After surgery, the patient experienced relief from pain during the remaining 3 months of his life. (B from W. D. Willis et al., 1999. *Proc Natl Acad Sci USA* 96: 7675–7679; drawing after H. J. W. Nauta et al., 1997. *J Neurosurg* 86: 538–542.)

■ BOX 13C | A Dorsal Column Pathway for Visceral Pain (*continued*)

13.4, left panel), while nociceptive information travels to the same (and additional) thalamic nuclei via the anterolateral system (see Figure 13.6A).

But how do painful signals that arise in the visceral organs of the pelvis, abdomen, and thorax enter the CNS and ultimately reach an individual's consciousness? The answer is via a component of the dorsal column–medial lemniscal pathway that conveys visceral nociception. Although Chapter 21 will present more information on the systems that receive and process visceral sensory information, at this juncture it is worth considering how a better understanding of this particular pathway has begun to affect clinical medicine.

Primary visceral afferents from the pelvic and abdominal viscera enter the spinal cord and synapse on second-order neurons in the dorsal horn of the lumbosacral spinal cord. As discussed in Box 13B and Chapter 21, some of these second-order neurons are cells that give rise to the anterolateral system and are thought to contribute to referred visceral pain sensations. However, other neurons synapse on neurons in the intermediate gray region of the spinal cord near the central canal. These neurons, in turn, send their axons not through the anterolateral white matter of the spinal cord (as might be expected for a pain pathway) but through the

dorsal columns in a position very near the midline (Figure A). Similarly, second-order neurons in the thoracic spinal cord that convey nociceptive signals from thoracic viscera send their axons rostrally through the dorsal columns along the dorsal intermediate septum, near the division of the gracile and cuneate tracts. These second-order axons then synapse on the dorsal column nuclei of the caudal medulla, where neurons give rise to arcuate fibers that form the contralateral medial lemniscus and eventually synapse on thalamocortical projection neurons in the ventral posterior thalamus.

This dorsal column visceral sensory projection appears to be the principal pathway by which painful sensations arising in the viscera are detected and discriminated. Several observations support this conclusion: (1) Neurons in the ventral posterior lateral nucleus, gracile nucleus, and near the central canal of the spinal cord all respond to noxious visceral stimulation; (2) responses of neurons in the ventral posterior lateral nucleus and gracile nucleus to such stimulation are greatly reduced by spinal lesions of the dorsal columns (Figure B), but not by lesions of the anterolateral white matter; and (3) infusion of drugs that block nociceptive synaptic transmission into the intermediate gray region of the sacral

spinal cord blocks the responses of neurons in the gracile nucleus to noxious visceral stimulation, but not to innocuous cutaneous stimulation.

The discovery of this visceral sensory component in the dorsal column–medial lemniscal system has helped explain why surgical transection of the axons that run in the medial part of the dorsal columns (a procedure termed *midline myelotomy*) generates significant relief from the debilitating pain that can result from visceral cancers in the abdomen and pelvis. Although the initial development of this surgical procedure preceded the elucidation of this visceral pain pathway, these new discoveries have renewed interest in midline myelotomy as a palliative neurosurgical intervention for cancer patients whose pain is otherwise unmanageable. Indeed, precise knowledge of the visceral sensory pathway in the dorsal columns has led to further refinements that permit a minimally invasive (*punctate*) surgical procedure that attempts to interrupt the second-order axons of this pathway within just a single spinal segment (typically, at mid- or lower-thoracic level; Figure C). In so doing, this procedure offers some hope to patients who struggle to maintain a reasonable quality of life in extraordinarily difficult circumstances.

---

have shown that cross-talk among these lines of sensory information exists and shapes somatosensory perception. Indeed, pain sensations are generated by a summation of inputs from various primary sensory afferents, and this information is modulated by spinal cord neurons as well as descending inputs from the brain.

Transmission of different qualities of pain sensations by parallel pain pathways

Second-order neurons in the anterolateral system project to several different structures in the brainstem and forebrain, making it clear that pain is processed by a diverse and distributed network of neurons. While the full significance of this complex pattern of connections remains unclear, these central destinations are likely to mediate

different aspects of the sensory and behavioral responses to a painful stimulus.

The spinothalamic tract, a part of the anterolateral system, consists of two adjacent pathways: lateral and anterior. The lateral spinothalamic tract conveys pain and temperature, while the anterior spinothalamic tract mediates the **sensory–discriminative** aspects of pain: the location, intensity, and quality of the noxious stimulation. These aspects of pain are thought to depend on information transmitted through the ventral posterior lateral nucleus (VPL) to neurons in the primary and secondary somatosensory cortex (Figures 13.5 and 13.6A). (The pathway for information from the face to the ventral posterior medial nucleus, or VPM, is considered later in this concept.) Although axons from the anterolateral system overlap with those from

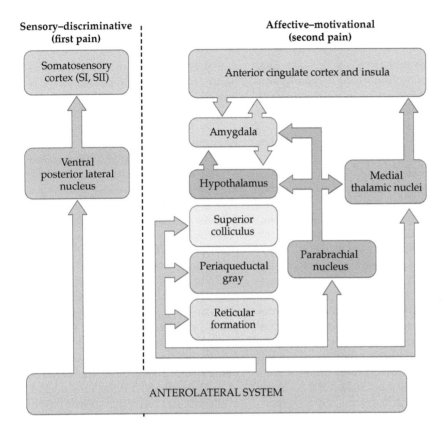

Sensory–discriminative (first pain)

Affective–motivational (second pain)

FIGURE 13.5 Two distinct aspects of the experience of pain
The anterolateral system supplies information to different structures in the brainstem and forebrain that contribute to different aspects of the experience of pain. The spinothalamic tract (left of dashed line) conveys signals that mediate the sensory discrimination of first pain. The affective-motivational aspects of second pain are mediated by complex pathways that reach integrative centers in the limbic forebrain.

the dorsal column system in the ventral posterior nuclei, these axons contact different classes of neurons. Consistent with mediating the sensory-discriminative aspects of pain, electrophysiological recordings from nociceptive neurons in the primary somatosensory cortex (SI) show that these neurons have small, localized receptive fields—properties commensurate with behavioral measures of pain localization.

Other parts of the anterolateral system convey information about the **affective–motivational** aspects of pain: the unpleasant feeling, the fear and anxiety, and the autonomic activation that accompany exposure to a noxious stimulus (the classic fight-or-flight response; see Chapter 21). Targets of these projections are numerous and include several subdivisions of the reticular formation, the periaqueductal gray, the deep layers of the superior colliculus, and the parabrachial nucleus in the rostral pons (see Figure 13.5). The parabrachial nucleus processes and transmits second pain signals to the amygdala, hypothalamus, and a distinct set of thalamic nuclei that lie medial to the ventral posterior nucleus, which we group together here as the medial

thalamic nuclei. These medial thalamic nuclei, which also receive input from anterolateral system axons, play an important role in transmitting nociceptive signals to both the anterior cingulate cortex and the insula. Together with the amygdala and hypothalamus, which are also interconnected with the cingulate cortex and insula, these limbic forebrain structures elaborate affective–motivational aspects of pain (see Chapter 32). For example, electrophysiological recordings in humans, which show that cingulate neurons respond to noxious stimuli, support the role of the anterior cingulate cortex in the perception of pain. Moreover, patients who have undergone cingulotomies report an attenuation of the unpleasantness that accompanies pain. Evidence from functional imaging studies in humans supports the view that different brain regions mediate the sensory–discriminative and affective–motivational aspects of pain. The presentation of a painful stimulus results in the activation of both primary somatosensory cortex and anterior cingulate cortex.

The full experience of pain involves the cooperative action of an extensive network of forebrain regions whose properties we are only beginning to understand. Indeed, brain-imaging studies frequently refer to the broad array of areas whose activity is associated with the experience of pain—including the somatosensory cortex, insular cortex, thalamus, and anterior cingulate cortex—as the **pain matrix**. In retrospect, the distributed nature of pain representation should not be surprising given that pain is a multidimensional experience with sensory, motor, affective, and cognitive effects. A distributed representation also explains why ablations of the somatosensory cortex do not usually alleviate chronic pain, even though they severely impair contralateral mechanosensory perception.

Information about noxious and thermal stimulation of the face originates from first-order neurons located in the trigeminal ganglion and from ganglia associated with cranial nerves VII, IX, and X (facial, glossopharyngeal, and vagal nerves, respectively) (Figure 13.6B). After entering the pons, these small myelinated and unmyelinated trigeminal fibers descend to the medulla, forming the spinal trigeminal tract and terminating in two subdivisions of the spinal trigeminal nucleus: the pars interpolaris and pars caudalis. Axons from the second-order neurons in these two trigeminal subdivisions cross the midline and terminate in a variety of targets in the brainstem and thalamus. Like their counterparts in the dorsal

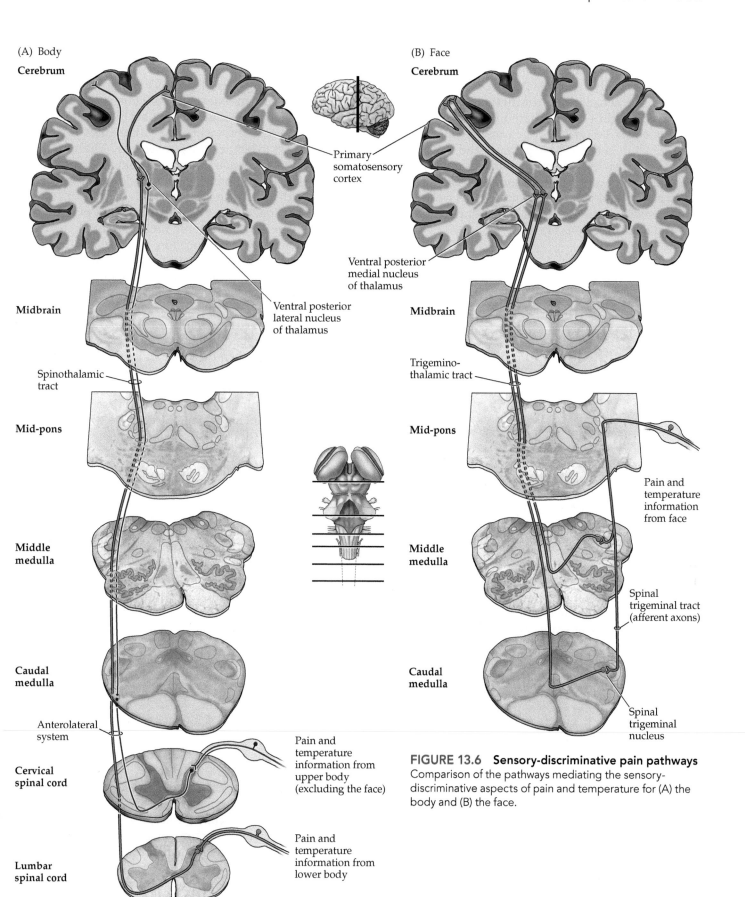

(A) Body
Cerebrum

Primary somatosensory cortex

(B) Face
Cerebrum

Ventral posterior medial nucleus of thalamus

Ventral posterior lateral nucleus of thalamus

Midbrain

Spinothalamic tract

Mid-pons

Midbrain

Trigemino-thalamic tract

Mid-pons

Pain and temperature information from face

Middle medulla

Middle medulla

Spinal trigeminal tract (afferent axons)

Caudal medulla

Caudal medulla

Spinal trigeminal nucleus

Anterolateral system

Cervical spinal cord

Pain and temperature information from upper body (excluding the face)

Lumbar spinal cord

Pain and temperature information from lower body

FIGURE 13.6 Sensory-discriminative pain pathways
Comparison of the pathways mediating the sensory-discriminative aspects of pain and temperature for (A) the body and (B) the face.

horn of the spinal cord, these targets can be grouped into those that mediate the sensory-discriminative aspects of pain and those that mediate the affective–motivational aspects. The sensory-discriminative aspects of facial pain are thought to be mediated by projections to the contralateral VPM (via the trigeminothalamic tract) and projections from the VPM to primary and secondary somatosensory cortex. Affective–motivational aspects are mediated by connections to various targets in the reticular formation and parabrachial nucleus, as well as by the medial nuclei of the thalamus, which supply the cingulate and insular regions of cortex.

CONCEPT 13.4 | Signals from a Variety of Innocuous Stimuli, as Well as From Internal Organs, Are Also Carried by Pain Pathways

LEARNING OBJECTIVES

13.4.1 Describe how temperature is detected by the nervous system.

13.4.2 Define what pruriceptors are and what they detect.

13.4.3 Define the term interoception and discuss how it differs from exteroception and proprioception.

13.4.4 Discuss the importance of interoception for homeostasis and how sensory signals from the internal organs are conveyed to the CNS.

Other modalities mediated by the anterolateral system

While the anterolateral system plays a critical role in mediating nociception, it is also responsible for transmitting a variety of other innocuous information to higher centers. For example, in the absence of the dorsal column system, the anterolateral system appears to be capable of mediating what is commonly called *non-discriminative touch*, a form of tactile sensitivity that lacks the fine spatial resolution that can be supplied only by the dorsal column system. The C-fiber low-threshold mechanoreceptors (C-LTMRs) mediate aspects of this non-discriminative (affective) touch (see Chapter 12). Thus, following damage to the dorsal column–medial lemniscal system, a "crude" form of tactile sensation remains, one in which two-point discrimination thresholds are increased and the ability to identify objects by touch alone (*stereognosis*) is markedly impaired.

As already noted in Concept 13.3, the anterolateral system is also responsible for mediating innocuous temperature sensation. The sensations of warmth and cold are thought to be subserved by separate sets of primary afferents. Neither of these afferent types responds to mechanical stimulation, and they are distinct from afferents that respond to temperatures that are considered painful (noxious heat, above 43°C; or noxious cold, below 15°C). Warmth-sensing fibers express TRPV3 and TRPV4, while cold-sensing fibers express TRPM8 (discussed in Concept 13.2); these channels endow each fiber type with the ability to respond with increasing spike discharges to increases or decreases in temperature, respectively. The information supplied by innocuous warm and cold afferents is transmitted to higher centers by distinct classes of spinal cord neurons that reside in lamina I.

In addition, subsets of C fibers, called pruriceptors, are activated by prurigenic (itch-inducing) chemicals such as histamine produced by mast cells, substance P produced by primary sensory neurons, and bile acid produced by the liver. Although there are many similarities between pain and itch sensations, emerging evidence indicates that itch is a distinct sensory modality that employs specific neural circuits. At least three subtypes of pruriceptors innervate the body and respond to different pruritogens that can then lead to itching behaviors. For example, an insectbite on the skin will cause mast cells to release histamine, which activates primary neurons that express histamine receptor type 1 (H1R). H1R-expressing pruriceptors make synaptic contacts with neurons in lamina I of the dorsal horn that contain gastrin releasing peptide (GRP). The third-order neuron expresses GRP receptors and excites pruriceptive neurons that ascend the spinothalamic tract and transmit the perception of itch stimuli to higher brain centers, including the thalamus, primary and secondary somatosensory cortex, prefrontal cortex, and anterior cingulate cortex, among others. Interestingly, itchiness is one of the most common side effects of drugs that target μ-opioid receptors (discussed in Concept 13.5), and this side effect seems to be related to cross-activation of GRP receptors in the spinal cord. Although itch is an unpleasant sensation, scratching an itch can produce a pleasant experience, which may in turn facilitate the scratching behavior. Neurons in the ventral tegmental area, a known reward center in the brain, are activated and release dopamine during itch-induced scratching behaviors, which is thought to promote the itch–scratch cycle.

As noted earlier, lamina I of the spinal cord consists of several distinct classes of modality-selective neurons that convey noxious and innocuous types of sensory information to the anterolateral system. These include individual classes of neurons that are sensitive to a variety of stimuli: sharp (first) pain, burning (second) pain, innocuous warmth, innocuous cold, the sense of itch, slow mechanical stimulation (non-discriminative, affective touch), and a class of inputs that innervate muscles and sense lactic acid and other metabolites that are released during muscle contraction. The last of which may contribute to the "burn" or ache that can accompany strenuous exercise.

Is lamina I merely an eclectic mixture of cells with different properties, or might a unifying theme account for this diversity? It has been proposed that the lamina I system functions as the sensory input to a network that is responsible for representing the physiological condition of the body—a modality that has been called **interoception**, to distinguish it from **exteroception** (sensing of stimuli external to the body and proprioception. These inputs drive the homeostatic mechanisms that maintain an optimal internal state. Some of these mechanisms are automatic, and the changes necessary to maintain homeostasis can be mediated by reflexive adjustment of the autonomic nervous system (see Chapter 21). For example, changes in temperature evoke autonomic reflexes (e.g., sweating or shivering) that counter a disturbance to the body's optimal temperature. Homeostatic disturbances are sometimes too great to be mediated by autonomic reflexes alone and require behavioral adjustments (e.g., putting on or taking off a sweater) to restore balance. In this conception, the sensations associated with the activation of the lamina I system—whether pleasant or noxious—motivate the initiation of behaviors appropriate to maintaining the physiological homeostasis of the body.

In addition to their roles already described, nociceptor neurons of the dorsal root ganglia are responsible for detecting painful stimuli from the internal organs (discussed in detail in Chapter 21). These afferents are predominantly unmyelinated, small-diameter C fibers and medium-diameter, thinly myelinated Aδ fibers. In the gastrointestinal tract alone, at least six classes of colon-innervating spinal afferents have been identified, defined by their termination locations within the gastrointestinal (GI) tract layers and by ex vivo responses to stimuli. Dorsal root ganglia neurons are necessary for conscious and unconscious sensations of the GI tract, including pain, stretch, mechanical force, and stroke. Critically, these neurons also play key roles in detecting the mechanical and microbial constituents within the GI tract and constitute the afferent limb of spinal and brainstem reflexes that enable long-range control of GI motility and secretion. Gastrointestinal pain is remarkably common and often associated with various diseases and disorders. Mechanisms of visceral pain are complex, including both peripheral and central sensitization mechanisms (see Concept 13.5 and also Box 13C). Understanding the complexities of interoception is an active area of research investigation.

CONCEPT 13.5 Sensitivity to Pain Is Subjective and Can Be Modified by a Variety of Factors

LEARNING OBJECTIVES

13.5.1 Explain what central and peripheral sensitization mechanisms are.

13.5.2 Describe how tissue damage contributes to increased pain sensitivity via both somatic and neural mechanisms.

13.5.3 Define allodynia and explain what mechanisms contribute to this pathogenic process.

13.5.4 Discuss what the placebo effect is and what is known about the physiological mechanisms that underlie it.

13.5.5 Explain how touch stimuli, descending pathways, and endogenous opioids modulate pain perception.

Sensitization: Peripheral mechanisms and treatments

Following a painful stimulus associated with tissue damage (e.g., cuts, scrapes, bruises, and burns), stimuli in the area of the injury and the surrounding region that would ordinarily be perceived as slightly painful are perceived as significantly more so, a phenomenon referred to as **hyperalgesia**. A good example of hyperalgesia is the increased sensitivity to temperature that occurs after sunburn. This effect is due to changes in neuronal sensitivity that occur at the level of peripheral receptors as well as their central targets.

Peripheral sensitization results from the interaction of nociceptors with the "inflammatory soup" of substances released when tissue is damaged. These substances arise from activated nociceptors or from non-neuronal cells that reside in, or migrate to, the injured area. Nociceptors release peptides and neurotransmitters such as substance P, calcitonin gene-related peptide (CGRP), and ATP, all of which further contribute to the inflammatory response (vasodilation, swelling, and the release of histamine from mast cells). The list of non-neuronal cells that contribute to this inflammatory soup includes mast cells, platelets, basophils, macrophages, neutrophils, endothelial cells, keratinocytes, and fibroblasts. These cells are responsible for releasing extracellular protons, arachidonic acid and other lipid metabolites, bradykinin, histamine, serotonin, prostaglandins, nucleotides, nerve growth factor (NGF), and numerous cytokines, chief among them interleukin-1β (IL-1β) and tumor necrosis factor α (TNF-α). Most of these substances interact directly with receptors or ion channels of nociceptive neurons, augmenting their responses (Figure 13.7). For example, the responses of the TRPV1 receptor to heat can be potentiated by direct interaction of the channel with extracellular protons or lipid metabolites. NGF and bradykinin also potentiate the activity of the TRPV1 receptor, but do so indirectly through the actions of separate cell surface receptors (TrkA and bradykinin receptors, respectively) and their associated intracellular signaling pathways. The prostaglandins are thought to contribute to peripheral sensitization by binding to G-protein-coupled receptors that increase levels of cyclic AMP within nociceptors. Prostaglandins also reduce the threshold depolarization required for generating action potentials via phosphorylation of a specific class of TTX-resistant sodium channels that are

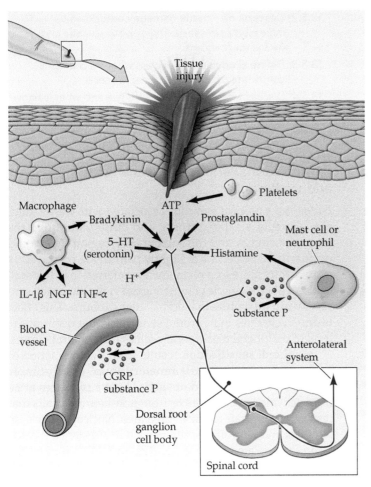

FIGURE 13.7 **Inflammatory response to tissue damage** Substances released by damaged tissues augment the response of nociceptive neurons. In addition, electrical activation of nociceptors causes the release of peptides and neurotransmitters that further contribute to the inflammatory response.

in the biosynthesis of prostaglandins. Interfering with neurotrophin or cytokine signaling has become a major strategy for controlling inflammatory disease and the resulting pain. Blocking the action of TNF-α with a neutralizing antibody has been significantly effective in the treatment of autoimmune diseases, including rheumatoid arthritis and Crohn's disease, leading to dramatic reductions in both tissue destruction and the accompanying hyperalgesia. Likewise, anti-NGF antibodies have been shown to prevent and to reverse the behavioral signs of hyperalgesia in animal models.

As already mentioned, peripheral sensitization and the associated hyperalgesia are initiated and maintained, in part, by the actions of CGRP. While CGRP administration alone does not cause nociception, injection of CGRP into the paw of a rodent will cause a reduction in the response threshold to a noxious stimulus through enhancement of sodium currents in nociceptor neurons. Activation of nociceptor endings can cause the release of CGRP from adjacent free nerve endings of the same axon, resulting in the release of additional proinflammatory mediators, which can further excite adjacent nerve endings. Thus, CGRP appears to facilitate nociceptive transmission and contributes to the development and maintenance of a sensitized state in nociceptor neurons. CGRP signaling has been implicated in a large number of pathogenic pain states, such as inflammatory bowel disease, rheumatoid arthritis, and migraines. CGRP-related therapies, such as CGRP receptor antagonists or monoclonal antibodies against CGRP or its receptor, are the first of their kind to be designed specifically to act on the trigeminal pain system for migraine treatments. These targeted therapies offer substantial improvements over existing drugs, as they are more specific and have few or no adverse effects, because of their peripheral site of action.

Sensitization: Central mechanisms and treatments

Central sensitization refers to a rapid onset, activity-dependent increase in the excitability of neurons in the dorsal horn of the spinal cord following high levels of activity in the nociceptive afferents. As a result, activity levels in nociceptive afferents that were subthreshold prior to the sensitizing event become sufficient to generate action potentials in dorsal horn neurons, contributing to an increase in pain sensitivity. This phenomenon typically occurs immediately after the painful event and can outlast the pain of the original stimulus by several hours. Although central sensitization is triggered in dorsal horn neurons by activity in nociceptors, the effects can generalize to other inputs that arise from low-threshold mechanoreceptors. Thus, stimuli that under normal conditions would be innocuous (such as brushing the surface of the

expressed in nociceptors. Cytokines can directly increase sodium channel activity, via activation of the MAP kinase signaling pathway, and can also potentiate the inflammatory response via increased production of prostaglandins, NGF, bradykinin, and extracellular protons.

The presumed purpose of the complex chemical signaling cascade arising from local damage is not only to protect the injured area, but also to promote healing and guard against infection by means of local effects such as increased blood flow and the migration of white blood cells to the site, and by the production of factors (e.g., resolvins) that reduce inflammation and resolve pain. Indeed, identifying the components of the inflammatory soup and their mechanisms of action is a fertile area of exploration in the search for potential analgesics (compounds that reduce pain intensity). For example, NSAIDs (*nonsteroidal anti-inflammatory drugs*), which include aspirin and ibuprofen, act by inhibiting cyclooxygenase (COX), an enzyme important skin) activate second-order neurons in the dorsal horn that receive nociceptive inputs, giving rise to a sensation of pain. The induction of pain by a normally innocuous stimulus that ordinarily does not elicit a painful response is referred to as **allodynia**.

As in peripheral sensitization, several different mechanisms contribute to central sensitization. One form of central sensitization, called *windup*, involves a progressive increase in the discharge rate of dorsal horn neurons in response to repeated low-frequency activation of nociceptive afferents. A behavioral correlate of the windup phenomenon has been studied by examining the perceived intensity of pain in response to multiple presentations of a noxious stimulus. Although the intensity of the stimulation is constant, the perceived intensity increases with each stimulus presentation. Windup lasts only during the period of stimulation and arises from the summation of the slow synaptic potentials evoked in dorsal horn neurons by nociceptive inputs. The sustained depolarization of the dorsal horn neurons results in part from the activation of voltage-dependent L-type calcium channels, and in part from the removal of the Mg^{2+} block of NMDA receptors. Removing the Mg^{2+} block increases the sensitivity of the dorsal horn neuron to glutamate, the neurotransmitter in nociceptive afferents.

Other forms of central sensitization that last longer than the period of sensory stimulation (e.g., allodynia) are thought to involve an LTP-like enhancement of postsynaptic potentials, much like that described for the hippocampus (see Chapter 8). These effects are dependent on NMDA receptor-mediated elevations of Ca^{2+} in spinal cord neurons postsynaptic to nociceptors. Reductions in the level of GABAergic or glycinergic inhibition in spinal cord circuits are also thought to contribute to persistent pain syndromes by increasing the excitability of dorsal horn projection neurons. One mechanism affecting GABA-mediated inhibition is the dysregulation of intracellular chloride. In conditions promoting central sensitization, the function or expression of a potassium–chloride co-transporter (KCC2) in dorsal horn neurons may become impaired, reducing the driving force for chloride and thus the strength of inhibitory transmission in the dorsal horn.

CGRP signaling has also been implicated in promoting central sensitization during pathological pain states. Early studies showed that stimulation of peripheral nociceptor axons evokes CGRP release from nociceptor central terminals in the spinal cord. While spinal administration of CGRP antisera did not alter baseline nociceptive behavioral responses, CGRP antisera were effective in blocking behavioral signs of hyperalgesia in rats with carrageenan-induced inflammation or experimental arthritis. These results suggest that CGRP does not mediate normal acute nociceptive signals, but rather promotes enhanced abnormal pain sensitivity in pathological states.

Microglia and astrocytes also contribute to the central sensitization process, especially when there is injury to the nerve, or in other chronic pain conditions associated with arthritis, chemotherapy, and cancer. For example, pro-inflammatory cytokines such as IL-1β released from microglia promote the widespread transcription of the COX-2 enzyme and ensuing production of prostaglandins in dorsal horn neurons. As described for nociceptive afferents, increased levels of prostaglandins in CNS neurons augment neuronal excitability. Thus, the analgesic effects of drugs that inhibit COX-2 transcription are due to actions in both the periphery and in the dorsal horn. Microglia also produce TNF-α and BDNF (*brain-derived neurotrophic factor*), which enhance excitatory synaptic transmission and suppress inhibitory synaptic transmission in nociceptive circuitry. Furthermore, astrocytes also produce chemokines such as CCL2 and CXCL1 to enhance pain transmission in the spinal cord. Finally, while microglia are activated after injury to a nerve in males and females, drugs that inhibit microglial activation are effective mainly in males, suggesting sex-specific effects of certain drugs after nerve damage. Indeed, sex differences are observed frequently in animal models for chronic pain, which may provide critical insight for the sex differences in pain often observed in humans.

As injured tissue heals, the sensitization induced by peripheral and central mechanisms typically declines and the threshold for pain returns to pre-injury levels. However, when the afferent nerve fibers or central pathways themselves are damaged—a frequent complication in pathological conditions, including diabetes, shingles, AIDS, multiple sclerosis, trauma, and stroke—these processes can persist. The resulting condition is referred to as **neuropathic pain**: a chronic, intensely painful experience that is difficult to treat with conventional analgesic medications and that presents with both allodynia and hyperalgesia. (Clinical Applications describes neuropathic pain associated with amputation of an extremity.)

Although acute pain occurs through the activation of nociceptors, light-touch-activated mechanosensory neurons can be recruited into the nociceptive network during neuropathic pain. One of the predominant concepts to explain this phenomenon is based on the gate control theory of pain (discussed in more detail later in this concept), which proposes that touch stimuli inhibit pain signaling through a feedforward inhibitory circuit in the superficial layers of the dorsal horn. During mechanical allodynia, an injury is proposed to impair this feedforward inhibitory circuit, which can cause light-touch neurons to engage with the nociceptive circuits in laminae I and II. Researchers found that in the presence of inhibitory receptor antagonists (to mimic injury-induced decrease in inhibition in the dorsal horn), Aβ-fiber input is sufficient to activate lamina I pain projection neurons through a polysynaptic

■ Clinical Applications

Phantom Limbs and Phantom Pain

Following the amputation of an extremity, nearly all patients have an illusion that the missing limb is still present. Although this illusion usually diminishes over time, it persists in some degree throughout the individual's life and can often be reactivated by injury to the stump or other perturbations. In some people, the perceptions may even increase over time. Such phantom sensations are not limited to amputated limbs; phantom breasts following mastectomy, phantom genitalia following castration, and phantoms of the entire lower body following spinal cord transection have all been reported (Figure A). Phantoms are also common after local nerve block for surgery. During recovery from brachial plexus anesthesia, for example, it is not unusual for the patient to experience a phantom arm, perceived as whole and intact, but displaced from the real arm. When the real arm is viewed, the phantom appears to "jump into" the arm and may emerge and reenter

intermittently as the anesthesia wears off. These sensory phantoms demonstrate that the central machinery for processing somatosensory information is not idle in the absence of peripheral stimuli; apparently, the central sensory processing apparatus continues to operate independently of the periphery, giving rise to these bizarre sensations.

Phantoms might simply be a curiosity —or a provocative clue about higher-order somatosensory processing— were it not for the fact that a substantial number of amputees also develop phantom pain. This common problem is usually described as a tingling or burning sensation in the missing part. Sometimes, however, the sensation becomes a more serious pain that individuals find increasingly debilitating. Phantom pain is, in fact, one of the more common causes of chronic pain syndromes and can be extraordinarily difficult to treat. Because of the widespread nature of central pain processing, ablation of the spinothalamic tract, portions of the

thalamus, or even primary sensory cortex does not generally relieve the discomfort felt by these individuals.

In recent years, it has become clearer that phantom sensations and phantom pain are probably a manifestation of maladaptive plasticity in neural circuits representing the sensation and actions of the body. Indeed, considerable functional reorganization of somatotopic maps in the primary somatosensory cortex occurs in individuals with limb loss and nerve injury. This reorganization starts immediately after an amputation and tends to evolve for several years. One of the effects of this process is that cortical neurons in affected regions acquire responses to previously silent inputs, typically mediated by long-range horizontal connections that span functional domains in somatotopic maps, with the potential to sprout new axonal collaterals that reinforce these newly functional inputs. Consequently, somatotopic domains in the postcentral gyrus (and subcortical

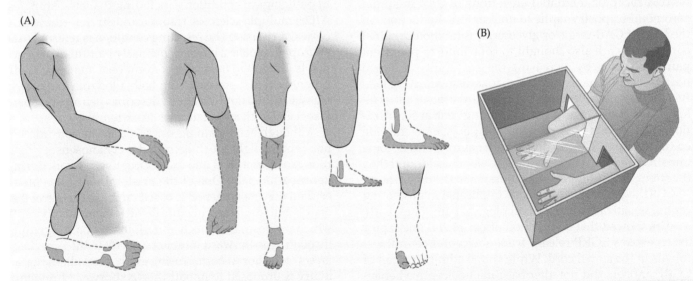

(A) Drawings of phantom arms and legs, based on patients' reports. The phantom is indicated by a dashed line, with the colored regions showing the most vividly experienced parts. Note that some phantoms are telescoped into the stump.
(B) Illustration of the mirror box designed by Ramachandran to relieve phantom pain with upper limb loss. The individual views his intact limb and its reflection in a mirror while commanding symmetrical movements of the remaining hand and the corresponding phantom. For some individuals, this experience immediately produces mobility of the phantom with a remarkable degree of relief from pain sensations. (A after K.A. Solonen, 1962. *Acta Orthop Scand Suppl* 54: 1–37.)

■ Clinical Applications (continued)

somatosensory centers) reorganize and neurons representing the missing or denervated body part begin responding to mechanical stimulation of other body parts. This is most common for body parts whose cortical representations are contiguous; thus, stimulation of the left side of the face, for example, can be experienced as if a missing left hand had been touched. Further evidence that the phenomenon of phantom limb is the result of a central representation is the experience of children born without limbs. Such individuals have rich phantom sensations, despite the fact that a limb never developed. This observation suggests that a full representation of the body exists independently of the peripheral elements that are mapped. Based on these results, Ronald Melzack proposed that

the loss of a limb generates an internal mismatch between the brain's representation of the body and the pattern of peripheral tactile input that reaches the neocortex. The consequence would be sensation that the missing body part is still present and functional.

Building on this conceptualization of phantom pain, V. S. Ramachandran has shown that "mirror box" therapy offers a low-tech form of virtual reality that may produce relief for individuals with phantom pain from limb loss (Figure B). Ramachandran reasoned that vision might normalize aberrant somatosensory and motor signals related to the missing limb if an individual is given visual feedback consistent with the intended movements of the missing limb. Thus, individuals view an intact limb and its reflection, while

"inserting" the phantom into the mirror-reversed visual percept of the intact limb. For some individuals at least, commanding symmetrical movements of the limbs in the mirror box gives rise to sensations of bilateral mobility with markedly diminished percepts of pain in the phantom.

The success of this simple intervention raises the intriguing possibility that visualization and virtual or augmented reality might prove to be a powerful means for promoting adaptive plasticity and neurorehabilitation. More generally, it reinforces the perspective that sensory perception, including pain, is generated actively in the brain and that the sensory cortices are not simply passive recipients of peripheral signals.

network. Multiple excitatory and inhibitory spinal cord neuron types across several dorsal horn laminae appear to be involved in the development and maintenance of allodynia. Thus, mechanical allodynia circuits are already in place under physiological conditions but are activated only in pathological conditions.

Neuropathic pain can arise spontaneously (i.e., without any stimulus), or it can be produced by mild stimuli that are common to everyday experience, such as the gentle touch and pressure of clothing, or warm and cool temperatures. People often describe their experience as a constant burning sensation interrupted by episodes of shooting, stabbing, or electric shock–like jolts. Because the disability and psychological stress associated with chronic neuropathic pain can be severe, much present research is being devoted to better understanding the mechanisms of peripheral and central sensitization, as well as glial activation and neuroinflammation, with the hope of developing more effective therapies for this debilitating syndrome.

The placebo effect

The word *placebo* means "I will please," and the **placebo effect** is defined as a beneficial effect (most often perceptual) following the administration of a pharmacologically inert "remedy." The placebo effect has a long history of use (and abuse) in medicine, but its reality is undisputed. In one classic study, medical students were given one of two different pills, one said to be a sedative and the

other a stimulant. In fact, both pills contained only inert ingredients. Of the students who received the "sedative," more than two-thirds reported feeling drowsy, and students who took two such pills felt sleepier than those who took only one. Conversely, a large fraction of the students who took the "stimulant" reported that they felt less tired. Moreover, about one-third of the entire group reported side effects ranging from headaches and dizziness to tingling extremities and a staggering gait. Only 3 of the 56 students in the group reported that the pills they took had no appreciable effect.

In another study of this general sort, 75% of patients suffering from postoperative wound pain reported satisfactory relief after an injection of sterile saline. The researchers who carried out this work noted that the responders were indistinguishable from the non-responders, both in the apparent severity of their pain and in their psychological makeup. Most tellingly, this placebo effect in postoperative patients could be blocked by naloxone, a competitive antagonist of opioid receptors, indicating that there is a substantial physiological basis for the pain relief experienced (see the next section). In addition, imaging studies show that the administration of a placebo with the expectation that it represents an analgesic agent is associated with activation of endogenous opioid receptors in cortical and subcortical brain regions that are part of the pain matrix, including the anterior cingulate and insular regions of cortex as well as the amygdala.

A common misunderstanding about the placebo effect is the view that individuals who respond to a therapeutically meaningless reagent are not suffering real pain, but only "imagining" it. This certainly is *not* the case. As discussed earlier, pain refers to an unpleasant sensory and emotional experience associated with actual or potential tissue damage, which is a subjective percept that can be elicited even in the absence of a nociceptive stimulus (e.g., during mechanical allodynia). Although the mechanisms by which the brain affects the perception of pain are only beginning to be understood, the effect is neither magical nor a sign of a suggestible intellect. In short, the placebo effect is quite real.

The physiological basis of pain modulation

Understanding the central modulation of pain perception was greatly advanced by the finding that electrical or pharmacological stimulation of certain regions of the midbrain produces relief of pain. This analgesic effect arises from activation of descending pain-modulating pathways that project to the dorsal horn of the spinal cord (as well as to the spinal trigeminal nucleus) and regulate the transmission of information to higher centers. One of the major brainstem regions that produce this effect is located in the periaqueductal gray of the midbrain. Electrical stimulation at this site in experimental animals not only produces analgesia by behavioral criteria, but also demonstrably inhibits the activity of nociceptive projection neurons in the dorsal horn of the spinal cord.

Further studies of descending pathways to the spinal cord that regulate the transmission of nociceptive information have shown that they arise from several brainstem sites, including the parabrachial nucleus, dorsal raphe, locus coeruleus, and medullary reticular formation (Figure 13.8A). The analgesic effects of stimulating the periaqueductal gray are mediated through these brainstem sites. These centers employ a wealth of different neurotransmitters (e.g., noradrenaline, serotonin, dopamine, histamine, acetylcholine) and can exert both facilitatory and inhibitory effects on the activity of neurons in the dorsal horn. The complexity of these interactions is made even greater by the fact that descending projections can exert their effects on a variety of sites within the dorsal horn, including the synaptic terminals of nociceptive afferents, excitatory and inhibitory interneurons, and the synaptic terminals of the other descending pathways, as well as by contacting the spinal cord projection neurons themselves. Although these descending projections were originally viewed as a mechanism that served primarily to inhibit the transmission of nociceptive signals, it is now evident that these projections provide a balance of facilitatory and inhibitory influences that ultimately determine the efficacy of nociceptive transmission. While more work is needed to fully elucidate these pathways, descending inputs from cortical regions—including secondary somatosensory cortex, insula, and anterior cingulate cortex—to subcortical regions and the spinal cord dorsal horn are demonstrated to modulate pain perception (see Figure 13.8A).

In addition to descending projections, local interactions between mechanoreceptive afferents and neural circuits within the dorsal horn can modulate the transmission of nociceptive information to higher centers (Figure 13.8B). These interactions are hypothesized to explain the ability to reduce the sensation of sharp pain through activating low-threshold mechanoreceptors. For example, if you knock your shin or stub a toe, a natural (and effective) reaction is to vigorously rub the site of injury for a minute or two. Such observations, supported by experiments in animals, led Ronald Melzack and Patrick Wall to propose that the flow of nociceptive information through the spinal cord is modulated by concomitant activation of low-threshold mechanoreceptors. Even though further investigation led to modifications of the original propositions in Melzack and Wall's "gate control theory of pain," the idea stimulated a great deal of work on pain modulation that continues today. This critical area of research has emphasized the importance of synaptic interactions within the dorsal horn for modulating the perception of pain intensity.

An exciting advance in this longstanding effort to understand central mechanisms of pain regulation has been the discovery of **endogenous opioids**. For centuries, opium derivatives such as morphine have been known to be powerful analgesics—indeed, they remain a mainstay of analgesic therapy today. In the modern era, animal studies have shown that a variety of brain regions are susceptible to the action of opioid drugs, particularly—and significantly—the periaqueductal gray matter and other sources of descending projections. In addition, there are opioid-sensitive neurons in the dorsal horn of the spinal cord. In other words, the areas that produce analgesia when stimulated are also responsive to exogenously administered opioids. It seems likely, then, that opioid drugs act at most or all of the sites shown in Figure 13.8 in producing their dramatic pain-relieving effects.

Several categories of endogenous opioids have been isolated from the brain and spinal cord and intensively studied. These agents are found in the same regions involved in the modulation of nociceptive afferents, although each of the families of endogenous opioid peptides has a somewhat different distribution. All three of the major groups—**enkephalins**, **endorphins**, and **dynorphins** (see Table 6.2)—are present in the periaqueductal gray matter. Enkephalins and dynorphins have also been found in the rostral ventral medulla and in those spinal cord regions involved in pain modulation.

One of the most compelling examples of the mechanism by which endogenous opioids modulate transmission of

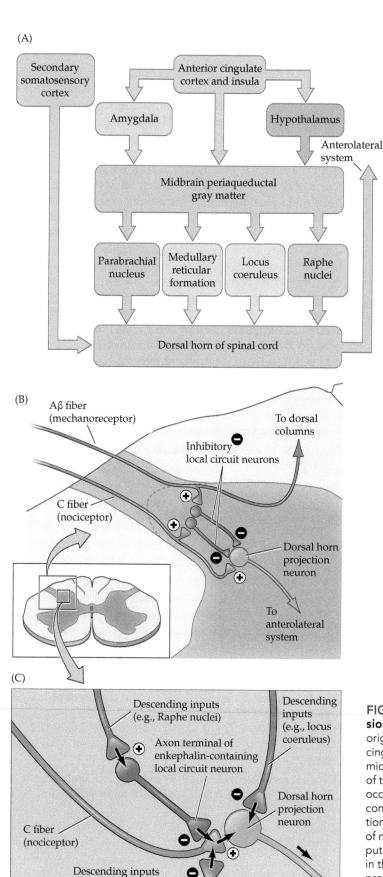

nociceptive information occurs at the first synapse in the pain pathway between nociceptive afferents and projection neurons in the dorsal horn of the spinal cord (Figure 13.8C). A class of enkephalin-containing interneurons within the dorsal horn synapses with the axon terminals of nociceptive afferents, which in turn synapse with dorsal horn projection neurons. The release of enkephalin onto the nociceptive terminals inhibits their release of neurotransmitter onto the projection neuron, thus reducing the level of activity that is passed on to higher centers. Enkephalin-containing interneurons are themselves the targets of descending projections, providing a powerful mechanism by which higher centers can decrease the activity transmitted by nociceptive afferents.

In a similar fashion, the analgesic effects of marijuana (*Cannabis*) led to the discovery of **endocannabinoids** (see Chapter 6). Exogenously administered cannabinoids are known to suppress nociceptive neurons in the dorsal horn of the spinal cord without altering the activity of non-nociceptive neurons. We now know that endogenous cannabinoids in the CNS act as neurotransmitters; they are released from depolarized neurons and activate cannabinoid receptors (CB_1) through a retrograde signaling mechanism. The actions of endocannabinoids are thought to *decrease* the release of neurotransmitters such as GABA and glutamate, thus modulating neuronal excitability. Evidence for a direct effect of endocannabinoids on the transmission of nociceptive signals comes from studies showing that analgesic effects induced by electrical stimulation of the periaqueductal gray can be blocked if CB_1 antagonists are administered. In addition, it appears that exposure to noxious stimuli increases the level of endocannabinoids in the periaqueductal gray matter, a finding that supports a major role for these molecules in the descending control of pain transmission. Cannabinoids also activate CB_2 receptors in microglia,

FIGURE 13.8 Descending systems modulate the transmission of ascending pain signals (A) These modulatory systems originate in the secondary somatosensory cortex, the anterior cingulate cortex and insula, the amygdala, the hypothalamus, the midbrain periaqueductal gray, the raphe nuclei, and other nuclei of the pons and rostral medulla. Complex modulatory effects occur at each of these sites, as well as in the dorsal horn. (B) Gate control theory of pain. One of the proposed mechanisms of action: Activation of mechanoreceptors modulates the transmission of nociceptive information to higher centers. (C) Descending inputs from the brainstem modulate the transmission of pain signals in the dorsal horn. Some inputs interact directly with dorsal horn projection neurons or the presynaptic terminals of C fibers. Others interact indirectly via enkephalin-containing local circuit neurons.

resulting in reduced activation of glial cells, which may further diminish pain transmission.

The story of endogenous anti-nociceptive compounds is impressive in its wedding of physiology, pharmacology, and clinical research to yield a richer understanding of the intrinsic modulation of pain. This information has finally begun to explain the subjective variability of painful stimuli and the striking dependence of pain perception on the context of experience. Many laboratories are exploring the precise mechanisms by which pain is modulated, motivated by the tremendous clinical benefits that would accrue from a deeper understanding of the pain system and its molecular underpinnings in the spinal cord and throughout the forebrain—wherever pain processing modifies cognition and behavior.

Summary

Whether studied from a structural or from a functional perspective, pain is an extraordinarily complex sensory modality. Because pain is an important means of warning an animal of dangerous circumstances, the mechanisms and pathways that subserve nociception are widespread and redundant. A distinct set of somatosensory afferents known as nociceptors transduces noxious stimulation of the body and conveys this information to neurons in the dorsal horn of the spinal cord. Pain is generated by an array of noxious-stimulus-detecting nociceptive neurons that sample both the internal environment and signals from the external world. This information is transmitted to a set of complex neural circuits in the spinal cord dorsal horn and subsequently to numerous brain regions, ultimately producing a diverse range of sensations and emotions. Pain can manifest with a range of different qualities, such as sharp pricking or dull burning, which highlights the complexity and heterogeneity of the underlying neural circuits. When tissue or neuronal damage persists, pain can become chronic and debilitating, and in some cases continues even after the wound has healed. Unraveling the biological basis of acute and chronic pain is a critical area of study because of the high prevalence of these conditions and lack of adequate treatment options.

■ Additional Reading

Reviews

Basbaum, A. I., D. M. Bautista, G. Scherrer and D. Julius (2009) Cellular and molecular mechanisms of pain. *Cell* 139: 267–284.

Bennet, D. L. H. and C. G. Woods (2014) Painful and painless channelopathies. *Lancet Neurol.* 13(6): 587–599.

Braz J., C. Solorzano, X. Wang and A. I. Basbaum (2014) Transmitting pain and itch messages: A contemporary view of the spinal cord circuits that generate gate control. *Neuron* 82: 522–536.

Di Marzo, V., P. M. Blumberg and A. Szallasi (2002) Endovanilloid signaling in pain. *Curr. Opin. Neurobiol.* 12: 372–379.

Fields, H. L. and A. I. Basbaum (1978) Brainstem control of spinal pain transmission neurons. *Annu. Rev. Physiol.* 40: 217–248.

Finn, D. P. and 5 others (2021) Cannabinoids, the endocannabinoid system, and pain: A review of preclinical studies. *Pain* 162: 5–25.

Gold, M. S. and G. F. Gebhart (2010) Nociceptor sensitization in pain pathogenesis. *Nat. Med.* 16: 1248–1257.

Hill, R. Z. and D. M. Bautista (2020) Getting in touch with mechanical pain mechanisms. *Trends Neurosci.* 43(5): 311–325.

Hunt, S. P. and P. W. Mantyh (2001) The molecular dynamics of pain control. *Nat. Rev. Neurosci.* 2: 83–91.

Koch, S., D. Acton and M. Goulding (2018) Spinal circuits for touch, pain, and itch. *Annu. Rev. Physiol.* 80: 189–217.

Lindsay, N. M. and 4 others (2021) Brain circuits for pain and its treatment. *Sci. Trans. Med.* 13(619): 7360.

Millan, M. J. (2002) Descending control of pain. *Prog. Neurobiol.* 66: 355–474.

Neugebauer, V., V. Galhardo, S. Maione and S. C. Mackey (2009) Forebrain pain mechanisms. *Brain Res. Rev.* 60: 226–242.

Patapoutian, A., A. M. Peier, G. M. Story and V. Viswanath (2003) ThermoTRP channels and beyond: Mechanisms of temperature sensation. *Nat. Rev. Neurosci.* 4: 529–539.

Peirs, C. and R. P. Seal (2016) Neural circuits for pain: Recent advances and current views. *Science* 354 (6312): 578–584.

Rainville, P. (2002) Brain mechanisms of pain affect and pain modulation. *Curr. Opin. Neurobiol.* 12: 195–204.

Scholz, J. and C. J. Woolf (2002) Can we conquer pain? *Nat. Rev. Neurosci.* 5 (Suppl): 1062–1067.

Tan, L. L. and R. Kuner (2021) Neocortical circuits in pain and pain relief. *Nat. Rev. Neurosci.* 22: 458–471.

Taves, S., T. Berta, G. Chen and R. R. Ji (2013) Microglia and spinal cord synaptic plasticity in persistent pain. *Neural Plast.* 2013: 753656.

Trang, T. and 5 others (2015) Pain and poppies: The good, the bad, and the ugly of opioid analgesics. *J. Neurosci.* 35: 13879–13888.

Zubieta, J.-K. and S. Christian (2009) Neurobiological mechanisms of placebo responses. *Ann. N.Y. Acad. Sci.* 1156: 198–210.

Important Original Papers

Basbaum, A. I. and H. L. Fields (1979) The origin of descending pathways in the dorsolateral funiculus of the spinal cord of the cat and rat: Further studies on the anatomy of pain modulation. *J. Comp. Neurol.* 187: 513–522.

Bautista, M. and 8 others (2006) TRPA1 mediates the inflammatory actions of environmental irritants and proalgesic agents. *Cell* 124: 1269–1282.

Blackwell, B., S. S. Bloomfield and C. R. Buncher (1972) Demonstration to medical students of placebo response and non-drug factors. *Lancet* 1: 1279–1282.

Caterina, M. J. and 5 others (1997) The capsaicin receptor: A heat-activated ion channel in the pain pathway. *Nature* 389: 816–824.

Caterina, M. J. and 8 others (2000) Impaired nociception and pain sensation in mice lacking the capsaicin receptor. *Science* 288: 306–313.

Choi, S. and 16 others (2020) Parallel ascending spinal pathways for affective touch and pain. *Nature* 587: 258–263.

Craig, A. D., E. M. Reiman, A. Evans and M. C. Bushnell (1996) Functional imaging of an illusion of pain. *Nature* 384: 258–260.

Huang, T. and 7 others (2018) Identifying the pathways required for coping behaviours associated with sustained pain. *Nature* 565: 86–90.

Hunt, S. P. and P. W. Mantyh (2001) The molecular dynamics of pain control. *Nat. Rev. Neurosci.* 2: 83–91.

LaMotte, R. H., X. Dong and M. Ringkamp (2014) Sensory neurons and circuits mediating itch. *Nat. Rev. Neurosci.* 15: 19–31.

Lavertu, G., S. L. Côté and Y. De Koninck (2014) Enhancing K–Cl co-transport restores normal spinothalamic sensory coding in a neuropathic pain model. *Brain* 137: 724–738.

Levine, J. D., H. L. Fields and A. I. Basbaum (1993) Peptides and the primary afferent nociceptor. *J. Neurosci.* 13: 2273–2286.

Murthy, S. E. and 10 others (2018) The mechanosensitive ion channel Piezo2 mediates sensitivity to mechanical pain in mice. *Sci. Trans. Med.* 10 (462): 9897.

Sorge, R. E. and 19 others (2015) Different immune cells mediate mechanical pain hypersensitivity in male and female mice. *Nat. Neurosci.* 18: 1081–1083.

Szczot, M. and 15 others (2018) PIEZO2 mediates injury-induced tactile pain in mice and humans. *Sci. Trans. Med.* 10 (462): 9892.

Zhang, K., D. Julius and Y. Cheng (2021) Structural snapshots of TRPV1 reveal mechanism of polymodal functionality. *Cell* 184(20): 5138–5150.

Books

Fields, H. L. (1987) *Pain*. New York: McGraw-Hill.

Fields, H. L. (Ed.) (1990) *Pain Syndromes in Neurology*. London: Butterworths.

Kolb, L. C. (1954) *The Painful Phantom*. Springfield, IL: Charles C. Thomas.

Skrabanek, P. and J. McCormick (1990) *Follies and Fallacies in Medicine*. New York: Prometheus Books.

Wall, P. D. and R. Melzack (2013) *Textbook of Pain*, 6th Edition. Philadelphia, PA: Elsevier Saunders.

14

Olfaction

Overview

Two sensory systems are associated with the nose: the olfactory (conscious smell) and vomeronasal (pheromone sensation) systems. Each system detects, encodes, relays, and represents information about airborne (also referred to as "volatile") chemical stimuli in the environment. Along with the gustatory, or taste, system (see Chapter 15), the olfactory and vomeronasal systems are collectively referred to as the chemosensory systems. The olfactory system is "tuned" to a subset of airborne molecules called odorants that help an individual organism identify other animals and plants, food, and noxious substances, as well as signals that distinguish self from others, thereby influencing a wide range of social behaviors, including reproduction and parenting. In most mammals—except, notably, humans and other primates where this system is diminished or vestigial—the vomeronasal system detects a more limited number of airborne stimuli released by predators, prey, potential mates, and offspring. Thus, the vomeronasal system is thought to influence more "innate" behavioral responses. For both olfaction and vomeronasal chemosensation, the initiation of sensory transduction relies on G-protein-coupled receptors (GPCRs) and second-messenger-mediated signaling in peripheral receptor neurons in the nose. In each system there are a large number of receptor genes that encode GPCRs with capacity to bind specific subsets of odorants or pheromones. After transduction by receptor neurons in sensory epithelia in the nose, olfactory and vomeronasal information is relayed to the main and accessory olfactory bulbs, respectively—forebrain structures where information is further processed and made available to multiple additional forebrain targets. The central representation of conscious olfactory perception, thought to be the primary function of the olfactory system, remains uncertain. Olfactory information is integrated with other sensory modalities to influence conscious perception of food, conspecifics, predators, prey, environmental attractants and hazards. Vomeronasal information activates amygdala and hypothalamic circuitry and influences predator avoidance, prey identification, aggression, and reproductive behaviors. Thus, the parallel chemosensory systems of olfactory and vomeronasal sensation construct representations of the molecular landscape of the air that organisms breathe, and extract key information that guides fundamental behaviors that can be broadly described as "food, family, friends, enemies, and sex."

Dr. Goran Bredberg/Science Photo Library

KEY CONCEPTS

14.1 The olfactory and vomeronasal systems process airborne molecules that influence a wide range of behaviors

14.2 Olfactory and vomeronasal transduction occurs via g-protein-coupled receptors

14.3 Olfactory and vomeronasal information is relayed directly to the main and accessory olfactory bulbs, and from there to multiple forebrain targets

14.4 In many species, including humans, olfactory capacity reflects the size and complexity of the olfactory system

CONCEPT
14.1

The Olfactory and Vomeronasal Systems Process Airborne Molecules That Influence a Wide Range of Behaviors

LEARNING OBJECTIVES

14.1.1 Identify the components of the olfactory and vomeronasal systems.

14.1.2 Define the range of behaviors influenced by olfaction.

The olfactory system

The olfactory system processes information about the identity, concentration, and quality of a wide range of airborne chemical stimuli called **odorants** and represents these stimuli as consciously perceived odors associated with a wide range of environmental sources. Odorants interact with **olfactory receptor neurons (ORNs)** found in an epithelial sheet, the **olfactory epithelium**, that lines the interior of the nose (Figure 14.1A,B). Axons arising from ORNs coalesce into the **olfactory nerve** (cranial nerve I). The bundles, or fascicles, of axons that compose the olfactory nerve extend through the **cribriform plate**, a thin

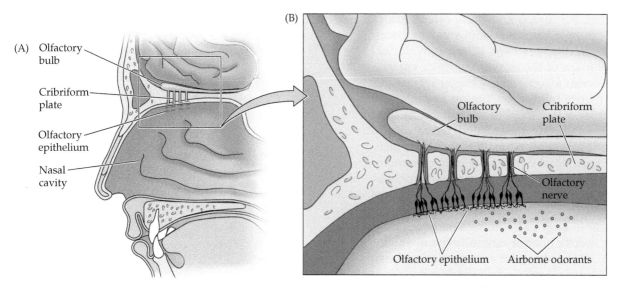

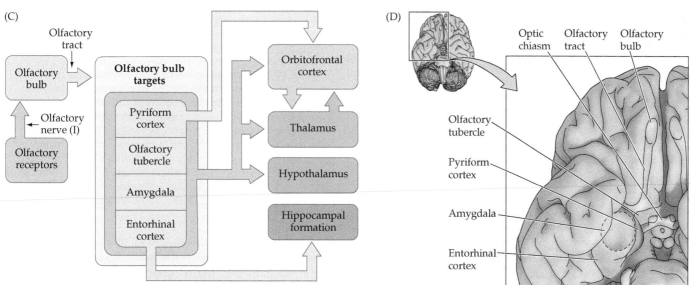

FIGURE 14.1 Organization of the human olfactory system (A) Peripheral and central components of the primary olfactory pathway. (B) Enlargement of region boxed in (A), showing the relationship between the olfactory epithelium (which contains the ORNs) and the olfactory bulb (the central target of ORNs). (C) The basic pathways for processing olfactory information. (D) Central components and basic connections of the olfactory system, seen in a ventral view of the human brain.

perforated region of the skull that separates the olfactory epithelium from the brain. These axons then project directly to neurons in the **olfactory bulb**, some of which have axons that in turn project to the pyriform cortex (see Concept 14.3) in the temporal lobe, as well as to other forebrain structures, via an axon pathway with multiple branches known as the **olfactory tract** (Figure 14.1C,D).

The olfactory system abides by the same principles that govern other sensory modalities: Sensory stimuli—in this case, airborne chemicals—interact with receptors at the periphery and are transduced and encoded into electrical signals, which are relayed via synaptic transmission to higher-order centers. Despite these basic similarities, far less is known about the neural representation of olfactory information in the CNS than is known for other sensory modalities. For example, the visual and somatosensory cortices (described in Chapters 9 and 12, respectively) feature topographic maps of the relevant receptor surface, and the auditory cortex (described in Chapter 10) features a computational map of frequencies, timing, and intensity of sound. The nature of the "map" of odorant information in the CNS remains uncertain. Indeed, until recently it was difficult to imagine how sensory qualities (e.g., odor identity, intensity, or behavioral significance) might be represented, or what features of chemosensory stimuli might be processed in parallel pathways as in other sensory systems (see below). The details of central conscious olfactory processing are unclear in large measure because of the difficulty of studying specific olfactory stimuli experimentally. It's challenging to identify key odor stimuli, and to present them to ORNs with consistent concentrations and without environmental "noise" from other molecules in the air that also are sensed as odorants. In addition, quantifying receptor responses and defining their precise influence on target neuron activities are daunting tasks. Nevertheless, it is clear that the olfactory system generates complex representations of odors that can influence information processing in a variety of forebrain regions, allowing olfactory perception to influence visceral, homeostatic, emotional, and cognitive behaviors.

The vomeronasal system

The **vomeronasal system** detects a distinct and more limited subset of airborne stimuli associated with behaviors that are essential for survival and reproduction. This parallel chemosensory pathway is found in most vertebrates. In mammals the vomeronasal system is prominent in carnivores (including dogs and cats) and rodents, and less robust or vestigial in primates (especially humans). The vomeronasal system encompasses a distinct receptor cell population in a separate compartment of the nasal epithelium called the **vomeronasal organ** (VNO), as well as a separate target region of the olfactory bulb—called the **accessory olfactory bulb**—where axons from chemosensory receptor cells in the VNO synapse (Figure 14.2).

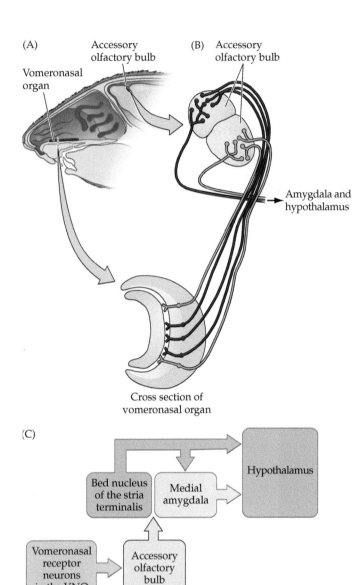

Figure 14.2 Organization of the vomeronasal pathway in non-primate mammals (A) The vomeronasal organ is the location of a unique population of vomeronasal chemosensory neurons that detect pheromones and kairomones. (B) The vomeronasal chemosensory neurons project specifically to the accessory olfactory bulb, a structure found in most non-primate mammals. (C) The relay of information from the accessory olfactory bulb is more limited than that for the main olfactory bulb, and primarily targeted to brain regions involved in homeostatic behaviors. (B after E. Pantages and C. Dulac, 2000. *Neuron* 28: 835–845.)

The projections of the accessory olfactory bulb are distinct from those of the remainder of the olfactory bulb (referred to as the "main" olfactory bulb in rodents and carnivores) and include subregions of the amygdala as their major target zones (see Figure 14.2C). This anatomical distinction provides an important clue to the primary function of the vomeronasal system: It encodes and processes chemical signals from conspecifics or predators to mediate sexual, reproductive, homeostatic, and aggressive responses—behaviors that also depend on circuits in the hypothalamus and amygdala (see Chapters 21 and 32). The specific stimuli detected and represented by the vomeronasal system that mediate **conspecific interactions** (e.g., mating, parental, and other social behaviors) are referred to as **pheromones**. The specific stimuli that mediate behavioral interactions with other animal species that are either predators (e.g., an owl, for a mouse) or prey (e.g., a mouse, for an owl) are referred to as **kairomones**.

The fate of the vomeronasal system in primates, especially humans, is mysterious. The vomeronasal organ is diminished in size in some New World primates (e.g., squirrel monkeys) and diminished or absent in Old World primates (e.g., rhesus monkeys), apes, and humans. In most primates, the genes for vomeronasal receptor proteins do not encode functional proteins (see Concept 14.2). Finally, a region that corresponds to the accessory olfactory bulb is not found in most primates. Nevertheless, primates, including humans, arguably have behavioral responses that can be attributed to stimuli similar to the pheromones recognized by the vomeronasal system in other animals. Thus, in some mammals the vomeronasal system provides a distinct chemosensory parallel pathway for detecting and processing chemosensory signals about reproduction, social interactions, predator threats, and prey opportunities. For primates, including humans, the representation of such information—if indeed it is represented specifically—remains uncertain.

Physiological and behavioral responses to airborne chemical stimuli

Consciously perceived odorants, as well as pheromones and kairomones, can elicit a variety of physiological responses. Examples of odorant responses generated by the olfactory system are the visceral motor responses to the aroma of appetizing food (salivation and increased gastric motility) or attractive scents (inhaling, sniffing), or to noxious smells (withdrawl, gagging, and in extreme cases, vomiting) or potentially hazardous or toxic odors (coughing, shortness of breath). The association of singular odors with attractive or aversive valences is a key aspect of conscious olfaction: The identification and behavioral significance of individual odors are tightly coupled. Olfaction, particularly in humans, encompasses a broad range of stimulus–association–behavioral response relationships. Some of these relationships are thought to drive **innate behaviors**, such as food seeking, aspects of mood, arousal, and social approach or avoidance. Others are thought to drive conscious choices based on odor detection and discrimination capacity. Thus, distinctly recognized odors can lead to the selection of the best smelling bread in a bakery, or the most attractive perfume at the cosmetics counter. In the case of highly trained "noses," these personal preferences based on odor grow into judgements of the most sophisticated bouquet among various wines, as well as the identity and sources of potentially harmful environmental odors

present in trace amounts. One example of the conscious association of an odor with a behavioral response is the human response to sulphurous chemicals that are added to natural gas, which is actually odorless. In this case, the association of the sulfur-laden "rotten egg" smell with the hazard–avoidance response to presumed natural gas in the air is based on a learned association between a consciously perceived olfactory stimulus and a specific response.

In animals with a vomeronasal system, pheromones—for example, from a receptive potential mate—can elicit a sequence of sexual behaviors. Detection of a kairomone from a predator or prey can result in "fear and flight" or "stay and fight" responses (see Chapters 21 and 32). These behaviors, once initiated, tend to have stereotypical, species-specific execution that usually does not reflect learning. For example, the detection of pheromones from her newborn pups will elicit lactation and nursing behaviors (see Chapter 25) in females of several mammalian species. Thus, the circuits for these behaviors, and their activation by pheromones or kairomones, are thought to be "hard-wired," with fairly invariant patterns of connectivity established during development. In primates, including humans, likely without a vomeronasal system, olfaction can also influence reproductive, parenting, and endocrine functions. Whether or not these responses are parallel to those regulated by vomeronasal sensation in other animals is uncertain.

| CONCEPT **14.2** | **Olfactory and Vomeronasal Transduction Occurs via G-Protein-Coupled Receptors** |

LEARNING OBJECTIVES

14.2.1 Establish the locations and cellular structure of olfactory and vomeronasal receptor neurons.

14.2.2 Define the central role of G-protein-coupled receptors in olfactory and vomeronasal transduction.

14.2.3 Discuss the distinct genetic and genomic features of olfactory and vomeronasal receptor genes.

14.2.4 Compare the signal transduction cascades and ion channels that transduce olfactory and vomeronasal stimuli into receptor potentials.

14.2.5 Describe the mechanism of ongoing loss and replacement of olfactory receptor neurons and their central connections.

The Olfactory Epithelium and Olfactory Receptor Neurons

The transduction and processing of olfactory information—a series of neural events that ultimately results in the conscious sense of smell—begins in the olfactory epithelium, the sheet of peripheral sensory receptor neurons and supporting cells that lines approximately half the surface of the nasal cavity (Figure 14.3A; see also Figure 14.1A). The

Figure 14.3 Cellular organization of the olfactory epithelium
(A) A summary of the major cell classes within the olfactory epithelium, as well as in the adjacent lamina propria. Neurons, supporting cells similar to glial cells, mucus secreting cells, and stem cells are found in the olfactory epithelium while the lamina propria is the sight of blood vessels, olfactory ensheathing cells that wrap the axons of olfactory receptor neurons (which are unmyelinated), and macrophages that mediate immune surveillance. (B) Distinctions between respiratory and olfactory (neural) epithelium and vomeronasal organ in the nasal cavity. From left to right: the nasal cavity of a juvenile mouse, composed of a fairly thin respiratory epithelium, the much thicker olfactory epithelium, the underlying lamina propria and the vomeronasal organ (VNO) encased within the nasal septum. The VNO is fully encased in the nasal septum and a large fascicle of axons exit the VNO in the septum to join the olfactory nerve. There is a sharp boundary between respiratory epithelium (labeled green here, based on expression of the transcription factor forkhead1) and olfactory epithelium. The remaining three panels show distinct cell classes in the olfactory epithelium. ORNs are labeled with olfactory marker protein (OMP; green), a molecule expressed uniquely in these neurons. Supporting (sustentacular) cells express another cell-class-specific molecule, Sus-4 (light brown). At far right, basal cells, the stem cells of the adult olfactory epithelium, are recognized by their expression of the filament protein cytokeratin 5 (dark brown). (C) Olfactory ensheathing cells (also called ensheathing glia; red) wrap bundles of unmyelinated axons from olfactory receptor neurons (green) as they extend through the lamina propria, toward the cribiform plate and into the olfactory bulb. (D) Macrophages (green) invade the lamina propria after an acute lesion to the olfactory epithelium that causes olfactory receptor neuron degeneration. (A after R. R. H. Anholt, 1987. *Trends Biochem Sci* 12: 58–62.)

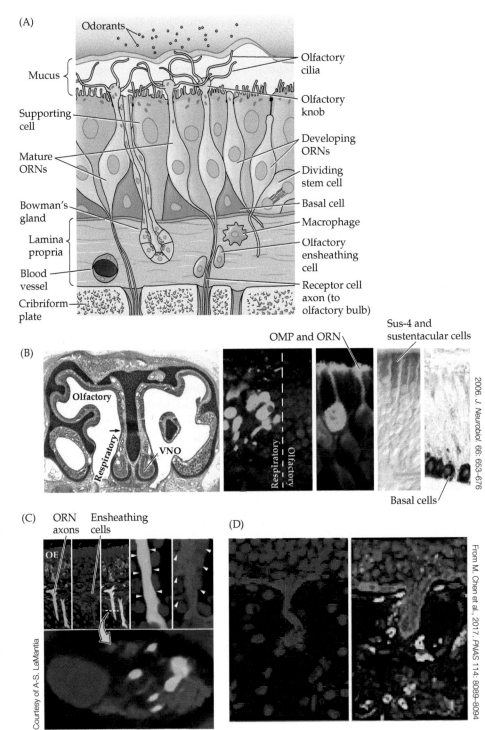

2006. *J. Neurobiol.* 66: 663–676

From M. Chen et al., 2017. *PNAS* 114: 8089–8094

Courtesy of A-S. LaMantia

primary neuron class of the olfactory epithelium is the olfactory receptor neuron (ORN), which transduces odorant stimuli into electrical signals that are then relayed to the CNS. The remaining intranasal surface is lined by respiratory epithelium similar to that in the trachea and lungs. Respiratory epithelium primarily maintains appropriate temperature and moisture for inhaled air (which may be

important for the presentation of odorants to ORNs) and provides an immune barrier that protects the nasal cavity from irritation and infection.

ORNs are bipolar neurons and have at their apical surface a single dendritic process that expands into a knoblike protrusion (Figure 14.3A) from which several microvilli, called **olfactory cilia**, extend into a thick layer of mucus that lines the nasal cavity at the air/epithelial interface. ORNs give rise to small-diameter, unmyelinated axons at their basal surface that relay olfactory information centrally. ORNs are the only neurons in the olfactory epithelium. There are no local circuit neurons (such as the bipolar or amacrine cells in the retina) that link receptors to cells that transduce stimuli with projection neurons (such as the retinal ganglion cells) that relay the transduced electrical signals to the brain. Instead, following transduction in the ORN apical dendrite, the relatively long axons of ORNs themselves extend toward the main olfactory bulb to relay chemosensory information directly to the forebrain. This distinguishes these chemosensory receptors from the sensory receptor cells in the cochlea, skin, and tongue, which make synapses onto bipolar peripheral sensory neurons that then relay information to the brainstem or spinal cord.

A layer of mucus lines the nasal cavity and protects the air-exposed neurons, respiratory epithelial cells, and supporting cells of the olfactory epithelium. The mucus also controls the ionic milieu of the olfactory cilia, the primary site of odorant transduction (see the end of this section). Mucus is produced by secretory specializations called **Bowman's glands** that are distributed throughout the olfactory epithelium. When the mucus layer thickens, usually in response to inflammation as during a cold, olfactory acuity decreases significantly. Two other cell classes, **basal cells** and **sustentacular** (supporting) **cells**, are also present in the olfactory epithelium. The mucus secreted by Bowman's glands also traps and neutralizes some potentially harmful agents. In both the respiratory and olfactory epithelium (Figure 14.3B), immunoglobulins and cytokines are secreted into the mucus where they provide an initial defense against harmful antigens or infectious agents such as bacteria or viruses (Clinical Applications). The sustentacular cells also have enzymes (cytochrome P-450s and others) that neutralize and catabolize organic chemicals

■ Clinical Applications

Only One Nose

Our other specialized sensory organs—the eyes and ears—are the subjects of safety campaigns that warn against risks to sight and hearing, and a robust industry manufactures protective devices such as goggles, sunglasses, and ear plugs. Despite its prominence, the nose is often overlooked when it comes to thinking about risks and consequences. This lack of attention to nasal peril has led to two eerily similar instances of medically induced anosmia resulting from exposure of the olfactory epithelium to zinc, which can be toxic to some tissues. In both cases, the reasonable desire to prevent a perceived "greater evil"—polio in the 1930s and the common cold in the 1990s—resulted in treatments that were not only ineffective in respect to their original intent but also in significant loss of smell for those who were exposed. In both instances, children were among those most affected.

In the late 1930s, the threat of polio was very real, and weapons to fight this devastating viral disease that preferentially afflicts children were limited. There was an impression that the poliovirus could be transmitted via the nasal mucosa (which did not turn out to be the primary route of infection). Many physicians thought that zinc-based nasal sprays could protect children from

(Continued)

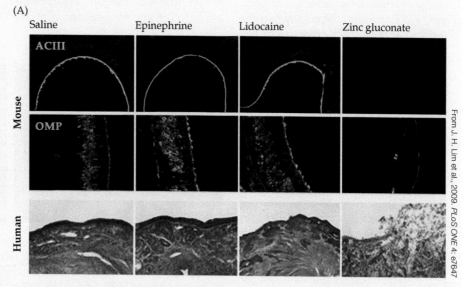

Figure A   The effects of nasal sprays or gels on ORNs in the mouse and on human olfactory epithelium (OE). The figure shows the consequences of saline, epinephrine, and lidocaine—all substances commonly found in preparations applied to OE—compared with those of zinc gluconate. In the mouse, adenylyl cyclase III labeling (ACIII) gauges the integrity of the signal transduction machinery for odorant detection at the ORN dendritic knob, while olfactory marker protein labeling (OMP) indicates the frequency of ORNs labeled by this particular molecular marker. In human OE, the effects of zinc gluconate can be seen in the diminished thickness and vacuolated appearance of the OE.

■ Clinical Applications (*continued*)

infection, even though there was little evidence for this hypothesis. Desperate times often lead to intemperate measures, however, and zinc nasal spray treatments were widely administered. Among the most ambitious was a 1937 trial involving 5000 children in Toronto who were given intranasal sprays of zinc sulfate. It soon became clear that the treatment did not prevent polio infection, and in a significant number of children, it had an unanticipated side effect—an irreversible loss of the sense of smell. Subsequent animal studies offered an explanation of this unfortunate consequence. Zinc ions cause dramatic, specific damage to ORNs, and elevated concentration or repeated exposure to zinc salts can almost completely destroy the olfactory epithelium (see figures)—including, presumably, the stem cells that regenerate ORNs throughout life.

Observations of zinc toxicity in the olfactory epithelium of animals (recorded in the literature as early as 1947), along with the unfortunate outcome of the 1937 Canadian trial, clearly raised a red flag against the further use of any zinc salt as an intranasal treatment. However, in the 1990s the notion that zinc was an effective antiviral agent (especially against rhinovirus, the cause of the common cold) led to the reemergence of zinc on the market, this time in unregulated dietary supplements and homeopathic remedies. Such was the case with Zicam nasal gel, in which the primary active ingredient was a zinc salt, zinc gluconate. Shortly after the introduction of this product, reports emerged of individuals losing their sense of smell after using Zicam gel. Not surprisingly, zinc gluconate, like zinc sulfate, causes significant cellular damage to the olfactory epithelium and can result in permanent disruption of olfactory sensation. On the basis of the clear risk of permanent damage to olfaction, in 2009 the FDA issued a warning to consumers to stop using zinc gluconate intranasal treatments.

This tale of the same mistake made twice holds two lessons. First, although less attended to than sight and hearing, olfaction is also vulnerable to peripheral insults that can seriously compromise sensation. Second, even seemingly innocuous compounds can pose significant risks that may be known but not recognized due to lack of rigorous oversight.

We have been reminded of the role of our noses as a first responder, and their vulnerability to major hazards, during the SARS-CoV2 pandemic that began in late 2019 and has continued through 2022, when this edition of *Neuroscience* was completed for publication. At the outset of the pandemic, one of the mysterious (to some) first symptoms of COVID-19, the name given to the disease caused by infection with the SARS-CoV2 corona virus variant, was a loss of smell (as well as taste, see Chapter 15). In some individuals, this COVID-related anosmia resolves as the additional symptoms of the disease recede, while in others, for reasons that are

unknown, it endures. The SARS-CoV2 virus does not appear to infect olfactory receptor neurons. Indeed, the available information indicates that the virus selectively infects subsets of sustentacular cells, the support cells of the olfactory epithelium. Sustentacular cells express the ACE-2 receptor, which is the cell surface protein used by SARS-CoV2 for entry into cells. The infected cells are thought to elicit a general inflammatory response in the olfactory epithelium. The inflammation plus the loss of sustentacular cells and the support they provide to olfactory receptor neurons is thought to secondarily lead to anosmia. There is no evidence that the virus directly infects olfactory receptor neurons. The extent to which COVID-19 impacts olfaction chronically, perhaps due to impaired regeneration in some individuals who have recovered from the disease, remains to be determined—like all of the other potential "long COVID" effects. Nevertheless, for COVID-19, we discovered once again the singularity of our sense of smell, and its susceptibility to environmental hazards.

(B)

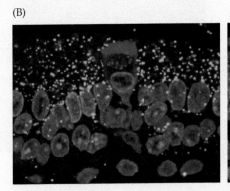 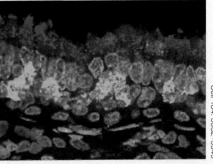

From M. Khan et al. 2021; *Cell* 184: 5932–5949

Figure B   Top: A sustentacular cell, identified by its position at the apical (lumenal/exposed directly to air in the nasal cavity) olfactory epithelium. Healthy sustentacular cells are labelled green for the sustentacular cell marker GPX3, but the infected cell, recognized by the presence of SARS-CoV2 RNA (red) and the expression of nucleocapsid protein encoded by the virus (blue) is not labelled for GPX3. This suggests that the infected cell is no longer functionally intact. Bottom: Sustentacular cell infection seen in the olfactory epithelium from a deceased individual with an active SARS-CoV2 infection, is extensive. Olfactory receptor neurons, labeled green for mRNAs encoding odorant receptor genes, are not infected by the virus. The disruption of the sustentacular cells, which regulate the mucus layer for odorant binding and maintain apical dendritic knobs and cilia of olfactory receptor neurons, is thought to disrupt olfaction in individuals with COVID-19.

and other potentially damaging molecules. Mucus also plays a role in the controlled diffusion and presentation of detectable odorants. Some odors are thought to form complexes with secreted odor-binding proteins in mucus that facilitate their diffusion and presentation to odorant receptors, while others may diffuse directly through the mucus layer at different rates. It is not clear how these mechanisms influence the final perception of any particular odor.

In addition to the passive protection provided by mucus, macrophages found throughout the nasal mucosa provide active surveillance of the respiratory and olfactory epithelium and isolate and remove harmful material. These immune cells also remove the remains of degenerating cells of the olfactory epithelium, including ORNs, which (unlike most other neurons) are continually lost and regenerated (see the next section). In the face of acute or sustained inflammation, leukocytes (also called white blood cells) are recruited via small blood vessels and capillaries to the **lamina propria**, a complex tissue between the olfactory epithelium and the cribriform plate. In addition to blood vessels, the lamina propria includes olfactory ensheathing cells that envelop sensory neuron axons (Figure 14.3C), the secretory region of mucus-secreting Bowman's glands, and several types of immune cells embedded in an extracellular matrix (Figure 14.3D). This entire apparatus—mucus layer, epithelium with neural and supporting cells, and lamina propria—is called the **nasal mucosa**. Given the unusual direct exposure of the olfactory epithelium to the external environment (outdoor or room air), immune protection is especially important to maintain the integrity of neuronal as well as supporting cells.

The Vomeronasal Organ and Vomeronasal Receptor Neurons

The vomeronasal organ (VNO) is located at the base of the nasal septum (see Figure 14.3B) and accesses air through small openings in the oral cavity beneath the upper lip rather than through the nasal cavity. Thus, the VNO acquires its airborne chemical stimuli from a different air stream than that inhaled into the nasal cavity. The VNO is lined with a sensory epithelium consisting of **vomeronasal receptor neurons (VRNs)** as well as supporting cells that are similar to those in the OE. The cellular appearance of VRNs is very similar to ORNs: VRNs have a single apical dendrite oriented toward the lumen of the VNO, and that dendrite ends in a dendritic knob with cilia like processes that are embedded in a mucus layer. The key distinctions between VRNs and ORNs, other than their location, are the identities of receptor proteins that bind pheromones and kairomones in VRNs versus those that bind odorants in ORNs (see below). In addition, the signal transduction pathways that transduce pheromone/kairomone versus odor binding are distinct, as are the ion channels that initiate receptor potentials that relay information to the CNS (see Concept 14.3).

Finally, VRNs extend their axons from the VNO. These axons join the olfactory nerve (cranial nerve 1), enter the CNS and then segregate to target the accessory olfactory bulb.

Degeneration and regeneration of ORNs and VRNs

As air is inspired through the nose, ORNs have constant and direct access (after diffusion through the mucus layer) to odorant molecules; however, this also exposes these neurons to airborne pollutants, allergens, microorganisms, viruses (see Clinical Applications), and other potentially harmful substances, subjecting them to more or less continual damage. Similarly, the constant exposure of VRNs to air inhaled via the oral cavity exposes these cells to a high risk of damage and degeneration. Some of this damage is minimized by local immune surveillance in the olfactory epithelium and vomeronasal organ. An optimal solution to the vulnerability of ORNs and VRNs in most animals is to maintain healthy populations of these cells by an ongoing cycle of degeneration and regeneration, analogous to that in other exposed epithelia (e.g., skin, intestine, lung). This constant process of ORN and VRN degeneration and regeneration occurs in all vertebrates, including mammals. ORN and VRN regeneration relies on maintaining a population of neural stem cells among basal cells in the olfactory epithelium and VNO. These stem cells divide and give rise to new receptor neurons (Figure 14.4A; see also Figure 14.3A and Chapter 26). The resident stem cells of the mature olfactory epithelium and VNO therefore retain the capacity to execute a program of division, transcriptional regulation, and differentiation (see Chapter 22) that recapitulates much of the initial developmental program that generates the earliest ORNs and VRNs (Figure 14.4B). Each newly generated ORN and VRN can then grow an axon that establishes new synaptic connections with appropriate targets. During the gradual loss and replacement of ORNs and VRNs, the new axons are remarkably faithful to specific targets (see the next section). Thus, ORNs and VRNs are the only classes of neurons in the adult mammalian nervous system that can be newly generated from neural stem cells, grow new axons for relatively long distances, and form new synaptic connections. Other vertebrates and invertebrates also replace ORNs or their equivalents throughout life; however, the mature nervous systems of some non-mammalian vertebrate species as well as of several invertebrates have additional regenerative capacity, including the genesis of new projection neurons in the retina that extend long axons and reestablish synaptic connections (see Chapter 26).

In rodents, most if not all olfactory neurons are lost and then renewed gradually every 6 to 8 weeks. This extended time period (6 to 8 weeks represents a significant portion of a mouse or rat's typical 1.5- to 2-year life span) suggests that neural regeneration is a gradual process, so that function can be maintained more or less continuously while small numbers of ORNs die and then are replaced. The period for complete turnover has not been defined in humans. Nevertheless, it is clear that ORNs will be regenerated if large populations of extant neurons, but not the neural stem cells that are maintained throughout life in the olfactory epithelium (see Figure 14.4A and Clinical Applications), are eliminated at one time. This can happen because of environmental exposure; viral or bacterial infection (see Clinical Applications); or traumatic head injuries such as whiplash that occurs in automobile accidents, when ORN axons can be sheared by the force of the impact because of differential movement of the neural tissue versus the cribriform plate. Unfortunately, this sort of large-scale regeneration after acute, massive injury does not fully restore normal function. In such individuals, after a period of complete **anosmia** (loss of the sense of smell), odor discrimination and identification as well as olfactory-guided behaviors often continue to be altered (see Clinical Applications).

In the mature olfactory system, many of the molecules that influence initial stem cell proliferation, neuronal differentiation, axon outgrowth, and synapse formation during development (see Chapters 22 and 23) are apparently retained or reactivated to perform similar functions for regenerating ORNs (see Figure 14.4B). Understanding how new ORNs differentiate, extend axons to the brain, and reestablish appropriate functional synaptic connections is obviously relevant to stimulating regeneration of functional connections elsewhere in the brain after injury or disease (see Chapter 26). Indeed, other specialized cell classes in the mature olfactory system are adapted to facilitate constant regeneration. In the mature olfactory nerve, glial cells called **olfactory ensheathing cells** surround axons as they grow toward and into the olfactory bulb (see Figure 14.3). These glial cells are believed to support the growth of new axons throughout the mature nervous system. In experimental therapies following damage to other CNS regions (e.g., the spinal cord), olfactory ensheathing cells have been used to construct cellular "bridges" across sites of axon damage to promote regeneration (see Chapter 26). Thus, the regenerative capacity of ORNs, with the assistance of other cell

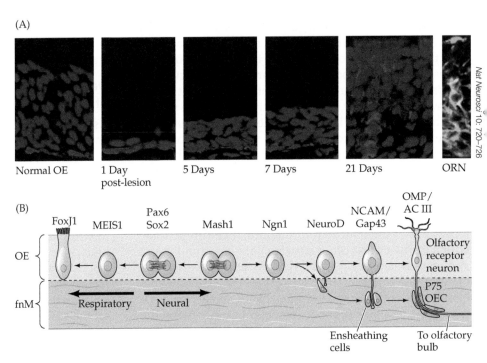

Normal OE | 1 Day post-lesion | 5 Days | 7 Days | 21 Days | ORN

Figure 14.4 Olfactory epithelium (OE) regeneration depends on basal cells
A reporter protein genetically labels OE basal cells and their descendants in a transgenic mouse line so that they appear red in the normal OE (left). When the OE is undisturbed, basal cells are seen in their appropriate position. Immediately after a lesion directly to the OE, however, basal cells begin to proliferate; their progeny are additional basal cells. Within 21 days, all cells in the regenerated epithelium have arisen from basal cells. Regenerated cells include ORNs, which are double-labeled for OMP (green) expression as well as the red reporter protein indicating the neuron's basal cell derivation (B) The likely lineage progression of basal cells in the OE. Several transcription factors (MEIS1, Pax6, Sox2, Mash1) distinguish basal cells and intermediate precursors. Other transcription factors (Ngn1, NeuroD), as well as neuronal markers (NCAM, GAP43) and ORN-specific molecules (OMP, ACIII), distinguish differentiating and mature ORNs. (B after C. W. Balmer and A.-S. LaMantia, 2005. *Dev Dynam* 234: 464–481, and J. E. Schwob et al., 2017. *J Comp Neurol* 525: 1034–1054.)

types in or adjacent to the olfactory epithelium, provides a potentially instructive model for understanding how regeneration of neurons or axons can be stimulated throughout the nervous system. This is a fundamental issue since in all mammals, including humans, neuronal replacement, regrowth, and related functional recovery after CNS damage do not occur to a useful degree (see Chapter 26).

Odorant detection

Odor transduction in the olfactory epithelium begins with odorant binding to specific odorant receptor proteins concentrated on the external surface of olfactory cilia that extend from the single apical dendrite of an ORN into the mucus layer of the olfactory epithelium. Air inspired through the nose in the course of breathing includes the volatile molecules that can be detected and transduced by ORNs to initiate the representation and perception of distinct smells. As with other sensory modalities, once a stimulus is initially detected by an ORN, the stimulus is thereafter represented by patterns of electrical activity generated in the ORN and the neural circuits to which it connects. Olfactory information is acquired phasically by briefly and repeatedly changing the volume of air inhaled into the nasal cavity by sniffing. For olfaction, the act of sniffing parallels eye movements for foveating high-contrast visual information or palpation (the repeated movement of fingertips over an object) for somatosensation. Sniffing is a cyclical act, and establishes transient changes in odorant concentration, as well as oscillatory detection and relay of odorant information that sharpens odorant representations in the olfactory bulb and subsequent processing in odorant recipient brain regions (see Figure 14.1).

Air inspiration and subsequent sniffing present odorants to the olfactory epithelium, where they are trapped and concentrated in the mucus layer so that they can bind to receptors on ORN or VRN cilia. The compartmental sensitivity of the cilia to odors was demonstrated in physiological experiments on isolated ORNs (Figure 14.5). Odorants presented to the *cilia* of an isolated ORN elicit a robust electrical response; those presented to the *cell body* do not. Cilia provide a greatly expanded cellular surface to which odorants can bind. ORN cilia have many of the same cytoskeletal features of all cilia—that is, the 9+2 array of microtubules anchored by a basal body at the origin of each cilium in

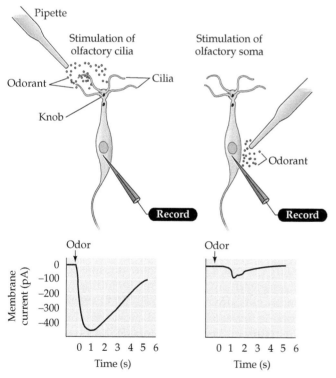

FIGURE 14.5 Receptor potentials are generated in the cilia of olfactory receptor neurons Odorants evoke a large inward (depolarizing) current when applied to the cilia (left), but only a small current when applied to the cell body (right). (After S. Firestein et al., 1991. *J Neurosci* 11: 3665–3572.)

the dendritc knob. The lack of dynein arms (cytoskeletal proteins that facilitate the "sliding" of microtubules on one another), however, makes ORN cilia *nonmotile*. Cytoskeletal and associated scaffolding proteins in ORN and VRN cilia localize odorant receptor proteins, associated signal transduction proteins, and ion channels that are crucial for odorant transduction to the cilium and dendritic knob.

Odorant receptor genes and proteins

Odorant receptor proteins embedded in the membrane of ORN and VRN cilia play a central role in encoding olfactory information. These receptor proteins were not discovered until the early 1990s (in contrast to the much earlier identification of photopigments in retinal photoreceptors) when Richard Axel and Linda Buck identified them based on potential homology to G-protein-coupled receptors (GPCRs). Unlike the limited number of opsins (see Chapter 9), olfactory receptor proteins are numerous (some species have more than 2000 distinct variants; see below) and they are encoded by a large family of **odorant receptor genes**, found in the genomes of most animals. Olfactory receptor proteins, like the photoreceptor opsins, are indeed homologous to GPCRs, a category that also includes muscarinic acetylcholine, dopamine, and norepinephrine (also called noradrenergic) receptors (see Chapter 6). Thus, in all invertebrates and vertebrates examined thus far, odorant receptor proteins have seven membrane-spanning hydrophobic domains, potential odorant binding sites in the extracellular domain of the protein, and the ability to interact with G-proteins at the carboxyl-terminal region of their cytoplasmic domain (Figure 14.6A). The amino acid sequences for several of the membrane-spanning regions of odorant receptor proteins vary substantially in most vertebrates, as do the extracellular and cytoplasmic domains. The specificity of odorant recognition and signal transduction is presumably the result of this molecular variability, and the extensive repertoire of detected odors is presumably due to the large number of odorant receptor proteins encoded by an equally large number of odorant receptor genes. Nevertheless, there is limited understanding of the molecular mechanisms by which these multiple receptors bind specific odorants and encode odor identity (see the next section).

The number of odorant receptor genes, though substantial in all species, varies widely, and their expression appears to be tightly controlled in spatially restricted subsets of ORNs. Thus, the questions of "how many" and "where" have driven the investigation of odorant receptor genes and their relationship to odor detection. Most of these genes are predicted to encode GPCRs (see Figure 14.6A); however, in some species, including the fruit fly *Drosophila melanogaster*, some molecular features shared by most GPCRs, particularly a G-protein binding domain, are not recognized in the amino acid sequences of the odorant receptor proteins

(A)

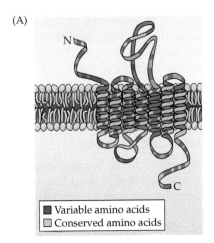

■ Variable amino acids
□ Conserved amino acids

FIGURE 14.6 (A) The generic structure of putative olfactory odorant receptor proteins. These proteins have seven transmembrane domains, plus a variable region that interacts with G-proteins, as is the case for all G-protein coupled receptors (GPCRs). The light blue regions of all three domains of the odorant receptor protein: extracellular (amino terminus: N), transmembrane, and intracellular (carboxy terminus:C) have amino acid sequences that are shared between odorant receptor proteins and other GPCRs. In contrast, the green regions are amino acid sequences in the odorant receptor proteins that are highly variable with sequences that distinguish the multiple ORs from other GPCRs as well as differentiating each OR protein. (B) The seven transmembrane domains common to all GPCRs are maintained in ORs throughout the animal kingdom. The proportionate sizes of the transmembrane domains (dark blue) and the cytoplasmic domains (light blue) vary in each species as do the presence and number of splice sites (red arrows) for differential splicing of mRNAs encoded by odorant receptor genes. Mammalian odorant receptor mRNAs are not differentially spliced. Those in *C. elegans* and *D. melanogaster* are, adding to the diversity in odorant receptor amino acid sequence and potentially function. (A after A. Menini, 1999. *Curr Op Neurobiol* 9: 419–426; B after L. Dryer, 2000. *BioEssays* 22: 803–810.)

(B)

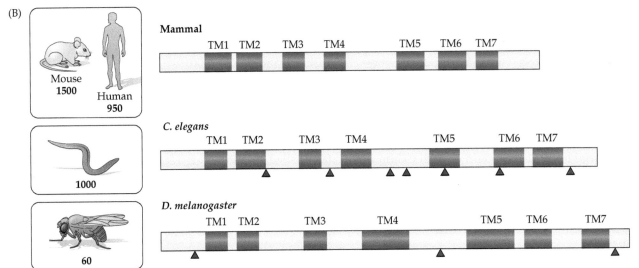

despite the retention of seven transmembrane domains (see paragraph on ionotropic and TAAR receptors below).

The number of individual odorant receptor genes in the genomes of most species, whether invertebrate or vertebrate, varies between 1000 and 2000, and these genes are homologous to additional members of the GPCR family. The worm *Caenorhabditis elegans* has approximately 1000 odorant receptor genes. In contrast, *Drosophila* has 61 odorant receptor genes. Although this remains a large number of genes for a single functional class of receptor proteins, the conseqeunces of the difference between 61 odorant receptor genes in the fly versus 1000 in the worm remain unclear. Most mammals have between 1000 and 2000 odorant receptor genes—approximately 3% to 5% of the protein-coding portion of their genomes (Figure 14.6B). Additional sequence analysis of apparent mammalian odorant receptor genes, however, suggests that many of these genes—about 60% in humans and chimpanzees versus 15% to 20% in mice and dogs—are not transcribed because of changes that have rendered them **pseudogenes**: DNA sequences that contain a promoter and transcription initiation site, but either cannot be transcribed into a

stable mRNA, or the transcript translated into a protein. Thus, the number of functional odorant receptor proteins encoded by stably transcribed genes and translatable mRNAs is estimated to be about 400 in humans and chimpanzees versus about 1200 in mice and 1000 in dogs (animals with apparently superior olfactory discriminatory ability; see Concept 14.4 and Box 14A). The relationship between these disparate numbers of odorant receptor proteins and distinctions in olfactory function is not fully understood.

The large odorant receptor gene families identified in most invertebrates and vertebrates, including humans and other mammals, are now referred to as *canonical* odorant receptors. In invertebrates and most vertebrates, additional, usually smaller, families of GPCRs have been associated with sensation of small subsets of volatile chemicals. These receptors—**ionotropic receptors (IRs)** in *Drosophila* and other insects, and **trace amine-associated receptors (TAARs)** in vertebrates—have little or no homology with canonical odorant receptors. Instead, IRs are thought to be related to ionotropic glutamate receptors, and TAARs are similar to aminergic GPCRs such as dopamine receptors (see Chapter 6). There are 16 ionotropic odorant receptors

■ BOX 14A | The "Dogtor" Is In

Conventional wisdom holds that having a pet, particularly a dog, is good for your health. Most of us assume that the primary benefits come from the companionship, as well as the daily exercise, that a dog provides. However, there may be more critical benefits of pet ownership that reflect the remarkable acuity of the canine olfactory system. One's family dog may in fact be a reliable source of early diagnosis for several cancers—albeit a diagnostician that likes to chew shoes and has a wet nose.

In the late 1980s, anecdotal reports emerged that claimed family dogs could use smell to identify moles and other skin blemishes on their owners that turned out to be malignant. In recounting this seemingly strange capacity of several dogs, H. Williams, one of the original discoverers, reported "a patient whose dog constantly sniffed at a mole on her leg. On one occasion, the dog even tried to bite the lesion off. ... [The] constant attention [of the dog] prompted her to seek medical advice. The lesion was excised and histology showed the lesion to be a malignant melanoma."

Subsequently, similar diagnoses by individual pets for their owners were reported, including a Labrador retriever that detected a basal cell carcinoma that had developed from an eczema lesion on its owner's skin. A slightly less anecdotal study relied on techniques used to train explosive-sniffing dogs for airport security. In this instance, George, a schnauzer, was trained to distinguish malignant melanomas in cell culture from their nonmalignant melanocyte counterparts. George was then introduced to a person who had several moles. One mole caused George to "go crazy"; a biopsy proved that the mole was indeed an early malignant melanoma.

Over the ensuing years, further anecdotal evidence suggested that dogs could recognize lung, breast, and bladder cancer using olfaction. These reports remained isolated anecdotes until 2006, when a truly systematic analysis of this apparent diagnostic capacity was published. In this study, five ordinary adult dogs were trained to distinguish exhaled breath samples from patients with lung or breast cancer versus controls who did not have cancer. The dogs were then tested for their ability to distinguish patients from controls in an entirely novel sample population. In this instance, the specificity and sensitivity of the dogs' ability to detect lung cancer from early to late stages was 99% as accurate as that of conventional biopsy diagnosis. The accuracy of breast cancer detection was slightly lower—approximately 90% that of conventional methods.

A similar study that challenged dogs to discriminate urine from patients with and without bladder cancer had parallel but somewhat less robust results. During the course of this study, however, the dogs consistently identified a presumed "control" sample as that from a patient with cancer. Clinicians were sufficiently alerted to perform further diagnostic tests, and in fact discovered a kidney carcinoma in this individual.

Aside from writing a new chapter in the saga of the salutatory relationship between humans and dogs, these observations have several implications for understanding the mechanisms and biological significance of olfactory acuity and selectivity. First, there is evidence that the concentration of alkanes and other volatile organic compounds is increased in air exhaled from patients with lung cancer. Thus, as indicated by preliminary studies of

odorant receptor molecule sensitivity, 7-transmembrane G-protein-coupled odorant receptors may be specialized to detect and discriminate a wide—and biologically significant—spectrum of volatile organic compounds at low concentrations. Second, the discrimination made between patients and controls, either by untrained individual dogs or the trained group of dogs, suggests that subtle distinctions in olfactory perception are clearly represented and can guide behavior. The apparent heightened olfactory ability in dogs may reflect a somewhat larger number of odorant receptors and/or relatively larger olfactory periphery that allow increased sensitivity, or specialized circuitry in the olfactory bulb, pyriform cortex, or other brain regions that assign cognitive significance to distinct olfactory stimuli. Whether this ability has adaptive significance for dogs or is just the ultimate smart pet trick is unclear.

Does this mean the term *pet scan* will soon take on a new meaning in clinical medicine? Clearly, the complexity of making critical diagnoses and the potential lack of reliability of dogs—however well trained—render routine use of diagnostic dogs difficult to imagine. Nevertheless, the remarkable olfactory capacity of these animals provides a starting point for an understanding of the molecular specificity of odorant receptors, as well as processing capacity and representations of olfactory information in the CNS. Such understanding may not only illuminate the functional characteristics of the olfactory system; it may provide a natural guide to specific molecules associated with disease states and the design of better diagnostic tools—or at least diagnoses that don't rely on cold, wet noses.

---

in *Drosophila* that are expressed in olfactory sensory neurons associated with antennae (a distinct set of IRs in *Drosophila* are believed to be taste receptor proteins; see Chapter 15). Terrestrial vertebrates, including mammals, have fewer active TAARs—6 TAAR genes in humans, 15 in mice—while some species of fish can have more than 100. The IRs and TAARs are apparently selective for specific, singular environmental stimuli. In *Drosophila*, select, nutritionally significant food odors as well as the insect repellent DEET activate IR-expressing ORNs, and their activation elicits a range of distinct behavioral responses. In vertebrates, TAARs are thought to recognize amines released during the decomposition of proteins, or those that may be secreted into bodily fluids. These signals are thought to mediate conscious repulsion and avoidance reactions, such as those that result from smelling foods that are spoiled and thus might be harmful to ingest.

Expression of canonical odorant receptor genes as well as IRs and TAARs in ORNs has been confirmed for only a limited subset of the substantial number of odorant receptor genes in most species. Localization of receptor expression has relied primarily on in situ hybridization (see Chapter 1) for mRNAs encoding the odorant receptor proteins, since generating antibodies against the proteins themselves has proved difficult. In addition, transgenic approaches have been used to detect odorant receptors

based on expression of genes encoding reporter proteins (see Chapter 1) driven by odorant receptor gene promoters (Figure 14.7). The most comprehensive mapping of odorant receptor gene expression has been completed in *Drosophila*, which not coincidentally has one of the smallest number of odorant receptor genes. In the fly, each of the ORNs (which are expressed on cells associated with the antennae and other facial structures, as flies and most other insects do not have "noses") expresses only one of the odorant receptor genes (see Figure 14.7A). In most vertebrates, including mammals, mRNAs for individual odorant receptor genes are also localized to single ORNs. Groups of these odorant receptor–specific ORNs occur in bilaterally symmetrical zones of olfactory epithelium. Additional evidence for restricted patterns of odorant receptor gene expression in spatially discrete subsets of ORNs comes from molecular genetic experiments (primarily in mice and flies) in which reporter proteins such as green fluorescent protein (GFP)

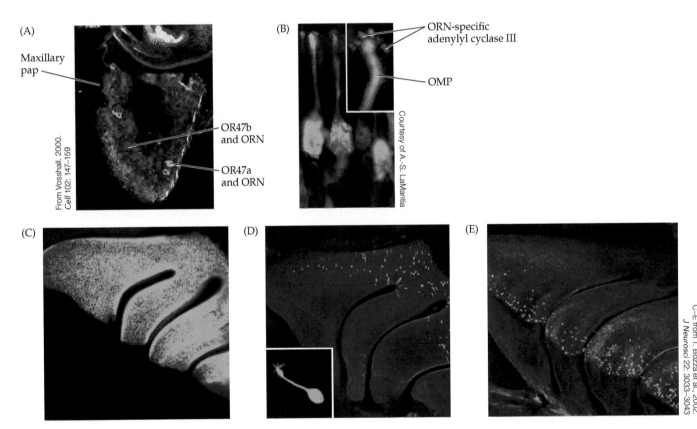

FIGURE 14.7 Odorant receptor gene expression (A) Olfactory receptor neurons in the distal domain of the *Drosophila* maxillary palp, a specialized olfactory organ on the fly's face (see also Figure 14.12), express distinct odorant receptor genes (OR47A, OR47B, labelled by in situ hybridization for the relevant mRNA). (B) Individual ORNs labeled immunohistochemically with olfactory marker protein (OMP, green; OMP is selective for *all* ORNs) and the ORN-specific adenylyl cyclase III (red) that is limited to olfactory cilia and is activated by odorant receptor proteins signaling (inset). The labels are in register with the segregation of signal transduction components to this domain. (C) The distribution of

OMP-expressing ORNs throughout the entire nasal epithelium of an adult mouse, demonstrated with an OMP-GFP reporter transgene. The protuberances oriented diagonally from left to right represent individual turbinates in the olfactory epithelium. The remaining bony and soft-tissue structures of the nose have been dissected away. (D) The distribution of ORNs expressing the I7 odorant receptor. These cells are restricted to a distinct domain or zone in the epithelium. The inset shows that odorant receptor-expressing cells are indeed cilia-bearing ORNs. (E) ORNs expressing the M71 odorant receptor are limited to a zone that is completely distinct from that of the I7 receptor.

or β-galactosidase are inserted into an odorant receptor gene locus (see Figure 14.7B–E; see also Figure 14.12 as well as Chapter 1 for a general summary of this approach). Finally, the mRNA that is translated to generate the odorant receptor protein uniquely expressed in each ORN is apparently transcribed from only one of the two allelic copies (one allele from maternal chromosomes, one allele from paternal chromosomes) of each odorant receptor locus. The mechanism of this **allelic silencing** is thought to reflect local chromatin confirmation changes in regions of the genome where odorant receptor genes are clustered. Local transcriptional feedback within each ORN reinforces the selection of one allele and the exclusion of the other so that this allelic choice is maintained. Remarkably, this allelic silencing must be recapitulated in all of the ORNs that are generated throughout an animal's lifetime (see the earlier section "Degeneration and regeneration of ORNs"). How this allelic exclusion mechanism is established and maintained in the ORN progeny of stem cells resident in the olfactory epithelium remains unknown.

Different odors probably activate spatially distinct subsets of ORNs specialized for binding and detection of a singular stimulus with a distinct molecular identity. It is unlikely, however, that most consciously perceived odors are encoded by binding of a single odorant to a single type of receptor protein. Furthermore, because only one of the two allelic copies of each odorant receptor gene is expressed in any particular receptor neuron (which effectively multiplies the number of distinct odorant receptor proteins by 2, since maternal and paternal alleles usually have some variation in sequence), there may be additional molecular resolution of odorants by subsets of ORNs. Thus, there is a likely consequence of the variable arithmetic and patterns of odorant receptor gene expression. Genomic and molecular diversity, along with complex cellular and spatial regulation of odorant receptor gene expression and the presumed selective activation of subsets of ORNs, probably contributes to the capacity of olfactory systems to detect and encode a wide range of complex and novel odors in the environment.

Molecular and physiological mechanisms of olfactory odorant transduction

Once an odorant is bound to an odorant receptor protein on the cilia of a particular ORN, several additional steps are required to generate a receptor potential that encodes the odorant's presence. This receptor potential converts the chemical information provided by the stimulus into electrical signals that can be interpreted by the brain. In mammals, the principal pathway for generating electrical activity in ORNs involves the activation of **cyclic nucleotide-gated ion channels** similar to those found in rod photoreceptors (see Chapter 9). ORNs express an olfactory-specific heterotrimeric G-protein, G_{olf}, whose α subunit dissociates upon odorant binding to

odorant receptor proteins and then activates **adenylyl cyclase III** (**ACIII**), an ORN-specific adenylate cyclase (Figure 14.8A; also see Figure 14.7B). Within individual ORNs, both G_{olf} and ACIII are restricted to the olfactory knob and cilia, consistent with the localization of odor transduction to this ORN cellular domain (see Figures 14.5 and 14.7). Stimulation of odorant receptor molecules by odorant binding leads to an increase in cyclic AMP (cAMP), which opens the cyclic nucleotide-gated channels and permits entry of Na^+ and Ca^{2+} (mostly Ca^{2+}), thus depolarizing the neuron (see Figure 14.8A). This depolarization, amplified by a Ca^{2+}-activated outward Cl^- current (resulting in further depolarization), is conducted passively from the cilia to the axon hillock region of the ORN, where action potentials are generated via voltage-regulated Na^+ channels and transmitted to the olfactory bulb.

There are also distinct signaling mechanisms for repolarization, recovery, and **adaptation** in response to odorants. Most of these mechanisms reflect increased Ca^{2+} that follows initial depolarization. In response to elevated Ca^{2+}, an Na^+/Ca^{2+} exchanger repolarizes the membrane. In addition, calcium/calmodulin kinase II reconstitutes the dissociated subunits of G_{olf} and diminishes cAMP levels via activation of phosphodiesterases. These events "reset" the receptor so that it can respond to the next odorant stimulus. Adaptation, the process by which sensory receptors adjust to constant stimulation by reducing their responsiveness, relies on cAMP-regulated phosphorylation of intracellular domains of the odorant receptor proteins, as well as engagement of β-arrestin (which serves a similar role in photoreceptor adaptation) to modify odorant receptor sensitivity. These mechanisms for adaptation probably play a role in perceived changes in sensitivity to smells, such as initially noticing, but later not sensing, the smell of cigarette smoke in a "smoking" hotel room. Thus, olfactory transduction parallels that in other sensory systems in the ability of the primary receptor neurons to modify their responses via altered channel or receptor activity in the sustained presence of specific stimuli.

In genetically engineered mice, elimination of any one of the major signal transduction elements associated with G-protein-coupled odorant receptors—G_{olf}, ACIII, or the cyclic nucleotide-gated channel—results in a loss of receptor response to odorants in ORNs (Figure 14.8B). In these mutant animals, the ORNs appear otherwise normal—they simply lack any response to odor. There is also complete loss of behavioral response to most odorants; in other words, the mice are anosmic. This common end point following loss of function of each signal transduction molecule demonstrates that each signaling step—receptor-mediated G-protein activation, adenylyl cyclase–mediated elevation of cAMP levels, and Ca^{2+}-mediated activation of the cyclic nucleotide-gated channel—contributes to the transduction of odorants. The use of a distinct set of G-proteins and channels in VRNs (see the next section) probably

(A)

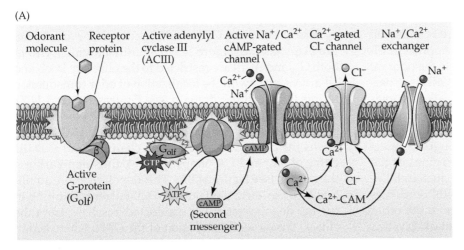

(B)

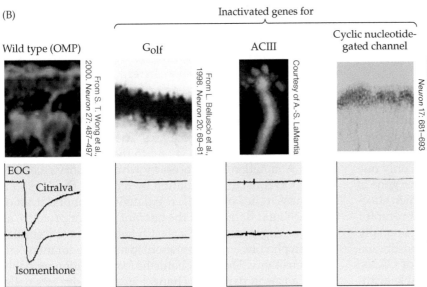

FIGURE 14.8 Molecular mechanisms of odorant transduction (A) Odorants in the mucus bind directly (or are shuttled via odorant binding proteins) to one of many receptor molecules located in the membranes of the cilia. This association activates an odorant-specific G-protein (G_{olf}) that, in turn, activates an adenylate cyclase (ACIII), resulting in the generation of cyclic AMP (cAMP). One target of cAMP is a cation-selective channel that, when open, permits the influx of Na^+ and Ca^{2+} into the cilia, resulting in depolarization. The ensuing increase in intracellular Ca^{2+} opens Ca^{2+}-gated Cl^- channels that provide most of the depolarization of the olfactory receptor potential. The receptor potential is reduced in magnitude when cAMP is broken down by specific phosphodiesterases to reduce its concentration. At the same time, Ca^{2+} complexes with calmodulin (Ca^{2+}-CAM) and binds to the channel, reducing its affinity for cAMP. Finally, Ca^{2+} is extruded through the Na^+/Ca^{2+} exchange pathway. (B) Consequences of inactivation of critical molecules in the odorant signal transduction cascade. The images of ORNs show expression of G_{olf}, ACIII, and the cyclic nucleotide-gated channel. The traces below show odorant-elicited electrical activity in the olfactory epithelium, measured extracellularly using the electro-olfactogram (EOG). In the wild type, a robust response results when either pleasant (citralva) or pungent (isomenthone) odors are presented. Inactivating any of the major signal transduction molecules linked to the 7-transmembrane odorant receptors abolishes these responses.(A after A. Menini, 1999. *Curr Opin Neurobiol* 9: 419–425.)

accounts for residual chemosensory behavior seen in these mice. Moreover, in invertebrates (including *Drosophila*), some odorant receptors may act as ion channels (like the IRs described in the previous section), directly influencing depolarization, and thus odor transduction, without activating G-protein-coupled signaling. Thus, although the overall molecular structure of odorant receptor proteins has been conserved, these proteins transduce odorants in diverse ways in different species.

Like other sensory receptor cells, individual ORNs are differentially sensitive to subsets of stimuli. Thus, there is ORN chemical **tuning** (Figure 14.9) similar to that of spectral tuning of photoreceptors in response to distinct portions of the visible light spectrum (see Chapter 9). The range of tuning for ORNs, however, is diverse. Understanding tuning is also complicated by uncertainty around how to classify molecules associated with specific types of perceived odors. There is no clear relationship between odorant molecules that parallels the relationship between the visible light spectrum, rods and cones.

ORNs can be responsive to several chemically distinct odorant molecules. Thus, it has been difficult to discern a code that might match groups of odorants with similar chemical structures to a particular ORN type or odorant receptor protein. Nevertheless, differences in odorant sensitivity detected in individual ORNs are probably parallel to expression of single odorant receptor genes in each cell. Most randomly selected single ORNs have broadly tuned responses to a variety of odorants (see Figure 14.9, neuron 1). For other ORNs, however, there may be more precise single odorant molecule specificity, possibly based on the binding affinity of distinct odorants for individual odorant receptor proteins (see below). Indeed, some ORNs exhibit marked selectivity to a single chemically defined odorant (compare neurons 2 and 3, Figure 14.9).

To determine whether there is a relationship between odorant responses and odorant receptors, ORNs that express single odorant receptors can be genetically labelled and isolated, and Ca^{2+} imaging of these cells (see Chapter 1) can be used to measure depolarization in response to

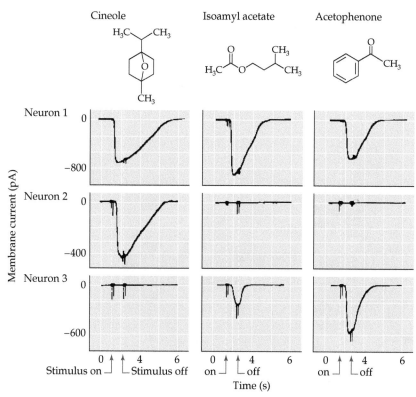

Cineole   Isoamyl acetate   Acetophenone

FIGURE 14.9 Responses of receptor neurons to selected odorants
Neuron 1 responds similarly to three different odorants. In contrast, neuron 2 responds to only one of these odorants. Neuron 3 responds to two of the three odorants. The responses of these receptor neurons were recorded by whole-cell patch clamp recording (see Box 4A); downward deflections represent inward currents measured at a holding potential of −55 mV. (After S. Firestein, et al., 1992. *Neuroreport* 3: 661–664.)

defined odorants in these molecularly identified individual ORNs (Figure 14.10). Consistent relationships exist between groups of odorants and the responses they elicit in individually isolated ORNs expressing the same odorant receptor. These cellular response profiles reflect chemical differences in subsets of odorant mixtures (defined by differences in carbon chain length of the molecular "backbone" of an odorant). These responses, like those described for narrower arrays of odors (see Figure 14.9), can be fairly diverse. For example, a single class of ORNs, defined by their expression of a single odorant receptor gene, responds relatively well to several different mixtures (Figure 14.10A). Apparently, there are ORNs that are "broadly tuned" to multiple chemical stimuli. ORNs expressing an odorant receptor protein that is known to bind with high affinity one or a few identified ligands (for example, the M71 odorant receptor, which is known to bind acetophenone and benzaldehyde; Figure 14.10B) can also be genetically labeled and their activity recorded. These ORNs respond specifically to the mixture that includes the limited odorant molecules that the receptor protein binds selectively, consistent with "narrow tuning". It is not known, however, whether any of these odorant molecules represent "best" stimuli (i.e., highest affinity) or most environmentally relevant odorants that each cell type can detect. For any given odorant receptor protein, the relationship between ORN response selectivity and consciously perceived odors that have behavioral significance remains uncertain.

Vomeronasal pheromone and kairomone transduction

In the late 1990s, the distinct identity of the vomeronasal system was confirmed at the genetic and molecular level with the cloning of a family of **vomeronasal receptors** (**VRs**). The vomeronasal receptors are genetically and molecularly distinct from odorant receptors. Thus, there is a

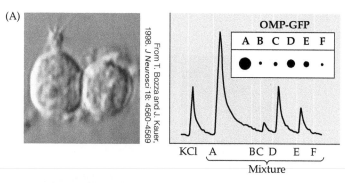

(A)

OMP-GFP

A B C D E F

KCl   A   BC D   E F
          Mixture

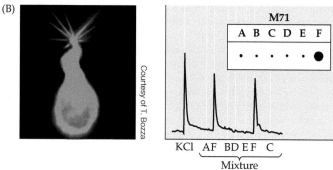

(B)

M71

A B C D E F

KCl AF BD EF C
        Mixture

FIGURE 14.10 Odorant receptor neuron selectivity
Odorant receptor neurons (ORNs) were isolated and tested for their responses to six different odorant molecule mixtures (indicated here as A–F). KCl was used as a control to demonstrate the capacity of the cell to generate an action potential. The size of the dots and magnitude of the spikes in the graphs indicate the strength of the electrical response to each odorant mixture. (A) Randomly chosen ORNs responded to several of the six mixtures. (B) The M71R-expressing cell was isolated by linking its gene to green fluorescent protein. M71-selected ORNs responded preferentially only to mixture F. (Graphs after T. Bozza et al., 2002. *J Neurosci* 22: 3033–3043.)

group of receptor proteins that are specifically expressed and have distinct transduction capacity in VRNs found in the VNO. The vomeronasal receptors are a separate large family (as many as 250 individual receptor genes in the mouse) of 7-transmembrane G-protein-coupled receptors expressed uniquely in VRNs. They fall into two major classes, **V1Rs** and **V2Rs** (Figure 14.11A). V1Rs have a limited extracellular domain, whereas V2Rs have an extensive extracellular domain. V1Rs and V2Rs use different G-protein-coupled cascades to activate signaling (Figure 14.11B). Thus, although VRNs in the VNO look much like ORNs, they are genetically, molecularly, structurally, and

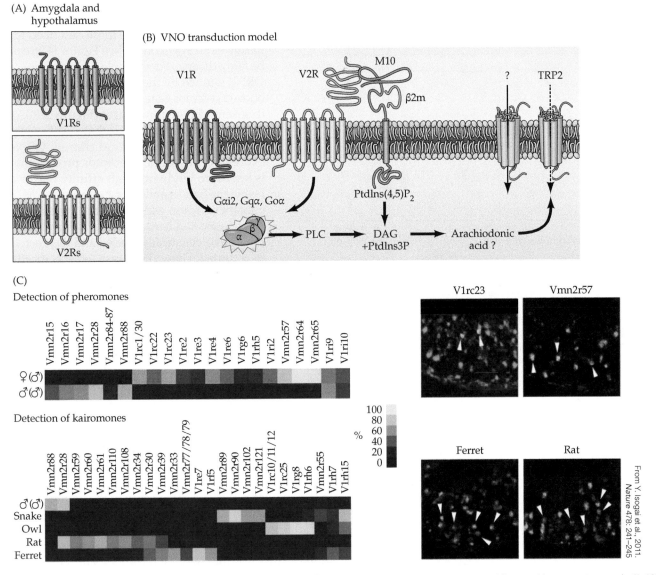

Figure 14.11 Vomeronasal receptor (VR) proteins and pheromone transduction. (A) V1Rs and V2Rs are distinct families of 7-transmembrane G-protein-coupled receptors. The amino terminus of the V1Rs is fairly simple, while that of the V2Rs includes an extensive extracellular domain. V1Rs and V2Rs are expressed uniquely by subsets of Vomeronasal receptor neurons. (B) The signal transduction mechanisms that activate the transient receptor potential (TRP) channel in the two classes of cells are distinct. V1Rs and V2Rs can activate several trimeric g-proteins. V2Rs also interact with additional transmembrane proteins to signal via the phosphotidyl inositol pathway. These transduction events lead to the activation of the Trp2 channel as well as additional channels whose identities remain uncertain (denoted by question mark, ?). (C) Selectivity of single VR proteins for sex-specific pheromonal cues (top) and species-specific kairomonal cues. The selectivity is not exclusive; however, based on the heat maps, it is clear that some receptor proteins (e.g., Vmn2r64 for sex-specific pheromones and V1rg8 for species-specific kairomones) respond dimorphically (Vmn2r64) or selectively (V1rg8), to distinct cues. These estimates of response selectivity are based on frequency of co-localization of single VRs (green) with expression of Egr1 protein (red) that is rapidly expressed in response to neuronal activity. (A after E. Pantages and C. Dulac, 2000. *Neuron* 28: 835–845; B after C. Dulac and A. T. Torello, 2003. *Nat Rev Neurosci* 4: 551–562.)

functionally different. Signal transduction in VRNs is accomplished via a different set of second messengers and cyclic nucleotide-gated ion channels. The V1Rs use the G-protein Gαi2, and the V2Rs use the G-protein Gαo (versus the olfactory-selective G_{olf} that is activated by odorant binding to odorant receptors in ORNs). The activity of **transient receptor potential (TRP)** channels, specifically TRP2, in VRNs is regulated by phospholipase C (PLC) and diacylglycerol (DAG) in response to G-protein stimulation via V1Rs and V2Rs. In contrast, the primary molecular mediators of excitability in ORNs are adenylate cyclase, cAMP, a cyclic nucleotide-gated ion channel, and the Ca^{2+}-gated Cl^- channel (see Figure 14.8). In addition, some V2Rs are coexpressed with and may interact with **major histocompatibility complex (MHC)** genes M10 and M1. The functional significance of the V2R–MHC interaction remains uncertain. The MHC protein may contribute to pheromone detection by subsets of V2Rs and also is thought to regulate the trafficking of the V2R to the vomeronasal receptor cell membrane.

The molecular map of pheromones (airborne molecular stimuli from conspecifics: e.g., mouse to mouse) and kairomones (airborne molecular stimuli from predators or prey) and its relationship to VRN activation is now fairly well understood. Based on cellular and physiological analyses, VRNs appear to be fairly broadly tuned (Figure 14.11C). However, some VRNs respond selectively to stimuli known to bind to distinct V1Rs and V2Rs. Several V1Rs and V2Rs respond robustly and differentially to pheromones as sex-specific or kairomones as predator–prey-specific cues (urine, etc.) in male versus female mice. Elimination of the TRP channels that mediate vomeronasal signal transduction (or elimination, replacement, or mutation of the genes encoding the stimulus-specific vomeronasal receptors) leads to changes in sexual or reproductive behavior, often in a **dimorphic** (see Chapter 25) manner, meaning the effects diverge in females versus males. For example, elimination of the TRPC2 VRN transduction channel in male mice increases affiliative (social) behaviors, including sexual behaviors, rather than the aggression usually seen with other male mice. Several V2Rs and fewer V1Rs respond robustly and differentially to predator cues. In mice, these cues include chemicals associated with urine or other bodily excretions from snakes, owls, rats, and ferrets. This selectivity is assumed to mediate avoidance responses so that mice can detect and evade these threatening species. V1Rs and V2Rs are segregated in the VNO: The V1R-expressing VRNs are found in the apical or lumenal region of the VNO sensory epithelium, while the VRNs expressing V2Rs are found more basally. These classes of receptors are also apparently functionally distinct: V1Rs participate more in pheromone sensing, and V2Rs in kairomone sensing.

CONCEPT 14.3 Olfactory and Vomeronasal Information Is Relayed Directly to the Main and Accessory Olfactory Bulbs, and from There to Multiple Forebrain Targets

LEARNING OBJECTIVES

14.3.1 Define the architecture of the main and accessory olfactory bulbs and the relay of peripheral odorant information via glomeruli.

14.3.2 Describe the relay and local processing of sensory information by main and accessory olfactory bulb projection neurons and interneurons.

14.3.3 Identify the forebrain targets of the main and accessory olfactory bulbs.

14.3.4 Evaluate the evidence for the representation and processing of odor identity in the pyriform cortex and other brain regions that are olfactory or vomeronasal targets.

The olfactory bulb

The transduction of odorants in the olfactory cilia and subsequent changes in electrical activity in ORNs are only the first steps in olfactory information processing. Unlike other primary "special" sensory receptor cells (e.g., photoreceptors in the retina, hair cells in the cochlea), ORNs have axons, and these axons relay odorant information directly to the brain via action potentials, without local processing after initial sensory transduction as in the retina, and without peripheral relay neuron intermediates as with audition, vestibular sense, somatosensation, and taste. As ORN axons leave the olfactory epithelium, they coalesce to form a large number of bundles that together make up the olfactory nerve (cranial nerve I). Each complete olfactory nerve (the aggregates of ORN axon bundles from the left or right olfactory epithelium) projects ipsilaterally to the olfactory bulb, which in humans lies on the ventral anterior aspect of the ipsilateral cerebral hemisphere. The most distinctive feature of the olfactory bulb is the array of **glomeruli**—more or less spherical accumulations of neuropil (dendrites, axons, glial processes) usually 100 to 200 μm in diameter. Glomeruli lie just beneath the surface of the bulb and are the synaptic target of the primary olfactory axons (Figure 14.12A-C). In vertebrates, ORN axons make excitatory glutamatergic synapses within the glomeruli. Remarkably, this relationship between the olfactory periphery—ORNs in the nose or those associated with insect sensory specializations—and glomeruli in the CNS is maintained across the animal kingdom (see Figure 14.12A, inset).

In mammals, including humans, within each glomerulus the axons of ORNs synapse on apical dendrites of

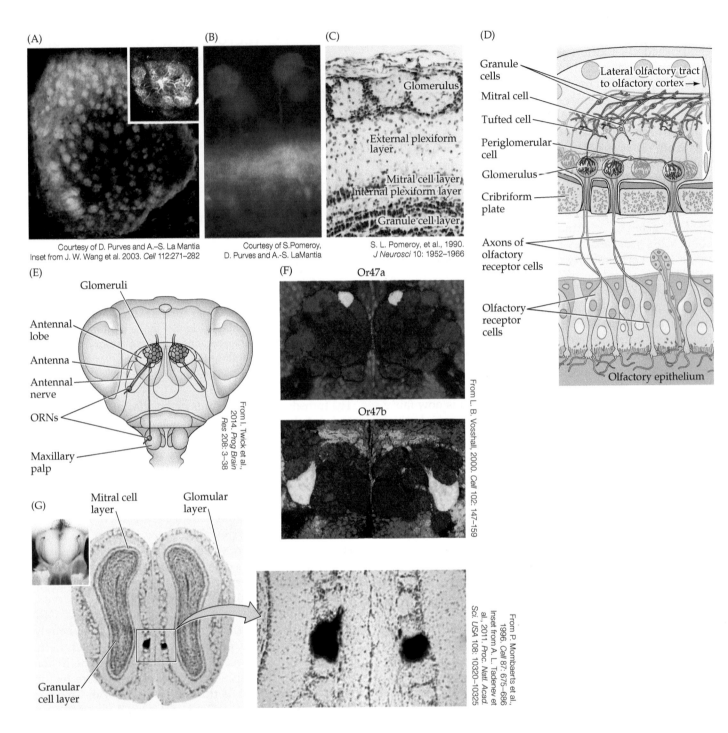

(A)

Courtesy of D. Purves and A.-S. La Mantia
Inset from J. W. Wang et al. 2003. *Cell* 112:271–282

(B)

Courtesy of S.Pomeroy,
D. Purves and A.-S. LaMantia

(C)

Glomerulus

External plexiform
layer

Mitral cell layer
Internal plexiform layer

Granule cell layer

S. L. Pomeroy, et al., 1990.
J Neurosci 10: 1952–1966

(D)

Granule
cells

Mitral cell

Tufted cell

Periglomerular
cell

Glomerulus

Cribriform
plate

Axons of
olfactory
receptor cells

Olfactory
receptor
cells

Lateral olfactory tract
to olfactory cortex →

Olfactory epithelium

(E)

Glomeruli

Antennal
lobe

Antenna

Antennal
nerve

ORNs

Maxillary
palp

From I. Twick et al.,
2014. *Prog Brain
Res* 208: 3–38

(F)

Or47a

Or47b

From L. B. Vosshall, 2000. *Cell* 102: 147–159

(G)

Mitral cell
layer

Glomular
layer

Granular
cell layer

From P. Mombaerts et al.,
1996. *Cell* 87: 675–686.
Inset from A. L. Tadenev et
al., 2011. *Proc. Natl. Acad.
Sci. USA* 108: 10320–10325

mitral cells, which are the principal projection neurons of the olfactory bulb. The cell bodies of the mitral cells are located in a distinct layer of the olfactory bulb deep within the glomeruli (Figure 14.12C,D). Mitral cells also use glutamate as their neurotransmitter, and extend axons to several additional forebrain targets to relay olfactory information for further processing (see later in this concept). Most mitral cells extend their primary dendrite into a single glomerulus, where the dendrite gives rise to an elaborate tuft of branches onto which the axons of ORNs

synapse (see Figure 14.12B,D). In addition, the lateral dendrites of mitral cells extend into a neuropil-dense region between the glomeruli and the mitral cell layer called the **external plexiform layer**. In the mouse, in which glomerular connectivity has been studied quantitatively, each glomerulus includes the apical dendrites of approximately 25 mitral cells, which in turn receive input from approximately 25,000 olfactory receptor axons.

Glomeruli across the animal kingdom share another feature: they receive input from multiple ORNs that

◀ **FIGURE 14.12 Organization of the primary olfactory pathway in insects and mammals** (A) When the bulb is viewed from its dorsal surface (visualized here in a living mouse in which the overlying bone has been removed), olfactory glomeruli can be seen. The dense accumulations of dendrites and synapses that constitutes glomeruli are stained here with a vital fluorescent dye that recognizes neuronal processes. (Inset) There is a similar arrangement of glomeruli in the antennal lobe (the equivalent of the olfactory bulb) in *Drosophila*. (B) Among the major neuronal components of each mammalian glomerulus are the apical tufts of mitral cells, which project to the piriform cortex and other olfactory bulb targets (see Figure 14.1C). In this image of a coronal section through the bulb, the mitral cells have been labeled retrogradely by placing the lipophilic tracer Di-I in the lateral olfactory tract. (C) The cellular structure of the mammalian (mouse) olfactory bulb, shown in a Nissl-stained coronal section. The five layers of the bulb are indicated. The glomerular layer includes the tufts of mitral cells, the axon terminals of ORNs, and periglomerular cells that define the margins of each glomerulus. The external plexiform layer is made up of lateral dendrites of mitral cells, cell bodies and lateral dendrites of tufted cells, and dendrites of granule cells that make dendrodendritic synapses with the other dendritic elements. The mitral cell layer is defined by the cell bodies of mitral cells, and mitral cell axons are found in the internal plexiform layer. Finally, granule cell bodies are densely packed into the granule cell layer. (D) Laminar and circuit organization of the mammalian olfactory bulb, shown diagrammatically in a cutaway view from its medial surface. Olfactory receptor cell axons synapse with mitral cell apical dendritic tufts and periglomerular cell processes within glomeruli. Granule cells and mitral cell lateral dendrites constitute the major synaptic elements of the external plexiform layer. (E) A schematic of the primary olfactory pathway in *Drosophila*. Olfactory receptor neurons are found in the antennae and the maxillary palps. Their axons project to the antennal lobe where the synapse in distinct glomeruli. (F) Axons from *Drosophila* olfactory receptor neurons expressing distinct odorant receptor proteins terminate in distinct glomeruli. (G) Axons from mammalian ORNs that express a particular odorant receptor gene converge on a small subset of bilaterally symmetrical glomeruli. These glomeruli, indicated in the boxed area in the upper panel, are shown at higher magnification in the lower panel. The projections from the olfactory epithelium have been labeled by a reporter transgene inserted by homologous recombination ("knocked in") into the genetic locus that encodes the particular receptor in the mouse.

express the same odorant receptor. This is the case for the limited number of glomeruli in the antennal lobe of *Drosophila* (the equivalent of the vertebrate olfactory bulb; Figure 14.12E,F). This relationship between molecular ORN identity and glomerular innervation is maintained for the more numerous glomeruli found in the olfactory bulbs of vertebrates. Thus, in the mouse, most, if not all, of the 25,000 axons that innervate each glomerulus come from ORNs that express the same, single odorant receptor gene (Figure 14.12G). This degree of convergence presumably increases the sensitivity of mitral cells to ensure maximum fidelity of encoding and relay of odorant identities. It may also maximize the signal strength from the convergent olfactory receptor neuron input by averaging out uncorrelated "background" noise.

Each glomerulus also includes dendritic processes from three other classes of local circuit neurons: **external tufted cells**, which are glutamatergic, **periglomerular cells**, which are GABAergic, and **short axon cells**, which are dopaminergic. These three cell classes make, respectively, excitatory, inhibitory, and modulatory synapses onto mitral cell dendrites within each glomerulus. In addition, a subset of periglomerular cells makes inhibitory synapses onto ORN axon terminals to provide local presynaptic inhibition. Finally, **granule cells**, whose cell bodies constitute the innermost layer of the vertebrate olfactory bulb, synapse primarily on the basal dendrites of mitral cells in the external plexiform layer (see Figure 14.12C,G). Granule cells lack an identifiable axon, and instead their dendrites make synapses onto mitral cell dendrites, primarily in the external plexiform layer. These dendrodendritic synapses made by olfactory granule cells, and the reciprocal dendrodendritic synapses made locally by the mitral cell dendrites, are thought to establish local lateral inhibitory circuits as well as participate in synaptic plasticity in the olfactory bulb. Olfactory granule cells and periglomerular cells are among the few classes of neurons in the forebrain that can be replaced throughout life in some mammals (see Chapter 26). In humans, however, the available evidence suggests that these cells, in contrast to ORNs (see Concept 14.2), are not lost and regenerated in adulthood.

The primary target of olfactory bulb projection neurons is the **pyriform cortex** and associated olfactory nuclei located in the lateral-basal region of cerebral cortex (see Figure 14.15). Mitral cells axons make synapses in all regions of the archicortical (three-layer) pyriform cortex, as well as in the amygdala and lateral entorhinal cortex, a parahippocampal neocortical (six-layer) area. Another class of mitral cell-like projection neurons called tufted cells have more limited projections. They innervate anterobasal olfactory nuclei proximal to the olfactory bulb. Thus, the olfactory bulb, via its projection neurons, relays odorant information processed by local glomerular circuits, to basal forebrain cortical regions that are probably essential for odor identification as well as to basal forebrain nuclei such as the amygdala that are more important for innate, homeostatic responses such as fear, and affiliative or aggressive social interactions to selected odors.

The accessory olfactory bulb

Axons from VRNs in the vomeronasal organ (VNO) project to the accessory olfactory bulb, which is usually recognized as a distinct structure located posterior, dorsal, and lateral to the main olfactory bulb. Thus, the primary projections of the VNO remain segregated from those of the main olfactory epithelium through the initial relay in the forebrain. VRNs expressing V1Rs versus V2Rs are segregated in the VNO, and their axons are segregated in the anterior versus posterior region of the accessory olfactory bulb (Figure 14.13). The cellular architecture of the accessory olfactory bulb is similar to that of the main olfactory bulb: There are glomeruli on the surface that include the same cell classes, and similar, although somewhat compressed, synaptic and cellular layers, including a distinct layer of mitral cells below the glomeruli. The output of accessory olfactory bulb mitral cells is more limited than that of mitral cells in the main olfactory bulb. The local processing of V1R and V2R information in individual glomeruli of the accessory olfactory bulb is thought to be similar to that in the main olfactory bulb; however, accessory olfactory bulb glomerular circuits and physiological responses have not been characterized in similar detail. Accessory olfactory bulb mitral cells project primarily to the medial amygdala as well as the bed nucleus of the stria terminalis (see Figure 14.2C), which serves as a relay to hypothalamic and brainstem circuits that regulate autonomic and neuroendocrine responses via vagal or thoracic motor neuron connections with peripheral sympathetic, parasympathetic, or enteric ganglia (see Chapter 21). These circuits use vomeronasal odorant information to influence emotional and innate responses essential for social interactions (see Chapter 32). Thus, the basal forebrain projections of the accessory olfactory bulb reinforce the conclusion that the vomeronasal system is a parallel pathway that encodes, relays, and represents pheromonal and kairomonal chemosensory signals that regulate homeostatic behaviors, independent of conscious volatile odor identification and perception, which are the province of the olfactory epitheium and main olfactory bulb.

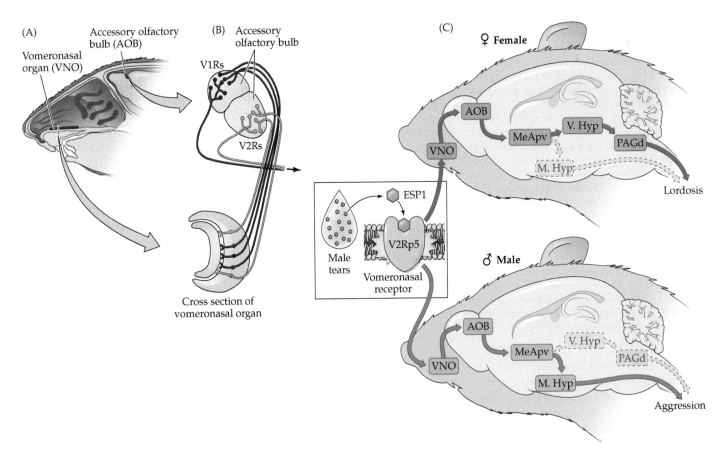

Figure 14.13 The vomeronasal system (A) A midsagittal section through the head of a mouse shows the location of the vomeronasal organ (VNO) in the nasal cavity, and the accessory olfactory bulb (AOB) located in the dorsal posterior region of the main olfactory bulb. (B) As diagrammed here, the two divisions of the AOB each have glomeruli (spherical units of neuropil where vomeronasal receptor neurons make synapses) that receive input from only one of two classes of vomeronasal receptor neurons, V1R or V2R (axons from vomeronasal receptor neurons expressing *vomeronasal receptor class 1*, dark blue, amd those expressing receptor class 2, light blue). (C) Activation of a specific V2R by a male pheromone found in tears leads to divergent relay of information through basal forebrain target nuclei and divergent behavioral responses: lordosis and reproductive receptivity in females; aggression in males. The female pathway includes target regions in the medial amygdala (Me A), the ventral hypothalamus (V. Hyp) and the midbrain (mid). The male pathway begins in the medial amygdala but then diverges to medial hypothalamus (M. Hyp). (B after E. Pantages and C. Dulac, 2000. *Neuron* 28: 835–845; C after J. Woodson, et al. 2017. *Neuron* 95: 1–2.)

Olfactory bulb circuits

The relationship between ORNs expressing one odorant receptor and small subsets of glomeruli (see Figure 14.12F,G) suggests that individual glomeruli respond specifically (or at least selectively) to distinct odorants. The selective (but not singular) responses of subsets of glomeruli to particular

odorants have been confirmed physiologically in invertebrates such as *Drosophila*, as well as in mice, using single and multiunit recordings, metabolic mapping, imaging with voltage-sensitive dyes, genetically encoded sensors of electrical activity, or intrinsic signals that depend on blood oxygenation. Some single odorants, or subsets of odorants with distinct chemical structures (e.g., length of the carbon chain in the backbone of the odorant molecule) can maximally activate one or a few glomeruli (Figure 14.14A). Increasing the odorant concentration available peripherally increases afferent and interneuron activity in individual glomeruli, as well as the number of ORN afferents active in subsets of glomeruli. Nevertheless, it appears that the output signal (mitral cell activation) generated from this peripheral concentration–sensitive activity is far less variable than the input responses to different concentrations. This suggests that a key aspect of the representation of olfactory information—**concentration-invariant coding** of odorant identity—is already established, at least in a rudimentary fashion, by glomerular circuit activity.

The neural circuitry responsible for establishing this concentration-invariant coding includes a balance of excitatory ORN afferent input modulated by local inhibition, as well as feedforward and feedback modulation of ORN excitatory inputs onto mitral cell dendrites (Figure 14.14B). The network of lateral interglomerular inhibitory and excitatory synapses sharpens the response of individual glomeruli to ORN stimulation. The innervation of mitral cells as well as of external tufted and periglomerular cells by ORN afferents provides a mechanism for feedforward influence so that the same afferent activity can influence mitral cells as well as the inhibitory and modulatory inputs onto the mitral cells. The network of granule cell inhibition in the external plexiform layer further shapes inter- and intraglomerular encoding of ORN afferent information. This local inhibitory network is activated by excitatory dendrodendritic synapses made by mitral or tufted cells onto granule cells, reciprocally inhibited by granule cell dendrodendritic synapses onto mitral and tufted cell

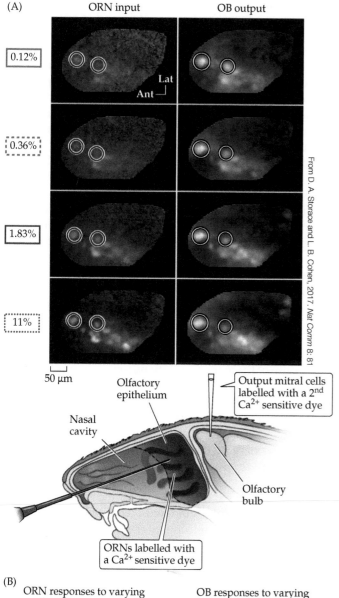

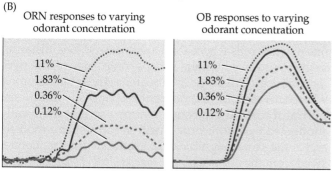

Figure 14.14   Differing sensitivity to odor concentration in the olfactory epithelium and bulb (A) Live imaging after loading two different calcium indicator dyes in the olfactory epithelium and bulb of the living mouse (lower inset). Using this approach, one can measure the activity of mouse odorant receptor neuron terminals that synapse within glomeruli (left, purple circles) versus that of mitral and tufted cells that project from glomeruli (right, gray circles) in response to four different concentrations (denoted by four different colors) of the same odor. (B) Representative traces showing the levels of activation of the olfactory receptor neuron afferents (left) and the olfactory bulb glomeruli target neurons (that project to the piriform cortex). The responses of the olfactory receptor neuron afferents are more sensitive to concentration changes than are those of the target neurons in the olfactory bulb. (B after D. A. Storace and L. B. Cohen, 2017. *Nat Comm* 8: 81.)

lateral dendrites. The local glomerular and extra-glomerular circuitry allows mitral cells to encode and relay odor information in register with odorant selectivity of ORNs whose projections converge on the glomerulus. This convergent information is further modulated by stimulus concentration and abundance as well as by activity in other unrelated glomeruli responding to different odorants. Mitral cell responses are also influenced by the active acquisition of odorant information via sniffing. Sniffing increases transiently the flow rate and concentration of odorants. The periodicity of the "sniff cycle" introduces time-dependent responses in the olfactory bulb that further shape stimulus selectivity of mitral cell responses to odorants.

It is still not clear how (or whether) odor identity and concentration are mapped across the entire array of glomeruli. The encoding of odor identity may reflect the activation of small subsets of glomeruli. Nevertheless, there is little evidence that single odorant receptor ligands correspond precisely to distinctly perceived smells. Given the response of small numbers of glomeruli to single odorant molecules, one might expect that complex natural odors such as those of coffee, fruits, cheeses, or spices—each of which includes more than 100 volatile compounds—would activate a very large number of olfactory glomeruli. Surprisingly, this is not the case. In mice, natural odorants presented at their normal concentrations activate a relatively small number of glomeruli (up to 20), each of which responds selectively to one or two molecules that characterize the complex odor. Thus, to solve the problem of representing complex odorants, the olfactory system appears to employ a **sparse coding** mechanism that extracts information from a small number of dominant chemicals within a complex odor, thus representing that complex odor via the maximal activation of a relatively small subset of glomeruli. One useful metaphor is to envision the sheet of glomeruli in the olfactory bulb as an array of lights on a movie marquee; the spatial distribution of the active and inactive glomeruli ("lit and unlit lights") produces a message that is unique for a given odorant at a particular concentration.

Pyriform cortical processing of odorant information

Mitral cell axons, as well as those from tufted cells, provide the only relay for olfactory information to the rest of the brain. The mitral cell axons from each olfactory bulb form the **lateral olfactory tract** which projects to the accessory olfactory nuclei, the olfactory tubercle, and the pyriform and entorhinal cortices, as well as to portions of the amygdala (Figure 14.15A; see also Figure 14.1). Most projections of the lateral olfactory tract are ipsilateral; however, a subset of mitral cell axons crosses the midline, presumably initiating bilateral processing of some aspects of olfactory information. In humans, the major target of the lateral olfactory tract is the three-layered pyriform cortex in the ventromedial aspect of the temporal lobe, near

the optic chiasm. Mitral cell inputs from glomeruli that receive odorant receptor–specific projections are distributed across the pyriform cortex (see Figure 14.15A). Accordingly, neurons in pyriform cortex respond to odors based on the relay of odorant information from ORNs through the olfactory bulb via mitral cell projections.

Recent work suggests that the segregation of projections based on the relationship between ORNs expressing a single odorant receptor protein and specific subsets of glomeruli in the olfactory bulb is far less constrained in the pyriform cortex. Thus, the projections of a single glomerulus focally innervated by ORNs expressing only one odorant receptor protein can be quite widespread in the pyriform cortex and additional basal forebrain targets of the olfactory bulb (see Figure 14.15A). Furthermore, pyriform cortical neurons have distinct responses to multiple versus single odors (Figure 14.15B). In fact, some individual pyriform cortical cells seem to be more broadly tuned to different odors than are cells in the olfactory bulb, while others are narrowly tuned to single odorant molecules presumed to bind to and maximally activate one or a few odorant receptors (see Figure 14.15B). Apparently, odorant receptor–based segregation of information in the olfactory bulb is not maintained in the pyriform cortex; however, there may be some segregation of inputs originating in distinct glomeruli in the amygdala. The anatomical redistribution of odorant information from the olfactory epithelium through the bulb to the pyriform cortex may influence an added level of complexity in the representations of olfactory sensation beyond odor identification; however, the nature of these representations remains somewhat difficult to discern based on patterns of connectivity alone—unlike the orderly topographic connectivity in the visual or somatosensory pathways that reflects key aspects of the functional organization of those systems.

The physiological correlate of the divergent projections of glomeruli to broad regions of the pyriform cortex is somewhat counterintuitive. Although the projections are anatomically broad, the responses of individual pyriform cortex neurons become more selective. Circuits in the olfactory bulb apparently encode information about both odor identity and abundance detected at the periphery, and relay this information in a temporal pattern established in part on the basis of the sniff cycle. In the pyriform cortex, the initial excitation in response to an odorant encoded and relayed by mitral cells in the olfactory bulb focally activates one or a few cells whose local axons extend long distances in the pyriform cortex and synapse onto a network of pyriform cortical GABAergic interneurons. These interneurons, when activated, inhibit odorant responses in additional pyriform cortical projection neurons so that the focally stimulated pyriform cortical neuron can respond robustly and at the same magnitude to the odorant information relayed from the olfactory bulb, *independent of changes in concentration.* Apparently, regardless of the relative strength of

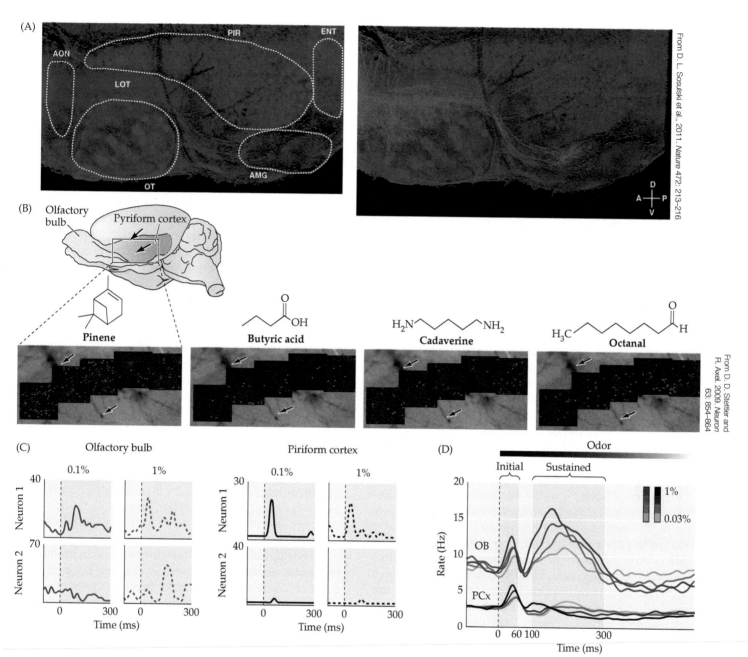

Figure 14.15 Relay and processing of odor information from the olfactory bulb (A) Anatomical tracing (using focal electroporation of TMR-dextran) of the projections of a single olfactory glomerulus via the lateral olfactory tract (LOT) to multiple olfactory bulb targets, including the piriform cortex (PIR), entorhinal cortex (ENT), amygdala (AMG), olfactory tubercle (OT), and accessory olfactory nucleus (AON). (B) Differential activation of widely distributed ensembles of neurons in piriform cortex by single odors. Lateral surface of the mouse brain showing the piriform cortex and outlining the region from which optical recordings of the electrical activity of single neurons were made. Four distinct odors recruited different subsets of cells across the piriform cortex. Activated cells (bright red) were recorded based on local change in fluorescence signal emitted by each cell. The fluorescence signal was due to a Ca^{2+}-sensitive dye introduced into all cells in the piriform cortex prior to the recording session. The arrows in each panel indicate a blood vessel, which provides a landmark to compare patterns of activation of the multiple odorants in this single animal. (C) Coincident responses of mitral cells (left) and piriform cortex neurons, whose responses have been recorded simultaneously, to differing concentrations of two odors. The olfactory bulb neurons still have somewhat differential responses based upon odor concentration; however, the pyriform cortical neurons generate the same magnitude responses to each concentration of each odor. (D) Temporal relationship between immediate and sustained responses of olfactory bulb (top traces) and piriform cortex (bottom traces) neurons in response to multiple odor concentrations. The piriform cortex neurons respond robustly to the initial presentation of the odor, independent of concentration, while the olfactory bulb neurons have both initial and sustained responses whose concentration sensitivity varies substantially. Neuron 1 in the pyriform cortex responds at approximately equal magnitude to the lower and higher odorant concentration at the periphery. Neuron 2 in the pyriform cortex has a minimal response to both concentrations. (C,D after K. E. Bolding and K. M. Franks. 2018. *Science* 361: eaat6904.)

connections of any one glomerular projection to a particular pyriform cortical neuron (Figure 14.15C), the identity of the stimulus relayed by that glomerulus to its pyriform cortex target neuron can be reliably represented by an invariant action potential response (Figure 14.15D). This odorant response would be of equal magnitude regardless of the concentration differences, and it would occur at the time of peak glomerular excitation, in register with the sniff cycle (see Figure 14.15D). This physiological mechanism, established by local circuits in the pyriform cortex, may provide a filter that diminishes "noise" in odor identification that is due to concentration. Thus, the information encoded and relayed from the pyriform cortex would reflect a high-fidelity representation of the most salient aspects of an odorant's identity.

Pyramidal neurons in the pyriform cortex project to a variety of forebrain targets. Thus, olfactory information—presumably about the essential aspects of odorant identity—is distributed broadly and influences a wide range of behaviors. Significant numbers of neurons in the pyriform cortex innervate several areas of the neocortex directly, including the orbitofrontal cortex in humans and other primates, where multimodal responses to complex stimuli—particularly food—include an olfactory component. Pyriform cortical neurons also project to the thalamus, hippocampus, hypothalamic nuclei, and amygdala. The connections between the pyriform cortex and the mediodorsal nucleus of the thalamus, a thalamic nucleus implicated in human memory (see Chapter 31), are thought to influence olfactory-guided declarative, or conscious, memory via mediodorsal connections with the frontal cortex. Projections to the hippocampus are similarly thought to play a role in olfactory-guided memory, but there is little indication of how olfaction and declarative memories are integrated. Finally, connections between the pyriform cortex, hypothalamus, and amygdala are thought to modulate visceral, appetitive, and sexual behaviors (these behaviors are also regulated by vomeronasal circuits in many animals).

CONCEPT
14.4

In Many Species, Including Humans, Olfactory Capacity Reflects the Size and Complexity of the Olfactory System

LEARNING OBJECTIVES

14.4.1 Compare olfactory behavioral capacity and olfactory system organization in multiple species, including humans.

14.4.2 Assess the role of vomeronasal sensation in guiding innate behaviors.

14.4.3 Define human olfactory system sensitivity and behavioral functions.

14.4.4 Evaluate the evidence for olfactory dysfunction in humans as a sign of neurodegenerative and psychiatric diseases.

Physiological and behavioral responses to odorant stimuli

All animals use odor information to guide a broad range of essential behaviors. Among the most essential are identification of appropriate, safe, and nutritive foods and detection of potentially toxic or hazardous odorants that emanate from equally hazardous sources. In addition, odorants influence social interactions that shape reproductive behaviors, predator and prey recognition, aggressive and affiliative social responses, and parenting. Odorants can elicit a variety of immediate physiological responses that reflect activation of the autonomic nervous system. Olfaction also has a *hedonic* perceptual component that determines whether a particular smell is pleasant and attractive or disagreeable and repulsive. The attractive versus repulsive valence of odors may also be influenced by learned associations. For example, the disagreeable odor, to some people, of certain cheeses is interpreted by other people as attractive based on learning and remembering the appetitive reward of what is experienced as a pleasant taste upon ingestion. Thus, neutral or even mildly aversive odors can elicit positive responses as a result of repeated exposure and pairing with behavioral rewards. This experience-dependent plasticity is thought to be mediated by synaptic and cellular mechanisms similar to those that underlie learning and memory for other modalities (see Chapters 8 and 30).

Physiological studies of olfactory stimuli in animals often assess responses associated with single odorant molecules or single odorant receptor protein activation. These single receptor–selective ligands, however, rarely represent salient environmental odors. Most behaviorally significant odors—for example, those that assist in distinguishing attractive entities such as baking bread or aversive or dangerous entities such as an electrical fire—are presumed to activate a wider range of odorant receptor proteins and ORNs. The identity of these odors reflects a combination of individual volatile molecules that elicit responses across multiple glomeruli. The capacity of the pyriform cortex to generate concentration-invariant representations of individual odorants probably contributes to the fidelity of identifying more complex odors. Some odors associated with molecules that have related chemical structures can be perceived as related. Furthermore, although animals have the capacity to discriminate among odors, either innately or as a learned response, accuracy diminishes rapidly with the presence of multiple "distracting" odors unless additional behavioral strategies, including sniffing and tracking (which can be considered a form of selective olfactory attention), are implemented. Thus, despite current understanding of the molecular logic of odorant receptor proteins, and insight into the circuitry of the main and accessory olfactory systems, the relationship between olfactory stimuli, central representations, and odor perception that guides behavior remains to be fully defined.

Olfactory perception in humans

In humans, olfaction is often considered the least acute of the senses, and many animals are believed to possess far superior olfactory abilities. This conclusion has been debated, however, based on comparisons made between us and other animals. A more accurate statement may be that humans like other species are specialized for and particularly skilled at the subset of olfactory tasks that are most significant for optimally adaptive human behaviors. The apparent chemosensory sophistication of some animals may be partly explained by those animals having increased numbers of olfactory receptor neurons and odorant receptor proteins (as discussed above) in an expanded olfactory epithelium, as well as a relatively larger portion of the forebrain devoted to olfaction (Figure 14.16). In a 70-kg human the surface area of the olfactory epithelium is approximately 10 cm²; in contrast, in a rat it is 15 cm², in a 3-kg cat it is about 20 cm², in dogs it is 150 to 170 cm², and in bloodhounds bred for their increased olfactory sensitivity it is 380 cm². With such a quantitative disadvantage, the human nose seems ill-suited for certain tasks, such as following a low-concentration scent to a specific target, that are second nature to a cat or dog. Nevertheless, humans can, when challenged, use their comparatively modest olfactory endowment to "sniff out" a scent trail. In this instance, the same relationship between sniffing and peak odorant responses in the olfactory bulb, as well as

concentration-invariant responses in the pyriform cortex, may enhance odor identification. Moreover, humans seem to use scent-tracking strategies that are similar to those of more olfactory-gifted animals: We pursue a tracking path that constantly bisects the linear scent trail (Figure 14.17A,B); we sniff frequently; our sniffing frequency increases as scent tracking is learned; and our performance improves with practice (Figure 14.17C). Thus, although humans do not routinely rely on olfaction as a major source of spatial information to locate objects, the human olfactory system has the capacity to use chemosensory information to track targets and locate items of interest in space.

Humans are also quite good at detecting and identifying individual airborne odorants with a wide range of aesthetic (unpleasant vs. pleasant) and behavioral (irritant vs. attractant) significance. The human olfactory system can detect ozone (the smell that accompanies lightning and electrical arcing) reliably at approximately 10 molecules per *billion* in room air. Similarly, humans can identify D-limonene, the major element of citrus smells, fairly reliably at 15 molecules per billion in room air (Figure 14.18A). Other molecules are detected only at much higher concentrations. For example, some estimates place human sensitivity to the odor of ethanol at 2000 molecules per billion.

A further complication in rationalizing the perception of odors is that their quality may change with odorant concentration. For example, at low concentrations the molecule indole has a "floral" odor that is reported as a pleasant perception, whereas at higher concentrations it smells "putrid" and is reported as an aversive perception (Figure 14.18B). The human olfactory system is also capable of making perceptual distinctions based on small changes in molecular structure; for example, the molecule D-carvone is identified as spearmint, whereas L-carvone smells like the caraway seeds found in rye bread (Figure 14.18C).

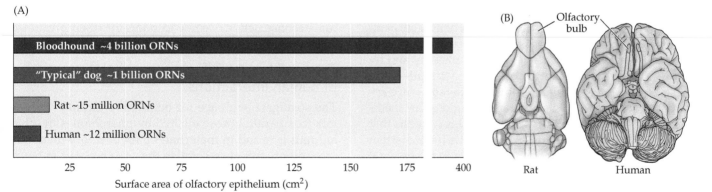

Figure 14.16 Odorant perception in mammals
(A) Comparison of the surface area (bars) of the olfactory epithelium and the number of ORNs in a human, a rat, a "typical" dog, and a bloodhound (bred for maximum olfactory discrimination). (B) Proportional sizes of the olfactory bulb (yellow) and piriform cortex (green) in rat and human brains; the bulbs and piriform cortex comprise relatively more of the forebrain in rats than they do in humans. (A, data from D. Shier et al., 2004. *Hole's Human Anatomy and Physiology*. Boston: McGraw-Hill.)

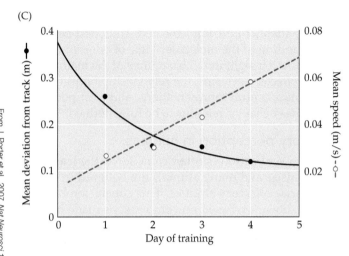

From J. Porter et al., 2007. Nat Neurosci 10: 27–29

FIGURE 14.17 Humans can track scents at low concentrations over long distances (A) The yellow line indicates the scent trail established by dragging a pheasant through a field (the pheasant, immobilized, is seen at the bottom of the picture). The red line indicates the path of a pointer tracking the scent. The dog's tracking includes several orthogonal digressions, which are common when a new scent is present at relatively low concentrations in a complex odor environment. (B) The yellow line shows a scent trail established with chocolate essential oil; the red line indicates the trail of a human tracking the scent. Like the dog, the human makes orthogonal digressions. (C) Learning curves for human scent tracking. Over a short training period, humans can acquire greater skill and accuracy tracking a scent at low concentrations (solid black trace). This improvement indicates that the olfactory system has the capacity for enhanced performance, perhaps by increased sensitivity to a learned signal over background "odor noise." Humans also acquire greater speed in tracking scents at low concentration over a small number of trials (dashed red trace). Apparently, olfactory sensation, like other sensory modalities, can be used in complex tasks in which performance speed as well as accuracy can be enhanced by repetition. (After J. Porter et al., 2007. *Nat Neurosci* 10: 27–29.)

There have been many attempts to classify human odor perception into categories based on chemical structure or broad identities (e.g., "floral") that parallel the division of the visible light spectrum into red, blue, and green (which correspond to the molecular specificity of photopigments in photoreceptors). Despite these efforts, there is no indication that any currently available scheme reflects the biological reality of odorant representation across the olfactory system. One of the most consistent aspects of olfactory perception, however, is the classification of odors as either pleasant and attractive versus unpleasant and repulsive. These basic properties of olfactory stimuli—their "aesthetic" or *hedonic* qualities (or lack thereof)—are apparently represented in distinct cortical regions that mediate olfactory perception (Figure 14.18D). This suggests that perceived hedonic properties of odorants have distinct representations in the forebrain, including in the cerebral cortex, perhaps to integrate the olfactory sensory information with subsequent motor planning and behavioral responses. It remains to be determined whether animals "map" odors or their attractive or repellent qualities based on single molecular or perceptual attributes. Most naturally occurring odors, regardless of their hedonic value, are blends of several odorant molecules, even though they

are typically perceived as a single smell (such as the scent of a particular perfume or the bouquet of a wine). The fragrance industry thrives on combining single odorant molecules into scents that can be marketed as distinctly identifiable. Similarly, the florid vocabulary that is sometimes used to describe wine or coffee—"notes of berries, with chocolate and a smoky finish"—are not only the stuff of parody, but in the presence of a trained "nose," a perceptual deconstruction of complex odors into their presumed simpler molecular components.

The social nose: Olfactory regulation of human interactions

The strongest evidence for odor-driven human behaviors that parallel those elicited by pheromones in other animals is found in maternal–infant olfactory-mediated recognition. Infants recognize their mother within hours after birth by smell, preferentially orienting toward their mother's breasts and showing increased rates of suckling when fed by their biological mother compared with being fed by other lactating females, or when presented experimentally with their mother's odor versus that of an unrelated female. There is evidence that the secretions from glands around the areolae of the maternal breast can cause

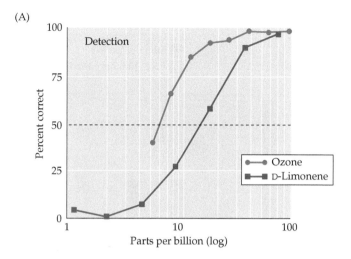

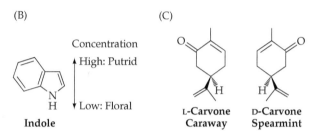

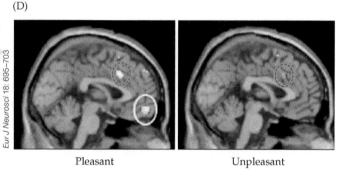

Eur J Neurosci 18: 695–703

| Pleasant | Unpleasant |

FIGURE 14.18 Human sensitivity to odors (A) In a controlled setting where room air is presented in precise mixtures with single odors, the threshold for detection reflects the concentration at which a human correctly identifies the presence of the odor above chance (50%). Humans can detect ozone, a somewhat unpleasant odor, at approximately 10 parts per billion. The pleasant and nutritionally significant odor of D-limonene can be identified at approximately 15 parts per billion. (B) Perception of the molecule indole is concentration-dependent. At low concentrations, it is perceived as a pleasant floral smell. At high concentrations, the same molecule (which is produced by bacteria in decomposing organic material) is experienced as putrid. (C) The D and L enantiomers of carvone produce very different olfactory perceptions (spearmint versus caraway) when present at similar concentrations. (D) Functional MRI analysis in typical humans indicates that odorants perceived as "pleasant" versus "unpleasant" elicit maximum activity in distinct regions of the orbitofrontal (white oval) and cingulate (red dashed-line ovals) cortex. (A, data from S. R. Cain et al., 2007. *Indoor Air* 17: 337–347.)

an infant to awaken, change her or his respiration, and increase suckling (Figure 14.19). This capacity of lactating females to produce, and of their infants to respond to, a potential pheromone-like cue seems to be independent of individual maternal–infant pairings. The areolar secretions of a non-familiar female are equally or more effective at eliciting infant responses that are those of the infant's own mother. A mother's recognition ability matches that of her infant, and mothers can reliably discriminate their own infant's scent from that of other infants of similar age, based on selecting otherwise indistinguishable clothing (e.g., simple white T-shirts) worn by a diverse group of individual infants, including their own. These observations parallel much more detailed analysis of olfactory and vomeronasal maternal–offspring bonding and subsequent maternal–offspring behavior in rodents, rabbits, and sheep—all species where the role of olfaction in parenting behaviors have been assessed in detail.

The influence of olfactory stimuli on human reproductive behaviors, if any exists, has been harder to securely identify. This uncertainty is clear from the variety of positive and negative results obtained in assessments of whether women living in single-sex dormitories tend to have synchronized menstrual cycles, potentially mediated by olfaction. There were a series of reports over 50 years ago that suggested this was the case. The apparent lack of conscious detection of olfactory cues that accompanied the reported synchronization suggests that the olfactory stimuli may be acting more as pheromones to directly signal a change in a specific, innate, homeostatic behavior in response to conspecifics. The original studies characterizing these responses have not been fully replicated, and there is continued contention about whether this particular phenomenon is robust and reproducible, or relevant to understanding human olfactory function. Studies of parallel responses in female animals, including rodents and non-human primates, are similarly inconclusive and controversial. Thus, the possibility of human sex-specific olfactory influence on this aspect of innate reproduction-related behavior remains uncertain.

It also remains unknown whether there is a representation of pheromone-like odorant information for social cues (i.e., airborne stimuli that elicit behavioral responses without conscious perception of a smell) in primates. As mentioned in Key Concept 14.1, vomeronasal peripheral and central structures, which process pheromones in rodents and carnivores, are diminished or absent in most primates, including humans, and the genes that encode vomeronasal receptor proteins are pseudogenes. Nevertheless, primates, including humans, have behavioral responses that can be attributed to stimuli similar to the pheromones that activate the vomeronasal system in other animals. There are controversial studies of male- and female-specific responses to odorants related

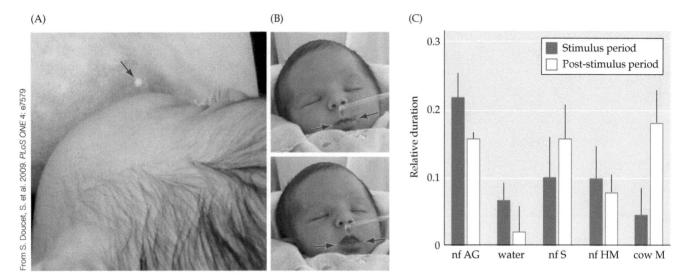

(A)

(B)

(C)

Figure 14.19 Maternal olfactory cues guide newborn humans to suckle (A) A nursing infant, when latched onto the nipple is also adjacent to glandular tissue (arrow) in the aerole (the thickened pigmented skin around the nipple) that produces a volatile signal. Infants can smell based upon the close proximity of their noses to the aerole. (B) Different odors can be presented to infants at distances that approximate those between the infant and the mother's areole, and oromotor behavioral responses, for example tongue protrusion (compare the infant's mouth at the red arrows in the upper and lower panel), recorded. (C) non-familiar (not from the infant's mother) aerolar gland secretions (nf AG) collected from volunteer nursing mothers elicits maximal oro-motor activity both during direct stimulation and for an extended post-stimulus period. Other stimuli including water, non-familiar sebaceous secretion (nf S), non-familiar human milk (nf HM) and cow's milk do not match the maximal stimulus profile both during and after stimulation. (After S. Doucet et al. 2009. *PLoS ONE* 4: e7579.)

chemically to androgens or estrogens in distinct regions of the hypothalamus (Figure 14.20), which presumably are relayed via the olfactory bulb because of the absence of the vomeronasal system. These odors are detected consciously, although individuals of both sexes report that they have little positive or negative hedonic value. There is also evidence that the location of these responses in the hypothalamus is reversed in homosexual versus heterosexual men and women (see Chapter 25). These studies all suggest that there may be an olfactory dimension to the sensory information that humans use to guide affliative and sexual behaviors. Nevertheless, in this instance the parallels between the data from animal studies and the more controversial data from human studies are far from clear.

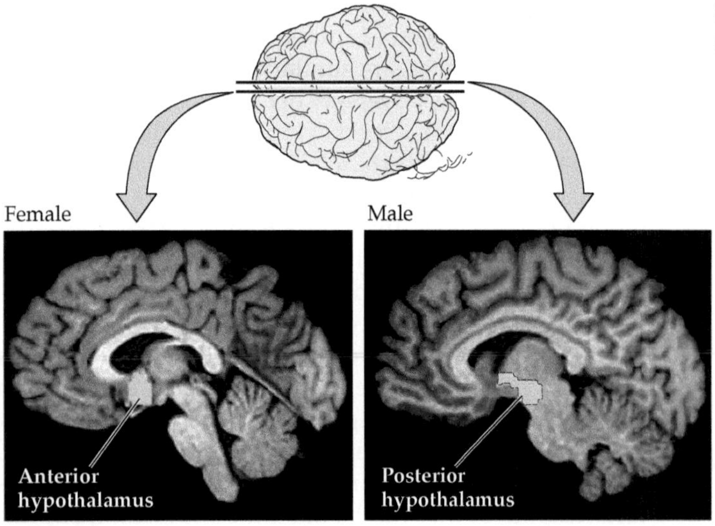

Assessing olfactory function in the laboratory or clinic

Most people are able to consistently identify a broad range of familiar yet distinct odorants, and they can distinguish different odors from one another. Indeed, many clinicians use uniquely scented "probes"—such as coffee grounds or soap—to test the functional integrity of the olfactory nerve (cranial nerve I) as part of the standard cranial nerve examination. Some individuals consistently fail to identify one or more common odors (Figure 14.21A). Such chemosensory deficits, known as selective anosmias, can be restricted to a single odorant. This may suggest that a specific element in the olfactory system—either an olfactory receptor gene or genes that control the expression or function of specific odorant receptor genes—is inactivated. Genetic analysis of anosmic individuals has

Female

Male

Anterior hypothalamus

Posterior hypothalamus

FIGURE 14.20 Differential patterns of activation in the hypothalamus of a typical human female (left) and male (right) after exposure to an estrogen- or androgen-containing odor mix.

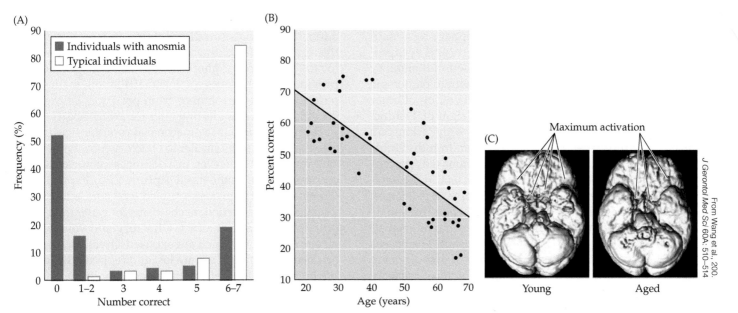

From Wang et al., 200. J Gerontol Med Sci 60A: 510–514

FIGURE 14.21 Loss of olfactory sensitivity (A) Anosmia is the inability to identify common odors. The majority of typical individuals presented with seven common odors (a test frequently used by neurologists) can identify all seven correctly (in this case, baby powder, chocolate, cinnamon, coffee, mothballs, peanut butter, and soap). Persons who are anosmic have difficulty identifying even these common scents. (B) The ability to identify 80 common odorants declines markedly between ages 20 and 70. Such loss of sensory acuity is normal. (C) Maximum activation (red) of orbitofrontal and medial (piriform cortex/amygdala) cerebral cortex by familiar odors in young and typical (i.e., without dementia) aged individuals. Areas of focal activation remain similar, but there is clearly diminished activity in the older individuals. (A after W. S. Cain and J. F. Gent, 1986. In *Clinical Measurement of Taste and Smell*, H. L. Meiselman and R. S. Rivlin [Eds.]. New York: Macmillan, pp. 170 -186; B after C. Murphy, 1986. In *Clinical Measurement of Taste and Smell*, H. L. Meiselman and R. S. Rivlin [Eds.]. New York: Macmillan, pp. 343–371.)

yet to confirm this possibility. Thus, unlike blindness and deafness, olfactory loss is difficult to classify as either peripheral or central in its origins.

Anosmias can be congenital, or they may be acquired following chronic sinus infection or inflammation, traumatic head injury, exposure to toxins, or specific infectious agents such as the SARS-CoV-2 virus associated with COVID-19 (see Clinical Applications). Olfactory loss is also a common consequence of aging (see Figure 14.21B). In some cases of olfactory loss, such disruption is not a source of great concern (e.g., the transient anosmia that occurs with a severe cold). If sustained, however, disruptions can lead to decreased appetite, weight loss, altered mood, and eventual malnutrition (especially in elderly individuals). If an anosmia is particularly specific and severe, it can affect a person's ability to identify and respond appropriately to potentially dangerous odors such as spoiled food, toxic chemicals, or smoke. For example, approximately 1 person in 1000 is insensitive to butyl mercaptan, the foul-smelling odorant released by skunks. More serious is the inability to detect hydrogen cyanide (1 person in 10), which can be lethal, or ethyl mercaptan, the chemical added to natural gas to enable people to detect gas leaks.

Like other sensory modalities, human olfactory capacity typically decreases with age. If otherwise healthy individuals are challenged to identify a large battery of common odorants, people 20 to 40 years of age can ordinarily identify 50% to 75% of the odors, whereas those between ages 50 and 70 correctly identify only 30% to 45% (Figure 14.21B,C). These changes may reflect either diminished peripheral sensitivity or altered activity of central olfactory structures in otherwise typical aging individuals. A more radically diminished or distorted sense of smell often accompanies neurodegenerative conditions associated with aging, especially Alzheimer's disease. In fact, odor discrimination (the ability to tell two odors apart, usually measured by a standardized "scratch and sniff" test known as the University of Pennsylvania Smell Identification Test) is often part of a battery of diagnostic tests administered at the early stages of age-related dementia and other neurodegenerative diseases.

In addition to pathological or degenerative age-related changes in olfaction, olfactory sensation and perception can be disrupted by brain tumors (e.g., gliomas or metastatic lesions from a non-neural source), chemotherapy, eating disorders, diabetes, neurological disorders (olfaction is often compromised early in the course of Parkinson's disease), and psychotic disorders (especially schizophrenia). Some of these olfactory deficits reflect disruption of specific central olfactory pathways (e.g., by brain tumors and some psychiatric disorders). Others probably reflect disrupted capacity of the olfactory periphery to sustain ongoing neurogenesis, ORN axon growth, and synaptogenesis. This may reflect early or increased vulnerability of ongoing ORN regeneration and regrowth to pathologic processes that eventually result in degeneration of other neuronal populations. One of the earliest signs of dysfunction in individuals ultimately diagnosed with Parkinson's disease is diminished olfactory function. In individuals with schizophrenia, olfactory hallucinations (i.e., perception of a stimulus that is not actually present in the environment) are among the earliest symptoms of psychosis. In individuals with autism, odorant detection thresholds can be lowered, and the experience of neutral or even attractive odors can be reported as unpleasant. The causes of olfactory deficits in this broad range of disorders are not known. Some dysfunction may reflect lost capacity of the olfactory epithelium to maintain neural stem cells or support the survival of newly generated neurons that normally replace damaged olfactory receptor neurons over the course of a lifetime (see Figure 14.3)—perhaps an early sign of more general pathogenic deficits in genesis or maintenance of optimally fuctioning neurons and circuits.

Summary

Two chemosensory systems—the olfactory and vomeronasal systems—detect and represent airborne molecules that provide information about essential aspects of life: food, friends, and sex. Olfactory sensory receptor neurons in the olfactory epithelium that lines the nasal cavity are bipolar neurons with an apical dendritic specialization where the transduction of odorants into neuronal activity occurs via binding to a large family of G-protein-coupled receptors. Activation of these receptors initiates second-messenger-mediated regulation of Na^+, Ca^{2+}, and Cl^- ion channels. These events ultimately generate action potentials in the afferent axons of the receptor neurons. The large number of G-protein-coupled odorant receptor molecules in most species is believed to establish sensitivity to the myriad odors that animals can discriminate. In most vertebrates, the vomeronasal pathway provides a parallel pathway for the detection of pheromones from conspecifics that provide attractive cues for social affiliation, mating, and parenting, and kairomones from predators or prey that provide aversive or alerting cues for avoidance, tracking, or capture. Peripheral vomeronasal receptor neurons are also bipolar neurons similar to olfactory receptor neurons. They are found, however, in a distinct structure, the vomeronasal organ, in the nasal septum. They express a different family of G-protein-coupled receptors, and their intercellular transduction mechanisms for generating action potentials are also distinct. Olfactory and vomeronasal receptor neurons, which are exposed directly to the environment via inspired air, are lost and replaced throughout life by neural stem cells maintained in both the olfactory epithelium and vomeronasal organ. Olfactory and vomeronasal information are relayed to distinct CNS targets. Olfactory receptor neurons project to the olfactory bulb, and the olfactory bulb projects primarily to the pyriform cortex. Vomeronasal receptor neurons project to the accessory olfactory bulb, which in turn projects to targets in the hypothalamus and amygdala. The processing and representation of olfactory and vomeronasal information in these target regions are not fully understood. In the initial steps, however, olfactory bulb and pyriform cortex neurons respond uniformly to distinct odors regardless of their peripheral concentration. These concentration-independent responses are thought to increase the fidelity of odor identification. The range of behaviors influenced by olfaction varies by species and reflects the proportions of sensory receptors and target regions devoted to olfaction versus other sensory modalities. Nevertheless, even in humans, where the olfactory system is proportionately diminished compared with other sensory modalities, olfaction is key for maternal–infant interactions, and is among the earliest targets for age-related or neurodegenerative disease–related decline.

■ Additional Reading

Reviews

Axel, R. (2005) Scents and sensibility: A molecular logic of olfactory perception (Nobel lecture). *Angew Chem., Int. Ed. (English)* 44 (38): 6110–6127.

Buck, L. B. (2000) The molecular architecture of odor and pheromone sensing in mammals. *Cell* 100: 611–618.

DuLac, C. and A. T. Torello (2003) Molecular detection of pheromone signals in mammals: from genes to behaviour. *Nat. Rev. Neurosci.* 4: 551–562.

Hildebrand, J. G. and G. M. Shepherd (1997) Mechanisms of olfactory discrimination: Converging evidence for common principles across phyla. *Annu. Rev. Neurosci.* 20: 595–631.

Knaup, U.B. (2010) Olfactory signaling in vertebrates and insects: differences and commonalities. *Nat. Rev. Neurosci.* 11: 188-200.

Mombaerts, P. (2004) Genes and ligands for odorant, vomeronasal and taste receptors *Nat. Rev. Neurosci.* 5: 263–278.

Schaal B, and four others (2020) Olfaction scaffolds the developing human from neonate to adolescent and beyond. *Phil. Trans. R. Soc. B* 375: 20190261.

Important Original Papers

Bolding, K.A. and K.M. Franks (2018) Recurrent cortical circuits implement concentration-invariant odor coding. *Science* 361: eaat 6904.

Bozza, T., P. Feinstein, C. Zheng and P. Mombaerts (2002) Odorant receptor expression defines functional units in the mouse olfactory system. *J. Neurosci.* 22: 3033–3043.

Buck, L. and R. Axel (1991) A novel multigene family may encode odorant receptors: A molecular basis for odor recognition. *Cell* 65: 175–187.

Graziadei, P. P. C. and G. A. Monti-Graziadei (1980) Neurogenesis and neuron regeneration in the olfactory system of mammals. III. Deafferentation and reinnervation of the olfactory bulb following section of the fila olfactoria in rat. *J. Neurocytol.* 9: 145–162.

Malnic, B., J. Hirono, T. Sato and L. B. Buck (1999) Combinatorial receptor codes for odors. *Cell* 96: 713–723.

Mombaerts, P. and 7 others (1996) Visualizing an olfactory sensory map. *Cell* 87: 675–686.

Sosulski, D. L., M. L. Bloom, T. Cutforth, R. Axel and S. R. Datta (2011) Distinct representations of olfactory information in different cortical centers. *Nature* 472: 213–216.

Stettler, D. D. and R. Axel (2009) Representations of odor in the piriform cortex. *Neuron* 63: 854–864.

Vassar, R. and 5 others (1994) Topographic organization of sensory projections to the olfactory bulb. *Cell* 79: 981–991.

Vosshall, L.B., A.M. Wong, and R. Axel (2000) An olfactory sensory map in the fly brain. *Cell* 102: 147-159.

Books

Barlow, H. B. and J. D. Mollon (1989) *The Senses*. Cambridge, UK: Cambridge University Press, chapters 17–19.

Doty, R. L. (ed.) (1995) *Handbook of Olfaction and Gustation.* New York: Marcel Dekker.

Farbman, A. I. (1992) *Cell Biology of Olfaction.* New York: Cambridge University Press.

Getchell, T. V., L. M. Bartoshuk, R. L. Doty and J. B. Snow, Jr. (1991) *Smell and Taste in Health and Disease.* New York: Raven Press.

Shier, D., J. Butler, and R. Lewis (2004) *Hole's Human Anatomy and Physiology.* Boston: McGraw-Hill.

15

Taste

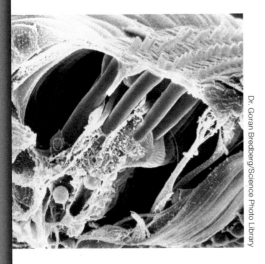

Dr. Goran Bredberg/Science Photo Library

KEY CONCEPTS

15.1 Detection of taste qualities is mediated by specialized cells in the taste buds and conveyed to the brain by three cranial nerves

15.2 Gustatory information flows to the brain through diverging and converging pathways and is encoded via the spatio-temporal firing patterns of neurons

15.3 Perception of food and beverages also involves olfaction, somatosensation, audition, and vision

15.4 Palatability or aversiveness of a food is mediated by the gustatory and limbic systems and can be modified by experience

Overview

In English, as in in many languages, *taste* has multiple meanings. In its strictest scientific definition, taste, or *gustation*, refers to a special sense that perceives and distinguishes the sweet, sour, bitter, umami, or salty quality and is mediated by taste buds in the mouth. In daily parlance, *taste* may also refer broadly to the integration of gustation, olfaction, and oral somatosensation. The most appropriate term for this is *flavor*, which emerges through the convergence of gustatory and other sensory inputs at many levels of the gustatory pathway. Yet another use of *taste* relates to preference and aesthetic quality. When Duke Ellington said "Create, and be true to yourself, and depend only on your own good taste," he was referring to this aesthetic quality. This connotation seems distant from chemosensation and other literal meanings. Yet gustation is linked to the most primal aesthetic quality and preference. Liking and disliking (termed *hedonic value*) are intimately associated with the taste of food and beverages. Sugar is sweet and palatable, while quinine is bitter and aversive. This intimate relationship between gustation and hedonic value is grounded in the interconnectedness between the gustatory and the limbic systems. In this chapter we touch on gustation, flavor, and hedonic value by exploring cellular mechanisms, anatomy, and the neural representation of percepts, as well as instances in which taste changes with experience and learning across mammals.

<table>
<tr><td>CONCEPT
15.1</td><td>**Detection of Taste Qualities Is Mediated by Specialized Cells in the Taste Buds and Conveyed to the Brain by Three Cranial Nerves**</td></tr>
</table>

LEARNING OBJECTIVES

15.1.1 Explain how the sense of taste is defined.

15.1.2 Identify the features of taste buds and their component cells.

15.1.3 Describe the molecular receptors and transduction pathways for the five different taste qualities.

The role of gustation

Taste in its simplest form relates to the perception of the basic taste qualities of **sweet**, **bitter**, **umami**, **salty**, and **sour**. Additional percepts—among them fatty, fizzy (carbonated), and metallic—may also be considered tastes in certain contexts. The commonality for taste is that the stimuli that elicit taste percepts are detected by membrane receptors on taste bud cells and recruit a generally common neural pathway from the periphery to the cortex (Figure 15.1). Curiously, some of these same receptors are also found in other tissues and

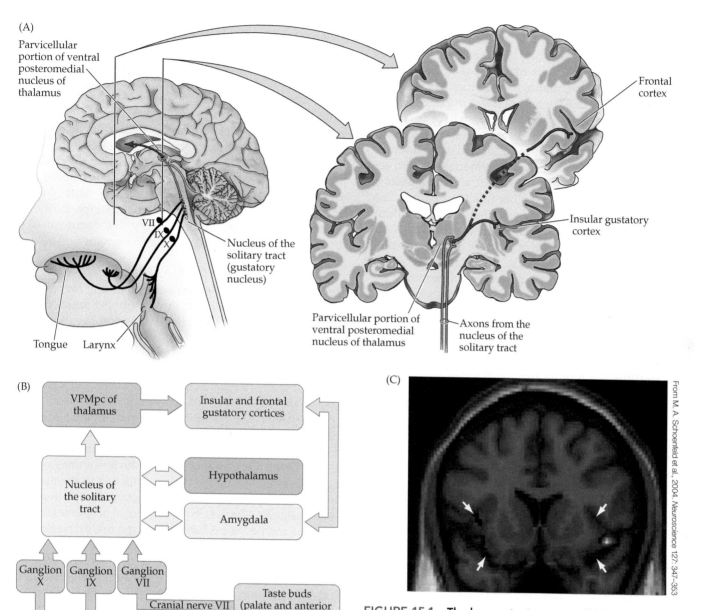

FIGURE 15.1 The human taste system (A) Taste buds in the mouth and upper alimentary canal are innervated by sensory neurons of three cranial nerves, terminating in the nucleus of the solitary tract in the medulla. Coronal sections show projections to the parvicellular portion of the ventral posteromedial nucleus of the thalamus (VPMpc) and subsequently to the gustatory and association cortices. (B) Organization of the gustatory system from taste buds to cortex. Several bidirectional inputs to gustatory cortex from other cortical and subcortical areas are omitted for clarity. (C) Functional MRI of the brain of a person consuming food. Note bilateral focal activation (red) in the gustatory portion of the insular cortex (demarcated by arrows).

may serve as important detectors that signal the presence of nutrients or toxins.

The number and types of chemicals that taste buds detect is quite large, ranging from sugars, amino acids, and salts to aliphatic, aromatic, and heterocyclic compounds, and even some small proteins. The most obvious and frequently stated role for taste is to identify nutritionally relevant compounds associated with calories, electrolytes, and proteins, such as sugars, salts, and amino acids, respectively. Yet this concept of a straightforward mapping between nutrients and taste often falls short in accounting for the complexity of taste. The taste system typically detects only relatively small molecules, while many nutrients can be macromolecules. While some macromolecules such as starch and fats may be partially digested into smaller units in the mouth, most are not. For instance, mono- and disaccharide sugars are sweet, but oligosaccharides and larger complex carbohydrates, which are nutritionally beneficial components of a natural diet, are not. Similarly, many individual amino acids and dipeptides are sweet or umami tasting, whereas polymerized amino acids (i.e., proteins) are usually tasteless (the taste of meat is conferred by amino acids and other small molecules). The taste system readily detects the essential nutrient sodium chloride (NaCl, or common table salt), but other less common salts may be bitter, sour, or metallic and readily distinguished from sodium chloride. Thus, taste may guide consumption both by directly detecting nutrients and by a learned association of particular perceptions with nutritionally and metabolically beneficial foods. The greatest diversity of chemical structures occurs among bitter-tasting compounds. Here the taste system is thought to serve as an early warning system to avoid ingesting toxins. Yet again, learning can modify the initial sensory signal; the many daily examples of bitter compounds that are sought and consumed include coffee, grapefruit, bitter vegetables, and beer. The significance of sour taste is not completely clear; perhaps it helps avoid consuming spoiled food or burdening metabolism with a large acid load.

The gustatory system is well tuned to detect nutrients and toxins at concentrations that are found in nature. Thus, sugars are detected in concentrations at and above tens of millimolar, i.e., at quantities necessary for nutritional value. Umami compounds, similarly, are detected by the taste system in the tens of millimolar, concentrations that are found in many foods. Sodium chloride, which must be consumed regularly to replace its loss in urine and to maintain electrolyte homeostasis, is preferred by animals and humans until it reaches the concentration found in blood and tissue (about 150 mM), beyond which it becomes progressively unpalatable. In contrast, detecting even small quantities of toxins to avoid ingesting them can be essential for survival. Accordingly, the taste system exhibits much higher sensitivity for most bitter stimuli. Bitter, toxic compounds, including alkaloids, glycosides, and aromatics, are often used by plants as defenses against predators, and toxins produced by ingested microorganisms may accumulate in animal tissues. Taste receptors detect many such compounds in the micromolar range. Indeed, it's possible that taste sensitivity of each bitter compound is calibrated to its toxicity. For example, the highly toxic compound strychnine is detected by human taste receptors at concentrations of 0.1 μM, whereas the related but less toxic compound brucine is detected only above 10 μM.

Taste sensitivities vary across individuals, a phenomenon attributed to anatomical as well as precise molecular heterogeneities. The term *super-taster* has been used to describe two seemingly disparate phenotypes. One phenotype includes individuals who have a much higher density of taste buds than average, and often perceive tastes as more intense. The other phenotype includes individuals who find phenylthiocarbamide and related compounds, which for most people are mildly bitter or tasteless, to be excruciatingly bitter. Both phenotypes can be traced to alleles of one taste receptor gene, and also to the development and density of fungiform papillae (which house taste buds) on the anterior tongue. The bitter intensity of many other compounds is often derived from multiple genes. For sweet taste, inherited differences are well documented in mice, but the story is less clear-cut in humans.

Taste buds and their cell types

Taste buds, the sensory end organs responsible for detecting taste stimuli, are distributed throughout the dorsal surface of the tongue, soft palate, pharynx, and upper part of the esophagus. On the tongue, taste buds cluster in specialized "bumps," or papillae, while on the soft palate, pharynx, and epiglottis they are flush with the epithelium. The papillae at the back of the tongue, termed *circumvallate* and *foliate*, are large and contain hundreds of taste buds each; other lingual papillae, termed *fungiform*, are small and contain just 1 to 20 taste buds, depending on the species (Figure 15.2A). In total, humans have between 1000 and 5000 taste buds, accounting in part for the wide range of taste sensitivities among people. Aging is often associated with a reduction in the number and size of taste buds and a concomitant decline in gustatory sensitivity. Changes in the functional properties of taste bud cells and nerves probably also play a role in this decline, although this is less extensively documented.

Each oral taste bud is a cluster of 40 to 80 cells, whereas taste buds in the pharynx and esophagus are quite small, with relatively few cells. Each taste bud contains both receptor and supporting elongated epithelial cells. Taste receptor cells are similar to auditory and vestibular hair cells, which are epithelial and unlike the neurons that are sensors for olfaction. With its base resting on the basement membrane, each taste receptor cell extends vertically toward the surface, terminating in fingerlike projections called microvilli (Figure 15.2B). The cells in a taste bud are packed together tightly. Their apical tips emerge through a small pore into the oral space, where they encounter tastants (i.e., chemicals that produce a taste sensation). Taste receptor proteins are located on the membranes of the apical microvilli and come in contact with dissolved taste stimuli. Thus, the microvilli are thought to be a principal site where taste transduction occurs. As with other sensory systems, transduction is the process by which recognition of the sensory stimulus on the outside of cells is converted to chemical or electrical intracellular signals. These signals eventually are conveyed to sensory nerve fibers for transmission to the brain.

(A)

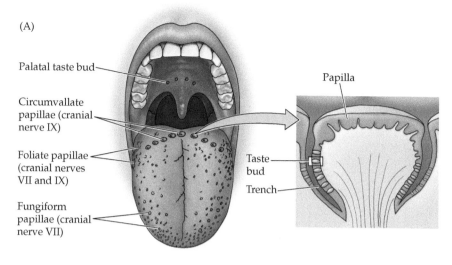

Palatal taste bud

Circumvallate papillae (cranial nerve IX)

Foliate papillae (cranial nerves VII and IX)

Fungiform papillae (cranial nerve VII)

Papilla

Taste bud

Trench

(B)

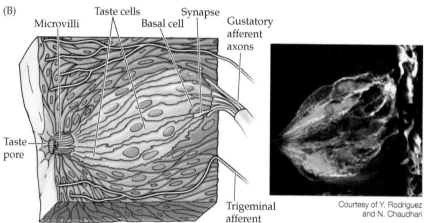

Microvilli

Taste cells

Basal cell

Synapse

Gustatory afferent axons

Taste pore

Trigeminal afferent axons

Courtesy of Y. Rodriguez and N. Chaudhari

FIGURE 15.2 Taste buds and taste papillae (A) Distribution of taste papillae on the dorsal surface of the tongue. The blowup to the right is a coronal section of a single circumvallate papilla, showing the placement of taste buds along the lateral walls of the papilla. (B) Diagram and light micrograph of a taste bud, showing different cell types. In the micrograph, type II and type III receptor cells are in green and red, respectively; type I glial-like cells are in white. The apical tips of these cells have microvilli projecting through the taste pore at left. Gustatory afferents innervate the receptor cells within the bud, while trigeminal afferents remain in the epithelium surrounding the taste bud.

Taste buds contain at least three distinct mature cell types. Type I cells are considered glial-like, and probably fulfill many functions similar to those of central glia, and supporting cells of other sensory epithelia. Such functions include general metabolic support and clearing transmitters via enzymes and transporters on the cell membrane. Some type I cells develop flat cytoplasmic winglike extensions that wrap around excitable chemosensory receptor cells. This arrangement may restrict the spread of neurotransmitter within the bud and may also limit which nerve fibers receive transmitter signals from any given receptor cell. Finally, some type I cells may also regulate the ionic environment within the taste bud, and thus control excitability and transmission to nerves.

Type II taste bud cells serve as chemosensory receptor cells for sweet, bitter, umami, or salty tastants (Figure 15.3A). Each type II cell contains only one or two categories of taste receptors, and thus is specialized for one or two taste submodalities. Taste receptors located in the cell membranes of these cells couple to a common set of signaling effector proteins that are found in the cytoplasm, as we discuss in more detail in the next section. An unusual characteristic of type II cells is that although they are excitable sensors and signal to nerves, they contain no aggregates of synaptic vesicles and no ultrastructurally visible pre- or postsynaptic membrane specializations. Instead, ATP, the afferent transmitter, is released by an unusual non-vesicular mechanism through large-pore membrane channels, an arrangement that is termed a *channel synapse*.

Type III taste bud cells serve as chemosensory receptor cells for sour tastants and some salts, and may also serve other functions (Figure 15.3B). Type III cells have well-developed synapses with transmitters released through conventional vesicles, although that does not preclude the possible use of a channel synapse similar to type II cells.

Although the cells of taste buds are specialized and excitable sensory detectors, they retain their capacity for renewal throughout adult life. In this regard they are similar to olfactory sensory neurons, and differ from auditory and vestibular hair cells. This property may reflect a key aspect of the biology of taste bud cells: They are exposed to water, to stimuli at a wide range of concentrations from hypo- to hypertonic, and potentially to toxic compounds, all of which represent environmental insults that limit their long-term viability. Individual taste bud cells have an average life span of 10 days, although type I cells often last less than a week, while type III cells may survive 2 months or longer. Both non-sensory epithelium and taste bud cells are regenerated from a common pool of basal cells, a situation resembling that of sensory neurons and supporting cells in olfactory epithelium. The proliferation of these basal progenitors and their differentiation into taste bud cells in adults is regulated both by afferent nerve–derived factors and by Wnt and Shh signaling pathways (see Chapter 23). Because inhibitors of these pathways are used as anti-cancer therapies, a common side effect of such treatments is a reduction in the

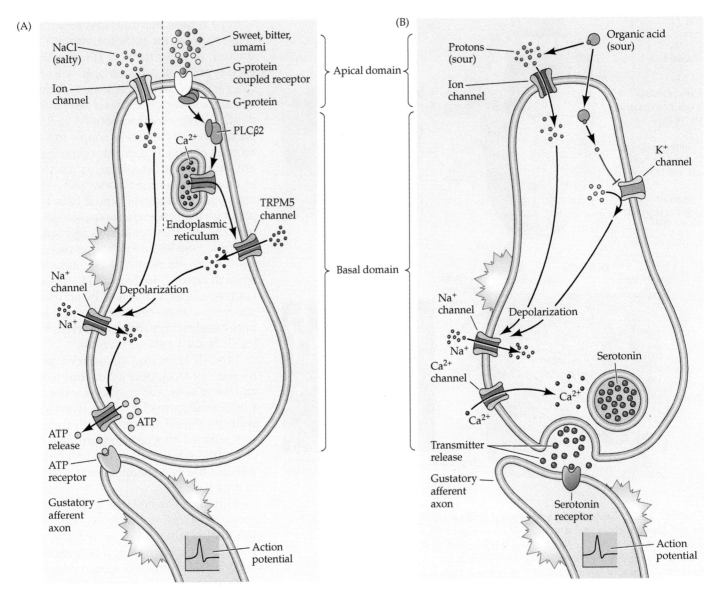

FIGURE 15.3 Polarized, excitable epithelial cells in taste buds are the receptor cells for taste (A) Type II cells house G-protein-coupled receptors selective for sweet, bitter, umami, or salty taste stimuli. The cell depicted has all the receptors, but typically only one or two receptor types are found in any one cell. When one of these taste receptors is activated, a transduction cascade is initiated, resulting in depolarization of the cell membrane and release of transmitter (ATP) on to the afferent nerve. This cascade is shown in greater detail in Figure 15.5. (B) Type III cells express ion channels for transducing sour taste stimuli. Here, transduction results in release of vesicles of neurotransmitter (serotonin) onto afferent nerve fibers.

size or number of taste buds and a consequent loss of taste sensory function. Radiotherapies for head and neck cancers also interfere with taste bud integrity and regeneration, and thus ageusia (complete loss of taste) or dysgeusia (taste perturbation) is often a pronounced side effect of head or neck radiation. Since the onset of the COVID-19 pandemic in 2020, dysgeusia has also been reported as a distinctive and common symptom of this novel coronavirus (Clinical Applications). Many other viral infections, mostly of taste nerves, are also reported to cause dysgeusias.

Taste receptors and transduction

To understand how the taste system is able to sense the wide range of chemical compounds that serve as taste stimuli, we must also examine the molecular receptors that bind these stimuli, the effectors with which they partner, and the cells that transform the detection into a signal that can be carried by nerves to the brain. These initial steps, termed sensory transduction, occur in type II and III taste bud cells.

Compared with those for vision and hearing, the stimuli and mechanisms for transducing gustatory stimuli are

■ Clinical Applications

Ageusia and Dysgeusia: Taste Loss and Taste Alterations from COVID-19

In early 2020 the world learned of a new threat, COVID-19, caused by the SARS-CoV-2 virus and producing respiratory and inflammatory conditions similar to those of other known coronaviruses. A perplexing symptom reported for COVID-19, however, was loss of smell and taste, or anosmia and ageusia. Researchers immediately began searching for a mechanism for this sensory phenotype. SARS-CoV-2 enters cells by binding to a specific receptor, angiotensin converting enzyme-2 (ACE2), on the surface of susceptible cells. The virus's spike glycoprotein (S glycoprotein) binds to ACE2, and the complex is then internalized. In the olfactory epithelium, ACE2 is expressed not on olfactory neurons but on the sustentacular, or supporting, cells that are intimately associated with and support olfactory neurons (see Chapter 14 Clinical Applications). So why the olfactory deficit? First, loss of supporting cells may affect olfactory neurons. Second, recent data show dramatic changes in chromatin organization in olfactory neurons, and downregulation of odorant receptor expression. Thus, olfactory deficits may derive from loss of the molecular detectors (receptors) themselves.

Are similar mechanisms at play in the case of taste? A simple early consensus was that self-reports of taste loss by individuals with COVID-19 may actually reflect loss of flavor. As discussed in this chapter, taste and smell converge at many locations in the cortex to produce flavor and preference. If smell is lost, an individual may register this as an impediment to recognizing and enjoying foods and report this as a loss of taste. Indeed, in the first several months of the pandemic, even many physicians conflated these two separate senses by asking patients about "loss of taste and smell." This was gradually rectified in more recent studies by applying rigorous assays for infection (such as PCR tests) and directly testing the concentrations at which individuals could identify taste stimuli and the reported intensity of such tastes. Meta-analyses of such population studies now show that taste loss associated with COVID-19 is not simply due to confusion with smell loss. Instead, a loss of taste sensitivity is apparent in one-third or more of COVID-19 patients. Much remains to decipher. There are suggestions, unverified as yet, that a subset of taste qualities (e.g., sweet) may be more severely degraded than others. Middle-aged individuals are more frequently and severely affected than young or elderly people. And curiously, the successive waves of viral variants (Delta, Omicron, etc.) may produce dysgeusia less frequently than earlier variants.

The cellular and neural mechanisms of taste loss remain widely debated but little understood. While some researchers have reported the presence of the virus receptor ACE2 on cells of taste buds, others have declared that it is primarily the surrounding non-sensory lingual epithelium, quite distant from taste buds, that bears ACE2 and proteases that would promote viral entry. Also, the virus is known to trigger broad pro-inflammatory responses, and some of these are associated with interruption of stem cell turnover and differentiation. In the absence of definitive evidence, the consensus as of this writing is that a combination of inflammation of lingual epithelium, which may block gustatory nerve signals or impair renewal of taste buds, along with specific virally induced cellular mechanisms in taste buds, may combine to produce taste loss during and after infection.

An interesting additional layer to COVID-related dysgeusia is the curious case of the antiviral drug Paxlovid (a combination of nirmatrelvir and ritonavir), which is effective at preventing severe illness after infection. Some individuals on this medication report what has been called "Paxlovid mouth," with metallic, bitter, sour, and other aversive taste sensations lasting throughout the course of the therapy. Perhaps the large number of diverse TAS2R bitter receptors in human taste buds produces a combined sensation that is both unfamiliar and impossibly horrid!

remarkably diverse. Thus, taste receptors themselves are diverse, and include several different types of ion channels and G-protein-coupled receptors (GPCRs) (Figure 15.4). How these various receptors interact with cellular machinery to achieve transduction also varies for the different taste qualities.

The main and best-studied molecular receptors for sweet and umami stimuli are GPCRs belonging to the three-member TAS1R family. These integral membrane proteins, residing on the surface of type II taste bud cells, are typical class C GPCRs. That is, they possess seven transmembrane helices, a large extracellular ligand-binding domain, and a cytoplasmic signaling domain that binds to and activates heterotrimeric G-proteins. TAS1R3 can dimerize with TAS1R1 to form umami-detecting receptors, or with TAS1R2 to form sweet-detecting receptors (see Figure 15.4C,D). The large extracellular domain of each TAS1R adopts a characteristic bilobed "venus flytrap" structure similar to that of metabotropic receptors for glutamate and GABA (see Chapter 6). The sweet taste receptors are unusual in having four or more ligand-binding sites which accommodate chemically diverse sweeteners. Sweet receptors have intrigued structural biologists because they can be activated by a variety of chemically dissimilar ligands. For instance, the monosaccharide fructose, several D-amino acids, the dipeptide

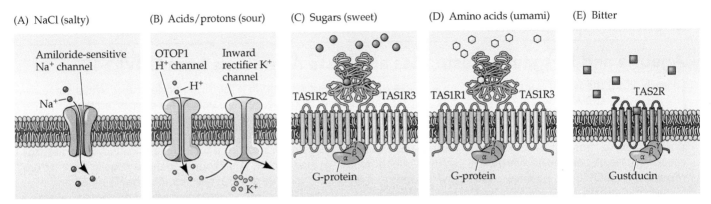

FIGURE 15.4 Receptors for taste Typically, only one or two classes of receptors are found in any cell. (A) NaCl is detected when Na⁺ permeates through the amiloride-sensitive Na⁺ channel, thereby depolarizing the taste bud cell and triggering action potentials. (B) Protons, dissociated from acids (sour stimuli), permeate into the cell through OTOP1 channels and depolarize the cell, a signal that is amplified by protons inhibiting an inward-rectifier K⁺ channel. (C–E) For sweet, umami (amino acid), and bitter tastants, different classes of G-protein-coupled receptors mediate transduction. These include a heterodimer of TAS1R2 and TAS1R3 that is activated by sweet stimuli, a dimer of TAS1R1 and TAS1R3 that is activated by certain amino acids, and several TAS2Rs that are activated by bitter stimuli.

aspartame, the thiazole saccharin, and several small proteins (e.g., brazzein) found in plants all bind TAS1R2+3 and elicit sweet taste. In contrast, umami stimuli are more homogeneous, including L-glutamate, L-aspartate, and oligopeptides rich in these two amino acids, as well as some nucleotide derivatives.

An interesting evolutionary note is that entire classes of animals have lost one or more of the TAS1Rs, while some species have repurposed existing TAS1Rs to detect nutrients relevant to their ecological niche. For instance, while most mammals use the TAS1R2+3 dimer as a sweet receptor, birds lost the *Tas1r2* gene early in their evolutionary lineage. Yet many species of songbirds as well as hummingbirds and other nectar feeders independently recovered sweet taste by evolving a modified *Tas1r1* gene that encodes a receptor not for amino acids but for sugars. Evolutionary loss of *Tas1r2* genes is also seen in many obligate carnivores that do not seek sweet foods, and an extreme example is the loss of all TAS1Rs in marine mammals that swallow food whole without tasting it. In addition to the TAS1Rs, there is also evidence of other membrane receptors, including GPCRs and transporters that bind and transduce sweet or umami stimuli, and these also may vary across species or feeding niches.

Bitter stimuli are detected via receptors that belong to the TAS2R family (see Figure 15.4E), which has 20 to 50 related members, depending on the species in question. Humans have 25 of these genes, encoding receptors of class A GPCRs. Such receptors, of which opsins in the retina (see Chapter 9) are a prime example, have the ubiquitous seven transmembrane regions, a short extracellular domain, and bind their ligands in or near the plane of the lipid membrane. This limited number of receptors, surprisingly, is able to recognize and bind thousands of bitter compounds,

from benign caffeine to lethal strychnine. Some of these receptors are activated by low concentrations of one or a few related compounds, whereas others are promiscuous and bind dozens of compounds. Conversely, any one compound may bind one or many distinct receptors. Thus, how bitter a compound tastes may derive from the number of receptors that bind it, the binding affinity of the receptor(s), and individual inherited variations. The TAS2Rs are expressed in various combinations in taste bud cells. In theory, this arrangement could allow the gustatory system to discriminate among different bitter tastants based on the combination of receptors activated. Nevertheless, there is only limited evidence that people and experimental animals are able to distinguish among most bitter tastants.

TAS1Rs and TAS2Rs are expressed in subsets of type II cells, and TAS2Rs are seldom in the same cells as TAS1Rs. Thus, TAS2R-expressing cells detect bitter but not sweet or umami taste stimuli. TAS1Rs exhibit a less discrete pattern of expression, resulting in some cells detecting only sweet or umami and others detecting both.

What cellular processes are triggered by a taste ligand binding to these taste GPCRs? The TAS1Rs and TAS2Rs share a common transduction process (Figure 15.5). When a tastant binds one of these receptors, a heterotrimeric G-protein on the cytoplasmic face of the receptor dissociates and activates a membrane-bound phospholipase C. As we saw in Chapter 7, this enzyme catalyzes the production of the second messenger inositol trisphosphate (IP₃), which in turn triggers the release of Ca²⁺ from intracellular stores. The cascade continues with Ca²⁺ activating transient receptor potential (TRP) channels in the cell membrane to depolarize the cell, which in turn elicits action potentials and elicits the release of the afferent transmitter ATP. An unusual feature of type II cells is that the transmitter is synthesized in

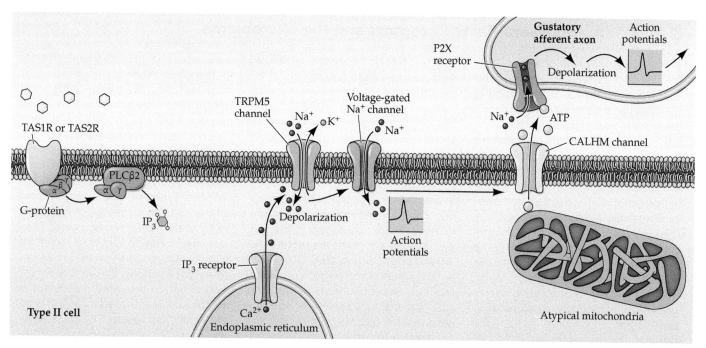

FIGURE 15.5   A shared sensory transduction cascade for G-protein-coupled taste receptors TAS2Rs and TAS1Rs appear to couple to different heterotrimeric G-proteins. Nevertheless, when these receptors are activated by their ligands, subsequent transduction steps are the same. The βγ subunits dissociate and activate PLCβ2, the enzyme that produces IP₃. The resulting elevation of cytosolic Ca²⁺ activates the TRPM5 channel, which depolarizes the cell and triggers action potentials. The strong and persistent depolarization then opens a voltage-gated large-pore calcium homeostasis modulator (CALHM) channel through which ATP is released and serves as an afferent transmitter. The ATP is produced locally by large mitochondria located near the cell membrane and is cytosolic, not packaged in vesicles. Released ATP activates P2X purinergic receptors (see Figure 6.20) to depolarize the afferent fiber. The final steps from action potentials to ATP release are similar for the transduction of salty tastants (NaCl).

specialized mitochondria, is not packaged in vesicles, and instead is released through voltage-gated large-pore channels. ATP, secreted into extracellular space, activates purinoceptors—ATP-gated ion channels on nerve fibers that are juxtaposed to the receptor cells (see Figure 15.5).

Ion-selective membrane channels serve as detectors for salty and sour stimuli. The cells that sense the most critical salt, NaCl, share many properties of type II cells, although the salt receptor itself is quite different from the taste GPCRs described earlier in this section. A salty solution in the mouth drives Na⁺ into the cytoplasm through epithelial sodium channels (called ENaCs; see Figure 15.4A), directly depolarizing the cell and eliciting action potentials. ATP release onto afferent fibers occurs via the same non-vesicular mechanism as for sweet, bitter, and umami transduction. The ENaC channel can be blocked by the diuretic amiloride, and in many animals the saltiness of NaCl is decreased by amiloride. An additional amiloride-insensitive pathway for detecting salts also exists in taste buds, likely in type III cells although the receptor(s) and transduction mechanism have yet to be defined. Human salty taste is mostly insensitive to amiloride.

In the case of acidic (sour) taste, protons are the stimulus. Extracellular protons that are dissociated from acidic stimuli enter type III taste cells through a newly discovered ion channel called OTOP1 (see Figure 15.4B). Organic acids may also passively diffuse through the cell membrane and dissociate in the cytoplasm to release protons. In both cases the cytoplasmic protons block an inward-rectifier K⁺ channel, resulting in a strong depolarization and action potentials. In these type III cells, serotonin is packaged into vesicles and transmitter release is thought to occur by the typical neuronal mechanisms examined in Chapter 5.

Beyond the canonical receptors for tastants, other stimuli exist in foods that may arguably be considered gustatory. Prominent among these is fat. Dietary triglycerides are large molecules that probably are recognized by texture rather than taste (Box 15A and Concept 15.3). However, salivary lipases are proposed to digest some of these molecules into long-chain fatty acids that are detected by free fatty acid receptors (FFARs) in taste bud cells. Another example is sodium–glucose co-transporters (SGLTs) repurposed as sweet taste receptors. Ingested glucose is transported into the same cells that house TAS1R sweet receptors. This dual sweet taste transduction thus produces a parallel redundant pathway for the metabolically important sugar.

Apart from detecting taste stimuli, many cells in the taste buds also participate in autocrine and paracrine signaling.

■ BOX 15A | Extraoral Taste Receptors and the Microbiome

Several receptors that function for taste detection, including the TAS1Rs and TAS2Rs, are found in the upper and distal airways, gastrointestinal (GI) tract, testes, pancreatic islets, kidneys, and choroid plexus of the brain. The cells that house these taste receptors also are of diverse types and include enteric neurons, adipocytes, numerous epithelia, and even cardiomyocytes. In these and many other locations, extraoral taste receptors serve as chemosensory detectors for signaling the presence of molecules of importance. Such molecules may represent toxins that were ingested or inhaled, pathogens within the body, or sometimes even intrinsic metabolites. As such, the extraoral taste receptors can be thought of as sentinels, alerting the body to the presence of chemicals that must be addressed metabolically or removed physically. Canonical downstream effectors such as PLCβ2 and TRPM5 often accompany the TAS1Rs and TAS2Rs in these extraoral locations. The functions these extraoral taste receptors perform are diverse, as a few examples here illustrate.

The dimeric receptor TAS1R2+3 detects sugars and sweeteners in the mouth and is also found in several locations in the GI tract. In the intestine, the binding of glucose or noncaloric sweeteners to this receptor is thought to increase the availability of sugar transporters, thus facilitating the absorption of glucose. The TAS1R2+3 dimer is also expressed on the surface of enteroendocrine cells in the intestinal epithelium. When activated by sugars in the gut, this receptor promotes the secretion of incretin peptides (e.g. GLP, GIP), which in turn enhance the secretion of insulin. Thus, sweet receptors on different cells in the gut may promote both the absorption of glucose from the gut and its clearance from the blood into tissues. An interesting genetic correlate is that certain polymorphisms in the TAS1R2 gene that decrease the affinity of sugar binding and reduce sweet taste perception also are associated with decreased glucose absorption. Individuals with these polymorphisms are reported to have a less acute rise and fall in blood glucose following meals, and may also

have abnormal glucose tolerance tests in the clinic. While the implication for high-potency sweeteners triggering metabolic misregulation of glucose through these mechanisms is apparent, this not been demonstrated clinically.

The large family of bitter taste receptors, the TAS2Rs, is expressed even more broadly across tissues than the TAS1Rs. Typically, only a small subset of the genes is expressed in individual cells. In the upper airway, a well-documented function for these receptors is as key players in a form of innate immunity. TAS2R38, one such receptor expressed on the ciliated cells in the epithelium, is activated by acyl-homoserine lactones secreted by gram-negative bacteria in sinonasal passages. Signaling downstream of these receptors accelerates ciliary beating, sweeping bacteria out of the airway. Additional bacterial metabolites are recognized by TAS2Rs expressed on adjacent solitary chemosensory cells (SCC), and these trigger the secretion of antimicrobial peptides. Both ciliary beating and secretion of antimicrobial peptides are

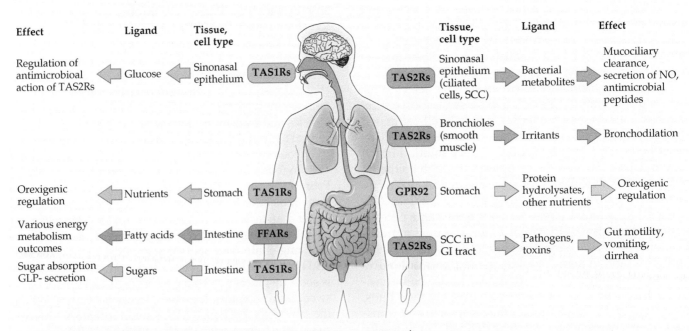

Examples of extraoral taste receptors, the tissues in which they are expressed, and possible functions. Many additional instances exist of receptors that serve roles in both taste and physiological regulation.

■ BOX 15A | **Extraoral Taste Receptors and the Microbiome** (*continued*)

innate immune responses. Cooperation among the cells of upper respiratory passages is thus orchestrated through the action of bitter taste receptors serving as early detectors of inhaled pathogens. Interestingly, many antibiotics used to treat sinonasal infections are bitter and may target upper airway cells in complementarity with their direct antimicrobial action.

Polymorphisms in TAS2R genes that result in altered bitter taste perceptions appear to also influence the ability of these genes to carry out their sentinel function. People who have decreased bitter perception for some foods have a greater predisposition for chronic rhinosinusitis (sinus infections), presumably because of disrupted ability to clear bacteria early. There are also suggestions that the airway microbiomes of people may differ in correlation with their different genetic complement of TAS2R alleles.

Activation of bitter receptors on smooth muscle cells of the lower airway leads to bronchodilation. Although the

ligands that trigger TAS2Rs in bronchioles in situ have not been identified, aerosolized bitter tastants may serve as therapeutic agents in certain conditions of asthma.

Another example of the sentinel role of TAS2Rs is seen in the case of solitary chemoreceptor cells in the gingival epithelium (gums). Here the receptors, activated by bacterial metabolites, secrete antimicrobial peptides that regulate the oral microbiome. Genetic or pathology-induced loss of proteins from the taste receptor signaling pathway is associated with increased numbers of bacteria in the periodontal space and altered balance between "good" and "pathogenic" bacteria, eventually producing periodontal disease and bone loss.

Finally, TAS2Rs are also present in the gut, on endocrine cells and smooth muscle cells. Their presence on endocrine cells mediates the release of a series of hormones involved in hunger and satiety such as cholecystokinin (CCK), glucagon-like peptide (GLP1), and ghrelin. TAS2Rs on smooth muscle cells

are responsible for modulating gastrointestinal motility. Together, these effects are coordinated to mediate the sensation of satiety and limit food intake. Deliberate activation of TAS2Rs in airway, gingiva, and the GI tract is being explored for therapeutic potential.

In addition to TAS1Rs and TAS2Rs, there are additional GPCRs that serve as both taste receptors and chemosensory sentinels in internal organs, notably the free fatty acid receptors (FFARs 1–4). When expressed in taste buds, such receptors detect dietary fatty acids. When expressed in epithelia of the lower digestive tract, the different FFARs detect long-chain fatty acids derived from ingested triglyceride fats, or short-chain fatty acids produced by fermentation of dietary fiber by the gut microbiome. Another example is GPR92, a GPCR that binds certain peptide products of protein digestion. Each of these types of molecules has distinct nutritional and metabolic significance, which the receptors may communicate to regulatory sites.

That is, a particular cell type, when stimulated by its appropriate stimulus, may secrete transmitters that modulate its own activity or the activity of other cells in the same taste bud. Thus, taste signals may be partly shaped within the taste bud. Some type III cells, in addition to sensing sour stimuli, may integrate signals from adjacent receptor cells. Coursing between the cells of each taste bud are fine nerve fibers, which we discuss in Concept 15.2.

CONCEPT **15.2**

Gustatory Information Flows to the Brain through Diverging and Converging Pathways and is Encoded via the Spatio-Temporal Firing Patterns of Neurons

LEARNING OBJECTIVES

15.2.1 Identify the components of the gustatory pathway and its general properties.

15.2.2 Identify and contrast labeled line and across-neuron pattern theories of gustatory neural coding.

15.2.3 Discuss the significance of cortical maps and temporal coding in the gustatory system.

Gustatory neural pathways

Chemosensory signals travel from taste buds to the brain via axons of gustatory afferent sensory neurons. The cell bodies of these neurons lie in one of three pairs of cranial sensory ganglia on the facial, glossopharyngeal, and vagus nerves (cranial nerves VII, IX, and X, respectively; see Figure 15.1A). These neurons are pseudounipolar, similar in appearance to somatosensory neurons in trigeminal and dorsal root ganglia. The peripheral processes of these neurons innervate receptor cells in taste buds. These gustatory afferent neurons are molecularly heterogeneous. However, it is not yet clear if a particular neuron type selectively innervates only one subset of taste bud cells (e.g., sweet-sensing type II cells expressing TAS1R2+3).

The central endings of peripheral gustatory afferents synapse onto brainstem neurons in the **nucleus of the solitary tract (NST)**, where second-order neurons reside. Some NST neurons receiving gustatory input project to other medullary centers and activate circuits related to ingestion (e.g., salivation, swallowing) or reflexes such as oromotor actions and cephalic phase reflex responses (e.g., rapid insulin release upon tasting sugars). Other gustatory neurons in the NST project in ascending fashion to the taste thalamus directly (in humans) or via an intermediate brainstem relay, the parabrachial nucleus (PBN in rodents; see Figure 15.1B). The thalamic target of gustatory brainstem nuclei (NST and PBN) is the parvicellular portion of the ventral posteromedial thalamic nucleus (VPMpc). Neurons in the VPMpc send projections to the gustatory cortex, which is located in a larger region known as the insular cortex. As we will see in Concept 15.3, the gustatory cortex plays a central role in many aspects of taste. Gustatory information does not stop in the gustatory cortex; it also reaches other areas, including the orbitofrontal and prefrontal cortices.

Multiple ascending pathways carry chemosensory information to the forebrain. Indeed, from the brainstem, signals diverge and reach a series of subcortical limbic regions such as the mediodorsal thalamus, the amygdala, and the hypothalamus, which process multisensory, affective, and metabolic information, respectively. From these regions, chemosensory signals reconverge onto the gustatory cortex.

While it is tempting to simplify neural pathways and think of them as unidirectional routes of information, in reality neural circuits are often recurrent. Many regions are reciprocally connected. This is the case for the gustatory system as well, where the gustatory cortex is bidirectionally connected with the thalamus, amygdala, other cortical areas, and brainstem nuclei. These reciprocal connections are important for generating complex activity patterns underlying temporal coding (see later in this concept) and for "top-down" modulation. Top-down modulation occurs when later processing stages—that is, in regions where higher-order neurons reside—shape the activity of earlier regions and nuclei. A classic example is the effect of projections from the gustatory cortex to the thalamus or to the NST. While the function of these top-down projections is not yet fully understood, they may play a role in mediating expectation—that is, they may be responsible for activating subcortical taste regions even before a taste stimulus is encountered.

Neural coding: Two contrasting theories

What happens in the brain that makes us perceive the sweetness of candy or the bitterness of coffee? Are there specialized neurons whose activation is necessary for the perception of each taste quality, or does perception emerge from the activity of large and distributed ensembles of multitasking neurons? In other words, what is the neural code for taste? Two opposing theories have contended against each other for decades (Figure 15.6). One theory, called **labeled line coding**, postulates that taste is mediated by specialized classes of neurons, each dedicated to a specific taste quality. According to this view, taste is carried from taste receptor cells to the cortex by sequential orders of

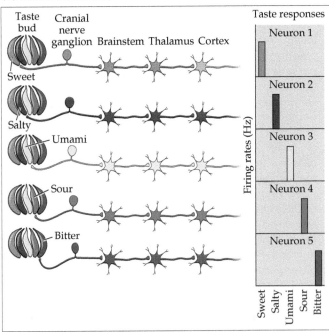

(A) Labeled Lines (LL)

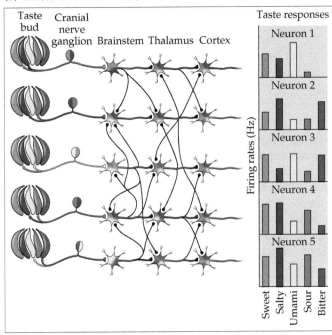

(B) Across Neuron Pattern (ANP)

FIGURE 15.6 Models of taste coding Schematic representation of the connectivity and responsiveness of neurons postulated by labeled line (LL) and across neuron pattern (ANP) coding theories. Individual taste qualities (sweet, bitter, etc.) are color-coded. Histograms on the right of each scheme exemplify the firing rate of five individual neurons. (A) In LL coding, narrowly tuned neurons selectively encode a single taste quality and connect preferentially to similarly tuned neurons, forming highways of single-quality information all the way to the cortex. (B) In ANP coding, broadly tuned neurons respond to multiple qualities and are interconnected along the axis.

neurons that are preferentially connected (hence forming a functional "line") and that are selectively excited by only one taste quality. Neurons that respond to only one taste quality are referred to as "narrowly tuned." The opposing theory, *across fiber pattern* or **across neuron pattern**, relies on a concept that in other sensory systems is described as distributed coding or population coding. The across neuron pattern theory posits that the perception of each taste quality is mediated by the combined activity of populations of neurons. Individual neurons do not need to be specialists; they can respond to multiple taste qualities with excitation or inhibition, as long as their responses are not the same for all stimuli. The across neuron pattern theory accepts that neurons coding for taste can be "broadly tuned."

Which one of these two theories is the correct one? Different experimental results support each one. At the level of taste buds, taste is mainly encoded by the activation of specific receptor cells, as expected from the segregated expression of many taste GPCRs. This means that taste buds may use primarily labeled line coding. However, recordings from nerve fibers and from neurons in the cranial nerve ganglia and in the brain provide mixed results. While some neurons appear to be narrowly tuned—a finding consistent with the labeled line theory—many others are broadly tuned and respond to multiple taste qualities—a result that refutes the labeled line theory and supports the across neuron pattern theory. To make matters even more complicated, the tuning of taste neurons can be significantly affected by changes in stimulus concentration as well as changes in the state of the animal (e.g., anesthetized versus awake). Regardless of the merits of each of the two theories, they both fall short in accounting for many of the fundamental features of gustatory coding. In their original formulation, neither one accounts for spatial and temporal coding, as we discuss next.

Cortical maps: Topographic versus distributed

As seen in some sensory systems, neurons can organize information according to a **topographic logic**. The somatosensory and visual cortices feature globally ordered topographic maps in which neurons representing a body part or stimulus location in the visual field are close to each other. Another form of topography is the spatially clustered representation of particular sensory features (e.g., orientation preference in visual cortex, sound frequency in auditory cortex). In contrast, in the rodent olfactory cortex, odorants evoke distributed patterns of activity with no discernible spatial clustering representing either stimulus location or chemical properties of the stimulus. What happens in the gustatory system? Is there such a thing as a taste map in either of these dimensions? Evidence from neural tracings and recordings shows that gustatory information from the tongue-tip to the pharynx appears in a rostral-to-caudal arrangement in the NST, although it is unclear whether

such a representation persists at later stages of gustatory processing. The question of whether neurons specifically encoding each of the five taste qualities (sweet, salty, bitter, sour, and umami) cluster next to each other has drawn much attention (Figure 15.7A). Early imaging studies in rodents and humans suggested that the gustatory cortex may feature a taste map, with neurons encoding a given quality sitting next to each other. However, more extensive recent evidence from calcium imaging in alert rodents and fMRI experiments in humans demonstrates that taste representations are distributed across the gustatory cortex, similar to what is seen in the olfactory cortex, which lacks chemotopic maps. Spatially distributed representations may be the signature of a system that is flexible, dynamic, plastic, and capable of integrating many types of stimuli.

Temporal coding

The studies that led to the formulation of the the labeled line and across-neuron theories of neural coding in the gustatory system measured neural responses to taste by averaging firing rates over multiple seconds. This procedure compresses the complexity of neural responses into a single number (the firing rate over a certain period) and hinders the ability to observe important temporal features in taste responses. More recent studies, which avoided this averaging and analyzed changes in firing activity over the course of a taste response, demonstrated that the gustatory system relies on temporal coding (Figure 15.7B). Time is important in two ways for representing taste qualities: (1) the specific timing at which action potentials occur and (2) the time course of multiphasic firing rate changes.

In the first case (spike timing), brainstem gustatory neurons encode taste not only by increasing or decreasing their firing rates, but also through precise control of the temporal pattern of action potentials. That is, a neuron could produce the same number of action potentials in 10 seconds in response to sweet or bitter stimuli, yet successfully encode the two qualities through different spike timings. As for the time course of firing rate changes, neurons in the gustatory cortex do not just produce sustained and tonic on–off responses to taste presentation. Instead, they dynamically modulate their firing rates. For instance, a neuron may respond to a sweet stimulus with a phasic increase in firing rates, followed by suppression, followed by a delayed wave of excitation. Other neurons may display different, yet equally rich, time-varying responses. Analysis of these temporal dynamics over populations of gustatory cortical neurons suggests that different aspects of a gustatory experience are encoded at different intervals. For instance, when a food is tasted, the initial response may reflect the contact (via somatosensory fibers); a next phase may represent taste quality; and a delayed response may correlate with the hedonic value of the stimulus. Such a temporal code can be flexible and context-dependent.

(A) Taste topography vs. distributed representation

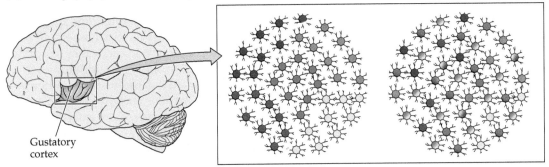

(B) Temporal coding

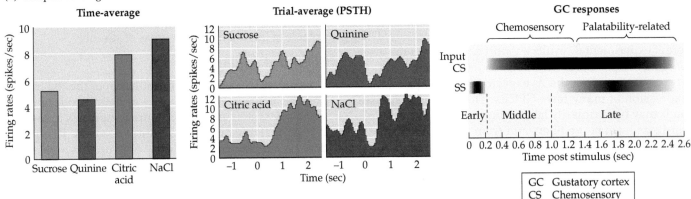

FIGURE 15.7 **Models of spatial and temporal coding of taste information in the gustatory cortex** (A) Spatial coding of gustatory information. The region of the brain in the black rectangle is the gustatory cortex. The inset at right represents two models for how neurons responding to a taste quality appear in the cortex. At left, neurons encoding the same taste quality (e.g., sweet–orange) are clustered in a topographic map. At right, narrowly tuned and broadly tuned neurons are interspersed without topographic clustering. (B) Temporal coding of gustatory information. The left panel is a histogram of responses to four tastants of a single neuron in the gustatory cortex of an alert rat. Firing rates are averaged over 2.5 seconds after stimulus delivery. The panels in the middle feature peristimulus time histograms (PSTHs) of the same neuron and show how firing rates for the four tastants vary over time. The right panel is a model for temporal coding of taste in the gustatory cortex (GC) with three temporal epochs (Early, Middle, Late) encoding distinct aspects of a gustatory experience. Somatosensory (SS) signals include Early touch on the tongue and Late signals associated with oromotor responses; chemosensory (CS) inputs drive responses in the Middle and Late epochs, including palatability processing (CS, chemosensory; SS, somatosensory). (A after J. A. Avery, 2021. *Curr Opin Physiol* 20: 23–28; B after L. M. Jones et al., 2007. *Proc Natl Acad Sci USA* 104: 18772–18777, © 2007 The National Academy of Sciences of the USA and D. B. Katz et al., 2001. *J Neurosci* 2001, 21: 4478-4489, © 2001 Society for Neuroscience.)

These rich dynamics are generated through neuronal interactions within each cortical area and across multiple areas. Furthermore, the rhythmicity of tongue and jaw movements in licking and mastication contributes to the temporal richness of spiking activity in gustatory circuits.

A unitary view of taste coding

Over the past decades, much of the discussion on taste coding has focused on labeled line and across neuron pattern coding as two alternative models. While the two may appear incompatible, experimental evidence lends partial support to both theories. Both narrowly and broadly tuned neurons have been observed at each processing stage along the gustatory pathway. Highly interconnected cortical areas may feature more broadly tuned neurons than lower relays such as those in the NST. Breadth of tuning may also depend on the state of the organism: Labeled line coding may be more prominent under anesthesia or low arousal, while across neuron pattern coding may predominate during alertness. Finally, the gustatory system may use the two coding strategies in different contexts. Labeled lines of narrowly tuned neurons may subserve fast recognition of taste and reflexive responses of rejection or acceptance; ensembles of broadly tuned neurons may favor integration of sensory, metabolic, and cognitive signals, modulation by different contexts, and plasticity of responses. Spatiotemporal dynamics may also be significant for encoding additional components of a tasting experience, including flavor (discussed in Concept 15.3) and hedonic value (discussed in Concept 15.4).

CONCEPT
15.3
Perception of Food and Beverages Also Involves Olfaction, Somatosensation, Audition, and Vision

LEARNING OBJECTIVES

15.3.1 Explain how gustation and olfaction are integrated in the brain.

15.3.2 Describe the sensations conveyed by trigeminal neurons and how they are integrated with gustatory signals to produce flavor.

15.3.3 Discuss how auditory and visual cues contribute to flavor perception and expectations.

Beyond gustation: Flavor

There is more to savoring food and beverages than detecting sweet, bitter, sour, salty, and umami tastants. Ice cream is not just sweet; it is also creamy and cold and has a distinctive flavor. Coffee is bitter but also warm and has a characteristic aroma. Chips are not just salty; they are also crunchy. In fact, the experience of eating (and drinking) engages all the senses. The unitary percept that emerges from the integration of the various sensory modalities is called flavor.

Integration of gustation and olfaction

The connection between gustation and olfaction is easy to experience: Just pinch your nose while eating jellybeans one at a time. They all taste similar when your nose is taken out of the picture: Cinnamon, vanilla, and licorice are sweet, while lemon, orange, and cherry are sweet–sour. The dramatic differences between the flavors are only apparent when you can both taste and smell. Odors related to food are detected either through orthonasal or retronasal olfaction (Figure 15.8A). Orthonasal olfaction is engaged when a food is smelled while it is still in the external environment, and is important for anticipating flavor. Retronasal olfaction is engaged during eating and drinking when odors travel from the mouth through the retronasal passage to the olfactory epithelium. Retronasal odors are perceived as integrated with the taste of the food or beverage being consumed.

Gustation and olfaction exert a mutual influence on each other, so much so that they are often confused. Retronasal odors can change the threshold for detecting gustatory stimuli. For instance, odors that are typically described as sweet (e.g., vanilla) enhance the perceived intensity of a sweet solution. Conversely, gustation can shape olfactory perception by changing sensory thresholds and by attributing gustatory qualities to odors, as in the case of vanilla, which is not actually sweet but is typically encountered in sweets. That is, the association of an odor with a tastant leads the odor to acquire a gustatory quality (Figure 15.8C).

Given the importance of the interaction between these two modalities, it is not surprising that much research

has concentrated on identifying the neural substrates of this interplay (Figure 15.8B). Early models assumed that the gustatory and olfactory systems carried separated streams of information (one originating from the taste buds and the other from the olfactory epithelium) to areas such as the orbitofrontal cortex, which would integrate the two modalities, creating the percept of flavor. More recent research, however, has shown that neurons in various stages of both the gustatory and olfactory pathways—and in regions shared by the two systems, such as the mediodorsal thalamus—can respond to stimuli of both modalities. Thus, gustation and olfaction are intertwined and in constant communication at multiple levels of their neural pathways, which explains the extent to which the two senses can modulate each other during food anticipation and consumption, as well as promote learning.

Touch, trigeminal sensations, and taste

Touch, temperature, texture, and even pain are intimately tied to taste. Chemesthesis is defined as the general somatic sensation elicited by exogenous chemical stimuli. This term refers to sensations such as the pungency, astringency, cooling, warmth, or burning produced by many spices and herbs. Think "hot" chilis, "cool" mint, and "warm" cinnamon. These oral sensations are mediated not by gustatory neurons but by neurons of cranial nerve V, the trigeminal nerve (see Figure 15.8B). Afferent endings of trigeminal neurons can be activated by certain chemicals, such as capsaicin and menthol (see Chapter 10). Across most skin, such chemicals do not gain access to the nerve endings under the epithelium. But the oral mucosa, thinner than skin, allows these compounds to penetrate and stimulate trigeminal nerve terminals in the mouth (see Figure 15.2B).

Some trigeminal nerve endings are similar to polymodal nociceptors and can be activated by capsaicin, leading some people to call chilis "hot" or "painful," depending on individual preference. Similarly, menthol and related compounds in mint permeate oral mucosa and activate TRP channels on cold-sensing neurons. Other compounds that produce mild to intense chemesthesis include alcohol, some organic acids, zingerone in ginger, eugenol in cloves, and allyl isothiocyanate in wasabi. The receptors for these compounds are diverse ion channels, including TRPV1-3, TRPA1, TRPM8, acid-sensing ion channels (ASICs), and several K^+ channels on the membranes of nerve fibers. Astringency, a sensation of local roughness or puckering, is commonly elicited by tannins in black and green tea and in red wine. Astringency is thought to reflect local dehydration or precipitation of proteins on the surface of the tongue, resulting in activation of stretch or other mechanosensory nerve fibers in the oral surface.

Trigeminal fibers are unevenly distributed in the oral epithelium and are plentiful near (but not in) taste buds. This intimate juxtaposition of gustatory and trigeminal fibers

(A) Ortho and retronasal olfaction

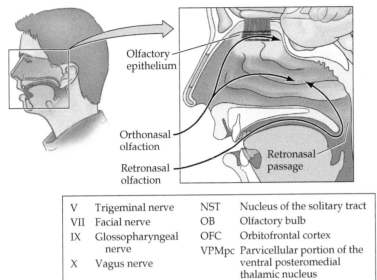

V	Trigeminal nerve	NST	Nucleus of the solitary tract
VII	Facial nerve	OB	Olfactory bulb
IX	Glossopharyngeal nerve	OFC	Orbitofrontal cortex
X	Vagus nerve	VPMpc	Parvicellular portion of the ventral posteromedial thalamic nucleus

(B) Pathways for taste, olfaction, somatosensation and their integration

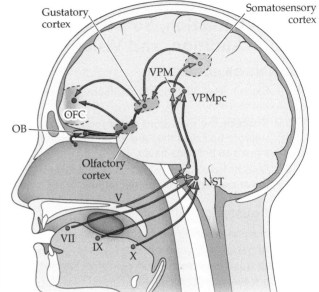

(C) Convergence of taste and olfaction in the gustatory insula

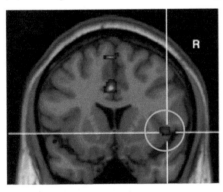
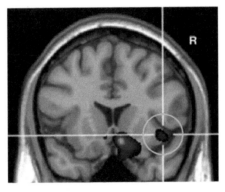
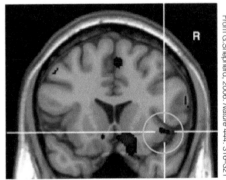

From G. Shepherd, 2006. *Nature* 444: 316–321

FIGURE 15.8 Anatomical and neural basis of integration of taste, odor and texture (A) Orthonasal and retronasal olfactory pathways. Odor molecules entering through the nostrils activate the orthonasal pathway; odor molecules from food and beverages in the mouth reach the olfactory epithelium through the nasopharynx, activating the retronasal pathway. (B) Simplified neural pathways for flavor. While olfactory, gustatory, and somatosensory signals originate from separate sensory epithelia and peripheral pathways, they converge at multiple stages in the nervous system. Roman numerals represent cranial nerves (NST, nucleus of the solitary tract; OB, olfactory bulb; OFC, orbitofrontal cortex; VPMpc, parvicellular portion of the ventral posteromedial thalamic nucleus). (C) Gustatory cortex: taste and olfaction. Functional MRI of the gustatory portion of the insular cortex of a person presented (from left to right) with retronasal strawberry odor, a sweet-tasting solution, and orthonasal fruit odors. Notice that the sweet taste and the retronasal strawberry odor both activate the same area of the insula. (A after J. M. Wolfe et al. 2018. *Sensation and Perception* 5th ed. Sunderland: Oxford University Press/Sinauer Associates; B after I. E. de Araujo et al. 2003. *Eur J Neurosci* 2003. 18: 2059–2068.)

suggests that the two types of sensory fibers may interact in peripheral tissues. Indeed, lingual trigeminal fibers stimulated by capsaicin release substance P, similarly to other polymodal nociceptors. If such neurotransmitters diffuse to nearby gustatory fibers, cross-talk between trigeminal and gustatory nerves may occur, and has been suggested to produce the distinctive flavors of many cuisines. Somatosensory fibers from the oral epithelium make their way to the trigeminal ganglion. The central axon projections of these oral trigeminal neurons enter the brainstem, producing an extensive terminal field in the principal trigeminal nucleus,

lateral to the NST, and also extending down to the spinal trigeminal nucleus. In addition to conveying direct somatosensory perception, branches of oral trigeminal neurons terminate directly onto second-order gustatory neurons in the NST. Temperature and pain signals, as well as electrical stimulation of trigeminal fibers, have been shown to modulate gustatory signals in the NST and PBN, and may be instrumental in producing a multisensory perception of ingested food and beverage.

Some oral trigeminal fibers are thermosensors; others are mechanosensors. Many tastes are directly affected when the stimulus is presented at different temperatures (e.g., melted ice cream is sweeter than frozen). The intensity of bitter and umami taste perceptions also increases as temperature rises. These effects may occur because of enhanced activation of taste buds or the temperature sensitivity of particular gustatory afferent fibers, or via interactions between gustatory and trigeminal fibers in the brainstem, pons, and cortical areas. The texture of food in the mouth, referred to as *mouthfeel*, is sensed by mechanoreceptive trigeminal neurons and is a critical part of how people and animals recognize foods and develop learned preferences. Oils and solid fats add textures such as smooth and slippery, for example, which are highly valued percepts. Soggy potato chips or a pizza blended to make a smoothie would be rejected immediately, even though all their gustatory stimuli remain unchanged.

Finally, it is interesting to note that self-reported dysgeusias are accompanied by diminished ability to perceive oral textures, and conversely, that patients with trigeminal disorders may exhibit decreased taste sensitivity. These are yet further examples pointing to extensive interactions between the gustatory and trigeminal sensory signals.

Other senses: Sounds and images

Olfaction and somatosensation are not the only senses that partner with gustation. Audition and vision do as well. A sound in the external environment can lead to the expectation of a specific flavor: Think of the pop of a can being opened, the splash of soda being poured, or a chef describing her dish. Each of these examples leads us to imagine and anticipate a taste sensation. Sounds in the oral cavity can be integrated with taste and contribute to flavor perception. In a classic study, participants were asked to eat potato chips while hearing their own mastication sounds through headphones. Then the volume and pitch of these sounds were manipulated. Participants reported the sounds as originating from the chips they were eating, not from the headphones, similarly to a ventriloquist's illusion. Furthermore, participants' ratings of the chips varied: Louder, higher-frequency crunching sounds were associated with the perception of more crispness. Similar effects have been reported for other foods (e.g., apples) and sounds (such as those related to carbonation).

Unlike smells, textures, and sounds, visual stimuli cannot be referred to intraoral events. Instead, images can be powerful cues that trigger expectations about imminent gustatory experiences. For instance, the color of a beverage was found to have a marked effect on flavor identification and enjoyment of the drink. Food labels and the presentation of dishes may exert the same kind of influence. Visual social communication, such as observing facial expressions related to food consumption (e.g., disgust or enjoyment) leads us to infer the quality of food.

The neural bases of the integration of visual, auditory, and gustatory stimuli are not yet fully understood. Evidence from rodents, non-human primates, and humans suggests that the gustatory cortex and adjacent portions of the insular cortex play a central role in the integration of auditory and visual signals associated with food and drinks. Auditory and visual information is likely to come from adjacent portions of the insular cortex involved in processing sounds and images and from regions such as the amygdala and the mediodorsal thalamus—a thalamic nucleus that is an integral part of the limbic system. The human insular cortex responds to images of food and to facial expressions of disgust. Similarly, rodents can learn to expect particular tastants following certain images or sounds. This occurs by associating auditory or visual stimuli with the subsequent availability of sweet, bitter, or complex solutions. As the predictive value of the auditory or visual cue is learned, neurons in the gustatory cortex begin to be recruited by these anticipatory stimuli. This is also true for cues from other sensory modalities (i.e., olfaction and taste).

Such anticipatory neural responses appear to have a dual function. First, they prepare the cortex to encode gustatory stimuli. While this preparatory activity can help an individual promptly recognize tastants, it can also erroneously bias perception toward an expected stimulus. For instance, human volunteers were trained to associate visual cues with the degree of bitterness of a solution. In some trials, individuals were given a misleading visual cue that suggested a mildly bitter solution, but instead they received a very bitter solution. In such trials, individuals rated the very bitter solution as mild, showing that expectations bias perception. And importantly, fMRI images showed that the gustatory cortex of volunteers responded to the very bitter stimulus as if it were a mild one. A second function of cue-evoked activity is to guide food-related actions based on expectations. Much like humans deciding to walk into a restaurant after reading the menu, rodents promptly enter a receptacle delivering pellets when an audiovisual signal predicts the availability of food. The changes of neural activity evoked in the gustatory cortex by the audiovisual cue causally contribute to the consummatory behavior. Experimentally silencing the cortex during the cue reduces the number of entrances into the receptacle.

As demonstrated throughout this concept, the gustatory system is involved in much more than just analyzing chemicals dissolved in saliva. The gustatory system integrates multisensory stimuli that are associated with eating and drinking—that is, it is inherently multimodal. The signals integrated into gustatory cortex can be from the external environment or from within the mouth. And such signals may produce expectations of taste and flavor or contribute to the perception of flavor.

CONCEPT
15.4

Palatability or Aversiveness of a Food Is Mediated by the Gustatory and Limbic Systems and Can Be Modified by Experience

LEARNING OBJECTIVES

15.4.1 Explain the concept of hedonic value and its behavioral expression.

15.4.2 Discuss the importance of the gustatory cortex and the amygdala in processing hedonic value.

15.4.3 Explain how perceived hedonic value can be affected by experience.

Taste and Hedonic Value

Taste and flavor are inherently linked to liking and disliking. The term used to define the pleasantness or aversiveness of a food or a drink is *hedonic value*. Nothing offers a clearer manifestation of the hedonic value associated with food than the facial expressions of infants and toddlers upon tasting ice cream, broccoli, or lemon for the first time. Sweets are highly palatable and evoke reactions such as tongue protrusion, lip licking, and lip smacking; bitter and sour tastes are aversive, at least in young and inexperienced tasters, and are associated with grimaces, gapes, and attempts to expel the undesired food (Figure 15.9A). These facial expressions are evolutionarily well conserved and are also observed in many other mammals. Rats and mice, much like human infants, protrude their tongues and gape if made to taste bitter fluids (Figure 15.9B). These orofacial reactions are used in research to infer the hedonic value of specific gustatory stimuli. Scientists can also gauge hedonic value of stimuli by monitoring consumption and avoidance behavior.

Processing hedonic value

According to a classic view, liking and disliking are determined by the activity of neurons in the limbic system. The medial prefrontal and orbitofrontal cortices, as well as the amygdala, ventral tegmental area (VTA), and ventral striatum, are well known for encoding the likability and desirability of foods. These variables are encoded within a distributed network of reward-processing areas.

Additional evidence, however, also supports a role for the gustatory system in encoding hedonic value. Recordings from multiple gustatory regions confirm the existence of neurons that respond similarly to tastants that are distinct but have matching hedonic value. For instance, responses to sucrose and NaCl (another pleasant tastant) are more similar, and they differ from responses to bitter or sour stimuli. More recently, analysis of the time course of firing responses in the gustatory cortex has revealed a neural signature of hedonic value. Likability or aversiveness of various taste solutions is encoded in a specific

temporal period following the presentation of the stimulus (see Figure 15.7B). That is, the gustatory cortex begins to elaborate the hedonic value of a taste only several hundreds of milliseconds after its presentation. Compared with taste, much less is known about how the hedonic value of flavor is encoded. We do know, however, that the gustatory cortex can encode the hedonic value of odors and plays a role in the development of odor preference.

How does the gustatory cortex acquire the information necessary to represent taste pleasantness or aversiveness? The processing of hedonic value in the gustatory cortex emerges through interactions with the limbic system, of

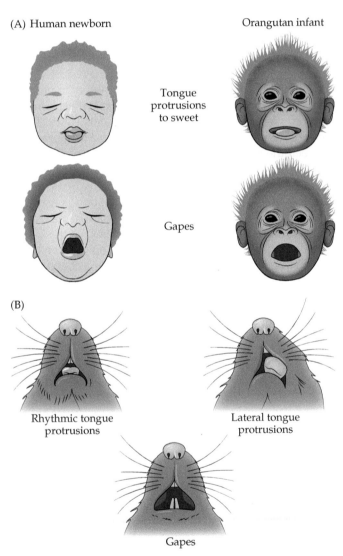

FIGURE 15.9 The hedonic value of taste is reflected in orofacial reactions. (A) Human babies and infant monkeys show characteristic orofacial reactions in response to rewarding and aversive tastants. Palatable tastants, such as sugar, evoke tongue protrusions, while aversive tastants, such as bitter, evoke gapes. (B) Tongue protrusions and gapes have also been observed in rodents. (After J. H. Grill and R. Norgren, 1978. *Brain Res* 143: 263–279, B after J. E. Steiner et al., 2001. *Neurosci Biobehav Rev* 25: 53–74.)

UNIT III
Movement and Its Central Control

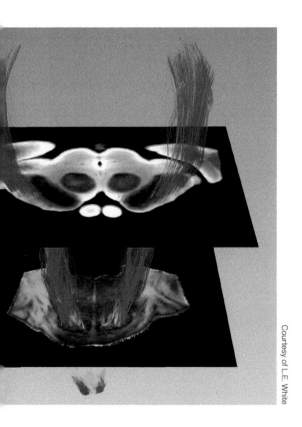

Courtesy of L.E. White

CHAPTER 16 **Lower Motor Neuron Circuits and Motor Control**

CHAPTER 17 **Upper Motor Neuron Control of the Brainstem and Spinal Cord**

CHAPTER 18 **Modulation of Movement by the Basal Ganglia**

CHAPTER 19 **Modulation of Movement by the Cerebellum**

CHAPTER 20 **Eye Movements and Sensorimotor Integration**

CHAPTER 21 **The Visceral Motor System**

Movements, whether voluntary or involuntary, are produced by spatial and temporal patterns of muscular contractions orchestrated by neural circuits in the brain and spinal cord. Analysis of these circuits is fundamental to an understanding of both typical behavior and the etiology of a variety of neurological disorders. This unit considers the brainstem and spinal cord circuitry that makes elementary reflex movements possible, as well as the circuits in the forebrain and cerebellum that organize the intricate patterns of neural activity responsible for more complex motor acts.

The "lower" motor neurons in the spinal cord and brainstem directly innervate skeletal muscles. These lower motor neurons are controlled directly by local circuits within the spinal cord and brainstem and indirectly by "upper" motor neurons in the cerebral cortex and brainstem. Circuits in the basal ganglia and cerebellum regulate upper motor neurons, facilitating the initiation and performance of movement with spatial and temporal precision. The autonomic divisions of the visceral motor system organize the innervation of visceral smooth muscles, cardiac muscle, and glandular secretions by a similar network of neurons in both lower and higher CNS centers. All of this circuitry works together to enable and coordinate complex sequences of body movements and ensure appropriate autonomic activity in support of these movements.

The various symptoms of movement disorders often signify damage to particular brain regions. For example, amyotrophic lateral sclerosis, Parkinson's disease, and Huntington's disease are the result of pathological changes in different parts of the motor system. Thus, knowledge of the various levels of motor control is essential for understanding, diagnosing, and treating these diseases.

On the previous page:

Demonstration of the corticospinal tracts (blue/green/magenta fibers) in a human brainstem constructed from MRI data obtained via diffusion tensor imaging (DTI). DTI measures the direction of water diffusion, which is highly constrained in white matter pathways by the organization of axons into parallel fascicles. Computational methods, called DTI tractography, are employed to generate images of fiber bundles that represent in 3D views the course of axonal projections in cerebral white matter. The upper slice is a diffusion-weighted image (DWI) taken from the level of the midbrain. The lower slice is a fractional anisotropy (FA) image color-coded for fiber orientation (see Atlas plate 5).

Lower Motor Neuron Circuits and Motor Control

Overview

Skeletal muscle contraction is initiated by "lower" motor neurons in the spinal cord and brainstem. The cell bodies of the lower neurons are located in the ventral horn of the spinal cord gray matter and in the motor nuclei of the cranial nerves in the brainstem. These neurons (also called α motor neurons) send axons directly to skeletal muscles via the ventral roots and spinal peripheral nerves or, in the case of brainstem motor nuclei, via cranial nerves. The spatial and temporal patterns of activation of lower motor neurons are determined primarily by local circuits located within the spinal cord and brainstem. The local circuit neurons receive direct input from sensory neurons and mediate sensorimotor reflexes; they also maintain precise interconnections that enable the coordination of a rich repertoire of rhythmical and stereotyped behaviors. The local circuit neurons also receive input from descending pathways from higher centers. These descending pathways comprise the axons of "upper" motor neurons that modulate the activity of lower motor neurons by influencing the local circuitry. The cell bodies of the upper motor neurons are located in brainstem centers, such as the vestibular nuclei, superior colliculus, and reticular formation, as well as in the cerebral cortex. These diverse sources of upper motor neurons initiate and guide a wide variety of both involuntary and voluntary movements. The axons of the upper motor neurons typically synapse on the local circuit neurons in the brainstem and spinal cord, which, via relatively short axons, make synaptic connections with the appropriate combinations of lower motor neurons. Lower motor neurons, therefore, are the final common pathway for transmitting information from a variety of sources to the skeletal muscles. Comparable circuits of interneurons and lower motor neurons may be recognized within the divisions of the visceral motor system, but consideration of these motor circuits will be reserved for Chapter 21. Until then, the principal context for our exploration of the central control of movement will be those movements that are executed by musculoskeletal systems.

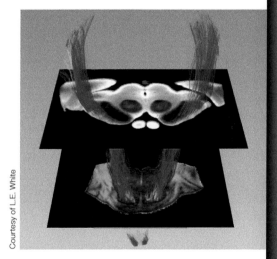

Courtesy of L.E. White

KEY CONCEPTS

16.1 Interacting subsystems in the CNS make essential and distinct contributions to motor control

16.2 Lower motor neurons in the spinal cord and brainstem map the body's musculature

16.3 Motor units of varying size produce appropriate movements

16.4 Local circuitry mediates reflexes that rapidly adjust muscle tension in response to sensory input

16.5 Local circuitry coordinates the output of lower motor neurons for rhythmic, stereotyped behavior

16.6 Damage to lower motor neurons results in "lower motor neuron syndrome"

CONCEPT
16.1

Interacting Subsystems in the CNS Make Essential and Distinct Contributions to Motor Control

LEARNING OBJECTIVES

16.1.1 Discuss the essential contributions of each of the four major subsystems within the CNS for the control of movement.

16.1.2 Describe the anatomical and functional relationships among the four major subsystems within the CNS for the control of movement.

16.1.3 State an anatomical definition of lower motor neurons.

Neural centers responsible for movement

The neural centers responsible for the control of movement can be divided into four distinct but highly interactive subsystems, each of which makes a unique contribution to motor control (Figure 16.1). The first of these subsystems is located within the gray matter of the spinal cord and the tegmentum of the brainstem. The relevant cells include the **lower motor neurons**, which send their axons out of the brainstem and spinal cord to innervate the skeletal muscles of the head and body, respectively, and the local circuit neurons, which are the major source of synaptic input to all lower motor neurons. Commands for movement, whether reflexive or voluntary, are ultimately conveyed to the muscles by the activity of the lower motor neurons; thus, these neurons comprise, in the words of the great British neurophysiologist Charles Sherrington, the "final common path" for initiating movement. The local circuit neurons that innervate the lower motor neurons receive sensory inputs as well as descending projections from higher centers. The circuits they form provide much of the coordination between different muscle groups that is essential for organized movement. Even after the spinal cord is disconnected from the brain in an experimental animal, appropriate stimulation of local circuits in the isolated spinal cord can elicit involuntary but highly coordinated limb movements that resemble walking.

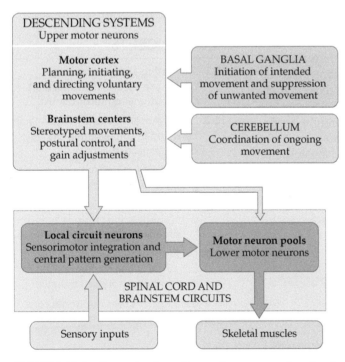

FIGURE 16.1 Organization of neural structures involved in the control of movement Four systems—local spinal cord and brainstem circuits, descending control centers in the cerebral cortex and brainstem, the cerebellum, and the basal ganglia—make essential and distinct contributions to motor control.

The second motor subsystem consists of the **upper motor neurons**, whose cell bodies lie in the brainstem or cerebral cortex, and whose axons descend to synapse with the local circuit neurons or (more rarely) with the lower motor neurons directly. The upper motor neuron pathways that arise in the cortex are essential for the initiation of voluntary movements and for complex spatiotemporal sequences of skilled movements. In particular, descending upper motor neuron pathways from cortical areas in the frontal lobe, including the **primary motor cortex** (Brodmann's area 4) and several divisions of the **premotor cortex** (mainly Brodmann's area 6), are essential for planning, initiating, and directing sequences of voluntary movements involving the head, trunk, and limbs. The frontal lobe also contains cortical areas that play a similar role in the control of eye movements (Brodmann's area 8). In addition, cortical areas in the anterior cingulate gyrus (Brodmann's area 24) govern the expression of emotions, especially with respect to the facial musculature. The posterior portion of the inferior frontal gyrus—typically in the left hemisphere (referred to as Broca's area or Brodmann's areas 44 and 45; see Chapter 31)—is a division of premotor cortex that plays a critical role in the production of speech. Upper motor neurons originating in the brainstem are responsible for regulating muscle tone and for orienting the eyes, head, and body with respect to vestibular, somatic, auditory, and visual sensory information. Their contributions also are critical for basic navigational movements and for the control of posture.

The third and fourth subsystems are massive, complex neural circuits with output pathways that have no direct access to either the local circuit neurons or the lower motor neurons. Instead, they control movement indirectly by regulating the activity of the upper motor neurons in the cerebral cortex and brainstem.

The larger of these latter two subsystems, the **cerebellum**, overlies the pons and fourth ventricle in the posterior cranium (see Chapter 19). The cerebellum functions via its efferent pathways to the upper motor neurons as a servomechanism, detecting and attenuating the difference, or "motor error," between an intended movement and the movement actually performed. The cerebellum mediates both real-time and long-term reductions in these inevitable motor errors (the latter being a form of motor learning). People or animals with cerebellar damage exhibit incoordination with persistent errors in controlling the direction and amplitude of ongoing movements.

Last, embedded in the depths of the forebrain, is a group of structures collectively referred to as the **basal ganglia**. The basal ganglia prevent upper motor neurons from initiating unwanted movements and prepare the motor circuits for the initiation of movements. The basal ganglia also play a role in habit formation and other forms of implicit learning. The problems with movements associated with disorders of basal ganglia, such as Parkinson's disease and Huntington's disease, attest to the importance of this subsystem in the regulation of transitions from one pattern of voluntary movements to another (see Chapter 18).

Despite much effort, the sequence of events that lead from thought and emotion to movement is still poorly understood. The picture is clearest, however, at the level of control of the skeletal muscles themselves. It therefore makes sense to begin an account of motor behavior by considering the anatomical and physiological relationships between lower motor neurons and the striated muscle fibers they innervate.

CONCEPT 16.2 | Lower Motor Neurons in the Spinal Cord and Brainstem Map the Body's Musculature

LEARNING OBJECTIVES

16.2.1 State an anatomical definition of a motor neuron pool.

16.2.2 Describe the general somatotopic organization of motor neuron pools in the ventral horn of the spinal cord and the spinal circuits that organize their output.

16.2.3 Differentiate the two principal types of lower motor neurons: α motor neurons and γ motor neurons.

Motor neuron–muscle relationships

An orderly relationship between the locations of motor neuron pools and the muscles they innervate is evident both along the length of the spinal cord and across the medial-to-lateral dimension of the cord, an arrangement that, in effect, provides a spatial map of the body's musculature. This map can be demonstrated in animal experiments by injecting individual muscle groups with visible tracers that are transported by the axons of the lower motor neurons in a retrograde direction from their terminals back to their cell bodies. The lower motor neurons that innervate each of the body's skeletal muscles can then be seen in histological sections of the ventral horns of the spinal cord. Each lower motor neuron innervates muscle fibers within a single muscle, and all the motor neurons innervating a single muscle, called the **motor neuron pool** for that muscle, are grouped together into a rod-shaped cluster that runs parallel to the long axis of the spinal cord for one or more spinal cord segments (Figure 16.2). For example, the motor neuron pools that innervate the arm

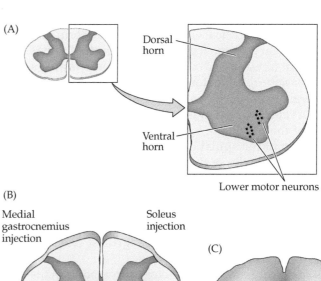

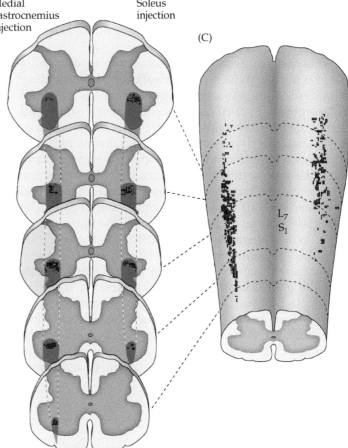

FIGURE 16.2   Distribution of lower motor neurons in the ventral horn of the spinal cord Motor neurons were identified by injecting a retrograde tracer into either the medial gastrocnemius or soleus muscle of the cat, thus labeling neuronal cell bodies and revealing their spatial distribution. A transverse section through the lumbar level of the spinal cord (A) shows lower motor neurons forming distinct, rod-shaped clusters (motor neuron pools) in the ipsilateral ventral horn. Spinal cord cross sections (B) and a reconstruction seen from the dorsal surface (C) illustrate the distribution of motor neurons innervating individual skeletal muscles in both axes of the cord. The rodlike shape and distinct distribution of different motor neuron pools are especially evident in the dorsal view of the reconstructed cord. The dashed lines in (C) represent the locations of individual lumbar and sacral spinal cord sections shown in (B). (After R. E. Burke et al., 1977. *J Neurophys* 40: 667–680.)

are located in the cervical enlargement of the cord, and those that innervate the leg are located in the lumbar enlargement (see Appendix Figure A3). There is also a map, or *topography*, of motor neuron pools in the medial-to-lateral dimension of the spinal cord. Motor neurons that innervate the axial musculature (i.e., the postural muscles of the trunk) are located most medially in the ventral horn of the spinal cord, whereas neurons that innervate the muscles of the shoulders (or pelvis in the lumbar spinal cord; see Figure 16.2) are lateral to the axial neurons. Lower motor neurons that innervate the proximal muscles of the arm are the next most lateral, while those that innervate the distal parts of the extremities, including the hands and fingers, lie farthest from the midline (Figure 16.3).

This spatial organization of motor neuron pools in the ventral horn provides a framework for understanding how descending projections of upper motor neurons and intersegmental spinal cord circuits control posture and modulate movement. Thus, medial lower motor neuron pools that govern postural control and the maintenance of balance receive input from upper motor neurons in the brainstem vestibular nuclei and reticular formation. They comprise long pathways that run in the medial and anterior (ventral) white matter of the spinal cord. The more lateral lower motor neuron pools that innervate the distal extremities are often concerned with the execution of skilled behavior; this is especially true of the lateral motor neurons of the cervical enlargement that innervate muscles of the forearm and hand in primates. These laterally placed lower motor neurons are governed by projections from motor divisions of the cerebral cortex that, in primates, run through the lateral white matter of the spinal cord. This same somatotopic plan is reflected in the location of local spinal cord circuits that interconnect

the lower motor neuron pools distributed along the longitudinal axis of the spinal cord (Figure 16.4). Thus, the patterns of connections made by local circuit neurons in the medial region of the intermediate zone are different from those made by local circuit neurons in the lateral region, and these differences are related to their respective functions. The medial local circuit neurons, which supply the lower motor neurons in the medial ventral horn, have axons that project to many spinal cord segments. Indeed, some projections run between the cervical and lumbar enlargements and participate in the coordination

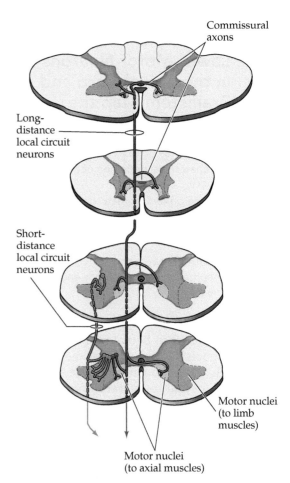

FIGURE 16.4 Local circuit neurons in the spinal cord gray matter Local circuit neurons that supply the medial region of the ventral horn are situated medially within the intermediate zone of the spinal cord gray matter. Their axons (red) extend over several spinal cord segments and terminate bilaterally. Those local circuit neurons that supply the lateral parts of the ventral horn are located more laterally; their axons (orange) extend over just a few spinal cord segments, always terminating on the same side of the cord as the cell body. Pathways that contact the medial parts of the spinal cord gray matter are involved primarily in the control of posture and locomotion; those that contact the lateral parts are involved in the fine control of the distal extremities.

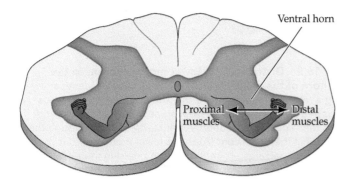

FIGURE 16.3 Somatotopic organization of lower motor neuron pools A cross section of the ventral horn at the cervical level of the spinal cord, illustrating that the motor neurons innervating the axial musculature are located medially, whereas those innervating the distal musculature are located more laterally.

of rhythmic movements of the upper and lower limbs (see Concept 16.5), while other axons terminate along the entire length of the cord and help mediate posture. Moreover, many of these neurons have axonal branches that cross the midline in the ventral commissure of the spinal cord to innervate lower motor neurons in the medial part of the contralateral hemicord. This arrangement ensures that groups of axial muscles on both sides of the body act in concert to maintain and adjust motor activity that requires synchronous bilateral coordination of muscles, such as maintenance of posture or breathing. In contrast, local circuit neurons in the lateral region of the intermediate zone have shorter axons that typically extend fewer than five segments and are predominantly ipsilateral. This more restricted pattern of connectivity provides the finer and more differentiated control that is exerted over the muscles of the distal extremities on one side, such as that required for the independent movement of individual fingers during typing, picking up small objects, or playing a musical instrument.

Two types of lower motor neurons are found in the motor neuron pools of the ventral horn. Large motor neurons are called **α motor neurons**; they innervate the striated muscle fibers that actually generate the forces needed for posture and movement. Interspersed among the α motor neurons are smaller **γ motor neurons**, which innervate specialized muscle fibers that, in combination with the nerve fibers that innervate them, are actually sensory receptors arranged in parallel with the force-generating striated muscle fibers. These specialized muscle fibers, called **muscle spindles** (see Chapter 12), are embedded within connective tissue capsules in the muscle and are thus referred to as intrafusal muscle fibers (*fusal* means capsular or spindle-shaped, in contrast to the surrounding unencapsulated striated muscle fibers, which are termed extrafusal). The intrafusal muscle fibers are innervated by sensory axons that send information to the spinal cord and brainstem about the length of the muscle. The function of the γ motor neurons is to regulate this sensory input by setting the intrafusal muscle fibers to an appropriate length (see Concept 16.4). The output of both types of lower motor neurons is coordinated to optimize movement, particularly when the lengths of active muscles change and the forces acting on the body are dynamic.

Comparable sets of motor neurons responsible for the control of muscles in the head, eyes, and neck are located in the brainstem. The lower motor neurons in the brainstem are distributed in the eight somatic and branchial motor nuclei of the cranial nerves that are located in the medulla, pons, and midbrain (see Appendix Figures A9 and A11). Their activity is controlled by analogous patterns of connections with local circuit neurons and upper motor neurons (see Chapter 17).

Motor Units of Varying Size Produce Appropriate Movements

LEARNING OBJECTIVES

16.3.1 State an anatomical definition of a motor unit.

16.3.2 Characterize in anatomical and physiological terms the three basic types of motor units: fast fatiguable; fast, fatigue-resistant; and slow.

16.3.3 Discuss the recruitment and activation of different types of motor units in the generation of varying levels of force during natural behaviors, such as the progression from standing to walking and then running.

The motor unit

Most extrafusal skeletal muscle fibers in mature mammals are innervated by only a single α motor neuron (immature muscle fibers are innervated by several α motor neurons; see Chapter 23). Since there are, by far, more muscle fibers than motor neurons, individual motor axons branch within muscles to synapse on multiple extrafusal fibers. These fibers are typically distributed over a relatively wide area within the muscle, presumably to ensure that the contractile force is spread evenly (Figure 16.5). In addition, this arrangement reduces the chance that damage to one or a few α motor neurons will significantly alter a muscle's action. Because an action potential generated by a motor neuron typically brings to contraction threshold all of the muscle fibers the neuron contacts, the single α motor neuron and its associated muscle fibers constitute the smallest unit of force that can be activated by the muscle. Sherrington was again the first to recognize this fundamental relationship between an α motor neuron and the muscle fibers it innervates, for which he coined the term **motor unit**.

Both motor units and the α motor neurons themselves vary in size. Small α motor neurons innervate relatively few muscle fibers to form motor units that generate small forces, whereas large motor neurons innervate larger, more powerful motor units. Motor units also differ in the types of muscle fibers they innervate. In most skeletal muscles, the smaller motor units comprise small "red" muscle fibers that contract slowly and generate relatively small forces; but because of their rich myoglobin content, plentiful mitochondria, and rich capillary beds, these small red fibers are resistant to fatigue. These small units are called **slow (S) motor units** and are especially important for activities that require sustained muscular contraction, such as maintaining an upright posture. Larger α motor neurons innervate larger, pale muscle fibers that generate more force; however, these fibers have sparse mitochondria and are therefore easily

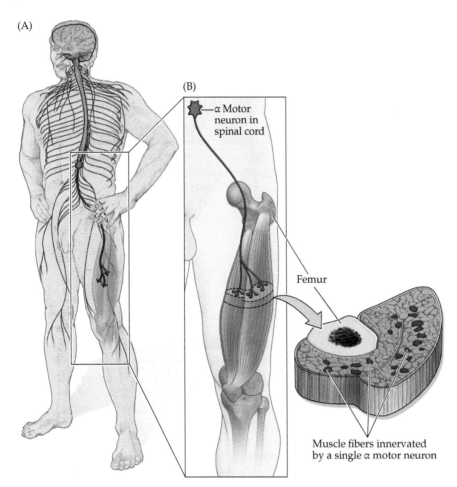

(A)

(B)

α Motor neuron in spinal cord

Femur

Muscle fibers innervated by a single α motor neuron

FIGURE 16.5 **The motor unit** (A) Diagram showing a lower motor neuron in the spinal cord and the course of its axon to its target muscle. (B) Each α motor neuron synapses with multiple fibers in the muscle. The α motor neuron and the muscle fibers it contacts define the motor unit. The cross section through the muscle shows the relatively diffuse distribution of muscle fibers (dark red) contacted by a single α motor neuron.

fast motor units are reached only during rapid movements requiring great force, such as jumping.

The functional distinctions between the various classes of motor units also explain some structural differences among muscles. For example, a motor unit in the soleus (a muscle important for posture that comprises mostly small motor units) has an average innervation ratio of 180 muscle fibers for each motor neuron. In contrast, the gastrocnemius, a muscle that comprises both small and larger motor units, has an innervation ratio of 1000 to 2000 muscle fibers per motor neuron and can generate forces needed for sudden changes in body position. Other differences are related to the highly specialized functions of particular muscles. For instance, the rotation of the eyes in the orbits requires rapid, precise movements that are generated by small forces; in consequence, extraocular muscle motor units are extremely small (with an average innervation ratio of only three fibers per unit) and have a very high proportion of muscle fibers capable of contracting with maximum velocity. More subtle motor unit variations are present in athletes on different training regimens; indeed, both the myofibril and neuronal properties of motor units are subject to use-dependent plasticity. This potential for change in part underlies neuromuscular adaptations to physical exercise and training (Box 16A). Thus, muscle biopsies show that sprinters have a larger proportion of powerful, but rapidly fatiguing, pale fibers in their leg muscles than do marathon runners.

Regulation of muscle force

Increasing or decreasing the number of motor units active at any one time changes the amount of force produced by a muscle. In the 1960s, Elwood Henneman and his colleagues at Harvard Medical School found that progressive increases in muscle tension could be produced by progressively increasing the activity of axons that provide input to the relevant pool of lower motor neurons. This gradual increase in tension results from the recruitment of motor units in a fixed order, according to their size. By stimulating either sensory nerves or upper motor pathways that project to a lower motor neuron pool while measuring the tension changes in the muscle, Henneman

fatigued. These units are called **fast fatigable (FF) motor units** and are especially important for brief exertions that require large forces, such as running or jumping. A third class of motor unit has properties in between those of the other two. These **fast fatigue-resistant (FR) motor units** are of intermediate size and are not quite as fast as FF motor units. They generate about twice the force of a slow motor unit and, as the name implies, are resistant to fatigue (Figure 16.6).

These distinctions among different types of motor units explain how the nervous system produces movements appropriate for different circumstances. In most muscles, small, slow motor units have lower thresholds for activation than do the larger units and are tonically active during motor acts that require sustained effort (standing, for instance). The thresholds for the large,

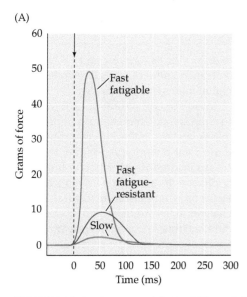

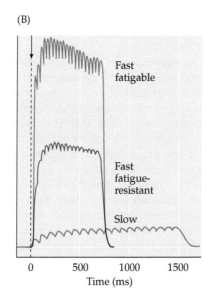

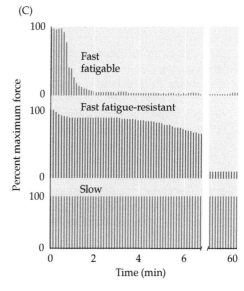

FIGURE 16.6 Force and fatigability of the three different types of motor units In each case, the response reflects stimulation of a single α motor neuron. (A) Change in muscle tension in response to a single action potential. (B) Tension in response to repetitive stimulation of each type of motor unit. (C) Response to repeated stimulation at a level that initially evokes maximum tension. The ordinate represents the force generated by each stimulus. Note the different timescales in the three panels and the strikingly different tensions generated and fatigue rates among motor units. (After R. E. Burke et al., 1973. *J Physiol* 234: 723–748.)

■ BOX 16A | Motor Unit Plasticity

Organisms with complex nervous systems demonstrate an astounding ability to acquire new motor skills and modify the strength and endurance of motor function. The neural basis of these abilities depends heavily on the operations of supraspinal motor centers (i.e., neural centers above the spinal cord) whose functions in volitional motor behavior and motor learning are described in Chapters 17–19. But what role—if any—do the motor units themselves play in the functional changes that underlie such abilities? Are motor units subject to use-dependent plasticity, and if so, how are the anatomical and physiological properties of motor units changeable? To address these questions, it is necessary to consider in more detail the range of phenotypes expressed by motor units.

When considering the structure and function of skeletal muscle, it is convenient to classify the constituent motor units into one of three categories: slow (S), fast fatigable (FF), or fast

fatigue-resistant (FR) (see Figure 16.6). However, with increasingly sophisticated means of characterizing the intrinsic architecture, biochemistry, and physiology of muscle fibers, it has become clear that most skeletal muscles possess a broader spectrum of fiber phenotypes that vary in speed of contraction, tension generation, oxidative capacity, and endurance. These variations among muscle fibers combine with corresponding variations in the morphological

and biophysical properties of α motor neurons to determine the physiological function of motor units (Figure A). Thus, the features of α motor neurons that serve small motor units explain why such neurons are easily depolarized to firing threshold but typically maintain only slow, steady rates of firing—properties that are well suited for the control of slow muscle fibers that mediate postural stability, for example. By

(Continued)

(A)

With increased motor unit size, α motor neurons exhibit:	
Increased	**Decreased**
Cell body size	Input resistance
Dendritic complexity	Excitability
Short-term EPSP potentiation with repeated activation	Ia EPSP amplitude
	PSP decay constant
Axonal diameter (i.e., faster conduction)	Duration of after-hyperpolarization
Number of axonal branches (i.e., more muscle fibers innervated)	

(A) Morphological and biophysical properties of α motor neurons that scale proportionally with the size of motor units

■ **BOX 16A** | **Motor Unit Plasticity** (*continued*)

contrast, α motor neurons that serve large motor units are more difficult to depolarize to threshold but are capable of achieving high frequencies of firing—properties consistent with the force-generating potential of FF muscle fibers that are recruited for production of maximum tension. Not surprisingly, muscle fibers that are intermediate in their functional properties are supplied by α motor neurons whose phenotypes are midway between these extremes.

An early clue to the nature of the mechanisms underlying motor unit plasticity came from a classic series of "cross-innervation" experiments performed by the Australian Nobel laureate J. C. Eccles and his colleagues, most notably A. J. Buller. The results demonstrated that the physiological properties of slow and fast muscle fibers could be reversed when the innervation to these fibers was surgically altered so that slow muscle fibers were innervated by a nerve that typically supplies fast fibers, and vice versa. Subsequent studies by other investigators demonstrated that the actual pattern of neural activity in a motor nerve, in addition to—or perhaps instead of—the molecular identity of the innervating motor neurons themselves, provides an instructive signal that can influence the expression of muscle fiber phenotype. For example, chronic electrical nerve stimulation transforms the metabolic and contractile properties of FF fibers to those consistent with S fibers (Figures B and C). Corresponding changes were also observed in the biophysical properties of the α motor neurons whose axons were stimulated. Although the effects were more subtle, stimulated α motor neurons were modified toward slow, fatigue-resistant motor units, with increased excitability, lengthened after-hyperpolarizations, and short-term depression of excitatory postsynaptic potential (EPSP) amplitudes following high-frequency activation.

It is much more difficult to control and interpret studies of exercising organisms; nevertheless, the same general principles of motor unit plasticity that were derived from nerve stimulation studies apply to neuromuscular adaptation in more naturalistic contexts, including resistance and endurance training. Thus, the nature and degree of muscle adaptation following exercise are functions of the tensions exerted by the muscle fibers and the duration of their increased activity. Most commonly, exercise regimes can "slow" the contractile properties of motor units while increasing the endurance and strength of muscle fibers. Moreover, the impact of exercise is distributed proportionally to motor units in order of their recruitment during training activities, with S motor units being affected most at low exertion levels, and FR and FF motor units being affected only if recruited by higher intensities of exercise.

Interestingly, neural contributions to exercise-induced changes in performance are not limited to alterations of motor unit phenotype. Indeed, the increase in strength achieved in the early phases of resistance training often exceeds what can be attributable to changes in the structure and function of muscle fibers, implying the operation of spinal and/or supraspinal neural mechanisms that mediate increased motor function. At the motor unit level, these neural adaptations include an increase in the instantaneous discharge rate, a reduction in discharge rate variability,

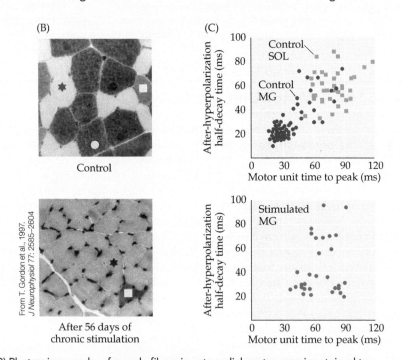

(B) Photomicrographs of muscle fibers in cat medial gastrocnemius stained to demonstrate the presence of myosin ATPase activity in alkaline conditions. In control muscle, fast fatigable fibers (circle) and fast fatigue-resistant fibers (square) stain darkly, but slow oxidative fibers (star) stain very lightly. Following 56 days of chronic electrical nerve stimulation, nearly all fibers acquired the histochemical phenotype of slow oxidative fibers. (C) The electrophysiological properties of the α motor neurons supplying the stimulated nerve also shifted toward those more characteristic of the slower motor units of the soleus muscle (SOL). The upper graph shows control data in which the faster motor units of the medial gastrocnemius (MG) are differentiated from the slower motor units of the soleus muscle by shorter neuronal after-hyperpolarizations and time-to-peak tension in the supplied muscle fibers. The lower graph shows the impact of chronic stimulation, which shifts the properties of MG motor neurons toward those seen in SOL motor neurons. (C after J. B. Munson et al., 1997. *J Neurophysiol* 77: 2605–2615.)

■ BOX 16A | Motor Unit Plasticity *(continued)*

and a marked decrease in interspike interval at the onset of contraction, all of which facilitate the rapid generation of tension (Figures D and E). Furthermore, studies of unilateral exercise (e.g., training one arm and not the other) have shown appreciable gains in the non-exercised limb, indicating the recruitment and adaptation of central neural circuits that have access to contralateral motor units. There are even documented gains in muscle strength with *imagined* exercise—a provocative result that may eventually have profound implications for athletic training and rehabilitation science.

There is still much to be learned about how motor units respond to alterations in strength and endurance training, and scientists are only beginning to probe the neurobiological and neuromuscular mechanisms that underlie skill acquisition. Pursuit of these aims will surely lead to a better understanding of how to maximize motor performance in exercising human (and non-human) individuals, as well as in the rehabilitation of patients coping with neuromuscular or other physical impairment.

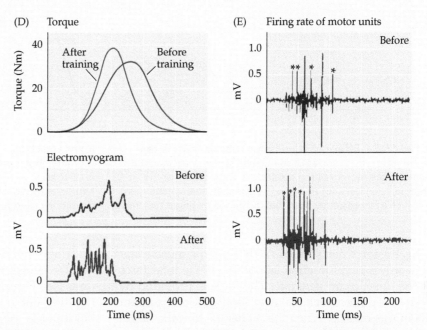

(D) Comparison of torque and electromyogram (EMG) activity during ballistic contractions of ankle dorsiflexor muscles in humans before and after dynamic training. Note the increased rate of tension development after training and the accompanying increase in rectified, surface EMG activity in the early phase of contraction. (E) These changes are associated with an increase in the instantaneous firing rate of motor units recorded from intramuscular electrodes; asterisks mark repetitive discharges of the same motor unit. (D,E after M. Van Cutsem et al., 1998. *J Physiol* 513: 295–305.)

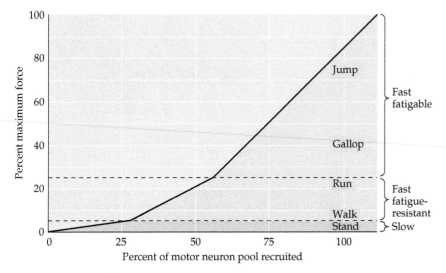

FIGURE 16.7 Motor neuron recruitment in the cat medial gastrocnemius muscle under different behavioral conditions Slow (S) motor units provide the tension required for standing. Fast fatigue-resistant (FR) motor units provide the additional force needed for walking and running. Fast fatigable (FF) motor units are recruited for the most strenuous activities, such as jumping. (After B. Walmsley et al., 1978. *J Neurophys* 41: 1203–1216.)

found that, in experimental animals, only the smallest motor units in the pool are activated by weak synaptic input. When synaptic input to the motor pool increases, progressively larger motor units that generate larger forces are recruited. Thus, as the synaptic activity driving a motor neuron pool increases, low-threshold S motor units are recruited first, then FR motor units, and finally, at the highest levels of activity, the FF motor units. Since these original experiments were performed, evidence for the orderly recruitment of motor units has been found in a variety of voluntary and reflexive movements, including exercise activities. This systematic relationship has come to be known as the **size principle**.

An illustration of how the size principle operates for the motor units of the medial gastrocnemius muscle in the cat is shown in Figure 16.7. When the animal is standing

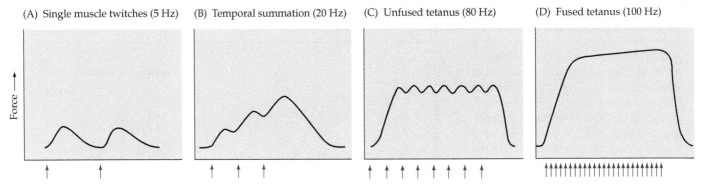

(A) Single muscle twitches (5 Hz) (B) Temporal summation (20 Hz) (C) Unfused tetanus (80 Hz) (D) Fused tetanus (100 Hz)

Force →

FIGURE 16.8 Effect of stimulation rate on muscle tension (A) At low frequencies of stimulation (arrows), each action potential in the motor neuron results in a single twitch of the related muscle fibers. (B) At higher frequencies, the twitches sum to produce a force greater than that produced by single twitches.

(C) At a still higher frequency of stimulation, the force produced is greater, but individual twitches are still apparent. This response is referred to as unfused tetanus. (D) At the highest rates of motor neuron activation, individual twitches are no longer apparent—a condition called fused tetanus.

quietly, the force measured directly from the muscle tendon is only a small fraction (about 5%) of the total force that the muscle can generate. The force is provided by the S motor units, which make up about 25% of the motor units in this muscle. When the cat begins to walk, larger forces are necessary. Locomotor activities that range from slow walking to fast running require up to 25% of the muscle's total force capacity. This additional need is met by the recruitment of FR motor units. Only movements such as galloping and jumping, which are performed infrequently and for short periods, require the full power of the muscle; such demands are met by the additional recruitment of the FF motor units. Thus, the size principle provides a simple solution to the problem of grading muscle force. The combination of motor units activated by such orderly recruitment optimally matches the physiological properties of different motor unit types with the range of forces required to perform different motor tasks.

The frequency of the action potentials generated by motor neurons also contributes to the regulation of muscle tension. The increase in force that occurs with increased firing rate reflects the temporal summation of successive muscle contractions. The muscle fibers are activated by the next action potential before they have time to completely relax, and so the forces generated by the temporally overlapping contractions are summed (Figure 16.8). The lowest firing rates during a voluntary movement are on the order of 8 Hz (Figure 16.9). As the firing rate of individual units rises (to a maximum of 20 to 25 Hz in the muscle being studied here), the amount of force produced increases. At the highest firing rates, individual muscle fibers are in a state of "fused tetanus"—that is, the tension produced in individual motor units no longer has peaks and troughs that correspond to the individual twitches evoked by the motor neuron's action potentials.

Under normal conditions, the maximum firing rate of motor neurons is less than that required for fused tetanus (see Figure 16.8). However, the asynchronous firing of different lower motor neurons provides a steady level of input to the muscle, which causes the contraction of a relatively constant number of motor units, and averages out the changes in tension due to contractions and relaxations of individual motor units. All this allows the resulting movements to be executed smoothly.

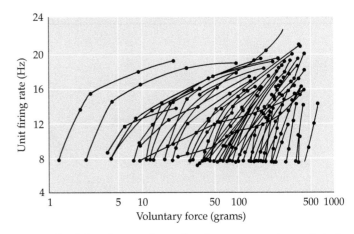

FIGURE 16.9 The number of active motor units and their rate of firing both increase with voluntary force Motor unit firing in a muscle of the human hand (each unit is represented here by a single trace) was recorded transcutaneously as the amount of voluntary force produced by the individual progressively increased. The lowest-threshold motor units generate the least amount of voluntary force and are recruited first. As the individual generates more and more force, both the number and the rate of firing of active motor units increase. (Note that all motor units fire initially at about 8 Hz.) (After A. W. Monster and H. Chan, 1977. *J Neurophys* 40: 1432–1443.)

Local Circuitry Mediates Reflexes That Rapidly Adjust Muscle Tension in Response to Sensory Input

LEARNING OBJECTIVES

16.4.1 Discuss the critical components of the myotatic reflex and how they interact to monitor and adjust muscle length.

16.4.2 Characterize the role of γ motor neurons in gain adjustment of muscle spindles.

16.4.3 Discuss the factors that account for muscle tone.

16.4.4 Discuss the critical components of the Golgi tendon circuit and how they interact to monitor and adjust the force of muscle contraction.

16.4.5 Compare and contrast the structure and function of muscle spindles and Golgi tendon organs—and their central modulation—in the regulation of muscle tension in various functional activities, such as yoga and weightlifting.

16.4.6 Discuss the critical components of the flexion–crossed extension reflex and how they interact to withdraw a limb from a harmful stimulus.

The spinal cord circuitry underlying muscle stretch reflexes

Local circuitry within the spinal cord mediates several sensorimotor reflexes. The simplest of these reflex arcs entails a sensory response to muscle stretch, which provides direct excitatory feedback to the motor neurons innervating the muscle that has been stretched. As already mentioned in Concept 16.2, the sensory signal for the stretch reflex originates in muscle spindles, the sensory receptors embedded within most muscles. The spindles comprise eight to ten intrafusal fibers arranged in parallel with the force-generating extrafusal fibers that make up the bulk of the muscle (Figure 16.10A).

Two classes of intrafusal fibers can be distinguished by differences in their structure and function: nuclear bag fibers and nuclear chain fibers (the nuclear bag fibers can be subdivided further into two subclasses, dynamic and static). The two classes differ in the arrangement of their nuclei (giving rise to their nomenclature, bag and chain fibers), the intrinsic architecture of their myofibrils, and their dynamic sensitivity to stretch. Most muscle spindles contain two or three nuclear bag fibers and at least twice that many nuclear chain fibers. Large-diameter sensory axons (group Ia afferents; see Table 12.1) are coiled around the middle region of each class of intrafusal fiber, forming so-called annulospiral primary endings (see Figure 16.10A). Nearly as large in diameter are the group II afferents, which form secondary endings, mainly on nuclear chain fibers; these are referred to as "flower-spray"

endings because of their short, petal-like contacts just outside the middle region of the fiber. Taken together, group Ia and group II afferents are the largest axons in peripheral nerves, and because action potential conduction velocity is a direct function of axon diameter (see Chapter 3), they mediate very rapid reflex adjustments when the muscle is stretched. The stretch imposed on the muscle deforms the intrafusal muscle fibers, which in turn initiates action potentials by activating mechanosensitive channels in the group I and II axon endings innervating the spindle.

Group Ia afferents tend to respond phasically to small stretches. This is because Ia afferent activity is dominated by signals transduced by the *dynamic* subtype of nuclear bag fiber whose biomechanical properties are sensitive to the *velocity* of fiber stretch. Group II afferents, which innervate *static* nuclear bag fibers and the nuclear chain fibers, signal the level of *sustained* fiber stretch by firing tonically at a frequency proportional to the degree of stretch, with less dynamic sensitivity. The centrally projecting branch of the sensory neuron forms monosynaptic excitatory connections with those α motor neurons in the ventral horn of the spinal cord that innervate the same (homonymous) muscle and, via intervening GABAergic local circuit neurons (called reciprocal-Ia-inhibitory interneurons), forms inhibitory connections with those α motor neurons that innervate antagonistic (heteronymous) muscles. This arrangement is an example of **reciprocal innervation** and results in rapid contraction of the stretched muscle and simultaneous relaxation of the antagonist muscle. This pattern of activity leads to especially rapid and efficient adjustments to changes in the length of the muscle (Figure 16.10B). The excitatory pathway from a spindle to the α motor neurons innervating the same muscle is unusual in that it is a monosynaptic reflex; in most cases, sensory neurons from the periphery do not contact lower motor neurons directly but instead exert their effects through local circuit neurons.

This monosynaptic reflex arc is variously referred to as the "stretch," "deep tendon," or "myotatic" reflex, and it is the basis of the knee, ankle, jaw, biceps, or triceps response tested in a routine physical examination. The tap of the reflex hammer on the tendon stretches the muscle, which evokes an afferent volley of activity in the Ia sensory axons that innervate the muscle spindles. The afferent volley is relayed to the α motor neurons in the brainstem or spinal cord, which then deliver an efferent volley to the same muscle. Since muscles are always under some degree of stretch, this reflex circuit, mediated largely by group II afferents, is typically responsible for the steady level of tension in muscles called **muscle tone**. Changes in muscle tone occur in a variety of pathological conditions, and these changes are assessed by examination of deep tendon reflexes (see Box 17D).

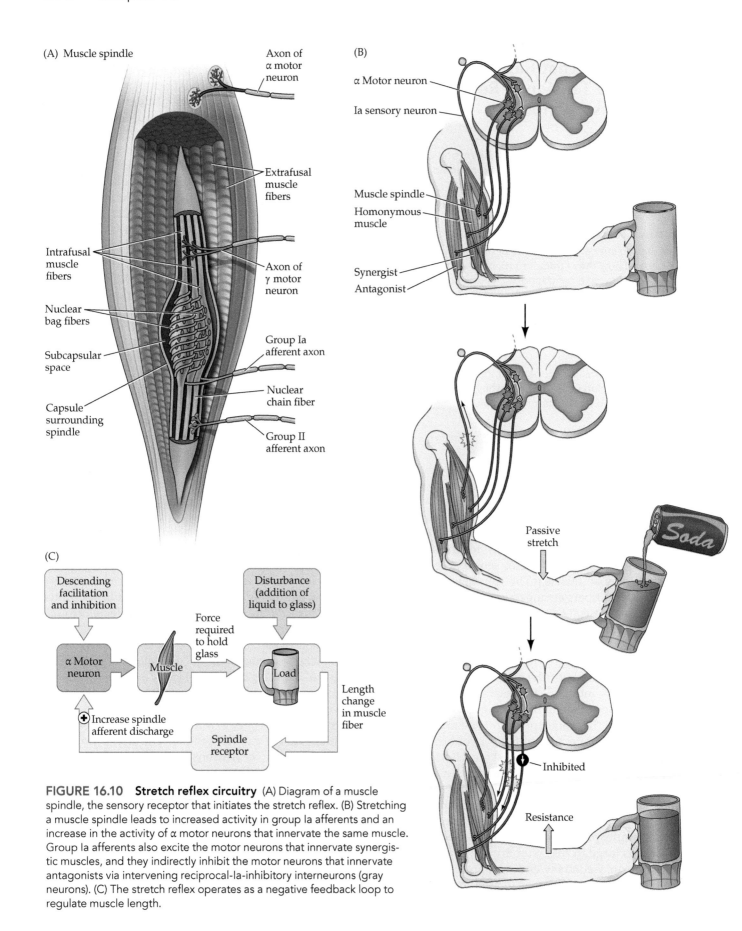

(A) Muscle spindle

Axon of α motor neuron

Extrafusal muscle fibers

Intrafusal muscle fibers

Axon of γ motor neuron

Nuclear bag fibers

Subcapsular space

Group Ia afferent axon

Nuclear chain fiber

Capsule surrounding spindle

Group II afferent axon

(B)

α Motor neuron

Ia sensory neuron

Muscle spindle

Homonymous muscle

Synergist

Antagonist

Passive stretch

Soda

Inhibited

Resistance

(C)

Descending facilitation and inhibition

Disturbance (addition of liquid to glass)

Force required to hold glass

α Motor neuron

Muscle

Load

Length change in muscle fiber

Increase spindle afferent discharge

Spindle receptor

FIGURE 16.10 Stretch reflex circuitry (A) Diagram of a muscle spindle, the sensory receptor that initiates the stretch reflex. (B) Stretching a muscle spindle leads to increased activity in group Ia afferents and an increase in the activity of α motor neurons that innervate the same muscle. Group Ia afferents also excite the motor neurons that innervate synergistic muscles, and they indirectly inhibit the motor neurons that innervate antagonists via intervening reciprocal-Ia-inhibitory interneurons (gray neurons). (C) The stretch reflex operates as a negative feedback loop to regulate muscle length.

In terms of engineering principles, the stretch reflex arc is a negative feedback loop used to maintain muscle length at a desired value (Figure 16.10C). In the context of motor control, the appropriate muscle length is specified by the activity of descending upper motor neuron pathways that influence the lower motor neuron pool. Deviations from the desired length are detected by the muscle spindles, since increases or decreases in the stretch of the intrafusal fibers alter the level of activity in the sensory axons that innervate the spindles. These changes lead, in turn, to adjustments in the activity of the α motor neurons, returning the muscle to the desired length by contracting the stretched muscle and relaxing the opposing muscle group, and by restoring the level of spindle activity and sensitivity.

Smaller γ motor neurons control the functional characteristics of the muscle spindles by modulating their level of excitability. As described earlier, when the muscle is stretched, the spindle is also stretched and the rate of discharge in the afferent fibers is increased. When the muscle shortens, the spindle is relieved of tension ("unloaded"), and the sensory axons that innervate the spindle might therefore be expected to fall silent, but in fact they remain active. The γ motor neurons terminate on the contractile poles of the intrafusal fibers, and the activation of these neurons causes intrafusal fiber contraction—in this way, maintaining the tension on the middle, or equatorial region, of the intrafusal fibers where the sensory axons terminate. Just as there are the dynamic and static functional classes of intrafusal muscle fibers, there are dynamic and static classes of γ motor neurons. When dynamic γ motor neurons fire, the dynamic response of the group Ia afferent is markedly enhanced. In contrast, when static γ motor neurons are activated, the dynamic response of the group Ia afferent is reduced and the static response is increased; the static response of the group II afferent likewise is enhanced under these conditions. Thus, co-activation of α and γ motor neurons allows spindles to function (i.e., send information centrally) at all muscle lengths during movement and postural adjustment.

Modifying the gain of muscle stretch reflexes

The level of γ motor neuron activity is often referred to as γ *bias*, or *gain*, and can be adjusted by upper motor neuron pathways as well as by local reflex circuitry. The gain of the myotatic reflex refers to the amount of muscle force generated in response to a given stretch of the intrafusal fibers. If the gain of the reflex is high, then a small amount of stretch applied to the intrafusal fibers will produce a large increase in the number of α motor neurons recruited and a large increase in their firing rates; this, in turn, will lead to a large increase in the amount of tension produced by the extrafusal fibers. If the gain is low, a greater stretch will be required to generate the same amount of tension in the extrafusal muscle fibers. In fact, the gain of the stretch reflex is continuously adjusted to meet different functional

requirements. For example, while standing in a moving bus, the gain of the stretch reflex can be modulated by upper motor neuron pathways to compensate for the variable changes that occur as the bus stops and starts or progresses relatively smoothly. During voluntary stretching, such as warming up for athletic performance, the gain of myotatic reflexes must be reduced to facilitate the lengthening of muscle fibers and other elastic elements of the musculotendinous system that is desirable under these circumstances. Thus, under the various demands of voluntary (and involuntary) movement, α and γ motor neurons are often co-activated by higher centers to prevent muscle spindles from being unloaded or overactivated (Figure 16.11).

In addition, the level of γ motor neuron activity can be modulated independently of α motor neuron activity to allow fine adjustments in movements. In general, the baseline activity level of γ motor neurons is high if a movement is relatively difficult and demands rapid and precise execution. For example, recordings from cat hindlimb muscles show that γ motor neuron activity is high when the animal has to perform a difficult movement, such as walking across a narrow beam. Unpredictable conditions, as when the animal is picked up or handled, also lead to marked increases in γ motor neuron activity and greatly increased spindle responsiveness.

However, γ motor neuron activity is not the only factor that sets the gain of the stretch reflex. The gain also depends on the level of excitability of the α motor neurons that serve as the principal efferent side of this reflex loop. Thus, in addition to the influence of descending upper motor neuron projections, local circuits in the spinal cord can change the gain of the stretch reflex by excitation or inhibition of either α or γ motor neurons. In addition, there are inhibitory interneurons that form axo-axonal synapses on the terminals of Ia afferents and are thus positioned to selectively suppress the transfer of excitatory drive to specific subpopulations of lower motor neurons. The activities of local circuits in the spinal cord are themselves influenced by the projections of upper motor neurons in the brainstem and cerebral cortex, as well as by neuromodulatory systems that originate in the brainstem reticular formation (see Chapter 17). Many of these neuromodulatory projections release biogenic amine neurotransmitters that bind to G-protein-coupled receptors and mediate long-lasting effects on the gain of segmental circuits in the spinal cord.

The spinal cord circuitry underlying the regulation of muscle force

Another sensory receptor that is important in the reflexive regulation of motor unit activity is the Golgi tendon organ. Golgi tendon organs are encapsulated afferent nerve endings located at the junction of a muscle and a tendon (Figure 16.12A). Each tendon organ is innervated by a single group Ib sensory axon (Ib axons are slightly smaller than

(A) α Motor neuron activation without γ

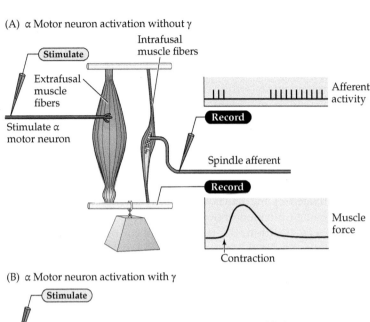

(B) α Motor neuron activation with γ

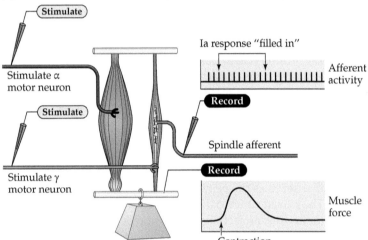

FIGURE 16.11 The role of γ motor neurons in regulating muscle spindle responses (A) When α motor neurons are stimulated without activation of γ motor neurons, the response of the Ia fiber decreases as the muscle contracts. (B) When both α and γ motor neurons are activated, there is no decrease in Ia firing during muscle shortening. Thus, the γ motor neurons can regulate the gain of muscle spindles so they can operate efficiently at any length of the parent muscle. (After C. C. Hunt and S. W. Kuffler, 1951. *J Physiol* 113: 298–314.)

the Ia axons that innervate the muscle spindles; see Table 12.1). In contrast to the parallel arrangement of extrafusal muscle fibers and spindles, Golgi tendon organs are in series with the extrafusal muscle fibers. When a muscle actively contracts, the force acts directly on the tendon, leading to an increase in the tension of the collagen fibrils in the tendon organ and consequent compression of the intertwined sensory nerve endings. Activation of the nonselective, cationic mechanosensitive ion channels in the nerve endings of the Golgi tendon organ results in a generator potential that, if suprathreshold, triggers generation of action potentials that are propagated along the group Ib axon to the spinal cord. The Ib axons from Golgi tendon

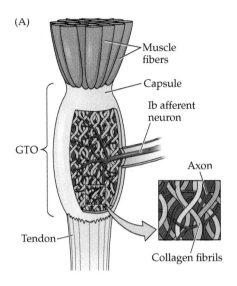

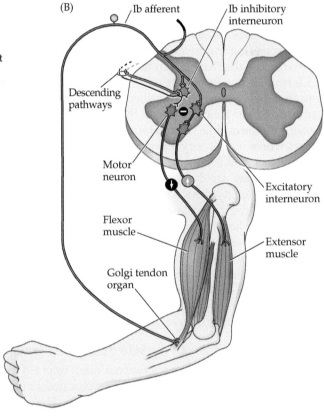

FIGURE 16.12 Golgi tendon organs and their role in the negative feedback regulation of muscle tension (A) Golgi tendon organs (GTO) are arranged in series with extrafusal muscle fibers because of their location at the junction of muscle and tendon. (B) The Ib afferents from tendon organs contact Ib inhibitory interneurons (gray neuron) that decrease the activity of α motor neurons innervating the same muscle. The Ib inhibitory interneurons also receive input from other sensory fibers (not illustrated), as well as from descending pathways. Ib afferents also contact excitatory interneurons (purple neuron) that activate α motor neurons innervating antagonistic muscles. This arrangement prevents muscles from generating excessive tension and helps maintain a steady level of tone during muscle fatigue.

organs contact GABAergic inhibitory local circuit neurons in the spinal cord (called Ib inhibitory interneurons) that synapse, in turn, with the α motor neurons that innervate the same muscle (Figure 16.12B). The Golgi tendon circuit is thus a negative feedback system that regulates muscle tension; it decreases the activation of a muscle when exceptionally large forces are generated and, in this way, protects the muscle. This reflex circuit also operates at lower levels of muscle force, counteracting small changes in muscle tension by increasing or decreasing the inhibition of α motor neurons. The same Ib afferents also make synaptic connections with excitatory interneurons that increase the excitability of α motor neurons that innervate the antagonistic muscle. Thus, at lower levels of muscle force, the Golgi tendon system tends to maintain a steady level of tension and a stable joint angle, counteracting effects that diminish muscle force (such as fatigue).

Like the muscle spindle system, the Golgi tendon organ system is subject to a variety of influences. The Ib inhibitory interneurons receive synaptic inputs from several other sources, including upper motor neurons, cutaneous receptors, muscle spindles, and joint receptors. The joint receptors comprise several types of receptors resembling Ruffini's and Pacinian corpuscles that are located in joint capsules (see Chapter 12). Joint receptors signal hyperextension or hyperflexion of the joint, thereby contributing to the protective functions mediated by Ib inhibitory interneurons when the risk of injury is markedly increased. Acting in concert, these diverse inputs regulate the responsiveness of Ib interneurons to activity arising in Golgi tendon organs.

Complementary functions of muscle spindles and Golgi tendon organs

From the preceding discussion, it should be evident that muscle spindles and Golgi tendon organs serve in a complementary fashion to help regulate motor performance through the operations of distinct spinal cord reflexes. Consider the circumstance of passively stretching a muscle. With passive stretch, most of the change in length occurs in the muscle fibers, since they are more elastic than the fibrils of the tendon. Thus, activity increases in spindle afferents with stretch while there is little change in the firing rate of Golgi tendon organ afferents (Figure 16.13A). Now, consider active muscle contraction. The generated force

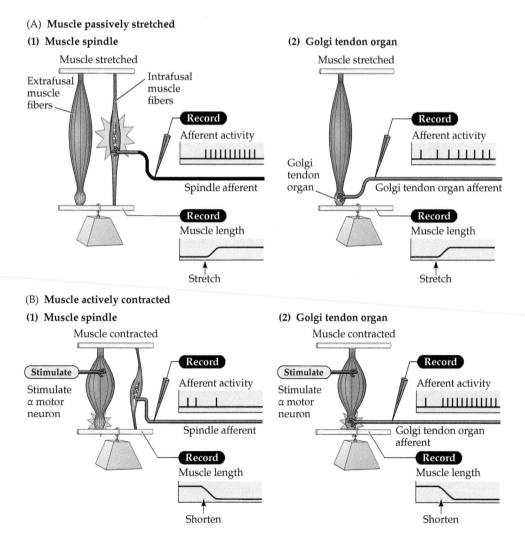

FIGURE 16.13 Muscle spindles and Golgi tendon organs The two types of muscle receptors, the muscle spindles (1) and the Golgi tendon organs (2), have different responses to passive muscle stretch (A) and active muscle contraction (B). Both afferents discharge in response to passively stretching the muscle, although the Golgi tendon organ discharge is much less than that of the spindle. When the extrafusal muscle fibers are made to contract by stimulation of their α motor neurons, however, the spindle is unloaded and its activity decreases, whereas the rate of Golgi tendon organ firing increases. (After H. D. Patton, 1965. In T. C. Ruch and H. D. Patton [Eds.], *Physiology and Biophysics*, 19th ed, Philadelphia: Saunders, pp. 181–206.)

(A) **Muscle passively stretched**

(1) **Muscle spindle**

Muscle stretched

Extrafusal muscle fibers

Intrafusal muscle fibers

Record
Afferent activity

Spindle afferent

Record
Muscle length

Stretch

(2) **Golgi tendon organ**

Muscle stretched

Golgi tendon organ

Record
Afferent activity

Golgi tendon organ afferent

Record
Muscle length

Stretch

(B) **Muscle actively contracted**

(1) **Muscle spindle**

Muscle contracted

Stimulate

Stimulate α motor neuron

Record
Afferent activity

Spindle afferent

Record
Muscle length

Shorten

(2) **Golgi tendon organ**

Muscle contracted

Stimulate

Stimulate α motor neuron

Record
Afferent activity

Golgi tendon organ afferent

Record
Muscle length

Shorten

is transmitted to the tendon and transduced by the Golgi tendon organ, leading to an increase in the firing rate of Ib afferents (Figure 16.13B). Thus, Golgi tendon organs are exquisitely sensitive to increases in muscle *tension* that arise from muscle contraction but, unlike spindles, are relatively insensitive to *passive stretch*. During active contraction, if it were not for a compensatory increase in the output of the relevant γ motor neurons, the muscle spindle would be unloaded and the activity of the associated Ia afferents would decrease (see Figure 16.13B). In short, the muscle spindle system is a feedback system that monitors and maintains muscle *length*, and the Golgi tendon system is a feedback system that monitors and maintains muscle *force*.

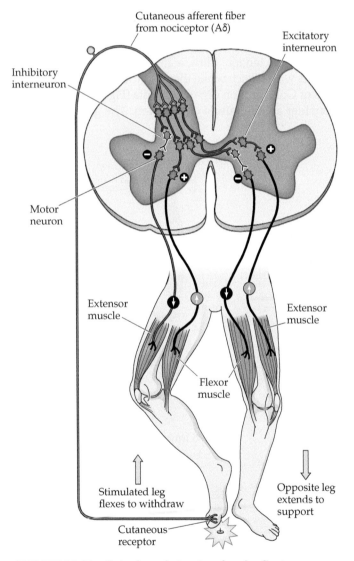

FIGURE 16.14 Spinal cord circuitry for the flexion–crossed extension reflex Stimulation of cutaneous receptors in the foot (by stepping on a tack, in this example) leads to activation of spinal cord local circuits that serve to withdraw (flex) the stimulated extremity and extend the other extremity to provide compensatory support.

Flexion reflex pathways

So far, this discussion has focused on reflexes driven by sensory receptors located within muscles or tendons. Other reflex circuitry mediates the withdrawal of a limb from a painful stimulus, such as a pinprick or the heat of a flame. Contrary to what might be imagined given the speed with which we are able to withdraw from a painful stimulus, this flexion reflex involves slowly conducting afferent axons and several synaptic links (Figure 16.14). As a result of activity in this circuitry, stimulation of nociceptive sensory fibers leads to withdrawal of the limb from the source of pain by activation of ipsilateral flexor muscles and reciprocal inhibition of the α motor neurons that innervate ipsilateral extensor muscles. Flexion of the stimulated limb is also accompanied by an opposite reaction in the contralateral limb (i.e., the contralateral extensor muscles are activated while flexor muscles are relaxed). This crossed extension reflex provides postural support during withdrawal of the affected limb from the painful stimulus.

As in the other reflex pathways, local circuit neurons in the flexion reflex pathway receive converging inputs from several different sources, including other spinal cord interneurons and upper motor neuron pathways. Although the functional significance of this complex pattern of connectivity is unclear, changes in the character of the reflex, following damage to descending pathways, provide some insight. Under normal conditions, a noxious stimulus is required to evoke the flexion reflex; following damage to descending pathways, however, other types of stimulation, such as squeezing a limb, can sometimes produce the same response. Alternatively, under some conditions, descending pathways can suppress the reflex withdrawal from a painful stimulus. These observations suggest that the descending projections to the spinal cord modulate the responsiveness of the local circuitry to a variety of sensory inputs.

CONCEPT **16.5** | ## Local Circuitry Coordinates the Output of Lower Motor Neurons for Rhythmic, Stereotyped Behavior

LEARNING OBJECTIVES

16.5.1 Describe the basic components of central pattern generators and their significance for locomotion and other forms of rhythmic, stereotyped behavior.

16.5.2 Discuss in broad terms how the four major subsystems of the motor system work together to activate and govern locomotion.

Spinal cord circuitry and locomotion

The contribution of local circuitry to motor control is not, of course, limited to reflexive responses to sensory inputs. Studies of rhythmic movements, such as locomotion and swimming in animal models (Box 16B), have demonstrated that local circuits in the spinal cord, called **central pattern generators**, are fully capable of controlling the timing and coordination of such complex patterns of movement, and of adjusting them in response to altered circumstances.

A good example is locomotion (walking, running, etc.). In quadrupeds and bipeds alike, the movement of a single limb during locomotion can be thought of as a cycle consisting of two phases: a *stance phase*, during which the limb is extended and placed in contact with the ground to propel the animal forward; and a *swing phase*, during which the limb is flexed to leave the ground and then brought forward to begin the next stance phase (Figure 16.15A). Increases in the speed of locomotion result from decreases in the amount of time it takes to complete a cycle, and most of the reduction in the cycle time is due to shortening of the stance phase; the swing phase remains relatively constant over a wide range of locomotor speeds.

In quadrupeds, changes in locomotor speed are accompanied by changes in the sequence of limb movements.

■ BOX 16B | Locomotion in the Leech and the Lamprey

All animals must coordinate body movements so they can navigate successfully in their environment. All vertebrates, including mammals, use local circuits in the spinal cord (central pattern generators) to control the coordinated movements associated with locomotion. The cellular basis of organized locomotor activity, however, is best understood in an invertebrate, the leech, and in a simple vertebrate, the lamprey.

Both the leech and the lamprey lack the peripheral appendages for locomotion possessed by many vertebrates (limbs, wings, flippers, fins, or their equivalent). Furthermore, their bodies comprise repeating muscle segments (as well as repeating skeletal elements in the lamprey). Thus, in order to move through the water, both animals must coordinate the movements generated by each segment. They do this by orchestrating a sinusoidal displacement of each body segment in sequence, so that the animal is propelled forward through the water.

The leech is particularly well suited for studying the circuitry responsible for coordinated movement. The nervous system in the leech consists of a series of interconnected segmental ganglia, each with motor neurons that innervate the corresponding segmental muscles (Figure A). These segmental ganglia facilitate electrophysiological studies of the circuitry because there is a limited number of neurons in each ganglion and each neuron has a distinct identity. Specific neurons can thus be recognized and studied from animal to animal and their electrical activity correlated with the sinusoidal swimming movements.

A central pattern generator circuit coordinates this undulating motion. In the leech, the circuit is an ensemble of sensory neurons, interneurons, and motor neurons repeated in each segmental ganglion that controls the local sequence of contraction and relaxation in each segment of the body wall musculature (Figure B). The sensory neurons detect the stretching and contraction of the body wall associated with the sequential swimming movements. Dorsal and ventral motor neurons in the circuit provide innervation to dorsal and ventral longitudinal muscles, whose phasic contractions propel the leech forward. The sensory and motor neuron signals are coordinated by interneurons that fire rhythmically, setting up phasic patterns of activity in the dorsal and ventral cells that lead to sinusoidal movement. The intrinsic swimming rhythm is established by a variety of membrane conductances that mediate periodic bursts of suprathreshold action potentials generated by depolarization, which are followed by well-defined periods of hyperpolarization.

The lamprey, one of the simplest vertebrates, is distinguished by its clearly segmented musculature and by its lack of bilateral fins or other appendages. In order to move through the water, the lamprey contracts and relaxes each muscle segment in sequence (Figure C),

(Continued)

(A) Leech

Posterior sucker

Head

Dorsal longitudinal muscle

Dorsoventral muscle

Segmental ganglion

Ventral longitudinal muscle

To muscle cells

(A) The leech propels itself through the water by sequential contraction and relaxation of the body wall musculature of each segment. The segmental ganglia in the ventral midline coordinate swimming, with each ganglion containing a population of identified neurons.

■ BOX 16B | Locomotion in the Leech and the Lamprey (continued)

which produces a sinusoidal motion, much like that of the leech. Again, a central pattern generator coordinates this sinusoidal movement.

Unlike the leech with its segmental ganglia, the lamprey has a continuous spinal cord that gives rise to nerves that connect each spinal level with the adjacent muscle segments. The lamprey spinal cord is simpler than that of other vertebrates, and several classes of identified neurons occupy stereotyped positions. This orderly arrangement again facilitates the identification and analysis of the neurons that constitute the central pattern generator circuit.

In the lamprey spinal cord, the intrinsic firing pattern of a set of interconnected sensory neurons, interneurons, and motor neurons establishes the pattern of undulating muscle contractions that underlie swimming (Figure D). The patterns of connectivity between neurons, the neurotransmitters used by each class of cell, and the physiological properties of the elements in the lamprey pattern generator are now known. One set of interneurons—known as excitatory premotor interneurons—release glutamate as their neurotransmitter and thereby excite one another, as well as nearby inhibitory interneurons. A local pool of these excitatory premotor interneurons generates a burst of activity for segmental motor output. One class of inhibitory interneurons makes reciprocal connections across the midline that coordinate the pattern-generating circuitry on each side of the spinal cord.

This circuitry in the lamprey provides a basis for understanding similar circuits that control locomotion in more complex vertebrates. Thus, investigations of pattern-generating circuits for locomotion in relatively simple animals have guided studies that have identified similar central pattern generators in the spinal cords of terrestrial mammals. Although different in detail, terrestrial locomotion ultimately relies on sequential movements similar to those that propel the leech and the lamprey through aquatic environments. Simple

aquatic organisms and more complex terrestrial vertebrates likely share many key features that facilitate central pattern generation, including the intrinsic physiological properties of spinal cord

neurons that establish rhythmicity and the modulation by descending monoaminergic input from the brainstem and hypothalamus that can alter rhythmic patterns and cycle frequency.

(B) Electrical recordings

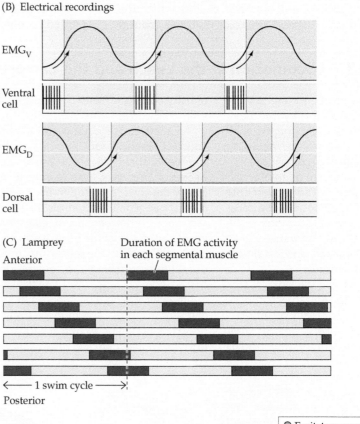

(C) Lamprey

(D) Lamprey

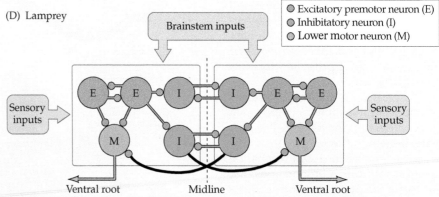

(B) Electromyographic recordings from the ventral (EMG$_V$) and dorsal (EMG$_D$) longitudinal muscles in the leech and the corresponding motor neurons show a reciprocal pattern of excitation for the dorsal and ventral muscles of a given segment.(C) In the lamprey, the pattern of activity across segments is also highly coordinated. (D) The elements of the central pattern generator in the lamprey have been worked out in detail, providing a guide to understanding homologous circuitry in more complex spinal cords. (E: excitatory premotor interneurons; I: inhibitory interneurons; M: lower motor neurons).

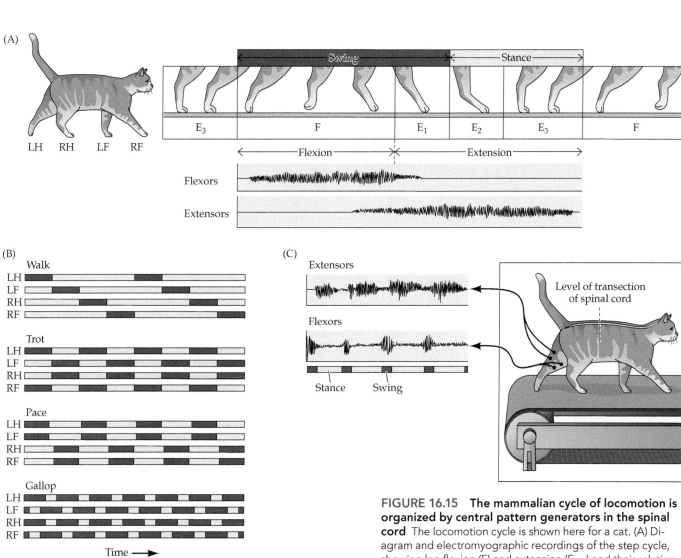

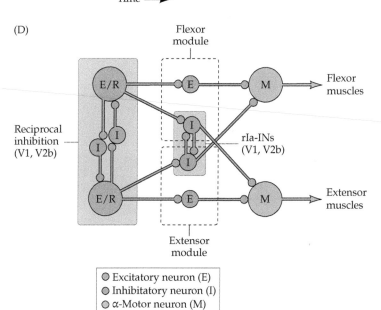

FIGURE 16.15   The mammalian cycle of locomotion is organized by central pattern generators in the spinal cord   The locomotion cycle is shown here for a cat. (A) Diagram and electromyographic recordings of the step cycle, showing leg flexion (F) and extension (E_{1-3}) and their relation to the swing and stance phases of locomotion. (B) Comparison of the stepping movements for different gaits. Red bars, foot lifted (swing phase); blue bars, foot planted (stance phase). (C) Transection of the spinal cord at the thoracic level isolates the hindlimb segments of the cord. After recovering from surgery, the hindlimbs are still able to walk on a treadmill, and reciprocal bursts of electrical activity can be recorded from flexors during the swing phase and from extensors during the stance phase of walking. (D) Schematic illustrating a circuit for central pattern generation of locomotion. Neuronal modules for flexion and extension antagonism (dashed boxes) comprise excitatory neurons (labelled E) and reciprocally connected Ia inhibitory interneurons (rIa-Ins; labelled I). These modules receive input from excitatory rhythm-generating interneurons (E/R), which are reciprocally inhibited by interneurons (labelled I) belonging to the V1 and V2b classes of spinal cord interneurons (expressing distinct transcription factors and derived from distinct embryonic lineages); rIa-INs also belong to the V1 and V2b neuronal classes. The output of this circuit is conveyed to flexor and extensor muscles by α motor neurons (labelled M). (A–C after K. Pearson, 1976. *Sci Am* 235: 72–86; D after O. Kiehn, 2016. *Nat Rev Neurosci* 17: 224–238.)

At low speeds, for example, there is a back-to-front progression of leg movements, first on one side and then on the other. As the speed increases to a trot, the movements of the right forelimb and left hindlimb are synchronized (as are the movements of the left forelimb and right hindlimb). At the highest speeds (a gallop), the movements of the two front legs are synchronized, as are the movements of the two hindlimbs (Figure 16.15B).

Given the precise timing of the movements of individual limbs and the necessity of coordinating these movements, it is natural to assume that locomotion is accomplished by higher centers that organize the spatial and temporal activity patterns of the individual limbs. Indeed, activation of centers in the brainstem, such as the **mesencephalic locomotor region** (also see Chapter 17), can trigger locomotion and change the speed and pattern of the movement by changing the level of activity delivered to the spinal cord. However, following transection of the spinal cord at the thoracic level, a cat's hindlimbs will still make coordinated locomotor movements if the animal is supported and placed on a moving treadmill (Figure 16.15C). Under these conditions, the speed of locomotor movements is determined by the speed of the treadmill, suggesting that the movement is nothing more than a reflexive response to the sensory input initiated by stretching the limb muscles. This possibility is ruled out, however, by experiments in which the dorsal roots are also sectioned. In this condition, locomotion still can be induced by the activation of local circuits either by the act of transecting the spinal cord, or by the intravenous injection of L-DOPA (a dopamine precursor), which may serve to release neurotransmitter from the axon terminals of the now-transected upper motor neuron pathways. Although the speed of walking is slowed and the movements are less coordinated than

under normal conditions, appropriate locomotor movements are still observed.

These and other observations in experimental animals show that the rhythmic patterns of limb movement during locomotion are not dependent on sensory input, nor are they wholly dependent on input from descending projections from higher centers. Rather, local circuitry provides for each limb a central pattern generator responsible for the alternating flexion and extension of the limb during locomotion. This central pattern generator comprises local circuit neurons that include excitatory glutameric neurons coupled to one another and a variety of inhibitory GABAergic and glycinergic neurons (Figure 16.15D). The mechanisms for generating different rhythms that characterize central pattern generator output for varying speeds of locomotion are not yet well understood. However, current evidence indicates that rhythmogenesis depends on both intrinsic membrane properties of excitatory local circuit neurons and network properties reflecting the distribution of connections within the circuit. The central pattern generators for the limbs are coupled to each other by additional modular circuits that coordinate left–right and forelimb–hindlimb activities (such as the long-distance local circuit neurons illustrated in Figure 16.4) in order to achieve the different sequences of movements that occur at different speeds.

Although locomotor movements can also be elicited in humans following damage to descending pathways, these are considerably less effective than the movements seen in the cat. The reduced ability of the transected spinal cord to mediate rhythmic stepping movements in humans presumably reflects an increased dependence of local circuitry on upper motor neuron pathways and the cortical and subcortical circuits that govern and modulate their output (Figure 16.16). Perhaps bipedal locomotion carries with it requirements for postural control greater than can be accommodated by

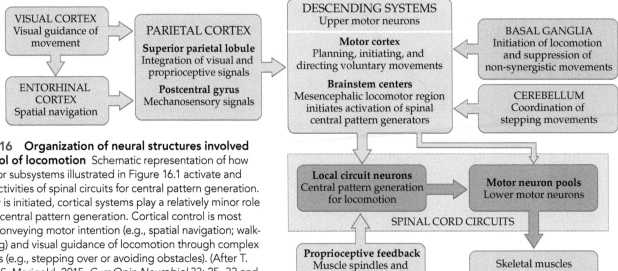

FIGURE 16.16 Organization of neural structures involved in the control of locomotion Schematic representation of how the four motor subsystems illustrated in Figure 16.1 activate and govern the activities of spinal circuits for central pattern generation. Once activity is initiated, cortical systems play a relatively minor role in sustaining central pattern generation. Cortical control is most relevant for conveying motor intention (e.g., spatial navigation; walking or running) and visual guidance of locomotion through complex environments (e.g., stepping over or avoiding obstacles). (After T. Drew and D. S. Marigold, 2015. *Curr Opin Neurobiol* 33: 25–33 and O. Kiehn, 2016. *Nat Rev Neurosci* 17: 224–238.)

spinal cord circuitry alone. Whatever the explanation, the basic oscillatory circuits that control such rhythmic behaviors as flying, walking, galloping, and swimming in many animals also play an important part in human locomotion.

| CONCEPT **16.6** | # Damage to Lower Motor Neurons Results in "Lower Motor Neuron Syndrome" |

LEARNING OBJECTIVE

16.6.1 Characterize the signs and symptoms associated with the lower motor neuron syndrome in terms of damage to the lower motor neurons themselves and secondary changes to denervated muscle.

The lower motor neuron syndrome

The complex of signs and symptoms that arise from damage to the lower motor neurons of the brainstem and spinal cord is referred to as the "lower motor neuron syndrome." In clinical neurology, this constellation of problems must be distinguished from the "upper motor neuron syndrome" that results from damage to the descending upper motor neuron pathways (see Table 17.1).

Damage to lower motor neuron cell bodies, or their peripheral axons, results in paralysis (loss of movement) or paresis (weakness) of the affected muscles, depending on the extent of the damage. In addition to paralysis and/or paresis, the lower motor neuron syndrome includes a loss of reflexes (areflexia) due to interruption of the efferent (motor) limb of the sensorimotor reflex arcs. Damage to lower motor neurons also entails a loss of muscle tone, since tone is dependent in part on the monosynaptic reflex arc that links the muscle spindles to the lower motor neurons (see Box 17D). The muscles involved may also exhibit fibrillations and fasciculations, which are spontaneous twitches characteristic of single denervated muscle fibers or motor units, respectively. These phenomena arise from changes in the excitability of single denervated muscle fibers in the case of fibrillation, and from pathological activity of injured α motor neurons (or their axons) in the case of fasciculations. These spontaneous contractions can be readily recognized in an electromyogram, providing an especially helpful clinical tool in diagnosing lower motor neuron disorders (Clinical Applications). A somewhat later effect of lower motor neuron damage is atrophy of the affected muscles due to long-term denervation and disuse.

■ Clinical Applications

Amyotrophic Lateral Sclerosis

Amyotrophic lateral sclerosis (ALS) is a neurodegenerative disease that affects an estimated 0.05% of the population in the United States. It is also called Lou Gehrig's disease, after the New York Yankees baseball player who died of the disorder in 1941. ALS is characterized by the slow but inexorable degeneration of α motor neurons in the ventral horn of the spinal cord and brainstem (lower motor neurons) and of neurons in the motor cortex (upper motor neurons). Affected individuals show progressive weakness due to upper and/or lower motor neuron involvement, wasting of skeletal muscles due to lower motor neuron involvement, and usually die within 5 years of onset. Sadly, these patients are condemned to watch their own demise, since cognitive faculties remain largely intact. No available therapy effectively prevents the progression of this disease, despite recent attention to this cause with the proliferation of "ice bucket challenges"

to raise public awareness and additional funds for research (Figure A).

Approximately 10% of ALS cases are familial, and several distinct familial forms have been identified. An autosomal dominant form of familial ALS (FALS) is caused by mutations of the gene that encodes the cytosolic antioxidant enzyme copper/zinc superoxide dismutase (SOD1). Mutations of *SOD1*

account for roughly 20% of families with FALS. A rare, autosomal recessive, juvenile-onset form is caused by mutations in a protein called alsin, a putative GTPase regulator. Another rare type of FALS consists of a slowly progressive, autosomal dominant, lower motor neuron disease without sensory symptoms, with onset in early adulthood. This form

(Continued)

Courtesy of Shawn Rocco, Duke Health News & Media

(A) The staff of the Duke ALS Clinic participate in an "ice bucket challenge" to raise awareness and financial support for research into the treatment and prevention of amyotrophic lateral sclerosis.

■ Clinical Applications (continued)

is caused by mutations of the protein dynactin, which binds to microtubules.

How these mutant genes lead to the phenotype of motor neuron disease is uncertain. Defects of axonal transport have long been hypothesized to cause ALS, perhaps because both upper and lower motor neurons give rise to some of the longest axonal projections in the nervous system and may be at greatest risk for injury secondary to impairments of intrinsic axonal structure and/or transport mechanisms. Evidence for this cause is that transgenic mice with mutant *SOD1* exhibit defects in slow axonal transport early in the course of the disease, and that mutant dynactin may modify fast axonal transport along microtubules. However, whether defective axonal transport is the cellular mechanism by which mutant protein and RNA aggregates lead to motor neuron disease remains to be clearly established. Nevertheless, recent human genetics studies have implicated *TANK-binding kinase 1 (TBK1)* mutations in ALS. TBK1 is a member of a family of kinases that are involved in innate immunity signaling pathways, and TBK1 seems to play a major role in autophagy—the process by which aberrant aggregates are cleared from cells.

In addition, recent studies have explored a variety of other pathogenic factors that may play a role in the majority of ALS cases, which are sporadic (i.e., non-familial). Among the plausible mechanisms are the activity of reactive oxygen species, the induction of apoptotic pathways, pro-inflammatory interactions between neurons and microglia, defective autophagy, mitochondrial dysfunction, and dysregulation of calcium homeostasis. Given that motor neurons are exceptionally vulnerable to disruption of mitochondrial function and that they tend to be weak in their capacity to buffer intracellular calcium, these last two factors are likely to contribute to the selective vulnerability of motor neurons in ALS. Additionally, researchers are increasingly investigating putative defects in the intracellular systems for clearing misfolded proteins,

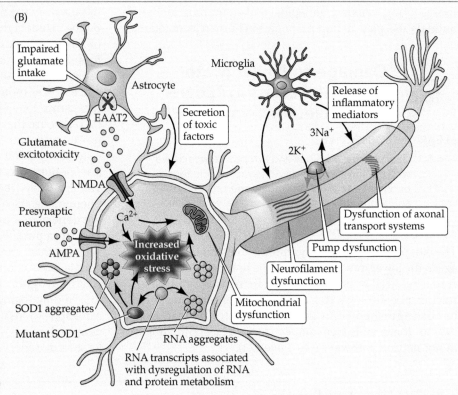

(B) Pathophysiological mechanisms in ALS. Among the mechanisms implicated in motor neuron degeneration in ALS are glutamate-mediated excitotoxicity, increased oxidative stress, mitochondrial dysfunction, accumulation of toxic protein and RNA aggregates, impaired axonal transport, and activation of pro-inflammatory microglia. (After N. Geevasinga et al., 2016. *Nat Rev Neurol* 12: 651–661.)

intracellular inclusions, and protein aggregates that typically accumulate in the motor neurons of patients and animal models of ALS.

There is now a growing body of evidence implicating hyperexcitability in cortical networks as an early feature of both sporadic and familial ALS, which may even precede and contribute to dysfunction and degeneration of lower motor neurons through glutamate-mediated excitotoxicity. Glutamate may accumulate to toxic levels in and around synaptic clefts if the excitatory amino acid transporter 2 (EAAT2) in astrocytes becomes impaired (Figure B). This would result in the excessive activation of ionotropic glutamate receptors (NMDA and AMPA), the induction of calcium-dependent second messenger systems associated with excitotoxicity, and the generation of destructive free radicals. The mechanisms of cortical

dysfunction and neuronal degeneration are likely to be multifactorial, with excitotoxicity being one of several pathways leading to neuronal loss (see Figure B). Another promising discovery in recent years is the association of a hexanucleotide expansion of the *C9orf72* gene as a significant risk factor for both ALS and frontotemporal dementia, suggesting that both diseases may share common pathophysiological mechanisms. At present, the function of this gene is not known, nor is it clear whether this genetic risk factor has a direct impact on cortical hyperexcitability.

Despite these uncertainties and the growing list of candidate mechanisms of neurodegeneration in sporadic cases of ALS, demonstration that specific mutations can cause familial ALS has given scientists valuable clues about the molecular pathogenesis of at least some forms of this tragic disorder.

Summary

Four distinct but highly interacting motor subsystems make essential contributions to motor control: lower motor neurons and their associated local circuits in the spinal cord and brainstem, descending upper motor neuron pathways that control these circuits, and the basal ganglia and cerebellum, which modulate the activity of upper motor neuron circuits. Alpha motor neurons in the spinal cord and in the cranial nerve nuclei of the brainstem directly link the nervous system and muscles, with each motor neuron and its associated muscle fibers constituting a functional entity called a motor unit. Motor units vary in size, amount of tension produced, speed of contraction, and degree of fatigability. Graded increases in muscle tension are mediated by both the orderly recruitment of different types of motor units and an increase in lower motor neuron firing frequency.

Local circuitry involving sensory inputs, local circuit neurons, and α and γ motor neurons is especially important in the reflexive control of muscle activity. The stretch reflex is a monosynaptic circuit with connections between sensory fibers arising from muscle spindles and the α motor neurons that innervate the same, or synergistic, muscles. The γ motor neurons regulate the gain of the stretch reflex by adjusting the level of tension in the intrafusal muscle fibers of the muscle spindle. This mechanism sets the baseline level of activity in α motor neurons and helps regulate muscle length and tone. Other reflex circuits provide feedback control of muscle tension and mediate essential functions, such as the rapid withdrawal of limbs from painful stimuli.

Much of the spatial organization and timing of muscle activation required for complex rhythmic movements such as locomotion is provided by specialized local circuits called central pattern generators. Because of the essential role of lower motor neurons in all of these circuits, damage to them results in paralysis or paresis of the associated muscle and other changes, including loss of reflex activity, loss of muscle tone, and eventual muscle atrophy.

■ Additional Reading

Reviews

Burke, R. E. (1981) Motor units: Anatomy, physiology and functional organization. In *Handbook of Physiology*, V. B. Brooks (Ed.). Section 1: *The Nervous System*, vol. 1, part 1. Bethesda, MD: American Physiological Society, pp. 345–422.

Grillner, S. and A. El Manira (2019) Current principles of motor control, with special reference to vertebrate locomotion. *Physiol. Rev.* 100: 271–320.

Henneman, E. (1990) Comments on the logical basis of muscle control. In *The Segmental Motor System*, M. C. Binder and L. M. Mendell (Eds.). New York: Oxford University Press, pp. 7–10.

Henneman, E. and L. M. Mendell (1981) Functional organization of the motoneuron pool and its inputs. In *Handbook of Physiology*, V. B. Brooks (Ed.). Section 1: *The Nervous System*, vol. 1, part 1. Bethesda, MD: American Physiological Society, pp. 423–507.

Kiehn, O. (2016) Decoding the organization of spinal circuits that control locomotion. *Nat. Rev. Neurosci.* 17: 224–238.

Lundberg, A. (1975) Control of spinal mechanisms from the brain. In *The Nervous System*, vol. 1: *The Basic Neurosciences*, D. B. Tower (Ed.). New York: Raven Press, pp. 253–265.

Nistri, A., K. Ostoumov, E. Sharifullina and G. Taccola (2006) Tuning and playing a motor rhythm: How metabotropic glutamate receptors orchestrate generation of motor patterns in the mammalian central nervous system. *J. Physiol.* 572: 323–334.

Patton, H. D. (1965) Reflex regulation of movement and posture. In *Physiology and Biophysics*, 19th Edition, T. C. Rugh and H. D. Patton (Eds.). Philadelphia: Saunders, pp. 181–206.

Prochazka, A., M. Hulliger, P. Trend and N. Durmuller (1988) Dynamic and static fusimotor set in various behavioral contexts. In *Mechanoreceptors: Development, Structure, and Function*, P. Hnik, T. Soulup, R. Vejsada and J. Zelena (Eds.). New York: Plenum, pp. 417–430.

Important Original Papers

Barker, D. (1948) The innervation of the muscle-spindle. *Q. J. Microsc. Sci.* 89: 143–186.

Burke, R. E., D. N. Levine, M. Salcman and P. Tsaires (1974) Motor units in cat soleus muscle: Physiological, histochemical, and morphological characteristics. *J. Physiol.* 238: 503–514.

Burke, R. E., D. N. Levine, P. Tsairis and F. E. Zajac III (1973) Physiological types and histochemical profiles in motor units of the cat gastrocnemius. *J. Physiol.* 234: 723–748.

Burke, R. E. and 4 others (1977) Anatomy of medial gastrocnemius and soleus motor nuclei in cat spinal cord. *J. Neurophysiol.* 40: 667–680.

Drew, T. and D. S. Marigold (2015) Taking the next step: Cortical contributions to the control of locomotion. *Curr. Opin. Neurobiol.* 33: 25–33.

Goetz, L. and 5 others (2016) On the role of the pedunculopontine nucleus and mesencephalic reticular formation in locomotion in nonhuman primates. *J. Neurosci.* 36: 4917–4929.

Henneman, E., E. Somjen, and D. O. Carpenter (1965) Excitability and inhibitability of motoneurons of different sizes. *J. Neurophysiol.* 28: 599–620.

Hunt, C. C. and S. W. Kuffler (1951) Stretch receptor discharges during muscle contraction. *J. Physiol.* 113: 298–315.

Liddell, E. G. T. and C. S. Sherrington (1925) Recruitment and some other factors of reflex inhibition. *Proc. R. Soc. London* 97: 488–518.

Lloyd, D. P. C. (1946) Integrative pattern of excitation and inhibition in two-neuron reflex arcs. *J. Neurophysiol.* 9: 439–444.

Monster, A. W. and H. Chan (1977) Isometric force production by motor units of extensor digitorum communis muscle in man. *J. Neurophysiol.* 40: 1432–1443.

Walmsley, B., J. A. Hodgson and R. E. Burke (1978) Forces produced by medial gastrocnemius and soleus muscles during locomotion in freely moving cats. *J. Neurophysiol.* 41: 1203–1215.

Books

Lieber, R. L. (2011) *Skeletal Muscle Structure, Function, and Plasticity*, 3rd Edition. Baltimore, MD: Lippincott Williams & Wilkins.

Sherrington, C. (1947) *The Integrative Action of the Nervous System*, 2nd Edition. New Haven, CT: Yale University Press.

Upper Motor Neuron Control of the Brainstem and Spinal Cord

Overview

The axons of upper motor neurons arise from cell bodies in higher centers and descend to influence the local circuits in the brainstem and spinal cord. These local circuits organize movements by coordinating the activity of the lower motor neurons that innervate different muscles. The sources of these upper motor neuron pathways include several brainstem centers and multiple cortical areas in the frontal lobe. The motor control centers in the brainstem are especially important in postural control, orientation toward sensory stimuli, locomotion, and orofacial behavior, with each center having a distinct influence. The mesencephalic locomotor region initiates locomotion. Two other centers, the vestibular nuclear complex and the reticular formation, make widespread contributions to the maintenance of body posture and position. The reticular formation also contributes to a variety of somatic and visceral motor circuits that govern the expression of autonomic and stereotyped somatic motor behavior. Also in the brainstem, the superior colliculus contains upper motor neurons that initiate orienting movements of the head and eyes. The primary motor cortex and a mosaic of "premotor" areas in the posterior frontal lobe, in contrast, are responsible for the planning, initiation, and control of complex sequences of voluntary movements, as well as mediating the somatic expression of emotional states. Most upper motor neurons, regardless of their source, influence the generation of movements by modulating the activity of the local circuits in the brainstem and spinal cord. Upper motor neurons in the cortex also control movement indirectly, via pathways that project to motor control centers in the brainstem, which in turn project to the local organizing circuits in the brainstem and spinal cord. These indirect pathways mediate the automatic adjustments in the body's posture that occur during cortically initiated voluntary movements. Injury or disease afflicting upper motor neurons compromises the ability of the brain to govern the activities of lower motor neuronal circuits, producing the upper motor neuron syndrome, which is characterized by weakness of voluntary movement with emergent muscle spasticity, increased muscle tone, and increased reflex activity.

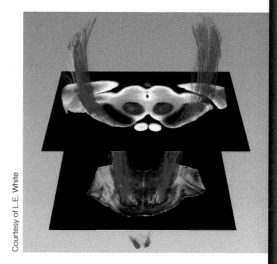

Courtesy of L.E. White

KEY CONCEPTS

17.1 Upper motor neurons give rise to lateral tracts in the spinal cord that govern skilled movements and medial tracts that influence posture, balance, and locomotion

17.2 Neurons in the primary motor cortex encode the intentions for body movements in central personal space

17.3 Neurons in the premotor cortex encode the intentions for body movements that are oriented toward extrapersonal space

17.4 Upper motor neurons in the brainstem help maintain balance, govern posture, initiate locomotion, and orient visual gaze

17.5 Damage to upper motor neurons produces "upper motor neuron syndrome"

CONCEPT
17.1

Upper Motor Neurons Give Rise to Lateral Tracts in the Spinal Cord that Govern Skilled Movements and Medial Tracts that Influence Posture, Balance, and Locomotion

LEARNING OBJECTIVES

17.1.1 Identify the cortical areas that give rise to lateral and medial descending projections to lower motor neurons.

17.1.2 Discuss the functional organization of lateral and medial descending projections from upper motor neurons in the motor cortex and brainstem to the spinal cord, with respect to the somatotopic organization of lower motor neurons.

17.1.3 Characterize the functional and anatomical organization of the corticospinal and corticobulbar tracts from cortex to lower motor circuits in the spinal cord and brainstem, respectively.

Organization of descending motor control

The spatial arrangement of the lower motor neurons and local circuit neurons within the spinal cord—the ultimate targets of the upper motor neurons—provides insight into the functions of different sets of upper motor neurons. As described in Chapter 16, lower motor neurons in the ventral horn of the spinal cord are distributed in a somatotopic fashion: The most medial part of the ventral horn contains lower motor neuron pools that innervate axial muscles or proximal muscles of the limbs, whereas the more lateral parts contain lower motor neurons that innervate the distal muscles of the limbs (Figure 17.1). The local circuit neurons, which lie primarily in the intermediate zone of the spinal cord and supply much of the direct input to the lower motor neurons, are also topographically arranged. Thus, the medial region of the intermediate zone of the spinal cord gray matter contains the local circuit neurons that synapse primarily with lower motor neurons in the medial part of the ventral horn, whereas the lateral regions of the intermediate zone contain local neurons that synapse primarily with lower motor neurons

(A)

Lateral white matter
(axons from motor cortex)

Proximal muscles ⟷ Distal muscles

Medial white matter
(axons from brainstem)

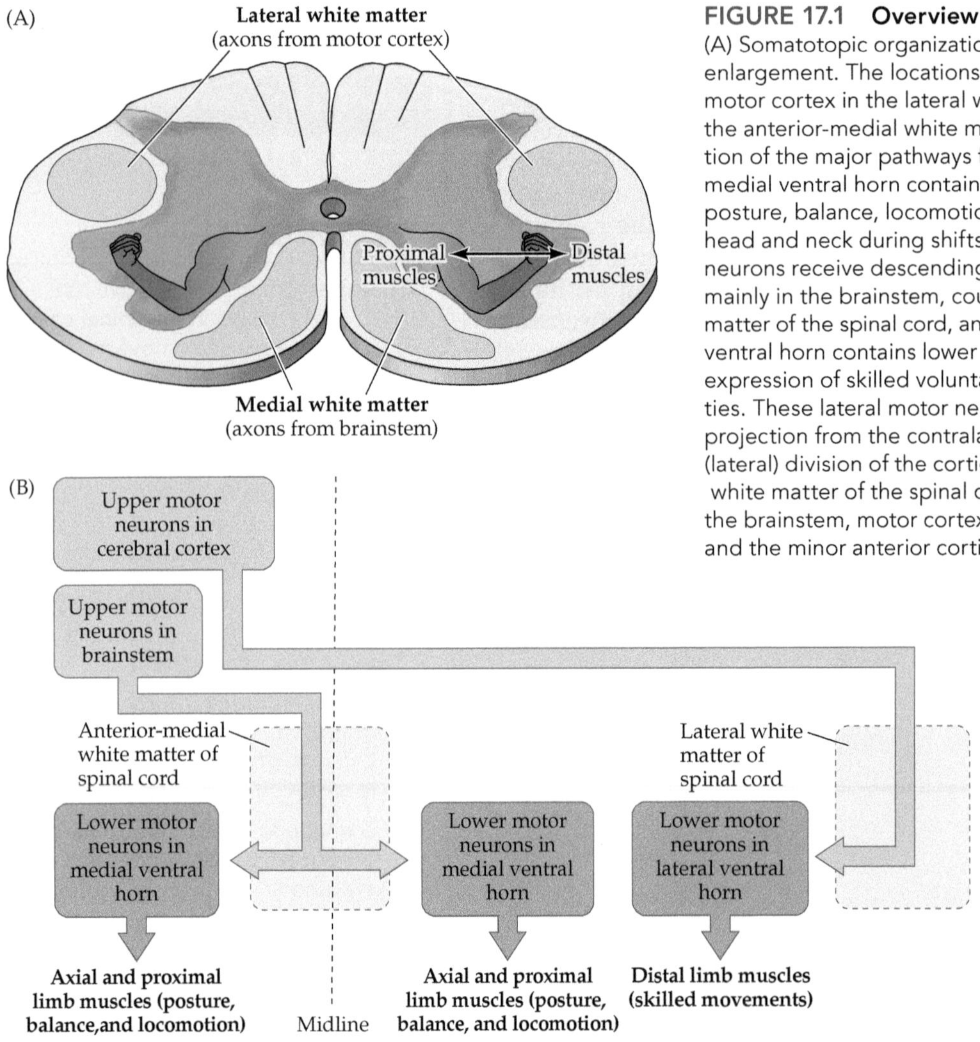

FIGURE 17.1 Overview of descending motor control
(A) Somatotopic organization of the ventral horn in the cervical enlargement. The locations of descending projections from the motor cortex in the lateral white matter and from the brainstem in the anterior-medial white matter are shown. (B) Schematic illustration of the major pathways for descending motor control. The medial ventral horn contains lower motor neurons that govern posture, balance, locomotion, and orienting movements of the head and neck during shifts of visual gaze. These medial motor neurons receive descending input from pathways that originate mainly in the brainstem, course through the anterior-medial white matter of the spinal cord, and then terminate bilaterally. The lateral ventral horn contains lower motor neurons that mediate the expression of skilled voluntary movements of the distal extremities. These lateral motor neurons receive a major descending projection from the contralateral motor cortex via the main (lateral) division of the corticospinal tract, which runs in the lateral white matter of the spinal cord. For simplicity, only one side of the brainstem, motor cortex, and lateral ventral horn is shown, and the minor anterior corticospinal tract is not illustrated.

(B)

Upper motor neurons in cerebral cortex

Upper motor neurons in brainstem

Anterior-medial white matter of spinal cord

Lateral white matter of spinal cord

Lower motor neurons in medial ventral horn

Lower motor neurons in medial ventral horn

Lower motor neurons in lateral ventral horn

Axial and proximal limb muscles (posture, balance, and locomotion)

Midline

Axial and proximal limb muscles (posture, balance, and locomotion)

Distal limb muscles (skilled movements)

in the lateral ventral horn. This somatotopic organization of the ventral and intermediate spinal cord gray matter provides an important framework for understanding the control of the body's musculature in posture and skilled movement, as well as for how the descending projections of different groups of upper motor neurons are organized to influence movements.

Differences in where upper motor neuron pathways from the cortex and brainstem terminate in the spinal cord conform to the functional distinctions between the local circuits that organize the activity of axial and distal muscle groups. Thus, most upper motor neurons that project to the medial part of the ventral horn also project to the medial region of the intermediate zone. The axons of these upper motor neurons course through the anterior-medial white matter of the spinal cord and give rise to collateral branches that terminate over many spinal cord segments among medial cell groups on both sides of the spinal cord. The sources of these projections are located primarily in the brainstem, and as their terminal zones in the medial spinal cord gray matter suggest, they are concerned primarily with proximal muscles that control posture, balance, orienting mechanisms, and the initiation and regulation of stereotyped, rhythmic behavior, including locomotion (see Figure 17.1B). In contrast, the large majority of axons that project from the motor cortex to the spinal cord course through the lateral white matter of the spinal cord and terminate in lateral parts of the ventral horn, with terminal fields that are restricted to only a few spinal cord segments. The major component of this corticospinal pathway is concerned with the voluntary expression of precise, skilled movements involving more distal parts of the limbs.

The corticospinal and corticobulbar tracts

The upper motor neurons in the cerebral cortex reside in several adjacent and highly interconnected areas in the posterior frontal lobe, which together mediate the planning and initiation of complex temporal sequences of voluntary movements. These cortical areas all receive regulatory input from the basal ganglia and cerebellum via relays in the ventrolateral thalamus (see Chapters 18 and 19), as well as inputs from sensory regions of the parietal lobe (see Chapter 12). Although the label "motor cortex" is sometimes used to refer to these frontal areas collectively, the term is more commonly restricted to the primary motor cortex located in the precentral gyrus and the paracentral lobule (Figure 17.2; see the Appendix and Plates 1–4 of the Atlas for annotated photographs of the gyral formations). The primary motor cortex can be distinguished from a complex mosaic of adjacent "premotor" areas both cytoarchitectonically (it is area 4 in Brodmann's nomenclature; see Figure 27.1) and by the low intensity of current necessary to elicit movements by electrical stimulation in this region. The low threshold for eliciting movements is

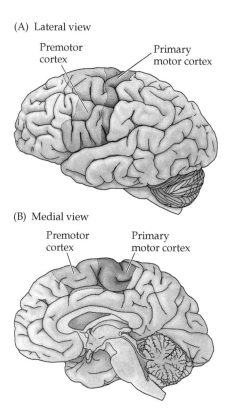

(A) Lateral view

Premotor cortex

Primary motor cortex

(B) Medial view

Premotor cortex

Primary motor cortex

FIGURE 17.2   Primary motor cortex and premotor areas in the human cerebral cortex   Seen here in lateral (A) and medial (B) views, the primary motor cortex is located in the precentral gyrus. The mosaic of premotor areas is more rostral.

an indicator of a relatively large and direct pathway from the primary area to the lower motor neurons of the brainstem and spinal cord.

The pyramidal cells of cortical layer 5 are the upper motor neurons of the primary motor cortex. Among these neurons are the conspicuous Betz cells, which are the largest neurons (by soma size) in the human CNS (Figure 17.3). Although it is often assumed that Betz cells are the principal upper motor neurons of the motor cortex, there are far too few of them to account for the number of axons that project from the motor cortex to the brainstem and spinal cord; indeed, in the human CNS they account for no more than 5% of the axons that project from the motor cortex to the spinal cord. Despite their small numbers, Betz cells play an important role in the activation of lower motor neurons that control muscle activities in the distal extremities. The remaining upper motor neurons are the smaller, non-Betz pyramidal neurons of layer 5 that are found in the primary motor cortex and in each division of the premotor cortex. The axons of these upper motor neurons descend in the **corticobulbar** and **corticospinal tracts**, terms that are used to distinguish axons that terminate in the brainstem ("bulbar" refers to the brainstem) or spinal cord. Along their course, these axons pass through the posterior limb of the internal capsule in the forebrain

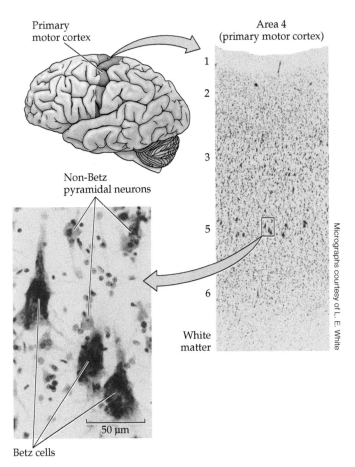

Primary motor cortex

Area 4 (primary motor cortex)

Non-Betz pyramidal neurons

Betz cells

50 μm

Micrographs courtesy of L. E. White

FIGURE 17.3 Cytoarchitectonic appearance of the primary motor cortex in the human brain Histological photomicrographs show Nissl-stained sections that demonstrate cell bodies; note the presence of Betz cells among the pyramidal neurons of cortical layer 5.

The consequence of the bilateral corticobulbar innervation is that damage to the corticobulbar fibers on only one side typically does not result in dramatic deficits in function.

There are three notable exceptions to the pattern of symmetrical, bilateral cortical innervation of the local circuits controlling cranial nerve motor nuclei. For each of these exceptions, corticobulbar inputs to the relevant local circuits arise from both cerebral hemispheres; but there is at least some bias in favor of inputs from the *contralateral* motor cortex. Specifically, the local circuits that organize the output of lower motor neurons in the hypoglossal nucleus (which governs tongue protrusion), the trigeminal motor nucleus (which governs chewing), and the part of the facial motor nucleus that innervates the lower face each receive corticobulbar input primarily from the contralateral motor cortex. The part of the facial motor nucleus that innervates the upper face is supplied more equally by the corticobulbar inputs from the two sides; this is an important clinical point that is also relevant for understanding facial expressions of emotion (Clinical Applications). Essentially, lower facial movements that may be performed unilaterally—such as pushing the tongue against one cheek, biting on one side of the mouth, or raising or lowering one corner of the mouth—are governed primarily by the contralateral motor cortex. Most other motor functions governed by cranial nerve nuclei in which the movements of the two sides are largely concurrent (e.g., vocalization, salivation, tearing, swallowing) are subject to more symmetrical, bilateral upper motor neuronal control.

Near the caudal end of the medulla, nearly all of the fibers in the medullary pyramids are corticospinal axons. Just before entering the spinal cord, about 90% of these axons cross the midline—*decussate*—to enter the lateral columns of the spinal cord on the opposite side, where they form the **lateral corticospinal tract**. The remaining 10% of the pyramidal tract fibers enter the spinal cord without crossing; these axons, which constitute the **ventral (anterior) corticospinal tract**, terminate bilaterally. Collateral branches of these axons cross the midline via the ventral white commissure of the spinal cord to reach the opposite ventral horn. The ventral corticospinal pathway arises primarily from dorsal and medial regions of the motor cortex that serve trunk and proximal limb muscles—the same divisions of the motor cortex that give rise to projections to the reticular formation (see Concept 17.4).

The lateral corticospinal tract forms a direct pathway from the cortex to the spinal cord and terminates primarily in the lateral portions of the ventral horn and intermediate gray matter. Some of these axons (including those derived from Betz cells) synapse directly on α motor neurons that govern the distal extremities (see Figures 17.1 and 17.4). This is particularly the case for corticospinal tract neurons localized to the anterior bank of the central sulcus where the contralateral upper extremity is represented. However, this privileged synaptic contact on lower motor neurons

to enter the cerebral peduncle at the base of the midbrain (Figure 17.4). They then pass through the base of the pons, where they are scattered among the transverse pontine fibers and nuclei of the basal pontine gray matter. They coalesce again on the ventral surface of the medulla, where they form the **medullary pyramids**. The components of this upper motor neuron pathway that innervate cranial nerve nuclei, the reticular formation, and the red nucleus (that is, the corticobulbar tract) leave the pathway at the appropriate levels of the brainstem (see Figure 17.4). There is also a massive corticobulbar projection that terminates among nuclei in the base of the pons that project in turn to the cerebellum; this projection is often called the corticopontine tract and will be discussed in Chapter 19.

Most corticobulbar axons that govern the cranial nerve motor nuclei (see the Appendix) terminate *bilaterally* on local circuit neurons embedded in the brainstem reticular formation (see Concept 17.4), rather than directly on the lower motor neurons in the motor nuclei. These local circuit neurons, in turn, coordinate the output of different groups of lower motor neurons in the cranial nerve motor nuclei.

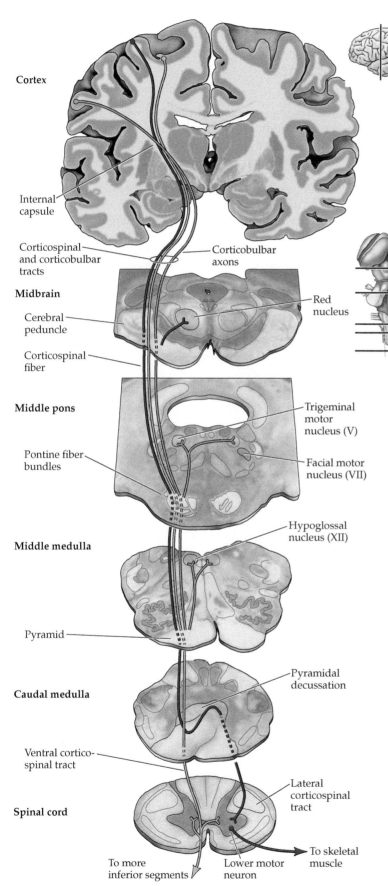

Cortex

Internal capsule

Corticospinal and corticobulbar tracts

Midbrain

Cerebral peduncle

Corticospinal fiber

Middle pons

Pontine fiber bundles

Middle medulla

Pyramid

Caudal medulla

Ventral cortico-spinal tract

Spinal cord

To more inferior segments

Corticobulbar axons

Red nucleus

Trigeminal motor nucleus (V)

Facial motor nucleus (VII)

Hypoglossal nucleus (XII)

Pyramidal decussation

Lateral corticospinal tract

Lower motor neuron

To skeletal muscle

FIGURE 17.4   The corticospinal and corticobulbar tracts Neurons in the motor cortex give rise to axons that travel through the internal capsule and coalesce on the ventral surface of the midbrain, within the cerebral peduncle. These axons continue through the pons and come to lie on the ventral surface of the medulla, giving rise to the medullary pyramids. As they course through the brainstem, cortico-bulbar axons (gold) give rise to bilateral collaterals that innervate brainstem nuclei (only collaterals to the trigeminal motor nuclei and the hypoglossal nuclei are shown). Most of the corticospinal fibers (dark red) cross in the caudal part of the medulla to form the lateral corticospinal tract in the spinal cord. Those axons that do not cross (light red) form the ventral corticospinal tract, which termi-nates bilaterally.

is restricted to a subset of α motor neurons that sup-ply the muscles of the forearm and hand. Most axons of the lateral corticospinal tract, in contrast, terminate among pools of local circuit neurons that coordinate the activities of the lower motor neurons in the lateral cell columns of the ventral horn that innervate different muscles. This difference in terminal distribution im-plies a special role for the lateral corticospinal tract in the control of the hands. Although selective damage to this pathway in humans is rarely seen, evidence from experimental studies in non-human primates indicates that direct projections from the motor cortex to the spi-nal cord are essential for the performance of discrete finger movements. This evidence helps explain the lim-ited recovery in humans after damage to the motor cor-tex or some component of this pathway. Immediately after such an injury, such patients are typically para-lyzed on the affected side. With time, however, some ability to perform voluntary movements reappears. These movements, which are presumably mediated by residual corticospinal inputs and by motor centers in the brainstem, are crude for the most part. The ability to perform fractionated finger movements, such as those required for writing, typing, playing a musical instru-ment, or buttoning clothes, typically remains impaired.

Finally, some components of the corticobulbar and corticospinal projections do not participate directly in upper motor control of lower motor neurons. These com-ponents are derived from layer 5 neurons in somatosen-sory regions of the anterior parietal lobe and terminate among local circuit neurons near the sensory trigeminal nuclei and dorsal column nuclei of the brainstem, and in

■ Clinical Applications

Patterns of Facial Weakness and Their Importance for Localizing Neurological Injury

The signs and symptoms pertinent to the cranial nerves and their nuclei are of special importance to clinicians seeking to pinpoint the neurological lesions that produce motor deficits. An especially instructive example is provided by the muscles of facial expression. It has long been recognized that the distribution of facial weakness provides important localizing clues indicating whether the underlying injury involves lower motor neurons in the facial motor nucleus (and/or their axons in the facial nerve) or the inputs that govern these neurons, which arise from upper motor neurons in the cerebral cortex. Damage to the facial motor nucleus or its nerve affects all the muscles of facial expression on the side of the lesion (lesion C in the figure); this is expected, given the intimate anatomical and functional linkage between lower motor neurons and skeletal muscles. A pattern of impairment that is more difficult to explain accompanies unilateral injury to the motor areas in the lateral frontal lobe (primary motor and premotor cortex), as occurs with strokes that involve the middle cerebral artery (lesion A in the figure). Most individuals with such injuries have difficulty controlling the contralateral muscles around the mouth but retain the ability to symmetrically raise their eyebrows, wrinkle their forehead, and squint.

Until recently, it was assumed that this pattern of inferior facial paresis with superior facial sparing could be attributed to (presumed) bilateral projections from the face representation in the lateral portion of the *primary motor cortex* to the facial motor nucleus; in this conception, the intact ipsilateral corticobulbar projections were considered sufficient to motivate the contractions of the superior muscles of the face. However, recent pathway-tracing studies in non-human primates have suggested a different explanation. These studies demonstrate two important facts that clarify the relations among the face representations in the cerebral cortex and the facial motor nucleus. First, the corticobulbar projections of the primary motor cortex are directed predominantly toward the lateral cell columns in the contralateral facial motor nucleus, which control the movements of the perioral musculature. Thus, the more dorsal cell columns in the facial motor nucleus that innervate superior facial muscles do not receive significant input from the primary motor cortex. Second, these dorsal cell columns are governed by premotor areas in the anterior cingulate gyrus, a cortical region that is associated with emotional processing (see Chapter 32). Therefore, strokes involving the middle cerebral artery spare the superior aspect of the face because the relevant upper motor neurons are in the cingulate gyrus, which is supplied by the anterior cerebral artery.

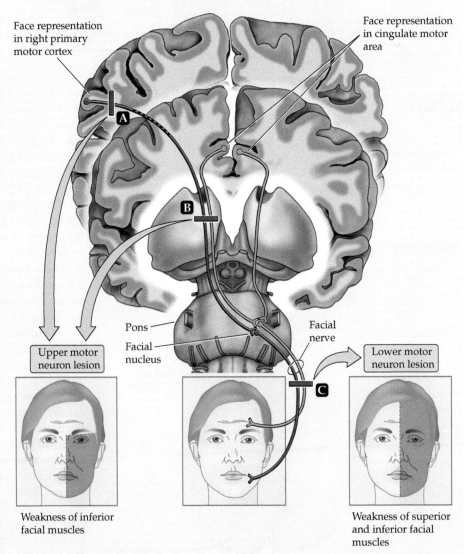

Face representation in right primary motor cortex

Face representation in cingulate motor area

Pons

Facial nucleus

Facial nerve

Upper motor neuron lesion

Lower motor neuron lesion

Weakness of inferior facial muscles

Weakness of superior and inferior facial muscles

Organization of projections from cerebral cortex to the facial motor nucleus and the effects of upper and lower motor neuron lesions

■ Clinical Applications (*continued*)

These studies have resolved an additional puzzle. Strokes involving the anterior cerebral artery or subcortical lesions that interrupt the corticobulbar projection (lesion B in the figure) seldom produce significant paresis of the superior facial muscles. Superior facial sparing in these situations may arise because these *cingulate motor areas* (see Figure 17.9) send descending projections through the corticobulbar pathway that bifurcate and innervate dorsal facial motor cell columns on both sides of the brainstem. Thus, the superior muscles of facial expression are controlled by symmetrical inputs from the cingulate motor areas in both hemispheres. These same cingulate motor areas also provide some measure of corticobulbar innervation to the dorsolateral facial motor nucleus, which governs the upper perioral musculature. This likely explains why individuals with injury to the lateral precentral gyrus (or the portion of the corticobulbar tract originating there) are often still able to express a genuine emotional smile despite voluntary weakness (see Box 33A).

the dorsal horn of the spinal cord. They are likely involved in modulating the transmission of proprioceptive signals and other mechanosensory inputs relevant to sensory perception and the monitoring of body movements. Interestingly, the corticospinal projection to the ventral horn is largest in vertebrates that have the most complex repertoire of fractionated movements with their hands or forepaws. In animals with little ability to execute skilled movements with their forepaws, the corticospinal projection is predominantly directed toward the dorsal horn, where it modulates sensory input to the brain and spinal cord.

CONCEPT
17.2

Neurons in the Primary Motor Cortex Encode Intentions for Body Movements in Central Personal Space

LEARNING OBJECTIVES

17.2.1 Discuss the functional organization of the primary motor cortex and its contributions to the control of volitional movement.

17.2.2 Characterize the representation of the body in the primary motor cortex and compare it to the representation of the body in the primary somatosensory cortex.

17.2.3 Discuss population coding in the primary motor cortex.

Functional organization of the primary motor cortex

Clinical observations and experimental work dating back more than 100 years provide a foundation for understanding the functional organization of the motor cortex. By the end of the nineteenth century, experimental work in animals by the German physiologists G. Theodor Fritsch and Eduard Hitzig had shown that electrical stimulation of the motor cortex elicits contractions of muscles on the contralateral side of the body. Around the same time, the British neurologist John Hughlings Jackson surmised that the motor cortex contains a complete spatial representation, or map, of the body's musculature. Jackson reached this conclusion from his observation that the movements accompanying certain types of epileptic seizures often began locally and proceeded to "march" systematically from one part of the body to another. For instance, partial motor seizures may begin with twitches and other non-purposeful movements of a finger, then involve the entire hand, and progressively affect the forearm, the arm, the shoulder, and finally the face.

This early evidence for motor maps in the cortex was confirmed shortly after the turn of the nineteenth century when the British neurophysiologist Sir Charles Sherrington published his classic maps of the organization of the motor cortex in great apes. These maps were created using focal electrical stimulation applied to the surface of the cortex. During the 1930s, one of Sherrington's students, the renowned neurosurgeon Wilder Penfield, extended this work by demonstrating that the human motor cortex also contains a spatial map of the contralateral body. By correlating the location of muscle contractions with the site of electrical stimulation on the surface of the motor cortex (the same method used by Sherrington), Penfield mapped the motor representation in the precentral gyrus in more than 400 neurosurgical patients. He found that this motor map shows the same general disproportions observed in the somatosensory maps in the postcentral gyrus (see Chapter 12). Thus, the musculature used in tasks requiring fine motor control (such as movements of the face and hands) is represented by a greater area of motor cortex than is the musculature requiring less precise motor control (such as that of the trunk) (Figure 17.5).

The introduction in the 1960s of intracortical microstimulation (a more refined method of cortical activation than cortical surface stimulation) allowed a more detailed understanding of motor maps. Microstimulation entails the delivery of brief electrical currents an order of magnitude smaller than those used by Sherrington and Penfield. By passing the current through the sharpened tip of a metal microelectrode inserted into the cortex, researchers could more focally stimulate the upper motor neurons in layer 5

(A)

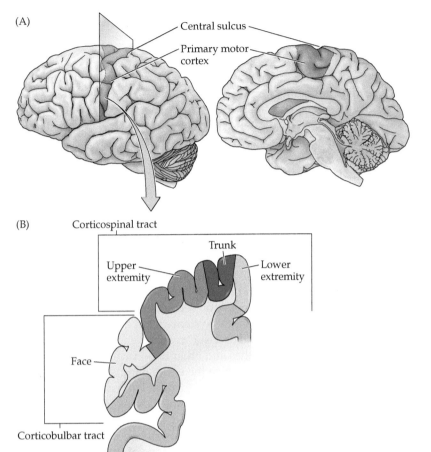

Central sulcus

Primary motor cortex

(B)

Corticospinal tract

Trunk

Upper extremity

Lower extremity

Face

Corticobulbar tract

FIGURE 17.5 Topographic map of movement in the primary motor cortex (A) Location of primary motor cortex in the precentral gyrus. (B) Section along the precentral gyrus, illustrating the somatotopic organization of the motor cortex. In contrast to the precise and detailed representation of the contralateral body in the primary somatosensory cortex (see Figure 12.12), the somatotopy of the primary motor cortex is much more coarse.

that project to lower motor neuron circuitry. Although intracortical stimulation generally confirmed Penfield's spatial map in the motor cortex, it also showed that the finer organization of the map is rather different from what most neuroscientists had imagined. For example, when microstimulation was combined with recordings of muscle electrical activity, even the smallest currents capable of eliciting a response initiated the excitation of several muscles (and the simultaneous relaxation of others), suggesting that organized movements rather than individual muscles are represented in the cortical map (Box 17A). Furthermore, within major subdivisions of the map (e.g., forearm or face regions), a particular movement could be elicited by stimulation of widely separated sites, supporting the argument that neurons in nearby regions are linked by local circuits in the cortex, brainstem or spinal cord to organize specific movements. This interpretation has been supported by the observation that the regions responsible for initiating different movements overlap substantially. The conclusion that movements—or action goals—are encoded in the cortex

(rather than the contractions of individual muscles) also applies to the motor areas of the frontal cortex that control eye movements, where focal stimulation elicits binocular shifts in the direction of gaze (see Chapter 20).

At about the same time that these microstimulation studies were undertaken, Edward Evarts and his colleagues at the National Institutes of Health were pioneering a technique in which implanted microelectrodes were used to record the electrical activity of individual motor neurons in awake, behaving monkeys. The monkeys were trained to perform a variety of motor tasks during the cortical recording, thus providing a means of correlating neuronal activity with voluntary movements. Evarts and his group found that the force generated by contracting muscles changed as a function of the firing rate of upper motor neurons. Moreover, the firing rates of the active neurons often changed *prior* to movements involving very small forces. Evarts therefore proposed that the primary motor cortex contributes to the initial phase of recruitment of the lower motor neurons that generate finely controlled movements. Additional experiments showed that the activity of neurons in the primary motor cortex is correlated not only with the magnitude but also with the direction of the force produced by muscles. Thus, some neurons show progressively less activity as the vector of the movement deviates from the neuron's "preferred direction."

A further advance was made in the 1970s with the introduction of *spike-triggered averaging* (Figure 17.6). By correlating the timing of a single cortical neuron's discharges with the onset times of the contractions generated by the various muscles used in a movement, this method provides an indirect means of measuring the influence of the single neuron on a population of lower motor neurons in the spinal cord. Recording such activity from different muscles as monkeys performed wrist flexion or extension demonstrated that the activity of multiple different muscles is directly facilitated by the discharges of a given upper motor neuron. This peripheral muscle group is referred to as the *muscle field* of the upper motor neuron. On average, the size of the muscle field in the wrist region is two or three muscles per upper motor neuron. These observations, which confirmed that single upper motor neurons contact several lower motor neuron pools, are consistent with the general conclusion that the activity of the upper motor neurons in the cortex controls *movements*, rather than individual muscles.

For the several reasons already discussed, the motor map in the precentral gyrus is much less precise than the somatotopic map in the postcentral gyrus, where the

■ BOX 17A | What Do Motor Maps Represent?

Electrical stimulation studies carried out in human patients by the neurosurgeon Wilder Penfield and his colleagues (and in experimental animals by Sherrington and, later, by Clinton Woolsey and his colleagues) clearly demonstrated a systematic motor map in the precentral gyrus (see text). The fine structure of this map and the nature of its representation, however, have been a continuing source of controversy. Is the map in the motor cortex a map of *musculature* that operates like a "piano keyboard" for the control of individual muscles? Is it a map of *movements*, in which specific sites control multiple muscle groups that contribute to the generation of particular actions? Is it a map of *intentions*, with the goal of the movement having preeminence over the means by which the goal is achieved?

Initial experiments implied that the map in the motor cortex is a fine-scale representation of individual muscles. Thus, stimulation of small regions of the map activated single muscles, suggesting that vertical columns of cells in the motor cortex were responsible for controlling the actions of particular muscles, much as columns in the somatosensory map are thought to analyze stimulus information from particular locations on the surface of the body (see Chapter 12).

More recent studies using anatomical and physiological techniques, however, have shown that the map in the motor cortex is far more complex than a columnar representation of particular muscles. Individual pyramidal tract axons are now known to terminate on sets of spinal motor neurons that innervate different muscles. This relationship is evident even for neurons in the hand representation of the motor cortex, the region that controls the most discrete, fractionated movements.

Furthermore, cortical microstimulation experiments have shown that contraction of a single muscle can be evoked by stimulation over a wide region of the motor cortex (about 2 to 3 mm in macaque monkeys) in a complex, mosaic fashion. It seems likely that horizontal connections within the motor cortex and local circuits in the spinal cord create ensembles of neurons that coordinate the pattern of firing in the population of ventral horn cells that ultimately generate a given movement.

Thus, while the somatotopic maps in the motor cortex generated by early studies are correct in their overall topography, the fine structure of the map is far more abstract. As discussed in the text, it is now widely accepted that the functional maps in the primary motor and premotor cortex are maps of movement. Although coarse somatotopy provides one means of understanding the organization of these motor maps (see Figure 17.5), Michael Graziano and his colleagues at Princeton University have proposed another scheme. Their microstimulation studies of awake, behaving monkeys suggest that the topographic representations of movement in the motor cortex are organized around ethologically relevant categories of motor behavior.

For example, microstimulation of sites in the arm region of the primary motor cortex often invoked movements of the arm that brought the monkey's hand into central space, where the animal might visually inspect and manipulate a held object (see figure; see also Figure 17.7B). Stimulation of more lateral regions (toward the face representation) often led to hand-to-mouth motions and mouth opening, whereas more medial stimulation sites (toward the trunk and leg representations) evoked climbing- or leaping-like postures. These observations suggest that the posterior regions of the motor cortex, including the primary motor cortex, are most concerned with manual and oral behaviors that occur in central personal space. Just anterior to this cortical region (in the premotor cortex) are sites that, when stimulated, elicit reaching motions and other outward arm movements directed away from the body. Other anterior sites may evoke coordinated, defensive postures as well, perhaps reflecting the integration of threatening sensory signals derived from extrapersonal space. New studies of the mirror motor system raise the intriguing possibility that what is actually represented in the motor cortex is the intention of movement or action goal, rather than movement per se.

Obviously, these are exciting times for investigators studying the cortical governance of movement. Unraveling the details of what is represented in motor maps still holds the key to understanding how patterns of activity in the primate motor cortex generate the rich repertoire of volitional movement.

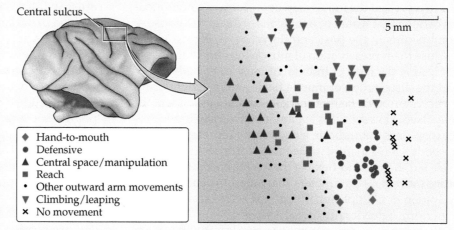

Central sulcus

♦ Hand-to-mouth
● Defensive
▲ Central space/manipulation
■ Reach
• Other outward arm movements
▼ Climbing/leaping
× No movement

5 mm

Topographic distribution of microstimulation sites that evoke behaviorally relevant movements in a macaque monkey. The rectangular region on the brain (left) shows the portion of the motor cortex under investigation. The shaded region in the map of stimulation sites indicates cortex folded into the anterior bank of the central sulcus. (After M. S. Graziano et al., 2005. *J Neurophysiol* 94: 4209–4223.)

(A) Detection of postspike facilitation

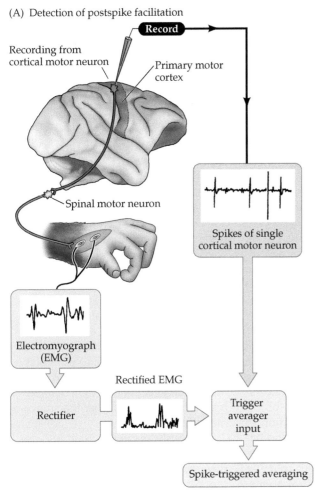

(B) Postspike facilitation by cortical motor neuron

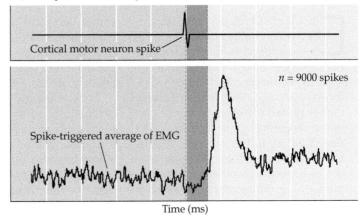

Time (ms)

FIGURE 17.6 The influence of single cortical upper motor neurons on muscle activity (A) Diagram illustrating the spike-triggered averaging method for correlating muscle activity with the discharges of single upper motor neurons. (B) The response of a thumb muscle (bottom trace) follows by a fixed latency the single spike discharge of a pyramidal tract neuron (top trace). This technique can be used to determine all the muscles that are influenced by a given motor neuron. (After R. Porter and R. Lemon, 1993. *Corticospinal Function and Voluntary Movement*. Oxford: Oxford University Press. © 1993 Oxford University Press.)

receptive fields of adjacent cortical neurons overlap in a smooth and continuous progression across the surface of the primary somatosensory cortex. Indeed, it is problematic to represent the motor map in the form of a homunculus cartoon that would be analogous to the somatosensory homunculus in the postcentral gyrus (see Figure 12.12), because the representation of muscle movement is not organized at the level of individual muscles or body parts, and the distribution of muscle fields among neighboring cortical neurons is neither spatially continuous nor temporally fixed. However, this apparent imprecision in the motor map does not indicate a degenerate representation of the body's musculature in the motor cortex. Rather, it suggests a dynamic and flexible means for encoding higher order movement parameters that entail the coordinated activation of multiple muscle groups across several joints to perform behaviorally useful actions.

This principle of upper motor neural control has been demonstrated by Michael Graziano and his colleagues at Princeton University, who extended the duration of cortical microstimulation in behaving monkeys to a timescale that more closely corresponds to the duration of volitional movements (from hundreds of milliseconds to several

seconds). When such stimuli are applied to the precentral gyrus, the resulting movements are sequentially distributed across multiple joints and are strikingly purposeful (Figure 17.7). Examples of motor patterns frequently elicited with prolonged microstimulation of the precentral gyrus are movements of the hand to the mouth as if to feed, movements that bring the hand to central space as if to inspect an object of interest, and defensive postures as if to protect the body from an impending collision. These findings reinforce the current view that purposeful movements are organized by the circuitry of the primary motor cortex and that their somatotopic organization is best understood in the context of ethologically relevant behaviors (see Box 17A and Concept 17.3).

Finally, the commands to perform precise movement patterns are encoded in the activity of a large population of upper motor neurons integrated by intracortical circuitry. One well-studied paradigm for exploring the nature of this "population code" involves recording from cortical neurons during visually guided reaching movements of the arm and hand. Using this paradigm, the direction of arm movements in monkeys could be predicted by calculating a "neuronal population vector" derived simultaneously from the discharges of a population of upper motor neurons that are "broadly tuned" in the sense that each neuron discharges prior to movements in many directions (Figure 17.8). These observations showed that the discharges of individual upper motor neurons cannot specify the direction of an arm movement, simply because they are tuned too broadly (likely reflecting the summed tuning of inputs from other upper motor neurons). Rather, each arm

FIGURE 17.7 Purposeful movements of the contralateral arm and hand in a macaque monkey Prolonged microstimulation of primary motor cortex sites near the middle of the precentral gyrus elicits coordinated movements of the hand and mouth (A) or movements of the arm that bring the hand toward central space, as if to visually inspect and manipulate a held object (B). The starting positions of the contralateral hand are indicated by the blue crosses, the elicited movements are illustrated with the curved black lines, and the final positions of the hand at the end of microstimulation are indicated by the red dots. (After M. S. Graziano et al., 2005. *J Neurophysiol* 94: 4209–4223.)

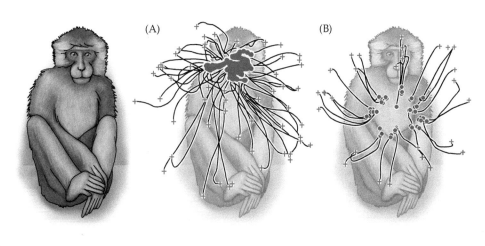

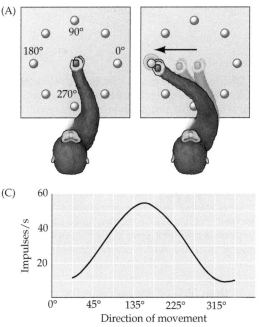

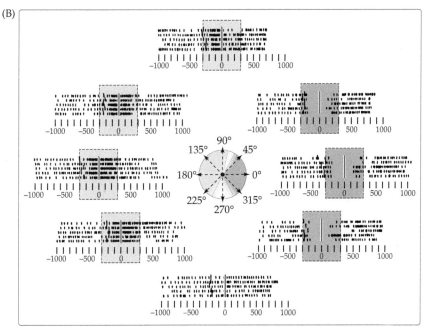

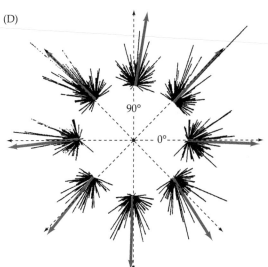

FIGURE 17.8 Directional tuning of an upper motor neuron in the primary motor cortex (A) A monkey is trained to move a joystick in the direction indicated by a light. (B) The activity of a single neuron was recorded during arm movements in each of eight different directions (0 indicates the time of movement onset; each short vertical line in this raster plot represents an action potential). The activity of the neuron increased before movements between 90° and 225° (yellow zone) but decreased in anticipation of movements between 45° and 315° (blue zone). (C) Plot showing that the neuron's discharge rate was greatest before movements in a particular direction, which defines the neuron's "preferred direction" in this experimental paradigm. (D) The black lines indicate the discharge rates of individual upper motor neurons prior to each direction of movement. By combining the responses of all the neurons in the recording session, a "population vector" (red arrows) can be derived that represents the movement direction encoded by the simultaneous activity of the entire population of recorded units. (After A. P. Georgeopoulos, et al., 1986. *Science* 233: 1416–1419.)

movement must be encoded by the concurrent discharges of a large population of such functionally linked neurons. The fact that the same site in the primary motor cortex can encode different trajectories of motion depending on the starting position of the limb (see Figure 17.7) suggests that multiple parameters of movement may be selected by the relevant ensemble of upper motor neurons to achieve a behaviorally useful action. Thus, microstimulation experiments use exogenous electrical currents to engage populations of upper motor neurons whose output encodes not simply the trajectory of arm motion, but also the final position of the hand in the context of an action goal.

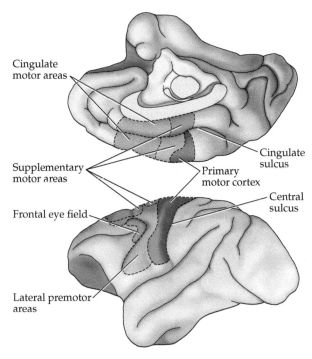

FIGURE 17.9   Divisions of the motor cortex in the macaque monkey brain As in humans, the primary motor cortex resides in the anterior bank of the central sulcus. Anterior to this region is a complex mosaic of premotor areas that extends from the frontal operculum on the lateral surface of the frontal lobe to the cingulate gyrus on the medial hemispheric surface. The lateral premotor and supplementary motor areas are involved in selecting and organizing purposeful movements of the limbs and face; the frontal eye fields organize voluntary gaze shifts (see Chapter 20), and the cingulate motor areas are involved in the expression of emotional somatic behavior (see Chapter 32). Current evidence supports the existence of comparable premotor areas in the human motor cortex. (After S. Geyer et al., 2000. *Anat Embryol* 202: 443–474.)

CONCEPT 17.3   Neurons in the Premotor Cortex Encode Intentions for Body Movements that Are Oriented toward Extrapersonal Space

LEARNING OBJECTIVES

17.3.1   Discuss the organization of the premotor cortex and the contributions of its lateral and medial divisions to the control of volitional movement.

17.3.2   Discuss the neurophysiological properties of mirror motor neurons and how their activities may contribute to action understanding and imitation learning.

The premotor cortex

A complex mosaic of interconnected frontal lobe areas that lie rostral to the primary motor cortex also contributes to motor functions (see Figure 17.2). This functional division of the motor cortex includes Brodmann's areas 6, 8, and 44/45 on the lateral surface of the frontal lobe and parts of areas 23 and 24 on the medial surface of the hemisphere. Although the organization of this premotor mosaic is best understood in macaque monkeys (Figure 17.9), recent functional brain imaging studies, as well as structural brain imaging studies in patients with frontal lobe injuries, suggest that a comparable distribution of premotor areas is present in humans. Each of the divisions of the premotor cortex receives extensive multisensory input from regions of the inferior and superior parietal lobules, as well as more complex signals related to motivation and intention from the rostral ("prefrontal") divisions of the frontal lobe. The upper motor neurons in this premotor cortex influence motor behavior both indirectly, through extensive reciprocal connections with the primary motor cortex, and directly, via axons that project through the corticobulbar and corticospinal pathways to influence the local circuits that organize the output of lower motor neurons in the brainstem and spinal cord.

Indeed, more than 30% of the axons in the corticospinal tract arise from neurons in the premotor cortex. Thus, past arguments that the premotor cortex occupies a higher position in a cortical hierarchy of motor control by operating through feedforward signals to the primary motor cortex are no longer tenable. Rather, a variety of experiments indicate that the premotor cortex uses information from other cortical regions to select movements appropriate to the context and goal of the action (see Chapter 33). The principal difference between the premotor cortex and primary motor cortex lies in the strength of their connections to lower motor neurons, with more upper motor neurons in the primary motor cortex making monosynaptic connections to α motor neurons, especially those in the ventral horn of the cervical spinal cord that control precise movements of the distal upper extremities. Recent evidence suggests that other differences may reflect the mapping of purposeful movements relative to personal and extrapersonal space and the nature of the signals that lead to the initiation of motor commands in the context of action goals. The action goals encoded by the primary motor cortex tend to be localized to personal space (within arm's length), while the action goals encoded by premotor cortex are more typically oriented toward extrapersonal space (beyond arm's length; see Box 17A). In the exciting new field of neuroengineering, the neural engrams of such action goals are now serving to drive machines and a variety of computerized systems through so-called brain–machine or brain-computer interfaces (Box 17B).

■ BOX 17B │ Minds and Machines

Science fiction has long imagined the melding of the human mind and mechanical or digital machines that would enact our thoughts without the obligatory actions of our evolved musculoskeletal effectors—that is, our physical bodies. In recent years, a consortium of neuroscientists, computer scientists, material scientists, and electrical, mechanical, and biomedical engineers have boldly envisioned the means for realizing what was once mere fantasy. In research laboratories and in some neurorehabilitation clinics around the world, such scientists are teaming up with neurologists, neurosurgeons, and physical therapists to translate *brain–machine interface* technology to clinical practice in the hopes of restoring function lost to neurological injury and disease.

(Continued)

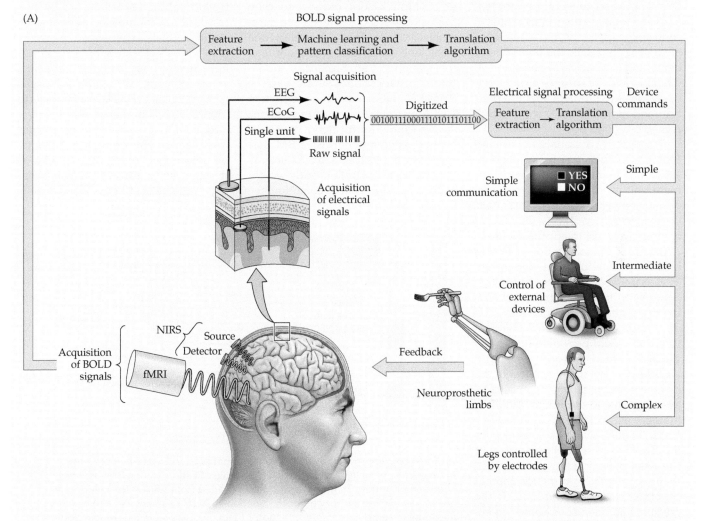

(A) General design of a brain–machine interface (BMI) system based on the invasive or noninvasive acquisition of brain-derived signals. Artificial neural networks decode brain activity and generate control signals that drive relatively simple, intermediate, or complex BMI systems. Visual, proprioceptive, or haptic feedback is provided to enhance brain control of BMI performance. (BOLD, blood oxygenation level–dependent; ECoG, electrocorticography; NIRS, near-infrared spectroscopy.) (After E. C. Leuthardt, Washington University School of Medicine.)

■ BOX 17B | Minds and Machines (continued)

Brain–machine interface (BMI; also known as brain–computer interface, BCI) refers to the systems and technologies that enable thought-controlled operation of virtual or real actuators for communication, movement, and the remote operation of a variety of computer-based systems to enable activities of daily living. The basic design of BMI systems involves (1) the acquisition of brain-generated signals that reflect information processing and the encoding of action goals; (2) the processing and decoding of brain signals using artificial neural networks to extract salient features and translate them into pragmatic control signals; (3) the implementation of control signals for the operation of digital and mechanical systems; and (4) the generation of sensory-based feedback signals to promote adaptive plasticity and improved brain control of BMI technology (Figure A).

Brain-derived signals for driving BMI technologies may be sampled invasively by methods such as single- or multi-unit recording of neuronal action potentials, local field potentials, or electrocorticography (ECoG); or noninvasively, using approaches such as electroencephalography (EEG) and blood oxygenation level–dependent (BOLD) fMRI or near-infrared spectroscopy (NIRS). Invasive means for acquiring brain-generated signals have the advantage of dense information content that would provide high-fidelity signals for processing and decoding in artificial neural networks, but with the obvious disadvantage of neurosurgical intervention and the risk of attending medical and postsurgical complications. Noninvasive means for acquiring brain-generated signals obviate the risks of neurosurgery; however, until recently the information content of signals recorded from outside the cranium had been considered too impoverished to be useful for driving BMI technologies. Improvements in signal processing and the performance of artificial neural networks have proved sufficient for the implementation of

BMI systems based on brain-generated signals acquired noninvasively.

One striking demonstration of the promise of such BMI systems was showcased in the opening ceremony for the 2014 FIFA (Fédération Internationale de Football Association) World Cup in São Paulo, Brazil. In a fleeting moment—amidst the pageantry and spectacle of the global celebration of the "beautiful game"—a 29-year-old Brazilian man who had suffered a complete spinal cord injury in the upper thoracic region 6 years earlier executed a simple kicking motion, sending the ball toward the referee in a ceremonial first kick. That was indeed "one small step for [a] man, one giant leap for mankind" (reminiscent of the interdisciplinary science supporting the Apollo 11 mission and the moment of Neil Armstrong's famous declaration), as this simple motor action was performed using a noninvasive BMI system driving a wearable exoskeleton to enable brain-controlled body-weight support, posture, and locomotion.

Quite unexpectedly, several members of the cohort of individuals who were training for this groundbreaking public demonstration of BMI technology experienced neurological improvement and some measure of clinically significant functional recovery. Eight individuals who were 3 to 13 years post spinal cord injury underwent 12 months of training with a multistage, BMI-based gait neurorehabilitation program that entailed immersive virtual reality training, enriched visual-tactile feedback, and extensive training with an EEG-controlled robotic exoskeleton. By the conclusion of the training period, all of the individuals experienced improvements in somatosensation across multiple dermatomes, and most regained some measure of voluntary muscle contraction below the level of the injury. Moreover, half of the individuals were upgraded to an incomplete paraplegia classification (Figure B). It remains to be determined which components of this complex, intensive training paradigm were most effective in promoting

(B)

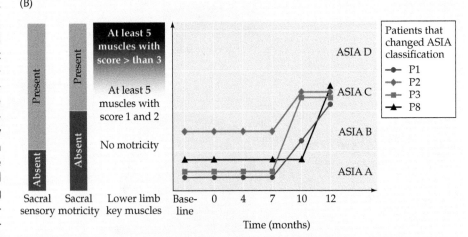

(B) Functional improvements in four of eight patients, indicated by upgrades in American Spinal Injury Association (ASIA) classification, during BMI-assisted neurorehabilitation. Three patients were upgraded from ASIA A (complete spinal cord injury with no sensory or motor function in sacral segments S4–S5) to ASIA C (incomplete spinal cord injury with motor function observable below neurologic level, with majority of affected muscles graded < 3 of 5). One patient was upgraded from ASIA B (incomplete spinal cord injury with sensory, but not motor, function preserved below neurologic level) to ASIA C. Four other patients (not shown) graded ASIA A showed functional gains in somatosensation and motor control but did not change ASIA classification. (After A. Donati et al., 2016. *Sci Rep* 6: 30383. CC BY 4.0.)

improved neurological function. Likewise, it is not known if these individuals have achieved a recovery "ceiling," or if ongoing BMI-assisted neurorehabilitation might promote even further functional gains years after injury.

The neurobiological mechanisms that underlie these functional improvements remain a matter of speculation. Perhaps some individuals classified as having complete spinal cord injury retain some latent corticospinal connections that may be "awakened" through long-term potentiation and synaptic sprouting (see Chapter 8) during intensive neurorehabilitation with virtual and real, brain-controlled BMI technologies. Ongoing synaptic and circuit plasticity at the cortical and spinal levels may further consolidate functional gains and promote ongoing adaptation to BMI-assisted neurorehabilitation.

Such dramatic demonstrations notwithstanding, BMI technology remains in its infancy, and the barriers to greater efficacy and widespread implementation are daunting. However, the pace of advancement in the multidisciplinary domains of science and technology that support BMI systems holds great promise for a future where minds and machines seamlessly integrate thought, feeling, and action.

The functions of the premotor cortex may be understood in terms of differences between the lateral and medial components of this region. As many as 65% of the neurons in the lateral premotor cortex have responses that are linked in time to the occurrence of movements; as in the primary motor area, many of these cells fire most strongly before and during movements made in a specific direction. However, these neurons are especially important in conditional ("closed-loop") motor tasks. For example, when a monkey is trained to reach in different directions depending on the nature of a visual cue, the appropriately tuned lateral premotor neurons begin to fire at the appearance of the cue, well before the monkey receives a signal to actually make the movement. As the animal learns to associate a new visual cue with the movement, appropriately tuned neurons begin to increase their rate of discharge in the interval between the cue and the onset of the signal to perform the movement. Rather than directly commanding the initiation of a movement, these neurons appear to encode the monkey's *intention* to perform a particular movement; thus, they seem to be particularly involved in the selection of movements based on external signals or events.

A ventrolateral subdivision of the premotor cortex has received considerable attention in recent years, after the discovery that a subset of its neurons responds not just in preparation for the execution of particular movements, such as a precision grip to retrieve a morsel of food, but also when the same action is *observed*, being performed by another individual (monkey or human). For example, these premotor neurons fire action potentials when a monkey observes the hand of a human trainer engaging in the same or similar action that would activate these same neurons during self-initiated movements (Figure 17.10). However, these so-called **mirror motor neurons** respond much less well when the same actions are pantomimed without the explicit presence of an action goal, such as an object to be grasped. Furthermore, they respond during the observation of goal-directed behavior even when the final stage of the action is hidden from view—for example, the grasping of an object known to the monkey to have been placed behind a small barrier. Recent studies have demonstrated that some mirror motor neurons show suppression of firing during action observation, even if the same neurons fire during action execution. Such neuronal activities may contribute to the suppression of imitation. Taken together, these findings suggest that the mirror motor system is involved in encoding the intention to make or suppress a specific movement based on the observation of the behaviorally relevant actions of others. Evidently, this system participates in an extended network of parietal and frontal regions that subserve action understanding and imitation learning, whether or not observed behavior is "mirrored" in one's own actions (Figure 17.11). The functions of the mirror motor system are among the most actively studied and debated domains of motor and cognitive neuroscience, but the full scope of this system's contributions to motor control, motor learning, and more complex brain functions such as social communication, language, theory of mind, and empathy remains to be elucidated.

Further evidence that the lateral premotor area is concerned with movement selection comes from the effects of cortical damage on motor behavior. Lesions in this region severely impair the ability of monkeys to perform visually cued conditional tasks, even though they can still respond to the visual stimulus and can perform the same movement in a different setting. Similarly, people with frontal lobe damage have difficulty learning to select a particular movement to be performed in response to a visual cue, even though they understand the instructions and can perform the movements. Individuals with lesions in the premotor cortex may also have difficulty performing movements in response to verbal commands.

Finally, a rostral division of the lateral premotor cortex in the human brain, especially in the left hemisphere, has evolved to play a special role in the production of speech sounds. This region, called **Broca's area** (which typically corresponds to Brodmann's areas 44 and 45, but may be localized to adjacent area 6 in some individuals), is critical for the production of speech and will be considered in detail in Chapter 31. The evolution of this premotor division in primates and its functional relation to semantic processing regions in the parietal and temporal lobe are areas of active investigation.

The medial division of the premotor cortex extends onto the medial aspect of the frontal lobe (including a division that has been referred to as the "supplementary motor area"). Like the lateral area, the medial premotor cortex mediates the selection of movements. However, this region appears to be specialized for initiating movements specified by *internal* rather than *external* cues ("open-loop" conditions). In contrast to lesions in the lateral premotor area, removal of the medial premotor area in a monkey reduces the number of self-initiated or "spontaneous" movements the animal makes, whereas the ability to execute movements in response to external cues remains largely intact. Imaging studies suggest that this cortical region in humans functions in much the same way. For example, functional brain imaging studies show that the medial region of the premotor cortex is activated when individuals perform motor sequences from memory (i.e., without

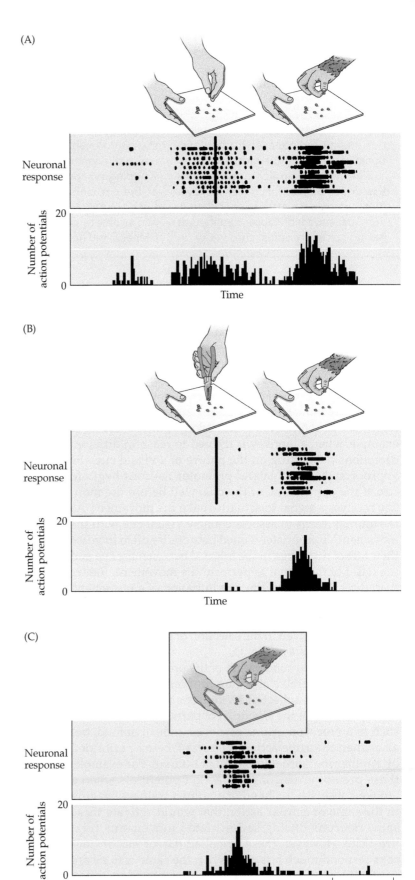

FIGURE 17.10 **Mirror motor neuron activity in a ventral sector of the lateral premotor cortex** In the panels, the upper graphics illustrate the monkey's view of the hand of the trainer placing a food morsel on a tray and the monkey's own hand extending to retrieve the morsel. The middle graphics illustrate raster plots that show the firing of the neuron relative to the observed and executed movements (each tick mark indicates an action potential, and each row represents one trial). The lower graphs are peristimulus response histograms aligned to the overlying raster plots. The mirror motor neuron fires during the passive observation of a human hand placing the morsel of food on the tray (A), as well as during the execution of a similar action to retrieve the food. (The vertical line in the raster plots indicates the time at which the food was placed on the tray; 1 to 2 s later, the monkey reaches to retrieve the morsel.) The same neuron does not respond when the food is placed with the aid of pliers (B), but it does fire during the monkey's reaching and retrieval movements when the monkey is allowed to observe its reach (B) and when the behavior is executed behind a barrier (C). These findings suggest that this division of the premotor cortex plays a role in encoding the observed actions of others. (After G. Rizzolatti et al., 1996. *Cogn Brain Res* 3: 131–141.)

(A)

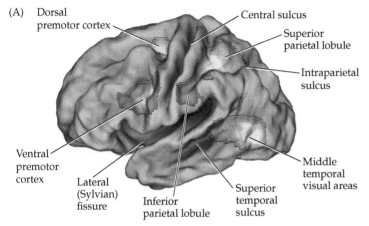

(B)

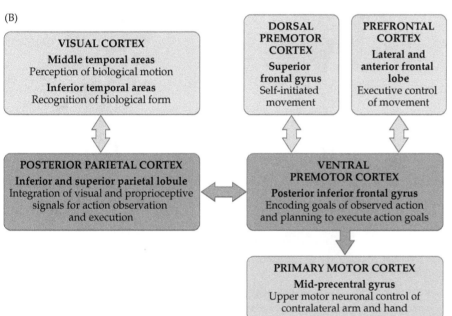

FIGURE 17.11 Cortical network for visually guided reach-to-grasp (A) Illustration of significant activation in humans during observation of reach-to-grasp. These same cortical areas are recruited during execution of reach-to-grasp with the right arm and hand, with the addition of activation in the mid-precentral gyrus (arm and hand region of the primary and premotor motor cortices). (B) Schematic summary of the cortical network for visually guided reach-to-grasp. Areas enriched in neurons with mirroring properties are colored red (mirror motor neurons are also found in other areas not illustrated). This network is proposed to encode the action goal of observed or executed reach-to-grasp. (A from G. Rizzolatti and C. Sinigaglia, 2016. *Cogn Brain Res* 3: 131–141 and S. Caspers et al., 2010. *Neuroimage* 50: 1148–1167.)

relying on an external instruction). In accord with this evidence, single-unit recordings in monkeys indicate that many neurons in the medial premotor cortex begin to discharge 1 to 2 s before the onset of a self-initiated movement. Among the areas of the medial premotor cortex are two divisions that will be considered in more detail elsewhere: a frontal eye field (see Figure 17.9) involved in directing visual gaze toward a location of interest (see also Chapter 20); and a set of areas in the depths of the cingulate sulcus (see Figure 17.9 and Clinical Applications) that plays a role in the expression of emotional behavior (see also Chapter 32).

In summary, both the lateral and medial areas of the premotor cortex are intimately involved in selecting a specific movement or sequence of movements from the repertoire of possible behaviorally relevant actions. The functions of the areas differ, however, in the relative contributions of external and internal cues to the selection process.

CONCEPT
17.4

Upper Motor Neurons in the Brainstem Help Maintain Balance, Govern Posture, Initiate Locomotion, and Orient Visual Gaze

LEARNING OBJECTIVES

17.4.1 Discuss the neural centers that give rise to medial descending projections from the brainstem to lower motor neurons.

17.4.2 Characterize the brainstem reticular formation in anatomical and functional terms.

17.4.3 Differentiate the motor control functions of spinal projections that arise from the vestibular complex from those that arise from the reticular formation.

17.4.4 Discuss the direct and indirect means by which upper motor neurons in the motor cortex influence spinal cord circuits.

Motor control centers in the brainstem

Several structures in the brainstem contain circuits of upper motor neurons whose activities serve to organize a variety of somatic movements involving the axial musculature of the trunk and the proximal musculature of the limbs. These movements include the maintenance of balance, the regulation of posture, the initiation and regulation of locomotion, and the orientation of visual gaze. They are governed by upper motor neurons in the nuclei of the vestibular complex, the reticular formation, and the superior colliculus (Figure 17.12). Such movements are usually necessary to support the expression of skilled motor behaviors involving the more distal parts of the extremities or, in the case of visual gaze, when attention is directed toward a particular sensory stimulus. Indeed, the relevant brainstem circuits are competent to direct many motor activities without supervision by higher motor centers in the cerebral cortex. However, these brainstem motor centers usually work in concert with divisions of the motor cortex that organize volitional movements, which always entail both skilled (voluntary) and supporting (reflexive) motor activities.

As described in Chapter 11, the vestibular nuclei are the major destination of the axons that form the vestibular division of the eighth cranial nerve; as such, they receive sensory information from the semicircular canals and the otolith organs that specifies the position of the head and its rotational and translational movements. Many of the cells in the vestibular nuclei that receive this information are upper motor neurons with descending axons that terminate in the medial region of the spinal cord gray matter, although some extend more laterally to contact the neurons that control the proximal muscles of the limbs. The projections from the vestibular nuclei that control axial muscles and those that influence proximal limb muscles originate from different cells and take somewhat different routes to the spinal cord (see Figure 17.12A).

Neurons in the medial vestibular nucleus give rise to a **medial vestibulospinal tract** that terminates bilaterally in the medial ventral horn, mainly in the cervical cord. There, the medial vestibulospinal tract regulates head position by reflex activation of neck muscles in response to the stimulation of the anterior semicircular canals resulting

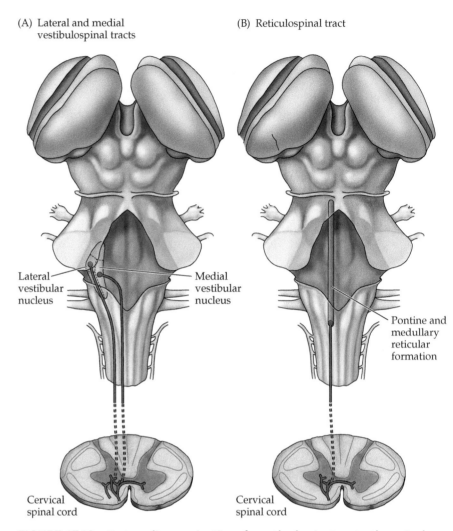

(A) Lateral and medial vestibulospinal tracts

(B) Reticulospinal tract

Lateral vestibular nucleus

Medial vestibular nucleus

Pontine and medullary reticular formation

Cervical spinal cord

Cervical spinal cord

FIGURE 17.12 **Descending projections from the brainstem to the spinal cord** Pathways that influence motor neurons in the medial part of the ventral horn originate in the vestibular nuclei (A) and the reticular formation (B).

from unexpected rapid, downward rotation of the head. For example, when an individual falls forward, the medial vestibulospinal tract mediates reflexive dorsiflexion of the neck as well as extension of the arms in an attempt to protect the upper body from injury. Neurons in the lateral vestibular nucleus are the source of the **lateral vestibulospinal tract**, which courses through the anterior white matter of the spinal cord in a slightly more lateral position relative to the medial vestibulospinal tract. Despite the modifier in its name, the lateral vestibulospinal tract terminates ipsilaterally among medial lower motor neuron pools that govern proximal muscles of the limbs. As discussed in more detail in Chapter 11, this tract facilitates the activation of limb extensor (antigravity) muscles when the otolith organs signal deviations from stable balance and upright posture. Other upper motor neurons in the vestibular nuclei project to local circuit neurons and lower motor neurons in the cranial nerve nuclei that control eye

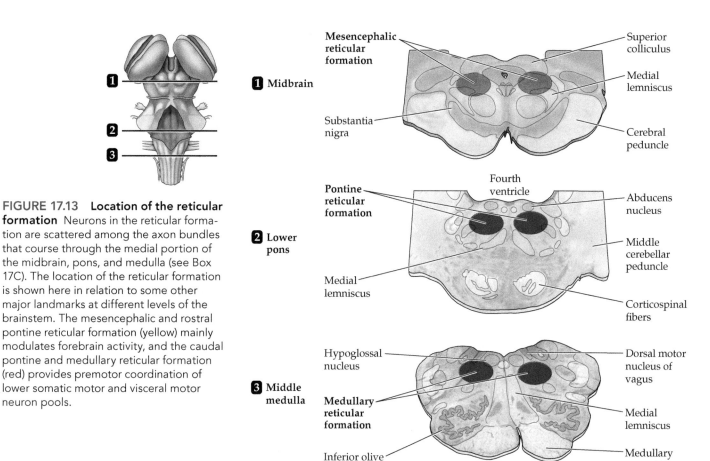

1 Midbrain

Mesencephalic reticular formation

Superior colliculus

Medial lemniscus

Substantia nigra

Cerebral peduncle

2 Lower pons

Fourth ventricle

Pontine reticular formation

Abducens nucleus

Middle cerebellar peduncle

Medial lemniscus

Corticospinal fibers

3 Middle medulla

Hypoglossal nucleus

Medullary reticular formation

Dorsal motor nucleus of vagus

Medial lemniscus

Inferior olive

Medullary pyramid

FIGURE 17.13 **Location of the reticular formation** Neurons in the reticular formation are scattered among the axon bundles that course through the medial portion of the midbrain, pons, and medulla (see Box 17C). The location of the reticular formation is shown here in relation to some other major landmarks at different levels of the brainstem. The mesencephalic and rostral pontine reticular formation (yellow) mainly modulates forebrain activity, and the caudal pontine and medullary reticular formation (red) provides premotor coordination of lower somatic motor and visceral motor neuron pools.

movements (the third, fourth, and sixth cranial nerve nuclei). This pathway produces the eye movements that maintain fixation while the head is moving (the vestibulo-ocular reflex; see Chapters 11 and 20).

The **reticular formation** is a complicated network of circuits in the core of the brainstem that extends from the rostral midbrain to the caudal medulla; it is similar in structure and function to the local circuitry in the intermediate gray matter of the spinal cord (Figure 17.13 and Box 17C). Unlike the well-defined sensory and motor nuclei of the cranial nerves, the reticular formation comprises numerous clusters of neurons scattered among a welter of interdigitating axon bundles; it is therefore difficult to subdivide anatomically. The neurons within the reticular formation serve a disparate variety of functions, including cardiovascular and respiratory control (see Chapter 21), governance of myriad sensorimotor reflexes (see Chapters 16 and 21), coordination of eye movements (see Chapter 20), regulation of sleep and wakefulness (see Chapter 28), and most important for the purpose of this discussion, the temporal and spatial coordination of limb and trunk movements, particularly those that control rhythmic, stereotypical behaviors such as locomotion. The descending motor control pathways

from the reticular formation to the spinal cord are similar to those of the vestibular nuclei; they terminate primarily in the medial parts of the gray matter, where they influence the local circuit neurons that coordinate axial and proximal limb muscles (see Figure 17.12B). With few exceptions, reticulospinal projections are distributed bilaterally to the medial ventral horns.

Both the vestibular nuclei and the reticular formation provide information to the spinal cord that maintains posture in response to environmental (or self-induced) disturbances of body position and stability. Direct projections from the vestibular nuclei to the spinal cord ensure a rapid compensatory *feedback* response to any postural instability detected by the vestibular labyrinth (see Chapter 11). In contrast, the motor circuits in the reticular formation are controlled largely by motor centers in the cerebral cortex, amygdala, hypothalamus, or brainstem. The relevant neurons in the reticular formation initiate *feedforward* adjustments that stabilize posture during ongoing movements.

The way neurons of the reticular formation maintain posture can be appreciated by analyzing their activity during voluntary movements. Even the simplest movements are accompanied by the activation of muscles that at first glance seem to have little to do with the primary

■ BOX 17C | The Reticular Formation

If one were to exclude from the structure of the brainstem the cranial nerve nuclei, the nuclei that provide input to the cerebellum, the long ascending and descending tracts that convey explicit sensory and motor signals, and the structures that lie dorsal and lateral to the ventricular system, what would be left is a central core region known as the *tegmentum* (Latin, "covering structure"), so named because it "covers" the ventral part of the brainstem. Scattered among the diffuse fibers that course through the tegmentum are small clusters of neurons that are collectively known as the reticular formation. With few exceptions, these clusters of neurons are difficult to recognize as distinct nuclei in standard histological preparations. Indeed, the modifying term *reticular* ("netlike") was applied to this loose collection of neuronal clusters because early histologists envisioned these neurons as part of a sparse network of diffusely connected cells that extends from the intermediate gray regions of the cervical spinal cord to the lateral regions of the hypothalamus and certain nuclei along the midline of the thalamus.

These early anatomical concepts were influenced by lesion experiments in animals and clinical observations in human patients made in the 1930s and 1940s. These studies showed that damage to the upper brainstem tegmentum produced coma, suggesting the existence of a neural system in the midbrain and rostral pons that supported typical conscious brain states and transitions between sleep and wakefulness. These ideas were articulated most influentially by G. Moruzzi and H. Magoun when they proposed a "reticular activating system" to account for these functions and the critical role of the brainstem reticular formation.

Current evidence generally supports the notion of an activating function of the rostral reticular formation; however, neuroscientists now recognize the complex interplay of a variety

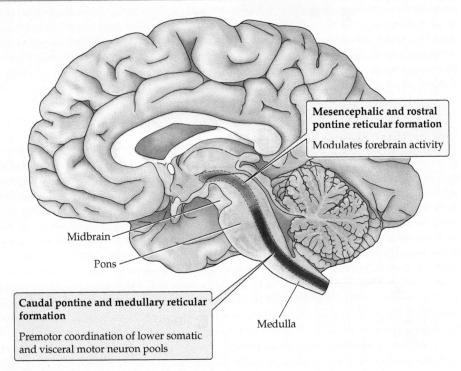

Mesencephalic and rostral pontine reticular formation

Modulates forebrain activity

Caudal pontine and medullary reticular formation

Premotor coordination of lower somatic and visceral motor neuron pools

Midbrain

Pons

Medulla

Midsagittal view of the brain showing the longitudinal extent of the reticular formation and highlighting the broad functional roles performed by neuronal clusters in its rostral (gold) and caudal (red) sectors.

of neurochemical systems (with diverse postsynaptic effects) comprising distinct cell clusters in the rostral tegmentum, and myriad other functions performed by neuronal clusters in more caudal parts of the reticular formation. Thus, with the advent of more precise means of demonstrating anatomical connections, as well as more sophisticated means of identifying neurotransmitters and the activity patterns of individual neurons, the concept of a "sparse network" engaged in a common function is now obsolete. Nevertheless, the term *reticular formation* remains, as does the daunting challenge of understanding the anatomical complexity and functional heterogeneity of this intricate brain region. Fortunately, two simplifying generalizations can be made. First, the functions of the different clusters of neurons in the reticular formation can be grouped into two broad categories: *modulatory*

functions and *premotor functions*. Second, the modulatory functions are found primarily in the rostral sector of the reticular formation, whereas most of the premotor functions are localized in more caudal regions.

Several clusters of large (magnocellular) neurons in the midbrain and rostral pontine reticular formation participate—together with certain diencephalic nuclei—in the modulation of conscious states (see Chapter 28). These effects are accomplished by long-range, diencephalic projections of cholinergic neurons near the superior cerebellar peduncle, as well as the more widespread forebrain projections of noradrenergic neurons in the locus coeruleus and serotonergic neurons in the raphe nuclei. Generally speaking, these biogenic amine neurotransmitters function as neuromodulators (see Chapter 6) that alter the membrane potential and thus the firing patterns

■ BOX 17C | The Reticular Formation (continued)

of thalamocortical and cortical neurons (the details of these effects are explained in Chapter 28). Also included in this category are the dopaminergic systems of the ventral midbrain that modulate corticostriatal interactions in the basal ganglia (see Chapter 18) and the responsiveness of neurons in the prefrontal cortex and limbic forebrain (see Chapter 32). However, not all modulatory projections from the rostral reticular formation are directed toward the forebrain. Although they are not typically considered part of the reticular formation, it is helpful to include in this functional group certain neuronal columns in the periaqueductal gray matter (surrounding the cerebral aqueduct) that project to the dorsal horn of the spinal cord and modulate the transmission of nociceptive signals (see Chapter 13).

Reticular formation neurons in the caudal pons and medulla oblongata generally serve a premotor function in the sense that they integrate feedback sensory signals with executive commands from upper motor neurons and deep cerebellar nuclei and, in turn,

organize the efferent activities of lower visceral motor and certain somatic motor neurons in the brainstem and spinal cord. Examples of this functional category include the smaller (parvocellular) neurons that coordinate a broad range of motor activities, including the gaze centers discussed in Chapter 20 and local circuit neurons near the somatic motor and branchiomotor nuclei that organize mastication, facial expressions, and a variety of reflexive orofacial behaviors such as sneezing, hiccupping, yawning, and swallowing. In addition, autonomic centers organize the efferent activities of specific pools of primary visceral motor neurons. Included in this subgroup are distinct clusters of neurons in the ventrolateral medulla that generate respiratory rhythms, and others that regulate the cardioinhibitory output of parasympathetic, preganglionic neurons in the nucleus ambiguus. Still other clusters organize more complex activities that require the coordination of both somatic motor and visceral motor outflow, such as gagging and vomiting, urination and defecation, and even laughing and crying.

One set of neuronal clusters that does not fit easily into this rostrocaudal framework is the set of neurons that give rise to the reticulospinal projections. As described in the text, these neurons are distributed in both rostral and caudal sectors of the reticular formation, and they give rise to long-range projections that innervate lower motor neuron pools in the medial ventral horn of the spinal cord. The reticulospinal inputs serve to modulate the gain of segmental reflexes involving the muscles of the trunk and proximal limbs and to relay initiation signals for certain stereotypical patterns of limb movement, such as locomotion.

In summary, the reticular formation is best viewed as a heterogeneous collection of distinct neuronal clusters in the brainstem tegmentum. These neuronal clusters either modulate the excitability of distant neurons in the forebrain and spinal cord or coordinate the firing patterns of more local lower motor neuron pools engaged in reflexive or stereotypical somatic motor and visceral motor behaviors.

purpose of the movement. For example, Figure 17.14 shows the pattern of muscle activity that occurs as an individual uses his arm to pull on a handle in response to an auditory tone. Activity in the biceps muscle begins about 200 ms after the tone. However, as the records show, the contraction of the biceps is accompanied by a significant increase in the activity of a proximal leg muscle, the gastrocnemius (as well as many other muscles not monitored in the experiment). In fact, contraction of the gastrocnemius muscle begins well before contraction of the biceps. These observations show that postural control during movement entails an anticipatory, or feedforward, mechanism (Figure 17.15). As part of the motor plan for moving the arm, the effect of the impending movement on body stability is predicted and used to generate a change in the activity of the gastrocnemius muscle. This change actually precedes and provides preparatory postural support for the movement of the arm. In the example given in Figure 17.14, contraction of the biceps would tend to pull the entire body forward, an action that is opposed by the contraction of the gastrocnemius muscle. In short, this feedforward mechanism

predicts the resulting disturbance in postural stability and generates an appropriate stabilizing response.

The importance of the reticular formation for feedforward mechanisms of postural control has been explored in more detail in cats trained to use a forepaw to strike an object. As expected, the forepaw movement is accompanied by feedforward postural adjustments in the other legs to maintain the animal upright. These adjustments shift the animal's weight from an even distribution over all four feet to a diagonal distribution pattern, in which the weight is carried mostly by the contralateral, nonreaching forelimb and the ipsilateral hindlimb. Lifting of the forepaw and postural adjustments in the other limbs can also be induced in an alert cat by electrical stimulation of the motor cortex. After pharmacological inactivation of the reticular formation, however, electrical stimulation of the motor cortex evokes only the forepaw movement, without the feedforward postural adjustments that normally accompany it.

The results of this experiment can be understood in terms of the fact that the upper motor neurons in the motor cortex influence the spinal cord circuits by two routes:

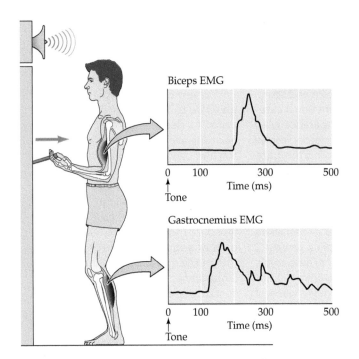

FIGURE 17.14 Anticipatory maintenance of body posture
At the onset of an audible tone, the individual pulls on a handle, contracting the biceps muscle. To ensure postural stability, contraction of the gastrocnemius muscle precedes that of the biceps. EMG refers to the electromyographic recording of muscle activity. (After L. M. Nashner, 1979. In *Progress in Brain Research, Vol. 50: Reflex Control of Posture and Movement*, R. Granit and O. Pompeiano [Eds.] Amsterdam: Elsevier/North Holland Biomedical Press, pp. 177–184.)

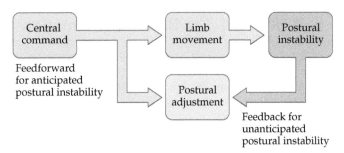

FIGURE 17.15 Feedforward and feedback mechanisms of postural control Feedforward postural responses are "preprogrammed" and typically precede the onset of limb movement (see Figure 17.14). Feedback responses are initiated by sensory inputs that detect postural instability.

direct projections to the spinal cord (as already discussed) and indirect projections to brainstem centers that in turn project to the spinal cord. The reticular formation is one of the major destinations of these latter projections from the motor cortex; thus, cortical upper motor neurons initiate both the reaching movement of the forepaw and also the postural adjustments in the other limbs necessary to maintain body stability. The forepaw movement is initiated by the direct pathway from the cortex to the spinal cord, whereas the postural adjustments are mediated via pathways from the motor cortex that reach the spinal cord indirectly, after an intervening relay in the reticular formation (the so-called cortico-reticulospinal pathway) (Figure 17.16).

Further evidence for the contrasting functions of the direct and indirect pathways from the motor cortex to the spinal cord comes from experiments carried out by the Dutch neurobiologist Hans Kuypers, who examined the behavior of rhesus monkeys that had the direct pathway to the spinal cord transected at the level of the medulla, leaving only the indirect pathways to the spinal cord via the brainstem centers intact. Immediately after the surgery, the animals were able to use axial and proximal

muscles to stand, walk, run, and climb, but they had great difficulty using the distal parts of their limbs (especially their hands) independently of other body movements. For example, the monkeys could cling to the cage but were unable to reach toward and pick up food with their fingers; rather, they used the entire arm to sweep the food toward them. After several weeks, the animals recovered some independent use of their hands and were again able to pick up objects of interest, but this action involved the concerted closure of all of the fingers. The ability to make independent, fractionated movements of the fingers, as in opposing the movements of the fingers and thumb to pick up an object, never returned.

These observations show that following damage to the direct projections from the motor cortex to the spinal cord at the level of the medulla, the indirect projections to the spinal cord from the motor cortex via the brainstem centers (or from brainstem centers alone) are capable of sustaining motor behavior that involves primarily the use of proximal muscles. In contrast, the direct projections from the motor cortex to the spinal cord provide the speed and agility of movements, and they enable a higher degree of precision in fractionated finger movements than is possible using the indirect pathways alone.

An additional brainstem structure, the **superior colliculus**, which is located in the dorsal midbrain, also contributes upper motor neuron pathways that govern lower motor neurons in the spinal cord. Although most mammals are likely to have direct projections from neurons in deep layers of the superior colliculus to the spinal cord (comprising a so-called colliculospinal or tectospinal tract), the major output of the superior colliculus to the spinal cord is mediated by the reticular formation. Thus, upper motor neurons in the superior colliculus innervate neural circuits in the reticular formation, which in turn give rise to reticulospinal projections that supply medial cell groups in the cervical cord. Functionally, this pathway plays a role in controlling axial musculature in the neck. These projections are particularly important in generating

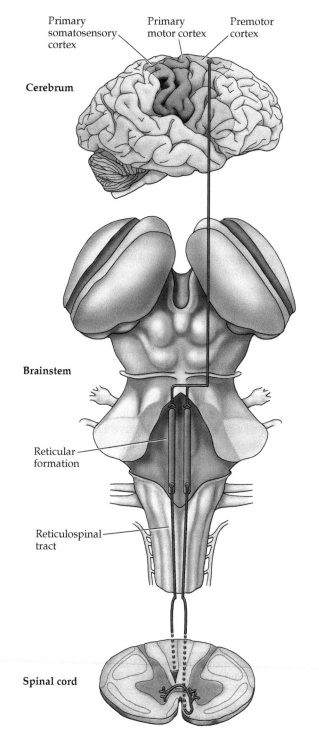

Primary somatosensory cortex

Primary motor cortex

Premotor cortex

Cerebrum

Brainstem

Reticular formation

Reticulospinal tract

Spinal cord

FIGURE 17.16 Indirect pathways from the motor cortex to the spinal cord Neurons in the motor cortex that supply the lateral part of the ventral horn to initiate movements of the distal limbs (see Figure 17.4) also terminate on neurons in the reticular formation to mediate postural adjustments that support the movement. The reticulospinal pathway terminates in the more medial parts of the ventral horn, where lower motor neurons that innervate axial and proximal muscles are located. Thus, the motor cortex can influence the activity of spinal cord neurons via both direct and indirect routes.

orienting movements of the head (Chapter 20 provides a detailed description of the role of the superior colliculus in the generation of head and eye movements).

Also in the midbrain is the mesencephalic locomotor region, which is involved in the initiation of locomotion (see Chapter 16). This region comprises a set of nuclei embedded in the reticular formation of the midbrain, just ventral and lateral to the periaqueductal gray matter. The mesenencephalic locomotor region projects to reticulospinal neurons in the medulla that, in turn, mediate the initiation and regulation of locomotion via connections with central pattern generators in the spinal cord.

In non-human primates and other mammals, a large nucleus in the tegmentum of the midbrain, termed the red nucleus, projects via the **rubrospinal tract** to the cervical level of the spinal cord (*rubro* [Latin, "red"] refers to the reddish color of this nucleus in fresh tissue, presumably due to the enrichment of its neurons with iron–protein complexes). Unlike the other projections from the brainstem to the spinal cord discussed thus far, the rubrospinal tract is located in the lateral white matter of the spinal cord; its axons terminate in lateral regions of the ventral horn and intermediate zone, where circuits of lower motor neurons governing the distal musculature of the upper extremities reside. Presumably, this projection participates together with the direct pathway from the motor cortex in the control of the arms (or forepaws). The limited distribution of rubrospinal projections may seem surprising, given the large size of the red nucleus in most mammals. However, the rubrospinal tract arises from especially large (magnocellular) neurons in the caudal pole of the red nucleus, which account for a relatively small fraction of the total number of neurons in the nucleus. In the human midbrain, there are few—if any—large neurons in the red nucleus; thus, if the rubrospinal tract exists in humans (which may not be the case in some individuals), its significance for motor control is dubious. Indeed, nearly all of the neurons in the red nucleus in humans are small (parvocellular) and do not project to the spinal cord at all; instead, many of these neurons relay information to the inferior olive, an important source of learning signals for the cerebellum (this role of the red nucleus is discussed in Chapter 19).

CONCEPT
17.5

Damage to Upper Motor Neurons Produces "Upper Motor Neuron Syndrome"

LEARNING OBJECTIVES

17.5.1 Discuss the signs and symptoms associated with damage to upper motor neurons.

17.5.2 Differentiate the upper motor neuron syndrome from the lower motor neuron syndrome.

support the individual's body weight, whereas those of an individual with damage at the cortical level often can. However, lesions that interrupt the descending pathways in the brainstem above the level of the vestibular nuclei but below the level of the red nucleus cause even greater extensor tone than that which occurs after damage to higher regions. Sherrington, who first described this phenomenon, called the increased tone **decerebrate rigidity**. In the cat, the extensor tone in all four limbs is so great after lesions that spare the vestibulospinal tracts that the animal can stand without support. Patients with severe brainstem injury at the level of the pons may exhibit similar signs of decerebration: arms and legs stiffly extended, jaw clenched, and neck retracted. The relatively greater hypertonia following damage to the nervous system above the level of the medulla oblongata is presumably explained by the remaining activity of the intact descending pathways from the vestibular nuclei and medullary reticular formation, which evidently have a net excitatory influence on the gain of segmental reflexes that contribute to posture and equilibrium in the context of impaired cortico-reticular regulation.

Summary

Two sets of upper motor neuron pathways make distinct contributions to the control of the local circuitry in the brainstem and spinal cord. One set originates from neurons in the frontal lobe and includes projections from the primary motor cortex and the nearby premotor areas. The premotor cortices are responsible for planning, initiating, and controlling complex sequences of voluntary movements, especially movements that are triggered by sensory cues or internal motivations, whereas the primary motor cortex is especially involved with the execution of skilled movements of the limb and facial musculature. The motor cortex influences movements *directly* by contacting lower motor neurons and local circuit neurons in the spinal cord and brainstem; and *indirectly* by innervating neurons in brainstem centers (mainly the reticular formation) that in turn project to lower motor neurons and circuits. The other major upper motor neuron pathways originate from brainstem centers—primarily the reticular formation and the vestibular nuclei—and are responsible for postural regulation. The reticular formation is especially important in *feedforward* control of posture (i.e., movements that occur in anticipation of changes in body stability). In contrast, the neurons in the vestibular nuclei that project to the spinal cord are especially important in *feedback* postural mechanisms (i.e., in producing movements that are generated in response to sensory signals that indicate an existing postural disturbance). Although the brainstem pathways can independently organize gross motor control, direct projections from the motor cortex to local circuit neurons in the brainstem and spinal cord are essential for the fine, fractionated movements of the face and the distal parts of the limbs that are especially important in activities of daily living and the expression of motor skill.

■ Additional Reading

Reviews

Dum, R. P. and P. L. Strick (2002) Motor areas in the frontal lobe of the primate. *Physiol. Behav.* 77: 677–682.

Gahery, Y. and J. Massion (1981) Coordination between posture and movement. *Trends Neurosci.* 4: 199–202.

Georgeopoulos, A. P., M. Taira and A. Lukashin (1993) Cognitive neurophysiology of the motor cortex. *Science* 260: 47–52.

Geyer, S., M. Matelli and G. Luppino (2000) Functional neuroanatomy of the primate isocortical motor system. *Anat. Embryol.* 202: 443–474.

Graziano, M. S. A. (2016) Ethological action maps: A paradigm shift for the motor cortex. *Trends. Cogn. Sci.* 20: 121–132.

Lemon R. (2019) Recent advances in our understanding of the primate corticospinal system. F1000Research 2019, 8(F1000 Faculty Rev): 274.

Nashner, L. M. (1979) Organization and programming of motor activity during posture control. In *Reflex Control of Posture and Movement*, R. Granit and O. Pompeiano (Eds.). *Prog. Brain Res.* 50: 177–184.

Nashner, L. M. (1982) Adaptation of human movement to altered environments. *Trends Neurosci.* 5: 358–361.

Rizzolatti, G. and 4 others (2021) The role of mirror mechanism in the recovery, maintenance, and acquisition of motor abilities. *Neurosci. Biobehav. Rev.* 127: 404–423.

Sherrington, C. and S. F. Grunbaum (1901) Observations on the physiology of the cerebral cortex of some of the higher apes. *Proc. R. Soc.* 69: 206–209.

Important Original Papers

BRAIN Initiative Cell Census Network (BICCN) (2021) A multimodal cell census and atlas of the mammalian primary motor cortex. *Nature* 598: 86–102.

Caspers, S., K. Zilles, A. R. Laird and S. B. Eickhoff (2010) ALE meta-analysis of action observation and imitation in the human brain. *NeuroImage* 50: 1148–1167.

Evarts, E. V. (1981) Functional studies of the motor cortex. In *The Organization of the Cerebral Cortex*, F. O. Schmitt, F. G. Worden, G. Adelman and S. G. Dennis (Eds.). Cambridge, MA: MIT Press, pp. 199–236.

Fetz, E. E. and P. D. Cheney (1978) Muscle fields of primate corticomotoneuronal cells. *J. Physiol. (Paris)* 74: 239–245.

Georgeopoulos, A. P., A. B. Swartz and R. E. Ketter (1986) Neuronal population coding of movement direction. *Science* 233: 1416–1419.

Graziano, M. S. A., T. N. S. Aflalo and D. F. Cooke (2005) Arm movements evoked by electrical stimulation in the motor cortex of monkeys. *J. Neurophysiol.* 94: 4209–4223.

Kuypers, H. G. J. M. (1958) Corticobulbar connexions to the pons and lower brain-stem in man. *Brain* 81: 364–388.

Lawrence, D. G. and H. G. J. M. Kuypers (1968) The functional organization of the motor system in the monkey. I. The effects of bilateral pyramidal lesions. *Brain* 91: 1–14.

Mitz, A. R., M. Godschalk and S. P. Wise (1991) Learning-dependent neuronal activity in the premotor cortex: Activity during the acquisition of conditional motor associations. *J. Neurosci.* 11: 1855–1872.

Rizzolatti, G., L. Fadiga, V. Gallese and L. Fogassi (1996) Premotor cortex and the recognition of motor actions. *Cogn. Brain Res.* 3: 131–141.

Roland, P. E., B. Larsen, N. A. Lassen and E. Skinhof (1980) Supplementary motor area and other cortical areas in organization of voluntary movements in man. *J. Neurophysiol.* 43: 118–136.

Schieber, M. H. and L. S. Hibbard (1993) How somatotopic is the motor cortex hand area? *Science* 261: 489–492.

Books

Graziano, M. S. A. (2009) The Intelligent Movement Machine: An Ethological Perspective on the Primate Motor System. Oxford, UK: Oxford University Press.

Nicolelis, M. A. L. (2011) Beyond Boundaries: The New Neuroscience of Connecting Brains with Machines—And How It Will Change Our Lives. New York: Times Books.

Passingham, R. (1993) *The Frontal Lobes and Voluntary Action*. Oxford, UK: Oxford University Press.

Penfield, W. and T. Rasmussen (1950) The Cerebral Cortex of Man: A Clinical Study of Localization of Function. New York: Macmillan.

Porter, R. and R. Lemon (1993) *Corticospinal Function and Voluntary Movement*. Oxford, UK: Oxford University Press.

Sherrington, C. (1947) *The Integrative Action of the Nervous System*, 2nd Edition. New Haven, CT: Yale University Press.

Sjölund, B. and A. Björklund (1982) *Brainstem Control of Spinal Mechanisms*. Amsterdam: Elsevier.

Modulation of Movement by the Basal Ganglia

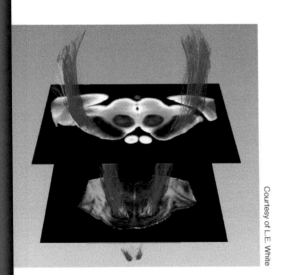

KEY CONCEPTS

18.1 The basal ganglia comprise a set of nuclei deep in the cerebral hemispheres

18.2 The basal ganglia influence movement by regulating the activity of upper motor neuronal circuits

18.3 Direct and indirect pathways regulate the initiation of voluntary movement and the suppression of unwanted movement

18.4 Dopamine modulates basal ganglia circuits by increasing or decreasing the excitability of medium spiny neurons

18.5 Hypokinetic movement disorders are associated with excessive inhibition of motor nuclei in the thalamus and brainstem

18.6 Hyperkinetic movement disorders are associated with insufficient inhibition of motor nuclei in the thalamus and brainstem

Overview

In contrast to upper motor neurons in the motor regions of the cerebral cortex and brainstem (discussed in Chapter 17), the basal ganglia and cerebellum do not directly influence lower motor neuronal circuitry; instead, these brain regions influence movement by regulating the activity of upper motor neuronal circuits. The term *basal ganglia* refers to a large and functionally diverse set of nuclei that lies deep within the cerebral hemispheres. The general function of the basal ganglia is to modulate *movement*, if that term may be taken broadly—that is, not only movement of the body, but also movement of thought and emotion or affect. In this unit, "Movement and Its Central Control," the focus is on the role of the basal ganglia in modulating the movements of the body. However, the general principles of anatomical organization and physiological operation apply throughout the basal ganglia to all of their explicitly motor and non-motor functions, some of which will be discussed in Unit V. The subset of these nuclei of the basal ganglia that is relevant to this account of motor function includes the caudate, the putamen, and the globus pallidus. Two additional structures, the substantia nigra in the base of the midbrain and the subthalamic nucleus in the ventral thalamus, are closely associated with the motor functions of these basal ganglia nuclei. The motor components of the basal ganglia, together with the substantia nigra and the subthalamic nucleus, comprise a subcortical loop that links most areas of the cerebral cortex with upper motor neurons in the motor cortex and in the brainstem. The neurons in this loop modulate their activity mainly at the beginning and ending of movement sequences, and their influences on upper motor neurons are required for functional regulation of voluntary movements. When one of these components of the basal ganglia or associated structures is compromised, the motor systems cannot switch smoothly between commands that initiate and maintain a movement and those that terminate the movement. The disordered movements that result can be understood as a consequence of maladaptive upper motor neuron activity that results from dysregulation of the control provided by the basal ganglia. Similar principles of function and dysfunction likely apply to parallel processing streams through the basal ganglia that serve other aspects of behavior, including cognition and emotional regulation.

The Basal Ganglia Comprise a Set of Nuclei Deep in the Cerebral Hemispheres

LEARNING OBJECTIVES

18.1.1 Identify the major components of the basal ganglia.

18.1.2 Identify important sources of input to basal ganglia circuits and discuss how those inputs are integrated by medium spiny neurons in the striatum.

Projections to the basal ganglia

The motor nuclei of the basal ganglia that modulate the movements of the body are divided into several functionally distinct groups. The first and largest of these groups is called the **striatum**, which includes two principal nuclei, the **caudate** and the **putamen** (Figure 18.1). There are other, more ventral divisions of the striatum, including the nucleus accumbens, that are associated with non-motor functions of the basal ganglia (Box 18A). An older term for the motor nuclei (plus additional components of the basal ganglia) is *corpus striatum*, which means "striped body," reflecting the fact that the caudate and the dorsal part of the putamen are joined by slender bridges of gray matter that extend through the internal capsule and confer a striped appearance in parasagittal sections through this area. These two subdivisions of the corpus striatum comprise the *input zone* of the basal ganglia, since their neurons are the destinations of most of the pathways that reach this complex from other parts of the brain (Figure 18.2). The

destinations of the incoming axons from the cerebral cortex are the dendrites of a class of cells in the corpus striatum called **medium spiny neurons** (Figure 18.3). The large dendritic trees of these neurons allow them to collect and integrate input from a variety of cortical, thalamic, and brainstem structures. The axons arising from the medium spiny neurons converge on neurons in the **pallidum**, which includes the **globus pallidus** and the **substantia nigra pars reticulata**. The globus pallidus and substantia nigra pars reticulata are the main sources of *output* from the basal ganglia complex to other parts of the brain.

Historically, the globus pallidus had been recognized as a component of the corpus striatum; however, given the important neurochemical, anatomical, and physiological distinctions between striatum and pallidum (see Concept 18.2), it is important to distinguish the globus pallidus from striatal divisions of the corpus striatum. Therefore, to avoid confusion that often attends this terminology, we will hereafter avoid the term *corpus striatum* in favor of more specific reference to components of the striatum or the pallidum.

Nearly all regions of the cerebral cortex project directly to the striatum, making the cortex the source of the largest input to the basal ganglia. The majority of these projections are from association areas in the frontal and parietal lobes, but substantial contributions also arise from the temporal, insular, and cingulate cortices, as well as from the amygdala and hippocampal formation. All of these projections, referred to collectively as the **corticostriatal pathway**, travel through the subcortical white matter on their way to the caudate and putamen (see Figure 18.2).

The cortical inputs to the caudate and putamen are not equivalent, however, and the differences in these inputs

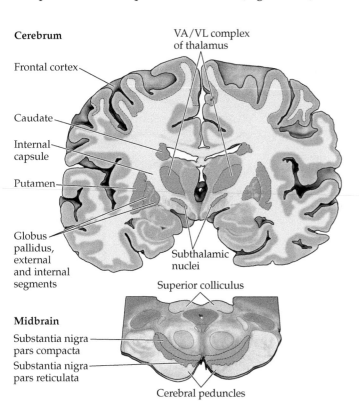

Cerebrum

Frontal cortex

VA/VL complex of thalamus

Caudate

Internal capsule

Putamen

Globus pallidus, external and internal segments

Subthalamic nuclei

Superior colliculus

Midbrain

Substantia nigra pars compacta

Substantia nigra pars reticulata

Cerebral peduncles

FIGURE 18.1   Motor components of the basal ganglia
The human basal ganglia comprise a set of gray matter structures, most of which are buried deep in the telencephalon, although some are found in the diencephalon and midbrain. The major components that receive and process movement-related signals are the striatum (caudate and putamen) and the pallidum (globus pallidus and substantia nigra pars reticulata). These structures border the internal capsule in the forebrain and midbrain (the cerebral peduncle is a caudal extension of the internal capsule). Smaller but functionally significant components of the basal ganglia system are the substantia nigra pars compacta and the subthalamic nucleus, which provide input to the striatum and pallidum, respectively. For the control of limb movements, output from the basal ganglia arises in the internal segment of the globus pallidus and is sent to the ventral anterior and ventral lateral nuclei (VA/VL complex) of the thalamus, which interact directly with circuits of upper motor neurons in frontal cortex. The substantia nigra pars reticulata projects to upper motor neurons in the superior colliculus and controls orienting movements of the eyes and head. Other divisions of the striatum and pallidum (not illustrated) participate in non-motor loops through the ventral portions of the basal ganglia.

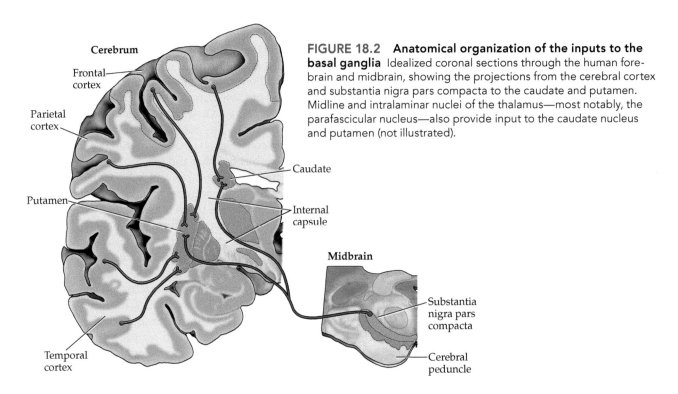

FIGURE 18.2 Anatomical organization of the inputs to the basal ganglia Idealized coronal sections through the human forebrain and midbrain, showing the projections from the cerebral cortex and substantia nigra pars compacta to the caudate and putamen. Midline and intralaminar nuclei of the thalamus—most notably, the parafascicular nucleus—also provide input to the caudate nucleus and putamen (not illustrated).

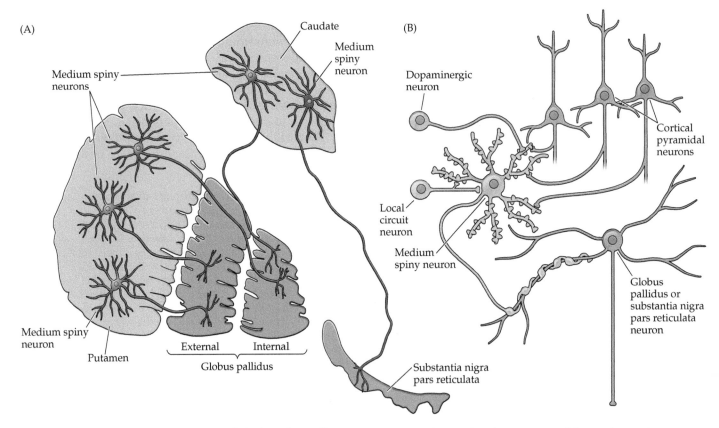

FIGURE 18.3 Neurons and circuits of the basal ganglia (A) Medium spiny neurons in the caudate and putamen. (B) Diagram showing convergent inputs onto a medium spiny neuron from cortical neurons, dopaminergic cells of the substantia nigra, and local circuit neurons within the striatum. The arrangement of these synapses indicates that the response of the medium spiny neurons to their principal input, derived from the cerebral cortex, can be modulated by dopamine and the inputs of local circuit neurons. The primary output of the medium spiny cells is to neurons in the globus pallidus and substantia nigra pars reticulata.

■ BOX 18A | Basal Ganglia Loops and Non-Motor Brain Functions

The basal ganglia traditionally have been regarded as motor structures that regulate the initiation of voluntary movements, such as those involving the limbs and eyes. However, if one is willing to apply the concept of "movement" figuratively as well as literally, then one may understand the functions of basal ganglia circuitry as modulating movements of body (limbs, trunk, eyes), mind (the content of thought), and emotion or affect (feeling or mood). Consider, for example, the wisdom of many cultures that use terms of movement (e.g., *moved*, *stirred*, *roused*) to account for the sublime impact of a glorious sunset or musical composition. Thus, the basal ganglia are central structures in anatomical circuits or loops that are involved in modulating aspects of behavior that are explicitly non-motor. These parallel loops originate in different regions of the cerebral cortex, engage specific subdivisions of the basal ganglia and thalamus, and ultimately affect areas of the frontal lobe outside the primary motor and premotor cortices. The most prominent of these non-motor loops are a *dorsolateral prefrontal loop*, involving the dorsolateral sector of the prefrontal cortex and the head of the caudate (see Chapter 27); and a *limbic loop* that originates in the orbitomedial prefrontal cortex, amygdala, and hippocampal formation and runs through ventral divisions of the striatum (see Chapter 32).

The anatomical, neurochemical, and neurophysiological similarity of these loops to the better understood motor loops suggests that the non-motor regulatory functions of the basal ganglia may be generally the same as the roles of basal ganglia in regulating the initiation and suppression of movement. For example, the prefrontal loop may regulate the initiation and termination of cognitive processes such as attention, working memory, and decision making.

(Continued)

Comparison of motor and non-motor basal ganglia loops.

■ **BOX 18A** | **Basal Ganglia Loops and Non-Motor Brain Functions** (*continued*)

Likewise, the limbic loop may regulate emotional and motivated behavior, as well as the transitions from one mood state to another. Indeed, the deterioration of cognitive and emotional function with disease progression in both Parkinson's and Huntington's diseases could be the result of the disruption of these non-motor loops.

In fact, a variety of other disorders are now thought to be caused, at least in part, by the dysfunction of non-motor components of the basal ganglia. For example, individuals with Tourette syndrome may produce inappropriate utterances and obscenities as well as unwanted vocal and motor tics and repetitive grunts. These manifestations may be a result of excessive activity in basal

ganglia loops that regulate the cognitive circuitry of the prefrontal speech areas. Another example is schizophrenia, which some investigators have argued is associated with aberrant activity within the limbic and prefrontal loops, resulting in hallucinations, delusions, disordered thoughts, and loss of emotional expression. In support of the argument for a basal ganglia contribution to schizophrenia, antipsychotic drugs are known to act on dopaminergic receptors, which are found in high concentrations in the striatum, as well as in the prefrontal cortex itself.

Still other psychiatric disorders, including obsessive-compulsive disorder, depression, and chronic anxiety, may also involve dysfunctions of

the limbic loop. Indeed, one particular component of the limbic loop in a ventral division of the striatum is the nucleus accumbens. This structure is implicated in both the neuropharmacology of addiction to drugs of abuse and of the expression of addictive reward-seeking behavior (see Chapter 32). A challenge for future research is to understand more fully the relationships between these clinical conditions and the functions of the basal ganglia. Nevertheless, extending the concepts of *hypokinesis* and *hyperkinesis* (see Concepts 18.5 and 18.6) to disordered movement of thought and emotion is proving to be a valuable framework for understanding mental health and a variety of mental illnesses.

---

reflect functional differences between the two nuclei. The caudate receives cortical projections primarily from multimodal association cortices and from motor areas in the frontal lobe that control eye movements and movements of thought, which are often conjoined. As their name implies, the association cortices do not process any one type of sensory information; rather, they receive input from several primary and secondary sensory cortices and their associated thalamic nuclei (see Chapter 27). The putamen, by contrast, receives input from the primary and secondary somatosensory cortices in the parietal lobe, the higher-order (extrastriate) visual cortices in the occipital and temporal lobes, the premotor and primary motor cortices in the frontal lobe, and the auditory association areas in the temporal lobe. The fact that different cortical areas project to different regions of the striatum implies that the corticostriatal pathway consists of multiple parallel pathways serving different functions. This interpretation is supported by the observation that the segregation is maintained in the output structures that receive projections from the striatum and in the output pathways that project from the basal ganglia to other brain regions (see the figure in Box 18A).

The distribution of parallel corticostriatal pathways within the striatum reflects the functional organization of the cerebral cortex. For example, visual and somatosensory cortical projections are topographically mapped within different regions of the putamen. Moreover, the cortical areas that are functionally interconnected at the level of the cortex give rise to projections that overlap extensively in the striatum. Anatomical studies by Ann Graybiel and

her colleagues at the Massachusetts Institute of Technology have shown that different cortical areas concerned with the hand (see Chapter 12) send projections that converge in specific rostrocaudal bands within the striatum; conversely, cortical areas concerned with the leg give rise to projections that converge in other striatal bands. These rostrocaudal bands therefore appear to be functional units concerned with the movement of particular body parts. Another study by the same group shows that the more extensive the interconnections of cortical areas by corticocortical pathways, the greater the overlap in their projections to the striatum. Thus, the specialization of functional units within the striatum reflects the specialization of the cortical areas that provide their input.

A further indication of functional subdivision within the striatum is evident when tissue sections obtained post mortem are stained for the presence of different neurotransmitters and their related enzymes. For example, when the striatum is stained for the enzyme acetylcholinesterase, which inactivates acetylcholine (see Chapter 6), a compartmental organization is revealed within the striatum. The compartments are defined by lightly stained regions, called *patches* or *striosomes*, surrounded by densely stained tissue, called *matrix* or *matrisomes*. Subsequent studies of the distributions of other neurochemicals, including peptide neurotransmitters, have cataloged a variety of neuroactive substances that localize to the patch or matrix compartments. Tract-tracing experiments in animals have likewise shown differences between these striatal compartments in the sources of their inputs from the cortex and in the destinations of their projections to other parts of the basal ganglia. For example, the matrix makes up the bulk of the striatum; it receives input from most areas of the cerebral cortex and sends projections to the globus pallidus and the substantia nigra pars reticulata. The patches in the caudate receive most of their input from the prefrontal cortex (see Chapter 27) and project preferentially to a different subdivision of the substantia nigra (the dopaminergic neurons of the pars compacta, discussed shortly). Distinct patterns of projection from medium spiny neurons in the patches and matrix further support the conclusion that functionally distinct pathways project in parallel from the cerebral cortex to the striatum.

The nature of the information transmitted to the caudate and putamen from the cerebral cortex is not understood. It is known, however, that collateral axons of corticocortical, corticothalamic, and corticospinal pathways all form excitatory glutamatergic synapses on the dendritic spines of medium spiny neurons (see Figure 18.3B). The number of contacts established between an individual cortical axon and a single medium spiny cell is very small, whereas the number of spiny neurons contacted by a single axon is extremely large. This divergence of the inputs from corticostriatal axons allows a single medium spiny neuron to integrate the influences of thousands of cortical cells.

The medium spiny cells also receive inputs from several sources besides the cerebral cortex, including other medium spiny neurons via their local axon collaterals, local circuit interneurons of the striatum, neurons in the midline and intralaminar nuclei of the thalamus (most notably, the parafascicular nucleus), and neurons in several nuclei of the brainstem that produce biogenic amine neurotransmitters. In contrast to the cortical inputs that synapse on the dendritic spines of the medium spiny neurons, the local circuit neuron and thalamic synapses are made on the dendritic shafts and close to the cell soma, where they can modulate the effectiveness of cortical synaptic activation of the more distal dendrites. One important set of brainstem inputs to the medium spiny neurons is dopaminergic, and it originates in a subdivision called the **substantia nigra pars compacta**, because of its densely packed cells. (The striatum also receives serotonergic inputs from the raphe nuclei; see Chapter 6.) The dopaminergic synapses are located on the base of the spine, in close proximity to the cortical synapses, where they selectively modulate cortical input (see Figure 18.3B). As a result, inputs from both the cortex and the substantia nigra pars compacta are relatively far from the initial segments of the medium spiny neurons' axons, where the nerve impulses are generated. Furthermore, medium spiny neurons express inward-rectifier potassium conductances that tend to remain open near resting membrane potentials, but close with depolarization. Accordingly, these neurons exhibit very little spontaneous activity and must simultaneously receive many excitatory inputs to overcome the stabilizing influence of this potassium conductance.

When the medium spiny neurons do become active, their firing is associated with the occurrence of a movement. Extracellular recordings show that these neurons typically increase their rate of discharge before an impending movement. Neurons in the putamen tend to discharge in anticipation of limb and trunk movements, whereas caudate neurons fire prior to eye movement. These anticipatory discharges are evidently part of a movement selection as well as a movement initiation process; in fact, they can precede the initiation of movement by as much as several seconds. Similar recordings have also shown that the discharges of some striatal neurons vary according to the location in space of the *destination* of a movement, rather than with the starting position of the limb relative to the destination. Thus, the activity of these cells may encode the *decision to move* toward a goal rather than the direction and amplitude of the actual movement necessary to reach the goal. Furthermore, medium spiny neurons increase their firing rate at the termination of a movement sequence, which routinely coincides with the initiation of a subsequent motor program (e.g., the reinitiation of stationary, stable posture following a sequence of steps). This temporal relationship between the firing of medium spiny neurons and the initiation and termination of movement sequences has implicated the basal ganglia in the selection of action plans and the instantiation of habitual patterns of movement (Box 18B).

■ BOX 18B | Making and Breaking Habits

To one degree or another, we are all "creatures of habit," which is to say that patterns of thought and movement often display repetitive stereotypes that may serve to increase the efficiency of goal-oriented behavior. Indeed, habitual patterns of movement become "second nature" as motivated behavior increasingly loses dependence on explicit outcomes (attainment of reward) and component movements are consolidated into stereotyped, automated sequences. It has been long suspected that this process of associative sensorimotor learning and action automation involves basal ganglia circuitry.

Presumably, one function of motor circuits in the basal ganglia is to acquire information related to stimulus-response associations and to initiate efficient patterns of movement driven by stimulus-response contingencies.

(Continued)

■ BOX 18B | Making and Breaking Habits (continued)

Research from the laboratory of Ann Graybiel at the Massachusetts Institute of Technology has shed considerable light on the contributions of striatal neurons and their through pathways in habit formation and the execution of habitual behavior. These studies show that the firing patterns of medium spiny neurons in the dorsolateral aspect of the striatum (primate putamen and head of the caudate nucleus) serve to "chunk" action sequences by accentuating the initiation and termination of overlearned patterns of movement. For example, as macaque monkeys freely viewed a visual display of possible targets for successive fixations (saccades; see Chapter 20), the

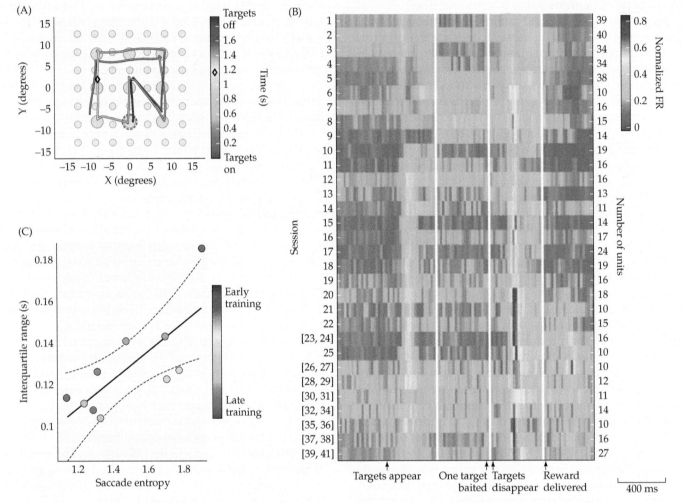

(A) Sample sequence of free visual scanning. Green targets appeared among an array of gray spots and a monkey began to scan the green targets until the scan path passed through a randomly chosen target (indicated by red dashed circle); the grid was then turned off and a reward provided. The monkey was not cued about the location of the baited target; it simply continued to scan across the green targets until they turned off at the conclusion of a rewarded trial. Black diamond indicates time (on color bar) and position (on grid) of monkey's gaze when the target became baited with reward. (B) All units recorded from the caudate nucleus in one monkey displayed across sessions. The firing rate (FR) of each unit was normalized and units were binned (20-ms bins), with each row representing the average activity of all units in a session. White vertical lines divide the phases of each session. Note the progressive increase in neuronal firing and the development of "sharpness" (increased tuning) of neuronal discharges across sessions at the start and especially at the termination of the visual scans. (C) Direct correlation between the tuning of neuronal activity in the striatum and the repetitiveness of visual scanning (saccade entropy). Smaller values of interquartile range (y-axis) indicate sharper tuning of neuronal activity, and lower values of saccade entropy (x-axis) indicate the formation of habitual patterns of scanning across the visual targets. Across sessions (color code), striatal neurons became more sharply tuned at the start and end of visual scans, and scan paths became more stereotyped. (After After T. M. Desrochers et al., 2015. *Neuron* 87: 853–868.)

discharges of medium spiny neurons in the head of the caudate nucleus became increasingly well tuned to the initiation and termination of stereotyped scan paths through the visual array as eye movements became more refined and habitual (see figure). This result suggests that striatal neurons encode an integrated cost–benefit signal by which reinforcement learning drives behaviors that minimize costs (in this case, the number of saccades necessary for completing a scan through the targets) and signal outcomes (completing a rewarded movement sequence).

Additional studies performed by Nicole Calakos, Henry Yin, and their colleagues at Duke University have teased apart the contributions of direct and indirect pathway striatal neurons in making and breaking habits (see Concept 18.3). These investigators used two-photon scanning laser microscopy to perform calcium imaging in mouse brain slices. Their goal was to record

simultaneously the evoked activities of both types of medium spiny neurons (and associated interneurons and glia) as a function of recent habitual behavior. The results indicated that plasticity mechanisms operating at the level of cortical inputs to striatal neurons are sufficient to drive habit formation, but with differential contributions of direct and indirect pathway projection neurons. As habits formed, there was a broadly distributed increase in the gain of striatal neuronal responses to cortical input, with a tendency for direct pathway striatal neurons to fire in advance of indirect pathway striatal neurons. Evidently, a timing competition between the direct and indirect pathways mediates the formation and expression of habitual patterns of movement. Plasticity advancing direct pathway activation would favor habit formation and reduce the probability of action cancellation associated with activation of the indirect pathway.

Interestingly, habit breaking was mediated by weakening the response of direct pathway neurons to cortical activation, rather than by strengthening indirect pathway connections. This implies that habit suppression is a manifestation of a reduced drive for volitional movement.

Taken together, these studies indicate that broadly distributed plastic changes in corticostriatal connections alter the propagation of activity through direct and indirect pathways and bias the output of the basal ganglia toward the consolidation of habitual movement. It remains to be determined how such mechanisms of circuit plasticity are related to the excessive or overly stereotyped movement routines that are commonly associated with certain neuropsychiatric conditions, including obsessive-compulsive disorder, autism spectrum disorders, and substance use disorders (see Chapters 32 and 33).

CONCEPT
18.2

The Basal Ganglia Influence Movement by Regulating the Activity of Upper Motor Neuronal Circuits

LEARNING OBJECTIVES

18.2.1 Identify the sources of major output from the basal ganglia to upper motor neuronal circuits.

18.2.2 Discuss the integration of input from medium spiny neurons on pallidal neurons.

18.2.3 Describe the principle of disinhibition and explain how it applies to the circuitry and functions of the basal ganglia.

Projections from the basal ganglia to other brain regions

The medium spiny neurons of the caudate and putamen give rise to inhibitory GABAergic projections that terminate in the globus pallidus and the substantia nigra pars reticulata in the pallidal nuclei of the basal ganglia (Figure 18.4). *Globus pallidus* means "pale body," a name that describes the appearance of the large number of myelinated axons in this nucleus; *pars reticulata* is so named because, unlike the pars compacta, axons passing through give it a netlike, or reticulated, appearance.

The globus pallidus and substantia nigra pars reticulata share the same types of neurons and perform comparable functions, albeit on the different types of signals they receive from the parallel streams of processing that flow through the basal ganglia. In fact, the pars reticulata may be understood as being a part of the globus pallidus that, during early brain development, became separated from the rest of the pallidum by the formation of the posterior limb of the internal capsule and cerebral peduncle. The striatal projections to these two nuclei resemble the corticostriatal pathways in that they terminate in rostrocaudal bands, the locations of which vary with the locations of sources in the striatum. A striking feature of these projections is the degree of convergence from the medium spiny neurons to the neurons of the globus pallidus and substantia nigra pars reticulata. In humans, for example, the striatum contains approximately 100 million neurons, about 75% of which are medium spiny neurons. In contrast, the main destination of their axons, the globus pallidus, comprises only about 700,000 neurons. Thus, on average, more than 100 medium spiny neurons innervate each neuron in the globus pallidus. However, despite this impressive degree of convergence, individual axons from the striatum sparsely contact many pallidal neurons before terminating densely on the dendrites of a particular neuron. Consequently, ensembles

(A)

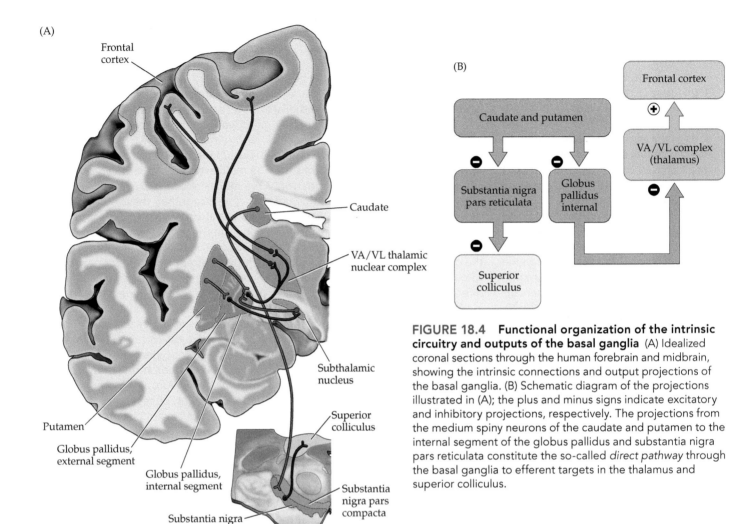

FIGURE 18.4 Functional organization of the intrinsic circuitry and outputs of the basal ganglia (A) Idealized coronal sections through the human forebrain and midbrain, showing the intrinsic connections and output projections of the basal ganglia. (B) Schematic diagram of the projections illustrated in (A); the plus and minus signs indicate excitatory and inhibitory projections, respectively. The projections from the medium spiny neurons of the caudate and putamen to the internal segment of the globus pallidus and substantia nigra pars reticulata constitute the so-called *direct pathway* through the basal ganglia to efferent targets in the thalamus and superior colliculus.

of medium spiny neurons exert a broad but functionally weak influence over many neurons, while at the same time strongly influencing a subset of neurons in the globus pallidus or substantia nigra pars reticulata. This pattern of innervation is important for understanding the role of the striatum in the selection and initiation of intended motor programs, as described in Concept 18.3.

The efferent neurons of the globus pallidus and substantia nigra pars reticulata together give rise to the major output pathways that allow the basal ganglia to influence the activity of upper motor neurons located in the motor cortex and in the brainstem (see Figure 18.4). The pathway to the cortex arises primarily in the medial division of the globus pallidus, called the **internal segment**, and reaches the motor cortex via a relay in the **ventral anterior nuclei** and **ventral lateral nuclei** of the dorsal thalamus. These thalamic nuclei project directly to motor areas of the cerebral cortex, thus completing a vast loop of circuitry that originates in multiple areas of the cortex and terminates in the motor areas of the frontal lobe, after successive stages of processing in the basal ganglia and thalamus. In contrast, many efferent

axons from substantia nigra pars reticulata have more direct access to upper motor neurons by synapsing on neurons in the superior colliculus that command head and eye movements, without an intervening relay in the thalamus. This difference between the globus pallidus and substantia nigra pars reticulata is not absolute, however, since many reticulata axons also project to the thalamus (mediodorsal and ventral anterior nuclei), where they contact relay neurons that project to the frontal eye fields of the premotor cortex (see Chapter 20). The thalamic relay is a mechanism for facilitating or suppressing inputs to circuits of upper motor neurons in the cortex—a level of organization that is not shared by the superior colliculus.

Because the efferent cells of both the globus pallidus and substantia nigra pars reticulata are GABAergic, the main output of the basal ganglia is *inhibitory*. In contrast to the quiescent medium spiny neurons, the neurons in both of these output structures have high levels of spontaneous activity that prevent unwanted movement by tonically inhibiting cells in the thalamus, superior colliculus, and other brainstem nuclei. Because the medium spiny neurons of the striatum also are GABAergic and inhibitory, the net effect of the phasic excitatory inputs that reach the striatum from the cortex is to open a physiological gate by inhibiting the tonically active inhibitory cells of the globus pallidus and substantia nigra pars reticulata (Figure 18.5). For example, in the absence of volitional body movements (and the intention to make movements), the globus pallidus neurons provide tonic inhibition to the relay cells in the ventral lateral and ventral anterior nuclei of the thalamus. When the pallidal cells are inhibited by activation of the medium spiny neurons (as signals for volitional movement converge on the striatum), the thalamic neurons are *disinhibited* and can trigger the activation of upper motor neurons in the cortex. This disinhibition allows the upper motor neurons to send commands to local circuit neurons and lower motor neurons that in turn initiate movement.

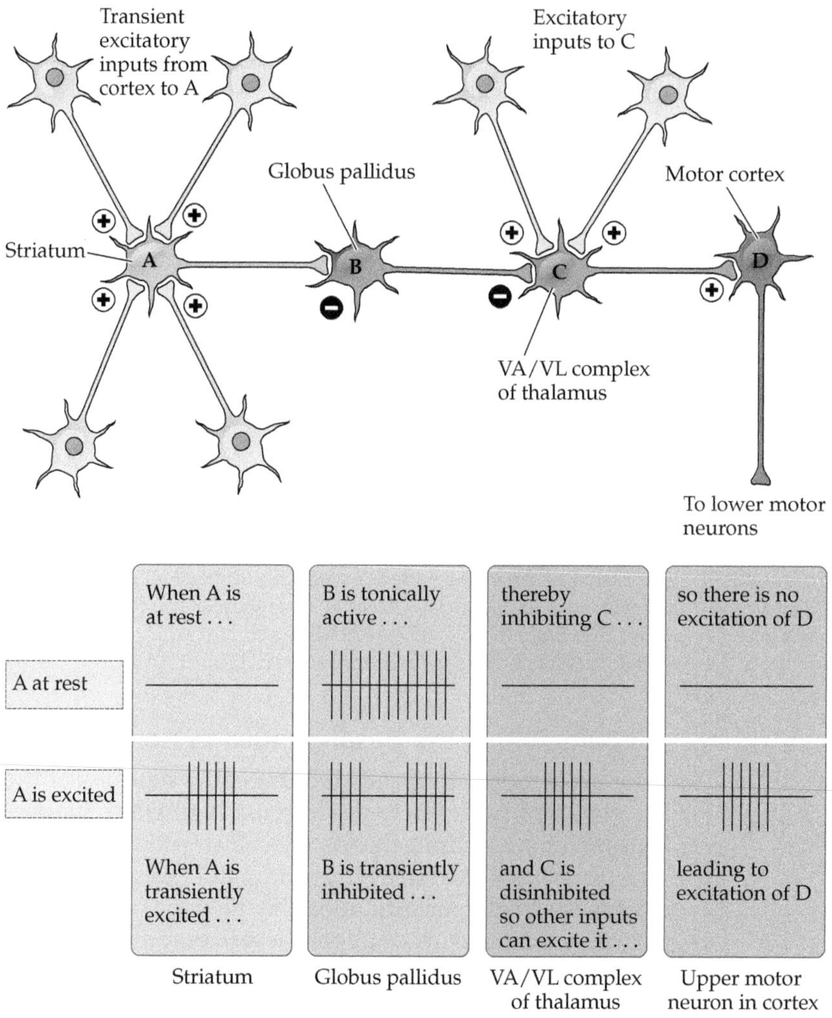

Evidence from studies of eye movements

The permissive, or gating, role of the basal ganglia in the initiation of movement is perhaps most clearly demonstrated by studies of eye movements carried out by Okihide Hikosaka and Robert Wurtz at the National Institutes of Health (Figure 18.6). As described in the previous section, the substantia nigra pars reticulata is part of the output circuitry of the basal ganglia. Instead of projecting to the ventral anterior and ventral lateral nuclei of the thalamus, however, it sends axons mainly to the deep layers of the superior colliculus. The upper motor neurons in these layers command the rapid, orienting movements of the eyes called *saccades* (see Chapter 20). When the eyes are fixating a visual target, these upper motor neurons are tonically inhibited by the spontaneously active reticulata cells, thus preventing unwanted saccades. Shortly before the onset of a saccade, the tonic discharge rate of the reticulata neurons is sharply reduced by input from the GABAergic medium spiny neurons of the caudate, which have been activated by signals from the cortex. The subsequent reduction in the tonic discharge from reticulata neurons disinhibits the upper motor neurons of the superior colliculus, allowing them to generate the bursts of action potentials that command the saccade. Thus, the projections from the substantia nigra pars reticulata to the upper motor neurons act as a physiological "gate" that must be "opened" to allow either sensory or other higher-order signals from cognitive centers to activate the upper motor neurons and initiate a saccade.

This brief account of the genesis of saccadic eye movements provides an important illustration of the principal functions of the basal ganglia in motor control: The basal ganglia facilitate the *initiation* of motor programs that express movement and the *suppression* of competing or non-synergistic motor programs that would otherwise interfere with the expression of sensory-driven or goal-directed behavior (see Box 18B). Chapter 20 provides a more complete account of sensorimotor integration and the origins of eye movements; the remainder of this chapter explains how the intrinsic and accessory circuits of the basal ganglia accomplish these principal functions in motor control and why disease that afflicts elements of these circuits can lead to devastating movement disorders.

FIGURE 18.5   A chain of nerve cells arranged in a disinhibitory circuit At the top is a diagram of the connections between neurons A and B and an excitatory neuron, C, which activates D, an upper motor neuron in the cortex. The colored boxes below diagram the pattern of action potential activity in A, B, C, and D both when neuron A is at rest and when neuron A fires transiently as a result of excitatory input. Such circuits are central to the gating operations of the basal ganglia.

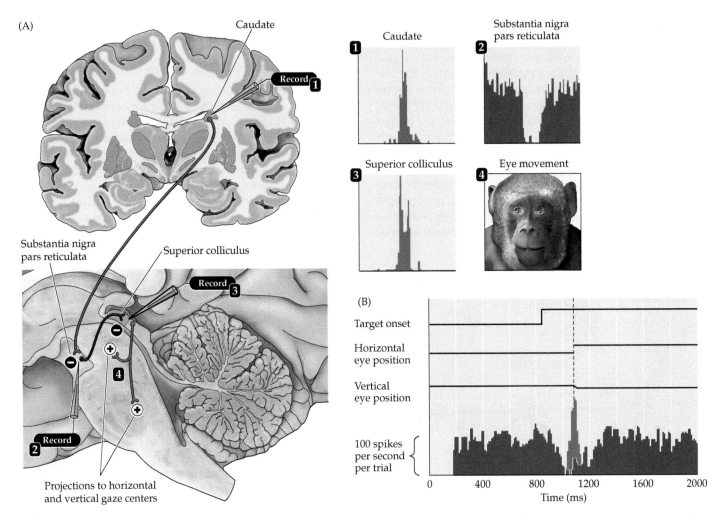

FIGURE 18.6 Role of basal ganglia disinhibition in the generation of saccadic eye movements (A) Medium spiny cells in the caudate respond with a transient burst of action potentials to an excitatory input from the cerebral cortex (1). Spiny cells inhibit tonically active GABAergic cells in the substantia nigra pars reticulata (2). As a result, the upper motor neurons in the deep layers of the superior colliculus are no longer tonically inhibited and can generate the bursts of action potentials that command a saccade (3, 4). (B) The graph shows the temporal relationship between inhibition in the substantia nigra pars reticulata (purple) and disinhibition in the superior colliculus (light blue) preceding a saccade to a visual target. (A1 after O. Hikosaka and R. H. Wurtz, 1986. *Exp Brain Res* 63: 659–662; A2–3 and B after O. Hikosaka and R. H. Wurtz, 1983. *J Neurophysiol* 49: 1285–1301.)

CONCEPT **18.3** | **Direct and Indirect Pathways Regulate the Initiation of Voluntary Movement and the Suppression of Unwanted Movement**

LEARNING OBJECTIVES

18.3.1 Identify the components of the direct and indirect pathways.

18.3.2 Discuss the neurophysiological means by which activation of the direct pathway facilitates the expression of voluntary movement.

18.3.3 Discuss the neurophysiological means by which activation of the indirect pathway facilitates the suppression of unwanted movement.

Circuits within the basal ganglia system

The projections from the medium spiny neurons of the caudate and putamen to the internal segment of the globus pallidus constitute the so-called *direct pathway* through the basal ganglia and, as illustrated in Figure 18.4, serve to release from tonic inhibition the thalamic neurons that drive cortical circuits of upper motor neurons. Thus, this direct pathway provides a means for the basal ganglia to facilitate the initiation of volitional movement.

Additional circuits of the basal ganglia constitute a so-called *indirect pathway* linking the caudate and putamen to the internal segment of the globus pallidus (Figure 18.7). This second pathway increases the level of tonic inhibition mediated by the projection neurons of the internal segment (and the substantia nigra pars reticulata). In the indirect pathway, a distinct population of medium spiny

Direct and indirect pathways

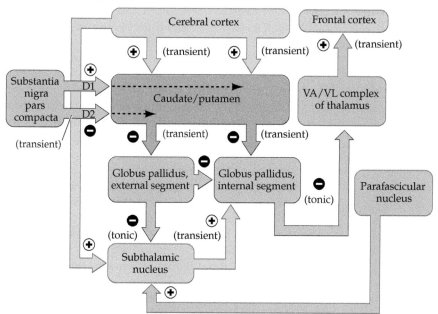

FIGURE 18.7 Disinhibition in the direct and indirect pathways through the basal ganglia In the direct pathway (beginning with the right downward arrow from Caudate/putamen), transiently inhibitory neurons in the caudate and putamen project to tonically active inhibitory neurons in the *internal* segment of the globus pallidus, which project in turn to the VA/VL complex of the thalamus. Transiently excitatory inputs to the caudate and putamen from the cortex are more efficacious when dopaminergic inputs from the substantia nigra pars compacta are co-activated and dopamine binds to D1 receptors. With activation of this direct pathway, excitatory inputs from the thalamus back to the cortex are transiently active. In the indirect pathway (beginning with the left downward arrow from the Caudate/putamen), transiently active inhibitory neurons from the caudate and putamen project to tonically active inhibitory neurons of the *external* segment of the globus pallidus. Note that the influence of nigral dopaminergic input to the D2-expressing striatal neurons in the indirect pathway is suppressive. Neurons in the external segment of the globus pallidus project to the subthalamic nucleus, which also receives a strong excitatory input from the cerebral cortex and the parafascicular nucleus of the thalamus. The subthalamic nucleus in turn projects to the internal segment of the globus pallidus, where its transiently excitatory drive serves to reinforce the tonic output of the internal segment. Thus, the functions of the direct and indirect pathways, which are otherwise in opposition to one another, become synergistic when dopamine is released onto the dendrites of medium spiny neurons in the striatum.

the output nuclei that provide the means by which the basal ganglia gain access to upper motor neurons. But as will become clear in the following discussion, *the indirect pathway antagonizes the activity of the direct pathway;* together, they function to open or shut the physiological gates that initiate and suppress movements.

The indirect pathway through the basal ganglia modulates the disinhibitory actions of the direct pathway. The subthalamic nucleus neurons that project to the internal segment of the globus pallidus and substantia nigra pars reticulata use glutamate as their neurotransmitter and are excitatory. When signals from the cortex activate the indirect pathway, the striatal medium spiny neurons discharge and inhibit the tonically active GABAergic neurons of the external globus pallidus. As a result of the removal of this tonic inhibition and the simultaneous arrival of excitatory inputs from the cerebral cortex and parafascicular nucleus, the subthalamic cells become more active, and by virtue of their excitatory synapses with the GABAergic cells of the internal segment of the globus pallidus and substantia nigra pars reticulata, they increase the inhibitory outflow of the basal ganglia. Concurrently, the inhibition of neurons in the external globus pallidus that project to the internal segment likewise serves to increase the activity of pallidal output neurons. In contrast to the direct pathway, which, when activated, releases thalamocortical and collicular circuits from tonic inhibition, the indirect pathway has the net effect of increasing the inhibitory influences of the basal ganglia on downstream motor centers. The balance of activity mediated by the direct and indirect pathways is the principal determinant of whether output from the pallidum to the thalamus or superior colliculus will select and facilitate the expression of the intended motor program.

These circuits not only facilitate the selection of a motor program; they also suppress competing motor programs that could interfere with the expression of sensory-driven or goal-oriented behavior. A concept called *focused selection* has increased understanding of this antagonistic interaction. According to this concept, the direct and indirect pathways are functionally organized in a center–surround fashion within the output nuclei of the basal ganglia (Figure 18.8). The influence of the direct

neurons projects to the lateral division of the globus pallidus, called the **external segment**. The external segment of the globus pallidus sends projections to both the adjacent internal segment and the **subthalamic nucleus** of the ventral thalamus (see Figure 18.1). The subthalamic nucleus also receives excitatory projections from the cerebral cortex (sometimes referred to as the *hyperdirect pathway*) and the parafascicular nucleus of the thalamus (recently termed the *super-direct pathway*). These excitatory inputs from the cortex and thalamus account for the ongoing activity of neurons in the subthalamic nucleus. In turn, the subthalamic nucleus projects diffusely back to the internal segment of the globus pallidus and to the substantia nigra pars reticulata. Thus, the indirect pathway feeds back onto pathway is tightly focused on particular functional units in the internal segment of the globus pallidus (and the substantia nigra pars reticulata), whereas the influence of the indirect pathway is much more diffuse, covering a broader range of functional units. Recall that individual axons from the striatum to the internal segment of the globus pallidus tend to synapse densely on single pallidal neurons while making sparse contacts on numerous pallidal cells; this provides a means for the direct pathway to focus its input on a "central" functional unit at the output stage of the basal ganglia. In contrast,

afferents from the subthalamic nucleus are distributed much more evenly throughout the internal segment, providing a means for the indirect pathway to suppress the activity of a broader "surrounding" set of functional units. Accordingly, when basal ganglia systems receive and process cortical signals, the suppression of competing or recently activated motor programs is reinforced, and simultaneously, the activation of the particular thalamocortical (or collicular) circuits that underlie the intended movement is facilitated. Recent studies in rodent models using optogenetic methods to selectivity activate or inhibit the striatal neurons giving rise to the direct and indirect pathways indicate that co-activation of both sets of striatal neurons is important for the smooth initiation and execution of new motor actions.

Precisely how these complex circuits of the basal ganglia interact to assist upper motor neuron systems in the execution of volitional behavior remains poorly understood, and this simplified description will undoubtedly be subject to revision as further anatomical and physiological details become available. Nevertheless, this account serves as a useful model for understanding the architecture and function of neural systems that achieve fine control of their output by an interplay between neural excitation and inhibition (recall, for example, the center–surround antagonism of ganglion cell receptive fields in the retina; see Chapter 9). Furthermore, this model provides an instructive framework for understanding disorders of movement that result from injury or disease that afflicts one or more components of the basal ganglia system (see Concepts 18.5 and 18.6).

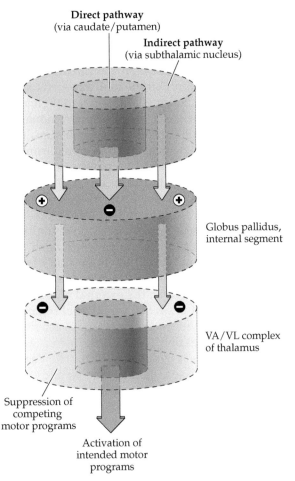

FIGURE 18.8 Center–surround functional organization of the direct and indirect pathways Integration of cortical input by the striatum leads to the co-activation of the direct and indirect pathways. With activation of the indirect pathway, neurons in a "surround" region of the internal segment of the globus pallidus are driven by excitatory inputs from the subthalamic nucleus; this reinforces the suppression of a broad set of competing motor programs. Simultaneously, activation of the direct pathway leads to the focal inhibition of a more restricted "center" cluster of neurons in the internal segment; this in turn results in the disinhibition (bottom arrow) of the VA/VL complex and the expression of the intended motor program.

CONCEPT 18.4 | Dopamine Modulates Basal Ganglia Circuits by Increasing or Decreasing the Excitability of Medium Spiny Neurons

LEARNING OBJECTIVES

18.4.1 Characterize the impacts of dopamine binding on the medium spiny neurons that differentially express D1 or D2 receptors.

18.4.2 Discuss the critical role of dopamine in shifting the balance between direct and indirect pathway activities in favor of the expression of movement.

Dopamine facilitates the expression of movement

As described in Concept 18.1, an important circuit within the basal ganglia system involves the dopaminergic cells in the pars compacta subdivision of the substantia nigra. Although this circuit derives from a relatively small pool of dopaminergic neurons, it exerts a profound influence over the integration of cortical input in the striatum. The medium spiny neurons of the striatum (especially from striosome compartments) project directly to the substantia nigra pars compacta, which in turn sends widespread dopaminergic projections back to the medium spiny neurons. The effects of dopamine on the spiny neurons are complex; they illustrate the principle that the action of a neurotransmitter is determined by the types of receptors expressed in postsynaptic neurons and by the downstream signaling pathways to which the receptors are linked (see Chapter 6). In this case, the same nigral neurons can *increase* the excitability of the spiny cells that project to the internal globus pallidus (the direct pathway) and *decrease* the excitability of the spiny cells that project to the external globus pallidus (the indirect pathway). This duality is achieved by the differential expression of two types of dopamine receptors—types D1 and D2—by the medium spiny neurons of the direct and indirect pathways, respectively.

Both D1 and D2 dopamine receptors are members of the 7-transmembrane G-protein-coupled family of cell surface receptors. The major functional difference between them is that the D1 receptors mediate the activation of G-proteins that *increase* cAMP, while D2 receptors act through different G-proteins that *decrease* cAMP. For both types of receptors, the dopaminergic synapses on medium spiny neurons tend to be located on the shafts of the spines that receive synaptic input from the cerebral cortex. This arrangement suggests that dopamine exerts its effects on the spiny neurons by modulating their responses to cortical input, with D1 receptors positioned to enhance the excitatory input from cortex and D2 receptors positioned to suppress this excitation. Since the actions of the direct and indirect pathways on the output of the basal ganglia are antagonistic, these different influences of dopamine on medium spiny neurons are functionally synergistic, producing a decrease in the inhibitory outflow of the basal ganglia, and the consequent release of projections from the thalamus to the frontal cortex or projections from the superior colliculus to circuits of lower motor neurons in the brainstem.

This dopaminergic input to the striatum may contribute to the modulation of motivated behavior. For example, in monkeys the latencies of saccades toward a target are shorter when the goal of the movement is associated with a larger reward. This effect is eliminated by caudate injections of the dopamine D1 receptor antagonist and enhanced by injections in the same site of the D2 receptor antagonist. These results suggest that the influence of motivation on motor performance may be modulated by circuits in the basal ganglia that recruit dopaminergic input from the midbrain. The role of dopamine in motivated behavior and the deleterious impact of addictive substances on dopaminergic modulation of basal ganglia function will be discussed in more detail in Chapter 32.

CONCEPT
18.5

Hypokinetic Movement Disorders Are Associated with Excessive Inhibition of Motor Nuclei in the Thalamus and Brainstem

LEARNING OBJECTIVES

18.5.1 Explain how the dysfunction of basal ganglia circuitry may lead to the expression of hypokinetic movement disorders.

18.5.2 Discuss the principal behavioral signs and symptoms associated with Parkinson's disease.

18.5.3 Discuss factors that may lead to the degeneration of midbrain dopaminergic neurons.

The modulatory influences of the dopaminergic circuit interconnecting the substantia nigra pars compacta and the striatum may also help explain many of the manifestations of basal ganglia disorders, especially those characterized by decreased voluntary movement (*hypokinesia*). For example, **Parkinson's disease** is the second most common degenerative disease of the nervous system (Alzheimer's disease being the leader). Described by James Parkinson in 1817, this disorder is characterized by slowness of movement (*bradykinesia*), rigidity of the extremities and neck, minimal facial expressions, and—somewhat paradoxically—tremor at rest, typically involving the fingers or head and neck. Walking entails short steps, stooped posture, and a paucity of associated movements such as arm swinging. In some individuals, these abnormalities of motor function are associated with dementia. Following a gradual onset, typically between the ages of 50 and 70, the disease progresses slowly and often culminates in death some 10 to 20 years later.

Unlike in many other neurodegenerative diseases (such as Alzheimer's disease and amyotrophic lateral sclerosis), the underlying degeneration in Parkinson's disease involves a single population of neurons, especially in the early to middle course of the disease: the dopaminergic neurons of the substantia nigra pars compacta and the adjacent ventral tegmental area of the midbrain. Thus, idiopathic Parkinson's disease is caused by the loss of the nigrostriatal dopaminergic neurons (Figure 18.9A). Although the cause of the progressive deterioration of these dopaminergic neurons is not known, genetic investigations provide clues to the etiology and pathogenesis. Whereas the majority of cases of Parkinson's disease are sporadic, there may be specific forms of susceptibility genes that confer increased risk of acquiring the disease, just as the *e4* allele of the *ApoE* gene increases the risk of Alzheimer's disease (see Chapter 30). Familial forms of Parkinson's caused by single gene mutations account for fewer than 10% of all cases; identification of these rare genes, however, is likely to provide insight into molecular pathways

(A)

Parkinson's   Without Parkinson's

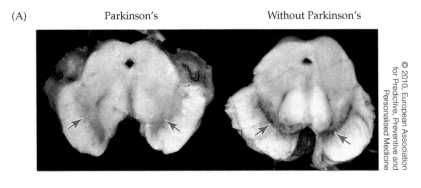

© 2010, European Association for Predictive, Preventive and Personalised Medicine

(B) Parkinson's disease (hypokinetic)

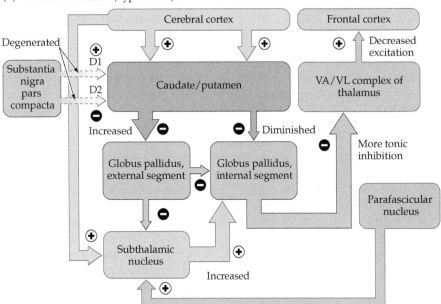

FIGURE 18.9 Degeneration of dopaminergic neurons reduces voluntary movement in Parkinson's disease (A) In the midbrain of an individual with Parkinson's disease, the substantia nigra pars compacta (heavily pigmented division) is largely absent in the region above the cerebral peduncles. The midbrain from an individual without Parkinson's disease shows the dark pigment (neuromelanin) that accumulates in the cell bodies of dopaminergic neurons of the substantia nigra pars compacta (cf. regions indicated with red arrows). (B) In Parkinson's disease, the loss of dopaminergic inputs provided by the substantia nigra pars compacta (dashed arrows) make it more difficult to generate the transient inhibition from the direct pathway from the caudate and putamen to the internal segment of the globus pallidus. The result of this change in the direct pathway is to sustain or increase the tonic inhibition from the internal segment of the globus pallidus to the thalamus (thicker arrow than corresponding arrow in Figure 18.7), making thalamic excitation of the motor cortex less likely (thinner arrow from thalamus to frontal cortex). (B after M. R. DeLong, 1990. *Trends Neurosci* 13: 281–285.)

that may underlie the disease. Mutations of three distinct genes—α-*synuclein*, *Parkin*, and *DJ-1*—have been implicated in rare forms of Parkinson's disease. Their identification provides an opportunity to generate transgenic mice carrying the mutant form of the human gene, potentially providing an animal model in which the pathogenesis can usefully be elucidated and therapies can be tested.

As described in Concept 18.4, activation of the nigrostriatal projection leads to opposite but synergistic effects on the direct and indirect pathways: The release of dopamine in the striatum increases the responsiveness of the direct pathway to corticostriatal input (a D1 effect) while decreasing the responsiveness of the indirect pathway (a D2 effect). Typically, both of these dopaminergic effects serve synergistically to decrease the inhibitory outflow of the basal ganglia and thus—by means of disinhibition of thalamic and brainstem nuclei—to increase the activation of upper motor neurons. In contrast, when the dopaminergic cells of the pars compacta are destroyed, as occurs in Parkinson's disease, the inhibitory outflow of the basal ganglia is abnormally high, and timely thalamic activation

of upper motor neurons in the motor cortex and brainstem is therefore less likely (Figure 18.9B).

In fact, many of the symptoms seen in Parkinson's disease and other hypokinetic movement disorders reflect a failure of the disinhibition normally mediated by the basal ganglia. Thus, individuals with Parkinson's tend to have diminished facial expressions and reduced amplitude of movement, such as diminished arm swinging during walking. Indeed, any movement is difficult to initiate and, once initiated, is often difficult to terminate. Disruption of the same circuits also increases the discharge rate of the inhibitory cells in the substantia nigra pars reticulata. The resulting increase in tonic inhibition reduces the excitability of upper motor neurons in the superior colliculus, thus reducing the frequency and amplitude of saccades.

Support for this explanation of hypokinetic movement disorders such as Parkinson's disease comes from studies of monkeys in which degeneration of the dopaminergic cells of the substantia nigra pars compacta has been induced by the neurotoxin 1-methyl-4-phenyl-1,2,3,6-tetrahydropyridine (MPTP). Monkeys (or humans) exposed to

MPTP develop symptoms that are very similar to those of individuals with Parkinson's disease. Furthermore, a second lesion placed in the subthalamic nucleus results in significant improvement in the ability of these animals to initiate movement, as would be expected based on the circuitry of the indirect pathway (see Figure 18.9B).

In humans, rather than creating lesions, neurologists and neurosurgeons are now increasingly using deep brain stimulation to normalize permissive patterns of neural activity in basal ganglia circuits, with the subthalamic nucleus most commonly targeted for neuromodulation (Clinical Applications).

■ Clinical Applications

Deep Brain Stimulation

Since the seminal demonstration by G. Fritsch and E. Hitzig in the nineteenth century that body movements could be induced when electrical currents were applied to cerebral tissue (in this case, motor cortex), clinicians have considered the possibility that certain neurological disorders affecting volitional movement could be treated with the application of acute or chronic electrical stimulation to key motor centers in the brain. However, it was not until well after the internally implanted cardiac pacemaker was introduced in the 1960s that technological advances made possible the implantation of a comparable device for the focal stimulation of brain structures. Such devices were introduced in the 1990s

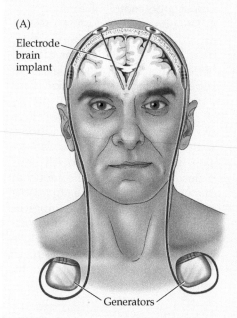

(A)

Electrode brain implant

Generators

(A) Illustration of a patient following implantation of a device for deep brain stimulation.

for the treatment of movement disorders, with their targets being components of the basal ganglia and thalamus located deep in the forebrain; hence the common term for this intervention, *deep brain stimulation.*

Prior to the 1990s, the treatment options for individuals with movement disorders were limited to pharmacological intervention, physical therapy, and in the most intractable cases, neurosurgical ablation of sites in the basal ganglia and thalamus that gate the initiation of movement. For intractable cases, the introduction of deep brain stimulation provided an obviously welcome alternative to the permanent destruction of brain structure. Currently, a combination of deep brain stimulation, pharmacological intervention, and physical therapy improves the health status of a significant number of individuals (more than would the use of just one approach).

Deep brain stimulation entails the implantation of battery-powered generator units, usually near the clavicles. These units produce electrical discharges that are passed through subcutaneous wires to electrodes implanted bilaterally into the brain (Figure A). (Recall that, with the exception of medial divisions of the premotor cortex, all of the cortical and subcortical neural circuitry in the forebrain that governs the activity of upper motor neurons is organized unilaterally; therefore, bilateral deep brain stimulation is necessary to achieve symmetrical results.) The placement of the electrodes requires careful stereotaxic surgery combined with radiological imaging of the patient's brain and electrophysiological recordings of

spontaneous and movement-related neuronal activity. The neuronal recordings are essential so that the neurosurgical team can recognize by sight and sound the characteristic discharge patterns of neurons in different nuclei of the basal ganglia and thalamus. This is done by displaying action potential waveforms on oscilloscopes and audio and computer monitors (see Figure 18.5). Once the target structures are localized, stimulation is tested to determine whether the desired clinical effect can be observed. After the patient has recovered from the implantation procedure, the generator units are activated and the parameters of stimulation are fine-tuned as various combinations of pulse widths, current amplitudes, and temporal patterns of pulse trains are employed and modified as needed.

A careful consideration of Figures 18.9B and 18.10B suggests several possible target sites of deep brain stimulation for individuals with hypokinetic or hyperkinetic movement disorders. In both diagnostic categories, neural activity in motor nuclei of the thalamus is abnormal; accordingly, the VA/VL complex of the thalamus is an appropriate target. However, abnormal activity in the thalamus is often the consequence of disturbances within the basal ganglia themselves, and it is arguably more desirable to manipulate the influence of the basal ganglia on thalamocortical circuits rather than directly altering activity in upper motor neuronal circuits by exogenous electrical stimulation. Thus, the two most common sites for deep brain stimulation in individuals with movement

(Continued)

■ Clinical Applications (continued)

disorders are the internal segment of the globus pallidus and the subthalamic nucleus. Preclinical studies in animal models are suggesting that the parafascicular nucleus of the thalamus, which gives rise to major excitatory projections to the subthalamic nucleus, may also serve as a strategic target for clinically efficacious deep brain stimulation. Regardless

of whether the neurological disorder to be rectified is hypokinetic (manifested as difficulty in expression of movement) or hyperkinetic (manifested as expression of unwanted movement), deep brain stimulation can be used to override intrinsic, pathological discharge patterns with stable and highly structured patterns of neural activity that better

facilitate the initiation and termination of volitional movement (Figure B).

Given the complexity of even a small volume of neural tissue (such as might be present near the tip of a stimulating electrode in the globus pallidus or subthalamic nucleus), it is not surprising that the exogenous induction of electrical currents can lead to complex patterns of activity and inactivity in the affected neural elements, as well as effects on non-neural elements (Figure C). Different intensities of deep brain stimulation may lead to the local release of neurotransmitters and neuromodulators. They may also lead to the generation of action potentials in afferent axons, neuronal cell bodies, efferent axons, and *fibers of passage* that originate elsewhere. However, the effects of electrical stimulation on certain intrinsic membrane properties, including voltage-dependent ion conductances, may block the generation of action potentials, thus silencing the affected neurons. Ideally, the net effect of these diverse changes is the amelioration of abnormal network activity that hinders the normal operation of upper motor neurons.

Despite ongoing uncertainties regarding its mechanisms of action, deep brain stimulation has offered hope to thousands of individuals who suffer neurological dysfunction ranging from the movement disorders discussed here to related disorders of non-motor basal ganglia loops, such as Tourette syndrome, depression, and obsessive-compulsive disorder (see Box 18A). The fact that stimulation protocols are adjustable provides clinicians with an unprecedented ability to manipulate the activities and functions of basal ganglia circuits whose operations are crucial to the typical expression of thought, emotion, and motor behavior.

(B) Prestimulation

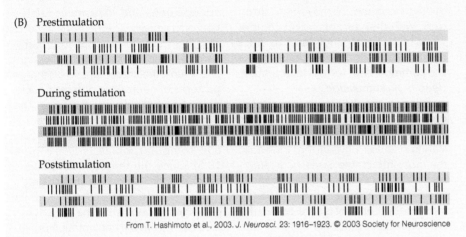

From T. Hashimoto et al., 2003. *J. Neurosci.* 23: 1916–1923. © 2003 Society for Neuroscience

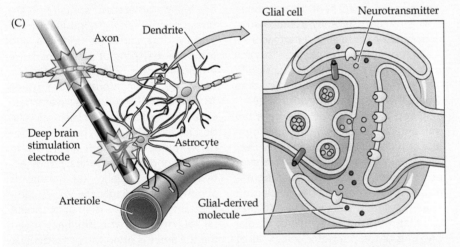

(B) Raster plots of action potentials recorded from a neuron in the internal segment of the globus pallidus in an awake rhesus monkey rendered Parkinsonian by the systemic administration of MPTP; each row lasts 1 s. The endogenous pattern of discharge is marked by irregular clusters of burst activity (top panel). Within seconds of the onset of subthalamic nucleus stimulation, the Parkinsonian symptoms abated, and the discharge of the pallidal neuron became much more regular (middle panel). (C) Deep brain stimulation induces myriad effects in brain tissue, including the direct modulation of action potential generation (green axon) and synaptic transmission (inset), the direct modulation of the release of local substances from glial cells (blue cell) that modify synaptic plasticity (inset), and the indirect modulation (mediated by astrocytic end feet; see Appendix) of arteriole tone affecting local blood flow. (C after C. C. McIntyre and R. W. Anderson, 2016. *J Neurochem* 139: 338–345.)

Other novel and promising therapeutic approaches include gene therapy and stem cell grafts. Gene therapy involves modifying a disease phenotype by introducing new genetic information into the affected organism. Although still in its infancy, this approach has the potential to revolutionize treatment of human disease. One such therapy proposed for Parkinson's disease would enhance release of dopamine in the caudate and putamen. In principle, this could be accomplished by implanting cells that have been genetically modified to express tyrosine hydroxylase, the enzyme that converts tyrosine to L-DOPA, which in turn is converted by a nearly ubiquitous decarboxylase into the neurotransmitter dopamine (see Figure 6.14). An alternative strategy involves "neural grafts" using stem cells. Stem cells are self-renewing, multipotent progenitors with broad developmental potential (see Chapters 22 and 26). This approach entails identifying and isolating stem cells, and identifying the growth factors needed to promote differentiation into the desired phenotype (i.e., dopaminergic neurons, in this application). The identification and isolation of multipotent mammalian stem cells have already been accomplished, and several factors likely to be important in differentiation of midbrain precursors into dopamine neurons have now been identified. Establishing the efficacy of this approach for individuals with Parkinson's disease would increase the possibility of its application to other neurodegenerative diseases.

CONCEPT 18.6	## Hyperkinetic Movement Disorders Are Associated with Insufficient Inhibition of Motor Nuclei in the Thalamus and Brainstem

LEARNING OBJECTIVES

18.6.1 Explain how the dysfunction of basal ganglia circuitry may lead to the expression of hyperkinetic movement disorders.

18.6.2 Discuss the principal behavioral signs and symptoms associated with Huntington's disease.

18.6.3 Discuss factors that may lead to the degeneration of D2-expressing medium spiny neurons.

Given the preceding discussion of how an overactive pallidum can suppress voluntary movement, it should not be surprising to learn that insufficient tonic output from the pallidum permits the expression of unwanted movement. Thus, knowledge of the architecture and neurophysiology of basal ganglia circuits also helps explain the motor abnormalities seen in *hyperkinetic* movement disorders, such as **Huntington's disease**.

In 1872, a physician named George Huntington described a group of patients seen by his father and grandfather in their practice in East Hampton, Long Island. The disease he defined, which became known as Huntington's disease (HD), is characterized by the gradual onset of deleterious changes in behavior, cognition, and movement beginning in the fourth and fifth decades of life. The disorder is inexorably progressive, resulting in death within 10 to 20 years. HD is inherited in an autosomal dominant pattern, a feature that has led to a much better understanding of its cause in genetic and molecular terms.

One of the more common inherited neurodegenerative diseases, HD usually presents as an alteration in mood (especially depression) or a change in personality that often takes the form of increased agitation, irritability, suspiciousness, and impulsive or eccentric behavior. Impairments of memory, attention, reasoning, and judgment may also occur. The hallmark of the disease, however, is a movement disorder consisting of unwanted rapid, rhythmical, or jerky motions of various body parts, with no clear purpose. These *choreiform* ("dancelike") movements may be confined to a finger or may involve a whole extremity, the facial musculature, or even the vocal apparatus. The movements themselves are involuntary, but the individual often incorporates them into apparently deliberate actions, presumably in an effort to obscure the condition. Typically, there is no associated weakness, ataxia, or sensory impairment.

A distinctive neuropathology is associated with these clinical manifestations: a profound but selective atrophy of the caudate and putamen, with some associated degeneration of the frontal and temporal cortices (Figure 18.10A). However, not all striatal neurons are equally susceptible, especially early in the course of disease. In individuals with HD, D2-expressing medium spiny neurons that project to the external segment of the globus pallidus degenerate. In the absence of their normal inhibitory input from these spiny neurons, the external globus pallidus cells become abnormally active; this activity reduces in turn the excitatory output of the subthalamic nucleus to the internal segment of the globus pallidus, and the inhibitory outflow of the basal ganglia is reduced (Figure 18.10B). Without the restraining influence of the basal ganglia, upper motor neurons can be activated by inappropriate signals, resulting in the undesired ballistic (flailing or jerking) and choreiform movements that characterize the disease.

The availability of extensive HD pedigrees has allowed geneticists to decipher the molecular cause of this disease. HD was one of the first human diseases in which DNA polymorphisms were used to localize the mutant gene, which in 1983 was mapped to the short arm of chromosome 4. This discovery led to an intensive effort to identify the HD gene within this region by positional cloning. Ten years later, these efforts culminated in identification of the gene (named *Huntingtin*) responsible for the disease. The *Huntingtin* mutation is an unstable triplet repeat present within the coding region of the gene consisting of a DNA

(A)

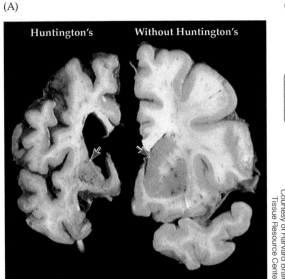

Huntington's   Without Huntington's

Courtesy of Harvard Brain
Tissue Resource Center

(B) Huntington's disease (hyperkinetic)

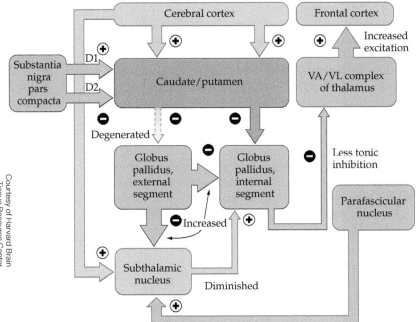

FIGURE 18.10 Degeneration of medium spiny neurons increases involuntary movement in Huntington's disease
(A) The size of the caudate and putamen (the striatum) is dramatically reduced in individuals with advanced Huntington's disease (cf. regions indicated with red arrows). In advanced disease (left slab), widespread neurodegeneration reduces gray matter and white matter throughout the cerebral hemisphere. (B) In Huntington's disease, medium spiny neurons of the indirect pathway degenerate (dashed arrow). This effect increases the tonic inhibition from the external segment of the globus pallidus to the subthalamic nucleus and the internal segment of the globus pallidus (thicker arrows than in Figure 18. 7). Consequently, the tonic inhibition of the VA/VL complex of the thalamus by the internal segment of the globus pallidus is insufficient to suppress thalamocortical activation (thinner arrow). Thus, thalamic excitation of the cortex is increased (thicker arrow), leading to the expression of unwanted motor activity. (B after M. R. DeLong, 1990. *Trends Neurosci* 13: 281–285.)

segment (CAG) that codes for the amino acid glutamine. In typical individuals, *Huntingtin* contains between 15 and 34 repeats, whereas in individuals with HD the gene contains 42 or more CAG repeats. The mechanism by which the increased number of polyglutamine repeats injures neurons is not clear. The leading hypothesis is that the increased numbers of glutamines alter protein folding, which somehow triggers a cascade of molecular events culminating in dysfunction and neuronal death.

Just as in Huntington's disease, imbalances in the fine control mechanism represented by the convergence of the direct and indirect pathways in the pallidum are apparent in other hyperkinetic movement disorders caused by diseases that affect primarily the subthalamic nucleus. The pathophysiology renders dysfunctional a source of excitatory input to the internal segment of the globus pallidus and substantia nigra pars reticulata, thus abnormally reducing the inhibitory outflow of the basal ganglia. A basal ganglia syndrome called **hemiballismus**, which is characterized by ballistic involuntary movements of the limbs, is the result of damage to the subthalamic nucleus. As in Huntington's disease, the involuntary movements of hemiballismus are initiated by the abnormal discharges of upper motor neurons that are receiving less than adequate governance via the tonic inhibition that the basal ganglia normally exert over the motor nuclei of the thalamus and brainstem.

As predicted by these accounts of hypokinetic and hyperkinetic movement disorders, GABA agonists and antagonists applied to substantia nigra pars reticulata of monkeys produce symptoms similar to those seen in human basal ganglia disease. For example, intranigral injection of bicuculline, which blocks the GABAergic inputs from the striatal medium spiny neurons to the reticulata cells, increases the amount of tonic inhibition on the upper motor neurons in the deep collicular layers. These animals exhibit fewer and slower saccades, reminiscent of individuals with Parkinson's disease. In contrast, injection of the GABA agonist muscimol into the substantia nigra pars reticulata decreases the tonic GABAergic inhibition of the upper motor neurons in the superior colliculus, with the result that the injected monkeys generate spontaneous, irrepressible saccades that resemble the involuntary movements characteristic of basal ganglia diseases such as hemiballismus and Huntington's disease (Figure 18.11).

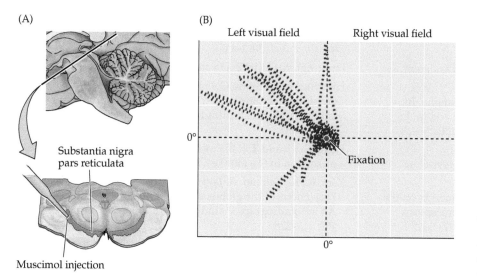

(A)

Substantia nigra
pars reticulata

Muscimol injection

(B)

Left visual field   Right visual field

0°

Fixation

0°

FIGURE 18.11   Infusing GABA agonist into the pallidum produces involuntary movements resembling hyperkinesia When the tonically active cells of the right substantia nigra pars reticulata are inactivated by an intranigral injection of the GABA agonist muscimol (A), the upper motor neurons in the deep layers of the right superior colliculus are disinhibited and the monkey generates spontaneous, irrepressible saccades (B). Because the cells in both the substantia nigra pars reticulata and the deep layers of the superior colliculus are arranged in spatially organized motor maps of saccade vectors (see Chapter 20), the direction of the involuntary saccades—in this case, mainly toward the upper left quadrant of the visual field—depends on the precise location of the injection site within the substantia nigra.

Such studies of the anatomical and physiological organization of the motor and oculomotor loops in health and disease provide a foundation for investigating the anterior and ventral circuitry of the basal ganglia that subserve a variety of non-motor functions in other aspects of behavior (see Box 18A). Thus, each functional loop through the basal ganglia is likely to exert a similar influence on the selection, initiation, and suppression of motor or non-motor programs that are instantiated at the level of the cerebral cortex. Moreover, the functional or clinical implications may be equally significant should injury, disease, or neurochemical imbalance impair the function of one or more components of the diverse loops that mediate the influence of the basal ganglia on cortical function and, ultimately, behavior.

Summary

The contributions of the basal ganglia to motor control are reflected in the deficits that result from damage to the component nuclei. Such lesions compromise the initiation and performance of voluntary movements, as exemplified by the paucity of movement typical of Parkinson's disease and in the inappropriate "release" of movements characteristic of Huntington's disease. The organization of the basic circuitry of the basal ganglia indicates how this constellation of nuclei modulates movement. With respect to motor function, the system forms a loop that originates in almost every area of the cerebral cortex and eventually terminates, after enormous convergence within the basal ganglia, on the upper motor neurons in the motor areas of the frontal lobe and in the superior colliculus. The efferent neurons of the basal ganglia influence the upper motor neurons in the cortex indirectly by gating the flow of information through relays in the ventral nuclei of the thalamus. The upper motor neurons in the superior colliculus that initiate saccadic eye movements are controlled by monosynaptic projections from the substantia nigra pars reticulata. In each case, the basal ganglia loops regulate movement by a process of disinhibition that results from the serial interaction within the basal ganglia circuitry of two sets of GABAergic neurons. Internal circuits within the basal ganglia system modulate the amplification of the signals that are transmitted through the loops.

■ Additional Reading

Reviews

Alexander, G. E. and M. D. Crutcher (1990) Functional architecture of basal ganglia circuits: Neural substrates of parallel processing. *Trends Neurosci.* 13: 266–271.

Cattaneo, E., C. Zuccato and M. Tartari (2005) Normal huntingtin function: An alternative approach to Huntington's disease. *Nat. Rev. Neurosci.* 6: 919–930.

DeLong, M. R. (1990) Primate models of movement disorders of basal ganglia origin. *Trends Neurosci.* 13: 281–285.

Gerfen, C. R. and C. J. Wilson (1996) The basal ganglia. In *Handbook of Chemical Neuroanatomy*, L. W. Swanson, A. Björklund and T. Hokfelt (Eds.). Vol. 12: *Integrated Systems of the CNS*, part III. New York: Elsevier Science Publishers, pp. 371–468.

Goldman-Rakic, P. S. and L. D. Selemon (1990) New frontiers in basal ganglia research. *Trends Neurosci.* 13: 241–244.

Graybiel, A. M. and C. W. Ragsdale (1983) Biochemical anatomy of the striatum. In *Chemical Neuroanatomy*, P. C. Emson (Ed.). New York: Raven Press, pp. 427–504.

Grillner, S. and 4 others (2005) Mechanisms for selection of basic motor programs: Roles for the striatum and pallidum. *Trends Neurosci.* 28: 364–370.

Hardy, J. (2010) Genetic analysis of pathways to Parkinson disease. *Neuron* 68: 201–206.

Hikosaka, O. and R. H. Wurtz (1989) The basal ganglia. In *The Neurobiology of Eye Movements*, R. H. Wurtz and M. E. Goldberg (Eds.). New York: Elsevier Science Publishers, pp. 257–281.

Kaji, R. (2001) Basal ganglia as a sensory gating devise for motor control. *J. Med. Invest.* 48: 142–146.

Ledonne, A. and N. B. Mercuri (2017) Current concepts on the physiopathological relevance of dopaminergic receptors. *Front. Cell. Neurosci.* 11: 27. doi: 10.3389/fncel.2017.00027

Mink, J. W. and W. T. Thach (1993) Basal ganglia intrinsic circuits and their role in behavior. *Curr. Opin. Neurobiol.* 3: 950–957.

Pollack, A. E. (2001) Anatomy, physiology, and pharmacology of the basal ganglia. *Neurol. Clin.* 19: 523–534.

Schapira, A. H. V., K. R. Chaudhuri and P. Jenner (2017) Non-motor features of Parkinson disease. *Nat. Rev. Neurosci.* 18: 435–450.

Shepherd, G. M. G. (2013) Corticostriatal connectivity and its role in disease. *Nat. Rev. Neurosci.* 14: 278–291.

Slaght, S. J. and 5 others (2002) Functional organization of the circuits connecting the cerebral cortex and the basal ganglia. Implications for the role of the basal ganglia in epilepsy. *Epileptic Disord.* Suppl 3: S9–S22.

Yin, H. H. (2017) The basal ganglia in action. *Neuroscientist* 23: 299–313.

Important Original Papers

Anden, N.-E. and 5 others (1966) Ascending monoamine neurons to the telencephalon and diencephalon. *Acta Physiol. Scand.* 67: 313–326.

Brodal, P. (1978) The corticopontine projection in the rhesus monkey: Origin and principles of organization. *Brain* 101: 251–283.

Crutcher, M. D. and M. R. DeLong (1984) Single cell studies of the primate putamen. *Exp. Brain Res.* 53: 233–243.

DeLong, M. R. and P. L. Strick (1974) Relation of basal ganglia, cerebellum, and motor cortex units to ramp and ballistic movements. *Brain Res.* 71: 327–335.

DiFiglia, M., P. Pasik and T. Pasik (1976) A Golgi study of neuronal types in the neostriatum of monkeys. *Brain Res.* 114: 245–256.

Hughes, R. N. and 6 others (2020) Ventral tegmental dopamine neurons control the impulse vector during motivated behavior. *Curr. Biol.* 30: 1–14.

Huntington, G. (1872) On chorea. *Med. Surg. Reporter* 26: 317.

Huntington's Disease Collaborative Research Group (1993) A novel gene containing a trinucleotide repeat that is expanded and unstable on Huntington's disease chromosomes. *Cell* 72: 971–983.

Kemp, J. M. and T. P. S. Powell (1970) The cortico-striate projection in the monkey. *Brain* 93: 525–546.

Kim, R., K. Nakano, A. Jayaraman and M. B. Carpenter (1976) Projections of the globus pallidus and adjacent structures: An autoradiographic study in the monkey. *J. Comp. Neurol.* 169: 217–228.

Kocsis, J. D., M. Sugimori and S. T. Kitai (1977) Convergence of excitatory synaptic inputs to caudate spiny neurons. *Brain Res.* 124: 403–413.

Mink, J. W. (1996) The basal ganglia: Focused selection and inhibition of competing motor programs. *Prog. Neurobiol.* 50: 381–425.

Nakamura, K. and O. Hikosaka (2006) Role of dopamine in the primate caudate nucleus in reward modulation of saccades. *J. Neurosci.* 26: 5360–5369.

Smith, Y., M. D. Bevan, E. Shink and J. P. Bolam (1998) Microcircuitry of the direct and indirect pathways of the basal ganglia. *Neuroscience* 86: 353–387.

Tecuapetla F., X. Jin, S. Q. Lima and R. M. Costa (2016) Complementary contributions of striatal projection pathways to action initiation and execution. *Cell* 166: 703–715.

Books

Bradley, W. G., R. B. Daroff, G. M. Fenichel and C. D. Marsden (Eds.) (1991) *Neurology in Clinical Practice.* Boston: Butterworth-Heinemann, chapters 29 and 77.

Donaldson, I., C. D. Marsden, K. P. Bhatia and S. A. Schneider (2012) *Marsden's Book of Movement Disorders.* Oxford, UK: Oxford University Press.

Klawans, H. L. (1989) *Toscanini's Fumble and Other Tales of Clinical Neurology.* New York: Bantam, chapters 7 and 10.

Steiner, H. and K. Tseng (Eds.) (2016) *Handbook of Basal Ganglia Structure and Function*, 2nd Edition. Amsterdam; Boston: Elsevier/Academic Press.

Modulation of Movement by the Cerebellum

Overview

In contrast to the upper motor neurons described in Chapter 17, the efferent cells of the cerebellum do not project directly to the local circuits of the brainstem and spinal cord that organize movement, nor do they directly contact the lower motor neurons that innervate muscles. Instead—like the basal ganglia—the cerebellum influences movements primarily by modifying the activity patterns of upper motor neurons. In fact, the cerebellum sends prominent projections to virtually all circuits that govern upper motor neurons. Anatomically, the cerebellum has two main gray matter structures: a laminated cortex on its surface and clusters of cells in nuclei buried deep in the white matter of the cerebellum. Pathways that reach the cerebellum from other brain regions (in humans, the largest contribution arises in the cerebral cortex) project to both components by means of afferent axons that send branches to both the deep nuclei and the cortex. Neurons in the deep nuclei are the main source of the output from the cerebellum. Their spatiotemporal patterns of activity are sculpted by descending input from the overlying cortex. In this way, the output of the cerebellum is integrated before being sent to circuits of upper motor neurons in the cerebral cortex, by way of thalamic relays, and in the brainstem. A primary function of the cerebellum is to detect the difference, or "motor error," between an intended movement and the actual movement, and through its influence over upper motor neurons, to reduce the error. These corrections can be made both during the course of the movement and as a form of motor learning when the correction is stored. When this feedback loop is damaged, as occurs in many cerebellar disorders and injuries, the afflicted individual makes persistent errors when executing behavior. The specific pattern of incoordination or dysmetria depends on the location of the damage, with impacts on body movement, thought, and/or affective regulation.

Courtesy of L.E. White

KEY CONCEPTS

19.1 The cerebellum comprises three major subdivisions: the cerebrocerebellum, spinocerebellum, and vestibulocerebellum

19.2 The cerebellar hemispheres coordinate movements of the ipsilateral body

19.3 Efferent output from the cerebellum to the brainstem and thalamus originates in the deep cerebellar nuclei and the vestibulocerebellum

19.4 Purkinje neurons integrate afferent input and modulate the output of deep cerebellar nuclei

19.5 The cerebellum coordinates ongoing movement by reducing motor error

19.6 Cerebellar injury compromises the coordination of movement, with or without impacts on cognitive or affective regulation

---

CONCEPT **19.1**

The Cerebellum Comprises Three Major Subdivisions: The Cerebrocerebellum, Spinocerebellum, and Vestibulocerebellum

LEARNING OBJECTIVE

19.1.1 Localize and discuss the major functional subdivisions of the cerebellar hemispheres.

19.1.2 Identify the major cerebellum peduncles and discuss the functional significance of each.

Organization of the cerebellum

The cerebellar hemispheres can be subdivided into three main parts based on differences in their sources of input (Figure 19.1A). By far, the largest

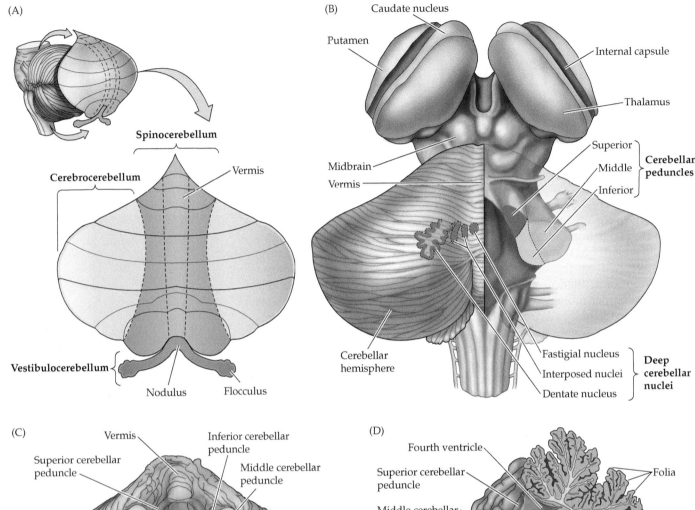

(A)

Cerebrocerebellum

Spinocerebellum

Vermis

Vestibulocerebellum {

Nodulus   Flocculus

(B)

Caudate nucleus

Putamen

Internal capsule

Thalamus

Midbrain

Vermis

Superior

Middle   Cerebellar peduncles

Inferior

Cerebellar hemisphere

Fastigial nucleus

Interposed nuclei   Deep cerebellar nuclei

Dentate nucleus

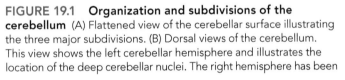

(C)

Vermis

Superior cerebellar peduncle

Inferior cerebellar peduncle

Middle cerebellar peduncle

Flocculus

Nodulus

Cerebellar hemisphere

(D)

Fourth ventricle

Superior cerebellar peduncle

Folia

Middle cerebellar peduncle

Inferior cerebellar peduncle

Flocculus

Nodulus

Cerebellar cortex

FIGURE 19.1   Organization and subdivisions of the cerebellum (A) Flattened view of the cerebellar surface illustrating the three major subdivisions. (B) Dorsal views of the cerebellum. This view shows the left cerebellar hemisphere and illustrates the location of the deep cerebellar nuclei. The right hemisphere has been removed to show the cerebellar peduncles. (C) Removal from the brainstem reveals the cerebellar peduncles on the anterior aspect of the inferior surface. (D) Paramedian sagittal section through the right cerebellar hemisphere showing the highly convoluted cerebellar cortex. The small gyri in the cerebellum are called *folia*.

subdivision in humans is the **cerebrocerebellum**. It occupies most of the lateral part of the cerebellar hemisphere and receives input, indirectly, from many areas of the cerebral cortex (Figure 19.2). This region of the cerebellum is especially well developed in primates and is particularly prominent in humans. The cerebrocerebellum is concerned with the regulation of highly skilled movements, especially the planning and execution of complex spatial and temporal sequences of movement (including speech).

Just medial to the cerebrocerebellum is the **spinocerebellum**. The spinocerebellum occupies the median and paramedian zones of the cerebellar hemispheres and is the only part that receives input directly from the spinal cord. The more lateral (paramedian) part of the spinocerebellum is concerned primarily with movements of distal muscles. The most median strip of cerebellar hemisphere lies along the midline and is called the **vermis**. The vermis is concerned primarily with movements of proximal

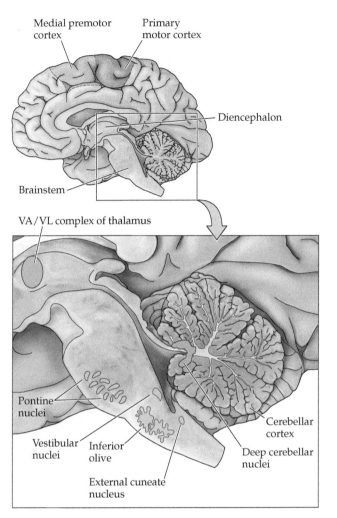

Medial premotor cortex

Primary motor cortex

Diencephalon

Brainstem

VA/VL complex of thalamus

Pontine nuclei

Vestibular nuclei

Inferior olive

External cuneate nucleus

Cerebellar cortex

Deep cerebellar nuclei

FIGURE 19.2 Components of the brainstem and diencephalon related to the cerebellum This sagittal section shows the major structures of the cerebellar system, including the cerebellar cortex, the deep cerebellar nuclei, and the ventral anterior and ventral lateral (VA/VL) complex (which is an important target of some deep cerebellar nuclei). Also shown are brainstem nuclei that provide input to the cerebellum.

TABLE 19.1 Major Components of the Cerebellum

Cerebellar cortex	Deep cerebellar nuclei	Cerebellar peduncles
Cerebrocerebellum	Dentate nucleus	Superior and middle peduncle
Spinocerebellum	Interposed nuclei	Inferior and middle peduncle
Vestibulocerebellum	Fastigial nucleus	Inferior peduncle

deep cerebellar nuclei (Table 19.1). Their axons project to the motor nuclei of the thalamus, which in turn relay signals to circuits of upper motor neurons in the primary motor and premotor divisions of the cerebral cortex. Efferent axons in the superior peduncle also project directly to upper motor neurons in the deep layers of the superior colliculus that control orienting movements of the head and eyes. In non-human species, neurons in the deep cerebellar nuclei also provide input to upper motor neurons in the caudal part of the red nucleus. The **middle cerebellar peduncle** (or **brachium pontis**) is an afferent pathway to the cerebellum; most of the cell bodies that give rise to this pathway are in the base of the contralateral pons, where they form the **pontine nuclei** (see Figure 19.2). Finally, the **inferior cerebellar peduncle** (or **restiform body**) is the smallest but most complex of the cerebellar peduncles, containing multiple afferent and efferent pathways. Afferent pathways in the inferior peduncle include axons from the vestibular nuclei, the spinal cord, and several regions of the brainstem tegmentum, while efferent pathways project to the vestibular nuclei and the reticular formation.

CONCEPT **19.2**

The Cerebellar Hemispheres Coordinate Movements of the Ipsilateral Body

LEARNING OBJECTIVES

19.2.1 Characterize the means by which afferent signals arising in the cortex of one cerebral hemisphere influence the opposite (contralateral) side of the cerebellum.

19.2.2 Characterize the means by which afferent sensory signals arising in the spinal cord and brainstem influence the same (ipsilateral) side of the cerebellum.

19.2.3 Characterize the means by which the inferior olivary nucleus influences the contralateral cerebellum.

Projections to the cerebellum

The cerebral cortex is by far the origin of the largest input to the cerebellum, and the major destination of this input is the cerebrocerebellum. However, these cortical axons do not project directly into the cerebellum. Rather, they

muscles; it also regulates certain types of eye movements (see Chapter 20). The third major subdivision is the **vestibulocerebellum**, the phylogenetically oldest part of the cerebellum. This portion comprises the caudal-inferior lobes of the cerebellum and includes the **flocculus** and **nodulus** (see Figure 19.1A). As its name suggests, the vestibulocerebellum receives input from the vestibular nuclei in the brainstem and is concerned primarily with the vestibulo-ocular reflex (see Chapter 11) and with the regulation of movements that maintain posture and equilibrium.

The connections between the cerebellum and other parts of the nervous system are made by three large pathways called **cerebellar peduncles** (Figure 19.1B–D; also see Figure 19.3). The **superior cerebellar peduncle** (or **brachium conjunctivum**) is almost entirely an efferent pathway. The neurons that give rise to this pathway are located in the

synapse on neurons in the ipsilateral pontine nuclei (i.e., on the same side of the brainstem as their hemisphere of origin). These pontine nuclei receive input from a wide variety of sources, including almost all areas of the cerebral cortex and the superior colliculus. The axons of the cells in the pontine nuclei, called **transverse pontine fibers** (or pontocerebellar fibers), cross the midline and enter the contralateral cerebellum via the middle cerebellar peduncle. Each of the two middle cerebellar peduncles contains approximately 20 million axons, making them among the largest pathways in the brain. (In comparison, the optic nerves each contain about 1 million axons, and the pyramidal tracts no more than 0.5 million each.) In fact, the size of the cerebral peduncles in the ventral portion of the human midbrain (each of which also contains about 20 million axons) is primarily due to the magnitude of the projection from the cerebral cortex that, via the pontine nuclei, provides afferent input to the cerebellum. (In comparison,

the corticospinal projection, which is often assumed incorrectly to account for the bulk of the cerebral peduncles, comprises 2% to 3% of the total number of axons in each cerebral peduncle.) This massive, crossed projection of transverse pontine fibers into the cerebellum via the middle cerebellar peduncle is the means by which signals originating in one *cerebral* hemisphere are sent to neural circuits in the opposite *cerebellar* hemisphere (Figure 19.3A).

Sensory pathways also project to the cerebellum (Figure 19.3B). Vestibular axons in the eighth cranial nerve as well as axons from the vestibular nuclei in the pons and medulla project to the vestibulocerebellum. In addition, somatosensory relay neurons in the **dorsal nucleus of Clarke** in the spinal cord and the **external** (or accessory) **cuneate nucleus** of the caudal medulla send their axons to the spinocerebellum (recall that these nuclei comprise groups of relay neurons innervated by proprioceptive axons from the lower and upper parts of the

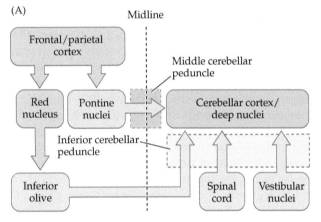

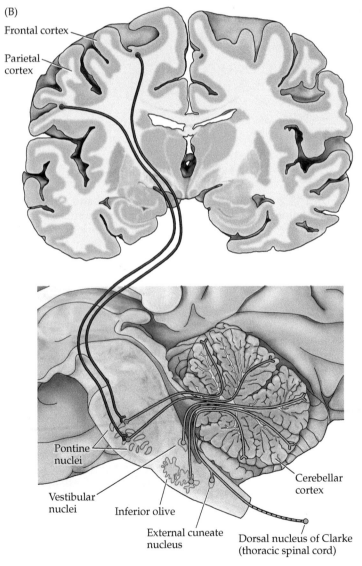

FIGURE 19.3 Functional organization of the inputs to the cerebellum (A) Diagram of the major inputs. (B) Idealized coronal and sagittal sections through the human brainstem and cerebrum, showing inputs to the cerebellum from the cerebral cortex, vestibular system, brainstem, and spinal cord. The cortical projections to the cerebellum are made via relay neurons in the pons. These pontine axons then cross the midline within the pons and project to the cerebellum via the middle cerebellar peduncle. Axons from the inferior olive, spinal cord, and vestibular nuclei enter via the inferior cerebellar peduncle.

body, respectively; see Chapter 12). Proprioceptive signals from the face are likewise relayed via the **mesencephalic trigeminal nucleus** to the spinocerebellum. The vestibular, spinal, and trigeminal inputs provide the cerebellum with information from the labyrinth in the ear, muscle spindles, and other mechanoreceptors that monitor the position and motion of the body. Visual and auditory signals are relayed via brainstem nuclei to the cerebellum; they provide the cerebellum with additional sensory signals that supplement the proprioceptive information regarding body position and motion.

The somatosensory input is topographically mapped in the spinocerebellum, providing the basis for orderly representations of the body within the cerebellum (Figure 19.4). However, these maps are "fractured"; that is, fine-grain electrophysiological analysis indicates that each small area of the body is represented multiple times by spatially separated clusters of cells, rather than by a specific site within a single continuous somatotopic map. The vestibular and spinal inputs remain ipsilateral as they pass through the inferior cerebellar peduncle and enter the cerebellum, having their origin on the same side of the brainstem and spinal cord (see Figure 19.3A). This arrangement ensures that the right cerebellum is concerned with the right half of the body and the left cerebellum with the left half. Thus, while many areas of the brain maintain *contralateral* representations (of the body and external space), the cerebellum maintains *ipsilateral* representations.

Finally, the entire cerebellum receives modulatory inputs from the **inferior olivary nucleus** (or **inferior olive**)

in the medulla oblongata. These inputs participate in the learning and memory functions served by cerebellar circuitry. The inferior olive receives input from a wide variety of structures, including the cerebral cortex (via a relay in the parvocellular, or small-celled, division of the red nucleus), the reticular formation, and the spinal cord. The so-called olivo-cerebellar axons exit medially from the inferior olive, cross the midline, and enter the cerebellum on the opposite side via the inferior cerebellar peduncle (see Figure 19.3A). Electrotonic gap junctions are abundant among neurons in the inferior olive, and these evidently play an important role in the timing and spatial distribution of cerebellar responses to olivary inputs.

CONCEPT
19.3

Efferent Output from the Cerebellum to the Brainstem and Thalamus Originates in the Deep Cerebellar Nuclei and the Vestibulocerebellum

LEARNING OBJECTIVES

19.3.1 Explain how the dentate nucleus primarily influences the premotor and associational cortices of the frontal lobe.

19.3.2 Distinguish open-loop from closed-loop circuits that engage portions of the cerebral cortex and the cerebellum.

19.3.3 Explain how the deep cerebellar nuclei influence brainstem centers that organize or govern the activities of lower motor neuron circuits.

19.3.4 Discuss the role of the red nucleus in motor control in human and non-human animals.

Projections from the cerebellum

The efferent neurons of the **cerebellar cortex** project to the deep cerebellar nuclei and to the vestibular complex; these structures project, in turn, to upper motor neurons in the brainstem and to thalamic nuclei that innervate upper motor neurons in the motor cortex (Figure 19.5). In each cerebellar hemisphere there are four major deep nuclei: the **dentate nucleus** (by far the largest in humans), two **interposed nuclei**, and the **fastigial nucleus**. Each receives input from a different region of the cerebellar cortex. Although the borders are not distinct, the cerebrocerebellum projects primarily to the dentate nucleus, and the spinocerebellum to the interposed and fastigial nuclei. The vestibulocerebellum projects directly to the vestibular complex in the brainstem. As discussed in Chapter 17, parts of the vestibular complex are sources of upper motor neurons that influence posture, equilibrium, and vestibulo-ocular eye movements.

Pathways originating in the dentate nucleus primarily influence the premotor and associational cortices of the frontal lobe, which function in planning and initiating

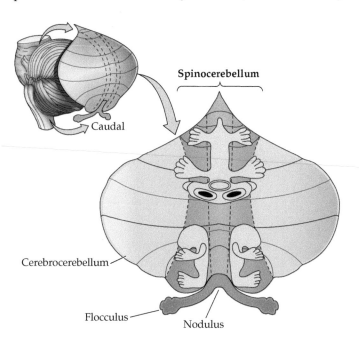

FIGURE 19.4 Somatotopic maps of the body surface in the cerebellum The spinocerebellum contains at least two maps of the body.

Spinocerebellum

Caudal

Cerebrocerebellum

Flocculus

Nodulus

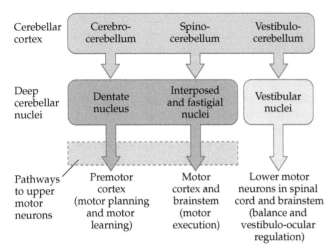

FIGURE 19.5 Functional organization of cerebellar outputs The three major functional divisions of the cerebellar hemispheres project to corresponding deep cerebellar nuclei and the vestibular nuclei, which in turn provide input to neural circuits that govern different aspects of motor control.

volitional movements. The pathways reach these cortical areas after a relay in the ventral lateral nuclear complex of the thalamus (Figure 19.6A). Since each cerebellar hemisphere is concerned with the ipsilateral side of the body, this pathway crosses the midline so the motor cortex in each hemisphere, which governs contralateral musculature, receives information from the appropriate cerebellar hemisphere. For this reason, the dentate axons that exit the cerebellum via the superior cerebellar peduncle cross the midline at the decussation of the superior cerebellar peduncle in the caudal midbrain, and then ascend to the contralateral thalamus. Along its course to the thalamus, this pathway also sends axons to eye-movement-related upper motor neurons in the superior colliculus and, in addition, sends collaterals to the parvocellular division of the **red nucleus** in the midbrain (which accounts for virtually the entire red nucleus in the human midbrain) (Figure 19.6B). This division of the red nucleus projects, in turn, to the inferior olive, thus providing a means for cerebellar output to feed back on a critical source of cerebellar input. This

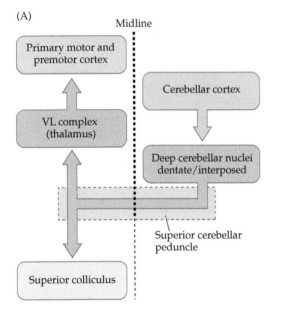

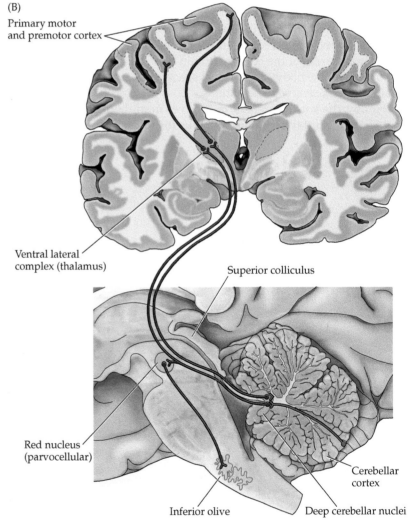

FIGURE 19.6 Functional organization of the major outputs from the cerebellum to cortical motor systems (A) Diagram of major outputs that affect upper motor neurons in the cerebral cortex. The axons of the deep cerebellar nuclei cross in the midbrain—in the decussation of the superior cerebellar peduncle—before reaching the thalamus and superior colliculus. (B) Idealized coronal and sagittal sections through the cerebrum and brainstem, showing the location of the structures and pathways diagrammed in (A) and a feedback circuit by which cerebellar output is directed to the inferior olive via the red nucleus.

feedback is crucial for the adaptive functions of cerebellar circuits in motor learning.

Anatomical studies using viruses to trace chains of connections between nerve cells have shown that, in addition to sending ascending projections to upper motor neurons concerned with the control of movement, large parts of the cerebrocerebellum form "closed loops" by sending information back to non-motor areas of the cortex. That is, a region of the cerebellum sends projections back to the same cortical areas (via thalamic projections) from which (via the pontine nuclei) its input signals originated. Such closed-loop cerebellar circuits provide a mechanism for the cerebellum to modulate its own input. In the case of cerebellar circuits that modulate the prefrontal cortex, these closed loops may also influence the coordination of non-motor programs—such as problem solving, language formation, and decision making—in a manner that is analogous to their modulation of movement-related signals. The closed loops run parallel to the more commonly recognized "open loops" that receive input from multiple cortical areas and funnel output back to upper motor neurons in specific regions of the motor and premotor cortices.

Spinocerebellar pathways are directed toward circuits of upper motor neurons that govern the execution of movement. The somatotopic organization of the spinal subdivision of the cerebellum is reflected in the organization of its efferent projections, both of which conform to the mediolateral organization of motor control in the spinal cord (see Chapter 16). Thus, the fastigial nuclei (which underlie the vermis near the midline of the cerebellum) project via the inferior cerebellar peduncle to nuclei of the reticular formation and vestibular complex that give rise to medial tracts governing the axial and proximal limb musculature (Figure 19.7). The more laterally positioned interposed nuclei (which underlie the paramedian subdivision of the spinocerebellum) send projections via the superior cerebellar peduncle to thalamic circuits that project to motor regions in the frontal lobe concerned with volitional movements of the limbs (see Figure 19.6). In non-human primates, axons from the interposed nuclei also send collaterals to a magnocellular (large-celled) division of the red nucleus that gives rise to the rubrospinal tract, a lateral tract of the spinal cord that functions synergistically with the lateral corticospinal tract. (As discussed in Chapter 17, this division of the red nucleus and its spinal projection are vestigial in humans relative to other primates and non-primate mammals.)

Most of the projection from the cerebellum to the eye-movement-related upper motor neurons in the superior colliculus arises in the dentate and interposed nuclei, which receive their input from the lateral portions of the cerebellar cortex. The output pathway travels in the superior cerebellar peduncle and crosses the midline to terminate on upper motor neurons in the deep layers of

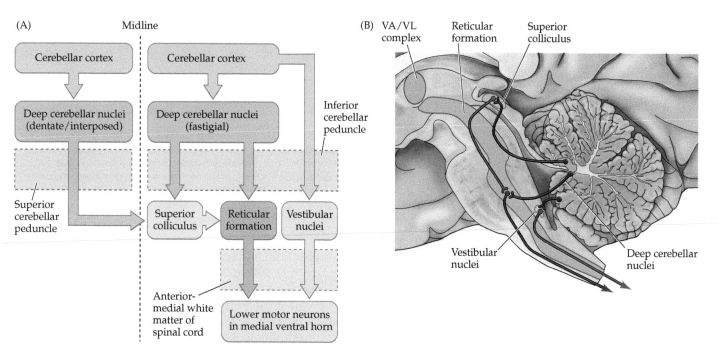

FIGURE 19.7   Functional organization of the major outputs from the cerebellum to brainstem motor systems (A) Diagram of major outputs that affect upper motor neurons in the brainstem. The axons of the deep cerebellar nuclei and the vestibulocerebellar cortex project to upper motor neurons that contribute to the control of axial and proximal limb musculature in the medial ventral horn of the spinal cord. (B) Idealized sagittal section through the brainstem, showing the location of the structures diagrammed in (A).

the superior colliculus on the contralateral side. So, for example, the right cerebellar hemisphere projects to the left superior colliculus, which in turn controls saccades toward the right half of the visual field (see Chapter 20).

The thalamic nuclei that receive projections from the cerebrocerebellum (dentate nuclei) and spinocerebellum (interposed nuclei) are segregated in two distinct subdivisions of the ventral lateral nuclear complex: the oral, or anterior, part of the posterolateral segment, and a region simply called *area X*. Both of these thalamic relays project directly to primary motor and premotor association cortices. Through these pathways, the cerebellum has access to the upper motor neurons that organize the sequence of muscular contractions underlying complex voluntary movements, as well as to circuits in frontal association cortex that exert executive control over planning movements (see Chapter 33).

Projections from the vestibulocerebellum course through the inferior cerebellar peduncle and terminate in nuclei of the vestibular complex in the brainstem. These nuclei govern the movements of the eyes, head, and neck that compensate for linear and rotational accelerations of the head (see Figure 19.5).

CONCEPT
19.4
Purkinje Neurons Integrate Afferent Input and Modulate the Output of Deep Cerebellar Nuclei

LEARNING OBJECTIVES

19.4.1 Differentiate the origins, functions, and terminations of mossy fibers and climbing fibers.

19.4.2 Describe the circuitry involved in the deep excitatory loop and the cortical inhibitory loop through the cerebellum.

19.4.3 Discuss the physiological role of Purkinje neurons in sculpting the spatiotemporal firing patterns of neurons in deep cerebellar nuclei.

19.4.4 Discuss the significance of long-term depression at the synapses between parallel fiber axons and the dendrites of Purkinje neurons.

Circuits within the cerebellum

The ultimate destination of the afferent pathways to the cerebellar cortex is a distinctive cell type called the Purkinje cell (Figures 19.8 and 19.9). The largest of these afferent pathways arises in widespread areas of the cerebral cortex and terminates in the pontine nuclei of the basal pons, as described in Concept 19.3. The pontine nuclei, in turn, project to the contralateral cerebellum. The axons from the pontine nuclei—and most other sources of cerebellar input from the brainstem and spinal cord—are called **mossy fibers** because of the appearance of their synaptic terminals. Mossy fibers send collateral branches that

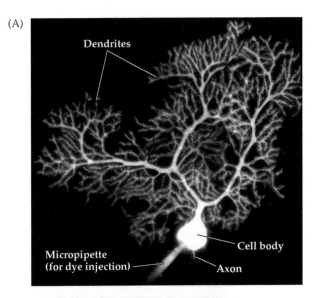

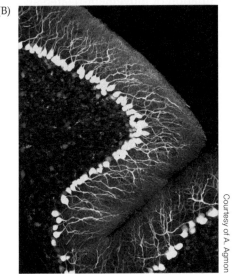

Courtesy of A. Agmon

FIGURE 19.8 Cerebellar cortical neurons (A) A cerebellar Purkinje neuron in a living slice from mouse cerebellum. The neuron has been visualized by infusing a fluorescent dye that indicates Ca²⁺ concentrations, via a micropipette inserted into the cell body. (B) Histological preparation of mouse cerebellar cortex. Purkinje neurons were engineered to fluoresce yellow, while granule cells fluoresce blue.

synapse both on neurons in the deep cerebellar nuclei and on granule cells in the granule cell layer of the cerebellar cortex. Cerebellar granule cells, which are widely held to be the most abundant class of neurons in the human brain, give rise to axons called **parallel fibers** that ascend to the outermost molecular layer of the cerebellar cortex. The parallel fibers bifurcate in the molecular layer to form T-shaped branches that extend for several millimeters parallel to the orientation of the small cerebellar gyri (called folia). There, they form excitatory synapses with the dendritic spines of the underlying Purkinje cells.

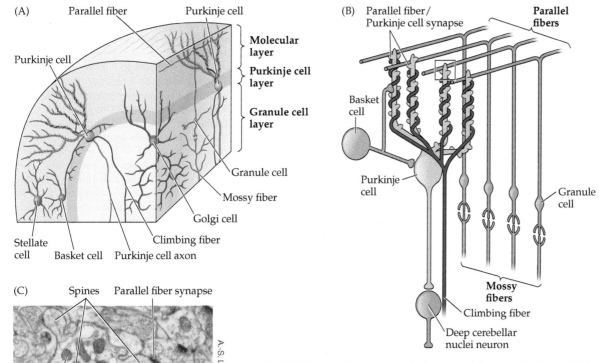

FIGURE 19.9 Neurons and circuits of the cerebellum (A) Neuronal types in the cerebellar cortex. Note that the various neuron classes are found in distinct layers. (B) Diagram showing convergent inputs onto the Purkinje cell from climbing and parallel fibers and from local circuit neurons (climbing and mossy fiber inputs to deep cerebellar neurons are omitted for clarity). The boxed region is shown at higher magnification in (C). The output of the Purkinje cells is to the deep cerebellar nuclei. (C) Electron micrograph showing a Purkinje cell dendritic shaft with spines contacted by parallel fibers in rhesus macaque cerebellum.

The Purkinje cells (see Figure 19.8A) are the most distinctive histological feature of the cerebellum. Their elaborate dendrites extend into the molecular layer from a single subjacent layer of giant Purkinje cell bodies (called the Purkinje cell layer) (see Figure 19.8B). In the molecular layer, the Purkinje cell dendrites branch extensively in a plane restricted at right angles to the trajectory of the parallel fibers (see Figure 19.9A). In this way, each Purkinje cell is in a position to receive input from a large number of parallel fibers (about 200,000), and each parallel fiber can contact a vast number of Purkinje cells (on the order of tens of thousands) (see Figure 19.9C). The Purkinje cells also receive a direct input on their dendritic shafts from **climbing fibers**, all of which arise in the contralateral inferior olive (see Figure 19.9B). Each Purkinje cell receives numerous synaptic contacts from a single climbing fiber. The climbing fibers provide a "training" signal that modulates the synaptic strength of the parallel fiber connection with the Purkinje cells.

The Purkinje cells project in turn to the deep cerebellar nuclei and comprise the only output cells of the cerebellar cortex. Since Purkinje cells are GABAergic, the output of the cerebellar cortex is wholly inhibitory. However, the neurons in the deep cerebellar nuclei also receive excitatory input from the collaterals of the mossy and climbing fibers. The inhibitory projections of Purkinje cells serve to sculpt the discharge patterns that deep nuclei neurons generate in response to their direct mossy and climbing fiber inputs (Figure 19.10).

Inputs from GABAergic interneurons modulate the inhibitory activity of Purkinje cells. The most powerful of these local inputs are inhibitory nests of synapses made with the Purkinje cell bodies by **basket cells** (see Figure 19.9). Another type of local circuit neuron, the **stellate cell**, receives input from the parallel fibers and provides an inhibitory input to the Purkinje cell dendrites. Finally, the molecular layer contains the apical dendrites of inhibitory interneurons called **Golgi cells**; these neurons have their cell bodies in the granule cell layer. Golgi cells receive input from the parallel fibers and provide an inhibitory feedback to the cells of origin of the parallel fibers (the granule cells). There are other classes of interneurons in the cerebellar cortex (including excitatory unipolar brush cells and inhibitory Lugaro cells; not illustrated in Figure 19.9), whose functions are less well understood.

This circuit module of excitatory and inhibitory cells is repeated over and over throughout every subdivision

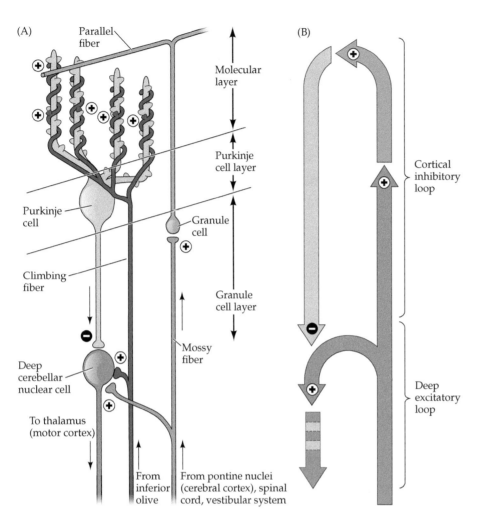

(A) The excitatory input from mossy fibers and climbing fibers to Purkinje cells and deep nuclear cells is basically the same. Additional convergent input onto the Purkinje cell from local circuit neurons (basket and stellate cells) and other Purkinje cells (not illustrated) establishes a basis for the comparison of ongoing movement and sensory feedback derived from it. The output of the Purkinje cell onto the deep cerebellar nuclear cell is inhibitory. (B) Conceptual diagram of the circuitry illustrated in (A). The deep cerebellar nuclei and their excitatory afferents constitute a *deep excitatory loop* whose output is shaped by a *cortical inhibitory loop* that inverts the "sign" of the input signals. The Purkinje neuron output to the deep cerebellar nuclear cell thus generates an error correction signal that can modify movements. The climbing fibers modify the efficacy of the parallel fiber–Purkinje cell connection, producing long-term changes in cerebellar output. (A after J. F. Stein, 1986. *Nature* 323: 217–220.)

of the cerebellum in all vertebrates, and suggests that—in spite of the differences in the sources of their inputs and in the destinations of their outputs—all of these subdivisions share a similar function. That is, in each subdivision, transformation of signal flow through these modules provides the basis for both the real-time regulation of movement and the long-term changes in regulation that underlie implicit learning and the acquisition of motor skill.

Description of the flow of signals through these complex modules may be simplified by distinguishing the two basic stages of cerebellar processing, beginning with the deep cerebellar nuclei. Mossy fiber and climbing fiber collaterals drive the activation of neurons in the deep cerebellar nuclei; this constitutes a *deep excitatory loop* in which input signals converge on the final output stage of cerebellar processing. However, as suggested earlier in this concept, the spatiotemporal patterns of the output activity are not simply faithful replications of the input patterns. The response patterns of the deep cerebellar nuclei to their direct inputs are modified by the descending inhibitory inputs of Purkinje cells, which are driven by these same two afferent pathways (i.e., the mossy and climbing fiber projections to the cerebellar cortex). For their part, Purkinje cells

integrate these principal inputs and invert their "sign" by responding to excitatory inputs with an inhibitory output (see Figure 19.10B). Thus, Purkinje cells convey the product of computations performed by a *cortical inhibitory loop* that comprises the circuitry of the cerebellar cortex, including the interneurons of the granule and molecular layers, as well as the Purkinje cells themselves. The interneurons control the flow of information through the cerebellar cortex. For example, the Golgi cells form an inhibitory feedback circuit that controls the temporal properties of the granule cell input to the Purkinje cells, whereas the basket cells provide lateral inhibition that may focus the spatial distribution of Purkinje cell activity.

The modulation of cerebellar output by the cerebellar cortex may be responsible for motor learning. According to a model proposed by Masao Ito and his colleagues at Tokyo University, the climbing fibers from the inferior olive relay the message of a motor error to the Purkinje cells, with motor error representing the difference between the precision and accuracy of an intended and an actual movement. This message is derived from inputs that the inferior olive receives from multiple structures (including the cerebral cortex and spinal cord) as well as from feedback signals from the cerebellum via the red nucleus, as described in Concept 19.3. The 1000 or so synapses made by a single climbing fiber with the proximal dendrites of a single Purkinje cell constitute one of the most powerful excitatory connections in the CNS. Activation of the climbing fibers induces a strong excitatory postsynaptic potential in Purkinje cells that generates an initial action potential followed by series of smaller "spikelets." This postsynaptic response is termed a *complex spike*; it typically occurs infrequently (1 to 2 Hz), depending on the context and demands of the concurrent movements. In contrast, the parallel fiber input to the Purkinje cells gives rise to individual action potentials, called *simple spikes*, that typically fire at a much higher rate (30 to 100 Hz). The impact of the climbing fiber input on Purkinje cell output is further enhanced by the gap junctions that electronically join and synchronize the activity of neurons in the inferior olive. These ensembles of olivary neurons both drive the activation of cerebellar circuits and also promote adaptive plasticity in the inhibitory output of the cerebellar cortex. The plasticity results from long-term reductions in the Purkinje cell responses to their parallel fiber inputs. The mechanism for this long-term depression is a complex chain of cellular events leading from the climbing fiber input to the Purkinje cells to the endocytosis of AMPA receptors at parallel fiber–Purkinje cell

synapses. (Recall that AMPA receptors mediate fast, excitatory responses at glutamatergic synapses; for an account of the cellular mechanisms responsible for this long-term reduction in the efficacy of the parallel fiber synapse on Purkinje cells, see Chapter 8.)

The reduction in the efficacy of parallel fiber input to Purkinje cells has the effect of increasing the response of neurons in the deep cerebellar nuclei to afferent activity (by weakening the influence of the inhibitory loop). Thus, the feedback signals from the cerebellum to circuits of upper motor neurons in the motor cortex and brainstem are altered as a consequence of climbing fiber activation. It is not yet understood at the circuit level how this alteration mediates a "correction" of movement error. Nevertheless, it is clear from studies of animal models and of humans with damage to the inferior olive that both short-term sensorimotor adaptation (error correction) and long-term motor learning require the modulation of cerebellar processing by climbing fiber activation.

Despite the consistency with which these basic structural and functional features of cerebellar circuitry are replicated throughout the cortex of the cerebellum, recent molecular, genetic, anatomical, and physiological studies have revealed longitudinal compartments in the cerebellar cortex (Figure 19.11). For example, subsets of Purkinje cells

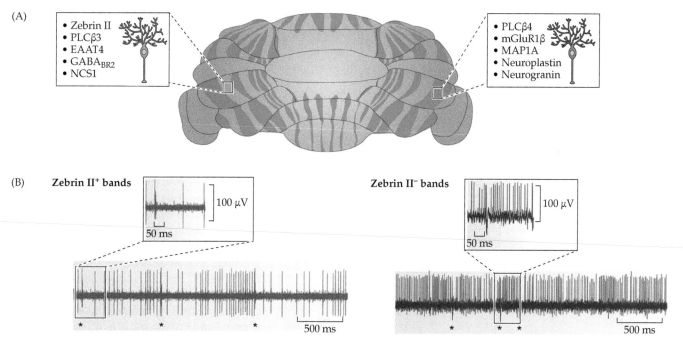

Figure 19.11 Compartments in the cerebellar cortex (A) Illustration of the mouse cerebellum showing alternating bands of zebrin II expression in Purkinje cells and co-localization of a variety of molecular markers. EAAT4, excitatory amino acid transporter 4; GABA$_{BR2}$, GABA$_B$ receptor subtype 2; MAP1A, microtubule-associated protein 1A; mGluR1β, metabotropic glutamate receptor 1β; NCS1, neuronal calcium sensor 1; PLCβ3, phospholipase Cβ3; PLCβ4, phospholipase Cβ4; neuroplastin is a member of the immunoglobulin superfamily that functions as a cell adhesion molecule, and neurogranin is a calmodulin-binding protein. (B) Extracellular recordings of Purkinje cells in zebrin II–positive and –negative bands of rat cerebellar cortex. Zebrin II–negative Purkinje cells show higher rates of sustained simple spike activity, while zebrin II–positive Purkinje cells show stronger suppression of simple spikes following complex spikes (asterisks). (From N. L. Cerminara et al., 2015. *Nat Rev Neurosci* 16: 79–93.)

show variable expression of zebrin II, which is an antigen localized to aldolase C (an enzyme in the glycolysis pathway). Purkinje cells that express zebrin II are clustered together into rostrocaudal bands that are interleaved with bands lacking zebrin II expression (reminiscent of modules found elsewhere in the CNS; see Box 12A). A variety of other molecular markers are co-localized to zebrin II–positive or –negative bands, including molecules related to glutamatergic neurotransmission and second-messenger systems within postsynaptic processes. One physiological consequence of this distinction is that long-term depression (as discussed earlier in this concept) is more prominent at synapses between parallel fibers and Purkinje neurons lacking zebrin II. These zebrin II–negative Purkinje neurons also tend to exhibit higher basal rates of simple spike activity. In contrast, long-term potentiation is more likely at parallel fiber–Purkinje cell synapses in zebrin II–positive bands where the rate of simple cell spikes is relatively low. These findings indicate that the canonical circuits of the cerebellar cortex may yet be further differentiated as the tools of contemporary neuroscience continue to reveal previously unrecognized order. It remains a challenge to determine how such regional variations in gene expression and physiological phenotype convey different capabilities in information processing across the cerebellar cortex.

(A) Purkinje cell

At rest

Wrist flexion and extension

(B) Deep nuclear cell

At rest

Wrist flexion and extension

FIGURE 19.12 Activity of Purkinje cells and cells of the deep cerebellar nuclei Neuronal activity is shown for a Purkinje cell (A) and a deep cerebellar nuclear cell (B) both at rest (upper records) and during movement of the ipsilateral wrist (lower records). The red traces represent wrist movement; up is flexion, down is extension. The durations of the wrist movements are indicated by the colored blocks. Both classes of cells are tonically active at rest. Rapid alternating movements result in the transient inhibition of the tonic activity of both cell types. (After W. T. Thach, 1968. *J Neurophys* 31: 785–797.)

CONCEPT
19.5

The Cerebellum Coordinates Ongoing Movement by Reducing Motor Error

LEARNING OBJECTIVES

19.5.1 Discuss how circuitry in the cerebellum generates signals that decrease motor error and increase the accuracy and precision of motor performance.

19.5.2 Characterize the signs and symptoms of cerebellar ataxia.

19.5.3 Discuss evidence from experiments on eye movements that demonstrate use-dependent corrections of motor error.

Cerebellar circuitry and the coordination of ongoing movement

As one would expect in a structure that monitors and adjusts motor behavior, neuronal activity in the cerebellum changes continually during the course of a movement. For instance, the execution of a relatively simple task such as flexing and extending the wrist back and forth elicits a dynamic pattern of activity in both the Purkinje cells and the deep cerebellar nuclear cells that closely follows the ongoing movement (Figure 19.12). Both types of cells are tonically active at rest and change their frequency of firing as movements occur. The neuronal responses are influenced by various aspects of movement, including relaxation or contraction of specific muscles, the position of the joints, and the direction of the next movement that will occur. All this information is encoded by changes in the discharge pattern of Purkinje cells, and these changes modulate the ongoing output of the deep cerebellar nuclear cells.

As these neuronal response properties predict, cerebellar injuries and disease tend to disrupt the modulation and coordination of ongoing movements, and the specific movements that are disrupted vary with the location of the damage (Clinical Applications). The hallmark of individuals with cerebellar damage is difficulty producing smooth, well-coordinated, multi-jointed movements. Instead, movements tend to be decomposed into jerky and imprecise elements, a condition referred to as **cerebellar ataxia**. Many of these difficulties in performing movements can be explained as disruption of the cerebellum's role in correcting errors in ongoing movements, since the cerebellar error correction mechanism normally ensures that movements are modified to cope with changing circumstances. As described in Concept 19.4, the Purkinje cells and deep cerebellar nuclear cells recognize potential

Prion Diseases

Creutzfeldt-Jakob disease (CJD) is a rare but devastating neurological disorder characterized by cerebellar ataxia, myoclonic jerks, seizures, and the fulminant progression of dementia. The onset is usually in middle age, and death typically follows within a year. The distinctive histopathology of the disease, termed *spongiform degeneration*, consists of neuron loss and extensive glial proliferation, mainly in the cortex of the cerebellum and cerebrum; the peculiar spongiform pattern is due to vacuoles in the cytoplasm of neurons and glia. CJD is the only human disease known to be transmitted by inoculation (either orally or into the bloodstream), or to be inherited through the germline. In contrast to other transmissible diseases mediated by microorganisms such as viruses or bacteria, the agent in this case is a protein called a prion.

Observations dating back half a century suggested that CJD was infectious. The major clue came from scrapie, a once-obscure disease of sheep that is also characterized by cerebellar ataxia, wasting, and intense itching. The ability to transmit scrapie from one sheep to another strongly suggested an infectious agent. Another clue came from the work of Carleton Gajdusek, a neurologist studying a peculiar human disease called kuru that occurred specifically in a group of New Guinea natives known to practice ritual cannibalism. Like CJD, kuru is a neurodegenerative disease characterized by devastating cerebellar ataxia and subsequent dementia, usually leading to death within a year. The striking similarities in the distinctive histopathology of scrapie and kuru—namely, spongiform degeneration—suggested a common pathogenesis and led to the successful transmission of kuru to apes and chimpanzees in the 1960s, confirming that CJD was indeed infectious. The prolonged period (months to years) between inoculation and disease onset led Gajdusek to suggest that the transmissible agent was what he called a *slow virus*.

These extraordinary findings spurred an intensive search for the infectious agent. The transmission of scrapie from sheep to hamsters by Stanley Prusiner at the University of California, San Francisco permitted biochemical characterization of partially purified fractions of scrapie agent from hamster brain. Oddly, Prusiner found that the infectivity was extraordinarily resistant to ultraviolet irradiation or nucleases—both treatments that degrade nucleic acids. It therefore seemed unlikely that a virus could be the causal agent. Conversely, procedures that modified or degraded proteins markedly diminished infectivity. In 1982, Prusiner coined the term *prion* to refer to the agent causing these transmissible spongiform encephalopathies (TSEs). He chose the term to emphasize that the agent was a proteinaceous infectious particle (and made the abbreviation a little more euphonious in the process). Since Prusiner's discoveries, a half-dozen more animal diseases—including the widely publicized bovine spongiform encephalopathy (BSE), or "mad cow disease"—and at least four more human diseases have been shown to be caused by prions.

Whether prions contain undetected nucleic acids or are really proteins remained controversial for some years. Prusiner strongly advocated a "protein-only" hypothesis, a revolutionary concept with respect to transmissible diseases. He proposed that the prion is a protein consisting of a modified (scrapie) form (PrP^{Sc}) of the normal host protein (PrP^C, for *prion protein control*), the propagation of which occurs by a conformational change of endogenous PrP^C to PrP^{Sc} autocatalyzed by PrP^{Sc}. That is, the modified form of the protein (PrP^{Sc}) transforms the normal form (PrP^C) into the modified form, much as crystals form in supersaturated solutions. Differences in the secondary structures of PrP^C and PrP^{Sc} seen with optical spectroscopy supported this idea. An alternative hypothesis, however, was that the agent is simply an unconventional nucleic acid–containing virus and that the accumulation of PrP^{Sc} is an incidental consequence of infection and cell death.

In the past two decades, a compelling body of evidence in support of the "protein-only" hypothesis has emerged. First, PrP^{Sc} and scrapie infectivity co-purify by several procedures, including affinity chromatography using an anti-PrP monoclonal antibody; no nucleic acid has been detected in highly purified preparations, despite intensive efforts. Second, TSEs can be inherited in humans, and the cause is now known to be a mutation (or mutations) in the gene coding for PrP. Third, transgenic mice carrying a mutant *PrP* gene equivalent to one of the mutations of inherited human prion disease develop a TSE. Thus, a defective protein is sufficient to account for the disease. Finally, transgenic mice carrying a null mutation for *PrP* do not develop a TSE when inoculated with scrapie agent, whereas wild-type mice do. These results argue convincingly that PrP^{Sc} must indeed interact with endogenous PrP^C to convert PrP^C to PrP^{Sc}, propagating the disease in the process. The protein is highly conserved across mammalian species, suggesting that it serves some essential function, although mice carrying a null mutation of *PrP* exhibit no detectable abnormalities.

These advances notwithstanding, many questions remain. What is the mechanism by which the conformational transformation of PrP^C to PrP^{Sc} occurs? How do mutations at different sites of the same protein culminate in the distinct phenotypes evident in diverse TSEs of humans? By what means do aggregates of transformed PrP^{Sc} pass from one neuron to another? Might there be selective vulnerabilities among subpopulations of neurons such that, in response to certain adverse conditions, protein transformation and aggregation ensue? Are conformational changes of proteins a common mechanism of other neurodegenerative diseases, such as Parkinson's disease, frontotemporal dementia, amyotrophic lateral sclerosis, and Alzheimer's disease? And do these findings suggest a therapy for the dreadful manifestations of TSEs?

Despite these unanswered questions, this work, which represents one of the most exciting chapters in modern neurological research, won Nobel Prizes in Physiology or Medicine for both Gajdusek (in 1976) and Prusiner (in 1997).

FIGURE 19.13 Contribution of the cerebellum to experience-dependent modification of saccadic eye movements Weakening of the lateral rectus muscle of the left eye causes the eye to undershoot the target (1). When the individual (in this case a monkey) is forced to use this eye by patching the right eye, multiple saccades must be generated to acquire the target (2). After 5 days of experience with the weak eye, the gain of the saccadic system has been increased, and a single saccade is now used to fixate the target (3). This adjustment of the gain of the saccadic eye movement system depends on an intact cerebellum. (After L. M. Optican and D. A. Robinson, 1980. *J Neurophys* 44: 1058–1076.)

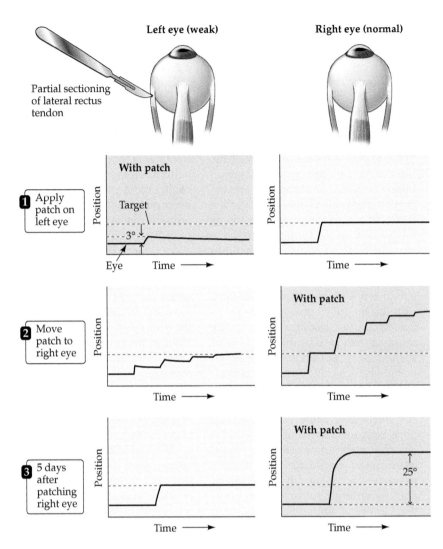

errors by comparing patterns of convergent activity that are concurrently available to both cell types. The deep nuclear cells then send corrective signals to the upper motor neurons in order to maintain or improve the accuracy and precision of the movement.

As in the case of the basal ganglia, studies of the oculomotor system (saccades in particular) have contributed greatly to understanding the contribution that the cerebellum makes to motor error reduction. For example, cutting part of the tendon to the lateral rectus muscles in one eye of a monkey weakens horizontal eye movements by that eye (Figure 19.13). When a patch is then placed over the normal eye to force the animal to use its weak eye, the saccades performed by the weak eye are initially *hypometric*—they fall short of visual targets. Over the next few days, however, the amplitude of the saccades gradually increases until they again become accurate. If the patch is then switched to cover the weakened eye, the saccades performed by the normal eye are now *hypermetric*. In other words, over a period of a few days the nervous system corrects the errors in the saccades made by the weak eye by increasing the gain in a region of the saccade motor system that controls both eyes (see Chapter 20). Lesions in the posterior vermis of the spinocerebellum (see Figure 19.1) eliminate this ability to reduce the motor error.

Similar evidence of the cerebellar contribution to movement has come from studies of the vestibulo-ocular reflex (VOR) in monkeys and humans. The VOR keeps the eyes trained on a visual target during head movements (see Chapter 11). The relative simplicity of this reflex has made it possible to analyze some of the mechanisms that enable motor learning as a process of error reduction. When the head moves, the eyes must move at the same velocity in the opposite direction to maintain a stable representation of the visual image on the retina. In these studies, the adaptability of the VOR to changes in the nature of incoming sensory information is challenged by fitting individuals (either monkeys or humans) with magnifying or minifying spectacles (Figure 19.14). Because the glasses alter the size of the visual image on the retina, the compensatory eye movements, which would normally maintain a stable image of an object on the retina, are either too large or too small. Over time, individuals learn to adjust the distance the eyes must move in response to head movements to accord with the artificially altered size of the visual field. Moreover, this change is retained for significant periods after the spectacles are removed and can be detected electrophysiologically in the responses that can be recorded from Purkinje cells and neurons in the deep cerebellar nuclei. Once again, if the cerebellum is damaged or removed, the ability of the VOR to adapt to the new conditions is lost. These observations support the conclusion that the cerebellum is critically important in error reduction during motor learning.

Normal vestibulo-ocular reflex (VOR)   VOR out of register   VOR gain reset

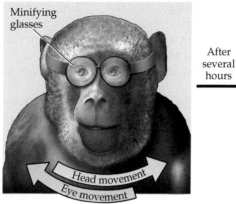

Head and eyes move in a coordinated manner to keep image on retina.

Eyes move too far in relation to image movement on the retina when the head moves.

Eyes move smaller distances in relation to head movement to compensate.

FIGURE 19.14 Learned changes in the vestibulo-ocular reflex in monkeys Normally, the VOR operates to move the eyes as the head moves, so that the retinal image remains stable. When the animal observes the world through minifying spectacles, the eyes initially move too far with respect to the "slippage" of the visual image on the retina. After some experience, however, the gain of the VOR is reset and the eyes move an appropriate distance in relation to head movement, thus compensating for the altered size of the visual image.

CONCEPT
19.6

Cerebellar Injury Compromises the Coordination of Movement, with or without Impacts on Cognitive or Affective Regulation

LEARNING OBJECTIVES

19.6.1 Discuss the signs and symptoms associated with the vestibulocerebellar syndrome.

19.6.2 Discuss the signs and symptoms associated with the cerebellar motor syndrome.

19.6.3 Discuss the emerging concept of the cerebellar cognitive-affective syndrome.

Motor and non-motor consequences of cerebellar lesions

A basic circuit module is present throughout all subdivisions of the cerebellum, and individuals with cerebellar damage, regardless of the cause or location, exhibit persistent errors in ongoing behavior. When the behavior in question involves volitional movements, movement errors are always on the same side of the body as the damage to the cerebellum, reflecting the cerebellum's unusual status as a brain structure in which sensorimotor information is represented ipsilaterally rather than contralaterally. Furthermore, somatic, visual, and other inputs are represented topographically within the cerebellum; as a result, the movement deficits following circumscribed cerebellar damage may be quite specific. For example, one of the most common cerebellar syndromes is caused by

degeneration in the anterior portion of the vermis (medial spinocerebellar cortex) in individuals with a long history of alcohol use disorder (Figure 19.15). Such damage specifically affects movement in the lower limbs, which are

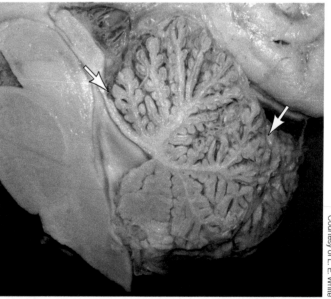

Courtesy of L. E. White

FIGURE 19.15 Pathological changes can provide insights about cerebellar function In this example, chronic alcohol use disorder has caused degeneration of the anterior vermis (arrows), while leaving other cerebellar regions intact. This localized degeneration produces difficulty coordinating walking, with little impairment of arm movements or speech. The orientation of this paramedian sagittal section is the same as in Figure 19.1D.

represented in the anterior spinocerebellum (see Figure 19.4). The consequences include a wide and staggering gait but little impairment of arm or hand movements. Thus, the topographical organization of the cerebellum allows cerebellar damage to disrupt the coordination of movements performed by some muscle groups but not others.

The implication of these pathologies is that the cerebellum is normally capable of integrating the moment-to-moment actions of muscles and joints throughout the body to ensure the smooth execution of a full range of motor behaviors. Thus, cerebellar lesions lead first and foremost to a lack of coordination of ongoing behavior (Box 19A).

■ BOX 19A | Genetic Analysis of Cerebellar Function

Since the early 1950s, investigators interested in motor behavior have identified and studied strains of mutant mice in which movement is compromised. These mutant mice are easy to spot: Following induced or spontaneous mutagenesis, the "screen" is simply to look for animals that have difficulty moving.

Genetic analysis suggested that some of these abnormal behaviors could be explained by single autosomal recessive or semidominant mutations, in which homozygotes are most severely affected. The strains were given names such as *reeler*, *weaver*, *lurcher*, *staggerer*, and *leaner* that reflected the nature of the motor dysfunction they exhibited (see table). The relatively large number of mutations that compromise movement suggested it might be possible to understand some aspects of motor circuits and function at the genetic level.

A common feature of the mutants is ataxia resembling that associated with cerebellar dysfunction in humans. Indeed, all the mutations are associated with some form of cerebellar malformation. The pathologies associated with the *reeler* and *weaver* mutations are particularly striking (see figure). In the *reeler* cerebellum, Purkinje cells, granule cells, and interneurons are all displaced from their usual laminar positions, and there are fewer granule cells than normal. In *weaver*, most of the granule cells are lost prior to their migration from the external granule layer (a proliferative region where cerebellar granule cells are generated during development), leaving only Purkinje cells and local interneurons to carry on the work of the cerebellum. Thus, these mutations causing deficits in motor behavior impair the development and final disposition of the neurons that make up the major processing circuits of the cerebellum (see Figures 19.8 and 19.9).

Early efforts to characterize the cellular mechanisms underlying these motor deficits were unsuccessful, and the molecular identity of the affected genes remained obscure. In the past few decades, however, both the *reeler* and *weaver* genes have been identified and cloned.

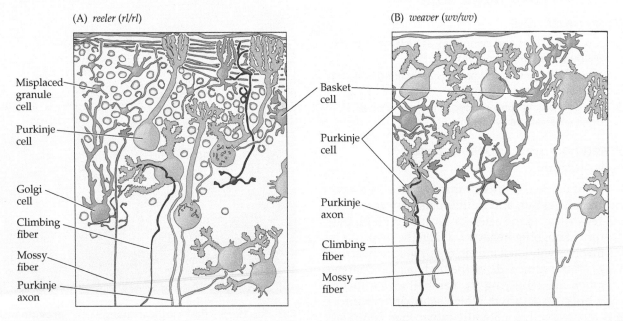

(A) *reeler (rl/rl)*   (B) *weaver (wv/wv)*

The cerebellar cortex is disrupted in both the *reeler* and *weaver* mutations. (A) The cerebellar cortex in homozygous *reeler* mice. The *reeler* mutation causes the major cell types of the cerebellar cortex to be displaced from their normal laminar positions. Despite the disorganization of the cerebellar cortex in *reeler* mutants, the major inputs—mossy fibers and climbing fibers—find appropriate targets. (B) The cerebellar cortex in homozygous *weaver* mice. The granule cells are missing, and the major cerebellar inputs synapse inappropriately on the remaining neurons. (After V. S. Caviness, Jr. and P. Rakic, 1978. *Annu Rev Neurosci* 1: 297–326.)

■ BOX 19A | Genetic Analysis of Cerebellar Function (continued)

The *reeler* gene was cloned through a combination of good luck and careful observation. In the course of making transgenic mice by inserting DNA fragments in the mouse genome, investigators in Tom Curran's laboratory created a new strain of mice that behaved much like *reeler* mice and had similar cerebellar pathology. This "synthetic" *reeler* mutation was identified by finding the position of the novel DNA fragment—which turned out to be on the same chromosome as the original *reeler* mutation. Further analysis showed that the same gene had indeed been mutated, and the *reeler* gene was subsequently identified. Remarkably, the protein encoded by this gene is homologous to known extracellular matrix proteins such as tenascin, laminin, and fibronectin (see Chapter 23). This finding makes sense, since the pathophysiology of the *reeler* mutation entails altered cell migration, resulting in misplaced neurons in the cerebellar cortex as well as the cerebral cortex and hippocampus.

Molecular genetic techniques have also led to cloning the *weaver* gene. Using linkage analysis and the ability to clone and sequence large pieces of mammalian chromosomes, Andy Peterson and his colleagues "walked" (i.e., sequentially cloned) several kilobases of DNA in the chromosomal region to find where the *weaver* gene mapped. By comparing normal and mutant sequences within this region, they determined *weaver* to be a mutation in an inward-rectifier K^+ channel (see Chapter 4). How this particular molecule influences the development of granule cells or causes their death in the mutants is not yet clear. Nevertheless, the story of the proteins encoded by the *reeler* and *weaver* genes indicates both the promise and the challenge of a genetic approach to understanding cerebellar function.

In recent years, such investigative opportunity has arisen with respect to the putative role of the cerebellum in the expression of disorders in the neurocognitive domain. Thus, genetic, behavioral, and clinical investigations have suggested that dysfunctional cerebellar circuits may contribute to the development of neurocognitive disorders, including autism spectrum disorders. Early disruption of the cerebellar circuitry has been shown to be positively correlated with autism; indeed, cerebellar injury conveys the largest nonheritable risk for the emergence of autism spectrum disorders. Among heritable factors, recent work using mouse models has shown that mutations in the genes that encode the SHANK (SH3 and multiple ankyrin repeat domains) family of postsynaptic scaffolding proteins produce impairments in cerebellar-dependent sensorimotor learning and alterations in Purkinje cell dendritic morphology. Perhaps such mutations induce synaptic modifications that alter the role of cerebellar circuitry in mediating neural activities that depend on precise temporal control. Whatever the explanation, it has become clear that early disturbances in cerebellar structure and function have broad implications for the ongoing (postnatal) construction and refinement of circuits in the cerebral hemispheres, including those that are involved in governing motor behavior, and also those involved in the regulation of cognitive processes and emotional expression.

Motor Mutations in Mice

Mutation	Inheritance	Chromosome affected	Behavioral and morphological characteristics
reeler (rl)	Autosomal recessive	5	Reeling ataxia of gait, dystonic postures, and tremors. Systematic malposition of neuron classes in the forebrain and cerebellum. Small cerebellum, reduced number of granule cells.
weaver (wv)	Autosomal recessive	?	Ataxia, hypotonia, and tremor. Cerebellar cortex reduced in volume. Most cells of external granule layer degenerate prior to migration.
leaner (tg1a)	Autosomal recessive	8	Ataxia and hypotonia. Degeneration of granule cells, particularly in the anterior and nodular lobes of the cerebellum. Degeneration of a few Purkinje cells.
lurcher (lr)	Autosomal semi-dominant	6	Homozygote dies. Heterozygote is ataxic with hesitant, lurching gait and has seizures. Cerebellum half normal size; Purkinje cells degenerate; granule cells reduced in number.
nervous (nr)	Autosomal recessive	8	Hyperactivity and ataxia. Ninety percent of Purkinje cells die between 3 and 6 weeks of age.
Purkinje cell degeneration (pcd)	Autosomal recessive	13	Moderate ataxia. All Purkinje cells degenerate between the fifteenth embryonic day and third month of age.
staggerer (sg)	Autosomal recessive	9	Ataxia with tremors. Dendritic arbors of Purkinje cells are simple (few spines). No synapses of Purkinje cells with parallel fibers. Granule cells eventually degenerate.

After V. S. Caviness, Jr. and P. Rakic. 1978. *Annu Rev Neurosci* 1: 297–326.

For example, damage to the vestibulocerebellum leads to the *vestibulocerebellar syndrome*, which is characterized by impairments in the ability to stand upright and maintain stable fixation. The eyes have difficulty maintaining fixation on stationary and moving visual objects; they drift from the target and then jump back to it with a corrective saccade, a phenomenon called **nystagmus**. Disruption of the pathways to the vestibular nuclei may also result in a reduction of muscle tone. In contrast, individuals with damage to the spinocerebellum, especially in the anterior lobe of the cerebellum and/or a particular division of the posterior lobe, are likely to exhibit the *cerebellar motor syndrome*, characterized by motor ataxia and **dysmetria**— over- and underreaching. Such persons tend to have difficulty controlling walking movements: They ambulate with a wide-based gait and small shuffling movements, which represent the inappropriate operation of groups of leg muscles that normally rely on sensory feedback to produce smooth, concerted actions. These individuals also tend to have difficulty performing rapid alternating movements with their hands and feet, a sign referred to as **dysdiadochokinesia**. Tremors during voluntary movements, known as **action** or **intention tremors**, accompany dysmetria due to disruption of the mechanism for detecting and correcting movement errors (Figure 19.16). Lesions of the cerebrocerebellum produce impairments in highly skilled sequences of learned movements, such as speech or playing a musical instrument, as well as in the acquisition of novel motor skills. Similar impairments may be seen with injury or disease that affects the afferent projections into the cerebellum or the efferent projections from the deep cerebellar nuclei to the brainstem and thalamus. The common feature of all of these signs, regardless of the site of the lesion, is the inability to perform smooth, precisely coordinated movements.

In recent years it has become clear that congenital malformation of cerebellar circuitry, altered expression of genes that organize and stabilize synaptic machinery, and/ or acquired injuries of the cerebellum may lead to atypical differences and impairments in cognitive or affective regulation (see Box 19A). Such impairments in non-motor domains of behavior may manifest even in the absence of motor ataxia—the more typical manifestation of cerebellar involvement in injury or disease. For example, recent studies of children and adults with acquired focal lesions of the cerebellum suggest a functional topography of ataxia, with this clinical concept expanded to include incoordination of cognitive and affective regulation (Figure 19.17). Thus, while lesions of the anterior lobe of the cerebellum or of a particular division of the posterior lobe are most likely to result in the cerebellar motor syndrome, injuries of the lateral hemispheres of the posterior lobe may leave motor coordination intact. Rather than producing the more familiar cerebellar motor syndrome, such lesions are more

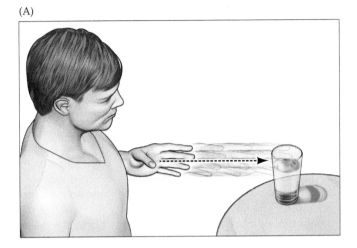

(A)

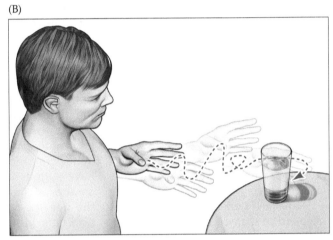

(B)

Figure 19.16 Illustration of appendicular ataxia with cerebellar damage (A) Smooth execution of a visually guided reach in an individual with typical cerebellum function. (B) Poorly coordinated visually guided reach (appendicular ataxia) in an individual with cerebellar damage. The hand takes a much less straight trajectory to the target, with irregular movements that typically overshoot or undershoot the visual target (dysmetria) and so require frequent corrective movements to execute the intended motor task.

likely to result in a kind of ataxia or dysmetria of thought or emotion, characterized by impaired executive functioning, visual spatial processing, linguistic skills, and regulation of affect (see Figure 19.17B). This constellation of non-motor impairments has become known as the *cerebellar cognitive-affective syndrome* or Schmahmann's syndrome (named for Jeremy D. Schmahmann, a neurologist at the Massachusetts General Hospital and Harvard Medical School who advanced a theory of dysmetria of thought and emotion). Furthermore, studies of major neurodevelopmental disorders, such as intellectual disability, autism spectrum disorders, attention deficit hyperactivity disorder, and Down syndrome have implicated atypicalities of cerebellar development. Moreover, many of the genes that

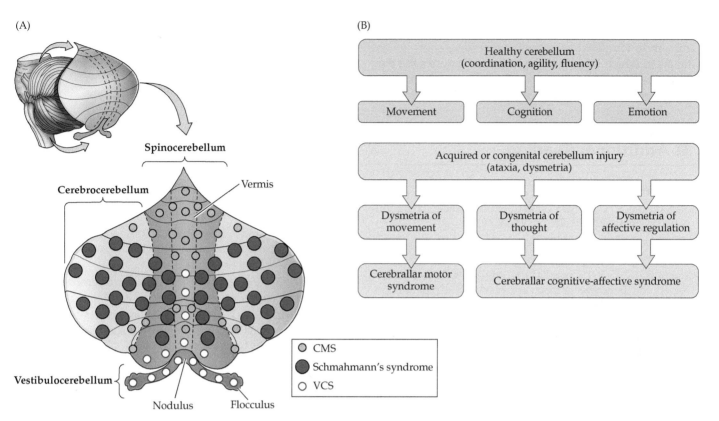

(A)

Spinocerebellum

Cerebrocerebellum

Vermis

Vestibulocerebellum

Nodulus   Flocculus

○ CMS
● Schmahmann's syndrome
○ VCS

(B)

Healthy cerebellum
(coordination, agility, fluency)

Movement   Cognition   Emotion

Acquired or congenital cerebellum injury
(ataxia, dysmetria)

Dysmetria of movement   Dysmetria of thought   Dysmetria of affective regulation

Cerebrallar motor syndrome   Cerebrallar cognitive-affective syndrome

Figure 19.17   Topography of motor and non-motor ataxias in cerebellar cortex (A) Flattened view of the cerebellar surface illustrating where acquired injury often results in the cerebellar motor syndrome (orange spots), the cerebellar cognitive-affective syndrome (blue spots), or the vestibulocerebellum syndrome (white spots). Large lesions affecting widespread cerebellar territories may produce combinations of syndromes, as predicted by this topography of cerebellar ataxias. (B) The healthy cerebellum coordinates the agile and fluent expression of movement, cognition, and emotion. Acquired injury of the cerebellum may result in a dysmetria of movement and the cerebellar motor syndrome, and/or a dysmetria of thought or affective regulation and the cerebellar cognitive-affective syndrome. (A after M. Manto and P. Mariën, 2015. *Cerebellum Ataxias* 2: 2; B after X. Guell et al., 2015. *Cerebellum* 14: 50–58.)

have been associated with autism spectrum disorders are expressed in the cerebellum during brain development in early life. These clinical associations are consistent with transsynaptic tract-tracing studies in non-human primates that show multi-neuronal connections between the posterolateral cerebrocerebellum and the prefrontal cortex, the sector of the cerebral mantle that is important for a variety of executive functions (see Chapter 33). Accordingly, clinicians, systems neuroscientists, and cognitive scientists alike are increasingly investigating the association of cerebellar function and dysfunction with the coordination of non-motor behavior, including the secondary consequences of the cerebellar motor syndrome for the organization and regulation of thought and emotion.

Summary

The cerebellum receives input from regions of the cerebral cortex that plan and initiate complex and highly skilled movements; it also receives input from sensory systems that monitor the course of movements. This arrangement enables a comparison of an intended movement with the actual movement and a reduction in the difference, or "motor error." The corrections of motor error produced by the cerebellum occur in real time and are stored over longer periods as a form of motor learning. Error correction is mediated by climbing fibers that ascend from the inferior olive to contact the dendrites of the Purkinje cells in the cerebellar cortex. Information provided by the climbing fibers modulates the effectiveness of the massively convergent input onto the Purkinje cells, which arrives via the parallel fibers from the granule cells. The granule cells receive information about the intended movement—and the actual performance of the movement—from the vast number of mossy fibers that enter the cerebellum from multiple sources. As might be expected, the output of the cerebellum from the deep cerebellar nuclei projects to circuits that govern all the major sources of upper motor neurons described in Chapter 17. The effects of cerebellar disease provide strong support for the idea that the cerebellum regulates the performance of movements. Thus, individuals with cerebellar disorders show severe ataxias in which the site of the lesion determines the particular movements affected, with the incoordination of movement on the same side of the body as the site of the lesion. Recent evidence supports a role for cerebellar circuitry in regulating performance in non-motor domains of behavior, including the organization and coordination of thought and emotion.

■ Additional Reading

Reviews

Allen, G. and N. Tsukahara (1974) Cerebrocerebellar communication systems. *Physiol. Rev.* 54: 957–1006.

Apps, R. and R. Hawkes (2009) Cerebellar cortical organization: A one-map hypothesis. *Nat. Rev. Neurosci.* 10: 670–681.

Cerminara, N. L., E. J. Lang, R. V. Sillitoe and R. Apps (2015) Redefining the cerebellar cortex as an assembly of non-uniform Purkinje cell microcircuits. *Nat. Rev. Neurosci.* 16: 79–93.

De Zeeuw, C. I., S. G. Lisberger and J. L. Raymond (2021) Diversity and dynamism in the cerebellum. *Nat. Neurosci.* 24:160–167.

Glickstein, M. and C. Yeo (1990) The cerebellum and motor learning. *J. Cogn. Neurosci.* 2: 69–80.

Lisberger, S. G. (1988) The neural basis for learning of simple motor skills. *Science* 242: 728–735.

Ohyama, T., W. L. Nores, M. Murphy and M. D. Mauk (2003) What the cerebellum computes. *Trends Neurosci.* 26: 222–227.

Robinson, F. R. and A. F. Fuchs (2001) The role of the cerebellum in voluntary eye movements. *Annu. Rev. Neurosci.* 24: 981–1004.

Sathyanesan, A. and 5 others (2019) Emerging connections between cerebellar development, behaviour and complex brain disorders. *Nat. Rev. Neurosci.* 20: 298–313.

Schmahmann, J. D. (2019) The cerebellum and cognition. *Neurosci. Lett.* 688: 62–75.

Thach, W. T. (2007) On the mechanism of cerebellum contributions to cognition. *Cerebellum* 6: 163–167.

Thach, W. T., H. P. Goodkin and J. G. Keating (1992) The cerebellum and adaptive coordination of movement. *Annu. Rev. Neurosci.* 15: 403–442.

Important Original Papers

Asanuma, C., W. T. Thach and E. G. Jones (1983) Distribution of cerebellar terminals and their relation to other afferent terminations in the ventral lateral thalamic region of the monkey. *Brain Res. Rev.* 5: 237–265.

Brodal, P. (1978) The corticopontine projection in the rhesus monkey: Origin and principles of organization. *Brain* 101: 251–283.

DeLong, M. R. and P. L. Strick (1974) Relation of basal ganglia, cerebellum, and motor cortex units to ramp and ballistic movements. *Brain Res.* 71: 327–335.

Eccles, J. C. (1967) Circuits in the cerebellar control of movement. *Proc. Natl. Acad. Sci. U.S.A.* 58: 336–343.

McCormick, D. A., G. A. Clark, D. G. Lavond and R. F. Thompson (1982) Initial localization of the memory trace for a basic form of learning. *Proc. Natl. Acad. Sci. U.S.A.* 79: 2731–2735.

Thach, W. T. (1968) Discharge of Purkinje and cerebellar nuclear neurons during rapidly alternating arm movements in the monkey. *J. Neurophysiol.* 31: 785–797.

Thach, W. T. (1978) Correlation of neural discharge with pattern and force of muscular activity, joint position, and direction of intended next movement in motor cortex and cerebellum. *J. Neurophysiol.* 41: 654–676.

Victor, M., R. D. Adams and E. L. Mancall (1959) A restricted form of cerebellar cortical degeneration occurring in alcoholic patients. *Arch. Neurol.* 1: 579–688.

Yang, Y. and S. G. Lisberger (2014) Purkinje-cell plasticity and cerebellar motor learning are graded by complex-spike duration. *Nature* 510: 529–532.

Books

Bradley, W. G., R. B. Daroff, G. M. Fenichel and C. D. Marsden (Eds.) (1991) *Neurology in Clinical Practice.* Boston: Butterworth-Heinemann, chapters 29 and 77.

Ito, M. (1984) *The Cerebellum and Neural Control.* New York: Raven Press.

Klawans, H. L. (1989) *Toscanini's Fumble and Other Tales of Clinical Neurology.* New York: Bantam, chapters 7 and 10.

Eye Movements and Sensorimotor Integration

Overview

Eye movements are easier to study than movements of other parts of the body. This fact arises in part from the relative simplicity of muscle actions on the eyeball. There are only six extraocular muscles, each of which has a specific role in adjusting eye position. Moreover, there is a limited set of stereotyped eye movements, and the central circuits governing each one are partially distinct. Eye movements have therefore been a useful model for understanding mechanisms of motor control. Indeed, much of what is known about the regulation of movements by the vestibular system, basal ganglia, and cerebellum has come from the study of eye movements (see Chapters 11, 18, and 19). In this chapter, the major features of eye movement control are used to illustrate principles of sensorimotor integration that also apply to more complex motor behaviors.

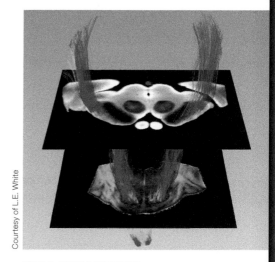

Courtesy of L.E. White

| CONCEPT 20.1 | **Eye Movements Are Necessary to Acquire and Foveate a New Visual Target and to Maintain Foveal Fixation** |

LEARNING OBJECTIVE

20.1.1 Discuss why eye movements are necessary to inspect a visual scene.

What eye movements accomplish

Eye movements are important in humans because high visual acuity is restricted to the fovea, the small circular region (about 1.2 mm in diameter) in the central retina that is densely packed with cone photoreceptors (see Chapter 9). Eye movements can direct the fovea to new objects of interest in the visual field—a process called **foveation**—or compensate for disturbances that cause the fovea to be displaced from an object already foveated.

Several decades ago, the Russian physiologist Alfred Yarbus demonstrated that eye movements reveal a good deal about the strategies used to inspect a scene. Yarbus used contact lenses with small mirrors on them to document (by the position of a reflected beam on photosensitive paper) the pattern of eye movements made while individuals examined a variety of objects and scenes. Figure 20.1 shows the changes in the direction of an individual's gaze while viewing a photograph. The thin, straight lines represent the quick, ballistic eye movements (**saccades**) used to align the foveae with particular parts of the scene. Little or no visual perception occurs during a saccade, which occupies only a few tens of milliseconds. The denser spots along these lines represent points of fixation where the observer paused for a variable period

KEY CONCEPTS

- **20.1** Eye movements are necessary to acquire and foveate a new visual target and to maintain foveal fixation

- **20.2** Eye movements are generated around three axes of rotation by three pairs of striated muscles

- **20.3** Conjugate eye movements rotate the eyes in the same direction, and disconjugate eye movements rotate the eyes in opposite directions

- **20.4** Neural circuits in the cerebral cortex and brainstem govern the amplitude, direction, and velocity of eye movements

From A. L. Yarbus, 1967. In *Eye Movements and Vision* (translated by B. Haigh), p. 181. New York: Plenum Press

 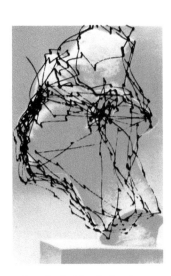

FIGURE 20.1 **Eye movements of an individual viewing a photograph** The individual was shown this photograph (left) of the famous bust of Queen Nefertiti. The diagram on the right shows the individual's eye movements over a 2-minute viewing period.

parts of the scene to examine especially interesting or informative features. The selection of locations of interest as targets of the saccades indicates that non-foveal areas of the retina have sufficient resolution to guide the foveae toward these locations for closer examination. In the figure, the spatial distribution of the fixation points is not random, and indicates that the individual spent much more time scrutinizing Nefertiti's eye, nose, mouth, and ear than examining the middle of her cheek or neck. Thus, eye movements allow us to scan the visual field, pausing to focus attention on the portions of the scene that convey the most significant information. It follows from Figure 20.1 that tracking eye movements can be used to determine which aspects of a scene are particularly arresting; in fact, today's corporate advertisers can use modern versions of Yarbus's method to determine which pictures and scene arrangements will best sell their products.

The importance of eye movements for visual perception has also been demonstrated by experiments in which a visual image is stabilized on the retina, either by paralyzing the extraocular eye muscles or by moving a scene in exact register with eye movements so that the different features of the image always fall on exactly the same parts of the retina (Box 20A). Such stabilized visual images rapidly disappear, for reasons that remain poorly understood. Nonetheless, observations on motionless images make it plain that eye movements are essential for visual perception.

to take in visual information from the location of interest. These results obtained by Yarbus, and subsequently by many others, showed that vision is an active process in which eye movements typically shift the view several times each second to direct the foveae toward selected

■ BOX 20A | The Perception of Stabilized Retinal Images

Visual perception depends critically on frequent changes of scene. Normally, our view of the world is changed by saccades, and tiny saccades that continue to move the eyes abruptly over a fraction of a degree of visual arc occur even when the observer stares intently at an object of interest. Moreover, continual drift of the eyes during fixation progressively shifts the image onto a nearby but different set of photoreceptors. As a consequence of these several sorts of eye movements (Figure A), our point of view changes more or less continually.

The importance of a continually changing scene for normal vision is dramatically revealed when the retinal image is stabilized. If a small mirror is attached to the eye by means of a contact lens and an image is reflected off the mirror onto a screen, then the

individual necessarily sees the same thing, whatever the position of the eye—every time the eye moves, the projected image moves by exactly the same amount (Figure B). Under these circumstances, the stabilized image actually disappears from perception within a few seconds!

A simple way to demonstrate the rapid disappearance of a stabilized retinal image is to visualize one's own retinal blood vessels. The blood vessels, which lie in front of the photoreceptor layer, cast a shadow on the underlying receptors. Although normally invisible, the vascular shadows can be seen by moving a source of light across the eye, a phenomenon first noted by J. E. Purkinje nearly two centuries ago. This perception can be elicited with an ordinary penlight pressed gently against the lateral side of the closed eyelid.

When the light is wiggled vigorously, a rich network of black blood vessel shadows (called a "Purkinje tree") appears against an orange background. (The vessels appear black because they are shadows.) By starting and stopping the movement, it is readily apparent that the image of the blood vessel shadows disappears within a fraction of a second after the light source is stilled.

The conventional interpretation of the rapid disappearance of stabilized images is retinal adaptation. In fact, the phenomenon is at least partly of central origin. Stabilizing the retinal image in one eye, for example, diminishes perception through the other eye, an effect known as interocular transfer. Although the explanation for these remarkable effects is not entirely clear, they emphasize the point that the visual system is most sensitive to dynamic stimuli.

■ BOX 20A | The Perception of Stabilized Retinal Images *(continued)*

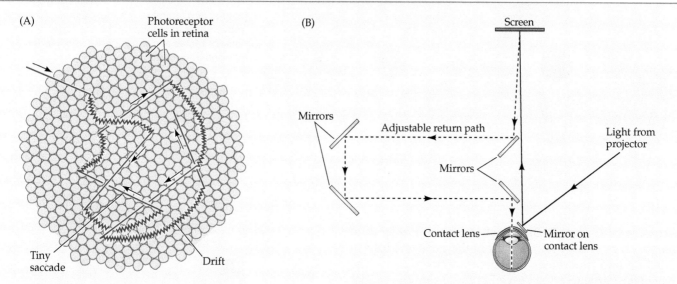

(A)   Photoreceptor cells in retina

Tiny saccade   Drift

(B)   Screen

Mirrors   Adjustable return path   Light from projector

Mirrors

Contact lens   Mirror on contact lens

(A) Diagram of the types of eye movements that continually change the retinal stimulus during fixation. The straight lines indicate microsaccades, and the zigzag lines indicate drift; the structures in the background are photoreceptor cells drawn approximately to scale. The normal scanning movements of the eyes (saccades) are much too large to be shown here but obviously contribute to the changes of view that we continually experience, as do slow tracking eye movements (although

the fovea tracks a particular object, the background scene nonetheless changes). (B) One means of producing stabilized retinal images. By attaching a small mirror to the eye, the scene projected onto the screen will always fall on the same set of retinal points, no matter how the eye is moved. (A after R. M. Pritchard, 1961. *Sci Am* 204: 72–78. B after L. A. Riggs et al., 1953. *J Opt Soc Am* 43: 495–501.)

CONCEPT
20.2

Eye Movements Are Generated around Three Axes of Rotation by Three Pairs of Striated Muscles

LEARNING OBJECTIVES

20.2.1 Discuss the actions of the three pairs of extraocular muscles.

20.2.2 Identify the cranial nerve that innervates each extraocular muscle.

Actions and innervation of extraocular muscles

Three antagonistic pairs of muscles in each orbit control eye movements: the **lateral** and **medial rectus muscles**; the **superior** and **inferior rectus muscles**; and the **superior** and **inferior oblique muscles**. These muscles are responsible for movements of the eye along three different axes: rotation around the vertical axis, either toward the nose (adduction) or away from the nose (abduction); rotation around the horizontal axis, either elevation or depression; and rotation around the optic axis, torsional movements that bring the top of the eye toward the nose (intorsion) or away from the nose (extorsion). Horizontal movements (rotations around the vertical axis) are controlled primarily

by the medial and lateral rectus muscles; the medial rectus muscle is responsible for adduction, the lateral rectus muscle for abduction (Figure 20.2). Vertical movements (rotations around the horizontal axis) require the coordinated action of the superior and inferior rectus muscles, as well as the oblique muscles. The relative contributions of the rectus and oblique muscles depend on the horizontal position of the eye. In the primary position (eyes straight ahead), both of these muscle groups contribute to vertical movements. Elevation is due to the action of the superior rectus and inferior oblique muscles, while depression is due to the action of the inferior rectus and superior oblique muscles. When the eye is abducted, the rectus muscles are the prime vertical movers; elevation is due to the action of the superior rectus, and depression is due to the action of the inferior rectus. When the eye is adducted, the oblique muscles are the prime vertical movers. In this position, elevation is due to the action of the inferior oblique muscle, while depression is due to the action of the superior oblique muscle. The oblique muscles are also primarily responsible for torsional movements.

The extraocular muscles are innervated by lower motor neurons in the pons and midbrain whose axons form three cranial nerves: the abducens, the trochlear, and the

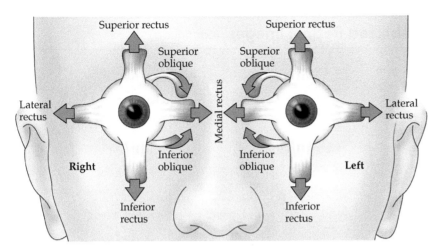

FIGURE 20.2   Extraocular muscles contribute to vertical and horizontal eye movements Horizontal movements are mediated by the medial and lateral rectus muscles, while vertical movements are mediated by the superior and inferior rectus and the superior and inferior oblique muscles.

oculomotor (Figure 20.3). The **abducens nerve** (cranial nerve VI) exits the brainstem from the pons–medullary junction and innervates the lateral rectus muscle. The **trochlear nerve** (cranial nerve IV) exits from the caudal midbrain and supplies the superior oblique muscle. The root of the trochlear nerve crosses the midline just before it exits from the dorsal surface of the brainstem and innervates the superior oblique muscle on the contralateral side—that is, contralateral relative to the location of the α motor neurons from which the axons in the nerve originated. This is the only somatic motor nerve—cranial or spinal—that wholly supplies muscle on the side of the body that is opposite to the origin of the nerve, and the only motor nerve to exit the dorsal aspect of the CNS. The **oculomotor nerve** (cranial nerve III), which exits from the rostral midbrain just medial to the cerebral peduncle, supplies the rest of the extraocular muscles. Although the oculomotor nerve governs several different muscles, each muscle receives its innervation from a separate pool of lower motor neurons within the nuclear complex that supplies the third nerve. For the lower motor neuron pools that supply the medial rectus, inferior rectus, and inferior oblique muscles, the axons from neurons in these pools grow through the ipsilateral oculomotor nerve to reach the muscles in the ipsilateral orbit. The lower motor neurons that govern the superior rectus muscle grow their axons across the midline of the midbrain and join the root of the contralateral oculomotor nerve. Thus, the superior rectus muscle is innervated by lower motor neurons whose cell bodies reside in the contralateral oculomotor nucleus (see Figure 20.3).

FIGURE 20.3   Innervation of the extraocular muscles by the cranial nerve nuclei governing eye movements The abducens nucleus innervates the ipsilateral lateral rectus muscle; the trochlear nucleus innervates the contralateral superior oblique muscle; and the oculomotor nucleus innervates the ipsilateral medial rectus, inferior rectus, and inferior oblique and the contralateral superior rectus.

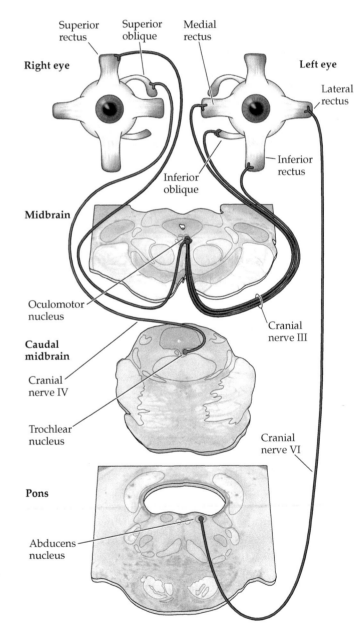

In addition to supplying the extraocular muscles, the oculomotor complex includes a distinct cell group that innervates the levator muscles of the ipsilateral eyelid; the axons from these neurons also travel in the third nerve. Finally, the third nerve carries preganglionic parasympathetic axons that are responsible for pupillary constriction from the nearby Edinger–Westphal nucleus (see Chapters 9 and 21). Thus, damage to the third nerve results in three characteristic deficits: impairment of eye movements, drooping of the eyelid (a clinical sign called **ptosis**), and pupillary dilation, because of the unopposed action of sympathetic inputs to the dilator muscles of the iris.

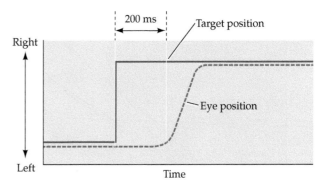

FIGURE 20.4 **Metrics of a saccadic eye movement**
The solid red line indicates the position of a fixation target, and the dashed blue line the position of the fovea. When the target moves suddenly to the right, there is a delay of about 200 ms before the eye begins to move to the new target position. (After A. F. Fuchs, 1967. *J Physiol* 191: 609–630.)

Conjugate Eye Movements Rotate the Eyes in the Same Direction, and Disconjugate Eye Movements Rotate the Eyes in Opposite Directions

CONCEPT
20.3

LEARNING OBJECTIVES

20.3.1 Discuss the four major types of conjugate eye movements and explain the functional purpose of each.

20.3.2 Discuss disconjugate eye movements and explain the functional purpose of inward and outward rotation of the eyes.

Types of eye movements and their functions

The five basic types of eye movements can be grouped into two functional categories: those that serve to *shift* the direction of gaze, and those that serve to *stabilize* gaze. Shifts in eye position are necessary to foveate new targets and to follow foveated targets as they move in visual space. Stabilizing movements of the eyes are used to maintain foveation when the head moves and when there are large-scale movements of the visual field. Thus, saccades, smooth pursuit movements, and vergence movements shift the direction of gaze, and vestibulo-ocular and optokinetic movements stabilize gaze. The functions of each type of eye movement are introduced here; the neural circuitry responsible for movements that shift the direction of gaze is presented in more detail in Concept 20.4 (see Chapters 11 and 19 for further discussion of the neural circuitry underlying gaze-stabilizing movements).

As noted in Concept 20.1, saccades are rapid, ballistic movements of the eyes that abruptly change the direction of fixation. They range in amplitude from the small movements made while reading to the much larger movements made while gazing around a room. Saccades can be elicited voluntarily, but they occur reflexively whenever the eyes are open. Indeed, so-called microsaccades occur even when the eyes are otherwise maintaining fixation on a target (see Box 20A). Saccades also occur during the rapid eye movement (REM) phase of sleep—a principal feature for which this phase of sleep is named (see Chapter 28).

Figure 20.4 shows the time course of a typical saccadic eye movement. After the onset of a target for a saccade (in this example, the stimulus was the movement of an already fixated target), it takes about 200 ms for eye movement to begin. During this delay, the position of the target with respect to the fovea (that is, how far the eye has to move) is computed, and the difference between the initial and intended position is converted into a motor command that activates the extraocular muscles to move the eyes the correct distance in the appropriate direction. Saccadic eye movements are said to be ballistic because the saccade-generating system usually does not respond to subsequent changes in the position of the target during the course of the eye movement. If the target moves again during this time (which is on the order of 15 to 100 ms), the saccade will miss the target, and a second saccade must be made to correct the error.

Smooth pursuit eye movements are much slower tracking movements of the eyes designed to keep a moving stimulus on the fovea once foveation is achieved. Such movements are under voluntary control in the sense that the observer can choose whether or not to track a moving stimulus (Figure 20.5). Surprisingly, only with task-specific training is it possible to make a smooth pursuit movement in the absence of a moving target. Most people who try to move their eyes in a smooth fashion without a moving target to track simply make a series of saccades along the imagined path through visual space.

Vergence movements align the fovea of each eye with targets located at different distances from the observer. Although vergence movements are required to track a visual target that may be moving closer or farther away, they are more commonly employed when abruptly shifting the direction of gaze, for example, from a near object to one that is more distant. Unlike other types of

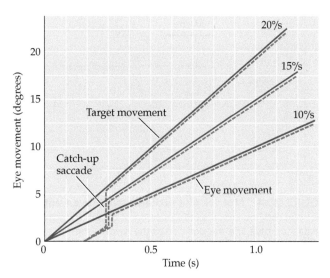

FIGURE 20.5 Metrics of smooth pursuit eye movements
These traces show eye movements (dashed blue lines) tracking a stimulus moving at one of three different velocities (solid red lines). After a quick saccade to foveate the target, the eye movement attains a velocity that matches the velocity of the target. (After A. F. Fuchs, 1967. *J Physiol* 191: 609–630.)

eye movements, in which the two eyes move in the same direction (**conjugate eye movements**), vergence movements are **disconjugate** (or **disjunctive**); they involve either a convergence or divergence of the lines of sight of each eye to foveate an object that is nearer or farther away. Convergence is one of the three reflexive visual responses elicited together to shift gaze from a distant to a near object. The other components of the so-called **near reflex triad** are accommodation of the lens, which by increasing the curvature of the lens brings the close object into focus, and pupillary constriction, which by reducing spherical aberration increases the depth of field and sharpens the image on the retina (see Chapter 9).

Vestibulo-ocular movements and **optokinetic eye movements** operate together to move the eyes and stabilize gaze relative to the external world, thus compensating for head movements. These reflexive responses prevent visual images from "slipping" on the surface of the retina as head position varies and, more rarely, when confronted with large-scale movements of the visual scene (such as a flowing river or a passing train).

The action of vestibulo-ocular movements can be appreciated by fixating an object and moving the head from side to side; the eyes automatically compensate for the head movement by moving the same distance and at the same velocity but in the opposite direction, thus keeping the image of the object at more or less the same place on the retina. The vestibular system detects brief, transient changes in head position and produces rapid, corrective eye movements using the pathways described in Chapter 11. Sensory information from the semicircular canals directs the eyes to move in a direction opposite to the head movement. Although the vestibular system operates effectively to counteract rapid movements of the head, it is relatively insensitive to slow movements (below 1 Hz) or to persistent rotation of the head. For example, if the vestibulo-ocular reflex is tested with continuous rotation of an individual and without visual cues about the movement of the image (i.e., with eyes closed or in the dark), the compensatory eye movements cease after only about 30 seconds of rotation. However, if the same test is performed with visual cues, eye movements persist. The compensatory eye movements in this case are due to the activation of another system that relies not on vestibular information, but on visual cues indicating motion of the visual field. This optokinetic system is especially sensitive to slow movements (below 1 Hz) of large areas of the visual field, and its response builds up slowly. These features complement the properties of the vestibulo-ocular reflex, especially as head movements slow down and vestibular signals decay (Figure 20.6). Thus, should a visual image slowly "slip" across the retina, the optokinetic system will respond by inducing compensatory movements of the eyes at the same speed and in the opposite direction.

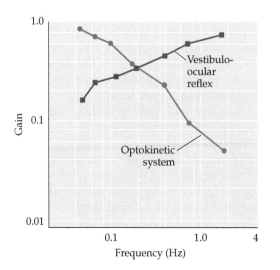

FIGURE 20.6 Operational ranges of the vestibulo-ocular and optokinetic systems Functions of the vestibulo-ocular and optokinetic systems were assessed independently in rabbits by rotating the animals with their eyes closed (to isolate the vestibulo-ocular reflex) or following recovery from bilateral labyrinthectomy (to isolate the optokinetic system). At low frequencies of movement (below 1 Hz or one back-and-forth cycle of stimulation per second), the gain of the vestibulo-ocular reflex (the ratio of eye movement to head movement) diminishes below unity. However, the gain of the optokinetic system (the ratio of eye movement to retinal slip) approaches unity at such low frequencies of stimulation. Thus, the vestibulo-ocular and optokinetic systems act in a complementary, frequency-dependent fashion to stabilize gaze over a broad range of stimulation frequencies. (After E. Baarsma and H. Collewijn, 1974. *J Physiol* 238: 603–625.)

The optokinetic system can be tested by seating an individual in front of a screen on which a series of horizontally moving vertical bars is presented. The eyes automatically track the stripes until the eyes reach the end of their excursion. Then there is a quick saccade in the direction opposite to the movement, followed once again by smooth pursuit of the stripes. This alternation of slow and fast movement of the eyes in response to such stimuli is called **optokinetic nystagmus**. Optokinetic nystagmus is a normal reflexive response of the visual and oculomotor systems in response to large-scale movements of the visual scene and should not be confused with the pathological nystagmus that can result from certain kinds of brain injury (for example, damage to the vestibular system or the cerebellum; see Chapters 11 and 19, respectively). Indeed, clinicians have long regarded eye movements to be key indicators of neurological function and dysfunction (Clinical Applications).

■ Clinical Applications

Eye Movements and Neurological Injury, Disease, and Disorder

Assessments of voluntary and involuntary eye movements have long been an important feature of neurological examinations by health professionals. Not only do such assessments test the integrity and functions of six cranial nerves (the paired oculomotor, trochlear, and abducens nerves), they also challenge central circuits that span nearly all major divisions of the CNS, except for the spinal cord (although spinal cord neurons do play a role in the sympathetic regulation of pupil diameter; see Chapter 21).

Just as any voluntary movement can be analyzed in terms of lower and upper motor neuronal control, so can the voluntary activity of the extraocular muscles that shift and stabilize visual gaze. Injuries that affect cranial nerves III, IV, or VI or the motor nuclei that supply them lead to lower motor neuronal signs and symptoms associated with the affected extraocular muscles and to predictable deficits in conjugate gaze (Figures A and B). A person afflicted with such an injury would experience *diplopia* (double vision), especially when gaze is cast in the direction of action for the affected muscle. For example, a sixth nerve palsy on the right (see lesion 1 in Figures A and B) or a third nerve palsy on the left (see lesion 2 in Figures A and B) would lead to severe diplopia with rightward gaze. (Complete injury to the oculomotor nerve would also impair the levator muscle of the upper eyelid; the ipsilateral eye would therefore be nearly closed and would need to be opened manually to facilitate vision and assessment of oculomotor function.) Damage to the fibers passing through the medial longitudinal fasciculus in the left side of the pons would likewise produce diplopia with rightward gaze (because of insufficient contraction of the left medial rectus muscle), with the added complication of nystagmus in the right eye (a condition that is not well explained,

(Continued)

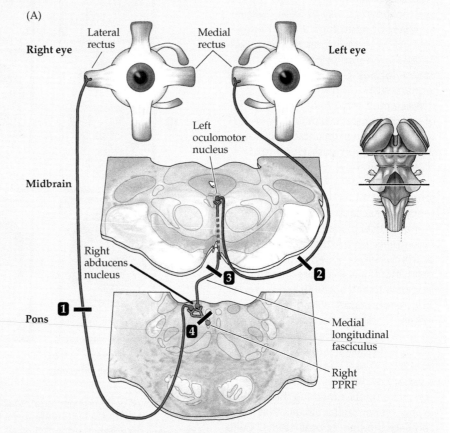

(A)

(A) Injuries to lower motor neurons and local brainstem circuits impair conjugate gaze and produce diplopia. Lesion 1 damages the right abducens nerve. Lesion 2 injures fibers in the left oculomotor nerve that innervate the left medial rectus muscle. Lesion 3 damages the left medial longitudinal fasciculus. Lesion 4 injures the local circuit neurons connecting the right paramedian pontine reticular formation (PPRF, or horizontal gaze center) to the right abducens nucleus. (See Figure 20.8 legend for additional detail.)

■ Clinical Applications (continued)

termed *internuclear ophthalmoplegia*; see lesion 3 in Figures A and B). Injury to the paramedian tegmentum of the pons may damage the horizontal gaze center (PPRF) and prevent conjugate gaze into the ipsilateral (right, in this case) visual field (see lesion 4 in Figures A and B). Similar deficits in conjugate gaze, but toward the contralateral visual field, are characteristic of damage to upper motor neurons in the frontal eye fields and superior colliculus, as described in Concept 20.4.

Given how much central circuitry in the cerebral hemispheres, cerebellum, and brainstem governs eye movements in natural viewing (see Figure 20.13), it is not surprising that eye movements may be disturbed in a variety of neurological and neuropsychiatric conditions that involve neurodegeneration or atypical neurodevelopment. For example, individuals with idiopathic Parkinson's disease may show mild impairments of smooth pursuit eye movements and decreased gain of voluntary saccades, while individuals with Huntington's disease may show difficulty initiating saccades in response to verbal instruction, with increased saccade latency and saccade slowing, especially in the vertical plane (see Chapter 18). Individuals with Alzheimer's disease or frontotemporal dementia typically show increases in errors when challenged with anti-saccade and saccade suppression tasks, with gaze-fixation instability and prolonged latency of voluntary and reflexive saccades. Other individuals with spinocerebellar ataxias (see Chapter 19) may show a variety of disturbances in voluntary eye movements, reflecting the degeneration of cerebellar circuits that help regulate the gain of brainstem circuits that govern conjugate eye movements, including the vestibulo-ocular reflex (see Chapter 11).

Finally, it is interesting to note how eye movements are often atypical in individuals with schizophrenia. Such

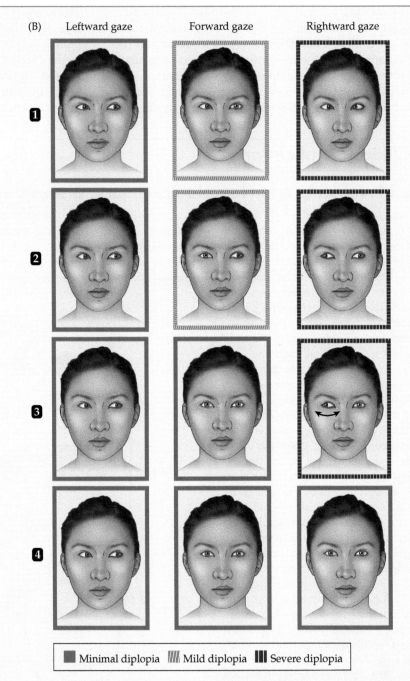

(B) Depiction of deficits in conjugate gaze associated with the four lesions shown in Figure A. Each of these lesions impairs a person's ability to make conjugate shifts of gaze toward their right. Lesion 3 typically results in internuclear ophthalmoplegia and nystagmus of the contralateral eye (double-headed curved arrow).

individuals show deficits in smooth pursuit eye movements (such as when following stimuli moving in sinusoidal trajectories; Figure C, left).

When free-viewing, individuals with schizophrenia often undersample visual scenes, restricting their gaze to a subset of available targets, even when

■ Clinical Applications (continued)

stimuli are inanimate and nonsocial (see Figure C, middle). Furthermore, people with schizophrenia often show marked instabilities when attempting to maintain fixation (see Figure C, right).

For individuals with neurodegenerative or neuropsychiatric disease, the co-occurrence of disturbances in eye movement suggests that the etiologies of these diverse conditions directly affect central circuits governing gaze shifting and stabilizing eye movements. It is also possible that disturbances in viewing behavior may exacerbate functional decline by altering perceptual and cognitive experience of the visual environment. In either case, it has become clear that clinical assessments of voluntary and involuntary eye movements are providing valuable diagnostic criteria and promising biomarkers of disease severity, progression, or regression.

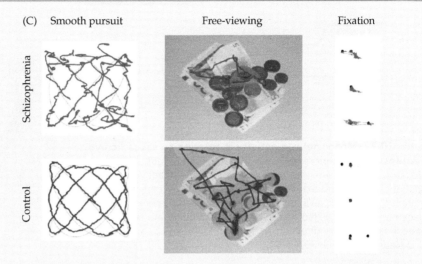

(C) Assessments of smooth pursuit, free-viewing, and fixation stability in individuals with schizophrenia (red traces) and in typical individuals (blue traces). Left: Smooth pursuit eye movements superimposed on sinusoidal patterns in two dimentions. Note the irregularities that degrade smooth pursuit in individuals with schizophrenia. Middle: Free-viewing reveals constrained patterns of visual sampling in individuals with schizophrenia (top) compared with typical individuals. Right: Assessment of fixation stability in three conditions: near distractor (top), single fixation target (center), and far distractor (bottom). Note the fixation instabilities in individuals with schizophrenia compared with typical individuals. (After P. J. Benson et al., 2012. *Biol Psychiat* 72: 716–724.)

CONCEPT
20.4

Neural Circuits in the Cerebral Cortex and Brainstem Govern the Amplitude, Direction, and Velocity of Eye Movements

LEARNING OBJECTIVES

20.4.1 Discuss how the amplitude and direction of eye movements are encoded in the firing of relevant lower motor neurons.

20.4.2 Discuss the neural circuits responsible for making saccadic eye movements.

20.4.3 Differentiate the roles of the frontal eye fields and the superior colliculus in directing gaze toward an object of interest in the visual field.

20.4.4 Discuss the complex neural circuits that govern and coordinate smooth pursuit movements.

Neural control of saccadic eye movements

Moving the eyes to fixate a new target in space (or indeed any other movement) involves two separate tasks: controlling the *amplitude* of movement (how far), and controlling the *direction* of movement (which way). The amplitude of a saccadic eye movement is encoded by the duration of neuronal activity in the lower motor neurons of the oculomotor nuclei. For instance, as shown in Figure 20.7, neurons in the abducens nucleus fire a burst of action potentials just prior to abducting the eye (by causing the lateral rectus muscle to contract) and are silent when the eye is adducted. The amplitude of the movement is correlated with the duration of the burst of action potentials in abducens neurons. Following each saccade, abducens neurons reach a new baseline level of discharge that is correlated with the position of the eye in the orbit. The steady baseline level of firing generates the muscle force needed to hold the eye in its new position.

The direction of the movement is determined by which eye muscles are activated. Although in principle any given direction of movement could be specified by independently adjusting the activity of individual eye muscles, the complexity of the task would be overwhelming. Instead, the direction of eye movement is controlled by local circuit neurons in two **gaze centers** in the reticular formation (see Box 17C), each of which is responsible for generating movements along a particular axis. The **paramedian pontine reticular formation** (PPRF), also called the horizontal gaze center, is a collection of local circuit neurons near the midline in the caudal pons. These neurons are responsible for generating horizontal eye movements. The **rostral interstitial nucleus**,

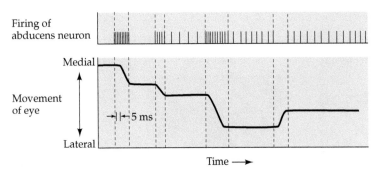

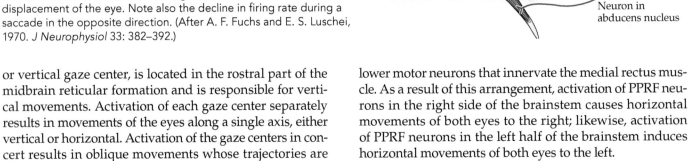

FIGURE 20.7 Motor neuron activity in relation to saccadic eye movements The experimental setup is shown on the right. In this example, an abducens lower motor neuron fires a burst of activity (upper trace) that precedes and extends throughout the movement (solid line). An increase in the tonic level of firing is associated with more lateral displacement of the eye. Note also the decline in firing rate during a saccade in the opposite direction. (After A. F. Fuchs and E. S. Luschei, 1970. *J Neurophysiol* 33: 382–392.)

or vertical gaze center, is located in the rostral part of the midbrain reticular formation and is responsible for vertical movements. Activation of each gaze center separately results in movements of the eyes along a single axis, either vertical or horizontal. Activation of the gaze centers in concert results in oblique movements whose trajectories are specified by the relative contribution of each center.

An example of how the PPRF works with the abducens and oculomotor nuclei to generate a horizontal saccade to the right is shown in Figure 20.8. Neurons in the PPRF innervate cells in the abducens nucleus on the same side of the brain. The abducens nucleus contains two types of neurons. One type comprises the lower motor neurons that innervate the lateral rectus muscle on the same side. The other type, called internuclear neurons, send their axons across the midline. These axons ascend in a fiber tract called the **medial longitudinal fasciculus** and terminate in the portion of the oculomotor nucleus that contains

lower motor neurons that innervate the medial rectus muscle. As a result of this arrangement, activation of PPRF neurons in the right side of the brainstem causes horizontal movements of both eyes to the right; likewise, activation of PPRF neurons in the left half of the brainstem induces horizontal movements of both eyes to the left.

Neurons in the PPRF also send axons to the medullary reticular formation, where they contact inhibitory local circuit neurons (not illustrated in Figure 20.8). These local

FIGURE 20.8 Synaptic circuitry responsible for horizontal movements of the eyes to the right This simplified diagram depicts how activation of local circuit neurons in the right horizontal gaze center (the right PPRF; orange) leads to increased activity of lower motor neurons and internuclear neurons in the right abducens nucleus. The lower motor neurons (green, far left) innervate the lateral rectus muscle of the right eye. The internuclear neurons (blue, center) project via the medial longitudinal fasciculus to the contralateral oculomotor nucleus, where they activate lower motor neurons (red, far right) that in turn innervate the medial rectus muscle of the left eye. The coordinated action of the right lateral rectus and left medial rectus muscles rotates the eyes to the right. Inhibitory local circuit neurons in the medullary reticular formation (not illustrated) inhibit activity in the left abducens nucleus, which has the effect of decreasing tone in the antagonistic muscles.

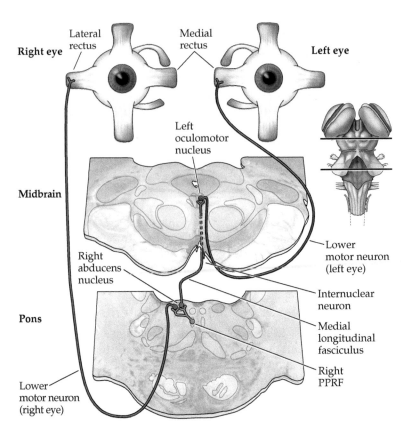

circuit neurons, in turn, project to the contralateral abducens nucleus, where they terminate on lower motor neurons and internuclear neurons. In consequence, activation of neurons in the PPRF on the right results in a reduction in the activity of the lower motor neurons in the left abducens nucleus, whose muscles would oppose movements of the eyes to the right (see Figure 20.7). Likewise, these inhibitory local circuit neurons in the medullary reticular formation inhibit the internuclear neurons that project from the left abducens nucleus to the right oculomotor nucleus, thus ensuring a commensurate reduction in the activity of lower motor neurons in the right oculomotor nucleus that innervate the right medial rectus. This inhibition of antagonists resembles the strategy used by local circuit neurons in the spinal cord to control limb muscle antagonists (see Chapter 16).

Although saccades can occur in complete darkness, they are often elicited when something in the visual field attracts attention and the observer directs the foveae toward the object of interest for more detailed examination. How, then, is sensory information about the location of a salient target in space transformed into an appropriate pattern of activity in the horizontal and vertical gaze centers? Two regions of the brain that project to the gaze centers are demonstrably important for the initiation and accurate targeting of saccadic eye movements: the **superior colliculus** of the midbrain (called the *optic tectum* in non-mammalian vertebrates) and several areas in the frontal and parietal cortex. Especially well studied is a region of the frontal lobe that lies in a rostral portion of the premotor cortex, known as the **frontal eye field** (classically, Brodmann's area 8, although in humans the frontal eye field may encroach posteriorly into Brodmann's area 6). Upper motor neurons in both the superior colliculus and frontal eye fields, each of which contains a topographical map of eye movement vectors, discharge immediately prior to saccades. Thus, activation of a particular site in the superior colliculus or in the frontal eye field elicits saccadic eye movements in a specified direction and for a specified distance. This movement is independent of the initial position of the eyes in the orbit. However, when the eyes are in the same initial position, the direction and distance of the elicited saccades are always the same for a given site of activation. Consistent with a topographic map of eye movement vectors, the direction and distance of the saccade change systematically when different sites in the frontal eye field are activated. Since each saccade is produced by the coordinated activity of all of the extraocular muscles, this arrangement is a good example of the principle that the activation of specific movements rather than individual muscles is encoded by the upper motor neurons.

Both the superior colliculus and the frontal eye field also contain cells that are activated by visual stimuli; however, the relationship between the sensory and motor responses of individual cells is better understood for the superior

colliculus. An orderly map of visual space is established by the topographical organization of the termination of retinal axons within the superior colliculus, as well as by inputs from cortical visual areas that participate in the dorsal spatial vision pathway (see Chapter 9). This sensory map is in register with the motor map that generates eye movements. Thus, neurons in a particular region of the superior colliculus are activated by visual stimuli in a limited region of visual space. This activation leads to the generation of a saccade by activating neighboring upper motor neurons that instigate movement of the eye by an amount just sufficient to align the foveae with the region of visual space that provided the stimulation (Figure 20.9).

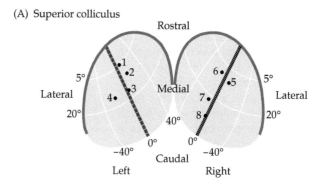

(A) Superior colliculus

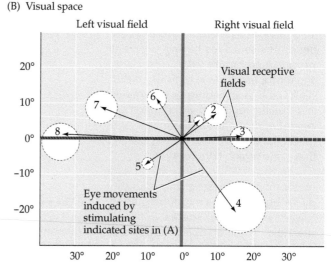

(B) Visual space

FIGURE 20.9 The sensory map of visual space in the superior colliculus is in register with the motor map that generates eye movements Evidence for this registration has been obtained from electrical recording and stimulation. (A) Surface views of the superior colliculus illustrating the location of eight separate electrode recording and stimulation sites. (B) Map of visual space showing the visual receptive field location of the sites in (A) (white circles), and the amplitude and direction of the eye movements elicited by stimulating these sites electrically (arrows). In each case, electrical stimulation results in eye movements that align the foveae with a region of visual space that corresponds to the visual receptive fields of neurons at that site. (After P. H. Schiller and M. Stryker, 1972. *J Neurophysiol* 35: 915–923.)

Neurons in the superior colliculus also respond to auditory and somatosensory stimuli. Indeed, location in space for these other modalities is mapped in register with the visual and motor maps in the colliculus. Topographically organized maps of auditory space and of the body surface in the superior colliculus can orient the eyes (and the head, via output projections from the superior colliculus to neurons that give rise to the reticulospinal tract; see Chapter 17) in response to a variety of different sensory stimuli. This registration of the sensory and motor maps in the colliculus illustrates an important principle of topographical maps: They provide an efficient mechanism for the transformation of sensory signals into the movements that are guided by these signals (in this case, the extraocular muscles and muscles of the posterior head and neck) (Box 20B). However, the motor map in the deep layers of the

---

■ **BOX 20B** | **Sensorimotor Integration in the Superior Colliculus**

The superior colliculus is a laminated structure in which differences between the layers provide clues about how sensory and motor maps interact to produce appropriate movements. As discussed in the main text, the superficial, or "visual," layer of the colliculus receives input from retinal axons that form a topographic map. Thus, each site in the superficial layer is activated maximally by the presence of a stimulus at a particular point of visual space. In contrast, neurons in the deeper, or "motor," layers generate bursts of action potentials that command saccades, effectively generating a motor map; thus, activation of different sites generates saccades having different vectors (see Figure 20.9). The visual and motor maps are *in register*, so that visual cells responding to a stimulus in a specific region of visual space are located directly above the motor cells that command eye movements toward that same region (Figure A).

The registration of the visual and motor maps suggests a simple strategy for how the eyes might be guided toward an object of interest in the visual field. When an object appears at a particular location in the visual field, it activates neurons in the corresponding part of the visual map. As a result, bursts of action potentials are generated by the subjacent motor cells to command a saccade that rotates the two eyes just the right amount to direct the foveae toward that same location in the visual field. This behavior is called *visual grasp* because successful sensorimotor integration results in the accurate foveation of a visual target.

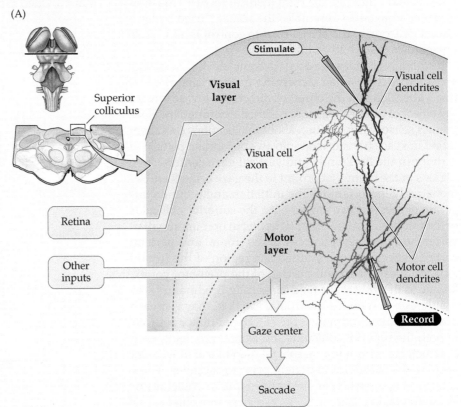

(A) The superior colliculus receives visual input from the retina and sends a command signal to the gaze centers to initiate a saccade. In the experiment illustrated here, a stimulating electrode activates cells in the visual layer, and a patch clamp pipette records the response evoked in a neuron in the subjacent motor layer. The cells in the visual and motor layers were subsequently labeled with a tracer called biocytin. This experiment demonstrates that the terminals of the visual neuron are located in the same region as the dendrites of the motor neuron.

This seemingly simple model, formulated in the early 1970s when the collicular maps were first found, assumes point-to-point connections between the visual and motor maps. In practice, however, these connections have been difficult to demonstrate. Neither the anatomical nor the physiological methods available at the time were sufficiently precise to establish these postulated synaptic connections. At about the same time, motor neurons were found to command saccades to nonvisual stimuli; moreover,

■ BOX 20B | Sensorimotor Integration in the Superior Colliculus (*continued*)

spontaneous saccades occur in the dark. Thus, it was clear that visual-layer activity is not always necessary for saccades. To confuse matters further, animals could be trained *not* to make a saccade when an object appeared in the visual field, showing that the activation of visual neurons is sometimes insufficient to command saccades. The fact that activity of neurons in the visual map is *neither necessary nor sufficient* for eliciting saccades led investigators away from the simple model of direct connections between corresponding regions of the two maps, toward models that linked the layers indirectly through pathways that detoured through the cortex.

Eventually, however, new and better methods resolved this uncertainty. Techniques for filling single cells with axonal tracers showed an overlap between descending visual-layer axons and ascending motor-layer dendrites, in accord with direct anatomical connections between corresponding regions of the maps. At the same time, in vitro whole-cell patch clamp recording (see Box 4A) permitted more discriminating functional studies that distinguished excitatory and inhibitory inputs to the motor cells. These experiments showed that the visual and motor layers do indeed have the functional connections required to initiate the command for a visually guided saccadic eye movement.

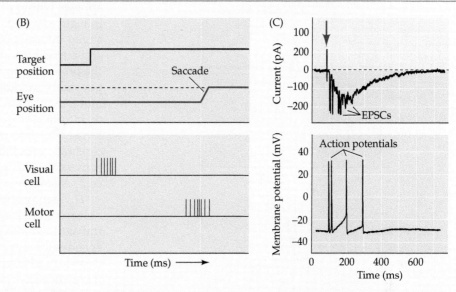

(B) The onset of a target in the visual field (top trace) is followed, after a short interval, by a saccade to foveate the target (second trace). In the superior colliculus, the visual cell responds shortly after the onset of the target, while the motor cell responds later, just before the onset of the saccade. (C) Bursts of excitatory postsynaptic currents (EPSCs) recorded from a motor-layer neuron in response to a brief (0.5 ms) current stimulus applied via a steel wire electrode in the visual layer (top; see arrow). These synaptic currents generate bursts of action potentials in the same cell (bottom). (B after R. H. Wurtz and J. E. Albano, 1980. *Annu Rev Neurosci* 3: 189–226; C after G. Ozen et al., 2000. *J Neurophysiol* 84: 460–471.)

A single brief electrical stimulus delivered to the superficial layer generates a prolonged burst of action potentials that resembles the command bursts that normally occur just before a saccade (Figures B and C).

These direct connections presumably provide the substrate for the very short-latency, reflex-like express saccades that are unaffected by destruction of the frontal eye fields. Other visual and nonvisual inputs to the deep layers probably explain why activation of the retina is neither necessary nor sufficient for the production of saccades.

---

superior colliculus is not simply organized in the framework established by the spatial distribution of the sensory inputs. Rather, the input signals must also be encoded in movement coordinates so that both sensory cues and cognitive signals can activate the motor responses required to move the eyes to the intended position in the orbits. Thus, the output of the superior colliculus specifies movement intention rather than movements to fixed positions in external space or on the body surface.

The organizing framework of this motor map was demonstrated in an ingenious series of studies performed by David Sparks and his colleagues at the University of Alabama. They showed that retinal error signals (i.e., the distance and direction of the retinal projection of the target

from the fovea) in retinotopic coordinates are often not sufficient to localize saccade targets. Using trained monkeys, the investigators cued a voluntary saccade with a brief flash of light, but before the saccade could be initiated, they stimulated a site in the deep layers of the superior colliculus that induced a saccade away from the point of fixation. They recorded eye movements to determine whether the change in eye position induced by the stimulation had an impact on the direction and distance of the cued saccade (Figure 20.10). If the saccade vectors were determined simply by the retinotopic coordinates of the target, then the monkey would be expected to make a saccade of the cued direction and distance (about 10° in the upward direction in this example). However, because of the deviated starting

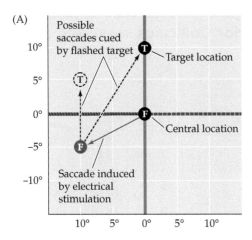

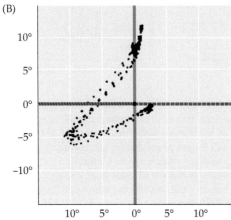

FIGURE 20.10 Saccades are encoded in movement coordinates, not retinotopic coordinates (A) Map of visual space illustrating experimental design. Monkeys were trained to fixate a central location (F, in black) and then perform a saccade to a remembered target location cued by a brief flash at a location 10° above the starting position (T, in black). After cueing, but before expression of the cued saccade, an electrical stimulus was applied to a site in the superior colliculus that induced a saccade down and to the left (to the location marked by the F in red). If the cued saccade was encoded in retinotopic coordinates, the monkey should move its eyes 10° above the stimulus-induced position of foveation (F, in red) to a location marked by the dash-encircled T. If the saccade was encoded in movement coordinates, then a compensatory saccade to the cued target location (T, in black) would be expected. (B) Consistent with the encoding of saccades in movement coordinates, the monkeys performed compensatory saccades upward and to the right, toward the location of the cued target. Dots represent eye movements sampled at 500 Hz. (After D. L. Sparks and L. E. Mays, 1983. *J Neurophysiol* 49: 45–63.)

position, the saccade should systematically miss the target position by the amount of the stimulation-induced deviation (indicated by the dashed arrow pointing upward to the dash-encircled T on the left side of Figure 20.10A). The results consistently showed, however, that this was not the case. The animals compensated for the stimulation-induced

shift by performing a compensatory saccade (a saccade indicated by the oblique, black dashed arrow to the T that appears within a black circle—the actual target location—in Figure 20.10A). This compensatory action was based on stored information about the location of the retinal image and current information about the position of the eyes in the orbit. The upper motor neurons that initiate the compensatory saccade are located at the expected site in the motor map of saccade vectors, but their activation depends on information in addition to the retinotopic location of the target. This information may be provided by circuits in the cerebral cortex that integrate this information and, in turn, activate the site in the superior colliculus that initiates the compensatory saccade (see Figure 20.10B).

This study and several that followed showed that signals from different sensory modalities are integrated and transformed into a common motor frame of reference that encodes the direction and distance of the eye movements necessary to foveate an intended target. This "place code" for intended eye position generated in the upper motor neurons of the superior colliculus is then translated into a "rate code" by downstream gaze centers in the reticular formation that can then direct the activity of lower motor neurons in the nuclei of cranial nerves III, IV, and VI (Box 20C).

The eye movement regions of the cerebral cortex collaborate with the superior colliculus in controlling saccades. Thus, the frontal eye field projects to the superior colliculus, and the superior colliculus projects to the PPRF on the contralateral side (Figure 20.11). (The superior colliculus also projects to the vertical gaze center, but for simplicity the discussion here is limited to the PPRF.) The frontal eye field can thus control eye movements by activating selected populations of upper motor neurons in the superior colliculus. This cortical area also projects directly to the contralateral PPRF; as a result, the frontal eye field can also control eye movements independently of the superior colliculus. The parallel inputs to the PPRF from the frontal eye field and superior colliculus are reflected in the different deficits that result from damage to these structures. Injury to the frontal eye field results in an inability to make saccades to the contralateral side and preferential looking toward the side of the lesion. These effects are transient, however; in monkeys with experimentally induced lesions of this cortical region, recovery is virtually complete in 2 to 4 weeks. Lesions of the superior colliculus increase the latency and decrease the accuracy, frequency, and velocity of saccades; yet saccades still occur, and the deficits also improve with time. These results suggest that the frontal eye fields and the superior colliculus provide complementary pathways for the control of saccades. Moreover, one of these structures appears to be able to compensate (at least partially) for the loss of the other. In support of this interpretation, combined lesions of the frontal eye field and the superior colliculus

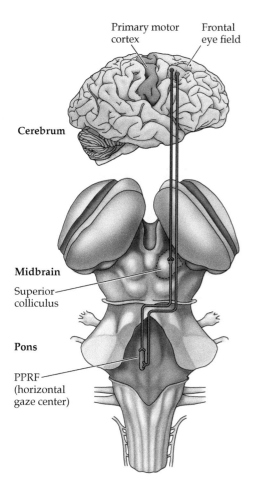

Primary motor cortex · Frontal eye field

Cerebrum

Midbrain

Superior colliculus

Pons

PPRF (horizontal gaze center)

FIGURE 20.11 Neurons in the frontal eye field collaborate with cells in the superior colliculus to control eye movements The projections shown here are from the frontal eye field in the right cerebral hemisphere (Brodmann's area 8) to the superior colliculus and the horizontal gaze center (PPRF). In humans, the frontal eye field can influence eye movements by either of two routes: indirectly, by projections to the ipsilateral superior colliculus, which in turn projects to the contralateral PPRF; and directly, by projections to the contralateral PPRF.

produce a dramatic and permanent loss in the ability to make saccadic eye movements.

These observations do not, however, imply that the frontal eye fields and the superior colliculus have the same functions. Superior colliculus lesions produce a permanent deficit in the ability to perform very short-latency, reflex-like eye movements called **express saccades**. Express saccades are evidently mediated by direct pathways to the superior colliculus from the retina or visual cortex that can access the upper motor neurons in the colliculus without extensive, and more time-consuming, processing in the frontal cortex (see Box 20B). In contrast, frontal eye field lesions produce permanent deficits in the ability to make saccades that are not guided by an external target. For example, people (or monkeys) with a lesion in the frontal eye field cannot voluntarily direct their eyes *away* from a stimulus in the visual field; this type of eye movement is called an *anti-saccade*. Such lesions also eliminate the ability to make a saccade to the remembered location of a target that is no longer visible.

■ BOX 20C | From Place Codes to Rate Codes

How does the pattern of activity in the superior colliculus get translated into a motor command that can be delivered to muscle fibers? Recall that neurons in the superior colliculus have "movement fields," discharging in conjunction with saccadic eye movements of a particular direction and amplitude. Movement fields are conceptually similar to the receptive fields that occur in various sensory areas of the brain. Across the entire population of collicular neurons, all possible saccade vectors are represented (Figure A). Because the movement fields are topographically organized, the superior colliculus forms a *motor map* of saccade vectors (or movement intentions; see main text).

(Continued)

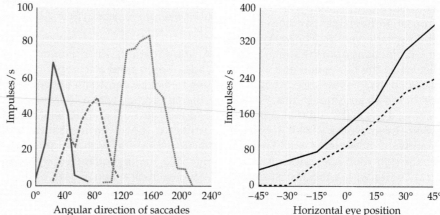

(A)

(B)

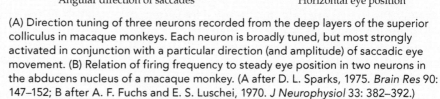

(A) Direction tuning of three neurons recorded from the deep layers of the superior colliculus in macaque monkeys. Each neuron is broadly tuned, but most strongly activated in conjunction with a particular direction (and amplitude) of saccadic eye movement. (B) Relation of firing frequency to steady eye position in two neurons in the abducens nucleus of a macaque monkey. (A after D. L. Sparks, 1975. *Brain Res* 90: 147–152; B after A. F. Fuchs and E. S. Luschei, 1970. *J Neurophysiol* 33: 382–392.)

■ BOX 20C | From Place Codes to Rate Codes (*continued*)

The direction and amplitude of eye movements are encoded quite differently by the extraocular muscles (Figure B). *Direction* is controlled by the ratio of activation of the different muscles, and *amplitude* is controlled by the magnitude of the activation of those muscles. In other words, to make a saccade go farther, the muscle pulling the eye must pull harder and longer than it would for a shorter saccade. Amplitude is therefore a *monotonic* function of muscle activation.

The pattern of activity must be transformed from a code in which collicular neurons are tuned for particular saccade amplitudes to a code in which most or all α motor neurons respond, regardless of saccade amplitude, but the level or duration of their activity varies monotonically with saccade amplitude. This transformation occurs before signals from the superior colliculus reach the α motor neurons that activate the extraocular muscles.

Various models have been proposed to explain this transformation. The basic idea, shared by all models, is that the saccade vector, as signaled by the locus of activity in the superior colliculus, is decomposed into two monotonic amplitude signals corresponding roughly to the horizontal and vertical components of the saccade vector. The weights of the projections from the superior colliculus to the horizontal and vertical gaze control centers are thought to be tuned to accomplish this. For example, a site in the superior colliculus where the movement fields encode 5° rightward movements would project to the rightward horizontal gaze control center with a modest strength.

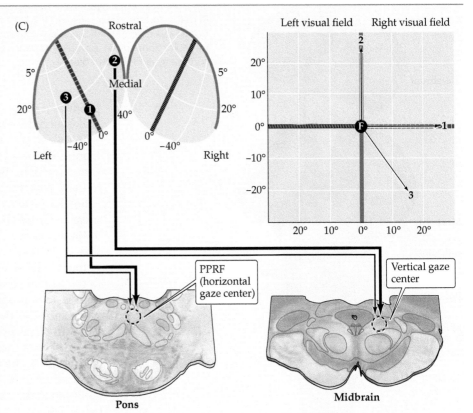

(C) Projections from the deep layers of the superior colliculus to the vertical and horizontal gaze centers in the mesencephalic and pontine reticular formation, respectively. Sites in the colliculus that encode horizontal movements (site 1) project mainly to the paramedian pontine reticular formation (PPRF, the horizontal gaze center), while sites that encode vertical movements (site 2) project mainly to the vertical gaze center in the mesencephalic reticular formation. Other sites that encode oblique saccades project to both gaze centers with weights proportional to the required horizontal and vertical displacements (thinner arrows projecting from site 3 to both gaze centers).

A site encoding 10° rightward saccades would send a stronger projection to that center. A site encoding an oblique saccade with a 10° horizontal and a 5° vertical component would project to both the horizontal and vertical centers, with weights proportional to the required contribution along each direction (Figure C).

This model is too simple to account for all the relevant experimental findings. However, it gives a general idea of how the brain might convert information encoded in one kind of format into another. This kind of transformation is a likely requirement of sensorimotor integration in many behavioral contexts where sensory cues guide movement.

Finally, the frontal eye fields are essential for systematically scanning the visual field to locate an object of interest within an array of distracting objects (see Figure 20.1). Figure 20.12 shows the responses of a frontal eye field neuron during a visual task in which a monkey was required to foveate a target located within an array of distracting objects. This frontal eye field neuron discharges at different levels to the same stimulus, depending on whether the stimulus is the target of the saccade or a "distractor," and on the location of the distractor relative to the actual target. For example, the differences between the middle and the left and right traces in Figure 20.12 demonstrate that the response to the distractor is much reduced if it is located close to the target in the visual field. Results such as

(A)

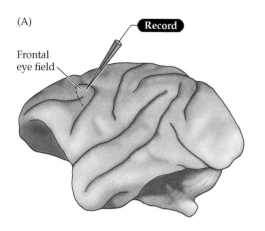

Record

Frontal
eye field

FIGURE 20.12 Responses of neurons in the frontal eye fields (A) Locus of the left frontal eye field on a lateral view of the rhesus monkey brain. (B) Activation of a frontal eye field neuron during visual search for a target. The vertical tick marks represent action potentials, and each row of tick marks is a different trial. The graphs below show the average frequency of action potentials as a function of time. The left-to-right change in color from beige to blue in each row indicates the time of onset of a saccade toward the target. In the left trace (1), the target (open red square) is in the part of the visual field "seen" by the neuron, and the response to the target is similar to the response that would be generated by the neuron even if no distractors (solid blue squares) were present (not shown). In the right trace (3), the target is far from the response field of the neuron. The neuron responds to the distractor in its response field. However, it responds at a lower rate than it would to exactly the same stimulus if the square were not a distractor but a target for a saccade (left trace). In the middle trace (2), the response of the neuron to the distractor has been sharply reduced by the presence of the target in a neighboring region of the visual field. (B after J. D. Schall, 1995. *Rev Neurosci* 6: 63–85.)

(B) **(1) Target in response field**

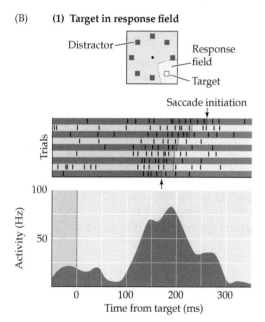

(2) Target adjacent to response field

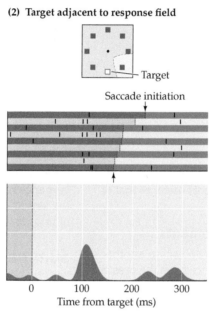

(3) Target distant from response field

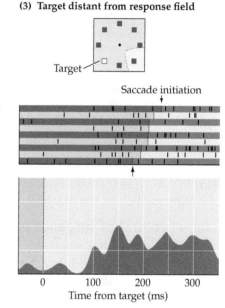

these suggest that lateral interactions within the frontal eye fields enhance the neuronal responses to stimuli that will be selected as saccade targets. They also suggest that such interactions suppress the responses to uninteresting and potentially distracting stimuli. These sorts of interactions presumably reduce the occurrence of unwanted saccades to distracting stimuli in the visual field.

Neural control of smooth pursuit movements

Until recently, smooth pursuit and saccades were considered to be mediated by different structures, but studies such as those carried out by Richard Krauzlis at the Salk Institute for Biological Studies indicate that these two types of eye movements involve many of the same structures. Not only are smooth pursuit movements mediated by neurons in the PPRF, they also are under the influence of motor control centers in the rostral superior colliculus and subareas within the frontal eye fields, both of which receive sensory input from the dorsal spatial vision pathway in the parietal

and temporal lobes. The exact routes by which visual information reaches the PPRF to generate smooth pursuit movements are not known, but pathways from the cortex to the superior colliculus and PPRF similar to those that mediate saccades may play a role; an indirect pathway through the cerebellum also has been suggested (Figure 20.13). It is clear, however, that neurons in the striate and extrastriate visual areas provide sensory information that is essential for the initiation and accurate guidance of smooth pursuit movements. In monkeys, neurons in the middle temporal area (which is largely concerned with the perception of moving stimuli; see Chapter 9) respond selectively to targets moving in a specific direction, and damage to this area disrupts smooth pursuit movements. In humans, damage to comparable areas in the parietal and occipital lobes also results in abnormalities of smooth pursuit movements. Finally, a pathway from the retina that detects movements of the visual stimulus on the retina (retinal drift) terminates in the cerebellum after relays in the pretectum and inferior olive

FIGURE 20.13 Sensory and motor structures and the connections that govern saccadic and smooth pursuit eye movements This illustration summarizes data from studies on the rhesus macaque brain. Although these two types of eye movements were once thought to be controlled by separate circuits in the forebrain and brainstem, it is now recognized that they are governed by similar networks of cortical and subcortical structures. Visual signals are processed by the dorsal spatial vision pathway, including the middle temporal and lateral intraparietal areas. Sensory and attentional signals then guide motor planning areas in the frontal eye field. These cortical areas interact with subcortical structures, including basal ganglia (caudate and substantia nigra pars reticulata) and pontine-cerebellar structures (pontine nuclei, cerebellar vermis, and vestibulocerebellum), that modulate the initiation and coordination of eye movements by the superior colliculus and downstream oculomotor centers in the reticular formation and vestibular nuclei. The eye movements regulated by this complex circuitry are guided by a variety of sensory and cognitive signals, including perception, attention, memory, and reward expectation. (After R. J. Krauzlis, 2005. *Neuroscientist* 11: 124–137.)

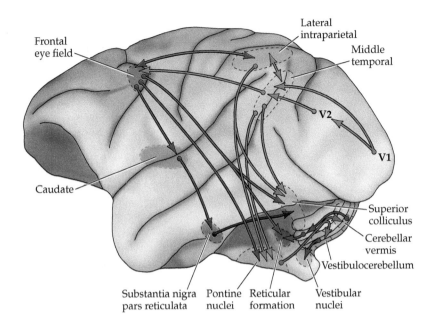

(see Chapter 19), and adjusts the gain of this system to ensure that the velocity of the eye movements matches that of the movement of the visual target.

Neural control of vergence movements

When a person wishes to look from one object to another object that is located at a different distance from the eyes, a saccade is made that shifts the direction of gaze toward the new object, and the eyes either diverge or converge until the object falls on the fovea of each eye. The structures and pathways responsible for mediating such vergence movements are not well understood, but they appear to include several extrastriate areas in the occipital lobe. Information about the location of retinal activity is relayed through the two lateral geniculate nuclei to the cortex, where the information from the two eyes is integrated. The appropriate command to diverge or converge the eyes, which is based largely on information from the two eyes about the amount of binocular disparity (see Chapter 9), is then sent from the occipital cortex to "vergence centers" in the brainstem. One such center is a population of local circuit neurons located in the midbrain near the oculomotor nucleus. These neurons generate a burst of action potentials that initiate a vergence movement, and the frequency of the burst determines its velocity. There is a division of labor within the vergence center, so that some neurons command convergence movements while others command

divergence movements. These neurons also coordinate vergence movements of the eyes with accommodation of the lens and pupillary constriction to maximize the clarity of images formed on the retina, as discussed in Chapter 9.

Summary

Despite their specialized function, the systems that control eye movements have much in common with the motor systems that govern movements of other parts of the body. Just as the spinal cord provides the basic intrinsic circuitry for coordinating the actions of muscles around a joint, the reticular formation of the pons and midbrain provides the basic circuitry that mediates movements of the eyes. Descending projections from upper motor neurons in the superior colliculus and the frontal eye field innervate gaze centers in the brainstem, providing a basis for integrating eye movements with sensory information that indicates the location of visual targets. The superior colliculus and the frontal eye field are organized in a parallel as well as a hierarchical fashion, enabling one of these structures to compensate for the loss of the other. Eye movements, like other movements, are also under the control of the basal ganglia and cerebellum; this control ensures the proper initiation and successful execution of these relatively simple motor behaviors, thus allowing observers to interact efficiently with the visual environment.

■ Additional Reading

Reviews

Borra, E. and G. Luppino (2021) Comparative anatomy of the macaque and the human frontal oculomotor domain. *Neurosci. Biobehav. Rev.* 126: 43–56.

Foulsham, T. (2015) Eye movements and their functions in everyday tasks. *Eye* 29: 196–199.

Fuchs, A. F., C. R. S. Kaneko and C. A. Scudder (1985) Brainstem control of eye movements. *Annu. Rev. Neurosci.* 8: 307–337.

Goettker, A. and K. R. Gegenfurtner (2021) A change in perspective: The interaction of saccadic and pursuit eye movements in oculomotor control and perception. *Vision Res.* 188: 283-296.

Hikosaka, O. and R. H. Wurtz (1989) The basal ganglia. In *The Neurobiology of Saccadic Eye Movements: Reviews of Oculomotor Research*, vol. 3, R. H. Wurtz and M. E. Goldberg (Eds.). Amsterdam: Elsevier, pp. 257–281.

Krauzlis, R. J., Laurent, G. and Z. M. Hafed (2017) Neuronal control of fixation and fixational eye movements. *Phil. Trans. R. Soc. B* 372: 20160205.

May, P. J. (2006) The mammalian superior colliculus: Laminar structure and connections. *Prog. Brain Res.* 151: 321–378.

Robinson, D. A. (1981) Control of eye movements. In *Handbook of Physiology*, V. B. Brooks (Ed.). Section 1: *The Nervous System*, vol. II: *Motor Control*, part 2. Bethesda, MD: American Physiological Society, pp. 1275–1320.

Schall, J. D. (1995) Neural basis of target selection. *Rev. Neurosci.* 6: 63–85.

Sparks, D. L. and L. E. Mays (1990) Signal transformations required for the generation of saccadic eye movements. *Annu. Rev. Neurosci.* 13: 309–336.

Spering, M. and M. Carrasco (2015) Acting without seeing: Eye movements reveal visual processing without awareness. *Trends Neurosci.* 38: 247–258.

Zee, D. S. and L. M. Optican (1985) Studies of adaption in human oculomotor disorders. In *Adaptive Mechanisms in Gaze Control: Facts and Theories*, A Berthoz and G. Melvill Jones (Eds.). Amsterdam: Elsevier, pp. 165–176.

Important Original Papers

Baarsma, E. and H. Collewijn (1974) Vestibulo-ocular and optokinetic reactions to rotation and their interaction in the rabbit. *J. Physiol.* 238: 603–625.

Fuchs, A. F. and E. S. Luschei (1970) Firing patterns of abducens neurons of alert monkeys in relationship to horizontal eye movements. *J. Neurophysiol.* 33: 382–392.

Optican, L. M. and D. A. Robinson (1980) Cerebellar-dependent adaptive control of primate saccadic system. *J. Neurophysiol.* 44: 1058–1076.

Schiller, P. H. and M. Stryker (1972) Single unit recording and stimulation in superior colliculus of the alert rhesus monkey. *J. Neurophysiol.* 35: 915–924.

Schiller, P. H., S. D. True and J. L. Conway (1980) Deficits in eye movements following frontal eye-field and superior colliculus ablations. *J. Neurophysiol.* 44: 1175–1189.

Sparks, D. L. and L. E. Mays (1983) Spatial localization of saccade targets. I. Compensation for stimulation-induced perturbations in eye position. *J. Neurophysiol.* 49: 45–63.

Sun, Z. and 4 others (2017) The same oculomotor vermal Purkinje cells encode the different kinematics of saccades and of smooth pursuit eye movements. *Sci. Rep.* 7: 40613; doi: 10.1038/srep40613

Books

Hall, W. C. and A. Moschovakis (Eds.) (2004) *The Superior Colliculus: New Approaches for Studying Sensorimotor Integration*. Methods and New Frontiers in Neuroscience Series. New York: CRC Press.

Leigh, R. J. and D. S. Zee (1983) *The Neurology of Eye Movements*. Contemporary Neurology Series. Philadelphia, PA: F. A. Davis.

Schor, C. M. and K. J. Ciuffreda (Eds.) (1983) *Vergence Eye Movements: Basic and Clinical Aspects*. Boston: Butterworth.

Yarbus, A. L. (1967) *Eye Movements and Vision* (trans. B. Haigh). New York: Plenum Press.

The Visceral Motor System

KEY CONCEPTS

21.1 The visceral (autonomic) motor system controls involuntary bodily functions

21.2 The sympathetic division prepares the body to mobilize resources in challenging situations

21.3 The parasympathetic division serves to increase metabolic resources and conserve energy

21.4 The enteric division is a semi-autonomous network of gastrointestinal neurons that promotes digestion

21.5 Visceral sensory signals serve local visceral motor reflexes and a central autonomic network

21.6 Visceral motor neurons use small molecule and neuropeptide neurotransmitters to mediate a variety of effects

Overview

The visceral, or autonomic, motor system controls involuntary functions mediated by the activity of smooth muscle fibers, cardiac muscle fibers, and glands. The system comprises two major divisions, the sympathetic and parasympathetic subsystems. The specialized innervation of the gut is a further, semi-independent division that is usually referred to as the enteric nervous system. Although these divisions are always active at some level, the sympathetic division mobilizes the body's resources for dealing with challenges of one sort or another. Conversely, parasympathetic activity predominates during states of relative quiescence, so that energy sources previously expended can be restored. This continuous neural regulation of the expenditure and replenishment of the body's resources is crucial for the overall physiological balance of bodily functions called homeostasis. Whereas the major controlling centers for somatic motor activity are the motor cortex in the frontal lobes and related subcortical nuclei, the major locus of central control in the visceral motor system is the hypothalamus, which in turn is modulated by circulating hormones and neural activity in the amygdala, hippocampus, insula, and other cortical regions in the ventral and medial aspects of the frontal lobes. The function of both principal divisions of the visceral motor system is governed by descending pathways from the hypothalamus and the reticular formation of the brainstem to preganglionic neurons in the brainstem and spinal cord, which in turn determine the activity of the primary, or lower, visceral motor neurons in autonomic ganglia located outside the CNS. The autonomic regulation of several organ systems of particular importance in clinical practice (including cardiovascular function, control of the bladder, and governance of the reproductive organs) is considered in more detail as specific examples of visceral motor control and the importance of central integration for the coordination of somatic motor and visceral motor function.

CONCEPT
21.1

The Visceral (Autonomic) Motor System Controls Involuntary Bodily Functions

LEARNING OBJECTIVES

21.1.1 Discuss the early studies that provided evidence for a visceral motor system.

21.1.2 Compare and contrast the distinctive features of the visceral motor system and the somatic motor system.

Early studies of the visceral motor system

Although humans must always have been aware of involuntary motor reactions to stimuli in the environment (e.g., narrowing of the pupil in response to bright light, constriction of superficial blood vessels in response to cold or fear, increased heart rate in response to exertion or anxiety), it was not until the late nineteenth century that the neural control of these and other visceral functions came to be understood in modern terms. The researchers who first rationalized the workings of the **visceral motor system** were Walter Gaskell and John Langley, British physiologists at Cambridge University. Gaskell's work preceded that of Langley and established the overall anatomy of the system as he carried out early physiological experiments that demonstrated some of its salient functional characteristics (e.g., that the heartbeat of an experimental animal is accelerated by stimulating the outflow of the upper thoracic spinal cord segments). Based on these and other observations, Gaskell concluded in 1866 that "every tissue is innervated by two sets of nerve fibers of opposite characters," and he further surmised that these actions showed "the characteristic signs of opposite chemical processes."

Using similar electrical stimulation techniques in experimental animals, Langley went on to establish the function of **autonomic ganglia** (which harbor the lower visceral motor neurons), defined the terms *preganglionic* and *postganglionic* (see Concept 21.2), and coined the phrase **autonomic nervous system**, which is commonly used as a synonym for visceral motor system (although some somatic motor actions related to emotion are closely tied to autonomic motor actions such as facial expressions; see Chapter 32). Langley's work on the pharmacology of the autonomic system initiated the classic studies indicating the roles of acetylcholine and the catecholamines in visceral motor function, and in neurotransmitter function more generally (see Chapter 6). In short, Langley's ingenious physiological and anatomical experiments established in detail the general proposition put forward by Gaskell on more circumstantial grounds.

The third major figure in the pioneering studies of the visceral motor system was Walter Cannon at Harvard Medical School, who during the early to middle 1900s devoted his career to understanding visceral motor functions in relation to homeostatic mechanisms, emotions, and other complex brain functions (see Chapter 32). Like Gaskell and Langley before him, Cannon based his work primarily on electrical stimulation in experimental animals, including activation of the hypothalamus, brainstem, and peripheral components of the system. He also established the effects of denervation in the visceral motor system, laying some of the foundation for current understanding of neuronal plasticity (see Chapters 8 and 24).

Distinctive features of the visceral motor system

Chapters 16 and 17 discussed in detail the organization of lower motor neurons in the CNS, their relationships to striated muscle fibers, and the means by which their activities are governed by higher motor centers. With respect to the efferent systems that govern the actions of smooth muscle fibers, cardiac muscle fibers, and glands, it is instructive to recognize the anatomical and functional features of the visceral motor system that distinguish it from the somatic motor system.

First, the lower motor neurons of the visceral motor system are located outside the CNS (Figure 21.1). The cell bodies of these primary visceral motor neurons are found in autonomic ganglia that are either close to the spinal cord (sympathetic division) or embedded in a neural **plexus**—a network of intersecting nerves—very near or in the target organ (parasympathetic and enteric divisions).

Second, the contacts between visceral motor neurons and the viscera are much less differentiated than the neuromuscular junctions of the somatic motor system. Visceral motor axons tend to be highly branched and give rise to many synaptic terminals at varicosities (swellings) along the length of the terminal axonal branch. Moreover, the surfaces of the visceral muscle usually lack the highly ordered structure of the motor end plates that characterizes postsynaptic target sites on striated muscle fibers. As a consequence, the neurotransmitters released by visceral motor terminals often diffuse for hundreds of microns before binding to postsynaptic receptors—a far greater distance than at the synaptic cleft of the somatic neuromuscular junction.

Third, whereas the principal actions of the somatic motor system are governed by motor cortical areas in the posterior frontal lobe (see Chapter 17), the activities of the visceral motor system are coordinated by a distributed set of cortical and subcortical structures in the ventral and medial parts of the forebrain and in the brainstem; collectively, these structures constitute a central autonomic network.

Finally, visceral motor terminals release a variety of neurotransmitters, including primary small-molecule neurotransmitters (which differ depending on whether the motor neuron in question is sympathetic or parasympathetic) and one or more of a variety of co-neurotransmitters that may be a different small-molecule neurotransmitter or a neuropeptide (see Chapter 6). These neurotransmitters interact with a diverse set of postsynaptic receptors that mediate myriad postsynaptic effects in smooth and cardiac muscle and glands. It should be clear, then, that whereas the major effect of somatic motor activation on striated muscle is nearly the same throughout the body, the effects of visceral motor activation are remarkably varied (Table 21.1). This fact should come as no surprise, given the challenge of maintaining homeostasis across the many organ systems of the body in the face of variable environmental conditions and ever-changing behavioral contingencies.

Table 21.1   Major Functions of the Visceral Motor System

Parasympathetic Division			
Target organ	**Location of preganglionic neurons**	**Location of ganglionic neurons**	**Actions**
Eye	Edinger–Westphal nucleus	Ciliary ganglion	Pupillary constriction, accommodation
Lacrimal gland	Superior salivatory nucleus	Pterygopalatine ganglion	Secretion of tears
Submandibular and sublingual glands	Superior salivatory nucleus	Submandibular ganglion	Secretion of saliva, vasodilation
Parotid gland	Inferior salivatory nucleus	Otic ganglion	Secretion of saliva, vasodilation
Head, neck (blood vessels, sweat glands, piloerector muscles)	None	None	None
Upper extremity	None	None	None
Heart	Nucleus ambiguus and dorsal motor nucleus of the vagus nerve	Cardiac plexus	Reduced heart rate
Bronchi, lungs	Dorsal motor nucleus of the vagus nerve	Pulmonary plexus	Bronchial constriction and secretion
Stomach	Dorsal motor nucleus of the vagus nerve	Myenteric and submucous plexus	Peristaltic movement and secretion
Pancreas	Dorsal motor nucleus of the vagus nerve	Pancreatic plexus	Secretion of insulin and digestive enzymes
Ascending small intestine, transverse large intestine	Dorsal motor nucleus of the vagus nerve	Ganglia in the myenteric and submucous plexus	Peristaltic movement and secretion
Descending large intestine, sigmoid, rectum	S3–S4	Ganglia in the myenteric and submucous plexus	Peristaltic movement and secretion
Adrenal gland	None	None	None
Ureter, bladder	S2–S4	Pelvic plexus	Contraction of bladder wall and inhibition of internal sphincter
Lower extremity	None	None	None

Table 21.1   Major Functions of the Visceral Motor System (*continued*)

Sympathetic Division			
Target organ	**Location of preganglionic neurons**	**Location of ganglionic neurons**	**Actions**
Eye			Pupillary dilation
Lacrimal gland			Protein secretion in tears
Submandibular and sublingual glands	Upper thoracic spinal cord (C8–T7)	Superior cervical ganglion	Vasoconstriction
Parotid gland			Vasoconstriction
Head, neck (blood vessels, sweat glands, piloerector muscles)			Sweat secretion, vasoconstriction, piloerection
Upper extremity	T3–T6	Stellate and upper thoracic ganglia	Sweat secretion, vasoconstriction, piloerection
Heart	Middle thoracic spinal cord (T1–T5)	Superior cervical and upper thoracic ganglia	Increased heart rate and stroke volume, dilation of coronary arteries
Bronchi, lungs		Upper thoracic ganglia	Vasodilation, bronchial dilation
Stomach		Celiac ganglion	Inhibition of peristaltic movement and gastric secretion, vasoconstriction
Pancreas	Lower thoracic spinal cord (T6–T10)	Celiac ganglion	Vasoconstriction, inhibition of insulin secretion
Ascending small intestine, transverse large intestine		Celiac, superior, and inferior mesenteric ganglia	Inhibition of peristaltic movement and secretion
Descending large intestine, sigmoid, rectum		Inferior mesenteric hypogastric and pelvic plexus	Inhibition of peristaltic movement and secretion
Adrenal gland	T9–L2	Cells of gland are modified neurons	Catecholamine secretion
Ureter, bladder	T10–L2	Hypogastric and pelvic plexus	Relaxation of bladder wall muscle and contraction of internal sphincter
Lower extremity	T10–L2	Lower lumbar and upper sacral ganglia	Sweat secretion, vasoconstriction, piloerection

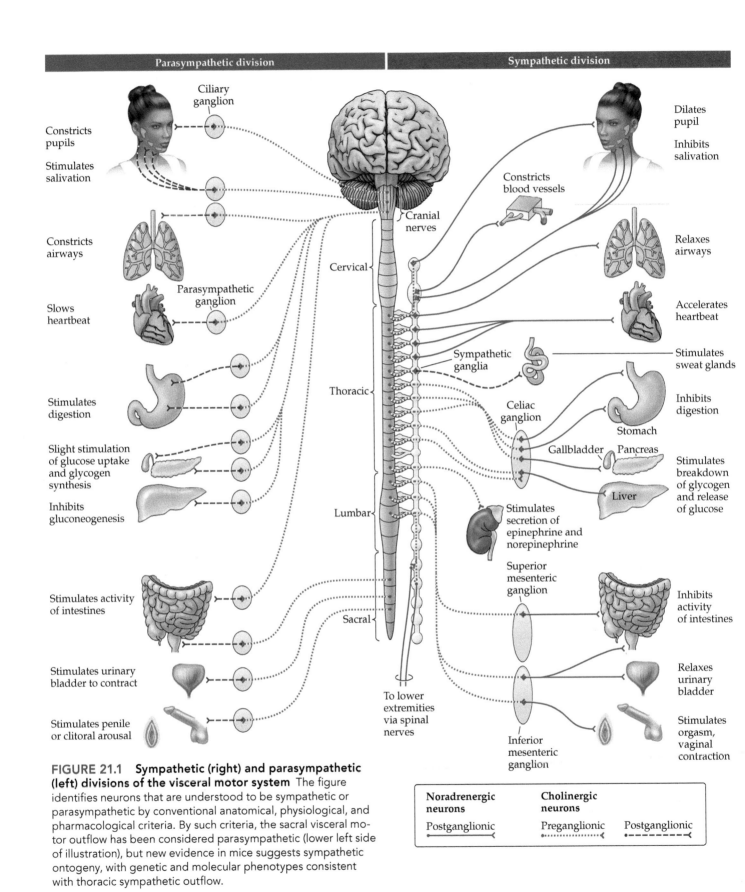

FIGURE 21.1 Sympathetic (right) and parasympathetic (left) divisions of the visceral motor system The figure identifies neurons that are understood to be sympathetic or parasympathetic by conventional anatomical, physiological, and pharmacological criteria. By such criteria, the sacral visceral motor outflow has been considered parasympathetic (lower left side of illustration), but new evidence in mice suggests sympathetic ontogeny, with genetic and molecular phenotypes consistent with thoracic sympathetic outflow.

	The Sympathetic Division
CONCEPT **21.2**	Prepares the Body to Mobilize Resources in Challenging Situations

LEARNING OBJECTIVES

21.2.1 Describe the anatomical organization of the sympathetic division of the visceral motor system, including the sources of preganglionic innervation and the location of postganglionic visceral motor neurons.

21.2.2 Characterize the major functions of the sympathetic division of the visceral motor system.

21.2.3 Compare and contrast the concepts of homeostasis and allostasis.

Anatomy and physiology of the sympathetic division

The activities of the neurons that make up the sympathetic division of the visceral motor system ultimately prepare individuals for "fight or flight," as Cannon famously put it. Cannon meant that, in extreme circumstances, heightened levels of sympathetic neural activity allow the body to make maximum use of its resources (particularly its metabolic resources), thereby increasing the chances of survival or success in threatening or otherwise challenging situations. Thus, during high levels of sympathetic activity, the pupils dilate and the eyelids retract (allowing more light to reach the retina and the eyes to move more efficiently); the blood vessels of the skin and gut constrict (rerouting blood to muscles, thus allowing them to extract a maximum of available energy); the hairs stand on end (which made our hairier ancestors look more fearsome); the bronchi dilate (increasing oxygenation); the heart rate accelerates and the force of cardiac contraction is enhanced (maximally perfusing skeletal muscles and the brain); and digestive and other vegetative functions become quiescent (thus diminishing activities that are temporarily unnecessary) (see Figure 21.1). At the same time, sympathetic activity stimulates the adrenal medulla to release epinephrine and norepinephrine into the bloodstream and causes the release of glucagon from the pancreas, further enhancing energy-mobilizing (or catabolic) functions. These coordinated responses illustrate an important principle of visceral motor function: There are circumstances that necessitate a departure from homeostatic set points in the regulation of the body's physiological systems (Box 21A), and such responses are coordinated by the sympathetic division of the visceral motor system. Thus, the short-term functional goal of autonomic activity is not always homeostasis (the maintenance of a constant internal state). Rather, the coordinated activity of visceral motor efferents may, for transient episodes,

impose *allostasis*—the restoration of homeostasis through physiological and behavioral change.

The neurons in the CNS that drive these effects are located in the spinal cord. They are arranged in a column of **preganglionic neurons** that extends from the uppermost thoracic to the upper lumbar segments (see Table 21.1) in a region of the spinal cord gray matter called the **intermediolateral cell column** in the **lateral horn** (Figure 21.2). The preganglionic neurons that control sympathetic outflow to the organs in the head and thorax are in the upper and middle thoracic segments, whereas those that control the abdominal and pelvic organs and targets in the lower extremities are in the lower thoracic and upper lumbar segments. The axons that arise from these spinal preganglionic neurons typically extend only a short distance, terminating in a series of **paravertebral** or **sympathetic chain ganglia**, which, as the name implies, are arranged in a chain that extends along most of the length of the vertebral column (see Figure 21.1). These preganglionic pathways to the ganglia are known as the *white communicating rami* because of the relatively light color imparted to the rami (singular, *ramus*) by the myelinated axons they contain (see Figure 21.2A). Roughly speaking, these preganglionic spinal neurons are comparable to somatic motor interneurons within the intermediate gray and ventral horns of the spinal cord (see Chapter 16).

The neurons in sympathetic ganglia are the primary or lower motor neurons of the sympathetic division in that they directly innervate smooth muscles, cardiac muscle, and glands. The **postganglionic axons** arising from these paravertebral sympathetic chain neurons travel to various targets in the body wall, joining the segmental spinal nerves of the corresponding spinal segments by way of the *gray communicating rami*. The gray rami are another set of short linking nerves, so named because the unmyelinated postganglionic axons give them a somewhat darker appearance than the myelinated preganglionic linking nerves (see Figure 21.2A).

In addition to innervating the sympathetic chain ganglia, the preganglionic axons that govern the viscera extend a longer distance from the spinal cord in the splanchnic nerves (nerves that innervate thoracic and abdominal viscera) to reach sympathetic ganglia that lie in the chest, abdomen, and pelvis. These **prevertebral ganglia** include sympathetic ganglia in the cardiac plexus; the celiac ganglion; the superior and inferior mesenteric ganglia; and sympathetic ganglia in the pelvic plexus. The postganglionic axons arising from the prevertebral ganglia provide sympathetic innervation to the heart, lungs, gut, kidneys, pancreas, liver, bladder, and reproductive organs; many of these organs also receive some postganglionic innervation from neurons in the sympathetic chain ganglia. Finally, a subset of thoracic preganglionic fibers in the splanchnic (visceral) nerves

(A)

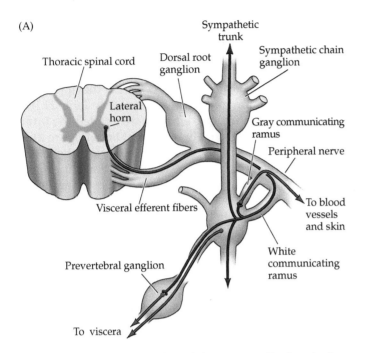

(B)

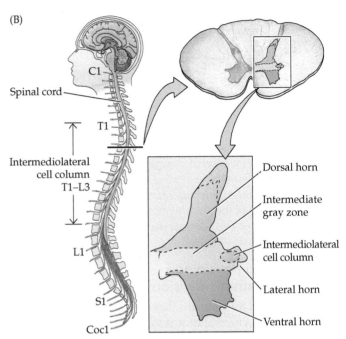

FIGURE 21.2 Organization of the preganglionic spinal outflow to sympathetic ganglia (A) General organization of the sympathetic division of the visceral motor system in the spinal cord and the preganglionic outflow to the sympathetic ganglia that contain the primary visceral motor neurons. (B) Cross section of the thoracic spinal cord at the level indicated, showing location of the sympathetic preganglionic neurons of the intermediolateral cell column in the lateral horn.

■ BOX 21A │ The Hypothalamus

The hypothalamus is located at the base of the forebrain, bounded by the optic chiasm rostrally and the midbrain tegmentum caudally. It forms the floor and ventral-lateral walls of the third ventricle and is continuous through the infundibular stalk with the posterior pituitary gland, as illustrated in Figure A. Given its central position in the brain and its proximity to the pituitary, it is not surprising that the hypothalamus integrates information from the forebrain, brainstem, spinal cord, and various intrinsic chemosensitive neurons.

What is surprising about this structure is the remarkable diversity of homeostatic and allostatic functions that are governed by this relatively small region of the forebrain. The diverse functions in which hypothalamic involvement is at least partially understood include: *the control of blood flow* (by promoting adjustments in cardiac output, vasomotor tone, blood osmolarity, and renal clearance, and

by motivating drinking and salt consumption); the *regulation of energy metabolism* (by monitoring blood glucose levels and regulating feeding behavior, digestive functions, metabolic rate, and temperature); the *regulation of reproductive activity* (by influencing gender identity, sexual orientation, and mating behavior, and in females, by governing menstrual cycles, pregnancy, and lactation); and the *coordination of responses to threatening conditions* (by governing the release of stress hormones, modulating the balance between sympathetic and parasympathetic tone, and influencing the regional distribution of blood flow).

Despite the impressive scope of hypothalamic control, the individual components of the hypothalamus use similar physiological mechanisms to exert their influence over these many functions (Figure B). Thus, hypothalamic circuits receive sensory and contextual information, compare that information

with biological set points, and activate relevant visceral motor, neuroendocrine, and somatic motor effector systems that restore homeostasis or elicit appropriate behavioral responses.

Like the overlying thalamus—and consistent with the scope of hypothalamic functions—the hypothalamus comprises a large number of distinct nuclei, each with its own specific pattern of connections and functions. The nuclei, most of which are intricately interconnected, can be grouped in three longitudinal regions referred to as *periventricular*, *medial*, and *lateral* (Figure C). The nuclei can also be grouped along the anterior–posterior dimension into the *anterior* (or preoptic), *tuberal*, or *posterior* region (see Figure A). The anterior-periventricular group contains the suprachiasmatic nucleus, which receives direct retinal input and drives circadian rhythms (see Chapter 28). More scattered neurons in the periventricular region (located along the wall of the third

■ BOX 21A | The Hypothalamus (continued)

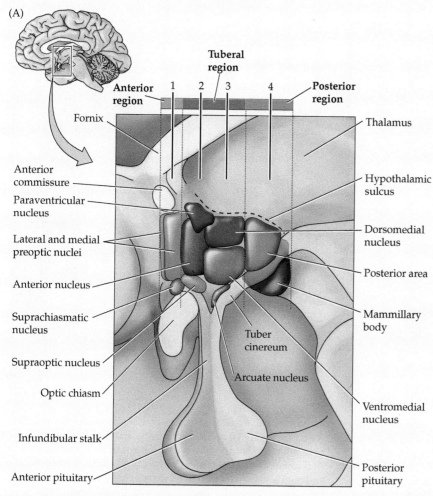

(A) Diagram of the human hypothalamus, illustrating its major nuclei.

Also in this region of the hypothalamus are the dorsomedial and ventromedial nuclei, which are involved in feeding, reproductive and parenting behavior, thermoregulation, and water balance. These nuclei receive inputs from structures in the limbic forebrain, as well as from visceral sensory nuclei in the brainstem (e.g., the nucleus of the solitary tract).

Finally, the lateral region of the hypothalamus is really a rostral continuation of the midbrain reticular formation (see Box 17C). Thus, the neurons of the lateral region are not grouped into nuclei, but are scattered among the fibers of the medial forebrain bundle, a prominent collection of axonal projections that run through the lateral hypothalamus. Cells in this lateral region control behavioral arousal and shifts of attention.

In summary, the hypothalamus regulates an enormous range of physiological and behavioral activities and serves as the key controlling center for visceral motor activity and homeostatic functions that are essential for survival.

(Continued)

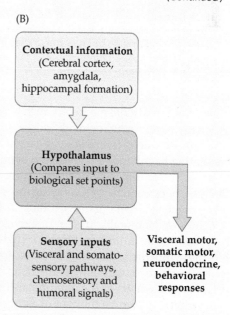

(B) Physiological mechanisms underlying hypothalamic function.

ventricle) manufacture peptides known as releasing or inhibiting factors, which control the secretion of a variety of hormones by the anterior pituitary. The axons of these neurons project to the median eminence, a region at the junction of the hypothalamus and pituitary stalk, where the peptides are secreted into the portal circulation that supplies the anterior pituitary.

Nuclei in the medial-tuberal region (*tuberal* refers to the tuber cinereum, the anatomical name given to the middle portion of the inferior surface of the hypothalamus) include the paraventricular and supraoptic nuclei, which contain the neurosecretory neurons whose axons extend into the posterior

pituitary. With appropriate stimulation, these neurons secrete oxytocin or vasopressin (antidiuretic hormone) directly into the bloodstream. Other neurons in the paraventricular nucleus project to autonomic centers in the reticular formation, as well as to preganglionic visceral motor neurons in the brainstem and spinal cord; these cells exert hypothalamic control over the visceral motor outflow throughout the body. The paraventricular nucleus receives inputs from other hypothalamic zones, which are in turn integrating input from cerebral cortex, hippocampus, amygdala, and other central structures, all of which are capable of influencing visceral motor function.

■ BOX 21A | The Hypothalamus (continued)

(C)
(1)

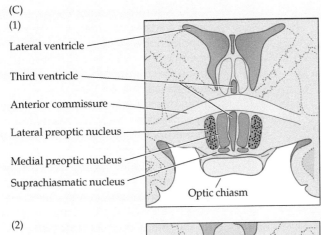

Lateral ventricle

Third ventricle

Anterior commissure

Lateral preoptic nucleus

Medial preoptic nucleus

Suprachiasmatic nucleus

Optic chiasm

(2)

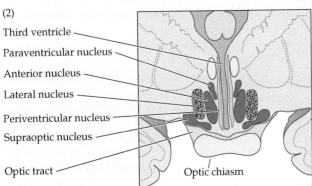

Third ventricle

Paraventricular nucleus

Anterior nucleus

Lateral nucleus

Periventricular nucleus

Supraoptic nucleus

Optic tract

Optic chiasm

(3)

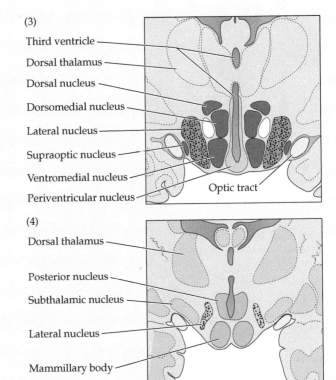

Third ventricle

Dorsal thalamus

Dorsal nucleus

Dorsomedial nucleus

Lateral nucleus

Supraoptic nucleus

Ventromedial nucleus

Periventricular nucleus

Optic tract

(4)

Dorsal thalamus

Posterior nucleus

Subthalamic nucleus

Lateral nucleus

Mammillary body

(C) Coronal sections through the human hypothalamus (see Figure A for location of sections 1–4). Color coding of the nuclei illustrates the two dimensions by which hypothalamic nuclei are subdivided (see text). Blue, red, and green illustrate nuclei in the anterior, tuberal, and posterior regions, respectively. The relative shading of these hues illustrates the three mediolateral zones: Lighter shading represents nuclei in the periventricular zone, whereas darker shades represent medial zone nuclei. Nuclei in the lateral zone are stippled to indicate the intermingled fibers of the median forebrain bundle. (1) Section through the anterior region illustrating the preoptic and suprachiasmatic nuclei. (2) Rostral tuberal region. (3) Caudal tuberal region. (4) Section through the posterior region illustrating the mammillary bodies.

---

innervates the adrenal medulla, which is generally regarded as a sympathetic ganglion modified for a specific endocrine function—namely, the release of catecholamines into the circulation to enhance a widespread sympathetic response to stress. In summary, sympathetic axons contribute to virtually all peripheral nerves, carrying innervation to an enormous range of target organs (see Table 21.1).

Cannon's memorable truism that the sympathetic activity prepares the animal for "fight or flight" notwithstanding, the sympathetic division of the visceral motor system is tonically active to maintain sympathetic function at appropriate levels, whatever the circumstances. Nor should the sympathetic system be thought of as responding in an all-or-none fashion; many sympathetic reflexes operate more or less independently, as might be expected from the obvious need to specifically control various organ functions (e.g., the heart during exercise, the bladder during urination, and the reproductive organs during sexual intercourse).

CONCEPT
21.3

The Parasympathetic Division Serves to Increase Metabolic Resources and Conserve Energy

LEARNING OBJECTIVES

21.3.1 Describe the anatomical organization of the parasympathetic division of the visceral motor system, including the sources of preganglionic innervation and the location of postganglionic visceral motor neurons.

21.3.2 Characterize the major functions of the parasympathetic division of the visceral motor system.

Anatomy and physiology of the parasympathetic division

The preganglionic outflow from the CNS to the ganglia of the parasympathetic division stems from neurons whose distribution is limited to the brainstem and (according to long-standing convention, but see later in this concept) the sacral part of the spinal cord (Figure 21.3; see also Figure 21.1). The cranial preganglionic

innervation arising from the brainstem, which is analogous to the preganglionic sympathetic outflow from the spinal cord, includes the **Edinger–Westphal nucleus** in the midbrain (which innervates the ciliary ganglion

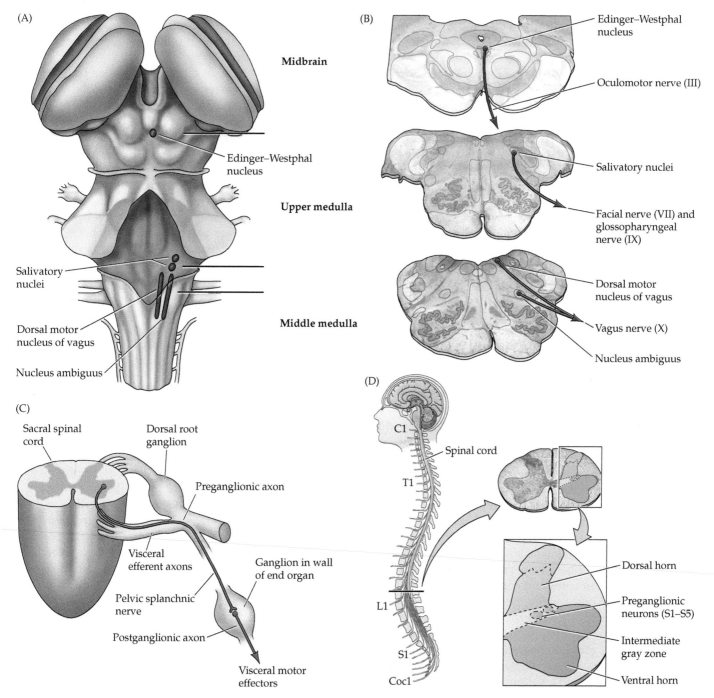

FIGURE 21.3   Organization of the cranial preganglionic outflow to parasympathetic ganglia and sacral visceral motor outflow (A) Dorsal view of brainstem showing the locations of the nuclei of the cranial part of the parasympathetic division of the visceral motor system. (B) Cross sections of the brainstem at the relevant levels [horizontal lines in (A)] showing the locations of these

parasympathetic nuclei. (C) Main features of the preganglionic visceral motor outflow in the sacral segments of the spinal cord. Until recently, this outflow has been considered parasympathetic (as in Figure 21.1); new evidence suggests sympathetic ontogeny (see text for details). (D) Cross section of the sacral spinal cord showing the location of sacral preganglionic neurons.

via the oculomotor nerve and mediates constriction of the pupil in response to increased light, as well as the accommodation reflex; see Chapter 9); the **superior** and **inferior salivatory nuclei** in the pons and medulla (which innervate the tear glands and salivary glands, mediating the production of tears and salivary secretion); a visceral motor division of the **nucleus ambiguus** in the medulla; and the **dorsal motor nucleus of the vagus nerve**, also in the medulla. Neurons in and adjacent to the ventrolateral part of the nucleus ambiguus provide an important source of cardio-inhibitory innervation to cardiac ganglia (a different division of the nucleus ambiguus provides branchiomotor innervation of striated muscle in the pharynx and larynx; see the Appendix). The more dorsal part of the dorsal motor nucleus of the vagus nerve primarily governs glandular secretion via the parasympathetic ganglia located in the viscera of the thorax and abdomen, whereas the more ventral part of the nucleus controls the motor responses of the lungs and gut elicited by the vagus nerve (e.g., constricting the bronchioles). The location of the parasympathetic brainstem nuclei is shown in Figure 21.3.

The sacral preganglionic innervation arises from neurons in the lateral gray matter of the sacral segments of the spinal cord, which are located in much the same position as the sympathetic preganglionic neurons in the intermediolateral cell column of the thoracic cord (see Figure 21.3C,D). The axons from these neurons travel in the splanchnic nerves to innervate ganglia in the lower third of the colon, rectum, bladder, and reproductive organs.

The **parasympathetic ganglia** innervated by preganglionic outflow from both cranial and sacral levels are in or near the end organs they serve. In this way they differ from the ganglionic targets of the sympathetic system (recall that both the paravertebral chain and prevertebral ganglia are located relatively far from their target organs; see Figure 21.1). An important anatomical difference between sympathetic and parasympathetic ganglia at the cellular level is that sympathetic ganglion cells tend to have extensive dendritic arbors and are, as might be expected from this arrangement, innervated by a large number of preganglionic fibers. Parasympathetic ganglion cells have few if any dendrites and consequently are each innervated by only one or a few preganglionic axons. This arrangement implies a greater diversity of converging influences on sympathetic ganglion neurons compared with parasympathetic ganglion neurons.

The overall function of the parasympathetic system, as Gaskell, Langley, and Cannon demonstrated, is generally opposite to that of the sympathetic system, serving to increase metabolic and other resources and to conserve energy during periods when the animal's circumstances allow it to "rest and digest." In contrast to the sympathetic functions enumerated in Concept 21.2, the activity of the parasympathetic system constricts the pupils, slows the heart rate, increases the peristaltic activity of the gut, and promotes emptying of urine from the bladder (*micturition* or *voiding*, as clinicians often call this process). At the same time, diminished activity in the sympathetic system allows the blood vessels of the skin and gut to dilate, the piloerector muscles to relax, and the outflow of catecholamines from the adrenal medulla to decrease.

Although most organs receive innervation from *both* the sympathetic and parasympathetic divisions of the visceral motor system (as Gaskell surmised), some receive only sympathetic innervation. These exceptional targets include the sweat glands, adrenal medulla, piloerector muscles of the skin, and most arterial blood vessels (see Table 21.1).

Until recently, there was little reason to challenge the conventional schema presented earlier in this concept for cranial-sacral parasympathetic outflow (see Figures 21.1 and 21.3). The standard classification of sacral visceral motor outflow as *parasympathetic* has been based on (1) anatomy—similarities in organization to vagal innervation (long preganglionic axons, short ganglionic axons); (2) physiology—presumed actions that oppose the effects mediated by the thoraco-lumbar (sympathetic) visceral motor outflow; and (3) pharmacology—general antagonism of end-organ action by blockade of muscarinic cholinergic receptors (see Concept 21.6). However, recent molecular and genetic analysis of the mouse nervous system has questioned this understanding of sacral visceral motor outflow. Studies by J.-F. Brunet and colleagues at the École Normale Supérieure in Paris suggest that the sacral division of the visceral motor system, which provides innervation to the pelvic organs, should now be considered *sympathetic*. The basis for this proposed reclassification is an analysis of 15 phenotypic and ontogenetic features of preganglionic and ganglionic elements showing that the sacral outflow and the thoraco-lumbar outflow share all 15 features. Furthermore, these shared features are distinct from features expressed by the parasympathetic outflow derived from the brainstem. Thus, this new work suggests that visceral motor outflow from the CNS may be understood in simple, bipartite terms comprising a cranial parasympathetic division and a spinal sympathetic division. While these molecular and genetic studies are compelling, it remains to be determined if these findings in mice generalize to all mammals, including humans. Furthermore, questions remain as to whether and how this "sacral sympathetic" concept might be reconciled with the anatomical, physiological, and pharmacological criteria that support the more conventional schema for cranial-sacral parasympathetic outflow.

CONCEPT
21.4

The Enteric Division Is a Semi-Autonomous Network of Gastrointestinal Neurons That Promotes Digestion

LEARNING OBJECTIVES

21.4.1 Describe the anatomical organization of the enteric division of the visceral motor system.

21.4.2 Differentiate the functions of the myenteric (or Auerbach's) plexus and the submucous (or Meissner's) plexus.

Anatomy and physiology of the enteric division

An enormous number of neurons are specifically associated with the gastrointestinal tract to control its many functions; indeed, it is likely that more neurons reside in the human gut than in the entire spinal cord. As already noted, the activity of the gut is modulated by both the sympathetic and parasympathetic divisions of the visceral motor system. However, the gut also has an extensive system of nerve cells in its wall (as do its accessory organs such as the pancreas and gallbladder) that do not fit neatly into the sympathetic or parasympathetic divisions of the visceral motor system (Figure 21.4A). To a surprising degree, these neurons and the complex enteric networks in which they are found operate more or less independently according to their own reflex rules;

as a result, many gut functions continue perfectly well without sympathetic or parasympathetic supervision (peristalsis, for example, occurs in isolated gut segments in vitro). Thus, most investigators prefer to classify the enteric nervous system as a unique, semi-autonomous division of the visceral motor system.

The neurons in the gut wall include local and centrally projecting sensory neurons that monitor mechanical and chemical conditions in the gut, local circuit neurons that integrate this information, and motor neurons that influence the activity of the smooth muscles in the wall of the gut and glandular secretions (e.g., of digestive enzymes, mucus, stomach acid, and bile). This complex arrangement of nerve cells intrinsic to the gut is organized into (1) the **myenteric** (or **Auerbach's**) **plexus**, which is specifically concerned with regulating the musculature of the gut; and (2) the **submucous** (or **Meissner's**) **plexus**, which is located, as the name implies, just beneath the mucus membranes of the gut and is concerned with chemical monitoring and glandular secretion (Figure 21.4B).

As already mentioned, the preganglionic parasympathetic neurons that influence the gut are primarily in the dorsal motor nucleus of the vagus nerve in the brainstem and the intermediate gray zone in the sacral spinal cord segments. The preganglionic sympathetic innervation that modulates the action of the gut plexuses derives from the thoraco-lumbar cord, primarily by way of the celiac, superior, and inferior mesenteric ganglia.

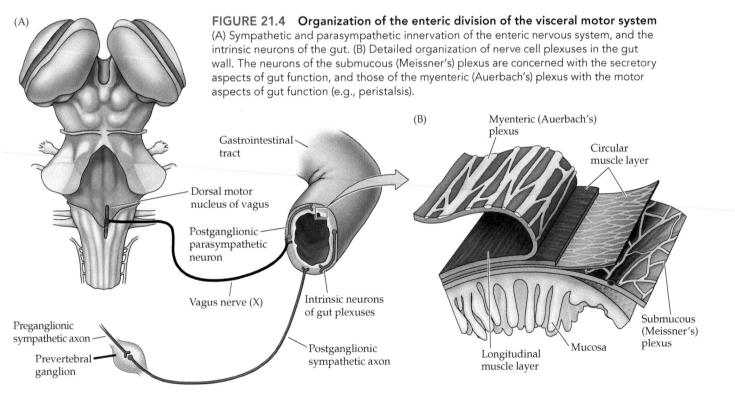

FIGURE 21.4 Organization of the enteric division of the visceral motor system
(A) Sympathetic and parasympathetic innervation of the enteric nervous system, and the intrinsic neurons of the gut. (B) Detailed organization of nerve cell plexuses in the gut wall. The neurons of the submucous (Meissner's) plexus are concerned with the secretory aspects of gut function, and those of the myenteric (Auerbach's) plexus with the motor aspects of gut function (e.g., peristalsis).

<table>
<tr><td>CONCEPT
21.5</td><td># Visceral Sensory Signals Serve Local Visceral Motor Reflexes and a Central Autonomic Network</td></tr>
</table>

LEARNING OBJECTIVES

21.5.1   Describe the sources of sensory signals derived from the viscera and the means by which such signals are conveyed to the CNS.

21.5.2   Discuss the functions of visceral sensory signals in local reflexes and higher integrative centers in the brain.

21.5.3   Identify and discuss the neural centers that integrate visceral sensory signals, distribute those signals to the forebrain, and regulate the outflow of efferent activity.

21.5.4   Discuss the importance of the hypothalamus as a key integrative center in the central autonomic network that organizes the expression of visceral motor activity.

Sensory components of the visceral motor system

Although the focus of this unit is movement and its central control, it is important to understand the sources of visceral sensory information and the means by which this input becomes integrated with visceral motor networks in the CNS. Generally speaking, afferent activity arising from the viscera serves two important functions. First, it provides feedback to local reflexes that modulate moment-to-moment visceral motor activity within individual organs. Second, it informs higher integrative centers of more complex patterns of stimulation that may signal potentially threatening conditions or require the coordination of more widespread

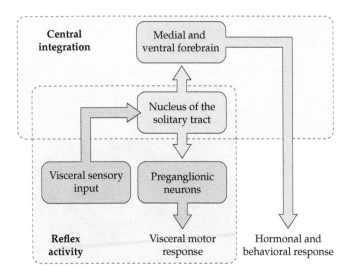

FIGURE 21.5   Distribution of visceral sensory information by the nucleus of the solitary tract Sensory information transduced via this pathway serves either local reflex responses or more complex hormonal and behavioral responses via integration within a central autonomic network. As expanded upon in Figure 21.7, forebrain centers also provide input to visceral motor effector systems in the brainstem and spinal cord.

visceral motor, somatic motor, neuroendocrine, and behavioral activities (Figure 21.5). The **nucleus of the solitary tract** in the medulla is a central structure in the brain that receives visceral sensory information and distributes it accordingly to serve both purposes.

The afferent fibers that provide these visceral sensory inputs arise from cell bodies in the dorsal root ganglia (as is the case of somatosensory modalities; see Chapters 12 and 13) and the sensory ganglia associated with the glossopharyngeal and vagus cranial nerves. However, far fewer visceral sensory neurons (by a factor of about 10) innervate their targets throughout the body in comparison with the number of mechanosensory neurons that innervate skin and deeper somatic structures. This relative sparseness of peripheral visceral sensory innervation accounts in part for why most visceral sensations are diffuse and difficult to localize precisely.

The spinal visceral sensory neurons in the dorsal root ganglia send axons peripherally, through sympathetic nerves, ending in sensory receptor specializations such as nerve endings that are sensitive to pressure or stretch (in the walls of the heart, bladder, and gastrointestinal tract); endings that innervate specialized chemosensitive cells (oxygen-sensitive cells in the carotid bodies); or nociceptive endings that respond to damaging stretch, ischemia, or the presence of irritating chemicals. The central axonal processes of these dorsal root ganglion neurons terminate on second-order neurons and local interneurons in the dorsal horn and in intermediate gray regions of the spinal cord. Some primary visceral sensory axons terminate near the lateral horn, where the preganglionic neurons of sympathetic and parasympathetic divisions are located; these terminals mediate visceral reflex activity in a manner not unlike that of the segmental sensorimotor reflexes described in Chapter 16.

In the dorsal horn, many of the second-order neurons that receive visceral sensory inputs are actually neurons of the anterolateral system, which also receives nociceptive or crude mechanosensory inputs from more superficial sources (see Chapter 13). As described in Box 13B, this is one means by which painful visceral sensations may be referred to more superficial somatic territories. Axons of these second-order visceral sensory neurons travel rostrally in the ventrolateral white matter of the spinal cord and the lateral sector of the brainstem and eventually reach the ventral posterior complex of the thalamus. However, the axons of other second-order visceral sensory neurons terminate before reaching the thalamus; the principal target of these axons is the nucleus of the solitary tract (Figure 21.6). Other brainstem targets of second-order visceral sensory axons are visceral motor centers in the medullary reticular formation (see Box 17C).

In the last few decades it has become clear that visceral sensory information, especially that related to painful visceral sensations originating in the lower abdomen, also ascends the CNS by another spinal pathway. Second-order neurons whose cell bodies are located near the central canal of the spinal cord send their axons through the dorsal columns to terminate in the dorsal column nuclei, where third-order neurons relay visceral nociceptive signals to the ventral posterior thalamus. Although the existence of this visceral pain pathway in the dorsal columns complicates the simplistic view of the dorsal column–medial lemniscal pathway as a discriminative mechanosensory projection and the anterolateral system as a pain pathway, mounting empirical and clinical evidence highlights the importance of this newly discovered dorsal column pain pathway in the central transmission of visceral nociception (see Box 13C).

In addition to these spinal visceral afferents, general visceral sensory inputs from thoracic and upper abdominal organs, as well as from viscera in the head and neck, enter the brainstem directly via the glossopharyngeal and vagus cranial nerves (see Figure 21.6). These glossopharyngeal and vagal visceral afferents terminate in the nucleus of the solitary tract. This nucleus, as described in the next section, integrates a wide range of visceral sensory information and transmits this information directly (and indirectly) to relevant visceral motor nuclei, to the brainstem reticular formation, and to several regions in the medial and ventral forebrain that coordinate visceral motor activity (see Figure 21.5).

Finally, unlike in the somatosensory system (where virtually all sensory signals gain access—albeit gated access—to conscious neural processing), sensory fibers related to

the viscera convey only limited information to consciousness. For example, most of us are completely unaware of the subtle changes in peripheral vascular resistance that raise or lower our mean arterial blood pressure, yet such covert visceral afferent information is essential for the functioning of autonomic reflexes and the maintenance of homeostasis. Typically, only painful visceral sensations enter conscious awareness (see Chapter 32).

Central control of visceral motor functions

The caudal part of the nucleus of the solitary tract is a key integrative center for reflexive control of visceral motor function and an important relay for visceral sensory information that reaches other brainstem nuclei and forebrain structures (Figure 21.7; see also Figure 21.5). The rostral part of this nucleus is a gustatory relay, receiving input from primary taste afferents (cranial nerves VII, IX,

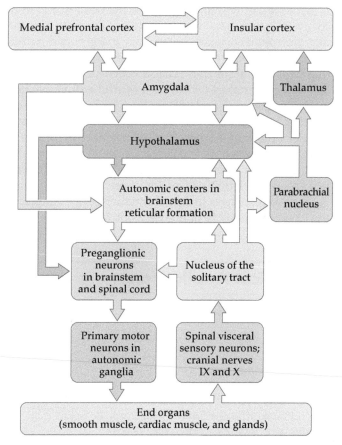

FIGURE 21.6 Organization of sensory input to the visceral motor system Afferent input from the cranial nerves relevant to visceral sensation (as well as afferent input ascending from second-order visceral afferents in the spinal cord) converges on the caudal division of the nucleus of the solitary tract (the rostral division is a gustatory relay; see Chapter 15).

FIGURE 21.7 A central autonomic network for the control of visceral motor function The distribution of visceral sensory information within this network is illustrated on the right side of the figure; the generation of visceral motor signals is shown on the left. However, extensive interconnections among autonomic centers in the forebrain (e.g., between the amygdala and associated cortical regions or hypothalamus) militate against a strict parsing of this network into afferent and efferent limbs. The hypothalamus, a key structure in this network, integrates visceral sensory input and higher-order visceral motor signals (see Box 21A).

and X) and sending projections to the gustatory nucleus in the ventral posterior thalamus (see Chapter 15). The caudal visceral sensory part of the nucleus of the solitary tract provides input to primary visceral motor nuclei, such as the dorsal motor nucleus of the vagus nerve and the nucleus ambiguus. It also projects to premotor autonomic centers in the medullary reticular formation and to higher integrative centers in the amygdala (specifically, the central group of amygdaloid nuclei; see Box 32B) and hypothalamus (see later in this section). In addition, the nucleus of the solitary tract projects to the **parabrachial nucleus** (so named because it envelopes the superior cerebellar peduncle, also known by the Latin name *brachium conjunctivum*). The parabrachial nucleus, in turn, relays visceral sensory information to the hypothalamus, amygdala, thalamus, and medial prefrontal and insular cortex (see Figure 21.7; for clarity, the cortical projections of the parabrachial nucleus are omitted).

Although one might argue that the posterior insular cortex is the primary visceral sensory area and the medial prefrontal cortex is the primary visceral motor area, it is more instructive to emphasize the interactions among these cortical areas and related subcortical structures. Taken together, they constitute a **central autonomic network**. This network accounts for the integration of visceral sensory information with input from other sensory modalities and from higher cognitive centers that process emotional experiences. Involuntary visceral reactions such as blushing in response to consciously embarrassing stimuli, vasoconstriction and pallor in response to fear, and autonomic responses to sexual situations are examples of the integrated activity of this network. Indeed, autonomic function is intimately related to emotional processing, as emphasized in Chapter 32.

The hypothalamus is a key component of this central autonomic network that deserves special consideration. The **hypothalamus** is a heterogeneous collection of nuclei in the base of the diencephalon that plays an important role in the coordination and expression of visceral motor activity (see Box 21A). The major outflow from the relevant hypothalamic nuclei is directed toward autonomic centers in the reticular formation; these centers can be thought of as dedicated premotor circuits that coordinate the efferent activity of preganglionic visceral motor neurons. They organize specific visceral functions such as cardiac reflexes, bladder control reflexes, sexual function reflexes, and critical reflexes underlying respiration and vomiting (see Box 17C).

In addition to these important projections to the reticular formation, hypothalamic control of visceral motor function is also exerted more directly by projections to the cranial nerve nuclei that contain parasympathetic preganglionic neurons, and to the sympathetic and parasympathetic preganglionic neurons in the spinal cord. Nevertheless, the autonomic centers of the reticular formation and the preganglionic visceral motor neurons that they control are competent to function autonomously should disease or injury impede the ability of the hypothalamus to govern the many bodily systems that maintain homeostasis. Figure 21.7 summarizes the general organization of this central autonomic control. Some important clinical manifestations of damage to this descending system are illustrated in Clinical Applications, and Box 21B discusses the relevance of this central control to the feeding system and obesity.

■ Clinical Applications

Horner's Syndrome

The characteristic clinical presentation of damage to the pathway that controls the sympathetic division of the visceral motor system to the head and neck is called Horner's syndrome, after the Swiss ophthalmologist who first described this clinical picture in the mid-nineteenth century. The main features, as illustrated in Figure A, are decreased diameter of the pupil on the side of the lesion (*miosis*), a droopy eyelid (*ptosis*), and a sunken appearance of the affected eye (*enophthalmos*). Less obvious signs are decreased sweating, increased skin temperature, and flushing of the skin

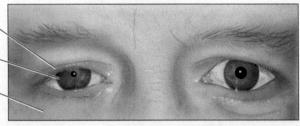

(A)

Drooping of eyelid (ptosis)

Ipsilateral pupillary constriction (miosis)

Apparent sinking of eyeball (enophthalmos)

(A) Major features of Horner's syndrome.

on the same side of the face and neck.

All these signs are explained by a loss of sympathetic tone due to damage somewhere along the pathway that connects visceral motor centers in the hypothalamus and reticular formation with sympathetic preganglionic neurons in the intermediolateral cell column of the thoracic spinal cord (Figure B). Lesions that interrupt these fibers often spare the descending parasympathetic pathways, which are located more medially in the brainstem and are more diffuse. The sympathetic

■ Clinical Applications (continued)

preganglionic targets that are affected by such lesions include the neurons in the intermediolateral cell column in spinal segments T1–T3 that control the dilator muscle of the iris and the tone in smooth muscles of the eyelid and globe, the paralysis of which leads to miosis, ptosis, and enophthalmos. The flushing and decreased sweating are likewise the result of diminished sympathetic tone, in this case governed by intermediolateral cell column neurons in somewhat lower thoracic segments. Damage to the descending sympathetic pathway in the brainstem will, of course, affect sweating and vascular tone in the rest of the body on the side of the lesion. However, if the damage is to the upper thoracic outflow (as is more typical), the upper thoracic chain, or the superior cervical ganglion, then the manifestations of Horner's

syndrome will be limited to the head and neck. Typical causes in these sites are traumatic injuries to the head and

neck, and tumors of the apex of the lung, thyroid, or cervical lymph nodes.

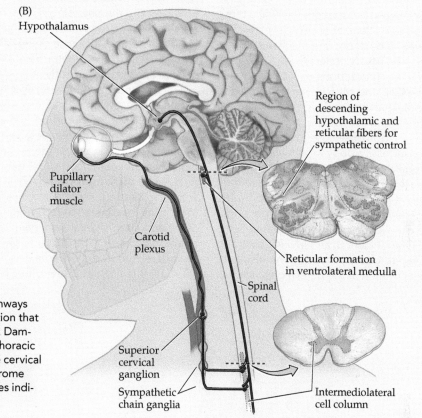

(B)
Hypothalamus
Pupillary dilator muscle
Carotid plexus
Superior cervical ganglion
Sympathetic chain ganglia
Spinal cord
Region of descending hypothalamic and reticular fibers for sympathetic control
Reticular formation in ventrolateral medulla
Intermediolateral cell column

(B) Diagram of the descending sympathetic pathways arising in the hypothalamus and reticular formation that can be interrupted to cause Horner's syndrome. Damage to the preganglionic neurons in the upper thoracic cord, to the superior cervical ganglion, or to the cervical sympathetic trunk can also cause Horner's syndrome (see also Figure 21.1). The transverse dashed lines indicate the level of the sections shown at right.

■ BOX 21B | Obesity and the Brain

Obesity—and its relationship to a broad range of diseases, including diabetes, cardiovascular disease, and cancer—has become a major public health concern in most countries, particularly the United States. Whereas the signature of obesity is an excess of body fat, the underlying cause or causes are generally thought to lie in abnormal regulation by the brain circuits that control appetite and satiety (the feeling of fullness following a meal). This fact makes weight loss particularly difficult for many individuals. Thus, understanding the CNS mechanisms that regulate food intake and metabolism is essential for developing effective strategies to

mitigate the health problems associated with this condition.

The brain regulates appetite and satiety via the neural activity that is modulated by chemical signals. These chemical signals are secreted into the circulation by fat-storing adipose tissues throughout the body. This feedback loop entails some of the central components of the visceral motor system, in addition to endocrine mechanisms via insulin, growth hormone, and a growing list of factors that signal metabolic state, adiposity (amount of body fat), and nutrient balances.

The peptide **ghrelin** is secreted by the stomach prior to feeding, presumably as a signal of hunger; adipocytes

(the cells that concentrate lipids in fatty tissues) increase their secretion of **leptin** into the circulation following feeding, presumably one of several signals for satiety (Figure A). The receptors for these peptides are concentrated in small groups of neurons in the ventrolateral and anterior hypothalamus (see Box 21A), which interact with additional hypothalamic neurons in the arcuate region. These ghrelin- and leptin-sensitive cells modulate the activity of neurons expressing the opiomelanocortin propeptide (POMC) and the subsequent secretion of α-melanocyte secreting hormone (α-MSH), one of the peptides encoded by the POMC transcript. This

(Continued)

■ BOX 21B | Obesity and the Brain (continued)

hormone evidently regulates appetite and satiety by acting on specific receptors (particularly the melanocortin receptor subtype called MCR-4) located on additional populations of hypothalamic and brainstem neurons (including those in the nucleus of the solitary tract), as well as by endocrine mechanisms that remain poorly understood.

The interactions of leptin, ghrelin, α-MSH, and MCR-4 were determined in animal models. Two recessive mutations in mice—obese (*ob/ob*) and diabetic (*db/db*)—were identified based on excessive body weight and failure to regulate food intake, respectively. When each mutation was cloned, the mutant gene in *ob* mice turned out to be the gene for leptin, and the *db* gene that for the leptin receptor. Mutations in the *POMC* (Figure B) and *MCR4* genes also lead to obesity in mice. The results of inactivation of the *ghrelin* gene are less

clear; however, pharmacological and physiological studies associate changes in ghrelin levels with altered feeding patterns and weight loss. Studies in mice have thus provided a solid framework for examining the physiological mechanisms regulating food intake in humans. Nonetheless, the relevance of the mice studies to morbid human obesity remained unclear until recently.

Genetic analysis of individuals in human pedigrees with extreme obesity (determined via measured body mass indices and weight/height ratios) has revealed mutations in one or more of the leptin, leptin receptor, or *MCR4* genes. As a result, these individuals have little sense of satiety after eating and thus fail to regulate food intake based on signals other than gastric distension, pain, and plasma osmolality. How this pathophysiology is related to less extreme degrees of obesity is not yet known, but it is

being intensely studied because of its implications for weight control in people with lower body mass indices.

The emerging understanding of body weight regulation by hypothalamic circuits that are modulated by feedback elicited by hormonal signals from fat tissues has provided new ways of thinking about pharmacological therapies for weight control. Although leptin mimetics have proved generally ineffective, leptin administration in people with leptin deficiencies does reduce food intake and obesity (Figure C). Currently, there is great interest in drugs that modulate α-MSH signaling via MCR-4. Although no effective pharmacological therapies presently exist, clinical investigators hope that such drugs, when combined with behavioral changes in dietary practices and physical activity, will effectively combat this often intractable and increasingly common health problem.

(A)

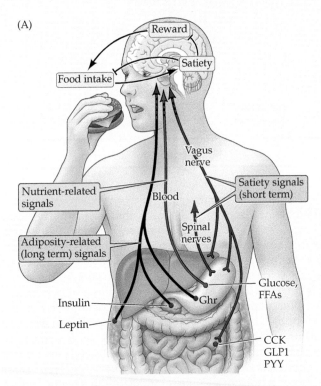

(B)

From L. Yaswen et al., 1999. *Nat Med* 5: 1066–1070

(C)

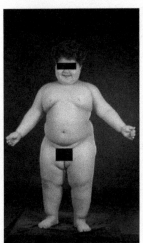
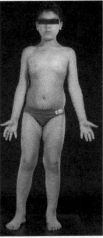

From S. O'Rahilly et al., 2003. *Endocrinology* 144: 3757–3764. Reprinted by permission of Oxford University Press on behalf of the Endocrine Society

(A) Body–brain dynamics in energy homeostasis. The CNS integrates longer-term, state-dependent signals (leptin, ghrelin [Ghr], insulin) and shorter-term, feeding-dependent signals related to nutrient content (glucose, free fatty acids [FFAs]), satiety (peptide YY [PYY], glucagon-like peptide 1 [GLP1], cholecystokinin [CCK]), and visceral motor activity of the gut. Central integration of such signals regulates food intake and energy expenditure. (After G. J. Morton et al., 2014, *Nature Rev Neurosci* 15: 367–378, adapted from J. Marx, 2003. *Science* 299: 846–849.) (B) A *POMC* knock-out mouse (left) and a wild-type littermate (right). (C) The effect of leptin treatment in a human. At age 3 years, the child weighed 42 kg (left); at age 7 years, following treatment, the same child weighed 32 kg (right).

<table>
<tr><td rowspan="1">CONCEPT
21.6</td><td>

Visceral Motor Neurons Use Small-Molecule and Neuropeptide Neurotransmitters to Mediate a Variety of Effects

</td></tr>
</table>

LEARNING OBJECTIVES

21.6.1 Characterize the principal neurotransmitters used by preganglionic and ganglionic neurons of the sympathetic and parasympathetic divisions of the visceral motor system.

21.6.2 Characterize the principal actions of the neurotransmitters used by preganglionic and ganglionic neurons of the sympathetic and parasympathetic divisions and the receptors that mediate them.

21.6.3 Discuss the autonomic regulation of cardiovascular function in terms of the relevant visceral sensory signals, their central integration, and the appropriate reflexive visceral motor responses.

21.6.4 Discuss the autonomic regulation of the bladder in terms of the necessary interplay between components of the somatic motor system and the sympathetic and parasympathetic divisions of the visceral motor system.

21.6.5 Discuss the autonomic regulation of sexual function in terms of the necessary interplay between components of the somatic motor system and the sympathetic and parasympathetic divisions of the visceral motor system.

Neurotransmission in the visceral motor system

The neurotransmitters used by the visceral motor system are of enormous importance in clinical practice, and drugs that act on the autonomic system are among the most important in the clinical armamentarium. Moreover, autonomic transmitters have played a major role in the history of efforts to understand synaptic function.

Acetylcholine is the primary neurotransmitter of both sympathetic and parasympathetic preganglionic neurons. Nicotinic receptors on autonomic ganglion cells are ligand-gated ion channels that mediate a fast EPSP (much like nicotinic receptors at the neuromuscular junction). In contrast, muscarinic acetylcholine receptors on ganglion cells are members of the 7-transmembrane G-protein-coupled receptor family (see Chapters 6 and 7), and they mediate slower synaptic responses. The primary action of muscarinic receptors in autonomic ganglion cells is to close K^+ channels, making the neurons more excitable and generating a prolonged EPSP. Acting in concert with the muscarinic activities are neuropeptides that serve as co-neurotransmitters at the ganglionic synapses. As described in Chapter 6, peptide neurotransmitters also tend to exert slowly developing but long-lasting effects on postsynaptic neurons. As a result of these two acetylcholine receptor types and

a rich repertoire of neuropeptide transmitters, ganglionic synapses mediate both rapid excitation and a slower modulation of autonomic ganglion cell activity.

The postganglionic effects of autonomic ganglion cells on their smooth muscle, cardiac muscle, or glandular targets are mediated by two primary neurotransmitters: norepinephrine (NE) and acetylcholine (ACh). For the most part, sympathetic ganglion cells release norepinephrine onto their targets (a notable exception is the cholinergic sympathetic innervation of sweat glands), whereas parasympathetic ganglion cells typically release acetylcholine. As might be expected from the foregoing account, these two neurotransmitters usually have opposing effects on their target tissue—contraction versus relaxation of smooth muscle, for example.

As described in Chapters 6 and 7, the specific effects of ACh and NE are determined by the type of receptor expressed in the target tissue and the downstream signaling pathways to which these receptors are linked. Peripheral sympathetic targets generally have two subclasses of noradrenergic receptors in their cell membranes, referred to as α and β receptors. Like muscarinic ACh receptors, both α and β receptors and their subtypes belong to the 7-transmembrane G-protein-coupled class of cell surface receptors. The different distribution of these receptors in sympathetic targets allows for a variety of postsynaptic effects mediated by norepinephrine released from postganglionic sympathetic nerve endings (Table 21.2).

The effects of acetylcholine released by parasympathetic ganglion cells onto smooth muscles, cardiac muscle, and glandular cells also vary according to the subtypes of muscarinic cholinergic receptors found in the peripheral target (Table 21.3). The two major subtypes are known as M_1 and M_2 receptors, M_1 receptors being found primarily in the gut and M_2 receptors in the cardiovascular system. Another subclass of muscarinic receptor, M_3, occurs in both smooth muscle and glandular tissues. Muscarinic receptors are coupled to a variety of intracellular signal transduction mechanisms that modify K^+ and Ca^{2+} channel conductances. They can also activate nitric oxide synthase, which promotes the local release of nitric oxide in some parasympathetic target tissues (see, for example, the section later in this concept that discusses autonomic control of sexual function).

In contrast to the relatively restricted responses generated by norepinephrine and acetylcholine released by sympathetic and parasympathetic ganglion cells, respectively, neurons of the enteric nervous system achieve an enormous diversity of effects by virtue of many different neurotransmitters, most of which are neuropeptides associated with specific cell groups in either the myenteric or submucous plexuses mentioned in Concept 21.4. The details of these agents and their actions are beyond the scope of this introductory account.

TABLE 21.2 Adrenergic Receptor Types and Some of Their Effects in Sympathetic Targets

Receptor	G-protein	Tissue	Response
α_1	G_q	Smooth muscle of blood vessels, iris, ureter, urethra, hairs, uterus	Contraction of smooth muscle
		Heart muscle	Positive inotropic effect ($\beta_1 \gg \alpha_1$)
		Salivary gland	Secretion
		Adipose tissue	Glycogenolysis, gluconeogenesis
		Sweat glands	Secretion
		Kidney	Na^+ reabsorbed
α_2	G_i	Adipose tissue	Inhibition of lipolysis
		Pancreas	Inhibition of insulin release
		Smooth muscle of blood vessels	Contraction
β_1	G_s	Heart muscle	Positive inotropic effect; positive chronotropic effect
		Adipose tissue	Lipolysis
		Kidney	Renin release
β_2	G_s	Liver	Glycogenolysis, gluconeogenesis
		Skeletal muscle	Glycogenolysis, lactate release
		Smooth muscle of bronchi, uterus, gut, blood vessels	Relaxation
		Pancreas	Stimulates insulin secretion
		Salivary glands	Thickened secretions
β_3	G_s	Adipose tissue	Lipolysis
		Smooth muscle of gut	Modulation of intestinal mobility
		Smooth muscle of bladder	Bladder filling

Many examples of specific autonomic functions could be used to illustrate in more detail how the visceral motor system operates. The three outlined here—control of cardiovascular function, control of the bladder, and control of sexual function—have been chosen primarily because of their importance in human physiology and clinical practice.

Autonomic regulation of cardiovascular function

The cardiovascular system is subject to precise reflex regulation so that an appropriate supply of oxygenated blood can reliably be provided to different body tissues under a wide range of circumstances. The sensory monitoring for this critical homeostatic process entails primarily mechanical (*barosensory*) information about pressure in the

TABLE 21.3 Cholinergic Receptor Types and Some of Their Effects in Parasympathetic Targets

Receptor	G-protein	Tissue	Response
Nicotinic	None (ionotropic receptor)	Most parasympathetic targets (and all autonomic ganglion cells)	Relatively fast postsynaptic response
Muscarinic (M_1)	Gq	Smooth muscles and glands of the gut	Smooth muscle contraction and glandular secretion (relatively slow response)
Muscarinic (M_2)	Gi	Smooth and cardiac muscle of cardiovascular system	Reduction in heart rate, smooth muscle contraction
Muscarinic (M_3)	Gq	Smooth muscles and glands of all targets	Smooth muscle contraction, glandular secretion

arterial system and, secondarily, chemical (*chemosensory*) information about the levels of oxygen and carbon dioxide in the blood. The parasympathetic and sympathetic activities relevant to cardiovascular control are determined by the information supplied by these sensors.

The mechanoreceptors, called baroreceptors, are located in the heart and major blood vessels; the chemoreceptors are located primarily in the carotid bodies, which are small, highly specialized organs located at the bifurcation of the common carotid arteries (some chemosensory tissue is also found in the aorta). The nerve endings in baroreceptors are activated by deformation as the elastic elements of the vessel walls expand and contract. The chemoreceptors in the carotid bodies and aorta respond directly to the partial pressure of oxygen and carbon dioxide in the blood. Visceral afferents from the aortic arch and carotid bifurcation reach the brainstem via the vagus nerve and glossopharyngeal

nerve, respectively. Both afferent systems convey their signals to the nucleus of the solitary tract, which relays this information to the hypothalamus and the relevant autonomic centers in the reticular formation (Figure 21.8).

The afferent information derived from changes in arterial pressure and blood gas levels reflexively modulates the activity of the relevant visceral motor pathways and, ultimately, smooth and cardiac muscles and other more specialized structures. For example, a rise in blood pressure activates baroreceptors that, via the pathway illustrated in Figure 21.8, inhibit the tonic activity of sympathetic preganglionic neurons in the spinal cord. In parallel, the pressure increase stimulates the activity of the parasympathetic preganglionic neurons in and adjacent to the nucleus ambiguus that influence heart rate and contractility. The carotid chemoreceptors also have some influence, but less than that stemming from the baroreceptors.

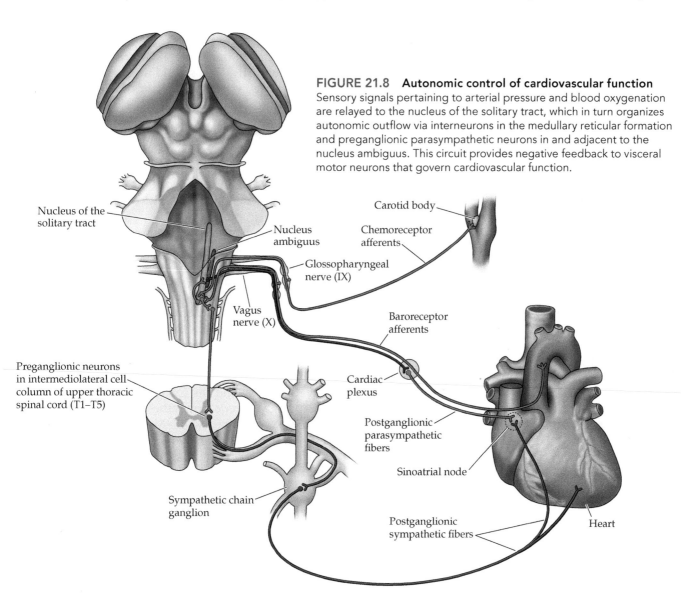

FIGURE 21.8 Autonomic control of cardiovascular function
Sensory signals pertaining to arterial pressure and blood oxygenation are relayed to the nucleus of the solitary tract, which in turn organizes autonomic outflow via interneurons in the medullary reticular formation and preganglionic parasympathetic neurons in and adjacent to the nucleus ambiguus. This circuit provides negative feedback to visceral motor neurons that govern cardiovascular function.

Nucleus of the solitary tract

Nucleus ambiguus

Carotid body

Chemoreceptor afferents

Glossopharyngeal nerve (IX)

Vagus nerve (X)

Baroreceptor afferents

Preganglionic neurons in intermediolateral cell column of upper thoracic spinal cord (T1–T5)

Cardiac plexus

Postganglionic parasympathetic fibers

Sinoatrial node

Sympathetic chain ganglion

Postganglionic sympathetic fibers

Heart

This shift in the balance of sympathetic and parasympathetic activity results in reduction of the stimulatory noradrenergic effects of postganglionic sympathetic innervation on the cardiac pacemaker and cardiac musculature. These effects are abetted by the decreased output of catecholamines from the adrenal medulla and the decreased vasoconstrictive effects of sympathetic innervation on the peripheral blood vessels. At the same time, activation of the cholinergic parasympathetic innervation of the heart decreases the discharge rate of the cardiac pacemaker in the sinoatrial node and slows the ventricular conduction system. These parasympathetic influences are mediated by an extensive series of parasympathetic ganglia in and near the heart, which release acetylcholine onto cardiac pacemaker cells and cardiac muscle fibers. As a result of this combination of sympathetic and parasympathetic effects, heart rate and the effectiveness of atrial and ventricular myocardial contraction are reduced and the peripheral arterioles dilate, thus lowering the blood pressure.

In contrast to this sequence of events in response to raised blood pressure, a fall in blood pressure (as might occur from blood loss) has the opposite effect—it attenuates parasympathetic activity while increasing sympathetic activity. As a result, norepinephrine is released from sympathetic postganglionic terminals, increasing the rate of cardiac pacemaker activity and enhancing cardiac contractility, at the same time increasing release of catecholamines from the adrenal medulla (which further augments these and many other sympathetic effects that enhance the response to this threatening situation). Norepinephrine released from the terminals of sympathetic ganglion cells also acts on the smooth muscles of the arterioles to increase the tone of the peripheral vessels, particularly those in the skin, subcutaneous tissues, and muscles, thus shunting blood away from these tissues to those organs where oxygen and metabolites are urgently needed to maintain function (e.g., brain, heart, and kidneys in the case of blood loss). If these reflex sympathetic responses fail to raise the blood pressure sufficiently (in which case the individual is said to be in shock), the vital functions of these organs begin to fail, often catastrophically.

A more mundane circumstance that requires a reflexive autonomic response to a fall in blood pressure is standing up. Rising quickly from a prone position produces a shift of some 300 to 800 mL of blood from the thorax and abdomen to the legs, resulting in a sharp (approximately 40%) decrease in the output of the heart. The adjustment to this normally occurring drop in blood pressure (called **orthostatic hypotension**) must be rapid and effective, as evidenced by the dizziness sometimes experienced in this situation. Indeed, anyone can briefly lose consciousness as a result of blood pooling in the lower extremities, which is the usual cause of fainting when standing still for exceptionally long periods.

The sympathetic innervation of the heart arises from the preganglionic neurons in the intermediolateral cell column of the spinal cord, extending from roughly the first through fifth thoracic segments (T1–T5; see Table 21.1). The primary visceral motor neurons are in the adjacent thoracic paravertebral and prevertebral ganglia of the cardiac plexus. The parasympathetic preganglionics, as already mentioned, are in and adjacent to the nucleus ambiguus (and to a lesser extent in the dorsal motor nucleus of the vagus nerve) and project to parasympathetic ganglia in and around the heart and great vessels.

Autonomic regulation of the bladder

The autonomic regulation of the bladder provides an especially instructive example of the interplay between components of the somatic motor system that are subject to volitional control (we usually have voluntary control over urination), and the sympathetic and parasympathetic divisions of the visceral motor system, which operate involuntarily. This should not be surprising given that, for many mammals, the act of urination (and of defecation) places the individual at increased risk for attack, since the capacity for immediate fight or flight is reduced. In addition, for many mammals, urine contains chemical signals (pheromones) that induce complex social behaviors. The neural control of bladder function therefore involves the coordination of relevant autonomic, somatic motor, and cognitive faculties that inhibit or promote urination.

The arrangement of afferent and efferent innervation of the bladder is shown in Figure 21.9A. The sympathetic innervation of the bladder originates in the lower thoracic and upper lumbar spinal cord segments (T10–L2), with preganglionic axons running to primary sympathetic neurons in the inferior mesenteric ganglion and the ganglia of the pelvic plexus. The postganglionic fibers from these ganglia travel in the hypogastric and pelvic nerves to the bladder, where sympathetic activity is believed to cause the smooth muscle of the bladder wall to relax and the internal urethral sphincter to close (postganglionic sympathetic fibers also innervate the blood vessels of the bladder). Stimulation of this sympathetic pathway in reflexive response to a modest increase in bladder pressure from the accumulation of urine thus allows the bladder to fill and prevents leakage of urine. At the same time, moderate distension of the bladder reflexively inhibits sacral outflow, which would otherwise cause contraction of the bladder musculature and bladder emptying. Contraction of the bladder musculature is promoted by preganglionic neurons in the sacral spinal cord segments (S2–S4) that innervate visceral motor neurons in ganglia in or near

(A)

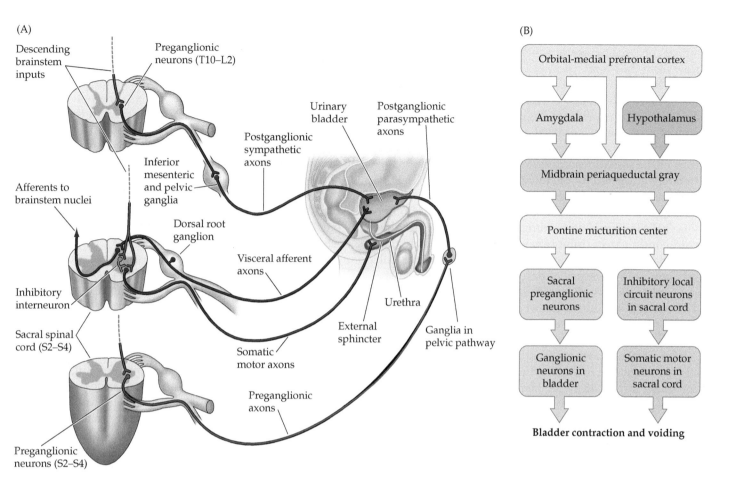

(B)

Descending brainstem inputs

Preganglionic neurons (T10–L2)

Afferents to brainstem nuclei

Inhibitory interneuron

Sacral spinal cord (S2–S4)

Preganglionic neurons (S2–S4)

Inferior mesenteric and pelvic ganglia

Dorsal root ganglion

Urinary bladder

Postganglionic sympathetic axons

Postganglionic parasympathetic axons

Visceral afferent axons

Urethra

External sphincter

Somatic motor axons

Preganglionic axons

Ganglia in pelvic pathway

Orbital-medial prefrontal cortex

Amygdala

Hypothalamus

Midbrain periaqueductal gray

Pontine micturition center

Sacral preganglionic neurons

Inhibitory local circuit neurons in sacral cord

Ganglionic neurons in bladder

Somatic motor neurons in sacral cord

Bladder contraction and voiding

FIGURE 21.9   Autonomic control of bladder function
(A) Organization of spinal control circuits. (B) Supraspinal control of micturition. The periaqueductal gray is an important integrative center that governs urination via its connections to the pontine micturition center, a component of the brainstem reticular formation that influences indirectly the lower visceral and somatic motor neurons that promote bladder contraction and voiding.

the bladder wall. (Conventionally, this is considered parasympathetic innervation of the bladder. However, as noted in Concept 21.3, recent molecular and genetic studies are suggesting that this sacral outflow should be reclassified as sympathetic.)

The afferent limb of this reflexive circuit is supplied by mechanoreceptors in the bladder wall that convey visceral afferent information to second-order neurons in the dorsal horn of the spinal cord. In addition to local connections within spinal cord circuitry, these neurons project to higher integrative centers in the periaqueductal gray of the midbrain. This midbrain region (which is also involved in the descending control of nociception; see Chapter 13) receives input from the hypothalamus, amygdala, and orbital-medial prefrontal cortex. These forebrain structures participate in limbic networks that evaluate risk and the emotional significance of contextual cues (see Chapter 32); in the context of bladder filling, they signal when it is safe and socially appropriate to urinate.

When the bladder is full, sacral visceral motor outflow increases and thoraco-lumbar motor outflow decreases, causing the bladder to contract and the internal sphincter muscle to relax. However, urine is held in check by the voluntary somatic motor innervation of the external urethral sphincter muscle. The voluntary control of the external sphincter is mediated by motor neurons of the ventral horn in sacral spinal cord segments (S2–S4), which cause the striated muscle fibers of the sphincter to contract. During bladder filling (and subsequently, until circumstances permit urination) these neurons are active, keeping the external sphincter closed and preventing voiding. During urination, this tonic activity is temporarily inhibited, leading to relaxation in the external sphincter muscle. Normally, this is only possible when integrative signals derived from the periaqueductal gray activate a collection of premotor neurons in the dorsal pontine reticular formation, known as the *pontine micturition center* (or Barrington's nucleus). The pontine micturition center projects to preganglionic neurons and inhibitory local circuit neurons in the sacral spinal cord; the net result is increased visceral motor outflow from the sacral cord (leading to stronger contraction of the bladder wall) and inhibition of the somatic lower motor neurons that innervate the external sphincter muscle (allowing for voiding) (Figure 21.9B). Thus, urination results from the coordinated activation of sacral visceral motor neurons and temporary inactivation of motor neurons of the somatic motor system; this coordination is ultimately governed by the integration of visceral sensory, emotional, social, and contextual cues.

Importantly, individuals who are paraplegic or otherwise have impaired descending control of the sacral spinal cord continue to exhibit reflexive, autonomic regulation of bladder function. Unfortunately, this reflex is not fully efficient in the absence of descending motor control, resulting in a variety of problems in people with paraplegia and others with diminished or defective central control of bladder function. The major difficulty in these cases is incomplete bladder emptying, which often leads to chronic urinary tract infections from the culture medium provided by retained urine, and thus the need for an indwelling catheter to ensure adequate drainage. Indeed, urinary tract morbidity is recognized as the second leading cause of death in individuals with spinal cord injury. In other individuals with urge incontinence and overactive bladder disorders, leakage of urine and an "absence of warning" are the problems. Mounting evidence obtained from functional and structural studies of the brains of such individuals implicates structural lesions or dysfunction that impair the integrative activity of the periaqueductal gray and its control over the pontine micturition center.

Autonomic regulation of sexual function

Much like control of the bladder, sexual responses are mediated by the coordinated activity of visceral motor and somatic motor innervation, both of which are governed by complex cognitive, emotional, and contextual cues processed in the limbic forebrain. Although these reflexes differ in detail in males and females, basic similarities, not only in humans but in mammals generally, allow the autonomic sexual responses of the two sexes to be considered together. These similarities include: (1) the mediation of vascular dilation, which causes penile or clitoral erection; (2) stimulation of prostatic or vaginal secretions; (3) smooth muscle contraction of the vas deferens during ejaculation in males or rhythmic vaginal contractions during orgasm in females; and (4) contractions of the somatic pelvic muscles that accompany orgasm in both sexes.

Like the urinary tract, the reproductive organs receive preganglionic visceral motor innervation from the sacral spinal cord and from the lower thoracic and upper lumbar spinal cord segments, and somatic motor innervation from α motor neurons in the ventral horn of the lower spinal cord segments (Figure 21.10). The sacral visceral motor pathway controlling the sexual organs in both males and females originates in the sacral segments S2–S4 and reaches the target organs via the pelvic nerves. Activity of the postganglionic neurons in the relevant ganglia causes dilation of penile or clitoral arteries, and a corresponding relaxation of the smooth muscles of the venous (cavernous) sinusoids, which leads to expansion of the sinusoidal spaces. As a result, the amount of blood in the tissue is increased, leading to a sharp rise in the pressure and an expansion of the cavernous spaces (i.e., erection). The mediator of the smooth muscle relaxation leading to erection is not acetylcholine (as in most postganglionic parasympathetic actions), but nitric oxide (NO; see Chapter 6). The drug sildenafil (Viagra) acts by inhibiting PDE-5, the predominant phosphodiesterase (PDE) expressed in erectile tissue, which leads to an increase in the intracellular concentration of cyclic GMP. This second messenger mediates the activity of endogenous NO; thus, PDE-5 inhibitors enhance the relaxation of the venous sinusoids and promote erection in males during sexual stimulation (when NO is released in erectile tissue). Sacral visceral motor outflow also provides excitatory input to the vas deferens, seminal vesicles, and prostate in males, or vaginal glands in females.

In contrast, visceral motor outflow from the thoraco-lumbar spinal cord causes vasoconstriction and loss of erection. This pathway to the sexual organs originates in spinal segments T11–L2 and reaches the target organs via the corresponding sympathetic chain ganglia and the inferior mesenteric and pelvic ganglia.

The afferent effects of genital stimulation are conveyed centrally from somatosensory endings via the dorsal roots of S2–S4, eventually reaching the somatosensory cortex (reflex sexual excitation may also occur by local stimulation, as is evident in individuals with spinal cord injuries sparing the sacral cord). The reflex effects of such stimulation are increased visceral motor outflow from the sacral cord, which, as noted, causes relaxation of the smooth muscles in the wall of the sinusoids and subsequent erection.

Finally, the somatic motor component of reflex sexual function arises from α motor neurons in the lumbar and sacral spinal cord segments. These neurons provide excitatory innervation to the bulbocavernosus and ischiocavernosus muscles, which are active during ejaculation in males and mediate the contractions of the perineal (pelvic floor) muscles that accompany orgasm in both males and females.

Sexual functions are governed centrally by the anterior-medial and medial-tuberal zones of the hypothalamus, which contain a variety of nuclei pertinent to visceral

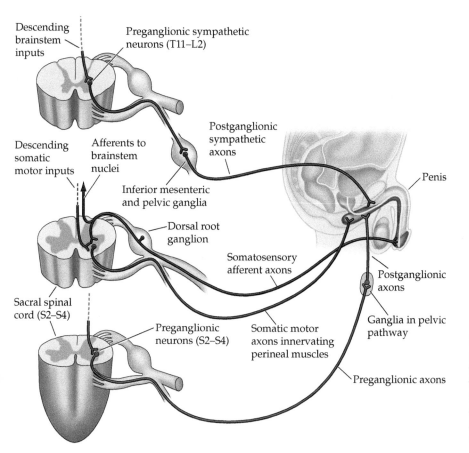

Descending brainstem inputs

Preganglionic sympathetic neurons (T11–L2)

Postganglionic sympathetic axons

Descending somatic motor inputs

Afferents to brainstem nuclei

Inferior mesenteric and pelvic ganglia

Penis

Dorsal root ganglion

Somatosensory afferent axons

Postganglionic axons

Sacral spinal cord (S2–S4)

Preganglionic neurons (S2–S4)

Somatic motor axons innervating perineal muscles

Ganglia in pelvic pathway

Preganglionic axons

FIGURE 21.10 Autonomic control of sexual function in the human male Similar to the central governance of the bladder, autonomic control of sexual function involves the integrative actions of forebrain and brainstem centers that regulate relevant visceral motor and somatic motor outflow from the thoracic, upper lumbar, and sacral spinal cord.

motor control and sexual and reproductive behavior (see Box 21A). Although they remain poorly understood, these nuclei appear to act as integrative centers for sexual responses and are also thought to be involved in more complex aspects of sexuality, such as sexual preference and gender identity (see Chapter 25). The relevant hypothalamic nuclei receive inputs from several areas of the brain, including—as one might imagine—the cortical and subcortical structures concerned with memory, emotion, and hedonic reward (see Chapters 30 and 32).

Summary

Sympathetic and parasympathetic ganglia, which contain the visceral lower motor neurons that innervate smooth muscles, cardiac muscle, and glands, are controlled by preganglionic neurons in the spinal cord and brainstem. The sympathetic preganglionic neurons that govern ganglion cells in the sympathetic division of the visceral motor system arise from neurons in the thoracic and upper lumbar segments of the spinal cord; parasympathetic preganglionic neurons, in contrast, are located in the brainstem. Preganglionic neurons in the sacral spinal cord have long been considered parasympathetic, but recent evidence shows that they share many features with preganglionic neurons in the thoraco-lumbar cord. Sympathetic ganglion cells are distributed in the sympathetic chain (paravertebral) and prevertebral ganglia, whereas the parasympathetic motor neurons are more widely distributed in ganglia that lie in or near the organs they control. Most visceral structures receive inputs from both the sympathetic and parasympathetic systems, which act in a generally antagonistic fashion. The diversity of autonomic functions is achieved primarily by different types of receptors for the two primary classes of postganglionic autonomic neurotransmitters, norepinephrine in the case of the sympathetic division and acetylcholine in the parasympathetic division. The visceral motor system is regulated by sensory feedback provided by dorsal root and cranial nerve sensory ganglion cells that make local reflex connections in the spinal cord or brainstem and project to the nucleus of the solitary tract in the brainstem. The visceral motor system is also regulated by descending pathways from the hypothalamus and brainstem reticular formation, the major control centers of homeostasis more generally. The importance of the visceral motor control of organs such as the heart, bladder, and reproductive organs—and the many pharmacological means of modulating autonomic function—have made visceral motor control a central theme in clinical medicine.

■ Additional Reading

Reviews

Brown, D. A. and 8 others (1997) Muscarinic mechanisms in nerve cells. *Life Sci.* 60 (13–14): 1137–1144.

Calabrò, R. S. and 8 others (2019) Neuroanatomy and function of human sexual behavior: A neglected or unknown issue? *Brain Behav.* 9: e01389.

Costa, M. and S. J. H. Brookes (1994) The enteric nervous system. *Am. J. Gastroenterol.* 89: S129–S137.

Craig, A. D. (2009) How do you feel—now? The anterior insula and human awareness. *Nat. Rev. Neurosci.* 10: 59–70.

Espinosa-Medina, I., O. Saha, F. Boismoreau and J.-F. Brunet (2018) The "sacral parasympathetic": Ontogeny and anatomy of a myth. *Clin. Auton. Res.* 28:13–21.

Fowler, C. J., D. Griffiths and W. C. de Groat (2008) The neural control of micturition. *Nat. Rev. Neurosci.* 9: 453–466.

Gershon, M. D. (1981) The enteric nervous system. *Annu. Rev. Neurosci.* 4: 227–272.

Guyenet, P. G. and 4 others (2020) Neuronal networks in hypertension: Recent advances. *Hypertension* 76: 300–311.

Holstege, G. (2005) Micturition and the soul. *J. Comp. Neurol.* 493: 15–21.

Silvani, A., G. Calandra-Buonaura, R. A. L. Dampney and P. Cortelli (2016) Brain–heart interactions: Physiology and clinical implications. *Philos. Trans. R. Soc. A* 374: 20150181.

Tish, M. M. and J. C. Geerling (2020) The brain and the bladder: Forebrain control of urinary (in)continence. *Front. Physiol.* 11: 658.

Important Original Papers

Espinosa-Medina, I. and 6 others (2016) The sacral autonomic outflow is sympathetic. *Science* 354: 893–897.

Jansen, A. S. P. and 4 others (1995) Central command neurons of the sympathetic nervous system: Basis of the fight or flight response. *Science* 270: 644–646.

Langley, J. N. (1894) The arrangement of the sympathetic nervous system chiefly on observations upon pilo-erector nerves. *J. Physiol.* 15: 176–244.

Langley, J. N. (1905) On the reaction of nerve cells and nerve endings to certain poisons chiefly as regards the reaction of striated muscle to nicotine and to curare. *J. Physiol.* 33: 374–473.

Lichtman, J. W., D. Purves and J. W. Yip (1980) Innervation of sympathetic neurones in the guinea-pig thoracic chain. *J. Physiol.* 298: 285–299.

Rubin, E. and D. Purves (1980) Segmental organization of sympathetic preganglionic neurons in the mammalian spinal cord. *J. Comp. Neurol.* 192: 163–174.

Books

Appenzeller, O. (1997) *The Autonomic Nervous System: An Introduction to Basic and Clinical Concepts*, 5th Edition. Amsterdam: Elsevier Biomedical Press.

Blessing, W. W. (1997) *The Lower Brainstem and Bodily Homeostasis.* New York: Oxford University Press.

Brading, A. (1999) *The Autonomic Nervous System and Its Effectors.* Oxford: Blackwell Science.

Burnstock, G. and C. H. V. Hoyle (1995) *The Autonomic Nervous System*, vol. 1: *Autonomic Neuroeffector Mechanism.* London: Harwood Academic.

Cannon, W. B. (1932) *The Wisdom of the Body.* New York: W. W. Norton.

Cardinali, D. P. (2018) *Autonomic Nervous System: Basic and Clinical Concepts.* Cham, Switzerland: Springer.

Furness, J. B. and M. Costa (1987) *The Enteric Nervous System.* Edinburgh: Churchill Livingstone.

Gabella, G. (1976) *Structure of the Autonomic Nervous System.* London: Chapman and Hall.

Jänig, W. (2006) *The Integrative Action of the Autonomic Nervous System: Neurobiology of Homeostasis.* Cambridge, UK: Cambridge University Press.

Langley, J. N. (1921) *The Autonomic Nervous System.* Cambridge, UK: Heffer & Sons.

Loewy, A. D. and K. M. Spyer (eds.) (1990) *Central Regulation of Autonomic Functions.* New York: Oxford University Press.

Pick, J. (1970) *The Autonomic Nervous System: Morphological, Comparative, Clinical, and Surgical Aspects.* Philadelphia, PA: J. B. Lippincott Co.

Randall, W. C. (ed.) (1984) *Nervous Control of Cardiovascular Function.* New York: Oxford University Press.

UNIT IV
The Changing Brain

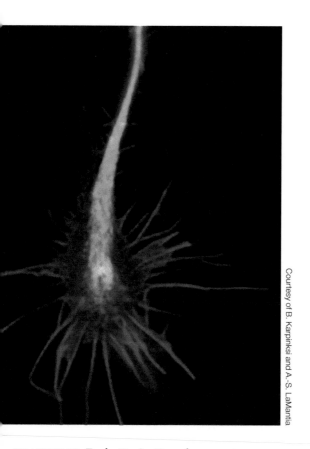

Courtesy of B. Karpinksi and A.-S. LaMantia

CHAPTER 22 **Early Brain Development**

CHAPTER 23 **Construction of Neural Circuits**

CHAPTER 24 **Experience-Dependent Plasticity in the Developing Brain**

CHAPTER 25 **Sex Differences and Neural Circuit Development**

CHAPTER 26 **Repair and Regeneration in the Nervous System**

The structure and functional capacity of the brain change dramatically over the human life span. As soon as the nervous system is established, coordinated gene expression, neuronal genesis, axonal and dendritic growth, and formation of connections yield a brain whose shape, size, and cellular architecture are continually transformed. These events rely on the transcriptional regulators, secreted signals and their receptors, adhesion and recognition molecules that determine appropriate neuronal identities, positions, and connections. The neural circuits that emerge eventually mediate a remarkably complex array of behaviors. Nevertheless, these circuits must be refined by subsequent postnatal experience to be maximally efficient for each individual. This refinement occurs via activity-dependent mechanisms that translate experience into altered synaptic efficacy that reflects changes in gene expression, neuronal growth, or in some cases, elimination of neuronal processes or synapses. Such changes are most pronounced during brief times in early life called critical periods. Finally, like any other organ, the brain is subject to damage and disease. Some injuries activate repair mechanisms that resemble those used to build the relevant circuits during development. Nevertheless, the capacity of the mature brain for full repair and regeneration is minimal. Limited numbers and types of new neurons can be generated in only a few brain regions and new connections made. These new neurons, however, cannot completely restore brain structure or function. Pathology in diseases such as amyotrophic lateral sclerosis, Parkinson's disease, and Alzheimer's disease disrupts mechanisms that typically contribute to neuronal development and subsequent maintenance of neural circuitry. The degeneration of brain circuits in these diseases ultimately accounts for the behavioral deficits seen in individuals with these disorders. Thus, from its initial construction, through experience-dependent use that elicits synaptic and functional plasticity, through challenges for repair in the face of damage and disease, the brain is a site of constant, lifelong change.

Early Brain Development

Overview

The elaborate architecture of the adult brain is the product of cell-to-cell signals, genetic instructions, and their consequences for stem cells that are set aside in the embryo during the earliest steps of development to generate the entire nervous system. These events include the establishment of the primordial central and peripheral nervous systems, the initial formation of the major brain regions, the generation of multiple classes of neurons and glial cells from undifferentiated neural stem or progenitor cells, and the migration of neurons or their immediate precursors from sites of generation to their final positions. These processes set the stage for the subsequent local differentiation of dendrites, axons, and synapses, as well as the growth of long-distance axon pathways and synaptic connections. When any of these processes goes awry—because of genetic mutation, disease, or exposure to drugs or other chemicals—the consequences can be disastrous. Indeed, many well-studied congenital brain defects result from interference with the typical mechanisms of early nervous system development, prior to the formation of synaptic connections. With the aid of cell biological, molecular, and genetic tools, investigators are beginning to understand the mechanisms underlying these extraordinarily complex events.

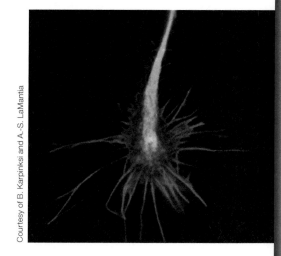

Courtesy of B. Karpinski and A.-S. LaMantia

KEY CONCEPTS

22.1 Neural stem cells, derived from pluripotent stem cells, generate the entire nervous system

22.2 Neural stem cells generate the central and peripheral nervous systems

22.3 Transcription factor patterning regulated by cell-cell signaling establishes distinct brain regions

22.4 Neurogenesis is an irreversible termination of the cell cycle that constrains neuron identity

22.5 Nerve cells often migrate from their site of neurogenesis to their final position

---

CONCEPT 22.1 | **Neural Stem Cells, Derived from Pluripotent Stem Cells, Generate the Entire Nervous System**

LEARNING OBJECTIVES

22.1.1 Define the capacity of stem cells to generate all cell types in an organism.

22.1.2 Describe the molecular and genetic manipulations that are required to make stem cells from somatic cells.

22.1.3 Discuss the properties of neural stem cells and their capacity to make all cell classes of the nervous system.

22.1.4 Assess the uses of stem cells for analysis of nervous system development, disease, and repair.

Stem cells

The nervous system, like every other organ system, is established by a subpopulation of fate-limited, tissue-specific stem cells derived from the **pluripotent stem cells** that form an embryo after fertilization. Thus, although stem cell biology has emerged as a promising field for establishing new therapies to repair degenerating or injured organs and to understand the pathogenesis of serious diseases such as several cancers, its foundations are found in understanding embryogenesis. Understanding key aspects of stem cell biology is critical for understanding the development of the nervous system as well as prospects (still quite speculative) for repair of degenerating or damaged neural tissue (Box 22A; see also Chapter 26). Thus, assessing the characteristics of stem cells in embryos, as well as those retained in maturing or mature tissues or in their in vitro counterparts, is vital for understanding vertebrate development and pathology, including that of the nervous system.

The foundational experiment that defined stem cell biology was carried out in the early 1960s. In this pioneering work (for which Sir John Gurdon shared the 2012 Nobel Prize in physiology or medicine with Shinya Yamanaka), a diploid nucleus with two copies of each chromosome (and thus two copies of each gene) from a somatic tissue in the frog—in this case the intestinal epithelium—was transplanted to an unfertilized frog oocyte whose nucleus had been removed (Figure 22.1A). The result, although obtained

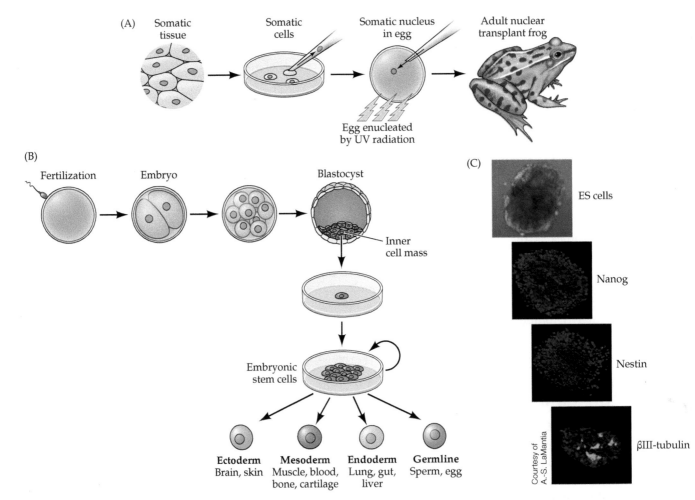

FIGURE 22.1 Stem cell specification accounts for development of all somatic tissues, including the nervous system (A) The first demonstration that the somatic nucleus of an adult tissue had the complete genomic information necessary to generate an entire organism, and thus could provide instruction to a stem cell to generate all the tissues in that organism. In this experiment, the nucleus of a somatic cell (from the intestine, in this example) was removed from its parent cell and then injected into a UV-irradiated oocyte (the UV irradiation completely inactivates the oocyte nucleus). This combination of somatic nucleus and oocyte was capable of generating an entire adult frog. Thus, the oocyte (in this case with a somatic nucleus, normally post-fertilization when the DNA content is restored to diploid from monoploid) had become a true stem cell capable of giving rise to all tissues in the organism. (B) In vitro experiments, in which cells from the inner cell mass of the mammalian blastocyst (the pluripotent stem cells generated by the first series of cell divisions of the fertilized oocyte) are isolated and maintained in cell culture conditions that preserve their capacity to generate all classes of somatic tissues. When injected into a host blastocyst, such cultured embryonic stem (ES) cells can integrate into an embryo and contribute to every tissue in the mature organism, including the germline. (C) Differentiation of ES cells in vitro to neurons. The ES cells initially express transcription factors associated with pluripotency, in this case Nanog (red). They then begin to express genes associated with neural stem cells, in this case Nestin (red). Finally, a subset of the progeny of the ES cell–derived neural stem cells become neurons, labeled with βIII-tubulin (green), a marker of neuron identity.

■ BOX 22A | Stem Cells: Promise and Peril

One of the most highly publicized issues in biology has been the potential use of stem cells to treat a variety of neurodegenerative conditions, including Parkinson's, Huntington's, and Alzheimer's diseases. Amidst the social, political, and ethical debate set off by the promise of stem cell therapies, the question of what exactly is a stem cell tends to get lost.

Neural stem cells are an example of a broader class of **somatic stem cells** found in various tissues, both during development and in the adult. All somatic stem cells share two fundamental characteristics: They are self-renewing; and upon terminal division and differentiation, they can give rise to the full range of cell classes within the relevant tissue. Thus, a neural stem cell can give rise to another neural stem cell, or to any of the differentiated cell types found in the central and peripheral nervous systems (i.e., inhibitory and excitatory neurons, astrocytes, and oligodendrocytes). A neural stem cell is distinct from a *neural progenitor cell*, which is incapable of continuous self-renewal and usually has the capacity to give rise to only one class of differentiated progeny. An oligodendroglial progenitor cell, for example, gives rise to oligodendrocytes until its

mitotic capacity is exhausted; a neural stem cell, in contrast, can generate more neural stem cells as well as a full range of differentiated neural and glial cell classes, presumably indefinitely.

Neural stem cells, and indeed all classes of somatic stem cells, are different from *embryonic* stem cells. **Embryonic stem cells (ES cells)** are derived from pre-gastrula embryos. Like somatic stem cells, ES cells have the potential for infinite self-renewal. However, ES cells can give rise to *all* tissue and cell types of the organism—including the germ cells that undergo meiosis and generate haploid gametes as well as neural and other somatic stem cells (Figure A). Somatic stem cells, by contrast, generate only diploid, tissue-specific cell types. In some cases, however, somatic cells such as skin cells or fibroblasts have been induced to acquire stem cell properties—including the ability to generate all tissues, as ES cells do. These *induced pluripotential stem cells*, or *IPSCs*, are produced in vitro by introducing the genes for several transcription factors associated with stem cells into somatic cells (e.g., fibroblasts). IPSCs hold out the possibility of generating stem cells from mature individuals for therapeutic uses in regeneration and tissue repair.

The ultimate therapeutic promise of stem cells—neural or other types—is their ability to generate newly differentiated cells and tissues to replace those that may have been lost because of disease or injury. Such therapies have been implemented in clinical trials for some forms of diabetes (replacement of islet cells that secrete insulin) and some hematopoietic diseases. In the nervous system, stem cell therapies have been suggested for replacement of dopaminergic cells lost to Parkinson's disease and replacing lost neurons in other degenerative disorders.

While intriguing, this projected use of stem cell technology raises some significant perils. These include ensuring the controlled division of stem cells when they are introduced into mature tissue, and identifying the appropriate molecular instructions to achieve differentiation of the desired cell class. Clearly, the latter challenge will need to be met with a fuller understanding of the signaling and transcriptional regulatory steps used during development to guide differentiation of relevant neuron classes in the embryo.

At present there is no clinically validated use of stem cells for human therapeutic applications in the nervous

(Continued)

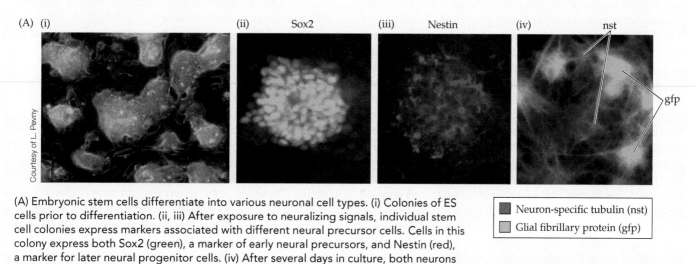

Courtesy of L. Pevny

(A) Embryonic stem cells differentiate into various neuronal cell types. (i) Colonies of ES cells prior to differentiation. (ii, iii) After exposure to neuralizing signals, individual stem cell colonies express markers associated with different neural precursor cells. Cells in this colony express both Sox2 (green), a marker of early neural precursors, and Nestin (red), a marker for later neural progenitor cells. (iv) After several days in culture, both neurons (red, labeled for neuron-specific tubulin) and astrocytes (green, labeled for glial fibrillary protein) have been generated from ES cells.

■ Neuron-specific tubulin (nst)
□ Glial fibrillary protein (gfp)

■ **BOX 22A** | **Stem Cells: Promise and Peril** (continued)

system. Nevertheless, some promising work in mice and other experimental animals indicates that both somatic and ES cells can acquire distinct identities if given appropriate instructions in vitro (i.e., prior to introduction into the host), and if delivered into a supportive host environment. For example, ES cells grown in the presence of platelet-derived growth factor, which biases progenitors toward glial fates, have generated oligodendroglial cells that can myelinate axons in myelin-deficient rats. Similarly, ES cells pretreated with retinoic acid matured into motor neurons when introduced into the developing spinal cord (Figure B). While such experiments suggest that a combination of proper instruction and correct placement can lead to appropriate differentiation of embryonic or somatic stem cells, there are still many issues to be resolved before the promise becomes reality.

(B)

Injection of stem cells
into spinal cord

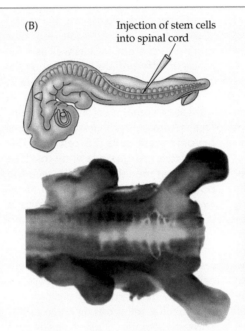

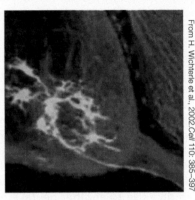

From H. Wichterle et al. 2002.Cell 110: 385–397

(B) Left top: Injection of fluorescently labeled embryonic stem cells into the spinal cord of a host chicken embryo shows that ES cells integrate into the host spinal cord and apparently extend axons. Below: The progeny of the grafted ES cells are seen in the ventral horn of the spinal cord. They have motor neuron–like morphologies, and their axons extend into the ventral root.

at low frequency (~10/750 attempts), showed that a single germline cell (in this case, an egg cell) with its haploid nucleus (thus, only one copy of each chromosome and each gene) removed and replaced by a somatic nucleus has the potential to generate a whole healthy frog, with all somatic tissues, including the central and peripheral nervous systems. Thus, these oocytes with a transplanted somatic nucleus were said to be **pluripotent**: capable of generating all cell types, tissues and organs, plus gametes that define the germline cells (oocytes and sperm) to make a complete organism capable of reproduction. This experiment suggests that somatic nuclei are not irrevocably altered over the course of differentiation through DNA excision or other mechanisms. They retain a full genome equivalent to that of a fertilized embryo with the full set of instructions necessary to make a new organism. The genesis of a whole frog from this hybrid cell—oocyte cytoplasm and somatic diploid nucleus—also defined a cell type, albeit created experimentally in this case, that can generate all other cells and tissues, including the nervous system, in the organism. This was the first demonstration of a true **embryonic stem (ES) cell** (Figure 22.1B). Indeed, the hybrid cell created by this experiment had the ploidy (chromosome number) of a somatic cell (two copies of each autosome and, variably,

two copies of the X chromosome in females, and one copy of the X and Y chromosomes in males) but not the additional influence of fertilization and nuclear fusion between oocyte and sperm. Nevertheless, the genomic information encoded by the nucleus could orchestrate the generation of an entire organism capable of reproduction. Thus, these oocyte–somatic nuclei hybrids were truly pluripotent—capable of giving rise to all cells in the organism, including gametes and the entire nervous system.

The next step for defining stem cell biology, including neural stem cell biology, was taken when Martin Evans and his colleagues isolated and propagated ES cells in vitro by harvesting cells from the inner cell mass of the blastocyst in mice and other mammals. (Other groups, using discarded human blastulae from in vitro fertilization procedures, isolated—with great controversy—human ES cells.) These ES cells, when maintained under rigorous conditions in vitro, can differentiate into all somatic cell types in the organism if given proper molecular instructions (see next section). This capacity is one of the major drivers of continued interest in stem cell biology: the prospect of creating and expanding the numbers of new cells in vitro that can replace damaged or dysfunctional cells in an otherwise intact tissue of a host individual in the context of degenerative disease or acute injury. Moreover, ES cells have become a key experimental tool for understanding the genetics underlying the development and function of the nervous system and other organs. When in vitro–propagated ES cells are injected into the blastocyst of an early-stage embryo (mouse embryo in most cases), they can create a chimeric mouse, including integrating into the germline. It is through this mechanism of germline integration that "knock-in" and "knock-out" mice carrying mutant genes are made in the laboratory from genetically modified ES cells (see Chapter 1). ES cells, under the appropriate culture conditions, can also generate neural stem cells (see later in this concept) and neurons (Figure 22.1C).

The molecular basis of stem cell pluripotency

The distinctive capacities of stem cells in the embryo and in vitro raise a central question for understanding development, including neural development. What is the molecular state that defines a stem cell? A corollary to that question provides an organizing principle for understanding neural development: How does a pluripotent stem cell in the embryo or in vitro access a subset of molecular instructions available from the complete genome

to become a neural stem cell? Shinya Yamanaka and his colleagues provided an essential molecular validation of the embryological and in vitro experiments that defined embryonic stem cells. Based on extensive studies of transcription factors (the proteins that bind to nuclear DNA to turn genes on and off) expressed early in developing embryos, Yamanaka and colleagues arrived at a combination of four—Sox2, Oct4, Kif4, and c-Myc—that could "reprogram" somatic cells to become pluripotent stem cells capable of generating an entire organism (in this case a mouse). These cells, called **induced pluripotent stem (iPS) cells**, like the hybrid somatic cells for the frog or ES cells from mice and other animals, are capable of generating a chimeric organism that is reproduction-competent (i.e., the progeny of iPS cells can integrate into the germline to become gametes) when injected into a host blastocyst (the cavity of a post-fertilization vertebrate embryo where ES cells that will make the new organism accumulate). These iPS cells, like ES cells, are pluripotent and can become all cell classes, including neurons (Figure 22.2A). The key role of transcription factors in inducing pluripotency suggests that the DNA in a somatic cell nucleus has to be reprogrammed by changing genes that are either

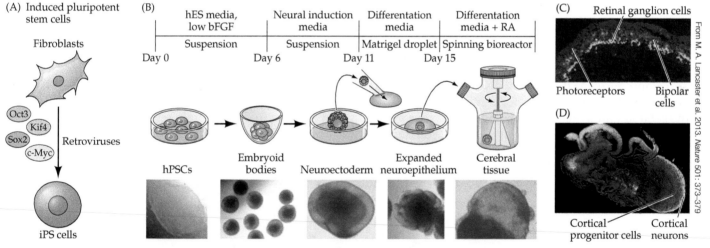

FIGURE 22.2 Changes in transcription factor regulation can return differentiated cells to stem cell identity (A) Somatic cells can be "reprogrammed" to become the equivalent of ES cells. Such cells are referred to as induced pluripotent stem (iPS) cells. The reprogramming relies on introducing four transcription factors (often using retroviruses with the mRNA encoding each of the transcription factors to deliver these factors) that are thought to be essential for establishing and maintaining the ES cell state from a somatic cell such as a fibroblast (a cell type found in the inner layers of the epidermis, among other places). In a small number of somatic cells in which these transcription factors are expressed, those cells will reacquire their ES cell properties, including the capacity to be introduced into a blastocyst and contribute to all tissues in the adult organism that originates from the chimeric embryo. (B) Human iPS cells, as well as those from other species can be grown under a

set of conditions that allows them to diversify and accumulate into aggregates (embryoid bodies) that then go on to make multicellular entities called organoids. The stem cells in organoids generate multiple cell classes found in specific tissues, depending on the combination of molecular signals provided to the iPS (or ES) cells during this process. bFGF, fibroblast growth factor beta; hES media, human embryonic stem cell media; hPSCs, human pluripotent stem cells; RA, retinoic acid. (C,D) Depending on the conditions, iPS- or ES-derived brain organoids can generate cells that resemble those in the retina (C) based on expression of retina-specific markers in subsets of cells (blue, photoreceptors; red, bipolar cells; green, retinal ganglion cells) or the forebrain (D), based on expression of forebrain-specific markers (red, cortical progenitor cells; green, cortical neurons). (After M. A. Lancaster et al. 2013. *Nature* 501: 373–379.)

on or off and changing the overall confirmation of the nuclear DNA to resemble that in the cells of a newly fertilized multicellular embryo at the blastula stage.

The capacity to create pluripotent stem cells from somatic cells also provides a new tool for understanding the molecular and cellular basis of human tissue functions as well as disease. iPS cells can be generated from humans, including those carrying mutations that cause congenital diseases. When given the right additional instructions in the form of developmental signaling molecules (see next section) or transcription factors that characterize a distinct class of cells in a specific organ, iPS cells can differentiate into cells that approximate the endogenous cell classes of that organ. Moreover, if conditions in vitro are varied to support iPS cell aggregation as well as differentiation, the differentiating cells can interact and form larger entities called **organoids** that often include multiple classes of cells from the tissue of interest (Figure 22.2B). These approaches are useful for generating samples of human neuronal classes at a scale sufficient for experiments not possible with human tissue samples, which are usually collected at autopsy. Indeed, using methods that support the generation of organoids, it is possible for neural stem cells derived from ES or iPS cells to give rise to aggregates that have multiple cell classes found in CNS regions such as the retina and forebrain (Figure 22.2C,D).

iPS cells also hold promise for more incisive analysis of pathogenesis in cell classes and tissues from humans with a broad range of genetic diseases. iPS cells can be generated from minimally invasive collection of skin cells and can then be instructed to differentiate into cell classes that are compromised by the relevant disease. This application of iPS cell biology may be particularly useful for defining the cellular and molecular pathology of nervous system disorders, especially because samples of live human nervous tissue from either typical individuals or those with brain disorders are nearly impossible to obtain.

Neural stem cells

Neural stem cells, identified either in an embryo or generated in vitro from ES cells, are proliferative precursor cells that have the capacity to differentiate into the three primary cell classes that define all neural tissue: neurons, astrocytes, and oligodendrocytes (Figure 22.3A–D). Neural stem cells have a distinct molecular identity from that of pluripotent ES cells, based on their expression of a subset of "marker" genes that distinguish them from stem cells of other organs and tissues such as the heart or skeletal muscles. One of the fundamental markers of neural stem cell identity is the basic helix-loop-helix (bHLH; see Concept 22.4) transcription factor Sox2 (Figure 22.3E), which is also one of the four factors necessary for reprogramming somatic cells into iPS cells (see previous section). In addition, neural stem cells can be identified based on genes that they inactivate early in their course of differentiation, both in the embryo and in in vitro experiments with ES or iPS cells. The neuron-restrictive silencing factor (NRSF, also known as REST, for repressor element 1-silencing transcription factor) is perhaps the best known of these repressors of neural precursor or neuron-specific genes. It can bind to specific upstream regulatory sequences (thus, repressor element) of multiple neuron-specific genes and prevent their transcription. Thus, its expression is highest in regions of the embryo where stem cells for non-neuronal tissues are found, and lowest where neurons are found.

The differentiation of neural stem cells into neurons and glia is regulated by subsets of secreted signals as well as by signaling domains of a wide range of cell surface proteins in neighboring cells (see Concept 22.3). The capacities of neural stem cells to differentiate into specific functional classes of neurons in distinct brain regions have been defined by embryonic transplantation experiments in which precursor cells from the rudimentary nervous system of an early-stage embryo or neural stem cells derived from ES cells in vitro are

(A) Neural stem cells   (B) Neurons   (C) Astrocytes   (D) Oligodendrocytes   (E) Embryo

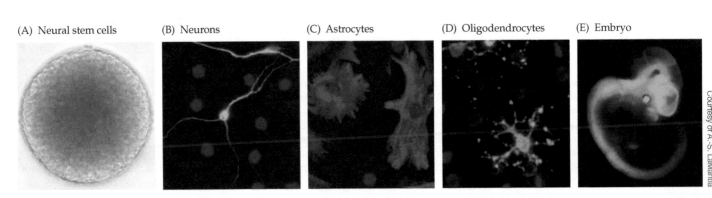

Courtesy of A.-S. LaMantia

FIGURE 22.3 **Neural stem cells have the capacity to give rise to the three major cell classes in the nervous system: neurons, astrocytes, and oligodendrocytes** (A) An aggregate of neural stem cells grown under in vitro conditions that support maintenance of neural stem cell identity. (B) Neural stem cells from aggregates like those shown in (A) can be cultured under conditions so that they generate postmitotic neurons that differentiate, grow dendrite- and axon-like processes called neurites, and even make synapses. (C) The same neural stem cells can also generate postmitotic, differentiated astrocytes, labeled here with an antibody against glial fibrillary acidic protein (GFAP), an astrocyte-selective marker. (D) Neural stem cells also generate oligodendrocytes, labeled here with the marker O4. (E) Neural stem cells in the embryo can be recognized by expression of one of the key reprogramming transcription factors, Sox2.

placed in a differentiating region of a more mature developing nervous system in an older host embryo. These transplanted early precursors, depending on the time they are grafted into the host region, can give rise to the same specific types of neurons and glia that are generated in the targeted region of the host. Neural stem cells in vitro, generated either from ES or iPS cells, can also be instructed by experimental manipulation of signaling molecules or transcription factors to differentiate into neurons and glia that share characteristics with neurons and glia in specific regions of the nervous system. These approaches are being developed with the hope of using engineered neural stem cells to replace degenerating neurons in the intact nervous system. The details of how such stem cell therapies might work reliably and safely in the nervous system—or any organ system—have yet to be fully established.

CONCEPT 22.2 | Neural Stem Cells Generate the Central and Peripheral Nervous Systems

LEARNING OBJECTIVES

22.2.1 Explain how gastrulation and neurulation in the embryo establish the entire nervous system.

22.2.2 Describe the derivation and differentiation of the major divisions of the peripheral nervous system.

22.2.3 Explain how the cranial placodes generate special sensory organs and peripheral sensory receptor neurons.

22.2.4 Describe the derivation and differentiation of the major divisions of the CNS.

Formation of the nervous system: Gastrulation and neurulation

The cells that will generate the nervous system—true neural stem cells within an embryo—become distinct early in the generation of a vertebrate embryo, concurrent with the establishment of the midline and the basic body axes: anterior–posterior (mouth–anus), dorsal–ventral (back–belly), and medial–lateral (midline–periphery). These axes are foundational for proper generation of every organ in the body, including the brain. In addition, the unique curvature of the human CNS generates a distinctive rostral–caudal (Latin, "nose–tail") axis in the developing brain (see Figure A1 in the Appendix). The axes, and thus the initiation of neural development, are critically dependent on the process of **gastrulation** and the molecular mechanisms that underlie this process. Gastrulation begins as the local invagination of a subset of cells in the very early embryo (which in mammals starts out as a single sheet of cells). By the time invagination is complete, the embryo consists of three layers of cells called the **germ layers**: an outer **ectoderm**; a middle **mesoderm** (the invagination of mesoderm cells initiates gastrulation); and an inner **endoderm** (Figure 22.4A).

Based on the position of the invaginating mesoderm and endoderm, gastrulation defines the midline as well as the anterior–posterior and dorsal–ventral axes of all vertebrate embryos. These axes then determine the position of all organ systems, including the peripheral and central nervous systems, as well as facial structures and appendages.

The formation of the **notochord** at the midline of the gastrulating embryo is a central event for the development of the nervous system. The notochord is a distinct cylinder of mesodermal cells that condenses at the midline as the mesoderm invaginates and extends from the mid-anterior to the posterior aspect of the embryo. It is generated at the site of a singular surface indentation called the **primitive node** or **pit**, which subsequently elongates to form the **primitive streak** or **groove**. As a result of these cell movements, the embryonic midline, and thus the axis of symmetry for the entire body forms based upon the position of the notochord. The ectoderm that lies immediately above the notochord, called **neuroectoderm**, gives rise to the entire nervous system (see Figure 22.4A). The notochord itself, however, is a transient structure that disappears once early development is complete. The notochord, by defining the midline and axis of symmetry in the embryo, determines the position of the nervous system, and is required for subsequent early neural differentiation. Along with cells that define the primitive pit, the notochord sends inductive signals (see Concept 22.3) to the overlying ectoderm that cause a subset of cells to differentiate into **neuroectodermal precursor cells**. During this process, called **neurulation**, the midline ectoderm that contains these cells thickens into a distinct columnar epithelium called the **neural plate** (Figure 22.4B). The lateral margins of the neural plate, referred to as the alar plate, then fold inward, transforming the neural plate into a tube called the **neural tube** (Figure 22.4C,D).

The portion of the neural plate at the midline, referred to as the basal plate, becomes the ventral region of the neural tube. The neural tube is not a uniform structure. The cells at the ventral midline of the neural tube differentiate into a specialized strip of epithelial-like cells called the **floorplate** (reflecting their position at the ventral and medial part of the neural tube, above the notochord). Molecular signals from the floorplate as well as from the notochord specify position and fate for the neuroectodermal precursors of the spinal cord and hindbrain. In the forebrain, non-floorplate ventral midline structures as well as neural crest–derived mesenchyme immediately adjacent to the prosencephalic vesicle provide similar signals. The multipotent neural stem cells within the neural tube subsequently give rise to the entire brain and spinal cord, as well as most of the peripheral nervous system, which is derived from a subset of these neuroectodermal precursors called the **neural crest** (see Figures 22.4B–D and 22.5). Thus, all of these cells are neural stem cells. They can divide to produce more neural stem cells (self-renewal being a hallmark of all stem cells) with the capacity to give rise to the full range of cell classes found in the brain, spinal cord, or peripheral nervous system. The unique positions of the self-renewing neural stem cells and limited capacity to make neurons and glial defines them as tissue-specific stem cells—their fate has been restricted to generate cells of a particular organ.

Neural stem cells of the neural tube and neural crest produce neurons, astrocytes, and oligodendrocytes (see Figure 22.3). Eventually, subsets of the neuroectodermal precursor cells, based upon their anterior-posterior, dorsal-ventral, or medial-lateral position, generate region- and fate-specified neural stem cells that differentiate into specific classes of neurons and glia in distinct brain structures. Molecular signals to the spinal cord and hindbrain from the floorplate and the somites, and to the forebrain from the cranial neural crest (see next section), instruct differentiation of cells in the ventral neural tube that eventually give rise to spinal and hindbrain motor neurons and related interneurons, closest to the ventral midline, or basal forebrain projection neurons as well as related interneurons, once again closest to the ventral midline. Precursor cells farther away from the ventral midline give rise to sensory relay neurons and related interneurons in more dorsal regions of the spinal cord and hindbrain. Similar precursors in the anterior region of the neural tube, further from the ventral midline, generate the entire range of dorsal forebrain regions, including the hippocampus and neocortex. The differentiation of these dorsal cell groups is also facilitated by a narrow strip of neuroepithelial cells at the dorsal midline of the neural tube referred to as the **roofplate** in the spinal cord. Roofplate cells are the source of secreted signals that influence dorsal neuronal identities in the spinal cord. In the forebrain, this signaling is mediated by two sources: the cortical "hem," an epithelial domain at the dorsal midline of the prosencephalic–telencephalic vesicles, and

the choroid plexus, which secretes signals directly into the developing ventricles to modulate forebrain stem cell proliferation and differentiation. Like the notochord, the floorplate and roofplate are transient structures that provide signals to the developing neural tube and all but disappear once initial nervous system development is complete.

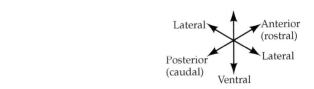

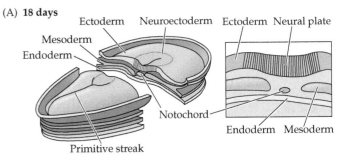

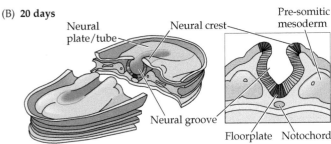

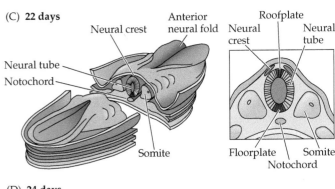

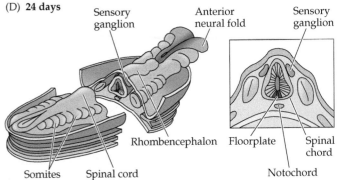

FIGURE 22.4 Neurulation in the mammalian embryo
On the left are dorsal views of a human embryo at several different stages of early development; each boxed view on the right is a midline cross section through the embryo at the same stage. (A) During late gastrulation and early neurulation, the notochord forms by invagination of the mesoderm in the region of the primitive streak. The ectoderm overlying the notochord becomes defined as the neural plate. (B) As neurulation proceeds, the neural plate begins to fold at the midline (adjacent to the notochord), forming the neural groove and, ultimately, the neural tube. The neural plate immediately above the notochord differentiates into the floorplate, whereas the neural crest emerges at the lateral margins of the neural plate (farthest from the notochord). (C) Once the edges of the neural plate meet in the midline, the neural tube is complete. The mesoderm adjacent to the tube then thickens and subdivides into structures called somites—the precursors of the axial musculature and skeleton. (D) As development continues, the neural tube adjacent to the somites becomes the rudimentary spinal cord, and the neural crest gives rise to sensory and autonomic ganglia (the major elements of the peripheral nervous system). Finally, the anterior ends of the neural plate (anterior neural folds) grow together at the midline and continue to expand, eventually giving rise to the brain.

At the lateral edges of the neural plate (the alar plate, which subsequently becomes the dorsal and medial aspect of the neural tube), a third population of precursor cells emerges: the neural crest. **Neural crest cells** arise from the region where the lateral edges (the alar regions) of the neural plate come together as it forms the neural tube (see Figure 22.4B). The neural crest cells undergo an essential cellular transformation called an epithelial-to-mesenchymal transition (see Concept 22.5), based on their location, the local availability of distinct signals in the dorsal neural tube, and the expression of transcription factors and downstream target genes that permit these cells to become migratory. The neural crest cells migrate away from the neural tube through a matrix of loosely packed mesenchymal cells that fill the spaces between the neural tube, embryonic epidermis, and somites (Figure 22.5A,B). Subsets

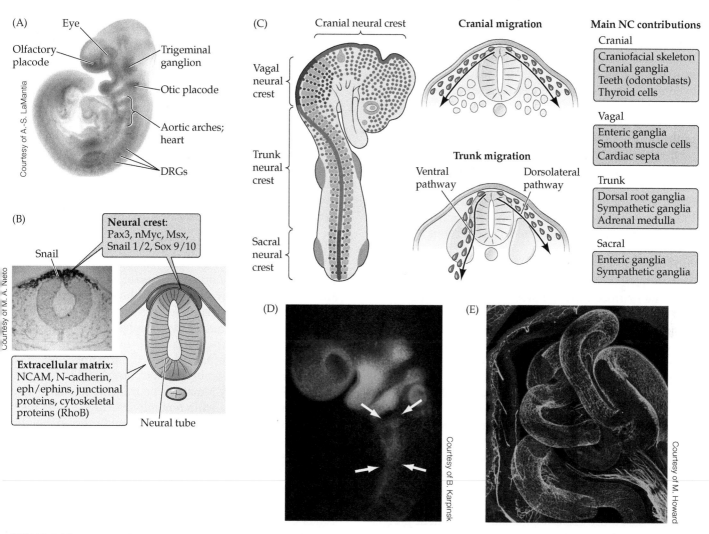

FIGURE 22.5   Neural crest in a developing embryo (A) This mid-gestation mouse embryo carries a reporter transgene that visualizes the migrating neural crest cells (blue) as they begin to accumulate in or near regions where sensory structures or major cardiac blood vessels will differentiate (e.g., the olfactory placode, eye, otic placode, and aortic arches) or at sites where sensory ganglia will form (e.g., the trigeminal ganglion, dorsal root ganglia [DRGs]). (B) The location of origin of the neural crest at the dorsal midline of the neural tube. This region is distinguished by a set of unique molecular determinants, including the transcription factors listed at the top and the adhesion molecules listed at the bottom. (C) Anterior–posterior (A–P) organization of four populations of neural crest cells defined based on their location in the neural tube, and the derivatives that each population gives rise to after the neural crest cells have migrated to cranial, vagal (referring to the territory in periphery innervated by the vagus nerve, cranial nerve X), trunk, or sacral regions of the embryo. In the head (top right), the cranial neural crest migrates immediately beneath the epidermis. In all other regions (bottom right), the neural crest migrates beneath the epidermis and between the neural tube and somites. (D) A mid-gestation mouse embryo carrying a reporter transgene that visualizes the migrating neural crest, including the migrating vagal neural crest (white arrows). (E) The vagal neural crest forms the enteric nervous system, which consists of a network of neurons and small ganglia (shown here by a green reporter transgene) that envelop the outer surface of the intestine. (C after M. Rothstein, 2018. *Dev Biol* 444: S170–S180.)

of neural crest cells follow different pathways, along which they are exposed to additional signals—from structures including the somites and notochord (of mesodermal origin) as well as the epidermis (derived from non-neural ectoderm)—that influence their specific differentiation (see Figure 22.5B). Thus, neural crest cells give rise to a variety of progeny, including the neurons and glia of the sensory and visceral motor (autonomic) ganglia, the neurosecretory cells of the adrenal gland, and the neurons of the enteric nervous system (Figure 22.5C–E). Neural crest cells also contribute to non-neural structures such as pigment cells beneath the epidermis, cartilage, and bone, particularly in the face and skull.

The past decades have witnessed an explosion of molecular biological knowledge about the inductive signaling events and their consequences for gene expression and differentiation that transform neuroectodermal precursors and neural stem cells into the diverse cell and tissue types of the nervous system, including the neural crest. The establishment of cellular diversity, outlined in detail later in this chapter, occurs in parallel with the formation of the anatomical structures that will define the gross subdivisions of the CNS (the spinal cord, brainstem, midbrain, and forebrain) as well as the peripheral nervous system (the sensory ganglia, autonomic ganglia, and enteric nervous system).

Formation of the peripheral nervous system

The peripheral nervous system is derived almost entirely from the neuroectodermal precursors of the neural plate that became neural crest cells, with the exception of a subset of special sensory neurons in the head that are generated from local regions of the head ectoderm called cranial placodes (see next section). Four populations of neural crest cells, distributed along the anterior–posterior axis of the neural tube, generate major subdivisions of the peripheral nervous system (see Figure 22.5C). The most anterior of these neural crest cells, composing the **cranial neural crest**, generate the cranial sensory ganglia (see Appendix), as well as non-neural cell classes that constitute the cranial skeleton, thyroid gland, and teeth. The **vagal neural crest** originates from the most anterior region of the spinal cord at its junction with the hindbrain. It gives rise to much of the enteric nervous system (see Figure 22.5C–E), the parasympathetic ganglia (see Chapter 21), and the smooth muscle cells in the aorta and other great vessels of the heart. The **trunk neural crest**, arising from most of the length of the spinal cord, gives rise to dorsal root sensory ganglia (see Chapters 12 and 13), the anterior sympathetic ganglia (see Chapter 21), and the chromaffin cells of the adrenal medulla. Finally, the **sacral neural crest**, found in the most posterior region of the spinal cord near the hindlimb buds, gives rise to additional neurons in the enteric nervous system and the posterior ganglia of the sympathetic chain. Neural crest cells from all four populations give rise to a variety of peripheral glial cells, including

the Schwann cells that form myelin sheaths around peripheral sensory, motor, and autonomic axons. They also give rise to melanocytes, the pigment- (melanin-) producing cells in the epidermis. Thus, these neuroectodermal-derived neural crest stem cells have a remarkably diverse spectrum of cell fates that include both neural and non-neural cell classes.

Several developmental disorders arise in part because of disrupted specification, migration, or differentiation of the neural crest. These neural crest–associated deficits are referred to as **neurochristopathies** and are often seen in syndromes that include disruptions of multiple organ systems. Neurochristopathies include craniofacial malformations, oropharyngeal malformations, sensory loss (deafness, visual impairment), cardiac malformations, pigment deficiencies or aberrations, gastrointestinal complications that are due to enteric nervous system loss or dysfunction, and several tumors, including neuroblastomas. Neurochristopathies define the pathology of several common genetic syndromes, including DiGeorge/22q11.2 Deletion syndrome, Down syndrome, Hirschsprung's disease, and Wardenburg syndrome. The breadth of organ systems and functions compromised by neurochristopathies indicates the essential role of the neural crest in establishing the peripheral nervous system as well as many of its targets.

Cranial placodes: Connecting the outside world and the brain

The specification and differentiation of the **cranial placodes** and **epibranchial placodes** from the head ectoderm—most of which will become skin—are essential for establishing additional components of the peripheral nervous system that are directly at the interface of the organism and the outside world. These patches of ectoderm are established on the surface of the embryo in the cranial and pharyngeal regions. They mostly remain as neurogenic ectoderm, and subsequently undergo differentiation, distinguishing them from the surrounding ectoderm that will presumably differentiate to epidermis. The placodes can make either peripheral sensory neurons or specializations that support sensory transduction in the nose, ears, and eyes (Figure 22.6). Thus, there are cranial or epibranchial placodes that give rise to the olfactory epithelium, the otic vesicle and organ of Corti, the lens and cornea (but not the neural part of the retina), and the mechanosensory neurons of the cranial sensory ganglia: the trigeminal, the geniculate, the petrosal, the superior vagal (jugular), and the nodose (see Figure 22.9A). Once the placodes are differentiated from the surface ectoderm, all of them undergo inductive interactions with neural crest cells, which drive further differentiation. In most cases, with the exception of the olfactory epithelium (which is derived entirely from placodal ectodermal precursors), the resulting cranial sensory structure will contain both placode- and neural crest–derived neurons. Each of these placode- or crest-derived sensory neuron populations

(A)

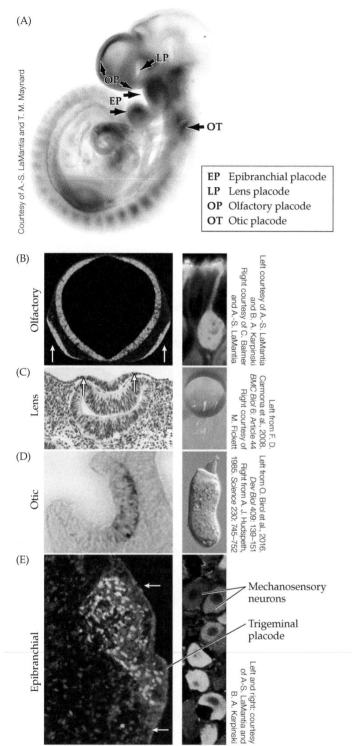

Courtesy of A.-S. LaMantia and T. M. Maynard

EP	Epibranchial placode
LP	Lens placode
OP	Olfactory placode
OT	Otic placode

(B) Olfactory

Left courtesy of A.-S. LaMantia and B. A. Karpinski
Right courtesy of C. Balmer and A.-S. LaMantia

(C) Lens

Left from F. D. Carmona et al., 2008. *BMC Biol* 6: Article 44
Right courtesy of M. Fickett

(D) Otic

Left from O. Birol et al., 2016. *Dev Biol* 409: 139–151
Right from A. J. Hudspeth, 1985. *Science* 230: 745–752

(E) Epibranchial

Mechanosensory neurons

Trigeminal placode

Left and right: courtesy of A.-S. LaMantia and B. A. Karpinski

FIGURE 22.6 Cranial placodes contribute to cranial sensory organs (A) The sites of the cranial placodes have been labeled in a mid-gestation mouse embryo by in situ hybridization for the transcript *Dgcr8*, a gene involved in microRNA processing. EP, epibranchial placode; LP, lens placode; OP, olfactory placode; OT, otic placode. (B) The olfactory placodes (arrows, left) are composed of neural precursors (labelled here by the early neural marker βIII-tubulin) that will give rise to olfactory receptor neurons (a single receptor is shown at right) as well olfactory epithelium supporting cells. (C) The lens placode (between the arrows, left) will give rise to the lens (shown at right) as eye development continues. (D) The otic placode, labeled by in situ hybridization for *Pax2* (left), will give rise to the hair cells of the inner ear (an auditory hair cell is shown at right). (E) The trigeminal placode, shown here labeled with an antibody for the transcription factor Six1 (red epithelial labelling between the arrows, left), will contribute mechanosensory neurons (red label, right) to the trigeminal ganglion.

The cranial placodes themselves are initially apparent as ectodermal thickenings during early development of the vertebrate head (see Figure 22.6). Adjacent to these thickenings are primarily populations of neural crest cells that remain as mesenchyme (loosely arrayed cells that are not organized into a "sheet" or epithelium; see Figure 22.5). The interaction of the placodes with the adjacent neural crest is key. In some cases (olfactory placode, lens placode, otic placode), the neural crest makes no contribution or only a negligible contribution to the sensory neuron populations. Nevertheless, there are essential interactions between the mesenchymal neural crest and the placodal ectoderm that drives morphogenesis and differentiation in all of these structures. In the case of the cranial ganglia, the neurons derived from placodal precursors and their axons are significant contributors to the relevant cranial nerves (trigeminal, vagal, etc.; see Appendix). These placodal precursors translocate into the nascent ganglia, become neural progenitors, and generate sensory—primarily mechanosensory—neurons (touch, pressure, and proprioception; see Chapter 9).

Formation of the CNS

Soon after the neural tube forms, the forerunners of the major brain regions become apparent as a result of morphogenetic movements that bend, fold, and constrict the tube. Initially, the anterior end of the tube forms a crook, or "cane handle" (Figure 22.7A). The end of the "handle" nearest the sharpest bend is the **cephalic flexure**, which balloons out to form the **prosencephalon**, which in turn gives rise to the forebrain. The midbrain (**mesencephalon**) forms as a bulge above the cephalic flexure. The hindbrain (**rhombencephalon**) forms in the long, relatively straight stretch between the cephalic flexure and the more caudal cervical flexure. Caudal to the cervical flexure, the neural tube forms the precursor of the spinal cord. This bending and folding diminish or enlarge the different regions of the lumen enclosed by the developing neural tube. These

has the unique capacity to transduce specific classes of sensory stimuli directly from the environment, leading to the initial encoding of information to be further processed in the brain. Accordingly, the result of placode formation is the establishment of the "lines into" the CNS from the head that carry information about the outside world.

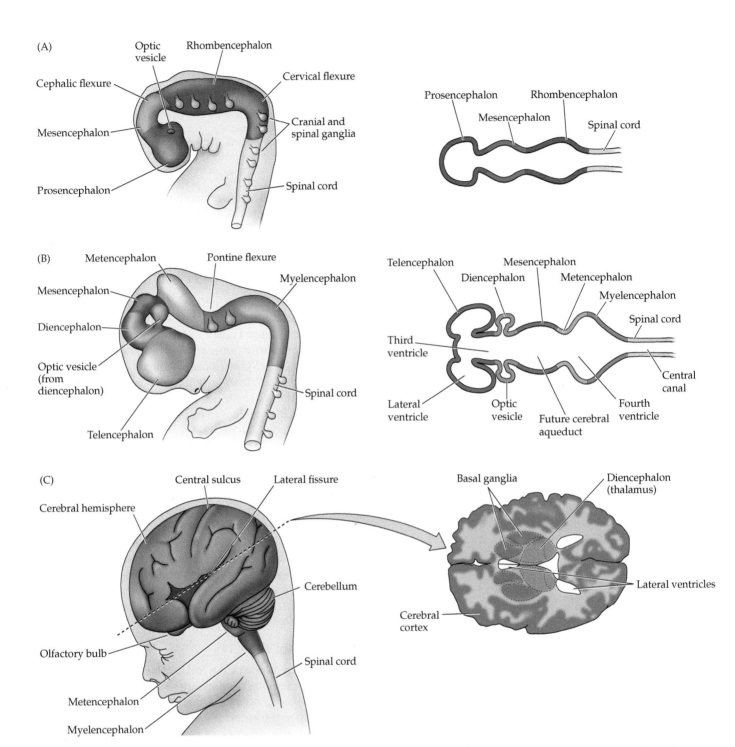

FIGURE 22.7 Regional specification of the developing brain (A) Early in gestation the neural tube becomes subdivided into the prosencephalon (at the anterior end of the embryo), mesencephalon, and rhombencephalon. The spinal cord differentiates from the more posterior region of the neural tube. The initial bending of the neural tube at its anterior end leads to a cane shape. At right is a longitudinal section of the neural tube at this stage, showing the position of the major brain regions. (B) Further development distinguishes the telencephalon and diencephalon from the prosencephalon; two other subdivisions —the metencephalon and myelencephalon—derive from the rhombencephalon. These subregions give rise to the rudiments of the major functional subdivisions of the brain, while the spaces they enclose eventually form the ventricles of the mature brain. At right is a longitudinal section of the embryo at the developmental stage shown at left. (C) The fetal brain and spinal cord are clearly differentiated by the end of the second trimester. Several major subdivisions, including the cerebral cortex and cerebellum, are clearly seen from the lateral surfaces. At right is a cross section through the forebrain at the level indicated showing the nascent sulci and gyri of the cerebral cortex, as well as the differentiation of the basal ganglia and thalamic nuclei.

lumenal spaces eventually become the ventricles of the mature brain (Figure 22.7B; also see the Appendix).

Once the primitive brain regions are established, they undergo at least two more rounds of partitioning, each of which elaborates the developing brain regions into forerunners of adult structures (Figure 22.7C). Thus, the lateral aspects of the rostral prosencephalon form the **telencephalon**. The two bilaterally symmetrical telencephalic vesicles include dorsal and ventral territories. The dorsal territory gives rise to the rudiments of the cerebral cortex and hippocampus, while the ventral territory gives rise to the basal ganglia (derived from embryonic structures called the **ganglionic eminences**), basal forebrain nuclei, and olfactory bulb. The more caudal portion of the prosencephalon forms the **diencephalon**, which contains the rudiments of the thalamus and hypothalamus, as well as a pair of lateral outpocketings—the **optic vesicles**—from which the neural portion of the retina will form. The dorsal portion of the mesencephalon gives rise to the superior and inferior colliculi, while the ventral portion gives rise to a collection of nuclei known as the midbrain tegmentum. The rostral part of the rhombencephalon becomes the **metencephalon** and gives rise to the adult cerebellum and pons. Finally, the caudal part of the rhombencephalon becomes the **myelencephalon** and gives rise to the adult medulla. Thus, the functional subdivisions of the mammalian brain and spinal cord reflect the morphogenetic processes that begin with a neural plate, followed by a neural tube that is further elaborated as development moves forward.

CONCEPT 22.3 | Transcription Factor Patterning Regulated by Cell-Cell Signaling Establishes Distinct Brain Regions

LEARNING OBJECTIVES

22.3.1 Discuss how regional development of the brain relies on restricted expression of genes in local domains of the developing neural tube.

22.3.2 Explain how patterned gene expression in the developing nervous system reflects the local action of cell-cell signaling molecules.

22.3.3 Explain how local cell-cell signaling and local transcription factor expression establish additional neuronal diversity.

The molecular basis of neural induction

How can a simple tube of neural stem cells produce such a variety of brain structures? At least part of the answer comes from an observation, made early in the twentieth century. After the initial morphogenetic development of the rudimentary brain regions described in the previous section, much of the neural tube is organized into repeating units called **neuromeres**. This discovery led to the idea that the process of **segmentation**—which in animal embryos (as well as in angiosperm plant embryos) establishes regional identity in the body by dividing the embryo into repeated units, or segments—might also establish regional identity in the developing brain. Enthusiasm for this hypothesis was stimulated by observations of the development of the body plan of the fruit fly *Drosophila* that linked these anatomically recognizable repeating units with molecular determinants and networks of genes that encode these molecules. In the fly, early expression of a class of genes called **homeotic** or **homeobox genes** guides the differentiation of the embryo into distinct segments that give rise to the head, thorax, and abdomen (Figure 22.8A,B). *Drosophila* homeobox genes code for DNA-binding proteins that modulate the expression of other genes that mediate morphogenesis. Similar genes have been identified in vertebrates, including mammals, and are referred to as **Hox genes**. Rather than having one copy of each segmentally essential homeobox gene, as seen in the fly, vertebrate Hox genes have undergone multiple duplications so that today's vertebrates have four "clusters" of homologous genes with similar functions. In most mammals, including humans, each Hox gene cluster is located on a distinct chromosome, and their anterior-to-posterior expression and function during regional development of the body and brain are reflected in their position $5' \rightarrow 3'$ on that chromosome (Figure 22.8C). In some cases in the developing mammalian nervous system, the pattern of Hox gene expression coincides with, or even precedes, the formation of morphological features—that is, the various bends, folds, and constrictions initially identified as neuromeres—that underlie the progressive regionalization of the developing neural tube, particularly in the hindbrain and spinal cord.

The relationship between Hox gene expression, initial morphogenetic distinctions in the neural tube, and the eventual organization of mature regions of the nervous system is best understood for the rhombencephalon, the region of the neural tube that will give rise to the hindbrain and the region where distinct repeating morphological domains (neuromeres) emerge during early neural tube differentiation. Very early during the morphogenetic events that distinguish the rhombencephalon from the rest of the brain, multiple Hox genes are expressed so that their boundaries coincide with local segments arrayed along the anterior–posterior axis called **rhombomeres** (Figure 22.9A). The rhombomeres are visible as local bulges, each bounded by a furrow or invagination of the neural tube. Each of the eight rhombomeres, either separately or in combination with an adjacent rhombomere, will give rise to the motor neurons that constitute each of the cranial motor nerves (see Appendix), as well as the neural crest that will contribute to each of

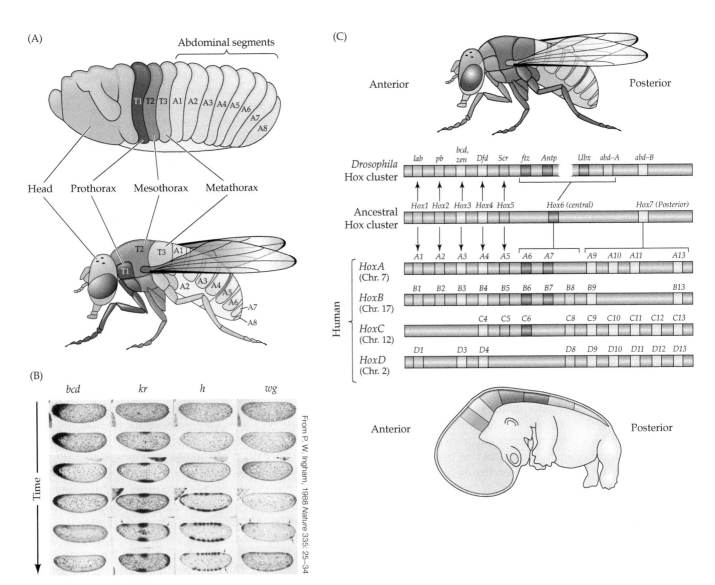

FIGURE 22.8 Sequential gene expression divides the embryo into regions and segments (A) The relationship of the embryonic segments in the *Drosophila* larva, defined by sequential gene expression, to the body plan of the mature fruit fly. (B) Temporal pattern of expression of four genes that influence the establishment of the body plan in *Drosophila*. A series of sections through the anterior–posterior midline of the embryo is shown from early to later stages of development (top to bottom in each row). Initially, expression of the gene *bicoid* (*bcd*) helps define the anterior pole of the embryo. Next, *krüppel* (*kr*) is expressed in the middle and then at the posterior end of the embryo, defining the anterior–posterior axis. Then *hairy* (*h*) expression helps delineate the domains that will eventually form the mature segmented body of the fly. Finally, the *wingless* (*wg*) gene is expressed, further refining the organization of individual segments. (C) Parallels between *Drosophila* segmental genes (the inferred "ancestral" homeobox genes from which invertebrate and vertebrate segmental genes evolved) and human Hox genes. Hox genes of humans (and of most mammals) have apparently been duplicated twice, leading to four independent groups, each on a distinct chromosome. The anterior-to-posterior pattern of Hox gene expression in both flies and mammals (including humans) follows the 3'-to-5' orientation of these genes on their respective chromosomes. (A,B after Gilbert, S. 1994. *Developmental Biology.* Sunderland: Sinauer; Lawrence, P. A. 1992. *The Making of a Fly: The Genetics of Animal Design.* Oxford: Blackwell Scientific Publicaitons; C after A. Veraksa et al., 2000. *Mol Genet Metab* 69: 85–100.)

the cranial sensory ganglia. Thus the anterior-posterior array of the cranial nerves reflects the initial subdivision of the hindbrain into rhombomeres. Expression of Hox genes and other transcription factors that define each rhombomere is maintained in the neural stem cells that generate each cranial motor nucleus, sensory relay nucleus, or peripheral sensory ganglion, thus providing unique identities for local neurogenesis and differentiation. The division of the rhombencephalon into segments with uniquely patterned transcription factor expression along the anterior–posterior axis and its direct relationship to the anterior–posterior organization of the cranial

(A)

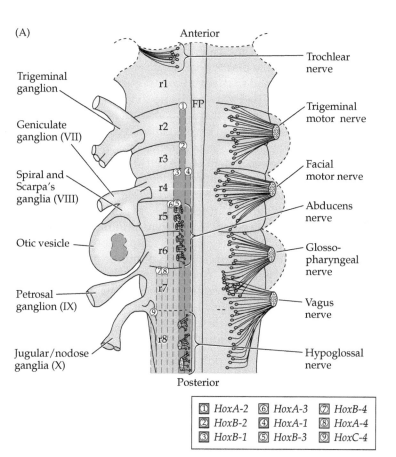

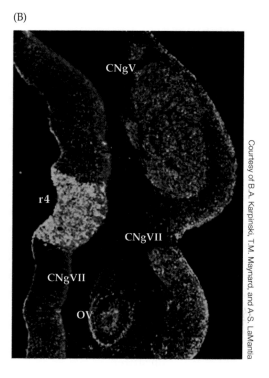

① HoxA-2	⑥ HoxA-3	⑦ HoxB-4
② HoxB-2	④ HoxA-1	⑧ HoxA-4
③ HoxB-1	⑤ HoxB-3	⑨ HoxC-4

FIGURE 22.9   The relationship between early segmental patterning in the hindbrain and the differentiation of cranial nerves and ganglia (A) In the vertebrate hindbrain, the regions of the neural tube that will give rise to the cranial nerves that innervate the muscles and skin of the head and neck are first identified by restricted expression of several Hox genes (colored bands) that give the neural stem cells found in a series of repeated units called rhombomeres a unique molecular identity. The subsequent specification of neural crest as well as neural progenitors that remain in the hindbrain guide the differentiation of the cranial sensory ganglia and nerves (left) as well as cranial motor or mixed nerves (right). FP, floorplate. (B) In a transverse section through the developing mid-gestation hindbrain in a mouse, rhombomere 4 (r4) is distinguished by expression of the *HoxB-1* gene. The relationship between this rhombomere, its neighbors (approximate boundaries, recognized by evagination and invagination of the hindbrain neuroepithelium, indicated by arrows), and adjacent cranial ganglia is indicated. CNgV, trigeminal ganglion; CNgVII, facial ganglion; OV, otic vesicle (A courtesy of A. Lumsden.)

nerves (Figure 22.9B) is perhaps the clearest example of how molecularly specified, genetically encoded embryonic segmentation and patterning influence the organization of the mature nervous system.

In vertebrates, Hox gene expression does not extend into the midbrain or forebrain (see Figure 22.8C); however, regional differences in expression of other transcription factors are seen in these subdivisions prior to and during the morphogenetic events that define them (see Figure 22.12). The genes that define early subdivisions of the vertebrate midbrain and forebrain (including members of the *distaless* [*DLX*] and *paired box* [*PAX*] families) are homologues of transcription factors in *Drosophila* that influence the development of body structures such as appendages, head and mouthparts, and sensory organs. As with the Hox genes in the early hindbrain, expression of these transcription factors confers identity upon neural stem cells in each local forebrain or midbrain region to influence their local fates and differentiation.

The patterned expression of Hox genes and the genes for other developmentally regulated transcription factors and signaling molecules does not by itself determine the fate of a group of embryonic neural precursors. Instead, this aspect of regionally distinct transcription factor expression during early brain development contributes to a broader series of cellular and molecular processes that further distinguish neural stem cells at specific positions in the neural tube. These neural stem cells, with divergent identities based on their position and history of gene expression, eventually produce distinct, fully differentiated brain regions with appropriate classes of neurons and glia.

Cell-cell signaling, neural induction, and patterning

Neural stem cells in the early neural plate and tube, and subsequently in each nascent brain region, must acquire instructions that establish their capacity to make neurons specific to each region. These instructions come from neighboring cells or tissues. During the first half of the twentieth century, the sources of cellular instructions needed for gastrulation and neurulation—those that give neural stem cells their identity during early embryonic development—were defined using a variety of experiments, now considered classic, based on the removal or transfer of embryonic tissues to assess the capacity of ectodermal, mesodermal, and endodermal cells to form organs composed of differentiated cell types. Cells that are moved either acquire the identity of the new region in which they are placed, having responded to instructions based on their new location, or they retain an identity that reflects their original position, presumably because they have already responded to immutable instructions at their original location.

Cells that are removed without replacement are either compensated by increased local cell proliferation, causing little noticeable disruption of subsequent development (a process referred to as embryonic regulation), or if they are already specified via signals no longer available in that region, their absence disrupts subsequent development. In some cases, relocation of cells causes a complete change of the local developmental program. For example, transplantation of the equivalent of the cells that define the primitive pit to another location on an early embryo can cause a second notochord to form and a second nervous system to develop (Figure 22.10) Similarly, the transplantation of a notochord to an ectopic location near more dorsal regions of the neural tube can cause an ectopic floorplate to form, resulting in local specification of motor neurons rather than sensory neurons (see below and Figure 22.12).

Taken together, these experiments show that *interactions* between cells in adjacent germ layers (e.g., mesoderm adjacent to ectoderm) are essential for regional and cellular identity in the developing embryo. By the early 1920s it

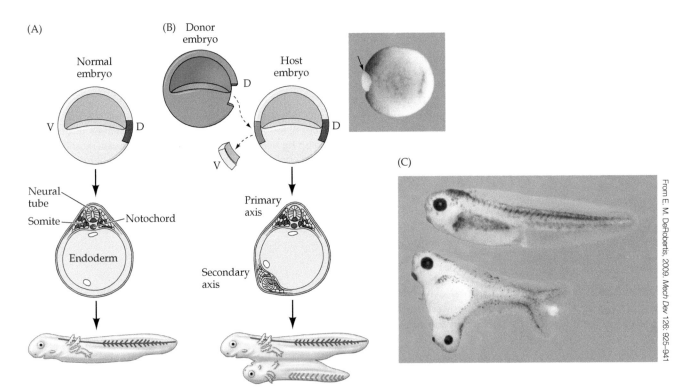

FIGURE 22.10   Induction of the nervous system from an undifferentiated blastula relies on a region of the early embryo called the organizer (A) The organizer is sometimes referred to as the dorsal lip of the blastopore (ventral [V] and dorsal [D] in this diagram are shown left to right) and consists of a small population of epithelial embryonic stem cells. This region of the embryo (in this case from a frog) gives rise to the notochord and contributes to other mesodermal derivatives such as the somites. The neural tube is established in the embryo based on its proximity to the organizer-derived notochord. (B) When the organizer (blue) from a donor embryo (often carrying a permanent marker to distinguish it from the tissue of the host) is dissected and transplanted into a host embryo as shown, a second notochord is generated from the donor organizer, and a second nervous system is induced in the host ectoderm (tan). The micrograph at right shows the organizer (arrow) from a non-pigmented embryo transplanted into a pigmented host. (C) A normal pigmented tadpole (top) compared with a two-headed tadpole (bottom) generated by transplanting an additional non-pigmented organizer as shown in the diagram in (B).

was clear that location of the primitive pit or its equivalent in non-mammalian embryos—collectively referred to as the embryonic organizer—or the apposition of notochord to overlying ectoderm is essential because these structures provide signals that establish neuroectodermal stem cells and thus the entire nervous system, a process known as **neural induction**. These experiments suggested strongly that neural induction relies on signals provided by adjacent cells or tissues; however, the ultimate proof of this conjecture did not emerge until the early 1990s.

The molecular basis of cell-cell signaling in developing brains

In the past 30 years, molecular and genetic approaches have demonstrated that the generation of cell identity and diversity—of which neural induction is only one example—results from the spatial and temporal control of different sets of genes by endogenous signaling molecules. Most of these molecular signals are secreted by one embryonic cell class or tissue and then diffuse through extracellular space to act on an adjacent cell class or tissue. These inductive signaling molecules are secreted by the embryonic structures that are critical for the morphogenesis and patterning of the CNS (as discussed in Concept 22.2 as well as in this Concept). These include the notochord, floorplate, roofplate, and neuroectoderm itself, as well as adjacent neural crest–derived mesenchymal cells or mesodermally derived tissues such as somites (Figure 22.11A). Different classes of receptors transduce the signals in the neuroectoderm to drive further cellular differentiation. In some cases the signals have graded effects based on the distance of target cells from the source. These effects may represent a diffusion gradient of the signal, or graded activity that is due to the distribution pattern of receptors or other signaling components. Other signals are more specific in their action, being most effective at the boundaries between distinct cell populations. The results of inductive signaling include changes in gene expression, shape, and motility in the target cells.

One of the first inductive signals to be identified was **retinoic acid (RA)**, a derivative of vitamin A and a member of the steroid and thyroid superfamily of hormones (Figure 22.11B). Retinoic acid is a small, lipophilic molecule synthesized via metabolic enzymes, similar to gonadal steroids and small-molecule neurotransmitters. RA activates a unique class of transcription factors that are also receptors for RA and related ligands—the **retinoid receptors**. There are multiple retinoid receptors, and they are expressed widely in the developing as well as adult brain. When activated by RA or related retinoids, the ligand–receptor complex modulates the expression of several target genes. The capacity of RA receptors, when bound by RA, to stimulate or repress gene expression depends on co-activators or co-repressors that form complexes with RA receptors when bound to nuclear

DNA. RA signaling drives cell proliferation as well as differentiation, regulating transitions between various classes of neural stem cells leading up to terminal neurogenesis. Excess retinoid signaling (caused by excessive or insufficient dietary intake of vitamin A or exposure to retinoid-based medications; see Clinical Applications) can cause severe birth defects, including incomplete neural tube closure and other disruptions of early brain morphogenesis.

Most inductive signaling molecules, however, are peptide hormones encoded by specific genes. The **fibroblast growth factor** (FGF) family of peptide hormones is among the largest sets of inductive signals. There are 22 different FGF ligands encoded by 22 genes in the human genome. All of these ligands bind to the same receptor tyrosine kinases that initiate a phosphorylation-based signaling cascade via the ras-MAP kinase pathway (Figure 22.11C). MAP kinase activation can lead to altered expression of several target genes, especially those that modulate cell proliferation and differentiation. Among mammals, including humans, FGF8 has emerged as a particularly important regulator of forebrain, midbrain, and cranial sensory neuron development. In addition, FGFs from the pre-somitic mesoderm (from which somites form) regulate spinal cord neurogenesis and differentiation.

The **bone morphogenetic proteins (BMPs)**, members of the TGF-β family of peptide hormones, are particularly important for a variety of events in neural induction and differentiation. The various BMPs play roles in the initial specification of the neural plate as well as the subsequent differentiation of the dorsal part of the spinal cord and hindbrain and the cerebral cortex. In humans and other mammals, six distinct genes encode six different BMP ligands. These ligands all activate a singular signaling pathway via the same receptor serine kinases, resulting in the phosphorylation and translocation to the nucleus of transcriptional regulators called SMADs (Figure 22.11D). Following BMP-dependent phosphorylation, three different phospho-SMADs—1, 5, and 8—translocate into the nucleus, bind to specific enhancer or supressor DNA sequences called BMP response elements (BMPre), and thereby influence transcription of several target genes.

In *mesodermal* cells, the BMPs (as their name suggests) elicit osteogenesis (bone cell differentiation). When *ectodermal* cells are exposed to BMPs, they assume an epidermal fate, forming structures associated with the skin. How, then, do ectodermal stem cells become or remain neuralized, given that BMPs are secreted by the somites and surrounding mesodermal tissue and instruct them to make bones and skin? The mechanism evidently relies on the local activity of additional secreted inductive signaling molecules, including **Noggin** and **Chordin**—two members of a broad class of **endogenous antagonists** that modulate signaling via the TGF-β family (including the BMPs). These antagonistic molecules can bind directly to

(A)

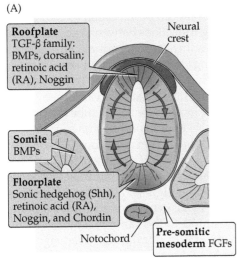

Roofplate
TGF-β family:
BMPs, dorsalin;
retinoic acid
(RA), Noggin

Neural
crest

Somite
BMPs

Floorplate
Sonic hedgehog (Shh),
retinoic acid (RA),
Noggin, and Chordin

Notochord

Pre-somitic
mesoderm FGFs

(B) Retinoic acid (RA)

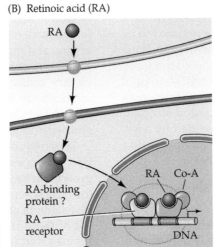

RA

RA-binding
protein ?

RA
receptor

RA   Co-A

DNA

(C) Fibroblast growth factor (FGF)

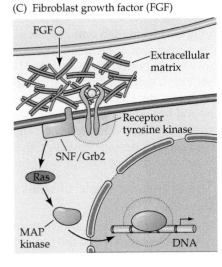

FGF

Extracellular
matrix

Receptor
tyrosine kinase

SNF/Grb2

Ras

MAP
kinase

DNA

(D) Bone morphogenetic protein (BMP)

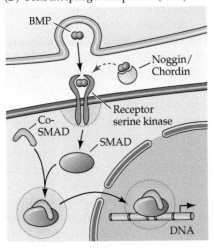

BMP

Noggin/
Chordin

Co-
SMAD

Receptor
serine kinase

SMAD

DNA

(E) Canonical Wnt

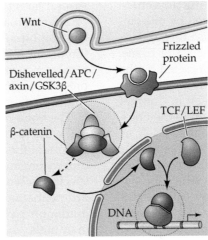

Wnt

Frizzled
protein

Dishevelled/APC/
axin/GSK3β

TCF/LEF

β-catenin

DNA

(F) Noncanonical Wnt

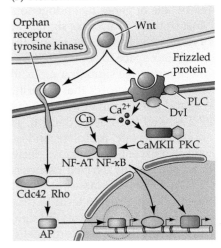

Orphan
receptor
tyrosine kinase

Wnt

Frizzled
protein

PLC

Ca²⁺

DvI

Cn

CaMKII PKC

NF-AT NF-κB

Cdc42 Rho

AP

(G) Sonic hedgehog (Shh)

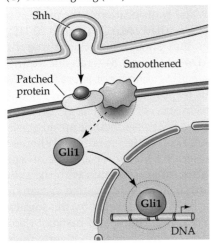

Shh

Smoothened

Patched
protein

Gli1

Gli1

DNA

FIGURE 22.11   Major inductive signaling pathways in vertebrate embryos
(A) The embryonic notochord, floorplate, and neural ectoderm, as well as adjacent tissues
such as somites, produce the molecular signals that induce cell and tissue differentiation
in the vertebrate embryo. (B–G) Schematics of ligands, receptors, and primary intracellular
signaling molecules for retinoic acid (RA); members of the FGF and TGF-β (BMP) super-
families of peptide hormones; the Wnt family of signals; and Sonic hedgehog (Shh). Each of
these pathways contributes to the initial establishment of the neural ectoderm, as well as to
the subsequent differentiation of distinct classes of neurons and glia throughout the brain.

BMPs, preventing their binding to BMP receptor proteins
(see Figure 22.11D). When BMPs are thus prevented from
binding to their "normal" receptors, neuroectoderm is
"rescued" from becoming epidermis and continues along

a path of neuralization. This negative regulation has reinforced speculation that becoming a neuron is the "default" fate for embryonic ectodermal cells. Once neuronal
precursor identity is firmly established, however, BMPs

can act on central and peripheral neural precursors or differentiating neurons to further influence their identity and fate.

Members of the **Wnt** family of secreted signals also modulate several aspects of nervous system morphogenesis and neuronal differentiation, including some aspects of neural crest differentiation. In contrast to several other inductive signaling pathways, the 19 human Wnt ligands (encoded by 19 separate genes) can activate two distinct signal transduction cascades, the "canonical" and "noncanonical" pathways. The canonical pathway was discovered first and is essential for regulating the identity of epithelial cells (cells held together by specific junctional complexes into a sheet or layer) versus mesenchymal cells (loosely associated, motile cells). In the developing nervous system, canonical Wnt signaling is especially important for initial neural crest migration. The noncanonical pathway is featured during early nervous system morphogenesis.

- The *canonical Wnt pathway* influences cell proliferation, adhesion, and differentiation after the initial morphogenesis of the nervous system (gastrulation and neurulation) is complete. This pathway relies on the activation of the Frizzled receptor in the presence of a co-receptor (LRP5/6) which leads to stabilization of β-catenin, a cellular messenger which is then translocated to the nucleus, where it influences gene expression via interactions with the TCF/LEF transcription factor complex (Figure 22.11E). Canonical Wnt signaling is a major regulator of the activity of the Snail transcription factors, which influence epithelial versus mesenchymal states in a variety of developing tissues, including the neural tube and neural crest (see also Figure 22.5B).

- The *noncanonical Wnt pathway*, also known as the planar cell polarity (PCP) pathway, regulates cell movements necessary for lengthening the neural plate and neural tube. In this transduction pathway, Wnt ligands activate receptor proteins (Frizzled), leading to changes in intracellular Ca^{2+} levels; alternatively, the Wnt ligands can bind an orphan receptor tyrosine kinase, leading to activation of a Jun kinase (Jnk) signaling pathway that can phosphorylate several intracellular targets, leading to changes in cell shape and polarity (Figure 22.11F).

Another peptide hormone essential for neural induction as well as progenitor specification and neuronal differentiation is **Sonic hedgehog (Shh)**. Sonic hedgehog is thought to be particularly important for three phases of neural development: (1) closing the neural tube, especially the anterior midline, (2) establishing the identity of neural precursors—particularly those for motor neurons—in the ventral portion of the spinal cord and hindbrain, and (3) modulating differentiation of postmitotic neurons, including axon growth and synapse formation. The transduction of signals via Shh (Figure 22.11G) requires the cooperative binding of two surface receptor proteins, Patched and Smoothened (the names are based on the appearances of their respective *Drosophila* mutants). In the absence of Shh, an inhibitory protein complex assembles that modulates a family of transcriptional regulators (Gli1, 2, and 3, originally discovered as oncogenes in *glio*mas). When this inhibitory complex is in place, only Gli3—which represses transcription of target genes—is available and active in the nucleus. When Shh is present, it binds to Patched and promotes the accumulation of Smoothened on the cell surface, causing disassembly of the inhibitory complex and allowing Gli1 (or Gli2) to be translocated into the nucleus, where it positively regulates the expression of genes that establish neural identity.

Integration of inductive signaling: Neuronal identity

Stem cells in the embryo are guided toward becoming neural stem cells, and ultimately neurons, via multiple signaling molecules that they encounter sequentially or simultaneously. These signaling molecules are either secreted by other cells nearby or are available on the cell surfaces of neighboring cells or in the extracellular matrix. The combined activities of retinoic acid, FGFs, BMPs (antagonized by Noggin and Chordin), Sonic hedgehog, and Wnts made available in this way specify a mosaic of transcription factor expression in subsets of precursor cells throughout the developing neuroectoderm, then the neural plate, and then the neural tube. The earliest signaling molecules are thought to be essential for distinguishing stem cells destined to become the neural crest at the margins of the neural plate from stem cells in the presumptive neural tube ectoderm that will give rise to neural stem cells that generate the brain and spinal cord.

The subsequent activity of inductive signals and their transcription factor targets for neural stem cell specification and differentiation in the CNS is best understood in the spinal cord, where these signals interact to establish differences in gene expression in the ventral, intermediate, and dorsal spinal cord. Sonic hedgehog from the floorplate (Figure 22.12A) and the BMP antagonist Noggin from the roofplate (Figure 22.12B), which antagonizes adjacent BMP signals that instruct the dorsal lateral ectoderm to become epidermis, are essential to establish cell identity based on ventral versus dorsal axes in the spinal cord. This mechanism of signaling and transcriptional control sequentially limits downstream gene expression in neural precursors, leading to specific patterns of gene expression in postmitotic neuronal progeny (Figure 22.12C). These specific patterns of transcription factor expression will provide a foundation for the differentiation of motor neurons, interneurons, and sensory relay neurons in the ventral, intermediate, and dorsal domains of the nascent spinal cord. Additional transcription factors, also influenced by local

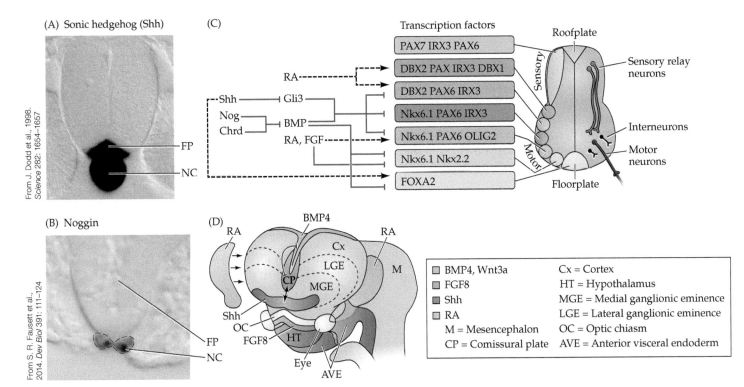

From J. Dodd et al., 1998. *Science* 282: 1654–1657

From S. R. Fausett et al., 2014. *Dev Biol* 391: 111–124

FIGURE 22.12 An integrated network of local signals specifies neural identity (A) A section through the embryonic chick spinal cord shows the distribution of the Sonic hedgehog signal (purple label) in the notochord (NC) and floorplate (FP). (B) The BMP antagonist Noggin (light blue label), which helps preserve the neural identity of ectoderm by preventing BMP signaling, is available from the roofplate (see part C and Figure 22.11D) as well as the floorplate and notochord. This image is a section through the embryonic mouse spinal cord; the mouse notochord is a somewhat smaller structure than that of the chick. (C) Interactions between Shh (via Gli3 repression), Noggin/Chordin, BMP, RA, and FGF lead to either expression (black dashed) or repression (red solid) of a set of transcription factors that distinguish different precursors. These distinct precursors, based on their dorsal-to-ventral position in the spinal cord, will go on to become sensory relay neurons (dorsal), interneurons (intermediate), or motor neurons (ventral). (D) A similar mechanism establishes identity for neuronal progenitors in the forebrain.

inductive signals, are thought to specify neuronal identity in immature postmitotic neuroblasts; they also support the position-specific final differentiation of motor neurons, interneurons, and sensory relay neurons in the mature ventral, intermediate, and dorsal spinal cord. Thus, the target of these transcription factors is the entire range of signals and effectors that influence neuronal migration, axon and dendrite growth, synapse formation, and neuronal identity—including neurotransmitter classes, receptor classes, and classes of ion channels for excitability.

This general mechanism—local signaling leading to local variation in transcription factor expression in distinct precursor cells or early postmitotic neuroblasts—operates throughout the developing central and peripheral nervous systems. The combination of transcription factors necessary to establish identity of specific neuron classes is often referred to as a *transcriptional code*. Transcriptional codes are thought to specify neuronal identity and facilitate neuronal diversity throughout the developing brain. Indeed, although some of the signaling molecules and transcription factors are different, a similar mechanism establishes

identity for neuronal progenitors in the forebrain (Figure 22.12D). In the forebrain, inductive signaling via molecules identical or analogous to those in the spinal cord establishes transcriptional differences among forebrain neural stem cells that prefigure morphogenesis and neuronal differentiation in ventral, dorsal, and intermediate forebrain subdivisions. These domains then give rise to the olfactory bulb, basal ganglia, basal forebrain (amygdala and other ventrolateral structures), hippocampus, and neocortex. Thus, locally available inductive signals, their receptors in adjacent cells and tissues, and the resulting regulation of gene expression (particularly of locally expressed transcription factors) specify cell identity as well as influence other aspects of neural development for the entire CNS.

Molecular and genetic disruptions of early neural development

The identification of the molecules involved in neural induction, neurogenesis, and the generation of neuronal diversity has led to a more informed way of thinking about the etiology of several congenital disorders of the nervous system. Anomalies such as *spina bifida* (failure of the posterior neural tube to close completely), *anencephaly* (failure of the anterior neural tube to close at all), *holoprosencephaly* (disrupted regional differentiation of the forebrain), and other brain malformations can result from environmental insults that disrupt inductive signaling, and from the mutation of genes that participate in this process. Some forms of *hydrocephalus* (a condition in which impedes flow of cerebrospinal fluid increases pressure and results in enlarged ventricles and eventual cortical atrophy from compression) can be traced to mutations of genes on the X chromosome, especially those for the L1 cell adhesion molecule (see Chapter 23).

Maternal dietary insufficiency of micronutrients (essential compounds needed at low concentrations) such as folic acid can disrupt embryonic neural tube formation by compromising cellular mechanisms essential for DNA replication and for cell division and motility. Conversely, excessive maternal intake of vitamin A—the metabolic precursor of retinoic acid—can impede neural tube closure and differentiation, or it can disrupt later aspects of neuronal differentiation because of an excess of ectopic RA signaling (Clinical Applications). Embryonic exposure to a variety of other drugs—alcohol and thalidomide are notable examples—can also elicit pathological differentiation of the embryonic nervous system by providing or blocking inductive signals at inappropriate times or places. Altered cholesterol metabolism can compromise Sonic hedgehog signaling because cholesterol molecules play a role in modulating the interaction of Shh with its

---

■ Clinical Applications

Inductive Signals and Neurodevelopmental Disorders

Retinoic acid: Teratogen and inductive signal

In the early 1930s, investigators noticed that vitamin A deficiency during pregnancy in animals led to a variety of fetal malformations. The most severe abnormalities affected the developing brain, which was often grossly malformed. At about the same time, experimental studies yielded the surprising finding that *excess* vitamin A caused similar defects. These observations suggested that a family of retinoid compounds—metabolic precursors or derivatives of vitamin A—are teratogenic. (*Teratogenesis* is the term for birth defects induced by exogenous agents.) The retinoids include the alcohol, aldehyde, and acid forms of vitamin A (retinol, retinal, and retinoic acid, respectively). The reasons for the adverse effects of retinoids on fetal development, however, remained obscure well into the late twentieth century.

The disastrous consequences of exposure to exogenous retinoids during human pregnancy were underscored in the early 1980s when the drug Accutane (the trade name for isotretinoin, or 13-*cis*-retinoic acid) was introduced

to treat severe acne. Women who used this drug during pregnancy had an increased number of spontaneous abortions and children born with a range of birth defects.

An important insight into the teratogenic potential of retinoids came when embryologists working on limb development in chicks found that RA mimics the inductive ability of tissues in the limb bud. Still, the mystery remained as to just what RA (or its absence) was doing to influence or compromise development. An essential clue came in the mid-1980s, when the receptors for RA were discovered. These receptors are members of the steroid and thyroid hormone receptor superfamily. When they bind RA or similar ligands, the receptors act as transcription factors to activate specific genes. Careful biochemical analysis showed that embryonic tissues synthesized RA, and subsequent studies showed that RA activates gene expression at several sites in the embryo, including the developing brain (Figures A and B). Among the most important targets for RA regulation are genes for other inductive signals, including Sonic hedgehog (see next section). Thus, an excess or deficiency of RA can disrupt typical development by

eliciting inappropriate patterns of retinoid-induced gene expression.

The role of RA as both a teratogen and an endogenous signaling molecule implies that the retinoids cause birth defects by mimicking or interfering with the typical signals that influence gene expression. The story provides a good example of how teratogenic, clinical, cellular, and molecular observations can be combined to explain seemingly bizarre developmental pathology.

Triple jeopardy: Diseases associated with Sonic hedgehog

Many molecules that are essential for early nervous system patterning and morphogenesis—such as the signaling molecule Sonic hedgehog, or SHH—have odd names and, at first glance, even more arcane functions. Nevertheless, mutations in the human genes for SHH and related signaling proteins are associated with at least three serious disorders. *Holoprosencephaly* disrupts the initial morphogenesis of two distinct cerebral hemispheres; *medulloblastoma* is the result of cancerous transformation of cerebellar granule neuron precursor cells; and

(Continued)

■ Clinical Applications (continued)

basal cell carcinoma is the most prevalent cancer of the skin, commonly seen in fair-skinned adults of middle age or older. While the three disorders affect distinct cellular targets (neural plate cells, cerebellar precursors, and basal epidermal cells, respectively), each of them can be caused by mutations in the *SHH* gene or in genes that encode the related receptors and signaling proteins Patched (PTC) and Smoothened (SMO).*

---

*Note that, per convention, gene names are italicized, while the names for the corresponding proteins are in standard type. Abbreviations for human genes and proteins are written in capital letters; when referring to other mammals, only the first letter of the abbreviation is capitalized.

From R. M. Anchan et al., 1997. *J Comp Neurol* 379: 171–184

From Linney and A.-S. LaMantia, 1994. *Advance Dev Dev Biol* 3: 73–114

(A) At left, retinoic acid activates gene expression in a subset of cells in the normal developing forebrain of a mid-gestation mouse embryo (blue areas indicate β-galactosidase reaction product, an indicator of gene expression in this experiment). At right, after maternal ingestion of a small quantity of retinoic acid (0.00025 mg/g of maternal weight), gene expression is ectopically activated throughout the forebrain. (B) At left, the brain of a normal mouse at term; at right, the grossly abnormal brain of a mouse whose mother ingested this same amount of retinoic acid at mid-gestation.

Holoprosencephaly

Holoprosencephaly is the most common malformation of the mammalian forebrain. Due to the varying severity, the incidence of holoprosencephaly across all human conceptuses is much higher—between 1/30 and 1/400 depending upon the gestational stage—than that seen for live births, which is estimated to be between 1/10,000 and 1/16,000. In a holoprosencephalic brain, the typical separation of the two hemispheres of the forebrain fails completely or partially. This collapse of the midline of the forebrain (Figure C) can secondarily disrupt development of midline facial structures; especially common is the failure of the eye primordium to split into two bilaterally symmetrical fields and the subsequent development of a single eye (cyclopia). Holoprosencephaly has a broad range of phenotypes, from mildly affected individuals (the 1/10,000 to 1/16,000 affected individuals who survive to birth) to embryonic lethality accompanied by early spontaneous abortion or stillbirth across the course of gestation (accounting for the greater prenatal incidence).

A small but consistent proportion of holoprosencephaly cases are associated with deletions or missense mutations in *SHH* on chromosome 7. Most of these mutations are sporadic (i.e., they arise spontaneously rather than being genetically inherited from a parent). Many cases reflect either microdeletions in the chromosomal region that includes *SHH*, or point mutations that cause missense transcripts of the gene, thus disrupting protein structure and function. Significant support for the association of the *SHH* genomic lesion with holoprosencephaly in humans comes from studies of mice in which the *Shh*

gene has been inactivated ("knocked out") and from zebrafish, in which several mutant alleles of *Shh* have been identified. In each instance, loss of Shh protein function results in the failure of forebrain hemisphere formation and in midline facial defects, including the cyclopean eye that characterizes the most severe forms of holoprosencephaly. Thus, one of the initial obligate functions of Sonic hedgehog—whose absence is not easily compensated for when the gene is inactivated—appears to be guiding the formation of the two cerebral hemispheres as well as bilaterally symmetrical facial structures.

Medulloblastoma

Even though it is the most common childhood brain tumor, medulloblastoma is fortunately rare, with an estimated frequency between 1 in 50,000 and 1 in 100,000 births. There is a 60% survival rate; however, surviving children are seriously compromised by the surgical and medical treatment necessary to prevent the growth of the tumor.

The pathogenesis of medulloblastomas subverts typical neurogenesis and cell migration in the cerebellum. Typically, a large number of cerebellar granule neurons are generated by precursors that migrate to the outside surface of the developing cerebellum, generating postmitotic neuroblasts that then migrate back past the Purkinje cells to their adult location (Figure D). Purkinje cells typically produce Sonic hedgehog, which acts as a mitogen to drive granule cell precursor division. This basic mechanism of granule cell genesis, along with the molecular pathology of medulloblastoma, led to the hypothesis that this devastating childhood tumor reflects altered SHH signaling. Most medulloblastoma cells have elevated levels of Gli1 (an oncogene product, so named because it is found at elevated levels in various *gliomas*, or glial cell tumors), and Gli1 levels are regulated by SHH signaling. Usually, binding of SHH to its receptor PTC (with the help of another SHH-regulated protein called SUFU) maintains low Gli1

■ Clinical Applications *(continued)*

levels, thus preventing the transcription of Gli1-regulated genes that drive cell proliferation. At least 9% of patients with medulloblastoma have loss-of-function mutations in *PTC*, and an additional 9% have mutations in *SUFU*. The likely contribution of both these genes to the pathogenesis of medulloblastomas was confirmed by studies in mutant mice in which either the *Ptc* or *Sufu* gene was inactivated. These mutations, particularly when accompanied by mutations in other major tumor-suppressor genes, result in mice with medulloblastoma with varying frequency, depending on the mutation.

Basal cell carcinomas

The final Sonic hedgehog–associated disorder, basal cell carcinoma of the skin, is by far the most common malady associated with this multifaceted developmental signaling pathway. In the United States alone there are at least 750,000 new basal cell carcinomas each year. A subset of basal cell carcinomas (which, because they are local neoplasms, are studied for somatic mutations in the tumor cells themselves rather than for heritable mutations in the individual) have mutations in either *SHH* (rare), *SMO* (rare), or *PTC* (very common). Once again, the use of mouse mutants confirmed the likely contribution of these genes to basal cell carcinoma pathogenesis. Moreover, basal carcinoma cells, when cultured, are responsive to manipulation of Shh signaling. Finally, and perhaps most intriguing, a rare autosomal dominant syndrome called nevoid basal carcinoma, or Gorlin syndrome, is caused by loss-of-function mutations in the *PTC* gene. Aside from having a high incidence of basal cell carcinoma, individuals with Gorlin syndrome also have a significantly increased incidence of medulloblastoma.

Taken together, these observations indicate the central contributions of seemingly obscure developmental signaling pathways to several disorders that, at first glance, might seem

completely unrelated. The contrast between the morphogenetic effects of excess or diminished RA signaling, via maternal diet or pharmacological exposure, or loss of SHH function in holoprosencephaly as well as disregulation of cell proliferation or differentiation (medulloblastoma and basal cell carcinoma), indicates how different tissue contexts can result in very different functions for the same molecules—and in very different pathologies when those functions are disrupted. The association of signaling with these four neurodevelopmental

and oncogenic disorders has led to new preventive and therapeutic approaches. For RA, there is now vigilant surveillance of prescriptions for RA-based medicines for acne treatment provided to women of child-bearing age (birth control prescriptions are provided at the same time). For SHH, small-molecule inhibitors of SMO, which typically promotes Gli1 stability and nuclear translocation unless inhibited by PTC in the absence of SHH, are currently in development. These inhibitors show some promise as therapeutic agents.

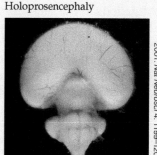

(C) Normal / Holoprosencephaly

(C) A typical late-gestation human brain, and a brain from a fetus with holoprosencephaly.

From E. S. Monuki and C. A. Walsh, 2001. *Nat Neurosci* 4: 1199–1206

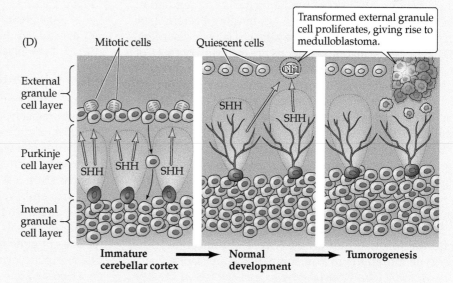

(D) The typical histogenesis of the cerebellum involves the migration of granule cell precursors (blue) to the external aspect of the cerebellum. The precursors divide in this location, and their postmitotic progeny cells migrate back into the cerebellum, where they differentiate into mature granule cells in the internal granule cell layer. In medulloblastoma, a subset of precursors transforms and divides uncontrollably (gray cells) because of a lack of SHH-mediated regulation of the Gli1 transcription factor.

receptor, Patched. Such metabolic disruptions, as well as rare mutations in the human *SHH* gene, are associated with a small proportion of the recorded cases of holoprosencephaly. Mutations in the genes for Shh and other proteins in the Shh pathway—particularly Patched—are also associated with the most prevalent childhood brain tumor, *medulloblastoma* (see Clinical Applications).

The X chromosome has emerged as a "hot spot" for single-gene mutations that result in significant disruption of brain development, in differing degrees for males and females. Males are effectively homozygous for X chromosome gene mutations since they have only one copy of the X chromosome (from their mothers). Thus if an X chromosome gene is lost in a male, that gene is completely absent in that individual. In contrast, X inactivation limits the amount of expression from the two copies of X chromosome genes in females. Thus, females are "mosaics" for these mutations: Some of their cells express only the mutant copy of the gene, while others express the normal copy. This makes females less susceptible to mutations on the X chromosome since a substantial number of cells in any organ will have the appropriate "dose" of X chromosome gene products, and presumably can compensate for the loss of those genes in adjacent cells where only the mutant allele is available. Fragile-X syndrome, the most common form of congenital mental retardation in males, is associated with triplet repeats in a subset of genes on the X chromosome, particularly the gene for fragile-X protein, which is involved in stabilizing dendritic processes and synapses. Most females who carry these triplet repeats heterozygously are either not phenotypic or only midly phenotypic. Mutations in additional single genes with distinct functions including regulation of DNA methylation (Methly CpG-binding Protein 2, or MECP2) are associated with rare X-linked genetic disorders. MECP2 loss-of-function mutations are the primary cause of an "autistic-like" disorder, Rett syndrome. This syndrome is seen nearly exclusively in girls, who are mosaic for a mutant (on one X chromosome) as well as functional MECP2 allele (on the other) because of X inactivation. Males with a mutant MECP2 gene do not survive through birth. Girls with Rett syndrome begin life similar to other typically developing children but then around age 2 to 3 begin to regress, losing language and cognitive function, and developing motor disabilities that are devastating for further intellectual and social behaviors.

Some disorders that compromise the nervous system reflect autosomal (not on the X or Y chromosome) single-gene mutations in homeobox-like transcription factors. *Aniridia* (characterized by loss of the iris in the eye and mild mental retardation) and *Waardenburg's syndrome* (characterized by craniofacial abnormalities, spina bifida, and hearing loss) are the result of homozygous or hemizygous (two different mutations that alter function; one maternal, the other paternal) mutations in the *PAX6* and *PAX3* genes, respectively, both of which encode hox-related transcription factors. Many additional rare single-gene mutations have been identified and correlated with behavioral deficits that also are seen in broader categories of neurodevelopmental disorders such as intellectual disability or autism. In addition, aneuploid disorders, which are defined by the deletion or duplication of a small region or large segment of an entire chromosome, are associated with neurodevelopmental disorders. Perhaps the best-known example of this class of disorders is *Down syndrome*, or *trisomy 21*, which is caused by the duplication of part or all of chromosome 21, most often due to failure of meiosis during the final stages of oogenesis. This duplication results in three copies of all the genes on chromosome 21, and parallel increases in the proteins encoded by these genes. Although the connections between aberrant gene dosage and the resulting anomalies of neural induction, patterning, and neurogenesis are not yet understood, such correlations provide a starting point for exploring the molecular pathogenesis of many congenital disorders of the nervous system.

CONCEPT 22.4 | Neurogenesis Is an Irreversible Termination of the Cell Cycle That Constrains Neuron Identity

LEARNING OBJECTIVES

22.4.1 Describe the different mechanisms that lead to the generation of postmitotic neurons versus glia.

22.4.2 Discuss how the time of final cell division influences the identity and fate of postmitotic neurons.

22.4.3 Identify the extrinsic and intrinsic factors that determine the fate of a newly generated neurons.

Initial differentiation of neurons and glia

The numbers of neurons versus glia in the human brain are not precisely known; estimates place both numbers at 86 *billion* or more of each class. Despite the uncertainty of numbers—one way or another there are a lot of cells—all of these cells must be generated over the course of a few months from a small population of neural stem cells in the early embryo. Neurogenesis begins after the initial patterning of the neural tube is complete. At this time, precursor cells in various brain regions have distinct signatures of gene expression that assign basic identities. These precursor cells are located in the neuroectoderm ventricular zone: the innermost layer of neuroectodermal cells surrounding the lumen of the neural tube, and a region of extraordinary proliferative activity during gestation. Except for a few specialized cases (see Chapter 26), the entire neuronal complement of the adult brain is produced during a time window that closes before birth;

thereafter, precursor cells mostly disappear, and in most brain regions, few if any new neurons can be added to replace those lost by age or injury. This is not the case for glial cells, which depending on the subtype—astrocytes, oligodendrocytes (or peripheral Schwann cells), and microglia—have varying degrees of turnover and replacement throughout life (see also Chapter 26).

Dividing neural precursor cells in the ventricular zone undergo a stereotyped pattern of cell movements as they progress through the cell cycle, leading to the formation of either new stem or precursor cells or postmitotic **neuroblasts** (immature nerve cells) that differentiate into neurons (Figure 22.13A). These movements, as well as the **polarity** of the precursors (thus molecular distinctions between the apical and basal domains; see also Chapters 1 and 23), are essential for regulating their proliferation and decisions to generate postmitotic neurons. The **apical domain** of early neural precursors faces the lumen of the neural tube (or the ventricles as development moves forward) and is where the **primary cilium** of each neural stem cell is found (Figure 22.13B). The primary cilium samples signals available from the amniotic fluid or cerebrospinal fluid that bathes the apical or ventricular surface and is essential for transducing these signals into instructions for progenitor proliferation or terminal neurogenic division. The extent and position of neural precursor **basal processes** are thought to influence modes of cell division as well as cell fate. Thus, the regulation of cell division based on local cell movements within the neuroepitheliym and neural precursor polarity is a key determinant of the fate of any cell in the developing nervous system.

New neural stem cells arise primarily from *symmetrical* divisions of neural stem cells in the developing neural tube (Figure 22.13C). These can renew themselves indefinitely. Perhaps counterintuitively, a substantial number of neural stem cells acquire and retain many of the molecular characteristics of glial cells. Thus, in the developing brain some multipotent neural precursors are called **radial glial cells** that also act as a substrate for migration of postmitotic neurons in the cerebral cortex (see Concept 22.5). These neural precursors have apical domains directly adjacent to the neural tube lumen of cerebral ventricles at later stages, and their basal processes extend to the outer surface of the neuroectoderm initially, or the pial surface of the developing brain and spinal cord as development proceeds. Postmitotic neurons, in contrast, are often generated from cells that divide *asymmetrically*: One of the two daughter cells will become a postmitotic neuroblast while the other reenters the cell cycle to give rise to yet another postmitotic progeny via asymmetrical division (see Figure 22.13A). Such asymmetrically dividing progenitors are molecularly distinct from more slowly dividing radial glial stem cells. They tend to divide more rapidly, but they have a limited capacity for division over time. In addition,

their apical domains are usually no longer in contact with the lumen or ventricle, and they can have basal processes of varying length that do not always reach the outer or pial surface of the developing brain. These neural precursors are sometimes called *transit amplifying cells* because they are a *transitional* form between stem cells and differentiated neurons, and they account for the *amplification* in numbers of differentiated cells due to their rapid mitotic kinetics and serial asymmetrical divisions. They are also referred to as *basal progenitors*, based on the retention of a basal process and their locations away from the ventricular or apical surface of the neuroepithelium.

The mode of cell division: symmetrical or asymmetrical; self-renewing (more progenitors) or neurogenic (one or both "daughter cells" are neurons); is an essential determinant of their future capacity of neural precursor cells to either generate more cells or yield the neurons that will constitute mature brain regions and circuits. Accordingly, there must be extensive molecular regulation of the proliferative and differentiation capacities of neural stem cells and newly generated neurons in the developing nervous system. Interactions between a family of transmembrane cell surface ligands, **Delta ligands**, and their **Notch cell surface receptors** are key regulators of neural stem cell decisions to generate either additional stem cells or postmitotic neurons (Figure 22.14A). These neurogenic decisions are made primarily on the basis of influences from immediately neighboring cells. Thus, signaling via Delta ligands, which occurs due to binding to Notch receptors, happens only between cells that are next to one another. Delta binding to Notch via apposition of the cell membranes of two neighboring cells leads to the cleavage of the intracellular domain of the receptor, liberating a protein fragment of the Notch receptor (the *Notch intracellular domain*, or *NICD*) into the cytoplasm, from whence it is transported into the nucleus. Once inside the nucleus, NICD binds to a transcriptional complex including the *recombining binding protein J (RBP-J)*, which is typically repressive. Binding the NICD, however, reverses RBP-J repression and results in the transcription of several genes, including a family of transcription factors genes called the *Hes* genes (named for their *Drosophila* counterparts, *hairless* and *enhancer-of-split*), which in turn influence the expression of transcription factors involved in the terminal differentiation of neural cells. Aside from the *Hes* genes themselves, the most important additional homologues of these factors in the vertebrate genome are referred to as the neurogenic **bHLH (basic helix-loop-helix)** transcription factors. These include a family of transcription factors called the neurogenins. Local Delta-Notch signaling among neighboring cells leads to downregulation of Delta in several cells (thus diminishing signaling capacity) as well as its upregulation in one or a few of the neighboring cells. In

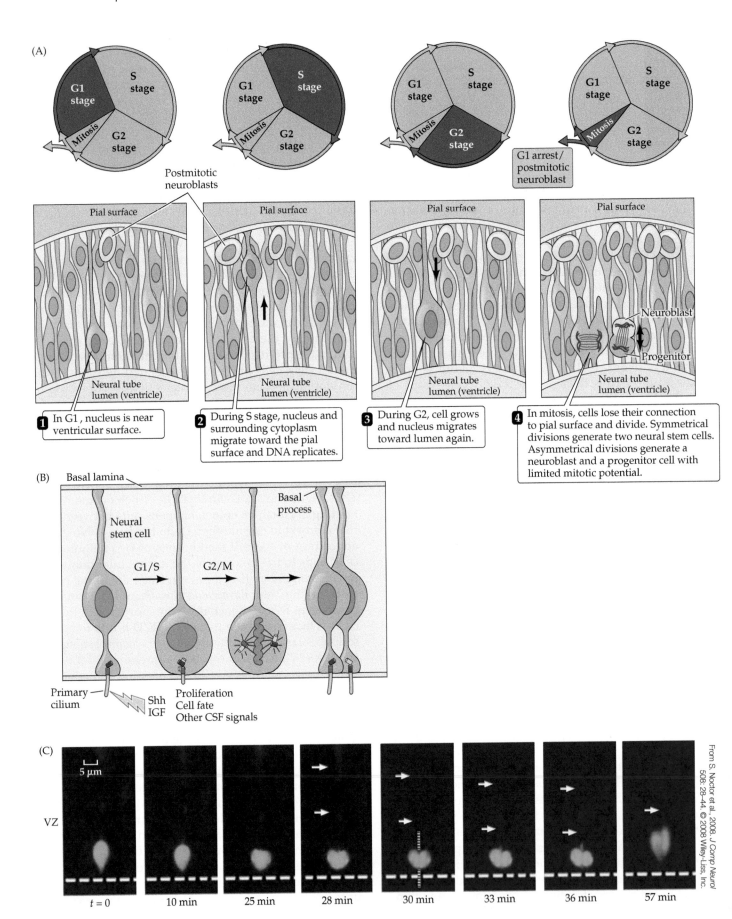

(A)

Postmitotic neuroblasts

Pial surface · Pial surface · Pial surface · Pial surface

G1 arrest/ postmitotic neuroblast

Neural tube lumen (ventricle)

Neuroblast

Progenitor

1 In G1, nucleus is near ventricular surface.

2 During S stage, nucleus and surrounding cytoplasm migrate toward the pial surface and DNA replicates.

3 During G2, cell grows and nucleus migrates toward lumen again.

4 In mitosis, cells lose their connection to pial surface and divide. Symmetrical divisions generate two neural stem cells. Asymmetrical divisions generate a neuroblast and a progenitor cell with limited mitotic potential.

(B) Basal lamina

Basal process

Neural stem cell

G1/S · G2/M

Primary cilium · Shh IGF · Proliferation Cell fate Other CSF signals

(C)

5 μm

VZ

$t = 0$ · 10 min · 25 min · 28 min · 30 min · 33 min · 36 min · 57 min

From S. Noctor et al., 2008. *J Comp Neurol* 508: 28–44. © 2008 Wiley-Liss, Inc.

◀ **FIGURE 22.13   Neural precursor cells undergo mitosis in the ventricular zone** (A) Precursor cells in the vertebrate neuroepithelium are attached both to the pial (outside) surface of the neural tube and to its ventricular (lumenal) surface. The nucleus of the cell translocates between these limits within a narrow cylinder of cytoplasm (the ventricular zone, VZ). When cells are closest to the outer surface of the tube, they enter the DNA synthesis phase (S stage) of the cell cycle. Once the nucleus moves back to the ventricular surface (G2 stage), the precursor cells lose their connection to the outer surface and enter mitosis. When mitosis is complete, the two daughter cells extend processes back to the outer surface of the neural tube, and the new precursor cells enter a resting (G1) phase of the cell cycle. At some point a precursor cell generates either another progenitor cell that will go on dividing and a daughter cell—a neuroblast—that will not divide further, or two postmitotic daughter cells. (B) Schematic of the apical (at the lumen of the neural tube) versus basal (at the outer surface of the neuroepithelium) polarity of neural stem cells in the developing neural tube. The apical domain is the site of the primary cilium, a motile organelle that is immersed in and samples soluble signals from the amniotic fluid (open neural tube) or cerebrospinal fluid (CSF; closed neural tube) including sonic hedgehog (SHH) and the insulin like growth factor (IGF). When the neural stem cell divides symmetrically, with the plane of division orthogonal to the apical surface, a new primary cilium is generated for the daughter cell. During symmetrical division that results in two self-renewing neural stem cells, the basal process is maintained and continues to interact with the basal lamina on the outer surface of the neuroepithelium. (C) Time-lapse microscopy permits visualization of symmetrical, vertically oriented division (red line) of a single radial glial stem cell in the cortex. The cell body is seen at the ventricular surface (dashed line); the arrows indicate the radially oriented process of the cell, which is mostly out of the focal plane necessary for visualizing the cell body. The radial processes are retained once the cell has divided. (B after S. Thomas et al., 2019. *Biol Cell* 111: 217–223.)

(A)

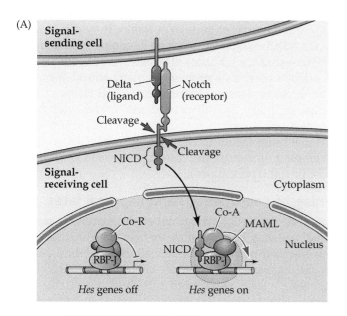

FIGURE 22.14   Delta-Notch signaling leads to neuronal differentiation Interaction between the Delta cell surface ligands and their Notch receptors on neural progenitor cells in close proximity to one another regulates transcription factors necessary for the generation of differentiated neurons. (A) Delta binding cleaves a protein fragment of the receptor (the Notch intracellular domain, or NICD). When the NICD is transported into the nucleus, it binds to RBP-J, a repressive transcription factor, inhibiting RBP-J transcriptional repression mediated via RPB-J binding with co-repressor proteins (Co-R). This results in association with co-activator proteins (Co-A) as well as the Notch-associated mastermind-like co-activator (MAML). This NICD-dependent signaling event leads to the transcription of (among others) the *Hes* genes and additional bHLH neurogenic factors responsible for neuronal differentiation. (B) Delta, Notch, and bHLH proteins are expressed at similar levels in a cluster of progenitor cells and neuroblasts. A stochastic increase in Delta ligands on a particular cell leads to downregulation of Delta in neighboring cells, while in the Delta-upregulated cell, bHLH gene expression is also upregulated, and the cell becomes primed for neuronal differentiation.

(B)

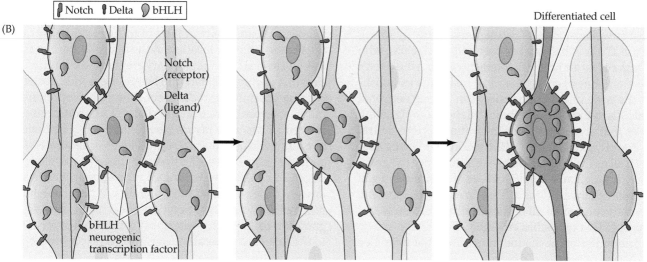

Delta-upregulated cells, bHLH gene expression is also upregulated, and the cell becomes primed for neuronal differentiation (Figure 22.14B). Cells whose Delta levels have been diminished remain as neural stem cells. Similar mechanisms, regulated by Notch signaling and bHLH neurogenic transcription factors, are thought to influence the generation of oligodendrocytes and astrocytes. Thus, the ultimate decision to exit the cell cycle and embark on terminal neurogenesis or gliogenesis appears to be regulated by local, cell contact–mediated interactions that depend initially on Delta-Notch signaling.

Generation of neuronal diversity

Neural precursor cells that look and act more or less the same give rise to postmitotic cells that are enormously diverse. On the most basic level, these precursors produce neurons and glial cells, two cell types with markedly different properties and functions. Neural stem cells then differentiate to produce the diverse neurons and glia of each of the brain regions specified during early morphogenetic events that guide initial steps of brain development. Thus, these molecular distinctions early in development profoundly influence morphology, neurotransmitter synthesis, cell surface molecules, and the types of synapses made or received. Together, variation in all of these characteristics distinguishes neuron classes of the spinal cord, brainstem, cerebellum, cerebral cortex, hippocampus, and subcortical nuclei, including the basal ganglia and thalamus.

The derivation of a neuron (the identity of the precursors that generate it, referred to as its lineage) or its history of interactions (sometimes generally called extrinsic factors) clearly influence neurogenesis and the subsequent acquisition of neuronal identities. Lineage does constrain the final differentiation of neurons and glial. Nevertheless, the bulk of evidence favors the view that essential aspects of neuronal differentiation are regulated primarily by local cell-cell interactions followed by distinct histories of transcriptional regulation via a "code" of transcription factors expressed in each cell following its terminal division, specified by diffusible as well as local cell-cell signals (Figure 22.15; see also Figure 22.12). A balance between cell lineage (i.e., who a cell's "parent" is) and cell-cell interactions (i.e., who a cell's neighbors are) has been invoked to explain the establishment of the wide range of diversity seen among neuronal and glial cells in the vertebrate brain. Many of the signaling molecules that are essential for the initial steps of neural induction and regionalization—BMPs, Sonic hedgehog, and Wnts, as well as Delta and Notch—also influence the genesis and differentiation of specific classes of neurons and glia via local cell-cell interactions at later times in development.

Among the targets of all of these pathways, a subset of neurogenic bHLH genes has emerged as central to subsequent differentiation of distinct neural or glial fates (see

Figure 22.15). There are multiple bHLH genes, and their restricted expression in distinct rudimentary brain regions exerts a powerful influence on cell identity in regions where they are found. Some of these bHLH genes (e.g., *Mash1*, *Mammalian achaete scute homolog 1*, now known as *Ascl1* for *Achaete Scute-like 1*) are homologues of genes originally discovered in developing fruit flies, while others have been identified based on their predicted amino acid sequences inferred from genomic sequences. These transcriptional regulators are used in diverse neuronal classes, presumably in a context-dependent fashion, to drive acquisition of distinct structural and functional characteristics.

These molecular details of cell signaling and subsequent lineage relationships provide an outline of how general cell classes are established. Nevertheless, there is presently no clear and complete explanation for the means by which any specific neuronal class achieves final mature identity. Neuronal identity must ultimately be defined not only by a history of signaling and transcriptional regulation but by current gene expression. In addition, each neuron acquires identity based on its connections and projections—thus the circuitry of which it is part. This gap in knowledge presents a problem in using neural stem cells to generate replacements for specific cell classes lost to disease or injury, as well as in efforts to understand how neural precursors can transform into the tumorogenic cells that cause medulloblastomas and other cancers of the developing or mature nervous system (see Concept 22.3 and Clinical Applications).

Timing of neurogenesis and neuronal fates

Different populations of spinal cord neurons, as well as retinal neurons, brainstem motor neurons, thalamic relay neurons, and projection neurons in the cerebral cortex, are distinguished by the time when they undergo their final cell division, referred to as their *birth date* (this does not mean, however, that developmental neuroscientists are "neuroastrologers"). Some of these distinctions are influenced by local differences in the signaling molecules and transcription factors that act on neural stem cells based on their location in the neural tube. Nevertheless, in several brain regions where there are multiple cell classes, often segregated into layers, the cells of each layer become postmitotic at different times. The vertebrate retina has several functionally distinct neuron classes arranged in discreet layers (see Figure 9.5). The majority of each cell class in each retinal layer is generated—becomes postmitotic—at different times in development (Figure 22.16A). The timing of retinal neurogenesis follows an order that is approximately, but not absolutely, in register with the final position of each cell class so that ganglion cells (in the outermost retinal cell layer closest to the lens) are generated first, and rods (photoreceptors

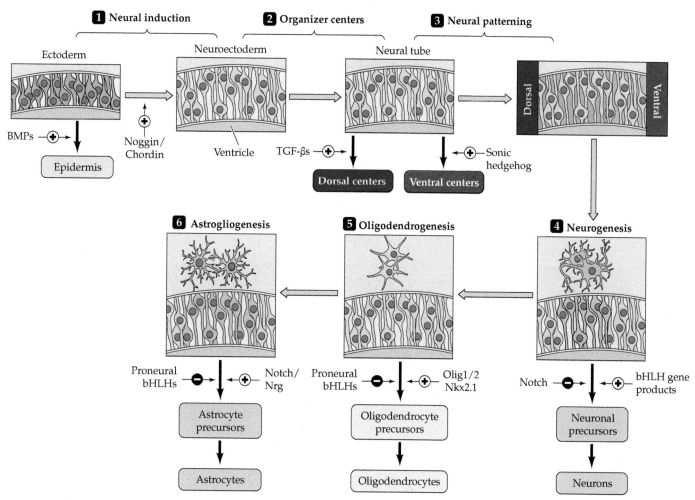

FIGURE 22.15   Molecular and cellular mechanisms that guide neuronal and glial differentiation (1–3) The steps by which ectoderm acquires neural identity. Neural precursors, or stem cells, are specified by a balance of BMP signaling (which favors epidermal fate for ectoderm) and signaling by endogenous antagonists (e.g., Noggin and Chordin) that leads to neural stem cell identities. Next, local sources of inductive signals, including TFG-β family members and Sonic hedgehog, establish gradients that influence subsequent neural precursor identities, as well as identifying local "organizers" (such as the floorplate and roofplate). (4–6) The specification of oligodendrocytes and astrocytes from multipotent neuronal precursors relies on local signals and the downstream transcription factors they engage. The balance of Notch signaling (see Figure 22.14) and transcriptional control by bHLH proneural transcription factors biases neural stem cells toward a differentiated neural fate. Similarly, antagonistic transcriptional regulation via either bHLH genes or three additional transcription factors—Olig1, Olig2, and Nkx2.1—influences the generation of oligodendrocytes. Continued antagonism between bHLH proteins, Notch signaling proteins, and the signaling molecule neuregulin (Nrg) is thought to influence the generation of mature astrocytes. Finally, in the adult brain, cells adjacent to the ventricles (which apparently have avoided becoming differentiated) remain as ependymal cells. These may include a subpopulation of neural stem cells (see Chapter 26). (After C. Kintner, 2002. *J Neurosci* 22: 639–643.)

in the innermost retinal cell layer, furthest from the lens) are generated last. This timing, the expression of transcription factors and downstream targets that accompany the process, and the subsequent timing of forming connections between cells in each retinal layer indicates that the schedule of neurogenesis can have a profound effect on subsequent circuit differentiation.

In the cerebral cortex, most neurons of the six cortical layers are generated in an inside-out manner; each layer consists of a cohort of cells "born"—and which therefore undergo their terminal cell divisions—at a distinct time. The firstborn cells are eventually located in the deepest layers; later generations of neurons migrate radially from the site of their final division in the ventricular zone, traveling through the older cells and coming to lie superficial to them (Figure 22.16B). These differences in time of cell origin are matched by differences in gene expression in distinct cohorts of cells. In the cortex, early-born neurons

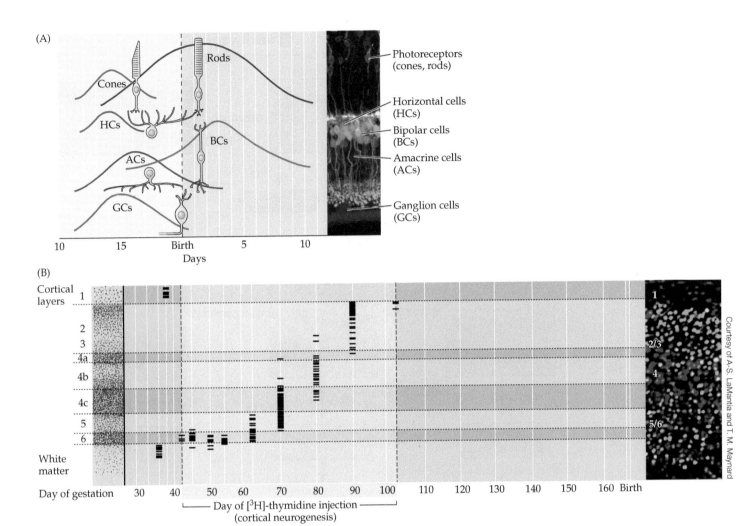

FIGURE 22.16 Relationship between timing of neurogenesis and post-mitotic neuron position in the developing CNS
(A) In the retina (micrograph at left), ganglion cells (GCs; bright red, are the projection neurons found in the outermost cell layer, closest to the lens). Amacrine (ACs; darker red), bipolar (BCs; green) and horizontal cells (HCs; orange) are all found in the middle layers of the retina. The photoreceptors (rods and cones; purple) are found in the innermost layer of the retina. These laminated cell classes are generated during staggered periods of neurogenesis (right). The peak rate of neurogenesis for each of these cell classes varies: the peak generation of ganglion cells, horizontal cells and cones is earliest; the amacrine cells follow, and the rods and bipolar cells reach peak neurogenesis at the latest time. (B) Generation of cortical neurons. The graph (left) covers a span of about 165 days during the gestation of a rhesus monkey. The final cell divisions of the neuronal precursors, determined by maximal incorporation of radioactive thymidine administered to the pregnant mother, occur primarily during the first half of pregnancy and are complete on or

about embryonic day 105. Each short horizontal line represents the position of a neuron heavily labeled by maternal injection of radiolabeled thymidine at the time indicated by the corresponding vertical line. The numerals on the left designate the cortical layers. The earliest-generated cells are found in a transient layer called the subplate (a few of these cells survive in the white matter) and in cortical layer 1 (the Cajal–Retzius cells). The image (right) shows molecular distinctions between newly generated neurons in the fetal mouse cerebral cortex. The green label is for the transcription factor Satb2 which is enhanced in layers 2/3 cortical projection neurons, and a subset of layers 5/6 neurons. The red label is for another transcription factor, Ctip2, that labels projection neurons in layers 4, 5 and 6. (A from N. Tian, 2012. Development of Retinal Ganglion Cell Dendritic Structure and Synaptic Connections. In Kolb, H., Fernandez, E., Nelson, R., [Eds.] *Webvision: The Organization of the Retina and Visual System.* Salt Lake City: University of Utah Health Sciences Center; B after P. Rakic, 1974. *Science* 183: 425–427.)

in lower cortical layers express transcription factors distinct from those of the later-born neurons in the upper cortical layers. Indeed, in most regions of the CNS where neurons are arranged into layered or laminar structures

(the hippocampus, cerebellum, superior colliculus, as well as the retina just described), there is a systematic relationship between the layers, time of cell origin, and additional molecular properties, including transcription factor expression. In these brain regions, neuroblasts from the ventricular zone either migrate or are passively displaced radially and outward, thus establishing a systematic relationship between the time of a cell's final division and its laminar position. The implication of this phenomenon is that organized periods of neurogenesis are important for the development of the cell types and their locations, and that these developmental distinctions prefigure or even thus influence the connections that will eventually characterize each distinct brain region.

CONCEPT **22.5**

Nerve Cells Often Migrate from Their Site of Neurogenesis to Their Final Position

LEARNING OBJECTIVES

22.5.1 Describe the basic mechanisms for cell motility and migration.

22.5.2 Define cell migration and discuss its consequences for the development of the peripheral and central nervous systems.

Cell motility and migration in the developing nervous system

Cell migration, a ubiquitous feature of all embryos, brings distinct classes of cells into appropriate spatial relationships within differentiating tissues. In the developing nervous system, migration provides opportunities for diverse cell classes to interact transiently during development, thereby constraining cell-cell signaling to specific times and places, and it ensures the appropriate final position of many postmitotic neurons. The capacity of a neural precursor or immature nerve cell to move and its path through a changing cellular environment are essential for its subsequent differentiation. The final location of a postmitotic neuron is especially critical, because neural function depends on precise connections made by these cells and their targets. Ultimately, a developing neuron must be in the right place at the right time so that it can be properly integrated into a functional circuit that can mediate behavior.

Neuronal migration is a special example of a more general mechanism shared across multiple tissues: **epithelial-to-mesenchymal transition**, or **EMT** (see Concept 22.2 and Figure 22.5), which initiates cell motility within multiple tissues. The converse of EMT, mesenchymal-to-epithelial transition (MET), marks the termination of cell motility and migration. Prior to EMT, polarized epithelial cells, each with a fairly simple cuboidal or columnar shape, have distinctive adhesion contacts mediated by a family of cell adhesion molecules called the cadherins (see Chapter 23). Related proteins that form tight

junctions hold cells together in an epithelial sheet. Additional cell surface proteins, including a family of molecules called integrins, interact with extracellular matrix proteins such as fibronectin and laminin that define the basement membrane or basal lamina that surrounds most epithelial tissues. During EMT, epithelial cells lose their junctional contacts, acquire an irregular "mesenchymal" shape, and become motile: they have membrane protrusions, make limited cell-cell contacts, and organize arrays of actin filaments that mediate force generation for cell movement. These mesenchymal cells either secrete or express proteases on the surfaces of their motile processes that degrade the basement membrane and allow free movement of the cells in a broader extracellular space. The generation of force occurs via additional cell-cell contact-mediated or secreted signaling that modifies the actin cytoskeleton to move the cell in a particular direction (Figure 22.17; see also Chapter 23).

The downregulation of cadherins and related proteins that maintain the epithelial state, and the upregulation of proteins necessary for mesenchymal cell identity and motility, depend on a subset of transcription factors, including Snail (see Figure 22.5B) and ZEB zinc finger transcription factors as well as the Twist bHLH transcription factor. All are upregulated in response to signals that promote EMT in epithelial tissues. These activators include a variety of the signaling molecules identified as inductive factors or neurogenic signals, including members of the TGF-β, Wnt, FGF, and Notch families (see Figure 22.11). These EMT transcription factors—Snail, ZEB, and Twist—bind a broad range of genes with E-box DNA binding elements in upstream regulatory sequences of genes that stabilize or disrupt the epithelial state. In some cases, E-box binding of Snail, Twist, or ZEB represses transcription of epithelial-promoting genes by recruiting additional repressive binding factors; for other downstream targets, binding of Snail, Twist, or ZEB drives transcription of genes necessary for the mesenchymal state and motility. During EMT (or MET when mesenchymal cells coalesce into epithelial tissues), Snail, ZEB, and Twist activity is modulated by phosphorylation (Snail, Twist) or sumoylation (ZEB) of each factor, nuclear retention or transport of each factor, as well as cytoplasmic degration of each factor. Subsets of the factors that initiate EMT and terminate it during MET regulate neuronal migration during initial formation of the peripheral and central nevous systems.

Neuronal migration in the peripheral nervous system

Several fundamental mechanisms that mediate EMT are necessary to initiate neural crest migration. All neural crest cells, regardless of their anterior–posterior origin, start out as neuroepithelial cells, with all of the intercellular

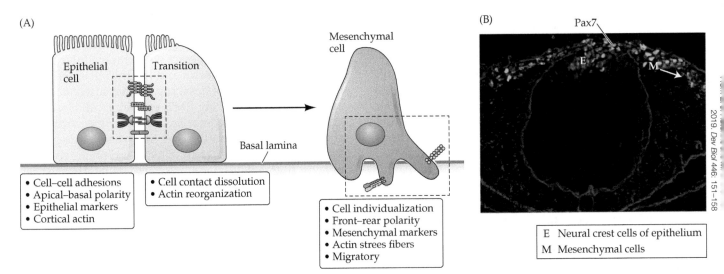

(A)

Epithelial cell

Transition

Mesenchymal cell

Basal lamina

- Cell–cell adhesions
- Apical–basal polarity
- Epithelial markers
- Cortical actin

- Cell contact dissolution
- Actin reorganization

- Cell individualization
- Front–rear polarity
- Mesenchymal markers
- Actin strees fibers
- Migratory

(B)

Pax7

E Neural crest cells of epithelium
M Mesenchymal cells

2019. *Dev Biol* 446: 151–158

FIGURE 22.17 Epithelial-to-mesenchymal transitions (EMTs) and their counterparts, mesenchymal to epithelial transitions (METs) are a key mechanism for neuronal migration (A) In the neuroepithelium of the neural tube, as well as in other epithelia throughout the body, individual cells are held together in the epithelial sheet by molecular junctions made via adhesion molecules and their receptors on the surfaces of adjacent cells (shown for the interface of the two epithelial cells, in the dotted box, left). Each epithelial cell retains a standard polarity across the epithelial sheet. The cellular junctions that maintain the epithelial organization and polarity are disrupted via transcriptional changes in response to local signals. This results in the dissolution of epithelial contacts between cells, the reorganization of the actin cytoskeleton, and a change in the state of the cell that supports motility and migration (shown at right). Mesenchymal cells have on their cell surfaces receptors and motor proteins that facilitate their capacity to interact with the extracellular matrix, other mesenchymal cells and to migrate (dotted box, right) (B) Neural crest cells in the embryonic chicken neural tube (E, labeled green by the neural crest–selective transcription factor Pax7), specified based on their position in the epithelium of the dorsal neural tube, undergo EMT and migrate laterally as mesenchymal cells (M). These migratory neural crest cells, however, still retain the molecular identity as neural crest (indicated by sustained Pax7 expression) established in the neuroepithelium of the neural tube. (A after S. Lamouille et al. 2014. *Mol Cell Biol* 15: 179.)

junctions and adhesive interactions that keep epithelial cells in place. To move away from the neural tube, neural crest cells must downregulate expression of these adhesive genes during a local EMT at the dorsomedial aspect of the neural tube. Presumptive neural crest cells in this region of the dorsal neural tube express several transcription factors, including Snail1 and Snail2 (see Figure 22.5B), that repress expression of intercellular junctional proteins and epithelial adhesion molecules. Neural crest motility depends upon the local generation of force via the actin cytoskeleton to move these cells. The extrinsic signals that mediate motility and attract neural crest cells to appropriate targets are those that serve similar functions for a broad range of cell-movement-related events. Similar signals mediate metastic migration of cancer cells as well as the outgrowth of axons and dendrites once neurons have acquired their final position (see Chapter 23). Indeed, some of the genes that modulate delamination in the neural crest—including the Snail genes and their downstream targets—can function as oncogenes if mutations arise that allow them to be expressed or activated in mature epithelial tissues. When motile neural crest cells reach their final destinations, they cease to express Snail and downstream mediators that induce the mesenchymal, migratory state.

This change is thought to reflect integration of several signals that neural crest cells encounter along their migratory route as well as at their final target. Understanding this normal event has implications for cancer biology, as inducing transformed, migratory cancer cells to revert to a stationary state would have clear therapeutic value.

The initial positional identity of neural crest cells in the neural tube is reflected in their final locations from various anterior–posterior levels in distinct parts of the body (Figure 22.18A,B). The final fates of neural crest cells—including becoming sensory, sympathetic, parasympathetic, and enteric neurons of the peripheral nervous system—are critically dependent on their proper EMT-mediated exit from specific anterior–posterior regions of the epithelium of the neural tube, and on their subsequent migration through terrain that provides instructive as well as tropic and trophic signals. Thus, as neural crest cells begin their journeys, they carry with them information about their point of origin, including expression of distinct Hox genes as well as other transcription factors that are limited to various spinal cord and hindbrain domains (see Figures 22.9 and 22.12). These initial molecular "identities" are crucial for these cells to "read" the terrain in the periphery through which they migrate and reach the appropriate destinations.

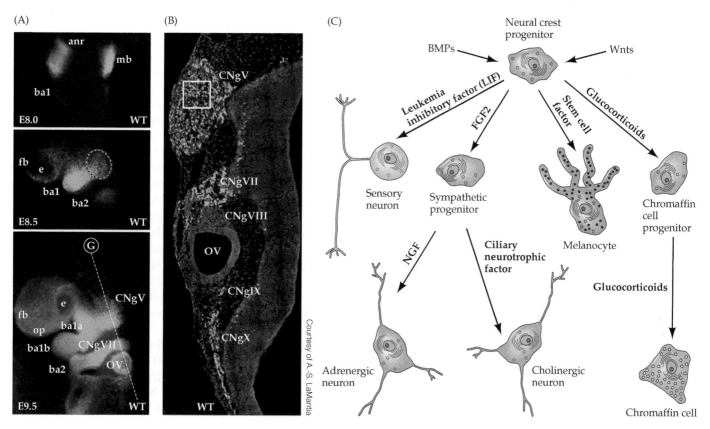

FIGURE 22.18   Derivatives of the neural crest become distinct peripheral structures (A) The cranial neural crest of a early midgestation mouse embryo migrates from the alar regions of the midbrain (top panel) into the mesenchyme between the optic cup and forebrain as well as into the branchial arches (middle panel). Adjacent to branchial arch 1 (dashed circle in middle panel), a population of neural crest cells coalesces adjacent to the trigeminal placode (see Figure 22.6) to begin to form the trigeminal ganglion. Subsequently (bottom panel), neural crest mesenchymal cells can be seen in the developing branchial arches (where they will contribute to cranial bones), around the eye and ear, and in the locations of the cranial ganglia. (B) Section through the hindbrain and periphery of a mid-gestation embryo at the level shown by the dashed line in the bottom panel of (A). The neural crest has been labeled by a green reporter transgene, and the placodal derivatives have been labeled with an antibody against Six1 (red cells), an established placode cell marker. The cranial somatosensory ganglia (V, trigeminal; VII, facial; IX, glossopharyngeal; X, vagal) are a mosaic of neural crest (green) and placode (red) cells. The otic vesicle epithelium (OV) and the associated acoustic ganglion (VIII), however, are completely derived from placodal precursors. (C) An array of signaling molecules available either from the migratory route or in the target destination of neural crest cells influences the differentiation of neural crest stem cells into sensory neurons, autonomic sympathetic or parasympathetic neurons, melanocytes, or adrenal chromaffin cells.

Neural crest cells in the trunk are largely guided along different migratory pathways by signals from non-neural peripheral structures such as somites (which eventually form the axial muscles and skeleton) and other rudimentary musculoskeletal or visceral tissues. Neural crest cells in the head are influenced by signals from the cranial placodes (see Figure 22.6) and the neural tube. The signals along these pathways can be secreted molecules (including some of the same peptide hormones used at earlier times for neural induction), cell surface ligands and receptors (adhesion molecules and other signals), or extracellular matrix molecules; most of these molecules also are used at later stages of development to guide axonal growth and targeting, as described in Chapter 23. Thus, the surfaces of cells in the embryonic periphery have specialized adhesion molecules such as neural cadherin (N-cadherin), neural cell adhesion molecule (NCAM), and Eph receptors and ephrin ligands (see Chapter 23), or these cells secrete distinct extracellular matrix molecules, including isoforms of laminin or fibronectin, to interact with migrating neural crest cells. In addition, signals secreted by targets including neurotrophic molecules (see Chapter 23) may influence neural crest migration. Finally, specific peptide hormone growth factors available in particular peripheral targets cause neural crest cells to differentiate into a broad range of functionally distinct cell classes (Figure 22.18C; see also Figure 22.5C). These cues modulate the expression of bHLH and other neurogenic and neuronal identity genes in neural crest cells during the transition from migratory precursor to postmitotic neuroblast. Thus, the balance of migratory capacity, instructive cues, and modification of gene expression seen during the transit of the neural crest from the neural tube to the periphery illustrates the influence of migration on the establishment of neuronal identity.

Neuronal migration in the CNS

Neuronal migration is not limited to the periphery. Neurons generated in several locations in the CNS must also move from the site of their initial genesis to a distant site, where they differentiate and are integrated into mature neural circuits. The mechanisms of central neuronal migration are diverse, and its successful completion is essential for many aspects of typical brain function.

A minority of nerve and glial cells in the CNS (and a few in the periphery) use recently generated axon pathways as migratory guides. These include subsets of cranial nerve motor neurons in the hindbrain; neurons that constitute nuclei in the pons that project to the cerebellum; and a small population of neurons that migrate during embryonic development from the olfactory epithelium in the nose to the hypothalamus, where they secrete gonadotropin-releasing hormone (GnRH), which is essential for regulating reproductive functions in the mature animal (see Chapters 15 and 25). In addition, some postmitotic neurons in the forebrain originate in the ventral structures called the ganglionic eminences (see Concept 22.2), acquire a mesenchymal migratory state (Figure 22.19), and move through the neuroepithelium to anterior or dorsal forebrain regions, including the olfactory bulb, hippocampus (see Chapter 26) and the cerebral cortex. Most of the migratory cells from the ventral forebrain are presumptive GABAergic interneurons that will populate specific layers of the olfactory bulb, hippocampus, and cerebral cortex.

The most prominent form of postmitotic neuronal migration in the CNS, however, is that guided by glial cells in a number of developing brain regions. Many neurons that migrate long distances in the CNS—particularly those in the cerebellum and cerebral cortex—are guided to their final destinations by glial processes. In the cerebellum, the glia that guide the migration of cerebellar granule cells, the most numerous neuron type in the adult brain, are called **Bergmann glia** (Figure 22.20A). In the mature cerebellum, granule cell bodies are found in a densely packed layer beneath the layer of single Purkinje cells, while granule cell axons extend past the monolayer of Purkinje cells into the molecular layer of the cerebellum where they bifurcate into branches called parallel fibers (see Chapter 20). The location and axonal arborizations of granule cells reflect their history of neurogenesis and migration. Granule cell precursors migrate to the outer surface (beneath the pia) of the cerebellum during very early hindbrain differentiation, where they establish a local neurogenic "zone" called the **external granule cell layer**. These granule cell precursors then divide rapidly, and their postmitotic neuronal progeny recognize the processes of Bergmann glia, which guide their migration into the internal granule cell layer. As granule cell bodies migrate, however, they acquire polarity that allows them to extend their parallel fiber axon branches into the plane parallel to the pial surface. The radially migrating process becomes the primary axon. Thus, as the granule cell body migrates, the migratory pathway supports the generation of its axon as well.

Glial guidance is also key for the migration of cortical projection neurons—the pyramidal neurons in the cortex that extend long axons to other cortical targets (those found in layers 2 and 3) and subcortical targets (those found in layers 5 and 6). Radial glial cells, which are the

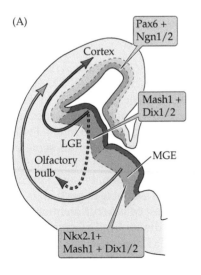

(A)

Pax6 +
Ngn1/2

Cortex

Mash1 +
Dix1/2

LGE

Olfactory bulb

MGE

Nkx2.1+
Mash1 + Dix1/2

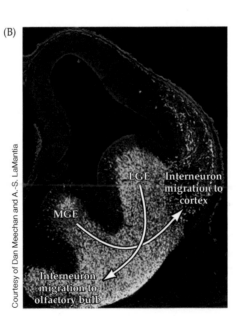

(B)

Courtesy of Dan Meechan and A.-S. LaMantia

LGE   Interneuron migration to cortex

MGE

Interneuron migration to olfactory bulb

FIGURE 22.19 Forebrain interneurons are generated in the basal forebrain and migrate into dorsal structures including the cerebral cortex, hippocampus and olfactory bulb. (A) The GABAergic interneurons of the cerebral cortex, hippocampus, and olfactory bulb are generated in the medial and lateral ganglionic eminences (MGE and LGE) of the developing forebrain. Once they become postmitotic, these interneurons undergo the equivalent of an epithelial-to-mesenchymal transition and migrate into the developing cortex and other distal forebrain targets. (B) A section through the forebrain of a mouse fetus during late gestation showing recently generated GABA-ergic interneurons, labeled with a reporter transgene (green) that identifies them at their origin in the ganglion eminences, migrating into the rudimentary cerebral cortex (red).

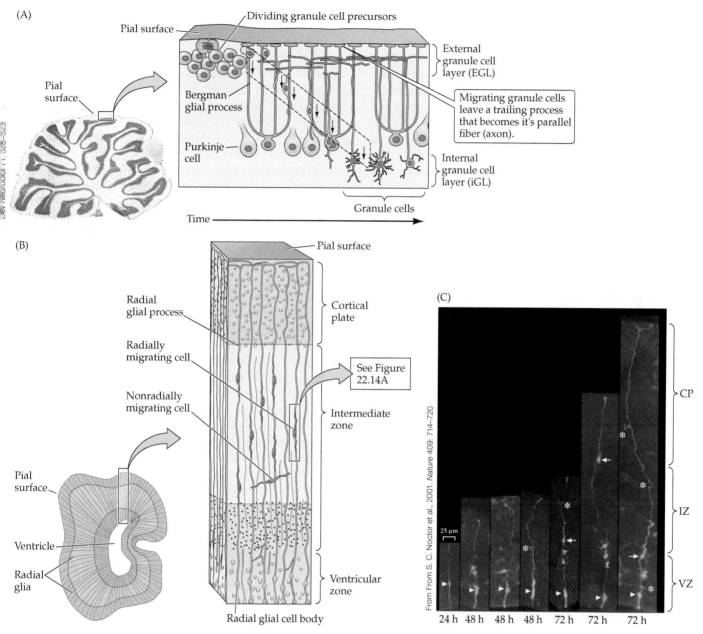

FIGURE 22.20   Migration guided by glial processes in the developing brain (A) Radial migration of postmitotic granule cell neurons (solid blue circles) from the external granule cell layer (EGL) of the developing cerebellum. Granule cell progenitors in the EGL are highly proliferative and generate large numbers of postmitotic granule cells. After becoming post-mitotic in the EGL, a postmitotic granule cell (GC) recognizes a Bergmann glial process and uses it as a substrate for movement downward in a roughly radial trajectory, past the Purkinje cells, toward the internal granule cell layer (IGL) (dashed circle). As the cell body is displaced radially along the Bergmann glial process, its basal (trailing) domain begins to elaborate two processes that extend orthogonally to the direction of the radial migration. These are the nascent parallel fiber branches of the newly generated granule cell. Finally, as radial migration of the cell body continues, its axonal process is generated. Thus, the migratory mechanism is responsible for translocating radially the granule cell body as well as generating the granule cell axon parallel to the pial surface. (B) Section through the developing cerebral cortex (left) showing radial glial processes from the ventricular to the pial surfaces. The magnified three-dimensional reconstruction (right) shows the radially oriented process of the radial glia, which newly generated cortical projection neurons adhere to selectively and use as a guide to migrate to their final position in specific cortical layers. Occasionally, a newly generated projection neuron may take a nonradial migratory route, which can lead to wide dispersion of neurons derived from the same precursor. Additional nonradial migrating neurons in the developing cortex are likely to be interneurons (see Figure 22.19). (C) Time-lapse microscopy of the cerebral cortex from a developing mouse showing radially migrating (arrows) and nonradially migrating (asterisks) neurons. (A after E. E. Govek et al., 2011. *Dev Neurobiol* 71: 528–523; B after W. M. Cowan, 1979. *Sci Am* 241: 124, based on Rakic, 1971.)

primary neural stem cells of the cortex, also act as guides for the inside-out migration of cortical projection neurons (see Figure 22.16B). Histological observations of embryonic brains made by Wilhelm His and Ramón y Cajal during the nineteenth and early twentieth centuries suggested that neuroblasts in the developing cerebral cortical hemispheres followed glial guides to their final locations. These light microscopic observations were supported by analyses of electron microscopic images of fixed tissue in the 1960s and 1970s as well as molecular labeling that identified the radial glial cells and migrating neurons as distinct cell classes (Figure 22.20B). More recent in vivo microscopic observation has confirmed these inferences (Figure 22.20C), and also has confirmed that radial glial cells serve dual functions: They are both stem cells and migration guides. The apparent scaffold for radial movement of postmitotic neurons established by cortical radial glia fits well with the orderly relationship between birth dates and final position of distinct cell types in the cerebral cortex (see Figure 22.16B). By adhering selectively to the glial process, newly generated cortical projection neurons can move past dividing cells in the ventricular and subventricular zones as well as already differentiating neurons in lower cortical layers (layers 5 and 6, for example, referred to during early cortical development as the **cortical plate**) toward the cortical surface, where they contact the end feet of radial glial cells (basal processes; see Figure 22.13) and disengage from the glial surface. As is the case with migrating cerebellar granule cells, migrating cortical neurons also generate a trailing process, in this case oriented toward the ventricular zone, that will then differentiate as the primary axon (see Chapter 23). Thus, glia-guided migration appears to ensure translocation of newly generated neurons past other neurons already in place in the developing nervous system, and may influence the ability of the migratory neurons to make connections once they have reached their final destination.

Molecular mechanisms of neuronal migration and cortical migration disorders

Increased understanding of molecular mechanisms and direct observation of migrating neurons and their glial guides in live developing brains indicate that the process of neuronal migration, particularly in the cortex, may be vulnerable to the effects of genetic mutations that disrupt either the ability of the nerve cell to move, the ability of radial glial cells to support migration, or both. This inference received initial support from the characterization of several single-gene mutants in the mouse that disrupted the orderly placement of neurons based on their time of origin. Subsequent work identified several proteins whose function is vital for normal migration along radial glia. Some of these proteins are found on the surface or in the cytoplasm of the migrating neuron itself; others are found on the glial cell surface (Figure 22.21A). When the function

of any of these proteins is disrupted, neurogenesis, migration, and cortical lamination can be compromised.

A particularly compelling demonstration of the reliance of cortical neurogenesis and migration on specific genes and their protein products has come from analysis of people with a variety of brain malformations that can be visualized using magnetic resonance imaging. Genetic analysis of these individuals has identified mutations in several genes that can disrupt cortical migration, including its impact on the pattern of sulci and gyri that emerge in a typical adult brain (Figure 22.21B). In some cases (such as mutations in the gene that encode the molecule Reelin, which is available near the glial end feet and is thought to influence the detachment of neurons from the radial glia), mutations in humans (Figure 22.21C) result in a cortical phenotype very similar to that seen in mice in which the analogous gene is experimentally deleted. The orderly "inside-out" position of cells in the cortical layers is altered, and in humans the pattern of sulci and gyri is changed substantially. Several genes that influence cortical neuronal migration, including *Lissencephaly 1* (*LIS1*) and *Doublecortin* (*DCX*), have been identified as causing lissencephaly ("smooth brain"), a condition in which the cortex has no sulci or gyri (Figure 22.21D). The LIS1 protein interacts with the cellular motor protein dynein, and mutations may disrupt dynein-mediated aspects of cell division in cortical progenitors or organelle transport in migrating cortical neurons. The DCX protein interacts with microtubules in migrating neurons and is thought to influence both the integrity of the cytoskeleton and the appropriate transport of organelles during initial differentiation.

Neuropathological observations as well as molecular and genetic studies have confirmed that several neurological problems, including epilepsy and some forms of intellectual disability, arise from the abnormal migration of cerebral cortical neurons. Several additional molecules that influence cell-cell signaling or cell adhesion are associated with disorders that are thought to reflect, in part, disrupted neuronal migration in the developing cerebral cortex. In particular, neuregulin (a secreted signal), NCAM, and *DISC1* (a gene mutated or deleted in a small number of cases of schizophrenia) are highly associated with risk for psychiatric diseases. Thus, disrupted cell migration (and its consequences for subsequent development of brain synapses and circuits) may underlie the pathology of several serious brain disorders.

Summary

The initial development of the nervous system depends on an intricate interplay of stem cell differentiation from ectodermal to neural stem cells, inductive signals that drive these events, cell proliferation to amplify the numbers of neurons, and cellular movements to ensure appropriate neuronal locations. In addition to the early

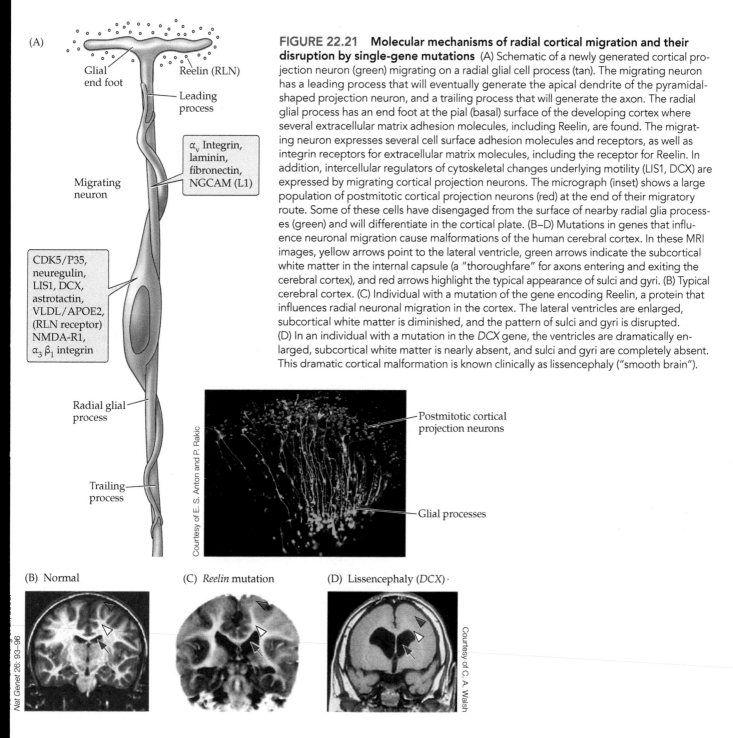

(A)

Glial end foot

Reelin (RLN)

Leading process

α_v Integrin, laminin, fibronectin, NGCAM (L1)

Migrating neuron

CDK5/P35, neuregulin, LIS1, DCX, astrotactin, VLDL/APOE2, (RLN receptor) NMDA-R1, $\alpha_3 \beta_1$ integrin

Radial glial process

Trailing process

Courtesy of E. S. Anton and P. Rakic

Postmitotic cortical projection neurons

Glial processes

(B) Normal

(C) *Reelin* mutation

(D) Lissencephaly (*DCX*)·

Courtesy of C. A. Walsh

FIGURE 22.21   Molecular mechanisms of radial cortical migration and their disruption by single-gene mutations (A) Schematic of a newly generated cortical projection neuron (green) migrating on a radial glial cell process (tan). The migrating neuron has a leading process that will eventually generate the apical dendrite of the pyramidal-shaped projection neuron, and a trailing process that will generate the axon. The radial glial process has an end foot at the pial (basal) surface of the developing cortex where several extracellular matrix adhesion molecules, including Reelin, are found. The migrating neuron expresses several cell surface adhesion molecules and receptors, as well as integrin receptors for extracellular matrix molecules, including the receptor for Reelin. In addition, intercellular regulators of cytoskeletal changes underlying motility (LIS1, DCX) are expressed by migrating cortical projection neurons. The micrograph (inset) shows a large population of postmitotic cortical projection neurons (red) at the end of their migratory route. Some of these cells have disengaged from the surface of nearby radial glia processes (green) and will differentiate in the cortical plate. (B–D) Mutations in genes that influence neuronal migration cause malformations of the human cerebral cortex. In these MRI images, yellow arrows point to the lateral ventricle, green arrows indicate the subcortical white matter in the internal capsule (a "thoroughfare" for axons entering and exiting the cerebral cortex), and red arrows highlight the typical appearance of sulci and gyri. (B) Typical cerebral cortex. (C) Individual with a mutation of the gene encoding Reelin, a protein that influences radial neuronal migration in the cortex. The lateral ventricles are enlarged, subcortical white matter is diminished, and the pattern of sulci and gyri is disrupted. (D) In an individual with a mutation in the *DCX* gene, the ventricles are dramatically enlarged, subcortical white matter is nearly absent, and sulci and gyri are completely absent. This dramatic cortical malformation is known clinically as lissencephaly ("smooth brain").

establishment of regional identity, cellular identity, and stem cell position within the brain, substantial migration of neuronal precursors or postmitotic neurons is necessary for the subsequent differentiation of classes of neurons and the eventual formation of specialized patterns of synaptic connections (see Chapters 8 and 23). The fate of individual neural precursors is not determined simply by their mitotic history; in addition, information required for differentiation arises from interactions between the developing cells, local signaling molecules, and the subsequent activity of distinct transcriptional regulators. All

of these events depend on the same categories of molecular and cellular phenomena: cell-cell signaling, changes in motility and adhesion, transcriptional regulation, and ultimately, cell-specific changes in gene expression. The molecules that participate in signaling during early brain development are the same as the signals used by mature cells: hormones, transcription factors, and second messengers (see Chapter 7), as well as cell adhesion molecules (see Chapter 23). Cell-cell signaling is essential for the progression from multipotent ectodermal stem cells to differentiation of neurons to build both the peripheral and central nervous systems. The identification and characterization of these molecules, the regions they specify, and the stem cells they influence in the developing brain have begun to provide insight into the genetic and cellular basis of several developmental disorders. These associations with brain pathologies reflect the vulnerability of signaling and transcriptional regulation during early neural development to the effects of genetic mutations, as well as to the actions of the many drugs and other chemicals that can compromise the elaboration of a typical nervous system.

■ Additional Reading

Reviews

Caviness, V. S., Jr. and P. Rakic (1978) Mechanisms of cortical development: A view from mutations in mice. *Annu. Rev. Neurosci.* 1: 297–326.

DeRobertis, E. M. (2009) Spemann's organizer and the self-regulation of embryonic fields. *Mech. Dev.* 126: 925–941.

Hatten, M. E. (1993) The role of migration in central nervous system neuronal development. *Curr. Opin. Neurobiol.* 3: 38–44.

Hu, W. F., Chahrour, M. H., and C. A. Walsh (2014) The diverse genetic landscape of neurodevelopmental disorders. *Ann. Rev. Genomics Hum. Genetics* 15: 195–213.

Ingham, P. (1988) The molecular genetics of embryonic pattern formation in *Drosophila*. *Nature* 335: 25–34.

Kessler, D. S. and D. A. Melton (1994) Vertebrate embryonic induction: Mesodermal and neural patterning. *Science* 266: 596–604.

Keynes, R. and R. Krumlauf (1994) Hox genes and regionalization of the nervous system. *Annu. Rev. Neurosci.* 17: 109–132.

Kintner, C. (2002) Neurogenesis in embryos and in adult neural stem cells. *J. Neurosci.* 22: 639–643.

LaMantia, A-S. (2020) Why does the face predict the brain? Neural crest induction, craniofacial morphogenesis and neural circuit development. *Front. Physiol.* 11: 610970.

Lamouille, S., Xu, J. and R. Derynck (2014) Molecular Mechanisms of epithelial-mesenchymal transition. *Nat. Rev. Mol. Cell Biol* 15: 178–196.

Peltjo, M. and H. Wichterle (2011) Programming embryonic stem cells to neuronal subtypes. *Curr. Opinion. Neurobiol.* 21: 43–51.

Rothstein, M., Bhattacharya, D., and M. Simoes-Costa (2018) The molecular basis of neural crest axial identity. *Dev. Biol.* 444: S170 -S180.

Important Original Papers

Angevine, J. B. and R. L. Sidman (1961) Autoradiographic study of cell migration during histogenesis of cerebral cortex in the mouse. *Nature* 192: 766–768.

Bulfone, A. and 5 others (1993) Spatially restricted expression of *Dlx-1, Dlx-2 (Tes-1), Gbx-2,* and *Wnt-3* in the embryonic day 12.5 mouse forebrain defines potential transverse and longitudinal segmental boundaries. *J. Neurosci.* 13: 3155–3172.

Eksioglu, Y. Z. and 12 others (1996) Periventricular heterotopia: An X-linked dominant epilepsy locus causing aberrant cerebral cortical development. *Neuron* 16: 77–87.

Hemmati-Brivanlou, A. and D. A. Melton (1994) Inhibition of activin receptor signaling promotes neuralization in *Xenopus. Cell* 77: 273–281.

Lancaster, M.A. and 9 others (2013) Cerebral organoids model human brain development and microcephaly. *Nature* 501: 373–379.

Liem, K. F., Jr., G. Tremml and T. M. Jessell (1997) A role for the roof plate and its resident TGFβ-related proteins in neuronal patterning in the dorsal spinal cord. *Cell* 91: 127–138.

Noden, D. M. (1975) Analysis of migratory behavior of avian cephalic neural crest cells. *Dev. Biol.* 42: 106–130.

Rakic, P. (1971) Neuron–glia relationship during granule cell migration in developing cerebral cortex: A Golgi and electron microscopic study in *Macacus rhesus. J. Comp. Neurol.* 141: 283–312.

Rakic, P. (1974) Neurons in rhesus monkey visual cortex: Systematic relation between time of origin and eventual disposition. *Science* 183: 425–427.

Sauer, F. C. (1935) Mitosis in the neural tube. *J. Comp. Neurol.* 62: 377–405.

Spemann, H. and H. Mangold (1924) Induction of embryonic primordia by implantation of organizers from a different species. Translated by V. Hamburger and reprinted in *Foundations of Experimental Embryology,* B. H. Willier and J. M. Oppenheimer (eds.) (1974). New York: Hafner Press.

Walsh, C. and C. L. Cepko (1992) Widespread dispersion of neuronal clones across functional regions of the cerebral cortex. *Science* 255: 434–440.

Yamada, T., M. Placzek, H. Tanaka, J. Dodd and T. M. Jessell (1991) Control of cell pattern in the developing nervous system. Polarizing activity of the floor plate and notochord. *Cell* 64: 635–647.

Zimmerman, L. B, J. M. De Jesus-Escobar and R. M. Harland (1996) The Spemann organizer signal Noggin binds and inactivates bone morphogenetic protein 4. *Cell* 86: 599–606.

Books

Gilbert, S. F. and M. Barresi (2016) *Developmental Biology,* 11th Edition, Chapters 9–15. Sunderland, MA: Sinauer Associates.

Lawrence, P. A. (1992) *The Making of a Fly: The Genetics of Animal Design.* Oxford: Blackwell Scientific Publications.

Moore, K. L. (1988) *The Developing Human: Clinically Oriented Embryology,* 4th Edition. Philadelphia: W. B. Saunders Company.

Construction of Neural Circuits

Overview

Once nerve cells have been generated, groups of neurons must become interconnected to form the neural circuits that mediate brain function. The first step in this process is to establish axons and dendrites in the newly generated neurons. The specification and growth of the characteristic single axon from a newly generated neuron, and the parallel specification and growth of dendrites, depend on cell polarity. Cell polarity in all cells relies on the movement of subsets of proteins and organelles to distinct domains. In the developing neuron, polarity reflects local signals that are available to one region or another of the developing neuron. These signals then elicit changes in the neuronal cytoskeleton that distinguish growing axons from dendrites. The subsequent growth of axons toward appropriate target cells, which can be local or distant, is critical for establishing neural circuits. The directed growth of axons and their recognition of appropriate synaptic targets depend on a leading process of the growing axon called a growth cone. Growth cones have a distinct cellular property, the capacity for motility. The dynamic behavior of growth cones depends in turn on adhesive, attractive, and repulsive molecular signals in the embryonic environment. Once axons find their way to appropriate targets and form synapses, molecular neurotrophic factors influence neuron survival. Dendrites do not usually grow as far from the neuronal cell body as axons. Nevertheless, their growth relies on some of the same cellular events: cytoskeletal reorganization, the extension of a growth cone–like leading process, and the recognition of signals that influence branching and formation of postsynaptic specializations. In addition to the regulation of individual neuron growth, the sizes and connections of entire populations of neurons must be regulated to match developing circuits to the targets they will control. The death of some neurons helps match the numbers of innervating neurons to the needs of their targets. Cell adhesion molecules, neurotrophic factors, and other signals also regulate the subsequent differentiation of axons and dendrites and the addition of synapses to match the numbers and strength of connections to target and circuit needs. As in other instances of intercellular communication, a variety of receptors and second-messenger molecules transduce adhesion and neurotrophic signals as synapses and circuits mature. Such signaling intermediates modify proteins that stabilize the molecular architecture of both pre- and postsynaptic specializations. These cellular mechanisms establish topographic maps and other orderly patterns of connectivity that allow animals to behave in increasingly sophisticated ways as they mature.

Courtesy of B. Karpinksi and A.-S. LaMantia

KEY CONCEPTS

- **23.1** Neural circuit construction relies on basic mechanisms of cell polarity

- **23.2** Neuronal growth cones are critical for establishing connections

- **23.3** Neuronal growth and synapse formation depend on signaling molecules

- **23.4** Axon, dendrite, and synapse development and numbers are regulated by trophic interactions

- **23.5** Axon, dendrite, and synaptic growth results in orderly patterns of connections, including topographic maps

| # Neural Circuit Construction Relies on Basic Mechanisms of Cell Polarity

LEARNING OBJECTIVES

23.1.1   Discuss how apical and basal polarity in epithelial cells relates to cell polarity in neurons.

23.1.2   Describe the roles of cytoskeletal organization and intracellular trafficking in neuronal polarization.

Neuronal polarization: The first step in neural circuit formation

Neurons are especially elaborate examples of **polarized epithelial cells**, a fundamental cell class found in most tissues. Epithelial cells assemble into sheets, held together by complex junctions that make one epithelial cell adhere to its neighbors. Polarized epithelial cells absorb molecules from the environment in the **apical domain** and secrete proteins and other cell products in the **basal domain** (Figure 23.1A–C). The tissues of the gut, lung, kidney, and pancreas are all sheets of polarized epithelial cells enclosing lumens from which molecules are absorbed and then released basally to adjacent tissues, especially blood vessels which then transport and distribute these molecules. The apical domain of an epithelial cell faces the lumen and may have specializations such as cilia or microvilli that increase the surface area for taking in and releasing specific molecules. Junctions that hold epithelial cells together are found on the basal domain, as are ion channels, small signaling molecules, and the machinery for intercellular protein exchange. The basolateral surface is typically specialized for intercellular communication. The apical and basal surfaces of many epithelial cells include sites for vesicular release of proteins; however, these secretory sites are distinguished by proteins specific to either the apical or basal domain. Apical and basolateral distinctions arise from the differential distribution of proteins that constitute the cellular cytoskeleton as well as additional junctional and scaffolding proteins that further establish polarity (see Figure 23.1A-C).

Neurons can be thought of as highly specialized epithelial cells. The fundamental polarity of most neurons reflects the distinction between the dendrites (specialized for signal transduction) and the axon (specialized for secretion). Thus, an essential first step in the differentiation of a neuron is cellular polarization. Once neurogenesis is complete and the neuroblast has entered a fully committed postmitotic state, the outgrowth of neuronal processes begins (Figure 23.1D,E). Initially, several apparently equivalent small extensions (referred to as *neurites*, since they have neither axonal nor dendritic identities at first) protrude from the immature neuron. Local cues disrupt the initial symmetry of the neuroblast, and after a process of amplification of these signals, a single process begins to differentiate as the axon. Soon after, microtubule and actin components of the cytoskeleton, as well as other proteins, are redistributed to further define the axon, and the remaining processes become dendrites (see Figure 23.2). Several studies, initially done in cell culture and confirmed in developing embryos, indicate that many proteins, particularly members of the PAR family, are distributed preferentially in the nascent axon (see Figure 23.1D,E). (PAR stands for "*partitioning-defective*"; PAR proteins were originally identified in the worm *Caenorhabditis elegans* based on their control of the axis of cell division and distribution of daughter cell proteins.) These proteins interact with cytoskeletal elements and signaling molecules, including Rho GTPase, related Rho kinases, and other protein kinases.

FIGURE 23.1   Cell polarity and the differentiation of axons and dendrites (A) Image of a simple polarized epithelial tube. The red labeling is actin filaments, which segregate to the apical domain, and the green is an adhesion molecule, e-cadherin, which is found at specialized adhesion contacts in the basolateral domain. (B) Molecular distinctions in the apical, basolateral, and basal domains of epithelial cells. In this image, epithelial cells from Drosophila express the Crumbs protein (red), which interacts with intracellular proteins to define the apical membrane; Discs large protein (blue), which influences cell adhesion, is limited to the basolateral domain; and Lachesin (green), another cell adhesion molecule, is seen in the basal domain. There is some overlap between Discs large and Lachesin in the basolateral domain (turquoise). The cell nucleus is purple. (C) Apical and basal distinctions in a simple epithelium. The apical domain has a distinctive actin cytoskeleton and membrane extensions (microvilli); there are tight junctions; the Golgi apparatus is oriented toward the apical membrane; secretory vesicles fuse to release contents and add membrane; and vesicular endocytosis internalizes membrane proteins as well as ligands bound to them and then targets endosomal compartments within the cell. The basal domain makes contact with the extracellular matrix; it has specialized adhesion contacts that bind the cell to the basement membrane, or basal lamina; and the plus ends of microtubules are oriented toward the basal lamina, which is a site of endosomal traffic. (D) Schematic of a newly generated postmitotic neuroblast extending processes and disrupting its symmetry to specify one of the processes as an axon. These events are accompanied by the polarized distribution of intracellular proteins such as the polarity scaffolding protein Par-3, which is localized to the tip of the growing axon (arrow). (E) Images of a neuroblast initiating neurite growth followed by axon specification. The result is a highly polarized neuron in which the dendrites (red; labeled for the microtubule-associated protein MAP2) are distinct from the axon (green; labeled for an axon-specific growth protein, GAP-43). (F) When Par-3 is functioning normally, one of the processes of the neuron (all labeled green) elongates and becomes an axon (top left). When Par-3 function is disrupted by overexpression, this difference in axonal growth is no longer seen—all of the processes have molecular characteristics of axons (top right). The axon-specific cytoskeleton protein tau (purple, bottom left) is usually seen only in the single developing axon. When Par-3 function is disrupted, tau is seen in all of the neurites (bottom right).

PAR proteins also interact with signal transduction pathways activated by secreted Wnts (see Chapter 22), cell surface–bound cell adhesion molecules, and neurotrophins (see Concept 23.4). When the function of PAR proteins or related signaling molecules is disrupted, the specification of a single axon does not occur (Figure 23.1F). PAR proteins and other polarity regulators also play a role in defining the regions of dendrites that receive synapses.

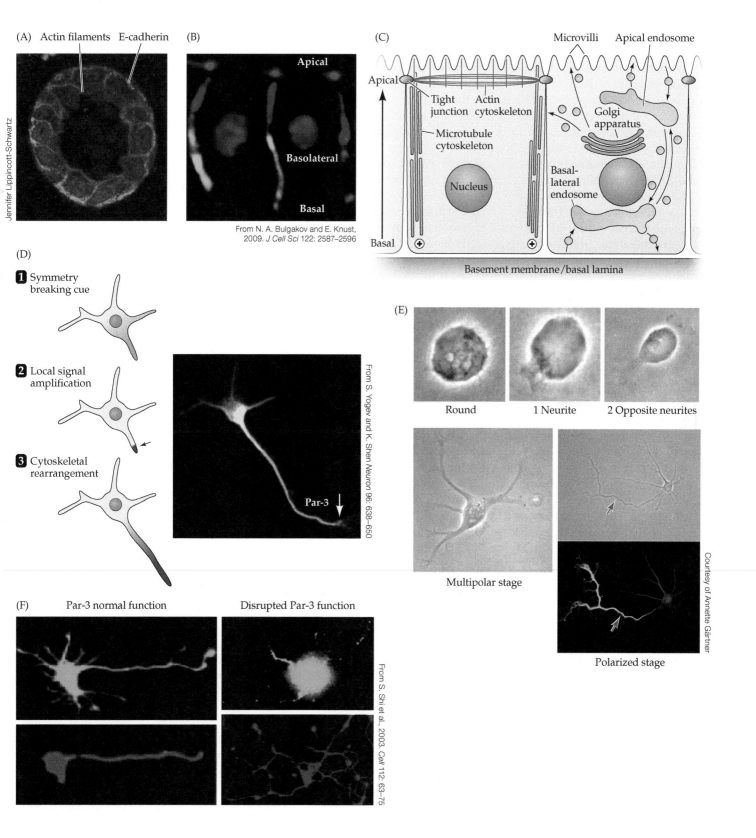

(A) Actin filaments   E-cadherin

Jennifer Lippincott-Schwartz

(B) Apical

Basolateral

Basal

From N. A. Bulgakov and E. Knust, 2009. *J Cell Sci* 122: 2587–2596

(C) Microvilli   Apical endosome

Apical

Tight junction   Actin cytoskeleton

Microtubule cytoskeleton

Golgi apparatus

Nucleus

Basal-lateral endosome

Basal

Basement membrane/basal lamina

(D)

1 Symmetry breaking cue

2 Local signal amplification

3 Cytoskeletal rearrangement

Par-3

From S. Yogev and K. Shen *Neuron* 96: 638–650

(E)

Round   1 Neurite   2 Opposite neurites

Multipolar stage

Polarized stage

Courtesy of Annette Gärtner

(F)   Par-3 normal function   Disrupted Par-3 function

From S. Shi et al. 2003. *Cell* 112: 63–75

Thus, in developing neurons, the molecular mechanisms that establish epithelial cell polarity are adapted to generate axons and dendrites, which continue to grow so that the connections that define a neural circuit can be made.

Protein and organelle trafficking in neurons

Neurons grow by elaborating axons and dendrites, which then make synaptic connections. The "raw materials" for constructing axons, dendrites, and synapses—proteins, lipids, and organelles synthesized or assembled primarily in the cell body—must be transported to the "construction sites" at the distal ends of developing axons or dendrites. This transport is achieved via a distinctive cell biological mechanism referred to as *trafficking*. To build an axon or dendrite, and subsequently the pre- and postsynaptic specializations that define connections in a neural circuit, multiple classes of cytoskeletal, cytoplasmic, and transmembrane proteins must be trafficked. In addition, organelles such as mitochondria, endosomes and lysosomes, endoplasmic reticulum, and Golgi cisternae must be distributed to sites where they are needed for growth or to establish mature functions. Finally, messenger RNAs (mRNAs) are often transported to axons and dendrites to facilitate local protein translation. Thus, the cellular mechanisms by which proteins, organelles, and mRNAs are moved into the growing processes of a neuron are a fundamental dimension of neural circuit construction.

Trafficking of proteins in axons and dendrites relies fundamentally on the availability of "highways" or substrates on which "cargos" of proteins and organelles can be moved. The elements of the axonal and dendritic cytoskeleton constitute these highways. Two families of cytoskeletal proteins—tubulin and actin—provide the primary substrates for movement of multiple cargos. The polymerization of both tubulin and actin relies on the hydrolysis of ATP and GTP, and thus requires substantial bioenergetic support. Polymerized tubulin forms neuronal microtubules that are arrayed parallel to the long axis of most axons and dendrites (Figure 23.2). In contrast, actin can be arranged in a variety of local patterns throughout the neuron. A third family of cytoskeletal proteins, the intermediate filaments, unique to vertebrates, can act as scaffolds for signaling and contributes to axon stability via formation of neurofilaments seen particularly in long axons.

Several regulatory proteins ensure the integrity of microtubule, actin, and intermediate filament proteins that constitute the developing cytoskeleton of a growing nerve cell. The most relevant proteins for the developing neuron include those that facilitate transport of specific cargos from the cell body to the sites of growth. These proteins that interact with microtubules are referred to as **motor proteins**. Motor proteins in axons include two families, the **kinesins** and the **dyneins** (see Figure 23.2A). Multiple kinesins regulate anterograde transport of diverse cargos into the distal

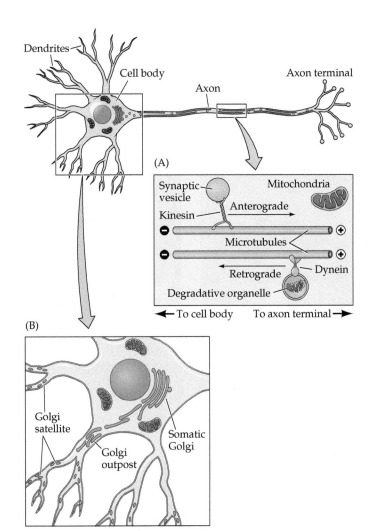

FIGURE 23.2 Membrane trafficking and protein transport are essential for neuron growth and synapse formation (A) Membrane vesicle and organelle transport in an axon. Microtubules with a consistent orientation (+ end toward the axon tip) provide a substrate for the binding of kinesin motor proteins, which also bind membrane vesicles as well as organelles such as mitochondria. The kinesins then mediate the movement of the vesicle cargo unidirectionally toward the axon terminal. In axons, the motor protein dynein is selective for endocytic vesicles that can transmit signals or are destined for lysosomal degradation in the cell body. (B) Dendritic membrane trafficking relies on the localization of a distinct domain of the Golgi apparatus, the Golgi outpost, to the proximal dendrites, as well as the distribution of Golgi satellites (smaller vesicular organelles that share functional properties with the Golgi apparatus localized furthest from the Golgi apparatus in the cell body) in the distal dendrite to facilitate differences in protein and membrane distribution in the dendrite versus axon. (A after M. T. Kelliher et al., 2019. *Curr Op Neurobiol* 57: 39–45; P. Guedes-Dias and E. Holzbaur, 2019. *Science* 366: 199; B after J. Wang et al., 2020. *Front Mol Neurosci* 13: 597391.)

aspect of an axon. Kinesins, in concert with microtubules, hydrolyze ATP to drive trafficking of cargo. The kinesin molecule, with cargo bound to one domain, uses this energy to move on the microtubule guides. There is some selectivity of kinesins for the type of cargo transported:

synaptic vesicles and organelles (particularly mitochondria) rely upon specific kinesins. Dyneins function as a multimeric complex of diverse cytoplasmic dynein subunits (each encoded by a distinct gene) that regulate retrograde transport of signals, phagocytic organelles (lysosomes, autophagosomes), and other degradation products from the distal axon toward the cell body. Finally, a large family of **myosin** motor proteins interacts with the actin cytoskeleton to facilitate local transport of proteins and mRNA, particularly in growth cones, dendrites, and dendritic spines. Protein trafficking is key for both axons and dendrites; however, because of their shorter size, proximity to the cell body, and distinct functions, dendrites also have specialized extensions of the Golgi apparatus, including vesicular bodies related to the Golgi called **Golgi outposts** (see Figure 23.2B). These extensions of the Golgi apparatus in dendrites are important for establishing dendritic growth and assembling postsynaptic sites during subsequent circuit development.

The complexity and stability acquired by axons and dendrites as their polarized microtubule/actin/neurofilament–based cytoskeleton is assembled are critical to the continuing directed growth and signaling capacity essential for circuit formation. Indeed, mutations in several genes that encode proteins essential for regulating trafficking result in serious developmental disorders including seizure disorders, spinal muscular atrophy, hereditary spastic paraplegias, and intellectual disability.

CONCEPT 23.2 | Neuronal Growth Cones Are Critical for Establishing Connections

LEARNING OBJECTIVES

23.2.1 Identify the key functional and cellular properties of growth cones.

23.2.2 Describe the mechanisms of growth cone motility.

23.2.3 Describe the basic mechanisms and cues for axon attraction and repulsion.

The axonal growth cone

Once an axon has been specified, it often must extend substantial distances from its parent neuronal cell body, navigating through complex embryonic terrain to find appropriate target regions and synaptic partners. In 1910, Ross G. Harrison first observed this phenomenon in a living tadpole and wrote, "The growing fibers are clearly endowed with considerable energy and have the power to make their way through the solid or semi-solid protoplasm of the cells of the neural tube. But we are at present in the dark with regard to the conditions which guide them to specific points." In the century that followed Harrison's fundamental observations, his description of the remarkable capacity of growing axons

to "make their way" through the embryo was confirmed using increasingly sophisticated microscopic imaging with improved resolution in living specimens, molecular labels, biochemical analysis, and mutations in key genes. This work defined many of the "conditions" that guide axons to "specific points." In addition, it is now clear that dendrites, especially primary dendrites such as those of cortical pyramidal cells or cerebellar Purkinje cells, must also extend over relatively long distances. The manner in which they do so is similar to that of axons—the growing tips of dendrites extend actively into the surrounding tissue, responding to signals that direct their growth and differentiation. Dendritic growth, however, is usually far more limited in distance than axonal growth.

Harrison recognized two fundamental features of axonal growth that are essential for both axons and dendrites during the initial phases of neural circuit development. First, the "considerable energy… and power" of growing axons reflect their capacity to rapidly add cellular membrane, cytoskeleton, and organelles and extend the axon via the **growth cone**, a specialized structure at the tip of the growing axon (Figure 23.3A). Growth cones are highly motile. They move relatively rapidly, explore the extracellular environment, determine the direction of growth, and then guide the extension of the axon in that direction. The primary morphological characteristic of a growth cone is a sheetlike expansion of the growing axon at its tip called a **lamellipodium**. When growth cones are examined in vitro, it is possible to visualize **filopodia**, spikelike processes that extend from each lamellipodium (Figure 23.3B,C). Filopodia rapidly form and disappear from the lamellipodium, like fingers reaching out to sense the environment. The lamellipodium and filopodia are distinguished from the axon shaft by different cytoskeletal molecules (see Figure 23.3C), particularly the presence of unpolymerized microtubules surrounded by nonfilamentous actin at the proximal aspect of the lamellipodium and the concentration of actin filaments and several actin-binding proteins in the filopodia. Lamellipodia and filopodia have the capacity to localize or concentrate receptors on their membrane surfaces to detect signaling molecules in the extracellular environment or on the surfaces of neighboring cells. ATP-dependent, force-generating interactions between cytoskeletal proteins in the lamellipodium and filopodia, mediated by signaling via cell surface receptors as well as ion channels, ultimately provide the "energy… and power" to propel the growth cone and its axon to its target. Growth cones in vivo appear to be similar, although their filopodia are more difficult to resolve. Nevertheless, as is the case in vitro, the growth cone extends away from the neuronal cell body, and as it does so, additional length is added to the axon (Figure 23.3D). Thus, the growth cone is a distinct, if transient, neuronal specialization whose activity is critical for establishing connections between nerve cells and their targets. Once a growth cone reaches

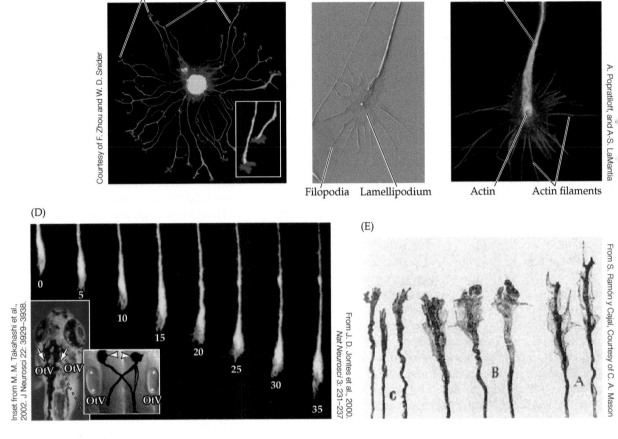

(A) Actin Microtubules

(B)

Filopodia Lamellipodium

(C) Microtubules

Actin Actin filaments

Courtesy of F. Zhou and W. D. Snider

A. Popratiloff, and A-S. LaMantia

(D)

0
5
10
15
20
25
30
35

OtV OtV

OtV OtV

Inset from M. M. Takahashi et al., 2002. J Neurosci 22: 3929–3938.

From J. D. Jontes et al., 2000. Nat Neurosci 3: 231–237

(E)

c B A

From S. Ramón y Cajal, Courtesy of C. A. Mason

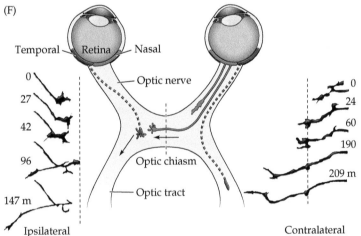

(F)

Temporal Retina Nasal

Optic nerve

0
27
42
96
147 m

Optic chiasm

Optic tract

Ipsilateral

0
24
60
190
209 m

Contralateral

and recognizes an appropriate target, it is gradually transformed into either a presynaptic ending for an axon or the terminal domain of a dendrite, since growing dendrites also have growth cones.

Santiago Ramón y Cajal, a contemporary of Harrison's, observed growth cones in several developing axon pathways using histological sections of fixed embryo specimens of a variety of mammals including cats, rabbits, and humans as well as amphibians and birds (rather than the tails of live tadpoles that Harrison viewed). These pathways appeared to be "pioneered" by one or a few early growing axons, whose growth cones are usually more complex than those of axons that follow the pioneers. These pioneer axons—the first axons to extend through a given region, thus establishing that terrain as hospitable to axon growth—also extend in new directions through territories not previously innervated. Cajal noted that when a pioneer axon's growth cone reaches a region where a choice must be made about which direction to take, its shape changes dramatically (Figure 23.3E). The lamellipodium of the growth cone expands as it encounters a potential guidance signal or target and extends numerous filopodia, actions that suggest an active search for appropriate cues to

direct subsequent growth. Indeed, growth cones become more complex at "decision points" in both the developing peripheral and central nervous systems (PNS and CNS).

In the PNS, the growth cones of both motor and sensory neurons change shape as they enter developing muscles of immature limbs or approach distinct targets in the differentiating dermis, presumably facilitating the selection of appropriate targets. Distinct molecules on or around the muscle surface—including cell surface adhesion molecules, receptors for secreted signals, and

◀ **FIGURE 23.3** **Growth cones guide axons in the developing nervous system** (A) A single dorsal root ganglion cell isolated in culture extends many processes. Each process has a long shaft in which microtubules (green) predominate, tipped by a growth cone in which actin (red) is the major molecular constituent. (B) A single growth cone from a trigeminal sensory neuron grown in culture visualized using a scanning electron microscope. Each filiopodium is distinct—they are not highly branched after extending from the lamellapodium. Instead, multiple filopodia extend from the lamellipodium. (C) An array of microtubules (green) extends into the lamellipodium, which includes a pool of both nonflilamentous and filamentous actin that forms a scaffold for numerous filopodia that define the growth cone. (D) A single Mauthner axon labeled with a fluorescent dye in a living zebrafish embryo, led through the spinal cord by a relatively simple growth cone. Over 35 minutes, the axon advances approximately 50 μm. Insets: At left, Mauthner neurons (arrows) in the hindbrain of a zebrafish embryo, adjacent to the otic vesicles (OtV), give rise to the sensory neurons of the inner ear. At right is a higher magnification of the hindbrain. The Mauthner neuron axons can be seen crossing the midline and extending down to the spinal cord. (E) In 1899, Ramon y Cajal recorded differences in growth cone shape in silver stained sections of the chick embryo. These differences were related systematically to their location in the gray matter (A), the ventral commissure of the spinal cord (B) and nascent spinal cord axon tracts (C). From these observations, he suggested that growth cone shape varies at "decision" regions. (F) Individual growth cones change dynamically at decision points. Variations in complexity for growth cones from retinal ganglion cell axons in the temporal (red) retina that will grow into the ipsilateral optic tract once they reach the optic chiasm and those from the nasal retina (green) that will cross the midline into the contralateral optic tract are summarized (center). At left are drawings of a single growth cone of a temporal retinal ganglion cell axon imaged over time showing its acquisition of complexity in the region of the optic chiasm (indicated by dotted line), followed by a return to a simpler shape once the "decision" to grow into the ipsilateral optic tract has been reached. At right is a similar series for a single growth cone from the nasal retina. (F after P. Bovolenta and C. Mason, 1987 *J Neurosci* 7: 1447–1460; P. Godement et al, 1994; *J Neurosci* 14: 7024–7039.)

specialized extracellular matrix molecules (see the next section)—mediate this change in shape for the axons of motor neurons. Similar molecules in the dermis (the layer under the epidermis) influence sensory neuron growth cones. In the CNS, growth cones in the optic nerve, especially after new axons are added as new ganglion cells are generated in the retina, remain somewhat simple. Similarly, in the CNS at the optic chiasm, where axons from the nasal retina must cross and those from the temporal retina must remain on the same side of the brain, the growth cones become far more complex, with an expanded lamellipodium and additional filopodia (Figure 23.3E,F). These changes are quite dynamic, and specific. Growth cones of retinal ganglion cell axons from the temporal retina reach the chiasm, explore, become complex with multiple branches and an expanded lamellipodium, and finally turn toward the ipsilateral optic tract. Those from the nasal retina reach the chiasm, explore, become complex and then cross the midline. Thus, growth cones from specific neuron classes dynamically explore peripheral targets such as muscles or developing CNS regions through which they grow. They alter shape and direction in real time by expanding their lamellipodium, adding filopodia, and ultimately reorienting the cytoskeletal elements that support axon extension in response to environmental cues. Selective, dynamic regulation and guidance of growth cones results in the extension of axons toward appropriate targets: an essential first step for optimal establishment of functional neural circuits.

The molecular basis of growth cone motility

Growth cone motility reflects rapid, controlled rearrangement of the cytoskeleton. The energy to move the axon is generated by ATP-dependent modification of the actin and microtubule cytoskeletons, and the resulting physical force is exerted by deformation of these proteins and the overlying membrane. The **actin cytoskeleton** regulates key changes in lamellipodial and filopodial shape for directed growth, while the **microtubule cytoskeleton** is primarily responsible for elongation of the axon itself (Figure 23.4A–C). The molecular composition of both the actin and microtubule cytoskeletons changes in distinct regions of the growth cone and axon, suggesting a great deal of dynamism within growing neural processes. Thus, defining the ways in which the actin and microtubule cytoskeletons are modified in growing axons is essential to understanding how growth cones and axons extend.

Actin is the primary molecular constituent of a network of cellular **filaments** found at the leading edge of the lamellipodium and within the core of the filopodia of a growth cone (see Figure 23.4A–C; see also Figure 23.3C). Tubulin is the primary molecular constituent of **microtubules** that run parallel to the axis of the axon and give it both structural integrity and a means for transporting proteins from the nerve cell body to the axon terminal, given that axons do not contain significant protein synthesis machinery (see Figure 23.2). Actin and tubulin are found in two forms in the growth cone and axon: freely soluble monomers in the cytoplasm, and protein polymers that form filaments (actin) or microtubules (tubulin) within filopodia or the axon itself (see Figure 23.4A). The dynamic polymerization and depolymerization of actin at the membrane of the lamellipodium, as well as within the filopodium, sets the direction of growth cone movement (see Figure 23.4B), in part by generating local forces that orient the growth cone toward or away from attractive or repulsive substrates (see Concept 23.3). Similarly, the polymerization and depolymerization of tubulin into microtubules consolidate the direction of axon extension by stabilizing the axon shaft in response to directional forces and underlying signaling mediated by the growth cone. These cytoskeletal elements, especially the actin filaments

FIGURE 23.4 Cell biological specializations underlie growth cone motility (A) Distinct types of actin and tubulin, the two key cytoskeletal proteins in all cells, are seen in discrete regions of the growth cone (shown schematically in the drawing, right). In a single growth cone grown in cell culture, filamentous actin (F-actin, red) is seen in the lamellipodium and filopodia. Tyrosinated microtubules (dynamic) are the primary tubular constituents of the lamellar region (green), and acetylated microtubules (stable) are restricted to the elongating axon itself (blue). (B) Growth cone turning reflects dynamic reorientation of the actin and tubulin cytoskeleton in the lamellipodium as well as filopodium (drawing at left). Dynamic actin filaments (red) extend into the boundaries of the lamelopodium and the filopodia. Dynamic microtubules (green) are assembled and disassembled as the growth cone establishes a direction of extension. Stable microtubules (blue) are seen in the axon shaft once the growth cone has continued to extend in a particular direction. This is illustrated by visualizing the dynamics of actin cytoskeleton (yellow, red) in a single growth cone imaged over an 8-hour interval. The distribution of filamentous actin, labeled red with a fluorescent actin-binding protein, changes in the region of the lamellipodium (yellow) as well as in the filopodia (red) as the direction of growth become oriented to the left in this series of images. (C) The distribution and dynamics of cytoskeletal elements in the growth cone. Globular actin (G-actin) can be incorporated into F-actin at the leading edge of a filopodium in response to attractive cues. Repulsive cues support disassembly and retrograde flow of G-actin toward the lamellipodium. Organized microtubules make up the cytoskeletal core of the axon, while more broadly dispersed microtubule subunits are found at the transition between the axon shaft and the lamellipodium. Actin- and tubulin-binding proteins regulate the assembly and disassembly of subunits to filaments or tubules. This process is influenced by changes in intracellular Ca^{2+} via voltage-regulated Ca^{2+} channels as well as transient receptor potential (TRP) channels. (D) A single growth cone (left) visualized in cell culture in which the actin cytoskeleton has been labelled (purple) as well as secretory vesicles that have been transported from the cell body (green). The secretory vesicles are associated with the actin cytoskeleton in individual filopodia, presumably to facilitate membrane addition via vesicular fusion as the growth cone extends. (E) Growth cones show rapid changes in Ca^{2+} concentration. In this example, a single filopodium (white arrowheads) undergoes a rapid increase in Ca^{2+}. (Schematics in A,B after Kahn and Baas, 2016. *Trends in Neurosci* 39: 433–440; C after A. B. Huber et al., 2003. *Ann Rev Neurosci* 26: 509–563.)

in the filopodia, provide a scaffold for the movement of membrane vesicles that will fuse with the growth cone membrane to extend the growth cone and axon (Figure 23.4D). The changes in cytoskeleton that facilitate directional growth also influence the responses to environmental cues (see Figure 23.4C). These cues can change local trafficking of cytoplasmic proteins that interact with the cytoskeleton. They can also facilitate the preferential insertion of membrane proteins, including cell surface receptors and ion channels, due to local changes in vesicle fusion (see Figure 23.4D) as the growth cone determines the direction of subsequent axon extension.

Several of the cytoplasmic proteins that influence local changes in growth cone motility and the direction of axon extension bind to actin and tubulin to regulate their polymerization and depolymerization. Actin-binding proteins are found throughout the growth cone cytoplasm. Most either bind actin directly or modify actin monomers by phosphorylation and other post-translational modifications. These molecules are particularly enriched at the inner surface of the growth cone membrane, where they mediate assembly and membrane anchoring of actin filaments to generate forces that direct the movement of the lamellipodium or extension of filopodia and promote the addition of membrane via vesicle fusion (see Figure 23.4D). Thus, these actin modifying proteins are central for the continued growth and extension of axons toward their appropriate targets. In addition, the actin cytoskeleton is the anchor for multiple "protein scaffolding" molecules that in turn localize or concentrate receptors and channels in the lamellipodial or filopodial membrane. These receptors and channels can further modulate the movement or choice of target of the growth cone. Microtubule-binding proteins concentrated in the axon shaft modulate post-translational modifications of monomeric and polymerized tubulin to stabilize the axon. The selectivity of motor proteins for transporting different cargos into or away from a growth cone, growing axon, or dendrite maintains the appropriate balance of cytoskeletal and membrane proteins for extension, target recognition, and synapse formation.

The constant flux between monomeric actin and tubulin versus polymerized actin filaments and microtubules is regulated via binding proteins that are activated by enzymatic cleavage of second messengers such as cAMP and cGMP, as well as influx of Ca^{2+} via plasma membrane ion channels or Ca^{2+} release from intracellular stores, usually in response to signaling via surface receptors. The energy demand of this signaling, and subsequent actin and tubulin polymerization and depolymerization, is also thought to explain the higher concentration of mitochondria in the lamellipodium of a growth cone as well as the growing axon. Some signals ultimately enhance mitochondrial ATP production to support energy-demanding cytoskeletal changes. The regulation of intracellular Ca^{2+} levels, either through voltage-regulated Ca^{2+} channels, transient receptor potential (TRP) channels activated by second messengers, or receptor-mediated second-messenger pathways that mobilize intracellular Ca^{2+} stores, is thought to be a major modulator of actin and microtubule dynamics in the growing axon. Fluctuations of Ca^{2+}, which acts as a second messenger, can be quite localized in the growth cone, sometimes in a single filopodium (Figure 23.4E). They are thought to influence decisions on direction of growth in response

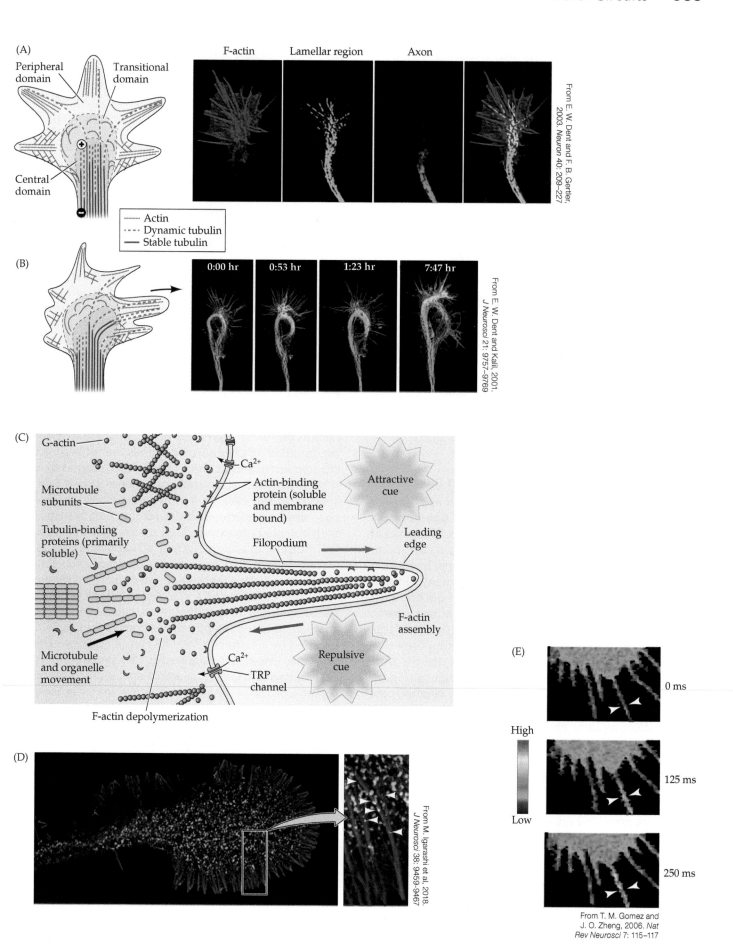

(A)

Peripheral domain

Transitional domain

Central domain

Actin
Dynamic tubulin
Stable tubulin

F-actin   Lamellar region   Axon

From E. W. Dent and F. B. Gertler, 2003. *Neuron* 40: 209–227

(B)

0:00 hr   0:53 hr   1:23 hr   7:47 hr

From E. W. Dent and Kalil, 2001. *J Neurosci* 21: 9757–9769

(C)

G-actin

Microtubule subunits

Tubulin-binding proteins (primarily soluble)

Microtubule and organelle movement

F-actin depolymerization

Ca^{2+}

Actin-binding protein (soluble and membrane bound)

Filopodium

Ca^{2+} TRP channel

Attractive cue

Leading edge

F-actin assembly

Repulsive cue

(D)

From M. Igarashi et al, 2018. *J Neurosci* 38: 9459–9467

(E)

High

Low

0 ms

125 ms

250 ms

From T. M. Gomez and J. O. Zheng, 2006. *Nat Rev Neurosci* 7: 115–117

to extrinsic cues. In parallel, cAMP, which can influence Ca^{2+} levels, is elevated in growth cones in response to extrinsic signals, including cell surface and extracellular matrix adhesion molecules. Thus, the conditions that Ross Harrison suggested "guide [growth cones] to specific points" are now understood to be changes in the cytoskeleton and local composition of membrane proteins of the growth cone and axon, mediated by intracellular signal transduction in response to adhesion molecules and diffusible signals available in the embryonic terrain through which the growth cone extends.

CONCEPT **23.3**

Neuronal Growth and Synapse Formation Depend on Signaling Molecules

LEARNING OBJECTIVES

23.3.1 Distinguish between diffusible and nondiffusible signals for axon guidance.

23.3.2 Identify the adhesion molecules and growth factors that direct axonal and dendritic growth.

23.3.3 Analyze the molecular mechanisms underlying axonal and dendritic guidance and growth.

23.3.4 Discuss the molecular mechanisms for selective synapse formation.

Diffusible signals: Chemotropic axon guidance

A growing axon must eventually find an appropriate target while avoiding inappropriate ones. **Chemotropic** signals, thought to be secreted by target cells located at a distance from the neurons that will innervate them, mediate this early selective growth of axons. In another instance of remarkable foresight (he had many of these instances!), Cajal proposed that such signals are most likely released by target cells to selectively attract growth cones to appropriate destinations. In addition to this *chemoattraction* predicted by Cajal, it was supposed that there might also be *chemorepellent* signals that discourage axon growth toward inappropriate regions (Figure 23.5A). Finally, it was assumed that these chemotropic factors would be available in a **gradient** (see Concept 23.5), reflecting their ability to diffuse freely through a tissue. Their concentration would therefore be high at the source, and lower farther from the source, to define the direction of growth. Despite the apparent importance of chemoattraction and chemorepulsion in constructing pathways and circuits, the identity of the signaling molecules remained uncertain until the early 1990s. One problem was the vanishingly small amounts of such factors expressed in developing embryos, which are also quite small. Another was that of distinguishing *tropic* molecules, which *guide* growing axons toward a source, from *trophic* molecules, which *support* the survival and

growth of neurons and their processes (see Concept 23.4). These problems were solved by analysis of attractive or repulsive activities in developing neural tissue collected from vertebrate (chick) embryos, followed by laborious biochemical purification as well as genetic analysis in both *C. elegans* and *Drosophila* that identified single-gene mutations associated with the misrouting of axons. Remarkably, the identity and function of chemoattractants and chemorepellents across phyla are highly conserved: What's chemotropic for the chick (and mouse, and humans) is also chemotropic for worms and flies.

The **netrins** and their receptors were one of the first families of guidance molecules with chemotropic properties to be identified in vertebrates (Figure 23.5B). In parallel, a related family called the **Uncoordinated** or **Unc** genes were one of the first families of chemotropic molecules to be identified based on genetic mutations in *C. elegans*. In chick embryos, netrins were identified as proteins with chemoattractant activity following biochemical purification from the developing spinal cord (see below). The first *Unc* gene was identified based on a behavioral phenotype: lack of proper movement in mutant worms. The cause was found to be misrouted axons, which resulted in misplaced or absent synapses, and disrupted circuits due to the loss of *Unc* function. These two "founding families" of chemotropic signals are related because chicken *netrin* and *C. elegans* Unc6 (also referred to as *netrin*) turned out to be orthologues: genes in two different species (in this case) or within a single genome that encode similar proteins with similar function. Netrins have high homology to extracellular matrix molecules such as laminin, and in some cases may actually interact with the extracellular matrix to influence directed axon growth. Netrin chemoattractant signals are transduced by specific receptors, including DCC (*deleted in colorectal cancer*), a transmembrane protein that binds netrin, and neogenin, a transmembrane receptor that is homologous to DCC but encoded by a distinct gene. DCC has a *C. elegans* Unc orthologue, Unc-40, thus suggesting the overall similarity of chemoattractant signaling across species. A different receptor, Unc5 (a *C. elegans* protein whose name is also used for its vertebrate orthologues), mediates netrin-dependent chemorepulsion. Thus, depending on the signal transduction capacity of the growing axon, which reflects the receptors on its surface, netrin can be either chemoattractive or chemorepulsive. While netrin was the first secreted signal identified with chemotropic activity, several other secreted factors have since been implicated in directed growth. Indeed, although netrin acts at the midline in the mammalian spinal cord, much of its guidance of spinal cord sensory neuron axons may be mediated by its local production by neuroepithelial cells that make it available along the pathway through which the axons grow before reaching the midline.

(A)

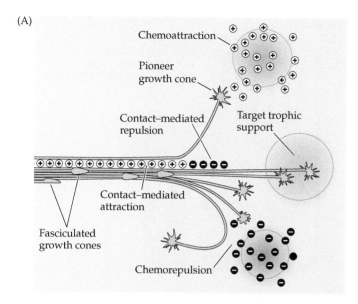

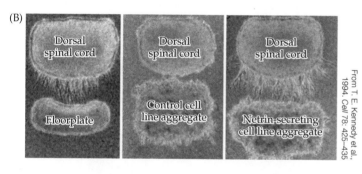

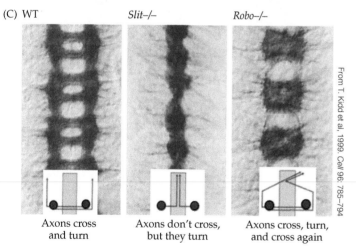

(C) WT   *Slit–/–*   *Robo–/–*

Axons cross and turn   Axons don't cross, but they turn   Axons cross, turn, and cross again

From T. E. Kennedy et al., 1994. *Cell 78*: 425–435.

From T. Kidd et al, 1999. *Cell 96*: 785–794

(D) Netrin/slit family

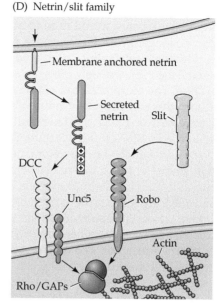

FIGURE 23.5 Attractive and repulsive signals provided by the embryonic environment influence growth cone behavior and axonal growth (A) Basic signal classes. Chemoattractant, or tropic, signals (pluses) can operate from a distance and reorient growth toward the source of the cue, often by acting on a pioneer growth cone that sets out a course distinct from that of the fasciculated followers. Adhesive cues acting on or near the surface of the axon shaft help maintain groups of axons as fascicles, which is essential for formation of coherent nerves and tracts. Chemorepellant signals (minuses) can also act from a distance, or at regions where axons must defasciculate from a nascent nerve in order to change their trajectory or avoid an inappropriate target. Trophic signals (gold) support survival, growth and differentiation of the axon and its parent nerve cell. (B) An in vitro assay from explanted tissue dissected from the dorsal spinal cord and floorplate of the chick embryo showing that the floorplate (left) releases soluble signaling molecules that act at a distance to attract growing dorsal spinal cord axons. An aggregate of cells from a cell line (middle) does not have this activity. An aggregate of cells from the same cell line transfected so that they express netrin (right), a chemoattractant signal produced in the floorplate, elicits directed growth similar to that elicited by the floorplate itself. (C) Slit and robo, two additional chemotropic cues, operate to guide axons across the midline (shown here in the *Drosophila* ventral nerve cord) and ensure that they do not cross back. Slit is essential for attracting axons across the midline; a loss-of-function mutation results in failure to cross. Robo is essential for preventing an axon after it has crossed the midline from crossing it again. WT, wild type. (D) Membrane anchored and secreted Netrin, as well as secreted Slit are available as tropic signals. The netrins bind either to the DCC receptor or Unc5 receptor to initiate Rho/GAP mediated signaling to modify the actin cytoskeleton of the growth cone. Slit binds to the Robo receptor, which also engages Rho/GAP signaling to modify the actin cytoskeleton. (A after A. B. Huber et al., 2003. *Annu Rev Neurosci* 26: 509–563.)

In the vertebrate spinal cord, netrin signaling can be downregulated by proteolytic cleavage of the DCC receptor once an axon crosses the midline, which probably contributes to ensuring that once axons cross the midline, they do not cross back. Nevertheless, it seems likely that there must be additional "active" chemorepulsive factors detected by specific receptors on the growth cone to further ensure that the axons do not cross back. These factors, like all chemorepulsive signals, would need to influence where axons can no longer grow. Furthermore, they would need to be "conditional": sensitivity to the repulsive factor can only be established immediately after the axon had crossed the midline. The secreted factor **slit** and its receptor **robo** function in this way to prevent ambiguity of direction of growth once an axon has crossed the midline. Slit and robo were identified first in *Drosophilia* mutants in which axons failed to cross the midline in the ventral nerve cord (Figure 23.5C,D). These two molecules are important for guiding vertebrate as well as invertebrate axons across the midline (slit) and then preventing them from crossing back once they have crossed (robo). Secreted slit protein diminishes directed axon growth when provided to neurons expressing the robo receptor. The robo receptor, like DCC, has no intrinsic intracellular signaling activity. Instead, robo is thought to engage non-receptor tyrosine kinases as well as GTPases after binding slit to transduce signals that lead to axon repulsion only after the axon has responded to chemoattractant signals while approaching and then crossing the midline (thus, it's conditionally activated). In the spinal cord, hindbrain, and other regions of the brain where axons cross, slit and robo are available immediately off the midline. Signaling via slit and robo is thought to reinforce the termination of the growth cone's sensitivity to netrin or other chemoattractant signals once the axon has crossed from one side of the CNS to the other. Thus, slit and robo (and probably several other molecules) and the signaling pathways they activate orchestrate the unidirectional crossing of axons at the midline via chemorepulsion in invertebrates as well as vertebrates.

Several other signals—some bound to cell surfaces, others secreted and available extracellularly—can act as chemotropic signals to influence the direction of axon growth via signaling at the growth cone. **Semaphorins** and their **plexin** receptors are essential mediators of chemotropic axon growth (Figure 23.6A). Semaphorins are a large and diverse set of signaling molecules. In vertebrates there are at least 20 semaphorins (abbreviated Sema3, Sema4, etc.) encoded by individual genes (rather than splice variants). Some semaphorins are transmembrane proteins and work in a way similar to that of the nondiffusible cell surface adhesion molecules described in detail in the next section. Others are associated with the extracellular side of a target cell's membrane via a membrane anchor that can be cleaved to release an active semaphorin into the extracellular space (see Figure 23.6A). Semaphorins can act as repulsive signals causing the collapse and retraction of growth cones as they approach an unfavorable target (Figure 23.6B), while others facilitate growth based on their availability from an optimal target. Semaphorin signaling through the plexin receptors influences cytoskeletal integrity and cell adhesion, often by destabilizing the actin

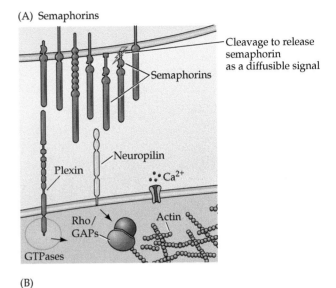

(A) Semaphorins

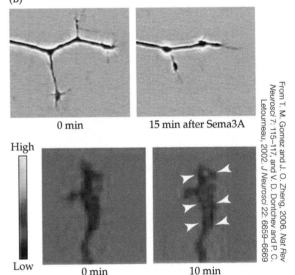

(B)

0 min   15 min after Sema3A

High

Low
0 min   10 min

From T. M. Gomez and J. O. Zheng, 2006. *Nat Rev Neurosci* 7: 115–117, and V. D. Dontchev and P. C. Letourneau, 2002. *J Neurosci* 22: 6659–6669

FIGURE 23.6 Semaphorins and their receptors mediate chemorepulsion (A) Many semaphorin ligands are either transmembrane proteins or membrane-associated proteins. Some can be proteolytically cleaved to release a soluble form of the ligand. There are two classes of semaphorin receptors, the plexins and the neuropilins. Both classes transduce semaphorin signals to influence GTPase activity and cause changes in the actin cytoskeleton. (B) Semaphorin signaling can collapse a growth cone, leading to process retraction (top panels). This activity is accompanied by Ca^{2+} influx into the growth cone upon exposure to semaphorin (bottom panels). The arrows indicate filopodial extensions that have begun to retract in response to semaphorin, and the subsequent increase in intracellular Ca^{2+}

cytoskeleton in an extending process. The plexins in vertebrates compose a large family of receptors that signal via their GTPase-activating domain, allowing GTPases to negatively regulate the cytoskeleton and cell adhesion. Plexins also act through the Rho GTPase–activating proteins (Rho/GAPs) to modulate the cytoskeleton via GTPase activity. A smaller set of semaphorin receptors, the neuropilins (there are two), interact primarily with a subset of semaphorins, the Sema3s. Neuropilins interact with protein scaffolds to ultimately regulate cell-cell adhesion. Thus, semaphorin signaling, based on diverse ligands, receptors, and signal transduction mechanisms, can provide substantial guidance information to growing axons.

Secreted molecules that are used as inductive and patterning signals at earlier stages of neural development (see Chapter 22) can also attract or repel growing axons. In particular, secreted Sonic hedgehog (Shh) and Wnts can act as chemoattractant or chemorepulsive signals. The attractive or repulsive influence of each depends on the capacity of the growing axon to "read" the Shh or Wnt signal via distinct signal transduction molecules expressed by the relevant growing axon. In addition, several other growth factors with growth-promoting activity in non-neural tissues act as secreted signals for directed axon growth. These include hepatocyte growth factor (HGF), initially characterized in the liver, and vascular endothelial growth factor (VEGF), which also promotes blood vessel differentiation and directed growth. These factors are expressed by targets to attract the growing axons of specific subsets of neurons. HGF is expressed in mesenchymal cells that will give rise to limb and craniofacial muscles, and it can mimic the chemoattractant properties of these cells for motor neurons that will innervate limb and facial muscles. VEGF can act as a chemoattractant for dorsal root ganglion neurons and autonomic neurons in peripheral targets where it also promotes vascular growth. VEGF can also be secreted by floorplate cells in the spinal cord, and it subsequently cooperates with netrin and other molecules to attract dorsal horn axons to the ventral midline. The broad array of secreted factors that act on growing axons as well as other non-neuronal cells indicates that these signals in the developing embryo coordinate axon growth perhaps to match it with target tissues where the same signals guide morphogenesis and differentiation.

Nondiffusible signals for axon guidance

The complex behavior of growth cones during axon extension suggests the presence of specific cues that are anchored to the chosen pathway to maintain a particular direction of axon growth, much like the fixed markers and blazes on a woodland trail. Unlike secreted, diffusible tropic signals, these cues would be stationary presumably to keep axons "on track" to a particular target that has been chosen rather than encourage them to grow in new directions. These nondiffusible signals include components of the extracellular matrix or ligands and receptors available on the surfaces of cells in regions through which axons grow. Some of these molecules can be found on the axons themselves, where they can coordinate the growth of multiple axons toward a single target along a pathway

established by a pioneer axon (see Concept 23.2)—a phenomenon known as **fasciculation** (see Figure 23.5A,B). Axon fasciculation relies on the capacity of growth cones of the follower axons to recognize the cell surface of a pioneer axon or their fellow axon as a more attractive substrate for growth. Fasciculation, as well as other aspects of directed axon growth, depends on families of nondiffusible cell adhesion molecules found either in the fairly stationary extracellular matrix upon which axons extend or anchored to the axon or growth cone membrane. These nondiffusible signals initiate intracellular signaling cascades that can alter the actin or microtubule cytoskeleton, promote membrane apposition, alter the complement of receptors and channels on the cell surface, or modify gene expression. The association of specific cell adhesion molecules with axon growth is based on experiments either in vitro, in which addition or removal of a particular molecule modifies the extent or direction of axon growth; or in vivo, in which genetic mutation, deletion, or manipulation disrupts the growth, guidance, or targeting of a particular axon pathway (Box 23A). Even though there are a daunting number of them, the molecules known to influence axon growth and guidance can be grouped into families of ligands and their receptors (Figure 23.7). The major classes of nondiffusible axon guidance signals are the extracellular matrix molecules and their integrin receptors; the Ca^{2+}-independent cell adhesion molecules (CAMs); the Ca^{2+}-dependent cell adhesion molecules, or cadherins; and the ephrins and Eph receptors.

The **extracellular matrix (ECM) cell adhesion molecules**, originally identified in the basal lamina (or basement membranes) of epithelial tissues (see the next paragraph and Figure 23.1C), were the first to be associated with axon growth. As their family name indicates, these adhesion molecules are found in the ECM, an adhesive macromolecular complex consisting of secreted proteins that aggregate outside cells. The most prominent members of this group are the **laminins**, the **collagens**, and **fibronectin** (see Figure 23.7C). The ECM's components can be secreted by a cell itself or by its neighbors; however, rather than diffusing away from the cell after secretion, they form polymers and create a durable local extracellular meshwork in the target region. A broad class of cell surface receptors known as **integrins** binds selectively to ECM molecules. Integrins are transmembrane proteins with intracellular domains; however, they do not have kinase activity or any other direct signaling capacity. Instead, the binding of laminin, collagen, or fibronectin to integrins triggers a signaling cascade via interactions between the cytoplasmic domains of integrins with cytoplasmic kinases or other signaling molecules that can stimulate axon growth and elongation. ECM–integrin signal transduction can occur via non-receptor cytoplasmic kinases such as the SRC kinases (SRC indicates the initial identification of this kinase

(A)

(B)

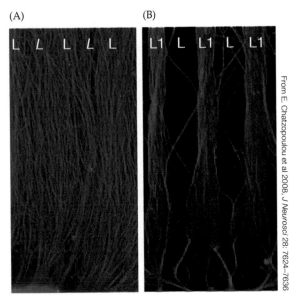

FIGURE 23.7 Cell adhesion molecules promote directed growth and fasciculation (A) Retinal axons grown in cell culture on a substrate of the extracellular matrix (ECM) molecule Laminin provided in alternating stripes of low (L) and high (*L*) concentration. The axons grow equally well on either concentration. (B) Retinal axons grown on alternating stripes of laminin (L) and the cell surface adhesion molecule L1. L1 elicits fasciculation, even though laminin can support axon guidance, suggesting that there are hierarchies of adhesive influences. (C) ECM molecules, including fibronectin and several isoforms of both laminin and collagen, serve as the ligands for multiple integrin receptors. Integrins transduce ECM signals by interacting with cytoplasmic protein kinases and activating Ca^{2+} channels. (D) Homophilic, Ca^{2+}-independent cell adhesion molecules (CAMs) are at once ligands and receptors. Homophilic binding activates intracellular kinases, leading to cytoskeletal changes. (E) Ca^{2+}-dependent adhesion molecules, or cadherins, are also capable of homophilic binding. They signal via activation of β-catenin, which influences gene expression. (F) Ephrins, which can be either transmembrane or membrane-associated, signal via the Eph receptors, which are receptor tyrosine kinases.

(C) Extracellular matrix molecules

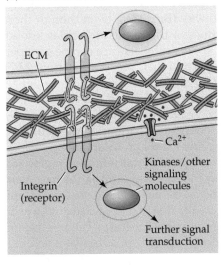

(D) CAMs

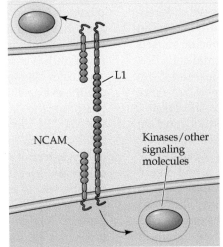

(E) Cadherins

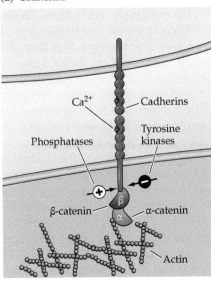

(F) Ephrins

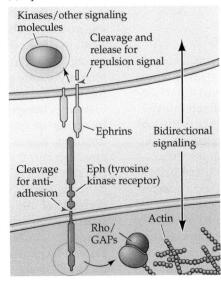

family in sarcoma tumor) as well as via second messengers, including cyclic nucleotides (cGMP and cAMP) and their downstream targets, or Ca^{2+}.

The role of ECM molecules in axon guidance is particularly clear in the embryonic PNS. Axons extending through peripheral sites such as the limbs grow through loosely arrayed mesenchymal cells (see Chapter 22) that fill the interstices of the embryo (between sheets of epithelial cells) before condensing to form muscles and bone. The spaces between these mesenchymal cells are rich in ECM molecules. Peripheral axons also grow along the interface of mesenchyme and epithelial tissues, including the boundaries between the neural tube and somites and between the mesenchyme and the epidermis. At these mesenchymal–epithelial interfaces, the **basal lamina**, which consists of organized sheets of ECM molecules and secreted proteins that bind to these ECM molecules, provides a supportive substrate for axon growth. The instructive capacity of ECM molecules is also essential for the directed regrowth of peripheral axons after injury in mature animals (see Chapter 26), suggesting that the mechanisms used for initial axon development can be reactivated for nervous system repair. In tissue culture as well as in the periphery of the embryo, different ECM molecules have

■ BOX 23A │ Choosing Sides: Axon Guidance at the Optic Chiasm

The functional requirement that a subset of axons from retinal ganglion cells in each eye must cross while the remaining axons project to the ipsilateral side of the brain was predicted based on optical principles—most notably by Sir Isaac Newton in the seventeenth century—and confirmed (much later) by neuroanatomists and neurophysiologists (see Chapter 12). The partial crossing, or *decussation*, of retinal axons is most striking in primates, including humans, in which approximately 50% of the axons cross and the other half do not. Although all other mammals also have crossed and uncrossed retinal projections, the percentage of uncrossed axons diminishes from 20%–30% in carnivores to less than 5% in most rodents. The frequency of uncrossed axons decreases even more in other vertebrates; thus, in amphibians, fish, and birds most or all of the retinal projection is crossed. For both functional and evolutionary reasons, the partial decussation of the retinal pathways and its variable extent in different species have engaged the imagination of biologists and others interested in vision for centuries.

For developmental neurobiologists, this phenomenon raises an obvious question: How do retinal ganglion cells "choose sides" such that some project contralaterally and others ipsilaterally? This question is central to understanding how the peripheral visual projection is organized to construct two accurate visual hemifield maps that superimpose points of space seen jointly by the two eyes (see Chapter 12). It also speaks to the more general issue in neural development of how axons distinguish between ipsilateral and contralateral targets in a bilaterally symmetric brain.

It is clear that the laterality of retinal axons is determined by initial cell identity and axon guidance mechanisms rather than by regressive processes that subsequently select or sculpt these projections. Thus, the distinction between the nasal and temporal retinal regions that project ipsilaterally and contralaterally is already apparent in the retina—as well as in axon trajectories at the midline and in the developing optic tract—long before the axons reach their targets. In the retina, this specificity is seen as a "line of decussation," or border, between ipsilaterally and contralaterally projecting retinal ganglion cells.

The line of decussation can be detected experimentally by injecting a retrograde tracer into the nascent optic tract of very young embryos. In the retinas of such embryos, there is a distinct boundary between the population of retinal ganglion cells projecting ipsilaterally in one eye (found in the temporal retina), as well as a complementary boundary for contralaterally projecting cells in the other eye (Figure). A molecular basis for this specificity was initially suggested by studies of albino mammals, including mice and humans. In albinos, in which single-gene mutation disrupts melanin synthesis throughout the animal, including in the pigment epithelium of the retina, the ipsilateral component of the retinal projection from each eye is dramatically reduced, the line of decussation in the retina is disrupted, and the distribution of glia and other cells in the vicinity of the optic chiasm is altered. These and other observations suggested that identity of retinal axons with respect to decussation is established in the retina, and further reinforced by axonal "choices" influenced by cues provided by cells within the optic chiasm.

Analysis of growth cone cell morphologies has shown that the chiasm is indeed a region where growth cones explore the molecular environment in a particularly detailed way, presumably to make choices pertinent to directed growth. Furthermore, molecular analysis reveals specialized neuroepithelial

(Continued)

(A) +/+

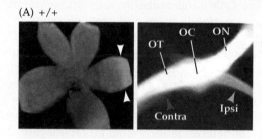

(B) *Zic2*kd/+

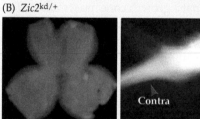

(C) *Zic2*kd/kd

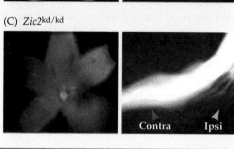

(A) At left, a small population of *Zic2*-expressing retinal ganglion cells (arrowheads) is seen in the ventrotemporal region of the normal retina (mounted flat by making several radial cuts). At right, the normal projection of one eye via the optic nerve (ON), through the optic chiasm (OC), and into the optic tract (OT) has been traced using a lipophilic dye placed in one eye. After the chiasm, labeled axons can be seen both in the contralateral (Contra) as well as the ipsilateral (Ipsi) optic tract. (B) When *Zic2* function is diminished in a mouse heterozygous for a *Zic2* knock-out mutation (in which expression of *Zic2* protein is diminished but not eliminated), the number of ipsilateral axons in the optic tract is similarly diminished. (C) When *Zic2* function is further diminished in homozygous *Zic2* knock-out mice, the ipsilateral projection can no longer be detected in the optic tract; thus, each optic tract consists of contralateral axons.

■ **BOX 23A** | **Choosing Sides: Axon Guidance at the Optic Chiasm** (*continued*)

cells in and around the chiasm that express several cell adhesion molecules associated with axon guidance. Interestingly, some of these molecules—particularly netrins, slits, and their robo receptors—do not influence decussation in the chiasm as they do in other regions of the nervous system. Instead, they are expressed in cells where the chiasm forms, apparently constraining its location on the ventral surface of the diencephalon. The establishment of ipsilateral versus contralateral identity is evidently more dependent on the zinc finger transcription factor Zic2, as well as on cell adhesion molecules of the ephrin family. Zic2, which is expressed specifically in the temporal retina, is associated with the expression of a distinct receptor, EphB1, in the axons arising from temporal retinal ganglion cells. The ephrin B2 ligand, which is recognized as a repellent of EphB1 axons, is found in midline glial cells in the optic chiasm. In support of the functional importance of these molecules, disrupting *Zic2*, *EphB1*, or *ephrin B2* gene function diminishes the degree of ipsilateral projection in developing mice; in accord with this finding, neither the *Zic2* nor the *ephrin B2* gene is expressed in vertebrate species that lack ipsilateral projections.

These observations provide a molecular framework for the identification of retinal ganglion cells and the sorting of their projections at the optic chiasm. How this sorting is related to the topography of tectal, thalamic, and cortical representations is not yet known. Most observations suggest that retinal topography is not faithfully preserved among axons in the optic tracts. The identity and position of axons from nasal and temporal retinas whose retinal ganglion cells "see" a common point in the binocular hemifield must therefore be restored in the thalamus and subsequently retained or reestablished in the thalamic projections to cortex. Choosing sides at the chiasm is only a first step in establishing maps of visual space.

---

different capacities to stimulate axon growth. The role of these molecules for axon growth in the CNS is less clear. Some of the same molecules are present in the extracellular spaces of the CNS, but they are not organized into a basal lamina as in the periphery, and therefore have been more difficult to study.

The **CAMs** and **cadherins** are transmembrane proteins found on growth cones and growing axons as well as on surrounding cells and targets (see Figure 23.7D,E). Moreover, both CAMs and cadherins have dual functions as ligands and receptors, usually via homophilic ("like with like") binding. Some of the CAMs, especially the L1 CAM, have been associated with fasciculation of groups of axons (see Figure 23.7B). Cadherins have been suggested as important determinants of final target selection during the transition that a growth cone must make to form a synapse (see later in this concept). There are several cadherins, all encoded by distinct genes, that influence axon growth at multiple locations in the developing nervous system. For both CAMs and cadherins, the unique ability of each class to function as both ligand and receptor (L1 CAM, for example, is its own receptor) may be important for recognition between specific sets of axons that fasciculate to form a coherent nerve or tract as well as their targets. This ligand–receptor duality encompasses both homophilic binding for like-recognition (potentially important for axon fasciculation) as well as heterophilic binding, which amplifies the potential for selective growth and guidance (potentially important for target discrimination). Both CAMs and cadherins rely on a somewhat indirect route of signal transduction because, like integrins, they have no intrinsic enzymatic activity. The Ca^{2+}-independent CAMs interact with cytoplasmic kinases to initiate cellular responses, whereas the Ca^{2+}-dependent cadherins engage the β-catenin pathway (also activated by Wnts; see Chapter 22).

A large family of **ephrin** ligands and their tyrosine kinase receptors (**Eph receptors**, or **Ephs**) constitute cell-cell recognition codes in a variety of tissues (see Figure 23.7F and Concept 23.5). Ephrins and Ephs function as nondiffusible axon guidance molecules in the developing nervous system. Growing axons use ephrins and Eph receptors to recognize appropriate pathways for growth as well as appropriate sites for synaptogenesis (see later in this concept). Although they are identified as ligands and receptors, the binding of ephrins with Ephs can initiate "reverse" signaling via the ephrins, which can interact with cytoplasmic protein kinases and cause changes in the cell expressing the ephrin ligand. This bidirectional signaling via ephrins and Ephs can alter the state of both axon and target. Ephrins and Ephs activate a variety of signaling pathways and, depending on the nature of signal transduction, can be either growth-promoting or growth-limiting. To limit axon growth, the extracellular domain of an ephrin ligand can be proteolytically cleaved, or Eph receptors can be removed via selective endocytosis, thus terminating signaling.

The dependence of axon growth and guidance on adhesive interactions and the signal transduction pathways activated by these interactions is underscored by the pathogenesis of several inherited human developmental or neurological disorders. These syndromes include X-linked hydrocephalus; MASA (an acronym for *m*ental retardation, *a*phasia, *s*huffling gait, and *a*dducted thumbs); Kallmann syndrome (which compromises reproductive and chemosensory function); X-linked spastic paraplegia; and several even rarer disorders (Clinical Applications). Some are consequences of mutations in genes encoding the Ca^{2+}-independent CAMs such as L1, while others compromise secreted guidance signals or additional mechanisms that influence axon growth. These mutations can also lead to the partial absence of the corpus callosum that connects the two cerebral hemispheres (referred to as *callosal agenesis*), or of the corticospinal tract, which carries cortical information to the spinal cord. These rare congenital anomalies are now understood to arise from mutations in genes that result in loss of function of cell surface adhesion molecules that support axon fasciculation and guidance.

Directed dendritic growth: Ensuring polarity

The establishment and maintenance of neuronal polarity—making one side of a neuron the origin of the axon, and the opposite side the origin of dendrites—was first described by Cajal at the outset of the twentieth century. (Cajal seems to have anticipated just about every aspect of modern neuroscience!) The maintenance of polarity is essential for building a nervous system that can consistently manage reception and integration of synaptic inputs and generation of synaptic outputs to define neural circuit function. The first step in neuronal polarization identifies the axon as the equivalent of the basolateral secretory domain of simpler polarized epithelial cells (see Concept 23.1) via activity of PAR proteins and subsequent stabilizers of the axon cytoskeleton (see Figures 23.1 and 23.2). This step, however, defines only the general site for synaptic outputs versus inputs onto a target neuron. The second step is the maintenance, guided growth, and local branching of dendrites that will provide appropriate post-synaptic sites for the terminals of growing axons. In the CNS, several classes of neurons, particularly projection neurons, have striking **dendritic polarization** that underlies their unique information-processing capacity due to the numbers and types of axonal inputs these dendritic arbors can accommodate (see Chapter 1). Retinal ganglion cells, olfactory

---

■ Clinical Applications

Axon Guidance Disorders

The disruption of axon guidance during human development has long been assumed to result in serious brain disorders. Until recently, however, it has been difficult to identify such disruptions and their specific anatomical and behavioral consequences. Parallel discoveries in human genetics and structural and functional brain imaging have made it possible to identify mutations in genes that regulate axon guidance during development, and to visualize the consequences for axon pathways in the mature human brain. There is now a growing list of rare but highly informative mutations in genes known to regulate growth cone pathfinding or axon extension during development (see table). These mutations cause distinct human brain diseases. The mutations can disrupt both CNS axon tracts and peripheral nerves (including cranial nerves) that control facial expression and eye movements. Their results include mild and specific motor or sensory deficits as well as more global difficulties, including intellectual disability and social and cognitive impairments. Four of these

deficits are particularly noteworthy based on their established functions in axon guidance and growth.

- The gene for the Ig-superfamily cell adhesion molecule **L1**, which is found on the X chromosome, is associated with a fairly severe syndrome (seen in males, since it is X-linked) that includes intellectual and language disabilities and severely diminished motor control.

- Mutations in **ROBO3**, a receptor for the slit family of secreted chemorepellant ligands, are associated with a rare disorder called *horizontal gaze palsy with progressive scoliosis*. Individuals with this disorder have limited or no ability to move their eyes horizontally, as well as postural and motor control difficulties that get worse with age.

- **KIF21A** is a member of the large family of kinesin motor proteins that mediate organelle and membrane traffic in the axon via interactions with the microtubule cytoskeleton. This particular kinesin is mutated in individuals with *congenital fibrosis of the extraocular muscles*, which

reflects diminished growth of the oculomotor and abducens nerves.

- The Ig-superfamily secreted adhesion signaling molecule **KAL1** is one of several mutant proteins involved in *Kallmann syndrome*, which disrupts olfactory nerve development peripherally and olfactory tract development centrally. Individuals with Kallmann syndrome are anosmic (thus, the KAL1 protein is also known as **anosmin**); they are also infertile because of the failure of gonadotropin-releasing hormone (GnRH) neurons to migrate from the olfactory placode to the hypothalamus (see Chapter 14).

In all four of these conditions, the mutated genes encode either cell adhesion molecules or cytoskeletal regulators of axon growth.

The advent of **diffusion tensor imaging (DTI)**, an imaging approach that allows visualization of axon pathways in the living human brain, has made it possible to see the presumed consequences of disrupted axon guidance. One of the most striking examples of axon guidance failure is agenesis of

(Continued)

■ Clinical Applications *(continued)*

(A) Midline   (B)   (C) Corticospinal tract   (D) Corticospinal tract   (E)

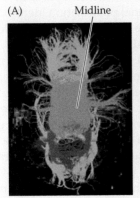 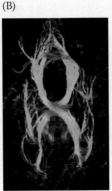 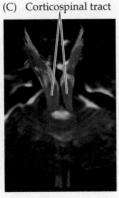  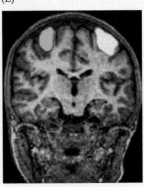

A,B from M. Wahl et al., 2009. Am J Neuroradiol 30: 282–289; C–E from S. Haller et al., 2008. Neuroradiology 50: 453–459

Diffusion tensor imaging (DTI) of a typical human compared with that of individuals who have pathologies of axonal guidance. DTI is a refinement of fMRI that applies algorithms to estimate the diffusion direction of water molecules in each voxel imaged. Because axon bundles tend to constrain the diffusion of water molecules in a single direction, DTI can identify major axon tracts. (A) In the control individual, the corpus callosum is seen as a corona of fibers crossing the midline (orange). (B) In an individual with partial agenesis of the corpus callosum, only a few fiber bundles cross the midline. (C,D) DTI images of an individual with horizontal gaze palsy with progressive scoliosis. The corticospinal tract (blue fibers) can be seen coursing through the pons (C) and the medulla (D). The tract does not cross at any brainstem level. (E) When the individual in (C,D) is asked to do a simple manual motor task, fMRI imaging reveals only ipsilateral activation in the motor cortex in each hemisphere rather than contralateral or bilateral activation.

the corpus callosum (Figures A and B), seen in a variety of developmental disorders, including L1 syndrome. Similarly, DTI has demonstrated that most axon pathways, including ascending sensory and descending motor tracts, fail to cross in the medulla in individuals with horizontal gaze palsy (Figures C and D). This anatomical failure is accompanied by functional changes, including anomalous, completely ipsilateral activation of the motor cortex during independent hand movements (Figure E).

The genetic and imaging observations described here reinforce the long-held view that axon guidance—because of its molecular complexity and the critical requirement of this process to achieve appropriate brain connections—is a likely target for diseases that alter brain function. The rare disorders associated with adhesion and related molecules indicate that specific central and peripheral axon pathways can be altered during development, with corresponding behavioral alterations seen throughout

the lifetime of the individual. It is also likely that disrupted axon guidance contributes to the pathogenesis of a wide range of clinically defined developmental disorders, including autism, attention deficit hyperactivity disorder (ADHD), and schizophrenia. Resolving the contribution of specific axon guidance molecules and disrupted axon pathways to these disorders will require additional genetic analysis and refinement of imaging techniques to observe subtle differences in connections in the brain.

Disorders Arising from Defects in Pathfinding or Axon Extension during Development[a]

Disorder	Mutant gene	Gene function	Axon pathways affected
L1 syndrome (also known as CRASH and MASA)	L1 (X-linked)	Ig-superfamily cell adhesion	Corpus callosum, corticospinal tract
Horizontal gaze palsy with progressive scoliosis	ROBO3	Adhesion receptor for chemo-repellant slit secreted proteins	Corticospinal tract, trochlear nerve
Congenital fibrosis of the extraocular muscles, type 1	KIF21A	Motor protein for microtubule-dependent organelle transport	Oculomotor, trochlear, abducens
Kallmann syndrome	KAL1 (anosmin)	Secreted Ig-superfamily adhesion signal	Olfactory nerve, olfactory tract, terminal nerve

[a]Developmental evidence for the function of these genes in the human disorders listed is based on studies of parallel (orthologous) genes in animals.

bulb mitral cells, cerebellar Purkinje cells, and cortical pyramidal neurons represent major neuron classes whose dendritic polarity and subsequent dendritic growth are both dramatic as well as essential for appropriate circuit function (see Chapters 1, 15, 19, and 33). Divergent cytological differentiation of dendritic versus axonal processes in these neurons requires molecular mechanisms that, quite literally, can tell one end of a neuron from the other. These include selective transport of organelles such as Golgi outposts (see Concept 23.1) and other targeted membrane vesicles that contribute to process growth and insertion of specific transmembrane proteins in growing dendrites versus axons (see Figure 23.2).

The growth and elaboration of dendrites, once they are established as distinct cellular domains, rely on nondiffusible adhesion-promoting and -inhibiting molecules that also influence axon growth and guidance: semaphorins, neuropilins, and related signaling intermediates (Figure 23.8). This mechanism has been best characterized for cerebral cortical neurons (see Figure 23.8A); however, similar molecular mechanisms are used by developing retinal ganglion cells and cerebellar Purkinje cells, which also must generate distinctive dendritic arbors (see Figure 23.8B). The chemotropic cell surface molecules Semaphorin 3A (Sema3A) and slit1 (see Figures 23.5 and 23.6) are central for directed growth of cerebral cortical projection neuron dendrites. Sema3A simultaneously repels the axons of developing cortical pyramidal neurons, as well as those of retinal ganglion and cerebellar Purkinje neurons, while acting as a chemoattractant for the dendrites of the same cells (see Figure 23.8C). Thus, in the developing cerebral cortex, secreted Sema3A is found at high concentrations in the marginal zone, the outermost layer of the nascent cortex, toward which the apical dendrites of cortical pyramidal neurons grow. This Sema3A *attracts* developing dendrites via neuropilin receptors. Polarized differences in the distribution of downstream signaling intermediates, including the local availability of soluble guanylyl cyclase (sGC) in the differentiating dendrite, convert the normally repulsive Sema3A/neuropilin signal into a chemoattractant signal. This downstream conversion does not happen in the axon, where sGC is absent, and thus the axon is repelled by the same Sema3A signal that attracts the dendrites, and it grows in the opposite direction of the dendrite. In parallel, the chemorepellant secreted signal slit1 also repels cortical projection neuron axon growth to ensure that the axon continues to grow toward the ventricular rather than pial surface of the developing cortex (see Figure 23.8A). Local Notch signaling reinforces the primary consequences of semaphorin signaling for dendritic polarization in the cortex. In addition, neurotrophins (see Concept 23.4 and Chapter 25), including brain-derived neurotrophic factor (BDNF), promote subsequent dendritic growth in the differentiating cortical neuropil (see Figure

23.8C. These adhesion and trophic signals are key for establishing and maintaining polarized growth of cortical pyramidal neurons as well as Purkinje cells and other dramatically polarized neuronal cell types such as retinal ganglion cells. These signals also influence growth and differentiation of local GABAergic interneurons, whose dendrites are often less polarized and for which axon growth is limited. Finally, although the overall directed growth of dendrites versus axons is clear, especially in highly polarized neurons, the local elaboration of dendrites can be dynamic (see Figure 23.8B). Thus, as they grow locally, dendritic branches are often retracted or remodeled. This dynamism probably reflects mechanisms that regulate the spacing and density of dendritic branches as well as axon terminal branches—referred to as **tiling**—in brain regions such as the cerebellum, cortex, and retina.

Dendritic and axonal tiling: Defining synaptic space

One last step in neuronal differentiation precedes the major phase of synapse formation and establishment of functional circuits. This step ensures proper modulation of dendritic and axonal growth so that each post- or presynaptic arbor occupies appropriate space to establish an optimal number or density of specific connections. This mechanism, tiling, was first observed in the PNS of *Drosophila* and has subsequently been analyzed in the developing CNS of mammals, especially mice. There are two key outcomes of this aspect of control of initial dendritic and axonal differentiation: First, developing dendrites and axons are regulated so they do not grow toward or become entangled with nearby dendrites or axons from the same neuron; and second, developing dendrites or axons from neighboring neurons are repelled from one another to a greater or lesser degree to ensure that each neuron's dendritic or axonal arbor provides adequate and orderly "coverage" for a particular region in a developing neural structure such as the retina or the cerebral cortex. Both dendrites and axons must be tiled. The mechanisms for dendritic versus axonal tiling, though similar in outcome, differ based upon the genes and proteins that mediate each.

The mechanisms for tiling must accomplish an intriguing feat: dual repulsion. First, each neuron must respond to molecular signals, presumably from its own dendrites or those of its nearest neighbors that are of the same cell type. These signals prevent dendrites of the same neuron from growing on top of or entangled with one another or crowding its nearest neighbor. Next, each neuron must respond to cues, either from the dendrites or axons of other neurons, that restrict the territories of each arbor within a local array of processes from similar neurons. A novel molecular mechanism has evolved to mediate this essential aspect of growth and avoidance of dendrites in space. The molecular basis for dendritic tiling was first defined

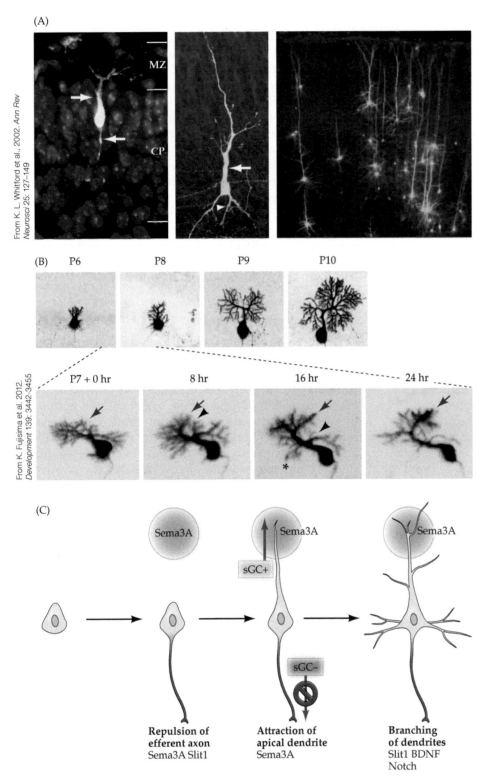

(A)

MZ

CP

From K. L. Whitford et al., 2002. *Ann Rev Neurosci* 25: 127–149

(B)   P6   P8   P9   P10

From K. Fujisima et al. 2012. *Development* 139: 3442-3455

P7 + 0 hr   8 hr   16 hr   24 hr

*

(C)

Sema3A   Sema3A   Sema3A

sGC+

sGC−

Repulsion of   Attraction of   Branching
efferent axon   apical dendrite   of dendrites
Sema3A Slit1   Sema3A   Slit1 BDNF
                                             Notch

FIGURE 23.8 Polarized dendritic growth relies on secreted signals (A) At left, the initial polarized differentiation of dendrite (upper arrow) and axon (lower arrow) of a cortical pyramidal or projection neuron. This neuron, which is migrating through the cortical plate (CP) on its way to its final destination at the outermost region of the developing cortex, has already begun to respond to directional cues in the marginal zone (MZ). Subsequently (middle image), the apical dendrite (oriented toward the MZ) begins to grow and branch. In addition, at this time the basal dendrites (arrowhead) of the neuron begin to differentiate, although they grow only in the local territory close to the cell body. (B) Top row: Dendritic branching in Purkinje cells during postnatal days 6–10 in different individual mice reflects extensive and rapid remodeling of the dendritic arbor. Bottom row: A single Purkinje cell from one mouse visualized over 24 hours on postnatal day 7. There is detectable remodeling of the dendritic arbor, Compare the branches indicated by the red arrows and black arrowheads, where differential growth (red) versus retraction (black) can be seen. The asterisk indicates a transient dendritic branch that appears and is retracted over the course of 16 hours. (C) A schematic of signals necessary for polarized, opposing growth of dendrites and axons. In this instance, signaling for a cortical pyramidal cell is outlined. Semaphorin 3A acts as both a chemoattractive and a chemorepulsive signal based on the activity of a soluble guanyl cyclase (sGC) that is available in dendrites but not in axons to convert the chemorepulsive activity to chemoattractive signaling This general mechanism also operates for retinal ganglion cells, cerebellar Purkinje cells, and several other classes of neurons with distinctly polarized axons and dendrites. (C after K. L. Whitford et al., 2002. *Ann Rev Neurosci* 25: 127–149.)

in the fly, based on the remarkable genomic structure of a single gene, *DSCAM1*, its close relative *DSCAM2* and the varied array of adhesion proteins these genes encode. In the fly, *DSCAM* has multiple splice acceptor and donor sites distributed over four exons (Figure 23.9A). This genomic structure yields a gene that theoretically can encode

at least 37,000 variants, and this flexibility is key for *Drosophila* dendritic tiling. When the *DSCAM* gene is unperturbed, dendrites grow in a normal distribution, avoiding their sibling dendrites and establishing reasonable spacing between dendrites from neighboring neurons (Figure 23.9B). When the *DSCAM* gene is mutated so that fewer splice variants are possible, or it is not expressed at all, dendritic tiling is impaired; sibling dendrites grow closer or on top of one another, and neighboring dendrites no longer keep their distance (Figure 23.9C). This has been seen for *Drosophila* sensory neurons whose dendrites cover the body wall, usually in an evenly distributed pattern, as well as for *Drosophila* retinal neurons whose connections must be precisely matched to the omatidial structure (columnar units defined by photoreceptors and adjacent processing neurons) of the fly's eye. The fundamental mechanism for dendritic tiling relies on distinctions between homophilic binding between DSCAM isoforms of the same identity, which leads, perhaps counterintuitively, to repulsion, so that dendrites from the same neuron recognize one another and don't grow on top of each other. Heterophilic binding, or the absence of a DSCAM variant on a nearby process, results in permissive growth that can lead to apposition, fasciculation, and other forms of dendritic contact.

Mammals also have a *DSCAM* gene. In fact, the acronym for *DSCAM* is derived from the location of the human orthologue, found on human chromosome 21, the chromosome duplicated in Down syndrome: *DSCAM* stands for *Down syndrome cell adhesion molecule*. The mammalian *DSCAM* gene, including that in humans, does not encode the large number of splice isoforms encoded by its *Drosophila* counterpart. Nevertheless, *DSCAM* does mediate some forms of dendritic tiling. This is particularly recognizable in the retinas of mice that carry a homozygous loss-of-function mutation in the *DSCAM* gene. In the retinas of wild-type mice, a key class of intraretinal relay neuron, amacrine cells (*amacrine* means "star-shaped"), is tiled across the retinal surface. In contrast, in the retinas of *DSCAM* mutant mice, dendrites of individual amacrine cells grow on top of one another, and the regular tiled distribution is completely disrupted (see Figure 23.9C). The sum of these repulsive and permissive growth mechanisms is to constrain neuronal differentiation so that the distribution of dendrites is optimized for the numbers of local and afferent synapses that will be made. These interactions presumably facilitate quantitative specificity of connections to ensure functional integrity. These mechanisms may also act to establish arrays of dendrites and axons whose initial patterns and numbers of connections are prepared for further activity-mediated specification and stabilization of synapses (see Chapter 24).

Axons and their terminal arbors are also tiled in target regions. This tiling is thought to underlie appropriate distribution of axon endings with distinct functions: for example, non-overlapping patterns of innervation of subclasses of somatosensory receptor axons in the skin (see Chapters 9 and 10), or the segregation of olfactory receptor axons so that they innervate only a subset of glomeruli in the olfactory bulb. This is also the case for multiple brainstem neurons—including dopaminergic, noradrenergic, and serotonergic neurons—whose axons are widely branched to provide broad modulatory inputs to the forebrain (see Chapter 6). Another class of alternatively spliced genes, the **protocadherins** (Figure 23.9D), are thought to regulate this example of tiling. The variable splicing of exons encoding the extracellular domain of the protocadherin proteins, combined with the invariant transcription of those encoding the intracellular domain, allows protocadherin isoforms to influence "self" versus "other" recognition for growing axons. The discrimination capacity of the protocadherins extends to variable expression from the paternal versus maternal alleles, allowing for even greater molecular diversity of recognition. The protocadherin "code" is complex; matching multiple or identical splice variants on the surfaces of two axons can result in recognition or avoidance. The axons of serotonergic neurons, located primarily in the dorsal raphe of the medulla, must distribute themselves in a non-overlapping pattern with fairly evenly spaced terminal fields throughout the striatum and the hippocampus, as well as several other targets. Serotonergic axons have a single isoform of protocadherin-α (Pcdh-α) on their surfaces that mediates homophobic repulsive interactions, resulting in axonal tiling. This regular distribution of serotonergic axon arbors and terminals is disrupted in mice that have a homozygous null mutation for Pcdh-α, or those in which both copies of Pcdh-α have been selectively eliminated by Cre-mediated recombination limited to serotonergic neurons. The serotonergic axons in these mutant mice lack the normally consistent spacing across forebrain targets such as the hippocampus (Figure 23.9E). This indicates that at least for these multiply branched axons that project broadly to several targets, a single, presumably repulsive, recognition mechanism prevents like axons from becoming entangled with one another or themselves, resulting in a continuous, appropriately spaced distribution of modulatory inputs.

Synapse formation

Once an axon reaches its target region, additional cell-cell interactions dictate which target cells to innervate from among a variety of potential synaptic partners. There are some absolute restrictions to synaptic associations. For example, neurons do not make synapses on nearby glial cells in the CNS or connective tissue cells in the PNS, even though glial cells can interact with pre-and postsynaptic domains to facilitate or stabilize synapses. Furthermore, instances have been described in which nerve and target cell types that are not normally interconnected show little or no inclination to establish synaptic partnerships with one another when confronted with the possibility by experimental manipulation, either in vivo or in vitro. When synaptogenesis does proceed, however, neurons and their targets in both the CNS and PNS appear

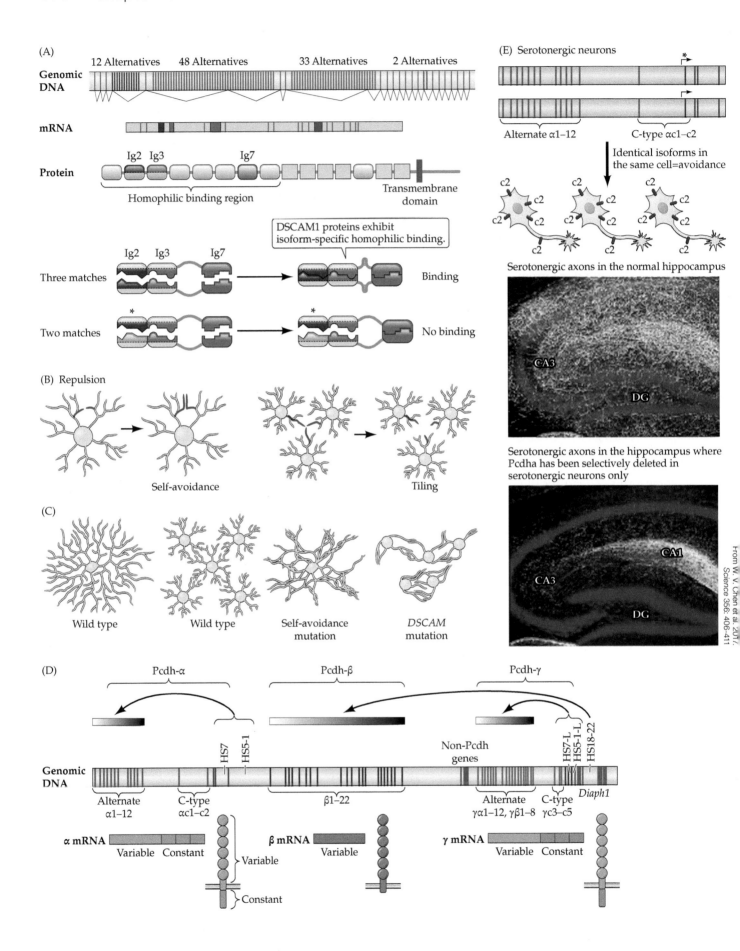

(A)

Genomic DNA — 12 Alternatives   48 Alternatives   33 Alternatives   2 Alternatives

mRNA

Protein — Ig2 Ig3   Ig7   Transmembrane domain

Homophilic binding region

DSCAM1 proteins exhibit isoform-specific homophilic binding.

Ig2 Ig3 Ig7

Three matches → Binding

Two matches → No binding

(B) Repulsion

Self-avoidance

Tiling

(C)

Wild type   Wild type   Self-avoidance mutation   *DSCAM* mutation

(D)

Pcdh-α   Pcdh-β   Pcdh-γ

HS7 HS5-1   Non-Pcdh genes   HS7-L HS5-1-L HS18-22

Genomic DNA

Alternate α1–12   C-type αc1–c2   β1–22   Alternate γα1–12, γβ1–8   C-type γc3–c5   *Diaph1*

α mRNA   Variable Constant   Variable Constant

β mRNA   Variable

γ mRNA   Variable Constant

Variable   Constant

(E) Serotonergic neurons

*

Alternate α1–12   C-type αc1–c2

Identical isoforms in the same cell=avoidance

c2

Serotonergic axons in the normal hippocampus

CA3   DG

Serotonergic axons in the hippocampus where Pcdha has been selectively deleted in serotonergic neurons only

CA3   CA1   DG

From W. V. Chen et al. 2017. *Science* 356: 406–411.

◀ **FIGURE 23.9   Splice variants, homophilic recognition or repulsion, dendritic and axon growth and tiling** (A) Top panel: In *Drosophila*, the *DSCAM1* gene has 12 alternative splice variants in exon 4, 48 variants in exon 6, 33 in exon 9, and 2 in exon 17. Middle panel: The variable mRNAs transcribed from this gene give rise to DSCAM proteins with variable homophilic immunoglobulin (Ig) adhesive domains. Bottom panel: If the three homophilic variable Ig domains are matched, there is recognition and binding. If the protein variability of one or more Ig domains is not matched (asterisk), recognition and binding do not occur. (B) A schematic of the consequences of repulsive interactions elicited by homophilic interactions of DSCAM1 variants. Within a single neuron (left panel), these repulsive interactions, based on homophilic binding, prevent branches of the same neuron from growing on top of one another, while the same mechanism for sets of different neurons (right panel) results in dendritic avoidance by each neuron of dendrites from other neurons, leading to dendritic tiling. (C) In the mouse retina, amacrine cells in a wild-type mouse are tiled, via murine DSCAM function. In the case of a *DSCAM* loss-of-function mutation, tiling fails; amacrine cell processes grow on top of one another, and cell bodies are no longer evenly spaced. (D) The clustered location of the three protocadherin genes α, β, and γ adjacent to one another on human

chromosome 5 . The orthologous protocadherins α, β, and γ are similarly clusted on mouse chromosome 18. Below each individual gene map, the mRNA and protein variation is shown. Exons in the alternative splice region of protocadherin α and γ, and all of the exons of protocadherin β can be combined to yield an array of distinct protocadherin transcripts and proteins from each single gene. For protocadherin α and γ there is also a constant region of exons that that encode an invariant intracellular domain. (E) In serotonergic neurons, whose axons project broadly throughout the forebrain, a single protocadherin α isoform consisting of the αc1 exon (purple) and the constant region (yellow) is expressed biallelically (thus there is equivalent mRNA transcribed from the maternal and paternal copy). This singular expression is key for repulsion between serotonergic axons and their terminal branches that results in axon tiling for broad uniform innervation in the mouse hippocampus. When the protocadherin α gene is selectively deleted in serotonergic neurons in the locus coeruleus in the mouse brainstem, their axons no longer broadly innervate the hippocampus and instead cluster in the CA1 subregion. (A,B,C [right] after D. Hattori et al., 2008. *Annu Rev Cell Dev Biol* 24: 597–620, C [left] J. L. Lefebvre et al., 2015. *Ann Rev Cell Dev Biol* 31: 741–777; D,E after M. Mountofaris et al. 2018. *Annu Rev Cell Dev Biol* 34: 471–493.)

to associate according to a continuously variable graded system of preferences. The target cells residing in muscles, autonomic ganglia, or the brain are certainly not equivalent—they have different locations along the body axes, and different functions. Nevertheless, they are not unique with respect to the innervation they can receive—neurons that might not normally innervate these targets can do so if provided an opportunity. This relative promiscuity most likely reflects the fact that potential pre- and postsynaptic sites may share many molecules that identify them as potential locations for making a connection. Thus, if there is a not a strong specific affinity, more generic recognition events occur, and anomalous connections will be made. This can result in functional disruption following neural injury and repair in the PNS and olfactory system, since regenerated patterns of innervation do not always faithfully follow original patterns of synaptic inputs (see Chapter 26).

Much of this imprecision may reflect the overlapping subsets of molecules that regulate general aspects of individual synapse formation as well as axonal and dendritic growth. Several observations show that many of the same adhesion molecules that participate in axon guidance contribute to the identification and stabilization of generic synaptic sites on target cells, as well as to the ability of a growing axon to recognize specific sites as optimal. Thus, in the first stages of synapse formation (Figure 23.10A), the ephrins, the Ca^{2+}-independent CAMs, and the cadherin families of Ca^{2+}-dependent adhesion molecules are all thought to influence a growth cone's recognition of suitable postsynaptic positions on dendrites, cell bodies, or other appropriate non-neural peripheral targets (i.e.,

muscle fibers). The presumptive presynaptic processes are derived by the conversion of a growth cone to an immature presynaptic terminal. In the next step, pre- and postsynaptic specializations must be elaborated for appropriate cellular specializations for synaptic communication (Figure 23.10B,C). Several soluble or secreted signals have been implicated in this process, including growth factors and neurotransmitters themselves. Subsequently, specialized synaptic adhesion molecules are expressed or recruited to the immature synaptic site that link pre- and postsynaptic domains so that the synapse emerges as a discrete, relatively stable intracellular specialization for local electrical or chemical signaling.

Among this list of molecules that initiate synaptogenesis, **neuregulin1 (Nrg1)**, a locally diffusible signaling protein, has emerged as an essential regulator of the expression and localization of postsynaptic receptors and other key proteins for the postsynaptic aspect of synaptic transmission. Nrg1 begins as a transmembrane protein usually made in presynaptic cells and can be released following proteolytic cleavage of the ectodomain (outside portion) of the protein. Cleaved, diffusible Nrg1, or the extracellular domain of its transmembrane form (particularly the region that contains an epidermal growth factor [EGF] repeat common to multiple signaling molecules), binds to specific receptors: the ErbB family of EGF-like receptors, found on the surfaces of many developing central neurons as well as on muscle cells and other targets of peripheral neurons. Nrg1 signaling is thought to elicit increased synthesis and insertion of neurotransmitter receptors at a nascent postsynaptic site. Intriguingly, the human *NRG1* gene is a site of multiple polymorphic changes (altered

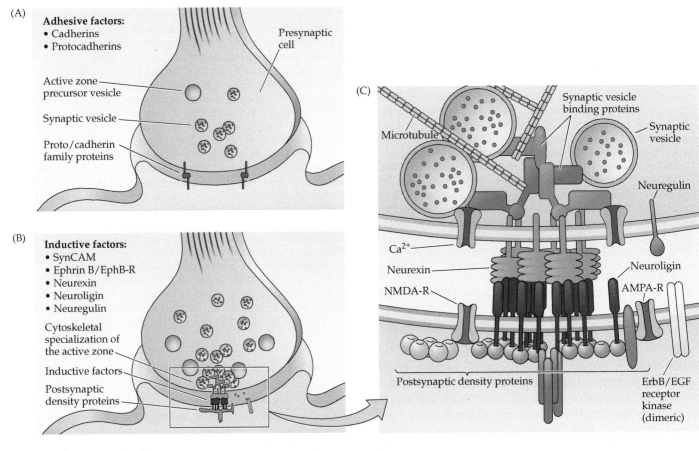

(A) **Adhesive factors:**
- Cadherins
- Protocadherins

Presynaptic cell

Active zone precursor vesicle

Synaptic vesicle

Proto/cadherin family proteins

(B) **Inductive factors:**
- SynCAM
- Ephrin B/EphB-R
- Neurexin
- Neuroligin
- Neuregulin

Cytoskeletal specialization of the active zone

Inductive factors

Postsynaptic density proteins

(C) Microtubule

Synaptic vesicle binding proteins

Synaptic vesicle

Neuregulin

Ca²⁺

Neurexin

Neuroligin

NMDA-R

AMPA-R

Postsynaptic density proteins

ErbB/EGF receptor kinase (dimeric)

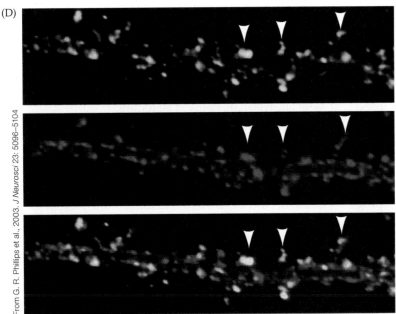

From G. R. Phillips et al., 2003. *J Neurosci* 23: 5096–5104

DNA sequences that differ from those of most individuals sequenced) associated with schizophrenia and other behavioral disorders thought to alter development or maintenance of synaptic connections. Individuals with these altered DNA sequences have a slightly (but significantly) increased probability of developing schizophrenia compared with individuals without the polymorphisms. The DNA sequence changes, however, do not result in altered amino acid sequences for Nrg1 protein (i.e., they occur in noncoding regions of the *NRG1* gene), and it remains

◄ **FIGURE 23.10   Molecular mechanisms involved in synapse formation** (A) Initiation of a synapse depends centrally on local recognition between the presumptive pre- and postsynaptic membranes mediated by members of the cadherin and protocadherin family of Ca^{2+} cell adhesion molecules (cadherins and protocadherins). This local recognition is accompanied by the initial accumulation of synaptic vesicles as well as transport vesicles that contain molecular components that contribute to the presynaptic active zone. (B) Once the initial specialization is established, additional adhesion molecules are recruited, including synaptic cell adhesion molecule (SynCAM), a member of the Ca^{2+}-independent, homophilic binding adhesion molecule family (like NCAM; see Figure 23.7B), neurexin and neuroligin (see panel C), and the ephrin B ligands and their EphB receptors. Adhesive signaling between these molecules initiates differentiation of the presynaptic active zone and the postsynaptic density. The presynaptic terminal also releases molecules (e.g., neuregulin) that influence the expression and clustering of postsynaptic receptors and associated proteins. (C) The interaction of neurexin (a presynaptic transmembrane adhesion protein) with neuroligin (a postsynaptic adhesion protein) is central for recruiting and retaining cytoskeletal elements that localize synaptic vesicles to the presynaptic terminal and mediate their fusion. In addition, neurexin is important for localizing voltage-gated Ca^{2+} channels to ensure local vesicle release. Neuroligin, upon binding neurexin, is essential for localizing neurotransmitter receptors and postsynaptic proteins to the postsynaptic specialization. Neuregulin is released via local proteolytic cleavage and binds to dimeric ErbB receptor kinases or to dimeric ErbB/epidermal growth factor (EGF) receptor kinases. (D) Distinct γ-protocadherin isoforms (green and red) are expressed at subsets of synaptic contacts on dendrites of hippocampal neurons in culture, suggesting that different synaptic sites may have different complements of adhesion molecules, perhaps conferring specificity to subsets of synapses. (A,B after C. L. Waites et al., 2005. *Annu Rev Neurosci* 28: 251–274; C after C. Dean and T. Dreshbach, 2006. *Trends Neurosci* 29: 21–29.)

uncertain how they might alter the gene's expression or the protein's activity.

Three families of adhesion molecules are particularly central to the construction of synapses: **neurexins**, found in the presynaptic membrane; their binding partners the **neuroligins**, found in the postsynaptic membrane; Neurexins and neuroligins bind one another and promote adhesion between the pre- and postsynaptic membranes (see Figure 23.10C). Neurexins have a specialized transmembrane domain that helps localize synaptic vesicles, docking proteins, and molecules that promote vesicle fusion in the presynaptic terminal. Neuroligins have similar functions for the postsynaptic site, where they interact with specialized postsynaptic proteins to promote the clustering of receptors and channels of the postsynaptic density as the synapse matures. Neurexins and neuroligins are shared by all developing synapses, perhaps explaining why some cells, when confronted by targets different from those they normally innervate, can make connections with the available, if unusual, target. There are additional synaptic cell adhesion molecules, including a small family of synCAMs, which includes three adhesion molecules (synCAMs 1, 2 and 3) that localize to pre-, post- and peri-synaptic (glial processes associated with synapses) sites and can elicit synapse formation. synCAMs also influence axonal and dendritic growth and guidance. These proteins belong to the larger family of Ca^{2+}-independent, immunoglobulin superfamily cell adhesions molecules that include NCAM and L1 (see Figure 23.7D). The extracellular domains of synCAMs can bind homophilically or heterophilically, while their intracellular domains interact with the same protein scaffolds in pre- and postsynaptic sites that also bind neurexin and neuroligin. The association of polymorphisms in neurexin, neuroligin, and synCAM genes with increased risk for behavioral disorders such as autism and schizophrenia has reinforced the hypothesis that these molecules are key for establishing appropriate connectivity and circuit function. There is some indication that different neuroligins as well as synCAMs are deployed at postsynaptic specializations in glutamatergic, excitatory synapses versus GABAergic, inhibitory synapses. Nevertheless, additional activity and experience dependent mechanisms are still required to sort out specific individual synapses made on the same cell from one another to achieve the ultimate precision required to establish functional neural circuits (see Chapter 24).

Many of the molecular mechanisms by which neighboring synapses are initially sorted out remain unclear; however, some common themes and compelling candidate molecules have emerged. First, the diversity of ephrin ligands and Eph receptors, along with their established roles in topographic map formation (see Concept 23.5), indicates that these molecules probably contribute to synapse specificity. Second, genes encoding additional candidates—all of which are cell surface or extracellular matrix cell adhesion molecules—have multiple sites for alternative splicing of transcripts, and thus can encode multiple variants of the same basic protein. Third, some of these variants tend to be distributed in different pre- and postsynaptic sites, sometimes in a single neuron. When these genes are mutated, patterns of connectivity are disrupted in subtle but informative ways. In the fly, the gene for the cell adhesion molecule DSCAM1 (see Figure 23.9) is also implicated in synapse formation, using mechanisms that are parallel to those underlying tiling (see the previous section). For synapse formation in the fly, as is the case for tiling, homophilic binding leads to repulsion, suggesting that an essential sorting rule prevents a neuron from making synapses with itself, or at two or more closely adjacent postsynaptic sites on other targets. When *DSCAM1* and its close relative *DSCAM2* (which has only two splice variants) are mutated in *Drosophila*, appropriate synapse sorting fails in the insect's eye, so that synapses of like rather than different cells cluster together. Thus, synapses may be sorted based on local variations of DSCAM-mediated repulsive recognition. In addition, once initial synapse formation is complete, there may be subsequent adhesion molecule–mediated interactions between pre-and postsynaptic domains and glial processes that stabilize synapses. In some cases, these interactions can result in the elaboration of extracellular chondroitin sulfate proteoglycan matrix proteins that complex with cell surface anchors to form specialized **perineuronal nets** that envelope the cell bodies and proximal dendrites of target neurons. Perineuronal nets influence the stability of some classes of synapses, including those made by subsets of inhibitory interneurons onto projection neurons in the cerebral cortex (see Chapters 24 and 28).

The protocadherins are also thought to underlie synapse specificity, in addition to their roles in axon tiling (see the previous section). There are multiple protocadherin genes. The protocadherin-α/-β/-γ cluster, from which the Pcdh-α isoform that is key for axon tiling is transcribed (see Figure 23.9D), has also been a focus of analysis for selective synapse formation. Protocadherin protein structure resembles that of the general cadherin family of cell adhesion molecules (see Figure 23.7E). The complex single gene locus or cluster that encodes Pcdh-α, -β, or -γ isoforms (see Figure 23.9D), however, is remarkably similar to *DSCAM* in the fly. Thus, there are three regions (α, β, and γ) consisting of multiple alternatively spliced exons that encode the extracellular and transmembrane domains of individual protocadherin variants (there are at least 50) and less variable domains that encode the intracellular portions of the protocadherin isoforms. As mentioned earlier, protocadherin isoforms on opposing cells bind to each other with varying affinity based on their degree of similarity (i.e., more homophilic; high binding) or divergence (i.e., more heterophilic; lower binding). Protocadherins are not uniformly expressed at neighboring synaptic sites in cultured neurons (Figure 23.10D) or in several neuron classes in the CNS. Thus, isoforms of protocadherins may invest both pre- and postsynaptic sites in the mammalian nervous system with distinct identities. Analysis of mutant mice in which protocadherin gene function is eliminated indicates that these genes are crucial for synapse formation generally. Synapse number and dendritic spine frequency in different neuronal classes and the distribution and function of glial processes at synapses are all disrupted by protocadherin mutations. Thus, there is probably substantial molecular specificity for initial synapse formation, particularly optimal sites where synapses of any sort (excitatory, inhibitory, dendritic shaft or spine) can form and where glial processes can be recruited to stabilize the nascent synapse. Further modulation of this specificity based on adhesion molecule diversity may further distinguish

synapse classes. It is still not clear, however, whether such molecular differences alone can explain the final patterns of synaptic connections that emerge in any particular neural circuit (see Concept 23.5 and Chapter 24). Indeed, there are additional families of molecules described in Concept 23.4 that regulate synapse number and specificity in a completely different fashion than the adhesion and recognition mechanisms described in this section.

CONCEPT
23.4

Axon, Dendrite, and Synapse Development and Numbers Are Regulated by Trophic Interactions

LEARNING OBJECTIVES

23.4.1 Define neurotrophic signal transduction, neurotrophins, and afferent–target interactions.

23.4.2 Discuss the relationship between trophic signaling, competition, and axon/dendrite/synapse elimination.

23.4.3 Explain the molecular consequences of neurotrophic signaling for neural circuit growth and stability.

Trophic interactions and the formation of neuronal connections

Once synaptic contacts are made between axons that have grown (variably) long distances and dendrites that have branched locally, neurons become dependent on optimal synaptic targets identified by the growth cone. This dependence determines the continued survival as well as further growth and differentiation of the afferent neuron and its connections. In the absence of synaptic targets, axons and dendrites of developing neurons typically atrophy and often die. This developmental dependency between neurons and their targets (and its continuation throughout the life span) is referred to as a **trophic interaction** (from the Greek *trophé*, meaning, roughly, "nourishment"). The nourishment provided by trophic interactions is not the sort derived from metabolites such as glucose or ATP. Instead, the dependence is based on signaling molecules provided by target cells, generally referred to as **neurotrophic factors** (also called **neurotrophins**) that bind to specific receptors on the afferent axon terminals. Neurotrophic factors, like some other intercellular signaling molecules (e.g., mitogens that promote cell proliferation and cytokines that regulate inflammation and immune responses), are secreted in small quantities from cells in target tissues. Neurotrophic factors are unique in that, unlike inductive signaling molecules and cell adhesion molecules that are found in multiple tissues, their expression is limited to neurons and a few non-neural neuronal targets such as muscles or glial cells. These factors are first detected after the initial populations of postmitotic neurons have been generated in the nascent

CNS and PNS and have begun to extend axons and dendrites. They help regulate the phase of neural development that begins once neurogenesis has concluded and process growth moves forward. Thus, neurotrophic signaling is essential for the formation, maintenance, and validation of connections that will define functional neural circuits in the mature brain.

Why do developing neurons depend so strongly on their targets, and what specific cellular and molecular interactions mediate this dependence? The answer to the first question lies in the changing scale of the developing nervous system and the body it serves. These changes of scale (a mouse and an elephant are both mammals with fundamentally similar nervous systems) need to precisely match the number of neurons in particular populations with the size and functional demands of their developing targets. A general—and surprising—strategy in vertebrate development is the production of an initial surplus of nerve cells (on the order of two- or threefold). The final number of neurons is subsequently established by the stabilization of those neurons that have selected a population of potential postsynaptic targets and have adequate access to target-derived trophic factors. The afferent neurons that do not find an appropriate target, or that fail to interact successfully with their intended targets, die. The elimination of supernumerary neurons, particularly the initiation of **apoptosis**—the highly regulated processes that result in cell death (see Chapter 26)—is the first major function in nervous system development that depends critically on the neurotrophins.

A series of studies dating from the early twentieth century showed that a target's size plays a major role in determining the size of the neuronal population that innervates it, presumably based on the target's limited provision of neurotrophic factors. The pioneering neuroembryologists Viktor Hamburger and Rita Levi-Montalcini made these seminal observations, first independently and then collaboratively, in the 1930s and 1940s. A critical finding was that the removal of a limb bud results, at later embryonic stages, in a striking reduction in the number of nerve cells, in this case lower, or α, motor neurons (see Chapter 17), in corresponding portions of the spinal cord (Figure 23.11A,B). This suggested that signals from the target cells in the limb bud influenced the survival of α motor neurons. This suggestion was supported by observations of typical embryos, in which an apparent surplus of motor neurons is generated before the limb differentiates and motor neuron axons grow into it. Initially, these surplus motor neurons innervate the immature limb, but then a large number of them die (see Figure 23.11)—presumably because of a lack of trophic support. Via this mechanism, motor neuron populations become matched to the size and functional requirements of the developing limb musculature. The original embryonic experiment suggests a mechanism for matching afferent neuron populations to their targets. The complete removal of the limb bud likely accelerates the loss of motor neurons by depriving them of a signal in the limb bud that normally supports the survival of some, but not all, of the motor neurons.

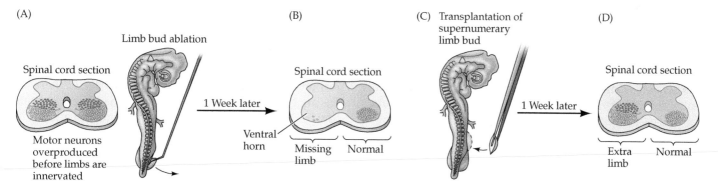

FIGURE 23.11 Target-derived trophic support regulates survival of related neurons (A) The chick spinal cord generates an excess of neurons (solid green dots) prior to the differentiation and innervation of the limb. Normally, some of these neurons are lost once the appropriate level of innervation is established in the developing limb bud. Limb bud amputation in a chick embryo at the appropriate stage of incubation (about 2.5 days) further depletes the pool of motor neurons that would have innervated the missing extremity. (B) Cross section of the lumbar spinal cord in an embryo that underwent this surgery about a week earlier. The motor neurons (open red dots) in the ventral horn that would have innervated the hindlimb have degenerated almost completely.

A normal complement of motor neurons is present on the other side; most of the normally supernumerary neurons have been lost. (C) Adding an extra limb bud before the normal period of cell death rescues early-generated neurons that normally would have died. (D) Such augmentation leads to an abnormally large number of limb motor neurons on the side related to the extra limb, and these neurons are recruited from the pool of cells overproduced at an earlier stage of development (green dots) rather than generated de novo through cell proliferation elicited by signals from the added target. (After V. Hamburger, 1958. *Amer J Anat* 102: 365–409; V. Hamburger, 1977. *Neurosci Res Prog Bull* 15, *Suppl. III*: 1–37; and M. Hollyday and V. Hamburger, 1976. *J Comp Neurol* 170: 311–320.)

Based on these observations, it seemed possible that when a limb bud is present, innervating neurons in the spinal cord *compete* with one another for a resource available in limited supply in the developing limb. In support of this idea, many neurons that would normally die can be rescued by augmenting the amount of target tissue available, presumably providing extra trophic support. Thus, in the developing chick embryo, experimentally adding a limb bud that can be innervated by the same spinal segments that innervate the normal limb results in an excess of α motor neurons in the corresponding regions of the spinal cord (Figure 23.11C,D). Careful monitoring of cell proliferation versus cell death shows that the extra cells are not generated from motor neuron precursors in response to a mitogenic signal from the extra target. Instead, they are "rescued" from a neuron population that is overproduced in early development and normally winnowed, presumably based on the limited trophic support, obtained via their axons, from the targets. Thus, the size of nerve cell populations in the adult is not fully determined by a rigid genetic program of cell proliferation followed by highly specified innervation of the target. The numbers of nerve cells innervating a target, and their connections with the target can be modified by interactions that match the degree of innervation to the amount of target available.

Competitive interactions and the formation of neuronal connections

Once the size of a neuronal population is established by initial trophic regulation, trophic interactions continue to modulate the formation of synaptic connections, beginning in embryonic life and extending far beyond birth. Certain problems must be solved during the establishment of innervation. One of these problems is ensuring that the "right" number of remaining axons innervates each target cell—"right" meaning the number must result in sufficient innervation to allow the target cell to integrate or process information and generate electrical output signals vital to the operation of its neural circuit. Another problem is ensuring that each individual axon innervates the "right" number of target cells. Getting these numbers correct—thus establishing appropriate convergence and divergence (see Chapter 1)—is another essential function of trophic interactions between developing neurons and target cells. Thus, trophic interactions optimize the quantitative aspects of connectivity in neural circuits during development, and also likely provide a mechanism to scale neural circuits so that they meet the demands of animals of different sizes throughout life.

Studying synaptic refinement and the role of trophic interactions in this process is a formidable challenge, especially in complex circuits of many regions of the CNS, including the cerebral cortex, The diverse inputs and outputs of these circuits makes it difficult to define precise relationships between afferent axons and targets as well as the trophic signaling that establishes these relationships. Thus, many fundamental ideas about the ongoing modification of developing neural circuitry have come from simpler, more accessible parts of the nervous system, most notably the vertebrate neuromuscular junction and autonomic ganglion cells. In both of these circuits, single classes of afferents and targets are precisely matched: α motor neurons for the neuromuscular junction (see Chapter 17), preganglionic motor neurons for autonomic ganglia (see Chapter 21). Each muscle fiber (the multinucleate single cells that define skeletal muscle) is ultimately innervated by a single α motor neuron. Similarly in many autonomic ganglia, each ganglion cell is innervated by a single preganglionic motor neuron. The mechanisms seen in the PNS also operate in the CNS. The most clear example of this sort of quantitatively precise matching of afferents and targets in the CNS is seen in the cerebellum, where each Purkinje cell is ultimately innervated by a single climbing fiber axon from neurons in the inferior olive (see Chapter 19).

The one-to-one matching of afferents and targets is not determined by precise, singular innervation during initial axon ingrowth and target recognition. Instead, each of these target cells, (muscle fiber, autonomic ganglion cell, or Purkinje cell), is innervated by axons from several α motor neurons, preganglionic motor neurons, or inferior olive neurons. This pattern of multiple axonal inputs, established as initial axon ingrowth, target recognition and synaptogenesis comes to an end, is called **polyneuronal innervation** (Figure 23.12). In such cases, axons from multiple neurons of each class—α motor neurons (see Figure 23.12A) preganglionic motor neurons (see Figure 23.12B), or inferior olive neurons (see Figure 23.12C)—make functional synapses on their targets that can elicit post-synaptic responses. The synapses made by all but one of these axons are gradually lost during early postnatal development until only one neuron innervates the target cell and elicits a change in its electrical activity. This process of loss is generally referred to as **synapse elimination**, although here "elimination" refers to a reduction in the number of physiologically measured or anatomically visualized subsets of axonal inputs to the target cells (i.e., how many different axons can elicit a postsynaptic response via synaptic endings made on the immature target cell). Perhaps counterintuitively, this process does not result in an overall reduction in the number of synaptic contacts (the individual specialized junctions between pre- and postsynaptic cells) made on postsynaptic cells (see Figure 23.13A). Indeed, as the synapses from the axons that withdraw from the target are eliminated, the single axon that remains establishes substantially larger numbers of individual synaptic contacts with the target cell —or a larger single territory for

(A) Muscle cells

(B) Ganglion cells

(C) Postnatal-day-7 cerebellum

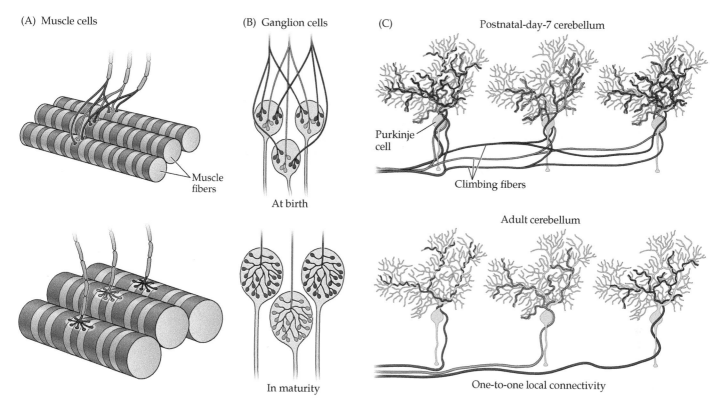

Muscle fibers

At birth

In maturity

Purkinje cell

Climbing fibers

Adult cerebellum

One-to-one local connectivity

FIGURE 23.12 Synapse number and pattern in the mammalian PNS are adjusted during early postnatal life In muscles (A), peripheral ganglia whose neurons have no dendrites (B), and Purkinje cells with extensive dendritic arbors (C), each axon innervates a larger number of target cells at birth than in maturity. Most of this rudimentary multiple innervation is eliminated shortly after birth. For muscles, ganglia, and Purkinje cells, however, the size and complexity of the terminal arbor that remains on each mature target cell increases. Thus, the stabilized single axon elaborates more and more terminal branches and synaptic endings on the target cell it will innervate in maturity. The common denominator of this process is not a net loss of synapses, but the removal of immature contacts from all but one or a few axons on each target, and the focus on fewer target cells by a progressively increasing amount of synaptic machinery for each axon that remains. (A,B after D. Purves and J. W. Lichtman, 1980. *Science* 210: 153–157; C after A. M. Wilson, et al., 2019. *Cell Reports* 29: 2849–2861.)

the single endplate synapse made by an α motor neuron and its muscle fiber target.

Patterns of electrical activity in the pre- and postsynaptic partners are thought to influence "competition" for target space and neurotrophic support that underlies synapse elimination. This electrical activity, however, does not necessarily depend on behavior or "experience" as later events in the final development of circuits do (see Chapter 24). Instead, it depends on the ongoing matufration of the neurons and targets as they acquire "excitability" and the cellular interactions that excitability makes possible. Thus, blocking action potential activity from the afferent motor neurons or depolarization via acetyl choline receptor activation in the target muscle cells (or their equivalent in autonomic ganglia) causes polyneuronal innervation to persist.

Many useful insights into the nature of input competition and subsequent rearrangement of the remaining synaptic connection during development have come from direct observations of competition between the presynaptic endings of two motor neuron axons for a single synaptic site on a developing muscle fiber. Using different colored fluorescent dyes or genetically encoded reporters that distinguish either each presynaptic terminal from different α motor neuron axons or the receptors that define the postsynaptic domain on the target muscle, the same multiply innervated neuromuscular junction can be followed over days, weeks, and even longer (Figure 23.13A). Competition between synapses arising from different motor neurons does not involve the active displacement of the "losing" input by the eventual "winner." Instead, it appears that the inputs of the two competitors initially occupy the same subregion of a nascent postsynaptic specialization but then gradually segregate (Figure 23.13B). The postsynaptic territory of the "losing" axon diminishes over time, and eventually the presynaptic ending atrophies and the axon retracts from the synaptic site. Neurotransmitter receptors beneath the terminal branches that eventually will be eliminated are also lost. This receptor loss occurs before the nerve terminal has withdrawn and presumably reduces the synaptic strength of the input, which results in further

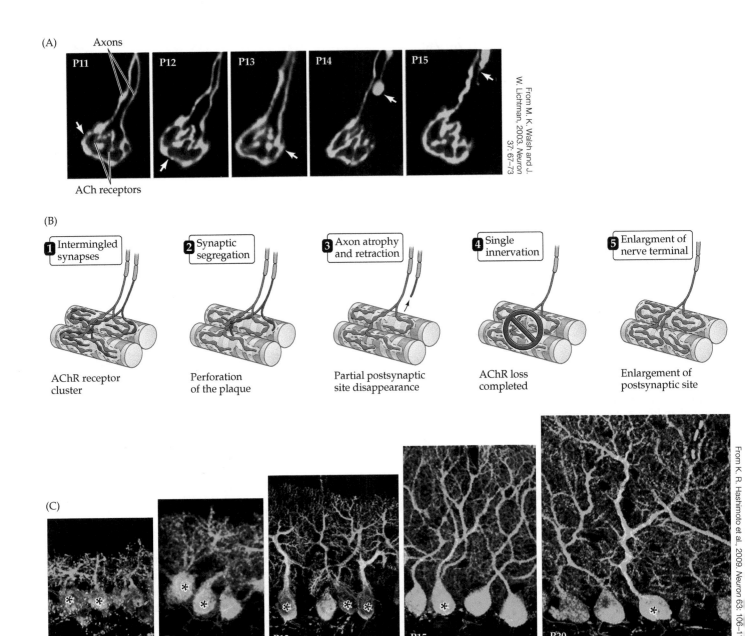

FIGURE 23.13 Elimination of multiple innervation in the PNS (neuromuscular junction) and CNS (cerebellum) (A) In this series of images, the same neuromuscular junction in a neonatal mouse has been imaged repeatedly, beginning on postnatal day 11 (P11). Initially, two axons (green and blue) innervate the muscle fiber (the local clustering of postsynaptic ACh receptors is shown in red). The arrow indicates the boundary between the postsynaptic territory of the green and blue axons. By P12, the proportion of territory occupied by the green and blue axons has begun to shift, with the blue axon terminal apparently losing postsynaptic space while the green axon terminal expands. This process continues on P13, and by P14 the blue axon has fully retreated, its synaptic terminal transformed into a large retraction bulb (arrow on P14). Within an additional day, the retracting axon is almost fully withdrawn from the synaptic site. (B) Schematic of the process by which two motor neurons innervate a single site of clustered ACh receptors (AChRs) on a muscle fiber and then segregate in parallel with the segregation of subsets of AChRs. Eventually only one axon will remain while a branch from the "unsuccessful" axon retracts and atrophies, and AChRs at the site from which the axon retracts are lost (indicated by the "do not" symbol). Finally, the single innervating axon terminal grows in register with the expansion of the postsynaptic clustered AChRs. (C) Single climbing fiber axons (red) in the developing cerebellum initially innervate multiple Purkinje cells (asterisks) during early postnatal development (P7). Over the course of approximately 2 weeks, individual climbing fibers retract immature axon branches from all but one Purkinje cell. The climbing fiber then elaborates an extensive axon that innervates most of the dendritic arbor of one (and only one) Purkinje cell. (B after K. Zito, 2000. *Neuron* 25: 269-278. © 2000 Cell Press.)

loss of postsynaptic receptors, leading to further reduction in the strength of the input. This downward spiral of selective synaptic efficacy for some inputs but not others and its molecular and cellular foundations, including access to target-derived trophic support, presumably results in withdrawal of the axon, and the elimination of its cellular and functional synaptic connection. The presynaptic terminals from the remaining axon continue to enlarge and strengthen as the end plate region expands during postnatal muscle growth. This elaboration of synaptic space (or addition of individual synapses in other instances outside the neuronmuscular junction) is thought to reflect exclusive access to trophic support from the target cell.

A similar process of polyneuronal innervation and input elimination occurs in a variety of other PNS and CNS regions. In autonomic ganglia, particularly ganglia whose neurons lack dendrites (Box 23B), multiple preganglionic motor neuron axons initially innervate each ganglion cell (see Figure 23.12B). In this instance of a central motor neuron making a synapse onto a target neuron (rather than non-neuronal muscle cells), a parallel process of eliminating all functional inputs but one occurs. Thus, neuronal targets, as well as muscle targets, apparently have the capacity to provide trophic support for which afferent axons compete to establish the appropriate number of inputs.

Similarly, in the cerebellum, each adult Purkinje cell is innervated by a single axon from neurons in the inferior olivary nucleus called a climbing fiber (see Chapter 20). During early development, however, each Purkinje cell receives multiple climbing fiber inputs, and each climbing fiber has branches that innervate multiple Purkinje cells (see Figure 23.12C). During postnatal life, each climbing fiber axon retracts branches from neighboring Purkinje cells and elaborates multiple branches and synaptic contacts on the apical dendrite of a single Purkinje cell target (Figure 24.13C). These interactions are thought to reflect the constraints of limited trophic support in the target Purkinje cell, and a differential capacity for one of the competing climbing fiber axons to establish access to that trophic support.

Thus, the pattern of synaptic connections that emerges in the adult is not simply a consequence of the biochemical identities of synaptic partners or other developmental rules that impose molecular specificity on synaptic connections. Rather, the mature wiring plan is the result of a much more flexible process in which neuronal connections are formed, removed, and remodeled according to local circumstances that reflect molecular constraints, the detailed structure and size of the target, the signals it can provide to the afferent neuron, and the capacity of the afferent neuron to respond to those signals, including those that reflect electrical activity. These interactions guarantee that every target cell is innervated—and continues to be innervated—by the "right" number of inputs and synapses, and that every innervating axon contacts the "right" number of target cells with an appropriate number of synaptic endings. Thus, the regulation of **convergence** (the number of inputs to a target cell) and **divergence** (the number of connections made by an afferent neuron) in the developing nervous system is another key consequence of trophic interactions between neurons and their targets. The regulation of convergence and divergence by neurotrophic interactions is also influenced by the shape and size of neurons, particularly the number and arborization of dendrites (see Box 23B). These adjustments shape the information processing capacity of neural circuits.

The molecular basis of trophic interactions

Three general assumptions help explain how trophic-mediated competition, at the molecular level, can result in an afferent neuron population whose size is appropriately matched to the target it innervates:

1. Neurons depend on target-derived trophic support for survival and for establishing appropriate innervation numbers and patterns.

2. Targets make limited quantities of trophic factors.

3. Accordingly, multiple inputs to any target must "compete" to bind the limited quantity of trophic signal and initiate downstream events for survival or growth.

Extensive studies of the neurotrophic protein **nerve growth factor** (NGF) provide support for these three assumptions. NGF was the first-discovered of an entire family of trophic factors. Characterization of its activity and regulation provides a model for understanding how neurotrophins provided by targets influence the survival and connections of the nerve cells that innervate them.

NGF was discovered as an "activity" that elicited robust growth of neuronal processes both in the animal and in cell culture. The idea that a specific molecule might elicit neuronal growth originated with experiments in which tumor cells secreting a substance that would be later identified as the NGF protein, encoded by the *NGF* gene (located on chromosome 1 in humans), were implanted into a host animal. These tumor cells survived, and they caused abnormal axon growth, particularly of axons from sympathetic ganglia, toward the implanted tumor cells. Subsequent experiments on cultured neurons grown in the presence or absence of what was later demonstrated to be NGF (Figure 23.14A–C) showed that this "activity" could enhance the survival of nerve cells in culture and support neurite growth. Support for the idea that NGF is important for neuronal survival in vivo emerged from several additional observations made initially in the sympathetic nervous system of rats and mice. Depriving developing mice of NGF (by the chronic administration of an NGF antiserum or by eliminating the gene for NGF selectively) resulted in adult mice that lacked most NGF-dependent

BOX 23B | Why Do Neurons Have Dendrites?

Perhaps the most striking feature of neurons is their diverse morphology. Some classes of neurons have no dendrites at all; others have a modest dendritic arborization; still others have an arborization that rivals the complex branching of a fully mature tree (see Figure 1.1). Why should this be? Although there are many reasons for this diversity, neuronal geometry influences the number of different inputs that a target neuron receives by modulating competitive interactions among the innervating axons.

Evidence that the number of inputs a neuron receives depends on its geometry has come from studies of the peripheral autonomic system, where it is possible to stimulate the full complement of axons innervating an autonomic ganglion and its constituent neurons. This approach is not usually feasible in the CNS because of the anatomical complexity of most central circuits. Because individual postsynaptic neurons can also be labeled via an intracellular recording electrode, electrophysiological measurements of the number of different axons innervating a neuron can routinely be correlated with target cell shape. In both parasympathetic and sympathetic ganglia, the degree of preganglionic convergence onto a neuron is proportional to its dendritic complexity. Thus, neurons that lack dendrites altogether are generally innervated by a single input, whereas neurons with increasingly complex dendritic arborizations are innervated by a proportionally greater number of different axons (Figure). This correlation of neuronal geometry and input number holds within a single ganglion, among different ganglia in a single species, and among homologous ganglia across a range of species. Since ganglion cells that have few or no dendrites are initially innervated by several different inputs (see text), confining inputs to the limited arena of the developing cell soma evidently enhances competition among them, whereas the addition of dendrites to a neuron allows multiple inputs to persist in peaceful coexistence. Importantly, the dendritic complexity of at least some classes of autonomic ganglion cells is influenced by neurotrophins.

A neuron innervated by a single axon will clearly be more limited in the scope of its responses than a neuron innervated by 100,000 inputs (1 to 100,000 is the approximate range of convergence in the mammalian brain). By regulating the number of inputs that neurons receive, dendritic form greatly influences function.

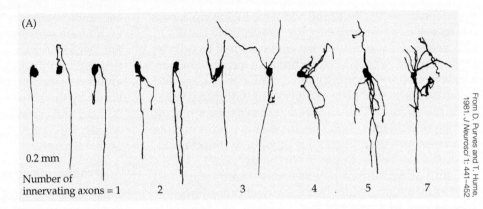

(A)

0.2 mm

Number of innervating axons = 1   2   3   4   5   7

From D. Purves and T. Hume, 1981. *J Neurosci* 1: 441–452

(A) Axons innervating ciliary ganglion cells in adult rabbits. Neurons studied electrophysiologically and then labeled by intracellular injection of a marker enzyme have been arranged in order of increasing dendritic complexity. The number of axons innervating each neuron is indicated.

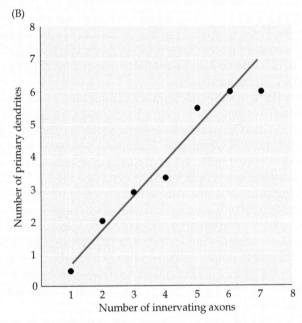

(B) This graph summarizes observations of the number of innervating axons on a large number of rabbit ciliary ganglion cells. There is a strong correlation between dendritic geometry and input number. (After D. Purves and T. Hume, 1981. *J Neurosci* 1: 441–452.)

(A)

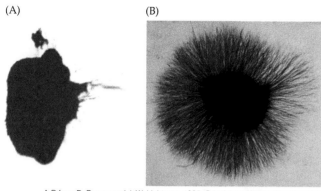

(B)

(C)

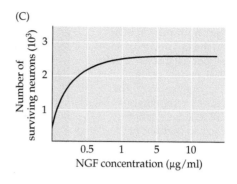

NGF concentration (µg/ml)

(D)

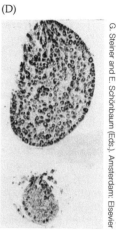

From R. Levi-Montalcini, 1972. The morphological effects of immunosympathectomy. In Immunosympathectomy, G. Steiner and E. Schönbaum (Eds.). Amsterdam: Elsevier

A,B from D. Purves and J. W. Lichtman, 1985. *Principles of Neural Development.* Sunderland, MA: Sinauer Associates. Courtesy of R. Levi-Montalcini

FIGURE 23.14   Effect of NGF on the outgrowth of neurites and survival of neurons (A) A chick sensory ganglion taken from an 8-day-old embryo and grown in organ culture for 24 hours in the absence of the neurotrophin NGF. Few, if any, neuronal branches grow out into the medium in which the explant is embedded. (B) A similar ganglion in identical culture conditions 24 hours after the addition of NGF to the medium. NGF stimulates a halo of neurite outgrowth from the ganglion cells. (C) NGF influences the survival of newborn rat sympathetic ganglion cells grown in culture for 30 days. Dose–response curves confirm the dependence of these neurons on the availability of NGF. (D) Cross section of a superior cervical ganglion from a normal 9-day-old mouse (top) compared with a similar section from a littermate that was injected daily since birth with NGF antiserum (bottom). The ganglion of the treated mouse shows marked atrophy, with obvious loss of nerve cells. (C after L. L. Chun and P. H. Patterson, 1977. *J Cell Biol* 75: 712–718.)

autonomic neurons (Figure 23.14D). Conversely, injection of exogenous NGF into newborn rodents caused enlargement of sympathetic ganglia (which are particularly NGF-dependent) because of additional cells as well as more extensive axonal and dendritic growth, an effect opposite to that of NGF deprivation. Finally, it was possible to show that NGF mRNA and protein are uniquely transcribed and translated by sympathetic neuron targets (a variety of smooth muscle, vascular, and glandular tissues; see Chapter 21). The dramatic influence of NGF on afferent neuron survival and its secretion from specific neuronal targets, together with what was known about the significance of neuronal death and regulation of neurite growth in development, suggested that NGF is indeed a target-derived neurotrophin that serves to match the number of nerve cells to the number of target cells.

From the outset, it was apparent that only certain classes of nerve cells respond to NGF. Thus, it was assumed that there must be other secreted proteins that served similar functions for trophic interactions between neurons and their targets. Furthermore, diversity of trophic factors would facilitate greater specificity between pre- and postsynaptic partners during development (or in regeneration; see Chapter 26). These factors, however, were presumed to be distinct from "chemotropic" molecules for initial axon guidance or synapse formation because—if they were like NGF—they would be expressed long after initial axon pathways were established, and their activity would be recognized only after connections had formed. In the 1980s and 1990s, work from several laboratories showed that NGF

belongs to a family of related trophic molecules called *neurotrophins*. Neurotrophins are functionally similar to a broader class of signaling molecules found throughout the organism and referred to generically as **growth factors**, which also have roles in neurogenesis, axonal and dendritic growth, and axon guidance as well as influencing cell proliferation, differentiation and survival in other tissues and organs. The expression and activity of the neurotrophins, however, are mostly limited to neurons and their targets. At present there are four well-characterized members of the neurotrophin family in addition to NGF: **brain-derived neurotrophic factor (BDNF)**, **neurotrophin-3 (NT-3)**, and **neurotrophins 4/5 (NT-4/5)** (Figure 23.15). In addition, two other ligand families, those that include the ciliary neurotrophic factor (CNTF) and the glial-derived neurotrophic factor (GDNF), also have neurotrophic signaling capacity (see later in this section).

Although the members of the neurotrophin family are homologous in amino acid sequence and structure, they are encoded by distinct genes and are very different in their specificity (see Figure 23.15A). NGF supports the survival of (and neurite outgrowth from) sympathetic neurons, while another family member—BDNF—does not. Conversely, BDNF, but not NGF, can support the survival of certain sensory ganglion neurons, which have a different embryonic origin. NT-3 supports both of these populations. Given the diverse systems whose growth and connectivity must be coordinated during neural development, this specificity is not surprising. Matching of target neurotrophic factors to neurons able to respond would reinforce axon targeting

FIGURE 23.15 Neurotrophins have distinct effects on different target neurons (A) Effect of NGF, BDNF, and NT-3 on the outgrowth of neurites from explanted dorsal root ganglia (DRG; left column), nodose ganglia (NG; middle column), and sympathetic ganglia (SG; right column). The specificities of these several neurotrophins are evident in the ability of NGF to induce neurite outgrowth from sympathetic and dorsal root ganglia, but not from nodose ganglia (which are cranial nerve sensory ganglia that have a different embryological origin from dorsal root ganglia); of BDNF to induce neurite outgrowth from dorsal root and nodose ganglia, but not from sympathetic ganglia; and of NT-3 to induce neurite outgrowth from all three types of ganglia. (B) Specific influence of neurotrophins in vivo. Distinct classes of peripheral somatosensory receptors and the dorsal root ganglion cells that give rise to these sensory endings depend on different trophic factors in specific target tissues. (B after M. Bibel and Y.-A. Barde, 2000. *Genes Dev* 14: 2919–2937.)

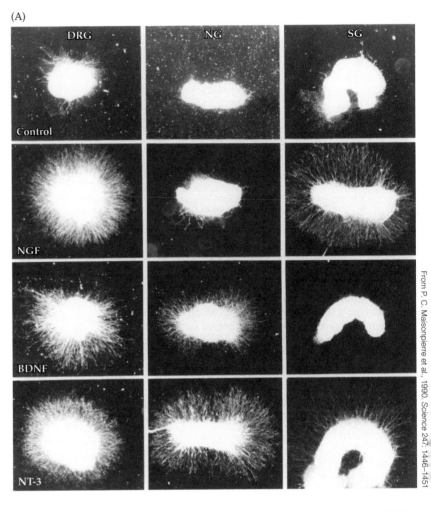

(A)

From P. C. Maisonpierre et al., 1990. *Science* 247: 1446–1451

and facilitate elaboration of appropriate connections. Indeed, different neurotrophins are selectively available in different targets. For example, the different receptor specializations in the skin that transduce somatosensory information (touch and proprioception versus pain and temperature) express different neurotrophins, and this specificity is matched by expression of neurotrophin receptors (see the next section) that distinguish the peripheral sensory neurons that innervate each specialized target (see Figure 23.15B). Subsequent work has shown that many distinct cell classes in the brain, including neurons in the olfactory bulb, cerebellum, hippocampus, and cerebral cortex, rely on distinct target-derived neurotrophins for their survival and differentiation. This supports the general conclusion that neurotrophic signaling detected by the axons of neurons that project to appropriate targets is an essential mechanism for establishing anatomically and functionally specific circuits in the developing nervous system.

In addition to the neurotrophins, other secreted molecules also have neurotrophic influences. These include CNTF, which is considered a cytokine because of its role in inflammation and immune responses beyond neurotrophic interactions; leukemia inhibitory factor (LIF), also a cytokine; and GDNF and related proteins (referred to as the GDNF family of ligands including neuturin, persephin, and artemin). GDNF ligands activate the RET tyrosine kinase receptor. GDNF/RET

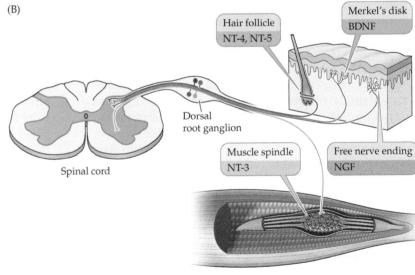

(B)

signaling can influence kidney and testis development, vascular growth, and also has been implicated in several cancers including thyroid and endocrine tumors. Thus, a variety of factors—some specific to the nervous system, others with neurotrophic activity that are also used for purposes beyond neural development—influence the survival and growth of developing nerve cells and the elaboration of neural circuits.

Neurotrophin signaling

Neurotrophic factors are clearly key regulators for three distinct cellular mechanisms: neuron survival or death, neural process growth or retraction, and initial synapse stabilization or elimination. The ways in which these ligands influence different aspects of neuronal differentiation reflect the details of signal transduction in response to neurotrophic molecules. The selective actions of neurotrophins also reflect their interactions with two classes of neurotrophin receptors: the **tyrosine kinase (Trk)** receptors and the **p75** receptor. There are three Trk receptors, each of which is a single transmembrane protein with a cytoplasmic tyrosine kinase domain. TrkA is primarily a receptor for NGF, TrkB is a receptor for BDNF and NT-4/5, and TrkC is a receptor for NT-3 (Figure 23.16A). The Trk receptors homodimerize, and in this dimeric state optimally bind their specific neurotrophin ligands. In addition, all neurotrophins activate the p75 receptor protein (Figure 23.16B). The interactions between neurotrophins, Trks, and p75 demonstrate another level of selectivity and specificity of neurotrophin signaling.

All neurotrophins are initially translated into "pro" peptides, which are proteolytically cleaved—in many cases intracellularly—to yield a maximally active neurotrophic ligand. In some cases, however, the neurotrophin propeptide is released and either cleaved extracellularly or maintained as a propeptide without proteolytic cleavage. The Trk receptors have high affinity for proteolytically cleaved ligand domains. In contrast, the p75 receptor has high affinity for unprocessed neurotrophins but low affinity for the processed ligands. The expression of a particular Trk receptor subtype or p75, as well as the local capacity for proteolytic cleavage of the pro-neurotrophins, therefore confers the capacity for distinct neurotrophin responses. Since Trk and p75 receptors are expressed only in subsets of neurons, and neurotrophins are available from different classes of targets, selective binding between ligand and receptor probably accounts for some of the specificity of neurotrophic interactions.

Cell biological experiments using a specialized culture system that isolates symmetrically growing neurites from their parent cell bodies, demonstrated that neurotrophin activity depended critically upon availability of neurotrophic ligands to axons and their terminals. When a neurotrophin was provided to only one of the two populations of axons growing from the same parent neurons in vitro—thus depriving cell bodies and the other neurites of access to the neurotrophin—only the axons with direct access to the neurotrophin responded with additional growth and branching (Figure 23.16C). This reinforced the conclusion that neurotrophins are detected in the target by afferent axons and their terminals. Additional experiments using this approach, in combination with other tools to visualize the binding of neurotrophins and subsequent cellular response, showed that the relay of the neurotrophic signal to the cell body leads to local changes in growth at the site where the signal is available. To accomplish this relay, neurotrophins bound to their transmembrane receptors (primarily the Trk subset) are selectively internalized by assembling a **signaling endosome** that also includes neurotrophins bound to the now-activated neurotrophin receptor/kinase, as well as an assembly of related signaling proteins (Figure 23.16D,E). This signaling endosome is transported back to the cell body (see Figure 23.2A), where its activated kinase domains can continue to transmit the neurotrophic signal transduced initially at the distal end of the axon. The binding of the neurotrophin at a distal target, and the subsequent endocytic internalization and retrograde transport of the ligand–receptor kinase complex (with the catalytic domain of the Trk receptor still facing the cytoplasm), facilitates a key aspect of neurotrophic signaling. In this way, a neurotrophic ligand can activate and maintain tyrosine kinase–related signaling from the site of availability to the neuronal cell body, which is often a substantial distance. The activated kinase domain, now protruding from the surface of the endosome that faces the cytoplasm, is available for catalytic activity as it is transported up the axon toward the cell body, and into the cell body itself. This signaling can modify existing cytoskeletal or cytoplasmic proteins, or it may influence transcriptional regulators that translocate to the nucleus upon neurotrophin receptor–dependent phosphorylation and modify gene expression. Such changes could influence the survival of the neuron as well as enhance its growth. The ongoing maintenance of trophic support provided to an afferent by a target neuron via this long-distance, endosome-mediated signaling may be a point of vulnerability for initiating pathology in several neurodegenerative diseases. Diminished endosomal signaling, in addition to diminished availability of neurotrophins from target cells, may lead to neuronal death or local retraction of axon or dendrite branches and the loss of synaptic connections.

The Trk neurotrophin receptors, via stimulation of their intercellular tyrosine kinase domains and subsequent phosphorylation of target proteins, engage three distinct second-messenger pathways that alter functions of target proteins (via phosphorylation and other post-translational modifications) or that change gene expression in the target cell (Figure 23.17). These receptors activate the small GTPase ras, which then activates a family of cytoplasmic protein kinases—the *mitogen activated protein* (MAP) kinases—to elicit a variety of cellular responses, including changes in gene expression. Trk receptors, upon binding of

(A)

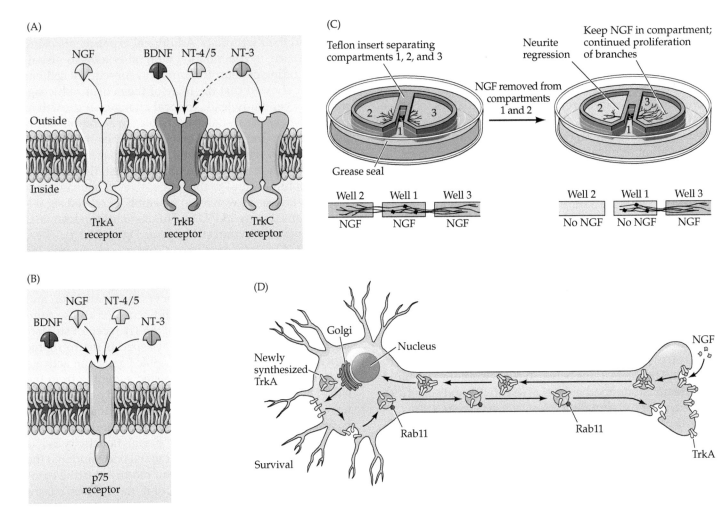

(B)

(C)

(D)

(E)

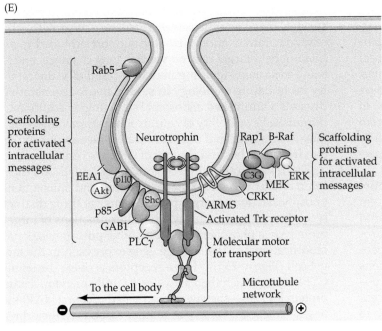

neurotrophins, also activate phospholipid second messengers via phospholipase C (PLC), and these second messengers then can activate downstream intracellular Ca^{2+} signaling. PLC-dependent signaling preferentially influences cellular responses that can lead to activity-dependent synapse plasticity (see Chapter 24). Finally, Trk receptors activate phosphoinositol 3 (PI3) kinase. PI3 kinase interacts with pathways that regulate the activity of Akt kinase, a cytoplasmic enzyme that modulates downstream signaling to either prevent or promote cell death, including via the mTOR pathway. The last of the neurotrophin receptors, p75, also engages three different intercellular signaling pathways. One of these is mediated by Rho GTPases (which function in much the same way as ras) that influence neurite growth. The second p75-dependent pathway leads to activation of the c-Jun transcription factor, which regulates the expression of genes involved in apoptosis. Finally,

◀ **FIGURE 23.16** **Neurotrophin receptors and neurotrophic signaling** (A) The Trk family of receptor tyrosine kinases for the neurotrophins. TrkA is primarily a receptor for NGF, TrkB a receptor for BDNF and NT-4/5, and TrkC a receptor for NT-3. Because of the high degree of structural homology among both the neurotrophins and the Trk receptors, there is some degree of cross-activation between factors and receptors. For example, NT-3 can bind to and activate TrkB under some conditions, as indicated by the dashed arrow. These distinct receptors allow various neurons to respond selectively to the different neurotrophins. (B) The p75 low-affinity neurotrophin receptor binds all neurotrophins at low affinities (as its name implies). This receptor confers the ability to respond to a broad range of neurotrophins on fairly broadly distributed classes of neurons in the PNS and CNS. (C) Three compartments ("wells") of a culture dish are separated by a Teflon divider sealed to the bottom of the dish with grease. A magnified view looking down on the wells is shown below. Isolated rat sympathetic ganglion cells plated in well 1 can grow through the grease seal into wells 2 and 3. Growth into a lateral well occurs as long as the well contains an adequate concentration of NGF. This local application does not influence the neurites in the other lateral well. Subsequent removal of NGF from a well causes local regression of neurites without affecting neurite survival in the other wells. (D) Neurotrophic signaling relies on retrograde transport of the endocytically internalized Trk receptors bound by their neurotrophin ligand. Following endocytic internalization, the endocytic vesicle is transported retrogradely to the cell body where it signals to the nucleus to elict changes in neuronal growth. The cytoplasmic domain of the activated Trk can also mediate phosphorylation of axonal proteins as it is transported from the axon terminal to the cell body. (E) Trk-mediated neurotrophin signaling at the axon is maintained and propagated by endocytic internalization of the ligand–receptor complex with several scaffolding proteins that bind one of three intracellular effectors: the Akt kinase, phospholipase C_γ (PLC_γ), or extracellular signal-regulated kinase (ERK). This "signaling endosome" can also bind molecular motors that engage the microtubule cytoskeleton. The signaling endosome, activated via neurotrophin binding, is then transported back to the cell body to activate downstream targets, including modifying gene expression. (C after R. B. Campenot, 1981. *Science* 214: 5 79–581. D after M. Ascano et al., 2012. *Trends Cell Biol* 22: 266–273; E after L. S. Zweifel et al., 2005. *Nat Rev Neurosci* 6: 615–625.)

p75 can also engage nuclear factor κB (NF-κB), which enhances expression of genes that promote cell survival. Thus, the final functional influence of any neurotrophin depends on (1) the receptor it encounters, and (2) that receptor's likelihood of engaging diverse intercellular signaling pathways that influence growth, gene expression, synaptic changes, or cell survival versus death.

Disruption of these neurotrophin-dependent processes, either during embryonic, fetal, or early postnatal life or in adulthood, probably contributes to neurodevelopmental or neurodegenerative conditions in which neurons fail to grow in the developing brain or die because of lack of appropriate trophic support in the mature brain. Such disruption has devastating consequences for the circuits that the neurons define and for the behaviors controlled by those circuits. Indeed, the pathogenic mechanisms of developmental disorders such as autism and schizophrenia, as well as neurodegenerative diseases as diverse as amyotrophic lateral sclerosis (ALS) and Parkinson's, Huntington's, and Alzheimer's diseases, may all reflect, at least in part, abnormalities of neurotrophic regulation.

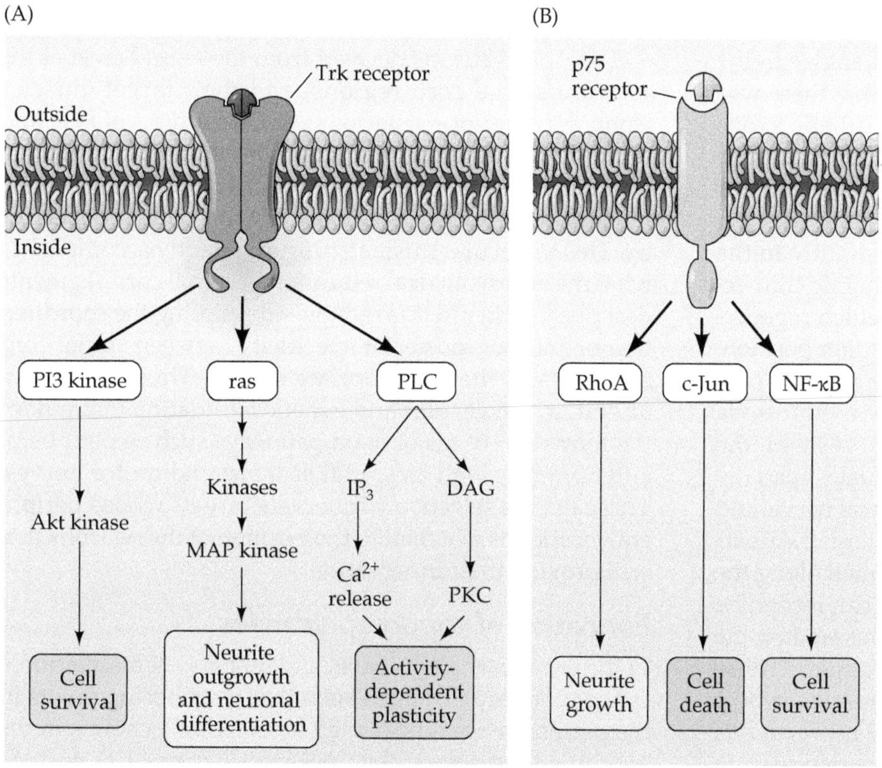

(A)

(B)

FIGURE 23.17 **Signaling via neurotrophins and their receptors** (A) Signaling via Trk dimers can lead to a variety of cellular responses, depending on the intracellular signaling cascade engaged by the receptor after binding to the ligand. The possibilities include cell survival (via the Akt pathway); neurite growth (via the MAP kinase pathway); and activity-dependent plasticity (via the Ca^{2+}/calmodulin and PKC pathways). (B) Signaling via the p75 pathway can lead to neurite growth via interaction with Rho kinases, or to cell cycle arrest and cell death via other distinct intracellular signaling cascades.

CONCEPT 23.5 | Axon, Dendrite, and Synaptic Growth Results in Orderly Patterns of Connections, Including Topographic Maps

LEARNING OBJECTIVES

23.5.1 Discuss pathway and target specificity in the developing nervous system.

23.5.2 Explain the role of gradients in topographic map formation.

23.5.3 Define the chemoaffinity hypothesis.

Pathway formation

Much of the detailed understanding of mechanisms underlying axon, dendrite, and synapse development focuses on interactions that happen in limited, small embryonic locations that engage subdomains (the growth cone, the axon, the dendrite) of a differentiating neuron shortly after neurogenesis is complete. There are, however, additional choices that growing axons in particular must make once entire populations of neurons have been generated in any location in the brain. These choices for growing axons must reflect the specificity of their intended, usually distal, targets. This aspect of axon growth and guidance, however, results in the establishment of larger axon pathways (often referred to as tracts in the CNS, like the spinothalamic tract; see Chapter 13), or nerves in the PNS, like the trigeminal nerve, that selectively connect one large population of neurons with another, usually distal, target. This "pathway specificity" in the developing or regenerating nervous system was first recognized by the British physiologist John Langley in the early twentieth century, when he observed that there was some precision in the regeneration of sympathetic axons to their peripheral targets. Langley recognized that if the anterior-to-posterior location of ganglion in the sympathetic chain was disrupted modestly, by experimentally altering the position of a posterior ganglion either slightly to the anterior or posterior, the specificity of autonomic connections would mirror those in an animal in which regeneration had occurred without a change in ganglion position. In contrast, when ganglion position was altered more substantially, regeneration to appropriate original targets via the relevant peripheral nerve was not seen. Instead, the regenerating axons grew into a different, usually spatially adjacent, target via a locally available peripheral nerve, and made anomalous connections with a novel target. Apparently, axons from subsets of neurons distributed along the body axes (anterior to posterior in this case) can recognize signals that coordinate their selective growth based on the parallel anterior–posterior location of their targets. Those signals, however, are not sufficient to dramatically reroute a regenerating axon to its original peripheral nerve and its original target if the parent neuron has been relocated to an extremely different location along the anterior–posterior axis. These experiments were subsequently repeated in developing animals, and performed in both the developing sympathetic chain as Langley had done in adults, and distinct segments of the spinal cord. The the same rule emerged for developing axon pathways as for regenerating axon pathways: Local cues direct entire populations of axons from neurons in autonomic ganglia or motor neurons in distinct segments of the developing spinal cord to appropriate targets based on anterior–posterior location. These cues, however, are not sufficient to maintain pathway specificity between autonomic or spinal cord projection neurons when they are translocated to dramatically distant positions.

A key mechanism for establishing this pathway specificity along the anterior–posterior axis of the spinal cord, and its targets in the periphery, is differential patterning of Hox transcription factors, which play a general role in establishing the anterior–posterior body axis in nearly every animal (see also Chapter 22). The basic anterior–posterior pattern of Hox gene expression is established by opposing gradients of FGF signals, which are highest anteriorly, and retinoic acid, which is highest posteriorly (see also Chapter 22). The consequence of this initial anterior–posterior signaling in the developing CNS of most vertebrates is to establish four domains of Hox gene expression that correspond approximately to the cervical (innervating the forelimbs), thoracic (innervating the axial muscles of the chest as well as the sympathetic chain ganglia), and lumbar (innervating the hindlimbs) spinal cord (Figure 23.18). Additional local signaling and transcriptional repression further refine these regional distinctions, and impose a "matching" between the motor neurons (and the peripheral sensory neurons derived from the neural crest at the relevant spinal cord regions) and their target muscles along the anterior–posterior axis. Disruptions of FGF and retinoic acid–dependent anterior–posterior patterning (see Chapter 22), as well as loss-of-function or mis-expression mutations of Hox genes, can disrupt this pathway specificity. This molecular mismatching apparently accomplishes what the embryological relocation of spinal cord segments described earlier in this section—disrupting the coordination of anterior–posterior specificity between spinal cord neurons and their appropriate targets. Thus, there is an underlying molecular and genetic foundation that guides the formation of major axon pathways such as peripheral autonomic as well as spinal sensory and motor nerves. These nerves selectively innervate targets whose peripheral location is matched to the position of the neurons that will provide their innervation.

Formation of topographic maps

In the somatosensory, visual, and motor systems, neuronal connections are arranged such that neighboring points in the periphery are represented by adjacent locations in the appropriate regions of the CNS (see Chapters 9, 11, and 16). In other systems (e.g., the auditory and olfactory systems),

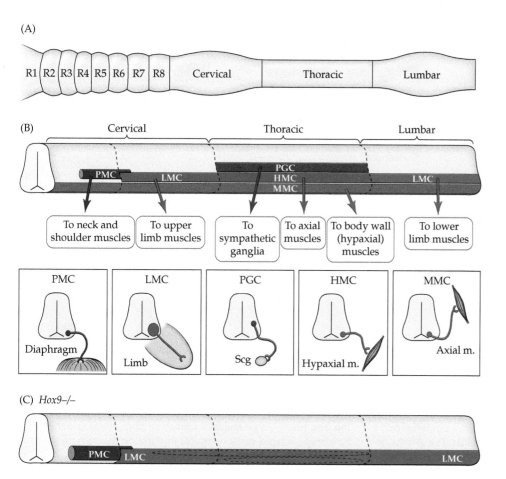

FIGURE 23.18   Matching CNS motor neurons to their peripheral targets (A) Based on 3′–5′ expression of subsets of the Hox clusters that parallels the anterior–posterior (A–P) axis of the spinal cord, the spinal cord, like the hindbrain, becomes specified into distinct regions that will generate motor neurons that innervate specific peripheral targets. (B) The division of the spinal cord motor neuron pools into "columns" that innervate the diaphragm (the phrenic motor column; PMC), the limbs (lateral motor column; LMC), the sympathetic ganglia (preganglionic column, PGC), the hypaxial muscles (body wall and abdomen, hypaxial motor column; HMC), and finally the axial musculature (medial motor column; MMC) relies on patterning each domain with a nested set of Hox genes. The overlapping transcriptional regulation established by these genes defines the cervical, thoracic and lumbar regions of the spinal cord and specifies the motor neurons in each region and column so that they can recognize signals from the appropriate target . The parallel A–P patterning in the target muscles or ganglia innervate guide and validate target choices of these specific classes of motor neurons. (C) When the Hox genes are mutated, the A–P identities of motor neurons are altered and motor neuron–target matching is disrupted. In this instance, complete loss of function of *Hox9C*, which regulates specification of the thoracic spinal cord, results in the loss of the PGC and HMC usually found in the thoracic cord, and an expansion of the lateral motor column. (After P. Philippidou and J. S. Dassen. 2013. *Neuron* 80: 12–34.)

there are also orderly representations of various stimulus attributes such as frequency or sensory receptor identity in at least some central regions (see Chapters 10 and 14). In all of these examples, there must be some sort of fairly precise recognition mechanisms during initial development to guide peripheral axons from distinct locations to appropriately "mapped" target locations in the brain. In the early 1960s, Roger Sperry (who also did pioneering work on the functional specialization of the cerebral hemispheres; see Chapter 33) proposed the **chemoaffinity hypothesis**. This hypothesis, based primarily on Sperry's work on the visual systems of frogs and goldfish but reminiscent of the rules established for pathway formation described in the previous section, provides a basic explanation for how topographic maps arise during development. In frogs and goldfish (and subsequently confirmed in mammals), the terminals of retinal ganglion cells form a precise topographic map in the optic tectum (the tectum is homologous to the mammalian superior colliculus) (Figure 23.19A). When Sperry crushed the optic nerve and allowed it to regenerate (fish and amphibians, unlike mammals, can regenerate CNS axon tracts; see Chapter 26), he found that retinal axons reestablished the original topographic pattern of connections in the tectum. Even if the eye was rotated 180 degrees, the regenerating axons grew back to their original tectal destinations (causing some behavioral confusion for the frog) (Figure 23.19B).

(A)

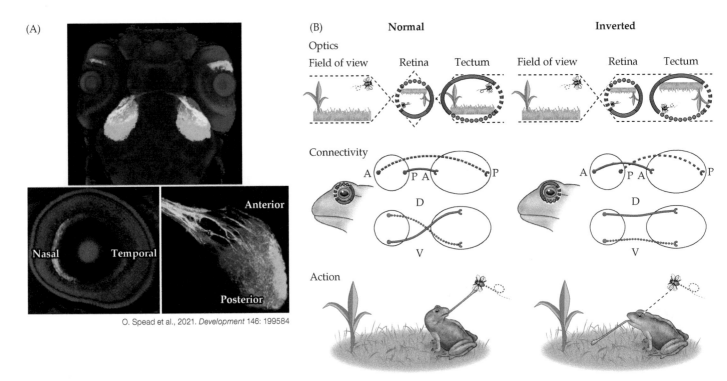

O. Spead et al., 2021. *Development* 146: 199584

FIGURE 23.19 Establishing topographic maps in the CNS: Mapping the retina onto the optic tectum (A) Mapping of the nasal and temporal retina onto the anterior versus posterior optic tectum visualized in a living, developing zebrafish. The affinity of axons from neurons in the temporal retina for the anterior tectum is substantial, and the capacity of axons from neurons in the nasal retina to ignore the anterior tectum target neurons and innervate the posterior tectum is similarly robust. (B) The significance of appropriate development of a retinal–tectal topographic map for the frog's capacity to detect and capture prey. Left: The optics of the normal visual system invert the image of the world, and the topographic mapping of the retinal image onto the tectum corrects that inversion so that the axes of the visual field and the retinal map in the tectum are aligned (top and middle panels). This allows the frog to use visual spatial information to orient to and then capture prey via rapid sensory–motor integration (bottom panel; see also Chapter 20). Right: When the retina is surgically rotated in the frog (in which axon regeneration after severing of the retinal axons is possible), the mapping of the retina onto the tectum still follows the original coordinates of the eye. This results in the frog perceiving the location of prey as being opposite to its actual position in space, and guiding its motor response to the incorrect location. (B after R. W. Sperry, 1963. *Proc. Natl. Acad. Sci. USA* 50: 703–10.)

Accordingly, Sperry proposed that each tectal cell carries a chemical "identification tag" and that the growing terminals of retinal ganglion cells have complementary tags such that the retinal cells seek out a specific location in the tectum. In a non-perturbed frog, such tags would therefore match the topography of the sensory surface, in this case the retina, with connections within the CNS target—the tectum. The "identification tags" were assumed to be cell adhesion or cell recognition molecules, and the "affinity" they engendered was presumed to be due to the selective binding of receptor molecules on retinal ganglion cell growth cones to complementary molecules on the tectal cells at appropriate relative positions. It is easy to imagine that such a mechanism might account for the full range of topographic maps seen in the nervous systems of many animals, including humans.

Further experiments in the amphibian and avian visual systems made the strictest form of the chemoaffinity hypothesis—that is, the labeling of each tectal location by a distinctive recognition molecule—untenable. Rather than displaying a precise "lock-and-key" affinity, the behavior of growing axons suggested that there are *gradients* of cell surface molecules to which growing axons respond. Normally, axons from the temporal region of the retina innervate the anterior pole of the tectum and avoid the posterior pole. Temporal retinal axons, when presented with a choice of cell membranes derived from anterior or posterior tectal regions as a substrate, grow exclusively on anterior membranes, avoiding membranes derived from the "wrong" region of the tectum (Figure 23.20A). A likely candidate for the negative-guidance signal for temporal axons in the posterior tectum was purified and its gene cloned. The protein—initially called RAGS (*repulsive axon guidance signal*)—turned out to belong to the Eph family of cell surface–bound adhesion and signaling molecules (see Figure 23.7F). The Ephs had been previously identified and characterized as cell-cell recognition molecules in tumor cells. In the eye and tectum, ephrins and Eph

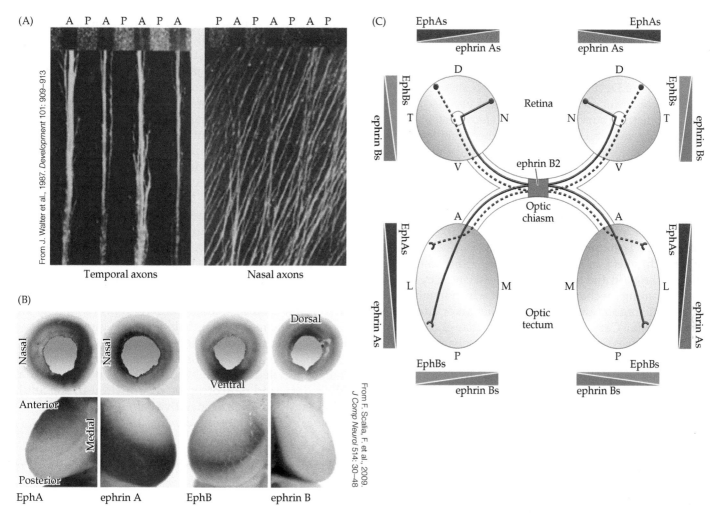

(A)

From J. Walter et al., 1987. *Development* 101: 909–913

Temporal axons   Nasal axons

(B)

Nasal   Nasal   Dorsal

Anterior   Ventral

Medial

Posterior

EphA   ephrin A   EphB   ephrin B

From F. Scalia, F. et al., 2009. *J Comp Neurol* 514: 30–48

(C)

EphAs   EphAs
ephrin As   ephrin As

EphBs   D   D   EphBs
ephrin Bs   Retina   ephrin Bs
T   N   N   T

ephrin B2
V   V

Optic chiasm

A   A

EphAs   EphAs
L   M   M   L
ephrin As   Optic   ephrin As
tectum

EphBs   P   P   EphBs
ephrin Bs   ephrin Bs

FIGURE 23.20 Molecular affinities determine mapping of the retina onto tectum (A) In vitro assay for cell surface molecules that contribute to topographic specificity in the optic tectum. A set of alternating, 90-µm-wide stripes of membranes from anterior (A) and posterior (P) optic tectum of chicks were laid down on a glass coverslip. The posterior membranes have fluorescent particles added to make the boundaries of the stripes apparent (top of panels). Explants of retina from either nasal or temporal retina were placed on the stripes. Temporal axons prefer to grow on anterior membranes and are repulsed by posterior membranes. In contrast, nasal retinal axons grow equally well on both stripes. (B) Distribution of Eph receptors and ephrin ligands in the retina and tectum of the

developing frog. The nasal retina is the site of elevated EphA and ephrin A expression, with EphA expressed in the anterior tectum and ephrin A in the posterior tectum (left). Similarly, the ventral retina is the site of highest expression of EphB, while the posterior tectum is a site of elevated EphB expression (right). The expression of ephrin B in the retina is reversed, so that it is highest in the dorsal retina and expressed modestly in the medial tectum. (C) The opposing gradients of Eph receptors and ligands across the retina and tectum establish a graded series of selective affinities for retinal ganglion cell axons so that the axons of neurons in the retina can recognize their optimal target territories upon arrival in the tectum. (C after T. Harada, et al., 2007. *Genes Dev* 21: 367–378.)

receptors are distributed in complementary gradients across the temporonasal and dorsoventral retina and the anterior–posterior and mediolateral tectum (Figure 23.20B). These gradients result in matching levels of specific ephrin ligands and receptors. Thus, ephrin and Eph binding and signaling facilitate the topographic mapping of the nasal and temporal retina along the anterior–posterior axis of the tectum. Subsequent work has associated several ephrins and Ephs with topographic mapping in numerous systems, including the mammalian visual

system (Figure 23.20C). These observations accord with the idea that chemoaffinity operates not by one-to-one or lock-and-key recognition, but by gradients of affinities that provide axons and their targets with markers of general position within a system of coordinates (like north, south, east, and west on a map). Additional sharpening of topography and other types of highly specific, behaviorally significant connections may rely on activity-dependent mechanisms (discussed in Chapter 24) that continue to modify patterns of synaptic connections.

Summary

Neurons in the developing brain must integrate a variety of molecular signals to determine where to send their axons, how to extend their dendrites, and what cells to form synapses on. Developing neurons also must determine whether to live or die as they either seek or fail to find targets. If a neuron survives the initial step of establishing a connection, molecular signals must then stabilize the target choice made by that neuron's axon and subseqeuntly regulate how many synapses are made. A remarkable transient cellular specialization, the growth cone, is responsible for axon growth and guidance. Growth cones explore the embryonic environment and determine the direction of axon growth as well as recognize appropriate targets. Their motile properties allow growth cones to approach, select, or avoid a target according to modulation of the actin and microtubule cytoskeleton by numerous signaling mechanisms, many of which involve local metabolic regulation via ATP production or changes in intracellular Ca^{2+}. The instructions that elicit growth cone responses come from chemotropic, chemorepellant, adhesive, and trophic molecules. These molecules are secreted into extracellular spaces, found on cell surfaces, or embedded in the extracellular matrix. Their cues ensure that coherent axon pathways are formed and prevent inappropriate connections. Initial growth of dendrites is influenced by similar adhesion and recognition signaling mechanisms, resulting in appropriate dendritic orientation, branching, and distribution. Attractive, repulsive and cell-cell recognition molecules also influence the differentiation of growth cones and developing dendritic domains into pre-and postsynaptic specializations that define a synapse. Further signals that specify synaptic partners, control afferent cell survival, and stabilize or destabilize nascent synapses, are transmitted by neurotrophins, molecules made by neuronal targets in small quantities that bind to a variety of receptors to elicit distinct cellular responses. Neurotrophic influences—cell survival or death, process growth, and modulation of synapse formation—help determine which neurons remain in a neural circuit, how they are connected, and how they continue to change. These mechanisms combine to generate major pathways that interconnect brain regions that will be functionally related, and they guide the initial establishment of orderly representations via topographic maps. Defects in the early guidance of axons, degeneration of dendrites, or subsequent regulation of synaptogenesis have been implicated in a variety of congenital neurological syndromes and developmental disorders, and neurotrophic dysfunction in the adult CNS may underlie degenerative pathologies such as Alzheimer's and Parkinson's diseases.

■ Additional Reading

Reviews

Heckman, E.L. and C.Q. Doe (2021) Establishment and maintenance of neural circuit architecture. *J. Neurosci.* 41: 1119—1129.

Huber, A. B., A. L. Kolodkin, D. D. Ginty and J. F. Cloutier (2003) Signaling at the growth cone: Ligand-receptor complexes and the control of axon growth and guidance. *Annu. Rev. Neurosci.* 26: 509–563.

Kolodkin, A. L. and M. Tessier-Lavigne (2010) Mechanisms and molecules of neuronal wiring: A primer. *Cold Spring Harb. Perspect. Biol.* 3(6): a001727. doi: 10.1101/cshperspect. a001727

Onesto, M.M. and 5 others (2021) Growth factors as axon guidance molecules: Lessons from in vitro studies. *Front. Neurosci.* 15: Article 678454.

Phillippidou, P. and J.S. Dasen (2013) Hox genes: choreographers in neural development, architects of circuit organization. *Neuron* 80: 12—34.

Reichardt, L. F. (2006) Neurotrophin-regulated signalling pathways. *Philos. Trans. R. Soc. Lond., B, Biol. Sci.* 361: 1545–1564.

Sanes, J. R. and J. W. Lichtman (1999) Development of the vertebrate neuromuscular junction. *Annu. Rev. Neurosci.* 22: 389–442.

Seiradake, E., Yvonne Jones, E., and R. Klein (2016) Structural perspectives on axon guidance. *Annu. Rev. Cell Dev. Biol.* 32:577—608.

Wiggin, G. R., J. P. Fawcett and T. Pawson (2005) Polarity proteins in axon specification and synaptogenesis. *Dev. Cell* 8 (6): 803–816.

Zipursky, S. L. and J. R. Sanes (2010) Chemoaffinity revisited: dscams, protocadherins, and neural circuit assembly. *Cell* 143: 343–353.

Important Original Papers

Baier, H. and F. Bonhoeffer (1992) Axon guidance by gradients of a target-derived component. *Science* 255: 472–475.

Brown, M. C., J. K. S. Jansen and D. Van Essen (1976) Polyneuronal innervation of skeletal muscle in new-born rats and its elimination during maturation. *J. Physiol.* 261: 387–422.

Campenot, R. B. (1977) Local control of neurite development by nerve growth factor. *Proc. Natl. Acad. Sci. U.S.A.* 74: 4516–4519.

Drescher, U. and 5 others (1995) In vitro guidance of retinal ganglion cell axons by RAGS, a 25 kDa tectal protein related to ligands for Eph receptor tyrosine kinases. *Cell* 82: 359–370.

Farinas, I. and 4 others (1994) Severe sensory and sympathetic deficits in mice lacking neurotrophin-3. *Nature* 369: 658–661.

Kaplan, D. R., D. Martin-Zanca and L. F. Parada (1991) Tyrosine phosphorylation and tyrosine kinase activity of the *trk* proto-oncogene product induced by NGF. *Nature* 350: 158–160.

Kennedy, T. E., T. Serafini, J. R. de la Torre and M. Tessier-Lavigne (1994) Netrins are diffusible chemotropic factors for commissural axons in the embryonic spinal cord. *Cell* 78: 425–435.

Kolodkin, A. L., D. J. Matthes and C. S. Goodman (1993) The *semaphorin* genes encode a family of transmembrane and secreted growth cone guidance molecules. *Cell* 75: 1389–1399.

Levi-Montalcini, R. and S. Cohen (1956) In vitro and in vivo effects of a nerve growth-stimulating agent isolated from snake venom. *Proc. Natl. Acad. Sci. U.S.A.* 42: 695–699.

Luo, Y., D. Raible and J. A. Raper (1993) Collapsin: A protein in brain that induces the collapse and paralysis of neuronal growth cones. *Cell* 75: 217–227.

Oppenheim, R. W., D. Prevette and S. Homma (1990) Naturally occurring and induced neuronal death in the chick embryo in vivo requires protein and RNA synthesis: Evidence for the role of cell death genes. *Dev. Biol.* 138: 104–113.

Serafini, T. and 6 others (1996) Netrin-1 is required for commissural axon guidance in the developing vertebrate nervous system. *Cell* 87: 1001–1014.

Domenici, C. and 8 others (2017) Floorplate-derived netrin is dispensable for commissural axon guidance. *Nature* 545: 350–354.

Sperry, R. W. (1963) Chemoaffinity in the orderly growth of nerve fiber patterns and connections. *Proc. Natl. Acad. Sci. U.S.A.* 50: 703–710.

Books

Loughlin, S. E. and J. H. Fallon (eds.) (1993) *Neurotrophic Factors*. San Diego, CA: Academic Press.

Purves, D. (1988) *Body and Brain: A Trophic Theory of Neural Connections*. Cambridge, MA: Harvard University Press.

Experience-Dependent Plasticity in the Developing Brain

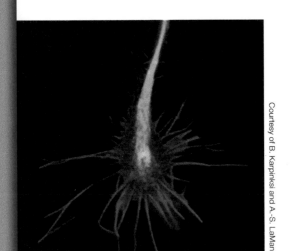

Courtesy of B. Karpinksi and A.-S. LaMantia

KEY CONCEPTS

- **24.1** Electrical activity in new neural circuits determines final numbers and patterns of functional connections

- **24.2** Electrical activity reflects initial experience and defines connections during critical periods

- **24.3** Relative levels of electrical activity across inputs determine final connections in neural circuits

- **24.4** Ion channels, neurotransmitters and their receptors, and neurotrophins regulate activity-dependent circuit development

Overview

Once brain regions and neuronal diversity have been established and axons, dendrites, and synapses have completed their initial growth, a final phase of developmental change begins. This phase of development—primarily during late prenatal and early postnatal life—relies on cellular changes driven by electrical activity in neurons, including synaptic and action potential activity, most often elicited by environmental stimuli that reflect the newborn's experience of their environment. The limited times of postnatal developmental change elicited by electrical activity in specific circuits are referred to as critical periods. As humans (and other mammals) continue to mature, the cellular mechanisms that modify neural connectivity in response to experience become less effective, and the dramatic changes in neural circuits and related behaviors seen during critical periods are no longer possible. Thus, critical periods are thought to optimize the brain of each individual so that circuitry is adapted early in life to specific demands that confront that individual throughout the balance of life. Changes in the numbers and organization of connections in many brain regions from birth through late adolescence, including the human cerebral cortex, indicate that experience-dependent critical periods are likely to influence connectivity and complex behaviors. Much of our knowledge of the influence of activity on developing neural circuits comes from studies of the mammalian visual system, where differences in input from each eye influence patterns of connections in the visual cortex that are essential for binocular vision. The cellular and molecular mechanisms that mediate activity-dependent developmental changes are in many ways similar to those that mediate synaptic modifications that underlie learning and memory. Many of the effects of activity during circuit development depend on secreted signals, including neurotransmitters and growth factors, transduced via second messengers and their effectors. These activity-elicited changes influence local gene expression and trophic interactions that lead to final adjustments of axon or dendrite growth, as well as synapse growth and stability. These mechanisms may be compromised in disorders that result in intellectual disability, developmental delay, autism spectrum disorder, or psychiatric diseases such as schizophrenia.

<table>
<tr><td>CONCEPT
24.1</td><td># Electrical Activity in New Neural Circuits Determines Final Numbers and Patterns of Functional Connections</td></tr>
</table>

LEARNING OBJECTIVES

24.1.1 Explain the role of activity in strengthening or weakening synaptic connections.

24.1.2 Discuss the association between postnatal brain growth and activity-dependent circuit changes.

24.1.3 Identify several brain regions and circuits that are influenced by early activity and experience.

Neural activity and brain development

Observations of humans and animals suggest that permanent behavioral changes can be encoded in the brain based on the environmental stimuli that define an individual human or animal's experience from birth onward. These observations were made more compelling by multiple studies in the first half of the twentieth century that stressed the importance of early experiences in the formation of a broad range of behaviors. There was an assumption that information acquired during early life had a disproportionate impact on subsequent behaviors, regardless of later experiences. The basis for this encoding of information, however, remained uncertain for most of the first half of the last century. In 1949, the psychologist D. O. Hebb hypothesized that the coordinated electrical activity of a presynaptic terminal and a postsynaptic neuron strengthens the synaptic connection between them. This simple

hypothesis defined a key principle to translate behavioral experience to altered neuronal connections: Electrical activity within the circuits can change the circuit. **Hebb's postulate**, as it has come to be known, was originally formulated to explain the cellular basis of learning and memory, but the concept has been generally applied to electrical activity in neural circuits that mediates modifications in synaptic strength or distribution, including those that occur during development. For developing neural circuits, Hebb's postulate implies that synaptic terminals strengthened by correlated activity during development will be retained by the target cells and that their parent axons will sprout new branches to make additional synaptic contacts. The synapses that are persistently weakened by uncorrelated activity will eventually lose their hold on the postsynaptic cell, leading to one or more of these outcomes:

1. The elimination of those synapses on the target cell and the withdrawal of the parent axon

2. The death of the neuron that gives rise to the weakened and eliminated synapses

3. The stabilization and growth of synapses from the less correlated afferent neuron on another target

The simplest statement of Hebb's postulate—that "neurons that fire together wire together"—provides a template for considering how activity leads to the retention or elimination of individual inputs onto a single target neuron when electrical activity occurs repetitively in a circuit (Figure 24.1). When applied to more complex circuits that engage manifold neurons and synapses, it is clear that numbers and patterns of connections can be changed by electrical

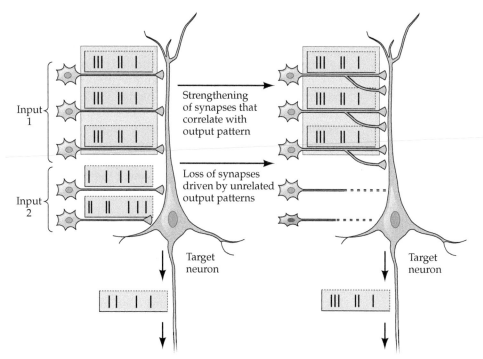

FIGURE 24.1 Hebb's postulate and the development of synaptic inputs In this drawing, a postsynaptic neuron is shown with two sets of presynaptic inputs, each with a different pattern of electrical activity. Activity patterns, corresponding to action potential frequency, are represented by the short vertical bars. In the example here, the three correlated inputs at the top (Input 1) are better able to activate the postsynaptic cell. These inputs cause the postsynaptic cell to fire a pattern of action potentials that follows the pattern seen in the input. As a result, the activity of the presynaptic terminals and the postsynaptic neuron is highly correlated. According to Hebb's postulate, these synapses are therefore strengthened. The two additional inputs (Input 2) relay a different pattern of activity that is less well correlated with the majority of the activity elicited in the postsynaptic cell. These synapses gradually weaken and are eventually eliminated (right-hand side of figure), while the correlated inputs form additional synapses.

activity across entire populations of synapses from diverse sources, thus influencing the final capacity of the circuit to integrate and process information.

The electrical activity initiated by exposure to and interaction with the environment following birth, and its consequences for making or breaking connections based on Hebb's initial formulation of **activity-dependent plasticity**, is a key element in the construction of any individual's nervous system. Intrinsic mechanisms controlled fairly rigidly by the genome—including those that specify neural progenitors and brain regions, constrain axon growth, generate the first synapses, and initiate the formation of topographic maps (see Chapters 22 and 23)—establish the general framework of circuitry required for most behaviors. These cellular and molecular mechanisms, however, do not yield final patterns of connectivity. The electrical activity generated by neurons as they differentiate, acquire excitable properties, and make initial synaptic connections establishes one driver of activity-dependent plasticity, usually prior to birth (in mammals) and the onset of environmentally generated experience. Subsequently, "typical" experiences driven by stimuli in an individual's environment, via the electrical activity and related synpatic changes they elicit, are thought to validate and enhance "optimal" behaviors and to preserve, amplify, or adjust initial numbers and arrangements of connections that are established by intrinsic developmental mechanisms.

"Diminished" or dramatically divergent experiences alter neural activity so that patterns of connections are changed to accommodate deficits or adaptive differences in behaviors and underlying synaptic and neural circuit function. These alterations can have some adaptive advantage—for example, when an infant's capacity to acquire information from the environment is changed because of visual or hearing impairments, the differences in experience and activity can enhance and modify neural circuits and amplify sensitivity to the remaining, intact sensory modalities. In contrast, sensory deprivation or early trauma caused by adverse circumstances for an otherwise "typical" infant or child may lead to long-lasting changes in numbers or patterns of connections and maladaptive behaviors that are not well matched to an environment in which the early trauma is no longer present. The eventual decline of the capacity to remodel connections using a mechanism that in outline is similar to that proposed by Hebb most likely explains changes in the capacity of the brain to acquire new information and generate completely new behaviors over a lifetime—a concept with obvious educational, clinical, and social implications.

Brain growth and activity-dependent circuit changes

The Hebbian formulation of the role of electrical activity in influencing the ongoing organization of neural circuits provides a framework for understanding the significance of an obvious phenomena: The brain continues to grow after birth (Figure 24.2A), roughly in parallel with the emergence and acquisition of increasingly complex behaviors. Since very few neurons are generated after birth in most mammalian brains (see Chapter 26), including the human brain, it is likely that the neurons that have fired and wired together continue to grow together to enhance the strength and effectiveness of circuits they have formed. Thus, the continued elaboration of dendrites, axons, and their branches, as well as the addition of synaptic connections, early in life must account for a significant portion of postnatal brain growth (Figure 24.2B). This growth perhaps provides a substrate for enhanced behavioral capacities, driven and then refined by Hebbian mechanisms. There is a dramatic addition of cytologically defined synaptic contacts (visualized with electron microscopy; see Chapter 1) during early postnatal life in the cerebral cortex (Figure 24.2C) as well as other brain regions. This early postnatal addition of synaptic contacts in the cortex is followed by a decline in synapse number during later childhood and adolescence. Accordingly, the events of progressive neuronal growth paralleled by synapse addition and then the elimination of a subset of synaptic connections probably reflect activity-dependent changes during early postnatal life. The genesis, growth, and interactions of several classes of glial cells (see Chapters 1 and 26) with developing synaptic connections probably facilitate the stabilization of activity-validated circuits as well as contribute to brain growth. Some of these glial cells—mostly astrocytes—are thought to further stabilize synaptic sites, while others—mostly microglia—are thought to help clear debris after synapses have been eliminated. In addition, oligodendroglial cells myelinate maturing axons. From birth through early adulthood, all of these neuronal and glial events occur in synchrony with the acquisition of sensory and motor abilities, the capacity for social interaction, and increasingly sophisticated cognitive behaviors, including spoken, signed, and written language in humans (see Concept 24.2 and Chapter 33). These coincidences suggest that the combination of activity-dependent modification of connections initially suggested by Hebb and corresponding brain growth and behavioral changes during early life must underlie how each individual's brain ultimately develops and grows to meet the challenges of adapting to a dynamic environment.

Systems, circuits, and activity-dependent development

Overall brain growth does not fully capture the essential contribution of activity-dependent changes to optimizing multiple functionally distinct circuits in an individual's brain, and thus the behaviors controlled by these circuits. There are multiple "systems" within the brain that are influenced by activity-dependent mechanisms that direct the growth, elaboration, and elimination of connections. Many

(A)   2 weeks   1 year   2 years

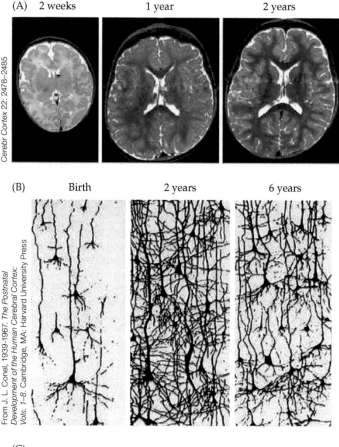

Cerebr Cortex 22: 2478–2485

(B)   Birth   2 years   6 years

From J. L. Conel, 1939-1967. *The Postnatal Development of the Human Cerebral Cortex,* Vols. 1–8. Cambridge, MA: Harvard University Press

(C)

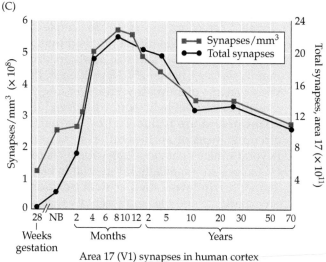

Area 17 (V1) synapses in human cortex

FIGURE 24.2   The human brain changes significantly during postnatal life (A) MRI images of a brain from the same child at 2 weeks, 1 year, and 2 years of age. During this period the brain increases substantially in size, the cortical sulci and gyri become more complex, and there is a significant increase in the subcortical white matter (lighter gray at 2 weeks, darker gray at 1 and 2 years). (B) Axons and dendrites in the human cerebral cortex at birth, 2 years, and 6 years of age. These drawings are based on Golgi-stained neurons in the cerebral cortex from individual postmortem samples of different ages. (C) Addition and then elimination of synapses in the human brain. These data are based on electron microscopic analysis of synapse density in the human primary visual cortex. (C graph after P. R. Huttenlocher et al., 1982. *Neurosci Lett* 33: 247–252.)

of these mechanisms are seen during early postnatal life, as the newborn and juvenile individual encounters both unique and repetitive environmental stimuli with more or less impact on adaptive behaviors. The result in each system is the matching of circuit organization, function, and behavior based on levels of neuronal synaptic transmission and action potential firing elicited by specific types of experience. Thus, experience and activity can "tune," or "sculpt," each system's connectivity to respond to an individual's needs based on their early environment and experience. Sensory pathways, and their constituent neurons and circuits, have proved to be the best systems thus far to study activity- and experience-dependent changes during early postnatal development. Circuits in the mammalian visual system, somatosensory system, and auditory system have been the most intensively studied models for understanding activity- and experience-dependent changes. In each case, a major target of activity or experience is topographic (or tonotopic, for the auditory system) maps of the sensory periphery established initially by selective, graded tropic and adhesion signaling (see Chapter 23). Thus, several aspects of functional visual capacities such as eye dominance, binocular vision, and orientation selectivity (see Concept 24.3) in the cerebral cortex are highly sensitive to activity-dependent changes that reflect either the overall functional integrity of the two eyes or the type of visual stimuli that the eyes detect (Figure 24.3A). Disproportionate exposure to auditory stimuli of a particular frequency facilitates expansion of the territory in the tonotopic map in the auditory cortex that is maximally responsive to that frequency, at the expense of diminished cortical space for other frequencies (Figure 24.3B). Similarly, the somatotopic map in the somatosensory cortex, as well as other relay nuclei in the brainstem and thalamus, can be shaped by the physical properties of the peripheral receptor surfaces and the type of somatosensory stimulation available during early life. This sensitivity has been most extensively analyzed in the map of individual whiskers that can be visualized in the brains of most rodents, including rats and mice, called the barrel field, reflecting the barrel-like shape of the discrete accumulations of nerve cells, processes, and glia that correspond to each of the whiskers in the periphery (Figure 24.3C). In addition to the influence that sensory experience has on the final arrangements of the connections that constitute topographic or computational maps in sensory systems, connections in hypothalamic or limbic circuits (see Chapters 19 and 29) that represent additional environmental stimuli such as stress, fear, and social interactions can be influenced by experience during early life. Thus, the transformation of environmental stimuli encountered by any individual to distinct patterns of neuronal activity during early life results in the refinement or change of topographic or functionally specific connections that reflect the unique experience of that individual as a result of developmental activity-dependent changes.

(A) Primary visual cortex

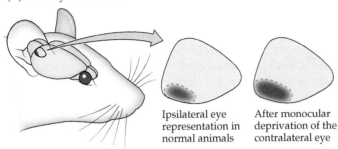

Ipsilateral eye representation in normal animals

After monocular deprivation of the contralateral eye

(B) Primary auditory cortex

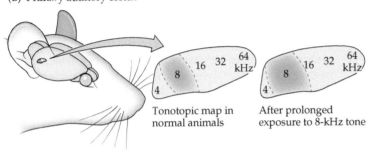

Tonotopic map in normal animals

After prolonged exposure to 8-kHz tone

(C) Primary somatosensory cortex

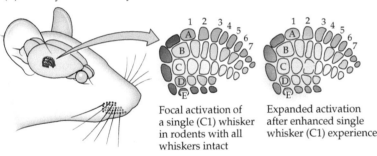

Focal activation of a single (C1) whisker in rodents with all whiskers intact

Expanded activation after enhanced single whisker (C1) experience

FIGURE 24.3 Mammalian sensory systems with significant activity-dependent plasticity during early life (A) The primary visual cortex is sensitive to experience-driven inputs from the two eyes for establishing the territory in each cortical hemisphere driven by each eye. After monocular deprivation, whereby visual activity in the cortex is driven by only one eye, the cortical territory responsive to stimulation in that eye—at least one synaptic relay from the eye—expands. (B) In the rodent primary auditory cortex, disproportionate exposure of a young animal to a tone of a single frequency results in the expansion of the territory in which auditory cortical neurons are responsive to that tone, at the expense of the representation of other tones with lower or higher frequencies. (C) Activation pattern of the barrel field in the rodent primary somatosensory cortex, which represents the whiskers in the whisker pad, when one whisker is stimulated. There is activation focused on the single barrel (C1) related to the whisker selected for preferential stimulation, as well as some activity in adjacent barrels. When whisker C1 is stimulated repeatedly and preferentially during early life, its representation expands substantially and the degree of responsiveness in the C1 barrel increases. (After K. Kole et al., 2018. *Neurosci Biobehav Rev* 84: 100–115.)

CONCEPT 24.2 | Electrical Activity Reflects Initial Experience and Defines Connections during Critical Periods

LEARNING OBJECTIVES

24.2.1 Define the term *critical periods*.

24.2.2 Identify at least three basic properties that all critical periods share.

24.2.3 Identify brain regions, circuits, and regions shaped by critical periods.

24.2.4 Describe the circuit changes elicited by critical period activity and experience.

Critical periods

For most animals, the behavioral repertoire for basic survival, including feeding, fighting, mating, and parenting (see Chapter 25), largely relies on patterns of connectivity established by intrinsic developmental mechanisms. Indeed, embryonic mechanisms and their developmental consequences are sufficient to create some remarkably sophisticated instinctual, or "innate," behaviors, including the complex repertoire for parental identification, food intake, and responses to predators seen in newborn birds and mammals (see Chapter 25). Nevertheless, the nervous systems of animals with increasingly complex repertoires of behaviors, including humans, adapt to and are influenced by the particular circumstances of an individual's environment beyond innate behavioral capacities. These environmental factors are especially influential in early life, during temporal windows called **critical periods**—the time when experience and the neural activity that reflects that experience have maximal effect on the acquisition or skilled execution of a particular behavior. Some behaviors, such as parental imprinting in hatchling birds (the process by which the hatchling recognizes its "parent" based on initial sensory experience after emerging from the egg), are seen only if animals encounter specific stimuli during a sharply restricted time (hours or days) in early postnatal (or posthatching) development (Box 24A). By contrast, critical periods for sensorimotor skills and complex behaviors that require learning over protracted periods of postnatal life end far less abruptly and provide far more time for environmentally acquired experience. In some cases, such as the acquisition of communication skills in young songbirds (see Chapter 25), or language in human infants (see later in this concept), detailed instructive influences from the

■ BOX 24A | Built-In Behaviors

The idea that animals possess an innate set of behaviors appropriate for a world not yet experienced has always been difficult to accept. However, the preeminence of instinctual responses is obvious to any biologist who looks at what animals actually do. Perhaps the most thoroughly studied examples occur in young birds. Hatchlings emerge from the egg with an elaborate set of innate behaviors. First, of course, is the complex behavior that allows the chick to escape from the egg. Once the chick has emerged, a variety of additional abilities indicate how much early behavior is "pre-programmed."

In a series of seminal observations based on his work with geese, Konrad Lorenz showed that goslings follow the first large, moving object they see and hear during their first day of life. Although this object is typically the mother goose, Lorenz found that goslings would imprint on a wide range of animate and inanimate objects presented during this period, including Lorenz himself. The window for imprinting in goslings is less than a day; if animals are not exposed to an appropriate stimulus during this time, they will never form the appropriate parental relationship. Once imprinting occurs, however, it is irreversible, and geese will continue to follow inappropriate objects (male conspecifics, people, or even inanimate objects).

In many mammals, auditory and visual systems are poorly developed at birth, and maternal imprinting relies on olfactory or gustatory cues. For example, during the first week of life (but not later), infant rats develop a lifelong preference for odors associated with their mother's nipples. As in birds, this filial imprinting plays a role in the rats' social development and later sexual preferences.

Imprinting is a two-way street, with parents (especially mothers) rapidly forming exclusive bonds with their offspring. This phenomenon is especially important in animals such as sheep that live in large groups or herds in which all the females produce offspring at about the same time of year. Ewes have a critical period 2 to 4 hours after giving birth during which they imprint on the scent of their own lamb. After about 4 hours, they rebuff approaches by other lambs.

Work by Harry Harlow and his colleagues at the University of Wisconsin in the 1950s underscored the relevance of these studies of avian imprinting to primates. Harlow isolated monkeys within a few hours of birth and raised them in the absence of either a natural mother or a human substitute. In the best known of these experiments, the baby monkeys had one of two maternal surrogates: a "mother" constructed of a wooden frame covered with wire mesh that supported a nursing bottle, or a similarly shaped object covered with soft terrycloth but without any source of nourishment for the young monkey. When presented with this choice, the baby monkeys preferred the terrycloth mother and spent much of their time clinging to it, even though the feeding bottle was with the wire mother. Harlow took this to mean that newborn monkeys have a built-in need for maternal care and have at least some innate idea of what a mother should feel like. Several other endogenous behaviors have been studied in infant monkeys, including a naïve monkey's fear reaction to the presentation of certain objects (e.g., a snake) and the "looming" response (fear elicited by the rapid approach of any formidable object). Most of these built-in behaviors have analogues in human infants.

Taken together, these observations make plain that many complicated behaviors, emotional responses, and other predilections are well established in the nervous system prior to any significant experience, and that the need for certain kinds of early experiences for normal development is predetermined. These built-in behaviors and their neural substrates have presumably evolved to give newborns a better chance of surviving in a reliably dangerous world.

Konrad Lorenz followed by imprinted geese.

environment (i.e., exposure to complex calls from fellow adult songbirds, or to words and sentences uttered by human caregivers) and opportunities for mimicry and repetition are required for an extended period to ensure normal development of the behavior. The availability of instructive experiences from the environment, as well as the neural capacity to respond to them, is key for successful completion of the critical period. These instructive influences are important for territorial and reproductive behaviors in a subset of non-human species. In some songbirds, young male birds acquire the capacity to produce species-specific song by mimicking adult "tutor" male birds during a limited period of postnatal life (see Chapter 25). If this essential instruction is withheld or disrupted, the birds are not effective in using communication to define their territories and compete for mates.

Although critical periods vary widely in both their duration and the behaviors they modify, they all share some basic properties. Each critical period encompasses the time during which a given behavior is especially susceptible to—indeed, requires—specific environmental influences in order to develop optimally. Environmental influence elicits neural activity in the relevant sensory pathway via stimulation of peripheral receptors, or motor pathways via rudimentary or initially uncoordinated movements. The acute nature of this activity—its frequency, amplitude, duration, and correlation, and the recurrence of the stimuli that elicit the activity—ultimately drives changes in synaptic connections. These influences of experience-driven neural activity can be as subtle as that of ongoing stimuli of light or sound encountered by an infant, or definitive as the precisely articulated instruction in one's native (or a foreign) language required to achieve fluent speech and accurate comprehension. Once critical periods end, the core features of behaviors that were once highly sensitive to experience-driven plasticity are largely unaffected by subsequent experience. This suggests that cellular and molecular mechanisms that are influenced by experience via neural activity must also change. Failure to be exposed to appropriate stimuli during the critical period is difficult or impossible to remedy subsequently—probably because the biological mechanisms needed to change connections during the critical period are no longer available in an older individual. In most mammals, including humans, critical periods seem to result in changes in organization and function of circuits in the cerebral cortex. Thus, much of our subsequent discussion of critical periods focuses on the consequences of experience and activity for influencing cerebral cortical growth, connections, and function from birth through young adulthood.

Critical periods in visual system development

Fundamental understanding of how changes in activity and connectivity might contribute to critical periods and behavioral capacity has come from studies of the developing visual system in animals with highly developed visual abilities, particularly cats and monkeys. The visual system is extremely amenable to the sorts of experimental manipulations necessary to test the relationship between experience, activity, and circuitry. It is relatively easy to either deprive or augment visual experience in an experimental animal; eyes can be sutured shut, or animals can be reared in illumination conditions ranging from total darkness to maximal light and patterns of light and dark. Such control

of sensory experience is almost impossible in any other modality—it is much harder to deprive an animal of auditory, somatosensory, olfactory, or taste stimuli. Moreover, the organization of the visual pathways provides ideal opportunities to evaluate how experience influences ongoing function and connections.

Information from the two eyes is first integrated in the primary visual (striate) cortex (see Chapters 11 and 12), where most afferents from the lateral geniculate nucleus (LGN) of the thalamus (where inputs from the right and left eye remain segregated after their partial crossing in the optic chiasm; see Figure 24.3A) terminate. In some mammals—carnivores, anthropoid primates, and humans—the afferent terminals form an alternating series of eye-specific domains in cortical layer 4 called **ocular dominance columns** (Figure 24.4A). In carnivores and primates, the ocular dominance columns are an interdigitated series of bands of synaptic terminals from LGN afferent cells driven by the right or left eye. Ocular dominance columns can be visualized by injecting tracers such as radioactive amino acids into one eye; the tracer is then transported along the visual pathway to specifically label the geniculo-cortical terminals (i.e., synaptic terminals in the visual cortex) corresponding to that eye. In the adult macaque, the domains representing the LGN input driven by one of the two eyes are stripes of about equal width (0.5 mm) that occupy roughly equal areas of layer 4 of the primary visual cortex. Electrical recordings confirm that the cells in layer 4 of macaques respond strongly or exclusively to stimulation of either the left or the right eye, while neurons in layers above and below layer 4 integrate inputs from both the left and right eyes and respond to visual stimuli seen by both eyes. Complete ocular dominance is thus seen in the domains (stripes) in cortical layer 4, where all neurons are driven exclusively by one eye or the other reflecting convergent information from LGN axons driven by one or the other eye. The same striped pattern is seen in awake humans using fMRI when the left versus the right eye is stimulated (Figure 24.4B). Ocular dominance can be seen beyond layer 4. It is measured based on the extent to which one or both eyes activate individual cortical neurons (i.e., in layers 2, 3, 5, and 6). The clarity of these patterns of anatomical and functional connectivity, and the precision by which experience via the two eyes can be manipulated, led to a series of experiments that defined the relationship between activity, experience, and circuitry (see Concept 24.3).

Evidence for critical periods in other sensory systems

Although the neural basis of critical periods has been most thoroughly studied in the mammalian visual system, similar phenomena exist in the auditory and somatosensory systems (see Figure 24.3B,C), olfactory and taste systems,

(A)

1 Radioactive amino acids injected in eye

2 Transsynaptic transport through the LGN terminates in layer 4 of the primary visual cortex

Cortical layers 1–3

Cortical layer 4

Optic radiation

Lateral geniculate nucleus

Optic nerve

Optic tract

3 Terminations are visible as bright bands on the autoradiogram

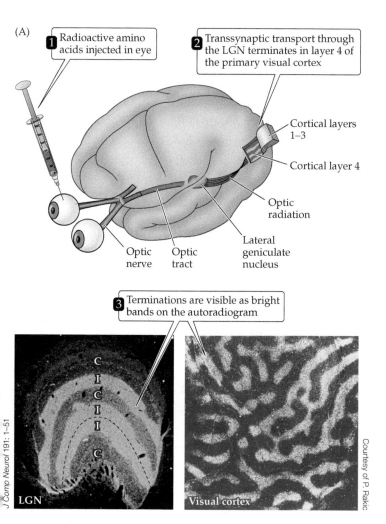

LGN

Visual cortex

(B)

R

L

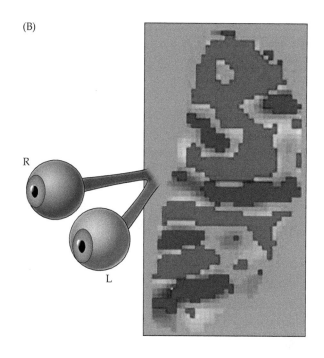

Courtesy of P. Rakic

FIGURE 24.4 **Ocular dominance columns in layer 4 of the primary visual cortex of an adult macaque** (A) The diagram illustrates transneuronal anterograde tracing of the pattern of single eye connections from the eye to the primary visual cortex. Following transsynaptic transport of the radioactive label, the distribution of ipsilateral (I) versus contralateral (C) retinal ganglion cell axon terminals is seen in the lateral geniculate nucleus (LGN; lower left). Geniculo-cortical terminals (from the labeled LGN layers) related to the injected eye are visible as a pattern of light stripes in an autoradiogram (lower right) of a section through layer 4 in the plane of the cortex (that is, as if looking down on the cortical surface). The dark areas are the zones occupied by geniculo-cortical terminals related to the unlabeled eye. (B) Ocular dominance columns in the human primary visual cortex measured using fMRI. Each eye (color-coded blue for right and red for left) was stimulated independently in this individual, and the regions of focal activation identified in the primary visual cortex. There is an interdigitating pattern of focal activation, consistent with segregation of right- and left-eye activation maintained in the primary visual cortex. (B after E. Yacoub et al., 2007. *Neuroimage* 37: 1161–1177.)

and primary motor pathways. Experiments on the role of auditory experience and neural activity in owls (which use auditory information to localize prey) indicate that neural circuits for auditory localization, as for mammalian visual circuits, are shaped by experience (Table 24.1). Thus, deafening an owl or altering its neural activity during early postnatal development compromises the bird's ability to localize sounds and accurately capture prey, and alters the neural circuits in the brain that mediate these capacities. In addition, this auditory critical period in the owl is coordinated with critical periods for vision, enabling the two sensory modalities to operate efficiently together to enhance detection and capture of prey. The development of song in many bird species provides another example of a critical period for auditory function, as well as critical periods for motor control of a complex behavior (see Chapter 25). In the somatosensory system, there is a critical period during which cortical maps can be changed by experience. In mice and rats, for instance, the anatomical patterns of "whisker barrels" in the somatosensory cortex (see Figure 24.3C and Chapter 9) can be altered by abnormal sensory experience (or by removing subsets of sensory receptors such as whiskers) during a narrow window in early postnatal life (see Concept 24.4). In the

TABLE 24.1 Critical Periods and Molecular Regulators for Some Neural Systems

System	Species[a]	Critical period (postnatal)[b]	Confirmed molecular regulators[c]
Neuromuscular junction	Mouse	Prior to day 12	ACh
Cerebellum	Mouse	Days 15–16	NMDA, mGluR1, Gq, PLCβ, PKCγ
Lateral geniculate	Mouse, ferret, cat	Prior to day 10	ACh, cAMP, MAOA, NO, MHC1, CREB nucleus layers
Ocular dominance	Cat, rat, mouse, ferret	3 weeks–months	GABA, NMDA, PKA, ERK, CaMKII, CREB, BDNF, tPA, protein synthesis, NE, ACh
Orientation bias	Cat, mouse	Prior to day 28	NR1, NR2A, PSD95
Somatosensory map	Mouse, rat	Prior to day 7–16	NR1, MAOA, 5-HT1B, cAMP, mGluR5, PLCβ, FGF8
Tonotopic map (cortex)	Rat	Days 16–50	ACh
Absolute pitch	Human	Before 7 years	Unknown
Taste, olfaction	Mouse	At birth	GABA, mGLuR2, NO, neurogenesis
Imprinting	Chick	14–42 hours	Catecholamines
Stress, anxiety	Rat, mouse	Prior to day 21	Hormones, 5-HT1A
Slow-wave sleep	Cat, mouse	Days 40–60	NMDA
Sound localization	Barn owl	Prior to day 200	GABA, NMDA
Birdsong	Zebra finch	Prior to day 100	GABA, hormones, neurogenesis
Language	Human	0–12 years	Unknown

[a]Primary research species for elucidation of molecular mechanisms.

[b]Although the details vary from system to system and from species to species, all critical periods are limited to a definite window of time during early postnatal (or posthatching) life and are complete before the onset of sexual maturation.

[c]Molecules known to regulate critical periods include neurotransmitters, their receptors, and related signaling proteins.

Source: T. K. Hensch, 2004. *Annu Rev Neurosci* 27: 549–579.

olfactory system, behavioral studies (outlined in Chapter 15) indicate that exposure to odors associated with maternal care or presence for a limited period can alter the ability to respond to such odorants—a change that can persist throughout life.

Critical periods are common in the development of sensory perception and related motor skills. This suggests that one imperative of postnatal brain development, perhaps reflected in the continued growth of the brain and its constituent neural and glial cells, as well as the malleability of synaptic connections during this time, is to adjust connectivity as precisely as possible to the environmental circumstances in which an individual will find themself. The robust acquisition of skills during distinct epochs of early life in a variety of species, followed by the decline of this capacity as life continues, provides one line of evidence for this conclusion. Additional evidence for critical periods in a variety of neural systems and species comes primarily from deprivation experiments (as well as selective augmentation experiments; see Figure 24.3B) analogous to those

done in the visual system, complemented by analysis using pharmacological approaches or genetically modified animals in which major neurotransmitter synthetic pathways are disabled, essential neurotransmitter receptors such as NMDA receptors are lost, or other major signaling molecules (e.g., Ca^{2+}/calmodulin kinases, BDNF, neurotrophin receptors [see Concept 24.4], or in the case of sexually dimorphic song behaviors in birds, gonadal steroids such as estrogen and testosterone [see Chapter 25]) have been disrupted. In each instance, these molecular modifications influence synaptic signaling and thus change the duration or efficiency of critical period–dependent plasticity.

Language development: A distinctly human critical period

Experimental and clinical observations of primary sensory systems and their sensitivity to activity and experience-dependent changes summarized earlier (and explored in greater detail in the following sections) provide biological explanations of mechanisms that initiate and terminate critical periods. These studies also raise the question of whether critical periods can be documented for even more complex behaviors, including cognitive function and, in humans, language, perhaps mediated by similar physiological mechanisms. While cellular and physiological evidence is not yet available, several behavioral observations, coupled with noninvasive imaging in young children, have defined an apparent critical period for language acquisition and production.

Exposure to language from birth onward is essential for the development of appropriate capacity to comprehend and produce meaningful communication. The various forms of early language exposure, including the "baby talk" that parents and other adults often use to communicate with infants and small children, may actually serve to emphasize important perceptual distinctions that facilitate proper language acquisition, production, and comprehension. To be effective, this linguistic experience must occur in early life. The requirement for perceiving and practicing language (as opposed to specific auditory, visual, or motor skills) during a critical period is apparent in studies of language acquisition in congenitally deaf children, whose nascent language ability (i.e., sign language) relies on seeing and moving the hands and fingers (the equivalent of spoken and heard language) rather than on listening and moving the lips, tongue, and larynx. Whereas most hearing and speaking babies begin producing speechlike sounds ("babbling") at about 7 months, congenitally deaf infants show obvious deficits in their early vocalizations and fail to develop language if not provided with an alternative form of symbolic expression such as sign language (see Chapter 33). If, however, deaf children are exposed to sign language at an early age (from approximately 6 months onward, which is particularly likely for the children of deaf, signing parents), they begin to "babble" with their hands, just as a hearing infant babbles audibly (Figure 24.5A). This manual babbling suggests that, regardless of the modality, early experience shapes language behavior. Children who have acquired speech but lose their hearing before puberty also suffer a substantial decline in spoken language, presumably because they are unable to hear themselves or others talk and thus lose the opportunity to refine their speech by auditory feedback during the final stages of the critical period for language.

The auditory details of the language an individual hears during early life shape both the perception and production of speech. Many of the thousands of human languages and dialects use appreciably different speech sounds called *phonemes* to produce spoken words (examples are the phonemes *ba* and *pa* in English; see Chapter 33). Very young human infants can perceive and discriminate between differences in *all* human speech sounds, and are not innately biased toward phonemes characteristic of any particular

language. However, this universal perceptual capacity does not persist. For example, adult Japanese speakers cannot reliably distinguish between the *r* and *l* sounds in English, presumably because this phonemic distinction is not made in Japanese and thus is not reinforced by experience during the critical period. Nonetheless, 4-month-old Japanese infants can make this discrimination as reliably as 4-month-olds raised in English-speaking households (as indicated by increased suckling frequency or head turning in the presence of a novel stimulus). By 6 months of age, however, infants begin to show preferences for phonemes in their native language over those in foreign languages, much as deaf infants do for moving digits that suggest signs. By the end of their first year, infants no longer respond robustly to phonetic elements that are peculiar to non-native languages. This can be considered—cautiously—comparable to very basic experiments in which single frequencies are presented during the auditory critical period in animals, and the changes in the tonotopic map favor response to that frequency at the expense of other frequencies (see Figure 24.3B). These sorts of observations provide additional evidence for the role of experience in shaping language capacity, as well as suggesting a critical period for the acquisition of phonetic perception and production.

The ability to perceive, learn, and produce distinct phonemes with clarity approximating, if not equalling, that of native speakers, as well as the ability to acquire a sense of the rules of grammar and usage in a language (see Chapter 33), persists for several more years, as evidenced by the fact that children can usually learn to speak a second language without accent and with fluent grammar until about age 7 or 8. After this age, however, performance gradually declines no matter what the extent of practice or exposure (Figure 24.5B). Changes in the patterns of activity in language regions of the brain in children versus adults suggest that the relevant neural circuits may undergo functional or structural modifications during the critical period for language (Figure 24.5C). Comparisons of patterns of activity in children ages 7 to 10 with the patterns of adults performing the same specific word-processing tasks suggest that different brain regions are activated for the same task in children versus adults. While the significance of such differences is not clear—they may reflect anatomical plasticity associated with critical periods, or distinct modes of performing language tasks in children versus adults—there is nevertheless an indication that brain circuits change to accommodate language function during early life, with very different patterns of activity seen in adulthood. Thus, although the cellular basis is difficult to study, it is likely that activity elicited by language experience leads to rearrangement of connectivity, analogous to the much better documented changes shaped by experience during the critical period in the visual cortex.

(A)

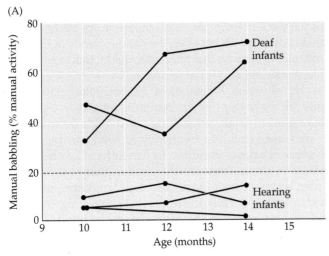

(B)

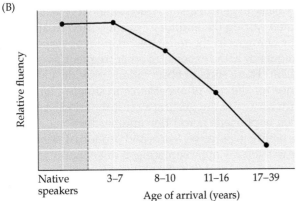

(C)

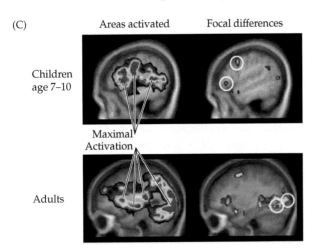

FIGURE 24.5 Evidence for a critical period for a complex human behavior: language (A) Critical periods in language are driven by learning and practicing the complex behavior. In this case, babbling is the precursor of language acquisition, and this "practice" that drives plastic change is modality-independent. In deaf children, babbling with the hands and fingers replaces the vocal babbling of hearing and speaking infants who have hearing and speaking parents. The two deaf infants studied here were raised by deaf, signing parents. Babbling was judged by scoring hand positions and shapes that showed some resemblance to the components of American Sign Language and comparing these scores with those of manual babble in three hearing infants. In these deaf infants with signing parents, meaningful hand shapes increased as a percentage of manual activity between ages 10 and 14 months. Hearing children raised by hearing, speaking parents do not produce similar hand shapes. (B) A critical period for learning language is demonstrated by the decline in language ability (fluency) of non-native speakers of English as a function of their age upon arrival in the United States. The ability to score well on tests of English grammar and vocabulary declines from approximately age 7 onward. (C) Maps derived from fMRI of adults and children performing visual word-processing tasks. Images are sagittal sections, with the front of the brain toward the left. The top row shows the range of active areas (left) and foci of activity based on group averages (right) for children ages 7 to 10. The bottom row shows analogous results for adults performing the same task. The differences in regions of maximal activation (shown in red in the images at left; highlighted by white circles in the right-hand images) indicate changes in either the circuitry or the mode of processing and performing the same task in children versus adults. (A after L. A. Petitto and P. F. Marentette, 1991. *Science* 251: 1493–1496; B after J. S. Johnson and E. I. Newport, 1989. *Cog Psychol* 21; C after B. L. Schlaggar et al., 2002. *Science* 296: 1476–1479.)

Critical periods and human brain growth

The advent of high-resolution, noninvasive imaging techniques has made it possible to reevaluate some basic aspects of the development of human brain structure and function in the context of physiological and behavioral understanding of critical periods and accompanying changes in neuronal growth that have been discerned in animal experiments. In the late 1980s, studies in multiple cortical areas in the rhesus monkey—which are very similar to human neocortical areas—demonstrated what had been suggested by less detailed analysis of synapses in the human visual cortex: The number of synapses throughout the primary sensory, motor, and association (including limbic; see Chapter 29) cortices increased during prenatal and a limited period of postnatal life, declined during a protracted period that included much of adolescence, and reached a steady state in early adulthood (Figure 24.6). This pattern of initial increase followed by a decline in synapse numbers indicated that critical periods may be mediated first by local growth of neural elements in an activity-dependent manner, followed by a subsequent elimination of some synapses and the selective growth and stabilization of other synapses —perhaps the cell biological equivalents of the functional consequences of Hebbian competition (see Figure 24.1). These quantitative anatomical observations suggest a cellular basis for activity-dependent plasticity and critical period phenomena throughout the cerebral cortex.

This suggestion received compelling support from a remarkable series of studies begun in the late 1990s and not completed until the end of the first decade of the new

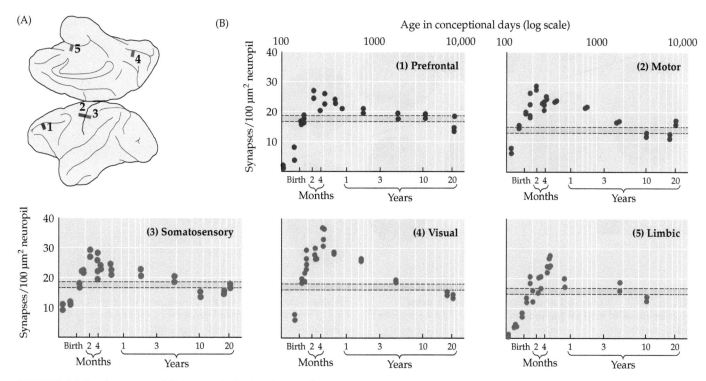

FIGURE 24.6 Synapse addition and elimination in rhesus monkey cortex (A) Location of brain regions where synapse density was measured between mid-gestation and 20 years of age. (B) Rapid addition followed by gradual decline of synapse density in the cerebral cortex. Age has been converted to a logarithmic scale of "conceptional days" in order to fit the entire life span onto one graph. Synapse addition apparently continues through early life, gradually declines throughout most of adolescence, and reaches a steady state (shaded horizontal bar) after puberty (between 2 and 3 years of age in the rhesus monkey). (After P. J. Rakic et al., 1986. *Science* 232: 232–235.)

millennium. A group of investigators at the National Institute of Mental Health did longitudinal MRI scans of the developing brains of 13 individual children starting at age 4 and ending at age 20 to measure the growth of their **gray matter** (the location of cell bodies, axon terminal branches, dendrites, and synapses) as well as **white matter** (mostly myelinated axons adjacent to gray matter) throughout the cortical mantle over time (see Figure 24.2A). This study's results paralleled those predicted from analyses of synapses in the developing human visual cortex (see Figure 24.2C) and the more comprehensive assessment in multiple cortical areas in rhesus monkeys (see Figure 24.6). Gray matter grows throughout the cortex during early life, then declines slightly over a protracted period of late childhood and early adolescence (Figure 24.7A). There are some important regional distinctions in this overall trajectory of early growth and subsequent loss of gray matter volume. Primary sensory cortices appear to have more robust early growth; however, decline is more prolonged in higher-order association cortices, including prefrontal, temporal, and parietal regions. These changes were confirmed in a larger study in which mean values were generated from multiple individuals at each age rather than for one individual

followed longitudinally. In these studies, males and females were analyzed separately (Figure 24.7B). Although the absolute size of the brain is distinct in the two sexes, the overall trajectory of gray matter volume growth and decline is parallel. This regressive process during late childhood and adolescence is specific to the gray matter. The white matter, where the multiple bundles of axons that interconnect cortical areas as well as those that connect the cortex with the rest of the brain are found, has a continual increase in both sexes. This is most likely due to the progressive addition of myelin during the same time period of postnatal life.

The elaboration followed by selective elimination of connections in the cerebral cortex—inferred from the increase and then decrease in gray matter volumes—may indeed underlie the remarkable capacity of the human brain to acquire and refine behaviors from birth through early adulthood. The cessation of this process is intriguingly coincident with the time in life when the process of learning new information and skills becomes increasingly difficult. Moreover, this prolonged process of experience-driven construction of cortical circuits seems to be altered in several disorders in which the development of cortical connections is thought to be the primary target for pathological change,

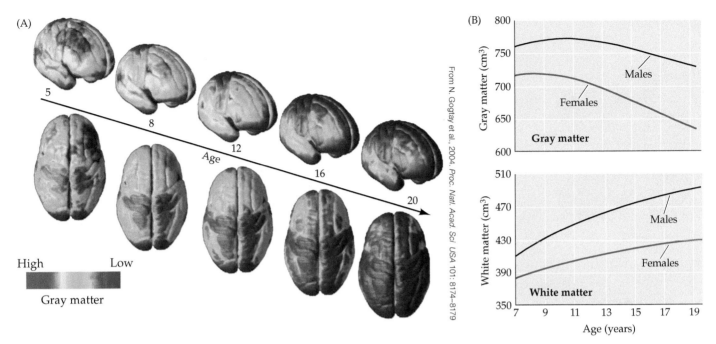

From N. Gogtay et al., 2004. *Proc. Natl. Acad. Sci. USA* 101: 8174–8179

FIGURE 24.7 Increased and decreased gray matter volumes parallel critical periods in humans (A) A composite map of cortical gray matter volume growth (red/yellow/green) and decline (blue/purple), based on longitudinal MRI scans of 13 typical individuals from 5 years of age until 20 years of age. There is initial growth of gray matter throughout the cortex, especially in primary sensory and motor regions, followed by gradual decline. There is some heterogeneity in the timing and rate of decline in primary sensory and motor versus association areas.

(B) Mean cortical gray matter (above) and white matter (below) volume in males and females from a cross-sectional study (i.e., means computed from multiple individuals rather than from the same individual at each age). Although the absolute growth of the male and female brains differs, gray matter volume increases and then decreases in roughly the same way. White matter volume, in contrast, increases throughout early childhood and adolescence. (B after R. K. Lenroot et al., 2007. *NeuroImage* 36: 1065–1073.)

including autism spectrum disorder, schizophrenia, and attention deficit hyperactivity disorder (ADHD). In children with ADHD, the rate of cortical growth during early postnatal life is delayed and the overall magnitude of growth is diminished compared with that of typically developing children (Figure 24.8A). These deficits in gray matter growth are greatest in association cortical areas that mediate cognitive, emotional, and social behaviors (see Chapter 27). Not only is growth diminished and slowed; decline in volume is enhanced. Thus, in children with ADHD, gray matter volume declines more dramatically, resulting in smaller cortical gray matter volumes in adulthood (Figure 24.8B).

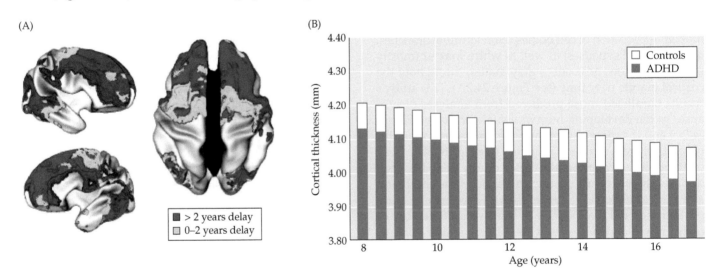

FIGURE 24.8 A behavioral disorder accompanied by altered addition of gray matter volume (A) Map of cortical regions in which gray matter volume increases more slowly in children with attention deficit hyperactivity disorder (ADHD) than in typically developing children. (B) The rate of decline of gray matter volume is equivalent for children with ADHD and typically developing children. The net result is lower gray matter volumes in adulthood for individuals with ADHD. (After P. Shaw et al., 2007. *Proc Natl Acad Sci USA* 104: 19649–19654.)

These observations of postnatal brain development in living humans allow one to infer (but do not prove) that experience- and activity-dependent mechanisms during critical periods may be primarily responsible for much of what is recognized as typical behavioral development, social development, and learning. They further suggest that the underlying substrate for these changes is the synaptic connections made by neurons, and whether they are retained or eliminated. This implies that the final endpoint (although not necessarily the initial cause) of several behavioral and psychiatric disorders may include the activity- and experience-dependent processes that shape and—under typical circumstances—optimize those connections that mediate complex behaviors.

CONCEPT 24.3 | Relative Levels of Electrical Activity across Inputs Determine Final Connections in Neural Circuits

LEARNING OBJECTIVES

24.3.1 Describe the influence of correlated activity on experience-driven critical periods.

24.3.2 Discuss the cellular and circuit consequences of altered activity during critical periods.

24.3.3 Identify the physiological mechanisms that help shape critical periods.

Experience, activity, and circuits: Lessons from visual system development

The behavioral and anatomical observations described in Concepts 24.1 and 24.2 all are consistent with the proposal that the electrical activity elicited by experience in the brain must somehow signal neural circuits to mature, resulting in an optimal (or when pathology strikes, suboptimal) pattern and number of connections and related behaviors. The pioneering, and to date, most complete analysis of the physiology underlying critical periods has been done in the mammalian visual system. This work began in the 1960s and has continued ever since. The visual system was ideal for testing the role of experience and activity in critical periods in part because of the clear distinction between activation of visual cortical neurons by one or the other eye via the retino-geniculo-cortical pathway. This feature meant that activity could be easily manipulated by closing one or both eyes, and that the responses of individual neurons could be measured to determine whether this change in "experience" resulted in a change in the responses of neurons and their underlying synaptic connections, perhaps restricted to early life—in other words, defining the critical period physiologically based on functionally relevant connections and their capacities for change in response to differences in sensory input.

As described in Chapter 9, if an electrode is passed at a shallow angle through the cortex while the responses of individual neurons in the primary visual cortex to light stimulation of one or the other eye are being recorded, detailed assessment of ocular dominance can be made at the level of individual cells that share, more or less, the same laminar location (see Figure 9.17). The original studies of visual cortical plasticity and critical periods were performed using a similar physiological approach in younger normal animals or those whose visual experience had been altered experimentally. Using extracellular electrode recording, the action potential responses of single cortical neurons in individual cortical layers were arbitrarily divided into seven "ocular dominance" groups based on their degree of response to either the contralateral or ipsilateral eye (Figure 24.9A). Group 1 cells are driven only by stimulation of the contralateral eye; group 7 cells are driven entirely by the ipsilateral eye; and neurons driven equally well by either eye are assigned to group 4. Based on this empirical measurement scheme, ocular dominance connectivity in all layers but layer 4 in primary visual cortex is normally distributed (approximately) across groups 1 through 7 in a normal adult (cats were used in these experiments). Most cortical neurons were activated to some degree by both eyes (distributed around a mean defined by "group 4" cells); however, a substantial minority were activated by either the contralateral or ipsilateral eye (see Figure 24.9A).

This normal distribution of ocular dominance responses—and thus the functional connectivity of individual visual cortical neurons—can be altered by visual experience. When one eye of a kitten was sutured closed early in life and the animal then matured to adulthood (which takes about 6 months), a remarkable change was observed. Once the eyelid was opened, electrophysiological recordings showed that very few cortical cells could be driven from the deprived (previously sutured) eye. Recordings from the retina and LGN layers in response to direct electrical stimulation in the deprived eye indicated that these more peripheral stations in the visual pathway remained interconnected and capable of transmitting information. Nevertheless, the ocular dominance distribution in the visual cortex had shifted; the eye that remained open was uniquely able to drive most cortical cells (Figure 24.9B, left). Thus, the absence of cortical cells that responded to stimulation of the closed eye was not a result of retinal degeneration or a loss of retinal connections to the thalamus. Rather, the deprived eye had been functionally disconnected from the visual cortex. Consequently, such animals are behaviorally blind in the deprived eye. This "cortical blindness," or *amblyopia*, is permanent (Clinical Applications; also see later in this concept). Even if the formerly deprived eye remains open, little or no recovery occurs.

The same manipulation—closing one eye—performed in adulthood has no effect on the responses of cells in the

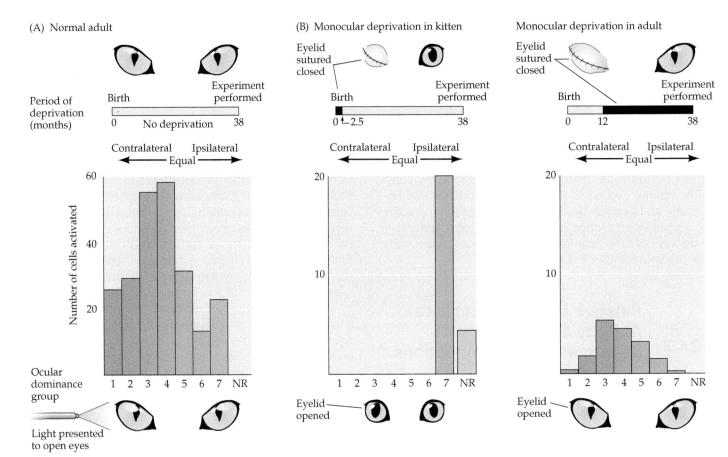

FIGURE 24.9 Manipulating peripheral visual experience by monocular deprivation defines a critical period for activity-dependent plasticity in the visual cortex (A–C) Histograms plot the number of cells that fall into one of the seven ocular dominance categories, defined based on the frequency of action potential activity elicited from visual cortical neurons following illumination in the relevant eye. (A) Ocular dominance distribution of single-unit recordings from a large number of neurons in the primary visual cortex of normal adult cats. Cells in group 1 were activated exclusively by the contralateral eye, cells in group 7 by the ipsilateral eye. There were no cells that were not responsive (NR) to light stimulation in one of the two retinas. (B) One eye of a newborn kitten (left) was closed from 1 week after birth until 2.5 months of age. After 2.5 months, the eye was opened and the kitten matured normally to 38 months. Note that the deprivation was relatively brief—the sutured eye had been open for 35.5 months of the cat's life. Even so, light presented to the open but transiently deprived eye elicited no electrical responses in visual cortical neurons. Visually responsive cells responded only to the ipsilateral (non-deprived) eye. In contrast, and as a control, a much longer period of monocular deprivation in an adult cat (right), beginning at 12 months of age, showed little effect on ocular dominance at 38 months, although overall cortical activity was diminished. (C) During the critical period, the connections in the visual cortex are sensitive to even brief changes in visual experience. Just 3 days of deprivation (left) produced a significant shift of cortical activation in favor of the non-deprived eye, and 6 days of deprivation (right) produced a shift in favor of the non-deprived eye almost as complete as that elicited by 2.5 months of deprivation. (D) In normal adult monkeys (left), ocular dominance columns are seen as alternating stripes of roughly equal width. After monocular deprivation (right), the columns related to the non-deprived eye (white stripes) are much wider than normal; those related to the deprived eye are shrunken. (E) After only a week of monocular deprivation during the critical period (left, short-term), axons terminating in layer 4 of the primary visual cortex from LGN neurons driven by the deprived eye have greatly reduced numbers of branches compared with those from the open eye. Deprivation for longer periods (right, long-term) does not result in appreciably larger changes in the arborization of geniculate axons. (A,B after D. H. Hubel and T. N. Wiesel, 1962. *J Physiol* 160: 106–154; T. N. Wiesel and D. H. Hubel, 1963. *J Neurophysiol* 26: 1003–1017; C D. H. Hubel and T. N. Wiesel, 1970. *J Physiol* 206: 419–436; E after A. Antonini and M P. Stryker, 1993. *Science* 260: 1819–1821.)

mature visual cortex. If one eye of an adult cat was closed for a year or more and then reopened, both the ocular dominance distribution measured across all cortical layers and the animal's visual behavior were indistinguishable from normal when tested through the reopened eye (see Figure 24.9B, right). Thus, sometime between when a kitten's eyes open (about a week after birth) and 1 year of age, visual

experience determines how the visual cortex is wired with respect to eye dominance. Additional experiments that varied the starting and stopping times for eye suturing and reopening identified the period of maximal sensitivity and malleability of visual cortical connections to altered visual experience. The timing of the critical period in the visual system was therefore first defined based on the potential

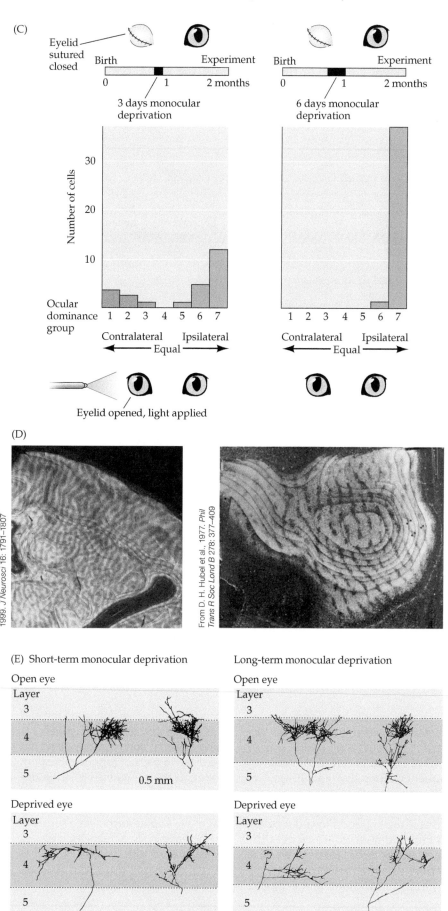

(C)

Eyelid sutured closed

Birth   Experiment   Birth   Experiment

0   1   2 months   0   1   2 months

3 days monocular deprivation   6 days monocular deprivation

Number of cells

30

20

10

Ocular dominance group   1 2 3 4 5 6 7   1 2 3 4 5 6 7

Contralateral   Ipsilateral   Contralateral   Ipsilateral

Equal   Equal

Eyelid opened, light applied

(D)

1999. J Neurosci 16: 1791–1807

From D. H. Hubel et al., 1977. Phil Trans R Soc Lond B 278: 377–409

(E) Short-term monocular deprivation   Long-term monocular deprivation

Open eye   Open eye

Layer   Layer
3   3

4   4

5   5

0.5 mm

Deprived eye   Deprived eye

Layer   Layer
3   3

4   4

5   5

of visual cortical neurons to respond optimally, suboptimally, or not at all to the deprived eye when it was reexposed to light. At the height of the visual critical period (about 4 weeks of age in a cat), as few as 3 or 4 days of eye closure profoundly alters the physiological ocular dominance profile of the striate cortex (Figure 24.9C). After this one-month peak, deprivation or manipulation has little or no permanent, detectable effect. In fact, eye closure alters noticeably eye-specific cortical responses only if the deprivation occurs during the first 3 months of a kitten's life. In keeping with the ethological observations described earlier in the chapter, David Hubel and Torsten Wiesel called this period of susceptibility to visual deprivation the critical period for ocular dominance development. Similar experiments in monkeys have shown that the same phenomenon occurs in primates, although the critical period is longer (up to about 6 months of age).

Visual deprivation during the critical period must result in some sort of cortical connectivity changes that influence the functional response properties of individual neurons—especially since neither retinal nor retino-geniculate activity or connections are altered. Indeed, the segregation of retinal afferents by eye of origin in the LGN, and the separation of the single eye recipient LGN layers, is present at birth and does not change in deprivation experiments. Subsequent anatomical studies established that physiological changes were due to changes in patterns of connections in the visual cortex, especially those made by geniculo-cortical axons. In monkeys, the stripelike pattern of geniculo-cortical axon terminals in layer 4 that defines ocular dominance columns is already present at birth, and this pattern reflects the functional segregation of synaptic terminals driven by one of the two eyes (Figure 24.9D; also see Figure 24.4). Indeed, the early formation of this pattern in layer 4 reflects a significant amount of segregation of the terminal branches of LGN axons that relay information from one eye or the other, and this early segregation occurs even in the absence of meaningful visual experience (see later in this concept). Subsequent observations have confirmed this initial experience-independent segregation, and there is some indication that specific molecular signals as well as pre-experience physiological activity may distinguish LGN cells innervated by one eye or the other. Thus, the visual cortex is clearly not a blank slate on which the effects of experience are inscribed. Nevertheless, animals deprived from birth of

■ Clinical Applications

Dancing in the Dark

In 1963, David Hubel and Torsten Wiesel demonstrated the fundamental properties of experience-guided, activity-dependent critical period plasticity in the visual cortex. Within the next decade, the implications of these foundational observations for diagnosis and early treatment of amblyopia were understood. Early intervention was possible only if early diagnosis was made. Unfortunately, then as now, early diagnoses were not always available before the critical period had closed for children with monocular eye alignment or eye occlusion anomalies. The challenge that remains, despite the capacity of pediatricians and ophthalmologists to detect and treat amblyopia before the close of the critical period, is whether binocular visual acuity and depth perception can be restored in individuals for whom uncorrected abnormal ocular competition early in life results in permanent visual impairment thereafter.

Within the last decade, several observations, primarily in animal models, have provided hope that a combination of noninvasive interventions in adults who experienced some form of monocular deprivation early in life (e.g., amblyopia, lens or corneal occlusion) can restore a great deal of lost visual capacity. In the key experiment, adult rats experienced monocular deprivation starting at the onset of the critical period (after eye opening, approximately 14 days of age) through adulthood (185 days, or approximately 6 months, of age). Half the rats were then placed in a completely light-free environment ("dark exposure"; see Figure A) for 10 days while the other half were maintained in standard

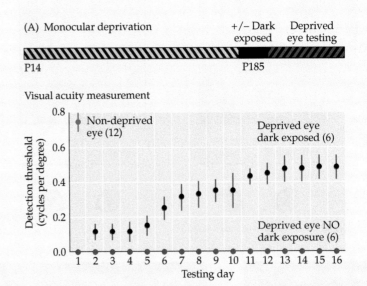

(A) Top panel: Experimental strategy for monocular deprivation of a rat pup at postnatal day 14 (P14), the onset of the visual experience–dependent critical period, through adulthood at P185, approximately 6 months of age. The rat pup can then be subjected to "dark exposure" (DE) for 10 days, or not. Subsequently, visual acuity can be tested to evaluate the function of the deprived eye, with and without DE. Bottom panel: Visual acuity measurement (based on the frequency of a line grating that can be resolved; i.e., cycles per degree) of the deprived eye with and without DE (6 animals measured in each treatment). Acuity in a deprived eye with no DE is 0 at the outset of the 16-day testing period. Acuity in a deprived eye with DE begins modestly higher than that without DE, and increases substantially over the subsequent days of testing. While this restoration of acuity is significant, it does not reach acuity levels measured in the non-deprived eye (6 animals measured/condition, error bars indicate standard error of the means). (After N. C. Eaton, et al., 2016. *Learn Mem* 23: 99–103.)

illumination conditions. Thereafter, all the rats were tested on a task that evaluated their visual acuity (the capacity to resolve linear patterns of increasing frequency). Rats that had 10 days of dark exposure following the monocular deprivation had modest acuity at the outset of testing, and showed substantial improvement in acuity over the course of the 16-day testing period. In contrast, rats that had been monocularly deprived without dark exposure (standard luminance) following the end of the deprivation period had no measurable visual acuity in the deprived eye after 16 days of repetitive testing. Although the acuity values in the deprived eye of dark-exposed rats did not reach those of the non-deprived eye, there was nevertheless a substantial increase in visual resolution capacity associated with dark exposure following the chronic disruption of binocular competition (Figure A).

vision in one eye develop abnormal patterns of ocular dominance stripes in the visual cortex, presumably because of the altered patterns of activity caused by deprivation (see Figure 24.9D, right). The stripes (demonstrated by tracer injection in one eye followed by transsynaptic transport as shown in Figure 24.4) related to the open eye are substantially wider, and the stripes representing the deprived eye are correspondingly diminished. The absence of cortical neurons that respond physiologically to the deprived eye is not simply a result of the relatively inactive inputs withering away. If this were the case, one would expect to see areas of layer 4 devoid of any thalamic innervation. Instead, inputs from the active (open)

eye take over some—but not all—of the territory that formerly belonged to the inactive (closed) eye. These inputs then dominate the physiological responses of the target cortical neurons.

These results suggest a **competitive interaction** for postsynaptic space between afferent axons driven by each of the two eyes during the critical period, reminiscent of Hebb's description of synaptic plasticity but in the context of development. In normal animals, an equivalent amount of synaptic territory occupied by axons from each eye is retained (and sharpened in terms of segregation of ocular dominance stripes in layer 4 of the cortex) if both eyes experience roughly comparable levels of visual stimulation.

■ Clinical Applications *(continued)*

The mechanisms by which dark exposure reactivates cortical synaptic plasticity and permits successful, repetition-mediated reacquisition of visual capacity remain uncertain. Earlier work indicated that dark exposure beginning in the critical period can prolong the critical period. These changes were accompanied by the upregulation of expression for genes that encode proteins essential for synaptic transmission and excitability or electrical activity in neurons. In addition, much of the retention of plasticity was thought to reflect delayed maturation of GABAergic inhibitory interneurons and their influence on distinct forms of synaptic plasticity. Cellular analysis in adult animals monocularly deprived from the critical period onward, and then visually "rescued" via dark exposure as adults as described above, shows that dark exposure increases the density of spines on visual cortical neuron dendrites toward levels that approximate those in non-deprived animals. In contrast, monocular deprivation over the same period without dark exposure results in substantial reduction in the density of visual cortical dendritic spines (Figure B).

Together, these results indicate that the adult visual cortex, after prolonged changes in binocular competition and subsequent binocular activity, retains measurable capacity for

(B)

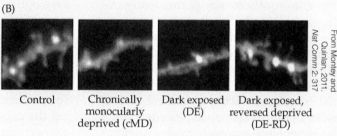

Control   Chronically monocularly deprived (cMD)   Dark exposed (DE)   Dark exposed, reversed deprived (DE-RD)

From Montey and Quinlan, 2011. Nat Comm 2: 317

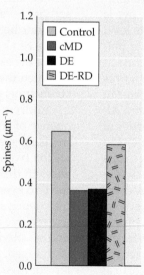

(B) The images at top (from left to right) show dendritic spines on a visual cortical neuron from a normal, non-deprived rat (control), a chronically monocularly deprived rat without dark exposure (cMD), a dark-exposed rat (DE), and a monocularly deprived then dark-exposed rat in which the previously deprived eye is given a competitive advantage by occluding the previously non-deprived eye (DE-RD). Reverse deprivation renders the chronically deprived eye maximally active. This activity, and the advantage of correlated synaptic drive it gives to neurons in the visual cortex driven by that eye, causes dendritic spines and presumably functional synapses to be added. Such changes may correct the cellular consequences of either cMD or DE. The graph below shows that the combination of DE plus RD increases spines on visual cortical neurons to control levels. DE or cMD alone diminishes spine numbers to a similar extent.

plasticity that can improve visual function, if certain conditions are put in place. Those conditions include dark exposure, which apparently restores cell biological and molecular mechanisms that can accommodate plasticity and functional recovery, and repeated

training to elicit maximal reacquisition of function. Thus, there is a new reason, hardly romantic but far more beneficial, for "Dancing in the Dark." With apologies to Fred Astaire, no more will one just "face the music," but can hope to "see the light."

However, when an imbalance in visual experience is induced by monocular deprivation, the active eye gains a competitive advantage and replaces many of the synaptic inputs from the closed eye. In this case, even though LGN axons arising from neurons innervated by the closed eye are retained in the cortex (albeit with much less extensive terminals and fewer functional synapses), few if any neurons fire action potentials when light is presented to the deprived eye. These observations in experimental animals have important implications for children with birth defects or ocular injuries that result in an imbalance of inputs from the two eyes. Unless the imbalance is corrected during the critical period, the child may ultimately have poor binocular fusion, diminished depth perception, and degraded acuity. In addition, recent work indicates that this mechanism is essential for organizing the orientation selectivity of cortical neurons that are driven by some degree by both eyes. Thus, a significant disruption of ocular competition and its consequences may permanently impair many aspects of a child's vision.

The idea that a competitive imbalance underlies the altered distribution of inputs after deprivation has been confirmed by suturing shut *both* eyes shortly after birth. This manipulation equally deprives all visual cortical neurons of normal experience during the critical period. The arrangement of ocular dominance recorded some months later is, by either electrophysiological or anatomical criteria, much closer to normal than if just one eye is closed. Thus, the balance of inputs, not the absolute level of activity, is a key feature for shaping the normal pattern of connections. This also reinforces the conclusion that modest correlation of activity between the two eyes prior to visual experience (i.e., retinal waves; see later in this concept) can influence initial ocular segregation. Although several peculiarities in the response properties of cortical cells deprived of normal light-dependent vision are apparent, roughly normal proportions of neurons responsive to the two eyes are found in binocularly deprived animals after the period of deprivation is finished. Because there is no added imbalance in visual activity favoring one eye or the other (both sets of cortical inputs being deprived), both eyes retain their territory in the cortex. If disuse atrophy of the closed-eye inputs were the main effect of monocular deprivation on visual cortical function, then binocular deprivation during the critical period would cause the visual cortex to be largely unresponsive; this is not the case. Furthermore, in dark-reared animals, even when the animals are reintroduced to light after the close of the typical critical period, visual activity and acuity return to levels that approximate those of animals reared in a typical environment.

Experiments using techniques that label individual axons from distinct layers in the LGN have shown in greater detail what happens to the arborizations of individual LGN neurons in layer 4 of visual cortex after visual deprivation (Figure 24.9E). At the level of single axons, loss of cortical territory related to the deprived eye and concomitant expansion of the open eye's territory are reflected in decreased size and complexity of the arborizations of LGN neurons related to the deprived eye, and increased growth and complexity of the arborizations related to the open eye (i.e., those arising from LGN cells innervated by the open eye). Individual axon arbors can be substantially altered after as little as 1 week of unequal deprivation. Presumably, the diminished extent of the axon arbors from an LGN cell driven by a deprived eye is accompanied by diminished frequency of individual synaptic endings made by that axon. The postsynaptic sites once occupied by these presynaptic terminals may be subsequently "claimed" by the more active axon, and then added to by that axon as postnatal development moves forward. This suggests that developing thalamic and cortical neurons can rapidly remodel their connections, presumably making, breaking, and replacing synapses, in response to environmental circumstances during the critical period.

Manipulating competition

To specifically test the role of correlated activity in driving the competitive postnatal rearrangement of cortical connections, it is necessary to create a situation in which activity levels in each eye remain the same but the correlations between the two eyes are altered. This circumstance can be created in experimental animals by cutting one of the extraocular muscles in one eye. This condition, in which the two eyes can no longer be aligned, is called **strabismus** (a condition that is also recognized clinically in children; see later in this concept). The major consequence of strabismus is that objects in the same location in visual space no longer stimulate corresponding points on the two retinas at the same time. As a result, differences in the visually evoked patterns of activity between the two eyes are far greater than normal. Unlike monocular deprivation, however, the overall amount of activity in each eye remains roughly the same; only the correlation of activity arising from corresponding retinal points is changed. Accordingly, the anatomical pattern of ocular dominance columns in layer 4 of cats in which input from both eyes remains active, but highly asynchronous, is *sharper* than normal. Apparently, the total independence of the two eyes further enhances the correlations between ipsilateral versus contralateral activity. In addition, ocular asynchrony prevents the binocular convergence that normally occurs in cells above and below layer 4; ocular dominance histograms from such animals show that cells in *all* layers are driven exclusively either by one eye or the other (Figure 24.10). Evidently, strabismus not only accentuates the competition between the two sets of thalamic inputs in layer 4, but also prevents binocular interactions in the other layers, which are mediated by local connections originating from cells in layer 4.

An intriguing test of Hebbian "fire together, wire together" stabilization and segregation of inputs was undertaken by making a vertebrate whose visual system is primarily monocular—the frog—anomalously binocular. Some frogs, including the species in which this experiment was performed, *Rana pipiens*, are remarkably receptive to ectopic, supernumerary transplanted peripheral structures, including a supernumerary eye. This "third eye" (the real thing, rather than the metaphorical one used by parents to make sure toddlers are not doing something they shouldn't) extends its axons into one of the optic tecta that are typically monocularly innervated. These two eyes are accordingly placed in "competition" for target space, and their activity, based on the very different parts of the visual world each eye captures, will be far more correlated *within* each eye than between the two eyes that innervate the single tectum. If the "fire together, wire together" hypothesis is correct, and the observations for ocular dominance columns in cats and monkeys described here have been interpreted reasonably, these two anti-correlated inputs should segregate from one another so that the within-eye terminals are grouped together, and separated from those of the other eye. This is indeed what happens (Figure 24.11). More striking, the segregation is not into two domains of equivalent size. Instead, it has the approximate periodicity and appearance of the ocular dominance column segregation seen in layer 4 of the primary visual cortex of carnivores and primates. This extreme (and extremely odd) experiment, which relies on a dramatic manipulation of ocular input and the independent light-driven activity of two eyes, provides essential mechanistic confirmation of the dominance of correlated activity in organizing and stabilizing synaptic connections. The striking similarity of segregated terminal fields in an anomalously binocularly innervated frog optic tectum to typically segregated mammalian ocular dominance columns suggests that a similar, perhaps singular, competitive Hebbian process may be at work to establish this anatomically distinct, functionally significant pattern of binocular connectivity.

Binocular competition and orientation tuning for binocular vision

The relationship between action potential activity correlated by eye of origin and the acquisition of other physiological properties that define neuronal identity and capacity for information processing in the primary visual cortex can be better defined by experiments that demonstrate the influence of binocular competition on

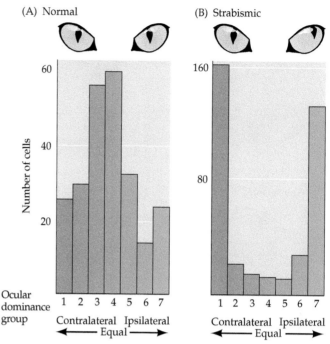

FIGURE 24.10 Ocular asynchrony prevents binocular convergence Ocular dominance histograms obtained by electrophysiological recordings in normal adult cats (A) and adult cats in which strabismus was induced during the critical period (B). The number of binocularly driven cells (groups 3, 4, and 5) is sharply decreased as a consequence of strabismus; most of the cells are driven exclusively by stimulation of one eye or the other. This enhanced segregation of the inputs presumably results from the greater discrepancy in the patterns of activity between the two eyes as a result of surgically interfering with normal conjugate vision. This pathological state is thought to enhance the relative degree of correlation within inputs from each eye and to decrease the possibility of correlation between eye inputs. (After D. H. Hubel and T. N. Wiesel, 1965. *J Neurophysiol* 28: 1041–1059.)

establishing orientation preference in individual cortical neurons (see Chapter 12). Prior to the onset of the critical period, there is little or no correlation between relatively broad orientation sensitivities in individual cortical neurons driven by both eyes: The maximal response to apparent preferred orientations is fairly low, and the orientations of the stimuli themselves are dissimilar. As visual system development progresses toward the start of the critical period for visual experience–evoked plasticity, the magnitude and frequency of the maximal responses to oriented stimuli in a single cell driven by both the right and left eyes increase dramatically. Their orientation preferences, however, remain dissimilar. The increased correlation of visually evoked stimuli, presumably correlated based on coincident activation of retinal ganglion cells by binocularly coherent stimuli with identical orientations, leads to the matching of orientation tuning of the right

(A)

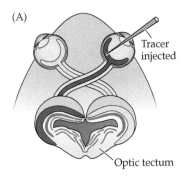

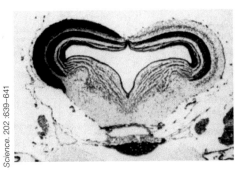

(B)

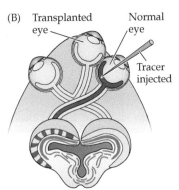

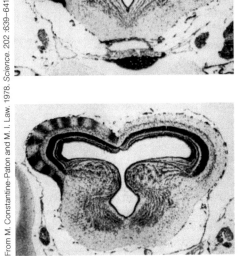

From M. Constantine-Paton and M. I. Law. 1978. *Science*. 202: 639–641

FIGURE 24.11 Hebb's postulate confirmed: Competition between more or less correlated inputs results in segregation of connections (A) In the normal frog, the retinal ganglion cell axons from the two eyes are mostly crossed, and their terminal fields in the optic tecta are continuously distributed. In this example, the diagram at left shows that the right eye has been injected with an anterograde tracer, and the micrograph at right shows that the left optic tectum has a continuous band of retinal ganglion cell terminals in the outermost layers. (B) When a third eye is grafted onto an early developing tadpole, the retinal ganglion cell axons from the third eye in the adult project to only one of the two tecta. In the tectum where these axons terminate, the axons establish stripes of terminal fields interdigitated with similar sized stripes from the other eye. Thus, it appears that since the inputs from each eye "fire together," they "wire together." Furthermore, the arrays of input stripes that parallel the size and interdigitation of ocular dominance columns in the carnivore and primate visual cortex (rather than two segregated domains) suggest that the nature of competing visual inputs favors periodic segregation. (After L. C. Katz and J. C. Crawley, 2002. *Nat Neurosci Rev* 3: 34–42.)

and left eye inputs to single cortical binocularly driven neurons (Figure 24.12). This process of shaping the circuitry for individual binocular neurons so that orientation information relayed by both eyes is in register can be modified by the same peripheral manipulations that change ocular dominance. Thus, if one eye or the other is closed during the critical period (monocular deprivation; see earlier in this concept), the matching of orientation tuning of binocular inputs does not occur, and cannot be restored if the closed eye is opened after the end of the critical period. In contrast, closing one eye for an extended time after the critical period is over has no effect on the matching of orientation tuning of the right and left eye inputs to single visual cortical binocular neurons. This suggests that correlated activity is a key determinant of not only general binocular circuitry, but of the circuitry that underlies feature detection (such as oriented bars or contrast and edges) in the visual system.

Amblyopia, strabismus, and critical periods for human vision

Developmental phenomena in the visual systems of experimental animals accord with clinical problems seen in children who have experienced similar deprivation. The loss of acuity, diminished stereopsis, and problems with fusion that arise from early deficiencies of visual experience are called **amblyopia** (Greek, "dim sight"). These functional difficulties are all believed to reflect the essential contribution made by normal binocular competition for defining, in

an experience-dependent manner, cortical circuitry necessary for binocular vision and depth perception during the critical period for visual cortical development.

In humans, amblyopia is most often the result of strabismus. Depending on the extraocular muscles affected, the misalignment can produce convergent strabismus, called **esotropia** ("crossed eyes"); or divergent strabismus, called **exotropia** ("wall eyes"). These alignment errors, both of which produce double vision (which describes a lack of proper binocular fusion as well as depth perception), are surprisingly common, affecting about 5% of children. In some of these individuals, the response of the visual system is to suppress input from one eye by mechanisms that are not completely understood but which are thought to reflect competitive interactions during the critical period. Presumably, the inputs to the LGN from the eye that is more optimally aligned are competitively advantaged. Thus, the corresponding LGN inputs are accorded more territory in the visual cortex. Functionally, the suppressed eye eventually comes to have very low acuity which may render an individual effectively blind in that eye. Early surgical correction of ocular misalignment in strabismic children (by adjusting the lengths of extraocular muscles during the critical period) has become an essential treatment to correct strabismus and preserve normal vision.

Another cause of visual deprivation in humans is cataracts, which can be caused by several congenital conditions and which render the lens or cornea opaque.

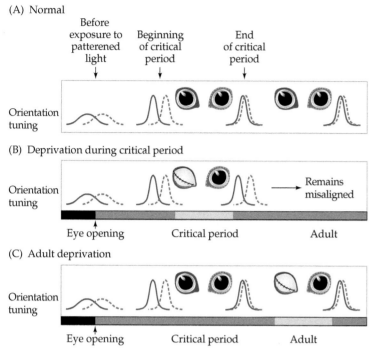

FIGURE 24.12 Binocular competition during the critical period aligns orientation tuning in binocularly innervated cortical neurons (A) Orientation tuning curves in a typically developing animal for a single neuron in the visual cortex that responds to oriented visual stimuli via a normal left eye (solid red) and right eye (dashed blue) at eye opening and the onset of visual experience. The responses are completely uncorrelated at the onset of vision. Thus, this binocularly innervated primary visual cortical neuron is sensitive to two very different orientations depending on which eye is stimulated. As postnatal development proceeds after eye opening, the responses of the cortical neuron increase in amplitude; however, the incongruence in their orientation tuning remains as the critical period begins. After the critical period, if binocular competition proceeds uninterrupted, the orientation tuning of binocularly driven single cortical neurons now matches. (B) When normal binocular competition is disrupted during the critical period (but not before or after) using transient monocular deprivation (MD, shown here as reversible lid closure of the left eye), the lack of binocular competition results in a failure to match orientation tuning of binocular inputs to single visual cortical neurons. (C) As a control, in response to the same transient monocular deprivation in an adult after the close of the critical period, the orientation congruence of binocular inputs to single cortical neurons does not change. (After B.-S. Wang et al., 2010. *Neuron* 65: 246–256.)

Diseases such as onchocerciasis ("river blindness," the result of infection by the parasitic nematode *Onchocerca volvulus*) and trachoma (caused by the parasitic bacterium *Chlamydia trachomatis*) affect millions of people in tropical regions. These diseases can lead to cataracts. A cataract or corneal occlusion in one eye is functionally equivalent to monocular deprivation in experimental animals. Left untreated in children, this ocular defect also results in irreversible damage to visual acuity in the deprived eye. If either the cataract or corneal occlusion is remedied by removing and replacing the biological lens with an artificial lens or removing the damaged cornea and providing a transplant before about 4 months of age, however, the consequences of this monocular deprivation are largely avoided. As expected from studies of bilateral visual deprivation in experimental animals, bilateral cataracts or corneal occlusions—which are similar to binocular deprivation in experimental animals—produce less dramatic deficits even if treatment is delayed. As predicted by detailed animal experiments, unequal competition during the critical period for normal vision is more deleterious than the complete disruption of visual input that occurs with binocular deprivation. Correction of unequal competitive interactions during the critical period has restored or preserved visual function in countless children.

In keeping with the findings in experimental animals, the visual abilities of individuals monocularly deprived of vision after the close of the critical period (for example, by late-arising cataracts or corneal scarring common in

adults) are much less compromised, even after decades of deprivation. When vision is restored in such individuals, there may be difficult psychological consequences, memorably described by the neurologist Oliver Sacks and then recounted in the play *Molly Sweeney* by the Irish playwright Brian Friel. Nor is there any evidence of anatomical change in this circumstance. For instance, an individual whose eye was surgically removed in adulthood showed normal ocular dominance columns when his brain was examined post mortem many years later. Thus, all of the predictions of decades of basic experiments defining critical period mechanisms and how they can be disrupted has led to a clearer understanding of several visual disorders and their prevention or treatment.

Oscillating electrical activity and critical periods

The evidence for critical periods initially relied on behavioral observations, then on electrophysiological data in postnatal animals that measured action potential activity elicited by relevant stimuli—especially visual stimuli—coupled with anatomical assessment of changes in patterns of connections. These approaches helped define the critical period for the visual system and demonstrated that "elicited" sensory experience—driven by external stimuli resulting in the transmission of action potential activity relayed from retina to LGN to visual cortex—is key. Nevertheless, there are additional forms of subthreshold as well as action potential–mediated physiological activity that occur prior to experience-driven activity. These forms of electrical activity establish a framework within which sensory experience can further influence patterns and function of connectivity. Local **oscillations**, or "waves," of electrical activity that are initially beneath the threshold for action potential generation (they reflect the fundamental excitable properties of developing neurons, including several classes of ion channels and some neurotransmitter receptors) are now known to be essential for shaping circuit networks so they are prepared for optimal experience-driven activity.

Most of the insights into the developmental importance of neuronal oscillations have come from analyses of the mammalian visual system. The initial evidence

for oscillatory activity came from analysis of activity in the retina that begins long before birth in most mammalian species and is referred to as **retinal waves** (Figure 24.13A,B). It was already known, based on anatomical tracing (see later in this concept), that an apparent scaffold of segregated inputs from the LGN, driven by one eye or the other, is present in the visual cortex of many animals, including rhesus monkeys, cats, and ferrets, either slightly before or after birth. It was unclear, however, how this segregation, prior to sustained experience-driven competition between the two eyes, was established. There are some molecular affinities reflecting the origin of ipsilateral-projecting versus contralateral-projecting retinal ganglion cells in the nasal versus temporal retina (see Box 23A) that bias them toward distinct target regions in the LGN. Nevertheless, these affinities alone are not sufficient to explain the initial pre- or perinatal segregation of the inputs from the two eyes. A substantial amount of this initial segregation is due to organized electrical activity in the retinas of individual animals before they are born or before their eyes have opened. Each retina in an individual fetus or newborn (prior to eye opening) generates independently a pattern of waves of electrical activity (usually measured as Ca^{2+} influx) that move across large populations of contiguous

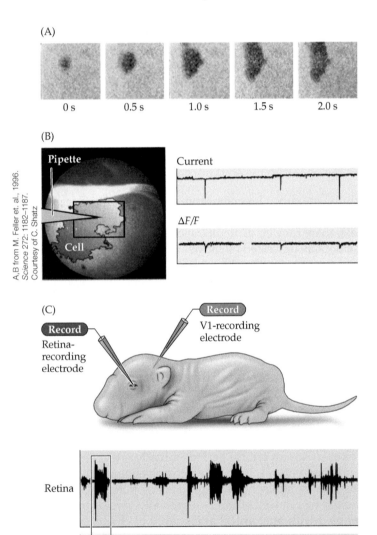

A,B from M. Feller et. al., 1996. *Science* 272: 1182–1187. Courtesy of C. Shatz

FIGURE 24.13 Spontaneous activity establishes rudimentary patterns of connectivity in the retino-geniculo-cortical pathway (A) Retinal waves, measured here by imaging calcium transients (proportional to action potential activity) in a flattened retina in vitro. The activity that defines a single wave, depicted using a gray scale that maps the local change of Ca^{2+} influx over time, measured by the change in fluorescence intensity ($\Delta F/F$) spreads across the retinal surface. The small gray spot in the upper left quadrant of the far left panel indicates the initiation of the wave at time 0 (0 s). The subsequent images, taken every 0.5 second, show the spreading excitation across a subregion of the retinal surface that defines the wave (which is ultimately relayed to the thalamus via action potential activity in retinal ganglion cells) until the wave abates (not shown). (B) The relationship between single cell excitation measured intracellularly (the downward deflections in the top trace indicate excitatory postsynaptic potentials) and calcium transients that define each retinal wave. Within the wave region outlined at left, the firing of a cell recorded at the position (shown by the pipette cartoon and box) is synchronized to the maximal depolarization events that define the wave, shown as the downward traces in the fluorescence intensity record (bottom trace). (C) Simultaneous extracellular recordings are made in the retina and primary visual cortex of an early postnatal rat pup, prior to eye opening and the capacity of the retina to relay light information from rods and cones to the ganglion cells. The spontaneous activity that defines retinal waves is shown as bursts of action potential activity. These bursts in the retina are highly correlated with bursts of action potential activity recorded in primary visual cortex in the same pup. (C after I. L. Hanganu et al., 2006. *J Neurosci* 26: 6728–6736.)

retinal cells in an orderly fashion. The waves are initiated in local retinal cells (amacrine cells), and this subthreshold activity leads to action potential firing by ganglion cells that is then relayed to the LGN. These waves, though coherent in each eye, are asynchronous between the two eyes. The lack of correlated activity establishes a modest competitive interaction—leading to Hebbian synaptic reinforcement—between the two eyes for target space in the LGN and via the LGN to the primary visual cortex. Thus, afferents driven by one eye are likely to segregate, at least partially, from those of the other eye. In experimental animals such as cats, ferrets, and rats, retinal waves can be abolished via pharmacological blockers of various neurotransmitters. Parallel experiments have been carried out in mice in which a variety of receptors or neurotransmitter transporters have been inactivated by mutation. These experiments suggest that concerted activity within a nascent circuit that includes acetylcholinergic, GABAergic, and glycinergic synaptic transmission from retinal amacrine cells combined with glutamate release from bipolar cells onto retinal ganglion cells (see Chapter 6) is responsible for the generation of retinal waves. In the absence of such synaptic signaling and the waves of electrical activity that result, segregation in the LGN by left eye versus right eye is substantially diminished or eliminated.

The excitatory drive established by retinal waves via the synaptic signals delivered to ganglion cells, and transmitted through LGN relay neurons that project to the primary visual cortex, establishes related oscillatory activity in the visual cortex. These oscillations can be recognized with multi-unit recordings from the neonatal primary visual cortex in animals. Such activity in developing cortical networks reflects voltage changes—synaptic as well as action potential activity—in large numbers of neurons and can be recorded using extracellular electrodes placed in the cortex or thalamus in early postnatal animals such as a rat or mouse (Figure 24.13C; also see Chapter 1), or can be detected using electroencephalography (EEG) in pre-term as well as full-term human infants. In early postnatal rat visual cortex, bursts of action potential activity in the cortex are highly correlated with wave-induced bursts of action potential activity from the retina. The absence of this cortical activity, induced experimentally using pharmacological agents, delays or abolishes appropriate maturation of visually evoked responses in the cortex. In humans, similar activity, detected as bursts of population electrical activity (generally called spindles) via EEG recordings, indicates that even before extensive visual experience, there is subthreshold oscillatory activity as well as bursts of action potential signaling that influence synapse and circuit maturation in the cortex. This activity is thought to be essential to prepare the cortex for visual experience–dependent critical period plasticity.

CONCEPT
24.4

Ion Channels, Neurotransmitters and Their Receptors, and Neurotrophins Regulate Activity-Dependent Circuit Development

LEARNING OBJECTIVES

24.4.1 Assess neurotransmitter regulation of experience dependent critical periods.

24.4.2 Define neurotrophic regulation of critical period plasticity.

24.4.3 Describe the mechanisms of signal transduction and transcriptional regulation that underlie critical periods.

Specific synaptic inputs and neurotransmitters mediate critical period plasticity

Electrical activity in the brain, usually elicited by environmental stimuli, during a critical period must be translated into molecular signals, primarily delivered via distinct synaptic inputs. Excitatory, inhibitory, and neuromodulatory neurotransmitters delivered to target neurons by inputs that use these synaptic signaling molecules are essential for initiating, maintaining, and terminating critical periods (Figure 24.14). Depending on the system, the afferent axons that innervate neurons that are targets of plastic change, several neurotransmitters, their receptors, and downstream signaling events (see Figure 24.14A) can influence the establishment, maintenance, or strength of specific connections. These influences reflect correlated activity in subsets of afferent inputs (as originally suggested by Hebb, see Figure 24.1), and they can further adjust connectivity for individual neurons when patterns of activity change, presumably reflecting altered peripheral inputs or experience. The convergent activity of a variety of inputs from distinct afferent classes, mediated by different neurotransmitters, can modify gene expression via transcription factors whose activity are sensitive to downstream consequences of neurotransmitter signaling. These include the transcription factor CREB (*cAMP response element binding* protein), which upon phosphorylation translocates to the nucleus to influence transcription. Finally, transcription can be affected during activity-dependent changes via chromatin-modifying mechanisms that alter DNA methylation state or histone modifications that influence more broadly the sets of genes available for transcription and those that are not.

The mechanisms underlying the influence of excitatory inputs on critical period plasticity in target neurons and circuits via glutamate reflect the capacity of NMDA receptors (NMDA-Rs) to detect local correlated synaptic activity. Thus, the correlated or non-correlated activity of glutamatergic inputs and its activation of

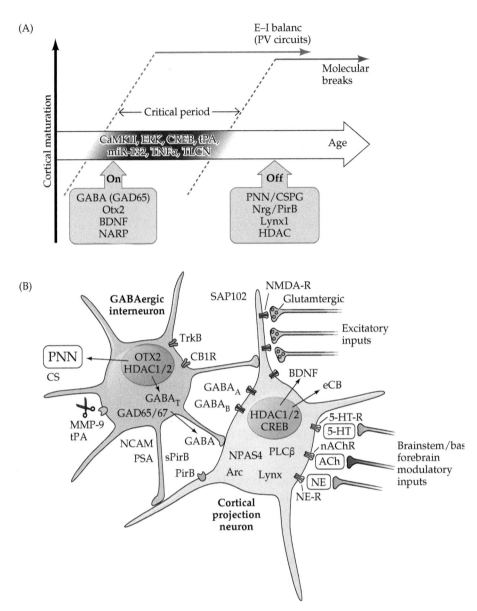

FIGURE 24.14 Molecular signals essential for opening and closing critical periods for cortical connections The key roles of the proteins summarized in this figure have been inferred primarily from studies in mutant mice, genetic manipulation in cultured neurons, and in some cases pharmacological manipulation in vivo or in vitro. (A) The schedule of opening and closing of the critical period in the mammalian visual cortex. At left, the availablility and activity of neurotransmitters (GABA), transcription factors (Otx2), neurotrophins (BDNF), and synapse-associated proteins (NARP) signal the onset of the critical period. As cortical circuit development progresses (the shaded horizontal arrow indicating increasing age of the developing animal) signal transduction genes (CaMKII, ERK, CREB, tPA), signaling molecules (TNFα), adhesion molecules (TLCN; telencephalin, a dendrite-associated cell adhesion molecule), and translational modulators (miR-132, a microRNA that regulates translation of target mRNAs) are critical to sustain plasticity of connections for the duration of the critical period. At right, several additional molecular mediators terminate the critical period, including chondroitin sulfate proteoglycans (CSPG), the major extracellular matrix proteins that comprise the perineuronal net (PNN), the signaling molecule neuregulin (Nrg), the adhesion molecule receptor PirB, the transcription factor Otx2, the membrane associated protein Lynx1, and histone deacetylases (HDAC) that modify the histone proteins that influence chromatin confirmation and transcriptional state. (B) Distinct molecular mediators of synapse formation and maintenance act in cortical projection neurons and interneurons to influence critical period plasticity. Key differences in the cortical projection neuron, the target of convergent inputs from multiple sources, include the transcriptional upregulation and secretion of BDNF and endocannabinoids (eCB) from cortical projection neurons, as well as their expression of several cell surface receptors and adhesion molecules (synaptic associated protein SAP102, Lynx1, PirB), excitatory and inhibitory neurotransmitter receptors (NMDA-R, GABAA, GABAB), intracellular signal transduction proteins (PLCβ), and the sensitivity of projection neurons to modulation by norepinephrine (NE), acetylcholine (ACh), and serotonergic (5-HT) afferent inputs. GABAergic interneurons influence critical periods by their sensitivity to BDNF via the TrkB receptor, endocannabinoids via the cannabinoid receptor 1 (CB1R), cell surface adhesion molecules such as NCAM and its polysialated forms (PSA), and the formation and modulation of the polyneuronal net of extracellular matrix proteins (PNN) and their controlled proteolytic cleavage by proteases such as matrix metalloproteinase 9 (MMP-9) and tissue plasminogen activating factor (tPA). (A after A. E. Takesian and T. Hesch, 2013. *Prog Brain Res* 207: 3–33; B from E. K. Choi et al., 2018, *Exp Mol Med* 50: 1–16.)

NMDA-Rs is central for critical period plasticity. This NMDA-R activation initiates Ca^{2+} influx at the postsynaptic specialization and thus modifying signaling in the postsynaptic cell (see Chapter 8). Additional ionotropic glutamate receptors—particularly the AMPA receptor (AMPA-R), which is permeable to Na^+ and K^+ but not Ca^{2+}, as well as members of the mGluR metabotropic glutamate receptors (see Chapter 8)—also establish or reinforce postsynaptic depolarization due to correlated activity. These glutamate receptor–mediated changes in the developing cerebral cortex, elicited by excitatory inputs from the thalamus as well as local and long-distance connections between other cortical projection neurons, can modify activation of additional voltage-gated channels, including L-type voltage-sensitive Ca^{2+} channels (L-VSCCs). L-VSCCs and other channels modify intracellular Ca^{2+} levels, and the activation of Ca^{2+}-dependent second-messenger cascades that alter cell adhesion signaling, influence the cytoskeleton, and ultimately change gene expression to stregthen or weaken a synaptic contact (see later in this concept). Thus, Hebb's postulate seems to be paralleled by the molecular mechanisms underlying excitatory neurotransmission from subsets of afferents via glutamate and its receptors.

Inputs made by inhibitory interneurons, mediated primarily by GABAergic neurotransmission, also influence how electrical activity in developing circuits shapes long-lasting changes in patterns, numbers, and strength of connections (see Figure 24.14). The key role for GABAergic neurotransmission appears to be coordinating the onset and cessation of peak critical period plasticity. Local GABAergic neurotransmission modulates the excitability of projection neurons that are often the primary targets of competing afferents (for example, LGN axons driven by each eye converging on a cortical pyramidal neuron). Thus, the activity of local inhibitory neurons during critical periods influences the effectiveness of correlated activity for modifying synaptic strength of competing inputs. The initial establishment of local GABAergic synaptic contacts appears to mark the onset of peak plasticity during critical periods. Differential influence of neurotrophins, including BDNF (see Chapter 23 and the next section), drives the onset of GABAergic synapse formation. The stimulation of GABAergic synapse formation indirectly influences the sensitivity of projection neuron targets to activity-dependent competition. Activation of these local GABAergic synapses can either reinforce or antagonize correlated activity of inputs and their targets. Thus, the **excitatory/inhibitory (E/1) balance** that is established by the onset of GABAergic neuron connectivity in several sensory cortical regions (visual, somatosensory, and auditory cortices) defines the onset of the critical period by shaping the context of correlated activity and competitive interaction for the target neuron. The further maturation of GABAergic interneurons, including the assembly of extracellular matrix proteins into a **perineuronal net** that ensheathes the GABAergic neuron and its processes—particularly the subset of GABAergic interneurons that are distinguished by expression of the Ca^{2+} binding protein parvalbumin—apparently influences the close of critical period plasticity (see Figure 24.14B). These changes all seem to be targeted at stabilizing synaptic organization and preventing malleability of patterns of connections via durable E/I balance once the critical period has closed.

Excitatory activity is essential for the formation in rodents of distinct whisker barrels in the somatosensory relay nuclei of the brainstem and thalamus as well as the somatosensory cortex (Figure 24.15A). When the excitatory input from one single row of whiskers is disrupted, the barrels driven by adjacent whiskers expand and reorganize the barrel cortical region that is no longer activated by the peripherally silenced row of whiskers (Figure 24.15B). If all of the peripheral excitatory input is eliminated during a critical period of early postnatal life by cutting the branch of the trigeminal nerve that innervates the whisker pad, the barrel pattern in the cortex is eliminated and the thalamic afferents are dispersed from their focal, segregated pattern (Figure 24.15C). Similarly, when excitation in the barrel cortex is diminished via mutations of specific glutamate receptors, the barrels are smaller and disordered, and the segregation of thalamic afferents is changed (Figure 24.15D).

The final neurotransmitters that influence critical period plasticity are those released by broadly projecting neuromodulatory inputs: noradrenergic inputs from the locus coeruleus, serotonergic inputs from the Raphe nucleus, cholinergic inputs from brainstem and basal forebrain nuclei, and dopaminergic inputs from the ventral tegmental area (VTA). Once again, the role of neuromodulators in shaping segregated inputs and their equally segregated targets is illustrated by the consequences for barrel cortex differentiation in mice carrying mutations in genes that encode proteins responsible for regulating local metabolism of serotonin (Figure 24.15E). In mice mutant for the serotonin transporter or the enzyme monoamine oxidase A, which degrades serotonin to an inactive molecule extracellularly, afferents that segregate to form individual barrel inputs or neurons that segregate to constitute whisker-selective targets fail to do so. Similarly, in mice lacking the serotonin transporter, which is responsible for serotonin reuptake from the synaptic cleft—thus limiting the availability of the transmitter—barrel afferents and targets also fail to segregate. This indicates that serotonergic modulation influences correlated activity and its consequences for selective connectivity in both afferents and targets that define the whisker barrels. Serotonergic, cholinergic, and noradrenergic neuromodulatory inputs (see Chapter 7) to other cortical areas, including visual cortex, also influence activity-dependent changes in connectivity during critical periods.

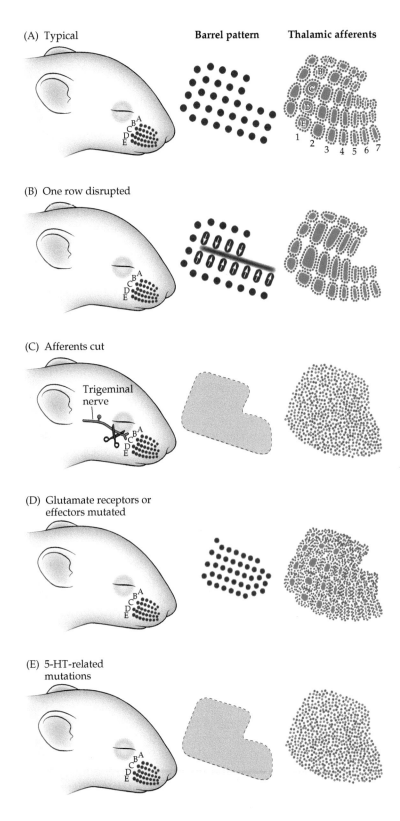

(A) Typical

Barrel pattern

Thalamic afferents

(B) One row disrupted

(C) Afferents cut

Trigeminal nerve

(D) Glutamate receptors or effectors mutated

(E) 5-HT-related mutations

FIGURE 24.15 Peripheral inputs and their influence on cortical representations and circuits The whiskers on the face of a mouse (left column) are referred to based on their location in rows (A through E). These whiskers are represented as "barrels" in the somatosensory cortex (middle column). The barrels are defined in layer 4 of the primary somatosensory cortex of most rodents by a ring of cortical neurons (gold) that form the barrel wall, and a cell sparse barrel center (blue) composed of dendrites, axon terminals, and synapses (right column). (A) In a typical mouse, each barrel receives segregated thalamic input in the barrel center that are separated from their neighbors by a ring of cortical neurons that comprise the barrel walls around the neuropil where the input terminates. (B) When one row of whisker inputs in the periphery is disrupted so that they no longer drive activity in response to peripheral stimulation (either by trimming to diminish activity or cauterization to eliminate the whiskers entirely), the barrels that correspond to that whisker are lost in the cortex, and the adjacent barrels, defined by the cells that constitute their barrel walls as well as the thalamic inputs in the barrel center expand into the functionally "silenced" territory once driven by the now damaged whiskers. (C) When all whisker inputs are silenced by cutting the afferent trigeminal nerve, the segregation of cortical cells that defines the barrels is lost, as is the segregation of thalamic afferents that apparently nucleates each barrel. (D) When a glutamate receptor is homozygously mutated selectively in the cerebral cortex (using conditional recombination; see Chapter 1), the segregation of barrels and thalamic afferents is altered. (E) When signal transduction molecules that mediate serotonergic signaling in the cortex are homozygously mutated, the cellular segregation defining barrels is also lost in the cortex, as is the segregation of thalamic afferents. (After R.S. Erzurmulu and P. Gaspar, 2012. *Eur J Neurosci* 35: 1540–1553; H. Li and M. C. Crair, 2011. *Ann NY Acad Sci* 1225: 119–129; M. Inan and M. C. Crair, 2007. *Neuroscientist* 12: 49–61.)

Transcriptional changes and activity-dependent plasticity

The influence of synaptic activity during critical periods, and the related changes in neurotransmitter receptor activation and electrical activity in afferent inputs and their target neurons, must ultimately be recorded as more or less permanent patterns, numbers, and strength of connections as the critical period draws to a close. The primary mechanisms for consolidating the activity-dependent changes of critical periods rely on intracellular signaling cascades initiated by the neurotransmitters described in the previous section that activate transcription factors. These transcription factors then modify expression of genes that influence neuronal growth or synaptic stability. These include genes for neurotrophic and other growth factors, as well as cytoskeletal and synaptic regulatory proteins that strengthen pre- and postsynaptic transmission.

Increased intracellular Ca^{2+} in target neurons, primarily due to input-mediated neurotransmitter signaling, results in the activation of multiple signal transduction cascades that ultimately modify gene expression. The Ca^{2+}-sensitive signaling intermediates of these cascades—protein kinases and phosphatases—are key mediators of cellular and molecular changes that lead to changes in synaptic connectivity during critical periods (Figure 24.16A). These intermediates include Ca^{2+}/calmodulin-dependent protein kinases (CaMKs) that phosphorylate the amino acids serine and threonine (thus, CaMKs are referred to as serine-threonine kinases) on target proteins at the postsynaptic site. These target proteins include the AMPA-R, and this modification further reinforces depolarization sensitivity and strengthens synapses via regulation of receptor trafficking at post-synaptic sites. In addition, the ras–mitogen-activated protein kinase (MAPK) pathway is activated by increases in intracellular Ca^{2+}, leading to changes in gene expression as well as alteration of the actin cytoskeleton. Finally, the protein phosphatase calcineurin is also activated by increased intracellular Ca^{2+} and also can modify signaling that leads to changes in gene expression. These signaling intermediates all converge on the regulation of transcription factor proteins that are resident in the nucleus and dependent on kinase activation or phosphatase deactivation to modulate their transcriptional activity (see Figure 24.16A). These include the transcription factor CREB (see Figure 24.14) and other modulators of gene expression. Multiple kinase- or phosphatase-regulated nuclear transcription factors, in the context of synaptic activity during the critical period (and beyond), regulate the transcription of a family of genes called **immediate early genes**, many (but not all) of which are transcription factors. These immediate early gene transcription factors, including the Fos transcription factor, define an initial transcriptional response to activity, and in turn activate a wave of additional gene expression that constitutes a "late response" (Figure 24.16B). The genes expressed during the late response include those that encode multiple proteins essential for dendritic growth and synaptic stability.

The ubiquitin ligase, Ube3a, plays an important role in degrading ARC, the *a*ctivity *r*egulated *c*ytoskeletal protein, which regulates, among other targets, membrane turnover and trafficking of the AMPA-R class of glutamate receptors. Cell adhesion signaling through neurexins (presynaptic) and neuroligins (postsynaptic) regulates the integrity of cytoskeletal scaffolding proteins, including postsynaptic density protein 95 (PSD95), the guanylate kinase protein GKAP, the scaffolding protein SHANK (*SH3*- and *ank*yrin repeat domains protein), and the activity-related cystokeletal protein HOMER (see Figure 24.15B). Cytoskeletal integrity at the pre- and postsynaptic membrane is essential for retaining the local accumulation of proteins for synaptic vesicle fusion (presynaptic) and receptor concentration (postsynaptic and presynaptic) at the synapse. If the synaptic cytoskeleton is disrupted, the efficiency and fidelity of synaptic transmission decline. Such changes can substantially diminish the effectiveness of activity

and experience in shaping maximally adaptive neural circuits and behaviors during a critical period. Finally, signaling mediated by metabotropic glutamate receptors (mGluRs, a group of glutamate-activated GPCRs) engages the tuberosclerosis complex genes 1 and 2 (*TSC1* and *TSC2*), originally identified as tumor suppressor genes. *TSC1* and *TSC2* ultimately regulate mTOR (mammalian target of rapamycin), a protein that controls bioenergetic homeostasis, and thus influences protein synthesis, stability, and turnover. The consequence of the activity of this combination of molecular mediators of excitatory synaptic transmission is to modify postsynaptic integrity and thus either strengthen the physical synaptic specialization or weaken it for eventual elimination.

A second key pathway for the molecular regulation of critical periods is via neurotrophins (see Chapter 23), particularly the brain-derived neurotrophin BDNF. BDNF in this instance is primarily available via secretion from the afferent synaptic input rather than the synaptic target (this can also be the case for other forms of neurotrophic signaling; see Chapter 23). BDNF, via the TrkB receptor tyrosine kinase, initiates a signaling cascade that depends on ras and Raf intermediates. The Raf serine-threonine kinase phosphorylates targets that lead to activation of the extracellular signal-regulated kinase (ERK) pathway, while ras (a small GTPase) binds and catalyzes the hydrolysis of GTP in its active form. Ultimately, ras and raf activate MEK and ERK kinases (MEK: mitogen-activated and ERK: extracellular signaling related kinases) that also target the CREB transcription factor as well as related activity-dependent transcriptional regulators in the nucleus. These include the RSK/MSK (RSK: ribosomal s6 kinase; MSK: mitogen and stress-activated protein kinase) family of kinases that translocate to the nucleus in response to ras/raf-dependent phosphorylation. This activation leads to changes in gene expression that record activity-dependent plasticity as an altered transcriptional state in the target cell. Mutations in these genes in mice can impact critical period plasticity, as well as additional forms of synaptic plasticity that remain once critical periods have ended.

The genes for many additional critical period–related signal transduction molecules as well as cytoskeletal and synaptic scaffold proteins (SHANK, neurexins and neuroligins, TSC1 and TSC2, as well as the Fragile X-mental retardation protein [FMRP]) are also, perhaps not surprisingly, targets for mutation associated with a range of neurodevelopmental disorders, including intellectual disability and autism spectrum disorders. In addition, local and global effects of modified neurotrophin signaling in genetically engineered mice (see Chapter 1) include altered synaptic plasticity during critical periods; however, a genetic association between neurotrophic signaling and neurodevelopmental disorders in humans has not been established. Evidence from studies in several animal models shows that when these signaling molecules

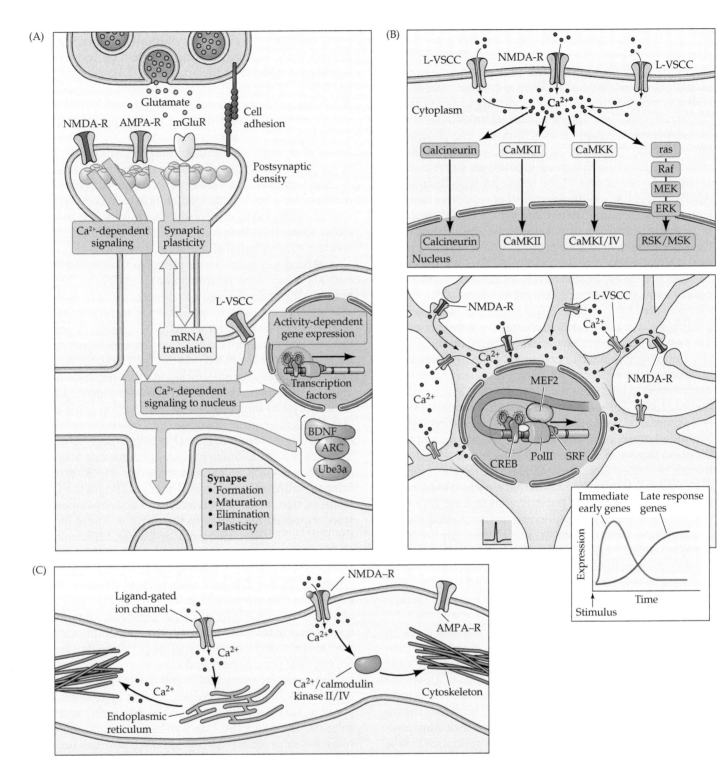

(A)

Glutamate

NMDA-R AMPA-R mGluR

Cell adhesion

Postsynaptic density

Ca²⁺-dependent signaling

Synaptic plasticity

L-VSCC

Activity-dependent gene expression

mRNA translation

Transcription factors

Ca²⁺-dependent signaling to nucleus

BDNF
ARC
Ube3a

Synapse
• Formation
• Maturation
• Elimination
• Plasticity

(B)

L-VSCC NMDA-R L-VSCC

Cytoplasm Ca²⁺

Calcineurin CaMKII CaMKK ras / Raf / MEK / ERK

Calcineurin CaMKII CaMKI/IV RSK/MSK

Nucleus

NMDA-R L-VSCC Ca²⁺

Ca²⁺

MEF2

NMDA-R

CREB PolII SRF

Immediate early genes Late response genes

Expression

Time

Stimulus

(C)

Ligand-gated ion channel

NMDA–R

AMPA–R

Ca²⁺

Ca²⁺

Ca²⁺/calmodulin kinase II/IV

Cytoskeleton

Ca²⁺

Endoplasmic reticulum

are manipulated genetically or pharmacologically, some critical period phenomena are altered. One major target of all these molecular signaling processes is apparently the network of local inhibitory connections made by GABAergic neurons (see Figure 24.13). Regulation of the number and placement of local inhibitory synapses, as well as the expression of GABA receptors at postsynaptic sites, seems to be exquisitely sensitive to changes in levels of electrical activity, and the signaling that these changes influence, during early postnatal life. Thus, the activity- and experience-dependent regulation of inhibitory connectivity, and E/I balance in developing neuronal circuits, has become a major focus for understanding the molecular and cellular mechanisms of activity- and experience-dependent changes in the brain and behavior during early post-natal life.

◀ **FIGURE 24.16 Transduction of electrical activity into cellular change via Ca²⁺ signaling** (A) Summary of the signal transduction essential for critical period synaptic plasticity, primarily at the postsynaptic specialization. Excitatory activity, which is experience-dependent after birth and relies on the appropriate maturation of peripheral sensory relay pathways, is proportional to the level of glutamate release from the presynaptic terminal. In turn, glutamate binds to the ionotropic receptors NMDA-R and AMPA-R, as well as to the metabotropic glutamate receptor, mGluR. The consequence of NMDA-R and AMPA-R activation via glutamate binding is depolarization that favors the influx of Ca²⁺ via the NMDA-R and the initiation of Ca²⁺-dependent signaling that can influence local cytoskeletal integrity and receptor distribution and stability. This aspect of structural modulation to translate electrical activity to cellular change also includes the modulation of Ca²⁺-dependent cell adhesion to either maintain or disrupt the relationship between pre- and postsynaptic sites. Signaling through mGluR activates second-messenger cascades that rely on mTOR activation to modulate mRNA translation into protein, or to activate ERK signaling, leading to altered nuclear gene transcription. (B) Correlated or sustained activity leads to increased Ca²⁺ via the NMDA receptor/channel (NMDAR) or the L type voltage gated Ca²⁺ channel (L-VSCC) and increased intracellular Ca²⁺ concentration, which results in activation of calcineurin (a protein phosphatase), CaMKII or CaMKIV, ERK, Rsk/Msk (Ribosomal S6 kinase, which phosphorylates ribosomal proteins and other signaling intermediates, also known as mitogen and stress activated kinase or Msk). Within the nucleus, these signaling intermediates then activate Ca²⁺-regulated transcription factors such as CREB, SRF (serum response factor) and MEF2 (myocyte enhancer factor) as well as other chromatin-binding proteins (not shown), to modify transcription. The Ca²⁺ dependent changes result in two phases of gene expression: an immediate early response and a late response, with different genes regulated for each phase. (C) Local increases in Ca²⁺ signaling in distal dendrites due to correlated or sustained activity may lead to local increases in Ca²⁺ concentration that modify cytoskeletal elements (actin- or tubulin-based structures), perhaps through the activity of kinases such as CaMKII/IV operating in the cytoplasm rather than the nucleus. Changes in these cytoskeletal elements lead to local changes in dendritic structure. In addition, increased local Ca²⁺ concentration may influence local translation of transcripts in the endoplasmic reticulum, including transcripts for neurotransmitter receptors and other modulators of postsynaptic responses. Increased Ca²⁺ may also influence the trafficking of these proteins, their insertion into the postsynaptic membrane, and their interaction with local scaffolds for cytoplasmic proteins (see Figure 23.10). (A after D. H. Ebert and M. E. Greenberg, 2013. *Nature* 493: 327–337; B after After E-L. Yap and M. E. Greenberg. 2018, *Neuron* 100: 330-348; C after R. O. Wong and A. Ghosh, 2002. *Nat Rev Neurosci* 3: 803–812.)

Summary

An individual animal's history of interaction with its environment—its "experience"—helps shape its neural circuitry and thus determines subsequent behavior. Experience during specific times in early life, referred to as critical periods, helps shape behaviors as diverse as maternal bonding and the acquisition of language. Correlated patterns of activity are thought to mediate critical periods by stabilizing concurrently active synaptic connections and weakening or eliminating connections whose activity is divergent. Some of this correlated activity depends on excitable properties of neurons that emerge prior to the influence of sensory-evoked, experience-dependent electrical signals, while other correlated activity is established by patterns of excitable changes elicited by sensory inputs or motor behaviors. The cellular and molecular mechanisms implicated in critical periods rely on the activity of several neurotransmitters, receptors, and intracellular signaling cascades that modify cytoskeletal integrity, receptor function and stability, and ultimately gene expression in response to changes in synaptic activity in a target cell. Genes encoding neurotrophins such as BDNF, extracellular matrix components, and neurotransmitter receptors are all targets for altered expression in response to synaptic activity during critical periods. Signaling via BDNF from the pre- to postsynaptic sites is especially important for the modification of gene expression that records activity-dependent change more permanently as transcriptional change. The most accessible and thoroughly studied example of a critical period is that responsible for the establishment of normal vision in mammals, including humans. When typical patterns of activity are disturbed during the critical period in early life (experimentally in animals or by pathology in humans), connectivity in the visual cortex is altered, as is visual function. If not reversed before the end of the critical period, these structural and functional alterations of brain circuitry are difficult or impossible to change. Observations of the addition and elimination of synapses throughout the cerebral cortex in animals, and parallel analysis of the increase and decrease of cortical gray matter volumes where such synapses are made in the brains of children and adolescents, indicate that a full range of human behaviors—including those compromised in conditions such as autism, schizophrenia, and ADHD—may be shaped by activity- and experience-dependent addition and subsequent selective elimination of synaptic connections during critical periods that begin at birth and end in early adulthood.

■ Additional Reading

Reviews

Giedd, J. N. and J. L. Rapoport (2010) Structural MRI of pediatric brain development: What have we learned and where are we going? *Neuron* 67: 728–734.

Hensch, T. K. (2004) Critical period regulation. *Annu. Rev. Neurosci.* 27: 549–579.

Katz, L. C. and C. J. Shatz (1996) Synaptic activity and the construction of cortical circuits. *Science* 274: 1133–1138.

Wiesel, T. N. (1982) Postnatal development of the visual cortex and the influence of environment. *Nature* 299: 583–591.

Wong, W. O. and A. Ghosh (2002) Activity-dependent regulation of dendritic growth and patterning. *Nat. Rev. Neurosci.* 10: 803–812.

Yap, E. L. and M. E. Greenberg (2018) Activity-regulated transcription: Bridging the gap between neural activity and behavior. *Neuron* 100: 330–348.

Important Original Papers

Antonini, A. and M. P. Stryker (1993) Rapid remodeling of axonal arbors in the visual cortex. *Science* 260: 1819–1821.

Cabelli, R. J., A. Hohn and C. J. Shatz (1995) Inhibition of ocular dominance column formation by infusion of NT-4/5 or BDNF. *Science* 267: 1662–1666.

Cases, O. and 5 others (1996) Lack of barrels in the somatosensory cortex of monoamine oxidase-deficient mice: Role of serotonin excess during the critical period. *Neuron* 16: 297–307.

Feller, M. B. and 4 others (1996) Requirement for cholinergic synaptic transmission in the propagation of spontaneous retinal waves. *Science* 272: 1182–1187.

Gogtay, N. and 11 others (2004) Dynamic mapping of human cortical development during childhood through early adulthood. *Proc. Natl. Acad. Sci. U.S.A.* 101: 8174–8179.

Hanganu, I. L., Y. Ben-Ari and R. Khazipov (2006) Retinal waves trigger spindle bursts in the neonatal rat visual cortex. *J. Neurosci.* 26: 6728–6736.

Horton, J. C. and D. R. Hocking (1999) An adult-like pattern of ocular dominance columns in striate cortex of newborn monkeys prior to visual experience. *J. Neurosci.* 16: 1791–1807.

Huang, Z. J. and 7 others (1999) BDNF regulates the maturation of inhibition and the critical period of plasticity in mouse visual cortex. *Cell* 98: 739–755.

Hubel, D. H. and T. N. Wiesel (1965) Binocular interaction in striate cortex of kittens reared with artificial squint. *J. Neurophysiol.* 28: 1041–1059.

Hubel, D. H. and T. N. Wiesel (1970) The period of susceptibility to the physiological effects of unilateral eye closure in kittens. *J. Physiol.* 206: 419–436.

Hubel, D. H., T. N. Wiesel and S. LeVay (1977) Plasticity of ocular dominance columns in monkey striate cortex. *Philos. Trans. R. Soc. Lond., B, Biol. Sci.* 278: 377–409.

Kuhl, P. K. and 4 others (1992) Linguistic experience alters phonetic perception in infants by 6 months of age. *Science* 255: 606–608.

LeVay, S., T. N. Wiesel and D. H. Hubel (1980) The development of ocular dominance columns in normal and visually deprived monkeys. *J. Comp. Neurol.* 191: 1–51.

Rakic, P. (1977) Prenatal development of the visual system in the rhesus monkey. *Philos. Trans. R. Soc. Lond., B, Biol. Sci.* 278: 245–260.

Rakic, P. and 4 others (1986) Concurrent overproduction of synapses in diverse regions of the primate cerebral cortex. *Science* 232: 232–235.

Wang, B.-S., R. Sarnaik and J. Cang (2010) Critical period plasticity matches binocular orientation preference in the visual cortex. *Neuron* 65: 246–256.

Wiesel, T. N. and D. H. Hubel (1965) Comparison of the effects of unilateral and bilateral eye closure on cortical unit responses in kittens. *J. Neurophysiol.* 28: 1029–1040.

Books

Curtiss, S. (1977) *Genie: A Psycholinguistic Study of a Modern-Day "Wild Child."* New York: Academic Press.

Hubel, D. H. (1988) *Eye, Brain, and Vision.* Scientific American Library Series. New York: W. H. Freeman.

Purves, D. (1994) *Neural Activity and the Growth of the Brain.* Cambridge, UK: Cambridge University Press.

Sex Differences and Neural Circuit Development

Overview

Several essential aspects of neuronal differentiation and circuit development rely on the activity of factors secreted systemically by the embryo or fetus itself. These factors elicit cellular changes and drive differentiation of specific brain regions and connections. The influences of these secreted factors, often from the fetus' other organs and delivered to the brain via fetal circulation, are key for establishing circuits that mediate some of the most fundamental and adaptive behaviors of any organism. The best example of systemic, secreted factors that shape neural circuit development are the steroid hormones produced by the gonadal tissues of an embryo or fetus of either sex. These factors are essential for establishing different (dimorphic) structures and circuits in the brains of females and males in both invertebrates and vertebrates, including mammals. Once the developing ovaries or testes have differentiated, the onset of secretion of gonadal steroids—primarily estrogen and testosterone—followed by changes in the level and temporal patterns influence multiple programs of brain differentiation. The best-established targets of gonadal steroid systemic regulation are circuits essential for fundamental reproductive and parenting behaviors. The sex-specific patterns of circulating, secreted signals are transduced via multiple steroid receptors. The receptors are selectively expressed in neural precursors or brain regions that then go on to develop differently in females versus males. These regions and their neuronal connections become specialized in register with different peripheral structures, particularly genitalia and mammary tissue. In addition, intrinsic signaling via gonadal steroids can mediate differentiation of sensory and motor pathways that serve male versus female behaviors for communication, mate identification or selection, and interactions with offspring. Thus, secreted, circulating signals can affect the development and maintenance of brain circuits that then influence integration of neural commands for key homeostatic behaviors.

Courtesy of B. Karpinksi and A.-S. LaMantia

KEY CONCEPTS

- **25.1** Systemic, secreted signals influence neural circuit development and maintenance

- **25.2** Sexual dimorphisms reflect systemic signaling in peripheral organs and related neural circuits

- **25.3** Systemic signals target neurons and circuits for reproductive and parenting behaviors

- **25.4** Complex human behaviors are difficult to associate with sex, gender, or early systemic signaling

CONCEPT
25.1

Systemic, Secreted Signals Influence Neural Circuit Development and Maintenance

LEARNING OBJECTIVES

- 25.1.1 Identify the sources of systemic signals that influence brain development.
- 25.1.2 Define the terms *chromosomal* and *phenotypic* sex, and discuss their roles in brain development.
- 25.1.3 Describe steroid hormone synthesis and receptor activity in brain development.

Organ-produced signals for the developing brain

The brain does not develop in a vacuum—its differentiation is accompanied by the development of the full set of tissues and organs that constitute a new individual. As these additional organs develop, some of them secrete signals, either locally or into the developing fetal circulation, that act on developing neurons and glia to influence neurogenesis, brain growth, and neural circuit differentiation. Some of these signals, including cortisol and estrogen and testosterone, are enzymatically synthesized hormones. At least one, serotonin, will also be synthesized by neurons as well, and used as a neurotransmitter as the brain matures. Other systemic signals are peptide hormones generated by transcription, translation, and sometimes proteolytic cleavage of genomically encoded loci. The intrinsic sources of these signals in the embryo include most developing organs: the liver, the vasculature, the adrenal glands, the pituitary gland, the thyroid gland, immature immune cells, and even the skin. The signals these organs secrete include corticotropin-releasing hormone (CRH), adrenocorticotropic hormone (ACTH), vascular endothelial growth factor (VEGF), and the insulin-like growth factors (IGFs). Additional signals include other members of the steroid–thyroid superfamily of hormones (see later in this concept) include cortisol, thyroid hormone, and retinoic acid (see Chapter 22). One particular class of systemic signals, however, estrogen and androgen produced by the developing gonads, will be the focus of much of this chapter.

Systemic signals that act on the developing mammalian embryonic or fetal brain in utero also come from maternal sources during the course of gestation. The placenta is a key source of these signals, both via its role in controlling the fetal–maternal vascular interface, and because the placenta itself secretes several signals that influence fetal brain development. These include serotonin, CRH, allopregnanolone, and IGFs. In addition, circulating signals from the maternal immune system—including several cytokines as well as hormones secreted by the maternal thyroid, pituitary, and adrenal glands—can pass through the placenta to influence several stages of prenatal brain development. The full consequences of these influences, and the specific neurons and circuits that are the most sensitive targets for many of these signals, remain uncertain. Placental-derived signals can influence overall brain growth as well as formation of circuits in the cerebral cortex. It is possible that disruption of this signaling can increase the risk for several neurodevelopmental disorders in a fetus if the integrity of its mother's placental secretions or function has been compromised. Thus, there can be little doubt that secreted, circulating signaling molecules from a variety of organs—both in the fetus and (in mammals) the mother—provide instructions for brain development. These signals ensure that an individual's brain will reflect the growth and differentiation of the body it will control as well as the maternal environment in which it developed.

Sex, gonads, bodies, and brains

Systemic signals produced by the nascent gonads are available to developing vertebrate embryos as soon as rudiments of these organs appear during embryogenesis and then differentiate during fetal development. Thus, understanding the divergent specification of female versus male gonadal tissues and their capacity to secrete gonadal steroid hormones provides perhaps the most informative example of how organogenesis beyond the brain influences brain development. To understand gonadal differentiation in the two sexes, it is necessary to define the genomic foundations of sex differences. *Chromosomal sex* is a biological term that refers specifically to an individual's sex chromosomes: the X and Y chromosomes in most animals, where several key genes for gonadal specification are found. Most species have two types of chromosomes: **autosomes**, which are identical in the two sexes; and **sex chromosomes**, the number or identity of which determine **chromosomal sex** (also referred to as *genotypic* sex). In some species, males have three copies of the sex chromosomes, whereas females have only two. In others, including humans, there are different chromosomal identities; most commonly, a male-specific chromosome is present or absent: Thus, typical females are XX and males are XY.

Not surprisingly, the genes critical for the development of **primary sex characteristics**—the gonadal tissues that support either male or female gametes in humans and most other mammals—are found on the sex chromosomes X and Y. The physical state of the gonads and external genitalia is the primary determinant of **phenotypic sex**—the fundamental physical attributes that define sex. A range of **secondary sex characteristics** further defines an individual's phenotypic sex; these include mammary glands in females, sex-specific hair patterns in males and females, musculoskeletal distinctions and organismal size differences. These phenotypic characteristics are more or less related to distinct reproductive and parenting functions in females and males and are regulated primarily by hormonal secretions from the gonads.

In humans, the letters X and Y identify the sex chromosomes, in contrast to the 22 pairs of autosomes, which are identified by numbers. With few exceptions (see the discussion of intersex conditions in Concept 25.4), individuals with two X chromosomes are biologically female, while those with one X and one Y chromosome are biologically male (Figure 25.1A). At this time, the biological sex of the embryo—its constituent cells and how they are organized— reflects the gonadal tissue that differentiates, and the subsequent anatomical structures that emerge based on divergent functional capacities of the developing gonad tissues in each sex. The correlations between genotype (i.e., chromosomal sex) and phenotype (i.e., phenotypic sex) for primary sex characteristics are best

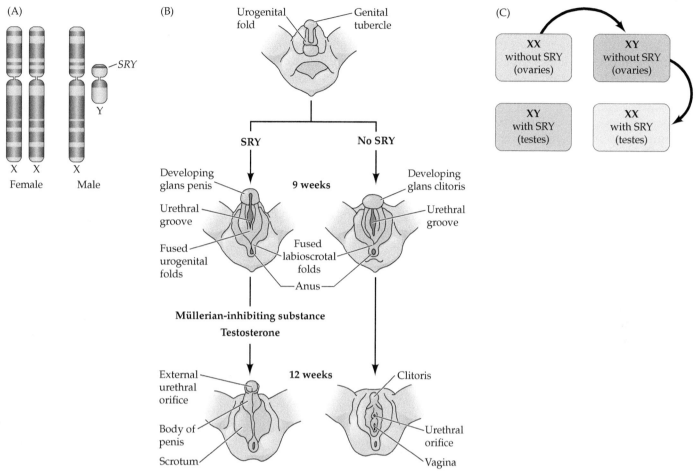

FIGURE 25.1   Chromosomal sex and primary sex determination in humans (A) The *SRY* gene, located on the short arm of the Y chromosome, initiates a cascade of gene expression and hormonal signaling that results in masculinization of human genitalia. (B) The genitalia of the early human embryo (weeks 4–7 of gestation) are sexually indifferent. Under the influence of *SRY* gene product, testes develop and produce hormones that result in male genitalia (left diagrams) between weeks 9 and 12 of gestation. Without the influence of the Y chromosome and its *SRY* gene, the human gonads become ovaries, whose hormonal cascade results in female external genitalia (right). (C) The consequences of translocation (in humans) or transgenic mis-expression (mice) of SRY activity in XY versus XX genotypes. (B, after K. L. Moore 1977. *The Developing Human*, 2nd Ed. Philadelphia: W. B. Saunders, p. 241; C after M. M. McCarthy et al., 2012, *J Neurosci* 32: 2241–2247.)

understood in mammals, including humans, for the male genotype, XY. The defining gene for the development of male-specific genotypic and phenotypic sex is found on the Y chromosome and is a single gene for a transcription factor known as **testis-determining factor (TDF)** or **SRY (sex-reversal gene on the Y chromosome)**. Remarkably, a translocation of the *SRY* gene from the Y chromosome to the X chromosome in a male can result in the complete masculinization of that male's XX offspring.

In most instances, the XY genotype leads to a male phenotype (hence, biological phenotypic sex is usually linked to genotypic sex) with testicles, epididymis, vas deferens, seminal vesicles, penis, and scrotum; the XX genotype results in the development of ovaries, oviducts, uterus, cervix, clitoris, labia, vagina, and mammary glands that define female phenotypic sex (Figure 25.1B). Rare apparently male individuals (and mutant mice, in which the *Sry*

gene was first cloned and sequenced) have two X chromosomes, but one of these carries the translocated *SRY* gene (transferred during spermatogenesis to the paternal X chromosome from the paternal Y chromosome during meiotic recombination). Despite the overwhelming majority of dual copies of X-chromosome genes in these individuals (and the corresponding lack of any additional Y genes), *SRY* alone is sufficient to completely masculinize them, and they become phenotypic males despite their otherwise genotypic female chromosomal identity (Figure 25.1C). In fact, whether *SRY* is translocated as an individual gene anywhere in the genome (autosomal chromosomes or sex chromosomes, inserted experimentally in transgenic mice) or is transferred to an XX individual along with the entire Y chromosome (resulting in an aneuploid XXY phenotype known as Klinefelter's syndrome; see Concept 25.4), the result is the same—a phenotypic male.

For much of recorded history, romantic love has been the province of poets, painters, and musicians. Scientists, it seemed, had little to add—and perhaps it was feared that scientists might diminish the charm of this intoxicating human experience. Nevertheless, over the past decade neuroscientists have added their interpretation to the countless couplets, cupids, and cantatas that celebrate coupling. The results seem to confirm the notion of songwriters and rock stars that love is a drug.

Beyond inspiring poetry and lyrics, the biological purpose of romantic love seems to be to reinforce mate selection and bonding for maximum parental effectiveness. Thus, the brain systems that are the most engaged by "being in love" are also associated with reward and reinforcement—the very same systems that are activated, often with disastrous consequences, by alcohol and drugs. In addition, in humans the state of romantic love (and also maternal love) relies on the activity of two peptide hormone neurotransmitters, oxytocin and vasopressin. These peptides are associated with social recognition and maternal–offspring bonding in diverse species, regardless of whether monogamy and romantic love (or its biological counterpart, pair bonding) are part of the behavioral repertoire. Finally, when a person is in love, activity in the brain regions that normally regulate social interactions—especially those in the cerebral cortex and basal forebrain that enhance social vigilance and caution—are diminished. Thus, love may actually be "blind" to the behavioral cues that in other social encounters inspire caution.

Monogamy and pair bonding were the first behaviors related to romantic love to be understood biologically. These two behaviors emerge in only a few mammalian species, including the prairie vole (*Microtus ochrogaster*). Prairie voles bond with a single mate and remain monogamous for life. In contrast, individuals of the closely related species montane vole (*Microtus montanus*) are promiscuous and do not form lifelong mating pairs. It turns out that these differences in mate selection and preference are matched by differences in the distribution of oxytocin and vasopressin receptors in the nucleus accumbens, caudate putamen, and ventral pallidum (Figures A and B)—all regions that are associated with reward, reinforcement, and addictive behaviors (see Chapter 31). These regions are also sites of enriched dopaminergic neurotransmission, which is known to contribute to reinforcement, including the maladaptive reinforcement seen in drug-taking behavior. Perhaps not surprisingly, vasopressin and dopamine both influence monogamous mate selection in prairie voles. Thus, when vasopressin antagonists are injected into the ventral pallidum of male prairie voles, monogamy is disrupted, and when dopamine antagonists are injected into the nucleus accumbens of female prairie voles, single-partner preference induced by mating and pair bonding is no longer seen (Figure C).

(A) Oxytocin receptors

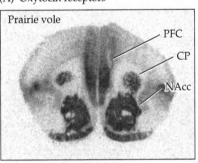

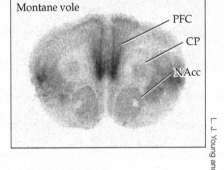

(B) Vasopressin receptors

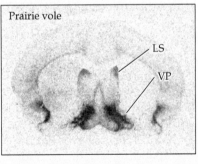

L. J. Young and Z. Wang, 2004, *Nat Neurosci* 7: 1048–1054

(A) Prairie voles (left), which form lifelong mating pair bonds, have a high density of oxytocin receptors in the nucleus accumbens (NAcc), caudate putamen (CP), and prefrontal cortex (PFC), all associated with reward and reinforcement. Promiscuous montane voles (right) lack a high density of receptors in the NAcc and CP. (B) Male prairie voles (left) have divergent densities of vasopressin receptors in their ventral pallidum (VP, part of the striatum) and lateral septal nuclei (LS). In the montane vole (right), there is a high density of receptors in the LS, but the density is diminished in the VP.

Other rare individuals carry a deletion or loss-of-function mutation of *SRY* on an otherwise intact Y chromosome and are phenotypic females. *SRY* is thus believed to be at the apex of a genetic network that mediates the differentiation of male primary and secondary sex characteristics (see Figure 25.1B,C). These genetic details of gonadal and secondary sex characteristic specification are crucial for brain development because, as described in detail in Concept 25.2, the organization of female versus male brains depends in large measure on the initial divergent differentiation of gonadal tissue. Surprisingly, however, *SRY* is not expressed in the brain. In most brains, sexual dimorphisms arise secondarily, in response to primary peripheral distinctions between males and females and the differences in signals available to the developing brain in each sex (Box 25A).

(C)

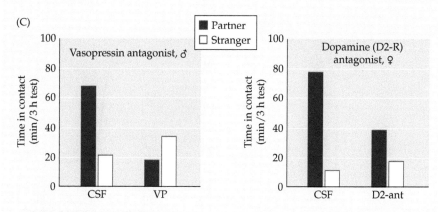

(C) Typical male prairie voles (left) presented with their pair-bonded partner versus a stranger prefer to mate with the partner, based on the amount of time spent in contact. If a vasopressin antagonist is infused into the ventral pallidum versus a cerebrospinal fluid (CSF) control, this preference diminishes substantially. Typical female prairie voles (right) also prefer to mate with their pair-bonded partner; however, if a dopamine antagonist is infused into the NAcc, this preference is decreased. (C after L. J. Young and Z. Wang, 2004. *Nat Neurosci* 7: 1048–1054.)

The contributions of the nucleus accumbens and of the ventral pallidum and its dopaminergic innervation from the ventral tegmental area (VTA) have led to a hypothesis for studying human romantic love: Individuals newly in love should have elevated brain activity in all of these regions in response to considering or seeing the object of their affections versus neutral stimuli. Indeed, when males and females monitored using fMRI were presented with images of their beloved versus images of friends or acquaintances, maximum brain activity was observed in the caudate putamen and VTA (Figure D). Moreover, since "love is blind," it was reasoned that the regions of the brain essential for social vigilance and caution, including several cortical regions and the amygdala, should be diminished in activation, which was also the case (Figure E). These observations placed love somewhat outside the brain systems that mediate emotion—particularly limbic regions (see

Chapter 31). Instead, brain regions essential for reward, risk taking, and social cognition seem to be featured.

Subsequent studies suggest that the ups and downs of love recruit additional brain circuits. Individuals recently rejected by a love interest retain maximum activation in the VTA and caudate nucleus in response to viewing an image of the lost love, but add several regions, including cortical regions also associated with motivation, calculation of gain and loss (such as in gambling or risk taking), emotional regulation, and even drug craving, that are not activated in the "in love" state. Conversely, when men and women who have been married an average of more than 20 years are presented with images of their spouse versus other familiar individuals, regions associated with attachment—particularly those activated by mother–child bonding—are differentially activated.

While one is unlikely to see Valentine cards reading "You've carried

away my VTA" any time soon, these results drain neither the passion nor mystery from love. Indeed, while we can identify the neural consequences of attraction and romance, the reasons why the beloved elicits these responses in the first place remain unknown. Nevertheless, these observations point to our subjective experience of romantic love as a key aspect of the overall biology of sex, reproduction, and parenting. The human experience of romantic love activates regions of the brain that favor reinforcement of a connection with a sexual partner, promote exclusivity in mating, and enhance cooperation in the rearing of offspring. These desires—passion, constancy, and domesticity—offer depth of feeling, whether expressed by poets or glimpsed by fMRI. Thus, love may be like a drug in the best sense, in that it can be highly therapeutic to those who find it.

(D)

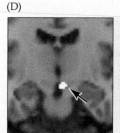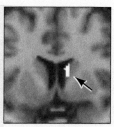

(E)

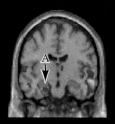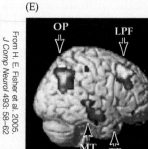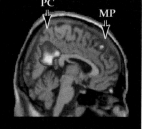

From H. E. Fisher et al. 2005. J Comp Neurol 493: 58–62

From A. Bartels and S. Zeki 2004. Neuroreport 11: 3829–3834

(D) Maximum focal activation in response to seeing a picture of one's beloved versus that of a friend or acquaintance is detected in the ventral tegmental area (left, arrow) and the caudate nucleus (right, arrow) in both young men and women in the early to middle stages of self-identified love affairs. (E) Brain regions that are deactivated upon seeing a picture of one's beloved versus that of a friend or acquaintance. Cortical areas in the occipital–parietal junction (OP), medial temporal (MT), temporal pole (TP), lateral prefrontal (LPF), posterior cingulate (PC), medial prefrontal/paracingulate (MP), and the amygdala (A) are all sites of diminished activation. These areas are also deactivated when mothers view pictures of their own children.

Gonadal steroid synthesis and signaling

The differentiation of male versus female gonadal tissue specified by chromosomal sex sets in motion a series of events that define major phenotypic dimorphisms both for peripheral sexually dimorphic structures, particularly gonads and genitalia, and in the brain. The steroid hormones **testosterone** and **estrogen** (Figure 25.2A), secreted by the developing testes and ovaries influence most aspects of sexual dimorphism. Steroids are a distinct class of circulating signaling molecules. In contrast to protein and peptide signals (see Chapter 22), steroids are not directly transcribed from genomically translated messenger RNAs. Instead, a metabolic precursor, cholesterol, is enzymatically modified to produce steroid variants with different activities and receptor specificities. Thus, the integrity of the genes that encode these gonadal steroid-generating enzymes is key for typical systemic signaling that influences sexually dimorphic development. Mutations in these genes can result in divergent developmental outcomes (see later in this concept).

The initial differentiation of male and female gonads is the central event for the translation of chromosomal sex into phenotypic sex, because the expression of the enzymes necessary to produce gonadal steroids is enhanced in ovaries and testes. Although the ovaries (XX) and testes (XY, dependent on a functional *SRY* gene) secrete both estrogens and androgens, the two gonadal tissue types secrete very different levels of these hormones at distinct developmental times. At critical junctures, differing hormone levels influence the undifferentiated primordial gonadal structures, resulting in divergent developmental paths of gonads themselves, the genitalia, and later, secondary sex characteristics of males and females. This early hormonal influence on the development of sexually dimorphic structures is sometimes referred to as the *organizational effects* of the gonadal steroids, reflecting their actions in guiding distinct male versus female differentiation in a variety of tissues, including the brain.

Genotypic males experience an early surge of testosterone, which, along with the peptide hormone Müllerian-inhibiting substance (MIS), masculinizes the genitalia (Figure 25.2B). Paradoxically, many of the effects of

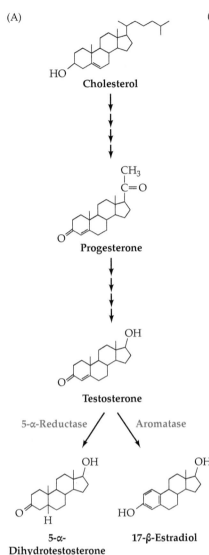

(A)

Cholesterol

Progesterone

Testosterone

5-α-Reductase   Aromatase

5-α-
Dihydrotestosterone   17-β-Estradiol

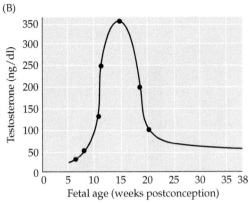

(B)

FIGURE 25.2 Gonadal steroids and their systemic availability during fetal development (A) All gonadal (sex) steroids are synthesized from cholesterol, which is converted to progesterone, the common precursor, by four enzymatic reactions (represented by four arrows). Progesterone can then be converted into testosterone via another series of enzymatic reactions; testosterone in turn is converted to 5-α-dihydrotestosterone (the active form of testosterone that binds best to testosterone receptors) via 5-α-reductase, or to 17-β-estradiol via an aromatase. 17-β-estradiol is the preferred ligand for the estrogen receptor subtypes (α, β, and X) and mediates most of the known gonadal steroid effects in the brains of both female and male rodents. (B) In human male fetuses, masculinization of the genitalia (and probably the brain) reflects increased secretion of testosterone by the immature testes between weeks 7 and 20 of gestation. Females do not have a parallel surge of gonadal steroids. (B after M. L. Gustafson and P. K. Donahoe, 1994. *Annu Rev Med* 45: 505–524.)

testosterone in the male brain are due to the conversion of testosterone to estrogen in the brain during mid-gestation (the second trimester of pregnancy in humans). Vertebrate neurons, especially those in brain regions that will develop divergently in males and females, express an enzyme called **aromatase** that converts testosterone to **estradiol**, an "active" form of estrogen that binds with high affinity to the estrogen receptor (see Figure 25.2A). Thus, a surge of testosterone secreted by the developing testes in a male fetus generates high levels of circulating testosterone that drives differentiation of the male genitalia peripherally. In parallel, this circulating testosterone surge also leads to increased estradiol-mediated signaling in neurons that express aromatase. Thus, estrogen and estradiol via local estrogen receptors in the developing male brain are assumed to mediate subsequent male sexual dimorphisms in the nervous system that reflect increased testosterone. There are, however, instances where testosterone acts directly (i.e., through its own receptors) on developing as well as mature neurons.

Estrogen, testosterone, and other steroids are highly lipophilic molecules and are usually transported via carrier proteins circulating in the blood. Thus, the fetuses of placental mammals are also exposed to estrogens generated by the maternal ovaries and placenta delivered via maternal circulation. This maternal estrogen, however, does not interfere with differentiation of phenotypic sex characteristics in the offspring. Developing fetuses have high levels of **α-fetoprotein**, a protein that binds circulating estrogens, including those from the maternal blood supply. In both sexes, the brain is apparently protected from early exposure because estrogens secreted by the maternal gonads into the bloodstream are bound by α-fetoprotein. Testosterone does not bind to α-fetoprotein, however, and is enzymatically converted to estradiol in neurons that express aromatase; therefore, the developing male brain especially is exposed to an early dose of masculinizing steroids—including 17-β-estradiol generated from testosterone by aromatase—from the testosterone surge generated by the embryonic testes (see Figure 25.2B). The consequences of this difference in steroid availability due to binding proteins, subsequent steroid signaling, and dimorphic brain development can be seen in mice in which the α-fetoprotein gene has been knocked out. Female mice that lack α-fetoprotein are infertile because of a failure of differentiation of the hypothalamic circuits that control ovulation, presumably the result of masculinizing overexposure to estrogen during embryonic development. Surprisingly, there is no information available on ovulation or fertility for the rare cases of human females that lack the gene for α-fetoprotein. Male mice and men lacking this gene are fully fertile.

Gonadal steroids act by binding to specific receptors for either testosterone or estrogen (Figure 25.3A). These receptors are distributed in fairly limited populations of cells in the mammalian brain and are particularly concentrated at sites where

reproductive and parenting functions are centrally represented and sexual dimorphisms are seen (Figure 25.3B). Their restricted pattern of expression is mostly independent of gonadal influence. Thus, the influence of gonadal steroids in the brains of developing females and males is sometimes referred to as being *activational* as well

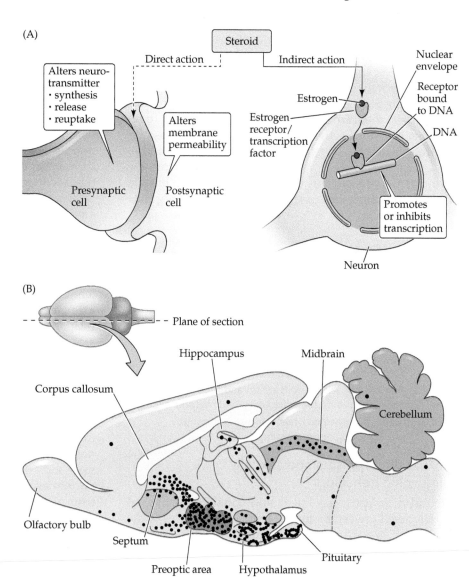

FIGURE 25.3 Effects of gonadal steroids on neurons (A) The left panel of this schematic lists several direct effects of steroid hormones on the pre- or postsynaptic membrane. These effects, which probably reflect the capacity of lipophilic gonadal steroids to diffuse through neuronal membranes and change their properties, can alter neurotransmitter release and influence neurotransmitter receptors. On the right are shown some indirect effects of these hormones, which bind to steroid receptor-transcription factors that act in the nucleus to influence gene expression. (B) Distribution of estradiol-sensitive neurons in a sagittal section of the rat brain. Animals were given radioactively labeled estradiol to identify sites where the ligand binds to presumed receptors; dots represent regions where the label accumulated. In the rat, most estradiol-sensitive neurons are located in the basal diencephalon and telencephalon, with a high concentration in the preoptic area, hypothalamus, and more laterally in the amygdala, which is not shown in this midsagittal section. (A after B. S. McEwen et al., 1979. *Annu Rev Neurosci* 2: 65–112; B after B. S. McEwen, 1976. *Sci Am* 235: 48–58.)

as organizational, since the steroid signals are eliciting responses in a subpopulation of previously sensitized neurons. Gonadal steroid receptors belong to a larger family of proteins called the **steroid–thyroid nuclear receptors**, which includes the receptors for vitamin A (retinoic acid; see Chapter 22), vitamin D, and glucocorticoids. Unlike most receptors for peptide hormones (or neurotransmitters), which are on the cell surface, most steroid–thyroid receptors are found either in the cytoplasm or in the nucleus (thus, they are sometimes called nuclear receptors). Their ligands, testosterone and estrogen in this case, must diffuse through the membrane to reach their receptors. Once inside the cell, steroid ligands often bind to proteins that protect them from degradation before binding to receptors. When the active forms of testosterone (as 5-α-dihydrotestosterone) or estrogen (as 17-β-estradiol) bind to their respective receptors, the receptors move from the cytoplasm to the nucleus, where they are able to bind to DNA recognition sites (response elements) and regulate gene expression.

CONCEPT
25.2

Sexual Dimorphisms Reflect Systemic Signaling in Peripheral Organs and Related Neural Circuits

LEARNING OBJECTIVES

25.2.1 Discuss relationships between dimorphic periphery and neural circuits in females and males.

25.2.2 Analyze gonadal steroid action and broader developmental mechanisms for sexually dimorphic circuits.

25.2.3 Identify dimorphic brain regions that control reproduction and parenting.

Sexual dimorphisms and sexually dimorphic behaviors

Sexual dimorphisms—clear and consistent physical differences between biological females and males of the same species—are seen throughout the animal kingdom and are usually associated with functional differences that promote reproduction or rearing of offspring. These physical differences are often recognized in peripheral structures found in one sex but not the other. Usually, the differences in female and male bodies are matched by different neural circuits in female and male brains. For example, in the tobacco hawk moth (*Manduca sexta*), the female and male antennae differ strikingly in size and structure: Female antennae are smaller and smooth, while male antennae are larger and lined with rows of ciliated structures (Figure 25.4A). These anatomical specializations are essential for female versus male reproductive behaviors; female antennae sense specific volatile odorants from the tobacco plants that are optimal egg-laying sites, while male antennae can detect extremely low concentrations of an airborne pheromone (see Chapter 14) secreted by a nearby female that identifies her as a potential mate.

This physical and related behavioral dimorphism is matched by dimorphic circuitry in the brain. Male and female antennae have olfactory receptor neurons whose axons project to glomerular structures in the antennal lobe (the equivalent of the vertebrate olfactory bulb; see Chapter 14), and the organization of these glomeruli in males versus females corresponds to the peripheral dimorphism (Figure 25.4B). The larger, more complex male antennae innervate a large accumulation of glomeruli in the male moth's antennal lobe called the macroglomerular complex, which has neural circuitry specialized for responses to female pheromone. Similar glomeruli in females are smaller, in register with the diminished size of female antennae, and their constituent cells respond more robustly to molecules released by tobacco plants. The development of these glomeruli depends on the antennae that project to them. If a male antenna is transplanted onto the head of a female at a late larval stage, a male macroglomerular complex develops in the female brain. The relationship of the dimorphic periphery, established based on the size and presumably number of sensory neurons innervating the male versus female antennal lobe, and the corresponding distinction in antennal lobe glomeruli, is reminiscent of the general quantitative relationships between afferents and targets that are mediated by trophic interactions during development (see the next section and Chapter 23).

Among the most thoroughly characterized sexual dimorphisms of body, behavior, and brain in non-mammalian vertebrates are those of several songbird species, including the zebra finch (*Taeniopygia guttata*) (Figure 25.5A,B), in which males produce complex songs and females do not. A male zebra finch's song is critical for attracting a mate and establishing territorial dominion. It is also an important aspect of parenting: Male hatchlings learn to sing from adult male "tutors" whose vocalizations they hear and mimic, thus determining their reproductive success. The peripheral pharyngeal structure that produces songs (the *syrinx*, a series of membranes around the trachea that vibrate based on air flow in the trachea and contraction of throat muscles) is larger and more differentiated in male birds than in females, an adaptation that accommodates the male's song production. Accordingly, in the songbird brain, the nuclei that control motor as well as sensory aspects of song production via the syrinx are larger in males. In both the body (in this instance, the syrinx) and the brain, many of these structural dimorphisms are under the control of circulating steroid hormones produced in differing quantities by the male and female gonads, particularly during early posthatching development. Thus, female zebra finches treated with male-specific levels of gonadal steroids during development acquire a highly differentiated, hypertrophic syrinx. In these experiments, females are

FIGURE 25.4 Sexual dimorphisms in the periphery and brain of the tobacco hawk moth (A) The antennae of male and female *Manduca sexta* are specialized for their different roles in courtship and mating behavior. The male antennae are larger and ribbed, maximizing receptor surface area for detection of low concentrations of female pheromone released into the air. (B) The physical dimorphism of the moths' antennae is matched by dimorphism in the olfactory glomeruli of their brain's antennal lobes, which are specialized for odorant-mediated, sex-specific behaviors. The male-specific macroglomerular complex is essential for processing the female pheromone.

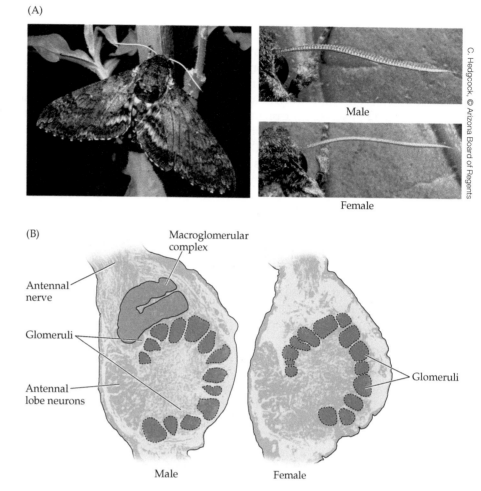

exposed to estradiol, even though in a male, the estradiol that masculinizes the hatchling is derived from endogenous gonadal-secreted testosterone via aromatase in sensitive neurons (see Concept 25.1). The song-control nuclei in the female zebra finch brain also become masculinized (Figure 25.5C), and these female birds—masculinized in peripheral sexually dimorphic structures (the syrinx) as well as in the brain—sing like their male counterparts.

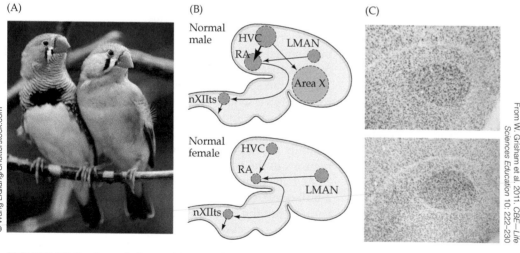

FIGURE 25.5 Sexual dimorphism in the zebra finch (A) The male zebra finch (left) is larger than the female (right) and has different feather patterns: an orange cheek patch and white-spotted chestnut flanks (both of which are gray in the female) and black upper chest feathers and white abdominal feathers (gray and tan, respectively, in the female). (B) Forebrain regions that control song in the zebra finch vary between sexes. The key areas shown here are both pallial (forebrain) structures: area X, which is similar to the mammalian medial striatum (but not exactly parallel), and the *hyperstriatum ventralis, pars caudalis* or more colloquially, the *higher vocal center* (both are abbreviated HVC), which is also a striatal motor control nucleus. In a male, area X can be easily recognized and the HVC is quite prominent. In the female, area X is absent and the HVC is substantially smaller than in the male. (C) The HVC in a female (bottom panel) can be masculinized by early posthatching exposure to estradiol (in males, estradiol is normally derived from testosterone and then aromatized to estradiol in the brain; see Figure 25.2) and results in an HVC that is similar in size to that of a male control (top panel). (B after A. P. Arnold, 1980. *Am Sci* 68: 165–173.)

The relationships between brain and peripheral sexual dimorphism seen in tobacco hawk moths and zebra finches illustrate a fundamental concept in the development of sex differences: Dimorphisms in body structure that are essential for reproductive and parenting behaviors are matched by sex-specific prenatal or early postnatal differentiation of distinct brain structures and circuits that ultimately control those peripheral specializations and behaviors. For the nervous systems of moths, birds, and several other species, including humans, these sex differences reflect differential growth of neurons and glia, often facilitated by distinct patterns of gonadal steroids acting on the dimorphic periphery, that leads to divergent organization of brain structures that are nevertheless present in both sexes. The divergent size and organization of these brain structures reflect both target size–based trophic interactions (see the next section) and the reinforcing actions of ongoing signaling via systemically available gonadal steroids on neurons in the dimorphic circuits.

Development of mammalian neural circuit dimorphisms

Testosterone and estrogen each have specific cellular targets in distinct tissues, both in the body and in the brain. The peripheral targets are often parts of the body directly involved in reproduction and parenting: the genitalia in both sexes and mammary glands in females. Many of the brain targets are neural structures that control the external genitalia, gonads, and other dimorphic structures (such as mammary glands) and mediate sex-specific behaviors. Most of the key brain regions are nuclei in the spinal cord and hypothalamus (see the next section). A well-studied example of CNS dimorphism related to motor control of sex-specific reproductive behavior is the difference in size of the **spinal nucleus of the bulbocavernosus (SNB)** primarily located in the fifth lumbar segment of the rodent spinal cord, which innervates muscles of the genitalia. The motor neurons of the SNB innervate two striated muscles of the male perineum (the bulbocavernosus and levator ani) that attach at the base of the penis and are involved in both penile erection and urination (Figure 25.6A). In female rats, the bulbocavernosus muscle is absent and the levator ani is dramatically reduced in size. The SNB, though present in both sexes, is significantly larger in males: its individual neurons are larger and there are more of them. Thus, the spinal cords of male and female rodents are sexually dimorphic in parallel with their genitalia.

The SNB dimorphism in rodents is established via the influence of testosterone around the time of birth; however, in this case the primary target of gonadal steroid hormone action is not in the nervous system, but the perineal muscles that control the genitalia. At birth, undeveloped bulbocavernosus muscles of similar size are present in male and female rodents (Figure 25.6B), and there are equivalent numbers of neurons in the SNB. The muscles in both males and females have testosterone receptors (as do all muscles, much to the chagrin, in the case of humans, of the governing bodies of major sports organizations), but only males have sufficient endogenous levels of testosterone to activate these receptors. The activation of testosterone receptors in the male perineal muscles spares these muscles from the apoptotic cell death for which the female bulbocavernosus muscle is destined shortly after birth. The male bulbocavernosus and levator ani continue to grow, driven by testosterone and other growth factors. Subsequently, the larger, dimorphic male peripheral target muscles support the survival and differentiation of significantly more SNB neurons in the male spinal cord, via trophic interactions (see Chapter 23) that are mostly independent of gonadal steroid influence. In contrast, the loss of a major target muscle in the female leads to target trophic factor deprivation, again largely independent of additional direct gonadal steroid influence, followed by cell atrophy and apoptosis in the SNB. If a female rodent is artificially exposed to testosterone during a sensitive period in postnatal development prior to the onset of cell death in the bulbocavernosus and levator ani, the muscles are rescued and the number of SNB neurons approaches that of the male. This mechanism for the development of a sexual dimorphism represents a special case of a more general rule: During development, structures in the CNS are matched to the periphery based on the level of trophic support provided by their targets. In addition to this trophic mechanism, SNB neurons in the male spinal cord express testosterone receptors, and activation of these receptors is thought to secondarily regulate the enhanced growth of SNB neurons in males.

Dimorphic differentiation of the human male versus female spinal cord motor neurons that innervate the genital musculature is considerably less clear than that in rodents. The human spinal cord structure that corresponds to the rodent SNB is **Onuf's nucleus**, which consists of two motor neuron clusters in the sacral spinal cord (the dorsomedial and ventrolateral groups). The dorsomedial group is not sexually dimorphic; however, human females have fewer neurons in the ventrolateral group than do males (Figure 25.6C). In contrast to rodents, adult human females retain a bulbocavernosus muscle (which serves to constrict the vagina), but the muscle is smaller than that in males. The difference in nuclear size and cell number in humans, as in rats, presumably reflects the matching of neuron number and size to the amount of target tissue, and therefore trophic support, from the male versus female genital muscles.

Neural development and reproductive behaviors

There are sexually dimorphic regions in the CNS that do not project directly to the genitalia. Nevertheless, their divergent cellular composition and size in females versus males are essential for differences in reproductive function.

(A) Male rat pelvis

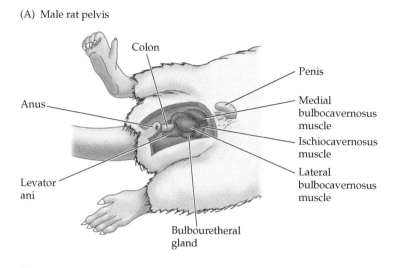

(B)

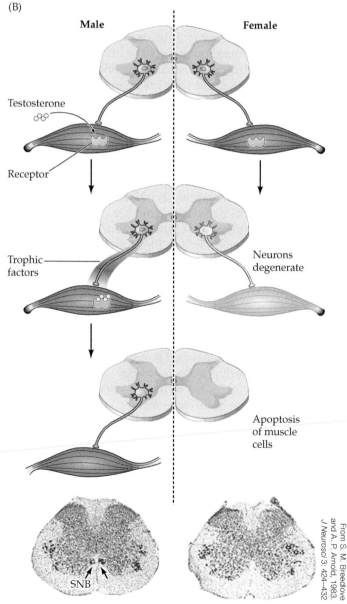

From S. M. Breedlove and A. P. Arnold, 1983. *J Neurosci* 3: 424–432

(C)

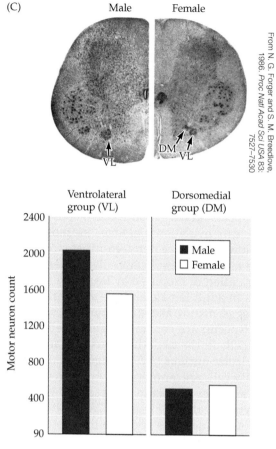

From N. G. Forger and S. M. Breedlove, 1986. *Proc Natl Acad Sci USA* 83: 7527–7530

FIGURE 25.6 The number of spinal motor neurons related to the perineal muscles is sexually dimorphic (A) The perineal region of a male rat. (B) Comparison of the developmental sequence for the bulbocavernosus muscle and the spinal motor neurons that innervate it, in male and female rats. Histological cross sections show the dimorphic spinal nucleus of the bulbocavernosus (SNB), found in the fifth lumbar spinal cord segment. Arrows point to the SNB in the male; there is no equivalent grouping of densely stained neurons in the female. (C) The micrograph shows Onuf's nucleus in the lumbar spinal cord of a human male and female. Histograms show motor neuron counts in the ventrolateral (VL) and dorsomedial (DM) groups of Onuf's nucleus in a human male and female. The dorsomedial group is not visible in the section through the male spinal cord. (A after S. M. Breedlove and A. P. Arnold, 1981. *Brain Res* 225: 297–307; B, diagram after J. A. Morris et al., 2004. *Nat Neurosci* 7: 1034–1039; C graphs after N. G. Forger and S. M. Breedlove, 1986. *Proc Natl Acad Sci USA* 83: 7527–7530.)

A major site of these dimorphisms is the hypothalamus, presumably because of its central role in the control of visceral motor function (see Chapter 21)—which includes the cellular, secretory, vascular, and smooth muscle control necessary for basic sexual function: gametogenesis and muscular and vascular control of the genitalia, in both males and females. The concentration of gonadal steroid receptors in the hypothalamus (see Figure 25.3B) reinforces this conclusion. Neurons in the anteroventral and medial preoptic areas of the anterior hypothalamus, where estrogen and androgen receptors are concentrated in the adult brain, mediate key sexual and reproductive behaviors. In most mammals there are sexually dimorphic differences in neuron number and size of subsets of hypothalamic nuclei.

The cycle for ovulation is perhaps the most distinctive fundamental difference between innate (controlled by dedicated circuits; not substantially influenced by cognitive regulation or modified by learning) reproductive behaviors in adult female and male mammals. Female gametes (eggs) mature at distinct intervals, whereas male gametes (sperm) are constantly generated. A specific group of cells in the hypothalamus, the **anteroventral paraventricular nucleus (AVPV)**, regulates the ovulatory cycle. AVPV cells are far more numerous in females than in males, and they project to different cell groups in the female hypothalamus (Figure 25.7A). The connections made by AVPV neurons regulate the systemic release of gonadotropin-releasing hormone by hypothalamic GnRH neurons (see Chapters 21 and 22) as well as prolactin secretion from the anterior pituitary, both of which are also key for optimal spermatogenesis as well as the cyclic control of ovulation. Sexual dimorphism in the AVPV arises during development due to the negative influence of elevated levels of testosterone in males. Transient elevated testosterone levels, converted to estradiol by aromatase in AVPV neurons, induce cell death in the developing male AVPV; the absence of similar levels of testosterone in females ensures AVPV cell survival. As is the case for many other sexually dimorphic cell groups, altering hormone levels during development can alter the AVPV dimorphism. Elevated testosterone at the critical stage of development in the female will cause AVPV cell death and lack of ovulation, and lack of testosterone signaling in males (via aromatized 17-β-estradiol binding to estrogen receptors) will rescue the AVPV cells that would normally die (see Figure 25.7A). In male mice in which the estrogen receptor has been knocked out, the AVPV neurons that would normally die are rescued as well.

In rodents, another nuclear group in the hypothalamus, the **sexually dimorphic nucleus of the preoptic area (SDN-POA)** in males is consistently larger and has more neurons than in females (Figure 25.7B). The size and cell number of this nucleus are regulated by testosterone during early postnatal development. In the male, estradiol (once again derived from testosterone via aromatase) influences anti-apoptotic genes and stabilizes SDN-POA neurons. In the female, the lack of significant levels of circulating testosterone leads to cell death. Females exposed to elevated levels of testosterone during early postnatal life develop an enlarged SDN-POA with more neurons, presumably because of diminished cell death. In contrast to what happens in the spinal cord nucleus of the bulbocavernosus, however, the effects of testosterone on cell survival in the SDN-POA are direct, via its conversion to estrogen in the brain and (presumably) the subsequent actions of estrogen via its receptors in SND-POA neurons. This dimorphism, like most others, can ultimately be linked to the differing ability of male and female gonadal tissues to provide distinct levels of testosterone and estrogen. Similar dimorphisms have been reported for several other nuclei in the preoptic area of the human hypothalamus; however, their consistency and relationship to sexual behavior remain controversial (see Concept 25.4).

In a range of laboratory animals, the hypothalamic preoptic area in general, including the SDN-POA, has been implicated in dimorphic sexual behaviors. In male rats, lesions of the entire preoptic area abolish all copulatory behavior, while more discrete lesions of the SDN-POA diminish the frequency of mounting and copulation. In female rats, such lesions yield individuals that avoid male partners and do not display female-specific copulatory behaviors. Thus, the preoptic area is thought to mediate mate selection and the preparatory behaviors for copulation, as well as some of the motor and visceral aspects of male intromission and ejaculation, and female copulatory responses. These behavioral differences are directly related to the sex differences in adult brain structure described here. This suggests that developmental mechanisms prepare female versus male brains for a set of "innate" behaviors that will only be performed once sexual maturity is reached far later in the life span.

This account of two hypothalamic nuclei with established sexual dimorphism in register with genotypic and phenotypic sex in males and females provides a starting point for considering a much broader set of observed sex differences in the brains of female versus male rodents (Table 25.1). These differences include volume differences in identified brain structures, including the amygdala (see Concept 25.3), and functionally distinct areas of the cerebral cortex. They also extend to cell biological distinctions in neuronal and glial numbers, as well as differentiation of dendrites, axons, and synapses. Finally, sex differences can be seen in molecular distinctions between cell classes, neuronal excitability, and circuit properties. The relationships between these neurobiological distinctions and the sex biases shown by each remain to be fully explained. Nevertheless, there is a substantial amount of evidence that the brains of male and female rodents differ when analyzed quantitatively by region, cell class, molecular identity, and circuit function.

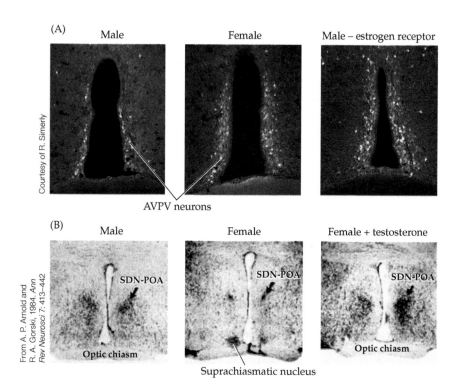

(A) Male   Female   Male – estrogen receptor

Courtesy of R. Simerly

AVPV neurons

(B) Male   Female   Female + testosterone

From A. P. Arnold and R. A. Gorski, 1984. *Ann Rev Neurosci* 7: 413–442

SDN-POA   SDN-POA   SDN-POA

Optic chiasm   Optic chiasm

Suprachiasmatic nucleus

FIGURE 25.7   Sexually dimorphic hypothalamic nuclei associated with sexual behaviors (A) The anteroventral paraventricular nucleus (AVPV) is a collection of dopaminergic neurons (labeled here with an antibody against the dopamine-synthetic enzyme tyrosine hydroxylase) that is larger in females than in males (the AVPVs from both female male and mice are shown here). It is likely that the larger size AVPV in females reflects the need for increased control of cyclic secretion of GnRH to regulate ovulation in females, as opposed to lower level chronic GnRH in males. When the estrogen receptor gene is inactivated in male mice (thus preventing the masculinizing effects of testosterone, converted in the brain to estradiol by aromatase), the AVPV is similar in size to that of the female (far right). (B) The sexually dimorphic nucleus of the preoptic area (SDN-POA) is larger in male rats than in female rats. This size difference can be approximated in genotypically female rats that are given testosterone perinatally (far right). (B from A. P. Arnold and R. A. Gorski, 1984. *Ann Rev Neurosci* 7: 413–442, and R. A. Gorski, 1983. In *Neuroendocrine Perspectives*, Vol. 2. E. E. Muller and R. M. MacLeod [Eds.], Amsterdam: Elsevier/North Holland.)

Table 25.1   Sexually Dimorphic Brain Regions, Neurons, Glia, and Circuits in Genotypic and Phenotypic Male and Female Rodents (and Other Mammalian Models)

Volume larger in males	Volume larger in females
mPOA (hypothalamus)	Orbital prefrontal cortex
VMN (hypothalamus)	AVPV (hypothalamus)
Bed nuc. stria terminalis	Locus coeruleus
Binocular area of the visual cortex	
Amygdala	

Cell number greater in males	Cell number greater in females
Activated microglia in mPOA	Radial glia in the mPOA/hypothalamus
Glia in visual cortex	Oligodendrocyte precursors
Glia in the globus pallidus and CA1	
Astrocytes in posterodorsal med. amygdala	

Synapses differ in males	Synapses differ in females
Hippocampal CA3 primary dendrites/spines	Arcuate nucleus axodendritic synapses
Arcuate nucleus axosomatic synapses	
VMN axodendritic synapses	

Source: McCarthy et al., 2017. *Nat Neurosci Rev* 18: 471–482.

CONCEPT

Systemic Signals Target Neurons and Circuits for Reproductive and Parenting Behaviors

CONCEPT 25.3

LEARNING OBJECTIVES

25.3.1 Define divergent gonadal steroid signaling for male versus female neural circuits and behaviors.

25.3.2 Explain how the targets and activities of gonadal steroids differ in the developing versus the adult brain.

25.3.3 Identify additional molecular and cellular mechanisms that contribute to sexual dimorphisms in the brain and in behavior.

Cellular and molecular bases of sexually dimorphic development

The establishment, maintenance, and plasticity of sexually dimorphic brain structures, neural circuits, and behaviors depend on **organizational effects** as well as **activational effects** of circulating levels of gonadal steroids (Figure 25.8). These distinctions refer primarily to two distinct periods of the life span when gonadal tissue differentiates or matures, and when levels of circulating gonadal steroids differ. The organizational effects of gonadal steroids occur as initial gonadal tissues differentiate in male and female fetuses, and are essential for establishing major differences in neuron numbers, glial classes and connections in sexually dimorphic brain structures. The activational effects occur later in life (they begin during puberty) when gonadal tissue matures further so that it can secrete constantly higher levels of gonadal steroids in males, or fluctuating higher levels in females.

Differing levels of estrogen (mostly as estradiol) and testosterone (as both estradiol and in some instances directly as testosterone) act during both the organizational and activational phases of sexually dimorphic brain development and emergence of related sex differences in behaviors (see Figure 25.8). In male rodents (and most other male mammals) there is a surge of testes-secreted testosterone during late fetal and early postnatal life that is absent in females (see Figure 25.8). This surge of testosterone, acting via either estrogen receptors after aromatization to estradiol or androgen receptors, is critical for selective cell death or cell survival in several nuclei of the hypothalamus (see Concept 25.2). Additional cellular and molecular distinctions emerge in the male brain that distinguish it from the female brain in response to substantial differences in late fetal and early postnatal gonadal steroid availability. Ultimately, the consequences of this organizational influence of the male gonadal steroid surge are to prepare circuits in the brain to execute male-specific reproductive behaviors, including mounting and subsequent copulatory behaviors (see Figure 25.8A).

In addition, circuits organized by the perinatal surge in testosterone in males will regulate the release of GnRH from the limited numbers of GnRH neurons in the preoptic area of the hypothalamus (see Chapter 21) to initiate and then maintain effective spermatogenesis. These behaviors and the gamete maturation necessary for effective reproduction, however, do not begin until a second substantial increase in steroid secretion from the testes occurs during puberty—the onset of the activational influence of gonadal steroids. This elevation of circulating gonadal steroids reaches a maximum by adulthood.

In contrast, the lack of change in gonadal steroid levels from a relatively low baseline secreted by the ovaries during late fetal and early postnatal development results in dimorphisms in the hypothalamic nuclei in the brain as well as additional molecular and cellular distinctions that characterize the neural circuits that regulate female gametogenesis and reproductive behaviors. Even though there is no noticeable change in gonadal steroid levels in females, the differential effects of the baseline levels (versus the elevated levels in males) nevertheless "organizes" the female pattern of neural circuits for reproduction. The lack of a late fetal and early postnatal surge in females is thought result in a "default" execution of developmental programs that result in distinctions such as increased cell numbers in the AVPV and decreased cell numbers in the SDN-POA (see Concept 25.2). Indeed, the mechanisms that pattern steroid hormone receptors and prepare sensitive areas of the developing brain for initial gonadal steroid signaling that leads to sexually dimorphic circuits are largely shared in the two sexes. This seems to be the case based on multiple experiments in which a surge in steroid-mediated signaling is induced experimentally in late fetal or early postnatal females. These experimental interventions result in the masculinization of relevant brain regions and parallel changes from female to male patterns of behavior, even though the chromosomal sex has not changed and (usually) secondary sex characteristics (external genitalia, etc.) retain their female pattern (with some exceptions, including for the differentiation of muscles associated with the genitalia; see Concept 25.2). The time that these steroid-driven changes are effective in chromosomal females is limited to approximately the same time as that of the endogenous testosterone surge in chromosomal males. Together, these observations indicate that there is a *critical period* (see Chapter 24) for sexually dimorphic circuit development, and during late fetal and early post-natal life, and that the mechanisms that distinguish male and female brains primarily reflect the consequences of the testosterone surge in males.

Molecular pathways that regulate cell survival and cell death are an essential target for estrogen and testosterone. Many sexually dimorphic nuclei achieve their distinct cell numbers and cell sizes through apoptotic cell death and subsequent growth of the surviving cells. It is not clear

(A) Influence of male steroids produced in testes

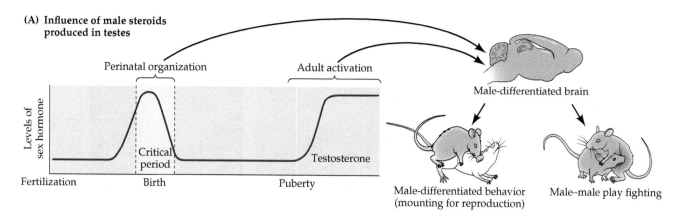

(B) Influence of female steroids produced in ovaries

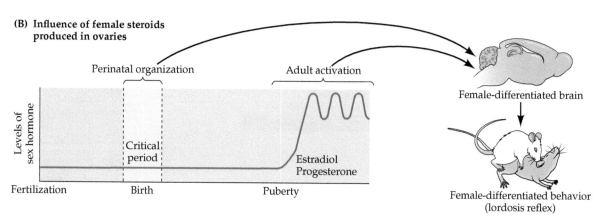

FIGURE 25.8 Organizational and activational influences of gonadal steroids on male and female brains and behaviors
(A) The time of release and the levels of circulating testosterone, secreted by the testes in male fetal, newborn, adolescent, and adult mice. The initial surge of testosterone during late fetal and early postnatal life organizes neural circuits so that they are adapted quantitatively and functionally to the masculinized periphery as well as male-specific behaviors. At puberty, the rapid increase and stable levels of circulating testosterone from the testes activate physiological functions (spermatogenesis) as well as support circuit activity for reproductive and social behaviors, including male–male "play fighting." (B) The developing female gonadal tissue does not produce a late fetal and early postnatal surge of gondal steroid (estrogen, primarily). This implies that the developmental mechanisms for female-specific brain circuitry that are sensitive to gonadal steroids are a "default" and do not require increased circulating systemic estrogen for their completion. Once puberty is reached, however, the rapid increase and then phasic surges of estrogen activate neural control of cyclic oogenesis as well as reproductive behaviors such as the lordosis reflex that signals female receptivity for copulation. Females do not typically acquire play-fighting behaviors. If a genotypic female (XX, no Sry) is exposed to a late fetal and early postnatal estradiol or testosterone surge, these behaviors are performed by the masculinized female (see also Figure 25.11). (After K. L. Meeh et al., 2021. *Dev Biol* 472: 75-84; S. M. Pellis and V. C. Pellis, 2017. *Learning Behav* 45: 355–366.)

how estrogen and testosterone initially stimulate mechanisms that favor either apoptosis or cell survival; however, recent evidence implies that genes that regulate apoptotic cell death are downstream of the initial activity of gonadal steroids in developing neurons. Male or female mice overexpressing the anti-apoptotic gene *Bcl2* have more neurons in the SNB. Mice in which the pro-apoptotic gene *Bax* has been inactivated do not display differences associated with neural circuits in the hypothalamus and basal forebrain (including the bed nucleus of the stria terminalis; see Chapter 14). Thus, the regulation of structural sexual dimorphisms that match phenotypic sex is most likely dependent on expression and activity of the genes that control apoptosis in response to gonadal steroid signaling, perhaps further modulated by trophic factors during brain development (see Chapters 23 and 26).

Besides influencing cell death, in certain instances gonadal steroids can act as mitogens for neurogenesis, or trophic factors, directly regulating neuronal size as well as process growth. During development, and to some extent throughout life, estradiol stimulates brain dimorphisms by influencing proliferation of both neurons and glia, cell size, dendrite length and branching, dendritic spine density, and synaptic connectivity of sensitive neurons, independent of cell survival or apoptosis (Figure 25.9A,B). Testosterone can also influence neuronal size and differentiation, at least in vitro, in neurons that express testosterone receptors (Figure 25.9C); however, the extent to

FIGURE 25.9 Estrogen and testosterone influence neuronal growth and differentiation
(A) A control explant (left) from the mousehypo-thalamus shows only a few silver-impregnated processes; an estradiol-treated explant (right) has many more neurites growing from its center. (B) Dendritic spine density in female rat hippocampal neurons in response to progesterone (a precursor of both estrogen and testosterone; see Figure 25.2A) and to estrogen. Recall that dendritic spines, which are small extensions from the dendritic shaft, are sites of synapses. Tracings at right are of representative apical dendrites from hippocampal pyramidal neurons: (1) after administration of progesterone and estrogen in high dosage; (2) after administration of progesterone and estrogen at basal levels; and (3) after administration of a progesterone receptor antagonist. (C) Effects of testosterone on embryonic rat pelvic ganglion in cell culture. In response to testosterone, processes become thicker and more highly branched, and the cell body (soma) grows in size. (B after C. S. Woolley and B. S. McEwen, 1993. *J Comp Neurol* 336: 293–306; C after S. M. Meusberger and J. R. Keast, 2001. *Neuroscience* 108: 331–340.)

(A)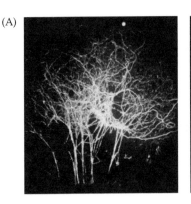

C. D. Toran-Allerand, 1976.
Brain Res 106: 407–412

(B) Dendritic spine density

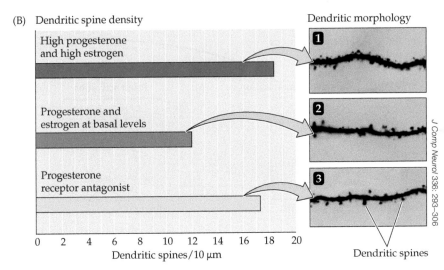

High progesterone and high estrogen

Progesterone and estrogen at basal levels

Progesterone receptor antagonist

0 2 4 6 8 10 12 14 16 18 20
Dendritic spines/10 μm

Dendritic morphology

Dendritic spines

J Comp Neurol 336: 293–306

(C)

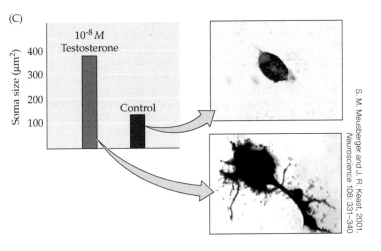

$10^{-8} M$ Testosterone

Control

Soma size (μm²)

S. M. Meusberger and J. R. Keast, 2001.
Neuroscience 108: 331–340

which these effects rely on the direct action of testosterone binding to its receptor versus aromatized 17-β-estradiol via estrogen receptors is unclear.

Signaling via estrogen and androgen in neurons and glia relies critically on the activation of estrogen and androgen receptors. Estrogen and testosterone also engage an extensive network of additional receptors, signal transduction molecules, and transcription factors. Estrogen and testosterone can bind to selective G-protein-coupled receptors to activate several kinases and their targets. Estrogen and androgen also have apparent influences on the number and activity of additional membrane receptors and channels, although the mechanisms of these actions are not fully understood. The consequences of this additional gonadal steroid influence include changes in electrical activity and altered downstream Ca^{2+} signaling. Finally, nuclear transcriptional co-factors (proteins that modulate the efficiency of transcription) are recruited to sites of estrogen or androgen nuclear receptors that have been activated ligand binding and are in turn bound to their specific DNA response elements. The assembly of larger complexes of gonadal steroid receptors and nuclear co-factors can influence histone modifications and transcription of specific estrogen- or androgen-responsive genes. The chromatin changes that result from histone modifications probably influence much broader networks of transcriptional regulation that can differ in steroid-responsive neurons in the developing and mature brain.

Gonadal steroid influences beyond the hypothalamus

Estrogens and androgens can influence neuronal and glial structure and function throughout life in brain regions beyond the hypothalamus. Aside from high concentrations

FIGURE 25.10 Estrogen and androgen receptor distribution in the brain is widespread Distribution in the rat brain of the three major receptor/transcription factors that bind estrogen (ERα and ERβ) and androgen/testosterone (AR), eliciting corresponding changes in gene expression. ERα, ERβ, and AR tend to be expressed in the same subsets of brain structures. However, these structures are not restricted to the hypothalamic nuclei that control gonadal function, sexual behavior, and parenting behavior; they also include large regions of the cerebral cortex, amygdala, hippocampus, thalamus, substantia nigra, and cerebellum. The significance of gonadal steroid receptor expression and activity at sites beyond the hypothalamus is less well understood than their reproduction-specific functions. Receptors in these brain structures may provide a substrate for the influence of these hormones on circuits and behaviors beyond those directly related to reproduction and parenting, including cognition (cortex), learning and memory (cortex, hippocampus, amygdala), aggression and stress (hippocampus, amygdala), pain sensation (thalamus, brainstem), and motor control (substantia nigra, cerebellum).

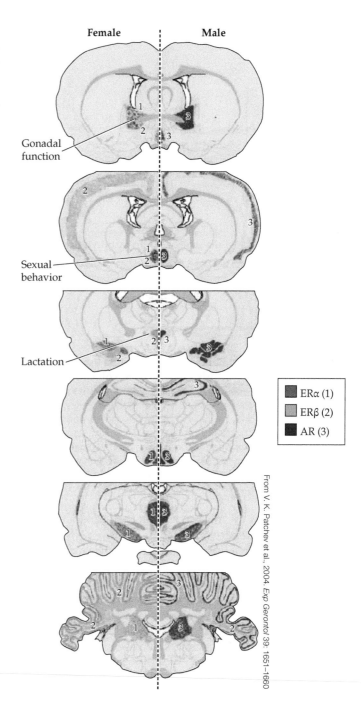

From V. K. Patchev et al., 2004. *Exp Gerontol* 39: 1651–1660

in the hypothalamus, there are significant numbers of both estrogen and androgen receptors in the developing and mature cerebral cortex, amygdala, and substantia nigra (Figure 25.10). All of these brain regions differ in size or cell number in females versus males (see also Table 25.1). The availability of these receptors, and presumably sex differences in either expression patterns or signaling capacity, in males and females can lead to divergent circuit differentiation in regions beyond the hypothalamus. These organizational differences during neural circuit development may be relevant for understanding male versus female behaviors that go beyond the direct regulation of reproductive functions.

The emergence of play fighting in male rodents, and its relative absence in females, provides an example of how organizational influences (those that depend on differences in gonadal steroid signaling during late fetal and early postnatal life; see the previous section) can distinguish additional brain regions and circuits beyond the spinal cord or hypothalamus in males versus females (Figure 25.11). In males, neural circuits in the amygdala are essential for play fighting: non-injurious grabbing and biting between juveniles that becomes a precursor for aggressive behaviors related to reproduction later in life. These behaviors are not seen in juvenile females; however, if juvenile females are exposed to testosterone early in life, they will engage in play fighting, similar to juvenile males. This difference apparently reflects distinctions in the proliferation, final numbers, and activity of astrocytes and microglial cells in the female versus male amygdala (see Figure 25.11). As early postnatal development draws to a close, there are more astrocytes in the female amygdala than in the male. In contrast, in the male amygdala there are more microglia, whose immune-related functions are thought to influence astrocyte survival or death, and thus subsequent synapse formation and plasticity. The testosterone surge in males, or excess testosterone exposure in females, results

in these differences in glial frequency and type by signaling that eventually activates endocannabinoid ligands and receptors (see Chapter 6). Thus, gonadal steroids clearly influence a behavioral sex difference beyond gametogenesis, genital function, or copulation, mediated by distinct cell classes and connections in the forebrain.

In the adult brain, gonadal steroids exert effects via estrogen and androgen nuclear receptor–mediated changes in gene expression that influence neuronal or glial function. In addition, they also can act to modify electrical signaling between neurons in a variety of mature brain

(A) Male

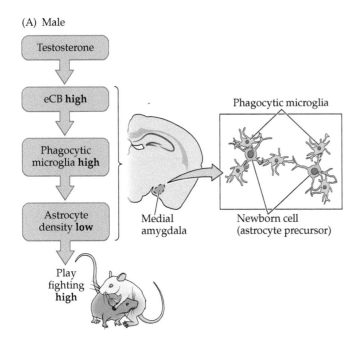

(B) Female

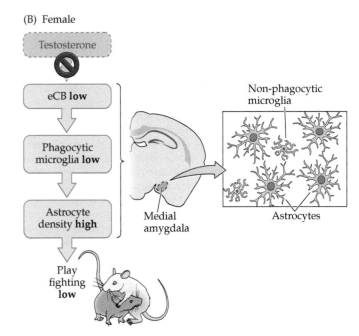

FIGURE 25.11 Cell biological distinctions established by systemic gonadal steroids result in dimorphic forebrain gene expression, cellular differentiation, circuits, and behaviors (A) In the presence of a late fetal and early postnatal surge in testosterone, typically in genotypic males, endocannabinoid (eCB) receptors are upregulated in the amygdala. Signaling via these receptors in phagocytic microglia activates these cells to engulf and kill astrocyte precursors. The relatively lower density of amygdala astrocytes is associated with male-specific behaviors, including play fighting, mediated in part by amygdala circuits. (B) In females, lower endocannabinoid receptor expression maintains the resident microglia in their non-phagocytic state, allowing astrocyte precursors to generate a larger population of astrocytes. If endocannabinoid signaling is stimulated in females (or an early gonadal steroid surge is provided experimentally), astrocyte numbers decline in the amygdala and male-typical play fighting behaviors are seen in genotypic (XX) females. (After K. L. Meeh et al., 2021, *Dev Biol* 472: 75-84.)

regions. Perhaps the most compelling example of this phenomenon is in the periventricular nucleus (PVN) of the hypothalamus, where fluctuating steroid levels facilitate the formation of gap junctions by regulating the transcription of relevant proteins. The resulting increase in gap junctions allows for neuronal synchrony in the PVN correlated with lactation and maternal behavior (Clinical Applications). In addition, the influences of gonadal steroids, particularly estrogen, on neuronal activity have been evaluated in the mature hippocampus. The hippocampus is an established site of neuronal plasticity (see Chapters 24 and 30) and is also sensitive to hormonal fluctuations, including those seen during estrous. Estrogen receptors are expressed by mature neurons and often are localized to the cytoplasm at synapses, as well as in the cytoplasm of the cell body (Figure 25.12A). Estrogen can modify excitable properties of hippocampal neurons, including K^+ and Ca^{2+} conductances and the rate of action potential firing. Estrogen can also influence hippocampal synaptic signaling and plasticity. Estrogen at relatively high concentrations (arguably higher than those seen physiologically) can increase the amplitude of excitatory postsynaptic currents (related to level of vesicular neurotransmitter release) over minutes to hours. When experimental estrogen exposure is coupled with high-frequency stimulation that elicits long-term potentiation (LTP; see Chapter 24), excitatory postsynaptic potentials are enhanced over baseline for sustained periods (Figure 25.12B,C). It is tempting to speculate that such changes during times of fluctuating gonadal steroid levels underlie the acquisition of learned behaviors and memories formed during these times. There is no solid evidence as yet that supports this speculation.

Estradiol can also stimulate an increase in the number of synaptic contacts in adult animals. For example, during periods of high circulating estrogen in the estrous cycle of female rodents (or after administration of exogenous estrogen), there is an increase in the density of spines (and presumably synapses) on the apical dendrites of pyramidal neurons in the hippocampus (see Figure 25.9B). These apparent changes in synaptic connectivity might contribute to differences in learning and memory during the course of the estrous cycle. Such differences have been observed in rodents using tests of spatial navigation and memory; however, the relevance of these laboratory behaviors to significant functional differences in reproductive behavior engendered by estrous-dependent hippocampal changes outside these experimental paradigms is not understood.

■ Clinical Applications

The Good Mother

In Hollywood movies, fairy tales, and myths, mothers have been portrayed as both saintly and evil. While these stories rarely address the source of maternal warmth and feeling (or lack thereof), recent observations suggest strongly that good mothers are made, not born.

The mothers in question are female rats, whose repertoire of motherly behaviors does not extend to either sacrificing her life for her children (as did Hollywood's Mildred Pierce—played by Joan Crawford in the 1945 film, but unlike Joan Crawford herself, as memorialized by Faye Dunaway in the 1981 film Mommie Dearest) or to sacrificing her children, like mythological mothers of old (according to Greek mythology, Medea killed her children as revenge against their father, Jason, for taking up with another woman). The sign of a good rat mother is the amount of time she spends licking and grooming her pups when she enters the nest for nursing, and her posture during nursing itself. A good rat mother arches her back distinctively (Figure), presumably allowing for better access for the pups without extreme spatial confinement. "Bad" rat mothers lick and groom much less frequently and do not arch their backs when nursing. Offspring reared by high *licking-grooming/arched back nursing* (LG/ABN) mothers grow up to have much greater adaptive response to stress and more modulated responses to fearful stimuli. When pups from low-LG/ABN mice are transferred to high-LG/ABN mothers, they acquire stress responses consistent with the maternal qualities of their new mother. Thus, "good" mothering, perhaps as much or even more than genetic predisposition (at least in rats), makes for much better adjusted offspring and ensures good mothering for the next generation.

Good maternal behavior apparently is essential for the health and well-being of offspring. Thus, the

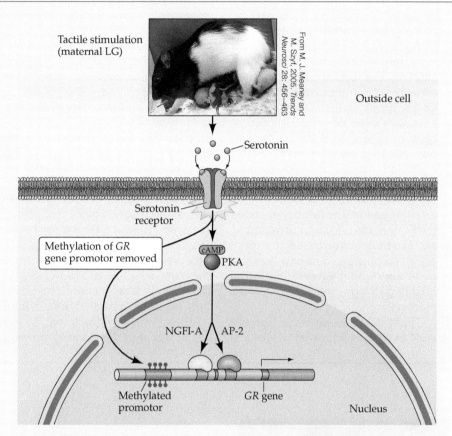

Tactile stimulation (maternal LG)

Outside cell

Serotonin

Serotonin receptor

Methylation of *GR* gene promotor removed

cAMP
PKA

NGFI-A   AP-2

Methylated promotor

GR gene

Nucleus

From M. J. Meaney and M. Szyf, 2005, *Trends Neurosci* 28: 456–463

The inset photo shows a "good" rat mother licking and grooming her pups; her back is arched to accommodate nursing. In the model diagrammed here, increased serotonin levels elicited by the increased tactile stimulation provided by a high-LG/ABN mother may lead to a signaling cascade that ultimately alters expression of the glucocorticoid receptor (*GR*) gene. In this model, adequate nursing eliminates methyl groups from the DNA in a brain-specific promoter region of *GR*, thus upregulating the *GR* gene in the brains of these pups.

acquisition of motherly skills becomes key in understanding the transmission of adaptive stress responses from generation to generation. Michael Meaney and his colleagues at McGill University asked whether good mother rats—high-LG/ABN females—were determined genetically or acquired their maternal behaviors based on their early experiences. By cross-fostering offspring of established low-LG/ABN mothers with high-LG/ABN mothers and vice versa, these researchers demonstrated that, regardless of genetic identity, maternal skills depended on the skills of the mother that reared the pups. High-LG/ABN mothers had

foster daughters that were also high-LG/ABN mothers—even if their birth mothers were low LG/ABN. Similarly, female offspring of high-LG/ABN mothers displayed low-LG/ABN maternal skills when reared by a low-LG/ABN foster mother. These observations suggest that good mothers are made, not born, and that the foundation for maternal skills is established early in life—in part by the maternal behaviors to which female pups are exposed.

Subsequent work from Meaney's laboratory as well as others has shown that one of the key biological targets of the effects of early maternal

(Continued)

■ Clinical Applications (continued)

behavior on offspring is the expression of glucocorticoid receptors in the hippocampus. These steroid–thyroid superfamily receptors (see Figure 25.2) are key regulators of the stress response throughout the organism. Offspring of high-LG/ABN mothers have high levels of glucocorticoid receptor expression in the hippocampus, and thus are presumably better equipped to deal with the deleterious effects of stress. But the maternal behaviors that establish these differences are apparently not strictly encoded by the genome (they are not heritable, and can be acquired based on experience), so how are differing levels of gene expression established in the offspring of high- versus low-LG/ABN mothers?

The tentative answer is that high-LG/ABN behavior elicits altered levels of serotonergic signaling in the offspring. This signaling, via a specific serotonin receptor and subsequent signaling cascade, establishes differential expression of the glucocorticoid receptor via genomic imprinting—local modification of DNA and chromatin that leads to long-lasting changes in gene expression. The details of this intriguing epigenetic mechanism for establishing essential differences in behavior remain to be determined. It appears that the glucocorticoid receptor gene has several regulatory regions that enable its transcription in different cell types. The signals initiated by maternal care alter the DNA methylation pattern and chromatin structure of a regulatory region that binds brain-specific transcription factors, including one called NGF1-A. Those rat pups that do not get adequate licking and grooming have methylated DNA in that region of the *GR* gene that would normally bind NGF1-A, which cannot bind to the methylated DNA and therefore does not activate transcription of the *GR* gene in the brain. As a result, the glucocorticoid receptor protein is not present to negatively regulate the stress response.

Whatever the details of the molecular mechanism may be, early experience can profoundly and irreversibly alter an entire lifetime of essential behaviors. Unfortunately, these studies also suggest that the happy endings enjoyed by fictional offspring who were fostered by wicked stepmothers—characters such as Snow White, Cinderella, and Hansel and Gretel—may need to be rewritten to project a far more grim ever-after than previously imagined.

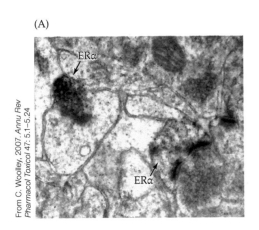

(A)

ERα

ERα

From C. Woolley, 2007. *Annu Rev Pharmacol Toxicol* 47: 5.1–5.24

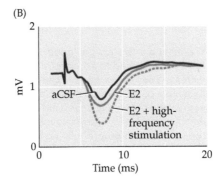

(B)

aCSF
E2
E2 + high-frequency stimulation

mV

Time (ms)

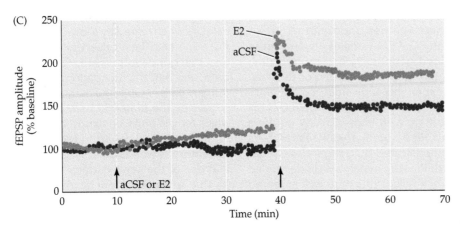

(C)

E2
aCSF

fEPSP amplitude (% baseline)

aCSF or E2

Time (min)

FIGURE 25.12 **Estrogen influences synaptic transmission** (A) Electron micrograph showing localization of the estrogen receptor α (ERα; the dark, "electron-dense" label) in postsynaptic processes (presumably spines) in the rat hippocampus. (B) Estrogen (E2) increases the amplitude of excitatory postsynaptic potentials in individual hippocampal neurons (the same physiological measurement done in artificial cerebrospinal fluid—aCSF—is shown as a control). High-frequency stimulation further enhances the effects of E2, suggesting that estrogen may modulate use-dependent plasticity in hippocampal synapses. (C) High-frequency stimulation in the presence of E2 in rat hippocampal slices (see Chapter 8) results in enhanced long-term potentiation consistent with a role for estrogen in synaptic and circuit plasticity in the hippocampus. The data shown here plot the frequency of EPSP values (fEPSP) over time. E2 alone results in a clear increase in EPSP values, and this effect is magnified and maintained after high-frequency stimulation, denoted by the second arrow. (From C. Woolley, 2007. *Annu Rev Pharmacol Toxicol* 47: 5.1–5.24.)

CONCEPT
25.4
Complex Human Behaviors Are Difficult to Associate with Sex, Gender, or Early Systemic Signaling

LEARNING OBJECTIVES

25.4.1 Distinguish sexual dimorphism from sex and gender differences for human brains and behaviors.

25.4.2 Describe human genetic syndromes that disrupt the relationship between chromosomal sex and phenotypic sex.

25.4.3 Discuss the evidence for human brain and behavioral differences associated with sex or gender.

The third rail of neuroscience: Human sex differences

Most of the current, robust characterizations of sexual dimorphisms and related sex differences in brains and behaviors reflect highly detailed observations of circuits and the peripheral organs they control for reproduction or parenting behaviors. These observations of clear sex differences are reinforced by experimental studies that selectively alter exposures to gonadal steroids or selectively delete key genes at critical developmental stages in animal models and reverse or alter these dimorphisms. The available evidence has shown that true neural dimorphisms—reliable differences in brain structures or circuits that parallel chromosomal and phenotypic sex—are seen in brain regions specifically associated with sexually dimorphic structures in the periphery. The key peripheral structures—that is, male and female genitalia and mammary glands—develop based on divergent gonadal steroid influences that are themselves established in register with the differentiation of male versus female gonadal tissue that in turn reflect chromosomal sex. They are directly associated with reproductive behaviors and are controlled by dimorphic circuits in the brain. The distinctions become far less clear, and arguments around the data that support "sex differences" in other brain regions and for other behaviors become far more heated. When such differences are evaluated for several human cognitive or social behaviors, as well as brain regions that may be related to complex dimensions of gender and human sexuality, the arguments become difficult to resolve with the available biological evidence.

This raises the question of whether there is a place for neuroscience—the sort done at the end of an electrode, via microscopic eyepieces, molecular characterization, or noninvasive human brain imaging—in addressing these issues. One can argue that despite the sensitive nature of such research, a more detailed understanding of these differences is essential for better diagnoses and treatment of several diseases of the nervous system. For example, several neurodevelopmental disorders, including autistic spectrum disorder and schizophrenia, have higher incidence in males, while others, such as major depression, subtypes of bipolar disorder, and anxiety disorder, are seen more frequently in females (see later in this concept). The key behavioral changes in these disorders are not easily related to the fundamental reproductive and parenting behaviors for which solid evidence of dimorphic neural circuitry is available. There are valid arguments that suggest that data addressing sex differences in these clinical disorders, and the related differences in brain structures that have been suggested, are incomplete at best and interpreted selectively, potentially leading to discrimination, mistreatment, and other harmful outcomes based on less than reliable information. This is especially true when applying this sort of information to questions of human gender, individual identity, and other social and psychological constructs. Perhaps the best statement of these fundamental concerns has come from M. M. McCarthy, a pioneer in the study of biologically defined brain and behavioral dimorphisms:

> *Gender is a uniquely human construct that combines self and societal awareness of one's sex, thereby including the influence of cultural norms, implicit bias and parental expectations (according to the World Health Organization definition of gender; see Further Information). This makes it difficult, if not impossible, to isolate a purely biological contribution to sex differences in human brain and behaviour. Indeed, some would argue that we should not even try as there is potential to do real harm by lending scientific credence to well-entrenched stereotypes.*
>
> *M.M. McCarthy et al., Nature Neuroscience Reviews 18, p. 471*

The issue of "real harm" caused by imprecise language, facile comparisons, or unsupported generalizations must be considered in reviewing and interpreting the following summary of current knowledge of sex-related differences in human brain structure and function. One must be cautious in assessing any data, particularly those that go beyond direct relationships between chromosomal sex, phenotypic sex, and reproduction-related behaviors. Most of the information in the following sections should be considered in this context. It is useful to ask how science might address these controversial questions with a variety of data, especially when such studies might provide insight into the pathogenesis or treatment of serious clinical disorders of the brain and behavior. The nature of the questions, however, must be thoughtfully matched to the resolution of the data available, and any conclusions considered in the context of assumptions and biases made in framing the questions. The biological basis of human sex, gender, and sexuality is beyond doubt one of the most freighted conversations that continues in science and more broadly in society, and everyone deserves a voice. Scientists—and students—should be part of the conversation. Nevertheless, all who consider these issues need to approach the data cautiously and critically. It is essential to recognize the ambiguities and limits of such studies, the biases they reflect, or the biases that emerge based upon one's interpretation of the data.

Human genetic disorders of genotypic and phenotypic sex

Chromosomal sex and phenotypic sex are not always aligned, and genetic variations in humans can challenge the usual definitions of female and male. Such genetic variation, called **intersexuality**, is apparent in 1% to 2% of all live births and can arise from a variety of mutations associated with sex chromosomes as well as autosomes (Table 25.2). The most common genomic variations result in misaligned chromosomal and phenotypic sex. XXY individuals (Klinefelter's syndrome) have male genitalia because of the presence of *SRY* on the Y chromosome (see Figure 25.1) but some female secondary sex characteristics (e.g., mammary tissue), presumably because of a "double dose" of the genes on the X chromosome. In XX females, one of each copy of most X-chromosome genes is inactivated via DNA modifications that ensure appropriate expression levels. The XXY genotype, and expression of Y chromosome genes, may disrupt the typical process of X inactivation. XO individuals (Turner syndrome) are small in stature, have rudimentary gonadal development, underdeveloped external genitalia (which usually appear female), and are sterile. XYY individuals are the least compromised in terms of accordance between chromosomal sex and phenotypic sex. Their gonadal tissues and external genitalia are male (although these individuals are sterile). Their primary identifying physical characteristic is slightly increased height.

Other genetic disorders that result in intersexuality are the result of mutations in genes that encode metabolic enzymes for steroid hormone production. One of the most prevalent examples is **congenital adrenal hyperplasia** (**CAH**), usually the result of mutations in the gene encoding 21-hydroxylase, an enzyme that, using testosterone as the precursor, synthesizes cortisol and aldosterone in the adrenal gland. In affected individuals, failure of cortisol and aldosterone synthesis leads to increased testosterone. XY individuals with CAH are dramatically masculinized, are often very tall at an early age, and undergo precocious puberty. In XX individuals, CAH leads to overactive adrenal secretion of testosterone during development, resulting in ambiguous, masculinized secondary sex characteristics.

Androgen insensitivity syndrome (**AIS**) (sometimes called *testicular feminization*) illustrates the results of genetic disruption of receptor-mediated responses to gonadal steroids. The best-studied cases of AIS are males who carry mutations in the gene encoding the receptors for testosterone and/or Müllerian-inhibiting substance (see Figure 25.1). In these XY individuals, testes form initially and secrete androgens; however, the testes rapidly become hypotrophic. The deficiency of testosterone receptors due to genetic mutation, and the subsequent failure of differentiation of testicular tissue capable of secreting elevated levels of testosterone, leads to the development of female external genitalia. XY individuals with AIS appear to be phenotypically female based upon their genitalia and secondary sex characteristics and self-identify as female, even though they have a Y chromosome and remnants of testicular tissues. Individuals with AIS

TABLE 25.2   Genetic Disorders Resulting in Lack of Registration of Genotypic and Phenotypic Sex in Humans

Syndrome	Mutation	Frequency	Phenotype
Klinefelter's	XXY	1/2500 live births	Male secondary sex characteristics
Turner	XO	1/10,000 live births	Incomplete female development
47-XYY	XYY	1/1000 live births	Male secondary sex characteristics
Congenital adrenal hyperplasia (CAH)	21-hydroxylase (chrom. 6)	1/5000 live births	Hypermasculinization of XY individuals; masculinization of XX individuals
Androgen insensitivity (AIS)	Testosterone receptor (X chrom.); Müllerian-inhibiting substance (MIS) (chrom. 19)	Rare (1/100,000 live births)	Hypotrophic testicular tissue, female syndrome secondary sex characteristics in XY individuals
Testes-at-12	5-α-reductase (chrom. 2)	Exceedingly rare	Incomplete male genital development prior to puberty; brain and behavior remain masculinized

present one of the strongest arguments that brain circuits in humans are masculinized mainly by the action of circulating testosterone.

Rare populations of XY individuals, particularly in geographically isolated regions where consanguinity is more frequent, carry two copies of a recessive mutation that makes them deficient in one of two forms of **5-α-reductase** encoded by two distinct genes found at different chromosomal locations. 5-α-reductase catalyzes the conversion of testosterone to the biologically active dihydrotestosterone during fetal and early postnatal life. Dihydrotestosterone is particularly important for the continued differentiation of male external genitalia. In these XY individuals, all of whom have intact copies of the *SRY* gene, the genitalia initially resemble those of females. Indeed, these individuals are perceived as females at birth and raised as females during their early lives. At puberty, however, testicular secretion of testosterone increases once again (long after the transient late fetal/early post-natal surge; see Figure 25.8). At this time, the second 5-α-reductase variant encoded by a separate gene regulates dihydrotestosterone synthesis from circulating testosterone. Accordingly, higher levels of dihydrotestosterone are available and in response the apparent clitoris enlarges into a penis and the testes descend. In the Dominican Republic, where this recessive syndrome has been thoroughly studied in particularly consanguineous families, the condition is referred to colloquially as *guevedoces*, roughly translated as "testes-at-12." Anecdotal reports indicate that most such individuals retain their masculine gender identity as well as a heterosexual orientation after puberty. Genetic testing can now identify affected children, most of whom are then raised from birth in a manner consistent with their genotypic sex. The rapid diagnosis of intersexuality based on chromosomal complement or established mutations is now considered standard clinical practice for at-risk individuals. Clinical interventions or support can include efforts to best align an individual's chromosomal sex with other phenotypic sex characteristics.

Sexual orientation and human brain structure

In the early 1990s, several high-profile studies of postmortem brain samples reported anatomical dimorphisms between the brains of homosexual and heterosexual men. This issue was approached primarily in males, most likely because of the increased availability of postmortem brain samples from self-identified homosexual men who died due to AIDS-related complications—a sad and serious confounding factor for interpreting these studies. Such analyses of anatomical dimorphisms were based on the notion that mechanisms resulting in a "homosexual brain" (if such a singular entity exists) would tend to make dimorphic structures more feminized in homosexual men and masculinized in homosexual women. Again, the validity of this assumption can be questioned.

Initial studies by Simon LeVay in an analysis of a sample of postmortem tissue from heterosexual and homosexual men that received substantial attention when it was published in the early 1990s indicated that such differences might exist. These observations, however, were only modestly significant and not absolutely predictive. They suggested that INAH3, a sexually dimorphic subnucleus of the **interstitial nuclei of the hypothalamus (INAH)**, the human homologue of the sexually dimorphic SDN-POA in rodents (see Figure 25.7B), was smaller on average in homosexual than in heterosexual men, and similar in size to that of heterosexual women. Nevertheless, the size of INAH3 alone was not a reliable predictor of sexual orientation in the sample reported by LeVay. Subsequent studies that have taken into account a significant occurrence of degenerative changes in brain tissue from individuals who died of AIDS-related illnesses (now far less common because of improved anti-retroviral treatment). These studies regardless of declared sexual orientation of the individuals from whom the post-mortem samples were obtained, have failed to replicate this suggested dimorphism. Indeed, analyses of several other anatomical dimorphisms of the hypothalamus in heterosexual and homosexual men (women remain largely unstudied) have failed to generate consistent results of reliable differences that either conform to or reverse the statistical associations of INAH size with chromosomal sex. Noninvasive anatomical imaging approaches such as MRI lack the resolution to identify these cellular differences in the hypothalamus. Given the current evidence in humans, it would seem that the volume of distinct hypothalamic nuclei and number of neurons do not reliably predict sexual orientation. Indeed, they may not even reliably indicate the chromosomal sex of any individual.

The application of fMRI has facilitated formulation of a different set of questions and provisional answers by mapping differences in activation of potentially functionally dimorphic regions in the brains of heterosexual and homosexual men and women in response to behaviorally relevant stimuli. These fMRI studies of the hypothalamus were undertaken based on the centrality of hypothalamic nuclei for sex differences in reproductive and parenting behaviors that are more thoroughly studied, with better biological resolution, in animal models. In addition, they reflect an assumption that olfactory stimuli are highly effective in activating hypothalamic circuits, and also in some way involved in influence reproductive behaviors. Again, these assumptions and their relevance for human behaviors can be questioned. In these studies, individuals have been carefully selected for comparable age, consistent sexual orientation (both heterosexual and homosexual), relationship status (a balanced percentage of both the heterosexual and homosexual individuals were in committed relationships), age, and HIV status (none of the individuals were HIV-positive). Heterosexual men and women show differential patterns of

(A) Androgen administered

Hypothalamus activation

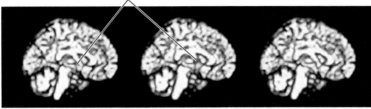

Heterosexual female   Homosexual male   Heterosexual male

(B) Estrogen administered

Cingulate cortex activation   Hypothalamus activation

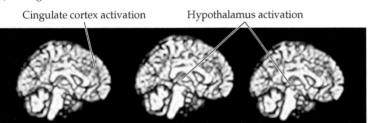

Heterosexual female   Homosexual female   Heterosexual male

H. Savic and P. Lindström, 2005. *Proc Nat Acad Sci USA* 102: 7356–7361

FIGURE 25.13 **Hypothalamic activation by estrogen and androgens in heterosexual and homosexual women and men** (A) Inhaling an androgen-related compound elicits focal activation of the hypothalamus (red) in heterosexual women and homosexual men; there is no activation in the hypothalamus of heterosexual men. (B) An inhaled estrogen-related compound activates the cingulate cortex, but not the hypothalamus, in heterosexual women. In homosexual women, estrogen elicits some activation in the hypothalamus but no in the cingulate cortex, similar to that seen in heterosexual men.

hypothalamic activation when presented with estrogen-related and androgen-related compounds as odorants (see Chapter 14). In homosexual men, androgens maximally activate the anterior hypothalamus as they do in heterosexual women; and in homosexual women, estrogens maximally activate the anterior hypothalamus, as in heterosexual men (Figure 25.13). The behavioral significance of this apparent reversal of a functional dimorphism in self-identified heterosexual or homosexual individuals is not clear. Several additional studies have examined statistical differences in the sizes and functional activation of other brain regions and structures using structural or functional MRI. While several individual studies suggest that areas of the cerebral cortex or subcortical nuclei may differ in size, shape, or activity, none of these differences appear robust enough to be predictive of sexual orientation based on brain morphology or patterns of functional activity.

A consistent relationship between brain structure, functional activation, and gender identity, if any, is even less clear for individuals whose chromosomal sex and phenotypic sex do not match their gender identity. Analyses that combine multiple studies with small sample sizes, and attempts to extract standardized data that permit more statistically robust conclusions (so-called meta-analyses) have identified potential mean differences in size, shape, or activation of the hypothalamus, striatum, cerebellum, and cerebral cortex. None of these differences are predictive of transgender identities; instead, they are modest statistical distinctions. Indeed, the conclusion reached from a recent meta-analysis by Mueller and colleagues published in 2021 suggests that "rather than being merely shifted towards either end of the male-female spectrum, transgender persons seem to present with their own unique brain phenotype."

Sex-based differences in cognitive functions

There is little scientific evidence that distinct cognitive abilities of men and women (if any consistent differences exist at all) differ in ways that correlate strictly with either genotypic or phenotypic sex. Thus, the conventional wisdom or prejudice that boys and girls or men and women have greater aptitudes for, or interest in, certain tasks or superior abilities for certain skills, is neither wise, nor judicious. Many apparent sex differences in cognitive tasks such as language, learning, memory, and visuospatial ability may reflect influences not directly related to genetically established sexual dimorphisms. Most of these distinctions, if they exist at all, are probably independent of the developmental history of differential exposure to gonadal steroids from testes or ovaries. Thus, statistically significant differences in the performance of men and women on a variety of tasks are at least as likely to represent social or cultural influences that result in different patterns of learned behaviors. The relationship of these behaviors to genotypic or phenotypic sex, if any, has not been established.

The issue of brain structural differences beyond quantitative cell biological dimorphisms detected via histological analysis in the spinal cord and hypothalamus has proved even more difficult to evaluate. This issue can be addressed with substantially greater quantitative resolution in experimental animals where issues of individual variation unrelated to genotypic sex can be controlled, cells can be counted, and additional histological assessment can provide reliable numbers. These analyses demonstrate differences in cell and synapse numbers, frequency of distinct cell classes, or specific neuronal features like dendritic spines in regions beyond the hypothalamus, including cerebral cortical areas, the hippocampus and the amygdala (Figure 25.14A). Parallel postmortem as well as structural MRI analyses in humans suggest that similar forebrain structures— the many cortical regions, the hippocampus, the amygdala, as well as cerebral commissures (axon tracts) such as the corpus callosum and anterior commissure that connect these regions in the two cerebral hemispheres —may differ in size or shape in men and women (Figure 25.14B). Thus, there is some agreement between animal studies and those in humans; the studies, however, differ in the approaches used, and the resolution possible to address differences that might have functional significane. Most of the studies in humans have been performed in small or highly heterogeneous samples and reflect highly derived analyses that reveal only small differences in size and shape. Finally, it has proven difficult to replicate many claims of sex differences in brain structure in additional independent samples.

Despite the difficulties studying sex differences in brain architecture beyond the spinal cord or hypothalamus in humans, a great deal of interest has focused on the amygdala as the most likely central site of sexual dimorphism. This interest reflects the established role of the amygdala in regulating the output of hypothalamic nuclei, robust gonadal steroid–sensitive sex differences in glial structure and related behaviors in male versus female rodents (see Figure 25.11), and the hypersexuality observed in animals and people with bilateral damage to the amygdala (see Clinical Applications, Chapter 30). Some MRI studies suggest that the male amygdala has a larger volume than the female; however, additional observations indicate that these differences are not significant when measurements are corrected for brain size, cranial volume or body weight that tend to differ in men and women, most likely independent of gonadally sexual dimorphism driven directly by gonadal steroid influences. Studies of intersex individuals (see Table 25.2) suggest that the amygdala may be sexually dimorphic, perhaps because of the influence of altered gonadal steroid levels in these conditions and their potential impact on glial and neuronal differentiation in the amygdala. These anatomical and imaging studies, however, are complicated by the generally smaller total brain size in genetically intersex individuals, and as yet have not been complimented by cellular analyses similar to those possible in animal models.

In parallel with these equivocal anatomical data, there have been similar functional studies, primarily based upon fMRI data, that suggest sex differences in information processing in regions beyond the hypothalamus (see Figure 25.13). One of the more apparently robust differences are the distinctions in the amygdala of men and women performing emotional memory tasks. In such tasks, individuals view aversive or frightening films or images that elicit an emotional response. Several weeks later, the individuals are evaluated for their memory of these images. In recalling emotionally charged content, the right amygdala is maximally activated in men, while the left is maximally activated in women. More recent work indicates that these differences are seen, statistically, in response to distinct aspects of facial cues indicating fearful or neutral emotional states. These functional differences suggest that *laterality of activation*, rather than differences in size of the nucleus, is the most robust sexual dimorphism in the human amygdala. The functional significance of this observation, however, is not known.

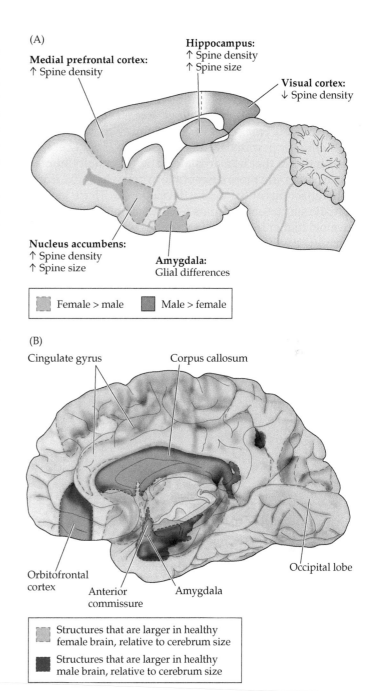

FIGURE 25.14 Potentially dimorphic brain regions beyond the hypothalamus (A) In the rodent brain, regions that differ in neuron or glial numbers, features, or proportion of cell types (identified based on cell counts or analysis of distinct cell classes and connections) include the frontal and cingulate cortex, the nucleus accumbens (part of the striatum), and the amygdala. These observations in animal models have motivated analysis of parallel brain regions, albeit without the cellular and physiological resolution possible in experimental animals, in the human brain. (B) In the human brain, the average size of these regions differs in females versus males. It is important to note, however, that these representations are based on mean estimates; individual variability makes it impossible to predict an individual's sex based solely on the sizes of the brain regions depicted here. (A after L. Cahill, 2006. *Nat Rev Neurosci* 7: 477–484; B after M. Uhl et al., 2022. *Front Mol Neurosci* 15: article 818390.)

Despite the paucity of reliable data that support anatomical or functional sex differences in the human brain beyond those in the spinal cord and hypothalamus directly related to external genitalia or reproduction, there are indications that some biological biases exist between the two genotypic sexes in vulnerability to neurodevelopmental or psychiatric diseases (Table 25.3). These biases indicate that human males, and potentially human male brains, are more susceptible to behavioral disorders that reflect disrupted neural development. These include autistic spectrum disorders (ASD), which have a four- to fivefold higher frequency in males, as well as biases in the sorts of social cognitive impairments seen in female versus male individuals with ASD. In addition, males have a two- to threefold higher incidence of attention deficit hyperactivity disorder (ADHD), which is thought to disrupt optimal function of neural circuits for attention and impulse control (see Chapters 29 and 33). Finally, there is a slightly elevated risk for schizophrenia in males. In contrast, females have generally increased frequency of adult-onset behavioral disorders, including major depression, distinct types of bipolar disorder, anxiety, obsessive-compulsive disorder (OCD), and post-traumatic stress disorder (PTSD).

It is difficult to assign these differences directly to the fundamental mechanisms for sexual differentiation of the brain that are currently well-established. Indeed, the causes of all of these disorders remain mostly undefined and, despite biases based upon chromosomal and phenotypic sex, all of these diseases are frequently diagnosed in both sexes. The existence of these sex biases nevertheless raises the question of whether developmental mechanisms that rely upon dimorphic systemic signals or ongoing gonadal steroid modulation of neural functions in some way distinguish some of the circuits for complex behaviors in human male versus female brains. The polar opposites of interpretation of all of these inconclusive data—uncertain anatomical and functional studies and statistically more robust differences in diagnostic frequency of several brain disorders—are reflected in titles of recent reviews articles published in the field: In 2022: "Sex is a defining feature of neuroimaging phenotypes in major brain disorders," and in 2021: "Dump the 'dimorphism': Comprehensive synthesis of human brain studies reveals few female-male differences beyond size." Thus, while questions of sex-related differences in human brain structure, function, and pathology remain a focus in neuroscience, the answers—if any definitive answers exist—remain a matter of debate that defies any singular, certain statement based upon rigorous scientific evidence.

TABLE 25.3 Sex Biases in the Incidence of Human Behavioral Disorders

Condition[a]	Sex differences in prevalence	Sex differences in onset	Sex differences in phenotype
Neuropsychiatric conditions with origins in development			
Autism spectrum disorder	Four to 5 times higher in males than in females	None	More social impairment in males; more affective symptoms in females
Conduct disorder oppositional defiance disorder	Three times higher in males than in females	Earlier onset in males	More externalizing symptoms in males; more affective symptoms in females
ADHD	Two to 3 times higher in males than in females	None	More hyperactivity, externalizing, and impulsivity in males; more internalizing, inattention, and intellectual impairment in females
Schizophrenia	1.42 times higher in males than in females	Earlier onset in males	More language disruption, positive symptoms, and severe course of illness in males; more affective symptoms in females
Neurological developmental conditions			
Dyslexia or reading impairment	Two to 3 times higher in males than in females	None	None known
Stuttering	2.3 times higher in males than in females	Adolescent onset 4 times higher in males	None known
Tourette syndrome	Three to 4 times higher in males than in females	Earlier onset in males	Greater tic severity in adulthood in females

(Continued)

TABLE 25.3   Sex Biases in the Incidence of Human Behavioral Disorders *(continued)*

Adult-onset neuropsychiatric conditions

Major depression	None before puberty; 2 times higher in females than in males post-puberty	None	None known
Bipolar disorder	None for bipolar I; bipolar II higher in females than in males	Earlier onset in males	Sex by genotype interaction
Generalized anxiety	Two times higher in females than in males	None	Higher chronicity and comorbidity with major depression in females
Panic disorder	2.5 times higher in females than in males	None	None known
OCD	1.5 times higher in females than in males	None	None known
PTSD	Two times higher in females than in males	None	More likely in females than in males following childhood trauma
Anorexia nervosa	Three times higher in females than in males	Unknown	None known
Bulimia	Three to 4 times higher in females than in males	Unknown	None known
Alcoholism or substance abuse	Higher in males than in females	Earlier in females	Females progress to addiction more quickly than males

Adult neurological conditions

Migraine	None pre-puberty, but 3 times higher in females than males post-puberty	None	None known
Stroke	Higher in males than females before age 85, but higher in females than males after age 85	Males 4 years before females	None known

Neurodegenerative diseases

MS (with exception of primary progressive MS)	Two times higher in females than in males	Earlier onset in females	More severe in males
Alzheimer's disease	1.5–2 times higher in females than in males, especially in those over 80 years	Earlier onset in females	More tangles and global pathology in females; pathology more highly correlated with clinical score in females
Parkinson's disease	1.5 times higher in males than in females	Males 2 years before females	None known
ALS	Three times higher in males than in females	Earlier onset in males	None known
Myasthenia gravis	Four times higher in females than in males	Earlier onset in females	None known

Summary

In most animals analyzed thus far, the brains of male and female individuals become specialized during pre- and postnatal development for the division of behavioral tasks that deal with reproduction and the rearing of offspring. These differences reflect the consequences of intrinsically produced signals provided primarily by the embryo itself based on its chromosomal and phenotypic sex: the genomic and physical characteristics typically associated with females versus males in the relevant species. In mammals, the strongest determinant of these differences is the initial differentiation of gonadal tissues, under the control of the masculinizing transcription factor, SRY. SRY determines an individual's genetic sex, and usually the phenotypic sex as well, however SRY is not expressed in the brain. The influence of SRY on the nervous system is thus indirect: SRY-mediated masculinization leads to differential generation of male versus female gonadal tissues during fetal development and thus to sex-specific levels of circulating gonadal hormones (estrogen and testosterone in particular over the course of development). These hormones profoundly influence the development of brain structures that subserve peripheral structures (genitalia, mammary glands) directly related to reproduction and parenting. Some dimorphisms reflect trophic regulation of cell survival and death based on parallel development of peripheral organs that these cells innervate or regulate (male and female genitalia, mammary glands in females, and related muscles). Differences in size, presence or absence of muscles, or glandular tissue result in distinct levels of target-derived trophic support. The origins, existence, and functional significance of dimorphisms related to distinctions in sexual orientation and gender identity remain controversial. Whether they exist at all, and whether they arise via learned behaviors that are used to define gender roles in society or intrinsic developmental mechanisms, is not known.

■ Additional Reading

Reviews

Eliot, L and 3 others. (2019) Dump the dimorphism: Comprehensive synthesis of human brain studies reveals few male-female differences beyond size. *Neurosci. and Biobehavioral Reviews* 125: 667-697.

Luders, E, and F. Kurth (2020) Structural differences between male and female brains. *Handbook of Clinical Neurology 175: (3rd Series) Sex Differences in Neurology and Psychiatry,* R. Lanzberger, G.S. Kranz, and I. Savic, editors.

McCarthy M. M. and 4 others (2012) Sex differences in the brain: The not so inconvenient truth. *J. Neurosci.* 32: 2241-2247.

McCarthy, M. M., B. M. Nugent, and K. M. Lenz (2017) Neuroimmunology and neuroepigenetics in the establishment of sex differences in the brain. *Nature Neuroscience Reviews* 18: 471-548

McEwen, B. S. (1999) Permanence of brain sex differences and structural plasticity of the adult brain. *Proc. Natl. Acad. Sci. U.S.A.* 96: 7128-7129.

Meeh, K. L. and 3 others (2021). The development of sex differences in the nervous system and behavior of flies, worms and rodents. *Dev. Biol.* 472: 75-84.

Morris, J. A., C. L. Jordan and S. M. Breedlove (2004) Sexual differentiation of the vertebrate nervous system. *Nat. Neurosci.* 7: 1034-1039.

Salminen, L. E. and 5 others (2022). Sex is a defining feature in neuroimaging phenotypes in major brain disorders. *Hum. Brain. Map.* 43: 500 -542.

Seaman S. J. and A.S. Kaufman (2010) Sexual differentiation and development of forebrain circuits. *Curr. Opinion in Neurobiol.* 20: 424-431.

Important Original Papers

Allen, L. S., M. Hines, J. E. Shryne and R. A. Gorski (1989) Two sexually dimorphic cell groups in the human brain. *J. Neurosci.* 9: 497-506.

Allen, L. S., M. F. Richey, Y. M. Chai and R. A. Gorski (1991) Sex differences in the corpus callosum of the living human being. *J. Neurosci.* 11: 933-942.

Beyer, C., B. Eusterschulte, C. Pilgrim and I. Reisert (1992) Sex steroids do not alter sex differences in tyrosine hydroxylase activity of dopaminergic neurons in vitro. *Cell Tissue Res.* 270: 547-552.

Breedlove, S. M. and A. P. Arnold (1981) Sexually dimorphic motor nucleus in the rat lumbar spinal cord: Response to adult hormone manipulation, absence in androgen-insensitive rats. *Brain Res.* 225: 297-307.

Byne, W. and 8 others (2002) The interstitial nuclei of the human anterior hypothalamus: An investigation of variation with sex, sexual orientation, and HIV status. *Horm. Behav.* 40: 86-92.

Cooke, B. M., G. Tabibnia and S. M. Breedlove (1999) A brain sexual dimorphism controlled by adult circulating androgens. *Proc. Natl. Acad. Sci. U.S.A.* 96: 7538-7540.

De Vries, G. J. and 9 others (2002) A model system for study of sex chromosome effects on sexually dimorphic neural and behavioral traits. *J. Neurosci.* 22: 9005-9014.

Forger, N. G. and S. M. Breedlove (1987) Motoneuronal death during human fetal development. *J. Comp. Neurol.* 264: 118-122.

Frederikse, M. E. and 4 others (1999) Sex differences in the inferior parietal lobule. *Cereb. Cortex* 9: 896-901.

Gorski, R. A., J. H. Gordon, J. E. Shryne and A. M. Southam (1978) Evidence for a morphological sex difference within the medial preoptic area of the rat brain. *Brain Res.* 143: 333–346.

Gron, G. and 4 others (2000) Brain activation during human navigation: Gender different neural networks as substrate of performance. *Nat. Neurosci.* 3: 404–408.

Lasco, M. S. and 4 others (2002) A lack of dimorphism of sex or sexual orientation in the human anterior commissure. *Brain Res.* 936: 95–98.

LeVay, S. (1991) A difference in hypothalamic structure between heterosexual and homosexual men. *Science* 253: 1034–1037.

Mueller, S. C. and 33 others (2021) The neuroanatomy of transgender identity: Mega-analytic findings from the ENIGMA transgender persons working group. *J. Sexual Med.* 18: 1122 – 1129.

Swaab, D. F. and E. Fliers (1985) A sexually dimorphic nucleus in the human brain. *Science* 228: 1112–1115.

Woolley, C. S. and B. S. McEwen (1992) Estradiol mediates fluctuation in hippocampal synapse density during the estrous cycle in the adult rat. *J. Neurosci.* 12: 2549–2554.

Books

Fausto-Sterling, A. (2000) *Sexing the Body*. New York: Basic Books.

Goy, R. W. and B. S. McEwen (1980) *Sexual Differentiation of the Brain*. Cambridge, MA: MIT Press.

LeVay, S. (1993) *The Sexual Brain*. Cambridge, MA: MIT Press.

LeVay, S. and J. Baldwin (2012) *Human Sexuality*, 4th Edition. Sunderland, MA: Sinauer/Oxford University Press.

Repair and Regeneration in the Nervous System

Courtesy of B. Karpinksi and A.-S. LaMantia

KEY CONCEPTS

26.1 Neural tissue has a distinct response to injury and limited potential for regeneration

26.2 The peripheral nervous system retains the capacity for axon regrowth and synaptic reinnervation

26.3 CNS axons and dendrites in most adult mammals lack the capacity for extensive regrowth

26.4 Adult vertebrate nervous systems retain some neural stem cells for limited replacement of neurons

Overview

The ability of brain tissue to alter, renew, or repair itself (beyond the ongoing molecular and cellular changes associated with synaptic plasticity) is limited. Unlike many other organs—notably the skin, lungs, intestine, and liver—that continuously generate new cells or extensively repair existing cells, mature human brains do not produce many new neurons once the initial complement is established during mid-gestation through early postnatal life, and existing neurons do not grow extensively once postnatal differentiation is complete. The modest recovery seen after most acute brain injuries—if there is recovery at all—is usually attributed to reorganization of function using remaining, intact circuits rather than repair of damaged brain tissue. The brain's capacity to repair itself in response to adult-onset neurodegenerative conditions such as Parkinson's, Huntington's, and Alzheimer's diseases is even more limited, leading to inexorable functional decline with little recovery. Nevertheless, some nervous system repair does occur after injury. The axons of neurons that project into the periphery (motor neurons, sensory ganglion neurons, and autonomic ganglion neurons) can regrow through vacated peripheral nerve sheaths. Using these sheaths as guides, central motor neuron axons can regrow through the periphery to reinnervate synaptic sites on muscles, and peripheral sensory neurons can eventually reinnervate sensory specializations in the skin. Few brain neurons can grow a new axon if the original axon is severed or injured, nor can neurons replace dendrites lost because of local tissue damage or degenerative disease. At least four barriers impede CNS regeneration. First, local injury of brain tissue or adult-onset degenerative diseases often lead to neuronal death. Second, several other cell classes, particularly glial cells, actively inhibit axon growth. Third, although neural stem cells are retained in the adult brain, most are constrained in their ability to divide, migrate, and differentiate. And fourth, immune responses in the nervous system, mediated by microglia, astrocytes, and oligodendrocytes, release cytokines that further inhibit extensive regrowth. Nevertheless, in some vertebrate species these impediments are circumvented in specific regions of the CNS. Neural stem cells are maintained in distinct niches over the life span, and they can generate new neurons with the capacity to engage in existing circuits or form entirely new circuits. Efforts to understand these exceptions provide a foundation for ongoing research into potential therapies for brain repair following traumatic injury, hypoxia or ischemia, stroke, or disease-related degeneration.

CONCEPT
26.1

Neural Tissue Has a Distinct Response to Injury and Limited Potential for Regeneration

LEARNING OBJECTIVES

26.1.1 Contrast the response to injury seen in brain tissue versus that in other tissues.

26.1.2 Distinguish damage and response to acute injury from that in neurodegenerative diseases.

26.1.3 Explain how functional recovery can occur without repair or regeneration.

26.1.4 Identify the three types of tissue repair that can occur in the nervous system after injury.

The damaged brain

Many organs are capable of repair and regeneration. Broken bones knit, and wounds heal. The epithelial cells of the epidermis, the intestinal lining, the airways, and the lung are constantly lost and replaced, as are blood cells. There are adaptive responses to transient or chronic disease states that respond to local or widespread, acute, or chronic degenerative loss of cells, for example in the gut epithelium or airways. These responses retain or restore reasonable levels of organ function despite disease-related pathology. The adult liver, an organ that confronts multiple toxins and metabolizes them to less harmful compounds, can repair itself through enhanced growth and function of remaining cells or regeneration of cells that have been lost, unless there is widespread acute or cumulative damage to the organ. In each of these cases, tissue repair depends on the retention in the adult of tissue-specific stem cells—maintained in protective environments called *niches*—that can proliferate and generate new differentiated cell classes to replace those that are damaged acutely or lost as a result of normal wear and tear. In contrast, the brain, especially in mammals, is generally refractory to repair. This deficiency was noted clinically at the dawn of medical history (Figure 26.1). The pessimistic warning of ancient Egyptian physicians that a vertebral displacement (presumably indicating spinal cord damage) resulting in paralysis of the arms and legs is "a disease one cannot treat" remains more or less true today. Although the hieroglyphics have changed, the message remains the same. Indeed, the understanding of the structure and function of nervous tissue gained in the intervening two millennia has not provided much encouragement.

Nervous tissue is made up of many classes of largely postmitotic, terminally differentiated, highly branched, interconnected nerve cells that communicate via cellular junctions to transmit electrical impulses. In addition, the lifelong storage of information via plastic changes at individual synapses throughout the brain (see Chapters 8, 24, and 30) presents an additional challenge for restoring function as well as cellular structure after injury or in the

FIGURE 26.1 A historically difficult problem: repairing the brain and spinal cord after injury This ancient Egyptian papyrus includes advice from a physician regarding spinal cord damage. The symbols in brown translate to: *When you examine a man with a dislocation of a vertebra of his neck, and you find him unable to move his arms, and his legs... Then you have to say: a disease one cannot treat.*

face of degenerative disease. Thus, repairing nervous tissue presents a far greater challenge than regenerating an organ such as the skin, gut epithelium, lung, or liver, all of which are far more limited in structure and in their range of cell classes. There is little evidence of wholesale regeneration in the brain, especially in vertebrates, including mammals: with few exceptions (see Chapter 14), neurons are not replaced by local neurogenesis after damage or loss due to degeneration. As postmortem analyses and histological studies of the brains of humans and of animal models progressed during the last century, it became clear that specifically localized lesions, devoid of neurons and instead populated with glial cells and lymphocytes, invariably accompany specific behavioral deficits after brain damage. These lesions remain visible even many years after the injury, and their histological appearance suggests they are sites of degeneration rather than regeneration. There appears to be little repair of damaged tissue toward a more typical appearance; the injured region is characterized either by a fluid-filled cyst or by "scar tissue" that lacks the histological integrity of neighboring brain regions. Thus, any adaptive changes in brain function following injury are unlikely to reflect wholesale replacement or remodeling of the cellular constituents of the injured brain.

Despite the apparent barriers to repair and regeneration, the view that the neural circuits, once damaged, can never be restored to their previous state has been challenged for more than a century. Key questions about brain regeneration and repair, as well as relevant data to address these questions, continue to be debated. Nevertheless, nearly all modern studies of functional recovery after brain injury indicate that improvement seen in some individuals following neural trauma primarily reflects the reorganization of intact circuits for functional compensation rather than the regrowth or replacement of damaged neurons. Indeed, some of the most successful efforts to restore behavioral function following damage to the nervous system have focused on engaging the neurons and circuits that remain after injury to compensate for those that have been lost (see later in this concept).

Neurodegenerative diseases: Different demands for repair

Acute injury to an otherwise healthy brain clearly presents multiple challenges for regeneration and repair of neurons and connections that have been lost. A different set of challenges is faced, however, when the brain is slowly damaged by neurodegenerative disease. Such diseases can lead either to widespread neuronal disconnection and loss, as in the case of Alzheimer's disease (Figure 26.2), or to synaptic damage and neuronal death that are limited to a particular cell class, region, or circuit, as occurs in Parkinson's disease, Huntington's disease, and amyotrophic lateral sclerosis (ALS). In each of these diseases, damage to the brain initially is focused on one subclass of nerve cell, often in a limited region. The degenerative changes in the targeted neurons usually begin slowly, long before any functional or behavioral changes can be detected (see Chapter 30). Nevertheless, connections are being lost during this time, and the resulting cellular changes render these damaged areas of the brain less likely to be repaired. The repair-opposing changes include abnormal glial cell growth and proliferation or infiltration of immune cells

that elicit a local response similar to that of inflammation in other tissues (see Concept 26.3).

Neurodegenerative diseases present several barriers to effective repair that seem even more daunting than those presented by acute localized brain damage. First, the early "silent" phase of damage must be detected in time to (perhaps) arrest further degenerative change that will result in even more widespread loss of neurons and connections. This early detection remains difficult because of a lack of reliable ways to identify individuals who are at risk for most of these diseases. There are some exceptions: Some fairly rare genetic variants of Alzheimer's and Parkinson's diseases, and most instances of Huntington's disease, which is caused primarily by mutations in the *Huntingtin* (*HTT*) gene, can be identified in individuals who have a family history of one these diseases. Despite such genetic diagnoses, however, arresting the degeneration, repairing the damage, and preventing further damage due to the disease remains more or less impossible. Once behavioral symptoms emerge, necessary repairs include restoring extensive networks of connections made by neurons beyond the primary pathologically targeted cell type. Thus, in instances where projection neurons in the cortex seem to be targeted (Alzheimer's disease) or motor neurons in the brainstem and spinal cord are lost (ALS), not only do these target neurons need to be stabilized or replaced, but all of the other neurons that connect to them, including interneurons that maintain an appropriate balance of local excitatory and inhibitory connections, must be instructed to reestablish damaged or lost connections. To be optimally effective, repair of already damaged neurons or circuits in neurodegenerative diseases must be accompanied by therapies that prevent new damage to circuits not initially in need of repair. Similarly, there must be strategies to prevent

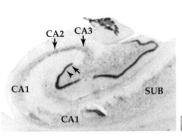
(A) Nissl stain of healthy human hippocampus

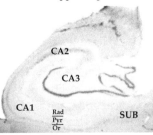
(B) Human Alzheimer's disease hippocampus

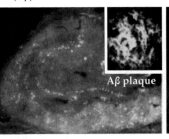
(C) Amyloid plaques (Aβ)

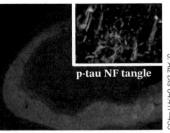
(D) Neurofibrillary tangles (phospho-tau)

From G. Fattepekar, et al., 2002. *AJNR Am J Neuroradiol* 23: 1313–1321

J. Alz Dis 64:417–435

FIGURE 26.2 Neurodegenerative diseases irreversibly damage cells and circuits, leading to diminished function
(A) Cytology of the hippocampus in a Nissl-stained specimen from a healthy human brain. CA1–4 indicates the position of the hippocampal CA fields (cornon ammu, the latin name for the hippocampus); Sub indicates the location of the subiculum, a transitional region between the hippocampus and the neocortex. (B) The density of Nissl-stained cells is diminished in the hippocampus from an individual with Alzheimer's disease, especially in the CA1 and CA2 regions. The layes of the hippocampal CA fields

are indicated at the bottom left: Rad defines the stratum radiatum, Pyr, the pyramidal cell layer, and Or, stratum oriens. (C) Another hippocampal section from same individual as in panel B, labeled immunohistochemically for amyloid protein Aβ. Aβ plaques, which are extracellular accumulations of misfolded amyloid protein, are distributed throughout CA1 and CA2 and disrupt the synaptic connectivity in those regions. (D) In parallel, accumulations of phosphorylated tau protein (phospo-tau) called neurofibrillary tangles in CA1 and CA2 neurons indicate cell pathology that leads to disconnection and disfunction.

renewed degenerative change in neurons or circuits that have been rescued or reconnected. Currently, none of these challenges have been adequately met. The barriers to doing so include understanding complex molecular pathogenesis in each disorder and devising strategies to reactivate the necessary steps of development that, at appropriate sites with appropriate frequency, originally made the lost connections. These issues will be the focus of several of the following sections of this chapter.

Functional reorganization without repair

Neurologists have long recognized that, over time, individuals who suffer focal strokes or sustain limited injuries to distinct brain regions can eventually recover some of the deficits seen immediately after the trauma. Movement in paralyzed limbs can improve (especially if physical therapy is included in the treatment plan), coordination and balance can be partially restored, and problems with verbal communication can diminish with intensive speech therapy. Such recovery is not thought to reflect significant regrowth or replacement of damaged neurons. Instead, the available evidence indicates that undamaged brain regions eventually become activated and reorganized to support, at least in part, functions whose primary representation was disrupted. The best understanding of functional recovery comes from studies of the primary motor cortex, where behaviors controlled by the motor cortex (generation of force and accuracy of movement) can be measured reliably over time after focal damage.

Observations made in animals have shown that circuits in the adult primary motor cortex retain some capacity for use-dependent plasticity, suggesting a biological mechanism for the reorganization and recovery seen in patients. The plasticity of the primary motor cortex reflects that region's rich array of horizontally spreading axonal connections. Thus, connections that might not be active when the system is intact can be "unmasked" when there is damage nearby, thus providing a functional "bridge" to relay and integrate information across the limited lesion. In addition, plastic changes that favor functional recovery—similar to long-term potentiation (LTP) or long-term depression (LTD; see Chapter 8)—may occur at synapses between intact excitatory or inhibitory neurons, perhaps resetting the excitatory–inhibitory balance to maximize circuit function and prevent seizures or other disruptions. There may also be some modest local growth of axon branches or dendrites as well as new synaptogenesis from intact neurons that further strengthens remaining connections; however, the extent to which this occurs remains uncertain. Finally, altered activity in the ipsilateral motor cortex may provide activation that can pattern the appropriate movement via spared contralateral pathways.

The consequences of motor cortex plasticity can be seen in stroke patients using fMRI in parallel with observing the patient's progress during rehabilitation. The most thorough studies have been of individuals with focal strokes in the subcortical white matter that result in specific deficits in hand movement and grip strength. In these individuals, the amount of cortical activity, particularly in the hand region of the motor map, is *increased* and broadened shortly after injury, and *declines* with improved function (Figure 26.3A). Indeed, those individuals with poorer outcomes did not show as extensive a decline in ectopic cortical activation. This surprising finding can be best appreciated by following differences in cortical activity in a single individual over time (Figure 26.3B,C). As hand movement improved in an individual with a focal pontine stroke (compromising axons from upper motor neurons in the hand representation), focal activity in the hand representation, which increased substantially shortly after the stroke, declined as functional recovery progressed. Ectopic or increased activity in primary and supplementary motor cortex, as well as in the primary and secondary visual cortices both ipsilateral and contralateral to the lesion, also diminished (see Figure 26.3B). Whether these changes represent diminished excitatory neuronal activity or increased inhibitory neuronal activity is not known since the fMRI signal does not resolve those classes of neuronal activity.

Similar observations have been made in individuals with strokes that compromise complex functions such as language. Activation of remaining brain circuits changes over time following rehabilitation and functional recovery; however, the magnitude, direction, and localization of these changes are more variable than changes that occur in the motor cortex. In sum, neural circuits that remain following focal brain damage can reorganize, based on changing patterns of activation, to accommodate functional recovery, even if rules for reorganization remain elusive. The limited nature of this reorganization and its absence in individuals with more profound impairments indicate the challenge of regaining normal function after the brain has been damaged.

These examples of functional restoration without repair are usually seen without substantial intervention beyond physical or behavioral therapy to support the reacquisition of lost movement or skills as reactivation or reorganization of existing neurons and circuits occurs and becomes stabilized. Over the last decade, however, strategies have emerged that use external electrical stimulation to activate remaining circuits in individuals with spinal cord injuries or with neurodegenerative diseases that compromise motor function. Epidural electrical stimulation, which does not require insertion of stimulating electrodes directly into neural tissue, has provided substantial functional recovery for small numbers of individuals. These individuals all had spinal cord injuries that eliminated descending cortical and brainstem motor control and coordination for motor neurons in the lumbosacral spinal cord (thus, the

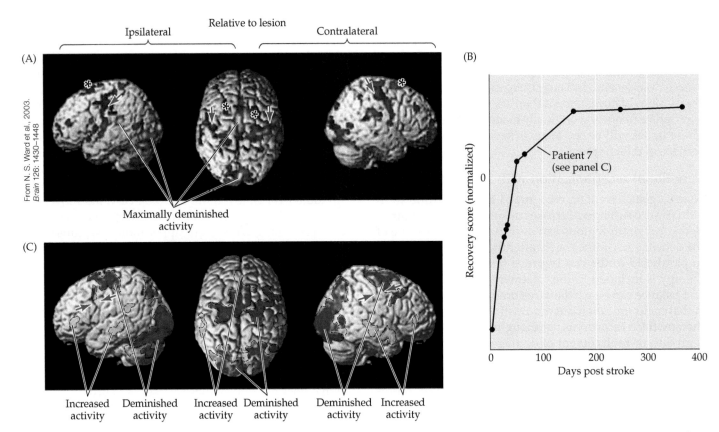

FIGURE 26.3   Altered cortical activity patterns are correlated with functional recovery after a focal stroke
(A) An fMRI map of *diminished* activity compared with age-matched control individuals following a middle cerebral artery (MCA) stroke. Regions of maximally diminished activity are red; orange and yellow reflect moderate changes. The data reflect the average of changes from six individuals (from a total of 20 studied—the others had unilateral infarcts either in the internal capsule or pons) with comparable MCA strokes projected onto an idealized brain surface. There is a statistically significant bilateral decline in activity in much of the precentral gyrus (primary motor cortex; blue arrows); the decline is slightly enhanced (red) contralateral to the lesion. There is also decline in activity in supplementary motor areas (asterisks). Finally, visual cortex activity is diminished. (B) Improvement of motor performance over time for a single individual. Improvement is seen as an increased "recovery score," which measures performance on a battery of tasks that require hand-grasping and controlled movement. (C) Areas of diminished (red) and increased (green) cortical activity over time in a single patient scanned multiple times. This individual (whose recovery is documented in panel B) had a pontine stroke that compromised axons carrying motor commands from the "hand" region of the primary motor cortex. This composite image shows bilateral diminished activity in the primary and supplementary motor areas, as well as in the visual cortex. The changes include diminished activity in the region of the primary motor cortex associated with the hand ipsilaterally and contralaterally (arrows). (A after N. S. Ward et al., 2003. *Brain* 126: 1430–1448.)

lesions were in the thoracic cord and left the distal spinal cord intact). The utility of this approach has relied on engineering and computational modeling to identify patterns and timing of stimulation most likely to restore coordinated muscle contractions so that the individuals can engage in walking and other activities that require controlled movement of the legs (Figure 26.4). Somewhat more invasive approaches have been used to stimulate intact basal ganglia circuits in which activity has been diminished because of disconnection following degeneration of dopaminergic inputs from the substantia nigra. In these approaches, bilateral electrodes are usually implanted into regions of the thalamus that control movement, the subthalamic nucleus (normally a "brake" for the activation of movement via the basal ganglia) or the globus pallidus (which influences the thalamus to initiate complex motor behaviors; see Chapter 18). In these instances, there is no actual repair of the damaged brain regions. Instead, the intact circuits adjacent to the damaged or degenerating brain region are engaged to approximate normal function.

Three types of neuronal repair

The damaged nervous system, either peripheral or central, can accomplish in varying degrees three types of cellular repair that can restore function The first type is the *regrowth of axons*, either from intact nerve cells in peripheral ganglia or intact central α motor neurons of the spinal cord and brainstem of which peripherally extending

(A)

(B)

(C)

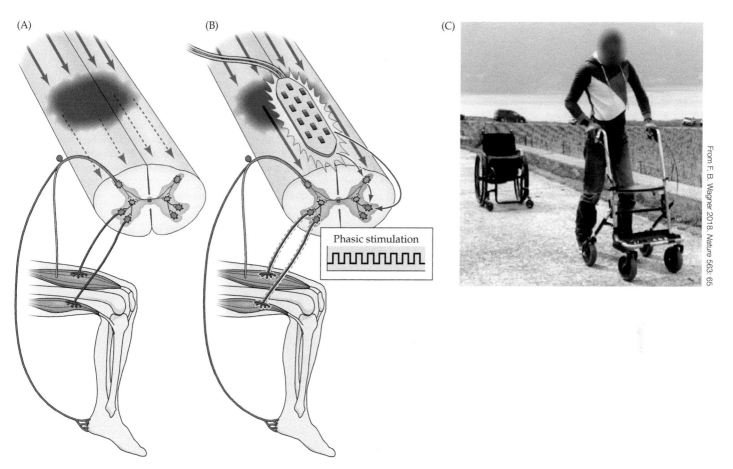

Phasic stimulation

From F. B. Wagner 2018, *Nature* 563: 65

FIGURE 26.4 Strategies to restore function without cellular repair in the nervous system (A) When there is a spinal cord injury (red shading), the connections between descending axons (solid green arrows) and the motor neurons and interneurons distal to the lesion (gray dashed arrows) are diminished or completely lost. Nevertheless, the local "reflex" circuitry (see Chapter 1)—sensory afferents (blue) and their connections with interneurons in the ventral horn as well as output motor neurons (red)—remains intact in the spinal cord distal to the lesion. (B) Epidural electrical stimulation (EES) via a surface electrode array placed on the dorsal spinal cord dura can stimulate the sensory afferents to control local patterning of activation among ventral horn interneurons and motor neurons for limb movement (walking; see Chapter 16). This control can be optimized by establishing phasic stimulation patterns that can be adapted for the needs of each individual based on their spinal cord injury. In addition, EES can recruit small numbers of descending axons that remain intact after a spinal cord "crush" injury (green arrows) as opposed to complete severing of the spinal cord (see Figure 26.7) to facilitate movement control. (C) An individual with a spinal cord lesion, assisted by EES, can walk independently with aid of a walker. (A,B after J. S. Calvert et al., 2019. *Neuromodulation* 22: 244–252.)

axons have been severed (Figure 26.5A). This scenario requires a reactivation of the developmental processes for axon growth and guidance into the periphery, as well as those for establishing sensory specializations in the skin or synapses at neuromuscular junctions. In addition, such repair entails activity-dependent competitive mechanisms similar to those used during development to ensure proper quantitative matching of newly regrown afferents to temporarily denervated targets. This first type of repair is seen primarily when sensory or motor nerves are damaged in the periphery, leaving the nerve cell bodies in the relevant sensory and autonomic ganglia or the spinal cord intact. This *peripheral nerve regeneration* is the most readily

accomplished type of repair in the nervous system, and the most clinically successful.

The second type of repair is *regrowth of damaged neurons*. Although the axons or dendrites of neurons may be injured, the neurons themselves often survive (Figure 26.5B). This type of repair requires first that nerve cells are capable of restoring their damaged processes and connections to some level of functional integrity. Thus, new dendrites, axons, and synapses must grow from an existing cell body—a phenomenon sometimes referred to as *sprouting*. To achieve this, several developmental mechanisms must be reactivated, including appropriate regulation of initial cell polarity to distinguish dendrites and axons; adhesion

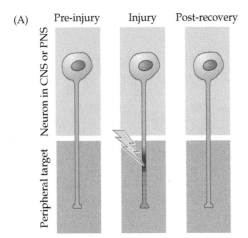

(A)

Pre-injury   Injury   Post-recovery

Neuron in CNS or PNS

Peripheral target

FIGURE 26.5 Three types of nervous system repair or regeneration
(A) In peripheral nerve regeneration, when peripheral axons are severed, the neuron, whether in a peripheral ganglion or in the CNS, regenerates the distal portion of the axon. (B) Repair of existing neurons at and around a site of injury in the CNS. Prior to injury, glial cells (dark purple) are quiescent. Immediately following the injury, the glial cells grow, axons and dendrites degenerate, and connections are lost. Following recovery, some modest axon and dendrite growth may be seen, but the hypertrophic glial cells remain to form a "scar" at the site of the tissue damage. (C) Neuronal replacement depends on the maintenance of neural stem cells (dark green) within or adjacent to the region where damage occurs. Following injury, this stem cell proliferates and gives rise to new neuroblasts that then differentiate and integrate into the damaged tissue. These new neurons (dark purple) make connections with existing cells.

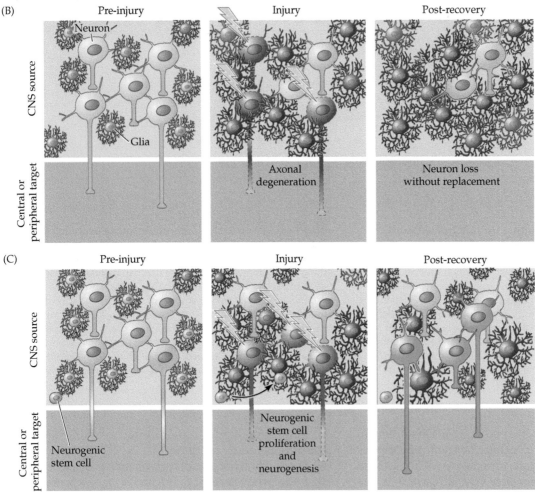

(B)   Pre-injury   Injury   Post-recovery

CNS source

Neuron

Glia

Central or peripheral target

Axonal degeneration

Neuron loss without replacement

(C)   Pre-injury   Injury   Post-recovery

CNS source

Central or peripheral target

Neurogenic stem cell

Neurogenic stem cell proliferation and neurogenesis

signaling to direct process extension; and trophic signaling to support growth and stable connections with appropriate targets (see Chapter 23). This type of repair also requires the cooperative regrowth of existing neuronal and glial elements in a more complex environment—the local mature neuropil—than that in the peripheral nervous system (PNS). It generally fails in the injured CNS except over limited distances, most likely because of local overgrowth of glial cells in response to injury and their production of signals that inhibit neuron growth. The subsequent loss of trophic support to damaged axons and dendrites due to inaccessibility of once innervated, now disconnected targets obscured by glial overgrowth, along with the action of inflammatory cytokines (released by glial cells, microglia, macrophages, and other immune cells in response to tissue damage; see Concept 26.3), may suppress reactivation of cellular mechanisms for dendritic and axonal regrowth as well as those for synapse formation. Thus, this second type of repair is quite limited.

The third type of repair is the *genesis of new neurons* to replace those that have been lost, whether through normal wear and tear or as a result of damage (Figure 26.5C). Such adult neuronal genesis occurs rarely, and its contributions to maintaining or restoring function are uncertain. The genesis of peripheral olfactory receptor neurons throughout adult life (see Chapter 14) is an example of this type of repair. It is ongoing and gradual; however, if there is acute traumatic damage to the olfactory epithelium or olfactory receptor neuron axons extending through the cribriform plate to the olfactory bulb, it also can occur. The growth of newly generated axons from new olfactory receptor neurons resembles peripheral nerve regeneration in some respects. The olfactory ensheathing cells that support ongoing olfactory receptor neuron axon growth to the olfactory bulb resemble peripheral Schwann cells more than central glia, even though their derivation from the neural crest (which generates Schwann cells) versus cranial placodes (which generate the neurons and supporting cells of the olfactory epithelium) remains controversial. The olfactory receptor neuron axons retain the capacity to make new synapses in the CNS, which is another fairly rare occurrence. If repair is accomplished by genesis of new neurons in other regions of the adult brain, several criteria must be met. First, nervous tissue must retain a population of *multipotent neural stem cells* (see Chapter 22) that can give rise to all of the cell types of the relevant brain region. Second, these neural stem cells must be maintained in a region that retains an appropriate environment for the genesis and differentiation of new nerve cells and glia—thus, a neural stem cell niche. Third, the regenerating tissue must preserve the capacity to recapitulate (or closely parallel) the local migration, process outgrowth, and synapse formation necessary to reconstitute functional networks of connections as well as long-distance connections.

Peripheral neurons seem to be quite capable of extensive axonal regrowth and reinnervation via newly remade synapses, leading to functional recovery. The extent of similar repair possible in the mammalian brain, however, is limited. Some other vertebrate species (and many more invertebrate species) exhibit the capacity for axon regrowth, neuron replacement, and wholesale regeneration of lost or damaged neural tissue. These examples in insects, fish, frogs, and birds suggest that a better understanding of neural stem cell biology, axon growth and guidance, synapse formation, and plasticity (see Chapters 22–24) might eventually indicate strategies to promote the repair of damaged neural tissue in humans.

| CONCEPT 26.2 | The Peripheral Nervous System Retains the Capacity for Axon Regrowth and Synaptic Reinnervation |

LEARNING OBJECTIVES

26.2.1 Explain how peripheral nerve regeneration relies on local guidance for axons to grow back to their original targets.

26.2.2 Identify the four major cellular elements that contribute to peripheral axon regrowth and the reinnervation of targets in the mature PNS.

26.2.3 Discuss the reliance of regenerating peripheral axons on adhesion and trophic signals from Schwann cells and target cells.

26.2.4 Compare the reinnervation versus developmental mechanisms for peripheral axon growth and synapse formation.

Peripheral nerve regeneration

In the early 1900s, the British neurologist Henry Head provided a dramatic account of repair in the PNS. By this time, it had become clear that damage to a peripheral nerve resulted in a gradual but usually incomplete restoration of sensory and motor function. The speed and precision of this recovery could be facilitated by the surgical reapposition of the two ends of the severed nerve. Apparently, if regeneration occurred in an environment that restored continuity of the existing nerve sheath, functional recovery could be far greater. Head's interest in this possibility culminated in a somewhat idiosyncratic approach to documenting the extent and functional precision of regeneration in damaged peripheral sensory and motor nerves. Rather than evaluating functional recovery in patients whose traumatic injuries were variable in location and extent of tissue damage, he chose to undergo a precise nerve transection and reapposition experiment *himself*, documenting the results as a personal narrative. In his 1905 paper, Head wrote:

> On April 25, 1903, the radial (ramus cutaneus radialis) *and external cutaneous nerves were divided (cut) in the neighborhood of my elbow, and after small portions had been excised, the ends were united with silk sutures. Before the operation the sensory condition of the arm and back of the hand had been minutely examined and the distances at which two points of the compass could be discriminated had been everywhere measured.*

Head, et al., 1905, *Brain* 28: 99–115

Head went on to carefully monitor the return of sensation to the parts of his hand that had been rendered insensitive by the lesion (Figure 26.6). His observations emphasized

(A)

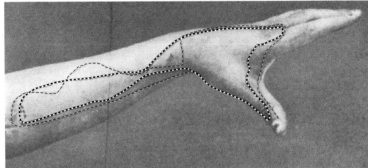

Site of cut

External cutaneous nerve

Site of cut

Superficial radial nerve

(B)

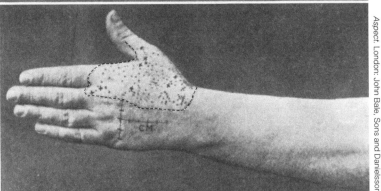

Aspect. London: John Bale, Sons and Danielsson

FIGURE 26.6 Henry Head's peripheral nerve regeneration experiment (A) Diagram of the human arm showing the location of the radial nerve (left), which was severed and the proximal and distal ends surgically reapposed in Head's experiment on his own arm; and the territory normally innervated by the radial nerve (purple, right). (B) Two photographs taken from Head's 1905 paper on his recovery from the peripheral nerve cut. The top image shows the outlines of regions of Head's lower arm and hand that were insensitive to painful stimuli (e.g., a pinprick), and the dotted line shows regions that were insensitive to light touch (e.g., a wisp of cotton). The bottom image shows the region of Head's hand and thumb that regained sensation after an initial period of recovery (2 to 6 months). The various marks within the resensitized region indicate "hot" and "cold" spots that were more or less sensitive to stimulation.

several important functional aspects of peripheral nerve regeneration in this one-of-a-kind experiment. The first indication of recovery was a difference in the return of general sensitivity to pressure and touch that was not well localized (a sensitivity he called "protopathic"), beginning at approximately 6 weeks and lasting about 13 weeks. Based on this observation, Head suggested that the "protopathic system regenerates more rapidly and with greater ease. It can triumph over want of apposition and the many disadvantages that are liable to follow traumatic division of a nerve." Head also experienced sensations that recovered more slowly and with less fidelity to his recollection of his original sensory state. These phenomena included sensitivity to light touch, temperature, pinprick, and two-point discrimination, as well as fine motor control, which he referred to as "epicritic" abilities. These faculties were not fully recovered over the 2 years between Head's surgical injury and publication of the paper he wrote about his observations of functional regeneration. He noted "the fibers of this system are more easily injured, and regenerate

more slowly, than those of the protopathic system. They are evidently more highly developed and approach more nearly to the motor fibres that supply voluntary muscle in the time required for their regeneration."

These observations—despite their unusual experimental source and limited number of participants in the study (only one!)—were the first to distinguish between the regenerative abilities of various classes of dorsal root ganglion cells and spinal motor neurons during the process of peripheral axon regrowth and reinnervation. They established the importance, and limits, of providing the denervated peripheral nerve sheath as a guide for regrowth. They also suggested, for the first time, that there may be a biological difference between sensory neurons that detect general, less well localized somatosensory information such as pressure and touch and more localized, specific information such as two-point discrimination and pain (see Chapters 12 and 13). Of course, the actual process of regeneration was inferred based on Head's perceptions of recovery of peripheral sensory and motor functions. The distinct schedules for regeneration of "protopathic" versus "epicritic" abilities are presumably related to differences in the initial specificity of different classes of dorsal root ganglion cell and motor neuron axons and the selective cues provided by peripheral glia as well as diverse targets during development. This specificity during regeneration is now known to rely on a variety of molecular signals, including several different neurotrophins that serve similar roles for initial formation of these pathways (see Figure 23.19). The importance of these cues, and their ability in the adult PNS to facilitate regeneration, is discussed in the next section.

The cellular basis of peripheral nerve repair

The cellular basis of peripheral nerve regeneration provides perhaps the clearest example of the relationship between the mechanisms that repair neural damage and those that promote initial axon growth and synapse formation during development. Even though the molecular mechanisms are similar, the environment for adult peripheral nerve repair (Figure 26.7A) is far different than that for initial peripheral axon growth in the embryo. Obviously, the distances that must be accommodated by the growing axon are larger, and the synaptic targets have already been specified. In

addition, in the embryo, these axons usually grow through undifferentiated mesenchymal cells that eventually become target muscles, skin and connective tissue. In the adult, these tissues are already differentiated, and their organization is far more complex. Four major cellular elements facilitate peripheral axon regrowth and the reinnervation of targets in the mature PNS (Figures 26.7B,C and 26.8). *Schwann cells*, the glial cells that myelinate peripheral axons, support the proximal live axon stump, provide a substrate for regrowth into the distal denervated nerve, and provide cell surface adhesion molecules as well as trophic factors to support regrowth. *Macrophages* clear the degenerating remains of severed axons and mediate immune responses that promote regrowth. *Fibroblasts* are connective tissue cells that provide a physical bridge between the proximal and distal nerve ends and produce local extracellular matrix (ECM) that provides a substrate for regrowth through the site of injury. Finally, local *endothelial cells* of the blood vessels associated with peripheral nerves provide a physical guide for regrowing axons as well as molecular cues that are essential for successful regeneration.

When axons in a peripheral nerve are severed (as in a clean laceration), the axon segments distal to the site of the cut degenerate, and macrophages clear the debris left at the site of injury as well as in the distal nerve sheath; the Schwann and connective tissue cells remain. When the axons are crushed (severely compressed, as in a traumatic

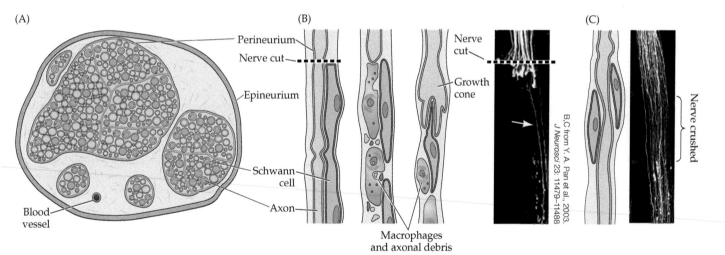

FIGURE 26.7 Regeneration in peripheral nerves (A) Cross section through a peripheral nerve showing the connective tissue sheath of the epineurium and the extracellular matrix–rich perineurium that immediately surrounds the axons and Schwann cells. (B) Degeneration and regeneration in an idealized single peripheral nerve "tube" of perineurium (the connective tissue membrane that ensheathes the entire nerve) and basal lamina. Once the axon is cut, the distal portion degenerates and is phagocytosed by macrophages. After the debris is mostly cleared, the proximal axon stump transforms into a growth cone, and this growth cone interacts with

the adjacent Schwann cells. The image next to the drawing shows this step in peripheral nerve regeneration imaged in a live mouse after peripheral nerve damage. The nascent growth cones can be seen at the site of the cut, and in a few cases (arrow) growing beyond it into the distal stump. (C) Regeneration is more efficient after crushing versus cutting a nerve. This panel shows a nerve adjacent to that which was cut in the same animal shown in panel B. The remaining axons have recovered far more rapidly, and there is extensive regeneration across the site of the crush.

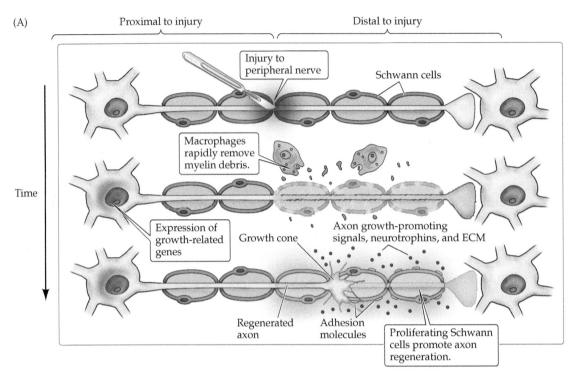

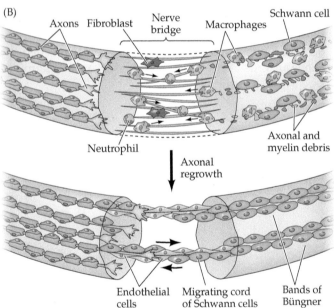

FIGURE 26.8 Molecular and cellular responses that promote adult peripheral nerve regeneration (A) After peripheral nerve injury, Schwann cells are essential for peripheral axon regrowth. Once the macrophages have cleared the debris from the degenerating peripheral stump, the Schwann cells proliferate, express adhesion molecules on their surface, and secrete neurotrophins and other growth-promoting signaling molecules. In parallel, the parent neuron of the regenerating axon expresses genes that restore it to a growth state. The gene products are often receptors, or signal transduction molecules, that allow the cell to respond to the factors provided by the Schwann cell. (ECM = extracellular matrix.) (B) Following peripheral nerve injury but prior to Schwann cell–mediated axon regrowth, several immune and connective tissue cells—including macrophages, neutrophils, and fibroblasts—as well as endothelial cells invade the damaged region of the nerve to create a "nerve bridge" region. The activity of these cells, including removal of debris and secretion of diffusible signals, facilitates organized extension of blood vessels and Schwann cells into linear bands called *bands of Büngner* that reconnect with the channels for axons in the distal nerve. (B after A.-C. Cattin and A. C. Lloyd. 2016. *Curr Op Neurobiol* 39: 38–46.)

impact injury) rather than severed, more rapid recovery occurs because the damaged segments have not fully degenerated (see Figure 26.7C). These intact segments and associated Schwann cells provide a guide for the regenerating proximal axons. In all cases, the ECM, including key proteins such as fibronectin and laminin, provides an essential scaffold for the regrowth of axons via signaling through integrin receptors on the growing axons as well as Schwann cells. When the nerve is fully severed, the space between the proximal and distal stumps becomes

organized initially as a "bridge" consisting of fibroblasts, fibrillar ECM, and macrophages (see Figure 26.8B). The invasion of macrophages into this space is essential because they secrete vascular endothelial growth factor (VEGF), which recruits endothelial cells into the damaged space to revascularize the territory. The endothelial cells and the blood vessel continuity they generate for the nerve provide a physical guide for axon growth across the bridge.

Following the initial bridging and the clearance of axon and myelin debris from the distal, denervated nerve (primarily by macrophages), Schwann cells begin to proliferate at the site of injury and extend across it (see Figure 26.8). This ordered migration of newly generated Schwann cells provides continuity for the proximal "live" axons by forming cellular "channels" in the distal peripheral nerve that constrain the regrowing axons. These channels, referred to as *bands of Büngner*, are composed of a relatively orderly array of Schwann cells, and the ECM components that support distal regrowth and axon guidance. The ECM associated with axon fascicles in the nerve before damage (also referred to as the basal lamina, although a peripheral nerve is not really an epithelium) surrounds the peripheral axons more or less continuously until they reach their targets. This is presumably why precise reapposition of distal and proximal nerve segments facilitates better recovery of function, especially of fine touch and movement (Henry Head's "epicritic" abilities; see earlier in this concept). Precise surgical reapposition is now done routinely using microscopic guidance to maximize recovery after peripheral nerve injury.

Both regenerating peripheral sensory and motor axons express *integrins* (see Figure 23.7A) that mediate recognition of ECM molecules and subsequent intracellular signaling that facilitates growth. The capacity of motor neurons, whose cell bodies are in the CNS, to reactivate the expression of growth-promoting factors suggests that mature central neurons can modify gene expression and respond to appropriate cues, at least in peripheral nerves, to accommodate axon regrowth during adulthood. If an injury is extensive and much of the nerve is lost, new Schwann cells can be generated by Schwann cell precursors remaining in the damaged proximal nerves. These new cells can then provide an appropriate environment to support the extension of a newly generated growth cone from the axon stump. Nevertheless, the absence of a well-aligned distal stump makes surgical repair and subsequent regeneration more difficult, limits the precision of regeneration, and dramatically diminishes the recovery of function.

Schwann cells and peripheral axon regrowth

The Schwann cell is the essential cellular mediator of peripheral axon regrowth through any conduit. Schwann cells provide physical support based upon their proximity to the growing axon, and molecular support that facilitates regeneration by recreating an environment similar to the milieu that supports axon guidance and growth during early development. Indeed, there is a close relationship between Schwann cells and growing axons at the interface of the proximal nerve and the bridge region that must be traversed to reach the distal portion of the nerve (Figure 26.9A). The myelin components normally made by Schwann cells are downregulated soon after injury (Figure 26.9B), presumably because these proteins and related transcriptional regulators might act to inhibit axon growth,

as they do for CNS neurons (see Concept 26.3). Schwann cells next secrete additional ECM molecules such as laminin, fibronectin, and collagens that provide a substrate for axon growth. These molecules initiate signaling that supports growth cone pathfinding and reextension of the axon through the area of injury and into the distal stump when possible. Schwann cells in an injured peripheral nerve also increase their expression of Sox2, a transcription factor associated with transcriptional activation in neural stem cells. They also increase expression of secreted signals, including VEGF and the cytokine TNF-α, that recruit vascular macrophages and other immune cells. Schwann cells also increase expression of several signaling intermediates, including intracellular kinases and related cell surface receptors or adhesion molecules (see Figure 26.9B), among them NCAM and L1and N-cadherin, and they re-express the adhesion receptor robo1 (see Chapter 23). Schwann cells also increase expression of Notch receptors, which can regulate cell identity as well as cell and axon migration (see Chapters 22 and 23). At the same time, regenerating axons, as well as vascular and immune cells that migrate to the site of injury and participate in forming the bridge between proximal and distal nerve segments (see Figures 26.8 and 26.9), express complementary cell surface adhesion molecules and co-receptors. These adhesion molecules probably mediate signaling that facilitates growth cone motility, force generation, and microtubule assembly in the newly generated portion of the axon.

Schwann cells near the site of injury in the distal end of the nerve increase expression and secretion of several neurotrophins, including NGF and BDNF (thought to be especially crucial for motor axon growth). p75 neurotrophin receptors are elevated on Schwann cells following injury as well as on the newly generated growth cones of regenerating peripheral axons (see Figure 26.9). As axons reinitiate growth during Schwann cell activation, there are some differences in timing of the availability of NGF, p75, and TrkA (all are primarily associated with nociceptive axon growth). These neurotrophins are expressed first in regenerating nerves, followed by expression of other neurotrophins and Trks that support mechanoreceptive and proprioceptive axon growth (see Chapter 23). The local availability of neurotrophins may promote a "growth" state (i.e., reactivate the capacity for *trophic* signaling) for the damaged axons, as well as attract the growing axons to appropriate local targets distal to the site of damage (i.e., *tropic* effects). Finally, once axons have traversed the site of injury and extended through the bands of Büngner maintained in the distal nerve, transcriptional regulators and proteins associated with myelination are upregulated in Schwann cells to remyelinate the regenerated axons (see Figure 26.9B). The specificity and timing of differential expression of neurotrophins and their receptors followed by remyelination provide a molecular explanation

(A)

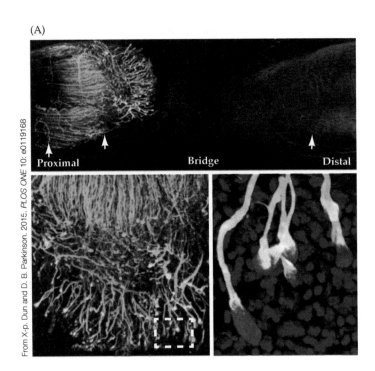

Proximal   Bridge   Distal

FIGURE 26.9 Schwann cell migration is mediated by sequential molecular signals (A) Top: Initiation of Schwann cell (red) migration and axon (green) regrowth at the interface of the proximal nerve and the nerve bridge region. In the distal nerve, Schwann cells remain aligned in parallel conduits that define the bands of Büngner. Bottom: Schwann cells migrate to the front of the injury site in register with the growth cones of small fascicles of axons extending into the nerve bridge, oriented parallel to the axis of the nerve. The dashed outline in the left image is magnified in the right image. (B) The sequence of molecular signaling that accompanies Schwann cell (SC) responses during axon regrowth and remyelination. First, injured proximal and distal axons are demyelinated and Schwann cells downregulate myelin proteins, secreted signals, and transcription factors. Next, a distinct subset of transcription factors, as well as neurotrophins, secreted signals, and cell surface ligands, receptors, and adhesion molecules, are upregulated transiently as Schwann cells facilitate axon regrowth. Finally, factors necessary for regrowth are downregulated, and signals necessary for remyelination of the regenerated axons by the Schwann cells are expressed. (B after G. Nocera and C. Jacob, 2020. *Cell Mol Life Sci* 77: 3977–3989).

(B)

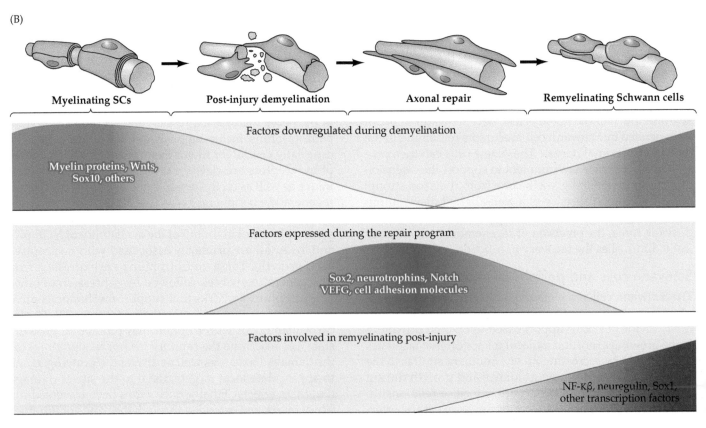

Myelinating SCs   Post-injury demyelination   Axonal repair   Remyelinating Schwann cells

Factors downregulated during demyelination

Myelin proteins, Wnts, Sox10, others

Factors expressed during the repair program

Sox2, neurotrophins, Notch VEFG, cell adhesion molecules

Factors involved in remyelinating post-injury

NF-κβ, neuregulin, Sox1, other transcription factors

of the early observations of Head and his colleagues that showed the initial recovery of nociception (Head's "protopathic" sense) followed by proprio- and mechanoreception (Head's "epicritic" sense).

Axon growth during regeneration

In anticipation of reactivation of nerve growth immediately after injury or in response to the changes orchestrated by the Schwann cells after debris is cleared and Schwann cells have begun to migrate into the distal stump, the regenerating peripheral neurons also change. The growth cones of regenerating peripheral axons are in direct contact with Schwann cells as they migrate into the site of nerve injury and retain this contact once they reenter the denervated distal nerve (see Figure 26.9A). Injured and actively regenerating sensory or motor neurons modify gene expression for cytoskeletal reassembly, signal transduction, and protein trafficking in the cell body to accommodate regrowth. Some of this modification relies on early reexpression of developmentally regulated transcription factors, including those in the bHLH family (see Chapter 22). The dynamics of the actin and microtubule cytoskeleton must be restored to a growth state so that growth cone navigation and axon extension can occur (see Chapter 23). Changes in gene expression enable growth cone extension and substrate recognition, and the genes that are switched on include several associated with proteins that modulate the cytoskeleton during developmental axon growth—actin- and microtubule-binding proteins, Ca^{2+}-regulated modulators, and molecular "motors" such as kinesins and dynein. In addition, genes selectively associated with the growth state, including *GAP43*, are reexpressed. The protein GAP43 (*g*rowth-*a*ssociated *p*rotein-*43*) is normally found in growing axons in the embryo (see Chapter 23) but is also present at high levels following axotomy. As regrowth reaches its conclusion, and synapses are formed, these growth-promoting genes are downregulated.

In some cases, damage is so extensive that the proximal and distal nerves cannot be directly rejoined, and recovery cannot be achieved without additional intervention. In such cases, surgeons attach both proximal and distal ends to a nearby intact nerve in the hope that growth will occur via the undamaged nerve and that denervated adjacent muscles or sensory specializations will be recognized once the redirected axons reach their terminal target. The other option is to take a length of nerve from another, less important nerve (e.g., a cutaneous nerve in the leg) and attach the proximal and distal ends to this heterologous graft. In experimental animals, the rate and magnitude of regrowth through such grafts can be accelerated by use-dependent mechanisms, such as treadmill running done by mice with peripheral nerve grafts following peripheral nerve damage (Figure 26.10A). Additional approaches have also been developed. In instances where surgical repair is possible (and fairly accurate), stimulation by implanted microelectrodes spanning the proximal and distal sites of injury enhances regeneration, apparently by enhancing neurotrophic secretion and gene expression that supports regrowth (see Figure 26.4). Finally, a broad range of engineered biomaterials that provide physical guidance, approximate ECM, or act as a delivery device for growth-promoting cells or growth factors may be able to substitute for the lost length of peripheral nerve (Figure

26.10B). These biomaterials serve as a scaffold for the migration and differentiation of Schwann cells and other necessary cell classes from the proximal nerve stump (see Figure 26.8) to support formation of a bridge to the distal nerve sheath. The success of such interventions relies on control of mechanical forces around the injured nerve after surgical repair (so that the conduit does not become misplaced, resulting in misrouting of regrowing axons). It also relies on stimulating cellular and molecular programs for regrowth in multiple cell classes in and around the nerve despite the insertion of a foreign substrate, and minimizing inflammation and infection and immune response to the engrafted biomaterial.

Regeneration of peripheral synapses

The extension of axons from mature neurons into peripheral nerve sheaths is only the first step in peripheral nerve regeneration. The next essential event in successful recovery of function is reinnervation of appropriate target tissues and reestablishment of synaptic connections. This process must occur for all three classes of peripheral axons (motor, sensory, and autonomic); however, it has been most thoroughly characterized at the neuromuscular junction (NMJ) and in the peripheral autonomic system (Box 26A). Because of the relative ease of identifying and visualizing synaptic sites on muscle fibers, the regenerating NMJ has been studied in great detail (Figure 26.11A,B). The ability to define major molecular constituents at the NMJ—individual synaptic terminals, ECM, postsynaptic receptors, and related proteins—gives insight into the stability of a denervated synaptic site after damage, as well as the changes that accompany reinnervation (Figure 26.11C,D). When skeletal muscle fibers are denervated, the denervated neuromuscular postsynaptic sites remain intact for weeks. Indeed, Schwann cells that aggregate around the NMJ, and even specialized components of the basal lamina that mark these synaptic sites, remain segregated in the absence of the synapse.

The muscle cells and Schwann cells at the denervated neuromuscular junction sites modify their secretion of signaling molecules, including neurotrophins, ECM components, and other growth factors that promote selective reinnervation by a regrowing axon from an α motor neuron. Presumably, the neurotrophins whose expression is increased at the denervated neuromuscular junction—NGF and BDNF—enhance the tropic and trophic signaling necessary to recapitulate target recognition and synaptogenesis. Those neurotrophins that are diminished—NT-3 and NT-4—may be more important for the maintenance of established synapses. Their decline may ensure that denervated synaptic sites accept innervation from regenerating axons.

Retaining the precise localization of postsynaptic acetylcholine receptors (AChRs) in the membrane of the denervated muscle at the postsynaptic site is essential for

(A)

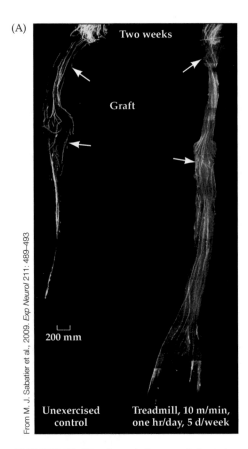

From M. J. Sabatier et al., 2009. *Exp Neurol* 211: 489–493

(B)

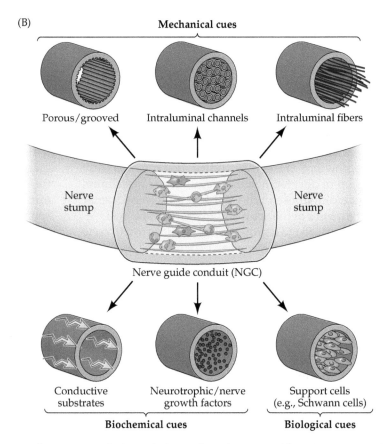

FIGURE 26.10 **Improving peripheral nerve regeneration**
(A) Activity and use can influence peripheral axon regrowth. The fibular nerve in the leg of a mouse was transected and a significant length of nerve removed. Subsets of axons have been labeled using a fluorescent reporter transgene. A heterologous graft of peripheral nerve from a donor mouse was positioned to establish a bridge region (between the arrows in each nerve shown). After the graft was complete, over the course of 2 additional weeks the mice were either left to recover while pursuing their normal cage activities (left) or given a regime of exercise on a treadmill (right). Exercised mice showed significantly more nerve growth through the graft and recovered function more rapidly. (B) Biocompatible materials can be used to make artificial conduits that bridge damaged regions in peripheral nerves. These conduits can use additional biocompatible materials to guide axon growth across the bridge region (top) or can be used to present adhesion molecules, neurotrophins, or Schwann cells to a regenerating peripheral nerve to improve the rate and extent of nerve repair (bottom). In all cases the artificial conduit is inserted into the damaged nerve to facilitate growth across the bridge region. (B after S. Vijayavenkataraman, 2020. *Acta Biomaterialia* 106: 54-69.)

regeneration. Thus, the AChR clustering that defines the postsynaptic membrane specialization remains following denervation, as does the local scaffold of proteins in the muscle cell cytoplasm that retain the AChRs at the synaptic specialization on the muscle fiber (see Figure 26.11D). The secreted factor neuregulin and its receptors, involved in the initiation of receptor clustering during initial synapse formation (see Chapter 23), are expressed at the denervated synapse. In addition, the ECM components that distinguish the synaptic portion of the muscle basal lamina are maintained when mature muscle fibers are denervated. This matrix includes specialized forms of laminin (synaptic, or S laminin) normally found at the neuromuscular synapse. In addition, the proteoglycan agrin is key for initiating or maintaining the synaptic site on muscle fibers by binding to the low-density lipoprotein receptor 4 (LRP4), a transmembrane receptor that then activates the muscle-specific receptor kinase MuSK. The activation of MuSK, via phosphorylation, leads to the recruitment of a transmembrane complex that includes the AChR itself (a multimeric ligand-gated ion channel/receptor), the transmembrane scaffolding protein dystrobrevin, the cytoplasmic scaffold dystrophin, and an additional cytoplasmic membrane–associated kinase, rapsyn. The immobilization of these AChrR complexes in the postsynaptic NMJ cluster, particularly at the top of the postsynaptic membrane foldings closest to the sites of motor neuron synaptic vesicle release, maximizes the initiation of neuromuscular synaptic transmission. The retention of the complexes following denervation helps maximize activity-dependent restoration of neuromuscular connectivity. The genes for many of the AChR scaffolding proteins are targets for causal mutations in muscular

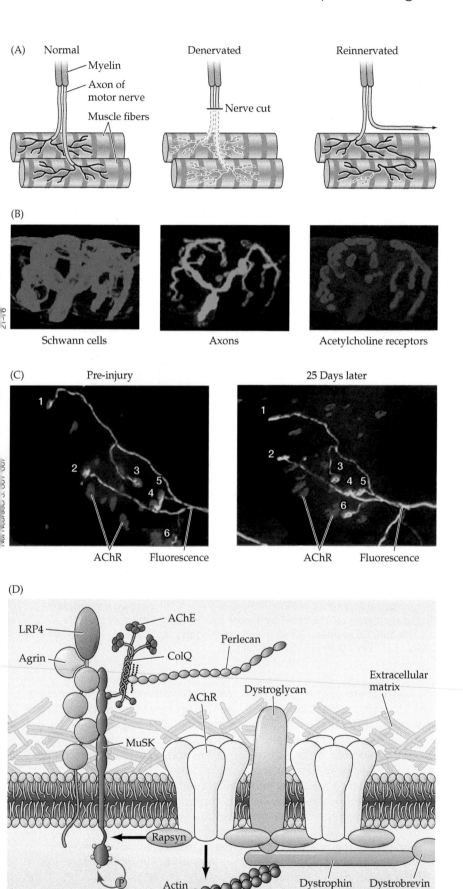

(A) Normal
Myelin
Axon of motor nerve
Muscle fibers

Denervated
Nerve cut

Reinnervated

(B)
Schwann cells
Axons
Acetylcholine receptors

(C)
Pre-injury
25 Days later
1
2   3   5
4
6
AChR   Fluorescence
1
2
3
4   5
6
AChR   Fluorescence

(D)
LRP4
Agrin
AChE
Perlecan
ColQ
AChR   Dystroglycan
Extracellular matrix
MuSK
Rapsyn
Actin   Dystrophin   Dystrobrevin
P

FIGURE 26.11   Reinnervation of muscles following peripheral motor nerve damage
(A) Schematic showing the degeneration of a distal motor axon, its regrowth, and the maintenance of the postsynaptic specialization on the muscle surface during the period of denervation. (B) The cellular components usually found at the neuromuscular junction (NMJ). When the axon degenerates, the Schwann cells and acetylcholine receptors (AChRs) remain in place. (C) The pattern of motor innervation in an isolated muscle before an injury and 25 days after, imaged in a single live mouse. The postsynaptic specializations of the NMJs on each individual muscle fiber are labeled in red (for AChR). The axon fluoresces green. This axon makes six synapses on six muscle fibers. Twenty-five days later, the axon has reinnervated all six sites, and the basic pattern is similar to that prior to the injury. (D) A specific molecular architecture in the synaptic cleft and postsynaptic membrane of the muscle preserves neuromuscular synaptic sites and facilitates reinnervation and functional recovery. The synaptic cleft is the site of several specific ECM molecules, including specific classes of laminin (e.g., the NMJ-specific variant S laminin) and special variants of collagen, including ColQ, which specifically binds perlecan, another ECM molecule found in the NMJ matrix. ColQ in turn binds and localizes acetylcholinesterase (AChE) to the synaptic cleft. The postsynaptic muscle membrane itself is key for maintaining ongoing NMJ repair and regeneration following denervation. Two key transmembrane receptor proteins, low-density lipoprotein receptor 4 (LRP4) and the muscle-specific tyrosine kinase MuSK, bind the protein agrin (secreted by the presynaptic terminal of the ingrowing motor neuron axon). The secretion of agrin into the ECM triggers the muscle cell to establish or maintain AChR clusters that rely on additional scaffolding proteins, including dystroglycan, dystrophin, dystrobrevin, and rapsyn. (D after M. L. Campanari et al., 2016. *Front Mol Neurosci* 9:160.)

The peripheral autonomic system has been studied in detail for more than a century with respect to neural regeneration. The accessibility and regenerative properties of peripheral axons allow a variety of investigations to be done with relative ease. Most of these studies have been done in the mammalian sympathetic system. Preganglionic sympathetic fibers, like other peripheral axons, regenerate when they are severed. Toward the end of the nineteenth century, the English physiologist John Langley, working at Cambridge University, found that sympathetic end-organ responses (e.g., blood vessel constriction, piloerection, and pupillary dilation) recovered a few weeks after cutting the preganglionic nerve to the superior cervical ganglion. As indicated in Figure A, the normal innervation of this and other sympathetic ganglia is selectively organized in that preganglionic axons arising from different spinal segments innervate particular functional classes of cells in the ganglion. Langley found that after reinnervation the end-organ responses were organized much as before; thus, stimulation of T1 elicited its particular constellation of largely nonoverlapping end-organ effects compared with stimulation of the preganglionic axons arising from T4.

Modern experiments confirmed Langley's observations and showed further that the normal pattern of innervation observed with intracellular recording is indeed reestablished following regeneration of the preganglionic axons. Selective reinnervation also occurs in parasympathetic ganglia. The chick ciliary ganglion has two functionally and anatomically distinct populations of ganglion cells: the ciliary cells and the choroidal cells. Because these ganglion cell types can be separately identified and are in turn innervated by preganglionic axons with different conduction velocities, one can ask whether these two populations are reinnervated by the same classes of axons that contacted them in the first place. As in the mammalian sympathetic

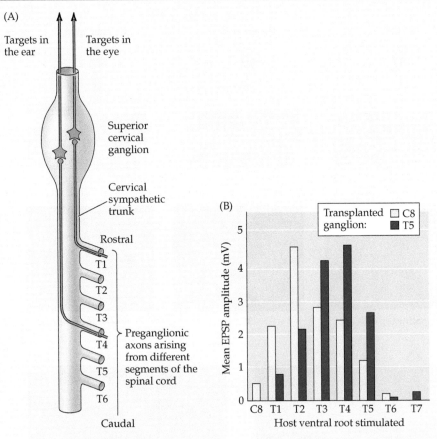

Evidence that synaptic connections between mammalian neurons form according to specific affinities between different classes of pre- and postsynaptic cells. (A) In the superior cervical ganglion, preganglionic neurons located in particular spinal cord segments (T1, for example) innervate ganglion cells that project to particular peripheral targets (the eye, for example). Establishment of these preferential synaptic relationships indicates that selective neuronal affinities are a major determinant of neural connectivity. (B) In a transplantation experiment, guinea pig donor ganglia C8 (superior cervical ganglion; control) and T5 (fifth thoracic ganglion) were transplanted into the superior cervical bed of a host animal. The graph shows the average postsynaptic response of neurons in the transplanted ganglia to stimulation of different spinal segments. Although there is overlap, the neurons in transplanted T5 ganglia are clearly reinnervated by a more caudal set of segments than are the transplanted C8 neurons. (EPSP = excitatory postsynaptic potential.) (B after D. Purves et al., 1981. *J. Physiol* 313: 49-64.)

system, appropriate contacts are reestablished during reinnervation.

The accurate reinnervation of different classes of sympathetic neurons is especially remarkable because the ganglion cells innervated by a particular spinal segment (and that innervate a particular target) are distributed more or less randomly through the ganglion. This arrangement implies that recognition of the pre- and postsynaptic elements must occur at the level of the target cells. One way to explore the implication that ganglion

cells bear some more or less permanent identity is to transplant different sympathetic chain ganglia from a donor animal to a host where the ganglia can be exposed to the same segmental set of preganglionic axons during reinnervation. One can then ask whether two different ganglia, normally innervated by different sets of axons, are selectively reinnervated by axons arising from different spinal segments.

As seen in Figure B, different sympathetic chain ganglia (in this case, the superior cervical ganglion and

■ **BOX 26A** | **Specific Regeneration of Synaptic Connections in Autonomic Ganglia** *(continued)*

fifth thoracic ganglion) are indeed distinguished by the preganglionic axons in the host cervical sympathetic trunk. A donor superior cervical ganglion transplanted to the cervical sympathetic trunk is reinnervated in a manner that approximates its original segmental innervation; the fifth thoracic ganglion transplanted to that position, however, is reinnervated by an overlapping but caudally shifted subset of the thoracic spinal cord segments that normally contribute to the cervical sympathetic trunk. This more caudal innervation approximates the original segmental innervation of the fifth thoracic ganglion.

These results indicate that ganglion cells carry with them a property that biases the innervation they receive, in confirmation of Langley's original concept of "chemoaffinity" as a basis for the selectivity of target cell innervation (see Chapter 23).

---

dystrophy, a disease in which neuromuscular junctions are denervated and muscle targets degenerate. Indeed, the first muscular dystrophy–causing mutation was found in the dystrophin gene (thus named for its association with muscular dystrophy) that encodes a key scaffolding protein for AChR clustering (see later in this section and Figure 26.11D).

Molecules found in the specialized basal lamina at the NMJ are also retained at the denervated postsynaptic site. These include molecules that are normally bound to specialized ECM proteins produced by perisynaptic Schwann cells at the synapse. In particular, these ECM proteins localize acetylcholinesterase (AchE), which mediates degradation of acetylcholine at the NMJ. AchE is concentrated in the synaptic cleft by the collagen-related protein ColQ. ColQ is localized to the synaptic cleft via binding to perlecan, another ECM protein. The ECM at the synapse also localizes and concentrates secreted growth factors—not only neurotrophins, but additional growth-promoting molecules such as fibroblast growth factors (see Chapter 22)—produced upon denervation. Accordingly, this highly organized region of the basement membrane, in concert with secreted factors, postsynaptic receptor proteins, and Schwann cells, defines the site of reinnervation. In fact, even if the muscle fiber is eliminated (which can be done experimentally in model organisms), motor axons recognize these specialized sites (maintained in "tubes" of basal lamina that normally surround muscle fibers) as optimal locations for reinnervation. Finally, a complex of transmembrane proteins in the muscle plasma membrane ensures the fidelity of synaptogenesis during regeneration. The transmembrane proteins agrin as well as the muscle-specific kinase MuSK link ColQ, AChE, and perlecan in the synapse-specific ECM with clustered nicotinic acetylcholine receptors (nAchRs) in the muscle postsynaptic membrane (see Figure 26.11D). This ECM–postsynaptic complex is particularly reinforced by two scaffolding proteins, dystroglycan and rapsyn, which anchor the complex to the actin cytoskeleton of the muscle cell. Many of these

molecules, if mutated or attacked by auto-antibodies or general inflammatory responses, can compromise ongoing maintenance and repair of the NMJ. Several neurodegenerative or autoimmune diseases, including muscular dystrophy, myasthenia gravis and other conditions that include muscle weakness and wasting as key symptoms, reflect failures of normal NMJ maintenance.

This molecular specificity, however, provides only one part of the instructions needed to reestablish synapses after peripheral nerve regeneration. Activity-dependent processes similar to those that eliminate polyneuronal innervation at neuromuscular synaptic sites during development (see Chapter 24) are also essential for restoring function after peripheral nerve damage. There is a fair degree of imprecision in the initial reinnervation of specific targets, as is evident in Henry Head's description of the slowness and imprecision of his recovery of fine motor and sensory function. This result has been confirmed by more recent studies in which reinnervation is observed over time in adult animals with experimentally damaged peripheral motor nerves (see Figure 26.11C). The subsequent regeneration can be fairly faithful to the original pattern, but there are often errors. Imprecision is due not only to a lack of fidelity of the regenerating axon to its original muscle target site. Polyneuronal innervation of neuromuscular synapses returns during regeneration and reinnervation. Much of this innervation is eventually eliminated, presumably via the same activity-dependent mechanisms that operate during the early postnatal period, when supernumerary axons are eliminated from the developing synapse (see Figure 23.13). If electrical activity is blocked during regeneration, either in the muscle fiber or in the afferent nerve, multiple innervation of the reinnervated neuromuscular junction site remains. This return of activity-dependent mechanisms for elimination of multiple innervation may be used to revalidate singly innervated motor units and thus optimize restoration of function.

CNS Axons and Dendrites in Most Adult Mammals Lack the Capacity for Extensive Regrowth

CONCEPT
26.3

LEARNING OBJECTIVES

26.3.1 Consider why neural regeneration is so limited in the CNS.

26.3.2 Relate cellular stress mechanisms caused by CNS injury to neuronal cell death.

26.3.3 Compare CNS glial and neuronal responses to injury with those in the PNS.

26.3.4 Identify the major cell classes involved in forming glial scars.

26.3.5 Describe the role of immune cells and inflammatory signaling in preventing CNS repair.

Regeneration in the CNS

There is very little long-distance axon growth or reestablishment of functional connections in the CNS following injury. This limited regrowth of damaged CNS axons, even those whose cell bodies remain intact, largely accounts for the relatively poor prognosis following brain or spinal cord damage. The only major exception to this unfortunate reality is for spinal cord and brainstem motor neurons. Although the cell bodies of these motor neurons are located in the CNS, their axons project via peripheral nerves to peripheral muscles and therefore have access to instructions for relatively successful peripheral regeneration. In addition, it is assumed that these motor neurons retain the capacity to reactivate developmental programs that support growth and reinnervation.

Damage to the CNS can occur in several ways. The brain or spinal cord can be injured acutely by physical trauma (Clinical Applications). Another type of damage is caused by a lack of oxygen created by locally diminished blood flow (ischemia) due to a vascular occlusion or local bleeding (stroke; see Clinical Applications in the Appendix). Damage can also arise from global deprivation of oxygen, as in drowning or cardiac arrest. A different type of damage arises from neurodegenerative diseases such as Alzheimer's and Parkinson's diseases and ALS, discussed at the outset of this chapter. In all of these cases, neuronal loss occurs for reasons that have not yet been fully explained. Because new central neurons are not generated in the adult brain and central axons have little ability to regenerate, however, the key to whatever recovery is possible after brain injury lies with complex cellular events that support the survival of those neurons that have not been killed outright and whose processes and synaptic connections remain relatively intact.

■ Clinical Applications

Casualties of War and Sports

We often use the same words to describe contests on the battlefield and playing field. One side "attacks" the other, defenses and offenses are "infiltrated," and in the end, one side might be declared to have "killed" the other—a terrible reality in war, an enthusiastic metaphor in sports. Until recently, however, it was not known that soldiers and athletes could sustain the same serious brain injuries on the battlefield and playing field. It appears that concussive injuries resulting from explosions and blunt trauma in a theater of war cause brain damage similar to that caused by punches and full body hits in the boxing ring, on the hockey rink, or on the football field. Recent efforts both by the U.S. military and by sports organizations acknowledge this dangerous parallel. These efforts have led to a better understanding of the relationship between closed-head traumatic brain injury (TBI), behavioral changes, and long-term pathology that results in mood, memory, and cognitive difficulties accompanied by degenerative changes in individuals' brains.

Single or repeated high-energy forces delivered to the head cause TBI. Such forces (whether from the detonation of an explosive device, the concussive force of a boxing punch, or a hockey or football hit) overwhelm the ability of the skull and the fluid-filled cushion created by the subarachnoid space to protect the brain from shearing forces. These forces can cause acute bleeding around the meninges or in brain tissue. In addition, blast waves, hard blows, and their sequelae cause distension of the long axons in white matter pathways, severing the axon from its parent cell body. Such immediate cellular damage may be accompanied by long-term neurodegenerative changes, especially in the brains of individuals who sustain repeated TBI.

While tissue damage and degeneration are thought to be the ultimate consequences of TBI, many primary symptoms are behavioral. Upon impact or immediately thereafter, individuals sustaining TBI experience loss

■ Clinical Applications (continued)

of consciousness, followed by loss of memory of events before or immediately after the injury, altered mental state (disorientation, confusion), and mild to moderate neurological deficits (diminished somatic or cranial nerve reflexes). Following these immediate symptoms, there may be sustained changes in the behavior of the individual, including altered mood (especially depression), cognitive impairment, impulsivity, and distorted sensory perception. Headaches and lethargy are often reported, and many individuals with TBI exhibit symptoms of post-traumatic stress disorder (PTSD). Of course, for soldiers, it is difficult to disentangle the direct consequences of a concussive force delivered to the head in an instant, and the physiological consequences of the ongoing stress of warfare.

Even more insidious, however, are the potential long-term consequences for victims of TBI. Depending on the severity and frequency of injuries, TBI victims may be at risk for recurrent headaches, depression, early-onset mild cognitive impairment, or more severe dementia. Indeed, there are anecdotal reports of athletes (especially boxers and football players) who have had such difficulties. John Grimsley, a linebacker for the Houston Oilers in the late 1980s and early 1990s (Figure A), had already begun experiencing memory loss and irritability before his death at age 45 of an accidental gunshot wound. Upon postmortem examination, Mr. Grimsley's brain was found to have numerous depositions of tau protein (Figure B), the primary constituent of neurofibrillary tangles that are one of the hallmarks of Alzheimer's disease pathology (see Figure 26.2D and Box 30C). In contrast, postmortem examination of a 65-year-old control individual with no history of TBI and no cognitive impairment showed no obvious tau

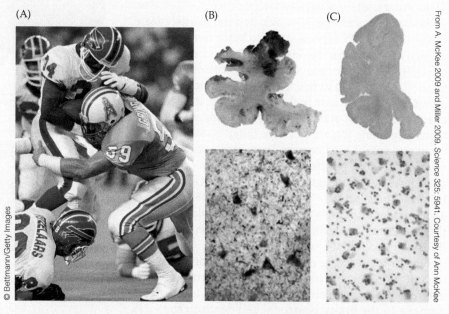

(A) John Grimsley (right) makes a head-on hit to a Buffalo Bills player. (B) Grimsley's brain on postmortem analysis after his death at age 45 from an accidental gunshot wound. Brown label indicates deposits of tau protein, usually associated with Alzheimer's pathology. (C) Brain of a 65-year-old man with no history of TBI or cognitive impairment.

© Bettmann/Getty Images

From A. McKee 2009 and Miller 2009. *Science* 325: 5941. Courtesy of Ann McKee

depositions (Figure C). Subsequent analysis of a small number of athletes with a history of TBI and signs of cognitive impairment, altered mood, and memory loss has revealed similar pathology. The neurological and behavioral deficits that follow TBI, paired with postmortem degenerative pathology, are now classified as a new syndrome, **chronic traumatic encephalopathy (CTE)**, which apparently occurs in athletes who sustain repeated TBI during the course of their careers. CTE may also emerge as a cause of post-service functional impairment for soldiers sustaining repeated TBI on the battlefield.

TBI and its potential long-term consequence CTE have become major foci for research into the casualties of war and sports. The U.S. military and professional sports organizations have taken steps to raise awareness and to

limit repeated instances of head injury that lead to TBI. In the military, there are now standard questionnaires and policies for any soldier who has been exposed to head trauma or explosive forces that do not result in an obvious head wound, and treatment or delayed return to battle is based on the answers. Professional sports organizations are supporting research to better understand why some players who sustain TBI progress to CTE later in life. These casualties of war and sport illustrate the vulnerability of the brain to serious, if initially unrecognized, injury. The long-term degenerative consequences point out the limited ability of the brain to repair itself once damage has occurred.

Cellular and molecular responses to injury in the CNS

There are at least three reasons for the differences between successful peripheral regeneration and the limited regeneration in the CNS. First, damage to brain tissue tends to engage mechanisms that lead to necrotic and apoptotic cell death for nearby neurons whose processes have been severed, rather than to reactivation of trophic and tissue repair mechanisms seen in the PNS. Second, the neuronal changes at the site of injury do not recapitulate developmental signaling that supports the initial establishment of brain and spinal cord circuits. Instead, a combination of glial growth and proliferation along with microglial activity (microglia have immune functions that lead to local inflammation) actively inhibits growth. Finally, there is an upregulation of growth-inhibiting molecules related to the chemorepellent factors that influence axon trajectories during development (see Figure 23.6A,B).

One of the striking differences in the consequences of central versus peripheral nerve cell damage is the extent of neuronal cell death that occurs after direct damage to the brain. Neuronal cell death in the CNS has been studied extensively in brains where hypoxia has occurred due to local vascular occlusion (a common form of stroke), as well as in instances where axons have sustained damage that disconnects their parent CNS cell bodies from their CNS targets. In response to hypoxia, neuron loss begins rapidly in the hypoxic region and continues for an extended period in the region immediately surrounding the initial site of hypoxia, referred to as the *penumbra* region. Thus, hypoxic events often eliminate most or all of the local neurons within and adjacent to sites of focal O_2 deprivation (Figure 26.12A). The consequences of severing axons that project to CNS targets (unlike α motor neurons whose CNS cell body gives rise to an axon that enters the PNS) for parent neuron death are similar; however, they may occur more gradually, depending on the extent of axon disconnection. The primary driver of adult neuronal cell death in the CNS is a cascade of events, similar to developmentally regulated apoptotic cell death (see Chapter 23), that activates mitochondrial-mediated apoptosis (Figure 26.12B). In mature neurons, cellular stress pathways can be activated by hypoxia, mitochondrial dysfunction and increased generation of reactive oxygen species, or DNA damage at the neuronal cell body. The transection of axons presumably leads to growth factor deprivation by disconnecting the parent neuron from its source of trophic support. Similar mechanisms of DNA damage and cellular stress (including changes in oxidative metabolism and damage from oxygen free radicals, which also trigger apoptosis) and trophic signaling disregulation and its impact on mitochondrial function have been suggested to underlie some neurodegenerative diseases, including ALS and Parkinson's disease.

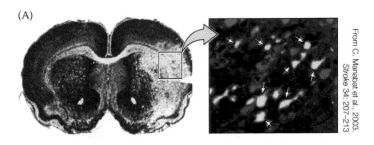

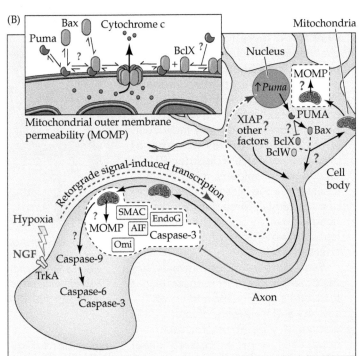

FIGURE 26.12 Consequences of hypoxia or ischemia in the mammalian brain (A) Section through the brain of a 7-day-old mouse in which the carotid artery was transiently constricted. Nissl stain (see Chapter 1) was used to visualize cell bodies. The lighter region (i.e., little or no staining) shows the extent of cell damage and loss caused by this brief deprivation of oxygen. In the higher-magnification image, cells were stained for the neuronal marker Neu-N (red-labelled cells) and for activated caspase-3 (yellow-labelled cells, indicated by white arrows), which identifies Neu-N expressing neurons undergoing apoptosis (the yellow color reflect the overlay of the green caspase-3 labeling with the red Neu-N labelling). (B) Schematic of the neuronal responses to hypoxia or growth factor withdrawal that result in apoptotic cell death. Signals initiated by low oxygen or loss of trophic support alter mitochondrial outer membrane permeability (MOMP). This change reflects interactions between the PUMA (*p53* upregulated modulator of apoptosis) protein and the Bax family of pro-apoptotic proteins as well as the Bcl-X and Blc-W anti-apoptotic proteins. The displacement of Bcl proteins by Bax results in permeability changes and the release of cytochrome *c*, a mitochondrial enzyme, into the cyptoplasm. Cytochrome *c* then activates a protein complex that results in caspase activity that initiates apoptosis.

In neurons, apoptosis in response to neuronal hypoxia, stress, or trophic factor deprivation is initiated by the upregulation of stress-responsive transcriptional regulators, including the tumor suppressor protein p53, the CCAAT enhancer binding protein homologue CHOP, and the fork-head transcription factor FOXO3a, among others. These transcription factors then influence the upregulation of the *p53 upregulated modulator of apoptosis* protein, or PUMA. PUMA interacts with members of the Bcl2 family of anti-apoptotic genes that normally antagonize the pro-apoptotic proteins Bak and Bax. When PUMA binds to anti-apoptotic proteins such as Bcl2 or BclX, it releases Bak and Bax, causing mitochondrial outer membrane permeability changes, which in turn increase oxidative stress and ultimately lead to the release of cytochrome *c*, a key enzyme in the electron transport chain. This initiates the activation of caspase-9 and -3 in the cytoplasm. These enzymes then begin the process of proteolysis, DNA fragmentation and membrane changes that ultimately cause the demise of the neuron (see Figure 26.12B).

In addition to apoptosis, **autophagy** has emerged as an intermediate response to cellular stress that can either prevent, or ultimately lead to, neuronal degeneration. Autophagic responses operate normally to target cellular contents for lysosomal degeneration. When neurons (or other cells) are under transient metabolic stress because of hypoxia or other bioenergetic fluctuations, the autophagic pathway can alter activity to accommodate dynamic metabolic demands, prevent oxidative stress, and move the bioenergetically deprived neuron as close as possible to a state that does not activate apoptosis. Disruptions of autophagy are implicated in several neurodegenerative disorders, including Parkinson's and Alzheimer's diseases. In addition, autophagic responses may enhance neuronal survival in the face of acute damage to the nervous system and subsequent neuroinflammatory activity.

A major source of celllular stress in neurons following CNS injury is glutamatergic overstimulation caused by bursts of abnormal activity arising after focal brain damage. Such overstimulation can also arise from uninhibited glutamatergic neuronal activity within epileptogenic foci or similar seizure activity generated secondarily at sites of brain damage. Elevated glutamatergic neuronal activity and its consequences are referred to as *excitotoxicity* and, if unchecked, can lead to neuron death, especially in the penumbra region. Following injury or seizure, large amounts of neurotransmitters are released. This enhanced signaling further modifies the effectiveness of members of the Bcl2 family of anti-apoptotic molecules, which normally oppose changes in mitochondrial function that reflect oxidative

stress (see Figure 26.12B). Thus, one of the key determinants of the long-term effects of damage to adult neural tissue is the extent to which the damage activates apoptosis, either directly because of immediate neuronal damage and subsequent cellular stress, or because of excitotoxicity following excess neurotransmitter release in response to brain injury.

Glial scar formation in the injured brain

As might be expected from cellular mechanisms of peripheral nerve injury, glial cells found at the site of injury contribute to the degenerative as well as limited regenerative processes that occur after brain damage; however, central glia differ from Schwann cells in their responses to injury and the initiating of neuronal regrowth. Brain injury elicits responses—that actively *oppose* neuronal regrowth from all three glial classes—astrocytes, oligodendrocytes, and microglia (Figure 26.13).

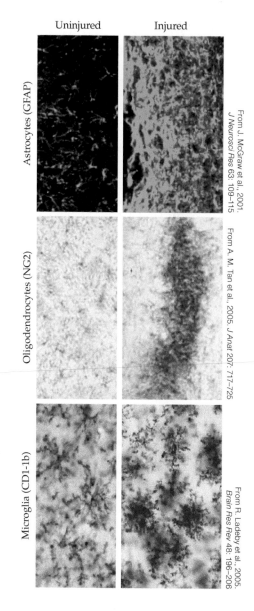

FIGURE 26.13 Reaction of the three major classes of glia in the CNS to local tissue damage In each case, there is growth and change in expression of molecules normally associated with each cell class. Top: Astrocytes labeled to visualize glial fibrillary acidic protein (GFAP) both before (left) and after (right) CNS injury. Center: The molecule NG2 identifies oligodendroglial precursors and immature oligodendrocytes in uninjured CNS tissue (left), whose expression is enhanced in oligodendroglial cells that proliferate at the site of an inury (right). Bottom: CD1-1b, a marker for microglia. In uninjured CNS tissue, microglial processes are somewhat sparce and slender (left). At a site of injury (right), microglia appear to have more, thicker processes.

Such cells, referred to as *activated glia,* are less susceptible to the stimuli that result in neuronal apoptosis after injury. Thus, survival and growth of glial cells are preserved while neighboring neurons die. Most brain lesions cause local proliferation of otherwise quiescent glial precursors, extensive growth of processes from existing glial cells in or around the site of injury, and some migration of glial cells or precursors to the site of injury. These reactions lead to the formation of a **glial scar**, which reflects a local overgrowth and sustained concentration of astrocytes, oligodendrocytes, and microglia and their processes (Figure 26.14; also see Figure 26.13). The glial scar also includes immune cell classes: macrophages, neutrophils, and lymphocytes. Both the glial and the immune cell classes invade the site of injury in waves. Within hours or a day, the immune cells arrive. The resident glial cells infiltrate the site of damage sequentially: Oligodendroglial precursors arrive first, microglia next, and then astrocytes. In addition, macrophages undergo significant proliferation and growth, adding to the physical barrier established by the glial scar. Glial scars are thought to be a major barrier for axon and dendrite regrowth in the CNS.

Several inhibitory molecules that prevent axon growth in the CNS are produced by the glial cells that contribute to glial scars (see Figure 26.13). The protein components of brain myelin produced by oligodendrocytes inhibit axon growth. Substrates enriched in myelin-associated proteins such as

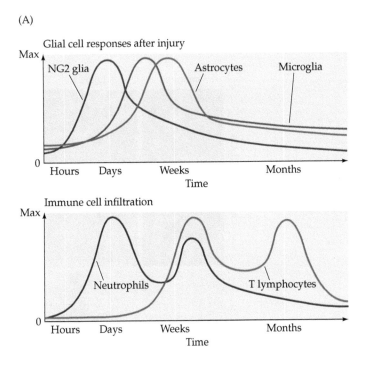

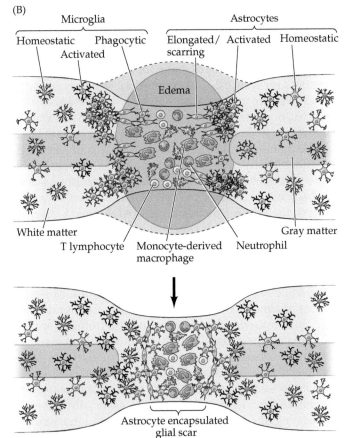

FIGURE 26.14 Glial cells and immune cells interact and prevent CNS regeneration by forming a glial scar
(A) Oligodendroglial precursors (NG2 glia), astrocytes, and microglia proliferate, invade, and reach peak activation at sites of CNS damage on different timescales: hours for NG2 glia, days for microglia, and weeks for astrocytes. Similarly, immune cells (neutrophils and T lymphocytes) respond within days (neutrophils) to weeks (T lymphocytes). (B) The glial scar reflects the sequential arrival of diverse glial and immune cell classes at the site of injury. As these cells accumulate, microglia transition from the homeostatic to the activated, phagocytic state, and subclasses of astrocytes differentiate into activated astrocytes, with significantly more branched processes, and scarring astrocytes. In addition, immune-derived macrophages as well as neutrophils and lymphocytes interact with the differentiating glial cells. Eventually, the scarring astrocytes form a capsule around the site of damage where cellular debris continues to be cleared and immune cells (but not astrocytes or NG2 glia) are retained. This mature glial scar presents a major barrier to the regrowth of axons across the site of CNS damage. (After J. C. Perez et al., 2021. *Front Aging Neruosci* 13.)

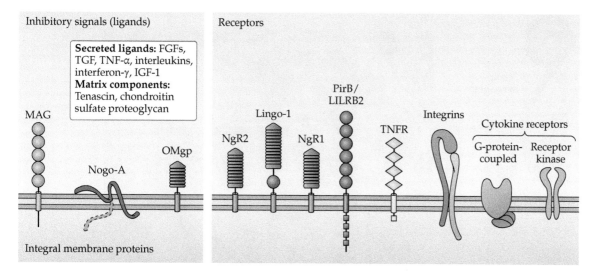

FIGURE 26.15 Molecular mediators of the CNS response to injury Glial cells either present on their cell surfaces or secrete inhibitory signals that limit axon growth after injury. Receptors for these signals are expressed on the newly generated axons of neurons whose original axons have been severed. Oligodendroglia in particular are sources of inhibitory signals, including the myelin-associated proteins MAG (*myelin-associated glycoprotein*) and OMgp (*oligodendrocyte myelin glycoprotein*). In addition, Nogo-A, a member of the reticulon family of endoplasmic reticulum membrane trafficking regulators, is thought to be expressed on the surfaces of glial cells as well as neurons. Lingo1 and PirB/LILRB2 are receptors for MAG and OMgp. NgR1 and 2 are receptors for Nogo-A. Tissue necrosis factor receptors (TNFRs) bind the inflammatory cytokine TNF-α. Integrins bind several ECM proteins, and two classes of cytokine receptors bind pro- and anti-inflammatory cytokines released by astrocytes, oligodendroglia, or macrophages at sites of injury. (After A. L. Kolodkin and M. Tessier-Lavigne, 2011. *Cold Spring Har Perspec Biol* 3: a001727.)

myelin-associated glycoprotein (MAG) diminish the growth capacity of CNS axons (Figure 26.15). This observation presents a puzzle because MAG is also produced by Schwann cells but does not impede peripheral regeneration most likely due to its transient down-reglation during PNS regeration (see Figure 26.9B). Several transmembrane receptors expressed on injured CNS axons interact with myelin proteins including MAG, as well as ECM molecules, secreted signals including tumor necrosis factor (TNF), and other cytokines to diminish axon regrowth (see the next section).

Another unfortunate consequence of glial scarring is secretion of molecules that act as chemorepulsive guidance cues during initial development. These include Semaphorin 3A (which causes growth cone collapse and axon withdrawal in the developing CNS), several ephrins, as well as slit, the ligand for the robo receptor (see Figures 23.5 and 23.7). The receptors for each of these molecules are upregulated on the growth cones of axons that approach the glial scar, resulting in local distortions in the direction of growth: The axons turn away from the glial scar and often form a tangled mass of endings at the scar's boundary. In addition, matrix components that inhibit axon growth (particularly tenascin and chondroitin sulfate proteoglycan) are enriched in the extracellular spaces within the glial scar. Thus, the proliferation and hypertrophy of glial cells, and their expression of chemorepellent or growth-inhibiting molecules, reinforce the limits of axon growth following focal brain injury.

Despite these multiple barriers to CNS axon regrowth, it is possible to provide a milieu that elicits growth of CNS axons from adult neurons. Facilitation of regrowth, however, depends on an experimental approach that is both impractical and limited in its efficacy. In experimental animals, severed axons in the optic nerve (recall that the optic nerve and retina are components of the CNS) or spinal cord can be provided with a peripheral nerve graft that offers Schwann cell, basal lamina, and connective tissue components that normally support peripheral nerve regeneration (Figure 26.16A). Central axons grow readily through the peripheral nerve graft. Some of these axons can even make synapses in the target territory to which the distal end of the graft is connected (Figure 26.16B). The numbers of synapses, and their functional capacity to restore vision, however, are very limited. Nevertheless, such experiments show that Schwann cells define an environment in the peripheral nerve sheath that is particularly well adapted to initiate and support the regrowth of damaged adult axons, whether they project to the periphery as do motor neurons during peripheral nerve repair or remain within the CNS. Perhaps more

(A)

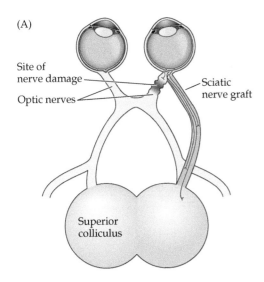

Site of nerve damage

Optic nerves

Sciatic nerve graft

Superior colliculus

(B)

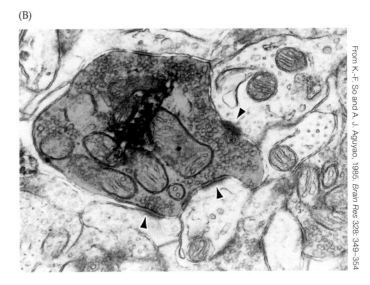

From K.-F. So and A. J. Aguyao, 1985. *Brain Res* 328: 349–354

FIGURE 26.16   Damaged axons in the CNS respond to growth-promoting properties of peripheral nerve sheaths and Schwann cells (A) Severed axons from the optic nerve are apposed to a peripheral nerve graft. The axons, which would normally not regenerate through the optic nerve (recall that the optic nerve, while physically in the periphery, is wholly within the CNS), now grow through the peripheral nerve graft to reach the superior colliculus, a normal target for retinal ganglion cells. (B) Regenerated axons make synapses with targets in the superior colliculus. The dark material is an electron-dense, intracellularly transported tracer that identifies specific synaptic terminals (arrowheads) as emanating from a regenerated retinal axon. (A after K.-F. So and A. J. Aguyao, 1985. *Brain Res* 328: 349–354.)

important, such experiments indicate that CNS neurons, beyond motor neurons when damaged, can return to a transcriptional state that will support growth. Without this very artificial solution to CNS regenerative axon growth, which circumvents the inhibitory environment of the glial scar, the net effect of CNS damage apparently overcomes any potential change in surviving neurons and prevents their regrowth.

Immune activation and inflammation following brain injury

Perhaps one of the most important challenges to CNS neuronal regeneration is the disruption of the barriers that separate CNS tissues from systemic immune cells. This disruption results in immune-inflammatory responses elicited by neutrophils, leukocytes, and glial cells. In some cases, the immune-inflammatory response is reinforced by additional cells of the monocyte lineage that enter the CNS after damage to the *blood-brain barrier* (see Appendix), a cellular interface between blood vessels and neural tissue that keeps most molecules and immune cells from entering the brain (Figure 26.17A). Acute physical damage to brain tissue, or the pathological changes that accompany neurodegenerative disorders, disrupt the blood-brain barrier: The tight junctions that link endothelial cells and thus prevent large molecules and circulating blood cells from directly entering

CNS tissue are diminished in their effectiveness or completely eliminated. This allows invasion of neutrophils and monocytes, the subsequent activation of microglia as well as astrocytes, and finally the invasion of T and B lymphocytes (Figure 26.17B). This inflammatory response is initiated by the release of damage-associated proteins (DAMPs) that include the contents of lysed neurons, mitochondrial enzymes, and non-protein signaling intermediates such as reactive oxygen species and purinergic (ATP, ADP) ligands. These proteins and small molecules initiate a cascade of cellular responses that elicit secretion of pro-inflammatory cytokines (Table 26.1), including cytokines that in turn attract additional neutrophils and other monocytes, cytokines that mediate T and B lymphocyte maturation, and perhaps most important, the pro-inflammatory cytokine interleukin-1 (IL-1). IL-1 modulates expression of several immune mediators that reinforce the inflammatory state. The end result of this neuroinflammatory response is once again local glial scarring that encapsulates the site of inflammation and forms a protective barrier between the inflamed site and adjacent healthy brain tissue. Because inflammatory mediators are targets of many drugs, this damage- or degeneration-induced inflammatory state could potentially be manipulated to minimize additional damage and improve the efficiency of repair, or at least limit the extent of neuronal damage or loss, after brain injury.

TABLE 26.1 **Major Immune Cell Classes, Cytokines, and Other Inflammatory Mediators Activated in Neuroinflammation Following Brain Injury**

Cell type	Mediator	Function
Neutrophils	CXCR2 (C-X-G motif chemokine receptor 2)	Chemokine that mediates neutrophil migration
	NE (neutrophil elastase)	Enzyme released by neutrophils to degrade extracellular matrix
Macrophages and microglia	CD1-1b (cluster of differentiation 1-1b)	Integrin that regulates migration of immune cells through tissues
	CCR2 (C-G motif chemokine receptor 2)	Chemokine receptor that coordinates monocyte chemotaxis
	CX3CR1 (C-X3-G motif chemokine receptor 1)	Chemokine receptor mediating macrophage and microglia migration
	IBA 1 (ionized calcium-binding adapter molecule 1)	Calcium-binding protein associated with microglia and macrophage activation
T cells	Rag 1 (recombination activating gene 1)	Enzyme required for B and T lymphocyte development
	IL-4 (inter1eukin-4)	Cytokine that aids in B and T lymphocyte proliferation and differentiation
Others	IL-1 (inter1eukin-1)	Pro-inflammatory cytokine that regulates transcription and production of multiple downstream inflammatory mediators
	Caspase-1	Enzyme that cleaves pro-IL-1β and pro-IL-18 to induce inflammation
	IL-18 (interleukin-18)	Pro-inflammatory cytokine that activates T lymphocytes
	IL-6 (interleukin-6)	Pleiotropic cytokine that induces a multitude of inflammatory responses
	GFAP (glial fibrillary acidic protein)	Intermediate filament protein expressed by astrocytes
	TNF-α (tumor necrosis factor α)	Pleotropic cytokine that can promote cell death, inflammatory cytokine production, and cell proliferation
	G-CSF (granulocyte colony-stimulating factor)	Stimulates proliferation and differentiation of hematopoietic cells as well as neural progenitors
	GM-CSF (granulocyte-macrophage colony-stimulating factor)	Promotes generation and activation of myeloid cells and neurons
	Type 1 IFN (type 1 interferon)	Regulates transcription of pro-inflammatory cytokines and chemokines
	IL-10 (interleukin-10)	Negatively regulates pro-inflammatory cytokine production
	TGF-β (transforming growth factor β)	Controls proliferation and differentiation of multiple immune cell types
	TREM2 (triggering receptor expressed on myeloid cells 2)	Stimulates proliferation and differentiation of hematopoietic cells as well as neural progenitors

Source: C. A. McKee and J. R. Lukens, 2016. *Front Immunol* Article 556

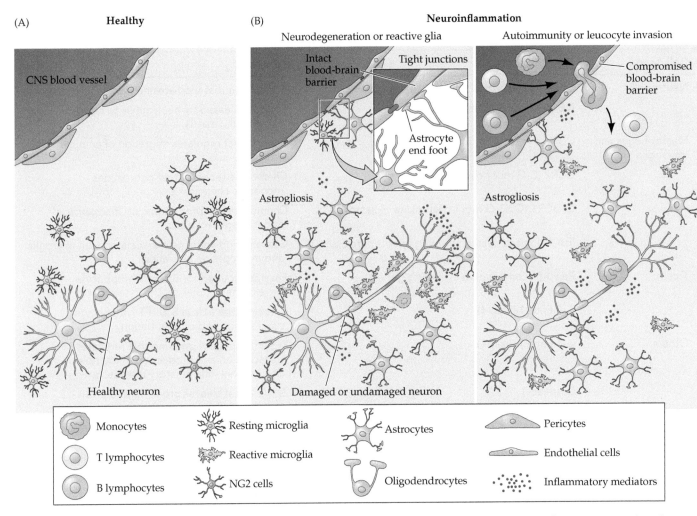

(A) Healthy

(B) Neuroinflammation

Neurodegeneration or reactive glia

Autoimmunity or leucocyte invasion

CNS blood vessel

Intact blood-brain barrier

Tight junctions

Compromised blood-brain barrier

Astrocyte end foot

Astrogliosis

Astrogliosis

Healthy neuron

Damaged or undamaged neuron

Monocytes | Resting microglia | Astrocytes | Pericytes
T lymphocytes | Reactive microglia | | Endothelial cells
B lymphocytes | NG2 cells | Oligodendrocytes | Inflammatory mediators

FIGURE 26.17 **Immune-mediated inflammatory responses that drive glial responses and resistance to neuronal regrowth and repair** (A) The distribution and state of glial cells in an undamaged brain in which the blood-brain barrier is intact. Quiescent microglia, a few activated microglia, astrocytes, oligodendrocytes, and presumed oligodendrocyte precursors (NG2 cells) are distributed throughout the tissue. (B) When local damage occurs in brain tissue (left) or when the blood-brain barrier is disrupted (right), cytokines and other signaling molecules activate microglia and astrocytes. In turn, astrocytic processes as well as the astrocytes themselves increase in frequency, providing the scaffold for a local glial scar. Once the blood-brain barrier is compromised, neutrophils and other monocytes rapidly infiltrate the brain tissue and release additional pro-inflammatory cytokines that elicit a more robust astrocyte response and reinforce local inflammation. This leads to an additional decrease in the potential for preserving neuronal survival, tissue integrity, and the possibility of even modest regrowth and repair. (A,B after A. Waisman et al., 2015. *Neurology* 14: 945–955.)

CONCEPT **26.4**

Adult Vertebrate Nervous Systems Retain Some Neural Stem Cells for Limited Replacement of Neurons

LEARNING OBJECTIVES

26.4.1 List the characteristics of adult neural stem cells and the factors that influence their neurogenic capacity.

26.4.2 Discuss two examples of ongoing neurogenesis in adult vertebrates and their functional consequences.

26.4.3 Analyze the sites and mechanisms of neurogenesis and migration in the adult mammalian CNS.

26.4.4 Discuss the possibilities and limits of adult neurogenesis for brain repair following injury or neurodegenerative disease.

Neurogenesis in the mature CNS

Few issues in modern neuroscience have engendered as much controversy as the ability of the adult CNS to generate new neurons in response to acute or degenerative damage to neural tissue. There was no reliable way to assess the extent of cell proliferation and mitotic activity in the mature brain until the advent of neuronal birth-dating techniques (see Chapter 22). These approaches use analogues of the DNA-specific nucleotide thymidine. When a cell divides in the presence of such analogues, including radioactive-labeled thymidine (H^3-T) or bromodeoxyuridine (BrdU) and other uracil variants, the analogue is incorporated into the nuclear DNA. If the division is a terminal division or that of a slowly dividing stem cell, the nuclear DNA will be heavily labeled, and detectable by autoradiography (H^3-T) or immunohistochemistry (BrdU). When these techniques became available in the late 1960s and early 1970s, the presence of labeled nuclei, suggesting cell division, in a subset of brain regions, including the olfactory bulb and hippocampus, was greeted with interest as well as skepticism. Indeed, the only unchallenged example of ongoing neurogenesis anywhere in the adult mammalian nervous system was that which occurs in the peripheral olfactory epithelium of all vertebrates (see Chapter 14). Clinical experience, buttressed by animal studies primarily in mammals, indicated that the mature brain was unlikely to undergo significant genesis of new functional nerve cells. Instead, much of the labeling of dividing cells in undamaged mature brains was attributed to the division of glial cells.

The ability to label cells undergoing DNA replication and presumably mitosis and to track their progeny in developing versus adult mammalian brains led to an overall discouraging assessment of the potential for generating new neurons in the adult mammalian brain. Quite simply, there did not seem to be any extensive addition of neurons after the completion of early postnatal development. While this general conclusion has had extensive support, some reservations were raised as early as the mid-1960s. Joseph Altman and his colleagues, then at the Massachusetts Institute of Technology, found that small numbers of apparent granule neurons in the hippocampus and olfactory bulb in guinea pigs and rats could be heavily labeled with H^3-T injected at adult ages. Their work suggested that these local inhibitory interneurons with short axons that remained within the hippocampus or olfactory bulb, respectively, might be added to the brain during adulthood, either replacing or augmenting the cohort generated during development. At the time, however, the lack of additional markers that could identify these cells securely as neurons versus glial cells, which are similar in size and appearance when stained with standard Nissl stains (see Chapter 1), led other investigators to conclude that many, if not all, of these cells were newly generated glia rather than neurons. This conclusion was reached based on evidence available at the time that glial cells continued to divide in the CNS throughout life. Moreover, it seemed unlikely that already differentiated neurons might divide to generate additional new neurons. Finally, the known properties of neural stem cells based upon observations in embryos, and the lack of tools to identify these cells securely in the adult brain, made it difficult to imagine which cells in a mature brain might have the capacity to generate new neurons.

Modern approaches have led to a much clearer understanding of the identity of mitotically active cells in the adult brain of many vertebrates, including mammals. It became clear that most adult tissues in animals as well as plants retain resident *stem cells* capable of generating the full range of differentiated cell classes in the tissue where they reside. Thus, it is reasonable to assume that neurogenesis in the adult brain most likely relies on similar resident stem cells. This conclusion was reinforced by clear evidence that mature, differentiated neurons do not de-differentiate and divide. For several decades, however, it was not entirely clear where adult neural stem cells might be found. A low level of glial cell proliferation—for both oligodendrocytes and astrocytes—does indeed continue in most mature brain regions throughout life. These committed glial precursors, however, are not neural stem cells capable of generating new neurons in mature brains. Instead, neural stem cells are maintained at a few distinct locations in the adult brains of several species, including humans. As discussed in Chapter 22, the neural stem cells of the embryonic nervous system give rise to the full complement of cell classes found in neural tissue—that is, diverse classes of differentiated neurons, astrocytes, and oligodendroglia (see Box 22A)—as well as to more neural stem cells. Apparently, specific regions in the mature CNS in several vertebrates provide an environment, or niche, that supports the maintenance of such multipotent neural precursors. These adult neural stem cells, like their embryonic counterparts, express primarily glial markers as well as developmentally regulated transcription factors (see Chapter 22). In the adult brain, however, these neural stem cells are distinguished by their limited locations, their proliferative characteristics, and their neurogenic capacity. The extent to which these stem cells give rise to neurons that continually replace or augment existing populations varies depending on the species, brain region, and conditions (e.g., growth, seasonal change, injury) that influence neurogenesis in the adult brain. In some cases, mostly in non-mammalian vertebrates, adult neurogenesis accommodates ongoing somatic growth of the mature individual—suggesting that bigger bodies might elicit bigger brains. In mammals, the issue of whether adult neurogenesis plays a significant role in normal brain function in many instances remains unresolved.

Adult neurogenesis in non-mammalian vertebrates

Observations in several non-mammalian vertebrate species, particularly in teleost fish and songbirds, make a strong case for the capacity of adult vertebrate brains to add new neurons and incorporate them into functional circuits that guide or even modify behavior. One of the first thoroughly characterized examples of ongoing vertebrate adult neurogenesis was that in the goldfish. Goldfish,

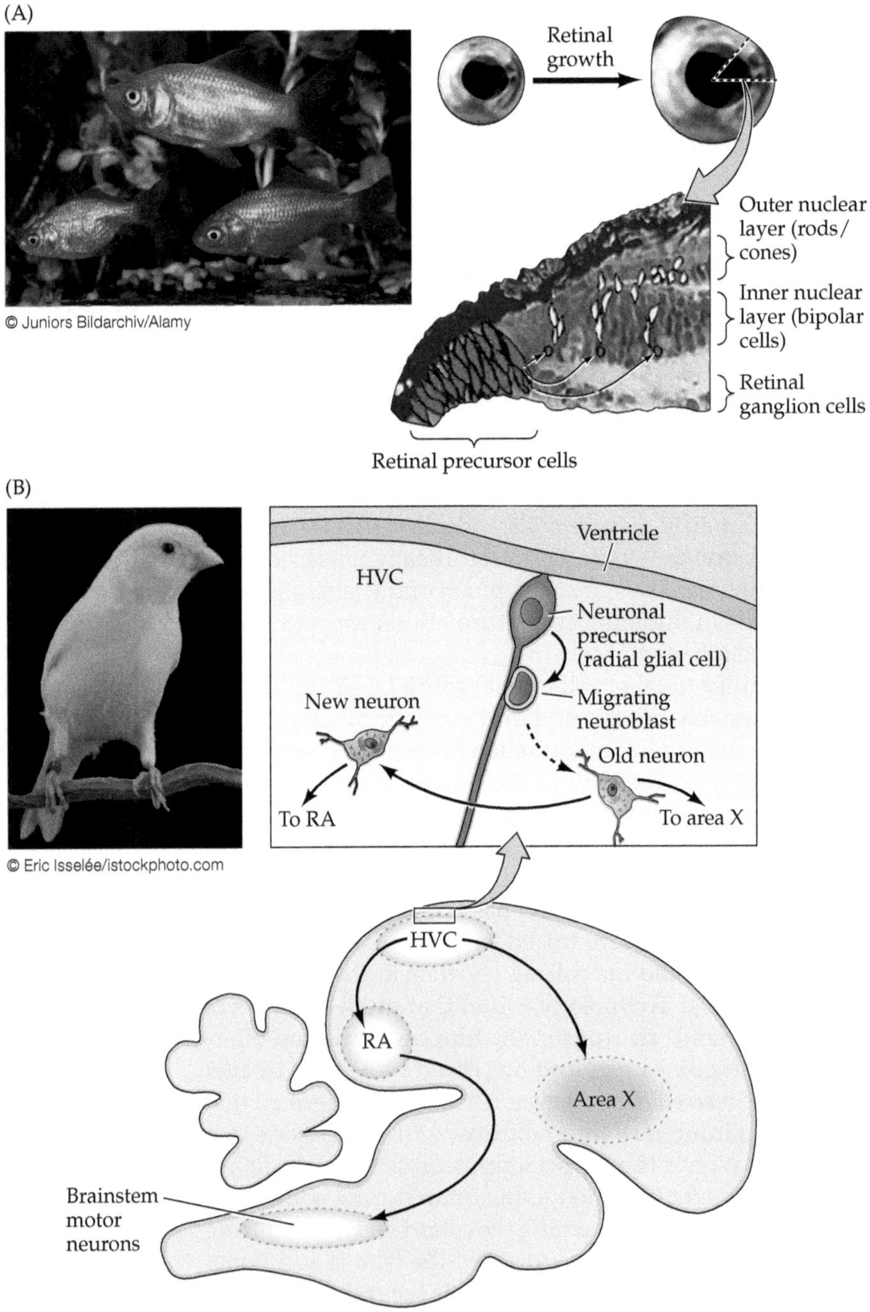

FIGURE 26.18 Adult neurogenesis in non-mammalian vertebrates (A) Given favorable environmental conditions, teleost fish such as goldfish grow throughout their entire adult life; the growth of the fish's body is matched by the growth of its eyes and brain. The retina grows by adding new neurons generated from a population of stem cells distributed in a ring at the very margin of the retina (red). These stem cells give rise to all retinal cell types except the rods (which are regenerated from precursors found in the existing differentiated region of the retina). (B) In a process of ongoing adult neurogenesis, male songbirds such as the canary lose and replace significant numbers of neurons in forebrain nuclei that control the acquisition, production and perception of song. These song-control centers include the HVC (higher vocal center), RA (robustus archistriatus), and area X, which is the equivalent of the caudate nucleus in the mammalian brain. In the HVC, a population of radial stem cells is maintained. The cell bodies are adjacent to the ventricular space, and their processes extend into the neuropil of the nucleus. A subset of neurons is retained in the nucleus as new ones are added. Neuroblasts migrate from the ventricular zone along the radial processes of the precursor cells and then integrate into circuits with existing neurons. (A after D. C. Otteson and P. F. Hitchcock, 2003. *Vis Res* 43: 927–936; B after S. A. Goldman, 1998. *J Neurobiol* 36: 267–286.)

like many other fish, continue to grow throughout their lifetimes. Fully adult fish can increase substantially in size after they reach maturity, depending on their environment and the availability of habitat and food. Their body growth is matched by the growth of sensory structures in the periphery, particularly the eye. By the early 1970s, several investigators had recognized that growth of the eye was accompanied by the generation of new retinal neurons—thus providing an example of adult CNS neurogenesis, given that the retina is part of the CNS. Subsequent work showed that these neurons are generated from a subset of stem cells that form a ring around the entire margin of the goldfish retina (Figure 26.18A). These cells are capable of generating all of the cell classes in the goldfish retina (with the exception of rod photoreceptors, which are regenerated by a distinct adult precursor cell; see Chapter 9 for a review of retinal cell classes). They retain many of the molecular characteristics of embryonic retinal stem cells. The new neurons generated by these adult retinal stem cells integrate into the existing retina in rings between

the precursor cells at the periphery and the existing differentiated retina—much like yearly growth rings added to tree trunks. Newly generated retinal local circuit neurons—bipolar and amacrine cells—make connections with new photoreceptors and new retinal ganglion cells. The axons of the new retinal ganglion cells enter and grow through the existing optic nerve and tract to reinnervate the optic tectum. Most of this axon growth happens along the ECM deposited at the glial-limiting membrane of the optic nerve. (The optic nerve is ensheathed by glial cells, which secrete matrix molecules that are the equivalent of the basal lamina conduits made by Schwann cells in the periphery.) Regeneration of all retinal cell classes in teleost fish can also happen in response to a local injury of existing retinal tissue; however, the details of how this limited, localized repair is completed are somewhat different than that for ongoing addition of cells. In parallel with neuronal addition in the eye and the growth of new retinal ganglion cell axons into the brain, the goldfish optic tectum—the primary central target for retinal axons in the fish—adds new neurons to accommodate the quantitative expansion of the retinal projection. These cells are not added in complementary rings; instead, populations of new neurons are added as crescents to the back of the tectum. Accordingly, the geometry of adult neurogenesis in the periphery and brain is mismatched. This divergence requires that new retinal inputs be constantly remapped along with existing retinal projections to maintain a faithful retino-tectal topographic map. Thus, there must be a great deal of dynamism in synaptic connections in the optic tectum of the adult goldfish. Connections must be made, broken, and remade to maintain the integrity of the retinotopic map as the fish and its retina grow throughout the fish's lifetime.

An equally striking example of adult neurogenesis is found in several species of songbirds, including the canary and the zebra finch. This ongoing neurogenesis occurs in several regions of the avian brain but has been most thoroughly studied in the structures that control vocalization and perception of song (Figure 26.18B; see also Figure 24.2). In most male songbirds, there is continual loss and addition of neurons in these regions. In some species, the cycle of loss and regeneration follows mating seasons and is under the control of gonadal steroids (see Chapter 25), while in others it occurs continually. Although it is tempting to speculate that the new neurons are a substrate for flexible acquisition, production, or perception of songs, this interpretation remains uncertain. Indeed, many birds that have significant amounts of adult neurogenesis in song-control regions show very little flexibility in their song after the critical period for song learning is complete in early life.

Regardless of the behavioral consequences, it is estimated that birds replace most of the neurons in several song-control centers of their brain several times over a lifetime. The new neurons for song-control centers are generated from neural stem cells found in a limited region of the neural tissue immediately adjacent to the forebrain lateral ventricles. Stem cell bodies are found in this zone, and their radial processes (much like the radial glia that are the stem cells of the developing mammalian neocortex; see Chapter 22) extend into song-control centers (see Figure 26.18B). These cells function as precursors, generating new neurons via asymmetrical divisions, and as migration guides that constrain the translocation of new neurons from the ventricular zone to the song-control nuclei (which can be considered comparable to regions of the striatum in the mammalian brain). Many of

the new neurons integrate themselves into behaviorally relevant circuits and have functional properties that are consistent with a contribution to song acquisition, production or perception. Nevertheless, a significant number die before they can fully differentiate, suggesting that there may be limits to the capacity of newly generated neurons to establish sufficient trophic support and activity-dependent validation to survive. A key feature of adult neurogenesis in the avian brain is that there is always a balance of existing, long-lived neurons and newly generated neurons. Thus, even this compelling example of adult neurogenesis occurs in the context of some stability in the mature brain.

Neurogenesis in the adult mammalian brain

Since the initial reports of potential neurogenesis in the adult mammalian brain in the late 1960s, the mechanisms and extent of neurogenesis in the adult mammalian brain have been continually reexamined in mice, rats, monkeys, and humans. Each successive re-examination was driven by the advent of new genetic and cell biological tools to distinguish more securely new neurons and neural stem cells as well as to trace their proliferation and cell lineages (see Chapters 1 and 22). Clearly, the capacity for ongoing neuronal replacement in specific CNS regions would provide a model for how regeneration might be elicited following brain injury or to combat neurodegenerative disease. Such approaches imagine stimulating resident, quiescent neural stem cells or using exogenous neural stem cells that have been instructed to generate the neuronal class that has been lost as a result of injury or disease. To date, the results of ongoing efforts to understand neurogenesis in the adult brain indicate that new nerve cells are generated reliably in just two regions in the mature mammalian CNS, the olfactory bulb and the hippocampus (Figure 26.19; see also Figure 14.3).

In these regions of the adult CNS, the newly generated nerve cells are primarily interneurons: granule cells and periglomerular cells in the olfactory bulb (see Chapter 14), or granule cells in the hippocampus (see Chapters 8 and 30). These newly generated olfactory or hippocampal interneurons are apparently the progeny of neural precursor or stem cells located in niches close to the surface of the lateral ventricles, relatively near either the bulb or hippocampus. For the mature olfactory bulb, neural stem cells are found in the anterior subventricular zone (SVZ) of the forebrain. In the hippocampus, they are found in the subgranular zone (SGZ) within the hippocampal formation. The neural stem cells in the SVZ and SGZ do not give rise to projection neurons with long axons. At least some of these new neurons become integrated into functional synaptic circuits (see Figure 26.19B,C); however, most new neurons generated in the adult brain die before being integrated into existing circuitry. No one has yet fully explained the functional significance of restricting

(A)

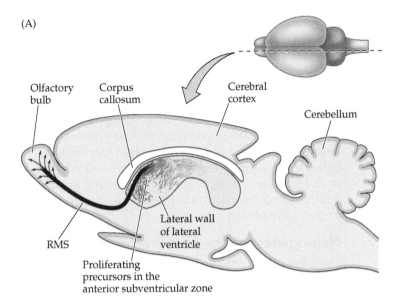

Olfactory bulb

Corpus callosum

Cerebral cortex

Cerebellum

RMS

Lateral wall of lateral ventricle

Proliferating precursors in the anterior subventricular zone

FIGURE 26.19 Neurogenesis in the adult mammalian brain (A) Neural precursors in the epithelial lining of the anterior lateral ventricles in the forebrain (a region called the anterior subventricular zone, or SVZ) give rise to postmitotic neuroblasts that migrate to the olfactory bulb via a distinctive pathway known as the rostral migratory stream (RMS). Neuroblasts that migrate to the bulb via the RMS become either olfactory bulb granule cells or periglomerular cells; both cell types function as interneurons in the bulb. (B) In the mature hippocampus, a population of neural precursors is resident in the basal aspect of the granule cell layer of the dentate gyrus (the subgranular zone, or SGZ). These precursors give rise to postmitotic neuroblasts that translocate from the basal aspect of the granule cell layer to more apical levels. In addition, some of these neuroblasts elaborate dendrites and a local axonal process and apparently become GABAergic interneurons within the dentate gyrus. (C) A newly generated granule neuron (labeled green using a genetic marker) is contacted by GABAergic synapses (labeled red) in the granule cell layer of the adult hippocampus. (A,B after F. H. Gage, 2000. *Science* 287: 5457.)

(B)

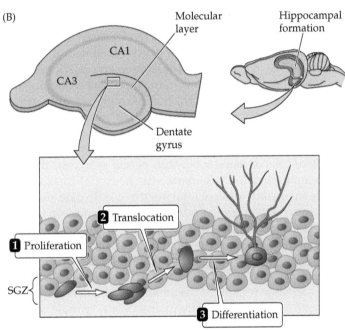

Molecular layer

Hippocampal formation

CA1

CA3

Dentate gyrus

2 Translocation

1 Proliferation

SGZ

3 Differentiation

(C)

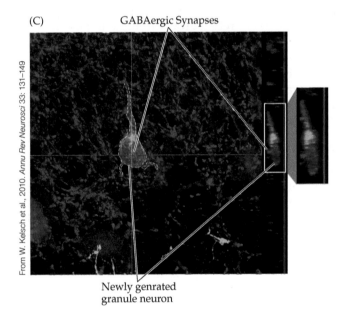

GABAergic Synapses

From W. Kelsch et al., 2010. *Annu Rev Neurosci* 33: 131–149

Newly genrated granule neuron

neurogenesis to just these few regions in the adult brain, the exclusive GABAergic interneuron identity of the newly generated neurons, or the ultimate behavioral consequences for the addition of such cells. Suggestions for functional significance include serving as a substrate for learning and memory in the hippocampus, and as a substrate for ongoing adjustment of olfactory representation to accommodate peripheral receptor turnover. The death of most and the integration of only a few of these newly generated neurons suggest that there may be a premium placed on stability in the mammalian brain—even in regions such as the olfactory bulb and hippocampus, which

are thought to be highly plastic—thus limiting opportunities for new neurons to join existing circuits. Nevertheless, the fact that any new neurons can be generated and incorporated in at least a few adult brain regions shows that this phenomenon can occur in the mammalian CNS.

Cellular and molecular mechanisms of adult neurogenesis

In regenerating tissues such as intestine or lung, stem cells are found in a distinct location. Presumably there is a local environment within the stem cell niche that is conducive to the maintenance of stem cells as well as to the division and

(A)

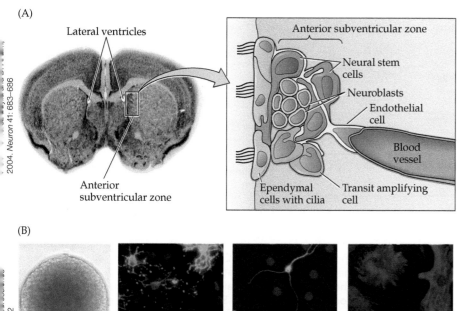

Lateral ventricles

Anterior
subventricular zone

Anterior subventricular zone

Neural stem
cells

Neuroblasts

Endothelial
cell

Blood
vessel

Ependymal
cells with cilia

Transit amplifying
cell

(B)

FIGURE 26.20   The forebrain's anterior subventricular zone provides a stem cell niche (A) Coronal section through the mouse brain indicating the location of the anterior subventricular zone in the anterior part of the lateral ventricles. The schematic shows the arrangement of cells at the ventricular surface. The ciliated ependymal cells form a tight epithelial boundary separating cerebrospinal fluid from brain tissue. Immediately adjacent are neural stem cells, whose processes are intercalated between the ependyma and all other cell types. These stem cells are also often seen in proximity to blood vessels. Transit amplifying cells are also found in this domain, and their progeny—the neuroblasts—are often clustered close by before adhering to glia that guide them into the rostral migratory stream (RMS; see Figure 26.21). (B) Isolation and propagation in vitro of neural stem cells from the anterior subventricular zone. At left is a *neurosphere*, a ball of neural stem and transit amplifying cells that has been generated clonally from a single founder dissociated from the adult subventricular zone epithelium. The adjacent three panels show differentiated cell types generated from the neurosphere. From left to right: oligodendroglia, neurons, and astrocytes. (A after A. Alvarez-Buylla and D. A. Lim, 2004. *Neuron* 41: 683–686.)

initial differentiation of cells that can reconstitute the adult tissue. The SVZ is a cell-dense region adjacent to the ventricular space found in the anterior aspect of the cortical hemispheres, adjacent to the ventricular space that is filled with cerebrospinal fluid (Figure 26.20A; also see Figure 26.19A). The SGZ is a region of cells immediately adjacent to the granule cell layer of the hippocampus (see Figure 26.19B). In addition, apparent adult neural stem cells can be isolated from subventricular regions of the cerebellum, midbrain, and spinal cord. These stem cells, whose neurogenic potential has been identified based on in vitro differentiation (Figure 26.20B), do not generate new neurons in the regions where they are found in the adult brain. Thus, the olfactory bulb and hippocampus remain the only regions of the adult mammalian brain where newly generated neurons are incorporated into existing circuits.

In both the SVZ and SGZ, neural stem cells have characteristics similar to those of astrocytes, including the expression of multiple molecules that are also seen in astrocytes. Thus, similar to the developing brain (as well as in avian brains; see Figure 26.18B), the multipotent neural stem cell has an apparent glial rather than neuronal identity (see Chapter 22). Furthermore, neural stem cells are often found in close proximity to blood vessels, suggesting that they may be regulated by circulating as well as local signaling molecules. Although the route of delivery is different, this proximity to signals suggests that these precursors, like their developing counterparts, rely on cell-cell signaling to guide proliferation and differentiation. It is also possible

that circulating signals relay information about the overall physiological state of the animal, thus linking broader homeostatic mechanisms with neurogenesis in the adult brain.

In order to generate differentiated neurons and glia, the neural stem cell must give rise to an intermediate precursor cell class, generally referred to as a **transit amplifying cell**. These cells retain the ability to divide; however, their cell cycles are much faster than those of stem cells, and they divide asymmetrically. After each cell division, a transit amplifying cell gives rise to a postmitotic daughter cell, plus another transit amplifying cell that reenters the cell cycle for an additional round of asymmetrical division. Transit amplifying cells are limited in their number of divisions, and eventually their potential for generating postmitotic blast cells is exhausted—thus yielding a terminal symmetric division. For neurogenesis to proceed constitutively, the transit amplifying cells must be replenished from the stem cell population. Newly generated, still undifferentiated neurons and glial cells—neuroblasts and glioblasts—are no longer competent to divide, and they move away from the SGZ or SVZ into regions of the olfactory bulb or hippocampus where mature neurons or glia are found. In the hippocampus, this distance is relatively small, and the cells undergo a modest local displacement to reach a final position relatively close to their site of

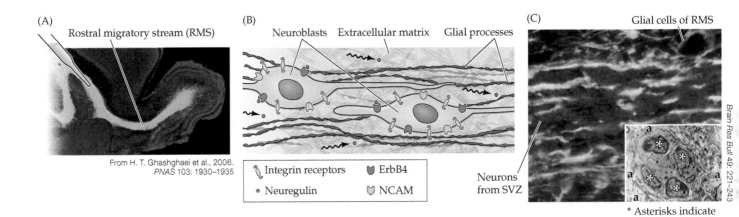

(A) Rostral migratory stream (RMS)

From H. T. Ghashghaei et al., 2006.
PNAS 103: 1930–1935

(B) Neuroblasts   Extracellular matrix   Glial processes

| ↳ Integrin receptors | ⬮ ErbB4 |
| • Neuregulin | ⬮ NCAM |

(C) Glial cells of RMS

Neurons from SVZ

Brain Res Bull 49: 221–243

* Asterisks indicate migrating neurons

FIGURE 26.21   New neurons in the adult brain migrate via a specific pathway (A) The rostral migratory stream (RMS) in the adult mouse brain can be demonstrated by injecting a tracer into the lateral ventricle. The cells in the SVZ take up the tracer (see Figure 26.19), and the labeled cells enter the forebrain tissue in a "stream" of migrating neurons. (B) Schematic of the mouse RMS. Glial processes (red) form conduits for migrating neurons. The ECM associated with these processes influences migration, mediated by integrin receptors for ECM components found on migrating neurons. Secreted neuregulin also influences motility of the migrating neurons in the RMS, via the ErbB4 neuregulin receptor. Finally, polysialylated NCAM on the surfaces of newly generated neurons facilitates migration through the RMS. (C) The glial cells of the RMS in a longitudinal view are labeled in green, and the neurons migrating from the SVZ via the RMS are labeled red. The neurons are constrained within apparent tubes composed of glial cells and processes. The inset shows a cross section of one such "tube" with glial processes (a) encapsulating the migrating neurons (asterisks). (B after H. T. Ghashghaei et al., 2007. *Nat Rev Neurosci* 8: 141–151.)

generation. For newly generated neurons destined for the olfactory bulb, however, the distance from the SVZ, which is in the anterior ventricular region adjacent to the cerebral cortical hemisphere, to the olfactory bulb (which has no recognizable ventricular space) is considerable. A specific migratory route, defined by a distinct subset of glial cells, facilitates migration of newly generated neurons from the anterior SVZ to the bulb. This route is referred to as the **rostral migratory stream**, or **RMS** (Figure 26.21); within it, neuroblasts move along channels defined by the surfaces of elongated glial cells (these glial cells do not have stem cell properties). In the RMS, as at other sites of cell migration or axon growth, an ECM, presumably secreted by the glial cells, facilitates migration. In addition, the migrating cells express the polysialylated form of the neural cell adhesion molecule NCAM, which promotes cell-cell interactions that facilitate migration. Secreted signaling molecules also influence migration in the RMS. In this case, neuregulin and its ErbB receptors, which also influence axon guidance and synapse formation in the periphery, facilitate motility, particularly via interaction with the ErbB4 neuregulin receptor. Thus, in the RMS, several developmentally regulated adhesion molecules and secreted signals mediate migration of new neurons through otherwise mature brain tissue.

Identification of molecular mediators of adult neurogenesis remains a major focus of current research. The most attractive hypothesis, and one that has support from the available data, is that the signaling molecules and transcriptional regulators used to define neural stem cells early in development are either retained or reactivated to facilitate neurogenesis in the adult. Accordingly, many of the inductive signaling molecules described in Chapter 22 as mediators of the specification of neural precursors and their progeny in the neural plate and neural tube are also active in adult SVZs. These include Sonic hedgehog (Shh), members of the fibroblast growth factor (FGF) family, TGF-β family members (including the bone morphogenetic proteins, or BMPs), and retinoic acid. The essential transcription factors associated with neural stem cells during initial CNS development, including Sox2, and those that identify transit amplifying cells and newly generated neurons in the embryonic brain, such as the bHLH neurogenic genes (see Chapter 22), are expressed in adult neural stem and transit amplifying cells. Finally, many of the adhesion molecules that influence cell migration and dendritic and axonal outgrowth during neural development (see Figures 23.3–23.5) also influence the migration and differentiation of newly generated neurons.

Adult neurogenesis, stem cells, and brain repair in humans

The addition or replacement of neurons in the adult brains of fish, birds, rats, and mice provide clear examples of how new neurons may be integrated into existing circuits, presumably preserving, replacing, or augmenting function. In most cases, neuron replacement is gradual and probably related to ongoing low-level neurogenesis rather than to wholesale reconstitution of brain tissue in response to injury. Nevertheless, the limited capacity to replace neurons in an adult brain has offered some promise that, under the right conditions, neuron replacement might be used to repair the injured brain. In humans, the entire subventricular zone of the cerebral hemispheres provides a supportive environment for neural stem cells. At present, however, there is no evidence that adult neurogenesis in humans occurs outside the hippocampus. The rostral migratory stream that facilitates the migration of newly generated neurons from the anterior subventricular zone to the olfactory bulb is absent in humans, suggesting that ongoing neurogenesis in the ventricular zone does not produce new neurons that migrate to the olfactory bulb. Moreover, the available evidence suggests that there is no neurogenesis in the adult cerebral cortex (Box 26B)—an observation that establishes a much higher threshold for the possibility of stem cell–mediated repair of cortical circuitry damaged by trauma, hypoxia, or neurodegenerative disease. Cell replacement therapies have been attempted in a relatively small number of individuals with Parkinson's disease (see Chapter 18), but the overall effectiveness of such treatments has been poor. Furthermore, the extent to which truly undifferentiated human neural stem cells can be made to acquire and maintain characteristics of dopaminergic neurons of the substantia nigra (or any other distinctive neuronal identity) as well as an appropriate pattern of synaptic connections is unclear. Thus, although there is some promise that understanding the maintenance of neural stem cells and their neurogenic capacity in limited regions of the adult brain has potential for repairing the damaged brain, the fulfillment of this promise may be extremely challenging.

■ BOX 26B | Nuclear Weapons and Neurogenesis

The presence of stockpiles of nuclear weapons in an ever-increasing number of nations has always been a heavy burden for world affairs. It therefore may come as a surprise that the nuclear weapons testing carried out during the height of the Cold War (from the early 1950s through 1963) might play a positive, if unanticipated, role in resolving a major conflict in neuroscience.

In the late 1990s, a consensus emerged that the hippocampus and olfactory bulb are sites of gradual, limited addition of new neurons in the adult brain of most mammalian species. This consensus, however, did not extend to the question of whether new neurons are added to the cerebral cortex, especially that of humans, in adulthood. If adults do indeed add new critical neurons in significant numbers, such a mechanism would demand revision of current notions of plasticity, learning, and memory; it would also offer new avenues for treating traumatic, hypoxic, and neurodegenerative cortical damage. Several reports in the mid-1990s, including some from experiments on non-human primates, suggested that there might be substantial addition of neurons to the adult cortex. Despite the provocative nature and exciting implications of these findings, several other laboratories had difficulty replicating the surprising and controversial results. The disparities engendered a polarized debate with no easy resolution. Clearly, what was needed was an independent means of assessing neurogenesis in the adult cortex, preferably in humans.

In a unique approach—and in one of the more successful searches for evidence of weapons of mass destruction in the last several years—Jonas Friesén and colleagues at the Karolinska Institute in Stockholm took advantage of fluctuations in environmental exposure to radioisotopes from nuclear weapons testing to evaluate when cortical neurons are indeed generated over an individual's lifetime. Their method relied on the recognition that the normally steady state of the isotope carbon-14 (^{14}C) in Earth's atmosphere had been dramatically altered for a brief period between the mid-1950s and early 1960s. During this time, many countries conducted multiple tests of nuclear weapons, introducing large amounts of ionizing radiation into the atmosphere and nearly doubling the atmospheric concentration of ^{14}C. The Nuclear Test Ban Treaty of 1963 (which was adhered to by most countries until fairly recently) put a fairly abrupt

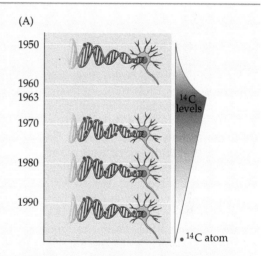

(A)

(A) Changes in atmospheric ^{14}C levels (right) and the availability of the isotope for incorporation into mitotic neurons at different times between 1955 (the start of frequent nuclear testing) and 1963 (when the Nuclear Test Ban Treaty was put in place). Neurons generated between 1963 and 1970, whether in adults or in individuals who underwent gestation and birth during that time frame, would incorporate significant amounts of ^{14}C into their nuclear DNA. (After E. Au and G. Fishell, 2006. *Nat Neurosci* 9: 1086–1088.)

end to this frightening period in human history, and atmospheric ^{14}C levels declined exponentially (Figure A).

This change in atmospheric ^{14}C concentration provided a natural version of experimental birth-dating techniques.

(Continued)

■ BOX 26B | Nuclear Weapons and Neurogenesis *(continued)*

Rather than injecting a bolus of tritiated thymidine or BrdU into the individual, people of varying ages had been naturally exposed to a "bolus" of ^{14}C that was incorporated into DNA being synthesized at the time. Thus, regardless of the age of the individual, cortical neurons generated between 1955 and 1963 should have a higher concentration of ^{14}C in their nucleus than those generated before or after that time frame.

To assess neurogenesis in this ingenious way, the researchers obtained autopsy samples from the cerebral cortices of seven individuals born between 1933 and 1973. The logic was that those born before 1955 would have significant numbers of ^{14}C-labeled neurons due to exposure as adults *only if there was indeed adult cortical neurogenesis*. If there was no adult neurogenesis, only those individuals born during (or shortly after) 1955–1963 should have ^{14}C-containing cortical neurons. To ensure that only cortical neurons were assessed, dissociated cortical cells were labeled fluorescently with a neuron-specific marker and then sorted so that a comparison could be made between fluorescent-tagged neurons and the non-fluorescent glia and supporting cells (non-neurons). ^{14}C levels were then measured using accelerator mass spectroscopy.

The results were clear: Individuals born before 1955 had no cortical neurons with elevated ^{14}C levels; thus, no neurons had been generated in their adult cortices (Figure B, top). In contrast, individuals born after 1955, but before the return of ^{14}C levels to baseline, had significant numbers of ^{14}C-labeled neurons; furthermore, the neuronal level of ^{14}C corresponded to the atmospheric level around the time of these individuals' gestations and births (see Figure B, bottom). The different values for

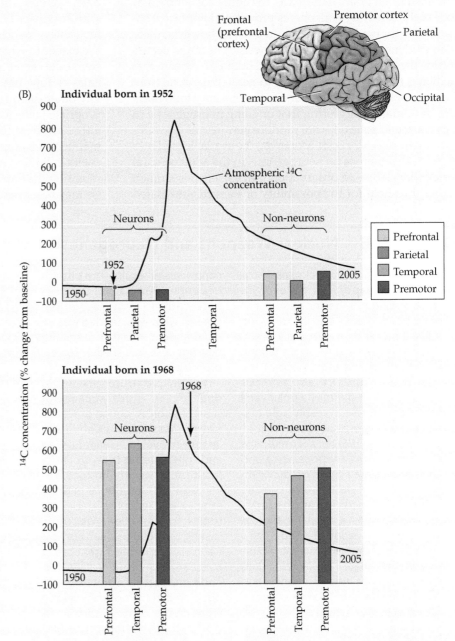

(B) Autopsy results for the individual born in 1952 showed no cortical neurons with elevated ^{14}C levels; thus, no neurons had been generated in the adult cortices. In contrast, the individual born in 1968 had significant numbers of ^{14}C-labeled neurons. Furthermore, in both cases the neuronal level of ^{14}C corresponded to the atmospheric level at around the time of the individuals' gestations and births. The slightly elevated levels for non-neural cells in the 1952 individual, and the lower levels in the 1968 individual, indicate that these cells turn over, and their ^{14}C content is altered by subsequent rounds of DNA synthesis and cell division. (After R. D. Bhardwaj et al., 2006. *PNAS* 103: 12564–12568.)

■ **BOX 26B** | **Nuclear Weapons and Neurogenesis** *(continued)*

the non-neural cells indicate that these cells turn over, so their ^{14}C content declines because of dilution by subsequent rounds of DNA synthesis and cell division.

To amplify this result, Friesén and colleagues studied a group of patients who had skin cancers and whose treatment included injections of thymidine analogues such as BrdU. They then examined cells heavily labeled with BrdU in the cortex post mortem. BrdU does not label neurons (recognized by staining for Neu-N, a neural marker) or neuron-specific neurofilaments; however, BrdU does label glial cells, recognized with antibodies against glial fibrillary protein (Figure C). This observation, especially when taken together with the ^{14}C study, argues strongly against significant neurogenesis in the adult cerebral cortex.

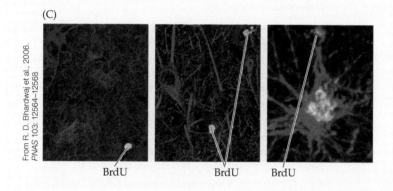

(C)

From R. D. Bhardwaj et al., 2006. *PNAS* 103: 12564–12568

BrdU   BrdU   BrdU

(C) The cortices from patients receiving BrdU during their lifetime were processed for BrdU histochemistry combined with neuronal and glial markers post mortem. Left: A BrdU nucleus (green) is distinct from cells labeled with Neu-N (red), a neuronal marker. Center: A similar distinction between neurons labeled for neurofilaments (red) and BrdU-labeled cells (green). Right: Cells labeled for GFAP are coincident with the BrdU-labeled nuclei, suggesting that only glia are generated in the adult brain.

Summary

There are three types of cellular repair in the adult nervous system, in addition to the functional reorganization of surviving neurons and circuits that typically follows brain damage. The first and most effective is the regrowth of severed peripheral axons either from peripheral sensory neurons or central motor neurons, usually via the peripheral nerve sheaths once occupied by their forerunners. After regrowth, these axons reestablish sensory and motor synapses on muscles or other targets. During this regeneration, mature Schwann cells provide many of the molecules that regulate axon regrowth and targeting; these molecules are mostly those used for the same purpose during initial development. A second, and far more limited, type of repair is local sprouting or longer extension of axons and dendrites at sites of traumatic damage or degenerative pathology in the brain or spinal cord. Major impediments to such local repair include: the death of damaged neurons because of trophic deprivation or other stress; the inhibition of axon growth by protein components of myelin; the inhibition of neuronal growth by cytokines released during the immune response to brain tissue damage, and the eventual formation of glial scars. The role of immune-mediated inflammation in establishing an anti-regenerative state in brain tissue is central. Molecular mediators of inflammation, including cytokines, their receptors, and related signaling intermediates, drive this process and establish barriers to neuronal regrowth. A third type of repair is generation of new neurons in the adult brain. Although there is no evidence for wholesale replacement of neurons and circuits in most vertebrate brains, the capacity for limited ongoing neuronal replacement exists in some species—sometimes in register with ongoing growth of the animal or because of seasonal variations. In most mammals, the olfactory bulb and the hippocampus are the only sites of adult neurogenesis in the CNS. In both of these brain regions, new neurons are generated by neural stem cells retained in specific restricted locations in the adult brain. Many of the molecules that regulate the maintenance, proliferation, and differentiation of adult neural stem cells and their progeny are used for similar purposes for neural stem cells in the embryonic brain. The challenge of developing this capacity to generate new neurons and circuits as a strategy for repair following brain injury or degenerative disease continues to capture the imagination of patients, physicians, and many neuroscientists.

■ Additional Reading

Reviews

Nocera, G. and C. Jacob (2020) Mechanisms of Schwann cell plasticity involved in peripheral nerve repair after injury. *Cell & Mol. Life Sciences* 77: 3977 – 3989.

Pemberton, J.M., Pogmore, J.P. and D.W. Andrews (2021) Neuronal cell life, death and axonal degeneration as regulated by the BCL -2 family protiens. *Cell Death and Differentiation* 28: 108 – 122.

Vijayavenkataraman, S. Nerve guide conduits for peripheral never injury repair: A review on design, materials and fabrication methods. *Acta Biomaterialia* 106: 54 – 69.

Perez, J-C., Gerber, Y.N., and F. Perrin (2021) Dynamic diversity of glial response among species in spinal cord. *Front. Aging Neurosci.* 13: 769548.

Obernier, K. and A. Alvarez-Buylla (2019) Neural stem cells: origin, heterogeneity, and regulation in the adult mammalian brain. *Development* 146: dev156059.

Case, L. C. and M. Tessier-Lavigne (2005) Regeneration of the adult central nervous system. *Curr. Biol.* 15: 749–753.

Gage, F. H. (2000) Mammalian neural stem cells. *Science* 287: 1433–1488.

Otteson, D. C. and P. F. Hitchcock (2003) Stem cells in the teleost retina: Persistent neurogenesis and injury-induced regeneration. *Vision Res.* 43: 927–936.

Rossini, P. M., C. Calautti, F. Pauri and J.-C. Baron (2003) Post-stroke plastic reorganisation in the adult brain. *Lancet Neurol.* 2: 493–502.

Song, Y., J. A. Panzer, R. M. Wyatt and R. J. Balice-Gordon (2006) Formation and plasticity of neuromuscular synaptic connections. *Int. Anesthesiol. Clin.* 44: 145–178.

Important Original Papers

Altman, J. and G. D. Das (1967) Postnatal neurogenesis in the guinea-pig. *Nature* 214: 1098–1101.

David, S. and A. J. Aguayo (1981) Axonal elongation into peripheral nervous system "bridges" after central nervous system injury in adult rats. *Science* 214: 931–933.

Easter, S. S. Jr. and C. A. Stuermer (1984) An evaluation of the hypothesis of shifting terminals in goldfish optic tectum. *J. Neurosci.* 4: 1052–1063.

Eriksson, P. S. and 6 others (1998) Neurogenesis in the adult human hippocampus. *Nat. Med.* 4: 1313–1317.

Goldman, S. A. and F. Nottebohm (1983) Neuronal production, migration, and differentiation in a vocal control nucleus of the adult female canary brain. *Proc. Natl. Acad. Sci. U.S.A.* 80: 2390–2394.

Graziadei, G. A. and P. P. Graziadei (1979) Neurogenesis and neuron regeneration in the olfactory system of mammals. II. Degeneration and reconstitution of the olfactory sensory neurons after axotomy. *J. Neurocytol.* 8: 197–213.

Head, H., W. H. R. Rivers and J. Sherren (1905) The afferent nervous system from a new aspect. *Brain* 28: 99–111.

Lois, C., J. M. Garcia-Verdugo and A. Alvarez-Buylla (1996) Chain migration of neuronal precursors. *Science* 271: 978–981.

Luskin, M. B. (1993) Restricted proliferation and migration of postnatally generated neurons derived from the forebrain subventricular zone. *Neuron* 11: 173–189.

Marshall, L. M., J. R. Sanes and U. J. McMahan (1977) Reinnervation of original synaptic sites on muscle fiber basement membrane after disruption of the muscle cells. *Proc. Natl. Acad. Sci. U.S.A.* 74: 3073–3077.

UNIT V
Complex Brain Functions and Cognitive Neuroscience

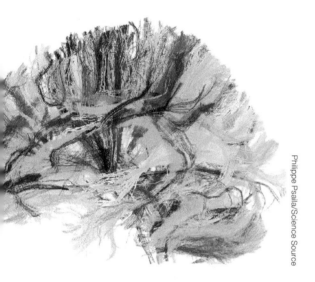

Philippe Psaila/Science Source

CHAPTER 27 **Cognitive Functions and the Organization of the Cerebral Cortex**

CHAPTER 28 **Cortical States**

CHAPTER 29 **Attention**

CHAPTER 30 **Memory**

CHAPTER 31 **Speech and Language**

CHAPTER 32 **Emotion**

CHAPTER 33 **Thinking, Planning, and Deciding**

The perception of physical and social circumstances, attending to those events that are especially important, remembering the past, planning for the future, experiencing emotions, and using language all rank among the most complex functions of the human brain. The intrinsic interest of these aspects of brain function is unfortunately equaled by the challenges—both technical and conceptual—involved in unraveling their neurobiological underpinnings. Nonetheless, much progress has been made in deciphering the basic psychological and computational properties of cognition, as well as in describing the structural and functional organization of many of the relevant brain regions. The older approach of clinical evaluation of individuals with brain damage and postmortem correlation is now complemented by noninvasive brain imaging and other techniques, allowing the study of cognitive processes in both atypical and typical individuals across the life span. At the same time, electrophysiological experiments in non-human primates and other animals have begun to elucidate the cellular correlates of many of these functions. This unit reviews the progress made in this relatively new field, now called *cognitive neuroscience*. The chapters that follow each explore a specific aspect of cognition in more depth, collectively demonstrating the remarkable complexity of the human brain.

Cognitive Functions and the Organization of the Cerebral Cortex

Overview

Given the importance of cognitive functions for human behavior and culture, it is hardly surprising that much of the human brain is devoted to the operations listed in the Unit Outline. The brain regions that support complex cognitive functions have historically been referred to as *association cortex* (or *cortices*), in reference to the fact that these regions associate (or integrate) sensory information derived from other brain regions to support adaptive behavior. Modern cognitive neuroscience research has greatly extended this simple framework, revealing how different regions of the cortex support very different sorts of computations—as reflected not only in their local processing but also in their global patterns of connection to other brain regions. This chapter advances a broad perspective on the organization of cognitive processes throughout the cortex, including how functions are supported by specific brain regions but are also embedded in distributed networks. The incredible complexity of cognitive functions means that any attempt to summarize their neural substrates will involve considerable simplification; new research continually identifies unexpected ways in which different brain regions contribute to thoughts and behavior.

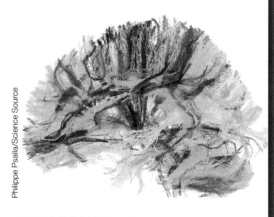

Philippe Psaila/Science Source

KEY CONCEPTS

27.1 The cerebral cortices are organized into subregions

27.2 The parietal cortex has many functions

27.3 The temporal cortex plays a critical role in object processing

27.4 The prefrontal cortex supports executive control, planning, and goal-directed action

---

CONCEPT **27.1** | **The Cerebral Cortices Are Organized into Subregions**

LEARNING OBJECTIVES

27.1.1 Explain how cytoarchitectonic areas are defined.

27.1.2 Describe the common features of the laminar structure of cortex.

27.1.3 Describe the overall patterns of connectivity within the cerebral cortex.

A primer on cortical structure

Before delving into a more detailed account of the functions of the different regions of the association cortices (Figure 27.1A), some general understanding of cortical structure and the organization of its canonical circuitry is useful. Most of the cortex that covers the cerebral hemispheres is *neocortex*, defined as cortex that has six cellular layers, or **laminae**. Each layer comprises more or less distinctive populations of cells based on their different densities, sizes, shapes, inputs, and outputs (Figure 27.1B). Despite an overall uniformity, regional differences based on these laminar features have long been apparent (Box 27A), allowing investigators to identify numerous subdivisions of the cerebral cortex. Such histologically defined subdivisions are referred to as **cytoarchitectonic areas**; the most common map includes subdivisions called *Brodmann's areas* (Figure 27.1C). Painstaking neuroanatomical work has revealed considerable structural variability in the organization of the cerebral

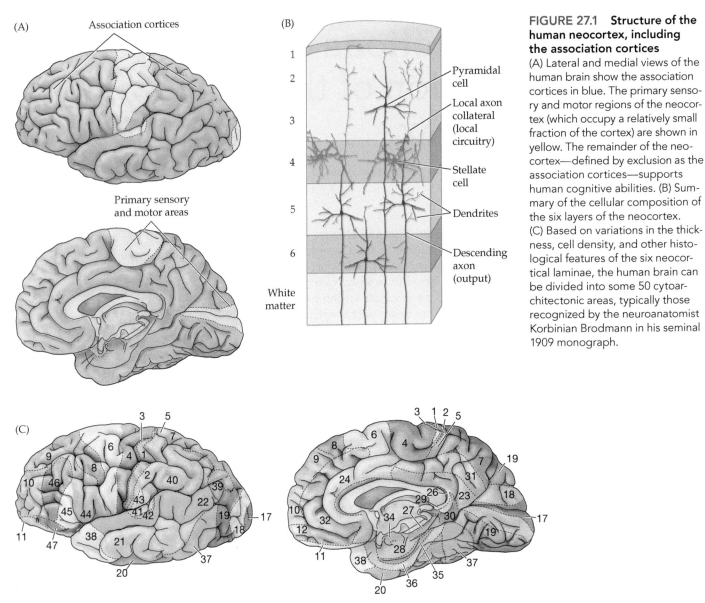

(A) Association cortices

Primary sensory and motor areas

(B)
1
2
3
4
5
6
White matter

Pyramidal cell

Local axon collateral (local circuitry)

Stellate cell

Dendrites

Descending axon (output)

FIGURE 27.1 Structure of the human neocortex, including the association cortices (A) Lateral and medial views of the human brain show the association cortices in blue. The primary sensory and motor regions of the neocortex (which occupy a relatively small fraction of the cortex) are shown in yellow. The remainder of the neocortex—defined by exclusion as the association cortices—supports human cognitive abilities. (B) Summary of the cellular composition of the six layers of the neocortex. (C) Based on variations in the thickness, cell density, and other histological features of the six neocortical laminae, the human brain can be divided into some 50 cytoarchitectonic areas, typically those recognized by the neuroanatomist Korbinian Brodmann in his seminal 1909 monograph.

(C)

cortex, not just across species but across individuals. For example, postmortem studies of the human frontal lobes have revealed that Brodmann's areas 9 and 46 vary dramatically in relative size and spatial position across individuals—a cellular variability that is not evident on large-scale measures of brain structure such as MRI.

Early in the twentieth century, cytoarchitectonically distinct regions were identified with little or no knowledge of their functional significance. Eventually, studies of individuals in whom one or more of these cortical areas had been damaged, supplemented by electrophysiological mapping in both laboratory animals and neurosurgical patients, showed that many of the regions that neuroanatomists had distinguished on histological grounds also support distinct cognitive functions. Other work examined the connectivity between different

cortical regions. Key techniques included tracing of white matter tracts in non-human primates, the limited pathway tracing that can be done in human brain tissue post mortem, and the newer noninvasive technique called **diffusion tensor imaging (DTI)** that can identify large bundles of axons connecting brain areas in living humans. DTI is one of several core techniques driving the Human Connectome Project, an ambitious effort to map the detailed functional and structural organization of the human brain. All of these techniques have shown that boundaries between cytoarchitectonic areas often accompany differences in connectivity. Thus, structural divisions in the cortex can be identified not only by the physiological response properties of their constituent cells but also by their patterns of local and long-distance connections.

■ BOX 27A | Cortical Lamination

Much knowledge about the cerebral cortex is based on descriptions of differences in cell number and density throughout the cortical mantle. Nerve cell bodies, because of their high metabolic rate, are rich in basophilic substances (RNA, for instance) and therefore tend to stain darkly with reagents such as cresyl violet acetate. These *Nissl stains* (named after Franz Nissl, who first described this technique when he was a medical student in nineteenth-century Germany) provide a dramatic picture of brain structure at the histological level. The most striking feature revealed in this way is the distinctive lamination of the cortex in humans and other mammals, as seen in the figure. Humans have three to six cortical layers, depending on the area of cortex. These layers, or *laminae*, are designated with the numerals 1–6 (or with Roman numerals I–VI). Laminar subdivisions are indicated with letters (layers 4a, 4b, and 4c of the visual cortex, for example).

Each of the cortical laminae in the so-called *neocortex* (which covers the bulk of the cerebral hemispheres and is defined by six layers) has characteristic functional and anatomical features (see Figures 27.1 and 27.2). For example, cortical layer 4 is typically rich in stellate neurons with locally ramifying axons; in the primary sensory cortices, these neurons receive input from the thalamus, the major sensory relay from the periphery. Layer 5, and to a lesser degree layer 6, contain pyramidal neurons whose axons typically leave the cortex. The generally smaller pyramidal neurons in layers 2 and 3 (which are not as distinct as their numerical assignments suggest) have primarily corticocortical connections, and layer 1 contains mainly neuropil. Early in the twentieth century, Korbinian Brodmann devoted his career to an analysis of brain regions distinguished in this way, describing some 50 distinct cortical regions, or *cytoarchitectonic areas* (see Figure 27.1C). These structural features of the cerebral cortex continue to figure importantly in discussions of the brain.

Not all cortex is six-layered neocortex. The hippocampal cortex, which lies deep in the temporal lobe and has been implicated in acquisition of declarative memories (see Chapter 30), has only three or four laminae. The hippocampal cortex is regarded as evolutionarily more primitive and is therefore called *archicortex* (*archi*, "first") to distinguish it from the six-layered neocortex. Another, presumably even more primitive, type of cortex is the *paleocortex* (*paleo*, "ancient"); paleocortex generally has three layers and is found on the ventral surface of the cerebral hemispheres and along the parahippocampal gyrus in the medial temporal lobe. The functional significance of different numbers of laminae in neocortex, archicortex, and paleocortex is not known, although it seems likely that the greater number of layers in neocortex reflects more complex information processing.

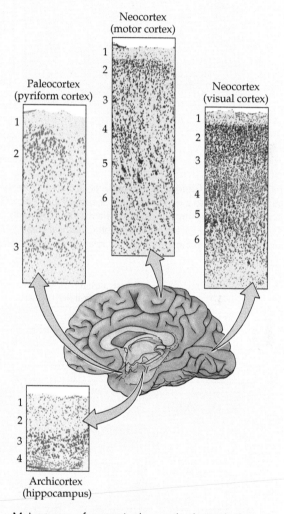

Major types of cortex in the cerebral mantle, based primarily on the different numbers of layers (laminae) apparent in histological sections.

Despite significant variations among different cytoarchitectonic areas, the circuitry of all cortical regions has some common features (Figure 27.2). First, each cortical layer has a primary source of inputs and a primary output target. Second, each area has connections in the vertical axis (called *columnar* or *radial* connections) and connections in the horizontal axis (called *lateral* or *horizontal* connections). Third, cells with similar functions tend to be arrayed in radially aligned groups that span all the cortical layers and receive inputs that are often segregated into radial bands or columns. Finally, interneurons within specific cortical layers give rise to extensive local axons that extend horizontally in the cortex, often linking functionally similar groups of cells. The particular circuitry of any cortical region is a variation on this canonical pattern of inputs, outputs, and vertical and horizontal patterns of connectivity.

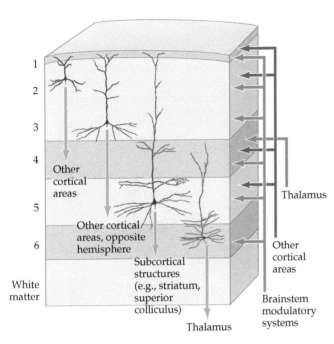

FIGURE 27.2 **Canonical neocortical circuitry** Green arrows indicate outputs to the major targets of each of the neocortical layers in humans; orange arrow indicates thalamic input (primarily to layer 4); purple arrows indicate input from other cortical areas; blue arrows indicate input from the brainstem modulatory systems to each layer.

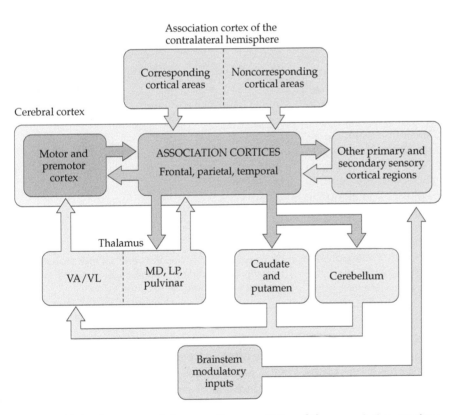

FIGURE 27.3 **Summary of the overall connectivity of the association cortices** VA = ventral anterior nucleus, VL = ventral lateral nucleus, MD = medial dorsal nucleus, LP = lateral posterior nucleus.

Connectivity patterns within the cerebral cortex

The general wiring plan for the association cortices is summarized in Figure 27.3. There are three major differences between the connectivity of association cortices and that of other cortical (or subcortical) brain regions. First, most inputs to association cortices are projections from other cortical areas, called **corticocortical connections**. Ipsilateral corticocortical connections arise from primary and secondary sensory and motor cortices, and from other association cortices within the same hemisphere. Corticocortical connections also arise from both corresponding and noncorresponding cortical regions in the opposite hemisphere via the corpus callosum and anterior commissure, which together are referred to as **interhemispheric connections**. In the association cortices of humans and other primates, corticocortical connections often form segregated radial bands or columns in which interhemispheric projection bands are interdigitated with bands of ipsilateral corticocortical projections.

Second, there are important differences in the patterns of thalamic input. Two thalamic nuclei that are not involved in relaying primary motor or sensory information provide much of the subcortical input to the association cortices: The pulvinar projects to the parietal association cortex, while the medial dorsal nuclei project to the frontal association cortex. In consequence, the signals coming into the association cortices via the thalamus reflect sensory and motor information that has *already* been processed in the primary sensory and motor areas of the cerebral cortex and is being fed back to the association regions. The primary sensory cortices, in contrast, receive thalamic information that is more directly related to peripheral sense organs (see Unit II). Similarly, much of the thalamic input to primary motor cortex is derived from the thalamic nuclei related to the basal ganglia and cerebellum rather than to other cortical regions (see Unit III).

Third, the association cortices receive important inputs from subcortical brain regions, including the dopaminergic nuclei in the midbrain, the noradrenergic and serotonergic nuclei in the brainstem, and cholinergic nuclei in the brainstem and basal forebrain. These diffuse inputs project to different cortical layers and, among other functions, contribute to learning, motivation, and arousal (the continuum of mental states that ranges from deep sleep to high alert; see Chapter 28). A variety of behavioral and psychiatric disorders, including addiction, depression, and attention deficit disorder, are associated with dysfunction in one or more of these neuromodulatory circuits. Current pharmacological therapies for these diseases rely on manipulation of signaling via modulatory inputs to the association cortices.

Concepts 27.2 through 27.4 provide overviews of the major association cortices, each of which supports the cognitive functions considered in detail in the rest of Unit V. This and the following chapters highlight both historical and current research that elucidates cortical function, including both specific studies that illustrate fundamental concepts as well as general conclusions that are increasingly drawn from aggregations of studies via meta-analyses (Box 27B).

■ BOX 27B | Large-Scale Neuroscience: Meta-Analyses and Consortium Studies

Cognitive neuroscience has grown dramatically over the past several decades. Every year thousands of new studies are published, spanning all of the research methods and cognitive functions described throughout this unit. The diversity of current research illustrates how cognitive neuroscience has grown into a vibrant, mature discipline. Yet this success points to an important challenge: How can researchers integrate results from many experiments to improve their inferences about cognitive functions? This box considers two such approaches: meta-analyses and research consortium studies.

Methods for combining information across multiple independent studies are called **meta-analysis** techniques (i.e., "analyses of analyses"). Generally, they improve the power and precision with which researchers can detect and specify conclusions. Nearly all studies in cognitive neuroscience involve relatively small samples; typical are electrophysiological studies of two or three monkeys, neuropsychological testing of a handful of individuals with brain lesions, and functional MRI or EEG recording from 50 or fewer research participants. By combining data across multiple studies, researchers increase the effective sample size for their analyses, allowing the identification of effects that replicate across studies. The advantages of meta-analyses go well beyond increased experimental power. Because different cognitive neuroscientists typically approach a research question in different ways—through different experimental tasks or different methods for data collection—the results of any single study might be

attributable to an idiosyncratic feature of its experiment. Those idiosyncratic features (e.g., the specific demographics of a subject sample) will average out over many studies, leaving only what is common to all the studies. Thus, meta-analyses may be especially important for understanding the neural basis of complex cognitive functions that can be evoked within a wide range of contexts.

In *qualitative meta-analyses*, a research team first identifies a comprehensive set of studies on the same cognitive function and then looks for similarities among their results. Most review articles approach the neuroscience literature in this manner: They identify a cognitive function (e.g., reward processing in adolescence) and then seek studies of that function that match some a priori criteria (e.g., using fMRI in a sample of 12- to 18-year-olds). Then, from the resulting set of studies, they derive conclusions about commonly observed results (e.g., increased striatal activity when peers are present). An early and influential example can be seen in the results of a review article that explored the major categories of cognitive functions discussed in this unit (Figure A).

(Continued)

(A)

Perception of Faces											
			Temporal						Occipito-temporal		
STUDY	38	Ins	42	22	21	20	Mt	37	19	18	17
Grady et al. 1990								x	x	x	
Haxby et al. 1994								x	x		
Haxby et al. 1995		x					x		x		
Sergent et al. 1992								x	x		
N. Kapur et al. 1995							x		x	x	x
Puce et al. 1995					x			x	x		
Clark et al. 1996								x			
Puce et al. 1996								x	x		
Kanwisher et al. 1997								x			
McCarthy et al. 1997								x			
Clark et al. 1998								x			

Source: Cabeza and Nyberg 2000.

(A) An early qualitative meta-analysis relied on the authors matching experimental tasks from specific studies (left column) to subregions of the brain, represented here by numbers corresponding to Brodmann's areas in the cerebral cortex and by abbreviations for other brain regions (Ins = insula, Mt = medial temporal/hippocampus). As shown, a variety of tasks involving the visual perception of human faces evoked activation in a region of the ventral temporal lobe (Brodmann's area 37, which includes the fusiform gyrus), although other regions were found to be activated in different studies. (After R. Cabeza and L. Nyberg, 2000. *J Cog Neurosci* 12: 1–47.)

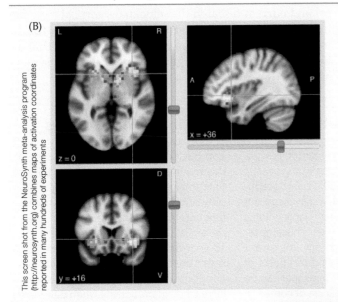

This screen shot from the NeuroSynth meta-analysis program (http://neurosynth.org) combines maps of activation coordinates reported in many hundreds of experiments

(B) Quantitative meta-analyses integrate data from many studies. When created from PET and fMRI data, such maps allow researchers to identify brain regions that have been consistently reported across experiments studying a single cognitive function or experimental task. This screen shot from the Neurosynth meta-analysis program (http://neurosynth.org) combines maps of activation coordinates reported in many hundreds of experiments, each of which was coded according to the specific cognitive functions it evoked. Shown are three cross sections (locations indicated by coordinates: axial, z; coronal, y; sagittal, x) through the meta-analytic map for "decision making," a process that consistently evokes activation in the insula (at crosshair in each image) and basal ganglia (visible in the axial image; upper left panel), as well as in a network of cortical regions.

Quantitative meta-analyses, in contrast, combine results from multiple studies into a single statistical framework. In a typical analysis technique, coordinates of fMRI activation found in many studies are extracted and used to create a probabilistic map of the combined data (Figure B). Brain regions that are reliably activated across many studies have high significance values and are highlighted on the overlaid color maps. Quantitative analyses of this sort not only can improve statistical power, but also can distinguish subtle functional differences within a brain region. In addition, their need for minimal human input increases their objectivity. Some cognitive neuroscience researchers have created automated systems that can evaluate thousands of studies, allowing a keyword-based search of the larger literature (e.g., see http://neurosynth.org). Other researchers have attempted to break down complex cognitive functions into their core processes, on the basis of prior research, and then determine patterns of brain activation associated with each of those processes (see http://cognitiveatlas.org for an example).

Finally, *consortium studies* involve coordinated data collection across many experimental sites, allowing

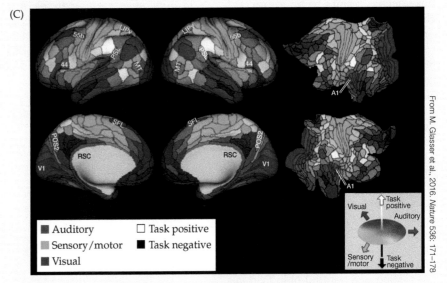

From M. Glasser et al., 2016. *Nature* 536: 171–178

(C) Consortium studies collect experimental data at many sites, increasing both the size and diversity of the subject sample. Researchers used data sets from the Human Connectome Project to partition the cerebral cortex into 180 regions that differed in their function, structure, or connectivity to other regions. Those regions are shown on inflated and flattened cortical surfaces, with black lines indicating boundaries between regions. The color of each region indicates to what extent it is associated with a set of functions: audition (red), vision (blue), sensory/motor (green), task positive (i.e., more active when performing a task; light gray), or task negative (i.e., less active when performing a task; dark gray). Labels indicate selected regions that correspond to commonly studied areas (e.g., V1 for primary visual cortex).

research to be conducted at very large scales. Suppose you want to understand how differences in brain structure and function (both across individuals and across the life span) interrelate with other predictors of personality and health (e.g., genomics). Drawing firm conclusions about such a complex topic —think of how many factors contribute to one's mental health!—would require

■ BOX 27B | Large-Scale Neuroscience: Meta-Analyses and Consortium Studies (continued)

data from very large and diverse samples, likely thousands of individuals of different ages and backgrounds. Conducting such a study would surely be beyond the capability of any one institution. Not only would it be extraordinarily difficult to recruit such a large sample at one location, but even if it were possible, the resulting sample would lack geographic diversity (and perhaps other sorts of diversity too).

More practical would be to collect the necessary sample via a consortium of institutions that agree to run the same experiments and to collect similar measures of brain structure and function. The Human Connectome Project, as a prototypic example, comprised functional and structural MRI data from more than 1100 participants collected by an international consortium of ten universities (Figure C). The ongoing Adolescent Brain Cognitive Development study targets an even larger sample: nearly 12,000 children (9 to 10 years old) who will participate in brain imaging and other studies for a decade. That study combines data from 21 sites spread throughout the United States, allowing for considerable geographic and socioeconomic diversity. Consortium studies are particularly important for questions about individual differences or disorders (e.g., understanding how brain structure might predict who develops autism or schizophrenia). Moreover, most such studies now adopt an *open-science* model by making their large-scale data available to the entire research community. Neuroscientists throughout the world use that data to ask and answer questions about the brain—increasing the diversity of the research community as well.

CONCEPT 27.2 | The Parietal Cortex Has Many Functions

LEARNING OBJECTIVES

27.2.1 Define contralateral neglect syndrome.

27.2.2 Explain why neglect syndrome may be associated with right-hemisphere but not left-hemisphere damage.

27.2.3 Provide an example of how neurons in parietal cortex may link stimuli to actions.

In 1941, the British neurologist W. R. Brain reported three individuals with unilateral parietal lobe lesions in whom the primary problem was varying degrees of difficulty paying attention to objects and events contralateral to the lesion. Brain described their peculiar deficiency in the following way:

Though not suffering from a loss of topographical memory or an inability to describe familiar routes, they nevertheless got lost in going from one room to another in their own homes, always making the same error of choosing a right turning instead of a left, or a door on the right instead of one on the left. In each case there was a massive lesion in the right parieto-occipital region, and it is suggested that this ... resulted in an inattention to or neglect of the left half of external space.

The patient who is thus cut off from the sensations which are necessary for the construction of a body scheme may react to the situation in several different ways. He may remember that the limbs on his left side are still there, or he may periodically forget them until reminded of their presence. He may have an illusion of their absence, i.e. they may "feel absent" although he knows that they are there;

he may believe that they are absent but allow himself to be convinced by evidence to the contrary; or, finally, his belief in their absence may be unamenable to reason and evidence to the contrary and so constitute a delusion.

W. R. Brain, 1941 (*Brain* 64, pp. 257 and 264)

This description is generally considered to be the first account of the link between parietal lobe lesions and deficits in attention or perceptual awareness. Based on a large number of individuals studied since Brain's pioneering work, these deficits are now referred to as **contralateral neglect syndrome**. The hallmark of contralateral neglect syndrome is an inability to attend to objects, or even one's own body, in a portion of space, despite the fact that visual acuity, somatic sensation, and motor ability remain intact. Affected individuals fail to report, respond to, or even orient to stimuli presented to the side of the body (or visual space) opposite the lesion. Automatic responses to stimuli are still present, however, revealing that the deficit is attentional and not sensory. For example, an individual with contralateral neglect syndrome may still withdraw her left arm in response to a pinprick.

Importantly, contralateral neglect syndrome is typically associated with damage to the *right* parietal cortex (see Chapter 29). The unequal distribution of this particular cognitive function between the hemispheres is thought to arise because the right parietal cortex mediates attention to both left and right halves of the body and extrapersonal space, whereas the left hemisphere mediates attention primarily to the right—a bias that may arise because of specialization of the *left* hemisphere for language. Thus, left parietal lesions tend to be compensated for by the intact right hemisphere. In contrast, when the right parietal cortex is damaged, there is little or no compensatory capacity in the left hemisphere to mediate attention to the left side of the body or extrapersonal space. This interpretation has been confirmed by noninvasive imaging of parietal lobe activity during specific attention tasks carried out by typical individuals. Such studies show that neural activity is increased in *both* the right and left parietal cortices when individuals are asked to perform tasks in the *right* visual field requiring selective attention to distinct aspects of a visual stimulus such as its shape, velocity, or color. However, when a similar challenge is presented in the *left* visual field, typically the right parietal cortex is activated, although left parietal cortical activity is often observed as well. There is also evidence of increased activity in the right frontal cortex during such tasks. This observation suggests that multiple regions contribute to attentive behavior, and perhaps to some aspects of the pathology of neglect syndromes. Overall, however, brain-imaging data are consistent with the clinical fact that contralateral neglect typically arises from a right parietal lesion, and these data endorse the broader idea of hemispheric specialization for attention, in keeping with the concept of hemispheric specialization for several other cognitive functions.

These clinical and brain-imaging observations do not, however, provide much insight into how the nervous system represents cognitive information in nerve cells and their interconnections. The apparent functions of the association cortices implied by clinical observations stimulated a wealth of informative electrophysiological studies in non-human primates, particularly macaque (usually rhesus) monkeys. As in humans, many cognitive abilities in monkeys are mediated by the association cortices of the parietal, temporal, and frontal lobes (Figure 27.4A). Moreover, these functions can be tested using behavioral paradigms that assess attention, identification, and planning capabilities—the broad functions assigned to the parietal, temporal, and frontal association cortices, respectively, in humans. By means of implanted electrodes, recordings from single neurons in the brains of awake, behaving monkeys are used to assess the activity of individual cells in the brain as various tasks are performed (Figure 27.4B).

Neurons in the parietal cortex of monkeys have been studied using this approach. The studies take advantage of the fact that monkeys can be trained to selectively attend to particular objects or events and report their experience in a variety of nonverbal ways, typically by looking at a response target (thus allowing their eye movements to be monitored) or manipulating a joystick. Attention-sensitive neurons can be identified by recording electrophysiological changes in neuronal activity associated with simultaneous changes in the attentive behavior of the animal. As might be expected from the clinical evidence in humans, neurons in specific regions of the parietal cortex of the rhesus monkey are activated when the animal attends to a target, but not when the same stimulus is ignored (Figure 27.5A). In another study, monkeys were rewarded with different

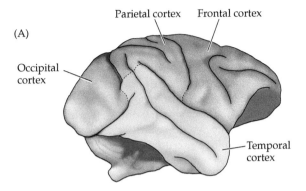

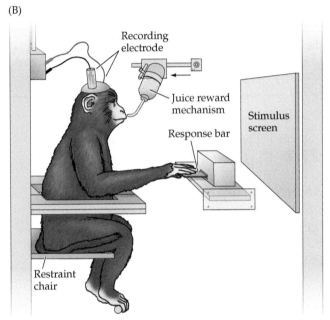

FIGURE 27.4 Recording from single neurons in the brain of an awake, behaving rhesus monkey (A) Lateral view of the rhesus monkey brain showing the parietal, temporal, occipital, and frontal cortices. (B) The animal is seated in a chair and gently restrained. Several weeks before data collection begins, a recording well is placed through the skull by means of a sterile surgical technique. For electrophysiological recording experiments, a microelectrode is inserted through the dura and arachnoid and into the cortex. The screen and response bar in front of the monkey are for behavioral testing. In this way, individual neurons can be monitored while the monkey performs specific cognitive tasks to gain a reward (e.g., fruit juice).

amounts of fruit juice for different target stimuli (Figure 27.5B). Not surprisingly, the frequency with which monkeys attended to each target varied with the amount of juice they expected for a correct response. Moreover, the activity of some neurons in parietal cortex varied systematically as a function of the amount of juice associated with each target, and therefore the amount of attention paid by the monkey to the target. Thus, the primate parietal cortex contains neurons that respond specifically when the animal attends to a behaviorally meaningful stimulus, and the vigor of the response reflects the amount of attention paid to the stimulus.

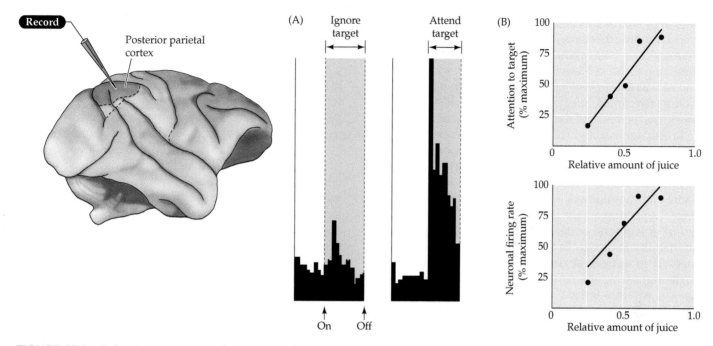

FIGURE 27.5 Selective activation of neurons in the monkey parietal cortex as a function of attention In this case, a rhesus monkey's attention is directed to a light associated with a fruit juice reward. (A) Although the baseline level of activity of the neuron being studied here remains unchanged when the monkey ignores a visual target (left), firing rate increases dramatically when the monkey attends to the same stimulus (right). The histograms indicate action potential frequency per unit of time. (B) When given a choice of where to attend, the monkey pays increasing attention to a particular visual target when more fruit juice reward can be expected for doing so (top), and the firing rate of the parietal neuron under study increases accordingly (bottom). (A after C. Colby et al., 1996. *J Neurophysiol* 76: 2841–2852; B after M. L. Platt and P. W. Glimcher, 1999. *Nature* 400: 233–238.)

CONCEPT
27.3

The Temporal Cortex Plays a Critical Role in Object Processing

LEARNING OBJECTIVES

27.3.1 Define agnosias and explain their etiology.

27.3.2 Explain the deficit associated with prosopagnosia.

27.3.3 Provide an example of population coding in object perception in temporal cortex.

Clinical evidence from individuals with lesions of the association cortex in the temporal lobe indicates that one of the major functions of this part of the brain is the recognition and identification of stimuli, particularly complex stimuli. Thus, damage to either temporal lobe can result in difficulty recognizing, identifying, and naming different categories of objects. These disorders, collectively called **agnosias** (Greek, "unknown"), are quite different from the neglect syndromes. As noted in Concept 27.2, individuals with right parietal lobe damage often deny awareness of sensory information in the left visual field (and are less attentive to the left sides of objects generally), despite the fact that the sensory systems are intact. Individuals with temporal lobe damage that leads to agnosia, by contrast, acknowledge the presence of a stimulus but are unable to report what it is. Agnosias have both a lexical aspect (a

mismatching of verbal or other cognitive symbols with sensory stimuli) and a mnemonic aspect (a failure to recall stimuli when confronted with them again).

One of the most thoroughly studied agnosias following damage to the temporal association cortex in humans is the inability to recognize and identify faces. This disorder, called **prosopagnosia** (Greek *prosopon*, "face" or "person"), was recognized by neurologists in the late nineteenth century and remains an area of intense investigation. After damage to the inferotemporal cortex, typically in the right hemisphere, individuals are often unable to identify familiar people by their facial characteristics, and in some cases cannot recognize a face at all. Nonetheless, such individuals are perfectly aware that some sort of visual stimulus is present and can describe particular aspects or elements of it without difficulty.

An example is the case of L. H., a 40-year-old minister and social worker who had sustained a severe head injury in an automobile accident when he was 18. (Note that the use of initials to identify neurological patients in published reports is standard practice.) After recovery, he could not recognize familiar faces, report that they were familiar, or answer questions about faces from memory. He could identify common objects, could discriminate subtle shape differences, and could recognize the sex, age, and even the likeability of faces. Moreover, he could identify particular people by nonfacial cues such as voice, body shape, and gait. The only other category of visual stimuli he had trouble recognizing was animals and their expressions, though these impairments were not as severe as for human faces, and he was able to lead a fairly normal and productive life. Noninvasive brain imaging showed that L. H.'s prosopagnosia was the result of damage to the right temporal lobe.

Prosopagnosia and related agnosias involving objects are specific instances of a broad range of functional deficits that have as their hallmark the inability to recognize a complex sensory stimulus as familiar, and to identify and name that stimulus as a meaningful entity in the environment. Depending on the laterality, location, and size of the lesion in temporal cortex, agnosias can be as specific as an inability to recognize human faces or as general as an inability to name most familiar objects. In general, lesions of the right temporal cortex lead to agnosia for faces and objects, whereas lesions of the corresponding regions of the left temporal cortex tend to result in difficulties with language-related material. The lesions that typically cause recognition deficits, particularly for faces, are in the inferotemporal cortex, in or near the fusiform gyrus; those that cause language-related problems in the left temporal lobe tend to be on the lateral surface of the cortex (see Chapter 31). Consistent with these conclusions, direct cortical stimulation in individuals whose temporal lobes are being mapped for neurosurgery (typically removal of a seizure focus) may induce transient prosopagnosia as a consequence of this abnormal activation of the relevant regions of the right temporal cortex.

More recently, brain imaging and direct electrophysiological recording studies have confirmed that the inferotemporal cortex, particularly the fusiform gyrus, mediates face recognition—and that nearby regions of the temporal and parietal cortices support processing of related social stimuli (e.g., biologically relevant motion). Recognizing these stimuli provides important clues to another person's emotional state and intentions, and the existence of brain regions specialized for processing this information has been taken as evidence for the important role of social behavior and cognition during human evolution.

In keeping with human deficits of recognition following temporal lobe lesions, neurons with responses that correlate with the recognition of specific stimuli are present in the temporal cortex of rhesus monkeys. The behavior of these neurons is generally consistent with one of the major functions ascribed to the human temporal cortex, namely, the recognition and identification of complex stimuli. For example, some neurons in the inferotemporal gyrus of the rhesus monkey cortex respond specifically to the presentation of a monkey face. These cells are often quite selective; some respond only to the frontal view of a face, others only to profiles. Furthermore, the cells are not easily deceived. When parts of faces or generally similar objects are presented, such cells typically fail to respond; in fact, the only things that confuse face-selective neurons are round or fuzzy objects such as apples, clock faces, or toilet brushes—all of which are vaguely facelike in appearance.

In principle, it is unlikely that specific faces or objects are encoded by the activity of single neurons in isolation. However, populations of neurons differently responsive to various features of faces or other objects can act in concert to enable the recognition of such complex sensory stimuli (Figure 27.6). The notion of such *population coding* of objects

(A)

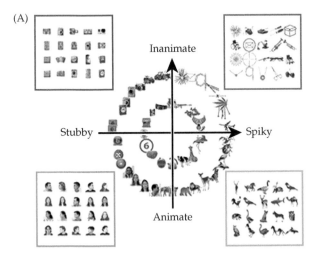

(B)

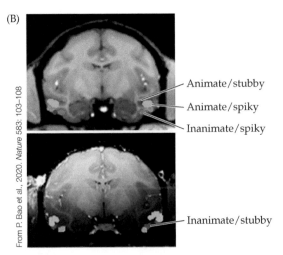

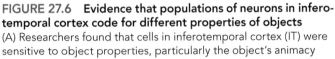

FIGURE 27.6 Evidence that populations of neurons in inferotemporal cortex code for different properties of objects
(A) Researchers found that cells in inferotemporal cortex (IT) were sensitive to object properties, particularly the object's animacy (e.g., a tool vs. a person) and general shape (e.g., concave/stubby vs. convex/spiky). Shown here are different objects that varied in those properties; each was shown to rhesus monkeys that were scanned using fMRI. (B) fMRI activation maps from two rhesus monkeys are shown on coronal structural MRI slices, with the areas of IT that were associated with each of the object categories indicated via color. Both rhesus monkeys show distinct patches of cortex sensitive to the different categories, with generally similar relative positions of those cortical patches. (After P. Bao et al., 2020. *Nature* 583: 103–108.)

From P. Bao et al., 2020. *Nature* 583: 103–108

is supported by the recent observation that face-selective neuronal responses in the temporal cortex of monkeys vary in intensity with respect to an average face. Both monkeys and humans are better at recognizing faces having extreme features—caricatures—than they are at recognizing less distinctive faces, suggesting that faces are identified by comparison with a mental standard or norm. Similarly, neurons in the inferotemporal cortex of monkeys respond much more strongly to caricatures of human faces than to an average human face. Such norm-based tuning has also been reported for neuronal responses to shapes in the inferotemporal cortex.

Recent studies suggest that such complex response properties may be based on a columnar anatomical arrangement similar to that in the primary visual cortex (see Chapter 9). Each column has been taken to represent different arrangements of complex features making up an object, the overall spatial pattern of neuronal activity representing the object in view. In keeping with this general idea, optical imaging of the surface of the temporal cortex shows that large populations of neurons are activated when monkeys view an object comprising several different geometric features. The locus of this activity in the upper layers of the cortex shifts systematically when object features, such as the orientation of a face, are systematically altered. Taken together, these further observations suggest that object identification relies on graded signals carried by a population of neurons rather than on the specific output of one or a few cells that are selective for a particular object.

CONCEPT 27.4 | The Prefrontal Cortex Supports Executive Control, Planning, and Goal-Directed Action

LEARNING OBJECTIVES

27.4.1 Explain the case of Phineas Gage, including the historical perspective on his changed personality and how modern research calls that perspective into question.

27.4.2 Provide examples of how clinicians use neuropsychological testing to identify deficits in cognition.

27.4.3 Describe how frontal lobe neurons support maintenance of rules for behavior.

Damage to the human frontal lobe, particularly to the regions anterior to motor cortex (i.e., **prefrontal cortex [PFC]**), can lead to any of a diverse set of functional deficits, often with devastating consequences for an individual's quality of life. This broad range of clinical effects stems from the fact that the frontal cortex has a wider repertoire of functions than any other neocortical region—consistent with the facts that the frontal lobes in humans and other primates are the largest of the brain's lobes, comprise a greater number of

cytoarchitectonic areas, and have the broadest pattern of interconnectivity with other brain regions.

The often dramatic nature of the behavioral deficits after frontal lobe damage reflects the role of this part of the brain in maintaining what is usually thought of as an individual's personality. The frontal cortex integrates complex information from sensory and motor cortices, as well as from the parietal and temporal association cortices. The result is an appreciation of self in relation to the world that allows behaviors to be planned and executed normally. When this ability is compromised, the afflicted individual often has difficulty carrying out complex behaviors that are appropriate to the circumstances. These deficiencies in the normal ability to match ongoing behavior to present or future demands are, not surprisingly, interpreted as a change in the individual's character.

The case that first called attention to the consequences of frontal lobe damage was that of Phineas Gage, who worked as a railroad crew foreman in mid-nineteenth-century Vermont (Figure 27.7). One day in 1848, Gage was tamping blasting powder into a hole using a heavy metal rod. A spark (likely triggered by the rod hitting metal) drove that rod through his left eye socket, destroying part of his frontal lobes. Gage was promptly taken to a local doctor who treated his wound and documented his miraculous survival from the extraordinary accident. While Gage eventually recovered from the injury and subsequent infection and illness, his personality was reported to have changed. After Gage's death, he was retrospectively described in evocative terms:

[Gage is] fitful, irreverent, indulging at times in the grossest profanity (which was not previously his custom), manifesting but little deference for his fellows, impatient of restraint or advice when it conflicts with his desires, at times pertinaciously obstinate, yet capricious and vacillating, devising many plans of future operations, which are no sooner arranged than they are abandoned in turn for others appearing more feasible. A child in his intellectual capacity and manifestations, he has the animal passions of a strong man. Previous to his injury, although untrained in the schools, he possessed a well-balanced mind, and was looked upon by those who knew him as a shrewd, smart businessman, very energetic and persistent in executing all his plans of operation. In this regard his mind was radically changed, so decidedly that his friends and acquaintances said he was "no longer Gage."

J. M. Harlow, 1868 (*Publications of the Massachusetts Medical Society* 2: 339–340)

This seemingly dramatic change in Gage's character became a textbook example of how the frontal lobes contribute to the control of impulsive, maladaptive behavior. Yet modern research provides a much more nuanced view of Gage (and, by extension, of others with similar frontal damage).

(A)

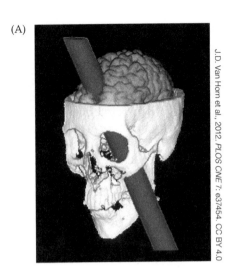

J.D. Van Horn et al. 2012. *PLOS ONE 7: e37454. CC BY 4.0*

(B)

From the collection of Jack and Beverly Wilgus, courtesy of Warren Anatomical Museum, Harvard Medical School

FIGURE 27.7 **The case of Phineas Gage** In 1848, Phineas Gage was a foreman on the team constructing a new railway in Vermont. While he was tamping down some blasting powder in a hole in the rock, the powder unexpectedly exploded, driving the tamping iron (a metal rod about 2 m in length and 3 cm in diameter) through his left cheek and out the top of his skull. (A) This modern reconstruction based on the damage to Gage's skull illustrates that the most likely path of the rod was through the middle of the frontal lobes, leading to significant damage to the orbitofrontal and medial prefrontal cortex. (B) The traditional narrative about Phineas Gage reports that this damage had dramatic effects on his personality; for instance, that he changed from being responsible and conscientious to being profane, reckless, and impulsive. Recent research suggests that this narrative is exaggerated. For example, Gage was hardly itinerant and unproductive; after the accident, he traveled independently to South America and remained employed there for several years. This photograph of Gage after his accident was recently discovered.

In contrast to the standard story—and even Harlow's postmortem description—Gage did not spend his post-accident years as an itinerant ne'er-do-well. Instead, he held gainful employment throughout the rest of his life, including an extended tenure as a stage-coach driver in Chile, a position that required considerable personal responsibility and for which contemporaneous accounts reported him in good health. Gage's deficits have been exaggerated over the years to fit current views about frontal lobe function, providing a cautionary tale about the overreliance on specific cases. An excellent summary of Gage's case and its changing interpretation over decades of neuroscience research can be found in the book by M. Macmillan listed in Additional Reading at the end of the chapter.

Recent studies of individuals with focal damage to particular regions of the frontal lobe—along with research using noninvasive brain imaging in unimpaired individuals—reveal how distinct cognitive processes are localized to different regions. Some of these functions can be clinically assessed using standardized tests such as the Wisconsin Card Sorting Task for planning (Box 27C), the delayed response task for short-term memory,

---

■ BOX 27C | Neuropsychological Testing

Long before PET scanning and fMRI were available to evaluate normal and abnormal cognitive brain function, clinicians and scientists used neuropsychological tests of behavior to evaluate the integrity of cognitive function. In turn, performance on those tests provided insights about the functional consequences of brain lesions and psychiatric disorders.

One of the most frequently used tests is the Wisconsin Card Sorting Task illustrated here. In this test, the examiner places four target cards with symbols that differ in number, shape, or color before an individual, who is given a deck of response cards with similar symbols on them. The individual must learn (by trial and error) the correct rule for matching the response cards to the target cards; for example, one rule might be "match the number of symbols on the card." As the individual places each card, the examiner then indicates whether the response is correct or incorrect, thus providing feedback that helps the individual learn the rule. After the individual learns the rule and makes ten consecutive correct responses, the examiner changes the sorting rule simply by saying "wrong." The individual must then learn the new sorting rule based on trial and error. This process of learning and then changing rules repeats until six cycles have been completed.

■ BOX 27C | Neuropsychological Testing *(continued)*

In 1963, the neuropsychologist Brenda Milner at the Montreal Neurological Institute showed that individuals with frontal lobe lesions have consistently poor performance in the Wisconsin Card Sorting Task. By comparing individuals with known brain lesions as a result of surgery for epilepsy or tumor, Milner was able to demonstrate that this impairment is fairly specific for frontal lobe damage. Particularly striking is the inability of individuals with frontal lobe damage to use previous information to guide subsequent behavior. A widely accepted explanation for the sensitivity of the Wisconsin Card Sorting Task to frontal lobe deficits is the planning aspect of this test. To respond correctly, the individual must retain information about the previous trial, which is then used to guide behavior on future trials. Processing this sort of information

is characteristic of normal frontal lobe function. In keeping with the conservation of neural function between humans and non-human primates, rhesus monkeys can be taught to perform a computerized variant of the Wisconsin Card Sorting Task. Lesion studies in monkeys confirm that the dorsolateral prefrontal cortex is crucial for performing this task, specifically for representing the rule (i.e., sort by shape, color, or number). Single-unit recordings in monkeys showing neurons whose activity specifically encodes particular rules endorse this conclusion.

A variety of other neuropsychological tests have been devised to evaluate the integrity of other cognitive functions. These include tasks in which an individual is asked to identify familiar faces in a series of pictures, and others in which distractor stimuli interfere with the

individual's ability to attend to salient stimulus features. An example of the latter is the Stroop Test (see Figure 33.10A), in which individuals are asked to name the ink color of words that might be semantically congruent (e.g., the word *green* printed in green ink), incongruent (e.g., the word *red* printed in blue ink), or neutral (e.g., the word *chair* printed in orange ink). Similarly, in the go/no-go task, individuals view a series of stimuli that typically generate the same motor response (i.e., press a button when you see an X) but must infrequently inhibit that response when another stimulus is presented (i.e., 10% of the time, a K is presented instead). Deficits in the ability to adapt behavior according to the rules of the task can indicate problems with prefrontal cortex function.

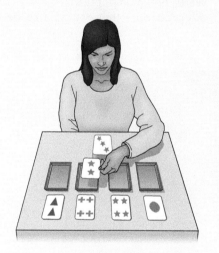

Sort by color

Sort by shape

Sort by number

and response inhibition tasks (e.g., the Stroop Test and go/no-go task). For example, the dorsal and lateral aspects of the frontal cortex are activated when typical individuals actively suppress a behavioral response when an expected pattern of events is violated in order to generate the appropriate behavior. By contrast, personal preferences for different types of rewards are correlated with individual differences in brain activation within the ventromedial prefrontal cortex, suggesting that this part of the brain signals the value individuals place on a reward independent of its physical attributes (Figure 27.8). All of these observations are consistent with the idea that

the common denominator of the cognitive functions subserved by the frontal cortex is the selection, planning, and execution of goal- and context-appropriate behavior (see Chapter 33).

Sadly, the effects of damage to the frontal lobes have also been documented by the many thousands of frontal lobotomies performed in the twentieth century as a means of treating mental illness (Clinical Applications). The rise and fall of lobotomies and related psychosurgeries provide compelling examples of the frailty of human judgment in medical practice and of the challenges inherent in treating mental disease.

(A)

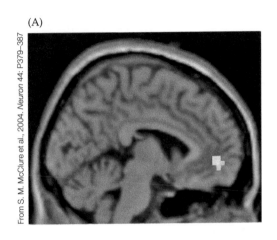

(B)

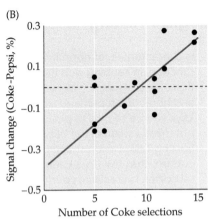

FIGURE 27.8 Activation in the ventromedial prefrontal cortex correlates with subjective preferences for soft drinks Individuals were given a taste test to determine whether they preferred Coke or Pepsi. They then underwent fMRI scans while they received squirts of each soft drink. Activation in the ventromedial prefrontal cortex in response to squirts of Coke compared with activation in response to squirts of Pepsi (A) was correlated with the frequency with which individuals selected Coke over Pepsi in the taste-test (B). (B after S. M. McClure et al., 2004. *Neuron* 44: P379–387.)

■ Clinical Applications

Psychosurgery

The consequences of frontal lobe destruction were all too well documented by a disturbing yet fascinating episode in medical practice. In the early twentieth century, there were few options for effective treatment of psychiatric conditions such as schizophrenia and depression. No psychotropic drugs had been yet identified, and individuals afflicted with many disorders were confined under custodial conditions that were dismal at best and brutal at worst. Many individuals (and their families) were desperate for any treatment that could alleviate their symptoms and restore some quality of life.

In the 1930s, Egas Moniz, a respected Portuguese neurologist, recognized that the frontal lobes were important in personality and behavior and concluded that interfering with frontal lobe function might alter the course of mental diseases such as schizophrenia and other chronic psychiatric disorders. He also recognized that destroying the frontal lobe would be relatively easy to do and, with the help of Almeida Lima, a neurosurgical colleague, introduced a simple surgical procedure for destroying most of the connections between the frontal lobe and the rest of the brain (Figure). This approach became known as a frontal lobotomy, or leukotomy.

The lobotomy technique was soon brought to the United States by the neurologist Walter Freeman, who became an equally strong advocate of this approach and devoted his life to treating a wide variety of mentally disturbed individuals in this way. In collaboration with neurosurgeon James Watts, Freeman popularized a form of the procedure that could be carried

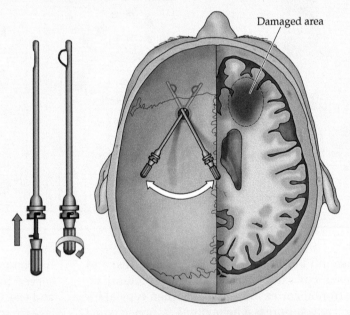

Damaged area

The surgical technique for frontal leukotomy under local anesthesia described and advocated by Egas Moniz and Almeida Lima. The "leukotome" was inserted into the brain at approximately the angles shown. When the leukotome was in place, a wire "knife" was extended and the handle rotated. The right side of the figure depicts a horizontal cutaway of the brain (parallel to the top of the skull) indicating Moniz's estimate of the extent of the damage done by the surgery. (After A. E. Moniz, 1936. *Tentatives Opératoires dans le Traitement de Certaines Psychoses.* Paris: Masson.)

■ Clinical Applications (continued)

out under local anesthesia, traveling widely across the country to demonstrate the technique and encourage its use. The technique became popular because it did relieve symptoms associated with some psychiatric conditions in many individuals—without obvious side effects in many cases—and more than 50,000 lobotomies were performed, mostly in the United States.

Rendering an individual relatively tractable, albeit permanently altered in personality, no doubt seemed the most humane of the difficult choices that faced psychiatrists and others

dealing with such individuals during that period—and Moniz was even awarded the 1949 Nobel Prize in Physiology or Medicine for this work. However, researchers and clinicians gradually recognized that lobotomies changed people's personalities—often in subtle but damaging ways. Many individuals reported not just dampened affect but also impaired motivation; that is, they lacked the goals and desires necessary for completing tasks and managing their lives. Others reported great difficulty in maintaining social relationships, both

professionally and personally. (To learn more about the personal costs of frontal lobotomies, and to hear the voices of some of the people who bore those costs, see the National Public Radio documentary series "My Lobotomy.") And in rare but striking cases such as that of Rosemary Kennedy, the operation left people essentially unable to function on their own. With the advent of increasingly effective psychotropic drugs in the late 1940s and early 1950s, frontal lobotomy as a psychotherapeutic strategy rapidly disappeared.

Prefrontal cortex contributors to planning adaptive behavior

In confirmation of the clinical evidence about the function of the frontal association cortices in neurological patients, neurons that appear to be specifically involved in planning have been identified in the frontal cortices of rhesus monkeys. One behavioral test used to study cells in the monkey frontal cortex is called the **delayed response task** (Figure 27.9A). Variants of this task are used to assess frontal lobe function in a variety of situations, including the clinical evaluation of frontal lobe function in humans. In the simplest version of the delayed response task, the monkey watches an experimenter place a food morsel in one of two wells; both wells are then covered. Subsequently, a screen is lowered for an interval of a few seconds to several minutes (the delay). When the screen is raised, the monkey gets only one chance to uncover the well containing food and retrieve the reward. Thus, the animal must decide that it wants the food, remember where it is placed, recall that the cover must be removed to obtain the food, and keep all this information available during the delay so that it can be used to get the reward. The monkey's ability to carry out this short-term memory task is diminished or abolished if the prefrontal cortex is destroyed bilaterally; this result is in accord with clinical findings in humans.

Some neurons in the prefrontal cortex, particularly those in and around the principal sulcus (Figure 27.9B), are activated when monkeys perform computerized variants of the delayed response task, and they are maximally active during the period of the delay, as if their firing represented information about the location of the food morsel maintained from the presentation part of the trial (i.e., the cognitive information needed to guide behavior when the screen is raised) (Figure 27.9C,D). Such neurons return to

a low level of activity during the actual motor phase of the task, suggesting that they represent short-term memory and planning (see Chapter 30) rather than the actual movement itself. Delay-specific neurons in the prefrontal cortex are also active in monkeys that have been trained to perform a variant of the delayed response task in which well-learned movements are produced in the absence of any cue. Evidently, these neurons are equally capable of using remembered information to guide behavior. Thus, if a monkey is trained to associate looking at a particular target with a delayed reward, the delay-associated neurons in the prefrontal cortex will fire during the delay, even if the monkey shifts its gaze to the appropriate region of the visual field in the absence of the target.

In addition to maintaining cognitive information during short delays, some neurons in prefrontal cortex also appear to participate directly in longer range planning of sequences of movements. When monkeys are trained to perform a motor sequence, such as moving a joystick to the left, then right, then left again, some neurons in prefrontal cortex fire at a particular point in the sequence (such as the third response), regardless of which movement (e.g., left or right) is made. Prefrontal neurons have also been found that are selective for each position in a learned motor sequence, thus ruling out the possibility that these neurons merely encode task difficulty or proximity to reward as the monkey nears the end of the series of responses. When these regions of prefrontal cortex are inactivated pharmacologically, monkeys lose the ability to execute sequences of movements from memory. These observations support the conclusion that the frontal lobe contributes specifically to the cognitive functions that use remembered information to plan and guide appropriate sequences of behavior.

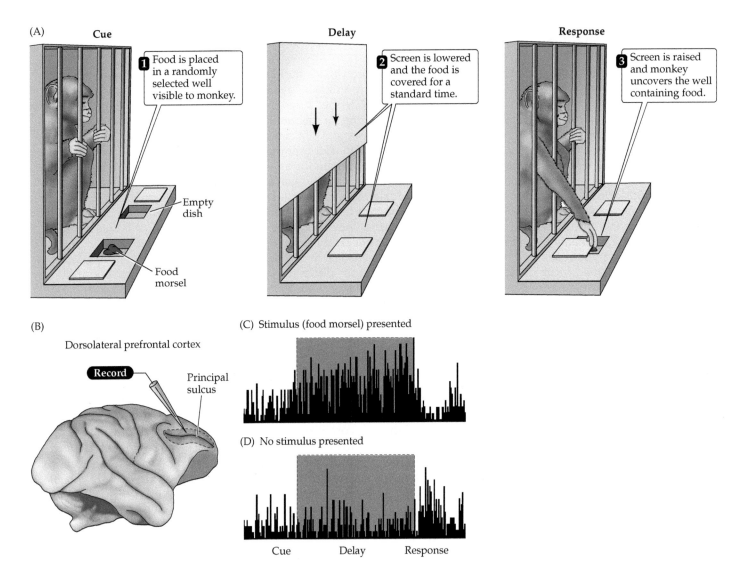

FIGURE 27.9 Activation of neurons near the principal sulcus of the frontal lobe during the delayed response task (A) Illustration of task. The experimenter randomly varies the well in which the food is placed. The monkey watches the morsel being covered, and then the screen is lowered for a standard time. When the screen is raised, the monkey is allowed to uncover only one well to retrieve the food. Typical monkeys learn this task quickly, usually performing at a level of 90% correct after fewer than 500 training trials, whereas monkeys with frontal lesions perform poorly. (B) Region of recording. (C) Activity of a delay-specific neuron in the prefrontal cortex of a rhesus monkey recorded during the delayed response task shown in (A). The histograms show the number of action potentials during the cue, delay, and response periods. The neuron begins firing more rapidly when the screen is lowered and remains active throughout the delay period. (D) When the screen is lowered and raised but no food is presented, the same neuron is less active. (After P. S. Goldman-Rakic, 1987. *Compr Physiol* 2011, Supplement 5: 373–417.)

Summary

The majority of the human cerebral cortex is devoted to tasks that transcend encoding primary sensations or commanding motor actions. Collectively, the association cortices mediate *cognitive* functions—broadly defined as the abilities that support and coordinate complex thoughts and behaviors in service of our goals. Descriptions of individuals with cortical lesions, functional brain imaging of typical individuals, and behavioral and electrophysiological studies of non-human primates show how different association cortices contribute to different functions.

Thus, parietal association cortex is involved in attention and awareness of the body and the stimuli that act on it; temporal association cortex is involved in the recognition and identification of highly processed sensory information; and frontal association cortex is involved in guiding complex behavior by planning responses to ongoing stimulation (or remembered information), matching such behaviors to the demands of a particular situation. The especially extensive association cortices in our species compared with those of other primates support the cognitive processes that define human culture.

■ Additional Reading

Reviews

Buschman, T. J. and S. Kastner (2015) From behavior to neural dynamics: An integrated theory of attention. *Neuron* 88: 127–144.

Carter, R. M. K. and S. A. Huettel (2013) A nexus model of the temporal–parietal junction. *Trends Cogn. Sci.* 17: 328–336.

Grill-Spector, K. and K. S. Weiner (2014) The functional architecture of the ventral temporal cortex and its role in categorization. *Nat. Rev. Neurosci.* 15: 536–548.

Hesse, J. K., and D. Y. Tsao (2020) The macaque face patch system: A turtle's underbelly for the brain. *Nat. Rev. Neurosci.* 21: 695–716.

Kanwisher, N. (2010) Functional specificity in the human brain: A window into the functional architecture of the mind. *Proc. Natl. Acad. Sci. U.S.A.* 107: 11163–11170.

Koechlin, E. (2016) Prefrontal executive function and adaptive behavior in complex environments. *Curr. Opin. Neurobiol.* 37: 1–6.

Szczepanski, S. M. and R. T. Knight (2014) Insights into human behavior from lesions to the prefrontal cortex. *Neuron* 83: 1002–1018.

Important Original Papers

Baldauf, S. and R. Desimone (2014) Neural mechanisms of object-based attention. *Science* 344: 424–427.

Brain, W. R. (1941) Visual disorientation with special reference to lesions of the right cerebral hemisphere. *Brain* 64: 224–272.

Crowe, D. A. and 6 others (2013) Prefrontal neurons transmit signals to parietal neurons that reflect executive control of cognition. *Nat. Neurosci.* 16: 1484–1491.

Desrochers, T. M., C. H. Chatham and D. Badre (2015) The necessity of rostrolateral prefrontal cortex for higher-level sequential behavior. *Neuron* 87: 1357–1368.

Elam, J. S. and 17 others (2021) The Human Connectome Project: A retrospective. *NeuroImage* 244: 118543.

Etcoff, N. L., R. Freeman and K. R. Cave (1991) Can we lose memories of faces? Content specificity and awareness in a prosopagnosic. *J. Cog. Neurosci.* 3: 25–41.

Funahashi, S., M. V. Chafee and P. S. Goldman-Rakic (1993) Prefrontal neuronal activity in rhesus monkeys performing a delayed anti-saccade task. *Nature* 365: 753–756.

Fuster, J. M. (1973) Unit activity in prefrontal cortex during delayed-response performance: Neuronal correlates of transient memory. *J. Neurophysiol.* 36: 61–78.

Geschwind, N. (1965) Disconnexion syndromes in animals and man. Parts I and II. *Brain* 88: 237–294.

Harlow, J. M. (1868) Recovery from the passage of an iron bar through the head. *Publications of the Massachusetts Medical Society* 2: 327–347.

Karcher, N. R., and D. M. Barch (2020). The ABCD study: Understanding the development of risk for mental and physical health outcomes. *Neuropsychopharmacology* 46: 131–142.

Kim, H. and 4 others (2016) Prefrontal parvalbumin neurons in control of attention. *Cell* 164: 208–218.

Platt, M. L. and P. W. Glimcher (1999) Neural correlates of decision variables in parietal cortex. *Nature* 400: 233–238.

Yarkoni, T. and 4 others (2011) Large-scale automated synthesis of human functional neuroimaging data. *Nat. Methods* 8: 665–670.

Books

Brickner, R. M. (1936) The Intellectual Functions of the Frontal Lobes. New York: Macmillan.

DeFelipe, J. and E. G. Jones (1988) Cajal on the Cerebral Cortex: An Annotated Translation of the Complete Writings. New York: Oxford University Press.

Garey, L. J. (1994) *Brodmann's "Localisation in the Cerebral Cortex."* London: Smith-Gordon. (Translation of K. Brodmann's 1909 book. Leipzig: Verlag von Johann Ambrosius Barth.)

Macmillan, M. (2000). *An Odd Kind of Fame: Stories of Phineas Gage.* Cambridge, MA: MIT Press.

Purves, D. and 5 others (2013) *Principles of Cognitive Neuroscience*, 2nd Edition. Sunderland, MA: Sinauer/Oxford University Press.

CHAPTER

28

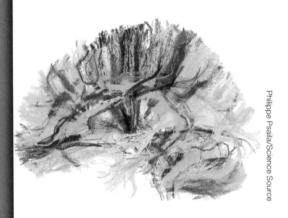

Philippe Psaila/Science Source

KEY CONCEPTS

28.1 Circadian cycles of function are regulated by neural circuits

28.2 Sleep supports physiological functions critical for health and behavior

28.3 Sleep progresses through stages of brain activity

28.4 Transitions between sleep and wakefulness rely on brain circuits

28.5 Selective impairments in cortical function can alter conscious experiences

28.6 A distributed set of brain regions becomes active when people disengage from active tasks

Cortical States

Overview

The brain varies in its activity as we change from sleep to wakefulness or from alert engagement with a task to disengaged self-reflection. Most progress has been made in understanding the variation in brain activity that follows a daily circadian pattern: the transitions between sleep and wakefulness. About a third of our lives is spent in various stages of sleep, which, except for periods of dreaming, is unconscious and fundamentally different from wakefulness. But even when we are fully awake, our awareness of the world around us and of our internal feelings and thoughts varies greatly from moment to moment. Understanding the distinctions between different cortical states and how the brain moves from one state to the next has become an increasingly important area of research in systems and cognitive neuroscience. This chapter reviews how the brain and the rest of the nervous system regulate changes across cortical states, and how this physiological regulation contributes to consciousness and awareness.

CONCEPT **28.1** | ## Circadian Cycles of Function Are Regulated by Neural Circuits

LEARNING OBJECTIVES

28.1.1 Understand the circuits that link environmental light levels to changes in brain function.

28.1.2 Describe the evidence that the suprachiasmatic nucleus serves as a master clock for the circadian cycle.

28.1.3 Summarize how genes and proteins contribute to the circuits that establish the circadian cycle.

The circadian cycle

Human physiology varies with **circadian** (Latin, "about a day") periodicity, and biologists have explored several questions about this 24-hour cycle. What happens, for example, when individuals are prevented from sensing the cues they normally use to distinguish night and day? Humans (and many other animals) have an internal "clock" that operates even in the absence of external information about the time of day, as shown by experiments in which volunteers spend days in an environment that lacks external time cues such as daily sunlight cycles, timekeeping devices, and exposure to digital media. In a typical experiment of this sort, participants undergo a 5- to 8-day period of acclimatization that includes normal social interactions, meals at the usual times, and temporal cues. During this period, the individuals typically arise and go to sleep at the usual times and maintain a 24-hour sleep–wake cycle. When the cues are removed, however, the volunteers awaken later each day (Figure 28.1A) but still show a cyclical pattern of physiological changes

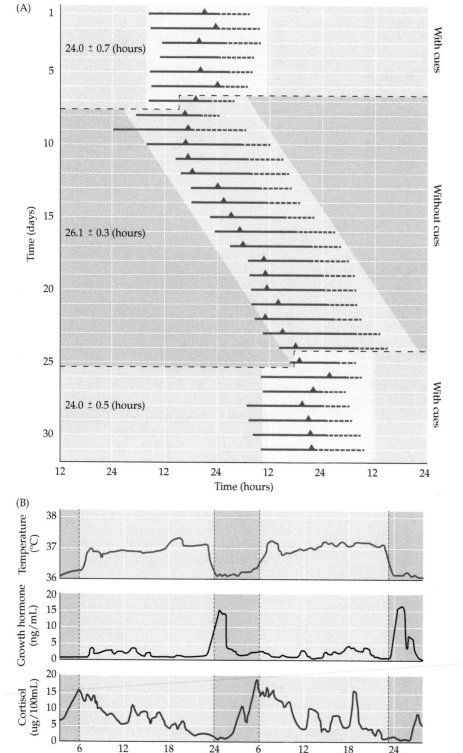

(A)

24.0 ± 0.7 (hours)

26.1 ± 0.3 (hours)

24.0 ± 0.5 (hours)

With cues

Without cues

With cues

Time (days)

Time (hours)

(B)

Temperature (°C)

Growth hormone (ng/mL)

Cortisol (ug/100mL)

Time of day (hours)

FIGURE 28.1 Circadian rhythmicity
(A) The illustration graphs the waking (blue) and sleeping (red) status of a volunteer in an isolation chamber with and without cues about the day–night cycle. Numbers represent the mean ± standard deviation of a complete wake–sleep cycle in each condition. Triangles represent times when the core body temperature was maximal. (B) Circadian rhythmicity of homeostatic regulation. Core body temperature and blood levels of growth hormone and cortisol all show a rhythmic pattern of roughly 24 hours. In the early evening, core temperature begins to decrease, whereas growth hormone begins to increase. The level of cortisol, which reflects stress, begins to increase toward morning and stays elevated for several hours. (A after J. Aschoff, 1965. *Science* 148: 1427–1432; B after J. A. Hobson, 1989. *Sleep.* New York: Scientific American Library.)

in different seasons and at different locations on the planet (Figure 28.1B). To photoentrain physiological processes with this day–night cycle, the biological clock must be able to detect variations in light levels. The receptors that sense these light changes are, not surprisingly, in the retina, as demonstrated by the fact that removing or covering the eyes abolishes photoentrainment. In mammals, however, the most important retinal detectors are not rod or cone cells, but neurons that lie within the ganglion cell layer of the retina. Unlike rods and cones, which are hyperpolarized when activated by light (see Chapter 11), these photosensitive ganglion cells contain a novel photopigment called **melanopsin** and are depolarized by light (Figure 28.2A,B). These unusual photoreceptors encode environmental illumination and thus reset the circadian clock, although rods and cones are still able to mediate some circadian entrainment in melanopsin-knockout mice.

The axons of these melanopsin-containing neurons run in the retinohypothalamic tract, which projects to the **suprachiasmatic nucleus (SCN)** of the anterior hypothalamus, the central site of the circadian control of homeostatic functions (Figure 28.2C). Activation of the SCN via this pathway evokes responses in the paraventricular nucleus of the hypothalamus and, ultimately, the preganglionic sympathetic neurons in the intermediolateral zone of the lateral horns

(e.g., body temperature)—albeit usually with a clock cycle now slightly longer than 24 hours in length.

Presumably, circadian clocks evolved to maintain appropriate daily rhythms of homeostatic functions in spite of the variable amount of daylight and darkness

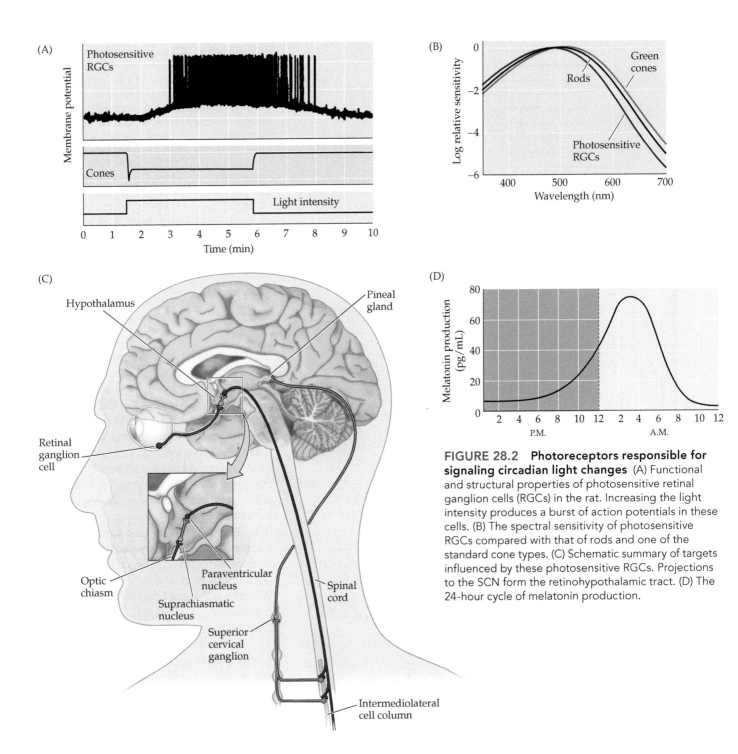

FIGURE 28.2 Photoreceptors responsible for signaling circadian light changes (A) Functional and structural properties of photosensitive retinal ganglion cells (RGCs) in the rat. Increasing the light intensity produces a burst of action potentials in these cells. (B) The spectral sensitivity of photosensitive RGCs compared with that of rods and one of the standard cone types. (C) Schematic summary of targets influenced by these photosensitive RGCs. Projections to the SCN form the retinohypothalamic tract. (D) The 24-hour cycle of melatonin production.

of the thoracic spinal cord. As described in Chapter 21, these preganglionic neurons modulate neurons in the superior cervical ganglia, some of whose postganglionic axons project to the **pineal gland** (*pineal* means "pine-cone-like") in the midline near the dorsal thalamus. The pineal gland synthesizes the sleep-promoting neurohormone **melatonin** (*N*-acetyl-5-methoxytryptamine) from tryptophan. When secreted into the bloodstream, melatonin modulates neural activity by interacting with

melatonin receptors on neurons in the SCN that in turn influence the sleep–wake cycle. Melatonin synthesis increases as light from the environment decreases, reaching a maximum between 2 and 4 A.M. (Figure 28.2D). In the elderly, the pineal gland produces less melatonin, perhaps explaining why older people sleep less at night. Melatonin supplements have been used to promote sleep in people with insomnia and to reduce the disruption of biological clocks that occurs with jet lag.

Most sleep researchers consider the suprachiasmatic nucleus to be a "master clock." Evidence for this conclusion is that removal of the SCN in experimental animals abolishes their circadian sleep–wake cycle. Furthermore, when SCN cells are placed in culture, they exhibit characteristic circadian rhythms of activity. Some other isolated cell types also show this rhythmicity, implying that the SCN is the apex of a hierarchy that governs physiological timing functions that are synchronized with the sleep–wake cycle, including body temperature, hormone secretion (e.g., cortisol), blood pressure, and urine output. In adults, urine production is reduced at night because of a circadian upregulation of antidiuretic hormone (ADH, also called vasopressin).

Molecular mechanisms of biological clocks

Virtually all animals—and many other organisms—adjust their physiology and behavior to the 24-hour day–night cycle under the influence of circadian variations. Recent studies have revealed much about the genes and proteins that constitute the molecular machinery underlying these effects. This work began in the early 1970s, with the discovery of three mutant strains of fruit flies whose circadian rhythms were abnormal. Analysis showed the mutations to be alleles that differed at a single locus, subsequently called the *period* or *per* gene. In the absence of normal environmental cues (that is, in constant light or dark), wild-type flies have periods of activity geared to a 24-hour cycle; per^S mutants have 19-hour rhythms, per^1 mutants have 29-hour rhythms, and per^0 mutants have no apparent circadian rhythm. The gene product Per, a nuclear protein, is found in many *Drosophila* cells pertinent to the production of the fly's circadian rhythms. Moreover, normal flies show a circadian variation in the amount of *per* mRNA and Per protein, whereas per^0 flies do not show circadian rhythmicity of gene expression.

Many of the genes and proteins responsible for circadian rhythms in fruit flies have now been discovered in mammals (Figure 28.3). In mice, the circadian clock arises from the temporally regulated activity of proteins (given here in capital letters) and genes (both abbreviations and full names in italics), including CRY (*Cry, cryptochrome*), CLOCK (*Clk, circadian locomotor output cycles kaput*), BMAL1 (*Bmal1, brain and muscle, ARNT-like*), REV-ERB (*NR1D1*), PER1 (*Per1, period1*), PER2 (*Per2, period2*), and PER3 (*Per3, period3*). These genes and the proteins they express give rise to regulatory transcription/translation feedback loops with both excitatory and inhibitory components (see Figure 28.3). The key points in this complex regulatory scheme are: (1) The concentrations of BMAL1 and the three PER proteins cycle in counterphase; (2) PER2 is a positive

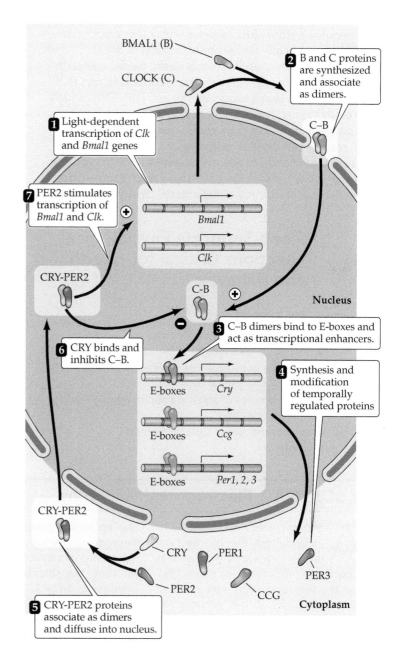

FIGURE 28.3   The molecular feedback loop believed to govern circadian clocks in mammals (After H. Okamura et al., 1999. *Science* 276: 2531–2534.)

regulator of the BMAL1 loop; and (3) CRY is a negative regulator of the period and cryptochrome loops. The two positive components of this scheme are influenced, albeit indirectly, by light and/or temperature. The cycle begins when CLOCK and BMAL1 bind to loci (called E-box elements) of target clock-controlled genes (CCGs) to drive transcription of *Cry* and the *Per* genes; REV-ERB proteins and CRY heterodimerize and translocate to the nucleus, where they inhibit the transcriptional activity of CLOCK and BMAL1. This cycle takes about 24 hours. In a second feedback loop, REV-ERB competes with a transcription factor ROR to cyclically repress transcription of *Bmal1*.

In the most recent update of this classic model, transcription of *Bmal1* is regulated by the transcription factors ROR (which activates *Bmal1* transcription) and REV-ERB (which represses transcription). CLOCK and BMAL1 proteins rhythmically drive direct transcriptional activation of REV-ERB, resulting in rhythmic expression of *Bmal1* that is antiphase to the *Per* and *Cry* genes. The collective action of these proteins—together with the genes that govern them—drives the molecular machinery of the internal circadian clock.

CONCEPT
28.2
Sleep Supports Physiological Functions Critical for Health and Behavior

LEARNING OBJECTIVES

28.2.1 Understand the evidence for sleep as a highly conserved behavior found in most animals.

28.2.2 Describe the consequences of sleep deprivation on physical and mental health.

28.2.3 Provide examples of atypical sleep in other mammalian species.

28.2.4 Summarize the role of sleep in memory consolidation.

Sleep: An overview

Sleep is defined behaviorally by the normal suspension of consciousness and electrophysiologically by specific brain wave criteria (Box 28A). It occurs in all mammals and nearly all vertebrates, with the notable exceptions of some animals that live in environments with no circadian cues (e.g., blind cave fish) or that maintain a high level of behavior continuously throughout the daily cycle (e.g., some species of sharks). We crave sleep when deprived of it, and continued sleep deprivation damages physical and mental health and can even be fatal. The clinical importance of sleep is obvious from the prevalence of sleep disorders; in the United States, about 25% of all adults suffer from some form of chronic or occasional sleep problem that interferes with their daily activities (Clinical Applications). Surprisingly, however, sleep is not the result of a simple diminution of brain activity. Indeed, in rapid eye movement (REM) sleep, the brain is about as active as it is when people are awake. Rather, sleep is a series of precisely controlled physiological states, the sequence of which is governed by a group of brainstem nuclei that project widely throughout the brain and spinal cord. Neuroscientists seek to understand both the basic mechanisms of sleep as well as some of its mysteries: Why is the brain so active during REM sleep? Why do we dream? And how does sleep restore our physical and mental health?

■ BOX 28A │ Electroencephalography

Although electrical activity recorded from the exposed cerebral cortex of a monkey was reported in 1875, it was not until 1929 that the first recordings were made of electrical activity on the human scalp. Since then, the electroencephalogram, or EEG, has provided a valuable tool to clinicians, particularly in the fields of sleep physiology and epilepsy, and to researchers seeking to understand the timing of brain activity, particularly of rapid processes such as attention.

EEG provides several advantages for neuroscientists. First, it is a simple way—at least compared with neuroimaging techniques such as fMRI—to collect data about brain function using relatively inexpensive equipment. Thus, it can be used in many institutions that might not have the resources for a neuroimaging laboratory. Second, it is noninvasive and thus can be readily

applied to any population of interest, even young babies. And third, it provides exceptional temporal resolution, sampling the electrical activity of the brain as rapidly as 1000 times a second. Its most serious limitation is poor spatial resolution. Data are typically collected from a set of electrodes placed in standard positions on the scalp (Figure A); while a modern EEG system may have 128 or more such electrodes, the electrical signals they measure are difficult to localize because of the electrical conductivity of the brain and scalp. Some researchers combine EEG with other techniques such as fMRI that can help localize the sources of fast electrical changes.

As noted in the first recordings, and still of interest today, the EEG signal contains periodicity within particular frequency bands (by convention, labeled with Greek letters) that correspond

to different cortical states. The alpha rhythm, 8 to 13 Hz, is typically recorded in awake individuals with their eyes closed; this was the first EEG signal recorded in humans. Beta activity is defined by frequencies of 14 to 35 Hz and is indicative of mental activity and attention. The theta and delta waves, which are characterized by frequencies of 4 to 7 Hz and less than 4 Hz, respectively, imply drowsiness, sleep, or one of a variety of pathological conditions; these slow waves in individuals with normal brain activity are the signature of stage IV non-REM sleep. The way these phenomena are generated is shown in Figures B and C.

Far and away the most obvious component of these various oscillations is the alpha rhythm. Its prominence in the occipital region—and its modulation by eye opening and closing—implies that it is somehow linked to

■ BOX 28A | Electroencephalography (continued)

visual processing. In fact, evidence from large numbers of individuals suggests that at least several different regions of the brain have their own characteristic rhythms; for example, within the alpha band, the classic alpha rhythm is associated with visual cortex, the mu rhythm is associated with the sensorimotor cortex around the central sulcus, and the kappa rhythm is associated with the auditory cortex.

These EEG rhythms depend in part on activity in the thalamus, as shown by the fact that thalamic lesions can reduce or abolish the oscillatory cortical discharge. The reticular activating system in the brainstem is also important in modulating EEG activity. For example, activation of the reticular formation changes the cortical alpha rhythm to beta activity, in association with greater behavioral alertness. Virtually all areas of the cortex participate in these oscillatory rhythms, which reflect a feedback loop between neurons in the thalamus and cortex (see Figure 28.9).

Animal studies that simultaneously recorded the scalp EEG and signals from electrodes within cortical layers revealed that a primary source of the EEG potential is primarily the pyramidal neurons and their synaptic connections in the deeper layers of the cortex (see Figures B and C). This conclusion was reached by noting the location of electrical field reversal upon passing an electrode vertically through the cortex
(*Continued*)

(B) An EEG electrode on the scalp measures the activity of a large number of neurons in the underlying regions of the brain, each of which generates a small electrical field that changes over time. The EEG signal is thought to represent primarily changes in dendritic potentials across large numbers of cells within the same brain region that respond to a given input in a synchronous manner. (After M. Bear et al., 2001. *Neuroscience Exploring the Brain*, 2nd ed. Philadelphia: William & Wilkins/Lippincott.)

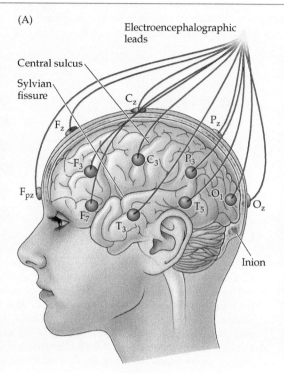

(A)

(A) The electroencephalogram represents the voltage difference between two electrodes applied to the scalp, typically by comparing each of many electrodes placed in standard positions to a reference electrode (e.g., at the mastoid behind the ear). In the example here, electrodes are placed in standard positions distributed over the head. Capital letters indicate position (FP = frontal pole, F = frontal, P = parietal, T = temporal, O = occipital, C = central); subscripts indicate left hemisphere (odd numbers), right hemisphere (even, not shown), or midline (z). The recording obtained from each electrode will differ depending on the activity of the population of neurons in the underlying brain region.

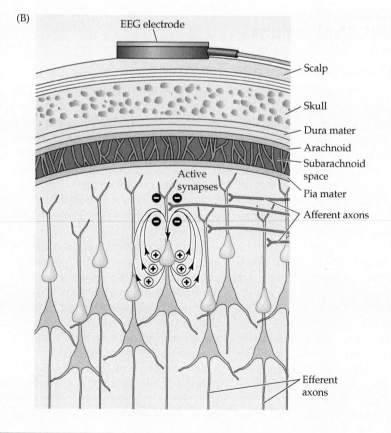

(B)

■ BOX 28A │ Electroencephalography *(continued)*

from surface to white matter. The oscillations observed in the EEG signal are thought to result from the reciprocal interaction of excitatory and inhibitory neurons in circuit loops. Despite these intriguing observations, the functional significance of these cortical rhythms remains an important area of research nearly a century after their discovery.

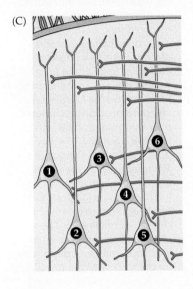 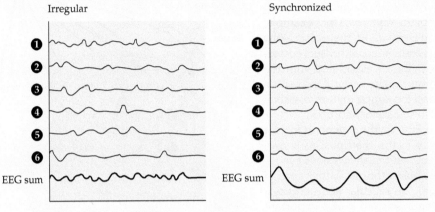

(C) Generation of the synchronous activity that characterizes deep sleep. In the pyramidal cell layer below the EEG electrode, each neuron receives thousands of synaptic inputs. If the inputs are irregular or out of phase, their algebraic sum will have a small amplitude, as occurs in the waking state. If, however, the neurons are activated at approximately the same time, then the EEG waves will tend to be in phase and their amplitude will be much greater, as occurs in the delta waves that characterize stage IV sleep. (After M. Bear et al., 2001. *Neuroscience Exploring the Brain*, 2nd ed. Philadelphia: William & Wilkins/Lippincott.)

■ Clinical Applications

Sleep Disorders and Their Treatment

A substantial fraction of the U.S. population experiences some kind of sleep disorder at one time or another. Sleep problems, which range from simply annoying to life-threatening, occur more frequently with advancing age but affect all segments of society, including young adults. The most significant problems are insomnia, sleep apnea, restless legs syndrome, and narcolepsy.

Insomnia is the inability to sleep for a sufficient length of time (or deeply enough) to produce a subjective sense of refreshment. This all-too-common problem has many causes. Short-term insomnia can arise from stress, jet lag, or simply drinking too much coffee. Another frequent cause is altered circadian rhythms associated with working night shifts. Improving sleep habits, avoiding stimulants such as caffeine, and in some cases, taking sleep-promoting medications can usually prevent these problems. Although minor insomnia may not always be reported as a serious medical problem, it may still have significant consequences for daily functioning, including decreased job efficiency and increased susceptibility to accidents.

More debilitating insomnia can be associated with major depression, potentially because of disruptions in the balance between the cholinergic, adrenergic, and serotinergic systems that control the onset and duration of the sleep cycles. Long-term insomnia is a particular problem in the elderly for reasons that are not well understood. Whatever the underlying physiology, the problem is aggravated by the fact that aged individuals are more prone to depression and often take medications that affect the neurotransmitter systems (and thus affect sleep patterns).

Sleep apnea refers to a pattern of interrupted breathing during sleep that affects many people, most often obese middle-aged males. People suffering from sleep apnea may wake up dozens or more times during the night, with the result that they experience little or no slow-wave sleep and spend less time in REM sleep. As a result, these individuals are continually tired and prone to depression, which can compound problems with poor sleep. Sleep apnea can even lead to death from respiratory arrest because the airway in susceptible individuals tends to collapse during sleep, thus blocking normal airflow. In sleep, breathing slows and muscle

■ Clinical Applications (continued)

tone decreases throughout the body, including the tone of the pharynx. If the output of the brainstem circuitry regulating commands to the chest wall or to pharyngeal muscles is decreased sufficiently, or if the airway is already compressed because of obesity, the pharynx tends to collapse as the muscles relax during the normal cycle of breathing. As a result, oxygen levels decrease and CO_2 levels rise. The rise in CO_2 causes an abrupt reflex inspiration, the force of which tends to wake up the affected individual. The most widely used remedy for sleep apnea is a positive-pressure mask that enhances airflow when worn during sleep.

Restless legs syndrome can affect people of any age but becomes increasingly prevalent with age. The disorder is characterized by unpleasant crawling, prickling, or tingling sensations in one or both legs and feet, and the urge to move them about to obtain relief. These sensations occur when the affected individual sits or lies down for prolonged periods. The result is constant leg movement during the day and fragmented sleep at night. The underlying cause of this syndrome is not understood, although it is more frequent in individuals with chronic diseases. In mild cases, a hot bath, massaging the legs, or eliminating caffeine may alleviate the problem. In more severe cases, drugs such as benzodiazepines may help.

Narcolepsy is a chronic problem that affects about 1 in 2000 people in the United States. Individuals with narcolepsy have frequent "REM sleep attacks" during the day, in which they enter REM sleep from wakefulness without going through non-REM sleep. These attacks can last from 30 seconds to 30 minutes or more. The onset of sleep in such individuals can be abrupt, with potentially disastrous consequences. This phenomenon entails a temporary loss of muscle control and is called *cataplexy*. Insights into the causes of narcolepsy have come from studies of dogs that suffer

from a genetic disorder similar to the human condition. In these animals, narcolepsy is caused by a mutation of the orexin-2 receptor gene (*Orx2*). (As mentioned in Concept 28.4, orexins are neuropeptides homologous to secretin and are found exclusively in cells in the tuberal region of the hypothalamus, which project to target nuclei responsible for wakefulness; see Figure 28.9.) Evidence from both dogs and mice suggests that the *Orx2* mutation causes hyperexcitability of the neurons that generate REM sleep and/or impairment of the circuits that inhibit REM sleep. Clinically, people with narcolepsy are treated with stimulants such as methylphenidate (Ritalin), modafinil, or amphetamines to increase their overall level of arousal.

Chronic fatigue syndrome (CFS) is not a sleep disorder per se but is believed to represent a related clinical problem. It is characterized by debilitating fatigue and other symptoms, such as unrefreshing sleep, memory and concentration problems, and a feeling of extraordinary tiredness after routine exercise. The usual definition includes 6 months or more of persistent or relapsing fatigue that impairs the individual's ability to function and is not relieved by sleep or rest. Other common symptoms can include tender lymph nodes, sore throat, muscle pain, joint pain without swelling or redness, and headaches of new type, pattern, or severity. The onset of CFS often follows a flu-like illness; it then tends to be chronic, with only a minority of individuals returning to their normal level of vigor. There are no specific tests for the disorder, which must be distinguished clinically from depression, Lyme disease, and other chronic diseases that can cause some of the same signs and symptoms. Treatment of CFS is primarily management of symptoms, and often includes cognitive behavioral therapy as well as a graduated exercise regimen that attempts to maximize the individual's stamina while not overexpending his or her

available energy. Rather than offering a cure, this approach focuses on improving a person's ability to cope with this illness.

Many drugs can affect sleep patterns, largely because so many neurotransmitters (e.g., acetylcholine, serotonin, norepinepherine, and histamine) are involved in regulating the various states of sleep (see Table 28.1). A simple but useful way of summarizing these effects is that in the waking state, the aminergic system is especially active. During non-REM sleep, both aminergic and cholinergic input decrease, but the decrease in aminergic activity is greater, so that cholinergic inputs become dominant. Thus, drugs alter the sleep pattern in one of two major ways: (1) by changing the relative activity of the inputs in any of the three states; or (2) by changing the time at which the different sleep states commence. For example, insomnia will ensue if, during the waking state, the aminergic input is increased relative to the cholinergic input. In contrast, hypersomnia occurs when there is increased cholinergic activity relative to the aminergic input.

Sleepiness. Falling asleep while doing a task, while not in itself a medical disorder, leads to enormous problems of work efficiency and even mortality. In the United States, the National Highway Traffic Safety Administration estimates that drowsy driving is responsible for tens of thousands of crashes and about 1000 deaths annually. Because of the large number of people who have sleep disorders, numerous drugs are available to treat these problems. One class of commonly used drugs is the benzodiazepines. These drugs decrease the time to onset of the deeper stages of sleep. Stimulant drugs that inhibit sleep are also commonly used, especially caffeine, which is an adenosine receptor antagonist (adenosine induces sleep). Compared with a placebo, benzodiazepines hasten the onset and depth of sleep. Caffeine has the opposite effect.

Healthy sleep and sleep deprivation

We spend a substantial fraction of our lives sleeping. To feel rested, most adults require about 7 to 8 hours of sleep, with considerable variation across individuals and over the life span (Figure 28.4A). The amount of sleep needed for optimal health is highest early in life and declines over the life span (Figure 28.4B). Infants sleep for most of the day–night cycle, including both at night and during the day. Children and teenagers need on average about 9 hours of sleep. Older adults often make up for shorter and lighter nightly sleep periods by napping during the day.

In mammals, sleep is necessary for successful behavior and even survival. Sleep-deprived rats lose weight despite increased food intake and progressively fail to regulate body temperature as it increases several degrees. They also develop infections, suggesting a compromised immune system. Rats completely deprived of sleep die within a few weeks. In humans, lack of sleep leads to impaired memory and reduced cognitive abilities and, if the deprivation persists, mood swings and, often, hallucinations. Individuals with a genetic disorder called fatal familial insomnia die within several years of onset. This rare disease, which appears in middle age, is characterized by hallucinations, seizures, loss of motor control, and the inability to enter a state of deep sleep.

Sleep deprivation need not be complete to harm physical and mental health. Even moderate amounts of reduced sleep create a sleep debt that must be repaid in the following days; otherwise judgment, reaction time, memory, and other cognitive functions will be impaired. Poor sleep has significant consequences for both individuals and society. In the United States, fatigue is estimated to contribute to more than 100,000 highway accidents each year, resulting in some 70,000 injuries and about 1000 deaths. The recognition that sleep deprivation can lead to errors in medical practice has resulted in new guidelines that limit consecutive work hours for medical students and residents, and many hospitals now take steps to counter sleep deprivation in their physicians and trainees.

Sleep in different mammalian species

A wide variety of animals have a rest–activity cycle that often (but not always) occurs in a circadian (i.e., daily) rhythm. Even among mammals, however, the organization of sleep depends on the species in question. As a general rule, predatory animals indulge in long, uninterrupted periods of wakefulness and sleep; such patterns are described as nocturnal or diurnal, depending whether they are active during the night or day. Like humans, predators have the luxury of using part of the daily cycle for life's necessities—acquiring food, caring for young—and recovering via sleep. The survival of animals that are preyed on depends much more critically on continued vigilance. Such species—as diverse as rabbits and giraffes—sleep during short intervals that usually last no more than a few minutes. Shrews, the smallest mammals, hardly sleep at all.

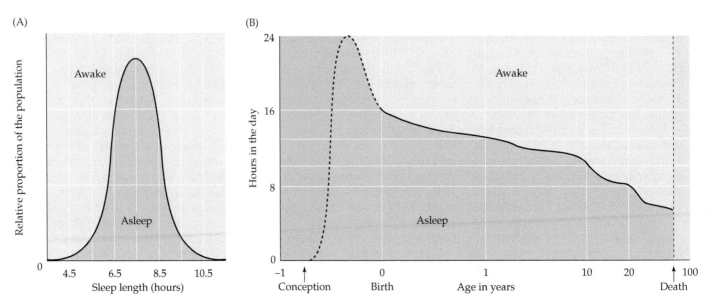

FIGURE 28.4   Duration of sleep (A) In adults, the duration of sleep each night is typically distributed with a mean of 7.5 hours and a standard deviation of about 1.25 hours. Thus, each night about two-thirds of the population sleeps between 6.25 and 8.75 hours. (B) Duration of daily sleep as a function of age. (After J. A. Hobson, 1989. *Sleep*. New York: Scientific American Library.)

Dolphins and seals show an especially remarkable solution to the challenge of maintaining vigilance during sleep. Their sleep alternates between the two cerebral hemispheres. EEG tracings taken simultaneously from left and right cerebral hemispheres of dolphins show that one hemisphere can exhibit the electroencephalographic signs of wakefulness while the other shows the characteristics of sleep. In short, although periods of rest are evidently essential to the proper functioning of the mammalian brain, and more generally to normal homeostasis, the manner in which rest is obtained depends on the particular needs of a species.

Why do we sleep?

Because an animal is particularly vulnerable while sleeping, there must be evolutionary advantages that outweigh this considerable disadvantage. For many diurnal species, one advantage is energy conservation: More energy must be expended to keep warm at night, when it is generally colder. Human body temperature has a 24-hour cycle (as do many other indices of activity and stress; see Figure 28.1B), reaching a minimum at night and thus reducing heat loss. Metabolism measured by oxygen consumption also decreases during sleep. Moreover, brain glycogen levels fall during the waking hours and are replenished at night. Because humans and many other animals that sleep at night are highly dependent on visual information to find food and avoid predators, prioritizing activity during the day increases energy efficiency. Another advantage of sleep is that its quiescence allows the brain (and body) to clear metabolic waste products produced during wakefulness. In effect, limiting activity during the night improves brain function during the day.

Another key advantage of sleep can be seen in its effects on memory. Substantial evidence from neuroscience and psychology demonstrates that periods of sleep are necessary for integrating experiences from the day into stable long-term representation in memory, a process called **consolidation**. The idea that memories are consolidated during sleep was first suggested by behavioral evidence that sleeping soon after learning improves subsequent recall. For example, performance in spatial memory tasks—such as when learning which objects are present in which locations—can actually improve following a nap.

Electroencephalographic studies further showed that REM sleep is more important for consolidation of nondeclarative memories (e.g., skills), whereas slow-wave sleep is more important for consolidation of declarative memories (e.g., facts, episodes from daily life). Other work has measured neuronal firing in the hippocampus and visual cortex both while rats ran in a figure-eight maze and during periods of slow-wave sleep before and after maze running. Researchers observed simultaneous

hippocampal–cortical reactivation that was thought to represent memory consolidation. And human neuroimaging studies have examined consolidation directly by measuring activation patterns while participants sleep inside the scanner. In one remarkable experiment, after participants learned associations between visual images and locations on a computer screen before sleeping, the spontaneous patterns of activation observed during sleep were more consistent with patterns of activation evoked during learning of those visual associations than with patterns that evoked by unfamiliar control stimuli with similar properties.

Additional evidence that memories are replayed and consolidated during sleep comes from experiments that present previously learned information while human or non-human animal participants are sleeping. A typical experimental paradigm begins with a learning task while the participant is awake; for example, a participant might learn object–location associations while a specific odor ("rose") is present in the room. Then, the same odor is presented during slow-wave sleep, during which it generates increased hippocampal activity that is thought to reflect the reactivation of the learned experience—and then leads to improved memory for object locations the next day.

CONCEPT **28.3** | **Sleep Progresses through Stages of Brain Activity**

LEARNING OBJECTIVES

28.3.1 Characterize the different stages of sleep and their physiological properties.

28.3.2 Describe the characteristic features of rapid eye movement (REM) sleep.

28.3.3 Summarize the relationship between REM sleep and dreaming.

The stages of sleep

The typical cycle of human sleep and wakefulness implies that, at specific times, various neural systems are active while others are turned off. For centuries—indeed, up until the 1950s—most researchers considered sleep a unitary phenomenon whose physiology was essentially passive and whose purpose was simply restorative. However, EEG recordings (see Box 28A) revealed that sleep actually comprises different stages that occur in a characteristic sequence.

Over the first hour after sleep onset, humans descend into successive stages of increasingly deep sleep (Figure 28.5). Initially, during "drowsiness," the frequency spectrum of the EEG shifts toward lower values and the amplitude of the cortical waves increases slightly. This drowsy

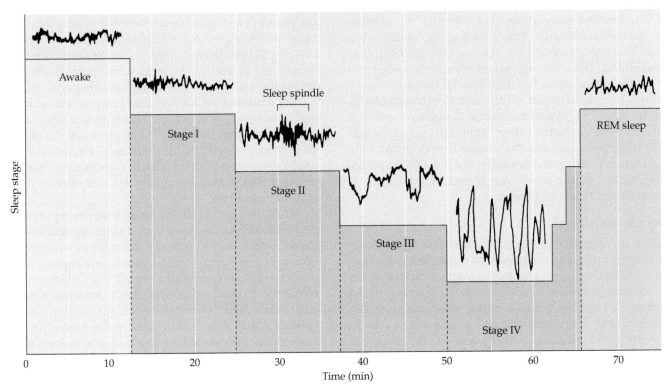

FIGURE 28.5 EEG recordings during the first hour of sleep in humans The waking state with the eyes open is characterized by high-frequency (15- to 60-Hz), low-amplitude (~30 μV) activity. This pattern is called beta activity. Descent into stage I non-REM sleep is characterized by decreasing EEG frequency (4 to 7 Hz) and increasing amplitude (50 to 100 μV), called theta activity. Descent into stage II non-REM sleep is characterized by 10- to 12-Hz oscillations (50 to 150 μV) called sleep spindles, which occur periodically and last for a few seconds. Stages III and IV of non-REM sleep are characterized by slower waves (also called delta activity) at 0.5 to 4 Hz (100 to 150 μV). After this level of deep sleep is reached, the sequence changes and a period of rapid eye movement, or REM, sleep ensues. REM sleep is characterized by low-voltage, high-frequency activity similar to the EEG activity of individuals who are awake. Although the diagram shows the stages as if they were of similar duration (for ease of comparison), they differ in duration and vary over the course of the night (see Figure 28.6). (After J. A. Hobson, 1989. *Sleep*. New York: Scientific American Library, 1989.)

period, called *stage I sleep*, eventually gives way to light *stage II sleep*, which is characterized by a further decrease in the frequency of the EEG waves and an increase in their amplitude, together with intermittent high-frequency spike clusters called **sleep spindles**. Sleep spindles are periodic bursts of activity at about 10 to 12 Hz that generally last 1 to 2 seconds and arise as a result of interactions between thalamic and cortical neurons. In *stage III sleep*, which represents moderate to deep sleep, the number of spindles decreases, whereas the amplitude of EEG activity increases further and the frequency continues to fall. In the deepest level of sleep, *stage IV sleep*, the predominant EEG activity consists of very low frequency (0.5 to 4 Hz), high-amplitude fluctuations called **delta waves** (or delta activity) the characteristic slow waves for which this phase of sleep is named; stages III and IV together are known as **slow-wave sleep**. (Note that delta waves can be thought of as reflecting synchronized electrical activity of cortical neurons.)

Taken together, sleep stages I through IV are called **non-rapid eye movement**, or **non-REM**, **sleep**. The most prominent feature of non-REM sleep is the slow-wave stage IV. It is more difficult to awaken people from slow-wave sleep, which is therefore considered to be the deepest stage of sleep. Following a period of slow-wave sleep, however, EEG recordings show that participants enter a quite different state called **rapid eye movement (REM) sleep**, for which EEG recordings appear remarkably similar to those of the awake state. After about 10 minutes in REM sleep, the brain typically cycles back through the four non-REM sleep stages. Slow-wave sleep is usually most pronounced early in an 8-hour sleep episode, although it occurs periodically during any given night (see Figure 28.6). On average, five periods of REM sleep occur each night, each having a successively longer duration.

In summary, the typical 8 hours of sleep experienced each night actually comprise several cycles interweaving non-REM and REM sleep, and the brain is quite active

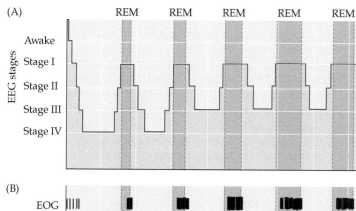

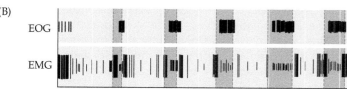

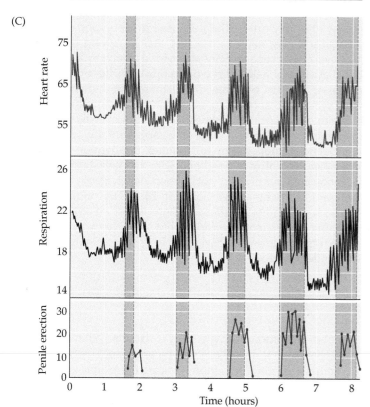

FIGURE 28.6 Physiological changes during the various sleep states Physiological status was tracked in a volunteer over a typical 8-hour sleep period. (A) The duration of REM sleep increases from 10 minutes in the first cycle to up to 50 minutes in the final cycle; note that slow-wave (stage IV) sleep is attained only in the first two cycles. (B) The electrooculogram (EOG, above) and movement of neck muscles, measured using an electromyogram (EMG, below). Other than the few slow eye movements approaching stage I sleep, all other eye movements evident in the EOG occur in REM sleep. The greatest EMG activity occurs during the onset of sleep and just before awakening. (C) The heart rate (beats per minute) and respiration (breaths per minute) slow in non-REM sleep, but increase almost to the waking levels in REM sleep. Penile erection (strain gauge units) occurs only during REM sleep. (After U. J. Jovanovic, 1971. *Normal Sleep in Man.* Stuttgart: Hippocrates Verlag.)

followed by decreases in muscle tone, body movements, heart rate, breathing, blood pressure, metabolic rate, and temperature. All these parameters reach their lowest values during stage IV sleep. Periods of REM sleep, in contrast, are accompanied by increases in blood pressure, heart rate, and metabolism to levels almost as high as those found in the awake state. REM sleep, as the name implies, is also characterized by rapid, ballistic eye movements, as well as by pupillary constriction, paralysis of many large muscle groups (although obviously not the diaphragm and other muscles used for breathing), and twitching of the smaller muscles in the fingers, toes, and middle ear. Spontaneous penile erection also occurs during REM sleep, a fact that can be clinically important in determining the course of treatment for a complaint of impotence. REM sleep has been observed in all mammals and in at least some birds; certain reptiles also have periods of increased brain activity during sleep that may be homologous to REM sleep in mammals.

Despite the similarity of EEG recordings obtained in REM sleep and in wakefulness, the two brain states are clearly not equivalent. For one thing, REM sleep is characterized by a greater prevalence of **dreaming** (Box 28B), a unique state of awareness that entails some features of memory and hallucinations (i.e., the experience of dreams is not related to corresponding sensory stimuli arising from the present environment). Because most muscles are inactive during REM sleep, the motor responses to dreams are relatively minor. (Sleepwalking, which is most common in children ages 4 to 12, and sleeptalking actually occur during non-REM sleep and are not usually accompanied or motivated by dreams.) Taken together, these observations have led to the aphorism that non-REM sleep is characterized by an inactive brain in an inactive body, whereas REM sleep is characterized by an active brain in an inactive body.

during much of this supposedly dormant, restful time. The amount of daily REM sleep decreases from about 8 hours at birth to 2 hours at 20 years to only about 45 minutes at 70 years of age.

The physiological changes in sleep states

A variety of additional physiological changes take place during the different stages of sleep (Figure 28.6). Stage I sleep is characterized by slow, rolling eye movements

■ BOX 28B | Dreaming

Despite the wealth of descriptive information about the stages of sleep and an intense research effort over more than a half-century of neuroscience research, the functional purposes of the various sleep states remain poorly understood. Whereas most sleep researchers accept the idea that the purpose of non-REM sleep is at least in part restorative, the function of REM sleep and of sleep in general remains a matter of considerable controversy.

A clue about additional purposes of sleep concerns dreams. The time of occurrence of dreams during sleep has been determined by waking volunteers during either non-REM or REM sleep and asking them if they were dreaming. Individuals awakened from REM sleep usually recall elaborate, vivid, and often emotional dreams; those awakened during non-REM sleep report fewer dreams, and when dreams do occur, they are more conceptual, less vivid, and tend to be less emotion-laden. Dreaming during light non-REM sleep tends to be more prevalent near the onset of sleep and before awakening. In any event, dreaming is not limited to REM sleep.

Dreams have been studied in a variety of ways, perhaps most notably within the psychoanalytic framework aimed at revealing unconscious thought processes considered to be at the root of neuroses. Sigmund Freud's *The Interpretation of Dreams*, published in 1900, speaks eloquently to the complex relationship between conscious and unconscious mentation. Freud thought that during dreaming the conscious "ego" relaxes its hold on the "id," or subconscious. These ideas have been out of fashion in recent decades, but to give Freud his due, at the time he made these speculations little was known about neurobiology of the brain in general and sleep in particular. In fact, some recent evidence supports Freud's idea that dreams often reflect events and conflicts of the day (the "day residue," in his terminology) and may play a role in memory. Several investigators have suggested that dreams help consolidate learned tasks, perhaps by further strengthening synaptic changes associated with recent experiences. The more general hypothesis that sleep is important in consolidating memories has been supported by studies of remembered spatial location in rodents in which ensembles of hippocampal neurons activated during a spatial memory task are reactivated during subsequent sleep, and by experiments in humans that show a sleep-dependent improvement in learning.

A quite different idea is that dreaming has evolved to dispose of unwanted memories that accumulate during the day by expunging parasitic modes of thought that would otherwise become overly intrusive, as occurs in compulsive thought disorders. Finally, some researchers have taken the more skeptical view that dream content may represent a relatively prosaic re-experiencing of recent episodes in our daily lives. In this view, dreams are just an epiphenomenal experience that accompanies the consolidation processes of events as our brain filters and organizes our memories.

Adding to the uncertainty about the purposes of sleep and dreaming is the fact that depriving humans of REM sleep for as much as 2 weeks has little or no obvious effect on their behavior. Furthermore, individuals taking serotonin reuptake inhibitors for depression have markedly less REM sleep. The apparent innocuousness of REM sleep deprivation contrasts markedly with the devastating effects of total sleep deprivation mentioned in Concept 28.2. The implication of these findings is that we can get along without REM sleep, but we need non-REM sleep in order to survive.

<table>
<tr><td>CONCEPT
28.4</td><td>**Transitions between Sleep and Wakefulness Rely on Brain Circuits**</td></tr>
</table>

LEARNING OBJECTIVES

28.4.1 Describe the role of the reticular activating system in governing the sleep–wake cycle.

28.4.2 Evaluate how the relative activity of monoaminergic and cholinergic systems differs in REM and non-REM sleep.

28.4.3 Understand how modulation of thalamocortical circuits generates EEG signatures of brain function.

The neural circuits governing sleep

The variation in physiological states that occurs during sleep is regulated by changes in the balance of excitation and inhibition in many neural circuits. What follows is a brief overview of these still incompletely understood circuits and the interactions among them that govern sleep and wakefulness.

The first evidence for circuits governing the sleep–wake cycle came from electrical stimulation of neurons near the junction of the pons and midbrain; that stimulation caused a state of wakefulness and arousal. This region of the brainstem was given the name **reticular activating system** (Figure 28.7A; see also Box 17D), under the assumption that wakefulness requires specialized activating circuitry—that is, wakefulness is not just the result of adequate sensory experience. In parallel, research on thalamic stimulation revealed that low-frequency pulses produced a slow-wave sleep (Figure 28.7B). These results indicated that the transitions between sleep and wakefulness follow from interactions between the brainstem, thalamus, and cortex.

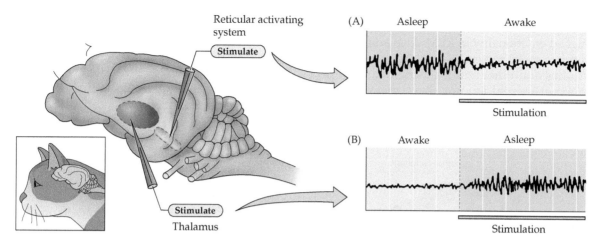

FIGURE 28.7 Activation of specific neural circuits triggers sleep and wakefulness (A) Electrical stimulation of the cholinergic neurons near the junction of the pons and midbrain (the reticular activating system) causes a sleeping cat to awaken. (B) Slow electrical stimulation of the thalamus causes an awake cat to fall asleep. Graphs show EEG recordings before and during stimulation. (After H. W. Magoun, 1952. *Arch Neurol Psychiatry* 67: 145–154.)

Further evidence that specific neural circuits govern the sleep–wake cycle was provided by work on the circuitry underlying REM sleep. The saccade-like eye movements (which occur in the absence of external visual stimuli) that define REM sleep are now known to arise because endogenously generated signals from the **pontine reticular formation** are transmitted to the motor region of the superior colliculus. As described in Chapter 20, collicular neurons project to the **paramedian pontine reticular formation (PPRF)** and the **rostral interstitial nucleus**, which coordinates timing and direction of eye movements. REM sleep is also characterized by EEG waves that originate in the pontine reticular formation and propagate through the lateral geniculate nucleus of the thalamus to the occipital cortex. These **pontine-geniculo-occipital (PGO) waves** provide a useful marker for the beginning of REM sleep; they also indicate yet another neural network by which brainstem nuclei can activate the cortex.

A further advance has come from fMRI and PET studies that have compared human brain activity in the awake state and in REM sleep. Activity in the amygdala, parahippocampal gyrus, pontine tegmentum, and anterior cingulate cortex increases during REM sleep, whereas activity in the dorsolateral prefrontal and posterior cingulate cortices decreases (Figure 28.8). The increase in activity of brain regions critical for emotional processing, coupled with a marked decrease in the influence of the prefrontal cortex,

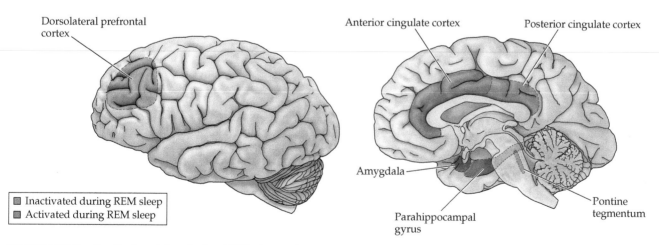

FIGURE 28.8 Cortical activity during REM sleep The diagram shows cortical regions whose activity is increased or decreased during REM sleep, as well as key subcortical structures described in the text. (After J. A. Hobson, 1989. *Sleep.* New York: Scientific American Library, 1989.)

presumably explains some characteristics of dreams (e.g., their heightened emotionality in REM sleep and their often inappropriate social content).

These studies and other studies using neuronal recording in experimental animals have revealed that a key component of the reticular activating system is a group of **cholinergic nuclei** near the pons–midbrain junction that project to thalamocortical neurons (Figure 28.9). The relevant neurons in the nuclei are characterized by high discharge rates during both waking and REM sleep and by quiescence during non-REM sleep. When stimulated, these nuclei cause desynchronization

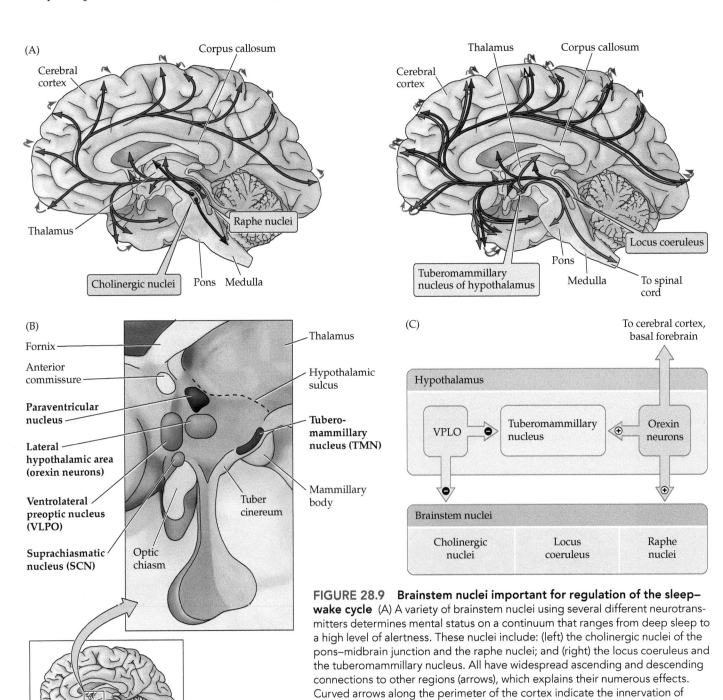

FIGURE 28.9 Brainstem nuclei important for regulation of the sleep–wake cycle (A) A variety of brainstem nuclei using several different neurotransmitters determines mental status on a continuum that ranges from deep sleep to a high level of alertness. These nuclei include: (left) the cholinergic nuclei of the pons–midbrain junction and the raphe nuclei; and (right) the locus coeruleus and the tuberomammillary nucleus. All have widespread ascending and descending connections to other regions (arrows), which explains their numerous effects. Curved arrows along the perimeter of the cortex indicate the innervation of lateral cortical regions not shown in this plane of section. (B) Location of hypothalamic nuclei involved in sleep. (C) Activation of VLPO neurons induces sleep. Orexin-containing neurons project to different nuclei and produce arousal.

TABLE 28.1   Summary of the Cellular Mechanisms That Govern Sleep and Wakefulness

Brainstem nuclei responsible	Neurotransmitter involved	Activity state of the relevant brainstem neurons
Wakefulness		
Cholinergic nuclei of pons–midbrain junction	Acetylcholine	Active
Locus coeruleus	Norepinephrine	Active
Raphe nuclei	Serotonin	Active
Tuberomammillary nuclei	Histamine	Active
Lateral hypothalamus	Orexin	Active
Non-REM sleep		
Cholinergic nuclei of pons–midbrain junction	Acetylcholine	Decreased
Locus coeruleus	Norepinephrine	Decreased
Raphe nuclei	Serotonin	Decreased
REM sleep		
Cholinergic nuclei of pons–midbrain junction	Acetylcholine	Active (PGO waves)
Raphe nuclei	Serotonin	Inactive
Locus coeruleus	Norepinephrine	Inactive

of the electroencephalogram: a shift in EEG activity from high-amplitude, synchronized waves to lower-amplitude, higher-frequency, desynchronized waves (see Box 28A). Activity of cholinergic neurons in the reticular activating system is a primary cause of wakefulness and REM sleep, whereas relative inactivity of those neurons produces non-REM sleep.

Activity of these cholinergic neurons is not, however, the only neuronal basis of wakefulness; also involved are the **noradrenergic neurons** of the locus coeruleus; the **serotonergic neurons** of the raphe nuclei; and the **histamine-containing neurons** in the tuberomammillary nucleus (TMN) of the hypothalamus (see Figure 28.9). The activation of these cholinergic and monoaminergic networks together produces the awake state. The locus coeruleus and raphe nuclei are modulated by the TMN neurons located near the tuberal region. The TMN is activated by neurons in the lateral hypothalamus that secrete the peptide **orexin** (also called **hypocretin**), which promotes waking. Conversely, antihistamines inhibit the histamine-containing TMN network, which is why they tend to make people drowsy.

These circuits responsible for the awake state are periodically inhibited by neurons in the ventrolateral preoptic nucleus (VLPO) of the hypothalamus. Thus, activation of VLPO neurons contributes to the onset of sleep, and lesions of VLPO neurons tend to produce insomnia. Recent work suggests that adenosine neurotransmission in the basal forebrain is also involved in the regulation of sleep. These complex interactions and effects are summarized in Table 28.1. Given that so many systems and transmitters are involved in the different phases of sleep, it is not surprising that a wide variety of drugs can influence the sleep cycle (see Clinical Applications).

Thalamocortical interactions in sleep

Neuronal activity (or its absence) in these brainstem nuclei affects sleep and wakefulness by modulating the interactions between the thalamus and the cortex. Specifically, the activity of several ascending systems from the brainstem decreases both the rhythmic bursting of the thalamocortical neurons and the related synchronized activity of cortical neurons (hence the diminution and ultimate disappearance of high-voltage, low-frequency slow waves during waking and during REM sleep).

To appreciate how different sleep states reflect modulation of thalamocortical activity, consider the electrophysiological responses of the relevant neurons. Thalamocortical neurons receive ascending projections from the locus coeruleus (noradregeneric), raphe nuclei (serotonergic), reticular activating system (cholinergic), and TMN (histaminergic) and project to cortical pyramidal cells. The primary characteristic of thalamocortical neurons is that they can be in one of two stable electrophysiological states: an intrinsic bursting (or oscillatory) state, or a tonically active state. The latter is generated when the thalamocortical neurons are depolarized, such as occurs when the reticular activating system generates wakefulness (Figure 28.10). In the tonically active state, thalamocortical neurons transmit information to the cortex that is correlated with the spike trains encoding peripheral stimuli. In contrast, when thalamocortical neurons are in the bursting state, the neurons in

FIGURE 28.10 Thalamocortical neurons and the sleep cycle Recordings from a thalamocortical neuron, showing the oscillatory mode corresponding to a sleep state and the tonically active mode corresponding to an awake state. An expanded view of oscillatory phase is shown at left. Bursts of action potentials are evoked only when the thalamocortical neuron is hyperpolarized sufficiently to activate low-threshold calcium channels. These bursts account for the spindle activity seen in EEG recordings in stage II sleep (see Figures 28.5 and 28.11). Depolarizing the cell either by injecting current or by stimulating the reticular activating system transforms this oscillatory activity into a tonically active mode. (After D. A. McCormick and H. C. Pape, 1990. *J Physiol* 431: 291–318.)

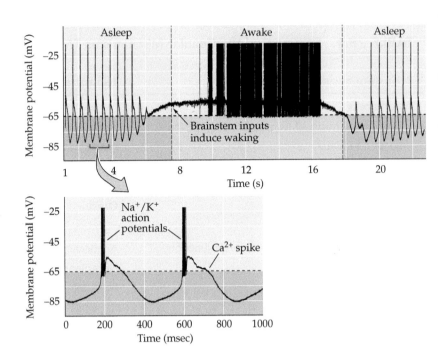

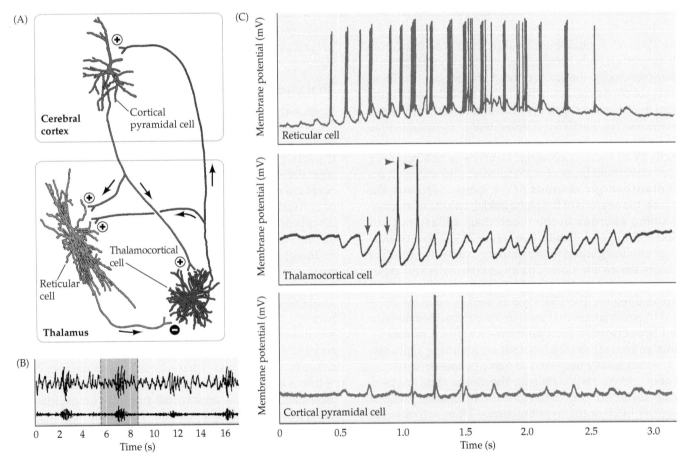

FIGURE 28.11 Thalamocortical feedback loop and the generation of sleep spindles (A) Diagram showing excitatory (+) and inhibitory (−) connections between thalamocortical cells, pyramidal cells in the cortex, and thalamic reticular cells, which provide the basis for sleep spindle generation. Inputs into thalamocortical and thalamic reticular cells are not shown. (B) EEG recordings illustrating sleep spindles (the bottom trace is filtered to accentuate the spindles). (C) Responses from individual thalamic reticular cells, thalamocortical cells, and cortical cells during generation of the middle spindle (boxed in panel B). The bursting behavior of the thalamocortical neurons elicits spikes in cortical cells, which are then evident as spindles in EEG recordings. (After M. Steriade et al., 1993. *Science* 262: 679–685.)

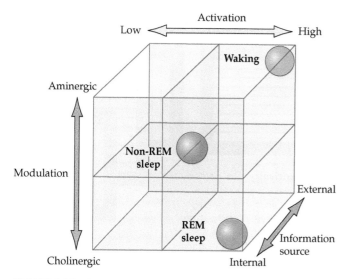

FIGURE 28.12   Summary of sleep–wake states In the waking state, activation is high, modulation is aminergic, and the information source is external. In REM sleep, activation is also high, modulation is cholinergic, and the information source is internal. The other states can likewise be remembered in terms of this general diagram. (After J. A. Hobson, 1989. *Sleep*. New York: Scientific American Library.)

the thalamus become synchronized with those in the cortex, as if to functionally disconnect the cortex from the outside world. The disconnection is maximal during slow-wave sleep, when EEG recordings show the lowest frequency and the highest amplitude.

The oscillatory, or bursting, state of thalamocortical neurons can be transformed into the tonically active state by activity in the cholinergic or monoaminergic projections from the brainstem nuclei (Figure 28.11). Moreover, the bursting state is stabilized by hyperpolarizing the relevant thalamic cells. Such hyperpolarization occurs as a consequence of stimulation by GABAergic neurons in the thalamic reticular nucleus. These neurons receive ascending information from the brainstem and descending projections from cortical neurons, and they contact the thalamocortical neurons. When neurons in the reticular nucleus undergo a burst of activity, they cause thalamocortical neurons to generate short bursts of action potentials, which in turn generate spindle activity in cortical EEG recordings (indicating a lighter sleep state). Figure 28.12 summarizes the relationships among the various sleep–wake states.

CONCEPT 28.5 | Selective Impairments in Cortical Function Can Alter Conscious Experiences

LEARNING OBJECTIVES

28.5.1 Understand how neuroscientists and philosophers define consciousness in terms of subjective experiences.

28.5.2 Provide examples of experimental paradigms that can be used to study consciousness.

28.5.3 Describe the functional deficits and capabilities exhibited by individuals with blindsight.

28.5.4 Characterize the consequences of surgical resection of the corpus callosum (split-brain surgery) on cognition and behavior.

28.5.5 Explain the loss of function seen in coma and describe the evidence that some aspects of consciousness are preserved.

Consciousness

Although being awake is clearly a prerequisite to being conscious in the sense of being normally aware of the world and the self, these functions are not equivalent. Most taxonomies of consciousness thus distinguish degrees of physiological *wakefulness* from aspects of the subjective *awareness* of the world that can vary based on brain states, neurological or psychiatric conditions, or experiences and behaviors. Understanding the links between brain states and subjective awareness has become a major topic for research in neuroscience.

Research on consciousness has been further complicated by the challenges of linking objective measures (e.g., of brain states) to subjective reports of awareness. Many non-human animals seem to behave as if they are aware of the surrounding world and its properties; they respond to stimuli in goal-appropriate ways (e.g., tracking prey), recognize conspecifics and coordinate behavior (e.g., pack hunting by wolves or dolphins), and even exert effort now for future benefit (e.g., shaping tools, caching). Yet even these capacities do not necessarily prove the existence of subjective mental states—or **phenomenological consciousness**—especially when considering that even organisms with very simple nervous systems (e.g., octopi) may exhibit complex behaviors. Nor would the demonstration that an organism exhibits a particular aspect of consciousness in one context necessarily generalize to other contexts. Even in the human nervous system, most neural processing is largely automatic, operating below the threshold of awareness. Think, for example, of all the homeostatic neural mechanisms that ensure your well-being in innumerable ways even while you are pondering the meaning of this sentence.

These daunting issues notwithstanding, neuroscientists have sought to address the basis of consciousness by ferreting out some signature of the neural processing that occurs when we are aware of something compared with when we are not. Philosophers refer to such situations as involving **access consciousness**—when the contents of our subjective experience can be reported through our speech or controlled behavior. Note that one can have phenomological consciousness without access consciousness, such as in cases where the nuances of a subjective experience (e.g., a spiritual epiphany) are not readily conveyed to others. Experiments typically measure neural activity while a particular sensory percept moves in and out of awareness. By asking the person (or experimental animal) to report these perceptual transitions (typically verbally or by a button press), investigators can compare neural activity during awareness of a stimulus with that when the individual is unaware of the stimulus. In such paradigms, the physical stimulus remains unchanged and thus serves as its own control. One paradigm for this purpose has been binocular rivalry. Binocular rivalry refers to the fact that when a particular stimulus is presented to one eye while a discordant stimulus is presented to the other, the visual percept is of either one stimulus or the other, and alternates back and forth every few seconds, rather than being a combination or blending of the views coming from the two eyes (Figure 28.13).

Using humans or monkeys trained to report what they are seeing at any given moment, electrophysiological methods and fMRI can assess changes in brain activity that occur when there is a switch of conscious content. For example, when the monocular inputs are faces and houses, recordings of fMRI activity show increases in the fusiform face area of the temporal lobe when a face is perceived, and in the parahippocampal place area when houses are seen. Similar paradigms in non-human primates reveal that neurons in different object-processing regions of visual cortex show increased activity depending on which stimulus category the animal perceives, consistent with the interpretation that activity in object-selective regions within visual cortex tracks perceptual experiences and not visual input.

By changing the experimental stimuli, neuroscientists can gain insight into the levels of processing necessary for subjective experience. One clever approach takes advantage of orientation aftereffects that can be induced by exposing participants to a series of lines (referred to as *gratings*) in a particular orientation (say, 45°) for a minute or two; following this exposure, "neutral" line stimuli (vertical lines) are perceived for several seconds as being slightly tilted in the direction opposite the angle of the inducing exposure. Researchers can run the same procedure without awareness of the inducing stimulus by masking it during its presentation, and then compare behavior and brain responses in the two conditions (with and without awareness). The orientation aftereffect remains present regardless, indicating that activity of neurons in primary visual cortex does not depend on an individual's awareness of the inducing stimulus. Similar results have been observed when looking at

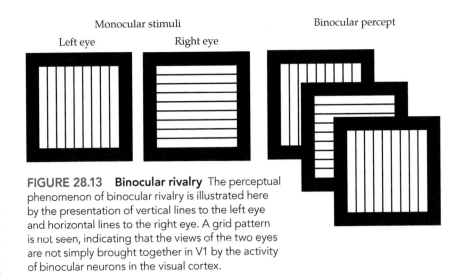

Monocular stimuli   Binocular percept
Left eye   Right eye

FIGURE 28.13 Binocular rivalry The perceptual phenomenon of binocular rivalry is illustrated here by the presentation of vertical lines to the left eye and horizontal lines to the right eye. A grid pattern is not seen, indicating that the views of the two eyes are not simply brought together in V1 by the activity of binocular neurons in the visual cortex.

depth-sensitive (i.e., disparity tuned to be active in response to stereoscopic stimuli) neurons in the primary visual cortex of monkeys. Again, little or no difference in activity is apparent when a monkey indicates behaviorally that it experienced a perception of depth compared with when it did not. Thus, unlike in later stages of visual processing, studies of early visual cortical processing reveal that neural activity is maintained even in the absence of visual awareness.

Visual awareness depends, at least in part, on the activity of some cortical regions not generally thought of as visual. For example, neuroimaging studies in humans have shown that perceptual changes in binocular rivalry and other bistable-image paradigms are associated with activity in frontal and parietal cortical regions (i.e., there is activity in these cortical regions that is time-locked to the individuals' reports of the perceptual changes). Consistent with this evidence, other perceptual effects such as "pop-out" of a visual stimulus, or the awareness of a previously missed stimulus feature, are also correlated with activity changes in frontal and parietal areas. Transient functional disruption of processing in these areas by transcranial magnetic stimulation (TMS) also disturbs perception. Disturbances in perception following damage to these higher-level brain regions further confirm that they are involved in perceptual awareness and consciousness, although their roles in these phenomena remain unclear.

One working hypothesis about the aspects of brain activity necessary for consciousness has been that subjective awareness requires a return of activity back to the relevant sensory processing regions—one that occurs much later in time than the initial feedforward cascade. This longer-latency activity, termed **reentrant neural activation**, has specifically been proposed as a mechanism leading to perceptual awareness. Because of the sluggishness of the BOLD signal, fMRI provides only limited information about reentrant activity. EEG studies, however, have suggested that reentrant activity in visual processing regions correlates with reported awareness of an attended visual stimulus.

Blindsight

Clinical conditions associated with damage to the brain have also contributed to understanding the neural basis of awareness and consciousness. A phenomenon of particular interest is **blindsight**. As described in Chapter 9, individuals with damage to the primary visual cortex are blind in the affected area of the contralateral visual field; for example, the individual known as DB experienced damage to most of V1 in the right hemisphere following a surgical operation, leading to near-complete blindness (a **scotoma**) in the left visual field. When tested using standard ophthalmologic approaches (e.g., flashing a light at different positions in space), individuals such as DB will attest that they see nothing—they have no subjective experience of any light, or anything else, in the region of blindness (Figure 28.14A). However, when forced to guess some feature of the (unseen) visual stimulus—such as where a light is positioned or whether a bar is oriented vertically or horizontally—their reports are highly accurate, leading to the evocative description of their experience as "blindsight." When DB was asked to point to the location of a target light flashed briefly within his left visual field, he could do so with near perfect accuracy (Figure 28.14B). Blindsight can also be simulated in typical individuals by transient inactivation of V1 by TMS applied over the occipital lobes. TMS creates a temporary scotoma for a specific region of the visual field, and again the features of simple stimuli presented in the unseen region tend to be guessed correctly at levels well above chance.

Functional neuroimaging and electrophysiological studies of individuals with blindsight show that the unseen stimuli elicit some activity in extrastriate regions beyond V1, implying that these cortical areas are needed for successful behavior in the absence of awareness. This evidence combined with the research described in the previous section of this concept leads to the conclusion that extrastriate activity may be necessary but not sufficient for awareness. One proposed explanation of blindsight is that subcortical visual processing of the stimulus, or coarse subcortical input to extrastriate cortex that bypasses V1, influences the individual's guesses.

Split-brain syndrome

The experience of individuals with *split-brain syndrome* (see also Figure 31.9) provides additional important data about the links between brain function and subjective experience. Beginning in the 1950s, neurosurgeons recognized that individuals with intractable epileptic seizures exhibited an expansion of their epileptic focus from one hemisphere to a homologous region in the other hemisphere, such as from the left temporal cortex to the right temporal cortex. To migitate the spread of the seizures in these individuals, they resected the major axonal tracts between the hemispheres, typically the corpus callosum (a procedure known as a callosotomy). This was a relatively rare procedure—one performed on perhaps a few hundred individuals over three decades before being phased out in favour of less drastic surgical procedures or drug administration—and many of the patients had suffered from such severe epilepsy that their cognitive or motor functions were already greatly

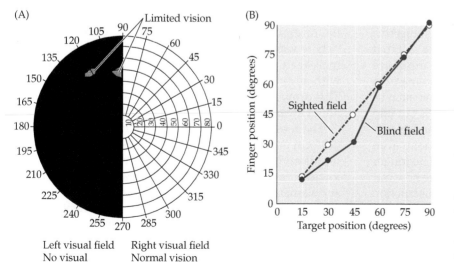

FIGURE 28.14 Blindsight (A) Following surgery to his occipital cortex, individual DB was left with near-complete blindness in his left visual field as measured using standard ophthalmologic methods. The center of this figure represents the center of fixation (i.e, when looking straight ahead), while the left and right sides represent the left and right visual fields, respectively. Areas shaded in white represented regions with normal vision (i.e., DB reported seeing lights flashed there); areas shaded in hatching had limited vision; and areas shaded in black had no reported perceptual experiences. (B) When lights were flashed in the blind left visual field, DB was asked to guess by pointing a finger at the location of the unseen lights—and could guess the location with great accuracy, even without any visual experience. (After After L. Weiskrantz et al., 1974. *Brain* 97: 709–728.)

undermined. However, about a dozen such patients had remarkable changes following surgery: Once their epilepsy was controlled, they reported improved quality of life. Their cognitive and motor functions seemed fully intact, they carried on conversations, walked and engaged in other coordinated actions, held steady jobs, and interacted socially with others. Yet because of their surgery, information could no longer flow directly between the cerebral hemispheres—leading to the concept of a "split brain."

Neuroscientists recognized that split-brain syndrome could provide important insights into how the brain supports conscious experience. Experiments showed that the divided hemispheres function relatively independently in split-brain syndrome, and that awareness generated by neural processing in one hemisphere is largely unavailable to the other. For example, when an individual with split-brain syndrome views simple written instructions such as "laugh" or "walk" in their left visual field—and thus the right brain hemisphere—they often have enough rudimentary verbal understanding in the right hemisphere to execute the commanded action. However, when asked to report *why* they laughed or walked, they typically confabulate a response using the superior language skills in the left hemisphere, saying, for instance, that something the experimenter said struck them as funny, or that they were tired of sitting and needed to walk a bit. Thus, the same individual would appear, under these circumstances, to harbor two relatively independent domains of awareness.

Whereas blindsight and split-brain patients both exhibit a lack of awareness of stimuli that nonetheless influence their behavior, it is also possible to be aware of something that doesn't actually exist. Perhaps the most striking illustration of this sort of phenomenon is the *phantom limb* experiences of amputees described in Clinical Applications, Chapter 13. Recall that a common experience following amputation is the patient's subjective awareness of the missing arm or leg, despite the fact that the physical limb and its peripheral sensory input are absent. Many such patients complain about extreme pain in the missing limb, reporting vivid sensations such as illusory fingers clenched so tightly that the nails cut into the palm. The awareness of the missing limb and sensations arising from it seem quite real to the patient, even though they fully recognize that the limb itself no longer exists. Collectively, these phenomena emphasize how our sense of self arises from an active process of construction orchestrated by our cortex.

Comatose states

An important pathological condition pertinent to consciousness is **coma** (Greek for "deep sleep"), a deeply

unconscious state defined by apparent unresponsiveness to sensory stimuli. The condition typically arises from injury or disease that compromises the function of the brainstem and other deep brain structures, such that the normal interaction of these structures with the cerebral cortex is interrupted.

Coma can arise from varying degrees of brain damage and thus is associated with significant variations in prognosis. Most comatose individuals recover consciousness within a few days or weeks as the compromised neurons and their associated circuits gradually regain their functions. Impairment can persist for much longer, however, if neural damage is more profound. Some patients recover consciousness after months, although typically with residual effects, and extremely rare cases have been reported in which consciousness is regained after some years. This variability has led to social, religious, and political controversy over the point at which an unresponsive patient should be considered to be in a **persistent vegetative state**, a diagnosis that raises ethical issues surrounding decisions about withholding life support. This sensitive issue means there has been, and will continue to be, interest in techniques that could contribute to a better understanding a given patient's prognosis. Electroencephalography has been fundamental in diagnosing *irreversible brain death*, which occurs when brain trauma is so severe that no EEG activity can be recorded (i.e., a flat electrical trace).

More recently, functional neuroimaging has also been used to evaluate persistent vegetative state, sometimes with surprising results. One 23-year-old female patient in a persistent vegetative state had suffered severe brain damage in an automobile accident 5 months previously, and had been unresponsive to external stimulation ever since. She was brought into an fMRI scanner to participate in several sorts of experiments, all involving passive listening to speech and simple instructions. When the researchers spoke to her in complex sentences (e.g., "There was milk and sugar in his coffee"), they observed increased fMRI activation in brain regions associated with auditory processing and sentence comprehension. This provided evidence of sensory system function, albeit not of awareness itself.

The researchers then verbally instructed the patient to perform particular mental imagery tasks, namely playing tennis and walking through her home. Remarkably, following the tennis instructions, there was activation in her supplementary motor area, a region associated with the coordination and preparation of motor movements. In contrast, following the home exploration instructions, there was activation in her parahippocampal and parietal cortices, regions associated with spatial processing. The fMRI activations the patient exhibited were nearly identical to

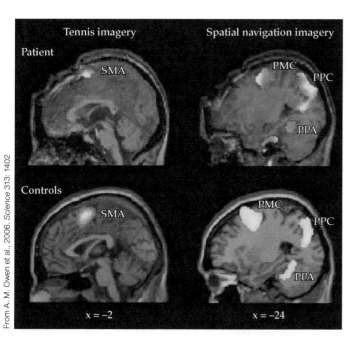

From A. M. Owen et al., 2006. *Science* 313: 1402

FIGURE 28.15 Functional MRI evidence for intact and purposeful brain function in a patient in a partial vegetative state Researchers used fMRI to scan a patient who made no overt responses to external stimuli. When she was asked to imagine playing tennis, the patient's brain exhibited increased fMRI activation in the supplementary motor area (SMA), just like that found in a group of typical control individuals. Similarly, when the patient was asked to imagine walking around her house, her brain showed activation in the premotor cortex (PMC), posterior parietal cortex (PPC), and parahippocampal place area (PPA), again just as in typical controls.

those of typical control individuals performing the same tasks (Figure 28.15). These results provided compelling evidence for aware, purposeful mental actions in an individual whose disorder precluded overt behavior.

CONCEPT **28.6**	# A Distributed Set of Brain Regions Becomes Active When People Disengage from Active Tasks

LEARNING OBJECTIVES

28.6.1 List the key brain regions in the default-mode network.

28.6.2 Describe potential purposes for default-mode activity in the human brain.

The default mode of brain function

In recent years, neuroscientists have observed something remarkable: increased activity in specific brain regions when people are awake and aware, but not pursuing any particular goal or task. On the basis of neuroimaging studies, neuroscientists have come to call this the *default* state of brain activity. Some brain regions, including the posterior cingulate cortex, the medial prefrontal cortex, and the temporal–parietal junction (TPJ), show consistently *greater* activity when people are in an alert resting state than when they are engaged with a demanding cognitive task. When the activity during a baseline or rest condition is subtracted from the activity during cognitive tasks, relative decreases of activity (generally called *deactivations*) are observed in these areas. Because cognitive tasks lead to activation increases in other brain areas carrying out the relevant processing, this finding led to the proposal that these regions constitute a network supporting a default mode of brain function that is engaged in the *absence* of any particular cognitive task (Figure 28.16). Further supporting the hypothesis that these areas constitute a **default mode network**, analysis of the functional connectivity of these areas (i.e., how much their activity covaries across time) has shown strong coupling during the resting state.

Single-unit recording in monkeys provides additional evidence about the characteristic features of the default-mode network. For example, neurons in the posterior cingulate cortex show decreases in firing rates during performance of active tasks that closely match the fMRI activation patterns observed in humans. In particular, their fluctuations are anticorrelated with activity in dorsal parietal regions associated with the attentional control network and predict lapses in attention and the ability to switch from one task to another. Moreover, neuroimaging studies also have shown abnormal activity patterns in the default-mode network in several major neurological and psychiatric disorders, including being less active in autism and more so in schizophrenia.

The obvious question is what purpose neural activity in a default-mode network serves—that is, why should these regions be active if and when the brain is doing nothing in particular? Although the default-mode network activity might be related to mental idling, another possibility is that this network is activated when attention is inwardly focused, the standard attentional control system being activated primarily when a person is focused on events and stimuli in the external environment (see Chapter 29). Whatever this network does, its activation pattern also occurs in monkeys, so it presumably evolved to carry out some relatively basic yet important function. The inverse relationship of activity in the default-mode network and in the dorsal frontal-parietal attentional control network during focused attention suggests that these complementary systems may play an interactive role in system-wide brain function related to engaging and disengaging attention and other cognitive functions.

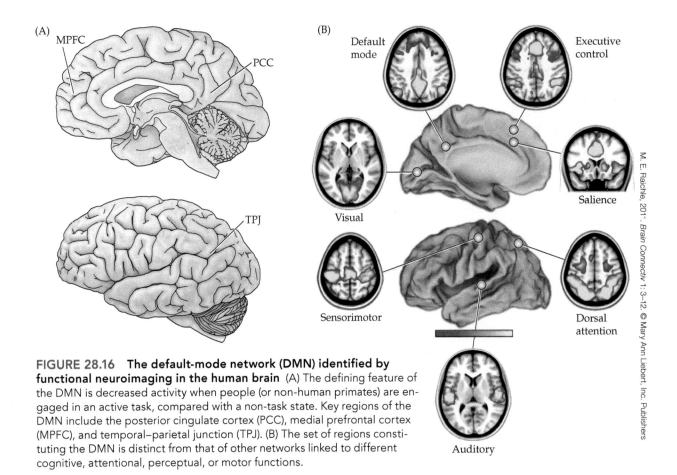

FIGURE 28.16 The default-mode network (DMN) identified by functional neuroimaging in the human brain (A) The defining feature of the DMN is decreased activity when people (or non-human primates) are engaged in an active task, compared with a non-task state. Key regions of the DMN include the posterior cingulate cortex (PCC), medial prefrontal cortex (MPFC), and temporal–parietal junction (TPJ). (B) The set of regions constituting the DMN is distinct from that of other networks linked to different cognitive, attentional, perceptual, or motor functions.

Labels in figure: (A) MPFC, PCC, TPJ. (B) Default mode, Executive control, Visual, Salience, Sensorimotor, Dorsal attention, Auditory. Source credit: M. E. Raichle, 201- . *Brain Connectiv* 1: 3–12. © Mary Ann Liebert, Inc. Publishers

Summary

The most obvious cortical states are sleep and wakefulness. Nearly all animals exhibit a restorative cycle of rest following daily activity, but only some animals (e.g., mammals and birds) organize the period of rest into distinct phases of non-REM and REM sleep. Why we need a restorative phase of suspended consciousness accompanied by decreased metabolism and lowered body temperature remains incompletely understood, although recent research has identified important brain functions (e.g., consolidation of information in memory) that take place during sleep. Even more mysterious is why the human brain is periodically active during sleep at levels not appreciably different from those of the waking state (that is, the neural activity during REM sleep). A complex physiological interplay involving the brainstem, thalamus, and cortex controls the degree of mental alertness on a continuum from deep sleep to wakeful attentiveness. A circadian clock located in the suprachiasmatic nucleus of the hypothalamus in turn influences these systems, adjusting cortical and other physiological states to appropriate durations during the 24-hour cycle of light and darkness that is fundamental to life on Earth. Recent research seeks to understand the puzzling question of why large regions of the cortex are more active at rest than when an individual is performing a task.

■ Additional Reading

Reviews

Brown, R. E. and 4 others (2012) Control of sleep and wakefulness. *Physiol. Rev.* 92 (3): 1087–1187. doi:10.1152/physrev.00032.2011

Green, C., J. Takahashi and J. Bass (2008) The meter of metabolism. *Cell* 134: 727–742.

Krueger, J. M., M. G. Frank, J. Wisor and S. Roy (2016) Sleep function: Toward elucidating an enigma. *Sleep Med. Rev.* 28: 46–54.

McCormick, D. A. (1992) Neurotransmitter actions in the thalamus and cerebral cortex. *J. Clin. Neurophysiol.* 9: 212–223.

Raichle, M. E. (2015) The brain's default mode network. *Annu. Rev. Neurosci.* 38: 433–447.

Rees, G., G. Kreiman and C. Koch (2002) Neural correlates of consciousness in humans. *Nat. Rev. Neurosci.* 3: 261–270.

Stoerig, P. and A. Cowey (1997) Blindsight in man and monkey. *Brain* 120: 535–559.

Storm, J. F. and 6 others (2017) Consciousness regained: Disentangling mechanisms, brain systems, and behavioral responses. *J. Neurosci.* 37: 10882–10893.

Weber, F. and Y. Dan (2016) Circuit-based interrogation of sleep control. *Nature* 538: 51–59.

Wolman, D. (2012) The split brain: A tale of two halves. *Nature* 483: 260–263.

Important Original Papers

Allison, T. and D. V. Cicchetti (1976) Sleep in mammals: Ecological and constitutional correlates. *Science* 194: 732–734.

Aschoff, J. (1965) Circadian rhythms in man. *Science* 148: 1427–1432.

Aserinsky, E. and N. Kleitman (1953) Regularly occurring periods of eye motility, and concomitant phenomena, during sleep. *Science* 118: 273–274.

Cashmore, A. R. (2003) Cryptochromes: Enabling plants and animals to determine circadian time. *Cell* 114: 537–543.

Colwell, C. S. and S. Michel (2003) Sleep and circadian rhythms: Do sleep centers talk back to the clock? *Nat. Neurosci.* 10: 1005–1006.

Czeisler, C. A. and 11 others (1999) Stability, precision, and near-24-hour period of the human circadian pacemaker. *Science* 274: 2177–2181.

Dueker, L. and 8 others (2013) Memory consolidation by replay of stimulus-specific neural activity. *J. Neurosci.* 33: 19373–19383.

Dunlap, J. C. (1993) Genetic analysis of circadian clocks. *Annu. Rev. Physiol.* 55: 683–727.

Fox, M. D. and 5 others (2005) The human brain is intrinsically organized into dynamic, anticorrelated function networks. *Proc. Natl. Acad. Sci. U.S.A.* 102: 9673–9678.

Hayden, B. Y., D. V. Smith and M. E. Platt (2009) Electrophysiological correlates of default-mode in macaque posterior cingulate cortex. *Proc. Natl. Acad. Sci. U.S.A.* 106: 5948–5953.

Hobson, J. A., R. Strickgold and E. F. Pace-Schott (1998) The neuropsychology of REM sleep and dreaming. *Neuroreport* 9: R1–R14.

Jovanovic, U. J. (1971) The recording of physiological evidence of genital arousal in human males and females. *Arch. Sex. Behav.* 1: 309–320.

King, D. P. and J. S. Takahashi (2000) Molecular mechanism of circadian rhythms in mammals. *Annu. Rev. Neurosci.* 23: 713–742.

Lu, J., D. Sherman, M. Devor and C. B. Saper (2006) A putative flip-flop switch for control of REM sleep. *Nature* 441: 589–594.

Magoun, H. W. (1952) An ascending reticular activating system in the brain stem. *AMA Arch. Neurol.* 67: 145–154.

McCormick D. A. and H. C. Pape (1990) Properties of a hyperpolarization-activated cation current and its role in rhythmic oscillation in thalamic relay neurones. *J. Physiol.* 431: 291–318.

Moruzzi, G. and H. W. Magoun (1949) Brain stem reticular formation and activation of the EEG. *Electroencephalogr. Clin. Neurophysiol.* 1: 455–473.

Okamura, H. and 8 others (1999) Photic induction of *mPer1* and *mPer2* in *Cry*-deficient mice lacking a biological clock. *Science* 286: 2531–2534.

Owen, A. M. and 5 others (2006) Detecting awareness in the vegetative state. *Science* 313: 1402.

Provencio, I. and 5 others (2000) A novel human opsin in the inner retina. *J. Neurosci.* 20: 600–605.

Raichle, M. E. and 5 others (2001) A default mode of brain function. *Proc. Natl. Acad. Sci. U.S.A.* 98: 676–682.

Shearman, L. P. and 10 others (2000) Interacting molecular loops in the mammalian circadian clock. *Science* 278: 1013–1019.

Steriade, M., D. A. McCormick and T. J. Sejnowski (1993) Thalamocortical oscillations in the sleeping and aroused brain. *Science* 262: 679–685.

Utevsky, A. V., D. V. Smith and S. A. Huettel (2014) Precuneus is a functional core of the default-mode network. *J. Neurosci.* 34: 932–940.

Vitaterna, M. H. and 9 others (1994) Mutagenesis and mapping of a mouse gene, *clock*, essential for circadian behavior. *Science* 264: 719–725.

Xie, L. and 12 others (2013) Sleep drives metabolite clearance from the adult brain. *Science* 342: 373–377. doi: 10.1126/science.1241224

Books

Hobson, J. A. (1989) *Sleep*. New York: Scientific American Library.

Hobson, J. A. (2002) *Dreaming*. New York: Oxford University Press.

McNamara, P., R. A. Barton and C. L. Nunn (2010) *Evolution of Sleep: Phylogenetic and Functional Perspectives*. Cambridge: Cambridge University Press.

Weiskrantz, L. (1986) *Blindsight: A Case Study and Its Implications*. Oxford: Oxford University Press.

Attention

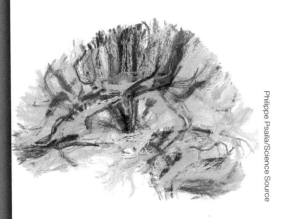

KEY CONCEPTS

29.1 Attention prioritizes some stimuli over others

29.2 Attention alters activity in brain regions associated with perception

29.3 Damage to key brain regions can disrupt attentional processes

29.4 A frontal-parietal network supports the allocation of attention

Overview

Attention is the cognitive function that focuses on some external or internal stimulus, prioritizing its processing over that of other stimuli. This idea can be traced back to William James, who stated:

Everyone knows what attention is. It is the taking possession by the mind, in clear and vivid form, of one out of what seem several simultaneously possible objects or trains of thought. ... It implies withdrawal from some things in order to deal effectively with others...

James, 1890 (*The Principles of Psychology*, pp. 403–404)

The mechanisms of attention allow humans and other organisms to focus on particular aspects of the flood of information available in the internal and external environments. This definition both connects attention to and distinguishes it from other cognitive processes considered in this unit. For example, as described in Chapter 28, wakefulness contains a continuum of brain states that range from inattentiveness to a fully alert and aroused state when one is directing attention to external stimuli. Attention similarly can be evoked unconsciously or consciously, but it differs from more general processes of arousal in that it focuses processing on specific stimuli. Attention also overlaps with executive functions (see Chapter 33) in that both support goal-directed behaviors through processes such as inhibition or selection. However, the targets of those processes differ, with attention targeting external stimuli or their internal representations and executive functions targeting rules for thought or action. Although early studies of attention were limited to behavioral measures such as reaction time and processing accuracy, or to investigating the consequences of brain damage on selective attention, over the last few decades neuroscientists have directly measured the influence of attention on brain activity. This chapter reviews the phenomenology of attention, its effects on sensory systems, and the neural systems that support its deployment.

CONCEPT **29.1** | **Attention Prioritizes Some Stimuli over Others**

LEARNING OBJECTIVES

29.1.1 Understand the distinction between exogenous and endogenous forms of attention.

29.1.2 Explain how covert attention differs from overt attention.

29.1.3 Define supramodal attention and distinguish its effects from those of unimodal attention.

Selective attention

Attentional processes allocate neural resources to prioritize the analysis of some information (e.g., the face of a friend) at the expense of other information (e.g., the surrounding visual environment). Such processes are often described as **selective attention**, as if the brain selects some stimuli for preferred processing. A classic demonstration of selective attention, experienced by everyone in daily life, has been evocatively labeled the *cocktail party effect*. Imagine you're at a crowded party where many small groups of people are carrying on conversations simultaneously. All of those conversations combine into a single pressure wave in the surrounding air, resulting in an extraordinarily complex signal for your auditory system to decode. Despite the cacophony, you can still focus on a single conversation—such as with a friend in front of you—and ignore the other conversations. Yet some particularly salient stimulus (e.g., someone shouting your name from across the room) could break through your focus and pull your attention away from that conversation.

Experiments that demonstrate the cocktail party effect typically present a person with different recorded messages in each ear at the same time (Figure 29.1). If that person is instructed to repeat one of the two messages, they can do so accurately—showing that they can attend selectively to one auditory stream and ignore the other. Remarkably, the nature of the unattended auditory message can change dramatically without a person always noticing. For example, the message in the unattended ear can change to a different story or switch from one language to another or even change from forward to backward speech (i.e., reversing the auditory signal); in all those cases, people typically do not notice the change. Other sorts of changes to the unattended message are usually noticed, such as a change in the gender of the speaker, a transition from speech to music, or the sudden onset of an alerting signal such as an alarm or hearing one's name called. From these and related studies, researchers concluded that attentional mechanisms filter unattended information at both relatively low levels (e.g., an entire auditory stream presented to one ear) and at relatively high levels (e.g., based on the content of information, such as one's name). In all cases, though, it is clear that attention has a selective role of prioritizing some information over other information based on the demands of the current task, the properties of the sensory environment, and one's current goals.

Endogenous versus exogenous attention

Voluntarily attending to a particular aspect of the environment, such as to an individual voice or a location in visual space, is called **endogenous attention**. In real-world settings such as a cocktail party, endogenous attention results from our goals: We want to hear what our friends say, and do not want to be distracted by other conversations. In

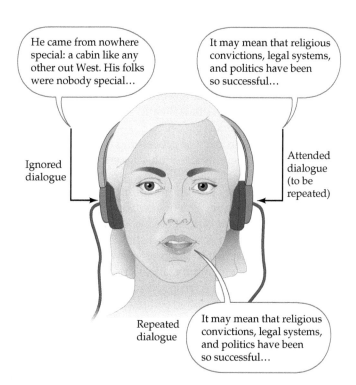

FIGURE 29.1   Attention as selective filtering An experiment using two voices speaking different dialogues presented separately to the left and right ears. See text for explanation.

laboratory experiments, however, researchers induce endogenous attention through explicit instructions (or implicit learning) about which stimuli are deserving of attention. In the Posner cueing paradigm (Figure 29.2A), individuals first maintain visual fixation on a central point in a visual display, and then an arrow appears on the screen (the *cue*) to indicate where an upcoming stimulus is most likely to occur (the *target*). In most trials (e.g., 80%), the target appears at the predicted location. When the target appears, individuals must perform either a detection task (e.g., indicating the side of the screen where the target was presented) or a discrimination task (e.g., indicating whether the target was a circle or an oval). To optimize their performance on the task, individuals typically shift the focus of their attention to the anticipated location of the target. Figure 29.2B shows the typical behavioral results for a detection task: Individuals respond faster to targets appearing at the cued location ("validly cued" trials) than those not appearing at the cued location ("invalidly cued" trials). In other versions of the paradigm, individuals show improved performance on discrimination tasks—such as distinguishing the properties of targets—when the targets appear at the cued location. This and related paradigms allow researchers to quantify the benefits of successfully shifting attention and the costs of attending to the wrong location.

In contrast, **exogenous attention** refers to the situation in which an unexpected noise, flash of light, movement, or other salient stimulus causes a shift away from a previous object of focus to that new stimulus. The shift in attention facilitates processing of information related to the unexpected stimulus—for example, an improved ability to identify a person's visual features when we orient to the location of their voice—while diminishing the efficacy of processing elsewhere. Like endogenous attention, exogenous attention has been studied in a variety of behavioral experiments. One approach has used trial-by-trial cueing in which a sensory cue such as a flash of light is presented at a particular location shortly before a target stimulus is presented either in that location or elsewhere (Figure 29.3). In such circumstances, individuals are again quicker to respond to a target presented in the cued location compared with an uncued location.

Endogenous and exogenous attention have important functional differences. In endogenous cueing, information about the likelihood that the target stimulus will occur in the cued location is provided by prior knowledge (e.g., being informed where the target is likely to occur). In contrast, exogenous attention is not driven by any explicit information about a likely target location. That is, even if an exogenous cue (e.g., a flash) is presented randomly in the two possible locations from trial to trial, and thus has no predictive value about where a target will occur, that cue nonetheless facilitates processing of targets presented at the same location, presumably because the cue automatically draws attention to its location.

Endogenous and exogenous attention also differ in the time courses of their influence on target processing. For endogenously cued attention, the improved processing of a cued target begins about 300 ms after the cue and can last for some seconds afterward, or longer if individuals

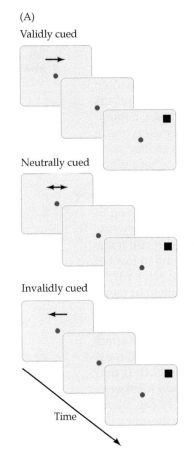

(A)
Validly cued

Neutrally cued

Invalidly cued

Time

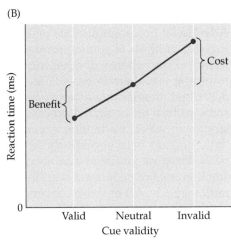

FIGURE 29.2 A cueing paradigm for studying endogenous visual spatial attention (A) In this paradigm, a centrally presented instructional cue indicates where a target will most likely be presented (validly cued), where it will be less likely to be presented (invalidly cued), and where the cue provides no information as to the likely target location (neutrally cued). (B) Typical results show the benefits and costs in the reaction time for target detection after valid and invalid cueing, relative to the neutral-cue condition. (After M. I. Posner et al., 1980. *J Exp Psychol* 109: 160–174.)

maintain their focus of attention on the instructed location. In contrast, exogenous cueing effects start earlier and are short-lived, beginning as early as 75 ms after the cue and lasting only a few hundred milliseconds or so. Moreover, at still longer intervals (~400–800 ms after the cue), the effect of exogenous cuing tends to reverse, with individuals actually being somewhat slower at responding to targets in the cued location. This *inhibition of return* (see Figure 29.3B) reflects the redeployment of attention to other locations when a target fails to appear within a short time at

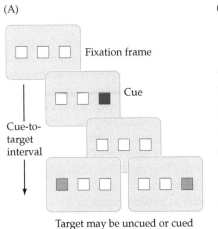

(A)

Fixation frame

Cue

Cue-to-target interval

Target may be uncued or cued

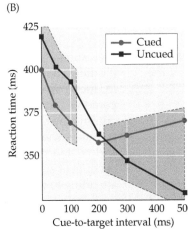

(B)

FIGURE 29.3 Exogenously triggered attention (A) In this paradigm a brief flash is presented in one of two possible target locations, serving as an exogenous cue for a target that might follow at that location or at the other location. The occurrence of a target at the cued versus uncued location is random, with the probability at each location being 50%. (B) Shortly after the exogenous cue (green-shaded time period), stimulus processing at that location is facilitated, as indicated by faster response times to cued relative to uncued targets. At longer intervals (purple-shaded time period), however, there is a decrement in performance for the cued targets, known as *inhibition of return*. (After R. M. Klein, 2000. *Trends Cogn Sci* 4: 138–147; data from M. Posner and Y. Cohen, 1984. In H. Bouma and D. G. Bouwhuis [eds.], *Attention and Performance X*, pp. 531–556. Hillsdale: Erlbaum.)

the cued location. In any event, it is clear that the pattern of effects on behavioral task performance differs between attentional shifts that are triggered endogenously and those that are driven by exogenous factors. One important question emerging from these findings is whether or not different neural systems mediate exogenous and endogenous attention—a topic taken up in more detail in Concept 29.2.

Overt versus covert attention

Another way of categorizing attention is whether it is overt or covert. **Overt attention** involves orienting the head and eyes to a stimulus, thereby aligning the area of maximum visual acuity with the target of perception. **Covert attention** involves directing attention to a stimulus without moving the head or eyes. The Russian psychologist and cyberneticist Alfred Yarbus first quantified overt attention by measuring people's patterns of gaze in response to viewing paintings and sculptures, using an ingenious system of small mirrors that were attached to the eyes and that redirected light to photo-tracing paper. He found that, in the absence of any instructions, people tended to look at the faces and eyes of individuals in the artwork. But when instructed to ascertain the ages or wealth of individuals depicted in a painting, people's gaze patterns shifted to focus on the individuals' bodies and clothing (i.e., information important for solving the task). Modern researchers use eye-tracking systems based on infrared cameras to monitor the position of gaze while people view pictures and make decisions.

An experimental example of covert attention was described by the German physicist and vision scientist Hermann von Helmholtz at the end of the nineteenth century (Figure 29.4). When Helmholtz briefly flashed arrays of letters on a screen and asked people to report the letter appearing at a particular location, he observed that if a person steadily fixated gaze on a particular point in the visual field but directed attention to another region of the field (that is, without moving the eyes), then the stimuli presented in the attended location were reported more accurately than were stimuli in the rest of the field. These and many related findings have established that attention to particular aspects of the environment—whether directed overtly or covertly—generally leads to improved processing of the attended stimuli, typically at the expense of the processing of other, simultaneously presented information.

Unimodal versus supramodal attention

The examples considered so far in this chapter reflect **unimodal attention**, the enhancement of processing within a single sensory modality. However, attentional processes can also act across sensory modalities, providing evidence for what is often called **supramodal attention**. When stimuli from two different modalities occur close together in time, attention to the stimulus in one of the modalities will tend to encompass concurrently occurring stimulation in

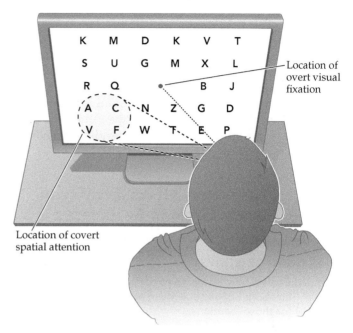

FIGURE 29.4   Studying visual spatial attention Selective attention to a subset of a visual scene enhances the processing of information from the attended portion at the expense of processing information from the rest of the scene. In the original experiment by Hermann von Helmholtz, individuals were briefly presented with an array of letters and asked afterward to recall the letters they had seen. Helmholtz observed that if a person was asked to covertly attend to a certain area of the visual field away from fixation, then the items in the attended portion of the letter array could be accurately reported, whereas items in unattended locations could not be.

another modality. This spreading of attention helps link simultaneously occurring stimulation from different modalities into a multisensory object, such as a barking dog, the look and voice of a friend, the odor and appearance of a pizza, or pretty much any complex event one can imagine.

Laboratory studies of supramodal attention typically take advantage of cross-modality congruence in properties of events, for example, by presenting visual and auditory stimuli at the same spatial location. Studies of event-related potentials (ERPs; see Concept 29.2) recorded from the scalp have shown that the electrophysiological responses elicited by auditory stimuli are enhanced when they occur in a visually attended location, even when they are task-irrelevant. Correspondingly, ERP responses to task-irrelevant visual stimuli are enhanced when they occur in a location being attended for auditory stimuli. Similar results are observed between the visual and tactile modalities, as well as between tactile and auditory modalities. Complementary studies using fMRI have indicated that these enhanced responses to stimuli in the task-irrelevant modality include increased activity at relatively low-level processing areas in the sensory cortices. Moreover, regions of frontal and parietal cortex linked to the control of spatial shifts in attention (see discussions in Concept 29.4) are similarly engaged regardless of whether the triggering cue is presented in the visual or auditory modality. Supramodal attention provides considerable biological value: If a particular spatial location carries important information via one sensory modality, stimulus information from another modality arising from the same location is also likely to be important for understanding the nature of the relevant object or event.

<table>
<tr><td>CONCEPT
29.2</td><td>## Attention Alters Activity in Brain Regions Associated with Perception</td></tr>
</table>

LEARNING OBJECTIVES

29.2.1 Explain how attention to particular spatial locations shapes activity in neurons selective for those locations.

29.2.2 Define event-related potentials and explain their value for studies of attention.

29.2.3 Describe how attention can be selective for stimulus features.

Effects of attention on sensory systems

The improvements in processing associated with attention (e.g., better detection of a visual target when it is presented in an attended spatial location) result from short-term changes in the activity of brain systems that support perception and other functions. To identify attentional effects on perception, neuroscientists take advantage of known features of brain organization, such as the mapping of different locations of visual space to different parts of visual cortex or the specialization of different neurons or brain regions for different categories of objects. Much of this research has been done with non-human primates.

Attention to spatial locations

Most research on spatial attention has examined the processing of visual stimuli, as we will consider in the examples that follow. However, similar effects have been observed for auditory stimuli, suggesting a more domain-general role for attention. Visual cortical neurons are typically spatially selective: They respond strongly only if a stimulus is presented within the cell's receptive field. The firing rate signals the optimal stimulus for that neuron (i.e., the characteristics to which the cell is tuned, such as a particular orientation, direction of movement, color, etc.). Once a cortical neuron is located and its receptive field is characterized, an animal's attention can be manipulated to investigate its effects on neuronal responsiveness. In studies with monkeys, when effective and ineffective stimuli (i.e., stimuli that matched or did not match the neuron's tuning curve) were presented together within a neuron's receptive field in visual area V4, the cell

fired strongly only when the effective stimulus was being attended. When a monkey attended to the ineffective stimulus, the neuron responded weakly, even though the visual input had not changed (Figure 29.5). These observations indicate that neuronal responses depend on the locus of attention within a neuron's receptive field, at least for cells in these areas of cortex. In the later stages of visual processing—that is, in the ventral pathway leading to the inferior temporal cortex—attention modulated the neuronal responses even if the ignored stimulus was relatively far away from the attended one, presumably because the receptive fields in this region are much larger.

Neuroscience techniques that have high temporal resolution, such as electroencephalography (EEG), can provide critical insights into attentional mechanisms. The presentation of a task-relevant visual stimulus (e.g., a shape flashed on a display) typically evokes a series of changes in the

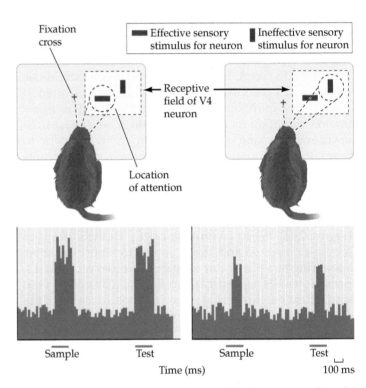

FIGURE 29.5 **Effects of attention on the firing rates of single neurons in the visual cortex** At the attended location (circled), two stimuli—sample and test—were presented sequentially; the monkey had to discriminate whether they were the same or different. Irrelevant stimuli were presented simultaneously with the sample and test but at a separate location in the receptive field. Stimuli could either be effective stimuli for the neuron (red bars in this example) or ineffective stimuli (purple bars). When both an effective stimulus and an ineffective stimulus were presented within the receptive field and the monkey attended to the effective stimulus, the neural responses were robust. When the monkey attended to the ineffective stimulus, however, the responses were much reduced, despite the presence of an effective stimulus in the receptive field. In short, the locus of attention has a clear effect on the activity of the relevant neurons. (After J. Moran and R. Desimone, 1985. *Science* 229: 782–784.)

electrical potential measured by EEG (Figure 29.6A); such changes are called **event-related potentials (ERPs)**. As examples, relatively rapid ERPs such as the P1 (so called because it is the first positive-polarity ERP in such tasks) have been linked to activity in early visual regions (e.g., V1–V4), while slightly slower ERPs such as the N1 have been associated with activity in extrastriate and parietal regions. By presenting those visual stimuli either inside or outside the focus of visual attention (Figure 29.6B), the effects of attention on the associated ERPs can be measured. In general, focused covert attention tends to increase the amplitude of these ERPs (Figure 29.6C), consistent with the results from

measurements of individual neuron firing rates shown in the previous figure. Neuroscientists can use modulations of ERPs (or lack thereof) as a marker for attentional effects—a technique with considerable importance for both basic science and clinical applications.

Our sensory systems are sensitive to other perceptual features besides stimulus location, and attention has been shown to modulate the processing of those features as well. Another non-human primate study assessed how the locus of spatial attention affects the orientation tuning curves of visual neurons after training monkeys to attend to one of two gratings. When a monkey attended

(A)

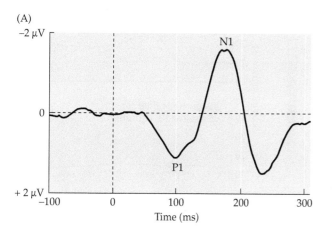

FIGURE 29.6 Effects of visual spatial attention on ERP responses to visual stimuli (A) Recorded over the left occipital lobe to a light flash in the right visual hemifield, this ERP shows the visual P1 and N1 components peaking at 100 and 180 ms, respectively. (B) In this paradigm for investigating the effects of visual spatial attention, the individual is attending to a location either in the right visual field or in the left visual field; they respond to unilaterally presented visual flash stimuli on that side and ignore stimuli on the other side. (C) Representative ERPs elicited by right-field stimuli when attended versus unattended are shown here, along with the corresponding topographical distributions on the scalp at the latency of the P1 peak. Attention enhances the amplitude of the sensory P1 component, with little change in waveform or scalp distribution. Such an effect is consistent with the conclusion that attention induces a gain enhancement in the responses to stimuli occurring in an attended region of space. μV = microvolts; by convention, negative-polarity changes in the ERP signal are indicated as upward deflections on the y-axis.

(B)

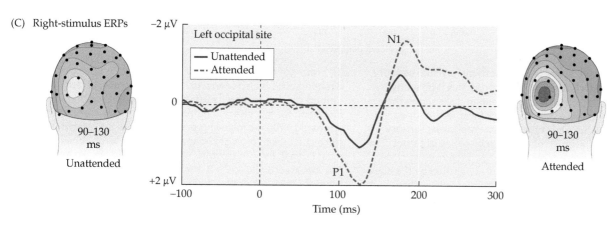

(C) Right-stimulus ERPs

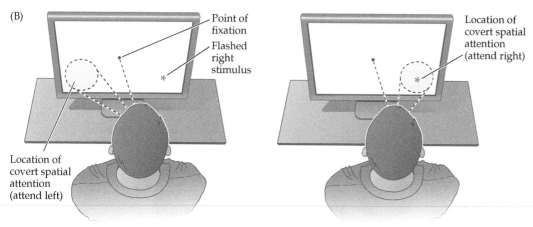

to the stimuli in the receptive field of the recorded neuron, responses were enhanced, as expected. Using this paradigm, however, allowed evaluation of how attention affected responses to gratings of different orientations. Although attention enhanced neural responses at all orientations, the effects were stronger for a neuron's preferred orientation (see Chapter 12).

Attention can also modulate activity in sensory regions in advance of stimulus presentation, as when we shift attention to a region of visual space in anticipation that a target will soon be presented there. Experimental studies investigate this preparatory activity by separating initial attention-directing cues from subsequent targets that are likely to be presented in the cued locations. Presentation of an initial cue increases activity in regions of visual cortex that are contralateral to the cued location. That is, cueing the individual to attend to the right side of visual space evokes activity in visual cortical regions in the left hemisphere—even before the target appears (Figure 29.7A). This preparatory activity matches well with the activity evoked

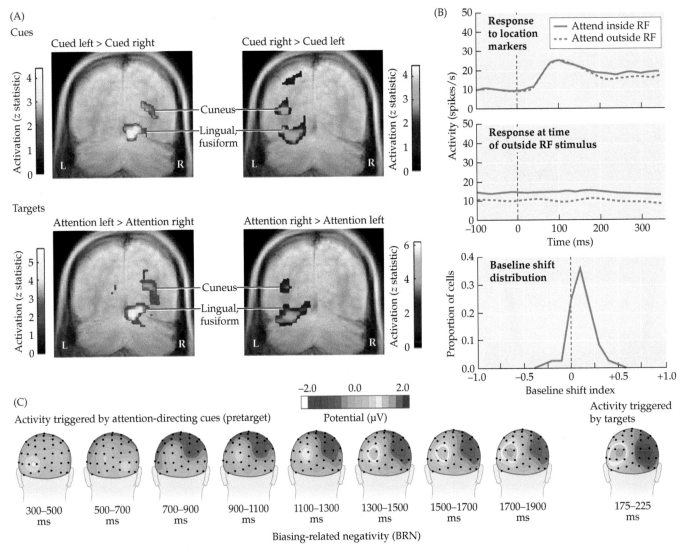

FIGURE 29.7 **Preparatory biasing activity in visual sensory cortex during spatial attention tasks** (A) This fMRI study of humans performing visual spatial attention tasks shows that engaging the attentional control network by attention-directing cues leads to enhanced activity in contralateral visual sensory cortex prior to—and even in the absence of—a visual target. The sensory cortex activity elicited by the targets themselves closely corresponds to the pretarget biasing activity elicited by the cues. (B) Single-unit recordings in awake, behaving monkeys show that attention to a location in space increases the background firing of V4 neurons that have receptive fields (RFs) in that location.

(C) Pretarget biasing of sensory cortex triggered by attention-directing cues is also seen with ERPs (a negative-polarity wave termed *biasing-related negativity*), which also provide timing information for this effect in humans. In this experiment, the target appeared 1900 ms following the onset of the cue; shown are times since cue onset (left eight images) and since target onset (rightmost image). μV = microvolts. (A from J. B. Hopfinger et al., 2000. *Nat Neurosci* 3: 284–291; B after S. J. Luck et al., 1997. *J Neurophysiol* 77: 24–42; C after T. Grent-'t-Jong and M. G. Woldorff, 2007. *PLOS Biol* 5: 114–126, CC BY 4.0.)

by the later target, and trials with larger preparatory activity tend to be associated with better performance (e.g., more rapid detection of targets). Neuroscience methods with better temporal resolution (such as EEG) reveal how preparatory activity builds following attention-directing cues (Figure 29.7B). Following an initial cue that directs attention to the right side of visual space, EEG signal amplitude builds over 1 to 2 sec until there is a strong biasing of activity toward right occipital cortex. That preparatory response bias again mirrors what is observed for targets themselves, as seen in the scalp distribution of the relatively rapid (100–200 ms latency) N1 ERP (Figure 29.7C).

Attention to features and objects

Many stimuli attract attention not because of their spatial locations but because of their identity (e.g., listening for one's name to be called, looking for a friend's face in a crowd). For such stimuli, attention helps prioritize particular stimulus features over other features that could be experienced but that are not as critical for the current task. Using single-unit recording approaches in monkeys and neuroimaging techniques in humans, neuroscientists have shown that attention to a visual feature such as color or motion increases activity in brain regions associated with that feature. For example, attending to a pattern of upward moving dots in one specific spatial location increases the firing rates of neurons that are sensitive to similar motion in other spatial locations. Why might the effects of attention spread to similar

features at different spatial locations in this manner? One possibility is that feature-based enhancements of neuronal activity could be particularly valuable for facilitating **visual search**—the process of finding a target within a complex scene of distracting information (e.g., seeking a particular item on a supermarket shelf).

Attention can also be directed at entire categories of stimuli. As discussed in Chapter 12, the visual system contains both regions whose neurons track relatively low-level properties of the visual display (e.g., lines, movement) and regions whose neurons are selective for particular stimulus categories (e.g., faces, scenes, places) and exemplars of those categories. Because some of the category-selective regions can be spatially distinguished from each other—such as the fusiform face area (FFA) versus the parahippocampal place area (PPA)—neuroimaging methods can readily distinguish large-scale effects of attention. Common paradigms involve presenting two such stimulus categories (e.g., faces and scenes) either as overlapping, partially transparent images or in an alternating sequence, and asking individuals to selectively attend to one category of objects and ignore the other (Figure 29.8). Object-selective attention increases activation in brain regions whose neurons show selectivity for that category (e.g., attending to faces increases activation in the FFA), and decreases activation in the ignored category to a level below that of passive viewing (e.g., attending to faces inhibits activation in the PPA). Consistent with these broad

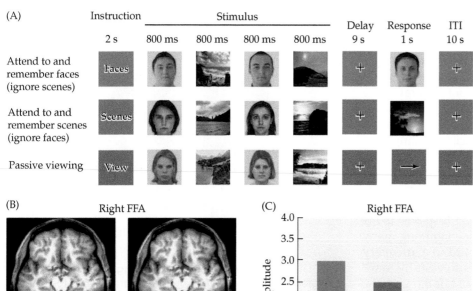

FIGURE 29.8 Attention can be selective to object categories (A) Depending on the condition, individuals were either instructed to attend to faces and ignore the scenes, attend to scenes and ignore faces, or passively view all images. They then viewed a rapidly presented series of faces and scenes (each 800 ms in duration) and, following a brief delay, judged whether a new image had been previously presented. (In the passive viewing condition, they just pressed a button consistent with the direction of an arrow.) After an inter-trial interval (ITI), the next trial began. (B,C) Activity in the fusiform face area (FFA), a brain region whose activity increases when individuals look at faces, was greater when the faces were being attended to than when they were being ignored—even though the physical stimuli were similar in the two conditions. (After A. Gazzaley, et al., 2005. *J Cogn Neurosci* 17: 507–517.)

neural effects, attention to one stimulus category improves the specificity of neuronal responses to its exemplars—and in turn improves people's ability to identify and respond to those stimuli from that category.

CONCEPT 29.3 | Damage to Key Brain Regions Can Disrupt Attentional Processes

LEARNING OBJECTIVES

29.3.1 Describe hemispatial neglect syndrome and the lesions that can cause it.

29.3.2 Provide two examples of lesions that can impair attentional control processes.

Disruptions in the control of attention caused by brain lesions

The previous concepts considered how attention can modulate the processing of stimuli, leading to faster responses or more accurate discriminations, and the accompanying changes in brain function within sensory cortices. In this and the next concept, we discuss research on brain mechanisms that support the control of attentional processes, enabling their deployment when needed to support effective behavior. Much of this research originated from observations about disorders of attention in people with brain damage; collectively, this work revealed that attentional control can be selectively impaired even if processes of sensation and perception remain largely intact (Clinical Applications).

Hemispatial neglect syndrome

Lesions to the right inferior parietal lobe and adjacent regions can impair attention to the left side of personal and extrapersonal space (i.e., the side contralateral to the lesion)—a collection of deficits called **hemispatial (or contralateral) neglect syndrome** (Figure 29.9). Individuals with hemispatial neglect show clear deficits orienting their attention to stimuli presented on the left side of visual space. For example, in the single-line bisection test (Figure 29.10A), individuals are asked to mark the center of a horizontal line. Individuals with neglect tend to ignore the left side of the line, and thus their estimate of the center is displaced to the right. If they are given a piece of paper containing many shapes (e.g., line segments) and are asked to mark every shape, individuals with neglect may successfully mark shapes on the right side of the paper but not those on the left (Figure 29.10B). This behavior occurs even though the individuals may freely move their eyes anywhere on the page, suggesting that the deficit lies in an inability to direct attention to objects in the neglected part of the visual field.

Research by neuroscientist Marlene Behrmann and her colleagues supports this interpretation: People with neglect begin attentional search tasks by gazing at stimuli within the ipsilesional visual field and much more rarely fixate on stimuli in the contralesional field. In addition, the left-sided neglect is not limited to ignoring objects in the left hemispace; individuals with this syndrome also tend to ignore the left sides of objects *wherever* they are in visual space. For example, if asked to draw a copy of an object, these individuals tend to draw only its right side (Figure 29.10C). Such individuals even tend to ignore the left side of their visual imagery and memory. So if asked to draw a clock from memory, they are likely to draw half a clock, sometimes remembering to include all 12 numbers in the drawing, but placing all of the numbers on the right (Figure 29.10D).

Even more striking evidence that these deficits are attentionally driven came from a clever paradigm in which individuals viewed a barbell shape consisting of two circles connected by a horizontal line and then detected the appearance of a target presented in one of the two circles. Consistent with what might be expected from the previous examples, individuals with hemispatial neglect were much faster and more accurate at detecting targets presented in the right circle. However, on some trials the barbell shape was presented on the screen and then rotated 180 degrees so that its former left side was now on the right side of visual space, and vice versa. Following the object rotation, individuals with hemispatial neglect were much faster and more accurate at detecting targets presented in the *left* circle—as if what they were neglecting was the left side of the object (as established by its initial presentation) and not the left side of visual space.

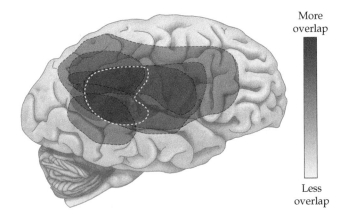

FIGURE 29.9 Cortical lesions leading to left hemispatial neglect syndrome This composite diagram shows the distribution of right-hemisphere damage in eight individuals with left hemispatial neglect. The degree of overlap of damaged brain areas across individuals is indicated by shading level. Although some of the lesions include parietal and frontal lobes, as well as parts of the temporal lobe, the region most commonly affected is in the right inferior parietal lobe (dashed line). (After K. M. Heilman et al., 1985. In *Clinical Neuropsychology*, 2nd ed., K. M. Heilman and E. Valenstein. New York: Oxford University Press.)

(A) "Bisect the line"

(B) "Cancel the lines"

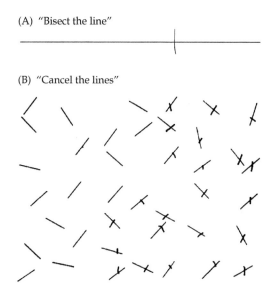

(C) "Copy this picture of a house"

Model   Patient's copy

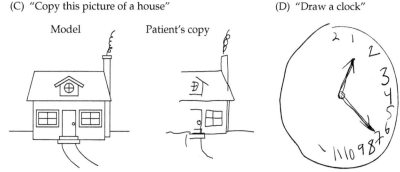

(D) "Draw a clock"

FIGURE 29.10 Clinical tests of left hemispatial neglect caused by damage to the right inferior parietal lobe The performances in the single-line bisection test (A) and the line cancellation test (B) shown here are characteristic of people with hemispatial neglect. (C) An example of a visual copying task as performed by an individual with hemispatial neglect. (D) A drawing of a clock face from memory by an individual with hemispatial neglect. (A,B from G. J. Luvizutto, 2020. Clinics [Sao Paulo, Brazil], 75, e1468. CC BY 4.0; C from V. W. Mark. 2003. *Front Biosci* 8: e172–189; D from P. Chen and K. M. Goedert, 2012. *J Neuropsych* 6: 270–289.)

■ Clinical Applications

Balint's Syndrome

A brain lesion that has striking effects on attention is bilateral damage to the dorsal posterior parietal and lateral occipital cortex, leading to a disorder known as **Balint's syndrome** (Figure A). Damage of this sort presents a distinct clinical picture and a very debilitating deficit.

First characterized by the Hungarian physician Rezsö Bálint, this syndrome has three defining characteristics: (1) simultanagnosia, the inability to attend to or perceive more than one visual object at a time; (2) optic ataxia, the impaired ability to reach for or point to an object in space under

visual guidance; and (3) oculomotor apraxia, difficulty voluntarily directing eye gaze toward objects in the visual field. Simultanagnosia is the deficit most closely associated with Balint's syndrome. If the neurologist holds up two different objects and *(Continued)*

(A)   Lateral views   Coronal MRI

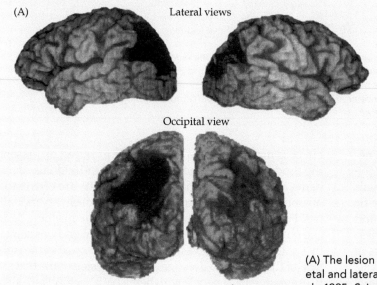

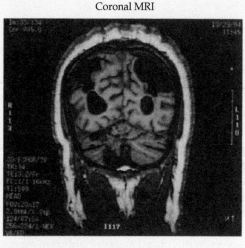

(A) The lesion in Balint's syndrome is typically located in posterior parietal and lateral occipital cortex bilaterally. (From S. R. Friedman-Hill et al., 1995. *Science* 269: 853–855.)

■ Clinical Applications (continued)

asks individuals with this syndrome what they see, they report seeing only one object or the other but not both, even if the objects are right next to each other. Unlike in hemispatial neglect syndrome, the relative positions of the objects within an individual's visual field do not matter. If the unseen object is jiggled to attract attention, individuals will then say they see it, but they will have lost the perception of the first object. Moreover, when people with Balint's syndrome are presented with an array of randomly distributed objects, half of which are one color and half another color, they report seeing just one color or the other, but not both (Figure B). But if the differently colored items are attached so that each object contains both colors, individuals then report seeing both colors in the array. Similarly, these individuals have trouble perceiving and comparing the lengths of two nearby rectangular bars unless they are connected as parts of the same overall object (Figure C). Thus, people with Balint's syndrome can attend to more than one stimulus quality or stimulus part, but only when the parts are embodied in the same object.

(B) "How many colors do you see?"

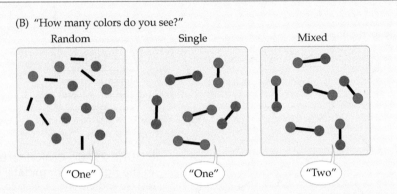

(B) The inability of people with Balint's syndrome to perceive or attend to more than one object at a time (simultanagnosia) prevents them from noticing more than one color in the displays shown in the left and middle panels. However, if the blue and red circles are connected to form single objects (right panel), the individual is able to report both colors, indicating that the deficit lies in the inability to attend to multiple objects rather than in a failure to attend to multiple qualities. (After G. W. Humphreys and M. J. Riddoch, 1993. In D. E. Meyer and S. Kornblum [eds.], *Attention and Performance XIV*, pp. 143–162. Cambridge, MA: The MIT Press.)

(C) "Are the two projections the same height?"

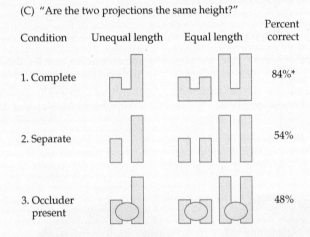

(C) Similarly, if individuals are asked to compare the lengths of two nearby rectangles, their performance is near chance (50% correct) if the objects are separated, whereas they perform much better when the components are connected as part of the same object. (After A. C. Cooper and G. W. Humphreys, 2000. *Neuropsychologia* 38: 723–733.)

The attentional deficits associated with hemispatial neglect are quite different from the perceptual deficits that follow lesions to the visual cortex (see Chapter 12). Individuals with visual cortical lesions are effectively blind in specific corresponding parts of the contralateral visual field—and that blindness is accompanied by a lack of responsiveness from neurons in the visual system (with rare exceptions in conditions such as blindsight; see Chapter 28). In individuals with neglect, however, objects presented to the left visual field can stimulate the visual system normally. Moreover, when an object in the left visual field is specifically pointed

out or is made particularly salient (for example, by presenting food to a hungry individual), some individuals report being able to see it. Although these deficits are often most obvious in vision, they are evident in other sensory modalities. For instance, many people with right hemisphere damage fail to attend to the left side of their own body, as shown by the tendency to shave or apply makeup on only the right side of the face, or to dress only the right side of the body.

Assessments of people with hemispatial neglect also elicit *extinction*. This phenomenon is revealed when the neurologist stands in front of the individual with arms outstretched and moves a finger on either the right or the left hand. If a finger on either side is moved by itself, the individual generally reports the presence of the moving finger correctly, presumably because a moving stimulus is a particularly strong attractor of attention, even for these individuals. If both fingers are moved at the same time, however, the individual typically reports seeing only the one on the right. This test suggests that the normal competition between the stimulus inputs from the two sides is now dominated by the right visual field, which "extinguishes" the input from the left. Extinction emphasizes again that the underlying problem is an attentional deficit, not a sensory one.

A possible reason why attentional deficits are most often associated with right parietal lesions is that this region influences mechanisms of attention in *both* hemispheres, whereas the corresponding left parietal area influences mainly those

on the right. There are, however, alternative explanations, and the anatomy is a good deal more complex than implied here. For instance, some investigators have suggested that the relatively greater importance of right parietal cortex in neglect syndromes reflects right hemispheric lateralization for vigilance or alertness. Interestingly, in non-human primates damage to the inferior parietal lobe on either the left or right side induces neglect of contralateral space.

Impairments of attentional control

Lesions in parts of the frontal cortex that are connected to the parietal cortex can also cause attentional deficits. In particular, lesions to the **frontal eye fields** disrupt the ability both to initiate eye movements to targets in the contralateral visual field and to direct attention toward that side (Box 29A). Moreover, lesions in these frontal regions can interfere

---

■ BOX 29A | Attention and the Frontal Eye Fields

Neuroscientists study the brain mechanisms of attention by examining the effects of attentional control regions on activity in sensory regions—and subsequently on behavior. The frontal eye fields (FEFs) generate saccadic eye movements to locations in visual space that warrant

attention. In the experiment shown here, a stimulating electrode was placed in the FEF in a locus that would evoke saccades to a given location with respect to the center of gaze; a second electrode recorded the activity of a visual cortical neuron responsive to the same location in visual space. The FEF

stimulation while the monkey attended the fixation point caused a saccade to the expected location and increased neuronal activity at the recoding site. The implication is that the FEF plays a key role in an attentional control network through its influence on extrastriate visual cortices.

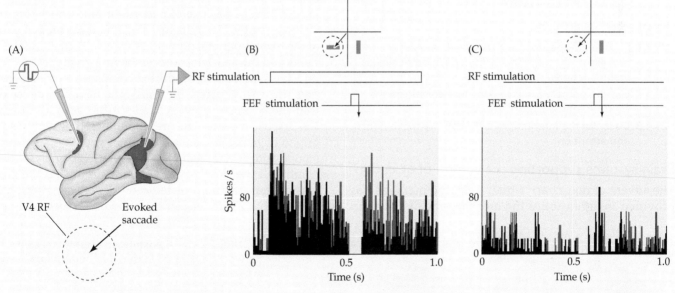

Microstimulation of sites within the FEF, below the threshold for eliciting a saccade, was carried out while the visual stimulus responses of single V4 neurons were recorded in monkeys performing a fixation task. (A) The stimulating electrode was positioned so that suprathreshold stimulation would evoke a saccade into the receptive field (RF) of the V4 cell under study. (B) This example shows the effect of subthreshold FEF micro-

stimulation on the response of a single V4 neuron to an oriented bar presented in the cell's receptive field. The mean response during control trials is shown in black; the enhanced response arising from the FEF microstimulation is shown in red. (C) On trials in which the visual stimulus was presented outside the receptive field of the V4 neuron, no enhancement is seen. (After T. Moore and K. M. Armstrong, 2003. *Nature* 421: 370–373.)

with other complex functions such as task switching and ignoring irrelevant information. As a rule, unilateral frontal lesions tend to have a greater effect on motor-related aspects of attention, compromising the ability to initiate or direct eye or limb movements toward contralateral space.

Brainstem lesions can also affect attentional control. The interactions between the superior colliculi and the parietal cortex are apparent in the so-called **Sprague effect**, in which the hemispatial neglect induced by a parietal lesion can be mostly compensated for by a lesion of the superior colliculus on the other side. The proposed explanation is that the parietal lesion-induced neglect results not from the cortical damage itself, but from an imbalance of activity between the two parietal lobes in attentional control. According to this theory, a lesion of the contralateral colliculus helps restore the appropriate balance, because of its connections to the parietal lobe on the same side. Regardless, this effect underscores an important role for the superior colliculus in attentional control, possibly via functional interactions with parietal cortex. Experimental studies in monkeys have demonstrated that these interactions are mediated by the pulvinar, the thalamic relay that connects the parietal cortex and the superior colliculus.

CONCEPT 29.4 | A Frontal-Parietal Network Supports the Allocation of Attention

LEARNING OBJECTIVES

29.4.1 Explain how neuroscientists can distinguish processes of attentional control from other similar brain functions.

29.4.2 List the primary brain regions that constitute the frontal-parietal control network.

29.4.3 Describe the interactions between eye movements and attention.

Brain systems supporting the control of attention

The advent of noninvasive brain-imaging methods has provided insights about the mechanisms of attention employed by typical individuals. In accord with studies of individuals with neglect, tasks involving attention reliably activate a set of brain regions in the dorsal parietal and dorsolateral frontal cortices that has come to be called the **frontal-parietal control network** (Figure 29.11). This network is activated by both endogenous and exogenous forms of attention, during which it modulates activity in the sensory cortices and other brain regions, resulting in more effective processing of some inputs and a less complete processing of others.

A key challenge in studying the brain mechanisms of attention comes from the need to distinguish effects specific to attention (e.g., moving the focus from one location to another) from other, nonspecific effects (e.g., processing a meaningful cue). And because many interesting aspects of attention occur covertly, without accompanying behavior, neuroscientists employ experimental and data analysis methods that compare attentionally demanding tasks with control tasks that omit key attentional processes. In one study of the frontal-parietal control network, research participants underwent parallel fMRI and EEG studies using the same attentional tasks. On each trial, they received either a cue that directed their attention to the left or right side of visual space ("attend cue") or a cue that instructed them not to shift their attention and just to look at the cue itself ("interpret cue"). By subtracting brain activity associated with the latter from that of the former (i.e., attend minus interpret), the researchers could eliminate activity common to cue evaluation and identify processes specific to attentional shifts (Figure 29.12A). The fMRI results revealed that more medial parts of the frontal-parietal network, specifically the frontal eye fields and intraparietal sulcus, showed greater activation on trials when attention shifted compared to trials when no shift was required. Similarly, the EEG data revealed that the two cue types evoked similar activity during the first 350 ms after cue presentation, followed by a sustained negative response specific to attentional shifts in frontal and parietal regions. Computational analyses that combined the fMRI and EEG data supported the conclusion that attention-related activity began in prefrontal cortex about 200 ms earlier than in parietal cortex (Figure

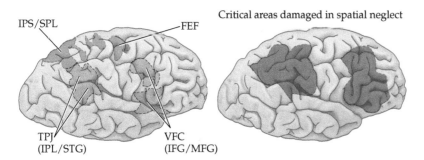

IPS/SPL — FEF

Critical areas damaged in spatial neglect

TPJ (IPL/STG)   VFC (IFG/MFG)

FIGURE 29.11 A postulated attentional control network, illustrated in the right hemisphere Areas in blue indicate the dorsal frontal-parietal regions that tend to be activated by endogenous stimuli; areas in yellow indicate the more ventral regions that tend to be activated during reorienting and by exogenous stimuli. IPS/SPL = intraparietal sulcus/superior parietal lobule; FEF = frontal eye fields; TPJ = temporal–parietal junction; IPL/STG = inferior parietal lobule/superior temporal gyrus; VFC = ventral frontal cortex; IFG/MFG = inferior frontal gyrus/middle frontal gyrus. (After M. Corbetta and G. L. Shulman, 2002. *Nat Rev Neurosci* 3: 201–215.)

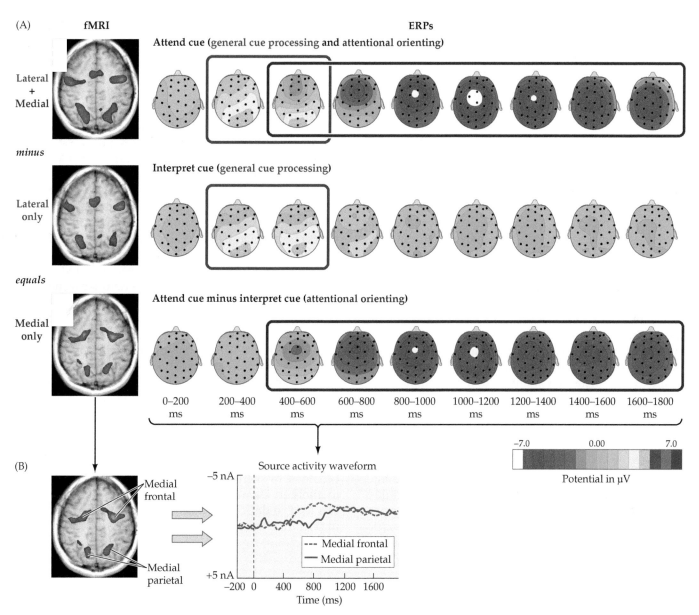

FIGURE 29.12 Temporal dynamics of cue-triggered activity in the frontal-parietal control network In this combined fMRI–ERP study, research participants were given a centrally presented instructional cue to shift attention to the left or right ("attend cue") to detect a possible upcoming target there, or a cue indicating that no shift of attention was required on that trial ("interpret cue"). (A) In the fMRI study, a contrast between the interpret-cue responses (second row) and the attend-cue responses (first row) revealed that the more medial portions of the frontal-parietal cortex were involved in attentional orienting and the more lateral areas with general cue processing (differences shown in the third row).

Corresponding contrasts of the ERP data showed that attend cues and interpret cues elicited similar general cue–processing activity in the first 350 ms, followed by a sustained negative wave over frontal, central, and parietal scalp lasting hundreds of milliseconds that was associated with attend cues only. (B) Using the fMRI activations to facilitate the analyses, source modeling of the ERP orienting-specific control activity showed that the frontal regions of the medial orienting network were activated 200 to 300 ms earlier than the parietal regions. nA = nanoamperes. (fMRIs from M. G. Woldorff et al., 2004. *J Cogn Neurosci* 16: 149–165; and after T. Grent-'t-Jong and M. G. Woldorff, 2007. *PLOS Biol* 5: 114–126.)

29.12B). These and other results have led cognitive neuroscientists to propose that the prefrontal cortex plays a particularly important role in initiating goal-directed shifts in attention, leading to subsequent changes in processing within parietal cortex.

Relationships between eye movements and attention

Other electrophysiological studies of attention in experimental animals have examined neurons in the lateral intraparietal area (LIP) of the posterior parietal cortex, as well as in the frontal eye fields (see Concept 29.3). These two regions of the monkey cortex are assumed to be homologous to parietal and frontal areas in humans where neuroimaging studies have shown activity related to attentional control. These particular areas may serve as integrating centers within a broader set of brain areas involved in attention. For instance, the firing rates of LIP neurons in response to a stimulus in their receptive field are greater when the task is to make a saccade to a target in the receptive field rather than simple fixation. Neuronal responses are also enhanced when a monkey attends to the stimulus in the receptive field but does not make a saccade, or when the saccade is delayed. These results suggest that enhanced neuronal responsiveness is not due to saccade preparation per se, but to the allocation of attention to the spatial location of the target in the neuron's receptive field, as occurs when the monkey plans to shift its gaze there.

A related question is how activity in these regions could lead to enhanced stimulus processing in the sensory cortices. Some relevant information had already been provided by fMRI studies in humans. When research participants direct sustained attention to a particular visual-field location expecting the onset of a visual stimulus there, increased activity is elicited not only in the frontal and parietal cortices but in extrastriate cortex as well. The implication is that the increased activity in visual cortex in the absence of visual stimulation reflects preparatory signals from the frontal-parietal network that favor the attended location. In accord with this idea, microstimulation of the frontal eye fields in monkeys improves performance in attentional tasks and simultaneously increases the activity of neurons in V4 with receptive fields aligned with the retinotopic locus of stimulation. Saccade-related activity in the frontal eye fields has also supported this "premotor theory" of attention, although other interpretations have also been suggested.

Summary

Attention research seeks to understand how processing resources are directed to deal effectively with ever-changing internal and external environments. Endogenous attention refers to the ability to voluntarily direct attention based on one's goals, expectations, or knowledge. Exogenous attention refers to involuntary shifts of attention triggered by salient stimuli in the environment. Both lead to enhanced processing of the information to which attention has been directed. Insight into both the psychological and the neural mechanisms of attention has been greatly advanced in recent years by combining older behavioral approaches with EEG, fMRI, and single-unit recording methods that can evaluate brain activity while humans or other animals are engaged in attentional tasks. Those studies have provided new insights into how directed attention alters activity in sensory cortices, based on both the location and content of the information to which attention is directed. Recent research has identified a frontal-parietal network whose activity is associated with the engagement of attentional processes, as when a stimulus indicates the need to shift attention from one location in space to another. Damage to cortical and subcortical regions can lead to deficits in attentional processing that have important clinical consequences.

■ Additional Reading

Reviews

Buxbaum, L. J. (2006) On the right (and left) track: Twenty years of progress in studying hemispatial neglect. *Cog. Neuropsych.* 23: 184–201. doi: 10.1080/02643290500202698

Corbetta, M. and G. L. Shulman (2002) Control of goal-directed and stimulus-driven attention in the brain. *Nat. Rev. Neurosci.* 3: 201–215.

Driver, J. (2001) A selective review of selective attention research from the past century. *Br. J. Psychol.* 92: 53–78.

Posner, M. I. and S. E. Petersen (1990) The attention system of the human brain. *Annu. Rev. Neurosci.* 13: 25–42.

Scolari, M., K. N. Seidl-Rathkopf and S. Kastner (2015) Functions of the human frontoparietal attention network: Evidence from neuroimaging. *Curr. Opin. Behav. Sci.* 1: 32–39.

Talsma, D., D. Senkowski, S. Soto-Faraco and M. G. Woldorff (2010) The multifaceted interplay between attention and multisensory integration. *Trends Cogn. Sci.* 14: 400–410.

Important Original Papers

Baldauf, D. and R. Desimone (2014) Neural mechanisms of object-based attention. *Science* 344: 424–427.

Behrmann, M., S. Watt, S. E. Black and J. J. S. Barton (1997). Impaired visual search in patients with unilateral neglect: an oculographic analysis. *Neuropsychologia* 35: 1445–1458.

Buschman, T. J. and E. K. Miller (2007) Top-down versus bottom-up control of attention in the prefrontal and posterior parietal cortices. *Science* 315: 1860–1862.

Cerf, M. and 6 others (2010) On-line voluntary control of human temporal lobe neurons. *Nature* 467: 1104–1108.

Cooper, A. A. and G. W. Humphreys (2000) Coding space within but not between objects: Evidence from Balint's syndrome. *Neuropsychologia* 38: 723–733.

Corbetta, M. and 10 others (1998) A common network of functional areas for attention and eye movements. *Neuron* 21: 761–773.

De Weerd, P., M. R. Peralta III, R. Desimone and L. G. Ungerleider (1999) Loss of attentional stimulus selection after extrastriate cortical lesions in macaques. *Nat. Neurosci.* 2: 753–758.

Friedman-Hill, S. R., L. C. Robertson and A. Treisman (1995) Parietal contributions to visual feature binding: Evidence from a patient with bilateral lesions. *Science* 269: 853–855.

Gazzaley, A. and 4 others (2005). Top-down enhancement and suppression of the magnitude and speed of neural activity. *J. Cogn. Neurosci.* 17: 507–517.

Grabowecky, M., L. C. Robertson and A. Treisman (1993) Preattentive processes guide visual search: evidence from patients with unilateral visual neglect. *J. Cogn. Neurosci.* 5: 288–302.

Green, J. J., S. M. Doesburg, L. M. Ward and J. J. McDonald (2011) Electrical neuroimaging of voluntary audiospatial attention: Evidence for a supramodal attention control network. *J. Neurosci.* 31: 3560–3564.

Grent-'t-Jong, T. and M. G. Woldorff (2007) Timing and sequence of brain activity in top-down control of visual-spatial attention. *PLOS: Biology,* https://doi.org/10.1371/journal.pbio.0050012

Heilman, H. and E. Valenstein (1985) *Clinical Neuropsychology,* 2nd Edition. New York: Oxford University Press.

Hillyard, S. A., R. F. Hink, V. L. Schwent and T. W. Picton (1973) Electrical signs of selective attention in the human brain. *Science* 182: 177–180.

Humphreys, G. W. and M. J. Riddoch (1993) Interactions between object and space systems revealed through neuropsychology. In *Attention and Performance,* vol. 14, *Synergies in Experimental Psychology, Artificial Intelligence, and Cognitive Neuroscience,* D. E. Meyer and S. Kornblum (eds.). Cambridge, MA: MIT Press, pp. 143–162.

Kastner, S. and 4 others (1999) Increased activity in human visual cortex during directed attention in the absence of visual stimulation. *Neuron* 22: 751–761.

Klein, R. M. (2000) Inhibition of return. *Trends Cogn. Sci.* 4:138–147.

McAdams, C. J. and J. H. R. Maunsell (1999) Effects of attention on orientation-tuning functions of single neurons in macaque cortical area V4. *J. Neurosci.* 19: 431–441.

Mesulam, M. M. (1981) A cortical network for directed attention and unilateral neglect. *Ann. Neurol.* 10: 309–325.

Moore, T., K. M. Armstrong and M. Fallah (2003) Visuomotor origins of covert spatial attention. *Neuron* 40: 671–683.

Moran, J. and R. Desimone (1985) Selective attention gates visual processing in the extrastriate cortex. *Science* 229: 782–784.

Posner, M. I. and Y. Cohen (1984) Components of visual orienting. In *Attention and Performance,* vol. 10, *Control of Language Processes,* H. Bouma and D. Bouwhuis (eds.). London: Erlbaum, pp. 531–556.

Posner, M. I., C. R. R. Snyder and B. J. Davidson (1980) Attention and the detection of signals. *J. Exp. Psychol. Gen.* 59: 160–174.

Ptak, R. and A. Schnider (2010) The dorsal attention network mediates orienting toward behaviorally relevant stimuli in spatial neglect. *J. Neurosci.* 30: 12557–12565.

Reynolds, J. H., T. Pasternak and R. Desimone (2000) Attention increases sensitivity of V4 neurons. *Neuron* 26: 703–714.

Smith, D. V., J. A. Clithero, C. Rorden and H.-O. Karnath (2013) Decoding the anatomical network of spatial attention. *Proc. Natl. Acad. Sci. U.S.A.* 110: 1518–1523.

Thompson, K. G., K. L. Biscoe and T. R. Sato (2005) Neuronal basis of covert spatial attention in the frontal eye field. *J. Neurosci.* 25: 9479–9487.

Tipper, S. P. and M. Behrmann (1996) Object-centered not scene-based visual neglect. *J. Exp. Psychol.: Hum. Percept. Perform.* 22: 1261–1278.

Treisman, A. and G. Gelade (1980) A feature integration theory of attention. *Cogn. Psychol.* 12: 97–136.

Woodman, G. F. and S. J. Luck (1999) Electrophysiological measurement of rapid shifts of attention during visual search. *Nature* 400: 867–869.

Books

James, W. (1890) *The Principles of Psychology.* New York: Henry Holt and Company.

Nobre, A. C. and S. Kastner (2014) *The Oxford Handbook of Attention.* Oxford: Oxford University Press.

Posner, M. I. and M. E. Raichle (1994) *Images of Mind.* New York: Scientific American Library.

Memory

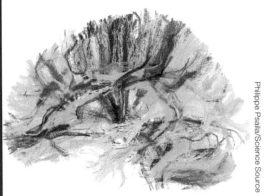

Philippe Psaila/Science Source

KEY CONCEPTS

30.1 Memory processes can be categorized by function

30.2 Memory encoding involves creating associations that support later recall

30.3 The medial temporal lobe supports declarative memory

30.4 Memories are stored in a distributed manner throughout the cerebral cortex

30.5 Nondeclarative memory relies on brain systems distinct from those supporting declarative memory

30.6 As humans age, changes in the brain alter memory processes

Overview

The abilities to store and retrieve information about past experiences are among the most important of the brain's complex functions. Memory is evident whenever we bring our past experiences into consciousness (e.g., reliving a vacation with family) or when past events or actions change our behaviors (e.g., playing a piece of music flawlessly after months of practice). Without memory, access to the past and imagination of the future would be lost. This chapter reviews the organization of human memory systems, surveys memory disorders and their implications, and considers some key questions about memory that remain unanswered. It builds upon research on *learning*, the processes by which information is acquired by the nervous system and is stored in neural circuits (see Chapter 8). Yet research on memory extends far beyond basic processes of learning to consider not only experience-dependent changes in behavior and the development of skills, but also conscious remembering of autobiographical experiences. Disorders of memory such as the pathological loss of previously stored information (retrograde amnesia) and the inability to store new information (anterograde amnesia) also provide important insights into brain function, while also highlighting the relevance of memory to a wide range of clinical disorders.

CONCEPT
30.1 | **Memory Processes Can Be Categorized by Function**

LEARNING OBJECTIVES

30.1.1 Understand the distinction between working memory and long-term memory.

30.1.2 Explain how nondeclarative memory differs from declarative memory.

30.1.3 Define and distinguish between classical and operant conditioning.

Memory: A Taxonomy of Functions

Modern research divides the larger concept of memory into distinct systems that can be distinguished in clinical cases (e.g., an individual may have a deficit in one system but intact abilities in another) and laboratory studies (e.g., two tasks that rely on different memory systems evoke activity in distinct neural circuits or brain regions). Figure 30.1 shows a schematic taxonomy of memory systems.

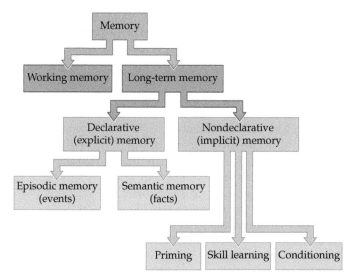

FIGURE 30.1   A taxonomy of memory systems Working (or short-term) memory keeps information available for seconds to minutes, after which time it decays and becomes unavailable. Long-term memory encodes information in a more stable form that can persist for hours, days, or even a lifetime. Within long-term memory, there is a distinction between declarative memory (for facts and events in one's life) and nondeclarative memory (for skills and behaviors learned implicitly).

Working memory

One key distinction among memory systems is the time-scale over which they operate. **Working memory** (also called short-term memory) maintains information for short periods of time (typically a few seconds to minutes) to achieve a particular goal or task. An everyday example is searching for a lost object; working memory allows the hunt to proceed efficiently by avoiding places already inspected. To experience a typical experimental paradigm for testing working memory, try to hold in memory the string of numbers "211776314911." That 12-digit string is longer than the typical person's working memory capacity (usually seven to nine items), which poses a challenge for working memory. You may have found yourself rehearsing the digits by repeating them in order (either vocally or subvocally) to refresh the information before it decays. An alternative approach is to minimize the working memory load by organizing the contents of memory into chunks that are individually easier to remember; in this example, the same string of numbers could be more easily kept in working memory as "21" + "1776" + "314" + "911." Intact working memory abilities are critical for a wide range of other cognitive functions and adaptive behaviors, as seen in the example of searching for a lost item. Tests of cognitive abilities designed to detect clinically relevant dysfunctions often include brief assessments of working memory; for

example, people with Alzheimer's disease show impairments in working memory compared with people with non-Alzheimer's mild cognitive impairment, who are themselves impaired compared with older adults who have no dementia or neurocognitive disease.

Working memory is closely related to executive and control processes (see Chapter 34). Indeed, it is sometimes considered to be a special category of executive function that operates on internal representations rather than on sensory input. Consistent with this view, neurons in the lateral prefrontal cortex (PFC) show continuing activity while information is maintained in working memory (e.g., the spatial location at which a stimulus was just seen). Yet those PFC neurons do not act alone, as seen both by measurements of activity in other regions of the brain and by examination of individuals with working memory deficits. For working memory tasks like that introduced in the previous paragraph (i.e., keeping a set of stimuli active in advance of a memory test), regions of the temporal–parietal cortex are critical. An evocative example can be seen in the case of an individual (known as K. F.) who had selective damage to that region (Figure 30.2A). K. F. was greatly impaired on tests of working memory; when presented with a list of seven numbers and asked to repeat its contents right away, he could typically remember only two numbers. But his abilities to encode and retrieve information from long-term memory were essentially normal; if he was repeatedly presented with the same ten-word list for later testing, he could learn that list just as well as similar-aged control participants. Conversely, individuals with damage to the medial temporal lobe (e.g., the case of H.M., see also *Clinical Applications*) often have normal working memory but great difficulty encoding and retrieving information from long-term memory (Figure 30.2B). These and other results indicate that successful maintenance of information in working memory relies on the influences of PFC neurons on information held in posterior sensory cortices.

Long-term memory

While information stored in working memory remains available for a relatively short time before decaying, other memories are more stable—and in some cases persist throughout one's lifetime. Information of particular significance in working memory can enter into **long-term memory** by conscious or unconscious rehearsal or practice. Long-term memory consists of at least two qualitatively different ways of storing information, generally referred to as *declarative memory* and *nondeclarative memory*. **Declarative memory** is the storage and retrieval of material that is available to consciousness and can be expressed by language (i.e., "declared"). Examples of declarative memory are the ability to remember facts about the world, the words to a song, or events from a recent vacation. **Nondeclarative memory**

(A) Patient K.F.: Impaired working memory
versus preserved declarative memory

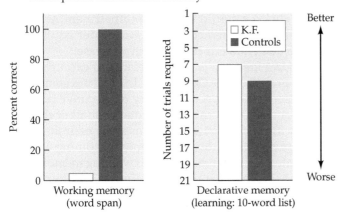

(B) Patient H.M. (amnesic): Preserved working memory
versus impaired declarative memory

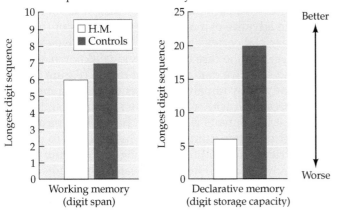

FIGURE 30.2 Working memory and declarative memory rely on different brain systems (A) The individual known as K. F., who had selective damage to temporal–parietal cortex, was greatly impaired on working memory tasks that required immediate recall, but showed abilities to encode and retrieve information into long-term memory that were normal for his age and education level. (B) The individual known as H. M., who had damage to medial temporal lobe and hippocampus, showed normal working memory but considerable impairments in encoding new information into long-term memory. (A after E. K. Warrington and T. Shallice, 1969. *Brain* 92: 885–896; B after D. A. Drachman and J. Arbit, 1966. *Arch Neurol* 15: 52–61.)

(also referred to as procedural, or implicit, memory) involves skills and associations that are generally acquired and retrieved at an unconscious level. Shooting a basketball and playing the piano are examples of nondeclarative memories; it is difficult to describe exactly how we do these things, and thinking about how to carry out such well-practiced activities may actually disrupt the ability to perform them efficiently. As discussed later in the chapter, the distinction between declarative and nondeclarative memory is well supported by anatomical, clinical, and other evidence.

There is general agreement that the so-called **engram**—the physical embodiment of any memory in neuronal machinery—depends on changes in the efficacy of synaptic connections and/or the actual growth and reordering of such connections. As discussed in Chapter 23, there is ample evidence that mechanisms of synaptic change can and do act over both short-term and long-term timescales. The term **consolidation** (Latin, "to make firm") refers to the progressive stabilization of memories following an initial encoding event. Consolidation involves changes in gene expression, protein synthesis, and other mechanisms of synaptic plasticity that allow the persistence of memories at the cellular level (see Chapters 8 and 24). Effective consolidation can be disrupted by interfering with these processes, as in the example of disordered sleep.

Although it makes good sense to divide human learning and memory into categories based on the accessibility of stored information to consciousness, this distinction becomes problematic when considering learning and memory processes in non-human animals. From an evolutionary point of view, it is of course unlikely that declarative memory arose de novo in humans with the development of language. Although some researchers favor different classification systems for humans as opposed to other animals, studies suggest that similar memory processes operate in all mammals and that these functions are carried out by homologous neural circuitry. In non-human mammals, declarative memory typically refers to information of which they are aware, and could be declared if the species in question had this ability. Another criterion of declarative memory in non-human animals is its dependence on the integrity of the medial temporal lobes, consistent with the evidence from studies of human memory, as described in Concept 30.3. Nondeclarative memory, in humans and other animals alike, can be thought of as the acquisition and storage of neural associations that are not available to consciousness and not dependent on the medial temporal lobes.

Nondeclarative memory

Priming is defined as a change in the processing of a stimulus due to a previous encounter with the same or a related stimulus with or without conscious awareness of the original encounter. For example, a list of words can be given with the instruction that participants identify some feature that is actually extraneous to the experiment (e.g., identifying the words as verbs, adjectives, or nouns). Sometime thereafter (often the next day), the same individuals are given a different test in which they are asked to fill in the missing letters of words with the letters of whatever words come to mind (Figure 30.3). The test list actually includes fragments of words that were presented in the first test, mixed among fragments of words that were not.

Participants tend to fill in the letters to make the words that were presented earlier at a higher rate than expected by chance, and fill them in more quickly than they do new words, even though they may have little or no conscious memory of seeing the words from the earlier list.

The information stored by priming, however, is not always reliable. Consider the list of words in Table 30.1A. If the list is read to a group of students who are immediately asked to identify which of several items were on the original list and which were not (Table 30.1B), the result is surprising. Typically, about half the students report that the word *sweet* was included in the original list; moreover, they are quite certain about it. The mechanism of such erroneous recognition is presumably the strong associations that have previously been made between the words on that list and the word *sweet*, which biases the students to think that *sweet* was a member of the original set. Clearly, memories, even those we feel quite confident about, are often false.

Priming is resistant to brain injury, aging, and dementia. As a result, its contributions are less obvious (and less easily studied) than other forms of memory that are compromised by specific brain insults, such as impaired declarative memory following damage to the medial temporal lobes (see Concept 30.3). Among other things, priming shows that previously presented information often influences subsequent behavior, even when we have no explicit awareness of that prior presentation. The significance of priming is well known—at least intuitively—to advertisers, teachers, spouses, and others who want to influence the way we think and act.

Psychologists and neuroscientists use the term *conditioning* to describe several processes by which an organism learns relationships between stimuli and actions. **Classical conditioning** occurs when an innate reflex is modified by associating its normal trigger with an unrelated stimulus; by virtue of the repeated association, the unrelated stimulus eventually triggers the original response. This type of conditioning was famously studied by the Russian psychologist Ivan Pavlov in experiments with dogs and other animals early in the twentieth century. The dogs' innate reflex was salivation (the *unconditioned response*) in reaction to the sight or smell of food (the *unconditioned stimulus*). The association was elicited in the animals by repeatedly pairing the sight and smell of food with the sound of a bell (the *conditioned stimulus*). The conditioned reflex was considered established when the conditioned stimulus (the sound of the bell) elicited salivation by itself (the *conditioned response*).

Operant conditioning alters the probability of a behavioral response by associating that

TABLE 30.1   The Fallibility of Human Memory[a]

(A) Initial list of words		(B) Subsequent test list
candy	honey	taste
sour	soda	point
sugar	chocolate	sweet
bitter	heart	chocolate
good	cake	sugar
taste	eat	nice
tooth	pie	
nice		

[a]After hearing list A read aloud, participants were asked to identify which items in list B had also been on list A. See text for the results.

response with a reward or punishment. In Edward Thorndike's original experiments during the 1890s, cats learned that pressing a lever opened a trap door that revealed a food reward. Although the cats initially pressed the lever only occasionally—and more or less by chance—they soon learned to associate this action with the reward, and subsequently became increasingly likely to press the lever. In Frederick Skinner's better-known experiments performed a few decades later, pigeons and rats learned to associate pressing a lever with receiving a food pellet; the experimental device (later known as a Skinner box; Figure 30.4)

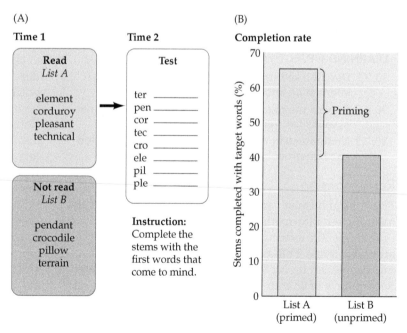

FIGURE 30.3 Priming (A) In a commonly used test, the participant is presented at time 1 with a list of words to study (list A) and is later tested using a word-stem completion task (time 2). The stems could also be completed from list B, which comprises words the subject did not see during the initial session. (B) Participants typically complete the stems with about 25% more studied than unstudied words; this percentage represents the effect of priming.

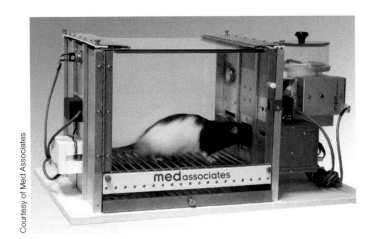

FIGURE 30.4 Modern example of a Skinner box
This apparatus is the most widely used method for studying operant conditioning.

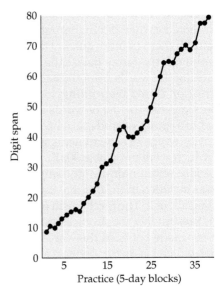

FIGURE 30.5 Increasing the digit span by practice and the development of associational strategies During many months involving 1 hour of practice each day for 3 to 5 days a week, a college student increased his digit memory span from 7 to 79 numbers. Random digits were read to him at the rate of 1 per second. If a sequence was recalled correctly, 1 digit was added to the next sequence. (After K. A. Ericsson et al., 1980. *Science* 208: 1181–1182.)

became widely used in psychological and neuroscience research. In both classical and operant conditioning, it takes several trials for the conditioning to become established. If the conditioned animal performs the desired response but the reward is no longer provided, the conditioning gradually disappears, a phenomenon called *extinction*.

CONCEPT 30.2 | Memory Encoding Involves Creating Associations That Support Later Recall

LEARNING OBJECTIVES

30.2.1 Describe key strategies for effectively encoding information into memory.

30.2.2 Explain the role of motivation and supporting brain systems for memory encoding.

30.2.3 Understand how the inability to forget information can benefit or impair everyday functioning.

Remembering

The typical human capacity for remembering relatively meaningless information is surprisingly limited (as noted in Concept 30.1, a string of seven to nine numbers or other arbitrary items). This stated capacity, however, is misleading. People can report 14 or 15 items in a briefly presented 5 × 5 matrix of 25 numbers or other objects if the experimenter points to specific boxes in the blank matrix during recall testing. Moreover, a person's digit memory span can be increased dramatically with practice. For example, a college student who for some months spent an hour each day being paid to successfully remember randomly presented numbers was able to recall a string of up to about 80 digits (Figure 30.5). He did this by making subsets of the string of numbers he was given signify dates or times

at track meets (he was a competitive runner)—in essence, giving meaningless items a meaningful context.

Competitive "mnemonists" who perform prodigious feats of memory—such as reciting the digits of π (3.1416...*n*) to more than 70,000 decimal places—use similar strategies of associating chunks of the larger sequence with meaningful elements. One such approach that can be used in everyday life was first identified in ancient Rome: the "method of loci" or "memory palace" strategy. Suppose you need to remember a list of unrelated words (e.g., those in Figure 30.3). Now imagine you are walking along a familiar path, perhaps from one building to another on a university campus or from room to room within your home. Visualize yourself at the first location on that path and link that location to the first word on the list; for example, if the first word is *pendant*, imagine you see a necklace suspended from a sign. At each stop along the path, build another association between the physical location and the next word in the list. This approach illustrates how a deep encoding process—here, active visualization of links between a spatial location and a to-be-remembered word—can help form strong associations that support later recall from memory.

The capacity of memory very much depends on what the information in question means to the individual and how readily it can be associated with information that has already been stored. A good chess player can remember the position of many more pieces on a briefly examined board than an inexperienced player, presumably because the positions have much more significance for individuals who understand the intricacies of the game (Figure 30.6). Arturo Toscanini, the late conductor of the NBC Philharmonic Orchestra, allegedly kept in his head the complete scores of more than 250 orchestral works, as well as the music and librettos for some 100 operas. Once, just before a concert in St. Louis, the first bassoonist approached Toscanini in some consternation, having just discovered that one of the keys on his bassoon was broken. After a minute or two of deep concentration, the story goes, Toscanini turned to the alarmed bassoonist and informed him that there was no need for concern, since that note did not appear in any of the bassoon parts for the evening's program. Such feats of memory are not achieved by rote learning but are a result of the fascination that aficionados bring to their special interests, sometimes in a pathological way (Box 30A).

Such examples indicate that motivation also plays an important role in memory. In one study of this issue, experimenters asked participants to examine a set of photographs that depicted either pieces of furniture or pieces of food (Figure 30.7). The participants were later tested with a much larger set of photographs that included images from the previously studied set along with new ones; the participants were asked to indicate whether a picture was "old" or "new." In one condition, the experimenters increased participants' hunger by depriving them of food for several hours. Predictably, participants were much more likely to

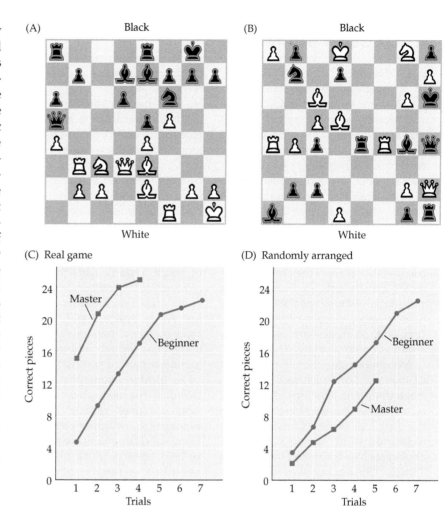

FIGURE 30.6 Retention of briefly presented information depends on past experience, context, and perceived importance (A) A sample of a chess position from a real game between grandmasters; shown is the board after white's 21st move in game 10 of the 1985 World Chess Championship between A. Karpov (white) and G. Kasparov (black). (B) A random arrangement of the same 28 pieces. (C) After briefly viewing chess boards drawn from real games, master players reconstruct the positions of the pieces with much greater efficiency than beginning players. (D) With a randomly arranged board, however, beginners perform as well as or better than accomplished players. (After W. G. Chase and H. A. Simon, 1973. *Cogn Psychol* 4: 55–81.)

FIGURE 30.7 Motivated memory
(A) Participants studied a set of pictures of food and non-food (i.e., furniture) items and were later tested for their ability to discriminate the pictures they had seen from a new set of pictures. In one condition, participants were made hungry by withholding food from them for several hours. (B) Memory for food items was significantly enhanced when participants were hungry, but there was no significant effect of hunger on memory for non-food pictures. Results such as these emphasize the importance of motivation and interest for memory performance.

■ BOX 30A | Savant Syndrome

A fascinating developmental anomaly of human memory is seen in rare individuals who until recently were referred to as *idiot savants*; the current literature tends to use the less pejorative phrase *savant syndrome*. Savants are people who, for a variety of poorly understood reasons (typically brain damage in the perinatal period), are severely restricted in most mental activities but extraordinarily competent in one particular domain. The grossly disproportionate skill compared with the rest of their limited mental life can be striking. Indeed, these individuals—whose special talent may be in memory, calculation, history, art, language, or music—are usually diagnosed as being severely impaired.

Many examples could be cited, but a summary of one such case suffices to make the point. The individual whose history is summarized here was given the fictitious name "Christopher" in a detailed study carried out by psychologists Neil Smith and Ianthi-Maria Tsimpli. Christopher was discovered to be severely brain damaged at just a few weeks of age (perhaps as the result of rubella during his mother's pregnancy or anoxia during birth; the record is uncertain in this respect). He had been institutionalized since childhood because he was unable to care for himself, could not find his way around, had poor hand–eye coordination, and had a variety of other deficiencies. Tests on standard IQ scales were low, consistent with his general inability to cope with daily life.

Despite his severe mental incapacitation, Christopher took an intense interest in books from the age of about 3, particularly those providing factual information and lists (e.g., telephone directories and dictionaries). At about age 6 or 7 he began to read technical papers that his sister sometimes brought home from work, and he showed a surprising proficiency in foreign languages. His special talent in the acquisition and use of language grew rapidly. As an early teenager, Christopher could translate from—and communicate in—a variety of languages in which his skills were described as ranging from rudimentary to fluent; these included Danish, Dutch, Finnish, French, German, modern Greek, Hindi, Italian, Norwegian, Polish, Portuguese, Russian, Spanish, Swedish, Turkish, and Welsh. This extraordinary level of linguistic accomplishment is all the more remarkable given that he had no formal training in language even at the elementary school level, and could not play tic-tac-toe or checkers because he was unable to grasp the rules needed to make moves in these games.

The neurobiological basis for such extraordinary individuals is not understood. It is fair to say, however, that savants are unlikely to have ability in their areas of expertise that exceeds the competency of typically intelligent individuals who focus passionately on a particular subject. Presumably, the savant's intense interest in a particular cognitive domain is due to one or more brain regions that continue to work reasonably well. Whether because of social feedback or self-satisfaction, savants spend a great deal of their mental time and energy practicing the skill they can exercise more or less normally. The result is that the relevant associations they make become especially rich, as Christopher's case demonstrates.

---

remember more pictures of food when they were hungry than when they were not. There was no effect of motivation on memory for pictures of furniture.

Neuroscientists have used extrinsic manipulations of motivation to understand how brain systems for reward learning interact with those that support memory. The delivery of unexpected rewards evokes activation in dopamine neurons within the ventral tegmental area (VTA); those neurons have broad projections throughout the brain (see Chapter 6). One important target of those neurons is the hippocampus, a key brain structure for organizing information in memory. In a seminal study, participants encoded visual scenes into long-term memory in advance of an expected test the next day; critically, some scenes were associated with relatively large rewards (e.g., $5) for accurate memory while others were associated with small rewards (e.g., 10¢) (Figure 30.8A). The large rewards evoked activation in the VTA, as measured by fMRI, which in turn generated greater activation in the hippocampus and better overall memory performance.

Why might reward-related signals in the VTA be so critical for memory? One intriguing possibility is that the VTA activation indicates that an unexpected stimulus was particularly important for subsequent memory (i.e., accurate memory would lead to greater future rewards), leading to an amplification of hippocampal responses that facilitate successful encoding (Figure 30.8B).

Forgetting

We often lament the transience of our memories. Yet it is critical to be able to forget; otherwise our brains would be impossibly burdened with a welter of useless and even maladaptive information that could be encoded in our memories. Many emotionally negative events diminish in memory over time—a feature of memory that supports our overall mental health. Fortunately, the human brain is very good at forgetting. Consistent with the unreliable performance on tests such as the one shown in Table 30.1, Figure 30.9A shows that the memory of a simple penny (or similar familiar coin in other countries)

(A)

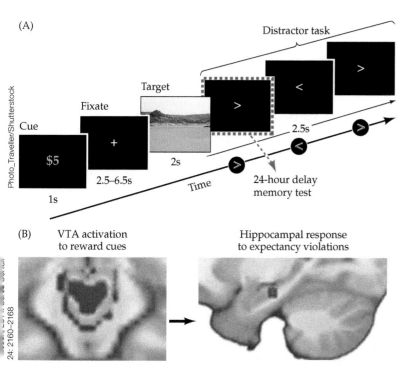

(B)   VTA activation to reward cues   Hippocampal response to expectancy violations

FIGURE 30.8 **Influence of reward information on memory** (A) An example task used to study reward effects via fMRI in humans. Participants were shown photographs of outdoor scenes and then asked to remember them in advance of a memory task 24 hours later. Photographs were associated with either a high or low reward. A control attention task (pressing buttons to indicate the directions of arrows) was included to provide a general measure of behavior and to minimize working memory contributions to learning. (B) fMRI activation in the ventral tegmental area (VTA), a primary site of dopamine neurons that signal unexpected rewards, predicts concurrent activation in the left hippocampus that is proportional to how much each reward differed from the participant's prior expectations. (A after R. A. Adcock et al., 2006. *Neuron* 50: 507–517.

is uncertain at best, in large part because keeping an accurate memory for a penny has little consequence for everyday life (i.e., we never encounter forged or inaccurate pennies). More generally, as time passes, people tend to gradually forget what they have encoded in long-term memory (Figure 30.9B).

Some rare individuals have difficulty forgetting. The best-known exemplar was studied over several decades by the Russian psychologist Alexander Luria, who referred to the individual simply as "S." Luria's description of an early encounter gives some idea why S, then a newspaper reporter, was so interesting:

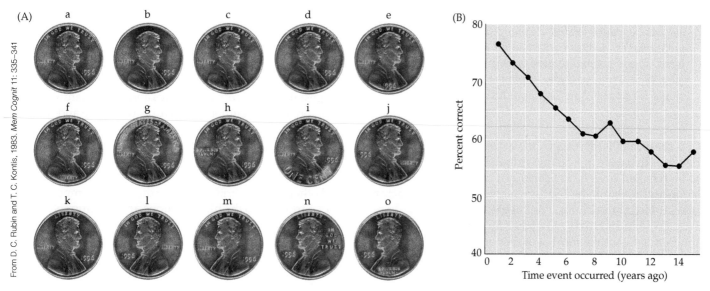

FIGURE 30.9 **Forgetting** (A) Different versions of the "heads" side of a penny. Despite innumerable exposures to this familiar design, few people are able to select (a) as the authentic version. Clearly, repeated information is not necessarily retained. (B) The deterioration of long-term memories was evaluated in this example by a multiple-choice test in which participants were asked to recognize the names of television programs that had been broadcast for only one season during the past 15 years. Forgetting of stored information that is no longer used evidently occurs gradually and progressively over the years (chance performance = 25%). (B after L. R. Squire, 1989. *J Exp Psychol: Learn Mem Cog* 15: 241–245.)

I gave S a series of words, then numbers, then letters, reading them to him slowly or presenting them in written form. He read or listened attentively and then repeated the material exactly as it had been presented. I increased the number of elements in each series, giving him as many as thirty, fifty, or even seventy words or numbers, but this too, presented no problem for him. He did not need to commit any of the material to memory; if I gave him a series of words or numbers, which I read slowly and distinctly, he would listen attentively, sometimes ask me to stop and enunciate a word more clearly, or, if in doubt whether he had heard a word correctly, would ask me to repeat it. Usually during an experiment he would close his eyes or stare into space, fixing his gaze on one point; when the experiment was over, he would ask that we pause while he went over the material in his mind to see if he had retained it. Thereupon, without another moment's pause, he would reproduce the series that had been read to him.

A. R. Luria (1987), *The Mind of a Mnemonist*, pp. 9–10

S's phenomenal memory did not always serve him well, however. He had difficulty ridding his mind of the trivial information on which he tended to focus, sometimes to the point of incapacitation. As Luria put it:

Thus, trying to understand a passage, to grasp the information it contains (which other people accomplish by singling out what is most important) became a tortuous procedure for S, a struggle against images that kept rising to the surface in his mind. Images, then, proved an obstacle as well as an aid to learning in that they prevented S from concentrating on what was essential. Moreover, since these images tended to jam together, producing still more images, he was carried so far adrift that he was forced to go back and rethink the entire passage. Consequently, a simple passage—a phrase, for that matter—would turn out to be a Sisyphean task.

Ibid., p. 113

S presumably represents one extreme of a continuum. Several otherwise typical individuals have what has come to be referred to as *hyperthymesia* or *highly superior autobiographical memory*, the best known of whom is the actor Marilu Henner. Although not negatively afflicted like S was, these individuals remember far more details about their daily lives than most of us. For example, when asked to remember days associated with a particular type of event (e.g., prior visits to a research laboratory to participate in experiments), these individuals can readily volunteer not only the dates and days of the week for those events but can also relate other events that they experienced on those particular days with great fidelity (e.g., their descriptions match diaries or laboratory records).

The Medial Temporal Lobe Supports Declarative Memory

LEARNING OBJECTIVES

30.3.1 Define anterograde amnesia and explain its typical causes.

30.3.2 Describe how experience might alter hippocampal structure and function.

Structures within the medial temporal lobe—the hippocampus in particular—are critical for establishing new declarative memories (Figure 30.10). Early evidence for this link came from extraordinary clinical cases of **amnesia** (Clinical Applications). Individuals with medial temporal damage retain the ability to remember events and other information acquired before the brain damage but are unable to form new memories. Thus, this sort of injury produces primarily **anterograde amnesia**, or difficulty forming and retrieving new memories.

Studies of animals with lesions of the medial temporal lobe have largely corroborated these findings. For example, one test of the presumed equivalence of declarative memory formation in animals involves placing rats into a pool filled with opaque water, thus concealing a submerged platform; this apparatus is now known as a Morris water maze. Surrounding the pool are prominent visual landmarks (Figure 30.11). Typical rats at first search randomly until they find the submerged platform. After repeated testing, however, they learn to swim directly to the platform no matter where they are initially placed in the pool by orienting to the landmarks. Rats with lesions of the hippocampus and nearby structures cannot learn to find the platform, suggesting that remembering its location relative to visual landmarks depends on the same neural structures critical to declarative memory formation in humans. Likewise, destruction of the hippocampus and parahippocampal gyrus in monkeys severely impairs their ability to perform delayed response tasks. These studies suggest that non-human primates and other mammals depend on medial temporal structures to encode and initiate the consolidation of memories of events, just as humans use these same brain regions for the initial encoding and consolidation of declarative memories.

Consistent with the evidence from studies of humans and other animals with lesions to the medial temporal lobe—in particular to the hippocampus and parahippocampal cortex—recent studies have shown that neurons in these areas are selectively recruited by tasks that involve declarative memory. For example, neuroimaging studies using PET show increased metabolism in the hippocampus of people studying information they will later be asked to recall. Studies using fMRI have also shown that the hippocampus and parahippocampal gyrus are activated in people studying a list of items to be remembered. Moreover, the amount of activity measured in these areas is higher

(A) Brain areas associated with declarative memory

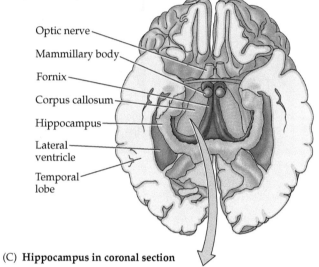

(B) Ventral view of hippocampus and related structures with part of temporal lobes removed

(C) Hippocampus in coronal section

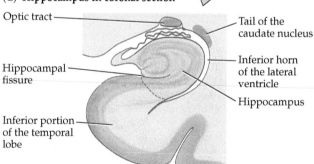

FIGURE 30.10   Brain areas that support declarative memory By inference from the results of damage to these structures, declarative memory is based on their physiological activity. (A) Studies of patients with amnesia have shown that the formation of declarative memories depends on the integrity of the hippocampus and its subcortical connections to the mammillary bodies and dorsal thalamus. (B) Location of the hippocampus as seen in a cutaway view in the horizontal plane. (C) The hippocampus as it would appear in a histological section in the coronal plane, at approximately the level indicated by the arrow in (B).

(A)

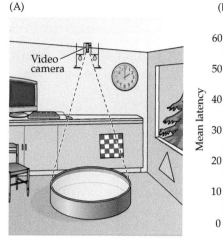

(B)

(C) Control rat

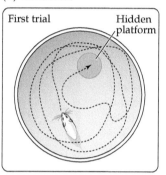
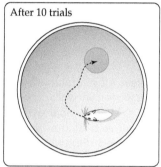

(D) Rat with hippocampal lesions

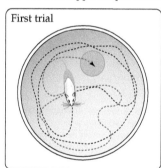
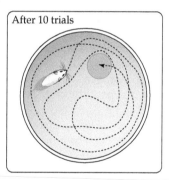

FIGURE 30.11   Spatial learning and memory in rodents depend on the hippocampus (A) Rats are placed in a circular tank about the size and shape of a child's wading pool, filled with opaque (milky) water. The surrounding environment contains visual cues such as windows, doors, a clock, and so on. A small platform is located just below the surface. As rats search for this resting place, the pattern of their swimming (indicated by the traces in C and D) is monitored by a video camera. (B) After a few trials, typical rats rapidly reduce the time required to find the platform, whereas rats with hippocampal lesions do not. Sample swim paths of typical (C) and hippocampus-lesioned (D) rats at the beginning of learning and after repeated attempts to learn the platform location. Rats with hippocampal lesions are unable to remember where the platform is located. (B after H. Eichenbaum, 2000. *Nat Rev Neurosci* 1: 41–50; C,D after F. Schenk and R. G. M. Morris, 1985. *Exp Brain Res* 58: 11–28.)

■ Clinical Applications

Clinical Cases That Illustrate the Neural Basis of Memory

H. M.

Henry Molaison, known to science only as "H. M." while he was still alive, suffered from severe epilepsy throughout his childhood and early adulthood that would eventually lead to an experimental surgery. A high school graduate, H. M. had been working as a technician in a small electrical business until his epileptic seizures became so severe and uncontrolled that he became unable to work. These attacks involved generalized convulsions with tongue biting, incontinence, and loss of consciousness (all typical of grand mal or tonic-clonic seizures). In 1953, at the age of 27, he underwent a bilateral medial temporal lobe resection in which the amygdala, uncus, hippocampal gyrus, and anterior two-thirds of the hippocampus were removed (Figures A–D). The surgery controlled H. M.'s epilepsy, but his life thereafter was radically changed.

The first formal psychological examination of H. M. was conducted nearly 2 years after the operation, at which time a profound memory deficit was obvious. Just before the exam, for instance, H. M. had been talking to the psychologist; yet a few minutes later he reported no recollection of this experience, denying that anyone had spoken to him. He gave the date as March 1953 and seemed oblivious to the fact that he had undergone an operation, or that he had become incapacitated as a result. Nonetheless, his score on the Wechsler–Bellevue Intelligence Scale was 112, a value not significantly different from his preoperative IQ. Various psychological tests failed to reveal any deficiencies in perception, abstract thinking, or reasoning; H. M. seemed highly motivated and, in the context of casual conversation, unremarkable. Importantly, he also performed well on tests of the ability to learn new skills, such as mirror writing or puzzle solving

MRI images of the brain of H. M., Henry Molaison. (A) Sagittal view of the right hemisphere; the area of the anterior temporal lobectomy is indicated by the white dotted line. The intact posterior hippocampus is indicated by the white arrow. (B–D) Coronal sections at approximately the levels indicated by the red lines in (A). Image (B) is the most rostral and is at the level of the amygdala. The amygdala and the associated cortex are entirely missing. Image (C) is at the level of the rostral hippocampus; again, this structure and the associated cortex have been removed. Image (D) is at the caudal level of the hippocampus; the posterior hippocampus appears intact, although somewhat shrunken. Outlines below give a clearer indication of the parts of H. M.'s brain that were ablated (black shading).

From S. Corkin et al., 1997, *J Neurosci* 17, 3964–3979. © 1997 Society for Neuroscience

(i.e., his ability to form nondeclarative memories was intact). Moreover, his early memories were easily recalled, showing that the structures removed during H. M.'s operation are not a permanent repository for such information. On the Wechsler Memory Scale (a specific test of declarative memory), however, he performed very poorly, and he could not recall a preceding test set once he had turned his attention to another part of the exam. These deficits, along with his obvious inability to recall events in his daily life,

■ Clinical Applications *(continued)*

all indicated a profound loss of short-term declarative memory function.

H. M. was studied extensively during the subsequent decades, primarily by Brenda Milner and her colleagues at the Montreal Neurological Institute. His memory deficiency continued unabated, and according to Milner, he had little idea who she was in spite of their acquaintance for nearly 50 years.

Sadly, he gradually came to appreciate his predicament. "Every day is alone," H. M. reported, "whatever enjoyment I've had and whatever sorrow I've had." H. M. died in 2008 at age 82.

N. A.

Born in 1938, N. A. grew up with his mother and stepfather, attending public schools in California. After a year of junior college, he joined the Air Force. In October of 1959 he was assigned to the Azores as a radar technician and remained there until December 1960, when a bizarre accident made him a celebrated neurological case.

N. A. was assembling a model airplane in his barracks room while, unbeknownst to him, his roommate was practicing thrusts and parries with a miniature fencing foil behind N. A.'s chair. N. A. turned suddenly and was stabbed through the right nostril. The foil penetrated the cribriform plate (the structure through which the olfactory nerve enters the brain) and took an upward course into the left forebrain. N. A. lost consciousness within a few minutes (presumably because of bleeding in the region of brain injury) and was taken to a hospital. There he exhibited right-side weakness and paralysis of the right eye muscles innervated by the third cranial nerve. Exploratory surgery was undertaken and the dural tear repaired. Gradually, N. A. recovered and was sent home to California. After some months, his only general neurological deficits were some weakness of upward gaze and mild double vision. He retained, however, a severe anterograde amnesia for declarative memories. MRI studies first carried out in 1986 showed extensive damage to the thalamus and nearby tracts, mainly on the left side; there was also damage to the right anterior temporal lobe. The exact extent of his lesion, however, is not known, as N. A. remains alive and well.

N. A.'s memory from the time of his injury (more than 60 years ago) to the present has remained impaired, and like H. M., he fails badly on formal tests of new learning ability. His IQ is 124, and he shows no defects in language skills, perception, or other measures of intelligence. He also learns new nondeclarative skills quite normally. His amnesia is not as dense as that of H. M. and is more verbal than spatial. He can, for example, draw accurate diagrams of material presented to him earlier. Nonetheless, he loses track of his possessions, forgets what he has done, and tends to forget who has come to visit him. He has only vague impressions of political, social, and sporting events that have occurred since his injury. When watching television, he tends to forget the story line during commercials. However, his memory for events prior to 1960 is extremely good; indeed, his lifestyle tends to reflect the 1950s.

R. B.

At the age of 52, R. B. suffered an ischemic episode during cardiac bypass surgery. Following recovery from anesthesia, a profound amnesic disorder was apparent. As in the cases of H. M. and N. A., R. B.'s IQ was normal (111) and he showed no evidence of cognitive defects other than memory impairment. R. B. was tested extensively for the next 5 years, and while his amnesia was not as severe as that of H. M. or N. A., he consistently failed standard tests of the ability to establish new declarative memories. When R. B. died of congestive heart failure in 1983, a detailed examination of his brain was carried out. The only significant finding was bilateral lesions of the hippocampus—specifically, cell loss in the CA1 region that extended the full rostral–caudal length of the hippocampus on both sides. The amygdala, thalamus, and mammillary bodies, as well as the structures of the basal forebrain, were normal. R. B.'s case is particularly important because it suggests that hippocampal lesions alone can result in profound anterograde amnesia for declarative memory.

K. C.

As a young man, K. C. had a motorcycle accident in which he sustained damage to several brain regions, including the hippocampus (Figure E). As with the other patients described here, K. C.'s intellectual abilities were well preserved; he is able to read, write, and play chess at much the same level as before his accident (Figure F). However, both his anterograde and retrograde episodic memory are severely impaired.

Unlike H. M.'s amnesia, K. C.'s retrograde amnesia covers his whole life and he can remember little, if any, personal history. Nonetheless, his memory for semantic information acquired before the accident is intact. He has a good vocabulary, and his knowledge of subjects such as mathematics, history, and geography is not greatly different from that of others with his educational background. K. C.'s case thus exemplifies how extensive medial temporal lobe damage can, in at least some cases, *(Continued)*

■ Clinical Applications (continued)

impair retrograde *episodic* memory while sparing retrograde *semantic* memory. It could be that K. C.'s general knowledge was acquired earlier than the episodic memories tested and hence was more consolidated and less dependent on the hippocampus and surrounding structures. However, he can readily retrieve semantic information (e.g., the meaning of highly technical terms) he acquired while working as a machinist, whereas he fails to remember events that occurred in the factory during the same time period.

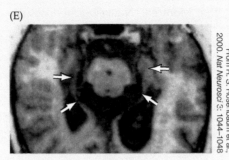

From R. S. Rosenbaum et al., 2000. *Nat Neurosci* 3: 1044–1048.

From E. Tulving, 2002. *Annu Rev Psychol* 53: 1–25.

(E) MRI image showing K. C.'s bilateral hippocampal and parahippocampal damage (arrows). (F) Although K. C. has severe episodic amnesia, his semantic memories remain largely intact.

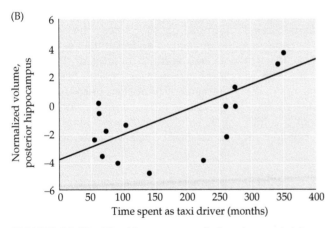

FIGURE 30.12　The hippocampus in London taxi drivers (A) Structural brain scans show that the posterior hippocampus, a region specialized for remembering spatial information, is larger in taxi drivers than in age-matched controls. (B) Hippocampus size scales positively with experience as a cabbie. (After E. A. Maguire et al., 2000. *Proc Natl Acad Sci USA* 97: 4398–4403. © 2000 National Academy of Sciences.)

for items that people subsequently remembered compared with the activity measured for items they later forgot.

Another example of the importance of medial temporal lobe structures in the formation and consolidation of declarative memories can be seen in studies of people whose careers require an exceptional memory. To become a taxi driver in London, one must pass a famously difficult test that challenges one's memory of the layout and contents of that labyrinthine city. A remarkable study showed that the posterior hippocampus, which appears to be particularly useful in remembering spatial information, is larger in London taxi drivers than in age-matched control individuals (Figure 30.12A). Confirming the role of experience in performance, the size of the posterior hippocampus in cab drivers scales positively with the number of months spent driving a cab (Figure 30.12B). Together, such findings support the idea that computations supported by the hippocampus and closely allied cortical areas of the medial temporal lobe assist in the transfer of declarative information into long-term memory, and that the robustness with which such memories are encoded depends on structural and functional changes of neural connections that occur as a result of experience.

CONCEPT 30.4

Memories Are Stored in a Distributed Manner throughout the Cerebral Cortex

LEARNING OBJECTIVES

30.4.1　Define retrograde amnesia and explain its typical causes.

30.4.2　Provide at least two examples of evidence supporting the conclusion that memories are stored in a distributed manner.

30.4.3　Describe key clinical cases that illustrate selective deficits in memory function.

While the regions of the medial temporal lobe are critical for memory retrieval processes, considerable research now supports the idea that declarative memories are stored throughout the cerebral cortex. The idea that memory traces are distributed over the cortex began with the work of the American neuroscientist Karl Lashley in the 1920s. Lashley removed regions of the cortex in rats (Figure 30.13A), performing this procedure either before or after the animals had learned to run mazes of varying difficulty. Lashley found that the location of the lesions did not matter much—only the extent of the tissue destruction and the difficulty of the task seemed consequential (Figure 30.13B,C). He summarized his findings in terms of what he called the *mass action principle*, which states that the degree of degradation in learning and memory depends on the amount of cortex destroyed—and that the more complex the learning task, the more disruptive the lesion. Only when the damage is widespread does network performance show a significant decline. These findings imply that, whereas acquiring declarative memories depends on the integrity of the medial temporal lobes, storing them over the long term depends on distributed cortical networks that are seriously impaired only when large portions of them are destroyed. Additional evidence for this perspective comes from examination of people with **retrograde amnesia**—the loss of previously stored memories of events preceding an injury or illness—which more typically follows broad damage throughout the cortex caused

by head trauma or neurodegenerative disorders such as Alzheimer's disease (Box 30B).

A second line of evidence supporting this interpretation comes from individuals with severe depression who undergo electroconvulsive therapy (ECT). The passage of enough electrical current through the brain causes the equivalence of a full-blown seizure. This remarkably useful treatment (which is performed under anesthesia in well-controlled circumstances) was discovered because depression in epileptics often remitted after a spontaneous seizure. However, ECT often causes both anterograde and retrograde amnesia. Patients typically do not remember the treatment itself or the events of the preceding days, and their recall of events over the previous 1 to 3 years can be affected. Animal studies (e.g., rats tested for maze learning) have confirmed the amnesic consequences of ECT. To mitigate this side effect (which may be the result of excitotoxicity and tends to resolve over a few months), ECT is often delivered to only one hemisphere at a time. The nature of amnesia following ECT supports the conclusion that long-term declarative memories are widely stored in the cerebral cortex.

Another sort of evidence comes from comparisons of the differential effects of damage to different cortical regions. Since cortical regions vary in their functions (see Chapter 28), it is not surprising that each region stores information that reflects its function. For example, the region that links speech sounds and their symbolic

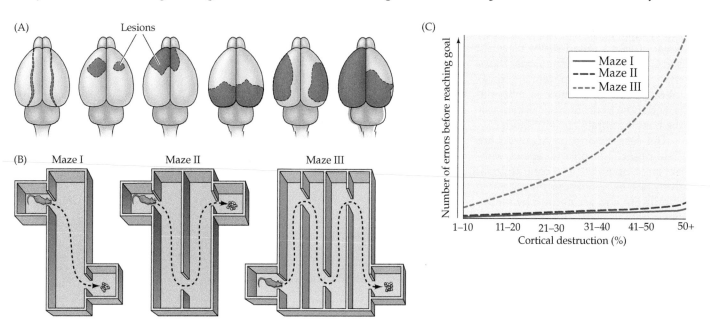

FIGURE 30.13 Lashley's experiments examining the effects of brain lesions on long-term memory (A) Lesions of varying size and location (red) were made in rat brains either before or after the animals had learned to run mazes (B) of varying complexity. (C) The reduction in learning that Lashley observed was proportional to the amount of tissue destroyed; the lesion

locations appeared inconsequential. The more complex the learning task, however, the more the lesions affected performance. (After K. S. Lashley and L. E. Wiley, 1933. *J Comp Neurol* 57: 3–55; K. S. Lashley, 1944. *J Comp Neurol* 80: 257–281; and the University of Rome Psychology Lab Website.)

■ BOX 30B | Alzheimer's Disease

ementia is a syndrome characterized by failure of recent memory and other intellectual functions. It is usually insidious in onset but tends to progress steadily. Alzheimer's disease (AD) is the most common dementia, accounting for 60% to 80% of cases in the elderly. This unfortunate condition afflicts about 10% of the U.S. population over the age of 65 and as much as 45% of the population over 85. The earliest signs are an impairment of recent memory function and attention, followed by failure of language skills, visual–spatial orientation, abstract thinking, and judgment. Alterations of personality inevitably accompany these defects.

A tentative diagnosis of AD is based on these characteristic clinical features and can be confirmed only by the distinctive cellular pathology evident on postmortem examination of the brain (Figure A). These histopathological changes consist of three principal features: (1) collections of intraneuronal cytoskeletal filaments called *neurofibrillary tangles*; (2) extracellular deposits of an abnormal protein (called amyloid) in so-called *senile plaques*; and (3) a diffuse loss of neurons. These changes are most apparent in the cerebral cortex, hippocampus, amygdala, and some brainstem nuclei (typically the basal forebrain nuclei) (Figure B).

The vast majority of AD cases are "late-onset," arising after age 60 without an obvious cause. In contrast, relatively rare early-onset forms appear in middle life and are caused by monogenic defects consistent with an autosomal dominant pattern of inheritance. Identification of the mutant genes in a few families with the early-onset form has provided considerable insight into the processes that go awry in AD.

Investigators long suspected that a mutant gene responsible for familial AD might reside on chromosome 21, primarily because clinical and neuropathologic features similar to AD often occur in individuals with Down syndrome (caused by an extra

copy of chromosome 21), but with a much earlier onset (about age 30 in most cases). A mutation of the gene encoding amyloid precursor protein (APP) emerged as an attractive candidate both because of the prominent amyloid deposits in AD together with isolation of a fragment of APP, Aβ peptide, from amyloid plaques. The gene that encodes APP was subsequently cloned by Dmitry Goldgaber and his colleagues and found to reside on chromosome 21. This discovery eventually led to the identification of mutations of the *APP* gene in almost 20 families with the early-onset, autosomal dominant form of AD. It should be noted, however, that only a few of the early-onset families (and none of the late-onset families) exhibited these particular mutations.

The mutant genes underlying two additional autosomal dominant forms

of AD have been subsequently identified as *presenilin 1* and *presenilin 2*. Mutations of these two genes modify processing of APP and result in increased amounts of a particularly toxic form of Aβ peptide, Aβ42. Thus, mutation of any one of several genes appears to be sufficient to cause a heritable form of AD, and these converge on abnormal processing of APP.

In the far more common late-onset form of AD, the disease is clearly not inherited in any simple sense (although the relatives of affected individuals are at a greater risk, for reasons that are not clear). The central role of APP in the families with early-onset forms of the disease nonetheless suggested that APP might be linked to the chain of events culminating in the sporadic forms of AD. Biochemists Warren Strittmatter and Guy Salvesen theorized that pathologic deposition of

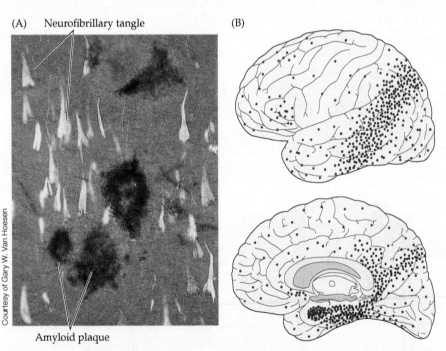

(A) Neurofibrillary tangle

(B)

Courtesy of Gary W. Van Hoesen

Amyloid plaque

(A) Histological section of the cerebral cortex from an individual with Alzheimer's disease, showing characteristic amyloid plaques and neurofibrillary tangles. (B) Distribution of pathologic changes (including plaques, tangles, neuronal loss, and gray matter shrinkage) in Alzheimer's disease. Dot density indicates severity of pathology. (B after H. Blumenfeld, 2002. *Neuroanatomy through Clinical Cases.* Sunderland, MA: Oxford University Press/Sinauer, based on A. Brun and E. Englund, 1981. *Histopathology* 5: 549–564.)

■ BOX 30B | Alzheimer's Disease *(continued)*

proteins complexed with Aβ peptide might be responsible.

To test this idea, Strittmatter and Salvesen immobilized Aβ peptide on nitrocellulose paper and searched for proteins in the cerebrospinal fluid of patients with AD that bound with high affinity. One of the proteins they detected was apolipoprotein E (ApoE), a molecule that normally chaperones cholesterol through the bloodstream. This discovery was especially provocative in light of another discovery, this one made by Margaret Pericak-Vance, Allen Roses, and their colleagues, who found that affected members of some families with the late-onset form of AD exhibited an association with genetic markers on chromosome 19. This finding was of particular interest because a gene encoding an isoform of ApoE is located in the same region of chromosome 19 implicated by the association studies. As a result, these researchers

began to explore the relationship of the different alleles of *ApoE* with individuals with a sporadic, late-onset form of AD.

There are three major alleles of *ApoE*: *e2*, *e3*, and *e4*. The frequency of allele *e3* in the general population is 0.78, and the frequency of allele *e4* is 0.14. The frequency of the *e4* allele in late-onset AD patients, however, is 0.52—almost four times higher than in the general population. Thus, the inheritance of the *e4* allele is a risk factor for late-onset AD. In fact, people homozygous for *e4* are about eight times more likely to develop AD compared with individuals homozygous for *e3*. Among individuals with no copies of *e4*, only 20% develop AD by age 75, compared with 90% of individuals with two copies of *e4*.

In contrast to the mutations of *APP* or *presenilin 1* and *presenilin 2* that cause early-onset familial forms of AD,

inheriting the *e4* form of ApoE is not sufficient to cause AD; rather, inheriting this gene simply increases the risk of developing AD. The cellular and molecular mechanisms by which the *e4* allele of ApoE increases susceptibility to late-onset AD are not understood, and elucidating these mechanisms is clearly an important goal.

Clearly, AD has a complex pathology and probably reflects a variety of related molecular and cellular abnormalities. So far, the most apparent common denominator seen in this complex disease is abnormal APP processing. In particular, accumulation of the toxic Aβ42 peptide is thought to be a key factor. This conclusion has led to efforts to develop therapies aimed at inhibiting formation or facilitating clearance of this toxic peptide. It is unlikely that this important problem will be understood without a great deal more research.

---

significance is located in the superior temporal lobe, and damage to this area typically results in an inability to link words and meanings (Wernicke's aphasia; see Chapter 32). This result supports the conclusion that the widespread connections of the hippocampus transfer declarative information to these and other language-related cortical sites (Figure 30.14). Similarly, the inability of individuals with temporal lobe lesions to remember and thus recognize objects or faces suggests that such memories are stored in these cortical sites.

One more line of support for the hypothesis that declarative memories are stored in cortical areas specialized for processing particular types of information comes from neuroimaging of people during the recollection of vivid memories. In one such study, participants first examined words paired with either pictures or sounds. Their brains were then scanned while they were asked to recall whether each test word was associated with either a picture or a sound. Functional images based on these scans showed that the cortical areas activated when participants

viewed pictures or heard sounds were reactivated when these percepts were vividly recalled. In fact, this sort of reactivation can be quite specific. Thus, different classes of visual images—for example, faces, houses, or chairs—tend to reactivate the same regions of the visual association

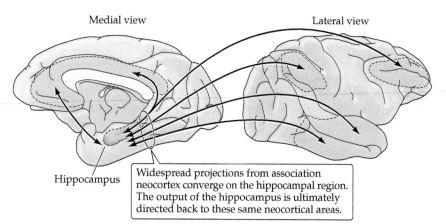

FIGURE 30.14 The hippocampus and possible declarative memory storage sites The rhesus monkey brain is shown because these connections are much better documented in non-human primates than in humans. Projections from numerous cortical areas converge on the hippocampus and the related structures known to be involved in human memory; most of these sites also send projections to the same cortical areas. Medial and lateral views are shown, the latter rotated 180° for clarity. (After G. W. Van Hoesen, 1982. *Trends Neurosci* 5: P345–350.)

FIGURE 30.15 Reactivation of visual cortex during vivid remembering of visual images (A) Participants were instructed either to view images of houses, faces, and chairs (left) or to imagine the objects in the absence of the stimulus (right). (B) At left, bilateral regions of ventral temporal cortex are specifically activated during perception of houses (yellow), faces (red), and chairs (blue). At right, when participants recall these objects, the same regions preferentially activated during the perception of each object class are reactivated. (A after R. L. Buckner and M. E. Wheeler, 2001. *Nature* 2: 624–634.)

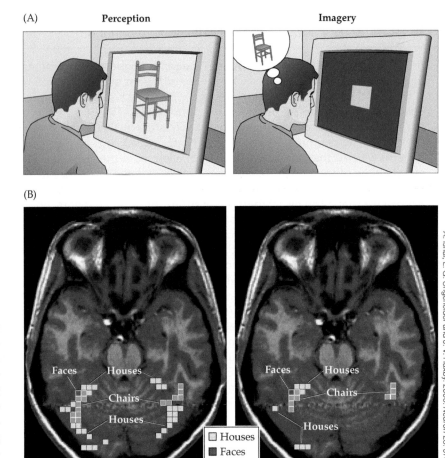

A. Ishai, L. G. Ungerleider and J. V. Haxby, 2000. *Neuron* 28: 979–990

cortex that were activated when the objects were actually perceived (Figure 30.15).

Finally, whereas the ability of individuals such as H. M. (see Clinical Applications) to remember facts and events from the period of their lives preceding their lesions demonstrates that the medial temporal lobe is not necessary for retrieving declarative information held in long-term memory, other studies suggest that these structures may be important for recalling declarative memories during the early stages of consolidation and storage in the cerebral cortex. This conclusion connects to remarkable work in animals showing that medial temporal lobe structures provide indexes—or a *cognitive map* (Box 30C)—for memories stored elsewhere in the brain.

■ BOX 30C | Place Cells and Grid Cells

Nearly 70 years ago psychologist Edward Tolman suggested that the brain must possess a cognitive map that represents remembered places in the environment. In 2014 the Nobel Prize in Physiology or Medicine was awarded to three neurophysiologists (John O'Keefe, May-Britt Moser, and Edvard Moser) for their studies of navigation in rodents that confirmed Tolman's idea.

This work began in the late 1960s with O'Keefe's observation that some neurons in the rat hippocampus fire robustly when and only when freely moving animals in an arena occupy a specific place (Figure A). Based on a series of studies by O'Keefe and his

colleagues, it became clear that the activity of different combinations of these *place cells* constituted the sort of learned cognitive map that Tolman had imagined. Although recording from neurons in behaving animals is now widely practiced, O'Keefe's group was one of the pioneers in this methodology.

The Mosers added to this understanding of animal navigation by further exploring the activity of neurons in the entorhinal cortex, a region adjacent to the hippocampus and whose neurons project to it. They found that cells in this area also code for place, but other neurons they named *grid cells* showed quite different patterns

of activity. Remarkably, each of these grid cells fired when the rat was in multiple loci that formed a hexagonal grid (Figure B), thus mapping every point in the arena over distances ranging from centimeters to a few meters. The implied significance of grid cells is providing measurements of the arena which place cells in the hippocampus then use to couple environmental cues with distance and direction.

Although the link from these findings in rats and mice to human hippocampal and entorhinal functions remains largely speculative, it is not difficult to see how the learning deficits seen in H. M. and other patients with medial temporal lobe damage are

■ BOX 30C | Place Cells and Grid Cells (*continued*)

related to these basic studies. Indeed, recordings from hippocampal neurons in patients undergoing surgery to relieve intractable epilepsy support this connection.

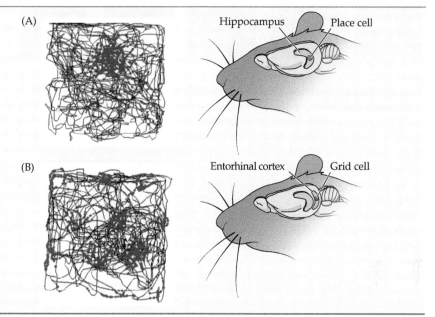

(A) Hippocampal place cells. The picture on the right shows the location of the hippocampus in a rat brain; the panel on the left illustrates place cells (orange dots) that fire only when the rat traverses a specific locus as it moves in an arena. (B) Entorhinal grid cells. The picture on the right shows the location of the entorhinal cortex adjacent to the hippocampus; the panel on the left represents the activity of a single grid cell as the rat traverses points in the arena that form a hexagonal grid. (From E. I. Moser et al., 2008. *Annu Rev Neurosci* 31: 69–89.)

CONCEPT
30.5

Nondeclarative Memory Relies on Brain Systems Distinct from Those Supporting Declarative Memory

LEARNING OBJECTIVES

30.5.1 Provide two examples of nondeclarative memory formation and describe their potential neural substrates.

30.5.2 Explain the role of the basal ganglia in motor skill learning.

Nondeclarative memory acquisition and storage

The fact that individuals such as H. M., N. A., and R. B. (see Clinical Applications) had no difficulty establishing or recalling nondeclarative memories implies that such information is laid down using a different anatomical substrate from that used in declarative memory formation. Nondeclarative memory apparently involves the basal ganglia, prefrontal cortex, amygdala, sensory association cortices, and cerebellum—but not the medial temporal lobe or midline diencephalon. In support of this view, perceptual priming (the unconscious influence of previously studied information on subsequent performance; see Concept 30.1) depends critically on the integrity of sensory association cortex. For example, lesions of the visual association cortex produce profound impairments in visual priming but leave declarative memory formation intact. Likewise, simple sensorimotor conditioning, such

as learning to blink following a tone that predicts a puff of air directed at the eye, relies on neural circuits in the cerebellum. Ischemic damage to the cerebellum following infarcts of the superior cerebellar artery or the posterior inferior cerebellar artery causes profound deficits in classical eye-blink conditioning but does not interfere with the ability to lay down new declarative memories. Evidence from such *double dissociations* endorses the idea that relatively independent brain systems govern the formation and storage of declarative and nondeclarative memories.

The connections between the basal ganglia and prefrontal cortex appear to be especially important for complex motor learning (see Chapter 19). Damage to either structure interferes with the ability to learn new motor skills. People with Huntington's disease, which causes atrophy of the caudate and putamen, perform poorly on motor skill learning tests such as manually tracking a spot of light, tracing curves using a mirror, or reproducing sequences of finger movements. Because the loss of dopaminergic neurons in the substantia nigra interferes with normal signaling in the basal ganglia, individuals with Parkinson's disease show similar deficits in motor skill learning (Figure 30.16), as do patients with prefrontal lesions caused by tumors or strokes. Neuroimaging studies have largely corroborated these findings, revealing activation of the basal ganglia and prefrontal cortex in typical people performing these same skill-learning tests. Activation of the basal ganglia and prefrontal cortex has also been observed in animals carrying out rudimentary motor learning and sequencing tasks.

(A)

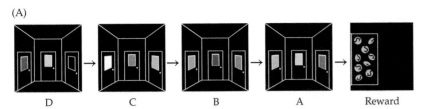

D   C   B   A   Reward

(B)

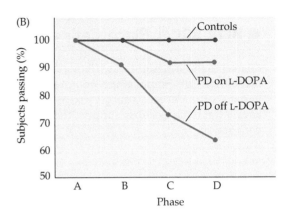

FIGURE 30.16 Parkinson's disease reveals a role for basal ganglia in nondeclarative memory (A) Participants performed a probabilistic learning task that had four levels. They first learned that selecting a door of one color (e.g., pink) in condition A led to a reward. Participants then learned that selecting a differently colored door (e.g., red) in condition B would permit them to proceed to condition A, where they could select the rewarded door. This procedure was continued until participants made choices from D → C → B → A → reward. (B) Participants with Parkinson's disease (PD) who were taking medication to replace depleted dopamine in the midbrain performed nearly as well as age-matched controls. However, participants with Parkinson's disease who were not on dopamine replacement were impaired in their ability to learn the task. (After D. Shohamy et al., 2005. *Behav Brain Res* 156: 191–199.)

In sum, a variety of evidence indicates that long-term memories, whether declarative or nondeclarative, are stored throughout the brain. This conclusion, however, does not imply that individual memory traces are randomly distributed over the cortex. The current view is that memories are stored primarily within the brain regions originally involved in processing each kind of information. That is, the striate and extrastriate visual cortices store memory traces for visual information, auditory cortices store memory traces for auditory information, and so on. Moreover, some inividuals with brain damage are impaired in very specific semantic or object categories, such as information about animals, and some forms of memory storage have been associated with memory mechanisms in restricted brain regions, such as the localization of eyeblink conditioning in the cerebellum, or fear conditioning in the amygdala.

CONCEPT
30.6

As Humans Age, Changes in the Brain Alter Memory Processes

LEARNING OBJECTIVES

30.6.1 Describe changes in brain structure that may influence memory processes as humans age.

30.6.2 Explain the limitations of brain-training exercises for minimizing or reversing age-related memory and cognitive declines.

Memory and aging

Although it is all too obvious that our outward appearance changes with age, most of us would like to believe that the brain is more resistant to the ravages of time.

Unfortunately, this optimistic view is not justified. The average weight of the typical human brain determined at autopsy steadily decreases from early adulthood onward (Figure 30.17). In elderly individuals, this effect can also be observed with noninvasive imaging as a slight but nonetheless significant shrinkage of the brain. Counts of synapses in the cerebral cortex generally decrease in old

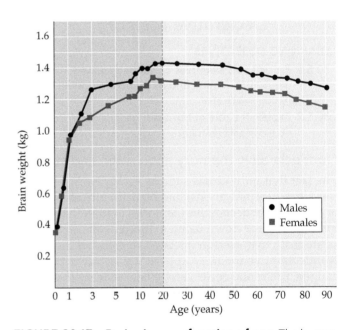

FIGURE 30.17 Brain size as a function of age The human brain reaches its maximum size (measured by weight in this case) in early adult life and decreases progressively thereafter. This decrease represents the gradual loss of neural circuitry in the aging brain presumably underlies the progressively diminished memory function in older individuals. (After A. S. Dekaban and D. Sadowsky, 1978. *Ann Neurol* 4: 345–356.)

NeuroImage 17: 1394–1402

Younger	Older, poor recall	Older, good recall

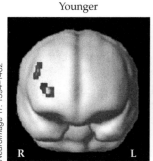 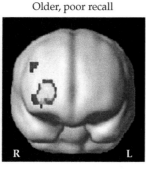 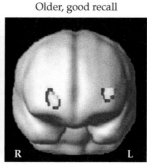

FIGURE 30.18 Compensatory activation of memory areas in high-functioning older adults During remembering, activity in prefrontal cortex was restricted to the right prefrontal cortex (following radiological conventions, the brain images are left–right reversed) in both young participants and older adult participants with poor recall. In contrast, older adult participants with relatively good memory showed activation in both right and left prefrontal cortex.

age (although the number of neurons probably does not change very much), suggesting that it is mainly the connections between neurons that are lost as humans grow old, consistent with the idea that the networks of connections that represent memories gradually deteriorate.

These observations accord with the difficulty older people have in making associations (e.g., remembering names, or the details of recent experiences) and with declining scores on tests of memory as a function of age. The normal loss of some memory function with age means that clinicians may have challenges in distinguishing individuals subject to normal aging from those who have Alzheimer's disease (see Box 30B). Although cognitive decline with age is ultimately inevitable, neuroimaging studies suggest that high-performing older adults may recruit additional neural resources to compensate for declines in processing efficiency (Figure 30.18).

Many commercial offerings today advertise that they forestall or reverse these effects of aging. Sometimes referred to as "exercising" or "training" the brain, such products often claim that the promised results are backed by "neuroscientific evidence." These products often involve a variety of games that test memory or other cognitive functions; people play the games via their phones or computers, and generally improve in their performance the more they play. Large-scale studies of such brain training have indicated that any benefits of the activities are marginal and transient at best. In most cases, performance improvements are limited to the specific games being played, and do not generalize to different sorts of cognitive challenges. For example, someone who spent many hours playing a word-clue matching game would probably become better at crossword puzzles but would be unlikely to show generalized improvements in long-term memory recall or executive function. Instead, considerable research indicates that improving overall cardiovascular health and building social connections are most critical for maintaining brain function—suggesting that time spent playing computerized brain-training games would be better spent going for a walk with friends.

Summary

Human memory entails many biological strategies and anatomical substrates. Primary among these are a system for memories that can be expressed by means of language and can be made available to the conscious mind (declarative memory) versus systems that concern skills and associations that are essentially nonverbal, operating at a largely unconscious level (nondeclarative, or procedural, memory). Evidence from people with amnesia indicates that the hippocampus and associated midline diencephalic and medial temporal lobe structures are critically important in acquiring and consolidating declarative memories, although not in storing them, which occurs primarily in the cerebral cortices. In contrast, the acquisition and consolidation of nondeclarative memories for motor and other unconscious skills depend on the integrity of the premotor cortex, basal ganglia, and cerebellum, and are not affected by lesions that impair the declarative memory system. The common denominator of stored information is generally thought to be alterations in the strength and number of the synaptic connections in the cerebral cortices that mediate associations between stimuli and the behavioral responses to them, which include perceptions, thoughts, and emotions as well as motor actions.

■ Additional Reading

Reviews

Biderman, N., A. Bakkour and D. Shohamy (2020). What are memories for? The hippocampus bridges past experience with future decisions. *Trends Cogn. Sci.* 24: 542–556.

Cabeza, R. and 12 others (2018) Maintenance, reserve and compensation: The cognitive neuroscience of healthy ageing. *Nat. Rev. Neurosci.* 19: 701–710.

Eichenbaum, H., A. R. Yonelinas and C. Ranganath (2007) The medial temporal lobe and recognition memory. *Annu. Rev. Neurosci.* 30: 123–152.

Gallistel, C. R. and L. D. Matzel (2013) The neuroscience of learning: Beyond the Hebbian synapse. *Annu. Rev. Psychol.* 64: 169–200. doi: 10.1146/anurev-psych-113011-143807

LaBar K. S. and R. Cabeza (2006) Cognitive neuroscience of emotional memory. *Nat. Rev. Neurosci.* 7: 54–64. doi: 10.1038/nrn1825

Lara, A. H. and J. D. Wallis (2015) The role of prefrontal cortex in working memory: A mini review. *Front. Syst. Neurosci.* 9: 173. https://doi.org/10.3389/fnsys.2015.00173

Schiller, D. and 6 others (2015) Memory and space: Towards an understanding of the cognitive map. *J. Neurosci.* 35: 13904–13911.

Squire, L. R. and J. T. Wixted (2011) The cognitive neuroscience of human memory since HM. *Annu. Rev. Neurosci.* 34: 259–288.

Zahodne, L. B. and P. A. Reuter-Lorenz (2019) Compensation and brain aging: A review and analysis of evidence. In *The Aging Brain: Functional Adaptation across Adulthood*, G. Samanez-Larkin (ed.). Washington, D.C.: American Psychological Association, pp. 185–216.

Important Original Papers

Adcock, R. A. and 4 others (2006) Reward-motivated learning: Mesolimbic activation precedes memory formation. *Neuron* 50: 507–517.

Cabeza, R., N. D. Anderson, J. K. Locantore and A. R. McIntosh (2002) Aging gracefully: Compensatory brain activity in high-performing older adults. *NeuroImage* 17: 1394–1402.

Dunsmoor, J. E., V. P. Murty, L. Davachi and E. A. Phelps (2015) Emotional learning selectively and retroactively strengthens memories for related events. *Nature* 520: 345–348.

Gobet, F. and H. A. Simon (1998) Expert chess memory: Revisiting the chunking hypothesis. *Memory* 6: 225–255.

Murty, V. P. and R. A. Adcock (2014) Enriched encoding: Reward motivation organizes cortical networks for hippocampal detection of unexpected events. *Cereb. Cortex* 24: 2160–2168.

Öztekin, I., B. McElree, B. P. Staresina and L. Davachi (2009). Working memory retrieval: contributions of the left prefrontal cortex, the left posterior parietal cortex, and the hippocampus. *J. Cogn. Neurosci.* 21: 581–593.

Scoville, W. B. and B. Milner (1957) Loss of recent memory after bilateral hippocampal lesions. *J. Neurol. Neurosurg. Psychiat.* 20: 11–21.

Squire, L. R. (1989) On the course of forgetting in very long-term memory. *J. Exp. Psychol.* 15: 241–245.

Talarico, J. M. and D. C. Rubin (2003) Confidence, not consistency, characterizes flashbulb memories. *Psychol. Sci.* 14: 455–461.

Wheeler, M. A., D. T. Stuss and E. Tulving (1995) Frontal lobe damage produces episodic memory impairment. *J. Int. Neuropsychol. Soc.* 1: 525–536.

Zola-Morgan, S. M. and L. R. Squire (1990) The primate hippocampal formation: Evidence for a time-limited role in memory storage. *Science* 250: 288–290.

Books

Baddeley, A., M. W. Eysenck and M. C. Anderson (2020) *Memory*, 3rd Edition. New York: Routledge.

Luria, A. R. (1987) *The Mind of a Mnemonist* (trans. L. Solotaroff). Cambridge, MA: Harvard University Press.

Neisser, U. (1982) *Memory Observed: Remembering in Natural Contexts*. San Francisco: W. H. Freeman.

Purves, D. and 5 others (2013) *Principles of Cognitive Neuroscience*, 2nd Edition. Sunderland, MA: Sinauer Associates.

Schacter, D. L. (2001) *The Seven Sins of Memory: How the Mind Forgets and Remembers*. Boston: Houghton Mifflin.

Speech and Language

Overview

Human culture, social networks, and personal relationships are supported by our capacity for language. Through language we imbue arbitrary symbols—expressed through speech, gesture, or writing—with meanings that convey ideas, feelings, desires, emotions, and more. Research on the neural mechanisms of language production and comprehension has become a central area of study in neuroscience, both because of language's centrality for our society and because of its importance for clinical practice. Understanding how linquistic processes are represented in the brain can provide new insights into developmental disorders, as seen in people who fail to develop a facility for language as a child and remain severely incapacitated throughout later life. Moreover, in advance of surgeries to remove dysfunctional brain tissue, clinicians and scientists often work together to identify and preserve cortical areas involved in language comprehension and production. Research using a variety of neuroscience methods has revealed that the linguistic abilities of humans depend on the integrity of several specialized areas located primarily in the association cortices of the left temporal, parietal, and frontal lobes. The linkage between speech sounds and their meanings is mainly represented in the left temporal and parietal cortices, and the circuitry for the motor commands that organize the production of meaningful speech is found mainly in the left frontal cortex. While the left hemisphere typically processes the lexical, grammatical, and syntactic aspects of language, the right hemisphere makes important contributions to processing the emotional content of speech. Studies of congenitally deaf individuals have shown that the cortical areas devoted to sign language are generally the same as those that organize spoken and heard communication. The regions of the brain devoted to language are therefore specialized for symbolic representation and communication rather than for heard and spoken language per se. Finally, many non-human animals exhibit communication skills that partially overlap with human language, leading to important work examining neural circuits that may have functional similarities across species.

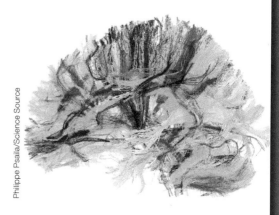

Philippe Psaila/Science Source

KEY CONCEPTS

31.1 Language production relies on both the vocal apparatus and cortical regions

31.2 Language comprehension relies on a distributed brain network

31.3 The right hemisphere makes important contributions to language

31.4 Language development includes a critical period during childhood

31.5 Nonhuman animals exhibit complex communicative abilities

CONCEPT 31.1 Language Production Relies on Both the Vocal Apparatus and Cortical Regions

LEARNING OBJECTIVES

31.1.1 Describe how speech is generated.

31.1.2 Understand the different elements of language and their properties.

31.1.3 Explain the evidence that led to the conclusion that language production relies on the left frontal lobe.

31.1.4 Provide an example of how genes contribute to language production.

The neural circuitry that supports human language production is distinct from, although related to, the circuitry concerned with the motor planning and control of the larynx, pharynx, mouth, and tongue—the structures that produce speech sounds. The regions of the brain that are specifically devoted to language transcend these more basic demands (Figure 31.1). The main concern of the areas of cortex that represent language is producing and interpreting symbols for communication—spoken and heard, written and read, or in the case of sign language, gestured and seen. Obedience to a set of rules for using these symbols (called grammar), ordering them to generate useful meanings (called syntax), and giving utterances the appropriate emotional valence by varying intensity, pitch, stress, and rhythm (called **prosody**) are all important and readily recognized features of language production, regardless of the particular mode of expression.

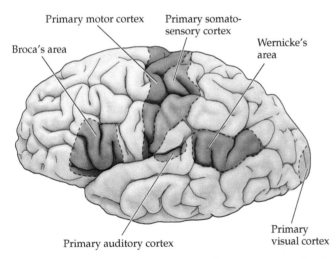

FIGURE 31.1 Diagram of the major brain areas involved in the comprehension and production of language The primary sensory, auditory, visual, and motor cortices are indicated to show the relation of Broca's and Wernicke's language areas to these other areas that are involved in the comprehension and production of speech, albeit in a less specialized way.

Elements of speech

The organs that produce speech include the lungs, which serve as a reservoir of air; the larynx, which is the source of the periodic quality of voiced sounds; and the pharynx, oral, and nasal cavities and their included structures (e.g., tongue, teeth, and lips), which modify (or filter) the speech sounds that eventually emanate from the speaker (Figure 31.2A). This generally accepted *source–filter model* of speech is an old one, having been proposed by Johannes Mueller in the nineteenth century (Figure 31.2B).

Although the physiological details are complex, the general operation of the vocal apparatus is simple. Air expelled from the lungs accelerates as it passes through a constricted opening between the **vocal folds** ("vocal cords") called the *glottis*, thus decreasing the pressure in the air stream according to Bernoulli's principle. As a result, the vocal folds come together until the pressure buildup from below forces them open again. The ongoing iteration of this process results in an oscillation of sound wave pressure, the frequency of which is determined primarily by the muscles that control the tension on the vocal folds. The fundamental frequencies of these oscillations range from about 100 to about 400 Hz, depending on the gender, size, and age of the speaker.

In any given language, the basic speech sounds are called *phones* and the percepts they elicit are called *phonemes*; different phones are produced as the muscles of the vocal tract change the tension on the vocal folds and the shape of the resonant cavities above the folds. Phonemes make up syllables in speech, which are used in turn to make up words, which are then strung together to create sentences. There are about 40 phonemes in English, and these are about equally divided between vowel and consonant speech sound percepts. Note that our subjective sense of the elements of speech (i.e., a succession of separate words and syllables) does not match the physical properties of the speech stimuli we produce. If one examines the auditory signal of a simple spoken phrase (Figure 31.3), there are evident pauses and breaks even within syllables themselves—yet what we hear seems integrated as one continuous flow of speech.

Vowel sounds are the periodic voiced elements of speech generated by the oscillation of the vocal cords. That oscillation can be modulated to produce different speech sounds, akin to how a musical instrument such as a clarinet can modulate oscillating air to produce different tones. And just as musical instruments have natural resonant properties that produce sounds at particular frequencies, the vocal tract produces acoustical energy within specific frequency bands called *formants* (i.e., peaks of power in the spectrum of a vocal sound stimulus; see Figure 31.2B). The power in the laryngeal source near the formant frequencies is reinforced, and power at other frequencies is, to varying degrees, filtered

(A)

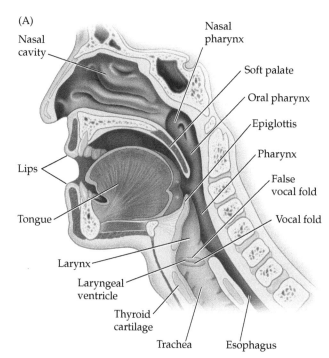

Nasal cavity

Nasal pharynx

Soft palate

Oral pharynx

Epiglottis

Pharynx

False vocal fold

Vocal fold

Lips

Tongue

Larynx

Laryngeal ventricle

Thyroid cartilage

Trachea

Esophagus

(B)

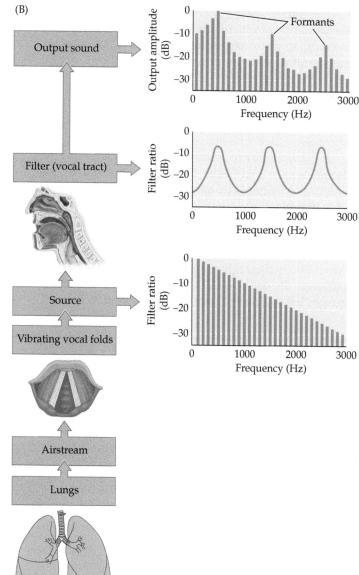

Output sound

Filter (vocal tract)

Source

Vibrating vocal folds

Airstream

Lungs

Formants

FIGURE 31.2   Generation of speech (A) The human vocal tract includes the vocal apparatus from the larynx to the lips. The structures above the larynx, including the pharynx, soft palate, and nasal cavity, shape and filter the harmonic series that are generated by the vocal folds when they vibrate. (B) The source–filter model of speech sound production. Using air expelled by the lungs, the vocal folds of the larynx are the source of the vibrations that become speech stimuli. Other components of the vocal tract, including the pharynx and structures of the oral and nasal cavities, filter the laryngeal harmonics by the superposition of their own resonances, thus creating the speech sound stimuli that we ultimately hear. (B after G. A. Miller, 1991. *The Science of Words.* New York: Scientific American Library.)

out. The resonance frequency of the first formant arises from the fact that the approximate length of the adult vocal tract in its relaxed state is about 17 cm, which is the quarter wavelength of a 68-cm sound wave; quarter wavelengths determine the resonances of pipes open at one end, which essentially describes the vocal tract. Since the speed of sound is about 33,500 cm/s, the lowest resonance frequency of an open tube or pipe of this length will be 33,500/68 or about 500 Hz; additional formants occur at about 1500 Hz and 2500 Hz. The changing shape of the

FIGURE 31.3   Recording of the spoken sentence "This is a glad time indeed" This plot shows how the intensity of the sound signal varies over time (about 2 seconds in total). Notably, the elements we perceive as speech sounds (e.g., words, syllables, and phones) do not map neatly to divisions of the auditory signal; for example, the breaks in intensity within words (e.g., the "g" in "glad") may be of longer duration than the breaks between words. (From D. A. Schwartz et al., 2003. *J Neurosci* 23: 7160–7168.)

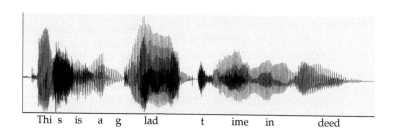

This   s   is   a   g   lad   t   ime   in   deed

vocal tract during speech changes the relative frequencies of the formants, producing different vowel sounds.

In contrast, consonant sounds involve rapid changes in the sound signal and are more complex. In English, consonants begin and/or end syllables, each of which includes a vowel sound. Consonant sounds are categorized according to the site in the vocal tract that determines them (the *place of articulation*), or the physical way they are generated (the *manner of articulation*). With respect to place, there are labial consonants (such as *p* and *b*), dental consonants (*f* and *v*), palatal consonants (*sh*), and glottal consonants (*h*, among many others). With respect to manner, there are plosive, fricative, nasal, liquid, and semi-vowel consonants. Plosives are produced by blocking the flow of air somewhere in the vocal tract, fricatives by producing turbulence, nasals by directing the flow of air through the nose, and so on. Another variation on the use of consonants is found in the "click languages" of southern Africa, of which about 30 survive today. Many of these languages have four or five different click sounds. They are produced by the tongue enclosing a pocket of air against the palate and then being sucked down from the roof of the mouth. The tongue release can occur at different positions, which creates different click sounds.

Cortical contributions to speech production

The distinction between language and the related motor and sensory functions on which it depends was first apparent in individuals with damage to specific brain regions. Clinical evidence from such cases showed that the ability to move the muscles of the larynx, pharynx, mouth, and tongue can be compromised without abolishing the ability to use language to communicate (even though a motor deficit may make communication difficult). Such individuals may still communicate through non-spoken language, such as the gestures used in American Sign Language (Box 31A), and through written language. Similarly, damage to the

---

■ BOX 31A │ Sign Language

The cortical organization of language does not simply reflect specializations for hearing and speaking; the language regions of the brain appear to be more broadly organized for processing symbols pertinent to social communication. Strong support for this conclusion has come from studies of sign language in individuals deaf from birth.

American Sign Language has all the components (grammar, syntax, and emotional tone) of spoken and heard language. Based on this knowledge, Ursula Bellugi and her colleagues at the Salk Institute for Biological Studies examined the cortical localization of sign language abilities in individuals who had suffered lesions of either the left or right hemisphere. All of these deaf individuals had never learned verbal language, had been signing throughout their lives, had deaf spouses, were members of the deaf community, and were right-handed. The individuals with left-hemisphere lesions, which in each case involved the language areas of the frontal and/or temporal lobes, had measurable deficits in sign production and comprehension when compared with typical

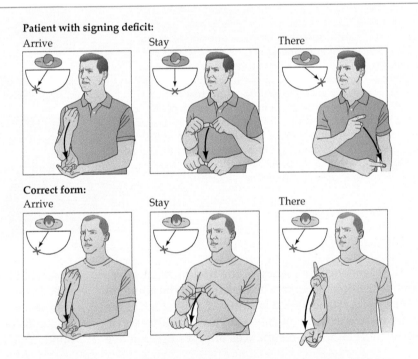

(A) Signing deficits in congenitally deaf individuals who learned sign language from birth and later suffered lesions of the language areas in the left hemisphere. Left-hemisphere damage produced signing problems in these individuals analogous to the aphasias seen after comparable lesions in hearing, speaking individuals. In this example, the individual (upper panels) is expressing the sentence "We arrived in Jerusalem and stayed there." Compared with a control (lower panels), the individual cannot properly control the spatial orientation of the signs. The direction of the correct signs and the inconsistent direction of the aphasic signs are indicated in the upper left-hand corner of each panel. (After U. Bellugi et al., 1989. *Trends Neurosci* 12: P380–388.)

■ BOX 31A | Sign Language *(continued)*

signers of similar age (Figure A). In contrast, the individuals with lesions in approximately the same areas in the right hemisphere did not have such signing aphasias. Instead, as predicted from hearing individuals with similar lesions, right-hemisphere abilities such as visuospatial processing, emotional processing, and the emotional tone evident in signing were impaired. Although the number of people studied was necessarily small—individuals who are both deaf and who have lesions of the language areas are understandably rare—the capacity for signed and seen communication was predominantly represented in the same areas

as spoken language in the left hemisphere. This evidence accords with the idea that the language regions of the brain are specialized for the representation of social communication by means of symbols, rather than for heard and spoken language per se.

The capacity for seen and signed communication, like its heard and spoken counterpart, emerges in early infancy. Careful observation of babbling in infants with normal hearing shows the production of a predictable pattern of sound production related to the ultimate acquisition of spoken language. Thus, babbling prefigures mature language, and indicates that an

innate capacity for language imitation is a key part of the process by which a full-blown language is ultimately acquired. The hearing offspring of deaf, signing parents babble with their hands in gestures that are apparently the forerunners of mature signs. As with verbal babbling, the amount of manual babbling increases with age until the child begins to form accurate, meaningful signs. These observations indicate that language development— whether verbal or gestural—relies on a similar process of acquiring the rudiments of symbolic communication from parental or other cues.

---

auditory pathways can impede the ability to hear without interfering with language functions per se (as is obvious in individuals who have become partially or wholly deaf later in life). Damage to specific brain regions, however, can compromise essential language functions while leaving

the motor and sensory infrastructure of verbal communication intact. These syndromes, collectively referred to as aphasias (Clinical Applications), diminish the capacity to recognize or employ the meaning of words, thus depriving such individuals of the linguistic understanding,

---

■ Clinical Applications

Clinical Presentations of Aphasia

In a classic Broca's aphasia, individuals cannot express themselves fluently because the organizational aspects of language (its grammar and syntax) have been disrupted, as shown in the following example reported by Howard Gardner (who is the interlocutor). This individual was a 39-year-old Coast Guard radio operator named Ford who had suffered a stroke that affected his left posterior frontal lobe.

"I am a sig...no...man...uh, well,... again." These words were emitted slowly, and with great effort. The sounds were not clearly articulated; each syllable was uttered harshly, explosively, in a throaty voice. With practice, it was possible to understand him, but at first I encountered considerable difficulty in this. "Let me help you," I interjected. "You were a signal..." "A sig-nal

man...right," Ford completed my phrase triumphantly. "Were you in the Coast Guard?" "No, er, yes,... ship...Massachu...chusetts...Coastguard...years." He raised his hands twice, indicating the number nineteen. "Oh, you were in the Coast Guard for nineteen years." "Oh... boy...right...right," he replied. "Why are you in the hospital, Mr. Ford?" Ford looked at me strangely, as if to say, Isn't it patently obvious? He pointed to his paralyzed arm and said, "Arm no good," then to his mouth and said, "Speech...can't say...talk, you see."

Howard Gardner, 1974
(The Shattered Mind: The Person after Brain Damage, pp. 60–61)

In contrast, the major difficulty in Wernicke's aphasia is putting together objects or ideas and the words that

signify them. Thus, in a Wernicke's aphasia, speech is fluent and well structured, but on closer inspection makes little or no sense, as is apparent in the following example (again from Gardner). The individual in this case was a 72-year-old retired butcher who had suffered a stroke affecting his left posterior temporal lobe.

Boy, I'm sweating, I'm awful nervous, you know, once in a while I get caught up, I can't get caught up, I can't mention the tarripoi, a month ago, quite a little, I've done a lot well, I impose a lot, while, on the other hand, you know what I mean, I have to run around, look it over, trebbin and all that sort of stuff. Oh sure, go ahead, any old think you want. If I could I would. Oh, I'm taking the word the wrong way to say, all of the barbers

(Continued)

■ Clinical Applications (continued)

*here whenever they stop you it's go-
ing around and around, if you know
what I mean, that is tying and tying for
repucer, repuceration, well, we were
trying the best that we could while
another time it was with the beds
over there the same thing...*

Ibid., p. 68

Despite the validity of Broca's and
Wernicke's original observations, the
classification of language disorders
is considerably more complex in
practice. For example, there is a third
broad category of language deficiency syndromes subsumed under
the general category of conduction
aphasia. These disorders arise from
lesions to the pathways connecting

Characteristics of Broca's and Wernicke's Aphasias

Broca's aphasia[a]	Wernicke's aphasia[b]
Halting speech	Fluent speech
Tendency to repeat phrases or words (perseveration)	Little spontaneous repetition
Disordered syntax	Syntax adequate
Disordered grammar	Grammar adequate
Disordered structure of individual words	Contrived or inappropriate words
Comprehension intact	Comprehension not intact

[a]Also called motor, expressive, or production aphasia.
[b]Also called sensory or receptive aphasia.

the relevant temporal and frontal regions, such as the arcuate fasciculus
that links Broca's and Wernicke's areas. Interruption of these pathways

may result in an inability to produce
appropriate responses to heard
communication, even though the
communication is understood.

---

grammatical and syntactic organization, or of appropriate
intonation that distinguish language from nonsense.

The first evidence for the localization of language function to a specific region of the cerebrum was provided by
the French neurologist Paul Broca in 1861. A 51-year-old
man had just been admitted to Broca's hospital because
of an acute infection; however, the man had longstanding
neurological problems that included epilepsy and, most
important for the history of science, a decade-long speech
impairment. Specifically, he had great difficulty producing
any coherent speech; for example, when trying to state his
name ("Leborgne"), he could only produce the nonsense
syllable "Tan," a nickname by which he became known
to posterity. Broca believed that language functions were
localized in the frontal lobes, and so he predicted that Tan's
language production deficits resulted from a lesion therein.
Less than a week later, Tan died—and an autopsy revealed
substantial damage to the inferior frontal lobe in the left
hemisphere, in line with Broca's prediction (Figure 31.4).

Tan's case provided landmark evidence for the general
idea of localization of function (i.e., that cortical regions
make distinct contributions to specific aspects of thought
and behavior, such as language and vision) and for the specific linking of language production to the frontal lobes.
It also led to a foundational rule about the localization of
language, which is that lesions of the left frontal lobe in a
region referred to as **Broca's area** affect the ability to *produce* language efficiently. This deficiency is called **motor** or
expressive aphasia, and is also known as **Broca's aphasia**.

(Expressive aphasias must be distinguished from *dysarthria*,
which is the impaired ability to move the muscles of the
mouth, tongue, and pharynx that mediate speech.) The deficient motor-planning aspects of expressive aphasias accord
with the complex motor functions of the posterior frontal
lobe and its proximity to the primary motor cortex.

However, Broca did not immediately recognize any
hemispheric asymmetry; only after the passage of a few

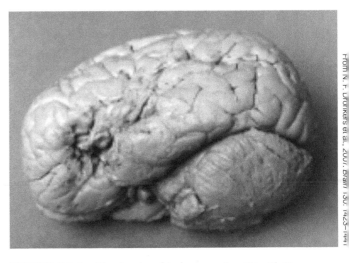

FIGURE 31.4   The brain of Leborgne (or "Tan") Shown
is the preserved brain of the patient whose inability to form coherent speech inspired Paul Broca's claims about the role of the
frontal lobes in language production.

years, and additional evidence from clinical cases, did he conclude that language production typically relies on the left frontal lobes. His subsequent claim that "we speak with the left hemisphere" has been amply confirmed by many more clinical cases and modern functional neuroimaging studies. Nor did Broca recognize the distinction between language production and comprehension, the latter of which relies on other other cortical regions. These important caveats—and the neuroscience research that provides a fuller picture of cortical contributions to language process—are considered later in this chapter.

Genes and language production

The search for genetic contributions to cognitive functions and their disorders is being pursued in many contexts. Because genes play some role in all phenotypic features, and because the propensity for language acquisition by infants is obvious, exploring the genes involved is plausible. Furthermore, the occurrence of language or reading problems that run in families makes plain that genetic anomalies can play a role in the normal development of these cognitive functions.

An inherited but quite rare disorder has more specifically raised the question of the genetic determination of language and, in popular accounts, whether there might be a "language gene." The gene of interest, called *FOXP2*, is located on human chromosome 7. It was discovered in 1990 in a family in which about half the members are afflicted. The affected individuals in the pedigree, known in the literature as the K.E. family, are unable to fluently select the movements of the vocal apparatus needed to make appropriate speech sounds. Thus, what they try to say is largely incomprehensible. The impairment, which is caused by a single autosomal recessive mutation, is thus one of motor organization, as it pertains to speech, rather than one of comprehension. The afflicted family members, however, also have lower IQs than their unafflicted relatives, indicating that the defect is not specific for language.

The mechanism by which the defective gene exerts these effects is not known, but the protein that *FOXP2* encodes is a transcription factor, meaning that the gene product is an agent that binds to the promoter regions of other genes to control their expression. The *FOXP2* gene is strongly expressed in other animals, including mice, where it affects many aspects of development, including the ultrasonic vocalization of these animals.

Interesting though this gene may be, reports about the discovery of a "language gene" were clearly unwarranted. Because it encodes a transcription factor, *FOXP2* affects many other genes with a range of developmental consequences, some evidently influencing the mechanisms that generate neural circuits in those parts of the brain that support the organization and expression of language.

CONCEPT 31.2 | Language Comprehension Relies on a Distributed Brain Network

LEARNING OBJECTIVES

31.2.1 Explain how receptive aphasia affects language comprehension.

31.2.2 Describe the McGurk effect and its relevance for understanding context effects in language comprehension.

31.2.3 Describe how the N400 response provides a marker for semantic violations.

Language comprehension

Considerable research supports the conclusion that language is not a unitary function localized in a single cortical region; instead, language production and comprehension rely on dissociable brain networks. Moreover, just as language production requires neural circuitry distinct from that supporting basic motor function, language comprehension relies on specialized brain systems that go beyond those supporting auditory perception of speech or visual perception of written words and gestures. Language comprehension, in particular, relies on the brain's ability to relate physical speech or written stimuli to representations stored in memory, building on the rich semantic structure that humans can perceive (Box 31B).

A milestone in the understanding of language comprehension came from work by the German neurologist Carl Wernicke in the late 1800s. Wernicke recognized that some people with aphasia who hear normally have great difficulty understanding speech; furthermore, these individuals often retain the ability to produce utterances with reasonable grammatical and syntactic fluency, but often without meaningful content. The underlying disorder is now referred to as **sensory aphasia** or **receptive aphasia**—or as **Wernicke's aphasia**, as a parallel to Broca's aphasia. Wernicke further observed that the brain damage that most commonly led to such aphasias involved lesions of the posterior and superior temporal lobe on the left side, regions now referred to as **Wernicke's area**. Deficits of reading and writing—*alexias* and *agraphias*, respectively—are separate disorders that can arise from damage to related but different brain areas; most people with aphasia, however, also have difficulty with these closely linked abilities. As an important reminder, the language comprehension deficits associated with damage to the posterior temporal lobe can be fully distinct from the production deficits associated with damage to the frontal lobe, leading to distinct profiles of dysfunctional behavior and clinical presentations (see Clinical Applications).

■ BOX 31B | Semantics: Extracting Meaning from Language

Research on language comprehension seeks to understand not just how people process the sounds and structure of language but also how they extract *meaning* from language. The study of how language and other symbolic systems generate meaning is known as semantics. Neuroscientists have long sought to understand semantic processing because of its relevance for many aspects of human thought and behavior. For example, memory researchers distinguish *episodic memory*, or representations of events from our past, from *semantic memory*, or representation of facts, concepts, and knowledge about the world. Understanding how the brain represents semantic concepts thus has important consequences not only for research on language, but also for research on memory, perception, and other processes of cognition.

As introduced in Chapter 30, the brain organizes episodic information in a distributed fashion, such that elements of our memories are stored throughout the brain. Semantic information seems to be similarly organized in a distributed fashion. Early evidence came from PET studies of naming images of different categories, such as photographs of the faces of well-known people, animals, or tools—each of which generated a distinct pattern of activation distributed throughout the cerebral cortex (Figure A). This observation helps explain the clinical finding that when a relatively limited region of the temporal lobe is damaged (usually but by no means always on the left side), language deficits are sometimes restricted to a particular *category* of objects. These studies are also consistent with electrophysiological studies that indicate that some aspects of language are organized according to categories of meaning rather than individual words.

More recent work combines higher-resolution fMRI with computational models of semantic relationships

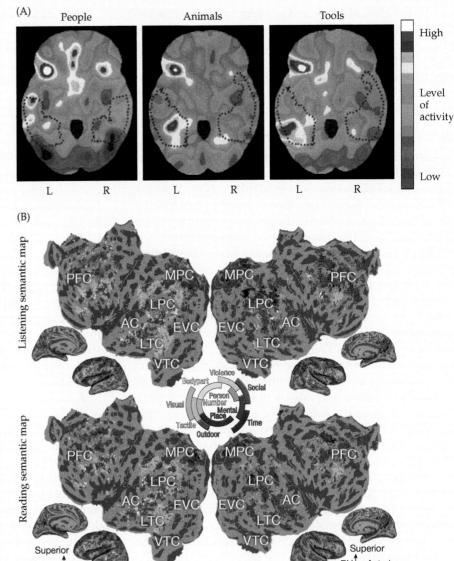

(A) Different regions in the temporal lobe are activated by different word categories. Dotted lines show the locations of the relevant temporal regions in these horizontal views. Note the different patterns of PET activity in the temporal lobe in response to each stimulus category. (From H. Damasio et al., 1996. *Nature* 380: 499–505.)

(B) Organization of semantic information throughout the cortex. Projected in color on the brain images at top and bottom are patterns of fMRI activation associated with the key components of a semantic map extracted from real-world stories; the color wheel at center provides labels for key categories that make up the semantic maps. Results indicate that semantic information is distributed throughout the cortex in a manner that is consistent across auditory and written language. Abbreviations: AC, auditory cortex; EVC, early visual cortex; LH/RH, left hemisphere/right hemisphere; LPC, lateral parietal cortex; LTC, lateral temporal cortex; MPC, medial parietal cortex; PFC, prefrontal cortex; VTC, ventral temporal cortex. (From F. Deniz et al., 2019. *J Neurosci* 39: 7722–7736. © 2019 Society for Neuroscience.)

■ BOX 31B | Semantics: Extracting Meaning from Language *(continued)*

in order to better understand the structure of meaning in the brain. In one evocative recent example, participants separately listened to and read stories taken from a popular radio program; each story presented a complex autobiographical narrative with many distinct semantic concepts. To construct a model of semantic content, the stories were transcribed and then analyzed for word co-occurrence within the text; concepts were considered related to the extent that they tended to co-occur in close proximity (e.g., in part of the same story) within this large corpus of natural speech. Then the relationships between the concepts were subjected to a data reduction technique that could extract a rough semantic space that described similarity relationships (see Figure B, center); for example, words describing mental states are more similar to words describing social relations than to words describing places or objects found outdoors. By examining the sensitivity of each brain location to the different features that defined that semantic space, the researchers could characterize the distribution of semantic information across the brain for both spoken language (see Figure B, top) and written language (see Figure B, bottom). The similarity in those distributions provided an elegant example of how semantic information is not only distributed broadly throughout the brain, but also is represented in a manner that seems largely independent of the manner by which that information is perceived.

Processing of speech: Phonology and meaning

More recent research has provided further insight into the different processes associated with language comprehension, particularly with regard to listening to and understanding human speech. Broadly considered, researchers have distinguished brain regions critical for phonological processing of speech stimuli from those that contribute to understanding the meanings conveyed.

Phonological processing relies largely on a distributed cortical network that includes both regions in temporal cortex that encode and disambiguate auditory signals and regions in frontal and parietal cortex that are associated with speech articulation (i.e., generation and rehearsal of speech sounds). Notably, how we perceive an auditory speech stimulus depends on its context. Straightforward examples can be seen in the cases of homonyms (i.e., words such as *bank* that have multiple distinct meanings) and homophones (i.e., words such as *dear* and *deer* that are represented by the same sound stimulus). The meaning of a sound stimulus containing one of those words depends on the surrounding speech—in the case of *bank*, for example, whether referring to a river or a building. We can also use visual cues to assist in speech comprehension, as when looking at the mouth movements of a conversation partner. The power of such contextual information can be seen in the visual–auditory illusion known as the **McGurk effect**, in which the auditory perception of a speech stimulus can be altered by the visual perception of a speaker's mouth movements (Figure 31.5A). Integration of visual and auditory information relies on processing in the superior temporal sulcus, as revealed by neuroimaging studies of the McGurk effect. In one clever example, researchers used fMRI to map the region of the superior temporal sulcus that contributes to audio–visual integration in the McGurk effect—and then disrupted the function of that region using transcranial magnetic stimulation (TMS). They found that TMS pulses delivered within about 100 ms of the onset of the auditory stimulus greatly attenuated the McGurk effect, such that the incongruent visual information no longer impaired speech comprehension (Figure 31.5B,C).

Given the rapidity of speech, techniques with relatively high temporal resolution provide important information about dynamic processes of comprehension. Electrophysiological techniques such as EEG have been especially important, particularly by identifying robust changes in the EEG signal that accompany key events of interest (i.e., event-related potentials, or ERPs). The most important such ERP for language comprehension is known as the N400 response, which is generated by the perception of a word (or other meaningful stimulus) that violates a semantic expectation. In the classic paradigm (Figure 31.6), participants read or listen to a series of words that form a grammatical sentence; each word is presented separately with a fixed interstimulus interval, so that the EEG response to each word can be isolated. Most sentences are semantically normal; that is, each word follows from the previous word largely as expected. However, a small number of sentences end with an unexpected word that is grammatically appropriate but semantically inappropriate (e.g., "I take coffee with cream and dog"). Those semantic deviants elicit a rapid brain response known as the N400, reflecting its timing (a peak at 400 ms) and negative amplitude, that has maximal amplitude over parietal cortex. Considerable research has used the N400 response as a marker of rapid language comprehension in the brain, allowing researchers to ask questions about the timing of key processes, the development of language comprehension over childhood, and even the relationship between language processing and other ways of extracting meaning from stimuli.

(A)

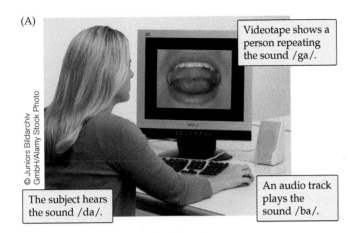

Videotape shows a person repeating the sound /ga/.

The subject hears the sound /da/.

An audio track plays the sound /ba/.

(B)

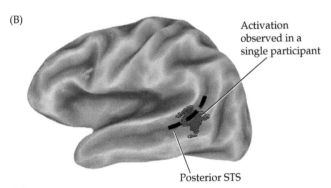

Activation observed in a single participant

Posterior STS

(C)

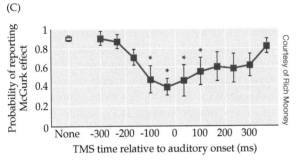

FIGURE 31.5 **The McGurk effect** (A) Participants listen to an audio recording that repeats a single syllable in isolation (e.g., /ba/) while watching a video of a speaker voicing a different syllable (e.g., /ga/). The resulting speech perception reveals that comprehension integrates both auditory and visual information, such that most people hear a speech sound (e.g., /da/) that, in visual terms, lies intermediate between the visible vocal tract configurations of the conflicting stimuli. If the participant closes their eyes, then they can readily report the correct auditory stimulus. (B) Using fMRI, researchers mapped the region of the superior temporal sulcus that responded to single words simultaneously in auditory and visual formats; shown in red is the activation observed in a single participant, with the posterior superior temporal sulcus indicated by the dashed line. (C) Individuals viewed an audio–visual stimulus that reliably generates a McGurk effect (see "None" for data in the absence of TMS) and received TMS pulses to the STS at various times before and after the onset of the stimulus. When a TMS pulse was delivered immediately before or after the onset of the stimulus, the likelihood of experiencing a McGurk effect was greatly attenuated. Error bars show standard error of the mean across subjects; asterisks show time points for which the effect was significantly attenuated ($p < 0.05$). (C after M. S. Beauchamp et al., 2010. *J Neurosci* 30: 2414–2417.)

(A)

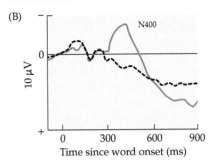

| ----- | IT | WAS | HIS | FIRST | DAY | AT | WORK |
| ——— | HE | SPREAD | THE | WARM | BREAD | WITH | SOCKS |

(B)

FIGURE 31.6 **Semantic processing of events that violate expectations** (A) Researchers interested in violations of semantic expectations have used simple stimuli in which words (or other meaningful stimuli) are presented sequentially. The key comparison is between words that are consistent with expectations and words that violate expectations; compare the last words of the two sentences here. (B) Words that violate a semantic expectation (blue line) evoke a negative potential in the ongoing EEG waveform that is known as the N400 response, reflecting its peak about 400 ms after the onset of the unexpected word. Words that are consistent with expectations (black line) do not evoke this response. (After M. Kutas and S. A. Hillyard, 1980. *Science* 207: 203–205.)

CONCEPT
31.3

The Right Hemisphere Makes Important Contributions to Language

LEARNING OBJECTIVES

31.3.1 Explain the role of the right hemisphere in supporting prosody.

31.3.2 Understand how and why clinicians seek to identify brain regions that support language processing.

31.3.3 Explain the clinical considerations that lead to split-brain syndrome and how that condition can be used to understand brain lateralization.

Hemispheric specialization in language processing

Much of the research on the neural basis of language—including that discussed so far in this chapter—has focused on regions in the left cerebral hemisphere. While lateralization can be seen in many of the processes discussed in earlier chapters of this unit (e.g., the unequal functions of the parietal lobes in attention and of the temporal lobes in recognizing different categories of objects), it is in language that the idea of lateralization has been most thoroughly documented. Because language is so important to humans, its lateralization has given rise to the misleading idea that the left hemisphere, in which the major capacity for verbal expression resides, is dominant over the right. But the true significance of lateralization for language (or for any other cognitive ability) lies in the efficient subdivision of complex functions between the two hemispheres rather than in any superiority of one hemisphere over the other.

Because the same gross anatomical and cytoarchitectonic areas exist in the cortex of both hemispheres, a puzzling issue remains. What do the comparable areas in the right hemisphere actually do? In fact, language deficits often do occur following damage to the right hemisphere, but those deficits are more subtle than, for comparison, the near-complete impairment in production seen in Broca's patient Tan (see Concept 31.1). For example, some individuals with damage to the right hemisphere can readily produce complex speech that includes appropriate vocabulary and follows grammatical rules but lacks the typical emotional and tonal components of language. When we speak, we stress different words, vary our intonation, and adjust our speaking rhythm. These prosodic aspects of speech provide critical information to listeners well beyond what is encoded in the words themselves; for example, changes in pitch over a sentence indicate whether an utterance is a question, statement, or command. In some languages, such as Mandarin, variations in tone are used to change the meaning of the word uttered. Deficiencies in speech prosody, referred to as **aprosodias**, are often associated with right-hemisphere damage to the cortical regions that correspond to Broca's and Wernicke's areas. Aprosodias emphasize that although the left hemisphere (or better put, specialized cortical regions in that hemisphere) figures prominently in the comprehension and production of language for most humans, other regions, including corresponding areas in the right hemisphere, are needed to generate the full richness of everyday speech.

Mapping language function in the brain

Understanding the location and lateralization of language processing has both scientific and clinical importance. Critically, not all individuals have the major language functions in the left hemisphere; some 1% to 5% of right-handed individuals and 30% of left-handed individuals exhibit right-hemisphere lateralization of language. An early method for assessing language lateralization, the **Wada test**, involves injecting a short-acting anesthetic (e.g., sodium amytal) into the left carotid artery of an individual who is performing a verbal task (e.g., counting out loud). If the left hemisphere is indeed dominant for language, then the individual becomes transiently aphasic for a few minutes before the anesthetic is diluted by the efficient circulatory system of the brain. Because this test is potentially dangerous, its use is typically limited to patients in advance of a neurosurgical procedure, such as before surgery to remove brain tissue that is generating epileptic seizures. Knowing the location of cortical regions critical for speech and language cortex is particularly important, to minimize risks of unwanted impairments following those surgeries. More recently, noninvasive techniques have become

increasingly popular for presurgical language mapping, including both approaches for disrupting brain function temporarily (e.g., TMS) and approaches for mapping regions active during language production (e.g., fMRI).

Once the dominant hemisphere has been identified—allowing a surgical plan that avoids language cortex—neurosurgeons typically map language areas more precisely by electrical stimulation of the cortex during surgery. By the 1930s, the neurosurgeon Wilder Penfield and his colleagues had already carried out a detailed localization of cortical capacities in a large number of patients. Penfield used electrical mapping techniques adapted from neurophysiological work in animals to delineate the language areas of the cortex prior to removing brain tissue in the treatment of tumors or epilepsy (Figure 31.7A). Such

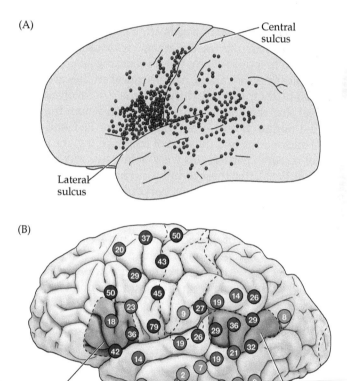

FIGURE 31.7 Evidence for the variability of language representation among individuals, determined by electrical stimulation during neurosurgery (A) Diagram from Penfield's original study illustrating sites in the left hemisphere at which electrical stimulation interfered with speech. (B) Diagram summarizing data from 117 patients whose language areas were mapped by electrical recording at the time of surgery. The number in each red circle indicates the percentage of patients who showed interference with language in response to stimulation at that site. Note that many of the sites that elicited interference fall outside the classic language areas (Broca's area, shown in purple; Wernicke's area, shown in blue). (A after W. Penfield and L. Roberts, 1959. *Speech and Brain Mechanisms.* Princeton, NJ: Princeton University Press; B after G. A. Ojemann et al., 1989. *Electroencephalo Clin Neurophys* 73: 453–463.)

intraoperative mapping guaranteed that the cure would not be worse than the disease and has been widely used ever since, with increasingly sophisticated stimulation and recording methods. More recent studies using electrophysiological recording methods during surgery have shown that a large region of the perisylvian cortex of the left hemisphere is clearly involved in language production and comprehension. A surprise, however, has been the variability in language localization (Figure 31.7B); the brain regions that support language only approximate those indicated by textbook treatments, with locations that differ unpredictably among individuals. Other conclusions from neurosurgical studies are equally unexpected. Bilingual individuals do not necessarily use the same bit of cortex for storing the names of the same objects in two different languages. Moreover, although neurons in the temporal cortex in and around Wernicke's area respond preferentially to spoken words, they do not show preferences for a particular word. Rather, a wide range of words can elicit a response at any given recording site.

Despite these advances, neurosurgical studies are complicated by their intrinsic difficulty, the risk involved, and the fact that the brains of the patients on whom they are carried out are not typical. The advent of PET and then fMRI allowed the investigation of language regions in volunteer research participants without invasive surgery (Figure 31.8). Recall that these techniques reveal the areas of the brain that are active during a particular task because the related electrical activity increases local metabolic activity and blood flow. These experiments have challenged excessively rigid views of the localization and lateralization of linguistic function. Although high levels of activity occur in the expected regions, large areas of both hemispheres are activated in word recognition or production tasks.

Split-brain syndrome: Evidence for lateralization

Until the 1960s, observations about language localization and lateralization were based primarily on individuals with brain lesions of varying severity, location, and etiology—or on then rare examples of language disruptions before or during neurosurgery (e.g., the Wada test discussed in the previous section). Up until that time, the inevitable uncertainties of clinical findings had allowed some skeptics to argue that language and other complex cognitive functions might not be localized or even lateralized in the brain. Definitive evidence supporting the inferences from neurological observations came from studies of individuals whose corpus callosum and anterior commissure were severed as a treatment for medically intractable epileptic seizures. (Recall from Chapter 8, Clinical Applications, that a small proportion of people with severe epilepsy are refractory to medical treatment, and that interrupting the connection between the two hemispheres was an effective way of treating epilepsy in highly selected

Passively viewing words

 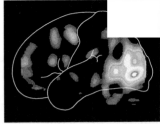

Listening to words

"Table"

 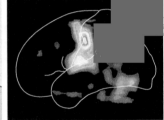

Speaking words

Table
"Table"

 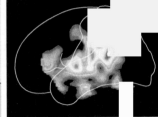

Generating word associations

Table
"Chair"

 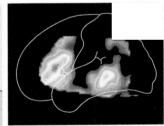

FIGURE 31.8 Language-related regions of the left hemisphere mapped by PET in intact individuals The left panels indicate the task being practiced prior to scanning. The PET scan images are shown on the right. Language tasks such as listening to words and generating word associations elicit activity in Broca's and Wernicke's areas, as expected. However, there is also activity in primary and association sensory and motor areas for both active and passive language tasks. These observations indicate that language processing involves many cortical regions in addition to the classic language areas. Warm colors indicate greater activation. (Illustrations after M. I. Posner and M. E. Raichle, 1994. *Images of Mind*. New York: Scientific American Library.)

From M. I. Posner and M. E. Raichle, 1994. *Images of Mind*. New York: Scientific American Library

individuals). In such individuals, investigators could assess the function of the two cerebral hemispheres independently because the major axon tracts that connect them had been interrupted. The first studies of these so-called split-brain patients—carried out by Roger Sperry, Michael Gazzaniga, and their colleagues—established the hemispheric lateralization of language beyond any doubt. Their work also demonstrated many other functional differences between the left and right hemispheres

(Figure 31.9) and stands as an extraordinary contribution to the understanding of brain organization.

To evaluate the functional capacity of each hemisphere in split-brain individuals, it is essential to provide information to one side of the brain only—for example, by

(A)

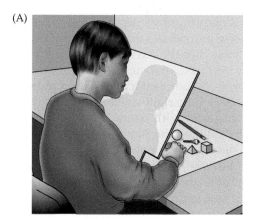

Left-hemisphere functions	Right-hemisphere functions
Analysis of right visual field	Analysis of left visual field
Stereognosis (right hand)	Stereognosis (left hand)
Lexical and syntactic language	Emotional coloring of language
Writing	Spatial abilities
Speech	Rudimentary speech

(B) Control individual

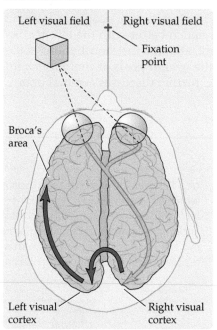

Split-brain individual

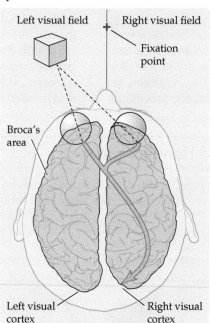

Split-brain individual

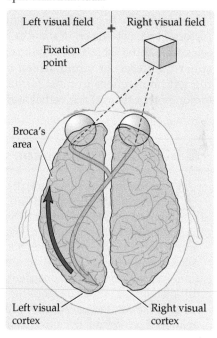

FIGURE 31.9 Confirmation of hemispheric specialization for language obtained by studying individuals in whom the connections between the right and left hemispheres have been surgically divided (A) Single-handed, vision-independent *stereognosis* can be used to evaluate the language capabilities of each hemisphere in individuals with split-brain syndrome. Objects held in the right hand, which provides somatosensory information to the left hemisphere, are easily named; objects held in the left hand, however, are not readily named by most of these individuals. The list shown here enumerates some of the different functional abilities of the left and right hemispheres, as deduced from a variety of behavioral tests in split-brain individuals. (B) Visual stimuli or simple instructions can also be given independently to the right or left hemisphere in intact and split-brain individuals. Since the left visual field is perceived by the right hemisphere (and vice versa; see Chapter 12), a briefly presented (*tachistoscopic*) instruction in the left visual field is appreciated only by the right brain (assuming that the individual maintains fixation on a mark in the center of the viewing screen). In typical control individuals, activation of the right visual cortex leads to hemispheric transfer of visual information via the corpus callosum to the left hemisphere. In split-brain individuals, information presented to the left visual field cannot reach the left hemisphere, and individuals are unable to produce a verbal report regarding the stimuli. However, such individuals are able to provide a verbal report of stimuli presented to the right visual field. A wide range of hemispheric functions can be evaluated using this method, even in control individuals.

asking the individual to use each hand independently to identify objects without any visual assistance (see Figure 31.9A). Recall from Chapter 9 that somatosensory information from the right hand is processed by the left hemisphere, and vice versa. Therefore, by asking the individual to describe an item being manipulated by one hand or the other, researchers can test the language capacity of the relevant hemisphere. Such studies show clearly that the two hemispheres differ in their language ability, in keeping with the postmortem correlations described earlier.

Split-brain individuals were able to name objects held in the right hand without difficulty, supporting the role of the left hemisphere in object naming and speech production. Yet most could not name objects held in the left hand, instead producing only an indirect description that relied on rudimentary words and phrases (for instance, "a round thing" instead of "a ball"); some could not provide any verbal account whatsoever. Observations using techniques to present visual information to the hemispheres independently (a method called *tachistoscopic presentation*; see Figure 31.9B) showed further that the left hemisphere responds to written commands, whereas the right hemisphere typically responds only to nonverbal stimuli (e.g., pictorial instructions or, in some cases, rudimentary written commands). These distinctions reflect broader hemispheric differences summarized by the statement that the left hemisphere in most humans is specialized for (among other things) the verbal and symbolic processing important in communication, whereas the right hemisphere is specialized for (among other things) visuospatial and emotional processing.

The ingenious studies of split-brain individuals put an end to the century-long controversy about language lateralization. In most people, the left hemisphere is unequivocally the seat of the explicitly verbal language functions. There is significant variation in the degree of lateralization among individuals, however, and it would be wrong to suppose that the right hemisphere has no language capacity. As noted, in some individuals the right hemisphere can produce rudimentary words and phrases, a few individuals have fully right-sided verbal functions, and even for the majority with strongly left-lateralized language semantic abilities, the right hemisphere is typically the source of our emotional coloring of language (see Chapter 33). Moreover, the right hemisphere in many split-brain individuals understands language to a modest degree, since these individuals can respond to simple visual commands presented tachistoscopically in the left visual field. Consequently, Broca's conclusion that we speak with our left brain is not strictly correct; it would be more accurate to say that most people understand language and speak much better with the left hemisphere than with the right, and that the contributions of the two hemispheres to the overall goals of communication are different. Box 31C further discusses cerebral dominance and handedness.

■ BOX 31C | Language and Handedness

Approximately nine out of ten people are right-handed, a proportion that appears to have been stable over thousands of years and across all cultures in which handedness has been examined. In addition, handedness, or its equivalent, is not peculiar to humans; many studies have demonstrated paw preference in animals ranging from mice to monkeys that is, at least in some ways, similar to human handedness. Unlike in humans, however, which hand is preferred varies about equally among individuals.

Handedness is usually assessed by having individuals answer a series of questions about preferred manual behaviors, such as "Which hand do you use to write?"; "Which hand do you use to throw a ball?"; and "Which hand do you use to brush your teeth?" Each answer is given a value, depending on the preference indicated, providing a quantitative measure of the inclination toward right- or left-handedness. Anthropologists have determined the incidence of handedness in ancient cultures by examining artifacts; the shape of a flint ax, for example, can indicate whether it was made by a right- or left-handed individual. Handedness in antiquity has also been assessed by noting the incidence of people in artistic representations who are using one hand

(A) Right-handed   Left-handed

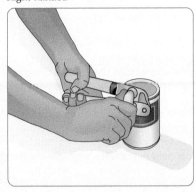

(A) A simple manual can opener is one example of the many common objects designed for use by the right-handed majority.

or the other. Based on this evidence, the human species appears always to have been primarily right-handed.

Whether an individual is right- or left-handed has several interesting consequences. As will be obvious to left-handers, the world of human artifacts is in many respects a right-handed one. Implements such as can openers, scissors, and power tools are constructed for the right-handed majority (Figure A). By the same token, the challenge of penmanship is different for English-speaking left- and right-handers by virtue of writing from left to right (Figure B). Perhaps as a consequence of such biases, the accident rate for left-handers in all categories (work, home, sports, traffic fatalities) is higher than for right-handers. However, there are also some advantages to being left-handed. For example, an inordinate number of international fencing champions have been left-handed. The reason for this fact is simply that the majority of any individual's opponents will be right-handed; therefore, the average fencer, whether right- or left-handed, is less practiced at parrying thrusts from left-handers.

Hotly debated over the years have been the related questions of whether being left-handed is in any sense "pathological," and whether being left-handed entails a diminished life expectancy. No one disputes the fact that there is currently a surprisingly small number of left-handers among the elderly (Figure C). These data have come from studies of the general population and have been supported by information gleaned from *The Baseball Encyclopedia* (in which longevity and other characteristics of a large number of left- and right-handers have been recorded because of interest in the U.S. national pastime).

Two explanations for this peculiar finding have been put forward. Stanley Coren and his collaborators at the University of British Columbia have argued that these statistics reflect a higher mortality rate among left-handers, partly as a result of increased accidents, but also because other data show left-handedness to be associated with a variety of pathologies (there is, for instance, a

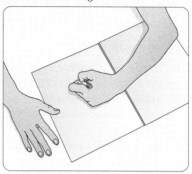

(B) Right-handed writing   Left-handed writing

(B) Writing techniques for right- and left-handed individuals.

higher incidence of left-handedness among individuals classified as mentally impaired). Coren and others have suggested that left-handedness may arise because of developmental problems in the pre- or perinatal period. If this were shown to be true, then a rationale for decreased longevity would have been identified that might combine with greater proclivity for accidents in a right-hander's world.

An alternative explanation, however, is that the diminished number of left-handers among the elderly at present is primarily a reflection of sociological factors—namely a greater acceptance of left-handed children today compared with earlier in the twentieth century. In this view, there are fewer older left-handers because in earlier generations parents, teachers, and other authority figures encouraged (and sometimes insisted on) right-handedness. The weight of the evidence favors the sociological explanation.

The relationship between handedness and other lateralized functions—language in particular—has long been a source of confusion. It is unlikely that there is any direct relationship between language and handedness, despite much speculation to the contrary. The most straightforward evidence on this point comes from the results of the Wada test described in Concept 31.3. The large number of such tests carried out for clinical purposes indicates that about 97% of humans, including the majority of left-handers, have verbal language functions represented in the left hemisphere (although that right-hemispheric

dominance for language is more common among left-handers). Since most left-handers don't have language function on the same side of the brain as the control of their preferred hand, it is hard to argue for any strict relationship between these two lateralized functions. In all likelihood, handedness, like language, is first and foremost an example of the advantage of having any specialized function in one hemisphere or the other to maximize wiring efficiency.

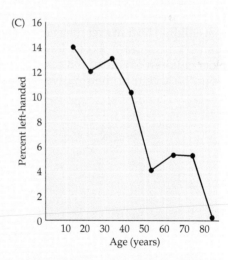

(C) Percentage of left-handers in the U.S. population as a function of age (based on more than 5000 individuals assessed in 1990). Taken at face value, these data indicate that right-handers live longer than left-handers. A more likely possibility is that the paucity of elderly left-handers simply reflects changes in the social pressures on children to become right-handed. (After S. Coren, 1992. *The Left-Hander Syndrome: The Causes and Consequences of Left-Handedness*. New York: The Free Press.)

CONCEPT 31.4 Language Development Includes a Critical Period During Childhood

LEARNING OBJECTIVES

31.4.1 Provide two different examples of evidence for a critical period in language development.

31.4.2 Explain how the absence of exposure to language during development can affect linguistic abilities later in life.

31.4.3 Define dyslexia and its effects on language abilities.

The development of speech and language abilities

Human linguistic abilities develop rapidly over early life. An influential theory of language development contends that infants and young children begin with an undifferentiated capacity for language learning that becomes refined following exposure to their native language. Newborn humans show a preference for speech and speechlike stimuli (e.g., animal vocal calls) over auditorily similar non-speech stimuli—but this general preference becomes increasingly specific to speech from their native language by about 3 months of age. Strikingly, while very young infants can readily distinguish speech sounds that aren't present in their native language (e.g., the /r/ and /l/ sounds that are absent in the Japanese language), that ability diminishes before about 1 year of age—replaced by improved discrimination of sounds that children hear in everyday native speech. Younger children often can learn second languages more rapidly than older children or adults, and can do so with less intrusion from the accent of their native language (Figure 31.10).

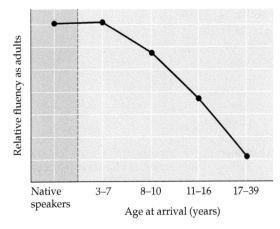

FIGURE 31.10 The critical period for language learning The critical period for fully fluent language learning is apparent in studies of the fluency of adult Chinese Americans as a function of the age of their arrival in the United States, marking the onset of significant exposure to English. Ultimate fluency starts to drop off when language learning begins after about age 7. (After J. S. Johnson and E. L. Newport, 1989. *Cogn Psychol* 21: 60–99.)

Collectively these facts illustrate two key principles of development: that neural circuitry is especially susceptible to modification during early life, and that this malleability gradually diminishes with maturation. The window for extensive neural modification supporting a behavior is referred to as a **critical period** (also called a sensitive period), and it is especially evident in the acquisition of language.

Studies of language acquisition in children who become deaf at different ages reveal the impact of experience on language development. Younger children who have acquired some speech but then lose their hearing suffer a substantial decline in the quality of spoken language (e.g., the ability to form words and sentences verbally). Without hearing their own speech during this critical period, they cannot refine that speech through auditory feedback. Losing hearing in adolescence or adulthood has a less marked effect on language skills. Consistent with these observations, neuroscience research supports the conclusion that the neural substrates of language production and comprehension differ in children and adults. When children and adults are tested in language tasks using fMRI, the brains of children tend to show activation in a broader set of regions well beyond the primary language-related structures identified in adults (Figure 31.11A). Other intriguing evidence for language development comes from diffusion tensor imaging (DTI), a variant of structural MRI that can map white matter tracts in the brain. Such work has shown that the primary pathways between Broca's and Wernicke's areas (e.g., the arcuate fasciculus), remain immature in young children (7 years old) and that communication between these regions may follow an alternative pathway called the extreme capsule fiber system (Figure 31.11B).

If experience with language during a critical period is necessary for its development, then the lack of such experience during childhood should lead to the inability to use language (or even to learn language) in adulthood. Consider what might happen if an otherwise typical child were never exposed to language. Would the child remain mute, or could he or she develop some ability to speak? If so, what sort of language would she or he have? The closest answers to these questions come from a handful of unfortunate cases in which children have been deprived of significant language exposure as a result of having been raised in conditions of complete social deprivation.

In the most fully documented case, a girl in a Los Angeles suburb was raised from infancy until age 13 under conditions of almost total isolation. Genie (as she is known in the scientific literature) was brought to the attention of social workers in 1970, who found her locked in a small room where she had been isolated and allegedly beaten if she made any noise. She was removed from these desperate conditions and taken to the children's hospital at the University of California, Los Angeles, where she was found to be in adequate general health. Given these highly unusual

(A)

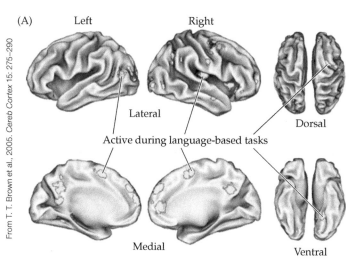

Left   Right

Lateral

Dorsal

Active during language-based tasks

Medial   Ventral

(B)

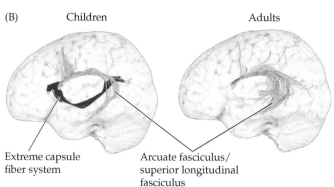

Children   Adults

Extreme capsule
fiber system

Arcuate fasciculus/
superior longitudinal
fasciculus

From T. T. Brown et al., 2005. *Cereb Cortex* 15: 275–290

FIGURE 31.11 Language-related differences between children and adults in brain function and structure (A) Areas in the brain that are differentially activated in children and adults during language-based tasks are shown in yellow. These differences provide a possible neural basis for the diminishing ability to learn a new language with increasing age, as the more diffuse set of brain regions observed in children crystallizes into a smaller, more functionally specific set of regions by adulthood. (B) By combining diffusion tensor imaging (DTI) measures of white matter tracts with fMRI measures of activation during auditory language comprehension, neuroscientists have identified the likely pathways that support language processing in children (left) and adults (right). Notably, children's language comprehension relies on two pathways connecting the frontal and temporal lobes: the arcuate fasciculus (green) and the extreme capsule fiber system (blue). By adulthood, only the former pathway is associated with functional connectivity between language-related regions. (B after J. Brauer et al., 2011. *Cereb Cortex* 21: 459–466.)

circumstances, a team of psychologists and linguists studied Genie's language and other cognitive skills during the subsequent 5 years. Although Genie had little or no language ability initially, the investigators found no evidence of brain damage or extreme general cognitive deficits, and they described her overall personality as rather docile and generally pleasant. As might be expected, Genie also received extensive remedial training to teach her the language skills that she had never learned as a child. Despite

these efforts, as well as daily life in more or less normal conditions in foster homes, Genie never acquired more than rudimentary language skills. Although she eventually learned a reasonable vocabulary, she could not put words together using typical grammatical rules, saying things like "Applesauce buy store?" when she wanted to ask whether she might buy some applesauce at the store. Genie's case and a few similar examples starkly define the importance of adequate early experience for successfully learning any language, in accord with the more abundant evidence of a critical period for learning a second language.

In summary, researchers agree that the normal acquisition of human language is subject to a critical period of approximately a decade; exposure and practice must occur within this time for a person to achieve full fluency. The increasing societal recognition of this critical period has led to investments in educational and family interventions that can support children's development, for example by exposing them to more spoken language (e.g., via reading out loud) at earlier ages. Some ability to learn language persists into adulthood, but at a reduced level of efficiency and ultimate performance. This generalization is consistent with much other evidence from neural development that underscores the special importance of early experience in the full development of cognitive abilities.

Reading and dyslexia

Problems with language production and comprehension during childhood can have significant negative consequences for educational outcomes and everyday life. Most common is dyslexia, which describes a broad category of disorders characterized by impairments related to written language, most notably in reading. People with dyslexia often have difficulty writing, are poor spellers, and are prone to errors arising from letter transposition. Dyslexia is not associated with impairments in general intelligence, memory, or most other cognitive processes, although it does commonly co-occur with disorders of attention such as ADHD. Because there is no specific diagnostic criterion for dyslexia, estimates of its prevalence vary widely; 5% to 15% of children are affected, with greater occurrence in boys.

For clinical diagnosis, the term *dyslexia* is subsumed under the clinical category Specific Learning Disorder, which characterizes neurodevelopmental disorders that affect children's performance in school and educational activities but that do not result from intellectual disabilities or neurological conditions. Dyslexia is by far the most common such disorder; others include dysgraphia (difficulty writing by hand) and dyscalculia (difficulty performing mathematical calculations). There is no accepted treatment for dyslexia, although identifying the problem early and implementing remediation through extra training and effort can ameliorate its impact, especially when interventions begin early (i.e., when the child is first learning to read). Consistent with the concept of a critical period discussed in the previous section, if children with dyslexia do not receive assistance until adolescence or later, then they may develop persistent reading challenges that impair their success in school and in life.

Several lines of evidence indicate that dyslexia has a neural basis, including studies of its heritability and genetics (i.e., it tends to run in families) and more recent work using brain imaging techniques. Researchers investigating the neural underpinnings of dyslexia have understandably focused on areas of the brain concerned with reading. fMRI studies indicate that a specific set of left-hemispheric brain areas is activated during reading. Some of these areas are also activated by spoken language, but one of them, the visual word form area (VWFA) located in the region of the left occipito-temporal sulcus, is selectively activated by written characters but not by spoken words or low-level visual stimuli (Figure 31.12). The organization of the VWFA appears to depend on experience, and activation levels in this area in children and adolescents predict word–phoneme decoding abilities. People with dyslexia tend to exhibit less fMRI activation in this general area compared with people who do not have dyslexia, as well as underdevelopment of the associated cortex and white matter tracts.

Evidence for a brain region functionally specific to reading may seem surprising, given that written language arose very late within the span of human evolution (about 5000 years ago) and that until recent centuries relatively few humans were literate. Thus, the VWFA could not have evolved to support reading per se, and may be better thought of as a brain region whose processing

characteristics could be adapted to the demands of reading rather than a brain region devoted specifically to words. An extension of this idea is that cultural inventions such as reading and writing may differ in expressions across cultures, but rely on similar brain circuits and thus share some similar constraints. For example, letters in all alphabets tend to be made up of a small number of strokes, which might in turn be related to the efficient use of receptive fields by neurons at successively higher levels of visual cortex. The argument is that by matching the appearance of letters to the inherent functions of neurons involved in recognizing elementary objects, writing and reading systems are similarly determined across cultures.

CONCEPT
31.5

Nonhuman Animals Exhibit Complex Communicative Abilities

LEARNING OBJECTIVES

31.5.1 Provide examples of flexible, context-dependent animal communication.

31.5.2 Describe the evidence that non-human primate communication relies on similar underlying brain systems as human language.

31.5.3 Explain the basic mechanisms of vocal learning and communication in birds.

Do other animals have language?

Over the centuries, theologians, natural philosophers, and a good many modern neuroscientists have argued that language abilities are uniquely human. Yet modern research reveals that animals as diverse as bees, birds, monkeys, and whales possess highly sophisticated systems of communication. One apparent difference between human language and animal communication lies in the range of meaning that can be conveyed. Human language has sufficient flexibility to associate any concept with a collection of arbitrary symbols; we create new words and phrases to describe events, elements of our environment, and even internal states. Non-human animal communication seems more restricted. When foraging honeybees return to their hive, they begin a dance whose direction and duration convey the angle (relative to the position of the sun) and the distance of a food source. Many prey species emit vocalizations that signal not only the presence of a predator but also other characteristics. For example, vervet monkeys have distinct alarm calls for different predators, each leading to a distinct type of behavior: A bark signaling an approaching leopard causes other monkeys to climb nearby trees, while a lower-pitched call signals that an eagle is circling, leading the group to hide in the low bushes. While these and many other examples provide impressive evidence of

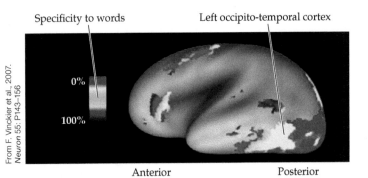

From F. Vinckier et al., 2007. Neuron 55: P143–156

Specificity to words | Left occipito-temporal cortex

0%

100%

Anterior | Posterior

FIGURE 31.12 Hierarchical organization across the extent of the VWFA in the left occipito-temporal cortex The figure shows an anterior–posterior gradient in stimulus specificity as stimuli being read are progressively changed from single letters to complete words. In typical individuals, but less so in those with dyslexia, increasingly closer approximations to the usual appearance of words (i.e., shifts from false fonts to regular fonts) lead to increasing anterior activation in the left but not the right occipito-temporal cortex. Regions shown in red respond equally to words and nonword stimuli, whereas regions shown in blue are selectively active for words.

adaptive communications, they nevertheless convey only a small set of potential concepts (e.g., aerial vs. terrestrial predator) and cannot generalize to new concepts in an arbitrarily complex fashion.

A series of studies in great apes, however, has claimed that the rudiments of language are evident in the behavior of our closest relatives. The most remarkable results have come from sophisticated experiments that train chimpanzees to use keyboards with a variety of symbols that can be arranged to express ideas in an interpretable manner (Figure 31.13) or to gesture via sign language to convey intentions. With appropriate training, chimpanzees can learn as many as 400 different symbols (or gestures) that can be combined into simple demands, questions, and even expressions of feeling, allowing them to have something resembling a rudimentary conversation with the experimenters. The more accomplished of these animals are alleged to have vocabularies of several thousand phrases (i.e., combinations of symbols) and to exhibit some hallmarks of language usage, including the creation of context-appropriate novel expressions. Even so, the way the animals use phrases is dramatically less impressive than that of human children with apparently similar vocabularies.

Given the challenge that claims of animal language capabilities present to long-held beliefs about human uniqueness, there has been continuing debate about those capabilities and their interpretation. Nonetheless, the issues raised deserve careful consideration by anyone interested in human language abilities and how our remarkable symbolic skills may have evolved from the communicative capabilities of our ancestors. The pressure for the evolution of some form of symbolic communication in great apes seems clear enough. Ethologists studying chimpanzees in the wild have described extensive social communication based on gestures, the manipulation of objects, and facial expressions. In addition, studies of monkeys have shown that some species typically use a variety of vocalizations in socially meaningful ways, and that these vocalizations may activate regions in the frontal and temporal lobes that are homologous to Broca's and Wernicke's areas in humans (Figure 31.14). This intricate social intercourse by gestures, facial expression, and limited vocalizations in non-human primates may rely on the same building blocks as human language; one need only think of the importance of gestures, facial expressions, and nonverbal human vocal sounds as ancillary aspects of our own speech to appreciate this point.

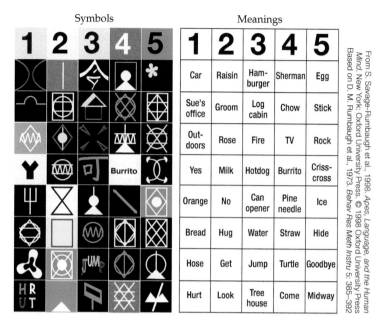

From S. Savage-Rumbaugh et al., 1998. *Apes, Language, and the Human Mind*. New York: Oxford University Press. © 1998 Oxford University Press. Based on D. M. Rumbaugh et al., 1973. *Behav Res Meth Instru* 5: 385–392

FIGURE 31.13 Rudiments of language in non-human primates Keyboard showing lexical symbols used to study symbolic communication in great apes.

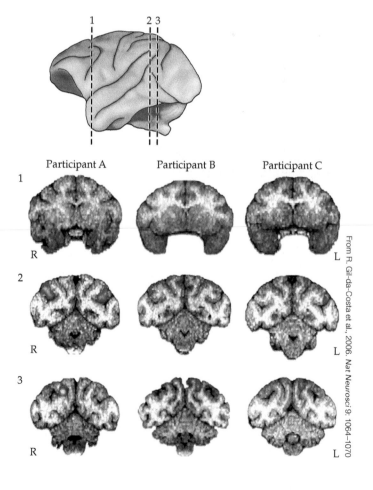

From R. Gil-da-Costa et al., 2006. *Nat Neurosci* 9: 1064–1070

FIGURE 31.14 Activation of areas in the frontal and temporal lobes of three rhesus monkeys responding to conspecific vocal calls The areas activated are arguably similar to the major language areas in the human brain.

Vocal communication in birds

Humans are not the only animals that learn to communicate during a critical period of development, and studies of that development in birds have added greatly to a better understanding of social communication and its neural basis. Some bird vocalizations are innate in the sense that they require no experience to be correctly produced and interpreted. For example, quails raised in isolation or deafened at hatching so that they never hear conspecific vocal stimuli nonetheless produce the full repertoire of species-specific vocalizations. Other species of birds, however, learn to communicate by vocal sounds via a developmental process that is in some respects similar to the way humans learn language. Particularly well characterized is vocal learning in song sparrows, canaries, and finches. These and other bird species use songs to define their territory and attract mates (Figure 31.15A,B).

As with human language, early sensory exposure and practice are key determinants of subsequent perceptual and behavioral capabilities. Furthermore, the developmental period for learning these vocal behaviors, as for learning language, is restricted to early life. (Canaries are exceptional in that they continue to build their song repertoire from season to season, which is one reason these birds have been such popular pets over the centuries.) Song learning in these species entails an initial stage of *sensory acquisition*, when the juvenile bird listens to and memorizes the song of an adult tutor (e.g., its father) of its own species. This period is followed by a stage of vocal learning through practice, when the young bird matches its song to the memorized tutor model by auditory feedback. This *sensory–motor learning* stage ends with the onset of sexual maturity, when songs become acoustically stable (called *crystallized song*) (Figure 31.15C).

In the species typically studied, young birds are especially impressionable during the first 2 months after hatching and become refractory to further exposure to tutor songs as they age, thus defining a critical (or sensitive) period for song learning. The early exposure to the tutor generates a memory that can remain intact for months (or longer) in some species before the onset of the vocal practice phase. Moreover, juveniles need to hear the tutor song only 10 to 20 times to vocally mimic it months later, and exposure to other songs after the sensory acquisition period does not affect this memory. The songs heard during this time, but not later, are the only ones that the young

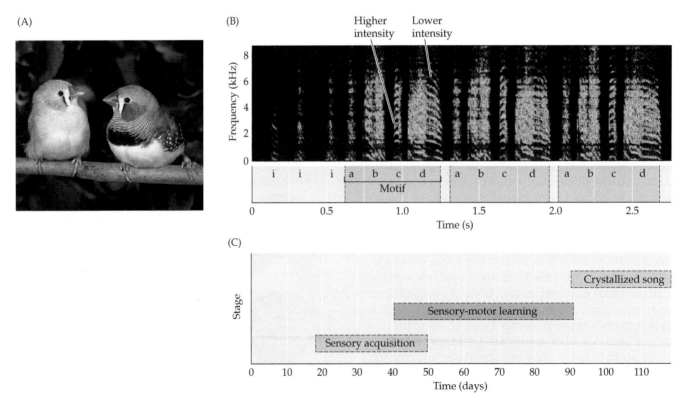

FIGURE 31.15   Birdsong learning (A) A pair of zebra finches, a model species for investigations of vocal learning. The male is on the right. (B) Spectrogram showing the song of an adult male zebra finch that is used in courting the female (note that considerable research shows that female songbirds also sing). The recording plots the frequency of the song against time, showing the syllables and motifs that characterize the song of this species. Color indicates the intensity of the vocal signal, with red representing higher intensity and blue lower. (C) The stages of song learning in the zebra finch (0 indicates the time of hatching). (Courtesy of Rich Mooney.)

bird mimics. Songbirds also exhibit learned regional dialects, much as human infants learn the language characteristic of the region in which they are raised.

Other studies indicate that birds have a strong intrinsic predisposition to learning the song of their own species. Thus, when presented during maturation with a variety of recorded songs that include their own species' song, together with that of other species, juvenile birds preferentially learn the song of their own species. This observation shows that juveniles are not really naïve, but are innately biased to learn the songs of their own species in preference to those of other species. Indeed, some evidence suggests that songbirds have a very rough template of their species song that is expressed in the absence of any exposure to that song or any other. Thus, birds raised in isolation produce highly abnormal "isolate" songs that have some characteristics of the song they would normally have learned (unlike Genie in the comparable human example). Such songs, however, are biologically ineffective in that they fail to attract mates.

Thus, the vocally relevant parts of the bird brain are already prepared during early life to learn the specific vocal sounds of the species, much as the brains of human infants are prepared at birth to learn language. Although the similarities with human language acquisition can be exaggerated, at least some aspects of human language have analogues in the vocal communicative abilities of other animals.

Summary

A variety of methods have been used to understand the organization of language in the human brain. This effort began in the nineteenth century by correlating clinical signs and symptoms with the location of brain lesions determined postmortem. In the twentieth and now twenty-first centuries, additional clinical observations together with studies of split-brain individuals, mapping prior to neurosurgery, transient anesthesia of a single hemisphere, and noninvasive imaging techniques such as PET and fMRI have greatly extended knowledge of the neural substrates of language. Together, these various approaches show that the perisylvian cortices of the left hemisphere are especially important for normal language in the vast majority of humans. The right hemisphere also contributes importantly to language, most obviously by giving it emotional tone. The similarity of the deficits after comparable brain lesions in congenitally deaf individuals and their speaking counterparts has shown further that the cortical representation of language is independent of the means of its expression or perception (spoken and heard versus gestured and seen). The specialized language areas that have been identified are the major components of a widely distributed set of brain regions that allow humans to communicate effectively by means of symbols that can be attached to objects, concepts, and feelings. Unlike social communication in other species, humans can manipulate and organize linguistic symbols to create an endless range of meanings, including combinations that convey novel concepts. Future research in neuroscience and related fields will continue to answer important questions about the origins of human language and the basis of its extraordinary development.

■ Additional Reading

Reviews

Binder, J. R. and 4 others (2009) Where is the semantic system? A critical review and meta-analysis of 120 functional neuroimaging studies. *Cereb. Cortex* 19: 3267–3296.

Bloomfield, T. C., T. Q. Gentner and D. Margoliash (2012) What birds have to say about language. *Nat. Neurosci.* 14: 947–948.

Evans, N. and S. C. Levinson (2009) The myth of language universals: Language diversity and its importance for cognitive science. *Behav. Brain Sci.* 32: 429–492.

Gazzaniga, M. S. (1998) The split brain revisited. *Sci. Am.* 329 (1): 50–55.

Gazzaniga, M. S. and R. W. Sperry (1967) Language after section of the cerebral commissures. *Brain* 90: 131–147.

Gibbon, J., C. Malapani, C. L. Dale and C. R. Gallistel (1997) Toward a neurobiology of temporal cognition: Advances and challenges. *Curr. Opin. Neurobiol.* 7: 170–184.

Kutas, M. and K. D. Federmeier. (2011). Thirty years and counting: Finding meaning in the N400 component of the event-related brain potential (ERP). *Annu. Rev. Psychol.* 62: 621–647.

Price, C. J. (2012) A review and synthesis of the first 20 years of PET and fMRI studies of heard speech, spoken language, and reading. *NeuroImage* 62: 816–847.

Seyfarth, D. M. and D. I. Cheney (1984) The natural vocalizations of non-human primates. *Trends Neurosci.* 7: 66–73.

Important Original Papers

Abe, K. and D. Watanabe (2012) Songbirds possess the spontaneous ability to discriminate syntactic rules. *Nat. Neurosci.* 14: 1067–1074.

Bagley, W. C. (1900–1901) The apperception of the spoken sentence: A study in the psychology of language. *Am. J. Psychol.* 12: 80–130.

Berwick, R. C., A. D. Friederici, N. Chomsky and J. J. Bolhuis (2013) Evolution, brain, and the nature of language. *Trends Cogn. Sci.* 17: 89–98.

Brauer, J., A. Anwader and A. D. Freiderici (2011) Neuroanatomical prerequisites for language functions in the maturing brain. *Cereb. Cortex* 21: 459–466.

Brown, T. T. and 5 others (2005) Developmental changes in human cerebral functional organization for word generation. *Cereb. Cortex* 15: 275–290.

Chang, E. F. and 4 others (2013) Human cortical sensorimotor network underlying feedback control of vocal pitch. *Proc. Nat. Acad. Sci. U.S.A.* 110: 2653–2658.

Deniz, F., A. O. Nunez-Elizalde, A. G. Huth and J. L. Gallant (2019). The representation of semantic information across human cerebral cortex during listening versus reading is invariant to stimulus modality. *J. Neurosci.* 39: 7722–7736.

Fromkin, V. and 4 others (1974) The development of language in Genie: A case of language acquisition beyond the "critical period." *Brain Lang.* 1: 81–107.

Gentner, T. Q., K. M. Fenn, D. Margoliash and H. C. Nusbaum (2006) Recursive syntactic pattern learning by songbirds. *Nature* 440: 1204–1207.

Gil-da-Costa, R. and 5 others (2006) Species-specific calls activate homologs of Broca's and Wernicke's areas in the macaque. *Nat. Neurosci.* 9: 1064–1070.

Kutas, M. and S. A. Hillyard (1980) Reading senseless sentences: Brain potentials reflect semantic incongruity. *Science* 207: 203–205.

Leonard, M. K. and E. F. Chang (2014) Dynamic speech representations in the human temporal lobe. *Trends Cogn. Sci.* 18: 472–479.

Miller, G. A. and J. C. R. Licklider (1950) The intelligibility of interrupted speech. *J. Acoust. Soc. Am.* 22: 167–173.

Pollick, A. S. and F. B. M. de Waal (2007) Ape gestures and language evolution. *Proc. Natl. Acad. Sci. U.S.A.* 104: 8184–8189.

Vinckier, F. and 5 others (2007) Hierarchical coding of letter strings in the ventral stream: Dissecting the inner organization of the visual word-form system. *Neuron* 55: 143–156.

Books

Anderson, S. (2012) *Languages: A Very Short Introduction.* Oxford: Oxford University Press.

Bloom, P. (2002) *How Children Learn the Meanings of Words.* Cambridge, MA: MIT Press.

Chomsky, N. (1957) *Syntactic Structures.* The Hague: Elsevier.

Darwin, C. (1872) *The Expression of Emotion in Man and Animals.* Reprint, Chicago: University of Chicago Press, 1965.

McNeil, D. (2000) *Language and Gesture.* Cambridge: Cambridge University Press.

Rogers, T. T. and J. L. McClelland (2004) *Semantic Cognition: A Parallel Distributed Processing Approach.* Cambridge, MA: MIT Press.

Tomasello, M. (2008) *Origin of Human Communication.* Cambridge, MA: MIT Press.

von Frisch, K. (1993) *The Dance Language and Orientation of Bees* (trans. Leigh E. Chadwick). Cambridge, MA: Harvard University Press.

Emotion

Overview

Emotions are critically important for human behavior. They combine subjectively experienced feelings with a diverse set of changes to our body state and behavior—from engagement of autonomic processes (e.g., altered heart rate and sweating) to motor responses (e.g., movements of the facial muscles and changes in posture). They can change our actions in the moment, as when anger leads to violence, or our tendencies over much longer time scales, as seen in disorders such as depression. And emotions are interrelated with the many other aspects of cognition discussed in this unit; for everyday examples, consider how happiness and sadness intertwine in our memories or how the anticipation of regret can change our decisions. Consistent with the broad role of emotion in cognition, research on the neural mechanisms of emotion has a similar breadth. Some core brain regions (e.g., the amygdala) contribute to many aspects of emotional processing. But emotion is also supported by a constellation of cortical and subcortical regions, including central components of the visceral motor system as well as regions in the forebrain and diencephalon that motivate lower motor neuron pools concerned with the somatic expression of emotional behavior. Dysfunction within brain structures that support emotional processing has been linked to a diverse set of clinical conditions, including drug abuse and psychiatric illnesses.

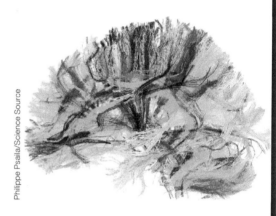

Philippe Psaila/Science Source

KEY CONCEPTS

32.1 Emotions integrate feelings, physiology, and behavior

32.2 The amygdala plays a central role in emotional processing

32.3 The cerebral cortices support emotional processing

32.4 Emotions interact with other cognitive processes

CONCEPT **32.1** | **Emotions Integrate Feelings, Physiology, and Behavior**

LEARNING OBJECTIVES

32.1.1 Provide an example of how neuroscientists can measure physiological changes that accompany emotion.

32.1.2 Distinguish among categorical, dimensional, and component process theories of emotion.

32.1.3 Describe the changes in brain function that lead to sham rage.

32.1.4 Explain the contributions of the visceral and somatic motor systems to emotion.

Defining emotion

In everyday use, the word *emotion* refers to conscious feelings as varied as happiness, anger, fear, surprise, disgust, jealousy, and much more. Yet considering emotions only in terms of subjective states is a very limiting perspective; for example, that view implies that non-human animals do not experience or use emotions similarly to humans. Researchers now conceptualize emotional states as a composite of subjective feelings, physiological responses, and behaviors that allow humans and other animals to react adaptively to internal and external stimuli (Figure 32.1).

Measurement of physiological changes that accompany emotions provides neuroscientists with important information about how subjective feelings are generated. Emotional arousal involves changes in the activity of the visceral (autonomic) motor system (see Chapter 21). Increases or decreases in heart rate, cutaneous blood flow (blushing or turning pale), skin temperature, sweating, piloerection, pupil size, and gut motility can all accompany various emotions. One physiological measure that has been particularly important for research on how different stimuli or situations elicit emotions is **skin conductance**, an index of sweating typically measured by electrodes on the palmar surface of the hands. Because fear and anxiety are linked to high arousal states, skin conductance provides information about moment-to-moment changes in these emotions. However, skin conductance is modulated by other emotions, such as sexual arousal, as well as by attentional orienting responses to novel stimuli. Skin conductance has also been used to identify stimulus-evoked arousal responses that arise unconsciously. For instance, individuals with prosopagnosia often show skin conductance responses to pictures of family members, despite being unable to consciously recognize the individuals by sight.

Physiological responses associated with emotions are brought about by changes in activity in the sympathetic, parasympathetic, and enteric components of the visceral motor system, which governs smooth muscle, cardiac muscle, and glands throughout the body. As discussed in Chapter 21, Walter Cannon argued that intense activity of the sympathetic division of the visceral motor system prepares the animal to fully utilize metabolic and other resources in challenging or threatening situations. Conversely, activity of the parasympathetic division promotes a building up of metabolic reserves. Cannon further suggested that the natural opposition of the expenditure versus the storage of resources is reflected in a parallel opposition of the emotions associated with these different physiological states. As he pointed out, "The desire for food and drink, the relish of taking them, all the pleasures of the table are naught in the presence of anger or great anxiety."

Classifying emotions

There are three primary perspectives on how emotions should be classified; each provides a different way of labeling emotions and characterizing their substrates. As implied by their name, **categorical theories** separate emotional experience into distinct categories that differ qualitatively in their subjective experiences, physiological states, and behavioral tendencies. Most of these theories argue that there is a small set of **basic emotions** that are universal across human societies and cultures; while the theories differ in exactly which emotions are contained in that set, those most commonly included are anger, disgust, fear, happiness, sadness, and surprise (Figure 32.2). Other subjective feelings (e.g., guilt, envy) are often considered to arise from combinations of these basic building blocks. While considerable research supports the claims that some emotional states are both universal across humans and are qualitatively different from other emotions (e.g., anger and sadness are always distinct), other emotional states are clearly context-dependent. For example, how pride is experienced and expressed depends on the setting and one's environment; cultures differ dramatically in whether expressions of pride (e.g., changes in language and body posture) are socially acceptable.

Dimensional theories argue that different emotions do not fall into discrete categories, but instead reflect different values along fundamental dimensions, most commonly **arousal** (i.e., high vs. low intensity) and **valence** (i.e., positive vs. negative affect). Many different emotional states can be characterized along these two dimensions (Figure 32.3). Highly arousing and highly positive subjective states tend to be characterized as "happy," while highly arousing but highly negative states are more associated with "angry." Note that dimensional theories do not consider our commonly labeled emotions to be fundamental elements of brain function; instead, lower-level processes that shape arousal and valence would be experienced as a particular emotion in a particular context.

A third perspective, collectively called **component process theories**, contends that differences among emotions

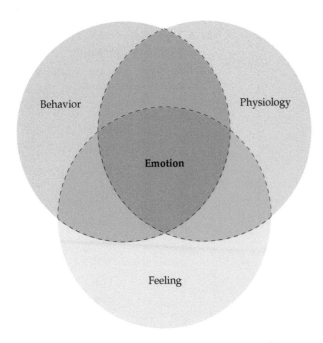

FIGURE 32.1 **Emotions** Emotions typically have three components: behavioral manifestations, a subjective feeling, and a physiological state.

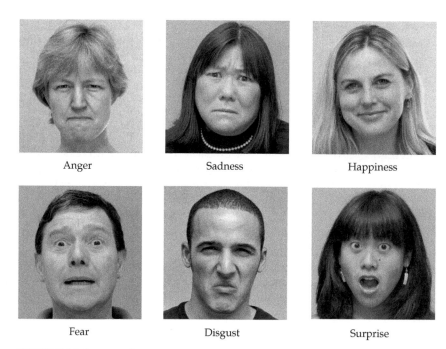

Anger   Sadness   Happiness

Fear   Disgust   Surprise

FIGURE 32.2   Facial expressions of emotions Categorical theories of emotion argue for a set of primary emotions, each of which represents a distinct combination of facial expressions and feelings, and potentiates distinct behaviors (e.g., anger leads to aggression, sadness to withdrawal). These six emotions are those most commonly included in categorical theories.

Circumplex model

Valence

Content
Satisfied   Pleased   Happy
Relaxed   Glad
Calm   Delighted   Excited
At ease   Serene

Sleepy   Aroused
Astonished

Tired

Alarmed

Droopy   Afraid   Tense

Annoyed   Angry
Bored   Frustrated
Sad   Miserable   Distressed
Gloomy   Depressed

Arousal

FIGURE 32.3   Dimensional theories of emotion An alternative perspective on emotions contends that different emotional states can be organized according to their values on fundamental dimensions. The circumplex model shown here has two primary dimensions: arousal and valence. Terms that describe similar emotions cluster within similar regions of the resulting space of possible emotions. (After J. A. Russell, 1980. *J Pers Soc Psychol* 39: 1161–1178.)

reflect differences in how people appraise a situation and respond to it (Table 32.1). Two emotions might share some appraisals but not others: Appraising a situation as novel might generate either joy (when the situation is conducive to one's goals) or anger (when the situation is obstructive). Considering emotions in terms of component processes fits well with current basic neuroscience research, as these processes can be readily linked to underlying brain regions or networks. Moreover, component process theories also can provide insights into clinical disorders. For example, the U.S. National Institute of Mental Health has adopted the Research Domain Criteria (RDoC) framework, which contends that mental disorders reflect profiles of dysfunction across a set of component processes (notably including emotion-related domains).

Integrating the components of emotion

These various physiological and behavioral components of emotions must be integrated in some way. In 1928, Philip Bard reported a series of experiments that pointed to the hypothalamus as a critical center for coordination of both the visceral and somatic motor components of emotional behavior (see Box 21A). Bard removed both cerebral hemispheres (including the cortex, underlying white matter, and basal ganglia) in a series of cats. When the anesthesia had worn off, the animals behaved as if they were enraged. The angry behavior occurred spontaneously and included the usual autonomic correlates of this emotion: increased blood pressure and heart rate, retraction of the nictitating membranes (the thin connective tissue sheets associated with feline eyelids), dilation of the pupils, and erection of the hairs on the back and tail. The cats also exhibited somatic motor components of anger, such as arching the back, extending the claws, lashing the tail, and snarling. This behavior was called **sham rage** because it had no obvious target. Bard showed that a complete response occurred as long as the caudal hypothalamus was intact (Figure 32.4). Sham rage could not be elicited, however, when the brain was transected at the junction of the hypothalamus and midbrain (although some uncoordinated components of the response were still apparent). Bard suggested that whereas the subjective experience of emotion might depend on an intact cerebral cortex, the expression of coordinated emotional behaviors does not necessarily entail cortical processes. The functional importance of emotions in all mammals is consistent with the involvement of phylogenetically older parts of the nervous

(A) No sham rage

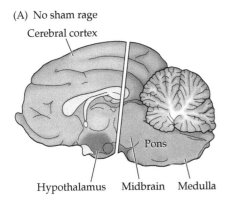

Cerebral cortex

Hypothalamus Midbrain Medulla

Pons

(B) Sham rage remains

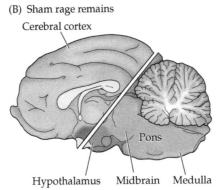

Cerebral cortex

Hypothalamus Midbrain Medulla

Pons

FIGURE 32.4 Midsagittal view of a cat's brain, illustrating the regions sufficient for the expression of emotional behavior (A) Transection through the midbrain, disconnecting the hypothalamus and brainstem, abolishes sham rage. (B) The integrated emotional responses associated with sham rage survive removal of the cerebral hemispheres as long as the caudal hypothalamus remains intact. (After J. E. LeDoux, 1987. In *Handbook of Physiology*, F. Blum et al. [Eds.]. Bethesda, MD: American Physiological Society, pp. 419–459.)

system. Bard also emphasized that emotional behaviors are often directed toward self-preservation, as Darwin had pointed out in his work on the evolution of emotion.

Complementary results were reported by Walter Hess, who showed that electrical stimulation of discrete sites in the hypothalamus of awake, freely moving cats could also lead to a rage response, and even to subsequent attack behavior. Moreover, stimulation of other sites in the hypothalamus caused a defensive posture that resembled fear. In 1949, a share of the Nobel Prize in Physiology or Medicine was awarded to Hess "for his discovery of the functional organization of the interbrain [hypothalamus] as a coordinator of the activities of the internal organs." Experiments like those of Bard and Hess led to the important conclusion that the basic circuits for organized behaviors accompanied by emotion are in the diencephalon and the

brainstem structures connected to it. Furthermore, their work emphasized that control of the involuntary motor system is not entirely separable from control of the voluntary pathways, an important consideration in understanding the motor aspects of emotion, as we discuss in the next section of this concept.

The routes by which the hypothalamus and other forebrain structures influence the visceral and somatic motor systems are complex. The major targets of the hypothalamus lie in the **reticular formation**, the tangled web of nerve cells and fibers in the core of the brainstem (see Box 17A). This structure contains more than 100 identifiable cell groups, including some of the nuclei that control the brain states associated with sleep and wakefulness described in Chapter 28. Other important circuits in the reticular formation control cardiovascular function, respiration, urination, vomiting, and swallowing. The reticular neurons receive hypothalamic input from and feed into both somatic and autonomic effector systems in the brainstem and spinal cord. Their activity can therefore produce widespread visceral motor and somatic motor responses, often overriding reflex functions and sometimes involving almost every organ in the body (as implied by Cannon's dictum about the sympathetic preparation of the animal for fight or flight). In addition to the hypothalamus, other sources of descending projections from the forebrain to the brainstem reticular formation contribute to the expression of emotional behavior.

The descending control of emotional expression thus entails two parallel systems that are anatomically and functionally distinct (Figure 32.5). The voluntary motor component described in detail in Unit III comprises the classic motor areas of the posterior frontal lobe and related

TABLE 32.1 Component Process Theories of Emotion[a]

Appraisal Criteria	Joy	Anger	Fear	Sadness
Novelty[b]	High	High	High	Low
Pleasantness	High	Variable	Low	Variable
Goal significance				
Outcome certainty	High	Very high	High	Very high
Conduciveness	Conducive	Obstructive	Obstructive	Obstructive
Urgency	Low	High	Very high	Low
Coping potential				
Agency	Self/other	Other	Other	Variable
Control	High	High	Variable	Very low
Power	High	High	Very low	Very low
Adjustment	High	High	Low	Medium

[a]Component process theories maintain that emotions arise from combinations of specific appraisals of and responses to an emotion-eliciting situation. Table adapted from Ellsworth & Scherer (2003).

[b]Both joy and anger can be elicited by novel situations. However, situations that induce anger tend to be obstructive to obtaining one's goals and are accompanied by a high sense of urgency to act, whereas those that induce joy tend to be conducive to one's goals and have a low sense of urgency. Different emotions resemble each other to the extent that they share some of these component cognitive processes.

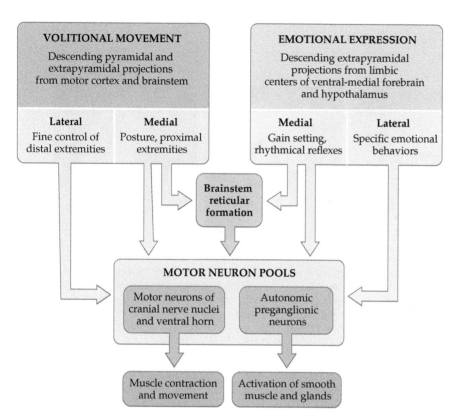

FIGURE 32.5   Components of the nervous system that organize the expression of emotional experience Diagram of the descending systems that control somatic and visceral motor effectors. Motor cortical areas in the posterior frontal lobe give rise to descending projections that, together with secondary projections arising in the brainstem, are organized into medial and lateral components. As described in Chapter 17, these descending projections account for volitional somatic movements. Functionally and anatomically distinct centers in the forebrain govern the expression of nonvolitional somatic motor and visceral motor functions, which are coordinated to mediate emotional behavior. Centers in the ventral-medial forebrain and hypothalamus also give rise to medial and lateral descending projections. For both systems of descending projections, the lateral components elicit specific behaviors (e.g., volitional digit movements and emotional facial expressions), while the medial components support and modulate the execution of such behaviors. The descending projections of both systems terminate in several integrative centers in the brainstem reticular formation, as well as the motor neuron pools of the brainstem and spinal cord. In addition, the forebrain centers innervate components of the visceral motor system that govern preganglionic autonomic neurons in the brainstem and spinal cord.

circuitry in the basal ganglia and cerebellum. The descending pyramidal and extrapyramidal projections from the motor cortex and brainstem ultimately convey the impulses responsible for voluntary somatic movements. Additionally, several cortical and subcortical structures in the medial frontal lobe and ventral parts of the forebrain, including related circuitry in the ventral part of the basal ganglia and hypothalamus, give rise to separate descending projections that run parallel to the pathways of the volitional motor system. These descending

projections of the ventral-medial forebrain terminate on visceral motor centers in the brainstem reticular formation, preganglionic autonomic neurons, and certain somatic premotor and motor neuron pools that also receive projections from volitional motor centers. The two types of facial paresis illustrated in Box 32A underscore this dual nature of descending motor control.

Emotions: Causes or effects?

The concerted action of the visceral and somatic motor systems in response to the diverse brain regions that control them is, in effect, an "emotional motor system." But do the subjective feelings of an emotion initiate this motor activity, or is it the other way around? Some evidence favors the latter view. For example, if individuals are given muscle-by-muscle instructions that result in facial expressions recognizable as anger, disgust, fear, happiness, sadness, or surprise without being told which emotion they are simulating, each pattern of facial muscle activity is accompanied by specific and reproducible differences in visceral motor activity (as measured by indices such as heart rate, skin conductance, and skin temperature). Moreover, autonomic responses are strongest when the facial expressions are judged to most closely resemble actual emotional expression and are often accompanied by the subjective experience of that emotion. One interpretation of these findings is that when voluntary facial expressions are produced, signals in the brain engage not only the motor cortex but also some of the circuits that produce emotional states. Perhaps this relationship helps explain how good actors can be so convincing, and why we are adept at recognizing the difference between a contrived facial expression and the spontaneous smile that accompanies a positive emotional state (see Box 32A).

This sort of evidence indicates that a major source of emotion (but certainly not the only source) is feedback from muscles and internal organs that are activated reflexively by external circumstances. However, physiological responses can also be elicited by complex and idiosyncratic stimuli mediated by the forebrain. For example, an anticipated tryst with a lover, a suspenseful episode in a

■ BOX 32A | Determination of Facial Expressions

(A) (1)

(2)

(3)

(4)

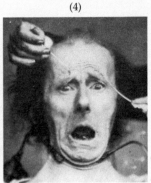

(A) Duchenne made use of early photography to study human facial expressions. (1) Duchenne with a Parisian shoemaker undergoing "faradization" of the facial muscles. (2) Bilateral electrical stimulation of the zygomaticus major mimicked a genuine expression of happiness, although closer examination shows insufficient contraction of the obicularis oculi (surrounding the eyes) compared with that evoked by spontaneous laughter (3). (4) Stimulation of the brow and neck produced an expression of "terror mixed with pain, torture … that of the damned"; however, the man reported no discomfort or emotional experience accompanying the evoked contractions.

In 1862, the French neurologist and physiologist G.-B. Duchenne published a remarkable treatise that linked emotions to facial expressions. He theorized that coordinated contractions of groups of muscles expressed distinct feelings that were common across individuals and cultures. To evoke those contractions, he pioneered the use of transcutaneous electrical stimulation (then called *faradization*, after the British chemist and physicist Michael Faraday) to activate single muscles and small groups of muscles in the face, dorsal surface of the head, and neck. He also documented the faces of the people he studied with another technological innovation of the time: photography (Figure A). His seminal contribution was the identification of muscles and muscle groups, such as the obicularis oculi, that are not easily controlled by force of the will, but are mainly "put into play by the sweet emotions of the soul." Duchenne concluded that the emotion-driven contraction of these muscle groups surrounding the eyes, together with the zygomaticus major, conveys the genuine experience of happiness, joy, and laughter. In recognition of these insights, psychologists sometimes refer to this facial expression as the "Duchenne smile."

In typical individuals, the difference between a forced smile (produced by voluntary contraction or electrical stimulation of facial muscles) and a spontaneous (or Duchenne) smile testifies to the convergence of descending motor signals from different forebrain centers onto premotor and motor neurons in the brainstem that control the facial musculature. The contrived smile of volition (sometimes called a *pyramidal smile*) is driven by the motor cortex, which communicates with the brainstem and spinal cord via the pyramidal tracts. The Duchenne smile is motivated by motor areas in the anterior cingulate gyrus (see Clinical Applications, Chapter 17) that access facial motor circuitry via multisyn-

(B)

Volitional motor paresis

Voluntary smile

Response to humor

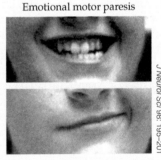

Emotional motor paresis

(B) Left panels: Mouth of an individual with a lesion that destroyed descending fibers from the right motor cortex displaying volitional facial paresis. When asked to show her teeth, the woman was unable to contract the muscles on the left side of her mouth (upper left), yet her spontaneous smile in response to a humorous remark was nearly symmetrical (lower left). Right panels: Face of a child with a lesion of the left forebrain that interrupted descending pathways from non-classic motor cortical areas, producing emotional facial paresis. When the child was asked to smile volitionally, the contractions of the facial muscles were nearly symmetrical (upper right). In spontaneous response to a humorous comment, however, the right side of the child's face failed to express emotion (lower right).

■ BOX 32A │ Determination of Facial Expressions (continued)

aptic, extrapyramidal pathways through the brainstem reticular formation.

Studies of individuals with specific neurological injury to these separate descending systems of control have further differentiated the forebrain centers responsible for control of the muscles of facial expression (Figure B). Individuals with unilateral facial paralysis due to damage of descending pathways from the motor cortex (upper motor neuron syndrome; see Chapter 17) have considerable difficulty moving their lower facial muscles on one side, either voluntarily or in response to commands, a condition called *volitional facial paresis* (see Figure B, left panels). Nonetheless, many such individuals produce symmetrical involuntary facial movements when they laugh, frown, or cry in response to amusing or distressing stimuli. In these individuals, pathways from regions of the forebrain other than the classic motor cortex in the posterior frontal lobe remain available to activate facial movements in response to stimuli with emotional significance.

A much less common form of neurological injury, called *emotional facial*

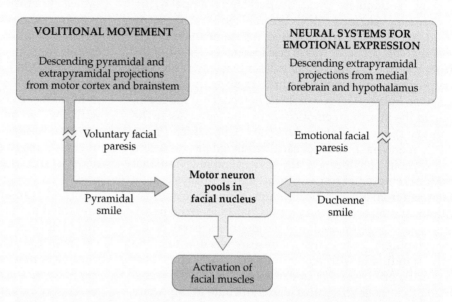

(C) The complementary deficits demonstrated in Figure B are explained by selective lesions of one of two anatomically and functionally distinct sets of descending projections that motivate the muscles of facial expression.

paresis, demonstrates the opposite set of impairments—that is, loss of the ability to express emotions by using the muscles of the face without loss of volitional control (see Figure B, right panels). These individuals are able to produce symmetrical pyramidal smiles, but fail to display spontaneous emotional expressions involving the facial musculature contralateral to the lesion. These two systems are diagrammed in Figure C.

---

novel or film, stirring patriotic or religious music, or accusations of dishonesty can all lead to autonomic activation and strongly felt emotions. The neural activity evoked by such complex stimuli is relayed from the forebrain to visceral and somatic motor nuclei via the hypothalamus and brainstem reticular formation, the major structures that coordinate the expression of emotional behavior.

In short, emotion and sensorimotor behavior are inextricably linked. As William James put it more than a century ago:

> What kind of an emotion of fear would be left if the feeling neither of quickened heart-beats nor of shallow breathing, neither of trembling lips nor of weakened limbs, neither of goose-flesh nor of visceral stirrings, were present, it is quite impossible for me to think … I say that for us emotion dissociated from all bodily feeling is inconceivable.
>
> William James, 1893 (*Psychology*, p. 379)

CONCEPT
32.2

The Amygdala Plays a Central Role in Emotional Processing

LEARNING OBJECTIVES

32.2.1 Describe the causes and consequences of Klüver–Bucy syndrome.

32.2.2 Explain how the amygdala contributes to fear conditioning.

32.2.3 Provide examples of how damage to the amygdala can affect perception and behavior.

32.2.4 Explain how emotional and body states can contribute to adaptive decision making.

The amygdala and its role in emotion

Neuroscientists have long sought to understand how regions of the forebrain support emotional experience and expression. Research by James Papez (pronounced *Papes*) in the mid-twentieth century proposed that specific brain circuits are devoted to emotion (much as the occipital cortex is devoted to vision, for instance). Papez focused on a set of midline and medial temporal lobe structures, including the **cingulate gyrus**, the **parahippocampal gyrus**, and the hippocampus, which became known as the *Papez circuit*. Yet as research progressed, it became increasingly clear that this proposal was at best incomplete: Some of the structures that Papez originally described (e.g., the hippocampus) now appear to have little to do with emotional behavior. Subsequent researchers, notably Paul MacLean, refined the idea of an emotional circuit to emphasize the primary role of the **amygdala** (Box 32B), which Papez hardly mentioned. To MacLean, the amygdala was a central player in a phylogenetically ancient **limbic system** that produces emotional responses through pathways largely separated from other cortical circuits. While the concept of a limbic system became very influential, modern neuroscientists have largely abandoned the idea of a separated, distinct emotional system. Instead, researchers now focus on understanding how different brain structures contribute to specific processes that underlie emotions, while recognizing that those structures also contribute to important non-emotional processes.

The single brain region most associated with emotion is undoubtedly the amygdala. Early evidence came from experiments in the 1930s by Heinrich Klüver and Paul Bucy, who removed the medial temporal lobes in rhesus monkeys. The resulting deficit led the monkeys to display a set of atypical behaviors—putting inappropriate objects into their mouths, hypersexuality, and excessive physical contact with objects in the environment—that are now known as Klüver–Bucy syndrome. Most important, the monkeys showed marked changes in emotional behavior. Because they had been caught in the wild, the monkeys had typically reacted with hostility and fear to humans before their surgery. Postoperatively, however, they were virtually tame. Motor and vocal reactions generally associated with anger or fear were no longer elicited by the approach of humans, and the animals showed little or no excitement when the experimenters handled them. Nor did they show fear when presented with a snake—a strongly aversive stimulus for a typical rhesus monkey. Subsequent work has shown that the emotional disturbances of Klüver–Bucy syndrome can be elicited by removal of the amygdala alone; in rare cases where there is damage to the amygdala and its cortical connections, humans exhibit similarly disordered motoric and emotional behavior.

Experiments first performed in the late 1950s by John Downer at University College London vividly demonstrated the importance of the amygdala in aggressive behavior. Downer removed one amygdala in rhesus monkeys, at the same time transecting the optic chiasm and the commissures that link the two hemispheres (principally, the corpus callosum and anterior commissure). In so doing, he produced animals with a single amygdala that had access only to visual inputs from the eye on the same side of the head. Downer found that the animals' behavior depended on which eye was used to view the world. When the monkeys were allowed to see with the eye on the side of the amygdala lesion, they behaved in some respects like those described by Klüver and Bucy (for example, they were relatively placid in the presence of humans). If, however, they were allowed to see only with the eye on the side of the intact amygdala, they reverted to their usual fearful and often aggressive behavior. Thus, in the absence of the amygdala, a monkey does not interpret the significance of the visual stimulus presented by an approaching human in the same way a normal animal does. Importantly, only visual stimuli presented to the eye on the side of the ablation produced this abnormal state; thus, if the animal was touched on either side, a full aggressive reaction occurred, implying that somatosensory information about both sides of the body had access to the remaining amygdala. These anecdotal data, taken together with what is now a rich trove of empirical results and clinical observations in both experimental animals and humans, show that the amygdala mediates neural processes that invest sensory experience with emotional significance.

Fear conditioning

To better understand the role of the amygdala in evaluating stimuli, and to define more precisely the specific circuits and mechanisms involved, several other animal models of emotional behavior have since been developed. One of the most useful models is based on conditioned fear responses in rats. Conditioned fear develops when an initially neutral stimulus is repeatedly paired with an inherently aversive one. Over time, the animal begins to respond to the neutral stimulus with behaviors similar to those elicited by the threatening stimulus (i.e., it learns to attach a new meaning to the neutral stimulus). Studies of the parts of the brain involved in the development of conditioned fear in rats have begun to shed some light on this process. After rats are trained to associate a tone with a mildly aversive foot shock delivered shortly afterward, the onset of that tone causes a marked increase in blood pressure and prolonged periods of behavioral freezing (a fearful reaction characterized by crouching without moving). Researchers demonstrated that the medial geniculate complex (MGC) is necessary for the development of the conditioned fear response (Figure 32.6). This result is not surprising, given that all auditory information that reaches the forebrain travels through the MGC of the dorsal thalamus (see Chapter 10). Yet the contributions of the MGC are not simply sensory. Even if the connections between the MGC and auditory cortex are severed, leaving only a direct projection between the MGC and the

■ BOX 32B | Anatomy of the Amygdala

The amygdala is a complex mass of gray matter buried in the anterior-medial portion of the temporal lobe, just rostral to the hippocampus (Figure A). It comprises multiple, distinct subnuclei and cortical regions that are richly connected to nearby cortical areas on the ventral and medial aspect of the hemispheric surface.

The amygdala (or amygdaloid complex, as it is often called) can best be thought of in terms of three major functional and anatomical subdivisions, each of which has a unique set of connections with other parts of the brain (Figures B and C). The medial group of subnuclei has extensive connections with the olfactory bulb and the olfactory cortex. The basolateral group, which is especially large in humans, has major connections with the cerebral cortex, especially the orbital and medial prefrontal cortex and the anterior temporal lobe. The central and anterior group of nuclei is characterized by connections with the hypothalamus and brainstem, including such visceral sensory structures as the nucleus of the solitary tract and the parabrachial nucleus (see Chapter 21).

The amygdala thus links cortical regions that process sensory information with hypothalamic and brainstem effector systems. Cortical inputs provide information about highly processed visual, somatosensory, visceral sensory, and auditory stimuli. These pathways from sensory cortical areas distinguish the amygdala from the hypothalamus, which receives relatively unprocessed sensory inputs. The amygdala also receives sensory input directly from some thalamic nuclei, the olfactory bulb, and visceral sensory relays in the brainstem. Thus, many neurons in the amygdala respond to visual, auditory, somatosensory, visceral sensory, gustatory, and olfactory stimuli.

Physiological studies have confirmed this convergence of sensory information. Moreover, highly complex stimuli are often required to evoke a neuronal response. For example, there are neurons in the basolateral group of nuclei that respond selectively to the sight of faces, very much like the responses observed in neurons in the inferior temporal cortex. In addition to sensory inputs (e.g., visual), the prefrontal and temporal cortical connections of the amygdala give it access to more overtly cognitive neocortical circuits, which integrate the emotional significance of sensory stimuli and guide complex behavior.

Finally, projections from the amygdala to the hypothalamus and brainstem (and possibly as far as the spinal cord) allow the amygdala to play an important role in the expression of emotional behavior by influencing activity in both the somatic and visceral motor efferent systems.

(A)

(B)

Central group

Medial group

Basolateral group

Courtesy of Joel Price

Amygdala

(C)

Orbital and medial prefrontal cortex

Amygdala (basolateral nuclei)

Ventral basal ganglia

Mediodorsal nucleus of the thalamus

(A) Coronal section through the forebrain at the level of the amygdala. (B) Histological section through the human amygdala (boxed area in Figure A), stained with silver salts to reveal the presence of myelinated fiber bundles. These bundles subdivide major nuclei and cortical regions within the amygdaloid complex. (C) The amygdala (specifically, the basolateral group of nuclei) participates in a "triangular" circuit linking the amygdala, the thalamic mediodorsal nucleus (directly and indirectly via the ventral parts of the basal ganglia), and the orbital and medial prefrontal cortex. These complex interconnections allow direct interactions between the amygdala and prefrontal cortex, as well as indirect modulation via the circuitry of the ventral basal ganglia.

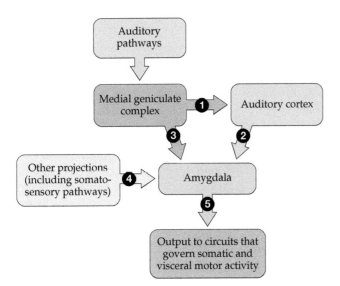

FIGURE 32.6 Pathways in the rat brain that mediate the association of auditory and aversive somatosensory stimuli Information processed by the auditory centers in the brainstem is relayed to the auditory cortex via the medial geniculate complex (1). The amygdala receives auditory information indirectly via the auditory cortex (2) and directly from one subdivision of the medial geniculate (3). The amygdala also receives sensory information about other sensory modalities, including pain (4). Thus, the amygdala is in a position to associate diverse sensory inputs, leading to new behavioral and autonomic responses to stimuli that were previously devoid of emotional content (5).

basolateral group of nuclei in the amygdala, fear conditioning remains intact. Conversely, if the part of the MGC that projects to the amygdala is also destroyed, the fear responses are abolished.

Because the amygdala is a site where neural activity produced by both tones and shocks can be processed, it is reasonable to suppose that the amygdala is also the site where learning about fearful stimuli occurs. This supposition led to the broader hypothesis that the amygdala participates in establishing associations between neutral sensory stimuli, such as a mild auditory tone or the sight of inanimate objects in the environment, and other stimuli that have some primary reinforcement value (Figure 32.7). The neutral sensory input can be stimuli in the external environment, stimuli communicated centrally via the special sensory afferent systems, or internal stimuli derived from activation of visceral sensory receptors. Stimuli with primary reinforcement value include both those with positive valence (e.g., the sight, smell, and taste of food) and those with negative valence (e.g., aversive tastes, loud sounds, or painful electrical shocks). Associative learning strengthens the connections relaying the information about the neutral stimulus, provided that the connections activate the postsynaptic neurons in the amygdala at the same time as inputs pertaining

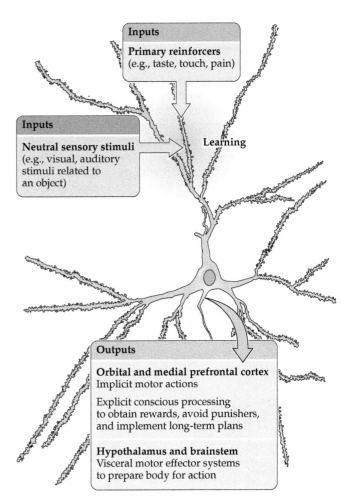

FIGURE 32.7 Model of associative learning in the amygdala relevant to emotional function Most neutral sensory inputs are relayed to neurons in the amygdala by projections from sensory processing areas that represent objects (e.g., faces). If these sensory inputs depolarize neurons at the same time as inputs that represent other sensations with primary reinforcing value, then associative learning would presumably occur by strengthening synaptic linkages between the previously neutral inputs and the neurons of the amygdala. The output of the amygdala would then inform a variety of integrative centers responsible for the somatic and visceral motor expression of emotion and for modifying behavior relevant to seeking rewards and avoiding punishment. (After E. T. Rolls, 1999. *The Brain and Emotion*. Oxford: Oxford University Press.)

to the primary reinforcer. The discovery that long-term potentiation (LTP) occurs in the amygdala provides further support for this hypothesis. Indeed, the acquisition of conditioned fear in rats is blocked by infusion into the amygdala of NMDA antagonists, which prevent the induction of LTP. Finally, the behavior of individuals with selective damage to the anterior-medial temporal lobe indicates that the amygdala plays a similar role in the human experience of fear (Box 32C).

■ BOX 32C | Fear and the Human Amygdala

Studies of fear conditioning in rodents show that the amygdala plays a critical role in the association of an innocuous auditory tone with an aversive mechanical sensation. Does this finding imply that the human amygdala is similarly involved in the experience of fear and the expression of fearful behavior? Reports of at least one extraordinary individual support the idea that the amygdala is indeed a key brain center for the experience of fear.

The individual (S. M.) suffers from a rare autosomal recessive condition called Urbach–Wiethe disease, a disorder that causes bilateral calcification and atrophy of the anterior-medial temporal lobes. As a result, both of S. M.'s amygdalas are extensively damaged, with little or no detectable injury to the hippocampal formation or nearby temporal neocortex (Figure A). She has no motor or sensory impairment, and no notable deficits in intelligence, memory, or language function. However, when asked to rate the intensity of emotion in a series of photographs of facial expressions, she could not recognize the emotion of fear (Figure B). Indeed, S. M.'s ratings of emotional content in fearful facial expressions were several standard deviations below the ratings of control individuals who had suffered brain damage outside the anterior-medial temporal lobe.

The investigators next asked S. M. (and control individuals with damage in other brain regions) to draw facial expressions of the same set of emotions from memory. Although the individuals obviously differed in artistic ability and the details of their renderings, S. M. (who has some artistic experience) produced skillful pictures of each emotion, except for fear (Figure C). At first, she could not produce a sketch of a fearful expression and, when prodded to do so, explained that she "did not know what an afraid face would look like." After several failed attempts, she

(Continued)

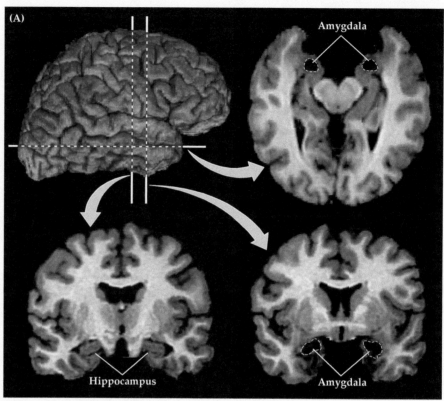

Courtesy of R. Adolphs

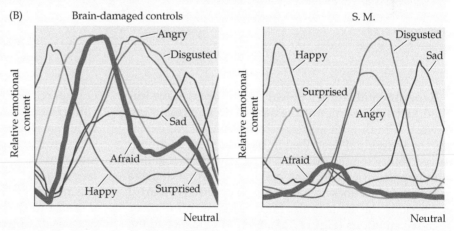

(A) MRI showing the extent of brain damage in S. M.; note the bilateral destruction of the amygdala and the preservation of the hippocampus. (B) Individuals with brain damage outside the anterior-medial temporal lobe (controls) and S. M. rated the emotional content of a series of facial expressions. Each colored line represents the intensity of the emotions judged in the face. S. M. recognized happiness, surprise, anger, disgust, sadness, and neutral qualities in facial expressions about as well as controls did. However, she failed to recognize fear (heavy orange lines). (B after R. Adolphs et al., 1995. *J Neurosci* 15: 5879–5891. © 1995 Society for Neuroscience.)

■ BOX 32C | Fear and the Human Amygdala (*continued*)

produced the sketch of a cowering figure with hair standing on end, evidently
because she knew these clichés about
the expression of fear. In short, S. M.
has a severely limited concept of fear
and, consequently, fails to recognize
the emotion of fear in facial expressions, in part because she fails to seek
out salient social information from the
eye regions of human faces. Studies of
other individuals with bilateral destruction of the amygdala are consistent
with this account. As might be expected, S. M.'s impairment also limits her
ability to experience fear in situations
where this emotion is appropriate.

Despite the admonition "have no
fear," to truly live without fear is to be
deprived of a crucial neural mechanism that facilitates appropriate social
behavior, helps make advantageous
decisions in critical circumstances, and
ultimately promotes survival.

(C) Sketches made by S. M. when asked to draw
facial expressions of emotion. (From R. Adolphs et al.,
1995. *J Neurosci* 15: 5879–5891. © 1995 Society for
Neuroscience.)

(C)

Happy

Sad

Surprised

Disgusted

Angry

Afraid

Experimental studies of fear conditioning have implied just such a role for the amygdala in associating
sensory stimuli with aversive consequences. For example, the individual described in Box 32C showed an impaired ability to recognize and experience fear, together
with impairment in decision making. Note that damage
to the amygdala does not only alter recognition of fearful
expressions; the damage also impairs judgments of other
negative emotions (Figure 32.8). Similar evidence of the
emotional influences on decision making have also come
from studies of people with lesions in the orbital and medial prefrontal cortex. These clinical observations imply
that the amygdala and prefrontal cortex, as well as their
striatal and thalamic connections, are not only involved
in processing emotions but participate in the complex
neural processing responsible for thinking and decidng.
These same neural networks are engaged by sensory
stimuli (e.g., facial expressions) that convey important

cues pertinent to appraising social circumstances and
conventions. Thus, when judging the trustworthiness
of human faces—a task of considerable importance for
successful interpersonal relations—neural activity in the
amygdala is specifically increased, especially when the
face in question is deemed untrustworthy (Figure 32.9).
It is not surprising, then, that individuals with bilateral
damage to the amygdala differ from controls in their
appraisals of trustworthiness; indeed, individuals with
such impairments often show inappropriately friendly
behavior toward strangers in real-life social situations.
Such evidence adds further weight to the idea that emotional processing is crucial for competent performance in
a wide variety of complex brain functions.

Emotional experience and decision making

The experience of emotion—even on a subconscious
level—has powerful influences on other brain functions,

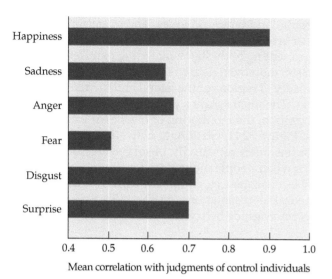

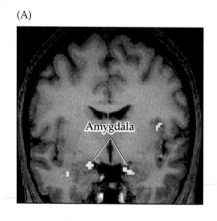

FIGURE 32.8 Damage to the amygdala affects emotional processing Individuals with damage to the amygdala viewed photographs of actors' faces expressing the six basic emotions shown here (see Figure 32.2 for similar examples) and then rated how intensely each image displayed the basic emotions. These ratings were then compared with those made by control individuals without brain damage. Among the individuals with damage to the amygdala, judgments about the happiness of facial expressions were very similar to those made by control individuals. However, judgments about negative emotions, particularly fear, were greatly impaired. (After R. Adolphs et al., 1999. *Neuropsychologia* 37: 1111–1117.)

including the neural faculties responsible for making decisions and for interacting with others (see Chapter 34 for additional examples). Evidence for such influences come principally from studies of individuals with damage to parts of the orbital and medial prefrontal cortex and of individuals with injury or disease involving the amygdala (see Clinical Applications). Such people often have impaired emotional processing, especially of emotions engendered by complex personal and social situations. As a result, they may have difficulty making advantageous decisions.

Adaptive decision making often entails the rapid evaluation of a set of possible outcomes with respect to the future consequences associated with each course of action. Emotional responses, whether experienced or anticipated, can provide important input to that evaluation process. Image that you face a difficult, high-stakes decision such as whether to remain in your current job or to uproot yourself and move across the country to begin a new career. Visualizing what your life would be like in the new position might generate a host of emotions—and whether those emotions convey excitement or fear might be a good guide to whether you should take the new position. Decision scientists call this the *affect heuristic*, meaning the emotions generated by the simulation of a decision can provide potential information about its consequences. Neuroscientists have argued, in turn, that the generation of conscious or subconscious mental representations of decision consequences involves activation

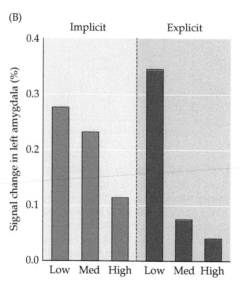

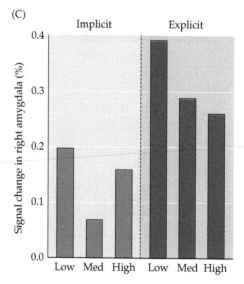

FIGURE 32.9 Activation of the amygdala during judgments of trustworthiness (A) Functional MRI shows increased neural activation bilaterally in the amygdala when typical individuals appraise the trustworthiness of human faces; activity is also increased in the right insular cortex. (B,C) The degree of activation is greatest when individuals evaluate faces that are considered untrustworthy (low, med, and high indicate ratings of trustworthiness; low = untrustworthy). The same effect was observed when individuals were instructed to evaluate the trustworthiness of the faces (explicit condition) or whether the faces were those of high school or university students (implicit condition). (After J. S. Winston et al., 2002. *Nat Neurosci* 5: 277–283.)

of brain circuits that themselves represent body states. Individuals with damage to key regions within those circuits—one of which is the ventromedial prefrontal cortex, or VMPFC—have difficulty simulating the consequences of their actions and thus are more likely to make maladaptive decisions (Figure 32.10).

It is important to emphasize that multiple brain regions contribute to the various components of emotional experience—and thus make distinct contributions to decision making. In addition to the amygdala and VMPFC, the insula has been shown to be particularly relevant for many types of decisions. As its Latin name implies ("insula" means "island"), the insular cortex lies hidden beneath the frontal and temporal lobes. The posterior insula receives a broad array of inputs from pathways important for representing body states, including pain, temperature, and taste; moreover, neurons in the anterior insula project broadly to regions of the lateral and medial prefrontal cortices. The insula plays a particularly important role in interoception, the monitoring of one's own internal body states (Figure 32.11). Strikingly, activation of the anterior insula has been repeatedly observed in human fMRI studies when people make decisions involving risky or aversive consequences, consistent with the interpretation that neural representations of potentially negative consequences may guide behavior toward safer choices.

(A)

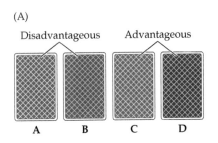

(B)

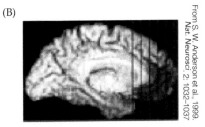

VMPFC damage overlap

From S. W. Anderson et al., 1999. *Nat. Neurosci.* 2: 1032–1037

(D)

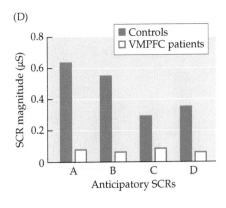

(C)

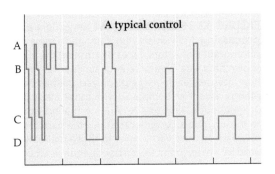

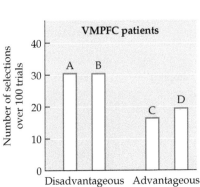

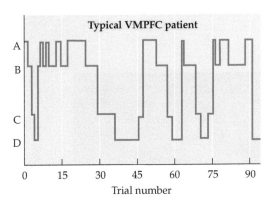

FIGURE 32.10 The Iowa Gambling Task (A) In this common experimental task, individuals select cards from four decks, each containing different combinations of winning and losing cards. Two of the decks contain many cards that lead to small monetary gains and a few cards that lead to large losses; choosing repeatedly from those disadvantageous decks leads to negative expected value. The other two decks have many cards leading to small losses and a few cards that lead to large gains; these advantageous decks have positive expected value. (B) Locations of VMPFC damage in the individuals tested in this experiment, with deeper colors of red indicating regions with more overlap across individuals. (C) Control individuals (i.e., without VMPFC damage; top graphs) typically sample all of the decks at the outset of the experiment and then gradually learn which decks are advantageous and which are disadvantageous. By contrast, individuals with VMPFC damage (bottom graphs) often fail to learn the differences between the decks, choosing the disadvantageous decks despite the negative feedback. (D) Choosing from the disadvantageous decks generates anticipatory skin conductance responses (SCRs) in control individuals but not in those with VMPFC damage. μS, microsiemens. (C,D after A. Bechara, 1994. *Cognition* 50: 7–15.)

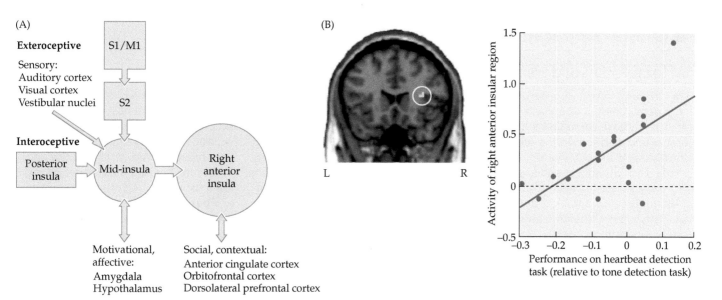

FIGURE 32.11   Interoception and the insula (A) The insula receives sensory inputs from a variety of sensory cortices, including information about the external world (exteroception) and about body states such as pain (interoception). These inputs are integrated in the mid-insula, which is interconnected with other brain regions that support affective states. (B) Activation of the insula has been linked to interoceptive processes; for example, fMRI studies show that individuals who are better able to monitor their own heartbeat show increased activity in the anterior insula. S1, primary somatosensory cortex; M1, primary motor cortex; S2, secondary somatosensory cortices. (A after A. D. Craig, 2007. In *Handbook of Emotions*, 3rd ed., M. Lewis et al. [Eds.]. New York: Guilford, pp. 395–408; B after H. D. Critchley et al., 2004. *Nat Neurosci* 7: 189–195.)

The Cerebral Cortices Support Emotional Processing

LEARNING OBJECTIVES

32.3.1 Describe the primary pathways through which the amygdala and cortical regions interact.

32.3.2 Explain the contributions of the right cerebral hemisphere to the emotional content of speech, mood, and facial expressions.

Cortical contributors to emotion processing

In animals such as the rat, most behavioral responses are highly stereotyped. In more complex brains, however, individual experiences often determine responses to stimuli. Thus, in humans a stimulus that evokes fear or sadness in one person may have little or no effect on the emotions of another. Although the pathways underlying such responses are not well understood, the amygdala and its interconnections with an array of neocortical areas in the prefrontal cortex and anterior temporal lobe, as well as several subcortical structures, appear to be especially important in the higher-order processing of emotion. In addition to its connections with the hypothalamus and brainstem centers that regulate visceral motor function, the amygdala has significant connections with several cortical areas in the orbital and medial aspects of the frontal lobe. These prefrontal cortical regions associate information from every sensory modality (including information

about visceral activities) and can thus integrate a variety of inputs pertinent to moment-to-moment experience. In addition, the amygdala projects to the thalamus (specifically, the mediodorsal nucleus), which projects in turn to these same cortical areas. Finally, the amygdala innervates neurons in the ventral portions of the basal ganglia that receive the major corticostriatal projections from the regions of the prefrontal cortex thought to process emotions. The amygdala thus can be considered a nodal point in a network that links together the cortical and subcortical brain regions involved in emotional processing.

Clinical evidence concerning the significance of this circuitry linked through the amygdala has come from functional imaging studies of individuals suffering from depression (Clinical Applications). In such individuals this set of interrelated forebrain structures shows atypical patterns of cerebral blood flow, especially in the left hemisphere. More generally, the amygdala and its connections to the prefrontal cortex and basal ganglia influence the selection and initiation of behaviors aimed at obtaining rewards and avoiding punishments (recall that the process of motor program selection and initiation is an important function of basal ganglia circuitry; see Chapter 18). The parts of the prefrontal cortex interconnected with the amygdala are also involved in organizing and planning future behaviors; thus, the amygdala may provide emotional input to overt (and covert) deliberations of this sort (see Concept 32.4).

■ Clinical Applications

Affective Disorders

Although some degree of disordered emotion is present in virtually all psychiatric problems, in affective (mood) disorders the essence of the disease is an abnormal regulation of the feelings of sadness and happiness. The most severe of these afflictions are major depression and bipolar disorder, which includes alternating episodes of depression and euphoria.)

Depression, the most common of the major psychiatric disorders, has a lifetime incidence of about 25% in women and about 15% in men. For clinical purposes, depression (as distinct from bereavement or neurotic unhappiness) is defined by a set of standard criteria. In addition to an abnormal sense of sadness, despair, and bleak feelings about the future (depression itself), these criteria include disordered eating and weight control, disordered sleeping, poor concentration, inappropriate guilt, and diminished sexual interest. The personally overwhelming quality of major depression has been compellingly described by afflicted authors such as William Styron and by afflicted psychologists such as Kay Jamison. The depressed individual's profound sense of despair has been nowhere better expressed than by Abraham Lincoln, who during a period of depression wrote:

> I am now the most miserable man living. If what I feel were equally distributed to the whole human family, there would not be one cheerful face on earth. Whether I shall ever be better, I cannot tell; I awfully forebode I shall not. To remain as I am is impossible. I must die or be better, it appears to me.

About half of all suicides occur in individuals with clinical depression.

When first identified, major depression and bipolar disorder were considered to arise from a neurotic inability to cope with external circumstances. It is now universally accepted that these conditions are neurobiological disorders. Among the strongest lines of evidence for this consensus are studies of the inheritance of these diseases. For example, the concordance of affective disorders is high in monozygotic compared with dizygotic twins. It has also become possible to study the brain activity of individuals suffering from affective disorders by noninvasive brain imaging. In at least one condition (unipolar depression), abnormal patterns of blood flow are apparent in the circuit interconnecting the amygdala, the mediodorsal nucleus of the thalamus, and the orbital and medial prefrontal cortex (see Box 32B). Of particular interest is the significant correlation of abnormal blood flow in the amygdala and the clinical severity of depression, as well as the observation that the abnormal blood flow pattern in the prefrontal cortex returns to normal when the depression has abated.

Despite evidence for a genetic predisposition and an increasing understanding of the brain areas involved, the cause of these conditions remains unknown. The efficacy of a large number of drugs that influence catecholaminergic and serotonergic neurotransmission strongly implies that the basis of the disease(s) is ultimately neurochemical (see Figures 6.15 and 6.16 for overviews of the projections of these neural systems). The majority of afflicted individuals (about 70%) can be effectively treated with one of a variety of drugs (including tricyclic antidepressants, monoamine oxidase inhibitors, and selective serotonin reuptake inhibitors, or SSRIs). Most successful are the SSRIs, which selectively block the uptake of serotonin without affecting the uptake of other neurotransmitters. Three such inhibitors—fluoxetine (Prozac), sertraline (Zoloft), and paroxetine (Paxil)—are especially effective in treating depression and have few of the side effects of the older, less specific drugs. Perhaps the best indicator of the success of these drugs has been their wide acceptance: Although the first SSRIs were approved for clinical use only in the late 1980s, they are now among the most prescribed pharmaceuticals.

Posttraumatic stress disorder (PTSD) typically emerges following exposure to a traumatic stressor, such as rape, robbery, or combat, that elicits feelings of fear, horror, or helplessness to forestall bodily injury or threat of death. Community-based studies in the United States estimate that 50% of people will have a traumatic experience during their lifetime, and an estimated 5% of men and 9% of women will develop PTSD as a result. Symptoms include persistently reexperiencing the traumatic event, avoiding reminders of the event, numbed responsiveness, and heightened arousal. PTSD is often accompanied by depression and substance abuse, each of which complicates treatment and recovery. While cognitive–behavioral therapies and antianxiety and antidepressant medications often help, there is no cure for this debilitating condition, which can persist for decades.

The following description gives some sense of what PTSD entails:

> For months after the attack, I couldn't close my eyes without envisioning the face of my attacker. I suffered horrific flashbacks and nightmares. For four years after the attack I was unable to sleep alone in my house. I obsessively checked windows, doors, and locks…I lost all ability to concentrate or even complete simple tasks. Normally social, I stopped trying to make friends or get involved in my community.
>
> P. K. Philips, www.adaa.org

Some of the structural abnormalities associated with PTSD are reductions

■ Clinical Applications (*continued*)

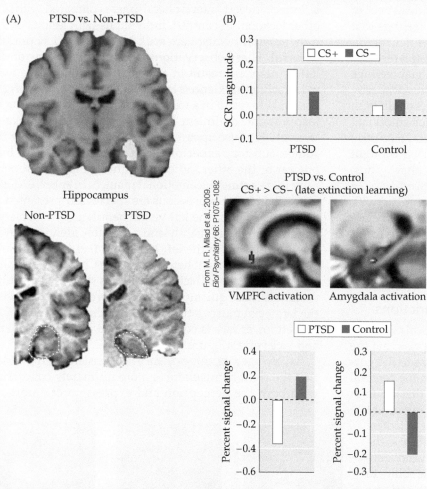

(A) PTSD vs. Non-PTSD

Hippocampus

Non-PTSD   PTSD

(B)

CS+ ■ CS−

PTSD vs. Control
CS+ > CS− (late extinction learning)

From M. R. Milad et al., 2009.
Biol Psychiatry 66: P1075–1082

VMPFC activation   Amygdala activation

□ PTSD ■ Control

(A) The volume of the hippocampus is often reduced in adults with PTSD and childhood abuse, compared with abused individuals who never developed PTSD (or with individuals without PTSD or abuse; not shown). Shown at top is a statistical map highlighting the region (in one hemisphere) where differences in volume were observed; shown below are brains of two representative individuals. (B) Compared with trauma-exposed control individuals, individuals with PTSD have a difficult time extinguishing fear response to cues that previously predicted a threat. During fear conditioning, one cue (CS+) predicted delivery of a mild shock whereas another cue (CS−) did not. Individuals then underwent an extinction procedure in which the shock was removed and the CS+ was now "safe." Despite the fact that the cue no longer predicted a shock, the individuals with PTSD continued to show greater skin conductance response (SCR), a measure of sympathetic arousal, and greater amygdala activity to the CS+ during the extinction test. In addition, they failed to engage the ventromedial prefrontal cortex (VMPFC) during the extinction test. This functional activity pattern is indicative of persistent hyperreactivity to threats and difficulty engaging executive control processes to suppress acquired fears when they are not appropriate to express. (A from R. G. Parsons and K. J. Ressler, 2013. *Nat Neurosci* 16: 146–153, after V. G. Carrión et al., 2010. *J Pediatr Psychol* 35: 559–569.)

in hippocampal and amygdala volume and altered dendritic remodeling in these structures (Figure A). Hippocampal atrophy has been linked to declarative memory deficits in some individuals with PTSD; in others, functional impairments in the amygdala are associated with hyperarousal symptoms and exaggerated responses to threats. Problems with fear reduction are further exacerbated by hyporesponsiveness in the rostral anterior cingulate and ventromedial prefrontal cortex, which provide inhibitory control over neurons in the amygdala (Figure B). Treatment with serotonin uptake inhibitors (e.g., Prozac) may partially reverse hippocampal volume differences and alleviate anxiety symptoms, but no single treatment cures this complex disorder. A major focus of ongoing research is to determine whether the brain alterations in PTSD are causal, or whether they are a consequence of the chronic stress associated with the syndrome.

Researchers and clinicians working with individuals who have PTSD face other challenges as well. Since it is generally considered unethical to induce physical or psychological trauma in the laboratory, the topic is difficult to approach experimentally. For instance, is it ethical to have individuals suffering from PTSD relive their painful past experiences for the purpose of studying these extreme emotions in the laboratory? As new treatments develop, additional dilemmas have emerged. For example, if a pharmacological agent selectively blocks emotional memories, should it be routinely administered to rape victims? If a genetic variant of a molecular marker is discovered to be a risk factor for developing PTSD, should military recruits be screened for it? Is it ethical to expose research animals to chronic stress to investigate the neurobiological mechanisms? Questions like these raise concerns not only for PTSD but for emotion research generally.

Finally, it is likely that interactions between the amygdala, the neocortex, and related subcortical circuits account for what is perhaps the most enigmatic aspect of emotional experience: the highly subjective feelings that attend most emotional states. Although the neurobiology of subjective experience remains incompletely understood, recent research argues that emotional feelings arise as a consequence of a more general cognitive capacity for self-awareness. In this conception, feelings entail both the immediate conscious experience of implicit emotional processing (arising from amygdala–neocortical circuitry) and explicit processing of semantically based thought (arising from hippocampal–neocortical circuitry). Thus, feelings can be plausibly conceived of as the product of an emotional working memory that sustains neural activity related to the processing of these various elements of emotional experience. Given the evidence for working memory functions in the prefrontal cortex, this portion of the frontal lobe—especially its orbital and medial aspects—is the likely substrate when such associations are conscious.

Cortical lateralization of emotional functions

Since functional asymmetries of complex cortical processes are commonplace, it should come as no surprise that the two hemispheres make different contributions to the governance of emotion. Emotion is lateralized in the cerebral hemispheres in at least two ways. First, as discussed in Chapter 31, the right hemisphere is especially important for the expression and comprehension of the affective aspects of speech. Thus, individuals with damage to the supra-Sylvian portions of the posterior frontal and anterior parietal lobes on the right side may lose the ability to express emotion by modulation of their speech patterns (this loss of emotional expression is referred to as *aprosody* or *aprosodia*; similar lesions in the left hemisphere give rise to Broca's aphasia). Individuals with aprosodia tend to speak in a monotone, no matter what the circumstances or meaning of what is said. For example, one such individual, a teacher, had trouble maintaining discipline in the classroom. Because her pupils (and even her own children) couldn't tell when she was angry or upset, she had to resort to adding phrases such as "I am angry and I really mean it" to indicate the emotional significance of her remarks. The wife of another individual with aprosodia felt her husband no longer loved her because he could not imbue his speech with cheerfulness or affection. Although such individuals cannot express emotion in speech, they nonetheless experience typical emotional feelings.

A second way in which the hemispheric processing of emotionality is asymmetrical concerns mood. Both clinical and experimental studies indicate that the left hemisphere is more involved with what can be thought of as positive emotions, whereas the right hemisphere is more involved with negative emotions. For example,

the incidence and severity of depression (see Clinical Applications) are significantly higher in individuals with lesions of the left anterior hemisphere compared with any other location. In contrast, individuals with lesions of the right anterior hemisphere are often described as unduly cheerful. These observations suggest that lesions of the left hemisphere result in a relative loss of positive feelings, facilitating depression, whereas lesions of the right hemisphere result in a loss of negative feelings, leading to inappropriate optimism. Hemispheric asymmetry related to emotion is also apparent in typical individuals. For instance, auditory experiments that introduce sound into one ear or the other indicate a right-hemisphere superiority in detecting the emotional nuances in speech. Moreover, when facial expressions are specifically presented to either the right or the left visual hemifield, the depicted emotions are more readily and accurately identified from the information in the left hemifield (i.e., the hemifield perceived by the right hemisphere; see Chapter 9). Finally, kinematic studies of facial expressions show that most individuals more quickly and fully express emotions with the left facial musculature than with the right (recall that the left lower face is controlled by the right hemisphere, and vice versa) (Figure 32.12).

Taken together, this evidence is consistent with the idea that the right hemisphere is more intimately concerned with both the perception and expression of emotions

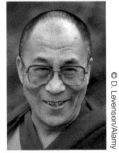

FIGURE 32.12 Asymmetrical smiles on some famous faces
Studies of typical individuals show that facial expressions are often more quickly and fully expressed by the left facial musculature than the right, as suggested by examination of these examples (try covering one side of the faces and then the other). Because the left lower face is governed by the right hemisphere, some psychologists have suggested that the majority of humans are "left-faced," in the same general sense that most of us are right-handed.

than is the left hemisphere. However, as in the case of other lateralized behaviors (language, for instance), both hemispheres participate in processing emotion.

From P. Vuilleumier and J. Driver, 2007, *Phil Trans Soc B Biol Sci 362*: 837–855; based on P. Vuilleumier et al., 2004, *Nat Neurosci 7*: 1271–1278

CONCEPT 32.4 Emotions Interact with Other Cognitive Processes

LEARNING OBJECTIVES

32.4.1 Describe the various pathways through which emotion-related activity in the amygdala can influence attention.

32.4.2 Explain how emotional content can contribute to memory consolidation.

32.4.3 Provide examples of the different methods for regulating emotions.

32.4.4 Characterize the interactions between the amygdala and lateral prefrontal cortex during emotion regulation.

Emotion–cognition interactions

One of the major lessons of recent neuroscience research is that higher-level cognitive functions are not isolated; emotional processes, for example, interact with processes of attention, memory, and executive function, among many others. Some emotion–cognition interactions arise through indirect pathways, such as emotion-induced engagement of the autonomic nervous system. As discussed in Concept 32.2, increased arousal following an emotional stimulus can potentiate processes of attention or decision making (e.g., identifying threats). Through what other pathways might emotion alter cognitive functions?

Emotions can shape perception by speeding the detection of important stimuli in the world and thus guiding us toward adaptive behaviors. Individuals with damage to key brain circuits for emotion, conversely, may have less sensitivity to emotional stimuli. Damage to the amygdala, for instance, does not in itself disrupt the ability to recognize faces. In one study (Figure 32.13), images of faces and houses were shown to individuals with different types of medial temporal lobe lesions. As discussed in Chapter 9, these different categories of visual stimuli engage different regions of extrastriate cortex: the fusiform gyrus for faces and the parahippocampal gyrus for houses. In people in whom the lesion was restricted to the hippocampus, there was a robust response in the fusiform gyrus to the faces—and that response was greater when the faces conveyed a fearful emotion. Individuals who had both hippocampus and amygdala damage also evinced fusiform activation to faces, but that activation did not depend on the emotion conveyed by the face. These results demonstrate that processing in the amygdala can alter responses in the visual system in an emotion-specific manner.

Neuroscientists have identified several pathways through which the amgydala influences processes of attention and goal-directed behaviors (Figure 32.14). Amygdala activity can cause the release of acetylcholine from the basal forebrain, leading to increased activity across many regions of cortex; this may bias attention toward stimuli whose emotional content makes them particularly relevant for behavior. Some clinical conditions are characterized by dysfunctional attention–emotion interactions. For example, people with major depressive disorder may become especially attentive to stimuli that evoke sad, stressful, or negative mood states—and which may also generate repetitive and intrusive thoughts about events or memories (a process called rumination). Recent neuroscience research links the cingulate cortex, in particular, to depression. Neuroimaging studies using fMRI have found that individuals with unmedicated active depression show higher functional connectivity between the cingulate cortex and orbitofrontal cortex; this connectivity diminishes when individuals take medication that alleviates depressive symptoms. Moreover, deep brain stimulation of the cingulate gyrus can alleviate treatment-resistant depression in at least some individuals.

Emotions also interact with the processes of memory. It has long been recognized that emotionally arousing events are often represented vividly in memory. Even if such events are not necessarily more accurately remembered (as shown in recent research on flashbulb memories), emotional content can make memories seem richer and more detailed. One potential explanation comes from the role of the amygdala in enhancing consolidation processes that occur

(A) People with hippocampus damage

Faces > Houses   Fearful > Neutral faces

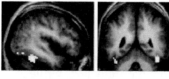

Fusiform gyrus

(B) People with hippocampus and amygdala damage

FIGURE 32.13 Amygdala damage alters the cortical processing of emotional faces Neuroscientists used fMRI to examine the cortical responses of individuals with hippocampus damage only (A) and with both hippocampus and amygdala damage (B) to faces with different emotional content. In both experimental groups, faces evoked robust responses in face-selective regions within the fusiform gyrus (when compared with houses, which were used as control visual stimuli). In the individuals with hippocampus damage only, the fusiform response was greater when the faces expressed a strong negative emotion, as would be found in individuals with no neurological damage. This emotional enhancement effect was absent in the individuals who also had amygdala damage.

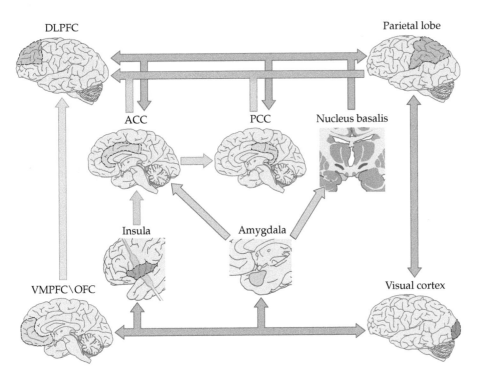

FIGURE 32.14 Schematic of neural pathways through which emotion and attention interact The amygdala influences attentional processing through multiple direct and indirect pathways. By stimulating the release of acetylcholine from the nucleus basalis, the amygdala can enhance attention throughout the cortex. Through projections to prefrontal cortex, notably the anterior cingulate cortex (ACC), the amygdala can shape processing in many other cortical regions (including the dorsolateral prefrontal cortex, DLPFC; posterior cingulate cortex, PCC; and ventromedial and orbitofrontal cortices, VMPFC and OFC). Feedback projections to sensory regions can enhance attention in sensory areas directly. (After K.S. LaBar, 2010. In *Encyclopedia of Behavioral Neuroscience*, G. F. Koob et al., [Eds.]. Academic Press: New York, pp. 469–476.)

elsewhere in the brain (Figure 32.15). Specifically, emotionally arousing information or experience may lead to the release of stress hormones (e.g., epinephrine) that modulate amygdala activity, which in turn alters functioning of other medial temporal lobe structures (e.g., the hippocampus) as well as regions of the cerebral cortex. Disruption of these processes by administration of the β-adrenergic blocker propranolol can inhibit the formation of memories, especially those involving emotional content. Such results have led to active investigation of such drugs as potential treatments for posttraumatic stress disorder (PTSD; see Clinical Applications), with mixed results. Some recent research suggests that delivery of propranolol immediately before therapeutic sessions that involve trauma reactivation may help reduce PTSD symptions, presumably by weakening the links between new stimuli and the traumatic memories.

Emotion regulation

No emotional experience is always good or always bad for mental or physical health. For example, some emotions that are often considered negative (e.g., sadness) may nevertheless be important to experience naturally (e.g., while grieving the loss of a loved one). In other situations, one's natural emotional experience might be maladaptive and interfere with the activities of daily life. The processes by which someone attempts to alter emotional experiences—by making them more or less intense, by shortening or lengthening their duration, or by changing one emotion to another—are known as **emotion regulation**.

Researchers have identified several distinct strategies for emotion regulation. Some occur prior to emotional experiences, as when people avoid events that might trigger unwanted emotions (i.e., situation selection). Others occur following the onset of an emotion; for example, when experiencing a sudden burst of anger, someone may try to downregulate their experiences and temper their behaviors (e.g., suppression). A particularly important strategy for both basic science and clinical applications alike is **cognitive reappraisal**, which involves reinterpreting an emotionally charged situation in order to alter its emotional impact. Imagine that you were viewing a scene of a young woman lying on a hospital bed. Such a scene might naturally elicit negative emotions such as sadness. However, you could reinterpret that situation to elicit positive emotions: The woman is now resting following a successful surgery (or childbirth, or organ donation to a family member) from which she'll fully recover. Reappraisal strategies are central to many forms of mental health treatments and therapies (e.g., cognitive–behavioral therapy).

Researchers interested in the neural basis of emotion regulation have examined the effects of different strategies (particularly cognitive reappraisal) on brain function using fMRI. A typical paradigm involves presenting visual images that evoke either emotional or neutral experiences (e.g., photographs of a war scene vs. a shopping center); on different trials, individuals are instructed to decrease their emotional response to the images via reappraisal or to experience their emotions naturally. Broadly summarized, engaging in active reappraisal leads to increased activity in dorsal and lateral prefrontal cortex and dorsal cingulate cortex, brain regions often associated with executive control of behavior, along with decreased activity in the amygdala and insula. Using techniques for measuring changes in the coactivity of different regions,

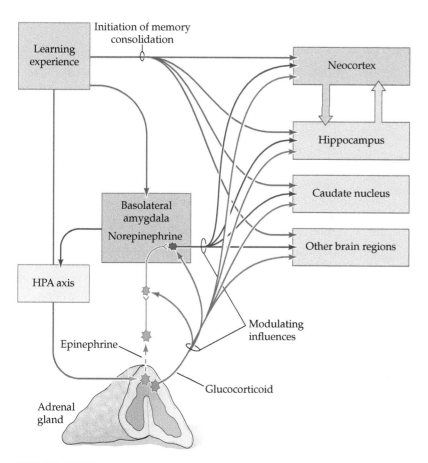

FIGURE 32.15 Enhancement of memories by emotions Hormonal releases associated with emotionally arousing experiences can alter memory consolidation. A primary pathway for those effects involves stress hormones that influence activity in the amygdala, which in turn influences storage processes in medial temporal lobe structures and elsewhere in the brain. HPA axis, Hypothalamic Pituitary Adrenal Axis. (After J. L. McGaugh, 2000. *Science* 287: 248–251.)

researchers have identified a consistent pattern by which emotion regulation increases the functional connectivity between regions (Figure 32.16): During regulation, fluctuations in the activity of lateral prefrontal cortex are mirrored by opposite-direction fluctuations in the amygdala (i.e., increases in prefrontal cortex activity co-occur with decreases in amygdala activity).

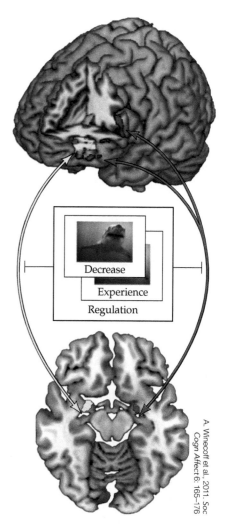

A. Winecoff et al., 2011. Soc Cogn Affect 6: 165–176

FIGURE 32.16 Interactions between the amygdala and lateral prefrontal cortex during emotion regulation In this fMRI study, individuals viewed images that had either emotional or neutral content. When the individuals were instructed to diminish their emotional reaction to the images via a process of cognitive reappraisal, they showed increased functional connectivity between the amygdala (left/right shown in red/yellow colors) and the lateral prefrontal cortex, compared with when they experienced their emotions naturally.

Summary

The word *emotion* covers a wide range of states that have in common the association of subjective feelings, body states, and behavior. Emotion-induced changes in body states are mediated by the visceral motor nervous system, which is itself regulated by inputs from many other parts of the brain. Behavioral changes associated with emotion are governed by a diverse set of brain structures, most notably the amygdala but also the hypothalamus and several regions of the cerebral cortex. Many clinical conditions are associated with dysfunctional emotional processing: the dampening of affect in depression, the disordered perception of emotional expressions following medial temporal damage, and the exaggerated reactions to stimuli by individuals with PTSD, among many others. Neuroscientists now see emotion as intertwined with cognitive processes such as attention, memory, and decision making. The prevalence and social significance of human emotions and their disorders ensure that the neurobiology of emotion will be an increasingly important theme in modern neuroscience.

■ Additional Reading

Reviews

Craig, A. D. (2007) Interoception and emotion: A neuro-anatomical perspective. In *Handbook of Emotions*, 3rd Edition, M. Lewis, J. M. Haviland-Jones and L. F. Barrett (Eds.). New York: Guilford Press, pp. 395–408.

Kragel, P. A. and K. S. LaBar (2016) Decoding the nature of emotion in the brain. *Trends Cog. Sci.* 20: 444–455.

LeDoux, J. E. (2012) Rethinking the emotional brain. *Neuron* 73(4): 653–676.

Lindquist, K. A. and 4 others (2012) The brain basis of emotion: A meta-analytic review. *Behav. Brain Sci.* 35: 121–202.

Mayberg, H. S. (1997) Limbic-cortical dysregulation: A proposed model of depression. *J. Neuropsychiatry Clin. Neurosci.* 9: 471–481.

Namburi, P. and 4 others (2015). Architectural representation of valence in the limbic system. *Neuropsychopharmacol.* 41: 1697–1715.

Phelps, E. A., K. M. Lempert and P. Sokol-Hessner (2014). Emotion and decision making: Multiple modulatory neural circuits. *Annu. Rev. Neurosci.* 37: 263–287.

Picó-Pérez, M. and 4 others (2017) Emotion regulation in mood and anxiety disorders: A meta-analysis of fMRI cognitive reappraisal studies. *Prog. Neuropsychopharmacol. Biol. Psychiatry* 79: 96–104.

Rolls, E. T. (2019) The cingulate cortex and limbic systems for emotion, action, and memory. *Brain Struct. Funct.* 224: 3001–3018.

Important Original Papers

Anderson, A. K. and E. A. Phelps (2001) Lesions of the human amygdala impair enhanced perception of emotionally salient events. *Nature* 411: 305–309.

Bard, P. (1928) A diencephalic mechanism for the expression of rage with special reference to the sympathetic nervous system. *Am. J. Physiol.* 84: 490–515.

Bremner, J. D. (2006) Traumatic stress: effects on the brain. *Dialogues Clin. Neurosci.* 8: 445–461.

Critchley, H. D. and 4 others (2004) Neural systems supporting interoceptive awareness. *Nat. Neurosci.* 7: 189–195.

Downer, J. L. de C. (1961) Changes in visual agnostic functions and emotional behaviour following unilateral temporal pole damage in the "split-brain" monkey. *Nature* 191: 50–51.

Dunsmoor, J. E., V. P. Murty, L. Davachi and E. A. Phelps (2015). Emotional learning selectively and retroactively strengthens memories for related events. *Nature* 520: 345–348.

Ekman, P., R. W. Levenson and W. V. Friesen (1983) Autonomic nervous system activity distinguishes among emotions. *Science* 221: 1208–1210.

Hayman, L. A. and 4 others (1998). Klüver-Bucy syndrome after bilateral selective damage of amygdala and its cortical connections. *Neuropsychiatry* 10: 354–358.

Klüver, H. and P. C. Bucy (1939) Preliminary analysis of functions of the temporal lobes in monkeys. *Arch. Neurol. Psychiat.* 42: 979–1000.

MacLean, P. D. (1949) Psychosomatic disease and the "visceral brain": Recent developments bearing on the Papez theory of emotion. *Psychosom. Med.* 11: 338–353.

Papez, J. W. (1937) A proposed mechanism of emotion. *Arch. Neurol. Psychiat.* 38: 725–743.

Phillips, R. G. and J. E. LeDoux (1992) Differential contribution of amygdala and hippocampus to cued and contextual fear conditioning. *Behav. Neurosci.* 106: 274–285.

Shackman, A. J. and 4 others (2010) Right dorsolateral prefrontal cortical activity and behavioral inhibition. *Psychol. Sci.* 20: 1500–1506.

Shin, L. M. and I. Liberzon (2010) The neurocircuitry of fear, stress, and anxiety. *Neuropsychopharmacol.* 35: 169–191.

Treadway, M. T., and 7 others (2014). Corticolimbic gating of emotion-driven punishment. *Nat. Neurosci.* 17: 1270–1275.

Vuilleumier, P. and 4 others (2004) Distant influences of amygdala lesion on visual cortical activation during emotional face processing. *Nat. Neurosci.* 7: 1271–1278.

Williams, M. A. and 4 others (2004) Amygdala responses to fearful and happy facial expressions under conditions of binocular suppression. *J. Neurosci.* 24: 2898–3004.

Winecoff, A. and 4 others (2011) Cognitive and neural contributors to emotion regulation in aging. *Soc. Cogn. Affect. Neurosci.* 6: 165–176.

Books

Barrett, L. F. (2017) *How emotions are made: The secret life of the brain.* New York: Houghton Mifflin Harcourt.

Barrett, L. F., M. Lewis and J. M. Haviland-Jones (2016) *Handbook of Emotions*, 4th Edition. New York: Guilford Press.

Damasio, A. R. (1994) *Descartes Error: Emotion, Reason, and the Human Brain.* New York: Avon Books.

Darwin, C. (1890) *The Expression of Emotion in Man and Animals*, 2nd Edition. In *The Works of Charles Darwin*, vol. 23, London: William Pickering (1989).

Ekman, P. and R. J. Davidson (1994) *The Nature of Emotions.* New York: Oxford University Press.

Gross, J. J. (2007) *Handbook of Emotion Regulation.* New York: Guilford Press.

James, W. (1890) *The Principles of Psychology*, vols. 1 and 2. New York: Dover Publications (1950).

LeDoux, J. (1998) *The Emotional Brain: The Mysterious Underpinnings of Emotional Life.* New York: Simon and Schuster.

Thinking, Planning, and Deciding

Overview

Our capacity for flexible, goal-directed cognition forms the foundation for much of what makes us human. We play chess, write novels, conduct scientific experiments, and complete many tasks that require organizing complex actions in service of some goal. The mental activities that allow such flexibility are often described as *cognitive control* (or *executive control*) *processes*: They select or modify other cognitive functions in response to changing environmental demands. Conversely, many neurological and psychiatric diseases are characterized by dysfunctions in processes such as thinking, planning, and deciding. For example, addiction, depression, schizophrenia, and obsessive-compulsive disorder have very different etiologies, but all are linked to impairments in cognitive flexibility.

Research on control processes has become one of the most important areas of research in neuroscience—as evidenced by the diversity of methods used by neuroscientists to understand complex cognitive functions. Studies of people with brain damage and advances in noninvasive brain imaging methods (e.g., fMRI) have provided insight into the brain networks associated with these key abilities. Experimental work examining non-human animals has revealed how neurons and circuits support flexible selection of actions. And computational models provide important links between measures of brain function and of behavior. Much of this research has converged on the conclusion that the prefrontal cortex (PFC) supports control processes, with different regions of the PFC making distinct contributions to complex behavior.

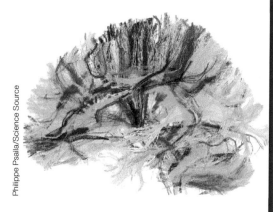

Philippe Psaila/Science Source

KEY CONCEPTS

33.1 The prefrontal cortex supports processes related to cognitive control

33.2 Lateral prefrontal cortex supports cognitive control

33.3 Orbitofrontal cortex supports the evaluation of the outcomes of behavior

33.4 Anterior cingulate cortex supports regulation of activity in other brain regions

33.5 The anterior insula incorporates information about body states into decision processes

33.6 Posterior cingulate cortex supports internally directed processes

---

CONCEPT **33.1** | **The Prefrontal Cortex Supports Processes Related to Cognitive Control**

LEARNING OBJECTIVES

33.1.1 Describe the key subregions of prefrontal cortex and their patterns of connectivity to other areas of the brain.

33.1.2 Provide examples of the effects of damage to prefrontal cortex.

33.1.3 Explain the differences between dysexecutive and disinhibition syndromes.

Prefrontal cortex: An overview

The localization of complex mental functions in the brain has a checkered history. At the beginning of the nineteenth century, many scientists already believed that the frontal lobes were critical for controlling behavior and for higher cognitive functions generally. This conviction was based on very limited evidence, such as the gross anatomical fact that the frontal lobes are relatively larger in humans and great apes than in other mammals (Figure 33.1).

FIGURE 33.1 Size of the cerebral cortex and prefrontal cortex in mammals (A) Among the seven species shown here, humans not only have the largest cerebral cortex but also have a larger PFC (blue) relative to the other (non-primate) species, even controlling for brain size. The porpoise brain is provided for size comparison; its PFC is not indicated because the borders of it are not known. (B) Within the order of primates, the size of the PFC is roughly proportional to that of the rest of the neocortex. Brodmann's work in the early twentieth century had suggested that humans and other great apes have a disproportionately large PFC, here measured as surface area of the cerebral cortex. (C) Later work has indicated that relative size of the PFC is roughly constant within the order of primates, measured here as volume of the cerebral cortex. (B after K. Brodmann, 1912. *Anat Anz* 41: 157–216; C after K. Semendeferi et al., 1997. *J Hum Evol* 32: 375–388 and K. Semendeferi et al., 2002. *Nat Neurosci* 5: 272–276.)

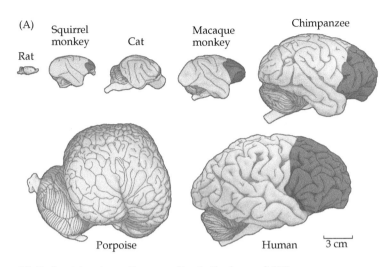

(A)

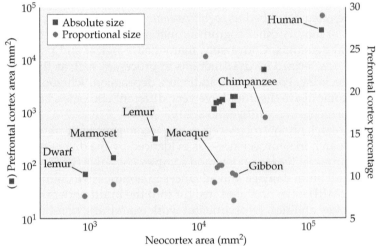

(B) Prefrontal cortex scaling according to Brodmann (1912)

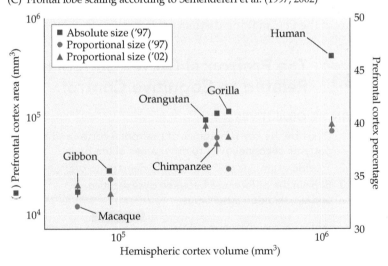

(C) Frontal lobe scaling according to Semendeferi et al. (1997, 2002)

Without experimental methods for altering or measuring brain function, however, the scientists of the day could not draw specific conclusions about how control processes might be represented in the brain.

The first studies of the frontal lobes in an animal were conducted by the German physiologist Eduard Hitzig and his colleague Gustav Fritsch in the late 1860s. When they electrically stimulated the posterior parts of the frontal lobes in a dog, the animal moved its limbs on the opposite side; conversely, damage to the posterior frontal lobe led to a lack of voluntary control over motor actions, although the animal could still engage in reflexive movements such as walking. Together, these results provided evidence that the more posterior portions of the frontal lobes are associated with motor function. Damage to the anterior frontal lobes, in contrast, caused neither paralysis nor any obvious sensory deficits; moreover, electrical stimulation of those regions evoked no discernible movements or muscle activity. From this limited evidence, Hitzig and Fritsch speculated that the anterior frontal lobes are associated with higher cognitive functions rather than sensation or motor control.

This experimental approach was extended and refined over the succeeding decades by the British physiologist David Ferrier and the Italian physiologist Leonardo Bianchi, each of whom created various frontal lobe lesions and followed up with careful observations of behavior. For example, Bianchi found that bilateral frontal damage caused deficits in a range of cognitive functions: failure to recognize known objects, inability to use past experience to guide behavior, deficits in initiative, loss of affective responses, and lack of coherent behavior. He also made the important observation that unilateral damage rarely caused these behavioral changes. These and other early studies provided the first clear evidence that the frontal lobes play a key role in the executive control of behavior. Although these early observations of Hitzig, Fritsch, Ferrier, and Bianchi remain valid today, more recent research has shown that the frontal lobes are part of a larger network that supports complex behaviors.

Organization and connectivity of PFC

The **prefrontal cortex** (**PFC**) is the portion of the frontal lobe anterior to the motor cortex in both humans and non-human primates (see Figure 33.1), and it is especially prominent in humans. Broadly considered, the PFC can be divided into regions that make distinct contributions to control processes: lateral prefrontal cortex, orbitofrontal cortex, and anterior cingulate cortex (Figure 33.2). (This chapter is organized around these divisions and the functions they support.) Unlike in other parts of the brain (e.g., visual cortex), the boundaries between divisions of the PFC do not clearly correspond to differences in cytoarchetectonics (e.g., Brodmann's areas) or electrophysiology (e.g., firing properties of neurons). Thus, it is difficult if not impossible to reliably identify the flow of information through the PFC; connections between different cortical regions are often reciprocal, and it is hard to differentiate which connections are feedback and which are feedforward. This robust connectivity is a conundrum that applies to the association cortices generally, as introduced in Chapter 27.

Despite this complexity, rough pathways from input to output can be identified (Figure 33.3). Information about sensory stimuli is conveyed to the orbitofrontal cortex (that is, the orbital portion of the PFC), where representations of the *values* of various options may be represented. Value signals, along with much other information, then flow rostrally and laterally to the lateral and medial PFC, where they are combined with information about goals and current environmental contexts to plan possible responses. From there, signals propagate to the premotor and parietal cortices, and finally to the motor and other cortical regions that give rise to behavior (remember that behavior is not limited to motor actions but also includes perception, attention, emotion, memory, and more). These pathways and their targets are influenced by neuromodulatory transmitters such as dopamine, serotonin, and acetylcholine; by specialized cortico-basal ganglia loops; and by emotional and memory processes in the amygdala and hippocampus, respectively.

Consequences of damage to prefrontal cortex

Someone with damage to prefrontal cortex—whether from stroke, trauma, surgery, or a degenerative disease—may appear unremarkable when first encountered. Perceptual, language, and motor abilities are usually unimpaired, and the ability to recall events or facts from memory is typically intact. Yet that same person may have profound difficulty carrying out simple activities and may have a greatly diminished quality of life. These subtle yet profound deficits suggest that a deeper examination of the functional properties of the prefrontal cortex is needed to understand its role in cognition, both normal and disordered.

Prefrontal cortex damage can lead to either of two general syndromes, depending on the affected region. Damage to (or degeneration of) lateral prefrontal cortex can lead to the constellation of symptoms called **dysexecutive syndrome**. Individuals with this syndrome do not present with obvious deficits in intelligence; they can use language normally and can remember events and facts. But they have great difficulty managing their daily lives. They fail to plan for the future, rarely initiate new projects or set long-term goals, leave tasks uncompleted if initiated, and have a limited span of attention. They may have difficulty interacting with others, both because of these deficits and because of difficulty in understanding the goals and thoughts of others. They show a lack of insight into their own and others' actions. They may even deny the existence of their problems, or create implausible

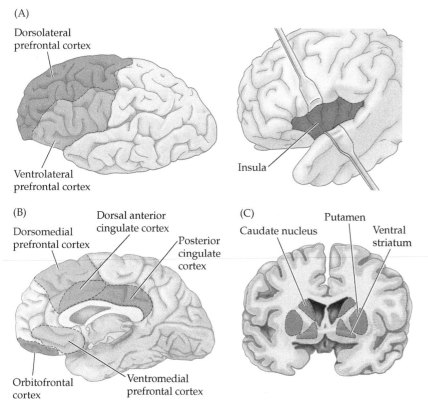

(A)

Dorsolateral prefrontal cortex

Ventrolateral prefrontal cortex

Insula

(B)

Dorsomedial prefrontal cortex

Dorsal anterior cingulate cortex

Posterior cingulate cortex

Orbitofrontal cortex

Ventromedial prefrontal cortex

(C)

Caudate nucleus

Putamen

Ventral striatum

FIGURE 33.2   Major brain regions involved in thinking, planning, and deciding Three views of the human brain: lateral (A), medial (B), and coronal (C). Highlighted are the dorsolateral prefrontal cortex, ventrolateral prefrontal cortex, insula, dorsomedial prefrontal cortex, ventromedial prefrontal cortex, dorsal anterior cingulate cortex, orbitofrontal cortex, and posterior cingulate cortex and their major targets in the striatum: the caudate nucleus, putamen, and ventral striatum.

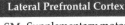

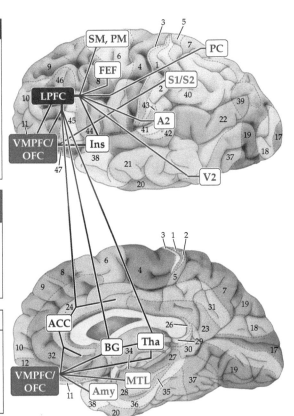

Lateral Prefrontal Cortex

SM: Supplementary motor cortex

PM: Premotor cortex

FEF: Frontal eye fields

PC: Parietal cortex

V2: Secondary visual cortex

A2: Secondary auditory cortex

Ventromedial Prefrontal Cortex/Orbitofrontal Cortex

Amy: Amygdala

MTL: Medial temporal lobe

S1/S2: Primary and secondary somatosensory cortices

Shared Regions of Connectivity

Tha: Thalamus

BG: Basal ganglia

ACC: Anterior cingulate cortex

Ins: Insula

FIGURE 33.3 Connectivity of the prefrontal cortex Neurons in the PFC project to and receive input from secondary sensory cortices, motor preparatory structures, and parietal cortex. This schematic diagram shows some of the major connections for the lateral prefrontal cortex (LPFC) and for the ventromedial prefrontal cortex and orbitofrontal cortex (VMPFC/OFC, combined here for simplicity). All indicated connections are bidirectional, with the important exception of a unidirectional projection from the LPFC to the basal ganglia (which project back to the LPFC via the thalamus). Numbers refer to Brodmann's areas.

explanations for those problems (i.e., confabulation) as their life deteriorates around them. As introduced in the chapter's upcoming concepts, these difficulties dealing with the real world are accompanied by impairments in specific tasks that can be carried out in a laboratory.

Damage to orbitofrontal and medial portions of the prefrontal cortex, in contrast, gives rise to what has been termed **disinhibition syndrome**. People with this syndrome, like people with dysexecutive syndrome, present with what generally appear to be typical cognitive functions. They tend to perform normally on laboratory tests of response selection and working memory, unlike people with dysexecutive syndrome. Nonetheless, their lives outside the laboratory are often chaotic. People with disinhibition syndrome often exhibit constant movement that is not channeled toward productive activities, and they may be euphoric or manic with an atypical sense of humor. Thus, they may laugh at inappropriate times in simple social situations, fail to respond to typical social cues, or reveal embarrassing personal information. Their outward expressiveness stands in sharp contrast to the quietness and apathy associated with lateral prefrontal damage. The famous case of Phineas Gage (see Chapter 27) has historically been considered an example of disinhibition syndrome, although recent investigation of Gage's life history suggests a more complex picture. Despite the accident that

destroyed much of his ventral frontal lobes, Gage nevertheless remained employed for most of his remaining life and failed to show some typical personality markers of this type of prefrontal damage (e.g., he was reported to be a good storyteller and could communicate with socially appropriate affect).

CONCEPT **33.2** | Lateral Prefrontal Cortex Supports Cognitive Control

LEARNING OBJECTIVES

33.2.1 Define dorsolateral prefrontal cortex and describe its pattern of connectivity to other brain regions.

33.2.2 Explain how dorsolateral prefrontal cortex supports the selection of behavioral rules.

33.2.3 Define ventrolateral prefrontal cortex and describe its pattern of connectivity to other brain regions.

33.2.4 Provide evidence for the conclusion that ventrolateral prefrontal cortex contributes to inhibition of inappropriate behaviors.

Planning and organization of behavior

Imagine driving along a regular route to work. If one day that route is blocked by an accident, you may quickly recognize the obstacle along your intended path and then turn the car down a side street that avoids the traffic jam. This is an example of *flexible* behavior: selecting an alternative action (instead of the one typically followed) that overcomes an immediate environmental challenge to reach an intended goal. Flexibility can be seen whenever behavior does not depend solely on learned habits or reflexes, but instead on processes that intervene between sensory input and motor (or other) output. In the example of driving, you might need to recognize the consequences of a road blockage, inhibit the actions you typically take, search your memory for alternatives, and then select a new course to take. Such flexibility is most commonly displayed by animals that have relatively large brains, such as primates, carnivores, and some cetaceans. The most flexible, complex, and future-oriented behaviors produced by humans and other mammals appear to be planned and organized, in part, by processes occurring in the **dorsolateral prefrontal cortex (DLPFC)**.

The DLPFC consists primarily of Brodmann's areas 9 and 46 (see Figure 33.3) and communicates broadly to cortical and subcortical regions: orbitofrontal cortex (OFC), anterior cingulate cortex, premotor cortex, and parietal cortex, among many others. Thus, it has connections well suited to serve as regulators of input-output pathways, using value and environmental context to shape this process. Accordingly, the function of the DLPFC is sometimes likened to switching in a railroad yard, rerouting connections between different tracks to align trains from their origins with their intended destinations (Figure 33.4). (This role is distinguished from that of OFC because it involves an active regulation of other circuits, not just a linkage of stimuli and their values.) In a similar manner,

the DLPFC may control the responses of other groups of neurons, making them more or less responsive to inputs and feedback, thereby producing different responses in different contexts.

A particularly important component of this type of control system is working memory (also called short-term memory), the ability to keep information in mind to guide behavior (see Chapter 30). Working memory is distinguished from long-term memory by its duration and purpose. If someone asks you to remember a phone number, you can maintain it in mind for tens of seconds (e.g., via repetition of its digits in your mind), but the memory can quickly fade if you become distracted. Specifically, firing rates of neurons in the DLPFC increase while information is maintained in short-term memory. For example, if a monkey is trained on a delayed response task, it will fixate a central spot and then covertly remember one of eight positions from a set of positions around the spot; the monkey must store that spatial information in this form of short-term memory. Neurons in the DLPFC show specific changes in their firing rate that depend on the position of the remembered dot; that is, their response depends on the contents of short-term memory. Consistent with a link between neuronal activity and short-term memory, damage to the DLPFC is associated with impairments in short-term memory capacity and duration. Such impairments are evident in experimental animals with ablations to the DLPFC as well as in humans who have had damage to this area.

Another important element in cognitive control is maintenance of rules and corresponding changes in behavior when the rules change. Neurons in the DLPFC show systematic patterns of activity that accord with specific rules, suggesting this area also maintains a representation of abstract information that guides complex behavior (Figure 33.5). Moreover, systematic changes in the firing rates of neurons in the DLPFC accompany changes in the rules that govern effective behavior in a particular context. In the **Wisconsin Card Sorting Task** (see Box 27C), an individual is shown a set of cards, each of which has a different number of distinct shapes of varying color. The individual must then place a new card according to an unstated rule, such as shape, color, or number. After a series of trials, the rule is surreptitiously switched. Individuals with DLPFC damage can learn to perform this task using the initial rule, but when the rule is changed they tend to continue with the old one and perform poorly. This impairment corresponds to the tendency for individuals with DLPFC lesions to become stuck in behavioral routines and not adapt to changing circumstances.

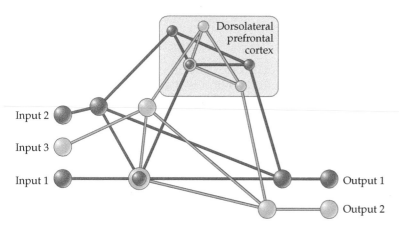

FIGURE 33.4 Schematic illustration of the function of the dorsolateral prefrontal cortex The brain can be thought of as a network that transforms inputs to outputs through information propagated along weighted connections. The path that information takes is, in turn, influenced by regulatory units thought to be housed in the DLPFC. These regulatory units receive information from other units in the system. (After E. K. Miller, 2000. *Nat Rev Neurosci* 1: 59–65.)

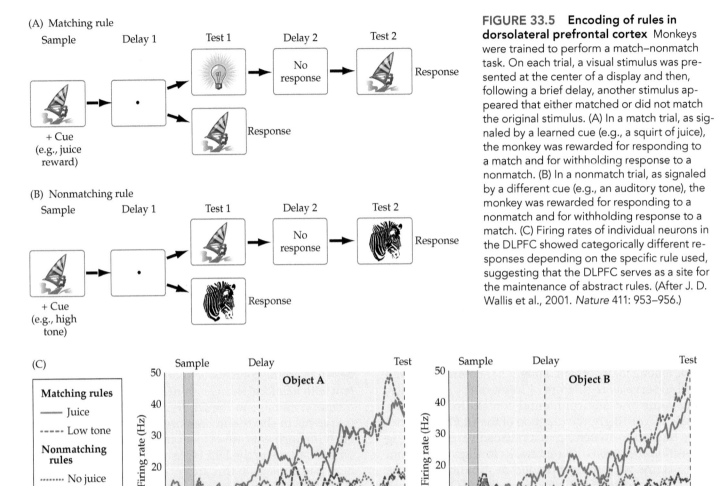

(A) Matching rule

Sample — Delay 1 — Test 1 — Delay 2 — Test 2 — Response

+ Cue
(e.g., juice
reward)

(B) Nonmatching rule

Sample — Delay 1 — Test 1 — Delay 2 — Test 2 — Response

+ Cue
(e.g., high
tone)

(C)

Matching rules
— Juice
----- Low tone

Nonmatching rules
.... No juice
---- High tone

FIGURE 33.5 Encoding of rules in dorsolateral prefrontal cortex Monkeys were trained to perform a match–nonmatch task. On each trial, a visual stimulus was presented at the center of a display and then, following a brief delay, another stimulus appeared that either matched or did not match the original stimulus. (A) In a match trial, as signaled by a learned cue (e.g., a squirt of juice), the monkey was rewarded for responding to a match and for withholding response to a nonmatch. (B) In a nonmatch trial, as signaled by a different cue (e.g., an auditory tone), the monkey was rewarded for responding to a nonmatch and for withholding response to a match. (C) Firing rates of individual neurons in the DLPFC showed categorically different responses depending on the specific rule used, suggesting that the DLPFC serves as a site for the maintenance of abstract rules. (After J. D. Wallis et al., 2001. *Nature* 411: 953–956.)

Impulsivity and self-control

After deciding to act and preparing to do so, individuals sometimes change their mind and countermand their planned actions. Individuals with damage to the **ventrolateral prefrontal cortex (VLPFC)** are impaired in this override function and respond impulsively. When tested, they respond more quickly but less accurately in timed tasks, and in life make poor decisions in various domains, including purchases, dietary choices, and social interactions. They may reach out and touch or even grab things that come into view or say the first thought that pops into their mind, even if it is inappropriate; and they continue to make bad choices even when they recognize that these actions are harmful.

The VLPFC is well positioned to govern the flow of information as it undergoes transformation from stimulus to behavior. It connects with sensory areas in the inferotemporal cortex and the auditory superior temporal gyrus, and has outputs that support a role in regulating the DLPFC. The VLPFC consists of Brodmann's areas 44, 45, and 12/47 (see Figure 33.3). The role of the VLPFC in inhibition is well documented (Figure 33.6). For example, disorders associated with reduced ability to inhibit unwanted actions and thoughts, including Tourette syndrome, obsessive-compulsive disorder (OCD), and clinical depression (habitual negative thoughts), are associated with damage to this region. When neuronal activity in the VLPFC is enhanced using transcranial magnetic stimulation (TMS), an individual's ability to suppress unwanted actions in laboratory tasks is improved. Consequently, TMS is currently being evaluated as a therapy to help individuals with OCD inhibit unwanted impulses.

Identifying the mechanisms that mediate behavioral inhibition involves using tests that target the ability to

(A)
Training phase

× 40 faces

Experimental phase

Think

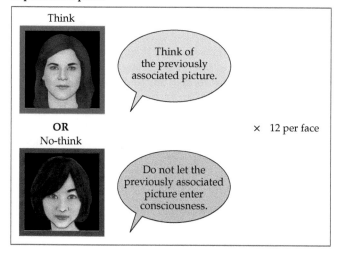

Think of the previously associated picture.

OR
No-think

Do not let the previously associated picture enter consciousness.

× 12 per face

(B)

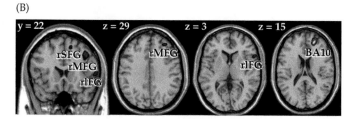

FIGURE 33.6 **Role of the ventrolateral prefrontal cortex in behavioral inhibition** (A) In one experiment, individuals learned to associate faces with particular scenes. Following training, presentation of the face naturally led to recall of the associated scene. On some trials, individuals were asked not to think of the associated scene—that is, to inhibit the thought process. (B) Deliberate inhibition of thought led to activation of the right inferior frontal gyrus (rIFG), a major component of the VLPFC. Colored areas indicate regions of hemodynamic activation; y/z measures indicate coronal and axial positions of brain images, respectively. Other regions activated, and thus potentially involved in inhibition, include Brodmann's area 10 (BA 10), the right medial frontal gyrus (rMFG), the right superior frontal gyrus (rSFG), and the right lateral frontal gyrus (rLFG). (From B. E. Depue et al., 2007. *Science* 317: 215–219.)

CONCEPT
33.3

Orbitofrontal Cortex Supports the Evaluation of the Outcomes of Behavior

LEARNING OBJECTIVES

33.3.1 Define orbitofrontal cortex and describe its pattern of connectivity to other brain regions.

33.3.2 Describe at least two sources of evidence for the role of orbitofrontal cortex in reward evaluation.

33.3.3 Define the common currency theory for orbitofrontal cortex function.

33.3.4 Explain the credit assignment problem in decision making.

Evaluation of options

Reward is a basic influence on the decisions individuals make. Generally speaking, the reward value of an option refers to the benefit it provides the decision maker, either in the short term (such as the relief provided by a cold drink on a hot day) or in the long term (such as a retirement annuity or the increased opportunities provided by enrolling one's child in a better school). Dysfunctions in reward processing are hallmarks of clinical disorders, including addiction (Clinical Applications). Estimating the value of an option involves perceiving its sensory properties, identifying the situational context, and retrieving information about past experiences with similar contexts from memory. For example, a diner might be attracted by a particular entrée but may recall a prior experience with food poisoning after eating that dish and decide to pass. Such details are combined to estimate a value that can be compared with the values of other options to guide decisions. **Evaluation**—the estimation of the value of an option based on both past and present information—has been linked to the **orbitofrontal cortex (OFC)**

countermand learned or habitual actions. The go/no-go task is an example. Individuals are told to perform a specific action when they see a stimulus—for example, press a button when a green light is illuminated. But occasionally a subsequent stimulus—say, a red light—supplants the first one, and the individual must ignore the green light and withhold the initially planned action. Another test entails task switching, in which individuals alternate between performing two different tasks. The role of inhibition in task switching may not be obvious at first, but performing one task for several trials tends to instill a particular response pattern. Switching from one task to another requires suppressing that practiced pattern, and failure to do so introduces confusion between tasks and conflict. Tasks that involve response inhibition reliably activate the VLPFC in fMRI studies; conversely, damage to the VLPFC impairs performance, such that individuals have difficulty inhibiting previously valid responses.

■ Clinical Applications

Addiction

Drug abuse and addiction show how neural circuits that typically support adaptive behaviors (e.g., learning from rewards) can become dysfunctional when chemicals modify their function. With this in mind, it is not surprising that most known drugs of abuse—including heroin and other opiates, cocaine, alcohol, marijuana, nicotine, amphetamines, and their synthetic analogues—act on the dopamine system.

Recall from Chapter 18 that the dorsal divisions of the basal ganglia (caudate, putamen, and globus pallidus) are instrumental in gating the activation of thalamocortical circuits that initiate volitional movement. Also mentioned briefly in Chapter 18 is the existence of other, parallel processing streams that similarly gate the activation of non-motor programs, including those that pertain to cognition and affect. The organization of these non-motor processing streams is fundamentally comparable to the "direct pathway" for volitional movement: There are major excitatory inputs from cortex to striatum, neuromodulatory projections from midbrain dopaminergic neurons to striatum, internuclear connections from striatum to pallidum, and output projections from pallidum to thalamus. What distinguishes the brain circuitry for reward and motivation from the motor circuitry discussed in Chapter 18 is the source and nature of the cortical input, the relevant divisions of striatum and pallidum that process this input, the source of dopaminergic projections from the midbrain, and the thalamic target of the pallidal output (Figure A).

Central to the processing of rewards are inputs from the amygdala, the subiculum (a ventral division of the hippocampal formation), and OFC that convey signals relevant to emotional reinforcement to ventral divisions of the anterior striatum, the largest component of which is called the *nucleus accumbens*. Like the caudate and putamen, the nucleus accumbens contains medium spiny neurons that integrate excitatory telencephalic inputs under the modulatory influence of dopamine. However, unlike the larger dorsal division of the striatum, the nucleus accumbens receives its dopaminergic projections from a collection of neurons that lies just dorsal and medial to the substantia nigra, in a region of the midbrain called the **ventral tegmental area (VTA)** (Figure B). The nucleus accumbens and the ventral tegmental area are primary sites where drugs of abuse interact with the processing of neural signals related to emotional reinforcement; they do so by prolonging the action of dopamine in the nucleus accumbens or by potentiating the activation of neurons in the ventral tegmental area and nucleus accumbens (Figures C and D).

Under normal conditions, these dopaminergic neurons are only phasically active; when they do fire a barrage of action potentials, however, dopamine is released in the nucleus accumbens, and medium spiny neurons are much more responsive to coincident excitatory input from telencephalic structures such as the amygdala and OFC. These activated striatal neurons in turn project to and inhibit pallidal neurons in a region just below the globus

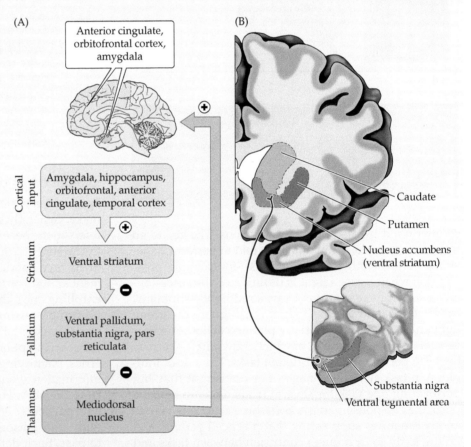

(A) The learning processes that are impaired during addiction rely on a circuit that comprises cortical inputs to striatum, internuclear projections from striatum to pallidum, pallidal output to thalamus, and thalamic projections back to cortex. (B) Coronal section through the rostral forebrain, showing the basal ganglia structures represented in Figure A and the dopaminergic projection from the ventral tegmental area of the midbrain to the nucleus accumbens, the principal component of the ventral striatum.

■ Clinical Applications (continued)

pallidus called the **ventral pallidum**, as well as in the pallidal division of the substantia nigra (pars reticulata). The suppression of tonic activity in the pallidum then disinhibits the mediodorsal nucleus of the thalamus, the nucleus that innervates cortical divisions of the forebrain. Activation of these cortical regions via the mediodorsal nucleus is reinforced by direct cortical projections of dopaminergic neurons in the ventral tegmental area and glutamatergic projections from the basolateral group of nuclei in the amygdala that target these same regions.

Unfortunately, the plastic potential of the dopamine system can be co-opted by chronic exposure to drugs of abuse, leading to cellular and molecular changes that promote abnormal regulation (see Figure D). In the ventral tegmental area of addicted individuals, the activity of the dopamine-synthesizing enzyme tyrosine hydroxylase increases, as does the

ability of VTA neurons to respond to excitatory inputs. The latter effect is secondary to increases in the activity of the transcription factor CREB (see Chapter 7) and the upregulation of GluR1, an important subunit of AMPA receptors for glutamate (see Chapter 6). In the nucleus accumbens, addiction is characterized by increases in another transcription factor, ΔFosB, in addition to induction of CREB with chronic exposure to at least some classes of addictive drugs. Activation of these molecular signaling pathways leads to a generalized reduction in the responsiveness of accumbens neurons to glutamate released by telencephalic inputs by regulation of different AMPA receptor subunits and/or changes in postsynaptic density proteins that alter the dynamics of receptor trafficking. However, during the phasic release of dopamine, the responsiveness of these striatal neurons is intensified, mediated in part by a shift in the expression of

D1 and D2 classes of dopamine receptors and a coordinated upregulation of cAMP–PKA signaling pathways.

The cellular and molecular maladaptations of ventral tegmental, striatal, and cortical neurons to chronic exposure of abusive drugs remain incompletely understood. Nevertheless, the net effect of these and other changes is that addiction dampens the response of this reinforcement circuitry to less potent natural rewards, while intensifying the response to addictive drugs. At a systems level, these changes are probably reflected in the hypofrontality —reduced baseline activity in OFC— commonly seen in addicts. Taken together, these alterations in neural processing could account for the waning influence of adaptive emotional signals in the operation of decision-making faculties as drug-seeking and drug-taking behaviors become habitual and eventually compulsive.

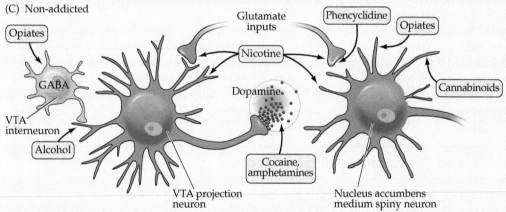

(C) Most drugs of abuse potentiate the activity of dopamine by interacting directly with dopamine synapses in the nucleus accumbens, or indirectly by modulating the activity of neurons in the ventral tegmental area (VTA). Other drugs may act directly on accumbens neurons to increase their responsiveness to telencephalic input. (D) Drug addiction (DA) is associated with cellular and molecular adaptations of this circuit. The net effect of addiction is a chronic decrease in basal activity and an increase in the intensity of phasic activity in the presence of abusive drugs. (After E. J. Nestler, 2005. *Nat Neurosci* 8: 1445–1449.)

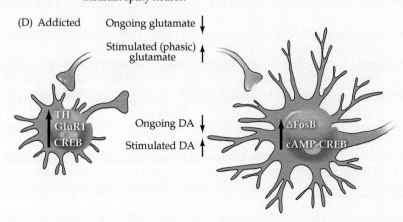

(Figure 33.7) by virtue of the OFC's anatomical connections, the behavioral disruptions that follow damage to this area, and its activation during learning and decision making. Note that neuroscientists sometimes use the alternative term **ventromedial prefrontal cortex (VMPFC)** to describe a broader region that encompasses parts of OFC as well as midline regions of the prefrontal cortex.

The OFC receives input from all of the major sensory modalities (vision, audition, somatic sensation, olfaction, and gustation), giving it access to the information necessary to identify decision options and estimate their values. Unlike other prefrontal regions, however, the OFC has few motor connections, consistent with the idea that the OFC provides inputs to systems that themselves inform the selection and execution of behavior. Furthermore, the OFC receives inputs from the hippocampus and adjacent regions in the medial temporal lobe that are involved in memory storage and retrieval (see Chapter 30). These inputs presumably provide information about prior experiences to improve estimates of value. Finally, the OFC receives inputs from reward-related dopamine neurons in the midbrain that help shape associations among objects, actions, and their consequences (Box 33A).

Direct evidence for the role of the OFC in evaluation is fairly strong. When monkeys choose among options varying in reward amount, reward type, and probability of reward delivery, firing rates of some neurons in the OFC track individual preferences for a particular option, a variable known as *subjective value*. For example, when a monkey that likes peanuts is fed them to satiety, the sensory properties of peanuts remain

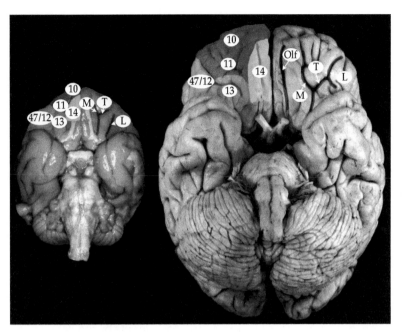

FIGURE 33.7 Gross anatomy of the orbitofrontal cortex The ventral surfaces of the brains of a rhesus monkey (left) and a human (right) are shown for comparison. In spite of the different sizes of the brains, the structure of the OFC (the colored regions) is largely preserved across the two species. Numbers refer to Brodmann's areas that are the major anatomical subdivisions of the OFC. The majority of research on the OFC in both species has focused on Brodmann's area 13. Olf = olfactory sulcus, M = medial orbital sulcus, T = transverse orbital sulcus, L = lateral orbital sulcus. (From J. D. Wallis, 2007. *Annu Rev Neurosci* 30: 31–56.)

the same but their value to the monkey is reduced. Satiety is accompanied by a reduction in the firing rates of OFC neurons in response to more peanuts, implying that the OFC encodes the subjective value of a food and not just its sensory properties.

■ BOX 33A | Dopamine and Reward Prediction Errors

Perhaps more than any other neurotransmitter, dopamine has permeated popular culture. References to dopamine as the "pleasure chemical" can be found on late-night TV, in newspaper editorials, and in casual conversation. Dopamine has risen to this prominence in part through its role in drug addiction. Nearly all addictive drugs exert their effects through their ability to alter dopamine release or reuptake from the relevant synapses—which leads to the inference that

dopamine signaling is the basis of pleasure. The popular view has some truth to it but is misleading.

Early links between dopamine and addiction came from the series of self-stimulation studies carried out in rats in the 1950s. James Olds and Peter Milner implanted electrodes in the medial forebrain bundle of rats; activation of this structure causes the release of dopamine in the nucleus accumbens of the striatum. Olds and Milner's innovation was allowing rats

to control the stimulation. Faced with the choice between self-stimulation and other activities such as eating and drinking, the rats chose to self-stimulate even to the point of death.

The most completely studied dopaminergic neurons are those whose cell bodies are in the ventral tegmental area (VTA) of the midbrain and the substantia nigra pars compacta (SNPc) (Figure A). These neurons project widely throughout the brain, but especially to the PFC and ventral striatum, where they are

■ BOX 33A │ Dopamine and Reward Prediction Errors (continued)

(A)

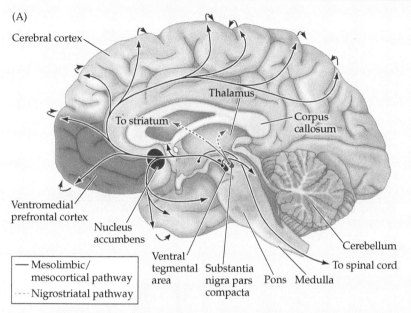

Cerebral cortex

Thalamus

To striatum

Corpus callosum

Ventromedial prefrontal cortex

Nucleus accumbens

Ventral tegmental area

Substantia nigra pars compacta

Pons   Medulla

Cerebellum

To spinal cord

— Mesolimbic/ mesocortical pathway

---- Nigrostriatal pathway

(B)

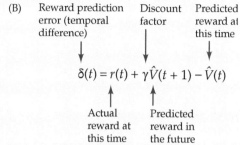

Reward prediction error (temporal difference)

Discount factor

Predicted reward at this time

$$\delta(t) = r(t) + \gamma \hat{V}(t+1) - \hat{V}(t)$$

Actual reward at this time

Predicted reward in the future

(A) Schematic of the dopamine system of the brain. Two of the major sources of dopamine in the brain are the ventral tegmental area (VTA) and the substantia nigra pars compacta (SNPc). These two regions house the cell bodies of dopamine neurons that project broadly throughout the brain. (B) The equation of the reward prediction error (RPE) signal—the hallmark of the response of the dopamine neuron. (After W. Schultz et al., 1997. *Science* 275: 1593–1599.)

(C)

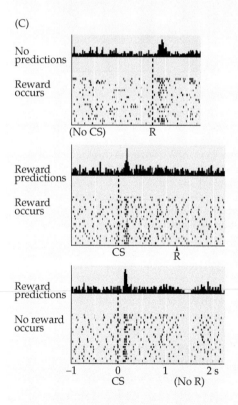

No predictions

Reward occurs

(No CS)   R

Reward predictions

Reward occurs

CS   R

Reward predictions

No reward occurs

−1   0   1   2 s
CS   (No R)

(C) In practice, the RPE signal is manifest as a systematic change in the brief response of dopamine neurons to reward. When the reward (R) is unexpected, it increases the baseline firing rate of the neurons. When the reward is paired with a cue (conditioned stimulus, CS), then following learning, the cue elicits a response while the reward itself no longer affects the neural response. Finally, when the cue is followed unexpectedly by a failure to provide a reward, the dopamine neurons briefly pause their firing, thus carrying a negative RPE signal. (From W. Schultz et al., 1997. *Science* 275: 1593–1599.)

thought to regulate neural activity. Recordings from dopamine neurons show that their responses can be described as a **reward prediction error (RPE)**. Whenever individuals execute some behavior, they predict the likely outcome (*reward*). RPE is simply the difference between what was predicted and what actually materialized (Figure B). If what materialized is better than predicted, the RPE is positive; if what materialized is worse than predicted, the RPE is negative. The RPE is important for learning: If one option is better than expected, individuals update their estimate of its value and choose it more often in the future—as predicted by formal learning models developed in psychology and computer science.

The phasic release of dopamine is also subject to experience-dependent plasticity (Figure C). During associative learning, for example, the activity of ventral tegmental neurons transitions from signalling the occurrence of a reward (see Figure C, top panel) to signalling a reward-predicting stimulus with diminished responsiveness to primary rewards themselves (see Figure C, middle panel). Interestingly, if a stimulus is not followed by delivery of the predicted reward, ventral tegmental neurons are suppressed at precisely the same time that an increased neuronal response would have signaled the presence of a reward (see Figure C, bottom panel). These observations suggest that the phasic release of dopamine signals the presence of reward relative to its prediction, rather than the unconditional presence of reward. The integration of such signals in the nucleus accumbens, OFC, and amygdala leads to the activation of behaviors directed at obtaining and consolidating the benefits of the rewarding event.

One way that dopamine might work is by regulating Hebbian learning (the idea that "neurons that fire together wire together"; see Chapter 25). If two

(Continued)

■ BOX 33A | Dopamine and Reward Prediction Errors *(continued)*

neurons fire in sequence, and if dopamine is also present, their connection may be strengthened. If dopamine is released when the outcome is better than expected, it would strengthen connections that are active right before that release occurred. Thus, dopamine may strengthen connections when the environment is better than expected and learning is favored. In the case of drug addiction, however, excess dopamine may hijack learning and lead to the formation of maladaptive habits.

These and many other studies studies emphasize the direct influence of dopamine circuitry over processes of learning that could lead to addiction—leading to the popular view that dopamine release may be the cause of pleasure. Yet the inference that dopamine supports pleasure turns out to be wrong. Subseqent research has distinguished the neuroanatomical substrates associated with the pleasure obtained from rewards (*liking*) from the motivation to seek those rewards (*wanting*). An individual can enjoy (*like*) things without having a drive to seek more of them, just as one can be driven (*want*) to pursue things, often with great vigor, without pleasure. Consider a smoker trying to quit who is miserable and hates stepping outside to smoke, but is compelled by his addiction to do it anyway. It is unlikely the smoker would describe the smoking experience as pleasurable. Or consider an individual with OCD, hand washing repeatedly; even though motivated to do it, the individual would not describe this process as enjoyable.

An elegant distinction between linking and wanting has been demonstrated in rats that have had lesions that damage neurons in the ventral striatum that receive dopaminergic imputs (Figure D). These rats lack the

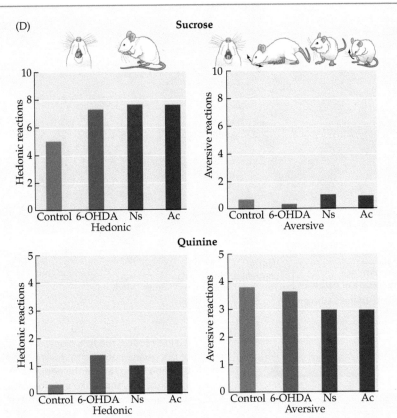

(D) Rats treated with the neurotoxin 6-hydroxydopamine HBr (6-OHDA) show dramatic decreases in the number of dopaminergic neurons in the striatum, compared with control animals. For example, in the experiment whose data are shown here, administration of 6-OHDA resulted in greater than 90% dopamine depletion in all rats (6-OHDA group), with subsamples showing greater than 98% depletion in the striatum (Ns) and greater than 99% depletion in the nucleus accumbens (Ac). These rats became aphagic, such that they would not seek out and eat food that was freely available in their cage. However, they showed normal hedonic ("liking") reactions (e.g., tongue protrusions) to a sucrose solution delivered directly to their mouths. They also showed normal aversive reactions (e.g., gaping mouths) when the sugar water was made bitter by the addition of quinine. (Adapted from K. C. Berridge and T. E. Robinson, 1998. *Brain Res Rev* 28: 309–36.)

motivation to obtain rewards (e.g., they will not press levers or even cross their cage to obtain food or drink). Yet if a pleasurable fluid is squirted into their mouths, they show typical facial expressions of pleasure; conversely, if a bitter fluid is squirted into their mouths, they show typical facial expressions of disgust. These results demonstrate that dopamine regulates the drive for seeking rewards but not pleasure. While the neural substrates of pleasure itself are not completely understood, neuroscientists have argued that it is supported by a different brain system, the μ-opioid system.

Evidence for the evaluative role of the OFC also comes from neuroimaging studies, which show a robust correlation between hemodynamic activity and the subjective value of options in decision-making tasks. This relationship is vividly illustrated in a study on the human enjoyment of wine. The investigators found that how much people enjoy the taste of wine depends on factors that do not affect its taste, such as how much the wine costs. When people believe they are drinking a more expensive wine, they tend to report that it tastes better. This change in appeal is reflected in changes in hemodynamic activity in the OFC, suggesting that the activity mediates the change in enjoyment associated with price. The relationship between activity in the OFC and personal preferences is so robust that measures of brain activity in the OFC can be used to predict purchasing behavior across a broad range of goods and experiences, including snack foods, beverages, and images that feature attractive people.

Indeed, a remarkable feature of the OFC is its apparent ability to contribute directly to decisions about so many different things, ranging from which soda to buy from a vending machine to which college to attend (see Figure 27.6). This flexibility has led to the proposal that the OFC computes value in a universal scale that allows comparison of diverse options. This **common currency theory** is supported by hemodynamic response patterns measured with fMRI. However, direct recordings from OFC neurons reveal a much greater diversity of response patterns—and understanding how the OFC may support processes beyond construction of a common currency has become a very active area of research.

Actually making a choice occurs after evaluation and requires active maintenance of the values of two or more options. This process seems likely to occur in the OFC, whose neurons show systematic changes in firing when the values of multiple options are maintained in short-term memory. Lesions to the OFC are associated with deficits in comparing the values of disparate options. For example, when a person chooses between options that differ along multiple dimensions, such as price, styling, and gas mileage of a new car, the brain must make separate comparisons and use the results to make a decision. Decisions that involve such integration of multiple sorts of information are particularly impaired by OFC damage.

This evidence extends the role of the OFC to include **credit assignment**, which is the process of identifying the one stimulus among many in the current context that is responsible for a reward or punishment. The idea of credit assignment was anticipated by the psychologist Edward Thorndike in his classic work on the *law of effect*. His idea was that "responses that produce a satisfying effect in a particular situation become more likely to occur again in that situation, and responses that produce a discomforting effect become less likely to occur again in that situation." More generally, stimuli associated with pleasant events gain pleasant associations themselves, and vice versa. This concept seems simple, but in many cases a large number of stimuli occur at the same time and their consequences may be delayed or provide both pluses and minuses. Brain mechanisms need to assign credit to the actual stimuli that predict rewards and

punishments, and a good deal of evidence supports the idea that the OFC contributes to this process. For example, lesions to the OFC selectively impair the ability to link rewards to events (Figure 33.8). Monkeys with OFC lesions assign positive value to stimuli that predict aversive events as long as the stimuli are surrounded by other positive events. It thus seems that the OFC is critical for accurately linking events with their values.

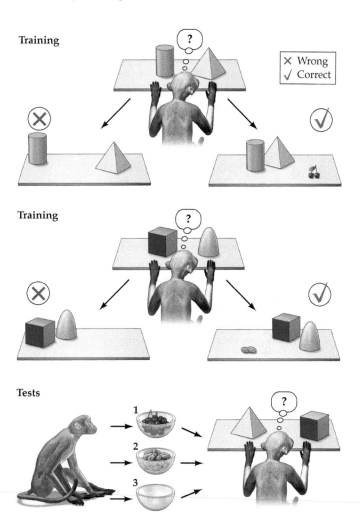

FIGURE 33.8   Effects of orbitofrontal cortex lesions on the reinforcer devaluation task Some of the most important insights into the role of the OFC come from the results of lesion tasks in primates. In the reinforcer devaluation task, a monkey is trained with pairs of stimuli that are associated through training with either a reward or no reward. The rewards themselves differ and are predicted by the stimuli. During the testing phase, preference for the stimulus is experimentally controlled by selective satiation. For example, to make the monkey more motivated to eat cherries, the monkey would be fed to satiation on peanuts (this process devalues the reinforcer). The monkey then chooses between stimuli associated with cherries or peanuts. Monkeys with OFC lesions do not adjust their preference toward the preferred food as well as non-lesioned control monkeys do. (After M. G. Baxter and E. A. Murray, 2002. *Nat Rev Neurosci* 3: 563–573.)

The OFC is not the only brain area in which values are computed, maintained, and compared. Other evidence supports roles for additional brain regions, including the ventral striatum and the dorsal anterior cingulate cortex (see Concept 33.4). Moreover, a good deal of evidence points to a more sophisticated role for the OFC in behavior. For example, one current theory holds that the OFC maintains a cognitive map of the set of currently relevant behavioral stimuli, their values, and potential outcomes. This idea suggests that the OFC operates much like a switchboard, linking the external world and internal states with the possible outcomes of choices. Given the importance of reward and motivation for behavior, elucidating the specific computations supported by OFC remains a critical area for modern neuroscience research.

CONCEPT **33.4**

Anterior Cingulate Cortex Supports Regulation of Activity in Other Brain Regions

LEARNING OBJECTIVES

33.4.1 Define the anterior cingulate cortex and describe its pattern of connectivity to other brain regions.

33.4.2 Provide examples of the potential contributions of anterior cingulate cortex to learning from feedback.

33.4.3 Explain the evidence for a role of anterior cingulate cortex in monitoring behavior.

Learning from the consequences of behavior

Humans and other intelligent animals are capable of learning from the consequences of their actions. Doing so requires circuitry that can evaluate the outcomes of decisions and update the control systems that regulate the connections between stimulus inputs and behavioral outputs (e.g., influencing the planning and rule selection processes described in the chapter's earlier concepts). Such cognitive control processes are most strongly associated with the **anterior cingulate cortex** (**ACC**), a region within the **dorsomedial prefrontal cortex** (**DMPFC**). As its name implies, the ACC includes the anterior portion of the cingulate gyrus along the brain's midline; it consists of Brodmann's area 24 as well as parts of areas 9, 6, and 32. Its anatomical connections with other brain regions make it well positioned to support cognitive control, since it has multiple inputs conveying information from a variety of systems, including perception, emotion, attention, and memory.

Activity in the ACC tracks the outcomes of actions and generates feedback signals useful for updating behavioral goals and for adopting new rules for action. Particularly important are outcomes that lead to rapid changes in behavior. Such outcomes include those that are unexpected or surprising, provide useful information and promote learning, or provide new information about other individuals or about pain, which signals what is or is not safe. For example, in laboratory studies the ACC is activated by errors that reduce rewards, which in real life would correspond to disappointment in not achieving a goal; studies using EEG have revealed that these error-related responses arise very rapidly, consistent with roles for the ACC in learning from negative outcomes and shaping behavior elsewhere in prefrontal cortex (Figure 33.9). Moreover, in tasks where the reward depends on the choices individuals make, ACC neurons signal the size of rewards that follow decisions. The ACC thus seems to track the values of outcomes when this information is used to guide future behavior.

The ACC also tracks counterfactual or fictive outcomes—in other words, the rewards or punishments associated with options that were not chosen. People and other animals not only monitor the consequences of their

FIGURE 33.9 Rapid prefrontal cortex responses to monetary gains and losses (A) By measuring brain functioning using EEG, cognitive neuroscientists can identify very rapid changes in brain activity associated with winning or losing money. Feedback about an unexpected monetary loss generates a negative-polarity event-related potential component within about 200–300 ms (black arrow in figure). (B) Source localization methods have linked this component to the anterior cingulate gyrus, a region that contributes to the engagement of control systems within the brain. μV = microvolts. (After W. J. Gehring and A. R. Willoughby 2002. *Science* 295: 2279–2292.)

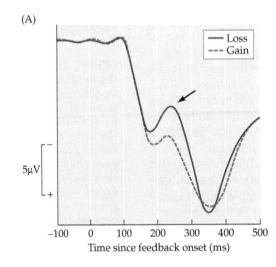

(A)

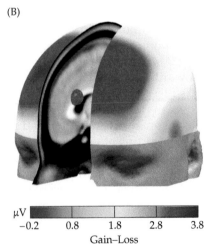

(B)

actions but also attend to what they might have experienced had they acted differently. There is evidence that the ACC monitors both types of outcomes simultaneously. For example, when stock traders play the market, they respond differently to financial returns depending on how other stocks they could have invested in perform—and fMRI studies using simulated stock markets have identified signals in the ACC that track counterfactual outcomes (e.g., what would have happened if an individual had sold their stock holdings). Across all of these examples, the role of the ACC can be regarded as complementary to that of the DLPFC: The ACC detects the need to change behavior, and the DLPFC implements that change.

Another influence on ACC activity is the **conflict** associated with different action plans activated at the same time. A classic example is the **Stroop effect** (Figure 33.10). In a typical Stroop scenario, people are faster at identifying the color of the word *red* when it is printed in red ink than they are when it is printed in green ink. In cases where the word label and ink color are incongruent, the two sources of information conflict. Resolving such conflicts based on task instructions relies on processing in the ACC, as shown both through direct electrophysiological recordings in human neurosurgical patients and functional neuroimaging studies in control individuals. Suppose you are instructed to press a left button when an arrow at the center of the computer screen points left and a right button when that arrow points right; if only one arrow is visible on the screen, there is no conflict and you can complete the task rapidly and without errors. But if the task-relevant arrow is flanked by other arrows that you must ignore, you will be faster and more accurate when the flanking arrows point in the same direction as the central one, and slower and less accurate when they point in the opposite direction. Adjudicating between the conflicting responses indicated by the differently oriented arrows requires control processes, leading to activity in neurons in the ACC and increased hemodynamic responses in this area as measured by fMRI.

Consistent with these observations, damage to the ACC leads to impairments in learning from the consequences of actions. An example is OCD, which is associated with atypical levels of activity in the ACC. Individuals with OCD are overly sensitive to stimuli that would generally be ignored, such as dirt and clutter. OCD is also associated with self-doubt (being unsure the oven is actually off, for example) and failures of self-control. Thus, these individuals feel compelled to perform acts such as counting and washing that they know are illogical and don't want to perform, but can't suppress. Cingulotomy—surgical ablation of the ACC—can be an effective treatment for the most severe cases of OCD, for reasons that are unclear. One possibility is that regulatory monitoring activity in the ACC is hyperactive in such individuals, leading to enhanced anxiety and maladaptive behavior. Suppressing this hyperactivity by severing the connections of the ACC may reduce individuals' sensitivity to minor errors, allowing them to move ahead with more adaptive responses. In any case, surgical intervention in OCD is an example where basic science and surgery have worked well together.

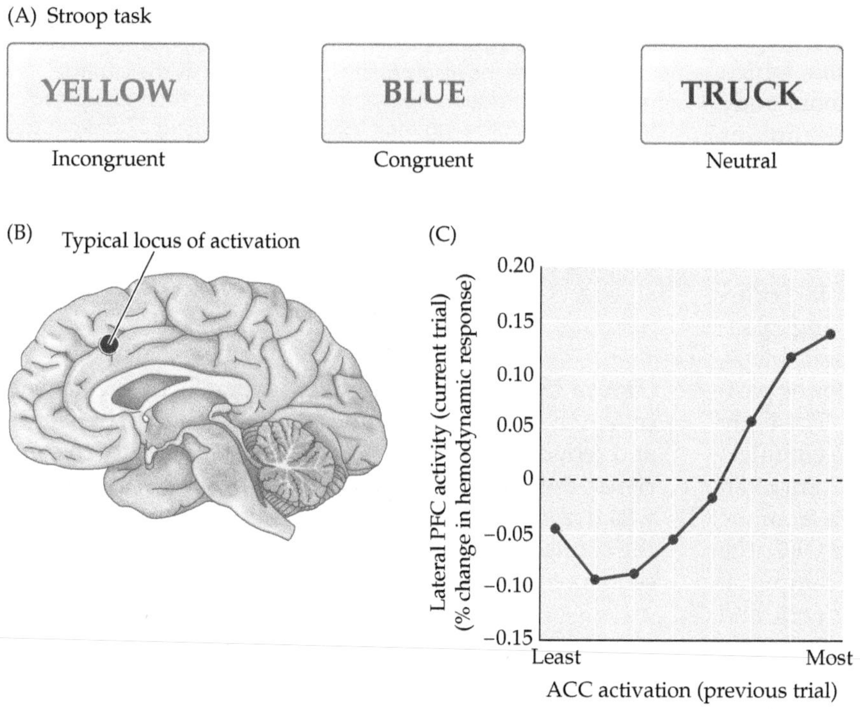

(A) Stroop task

YELLOW	BLUE	TRUCK
Incongruent	Congruent	Neutral

(B) Typical locus of activation

(C)

FIGURE 33.10 **Anterior cingulate cortex and the Stroop effect** (A) In the Stroop Task, individuals are asked to say a word printed in a color on a card or computer screen. When the word is incongruent—a different color than the color of the ink—individuals are slower to respond. When the word is neutral (not color related) or congruent (same color), individuals have no difficulty responding. It is thought that the competition between saying the word and saying the ink color come into conflict. (B) During cognitive conflict, responses in the anterior cingulate cortex (ACC) are greatly enhanced. (C) Following conflicting trials, individuals adjust their behavior to improve performance. This adjustment is thought to reflect changes driven by inputs to the dorsolateral prefrontal cortex from the ACC. Consistent with this interpretation, increased activation of the ACC on the previous trial, as measured by fMRI, led to increased activation in the DLPFC on the current trial. (B after G. Bush et al., 1998. *Hum Brain Mapp* 6: 270–282; C after J. G. Kerns et al. 2004. *Science* 303: 1023–1026.)

CONCEPT
33.5

The Anterior Insula Incorporates Information about Body States into Decision Processes

LEARNING OBJECTIVES

33.5.1 Define the insular cortex and describe its pattern of connectivity to other brain regions.

33.5.2 Explain why damage to insular cortex might reduce cravings associated with addiction.

The internal milieu

Just as individuals use cognitive information to deal with information about reward, error, memory, and surprise, they also use information about basic body states to regulate behavior. These states include hunger, thirst, temperature, pain, itch, fatigue, heart rate, and many more. Although these sensations are often unconscious, they nonetheless affect decision making. These processes are most closely associated with the **anterior insula**, which receives visceral inputs—that is, information about body states, largely via subcortical brain areas. The anterior insula is also associated with emotional awareness and expression, relaying this information to the rest of the cerebral cortex as well as subcortical areas involved in the expression of emotions (see Chapter 32).

The anterior insula is not part of the prefrontal cortex but is discussed here because of its role in thinking, planning, and deciding. It lies buried within the lateral sulcus, which separates the temporal lobe from the inferior parietal and frontal lobes (see Figure 33.2A, right). The body states that activate it are also associated with the cingulate cortex; indeed, the anterior insula and the ACC are often co-activated. The anterior insula receives inputs from the ACC, the inferior temporal lobe, the PFC, the OFC, the central nucleus of the amygdala, and the hippocampus. It is also linked to attention, time perception, romantic and maternal love, mood, speech, and music—presumably because at least some of these states are accompanied by body states often linked to emotion. Thus, the anterior insula can be seen as part of a larger system that uses information about internal states to help make decisions and guide behavior.

The insula figures in several disorders, the most prominent being compulsive gambling and drug addiction, both of which affect decision making. Indeed, there is some evidence that drug addiction and compulsive behaviors such as gambling may be the outcome of a vicious cycle in which the addictive substance or behavior begins to take over decision making, and push the addict toward more of that substance or behavior. One study examined individuals who had experienced acute damage to the insula (typically through stroke) and then tried to quit smoking

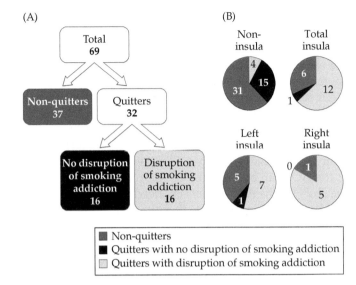

FIGURE 33.11 Insula damage is associated with an improved ability to quit smoking (A) Among all individuals studied, 19 had insula damage and 50 had other damage that did not involve the insula (that is, they served as the control group). All individuals were smokers; 32 of the total group had tried quitting following their lesion; the remaining 37 did not try. Of the 32 who tried to quit, half experienced a disruption in their smoking addiction; that is, quitting was relatively effortless and the urge to smoke again was low. (B) Critically, the individuals who experienced disrupted smoking addiction were more likely to be the individuals with insula damage. (After N. H. Naqvi et al., 2007. *Science* 315: 531–534.)

(Figure 33.11). Those individuals experienced greater success in their attempts to quit, a lower likelihood of relapse, and reduced cravings for cigarettes. An inability to experience the body states typically associated with nicotine withdrawal may have allowed these individuals to lose the connection between smoking and its immediate consequences, thereby allowing them to break the habit.

CONCEPT
33.6

Posterior Cingulate Cortex Supports Internally Directed Processes

LEARNING OBJECTIVES

33.6.1 Define the posterior cingulate cortex and describe its pattern of connectivity to other brain regions, including within the default-mode network.

33.6.2 Explain the functions associated with the default-mode network.

33.6.3 Describe the effects of damage to posterior cingulate cortex on behavior.

33.6.4 Characterize key conclusions from neuroscience research on awareness and volition.

Self-awareness

In addition to an awareness of body states, humans (and presumably many other animals) are aware of themselves as part of a larger world. We are creative and unpredictable, and we ruminate and muse. These ineffable, but important, aspects of mental life are at best difficult to study (Box 33B). But as neuroscientists have begun to explore these issues, a region that appears to be especially important is the **posterior cingulate cortex** (**PCC**), a midline cortical structure lying at the caudal end of the cingulate sulcus that includes Brodmann's areas 31 and 23.

The PCC, like the anterior insula, it is not part of the prefrontal cortex, and until recently it received relatively little attention from cognitive neuroscientists. When examined

---

■ BOX 33B | What Does Neuroscience Have to Say about Free Will?

Any discussion of thinking, planning, and decision making is bound at some point to come up against the philosophical question of "free will." The debate over free will, which like many philosophical arguments goes back millennia, asks whether individuals govern their own actions, or whether all behaviors are in fact determined.

Neuroscience has been a player in the free will debate since a study by neurophysiologist Benjamin Libet in 1984. Libet took advantage of a classic finding from EEG: that an individual's actions are preceded by an elevation in neural activity (Figure A) that likely reflects the responding of premotor cortex neurons, which are thought to directly influence motor neurons through feedforward projections. Libet wondered whether this signal, which became known as the **readiness potential** (Figure B), reflects brain events that occurred before conscious awareness, which would indicate that individuals only become aware of their intentions after the action has already been planned. He asked people to look at a specially designed clock and note, privately, the time on the clock when they first felt the conscious urge to act. (Their task was simply to move their arm when they felt the urge.) Individuals were then asked to indicate the time when the event occurred.

Libet found that the readiness potential was statistically detectable up to 300 ms before individuals reported the conscious intention to act. More modern methods using more sophisticated brain measures show that brain activity can predict choices as

(Continued)

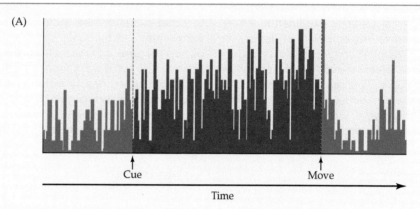

(A) When we move our limbs, our action is preceded by specific patterns of brain activity that seem to have a direct causative role in preparing for the action. One of the most prominent such signals occurs in the premotor cortex, a region of the brain located rostrally to the motor cortex. The intention period that precedes the movement is associated with a gradual increase in firing rate of neurons in the premotor cortex, a region that provides direct inputs to the motor cortex. Once the movement begins, neurons in this region return to baseline level, suggesting that their major contribution is to promote motor intention.

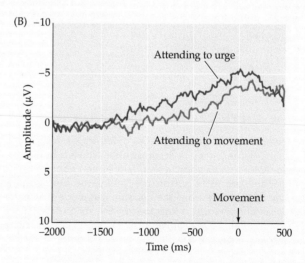

(B) The readiness potential. When EEG electrodes are placed over the premotor cortex, they show a consistent ramping up of activity that precedes the performance of the action (the black arrow at time 0). This ramping up is greater when individuals attend to the urge to move (red line) rather than attending to the movement itself (blue line), but it is observed in both cases. (After A. Sirigu et al., 2003. *Nat Neurosci* 7: 80–84.)

much as 20 seconds in advance (Figures C and D). These findings suggest that consciousness is an effect, not a cause, of the decision to act, which occurs unconsciously. This in turn would imply that one's actions are not generated consciously but instead are elicited by other processes, and that consciousness is simply a passive observer. An alternative explanation—now supported by empirical evidence—is that the brain continually develops potential plans for action and that signals for those plans can be observed in advance of their executions. According to this model, individuals don't freely select their actions, but rather reject some and permit others, and this veto process results in the indeterminacy or unpredictability of behavior. Not surprisingly, this interpretation has attracted a great deal of debate among philosophers.

Other pertinent evidence comes from some very simple psychological experiments looking at the factors that influence free choice. In one experiment, individuals chose one of two pictures of faces as more attractive (the faces were monochrome and were rated nearly equal in attractiveness by independent raters). The experimenter then handed the participants the picture and asked them to explain why they chose the one they did. Unbeknownst to the individuals, however, the experimenter used sleight of hand to change which picture was handed to the individuals. In most cases, participants did not notice the switch. Then when faced with the need to explain their choice, most individuals made up plausible explanations. This finding suggests that the human brain will often fabricate beliefs and memories to support what it thinks occurred. Perhaps our sense of free will is a similar confabulation that we employ to make sense of a determined reality we can't accept.

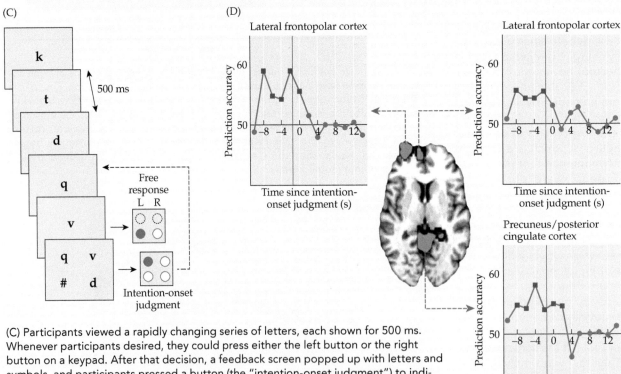

(C) Participants viewed a rapidly changing series of letters, each shown for 500 ms. Whenever participants desired, they could press either the left button or the right button on a keypad. After that decision, a feedback screen popped up with letters and symbols, and participants pressed a button (the "intention-onset judgment") to indicate which letter had been visible when the decision was made. The mean time reported for that initial sense of awareness is indicated by the solid vertical lines in the next panel. (After C. S. Soon et al., 2008. *Nat Neurosci* 11: 543–545.) (D) The researchers applied machine learning analyses to fMRI data collected before the reported onset of the decision. In a striking result, activation in the anterior part of the prefrontal cortex (i.e., lateral and medial frontopolar cortex) and in the medial parietal cortex (i.e., precuneus and posterior cingulate cortex) predicted the subsequent decision as much as 8 seconds in advance. Because the fMRI signal lags neuronal activity by about 5 seconds, this result means that activity predictive of movements was present in the brain at least 12 seconds before the movements themselves. Red squares indicate time points with significant predictive power. (After C. S. Soon et al., 2008. *Nat Neurosci* 11: 543–545.)

with recording electrodes in experimental animals, PCC neurons do not generally respond to sensory stimuli. Instead, they tend to exhibit slow, long-lasting fluctuations in firing rates. Neuroimaging during standard cognitive laboratory tasks further reveals that this region shows reduced activity during performance of most tasks and returns to a high baseline activity level during the delay between trials. When individuals are distracted or daydreaming, activity in the PCC is elevated. By some measures, the PCC is the most metabolically active part of the brain, the implication being that it must be doing something vital. But precisely what that is remains a mystery.

These facts are puzzling. Why would the brain include a region whose activity is—or seems to be—antagonistic to the sorts of functions probed in laboratory tasks? One hypothesis is that some brain functions relevant to cognition happen at rest or between the trials of a test; those functions are supported by a brain system called the default-mode network (DMN) (Figure 33.12; also see Chapter 28). What might these default functions be? Key candidates involve regulating information relevant to the self, such as retrieval of information from autobiographical memory. Humans and presumably other animals are especially interested in information about their past, perhaps subconsciously and even during sleep. This focus would help individuals evaluate past actions so they could make better plans for the future. Other candidate

functions involve reflection on one's internal states, both physiological and cognitive. When people stop performing some active task, they may disengage from the outside world and focus on their own thoughts, experiences, and feelings—and those transitions from external to internal may be identifiable as similar state changes in the brain.

The PCC has been the most reliably and strongly modulated node in the DMN, but it is not the only region important for these self-directed processes. Other key regions include midline aspects of the PFC and the temporal–parietal junction. All of these regions typically show concomitant decreases in activity during active engagement with a task, and increases in activity once a task is completed—although each region can be separately engaged by distinct cognitive demands (e.g., reward evaluation, as noted for the OFC in Concept 33.3). The relationship among these regions can be explored using neuroimaging, which can provide insights into correlations between activity in different brain regions. Two regions with similar activity patterns over time are assumed to be functionally connected, though it should be emphasized that this measure is indirect and does not necessarily imply direct structural connections; for example, two regions might exhibit a pattern of functional connectivity because both are influenced similarly by a third region. Functional connectivity studies have been particularly important in understanding the DMN and its involvement in psychiatric disorders. Disruptions of

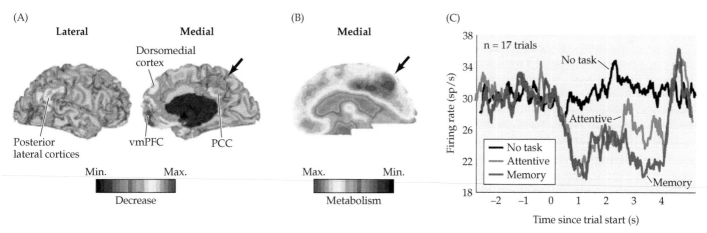

FIGURE 33.12 **Default-mode function and the posterior cingulate cortex** (A) Images of the human brain showing metabolic responses as measured by PET. Individuals either performed standard laboratory tasks or viewed the same stimuli passively—that is, without any task requirements. Active task performance led to consistent decreases in activity in the default-mode network (DMN), a set of regions whose responses are enhanced at cognitive rest. Prominent among the DMN regions are the posterior cingulate cortex, the VMPFC, and the temporal–parietal junction. (B) Resting metabolic rates in the human brain are high in the DMN, especially in the posterior cingulate cortex, which is among the most metabolically active regions of the brain. (C) Responses of single neurons in the monkey PCC show similar patterns as the hemodynamic response. That is, they show sustained reductions in activity during the cognitively demanding parts of rapid laboratory tasks and a return to high baseline levels during the delay between trials. This figure comes from a study in which rhesus monkeys performed either a difficult attention task (blue line) or short-term memory task (red line), or else performed no task (black line). In the task, tonic firing rates during the delay were reduced during the difficult part of the task relative to the rest condition. When information was available in short-term memory, responses were further reduced. (A,B from D. A. Gusnard and M. E. Raichle, 2001. *Nat Rev Neurosci* 2: 685–694. Part A Courtesy of D. Van Essen and A. Z. Snyder, Washington University; C after B. Y. Hayden et al., 2009. *Proc Natl Acad Sci USA* 106: 5948–5953.)

functional connectivity within the DMN have been linked to schizophrenia, autism, and OCD. Studies of functional connectivity indicate that the PCC represents a hub region within the DMN, in that its connectivity with other DMN regions increases during rest periods while its connectivity with control-related regions in the frontal and parietal cortices increases during active tasks.

Brain imaging studies have linked the PCC to autobiographical memory, or thinking about one's personal experiences. The anatomical connections between the PCC and the medial temporal lobe—which plays an important role in memory (see Chapter 30)—are consistent with the idea that PCC activation reflects these processes. Strikingly, activation of the PCC is also associated with thinking about oneself in the future and considering others in relation to oneself. These processes may involve complex representations of how the self relates to the broader environment, which includes the future and the roles of other people in it. To construct such representations, one would have to draw on memories to formulate a best guess about the present and derive a reasonable prediction about the future. The idea that the PCC helps individuals situate themselves in the larger world can potentially explain the relationship between PCC activity and reward value in economic choice tasks. Economic choices involve evaluation of past encounters with the options available. Such memory retrieval could also explain the antagonistic relationship between PCC activity and task performance: To perform an attentionally demanding task, the process of retrieving past memories must be inhibited to focus on the present.

Some of the most compelling indications of PCC function come from the effects of damage to this area. The PCC is one of the first regions affected in Alzheimer's disease (see Box 30B), and the progression of the disease is associated with degeneration of the PCC as measured after death. Given the close association between the progression of Alzheimer's disease and the loss of autobiographical memories, this finding provides additional evidence that the PCC in particular, and default-mode function in general, is linked with the regulation of autobiographical memory.

One recent study highlights this linkage directly by examining the activity of single neurons in the brains of monkeys performing a difficult learning task. The task required that monkeys learn to pair scenic photographs with eye movements to the left or right. Once an association was learned, the researchers added new picture–action associations for the monkeys to learn. Activation of PCC neurons

tracked these associations and was particularly high following errors and when the monkeys were confronted with new scenes. When the investigators inactivated the PCC by injecting a small amount of the GABA agonist muscimol, learning was impaired, providing direct evidence that the PCC plays a causal role in regulating learning.

Given the contexts that activate and deactivate the PCC, it seems likely that its function is much more complex than simply regulating memory retrieval. To give just one example, the PCC appears to play a key role in regulating exploratory behavior. Individuals often must make complex decisions in which their choice affects the range of options available to them in the future. Such decisions are particularly common in complex foraging situations and in social interactions. Activation of the PCC is higher during such strategic decisions. Such decisions involve memory, but also emotion, prospection, error monitoring, the delicate trade-off between exploration and exploitation, and the integration of all these factors into a decision.

Summary

Humans and other intelligent animals stand out because of the number and complexity of the associations that can be made between sensory input and behavioral responses. This flexibility depends, at least in part, on the prefrontal cortex, a region that consists of many areas that work together to produce rich, sophisticated, and even creative behaviors. These functions are not mediated by the prefrontal cortex alone, but recruit an extended network of structures with sometimes overlapping and sometimes conflicting roles. These areas acquire information about both the state of the world and the state of the body, and elicit further associations that can then be used to evaluate options, deal with conflicting possibilities, and regulate the allocation of cognitive resources accordingly. This network incorporates rapidly changing contextual information to generate an effective plan of action, inhibit unwanted or maladaptive plans, and monitor the consequences of whatever an individual ultimately chooses to do. Insults to this extended network, whether through stroke, degenerative neurological disorders, trauma, or drug abuse, compromise these functions. Research that improves our understanding of how this mosaic of brain regions contributes to thinking, planning, and deciding will in turn provide insights into treatments for many neurological and psychiatric disorders.

■ Additional Reading

Reviews

Banich, M. T. and B. E. Depue (2015) Recent advances in understanding neural systems that support inhibitory control. *Curr. Opin. Behav. Sci.* 1: 17–22.

Delgado, M. R. and 6 others (2016) Viewpoints: Dialogues on the functional role of the ventromedial prefrontal cortex. *Nat. Neurosci.* 19: 1545–1552.

Domenech, P. and E. Koechlin (2015) Executive control and decision-making in the prefrontal cortex. *Curr. Opin. Behav. Sci.* 1: 101–106.

Kolling, N., T. E. J. Behrens, M. K. Wittmann and M. F. S. Rushworth (2016) Multiple signals in anterior cingulate cortex. *Curr. Opin. Neurobiol.* 37: 36–43.

Miller, E. K. (2000) The prefrontal cortex and cognitive control. *Nat. Rev. Neurosci.* 1: 59–65.

Ott, T. and A. Nieder (2019) Dopamine and cognitive control in prefrontal cortex. *Trends Cogn. Sci.* 23: 213–234.

Pearson, J. M. and 4 others (2011) Posterior cingulate cortex: Adapting behavior to a changing world. *Trends Cogn. Sci.* 15: 143–151.

Important Original Papers

Bechara, A., A. R. Damasio, H. Damasio and S. W. Anderson (1994) Insensitivity to future consequences following damage to the human prefrontal cortex. *Cognition* 50: 7–15.

Depue, B. E., T. Curran and M. T. Banich (2007) Prefrontal regions orchestrate suppression of emotional memories via a two-phase process. *Science* 317: 215–219.

Dixon, M. L., K. C. R. Fox and K. Christoff (2014) A framework for understanding the relationship between externally and internally directed cognition. *Neuropsychologia* 62: 321–330.

Fellows, L. K. and M. J. Farah (2005) Is anterior cingulate cortex necessary for cognitive control? *Brain* 128: 788–796.

Funahashi, S., C. J. Bruce and P. S. Goldman-Rakic (1989) Mnemonic coding of visual space in the monkey's dorsolateral prefrontal cortex. *J. Neurophysiol.* 61: 331–349.

Hayden, B. Y., D. Smith and M. L. Platt (2009) Electrophysiological correlates of default-mode processing in macaque posterior cingulate cortex. *Proc. Natl. Acad. Sci. U.S.A.* 106: 5948–5953.

Kerns, J. G. and 5 others (2004) Anterior cingulate conflict monitoring and adjustments in control. *Science* 303: 1023–1026.

Naqvi, N. H., D. Rudrauf, H. Damasio and A. Bechara (2007) Damage to the insula disrupts addiction to cigarette smoking. *Science* 315: 531–534.

Schultz, W., P. Dayan and P. R. Montague (1997) A neural substrate of prediction and reward. *Science* 275: 1593–1599.

Utevsky, A. V., D. V. Smith, and S. A. Huettel (2014) Precuneus is a functional core of the default-mode network. *J. Neurosci.* 34: 932–940.

Wallis, J. D., K. C. Anderson and E. K. Miller (2001) Single neurons in prefrontal cortex encode abstract rules. *Nature* 411: 953–956.

Books

Damasio, A. (2005) *Descartes' Error: Emotion, Lesion, and the Human Brain*. New York: Penguin Books.

Fuster, J. (2015) *The Prefrontal Cortex*, 5th Edition. New York: Academic Press.

Glimcher, P. and E. Fehr (2013) *Neuroeconomics: Decision Making and the Brain*, 2nd Edition. London: Academic Press.

Passingham, R. E. and S. P. Wise (2012) *The Neurobiology of the Prefrontal Cortex: Anatomy, Evolution, and the Origin of Insight*. London: Oxford University Press.

APPENDIX

Survey of Human Neuroanatomy

Overview

Perhaps the major reason that neuroscience remains such an exciting field is the wealth of unanswered questions about the fundamental organization and function of the human brain. To understand this remarkable organ (and its interactions with the body that it governs), the myriad cell types that constitute the nervous system must be identified, their mechanisms of excitability and plasticity characterized, their interconnections traced, and the physiological role of the resulting neural circuits defined in behaviorally meaningful contexts. These challenges have been at the forefront of the five units of this textbook, where a broad range of questions about how nervous systems are organized and how they generate behavior has been addressed (albeit leaving many questions unanswered, especially those that pertain to distinctly human behaviors). This appendix provides an anatomical framework for the integration of this knowledge and its application to the human nervous system. It reviews the basic terms and anatomical conventions used in discussing human neuroanatomy, and provides a general picture of the organization of the human forebrain, brainstem, and spinal cord. The appendix is followed by an atlas of surface and sectional images of the human central nervous system (CNS) on which relevant neuroanatomical structures are identified.

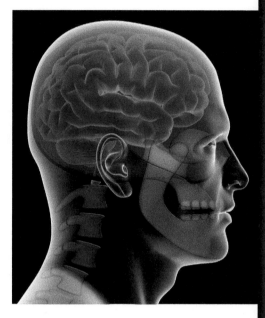

KEY CONCEPTS

A.1 The regional organization of the CNS may be understood embryologically and discussed with standard anatomical terms that are used throughout the body

A.2 The spinal cord and spinal nerves mediate sensation and motor control below the head

A.3 The brainstem mediates sensation and motor control for the head and influences many neurological and physiological functions

A.4 Despite the complexity of its appearance, the surface of the human brain conforms to a basic organizational plan

A.5 Sectional views of the forebrain reveal cortical variation and deep gray and white matter structures

A.6 The brain and spinal cord are critically dependent on a constant supply of blood and the integrity of the blood-brain barrier

A.7 The meninges and ventricular system protect the CNS and facilitate the production and circulation of cerebral spinal fluid

| CONCEPT A.1 | The Regional Organization of the CNS May Be Understood Embryologically and Discussed with Standard Anatomical Terms that Are Used throughout the Body |

LEARNING OBJECTIVES

A.1.1 Discuss position in various divisions of the CNS using the following pairs of directional terms: anterior/posterior; rostral/caudal; superior/inferior; dorsal/ventral; and medial/lateral.

A.1.2 Differentiate the three standard orthogonal planes and the transverse and longitudinal planes that are used to image or section the CNS.

A.1.3 Identify each of the major subdivisions of the adult nervous system, and relate them to their embryological precursors and associated ventricular spaces.

Neuroanatomical terminology

The terms used to specify *location* in the CNS are the same as those used for the gross anatomical description of the rest of the body (Figure A1A). Thus, *anterior* and *posterior* indicate, respectively, front and behind; *rostral* and *caudal*, nose and "tail" (i.e., the lower spinal region); *dorsal* and *ventral*, top

(A)

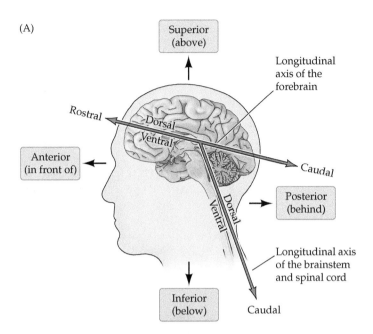

(B)

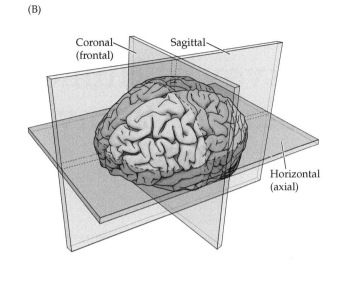

FIGURE A1 **Axes of the human nervous system**
(A) A flexure in the long axis of the nervous system arose as humans evolved upright posture, leading to an approximately 120° angle between the long axis of the brainstem and that of the forebrain. The consequences of this flexure for anatomical terminology are indicated here. The terms *anterior*, *posterior*, *superior*, and *inferior* refer to the long axis of the body, which is straight. Therefore, these terms indicate the same direction for both the forebrain and the brainstem. In contrast, the terms *dorsal*, *ventral*, *rostral*, and *caudal* refer to the long axis of the CNS. The dorsal direction is toward the back for the brainstem and spinal cord, but toward the top of the head for the forebrain. The opposite direction is ventral. The rostral direction is toward the top of the head for the brainstem and spinal cord, but toward the face for the forebrain. The opposite direction is caudal. (B) The major planes of section used in cutting or imaging the brain.

and bottom (back and belly); and *medial* and *lateral*, at the midline or to the side. But the use of these coordinates in the body versus their use to describe position in the brain can be confusing, especially as these terms are applied to humans. For the entire body, these anatomical terms refer to the long axis, which is straight. The long axis of the human CNS, however, has a bend in it. In humans (and other bipeds), the rostral–caudal axis of the forebrain is tilted forward (because of the cephalic flexure that forms in embryogenesis; see Chapter 22) with respect to the long axis of the brainstem and spinal cord (see Figure A1A). Once this forward tilt is appreciated, the other terms that describe position in the brain and the terms used to identify planes of section can be easily assigned.

The proper assignment of the anatomical axes dictates the standard planes for histological sections or live images that are used to study the internal anatomy of the brain and to localize function (Figure A1B). **Horizontal sections** (also referred to as **axial sections**) are taken parallel to the rostral–caudal axis of the brain; thus, in an individual standing upright, such sections are parallel to the ground. Sections taken in the plane dividing the two hemispheres are **sagittal** and can be further categorized as **midsagittal** or **parasagittal**, according to whether the section is at the midline (midsagittal) or is more lateral (parasagittal). Sections in the plane of the face are called **coronal** or **frontal**.

Different terms are usually used to refer to sections of the brainstem and spinal cord. The plane of section orthogonal to the long axis of the brainstem and spinal cord is the **transverse**, whereas sections parallel to this axis are **longitudinal**. In a transverse section through the human brainstem and spinal cord, the dorsal–ventral axis and the posterior–anterior axis indicate the same directions (see Figure A1A). This terminology is essential for understanding the basic subdivisions of the nervous system and for discussing the locations of brain structures in a common frame of reference.

Basic subdivisions of the CNS

As detailed in Chapter 22, the four embryological divisions of the CNS arise in early brain development after neurulation, as three swellings appear at the cephalic end of the neural tube (see Figure 22.7); together, these swellings develop into the brain, while the rest of the neural tube gives rise to the spinal cord. The most rostral of the three swellings, the **prosencephalon** ("forward brain" or "front brain"), divides into two parts: the **telencephalon** ("end brain" or "outer brain"), which gives rise to the cerebral hemispheres, and the **diencephalon** ("between brain" or "through brain"), from which are derived the thalamus, the hypothalamus, and also the retina (via the optic vesicle). These structures together make up the adult **forebrain**. The **mesencephalon** is the middle swelling in the embryonic brain, and it does not divide further; this division becomes the **midbrain** of

Embryonic brain		Adult brain derivatives	Associated ventricular space
Prosencephalon (forebrain)	Telencephalon	Cerebral cortex	Lateral ventricles
		Cerebral nuclei (basal ganglia, amygdala, basal forebrain)	
	Diencephalon	Thalamus	Third ventricle
		Hypothalamus	
		Retina	
Mesencephalon (midbrain)		Superior and inferior colliculi Red nucleus Substantia nigra	Cerebral aqueduct
Rhombencephalon (hindbrain)	Metencephalon	Cerebellum	Fourth ventricle
		Pons	
	Myelencephalon	Medulla oblongata	Fourth ventricle
Spinal cord		Spinal cord	Central canal

FIGURE A2 Representative relationships between the embryonic and adult forms of the central nervous system See Chapter 22 for an account of brain development that more fully explains the formation of regional identity in the developing CNS, including the origin of the ventricular spaces.

the adult. The **rhombencephalon** is also known as the **hindbrain**; it is the third of the three cephalic swellings, and it develops just caudal to the mesencephalon. The rhombencephalon further divides into the **metencephalon**, which becomes the pons and the overlying cerebellum, and the **myelencephalon**, which becomes the medulla oblongata (or simply the medulla). The term **brainstem** is used commonly to refer to the midbrain, pons, and medulla as a collective structure, despite their distinct embryological origins. The neural tube caudal to the three cephalic swellings becomes the spinal cord.

Because the nervous system starts out as a simple tube, the lumen of the tube remains in the adult brain as a series of connected, fluid-filled spaces. These spaces, known as the **ventricles**, are filled with **cerebrospinal fluid (CSF)** and provide important landmarks on sectional images of the nervous system. As the brain grows, the shape of the ventricles changes from that of a simple tube to that of the complex adult form (see the section "The Ventricular System" in Concept A.7). The ventricles, although continuous, acquire different names in each of the embryological subdivisions of the CNS. Thus, the spaces inside the hemispheres are known as the lateral ventricles, and the space inside the diencephalon is the third ventricle. The space inside the midbrain is called the cerebral aqueduct. The space inside the developing rhombencephalon (between the cerebellum and the pons and rostral medulla) is called the fourth ventricle. In embryos and young children, the opening in the spinal cord is patent and is known as the central canal.

Figure A2 accounts for the conserved relationships among the parts of the developing brain and their adult brain derivatives, including the components of the ventricular system. Figure A3 shows the subdivisions of the CNS as they are situated in the human body, including illustration of the relationship among the spinal cord, the

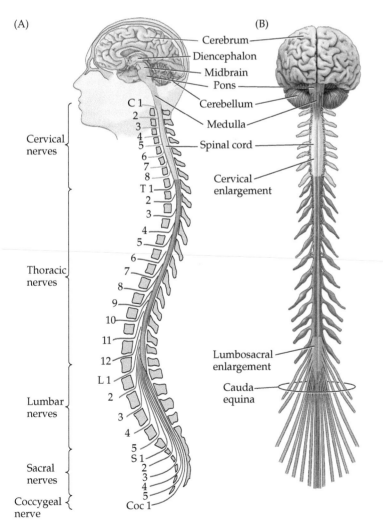

(A)

(B)

Cerebrum
Diencephalon
Midbrain
Pons
Cerebellum
Medulla
Spinal cord

Cervical enlargement

Lumbosacral enlargement

Cauda equina

Cervical nerves

Thoracic nerves

Lumbar nerves

Sacral nerves

Coccygeal nerve

C 1
2
3
4
5
6
7
8
T 1
2
3
4
5
6
7
8
9
10
11
12
L 1
2
3
4
5
S 1
2
3
4
5
Coc 1

FIGURE A3 Subdivisions and components of the central nervous system (A) A lateral view indicating the major subdivisions and components of the CNS. (Note that the position of the brackets on the left side of the figure refers to the location of the spinal nerves as they exit the intervertebral foramina, not the position of the corresponding spinal cord segments.) (B) The CNS in ventral view, indicating the emergence of the spinal nerves, the cervical and lumbar enlargements, and the cauda equina.

spinal nerves, and the vertebrae. The same embryonic relationships shown in Figure A2 should be discoverable in the adult form in Figure A3, although the relatively greater growth of the **cerebral hemispheres** makes some of these relationships difficult to appreciate since the hemispheres are the largest and most prominent feature of the human brain (Figure A4).

In humans, the cerebral hemispheres (the outermost portions of which are continuous, highly folded sheets of cortex, with each sheet roughly the size of a medium pizza) are characterized by **gyri** (singular: *gyrus*), or crests of folded cortical tissue, and by **sulci** (singular: *sulcus*), which are the grooves or spaces that divide gyri from one another. Although gyral and sulcal patterns vary among individuals, several consistent landmarks divide the **cerebral cortex** in each hemisphere into four **lobes**. The names of the lobes are derived from the cranial bones that overlie them: **occipital** , **temporal** , **parietal**, and **frontal**. A key feature of the surface anatomy of the cerebrum is the **central sulcus** located roughly halfway between the rostral and caudal poles of the hemispheres on the cerebrum's lateral surface (see Figure A4). This prominent sulcus divides the frontal lobe in the rostral half of the hemisphere from the more caudal parietal lobe. Other prominent landmarks that divide the cerebral lobes are the **lateral fissure** (also called

the **Sylvian fissure**), which divides the temporal lobe inferiorly from the overlying frontal and parietal lobes, and the **parieto-occipital sulcus**, which separates the parietal lobe from the occipital lobe on the medial surface of the hemisphere. The remaining major subdivisions of the forebrain are not visible from the surface; they comprise gray matter and white matter structures that lie deeper in the cerebral hemispheres and can be seen only in sectional views.

Next we describe the characteristic superficial features of these major subdivisions of the human CNS and their internal organization in more detail from the caudal to rostral direction, beginning with the spinal cord.

CONCEPT A.2 | The Spinal Cord and Spinal Nerves Mediate Sensation and Motor Control below the Head

LEARNING OBJECTIVES

A.2.1 Characterize the longitudinal organization of the spinal cord.

A.2.2 Differentiate the dorsal roots from the ventral roots.

A.2.3 Discuss the organization and composition of a typical spinal nerve.

A.2.4 Discuss the organization of gray matter in the spinal cord and the general functions associated with the dorsal, ventral, and lateral horns.

A.2.5 Discuss the organization of white matter in the spinal cord and the general functions associated with each column.

External anatomy of the spinal cord

The spinal cord extends caudally from the brainstem, running from the medullary-spinal junction at about the level of the first cervical vertebra to about the level of the first lumbar vertebra in adults (see Figure A3). The vertebral column (and the spinal cord within it) is divided into **cervical**, **thoracic**, **lumbar**, **sacral**, and **coccygeal** regions. The peripheral nerves (called the **spinal** or **segmental nerves**) that innervate much of the body arise from the spinal cord's 31 pairs of spinal nerves. On each side of the midline, the cervical region of the cord gives rise to 8 cervical nerves (C1–C8), the thoracic region to 12 thoracic nerves (T1–T12), the lumbar region to 5 lumbar nerves (L1–L5), the sacral region to 5 sacral nerves (S1–S5), and the coccygeal region to a single coccygeal nerve. The spinal nerves leave the vertebral column through the intervertebral foramina that lie adjacent to the respectively numbered vertebral body. Sensory information carried by the afferent axons of the spinal nerves enters the cord via the **dorsal roots**, and motor commands carried by the efferent axons leave the cord via the **ventral roots** (Figure A5). Once the dorsal and ventral roots join, sensory and motor axons (with some exceptions) travel together in the spinal nerves.

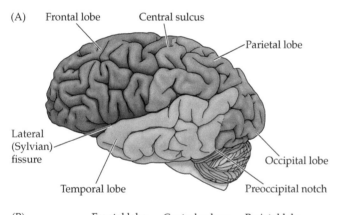

(A)
Frontal lobe   Central sulcus
Parietal lobe
Lateral (Sylvian) fissure
Temporal lobe   Occipital lobe
Preoccipital notch

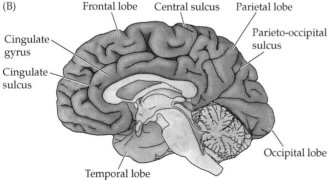

(B)
Frontal lobe   Central sulcus   Parietal lobe
Cingulate gyrus
Parieto-occipital sulcus
Cingulate sulcus
Occipital lobe
Temporal lobe

FIGURE A4 Surface anatomy of the cerebral hemisphere These depictions show the four lobes of the brain and the major fissures and sulci that help define their boundaries. (A) Lateral view. (B) Midsagittal view.

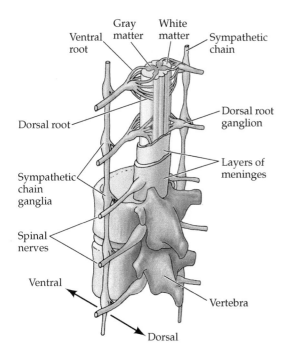

FIGURE A5   Relationship of the spinal cord and spinal nerves in the vertebral column Sensory information carried by the spinal nerves enters the cord via the dorsal roots; motor commands leave the cord via the ventral roots. Once the dorsal and ventral roots join, sensory and motor axons travel together in the spinal nerves.

Two regions of the spinal cord are enlarged to accommodate the greater number of nerve cells and connections needed to process information related to the upper and lower limbs. The spinal cord expansion that corresponds to the arms is called the **cervical enlargement** and includes spinal segments C3–T1; the expansion that corresponds to the legs is called the **lumbosacral enlargement** and includes spinal segments L1–S2 (see Figure A3B). Because the spinal cord is considerably shorter than the vertebral column in adults (see Figure A3A), lumbar and sacral nerves run for some distance in the vertebral canal before emerging, thus forming a collection of nerve roots known as the **cauda equina**. The space surrounding the cauda equina is the target for an important clinical procedure—the *lumbar puncture*—that allows for the collection of CSF by placing a needle into this lumbar cistern to withdraw fluid for analysis. In addition, local anesthetics can be safely introduced into the cauda equina, producing spinal anesthesia; at this level, the risk of damage to the spinal cord with insertion of a needle is minimal.

Internal anatomy of the spinal cord

The arrangement of gray and white matter in the spinal cord is relatively simple: The interior of the cord is formed by gray matter, which is surrounded by white matter (Figure A6A). In transverse sections, the gray matter is conventionally divided into dorsal (posterior) and ventral (anterior) "horns." The neurons of the **dorsal horns** receive sensory information that enters the spinal cord via the dorsal roots of the spinal nerves (Figure A6B). The **lateral horns** (see Chapter 21) are present primarily in the thoracic region and contain the preganglionic visceral motor neurons that project to the sympathetic ganglia (illustrated in Figure A5). The **ventral horns** contain the cell bodies of motor neurons

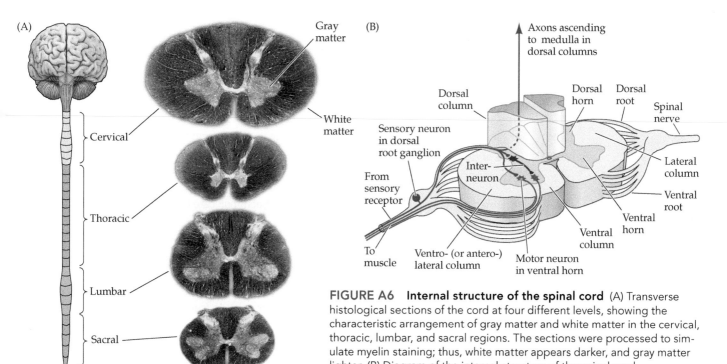

FIGURE A6   Internal structure of the spinal cord (A) Transverse histological sections of the cord at four different levels, showing the characteristic arrangement of gray matter and white matter in the cervical, thoracic, lumbar, and sacral regions. The sections were processed to simulate myelin staining; thus, white matter appears darker, and gray matter lighter. (B) Diagram of the internal structure of the spinal cord.

(A)

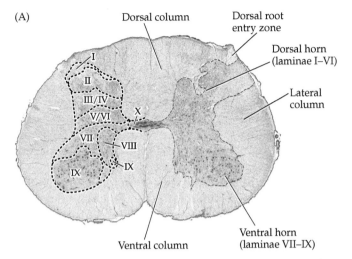

(B)

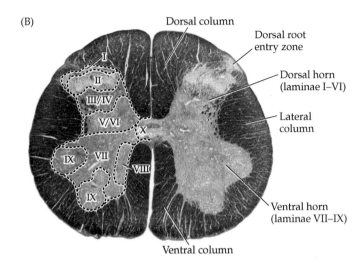

FIGURE A7 Internal histology of the human spinal cord in a lumbar segment (A) Photomicrograph of a section stained for the demonstration of Nissl substance (showing cell bodies in a blue stain). (B) Photomicrograph of a section that was acquired and processed to simulate myelin staining. On the left side of both images, dotted lines indicate the boundaries between cytoarchitectonic divisions of spinal cord gray matter, known as Rexed's laminae (see Table A1). Among the more conspicuous divisions are lamina II, which corresponds to the substantia gelatinosa and is important in pain transmission (see Chapter 13), and lamina IX, which contains the columns of lower motor neurons that innervate skeletal muscle (see Chapter 16).

that send axons via the ventral roots of the spinal nerves to terminate on striated muscles. These major divisions of gray matter have been further subdivided according to the distribution of neurons in the dorsal–ventral axis.

The Swedish neuroanatomist Bror Rexed recognized that neurons in the dorsal horn are organized into layers, and that neurons in the ventral horns (especially in the enlargements) are arranged into longitudinal columns

(Figure A7). Rexed proposed a scheme for naming these subdivisions (since termed *Rexed's laminae*) that is still used by neuroanatomists and clinicians, although more descriptive terms are also applied (Table A1).

The white matter of the spinal cord is subdivided into dorsal (or posterior), lateral, and ventral (or anterior) columns, each of which contains axon tracts related to specific functions. The **dorsal columns** carry ascending sensory

TABLE A1 Subdivisions of Spinal Cord Gray Matter

Division	Rexed's lamina	Descriptive term	Significance
Dorsal horn	I	Marginal zone	Projection neurons that receive input from small-diameter afferents; one source of anterolateral system projections
	II	Substantia gelatinosa	Interneurons that receive input mainly from small-diameter afferents; integrates feedforward and feedback (descending) inputs that modulate pain transmission (see Chapter 13)
	III/IV	Nucleus proprius	Interneurons that integrate inputs from small- and large-diameter afferents
	V/VI	Base of dorsal horn	Projection neurons that receive input from both large- and small-diameter afferents and spinal interneurons; another source of anterolateral system projections
Intermediate zone (lateral horn)	VII	Intermediate gray	Mainly interneurons that communicate between dorsal and ventral horns; in the thoracic cord, also contains projection neurons of the dorsal nucleus of Clarke, a spinocerebellar relay (see Chapter 19), and the sympathetic preganglionic visceral motor neurons of the intermediolateral cell column (underlying the lateral horn); in the sacral cord, also contains preganglionic visceral motor neurons (see Chapter 21)
Ventral horn	VIII	Motor interneurons	Interneurons in the medial aspect of ventral horn that coordinate the activities of lower motor neurons (see Chapter 16)
	IX	Motor neuron columns	Columns of lower motor neurons that govern limb musculature (see Chapter 16)
Central zone	X	Central gray	Interneurons surrounding the rudiment of the central canal

information mainly from somatic mechanoreceptors (see Figure A6B). The **lateral columns** include axons that extend from the cerebral cortex to interneurons and motor neurons in the ventral horns; this important pathway is called the **lateral corticospinal tract** (see Chapter 17). The lateral columns also convey proprioceptive signals from spinal cord neurons to the cerebellum (see Chapter 19). The **ventral** (and **ventrolateral**, or **anterolateral**) **columns** carry both ascending information about pain and temperature, and descending motor information from the brainstem and motor cortex concerned with postural control and gain adjustment.

CONCEPT A.3 | The Brainstem Mediates Sensation and Motor Control for the Head and Influences Many Neurological and Physiological Functions

LEARNING OBJECTIVES

A.3.1 Discuss five fundamental functions associated with the brainstem.

A.3.2 Describe and localize the principal features of the brainstem as seen from the surface, including the attachments of the cranial nerves.

A.3.3 Describe the sensory and motor signals conveyed by each cranial nerve.

A.3.4 Identify the major subdivisions of the brainstem and spinal cord, as seen in representative transverse cross sections.

A.3.5 Discuss the relationship between the cranial nerves and the corresponding cranial nerve nuclei.

Brainstem and cranial nerves

The brainstem is one of the most complex regions of the CNS. It comprises the midbrain, pons, and medulla and is continuous rostrally with the diencephalon (thalamus and hypothalamus); caudally it is continuous with the spinal cord. Although the medulla, pons, and midbrain participate in myriad specific functions that are beyond the scope of this Appendix, the integrated actions of these brainstem subdivisions give rise to five fundamental functions. First, the brainstem is the target or source for the **cranial nerves** that deal with sensory and motor function in the head and neck, and it provides for local circuits that integrate afferent signals and coordinate or organize efferent signals (Table A2). Second, the brainstem provides a "throughway" for all of the ascending sensory tracts from the spinal cord; the sensory tracts for the head and neck; the descending motor tracts from the forebrain; and local pathways that link eye movement centers. Third, the brainstem gives rise to and receives important connections from the overlying cerebellum. Fourth, the brainstem governs a variety of autonomic functions, such as breathing and cardiovascular regulation (see Chapter 21). Finally, the brainstem is involved in

regulating the level of consciousness, primarily though the extensive forebrain projections of key modulatory centers in the brainstem core that form a complicated network of circuits known as the **reticular formation** (see Box 17C).

Understanding the internal anatomy of the brainstem is generally regarded as essential for neurological diagnosis and the practice of clinical healthcare. Brainstem structures are compressed into a relatively small volume that has a regionally restricted vascular supply. Thus, vascular accidents in the brainstem—which are common—result in distinctive, and often devastating, combinations of functional deficits (see later in this concept). These deficits can be used both for diagnosis and for better understanding the intricate anatomy of the medulla, pons, and midbrain.

Unlike the surface appearance of the spinal cord, which is relatively homogeneous along its length, the surface appearance of each brainstem subdivision is characterized by unique bumps and bulges formed by the underlying gray matter (nuclei) or white matter (tracts) (Figure A8). A series of swellings on the dorsal and ventral surfaces of the medulla reflects many of the major structures in this caudal part of the brainstem. One prominent landmark that can be seen laterally is the **inferior olive**. Just medial to the inferior olives are the **medullary pyramids**, prominent swellings on the ventral surface of the medulla formed by the underlying descending corticospinal tracts (see Chapter 17).

The **pons** (Latin, "bridge") is rostral to the medulla and is easily recognized by the mass of decussating fibers that cross (bridge) the midline on its ventral surface, giving rise to the name of this subdivision. The **cerebellum** is attached to the dorsal aspect of the pons by three large white matter tracts: the **superior**, **middle**, and **inferior cerebellar peduncles**. Each of these tracts contains either efferent (superior and inferior) or afferent (inferior and middle) axons from or to the cerebellum (see Chapter 19).

The midbrain contains the **superior** and **inferior colliculi** defining its dorsal surface, or *tectum* (Latin, "roof"). Several midbrain nuclei lie in the ventral portion of the midbrain, including the bipartite structure the **substantia nigra** (pars reticulata and pars compacta; see Chapter 18), and the **red nucleus** (see Chapter 19). Another noteworthy anatomical feature of the midbrain is the presence of the prominent **cerebral peduncles** that are visible from the ventral surface; these structures are formed by massive projections from the cerebral cortex to targets in the brainstem and spinal cord.

The surface features of the midbrain, pons, and medulla can be used as landmarks for locating the source and termination of the majority of cranial nerves in the brainstem. Unlike for the spinal nerves, the entry and exit points of the cranial nerves are not regularly arrayed along the length of the brainstem. Two cranial nerves, the **olfactory nerve** (I) and the **optic nerve (II)**, enter the forebrain directly. The remaining ten pairs of cranial nerves enter and exit at distinct regions of the ventral (and in one case, the dorsal) surface of the midbrain, pons, and medulla (see Figure A8). The

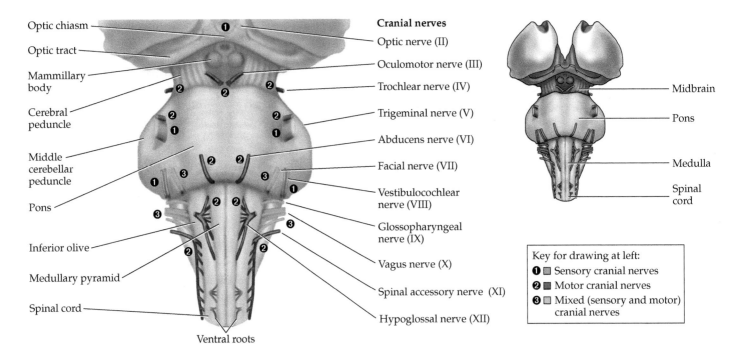

FIGURE A8   Cranial nerves of the brainstem This ventral view of the brainstem shows the locations of the cranial nerves as they enter or exit each of the brainstem subdivisions (midbrain, pons, and medulla, shown at the right).

TABLE A2   The Cranial Nerves and Their Primary Functions

Cranial nerve	Name	Sensory and/or motor	Major function
I	Olfactory nerve	Sensory	Sense of smell
II	Optic nerve	Sensory	Vision
III	Oculomotor nerve	Motor	Eye movements; pupillary constriction and accommodation; muscle of uppe (elevation, depression, extorsion, adduction, or medial movements) r eyelid
IV	Trochlear nerve	Motor	Eye movements (intorsion, downward gaze)
V	Trigeminal nerve	Sensory and motor	Somatic sensation from face, mouth, cornea; muscles of mastication
VI	Abducens nerve	Motor	Eye movements (abduction or lateral movements)
VII	Facial nerve	Sensory and motor	Controls the muscles of facial expression; taste from anterior tongue; lacrimal and salivary glands; somatic sensation from a small part of ear
VIII	Vestibulocochlear (auditory) nerve	Sensory	Hearing; sense of balance
IX	Glossopharyngeal nerve	Sensory and motor	Swallowing muscle; sensation from posterior tongue and pharynx; taste from posterior tongue; carotid baroreceptors and chemoreceptors; salivary gland; somatic sensation from a small part of ear
X	Vagus nerve	Sensory and motor	Autonomic functions of gut; cardiac inhibition; sensation from larynx and pharynx; muscles of vocal cords; swallowing; taste from posterior pharynx; somatic sensation from a small part of ear
XI	Spinal accessory nerve	Motor	Shoulder and neck muscles
XII	Hypoglossal nerve	Motor	Movements of tongue

oculomotor nerve (III) exits into the space between the two cerebral peduncles on the ventral surface of the midbrain. The **trochlear nerve (IV)** associated with the caudal midbrain is the only cranial nerve to exit on the dorsal surface of the brainstem; it is also the only nerve (cranial or spinal) that wholly supplies muscle that is situated contralateral to the cell bodies of the relevant lower motor neurons. The **trigeminal nerve (V)**—the thickest cranial nerve—exits the ventrolateral pons by traversing the fibers of the middle cerebellar peduncle. The **abducens nerve (VI)**, **facial nerve (VII)**, and **vestibulocochlear nerve (VIII)** emerge in a medial to lateral manner, respectively, at the junction of the pons and medulla, with the abducens nerve emerging most medially and the vestibulocochlear nerve emerging most laterally. The **glossopharyngeal nerve (IX)** and the **vagus nerve (X)** are associated with the lateral medulla, whereas the **hypoglossal nerve (XII)** exits the ventromedial medulla between the medullary pyramid and the inferior olive. The **spinal accessory nerve (XI)** does not originate in the brainstem but, as its name implies, exits the lateral portion of the upper cervical spinal cord.

Despite the somewhat irregular distribution of these cranial nerves relative to their points of attachment to the brainstem, there is an orderly arrangement of certain groups of nerves. For example, the nerves that convey motor signals to muscles derived from embryological somitomeres—cranial nerves III, IV, VI, and XII—all attach to the brainstem along the same parasagittal plane (see Figure A8). Likewise, the nerves that supply muscles derived from the embryological pharyngeal arches (also known as branchiomeres)—the motor root of cranial nerve V and cranial nerves VII, IX, X, and XI—also emerge from the same parasagittal plane through the brainstem (with some distortion to account for the lateral expansion of the pons). The sensory root of cranial nerve V and cranial nerve VIII attach to the brainstem in a similar but slightly more lateral parasagittal plane; these are the cranial nerves that convey predominantly sensory signals to the brainstem. Table A2 presents a more complete description of the major functions of the cranial nerves, the locations of the cell bodies whose axons form the nerves, and simple means for testing major functions associated with each nerve.

Cranial nerve nuclei within the brainstem are the targets of cranial sensory nerves or the source of cranial motor nerves (Table A3, Figure A9). These nuclei are located in the **tegmentum** or central core of the brainstem, between the

TABLE A2 The Cranial Nerves and Their Primary Functions (*continued*)

Location of cells whose axons form the nerve	Clinical test of function
Nasal epithelium	Test sense of smell with standard odor
Retina	Assess acuity, pupillary light reflex, and integrity of visual field
Oculomotor nucleus in midbrain; Edinger–Westphal nucleus in midbrain	Test eye movements (individual can't look up, down in neutral position, or medially if nerve involved); look for ptosis and pupillary dilation; assess pupillary light reflex
Trochlear nucleus in midbrain	Test for downward eye movement when eye adducted
Trigeminal sensory (gasserian or semilunar) ganglion; trigeminal motor nucleus in pons	Test sensation on face; test ability to clamp jaw tightly; palpate masseter and temporalis muscles
Abducens nucleus in pons	Test for lateral eye movement
Facial motor nucleus in pons; geniculate ganglion; superior salivatory nuclei in pons	Test facial expressions of upper and lower face; test taste on anterior tongue
Spiral ganglion; vestibular (Scarpa's) ganglion	Test audition with tuning fork; test vestibular function by assessing gaze fixation during head rotation and balance during perturbation; perform caloric test
Nucleus ambiguus in rostral medulla; superior and inferior glossopharyngeal nerve ganglia; inferior salivatory nucleus in pons and rostral medulla	Test swallowing and pharyngeal gag reflex
Dorsal motor nucleus of vagus; nucleus ambiguus; superior and inferior vagal nerve ganglion	Test swallowing and pharyngeal gag reflex; assess hoarseness; observe uvula and posterior pharynx at rest and during phonation
Spinal accessory nucleus in superior cervical cord	Test shrugging shoulders and turning head side to side
Hypoglossal nucleus in rostral medulla	Test tongue protrusion (if deviated, points to side of lesion) and symmetry of force when pushing tongue against cheek

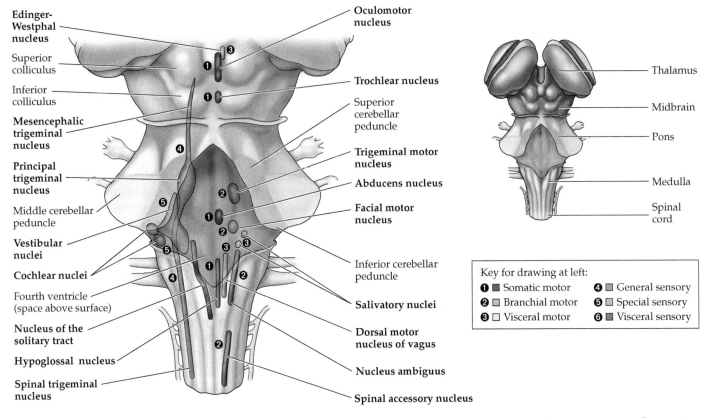

FIGURE A9   Cranial nerve nuclei of the brainstem
This "phantom" view of the brainstem's dorsal surface shows the locations of the brainstem nuclei that are either the target or the source of the cranial nerves. (See Table A2 for the relationship between each cranial nerve and cranial nerve nuclei, and Table A3 for a functional scheme that localizes cranial nerve nuclei with respect to brainstem subdivision and sensory or motor function.) With the exception of the nuclei associated with the trigeminal nerve, there is close correspondence between the location of the cranial nerve nuclei in the midbrain, pons, and medulla and the location of the associated cranial nerve roots. At right, the territories of the major brainstem subdivisions are indicated on the dorsal surface.

TABLE A3   Classification and Location of the Cranial Nerve Nuclei[a]

LOCATION	SOMATIC MOTOR	BRANCHIAL MOTOR	VISCERAL MOTOR	GENERAL SENSORY	SPECIAL SENSORY	VISCERAL SENSORY
Midbrain	Oculomotor nucleus (III) Trochlear nucleus (IV)		Edinger–Westphal nucleus (III)	Trigeminal sensory: mesencephalic nucleus (V, VII)		
Pons	Abducens nucleus (VI)	Trigeminal motor nucleus (V) Facial nucleus (VII)	Superior salivatory nucleus (VII) Inferior salivatory nucleus (IX)	Trigeminal sensory: principal nucleus (V, VII, IX, X) Trigeminal sensory: spinal nucleus (V, VII, IX, X)	Vestibular nuclei (VIII) Cochlear nuclei (VIII)	
Medulla	Hypoglossal nucleus (XII)	Nucleus ambiguus (IX, X) Spinal accessory nucleus (XI)	Nucleus ambiguus (X) Dorsal motor nucleus of vagus (X)		Nucleus of the solitary tract, rostral division (VII, IX, X)	Nucleus of the solitary tract caudal division (IX, X)

[a]Associated cranial nerves are shown in parentheses.

ventricular system dorsally and the division-specific structures and long motor pathways located ventrally. Cranial nerve nuclei that receive sensory input (analogous to the dorsal horns of the spinal cord) are located separately from those that give rise to motor output (which are analogous to the ventral horns). The primary sensory neurons that innervate these nuclei are found in ganglia associated with the cranial nerves—similar to the relationship between dorsal root ganglia and the spinal cord. In general, sensory nuclei are found laterally in the brainstem, whereas motor nuclei are located more medially (see Figure A9).

A more precise accounting of the location of cranial nerve nuclei should be appreciated in relation to the embryological origins of the nuclei and the target tissues they innervate (Figure A10). Early in the development of the CNS, the neural tube establishes regional identity in the rostral–caudal axis (the divisions of the brainstem and spinal cord discussed in Concept A.1). The neural tube also gives rise to an important differentiation of dorsal and ventral identity, with the dorsal gray matter establishing an **alar plate** and the ventral gray matter establishing a **basal plate** (see Chapter 22). The alar and basal plates are separated by a shallow longitudinal groove called the **sulcus limitans**, which extends the length of the spinal cord through the mesencephalon. The alar plates gives rise to the dorsal horn of the spinal cord and the sensory nuclei of the cranial nerves, and the basal plate gives rise to the ventral horn of the spinal cord and the motor nuclei of the cranial nerves.

In the spinal cord, the division of sensory (dorsal) and motor (ventral) gray matter is straightforward and easy

to appreciate in transverse sections. In the brainstem, the enlargement of the ventricular system that generates the fourth ventricle (see Concept A.7) contributes to the lateral displacement of the alar plate (see Figure A10B). Thus, in the tegmentum of the metencephalon and myelencephalon, the derivatives of the alar plate (the sensory nuclei) are located lateral to the derivatives of the basal plate (the motor nuclei). The derivatives of the basal plate differentiate further into three types of motor nuclei that are arranged in a medial to lateral progression, with respect to the embryological derivatives they innervate. Along the dorsal midline of the tegmentum are the **somatic motor nuclei** that project to striated muscles derived from somitomeres (extraocular muscles and extrinsic muscles of the tongue). Next, in a slightly more lateral column, are the **visceral motor nuclei** that project to peripheral ganglia innervating smooth muscle or glandular targets, similar to preganglionic motor neurons in the spinal cord that innervate autonomic ganglia. The **branchial motor nuclei** project to muscles derived from the pharyngeal or branchial arches, which give rise to the muscles (and bones) of the jaws, larynx, pharynx, and other craniofacial structures. The branchial motor nuclei migrate away from the dorsal aspect of the tegmentum and occupy a more central position just lateral to the visceral motor column, but still medial to the sensory nuclei (see curved arrows in Figure A10B).

The rostrocaudal organization of the cranial nerve nuclei (all of which are bilaterally symmetrical) reflects the rostrocaudal distribution of head and neck structures (Figure A11). The more caudal the nucleus, the more

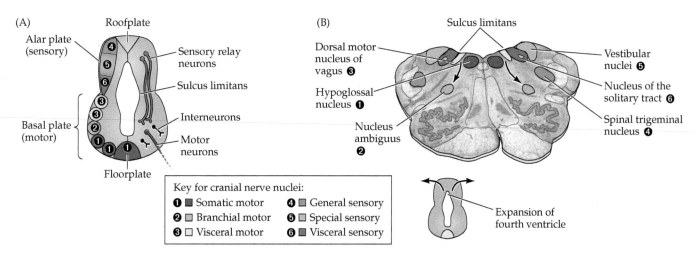

FIGURE A10   Embryological derivation of internal structure in the brainstem (A) Illustration of a transverse section through the developing neural tube demonstrating the division of the alar plate from the basal plate by the sulcus limitans. The alar plate differentiates into the dorsal horn of the spinal cord and the sensory nuclei of the brainstem. The basal plate differentiates into the ventral horn of the spinal cord and the motor nuclei of the brainstem. (B) Representative transverse section from the brainstem

(middle medulla) illustrating the location and identity of alar and basal plate derivatives. With expansion of the fourth ventricle, the alar plate derivatives develop lateral to the basal plate derivatives, like the opening of a book, with the floorplate in the position of the binding of the book (see inset). Note the secondary migration (curved arrows in main illustration) of branchial motor nuclei, such as the nucleus ambiguus, to an intermediate position in the brainstem tegmentum.

FIGURE A11 Internal organization of the brainstem Transverse sections through the brainstem along the rostral–caudal axis show the locations of the cranial nerve nuclei in six representative sections. The vascular territories for these brainstem sections are illustrated in Figure A21.

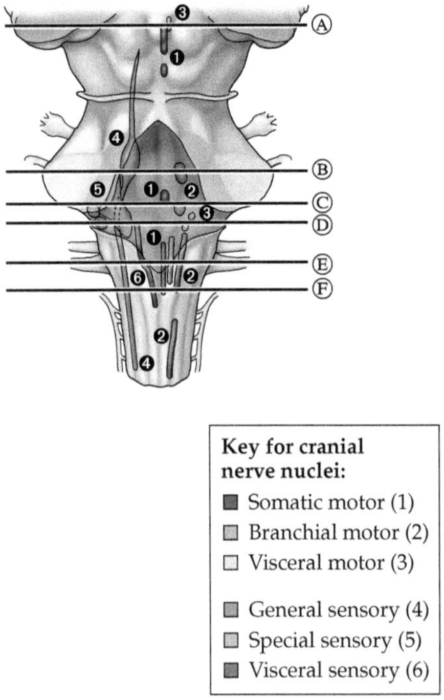

Ⓐ Midbrain

Superior colliculus
Substantia nigra
Edinger-Westphal nucleus (3)
Oculomotor nucleus (1)

Ⓑ Middle pons

Fourth ventricle
Trigeminal motor nucleus (2)
Pyramidal tract
Superior cerebellar peduncle
Principal trigeminal nucleus (4)
Middle cerebellar peduncle

Ⓒ Lower pons

Vestibular nuclei (5)
Facial nucleus (2)
Medial lemniscus
Abducens nucleus (1)
Spinal trigeminal nucleus (4)
Middle cerebellar peduncle

Ⓓ Upper medulla

Vestibular nuclei (5)
Nucleus of the solitary tract (5, 6)
Inferior olivary nucleus
Cochlear nuclei (5)
Spinal trigeminal nucleus (4)
Inferior cerebellar peduncle

Ⓔ Middle medulla

Vestibular nuclei (5)
Nucleus of the solitary tract (6)
Hypoglossal nucleus (1)
Dorsal motor nucleus of vagus (3)
Spinal trigeminal nucleus (4)
Nucleus ambiguus (2)
Medullary pyramid

Ⓕ Caudal medulla

Nucleus of the solitary tract (6)
Cuneate nucleus
Spinal trigeminal nucleus (4)
Hypoglossal nucleus (1)
Gracile nucleus
Dorsal motor nucleus of vagus (3)
Medial lemniscus

Key for cranial nerve nuclei:
- ◼ Somatic motor (1)
- ◻ Branchial motor (2)
- ◻ Visceral motor (3)
- ◼ General sensory (4)
- ◻ Special sensory (5)
- ◼ Visceral sensory (6)

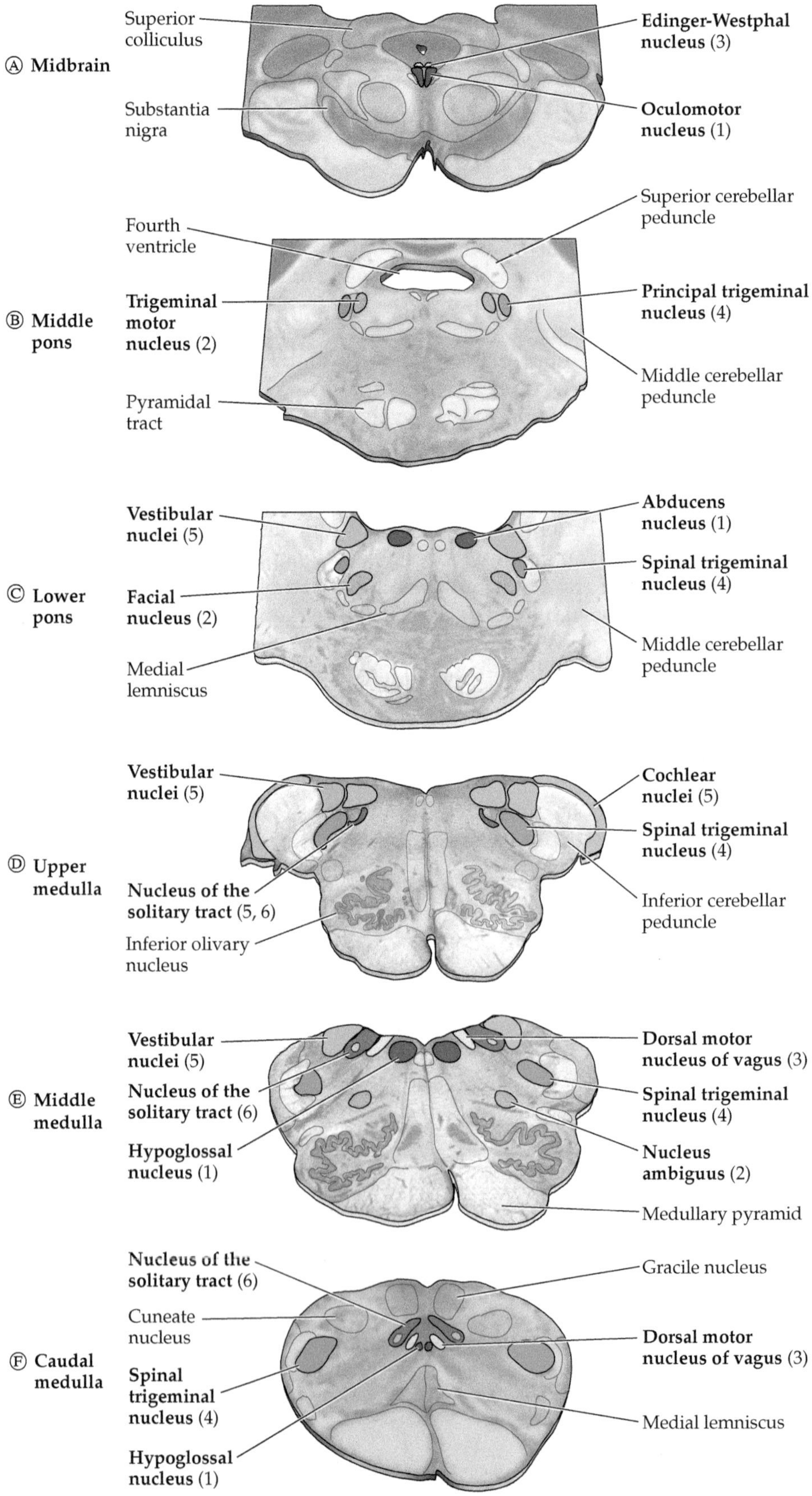

caudally located are the target structures in the periphery. For example, the spinal accessory nucleus in the cervical spinal cord and caudal medulla provides branchial motor innervation for neck and shoulder muscles, and the motor nucleus of the vagus nerve provides preganglionic (parasympathetic) innervation for many enteric and visceral targets. In the pons, the sensory and motor nuclei are concerned primarily with somatic sensation from the face (the principal trigeminal nuclei) as well as movement of the jaws and the muscles of facial expression (the trigeminal motor and facial nuclei). Farther rostrally, in the mesencephalic portion of the brainstem, are nuclei concerned primarily with eye movements (the oculomotor and trochlear nuclei) and preganglionic parasympathetic innervation of the iris (the Edinger–Westphal nucleus). While this list is not complete, it indicates the basic order of the rostrocaudal organization of the brainstem.

Healthcare professionals assess combinations of cranial nerve deficits to infer the location of brainstem lesions, or to place the source of brain dysfunction either in the spinal cord or forebrain. The most common brainstem lesions reflect the vascular territories that supply subsets of cranial nerve nuclei as well as ascending and descending tracts, which are located generally in the tegmentum (sensory) or basal (motor) regions of the brainstem (see Concept A.6). For example, an occlusion of the posterior inferior cerebellar artery (PICA), a branch of the vertebral artery that supplies the dorsolateral region of the middle and rostral medulla, results in damage to several cranial nerve nuclei and tracts (see the "Upper medulla" section in Figure A11). Accordingly, there are functional deficits that reflect the loss of the spinal trigeminal nucleus, the vestibular and cochlear nuclei, and the nucleus ambiguus (which contains branchial motor neurons that project to the larynx and pharynx) on the same side as the lesion. In addition, ascending pathways from the spinal cord that relay pain and temperature from the contralateral body surface are disrupted, leading to a contralateral loss of these functions (see Chapter 13). Finally, the inferior cerebellar peduncle, which contains projections that relay information about body position to the cerebellum for postural control, is damaged. This loss results in ataxia (incoordination) on the side of the lesion (see Chapter 19).

Anatomical relationships and shared vascularization, rather than any functional principle, unite these deficits and allow clinical localization of brainstem damage. For both clinicians and neurobiologists, understanding the brainstem requires integrating regional anatomical information with knowledge about functional organization, physiology, and pathology.

CONCEPT A.4 | Despite the Complexity of Its Appearance, the Surface of the Human Brain Conforms to a Basic Organizational Plan

LEARNING OBJECTIVES

A.4.1 Identify the four paired lobes of the cerebral cortex and describe the boundaries of each.

A.4.2 Sketch the major features of each cerebral lobe, as seen from the lateral view, identifying major gyri and sulci that characterize each lobe.

A.4.3 Identify and discuss distinct features of the forebrain and hindbrain that are visible on the ventral surface of the brain.

A.4.4 Identify and discuss distinct features of the forebrain and hindbrain that are visible in midsagittal section of the brain.

Lateral surface of the brain

A lateral view of the human brain is the best perspective from which to appreciate all four lobes of the cerebral hemisphere (see Figure A4A). In this view, the two most salient landmarks are the deep lateral fissure that separates the temporal lobe from the overlying frontal and parietal lobes, and the central sulcus, which serves as the boundary between the frontal and parietal lobes (Figure A12). A particularly important feature of the frontal lobe is the **precentral gyrus**. (The prefix *pre-*, when used anatomically, refers to a structure that is in front of or anterior to another.) The cortex of the precentral gyrus is referred to as the **motor cortex** and contains neurons whose axons project to the lower somatic and branchial motor neurons in the brainstem and spinal cord (see Chapter 17). Anterior to the precentral gyrus are three long parallel gyri called the **superior, middle**, and **inferior frontal gyri**. The posterior portion of the left inferior frontal gyrus is the typical localization of Broca's area, which is involved in the expression of language (see Chapter 31).

The lateral surface of the temporal lobe, like the frontal lobe, features three long parallel gyri termed the **superior, middle**, and **inferior temporal gyri**. The superior aspect of the temporal lobe contains cortex concerned with audition and language reception (see Chapter 10), and inferior portions of the lobe deal with highly processed visual information (see Chapter 9). Hidden beneath the frontal and temporal lobes, the **insular cortex**, or **insula**, can be seen only if these two lobes are pulled apart or removed (see Figure A12B). The posterior portion of the insula is concerned largely with visceral and autonomic function, including taste. More rostral portions of the insula are involved in implicit feelings and their impact on social cognition (see Chapter 32). In the anterior parietal

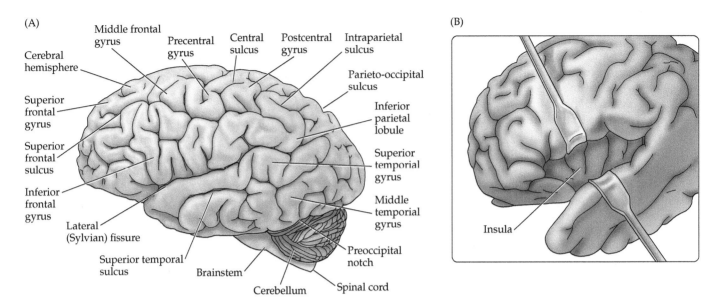

FIGURE A12   Lateral view of the human brain (A) Illustration of some of the major gyri and sulci from this perspective. (B) The banks of the lateral (Sylvian) fissure have been retracted to expose the insula.

lobe just posterior to the central sulcus is the **postcentral gyrus**; this gyrus harbors cortex that is concerned with somatic (bodily) sensation and is therefore referred to as the **somatosensory cortex** (see Chapter 12). Posterior to the postcentral gyrus are two gyral formations called the **superior** and **inferior parietal lobules**, which are separated by the **intraparietal sulcus**. These cortical regions associate somatosensory, visual, auditory, and vestibular signals and generate a neural construct of the body, the position of its parts, and its movements relative to the world around it (body image or body schema).

The boundary between the parietal lobe and the occipital lobe, the most posterior of the hemispheric lobes, is a somewhat arbitrary line from the parieto-occipital sulcus to the preoccipital notch. The occipital lobe, only a small part of which is apparent from the lateral surface of the brain, is concerned primarily with vision and visualization (even when the eyelids are closed). In addition to its role in primary and sensory processing, each cortical lobe participates in complex brain functions related to one or more dimensions of cognition (see Chapter 27). Thus, the parietal lobe is critical for attending to stimuli, the temporal lobe is used to recognize stimuli, and the frontal lobe is critical in planning responses to stimuli and in the organization of future behavior; the occipital lobe is involved in all aspects of visual perception and may also participate in multimodal sensory processing.

Dorsal and ventral surfaces of the brain

Although the primary subdivisions of the cerebral hemispheres can be appreciated from a lateral view, other key landmarks are better seen from the dorsal and ventral surfaces (Figure A13). When viewed from the dorsal surface,

the approximate bilateral symmetry of the cerebral hemispheres is apparent. Defining the axis of symmetry in this view of the brain is a deep space, called the **longitudinal fissure**, that separates the cerebral cortex of the two hemispheres. Although there is some variation (no two brains are identical, and even the two hemispheres of the same brain are not mirror copies of one another), major landmarks such as the central sulci and intraparietal sulci are usually very similar in arrangement on the two sides. If the cortical hemispheres are spread slightly apart from the longitudinal fissure, another major structure, the **corpus callosum**, can be seen bridging the two hemispheres (see Figure A13C). This structure contains axons that originate from pyramidal neurons in the cerebral cortex of both hemispheres. Callosal axons interconnect neurons in opposite (homotypical) cortical regions.

The external features of the brain that are best seen on its ventral surface are shown in Figure A13B. Extending along the inferior surface of the frontal lobe near the midline are the **olfactory tracts**, which arise from enlargements at their anterior ends called the **olfactory bulbs**. The olfactory bulbs receive input from neurons in the epithelial lining of the nasal cavity whose axons make up the first cranial nerve (cranial nerve I is therefore called the olfactory nerve; see Table A2 and Chapter 14). The olfactory bulbs and tracts lie on the medial margins of the **orbital gyri**, so named because these complex and highly variable gyri of the ventral frontal lobe are situated just superior to the orbits of the skull. On the ventromedial surface of the temporal lobe, the **parahippocampal gyrus** conceals the amygdala and also the **hippocampus**, a highly convoluted cortical structure that contributes to spatial navigation and the generation of cognitive maps for the acquisition of new information, such

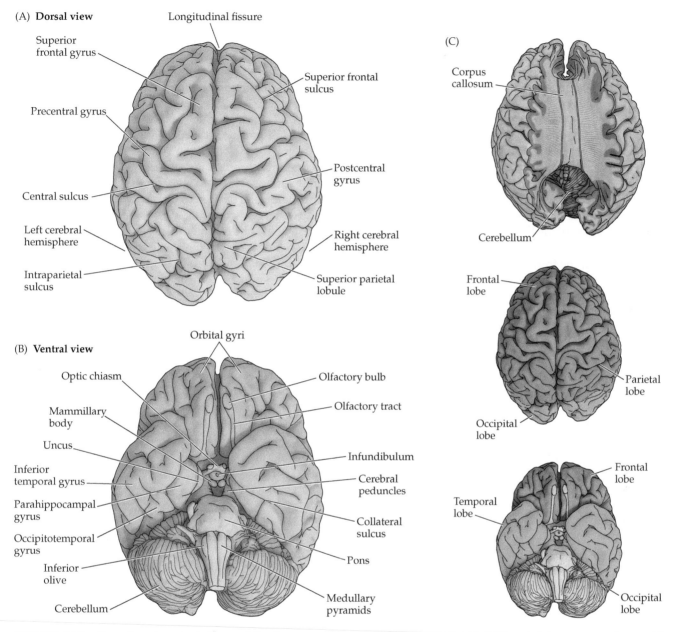

(A) **Dorsal view**

Longitudinal fissure

Superior frontal gyrus

Superior frontal sulcus

Precentral gyrus

Postcentral gyrus

Central sulcus

Left cerebral hemisphere

Right cerebral hemisphere

Intraparietal sulcus

Superior parietal lobule

(B) **Ventral view**

Orbital gyri

Optic chiasm

Olfactory bulb

Mammillary body

Olfactory tract

Uncus

Infundibulum

Inferior temporal gyrus

Cerebral peduncles

Parahippocampal gyrus

Occipitotemporal gyrus

Collateral sulcus

Inferior olive

Pons

Cerebellum

Medullary pyramids

(C)

Corpus callosum

Cerebellum

Frontal lobe

Parietal lobe

Occipital lobe

Temporal lobe

Frontal lobe

Occipital lobe

FIGURE A13 **Dorsal and ventral views of the human brain** (A) Dorsal view. (B) Ventral view. Both illustrations highlight some of the major features visible from these perspectives. (C) In the upper image (dorsal view), the cerebral cortex has been dissected away to reveal the underlying corpus callosum. The two lower images highlight the four lobes of the cerebral cortex. (C after J. W. Rohen and C. Yokochi, 1993. *Color Atlas of Anatomy.* New York: Igaku-Shoin.)

as episodic and declarative memory (see Chapter 30). A prominent medial protrusion of the parahippocampal gyrus is the **uncus**, which includes the cortical divisions of the amygdala. Between the parahippocampal and the inferior temporal gyri is the **occipitotemporal gyrus**, the posterior portion of which is sometimes called the **fusiform gyrus** (hidden by the cerebellum in Figure A13B). At the most central aspect of the ventral surface of the forebrain is the **optic chiasm**, and immediately posterior, the ventral surface of the **hypothalamus**, including the **infundibulum** (also called the **pituitary stalk**, at the base of the pituitary gland) and the **mammillary bodies**. Posterior to the hypothalamus, the paired cerebral peduncles are located on either side of the ventral midline of the midbrain. Finally, the ventral surfaces of the pons, medulla, and cerebellar hemispheres (see Figure A8) can be seen in this ventral view.

Midsagittal surface of the brain

When the brain is hemisected in the midsagittal plane, all of its major subdivisions plus several additional structures are visible on the cut surface. In this view, the cerebral hemispheres, because of their greater size, are still the most obvious structures. The frontal lobe of each hemisphere extends forward from the central sulcus, the medial end of which can just be seen terminating in the **paracentral lobule** (Figure A14A,B). The parieto-occipital sulcus, running from the superior to the inferior aspect of the hemisphere, is most obvious in this view of the hemisphere as it separates the **precuneus gyrus** in the parietal lobe from two major gyri in the occipital lobe. The **calcarine sulcus** divides the medial surface of the occipital lobe, running at nearly a right angle from the parieto-occipital sulcus and marking the location of the **primary visual cortex** (see Chapter 9). The upper bank of the calcarine sulcus is formed by the **cuneus gyrus** and the lower bank by the **lingual gyrus**. A long sulcus that follows the curvature of the corpus callosum, the **cingulate sulcus**, extends across the medial surface of the frontal and parietal lobes, ending in a dorsal ramus that marks the posterior boundary of the paracentral lobule. Below the cingulate sulcus is the **cingulate gyrus**, a prominent component of the **limbic forebrain** (*limbic* means "border" or "rim"), which comprises cortical

and subcortical structures in the frontal and temporal lobes that form a medial rim of cerebrum roughly encircling the corpus callosum and diencephalon. The limbic forebrain is important in the experience and expression of emotion, as well as the regulation of attending visceral motor activity (see Chapter 32). Finally, ventral to the cingulate gyrus is the cut, midsagittal surface of the corpus callosum.

Although parts of the diencephalon, brainstem, and cerebellum are visible at the ventral surface of the brain, their overall structure is especially clear from the midsagittal surface (Figure A14C). From this perspective, the diencephalon can be seen to consist of two parts. The **thalamus**, the largest component of the diencephalon, comprises several subdivisions, all of which relay information to the cerebral cortex from other parts of the brain (Box A). The hypothalamus—a small but crucial part of the diencephalon—is devoted to the control of homeostatic and reproductive functions, among other diverse activities (see Box 21A). The hypothalamus is intimately related, both structurally and functionally, to the pituitary gland,

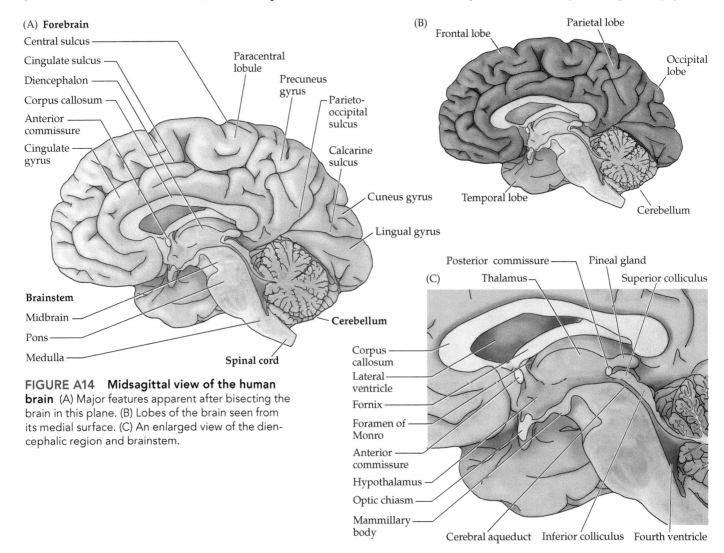

FIGURE A14 Midsagittal view of the human brain (A) Major features apparent after bisecting the brain in this plane. (B) Lobes of the brain seen from its medial surface. (C) An enlarged view of the diencephalic region and brainstem.

■ BOX A | Thalamus and Thalamocortical Relations

With one notable exception, studies of the sensory and motor systems in Units II and III have all included descriptions of important connections between the thalamus and some circumscribed division of the cerebral cortex. Here we draw together these descriptions into a brief consideration of the thalamus and its anatomical relations to the cerebral cortex.

The thalamus is a large mass of gray matter in the dorsal aspect of the diencephalon, superior to the hypothalamus (the hypothalamus is described in Box 21A) and medial to the massive collections of fibers that form the genu and posterior limb of the internal capsule. Conventionally, the thalamus comprises three main parts: the *epithalamus*, a small strip of tissue on the dorsomedial aspect of the thalamus to which the pineal gland is attached; the *subthalamus*, a region just above and slightly posterior to the hypothalamus containing nuclei that modulate basal ganglia output (including the subthalamic nucleus discussed in Chapter 18); and the *dorsal thalamus*, which is the largest and most complex of the three parts. The dorsal thalamus is the part that is now simply called the *thalamus* (Greek,

"inner chamber" or "marriage bed," a Galenic reference to the thalamus's central position in the brain and possibly also a sexual reference implying the regeneration or procreation of sensory signals). It is this part of the dorsal diencephalon that is most closely associated anatomically and functionally with the circuitry of the cerebral cortex.

The mammalian thalamus is a complex (in the sense that the amygdala is a complex; see Box 32B) comprising some 50 or so nuclear subdivisions that maintain distinct connections of inputs and outputs (Figure A). Despite the complexity of the thalamus, it is possible to understand its structure in broad terms based on the locations of its subdivisions and the patterns of their projections to the cerebral cortex. Broadly speaking, the thalamus is divided into medial, lateral, and anterior sectors by a Y-shaped band of white matter called the internal medullary lamina. Thus, there are groups of medial nuclei, lateral nuclei, and anterior nuclei; there are nuclei embedded in the internal medullary lamina itself; and there are nuclei along the midline of the thalamus. There is also a thin, shell-shaped nucleus that envelops

the thalamus laterally, called the reticular nucleus (not to be confused with the brainstem reticular formation), which has a profound influence over the firing patterns of thalamocortical projection neurons (see Chapter 28).

The thalamus is usually considered a major relay station for sending sensory signals and signals that modulate motor control to specific areas of the cerebral cortex (with the exception of the olfactory cortex, which receives sensory signals directly from the olfactory bulb). These sensory and motor signals arise mainly from nuclei in the lateral group. The thalamus also conveys less well understood but nonetheless specific signals to association-al areas of the cerebral cortex; these arise from medial and anterior nuclei and from certain nuclei in the posterior pole of the thalamus. Thus, every area of the cerebral cortex receives incoming signals that are topographically organized from some particular subdivision of the thalamus. Indeed, the degree to which the functional—and even the anatomical—identify of cortical areas is determined by specific connections with thalamic nuclei

(Continued)

(A)

Anterior nuclear group

Thalamic reticular nucleus

Lateral nuclear group (all others):

Ventral anterior

Lateral dorsal

Ventral lateral

Lateral posterior

Ventral posterior lateral

Ventral posterior medial

Midline thalamic nuclei

Medial nuclear group (mediodorsal nucleus)

Intralaminar nuclei

Internal medullary lamina

Pulvinar

Medial geniculate complex

Lateral geniculate nucleus

(A) Subdivisions of the thalamus in the human brain.

From After H. Blumenfeld, 2022. *Neuroanatomy through Clinical Cases*, 3rd ed. Sunderland, MA: Oxford University Press

■ BOX A | Thalamus and Thalamocortical Relations (continued)

remains an active area of research and debate in developmental and systems neuroscience.

These so-called specific thalamic projections are targeted to the middle layers of the cerebral cortex, where they serve to drive or sharply modulate activity in local columnar circuits of cortical neurons (Figure B). Examples include thalamocortical projections from the *visual thalamus* (lateral geniculate nucleus projections to the lingual and cuneus gyri; see Chapter 9), the *auditory thalamus* (medial geniculate complex projections to superior, transverse temporal gyri; see Chapter 10), the *somatosensory thalamus* (ventral posterior lateral and ventral posterior medial nuclei to the postcentral gyrus; see Chapters 12 and 13), or the *motor thalamus* (ventral anterior and ventral lateral nuclei projections to the precentral gyrus; see Chapters 18 and 19). However, there are prominent thalamic nuclei—such as the pulvinar in the posterior pole of the thalamus—that are themselves driven primarily by inputs from layer 5 of the cerebral cortex rather than by lower-order sensory or motor centers. Such thalamic nuclei, in turn, provide higher-order input that drives activity in other (non-primary) cortical areas (see Figure B). Thus, the thalamus may serve as both a relay of first-order sensory and motor input signals to relevant primary cortical areas, and a distributer of higher-order output signals from one cortical area to another. Evidently, it is the timely activation of such thalamocortical and cortico-thalamocortical projections that triggers sensory representation and the execution of behavioral programs that move body, mind, and emotion.

In contrast to these "specific" thalamocortical relations, there are much more diffuse projections arising from intralaminar and midline nuclei that terminate diffusely in the upper layers of the cerebral cortex (ascending dashed arrows in Figure B). Rather than conveying specific sensory or motor signals, these so-called nonspecific thalamic projections have widespread modulatory influences over vast networks of cortical neurons—the sorts of influences that could mediate attention, mood change, behavioral arousal, and transitions in sleep and wakefulness. Unfortunately, it is also these nonspecific projections of the thalamus that synchronize paroxysmal activity in generalized seizures, accounting for the nearly simultaneous and rhythmical discharge of cortical neurons across the cerebral hemispheres (see Clinical Applications, Chapter 8).

Finally, thalamocortical projections are reciprocated by massive systems of inputs from layer 6 of the cerebral cortex that appear morphologically and physiologically to serve as feedback modulators of the same thalamic neurons that provide driving signals to cortical networks (descending dashed arrows in Figure B). In fact, for certain thalamic nuclei such as the lateral geniculate nucleus, the number of presumptive modulatory inputs derived from the cerebral cortex is several-fold larger than the number of synaptic connections received from lower-order processing centers (the retina, in the case of the lateral geniculate nucleus). Despite the preeminence of these corticothalamic inputs, the precise role of feedback modulation in thalamic function remains poorly understood. Clearly, there remains much to be discovered concerning the neural computations hosted by this "inner chamber."

(B)

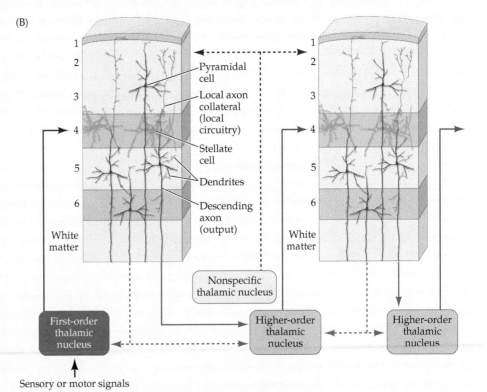

(B) Thalamocortical relations. Some specific thalamic nuclei are "first-order" relays of sensory or motor signals to the middle layers of primary cortex; other, "higher-order" nuclei distribute output signals via cortico-thalamocortical circuits (solid arrows). The thalamus also receives and distributes modulatory signals (dashed arrows). (After S. M. Sherman and R. W. Guillery, 2011. *J Neurophysiol* doi:10.1152/jn.00429.2011.)

a critical endocrine organ whose posterior part is an extension of the hypothalamus through the infundibulum.

The midbrain lies caudal to the thalamus, and the pons is caudal to the midbrain. The cerebellum lies over the pons and rostral medulla just beneath the occipital lobe of the cerebral hemispheres. From the midsagittal surface, the most visible feature of the cerebellum is the **cerebellar cortex**, a continuous layered sheet of cells folded into small convolutions called **folia** (early anatomists called this view of the cerebellum an *arbor vitae*—Latin, "tree of life"). The most caudal structure seen from the midsagittal surface of the brain is the medulla, which merges into the spinal cord.

CONCEPT A.5 | Sectional Views of the Forebrain Reveal Cortical Variation and Deep Gray and White Matter Structures

LEARNING OBJECTIVES

A.5.1 Identify the major white matter and gray matter structures that are apparent in sectional views of the forebrain, including the corpus callosum, anterior commissure, internal capsule, fornix, basal ganglia, thalamus, hypothalamus, and amygdala and hippocampus.

A.5.2 Sketch the relationships of different components of the basal ganglia and the thalamus relative to the internal capsule in coronal and axial sections through the forebrain.

Internal anatomy of the forebrain

A much more detailed neuroanatomical picture of the forebrain is apparent in gross or histological slices. In these slices (or sections), deep structures that are not visible from any brain surface can be identified. In addition, relationships between brain structures seen from the

surface can be appreciated more fully. The major challenge to understanding the internal anatomy of the brain is to integrate the rostral–caudal, dorsal–ventral, and medial–lateral landmarks seen on the brain surface with the position of structures seen in brain sections taken in the horizontal (axial), frontal, and sagittal planes. This challenge is not only important for understanding brain function; it is essential for interpreting noninvasive images of the brain, most of which are displayed as sections (see Atlas).

In any plane of section through the forebrain, the cerebral cortex is evident as a thin layer of neural tissue that covers the entire cerebrum. Most cerebral cortex is made up of six layers and is referred to as **neocortex** (see Box 27A). Phylogenetically older cortex (**paleocortex**) with fewer cell layers occurs on the inferior and medial aspect of the temporal lobe within the parahippocampal gyrus and in the pyriform cortex (a major division of the olfactory cortex near the junction of the temporal and frontal lobes). The simplest and most primitive division of the cortex, the **archicortex**, occurs in the hippocampus. The hippocampal cortex is folded into the medial aspect of the temporal lobe and therefore is visible only in dissected brains or in sections (Figure A15).

Embedded within the cerebral hemispheres are the **cerebral nuclei**, the largest of which are the components of the **basal ganglia**: the **caudate** and **putamen nuclei** (together referred to as **striatum**) and the **globus pallidus** (Figure A16). (The term *ganglia* does not usually refer to nuclei in the brain; the usage here is an exception.) The basal ganglia are visible in sections through the forebrain that also contain the lateral ventricles. The anterior "head" and central "body" of the caudate nucleus form the lateral wall of the lateral ventricle, and the tail of the caudate may be found in the temporal lobe in the roof of the temporal horn of the lateral ventricle. The neurons of these large nuclei receive input from the cerebral cortex and participate in the organization and guidance of complex motor

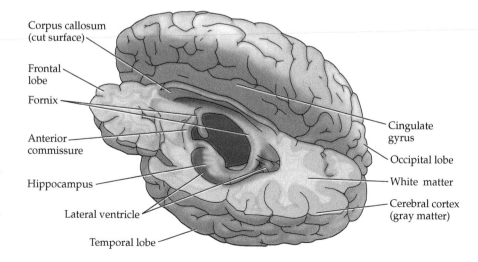

FIGURE A15 Major internal structures of the brain In this view, the upper half of the left hemisphere has been dissected away, revealing the temporal horn of the lateral ventricle, the hippocampus, the fornix, and the anterior commissure.

Corpus callosum (cut surface)
Frontal lobe
Fornix
Anterior commissure
Hippocampus
Lateral ventricle
Temporal lobe
Cingulate gyrus
Occipital lobe
White matter
Cerebral cortex (gray matter)

(A)

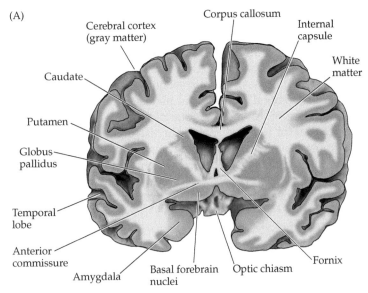

(C)

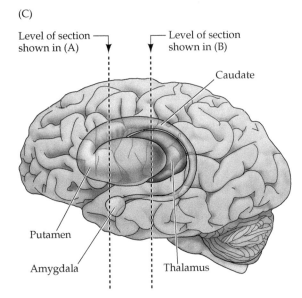

(B)

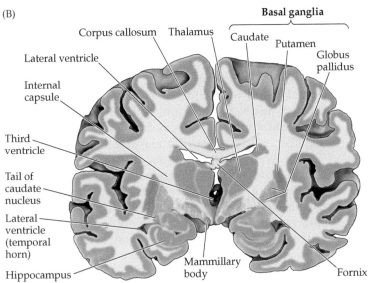

FIGURE A16   Internal structures of the brain seen in coronal section (A) This section passes through the breadth of the basal ganglia. (B) A more posterior section also includes the thalamus. (C) A transparent view of the cerebral hemisphere showing the approximate locations of the sections in (A) and (B) relative to deep gray matter (the basal ganglia, thalamus, and amygdala are represented). Notice that because the caudate nucleus has a "tail" that arcs into the temporal lobe, it appears twice in section (B); the same is true of other brain structures, including the lateral ventricle. (Also see Figure A20 and Atlas Plate 2.) (After H. Blumenfeld, 2022. *Neuroanatomy through Clinical Cases*, 3rd ed. Sunderland, MA: Oxford University Press.)

functions (see Chapter 18). In the base of the forebrain, ventral and medial to the basal ganglia are several smaller clusters of nerve cells known as the **septal** or **basal forebrain nuclei**. These nuclei are of particular interest because they modulate neural activity in the cerebral cortex and hippocampus, and they are among the forebrain systems that degenerate in Alzheimer's disease. The other clearly discernible structure visible in sections through the cerebral hemispheres at the level of the uncus is the **amygdala**, a complex of nuclei and cortical divisions that lies in front of (anterior to) the hippocampus, just anterior and dorsal to the anterior pole of the temporal lobe (see Box 32B).

In addition to these cortical and nuclear structures, several important axon tracts are localized to the internal anatomy of the forebrain. As already mentioned in Concept A.4, cerebral cortex in the two cerebral hemispheres is interconnected by the corpus callosum; in some anterior

sections, the smaller **anterior commissure** that interconnects cortex in the anterior temporal lobes and ventral frontal lobes can also be seen (see Figure A16A). Axons descending from, and others that are ascending to, the cerebral cortex assemble into another large fiber bundle tract called the **internal capsule** (see Figure A16A,B). The internal capsule lies just lateral to the diencephalon (forming a "capsule" around it), and many of its axons arise from or terminate in the thalamus. The internal capsule is seen most clearly in frontal sections through the middle one-third of the rostrocaudal extent of forebrain, or in horizontal sections through the level of the thalamus. Other axons descending from the cortex in the internal capsule continue past the diencephalon to enter the cerebral peduncles of the midbrain. Axons in these corticobulbar and corticospinal tracts project to several targets in the brainstem and spinal cord, respectively (see Chapter 17).

Thus, the internal capsule is the major pathway linking the cerebral cortex to the rest of the brain and spinal cord. Injury to this structure interrupts the flow of ascending and descending nerve impulses, often with devastating consequences. The internal capsule is also a useful landmark for understanding the distribution of major cerebral nuclei, including the basal ganglia and the thalamus. The caudate nucleus and the thalamus lie on the medial aspect of the internal capsule, while the globus pallidus and putamen are found on its lateral aspect. Finally, a smaller fiber bundle within each of the hemispheres, the **fornix**, interconnects the hippocampus and the hypothalamus and the septal region of the basal forebrain.

| CONCEPT **A.6** | The Brain and Spinal Cord Are Critically Dependent on a Constant Supply of Blood and the Integrity of the Blood-Brain Barrier |

LEARNING OBJECTIVES

A.6.1 Identify the major blood vessels that contribute to the anterior and posterior circulation of the brain.

A.6.2 Sketch the anastomotic ring of blood vessels (the circle of Willis) at the base of the brain.

A.6.3 Describe the system of vessels for venous drainage of blood from the brain into the jugular veins.

A.6.4 Identify the major blood vessels that supply the spinal cord.

A.6.5 Discuss the principal elements that form the blood-brain barrier.

A.6.6 Discuss the importance of the blood-brain barrier for protecting the brain and maintaining homeostasis of its interstitial fluids.

Blood supply of the brain and spinal cord

Understanding the blood supply of the brain and spinal cord is crucial for neurological diagnoses and the practice of medicine, particularly for neurology and neurosurgery. Damage to major blood vessels by trauma or stroke results in combinations of functional deficits that reflect both local cell death and the disruption of axons passing through the region compromised by the vascular damage. Thus, a firm knowledge of the major cerebral blood vessels and the neuroanatomical territories they perfuse facilitates the initial diagnoses of a broad range of brain damage and disease.

The entire blood supply of the brain and spinal cord depends on two sets of branches from the dorsal aorta (Figures A17 and A18). The **internal carotid arteries** are branches of the common carotid arteries, while the **vertebral arteries** arise from the subclavian arteries. These

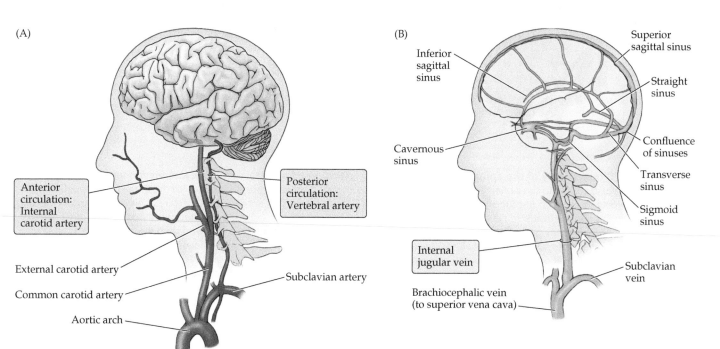

FIGURE A17 Anterior and posterior circulation and venous drainage of the brain (A) Arterial supply to the brain and upper spinal cord is derived from the internal carotid arteries (anterior circulation) and the vertebral arteries (posterior circulation). (B) Drainage of venous blood is through sinuses where the inner and outer layers of dura mater separate to create vascular channels that finally supply the internal jugular vein. (After H. Blumenfeld, 2022. *Neuroanatomy through Clinical Cases*, 3rd ed. Sunderland, MA: Oxford University Press.)

(A)

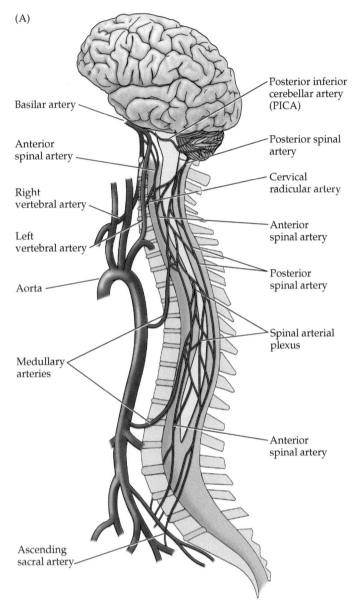

(B)

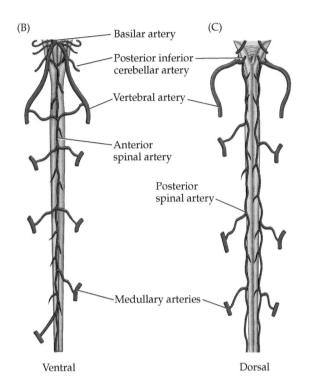

Ventral

(C)

Dorsal

(D)

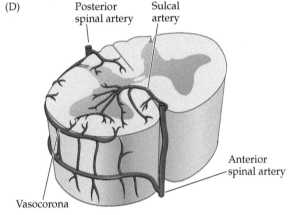

FIGURE A18 **Blood supply of the spinal cord** (A) View of the left side of blood supply to the brainstem and spinal cord in relation to the aorta from which the supply is derived. (B) View of the ventral (anterior) surface of the spinal cord. At the level of the medulla, the vertebral arteries give off branches that merge to form the anterior spinal artery. Approximately 8 to 12 segmental arteries (which arise from various branches of the aorta) supply the anterior and posterior spinal arteries along their course. These segmental arteries are also known as medullary arteries. (C) Typically, the posterior inferior cerebellar arteries give rise to paired posterior spinal arteries that run along the dorsal (posterior) surface of the spinal cord. (D) Cross section through the spinal cord, illustrating the distribution of the anterior and posterior spinal arteries. The anterior spinal artery gives rise to numerous sulcal branches that supply the anterior two-thirds of the spinal cord in alternating fashion (one side and then the other) along the length of the cord. The posterior spinal arteries supply much of the dorsal horn and the dorsal columns. A network of vessels known as the vasocorona connects these two sources of supply and sends branches into the white matter around the margin of the spinal cord. (A after H. Blumenfeld, 2022. *Neuroanatomy through Clinical Cases*, 3rd ed. Sunderland, MA: Oxford University Press.)

two major sets of arterial branches and the vascular distributions supplied by them are often conceptualized as the **anterior** and **posterior circulation**, with the anterior circulation derived from the internal carotid arteries and the posterior circulation from the vertebral arteries (see Figure A17A). Generally speaking, the anterior circulation supplies the forebrain (the cerebral hemispheres and diencephalon), while the posterior circulation supplies the brainstem, cerebellum, and upper portion of the spinal cord. However, the posterior cerebral artery supplies the posterior forebrain, including some deep structures. Thus, as indicated by listing this artery twice in Table A4, the

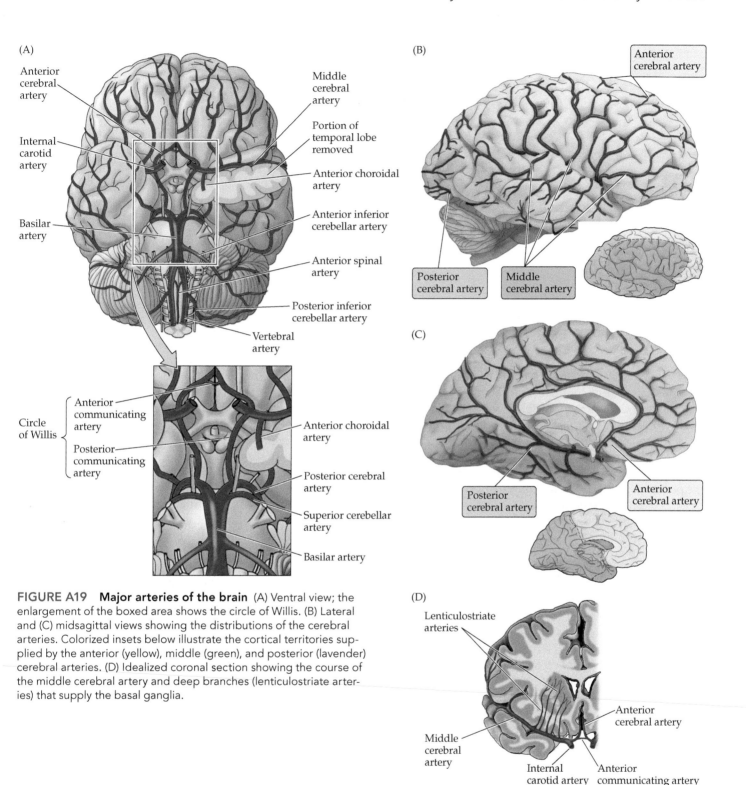

FIGURE A19 Major arteries of the brain (A) Ventral view; the enlargement of the boxed area shows the circle of Willis. (B) Lateral and (C) midsagittal views showing the distributions of the cerebral arteries. Colorized insets below illustrate the cortical territories supplied by the anterior (yellow), middle (green), and posterior (lavender) cerebral arteries. (D) Idealized coronal section showing the course of the middle cerebral artery and deep branches (lenticulostriate arteries) that supply the basal ganglia.

posterior cerebral artery contributes to both the anterior and posterior circulations supplying the forebrain and hindbrain, respectively.

Figure A19 shows the major arteries of the brain. Anterior to the spinal cord and brainstem, the internal carotid arteries branch to form two major cerebral arteries, the **anterior**

and **middle cerebral arteries**. The right and left vertebral arteries come together at the level of the pons on the ventral surface of the brainstem to form the midline **basilar artery**. The basilar artery joins the blood supply from the internal carotids in an arterial ring at the base of the brain (in the vicinity of the hypothalamus and cerebral peduncles) called

TABLE A4   Organization of Blood Supply to Brain and Spinal Cord

Circulation	Supply	Cerebrospinal artery
Anterior	Internal carotid arteries	Anterior cerebral arteries
		Middle cerebral arteries
		Anterior choroidal arteries
		Posterior communicating arteries
		Posterior cerebral arteries
Posterior	Vertebral/basilar arteries	Posterior cerebral arteries
		Superior cerebellar arteries
		Anterior inferior cerebellar arteries
		Posterior inferior cerebellar arteries
		Anterior spinal artery (upper portion)
		Posterior spinal arteries (upper portion)

the **circle of Willis** (see Figure A19A). The **posterior cerebral arteries** arise at this confluence, as do three small bridging arteries, the single **anterior** and paired **posterior communicating arteries**. In most humans, the posterior cerebral artery receives its blood supply from the vertebral/basilar system. In some people, the posterior communicating artery is quite large, and the posterior cerebral artery may be perfused primarily by the carotid artery. Conjoining the major sources of cerebral vascular supply via the circle of Willis presumably improves the chances of any region of the brain continuing to receive blood if one of the major arteries becomes occluded.

Each of the major arterial branches that make up the anterior circulation (i.e., those branches derived from the internal carotid artery plus the posterior cerebral artery) gives rise to superficial vessels that supply cortical structures and deep vessels that penetrate the surface of the brain and supply internal structures. An extensive region of the central and lateral cerebral hemispheres is supplied by the middle cerebral artery (shaded green in Figure A19B). Included in this region are the sensorimotor areas that govern the upper extremities and face, and the language areas of the left hemisphere (Broca's area and Wernicke's area; see Chapter 31). The anterior cerebral artery supplies regions in the medial aspect and dorsal and orbital margins of the frontal lobe, and the medial aspect and dorsal margin of the anterior parietal lobe (yellow area in Figure A19B). Included in this extended territory are sensorimotor areas in the paracentral lobule that govern the lower extremity, accessory motor areas in the cingulate gyrus that govern the upper face (see Clinical Applications, Chapter 17), and limbic areas in the medial frontal lobe. The posterior cerebral artery supplies regions in the posterior parietal lobe, inferior temporal lobe, and occipital lobe (lavender in Figure A19B). Included in this region are primary and associational (higher-order) visual areas in each lobe and limbic regions in the posterior cingulate and parahippocampal gyri.

As Figure A20 illustrates, the internal structures of the forebrain are supplied by an anterior-to-posterior pattern of deeply penetrating branches that arise from the proximal segments of the major cerebral arteries. The anterior cerebral artery supplies the anterior caudate and putamen and the anterior limb of the internal capsule. The middle cerebral artery supplies the body of the caudate and most of the putamen, most of the globus pallidus, the middle part (or genu) of the internal capsule, and the anterior hypothalamus. These deep-penetrating branches of the middle cerebral artery are usually called the **lenticulostriate arteries** (see Figure A19D). The **anterior choroidal artery**, which arises from the middle cerebral artery just distal to the circle of Willis, supplies the amygdala, hippocampus, anterior part of the thalamus, part of the globus pallidus, posterior limb of the internal capsule, and choroid plexus of the lateral ventricle. The posterior communicating and posterior cerebral arteries supply the posterior hypothalamus, most of the thalamus, and the choroid plexus of the third ventricle. Thus, the deep structures of the forebrain are divided approximately into four sectors progressing from anterior to posterior, and each sector is perfused by a different artery.

Blood from the anterior circulation makes its passage through the brain from the arterial vasculature back to the heart via the internal **jugular veins** through a series of venous sinuses (see Figure A17B). The major venous sinuses inside the cranium are formed by a separation of the two layers of dura mater, the tough outer component of the meninges that surrounds the brain (see Concept A.7).

(A)

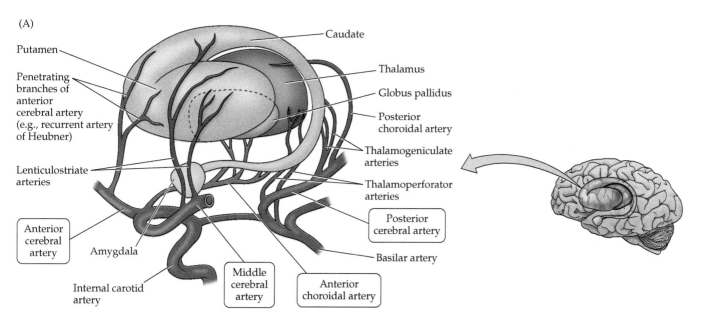

Putamen

Penetrating branches of anterior cerebral artery (e.g., recurrent artery of Heubner)

Lenticulostriate arteries

Anterior cerebral artery

Amygdala

Internal carotid artery

Middle cerebral artery

Caudate

Thalamus

Globus pallidus

Posterior choroidal artery

Thalamogeniculate arteries

Thalamoperforator arteries

Posterior cerebral artery

Basilar artery

Anterior choroidal artery

(B)

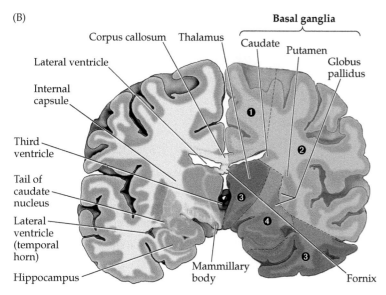

Basal ganglia

Corpus callosum   Thalamus   Caudate   Putamen

Lateral ventricle

Internal capsule

Third ventricle

Tail of caudate nucleus

Lateral ventricle (temporal horn)

Hippocampus

Globus pallidus

Mammillary body

Fornix

❶ ⬜ Anterior cerebral artery   ❸ ⬛ Posterior cerebral artery
❷ ⬜ Middle cerebral artery   ❹ ⬛ Anterior choroidal artery

FIGURE A20   Blood supply of deep cerebral structures (A) Deeply penetrating branches of the anterior and posterior circulation that supply the basal ganglia and thalamus. (B) Distribution of deep branches of the major cerebral arteries in a representative coronal section. (After H. Blumenfeld, 2022. *Neuroanatomy through Clinical Cases*, 3rd ed. Sunderland, MA: Oxford University Press.)

Thus, most of the superficial veins of the cerebrum drain into the **superior sagittal sinus** along the dorsal midline of the hemisphere, or the **cavernous sinus** in the base of the cranium. The superior sagittal sinus and the deeper sinuses drain into the **confluence of sinuses** at the posterior end of the longitudinal fissure, before giving rise to the **transverse sinuses**, which are oriented roughly in the horizontal plane. Venous blood then passes a short distance in the anterior direction before the transverse sinuses turn in the inferior direction, curving into a sigmoid shape. Finally, venous blood exits the cranial vault as the **sigmoid sinuses** pass through the skull base and supply the internal jugular veins.

The posterior circulation of the brain supplies the posterior cerebral cortex, thalamus, and brainstem. The pattern of arterial distribution is similar for all the subdivisions of the brainstem: Midline arteries supply medial structures, lateral arteries supply the lateral brainstem, and dorsolateral arteries supply dorsolateral brainstem structures and the cerebellum (Figure A21). Among the most important dorsolateral arteries (also called **long circumferential arteries**) are the cerebellar arteries: the **posterior inferior cerebellar artery (PICA)**, the **anterior inferior cerebellar artery (AICA)**, and the **superior cerebellar artery (SCA)**, each of which supplies distinct regions of the medulla or pons along its way to the cerebellum. These arteries, as well as branches of the posterior cerebral and basilar arteries that penetrate the brainstem from its ventral and lateral surfaces—the **paramedian** and **short circumferential arteries**—are especially common sites of occlusion and result in specific

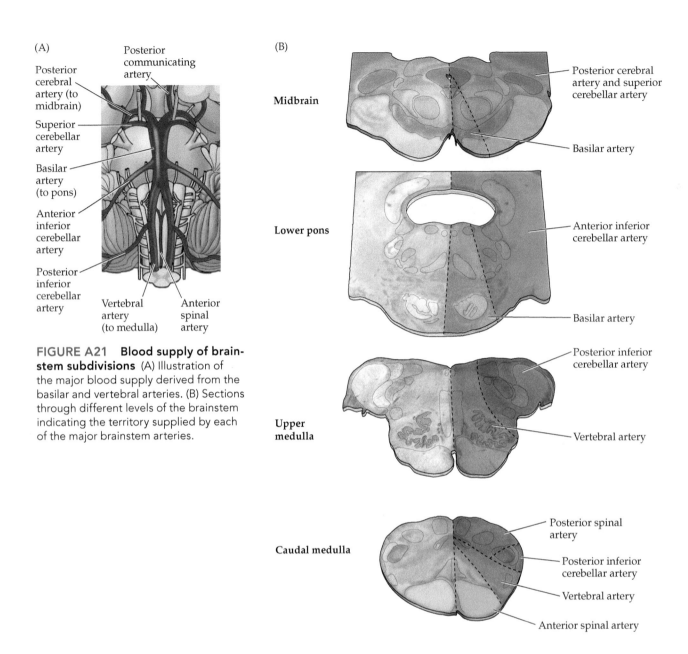

(A)

Posterior communicating artery

Posterior cerebral artery (to midbrain)

Superior cerebellar artery

Basilar artery (to pons)

Anterior inferior cerebellar artery

Posterior inferior cerebellar artery

Vertebral artery (to medulla)

Anterior spinal artery

(B)

Midbrain

Posterior cerebral artery and superior cerebellar artery

Basilar artery

Lower pons

Anterior inferior cerebellar artery

Basilar artery

Upper medulla

Posterior inferior cerebellar artery

Vertebral artery

Caudal medulla

Posterior spinal artery

Posterior inferior cerebellar artery

Vertebral artery

Anterior spinal artery

FIGURE A21   Blood supply of brainstem subdivisions (A) Illustration of the major blood supply derived from the basilar and vertebral arteries. (B) Sections through different levels of the brainstem indicating the territory supplied by each of the major brainstem arteries.

functional deficits of cranial nerve, somatosensory, and motor function. Most vascular lesions of the brainstem are unilateral, since each side of the brainstem is supplied by different sets of circumferential vessels. However, this may not be the case if the basilar artery itself is blocked, since it gives rise to vessels that supply both sides.

Blood is supplied to the spinal cord by the vertebral arteries and, typically, 8 to 12 **medullary arteries** that arise from segmental branches of the aorta (see Figure A18). These medullary arteries supply a single **anterior spinal artery** and a pair of **posterior spinal arteries**. An anastomotic network of vessels known as the **vasocorona** connects these two sources of supply and sends branches into a narrow zone of white matter around the margin of the spinal cord. The vasocorona may be sufficient to

supply the most lateral white matter in cases in which the anterior spinal artery is occluded. Nevertheless, if any of the medullary arteries are obstructed or damaged (during abdominal surgery, for example), the blood supply to specific parts of the spinal cord may be compromised. The pattern of resulting neurological damage differs according to whether supply to a posterior spinal artery or the anterior spinal artery is interrupted. As might be expected from the arrangement of ascending and descending neural pathways in the spinal cord, loss of the posterior supply generally leads to loss of sensory functions, whereas loss of the anterior supply causes motor deficits.

The physiological demands on the brain's blood supply are particularly significant because neurons are more sensitive to oxygen and glucose deprivation than are cells with lower rates of metabolism. The high metabolic rate of neurons means that brain tissue deprived of oxygen and glucose as a result of compromised blood supply is likely to sustain transient or permanent damage. Even brief loss of blood supply (referred to as ischemia) can cause cellular changes that, if not quickly reversed, can lead to cell death through the mechanisms of excitotoxicity. Sustained loss of blood supply leads much more directly to death and degeneration of the deprived cells. Stroke—an anachronistic term that refers to the death or dysfunction of brain tissue due to vascular disease—often follows the occlusion of (or hemorrhage from) the brain's arteries (Clinical Applications). Historically, studies of the functional consequences of strokes, and their relation to vascular territories in the brain and spinal cord, provided information about the location of various brain functions. The location of the major language functions in the left hemisphere, for instance, was discovered in this way in the latter part of the nineteenth century (see Chapter 31). Now, noninvasive functional imaging techniques based on blood flow have largely supplanted the correlation of clinical signs and symptoms with the location of tissue damage observed at autopsy (see Chapter 1).

The Blood-Brain Barrier

In addition to their susceptibility to oxygen and glucose deprivation, brain cells are at risk from toxins circulating in the bloodstream. The brain is specifically protected in this respect, however, by the **blood-brain barrier**. The interface

---

■ Clinical Applications

Stroke

Stroke is the most common neurological cause for hospital admission and in 2020 was the fifth leading cause of death in the United States (after heart disease, cancer, COVID-19, and accidents). The term stroke refers to the sudden appearance of a limited neurological deficit, such as weakness or paralysis of a limb, or the sudden inability to speak. The onset of the deficit within seconds, minutes, or a few hours typically implicates a vascular etiology. Brain function is exquisitely dependent on a continuous supply of oxygen and glucose, as evidenced by the onset of unconsciousness within about 10 seconds of blocking the brain's blood supply (by cardiac arrest, for instance). The damage to neurons is at first reversible but becomes permanent if the blood supply is not restored.

Strokes can be subdivided into three main types: *thrombotic, embolic,* and *hemorrhagic.* The thrombotic variety is caused by a local reduction of blood flow arising from a blood clot (thrombus) that builds up around an atherosclerotic plaque in one of the cerebral blood vessels, eventually occluding it. A reduction of blood flow also can arise when an embolus (an object loose in the bloodstream, such as a thrombus) dislodges from the heart (or from an atherosclerotic plaque in the carotid or vertebral artery) and travels to a cerebral artery (or arteriole), where it forms a plug and causes an embolic stroke. A hemorrhagic stroke occurs when a cerebral blood vessel ruptures, as can occur as a result of hypertension, a congenital aneurysm (bulging of a vessel), a congenital arteriovenous malformation, or a traumatic injury involving the meninges or the brain itself. Approximately 50% of strokes are thrombotic, 30% embolic, and 20% hemorrhagic.

The diagnosis of stroke relies primarily on an accurate history and a competent neurological examination. Indeed, the neurologist C. Miller Fisher, a master of bedside diagnosis, remarked that medical students and residents should learn neurology "stroke by stroke." Understanding the portion of the brain supplied by each of the major arteries (see text) enables an astute clinician to identify the occluded blood vessel.

More recently, imaging techniques such as CT scans and MRI (see Chapter 1) have greatly facilitated the clinician's ability to identify and localize small hemorrhages and regions of permanently damaged tissue. Moreover, Doppler ultrasound, magnetic resonance angiography, and imaging of blood vessels by direct infusion of radio-opaque dye can now pinpoint atherosclerotic plaques, aneurysms, and other vascular abnormalities.

Several therapeutic approaches to strokes are feasible. Dissolving a thrombotic plug by tissue plasminogen activator (TPA) and other compounds, or by breaking through a clot with endovascular devices, is now standard clinical practice for selected stroke victims. Recent understanding of some of the mechanisms by which ischemia injures brain tissue has made pharmacological strategies to minimize neuronal injury after stroke a potentially effective intervention. Hemorrhagic strokes may be treated neurosurgically by finding and stopping the bleeding from the defective vessel (when that is technically possible).

Although all of these approaches can minimize functional loss, stroke remains a serious health risk from which there is never full recovery. The inability of the mature brain to replace large populations of dead or damaged neurons, or to repair long axon tracts once they have been compromised, invariably prevents the complete restoration of lost functions. Despite these seemingly intractable limitations, novel strategies for neurorehabilitation continue to be investigated and introduced into clinical practice, offering a measure of hope to those afflicted with stroke and the disabilities that accompany cerebrospinal injury.

between the walls of capillaries and the surrounding tissue is important throughout the body, as it keeps vascular and extravascular concentrations of ions and molecules at appropriate levels in these two compartments. In the brain, this interface is especially significant—hence its unique and alliterative name. The special properties of the blood-brain barrier were first observed by the nineteenth-century bacteriologist Paul Ehrlich, who noted that intravenously injected dyes leaked out of capillaries in most regions of the body to stain the surrounding tissues; brain tissue, however, remained unstained. Ehrlich wrongly concluded that the brain had a low affinity for the dyes. It was his student, Edwin Goldmann, who showed that in fact such dyes do not traverse the specialized walls of brain capillaries.

The restriction of large molecules such as Ehrlich's dyes (and many smaller molecules) to the vascular space is the result of tight junctions between neighboring capillary endothelial cells in the brain (Figure A22). Such junctions are not found in capillaries elsewhere in the body, where the spaces between adjacent endothelial cells allow much more

ionic and molecular traffic. The structure of tight junctions was first demonstrated in the 1960s by Tom Reese, Morris Karnovsky, and Milton Brightman. Using electron microscopy after the injection of electron-dense intravascular agents such as lanthanum salts, they showed that the close apposition of the endothelial cell membranes prevented such ions from passing (see Figure A22B). Substances that traverse the walls of brain capillaries must move *through* the endothelial cell membranes. Accordingly, molecular entry into the brain should be determined by an agent's solubility in lipids, the major constituent of cell membranes. Nevertheless, many ions and molecules not readily soluble in lipids *do* move quite efficiently from the vascular space into brain tissue. A molecule such as glucose, the primary source of metabolic energy for neurons and glial cells, is an obvious example. This paradox is explained by the presence of specific transporters in the endothelial cell membrane for glucose and other critical molecules and ions.

In addition to tight junctions, astrocytic *end feet* (the terminal regions of astrocytic processes) surround the outside of capillary endothelial cells (see Figure A22A). The reason for this endothelial–glial allegiance is unclear, but may reflect an influence of astrocytes on the formation and maintenance of the blood-brain barrier or the passage of cerebrospinal fluid from perivascular space through aqueous channels in the astrocytic end feet (see Concept A.7).

The brain, more than any other organ, must be carefully shielded from abnormal variations in its ionic milieu, as well as from the potentially toxic molecules that find their way into the vascular space by ingestion, infection, or other means. The blood-brain barrier is thus crucial for protection and homeostasis. It also presents a significant problem for the delivery of drugs to the brain. Large (or lipid-insoluble) molecules can be introduced to the brain only by transiently disrupting the blood-brain barrier with hyperosmotic agents such as the sugar mannitol.

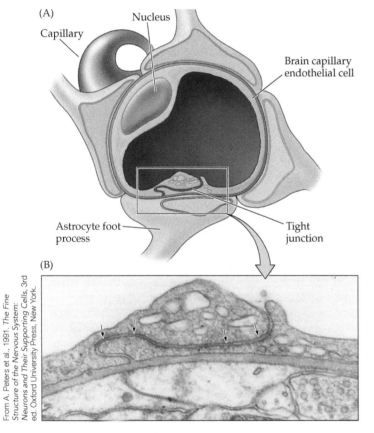

(A)

Capillary

Nucleus

Brain capillary endothelial cell

Astrocyte foot process

Tight junction

(B)

From A. Peters et al. 1991. *The Fine Structure of the Nervous System: Neurons and Their Supporting Cells*, 3rd ed. Oxford University Press, New York.

FIGURE A22   Cellular basis of the blood-brain barrier
(A) Diagram of a brain capillary in cross section and reconstructed views, showing endothelial tight junctions and the investment of the capillary by astrocytic end feet. (B) Electron micrograph of boxed area in (A), showing the appearance of tight junctions between neighboring endothelial cells (arrows). (A after G. W. Goldstein and A. L. Betz, 1986. *Sci Am* 255: 74–83.)

CONCEPT A.7   The Meninges and Ventricular System Protect the CNS and Facilitate the Production and Circulation of Cerebral Spinal Fluid

LEARNING OBJECTIVES

A.7.1   Identify the meningeal layers and discuss the significance of each one.

A.7.2   Describe the production and distribution of cerebrospinal fluid through the ventricular spaces in the forebrain and brainstem and throughout the subarachnoid space.

A.7.3   Describe the intracranial glymphatic system and discuss its significance for neurological health and disease.

The meninges

The cranial cavity is conventionally divided into three regions called the anterior, middle, and posterior cranial fossae. Surrounding and supporting the brain within this cavity are three protective tissue layers, which also extend down the brainstem and the spinal cord. Together these layers are called the **meninges** (Figure A23). The outermost layer of the meninges is called the **dura mater** ("tough mother" or "strong mother," referring to its thick and tough qualities). Most prominently along the dorsal and posterior midline and along the lateral aspects of the upper margin of the posterior fossa, the two layers of the dura mater separate to form the dural venous sinuses mentioned in Concept A.6 (the superior sagittal and transverse sinuses, respectively). The middle layer is called the **arachnoid mater** because of spiderweb-like processes called arachnoid trabeculae, which extend from it toward the third layer, the **pia mater** ("tender mother" or "delicate mother"), a delicate layer of cells that envelopes subarachnoid vessels and apposes the basement membrane on the outer glial surface of the brain. The subarachnoid space in between the arachnoid and pia is filled with cerebrospinal fluid (CSF), the fluid that fills the brain's ventricular spaces (see the next section), as well as branches of the major cerebral and spinal arteries and veins that course across the surface of the brain and spinal cord. Because the pia closely adheres to the brain as its surface curves and folds whereas the arachnoid does not, there are places, called **cisterns**, where the subarachnoid space enlarges to form significant collections of CSF. The subarachnoid space is also a frequent site of bleeding following trauma or in cerebrovascular disease, such as during rupture of an aneurysm (local outward bulge) on a cerebral artery. A collection of blood between the meningeal layers is referred to as a subdural or subarachnoid hemorrhage (or hematoma), as distinct from bleeding within the brain itself, which is called an intraparenchymal hemorrhage.

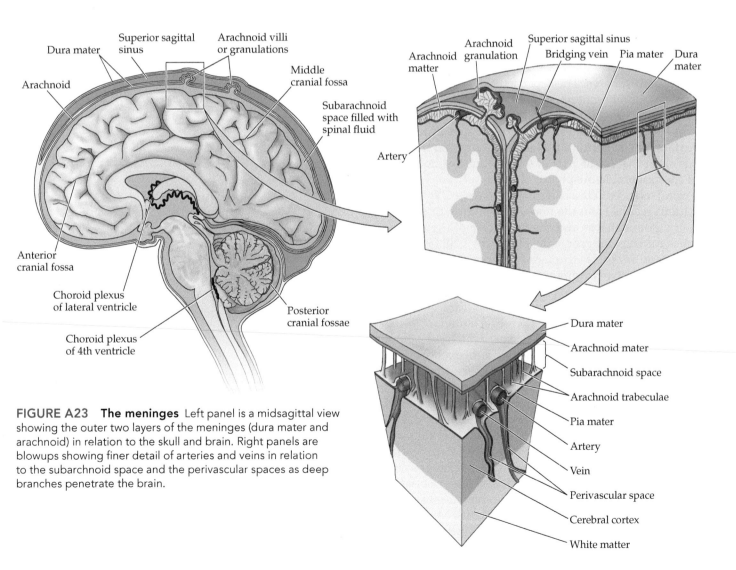

FIGURE A23 **The meninges** Left panel is a midsagittal view showing the outer two layers of the meninges (dura mater and arachnoid) in relation to the skull and brain. Right panels are blowups showing finer detail of arteries and veins in relation to the subarachnoid space and the perivascular spaces as deep branches penetrate the brain.

The ventricular system

The cerebral ventricles are a series of interconnected, fluid-filled spaces that lie in the core of the forebrain and brainstem (Figures A24 and A25). These spaces are filled with CSF produced by a modified vascular structure called the **choroid plexus**, which is present in each ventricle. CSF percolates through the ventricular system and flows into the subarachnoid space through perforations in the thin covering of the fourth ventricle (midline foramen of Magendie and two smaller lateral foramina of Luschka; see Figure A24); it is eventually passed through specialized structures called **arachnoid villi** or **granulations** along the dorsal midline of the forebrain (see Figure A23) and returned to the venous circulation mainly via the superior sagittal sinus.

The presence of ventricular spaces in the various subdivisions of the brain reflects the fact that the ventricles are the adult derivatives of the open space, or lumen, of the embryonic neural tube (see Chapter 22). Although they have no unique function, the ventricular spaces present in sections through the brain provide another useful guide to location (see Figure A2). The largest of these spaces are the **lateral ventricles** (considered the first and second ventricles), one within each of the cerebral hemispheres. These particular ventricles are best seen in frontal sections, where their ventral and lateral surfaces are usually defined by the basal ganglia, their dorsal surface by the corpus callosum, and their medial surface by the **septum pellucidum**, a membranous tissue sheet that forms part of the midline sagittal surface of the cerebral hemispheres. The lateral ventricles, like several telencephalic structures, possess a C shape. This pattern results from the non-uniform growth of the cerebral hemispheres and the formation of the temporal lobes during embryonic development. CSF flows from the lateral ventricles through small openings (called the **interventricular foramina**, or the **foramina of Monro**) into a narrow midline space between the right and left diencephalon, the **third ventricle**. The third ventricle is continuous caudally with the **cerebral aqueduct** (also referred to as the **aqueduct of Sylvius**), which runs though the midbrain. At its caudal end, the aqueduct opens into the **fourth ventricle**, a larger space dorsal to the pons and medulla. The fourth ventricle, covered on its dorsal aspect by the cerebellum, narrows caudally to form the central canal of the spinal cord, which normally does not remain patent beyond the early postnatal period.

Recent studies have demonstrated that, in addition to the bulk flow of CSF through the ventricular system, the subarachnoid space, and into the superior sagittal sinus, CSF also passes through the interstitial spaces of brain tissue itself (i.e., brain parenchyma). Maiken Nedergaard and Steven Goldman and their colleagues at the

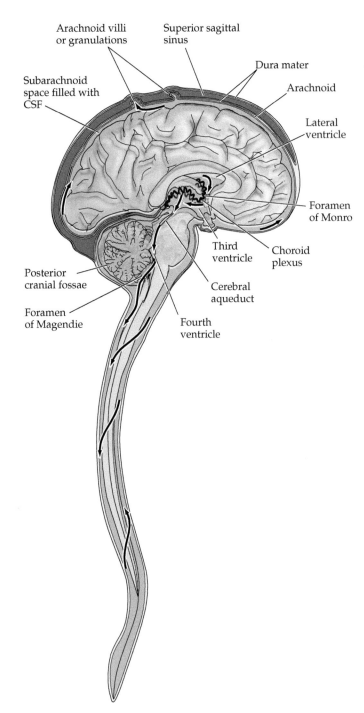

FIGURE A24   Circulation of cerebrospinal fluid
Cerebrospinal fluid (CSF) is produced by the choroid plexus and flows from the lateral ventricles through the paired interventricular foramina (singular: foramen; foramina of Monro) into the third ventricle, through the cerebral aqueduct, and into the fourth ventricle. CSF exits the ventricular system through several foramina associated with the fourth ventricle (e.g., foramen of Magendie along the midline) into the subarachnoid space surrounding the CNS. CSF is eventually passed through the arachnoid granulations and returned to the venous circulation in the superior sagittal sinus.

FIGURE A25 **Ventricular system of the human brain** (A) Location of the ventricles as seen in a transparent left lateral view. (B) Dorsal view of the ventricles.

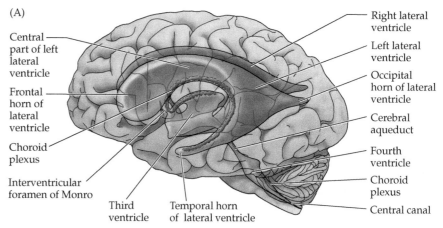

(A)

Central part of left lateral ventricle

Frontal horn of lateral ventricle

Choroid plexus

Interventricular foramen of Monro

Third ventricle

Temporal horn of lateral ventricle

Right lateral ventricle

Left lateral ventricle

Occipital horn of lateral ventricle

Cerebral aqueduct

Fourth ventricle

Choroid plexus

Central canal

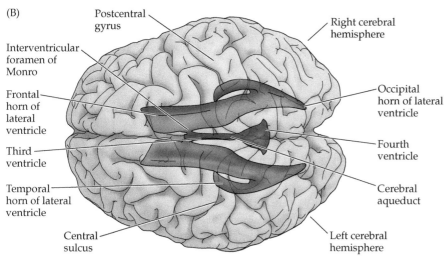

(B)

Postcentral gyrus

Interventricular foramen of Monro

Frontal horn of lateral ventricle

Third ventricle

Temporal horn of lateral ventricle

Central sulcus

Right cerebral hemisphere

Occipital horn of lateral ventricle

Fourth ventricle

Cerebral aqueduct

Left cerebral hemisphere

University of Rochester and the University of Copenhagen used chemical dyes and advanced in vivo microscopy to observe the passage of CSF through the parenchyma. Some quantity of CSF enters the perivascular space that surrounds the arterial branches penetrating deep into the brain from the subarachnoid compartment. Propelled by the pumping of arterial blood, this CSF moves into brain tissue by passing through water channels comprising aquaporin-4 proteins in the astrocytic end feet. As the CSF passes through the parenchyma and mixes with extracellular fluid in the interstitial spaces, metabolic waste and discarded proteins are carried away (Figure A26). This fluid eventually passes into the perivascular spaces surrounding small veins and flows back into the subarachnoid space or into newly discovered lymphatic vessels that course along the superior sagittal sinus. It is estimated that this system, termed the brain's **glymphatic system** because of the participation of glial cells in a lymphatic-like system, is responsible for removing nearly the brain's own weight in waste material over the course of a year.

Not surprisingly, the discovery of this glymphatic system has led to keen interest in its role in brain health and neurological disease. One intriguing observation is that the rate of glymphatic flow increases during slow-wave sleep (see Chapter 28), when the brain's interstitial spaces are thought to expand by some 50% or so. This expansion helps create convective flow of interstitial fluids through the parenchyma and a significant increase in the efficiency of waste removal. The finding of beta-amyloid and synuclein proteins (proteins implicated in Alzheimer's disease and Parkinson's disease, respectively) in fluids flowing through the glymphatic system suggests that this system may serve to remove potentially toxic substances from the brain. Furthermore, it raises the possibility that disruption of this cleansing function might contribute to the onset or progression of neurological disease. Perhaps this circadian mechanism of waste removal is responsible for the association of poor sleep in middle age and an increased risk of cognitive decline in later years. It may also help rationalize what would seem to be an excessively high rate of daily CSF production: The normal total volume of

FIGURE A26 Glymphatic system of the brain Cerebrospinal fluid (CSF) passes from arterial perivascular space through the substance of the brain. As it does so, metabolic wastes and discarded proteins are rinsed from the parenchyma and pass out of the brain via the perivascular spaces surrounding veins. This convective flow of CSF and interstitial fluid increases during sleep, when extracellular spaces expand. (After M. Nedergaard and S. A. Goldman, 2016. *Sci Am* 314: 44–49.)

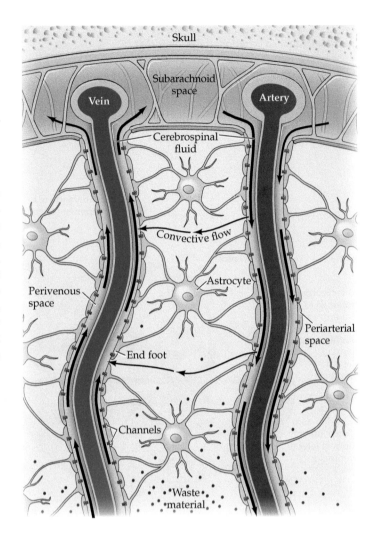

CSF in the ventricular system and subarachnoid space is approximately 150 mL, while the choroid plexus produces approximately 500 mL of CSF per day. Thus, the entire volume of CSF present in the ventricular system is turned over several times a day. However, this high rate of CSF production and clearance poses a risk if there is a blockage of CSF flow through the ventricular spaces or the arachnoid granulations. Obstruction results in an excess of CSF in the intracranial cavity, a dangerous condition called **hydrocephalus** (literally, "water head") that can lead to enlargement of the ventricles and compression of the brain.

■ References

Blumenfeld, H. (2022) *Neuroanatomy through Clinical Cases*, 3rd Edition. NewYork: Oxford University Press.

Brightman, M. W. and T. S. Reese (1969) Junctions between intimately opposed cell membranes in the vertebrate brain. *J. Cell Biol.* 40: 648–677.

England, M. A. and J. Wakely (1991) *Color Atlas of the Brain and Spinal Cord: An Introduction to Normal Neuroanatomy*. St. Louis, MO: Mosby Yearbook.

Goldstein, G. W. and A. L. Betz (1986) The blood–brain barrier. *Sci. Am.* 255: 74–83.

Hablitz, L. M. and M. Nedergaard (2021) The glymphatic system: A novel component of fundamental neurobiology. *J. Neurosci.* 41:7698–7711.

Haines, D. E. (2007) *Neuroanatomy: An Atlas of Structures, Sections, and Systems*, 7th Edition. Baltimore, MD: Lippincott Williams & Wilkins.

Mai, J. K. and G. Paxinos (2012) *The Human Nervous System*, 3rd Edition. New York: Elsevier.

Martin, J. H. (2012) *Neuroanatomy: Text and Atlas*, 4th Edition. New York: McGraw-Hill Medical.

Nedergaard, M. and S. A. Goldman (2016) Brain drain. *Sci. Am.* 314: 44–49.

Netter, F. H. (1983) *The CIBA Collection of Medical Illustrations*, vols. I and II. West Caldwell, NJ: CIBA Pharmaceutical Co.

Parent, A. and M. B. Carpenter (1996) *Carpenter's Human Neuroanatomy*, 9th Edition. Baltimore, MD: Williams & Wilkins.

Peters, A., S. L. Palay and H. deF. Webster (1991) *The Fine Structure of the Nervous System: Neurons and Their Supporting Cells*, 3rd Edition. New York: Oxford University Press.

Reese, T. S. and M. J. Karnovsky (1967) Fine structural localization of a blood–brain barrier to exogenous peroxidase. *J. Cell Biol.* 34: 207–217.

Rexed, B. (1952) The cytoarchitectonic organization of the spinal cord of the cat. *J. Comp. Neurol.* 96: 414–495.

ATLAS

The Human Central Nervous System

This series of seven plates presents labeled images of the human brain and spinal cord. The surface features of the brain are shown in photographs of a postmortem specimen after removal of the meninges and superficial blood vessels (Plate 1). Sectional views of the forebrain in each of three standard anatomical planes (see Figure A1) are derived from T1-weighted magnetic resonance imaging of a living subject (Plates 2–4). In these images, compartments filled with aqueous fluids, such as the ventricles and subarachnoid spaces, appear dark; tissues that are enriched with lipids, such as white matter, appear bright; and tissues that are relatively poor in lipid (myelin) and high in water content, such as gray matter, appear in intermediate shades of gray. Thus, the appearance of gray matter and white matter in the T1-weighted series is similar to what would be observed when dissecting a brain specimen obtained postmortem. Plate 5 displays computed representations of fiber tracts in the white matter of the human brain in a living subject obtained by means of diffusion tensor imaging (see Chapter 1 and Chapter 23, Clinical Applications). The final images are transverse sections obtained from the major subdivisions of the brainstem (Plate 6) and spinal cord (Plate 7). Each of these histological images was acquired and processed to simulate myelin staining; thus, white matter appears dark, while gray matter and poorly myelinated fibers appear light. Note the small insets that show the actual, typical size of cross sections through the human brainstem and spinal cord.

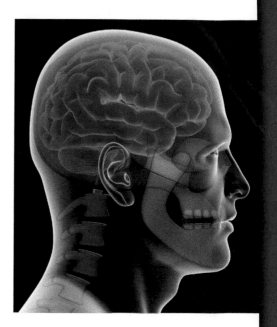

The brain and spinal cord images in these plates are also present in the expanded atlases featured in the neuroanatomical media that accompanies this volume, Sylvius 4 Online: Interactive Atlas and Visual Glossary of the Human Central Nervous System. Sylvius provides a unique computer-based learning environment for exploring and understanding the structure of the human central nervous system. The program features annotated surface views of the human brain, as well as interactive tools for dissecting the central nervous system and viewing annotated cross sections of these specimens. Sylvius also incorporates a comprehensive, searchable database of more than 500 neuroanatomical terms, all concisely defined and visualized in photographs; T1- weighted images; and illustrations from this text. Sylvius 4 Online access cards can be found on Oxford Learning Link or bundled with your e-Book.

PLATE 1   PHOTOGRAPHIC ATLAS: BRAIN SURFACE

(A)

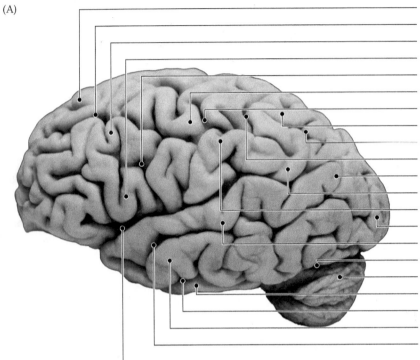

Superior frontal gyrus
Superior frontal sulcus
Middle frontal gyrus
Inferior frontal gyrus
Inferior frontal sulcus
Precentral gyrus
Central sulcus
Superior parietal lobule
Intraparietal sulcus
Postcentral sulcus
Angular gyrus
Supramarginal gyrus
Postcentral gyrus
Lateral occipital gyri
Superior temporal gyrus
Preoccipital notch
Cerebellar hemisphere
Inferior temporal gyrus
Inferior temporal sulcus
Middle temporal gyrus
Superior temporal sulcus
Lateral (Sylvian) fissure

(B)

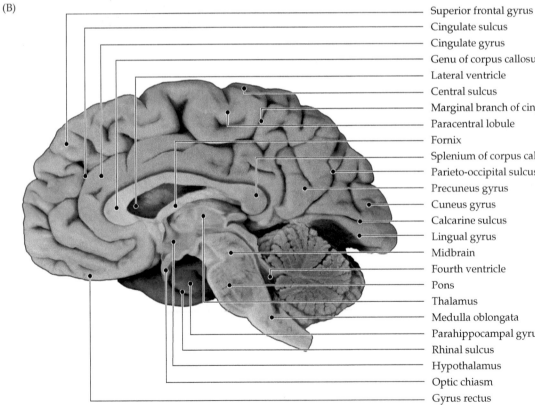

Superior frontal gyrus
Cingulate sulcus
Cingulate gyrus
Genu of corpus callosum
Lateral ventricle
Central sulcus
Marginal branch of cingulate sulcus
Paracentral lobule
Fornix
Splenium of corpus callosum
Parieto-occipital sulcus
Precuneus gyrus
Cuneus gyrus
Calcarine sulcus
Lingual gyrus
Midbrain
Fourth ventricle
Pons
Thalamus
Medulla oblongata
Parahippocampal gyrus
Rhinal sulcus
Hypothalamus
Optic chiasm
Gyrus rectus

Surface features of a human brain specimen. (A) Lateral view of the left hemisphere. (B) Midsagittal view of right hemisphere. (C) Dorsal view. (D) Ventral view.

(C)

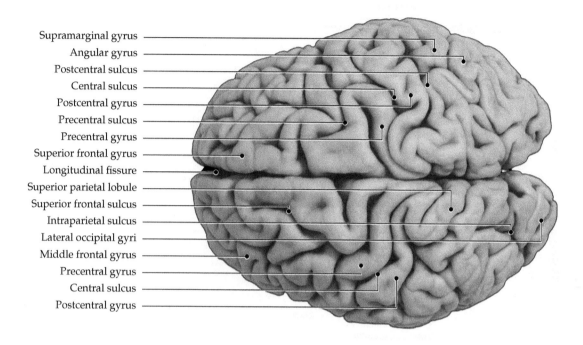

- Supramarginal gyrus
- Angular gyrus
- Postcentral sulcus
- Central sulcus
- Postcentral gyrus
- Precentral sulcus
- Precentral gyrus
- Superior frontal gyrus
- Longitudinal fissure
- Superior parietal lobule
- Superior frontal sulcus
- Intraparietal sulcus
- Lateral occipital gyri
- Middle frontal gyrus
- Precentral gyrus
- Central sulcus
- Postcentral gyrus

(D)

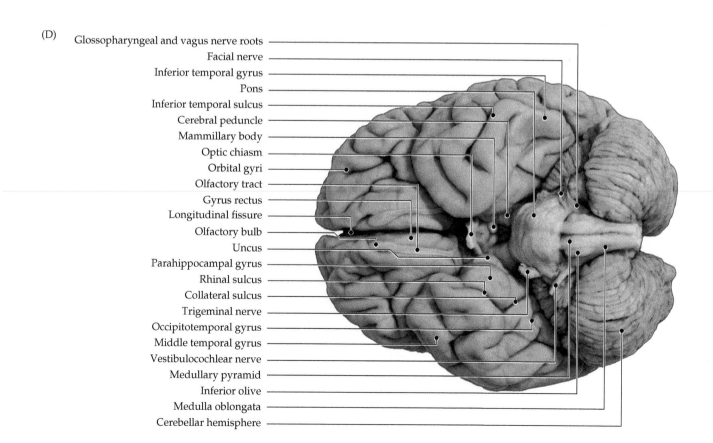

- Glossopharyngeal and vagus nerve roots
- Facial nerve
- Inferior temporal gyrus
- Pons
- Inferior temporal sulcus
- Cerebral peduncle
- Mammillary body
- Optic chiasm
- Orbital gyri
- Olfactory tract
- Gyrus rectus
- Longitudinal fissure
- Olfactory bulb
- Uncus
- Parahippocampal gyrus
- Rhinal sulcus
- Collateral sulcus
- Trigeminal nerve
- Occipitotemporal gyrus
- Middle temporal gyrus
- Vestibulocochlear nerve
- Medullary pyramid
- Inferior olive
- Medulla oblongata
- Cerebellar hemisphere

PLATE 2 CORONAL MR ATLAS

(A)

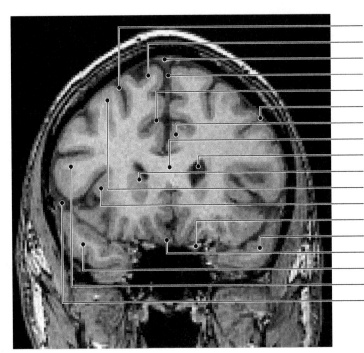

Superior frontal sulcus
Superior frontal gyrus
Superior sagittal sinus
Longitudinal fissure
Cingulate sulcus
Inferior frontal sulcus
Cingulate gyrus
Corpus callosum, genu
Lateral ventricle, anterior horn
Caudate
Middle frontal gyrus
Insular gyri
Optic nerve
Temporal pole
Gyrus rectus
Middle temporal gyrus
Inferior frontal gyrus
Lateral (sylvian) fissure

(B)

Inferior frontal gyrus
Middle frontal gyrus
Superior frontal gyrus
Superior sagittal sinus
Superior frontal sulcus
Longitudinal fissure
Cingulate sulcus
Inferior frontal sulcus
Cingulate gyrus
Corpus callosum, body
Lateral ventricle
Internal capsule
Anterior commissure
Third ventricle
Optic tract
Superior temporal sulcus
Caudate
Inferior temporal sulcus
Globus pallidus
Amygdala
Putamen
Insular gyri
Inferior temporal gyrus
Superior temporal gyrus
Middle temporal gyrus
Lateral (sylvian) fissure

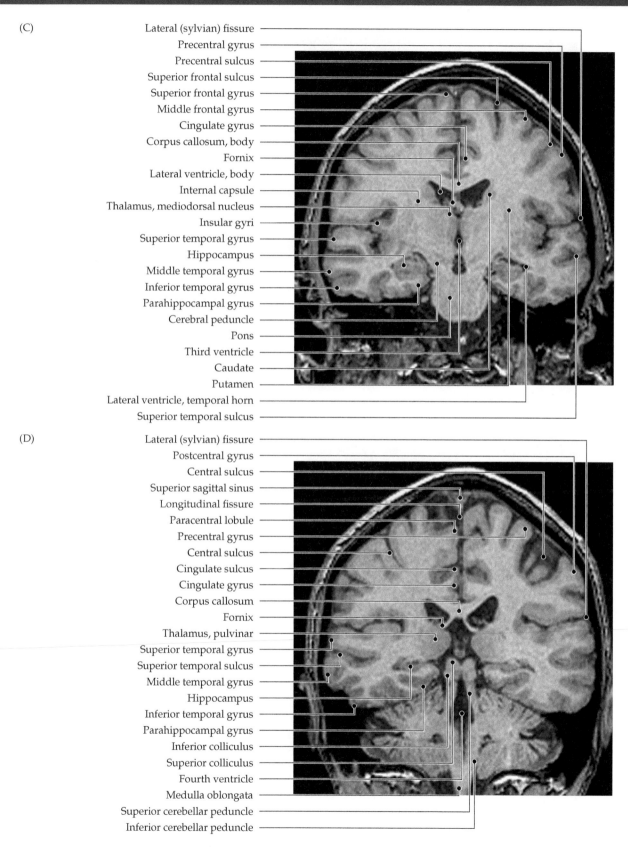

Coronal sections of the human brain demonstrating internal forebrain structures in magnetic resonance images; images in (A)–(D) are arranged from rostral to caudal.

(C)

Lateral (sylvian) fissure
Precentral gyrus
Precentral sulcus
Superior frontal sulcus
Superior frontal gyrus
Middle frontal gyrus
Cingulate gyrus
Corpus callosum, body
Fornix
Lateral ventricle, body
Internal capsule
Thalamus, mediodorsal nucleus
Insular gyri
Superior temporal gyrus
Hippocampus
Middle temporal gyrus
Inferior temporal gyrus
Parahippocampal gyrus
Cerebral peduncle
Pons
Third ventricle
Caudate
Putamen
Lateral ventricle, temporal horn
Superior temporal sulcus

(D)

Lateral (sylvian) fissure
Postcentral gyrus
Central sulcus
Superior sagittal sinus
Longitudinal fissure
Paracentral lobule
Precentral gyrus
Central sulcus
Cingulate sulcus
Cingulate gyrus
Corpus callosum
Fornix
Thalamus, pulvinar
Superior temporal gyrus
Superior temporal sulcus
Middle temporal gyrus
Hippocampus
Inferior temporal gyrus
Parahippocampal gyrus
Inferior colliculus
Superior colliculus
Fourth ventricle
Medulla oblongata
Superior cerebellar peduncle
Inferior cerebellar peduncle

PLATE 3   AXIAL MR ATLAS

(A)

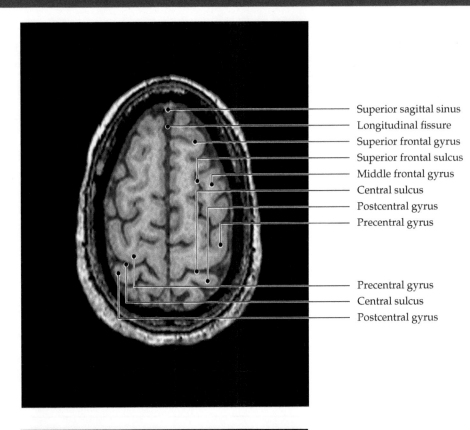

Superior sagittal sinus
Longitudinal fissure
Superior frontal gyrus
Superior frontal sulcus
Middle frontal gyrus
Central sulcus
Postcentral gyrus
Precentral gyrus

Precentral gyrus
Central sulcus
Postcentral gyrus

(B)

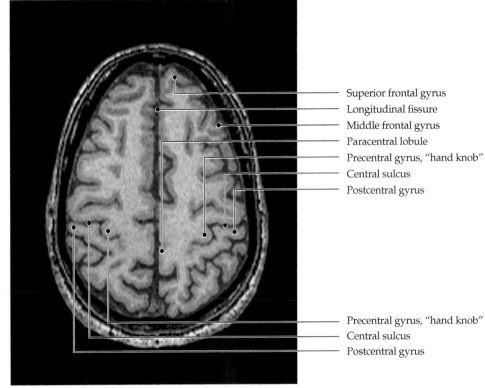

Superior frontal gyrus
Longitudinal fissure
Middle frontal gyrus
Paracentral lobule
Precentral gyrus, "hand knob"
Central sulcus
Postcentral gyrus

Precentral gyrus, "hand knob"
Central sulcus
Postcentral gyrus

Axial sections of the human brain demonstrating internal forebrain structures in T1-weighted magnetic resonance images; images in (A)–(H) are arranged from superior to inferior.

(C)

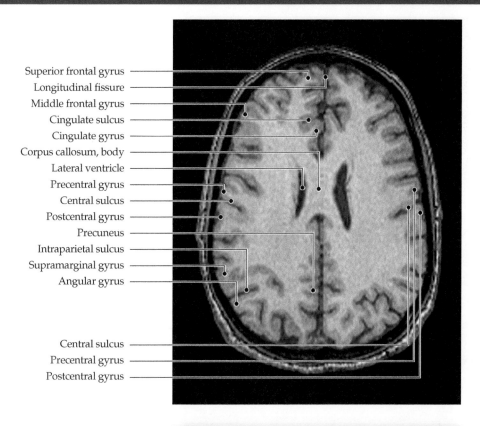

Superior frontal gyrus
Longitudinal fissure
Middle frontal gyrus
Cingulate sulcus
Cingulate gyrus
Corpus callosum, body
Lateral ventricle
Precentral gyrus
Central sulcus
Postcentral gyrus
Precuneus
Intraparietal sulcus
Supramarginal gyrus
Angular gyrus

Central sulcus
Precentral gyrus
Postcentral gyrus

(D)

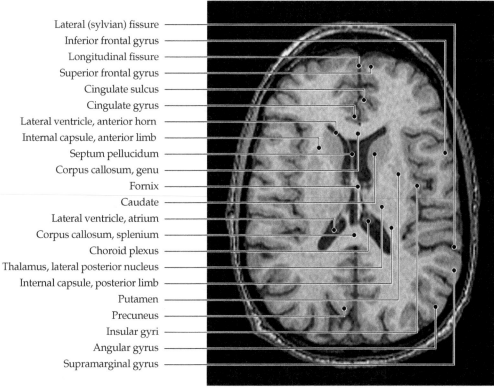

Lateral (sylvian) fissure
Inferior frontal gyrus
Longitudinal fissure
Superior frontal gyrus
Cingulate sulcus
Cingulate gyrus
Lateral ventricle, anterior horn
Internal capsule, anterior limb
Septum pellucidum
Corpus callosum, genu
Fornix
Caudate
Lateral ventricle, atrium
Corpus callosum, splenium
Choroid plexus
Thalamus, lateral posterior nucleus
Internal capsule, posterior limb
Putamen
Precuneus
Insular gyri
Angular gyrus
Supramarginal gyrus

PLATE 3 AXIAL MR ATLAS (CONTINUED)

(E)

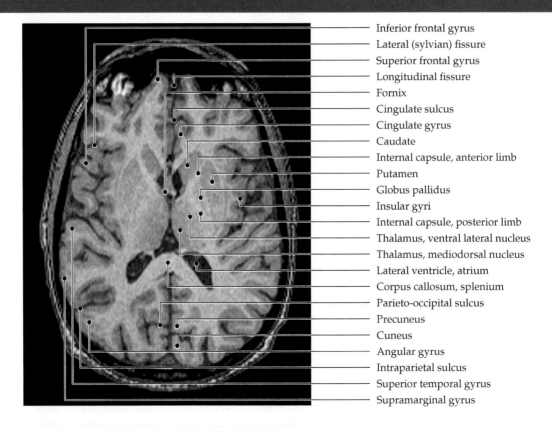

Inferior frontal gyrus
Lateral (sylvian) fissure
Superior frontal gyrus
Longitudinal fissure
Fornix
Cingulate sulcus
Cingulate gyrus
Caudate
Internal capsule, anterior limb
Putamen
Globus pallidus
Insular gyri
Internal capsule, posterior limb
Thalamus, ventral lateral nucleus
Thalamus, mediodorsal nucleus
Lateral ventricle, atrium
Corpus callosum, splenium
Parieto-occipital sulcus
Precuneus
Cuneus
Angular gyrus
Intraparietal sulcus
Superior temporal gyrus
Supramarginal gyrus

(F)

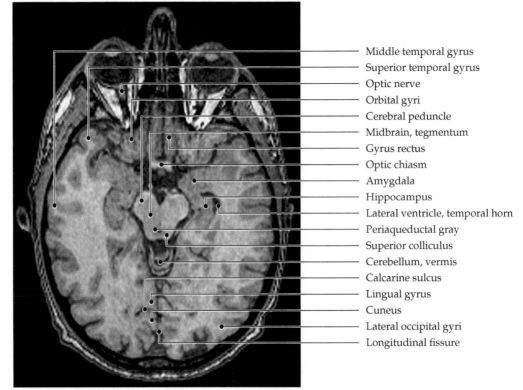

Middle temporal gyrus
Superior temporal gyrus
Optic nerve
Orbital gyri
Cerebral peduncle
Midbrain, tegmentum
Gyrus rectus
Optic chiasm
Amygdala
Hippocampus
Lateral ventricle, temporal horn
Periaqueductal gray
Superior colliculus
Cerebellum, vermis
Calcarine sulcus
Lingual gyrus
Cuneus
Lateral occipital gyri
Longitudinal fissure

Axial sections of the human brain demonstrating internal forebrain structures in T1-weighted magnetic resonance images; images in (A)–(H) are arranged from superior to inferior.

(G)

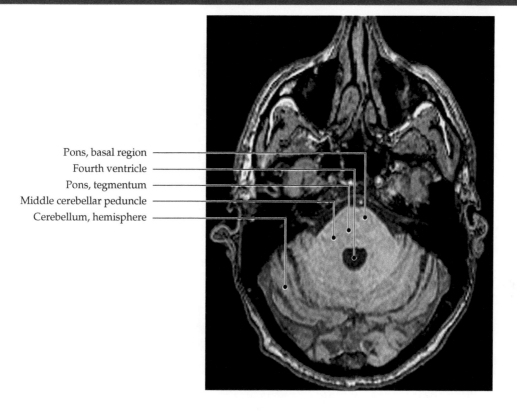

Pons, basal region

Fourth ventricle

Pons, tegmentum

Middle cerebellar peduncle

Cerebellum, hemisphere

(H)

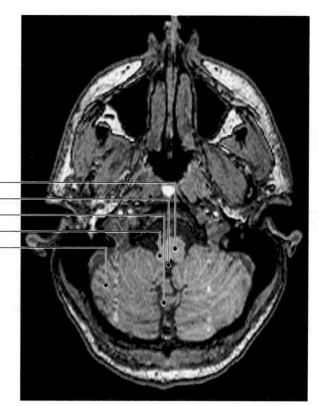

Medulla oblongata, tegmentum

Fourth ventricle

Cerebellum, vermis

Inferior cerebellar peduncle

Cerebellum, hemisphere

PLATE 4   SAGITTAL MR ATLAS

(A)

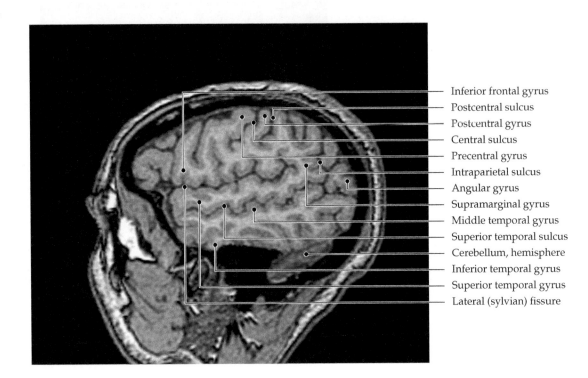

— Inferior frontal gyrus
— Postcentral sulcus
— Postcentral gyrus
— Central sulcus
— Precentral gyrus
— Intraparietal sulcus
— Angular gyrus
— Supramarginal gyrus
— Middle temporal gyrus
— Superior temporal sulcus
— Cerebellum, hemisphere
— Inferior temporal gyrus
— Superior temporal gyrus
— Lateral (sylvian) fissure

(B)

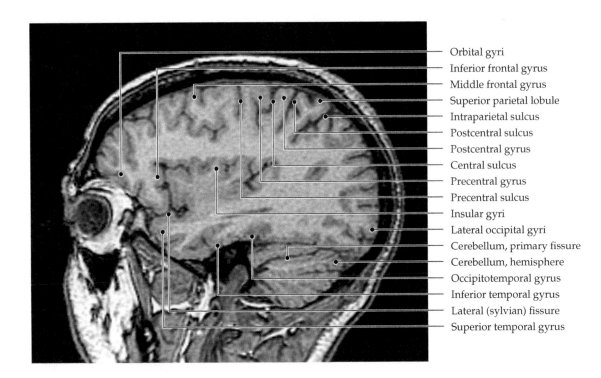

— Orbital gyri
— Inferior frontal gyrus
— Middle frontal gyrus
— Superior parietal lobule
— Intraparietal sulcus
— Postcentral sulcus
— Postcentral gyrus
— Central sulcus
— Precentral gyrus
— Precentral sulcus
— Insular gyri
— Lateral occipital gyri
— Cerebellum, primary fissure
— Cerebellum, hemisphere
— Occipitotemporal gyrus
— Inferior temporal gyrus
— Lateral (sylvian) fissure
— Superior temporal gyrus

Sagittal sections of the human brain demonstrating internal forebrain structures in T1-weighted magnetic resonance images; images in (A)–(D) are arranged from lateral to medial.

(C)

Precuneus
Parieto-occipital sulcus
Postcentral sulcus
Superior frontal gyrus
Precentral sulcus
Precentral gyrus
Central sulcus
Postcentral gyrus
Caudate, head
Caudate, body
Nucleus accumbens
Orbital gyri
Globus pallidus
Hippocampus
Parahippocampal gyrus
Internal capsule
Thalamus
Lateral ventricle, atrium
Fornix
Cerebellum, hemisphere
Lingual gyrus
Cuneus

(D)

Parieto-occipital sulcus
Precuneus gyrus
Marginal branch of cingulate sulcus
Central sulcus
Paracentral lobule
Superior frontal gyrus
Fornix
Cingulate sulcus
Cingulate gyrus
Lateral ventricle
Corpus callosum, genu
Orbital gyri
Hypothalamus
Thalamus
Midbrain
Pons
Fourth ventricle
Medulla oblongata
Corpus callosum, splenium
Cerebellum, vermis
Lingual gyrus
Calcarine sulcus
Spinal cord
Cuneus gyrus

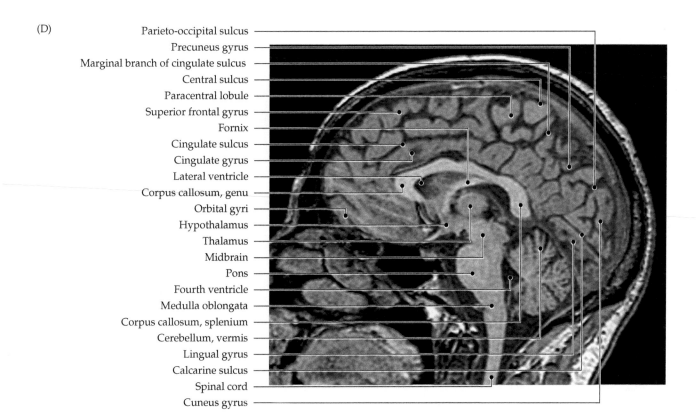

PLATE 5 DIFFUSION TENSOR IMAGING

(A)

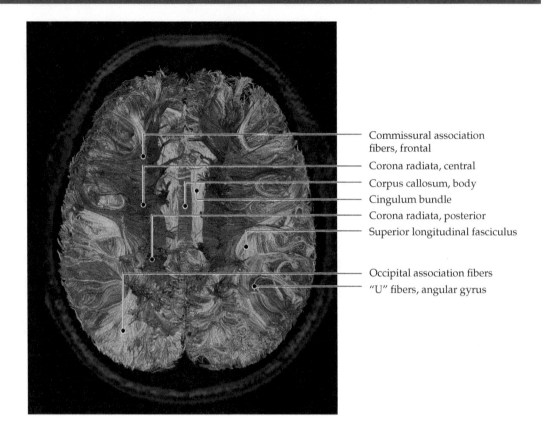

Commissural association fibers, frontal

Corona radiata, central

Corpus callosum, body

Cingulum bundle

Corona radiata, posterior

Superior longitudinal fasciculus

Occipital association fibers

"U" fibers, angular gyrus

(B)

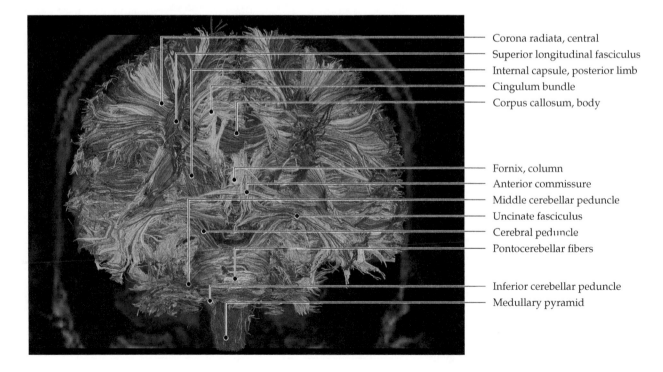

Corona radiata, central

Superior longitudinal fasciculus

Internal capsule, posterior limb

Cingulum bundle

Corpus callosum, body

Fornix, column

Anterior commissure

Middle cerebellar peduncle

Uncinate fasciculus

Cerebral peduncle

Pontocerebellar fibers

Inferior cerebellar peduncle

Medullary pyramid

(A) Axial, (B) coronal, and (C) sagittal sections through a diffusion tensor imaging dataset used to compute fiber tracts, which represent the structure of white matter fibers in a living human brain. (D) shows color code for spatial orientation of fiber tracts. (Images courtesy of Allen W. Song and Iain Bruce, Duke-UNC Brain Imaging and Analysis Center.)

(C)

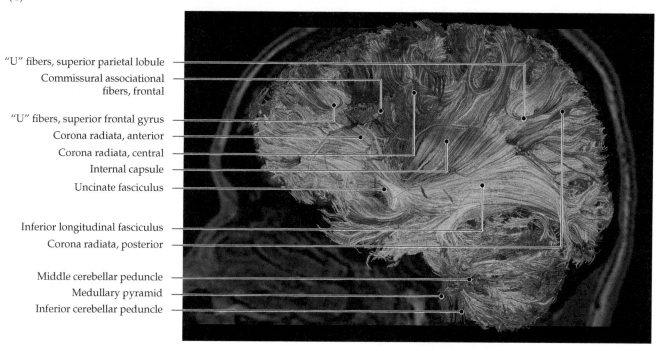

"U" fibers, superior parietal lobule
Commissural associational fibers, frontal
"U" fibers, superior frontal gyrus
Corona radiata, anterior
Corona radiata, central
Internal capsule
Uncinate fasciculus
Inferior longitudinal fasciculus
Corona radiata, posterior
Middle cerebellar peduncle
Medullary pyramid
Inferior cerebellar peduncle

(D)

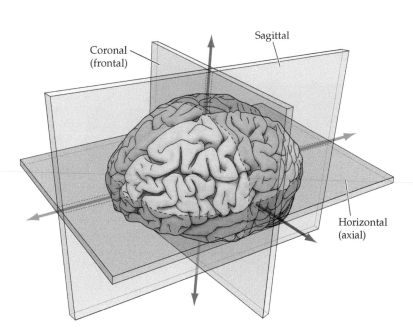

Coronal (frontal)
Sagittal
Horizontal (axial)

PLATE 6   BRAINSTEM ATLAS

(A)

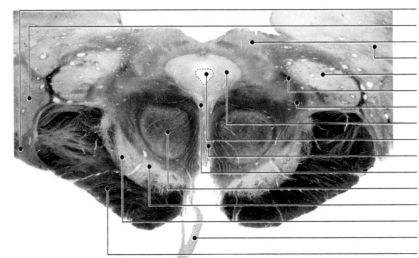

- Optic tract
- Lateral geniculate nucleus
- Superior colliculus
- Pulvinar
- Medial geniculate nucleus
- Anterolateral system
- Medial lemniscus
- Periaqueductal gray
- Cerebral aqueduct
- Raphe nuclei
- Oculomotor complex
- Red nucleus
- Substantia nigra, pars compacta
- Substantia nigra, pars reticulata
- Oculomotor nerve
- Cerebral peduncle

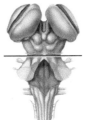

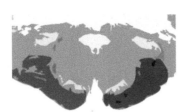

(B)

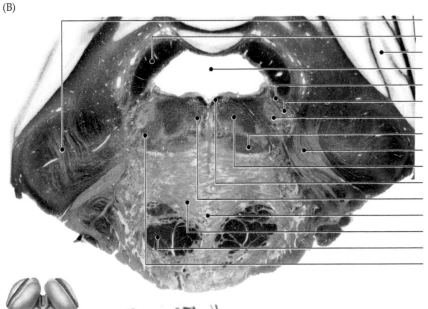

- Middle cerebellar peduncle
- Superior cerebellar peduncle
- Cerebellum, cortex
- Fourth ventricle
- Mesencephalic trigeminal tract and nucleus
- Chief sensory nucleus of the trigeminal complex
- Trigeminal motor nucleus
- Medial lemniscus
- Trigeminal nerve roots
- Central tegmental tract
- Medial longitudinal fasciculus
- Tectospinal fibers
- Pontine nuclei
- Pontocerebellar fibers
- Corticobulbar and corticospinal fibers
- Anterolateral system

Transverse sections of the human brainstem acquired and prepared to simulate myelin staining. (A) Midbrain. (B) Pons. (C) Rostral medulla oblongata. (D) Caudal medulla oblongata. Sections in insets printed at actual size.

(C)

Inferior cerebellar peduncle
External cuneate nucleus
Dorsal motor nucleus of vagus
Solitary tract
Nucleus of the solitary tract
Hypoglossal nucleus
Nucleus ambiguus
Medial longitudinal fasciculus
Tectospinal tract
Medial lemniscus
Medial vestibular nucleus
Spinal vestibular nucleus
Inferior olivary nucleus
Medullary pyramid
Spinal trigeminal nucleus
Anterolateral system
Spinal trigeminal tract

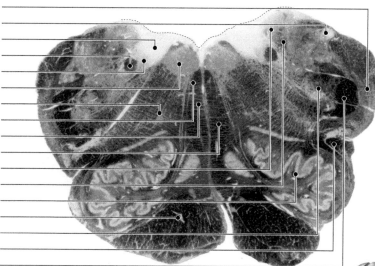

(D)

Gracile tract
Cuneate tract
Cuneate nucleus
Gracile nucleus
Pyramidal decussation
Spinal accessory nucleus
Anterolateral system
Spinal trigeminal nucleus, magnocellular layer
Spinal trigeminal tract
Spinal trigeminal nucleus, gelatinosa layer
Dorsal spinocerebellar tract

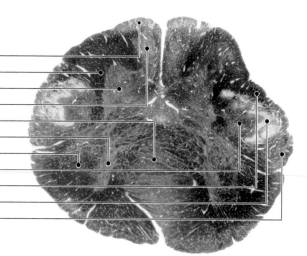
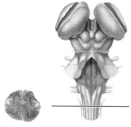

PLATE 7   SPINAL CORD ATLAS

(A)

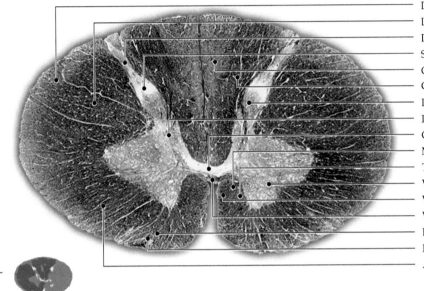

Dorsal spinocerebellar tract
Lateral corticospinal tract
Dorsolateral fasciculus
Substantia gelatinosa
Gracile tract
Cuneate tract
Dorsal horn
Intermediate gray
Central gray
Medial longitudinal fasciculus
Tectospinal tract
Ventral horn
Ventral corticospinal tract
Ventral white commissure
Lateral vestibulospinal tract
Reticulospinal tract
Anterolateral system

(B)

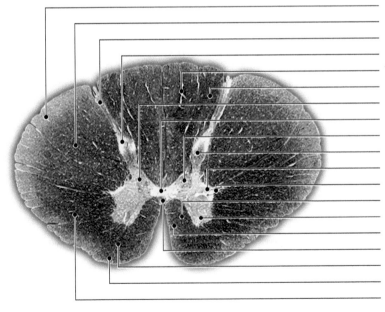

Dorsal spinocerebellar tract
Lateral corticospinal tract
Dorsolateral fasciculus
Substantia gelatinosa
Gracile tract
Cuneate tract
Intermediate gray
Central gray
Clarke's nucleus
Dorsal horn
Intermediolateral cell column
Lateral horn
Medial longitudinal fasciculus
Ventral horn
Ventral corticospinal tract
Ventral white commissure
Lateral vestibulospinal tract
Reticulospinal tract
Anterolateral system

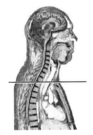

Transverse sections of the human spinal cord acquired and prepared to simulate myelin staining. (A) Cervical segment. (B) Thoracic segment. (C) Lumbar segment. (D) Sacral segment. Sections in insets printed at actual size.

(C)

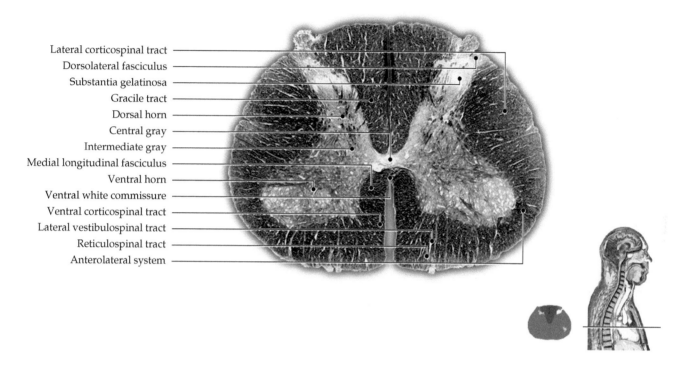

Lateral corticospinal tract
Dorsolateral fasciculus
Substantia gelatinosa
Gracile tract
Dorsal horn
Central gray
Intermediate gray
Medial longitudinal fasciculus
Ventral horn
Ventral white commissure
Ventral corticospinal tract
Lateral vestibulospinal tract
Reticulospinal tract
Anterolateral system

(D)

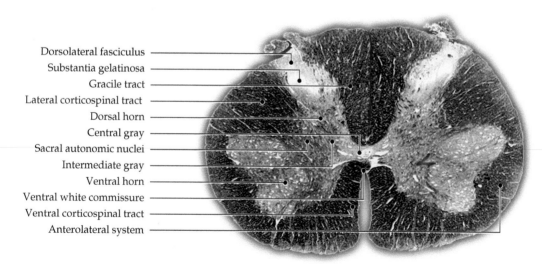

Dorsolateral fasciculus
Substantia gelatinosa
Gracile tract
Lateral corticospinal tract
Dorsal horn
Central gray
Sacral autonomic nuclei
Intermediate gray
Ventral horn
Ventral white commissure
Ventral corticospinal tract
Anterolateral system

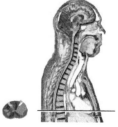

Glossary

A

5-α-reductase An enzyme that can convert testosterone, secreted by gonadal or adrenal tissues, to dihydrotestosterone (DHT), a form of testosterone that cannot be converted to estrogen, and with a high affinity for the testosterone receptor. Many masculinizing effects of testosterone, especially on the development of secondary sex characteristics (development of male genitalia at puberty, male-specific hair patterns) are thought to reflect the activity of 5-α-reductase converting testosterone into DHT.

Aδ fiber group Myelinated, faster conducting pain fibers.

abducens nerve Cranial nerve VI, an efferent nerve that controls the lateral rectus muscle of the eye.

access consciousness The ability to report the contents of subjective experience through speech or action.

accessory olfactory bulb The target of axons from the VNO, adjacent to the main olfactory bulb that relays vomeronasal information to the hypothalamus and other basal forebrain regions.

accommodation 1. Dynamic changes in the lens of the eye that enable the viewer to focus. When viewing distant objects, the lens is made relatively thin and flat; for near vision, the lens becomes thicker and rounder and has more refractive power. 2. Term used by Piaget (along with assimilation) to describe how children might react to a new person, event, or object by modifying their scheme of thought.

across neuron pattern Theory that relies on a concept that in other sensory systems is described as distributed coding or population coding. Contrast with *labeled line coding*.

acetylcholine (Ach) Neurotransmitter at motor neuron synapses, in autonomic ganglia, and in a variety of central synapses. Binds to two types of acetylcholine receptors (AChRs), either ligand-gated ion channels (nicotinic receptors) and G-protein-coupled receptors (muscarinic receptors).

acetylcholinesterase (AChE) Enzyme in the synaptic cleft that clears the cleft of acetylcholine released by the presynaptic cell. AChE hydrolyzes ACh into acetate and choline; the choline is then transported back into nerve terminals, where it is used to resynthesize ACh.

actin A cytoskeletal protein involved in maintaining cell shape and organelle movement.

actin cytoskeleton The meshwork of polymers of the fibrillary protein actin that forms a flexible, but strong scaffold to allow for localization of key proteins and organelles in the cytoplasm, and maintains the integrity of the cell membrane so that the cell retains its volume. The actin cytoskeleton is also essential for force generation in motile cells or motile cell extensions like axonal or dendritic growth cones.

action potential The electrical signal generated and conducted along axons (or muscle fibers) by which information is conveyed from one place to another in the nervous system (or within muscle fibers).

action tremors See *intention tremors*.

activation The time-dependent opening of ion channels in response to a stimulus, typically membrane depolarization.

activational effects The effects of gonadal steroids on circuits in the brain, mostly from puberty onward, that elicit changes in reproduction-related or other sex-specific behaviors. Activational effects can initiate behaviors that are then sustained in a sex-specific way, or they can be transient, like the cyclic levels of estrogen in females, that elicit periodic ovulation.

active transporters Transmembrane proteins that actively move ions into or out of cells against their concentration gradients. Their source of energy may be ATP or the electrochemical gradients of various ions. See *co-transporters; ion exchangers*.

active zone The location within the presynaptic terminal where synaptic vesicles fuse with the presynaptic plasma membrane to discharge their neurotransmitters into the synaptic cleft.

activity-dependent plasticity Changes in the location, strength, or efficiency of synaptic connections that are elicited by changes in voltage—receptor potentials, action potentials, and other depolarizing or hyperpolarizing events in neurons. The electrical activity of these events, especially if they are repeated at similar frequency, is transduced via second messenger systems including varying Ca^{2+} levels into molecular signals that modify synaptic position and structure.

adaptation In the context of evolution, moving animal phenotypes closer to the demands of their environments. Roughly synonymous with evolving "fitness."

adenylyl cyclase III (ACIII) Membrane-bound enzyme that can be activated by G-proteins to catalyze the synthesis of cyclic AMP from ATP.

adrenaline See *epinephrine*.

affective–motivational The fear, anxiety, and autonomic nervous activation that accompany exposure to a noxious stimulus.

afferent neurons Neurons or axons that conduct action potentials from the periphery toward the central nervous system.

age-related macular degeneration A group of age-related disorders in which blurred vision or central vision loss results from damage to the macula.

agnosias The inability to name objects; literally means "not knowing."

alar plate Embryonic gray matter structure in the brainstem and spinal cord that gives rise to sensory nuclei of the brainstem and the dorsal horn of the spinal cord.

allelic silencing A genomic mechanism that renders one of the two copies of each gene (paternal and maternal copies on each chromosome) unable to be expressed. Allelic silencing is thought to be due to a combination of direct binding of transcription regulators to regulatory sequences of one but not the other allele, as well as local histone modifications. Allelic silencing is a more general example of specific forms of transcriptional selection of one copy versus another of a particular gene. Other examples include parental imprinting when either the paternal or maternal allele of a gene is selectively expressed and the other suppressed, and X-inactivation in females when one of two copies of identical genes on each X chromosome is silenced to maintain appropriate gene dosage.

allodynia The induction of pain by a normally innocuous stimulus.

α-fetoprotein A protein that actively sequesters circulating estrogens, preventing maternal estrogen from affecting the sexual differentiation of the fetus.

alpha (α) motor neurons Neurons in the ventral horn of the spinal cord that innervate force-generating, extrafusal fibers of skeletal muscle.

amacrine cells Retinal neurons that mediate lateral interactions between bipolar cell terminals and the dendrites of ganglion cells.

amblyopia Diminished visual acuity as a result of the failure to establish appropriate visual cortical connections in early life.

ametropia A group of visual disorders in which structures in the eye are not capable of properly focusing light rays onto the retina. This results in blurred vision.

amnesia The pathological inability to remember or establish memories; retrograde amnesia is the inability to recall existing memories, whereas anterograde amnesia is the inability to lay down new memories.

AMPA receptors See *ionotropic glutamate receptors.*

ampullae The juglike swellings at the base of the semicircular canals that contain the hair cells and cupulae (plural). See *cupula* (singular).

amygdala A nuclear and corticoid complex in the anterior-medial temporal lobe that forms part of the limbic forebrain; its major functions concern implicit processing with respect to autonomic, emotional, and sexual behavior.

androgen insensitivity syndrome (AIS) A condition in which, due to a defect in the gene that codes for the androgen receptor, testosterone cannot act on its target tissues. Also called *testicular feminization.*

anopsias A large deficit in the visual field resulting from pathological changes in some component of the primary visual pathway.

anosmia Loss of the sense of smell; can be total or restricted to a single odorant.

anosmin The protein encoded by the gene mutated in Kallman's syndrome, also known as KAL1. Anosmin is a transmembrane Ca^{2+} independent cell adhesion molecule. In the absence of anosmin protein, olfactory receptor neurons do not extend their axons to the olfactory bulb nor do GNRH neurons migrate to the hypothalamus. Thus, individual lacking anosmin because of a specific gene mutation are both unable to smell things (anosmia) and also sterile due to a lack of gonadotrophin releasing hormone secretion.

anterior cerebral arteries Major vessels derived from the internal carotid arteries that supply the anterior and medial aspects of the frontal and parietal lobes, including associated deep structures.

anterior choroidal artery Branch of the proximal middle cerebral artery that supplies blood to the medial temporal lobe and deep white matter and gray matter, including parts of the basal ganglia and the internal capsule.

anterior cingulate cortex (ACC) The portion of the midline frontal lobe comprising the anterior extent of the cingulate gyrus and adjacent cortex; its dorsal regions are associated with executive functions.

anterior circulation Vasculature derived from the internal carotid arteries that supplies blood to the forebrain.

anterior commissure A small commissural fiber tract that lies anterior to the third ventricle and inferior to the genu of the corpus callosum; like the callosum, it serves to connect the two hemispheres, but its origins and terminations are mainly in the ventral frontal lobe, olfactory bulb, and anterior temporal lobe.

anterior communicating arteries Small vessels that cross the midsagittal plane joining the two anterior cerebral arteries, forming the anterior aspect of the circle of Willis.

anterior inferior cerebellar artery (AICA) Long circumferential branch of the basilar artery that supplies dorsolateral aspects of the caudal pons and anterior and inferior aspects of the cerebellum

anterior insula A functional division of the cerebral cortex located within the lateral sulcus between the temporal lobe and the frontal and parietal lobes. It is associated with emotion and homeostatic regulation.

anterior spinal artery Principal artery on the anterior aspect of the spinal cord supplied by the vertebral arteries and the medullary arteries; gives rise to some 200 unilateral sulcal branches that alternate left and right along the length of the spinal cord supplying blood to the anterior two-thirds of the cord.

anterograde Signals or impulses that travel "forward," e.g., from the cell body to the axon terminal, from the presynaptic terminal to the postsynaptic cell, or from the CNS to the periphery.

anterograde amnesia The inability to lay down new memories.

anterolateral columns See *ventral columns.*

anterolateral system Ascending sensory pathway in the spinal cord and brainstem that carries information about pain and temperature to the thalamus.

anteroventral paraventricular nucleus (AVPV) A sexually dimorphic hypothalamic nucleus that regulates cyclical ovulation in female mammals.

antiporters Active transporters that use the energy from ionic gradients to carry multiple ions across the membrane in opposite directions.

apical domain The region of an epithelial cell in an epithelium that surrounds a lumen (an open space in a developing embryo or mature organ that is usually fluid-filled) that is closest to the lumenal surface. The apical domain of an epithelial cells is usually specialized for vesicular fusion-mediated secretion and subsequent recycling of vesicle membrane as well as signaling via the cilium that is associated with the apical domain of all epithelial cells.

apoptosis Cell death resulting from a programmed pattern of gene expression; also known as *programmed cell death*.

aprosodias The inability to infuse language with its normal emotional content. See *prosody*.

aqueduct of Sylvius See *cerebral aqueduct*.

aqueous humor A clear, watery liquid that supplies nutrients to the cornea and lens of the eye.

arachnoid mater One of the three coverings of the brain that make up the meninges; lies between the dura mater and the pia mater and forms a spiderweb-like network of trabeculae (arachnoid mater means "spiderweb-like mother) that allows for the flow of cerebrospinal fluid and the distribution of superficial blood vessels in the subarachoid space.

arachnoid villi Protrusions of arachnoid mater into the superior sagittal sinus that allow for the passage of cerebrospinal fluid from the subarachnoid space into the venous drainage.

arbor The network of branches issued by a dendrite or an axon. The pattern of branching in both dendrites and axons, visualized using histological approaches and microscopy, resembles the branching patterns of trees, giving rise the descriptive term "arbor" for these processes.

archicortex Phylogenetically, the simplest and most primitive division of the cerebral cortex, which occurs in the hippocampus.

aromatase The enzyme that converts testosterone, and some additional steroids, to the active form of estrogen, 17β estradiol.

arousal 1. A global state of the brain (or the body) reflecting an overall level of responsiveness. 2. The degree of intensity of an emotion.

associational systems Neural cell circuits that are not part of the relatively defined sensory (input) and motor (output) systems; they mediate the most complex and least well-defined brain functions that require the integration or association of signals from multiple sensory and/or motor systems.

associativity A mechanism that serves to link together two or more independent processes. For example, associative learning results from the pairing of unconditioned and conditioned stimuli presented to an experimental subject. In the hippocampus, associativity allows a weakly activated group of synapses to undergo long-term potentiation when a nearby group of synapses is strongly activated.

astigmatism A condition in which imperfections in the curvature of the cornea or lens prevents light from being properly focussed onto the retina.

astrocytes One of the three major classes of glial cells found in the central nervous system; important in maintaining and regulating, in a variety of ways, an appropriate chemical environment for neuronal signaling; also involved in the formation of the blood-brain barrier, the secretion of substances that influence the construction of new synaptic connections, and the proliferation of new cells in the adult brain that retain characteristics of stem cells.

ATP1A gene Gene encoding alpha subunit of Na$^+$/K+ ATPase pump.

ATP1B gene Gene encoding beta subunit of Na$^+$/K$^+$ ATPase pump.

ATPase pumps Membrane pumps that use the hydrolysis of ATP to translocate ions against their electrochemical gradients.

auditory area 1 (A1) The cortical target of the neurons in the medial geniculate nucleus; the terminus of the primary auditory pathway.

auditory meatus Opening of the external ear canal.

Auerbach's plexus See *myenteric plexus*.

augmentation An activity-dependent form of short-term synaptic plasticity that enhances synaptic transmission over a time course of a few seconds. Augmentation is caused by an increase in the amount of neurotransmitter released in response to presynaptic action potentials and results from persistent calcium signaling within presynaptic terminals, perhaps due to actions on the SNARE-regulatory protein, munc13.

autonomic ganglia Collections of autonomic motor neurons outside the central nervous system that innervate visceral smooth muscles, cardiac muscle, and glands.

autonomic motor division See *visceral motor division*.

autonomic nervous system The components of the nervous system (peripheral and central) concerned with the regulation of smooth muscle, cardiac muscle, and glands. Also known as the visceral motor system; sometimes called the "involuntary" nervous system. Consists of sympathetic and parasympathetic divisions, and a semi-autonomous division in the gut, the enteric nervous system.

autophagy A cellular state in which proteins and other macromolecules within a cell are isolated and eventually trafficked to lysosomes for degradation. Autophagy is regulated by a number of genes, many of which respond to cellular damage, infection, or oxidative stress.

autosomes Any chromosome other than the X and Y sex chromosomes.

auxilin An accessory protein that promotes the actions of HSC70 during vesicle uncoating after endocytosis.

axial sections See *horizontal sections*.

axon The neuronal process (typically, much longer than any dendrite) that conveys the action potential from the nerve cell body to its terminals.

B

bacteriorhodopsin A protein that, in response to light of the proper wavelength, acts as a proton pump transporting protons from inside the cell to outside; in its native host, the resulting proton gradient is subsequently converted into chemical energy; when engineered into a neuron for optogenetics, it hyperpolarizes the neuron when exposed to light.

Balint's syndrome A neurological syndrome, caused by bilateral damage to the posterior parietal and lateral occipital cortex, that has three hallmark symptoms: (1) *simultanagnosia*, the inability to attend to and/or perceive more than one visual object at a time; (2) *optic ataxia*, the impaired ability to reach for or point to an object in space under visual guidance; and (3) *oculomotor apraxia*, difficulty voluntarily directing the eye gaze toward objects in the visual field with a saccade. Simultanagnosia is the sign most closely associated with the syndrome, and the one most studied from a cognitive neuroscience standpoint.

basal cells Basal cells are found in the region of the olfactory epithelium adjacent to the lamina propria, where blood vessels and connective tissue that support the olfactory epithelium are found. They retain neural stem cell identity and can generated new olfactory receptor neurons throughout life.

basal domain The region of a polarized epithelial cell that is specialized for local secretion and local endocytic trafficking. The basal domain is also the domain that is adjacent and adherent to the basal lamina, the layer of polymerized extracellular matrix proteins that defines the base of an epithelial sheet.

basal forebrain nuclei Cerebral nuclei anterior to the hypothalamus and ventral to the basal ganglia; give rise to widespread modulatory projections to diverse targets in the cerebral hemispheres.

basal ganglia Cerebral nuclei lying deep in the subcortical white matter of the cerebral hemispheres lateral and central to the lateral ventricle. The caudate, putamen, and globus pallidus are the major components of the basal ganglia; together with the subthalamic nucleus and substantia nigra, these structures modulate the initiation and suppression of behavior.

basal lamina A thin layer of extracellular matrix material (primarily collagen, laminin, and fibronectin) that surrounds muscle cells and Schwann cells. Also underlies all epithelial sheets. Also called the basement membrane.

basal plate Embryonic gray matter structure in the brainstem and spinal cord that gives rise to motor nuclei of the brainstem and the ventral horn of the spinal cord.

basal processes The processes of cells in a complex epithelium that extend toward the outer surface of the epithelium, away from the lumen. Basal process of epithelial cells often make contact with the basement membrane or basal lamina that defines the extracellular region immediately adjacent to the outer surfaces of most epithelial tissues.

basic emotions In categorical theories, the set of fundamental emotions thought to be expressed similarly across cultures; they are defined by specific combinations of neural, physiological, and subjective features.

basilar artery Major vessel formed by the fusion of the two vertebral arteries that lies along the ventral midline of the pons; gives rise to the anterior inferior and superior cerebellar arteries before bifurcating and giving rise to the paired posterior cerebral arteries at the midbrain.

basilar membrane The membrane that forms the floor of the cochlear duct, on which the cochlear hair cells are located.

basket cells Inhibitory interneurons in the cerebellar cortex whose cell bodies are located within the Purkinje cell layer and whose axons make basket-like terminal arbors around Purkinje cell bodies, providing lateral inhibition that focuses the spatial distribution of Purkinje cell activity.

belt and parabelt regions Regions of the auditory cortex that surround the core region.

Bergmann glia The resident glial cells of the cerebellum that have cell bodies intermixed with those of the Purkinje cells, the primary cerebellar projection neuron. The Bergman glia extend radially oriented processes toward the pial surface of the cerebellum. During the genesis of granule cells, the excitatory interneuron of the cerebellum, the processes of Bergmann glia act as guides for the post-mitotic granule cells that have been generated near the surface of the developing cerebellum and must migrate back into the cerebellar tissue to establish their final position and make appropriate connections.

bHLH (basic helix-loop-helix) Neurogenic transcription factors (named for a shared *basic helix-loop-helix* amino acid motif that defines their DNA-binding domain) that have emerged as central to the differentiation of distinct neural and glial fates.

binocular disparity Since human eyes are displaced from each other, they generate slightly different views of the world. Binocular disparity is the difference in object location detected by each eye and is used for stereopsis.

binocular field The two symmetrical, overlapping visual hemifields. The left hemifield includes the nasal visual field of the right eye and the temporal visual field of the left eye; the right hemifield includes the temporal field of the right eye and the nasal field of the left eye.

biogenic amines Category of small-molecule neurotransmitters; includes the catecholamines (epinephrine, norepinephrine, dopamine), serotonin, and histamine.

bipolar cells Retinal neurons that provide a direct link between photoreceptor terminals and ganglion cell dendrites.

bitemporal hemianopsia A loss of vision confined to the temporal visual field of each eye. Due to lesions at the optic chiasm. Also called *heteronomous hemianopsia*.

bitter One of the five basic tastes; the taste quality, generally considered unpleasant, produced by substances like quinine or caffeine. Compare *salt*, *sour*, and *sweet*. Bitter is transduced by taste cells via the T2R G-protein-coupled taste receptors.

blindsight A pathological phenomenon in which patients with damage to the primary visual cortex are blind in the affected area of the contralateral visual field.

blood oxygenation level-dependent (BOLD) Endogenous signals reflecting the oxygenation of hemoglobin in blood that are modulated by changes in the local level of neural activity; for example, when neural activity in a local brain region increases, more oxygen is consumed and within seconds the local microvasculature responds by increasing the flow of oxygen-rich blood to the active region, thus constituting a BOLD signal that may be detected by fMRI.

blood–brain barrier A diffusion barrier between the cerebral vasculature and the substance of the brain formed by tight junctions between capillary endothelial cells and the surrounding astrocytic endfeet.

bone morphogenetic proteins (BMPs) Peptide hormones that play important roles in neural induction and differentiation.

Bowman's glands Mucous producing specializations composed of secretory cells surrounding a lumen that is continuous with the surface of the olfactory epithelium.

brachium conjunctivum See *cerebellar peduncles.*

brachium pontis See *cerebellar peduncles.*

brain The rostral (supraspinal) portion of the central nervous system comprised of the cerebral hemispheres, diencephalon, cerebellum, and brainstem.

brain-derived neurotrophic factor (BDNF) One member of a family of neutrophic factors, the best-known constituent of which is nerve growth factor (NGF).

brainstem The portion of the brain that lies between the diencephalon and the spinal cord; comprises the midbrain, pons, and medulla.

branchial motor nuclei Brainstem nuclei (derived from the basal plate) that give rise to efferent fibers innervating striated muscle fibers derived from embryonic branchial (pharyngeal) arches; located in an intermediate position in the tegmentum after embryonic migration from a more dorsal location.

Broca's aphasia Difficulty producing speech as a result of damage to Broca's area in the left frontal lobe. Also called *motor, expressive,* or *production aphasia.*

Broca's area An area in the left inferior frontal lobe specialized for the production of speech and the expression of language in non-vocal forms.

C

11-*cis*-retinal A light-absorbing chromophore that binds opsin proteins and triggers phototransduction. When 11-*cis*-retinal absorbs a photon of light, its configuration changes to an all-*trans* isomer.

C fiber group Unmyelinated, slower conducting pain fibers.

c-fos A transcription factor, originally isolated from *cellular feline osteosarcoma* cells, that binds as a heterodimer, thus activating gene transcription.

Ca²⁺/calmodulin-dependent protein kinase, type II A protein kinase that is activated by the second messenger, calcium ions, binding to the calcium-binding protein calmodulin. Once activated by calcium and calmodulin, this protein kinase can phosphorylate numerous substrate proteins to alter their signaling properties.

CACNA genes Ca^{2+} channel genes.

cadherins A family of calcium-dependent cell adhesion molecules found on the surfaces of growth cones and the cells over which they grow.

calbindin A protein that slows transient changes in intracellular calcium concentration by reversibly binding calcium ions.

calcarine sulcus Major sulcus on the medial aspect of the occipital lobe that divides the cuneus and lingual gyri; the primary visual (striate) cortex lies largely within this sulcus.

calcium imaging Method of monitoring by optical means the levels of calcium within cells using calcium-sensitive fluorescent dyes; calcium dynamics within the cytoplasm of neurons reflect the integration of synaptic inputs and the generation of postsynaptic electrical activity.

calcium pump ATPases that remove calcium ions from the cytoplasm of cells. Calcium pumps are found both on the plasma membrane and intracellular membranes, such as the endoplasmic reticulum.

calmodulin A calcium-binding protein that serves as a sensor for many calcium-regulated intracellular signaling processes.

cAMP response element binding protein (CREB) A protein activated by cyclic AMP that binds to specific regions of DNA, thereby increasing the transcription rates of nearby genes.

CAMs The general abbreviation for all *cell adhesion molecules.*

cataracts Opacities in the lens of the eye that cause a loss of transparency and, ultimately, degrading vision.

catecholamines A term referring to molecules containing a catechol ring and an amino group; examples are the neurotransmitters epinephrine, norepinephrine, and dopamine.

categorical theories (of emotion). Theoretical frameworks contending that humans experience a small set of discrete basic emotions (e.g., anger, fear) that can combine into a larger set of complex emotions (e.g., guilt, frustration).

cauda equina The collection of segmental ventral and dorsal roots that attach to the lumbosacral enlargement through the caudal segments of the spinal cord and pass out of the spinal canal through caudal intervertebral foramina and the sacrum.

caudate One of the three major components of the striatum (the other two are the putamen and nucleus accumbens)

cavernous sinus Dural venous sinus in the anterior middle cranial fossa; drains venous blood from ventral cerebral hemisphere and the face. Cranial nerves III, IV, two divisions of V, and VI, and the internal carotid artery all pass through the cavernous sinus.

cell-impermeant molecules Chemical signals that are incapable of permeating the plasma membrane, either because they are too hydrophobic or are attached to the membranes of other nearby cells. Such molecules activate intracellular signaling by activating receptors on the plasma membrane of their target cells.

cell-permeant molecules Hydrophobic chemical signals that are capable of permeating the plasma membranes of their target cells. Such molecules activate intracellular signaling by activating receptors in the cytoplasm or nucleus of their target cells.

central autonomic network Collection of nuclei and cortical regions in the brain that integrate visceral sensory signals, distribute those signals to more widespread brain regions, and give rise to signals that govern visceral motor activity.

central nervous system (CNS) The brain and spinal cord of vertebrates (by analogy, the central nerve cord and ganglia of invertebrates).

central pattern generators Oscillatory spinal cord or brainstem circuits responsible for programmed, rhythmic movements such as locomotion.

central sensitization Increased excitability of neurons in the dorsal horn following high levels of activity in nociceptive afferents and that can result in hyperalgesia and allodynia.

central sulcus A major sulcus on the dorsolateral aspect of the cerebral hemispheres that forms the boundary between the frontal and parietal lobes. The anterior bank of the sulcus contains the primary motor cortex; the posterior bank contains the primary sensory cortex.

cephalic flexure Sharp bend in the neural tube which during early neurulation balloons out to form the prosencephalon, which in turn will give rise to the forebrain and later to the cerebral hemispheres.

cerebellar ataxia A pathological inability to make coordinated movements, associated with lesions or congenital malformation of the cerebellum.

cerebellar cortex Laminated, superficial gray matter of the cerebellum.

cerebellar peduncles Three bilateral pairs of tracts that convey axons to and from the cerebellum. The **superior cerebellar peduncle**, or **brachium conjunctivum**, is primarily an efferent motor pathway; the **middle cerebellar peduncle**, or **brachium pontis**, is an afferent pathway arising from the pontine nuclei. The smallest but most complex is the **inferior cerebellar peduncle**, or **restiform body**, which encompasses multiple afferent and efferent pathways.

cerebellum Prominent hindbrain structure concerned with motor coordination, posture, and balance; derived from the embryonic metencephalon. Composed of a three-layered cortex and deep nuclei; attached to the brainstem by the cerebellar peduncles.

cerebral achromatopsia Loss of color vision as a result of damage to extrastriate visual cortex.

cerebral akinetopsia A rare disorder in which one is unable to appreciate the motion of objects.

cerebral aqueduct Narrow channel derived from the lumen of the neural tube in the dorsal mesencephalon that connects the third and fourth ventricles. Also called the aqueduct of Sylvius.

cerebral cortex The superficial gray matter of the cerebral hemispheres derived from the outer aspect of the telencephalic vesicles.

cerebral hemispheres The two symmetrical halves of the cerebrum derived from the telencephalic vesicles.

cerebral nuclei Masses of gray matter located in the deep or basal regions of the cerebral hemispheres, including the basal ganglia, basal forebrain nuclei, septal nuclei, and nuclear components of the amygdala.

cerebral peduncles Paired "stalks" (peduncle means stalk) of white matter that define the ventral aspect of the midbrain; contain major axon tracts that originate in the cerebral cortex and terminate in the brainstem and spinal cord, including the corticopontine, corticobulbar and corticospinal tracts.

cerebrocerebellum Lateral part of the cerebellar hemisphere, greatly expanded in humans, that receives input from the cerebral cortex via axons from pontine relay nuclei and sends output to the premotor and prefrontal cortex via the thalamus; concerned with the planning and execution of complex spatial and temporal sequences of skilled movement.

cerebrospinal fluid (CSF) Clear and cell-free fluid that fills the ventricular system of the central nervous system produced by choroid plexus in the ventricles.

cervical Rostral region of the spinal cord related to the upper trunk and upper extremities.

cervical enlargement The spinal cord expansion that relates to the upper extremities; includes spinal segments C3–T1.

channel-linked receptors Receptors that are ligand-gated ion channels; binding of ligands leads to channel opening.

channelrhodopsin Typically, a protein that, in response to light of the proper wavelength, opens a channel that is permeable to cations; when engineered into a neuron for optogenetics, it depolarizes the neuron when exposed to light; anion-conducting channelrhodopsins have also been discovered, which would have inhibitory effects when activated in mature neurons.

chemoaffinity hypothesis The idea that nerve cells bear chemical labels that determine their connectivity.

chemotropic Molecules in developing tissues, primarily secreted, that can attract migration cells or in the case of the nervous system, growing axons and dendrites toward the source of the relevant molecule.

cholinergic nuclei Nuclei in which synaptic transmission is mediated by acetylcholine.

Chordin An endogenous antagonist of Bmps that acts in combination with Noggin.

choroid plexus Specialized, highly vascularized epithelium in the ventricular system that produces cerebrospinal fluid.

chromosomal sex The sex of an individual organism based upon the sex chromosomes in its genome.

chronic traumatic encephalopathy (CTE) A neurological syndrome that is caused by repeated concussive force delivered to the head. This syndrome is common in athletes who sustain constant forceful blows to the head, including boxers, hockey players and football players. In addition, soldiers exposed to repeated explosive blasts at close range are at risk. While alive, individuals with CTE suffer a gradual dementia like decline in cognition and social behaviors. At autopsy, their brains have several signs of neurodegenerative and neuroinflammatory damage, and resemble those of individuals, often far older in age, with Alzheimer's disease or other age-related neurodegenerative disorders.

ciliary body Two-part ring of tissue encircling the lens of the eye. The muscular component is important for adjusting the refractive power of the lens. The vascular component produces the fluid that fills the front of the eye.

ciliary ganglion A parasympathetic ganglia located behind the eye.

cingulate gyrus Prominent gyrus on the medial aspect of each cerebral hemisphere, lying just superior to the corpus callosum; a major component of the limbic forebrain.

cingulate sulcus Prominent sulcus on the medial aspect of each cerebral hemisphere.

circadian Refers to variations in physiological functions that occur on a daily basis.

circle of Willis Arterial anastomosis on the ventral aspect of the midbrain; connects the posterior and anterior cerebral circulation.

circumferential endings Specialized endings of mechanosensory afferent neurons in hairy skin arranged as palisades around hair follicles.

cisterns Large, cerebrospinal fluid-filled spaces that lie within the subarachnoid space.

Clarke's nucleus A group of relay neurons (also called the dorsal nucleus of Clarke) located in the medial aspect of the intermediate gray matter of the spinal cord (lamina VII) in spinal levels T1 through L2–3; conveys proprioceptive signals originating in the lower body to the ipsilateral cerebellum and the dorsal column nuclei via the dorsal spinocerebellar tract.

classical conditioning Also called *conditioned reflex*. The modification of an innate reflex by associating its normal triggering stimulus with an unrelated stimulus. The unrelated stimulus comes to trigger the original response by virtue of this repeated association. Compare *operant conditioning*.

clathrin The most important protein for endocytotic budding of vesicles from the plasma membrane; its three-pronged "triskelia" attach to the vesicular membrane to be retrieved.

CLCN genes Cl⁻ channel genes.

climbing fibers Axons that originate in the inferior olivary nuclei, ascend through the inferior cerebellar peduncle, and make terminal arborizations that invest the proximal dendritic trees of Purkinje cells; induce complex spikes and long-term depression in cerebellar Purkinje neurons.

co-transmitters Two or more types of neurotransmitters within a single synapse; may be packaged into separate populations of synaptic vesicles or co-localized within the same synaptic vesicles.

co-transporters Active transporters that use the energy from ionic gradients to carry multiple ions across the membrane in the same direction.

coccygeal Most caudal region of the spinal cord.

cochlea The coiled structure in the inner ear where vibrations caused by sound are transduced into neural impulses.

cognitive neuroscience The field of neuroscience devoted to studying and understanding cognitive functions.

cognitive reappraisal The process through which someone reinterprets a situation that generates an emotional experience, so as to reduce or alter its impact.

coincidence detector A device that detects the simultaneous presence of two or more signals. In the context of long-term synaptic plasticity, a mechanism for detecting the coincidence of two or more synaptic signals; for example, NMDA receptors detect the simultaneous occurrence of presynaptic glutamate release and postsynaptic depolarization during long-term synaptic potentiation.

collagens Fibrillary extracellular matrix proteins with multiple binding domains for a variety of cell surface receptors.

color blindness The decreased ability to detect differences in certain shades of color. This most commonly involves the impaired ability to detect differences in shades of red and green.

columns Term used to describe an elongated gray matter structure (e.g., the motor neuronal pool in the ventral horn of the spinal cord that innervates a muscle) or a subdivision of white matter (e.g., a region of white matter in the spinal cord containing long axon tracts).

coma A pathological state of profound and persistent unconsciousness.

commissures Axon tracts that cross the midline of the brain or spinal cord.

common currency theory A hypothesis about the function of the reward system. In this theory, the brain uses a single scale to compare the values of all goods; the scale is consistent for different items regardless of their type.

competitive interaction The struggle among nerve cells, or nerve cell processes, for limited resources essential to survival or growth.

component process theories (of emotion). Theoretical frameworks that define emotions based on the specific processes that are elicited when people appraise the meaning of events and determine their responses to those events.

computational map An assembly of neural circuits in a specific brain region that represent inputs that do not have a direct correspondence to a topographic map, such as those in the somatosensory or visual systems. Some cognitive capacities, including language and declarative memory, are thought to depend on computational maps.

computerized tomography (CT) Radiographic procedure in which a three-dimensional image of a body structure is constructed by computer from a series of cross-sectional X-ray images.

concentration-invariant coding Robust responses of constant magnitude in specifically tuned target neurons in a sensory relay pathway that are independent of variation of the amount of peripheral stimulation above a threshold. Thus, while the peripheral receptor neuron that directly encounters the stimulus may increase its firing in response to greater availability of the stimulus, the response of the target neuron responds equally to the threshold concentration of the stimulus as well as far greater concentrations of the same stimulus.

concha A component of the external ear.

cone opsins Three different opsins (also called photopsins) that are generated in cone photoreceptors and that respond to different wavelengths of visible light to promote color vision.

conditional mutations A genetic engineering approach, typically reliant upon the Cre/lox system, whereby an exogenous recombinase enzyme recognizes unique DNA excision sequences (loxP sequences) introduced at the 5′ and 3′ ends of an endogenous gene and eliminates the intervening sequence.

conductive hearing loss Diminished sense of hearing due to the reduced ability of sounds to be mechanically transmitted to the inner ear. Common causes include occlusion of the ear canal, perforation of the tympanic membrane, and arthritic degeneration of the middle ear ossicles. Contrast with sensorineural hearing loss.

cones Photoreceptor cells specialized for high visual acuity and the perception of color.

conflict A psychological process that arises when multiple competing demands compete for control of behavior or attention. It is usually associated with increased error rates and/or reaction times.

confluence of sinuses Dural venous sinus formed by the junction of the superior sagittal sinus with the transverse sinuses.

congenital adrenal hyperplasia (CAH) Genetic deficiency that leads to overproduction of androgens and a resultant masculinization of external genitalia in genotypic females.

conjugate eye movements The paired movements of the two eyes in the same direction, as occurs in saccades, smooth pursuit eye movements, optokinetic movements, and the vestibulo-ocular reflex.

connexins Transmembrane proteins that serve as the subunits of connexons, the transcellular channels that permit electrical and metabolic coupling between cells at electrical synapses.

connexons Precisely aligned, paired transmembrane channels that form gap junctions between cells. They are formed from *connexins*, members of a specialized family of channel proteins.

consolidation The process through which memory traces are strengthened some time after initial encoding (e.g., during sleep).

conspecific interactions Social interactions between individuals of the same species. These interactions can be between adults of the same sex, often for competition for dominance or mating. They can also be between adults of opposite sexes for mating. Finally, these interactions can be between adults and newborn or young offspring of the same species to provide food, protection, and behavioral guidance for newly born individuals.

contralateral neglect syndrome Neurological condition in which the patient does not acknowledge or attend to the left visual hemifield or the left half of the body. The syndrome typically results from lesions of the right parietal cortex.

contrast The difference in luminance between objects. The human visual system is more sensitive to detecting differences in contrast than in measuring absolute values of light.

convergence Innervation of a target cell by axons from more than one neuron. In vision refers specifically to the convergence of both rod and cone photoreceptor cells onto retinal ganglion cells.

cornea The transparent surface of the eyeball in front of the lens; the major refractive element in the optical pathway.

coronal Standard anatomical planes of section; any vertical plane passing parallel to the medial-to-lateral axis through the head (in humans, parallel to the face), dividing the head into anterior (front) and posterior (rear) segments. Also known as *frontal*.

corpus callosum The large medial fiber bundle that connects the cortices of the two cerebral hemispheres.

copy number variants A genetic change that does not reflect alteration of a single gene (based upon divergent DNA sequence and resulting divergent trait), but loss or gain of one of the two copies of a wild type gene or multiple genes. The losses (deletion copy number variants) or gains (duplication copy number variants) result in a change in gene dosage—the amount of messenger RNA transcribed and thus protein translated of the deleted or duplicated genes. Copy number variants can be benign and inherited across generations with no observable phenotypic variation, or deleterious leading to sub-optimal outcomes or disease. A common copy number variant that leads to a syndromic disease (caused by a genetic change with multiple divergent traits) is Trisomy 21, duplication of multiple genes on Chromosome 21 that results in Down Syndrome.

cortex (pl. cortices) The superficial mantle of gray matter (a sheet-like array of nerve cells) covering the cerebral hemispheres and cerebellum, where most of the neurons in the brain are located.

cortical column A vertical arrangement of neurons within the cerebral cortex, in which neurons exhibit similar tuning to receptive field attributes.

cortical plate The accumulation of initially post-mitotic cortical projection neurons generated from the ventricular and subventricular zone that migrate via radial glial processes outward toward the basal/pial surface of the developing cortex.

corticobulbar tract Pathway carrying motor information from the motor cortex to brainstem nuclei.

corticocortical connections Connections made between cortical areas in the same hemisphere, or between corresponding areas in the two hemispheres via the cerebral commissures.

corticospinal tract Pathway carrying motor information from the motor cortex to the spinal cord. Essential for the performance of discrete voluntary movements, especially of the hands and feet.

corticostriatal pathway Excitatory (glutameric) projections from deep layers of the cerebral cortex to the striatum. Projections are organized topographically with distinct cortical areas projecting to distinct divisions of the striatum.

covert attention The focusing of visual attention toward a location or item in the visual field without shifting the direction of gaze. Can apply to other sensory modalities or to attentional paradigms. Compare *overt attention*.

cranial nerve ganglia The sensory ganglia associated with the cranial nerves; these correspond to the dorsal root ganglia of the segmental nerves of the spinal cord.

cranial nerves The 12 pairs of sensory, motor, and mixed sensorimotor nerves that innervate targets in the head and neck.

cranial neural crest The portion of the neural crest, a migratory population of neural tube-derived progenitors, that moves into the interstitial space between the surface ectoderm and the neural tube ectoderm in the most anterior portion of a vertebrate embryo that becomes the forebrain and the head.

cranial placodes The local thickening of the non-neural surface ectoderm of the mid-gestation embryo that undergo a form of neural induction so that they can generate cranial peripheral sensory neurons (mechanoreceptor cells in the cranial sensory ganglia, olfactory and vomeronasal receptor neurons in the nose, and hair cells in the inner ear) as well as the lens in the eye.

Cre recombinase A viral DNA cutting enzyme used to excise a floxed exons. See *Cre/lox*.

Cre/lox A genetic engineering system for achieving conditional mutations of endogenous mammalian genes using introduced loxP sequences, which are not found in mammalian genomes but occur in bacterial genomes and are targeted by certain viruses, and a viral DNA cutting enzyme, Cre recombinase. With expression of the Cre DNA introduced into host genome, the resulting Cre recombinase engages the loxP binding sites, and the intervening endogenous exon targeted for elimination (the so-called floxed sequence) is excised.

credit assignment When multiple events occur and are associated with different values, the decision-maker must determine which event produces which value. This process, known as credit assignment, is trivially easy in many cases, but in other cases, can be quite difficult.

cribiform plate A bony structure of the facial portion of the skull comprising many small fensetra (tiny holes), at the level of the eyebrows, that separates the olfactory epithelium from the brain. The axons from the olfactory sensory neurons pass through the tiny holes of the cribiform plate to enter the brain.

crista The hair cell-containing sensory epithelium of the semicircular canals.

critical periods Restricted developmental periods during which the nervous system is particularly sensitive to the effects of experience.

cuneate nucleus Somatosensory relay nucleus in the lower medulla containing second-order sensory neurons that relay mechanosensory information originating in peripheral receptors in the upper body (excluding the face) to the contralateral thalamus.

cuneate tract Lateral division of the dorsal column in the upper half of the spinal cord containing the central processes of first-order afferents and the postsynaptic dorsal column projection; conveys mechanosensory signals derived from the upper body excluding the face.

cuneus gyrus Gyral structure on the superior aspect of the medial occipital lobe forming the upper bank of the calcarine sulcus; portion of the primary visual cortex that represents the inferior quadrant of the contralateral visual hemifields.

cupula Gelatinous structure in the semicircular canals in which the hair cell bundles are embedded.

cyclic nucleotide-gated channels A class of ion channels that are activated and inactivated by second messenger cascades. These second messenger cascades usually involve the activation of a G-protein that is coupled to a G-protein-coupled receptor leading to increased phosphorylation capacity of adenylyl or guanyl cyclases: enzymes that can phosphorylate the channels and modify their permeability to ions.

cytoarchitectonic areas See *cytoarchitecture*.

cytoarchitecture Distinct regions of the neocortical mantle identified by differences in cell size, packing density, and laminar arrangement (layering). Most prominent in humans is the 6-layered neocortex. The evolutionary older archicortex (or hippocampal cortex) has 3–4 layers, and the ancient paleocortex has 3 layers.

D

decerebrate rigidity Excessive tone in extensor muscles as a result of damage to descending motor pathways at the level of the brainstem.

declarative memory Memories available to consciousness that can be expressed by language.

decussation Midline crossing of axons.

deep cerebellar nuclei Subcortical nuclei at the base of the cerebellum that give rise to output from the cerebellum to the thalamus and brainstem; integrate afferent signals to the cerebellum and cortical processing conveyed by Purkinje neurons.

default mode network A collection of brain regions whose activity tends to decrease when people engage in an active task but increase when people engage in undirected thinking, reflection, mind-wandering, or similar processes.

delayed response genes Genes whose protein products are not produced rapidly after a triggering stimulus. The delay in gene expression is caused by the requirement for transcriptional regulators that must first be synthesized in response to the stimulus.

delayed response task A behavioral paradigm used to test cognition and memory.

delta ligands Transmembrane proteins whose ectodomain (the region of the protein that extends beyond the cell's outer membrane) binds to receptors (notch proteins) on the surfaces of adjacent cells to initiate a signaling cascade that promotes local cellular differentiation.

delta waves Slow (<4 Hz) electroencephalographic waves that characterize stage IV (slow-wave) sleep.

dendrites Neuronal processes (typically, much shorter than the axon) arising from the nerve cell body that receive synaptic input.

dendritic polarization The process by which the dendrite of a nerve cell, which is the primary site for synaptic input at post-synaptic specializations becomes distinguished from the axon, which is the primary site for synaptic output at presynaptic terminals.

dentate nucleus Largest and most lateral of the deep cerebellar nuclei; source of output from the cerebrocerebellum to the premotor and prefrontal cortex via the thalamus and to the parvocellular red nucleus.

depolarization Displacement of a cell's membrane potential toward a less negative value.

dermatomes The area of skin supplied by the sensory axons of a single dorsal root ganglion.

deuteranopia A type of color blindness in which the ability to detect medium wavelengths of visible light (such as shades of green) is impaired.

dichromatic The color vision that arises when animals have only two cone types.

diencephalon Portion of the brain derived from the posterior part of the embryonic forebrain vesicle that lies just rostral to the midbrain; comprises the thalamus and hypothalamus.

diffusion tensor imaging (DTI) A type of magnetic resonance imaging used in live humans that allows for the selective visualization of large axon tracts in the brain based upon the alignment of the water molecules in myelinated axons bundled together and extending in the same direction.

dimensional theories (of emotion). Theoretical frameworks that represent emotional experiences and expressions as points within a continuous space of two or more dimensions (e.g., arousal and valence).

dimorphic Having two different forms depending on genotypic or phenotypic sex.

diplopia Double vision.

disconjugate eye movements Movements of the two eyes in opposite directions, such as in vergence eye movements.

disinhibition syndrome Also called *frontal disinhibition syndrome*. A collection of behavioral signs and symptoms, typically caused by damage to the ventral prefrontal cortex; manifested by a loss of control, inappropriate outbursts, and a lack of inhibition in social settings. Compare *dysexecutive syndrome*.

dissociated sensory loss Loss of mechanosensation on one side of the body accompanied by pain and temperature deficits on the other side of the body, often caused by lateral hemisection of the spinal cord.

divergence The branching of a single axon to innervate multiple target cells.

dominant If a trait differs between parents, and the offspring inherits that trait as an exact replica of the trait in one of the two parents, then at least one the two copies of the gene encoding that trait in the parent as well as its replica offspring is said to be dominant; the offspring must inherit at least one dominant copy.

dopamine A catecholamine neurotransmitter that is involved many brain functions, including motivation, reward and motor control.

dorsal columns Major ascending tracts in the spinal cord that carry mechanosensory information from first-order sensory neurons in dorsal root ganglia and second-order, postsynaptic dorsal column projection neurons to the dorsal column nuclei; also called the posterior funiculi.

dorsal horns The dorsal portions of the spinal cord gray matter derived from the alar plate; populated by neurons that process somatosensory information.

dorsal motor nucleus of the vagus nerve Visceral motor nucleus of the rostral medulla containing parasympathetic preganglionic neurons that innervate thoracic and upper abdominal visceral, mediating a range of autonomic functions.

dorsal nucleus of Clarke Column of relay neurons in the medial aspect of the intermediate gray matter of the spinal cord from T1–L3; receives first-order sensory signals from proprioceptors that supply the lower body and gives rise to ipsilateral dorsal spinocerebellar tract. Also called Clarke's nucleus.

dorsal root ganglia The segmental sensory ganglia of the spinal cord; they contain the cell bodies of the first-order neurons of all somatic sensory and visceral sensory pathways arising in the spinal cord.

dorsal roots The bundle of axons that runs from the dorsal root ganglia to the dorsal horn of the spinal cord, carrying somatosensory information from the periphery.

dorsal spinocerebellar tract Axonal projection arising from Clarke's nucleus to the ipsilateral cerebellum and the dorsal column nuclei; conveys proprioceptive signals originating in the lower body.

dorsal stream Cortical visual pathways that originate in primary visual cortex and extend into the parietal cortex. This pathway processes information important for object spatial location and movement.

dorsolateral prefrontal cortex (DLPFC) A functional division of the prefrontal cortex roughly corresponding to the middle and superior frontal gyri, as located anterior to motor cortex and the frontal eye fields. Compare *ventrolateral prefrontal cortex.*

dorsolateral tract of Lissauer A small bundle of mostly unmyelinated axons that is situated on the posterior margin of the dorsal horn and that conveys pain information to second order neurons in Rexed's laminae 1, 2, and 5.

dorsomedial prefrontal cortex (DMPFC) A functional division of the prefrontal cortex roughly corresponding to the medial surface dorsal to the corpus callosum.

dreaming A unique state of awareness that entails some features of memory and hallucinations in the sense that the experience of dreams is not related to corresponding sensory stimuli arising from the present environment.

DSCAM A class of transmembrane cell adhesion molecules encoded by the DSCAM genes. In *Drosophila*, the DSCAM gene structure allows for a total of 37,000 possible splice variants based upon its complex structure of exons for alternative splicing. The DSCAMs can either bind homophilically (binding between the same isoforms) or heterophilically (binding between different isoforms) to initiate avoidance of dendrites or axon branches from the same neuron, or recognition and apposition of dendrites or axons from different cells.

dura mater The thick external covering (dura mater means "tough mother") of the brain and spinal cord; one of the three components of the meninges, the other two being the pia mater and arachnoid mater

dynamin A GTP-hydrolyzing enzyme involved in the fission of membranes during endocytosis.

dyneins A family of multimeric motor proteins that generate force for trafficking cargo via ATP hydrolysis. Dynein multimers have two microtubule binding domains, and several globular domains that support tail domains for binding specific cargos. Dyneins are primarily used for retrograde transport: from the periphery to the cell body.

dynorphins A class of endogenous opioid.

dysdiadochokinesia Difficulty performing rapid alternating movements.

dysexecutive syndrome Also called *frontal dysexecutive syndrome*. A collection of behavioral signs and symptoms, typically caused by damage to the dorsolateral prefrontal cortex; manifested by an inability to change behavior willfully and flexibly according to context. Compare *disinhibition syndrome*.

dysmetria Inaccurate movements due to faulty judgment of distance, especially over- or underreaching; characteristic of cerebellar pathology.

E

ectoderm The most superficial of the three embryonic germ layers; gives rise to the nervous system and epidermis.

Edinger–Westphal nucleus Midbrain visceral motor nucleus containing the parasympathetic preganglionic neurons that constitute the efferent limb of the pupillary light reflex.

efferent neurons Neurons or axons that conduct information away from the central nervous system toward the periphery.

electrical synapse Synapses that transmit information via the direct flow of electrical current at gap junctions.

electrochemical equilibrium The condition in which no net ionic flux occurs across a membrane because ion concentration gradients and opposing transmembrane potentials are in exact balance.

electroencephalography (EEG) The study of electrical potentials generated in the brain recorded from electrodes placed on the scalp.

electrophysiological recording Measure of the electrical activity across the membrane of a nerve cell by use of electrodes.

embryonic stem cells (ES cells) Cells derived from pre-gastrula embryos that have the potential for infinite self-renewal and can give rise to *all* tissue and cell types of the organism. See *neural stem cells.*

emotion regulation The process through which someone alters their emotional experiences, such as through suppression or reappraisal.

end plate current (EPC) A macroscopic postsynaptic current resulting from the summed opening of many ion channels; produced by neurotransmitter release and binding at the motor end plate.

end plate potential (EPP) Depolarization of the membrane potential of skeletal muscle fiber, caused by the action of the transmitter acetylcholine at the neuromuscular synapse.

end plates The complex postsynaptic specializations at the site of nerve contact on skeletal muscle fibers.

endocannabinoids A family of endogenous signals that participate in several forms of synaptic transmission, interacting cannabinoid receptors. These receptors are the molecular targets of the psychoactive component of the marijuana plant, *Cannabis.*

endocrine Referring to the release of signaling molecules whose effects are made widespread by distribution in the general circulation.

endocytosis A process that brings materials into a cell. At chemical synapses, endocytosis is responsible for the retrieval of synaptic vesicle components delivered to the presynaptic membrane via exocytosis.

endoderm The innermost of the three embryonic germ layers. Gives rise to the digestive and respiratory tracts and the structures associated with them.

endogenous antagonists The endogenous antagonists are secreted proteins from the notochord and other sources that bind to Bmp ligands and inactivate them. The specification of the neuroectoderm relys upon the activity of the endogenous antagonists to prevent the undifferentiated ectoderm from becoming epidermal ectoderm.

endogenous attention A form of attention in which processing resources are directed voluntarily to specific aspects of the environment; typically prompted by experimental instructions or, more normally, by an individual's goals, expectations, and/or knowledge. Compare *exogenous attention.*

endogenous opioids Peptide neuritransmitters in the central nervous system that have the same pharmacological effects as morphine and other derivatives of opium, being agonists at opioid receptors, virtually all of which contain the sequence Tyr-Gly-Gly-Phe. There are three classes: dynorphins, endorphins, and enkephalins.

endolymph The potassium-rich fluid filling both the cochlear duct and the membranous labyrinth; bathes the apical end of the hair cells.

endorphins A family of neuropeptides first identified as endogenous mimics of morphine. These neuropeptides, as well as morphine, act by activating opioid receptors.

engram A term used to describe the physical basis of a stored memory.

enkephalins A class of endogenous opioid.

enteric system A subsystem of the visceral motor system, made up of small ganglia and individual neurons scattered throughout the wall of the gut; influences gastric motility and secretion. Also called the *enteric nervous system.*

enzyme-linked receptors Receptors that activate intracellular signaling processes via their enzymatic activity. Most of these receptors are protein kinases that phosphorylate intracellular target proteins.

ephrin A large family of transmembrane cell surface adhesion molecules, also referred to as *Eph* ligands that can bind to and activate the protein kinase activity of ephrin receptors. Ephrins also can initiate retrograde signaling in the cell whose membrane they are embedded in.

Eph receptors (Ephs) Ephrin or Eph receptors are a large family of transmembrane receptor tyrosine kinases (RTKs). Upon binding of an eprhin (also known as an eph ligand) embedded in the membrane of a neighboring cell or process, the tyrosine kinase domain of the receptor is activated leading to phosphorylation of multiple protein targets and initiation of signaling that changes the cytoskeleton, cell motility and gene expression.

epibranchial placodes The cranial placodes found in the branchial arches that contribute neural precursors to cranial sensory ganglia to generate mechanoreceptor sensory neurons.

epinephrine Catecholamine hormone and neuro-transmitter that binds to adrenergic G-protein-coupled receptors.

epithelial-to-mesenchymal transition (EMT) The dissolution of tight junctions and other molecular specializations that maintain cells in a sheet like (epithelial) arrangement so that the liberated cells can acquire motile capacity and migrate to distal locations. The delamination and migration of the neural crest is the best-known example of an epithelial-to-mesenchymal transition in the embryo. These changes, resulting in migratory cells, can also be pathological in mature tissues, leading to metastatic tumor formation when transformed cells escape epithelial constraints.

equilibrium potential The membrane potential at which a given ion is in electrochemical equilibrium.

esotropia A type of ocular misalignment in which one or both eyes of an individual turn inward (toward the nose).

estradiol The principal estrogen, secreted by ovarian follicles.

estrogen A steroid hormone (including estradiol) that affects sexual differentiation during development and reproductive function and behavior in mature adults.

evaluation Assigning a specific value to an option or possible action. Usually based on learned associations with past experiences with similar options or actions.

event-related potentials (ERPs) Averaged EEG recordings measuring time-locked brain responses to repeated presentations of a stimulus or repeated execution of a motor task.

excitatory Postsynaptic potentials that increase the probability of firing a postsynaptic action potential (EPSPs).

excitatory/inhibitory (E/1) balance The quantitative relationship between the amount of glutamatergic, depolarizing input and GABAergic (or glycinergic in some cases) hyperpolarizing input to an individual target neuron and how this relationship influences the probability of firing, especially in the context of correlated activity that drives synaptic plasticity or stability.

excitatory postsynaptic potentials (EPSPs) Postsynaptic potentials that increase the probability of firing a postsynaptic action potential.

exocytosis A form of cell secretion resulting from the fusion of the membrane of a storage organelle, such as a synaptic vesicle, with the plasma membrane.

exogenous attention Also called *reflexive attention*. A form of attention in which processing resources are directed to specific aspects of the environment in response to a sudden stimulus change, such as a loud noise or sudden movement, that attracts attention automatically. Compare *endogenous attention*.

exotropia A type of ocular misalignment in which one or both eyes of an individual turn outward (toward the temporal bones, or the ears).

express saccades A reflexive type of saccade in response to the sudden appearance of a sensory stimulus; mediated by a direct pathway from the retina (and auditory and somatosensory centers in the brainstem) to the superior colliculus.

expressive aphasia See *Broca's aphasia*.

external cuneate nucleus A group of relay neurons just lateral to the cuneate nucleus in the caudal medulla; conveys proprioceptive signals originating in the upper body, excluding the face, to the ipsilateral cerebellum via the cuneocerebellar tract.

external granule cell layer A transient accumulation of neuronal stem cells on the surface of the developing cerebellum that generates large numbers of granule cell neurons, the primary interneuron of the cerebellum.

external plexiform layer A layer of neural tissue composed primarily of neuronal and glial processes as well as synapses located toward the outside surface of the nervous system. The external plexiform layer in the retina is the synaptic layer adjacent to photoreceptor cell bodies (the outer limit of the retina) where synapses are made between photoreceptors and bipolar cells. The external plexiform layer of the olfactory bulb, located underneath the outermost glomerular layer, is the site of synapses between lateral dendrites of mitral cells and granule cell processes.

external segment A lateral subdivision of the globus pallidus.

external tufted cells A distinct class of excitatory projection neuron in the olfactory bulb whose cell body is found in the outermost region of the external plexiform layer. The dendrites of external tufted cells contribute to glomeruli, and their axons extend from the olfactory bulb into the olfactory tract to basal forebrain target nuclei—primarily the accessory olfactory nucleus and the olfactory tubercle. External tufted cells, and their close relatives middle tufted cells (found in the mid-region of the external plexiform layer), do not project to the pyriform cortex.

exteroception The modality of touch and pressure.

extracellular matrix (ECM) cell adhesion molecules Fibrillar proteins usually secreted by epithelial cells that form a macromolecular meshwork to provide a substrate for binding and stability of epithelial cells (via the basal lamina, which is composed of extracellular matrix adhesion molecules) or signals for cell motility and force generation. Fibronectin, laminins, and collagens are the main subtypes of extracellular matrix (ECM) cell adhesion molecules.

extracellular recording Recording the electrical potentials in the extracellular space near active neurons. Compare *intracellular recording*.

extrafusal fibers Fibers of skeletal muscles that generate primary biomechanical force during muscle contraction; a term that distinguishes ordinary muscle fibers from the specialized intrafusal fibers associated with muscle spindles.

extraocular muscles Six extrinsic eye muscles (and one accessory extraocular muscle) that control movement of the eye and eyelid.

extrastriate visual areas All regions of cortex that process visual information other than striate cortex (also called primary visual cortex). Extrastriate visual cortex does not receive strong direct input from the lateral geniculate nucleus.

F

facial nerve (VII) Cranial nerve VII, a mixed sensorimotor nerve that conveys visceral afferents (taste from anterior two-thirds of tongue) and somatosensory afferents (mechanosensation from skin on or near pinna) to the brainstem, and branchial motor efferents from the brainstem to the muscles of mastication and visceral motor efferents to lacrimal and salivary glands. Middle of three cranial nerves that attaches to the brainstem at the junction of the pons and medulla.

fasciculation The process by which growth cones and growing axons use adhesion signaling via homophilic or heterophilic cell surface molecules to grown adjacent to one another for substantial distances, forming a bundle that will eventually be part of either a nerve in the periphery or a tract in the central nervous system. Dendrites can also be bundled using similar mechanisms.

fast fatigable (FF) motor units Large motor units comprising large, pale muscle fibers that generate large amounts of force; however, these fibers have sparse mitochondria and are therefore easily fatigued. FF motor units are especially important for brief exertions that require large forces, such as running or jumping.

fast fatigue-resistant (FR) motor units Motor units of intermediate size comprising muscle fibers that are not as fast as FF motor units, but generate about twice the force of S motor units and are resistant to fatigue.

fastigial nucleus Most medial of the deep cerebellar nuclei; source of output from the median spinocerebellum to brainstem upper motor neurons.

fibroblast growth factor (FGF) A peptide growth factor, originally defined by its mitogenic effects on fibroblasts; also acts as an inducer during early brain development.

fibronectin A large cell adhesion molecule that binds integrins.

filaments Cytoskeletal elements formed from polymers of actin as well as intermediate filament proteins that interact with other cytoskeletal elements to provide support for membrane extensions and help stabilize mature processes.

filopodia Slender protoplasmic projections, arising from the growth cone of an axon or a dendrite, that explore the local environment.

first pain A category of pain perception described as sharp.

fixations Brief maintenance of gaze on an object or location.

flocculus Lateral portion of the vestibulocerebellum that receives input from the vestibular nuclei in the brainstem and the vestibular nerve; coordinates the vestibulo-ocular reflex and movements that maintain posture and equilibrium.

floorplate A specialized region of columnar neuroepithelial cells at the midline of the ventral neural tube, just above the notochord, that becomes a source for secreted signals, particularly Sonic hedgehog (Shh) that establish the ventral-dorsal pattern of the neural tube.

folia The gyral formations of the cerebellar cortex.

foramina of Monro See *interventricular foramina*.

forebrain The anterior portion of the brain derived from the prosencephalon that includes the diencephalon and telencephalon.

fornix Axon tract, best seen from the medial surface of the divided brain, that interconnects the septal nuclei and hypothalamus with the hippocampus.

fourth ventricle The ventricular space derived from the lumen of the neural tube that lies between the pons and rostral medulla and the cerebellum.

fovea Area of the retina specialized for high acuity in the center of the macula; contains a high density of cones and few rods.

foveation Aligning the foveae with a visual target and maintaining fixation.

free nerve endings Afferent fibers that lack specialized receptor cells; they are especially important in the sensation of pain and tempertaure.

frontal See *coronal*.

frontal eye fields A region of the frontal lobe that lies in a rostral portion of the premotor cortex and that contains cells that respond to visual and motor stimuli.

frontal lobe The hemispheric lobe that lies anterior to the central sulcus and superior to the lateral fissure.

frontal-parietal control network A set of brain regions that tend to be jointly activated when people are engaged in an attentionally-demanding task; it is thought to modulate activity in sensory cortices in the service of task goals.

functional magnetic resonance imaging (fMRI) Magnetic resonance imaging that detects changes in blood flow and therefore identifies regions of the brain that are particularly active during a given task.

fusiform gyrus See *occipitotemporal gyrus*.

G

G-protein-coupled receptors A large family of neurotransmitter or hormone receptors, characterized by seven transmembrane domains; the binding of these receptors by agonists leads to the activation of intracellular G-proteins. See *metabotropic receptors*.

G-proteins Proteins that are activated by exchanging bound GDP for bound GTP (and thus also known as GTP-binding proteins).

gamma (γ) motor neurons Class of spinal motor neurons specifically concerned with the regulation of muscle spindle length; these neurons innervate the contractile elements of intrafusal muscle fibers in muscle spindles.

ganglia (sing. ganglion) Collection of hundreds to thousands of neurons found outside the brain and spinal cord along the course of peripheral nerves.

ganglionic eminences The bilaterally symmetric accumulations of forebrain neural precursors at the ventromedial aspect of the telencephalon that give rise to the nuclei of the basal ganglia (caudate, putamen, globus pallidus, subthalamic nucleus) as well as the majority of GABAergic interneurons that migrate into the cerebral cortex.

gap junctions Specialized intercellular contacts formed by channels that directly connect the cytoplasm of two cells.

gastrulation The cell movements (invagination and spreading) that transform the embryonic blastula into the gastrula.

gaze centers Collections of local circuit neurons in the reticular formation that organize the output of cranial nerves III, IV, and VI to control eye movements along the horizontal or vertical axis.

gene Hereditary unit located on the chromosomes; genetic information is carried by linear sequences of nucleotides in DNA that code for corresponding sequences of amino acids.

generator potential See *receptor potentials*.

genetic analysis The analysis of the relationship between single genes and the phenotypes to which each gene contributes.

genetic engineering A methodological means for inducing mutations in genes or otherwise editing or altering the structure and/or the function of targeted genes for experimental or therapeutic benefit. Also called *reverse genetics*.

genome-wide association studies (GWAS) A statistical correlation of likely associated genes drawn from analyses of large cohorts of individuals with the same phenotype or clinical diagnoses.

genomics The comprehensive analysis of nuclear DNA sequences within or between species or individuals. Genomic analyses include "whole exome" evaluation in which all nuclear regions that code RNA transcripts are assessed or compared, and "whole genome" evaluation in which the entire DNA sequence of an individual or organism is assessed or compared.

germ layers The three layers—ectoderm, mesoderm, and endoderm—of the developing embryo from which all adult tissues arise. Neural cells and structures arise from the ectoderm and from the mesodermally-generated notochord.

glaucoma Condition in which the eye's aqueous humor is not adequately drained, resulting in increased intraocular pressure, reduced blood supply to the eye, and eventual damage to the retina.

glial cells (glia) The support cells associated with neurons (astrocytes, oligodendrocytes, and microglial cells in the central nervous system; Schwann cells in peripheral nerves; and satellite cells in ganglia).

glial scar Local proliferation of glial precursors and extensive growth of processes from existing glia within or around the site of a brain injury.

globus pallidus Principal component of the pallidum. External division is involved with the indirect pathway from striatum to pallidum. Internal division provides major output from the basal ganglia to motor circuits in the thalamus and brainstem.

glomeruli Characteristic collections of neuropil in the olfactory bulb; formed by dendrites of mitral cells and terminals of olfactory receptor cells, as well as processes from local interneurons.

glossopharyngeal nerve (IX) Cranial nerve IX, a mixed sensorimotor nerve that conveys visceral afferents (taste from posterior one-third of tongue and sensation from oropharynx and middle ear) and somatosensory afferents (mechanosensation from skin on or near pinna) to the brainstem, and branchial motor efferents from the brainstem to muscles of the soft palate and pharynx and visceral motor efferents to a salivary (parotid) gland. Attaches to the rostral medulla in a cleft between the inferior olive and the inferior cerebellar peduncle, just superior to the roots of the vagus nerve.

glutamate–glutamine cycle A metabolic cycle of glutamate release and resynthesis involving both neuronal and glial cells.

glymphatic system Vascular, glial, and lymphatic system that allows for the passage of cerebrospinal fluid (CSF) from arterial perivascular space through the substance of the brain and back into venous perivascular space; with this flow of CSF, metabolic wastes and discarded proteins are rinsed from the parenchyma and pass out of the brain, especially during sleep when extracellular spaces in the brain expand.

Goldman equation Mathematical formula that permits membrane potential to be calculated for case where a membrane is permeable to multiple ions.

G_olf A G-protein found uniquely in olfactory receptor neurons.

Golgi cells Inhibitory interneurons in the granular cell layer of the cerebellar cortex that provide inhibitory feedback from parallel fibers to granule cells, regulating the temporal properties of the granule cell input to the Purkinje cells.

Golgi outpost A specialized membranous organelle found distant from the cell body that resembles the Golgi apparatus and acts as a microtubule organizing center in some neuronal processes, as well as facilitate the addition of membrane and membrane proteins for dendritic growth.

Golgi tendon organs Receptors at the interface of muscle and tendon that provide mechanosensory information to the central nervous system about muscle tension.

gracile nucleus Somatosensory nucleus in the caudal medulla containing second-order sensory neurons that relay mechanosensory information originating in peripheral receptors in the lower body to the contralateral thalamus via the medial lemniscus.

gracile tract Medial division of the dorsal column containing the central processes of first-order afferents and the postsynaptic dorsal column projection; conveys mechanosensory signals derived from the lower body (also conveys visceral pain signals from the lower abdominal viscera).

gradient A discontinuous distribution of secreted or cell surface bound molecules in which concentration of the molecule in question is highest at either a secretory source or a distinct region of cells within a tissue for cell surface molecules, and then diminishes. Gradients can be steep: where concentration of the molecule in question is quite high and over a relatively short distance declines substantially, or shallow, where concentration declines slowly over a longer distance.

granulations See *arachnoid villi*.

granule cells A class of olfactory bulb interneurons with cell bodies located in the granule cell layer, the innermost layer of the olfactory bulb. Granule cells do not have an axon; instead, their dendrites extend into the external plexiform layer and make inhibitory GABAergic dendro-dendritic synapses primarily onto the lateral dendrites of mitral cells.

gray matter General term that describes regions of the central nervous system rich in neuronal cell bodies and neuropil; includes the cerebral and cerebellar cortices, the nuclei of the brain, and the central portion of the spinal cord.

groove A furrow or invagination seen from the surface of the developing embryo that is usually the site of additional morphogenetic movement and tissue-tissue interactions.

group Ib afferents Axons of primary sensory neurons that innervate Golgi tendon organs.

growth cone The specialized end of a growing axon (or dendrite) that generates the motive force for elongation.

growth factors A broad class of genetically encoded peptide and enzymatically synthesized molecules that elicit cell proliferation, process growth and can also determine the survival or death of a differentiated cell.

GTP-binding protein (G-protein) Proteins that are activated by exchanging bound GDP for bound GTP.

gyri (sing. gyrus) Folds in the cerebral cortex. Also called *convolutions*.

H

habituation Reduced behavioral responsiveness to the repeated occurrence of a sensory stimulus.

hair cells The sensory cells in the inner ear that transduce mechanical displacement into neural impulses.

halorhodopsin A protein that, in response to light of the proper wavelength, opens a channel that is selectively permeable to chloride ions; when engineered into a mature neuron for optogenetics, it inhibits the neuron when exposed to light.

haptic touch The exploration and perception of somatosensory stimuli using active touching and proprioception.

Hebb's postulate The idea that when pre- and postsynaptic neurons fire action potentials at the same time, the synaptic association between those cells strengthens. Sometimes phrased as "cells that fire together wire together," the postulate provides one explanation for the formation of certain neural networks.

helicotrema The opening at the apex of the cochlea that joins the perilymph-filled cavities of the scala vestibuli and scala tympani.

hemiballismus A basal ganglia syndrome resulting from damage to the subthalamic nucleus and characterized by involuntary, ballistic movements of the limbs.

hemispatial (or contralateral) neglect syndrome) Neurological condition in which the patient does not acknowledge or attend to the left visual hemifield or the left half of the body. The syndrome typically results from lesions of the right parietal cortex.

heterotrimeric G-proteins A large group of proteins consisting of three subunits (α, β, and γ) that can be activated by exchanging bound GDP for GTP, resulting in the liberation of two signaling molecules—αGTP and the $\beta\gamma$ dimer.

hierarchical processing A type of neural processing in which higher brain regions in a network (or superordinate brain regions) can modulate or control the activity of prior (or subordinate) brain regions.

hindbrain See *rhombencephalon*.

hippocampus A cortical structure in the dorsomedial margin of the parahippocampal gyrus; in humans, concerned with short-term declarative memory, among many other functions.

histamine A biogenic amine neurotransmitter, derived from the amino acid histidine, that is involved in arousal, attention, and other central and peripheral functions.

histamine-containing neurons Neurons in the tuberomammillary nucleus (TMN) of the hypothalamus that contribute to the neuronal basis of wakefulness.

homeobox genes The family of genes that encode transcription factor proteins with a homeodomain DNA binding domain (the homeobox). The homeobox genes are essential for early anterior-posterior patterning in most animal embryos.

homeotic genes Genes that determine the developmental fate of an entire segment of an animal. Mutations in these genes drastically alter the characteristics of the body segment (as when wings grow from a fly body segment that should have produced legs).

homologous recombination An endogenous cellular mechanism for DNA replication and repair involving DNA polymerases and ligases; may be used in genetic engineering to replace ("recombine") a native sequence of nucleotides in a gene with an exogenous sequence.

homonymous hemianopsia A loss of vision in both left and right hemifields due to lesions of the optic tract.

homonymous quadrantanopsia A loss of vision in both left of right quadrants of the visual field due to lesions of the optic radiation.

homozygosity mapping Statistical assessment of specific traits or phenotypes, especially those thought to represent mutations. In homozygosity mapping, one first determines whether a presumed trait is inherited in a dominant (one parent/one offspring share the trait) or recessive pattern (neither parent have the trait seen in an offspring) for a single generation. Using a pedigree across as many generations as possible one can assess whether a trait is predictably inherited from parents to offspring in a dominant or recessive pattern, and use a variety of methods to assess that gene's identity based upon the pattern of inheritance.

horizontal cells Retinal neurons that mediate lateral interactions between photoreceptor terminals and the dendrites of bipolar cells.

horizontal sections Standard anatomical planes of section; when standing, horizontal sections are parallel to the ground. Also known as *axial sections*.

Hox genes A group of conserved genes characterized by a specific DNA sequence—the homeobox—and that specify body axis (particularly the anterior–posterior axis) and regional identity in the developing vertebrate embryo.

Hsc70 An ATP-hydrolyzing enzyme involved in dissociation of clathrin-coats on vesicles following endocytosis.

Huntington's disease An autosomal dominant genetic disorder in which a single gene mutation results in personality changes, progressive loss of control of voluntary movement, and eventually death. Primary target early in the disease is medium spiny neurons of the striatum that participate in the indirect pathway.

hydrocephalus Enlarged ventricles that can compress the brain and expand the entire cranium as a result of increased cerebrospinal fluid pressure (typically due to a mechanical outflow blockage).

hyperacusis A painful sensitivity to moderate or even low-intensity sounds.

hyperalgesia Increased perception of pain.

hyperopia Far sightedness. The condition where a weak refracting system or an eyeball that is too short causes light to be focused behind the retina, and these individuals cannot focus on near objects.

hyperpolarization The displacement of a cell's membrane potential toward a more negative value.

hypocretin Another name for orexin.

hypoglossal nerve (XII) Cranial nerve XII, a somatic motor nerve that conveys efferents from the brainstem (hypoglossal nucleus) to the extrinsic muscles of the tongue. Attaches to the rostral medulla in a cleft between the medullary pyramid and the inferior olive.

hypothalamus Heterogeneous collection of small nuclei in the base of the diencephalon that plays an important role in the coordination and expression of visceral motor activity, as well as neuroendocrine and somatomotor activities that promote homeostasis and allostasis. The diverse functions in which hypothalamic involvement is at least partially understood include: the control of blood flow, the regulation of energy metabolism, the regulation of reproductive activity, and the coordination of responses to threatening conditions.

I

immediate early gene A gene whose protein product is produced rapidly, within 30–60 minutes, after a triggering stimulus. These often serve as transcriptional activators for delayed response genes. *C-fos* is one of the best-known examples of an immediate early gene.

inactivation   The time-dependent closing of ion channels in response to a stimulus, typically membrane depolarization.

induced pluripotent stem (iPS) cells   iPS cells are stem cells capable of generating all tissues, including gametes, in a mature organism that have been produced via the reprogramming of the transcriptional capacity of a differentiated somatic cell (therefore a cell in a mature tissue like the skin). The reprogramming relies upon the introduction of a set of transcription factors that are thought to reset the capacity of the nuclear genome and thus the developmental potential of the iPS cell.

inferior cerebellar peduncles   See *cerebellar peduncles.*

inferior colliculi (sing. colliculus)   Paired hillocks on the dorsal surface of the midbrain; concerned with auditory processing.

inferior division of the retina   Referring to the region of the visual field of each eye that corresponds to the bottom half of the retina.

inferior frontal gyri   Inferior of three, parallel longitudinal gyri anterior to the precentral sulcus; part of the prefrontal cortex and anterior premotor cortex, including Broca's area in the left hemisphere.

inferior oblique muscles   Extraocular muscles that extort the eyeballs when in the primary position (eyes straight ahead) and rotate upward when in adduction.

inferior olivary nucleus   Prominent nucleus in the ventral medulla; source of climbing fiber input to the contralateral cerebellum; induces complex spikes and long-term depression in cerebellar Purkinje neurons. Also called *inferior olive.*

inferior olive   See *inferior olivary nucleus.*

inferior parietal lobules   Gyral formations of the lateral (inferior) parietal lobes that are involved in associating somatosensory, visual, auditory, and vestibular signals and generating a neural construct of the body, the position of its parts, and its movements (body image or schema).

inferior rectus muscles   Extraocular muscles that rotate the eyeballs downward.

inferior salivatory nuclei   Visceral motor nucleus of the caudal pons containing parasympathetic preganglionic neurons that mediate salivation.

inferior temporal gyri   Inferior of three, parallel longitudinal gyri that define the inferior-lateral margin of the temporal lobe; involved in associational functions pertaining to stimulus recognition.

infundibulum   The connection between the hypothalamus and the pituitary gland; also known as the *pituitary stalk.*

inhibitory   Postsynaptic potentials that decrease the probability that a postsynaptic cell will generate an action potential (IPSPs).

inhibitory postsynaptic potentials (IPSPs)   Postsynaptic potentials that decrease the probability that a postsynaptic cell will generate an action potential (IPSPs).

innate behaviors   Behaviors that can be executed by an individual with minimal or no learning. These behaviors include several sexual behaviors in rodents, as well as aggression, seeking of food and water, ingestion, chewing and swallowing. Other examples include regulation of blood pressure, body temperature, and water balance for homeostasis.

inner nuclear layer   The neuronal layer in the retina that contains the cell bodies of the retinal interneurons (i.e., horizontal cells, bipolar cells, and amacrine cells).

inner plexiform layer   The synaptic layer in the retina that separates retinal interneurons (i.e., horizontal cells, bipolar cells, and amacrine cells) from retinal ganglion cells.

inositol trisphosphate (IP3) receptor   A ligand-gated ion channel in the endoplasmic reticulum membrane. This receptor binds to the second messenger, IP$_3$, and elevates cytoplasmic calcium concentration by mediating flux of calcium out of the lumen of the endoplasmic reticulum.

insomnia   A clinical disorder characterized by difficulty in falling asleep and/or remaining asleep.

insular cortex (insula)   The portion of the cerebral cortex that is buried within the depths of the lateral fissure by the growth inferior frontal and parietal lobes and the superior temporal lobe. The posterior portion of the insula is concerned largely with visceral and autonomic function, including taste, while more rostral portions are involved in implicit feelings and their impact on social cognition. Also called the insula.

integrins   A family of receptor molecules found on growth cones that bind to cell adhesion molecules such as laminin and fibronection.

intention tremors   Tremor that occurs while performing a voluntary motor act; characteristic of cerebellar pathology. Also called *action tremor.*

interaural level differences   The loudness difference of a sound in one ear vs. the other, based on the acoustic shadow cast by the head.

interaural timing differences   The time delay of a sound to arrive at one ear versus the other, based on its angle of incidence with respect to the head.

interhemispheric connections   The corpus callosum and anterior commissure, together. Mediate corticocortical connections between cortical regions in the opposite hemispheres.

intermediolateral cell column   Rod-shaped distribution of sympathetic preganglionic neurons in the lateral, intermediate gray matter of the spinal cord; in thoracic segments, accounts for a lateral protrusion of gray matter into the white matter known as the lateral horn.

internal arcuate fibers   Axons of dorsal column nuclei in the caudal brainstem that sweep across the midline and turn in the rostral direction forming the medial lemniscus.

internal capsule   Large, fan-shaped white matter tract that lies between the diencephalon and the basal ganglia, formed by the growth of axons supplying the cerebral cortex (mainly from the thalamus) and axons originating in the cerebral cortex and terminating in subcortical targets; features an anterior limb, a genu, and a posterior limb.

internal carotid arteries   Large arteries, one on each side of the head, that carry blood to the head. They divide into an external branch (supplying the neck and face), and an internal branch (supplying the brain and eye).

internal segment   A medial subdivision of the globus pallidus.

interneurons   Technically, a neuron in the pathway between primary sensory and primary effector neurons; more generally, a neuron whose relatively short axons branch locally to innervate other neurons. Also known as *local circuit neuron.*

interoception The sense of the internal state of the organism.

interposed nuclei Intermediate deep cerebellar nuclei; source of output from the paramedian spinocerebellum to the motor cortex via the thalamus and to brainstem upper motor neurons.

intersexuality Having a biologically ambiguous or intermediate sex based either upon indeterminate chromosomal sex (XXY males), primary sex characteristics (differentiation of gonads, genitalia), or secondary sex characteristics.

interstitial nuclei of the hypothalamus (INAH) Four cell groups located slightly lateral to the third ventricle in the anterior hypothalamus of primates; thought to play a role in sexual behavior.

interventricular foramina Narrow channels that allow for the passage of cerebrospinal fluid from the paired lateral ventricles into the single third ventricle; form in each hemisphere between the fornix and medial aspect of the anterior thalamus. Also known as the *foramina of Monro*.

intracellular receptors Receptors that participate in signal transduction by binding to cell-permeant chemical signals. The activate form of these receptors typically interact with nuclear DNA and produce new mRNA and protein within target cells.

intracellular recording Recording the potential between the inside and outside of a neuron with a microelectrode. Compare *extracellular recording*.

intracellular signal transduction A process that converts binding of ligands to plasma membrane receptors to intracellular signaling processes.

intrafusal muscle fibers Specialized muscle fibers found in muscle spindles.

intraparietal sulcus Prominent longitudinal sulcus of the posterior parietal lobe that divides the superior and inferior parietal lobules.

intrinsically photosensitive ganglion cells A small subset of retinal ganglion cells that generate the photopigment melanopsin. These cells are capable of phototransduction and depolarize in response to light.

ion channels Integral membrane proteins possessing pores that allow only certain ions to diffuse across cell membranes, thereby conferring selective ionic permeability.

ion exchangers Membrane transporters that exchange intracellular and extracellular ions against their concentration gradient by using the electrochemical gradient of other ions as an energy source. See *antiporters* and *co-transporters*.

ion selectivity The ability of channels to discriminate between different ions.

ionotropic glutamate receptors Ligand-gated ion channels that are activated by the neurotransmitter glutamate.

ionotropic receptors (IRs) Receptors in which the ligand binding site is an integral part of the receptor molecule.

J

joint receptors Mechanoreceptors found in and around joints; especially important for monitoring finger movements during fine manual manipulations.

jugular veins Principal means for draining venous blood from the cranium; arise from the sigmoid sinuses as they pass through the jugular foramina in the base of the skull.

K

kainate receptors See *ionotropic glutamate receptors*.

kairomones Volatile chemicals from other species indicating predator, prey, or symbiotic status that bind specifically to vomeronasal receptor proteins localized on subsets of vomeronasal sensory receptor cells.

KAL1 See *anosmin*.

KCN genes K$^+$ channel genes.

KIF21A A member of the Kinesin family of molecular motors that interact with the microtubule cytoskeleton as well as with macromolecules and organelles referred to as cargo so that the cargo is transported from the cell body to the periphery of the cell. KIF21A and related kinesins are essential for moving proteins and organelles from the cell body to the axon as well as transporting cargos back to the cell body, often as endosomes for signaling to the cell body, or as membrane bound compartments for protein turnover and degradation.

kinesins A family of multimeric motor proteins whose structure include a dual head domain that interacts with the microtubule cytoskeleton to facilitate trafficking, and tail domains that bind specific cargo. Kinesins are primarily engaged for anterograde transport of cargos: from the cell body to peripheral sites in the cell, like the axon, growth cone, or presynaptic terminal.

knock-in A type of transgenic mouse in which a specific gene has been selectively added to the genome, either to change gene dosage (thus a transgenic duplication copy number variant) or to substitute a mutant gene for the endogenous, wild type gene. Gene knock-ins also rely (usually) upon homologous recombination approaches.

knock-out A type of transgenic mouse in which a specific gene or multiple genes have been targeted for inactivation or excision from the genome. The most common approach to knocking out a gene or creating a deletion copy number variant relies upon homologous recombination (some more recent approaches use CRISPR gene editing as well). An exogenous non-coding DNA sequence can be substituted for the coding sequence found in the genome based upon the homology of the flanking regions of the sequence and the likelihood for the exogenous sequence to be recombined (encorporated) in to the host DNA in place of the endogenous sequence.

koniocellular pathway A third poorly understood pathway from retina to cortex characterized by the anatomical location of it cells in the lateral geniculate nucleus that process of short wavelength light.

L

L1 A member of the Ca^{2+} independent family of transmembrane cell surface adhesion molecules. L1 binds homophilically (thus to other L1 molecules on adjacent cells) It signals primarily through non-receptor tyrosine kinases Fyn and Src. It is particularly essential for the fasciculation (binding together) of axons within growing nerves.

labeled line coding   A theory that postulates that taste is mediated by specialized classes of neurons, each dedicated to a specific taste quality.

labyrinth   A set of interconnected chambers in the internal ear, comprising the cochlea, vestibular apparatus, and the bony canals in which these structures are housed.

lamellipodium   A sheetlike extension, rich in actin filaments, on the leading edge of a motile cell or growth cone.

lamina propria   A lamina propria includes loose connective tissue as well as small blood vessels that underlie a mucosal epithelium like that of the gut, the lungs, or the lining of the nasal cavity (the olfactory epithelium). The apposition of small blood vessels to the mucosa via the connective tissue scaffold provided by the lamina propria allows for the passage of molecules between the mucosa and circulatory system.

laminae (sing. **lamina**)   Cell layers that characterize the neocortex, hippocampus, and cerebellar cortex. The gray matter of the spinal cord is also arranged in laminae.

laminins   Large cell adhesion molecules that bind integrins. Laminin is a major component of the extracellular matrix.

lateral columns   The lateral regions of spinal cord white matter that convey motor information from the brain to the ventral horn via the lateral corticospinal tract and convey proprioceptive signals from spinal cord neurons to the cerebellum via the spinocerebellar tracts.

lateral corticospinal tract   Spinal portion of the corticospinal tract in the lateral column of the spinal cord derived from the contralateral motor cortex; governs skilled movements of the extremities.

lateral fissure   The cleft on the lateral surface of the brain that separates the temporal lobe below from the frontal and parietal lobes above. Also called the *Sylvian fissure*.

lateral geniculate nucleus   The portion of the thalamus that sends (and receives) axons to the cerebral cortex via the internal capsule.

lateral horns   Lateral protrusion of intermediate gray matter into the adjacent white matter known; characteristic feature of the thoracic segments of the spinal cord. See *intermediolateral cell column*.

lateral olfactory tract   The bundle of mitral and tufted cell axons that relay olfactory information to the accessory olfactory nuclei, the olfactory tubercle, the pyriform and entorhinal cortices, and portions of the amygdala.

lateral rectus muscles   Extraocular muscles that rotate the eyeballs laterally.

lateral ventricles   Largest of the ventricles derived from the lumen of the neural tube that expanded in the formation of the paired telencephalic vesicles; components of the lateral ventricles are present in each lobe of the cerebral hemisphere: anterior (frontal) horn, body, atrium, posterior (occipital) horn, and temporal horn.

lateral vestibulospinal tract   Ipsilateral projection of the anterior white matter of the spinal cord from the lateral vestibular nuclei to the medial ventral horn; mediates reflexes that activate extensor (anti-gravity) muscles with rapid lateral roll of the head, as when jostled to one side when standing on a moving train.

lens   A biconvex, transparent structure in the anterior segment of the eye that refracts light onto the retina.

lenticulostriate arteries   Numerous small branches of the middle and anterior cerebral arteries that penetrate deep into the anterior cerebral hemispheres supplying blood to deep white matter and cerebral nuclei, including the basal ganglia.

lesion studies   The method of observing and documenting change in function following damage (lesion) of a distinct brain region, nerve, or tract; damage may be acquired in humans or induced experimentally in non-human models; predominant method of studying the human nervous system prior to the advent of modern neurophysiological and brain imaging tools.

ligand-gated ion channels   Ion channels that respond to chemical signals rather than to the changes in membrane potential generated by ionic gradients. The term covers a large group of neurotransmitter receptors that combine receptor and ion channel functions into a single molecule.

limbic forebrain   Constellation of cortical and subcortical structures in the frontal and temporal lobes that form a medial rim of cerebrum roughly encircling the corpus callosum and diencephalon (limbic means "border" or "rim"). Comprises an olfactory division that processes olfactory cues; a parahippocampal division that generates cognitive maps in spatial frameworks that facilitate the acquisition of episodic and declarative memory; and an amygdaloid/orbital cortical division that is important in the experience and expression of emotion.

limbic system   Term that refers to those cortical and subcortical structures concerned with the emotions; the most prominent components are the cingulate gyrus, the hippocampus, and the amygdala.

lingual gyrus   Gyral structure on the inferior aspect of the occipital lobe forming the lower bank of the calcarine sulcus; portion of the primary visual cortex that represents the superior quadrant of the contralateral visual hemifields.

lobes   The four major divisions of the cerebral hemispheres named for the overlying cranial bones (frontal, parietal, occipital, and temporal).

local circuit neurons   General term referring to a neuron whose activity mediates interactions among other neurons in the CNS; exemplified by short-axon neurons in the spinal cord that mediate transmission of signals from sensory neurons to motor neurons. Interneuron is often used as a synonym.

long circumferential arteries   Long branches of the vertebral and basilar arteries that supply dorsolateral aspects of the brainstem and cerebellum.

long-term depression (LTD)   A form of long-term synaptic plasticity that produces a persistent, activity-dependent weakening of synaptic transmission.

long-term memory   Memories that last days, weeks, months, years, or a lifetime.

long-term potentiation (LTP)   A form of long-term synaptic plasticity that produces a persistent, activity-dependent strengthening of synaptic transmission.

longitudinal   Standard anatomical planes of section that pass through the CNS parallel to its long axis.

longitudinal fissure   prominent long, deep, sagittally oriented cleft that separates the two cerebral hemispheres on the dorsal midline; also called the *sagittal fissure*.

longitudinal lanceolate endings   Specialized endings of mechanosensory afferent neurons in hairy skin that form collars or rings around hair follicles.

loudness   The sensory quality elicited by the intensity of sound stimuli.

lower motor neurons   Neurons that send their axons out of the brainstem and spinal cord to innervate the skeletal muscles of the head and body; includes alpha and gamma motor neurons.

lumbar   Caudal region of the spinal cord between the thoracic and sacral regions related to the lower extremities.

lumbosacral enlargement   The spinal cord expansion that relates to the lower extremities; includes spinal segments L1–S2.

M

macroscopic currents   Ionic currents flowing through large numbers of ion channels distributed over a substantial area of membrane.

macula   The sensory epithelium of the otolith organs, comprising hair cells and associated supporting cells.

macula lutea   The central region of the retina that contains the fovea (the term derives from the yellowish appearance of this region in ophthalmoscopic examination); also, the sensory epithelia of the otolith organs.

macular sparing   The loss of vision throughout wide areas of the visual field, with the exception of foveal vision.

magnetic resonance imaging (MRI)   A noninvasive technique that uses magnetic energy and radiofrequency pulses to generate images that reveal structural and/or functional information in the living brain.

magnetic source imaging (MSI)   A non-invasive means for localizing brain activity that combines magnetoencephalography with structural magnetic resonance imaging.

magnetoencephalography (MEG)   A passive and noninvasive functional brain-imaging technique that measures the tiny magnetic fields produced by active neurons, in order to identify regions of the brain that are particularly active during a given task.

magnocellular layers   A component of the primary visual pathway specialized for the perception of motion; so named because of the relatively large ("magno") cells involved.

major histocompatibility complex (MHC)   A macromolecular assembly of immune proteins that mediate recognition of antigens derived from an individual organism itself ("self") and those derived from external sources ("other"). The MHC provides a unique molecular signature for an individual, and is thought to act as a stimulus in for the vomeronasal system.

mammillary bodies   Small prominences on the ventral surface of the posterior diencephalon; anatomically and functionally part of the caudal hypothalamus that is related to the hippocampus and its role in memory formation.

McGurk effect   The misperception of speech sounds caused by conflicting visual stimuli.

mechanoreceptors   Receptors specialized to sense mechanical forces.

mechanosensitive   Ion channels that respond to mechanical distortion of the plasma membrane.

medial lemniscus   Axon tract in the brainstem that carries mechanosensory information from the dorsal column nuclei to the ipsilateral thalamus.

medial longitudinal fasciculus   Axon tract that carries excitatory projections from the abducens nucleus to the contralateral oculomotor nucleus; important in coordinating conjugate eye movements.

medial rectus muscles   Extraocular muscles that rotate the eyeballs medially.

medial vestibulospinal tract   Bilateral projection of the anterior-medial white matter of the spinal cord from the medial vestibular nuclei to the cervical cord; mediates reflexes that extend the arms and dorsiflex the neck with rapid downward pitch of the head, as when falling forward.

medium spiny neurons   The principal projection neurons of the striatum.

medullary arteries   Segmental branches of the descending aorta that supply blood to the vertebral column and the spinal cord.

medullary pyramids   Longitudinal bulges on the ventral aspect of the medial medulla formed by the corticospinal tract and a small remnant of the corticobulbar tract.

Meissner corpuscles   Encapsulated cutaneous mechanosensory receptors in the tips of dermal papillae specialized for the detection of fine touch and pressure.

Meissner's plexus   See *submucous plexus*.

melanopsin   A photopigment located in the retinal ganglion cells that help to set the biological clock.

melatonin   Sleep-promoting neurohormone produced in the pineal gland.

membrane conductance   The reciprocal of membrane resistance. Changes in membrane conductance result from, and are used to describe, the opening or closing of ion channels.

Mendelian analysis   Assessment of the inheritance of traits—single observable characteristics of any organism—and the presumed single genes that control these traits. Mendelian analysis is based upon the probability of transmission of a single trait observed in offspring from parents with or without that trait. This approach is named for the pioneering geneticist Gregor Mendel, who identified the association between inherited traits and the units of inheritance he inferred to be single genes based upon the mathematics of probability of passing of traits from one generation to the next.

meninges   The external covering of the brain and spinal cord; includes the pia, arachnoid, and dura mater.

Merkel cells   Specialized cells in the basal epidermis that contact Merkel afferents forming Merkel cell-neurite complexes; Merkel cells signal the static aspect of a touch stimulus, such as light pressure, and release peptide neurotransmitters onto the terminals of Merkel afferents.

mesencephalic locomotor region   Collection of neurons in the reticular formation of the midbrain tegmentum that can trigger locomotion and change the speed and pattern of the movement by changing the level of activity delivered to the spinal cord through reticulospinal projections originating in the pons and medulla.

mesencephalic trigeminal nucleus   Array of pseudounipolar proprioceptive neurons in the rostral pons and midbrain (mesencephalon) on the ventral-lateral margin of the periaqueductal gray; mediates sensory limb of myotatic reflexes for jaw muscles and other striated muscles of the anterior cranium.

mesencephalon   See *midbrain*.

mesoderm   The middle of the three embryonic germ layers; gives rise to muscle, connective tissue, skeleton, and other structures.

meta-analysis   A method for combining information across multiple independent studies.

metabotropic receptors   Receptors that are indirectly activated by the action of neurotransmitters or other extracellular signals, typically through the aegis of G-protein activation. Also called *G-protein-coupled receptors*.

metencephalon   The part of the embryonic hindbrain (the entire rhombencephalon at earliest stages) that generates the pons and the cerebellum, and the trigeminal (V), abducens (VI), facial (VII) and vestibulocochlear (VIII) cranial nerves and surrounds the fourth ventricle in the mature brainstem.

Meyer's loop   That part of the optic radiation that runs in the caudal portion of the temporal lobe.

microglial cells   One of the three major classes of glial cells found in the central nervous system derived primarily from hematopoietic precursor cells; function as scavenger cells that remove cellular debris from sites of injury or normal cell turnover, and secrete signaling molecules that modulate local inflammatory responses.

microscopic currents   Ionic currents flowing through single ion channels.

microtubule cytoskeleton   The polymers of a variety of tubulin proteins that form a framework of parallel tubes that provide stability and resilience to long cellular processes, especially dendrites and axons. Polymerization of tubulins occurs as an axon or dendrite extends from the cell body of a differentiating neuron. Once the process is stable, this arrangement of parallel microtubules maintains the volume of the process and also acts as a set of tracks upon which molecules or organelles are transported from the cell body to distal domains of the processes using the molecular "motor" proteins kinesin or dynein.

microtubules   Polymerized macromolecules that form extended tubular arrays to define the cytoskeleton, especially in axons in the nervous system. Microtubules are assembled from tubulin protein subunits, and there is a substantial number or distinct tubulins, encoded by different genes.

midbrain   The most rostral portion of the brainstem; identified by the superior and inferior colliculi on its dorsal surface, and the cerebral peduncles on its ventral aspect. Also known as the *mesencephalon*.

middle cerebellar peduncles   See *cerebellar peduncles*.

middle cerebral arteries   Major vessels derived from the internal carotid arteries that supply the lateral aspects of the frontal, parietal and temporal lobes, including associated deep structures.

middle frontal gyri   Middle of three, parallel longitudinal gyri anterior to the precentral sulcus; part of the prefrontal cortex and anterior premotor cortex.

middle temporal area   Region of the extrastriate cortex in which neurons respond mainly to movement without regard to color.

middle temporal gyri   Middle of three, parallel longitudinal gyri that define the lateral aspect of the temporal lobe; part of the lateral temporal network that encodes language content and, at its posterior margin, contains visual areas involved in motion discrimination.

midsagittal   Standard anatomical plane of section; the vertical plane passing from anterior to posterior through the midline dividing the body (and brain) into right and left sections.

miniature end plate potentials (MEPPs)   Small, spontaneous depolarization of the membrane potential of skeletal muscle cells, caused by the release of a single quantum of acetylcholine.

miosis   Excessive constriction of the pupil.

mirror motor neurons   Neurons in the posterior frontal and inferior parietal lobes that respond during the execution of goal-oriented action and the observation of the same actions, even when such actions are not executed.

mitral cells   The major output neurons of the olfactory bulb.

mixture suppression   Mixing two tastants leads to a reduction in the perceived intensity of each component.

modulatory   Many secreted signals that act on neurons or other neuron-related targets (glial cells, muscle cells, glandular cells) do not cause changes in the excitability of the target cells that result in the initiation of an all or none electrical signal that can be transmitted to additional cells (an action potential). Instead, these modulators change the membrane properties and excitability of the target cell making it more or less likely that the target cell can reach the voltages necessary to elicit an action potential.

monomeric G-proteins   GTP-binding proteins, also called small G-proteins, that relay signals from activate cell surface receptors to intracellular targets. In contrast to heterotrimeric G-proteins, monomeric G-proteins consist of a single protein. Also called *small G-proteins*.

mossy fibers   Afferent axons to the cerebellum from all sources except for the inferior olivary nuclei; the vast majority enter the cerebellum via the inferior and middle cerebellar peduncles.

motor aphasia   See *Broca's aphasia*.

motor cortex   Region of the cerebral cortex in the posterior frontal lobe that gives rise to corticobulbar and corticospinal projections and is concerned with motor behavior. Includes the primary motor cortex in the anterior bank of the central sulcus that is essential for the voluntary control of movement, and the premotor cortex (anterior to the primary motor cortex) that is involved in planning and programming voluntary movements.

motor neurons   By common usage, nerve cells that innervate and send efferent signals to skeletal muscle.

motor neuron pool   All the lower motor neurons innervating a single muscle.

motor proteins   Proteins that bind microtubules as well as several additional proteins to facilitate movement of a variety of cellular cargos including proteins, vesicles, or organelles.

motor systems A broad term used to describe all the central and peripheral structures that support motor behavior.

motor unit A motor neuron and the skeletal muscle fibers it innervates.

muscarinic ACh receptors (mAChRs) Metabootropic ACh receptors that can be pharmacologically identified by their selective activation by muscarine.

muscle spindles Highly specialized mechanosensory organs found in most skeletal muscles; provide proprioceptive information about muscle length.

muscle tone The normal, ongoing tension in a muscle; measured by resistance of a muscle to passive stretching.

mutations If a trait is transmitted without change, it is assumed that both copies of the gene encoding that trait are "wild type": the most common and optimally adaptive variant, shared by both parents, and unchanged in the offspring. If a trait in offspring diverges from the most common one, either with increased, decreased or no observable change in adaptive advantage or heritability, the gene for that trait is a variant of the wild type gene that encodes the related trait, and considered changed or mutated from the wild type gene.

mydriasis Excessive dilation or widening of the pupil.

myelencephalon The part of the hindbrain that gives rise to the medulla and the glossopharyngeal (IX), vagal (X), spinal accessory (XI), and hypoglossal (XII) cranial nerves (motor neurons and neural crest that will form related cranial ganglia).

myelin The multilaminated wrapping around many axons formed by oligodendrocytes or Schwann cells.

myelination Process by which glial cells (oligodendrocytes or Schwann cells) wrap around axons to form multiple layers of glial cell membrane, thus insulating the axonal membrane and increasing conduction velocity.

myenteric plexus Network of neurons in the enteric division of the visceral motor system concerned with regulating the musculature of the gut. Also called *Auerbach's plexus*.

myopia Near sightedness. The condition where too much refractive power of the eye or an eyeball that is too long causes light to be focused in front of the retina, and such individuals cannot focus on distant objects.

myosin A family of proteins that form fibrillary multimers, or more limited complexes to interact with actin and generate motility in some cells.

myotatic reflex A fundamental spinal reflex that is generated by the motor response to afferent sensory information arising from muscle spindles; also called a "stretch" or "deep tendon" reflex. The knee jerk reaction is a common example.

N

Na⁺/Ca2⁺ exchanger An active transport protein that removes calcium from the cytoplasm of cells by exchanging intracellular calcium ions for extracellular sodium ions.

nasal division of the retina Referring to the region of the visual field of each eye in the direction of the nose. See *binocular field*.

nasal mucosa The nasal mucosa is the general term for the entire epithelial lining of the nasal cavities. It includes both the non-neural respiratory epithelium and the neural olfactory sensory epithelium and their constituent cells. It is named based upon the layer of mucous that coats the entire outer surface of the respiratory and olfactory epithelia in the nose.

near reflex triad Reflexive response induced by changing binocular fixation to a closer target; comprises convergence, accommodation, and pupillary constriction.

near response A triad of events that occurs when viewing close objects with high acuity. The near response includes convergent eye movements, accommodation of the lens, and pupil constriction.

neocortex The six-layered cortex that forms the surface of most of the cerebral hemispheres (all cerebral cortex lateral and dorsal to the rhinal sulcus).

neophobia A general dislike of unfamiliar and novel foods.

Nernst equation A mathematical formula that predicts the electrical potential generated ionically across a membrane at electrochemical equilibrium.

nerve growth factor (NGF) A neurotrophic protein factor required for survival and differentiation of sympathetic ganglion cells and certain sensory neurons. Preeminent member of the neurotrophin family of growth factors.

nerves A collection of peripheral axons that are bundled together and travel a common route.

netrins A family of diffusible molecules that act as attractive or repulsive cues to guide growing axons.

neural circuits A collection of interconnected neurons mediating a specific function.

neural crest cells Cells that migrate to become a variety of cells types and structures, including peripheral sensory neurons, enteric neurons, and glial cells.

neural crest A transient region where the edges of the folded neural plate come together, at the dorsalmost limit of the neural tube. Gives rise to **neural crest cells** that migrate to become a variety of cells types and structures, including peripheral sensory neurons, enteric neurons, and glial cells as well as facial bones, teeth, and melanocytes in the skin.

neural induction The mechanism by which ectodermal cells, in response to local signals available in the embryo, acquire neural stem cell identity.

neural systems Groups of neural circuits that are dedicated to transducing, encoding, relaying, and processing one particular type of information. The neurons and circuits that comprise a neural system are often distributed throughout the brain or periphery, interconnected by axon tracts in the brain or nerves in the periphery. An example of a neural system is the visual system in most animals: the collection of neurons and the circuits they constitute throughout the animal's brain and body that process information about the visible electromagnetic spectrum (light).

neural stem cells The neuroectodermal cells, established immediately after gastrulation via signals from the notochord, that have the capacity to give rise to all neuronal and glial cell types of the CNS and PNS, plus the non-neural derivatives of the neural crest.

neural tube   The primordium of the brain and spinal cord; derived from the neural ectoderm.

neuregulin1 (Nrg1)   Neuregulin 1 is a member of the broader class of secreted neuregulin ligands. Via binding to the Erb family of neuregulin receptors, neuregulin1 and related neuregulins can mediate cell motility, local receptor clustering or a variety of growth responses.

neurexins   Adhesion molecules of the presynaptic membrane in developing synapses. Neurexins bind to neuroligin in the postsynaptic membrane, promoting adhesion, and help localize synaptic vesicles, docking proteins, and fusion molecules.

neuroblasts   Dividing cells, the progeny of which develop into neurons; immature nerve cells.

neurochristopathies   Neurocristopathies are morphogenetic disorders that reflect the disruption of specification, migration or differentiation of the neural crest. Craniofacial disorders as well as disorders of the developing enteric nervous system like Hirschprung's disease and some cardiac malformations that reflect anomalies of the neural crest that contributes to the outflow tract of the heart are all considered neurocristopathies.

neuroectoderm   The portion of the outermost embryonic germ layer (the ectoderm) that based upon its proximity to the mesodermally derived notochord differentiates into a field of multipotent neural stem cells that give rise to the entire nervous system (CNS and PNS).

neuroectodermal precursor cells   The cells in the neuroectoderm that can give rise via proliferation to a variety of neural and glial cell classes.

neuroglia   See *glial cells*.

neuroligins   Postsynaptic binding partners of the presynaptic adhesion molecule neurexin. Promote the clustering of receptors and channels of the postsynaptic density as the synapse matures.

neuromeres   The repeating units of the neural tube.

neuron doctrine   The fundamental concept that the brain and peripheral nervous system are composed of discrete cells that interact through specialized, non-continuous junctions. The neuron doctrine received its foundational support in the early 1900s via the histological observations of the pioneering neuroanatomist S. Ramon y Cajal and the pioneering neurophysiologist C. Sherrington. The neuron doctrine is an extension of the cell theory proposed by M. Schleiden and T. Schwann in the 1830s.

neurons   Also called *nerve cells*. Cells specialized for the generation, conduction, and transmission of electrical signals in the nervous system.

neuropathic pain   A chronic, intensely painful experience that is difficult to treat with conventional analgesic medications.

neuropeptides   A general term describing a large number of peptides that are synthesized by neurons and function as neurotransmitters or neurohormones.

neuropil   The dense tangle of axonal and dendritic branches, the synapses between them, and associated glia cell processes that lies between neuronal cell bodies in the gray matter of the brain and spinal cord.

neurotransmitter receptors   The transmembrane proteins whose extracellular regions have distinct domains for the binding of signaling molecules secreted by neurons or other cells (subsets of glial, glandular and immune cells). The binding of these secreted signaling molecules, referred to as neurotransmitters, results in a change in protein confirmation or activity. In some cases, neurotransmitter receptors are also ion channels, and the binding of the secreted signal results in direct activation of the channel and a change in voltage across the membrane of the cell where the receptors are found. In other cases, binding of the signal activates a catalytic domain that facilitates intracellular signaling that also leads to a change in excitability of the target cell.

neurotransmitters   Substances released by synaptic terminals for the purpose of transmitting information from one cell (the **presynaptic cell**) to another (the **postsynaptic cell**).

neurotrophic factors   Chemical substances, secreted by cells in a target tissue, that promote the growth and survival of neurons.

neurotrophin 4/5 (NT-4/5)   NT-4 and NT-5 are names used for the same member of the neurotrophin family of secreted growth and survival signaling molecules. NT-4/5 signals through the TrkB receptor as the LNGFR/P75.

neurotrophin-3 (NT-3)   A member of the neurotrophin family of secreted growth and survival signaling molecules. NT-3 binds with high affinity to both the TrkB and TrkC neurotrophin receptor-kinases, as well as with low affinity to the low affinity neurotrophin receptor (LNGFR, also known as P75). NT-3 has trophic effects on a variety of peripheral and central neurons.

neurotrophins   A family of trophic factor molecules that promote the growth and survival of several different classes of neurons.

neurulation   The process by which the neuroectoderm rounds into the neural tube. This process establishes the midline of the neural plate as the ventral midline of the CNS and the lateral edges of the neural plate as the dorsal, or alar regions that give rise to the neural crest at the margins, or the dorsal CNS.

nicotinic ACh receptor (nAChR)   Ionotropic ACh receptors that can be pharmacologically identified by their selective activation by nicotine.

NMDA receptors   See *ionotropic glutamate receptors*.

nociceptors   Cutaneous and subcutaneous receptors (especially free nerve endings) specialized for the detection of harmful (noxious) stimuli.

nodes of Ranvier   Periodic gaps in the myelination of axons where action potentials are generated.

nodulus   Medial portion of the vestibulocerebellum that receives input from the vestibular nuclei in the brainstem and the vestibular nerve; coordinates the vestibulo-ocular reflex and movements that maintain posture and equilibrium.

Noggin   An endogenous antagonist of the Bmp family of Tgfβ ligands that binds secreted Bmps and inactivates them, preventing the acquisition of epidermal fate for ectodermal cells that go on to become neural plate neural stem cells. Noggin is secreted by the notochord, in combination with positive signals like Shh that drive neuronal differentiation.

non-rapid eye movement (non-REM) sleep Collectively, those phases of sleep (stages I–IV) characterized by the absence of rapid eye movements.

nondeclarative memory Unconscious memories such as motor skills and associations. Also called *procedural memory*.

noradrenaline See *norepinephrine*.

noradrenergic neurons Neurons that contribute to the neuronal basis of wakefulness.

norepinephrine Catecholamine hormone and neurotransmitter that binds to α- and β-adrenergic receptors, both of which are G-protein-coupled receptors. Also known as *noradrenaline*.

Notch cell surface receptors These transmembrane proteins transduce signals upon binding a delta ligand on the surface of a neighboring cell. Delta binding to Notch activates a local cascade at the inner surface of the cell membrane that recruits a protease to cleave the intracellular domain of the Notch protein. This Notch intracellular domain (NICD) then complexes with other cytoplasmic proteins and is translocated to the nucleus to regulate gene expression.

notochord A transient, cylindrical structure of mesodermal cells underlying the neural plate (and later the neural tube) in vertebrate embryos. Source of important inductive signals for neural development.

NSF NEM-sensitive *fusion* protein. An enzyme responsible for dissociating complexes of SNARE proteins.

nucleus (pl. nuclei) Collection of nerve cells in the brain and spinal cord that are anatomically discrete, and which typically serve a particular function.

nucleus ambiguus Branchial motor nucleus of the rostral medulla containing somatic motor neurons that innervate striated muscles of the larynx and pharynx; also contains a visceral motor division with parasympathetic preganglionic neurons that mediate slowing of the heart rate.

nucleus of the solitary tract (NST) Nucleus of the caudal pons and rostral medulla that contains a rostral gustatory division and a caudal visceral sensory division; integrates inputs relayed from the rostral division and several primary and secondary visceral sensory afferents that are relevant to the autonomic control of the gut, the cardiovascular system, and other target organs; receives visceral and taste information via several cranial nerves and relays this information to the thalamus.

nystagmus Repetitive rotational movements of the eyes normally elicited by large-scale motion of the visual field (optokinetic nystagmus), with each cycle involving a slower phase driven by central circuits in the brainstem and higher brain centers and a faster, reflexive phase resetting the position of the eye in the orbit; in the absence of physiological visual or vestibular stimuli, nystagmus may indicate brainstem or cerebellar pathology.

O

occipital lobe The posterior lobe of the cerebral hemisphere; primarily devoted to vision.

occipitotemporal gyrus Longitudinal gyrus of the inferior temporal lobe between the parahippocampal gyrus and the inferior temporal gyrus; part of the ventral "what" visual processing stream concerned with object recognition. Posterior portion also known as the *fusiform gyrus*.

ocular dominance columns The segregated termination patterns of thalamic inputs representing the two eyes in the primary visual cortex of some mammalian species.

oculomotor nerve (III) Cranial nerve III, a mixed efferent nerve with somatic motor components that controls the superior rectus, the inferior rectus, the medial rectus and the inferior oblique eye muscles, as well as the levator palpebrae superioris—a muscle that retracts the upper eyelid; also contains a parasympathetic component that constricts the pupil.

odorant receptor proteins The G-protein-coupled, seven transmembrane domain proteins encoded by multiple related odorant receptor genes (as many as 2000) in the genomes of all animals. The extracellular domains of these odorant receptor proteins bind airborne molecules, called odorants, that are eventually detected as odors by the olfactory systems of all animals.

odorants Molecules capable of eliciting responses from receptors in the olfactory mucosa.

OFF bipolar cell A type of retinal bipolar cell that depolarizes in response to decreases of light within the receptive field.

OFF ganglion cell A type of retinal ganglion cell that intensifies its discharge rate (i.e., its activity) to decreases of light within the receptive field.

OFF-center ganglion cell A visual neuron whose receptive field center is inhibited by light.

olfactory bulbs Telencephalic structure that lies on the orbital surface of the frontal lobe and receives axons from cranial nerve I; contains local circuit neurons and project neurons that transmit olfactory signals to the olfactory cortex via the olfactory tracts.

olfactory cilia Actin based protrusions from the apical domain of an olfactory sensory receptor neuron. The olfactory cilia are the site of concentration of odorant receptor molecules and the cytoplasmic signaling intermediates necessary for odor transduction and the initiation of odor processing in the olfactory pathway.

olfactory ensheathing cells The glial cells that surround the unmyelinated axons of the peripheral portion of the olfactory nerve. These cells have several properties of Schwann cells that fulfill a similar function for peripheral somatosensory, motor and autonomic axons in peripheral nerves throughout the body.

olfactory epithelium Pseudostratified epithelium that contains olfactory receptor cells, supporting cells, and mucus-secreting glands.

olfactory nerve (I) The set of bundles of unmyelinated axons originating from the olfactory receptor neurons in the olfactory epithelium of the nose that project through the cribiform plate and terminate in the olfactory bulb.

olfactory receptor neurons (ORNs) Bipolar neurons in olfactory epithelium that contain receptors for odorants.

olfactory tracts The projection from the olfactory bulb to the various divisions of the olfactory cortex in the ventromedial forebrain. See *lateral olfactory tract*.

oligodendrocytes One of the three major classes of glial cells found in the central nervous system; their major function is to lay down myelin, which facilitates the efficient generation and rapid conduction of action potentials; also produce signaling molecules that modulate growth cone activity in regenerating axons.

olivary pretectal nucleus A region of the midbrain that is innervated by retinal ganglion cells and is responsible for mediating pupillary light reflexes.

ON bipolar cell A type of retinal bipolar cell that depolarizes in response to increases of light within the receptive field.

ON-center ganglion cell A visual neuron whose receptive field center is excited by light.

ON ganglion cell A type of retinal ganglion cell that increases the intensify of its discharge rate (i.e., its activity) to increases of light within the receptive field.

ON/OFF ganglion cells A type of retinal ganglion cell whose activity can respond transiently to both increases and decreases in light

Onuf's nucleus Sexually dimorphic nucleus in the human spinal cord that innervates striated perineal muscles mediating contraction of the bladder in males, and vaginal constriction in females.

operant conditioning A form of conditioning shaped by reward rather that pairing a reflex response with an arbitrary signal.

optic chiasm The junction of the two optic nerves on the ventral aspect of the diencephalon, where axons from the nasal divisions of each retina cross the midline.

optic flow The pattern of global visual motion across the retina as an observer moves through space.

optic nerve(II) The nerve (cranial nerve II) containing the axons of retinal ganglion cells; extends from the eye to the optic chiasm.

optic tract The axons of retinal ganglion cells after they have passed through the region of the optic chiasm en route to the lateral geniculate nucleus of the thalamus.

optic vesicles The evagination of the forebrain vesicles that generates the retina and induces lens formation in the overlying ectoderm.

optogenetics The use of genetic tools to induce neurons to become sensitive to light, such that experimenters can excite or inhibit a cell by exposing it to light.

optokinetic eye movements Movements of the eyes that compensate for head movements; the stimulus for optokinetic movements is large-scale motion of the visual field.

optokinetic nystagmus Repeated reflexive responses of the eyes to ongoing large-scale movements of the visual scene.

orbital gyri Gyral formations on the ventral aspect of the frontal lobes that lie superior to the orbits in the anterior cranial fossae; part of the prefrontal cortex that is involved in implicit processing, including emotion, bodily feeling, and related aspects of cognition.

orbitofrontal cortex (OFC) The division of the prefrontal cortex that lies above the orbits in the most rostral and ventral extension of the sagittal fissure; important in emotional processing and decision making.

orexin A peptide secreted by the hypothalamus, which promotes waking. Also called *hypocretin*.

organizational effects The actions of gonadal steroids on the developing brains of females and males that define sexually dimorphic patterns numbers and patterns of neurons and circuits that will eventually mediate reproduction-related and potentially other sex-specific behaviors.

organoids An organoid is an aggregate of ES or iPS cell derivatives produced in cell culture conditions that sequentially support the differentiation of multiple cell types found in a mature tissue in vivo. Thus, there are brain organoids that approximate regions like the retina and cerebral cortex. These organoids have cell classes found in these brain regions in vivo, and the cells within the organoid apparently organize themselves into layers that resemble those in the retina or cortex.

orthologous genes Genes expressed in model organisms that are identical or similar to target genes (typically expressed in humans and associated with disease) based on sequence and chromosomal location.

orthostatic hypotension Fall in blood pressure upon standing up as a result of blood pooling in the lower extremities.

oscillations Rhythmic patterns of either sub-threshold or spike related electrical activity in the brain that continues over extended periods of time, and can influence the capacity for neurons to engage in plastic changes in synaptic connections.

oscillopsia Inability, as a result of vestibular damage, to fixate visual targets while the head is moving.

ossicles The bones of the middle ear.

otoacoustic emissions Sounds generated or modified by active mechanisms in the ear.

otoconia The calcium carbonate crystals that rest on the otolithic membrane overlying the hair cells of the sacculus and utricle.

otolith organs The two organs in the labyrynth of the inner ear—the utricle and saccule—that respond to linear accelerations of the head and static head position relative to the gravitational axis.

otolithic membrane The gelatinous and fibrous membrane on which the otoconia lie and in which the tips of the hair bundles are embedded.

outer nuclear layer The neuronal layer in the retina that contains the cell bodies of canonical photoreceptors (i.e., rods and cones).

outer plexiform layer The synaptic layer in the retina that separates the photoreceptors from the retinal interneurons (i.e., horizontal cells, bipolar cells, and amacrine cells).

oval window Site where the middle ear ossicles transfer vibrational energy to the cochlea.

overshoot phase   The peak, positive-going phase of an action potential, caused by high membrane permeability to a cation such as Na⁺ or Ca²⁺.

overt attention   The focusing of attention (typically visual) by voluntarily shifting gaze. Compare *covert attention*.

P

P2X receptors   A family of ionotropic purinergic neurotransmitter receptors.

p75   The alternate name for the low affinity neurotrophin receptor (LNGFR). P75/LNGFR interacts with the AKT kinase pathway, and is particularly important for neurotrophin signaling that influences neuronal survival or death.

Pacinian corpuscles   Encapsulated cutaneous mechanosensory receptors in the deep dermis (also found in other tissues) specialized for the detection of high-frequency vibrations.

pain matrix   A broad array of brain areas, including the somatosensory cortex, insular cortex, amygdala, and anterior cingulate cortex, whose activity is associated with the experience of pain.

paleocortex   Phylogenetically primitive cortex with few cell layers on the inferior and medial aspect of the temporal lobe within the parahippocampal gyrus and the junction of the temporal and frontal lobes.

pallidum   Division of the basal ganglia that receives input from the striatum and provides inhibitory (GABAergic) output to the thalamus and brainstem.

parabrachial nucleus   Nucleus of the rostral pons that relays visceral sensory information to the hypothalamus, amygdala, thalamus, and medial prefrontal and insular cortex.

paracentral lobule   Gryal formation on the medial aspect of the cerebral hemisphere formed by the fusion of the pre- and post-central gyri surrounding the medial termination of the central sulcus; comprises the somatic sensorimotor representation of the contralateral foot.

paracrine   Referring to the secretion of hormone-like agents whose effects are mediated locally rather than by the general circulation.

parahippocampal gyrus   Medial-most gyral structure in the inferior temporal lobe; part of the medial temporal lobe memory system that generates cognitive maps in spatial frameworks that facilitate the acquisition of episodic and declarative memory.

parallel fibers   The bifurcated axons of cerebellar granule cells that extend along the length of the folia in the molecular layer of the cerebellar cortex where they synapse on dendritic spines of Purkinje cells.

parallel pathways   Afferent pathways that carry distinct submodalities of sensory information centrally at the same time along anatomically distinct projections.

paramedian circumferential arteries   Shorter branches of the vertebral and basilar arteries that supply lateral aspects of the brainstem.

paramedian pontine reticular formation (PPRF)   Neurons in the reticular formation of the pons that coordinate the actions of motor neurons in the abducens and oculomotor nuclei to generate horizontal movements of the eyes; also called the horizontal gaze center.

parasagittal   Standard anatomical planes of section; any vertical plane passing from anterior to posterior that is parallel to the sagittal plane.

parasympathetic ganglia   Locus of primary parasympathetic motor neurons; unlike sympathetic ganglia, which are relatively close to the spinal column, parasympathetic ganglia are further removed and typically embedded in or very near the end organs they innervate.

paravertebral sympathetic chain ganglia   Chain of cervical and thoracic sympathetic ganglia located lateral to the spinal column.

parietal lobe   The hemispheric lobe that lies between the frontal lobe anteriorly, and the occipital lobe posteriorly.

parieto-occipital sulcus   Prominent sulcus on the medial surface of the cerebral hemisphere that divides the parietal and occipital lobes.

Parkinson's disease   Progressive neurodegenerative disease of the substantia nigra pars compacta that results in a characteristic tremor at rest and a general paucity of movement.

parvocellular layers   A component of the primary visual pathway specialized for the detection of detail and color; so named because of the relatively small cells involved.

passive electrical responses   Responses to applied electrical currents that do not require activation of voltage-gated ion channels.

patch clamp recording   An extraordinarily sensitive voltage clamp method that permits the measurement of ionic currents flowing through individual ion channels.

peptide neurotransmitters   See *neuropeptides*.

periglomerular cells   A class of olfactory bulb interneurons whose cell bodies surround individual glomeruli and whose dendrites extend into the glomerular neuropil where they make inhibitory GABAergic dendro-dendritic synapses on mitral and tufted cell dendrites.

perilymph   The potassium-poor fluid that bathes the basal end of the cochlear hair cells.

perineuronal nets   Extracellular matrix proteins that combine with cell surface anchors to form specialized nets that envelope the cell bodies and proximal dendrites of target neurons. Perineuronal nets influence the stability of some classes of synapses, including those made by subsets of inhibitory interneurons onto projection neurons in the cerebral cortex.

peripheral nervous system (PNS)   All nerves and neurons that lie outside the brain and spinal cord; the network of neurons and axon pathways distributed throughout an animal's body.

peripheral sensitization   Increased responsiveness of peripheral pain-sensing neurons following tissue damage that is one source of hyperalgesia.

persistent vegetative state   A state that results from profound damage to the brain, perhaps by injury or disease, that is characterized by a lack of awareness. A patient with persistant vegetative state typically can still react to stimuli and exhibit degrees of wakefulness and quiescence.

phase locking   Temporal synchrony of neural response to the phase of a periodic signal. **phenomenological consciousness**   The subjective experience of being a particular state, such as what it feels like to see a color or feel frustration.

phenotype   The specific characteristics that define a single trait, or sometimes a group of related traits. For example, in mammals, the colors of fur are considered phenotypes: the phenotype of fur can be defined by one color (black fur on a cat) or multiple colors (a calico cat with black, brown, and white fur intermixed).

phenotypic sex   The visible somatic features that define one sex from the other. Phenotypic sex can variably include body hair patterns, body size and musculature, voice, as well as the primary sex characteristics associated with males versus females.

pheromones   Species-specific odorants that play important roles in behavior in some animals, including many mammals.

photopic vision   Vision at high light levels, which is mediated almost entirely by cone cells. Contrast with *scotopic vision*.

photoreceptors   The specialized neurons in the eye—rods and cones—that are sensitive to light.

phototransduction   The process by which light is converted in electrical signals in the retina.

pia mater   The innermost of the three layers of the meninges; a delicate layer (pia mater means "tender or delicate mother") closely applied to the surface of the brain.

pineal gland   Midline neural structure lying on the dorsal surface of the midbrain; important in the control of circadian rhythms (and, incidentally, considered by Descartes to be the seat of the soul).

pinhole effect   An optical concept in which light passing through a small opening can act to focus light. This occurs by preventing out of focus light from passing through the opening.

pinna   A component of the external ear.

pitch   The sensory quality roughly corresponding to periodic vibrations of sound stimuli.

pituitary stalk   See *infundibulum*.

placebo effect   The physiological effects resulting from administration of an inert substance (a placebo).

plexin   A family of transmembrane receptors for the semaphorin adhesion signaling molecules. Plexins interact with GTPases that in turn modulate the cytoskeleton to alter cell-cell contact or motility.

plexus   A complex network of nerves, blood vessels, or lymphatic vessels.

pluripotent   Pluripotency is the capacity of a stem cell to generate an entire organism that can reproduce via generation of gametes.

pluripotent stem cells   Mitotically active progenitor cells that have the capacity to give rise to stem cells that generate all of the organs in a mature organism, plus the precursors of the gametes that allow the organism to reproduce another generation.

PMCA   *Plasma membrane calcium ATPase*, which is responsible for translocating Ca^{2+} from cytoplasm to the extracellular medium.

polarity   The molecular and cell biological distinctions that distinguish apical (top) versus basal (bottom) domains of cells. In simple epithelial cells, the apical surface receives signals while the basal surface transmits signals.

polarized epithelial cells   Cells arranged in sheets or layers, that have an apical (top) and basal (bottom) domain that are distinguished by molecular differences. The apical domain of an epithelial cell is specialized for interactions and transduction of signals from the environment. The basal domain is specialized for secretion.

polyneuronal innervation   A state in which neurons or muscle fibers receive synaptic inputs from multiple, rather than single, axons.

pons   One of the three major divisions of the brainstem, lying between the midbrain rostrally and the medulla oblongata caudally; derived from the embryonic metencephalon.

pontine nuclei   Collections of neurons in the base of the pons that receive input from the ipsilateral cerebral cortex and send their axons across the midline to the contralateral cerebellum via the middle cerebellar peduncle.

pontine reticular formation   Collections of neurons in the pons that receive input from the cerebral cortex and send their axons across the midline to the cerebellar cortex via the middle cerebellar peduncle.

pontine-geniculo-occipital (PGO) waves   Characteristic encephalographic waves that signal the onset of rapid eye movement sleep.

pore   Structural feature of an ion channel that allows ions to diffuse through the channel.

pore loop   An extracellular domain of amino acids, found in certain ion channels, that lines the channel pore and allows only certain ions to pass.

positron emission tomography (PET)   A technique for examining brain function following injection of unstable, positron-emitting isotopes that are then incorporated into bioactive molecules or metabolites; the emission of positrons are detected by gamma ray detectors and tomographic images are computed that indicate the localization and concentration of the isotopes.

post-tetanic potentiation (PTP)   An enhancement of synaptic transmission resulting from high-frequency trains of action potentials. See *synaptic potential*.

postcentral gyrus   The gyrus that forms the posterior bank of the central sulcus; contains the primary somatic sensory cortex.

posterior cerebral arteries   Major vessels derived from the basilar artery that supply the ventral midbrain and the posterior, inferior, and medial aspects of the occipital, parietal, and temporal lobes, including associated deep structures.

posterior cingulate cortex (PCC)   A functional division of the cerebral cortex located on its midline surface caudal to the central sulcus that surrounds the corpus callosum. It is associated with task-negative cognition, including mind-wandering, and reward.

posterior circulation   Vasculature derived from the vertebral and basilar arteries that supplies blood to the hindbrain and posterior forebrain.

posterior communicating arteries Small vessels that join the internal carotid arteries to the posterior cerebral arteries, forming the lateral aspects of the circle of Willis.

posterior funiculi Major ascending tracts in the spinal cord that carry mechanosensory information from first-order sensory neurons in dorsal root ganglia and second-order, postsynaptic dorsal column projection neurons to the dorsal column nuclei; also called *dorsal columns*.

posterior inferior cerebellar artery (PICA) Long circumferential branch of the vertebral artery that supplies dorsolateral aspects of the medulla and posterior and inferior aspects of the cerebellum.

posterior spinal arteries Principal arteries on the posterior aspect of the spinal cord supplied by the vertebral or posterior inferior cerebellar arteries; supplies blood to the posterior one-third of the spinal cord.

postganglionic axons Axons that link visceral motor neurons in autonomic ganglia to their targets.

postsynaptic Referring to the compartment of a neuronal process (typically, a dendritic spine or shaft) or a location on a cell body that is specialized for transmitter reception; downstream at a synapse.

postsynaptic current (PSC) The current produced in a postsynaptic neuron by the binding of neurotransmitter released from a presynaptic neuron.

postsynaptic density A cytoskeletal junction in developing synapses that may serve to organize postsynaptic receptors and speed their response to neurotransmitter.

postsynaptic potential (PSP) The potential change produced in a postsynaptic neuron by the binding of neurotransmitter released from a presynaptic neuron.

posttraumatic stress disorder (PTSD) A clinical condition that emerges following the experience of one or more traumatic, stressful events. Symptoms include heightened arousal, emotional numbness, avoidance of event reminders, and persistent reexperiencing of the traumatic event(s).

potentiation An activity-dependent form of short-term synaptic plasticity that enhances synaptic transmission. Potentiation is caused by an increase in the amount of neurotransmitter released in response to presynaptic action potentials and results from persistent calcium actions within presynaptic terminals. Because potentiation acts over a time course of seconds to minutes, it often outlasts the high-frequency trains of action potentials that evoke it, leading to the phenomenon of post-tetanic potentiation.

power spectrum The distribution of power across frequency components of a sound, as computed via Fourier decomposition.

pre-propeptides The first protein translation products synthesized in a cell. These polypeptides are usually much larger than the final, mature peptide and often contain signal sequences that target the peptide to the lumen of the endoplasmic reticulum.

precentral gyrus The gyrus that forms the anterior bank of the central sulcus; contains the primary motor cortex.

precuneus gyrus Gyral structure on the medial aspect of the parietal lobe between the dorsal ramus of the cingulate sulcus and the parieto-occipital sulcus; a component of the

prefrontal cortex (PFC) Cortical regions in the frontal lobe that are anterior to the primary and association motor cortices; thought to be involved in planning complex cognitive behaviors and in the expression of personality and appropriate social behavior.

preganglionic neurons Visceral motor neurons in the spinal cord and brainstem that innervate autonomic ganglia.

preganglionic parasympathetic neurons First order neurons in the parasympathetic branch of the autonomic nervous system. These neurons reside within the CNS and extend long axons into the periphery to innervate second order neurons (or postganglionic parasympathetic neurons) in the parasympathetic ganglia associated with various target organs.

premotor cortex Motor association areas in the frontal lobe anterior to the primary motor cortex; involved in planning or programming of voluntary movements and a source of descending projections to motor neurons in the spinal cord and cranial nerve nuclei.

presbyopia The condition in which aging affects the accommodative ability of the eye.

presynaptic Referring to the compartment of a neuronal process (typically, a terminal of an axon) at a synapse specialized for transmitter release; upstream at a synapse.

prevertebral ganglia Sympathetic ganglia that lie anterior to the spinal column (distinct from the sympathetic chain ganglia).

primary cilium A singular microtubule-based structure that extends from the apical surface of al cells, including neurons, and acts as a sensor and transducer of signals.

primary motor cortex A major source of descending projections to motor neurons in the spinal cord and cranial nerve nuclei; located in the precentral gyrus (Brodmann's area 4) and essential for the voluntary control of movement.

primary sex characteristics The distinguishing body features related to chromosomal sex: male versus female gonads and genitalia.

primary visual cortex Brodmann's area 17 in the medial occipital lobe; major cortical target of the thalamic lateral geniculate nucleus. Also called *striate cortex* because of the prominence of a heavily myelinated stripe in layer 4 (called the stria of Gennari) that gives this region a striped (striated) appearance.

priming A change in the processing of a stimulus due to a previous encounter with the same or a related stimulus, with or without conscious awareness of the original encounter.

primitive node An important source of neural inductive signals during gastrulation.

pit See *primitive node*.

principal nucleus Main nuclear division of the trigeminal nuclear complex of the brainstem in the pons (also known as the chief *sensory nucleus*) that receives mechanosensory afferents from the trigeminal nerve; gives rise to the trigeminal lemniscus the supplies the contralateral ventral posterior medial nucleus.

See **projection neurons** Neurons with long axons that project to distant targets.

promoter DNA sequence (usually within 35 nucleotides upstream of the start site of transcription) to which the RNA polymerase and its associated factors bind to initiate transcription.

propeptide Partially processed forms of proteins containing peptide sequences that play a role in the correct folding of the final protein.

prosencephalon The part of the brain that includes the diencephalon and telencephalon derived from the embryonic forebrain vesicle.

prosody The normal rhythm, stress, and tonal variation of speech that give it emotional meaning.

prosopagnosia The inability to recognize faces; usually associated with lesions to the right inferior temporal cortex.

protanopia A type of color blindness in which the ability to detect longer wavelengths of visible light (such as shades of red) is impaired.

protein kinases Enzymes that participate in intracellular signal transduction by phosphorylating their target proteins, thereby altering the function of these targets.

protein phosphatases A family of enzymes that participate in intracellular signal transduction by removing phosphate groups from their target proteins, thereby altering the function of these targets.

protocadherins A large family of the cadherin class of Ca^{2+} dependent cell adhesion molecules. There are at least 50 or more genes in the mammalian genome that encode protocadherins, and most protocadherin genes are organized so that their exons can be alternatively spliced to generate multiple protocadherin isoforms from the same gene.

pseudogenes A sequence in the genome of an animal or plant that has features of an open reading frame—promoter and repressor regions, sequences for binding of the basal transcriptional complex—that can be transcribed to encode a protein or a functional non-coding RNA. Nevertheless, pseudogenes are not transcribed to messenger RNA. Several apparent odorant receptor genes and vomeronasal receptor genes in humans are pseudogenes that are thought to have been silenced via mutations over the course of evolution.

ptosis A drooping of the upper eyelid.

pupil The perforation in the center of the iris that allows light to enter the eye. The pupillary light reflex mediates pupillary constriction in full light and expansion (dilation) in dim light; these responses can also be induced by chemicals and by certain emotional states, and thus can be clinically important.

pupillary light reflex The reduction in the diameter of the pupil that occurs when sufficient light falls on the retina.

putamen One of the three major components of the striatum (the other two are the caudate and the nucleus accumbens). Also called *putamen nuclei*.

putamen nuclei See *putamen*.

pyriform cortex Component of cerebral cortex in the temporal lobe pertinent to olfaction; so named because of its pearlike shape.

R

radial glial cells Glial cells that contact both the luminal and pial surfaces of the neural tube, providing a substrate for neuronal migration.

rapid eye movement (REM) sleep The phase of sleep characterized by low-voltage, high-frequency electroencephalographic activity accompanied by rapid eye movements.

rapidly adapting afferents Afferents that fire transiently in response to stimulus onset or offset.

ras The first monomeric G-protein discovered. It is involved in many types of neuronal signaling and also controls differentiation and proliferation of non-neuronal cells.

readiness potential An electrical potential, recorded from the motor and premotor cortices with EEG electrodes, that signals the intention to initiate a voluntary movement well in advance of actual production of the movement.

receptive aphasia See *Wernicke's aphasia*.

receptive field The region of a receptive surface (e.g., the body surface, or a specialized structure such as the retina) within which a specific stimulus elicits the greatest action potential response from a sensory cell in a sensory ganglion or within the CNS.

receptor molecules Molecules that binds to chemical signals and transduce these signals into a cellular response.

receptor potentials The membrane potential change elicited in receptor neurons during sensory transduction. Also called *generator potentials*. Compare *synaptic potential*.

recessive If a trait is not seen in either parent, but emerges in the offspring, then one of the two copies of that gene in each parent is said to be recessive, and the offspring must inherit both recessive copies to exhibit the trait.

reciprocal innervation Pattern of connectivity in local circuits of the spinal cord involving excitatory and inhibitory interneurons arranged to ensure that contraction of agonistic muscles produce forces that are opposite to those generated by contraction of antagonistic muscles; thus, reciprocal innervation mediates the simultaneous relaxation of antagonists during contraction of agonists.

red nucleus Prominent parvocellular nucleus of the midbrain tegmentum involved in regulating activity in the inferior olivary nucleus; integrates input from the cerebral cortex and the contralateral dentate nucleus of the cerebrocerebellum. In non-human mammals, also features a magnocellular division that gives rise to the rubrospinal tract, which participates in upper motor neuronal control of the distal musculature of the upper limbs or forelimbs.

reentrant (or recurrent) neural activation Following a stimulus or event, a process in which neural activity is fed back to the same brain region activated earlier in the processing sequence.

refractory period The brief period after the generation of an action potential during which a second action potential is difficult or impossible to elicit.

regenerative A process that is self-sustaining. For example, action potential propagation is regenerative because an action potential produced at one location depolarizes downstream regions, thereby activating voltage-gated ion channels to generate an action potential in these regions.

reporter protein   A protein from one species, encoded by a gene not found in the genome of another, that can be inserted into the genome of a host (that does not have the gene in its genome) and expressed under the control of the host's regulatory gene sequences to identify cells in which the regulator sequences normally control expression of the endogenous gene.

restiform body   See *cerebellar peduncles*.

resting membrane potential   The inside-negative electrical potential that is normally recorded across all cell membranes.

reticular activating system   Region in the brainstem tegmentum that, when stimulated, causes arousal; involved in modulating sleep and wakefulness.

reticular formation   A network of neurons and axons that occupies the core of the brainstem, giving it a reticulated ("net-like") appearance in myelin-stained material; major functions are modulatory (e.g., regulating states of consciousness) and premotor (e.g., coordinating eye movements, posture, and the regulation of respiratory and cardiac rhythms).

retina   Laminated neural component of the eye that contains the photoreceptors (rods and cones) and the initial processing machinery for the primary (and other) visual pathways.

retinal ganglion cells   The projection neurons of the retina that transmit visual information from the retina to retino-recipient regions of the hypothalamus, thalamus, and midbrain.

retinal waves   A type of oscillatory activity established in the developing mammalian retina, usually prenatally or before eye opening, that is independent of visual input. These oscillations are established by subthreshold activity of subsets of amacrine cells leading to rhythmic firing of subsets of retinal ganglion cells in a spatially distinct manner.

retino-recipient brain regions   Regions of the brain that are directly innervated by retinal ganglion cells.

retino-geniculo-cortical pathway   Another term for the primary visual pathway.

retinoic acid (RA)   A derivative of vitamin A that acts as an inductive signal during early brain development.

retinoid receptors   The family of nuclear transcription factor proteins that form heterodimers and bind isomers of retinoic acid to initiate recognition of response element DNA binding sequences and subsequently influence gene expression. Retinoid receptors are a subclass of the broader steroid/thyroid receptor-transcription factor family.

retinotopic maps   Visual field maps that contain the topographic mapping of visual information from the retina to the brain.

retrograde   Signals or impulses that travel "backward," e.g., from the axon terminal toward the cell body, or from the postsynaptic cell to the presynaptic terminal, or from the periphery to the CNS.

retrograde amnesia   The inability to recall existing memories.

reversal potential   Membrane potential of a postsynaptic neuron (or other target cell) at which the action of a given neurotransmitter causes no net current flow.

reward prediction error (RPE)   A quantity given by the difference between the reward that was expected and what actually occurs; the activity of some dopaminergic neurons seems to convey this quantity.

reward   A poorly defined term that generally refers to a sense of pleasure following a successful response to some challenge. Often taken to entail dopaminergic neural circuitry.

rhodopsin   The photopigment found in rods.

rhombencephalon   The caudal part of the brain between the mesencephalon and the spinal cord derived from the embryonic hindbrain vesicle; includes the pons, cerebellum, and medulla. Also known as the hindbrain.

rhombomeres   A repeated unit or segment of the developing hindbrain that is distinguished by a distinct signature of transcription factors expression. Individual rhombomeres go on to generate cranial nerves along the anterior–posterior axis.

rising phase   The initial, depolarizing, phase of an action potential, caused by the regenerative, voltage-dependent influx of a cation such as Na^+ or Ca^{2+}.

robo   The transmembrane receptor for Slit that activates signaling via interactions with non-receptor tyrosine kinases, RhoGTPases and other cytoplasmic signaling molecules to initiate the repulsion/avoidance of an axon or dendrite for a specific direction or location.

ROBO3   An additional ROBO receptor, encoded by a separate gene. ROBO3 is particularly essential for the guidance of axons through the ventral commissure at the midline spinal cord, and ensuring that these axons do not turn around and cross back at the midline.

rods   Photoreceptor cells specialized for operating at low light levels.

roofplate   The thinned dorsal-most medial region of the neural tube, where the two edges of the lateral/alar neuroectoderm fuse during neurulation. This neuroectodermal cells of this region, like the floorplate at the ventral midline, provide secreted signals to specify the neural crest as well as dorsal cell types in the neural tube.

rostral (R)   Anterior, or "headward."

rostral interstitial nucleus   Cluster of neurons in the reticular formation that coordinates the actions of neurons in the oculomotor nuclei to generate vertical movements of the eye; also called the *vertical gaze center*.

rostral migratory stream (RMS)   A specific migratory route, defined by a distinct subset of glial cells, that facilitates migration of newly generated neurons from the stem cell niche of the anterior subventricular zone to the olfactory bulb.

rostrotemporal (RT)   One of the divisions of the core region of the auditory cortex in non-human primates.

rotational movements   Angular or turning motions around the X, Y, or Z axes.

round window   Along with the oval window, a region at the base of the cochlea where the overlying bone is absent.

rubrospinal tract   In non-human mammals, the pathway from the magnocellular divisions of the red nucleus of the midbrain to the spinal cord; participates with the lateral corticospinal tract in governing the distal extremities. In humans, however, the corticospinal tract serves this function and the rubrospinal tract is vestigial (perhaps even nonexistent).

Ruffini corpuscles   Encapsulated cutaneous mechanosensory receptors in the deep dermis (also found in other tissues) specialized for the detection of cutaneous stretching produced by digit or limb movements.

ryanodine receptor   A ligand-gated ion channel in the endoplasmic reticulum membrane. This receptor binds to the drug ryanodine and elevates cytoplasmic calcium concentration by mediating flux of calcium out of the lumen of the endoplasmic reticulum.

S

saccades   See *saccadic eye movements.*

saccadic eye movements   Ballistic, conjugate eye movements that change the point of foveal fixation.

saccadic suppression   The phenomenon that humans do not perceive motion of the visual field during saccadic eye movements.

saccule   The otolith organ that detects linear accelerations and head tilts in the vertical plane.

sacral   Caudal region of the spinal cord between the lumbar and coccygeal regions related to the lower extremities and pelvic visceral motor outflow.

sacral neural crest   The portion of the neural crest , a migratory population of neural tube-derived progenitors, that along with the vagal neural crest, migrates primarily into the gut to constitute the enteric nervous system

sagittal   Standard anatomical plane of section; the vertical plane passing from anterior to posterior through the midline dividing the body (and brain) into right and left sections.

salt   One of the five basic tastes; the taste quality produced by the cations of salts (e.g., the sodium in sodium chloride produces the salty taste). Some cations also produce other taste qualities (e.g., potassium tastes bitter as well as salty). The purest salty taste is produced by sodium chloride (NaCl), common table salt. Salt taste is transduced by taste cells via an ameloride sensitive Na$^+$ channel.

saltatory   Mechanism of action potential propagation in myelinated axons; so named because action potentials "jump" from one node of Ranvier to the next due to generation of action potentials only at these sites.

scala media   The fluid-filled chamber within the cochlea that sits on the basilar membrane and that lies between the scala vestibuli and the scala tympani.

scala tympani   The fluid-filled chamber within the cochlea at the base of which is located the round window.

scala vestibuli   The fluid-filled chamber within the cochlea at the base of which is located the oval window.

Scarpa's ganglion   See *vestibular nerve ganglion.*

Schwann cells   Glial cells in the peripheral nervous system that lay down myelin, which facilitates the efficient generation and rapid conduction of action potentials; also facilitate axon regeneration in damaged nerves (named after the nineteenth-century anatomist and physiologist Theodor Schwann).

sclera   The external connective tissue coat of the eyeball.

SCN genes   Na$^+$ channel genes. These genes produce proteins that differ in their structure, function, and distribution in specific tissues.

scotoma   A small deficit in the visual field resulting from pathological changes in some component of the primary visual pathway.

scotopic vision   Vision in dim light, where the rods are the operative receptors.

second pain   A category of pain perception described as more delayed, diffuse, and longer-lasting than first pain.

secondary sex characteristics   Anatomical characteristics, such as breasts and facial hair, that generally differ between the sexes but are not necessarily concordant with the chromosomal sex of the individual

segmental nerves   See *spinal nerves.*

segmentation   The anterior–posterior division of animals into roughly similar repeating units.

selective attention   The allocation of processing resources toward one stimulus or spatial location in order to facilitate its evaluation over that of other stimuli or locations.

selective serotonin reuptake inhibitors (SSRIs)   A class of drugs that work by inhibiting the ability of the SERT serotonin transporter to take serotonin up into presynaptic terminals.

selectivity filter   Structure within an ion channel that allows selected ions to permeate, while rejecting other types of ions.

semaphorins   A family of diffusible, growth-inhibiting molecules.

semicircular canals   The vestibular end organs in the inner ear that sense rotational accelerations of the head.

sensitization   Increased sensitivity to stimuli in an area surrounding an injury. Also, a generalized aversive response to an otherwise benign stimulus when it is paired with a noxious stimulus.

sensorineural hearing loss   Diminished sense of hearing due to damage of the inner ear or its related central auditory structures. Contrast with *conductive hearing loss.*

sensory aphasia   See *Wernicke's aphasia.*

sensory systems   Term sometimes used to describe all the components of the central and peripheral nervous system concerned with sensation.

sensory transduction   Process by which the energy of a stimulus is converted into electrical signals by peripheral sensory receptors and then processed by the central nervous system.

sensory–discriminative   The aspect of pain that allows one to distinguish the location, intensity, and quality of a noxious stimulation.

septal forebrain nuclei   Cerebral nuclei at the anterior base of the septum pellucidum; give rise to widespread modulatory projections to diverse targets in the cerebral hemispheres.

septum pellucidum   Non-neural tissue that forms the medial wall along the anterior horns, bodies, and atria of the paired lateral ventricles.

SERCA   *Sarcoplasmic/endoplasmic reticulum calcium ATPase,* responsible for translocating Ca^{2+} from cytoplasm to these intracellular storage organelles.

serotonergic neurons   Neurons that contribute to the neuronal basis of wakefulness.

serotonin   A biogenic amine neurotransmitter, derived from the amino acid tryptophan, that is involved in a wide range of behaviors, including emotional states and mental arousal.

sex chromosomes Either of a pair of chromosomes (XX in female or XY in male mammals) that differ between the sexes.

sexually dimorphic nucleus of the preoptic area (SDN-POA) A hypothalamic nucleus that in humans and several other mammals differs in size in males versus females: usually the male SDN-POA is larger than that of the female. The SDN-POA is thought to regulate sexual behaviors directly involved with reproduction as well as partner selection.

sham rage An emotional reaction elicited in cats by electrical stimulation of the hypothalamus, characterized by hissing, growling, and attack behaviors directed randomly toward innocuous targets.

short axon cells A class of olfactory bulb interneurons whose cell bodies are found in the glomerular layer in extra-glomerular spaces defined by the periglomerular cells as well as in the external plexiform layer. Some of these GABAergic interneurons also use dopamine as a co-neurotransmitter. Short axon cells have axons that interconnect multiple glomeruli and provide additional lateral inhibition between glomeruli.

short circumferential arteries The shortest branches of the vertebral and basilar arteries that supply medial aspects of the brainstem.

sigmoid sinuses Dural venous sinuses that convey venous blood from the transverse sinuses through the jugular foramina and into the jugular veins.

signal amplification A consequence of intercellular or intracellular signal transduction, resulting from the involvement of reactions that generate a much larger number of products than the number of molecules required to initiate the process. Signal amplification is one of the most important advantages of chemical signaling.

signaling endosome A vesicular structure, internalized into the cell, usually at an axonal or dendritic process, via the invagination and then constriction of the plasma membrane. These vesicles include transmembrane receptor kinases that have bound their activating ligand while still on the cell surface. The ligand remains bound to the receptor, but as an internalized vesicle the ligand activated kinase domain faces the cytoplasm of the cell. The vesicle can then be transported retrogradely to the cell body and continue to transduce the signal detected at its point of origin throughout the entire cell.

size principle The orderly recruitment of motor neurons by size to generate increasing amounts of muscle tension.

skin conductance A stimulus-induced increase in the electrical conductance of the skin due to increased hydration.

SLC8A gene A family of 3 genes encoding Na^+/Ca^{2+} exchangers, which are responsible for translocating Ca^{2+} from cytoplasm to the extracellular medium.

sleep spindles Periodic bursts of activity at about 10 to 12 Hz that generally last 1 to 2 seconds and arise as a result of interactions between thalamic and cortical neurons; intermittent high-frequency EEG spike clusters characteristic of stage II sleep.

slit A secreted signaling molecule that acts a repulsive cue for growing axons or dendrites.

slow (S) motor units Small motor units comprising small muscle fibers that contract slowly and generate relatively small forces; but because of their rich myoglobin content, plentiful mitochondria, and rich capillary beds, these small red fibers are resistant to fatigue. S motor units are especially important for activities that require sustained muscular contraction, such as maintaining an upright posture.

slow-wave sleep The component of sleep characterized by delta waves.

slowly adapting afferents Afferents that continue to fire, with only modest decrement in firing, in response to the sustained presence of a stimulus.

small G-proteins See *monomeric G-proteins*.

small-molecule neurotransmitters The non-peptide neurotransmitters such as acetylcholine, the amino acids glutamate, aspartate, GABA, and glycine, as well as the biogenic amines.

smooth pursuit eye movements Slow, tracking movements of the eyes designed to keep a moving object aligned with the foveae.

SNAP-25 A SNARE associated with the plasma membrane. This protein forms a SNARE complex with synaptobrevin and syntaxin that mediates fusion of synaptic vesicles with the presynaptic plasma membrane.

SNAPs Soluble *NSF-attachment proteins*. A protein that attaches the enzyme NSF to SNARE complexes, to allow NSF to dissociate the SNARE complexes.

SNAREs *SNAP receptors*. Proteins that are found on two membranes and are responsible for fusing the two membranes together.

somatic motor division The components of the motor system that support skeletal movements mediated by the contraction of skeletal muscles that are derived from embryonic somites or somitomeres.

somatic motor nuclei Brainstem nuclei (derived from the basal plate) that give rise to efferent fibers innervating striated muscle fibers derived from embryonic somitomeres; located in the dorsal tegmentum alongside the midline.

somatic stem cells Cells that can divide to give rise to more cells like itself, but also can divide to give rise to a new stem cell plus one or more differentiated cells of the relevant tissue type (e.g., a hematopoietic stem cell can give rise to all types of blood cells, neural stem cells give rise to all neuronal types, and glial stem cells to glia). Contrast with *embryonic stem cell*.

somatosensory cortex Functional division of the cerebral cortex in the postcentral gyrus and anterior parietal lobe that receives somatosensory projections from the ventral posterior complex of thalamic nuclei; processes somatosensory information from the body surface, subcutaneous tissues, muscles, and joints.

somatotopic maps Cortical or subcortical arrangements of sensory inputs and local circuits that reflect the topological organization of the body.

Sonic hedgehog (Shh) An inductive signaling hormone essential for development of the mammalian nervous system; believed to be particularly important for establishing the identity of neurons in the ventral portion of the developing spinal cord and hindbrain.

sour   One of the five basic tastes; the taste quality produced by the hydrogen ion in acids. Sour tastes are transduced by taste cells via a H⁺ selective TRP channel.

sparse coding   A pattern of electrical activation across a large set of neurons, all of which can respond at varying levels to a stimulus, in which only a very small subset of the neurons is activated strongly in response to that specific stimulus. The strong responses in such a subset of the neurons are thought to efficiently represent the identity of the stimulus, extracting this information from the broader population of responsive neurons that introduce a high degree of "noise" in the system based upon less ordered firing in response to the same stimulus.

spectral cues   The characteristic pattern of frequency filtering based on the angle of incidence of a sound with respect to the ear.

spectrogram   The frequency components of a sound across time, as computed by Fourier decomposition.

spike timing-dependent plasticity (STDP)   Changes in synaptic transmission that depend upon the precise temporal relationship between presynaptic action potentials and postsynaptic responses.

spinal accessory nerve (XI)   Cranial nerve XI, a branchial motor nerve that conveys efferents from the rostral cervical spinal cord (spinal accessory nucleus) to the upper trapezius and sternocleidomastoid muscles. Attaches to the rostral cervical cord in a cleft medial to the inferior cerebellar peduncle; enters the cranial vault through the foramen magnum and exits via the jugular foramen.

spinal cord   The caudal (post cranial) portion of the central nervous system (CNS) that extends from the lower end of the brainstem (the medulla) to the cauda equina; mediates the transmission of afferent and efferent neural signals between the CNS and the body.

spinal nerves   Mixed sensory and motor nerves that arise in bilaterally symmetrical pairs from each of 31 segments of the spinal cord. Also called *segmental nerves*.

spinal nucleus   Component of the trigeminal nuclear complex of the brainstem in the caudal pons and medulla oblongata that receives afferents from the trigeminal nerve concerning pain and temperature (also receives collateral of mechanosensory afferents); comprises several subdivisions each of which gives rise to the trigeminothalamic tract that supplies the contralateral ventral posterior medial nucleus.

spinal nucleus of the bulbocavernosus (SNB)   Sexually dimorphic collection of neurons in the lumbar region of the rodent spinal cord that innervate striated perineal muscles.

spinal shock   The initial, short-lived period of flaccid paralysis that accompanies damage to upper motor neurons or their descending pathways to lower motor neurons.

spinocerebellum   Medial part of the cerebellum that receives proprioceptive input from the spinal cord and sends output to the motor cortex via the thalamus and to brainstem upper motor neurons; includes paramedian zones that coordinate movements of distal muscles, and a median zone, called the vermis, that coordinates movements of proximal muscles, including eye movements.

spiny stellate cells   A class of excitatory neuron that resides within layer 4 of the cerebral cortex.

splice variants   Variable messenger RNA transcripts derived from the same gene that are typically produced by including or excluding certain exons from a gene; the result such alternative splicing is the production of a diverse set of related protein products.

split-brain patients   Individuals who have had the cerebral commissures divided in the midline to control epileptic seizures.

Sprague effect   Hemispatial neglect induced by a parietal lesion in humans can be mostly compensated by a lesion of the superior colliculus on the other side.

SRY (sex-reversal gene on the Y chromosome)   Gene on the Y chromosome whose expression triggers a signaling and transcriptional regulatory cascade that masculinizes the developing fetus.

stellate cells   Inhibitory interneurons in the cerebellar cortex that receive parallel fiber input and provide inhibitory output to Purkinje cell dendrites.

stereocilia   The actin-rich processes that, along with the kinocilium, form the hair bundle extending from the apical surface of the hair cell; site of mechanotransduction.

stereopsis   The perception of depth that results from the fact that the two eyes view the world from slightly different angles.

steroid–thyroid nuclear receptors   A large class of receptor transcription factors that selectively bind different members of the steroid-thyroid family of hormones. This class of receptor-transcription factors includes estrogen and androgen receptors.

strabismus   Developmental misalignment of the two eyes; may lead to binocular vision being compromised.

stria vascularis   Specialized epithelium lining the cochlear duct that maintains the high potassium concentration of the endolymph.

striate cortex   See *primary visual cortex*.

striatum   General term applied to the caudate, putamen, nucleus accumbens and other minor divisions of the ventral basal forebrain. The name derives from the bridges ("striations") of gray matter that unite the caudate and putamen around which course fibers of the anterior limb of the internal capsule. Principal component of the corpus striatum, an historical term that has been used collectively to refer to the striatum and the globus pallidus.

striola   A line found in both the sacculus and utricle that divides the hair cells into two populations with opposing hair bundle polarities.

Stroop effect   A slowing of response time when a stimulus harbors inherently conflicting information (e.g., responding to the word *red* printed in green ink versus the word printed in red ink).

submucous plexus   Network of neurons in the enteric division of the visceral motor system just beneath the mucus membranes of the gut and is concerned with chemical monitoring and glandular secretion. Also called *Meissner's plexus*.

substance P   An 11-amino acid neuropeptide; the first neuropeptide to be discovered.

substantia nigra Bipartite gray matter structure in the ventral midbrain; contains a pallidal division, called the **pars reticulata**, defined by a network (reticulum) of cells that provide inhibitory (GABAergic) output from the basal ganglia to the thalamus and brainstem, and a compact division, called the **pars compacta,** comprising densely packed neurons that synthesize and release dopamine in the caudate and putamen.

substantia nigra pars compacta Compact cell division of the substantia nigra in the ventral midbrain featuring densely packed neurons that synthesize and release dopamine in the caudate and putamen.

substantia nigra pars reticulata Pallidal division of the substantia nigra in the ventral midbrain featuring a network (reticulum) of cells that provide inhibitory (GABAergic) output from the basal ganglia to the thalamus and brainstem.

subthalamic nucleus A nucleus in the ventral thalamus that receives input from the cerebral cortex and external segment of the globus pallidus and sends excitatory (glutamatergic) projections to the internal segment of the globus pallidus. A component of the indirect pathway from striatum to pallidum.

sulci (sing. sulcus) Spaces between gyri; the largest of these spaces are called fissures.

sulcus limitans Shallow longitudinal groove in the lateral wall of the lumen of the neural tube that defines a boundary between the alar and basal plates.

summation The addition in space and time of sequential synaptic potentials to generate a postsynaptic response larger than that produced by a single synaptic potential.

superior cerebellar artery (SCA) Superior of the three pairs of arteries that supply blood to the cerebellum and the dorsolateral aspect of the brainstem; originates from the distal basilar artery just caudal to the posterior cerebral artery along the ventral aspect of the rostral pons.

superior cerebellar peduncle See *cerebellar peduncles.*

superior colliculus (pl. colliculi) Laminated gray matter structure that forms part of the roof of the midbrain; plays an important role in orienting movements of the head and eyes.

superior division of the retina Referring to the region of the visual field of each eye that corresponds to the top half of the retina.

superior frontal gyri Superior of three, parallel longitudinal gyri that define the dorsomedial margin of frontal lobe anterior to the precentral sulcus; part of the prefrontal cortex and anterior premotor cortex.

superior oblique muscles Extraocular muscles that intort the eyeballs when in the primary position (eyes straight ahead) and rotate downward when in adduction.

superior parietal lobules Gyral formations of the dorsal (superior) parietal lobes that are involved in associating somatosensory, visual, auditory, and vestibular signals and generating a neural construct of the body, the position of its parts, and its movements (body image or schema).

superior rectus muscles Extraocular muscles that rotate the eyeballs upward.

superior sagittal sinus Large dural venous sinus along the dorsal aspect of the longitudinal fissure; provides for the drainage of venous blood from the dorsal cerebral hemisphere and the return of cerebrospinal fluid via the arachnoid granulations.

superior salivatory nuclei Visceral motor nucleus of the rostral pons containing parasympathetic preganglionic neurons that mediate tearing and salivation.

superior temporal gyri Superior of three, parallel longitudinal gyri that define the dorsolateral margin of temporal lobe; part of the auditory cortex and lateral temporal network that encodes language content.

suprachiasmatic nucleus (SCN) Hypothalamic nucleus lying just above the optic chiasm that receives direct input from the retina; involved in light entrainment of circadian rhythms.

supramodal attention The focusing of attention on stimulus information across multiple modalities at the same time.

sustentacular cells The primary support cells of the olfactory epithelium. Sustentacular cells help to maintain appropriate ionic milieu and epithelial integrity for the olfactory sensory neurons and their basal cell precursors throughout life.

sweet One of the five basic tastes; the taste quality produced by some sugars, such as glucose, fructose, and sucrose. These three sugars are particularly biologically useful to us, and our sweet receptors are tuned to them. Some other compounds (e.g., saccharin, aspartame) are also sweet. Sweet is transduced by taste cells via the T1R class of G-protein-coupled taste receptors.

Sylvian fissure See *lateral fissure.*

sympathetic A division of the visceral motor system (division) in vertebrates comprising, for the most part, adrenergic ganglion cells located relatively far from the related end organs and the central preganglion neurons that innervate them.

See **synapse elimination** The developmental process by which the number of axons innervating some classes of target cells is diminished. Also called input elimination.

synapses The junctions between neurons where information is passed from one to the other; typically refers to chemical synapses where a physical cleft exists between communicating neurons, but could also refer to electrical synapses mediated by gap junctions.

synapsin A protein which reversibly binds to synaptic vesicles and is responsible for tethering these vesicles within a reserve pool.

synaptic cleft The space that separates pre- and postsynaptic neurons at chemical synapses.

synaptic delay The time interval between signals in presynaptic and postsynaptic neurons. The synaptic delay is typically much briefer for electrical synapses than chemical synapses.

synaptic depression A short-term decrease in synaptic strength resulting from the depletion of synaptic vesicles at active synapses.

synaptic facilitation An increase in synaptic strength that occurs when two or more action potentials invade the presynaptic terminal within a few milliseconds of each other. Facilitation is typically caused by an increase in the amount of neurotransmitter released by a presynaptic action potential and lasts for a fraction of a second.

synaptic potential A membrane potential change (or a conductance change) generated by the action of a chemical transmitter agent. Synaptic potentials allow the transmission of information from one neuron to another. Compare *receptor potentials*.

synaptic vesicle cycle Sequence of budding and fusion reactions that occurs in presynaptic terminals to maintain the supply of synaptic vesicles.

synaptic vesicles Spherical, membrane-bound organelles in presynaptic terminals that store neurotransmitter molecules and associated molecular machinery that facilitates exocytosis.

synaptobrevin A SNARE protein located in the membrane of synaptic vesicles. This protein forms a SNARE complex with syntaxin and SNAP-25 that mediates fusion of synaptic vesicles with the presynaptic plasma membrane.

synaptojanin A protein involved in uncoating of synaptic vesicles. It works by modifying a vesicular lipid, which serves as a cue for vesicle uncoating by Hsc70.

synaptotagmins A family of calcium-binding proteins found in the membrane of synaptic vesicles and elsewhere. Synaptotagmins 1 and 2 serve as the calcium sensors that trigger the rapid release of neurotransmitters.

syntaxin A SNARE protein found primarily in the plasma membrane. This protein forms a SNARE complex with synaptobrevin and SNAP-25 that mediates fusion of synaptic vesicles with the presynaptic plasma membrane.

T

taste buds Onion-shaped structures in the mouth and pharynx that contain taste cells.

tectorial membrane The fibrous sheet overlying the apical surface of the cochlear hair cells; produces a shearing motion of the stereocilia when the basilar membrane is displaced.

tegmentum A general term that refers to the central core of the brainstem.

telencephalon The part of the brain derived from the anterior part of the embryonic forebrain vesicle; includes the cerebral hemispheres (cerebral cortex and cerebral nuclei).

temporal division of the retina Referring to the region of the visual field of each eye in the direction of the temple.

temporal lobe The hemispheric lobe that lies inferior to the lateral fissure.

terminal field The spatial and cellular extent of synaptic endings made by an axon once it reaches it targets. The terminal field of an axon can include multiple neuronal or peripheral (muscle/gland) target cells, or it can be restricted to a single target cell.

testis-determining factor (TDF) The original name for the gene product of the *SRY* gene (See *SRY*).

testosterone The principal androgen, synthesized in the testes and, in lesser amounts, in the ovaries and adrenal glands.

thalamocortical relay cells Excitatory neurons within the dorsal thalamus that receive sensory input and project axons to the cerebral cortex.

thalamus A collection of nuclei that forms the major component of the dorsal diencephalon. Although its subdivisions and functions are many, a primary role of the thalamus is to interact with neural circuits in the cerebral cortex through reciprocal, topographically organized interconnections.

thermosensitive Ion channels that respond to heat.

third ventricle Narrow, slit-like ventricle derived from the lumen of the neural tube between the paired diencephalon, which form the lateral wall of the third ventricle.

thoracic Intermediate region of the spinal cord related to the trunk and sympathetic outflow.

threshold potential The level of membrane potential at which an action potential is generated.

tiling The optimal spacing of dendritic or axonal arbors from neighboring cells. Appropriate tiling is essential for quantitative matching of afferent and target populations.

timbre The perceptual quality of sound based on the totality of its component frequencies (as distinct from pitch).

tinnitus A pathological condition characterized by spontaneous ringing or rushing noises, which can be either peripheral or central in origin.

tonotopic organization Topographic organization according to the progression of sound frequencies to which neurons are most sensitive.

tonotopy The topographic mapping of sound frequency across the surface of a structure, which originates in the cochlea and is preserved in ascending auditory structures, including the auditory cortex.

topographic logic A system by which neurons organize information. The somatosensory and visual cortices feature globally ordered topographic maps in which neurons representing a body part or stimulus location in the visual field are close to each other.

topographic maps Point-to-point correspondence between neighboring regions of the sensory periphery (e.g., the visual field or the body surface) and neighboring neurons within the central components of the system (e.g., in the brain and spinal cord).

touch domes Specialized epidermal structures containing Merkel cells.

trace amine-associated receptors (TAARs) A distinct family of G-protein coupled odorant receptor proteins expressed in limited numbers of olfactory receptor neurons. These receptors have higher homology to biogenic amine neurotransmitter receptors (e.g., dopamine, noradrenaline, serotonin) than the larger family of canonical odorant receptor proteins. They bind volatile amines, aminergic molecules released by the decomposition of proteins. It is thought that the TAARs are essential for detection of rotting or decaying biological materials including food to ensure avoidance of potentially toxic substances.

tracrRNA Small, trans-encoding RNA that combines with a specific guide RNA species to form an RNA duplex, which then acts to guide a bacterial excision/repair enzyme (endonuclease Cas9) to a genomic location targeted for excision. Following Following Cas9 excision, the DNA may be repaired by non-homologous end joining, yielding a microdeletion mutation; alternatively, a donor DNA sequence can be inserted following Cas9 cleavage via a mechanism similar to homologous recombination.

tracts Bundles of fasciculated axons in the central nervous system that are gathered into compact structures and typically share a common origin and termination; more or less analogous to nerves in the periphery.

transcranial magnetic stimulation (TMS) Localized, noninvasive stimulation of cortical neurons through the induction of electrical current by the application of strong, focal magnetic fields.

transcription factors See *transcriptional activator proteins*.

transcriptional activator proteins DNA-binding proteins that attach near the start site of a gene, thereby activating transcription of the gene. Also called *transcription factors*.

transducin G-protein involved in the phototransduction cascade.

transgenic mice Mice in which a gene or genes has been artificially modified using genetic engineering approaches. Transgenic mice can have a variety of genetic changes introduced by experimental methods that rely upon the rules by which DNA is recombined and replicated.

transient receptor potential (TRP) The transient receptor potential (TRP) family of ion channels constitute approximately 28 individual genes and the proteins that they encode. All are transmembrane cation selective channels that mediate depolarization in response primarily to various sensory stimuli. These include taste stimuli that interact with the T1R/T2R GPCR taste receptors to activate the TRPM5 channel. There also TRP channels involved in transducing mechanical displacement/stretch across cellular membranes.

transit amplifying cell A precursor cell, capable of rapid asymmetric divisions, descended from a stem cell. Transit amplifying cells can generate a large number of post-mitotic cells via a series of asymmetric divisions that yield one post-mitotic cell that goes on to differentiate and one precursor. These cells, however, are not self-renewing like stem cells—eventually, there is a terminal symmetric division in which both progeny cease to divide.

translational movements Linear motion along the X, Y, and Z axes.

transverse Standard anatomical planes of section that pass through the CNS orthogonal to its long axis.

transverse pontine fibers Axons of pontine nuclei that cross the midline and form the middle cerebellar peduncles.

transverse sinuses Dural venous sinuses that convey venous blood in the anterior direction from the confluence of sinuses at the back of the cranium to the sigmoid sinuses.

traveling wave The sound-evoked propagation of motion from the basal toward the apical end of the basilar membrane.

trichromatic Referring to the presence of three different cone types in the human retina, which generate the initial steps in color vision by differentially absorbing long, medium, and short wavelength light.

trigeminal brainstem complex Nuclei of the brainstem that receive or give rise to sensory or motor axons in the trigeminal nerve; comprises the mesencephalic trigeminal nucleus in the midbrain and rostral pons, the principal (chief sensory) nucleus in the pons, the trigeminal motor nucleus in the pons, and the spinal trigeminal nucleus, which itself contains several subdivisions, in the caudal pons and medulla oblongata.

trigeminal lemniscus Axonal projection arising from the principal nucleus of the trigeminal nuclear complex of the brainstem and terminating in the contralateral ventral posterior medial nucleus of the thalamus; conveys mechanosensory signals derived from the face.

trigeminal nerve (V) Cranial nerve conveying somatosensory information from the face to the trigeminal nuclear complex of the brainstem; also conveys motor signals from the brainstem to the muscles of mastication.

tripartite synapse A three-way junction involving a presynaptic terminal, a postsynaptic process, and neighboring glia.

trochlear nerve (IV) Cranial nerve IV, an efferent motor nerve that controls the superior oblique muscle of the eye.

trophic interaction Referring to the long-term interdependence between nerve cells and their targets.

trunk neural crest The portion of the neural crest, a migratory population of neural tube-derived progenitors, that moves into the space between the developing vertebrae and somites to become primarily the dorsal root sensory ganglia.

tuning The extent of selective electrical response of a particular sensory receptor or central neuron to a specific stimulus or stimulus attribute. When a peripheral receptor or central neuron responds maximally to a certain stimulus: for example, neurons in the visual cortex to edges with distinct angles of orientation, or cells in the olfactory bulb or pyriform cortex in response to a particular odorant, those cells are said to be "tuned" to that stimulus.

tuning curve The function obtained when a neuron's receptive field is tested with stimuli at different orientations; its peak defines the maximum sensitivity of the neuron in question.

two-point discrimination Distance between caliper tips needed to distinguish one versus two points of stimulation.

tympanic membrane The eardrum.

tyrosine kinase (Trk) Tyrosine kinases are a broad and large class of enzymes that catalyze the phosphorylation of specific tyrosine residues in multiple protein targets. Tyrosine kinases can also be receptors, especially for multiple growth factors including neurotrophins that regulate neuronal growth and survival.

U

umami The last of five basic tastes: umami is taste detected in response to amino acids in proteins like meat. Umami is also referred to as "savory" taste. It is transduced via the T1R class of G-protein- coupled taste receptors.

Uncoordinated (Unc) A family of mutations in the nematode worm *C. elegans* that are named for the changes in movement that can be observed visibly in the individual mutant worms. Several *Unc* mutants were eventually found to have mutations targeted to cell surface adhesion molecules or secreted signals that influence axon growth and guidance.

uncus Medial protrusion of the anterior parahippocampal gyrus formed by cortical division of the amygdala.

undershoot The final, hyperpolarizing phase of an action potential, typically caused by the voltage-dependent efflux of a cation such as K^+.

unimodal attention The enhancement of processing within one sensory modality (e.g., vision).

upper motor neuron syndrome Signs and symptoms that result from damage to descending motor pathways; these include weakness, spasticity, clonus, hyperactive reflexes, and a positive Babinski sign.

utricle The otolith organ that senses linear accelerations and head tilts in the horizontal plane.

V

V1Rs A sub-class of vomeronasal receptors that interact with the G-protein Gαi2 to transduce vomeronasal sensory stimuli.

V2Rs A sub-class of vomeronasal receptors that interact with the G-protein Gαo to transduce vomeronasal signals.

vagal neural crest The region of the neural crest that arises from the posterior rhombencephalon, migrates into the visceral endoderm and eventually gives rise to the enteric nervous system as well as most of the parasympathetic ganglia.

vagus nerve (X) Cranial nerve X, a mixed sensorimotor nerve that conveys visceral afferents (taste from posterior oropharynx) and somatosensory afferents (mechanosensation from skin on or near pinna) to the brainstem, and branchial motor efferents from the brainstem to muscles of the larynx and pharynx and visceral motor efferents to widely distributed targets in the thorax and upper abdomen. Attaches to the rostral medulla in a cleft between the inferior olive and the inferior cerebellar peduncle, just inferior to the glossopharyngeal nerve.

valence The degree of pleasantness of a stimulus.

vasocorona A network of blood vessels on the lateral and ventrolateral margins of the spinal cord connecting circumferential branches of the posterior and anterior spinal arteries.

vection The sense of self-motion created by visual flow.

ventral (anterior) corticospinal tract Spinal portion of the corticospinal tract in the anterior–medial column of the spinal cord derived from the ipsilateral motor cortex; contributes to postural control.

ventral anterior nuclei Nuclei in the ventral tier of the thalamus that receive input from the basal ganglia and cerebellum and project to the motor cortex.

ventral columns The ventral (anterior) and ventrolateral (anterolateral) regions of spinal cord white matter that convey both ascending information about pain and temperature, and descending motor information from the brainstem and motor cortex concerned with postural control and gain adjustment. Also known as *ventrolateral columns* or *anterolateral columns*.

ventral horns The ventral portion of the spinal cord gray matter derived from the basal plate; populated by interneurons and primary motor neurons.

ventral lateral nuclei Nuclei in the ventral tier of the thalamus that receive input from the basal ganglia and cerebellum and project to the motor cortex.

ventral pallidum A structure within the basal ganglia whose fibers project to thalamic nuclei, such as the mediodorsal nucleus.

ventral posterior medial (VPM) nucleus A component of the ventral posterior complex of thalamic nuclei that receives brainstem projections carrying somatic sensory information

from the face including the inputs from the facial, glossopharyngeal and vagal nerve that innervate the taste buds in the tongue peripherally and the gustatory nucleus portion of the solitary nucleus in the brainstem.

ventral roots The collection of nerve fibers containing motor axons that exit ventrally from the spinal cord and contribute the motor component of each segmental spinal nerve.

ventral stream Cortical visual pathways that originate in primary visual cortex and extend into the temporal cortex. This pathway processes information important for object recognition and visual perception.

ventral tegmental area (VTA) A part of the midbrain that contains many dopaminergic neurons and is important for reward and learning.

ventricles The spaces in the vertebrate brain that are filled with cerebrospinal fluid and represent the lumen of the embryonic neural tube.

ventrolateral columns See *ventral columns*.

ventrolateral prefrontal cortex (VLPFC) A functional division of the prefrontal cortex roughly corresponding to the inferior frontal gyrus and surrounding sulci, as located anterior to motor cortex. Compare *dorsolateral prefrontal cortex*.

ventromedial prefrontal cortex (VMPFC) The ventral portion of the prefrontal cortex surrounding the hemispheric midline; plays a key role in the control of emotions and social behavior.

vergence movements Disjunctive movements of the eyes (convergence or divergence) that align the fovea of each eye with targets located at different distances from the observer.

vermis Median zone of the spinocerebellum that receives proprioceptive input from the spinal cord and sends output to brainstem upper motor neurons; coordinates movements of proximal muscles, including eye movements.

vertebral arteries Major source of posterior circulation to hindbrain and posterior forebrain.

See **vestibular nerve ganglion** Contains the bipolar afferent neurons that innervate the semicircular canals and otolith organs of the auditory vestibule. Also called *Scarpa's ganglion*.

vestibular nuclei Clusters of neurons in the medulla that receive direct innervation from the vestibular nerve.

vestibulo-ocular movements Involuntary movement of the eyes in response to displacement of the head; this reflex allows retinal images to remain stable during head movement.

vestibulo-ocular reflex (VOR) Involuntary movement of the eyes in response to displacement of the head. This reflex allows retinal images to remain stable while the head is moved.

vestibulocerebellum Caudal-inferior lobes of the cerebellum, including the flocculus and nodulus, that receives input from the vestibular nuclei in the brainstem and the vestibular nerves; concerned with the vestibulo-ocular reflex and the coordination of movements that maintain posture and equilibrium.

vestibulocochlear nerve (VIII) Cranial nerve VIII, a sensory nerve that conveys vestibular afferents from the various components of the vestibular labyrinth and auditory afferents from the cochlea to the vestibular and cochlear nuclei, respectively, in the dorsolateral caudal pons and rostral medulla. Lateral-most of three cranial nerves that attaches to the brainstem at the junction of the pons and medulla.

visceral motor division The components of the nervous system (peripheral and central) concerned with the regulation of smooth muscle, cardiac muscle, and glands; organized anatomically and physiologically into sympathetic, parasympathetic, and enteric divisions. Also known as the *autonomic nervous system* or *autonomic motor division*.

visceral motor nuclei Nuclei (derived from the basal plate) in the brainstem and spinal cord that give rise to efferent fibers innervating smooth muscle, cardiac muscle or glands. In the brainstem, located in the dorsal tegmentum just lateral to somatic motor nuclei; in the spinal cord, located in the intermediolateral cell column of thoracic and sacral segments.

visceral motor system See *autonomic nervous system*.

visible light The portion of the electromagnetic radiation that is detectable by the human eye. The human eye can respond to electromagnetic radiation whose wavelength falls between 380 and 750 nanometers.

visual acuity The measure of an eye (or an individual) to be able to recognize small details and distinguish objects within the center of the visual field.

visual field The area in the external world normally seen by one or both eyes (referred to, respectively, as the monocular and binocular fields).

visual receptive field A region of visual space in retinal neurons and the brain's visual centers where both the presence and properties of light will alter its activity or firing pattern.

visual search The controlled allocation of attention across elements within a visual display, with the goal of identifying an element that matches some target property.

vitreous humor A gelatinous substance that fills the space between the back of the lens and the surface of the retina.

vocal folds Source of vocal vibration in the larynx. Synonymous with *vocal cords*.

voltage clamp method A technique that uses electronic feedback to simultaneously control the membrane potential of a cell and measure the transmembrane currents that result from the opening and closing of ion channels.

voltage gated Term used to describe ion channels whose opening and closing is sensitive to membrane potential.

voltage sensor Charged structure within a membrane-spanning domain of an ion channel that confers the ability to sense changes in transmembrane potential.

vomeronasal organs (VNO) A pair of chemical sensing organs in the septum (medial process) of the olfactory epithelium that are specialized for the detection and transduction of specific classes of volatile chemicals, pheromones and kairomones. The sensory neurons of the VNO are bipolar vomeronasal sensory receptor neurons that resemble olfactory receptor neurons in the olfactory epithelium. The VNO is the site of expression of a distinct family of GPCR chemosensory receptors that specifically bind pheromones and kairomones.

vomeronasal receptor neurons (VRNs) A class of bipolar chemosensory neurons found in the vomeronasal organ that uniquely express vomeronasal receptors, and whose axons project to the accessory olfactory bulb.

vomeronasal receptors (VRs) A large class of 7-transmembrane G-protein coupled receptor proteins (GPCRs) that bind and transduce phermonal and kairomonal signals. Vomeronasal receptor proteins are localized only to vomeronasal receptor neurons in the vomeronasal organ.

vomeronasal system A specialized chemical detection system that detects pheromones—volatile chemicals released into the air from conspecifics to regulate social interactions—or kairomones—volatile chemicals from other species indicating predator, prey, or symbiotic status. The vomeronasal system includes the peripheral sensory organ, the vomeronasal organ adkacent to the olfactory epithelium in the nose, and its primary target in the forebrain, the accessory olfactory bulb.

W

Wada test A procedure sometimes used as a diagnostic tool to determine the location of the speech and language cortex in preparation for neurosurgery. Involves carotid injection of an anesthetic agent.

Wernicke's aphasia Difficulty comprehending speech as a result of damage to Wernicke's language area. Also called *sensory* or *receptive aphasia*.

Wernicke's area Region of cortex in the superior and posterior region of the left temporal lobe that helps mediate language comprehension. Named after the nineteenth-century neurologist Carl Wernicke.

white matter A general term that refers to regions of the brain and spinal cord containing large axonal tracts; the phrase derives from the fact that axonal tracts have a whitish cast when viewed in the freshly cut material due to the abundance of myelin.

wide-dynamic-range neurons Multimodal lamina V neurons that receive converging inputs from nociceptive and non-nociceptive afferents, a quality which makes them a likely substrate for referred pain.

Wisconsin Card Sorting Task A cognitive test that involves classifying a set of cards, each showing one or more images of a simple shape, into categories based on rules that periodically change throughout the session.

Wnt A large family of secreted ligands that regulate stem and precursor cell proliferation, transcriptional activation/repression, and differentiation within and beyond the nervous system.

working memory Memory for information held briefly in mind, typically to help accomplish some specific task.

Box References

CHAPTER 1 Studying the Nervous System

BOX 1A Model Organisms in Neuroscience

Bockamp, E. and 7 others (2002) Of mice and models: Improved animal models for biomedical research. *Physiol. Genomics* 11: 115–132.

Muquit, M. M. and M. B. Feany (2002) Modelling neurodegenerative diseases in *Drosophila*: A fruitful approach? *Nat. Rev. Neurosci.* 3: 237–243.

Rinkwitz, S., P. Mourrain and T. S. Becker (2011) Zebrafish: An integrative system for neurogenomics and neurosciences. *Prog. Neurobiol.* 93: 231–243.

Sengupta, P. and A. D. Samuel (2009) *Caenorhabditis elegans*: A model system for systems neuroscience. *Curr. Opin. Neurobiol.* 19: 637–643.

CHAPTER 2 Electrical Signals of Nerve Cells

CLINICAL APPLICATIONS Anesthesia and Neuronal Electrical Signaling

Franks, N. P. (2006) Molecular targets underlying general anaesthesia. *Br. J. Pharmacol.* 147: S72–S81.

Heurteaux, C. and 10 others (2004) TREK-1, a K$^+$ channel involved in neuroprotection and general anesthesia. *EMBO J.* 23: 2684–2695.

Hille, B. (1977) Local anesthetics: Hydrophilic and hydrophobic pathways for the drug-receptor reaction. *J. Gen. Physiol.* 69: 497–515.

Kopp Lugli, A., C. S. Yost and C. H. Kindler (2009) Anaesthetic mechanisms: Update on the challenge of unravelling the mystery of anaesthesia. *Eur. J. Anaesthesiol.* 26: 807–820.

Lirk, P., S. Picardi and M. W. Hollmann (2014) Local anaesthetics: 10 essentials. *Eur. J. Anaesthesiol.* 31: 575–585.

Lizarraga, I., J. P. Chambers and C. B. Johnson (2008) Synergistic depression of NMDA receptor-mediated transmission by ketamine, ketoprofen and L-NAME combinations in neonatal rat spinal cords in vitro. *Br. J. Pharmacol.* 153: 1030–1042.

Magorian, T., K. B. Flannery and R. D. Miller (1993) Comparison of rocuronium, succinylcholine, and vecuronium for rapid-sequence induction of anesthesia in adult patients. *Anesthesiology* 79: 913–918.

Pavel, M. A., E. N. Petersen, H. Wang, R. A. Lerner and S. B. Hansen (2020) Studies on the mechanism of general anesthesia. *Proc. Natl. Acad. Sci U.S.A.* 117: 13757–13766.

Schaller, S. J. and H. Fink (2013) Sugammadex as a reversal agent for neuromuscular block: An evidence-based review. *Core Evid.* 8: 57–67.

Scholz, A. (2002) Mechanisms of (local) anaesthetics on voltage-gated sodium and other ion channels. *Br. J. Anaesth.* 89: 52–61.

Sirois, J. E., J. J. Pancrazio, C. Lynch and D. A. Bayliss (1998) Multiple ionic mechanisms mediate inhibition of rat motoneurons by inhalation anaesthetics. *J. Physiol.* 512: 851–862.

Thomson, A. M., A. P. Bannister, D. I. Hughes and H. Pawelzik (2000) Differential sensitivity to Zolpidem of IPSPs activated by morphologically identified CA1 interneurons in slices of rat hippocampus. *Eur. J. Neurosci.* 12: 425–436.

Zanos, P. and T. D. Gould (2018) Mechanisms of ketamine action as an antidepressant. *Mol. Psychiatry* 23: 801–811.

BOX 2A The Remarkable Giant Nerve Cells of Squid

Llinás, R. (1999) *The Squid Synapse: A Model for Chemical Transmission.* Oxford, UK: Oxford University Press.

Young, J. Z. (1939) Fused neurons and synaptic contacts in the giant nerve fibres of cephalopods. *Philos. Trans. R. Soc. Lond. B* 229: 465–503.

BOX 2B Action Potential Form and Nomenclature

Barrett, E. F. and J. N. Barrett (1976) Separation of two voltage-sensitive potassium currents, and demonstration of a tetrodotoxin-resistant calcium current in frog motoneurones. *J. Physiol.* 255: 737–774.

Chen, S., G. J. Augustine and P. Chadderton (2016) The cerebellum linearly encodes whisker position during voluntary movement. *eLife* 5: e10509.

Dodge, F. A. and B. Frankenhaeuser (1958) Membrane currents in isolated frog nerve fibre under voltage clamp conditions. *J. Physiol.* 143: 76–90.

Hodgkin, A. L. and A. F. Huxley (1939) Action potentials recorded from inside a nerve fibre. *Nature* 144: 710–711.

Llinás, R. and Y. Yarom (1981) Electrophysiology of mammalian inferior olivary neurones *in vitro*. Different types of voltage-dependent ionic conductances. *J. Physiol.* 315: 549–567.

CHAPTER 3 Voltage-Dependent Membrane Permeability

BOX 3A The Voltage Clamp Method

Cole, K. S. (1968) *Membranes, Ions and Impulses: A Chapter of Classical Biophysics.* Berkeley, CA: University of California Press.

CLINICAL APPLICATIONS   Multiple Sclerosis

Bhat, R. and L. Steinman (2009) Innate and adaptive autoimmunity directed to the central nervous system. *Neuron* 64: 123–132.

Derfuss, T. and 18 others (2009) Contactin-2/TAG-1-directed autoimmunity is identified in multiple sclerosis patients and mediates gray matter pathology in animals. *Proc. Natl. Acad. Sci. U.S.A.* 106: 8302–8307.

Mahad, D. H., B. D. Trapp and H. Lassmann (2015) Pathological mechanisms in progressive multiple sclerosis. *Lancet Neurol.* 14: 183–193.

Ransohoff, R. M. (2007) Natalizumab for multiple sclerosis. *New Engl. J. Med.* 356: 2622–2629.

Trapp, B. D and P. K. Stys (2009) Virtual hypoxia and chronic necrosis of demyelinated axons in multiple sclerosis. *Lancet Neurol.* 8: 280–291.

Trapp, B. D. et al. (1998) Axonal transection in the lesions of multiple sclerosis. *N. Engl. J. Med.* 338: 278–285.

Waxman, S. G. (2006) Ions, energy, and axonal injury: Towards a molecular neurology of multiple sclerosis. *Trends Mol. Med.* 12: 192–195.

Zanvil, S. S. and L. Steinman (2003) Diverse targets for intervention during inflammatory and neurodegenerative phases of multiple sclerosis. *Neuron* 38: 685–688.

CHAPTER 4   Ion Channels and Transporters

BOX 4A   The Patch Clamp Method

Dunlop, J., M. Bowlby, R. Peri, D. Vasilyev and R. Arias (2008) High-throughput electrophysiology: An emerging paradigm for ion-channel screening and physiology. *Nat. Rev. Drug Discov.* 7: 358–368.

Hamill, O. P., A. Marty, E. Neher, B. Sakmann and F. J. Sigworth (1981) Improved patch-clamp techniques for high-resolution current recording from cells and cell-free membrane patches. *Pflügers Arch.* 391: 85–100.

Levis, R. A. and J. L. Rae (1998) Low-noise patch-clamp techniques. *Meth. Enzym.* 293: 218–266.

Sakmann, B. and E. Neher (1995) *Single-Channel Recording*, 2nd Edition. New York: Plenum Press.

BOX 4B   Toxins That Poison Ion Channels

Cahalan, M. (1975) Modification of sodium channel gating in frog myelinated nerve fibers by *Centruroides sculpturatus* scorpion venom. *J. Physiol.* 244: 511–534.

Catterall, W. A. and 5 others (2007) Voltage-gated ion channels and gating modifier toxins. *Toxicon* 49: 124–141.

Dutertre, S. and R. J. Lewis (2010) Use of venom peptides to probe ion channel structure and function. *J. Biol. Chem.* 285: 13315–13320.

Green, B. R. and 9 others (2016) Structural basis for the inhibition of voltage-gated sodium channels by conotoxin μO§-GVIIJ. *J. Biol. Chem.* 291: 7205–7220.

Narahashi, T. (2008) Tetrodotoxin: A brief history. *Proc. Jpn. Acad. Ser. B Phys. Biol. Sci.* 84: 147–154.

Schmidt, O. and H. Schmidt (1972) Influence of calcium ions on the ionic currents of nodes of Ranvier treated with scorpion venom. *Pflügers Arch.* 333: 51–61.

CLINICAL APPLICATIONS   Neurological Diseases Caused by Altered Ion Channels

Baig, S. M. and 16 others (2011) Loss of Ca$_v$1.3 (*CACNA1D*) function in a human channelopathy with bradycardia and congenital deafness. *Nat. Neurosci.* 14: 77–84.

de Lera Ruiz, M. and R. L. Kraus (2015) Voltage-gated sodium channels: Structure, function, pharmacology, and clinical indications. *J. Med. Chem.* 58: 7093–7118.

Escayg, A. and A. L. Goldin (2010) Sodium channel SCN1A and epilepsy: Mutations and mechanisms. *Epilepsia* 51: 1650–1658.

Hoeijmakers, J. G. et al. (2012) Small nerve fibres, small hands and small feet: A new syndrome of pain, dysautonomia and acromesomelia in a kindred with a novel Nav1.7 mutation. *Brain* 135: 345–358.

Shieh, C.-C., M. Coghlan, J. P. Sullivan and M. Gopalakrishnan (2000) Potassium channels: Molecular defects, diseases, and therapeutic opportunities. *Pharmacol. Rev.* 52: 557–593.

Spillane, J., D. M. Kullmann and M. G. Hanna (2016) Genetic neurological channelopathies: Molecular genetics and clinical phenotypes. *J. Neurol. Neurosurg. Psychiatry* 87: 37–48.

Waxman, S. G and G. W. Zamponi (2014) Regulating excitability of peripheral afferents: Emerging ion channel targets. *Nat. Neurosci.* 17: 153–163.

Zamponi, G. W., J. Striessnig, A. Koschak and A. C. Dolphin (2015) The physiology, pathology, and pharmacology of voltage-gated calcium channels and their future therapeutic potential. *Pharmacol. Rev.* 67: 821–870.

CHAPTER 5   Synaptic Transmission

CLINICAL APPLICATIONS   Diseases That Affect the Presynaptic Terminal

Chen, J., S. Yu, Y. Fu and X. Li (2014) Synaptic proteins and receptors defects in autism spectrum disorders. *Front. Cell. Neurosci.* 8: 276.

Engel, A. G. (1994) Congenital myasthenic syndromes. *Neurol. Clin.* 12: 401–437.

Gart, M. S. and K. A. Gutowski (2016) Overview of botulinum toxins for aesthetic uses. *Clin. Plast. Surg.* 43: 459–471.

Grumelli, C. and 5 others (2005) Internalization and mechanism of action of clostridial toxins in neurons. *Neurotoxicology* 26: 761–767.

Hülsbrink, R. and S. Hashemolhosseini (2014) Lambert-Eaton myasthenic syndrome—diagnosis, pathogenesis and therapy. *Clin. Neurophysiol.* 125: 2328–2336.

Humeau, Y., F. Doussau, N. J. Grant and B. Poulain (2000) How botulinum and tetanus neurotoxins block neurotransmitter release. *Biochimie* 82: 427–446.

Maselli, R. A. (1998) Pathogenesis of human botulism. *Ann. N. Y. Acad. Sci.* 841: 122–139.

Silva, J. P., J. Suckling and Y. Ushkaryov (2009) Penelope's web: Using α-latrotoxin to untangle the mysteries of exocytosis. *J. Neurochem.* 111: 275–290.

Südhof, T. C. (2008) Neuroligins and neurexins link synaptic function to cognitive disease. *Nature* 455: 903–911.

Sutton, R. B., D. Fasshauer, R. Jahn and A. T. Brünger (1998) Crystal structure of a SNARE complex involved in synaptic exocytosis at 2.4 Å resolution. *Nature* 395: 347–353.

Vincent, A. (2010) Autoimmune channelopathies: Well-established and emerging immunotherapy-responsive diseases of the peripheral and central nervous systems. *J. Clin. Immunol.* 30: S97–S102.

BOX 5A The Tripartite Synapse

Buchanan, J. et al. (2022) Oligodendrocyte precursor cells ingest axons in the mouse neocortex. *Proc. Natl. Acad. Sci. USA* 119 (48) e2202580119..

Cornell-Bell, A. H., S. M. Finkbeiner, M. S. Cooper and S. J. Smith (1990) Glutamate induces calcium waves in cultured astrocytes: Long-range glial signaling. *Science* 247: 470–473.

Fiacco, T. A., C. Agulhon and K. D. McCarthy (2009) Sorting out astrocyte physiology from pharmacology. *Annu. Rev. Pharmacol. Toxicol.* 49: 151–174.

Han, X. and 13 others (2013) Forebrain engraftment by human glial progenitor cells enhances synaptic plasticity and learning in adult mice. *Cell Stem Cell* 12: 342–353.

Haydon, P. G. and M. Nedergaard (2014) How do astrocytes participate in neural plasticity? *Cold Spring Harb. Perspect. Biol.* 7: a020438.

Jahromi, B. S., R. Robitaille and M. P. Charlton (1992) Transmitter release increases intracellular calcium in perisynaptic Schwann cells in situ. *Neuron* 8: 1069–1077.

Lee, J. H. and 7 others (2021) Astrocytes phagocytose adult hippocampal synapses for circuit homeostasis. *Nature* 590: 612–617.

Lee, S. and 7 others (2010) Channel-mediated tonic GABA release from glia. *Science* 330: 790–796.

Olsen, M. L. and 5 others (2015) New insights on astrocyte ion channels: Critical for homeostasis and neuron-glia signaling. *J. Neurosci.* 35: 13827–13835.

Paolicelli, R. C. and 11 others (2011) Synaptic pruning by microglia is necessary for normal brain development. *Science* 333: 1456–1458.

Perea, G. and A. Araque (2007) Astrocytes potentiate transmitter release at single hippocampal synapses. *Science* 317: 1083–1086.

Perea, G., M. Navarrete and A. Araque (2009) Tripartite synapses: Astrocytes process and control synaptic information. *Trends Neurosci.* 32: 421–431.

Wang, C. and 11 others (2020) Microglia mediate forgetting via complement-dependent synaptic elimination. *Science* 367: 688–694.

Witcher, M. R., S. A. Kirov and K. M. Harris (2007) Plasticity of perisynaptic astroglia during synaptogenesis in the mature rat hippocampus. *Glia* 55: 13–23.

CHAPTER 6 Neurotransmitters and Their Receptors

BOX 6A Neurotoxins That Act on Neurotransmitter Receptors

Han, T. S., R. W. Teichert, B. M. Olivera and G. Bulaj (2008) Conus venoms: A rich source of peptide-based therapeutics. *Curr. Pharm. Des.* 14: 2462–2479.

Lebbe, E. K. M., S. Peigneur, I. Wijesekara and J. Tytgat (2014) Conotoxins targeting nicotinic acetylcholine receptors: An overview. *Mar. Drugs* 12: 2970–3004.

Lewis, R. L. and L. Gutmann (2004) Snake venoms and the neuromuscular junction. *Semin. Neurol.* 24: 175–179.

Tsetlin, V. I. (2015) Three-finger snake neurotoxins and Ly6 proteins targeting nicotinic acetylcholine receptors: Pharmacological tools and endogenous modulators. *Trends Pharmacol. Sci.* 36: 109–123.

CLINICAL APPLICATIONS Myasthenia Gravis: An Autoimmune Disease of Neuromuscular Synapses

Elmqvist, D., W. W. Hofmann, J. Kugelberg and D. M. J. Quastel (1964) An electrophysiological investigation of neuromuscular transmission in myasthenia gravis. *J. Physiol.* 174: 417–434.

Farrugia, M. E. and A. Vincent (2010) Autoimmune mediated neuromuscular junction defects. *Curr. Opin. Neurol.* 23: 489–495.

Gilhus, N. E. (2016) Myasthenia gravis. *New Engl. J. Med.* 375: 2570–2581.

Harvey, A. M. and J. L. Lilienthal (1941) Observations on the nature of myasthenia gravis: The intra-arterial injection of acetylcholine, prostigmine, and adrenaline. *Bull. Johns Hopkins Hosp.* 69: 566–577.

Harvey, A. M., J. L. Lilienthal Jr. and S. A. Talbot (1942) Observations on the nature of myasthenia gravis: The effect of thymectomy on neuro-muscular transmission. *J. Clin. Invest.* 21(5): 579–588.

Patrick, J. and J. Lindstrom (1973) Autoimmune response to acetylcholine receptor. *Science* 180: 871–872.

Vincent, A. (2002) Unravelling the pathogenesis of myasthenia gravis. *Nat. Rev. Immunol.* 2: 797–804.

BOX 6B Excitatory Actions of GABA in the Developing Brain

Berglund, K. and 8 others (2006) Imaging synaptic inhibition in transgenic mice expressing the chloride indicator, Clomeleon. *Brain Cell Biol.* 35: 207–228.

Cherubini, E., J. L. Gaiarsa and Y. Ben-Ari (1991) GABA: An excitatory transmitter in early postnatal life. *Trends Neurosci.* 14: 515–519.

Glykys, J. and 7 others (2009) Differences in cortical versus subcortical GABAergic signaling: A candidate mechanism of electroclinical uncoupling of neonatal seizures. *Neuron* 63: 657–672.

Obata, K., M. Oide and H. Tanaka (1978) Excitatory and inhibitory actions of GABA and glycine on embryonic chick spinal neurons in culture. *Brain Res.* 144: 179–184.

Owens, D. F. and A. R. Kriegstein (2002) Is there more to GABA than synaptic inhibition? *Nat. Rev. Neurosci.* 3: 715–727.

Payne, J. A., C. Rivera, J. Voipio and K. Kaila (2003) Cation-chloride co-transporters in neuronal communication, development and trauma. *Trends Neurosci.* 26: 199–206.

Rivera, C. and 8 others (1999) The K$^+$/Cl$^-$ co-transporter KCC2 renders GABA hyperpolarizing during neuronal maturation. *Nature* 397: 251–255.

BOX 6C Marijuana and the Brain

Adams, A. R. (1941) Marihuana. *Harvey Lect.* 37: 168.

Freund, T. F., I. Katona and D. Piomelli (2003) Role of endogenous cannabinoids in synaptic signaling. *Physiol. Rev.* 83: 1017–1066.

Gerdeman, G. L., J. G. Partridge, C. R. Lupica and D. M. Lovinger (2003) It could be habit forming: Drugs of abuse and striatal synaptic plasticity. *Trends Neurosci.* 26: 184–192.

Howlett, A. C. (2005) Cannabinoid receptor signaling. *Handb. Exp. Pharmacol.* 168: 53–79.

Iversen, L. (2003) *Cannabis* and the brain. *Brain* 126: 1252–1270.

Mechoulam, R. (1970) Marihuana chemistry. *Science* 168: 1159–1166.

Onaivi, E. S. (2009) Cannabinoid receptors in brain: Pharmacogenetics, neuropharmacology, neurotoxicology, and potential therapeutic applications. *Int. Rev. Neurobiol.* 88: 335–369.

Shao, Z. and 6 others (2016) High-resolution crystal structure of the human CB1 cannabinoid receptor. *Nature* 540: 602–606.

CHAPTER 7   Molecular Signaling within Neurons

BOX 7A   Dynamic Imaging of Intracellular Signaling

Chalfie, M., Y. Tu, G. Euskirchen, W. W. Ward and D. C. Prasher (1994) Green fluorescent protein as a marker for gene expression. *Science* 263: 802–805.

Connor, J. A. (1986) Digital imaging of free calcium changes and of spatial gradients in growing processes in single mammalian central nervous system cells. *Proc. Natl. Acad. Sci. U.S.A.* 83: 6179–6183.

Finch, E. A. and G. J. Augustine (1998) Local calcium signaling by IP$_3$ in Purkinje cell dendrites. *Nature* 396: 753–756.

Grynkiewicz, G., M. Poenie and R. Y. Tsien (1985) A new generation of Ca^{2+} indicators with greatly improved fluorescence properties. *J. Biol. Chem.* 260: 3440–3450.

Livet, J. and 7 others (2007) Transgenic strategies for combinatorial expression of fluorescent proteins in the nervous system. *Nature* 450: 56–62.

Rodriguez, E. A. and 8 others (2017) The growing and glowing toolbox of fluorescent and photoactive proteins. *Trends Biochem. Sci.* 42: 111–129.

Shimomura, O. (2009) Discovery of green fluorescent protein (GFP) (Nobel Lecture). *Angew Chem. Int. Ed. Engl.* 48: 5590–5602.

Tsien, R. Y. (2010) Nobel lecture: Constructing and exploiting the fluorescent protein paintbox. *Integr. Biol. (Camb.)* 2: 77–93.

Vidal, G. S., M. Djurisic, K. Brown, R. W. Sapp and C. J. Shatz (2016) Cell-autonomous regulation of dendritic spine density by PirB. *eNeuro* 3: 1–15.

CLINICAL APPLICATIONS   Molecular Basis of Psychiatric Disorders

Charnet, D. S., J. D. Buxbaum, P. Sklar and E. J. Nestler (2014) *Neurobiology of Mental Illness*, 4th Edition. New York: Oxford University Press.

Craddock, N. and L. Forty (2006) Genetics of affective (mood) disorders. *Eur. J. Human Gen.* 14: 660–668.

Duric, V. and 7 others (2010) A negative regulator of MAP kinase causes depressive behavior. *Nat. Med.* 16: 1328–1332.

Howe, A. S. and 22 others (2016) Candidate genes in panic disorder: Meta-analyses of 23 common variants in major anxiogenic pathways. *Mol. Psychiatry* 21: 665–679.

Kambeitz, J., A. Abi-Dargham, S. Kapur and O. D. Howes (2014) Alterations in cortical and extrastriatal subcortical dopamine function in schizophrenia: Systematic review and meta-analysis of imaging studies. *Br. J. Psychiatry* 204: 420–429.

Karam, C. S. and 8 others (2010) Signaling pathways in schizophrenia: Emerging targets and therapeutic strategies. *Trends Pharmacol. Sci.* 31: 381–390.

Krishnan, V. and E. J. Nestler (2008) The molecular neurobiology of depression. *Nature* 457: 894–902.

Lakhan, S. E., M. Caro and N. Hadzimichalis (2013) NMDA receptor activity in neuropsychiatric disorders. *Front. Psychiatry* 4: 52.

Margolis, R. L. (2009) Neuropsychiatric disorders: The choice of antipsychotics in schizophrenia. *Nat. Rev. Neurosci.* 5: 308–310.

Marshall, M. (2020) The hidden links between mental disorders. *Nature* 581: 19–21.

Sharp, T. and P. J. Cowen (2011) 5-HT and depression: Is the glass half-full? *Curr. Opin. Pharmacol.* 11: 45–51.

Shin, J. K., D. T. Malone, I. T. Crosby and B. Capuano (2011) Schizophrenia: A systematic review of the disease state, current therapeutics, and their molecular mechanisms of action. *Curr. Med. Chem.* 18: 1380–1404.

Sokolowska, E. and I. Hovatta (2013) Anxiety genetics—findings from cross-species genome-wide approaches. *Biol. Mood Anxiety Disord.* 3: 9.

Stein, K., A. A. Maruf, D. J. Müller, J. R. Bishop and C. A. Bousman (2021) Serotonin transporter genetic variation and antidepressant response and tolerability: A systematic review and meta-analysis. *J. Pers. Med.* 11: 1334.

BOX 7B   Dendritic Spines

Bhatt, D. H., S. Zhang and W. B. Gan (2009) Dendritic spine dynamics. *Ann. Rev. Physiol.* 71: 261–282.

Cornejo, V. H., N. Ofer and R. Yuste (2022) Voltage compartmentalization in dendritic spines in vivo. *Science* 375: 82–86.

Goldberg, J. H., G. Tamas, D. Aronov and R. Yuste (2003) Calcium microdomains in aspiny dendrites. *Neuron* 40: 807–821.

Harnett, M. T., J. K. Makara, N. Spruston, W. L. Kath and J. C. Magee (2012) Synaptic amplification by dendritic spines enhances input cooperativity. *Nature* 491: 599–602.

Harris, K. M. and R. J. Weinberg (2012) Ultrastructure of synapses in the mammalian brain. *Cold Spring Harb. Perspect. Biol.* 4: a005587.

Nishiyama, J. and R. Yasuda (2015) Biochemical computation for spine structural plasticity. *Neuron* 87: 63–75.

Noguchi, J., M. Matsuzaki, G. C. Ellis-Davies and H. Kasai (2005) Spine-neck geometry determines NMDA receptor-dependent Ca^{2+} signaling in dendrites. *Neuron* 46: 609–622.

Penzes, P., M. E. Cahill, K. A. Jones, J. E. VanLeeuwen and K. M. Woolfrey (2011) Dendritic spine pathology in neuropsychiatric disorders. *Nat. Neurosci.* 14: 285–293.

Popovic, M. A., N. Carnevale, B. Rozsa and D. Zecevic (2015) Electrical behaviour of dendritic spines as revealed by voltage imaging. *Nat. Commun.* 6: 8436.

Sabatini, B. L., T. G. Oertner and K. Svoboda (2002) The life cycle of Ca^{2+} ions in dendritic spines. *Neuron* 33: 439–452.

Santamaria, F., S. Wils, E. De Schutter and G. J. Augustine (2006) Anomalous diffusion in Purkinje cell dendrites caused by spines. *Neuron* 52: 635–648.

Sheng, M. and E. Kim (2011) The postsynaptic organization of synapses. *Cold Spring Harb. Perspect. Biol.* 3: a00567.

Spacek, J., MUDr., DrSc., FRMS, Professor of Pathology; Charles University Prague University Hospital, Hradec Kralove, Czech Republic; spacek@lfhk.cuni.cz.

Vallés, A. S. and F. J. Barrantes (2021) Nanoscale sub-compartmentalization of the dendritic spine compartment. *Biomolecules* 11: 1697.

CHAPTER 8   Synaptic Plasticity

BOX 8A   Genetics of Learning and Memory in the Fruit Fly

Androschuk, A., B. Al-Jabri and F. V. Bolduc (2015) From learning to memory: What flies can tell us about intellectual disability treatment. *Front. Psychiatry* 6: 85.

Davis, R. L. (2004) Olfactory learning. *Neuron* 44: 31–48.

Liao, D., N. A. Hessler and R. Malinow (1995) Activation of postsynaptically silent synapses during pairing-induced LTP in CA1 region of hippocampal slice. *Nature* 375: 400–404.

Petralia, R. S. and 6 others (1999) Selective acquisition of AMPA receptors over postnatal development suggests a molecular basis for silent synapses. *Nat. Neurosci.* 2: 31–36.

Quinn, W. G., W. A. Harris and S. Benzer (1974) Conditioned behavior in *Drosophila melanogaster. Proc. Natl. Acad. Sci. U.S.A.* 71: 708–712.

Tully, T. (1996) Discovery of genes involved with learning and memory: An experimental synthesis of Hirshian and Benzerian perspectives. *Proc. Natl. Acad. Sci. U.S.A.* 93: 13460–13467.

Waddell, S. and W. G. Quinn (2001) Flies, genes, and learning. *Annu. Rev. Neurosci.* 24: 1283–1309.

Weiner, J. (1999) *Time, Love, Memory: A Great Biologist and His Quest for the Origins of Behavior.* New York: Knopf.

BOX 8B   Silent Synapses

Derkach, V. A., M. C. Oh, E. S. Guire and T. R. Soderling (2007) Regulatory mechanisms of AMPA receptors in synaptic plasticity. *Nat. Rev. Neurosci.* 8: 101–113.

Gomperts, S. N., A. Rao, A. M. Craig, R. C. Malenka and R. A. Nicoll (1998) Postsynaptically silent synapses in single neuron cultures. *Neuron* 21: 1443–1451.

Huang, Y. H. and 12 others (2009) In vivo cocaine experience generates silent synapses. *Neuron* 63: 40–47.

Liao, D., N. A. Hessler and R. Malinow (1995) Activation of postsynaptically silent synapses during pairing-induced LTP in CA1 region of hippocampal slice. *Nature* 375: 400–404.

Luscher, C., R. A. Nicoll, R. C. Malenka and D. Muller (2000) Synaptic plasticity and dynamic modulation of the postsynaptic membrane. *Nat. Neurosci.* 3: 545–550.

Petralia, R. S. and 6 others (1999) Selective acquisition of AMPA receptors over post–natal development suggests a molecular basis for silent synapses. *Nat. Neurosci.* 2: 31–36.

CLINICAL APPLICATIONS   Epilepsy: The Effect of Pathological Activity on Neural Circuitry

Dyro, F. M. (1989) *The EEG Handbook.* Boston: Little, Brown.

Engel, J., Jr. and T. A. Pedley (eds.) (2008) *Epilepsy: A Comprehensive Textbook,* 2nd Edition. Philadelphia: Lippincott-Raven.

McNamara, J. O., Y. Z. Huang and A. S. Leonard (2006) Molecular signaling mechanisms underlying epileptogenesis. *Sci. STKE* 356: re12.

Scheffer, I. E. (2014) Epilepsy genetics revolutionizes clinical practice. *Neuropediatrics* 45: 70–74.

CHAPTER 9   Vision

BOX 9A   The Importance of Context in Color Perception

Land, E. (1986) Recent advances in Retinex theory. *Vision Res.* 26: 7–21.

Purves, D. and R. B. Lotto (2011) *Why We See What We Do Redux: An Empirical Theory of Vision.* Sunderland, MA: Sinauer Associates, chapters 2 and 3, pp. 15–91.

BOX 9B   The Perception of Light Intensity

Adelson, E. H. (1999) Light perception and lightness illusions. In *The Cognitive Neurosciences,* 2nd Edition, M. Gazzaniga (ed.). Cambridge, MA: MIT Press, pp. 339–351.

Purves, D. and R. B. Lotto (2011) *Why We See What We Do Redux: An Empirical Theory of Vision.* Sunderland, MA: Sinauer Associates, chapters 2 and 3, pp. 15–91.

Purves, D., Y. Morgenstern and W. T. Wojtach (2015) Perception and reality: Why a wholly empirical paradigm is needed to understand vision. *Front. Syst. Neurosci.* 9: 156.

CHAPTER 10   Hearing

CLINICAL APPLICATIONS   Hearing Loss: Causes and Treatments

Kral, A. and G. M. O'Donoghue (2010) Profound deafness in childhood. *New Engl. J. Med.* 363: 1438–1450.

Lasak, J. M., P. Allen, T. McVay and D. Lewis (2014) Hearing loss: Diagnosis and management. *Prim. Care* 41: 19–31.

Litovsky, R. Y. and K. Gordon. (2016) Bilateral cochlear implants in children: Effects of auditory experience and deprivation on auditory perception. *Hear. Res.* 338: 76–87.

Moore, D. R. and R. V. Shannon (2009) Beyond cochlear implants: Awakening the deafened brain. *Nat. Neurosci.* 12: 686–691.

Niparko, J. K., and 6 others. (2010) Spoken language development in children following cochlear implantation. *JAMA* 303: 1498–1506.

Tremblay, K. L. and C. W. Miller (2014) How neuroscience relates to hearing aid amplification. *Int. J. Otolaryngol.* 2014: article 641652.

Wilson, B. S. (2015) Getting a decent (but sparse) signal to the brain for users of cochlear implants. *Hear. Res.* 322: 24–38.

Wilson, B. S. and M. F. Dorman (2008) Cochlear implants: A remarkable past and a brilliant future. *Hear. Res.* 242: 3–21.

BOX 10A   The Sweet Sound of Distortion

Jaramillo, F., V. S. Markin and A. J. Hudspeth (1993) Auditory illusions and the single hair cell. *Nature* 364: 527–529.

Planchart, A. E. (1960) A study of the theories of Giuseppe Tartini. *J. Music Theory* 4: 32–61.

Robles, L., M. A. Ruggero and N. C. Rich (1991) Two-tone distortion in the basilar membrane of the cochlea. *Nature* 439: 413–414.

CHAPTER 11 The Vestibular System

CLINICAL APPLICATIONS Clinical Evaluation of the Vestibular System

Bárány, R. (1916) Some new methods for functional testing of the vestibular apparatus and the cerebellum. Nobel Lecture, September 11, 1916. In *Nobel Lectures, Physiology or Medicine 1901–1921*. Amsterdam: Elsevier, 1967, pp. 500–511.

Ishiyama, G., I. A. Lopez, A. R. Sepahdari and A. Ishiyama (2015) Meniere's disease: Histopathology, cytochemistry, and imaging. *Ann. N. Y. Acad. Sci.* 1343: 49–57.

Mann, W. and H. T. Gouveris (2009) Diagnosis and therapy of vestibular schwannoma. *Expert Rev. Neurother.* 9(8): 1219–1232.

Sajjadi, H. and M. M. Paparella (2008) Meniere's disease. *Lancet* 372: 406–414.

BOX 11A Mauthner Cells in Fish

Eaton, R. C., R. A. Bombardieri and D. L. Meyer (1977) The Mauthner-initiated startle response in teleost fish. *J. Exp. Biol.* 66: 65–81.

Furshpan, E. J. and T. Furukawa (1962) Intracellular and extracellular responses of the several regions of the Mauthner cell of the goldfish. *J. Neurophysiol.* 25: 732–771.

Jontes, J. D., J. Buchanan and S. J. Smith (2000) Growth cone and dendrite dynamics in zebrafish embryos: Early events in synaptogenesis imaged in vivo. *Nat. Neurosci.* 3: 231–237.

Korn, H. and D. S. Faber (2005) The Mauthner cell half a century later: A neurobiological model for decision-making? *Neuron* 47: 13–28.

O'Malley, D. M., Y. H. Kao and J. R. Fetcho (1996) Imaging the functional organization of zebrafish hindbrain segments during escape behaviors. *Neuron* 17: 1145–1155.

CHAPTER 12 Touch and Proprioception

CLINICAL APPLICATIONS Dermatomes

Haymaker, W. and B. Woodhall (1967) *Peripheral Nerve Injuries: Principles of Diagnosis*. New York: American Association of Neurological Surgeons.

Rosenzweig, M. R., S. M. Breedlove and A. L. Leiman (2005) *Biological Psychology*, 3rd Edition. Sunderland, MA: Sinauer Associates.

BOX 12A Specialized Mechanosensation in Animals

Catania, K. C. and J. H. Kaas (1996) The unusual nose and brain of the star-nosed mole. *BioScience* 46: 578–586.

Sterbing-D'Angelo, S. J., M. Chadha, K. L. Marshall and C. F. Moss (2017) Functional role of airflow-sensing hairs on the bat wing. *J. Neurophysiol.* 117: 705–712.

CHAPTER 13 Pain

BOX 13A Capsaicin

Caterina, M. J. and 5 others (1997) The capsaicin receptor: A heat-activated ion channel in the pain pathway. *Nature* 389: 816–824.

Caterina, M. J. and 8 others (2000) Impaired nociception and pain sensation in mice lacking the capsaicin receptor. *Science* 288: 306–313.

Szallasi, A. and P. M. Blumberg (1999) Vanilloid (capsaicin) receptors and mechanisms. *Pharm. Rev.* 51: 159–212.

Tominaga, M. and 8 others (1998) The cloned capsaicin receptor integrates multiple pain-producing stimuli. *Neuron* 21: 531–543.

Zygmunt, P. M. and 7 others (1999) Vanilloid receptors on sensory nerves mediate the vasodilator action of anandamide. *Nature* 400: 452–457.

BOX 13B Referred Pain

Capps, J. A. and G. H. Coleman (1932) *An Experimental and Clinical Study of Pain in the Pleura, Pericardium, and Peritoneum*. New York: Macmillan.

Head, H. (1893) On disturbances of sensation with special reference to the pain of visceral disease. *Brain* 16: 1–32.

Kellgren, J. H. (1939–1942) On the distribution of pain arising from deep somatic structures with charts of segmental pain areas. *Clin. Sci.* 4: 35–46.

BOX 13C A Dorsal Column Pathway for Visceral Pain

Al-Chaer, E. D., N. B. Lawand, K. N. Westlund and W. D. Willis (1996) Visceral nociceptive input into the ventral posterolateral nucleus of the thalamus: a new function for the dorsal column pathway. *J. Neurophysiol.* 76: 2661–2674.

Al-Chaer, E. D., N. B. Lawand, K. N. Westlund and W. D. Willis (1996) Pelvic visceral input into the nucleus gracilis is largely mediated by the postsynaptic dorsal column pathway. *J. Neurophysiol.* 76: 2675–2690.

Becker, R., S. Gatscher, U. Sure and H. Bertalanffy (2001) The punctate midline myelotomy concept for visceral cancer pain control—case report and review of the literature. *Acta Neurochir.* (Suppl.) 79: 77–78.

Hirshberg, R. M., E. D. Al-Chaer, N. B. Lawand, K. N. Westlund and W. D. Willis (1996) Is there a pathway in the posterior funiculus that signals visceral pain? *Pain* 67: 291–305.

Hitchcock, E. R. (1970) Stereotactic cervical myelotomy. *J. Neurol. Neurosurg. Psychiatry* 33: 224–230.

Kim, Y. S. and S. J. Kwon (2000) High thoracic midline dorsal column myelotomy for severe visceral pain due to advanced stomach cancer. *Neurosurgery* 46: 85–90.

Nauta, H. J. W., E. Hewitt, K. N. Westlund and W. D. Willis Jr. (1997) Surgical interruption of a midline dorsal column visceral pain pathway: Case report and review of the literature. *J. Neurosurg.* 86(3): 538–542.

Nauta, H. and 8 others (2000) Punctate midline myelotomy for the relief of visceral cancer pain. *J. Neurosurg.* (*Spine 2*) 92: 125–130.

Willis, W. D., E. D. Al-Chaer, M. J. Quast and K. N. Westlund (1999) A visceral pain pathway in the dorsal column of the spinal cord. *Proc. Natl. Acad. Sci. U.S.A.* 96: 7675–7679.

CLINICAL APPLICATIONS Phantom Limbs and Phantom Pain

Barbin, J., V. Seetha, J. M. Casillas, J. Paysant and D. Pérennou (2016) The effects of mirror therapy on pain and motor control of phantom limb in amputees: A systematic review. *Ann. Phys. Rehabil. Med.* 59: 270–275. doi: 10.1016/j.rehab.2016.04.001

Melzack, R. (1990) Phantom limbs and the concept of a neuromatrix. *Trends Neurosci.* 13: 88–92.

Nashold, B. S., Jr. (1991) Paraplegia and pain. In *Deafferentation Pain Syndromes: Pathophysiology and Treatment*, B. S. Nashold, Jr. and J. Ovelmen-Levitt (eds.). New York: Raven Press, pp. 301–319.

Pons, T. P. and 5 others (1991) Massive reorganization of the primary somatosensory cortex after peripheral sensory deafferentation. *Science* 252: 1857–1860.

Ramachandran, V. S. and S. Blakeslee (1998) *Phantoms in the Brain.* New York: William Morrow & Co.

Solonen, K. A. (1962) The phantom phenomenon in amputated Finnish war veterans. *Acta. Orthop. Scand. Suppl.* 54: 1–37.

CHAPTER 14 Olfaction

CLINICAL APPLICATIONS Only One Nose

Federal Drug Administration News Release (2009) FDA advises consumers not to use certain Zicam cold remedies: Intranasal zinc product linked to loss of sense of smell. www.fda.gov/newsevents/newsroom/pressannouncements/ucm167065.htm.

Jafek, B. W., M. R. Linschoten and B. W. Murrow (2004) Anosmia after intranasal zinc gluconate use. *Amer. J. Rhinol.* 18: 137–141.

Khan, M., S.-J. Yoo, M. Clijsters, W. Backaert et al. (2021) Visualizing in deceased COVID-19 patients how SARS-CoV-2 attacks the respiratory and olfactory mucosae but spares the olfactory bulb. *Cell* 184: 5932–5949.

Lim, J. H. and 6 others (2009) Zicam-induced damage to mouse and human nasal tissue. *PLoS One* 4: e7647.

Rutty, C. J. (1996) The middle-class plague: Epidemic polio and the Canadian state, 1936–1937. *Can. Bull. Med. Hist.* 13: 277–314.

BOX 14A The "Dogtor" Is In

Church, J. and H. Williams (2001) Another sniffer dog for the clinic? *Lancet* 358: 930.

McCulloch, M. and 5 others (2006) Diagnostic accuracy of canine scent detection in early- and late-stage lung and breast cancers. *Integ. Cancer Therap.* 5: 30–39.

Phillips, M. and 7 others (2003) Detection of lung cancer with volatile markers in the breath. *Chest* 123: 2115–2123.

Willis, C. M. and 7 others (2004) Olfactory detection of human bladder cancer by dogs: Proof of principle study. *BMJ* 329: 712.

CHAPTER 15 Taste

BOX 15A Extraoral Taste Receptors and the Microbiome

D'Urso, O. and F. Drago (2021) Pharmacological significance of extra-oral taste receptors. *Eur. J. Pharmacol.* 910: 174480.

Harmon, C. P. et al. (2021) Bitter taste receptors (T2Rs) are sentinels that coordinate metabolic and immunological defense responses. *Curr. Opin. Physiol.* 20: 70–76.

Xi, R., X. Zheng and M. Tizzano (2022) Role of taste receptors in innate immunity and oral health. *J. Dent. Res.* 101(7): 759–768.

CLINICAL APPLICATIONS Ageusia and Dysgeusia: Taste Loss and Taste Alterations from COVID-19

Cooper, K. W. et al. (2020) COVID-19 and the chemical senses: Supporting players take center stage. *Neuron* 107(2): 219–233.

Hannum, M. E. et al. (2022) Taste loss as a distinct symptom of COVID-19: A systematic review and meta-analysis. *Chem. Senses* 47: bjac001. https://doi.org/10.1093/chemse/bjac001.

Zazhytska, M. et al. (2022) Non-cell-autonomous disruption of nuclear architecture as a potential cause of COVID-19-induced anosmia. *Cell* 185(6): 1052–1064.e12.

CHAPTER 16 Lower Motor Neuron Circuits and Motor Control

BOX 16A Motor Unit Plasticity

Aagaard, P., J. Bojsen-Møller and J. Lundbye-Jensen (2020) Assessment of neuroplasticity with strength training. *Exerc. Sport Sci. Rev.* 48(4): 151–162.

Brownstone, R. M., T. V. Bui and N. Stifani (2015) Spinal circuits for motor learning. *Curr. Opin. Neurobiol.* 33: 166–173.

Buller, A. J., J. C. Eccles and R. M. Eccles (1960a) Differentiation of fast and slow muscles in the cat hind limb. *J. Physiol.* 150: 399–416.

Buller, A. J., J. C. Eccles and R. M. Eccles (1960b) Interactions between motoneurones and muscles in respect of the characteristic speeds of their responses. *J. Physiol.* 150: 417–439.

Button, D. C. and J. M. Kalmar (2019) Understanding exercise-dependent plasticity of motoneurons using intracellular and intramuscular approaches. *Appl. Physiol. Nutr. Metab.* 44(11): 1125–1133.

Close, R. (1965) Effects of cross-union of motor nerves to fast and slow skeletal muscles. *Nature* 206: 831–832.

Duchateau, J., J. G. Semmler and R. M. Enoka (2006) Training adaptations in the behavior of human motor units. *J. Appl. Physiol.* 101: 1766–1775.

Gordon, T., N. Tyreman, V. F. Rafuse and J. B. Munson (1997) Fast-to-slow conversion following chronic low-frequency activation of medial gastrocnemius muscle in cats. I. Muscle and motor unit properties. *J. Neurophysiol.* 77: 2585–2604.

Lieber, R. L. (2002) *Skeletal Muscle Structure, Function, and Plasticity,* 3rd Edition. Baltimore, MD: Lippincott Williams & Wilkins.

Munson, J. B., R. C. Foehring, L. M. Mendell and T. Gordon (1997) Fast-to-slow conversion following chronic low-frequency activation of medial gastrocnemius muscle in cats. II. Motoneuron properties. *J. Neurophysiol.* 77: 2605–2615.

Van Cutsem, M., J. Duchateau and K. Hainaut (1998) Changes in single motor unit behaviour contribute to the increase in contraction speed after dynamic training in humans. *J. Physiol.* 513: 295–305.

BOX 16B Locomotion in the Leech and the Lamprey

Alford, S. T. and M. H. Alpert (2014) A synaptic mechanism for network synchrony. *Front. Cell. Neurosci.* 8: 290. https://doi.org/10.3389/fncel.2014.00290.

Grillner, S. and A. El Manira (2019) Current principles of motor control, with special reference to vertebrate locomotion. *Physiol. Rev.* 100: 271–320.

Kristan, W. B., Jr., R. L. Calabrese and W. O. Friesen (2005) Neuronal control of leech behavior. *Prog. Neurobiol.* 76: 279–327.

Marder, E. and R. L. Calabrese (1996) Principles of rhythmic motor pattern generation. *Physiol. Rev.* 76: 687–717.

Mullins, O. J., J. T. Hackett, J. T. Buchanan and W. D. Friesen (2011) Neuronal control of swimming behavior: Comparison of vertebrate and invertebrate model systems. *Prog. Neurobiol.* 93: 244–269.

Sharples, S. A., K. Koblinger, J. M. Humphreys and P. J. Whelan (2014) Dopamine: A parallel pathway for the modulation of spinal locomotor networks. *Front. Neural Circuits* 8: 55. https://doi.org/10.3389/fncir.2014.00055.

CLINICAL APPLICATIONS Amyotrophic Lateral Sclerosis

Amin, A., N. D. Perera, P. M. Beart, B. J. Turner and F. Shabanpoor (2020) Amyotrophic lateral sclerosis and autophagy: Dysfunction and therapeutic targeting. *Cells* 9: 2413.

Balendra, R. and A. M. Isaacs (2018) *C9orf72*-mediated ALS and FTD: Multiple pathways to disease. *Nat. Rev. Neurol.* 14(9): 544–558.

Boillee, S., C. Vande Velde and D. W. Cleveland (2006) ALS: A disease of motor neurons and their nonneuronal neighbors. *Neuron* 52: 39–59.

Geevasinga, N., P. Menon, P. H. Özdinler, M. C. Kiernan and S. Vucic (2016) Pathophysiological and diagnostic implications of cortical dysfunction in ALS. *Nat. Rev. Neurol.* 12: 651–661.

Hadano, S. and 20 others (2001) A gene encoding a putative GTPase regulator is mutated in familial amyotrophic lateral sclerosis 2. *Nat. Genet.* 29: 166–173.

Oakes, J. A., M. C. Davies and M. O. Collins (2017) TBK1: A new player in ALS linking autophagy and neuroinflammation. *Mol. Brain* 10: 5. https://doi.org/10.1186/s13041-017-0287-x.

Puls, I. and 13 others (2003) Mutant dynactin in motor neuron disease. *Nat. Genet.* 33: 455–456.

Taylor, J. P., R. H. Brown, Jr. and D. W. Cleveland (2016) Decoding ALS: From genes to mechanism. *Nature* 539: 197–206.

CHAPTER 17 Upper Motor Neuron Control of the Brainstem and Spinal Cord

CLINICAL APPLICATIONS Patterns of Facial Weakness and Their Importance for Localizing Neurological Injury

Jenny, A. B. and C. B. Saper (1987) Organization of the facial nucleus and corticofacial projection in the monkey: A reconsideration of the upper motor neuron facial palsy. *Neurology* 37: 930–939.

Kuypers, H. G. (1958) Corticobulbar connexions to the pons and lower brainstem in man. *Brain* 81: 364–489.

Morecraft, R. J., J. L. Louie, J. L. Herrick and K. S. Stilwell-Morecraft (2001) Cortical innervation of the facial nucleus in the non-human primate: A new interpretation of the effects of stroke and related subtotal brain trauma on the muscles of facial expression. *Brain* 124: 176–208.

Morecraft, R. J., K. S. Stilwell-Morecraft, and W. R. Rossing (2004) The motor cortex and facial expression: New insights from neuroscience. *Neurologist* 10: 235–249.

BOX 17A What Do Motor Maps Represent?

Barinaga, M. (1995) Remapping the motor cortex. *Science* 268: 1696–1698.

Graziano, M. S. A. (2016) Ethological action maps: A paradigm shift for the motor cortex. *Trends. Cogn. Sci.* 20: 121–132.

Graziano, M. S. A., T. N. S. Aflalo and D. F. Cooke (2005) Arm movements evoked by electrical stimulation in the motor cortex of monkeys. *J. Neurophysiol.* 94: 4209–4223.

Lemon, R. (1988) The output map of the primate motor cortex. *Trends Neurosci.* 11: 501–506.

Penfield, W. and E. Boldrey (1937) Somatic motor and sensory representation in the cerebral cortex of man studied by electrical stimulation. *Brain* 60: 389–443.

Schieber, M. H. and L. S. Hibbard (1993) How somatotopic is the motor cortex hand area? *Science* 261: 489–491.

Woolsey, C. N. (1958) Organization of somatic sensory and motor areas of the cerebral cortex. In *Biological and Biochemical Bases of Behavior*, H. F. Harlow and C. N. Woolsey (eds.). Madison: University of Wisconsin Press, pp. 63–81.

Yokoi, A. and J. Diedrichsen (2019) Neural organization of hierarchical motor sequence representations in the human neocortex. *Neuron* 103: 1178–1190.

BOX 17B Minds and Machines

Chaudhary, U., N. Birbaumer and A. Ramos-Murguialday (2016) Brain–computer interfaces for communication and rehabilitation. *Nat. Rev. Neurol.* 12: 513–525.

Donati, A. R. C. and 19 others (2016) Long-term training with a brain-machine interface-based gait protocol induces partial neurological recovery in paraplegic patients. *Sci. Rep.* 6: 30383. https://doi.org/10.1038/srep30383.

Lebedev, M. A. and M. A. L. Nicolelis (2017) Brain-machine interfaces: From basic science to neuroprostheses and neurorehabilitation. *Physiol. Rev.* 97: 767–837.

Saha, S., K. A. Mamun, K. Ahmed, R. Mostafa, G. R. Naik, S. Darvishi, A. H. Khandoker and M. Baumert (2021) Progress in brain computer interface: Challenges and cpportunities. *Front. Syst. Neurosci.* 15: 578875.

Figure A After Dr. Eric C. Leuthardt, Professor of Neurological Surgery, Washington University School of Medicine, Director of The Center for Innovation in Neuroscience and Technology.

BOX 17C The Reticular Formation

Blessing, W. W. (1997) Inadequate frameworks for understanding bodily homeostasis. *Trends Neurosci.* 20: 235–239.

Holstege, G., R. Bandler and C. B. Saper (eds.) (1996) *Progress in Brain Research*, Volume 107. Amsterdam: Elsevier.

Loewy, A. D. and K. M. Spyer (eds.) (1990) *Central Regulation of Autonomic Functions*. New York: Oxford University Press.

Mason, P. (2001) Contributions of the medullary raphe and ventromedial reticular region to pain modulation and other homeostatic functions. *Annu. Rev. Neurosci.* 24: 737–777.

Moruzzi, G. and H. W. Magoun (1949) Brain stem reticular formation and activation of the EEG. *EEG Clin. Neurophys.* 1: 455–476.

CHAPTER 18 Modulation of Movement by the Basal Ganglia

BOX 18A Basal Ganglia Loops and Non-Motor Brain Functions

Alexander, G. E., M. R. DeLong and P. L. Strick (1986) Parallel organization of functionally segregated circuits linking basal ganglia and cortex. *Annu. Rev. Neurosci.* 9: 357–381.

Desrochers, T. M., K. Amemori and A. M. Graybiel (2015) Habit learning by naive macaques is marked by response sharpening of striatal neurons representing the cost and outcome of acquired action sequences. *Neuron* 87: 853–868.

Drevets, W. C. and 6 others (1997) Subgenual prefrontal cortex abnormalities in mood disorders. *Nature* 386: 824–827.

Jahanshahi, M., I. Obeso, J. C. Rothwell and J. A. Obeso (2015) A fronto–striato–subthalamic–pallidal network for goal-directed and habitual inhibition. *Nat. Rev. Neurosci.* 16: 719–732.

Middleton, F. A. and P. L. Strick (2000) Basal ganglia output and cognition: Evidence from anatomical, behavioral, and clinical studies. *Brain Cogn.* 42: 183–200.

Shepherd, G. M. G. (2013) Corticostriatal connectivity and its role in disease. *Nat. Rev. Neurosci.* 14: 278–291.

Smith, K. S. and A. M. Graybiel (2016) Habit formation. *Dialog. Clin. Neurosci.* 18: 33–43.

BOX 18B   Making and Breaking Habits

Desrochers, T. M., K. Amemori and A. M. Graybiel (2015) Habit learning by naive macaques is marked by response sharpening of striatal neurons representing the cost and outcome of acquired action sequences. *Neuron* 87: 853–868.

O'Hare, J. K. and 6 others (2016) Pathway-specific striatal substrates for habitual behavior. *Neuron* 89: 472–479.

Smith, K. S. and A. M. Graybiel (2016) Habit formation. *Dialog. Clin. Neurosci.* 18: 33–43.

CLINICAL APPLICATIONS   Deep Brain Stimulation

Faggiani, E., and A. Benazzouz (2016) Deep brain stimulation of the subthalamic nucleus in Parkinson's disease: From history to the interaction with the monoaminergic systems. *Prog. Neurobiol.* https://doi.org/10.1016/j.pneurobio.2016.07.003.

Hashimoto, T., C. M. Elder, M. S. Okun, S. K. Patrick and J. L. Vitek (2003) Stimulation of the subthalamic nucleus changes the firing pattern of pallidal neurons. *J. Neurosci.* 23: 1916–1923.

Kringelbach, M. L., N. Jenkinson, S. L. F. Owen and T. Z. Aziz (2007) Translational principles of deep brain stimulation. *Nat. Rev. Neurosci.* 8: 623–635.

McIntyre, C. C. and R. W. Anderson (2016) Deep brain stimulation mechanisms: The control of network activity via neurochemistry modulation. *J. Neurochem.* 139 (Suppl. 1): 338–345.

Rosenbaum, R. and 6 others (2014) Axonal and synaptic failure suppress the transfer of firing rate oscillations, synchrony and information during high frequency deep brain stimulation. *Neurobiol. Dis.* 62: 86–99.

Stefani, A., V. Trendafilov, C. Liguori, E. Fedele and S. Galati (2017) Subthalamic nucleus deep brain stimulation on motor-symptoms of Parkinson's disease: Focus on neurochemistry. *Prog. Neurobiol.* https://doi.org/10.1016/j.pneurobio.2017.01.003.

Watson, G. D. R., R. N. Hughes, E. A. Petter, I. P. Fallon, N. Kim, F. P. U. Severino and H. H. Yin (2021) Thalamic projections to the subthalamic nucleus contribute to movement initiation and rescue of parkinsonian symptoms. *Sci. Adv.* 7: eabe9192.

Wichmann, T. and M. R. DeLong (2006) Deep brain stimulation for neurologic and neuropsychiatric disorders. *Neuron* 52: 197–204.

CHAPTER 19   Modulation of Movement by the Cerebellum

CLINICAL APPLICATIONS   Prion Diseases

Carlson, G. A. and S. B. Prusiner (2021) How an infection of sheep revealed prion mechanisms in Alzheimer's disease and other neurodegenerative disorders. *Int. J. Mol. Sci.* 22: 4861.

Gajdusek, D. C. (1977) Unconventional viruses and the origin and disappearance of kuru. *Science* 197: 943–960.

Gibbs, C. J., D. C. Gajdusek, D. M. Asher and M. P. Alpers (1968) Creutzfeldt-Jakob disease (spongiform encephalopathy): Transmission to the chimpanzee. *Science* 161: 388–389.

Harris, D. A. and H. L. True (2006) New insights into prion structure and toxicity. *Neuron* 50: 353–357.

Prusiner, S. B. (1982) Novel proteinaceous infectious particles cause scrapie. *Science* 216: 136–144.

Rhodes, R. (1997) *Deadly Feasts: Tracking the Secrets of a Terrifying New Plague.* New York: Simon and Schuster.

Walsh, D. M. and D. J. Selkoe (2016) A critical appraisal of the pathogenic protein spread hypothesis of neurodegeneration. *Nat. Rev. Neurosci.* 17: 251–260.

BOX 19A   Genetic Analysis of Cerebellar Function

Caviness, V. S., Jr. and P. Rakic (1978) Mechanisms of cortical development: A view from mutations in mice. *Annu. Rev. Neurosci.* 1: 297–326.

D'Arcangelo, G. and 5 others (1995) A protein related to extracellular matrix proteins deleted in the mouse mutation *reeler. Nature* 374: 719–723.

Kloth, A. D. and 16 others (2015) Cerebellar associative sensory learning defects in five mouse autism models. *eLife* 4: e06085.

Monteiro, P. and G. Feng (2017) SHANK proteins: Roles at the synapse and in autism spectrum disorder. *Nat. Rev. Neurosci.* 18: 147–157.

Patil, N. and 5 others (1995) A potassium channel mutation in *weaver* mice implicates membrane excitability in granule cell differentiation. *Nat. Genet.* 11: 126–129.

Rakic, P. (1977) Genesis of the dorsal lateral geniculate nucleus in the rhesus monkey: Site and time of origin, kinetics of proliferation, routes of migration and pattern of distribution of neurons. *J. Comp. Neurol.* 176: 23–52.

Rakic, P. and V. S. Caviness Jr. (1995) Cortical development: A view from neurological mutants two decades later. *Neuron* 14: 1101–1104.

Taroni, F. and S. DiDonato (2004) Pathways to motor incoordination: The inherited ataxias. *Nat. Rev. Neurosci.* 5: 641–655.

CHAPTER 20   Eye Movements and Sensorimotor Integration

BOX 20A   The Perception of Stabilized Retinal Images

Barlow, H. B. (1963) Slippage of contact lenses and other artifacts in relation to fading and regeneration of supposedly stable retinal images. *Q. J. Exp. Psychol.* 15: 36–51.

Coppola, D. and D. Purves (1996) The extraordinarily rapid disappearance of entopic images. *Proc. Natl. Acad. Sci. U.S.A.* 96: 8001–8003.

Heckenmueller, E. G. (1965) Stabilization of the retinal image: A review of method, effects and theory. *Psychol. Bull.* 63: 157–169.

Krauskopf, J. and L. A. Riggs (1959) Interocular transfer in the disappearance of stabilized images. *Amer. J. Psychol.* 72: 248–252.

Martinez-Conde, S., J. Otero-Millan and S. L. Macknik (2013) The impact of microsaccades on vision: Towards a unified theory of saccadic function. *Nat. Rev. Neurosci.* 14: 83–96.

Pritchard, R. M. (1961) Stabilized images on the retina. *Sci. Amer.* 204(6): 72–78.

Riggs, L. A., F. Ratliff, J. C. Cornsweet and T. N. Cornsweet (1953) The disappearance of steadily fixated visual test objects. *J. Opt. Soc. Am.* 43: 495–501.

Rucci, M. and J. D. Victor (2014) The unsteady eye: An information-processing stage, not a bug. *Trends Neurosci.* 38: 194–206.

CLINICAL APPLICATIONS Eye Movements and Neurological Injury, Disease, and Disorder

Anderson, T. J. and M. R. MacAskill (2013) Eye movements in patients with neurodegenerative disorders. *Nat. Rev. Neurol.* 9: 74–85.

Benson, P. J. and 5 others (2012) Simple viewing tests can detect eye movement abnormalities that distinguish schizophrenia cases from controls with exceptional accuracy. *Biol. Psychiatry* 72: 716–724.

Ivleva, E. I. and 8 others (2014) Smooth pursuit eye movement, prepulse inhibition, and auditory paired stimuli processing endophenotypes across the schizophrenia-bipolar disorder psychosis dimension. *Schizophr. Bull.* 40: 642–652.

Ma, Y. and 8 others (2015) Association of chromosome 5q21.3 polymorphisms with the exploratory eye movement dysfunction in schizophrenia. *Sci. Rep.* 5: 10299. https://doi.org/10.1038/srep10299.

BOX 20B Sensorimotor Integration in the Superior Colliculus

Isa, T. and W. C. Hall (2009) Exploring the superior colliculus in vitro. *J. Neurophysiol.* 102: 2581–2593.

Lee, P. H., M. C. Helms, G. J. Augustine and W. C. Hall (1997) Role of intrinsic synaptic circuitry in collicular sensorimotor integration. *Proc. Natl. Acad. Sci. U.S.A.* 94: 13299–13304.

Ozen, G., G. J. Augustine and W. C. Hall (2000) Contribution of superficial layer neurons to premotor bursts in the superior colliculus. *J. Neurophysiol.* 84: 460–471.

Sparks, D. L. and J. S. Nelson (1987) Sensory and motor maps in the mammalian superior colliculus. *Trends Neurosci.* 10: 312–317.

Wurtz, R. H. and J. E. Albano (1980) Visual-motor function of the primate superior colliculus. *Annu. Rev. Neurosci.* 3: 189–226.

BOX 20C From Place Codes to Rate Codes

Fuchs, A. F. and E. S. Luschei (1970) Firing patterns of abducens neurons of alert monkeys in relationship to horizontal eye movement. *J. Neurophysiol.* 33: 382–392.

Groh, J. M. (2001) Converting neural signals from place codes to rate codes. *Biol. Cybern.* 85: 159–165.

Sparks, D. L. (1975) Response properties of eye movement-related neurons in the monkey superior colliculus. *Brain Res.* 90: 147–152.

CHAPTER 21 The Visceral Motor System

BOX 21A The Hypothalamus

Saper, C. B. (2012) Hypothalamus. In *The Human Nervous System*, 3rd Edition, J. K. Mai and G. Paxinos (eds.). Amsterdam: Elsevier, pp. 548–583.

Swanson, L. W. and P. E. Sawchenko (1983) Hypothalamic integration: Organization of the paraventricular and supraoptic nuclei. *Annu. Rev. Neurosci.* 6: 269–324.

BOX 21B Obesity and the Brain

Horvath, T. L. and S. Diano (2004) The floating blueprint of hypothalamic feeding circuits. *Nat. Rev. Neurosci.* 5: 662–667.

Huxing, C., M. López and K. Rahmouni (2017) The cellular and molecular bases of leptin and ghrelin resistance in obesity. *Nat. Rev. Endocrinol.* 13(6): 338–351.

Kaye, W. H., J. L. Fudge and M. Paulus (2009) New insights in symptoms and neurocircuit function of anorexia nervosa. *Nat. Rev. Neurosci.* 10: 573–584.

Loos, R. J. F. and G. S. H. Yeo (2022) The genetics of obesity: From discovery to biology. *Nat. Rev. Genet.* 23: 120–133.

Marx, J. (2003) Cellular warriors at the battle of the bulge. *Science* 299: 846–849.

Morton, G. J., T. H. Meek and M. W. Schwartz (2014) Neurobiology of food intake in health and disease. *Nat. Rev. Neurosci.* 15: 367–378.

O'Rahilly, S., I. S. Farooqi, G. S. H. Yeo and B. G. Challis (2003) Human obesity—lessons from monogenic disorders. *Endocrinology* 144: 3757–3764.

Saper, C. B., T. C. Chou and J. K. Elmquist (2002) The need to feed: Homeostatic and hedonic control of eating. *Neuron* 36: 199–201.

Schwartz, M. W., S. C. Woode, D. Porte, R. J. Seely and D. G. Baskin (2000) Central nervous system control of food intake. *Nature* 404: 661–671.

Yaswen, L., N. Diehl, M. B. Brennan and U. Hochgeschwender (1999) Obesity in the mouse model of pro-opiomelanocortin deficiency responds to peripheral melanocortin. *Nat. Med.* 5: 1066–1070.

Ziauddeen, H., I. S. Farooqi and P. C. Fletcher (2012) Obesity and the brain: How convincing is the addiction model? *Nat. Rev. Neurosci.* 13: 279–286.

CHAPTER 22 Early Brain Development

BOX 22A Stem Cells: Promise and Peril

Brustle, O. and 7 others (1999) Embryonic stem cell derived glial precursors: A source of myelinating transplants. *Science* 285: 754–756.

Castro, R. F., K. A. Jackson, M. A. Goodell, C. S. Robertson, H. Liu and H. D. Shine (2002) Failure of bone marrow cells to transdifferentiate into neural cells in vivo. *Science* 297: 1299.

Dolmetsch, R. and D. H. Geschwind (2011) The human brain in a dish: The promise of iPSC-derived neurons. *Cell* 145: 831–834.

Seaberg, R. M. and D. Van Der Kuoy (2003) Stem and progenitor cells: The premature desertion of rigorous definition. *Trends Neurosci.* 26: 125–131.

Wichterle, H., I. Lieberam, J. A. Porter and T. M. Jessell (2002) Directed differentiation of embryonic stem cells into motor neurons. *Cell* 110: 385–397.

Wu, S. M. and K. Hochedlinger (2011) Harnessing the potential of induced pluri-potent stem cells for regenerative medicine. *Nat. Cell Biol.* 13: 497–505.

CLINICAL APPLICATIONS Inductive Signals and Neurodevelopmental Disorders

Anchan, R. M., D. P. Drake, C. F. Haines, E. A. Gerwe and A. S. LaMantia (1997) Disruption of local retinoid-mediated gene expression accompanies abnormal development in the mammalian olfactory pathway. *J. Comp. Neurol.* 379: 171–184.

Deya-Grosjean, L. and S. Couve-Privat (2005) Sonic hedgehog signaling in basal cell carcinomas. *Cancer Lett.* 225: 181–192.

Evans, R. M. (1988) The steroid and thyroid hormone receptor superfamily. *Science* 240: 889–895.

Johnson, R. L. and C. J. Tabin (1997) Molecular models for vertebrate limb development. *Cell* 90: 979–990.

LaMantia, A.-S., M. C. Colbert and E. Linney (1993) Retinoic acid induction and regional differentiation prefigure olfactory pathway formation in the mammalian forebrain. *Neuron* 10: 1035–1048.

Lammer, E. J. and 11 others (1985) Retinoic acid embryopathy. *N. Engl. J. Med.* 313: 837–841.

Linney, E. and A. S. LaMantia (1994) Retinoid signaling in mouse embryos. In *Advances in Developmental Biology*, Volume 3, P. Wassarman (ed.). Greenwich, CT: JAI Press/Elsevier Science, pp. 73–114.

Marino, S. (2005) Medulloblastoma: Developmental mechanisms out of control. *Trends Molec. Med.* 11: 17–22.

Maynard, T. M. and 5 others (2013) 22q11 gene dosage establishes a dynamic range for sonic hedgehog and retinoid signaling during early cardiovascular and brain development. *Hum. Mol. Genetics* 22: 300–312.

Monuki, E. S. and C. A. Walsh (2001) Mechanisms of cerebral cortical patterning in mice and humans. *Nat. Neurosci.* 4: 1199–1206.

Muenke, M. and P. A. Beachy (2000) Genetics of ventral brain development and holoprosencephaly. *Curr. Opin. Genet. Devel.* 10: 262–269.

Roymer, J. and T. Curran (2005) Targeting medulloblastoma: Small molecule inhibitors of the Sonic hedgehog pathway as potential cancer therapeutics. *Cancer Res.* 65: 4975–4978.

Schardein, J. L. (1993) *Chemically Induced Birth Defects*, 2nd Edition. New York: Marcel Dekker.

Thaller, C. and G. Eichele (1987) Identification and spatial distribution of retinoids in the developing chick limb bud. *Nature* 327: 625–628.

Tickle, C., B. Alberts, L. Wolpert and J. Lee (1982) Local application of retinoic acid to the limb bud mimics the action of the polarizing region. *Nature* 296: 564–565.

Warkany, J. and E. Schraffenberger (1946) Congenital malformations induced in rats by maternal vitamin A deficiency. *Arch. Ophthalmol.* 35: 150–169.

CHAPTER 23 Construction of Neural Circuits

BOX 23A Choosing Sides: Axon Guidance at the Optic Chiasm

Guillery, R. W. (1974) Visual pathways in albinos. *Sci. Amer.* 230: 44–54.

Guillery, R. W., C. A. Mason and J. S. Taylor (1995) Developmental determinants at the mammalian optic chiasm. *J. Neurosci.* 15: 4727–4737.

Herrera, E. and 8 others (2003) Zic2 patterns binocular vision by specifying the uncrossed retinal projection. *Cell* 114: 545–557.

Purves, D. and R. I. Hume (1981) The relation of postsynaptic geometry to the number of presynaptic axons that innervate autonomic ganglion cells. *J. Neurosci.* 1: 441–452.

Rasband, K., M. Hardyv and C. B. Chien (2003) Generating X: Formation of the optic chiasm. *Neuron* 39: 885–888.

Williams, S. E. and 9 others (2003) Ephrin-B2 and EphB1 mediate retinal axon divergence at the optic chiasm. *Neuron* 39: 919–935.

CLINICAL APPLICATIONS Axon Guidance Disorders

Engle, E. C. (2010) Human genetic disorders of axon guidance. *Cold Spring Harb. Perspect. Biol.* 2: a001784.

Haller, S., S. G. Wetzel and J. Lütschg (2008) Functional MRI, DTI and neurophysiology in horizontal gaze palsy with progressive scoliosis. *Neuroradiology* 50: 453–459.

Jen, J. C. and 34 others (2004) Mutations in a human *ROBO* gene disrupt hindbrain axon pathway crossing and morphogenesis. *Science* 304: 1509–1513.

Jouet, M. and 9 others (1994) X-linked spastic paraplegia (SPG1), MASA syndrome, and X-linked hydrocephalus result from mutations in the *L1* gene. *Nat. Genet.* 7: 402–407.

Legouis, R. and 14 others (1991) The candidate gene for the X-linked Kallmann syndrome encodes a protein related to adhesion molecules. *Cell* 67: 423–435.

Vits, L. and 12 others (1994) MASA syndrome is due to mutations in the neural cell adhesion gene *L1CAM*. *Nat. Genet.* 7: 408–413.

Wahl, M. et al. (2009) Variability of homotopic and heterotopic callosal connectivity in partial agenesis of the corpus callosum: A 3T diffusion tensor imaging and Q-ball tractography study. *Am. J. Neuroradiol.* 30: 282–289.

Yamada, K., D. G. Hunter, C. Andrews and E. C. Engle (2005) A novel *KIF21A* mutation in a patient with congenital fibrosis of the extraocular muscles and Marcus Gunn jaw-winking phenomenon. *Arch. Ophthalmol.* 123: 1254–1259.

BOX 23B Why Do Neurons Have Dendrites?

Hume, R. I. and D. Purves (1981) Geometry of neonatal neurons and the regulation of synapse elimination. *Nature* 293: 469–471.

Purves, D. and R. I. Hume (1981) The relation of postsynaptic geometry to the number of presynaptic axons that innervate autonomic ganglion cells. *J. Neurosci.* 1: 441–452.

Purves, D. and J. W. Lichtman (1985) Geometrical differences among homologous neurons in mammals. *Science* 228: 298–302.

Purves, D., E. Rubin, W. D. Snider and J. W. Lichtman (1986) Relation of animal size to convergence, divergence and neuronal number in peripheral sympathetic pathways. *J. Neurosci.* 6: 158–163.

Snider, W. D. (1988) Nerve growth factor promotes dendritic arborization of sympathetic ganglion cells in developing mammals. *J. Neurosci.* 8: 2628–2634.

CHAPTER 24 Experience-Dependent Plasticity in the Developing Brain

BOX 24A Built-In Behaviors

Bartels, A. and S. Zeki (2000) The neural basis of romantic love. *NeuroReport* 11: 3829–3834.

Fisher, H. E., A. Aron, and L. L. Brown (2005) Romantic love: An fMRI study of neural mechanisms for mate choice. *J. Comp. Neurol.* 493: 58–62.

Harlow, H. F. (1959) Love in infant monkeys. *Sci. Amer.* 2: 68–74.

Harlow, H. F. and R. R. Zimmerman (1959) Affectional responses in the infant monkey. *Science* 130: 421–432.

Lorenz, K. (1970) *Studies in Animal and Human Behaviour* (translated by R. Martin). Cambridge, MA: Harvard University Press.

Macfarlane, A. J. (1975) Olfaction in the development of social preferences in the human neonate. *Ciba Found. Symp.* 33: 103–117.

Schaal, B. E. and 5 others (1980) Les stimulations olfactives dans les relations entre l'enfant et la mère. *Reprod. Nutr. Dev.* 20: 843–858.

Tinbergen, N. (1953) *Curious Naturalists*. Garden City, NY: Doubleday.

Young, L. J. and Z. Wang (2004) The neurobiology of pair bonding. *Nat. Neurosci.* 7: 1048–1054.

CLINICAL APPLICATIONS   Dancing in the Dark

Eaton, N. C., H. M. Sheehan and E. M. Quinlan (2016) Optimization of visual training for full recovery from severe amblyopia in adults. *Learn. Mem.* 23: 99–103.

Meaney, M. J. and M. Szyf (2005) Maternal care as a model for experience-dependent chromatin plasticity? *Trends Neurosci.* 28: 456–463.

Montey, K. L. and E. M. Quinlan (2011) Recovery from chronic monocular deprivation following reactivation of thalamocortical plasticity by dark exposure. *Nat. Comm.* 2: 317.

Tropea, D. and 6 others (2006) Gene expression changes and molecular pathways mediating activity dependent plasticity in the visual cortex. *Nat. Neurosci.* 9: 660–668.

CHAPTER 25   Sex Differences and Neural Circuit Development

BOX 25A   The Science of Love (or, Love As a Drug)

Acevedo, B. P., A. Aron, H. E. Fisher and L. L. Brown (2011) Neural correlates of long-term intense romantic love. *Soc. Cogn. Affect. Neurosci.* 7: 145–159.

Aron, A. and 5 others (2005) Reward, motivation, and emotion systems associated with early-stage intense romantic love. *J. Neurophysiol.* 94: 327–337.

Bartels, A. and S. Zeki (2000) The neural basis of romantic love. *NeuroReport* 11: 3829–3834.

Bartels, A. and S. Zeki (2004) The neural correlates of maternal and romantic love. *NeuroImage* 21: 1155–1166.

Fisher, H. E., A. Aron, and L. L. Brown (2005) Romantic love: An fMRI study of neural mechanisms for mate choice. *J. Comp. Neurol.* 493: 58–62.

Fisher, H. E., L. L. Brown, A. Aron, G. Strong and D. Mashek (2010) Reward, addiction, and emotion regulation systems associated with rejection in love. *J. Neurophysiol.* 104: 51–60.

Insel, T. R. and L. J. Young (2001) The neurobiology of attachment. *Nat. Rev. Neurosci.* 2: 129–136.

Young, L. J. and Z. Wang (2004) The neurobiology of pair bonding. *Nat. Neurosci.* 7: 1048–1054.

Zeki, S. (2007) The neurobiology of love. *FEBS Lett.* 581: 2575–2579.

CLINICAL APPLICATIONS   The Good Mother

Belay, H. and 5 others (2011) Early adversity and serotonin transporter genotype interact with hippocampal glucocorticoid receptor mRNA expression, corticosterone, and behavior in adult male rats. *Behav. Neurosci.* 125: 150–160.

McGowan, P. O. and 6 others (2011) Broad epigenetic signature of maternal care in the brain of adult rats. *PLoS ONE* 6(2): e14739.

Meaney, M. J. (2001) Maternal care, gene expression, and the transmission of individual differences in stress reactivity across generations. *Annu. Rev. Neurosci.* 24: 1161–1192.

Meaney, M. J. and M. Szyf (2005) Maternal care as a model for experience-dependent chromatin plasticity? *Trends Neurosci.* 28: 456–463.

CHAPTER 26   Repair and Regeneration in the Nervous System

BOX 26A   Specific Regeneration of Synaptic Connections in Autonomic Ganglia

Landmesser, L. and G. Pilar (1970) Selective reinnervation of two cell populations in the adult pigeon ciliary ganglion. *J. Physiol.* 211: 203–216.

Langley, J. N. (1897) On the regeneration of pre-ganglionic and post-ganglionic visceral nerve fibres. *J. Physiol.* 22: 215–230.

Purves, D. and J. W. Lichtman (1983) Specific connections between nerve cells. *Annu. Rev. Physiol.* 45: 553–565.

Purves, D., W. Thompson and J. W. Yip (1981) Re-innervation of ganglia transplanted to the neck from different levels of the guinea-pig sympathetic chain. *J. Physiol.* 313: 49–63.

CLINICAL APPLICATIONS   Casualties of War and Sports

DeKosky, S. T., M. D. Ikonomovic and S. Gandy (2010) Traumatic brain injury: Football, warfare, and long-term effects. *N. Engl. J. Med.* 363: 1293–1296.

Meyer, K. S., D. W. Marion, H. Coronel and M. S. Jaffee (2010) Combat-related traumatic brain injury and its implications to military healthcare. *Psychiatr. Clin. N. Am.* 33: 783–796.

Miller, G. (2009) A late hit for pro football players. *Science* 325: 670–672.

McKee, A. C. and 9 others (2009) Chronic traumatic encephalopathy in athletes: Progressive tauopathy after repetitive head injury. *J. Neuropathol. Exp. Neurol.* 68: 709–735.

BOX 26B   Nuclear Weapons and Neurogenesis

Au, E. and G. Fishell (2006) Adult cortical neurogenesis: Nuanced, negligible, or nonexistent? *Nat. Neurosci.* 9: 1086–1088.

Bhardwaj, R. D. and 10 others (2006) Neocortical neurogenesis in humans is restricted to development. *Proc. Natl. Acad. Sci. U.S.A.* 103: 12564–12568.

Gould, E., A. J. Reeves, M. S. Graziano and C. G. Gross (1999) Neurogenesis in the neocortex of adult primates. *Science* 286: 548–552.

Koketsu, D., A. Mikami, Y. Miyamoto and T. Hisatsune (2003) Nonrenewal of neurons in the cerebral neocortex of adult macaque monkeys. *J. Neurosci.* 23: 937–942.

Kornack, D. R. and P. Rakic (2001) Cell proliferation without neurogenesis in adult primate neocortex. *Science* 294: 2127–2130.

Rakic, P. (2006) No more cortical neurons for you. *Science* 313: 928–929.

CHAPTER 27 Cognitive Functions and the Organization of the Cerebral Cortex

BOX 27B Large-Scale Neuroscience: Meta-Analyses and Consortium Studies

Cabeza, R. and L. Nyberg (2000) Imaging cognition II: An empirical review of 275 PET and fMRI studies. *J. Cogn. Neurosci.* 12(1): 1–47.

Glasser, M., T. Coalson and E. Robinson et al. (2016) A multimodal parcellation of human cerebral cortex. *Nature* 536: 171–178.

BOX 27C Neuropsychological Testing

Berg, E. A. (1948) A simple objective technique for measuring flexibility in thinking. *J. Gen. Psychol.* 39: 15–22.

Lezak, M. D. (1995) *Neuropsychological Assessment*, 3rd Edition. New York: Oxford University Press.

Milner, B. (1963) Effects of different brain lesions on card sorting. *Arch. Neurol.* 9: 90–100.

Milner, B. and M. Petrides (1984) Behavioural effects of frontal-lobe lesions in man. *Trends Neurosci.* 4: 403–407.

Stoet, G. and L. H. Snyder (2009) Neural correlates of executive control functions in the monkey. *Trends Cogn. Sci.* 13: 228–234.

CLINICAL APPLICATIONS Psychosurgery

Brickner, R. M. (1932) An interpretation of function based on the study of a case of bilateral frontal lobectomy. *Proceedings of the Association for Research in Nervous and Mental Disorders* 13: 259–351.

Brickner, R. M. (1952) Brain of patient A after bilateral frontal lobectomy: Status of frontal lobe problem. *Arch. Neurol. Psychiatry* 68: 293–313.

Freeman, W. and J. Watts (1942) *Psychosurgery: Intelligence, Emotion and Social Behavior Following Prefrontal Lobotomy for Mental Disorders.* Springfield, IL: Charles C. Thomas.

Moniz, E. (1937) Prefrontal leukotomy in the treatment of mental disorders. *Amer. J. Psychiatry* 93: 1379–1385.

Valenstein, E. S. (1986) *Great and Desperate Cures: The Rise and Decline of Psychosurgery and Other Radical Treatments for Mental Illness.* New York: Basic Books.

CHAPTER 28 Cortical States

BOX 28A Electroencephalography

Adrian, E. D. and K. Yamagiwa (1935) The origin of the Berger rhythm. *Brain* 58: 323–351.

Andersen, P. and S. A. Andersson (1968) *Physiological Basis of the Alpha Rhythm.* New York: Appleton-Century-Crofts.

Bear, M., M. A. Paradiso and B. Connors (2001) *Neuroscience: Exploring the Brain*, 2nd Edition. Philadelphia: Williams & Wilkins/Lippincott.

Caton, R. (1875) The electrical currents of the brain. *Br. Med. J.* 2: 278.

Da Silva, F. H. and W. S. Van Leeuwen (1977) The cortical source of the alpha rhythm. *Neurosci. Lett.* 6: 237–241.

Dempsey, E. W. and R. S. Morrison (1943) The electrical activity of a thalamocortical relay system. *Amer. J. Physiol.* 138: 273–296.

Niedermeyer, E. and F. L. Da Silva (1993) *Electroencephalography: Basic Principles, Clinical Applications, and Related Fields.* Baltimore: Williams & Wilkins.

Nuñez, P. L. (1981) *Electric Fields of the Brain: The Neurophysics of EEG.* New York: Oxford University Press.

CLINICAL APPLICATIONS Sleep Disorders and Their Treatment

Anderson, J. S. and C. E. Ferrans (1997) The quality of life of persons with chronic fatigue syndrome. *J. Nerv. Ment. Dis.* 1 85: 359–367.

Friedberg, F. and L. A. Jason (1998) *Understanding Chronic Fatigue Syndrome: An Empirical Guide to Assessment and Treatment.* Washington, D.C.: American Psychological Association.

Holmes, G. P. and 15 others (1988) Chronic fatigue syndrome: A working case definition. *Ann. Intern. Med.* 108: 387–389.

Komaroff, A. L. (2000) The biology of chronic fatigue syndrome. *Amer. J. Med.* 108: 169–171.

BOX 28B Dreaming

Foulkes, D. (1999) *Children's Dreaming and the Development of Consciousness.* Cambridge, MA: Harvard University Press.

Hobson, J. A. (1990) Sleep and dreaming. *J. Neurosci.* 10: 371–382.

Hobson, J. A. (2002) *Dreaming.* New York: Oxford University Press.

Hobson, J. A., R. Strickgold and E. F. Pace-Schott (1998) The neuropsychology of REM sleep and dreaming. *NeuroReport* 9: R1–R14.

CHAPTER 29 Attention

CLINICAL APPLICATIONS Balint's Syndrome

Cooper, A. A. and G. W. Humphreys (2000) Coding space within but not between objects: Evidence from Balint's syndrome. *Neuropsychologia* 38: 723–733.

Friedman-Hill, S. R., L. C. Robertson and A. Treisman (1995) Parietal contributions to visual feature binding: Evidence from a patient with bilateral lesions. *Science* 269: 853–855.

Humphreys, G. W. and M. J. Riddoch (1993) Interactions between object and space systems revealed through neuropsychology. In *Attention and Performance, Volume 14, Synergies in Experimental Psychology, Artificial Intelligence, and Cognitive Neuroscience*, D. E. Meyer and S. Kornblum (eds.). Cambridge, MA: MIT Press, pp. 143–162.

BOX 29A Attention and the Frontal Eye Fields

Moore, T., K. M. Armstrong (2003) Selective gating of visual signals by microstimulation of frontal cortex. *Nature* 421: 370–373.

Moore, T., K. M. Armstrong and M. Fallah (2003) Visuomotor origins of covert spatial attention. *Neuron* 40: 671–683.

Thompson, K. G., K. L. Biscoe and T. R. Sato (2005) Neuronal basis of covert spatial attention in the frontal eye field. *J. Neurosci.* 25: 9479–9487.

CHAPTER 30 Memory

BOX 30A Savant Syndrome

Blumenfeld, H. (2002) *Neuroanatomy through Clinical Cases*. Sunderland, MA: Sinauer Associates, based on Brun, A. and E. Englund (1981) Regional pattern of degeneration in Alzheimer's disease: Neuronal loss and histopathological grading. *Histopathology* 5: 549–564.

Howe, M. J. A. (1989) *Fragments of Genius: The Strange Feats of Idiots Savants*. New York: Routledge.

Miller, L. K. (1989) *Musical Savants: Exceptional Skill in the Mentally Retarded*. Hillsdale, NJ: Lawrence Erlbaum Associates.

Moser, E. I., E. Kropff and M.-B. Moser (2008) Place cells, grid cells, and the brain's spatial representation system. *Annu. Rev. Neurosci.* 31: 69–89.

Smith, N. and I.-M. Tsimpli (1995) *The Mind of a Savant: Language Learning and Modularity*. Oxford, UK: Basil Blackwell.

Smith, S. B. (1983) *The Great Mental Calculators: The Psychology, Methods, and Lives of Calculating Prodigies, Past and Present*. New York: Columbia University Press.

CLINICAL APPLICATIONS Clinical Cases That Illustrate the Neural Basis of Memory

Corkin, S. et al. (1997) H. M.'s medial temporal lobe lesion: Findings from magnetic resonance imaging. *J Neurosci* 17: 3964–3979.

Corkin, S., D. G. Amaral, R. G. González, K. A. Johnson and B. T. Hyman (1997) H.M.'s medial temporal lobe lesion: Findings from MRI. *J. Neurosci.* 17: 3964–3979.

Hilts, P. J. (1995) *Memory's Ghost: The Strange Tale of Mr. M. and the Nature of Memory*. New York: Simon and Schuster.

Rosenbaum, R. S. and 6 others (2000) Remote spatial memory in amnesic person with extensive bilateral hippocampal lesions. *Nat. Neurosci.* 3: 1044–1048.

Scoville, W. B. and B. Milner (1957) Loss of recent memory after bilateral hippocampal lesions. *J. Neurol. Neurosurg. Psychiatry* 20: 11–21.

Squire, L. R., D. G. Amaral, S. M. Zola-Morgan, M. Kritchevsky and G. Press (1989) Description of brain injury in the amnesic patient N.A. based on magnetic resonance imaging. *Exp. Neurol.* 105: 23–35.

Teuber, H. L., B. Milner and H. G. Vaughn (1968) Persistent anterograde amnesia after stab wound of the basal brain. *Neuropsychologia* 6: 267–282.

Tulving, E. (2002) Episodic memory: From mind to brain. *Annu. Rev. Psychol.* 53: 1–25.

Zola-Morgan, S., L. R. Squire and D. Amaral (1986) Human amnesia and the medial temporal region: Enduring memory impairment following a bilateral lesion limited to the CA1 field of the hippocampus. *J. Neurosci.* 6: 2950–2967.

BOX 30B Alzheimer's Disease

Blumenfeld, H. (2002) *Neuroanatomy through Clinical Cases*. Sunderland, MA: Sinauer Associates, based on Brun, A. and Englund, E. (1981) Regional pattern of degeneration in Alzheimer's disease: Neuronal loss and histopathological grading. *Histopathology* 5: 549–564.

Citron, M. and 8 others (1992) Mutation of the β-amyloid precursor protein in familial Alzheimer's disease increases β-protein production. *Nature* 360: 672–674.

Corder, E. H. and 8 others (1993) Gene dose of apolipoprotein E type 4 allele and the risk of Alzheimer's disease in late-onset families. *Science* 261: 921–923.

Goldgaber, D., M. I. Lerman, O. W. McBride, U. Saffiotti and D. C. Gajdusek (1987) Characterization and chromosomal localization of a cDNA encoding brain amyloid of Alzheimer's disease. *Science* 235: 877–880.

Gotz, J. and L. M. Ittner (2008) Animal models of Alzheimer's disease and frontotemporal dementia. *Nat. Rev. Neurosci.* 9: 532–534.

Murrell, J., M. Farlow, B. Ghetti and M. D. Benson (1991) A mutation in the amyloid precursor protein associated with hereditary Alzheimer's disease. *Science* 254: 97–99.

Rogaev, E. I. and 20 others (1995) Familial Alzheimer's disease in kindreds with missense mutations in a gene on chromosome 1 related to the Alzheimer's disease type 3 gene. *Nature* 376: 775–778.

Rosenbaum, R. S. and 6 others (2000) Remote spatial memory in amnesic person with extensive bilateral hippocampal lesions. *Nat. Neurosci.* 3: 1044–1048.

Sherrington, R. and 33 others (1995) Cloning of a gene bearing missense mutations in early-onset familial Alzheimer's disease. *Nature* 375: 754–760.

Whitehouse, P. J. and D. George (2008) *The Myth of Alzheimer's*. New York: St. Martin's Press.

BOX 30C Place Cells and Grid Cells

Brun, V. H. and 6 others (2002) Place cells and place recognition maintained by direct entorhinal-hippocampal circuitry. *Science* 296: 2243–2246.

Fyhn, M., S. Molden, M. P. Witter, E. I. Moser and M. B. Moser (2004) Spatial representation in the entorhinal cortex. *Science* 305: 1258–1264.

Jacobs, J. and 10 others (2013) Direct recordings of grid-like neuronal activity in human spatial navigation. *Nat. Neurosci.* 6: 1188–1190.

Moser, E. I., E. Kropff and M.-B. Moser (2008) Place cells, grid cells, and the brain's spatial representation system. *Ann. Rev. Neurosci.* 31: 69–89.

O'Keefe, J. (1976) Place units in the hippocampus of the freely moving rat. *Exp. Neurol.* 51: 78–109.

O'Keefe, J. and L. Nadel (1978) *The Hippocampus as a Cognitive Map*. Oxford, UK: Oxford University Press.

Tolman, E. C. (1948) Cognitive maps in rats and men. *Psychol. Rev.* 55: 189–208.

CHAPTER 31 Speech and Language

BOX 31A Sign Language

Bellugi, U., H. Poizner and E. S. Klima (1989) Language, modality, and the brain. *Trends Neurosci.* 12: 380–388.

CLINICAL APPLICATIONS Clinical Presentations of Aphasia

Gardner, H. (1974) *The Shattered Mind: The Person after Brain Damage*. New York: Vintage.

Sandrone, S. (2013) Norman Geschwind (1926–1984). *J. Neurol.* 260: 3197–3198.

BOX 31B Semantics: Extracting Meaning from Language

Damasio, H., T. J. Grabowski, D. Tranel, R. D. Hichwa and A. Damasio (1996) A neural basis for lexical retrieval. *Nature* 380: 499–505.

Deniz, F. et al. (2019) The representation of semantic information across human cerebral cortex during listening versus reading is invariant to stimulus modality. *J. Neurosci.* 39(39): 7722–7736.

BOX 31C Language and Handedness

Bakan, P. (1975) Are left-handers brain damaged? *New Sci.* 67: 200–202.

Coren, S. (1992) *The Left-Hander Syndrome: The Causes and Consequence of Left-Handedness.* New York: Free Press.

Davidson, R. J. and K. Hugdahl (eds.) (1995) *Brain Asymmetry.* Cambridge, MA: MIT Press.

Salive, M. E., J. M. Guralnik and R. J. Glynn (1993) Left-handedness and mortality. *Am. J. Pub. Health* 83: 265–267.

CHAPTER 32 Emotion

BOX 32A Determination of Facial Expressions

Duchenne, G.-B. (1876) Mécanisme de la physionomie humaine. In *Atlas*, Deuxième édition. Paris: J.-B. Bailliere et Fils, p. 1.

Duchenne de Boulogne, G.-B. (1862) *Mecanisme de la Physiomonie Humaine.* Paris: Editions de la Maison des Sciences de l'Homme. Edited and translated by R. A. Cuthbertson (1990) Cambridge, UK: Cambridge University Press.

Holstege, G. et al. (1996) The emotional motor system. In *Progress in Brain Research, Volume 107, The Emotional Motor System*, G. Holstege et al. (eds.). Amsterdam: Elsevier, pp. 3–6.

Hopf, H. C., W. Müller-Forell and N. J. Hopf (1992) Localization of emotional and volitional facial paresis. *Neurology* 42: 1918–1923.

Trosch, R. M., G. Sze, L. M. Brass and S. G. Waxman (1990) Emotional facial paresis with striatocapsular infarction. *J. Neurol. Sci.* 98: 195–201.

Waxman, S. G. (1996) Clinical observations on the emotional motor system. In *Progress in Brain Research*, Volume 107, G. Holstege, R. Bandler and C. B. Saper (eds.). Amsterdam: Elsevier, pp. 595–604.

Wellcome Library, London. Wellcome Images. images@wellcome.ac.uk. http://wellcomeimages.org.

BOX 32B Anatomy of theAmygdala

Price, J. L., F. T. Russchen and D. G. Amaral (1987) The limbic region II: The amygdaloid complex. In *Handbook of Chemical Neuroanatomy, Volume 5, Integrated Systems of the CNS, Part I, Hypothalamus, Hippocampus, Amygdala, Retina*, A. Björklund and T. Hökfelt (eds.). Amsterdam: Elsevier, pp. 279–388.

Phelps, E. A. and P. J. Whalen (2009) *The Human Amygdala.* New York: Guilford Press.

BOX 32C Fear and the Human Amygdala

Adolphs, R., D. Tranel, H. Damasio and A. R. Damasio (1995) Fear and the human amygdala. *J. Neurosci.* 15: 5879–5891.

Adolphs, R. and 5 others (2005) A mechanism for impaired fear recognition after amygdala damage. *Nature* 433: 68–72.

Bechara, A., H. Damasio, A. R. Damasio and G. P. Lee (1999) Differential contributions of the human amygdala and ventromedial prefrontal cortex to decision-making. *J. Neurosci.* 19: 5473–5481.

CLINICAL APPLICATIONS Affective Disorders

Breggin, P. R. (1994) *Talking Back to Prozac: What Doctors Won't Tell You about Today's Most Controversial Drug.* New York: St. Martin's Press.

Carrión, V. G., B. W. Haas, A. Garrett, S. Song and A. L. Reiss (2010) Reduced hippocampal activity in youth with posttraumatic stress symptoms: An fMRI study. *J. Pediatr. Psychol.* 35(5): 559–569.

Drevets, W. C. and M. E. Raichle (1994) PET imaging studies of human emotional disorders. In *The Cognitive Neurosciences*, M. S. Gazzaniga (ed.). Cambridge, MA: MIT Press, pp. 1153–1164.

Freeman, P. S., D. R. Wilson and F. S. Sierles (1993) Psychopathology. In *Behavior Science for Medical Students*, F. S. Sierles (ed.). Baltimore: Williams and Wilkins, pp. 239–277.

Greenberg, P. E., L. E. Stiglin, S. N. Finkelstein and E. R. Berndt (1993) The economic burden of depression in 1990. *J. Clin. Psychiatry* 54: 405–424.

Jamison, K. R. (1995) *An Unquiet Mind.* New York: Alfred A. Knopf.

Jefferson, J. W. and J. H. Griest (1994) Mood disorders. In *Textbook of Psychiatry*, J. A. Talbott, R. E. Hales and S. C. Yudofsky (eds.). Washington, DC: American Psychiatric Press, pp. 465–494.

Milad, M. R. and 9 others (2009) Neurobiological basis of failure to recall extinction memory in posttraumatic stress disorder. *Biol. Psychiatry* 66(12): 1075–1082.

Parsons, R. G. and K. J. Ressler (2013) Implications of memory modulation for post-traumatic stress and fear disorders. *Nat. Neurosci.* 16: 146–153.

Robins, E. (1981) *The Final Months: A Study of the Lives of 134 Persons Who Committed Suicide.* New York: Oxford University Press.

Styron, W. (1990) *Darkness Visible: A Memoir of Madness.* New York: Random House.

Wong, D. T. and F. P. Bymaster (1995) Development of antidepressant drugs: Fluoxetine (Prozac) and other selective serotonin uptake inhibitors. *Adv. Exp. Med. Biol.* 363: 77–95.

Wong, D. T., F. P. Bymaster and E. A. Engleman (1995) Fluoxetine (Prozac), the first selective serotonin uptake inhibitor and an antidepressant drug: Twenty years since its first publication. *Life Sciences* 57: 411–441.

Wurtzel, E. (1994) *Prozac Nation: Young and Depressed in America.* Boston: Houghton-Mifflin.

CHAPTER 33 Thinking, Planning, and Deciding

CLINICAL APPLICATIONS Addiction

Nestler, E. J. (2005) Is there a common molecular pathway for addiction? *Nat. Neurosci.* 8: 1445–1449.

BOX 33A Dopamine and Reward Prediction Errors

Berridge, K. C. and T. E. Robinson (1998) What is the role of dopamine in reward: Hedonic impact, reward learning, or incentive salience? *Brain Res. Rev.* 28(3): 309–369.

Murayama, K., M. Matsumoto, K. Izuma and K. Matsumoto (2010) Neural basis of the undermining effect of monetary reward on intrinsic motivation. *Proc. Natl. Acad. Sci. U.S.A.* 107: 20911–20916.

Schultz, W., P. Dayan and P. R. Montague (1997) A neural substrate of prediction and reward. *Science* 275: 1593–1599.

BOX 33B   What Does Neuroscience Have to Say about Free Will?

Eagleman, D. M. (2004) The where and when of intention. *Science* 303: 1144–1146.

Pearson, J. M., S. R. Heilbronner, D. L. Barack, B. Y. Haden and M. L. Platt (2011) Posterior cingulate cortex: Adapting behavior to a changing world. *Trends Cogn. Sci.* 15: 143–151.

Sirigu, A., E. Daprati, S. Ciancia, P. Giraux, N. Nighoghossian, A. Posada and P. Haggard (2003) Altered awareness of voluntary action after damage to the parietal cortex. *Nat. Neurosci.* 7: 80–84.

Soon, C., M. Brass, H. J. Heinze, et al. (2008) Unconscious determinants of free decisions in the human brain. *Nat. Neurosci.* 11: 543–545.

APPENDIX   Survey of Human Neuroanatomy

BOX A   Thalamus and Thalamocortical Relations

Blumenfeld, H. (2022) *Neuroanatomy through Clinical Cases*, 3rd Edition. New York: Oxford University Press/Sinauer Associates.

Jones, E. G. (2007) *The Thalamus*, 2nd Edition. Cambridge, UK: Cambridge University Press.

Sherman, S. M. (2016) Thalamus plays a central role in ongoing cortical functioning. *Nature Neurosci.* 19: 533–541.

Sherman, S. M. and R. W. Guillery (2011) Distinct functions for direct and transthalamic corticocortical connections. *J. Neurophysiol.* https://doi.org/10.1152/jn.00429.2011.

CLINICAL APPLICATIONS   Stroke

Belov Kirdajova, D., J. Kriska, J. Tureckova and M. Anderova (2020) Ischemia-triggered glutamate excitotoxicity from the perspective of glial cells. *Front. Cell. Neurosci.* 14: 51.

Carmichael, S. T. (2016) The 3 Rs of stroke biology: Radial, relayed, and regenerative. *Neurotherapeutics* 13: 348–1359.

Krakauer, J. W. and S. T. Carmichael (2017) *Broken Movement: The Neurobiology of Motor Recovery after Stroke*. Cambridge, MA: MIT Press.

Murphy, T. H. and D. Corbett (2010) Plasticity during stroke recovery: From synapse to behavior. *Nat. Rev. Neurosci.* 10: 861–872.

Ropper, A. and M. Samuels (2009) *Adams and Victor's Principles of Neurology*, 9th Edition. New York: McGraw-Hill, pp. 746–84.

Illustration Credits

UNIT OPENERS

UNIT 1
Opening Image, Unit I: Structure Of A Chemical Synapse Within The Cerebral Cortex. A Presynaptic Terminal (Top) Forms A Synapse With A Dendritic Spine Of The Postsynaptic Neuron (Bottom). Colors Indicate Different Organelles Found Within These Structures. Courtesy Of Alain Burette And Richard Weinberg.

UNIT 2
Opening Image, Unit II: Scanning electron micrograph of outer hair cells (stereocilia) in the inner ear. Dr. Goran Bredberg/Science Source.

UNIT 3
Opening Image, Unit III: Demonstration of the corticospinal tracts (blue/green/magenta fibers) in a human brainstem constructed from MRI data obtained via diffusion tensor imaging (DTI). Courtesy of L. E. White.

UNIT 4
Opening Image, Unit IV: A trigeminal ganglion from an E10.5 mouse embryo grown in cell culture for 24 hours. The green is neuron-selective label (b-tubulin III), the red is a transcription factor marker (Six1) and the blue is a nuclear DNA label (DAPI). Courtesy of Anthony-Samuel LaMantia.

UNIT 5
Opening Image, Unit V: Advanced MRI brain scan. Philippe Psaila/Science Source.

CHAPTER 1   Studying the Nervous System

Opening Image: Diffusion MRI, also referred to as diffusion tensor imaging or DTI, of the human brain. Courtesy of Allen W. Song, Duke–UNC Brain Imaging and Analysis Center. Tainaka, K., T. C. Murakami, E. Susaki and C. Shimizu (2018) Chemical landscape for tissue clearing based on hydrophilic reagents. *Cell Rep.* 24: 2196–2210. **Figure 1.2A**: Feirreira, F. R. M., M. I. Nogieira and J. DeFelipe (2014) The influence of James and Darwin on Cajal and his research into the neuron theory and evolution of the nervous system. *Front. Neuroanat.* 8: article 1. **Figure 1.2B**: Navarrete, M. and A. Araque (2014) The Cajal school and the physiological role of astrocytes: A way of thinking. *Front. Neuroanat.* 8: article 33. **Figure 1.2C right**: Kalil, K., G. Szebenyi and E. W. Dent (2000) Common mechanisms underlying growth cone guidance and axon branching. *Dev. Neurobio.* 44: 145–158. **Figure 1.3B**: Peters, A.,

S. L. Palay and H. deF. Webster (1991) *The Fine Structure of the Nervous System: Neurons and Their Supporting Cells*, 3rd Edition. New York: Oxford University Press. **Figure 1.3C**: Peters, A., S. L. Palay and H. deF. Webster (1991) *The Fine Structure of the Nervous System: Neurons and Their Supporting Cells*, 3rd Edition. New York: Oxford University Press. **Figure 1.3D**: Peters, A., S. L. Palay and H. deF. Webster (1991) *The Fine Structure of the Nervous System: Neurons and Their Supporting Cells*, 3rd Edition. New York: Oxford University Press. **Figure 1.3E**: Peters, A., S. L. Palay and H. deF. Webster (1991) *The Fine Structure of the Nervous System: Neurons and Their Supporting Cells*, 3rd Edition. New York: Oxford University Press. **Figure 1.3F**: Peters, A., S. L. Palay and H. deF. Webster (1991) *The Fine Structure of the Nervous System: Neurons and Their Supporting Cells*, 3rd Edition. New York: Oxford University Press. **Figure 1.3G**: Peters, A., S. L. Palay and H. deF. Webster (1991) *The Fine Structure of the Nervous System: Neurons and Their Supporting Cells*, 3rd Edition. New York: Oxford University Press. **Figure 1.4A–C**: Jones, E. G. and M. W. Cowan (1983) The nervous tissue. In *The Structural Basis of Neurobiology*, E. G. Jones (ed.). New York: Elsevier, chapter 8. **Figure 1.4D,E**: Nishiyama, A., M. Komitova, R. Suzuki and X. Zhu (2009) Polydendrocytes (NG2 cells): Multifunctional cells with lineage plasticity. *Nat. Rev. Neurosci.* 10: 9–22. **Figure 1.4I**: Bhat, M. A. and 11 others (2001) Axon-glia interactions and the domain organization of myelinated axons requires Neurexin IV/Caspr/Paranodin. *Neuron* 30: 369–383. **Figure 1.8A**: Wertz, A. and 12 others (2015) Single-cell–initiated monosynaptic tracing reveals layer-specific cortical network modules. *Science* 349: 70–74. **Figure 1.8B**: Mank, M. and 12 others (2008) A genetically encoded calcium indicator for chronic in vivo two-photon imaging. *Nat. Meth.* 5: 805–811. **Figure 1.8C,D**: Ohki, K., S. Chung, Y. H. Ch'ng, P. Kara and R. C. Reid (2005) Functional imaging with cellular resolution reveals precise micro-architecture in visual cortex. *Nature* 433: 597–603. **Figure 1.9A**: Zhang, F. and 12 others (2011) The microbial opsin family of optogenetic tools. *Cell* 147: 1446–1457. **Figure 1.9C**: Kravitz, A. V. and 6 others (2010) Regulation of parkinsonian motor behaviours by optogenetic control of basal ganglia circuitry. *Nature* 466: 622–626. **Figure 1.13A**: Ramsköld, D., E. T. Wang, C. B. Burge and R. Sandberg (2009) An abundance of ubiquitously expressed genes revealed by tissue transcriptome sequence data. *PLoS Comput. Biol.* 5(12): e1000598. **Figure 1.13C**: Bond, J. and 11 others (2002) ASPM is a major determinant of cerebral cortical size. *Nat. Genet.* 32: 316–320. **Figure 1.14A**: Stewart, T. A. and B. Mintz (1981) Successive generations of mice produced from an established culture line of euploid teratocarcinoma cells. *Proc. Natl. Acad. Sci. U.S.A.* 78:

6314–6318. **Figure 1.15B**: Jung, T.-P., S. Makeig, M. Westerfield, J. Townsend et al. (2001) Analysis and visualization of single-trial event-related potentials. *Hum. Brain Mapp.* 14: 166–185. **Figure 1.16C**: Khairy, S. and 5 others (2015) Duodenal obstruction as first presentation of metastatic breast cancer. *Case Rep. Surg.* 2015: article 605719. **Figure 1.16D**: Pagano, G., F. Niccolini and M. Politis (2016) Imaging in Parkinson's disease. *Clin. Med.* 16: 371–375. **Figure 1.17B**: Seiger, R. and 10 others (2015) Voxel-based morphometry at ultra-high fields: A comparison of 7T and 3T MRI data. *NeuroImage* 113: 207–216. **Figure 1.17D**: Goodyear, B., E. Liebenthal and V. Mosher (2014) Active and passive fMRI for presurgical mapping of motor and language cortex. In *Advanced Brain Neuroimaging Topics in Health and Disease— Methods and Applications*, D. Duric (ed.). InTech. https://doi.org/10.5772/58269. **Figure 1.18A**: National Institute of Mental Health, National Institutes of Health, Department of Health and Human Services. Figure 1.18B: Judith Schaechter, PhD/MGH Martinos Center for Biomedical Imaging.

CHAPTER 2 Electrical Signals of Nerve Cells

Figure 2.3: Hodgkin, A. L. and A. W. Rushton (1946) The electrical constants of a crustacean nerve fibre. *Proc. R. Soc. Lond. B* 133: 444–478. **Clinical Applications**: Scholz, A. (2002) Mechanisms of (local) anaesthetics on voltage-gated sodium and other ion channels. *Br. J. Anaesth.* 89: 52–61; Thomson, A. M., A. P. Bannister, D. I. Hughes and H. Pawelzik (2000) Differential sensitivity to Zolpidem of IPSPs activated by morphologically identified CA1 interneurons in slices of rat hippocampus. *Eur. J. Neurosci.* 12: 425–436; Lizarraga, I., J. P. Chambers and C. B. Johnson (2008) Synergistic depression of NMDA receptor-mediated transmission by ketamine, ketoprofen and L-NAME combinations in neonatal rat spinal cords in vitro. *Br. J. Pharmacol.* 153: 1030–1042; Sirois, J. E., J. J. Pancrazio, C. Lynch and D. A. Bayliss (1998) Multiple ionic mechanisms mediate inhibition of rat motoneurones by inhalation anaesthetics. *J. Physiol.* 512: 851–862. **Figure 2.7**: Hodgkin, A. L. and B. Katz (1949) The effect of sodium ions on the electrical activity of the giant axon of the squid. *J. Physiol.* 108: 37–77. **Figure 2.9**: Hodgkin, A. L. and B. Katz (1949) The effect of sodium ions on the electrical activity of the giant axon of the squid. *J. Physiol.* 108: 37–77. **Box 2B, part A**: Hodgkin, A. L. and A. F. Huxley (1939) Action potentials recorded from inside a nerve fibre. *Nature* 144: 710–711. **Box 2B, part B**: Dodge, F. A. and B. Frankenhaeuser (1958) Membrane currents in isolated frog nerve fibre under voltage clamp conditions. *J. Physiol.* 143: 76–90. **Box 2B, part C**: Barrett, E. F. and J. N. Barrett (1976) Separation of two voltage-sensitive potassium currents, and demonstration of a tetrodotoxin-resistant calcium current in frog motoneurones. *J. Physiol.* 255: 737–774. **Box 2B, part D**: Llinás, R. and Y. Yarom (1981) Electrophysiology of mammalian inferior olivary neurones in vitro. Different types of voltage-dependent ionic conductances. *J. Physiol.* 315: 549–567. **Box 2B, part E**: Chen, S., G. J. Augustine and P. Chadderton (2016) The cerebellum linearly encodes whisker position during voluntary movement. *eLife* 5: e10509.

CHAPTER 3 Voltage-Dependent Membrane Permeability

Figure 3.1: Hodgkin, A. L., A. F. Huxley and B. Katz (1952) Measurements of current-voltage relations in the membrane of the giant axon of *Loligo*. *J. Physiol.* 116: 424–448. **Figure 3.2**: Hodgkin, A. L., A. F. Huxley and B. Katz (1952) Measurements of current-voltage relations in the membrane of the giant axon of Loligo. *J. Physiol.* 116: 424–448. **Figure 3.3**: Hodgkin, A. L., A. F. Huxley and B. Katz (1952) Measurements of current-voltage relations in the membrane of the giant axon of Loligo. *J. Physiol.* 116: 424–448. **Figure 3.4**: Hodgkin, A. L. and A. F. Huxley (1952a) Currents carried by sodium and potassium ions through the membrane of the giant axon of Loligo. *J. Physiol.* 116: 449–472. **Figure 3.5**: Moore, J. W., M. P. Blaustein, N. C. Anderson and T. Narahashi (1967) Basis of tetrodotoxin's selectivity in blockage of squid axons. *J. Gen. Physiol.* 50: 1401–1410; Armstrong, C. M. and L. Binstock (1965) Anomalous rectification in the squid giant axon injected with tetraethylammonium chloride. *J. Gen. Physiol.* 48: 859–872. **Figure 3.6**: Hodgkin, A. L. and A. F. Huxley (1952b) The components of membrane conductance in the giant axon of Loligo. *J. Physiol.* 116: 473–496. **Figure 3.7**: Hodgkin, A. L. and A. F. Huxley (1952b) The components of membrane conductance in the giant axon of Loligo. *J. Physiol.* 116: 473–496. **Figure 3.8**: Hodgkin, A. L. and A. F. Huxley (1952d) A quantitative description of membrane current and its application to conduction and excitation in nerve. *J. Physiol.* 116: 507–544. **Figure 3.11B**: Chen, C. and 17 others (2004) Mice lacking sodium channel beta1 subunits display defects in neuronal excitability, sodium channel expression, and nodal architecture. *J. Neurosci.* 24: 4030–4042. © 2004 Society for Neuroscience. **Clinical Applications**: Trapp, B. D. et al. (1998) Axonal transection in the lesions of multiple sclerosis. *N. Engl. J. Med.* 338: 278–285.

CHAPTER 4 Ion Channels and Transporters

Figure 4.1B,C: Bezanilla, F. and A. M. Correa (1995) Single-channel properties and gating of Na^+ and K^+ channels in the squid giant axon. In *Cephalopod Neurobiology*, N. J. Abbott et al. (eds.). New York: Oxford University Press, pp. 131–151. © 1995 Oxford University. **Figure 4.1D**: Vanderberg, C. A. and F. Bezanilla (1991) A sodium channel model based on single channel, macroscopic ionic, and gating currents in the squid giant axon. *Biophys. J.* 60: 1511–1533. **Figure 4.1E**: Correa, A. M. and F. Bezanilla (1994) Gating of the squid sodium channel at positive potentials. II. Single channels reveal two open states. *Biophys. J.* 66: 1864–1878. **Figure 4.2A–C**: Hille, B. (2001) *Ion Channels of Excitable Membranes*. Sunderland, MA: Sinauer Associates, pp. 61–93. Courtesy of C. K. Augustine and F. Bezanilla. **Figure 4.2D**: Augustine, C. K. and F. Bezanilla (1990) Phosphorylation modulates potassium conductance and gating current of perfused giant axons of squid. *J. Gen. Physiol.* 95: 245–271. **Figure 4.2E**: Perozo, E., D. S. Jong and F. Bezanilla (1991) Single-channel studies of the phosphorylation of K^+ channels in the squid giant axon. II. Nonstationary conditions. *J. Gen. Physiol.* 98: 19–34. **Figure 4.4A**: Kitano, Y. and 8 others (2019) Effects of mirogabalin, a novel ligand for the $\alpha2\delta$ subunit of voltage-gated calcium channels, on N-type calcium channel currents of rat dorsal root ganglion culture neurons. *Pharmazie* 74: 147–149. **Figure 4.4B**: Watanabe, J., A. Rozov and L. P. Wollmuth (2005) Target-specific regulation of synaptic amplitudes in the neocortex. *J. Neurosci.* 25: 1024–1033. **Figure 4.4C**: Ingram, N. T, G. L. Fain and A. P. Sampath (2020) Elevated energy requirement of cone photoreceptors. *Proc. Natl. Acad. Sci. U.S.A.* 117: 19599–19603. **Figure**

4.4D: Cesare, P., L. V. Dekker, A. Sardini, P. J. Parker and P. A. McNaughton (1999) Specific involvement of PKC-epsilon in sensitization of the neuronal response to painful heat. *Neuron* 23(3): 617–624. **Figure 4.4E**: Ricci, A. J., H. J. Kennedy, A. C. Crawford and R. Fettiplace (2005) The transduction channel filter in auditory hair cells. *J. Neurosci.* 25: 7831–7839. **Figure 4.5**: Wang, H. and J. Siemens (2015) TRP ion channels in thermosensation, thermoregulation and metabolism. *Temperature* 2: 178–187. **Figure 4.6A,B**: Doyle, D. A. and 7 others (1998) The structure of the potassium channel: Molecular basis of K^+ conduction and selectivity. *Science* 280: 69–77. **Figure 4.6D,E**: Ahuja, S. and 34 others (2015) Structural basis of Nav1.7 inhibition by an isoform-selective small-molecule antagonist. *Science* 350: aac5464. **Figure 4.6G**: Dutzler, R. et al. (2002) X-ray structure of a ClC chloride channel at 3.0 Å reveals the molecular basis of anion selectivity. *Nature* 415: 287–294. **Figure 4.6H**: Park, E., E. B. Campbell and R. MacKinnon (2017) Structure of a CLC chloride ion channel by cryo-electron microscopy. *Nature* 541: 500–505. **Figure 4.7A,B**: Long, S. B., E. B. Campbell and R. Mackinnon (2005b) Voltage sensor of Kv1.2: Structural basis of electromechanical coupling. *Science* 309: 903–908. **Figure 4.8A**: Traynelis, S. F. and 9 others (2010) Glutamate receptor ion channels: Structure, regulation, and function. *Pharmacol. Rev.* 62: 405–496. **Figure 4.8B,C**: Sobolevsky, A. I., M. P. Rosconi and E. Gouaux (2009) X-ray structure, symmetry and mechanism of an AMPA-subtype glutamate receptor. *Nature* 462: 745–756. **Figure 4.8D,E**: Hansen, K. B. and 18 others (2021) Structure, function, and pharmacology of glutamate receptor ion channels. *Pharmacol. Rev.* 73(4): 298–487. **Figure 4.9A–C**: Napolitano, L. M. R., V. Torre and A. Marchesi (2021) CNG channel structure, function, and gating: A tale of conformational flexibility. *Pflügers Arch.* 473: 1423–1435. **Figure 4.9D**: Xue, J. et al. (2022) Structural mechanisms of assembly, permeation, gating, and pharmacology of native human rod CNG channel. *Neuron* 110(1): 86–95. **Figure 4.10A,B**: Gao, Y., E. Cao, D. Julius and Y. Cheng (2016) TRPV1 structures in nanodiscs reveal mechanisms of ligand and lipid action. *Nature* 534: 347–351. **Figure 4.11A**: Ge, J. and 9 others (2015) Architecture of the mammalian mechanosensitive Piezo1 channel. *Nature* 527: 64–69. **Figure 4.11B**: Kefauver, J. M., A. B. Ward and A. Patapoutian (2020) Discoveries in structure and physiology of mechanically activated ion channels. *Nature* 587: 567–576. **Figure 4.12A**: Shinoda, T., H. Ogawa, F. Cornelius and C. Toyoshima (2009) Crystal structure of the sodium-potassium pump at 2.4 Å resolution. *Nature* 459: 446–450. **Figure 4.12B**: Toyoshima, C., H. Nomura and T. Tsuda (2004) Luminal gating mechanism revealed in calcium pump crystal structures with phosphate analogues. *Nature* 432: 361–368. **Figure 4.13**: Hodgkin, A. L. and R. D. Keynes (1955) Active transport of cations in giant axons from Sepia and Loligo. *J. Physiol.* 128: 28–60. **Figure 4.14A**: Lingrel, J. B., J. Van Huysse, W. O'Brien, E. Jewell-Motz, R. Askew and P. Schultheis (1994) Structure-function studies of the Na,K-ATPase. *Kidney Int. Suppl.* 44: S32–S39. **Figure 4.14B**: Nyblom, M. and 7 others (2013) Crystal structure of Na^+, K^+-ATPase in the Na^+-bound state. *Science* 342: 123–127. **Figure 4.16A**: Liao, J. et al. (2016) Mechanism of extracellular ion exchange and binding-site occlusion in a sodium/calcium exchanger. *Nat. Struct. Mol. Biol.* 23(6): 590–599. **Figure 4.16B**: Iwaki, M. et al. (2020) Structure-affinity insights into the Na^+ and Ca^{2+} interactions with multiple sites of a sodium-calcium exchanger. *FEBS J.* 287: 4678–4695.

Box 4B, part A: Schmitt, O. and H. Schmidt (1972) Influence of calcium ions on the ionic currents of nodes of Ranvier treated with scorpion venom. *Pflügers Arch.* 333: 51–61. **Box 4B, part B**: Cahalan, M. (1975) Modification of sodium channel gating in frog myelinated nerve fibers by *Centruroides sculpturatus* scorpion venom. *J. Physiol.* 244(2): 511–534. **Clinical Applications, Figures A,B**: Waxman, S. G. and G. W. Zamponi (2014) Regulating excitability of peripheral afferents: Emerging ion channel targets. *Nat. Neurosci.* 17: 153–163; Hoeijmakers, J. G. et al. (2012) Small nerve fibres, small hands and small feet: A new syndrome of pain, dysautonomia and acromesomelia in a kindred with a novel Nav1.7 mutation. *Brain* 135: 345–358. **Clinical Applications, Figure C**: Baig, S. M. and 16 others (2011) Loss of Cav1.3 (CACNA1D) function in a human channelopathy with bradycardia and congenital deafness. *Nat. Neurosci.* 14: 77–84.

CHAPTER 5 Synaptic Transmission

Figure 5.2A,B: Sotelo, C., R. Llinas and R. Baker (1974) Structural study of inferior olivary nucleus of the cat: Morphological correlates of electrotonic coupling. *J. Neurophysiol.* 37: 541–559. **Figure 5.2D**: Maeda, S. and 6 others (2009) Structure of the connexin 26 gap junction channel at 3.5 Å resolution. *Nature* 458: 597–602. **Figure 5.3A**: Furshpan, E. J. and D. D. Potter (1959) Transmission at the giant motor synapses of the crayfish. *J. Physiol.* 145: 289–324. **Figure 5.3B**: Beierlein, M., J. R. Gibson and B. W. Connors (2000) A network of electrically coupled interneurons drives synchronized inhibition in neocortex. *Nat. Neurosci.* 3: 904–910. **Figure 5.4A,B**: Burette, A. C. and 6 others (2012) Electron tomographic analysis of synaptic ultrastructure. *J. Comp. Neurol.* 520: 2697–2711. **Figure 5.5A–D**: Fatt, P. and B. Katz (1952) Spontaneous subthreshold activity at motor nerve endings. *J. Physiol.* 117: 109–127. **Figure 5.6A,B**: Boyd, I. A. and A. R. Martin (1956) The end-plate potential in mammalian muscle. *J. Physiol.* 132: 74–91. **Figure 5.7B**: Augustine, G. J. and R. Eckert (1984) Divalent cations differentially support transmitter release at the squid giant synapse. *J. Physiol.* 346: 257–271. **Figure 5.8A**: Smith, S. J., J. Buchanan, L. R. Osses, M. P. Charlton and G. J. Augustine (1993) The spatial distribution of calcium signals in squid presynaptic terminals. *J. Physiol.* 472: 573–593. **Figure 5.8B**: Miledi, R. (1973) Transmitter release induced by injection of calcium ions into nerve terminals. *Proc. R. Soc. Lond. B* 183: 421–424. **Figure 5.8C**: Adler, E. M., G. J. Augustine, M. P. Charlton and S. N. Duffy (1991) Alien intracellular calcium chelators attenuate neurotransmitter release at the squid giant synapse. *J. Neurosci.* 11: 1496–1507. **Figure 5.9C**: Heuser, J. E., T. S. Reese, M. J. Dennis, Y. Jan, L. Jan and L. Evans (1979) Synaptic vesicle exocytosis captured by quick freezing and correlated with quantal transmitter release. *J. Cell Biol.* 81: 275–300. **Figure 5.10A–E**: Heuser, J. E. and T. S. Reese (1973) Evidence for recycling of synaptic vesicle membrane during transmitter release at the frog neuromuscular junction. *J. Cell Biol.* 57: 315–344. **Figure 5.11A**: Takamori, S. and 21 others (2006) Molecular anatomy of a trafficking organelle. *Cell* 127: 831–846. **Figure 5.12A**: Sutton, R. B., D. Fasshauer, R. Jahn and A. T. Brünger (1998) Crystal structure of a SNARE complex involved in synaptic exocytosis at 2.4 Å resolution. *Nature* 395: 347–353; Madej, T. et al. (1998) MMDB and VAST+: Tracking structural similarities between macromolecular complexes. *Nucleic Acids Res.* 42(D1): D297–303. **Figure 5.12B**: Radhakrishnan, A. et al. (2021) Symmetrical arrangement of proteins under release-ready vesicles in presynaptic terminals. *Proc. Natl. Acad. Sci. U.S.A.* 118(5): e2024029118. **Figure 5.12C**: Zhou, Q. et al. (2015) Architecture of the synaptotagmin-SNARE machinery for neuronal exocytosis. *Nature* 525: 62–67. **Clinical Applications, Figure B**: Sutton, R. B., D. Fasshauer, R. Jahn and A. T. Brünger (1998) Crystal structure of a SNARE complex involved in synaptic exocytosis at 2.4 Å resolution. *Nature* 395: 347–353. **Figure 5.13A**: Fotin, A. and 6 others (2004) Molecular model for a complete clathrin lattice from electron cryomicroscopy. *Nature* 432: 573–579. **Figure 5.13B**: Reubold, T. F. and 12 others (2015) Crystal structure of the dynamin tetramer. *Nature* 525: 404–408. **Figure 5.13C**: Shupliakov, O. and L. Brodin (2010) Recent insights into the building and cycling of synaptic vesicles. *Exp. Cell Res.* 316, 1344–1350. **Figure 5.16A–C**: Takeuchi, A. and N. Takeuchi (1960) On the permeability of end-plate membrane during the action of transmitter. *J. Physiol.* 154: 52–67. **Figure 5.17A,B**: Takeuchi, A. and N. Takeuchi (1960) On the permeability of end-plate membrane during the action of transmitter. *J. Physiol.* 154: 52–67. **Box 5A, part A**: Buchanan, J. et al. (2021) Oligodendrocyte precursor cells prune axons in the mouse neocortex. *bioRxiv* 2021.05.29.446047. https://doi.org/10.1101/2021.05.29.446047. **Box 5A, parts B,C**: Witcher, M. R., S. A. Kirov and K. M. Harris (2007). Plasticity of perisynaptic astroglia during synaptogenesis in the mature rat hippocampus. *Glia* 55: 13–23. **Box 5A, part D**: Cornell-Bell, A. H., S. M. Finkbeiner, M. S. Cooper and S. J. Smith (1990). Glutamate induces calcium waves in cultured astrocytes: Long-range glial signaling. *Science* 247: 470–473. **Box 5A, part E**: Perea, G. and A. Araque (2007) Astrocytes potentiate transmitter release at single hippocampal synapses. *Science* 317: 1083–1086.

CHAPTER 6   Neurotransmitters and Their Receptors

Box 6A, part B: Tsetlin, V. I. (2015) Three-finger snake neurotoxins and Ly6 proteins targeting nicotinic acetylcholine receptors: Pharmacological tools and endogenous modulators. *Trends Pharmacol. Sci.* 36: 109–123. **Box 6A, part D**: Lebbe, E. K. M., S. Peigneur, I. Wijesekara and J. Tytgat (2014) Conotoxins targeting nicotinic acetylcholine receptors: An overview. *Mar. Drugs* 12: 2970–3004. **Figure 6.3A–C**: Unwin, N. (2005) Refined structure of the nicotinic acetylcholine receptor at 4 Å resolution. *J. Mol. Biol.* 346: 967–989. **Figure 6.3D,E**: Miyazawa, A., Y. Fujiyoshi and N. Unwin (2003) Structure and gating mechanism of the acetylcholine receptor pore. *Nature* 423: 949–955. **Figure 6.4A,B**: Haga, K. and 10 others (2012) Structure of the human M2 muscarinic acetylcholine receptor bound to an antagonist. *Nature* 482: 547–551. **Clinical Applications, Figure A**: Harvey, A. M., J. L. Lilienthal Jr. and S. A. Talbot (1942) Observations on the nature of myasthenia gravis: The effect of thymectomy on neuro-muscular transmission. *J. Clin. Invest.* 21(5): 579–588. **Clinical Applications, Figure B**: Elmqvist, D., W. W. Hofmann, J. Kugelberg and D. M. J. Quastel (1964) An electrophysiological investigation of neuromuscular transmission in myasthenia gravis. *J. Physiol.* 174: 417–434. **Figure 6.6A**: Watanabe, J., A. Rozov and L. P. Wollmuth (2005) Target-specific regulation of synaptic amplitudes in the neocortex. *J. Neurosci.* 25(4): 1024–1033. © 2005 Society for Neuroscience. **Figure 6.6B**: Mott, D. D., M. Benveniste and R. J. Dingledine (2008) pH-dependent inhibition of kainate receptors by zinc. *J. Neurosci.* 28: 1659–1671. © 2008 Society for Neuroscience. **Figure 6.7A**: Traynelis, S. F. and 9 others (2010) Glutamate receptor ion channels: Structure, regulation, and function. *Pharmacol. Rev.* 62: 405–496. **Figure 6.7B,C**: Sobolevsky, A. I., M. P. Rosconi and E. Gouaux (2009) X-ray structure, symmetry and mechanism of an AMPA-subtype glutamate receptor. *Nature* 462: 745–756. **Figure 6.7D,E**: Hansen, K. B. and 18 others (2021) Structure, function, and pharmacology of glutamate receptor ion channels. *Pharmacol. Rev.* 73(4): 298–487. **Figure 6.8C–E**: Karakas, E. and H. Furukawa (2014) Crystal structure of a heterotetrameric NMDA receptor ion channel. *Science* 344(6187): 992–997. **Figure 6.8F**: Zhu, S. and 6 others (2016) Mechanism of NMDA receptor inhibition and activation. *Cell* 165(3): 704–714. **Figure 6.9**: Pin, J.-P. and B. Bettler (2016) Organization and functions of mGlu and GABAB receptor complexes. *Nature* 540: 60–68. **Figure 6.11A**: Chavas, J. and A. Marty (2003) Coexistence of excitatory and inhibitory GABA synapses in the cerebellar interneuron network. *J. Neurosci.* 23(6): 2019–2030. **Figure 6.11B,C**: Miller, P. S. and A. R. Aricescu (2014) Crystal structure of a human GABAA receptor. *Nature* 512: 270–275. **Figure 6.11D**: Puthenkalam, R. and 6 others (2016) Structural studies of GABAA receptor binding sites: Which experimental structure tells us what? *Front. Mol. Neurosci.* 9: 44. **Figure 6.12**: Pin, J.-P. and B. Bettler (2016) Organization and functions of mGlu and GABAB receptor complexes. *Nature* 540: 60–68. **Box 6B, part B (graph and images)**: Berglund, K. and 8 others (2006) Imaging synaptic inhibition in transgenic mice expressing the chloride indicator, Clomeleon. *Brain Cell Biol.* 35: 207–228. **Box 6B, part C**: Obata, K., M. Oide and H. Tanaka (1978) Excitatory and inhibitory actions of GABA and glycine on embryonic chick spinal neurons in culture. *Brain Res.* 144(1): 179–184. **Figure 6.13**: Du, J. et al. (2015) Glycine receptor mechanism elucidated by electron cryo-microscopy. *Nature* 526: 224–229. **Figure 6.16A**: Chien, E. Y. and 10 others (2010) Structure of the human dopamine D3 receptor in complex with a D2/D3 selective antagonist. *Science* 330: 1091–1095. **Figure 6.16B/inactive_left**: Betke, K. M., C. A. Wells and H. E. Hamm (2012) GPCR mediated regulation of synaptic transmission. *Prog. Neurobiol.* 96: 304–321. **Figure 6.16B/active_right**: Rasmussen, S. G. and 19 others (2011) Crystal structure of the m2 adrenergic receptor–Gs protein complex. *Nature* 477: 549–555. **Figure 6.19A**: Wacker, D. and 12 others (2017) Crystal structure of an LSD-bound human serotonin receptor. *Cell* 168: 377–389. **Figure 6.19B**: Hassaine, G. and 14 others (2014) X-ray structure of the mouse serotonin 5-HT3 receptor. *Nature* 512(7514): 276–281. **Figure 6.20A–C**: Kawate, T., J. C. Michel, W. T. Birdsong and E. Gouaux (2009) Crystal structure of the ATP-gated P2X(4) ion channel in the closed state. *Nature* 460: 592–598. **Figure 6.20D**: Jaakola, V. P. and A. P. Ijzerman (2010) The crystallographic structure of the human adenosine A2A receptor in a high-affinity antagonist-bound state: Implications for GPCR drug screening and design. *Curr. Opin. Struct. Biol.* 20: 401–414. **Box 6C, part B**: Iversen, L. (2003) Cannabis and the brain. *Brain* 126: 1252–1270. **Box 6C, part C**: Shao, Z. and 6 others (2016) High-resolution crystal structure of the human CB1 cannabinoid receptor. *Nature* 540: 602–606. **Figure 6.23A,B**: Freund, T. F., I. Katona and D. Piomelli (2003)

Role of endogenous cannabinoids in synaptic signaling. *Physiol. Rev.* 83: 1017–1066. **Figure 6.23C**: Iversen, L. (2003) Cannabis and the brain. *Brain* 126: 1252–1270. **Figure 6.24B,C**: Ohno-Shosaku, T., T. Maejima and M. Kano (2001) Endogenous cannabinoids mediate retrograde signals from depolarized postsynaptic neurons to presynaptic terminals. *Neuron* 29: 729–738.

CHAPTER 7   Molecular Signaling within Neurons

Box 7A, part A: Grynkiewicz, G., M. Poenie and R. Y. Tsien (1985) A new generation of Ca^{2+} indicators with greatly improved fluorescence properties. *J. Biol. Chem.* 260: 3440–3450. **Box 7A, part B**: Finch, E. A. and G. J. Augustine (1998) Local calcium signaling by IP3 in Purkinje cell dendrites. *Nature* 396: 753–756. **Box 7A, part D**: Vidal, G. S., M. Djurisic, K. Brown, R. W. Sapp and C. J. Shatz (2016) Cell-autonomous regulation of dendritic spine density by PirB. *eNeuro* 3(5): ENEURO.0089-16.2016. https://doi.org/10.1523/ENEURO.0089-16.2016. **Box 7A, part E**: Livet, J. and 7 others (2007) Transgenic strategies for combinatorial expression of fluorescent proteins in the nervous system. *Nature* 450: 56–62. **Figure 7.9A**: Berman, H. M., J. Westbrook, Z. Feng, G. Gilliland, T. N. Bhat, H. Weissig, I. N. Shindyalov and P. E. Bourne (2000) The Protein Data Bank. *Nucleic Acids Res.*, 28: 235–242. **Figure 7.9B**: Craddock, T. J. A., J. A. Tuszynski and S. Hameroff (2012) Cytoskeletal signaling: Is memory encoded in microtubule lattices by CaMKII phosphorylation? *PLoS Comput. Biol.* 8: e1002421. **Figure 7.9C**: Leonard, T. A., B. Różycki, L. F. Saidi, G. Hummer and J. H. Hurley (2011) Crystal structure and allosteric activation of protein kinase C βII. *Cell* 144: 55–66. **Figure 7.9D**: Turk, B. E. (2007) Manipulation of host signalling pathways by anthrax toxins. *Biochem. J.* 402: 405–417. **Figure 7.10A**: Bollen, M., W. Peti, M. J. Ragusa and M. Beullens (2010) The extended PP1 toolkit: Designed to create specificity. *Trends Biochem. Sci.* 35: 450–458. **Figure 7.10B**: Cho, U. S. and W. Xu (2007) Crystal structure of a protein phosphatase 2A heterotrimeric holoenzyme. *Nature* 445: 53–57. **Figure 7.10C**: Li, H., A. Rao and P. G. Hogan (2011) Interaction of calcineurin with substrates and targeting proteins. *Trends Cell Biol.* 21: 91–103. **Box 7B, part B**: Harris, K. M. and R. J. Weinberg (2012) Ultrastructure of synapses in the mammalian brain. *Cold Spring Harb. Perspect. Biol.* 4: a005587. **Box 7B, part C**: Spacek, J., MUDr., DrSc., FRMS, Professor of Pathology; Charles University Prague University Hospital, Hradec Kralove, Czech Republic; spacek@lfhk.cuni.cz. **Box 7B, part D**: Sabatini, B. L., T. G. Oertner and K. Svoboda (2002) The life cycle of Ca^{2+} ions in dendritic spines. *Neuron* 33: 439–452. **Box 7B, part E**: Sheng, M. and E. Kim (2011) The postsynaptic organization of synapses. *Cold Spring Harb. Perspect. Biol.* 3: a005678.

CHAPTER 8   Synaptic Plasticity

Figure 8.1A,B: Charlton, M. P. and G. D. Bittner (1978) Presynaptic potentials and facilitation of transmitter release in the squid giant synapse. *J. Gen. Physiol.* 72: 487–511. **Figure 8.1C**: Swandulla, D., M. Hans, K. Zipser and G. J. Augustine (1991) Role of residual calcium in synaptic depression and posttetanic potentiation: Fast and slow calcium signaling in nerve terminals. *Neuron* 7: 915–926. **Figure 8.1D**: Betz, W. J. (1970) Depression of transmitter release at the neuromuscular

junction of the frog. *J. Physiol.* 206: 629–644. **Figure 8.1E**: Lev-Tov, A., M. J. Pinter and R. E. Burke (1983) Posttetanic potentiation of group Ia EPSPs: Possible mechanisms for differential distribution among medial gastrocnemius motoneurons. *J. Neurophysiol.* 50: 379–398. **Figure 8.2A**: Katz, B. (1966) *Nerve, Muscle, and Synapse*. New York: McGraw-Hill. **Figure 8.2B**: Malenka, R. C. and S. A. Siegelbaum (2001) Synaptic plasticity: Diverse targets and mechanisms for regulating synaptic efficacy. In *Synapses*, W. M. Cowan, T. C. Sudhof and C. F. Stevens (eds.). Baltimore: Johns Hopkins University Press, pp. 393–413. **Figure 8.3A–E**: Squire, L. R. and E. R. Kandel (1999) *Memory: From Mind to Molecules*. New York: Scientific American Library. **Figure 8.4A–C**: Squire, L. R. and E. R. Kandel (1999) *Memory: From Mind to Molecules*. New York: Scientific American Library. **Figure 8.5A,B**: Squire, L. R. and E. R. Kandel (1999) *Memory: From Mind to Molecules*. New York: Scientific American Library. **Box 8A**: Tully, T. (1996) Discovery of genes involved with learning and memory: An experimental synthesis of Hirshian and Benzerian perspectives. *Proc. Natl. Acad. Sci. U.S.A.* 93: 13460–13467. **Figure 8.7A–C**: Malinow, R., H. Schulman and R. W. Tsien (1989) Inhibition of postsynaptic PKC or CaMKII blocks induction but not expression of LTP. *Science* 245: 862–866. **Figure 8.7D**: Abraham, W. C., B. Logan, J. M. Greenwood and M. Dragunow (2002) Induction and experience-dependent consolidation of stable long-term potentiation lasting months in the hippocampus. *J. Neurosci.* 22: 9626–9634. **Figure 8.8**: Gustafsson, B., H. Wigstrom, W. C. Abraham and Y. Y. Huang (1987) Long-term potentiation in the hippocampus using depolarizing current pulses as the conditioning stimulus to single volley synaptic potentials. *J. Neurosci.* 7: 774–780. **Figure 8.10**: Nicoll, R. A., J. A. Kauer and R. C. Malenka (1988) The current excitement in long-term potentiation. *Neuron* 1: 97–103. **Figure 8.11A,B**: Matsuzaki, M., N. Honkura, G. C. Ellis-Davies and H. Kasai (2004) Structural basis of long-term potentiation in single dendritic spines. *Nature* 429: 761–766. **Figure 8.11C**: Liao, D., N. A. Hessler and R. Malinow (1995) Activation of postsynaptically silent synapses during pairing-induced LTP in CA1 region of hippocampal slice. *Nature* 375: 400–404. **Box 8B, part A**: Liao, D., N. A. Hessler and R. Malinow (1995) Activation of postsynaptically silent synapses during pairing-induced LTP in CA1 region of hippocampal slice. *Nature* 375: 400–404. **Box 8B, part C**: Petralia, R. S. and 6 others (1999) Selective acquisition of AMPA receptors over postnatal development suggests a molecular basis for silent synapses. *Nat. Neurosci.* 2: 31–36. **Figure 8.12A,B**: Lee, S. J., Y. Escobedo-Lozoya, E. M. Szatmari and R. Yasuda (2009) Activation of CaMKII in single dendritic spines during long-term potentiation. *Nature* 458: 299–304. **Figure 8.14A,B**: Frey, U. and R. G. Morris (1997) Synaptic tagging and long-term potentiation. *Nature* 385: 533–536. **Figure 8.15A**: Squire, L. R. and E. R. Kandel (1999) *Memory: From Mind to Molecules*. New York: Scientific American Library. **Figure 8.15B,C**: Engert, F. and T. Bonhoeffer (1999) Dendritic spine changes associated with hippocampal long-term synaptic plasticity. *Nature* 399: 66–70. **Figure 8.16B**: Mulkey, R. M., C. E. Herron and R. C. Malenka (1993) An essential role for protein phosphatases in hippocampal long-term depression. *Science* 261: 1051–1055. **Figure 8.17B**: Sakurai, M. (1987) Synaptic modification of parallel fibre-Purkinje cell transmission in in vitro guinea-pig cerebellar slices. *J. Physiol.* 394: 463–480.

Figure 8.18A–C: Bi, G. Q. and M. M. Poo (1998) Synaptic modifications in cultured hippocampal neurons: Dependence on spike timing, synaptic strength, and postsynaptic cell type. *J. Neurosci.* 18: 10464–10472. **Clinical Applications:** Dyro, F. M. (1989) *The EEG Handbook.* Boston: Little, Brown.

CHAPTER 9 Vision

Figure 9.2A–C: Westheimer, G. (1974) The eye. In *Medical Physiology*, 13th Edition, V. B. Mountcastle (ed.). St. Louis: Mosby. **Figure 9.6:** Baylor, D. A., A. L. Hodgkin and T. D. Lamb (1974) The electrical response of turtle cones to flashes and steps of light. *J. Physiol.* 242: 685–727. **Figure 9.7A:** Dratz, E. A. and P. A. Hargrave (1983) The structure of rhodopsin and the rod outer segment disk membrane. *Trends Biochem. Sci.* 8(4): 128–131. **Figure 9.7B:** Stryer, L. (1987) The molecules of visual excitation. *Sci. Am.* 257(1): 42–50. **Figure 9.9A:** Curcio, C. A., K. R. Sloan, R. E. Kalina and A. E. Hendrickson (1990) Human photoreceptor topography. *J. Comp. Neurol.* 292(4): 497–523; Purves, D. and R. B. Lotto (2011) *Why We See What We Do Redux: An Empirical Theory of Vision.* Sunderland, MA: Sinauer Associates, Figure A.3. **Figure 9.10B:** Baylor, D. A. (1987) Photoreceptor signals and vision. *Invest. Ophthalmol. Vis. Sci.* 28: 34–49. **Figure 9.17A:** Watanabe, M. and R. W. Rodieck (1989) Parasol and midget ganglion cells of the primate retina. *J. Comp. Neurol.* 289: 434–454. **Figure 9.19A:** Hubel, D. H. (1988) *Eye, Brain, and Vision.* New York: Scientific American Library. **Figure 9.22B,C:** Sereno, M. I. and 7 others (1995) Borders of multiple visual areas in humans revealed by functional magnetic resonance imaging. *Science* 268: 889–893. **Box 9A, parts A,B:** Purves, D. and R. B. Lotto (2011) *Why We See What We Do Redux: An Empirical Theory of Vision.* Sunderland, MA: Sinauer Associates, Figure 3.18.

CHAPTER 10 Hearing

Figure 10.2B,C: Kikuchi, Y. et al. (2014) Processing of harmonics in the lateral belt of macaque auditory cortex. *Front. Neurosci.* 8: 204. **Figure 10.5 micrograph:** Counter, S. A. et al. (1991) Acoustic trauma in extracranial magnetic brain stimulation. *Electroencephalog. Clin. Neurophysiol.* 78: 173–184. **Figure 10.6 drawing:** P. Dallos (1992) The active cochlea. *J. Neurosci.* 12: 4575–4585. **Figure 10.6 graphs:** Von Békésy, G. (1960) *Experiments in Hearing.* New York: McGraw-Hill. **Figure 10.8B:** Kachar, B. et al. (2000) High-resolution structure of hair-cell tip links. *Proc. Natl. Acad. Sci. U.S.A.* 97: 13336–13341. © 2000 National Academy of Sciences, U.S.A. **Figure 10.9B:** Lewis, R. and A. Hudspeth (1983) Voltage- and ion-dependent conductances in solitary vertebrate hair cells. *Nature* 304: 538–541. **Figure 10.10:** Palmer, A. R. and I. J. Russell (1986) Phase-locking in the cochlear nerve of the guinea-pig and its relation to the receptor potential of inner hair-cells. *Hear. Res.* 24: 1–15. **Figure 10.11A,B:** Groh, J. M. (2014) *Making Space: How the Brain Knows Where Things Are.* Cambridge, MA: Harvard University Press. © 2014 Jennifer M. Groh. **Figure 10.11C:** Zheng, L. et al. (1999) Synthesis and decomposition of transient-evoked otoacoustic emissions based on an active auditory model. *IEEE Trans. Biomed. Eng.* 46(9): 1098–1106. **Figure 10.14A:** Pierce, J. R. (1983) *The Science of Musical Sound.* New York: Scientific American Library, distributed by W.H. Freeman. **Figure 10.14B:** Bulkin,

D. A. and J. M. Groh (2011) Systematic mapping of the monkey inferior colliculus reveals enhanced low frequency sound representation. *J. Neurophysiol.* 105(4): 1785–1797. **Figure 10.14C:** Javel, E. (1994) Shapes of cat auditory nerve fiber tuning curves. *Hear. Res.* 81(1–2): 167–188. **Figure 10.15A–C:** Machens, C. K. (2004) Linearity of cortical receptive fields measured with natural sounds. *J. Neurosci.* 24(5): 1089–1100. **Figure 10.16A–D:** Groh, J. M. (2014) *Making Space: How the Brain Knows Where Things Are.* Cambridge, MA: Harvard University Press. © 2014 Jennifer M. Groh. **Figure 10.17:** Jeffress, L. A. (1948) A place theory of sound localization. *J. Comp. Physiol. Psychol.* 41: 35–39.

CHAPTER 11 The Vestibular System

Figure 11.3: Dickman, J. D., D. Huss and M. Lowe (2004) Morphometry of otoconia in the utricle and saccule of developing Japanese quail. *Hear. Res.* 188: 89–103. **Figure 11.6A,B:** Fernández, C. and J. M. Goldberg (1976) Physiology of peripheral neurons innervating otolith organs of the squirrel monkey, Parts 1, 2, 3. *J. Neurophysiol.* 39: 970–1008. **Figure 11.9:** Fernández, C. and J. M. Goldberg (1976) Physiology of peripheral neurons innervating otolith organs of the squirrel monkey, Parts 1, 2, 3. *J. Neurophysiol.* 39: 970–1008. **Figure 11.10A:** Gibson, J. J. (1947) Motion picture testing and research, Report No 7, Army Air Force Aviation Psychology Program Research Reports. Washington, DC: US Government Printing Office. **Figure 11.10B:** Groh, J. M. (2014) *Making Space: How the Brain Knows Where Things Are.* Cambridge, MA: Harvard University Press. © 2014 Jennifer M. Groh. **Figure 11.10C:** Groh, J. M. (2014) *Making Space: How the Brain Knows Where Things Are.* Cambridge, MA: Harvard University Press. © 2014 Jennifer M. Groh. **Box 11A, part A:** Eaton, R. C., R. A. Bombardieri and D. L. Meyer (1977) The Mauthner-initiated startle response in teleost fish. *J. Exp. Biol.* 66: 65–81. **Box 11A, parts B,C:** Furshpan, E. J. and T. Furukawa (1962) Intracellular and extracellular responses of the several regions of the Mauthner cell of the goldfish. *J. Neurophysiol.* 25: 732–771.

CHAPTER 12 Touch and Proprioception

Figure 12.2C: Weinstein, S. (1968) Intensive and extensive aspects of tactile sensitivity as a function of body part, sex, and laterality. In *The Skin Senses*, D. R. Kenshalo (ed.). Springfield, IL: Charles C. Thomas, pp.195–222. **Figure 12.4A:** Johansson, R. S. and A. B. Vallbo (1983) Tactile sensory coding in the glabrous skin of the human. *Trends Neurosci.* 6: 27–32. **Figure 12.4B:** Abraira, V. E. and D. D. Ginty (2013) The sensory neurons of touch. *Neuron* 79: 618–639. **Figure 12.5:** Phillips, J. R., R. S. Johansson and K. O. Johnson (1990) Representation of braille characters in human nerve fibres. *Exp. Brain Res.* 81: 589–592. **Figure 12.6A:** Iggo, A. and A. R. Muir (1969) The structure and function of a slowly adapting touch corpuscle in hairy skin. *J. Physiol.* 200: 763–796. **Clinical Applications, Figure A:** Rosenzweig, M. R., S. M. Breedlove and A. L. Leiman (2005) *Biological Psychology*, 3rd Edition. Sunderland, MA: Sinauer Associates. **Clinical Applications, Figure B:** Haymaker, W. and B. Woodhall (1967) *Peripheral Nerve Injuries: Principles of Diagnosis.* New York: American Association of Neurological Surgeons, based on Foerster, O. (1933) The dermatomes in man. *Brain* 56(1): 1–39. **Clinical Applications, Figure C:** Haymaker,

W. and B. Woodhall (1967) *Peripheral Nerve Injuries: Principles of Diagnosis*. New York: American Association of Neurological Surgeons. **Figure 12.7A**: Matthews, P. B. C. (1964) Muscle spindles and their motor control. *Physiol. Rev.* 44: 219–289. **Figure 12.8A glabrous inset**: Abraira, V. and D. D. Ginty (2013) The sensory neurons of touch. *Neuron* 79: 618—639; Johansson, R. S. and A. B. Vallbo (1983) Tactile sensory coding in the glabrous skin of the human. *Trends Neurosci.* 6: 27–32. **Figure 12.8A hairy skin inset**: Abraira, V. and D. D. Ginty (2013) The sensory neurons of touch. *Neuron* 79: 618–639; Johansson, R. S. and A. B. Vallbo (1983) Tactile sensory coding in the glabrous skin of the human. *Trends Neurosci.* 6: 27–32. **Figure 12.9A**: Abraira, V. and D. D. Ginty (2013) The sensory neurons of touch. *Neuron* 79: 618–639. **Figure 12.9B–D**: Abraira, V. and D. D. Ginty (2013) The sensory neurons of touch. *Neuron* 79: 618–639. **Figure 12.11**: Brodal, P. (1992) *The Central Nervous System: Structure and Function*. New York: Oxford University Press, p. 151; Jones, E. G. and D. P. Friedman (1982) Projection pattern of functional components of thalamic ventrobasal complex on monkey somatosensory cortex. *J. Neurophysiol.* 48: 521–544. **Figure 12.12A–C**: Penfield, W. and T. Rasmussen (1950) *The Cerebral Cortex of Man: A Clinical Study of Localization of Function*. New York: Macmillan; Corsi, P. (ed.) (1991) *The Enchanted Loom: Chapters in the History of Neuroscience*. New York: Oxford University Press. **Figure 12.14**: Merzenich, M. M., R. J. Nelson, M. P. Stryker, M. S. Cynader, A. Schoppmann and J. M. Zook (1984) Somatosensory cortical map changes following digit amputation in adult monkeys. *J. Comp. Neurol.* 224: 591–605. **Figure 12.15A–C**: Jenkins, W. M. et al. (1990) Functional reorganization of primary somatosensory cortex in adult owl monkeys after behaviorally controlled tactile stimulation. *J. Neurophysiol.* 63: 82–104. **Box 12A, parts B,C**: Catania, K. C. and J. H. Kaas (1996) The unusual nose and brain of the star-nosed mole. *BioScience* 46: 578–586. **Box 12A, part E**: Sterbing-D'Angelo, S. J., M. Chadha, K. L. Marshall and C. F. Moss (2017) Functional role of airflow-sensing hairs on the bat wing. *J. Neurophysiol.* 117: 705–712. **Table 12.1**: Handler, A. and D. D. Ginty (2021) The mechanosensory neurons of touch and their mechanisms of activation. *Nat. Rev. Neurosci.* 22: 521–537; Abraira, V. and D. D. Ginty (2013) The sensory neurons of touch. *Neuron* 79: 618–639.

CHAPTER 13   Pain

Figure 13.1B,C: Fields, H. L. (1987) *Pain*. New York: McGraw-Hill, p. 19. **Figure 13.2A–C**: Fields, H. L. (ed.) (1990) *Pain Syndromes in Neurology*. London: Butterworths. **Box 13C, part B**: Willis, W. D., E. D. Al-Chaer, M. J. Quast and K. N. Westlund (1999) A visceral pain pathway in the dorsal column of the spinal cord. *Proc. Natl. Acad. Sci. U.S.A.* 96: 7675–7679. **Box 13C, part C (micrograph)**: Hirshberg, R. M., E. D. Al-Chaer, N. B. Lawand, K. N. Westlund and W. D. Willis (1996) Is there a pathway in the posterior funiculus that signals visceral pain? *Pain* 67: 291–305. **Box 13C, part C (drawing)**: Nauta, H. J. W., E. Hewitt, K. N. Westlund and W. D. Willis Jr. (1997) Surgical interruption of a midline dorsal column visceral pain pathway: Case report and review of the literature. *J. Neurosurg.* 86(3): 538–542. **Clinical Applications, Figure A**: Solonen, K. A. (1962) The phantom phenomenon in amputated Finnish war veterans. *Acta Orthop. Scand. Suppl.* 54: 1–37.

CHAPTER 14   Olfaction

Figure 14.2B: Pantages, E. and C. Dulac (2000) A novel family of candidate pheromone receptors in mammals. *Neuron* 28: 835–845. **Figure 14.3D**: Chen, M., R. R. Reed and A. P. Lane (2017) Acute inflammation regulates neuroregeneration through the NF-kB pathway in olfactory epithelium. *Proc. Natl. Acad. Sci. U.S.A.* 114: 8089–8094. **Figure 14.4**: Leung, C., P. A. Coulombe and R. R. Reed (2007) Contribution of olfactory neural stem cells to tissue maintenance and regeneration. *Nat. Neurosci.* 10: 720–726. **Figure 14.4B**: Balmer, C. W. and A.-S. LaMantia (2005) Noses and neurons: Induction, morphogenesis, and neuronal differentiation in the peripheral olfactory pathway. *Dev. Dyn.* 234: 464–481; Schwob, J. E., W. Jang, E. H. Holbrook, B. Lin et al. (2017) Stem and progenitor cells of the mammalian olfactory epithelium: Taking poietic license. *J. Comp. Neurol.* 525: 1034–1054. **Figure 14.5**: Firestein, S., F. Zufall and G. M. Shepherd (1991) Single odor-sensitive channels in olfactory receptor neurons are also gated by cyclic nucleotides. *J. Neurosci.* 11: 3665–3572. **Figure 14.6A**: Menini, A. (1999) Calcium signalling and regulation in olfactory neurons. *Curr. Opin. Neurobiol.* 9: 419–426. **Figure 14.6B**: Dryer, L. (2000) Evolution of odorant receptors. *BioEssays* 22: 803–810. **Figure 14.7A**: Vosshall, L. B. (2000) An olfactory sensory map in the fly brain. *Cell* 102: 147–159. **Clinical Applications, Figure A:** Lim, J. H., G. E. Davis, Z. Wang, V. Li, Y. Wu, T. C. Rue et al. (2009) Zicam-induced damage to mouse and human nasal tissue. *PLoS ONE* 4(10): e7647. **Clinical Applications, Figure B**: Khan, M., S.-J. Yoo, M. Clijsters, W. Backaert et al. (2021) Visualizing in deceased COVID-19 patients how SARS-CoV-2 attacks the respiratory and olfactory mucosae but spares the olfactory bulb. *Cell* 184: 5932–5949. **Figure 14.8A**: Menini, A. (1999) Calcium signalling and regulation in olfactory neurons. *Curr. Opin. Neurobiol.* 9: 419–425. **Figure 14.8B (wild type OMP)**: Wong, S. T. and 8 others (2000) Disruption of the type III adenylyl cyclase gene leads to peripheral and behavioral anosmia in transgenic mice. *Neuron* 27: 487–497. **Figure 14.8B (Golf)**: Belluscio, L., G. H. Gold, A. Nemes and R. Axel (1998) Mice deficient in Golf are anosmic. *Neuron* 20: 69–81. **Figure 14.8B (CNG)**: Brunet, L., G. H. Gold and J. Ngai (1996) General anosmia caused by a targeted disruption of the mouse olfactory cyclic nucleotide–gated cation channel. *Neuron* 17: 681–693. **Figure 14.9**: Firestein, S. and G. M. Shepherd (1992) Neurotransmitter antagonists block some odor responses in olfactory receptor neurons. *NeuroReport* 3: 661–664. **Figure 14.10A,B graphs**: Bozza, T., P. Firestein, C. Zheng and P. Mombaerts (2002) Odorant receptor expression defines functional units in the mouse olfactory system. *J. Neurosci.* 22: 3033–3043. **Figure 14.10A**: Bozza, T. and J. Kauer (1998) Odorant response properties of convergent olfactory receptor neurons. *J. Neurosci.* 18(12): 4560–4569. **Figure 14.11A**: Pantages, E. and C. Dulac (2000) A novel family of candidate pheromone receptors in mammals. *Neuron* 28: 835–845. **Figure 14.11B**: Dulac, C. and A. T. Torello (2003) Molecular detection of pheromone signals in mammals: From genes to behaviour. *Nat. Rev. Neurosci.* 4: 551–562. **Figure 14.11C**: Isogai, Y., S. Si, L. Pont-Lezica, T. Tan et al. (2011) Molecular organization of vomeronasal chemoreception. *Nature* 478: 241–245. **Figure 14.12A inset**: Wang, J. W. et al. (2003) Two-photon calcium imaging reveals an odor-evoked map of activity in the fly brain. *Cell*

112: 271–282. **Figure 14.12C**: Pomeroy, S. L., A.-S. LaMantia and D. Purves (1990) Postnatal construction of neural activity in the mouse olfactory bulb. *J. Neurosci.* 10: 1952–1966. **Figure 14.12E**: Twick, I., J. A. Lee and M. Ramaswami (2014) Olfactory habituation in Drosophila-odor encoding and its plasticity in the antennal lobe. *Prog. Brain Res.* 208: 3–38. **Figure 14.12F**: Vosshall, L. B. (2000) An olfactory sensory map in the fly brain. *Cell* 102: 147–159. **Figure 14.12G**: Mombaerts, P. and 7 others (1996) Visualizing an olfactory sensory map. *Cell* 87: 675–686; Tadenev, A. L. D., H. M. Kulaga, H. L. May-Simera, M. W. Kelley, N. Katsanis and R. R. Reed (2011) Loss of Bardet-Biedl syndrome protein-8 (BBS8) perturbs olfactory function, protein localization, and axon targeting. *Proc. Natl. Acad. Sci. U.S.A.* 108: 10320–10325. **Figure 14.13B**: Pantages, E. and C. Dulac (2000) A novel family of candidate pheromone receptors in mammals. *Neuron* 28: 835–845. **Figure 14.13C**: Woodson, J., A. Niemeyer and J. Bergan (2017) Untangling the neural circuits for sexual behavior. *Neuron* 95: 1–2. **Figure 14.14A**: Storace, D. A. and L. B. Cohen (2017) Measuring the olfactory bulb input-output transformation reveals a contribution to the perception of odorant concentration invariance. *Nat. Commun.* 8: 81. **Figure 14.14B**: Storace, D. A. and L. B. Cohen (2017) Measuring the olfactory bulb input-output transformation reveals a contribution to the perception of odorant concentration invariance. *Nat. Commun.* 8: 81. **Figure 14.15A**: Sosulski, D. L., M. L. Bloom, T. Cutforth, R. Axel and S. R. Datta (2011) Distinct representations of olfactory information in different cortical centers. *Nature* 472: 213–216. **Figure 14.15B**: Stettler, D. D. and R. Axel (2009) Representations of odor in the piriform cortex. *Neuron* 63: 854–864. **Figure 14.15C,D**: Bolding, K. E. and K. M. Franks (2018) Recurrent cortical circuits implement concentration-invariant odor coding. *Science* 361: eaat6904. **Figure 14.16A**: Shier, D., J. Butler and R. Lewis (2004) *Hole's Human Anatomy and Physiology*. Boston: McGraw-Hill. **Figure 14.17A,B**: Porter, J. and 8 others (2007) Mechanisms of scent-tracking in humans. *Nat. Neurosci.* 10: 27–29. **Figure 14.17C**: Porter, J. and 8 others (2007) Mechanisms of scent-tracking in humans. *Nat. Neurosci.* 10: 27–29. **Figure 14.18A**: Cain, W. S., R. Schmidt and P. Wolkoff (2007) Olfactory detection of ozone and d-limonene: Reactants in indoor spaces. *Indoor Air* 17: 337–347. **Figure 14.18D**: Rolls, E. T., M. L. Kringelbach and I. E. T. de Araujo (2003) Different representations of pleasant and unpleasant odours in the human brain. *Eur. J. Neurosci.* 18: 695–703. **Figure 14.19A,B**: Doucet, S., R. Soussignan, P. Sagot and B. Schaal (2009) The secretion of areolar (Montgomery's) glands from lactating women elicits selective, unconditional responses in neonates. *PLoS ONE* 4: e7579. **Figure 14.19C**: Doucet, S., R. Soussignan, P. Sagot and B. Schaal (2009) The secretion of areolar (Montgomery's) glands from lactating women elicits selective, unconditional responses in neonates. *PLoS ONE* 4: e7579. **Figure 14.20**: Savic, I., H. Berglund, B. Gulyas and P. Roland (2001) Smelling of odorous sex hormone-like compounds causes sex-differentiated hypothalamic activations in humans. *Neuron* 31: 661–668. **Figure 14.21C**: Wang, J., P. Eslinger, M. B. Smith and Q. X. Yang (2005) Functional magnetic resonance imaging study of human olfaction and normal aging. *J. Gerontol. A Biol. Sci. Med. Sci.* 60A: 510–514.

CHAPTER 15 Taste

Figure 15.1C: Schoenfeld, M. A. and 6 others (2004) Functional magnetic resonance tomography correlates of taste perception in human primary taste cortex. *Neuroscience* 127: 347–353. **Figure 15.7A**: Avery, J. A. (2021) Against gustotopic representation in the human brain: There is no Cartesian restaurant. *Curr. Opin. Physiol.* 20: 23–28. **Figure 15.7B (a,b)**: Jones, L. M., A. Fontanini, B. F. Sadacca, P. Miller and D. B. Katz (2007) Natural stimuli evoke dynamic sequences of states in sensory cortical ensembles. *Proc. Natl. Acad. Sci. U.S.A.* 104(47): 18772–18777. **Figure 15.7B (c)**: Katz, D. B., S. A. Simon and M. A. Nicolelis (2001) Dynamic and multimodal responses of gustatory cortical neurons in awake rats. *J. Neurosci.* 21: 4478–4489. **Figure 15.8A**: Wolfe, J. M. et al. (2018) *Sensation & Perception*, 5th Edition. New York: Oxford University Press/Sinauer Associates. **Figure 15.8B**: de Araujo, I. E., E. T. Rolls, M. L. Kringelbach, F. McGlone and N. Phillips (2003) Taste-olfactory convergence, and the representation of the pleasantness of flavour, in the human brain. *Eur. J. Neurosci.* 18(7): 2059–2068. **Figure 15.8C**: Shepherd, G. (2006) Smell images and the flavour system in the human brain. *Nature* 444: 316–321. **Figure 15.9A**: Grill, H. J. and R. Norgren (1978) The taste reactivity test. I. Mimetic responses to gustatory stimuli in neurologically normal rats. *Brain Res.* 143(2): 263–279. **Figure 15.9B**: Steiner, J. E., D. Glaser, M. E. Hawilo and K. C. Berridge (2001) Comparative expression of hedonic impact: Affective reactions to taste by human infants and other primates. *Neurosci. Biobehav. Rev.* 25(1): 53–74. **Figure 15.10A**: Leblanc, H. and S. Ramirez (2020) Linking social cognition to learning and memory. *J. Neurosci.* 40: 8782–8798. **Figure 15.10B left**: Nakai, J. et al. (2020) Another example of conditioned taste aversion: Case of snails. *Biology* 9(12): 422. **Figure 15.10B right**: Lavi, K. et al. (2018) Encoding of conditioned taste aversion in cortico-amygdala circuits. *Cell Rep.* 24(2): 278–283. **Quote, opening page**: Quotetab.com, https://www.quotetab.com/quote/by-duke-ellington/create-and-be-true-to-yourself-and-depend-only-on-your-own-good-taste.

CHAPTER 16 Lower Motor Neuron Circuits and Motor Control

Figure 16.2A–C: Burke, R. E., P. L. Strick, K. Kanda, C. C. Kim and B. Walmsley (1977) Anatomy of medial gastrocnemius and soleus motor nuclei in cat spinal cord. *J. Neurophysiol.* 40: 667–680. **Figure 16.6A–C**: Burke, R. E., D. N. Levine, P. Tsairis and F. E. Zajac III (1973) Physiological types and histochemical profiles in motor units of the cat gastrocnemius. *J. Physiol.* 234: 723–748. **Box 16A, part B**: Gordon, T., N. Tyreman, V. F. Rafuse and J. B. Munson (1997) Fast-to-slow conversion following chronic low-frequency activation of medial gastrocnemius muscle in cats. I. Muscle and motor unit properties. *J. Neurophysiol.* 77: 2585–2604. **Box 16A, part C**: Munson, J. B., R. C. Foehring, L. M. Mendell and T. Gordon (1997) Fast-to-slow conversion following chronic low-frequency activation of medial gastrocnemius muscle in cats. II. Motoneuron properties. *J. Neurophysiol.* 77: 2605–2615. **Box 16A, parts D,E**: Van Cutsem, M., J. Duchateau and K. Hainaut (1998) Changes in single motor unit behaviour contribute to the increase in contraction speed after dynamic training in humans. *J. Physiol.* 513: 295–305. **Figure 16.7**:

Walmsley, B., J. A. Hodgson and R. E. Burke (1978) Forces produced by medial gastrocnemius and soleus muscles during locomotion in freely moving cats. *J. Neurophysiol.* 41: 1203–1216. **Figure 16.9**: Monster, A. W. and H. Chan (1977) Isometric force production by motor units of extensor digitorum communis muscle in man. *J. Neurophysiol.* 40: 1432–1443. **Figure 16.11A,B**: Hunt, C. C. and S. W. Kuffler (1951) Stretch receptor discharges during muscle contraction. *J. Physiol.* 113: 298–314. **Figure 16.13A,B**: Patton, H. D. (1965) Reflex regulation of movement and posture. In *Physiology and Biophysics*, 19th Edition, T. C. Ruch and H. D. Patton (eds.). Philadelphia: Saunders, pp. 181–206. **Figure 16.15A–C**: Pearson, K. (1976) The control of walking. *Sci. Amer.* 235(6): 72–86. **Figure 16.15D**: Kiehn, O. (2016) Decoding the organization of spinal circuits that control locomotion. *Nat. Rev. Neurosci.* 17: 224–238. **Figure 16.16**: Kiehn, O. (2016) Decoding the organization of spinal circuits that control locomotion. *Nat. Rev. Neurosci.* 17: 224–238; Drew, T. and D. S. Marigold (2015) Taking the next step: Cortical contributions to the control of locomotion. *Curr. Opin. Neurobiol.* 33: 25–33. **Clinical Applications, Figure B**: Geevasinga N., P. Menon, P. H. Özdinler, M. C. Kiernan and S. Vucic (2016) Pathophysiological and diagnostic implications of cortical dysfunction in ALS. *Nat. Rev. Neurol.*12: 651–661.

CHAPTER 17   Upper Motor Neuron Control of the Brainstem and Spinal Cord

Box 17A: Graziano, M. S. A., T. N. S. Aflalo and D. F. Cooke (2005) Arm movements evoked by electrical stimulation in the motor cortex of monkeys. *J. Neurophysiol.* 94: 4209–4223. **Figure 17.6**: Porter, R. and R. Lemon (1993) *Corticospinal Function and Voluntary Movement*. Oxford: Oxford University Press. © 1993 Oxford University Press. **Figure 17.7**: Graziano, M. S. A., T. N. S. Aflalo and D. F. Cooke (2005) Arm movements evoked by electrical stimulation in the motor cortex of monkeys. *J. Neurophysiol.* 94: 4209–4223. **Figure 17.8A–D**: Georgeopoulos, A. P., A. B. Swartz and R. E. Ketter (1986) Neuronal population coding of movement direction. *Science* 233: 1416–1419. **Figure 17.9**: Geyer, S., M. Matelli and G. Luppino (2000) Functional neuroanatomy of the primate isocortical motor system. *Anat. Embryol.* 202: 443–474. **Box 17B, part A**: Dr. Eric C. Leuthardt, Professor of Neurological Surgery, Washington University School of Medicine, Director of The Center for Innovation in Neuroscience and Technology. **Box 17B, part B**: Donati, A. et al. (2016) Long-term training with a brain-machine interface-based gait protocol induces partial neurological recovery in paraplegic patients. *Sci. Rep.* 6: 30383. **Figure 17.10A–C**: Rizzolatti, G., L. Fadiga, V. Gallese and L. Fogassi (1996) Premotor cortex and the recognition of motor actions. *Cogn. Brain Res.* 3: 131–141. **Figure 17.11A**: Rizzolatti, G. and C. Sinigaglia (2016) The mirror mechanism: A basic principle of brain function. *Nat. Rev. Neurosci.* 17: 757–765; Caspers, S. et al. (2010) ALE meta-analysis of action observation and imitation in the human brain. *NeuroImage* 50(3): 1148–1167. **Figure 17.14**: Nashner, L. M. (1979) Organization and programming of motor activity during posture control. In *Progress in Brain Research, Volume 50, Reflex Control of Posture and Movement*, R. Granit and O. Pompeiano (eds.). Amsterdam: Elsevier/North Holland Biomedical Press, pp. 177–184.

CHAPTER 18   Modulation of Movement by the Basal Ganglia

Figure 18.6A1: Hikosaka, O. and R. H. Wurtz (1986) Cell activity in monkey caudate nucleus preceding saccadic eye movements. *Exp. Brain Res.* 63: 659–662. **Figure 18.6A2–3**: Hikosaka, O. and R. H. Wurtz (1983) Visual and oculomotor functions of monkey substantia nigra pars reticulata. IV. Relation of substantia nigra to superior colliculus. *J. Neurophysiol.* 49: 1285–1301. **Figure 18.6B**: Hikosaka, O. and R. H. Wurtz (1983) Visual and oculomotor functions of monkey substantia nigra pars reticulata. IV. Relation of substantia nigra to superior colliculus. *J. Neurophysiol.* 49: 1285–1301. **Figure 18.9B**: DeLong, M. R. (1990) Primate models of movement disorders of basal ganglia origin. *Trends Neurosci.* 13: 281–285. **Clinical Applications, Figure B**: Hashimoto, T., C. M. Elder, M. S. Okun, S. K. Patrick and J. L. Vitek (2003) Stimulation of the subthalamic nucleus changes the firing pattern of pallidal neurons. *J. Neurosci.* 23: 1916–1923. **Clinical Applications, Figure C**: McIntyre, C. C. and R. W. Anderson (2016) Deep brain stimulation mechanisms: The control of network activity via neurochemistry modulation. *J. Neurochem.* 139 (Suppl. 1): 338–345. **Figure 18.10B**: DeLong, M. R. (1990) Primate models of movement disorders of basal ganglia origin. *Trends Neurosci.* 13: 281–285. **Box 18B, parts A–C**: Desrochers, T. M., K. Amemori and A. M. Graybiel (2015) Habit learning by naive macaques is marked by response sharpening of striatal neurons representing the cost and outcome of acquired action sequences. *Neuron* 87: 853–868.

CHAPTER 19   Modulation of Movement by the Cerebellum

Figure 19.10A: Stein, J. F. (1986) Role of the cerebellum in the visual guidance of movement. *Nature* 323: 217–220. **Figure 19.11A,B**: Cerminara, N. L., E. J. Lang, R. V. Sillitoe and R. Apps (2015) Redefining the cerebellar cortex as an assembly of non-uniform Purkinje cell micro-circuits. *Nat. Rev. Neurosci.* 16: 79–93. **Figure 19.12A,B**: Thach, W. T. (1968) Discharge of Purkinje and cerebellar nuclear neurons during rapidly alternating arm movements in the monkey. *J. Neurophysiol.* 31: 785–797. **Figure 19.13**: Optican, L. M. and D. A. Robinson (1980) Cerebellar-dependent adaptive control of primate saccadic system. *J. Neurophysiol.* 44: 1058–1076. **Box 19A, parts A,B**: Caviness, V. S., Jr. and P. Rakic (1978) Mechanisms of cortical development: A view from mutations in mice. *Annu. Rev. Neurosci.* 1: 297–326. **Figure 19.17A**: Manto, M. and P. Mariën (2015) Schmahmann's syndrome: Identification of the third cornerstone of clinical ataxiology. *Cerebellum & Ataxias* 2: 2. **Figure 19.17B**: Guell, X., F. Hoche and J. D. Schmahmann (2015) Metalinguistic deficits in patients with cerebellar dysfunction: Empirical support for the dysmetria of thought theory. *Cerebellum* 14(1): 50–58.

CHAPTER 20   Eye Movement and Sensorimotor Integration

Figure 20.1: Yarbus, A. L. (1967) Eye movements during perception of complex objects. In *Eye Movements and Vision* (translated by B. Haigh). New York: Plenum Press, p. 181. **Box 20A, part A**: Pritchard, R. M. (1961) Stabilized images on the retina.

Sci. Amer. 204(6): 72–78. **Box 20A, part B**: Riggs, L. A., F. Ratliff, J. C. Cornsweet and T. N. Cornsweet (1953) The disappearance of steadily fixated visual test objects. *J. Opt. Soc. Am.* 43: 495–501. **Figure 20.4**: Fuchs, A. F. (1967) Saccadic and smooth pursuit eye movements in the monkey. *J. Physiol.* 191: 609–630. **Figure 20.5**: Fuchs, A. F. (1967) Saccadic and smooth pursuit eye movements in the monkey. *J. Physiol.* 191: 609–630. **Figure 20.6**: Baarsma, E. and H. Collewijn (1974) Vestibulo-ocular and optokinetic reactions to rotation and their interaction in the rabbit. *J. Physiol.* 238: 603–625. **Clinical Applications, Figure C**: Benson, P. J. and 5 others (2012) Simple viewing tests can detect eye movement abnormalities that distinguish schizophrenia cases from controls with exceptional accuracy. *Biol. Psychiatry* 72(9): 716–724. **Figure 20.7**: Fuchs, A. F. and E. S. Luschei (1970) Firing patterns of abducens neurons of alert monkeys in relationship to horizontal eye movements. *J. Neurophysiol.* 33: 382–392. **Figure 20.9A,B**: Schiller, P. H. and M. Stryker (1972) Single unit recording and stimulation in superior colliculus of the alert rhesus monkey. *J. Neurophysiol.* 35: 915–923. **Box 20B, part B**: Wurtz, R. H. and J. E. Albano (1980) Visual-motor function of the primate superior colliculus. *Annu. Rev. Neurosci.* 3: 189–226. **Box 20B, part C**: Ozen, G., G. J. Augustine and W. C. Hall (2000) Contribution of superficial layer neurons to premotor bursts in the superior colliculus. *J. Neurophysiol.* 84: 460–471. **Figure 20.10A,B**: Sparks, D. L. and L. E. Mays (1983) Spatial localization of saccade targets. I. Compensation for stimulation-induced perturbations in eye position. *J. Neurophysiol.* 49: 45–63. **Box 20C, part A**: Sparks, D. L. (1975) Response properties of eye movement-related neurons in the monkey superior colliculus. *Brain Res.* 90: 147–152. **Box 20C, part B**: Fuchs, A. F. and E. S. Luschei (1970) Firing patterns of abducens neurons of alert monkeys in relationship to horizontal eye movements. *J. Neurophysiol.* 33: 382–392. **Figure 20.12B**: Schall, J. D. (1995) Neural basis of saccade target selection. *Rev. Neurosci.* 6: 63–85. **Figure 20.13**: Krauzlis, R. J. (2005) The control of voluntary eye movements: New perspectives. *Neuroscientist* 11: 124–137.

CHAPTER 21   The Visceral Motor System

Box 21B, part A: Marx, J. (2003) Cellular warriors at the battle of the bulge. *Science* 299: 846–849; Morton, G. J., T. H. Meek and M. W. Schwartz (2014) Neurobiology of food intake in health and disease. *Nat. Rev. Neurosci.* 15: 367–378. **Box 21B, part B**: Yaswen, L., N. Diehl, M. B. Brennan and U. Hochgeschwender (1999) Obesity in the mouse model of pro-opiomelanocortin deficiency responds to peripheral melanocortin. *Nat. Med.* 5: 1066–1070. **Box 21B, part C**: O'Rahilly, S., S. Farooqi, G. S. H. Yeo and B. G. Challis (2003) Human obesity: Lessons from monogenic disorders. *Endocrinology* 144: 3757–3764. Reprinted by permission of Oxford University Press on behalf of the Endocrine Society.

CHAPTER 22   Early Brain Development

Figure 22.2: Lancaster, M. A., M. Renner, C.-A. Martin, W. Wenzel and 6 others (2013) Cerebral organoids model human brain development and microcephaly. *Nature* 501: 373–379. **Figure 22.5C**: Rothstein, M. (2018) The molecular basis of neural crest axial identity. *Dev. Biol.* 444: S170–S180. **Figure 22.6C, left**: Carmona, F. D., R. Jimenez and J. M. Collinson (2008) *BMC Biol.* 6: article 44. **Figure 22.6D, left**: Birol, O. et al. (2016) The mouse Foxi3 transcription factor is necessary for the development of posterior placodes. *Dev. Biol.* 409: 139–151. **Figure 22.6D, right**: Hudspeth, A. J. (1985) The cellular basis of hearing: The biophysics of hair cells. *Science* 230: 745–752. **Figure 22.8A**: Gilbert, S. (1994) *Developmental Biology.* Sunderland, MA: Sinauer Associates; Lawrence, P. A. 1992. *The Making of a Fly: The Genetics of Animal Design.* Oxford: Blackwell Scientific Publications. **Figure 22.8B**: Gilbert, S. (1994) *Developmental Biology.* Sunderland, MA: Sinauer Associates; Lawrence, P. A. 1992. *The Making of a Fly: The Genetics of Animal Design.* Oxford: Blackwell Scientific Publications. **Figure 22.8B, photo**: Ingham, P. W. (1988) The molecular genetics of embryonic pattern formation in *Drosophila. Nature* 335: 25–34. **Figure 22.8C**: Veraska, A., M. Del Campo and W. McGinnis (2000) Developmental patterning genes and their conserved functions: From model organisms to humans. *Mol. Genet. Metab.* 69: 85–100. **Figure 22.10C**: De Robertis, E. M. and H. Kuroda (2004) Dorsal-ventral patterning and neural induction in Xenopus embryos. *Annu. Rev. Cell Dev. Biol.* 20: 285–308. **Figure 22.12A**: Dodd, J. et al. (1998) The when and where of floor plate induction. *Science* 282: 1654–1657. **Figure 22.12B**: Fausett, S. R. et al. (2014) BMP antagonism by Noggin is required in presumptive notochord cells for mammalian foregut morphogenesis. *Dev. Biol.* 391: 111–124. **Figure 22.13B**: Thomas, S., L. Boutaud, M. L. Reillyu and A. Benmerah (2019) Cilia in hereditary cerebral anomalies. *Biol. Cell* 111: 217–231. **Figure 22.13C**: Noctor, S. et al. (2008) Distinct behaviors of neural stem and progenitor cells underlie cortical neurogenesis. *J. Comp. Neurol.* 508: 28–44. **Figure 22.15**: Kintner, C. (2002) Neurogenesis in embryos and in adult neural stem cells. *J. Neurosci.* 22: 639–643. **Figure 22.16A**: Tian, N. (2012) Development of retinal ganglion cell dendritic structure and synaptic connections. In *Webvision: The Organization of the Retina and Visual System*, H. Kolb, E. Fernandez and R. Nelson (eds.). Salt Lake City: University of Utah Health Sciences Center. **Figure 22.16B**: Rakic, P. (1974) Neurons in rhesus monkey visual cortex: Systematic relation between time of origin and eventual disposition. *Science* 183: 425–427. **Figure 22.17A**: Lamouille, S., J. Xu and R. Derynck (2014) Molecular mechanisms of epithelial mesenchymal transition. *Mol. Cell Biol.* 15: 179. **Figure 22.17B**: Hutchins, E. J. and M. E. Bronner (2019) Draxin alters laminin organization during basement membrane remodeling to control cranial neural crest EMT. *Dev. Biol.* 446: 151–158. **Figure 22.20A (bottom)**: Govek, E.-E., M. E. Hatten and L. Van Aelst (2011) The role of rho GTPase proteins in CNS neuronal migration. *Dev. Neurobiol.* 71: 528–523. **Figure 22.20B**: Cowan, W. M. (1979) The development of the brain. *Sci. Am.* 241: 124. **Figure 22.20C**: Noctor, S. C. et al. (2001) Neurons derived from radial glial cells establish radial units in neocortex. *Nature* 409: 714–720. **Figure 22.21B,C**: Hong, S. E. and 7 others (2000) Autosomal recessive lissencephaly with cerebellar hypoplasia is associated with human RELN mutations. *Nat. Genet.* 26: 93–96. **Box 22A, part B**: Wichterle, H. et al. (2002) Directed differentiation of embryonic stem cells into motor neurons. *Cell* 110: 385–397. **Clinical Applications, Figure A**: Anchan, R. M., D. P. Drake, C. F. Haines and E. A. Gerwe (1997) Disruption of local retinoid-mediated gene expression accompanies abnormal development in the mammalian olfactory pathway. *J. Comp. Neurol.* 379: 171–184. **Clinical Applications, Figure**

B: Linney, E. and A.-S. LaMantia (1994) Retinoid signaling in mouse embryos. In *Advances in Developmental Biology*, Volume 3, P. Wassarman (ed.). Greenwich, CT: JAI Press/Elsevier Science, pp. 73–114. **Clinical Applications, Figure C**: Monuki, E. S. and C. A. Walsh (2001) Mechanisms of cerebral cortical patterning in mice and humans. *Nat. Neurosci.* 4: 1199–1206.

CHAPTER 23 Construction of Neural Circuits

Figure 23.1B: Bulgakov, N. A. and E. Knust (2009) A conserved di-basic motif of Drosophila Crumbs contributes to efficient ER export. *J. Cell Sci.* 122: 2587–2596. **Figure 23.1D**: Yogev, S. and K. Shen. (2017) Establishing neuronal polarity with environmental and intrinsic mechanisms. *Neuron* 96: 638–650. **Figure 23.1F**: Shi, S., L. Jan and Y. Jan (2003) Hippocampal neuronal polarity specified by spatially localized mPar3/mPar6 and PI 3-kinase activity. *Cell* 112: 63–75. **Figure 23.2A**: Kelliher, M. T., H. A. J. Saunders and J. Wildonger (2019) Microtubule control of functional architecture in neurons. *Curr. Opin. Neurobiol.* 57: 39–45; Guedes-Dias, P. and E. L. F. Holzbauer (2019) Axonal transport: Driving synaptic function. *Science* 366: 199. **Figure 23.2B**: Wang, J., L. Fourriere and P. A. Gleeson (2020) Local secretory trafficking pathways in neurons and the role of dendritic Golgi outposts in different cell models. *Front. Mol. Neurosci.* 13: 597391. **Figure 23.3D inset; whole zebrafish embryo**: Takahashi, M., M. Narushima and Y. Oda (2002) In vivo imaging of functional inhibitory networks on the Mauthner cell of larval zebrafish. *J. Neurosci.* 22: 3929–3938. **Figure 23.3D inset & large image; Mauthner cell axons and growth cone**: Jontes, J. D., J. Buchanan and S. Smith (2000) Growth cone and dendrite dynamics in zebrafish embryos: Early events in synaptogenesis imaged in vivo. *Nat. Neurosci.* 3: 231–237. **Figure 23.3F**: Bovolenta P. and C. Mason (1987) Growth cone morphology varies with position in the developing mouse visual pathway from retina to first targets. *J. Neurosci.* 7: 1447–1460; Godement, P., L. C. Wang and C. A. Mason (1994) Retinal axon divergence in the optic chiasm: Dynamics of growth cone behavior at the midline. *J. Neurosci.* 14: 7024–7039. [Erratum in *J. Neurosci.* 1 March 1995, 15(3) np; https://doi.org/10.1523/JNEUROSCI.15-03-j0002.1995.] **Figure 23.4A,illus**: Kahn, O. I. and P. W. Baas (2016) Microtubules and growth cones: Motors drive the turn. *Trends Neurosci.* 39: 433–440; Dent, E. W. and F. B. Gertler (2003) Cytoskeletal dynamics and transport in growth cone motility and axon guidance. *Neuron* 40: 209–227. **Figure 23.4B illus**: Kahn, O. I. and P. W. Baas (2016) Microtubules and growth cones: Motors drive the turn. *Trends Neurosci.* 39: 433–440. **Figure 23.4B**: Dent, E. W. and K. Kalil (2001) Axon branching requires interactions between dynamic microtubules and actin filaments. *J. Neurosci.* 21: 9757–9769. **Figure 23.4C**: Huber, A. B., A. L. Kolodkin, D. D. Ginty and J.-F. Cloutier (2003) Signaling at the growth cone: Ligand-receptor complexes and the control of axon growth and guidance. *Annu. Rev. Neurosci.* 26: 509–563. **Figure 23.4D**: Igarashi, M. et al. (2018) New observations in neuroscience using superresolution microscopy. *J. Neurosci.* 38: 9459–9467. **Figure 23.4E**: Gomez, T. M. and J. O. Zheng (2006) The molecular basis for calcium-dependent axon pathfinding. *Nat. Rev. Neurosci.* 7: 115–117. **Figure 23.5A**: Huber, A. B., A. L. Kolodkin, D. D. Ginty and J. F. Cloutier (2003) Signaling at the growth cone: Ligand-receptor

complexes and the control of axon growth and guidance. *Annu. Rev. Neurosci.* 26: 509–563. **Figure 23.5B**: Kennedy, T. E., T. Serafini, J. R. de la Torre and M. Tessier-Lavigne (1994) Netrins are diffusible chemotropic factors for commissural axons in the embryonic spinal cord. *Cell* 78: 425–435. **Figure 23.5C**: Kidd, T., K. S. Bland and C. S. Goodman (1999) Slit is the midline repellent for the robo receptor in *Drosophila*. *Cell* 96: 785–794. **Figure 23.6B, upper**: Gomez, T. M. and J. O. Zheng (2006) The molecular basis for calcium-dependent axon pathfinding. *Nat. Rev. Neurosci.* 7: 115–117; Dontchev, V. D. and P. C. Letourneau (2002) Nerve growth factor and semaphorin 3A signaling pathways interact in regulating sensory neuronal growth cone motility. *J. Neurosci.* 22: 6659–6669. **Figure 23.6B, lower**: Gomez, T. M. and J. O. Zheng (2006) The molecular basis for calcium-dependent axon pathfinding. *Nat. Rev. Neurosci.* 7: 115–117; Dontchev, V. D. and P. C. Letourneau (2002) Nerve growth factor and semaphorin 3A signaling pathways interact in regulating sensory neuronal growth cone motility. *J. Neurosci.* 22: 6659–6669. **Figure 23.7A,B**: Chatzopoulou, E., A. Miguez, M. Savvaki, G. Levasseur et al. (2008) Structural requirement of TAG-1 for retinal ganglion cell axons and myelin in the mouse optic nerve. *J. Neurosci.* 28: 7624–7636. **Figure 23.8A**: Whitford, K. L., P. Dijkhuizen, F. Polleux and A. Ghosh (2002) Molecular control of cortical dendrite development. *Annu. Rev. Neurosci.* 25: 127–149. **Figure 23.8B**: Fujishima, K., R. Horie, A. Mochizuki and M. Kengaku (2012) Principles of branch dynamics governing shape characteristics of cerebellar Purkinje cell dendrites. *Development* 139: 3442–3455. **Figure 23.9A**: Hattori, D., S. S. Millard, W. M. Wojtowicz and S. L. Zipursky (2008) Dscam-mediated cell recognition regulates neural circuit formation. *Annu. Rev. Cell Dev. Biol.* 24: 597–620. **Figure 23.9B**: Hattori, D., S. S. Millard, W. M. Wojtowicz and S. L. Zipursky (2008) Dscam-mediated cell recognition regulates neural circuit formation. *Annu. Rev. Cell Dev. Biol.* 24: 597–620. **Figure 23.9C, fly**: Lefebvre, J. L., J. R. Sanes and J. N. Kay (2015) Development of dendritic form and function. *Annu. Rev. Cell Dev. Biol.* 31: 741–777. **Figure 23.9C, mouse**: Hattori, D., S. S. Millard, W. M. Wojtowicz and S. L. Zipursky (2008) Dscam-mediated cell recognition regulates neural circuit formation. *Annu. Rev. Cell Dev. Biol.* 24: 597–620. **Figure 23.9D**: Mountofaris, M., D. Canzio, C. L. Nwakeze, W. V. Chen et al. (2018) Writing, reading, and translating the clustered protocadherin cell surface recognition code for neural circuit assembly. *Annu. Rev. Cell Dev. Biol.* 34: 471–493. **Figure 23.9E**: Chen, W. V. et al. (2017) Pcdhαc2 is required for axonal tiling and assembly of serotonergic circuitries in mice. *Science* 356: 406–411. **Figure 23.10A,B**: Waites, C. L., A. M. Craig and C. C. Garner (2005) Mechanisms of vertebrates synaptogenesis. *Annu. Rev. Neurosci.* 28: 251–274. **Figure 23.10C**: Dean, C. and T. Dresbach (2006) Neuroligins and neurexins: Linking cell adhesion, synapse formation, and cognitive function. *Trends Neurosci.* 29: 21–29. **Figure 23.10D**: Phillips, G. R. and 6 others (2003) Gamma-protocadherins are targeted to subsets of synapses and intracellular organelles in neurons. *J. Neurosci.* 23: 5096–5104. **Figure 23.11A–C**: Hamburger, V. (1958) Regression versus peripheral controls of differentiation in motor hypoplasia. *Amer. J. Anat.* 102: 365–409; Hamburger, V. (1977) The developmental history of the motor neuron. *Neurosci. Res. Program Bull.* 15(Suppl.): iii–37; Hollyday, M. and V. Hamburger (1976) Reduction of the naturally occurring motor neuron loss by enlargement of the periphery. *J. Comp. Neurol.* 170: 311–320. **Figure 23.12A,B**: Purves, D. and J. W. Lichtman (1980) Elimination of synapses in the developing nervous system. *Science* 210: 153–157. **Figure 23.12C**: Wilson, A. M. and 6 others (2019) Developmental re-wiring between cerebellar climbing fibers and Purkinje cells begins with positive feedback synapse addition. *Cell Rep.* 29: 2849–2861. **Figure 23.13A**: Walsh, M. K. and J. W. Lichtman (2003) In vivo time-lapse imaging of synaptic takeover associated with naturally occurring synapse elimination. *Neuron* 37: 67–73. **Figure 23.13B**: Zito, K. (2000) The flip side of synapse elimination. *Neuron* 25: 269–278. © 2000 Cell Press. **Figure 23.13C**: Hashimoto, K., R. Ichikawa, K. Kitamura, M. Watanabe and M. Kano (2009) Translocation of a "winner" climbing fiber to the Purkinje cell dendrite and subsequent elimination of "losers" from the soma in developing cerebellum. *Neuron* 63: 106–118. **Figure 23.14A,B**: Purves, D. and J. W. Lichtman (1985) *Principles of Neural Development.* Sunderland, MA: Sinauer Associates. **Figure 23.14C**: Chun, L. L. and P. H. Patterson (1977) Role of nerve growth factor in the development of rat sympathetic neurons in vitro. III; Effect on acetylcholine production. *J. Cell Biol.* 75: 712–718. **Figure 23.14D**: Levi-Montalcini, R. (1972) The morphological effects of immuno-sympathectomy. In *Immonosympathectomy*, G. Steiner and E. Schönbaum (eds.). Amsterdam: Elsevier. **Figure 23.15A**: Maisonpierre, P. C. and 6 others (1990) Neurotrophin-3: A neurotrophic factor related to NGF and BDNF. *Science* 247: 1446–1451. **Figure 23.15B**: Bibel, M. and Y.-A. Barde (2000) Neurotrophins: Key regulators of cell fate and cell shape in the vertebrate nervous system. *Genes Dev.* 14: 2919–2937. **Figure 23.16C**: Campenot, R. B. (1981) Regeneration of neurites on long-term cultures of sympathetic neurons deprived of nerve growth factor. *Science* 214: 579–581. **Figure 23.16D**: Ascano, M., D. Bodmer and R. Kuruvilla (2012) Endocytic trafficking of neurotrophins in neural development. *Trends Cell Biol.* 22: 266–273. **Figure 23.16E**: Zweifel, L. S., R. Kuruvilla and D. D. Ginty (2005) Functions and mechanisms of retrograde neurotrophin signalling. *Nat. Rev. Neurosci.* 6: 615–625. **Figure 23.18**: Philippidou, P. and J. S. Dassen (2013) Hox genes: Choreographers in neural development, architects of circuit organization. *Neuron* 80: 12–34. **Figure 23.19A**: Spead, O., C. J. Weaver, T. Moreland and F. E. Poulain (2021) Live imaging of retinotectal mapping reveals topographic map dynamics and a previously undescribed role for Contactin 2 in map sharpening. *Development* 148(22): dev199584. https://doi.org/10.1242/dev.199584. **Figure 23.19B**: Sperry, R. W. (1963) Chemoaffinity in the orderly growth of nerve fiber patterns and connections. *Proc. Natl. Acad. Sci. U.S.A.* 50: 703–710. **Figure 23.20A**: Walter, J., S. Henke-Fahle and F. Bonhoeffer (1987) Avoidance of posterior tectal membranes by temporal retinal axons. *Development* 101: 909–913. **Figure 23.20B**: Scalia, F., J. R. Currie and D. A. Feldheim (2009) Eph/Ephrin gradients in the retinotectal system of *Rana pipiens*: Developmental and adult expression patterns. *J. Comp. Neurol.* 514: 30–48. **Figure 23.20C**: Harada, T., C. Harada and L. F. Parada (2007) Molecular regulation of visual system development: More than meets the eye. *Genes Dev.* 21: 367–378. **Box 23A, parts A–C**: Herrera, E. and 8 others (2003) Zic2 patterns binocular vision by specifying the uncrossed retinal projection. *Cell* 114: 545–557. **Box 23B, part A**: Purves, D.

and R. I. Hume (1981) The relation of postsynaptic geometry to the number of presynaptic axons that innervate autonomic ganglion cells. *J. Neurosci.* 1: 441–452. **Box 23B, part B**: Purves, D. and R. I. Hume (1981) The relation of postsynaptic geometry to the number of presynaptic axons that innervate autonomic ganglion cells. *J. Neurosci.* 1: 441–452. **Clinical Applications, Figures A,B**: Wahl, M. et al. (2009) Variability of homotopic and heterotopic callosal connectivity in partial agenesis of the corpus callosum: A 3T diffusion tensor imaging and Q-ball tractography study. *Am. J. Neuroradiol.* 30: 282–289. **Clinical Applications, Figures C,D**: Haller, S., S. G. Wetzel and J. Lütschg (2008) Functional MRI, DTI and neurophysiology in horizontal gaze palsy with progressive scoliosis. *Neuroradiology* 50: 453–459. **Clinical Applications, Figure E**: Haller, S., S. G. Wetzel and J. Lütschg (2008) Functional MRI, DTI and neurophysiology in horizontal gaze palsy with progressive scoliosis. *Neuroradiology* 50: 453–459.

CHAPTER 24   Experience-Dependent Plasticity in the Developing Brain

Figure 24.2A: Gilmore, J. H. et al. (2012) Longitudinal development of cortical and subcortical gray matter from birth to 2 years. *Cereb. Cortex* 22: 2478–2485. **Figure 24.2B**: Conel, J. L. (1939–1967) *The Postnatal Development of the Human Cerebral Cortex*, Volumes 1–8. Cambridge, MA: Harvard University Press. **Figure 24.2C**: Huttenlocher, P. R., C. De Courten, L. J. Garey and H. Van der Loos (1982) Synaptogenesis in human visual cortex: Evidence for synapse elimination during normal development. *Neurosci. Lett.* 33: 247–252. **Figure 24.2 inset**: Shapson-Coe, A., M. Januszewski, D. R. Berger and A. Pope (2021) A connectomic study of a petascale fragment of human cerebral cortex. *bioRxiv* 2021. https://doi.org/10.1101/2021.05.29.446289. **Figure 24.3**: Kole, K., W. Scheenen, P. Tiesinga and T. Celikel (2018) Cellular diversity of the somatosensory cortical map plasticity. *Neurosci. Biobehav. Rev.* 84: 100–115. **Figure 24.4A**: LeVay, S., T. N. Wiesel and D. H. Hubel (1980) The development of ocular dominance columns in normal and visually deprived monkeys. *J. Comp. Neurol.* 191: 1–51. **Figure 24.4B**: Yacoub, E. et al. (2007) Robust detection of ocular dominance columns in humans using Hahn Spin Echo BOLD functional MRI at 7 Tesla. *NeuroImage* 37: 1161–1177. **Figure 24.5A**: Petitto, L. A. and P. F. Marentette (1991) Babbling in the manual mode: Evidence for the ontogeny of language. *Science* 251: 1493–1496. **Figure 24.5B**: Johnson, J. S. and E. I. Newport (1989) Critical period effects in second language learning: The influences of maturational state on the acquisition of English as a second language. *Cogn. Psychol.* 21: 60–99. **Figure 24.5C**: Schlaggar, B. L. and 5 others (2002) Functional neuroanatomical differences between adults and school-age children in the processing of single words. *Science* 296: 1476–1479. **Figure 24.6**: Rakic, P., J. P. Bourgeois, M. F. Eckenhoff, N. Zecevic and P. S. Goldman-Rakic (1986) Concurrent overproduction of synapses in diverse regions of the primate cerebral cortex. *Science* 232: 232–235. **Figure 24.7A**: Gogtay, N. and 11 others (2004) Dynamic mapping of human cortical development during childhood through early adulthood. *Proc. Natl. Acad. Sci. U.S.A.* 101: 8174–8179. **Figure 24.7B**: Lenroot, R. K. and 11 others (2007) Sexual dimorphism of brain development trajectories during childhood and adolescence. *NeuroImage* 36: 1065–1073. **Figure 24.8A,B**: Shaw, P. and 9 others (2007) Attention-deficit/hyperactivity disorder is characterized by a delay in cortical maturation. *Proc. Natl. Acad. Sci. U.S.A.* 104: 19649–19654. © 2007 National Academy of Sciences, U.S.A. **Figure 24.8B**: Shaw, P. and 8 others (2006) Longitudinal mapping of cortical thickness and clinical outcome in children and adolescents with attention-deficit/hyperactivity disorder. *Arch. Gen. Psychiatry* 63: 540–549. **Figure 24.9A,B**: Hubel, D. H. and T. N. Wiesel (1962) Receptive fields, binocular interaction and functional architecture in the cat's visual cortex. *J. Physiol.* 160: 106–154. **Figure 24.9B, left**: Wiesel, T. N. and D. H. Hubel (1963) Single-cell responses in striate cortex of kittens deprived of vision in one eye. *J. Neurophysiol.* 26: 1003–1017. **Figure 24.9B, right**: Hubel, D. H. and T. N. Wiesel (1970) The period of susceptibility to the physiological effects of unilateral eye closure in kittens. *J. Physiol.* 206: 419–436. **Figure 24.9C**: Hubel, D. H. and T. N. Wiesel (1970) The period of susceptibility to the physiological effects of unilateral eye closure in kittens. *J. Physiol.* 206: 419–436. **Figure 24.9D, left**: Horton, J. C. and D. R. Hocking (1999) An adult-like pattern of ocular dominance columns in striate cortex of newborn monkeys prior to visual experience. *J. Neurosci.* 16: 1791–1807. **Figure 24.9D, right**: Hubel, D. H., T. N. Wiesel and S. LeVay (1977) Plasticity of ocular dominance columns in monkey striate cortex. *Philos. Trans. R. Soc. Lond. B* 278: 377–409. **Figure 24.9E**: Antonini, A. and M. P. Stryker (1993) Rapid remodeling of axonal arbors in the visual cortex. *Science* 260: 1819–1821. **Figure 24.10A,B**: Hubel, D. H. and T. N. Wiesel (1965) Binocular interaction in striate cortex of kittens reared with artificial squint. *J. Neurophysiol.* 28: 1041–1059. **Figure 24.11**: Katz, L. C. and J. C. Crawley (2002) Development of cortical circuits: Lessons from ocular dominance columns. *Nat. Rev. Neurosci.* 3: 34–42; Constantine-Paton, M. and M. I. Law (1978) Eye-specific termination bands in tecta of three-eyed frogs. *Science* 202: 639–641. **Figure 24.12**: Wang, B.-S., R. Sarnaik and J. Cang (2010) Critical period plasticity matches binocular orientation preference in the visual cortex. *Neuron* 65: 246–256. **Figure 24.13A,B**: Feller, M. B., D. P. Wellis, D. Stellwagen, F. S. Werblin and C. J. Shatz (1996) Requirement for cholinergic synaptic transmission in the propagation of spontaneous retinal waves. *Science* 272: 1182–1187. **Figure 24.13C**: Hanganu, I. L., Y. Ben-Ari and R. Khazipov (2006) Retinal waves trigger spindle bursts in the neonatal rat visual cortex. *J. Neurosci.* 26: 6728–6736. **Figure 24.14A**: Takesian, A. E. and T. K. Hesch (2013) Balancing plasticity/stability across brain development. *Prog. Brain Res.* 207: 3–33. **Figure 24.14B**: Choi, E. K. et al. (2018) Cyclin B1 stability is increased by interaction with BRCA1, and its overexpression suppresses the progression of BRCA1-associated mammary tumors. *Exp. Mol. Med.* 50: 1–16. **Figure 24.15**: Erzurmulu, R. S. and P. Gaspar (2012) Development and critical period plasticity of the barrel cortex. *Eur. J. Neurosci.* 35: 1540–1553; Li, H. and M. C. Crair (2011) How do barrels form in somatosensory cortex? *Ann. N. Y. Acad. Sci.* 1225: 119–129; Inan, M. and M. C. Crair (2007) Development of cortical maps: Perspectives from the barrel cortex. *Neuroscientist* 12: 49–61. **Figure 24.16A**: Ebert, D. H. and M. E. Greenberg (2013) Activity-dependent neuronal signaling and autism spectrum disorder. *Nature* 493: 327–337. **Figure 24.16B**: Yap, E.-L. and M. E. Greenberg (2018)

Activity-regulated transcription: Bridging the gap between neural activity and behavior. *Neuron* 100: 330–348. **Figure 24.16C**: Wong, R. O. and A. Ghosh (2002) Activity-dependent regulation of dendritic growth and patterning. *Nat. Rev. Neurosci.* 3: 803–812. **Clinical Applications, Figure A**: Eaton, N. C., H. M. Sheehan and E. M. Quinlan (2016) Optimization of visual training for full recovery from severe amblyopia in adults. *Learn. Mem.* 23: 99–103. **Clinical Applications, Figure B**: Montey, K. L. and E. M. Quinlan (2011) Recovery from chronic monocular deprivation following reactivation of thalamocortical plasticity by dark exposure. *Nat. Commun.* 2: 317.

CHAPTER 25   Sex Differences and Neural Circuit Development

Figure 25.1B: Moore, K. L. (1977) *The Developing Human*, 2nd Edition. Philadelphia: W. B. Saunders, p. 241. **Figure 25.1C**: McCarthy, M. M., A. P. Arnold, G. F. Ball, J. D. Blaustein et al. (2012) Sex differences in the brain: The not so inconvenient truth. *J. Neurosci.* 32: 2241–2247. **Figure 25.2B**: Gustafson, M. L. and P. K. Donahoe (1994) Male sex determination: Current concepts of male sexual differentiation. *Annu. Rev. Med.* 45: 505–524. **Figure 25.3A**: McEwen, B. S., P. G. Davis, B. Parsons and D. W. Pfaff (1979) The brain as a target for steroid hormone action. *Annu. Rev. Neurosci.* 2: 65–112. **Figure 25.3B**: McEwen, B. S. (1976) Interactions between hormones and nerve tissue. *Sci. Am.* 235: 48–58. **Figure 25.5B**: Arnold, A. P. (1980) Sexual differences in the brain. *Am. Sci.* 68: 165–173. **Figure 25.5C**: Grisham, W. and 6 others (2011) Using digital images of the zebra finch song system as a tool to teach organizational effects of steroid hormones: A free downloadable module. *CBE Life Sci. Educ.* 10: 222–230. **Figure 25.6A**: Breedlove, S. M. and A. P. Arnold (1981) Sexually dimorphic motor nucleus in the rat lumbar spinal cord: Response to adult hormone manipulation, absence in androgen-insensitive rats. *Brain Res.* 225: 297–307. **Figure 25.6B illustration**: Morris, J. A., C. L. Jordan and S. M. Breedlove (2004) Sexual differentiation of the vertebrate nervous system. *Nat. Neurosci.* 7: 1034–1039. **Figure 25.6B micrographs**: Breedlove, S. M. and A. P. Arnold (1983) Hormonal control of a developing neuromuscular system. II. Sensitive periods for the androgen-induced masculinization of the rat spinal nucleus of the bulbocavernosus. *J. Neurosci.* 3: 424–432. **Figure 25.6C**: Forger, N. G. and S. M. Breedlove (1986) Sexual dimorphism in human and canine spinal cord: Role of early androgen. *Proc. Natl. Acad. Sci. U.S.A.* 83: 7527–7530. **Figure 25.7B**: Arnold, A. P. and R. A. Gorski (1984) Gonadal steroid induction of structural sex differences in the central nervous system. *Ann. Rev. Neurosci.* 7: 413–442; Gorski, R. A. (1983) Steroid-induced sexual characteristics in the brain. In *Neuroendocrine Perspectives*, Volume 2, E. E. Muller and R. M. MacLeod (eds.). Amsterdam: Elsevier/North Holland. **Figure 25.8A,B**: Meeh, K. L., C. T. Rickel, A. J. Sansano and T. R. Shirangi (2021) The development of sex differences in the nervous system and behavior of flies, worms, and rodents. *Dev. Biol.* 472: 75–84. **Figure 25.8 playfighting illustration**: Pellis, S. M. and V. C. Pellis (2017) What is play fighting and what is it good for? *Learn. Behav.* 45: 355–336. **Figure 25.9A**: Toran-Allerand, C. D. (1976) Sex steroids and the development of the newborn mouse hypothalamus and preoptic area in vitro: Implications for sexual differentiation. *Brain Res.* 106: 407–412.

Figure 25.9B: Woolley, C. S. and B. S. McEwen (1993) Roles of estradiol and progesterone in regulation of hippocampal dendritic spine density during the estrous cycle in the rat. *J. Comp. Neurol.* 336: 293–306. **Figure 25.9C**: Meusburger, S. M. and J. R. Keast (2001) Testosterone and nerve growth factor have distinct but interacting effects on neurotransmitter expression of adult pelvic ganglion cells in vitro. *Neuroscience* 108: 331–340. **Figure 25.10**: Patchev, V. K., J. Schroeder, F. Goetz, W. Rhode and A. V. Patchev (2004) Neurotropic actions of androgens: Principles, mechanisms and novel targets. *Exp. Gerontol.* 39: 1651–1660. **Figure 25.11**: Meeh, K. L., C. T. Rickel, A. J. Sansano and T. R. Shirangi (2021) The development of sex differences in the nervous system and behavior of flies, worms, and rodents. *Dev. Biol.* 472: 75–84. **Figure 25.12A–C**: Wooley, C. (2007) Acute effects of estrogen on neuronal physiology. *Annu. Rev. Pharmacol. Toxicol.* 47: 5.1–5.24. **Box 25A, part A–C**: Young, L. J. and Z. Wang (2004) The neurobiology of pair bonding. *Nat. Neurosci.* 7: 1048–1054. **Box 25A, part D**: Fisher, H. E., A. Aron and L. L. Brown (2005) Romantic love: An fMRI study of neural mechanisms for mate choice. *J. Comp. Neurol.* 493: 58–62. **Box 25A, part E**: Bartels, A. and S. Zeki (2000) The neural basis of romantic love. *NeuroReport* 11: 3829–3834. **Figure 25.13**: Savic, I., H. Berglund and P. Lindström (2005) Brain response to putative pheromones in homosexual men. *Proc. Natl. Acad. Sci. U.S.A.* 102: 7356–7361. **Figure 25.14A**: Cahill, L. (2006) Why sex matters for neuroscience. *Nat. Rev. Neurosci.* 7: 477–484. **Figure 25.14B**: Uhl, M., M. J. Schmeisser and S. Schumann (2022) The sexual dimorphic synapse: From spine density to molecular composition. *Front. Mol. Neurosci.* 15: article 818390. **Clinical Applications**: Meaney, M. J. and M. Szyf (2005) Maternal care as a model for experience-dependent chromatin plasticity? *Trends Neurosci.* 28: 456–463. **Table 25.1**: McCarthy, M. M., B. M. Nugent and K. M. Lenz (2017) Neuroimmunology and neuroepigenetics in the establishment of sex differences in the brain. *Nat. Rev. Neurosci.* 18: 471–548.

CHAPTER 26   Repair and Regeneration in the Nervous System

Figure 26.1: Case, L. C. and M. Tessier-Lavigne (2005) Regeneration of the adult central nervous system. *Curr. Biol.* 15(18): R749–R753. Courtesy of the New York Academy of Medicine Library. **Figure 26.2A**: Fatterpekar, G. M., T. P. Naidich, B. N. Delman, J. G. Aguinaldo et al. (2002) Cytoarchitecture of the human cerebral cortex: MR microscopy of excised specimens at 9.4 Tesla. *Am. J. Neuroradiol.* 23: 1313–1321. **Figure 26.2B–D**: Furcila, D., J. DeFelipe and L. Alonso-Nanclares (2018) A study of amyloid-β and phosphotau in plaques and neurons in the hippocampus of Alzheimer's disease patients. *J. Alzheimer's Dis.* 64: 417–435. **Figure 26.3A**: Ward, N. S. et al. (2003) Neural correlates of outcome after stroke: A cross-sectional fMRI study. *Brain* 126: 1430–1448. **Figure 26.3B,C**: Ward, N. S. et al. (2003) Neural correlates of outcome after stroke: A cross-sectional fMRI study. *Brain* 126: 1430–1448. **Figure 26.4A,B**: Calvert, J. S. et al. (2019) Emergence of epidural electrical stimulation to facilitate sensorimotor network functionality after spinal cord injury. *Neuromodulation* 22: 244–252. **Figure 26.4C**: Wagner, F. B. (2018) Targeted neurotechnology restores walking in humans with spinal cord injury. *Nature* 563: 65. **Figure 26.6B**: Head, H. et al.

(1905) *The Afferent Nervous System from a New Aspect*. London: John Bale, Sons and Danielsson. **Figure 26.7B,C Photos**: Pan, Y. A., T. Misgeld, J. W. Lichtman and J. R. Sanes (2003) Effects of neurotoxic and neuroprotective agents on peripheral nerve regeneration assayed by time-lapse imaging in vivo. *J. Neurosci.* 23: 11479–11488. **Figure 26.8B**: Cattin, A.-C. and A. C. Lloyd (2016) The multicellular complexity of peripheral nerve regeneration. *Curr. Opin. Neurobiol.* 39: 38–46. **Figure 26.9A**: Dun, X.-p. and D. B. Parkinson (2015) Visualizing peripheral nerve regeneration by whole mount staining. *PLoS ONE* 10: e0119168. **Figure 26.9B**: Nocera, G. and C. Jacob (2020) Mechanisms of Schwann cell plasticity involved in peripheral nerve repair after injury. *Cell Mol. Life Sci.* 77: 3977–3989. **Figure 26.10A**: Sabatier, M. J. et al. (2009) Treadmill training promotes axon regeneration in injured peripheral nerves. *Exp. Neurol.* 211: 489–493. **Figure 26.10B**: Vijayavenkataraman, S. (2020) Nerve guide conduits for peripheral nerve injury repair: A review on design, materials and fabrication methods. *Acta Biomater.* 106: 54–69. **Figure 26.11B**: Pitts, E. V., S. Potluri, D. M. Hess and R. J. Balice-Gordon (2006) Neurotrophin and Trk-mediated signaling in the neuromuscular system. *Int. Anesthesiol. Clin.* 44: 21–76. **Figure 26.11C**: Nguyen, Q. T., J. R. Sanes and J. W. Lichtman (2002) Pre-existing pathways promote precise projection patterns. *Nat. Neurosci.* 5: 861–867. **Figure 26.11D**: Campanari, M.-L. et al. (2016) Neuromuscular junction impairment in amyotrophic lateral sclerosis: Reassessing the role of acetylcholinesterase. *Front. Mol. Neurosci.* 9: 160. **Box 26A, part B**: Purves, D. et al. (1981) Re-innervation of ganglia transplanted to the neck from different levels of the guinea-pig sympathetic chain. *J. Physiol.* 313: 49–64. **Clinical Applications, Figure B top**: McKee, A. C. and 9 others (2009) Chronic traumatic encephalopathy in athletes: Progressive tauopathy after repetitive head injury. *J. Neuropathol. Exp. Neurol.* 68: 709–735; Miller, G. (2009) A late hit for pro football players. *Science* 325: 670–672. **Clinical Applications, Figure B bottom, C**: McKee, A. C. and 9 others (2009) Chronic traumatic encephalopathy in athletes: Progressive tauopathy after repetitive head injury. *J. Neuropathol. Exp. Neurol.* 68: 709–735; Miller, G. (2009) A late hit for pro football players. *Science* 325: 670–672. **Figure 26.12A**: Manabat, C. and 8 others (2003) Reperfusion differentially induces caspase-3 activation in ischemic core and penumbra after stroke in immature brain. *Stroke* 34: 207–213. **Figure 26.13 (top)**: McGraw, J., G. W. Hiebert and J. D. Stevens (2001) Modulating astrogliosis after neurotrauma. *J. Neurosci. Res.* 63: 109–115. **Figure 26.13 (center)**: Tan, A. M., W. Zhang and J. M. Levine (2005) NG2: A component of the glial scar that inhibits axon growth. *J. Anat.* 207: 717–725. **Figure 26.13 (bottom)**: Ladeby, R. and 6 others (2005) Microglial cell population dynamics in the injured adult CNS. *Brain Res. Rev.* 48: 196–206. **Figure 26.14A,B**: Perez, J. C., Y. N. Gerber and F. E. Perrin (2021) Dynamic diversity of glial response among species in spinal cord injury. *Front. Aging Neurosci.* 13: 769548. https://doi.org/10.3389/fnagi.2021.769548. **Figure 26.15**: Kolodkin, A. L. and M. Tessier-Lavigne (2011) Mechanisms and molecules of neuronal wiring: A primer. *Cold Spring Harb. Perspect. Biol.* 3(6): a001727. **Figure 26.16A**: So, K.-F. and A. J. Aguyao (1985) Lengthy regrowth of cut axons from ganglion cells after peripheral nerve transplantation into the retina of adult rats. *Brain Res.* 328: 349–354. **Figure 26.16B**: So, K.-F. and A. J. Aguyao (1985) Lengthy regrowth of cut axons from ganglion cells after peripheral nerve transplantation into the retina of adult rats. *Brain Res.* 328: 349–354. **Figure 26.17A,B:** Waisman, A. et al. (2015) Innate and adaptive immune responses in the CNS. *Lancet Neurol.* 14: 945–955. **Figure 26.18A (Illustration):** Otteson, D. C. and P. F. Hitchcock (2003) Stem cells in the teleost retina: Persistent neurogenesis and injury-induced regeneration. *Vis. Res.* 43: 927–936. **Figure 26.18B (Illustration):** Goldman, S. A. (1998) Adult neurogenesis: From canaries to the clinic. *J. Neurobiol.* 36: 267–286. **Figure 26.19A,B:** Gage, F. H. (2000) Mammalian neural stem cells. *Science* 287: 1433–1438. **Figure 26.19C:** Kelsch, W. et al. (2010) Watching synaptogenesis in the adult brain. *Annu. Rev. Neurosci.* 33: 131–149. **Figure 26.20A:** Alvarez-Buylla, A. and D. A. Lim (2004) For the long run: Maintaining germinal niches in the adult brain. *Neuron* 41: 683–686. **Figure 26.20B:** Councill, J. H. et al. (2006) Limited influence of olanzapine on adult forebrain neural precursors in vitro. *Neuroscience* 140: 111–122. **Figure 26.21A:** Ghashghaei, H. T. et al. (2006) The role of neuregulin–ErbB4 interactions on the proliferation and organization of cells in the subventricular zone. *Proc. Natl. Acad. Sci. U.S.A.* 103: 1930–1935. **Figure 26.21B:** Ghashghaei, H. T. et al. (2007) Neuronal migration in the adult brain: Are we there yet? *Nat. Rev. Neurosci.* 8: 141–151. **Figure 26.21C:** Peretto, P. et al. (1999) The subependymal layer in rodents: A site of structural plasticity and cell migration in the adult mammalian brain. *Brain Res. Bull.* 49: 221–243. **Box 26B, part A:** Au, E. and G. Fishell (2006) Adult cortical neurogenesis: Nuanced, negligible, or nonexistent? *Nat. Neurosci.* 9: 1086–1088. **Box 26B, part B:** Bhardwaj, R. D. et al. (2006) Neocortical neurogenesis in humans is restricted to development. *Proc. Natl. Acad. Sci. U.S.A.* 103: 12564–12568. © 2006 National Academy of Sciences, U.S.A. **Box 26B, part C:** Bhardwaj, R. D. et al. (2006) Neocortical neurogenesis in humans is restricted to development. *Proc. Natl. Acad. Sci. U.S.A.* 103: 12564–12568. © 2006 National Academy of Sciences, U.S.A.

CHAPTER 27 Cognitive Functions and the Organization of the Cerebral Cortex

Box 27B, part A: Cabeza, R. and L. Nyberg (2000) Imaging cognition II: An empirical review of 275 PET and fMRI studies. *J. Cogn. Neurosci.* 12(1): 1–47. **Box 27B, part C:** Glasser, M., T. Coalson and E. Robinson et al. (2016) A multi-modal parcellation of human cerebral cortex. *Nature* 536: 171–178. **Figure 27.5A:** Colby, C. L, J. R. Duhamel and M. E. Goldberg (1996) Visual, presaccadic, and cognitive activation of single neurons in monkey lateral intraparietal area. *J. Neurophysiol.* 76(5): 2841–2852. **Figure 27.5B:** Platt, M. L. and P. W. Glimcher (1999) Neural correlates of decision variables in parietal cortex. *Nature* 400: 233–238. **Figure 27.6:** Bao, P., L. She, M. McGill et al. (2020) A map of object space in primate inferotemporal cortex. *Nature* 583: 103–108. **Figure 27.7A:** Van Horn, J. D., A. Irimia, C. M. Torgerson, M. C. Chambers, R. Kikinis and A. W. Toga (2012) Mapping connectivity damage in the case of Phineas Gage. *PLoS ONE* 7: e37454. https://doi.org/10.1371/journal.pone.0037454. **Figure 27.7B:** Originally from the collection of Jack and Beverly Wilgus, and now in the Warren Anatomical Museum, Harvard Medical School. **Figure 27.8A,B:** McClure, S. M. and 5 others (2004) Neural correlates of behavioral preference for culturally familiar drinks. *Neuron* 44(2): 379–387. **Figure 27.9A:** Goldman-Rakic, P. S. (1987) Circuitry of the prefrontal cortex and the regulation of behavior by representational memory. In *Handbook of Physiology: Section 1 (The Nervous System) (Volume 5: Higher Functions of the Brain, Part I),* F. Plum (ed.). Bethesda, MD: American Physiological Society, pp. 373–417. **Figure 27.9B:** Goldman-Rakic, P. S. (1987) Circuitry of the prefrontal cortex and the regulation of behavior by representational memory. In *Handbook of Physiology: Section 1 (The Nervous System) (Volume 5: Higher Functions of the Brain, Part I),* F. Plum (ed.). Bethesda, MD: American Physiological Society, pp. 373–417. **Figure 27.9C,D:** Goldman-Rakic, P. S. (1987) Circuitry of the prefrontal cortex and the regulation of behavior by representational memory. In *Handbook of Physiology: Section 1 (The Nervous System) (Volume 5: Higher Functions of the Brain, Part I),* F. Plum (ed.). Bethesda, MD: American Physiological Society, pp. 373–417.

CHAPTER 28 Cortical States

Figure 28.1A: Aschoff, J. (1965) Circadian rhythms in man. *Science* 148: 1427–1432. **Figure 28.1B:** Hobson, J. A. (1989) *Sleep.* New York: Scientific American Library. **Figure 28.3:** Okamura, H. and 8 others (1999) Photic induction of mPer1 and mPer2 in Cry-deficient mice lacking a biological clock. *Science* 286: 2531–2534. **Box 28A, parts B,C:** Bear, M., M. A. Paradiso and B. Connors (2001) *Neuroscience: Exploring the Brain,* 2nd Edition. Philadelphia: Williams & Wilkins/ Lippincott. **Figure 28.4A,B:** Hobson, J. A. (1989) *Sleep.* New York: Scientific American Library. **Figure 28.5:** Hobson, J. A. (1989) *Sleep.* New York: Scientific American Library. **Figure 28.6A–C:** Jovanovic, U. J. (1971) *Normal Sleep in Man.* Stuttgart: Hippokrates Verlag. **Figure 28.7A,B:** Magoun, H. W. (1952) An ascending reticular activating system in the brain stem. *AMA Arch. Neurol. Psychiatry* 67: 145–154. **Figure 28.8:** Hobson, J. A. (1989) *Sleep.* New York: Scientific American Library. **Figure 28.10:** McCormick, D. A. and H. C. Pape (1990) Properties of a hyperpolarization-activated cation current and its role in rhythmic oscillation in thalamic relay neurones. *J. Physiol.* 432: 291–318. **Figure 28.11A:** Steriade, M., D. A. McCormick and T. J. Sejnowski (1993) Thalamocortical oscillations in the sleeping and aroused brain. *Science* 262: 679–685. **Figure 28.11B,C:** Steriade, M., D. A. McCormick and T. J. Sejnowski (1993) Thalamocortical oscillations in the sleeping and aroused brain. *Science* 262: 679–685. **Figure 28.12:** Hobson, J. A. (1989) *Sleep.* New York: Scientific American Library. **Figure 28.14:** Weiskrantz, L. et al. (1974) Visual capacity in the hemianopic field following a restricted occipital ablation. *Brain* 97(1): 709–728. **Figure 28.15:** Owen, A. M., M. R. Coleman, M. Boly, M. H. Davis et al. (2006) Detecting awareness in the vegetative state. *Science* 313: 1402. **Figure 28.16B:** Raichle, M. E. (2011) The restless brain. *Brain Connect.* 1(1): 3–12.

CHAPTER 29 Attention

Figure 29.2A,B: Posner, M. I., C. R. R. Snyder and B. J. Davidson (1980) Attention and the detection of signals. *J. Exp. Psychol. Gen.* 59: 160–174. **Figure 29.3A,B:** Klein, R. M. (2000) Inhibition of return. *Trends Cogn. Sci.* 4: 138–147; Posner, M. I. and Y. Cohen (1984) Components of visual orienting. In *Attention and Performance, Volume 10, Control of Language Processes,* H. Bouma and D. Bouwhuis (eds.). London: Erlbaum, pp. 531–556. **Figure 29.5:** Moran, J. and R. Desimone (1985) Selective attention gates visual processing in the extrastriate cortex. *Science* 229: 782–784. **Figure 29.7B**: Luck, S. J. et al. (1997) Neural mechanisms of spatial selective attention in areas V1, V2, and V4 of macaque visual cortex. *J. Neurophysiol.* 77: 24–42. **Figure 29.7C**: Grent-'t-Jong, T. and M. G. Woldorff (2007) Timing and sequence of brain activity in top-down control of visual-spatial attention. *PLoS Biol.* 5(1): e12. https://doi.org/10.1371/journal.pbio.0050012. **Figure 29.8A–C**: Gazzaley, A. et al. (2005) Top-down enhancement and suppression of the magnitude and speed of neural activity. *J. Cogn. Neurosci.* 17(3): 507–517. **Figure 29.9**: Heilman, K. M., R. T. Watson and E. Valenstein (1985) Neglect and related disorders. In *Clinical Neuropsychology*, 2nd Edition, K. M. Heilman and E. Valenstein (eds.). New York: Oxford University Press, pp. 243–293. **Figure 29. 10A,B**: Luvizutto, G. J. et al. (2020) Norm scores of cancelation and bisection tests for unilateral spatial neglect: Data from a Brazilian population. *Clinics (Sao Paulo)* 75: e1468. https://doi.org/10.6061/clinics/2019/e1468. **Figure 29.10C**: Mark, V. W. (2003) Acute versus chronic functional aspects of unilateral spatial neglect. *Front. Biosci.* 8: e172–189. **Figure 29.10D**: Chen, P. and K. M. Goedert (2012) Clock drawing in spatial neglect: A comprehensive analysis of clock perimeter, placement, and accuracy. *J. Neuropsychol.* 6(2): 270–289. **Figure 29.11**: Corbetta, M. and G. L. Shulman (2002) Control of goal-directed and stimulus-driven attention in the brain. *Nat. Rev. Neurosci.* 3: 201–215. **Figure 29.12A,B fMRIs**: Woldorff, M. G., C. J. Hazlett, H. M. Fichtenholtz, D. H. Weissman, A. M. Dale et al. (2004) Functional parcellation of attentional control regions of the brain. *J. Cogn. Neurosci.* 16: 149–165. **Figure 29.12A,B illustration and graph**: Grent-'t-Jong, T. and M. G. Woldorff (2007) Timing and sequence of brain activity in top-down control of visual-spatial attention. *PLoS Biol.* 5(1): e12. https://doi.org/10.1371/journal.pbio.0050012. **Box 29A, parts A–C**: Moore, T. and K. M. Armstrong (2003) Selective gating of visual signals by microstimulation of frontal cortex. *Nature* 421: 370–373. **Clinical Applications, Figure A**: Friedman-Hill, S. R., L. C. Robertson and A. Treisman (1995) Parietal contributions to visual feature binding: Evidence from a patient with bilateral lesions. *Science* 269: 853–855. **Clinical Applications, Figure B**: Humphreys, G. W. and M. J. Riddoch (1993) Interactions between object and space systems revealed through neuropsychology. In *Attention and Performance, Volume 14: Synergies in Experimental Psychology, Artificial Intelligence, and Cognitive Neuroscience*, D. E. Meyer and S. Kornblum (eds.). Cambridge, MA: MIT Press, pp. 143–162. **Clinical Applications, Figure C**: Cooper, A. C. and G. W. Humphreys (2000) Coding space within but not between objects: Evidence from Balint's syndrome. *Neuropsychologia* 38: 723–733.

CHAPTER 30 Memory

Figure 30.2A: Warrington, E. K. and T. Shallice (1969) The selective impairment of auditory-verbal short-term memory. *Brain* 92: 885–896. **Figure 30.2B**: Drachman, D. A. and J. Arbit (1966) Memory and the hippocampal complex. II. Is memory a multiple process? *Arch. Neurol.* 15: 52–61. **Figure 30.5**: Ericsson, K. A., W. G. Chase and S. Faloon (1980) Acquisition of a memory skill. *Science* 208: 1181–1182. **Figure 30.6A–D**: Chase, W. G. and H. A. Simon (1973) *The mind's eye in chess. In Visual Information Processing*, W. G. Chase (ed.). New York: Academic Press, pp. 215–281. **Figure 30.7A,B**: Morris, J. S. and R. J. Dolan

(2001) Involvement of human amygdala and orbitofrontal cortex in hunger-enhanced memory for food stimuli. *J. Neurosci.* 21: 5304–5310. **Figure 30.8A**: Adcock, R. A. et al. (2006) Reward-motivated learning: Mesolimbic activation precedes memory formation. *Neuron* 50(3): 507–517. **Figure 30.8B**: Murty, V. P. and R. A. Adcock (2014) Enriched encoding: Reward motivation organizes cortical networks for hippocampal detection of unexpected events. *Cereb. Cortex* 24(8): 2160–2168. **Figure 30.9A**: Rubin, D. C. and T. C. Kontis (1983) A schema for common cents. *Mem. Cog.* 11: 335–341. **Figure 30.9B**: Squire, L. R. (1989) On the course of forgetting in very long-term memory. *J. Exp. Psychol.* 15: 241–245. **Clinical Applications, Figure A–D**: Corkin, S., D. G. Amaral, R. G. Gonzalez, K. A. Johnson and B. T. Hyman (1997) H.M.'s medial temporal lobe lesion: Findings from MRI. *J. Neurosci.* 17: 3964–3979. **Clinical Applications, Figure E**: Rosenbaum, R. S. and 6 others (2000) Remote spatial memory in amnesic person with extensive bilateral hippocampal lesions. *Nat. Neurosci.* 3: 1044–1048. **Clinical Applications, Figure F**: Tulving, E. (2002) Episodic memory: From mind to brain. *Annu. Rev. Psychol.* 53: 1–25. **Figure 30.11B**: Eichenbaum, H (2000) A cortical-hippocampal system for declarative memory. *Nat. Rev. Neurosci.* 1: 41–50. **Figure 30.11C,D**: Schenk, F. and R. G. Morris (1985) Dissociation between components of spatial memory in rats after recovery from the effects of retrohippocampal lesions. *Exp. Brain Res.* 58: 11–28. **Figure 30.12A,B**: Maguire, E. A. and 6 others (2000) Navigation-related structural change in the hippocampi of taxi drivers. *Proc. Natl. Acad. Sci. U.S.A.* 97: 4398–4403. **Box 30B, part B**: Blumenfeld, H. (2002) *Neuroanatomy through Clinical Cases.* Sunderland, MA: Sinauer Associates, based on Brun, A. and Englund, E. (1981) Regional pattern of degeneration in Alzheimer's disease: Neuronal loss and histopathological grading. *Histopathology* 5: 549–564. **Box 30C, parts A,B**: Moser, E. I., E. Kropff and M.-B. Moser (2008) Place cells, grid cells, and the brain's spatial representation system. *Annu. Rev. Neurosci.* 31: 69–89. **Figure 30.13A–C**: Lashley, K. S. and L. E. Wiley (1933) Studies of cerebral function in learning. IX. Mass action in relation to the number of elements in the problem to be learned. *J. Comp. Neurol.* 57: 3–55; Lashley, K. S. (1944) Studies of cerebral function in learning. XIII. Apparent absence of transcortical association in maze learning. *J. Comp. Neurol.*, 80: 257–281. **Figure 30.14**: Van Hoesen, G. W. (1982) The parahippocampal gyrus. *Trends Neurosci.* 5: 345–350. **Figure 30.15A**: Buckner, R. L. and M. E. Wheeler (2001) The cognitive neuroscience of remembering. *Nat. Rev. Neurosci.* 2: 624–634. **Figure 30.15B**: Buckner, R. L. and M. E. Wheeler (2001) The cognitive neuroscience of remembering. *Nat. Rev. Neurosci.* 2: 624–634; Ishai, A., L. G. Ungerleider and J. V. Haxby (2000) Distributed neural systems for the generation of visual images. *Neuron* 28: 979–990. **Figure 30.16A,B**: Shohamy, D., C. E. Myers, S. Grossman, J. Sage and M. A. Gluck (2005) The role of dopamine in cognitive sequence learning: Evidence from Parkinson's disease. *Behav. Brain Res.* 156: 191–199. **Figure 30.17**: Dekaban, A. S. and D. Sadowsky (1978) Changes in brain weights during the span of human life: Relation of brain weights to body heights and body weights. *Ann. Neurol.* 4: 345–356. **Figure 30.18**: Cabeza, R., N. D. Anderson, J. K. Locantore and A. R. McIntosh (2002) Aging gracefully: Compensatory brain activity in high-performing older adults. *NeuroImage* 17: 1394–1402.

CHAPTER 31 Speech and Language

Figure 31.2B: Miller, G. A. (1991) The spoken word. In *The Science of Words*. New York: Scientific American Library, chapter 4, p. 69. **Figure 31.3**: Schwartz, D. A., C. Q. Howe and D. Purves (2003) The statistical structure of human speech sounds predicts musical universals. *J. Neurosci.* 23(18): 7160–7168. **Figure 31.4**: Dronkers, N. F. et al. (2007) Paul Broca's historic cases: High resolution MR imaging of the brains of Leborgne and Lelong. *Brain* 130(5): 1432–1441. **Figure 31.5B**: Beauchamp, M. S., A. R. Nath and S. Pasalar (2010) fMRI-guided transcranial magnetic stimulation reveals that the superior temporal sulcus is a cortical locus of the McGurk effect. *J. Neurosci.* 30(7): 2414–2417. **Figure 31.5C**: Beauchamp, M. S., A. R. Nath and S. Pasalar (2010) fMRI-guided transcranial magnetic stimulation reveals that the superior temporal sulcus is a cortical locus of the McGurk effect. *J. Neurosci.* 30(7): 2414–2417. **Figure 31.6A,B**: Kutas, M. and S. A. Hillyard (1980) Reading senseless sentences: Brain potentials reflect semantic incongruity. *Science* 207(4427): 203–205. **Figure 31.7A**: Roberts, L. (1959) Evidence from cortical mapping. In *Speech and Brain Mechanisms*, W. Penfield and L. Roberts (eds.). Princeton, NJ: Princeton University Press, pp. 119–136. **Figure 31.7B**: Ojemann, G. A., I. Fried and E. Lettich (1989) Electrocorticographic (EcoG) correlates of language. *Electroencephalogr. Clin. Neurophysiol.* 73(5): 453–463. **Figure 31.8 scans**: Posner, M. I. and M. E. Raichle (1994) Interpreting words. In *Images of Mind*. New York: Scientific American Library, chapter 5. **Figure 31.8 illustration**: Posner, M. I. and M. E. Raichle (1994) Interpreting words. In *Images of Mind*. New York: Scientific American Library, chapter 5. **Figure 31.10**: Johnson, J. S. and E. L. Newport (1989) Critical period effects in second language learning: The influence of maturational state on the acquisition of English as a second language. *Cogn. Psychol.* 21: 60–99. **Figure 31.11A**: Brown, T. T., H. M. Lugar, R. S. Coalson, F. M. Miezin, S. E. Petersen and B. L. Schlagger (2005) Developmental changes in human cerebral functional organization for word generation. *Cereb. Cortex* 15: 275–290. **Figure 31.11B**: Brauer, J., A. Anwander and A. D. Friederici (2011) Neuroanatomical prerequisites for language functions in the maturing brain. *Cereb. Cortex* 21(2): 459–466. **Figure 31.12**: Vinckier, F. and 5 others (2007) Hierarchical coding of letter strings in the ventral stream: Dissecting the inner organization of the visual word-form system. *Neuron* 55: 143–156. **Figure 31.13**: Savage-Rumbaugh, S., S. G. Shanker and T. J. Taylor (1998) *Apes, Language, and the Human Mind*. New York: Oxford University Press. **Figure 31.14**: Gil-da-Costa, R. and 5 others (2006) Species-specific calls activate homologs of Broca's and Wernicke's areas in the macaque. *Nat. Neurosci.* 9: 1064–1070. **Box 31A**: Bellugi, U., H. Poizner and E. S. Klima (1989) Language, modality, and the brain. *Trends Neurosci.* 12: 380–388. **Box 31B, part A**: Damasio, H., T. J. Grabowski, D. Tranel, R. D. Hichwa and A. Damasio (1996) A neural basis for lexical retrieval. *Nature* 380: 499–505. **Box 31B, part B**: Deniz, F. et al. (2019) The representation of semantic information across human cerebral cortex during listening versus reading is invariant to stimulus modality. *J. Neurosci.* 39(39): 7722–7736. **Box 31C, parts A–C**: Coren, S. (1992) *The Left-Hander Syndrome: The Causes and Consequence of Left-Handedness*. New York: The Free Press. **Clinical Applications Extract**: Gardner, H. (1974) *The Shattered Mind: The Person after Brain Damage*. New York: Knopf, pp. 60–61.

CHAPTER 32 Emotion

Figure 32.3: Russell, J. A. (1980) A circumplex model of affect. *J. Pers. Soc. Psychol.* 39: 1161–1178. **Figure 32.4**: LeDoux, J. E. (1987) Emotion. In *Handbook of Physiology, Supplement 5: The Nervous System, Higher Functions of the Brain*, F. Blum et al. (eds.). Bethesda, MD: American Physiological Society, pp. 419–459. **Box 32A, part A1**: Duchenne, G.-B. (1876) *Mécanisme de la physionomie humaine. In Atlas*, Deuxième édition. Paris: J.-B. Bailliere et Fils, p. 1. **Box 32A, part A2**: Wellcome Library, London. Wellcome Images. images@wellcome.ac.uk. http://wellcomeimages.org. **Box 32A, part A3**: Wellcome Library, London. Wellcome Images. images@wellcome.ac.uk. http://wellcomeimages.org. **Box 32A, part A4**: Wellcome Library, London. Wellcome Images. images@wellcome.ac.uk. http://wellcomeimages.org. **Box 32A, part B left**: Holstege, G. et al. (1996) The emotional motor system. In *Progress in Brain Research, The Emotional Motor System*, volume 107, G. Holstege et al. (eds.). Amsterdam: Elsevier, pp. 3–6. **Box 32A, part B right**: Trosch, R. M., G. Sze, L. M. Brass and S. G. Waxman (1990) Emotional facial paresis with striatocapsular infarction. *J. Neurol. Sci.* 98: 195–201. **Figure 32.7**: Rolls, E. T. (1999) *The Brain and Emotion*. Oxford: Oxford University Press. **Figure 32.8**: Adolphs, R. et al. (1999) Recognition of facial emotion in nine individuals with bilateral amygdala damage. *Neuropsychologia* 37: 1111–1117. **Figure 32.9A–C**: Winston, J. S., B. A. Strange, J. O'Doherty and R. J. Dolan (2002) Automatic and intentional brain responses during evaluation of trustworthiness of faces. *Nat. Neurosci.* 5: 277–283. **Box 32C, parts B,C**: Adolphs, R., D. Tranel, H. Damasio and A. R. Damasio (1995) Fear and the human amygdala. *J. Neurosci.* 15(9): 5879–5891. **Clinical Applications, Figure A**: Parsons, R. G. and K. J. Ressler (2013) Implications of memory modulation for post-traumatic stress and fear disorders. *Nat. Neurosci.* 16: 146–153; Carrión, V. G., B. W. Haas, A. Garrett, S. Song and A. L. Reiss (2010) Reduced hippocampal activity in youth with posttraumatic stress symptoms: An fMRI study. *J. Pediatr. Psychol.* 35(5): 559–569. **Clinical Applications, Figure B**: Milad, M. R. and 9 others (2009) Neurobiological basis of failure to recall extinction memory in posttraumatic stress disorder. *Biol. Psychiatry* 66(12): 1075–1082. **Figure 32.10B**: Anderson, S. W. et al. (1999) Impairment of social and moral behavior related to early damage in the human prefrontal cortex. *Nat. Neurosci.* 2: 1032–1037. **Figure 32.10C,D**: Bechara, A. et al. (1994) Insensitivity to future consequences following damage to the human prefrontal cortex. *Cognition* 50: 7–15. **Figure 32.11A**: Craig, A. D. (2007) Interoception and emotion: A neuroanatomical perspective. In *Handbook of Emotions*, 3rd Edition, M. Lewis, J. M. Haviland-Jones and L. F. Barrett (eds.). New York: Guilford, pp. 395–408. **Figure 32.11B**: Critchley, H. D. et al. (2004) Neural systems supporting interoceptive awareness. *Nat. Neurosci.* 7: 189–195. **Figure 32.13**: Vuilleumier, P. and J. Driver (2007) Modulation of visual processing by attention and emotion: Windows on causal interactions between human brain regions. *Philos. Trans. R. Soc. B* 362: 837–855; Vuilleumier, P. et al. (2004) Distant influences of amygdala lesion on visual cortical activation during emotional face processing. *Nat. Neurosci.* 7:

1271–1278. **Figure 32.14**: LaBar, K. S. (2010) Emotion–cognition interactions. In *Encyclopedia of Behavioral Neuroscience*, G. F. Koob et al. (eds.). New York: Academic Press, pp. 469–476. https://doi.org/10.1016/B978-0-08. **Figure 32.15**: McGaugh, J. L. (2000) Memory: A century of consolidation. *Science* 287(5451): 248–251. **Figure 32.16**: Winecoff, A., K. S. Labar, D. J. Madden, R. Cabeza and S. A. Huettel (2011) Cognitive and neural contributors to emotion regulation in aging. *Soc. Cogn. Affect. Neurosci.* 6(2): 165–176. https://sites.duke.edu/huettellab/files/2013/02/2010_Winecoff_SCAN.pdf.

CHAPTER 33   Thinking, Planning, and Deciding

Figure 33.1B: Brodmann, K. (1912) Neue Ergebnisse über die vergleichende histologische Lokalisation der Grosshirnrinde mit besonderer Berücksichtigung des Stirnhirns. *Anat. Anz.* 41: 157–216. **Figure 33.1C**: Semendeferi, K., H. Damasio, R. Frank and G. W. Van Hoesen (1997) The evolution of the frontal lobes: A volumetric analysis based on three-dimensional reconstructions of magnetic resonance scans of human and ape brains. *J. Hum. Evol.* 32: 375–388; Semendeferi, K., A. Lu, N. Schenker and H. Damasio (2002) Humans and great apes share a large frontal cortex. *Nat. Neurosci.* 5: 272–276. **Figure 33.4**: Miller, E. K. (2000) The prefrontal cortex and cognitive control. *Nat. Rev. Neurosci.* 1: 59–65. **Figure 33.5A–C**: Wallis, J. D., K. C. Anderson and E. K. Miller (2001) Single neurons in prefrontal cortex encode abstract rules. *Nature* 411: 953–956. **Figure 33.6A**: Depue, B. E., T. Curran and M. T. Banich (2007) Prefrontal regions orchestrate suppression of emotional memories via a two-phase process. *Science* 317: 215–219. **Figure 33.6B**: Depue, B. E., T. Curran and M. T. Banich (2007) Prefrontal regions orchestrate suppression of emotional memories via a two-phase process. *Science* 317: 215–219. **Figure 33.7**: Wallis, J. D. (2007) Orbitofrontal cortex and its contribution to decision-making. *Ann. Rev. Neurosci.* 30: 31–56. **Figure 33.8**: Baxter, M. G. and E. A. Murray (2002) The amygdala and reward. *Nat. Rev. Neurosci.* 3: 563–573. **Figure 33.9A,B**: Gehring, W. J. and A. R. Willoughby (2002) The medial frontal cortex and the rapid processing of monetary gains and losses. *Science* 295: 2279–2292. **Box 33A, part B**: Schultz, W., P. Dayan and P. R. Montague (1997) A neural substrate of prediction and reward. *Science* 275: 1593–1599. **Box 33A, part C**: Schultz, W., P. Dayan and P. R. Montague (1997) A neural substrate of prediction and reward. *Science* 275: 1593–1599. **Box 33A, part D**: Berridge, K. C. and T. E. Robinson (1998) What is the role of dopamine in reward: Hedonic impact, reward learning, or incentive salience? *Brain Res. Rev.* 28(3): 309–369. **Figure 33.10B**: Bush, G., P. J. Whalen, B. R. Rosen, M. A. Jenike, S. C. McInerney and S. L. Rauch (1998) The counting Stroop: An interference task specialized for functional neuroimaging—Validation study with functional MRI. *Hum. Brain Mapp.* 6: 270–282. **Figure 33.10C**: Kerns, J. G., J. D. Cohen, A. W. MacDonald III, R. Y. Cho, V. A. Stenger and C. S. Carter (2004) Anterior cingulate conflict monitoring and adjustments in control. *Science* 303: 1023–1026. **Figure 33.11A,B**: Naqvi, N. H., D. Rudrauf, H. Damasio and A. Bechara (2007) Damage to the insula disrupts addition to cigarette smoking. *Science* 315: 531–534. **Box 33B, part B**: Sirigu, A., E. Daprati, S. Ciancia, P. Giraux, N. Nighoghossian, A. Posada and P. Haggard (2003) Altered awareness of voluntary action after damage to the parietal cortex. *Nat. Neurosci.* 7: 80–84. **Box 33B, parts C,D**: Soon, C., M. Brass, H. J. Heinze et al. (2008) Unconscious determinants of free decisions in the human brain. *Nat. Neurosci.* 11: 543–545. **Figure 33.12A**: Gusnard, D. A. and M .E. Raichle (2001) Searching for a baseline: Functional imaging and the resting human brain. *Nat. Rev. Neurosci.* 2: 685–694. **Figure 33.12B**: Gusnard, D. A. and M. E. Raichle (2001) Searching for a baseline: Functional imaging and the resting human brain. *Nat. Rev. Neurosci.* 2: 685–694. **Figure 33.12C**: Hayden, B. Y., D. Smith and M. L. Platt (2009) Electrophysiological correlates of default-mode processing in macaque posterior cingulate cortex. *Proc. Natl. Acad. Sci. U.S.A.* 106: 5948–5953. **Clinical Applications, Figures C,D**: Nestler, E. J. (2005) Is there a common molecular pathway for addiction? *Nat. Neurosci.* 8: 1445–1449.

APPENDIX

Figure A13C: Rohen, J. W. and C. Yokochi (1993) *Color Atlas of Anatomy*. New York: Igaku-Shoin. **Figure A16**: Blumenfeld, H. (2022) *Neuroanatomy through Clinical Cases*, 3rd Edition. New York: Oxford University Press/Sinauer Associates. **Figure A17A,B**: Blumenfeld, H. (2022) *Neuroanatomy through Clinical Cases*, 3rd Edition. New York: Oxford University Press/Sinauer Associates. **Figure A18A**: Blumenfeld, H. (2022) *Neuroanatomy through Clinical Cases*, 3rd Edition. New York: Oxford University Press/Sinauer Associates. **Figure A20A,B**: Blumenfeld, H. (2022) *Neuroanatomy through Clinical Cases*, 3rd Edition. New York: Oxford University Press/Sinauer Associates. **Figure A22A**: Goldstein, G. W. and A. L. Betz (1986) The blood–brain barrier. *Sci. Am.* 255: 74–83. **Figure A22B**: Peters, A., S. L. Palay and H. deF. Webster (1991) *The Fine Structure of the Nervous System: Neurons and Their Supporting Cells*, 3rd Edition. New York: Oxford University Press. **Figure A26**: Nedergaard, M. and S. A. Goldman (2016) Brain drain. *Sci. Am.* 314: 44–49. **Box A, part A**: Blumenfeld, H. (2022) *Neuroanatomy through Clinical Cases*, 3rd Edition. New York: Oxford University Press/Sinauer Associates. **Box A, part B**: Sherman, S. M. and R. W. Guillery (2011) Distinct functions for direct and transthalamic corticocortical connections. *J. Neurophysiol.* https://doi.org/10.1152/jn.00429.2011.

Index

Page references followed by a *t* indicate table, *f* indicate figure, and *b* indicate box.

A

Abducens nerve, 270*f*
  anatomy of, A–9, A–10*f*
  eye movements and, 474
  function of, A–8*t*
  injuries to, 477, 477*f*
Abducens nucleus, 269, 270*f*
Abnormal SPindle-like
  Microcephaly-associated (*ASPM*)
  gene, 21*f*, 22
Absolute pitch, 606*t*
ACC. *See* Anterior cingulate cortex
Acceleration, perception of, 268, 268*f*
Access consciousness, 731–32
Accessory olfactory bulb, 326–27,
  326*f*, 343–44, 344*f*
Accommodation, 207, 207*f*, 233
Accutane (isotretinoin), 537
ACE2 (angiotensin converting
  enzyme-2), 361
Acetylcholine (ACh)
  amygdala and release of, 815
  description of, 124–31
  electrical actions of, 112
  functional features of, 124*t*
  mAChRs and, 126*b*, 130–31, 130*f*
  nAChR and, 125, 127–28, 128*f*
  neurotransmission, 507
  postsynaptic currents and
    receptors of, 111–12, 112*f*, 584*f*
  release of, 98
  structure of, 123*f*
  in synaptic vesicles, 102
  synthesis and metabolism of, 125,
    125*f*
Acetylcholine receptors (AChRs),
  671–72
Acetylcholinesterase (AChE), 124*t*,
  125, 129, 129*f*
ACh. *See* Acetylcholine
AChE (acetylcholinesterase), 124*t*,
  125, 129, 129*f*
AChRs (acetylcholine receptors),
  671–72
Acid-sensing ion (ASIC) channels,
  307
ACIII (adenylyl cyclase III), 337
Aconitine, 75*b*
Acoustical trauma, 238*b*

Acromelic acid, 127*b*
Across neuron pattern coding, 366*f*,
  367
ACTH (adrenocorticotropic
  hormone), 148*f*, 630
Actin
  distribution of, 556
  function of, 109
Actin-binding proteins, 562, 563*f*
Actin cytoskeleton, 561–62, 563*f*
Action potentials
  in auditory nerve, 245
  blocking, 583
  conduction of, 62–64, 63*f*, 389
  description of, 38, 38*f*
  falling phase of, 52*b*
  feedback loops and generation of,
    62, 62*f*
  form and nomenclature of, 52*b*
  frequency of, 39
  generation of, 281
  information transfer via, 41, 44
  ion channels and, 50–51, 50*f*
  long-distance signaling and,
    62–64, 63*f*
  mathematical simulation of, 60–61,
    61*f*
  membrane permeability and,
    50–51, 50*f*
  myelination and conduction
    velocities of, 64–67, 64*t*, 65*f*, 66*f*
  Na⁺ channels and generation of,
    51–53, 51*f*
  overshoot phase of, 52, 52*b*
  refractoriness and, 64
  regenerative, 62
  rising phase of, 52, 52*b*
  saltatory propagation and, 65*f*, 67
  undershoot phase of, 52*b*, 53
Action tremors, 468, 468*f*
Activated glia, 680
Activational effects, of gonadal
  steroids, 642, 643*f*
Active transporters, 44, 44*f*
  functions of, 90
  ion gradients and, 86–90
Active zone, 96, 97*f*
Activity-dependent plasticity, 600–
  601, 602*f*, 612*f*, 624–26, 626*f*
Acute acoustical trauma, 238*b*

Adaptor proteins, 109, 110*f*
Addiction, 826–27, 826*f*, 827*f*
Adenosine triphosphate (ATP)
  adenylyl cyclase conversion of,
    163
  as co-transmitters, 145–46
  functional features of, 124*t*
  hydrolysis of, 87–89
  structure of, 123*f*
Adenylyl cyclase, ATP conversion
  and, 163
Adenylyl cyclase III (ACIII), 337
ADHD (attention deficit
  hyperactivity disorder), 610, 610*f*,
  654, 654*t*, 717
Adipose tissues, 505*b*
Adolescent Brain Cognitive
  Development study, 703*b*
Adrenal glands, 650
Adrenaline. *See* Epinephrine
Adrenal medulla, 498, 500
Adrenergic receptors, 507, 508*t*
Adrenocorticotropic hormone
  (ACTH), 148*f*, 630
Aequorea victoria (jellyfish), 164*b*
Affect heuristic, 809
Affective disorders, 812–13
Affective-motivational aspects of
  pain, 312, 312*f*, 314
Afferent neurons, 9–10
2-AG (arachidonoylglycerol), 149,
  151*f*
Age-related macular degeneration,
  207
Ageusia, 360, 361
Aging
  brain size and, 772, 772*f*
  memory and, 772–73, 772*f*, 773*f*
Agnosias, 705–6
Agraphias, 781
AICA (anterior inferior cerebellar
  artery), A–25, A–26*f*
AIS (androgen insensitivity
  syndrome), 650–51, 650*t*
A kinase anchoring proteins
  (AKAPs), 165
AKT1 (AKT1 protein kinase) gene,
  169*b*
Akt kinase, 590, 590*f*, 591*f*
Alar plate, A–11, A–11*f*

Alcoholism, 655*t*
Aldolase C, 462
Alexias, 781
Allelic silencing, 337
Allodynia, 317, 319
α-Fetoprotein, 635
ALS (amyotrophic lateral sclerosis),
  399–400, 400*f*, 655*t*, 660
Alzheimer's disease
  autophagy in, 679
  diagnosis of, 768*b*, 768*f*
  eye movement impairment and,
    478
  hallmarks of, 677, 677*f*
  heritability of, 768*b*
  PCC and, 838
  repair for damage from, 660, 660*f*
  sex-based differences in, 655*t*
Ambien (zolpidem), 42, 42*f*, 137
Amblyopia, 618
American Sign Language, 778*b*, 778*f*
Ametropia (refractive errors), 206–7,
  207*f*
Amino acids
  neuropeptides and sequences of,
    148*f*
  opioid peptides, 147–48, 148*f*, 149*t*
  substance P and, 147, 148*f*
Amino acids, structures of, 123*f*
Amnesia
  anterograde, 762, 765, 767
  brain areas in, 763*f*
  clinical cases of, 765–66
  ECT and, 767
  retrograde, 765–67
Amnesiac mutation, fruit fly, 186*b*
AMPA receptors, 626*f*, 827
  activation of, 176, 176*f*
  endocytosis of, 461
  EPSP mediation and, 190
  in hippocampus, 193*f*
  ligand-gated ion channels and, 81,
    84, 84*f*, 85*f*
  loss of, 194*b*
  in LTP, 190–91, 191*f*
  nomenclature of, 131

postsynaptic responses and, 132, 132*f*
structure of, 132, 133*f*
Amphetamines
 DAT inhibition and, 142
 function of, 169*b*
 for narcolepsy, 721
Amphibians
 studies on, 24*b*
 visual systems of, 594
Amphiphysin, 109
Ampullae, 262, 262*f*, 266–67, 267*f*
Amputations, 298, 298*f*
Amygdala
 ACh release and, 815
 anatomy of, 805*b*, 805*f*, A–20, A–20*f*
 associative learning in, 806, 806*f*
 astrocytes in, 645, 646*f*
 damage to, 808, 809*f*, 813, 815, 815*f*
 depression and, 811
 emotional memory tasks and, 653
 emotions and, 803–10
 fear conditioning and, 804–8, 806*f*, 807*b*, 807*f*
 innervation of, 811
 Klüver-Bucy syndrome and, 804
 location of, 727*f*
 LTP in, 806
 secondary somatosensory cortex projections to, 297
 sex-based differences in, 652–53, 653*f*
 trustworthiness judgments and, 808, 809*f*
Amyloid plaques, 768*b*, 768*f*
Amyloid precursor protein (APP), 768*b*
Amyotrophic lateral sclerosis (ALS), 399–400, 400*f*, 655*t*, 660
Anandamide, 149, 151*f*
Andersen, Per, 187
Androgen insensitivity syndrome (AIS), 650–51, 650*t*
Androgen receptors, distribution of, 644–45, 645*f*
Androgens, 634, 644–45, 645*f*, 652, 652*f*
Anesthesia
 general, 42–43
 local, 41, 42*f*
 neuronal electrical signals and, 41–43, 42*f*, 43*f*
 regional, 41–42
Anesthetics, 41–43
 Na⁺ channels and, 307
Anginal pain, 308, 309*b*
Angiotensin converting enzyme-2 (ACE2), 361
Angiotensin II, 148*f*
Aniridia, 540
Anopsias, 223*f*
Anorexia nervosa, 655*t*
Anosmias, 329–30, 332, 353
Anosmin, 571, 572*t*
Anterior, definition of, A–1, A–2*f*

Anterior cerebral arteries, A–23, A–23*f*
Anterior cerebral artery stroke, 409
Anterior chambers, 206, 206*f*
Anterior choroidal artery, A–24, A–25*f*
Anterior cingulate cortex (ACC)
 location of, 727*f*
 monitoring, 832–33
 OCD and, 833
 pain perception and, 312
Anterior cingulate gyrus, 380
Anterior circulation, A–21*f*, A–22–A–23, A–24*t*
Anterior commissure, 497*f*, 498*f*, 652, 653*f*, A–20, A–20*f*
Anterior communicating arteries, A–23*f*, A–24
Anterior inferior cerebellar artery (AICA), A–25, A–26*f*
Anterior insula, 834, 834*f*
Anterior nuclei, 496*b*, 497*f*, 498*f*
Anterior spinal arteries, A–22*f*, A–26
Anterograde, definition of, 16
Anterograde amnesia, 762, 765, 767
Anterolateral columns, A–6
Anterolateral system
 modalities mediated by, 314–15
 pain pathways of, 308, 308*f*, 309*f*
Anteroventral paraventricular nucleus (AVPV), 640, 641*f*, 642
Antibody stains, 16, 17*f*
Antihistamines, 144
Antiporters, 89, 89*f*
Antipsychotic drugs, 169*b*
Anxiety disorders
 molecular basis of, 170*b*
 sex-based differences in, 655*t*
Anxiolytic drugs, 170*b*
Apamin, 75*b*
Aphasias, 779–81
Apical domain
 of neural precursor cells, 541
 of polarized epithelial cells, 556
Aplysia californica (sea slug), 24*b*, 182–85, 183*f*
ApoE (apolipoprotein E), 769*b*
ApoE gene, 443
Apolipoprotein E (ApoE), 769*b*
Apoptosis
 hypoxia/ischemia injury and, 679*f*
 initiation of, 581
 mechanisms of, 679–80
 sexually dimorphic development and, 642–43
APP (amyloid precursor protein), 768*b*
Appendicular ataxia, 468*f*
Appendix, referred pain from inflamed, 309*b*
Aprosodias, 785, 814
Aprosody, 814
Aqueduct of Sylvius, A–30, A–30*f*, A–31*f*
Aqueous humor, 206, 206*f*
Arachidonic acid, 315

Arachidonoylglycerol (2-AG), 149, 151*f*
Arachnoid granulations, A–30, A–30*f*
Arachnoid mater, A–29, A–29*f*
Arachnoid villi, A–30, A–30*f*
Archicortex, 699*b*, A–19
Area X, 458
Areca catechu (betel nuts), 126*f*, 127*b*
Arecoline, 126*f*, 127*b*
Areflexia, 399
Armadillo repeat gene deleted in Velo-Cardio-Facial syndrome (ARVCF) gene, 169*b*
Aromatases, 634, 634*f*
Arousal, emotions and, 798
Arterial blood vessels, innervation of, 500
Artificial neural networks, 416*b*
ARVCF (Armadillo repeat gene deleted in Velo-Cardio-Facial syndrome) gene, 169*b*
ASD. *See* Autism spectrum disorder
ASIC (acid-sensing ion) channels, 307
Aspartate, structure of, 123*f*
Aspirin, 316
ASPM (Abnormal SPindle-like Microcephaly-associated) gene, 21*f*, 22
Associational systems, organization of, 15
Association cortices
 connectivity of, 700, 700*f*
 damage to, 700–701
 function of, 697
 structure of, 697–99, 698*f*
Associative learning, in amygdala, 806, 806*f*
Associativity, LTP and, 189, 189*f*
Astigmatism, 207
Astringency, 369
Astrocytes
 in amygdala, 645, 646*f*
 central sensitization process and, 317
 end feet, A–28, A–28*f*
 neural stem cells giving rise to, 522, 522*f*
 neuronal growth and, 679–80, 679*f*
 overview of, 6, 7*f*
Asymmetrical smiles, 814*f*
Ataxia, 82
 appendicular, 468*f*
 in cerebellar cortex, 468, 469*f*
Athletes, training regimens and, 384
ATP. *See* Adenosine triphosphate
ATP1A gene, 87
ATP1B gene, 87
ATPase pumps, 87–89, 87*f*, 88*f*
Attend cue, 750
Attention
 brain lesions and, 746–48, 746*f*, 747*f*
 brain systems supporting control of, 750–51, 750*f*, 751*f*
 definition of, 738

endogenous compared to exogenous, 739–41, 740*f*
eye movements and, 751–52
to features and objects, 745–46, 745*f*
frontal eye fields and, 749–50, 749*b*, 749*f*, 752
hemispatial neglect syndrome and, 746, 746*f*, 747*f*, 748–49
mediation of, 703–4, 704*f*
overt compared to covert, 741, 741*f*
parietal cortex and, 705*f*
selective, 739, 739*f*
to spatial locations, 742–45, 742*f*, 743*f*, 744*f*
unimodal compared to supramodal, 741–42
Attention deficit hyperactivity disorder (ADHD), 610, 610*f*, 654, 654*t*, 717
Attention-sensitive neurons, 704
Audible spectrum, 238
Auditory area 1, 250
Auditory cortex, 250, 251*f*, 252, 252*f*, 254–55
Auditory cues, taste and, 371
Auditory meatus, 240, 241*f*
Auditory nerve
 action potentials in, 245
 anatomy of, 241*f*, 262*f*, A–9
 cochlea and, 242*f*
 damage to, 239*b*
 function of, A–8*t*
 response properties of, 254*f*
 sound frequency information in, 252–55, 253*f*, 254*f*
 tuning of, 249, 253*f*, 254, 254*f*
Auditory pathways, 250, 251*f*, 252, 254
Auditory perception, 252
Auditory system, 236–58
Auditory thalamus, A–18*b*
Auerbach's plexus, 501, 501*f*
Augmentation, synaptic plasticity and, 180*f*, 181
Autism spectrum disorder (ASD)
 cortical connections in, 610
 default-mode networks and, 735
 sex-based differences in, 654, 654*t*
 somatosensory system in, 300
Autoimmune diseases, 129, 129*f*
Autonomic ganglia, 19, 491
Autonomic motor system. *See* Visceral motor system
Autonomic nervous system, 491
Autonomic reflexes to temperature, 315
Autophagy, 679
Autosomes, 630
Auxilin, 109, 110*f*
Aversive tastants, 372, 372*f*
Avians
 studies on, 24*b*
 visual systems of, 594

AVPV (anteroventral paraventricular nucleus), 640, 641*f*, 642
Axel, Richard, 333
Axial sections, A–2
Axon guidance
    diffusible, chemotropic, 564–67, 565*f*, 566*f*
    disorders, 571–72, 572*f*, 572*t*
    nondiffusible, 567–68, 568*f*, 570–71
    at optic chiasm, 569*b*, 569*f*
Axons
    action potential conduction and, 389
    biological features of, 4
    after brain injury, 681
    of corticocortical, corticothalamic, corticospinal pathways, 435
    of corticospinal tracts, 405–6, 435
    current flow in, 40*f*, 41
    decussation and, 569*b*
    differentiation of, 557*f*
    electron microscopy of, 5*f*
    fascicles of, 669, 670*f*
    fasciculation and, 567, 568*f*, 570
    geniculo-cortical, 613
    gradients and, 564, 594–955
    growth cones, 559–61, 560*f*
    labeling, 616
    MS and loss of, 67
    myelination of, 64–67, 64*t*, 65*f*, 66*f*
    neuronal divergence and, 6
    NMJ and peripheral, 671, 672*f*
    nociceptive, 304
    of ORNs, 341–43, 342*f*
    pathway formation of, 592, 593*f*
    postganglionic, 495, 499*f*
    preganglionic, 499*f*, 500
    protein and organelle trafficking in, 558–59, 558*f*
    regrowth of, 662–63, 664*f*, 668–70
    from retinal ganglion cells, 222
    Schwann cells and regrowth of, 669–70, 670*f*
    terminal field of, 4, 6
    tiling, 573–75, 576*f*
    of upper motor neurons, 403
Aβ fibers, 290–91
Aδ fiber group, 304–6

B

Babbling, 607, 608*f*
Babinski sign, 426, 426*f*
"Baby talk," 607
Baclofen, 127*b*
Bacteriorhodopsin, 12, 14*f*
Balance, upper motor neurons help maintaining, 419–25
Bálint, Rezsö, 747
Balint's syndrome, 747–48, 747*f*, 748*f*
Ballistic eye movements, 471
Banded krait (*Bungarus multicinctus*), 126*b*, 126*f*
Bands of Büngner, 668*f*, 669, 670*f*
Barbiturates, 42, 137

Bard, Philip, 799–800
Barn owls, 257
Baroreceptors, 509–10
Barosensory information, 509
Barrington's nucleus (pontine micturition center), 511–12, 511*f*
Bartter syndrome type IV, 83
Basal cell carcinomas, 335*b*, 538–39
Basal cells, 329
Basal domain, of polarized epithelial cells, 556
Basal forebrain nuclei, A–20, A–20*f*
Basal ganglia
    anatomical loops of, 433*b*, 433*f*
    circuits of, 432*f*, 440–42
    cogwheel rigidity in, 427*b*
    components of, A–19, A–20*f*
    corticostriatal interactions, 423*b*
    deep brain stimulation targeting, 445–46, 446*f*
    development of, 528*f*
    direct pathway through, 440, 441*f*, 442*f*
    dopamine circuits modulated by, 442–43
    eye movements and, 439, 440*f*
    habitual patterns and, 435*b*, 436*f*
    indirect pathway through, 440–41, 441*f*, 442*f*
    inputs to, 431, 432*f*
    intrinsic circuitry and outputs of, 438*f*
    in locomotion control, 398*f*
    motor components of, 431*f*
    movement modulated by, 380–81, 380*f*, 430–49
    neurons of, 432*f*
    nondeclarative memory and, 771, 772*f*
    non-motor functions of, 433*b*, 433*f*
    projections to, 431, 434–35
    tremor at rest in, 427*b*
Basal lamina, 568, 570
Basal plate, 523, A–11, A–11*f*
Basal processes, neural precursor, 541
Basal progenitors, 541
Basic emotions, 798
Basic helix-loop-helix (bHLH), 541, 543*f*, 544, 545*f*
Basilar artery, A–23, A–23*f*
Basilar membrane
    anatomy of, 243
    cochlea and, 242*f*, 243*f*
    phase locking and, 253
    sound distortion and, 247*b*
    traveling waves and, 244*f*
Basket cells, 459–60, 459*f*
Basophils, 315
Batrachotoxin, 75*b*
Bats, 238, 296*f*, 297*b*
Bax gene, 643
B cells, 682
Bcl2 gene, 643
Bcl2 gene family, 679
BDNF. *See* Brain-derived neurotrophic factor

Behavioral plasticity, 182–85, 183*f*, 184*f*, 185*f*
Behavioral sensitization, 185*f*
Behaviors
    addiction and, 826–27, 826*f*, 827*f*
    built-in, 603*b*
    flexible, 822–23
    gray matter volumes and, 610*f*
    inhibition of, 824–25
    instinctual, 602
    learning from consequences of, 832–33, 832*f*, 833*f*
    planning and organizing of, 822–23
    sexually dimorphic, 636–38
Behrmann, Marlene, 746
Békésy, Georg von, 244, 248
Bellugi, Ursula, 778*b*
Belladonna, 127*b*
Bell's palsy, 243
Bellugi, Ursula, 778*b*
Belt and parabelt regions, of auditory cortex, 252, 252*f*
Benadryl (diphenhydramine), 144
Benign familial neonatal convulsion (BFNC), 82
Benzer, Seymour, 186*b*
Benzodiazepines, 42, 137, 170*b*, 721
Bergmann glia, 550, 551*f*
Beta-amyloid proteins, A–31
β-Arrestin, 337
Betel nuts (*Areca catechu*), 126*f*, 127*b*
Betz cells, 405, 406*f*
BFNC (benign familial neonatal convulsion), 82
bHLH (basic helix-loop-helix), 541, 543*f*, 544, 545*f*
Bianchi, Leonardo, 820
Bicuculline, 127*b*
Binocular competition, 617–18, 619*f*
Binocular convergence, 616–17, 617*f*
Binocular disparity, 227
Binocular rivalry, 732, 732*f*
Binocular vision
    combined inputs from, 227
    field of view, 208*f*, 209
    orientation tuning in, 617–18, 618*f*
Biogenic amines, 124, 140–45
Biological clocks
    circadian cycle and, 714–17
    molecular mechanisms of, 717–18, 717*f*
Bipolar cells
    ON/OFF pathways in, 217–18, 218*f*
    retinal, 209, 210*f*
Bipolar disorder, 170*b*, 655*t*, 812
Birdsong, 602–4, 606*t*, 794–95, 794*f*
Bitemporal hemianopsia, 223*f*, 224*b*
Bitter taste, 356, 358
    as aversive, 372, 372*f*
    receptors detecting, 362, 362*f*
Bladder
    autonomic regulation of, 510–12, 511*f*
    cancer detection, 335*b*
    innervation of, 495
    referred pain from, 309*b*

visceral motor system functions and, 492*t*
Blastocysts, 520–21
Blindness, 83
Blindsight, 733, 733*f*
Blind spot (scotoma), 208, 223*b*, 733
Bliss, Timothy, 187
Blood-brain barrier, 6, 682, 684*f*, A–27–A–28, A–28*f*
Blood flow
    emotions and, 798
    hypothalamic control of, 496*b*
Bloodhounds, 349, 349*f*
Blood oxygenation level-dependent (BOLD) MRI, 30, 416*b*
Blood pressure, responses to, 510
Bmal1 gene, 717–18, 717*f*
BMAL1 protein, 717–18, 717*f*
BMI (brain-machine interface), 415*b*, 415*f*, 416*f*
BOLD (blood oxygenation level-dependent) MRI, 30, 416*b*
Bone morphogenetic proteins (BMPs)
    inductive signaling pathways of, 533–35, 534*f*
    neuronal differentiation and, 545*f*
    neuron development and, 535
    neuron migration and, 690
    Noggin antagonism of, 536*f*
Bony labyrinth, 263
Botulinum toxins, 108
Botulism, presynaptic terminals in, 108
Bouncing vision (oscillopsia), 269
Bovine spongiform encephalopathy (BSE), 463
Bowman's glands, 329
Brachium conjunctivum (superior cerebellar peduncle), 452*f*, 453, A–10*f*, A–12*f*
Brachium pontis (middle cerebellar peduncle), 452*f*, 453, A–10*f*, A–12*f*
Bradykinesia, Parkinson's disease and, 443
Bradykinin, 315
Brain
    activity-dependent development of, 600–601, 602*f*
    aging and size of, 772, 772*f*
    amnesia and, 763*f*
    anatomy of, 19
    androgen receptor distribution in, 644–45, 645*f*
    arteries of, A–23–A–24, A–23*f*
    attentional control supported by systems of, 750–51, 750*f*, 751*f*
    blood supply of, A–21–A–27, A–21*f*, A–24*t*
    BOLD MRI for, 30
    catecholamine neurotransmitters in, 140–43, 141*f*
    CNS and, 18, 19*f*
    complexity of, 31–32
    coronal view of, 821*f*
    cranial placodes and, 526–27, 527*f*

critical periods in, 602–4, 608–11, 609f, 610f
CT imaging of, 27–28, 28f
damaged, 659–60
decision making, thinking and planning regions of, 821f
declarative memory and, 756f
default state of, 735–36, 736f
development, 528f, 598–628
dopamine in, 140
dorsal surface of, A–14–A–15, A–15f
DTI for, 29f, 30
early development of, 517–54
embryonic subdivisions of, 526–29
epinephrine in, 143
estrogen receptor distribution in, 644–45, 645f
experience-dependent plasticity in developing, 598–628
fMRI for, 29f, 30
functional reorganization of, 661–62, 662f, 663f
GABA actions and development of, 138b, 138f
genetic models of function and disease of, 22–25, 23f
glial scar formation in, 679–82, 679f, 680f
glymphatic system of, A–31, A–32f
histamine levels in, 143, 143f
hypoxia/ischemia effects in, 679f
imaging of living, 26
intracellular signaling in, 164b
language representation in, 776, 776f, 785–86, 785f, 786f
lateral surface of, 821f, A–13–A–14, A–14f
lesions, attention and, 746–48, 746f, 747f
marijuana and, 150b, 150f
medial view of, 821f
MEG for, 30–31, 31f
midsagittal surface of, A–15–A16, A–16f, A19
MRI for, 29–30, 29f
Na⁺ pumps and function of, 87
neurogenesis in adult mammalian, 687–88, 688f
noninvasive neurophysiological imaging of, 26–27, 27f
norepinephrine in, 142
obesity and, 505b, 506f
organization of, 1
organ-produced signals for, 630
PET imaging of, 28, 28f
postnatal growth of, 600, 601f
repair and regeneration of, 658–93
resting state of, 735
retino-recipient regions of, 221–23, 222f
sexual dimorphisms and, 638–40, 641f, 641t, 653f
sexual orientation and, 651–52, 652f
silent synapses and development of, 193b

ventral surface of, A–14–A–15, A–15f
working memory and, 756f
Brain, W. R., 703
Brain death, 734
Brain-derived neurotrophic factor (BDNF), 173
    critical period regulation and, 625
    dendritic growth promotion by, 573
    microglia production of, 317
    NMJ expression of, 671
    secretion of, 669
    targets, 587–88, 588f
Brain injury
    axon growth after, 681
    excitotoxic, 131
    immune activation and inflammation after, 682, 683t, 684f
    traumatic, 676–77, 677f
Brain-machine interface (BMI), 415b, 415f, 416f
Brain repair, neurogenesis and, 690–91
Brainstem, A–3
    anatomy of, 453f, A–7–A–13
    assessment of, 271–72, 271f, 272f
    blood supply of, A–25–A–26, A–26f
    cerebellum and components of, 453f
    cerebellum projections to, 455–58, 457f
    cranial nerve nuclei of, A–10f
    internal organization of, A–12f
    in locomotion control, 398f
    lower motor neuron cell bodies in, 379
    motor control centers in, 420–25, 420f
    in movement control, 380, 380f
    projections to, 291
    projections to spinal cord, 420f
    sensory inputs to, 503
    upper motor neuron control of, 379, 380, 403–28, 455
Branchial arches, 549f, A–11
Branchial motor nuclei, A–11, A–11f
BrdU histochemistry, 692b, 693f
Breast cancer detection, 335b
Brightman, Milton, A–28
Brightness, 220b, 220f
Broca's aphasia, 779–81
Broca's area, 418, 780, 785
Brodmann, Korbinian, 699b
Brodmann's cytoarchitectonic areas
    area 2v, 276
    area 3a, 276
    area 4, 380, 405
    area 6, 380, 414
    area 8, 380, 414
    area 22, 255
    area 23, 414, 835
    area 24, 380, 414
    area 31, 835
    area 41, 250

area 42, 250
    area 44, 380, 414
    area 45, 380, 414
    structure of, 697–98, 699b
Bronchi, 492t
BSE (bovine spongiform encephalopathy), 463
Buck, Linda, 333
Bucy, Paul, 804
Bulbocavernosus muscle dimorphism, 638
Bulimia, 655t
Buller, A. J., 386b
α-Bungarotoxin, 125, 126b, 126f, 127
Bungarus multicinctus (banded krait), 126b, 126f
Bupivacaine, 41, 42f

C

C4A (C4 complement factors) gene, 169b
C4B (C4 complement factors) gene, 169b
C9orf72 gene, 400
CA (cornu Ammonis), 187, 187f
Ca2⁺ ion channels. See Calcium ion channels
Ca2⁺ ions. See Calcium ions
CACNA1A mutations, 82
CACNA1D genes, 83
CACNA1H mutations, 82
CACNA genes, 76, 77f
CACNG2 gene, 170b
E-Cadherin, imaging of, 557f
Cadherins, 547, 549, 568f, 570, 669
Caenorhabditis elegans (nematode worm), 24b, 164b, 334, 334f, 556, 557f, 564
Caffeine, 146, 721
Cajal, Santiago Ramón y, 3–4, 3f, 8, 171b, 552, 560, 564, 571
Calakos, Nicole, 437b
Calbindin, 161
Calcarine sulcus, A–16, A–16f
Calcitonin gene-related peptide (CGRP), 315
Calcium/calmodulin-dependent protein kinase II (CaMKII), 104, 191, 191f, 622f
    activation of, 166f
    function of, 165, 167
Calcium Green dye, 163b, 164f
Calcium homeostasis, 400
Calcium imaging, of neural circuits, 12, 13f
Calcium (Ca2⁺) ion channels
    blockage of, 127b
    neurological diseases caused by altered, 82–83
    presynaptic, 100
    selectively filters of, 80
    voltage-gated, 76, 77f, 100–101
Calcium (Ca2⁺) ions
    chelators, 190
    cytosolic concentration of, 161

intracellular messaging by, 161, 162f, 163
    LTD and release of, 176, 176f
    neurotransmitter secretion and, 99–101, 100f, 101f
    postsynaptic signals, 171b, 172f
    postsynaptic signal transduction and, 191–92, 191f
    in presynaptic terminal, 180–81
    Purkinje cell concentration of, 164b, 164f
    removal mechanisms of, 162f
    sources of, 162f
Calcium pumps, 87, 87f, 161
Calmodulin, 161
Caloric testing, of vestibular function, 272, 272f
CAMK2B gene, 169b
CaMKII. See Calcium/calmodulin-dependent protein kinase II
cAMP (cyclic adenosine monophosphate), 162f, 163
cAMP-dependent protein kinase (PKA), 163, 165, 166f, 184–85, 185f
cAMP response element binding protein (CREB)
    cerebellar LTD and, 197
    CRE binding, 184–85
    drug abuse and, 827
    transcriptional regulation of, 173–74, 174f
cAMP response elements (CREs), 184–85
CAMs. See Cell adhesion molecules
Canal reuniens, 262f
Canaries, 686f, 687
Cancer, dogs detecting, 335b
Cannabinoid receptors (CB₁), 321
Cannabis, 149, 150b, 150f
Cannon, Walter, 491, 495, 798
Canonical Wnt family signaling, 534f, 535
Capsaicin, 306b, 306f
Caraway seeds, 349, 351f
Carbamazepine, 199
Carbon-14, 691b
Cardiac pain, ASIC3 channels and, 307
Cardiovascular function, autonomic regulation of, 508–10, 509f
Carotid bodies, chemoreceptors in, 509–10
Caspase-1, after brain injury, 683t
Caspase-3, 679f
Castillo, Jose del, 99
Cataplexy, 721
Cataracts, 206, 618–19
Catecholamine neurotransmitters, 140–43, 141f, 142f
Catecholamines
    functional features of, 124t, 140
    structure of, 123f
Catecholamine synthesis, 177, 177f
Catechol-O-methyltransferase (COMT) gene, 142, 169b

Categorical theories of emotions, 798, 799f
Cats, 381f, 387–88, 387f, 397f
Cauda equina, A–3f, A–5
Caudal, definition of, A–1, A–2f
Caudal medulla, 290f, 313f
Caudate
  anatomy of, 431, 431f, A–19, A–20f
  inhibitory neurons of, 441f
  inputs to, 435
  medium spiny neurons of, 437
Caudate putamen (CP), 632b, 632f
Cavernous, A–21f, A–25
CB₁ (cannabinoid receptors), 321
CCL2, pain transmission and, 317
Cell adhesion molecules (CAMs)
  extracellular matrix, 567–68, 568f
  growth and fasciculation promoted by, 568f, 570
  Schwann cells and expression of, 668f, 669
Cell-associated signaling molecules, 157–58, 157f
Cell-attached recording, 72b
Cell-cell signaling, 532–35, 532f, 534f
Cell-impermeant signaling molecules, 157, 157f
Cell-permeant signaling molecules, 157–58, 157f
Cell polarity, differentiation and, 541, 557f
Cellular receptors, categories of, 158–59, 158f
Central autonomic network, 503f, 504
Central canal development, 528f
Central nervous system (CNS), 4, 6
  adult forms of, A–3f
  axes of, A–1, A–2f
  axonal growth cones in, 561
  dendritic polarization in, 571, 573
  formation of, 527, 528f, 529
  injury responses in, 678–79, 678f, 681, 681f
  input elimination in, 584f
  local tissue damage in, 679f
  motor control and subsystems of, 379–81
  neural induction in, 532–33, 532f
  neurogenesis in mature, 684–85
  neuronal migration in, 550, 550f, 551f, 552
  organization of, 18–20, 19f
  patterning of, 533
  regeneration in, 676
  sexual dimorphisms in, 638
  somatosensory information to, 279–81
  subdivisions of, A–2–A–4, A2f
  synapse formation in, 575, 577
  topographic map formation in, 592–93, 594f
  visualizing neurons in, 8–9, 8f
Central pattern generators, 395, 395b, 397f, 398, 425
Central sensitization, 316–17, 319
Central zone, A–6t

Cephalic flexure, 527, 528f
Cerebellar ataxia, 462, 463
Cerebellar cognitive-affective syndrome, 468
Cerebellar cortex
  anatomy of, A–19
  ataxias in, 468, 469f
  compartments in, 461f
  connections of, 460, 460f
  layers of, 459, 459f
  projections from, 455, 456f
Cerebellar motor syndrome, 468
Cerebellar peduncles, 452f, 453, A–7
Cerebellum
  anatomy of, A–19
  circuits within, 458–62, 459f
  critical periods and, 606t
  development of, 528f
  eye movement and, 464f
  genetic analysis of function, 466b, 466f, 467t
  innervation of, 584f, 585
  location of, A–3f
  in locomotion control, 398f
  LTD in, 195–97, 196f
  in movement control, 380, 380f
  movement modulated by, 451–70
  occipital lobe of, 222, 653f
  organization and subdivisions of, 451–53, 452f, 453f, 453t
  pathological changes to, 465, 465f, 468–69
  projections from, 455–58, 456f, 457f
  projections to, 453–55, 454f
  somatic maps of body surface in, 455, 455f
  spongiform degeneration of, 463
  tracts of, A–7
Cerebral achromatopsia, 231
Cerebral akinetopsia, 230–31
Cerebral aqueduct, A–30, A–30f, A–31f
Cerebral arteries, A–23, A–23f
Cerebral cortex
  anatomy of, A–4
  development of, 528f
  hyperdirect pathway and, 441
  layers of, 545–46
  lobes of, A–4, A–4f
  memory storage in, 766–70
  motor areas of, 438
  Nissl-stained layers of, 8f
  organization of, 697–712
  plasticity in, 298–99, 298f, 299f
  size of, 820f
  upper motor neuron cell bodies in, 379
  visual centers in, 227–31
Cerebral hem, 814
Cerebral hemispheres, A–4, A–4f
Cerebral nuclei, A–19
Cerebrocerebellum
  anatomy of, 452, 452f, 453t
  closed loops of, 457
  projections from, 455, 456f

Cerebrospinal fluid (CSF)
  circulation of, A–30, A–30f
  cisterns and, A–29
  daily production of, A–31–A–32
  ventricles in, A–3
Cerebrum
  anatomy of, 431f, 432f
  blood supply of, A–25, A–25f
  in discriminative pain pathways, 313f
  location of, A–3f
  spongiform degeneration of, 463
  in touch pathways, 290f
Cervical enlargement, A–3f, A–5
Cervical nerves, A–3f
Cervical spinal cord, 290f, 313f
Cervical vertebrae, A–4, A–5f
C fiber group
  lanceolate endings and, 286
  low-threshold mechanoreceptors, 314
  pain and, 304–5
  substance P release from, 147
  termination of, 308f
c-fos gene, 173, 175
CFS (chronic fatigue syndrome), 721
cGMP (cyclic guanosine monophosphate), 162f, 163, 211–12, 212f
CGRP (calcitonin gene-related peptide), 315
Chalfie, Martin, 164b
Channel-linked receptors, 158, 158f
Channelopathies, 82–83
Channelrhodopsin, 12, 14f
Channel synapse, 359
Charybdotoxin, 75b
ChAT (choline acetyltransferase), 125
Chemesthesis, 369
Chemical irritants, pain transduction and, 307
Chemical signaling
  classes of molecules in, 157–58, 157f
  forms of, 155–56, 156f
  intracellular signal transduction and, 156
  membrane-bound molecules in, 158
  receptor types in, 158–59, 158f
  signal amplification and, 156, 157f
Chemical synapses
  function of, 93
  signaling at, 96, 97f, 98
  structure of, 94–95, 94f, 97f
Chemical warfare, neurotoxins and, 127b
Chemoaffinity hypothesis, 593
Chemoattraction, 564, 567
Chemokines, 683t
Chemoreceptors, 509–10
Chemorepulsion, 564, 567
Chemosensory information, 509
Chemosensory systems, 324–54. See also Olfactory system; Vomeronasal system

Chemotropic axon guidance, 564–67, 565f, 566f
Chimpanzees, 793
Chlamydia trachomatis, 619
Chlordiazepoxide (Librium), 137, 170b
Chloride (Cl⁻) co-transporters, 138b, 138f
Chlorpromazine, 169b
Cholecystokinin, 148f
Cholesterol, 634, 634f
Choline, ACh hydrolysis and, 125
Choline acetyltransferase (ChAT), 125
Cholinergic nuclei, 728–29, 728f, 729t
Chondrodendron tomentosum, 126b
Chordin, 533, 534f, 535, 536f
Choreiform movements, 447
Choroid, anatomy of, 206f
Choroidal cells, 674b
Choroid plexus, A–30, A–30f, A–31f
CHRNA7 gene, 169b
Chromosomal sex (genotypic sex), 630–31, 631f
Chronic fatigue syndrome (CFS), 721
Chronic traumatic encephalopathy (CTE), 677
Ciliary body, anatomy of, 206
Ciliary cells, 674b
Ciliary ganglion, 232, 232f
Ciliary ganglion cells, 586b, 586f
Ciliary muscles, 206f
Ciliary neurotrophic factor (CNTF), 587–88
Cingulate gyrus, 653f, 804, A–16, A–16f
Cingulotomy, 833
Circadian cycle, 714–17
Circadian rhythmicity, 715, 715f
Circle of Willis, A–23f, A–24
Circumferential endings, 283t, 284f, 286
Circumvallate papillae, 358, 359f
Cisterns, A–29
Citrulline, 152, 153f
CJD (Creutzfeldt-Jakob disease), 463
c-Jun transcription factor, 590
Clarke's nucleus, 293, 293f
Classical conditioning, 757
Clathrin, 109, 110f
CLCN genes, 76
Cl⁻ (chloride) co-transporters, 138b, 138f
Click languages, 778
Climbing fiber inputs, in Purkinje cells, 584f, 585
Climbing fibers, 459–61, 459f, 460f
Clk gene, 717, 717f
Clock drawing test, 746, 747f
Clonus, 427b
Clostridium bacteria, 108
Clozapine, 169b
c-Myc transcription factor, 521
CNS. See Central nervous system
CNTF (ciliary neurotrophic factor), 587–88
Coated vesicles, 103, 109, 110f

Cocaine, DAT inhibition and, 142
Coccygeal nerve, A–3f
Coccygeal vertebrae, A–4, A–5f
Cochlea, 242f, 243–45, 243f, 244f, 246f, 250, 251f, 262f
Cochlear hair cells
  cochlea and, 242f
  damage to, 239b
  death of, 239b
  hair bundles in, 245, 245f
  mechanoelectrical transduction and, 245–48, 246f
  receptor potentials in, 247, 248f
  sound transduction by inner, 245–48
  sound transduction by outer, 248–50
  structure of, 245, 245f, 246f
Cochlear implants, 239b, 240f
Cochlear microphonics, 249
Cochlear nuclei, A–10f, A–12f
Cocktail party effect, 739
Cognitive control, 819
Cognitive function
  age and decline in, 773
  consortium studies and, 702b, 702f
  emotions and, 815–16
  importance of, 697
  meta-analysis techniques and, 701b, 701f, 702f
  neurexins and, 109
  sex-based differences in, 652–54
  TBIs in, 676–77
Cognitive neuroscience, 32
Cognitive reappraisal, 816
Coincidence detectors, 187, 189–92, 190f
Cole, Kenneth, 49, 55b
Collagens, 567, 568f
Colliculospinal tract, 424
Color constancy, 215b, 216f
Color contrast, 215b, 216f
Color opponent, 215
Color vision, 213–16, 215b, 215f, 216f
ColQ, 673f, 675
Columnar connections, 699
Columns, nervous system, 19
Coma, 734–35, 735f
Combination tones, 247b
Commissures, nervous system, 19
Common currency theory, 831
Communication, of songbirds, 602–4, 794–95, 794f
Competitive interactions, 615–16
Complex spikes, 52, 461
Component process theories of emotions, 798–99, 800t
Computational maps, 16
Computerized tomography (CT), 27–28, 28f
COMT (catechol-O-methyltransferase) gene, 142, 169b
Concentration-invariant coding, 345
Concha, 240, 241f
Conditional mutations, 25

Conditioned response, 757
Conditioned stimulus, 757
Conditioned taste aversion (CTA), 373–74
Conditioning, 757
Conduct disorder, 654t
Conduction velocities, 64–67, 64t, 65f, 66f
Conductive hearing loss, 239b
Cone photoreceptors, 209, 210f
  anatomical distribution of, 213f
  color vision and, 213–16, 215b, 215f, 216f
  convergence in, 212, 214f
  opsins in, 211
  phototransduction in, 211
  specialization of, 212–13
  types of, 214, 215f
Cone snails (Conus sp.), 126f, 127b
Conflicts, action plans and, 833, 833f
Confluence of sinuses, A–21f, A–25
Congenital adrenal hyperplasia (CAH), 650, 650t
Congenital fibrosis of the extraocular muscles, 571, 572t
Congenital insensitivity to pain, 83
Congenital myasthenic syndromes, 107
Congenital stationary night blindness (CSNB), 83f
Conjugate eye movement, 233, 476
Conjugate gaze deficits, 477–78, 477f, 478f
Connexins, 95, 95f
Connexons, 94–96, 95f
α-Conotoxins, 126f, 127b
Consciousness
  access, 731–32
  defining, 731
  memory and, 755–56
  phenomenological, 731
  reentrant neural activation and, 732–33
Consequences, learning from, 832–33, 832f, 833f
Consolidation, memory and, 723, 756
Consonant sounds, 778
Consortium studies, 702b, 702f
Conspecific interactions, 327
Contactin-2, 68
Contralateral neglect syndrome, 703–4, 746, 746f, 747f, 748–49
Contrast, retinal ganglion cells and, 219–20, 219f
Control systems, 819
Conus sp. (cone snails), 126f, 127b
Convergence
  neuronal, 6
  regulation of, 585
  in rod and cone photoreceptors, 212, 214f
Copy number variants, 22
Coren, Stanley, 789b
Cornea, anatomy of, 206, 206f
Corneal scarring, 619
Cornu Ammonis (CA), 187, 187f

Coronal, definition of, A–2
Corpus callosum, 652, 653f, A–14, A–15f
Corpus striatum, 140, 431
Corrective lenses, VOR and, 270f, 273
Cortex
  corticospinal and corticobulbar tracts in, 407f
  sleep interactions of thalamus and, 729, 730f, 731
  speech and, 778–81
  vestibular pathways to, 276, 276f
Cortical columns, 227–28
Cortical connections
  developmental disorders and, 609–10
  remodeling, 600
Cortical inhibitory loops, 460, 460f
Cortical neurons, 546f, 611, 612f, 616
Cortical plate, 552
Cortical pyramidal cells, 2f. See also Pyramidal neurons
Cortical states, 714–36
Cortices, nervous system, 19
Corticobulbar tracts, 405–9, 407f, 414
Corticocortical connections, 700
Corticocortical pathway, axons of, 435
Corticospinal tracts
  anatomy of, 407f
  axons of, 405–6, 435
  lateral, 406–7, 407f
  in primary motor cortex, 414
  ventral, 406–7, 407f
Corticostriatal pathway, 431, 434
Corticothalamic pathway, axons of, 435
Corticotropin-releasing hormone (CRH), 630
Co-transmitters, 122
Co-transporters, 89, 89f
Covert attention, 741, 741f
COVID-19 (SARS-CoV2) pandemic, 330, 330f, 360, 361
COX (cyclooxygenase), 316
COX-2 (cyclooxygenase 2), 317
CP (caudate putamen), 632b, 632f
Cranial nerve ganglia, 9–10
Cranial nerves. See also specific nerves
  anatomy of, A–7–A–13, A–8f
  differentiation of, 529–31, 531f
  function of, 242–43, A–8t
  nuclei, A–9–A–13, A–10f, A–10t
  segmental patterning of, 529–31, 531f
Cranial neural crest, 525f, 526
Cranial placodes, 526–27, 527f
CRASH, 572t
CREB. See cAMP response element binding protein
Credit assignments, 831
Cre/lox system, 25
Cre recombinase, 25
CREs (cAMP response elements), 184–85
Creutzfeldt-Jakob disease (CJD), 463

CRH (corticotropin-releasing hormone), 630
Cribriform plate, 325–26, 325f
CRISPR-Cas9 DNA editing, 25
Crista, 266–67, 267f
Critical periods
  activity-dependent plasticity and, 612f, 624–26, 626f
  in brain development, 602–4, 608–11, 609f, 610f
  gray matter volume and, 609–10, 610f
  for language acquisition, 790–91, 790f
  in language development, 606–7, 606t, 608f
  in neural systems, 604–6, 606t
  neurotransmitters mediating, 621, 622f, 623
  oscillating electrical activity and, 619–21, 620f
  properties of, 602–3, 604
  regulation of, 621–26
  in visual system development, 604, 605f, 618–19
Crocodiles, 296f, 297b
Crohn's disease, 316
Crossed eyes, 618
Crustaceans, 24b
Cry gene, 717–18, 717f
CRY protein, 717–18, 717f
CSF. See Cerebrospinal fluid
CSNB (congenital stationary night blindness), 83f
CT (computerized tomography), 27–28, 28f
CTA (conditioned taste aversion), 373–74
CTE (chronic traumatic encephalopathy), 677
Cuneate nucleus, 290f, 291, A–12f
Cuneate tract, 290f, 291
Cuneus gyrus, A–16, A–16f
Cupula, 267, 267f
Curare, 126b
Curran, Tom, 467b
Curtis, Howard, 49
CXCL1, pain transmission and, 317
Cyclic adenosine monophosphate (cAMP), 162f, 163
Cyclic guanosine monophosphate (cGMP), 162f, 163, 211–12, 212f
Cyclic nucleotide-gated ion channels, 76, 77f, 84, 85f, 337
Cyclic nucleotides, 163
Cyclooxygenase (COX), 316
Cyclooxygenase 2 (COX-2), 317
Cytoarchitectonic areas. See Brodmann's cytoarchitectonic areas
Cytoarchitecture, 9
Cytochrome c, 679f, 680
Cytokines
  after brain injury, 683t
  release of, 315

Cytoskeleton
   actin, 561–62, 563*f*
   components of, 556
   microtubule, 561, 563*f*

D

D3 dopamine receptors, 142, 142*f*
DAG (diacylglycerol), 151*f*, 162*f*, 163, 196, 196*f*, 197
Damage-associated proteins (DAMPs), 682
Danio rerio (zebrafish), 24*b*, 560*f*
DAOA (D-amino acid oxidase activator) gene, 169*b*
DAO (D-amino acid oxidase) gene, 169*b*
DARPP-32 gene, 169*b*
Darwin, Charles, 800
DAT (dopamine co-transporters), 142
dB (decibels), 237
db/db (diabetic) mutations, 506*b*
DBS (deep brain stimulation), 445–46, 446*f*
DCC receptors, 564–66, 565*f*
DCX (doublecortin) protein, 552
Deafness, 83, 778*b*
   language development and, 607, 608*f*
Decerebrate rigidity, 273, 428
Decibels (dB), 237
Decision making
   addiction and, 826–27, 826*f*, 827*f*
   anterior insula and, 834, 834*f*
   brain regions in, 821*f*
   common currency theory and, 831
   credit assignments and, 831
   emotions and, 808–10, 810*f*, 811*f*
   free will and, 835*b*, 835*f*, 836*f*
   impulsivity and self-control in, 823–25
   reinforcer devaluation task and, 831*f*
   rewards and, 825–32
   thinking, planning and, 819–38
Declarative memory
   brain systems in, 756*f*
   definition of, 755–56
   hippocampus and storage of, 769, 769*f*
   medial temporal lobe supporting, 762–66, 770
Decussation, 226, 569*b*
Deep brain stimulation (DBS), 445–46, 446*f*
Deep cerebellar nuclei
   activity of, 462*f*
   anatomy of, 452*f*, 453, 453*t*, 456*f*, 457*f*
   connections of, 460, 460*f*
   error recognition by, 462, 464
   inputs to, 455
Deep excitatory loops, 460, 460*f*
Default mode networks, 735–36, 736*f*, 837–38, 837*f*
Defecation, regulation of, 510

Delayed response genes, 175
Delayed response tasks, 711, 712*f*
Delta ligands, 541, 543*f*
Delta-Notch signaling, 541, 543*f*, 544
Delta-waves, 724
Dementia, 768*b*. *See also* Alzheimer's disease
Dendrites
   arborization of, 4, 6
   branching of, 4, 5*f*
   differentiation of, 557*f*
   directed growth of, 571, 573, 574*f*
   function of, 586*b*, 586*f*
   neuronal convergence and, 6
   pathway formation of, 592, 593*f*
   protein and organelle trafficking in, 558–59, 558*f*
   Purkinje cell, 459, 459*f*
   sprouting of, 663–64, 664*f*
   tiling, 573–75, 576*f*
Dendritic polarization, 571, 573, 574*f*
Dendritic spines, 168, 171*b*, 171*f*, 643, 644*f*
Dendrotoxin, 75*b*
Dental consonants, 778
Dentate nucleus, 455, 456*f*
Depolarization, 39, 39*f*
   Na$^+$ and K$^+$ conductances and, 69, 69*f*
   voltage-dependent ion currents activated by, 56–58, 57*f*
Depression
   amygdala and, 811
   definition of, 812
   ECT and, 767
   incidence of, 812
   insomnia and, 720
   medications for treating, 812, 815
   molecular basis of, 169*b*
   sex-based differences in, 654, 655*t*
   VLPFC and, 824
Dermatomes, 280
Desflurane, 43
Desipramine, 169*b*
Deuteranopia, 215*f*, 216
Deuterocebic (db/db) mutations, 506*b*
Diabetic (*db/db*) mutations, 506*b*
Diacylglycerol (DAG), 151*f*, 162*f*, 163, 196, 196*f*, 197
Diazepam (Valium), 42*f*, 137, 170*b*
Dichromatic vision, 215*f*, 216
Diencephalon, A–2
   anatomy of, 453*f*, A–16, A–16*f*
   cerebellum and components of, 453*f*
   formation of, 528*f*, 529
   location of, A–3*f*
Diethyl ether, 43
Diffusion tensor imaging (DTI), 29*f*, 30, 571–72, 572*f*, 698
Digenea simplex, 127*b*
Digit memory span, 758, 758*f*
Dihydrotestosterone, 634*f*, 651
Dihydroxyphenylalanine (DOPA), 140, 141*f*
Dilantin (phenytoin), 199

Dimensional theories of emotions, 798, 799*f*
Diphenhydramine (Benadryl), 144
Diplopia, 227, 479, 479*f*, 480*f*
DISC1 (Disrupted in Schizophrenia 1) gene, 169*b*, 552
Disconjugate eye movement, 233, 476
Disinhibition syndrome, 822
Disrupted in Schizophrenia 1 (*DISC1*) gene, 169*b*, 552
Dissociated sensory loss, 308
Distaless mutations, 531
Divergence
   neuronal, 6
   regulation of, 585
DJ-1 mutations, 444
DLG3 gene, 169*b*
DLPFC (Dorsolateral prefrontal cortex), 728, 728*f*, 823–24, 823*f*, 824*f*
DNA
   coding sequences, 20
   Cre recombinase, 25
   CRISPR-Cas9 DNA editing, 25
   labelled, 685
   regulatory sequences, 20
   transcription to RNA, 172–73, 173*f*
Dogs, 334, 335*b*, 349, 349*f*, 350*f*
Dolphins, audible spectrum of, 238
Dominant genes, 22
DOPA (dihydroxyphenylalanine), 140, 141*f*
Dopamine
   basal ganglia circuits modulated by, 442–43
   in brain regions, 140
   functional features of, 124*t*
   habits and, 830*b*
   metabotropic receptors and, 142, 142*f*
   in popular culture, 828*b*
   production of, 140, 141*f*
   RPEs and, 828*b*, 829*f*, 830*f*
   structure of, 123*f*
   system schematic of, 829*f*
Dopamine co-transporters (DAT), 142
Dopamine receptors
   activation of, 142, 142*f*
   antagonists of, 169*b*
   D1 and D2, 443
Dopaminergic neurons, in Parkinson's disease, 140, 443–44, 444*f*
Dorsal, definition of, A–1, A–2*f*
Dorsal column nuclei, 291–93, 293*f*
Dorsal columns, A–6
Dorsal horn neurons, 316–17, 499*f*, 502
Dorsal horns, A–5, A–5*f*, A–6*t*
Dorsal motor nucleus of the vagus nerve, 499*f*, 500, 510, A–10*f*, A–12*f*
Dorsal nucleus of Clark, 454, 454*f*
Dorsal raphe, 320, 321*f*
Dorsal root ganglia, 9–10, 280
   anatomy of, 499*f*

spinal visceral sensory neurons in, 502
Dorsal roots, A–4, A–5*f*
Dorsal spinocerebellar tract, 293, 293*f*
Dorsal stream, 230–31, 231*f*
Dorsal thalamus, 222, 222*f*, 804, A–17*b*
   ventral anterior nuclei, ventral lateral nuclei of, 438
Dorsiflexor muscles, 387*f*
Dorsolateral prefrontal cortex (DLPFC), 728, 728*f*, 823–24, 823*f*, 824*f*
Dorsolateral prefrontal loop, 433*b*, 433*f*
Dorsolateral tract of Lissauer, 307, 308*f*
Dorsomedial nucleus, 497*b*, 497*f*, 498*f*
Doublecortin (DCX) protein, 552
Double dissociations, 771
Downer, John, 804
Down syndrome, 540, 575
DRD2 (type 2 dopamine receptor) gene, 169*b*
Dreaming, 725, 726*b*
Drosophila melanogaster (fruit flies)
   axonal and dendritic tiling in, 573–75, 576*f*
   embryonic development of, 529–31, 530*f*
   learning and memory in, 186*b*
   odorant receptor genes of, 333–34, 334*f*
   olfactory bulbs of, 342*f*, 343
   slit factor and robo receptors in, 565*fm* 566
   studies on, 24*b*
Drug abuse, 826–27, 826*f*, 827*f*
DSCAM1 gene, 574–75, 576*f*, 579
DSCAM2 gene, 574, 579
D-Serine, metabolism of, 169*b*
DTI (diffusion tensor imaging), 29*f*, 30, 571–72, 572*f*, 698
Duchenne, G.-B., 802*b*, 802*f*
Duchenne smile, 802*b*
Dunce mutation, fruit fly, 186*b*
Dura mater, A–29, A–29*f*
Dutchman's breeches, 127*b*
Dynactin mutations, 400
Dynamin, 109, 110*f*
Dyneins, 558–59, 558*f*
Dynorphin A, 148*f*, 149*t*
Dynorphin B, 149*t*
Dynorphins, 148, 148*f*, 149*t*, 320
Dysarthria, 780
Dysdiadochokinesia, 468
Dysexecutive syndrome, 821
Dysgeusia, 360, 361
Dyslexia, 654*t*, 791–92, 792*f*
Dysmetria, 468
Dystrobrevin, 672
Dystroglycan, 672, 675

E

E/1 (excitatory/inhibitory) balance, 623
EA1 (episodic ataxia type 1), 82
EA2 (episodic ataxia type 2), 82
EAAT2 (excitatory amino acid transporter 2), 400
EAAT4 (excitatory amino acid transporter 4), 461f
EAATs (excitatory amino acid transporters), 131, 131f
Ear canal occlusion, 239b
Ears
  external, 240–41, 241f
  inner, 241f, 243–45
  middle, 239b, 241–43, 241f
  sound generated by, 249–50, 249f
Eccles, J. C., 386b
ECM. See Extracellular matrix
ECoG (electrocorticography), 416b
ECT (electroconvulsive therapy), 767
Ectoderm, 523, 524f, 533, 534f
Ectodermal cells, 533, 534f
Edinger-Westphal nucleus, A–10f, A–12f
  anatomy of, 499–500, 499f
  projections to, 232
  pupillary constriction and, 475
EEG. See Electroencephalography
Effector pathways, G-proteins and, 160f, 161
Efferent neurons, 9–10
Ehrlich, Paul, A–28
Eimer's organs, 296b
Electrical signals, neuronal
  anesthesia and, 41–43, 42f, 43f
  current flow, 39, 40f, 41
  encoding information with, 37–39
  ion movements producing, 44–45, 44f
  long-distance transmission of, 39, 40f, 41, 44
  in membrane potential creation, 45–46
  passive and active, 38–39, 39f
  types of, 37–38, 38f
Electrical synapses, 4
  advantages of, 95–96
  function of, 93
  gap junctions at, 94, 94f, 96f
  signaling at, 95–96, 96f
  structure of, 94, 94f, 95f
Electrochemical equilibrium, 45–46, 45f
Electroconvulsive therapy (ECT), 767
Electrocorticography (ECoG), 416b
Electroencephalography (EEG)
  brain-machine interfaces and, 416b
  on brain systems supporting attentional control, 750–51, 751f
  electrodes in, 719b, 719f
  ERPs and, 26, 27f
  N400 response and, 783, 784f
  oscillations in, 621

rhythms in, 718b
on seizures, 199f
sleep and, 718b, 720f, 723–24, 724f
Electromyograms (EMGs)
  of dorsiflexor muscles, 387f
  in myasthenia gravis, 129f
  in sleep, 725f
Electron microscopy, 4, 5f
Electrooculogram (EOG), 725f
Electro-olfactogram (EOG), 338f
Electrophysiological recordings, 10 11, 11f, 18, 18f
11-cis-retinal, 211
Ellington, Duke, 356
Embolic stroke, A–27
Embryonic organizer, 533
Embryonic stem cells, 519b, 519f, 520–21, 520f, 535
Embryos
  CNS formation in, 527, 528f, 529
  cranial placodes in, 526–27, 527f
  inductive signaling in, 533–35, 534f
  nervous system formation in, 523–26, 524f, 525f
  neural crest in, 523–24, 525f
  neurulation in, 523, 524f
  PNS formation in, 526
EMGs. See Electromyograms
Emmetropia (normal vision), 207f
Emotional facial paresis, 803b, 803f
Emotional memory tasks, 653
Emotions
  amygdala and, 803–10
  arousal and, 798
  basic, 798
  blood flow and, 798
  categorical theories of, 798, 799f
  classifying, 798–99
  cognitive function and, 815–16
  component process theories of, 798–99, 800t
  components of, 798f
  decision making and, 808–10, 810f, 811f
  defining, 797–98
  dimensional theories of, 798, 799f
  facial expressions and, 802b, 808b, 809f, 813–15, 814f, 815f
  fear conditioning and, 804–8, 806f, 807b, 807f
  hemispheric asymmetry and, 814
  integrating components of, 799–801, 800f, 801f
  memory and, 815–16, 817f
  motor activity and, 801
  nervous system and expression of, 801f
  PFC and processing of, 811, 814–15
  physiological changes and, 798
  regulation of, 816–17
  reticular formation and, 800
  skin conductance and, 798
  stress hormones and, 816
  subjectivity of, 814
  valence and, 798

EMT (epithelial-to-mesenchymal transition), 547, 548f
Endocannabinoids
  analgesic effects of, 321–22
  discovery of, 150b
  functional features of, 124t
  GABA release and, 152, 152f
  synaptic transmission of, 149, 151f, 152
Endocochlear potential, 247–48
Endocrine signaling, 156, 156f
Endocytosis
  of AMPA receptors, 461
  local recycling of synaptic vesicles and, 103–4, 103f
  mechanisms of, 109, 110f
  neurological diseases and defects in, 107–9
  after neurotransmitter release, 110f
Endoderm, 523, 524f
Endogenous antagonists, 533, 534f, 535, 536f
Endogenous attention, 739–41, 740f
Endogenous opioids, 320–21
Endolymph, 247
Endolymphatic ducts, 262f
Endolymphatic fluid, 267
Endoplasmic reticulum, 146–47
Endorphins, 147–48, 148f, 149t, 320
Endothelial cells, 315, A–28, A–28f
  peripheral nerve regeneration and, 667, 667f
End plate currents (EPCs)
  EPPs and, 114–15, 115f
  postsynaptic, 112–13, 113f
  production of, 112, 112f
  reversal potentials and, 113–14, 113f, 114f
End plate potentials (EPPs), 98–99, 99f, 100f
  EPCs and, 114–15, 115f
End plates, 98
Engram, 756
Enkephalins, 148, 148f, 149t, 320
Enophthalmos, 504–5, 504f
Enteric division, of visceral motor system, 20, 501, 501f
Entorhinal cortex, 398f
Enzyme-linked receptors, 158–59, 158f
EOG (electro-olfactogram), 338f
EOG (electrooculogram), 725f
EPCs. See End plate currents
EphB1 gene, 570b
Eph receptors (Ephs), 570, 594–95
Ephrin B2 gene, 570b
Ephrin ligands, 570
Ephrins, 681
Epibranchial placodes, 526, 527f
Epidural electrical stimulation, 661–62, 663f
Epilepsy
  altered ion channels and, 82
  cure or prevention for, 199
  description of, 198
  neural circuitry and, 198–99

split-brain patients and, 786–88, 787f
Epinephrine (adrenaline)
  brain levels of, 143
  functional features of, 124t
  structure of, 123f, 141f
  synthesis of, 143
Episodic ataxia type 1 (EA1), 82
Episodic ataxia type 2 (EA2), 82
Episodic memory, 782b
Epithalamus, A–17b
Epithelial-to-mesenchymal transition (EMT), 547, 548f
EPPs (end plate potentials), 98–99, 99f, 100f
Eps-15, 109
Epsin, 109
EPSPs. See Excitatory postsynaptic potentials
Equilibrium, vestibular complex and, 455
ERK (extracellular-regulated kinase) pathway, 625, 626f
ERPs (event-related potentials), 26, 27f, 741, 743, 743f
Esophagus, referred pain from, 309b
Esotropia, 619
Estradiol, 634–36, 634f, 635f, 640, 643, 646
Estrogen
  brain distribution of, 644–45, 645f
  hypothalamic activation by, 652, 652f
  lipophilic structure of, 635
  neuronal growth/differentiation and, 644f
  sexual development and, 634–35
  sexually dimorphic development and, 642–44
  synaptic transmission influenced by, 646, 648f
Estrogen receptors, 634, 635, 644–45, 645f
Ethyl mercaptan, 353
Eustachian tube, 241f
Evaluation, OFC and, 825, 828, 830–32
Evans, Martin, 520
Evarts, Edward, 410
Event-related potentials (ERPs), 26, 27f, 741, 743, 743f
Excitatory amino acid transporter 2 (EAAT2), 400
Excitatory amino acid transporter 4 (EAAT4), 461f
Excitatory amino acid transporters (EAATs), 131, 131f
Excitatory/inhibitory (E/1) balance, 623
Excitatory neurotransmitters, 9
Excitatory postsynaptic potentials (EPSPs), 116–17, 117f
  AMPA receptor mediation and, 190
  AMPA receptor-produced, 132, 132f
  causes of, 198f

NMDA receptor-produced, 132, 132f
presynaptic action potentials and, 180f
production of, 187, 188f, 189f
Excitotoxic brain injury, 131
Excitotoxicity, 679
Executive control, 819
Exercise, motor unit firing during, 387–88, 387f
Exocytosis
  mechanisms of, 104, 106f, 109
  neurological diseases and defects in, 107–9
  neurotransmitter release and, 98, 102f
  process of, 98
Exogenous attention, 739–41, 740f
Exons, 20
Exotropia, 619
Expressive aphasia, 780
Express saccades, 485
External cuneate nucleus, 292, 293f, 454, 454f
External ear, 240–41, 241f
External granule cell layers, 550
External tufted cells, 342f, 343
Exteroception, 315
Extinction, hemispatial neglect and, 748–49
Extracellular matrix (ECM)
  axon regrowth and, 668, 669
  cell adhesion molecules, 567–68, 568f
  CNS response to injury and, 681, 681f
  fibroblasts producing, 667
  integrins and recognition of, 669
  molecules bound to, 675
Extracellular recordings, 10–11
Extracellular-regulated kinase (ERK) pathway, 625, 626f
Extrafusal fibers, 288
Extrafusal muscle fibers, 288, 288f
Extraocular muscles, 233
  actions and innervation of, 473–75, 474f
  congenital fibrosis of, 571, 572t
  eye movements and, 474f
  spindle density and, 289
Extraoral taste receptors, 364b, 364f
Extrastriate visual areas, 230–31, 231f
Eye movements
  amplitude of, 479, 480f
  attention and, 751–52
  axes of rotation in, 473–75
  ballistic, 471
  basal ganglia and, 439, 440f
  Brodmann's area 8 and, 380
  cerebellar contribution to, 464f
  conjugate, 233, 476
  direction of, 479, 480f
  disconjugate, 233, 476
  extraocular muscles and, 474f
  functions of, 475–77

horizontal, 480, 480f
hypermetric, 464
hypometric, 464
importance of, 471–72, 472f
injury, disease, disorder and, 477–79, 477f, 478f, 479f
optokinetic, 476, 476f
saccadic, 233, 257, 272, 464f, 475, 475f, 479–87, 480f, 481, 484f
semicircular canals directing, 476
sensorimotor integration and, 471–88
smooth pursuit, 233, 475, 476f, 478–79, 479f, 487–88, 488f
spindle density and, 289
upper motor neurons and, 457–58
vergence, 475–76, 488
vestibulo-ocular, 269, 270f, 273, 455, 476, 476f
in viewing photographs, 472f
Eyes. See also specific anatomy
  accommodation in, 207, 207f, 233
  anatomy of, 206, 206f
  blindness and, 83
  crossed, 618
  lateral vestibulospinal tract and, 420–21
  projections specific to individual, 226–27
  refractive errors in, 206–7, 207f
  sound localization in relation to, 257–58, 258f
  strabismus and, 616–17, 617f
  visceral motor system functions with, 492t
  visible light and, 205–6
  visual deprivation and, 611–21
  wall, 618

F

Face
  identification of, 705–6
  pain pathway to, 312, 313f, 314
  proprioceptive pathways, 292, 455
  temperature pathway to, 312, 313f, 314
Facial expressions
  asymmetrical, 814f
  determination of, 802b, 802f, 803f
  emotions and, 802b, 808b, 809f, 813–15, 814f, 815f
  in Parkinson's disease, 443
  trustworthiness of, 808, 809f
  voluntary, 801
Facial motor nucleus, 408, 408f, A–10f
Facial nerve
  anatomy of, 262f, 499f, A–9
  facial pain and, 312
  function of, A–8t
Facial nucleus, A–12f
Facial paresis, 801, 803b, 803f
Facial weakness patterns, 408–9, 408f
Falling phase, of action potentials, 52b

Familial amyotrophic lateral sclerosis (FALS), 399–400
Familial hemiplegic migraine type 1 (FHM1), 82
Familial infantile myasthenia, 107–8
Faraday, Michael, 802b
Faradization, 802b
Farsightedness (hyperopia), 207, 207f
Fasciculation, 567, 568f, 570
Fast fatigable (FF) motor units
  force provided by, 387–88, 387f
  function of, 384, 385f, 386b
  staining of, 386b, 386f
Fast fatigue-resistant (FR) motor units
  force provided by, 387–88, 387f
  function of, 384, 385f, 386b
Fastigial nucleus, 455, 456f
Fatt, Paul, 98
Fatty acid hydrolase, 149
Fear conditioning, 804–8, 806f, 807b, 807f
FEFs. See Frontal eye fields
Female sexual function, 512–13
Fentanyl, 42, 43f
Ferrier, David, 820
FFA (fusiform face area), 745, 745f
FFARs (free fatty acid receptors), 363, 365b
FF motor units. See Fast fatigable motor units
FGFs. See Fibroblast growth factors
FHM1 (familial hemiplegic migraine type 1), 82
Fibroblast growth factors (FGFs)
  denervation and production of, 675
  inductive signaling and, 534f
  neural induction in, 533
  neuron development and, 535
  neuron migration and, 690
Fibroblasts, 315, 667, 667f
Fibronectin, 547, 549, 567, 568f
Fight or flight response, 495
Filopodia, 559–62, 560f
"Fire together, wire together" hypothesis, 617
First pain, 305, 305f
Fish, Mauthner cells in, 274b, 275f
Fisher, C. Miller, A–27
5-HT. See Serotonin
5-HT (serotonin) receptors, 144–45, 145f
5-α-reductase, 650t, 651
Fixation points, visual, 233
Flavor, 356, 369
Flexible behavior, 822–23
Flexion-crossed extension reflex, 394, 394f
Flexion reflex, 394, 394f
Flocculus, 452f, 453
Floorplate, 523, 524f, 534f
"Flower-spray" endings, 389
Fluoxetine (Prozac), 169b, 812
Fluphenazine, 169b

fMRI (functional MRI), 29f, 30, 604, 605f, 651–52, 702b, 702f, 750–51, 751f
FMRP (fragile X-mental retardation protein), 625
Focused selection concept, 441–42
Folia, 452f, 458, A–19
Foliate papillae, 358, 359f
Foramen of Magendie, A–30, A–30f
Foramina of Luschka, A–30
Foramina of Monro, A–30, A–30f, A–31f
Forebrain, A–2
  anterior SVZ, 689f
  emotions and, 800–801, 801f
  facial expressions and, 802b
  internal anatomy of, A–19–A–21, A–19f, A–20f
Forgetting, 760–62, 761f
Formants, 776
Fornix, A–20f, A–21
47-XXY syndrome, 650, 650t
Fourier decomposition, 237, 237f
Fourth ventricle, 528f, A–10f, A–12f, A–30, A–30f, A–31f
Foveae
  anatomy of, 206f, 207
  image maintenance and, 233
Foveation, 471, 482b, 483f
FOXP2 gene, 781
Fragile X-mental retardation protein (FMRP), 625
Fragile-X syndrome, 540
Frataxin (FXN), 299
Free fatty acid receptors (FFARs), 363, 365b
Freeman, Walter, 710b
Free nerve endings, 283t, 284, 284f, 286, 304
Free visual scanning, 436b, 436f
Free will, neuroscience on, 835b, 835f, 836f
Frequency, of sound, 252–55, 253f, 254f
Freud, Sigmund, 726b
Friedreich's ataxia, 299
Friel, Brian, 619
Friesén, Jonas, 691b
Fritsch, Gustav, 409, 445, 820
Frizzled receptor, 535
FR motor units. See Fast fatigue-resistant motor units
Frogs, 518, 518f, 520
Frontal, definition of, A–2
Frontal eye fields (FEFs)
  anatomy of, 485f
  attention and, 749–50, 749b, 749f, 752
  function of, 485–87
  neuron responses in, 487f
  saccadic eye movements and, 481
Frontal lobes
  anatomy of, A–4, A–4f
  in movement control, 380
  neuron activation in, 711, 712f
  size of, 820f
  studies on, 819–20

Frontal lobotomies, 709, 710b, 710f
Frontal-parietal control network, 750–51, 750f, 751f
Frontotemporal dementia, 478
Fruit flies. See Drosophila melanogaster
Functional MRI (fMRI), 29f, 30, 604, 605f, 651–52, 702b, 702f, 750–51, 751f
Fungiform papillae, 358, 359f
Fura-2, 101f, 163b, 164f
Fused tetanus, 388, 388f
Fusiform face area (FFA), 745, 745f
Fusiform gyrus, 230, A–15, A–15f
FXN (frataxin), 299

G

GABA (γ-aminobutyric acid)
   brain development and actions of, 138b, 138f
   endocannabinoids and release of, 152, 152f
   functional features of, 124t
   glycine and, 135–39, 136f
   removal of, 136
   structure of, 123f
   synthesis and metabolism of, 136, 136f
GABA_{BR2}, 461f
GABAergic agonists, 448, 449f
GABAergic inhibitory interneurons, 615
GABAergic interneurons, 459
GABAergic neurons, 137f, 438–39
GABAergic synapses, 136–37
GABA receptors
   ionotropic, 136–37, 137f, 139
   metabotropic, 139, 139f
GABA transaminase, 136
GAD2 (glutamatic acid decarboxylase) gene, 170b
GAD (glutamatic acid decarboxylase) gene, 136
Gage, Phineas, 707, 708f, 822
Gajdusek, Carleton, 463
Galef, Bennett, 373
γ-aminobutyric acid. See GABA
γ-Bias, 389
Ganglia, PNS, 18, 19f
Ganglion cells. See also Retinal ganglion cells
   ciliary, 586b, 586f
   identification of, 674b
   intrinsically photosensitive, 210–11
   midget, 225–26, 225f
   ON/OFF, 217–18, 217f, 218f, 219f
   parasol, 225–26, 225f
Ganglionic eminences, 529, 550, 550f
GAP43 (growth-associated protein-43) gene, 671
Gap junctions, 4, 94, 94f, 96f
   Purkinje cell output and, 461
GAPs (GTPase-activating proteins), 160
Garcia, John, 373
Gardner, Howard, 779

Gaskell, Walter, 491
Gastrocnemius muscle, 384, 387–88, 387f, 423, 424f
Gastrulation, 523
Gate theory of pain, 320, 321f
Gaze
   active control of, 231–32
   orientation, 419–25
   shifting and stabilizing, 475
Gaze centers, 478–81
Gazzaniga, Michael, 787
G-CSF (granulocyte colony-stimulating factor), 683t
GDNF (glial-derived neurotrophic factor), 587–88
Geese, 603b
GEFS (generalized epilepsy with febrile seizures), 82
GEFs (guanine nucleotide exchange factors), 160
Gender, brain size and, 772f
General anesthesia, 42–43
Generalized anxiety disorder, 170b
Generalized epilepsy with febrile seizures (GEFS), 82
Generator potentials. See Receptor potentials
Genes. See also specific types
   components of, 20
   dominant, 22
   ion channels encoded by, 76–78, 77f
   language production and, 781
   mutations in, 20–22, 21f
   orthologous, 22
   recessive, 22
   splice variants of, 20
Genetic analysis, 20–22
Genetic engineering, 22, 23f, 25
Genetics
   brain function and disease models in, 22–25, 23f
   genomics and, 20–21, 21f
Geniculo-cortical axons, 613
Genome-wide association studies (GWAS), 22
Genomics, 20–21, 21f
Genotypic males, 634–35
Genotypic sex (chromosomal sex), 630–31, 631f
Germ layers, 523, 532
GFAP (glial fibrillary acidic protein), 683t
GFP (green fluorescent protein), 164b, 164f
Ghrelin, 505b
Gill withdrawal reflex, Aplysia, 182–83, 183f
Glabrous skin, 279, 284f
Glaucoma, 206
Gli3, 535, 536f
Glial cells
   activated, 680
   calcium elevation in, 119f, 120b
   classes of, 679f
   differentiation of, 540–44, 545f
   discovery of, 2–4, 3f

generation of, 544
   neuronal migration and, 550, 551f, 552
   overgrowth of, 664–65
   radial, 541, 550, 551f, 552
   support cell theory of, 4
   types of, 6–8, 7f
Glial-derived neurotrophic factor (GDNF), 587–88
Glial fibrillary acidic protein (GFAP), 683t
Glial scar formation, 679–82, 679f, 680f
Gliotransmitters, 119f, 120b
Globus pallidus
   anatomy of, 431f, 437, 438f, A–19, A–20f
   efferent cells of, 438–39
   external segment of, 441, 441f
   internal segment of, 438, 438f, 440, 441f
   neurons of, 432f
Glomeruli, 341–46, 342f, 344f
Glossopharyngeal nerve
   anatomy of, A–9
   facial pain and, 312
   function of, A–8f
   sensory inputs from, 503
Glottal consonants, 778
Glottis, 776
Glucose
   GABA synthesis and, 136
   neuron metabolism of, 131
Glutamate
   accumulation of, 400
   cycling of, 131f
   functional features of, 124t
   neurotransmission by, 131–35
   photoreceptors releasing, 220
   release of, 196–97, 196f
   silent synapses releasing, 193b
   structure of, 123f
   synthesis of, 131, 131f
Glutamate-glutamine cycle, 131, 131f
Glutamate receptors, 127b
   ionotropic, 131–35, 132f, 133f, 134f
   metabotropic, 135, 135f
Glutamatic acid decarboxylase (GAD) gene, 136
Glutamatic acid decarboxylase (GAD2) gene, 170b
Glutaminase, 131, 131f
Glycine
   functional features of, 124t
   GABA and, 135–39, 136f
   structure of, 123f
   synthesis of, 136f, 139
Glycine receptors, 139, 140f
Glymphatic system, A–31, A–32f
GM-CSF (granulocyte-macrophage colony-stimulating factor), 683t
GnRH (gonadotropin-releasing hormone), 550, 571, 642
Goldfish, neurogenesis in, 685–86, 686f

Goldgaber, Dmitry, 768b
Goldman, Steven, A-30–A-31
Goldman equation, 49–50
Goldmann, Edwin, A–28
G_{olf} (olfactory-specific heterotrimeric g-protein), 337, 338f
Golgi, Camillo, 3, 8, 8f
Golgi cells, 459, 459f
Golgi outposts, 558f, 559
Golgi tendon organs, 288f, 289
   muscle spindles and, 393–94, 393f
Gonadal sex steroids, 634–36, 635f, 636f, 642, 643f, 644–46. See also specific hormones
Gonadotropin-releasing hormone (GnRH), 550, 571, 642
G-protein-coupled receptors (GPCRs), 111, 111f, 143, 324
   function of, 159
   odorant receptor proteins and, 333–34, 334f
   stimulation of, 145–46
   structure of, 158f
   taste transduction and, 361–63, 363f
G-proteins. See GTP-binding proteins
Gracile nucleus, 290f, 291, A–12f
Gracile tract, 290f, 291
Gradients, axons and, 564, 594–955
Grammar, 776
Granisetron (Kytril), 145
Granule cells, 342f, 343, 459f
Granulocyte colony-stimulating factor (G-CSF), 683t
Granulocyte-macrophage colony-stimulating factor (GM-CSF), 683t
Graybiel, Ann, 434, 436b
Gray communicating rami, 495
Gray matter
   critical periods and volume of, 609–10, 610f
   definition of, 19
   motor, A–11
   sensory, A–11
   spinal cord, 382, 382f, A–6f
Graziano, Michael, 411b, 412
Green fluorescent protein (GFP), 164b, 164f
Grid cells, 770b, 771f
Grimsley, John, 677, 677f
GRIN2B gene, 170b
GRK3 gene, 170b
Group Ia afferents, 288, 288f
Group Ib afferents, 288f, 289
Growth-associated protein-43 (GAP43) gene, 671
Growth cones
   axonal, 559–61, 560f
   motility of, 561–62, 563f, 564
   properties of, 559
Growth factors, 587
GTPase-activating proteins (GAPs), 160
GTP-binding proteins (G-proteins)
   activation of, 111, 111f
   binding of, 143

effector pathways associated with, 160f, 161
heterotrimeric, 159–60, 159f
monomeric, 159f, 160
signal transduction and, 159
Guanine nucleotide exchange factors (GEFs), 160
Guevedoces, 650t, 651
Gurdon, John, 518
Gustation, 356. *See also* Taste
Gustatory brainstem nuclei, 366
Gustatory cortex, 366, 367, 368f
Gut, innervation of, 495
GWAS (genome-wide association studies), 22
Gyri, A–4

H

Habits
dopamine and, 830b
making and breaking, 435b, 436f
neuron plasticity driving, 437b
Habituation, process of, 182, 183f
Hair cells. *See* Cochlear hair cells; Vestibular hair cells
Hairy mutation, 530f
Hairy skin, 279, 284f, 286
Haloperidol, 169b
Halorhodopsin, 12, 14f
Halothane, 43
Hamburger, Viktor, 581
Handedness, language and, 788b, 788f, 789f
Haptic touch, 299
Harlow, Harry, 603b
Harlow, J. M., 707–8
Harris, Bill, 186b
Harrison, Ross G., 559, 564
Head, 492t, 503, 527, 527f. *See also* Face
Head, Henry, 665–67, 666f, 669
Headaches, 82
Head movements, 265–69
Hearing, 236–58
Hearing, human, 238
Hearing aids, 239b, 239f
Hearing loss, 83, 238b, 239f, 240f, 248
Heart
influences on rates of, 510
innervation of, 495, 510
referred pain from, 309b
visceral motor system functions and, 492t
Hebb, Donald O., 187, 599–600
Hebbian competition, 608
Hebbian learning, 829b
Hebb's postulate, 599–600, 599f
Hedonic learning, 373–74, 374f
Hedonic value, 356, 372–73, 373f
Helicotrema, 243
Helmholtz, Hermann von, 741
Hemianopsias, 223f, 224b
Hemiballismus, 448
Hemispatial neglect syndrome, 746, 746f, 747f, 748–49

Hemispheres, anatomical differences, 787–88, 787f
Hemispheric asymmetry, emotion and, 814
Hemorrhagic stroke, A–27
Henbane, 127b
Henneman, Elwood, 384
Hepatocyte growth factor (HGF), 567
Hertz (Hz), 237
Heschl's gyri, 250
Hes genes, 541
Hess, Walter, 800
Heteronomous hemianopsia, 223f, 224b
Heterosexual individuals, 651–52, 652t
Heterotrimeric G-proteins, 159–60, 159f
Heuser, John, 102, 103
HGF (hepatocyte growth factor), 567
Hierarchical processing, in extrastriate visual areas, 230
Hikosaka, Okihide, 439
Hindbrain, 527, 530, 531f, 560f, A–3
Hippocampal atrophy, 813
Hippocampal cortex, 699b
Hippocampus
adult neurogenesis and, 691b
AMPA receptors in, 193f
anatomy of, A–14, A–15f
CA1 and CA3 pyramidal neurons in, 187, 187f
cognitive maps and, 770b
declarative memory storage and, 769, 769f
excitatory synapses in, 171f
in London taxi drivers, 766, 766f
long-term synaptic plasticity at, 186–89, 187f, 188f, 189f
LTD in, 186–87, 194–95, 195f
memory and, 762, 763f, 766, 766f
neuronal plasticity in, 646
NMDA receptors in, 193f
secondary somatosensory cortex projections to, 297
sex-based differences in, 652–53, 653f
trisynaptic circuit of, 187f
His, Wilhelm, 552
Histamine
brain levels of, 143, 143f
functional features of, 124t
release of, 315
structure of, 123f
synthesis of, 143–44, 144f
Histamine-containing neurons, 729
Histidine, 143, 144f
Histology, 8
Hitzig, Eduard, 409, 445, 820
HMC (hypaxial motor column), 593f
Hodgkin, Alan, 49, 51, 53, 56–62, 70–71
Holoprosencephaly, 537–38, 539f
Homeobox genes, 529
Homeostasis, 490, 506f
Homeotic genes, 529

HOMER protein, 625
Homologous recombination, 24b, 25
Homonymous hemianopsia, 223f, 224b
Homonymous quadrantanopsia, 224b
Homosexual individuals, 651–52, 652t
Homozygosity mapping, 22
Homunculus, 294, 295f
Horizontal cells, retinal, 209, 210f, 219–20, 219f
Horizontal connections, 699
Horizontal gaze center, 478–81
Horizontal gaze palsy with progressive scoliosis, 571, 572t
Horizontal sections, A–2
Horner's syndrome, 504–5, 504f, 505f
Horseradish peroxidase (HRP), 103, 103f
Hox genes, 529–31, 530f, 531f
HRP (horseradish peroxidase), 103, 103f
Hsc70, 109, 110f
HTR2A (5-HT$_{2A}$ receptor) gene, 169b
5HTT gene, 170b
Hubel, David, 613, 614
Human Connectome Project, 702f, 703b
Human sex differences, 649–50
Huntingtin mutation, 447–48, 660
Huntington, George, 447
Huntington's disease
basal ganglia and, 381, 434b, 771
eye movement impairment and, 478
hyperkinetic movements in, 447–49, 448f, 449f
repair for damage from, 660
Huxley, Andrew, 56–62, 70–71
Hydrocephalus, 537, 570, A–32
Hydrogen cyanide, 353
γ-Hydroxybutyrate, 136
Hypaxial motor column (HMC), 593f
Hyperacusis, 243
Hyperalgesia, 315, 316
Hyperglycinemia, 139
Hyperkinesis, 434b
Hyperkinetic movement disorders, 447–49, 448f, 449f
Hypermetric eye movements, 464
Hyperopia (farsightedness), 207, 207f
Hyperpolarization
description of, 38, 39f
of horizontal cells, 219f, 220
of photoreceptors, 209–10, 211f
Hypersomnia, 721
Hypertonia, 427b
Hypnotic agents, 42
Hypocretin, 729
Hypoglossal nerve, A–8t, A–9, A–12f
Hypoglossal nucleus, A–10f
Hypokinesis, 434b
Hypokinetic movement disorders, 443–47, 444f

Hypometric eye movements, 464
Hypothalamic nucleus, sexually dimorphic, 640, 641f
Hypothalamic sulcus, 497f
Hypothalamus, 222, 222f
anatomy of, 496b, 497f, 498f, 504, A–15, A–16
function of, 496b, 497f
Horner's syndrome and, 504–5, 504f, 505f
interstitial nuclei of, 651
periventricular nucleus of, 646
sex hormones activating, 651–52, 652f
sexual dimorphisms and, 638–40, 641f
visceral motor system and, 490
Hypotonia, 426, 427b
Hypoxia, 678–79, 678f
Hz (hertz), 237

I

Ibotenic acid, 127b
Ibuprofen, 316
Idiot savants, 760b
IEM (inherited erythromelalgia), 83
IGFs (insulin-like growth factors), 630
IL-1β (interleukin-1β), 315, 317
Imidazoleamine, 123f
Immediate early genes, 173, 175, 625
Immune system, after brain injury, 682, 683t, 684f
Implicit memory, 756
Imprinting, 603b, 606t, 648
Impulsivity, 823–25
INAH (interstitial nuclei of the hypothalamus), 651
Incus, 241f
Inderol (propranolol), 143, 816
Indoleamine, 123f
Induced pluripotential stem cells (IPSCs), 519b, 521–22, 521f
Infections, 316
Inferior, definition of, A–2f
Inferior cerebellar peduncle (restiform body), 452f, 453, A–10f, A–12f
Inferior colliculi, A–7, A–10f
auditory information and, 250
Inferior frontal gyrus, A–13, A–14f
Inferior oblique muscles, 473
Inferior olivary nucleus (inferior olive), 454f, 455, 461, A–7, A–12f
Inferior parietal lobules, A–14, A–14f
Inferior peduncles, A–7
Inferior rectus muscles, 473
Inferior salivatory nuclei, 500
Inferior temporal gyrus, A–13, A–14f
Inflamed appendix, referred pain from, 309b
Inflammation
after brain injury, 682, 683t, 684f
tissue damage and, 316f
Inflammatory bowel disease, 316

Information storage
  association and, 758, 758f
  neural mechanisms for, 187–88
  priming and, 756–57, 757f, 771
  retention in, 759f
Infundibular stalk, 496b, 497f
Infundibulum, A–15, A–15f
Inhalation anesthetics, 43
Inherited erythromelalgia (IEM), 83
Inhibitory neurotransmitters, 9
Inhibitory postsynaptic potentials
  (IPSPs), 116–17, 117f
Innate behaviors, 327
Inner ear, 241f, 243–45
Inositol triphosphate (IP₃)
  calcium release and, 176, 176f
  diffusion into dendritic shafts, 171b
  production of, 196–97, 196f
  second messenger function of,
    161, 162f
Inositol triphosphate (IP₃) receptors,
  161, 162f
Inside-out patch recording, 72b
In situ hybridization, 16, 17f
Insomnia, 720
Insula, 810, 811f, 834, 834f, A–13,
  A–14f
Insular taste cortex, 366
Insulin-like growth factors (IGFs),
  630
Integrins, 158, 567, 568f, 669
Intelligence. See Cognitive function
Intention tremors, 468, 468f
Interaural level differences, 255–57,
  255f, 257f
Interaural timing differences, 255–
  56, 255f, 256f
Interferons, 683t
Interhemispheric connections, 700
Interleukin-1β (IL-1β), 315, 317
Interleukins, after brain injury, 683t
Intermediate zone, A–6t
Intermediolateral cell column, 495
Internal arcuate fibers, 290f, 292
Internal capsule, A-20–A-21, A–20f
Internal carotid arteries, A–21, A–21f
Interneurons, 9
  excitatory synapses and, 182–83,
    184f
  GABAergic, 459
  information flow and, 460
  spinal cord, 10
  tetanus toxin targeting, 108
Internuclear ophthalmoplegia, 478
Interoception, 315
  insula and, 810, 811f
Interposed nucleus, 455, 456f
The Interpretation of Dreams (Freud),
  726b
Interpret cue, 750
Intersexuality, 650, 650t
Interstitial nuclei of the
  hypothalamus (INAH), 651
Interventricular foramina, A–30,
  A–31f
Intracellular receptors, 158f, 159

Intracellular recordings, 10–11
Intracellular signaling, dynamic
  imaging of, 163b, 164f
Intracellular signal transduction, 156
Intrafusal muscle fibers, 288, 288f,
  383, 389, 390f, 391
Intraparietal sulcus, A–14, A–14f
Intrinsically photosensitive ganglion
  cells, 210–11
Introns, 20
Ion channels. See also specific ions
  action potentials and, 50–51, 50f
  active transporters and, 44, 44f
  cyclic nucleotide-gated, 76, 77f, 84,
    85f, 337
  electrochemical equilibrium and,
    45, 45f, 46
  functions of, 90
  genes encoding varieties of, 76–78,
    77f
  ligand-gated, 76, 77f, 81, 84, 84f,
    111f, 128, 158
  mechanosensitive, 77–78, 77f,
    85–86, 86f
  membrane permeability and,
    44–46, 44f, 45f, 46f, 74
  microscopic/macroscopic currents
    flowing through, 71f, 72–73
  neurological diseases by altered,
    82–83
  patch clamp measurements of
    currents flowing through,
    70–74
  pores of, 78–80, 79f
  structure of, 79f
  thermosensitive, 76–77, 77f, 84–85,
    85f
  toxins that poison, 75b
  voltage-gated, 74, 74f, 80, 81f,
    100–101, 247, 263
Ion concentration gradients, across
  neural membranes, 47–49, 47t, 49f
Ion currents
  flow across membranes, 54–56, 56f
  membrane conductance and,
    59–62, 59f, 60f
  membrane potentials and
    amplitude of, 57, 57f
  voltage-dependent, 56–62
  voltage-sensitive, 56, 56f
Ion exchangers, 89–90, 89f, 90f
Ion fluxes, 46, 47f, 114–16, 115f
Ion gradients, active transporters
  and, 86–90
Ionotropic GABA receptors, 136–37,
  137f, 139
Ionotropic glutamate receptors,
  131–35, 132f, 133f, 134f
Ionotropic glycine receptors, 139,
  140f
Ionotropic purinergic receptors,
  145, 146f
Ionotropic receptors, 110, 111f,
  334–36
Ion selectivity, 74
  in channel pores, 78–80, 79f
Iowa Gambling Task, 810f
IP₃. See Inositol triphosphate

IP₃ (inositol triphosphate) receptors,
  161, 162f
IPSCs (induced pluripotential stem
  cells), 519b, 521–22, 521f
IPSPs (inhibitory postsynaptic
  potentials), 116–17, 117f
Irises, 206f
Irreversible brain death, 734
Ischemia, 676, 678f
Isotretinoin (Accutane), 537
Itch-inducing chemicals, 314
Ito, Masao, 195, 460

J

Jackson, John Hughlings, 409
James, William, 738, 803
Jamison, Kay, 812
Jellyfish (Aequorea victoria), 164b
Jet lag, 720
Jnk (jun kinase) activation, 535
Joint receptors, 289, 393
Joro spiders, 127b
Jugular veins, A–24
Jun kinase (Jnk) activation, 535

K

Kaas, Jon, 298
Kainate, 127b
Kainate receptors, 131–32, 132f, 133f
Kairomones, 327, 340f, 341
KAL1 mutations, 571, 572t
Kallmann syndrome, 570, 571, 572t
Kandel, Eric, 182
Kano, Masanobu, 195
Karnovsky, Morris, A–28
Katz, Bernard, 49, 51, 53, 98–102
KCC2 (potassium-chloride co-
  transporter), 317
K⁺ channels. See Potassium ion
  channels
KCNA1 genes, 82
KCN genes, 76
KCNQ2 genes, 82
KCNQ3 genes, 82
KCNQ4 genes, 83
K⁺ (potassium) co-transporters, 138b
Kennedy, Rosemary, 711b
Keratinocytes, 315
Ketamine, 42–43, 43f, 139
Keynes, Richard, 87
Kidneys
  development of, 588
  innervation of, 495
Kif4 transcription factor, 521
KIF21A mutations, 571, 572t
Kindling, 198–99
Kinesins, 558–59, 558f
K⁺ ions. See Potassium ions
Klinefelter's syndrome (XXY
  phenotype), 631, 650, 650t
Klüver, Heinrich, 804
Klüver-Bucy syndrome, 804
Knee-jerk reflex. See Myotatic reflex
Knock-in mutations, 22, 25
Knock-out mutations, 22, 25

Koniocellular pathway, 225f, 226
Krauzlis, Richard, 487
Krüppel mutation, 530f
Kuffler, Stephen, 18
Kuru, 463
Kuypers, Hans, 424
Kytril (granisetron), 145

L

L1, cell adhesion molecule, 669
L1 syndrome, 571–72, 572t
Labeled-line coding, 253, 366–67,
  366f
Labial consonants, 778
Labyrinth, vestibular, 261–62, 262f
Lacrimal gland, 492t
Lambert-Eaton myasthenic
  syndrome (LEMS), 107
Lamellipodium, 559–62, 560f
Laminae, cortical, 697, 699b
Lamina propria, 331
Laminins, 547, 549, 567, 568f, 672
Lamprey eels, locomotion in, 395b,
  396f
Lanceolate endings, longitudinal,
  283t, 284f, 286
Langley, John N., 110, 491, 592, 674b
Language
  of birds, 794–95, 794f
  brain areas involved in, 776, 776f,
    785–86, 785f, 786f
  comprehension, 781, 782b
  critical periods in development of,
    606–7, 606t, 608f
  genes and production of, 781
  handedness and, 788b, 788f, 789f
  hemispheric specialization in
    processing, 784–85
  lateralization of, 784–85
  learning, 790–91, 790f
  N400 response and
    comprehension of, 783, 784f
  nonhuman animals and, 792–93,
    793f
  phonemes and, 607
  reading, dyslexia and, 791–92, 792f
  semantics and, 782b, 782f
  sign, 778b, 778f
  speech and, 775–95
  split-brain syndrome and, 786–88,
    787f
Large intestine, 492t
Larynx, 776, 777f
Laser-assisted in situ keratomileusis
  (LASIK) corneal surgery, 207
Lashley, Karl, 767
LASIK (Laser-assisted in situ
  keratomileusis) corneal surgery,
  207
Lateral, definition of, A–2
Lateral cervical nucleus (LCN),
  290f, 292
Lateral columns, A–6
Lateral connections, 699
Lateral corticospinal tract, 406–7, A–6
Lateral (Sylvian) fissure, A–4, A–4f

Lateral geniculate nucleus (LGN)
  anatomy of, 222, 222f
  arborization of, 612f, 616
  critical periods and, 606t
  inputs from, 604, 605f
Lateral horns, 495, A–5, A–5f, A–6t
Lateral intraparietal area (LIP), 751–52
Laterality of activation, 653
Lateral motor column (LMC), 593f
Lateral olfactory tract, 346, 347f
Lateral preoptic nuclei, 496b, 497f, 498f
Lateral rectus muscles, 464, 464f, 473
Lateral sulcus, 250, 252f
Lateral superior olive (LSO), 250, 251f, 257, 257f
Lateral ventricles, 498f, A–30, A–30f, A–31f
  development of, 528f
Lateral vestibular nuclei, 273, 274f
Lateral vestibulospinal tracts, 273, 274f, 420–21, 420f
Latrodectism, 108–9
α-latrotoxin, 108–9
Law of effect, 831
LCN (lateral cervical nucleus), 290f, 292
Leaner mutations, 466b, 467t
Learning
  associative, in amygdala, 806, 806f
  conditioned, 757
  from consequences of behaviors, 832–33, 832f, 833f
  definition of, 754
  in fruit flies, 186b
  Hebbian, 829b
  hedonic, 373–74, 374f
  language and speech, 790–91, 790f
Leeches, locomotion in, 395b, 395f
Lefkowitz, Robert, 143
Left-handedness, 788b, 788f, 789f
LEMS (Lambert-Eaton myasthenic syndrome), 107
Lens, eye
  accommodation of, 476
  elasticity of, 207, 207f
  near reflex triad and, 476
Lenticulostriate arteries, A–24, A–25f
Leptin, 505b, 506f
Lesion studies, 16
Leucine enkephalin, 148f
Leucotomy, 710b
Leu-enkephalin, 149t
Leukemia inhibitory factor (LIF), 588
LeVay, Simon, 651
Levi-Montalcini, Rita, 581
LGN. See Lateral geniculate nucleus
LHFPL5 mutations, 248
LHRH (luteinizing hormone releasing hormone), 148f
Libet, Benjamin, 835b
Librium (chlordiazepoxide), 137, 170b
Lidocaine, 41, 42f
LIF (leukemia inhibitory factor), 588
Ligand-gated ion channels, 111f

activation of, 77f
chemical signals and, 76, 81, 84, 84f
nACh receptor as, 128
receptor and transducing functions of, 158
Light intensity, 220b, 220f
Lima, Almeida, 710b
Limb buds, 581–82, 581f
Limbic forebrain, A–16
Limbic loop, 433b, 433f
Limbic system, 804
Limb position, perception of, 288–89, 288f
Lincoln, Abraham, 812
Line cancellation test, 747f
Lingual gyrus, A–16, A–16f
LIP (lateral intraparietal area), 751–52
LIS1 (lissencephaly 1), 552
Lissencephaly, 552, 553f
Lissencephaly 1 (LIS1), 552
Liver, innervation of, 495
Llinás, Rodolfo, 100
LMC (lateral motor column), 593f
Lobotomies, 709, 710b, 710f
Local anesthesia, 41, 42f
Local circuit neurons, 9
Locomotion
  in leeches and lamprey eels, 395b, 395f, 396f
  neural structures in control of, 398–99, 398f
  phases of, 395, 397f, 398
  upper motor neurons initiating, 419–25
Locus coeruleus, 320, 321f
Loewi, Otto, 98
Lomo, Terje, 187
London taxi drivers, hippocampus in, 766, 766f
Long circumferential arteries, A–25, A–26f
Longitudinal, definition of, A–2
Longitudinal fissure, A–14, A–15f
Longitudinal lanceolate endings, 283t, 284f, 286
Long-term depression (LTD), 177
  Ca²⁺ release and, 176, 176f
  in cerebellum, 195–97, 196f
  in hippocampus, 186–87, 194–95, 195f
  synaptic plasticity and, 176
Long-term memory, 755–56, 767, 767f
Long-term potentiation (LTP)
  AMPA receptors in, 190–91, 191f
  in amygdala, 806
  associativity and, 189, 189f
  CaMKII activity in CA1 neuron during, 191, 191f
  expression of, 190–92, 191f, 192f, 193f
  at hippocampal synapse, 186–87, 188f, 189f
  induction of, 189–90, 190f
  NMDA receptors as coincidence detectors for, 189–92, 190f
  properties of, 187–89, 189f

signaling mechanisms underlying, 192, 192f
Lorenz, Konrad, 603b, 603f
Loudness, 237, 252
Love, science of, 632b, 632f
Lower extremities, 492t
Lower motor neurons
  lesions, signs and symptoms of, 426t
  motor control and, 379–401
  in movement control, 380, 380f
  somatotopic organization of, 382, 382f
  sources of, 379
Lower motor neuron syndrome, 399, 426t
LRP5/6, 535
LSD (lysergic acid diethylamide), 144, 145f
LSO (lateral superior olive), 250, 251f, 257, 257f
LTD. See Long-term depression
LTP. See Long-term potentiation
Lugaro cells, 459
Lumbar nerves, A–3f
Lumbar puncture, A–5
Lumbar spinal cord, 290f, 313f
Lumbar vertebrae, A–4, A–5f
Lumbosacral enlargement, A–3f, A–5
Luminance, 214f, 220b, 220f
Lungs, 492t, 495
Luria, Alexander, 761–62
Lurcher mutations, 466b, 467t
Luteinizing hormone releasing hormone (LHRH), 148f
Lysergic acid diethylamide (LSD), 144, 145f
Lysophosphatidylinositol, 151f

M

M₁ receptors, 507, 508t
M₂ receptors, 507, 508t
mAChRs (muscarinic ACh receptors), 126b, 130–31, 130f
MacKinnon, Rod, 78, 80
MacLean, Paul, 804
Macmillan, M., 708
Macrophages
  after brain injury, 683t
  inflammatory responses and, 315
  peripheral nerve regeneration and, 667, 667f
Macroscopic currents, 71f, 72–73
Macula, of otolith organs, 263, 264f, 265
Macula lutea, 207
Macular degeneration, age-related, 207
Macular sparing, 223f, 224b
Mad cow disease, 463
Magnetic resonance imaging (MRI), 29–30, 29f
Magnetic source imaging (MSI), 31
Magnetoencephalography (MEG), 30–31, 31f
Magoun, H., 422b

MAGs (myelin-associated glycoproteins), 681, 681f
Major depressive disorders, 169b, 654, 655t, 812, 815
Major histocompatibility complex (MHC) genes, 341
Male sexual function, 512–13, 512f
Malignant melanomas, 335b
Malleus, 241f
Mammals
  audible spectrum of, 238
  olfactory bulb organization in, 341–43, 342f
  olfactory perception in, 349, 349f
Mammillary bodies, 497f, 498f, A–15, A–15f
Manduca sexta (tobacco hawk moth), 636, 637f, 638
MAO (monoamine oxidase), 142, 169b
MAOA gene, 170b
MAP1A, 461f
MAPKs (mitogen-activated protein kinases), 166f, 167, 589
Marijuana
  analgesic effects of, 321–22
  brain and, 150b, 150f
Marmosets, 254
MASA (mental retardation, aphasia, shuffling gait, and adducted thumbs), 570, 572t
Mass action principle, 767
Mast cells, 315
Maternal-offspring bonding, 632b
Maternal olfactory cues, 350–51, 352f
Mauthner cells, 274b, 275f, 560f
Max Planck Institutes, 71
McCarthy, M. M., 649
McGurk effect, 783, 784f
MCR-4 (melanocortin receptor subtype 4), 506b
Meaney, Michael, 647
Mechanoelectrical transduction (MET), 245–48, 246f, 263
Mechanoreceptors. See Somatosensory afferents
Mechanosensitive ion channels, 77–78, 77f, 85–86, 86f
Medial, definition of, A–2
Medial geniculate complex (MGC), 250, 251f, 252, 804, 806, 806f
Medial lemniscus, 290f, 292–93, 293f, A–12f
Medial longitudinal fasciculus, 269, 270f, 477, 480, 480f
Medial motor column (MMC), 593f
Medial nucleus of the trapezoid body (MNTB), 250, 251f
Medial prefrontal cortex (MPFC), 735, 736f
Medial premotor cortex, 418–19, 453f
Medial preoptic nuclei, 496b, 497f, 498f
Medial rectus muscles, 473
Medial superior olive (MSO), 250, 251f, 256, 256f
Medial temporal lobe, declarative memory and, 762–66, 770

Medial vestibular nuclei, 269, 270f, 273, 274f

Medial vestibulospinal tracts, 273, 274f, 420, 420f

Medium spiny neurons, 431, 432f, 435, 437, 440–41

Medulla
  anatomy of, A–7, A–8f
  blood supply in, A–22f, A–26
  corticospinal and corticobulbar tracts in, 407f
  cranial nerve nuclei in, A–10t
  in discriminative pain pathways, 313f
  location of, A–3f

Medulla oblongata, 455

Medullary arteries, A–22f, A–26

Medullary pyramids, 406, A–7, A–12f

Medullary reticular formation, 320, 321f

Medulloblastoma, 537–40, 539f

MEG (magnetoencephalography), 30–31, 31f

Meissner corpuscles, 283t, 284f, 285, 286f

Meissner's plexus, 501, 501f

Melanin, 206

Melanocortin receptor subtype 4 (MCR-4), 506b

Melanopsin, 210, 222, 715

Melatonin, 716, 716f

Melperone, 169b

Melzack, Ronald, 319b, 320

Membrane conductance, 59–62, 59f, 60f

Membrane permeability, 111–16
  action potentials and, 50–51, 50f
  Goldman equation for, 49–50
  ion channels and, 44–46, 44f, 45f, 46f, 74
  to K+ ions, 54
  to Na+ ions, 54–55
  resting membrane potentials and, 49–50, 50f
  voltage-dependent, 54–68

Membrane potentials. See also Resting membrane potentials
  creation of, 45–46
  depolarizing, 39, 39f
  hyperpolarizing, 38, 39f
  ion current amplitude and, 57, 57f
  Nernst equation and, 45–46

Membranes
  ion concentration gradients across, 47–49, 47t, 49f
  ion current flow across, 54–56, 56f

Membranous labyrinth, 263

Memory, 754–73
  aging and, 772–73, 772f, 773f
  capacity of, 758–59
  cerebral cortex storage of, 766–70
  clinical cases illustrating neural basis of, 764–66, 764f, 766f
  cognitive maps and, 770b
  consolidation and, 723, 756

declarative, 755–56, 756f, 762–66, 769–70, 769f
  ECT and, 767
  emotions and, 815–16, 817f
  episodic, 782b
  facial identification and, 705–6
  fallibility of, 757t
  forgetting and, 760–62, 761f
  formation of, 188–89
  in fruit flies, 186b
  functions of, 754, 755f
  hippocampus and, 762, 763f, 766, 766f
  implicit, 756
  long-term, 755–56, 767, 767f
  motivated, 759–60, 759f
  nondeclarative, 755–58, 771–72, 772f
  parahippocampal gyrus and, 762
  priming and, 756–57, 757f, 771
  procedural, 756
  remembering in, 758–60
  semantic, 782b
  sex-based differences in, 653
  short-term, 755
  sleep and, 723
  spatial, 723, 766, 766f
  TBIs and, 677
  visual cortex and, 769–70, 770f
  working, 755, 756f

Mendel, Gregor, 21

Mendelian analysis, 21–22

Meniere, Prosper, 271

Meniere's disease, 271

Meninges, A–29, A–29f

Mental retardation, aphasia, shuffling gait, and adducted thumbs (MASA), 570, 572t

Meperidine, 147

MEPPs (miniature end plate potentials), 99, 99f, 100f

Merkel cell-neurite complexes (touch domes), 283t, 284f, 286

Merkel cells, 283t, 284–87, 284f, 286f, 297b

Merzenich, Michael, 298

Mesencephalic locomotor region, 398, 403

Mesencephalic trigeminal nucleus, 293, 455, A–10f

Mesencephalon, 527, 528f, 529, A-2–A-3

Mesenchymal-to-epithelial transition (MET), 547, 548f

Mesoderm, 523, 524f, 533

Mesodermal cells, 533, 534f

Messenger RNAs (mRNAs), 558

MET (mechanoelectrical transduction), 245–48, 246f, 263

MET (mesenchymal-to-epithelial transition), 547, 548f

Meta-analysis techniques, 701b, 701f, 702f

Metabotropic GABA receptors, 139, 139f

Metabotropic glutamate receptors, 135, 135f

Metabotropic purinergic receptors, 145–46, 146f

Metabotropic receptors, 110–11, 111f

Metencephalon, 528f, 529, A–3

Met-enkephalin, 149t

Methadone, 147

Methionine enkephalin, 123f

1-methyl-4-phenyl-1,2,3,6-tetrahydropridine (MPTP), 444–45, 446f

Methylphenidate (Ritalin), 721

Meyer's loop, 224b, 224f

MGC (medial geniculate complex), 250, 251f, 252, 804, 806, 806f

MHC (major histocompatibility complex) genes, 341

Mice, 467t, 605
  DSCAM1 gene in, 575
  hedonic learning in, 373–74, 374f
  motor mutations in, 467t
  odorant receptor genes of, 334
  olfactory bulbs of, 342f, 343, 345f
  transgenic, 22, 23f, 24b, 25
  whisker barrels of, 601, 602f, 623, 624f

Microelectrodes, 37

Microglial cells
  after brain injury, 683t
  central sensitization process and, 317
  neuronal growth and, 679–80, 679f
  overview of, 6–8, 7f

Microscopic currents, 71f, 72–73

Microtubule cytoskeleton, 561, 563f

Microtubules, 561–62, 563f

Microtus montanus (montane voles), 632b

Microtus ochrogaster (prairie voles), 632b

Micturition, 511–12, 511f

Midbrain, A–3
  anatomy of, A–19
  corticospinal and corticobulbar tracts in, 407f
  cranial nerve nuclei in, A–10t
  in discriminative pain pathways, 313f
  location of, A–3f
  locomotor region of, 425
  olivary pretectal nucleus of, 222, 222f
  pain modulation and, 320, 321f
  in touch pathways, 290f

Middle cerebellar peduncle (brachium pontis), 452f, 453, A–10f, A–12f

Middle cerebral arteries, A–23, A–23f

Middle cerebral artery stroke, 408

Middle ear, 239b, 241–43, 241f

Middle frontal gyrus, A–13, A–14f

Middle medulla, 313f

Middle peduncles, A–7

Middle temporal area, 230–31, 231f

Middle temporal gyrus, A–13, A–14f

Midline myelotomy, 310f, 311b

Mid-pons
  in discriminative pain pathways, 313f
  in touch pathways, 290f

Midsagittal, definition of, A–2

Migraine headaches, 82

Migraines, 316, 655t

Miledi, Ricardo, 100

Milner, Brenda, 709b, 765

Milner, Peter, 828b

Mimicry, development and, 603

Miniature end plate potentials (MEPPs), 99, 99f, 100f

Miosis, 232, 504–5, 504f

Mirror box therapy, 318f, 319b

Mirror motor neurons, 417, 418f

MIS (Müllerian-inhibiting substance), 634, 650

Mitogen-activated protein kinases (MAPKs), 166f, 167, 589

Mitral cells, 342–43, 342f

Mixture suppression, 373

MMC (medial motor column), 593f

Mnemonists, 758

MNTB (medial nucleus of the trapezoid body), 250, 251f

Modafinil, 721

Modulatory neurotransmitters, 9

Molaison, Henry, 764–65, 764f

Molecular signaling
  in neurons, 155–77
  strategies of, 155–57

Moles, 296b, 296f

Molly Sweeney (Friel), 619

Monaural hearing deficits, 239b

Moniz, Egas, 710b

Monkeys, 24b
  attention in, 704, 704f, 705f
  auditory cortex divisions in, 250
  cerebral cortex plasticity in, 298–99, 299f
  language and, 793, 793f
  motor maps and, 411b
  with OFC lesions, 831, 831f
  otolith organ responses in, 266, 266f
  premotor cortex divisions in, 414, 414f
  purposeful movements of contralateral arm and hand in, 413f
  sound frequency perception in, 253f, 254
  synapse addition and elimination in, 609f
  vestibular nerve axonal responses in, 268, 268f
  VOR in, 465f

Monoamine oxidase (MAO), 142, 169b

Monocular deprivation, 612f, 614–16, 615f

Monocytes after brain injury, 682

Monogamy, pair bonding and, 632b

Monomeric G-proteins (small G-proteins), 159f, 160

Montane voles (*Microtus montanus*), 632*b*

Mood disorders, molecular basis of, 169*b*

Morphine, 42, 43*f*, 147

Morris water maze, 762, 763*f*

Moruzzi, G., 422*b*

Moser, Edvard, 770*b*

Moser, May-Britt, 770*b*

Mossy fibers, 458–60, 459*f*

Mother-child bonding, 633*b*

Motherly behaviors, 647–48, 647*f*

Motivated memory, 759–60, 759*f*

Motor activity, emotions and, 801

Motor aphasia, 780

Motor control
    CNS subsystems and, 379–81
    descending, 404–5, 404*f*
    lower motor neurons and, 379–401

Motor cortex
    anatomy of, A–13, A–14*f*
    circuits, 661
    divisions of, 414, 414*f*
    in locomotion control, 398*f*
    microstimulation experiments in, 411*b*
    pathways to spinal cord, 425*f*
    plasticity of, 661

Motor loops, 433*b*, 433*f*

Motor maps, representation by, 411*b*

Motor neuron pools, 381–82, 381*f*, 382*f*

Motor neurons
    botulinum toxins targeting, 108
    efferent, 10
    gamma, 289
    innervation of, 582
    mirror, 417, 418*f*
    muscle relationships with, 381–83, 381*f*, 382*f*

Motor nuclei, A–11

α-motor neurons
    function of, 383
    muscle spindle regulation by, 391, 392*f*
    muscle tone and, 427*b*
    size and, 385*b*, 385*t*

γ-motor neurons, 289
    activity levels of, 391
    function of, 383
    muscle spindle regulation by, 391, 392*f*
    stretch reflex and, 391

Motor proteins, 558–59, 558*f*

Motor systems
    organization of, 15
    somatosensory system interacting with, 299

Motor thalamus, A–18*b*

Motor units
    anatomy of, 383, 384*f*
    classes of, 383–84, 385*f*
    firing rates of, 388, 388*f*
    plasticity of, 385*b*, 386*f*, 487*f*

Mountcastle, Vernon, 18, 295

Mouthfeel, 371

Movement disorders
    deep brain stimulation for, 445–46, 446*f*
    hyperkinetic, 447–49, 448*f*, 449*f*
    hypokinetic, 443–47, 444*f*

Movements, 377
    basal ganglia in modulation of, 430–49
    cerebellum in modulation of, 451–70
    choreiform, 447
    coordination of, 382–83, 398, 462–64
    dopamine facilitating expression of, 442–43
    fine control of, 427
    intentional, 417
    motor loops and, 433*b*, 433*f*
    neural centers and, 380–81, 380*f*
    purposeful, 412, 413*f*
    rhythmic, 383

MPFC (medial prefrontal cortex), 735, 736*f*

MPTP (1-methyl-4-phenyl-1,2,3,6-tetrahydropridine), 444–45, 446*f*

MRI (magnetic resonance imaging), 29–30, 29*f*

mRNAs (messenger RNAs), 558

MS (multiple sclerosis), 67–68, 655*t*

MSI (magnetic source imaging), 31

MSO (medial superior olive), 250, 251*f*, 256, 256*f*

Mucus, production of, 329

Mueller, Johannes, 776

Müllerian-inhibiting substance (MIS), 634, 650

Multiple sclerosis (MS), 67–68, 655*t*

Muscarinic ACh receptors (mAChRs), 126*b*, 130–31, 130*f*

Muscarinic cholinergic receptors, 507, 508*t*

Muscimol, 127*b*

Muscle fibers, 113*f*, 386*b*, 386*f*

Muscle fields, of upper motor neurons, 410

Muscles
    motor neuron relationships with, 381–83, 381*f*, 382*f*
    in movement coordination, 383
    peripheral nerve reinnervation of, 673*f*
    regulation of force, 384, 387–88, 387*f*, 388*f*, 391–93, 392*f*
    relaxation of, 389
    stimulation rates and tension of, 388, 388*f*
    stretch reflex and, 389, 390*f*, 391
    twitches of, 388, 388*f*

Muscle spindles
    anatomy of, 383
    density of, 289
    Golgi tendon organs and, 393–94, 393*f*
    proprioception and, 288–89, 288*f*
    stretch reflex and, 389, 390*f*, 391
    γ-motor neuron regulation of, 391, 392*f*

Muscle tone, 389, 427*b*

Musculoskeletal system, proprioceptors in, 288*f*

Music, perception of, 237, 238*f*, 252, 254

MuSK, activation of, 672

Mutations. *See also specific mutations*
    conditional, 25
    in genes, 20–22, 21*f*
    knock-in, knock-out, 22, 25

Myasthenia gravis, 129, 129*f*, 655*t*

Myasthenic syndromes, 107–8

Mydriasis, 232

Myelencephalon, 528*f*, 529, A–3

Myelin, 6
    action potential conduction speed and, 64, 66, 66*f*
    MS and loss of, 67–68

Myelin-associated glycoproteins (MAGs), 681, 681*f*

Myelination
    axon regrowth and, 669, 670*f*
    conduction velocities and, 64–67, 64*t*, 65*f*, 66*f*
    somatosensory afferents and, 281

Myenteric plexus, 501, 501*f*

Myofibrils, training regimens and, 384

Myopia (nearsightedness), 206–7, 207*f*

Myosin, 559

Myosin ATPase, 386*f*

Myotatic (knee-jerk) reflex
    mediation of, 292
    neural circuits controlling, 9–11, 10*f*, 11*f*

N

N400 response, 783, 784*f*

Na$^+$/Ca^{2+} (sodium/calcium) exchangers, 89, 89*f*, 90*f*, 161

Na$^+$ channels. *See* Sodium ion channels

nAChR. *See* Nicotinic ACh receptor

Na$^+$ (sodium) co-transporters, 138*b*

Na$^+$ ions. *See* Sodium ions

Na$^+$ (sodium) pumps, 87–89, 87*f*, 88*f*

Narcolepsy, 721

Nasal cavity, 327–29, 328*f*

Nasal mucosa, 329, 331

Nav1.7 gene, 307

Nav1.8 gene, 307

Nav1.9 gene, 307

NCAMs (neural cell adhesion molecules), 158, 549, 669, 690

NCS1 (neuronal calcium sensor 1), 461*f*

NE. *See* Norepinephrine

Near-infrared spectroscopy (NIRS), 416*b*

Near reflex triad, 476

Near response, 233

Nearsightedness (myopia), 206–7, 207*f*

Neck
    sensory inputs from, 503
    spindle density and, 289

visceral motor system functions and, 492*t*

Nedergaard, Maiken, A–30–A–31

Neher, Erwin, 71, 112

Nematode worm (*Caenorhabditis elegans*), 24*b*, 164*b*, 334, 334*f*, 556, 557*f*, 564

NEM-sensitive fusion protein (NSF), 104

Neocortex, A–19. *See also* Primary visual cortex
    canonical circuitry of, 700*f*
    cortical laminae in, 699*b*
    structure of, 697, 698*f*

Neoendorphins, 149*t*

Neophobia, 373

Neostigmine, 129

Nernst equation, 45–46, 49

Nernst equilibrium potential, 248

Nerve cells. *See* Neurons

Nerve growth factor (NGF)
    activation of, 175*f*
    blockade of, 316
    catecholamine synthesis and, 177
    NMJ expression of, 671
    release of, 315
    secretion of, 669
    signaling by, 175–76
    trophic interactions and, 585, 587, 587*f*

Nervous mutations, 467*t*

Nervous system. *See also* Central nervous system; Peripheral nervous system
    cellular components of, 2–4, 2*f*, 3*f*
    emotional expression and, 801*f*
    formation of, 523–26, 524*f*, 525*f*
    genetic engineering modeling of, 22, 23*f*, 25
    genome and, 20, 21*f*
    organization of, 18–20, 19*f*
    stem cells in development of, 517–23, 518*f*
    studying, 1–32

NET (norepinephrine transporter), 142

Netrins, 564–66, 565*f*

Neu-N marker, 678*f*

Neural cell adhesion molecules (NCAMs), 158, 549, 669, 690

Neural circuits
    brain imaging for analysis of, 26–32
    construction of, 555–96
    influences on, 600
    myotatic reflex controlled by, 9–11, 10*f*, 11*f*
    organization and function of, 9–11, 10*f*, 11*f*
    overview of, 9
    visualizing and analyzing, 11–13, 13*f*, 14*f*

Neural crest, 523–24, 525*f*, 526

Neural crest cells, 525–26, 525*f*, 548–49, 549*f*

Neural induction, 529–33, 532*f*

Neural plate, 523, 524*f*

Neural precursor cells, 541, 542*f*

Neural progenitor cells, 519*b*
Neural regeneration, 674*b*
Neural stem cells
    cell classes derived from, 522–23, 522*f*
    neurogenesis in mature CNS and, 685
    promise and peril of, 519*b*
    in SVZ and SGZ, 687, 688*f*, 689
Neural systems
    brain imaging for analysis of, 26–32
    computational maps and, 16
    critical periods in, 604–6, 606*t*
    functional analysis of, 16–18, 18*f*
    genetic analysis of, 21–22
    overview of, 15–16
    structural analysis of, 16, 17*f*
    topographic maps and, 15
Neural tubes, 523–25, 524*f*, 525*f*, A–11
Neuregulin, 552, 553*f*
Neuregulin1 (Nrg1), 577–78
Neuregulin (NRG1) gene, 169*b*
Neurexins
    cognitive disorders and, 109
    critical period regulation and, 625
    function of, 579
    interactions of, 578*f*
Neurites, 556, 587*f*
Neuroanatomy
    human, A–1–A–32
    terminology, A–1–A–2
Neuroblasts, 536, 541, 542*f*, 556
Neurochristopathies, 526
Neurodegenerative diseases, 660–61, 660*f*
Neurodevelopmental disorders, 537–39, 538*f*, 539*f*
Neuroectoderm, 523–24, 524*f*
Neuroectodermal precursor cells, 523–25
Neuroengineering, 415, 415*b*, 415*f*
Neurofibrillary tangles, 768*b*, 768*f*
Neurogenesis
    in adult mammalian brain, 687–88, 688*f*
    in adult non-mammalian vertebrates, 685–87, 686*f*
    brain repair and, 690–91
    in mature CNS, 684–85
    molecular mechanisms of adult, 688–90, 690*f*
    neuroblasts and, 556
    nuclear weapons and, 691*b*
    process of, 540
    sexually dimorphic development and, 643–44
    timing of, 544–47, 546*f*
Neurogenic stem cells, 689
Neuroglia. *See* Glial cells
Neurogranin, 461*f*
Neurokinin A, 147
Neuroligins, 578*f*, 579, 625
Neuromeres, 529
Neuromuscular junctions (NMJs)
    cellular components at, 673*f*
    critical periods and, 606*t*

innervation of, 582–83, 584*f*
    peripheral axons and, 671, 672*f*
    repair of, 675
Neuromuscular synapses, plasticity, 181*f*
Neuronal calcium sensor 1 (NCS1), 461*f*
Neuronal migration
    in CNS, 550, 550*f*, 551*f*, 552
    disorders of, 552, 553*f*
    molecular mechanisms of, 552, 553*f*
    in PNS, 547–50, 549*f*
Neuronal plasticity, in hippocampus, 646
Neuronal repair, types of, 662–65, 664*f*
Neuron doctrine, 3–4
Neuron-restrictive silencing factor (NRSF), 522
Neurons. *See also* Electrical signals; *specific types*
    activity-dependent changes to, 600
    afferent, 9–10
    anatomy of, 4, 6
    attention-sensitive, 704
    of basal ganglia, 432*f*
    connections of, 582–85
    convergence of, 6
    cortical, 546*f*
    dendrite function and, 586*b*, 586*f*
    developmental disruptions in, 536–40
    differentiation of, 2, 2*f*, 540–44, 543*f*, 545*f*
    discovery of, 2–4, 3*f*
    divergence of, 6
    dorsal horn, 316–17, 502
    efferent, 9–10
    electrical signals of, 37–53
    electron microscopy visualizing, 4, 5*f*
    GABAergic, 137*f*
    genesis of new, 664*f*, 665
    glucose metabolized by, 131
    Golgi-stained, 8*f*
    histamine-containing, 729
    identity of, 535–36, 536*f*
    immature, 556
    local circuit, 9
    migration of, 690, 690*f*
    molecular signaling in, 155–77
    neural stem cells giving rise to, 522, 522*f*
    noradrenergic, 729
    polarization of, 556–58, 557*f*
    postsynaptic, 4, 94, 94*f*
    postsynaptic dorsal column, 292
    preferred orientation of, 228, 229*f*
    preganglionic, 232, 495
    presynaptic, 4, 94, 94*f*
    projection, 9
    protein and organelle trafficking in, 558–59, 558*f*
    pseudounipolar, 279*f*, 281
    receptive fields of, 16–17, 17*f*
    regrowth of, 663–65, 664*f*

second messengers used by, 161, 162*f*, 163
    serotonergic, 575, 729
    signals of, 35
    spiny stellate, 227, 228*f*
    squid and giant, 47*t*, 48*b*, 49, 49*f*, 50*b*, 51*f*
    steroid hormones' effects on, 635*f*
    survival of, 581*f*, 587*f*
    synaptic connections between, 179
    visual fields of, 216–17
    visualizing in CNS and PNS, 8–9, 8*f*
    vomeronasal receptor, 331–33, 340–41
    wide-dynamic-range, 308
Neuropathic pain, 317, 319
Neuropeptide K, 147
Neuropeptide receptors, 148–49
Neuropeptides
    amino acid sequences of, 148*f*
    description of, 124
    functional features of, 124*t*
    receptors of, 148–49
    structure of, 123*f*
    synthesis and packaging of, 146–47, 147*f*
Neuropeptide γ, 147, 148*f*
Neuropils, 9
Neuroplastin, 461*f*
Neuropsychological testing, 708*b*, 709*f*
Neurorehabilitation, BMI-assisted, 416*b*, 416*f*
Neuroscience
    advances in, 31–32
    cognitive, 32
    on free will, 835*b*, 835*f*, 836*f*
    human sex differences and, 649–50
    model organisms in, 24*b*
Neurotensin, 148*f*
Neurotoxins, 126*b*, 126*f*. *See also specific toxins*
Neurotransmitter receptors, 4
    classes of, 93
    neurotoxins acting on, 126*b*, 126*f*
    neurotransmitters and, 122–53
    types of, 110–11, 111*f*
Neurotransmitters. *See also specific types*
    Ca²⁺ ions and secretion of, 99–101, 100*f*, 101*f*
    catecholamine, 140, 141*f*, 142*f*
    categories of, 123*f*, 124
    in critical period mediation, 621, 622*f*, 623
    endocytosis after release of, 110*f*
    excitatory, 9
    exocytosis and, 98, 102*f*
    functional features of, 124*t*
    inhibitory, 9
    modulatory, 9
    presynaptic terminal release of, 101, 101*f*
    quantal release of, 98–99, 99*f*, 100*f*
    receptors and, 122–53

secretion of, 93
    synaptic vesicles and, 4, 94–95, 94*f*, 98, 101–2, 102*f*
    use cycle of, 93
    visceral motor system releasing, 491, 507–8, 508*t*
    voltage-gated ion channel release of, 100–101, 100*f*
Neurotrophic factors, 580–81
Neurotrophins
    critical period regulation and, 625
    dendritic growth promotion by, 573
    description of, 580–81
    disruption of, 591
    NMJ expression of, 671
    NT-3, 587, 588*f*, 671
    NT-4, 671
    NT-4/5, 587, 588*f*
    signaling of, 589–91, 590*f*, 591*f*
    targets of, 587–88, 588*f*
Neurulation, 523, 524*f*
Neutrophils
    after brain injury, 682, 683*t*
    inflammatory responses and, 315
Newton, Isaac, 569*b*
NF-κB (nuclear factor κB), 591
NGF. *See* Nerve growth factor
Nicotinia tabacum (tobacco plant), 126*b*
Nicotinic ACh receptor (nAChR)
    antibodies to, 129
    function of, 125
    structure of, 128*f*
    subunits of, 127–28, 128*f*
Nicotinic cholinergic receptors, 507, 508*t*
Night blindness, 83
Nirmatrelvir, 361
NIRS (near-infrared spectroscopy), 416*b*
Nissl, Franz, 699*b*
Nissl staining, 8*f*, 9, 699*b*
Nitric oxide (NO)
    functional features of, 124*t*
    synthesis, release, and termination of, 152, 153*f*
NMDA receptors
    as coincidence detectors for LTP, 189–92, 190*f*
    critical period regulation and, 621, 623
    function of, 134*f*
    gating of, 135
    in hippocampus, 193*f*
    nomenclature of, 131
    postsynaptic responses and, 132, 132*f*
    structure of, 132–33, 134*f*, 135
    subunits of, 133, 135
NMJs. *See* Neuromuscular junctions
NO. *See* Nitric oxide
Nociception, transmission of, 320, 321*f*
Nociceptive signals, 305–7
Nociceptors, 281, 303–5, 304*f*
Nodes of Ranvier, 65*f*, 66–67

Nodulus, 452f, 453
Noggin, 533, 534f, 535, 536f
Noncanonical Wnt family signaling, 534f, 535
Nondeclarative memory, 755–58, 771–72, 772f
Non-discriminative touch, 314
Non-REM sleep, 724, 726b
Nonsteroidal anti-inflammatory drugs (NSAIDs), 316
Nonsyndromic sensorineural deafness type 2, 83
Noradrenergic neurons, 729
Noradrenergic receptors, 507
Norepinephrine (NE)
    brain levels of, 142
    functional features of, 124t
    neurotransmission of, 143, 507
    release of, 510
    signaling by, 156
    structure of, 123f, 141f
    synthesis of, 142
Norepinephrine transporter (NET), 142
Normal vision (emmetropia), 207f
Nose safety, 329–30, 329f, 330f
Notch cell surface receptors, 541, 543f
Notochord, 523, 524f, 532, 534f
Nrg1 (neuregulin1), 577–78
NRG1 (neuregulin) gene, 169b
NRSF (neuron-restrictive silencing factor), 522
NSAIDs (nonsteroidal anti-inflammatory drugs), 316
NSF (NEM-sensitive fusion protein), 104
NST (nucleus of the solitary tract), 365–66, 502–4, 502f, 503f, A–10f, A–12f
NSTX-3, 127b
NT-3 (neurotrophin-3), 587, 588f, 671
NT-4 (neurotrophin-4), 671
NT-4/5 (neurotrophins 4/5), 587, 588f
N transporter 1 (SN1), 131
Nuclear bag fibers, 389, 390f
Nuclear chain fibers, 389, 390f
Nuclear factor κB (NF-κB), 591
Nuclear receptors, 174
Nuclear Test Ban Treaty of 1963, 691b
Nuclear weapons, neurogenesis and, 691b
Nuclei, nervous system, 19
Nuclei of the lateral lemniscus, 251f
Nucleotide release, 315
Nucleus accumbens
    neurons activated in, 826
    oxytocin receptors in, 632b, 632f
Nucleus ambiguus, 499f, 500, A–10f, A–12f
Nucleus gracilis, 291
Nucleus of the solitary tract (NST), 365–66, 502–4, 502f, 503f, A–10f, A–12f
Nystagmus, 468, 477

O

Obese (ob/ob) mutations, 506b
Obesity, brain and, 505b, 506f
Object processing, 705–7, 706f
ob/ob (obese) mutations, 506b
Obsessive-compulsive disorder (OCD), 654, 655t, 824, 833
Occipital lobe, 222, 653f, A–4, A–4f
Occipitotemporal gyrus, A–15, A–15f
OCD (obsessive-compulsive disorder), 654, 655t, 824, 833
Oct4 transcription factor, 521
Octapeptide, 148f
Ocular asynchrony, 616–17, 617f
Ocular dominance
    critical periods and, 606t
    visual deprivation and, 611–16, 612f
Ocular dominance columns, 225f, 227, 604, 605f, 606t
Oculomotor nerve, 270f, 474
    anatomy of, 475, 499f, 500, A–9
    function of, A–8t
    injuries to, 477, 477f
Oculomotor nucleus, 269, 270f, 480–81, A–10f, A–12f
Odorant receptor proteins and genes, 333–37, 334f, 336f
Odorants. See also Olfactory receptor neurons
    description of, 325
    detection of, 332–33, 333f
    human perception of, 349–50, 349f, 350f
    human sensitivity to, 350–52, 351f
    physiological and behavioral responses to, 348–49
    receptor genes and proteins, 333–37, 334f, 336f
    sparse coding of, 346
    taste integration with, 369, 370f
    transduction mechanisms of, 332–33, 337–39, 338f, 339f
OFC. See Orbitofrontal cortex
OFF bipolar cells, 217–18, 218f
OFF-center ganglion cells, 217–18, 217f, 218f, 219f
Ohm's Law, 59
O'Keefe, John, 770b
Olanzapine, 169b
Olds, James, 828b
Olfaction, 325–54
Olfactory bulbs
    accessory, 326–27, 326f, 343–44, 344f
    adult neurogenesis and, 691b
    anatomy of, 325f, 326, 805b, A–14, A–15f
    circuits in, 345–46, 345f
    development of, 528f
    organization of, 341–43, 342f
    pyriform cortex projections from, 343, 346, 347f
Olfactory cilia, 328–29
Olfactory cortex, 805b
Olfactory deficits, 361
Olfactory ensheathing cells, 328f, 332
Olfactory epithelium
    olfactory receptor neurons and, 327–29, 331
    organization of, 325, 325f
    regeneration of, 331–32, 332f
    structure and function of, 328f
    zinc toxicity in, 330
Olfactory function, assessment of, 352–54, 353f
Olfactory loss, 352–54, 353f
Olfactory marker proteins (OMPs), 328f
Olfactory nerve, 325, 326f, A–7, A–8t
Olfactory odorant transduction, 332–33, 337–39, 338f, 339f
Olfactory perception, in humans, 349–50, 349f, 350f
Olfactory receptor neurons (ORNs), 325
    axons of, 341–43, 342f
    concentration-invariant coding and, 345
    degeneration and regeneration of, 331–32, 332f
    genesis of, 665
    odorant receptor proteins and genes in, 333–37, 334f, 336f
    odor detection and, 332–33, 333f
    olfactory epithelium and, 327–29, 331
    olfactory odorant transduction mechanisms and, 332–33, 337–39, 338f, 339f
    selectivity of, 339, 339f
    structure and function of, 328f
Olfactory-specific heterotrimeric G-protein (Golf), 337, 338f
Olfactory system
    function of, 324
    organization of, 325–26, 325f
    transduction mechanisms in, 327–29, 331–39
Olfactory tract, 325f, 326, A–14, A–15f
Oligodendrocytes
    neural stem cells giving rise to, 522, 522f
    neuronal growth and, 679–80, 679f
    overview of, 6, 7f
Oligodendroglia, 4, 5f
Olivary pretectal nucleus, of midbrain, 222, 222f
OMPs (olfactory marker proteins), 328f
ON bipolar cells, 217–18, 218f
ON-center ganglion cells, 217–18, 217f, 218f, 219f
Onchocerca volvulus, 619
Onchocerciasis (river blindness), 619
Ondansetron (Zofran), 145
Onuf's nucleus, 638, 639f
Open-science model, 703b
Operant conditioning, 757
Opioid peptides, 147–48, 148f, 149t
Opiomelanocortin propeptide (POMC), 505b
Opium, 147

Oppositional defiance disorder, 654t
Opsins, 12, 14f, 211
Optic chiasm, 226, 496b, 497f, 498f, 569b, 569f, A–15
Optic disk, 206f, 208
Optic flow, 269, 270f
Optic nerve, 208, 209, 210f, 226, A–8t, A–9
Optic tectum, 481
Optic tract, 226, 498f
Optic vesicles, 528f, 529
Optogenetics, 12–13, 14f
Optokinetic eye movements, 476, 476f
Optokinetic nystagmus, 477
Orbital gyri, A–14, A–15f
Orbitofrontal cortex (OFC), 653f
    anatomy of, 828f
    credit assignments and, 831
    evaluation and, 825, 828, 830–32
    lesions, 831, 831f
    reinforcer devaluation task and, 831f
Orb weaver spiders, 127b
Orexin, 729
Orexin-2 receptor gene (Orx2), 721
Organelle trafficking, 558–59, 558f
Organizational effects, of gonadal steroids, 642, 643f
Organ of Corti, 242f
Organoids, 25, 522
Organophosphates, 125
Orgasm, regulation of, 512–13
Orientation, perception of, 276
Orientation bias, 606t
Orientation tuning, in binocular vision, 617–18, 618f
ORNs. See Olfactory receptor neurons
Orthologous genes, 22
Orthostatic hypotension, 510
Oscillations, critical periods and, 619–21, 620f
Oscillopsia (bouncing vision), 269
Ossicles, 242–43
Ossification, of middle ear, 239b
Otoacoustic emissions, 249, 249f
Otoconia, 263, 264f
Otolithic membrane, 263–66, 264f, 265f
Otolith organs, 262–66, 264f, 265f, 266f
Outside-out patch recording, 72b
Oval window, 241f, 242–43, 242f
Ovaries, hormone secretion by, 634, 643f
Overshoot phase, of action potentials, 52, 52b
Overt attention, 741, 741f
Owls, 605
Oxytocin, 148f
Oxytocin receptors, 632b, 632f

P

P2X receptors, 145, 146f
p75 receptors, 589–91, 590f, 669
Pacemaker, cardiac, 510

Pacinian corpuscles, 283*t*, 284*f*, 285–86, 286*f*, 297*b*
Pain. *See also* Proprioception
  central pathways of, 307–8, 308*f*, 309*f*, 311
  channelopathies and, 82–83
  dermatomes and location of, 280
  experience of, 312*f*
  first, 305, 305*f*
  gate theory of, 320, 321*f*
  intensity of, 317
  neuropathic, 317, 319
  nociceptors and basis of, 303–5, 304*f*
  pathways to face, 312, 313*f*, 314
  perception of, 303
  persistent syndromes, 317
  phantom, 318*b*, 318*f*
  physiological basis of modulating, 320–21, 321*f*
  placebo effect and, 319–20
  referred, 309*b*, 309*f*
  second, 305, 305*f*
  sensitization to, 315–17, 319
  sensory-discriminative aspects of, 311–14, 312*f*, 313*f*
  transmission of, 321*f*
  visceral, 310*b*, 310*f*, 502–3
Pain matrix, 312, 319
Pair bonding, 632*b*
Paired box mutations, 531
Palatable tastants, 372, 372*f*
Palatal consonants, 778
Paleocortex, 699*b*, A–19
Pallidum, neurons of, 431, 432*f*
Pancreas, 492*t*, 495
Panic disorder, 170*b*, 655*t*
Papez, James, 803–4
Papez circuit, 804
Par-3, 557*f*
Parabrachial nucleus (PBN), 312, 314, 320, 321*f*, 366, 504
Paracentral lobule, A–16, A–16*f*
Paracrine signaling, 156, 156*f*
Parahippocampal gyrus, 699*b*, 804, A–14, A–15*f*
  location of, 727*f*
  memory and, 762
Parahippocampal place area (PPA), 745
Parallel fibers
  anatomy of, 458–59, 459*f*
  glutamate release by, 196*f*
  Purkinje cell inputs and, 461
  synaptic plasticity and, 176
Paramedian circumferential arteries, A–25
Paramedian pontine reticular formation (PPRF), 478–81, 727
Parasagittal, definition of, A–2
Parasol ganglion cells, 225–26, 225*f*
Parasympathetic division, of visceral motor system, 490
  anatomy and physiology of, 498–500, 499*f*
  major functions of, 492*t*

neurons of, 491, 494*f*
  organization of, 20
Parasympathetic ganglia, 499*f*, 500
Paraventricular nucleus, 497*b*, 497*f*, 498*f*, 715–16, 716*f*
Paravertebral chain ganglia, 495
Parenchyma, A–31
Parietal association cortex, 700, 704
Parietal cortex
  attention and, 705*f*
  functions of, 703–4, 704*f*, 705*f*
  in locomotion control, 398*f*
  visually-guided reach-to-grasp and, 419*f*
Parietal lobe
  anatomy of, A–4, A–4*f*
  lesions of, 703–4
  Sprague effect and, 750
Parietoinsular vestibular cortex (PIVC), 276, 276*f*
Parieto-occipital sulcus, A–4, A–4*f*
Parkin mutations, 444
Parkinson, James, 443
Parkinson's disease
  antipsychotic drugs and, 169*b*
  autophagy in, 679
  basal ganglia and, 381, 434*b*, 771, 772*f*
  cogwheel rigidity in, 427*b*
  dopaminergic neurons in, 140, 443–44, 444*f*
  eye movement impairment and, 478
  hypokinetic movement in, 443–45, 444*f*
  repair for damage from, 660
  sex-based differences in, 655*t*
  somatosensory system in, 299–300
  tremor at rest in, 427*b*
Parotid gland, 492*t*
Paroxetine (Paxil), 812
Paroxysmal extreme pain disorder (PEPD), 83
PAR proteins, 556, 557*f*
Pars caudalis, 312
Pars interpolaris, 312
Pars reticulata, 827
Parvocellular neurons, 423*b*
Passive electrical responses, 38–39, 39*f*
Patch clamp method
  configurations of, 72*b*
  ion channel current flow measured with, 70–74
  K+ ion channel measured with, 73–74, 73*f*
  Na+ ion channel measured with, 71–73, 71*f*
  single ACh receptor currents and, 112, 112*f*
Patched proteins, 535, 538, 540
Pathology, 8
Pavlov, Ivan, 757
PAX3 mutations, 540
PAX6 mutations, 540
Paxil (paroxetine), 812

Paxlovid, 361
PBN (parabrachial nucleus), 312, 314, 320, 321*f*, 366, 504
PCC. *See* Posterior cingulate cortex
PCP (planar cell polarity) pathway, 535
PDEs (phosphodiesterases), 211, 211*f*, 513
Penfield, Wilder, 276, 409, 411*b*, 785
Pentobarbital, 42, 42*f*, 137
PEPD (paroxysmal extreme pain disorder), 83
Peppers, capsaicin and, 306*b*, 306*f*
Peptide hormones, 534*f*
Peptide neurotransmitters. *See* Neuropeptides
Perception, disturbances in, 732–33
Per gene, 717–18, 717*f*
Periaqueductal gray matter, 293, 320–21, 321*f*, 423*b*
Pericak-Vance, Margaret, 769*b*
Periglomerular cells, 342*f*, 343
Perilymph, 243
Perineal muscles, dimorphism in, 639*f*
Perineuronal nets, 580, 623
Peripheral autonomic system, 674*b*
Peripheral nerve regeneration
  axon regrowth in, 664*f*
  cellular basis of, 667–69, 667*f*, 668*f*
  efficacy of, 663
  grafts in, 671, 672*f*
  Head's experiments on, 665–67, 666*f*
  muscle reinnervation and, 673*f*
  Schwann cells and axon regrowth, 669–70, 670*f*
Peripheral nervous system (PNS), 4, 6
  axonal growth cones in, 560–61
  ECM molecules in axon guidance in, 568
  EMT in, 547, 548*f*
  formation of, 526
  input elimination in, 584*f*
  MET in, 547, 548*f*
  neuronal migration in, 547–50, 549*f*
  organization of, 18–20, 19*f*
  synapse formation in, 575, 577
  synapses in early life in, 583*f*
  visualizing neurons in, 8–9, 8*f*
Peripheral sensitization, 315–16
Peripheral synapse regeneration, 671–75, 673*f*, 674*b*
Periventricular nucleus (PVN), 496*b*, 498*f*, 646
Perlecan, 675
PER protein, 717, 717*f*
Persistent pain syndromes, 317
Persistent vegetative state, 734
Personality, PFC damage and, 707, 708*f*
PET (positron emission tomography), 28, 28*f*, 782*b*, 782*f*, 786*f*
Peterson, Andy, 467*b*
Pets, benefits of, 335*b*

PFC. *See* Prefrontal cortex
PGC (preganglionic column), 593*f*
PGO (pontine-geniculo-occipital) waves, 727
Phantom limbs, 318*b*, 318*f*, 734
Phantom pain, 318*b*, 318*f*
Pharyngeal arches, A–11
Pharynx, 776, 777*f*
Phase locking, 253
Phenelzine, 142
Phenobarbital, 137, 199
Phenomenological consciousness, 731
Phenotypes, 22
Phenotypic sex, 630
Phenytoin (Dilantin), 199
Pheromones, 327, 340*f*, 350–52
β-philanthotoxin, 127*b*
Philips, P. K., 812
Phonemes, 607, 776
Phonological processing of speech, 783, 784*f*
Phorbol esters, 167
Phosphatidylethanolamine, 151*f*
Phosphatidylinositol, 151*f*
Phosphatidylinositol bisphosphate (PIP$_2$), 162*f*, 163
Phosphodiesterases (PDEs), 211, 211*f*, 513
Phosphoinositol 3 (PI3) kinase, 590
Phospholipase C (PLC)
  PIP$_2$ splitting by, 162*f*, 163
  signaling, 590, 590*f*
  stimulation of, 176
  structure of, 151*f*
Phospholipase Cβ3 (PLCβ3), 461*f*
Phospholipase Cβ4 (PLCβ4), 461*f*
Phospholipase D, 151*f*
Photopic vision, 213, 214*f*
Photoreceptors, 206. *See also* Cone photoreceptors; Rod photoreceptors
  circadian light changes and, 715, 716*f*
  electrical activity of, 212, 212*f*
  glutamate release from, 220
  hyperpolarization of, 209–10, 211*f*
  retinal, 209, 210*f*
  retinal distribution of, 213*f*
  visual fields of, 216–17
Phototransduction, 209–12, 211*f*, 212*f*
Phrenic motor column (PMC), 593*f*
PI3 (phosphoinositol 3) kinase, 590
Pia mater, A–29, A–29*f*
PICA (posterior inferior cerebellar artery), A–13, A–25, A–26*f*
Piezo channels
  Piezo1, 287
  Piezo2, 287, 288, 307
  structure and gating of, 85, 86*f*
Pigment epithelium, 206, 209, 210*f*
Piloerector muscle innervation, 500
Pineal gland, 716, 716*f*
Pinhole effect, 207
Pinna, 240, 241*f*

PIP₂ (phosphatidylinositol bisphosphate), 162*f*, 163
Pitch, 237, 254
Pituitary stalk, A–15, A–15*f*
PIVC (parietoinsular vestibular cortex), 276, 276*f*
PKA (cAMP-dependent protein kinase), 163, 165, 166*f*, 184–85, 185*f*
PKC (protein kinase C), 166*f*, 167, 176, 176*f*, 197
Placebo effect, pain and, 319–20
Place cells, 770*b*, 771*f*
Place codes, 485*b*
Placenta, hormone secretion by, 634
Planar cell polarity (PCP) pathway, 535
Planning
    PFC and, 711, 712*f*
    thinking, decision making and, 819–38
Plasma membranes, presynaptic, 102–4, 102*f*, 103*f*, 106*f*
Plasticity. *See also* Synaptic plasticity
    activity-dependent, 600–601, 602*f*, 612*f*, 624–26, 626*f*
    cerebral cortex, 298–99, 298*f*, 299*f*
    experience-dependent, 598–628
    habits and neuron, 437*b*
    of motor units, 385*b*, 386*f*, 487*f*
    taste, 373
    vestibulo-ocular reflex, 273
Platelets, 315
PLC. *See* Phospholipase C
PLCβ3 (phospholipase Cβ3), 461*f*
PLCβ4 (phospholipase Cβ4), 461*f*
Plexins, 566
Plexus, 491
Plosives, 778
Pluripotent stem cells, 517, 520
PMC (phrenic motor column), 593*f*
PMCA calcium pump, 87, 87*f*
PNS. *See* Peripheral nervous system
Polarity, differentiation and, 541, 557*f*
Polarized epithelial cells, 556
Poliovirus transmission, 329–30
Polyglutamine expansion, 82
Polyneuronal innervation, 582–85, 583*f*
POMC (opiomelanocortin propeptide), 505*b*
Pons
    anatomy of, 485*f*, A–7
    corticospinal and corticobulbar tracts in, 407*f*
    cranial nerve nuclei in, A–10*t*
    location of, A–3*f*
Pontine-geniculo-occipital (PGO) waves, 727
Pontine micturition center (Barrington's nucleus), 511–12, 511*f*
Pontine nuclei, 453, 453*f*
Pontine reticular formation, 727
Pontine tegmentum, 727*f*
Pontocerebellar fibers (transverse pontine fibers), 454
Population coding of objects, 706–7, 706*f*

Pore loops, 78–80, 79*f*
Pores, ion channel, 78–80, 79*f*
Positron emission tomography (PET), 28, 28*f*, 782*b*, 782*f*, 786*f*
Postcentral gyrus, 292
    anatomy of, A–13–A–14, A–14*f*
    cortical targets in, 294, 294*f*
    in locomotion control, 398*f*
    somatotopic map of, 410, 412
Posterior, definition of, A–1
Posterior cerebral arteries, A–23*f*, A–24
Posterior chambers, 206*f*
Posterior cingulate cortex (PCC)
    Alzheimer's disease and, 838
    default mode network and, 837–38, 837*f*
    default-mode networks and, 735, 736*f*
    location of, 728, 728*f*
    self-awareness and, 835–38
Posterior circulation, A–21*f*, A-22–A–23, A–24*t*
Posterior communicating arteries, A–23*f*, A–24
Posterior funiculi, 291
Posterior inferior cerebellar artery (PICA), A–13, A–25, A–26*f*
Posterior spinal arteries, A–22*f*, A–26
Postganglionic axons, 495, 499*f*
Postnatal period, brain growth during, 600, 601*f*
Postsynaptic cells, 4
Postsynaptic currents (PSCs), 115
Postsynaptic density, 98
Postsynaptic density protein 95 (PSD95), 625
Postsynaptic dorsal column (PSDC) neurons, 292
Postsynaptic neurons, 94, 94*f*
    presynaptic inputs and, 599*f*
Postsynaptic potentials (PSPs), 115–16
Postsynaptic signals, 111–12, 118*f*, 120
Post-tetanic potentiation (PTP), 181
Post-traumatic stress disorder (PTSD), 654, 655*t*, 677, 812–13, 813*f*
Posture, 403
    anticipatory maintenance of, 423, 424*f*
    reticular formation in control of, 423–24, 424*f*
    upper motor neurons governing, 419–25
    vestibular complex and, 455
Potassium-chloride co-transporter (KCC2), 317
Potassium (K⁺) co-transporters, 138*b*
Potassium (K⁺) ion channels
    Ca²⁺-dependent, 247
    inhibition of, 143
    in late current, 58, 58*f*
    membrane permeability and, 44–46, 45*f*, 46*f*
    neurological diseases caused by altered, 82–83

patch clamp measurements of, 73–74, 73*f*
    resting membrane potentials generated from, 47, 47*t*, 49, 49*f*
    structure of, 79*f*
    toxins that poison, 75*b*
    voltage-gated, 74, 74*f*, 80, 81*f*
Potassium (K⁺) ions
    conductance, 59–62, 59*f*, 60*f*, 61*f*
    EPC to EPP changes and, 114–15, 115*f*
    hair cell signal transduction and, 246–47, 246*f*
    membrane permeability to, 54
    Nernst equilibrium potential for, 248
    translocation by Na⁺ pumps, 88–89, 88*f*
Potentiation, synaptic plasticity and, 181
Power spectrum, 237, 237*f*
PP1 (protein phosphatase 1), 167, 168*f*
PP2A (protein phosphatase 2A), 167–68, 168*f*
PP2B (protein phosphatase 2B), 168, 168*f*
PPA (parahippocampal place area), 745
PPRF (paramedian pontine reticular formation), 478–81, 727
Prairie voles (*Microtus ochrogaster*), 632*b*
Precentral gyrus, 409–10, 410*f*, 412, A–13, A–14*f*
Precuneus gyrus, A–16, A–16*f*
Predators, vomeronasal system and, 327
Preferred orientation, of neuron, 228, 229*f*
Prefrontal cortex (PFC)
    caudate input from, 435
    connectivity of, 821, 822*f*
    damage to, 707–9, 708*f*, 821–22
    emotion processing and, 811, 814–15
    nondeclarative memory and, 771
    organization of, 821, 821*f*
    oxytocin receptors in, 632*b*, 632*f*
    planning and, 711, 712*f*
    size of, 820*f*
    studies on, 819–20
    visually-guided reach-to-grasp and, 419*f*
Prefrontal neurons, selectivity of, 711
Preganglionic axons, 499*f*, 500
Preganglionic column (PGC), 593*f*
Preganglionic neurons, 232, 495, 499*f*, 500
Premotor cortex
    anatomy of, 405, 405*f*
    divisions of, 414–19, 414*f*
    mirror motor neurons in, 417, 418*f*
    in movement control, 380
    speech and, 380
    visually-guided reach-to-grasp and, 419*f*

Preoptic area, sexual behaviors and, 640, 641*f*
Pre-propeptides, 146–47, 147*f*
Presbyopia, 207
Presenilin 1 gene, 768*b*
Presenilin 2 gene, 768*b*
Presynaptic cells, 4
Presynaptic inputs, postsynaptic neurons and, 599*f*
Presynaptic neurons, 94, 94*f*
Presynaptic terminals
    disorders affecting, 107–9
    electron microscopy of, 119*f*
    local recycling of synaptic vesicles in, 103–4, 103*f*
    neurotransmitter release from, 101, 101*f*
Prevertebral ganglia, 495
Prey, vomeronasal system and, 327
Primary motor areas, 698*f*
Primary motor cortex
    anatomy of, 405, 405*f*, 414, 414*f*, 453*f*
    corticobulbar and corticospinal tracts in, 414
    cytoarchitectonic appearance of, 406*f*
    facial motor nucleus and, 408
    functional organization of, 409–14, 410*f*
    in movement control, 380
    visually-guided reach-to-grasp and, 419*f*
Primary sensory areas, 698*f*
Primary sex characteristics, 630–31, 631*f*
Primary somatosensory cortex
    amputation and functional changes in, 298, 298*f*
    anatomy of, 294–95, 295*f*
    functional hierarchies in, 295, 295*f*, 297
    phantom limb and, 318*b*
    repetitive behavior and changes in, 298–99, 299*f*
    in touch pathways, 290*f*
Primary visual cortex
    anatomy of, 222, 222*f*, A–16
    columnar anatomical arrangement in, 707
    depth-sensitive neurons in, 732
    edge detection and orientation preference in, 228–30, 229*f*, 230*f*
    extrastriate visual areas and, 230
    ocular dominance columns in, 604, 605*f*
    organization of, 227–28, 228*f*
    retinotopic maps of, 226, 226*f*
Primates. *See also* Monkeys
    odorant receptor genes of, 334
    studies on, 24*b*
Priming, 756–57, 757*f*, 771
Primitive node/pit, 523, 533
Primitive streak/groove, 523
Principal nucleus, 292
Principal sulcus, 711, 712*f*

Principal trigeminal nucleus, A–10*f*, A–12*f*
Prion diseases, 463
Procedural memory, 756
Progesterone, 634*f*, 644*f*
Progesterone receptor antagonists, 644*f*
Projection neurons, 9
Promoters, 20, 173–75
Propeptides, 147, 147*f*
Propofol, 42, 42*f*
Propranolol (Inderol), 143, 816
Proprioception, 278. *See also* Pain
  central pathways of, 292–93, 293*f*
  dermatomes and location of, 280
  facial pathways in, 292
  lower and upper body pathways in, 292–93, 293*f*
  mechanoreceptors for, 288–89, 288*f*
  in musculoskeletal system, 288*f*
Prosencephalon, 527, 528*f*, 529, A–2
Prosody, 776
Prosopagnosia, 231, 705–6
Prostaglandins, 315–17
Prostate gland, referred pain from, 309*b*
Protanopia, 215*f*, 216
Protein kinases
  activation of, 166*f*
  function of, 165, 167
  PKA, 163, 165, 166*f*, 184–85, 185*f*
  PKC, 166*f*, 167, 176, 176*f*, 197
  receptor function of, 158–59
Protein phosphatases, 165, 167–68, 168*f*
Protein phosphorylation, 165, 165*f*, 177, 177*f*
Protein trafficking, 558–59, 558*f*
Protein tyrosine kinases, 167
Protocadherins, 575, 576*f*, 580
Prozac (fluoxetine), 169*b*, 812
Prurigenic chemicals, 314
Prusiner, Stanley, 463
PSCs (postsynaptic currents), 115
PSD95 (postsynaptic density protein 95), 625
PSDC (postsynaptic dorsal column) neurons, 292
Pseudogenes, 334
Pseudounipolar neurons, 279*f*, 281
PSPs (postsynaptic potentials), 115–16
Psychiatric disorders, molecular basis of, 169*b*
Psychoses, molecular basis of, 169*b*
Psychosurgery, 710*b*, 710*f*
PTC mutations, 539
Ptosis, 475, 504–5, 504*f*
PTP (post-tetanic potentiation), 181
PTSD (post-traumatic stress disorder), 654, 655*t*, 677, 812–13, 813*f*
Punctate midline myelotomy, 310*f*, 311*b*
Pupils
  anatomy of, 206, 206*f*

constriction of, 475, 476
light reflex of, 222, 232, 232*f*
pinhole effect and, 207
size changes of, 232–33
Purinergic receptors, 145–46, 146*f*
Purines
  catabolism of, 145
  receptors of, 145–46, 146*f*
  structure of, 123*f*
Purkinje, J. E., 472*b*
Purkinje cell degeneration mutations, 467*t*
Purkinje cells
  activity of, 462*f*
  calcium concentration, 164*b*, 164*f*
  climbing fiber inputs in, 584*f*, 585
  cortical inhibitory loops and, 460
  dendrites, 459, 459*f*
  description of, 458–62, 460*f*
  error recognition by, 462, 464
  gap junctions and output of, 461
  Golgi-stained, 8*f*
  imaging of, 164*f*
  innervation of, 176, 176*f*
  parallel fibers and, 461
  signal integration of, 274
  visualization of, 458*f*
Purkinje trees, 472*b*
Putamen
  anatomy of, 431, 431*f*
  inhibitory neurons of, 441*f*
  medium spiny neurons of, 437
Putamen nuclei, A–19, A–20*f*
PVN (periventricular nucleus), 496*b*, 498*f*, 646
Pyramidal neurons
  CaMKII activity during LTP in, 191, 191*f*
  cortical layer of, 699*b*
  in hippocampus, 187, 187*f*
  imaging of, 164*f*
  visual cortex, 227, 228*f*
Pyramidal smiles, 802*b*
Pyramidal tract, A–12*f*
Pyridoxal phosphate, 136
Pyriform cortex
  anatomy of, 325*f*, 326
  olfactory bulb projections to, 343, 346, 347*f*
  olfactory processing of, 346, 347*f*, 348

Q

Quadrantanopsia, homonymous, 224*b*
Quails, 794
Qualitative meta-analysis techniques, 701*b*, 701*f*
Quantitative meta-analysis techniques, 702*b*, 702*f*
Quinn, Chip, 186*b*
Quisqualate, 127*b*
Quisqualis indica, 127*b*

R

RA. *See* Retinoic acid
Rabbits, 187, 586*b*, 586*f*
Radial connections, 699
Radial glial cells, 541, 550, 551*f*, 552
RAGS protein, 594
Ramachandran, V. S., 318*f*, 319*b*
Rapid eye movement (REM) sleep, 475, 718, 721, 723–25, 725*f*, 726*b*, 727*f*. *See also* Sleep
Rapidly adapting afferents, 282, 283*f*
Rapsyn, 672, 675
Ras protein, 160
Rate codes, 485*b*
Rats, 605
  activity-dependent plasticity and, 602*f*
  declarative memory in, 762, 763*f*
  fear conditioning in, 804, 806*f*
  monocular deprivation in, 614, 615*f*
  motherly behaviors in, 647–48, 647*f*
  sleep deprivation in, 722
RDoC (Research Domain Criteria) framework, 799
Reach-to-grasp, visually-guided, 419*f*
Readiness potential, 835*b*, 835*f*
Reading, dyslexia and, 791–92, 792*f*
Reading deficits, 781
Receptive aphasia, 781
Receptive fields. *See also* Visual fields
  circuitry for responses for, 219*f*
  neuronal, 16–17, 17*f*
  tuning of, 16–17, 17*f*
  visual, 216–17
Receptive fields, mechanoreceptor, 282, 283*f*
Receptor molecules, 110
Receptor potentials, 11
  description of, 38, 38*f*
  generation of, 287, 287*f*
  hair cell displacement and, 245
  in hair cells, 247, 248*f*
Recessive genes, 22
Reciprocal innervation, 389
Rectum, 492*t*
Red nucleus, 456–57, 456*f*, 461, A–7
Reeler mutations, 466*b*, 466*f*, 467*t*
Reelin protein, 552
Reentrant neural activation, 732–33
Reese, Tom, 102, 103, A–28
Referred pain, 309*b*, 309*f*
Reflex arcs, 389, 391
Refractive errors (ametropia), 206–7, 207*f*
Refractory periods, 61
Regenerative action potentials, 62
Regional anesthesia, 41–42
Reinforcer devaluation task, 831*f*
Reissner's membrane rupture, 248
Remembering, 758–60
REM (rapid eye movement) sleep, 475, 718, 721, 723–25, 725*f*, 726*b*, 727*f*. *See also* Sleep

Repair and regeneration in nervous system, 658–93
Repetition, development and, 603
Reporter proteins, 25
Reproductive behaviors, dimorphism in, 638–40
Research Domain Criteria (RDoC) framework, 799
Restiform body (inferior cerebellar peduncle), 452*f*, 453, A–10*f*, A–12*f*
Resting membrane potentials
  description of, 37–38
  K⁺ channels generating, 47, 47*t*, 49, 49*f*
  membrane permeability and, 49–50, 50*f*
Restless legs syndrome, 721
Reticular, definition of, 422*b*
Reticular activating system, 726, 727*f*
Reticular formation
  affective-motivational aspects of pain and, 314
  anatomy of, 422*b*, 422*f*, 457*f*, A–7
  emotions and, 800
  functions of, 421
  Horner's syndrome and, 504–5, 504*f*, 505*f*
  location of, 421*f*
  modulatory functions of, 422*b*
  postural control and, 423–24, 424*f*
  premotor functions of, 423*b*
  sensory inputs to, 503
  upper motor neurons and, 379, 382, 420
Reticular theory, 3
Retinal amacrine cells, 2*f*
Retinal bipolar cells, 2*f*
Retinal drift, 487–88
Retinal error signals, 483
Retinal ganglion cells, 2*f*, 209–11, 210*f*. *See also* Ganglion cells
  axons from, 222
  circuitry for responses of, 218*f*
  contrast and, 219–20, 219*f*
  decussation of, 569*b*
  ON/OFF, 217–18, 217*f*, 218*f*, 219*f*
  receptive field responses of, 219*f*
  types and functions of, 221–23, 222*f*
Retinal images, stabilized, 472*b*, 473*f*
Retinal waves, recording of, 620–21, 620*f*
Retinas
  anatomy of, 206, 206*f*
  blind spot and, 208
  cell types in, 209, 210*f*
  ganglion cell layer of, 715, 716*f*
  inferior division of, 208*f*, 209
  inner nuclear layer of, 209, 210*f*
  inner plexiform layer of, 209, 210*f*
  lateral interactions in, 219–20, 219*f*
  light intensity and, 220*b*, 220*f*
  nasal division of, 208*f*, 209
  neural circuitry in, 209, 210*f*
  non-foveal areas of, 472
  ON/OFF pathways in, 217–18, 217*f*, 218*f*, 219*f*

optic flow across, 269, 270f
outer nuclear layer of, 209, 210f
outer plexiform layer of, 209, 210f
photoreceptor distribution in, 213f
phototransduction in, 209–12, 211f, 212f
projections of visual field onto, 207–9, 208f
spatial details from, 226, 226f
superior division of, 208f, 209
temporal division of, 208f, 209
topographic mapping of, 594f, 595f
ventrotemporal region of, 569f
Retino-geniculo-cortical pathway, 223, 225–26, 225f
Retinoic acid (RA)
    inductive signaling pathways and, 534f, 537
    neural induction and, 533
    neurodevelopmental disorders and, 537, 538f
    neuron migration and, 690
Retinoid receptors, 533
Retinoids, 158
Retino-recipient brain regions, 221–23, 222f
Retinotopic maps, 226, 226f
Retrograde, definition of, 16
Retrograde amnesia, 765–67
Rett syndrome, 540
RET tyrosine kinase receptors, 588
Reversal potentials, 113–14, 113f, 114f, 117f
Reward prediction errors (RPEs), 828b, 829f, 830f
Rewards, 825–32
Rexed, Bror, A–6
Rexed's laminae, 289, 307, 308f, A–6
RGS4 gene, 169b
Rhesus monkeys. See Monkeys
Rheumatoid arthritis, 316
Rhodopsin, 211
Rho/GAP proteins, 566
Rho GTPases, 566, 590
Rhombencephalon, 524f, 527–30, 528f, A–3
Rhombomeres, 529–30, 531f
Right-handedness, 788b, 788f, 789f
Rising phase, of action potentials, 52, 52b
Ritalin (methylphenidate), 721
Ritonavir, 361
River blindness (onchocerciasis), 619
RMS (rostral migratory system), 690, 690f
RNA, DNA transcription to, 172–73, 173f
RNA polymerase, 173
ROBO3 mutations, 571, 572t
Robo receptors, 565f, 566
Robotic exoskeletons, 416b
Rocuronium, 43
Rodents. See Mice; Rats
Rod photoreceptors, 209, 210f
    anatomical distribution of, 213f
    convergence in, 212, 214f

phototransduction in, 211, 211f
rhodopsin in, 211
specialization of, 212–13
Romantic love, science of, 632b
Roofplate, 524, 524f, 534f
Ropivacaine, 41, 42f
Roses, Allen, 769b
Rostral, definition of, A–1
Rostral area, 250
Rostral interstitial nucleus, 479–80, 727
Rostral medulla, in touch pathways, 290f
Rostral migratory system (RMS), 690, 690f
Rostrotemporal area, 250
Rotational movements, 262
Rough endoplasmic reticulum, 146–47
Round window, 241f, 243, 243f
RPEs (reward prediction errors), 828b, 829f, 830f
Rubrospinal tract, 425
Ruffini corpuscles, 283t, 284f, 285–86, 286f
Ruggero, M., 247b
Rutabaga mutation, fruit fly, 186b
Ryanodine receptors, 163

S

Saccades, 233, 257
    amplitude of, 475, 475f
    description of, 471
    entropy, 436f
    express, 485
    frontal eye fields and, 481
    hemispheric dysfunction and, 272
    motor neuron activity in, 480f
    movement coordinates and, 483, 484f
    neural control of, 479–87, 480f, 484f
    studies of, 464, 464f
Saccadic suppression, 233
Saccular maculae, 264f, 265
Saccules, 262–66, 262f, 264f
Sacks, Oliver, 619
Sacral nerves, A–3f
Sacral neural crest, 525f, 526
Sacral vertebrae, A–4, A–5f
Sagittal, definition of, A–2
Sakmann, Bert, 71, 112
Salivary gland innervation, 500
Salivatory nuclei, 499f, 500, A–10f
Saltatory propagation, 65f, 67
Salty taste, 356
    receptors detecting, 362f, 363
Salvesen, Guy, 768b
SANDD (sinoatrial node dysfunction and deafness), 83
SARS-CoV2 (COVID-19) pandemic, 330, 330f, 360, 361
SAT2 (System A transporter 2), 131
Savant syndrome, 760b
Saxitoxin, 75b

SCA (superior cerebellar artery), A–25, A–26f
SCA6 (spinocerebellar ataxia type 6), 82
Scala media, 242f, 243
Scala tympani, 242f, 243
Scala vestibuli, 243
Scarpa's ganglion, 268
SCCs (solitary chemosensory cells), 364b
Schaffer collateral-CA1 synapses
    LDP at, 194–95
    LTP at, 187, 188f, 189f, 191–92
Schizophrenia
    antipsychotic drugs and, 169b
    basal ganglia and, 434b
    cortical connections in, 610
    default-mode networks and, 735
    eye movements in, 478–79, 479f
    neurexins and, 109
    Nrg1 and, 578
    sex-based differences in, 654, 654t
    susceptibility genes in, 169b
Schmahmann, Jeremy D., 468
Schmahmann's syndrome, 468
Schwann cells
    axon regrowth and, 669–70, 670f
    CAM expression and, 668f, 669
    overview of, 6, 7f
    peripheral nerve regeneration and, 667, 667f
    secretion by, 669, 671
Sclera, 206, 206f, 233
SCN (suprachiasmatic nucleus), 222, 222f, 496b, 497f, 498f, 715–17, 716f
SCN9A gene, 307
SCN genes, 76, 82
Scopolamine, 127b
Scotoma (blind spot), 208, 223b, 733
Scotopic vision, 213, 214f
Scrapie, 463
SCT (spinocervical tract), 290f, 292
SDN-POA (sexually dimorphic nucleus of the preoptic area), 640, 641f, 642
Sea slug (Aplysia californica), 24b, 182–85, 183f
Secondary sex characteristics, 630
Secondary somatosensory cortex, 292, 294f, 297
Second messengers
    calcium-dependent cascades, 623
    effector pathways and, 161
    neural function changed by, 172–75
    neurons using, 161, 162f, 163
    protein phosphorylation and, 165, 165f
    signal amplification and, 156
Second pain, 305, 305f
Sedatives, 42, 42f
Segmental nerves, A–4, A–5f
Segmentation process, 529
Seizures, 82, 198–99, 199f, 679
Selective attention, 739, 739f
Selective serotonin reuptake inhibitors (SSRIs), 144, 812

Self-awareness, PCC and, 835–38
Self-control, 823–25
Semantic memory, 782b
Semantics, 782b, 782f
Semaphorins, 566–67, 566f, 681
Semicircular canals
    anatomy of, 241f, 262, 262f, 266–67, 267f
    eye movement impact of, 476
    functional organization of, 267f
    head rotations sensed in, 266–69
    sensory information from, 420
Senile plaques, 768b
Sensitization
    behavioral, 185f
    to pain, 315–17, 319
    process of, 182
    short-term, 183–85, 183f, 184f, 185f
Sensorimotor integration, eye movements and, 471–88
Sensorimotor learning, 435b
Sensorineural hearing loss, 239b, 248
Sensory aphasia, 781
Sensory-discriminative aspects of pain, 311–14, 312f, 313f
Sensory endings, "flower spray," 389
Sensory nuclei, A–11
Sensory systems, organization of, 15
Sensory transduction, mechanisms of, 287, 287f
Septal forebrain nuclei, A–20, A–20f
Septum pellucidum, A–30
SERCA calcium pump, 87, 87f
Ser kinases, 165
Serotonergic neurons, 575, 729
Serotonin (5-HT)
    functional features of, 124t, 144–45
    interneurons releasing, 183, 184f
    release of, 315
    structure of, 123f
    synthesis of, 144, 144f
Serotonin (5-HT) receptors, 144–45, 145f
Serotonin transporter (SERT), 144
Sertraline (Zoloft), 812
Severe myoclonic epilepsy of infancy, 82
Sevoflurane, 43
Sex characteristics, 630
Sex chromosomes, 630
Sex differences, circuit development and, 629–56
Sex-reversal gene on the Y chromosome (SRY), 631–32, 634, 650
Sex steroids. See Steroid hormones
Sexual dimorphisms
    brain and, 638–40, 641f, 641t, 653f
    description of, 636–38
    hormone influences on, 634
Sexual function, autonomic regulation of, 512–13, 512f
Sexually dimorphic behaviors, 636–38
Sexually dimorphic development, 634–36, 642–44

Sexually dimorphic nucleus of the preoptic area (SDN-POA), 640, 641f, 642
Sexual orientation, 651–52, 652f
SGLTs (sodium-glucose co-transporters), 363
SGZ (subgranular zone), 687, 688f, 689
Sham rage, 799
SHANK protein, 467b, 625
Sheep, 463
Sherrington, Charles, 3, 4, 380, 409, 428
Shh. *See* Sonic hedgehog
Shimomura, Osamu, 164b
Short axon cells, 343
Short circumferential arteries, A–25
Short-term memory, 755
Sigmoid, 492t
Sigmoid sinuses, A–21f, A–25
Signal amplification, chemical signaling, 156, 157f
Signaling endosomes, 589, 590f
Signal transduction
 amplification in, 156–57, 157f
 components of, 184
 in hair cells, 246–48, 246f
 neuronal, 175–77
 olfactory odor, 337–39, 338f, 339f
 pathways, 157–58, 157f
 postsynaptic, 191–92, 191f
Sign-inverting synapses, 218f, 219f
Sign language, 778b, 778f
Sildenafil (Viagra), 513
Silent synapses, 193b, 193f, 194f
Simple spikes, 461
Simultanagnosia, 747–48
Sine waves, 236–37, 237f
Single-line bisection test, 746, 747f
Sinoatrial node dysfunction and deafness (SANDD), 83
Size principle, 387–88, 387f
Skin
 glabrous, 279, 284f
 hairy, 279, 284f, 286
 mechanical forces on, 278–79
 piloerector muscles of, 500
 specialized mechanoreceptors and endings in, 284–86, 284f, 286f
Skin conductance, emotions and, 798
Skinner, Frederick, 757
Skinner box, 757–58, 758f
Skou, Jens Christian, 87
SLC8A genes, 89
Sleep
 definition of, 718
 deprivation, 721
 disorders, 720–21
 dreaming and, 725, 726b
 duration of, 722f
 EEG and, 718b, 720f, 723–24, 724f
 healthy, 721
 memory and, 723
 neural circuits governing, 726–29, 727f, 728f, 729t
 non-REM, 724, 726b

physiological changes during states of, 725, 725f
 purpose of, 723
 REM, 475, 718, 721, 723–25, 725f, 726b, 727f
 slow-wave, 606t, 724, A–31
 stages of, 723–25, 724f
 thalamocortical interactions in, 729, 730f, 731
 in varieties of animals, 721–22
Sleep apnea, 720–21
Sleepiness, 721
Sleep spindles, 724, 730f
Sleep-wake cycles, 714–17, 715f, 731f
Sleepwalking, 725
Slit factor, 565f, 566, 681
Slowly adapting afferents, 282, 283f
Slow (S) motor units
 force provided by, 387–88, 387f
 function of, 383–84, 385f, 386b
 staining of, 386b, 386f
Slow-wave sleep, 606t, 724, A–31. *See also* Sleep
SMADs, 533, 534f
Small G-proteins (monomeric G-proteins), 159f, 160
Small intestine, 492t
Small-molecule neurotransmitters, 123f, 124
Smith, Neil, 760b
Smoking cessation, 834, 834f
SMO mutations, 539
Smoothened proteins, 535, 538
Smooth pursuit eye movement, 233, 475, 476f, 478–79, 479f, 487–88, 488f
S motor units. *See* Slow motor units
SN1 (N transporter 1), 131
Snail genes, 547–48
SNAP-25, 104, 105f, 106f, 107
SNAP receptors (SNAREs), 104, 105f, 106f, 108, 109
SNAPs (soluble NSF-attachment proteins), 104, 105f, 106f, 107
SNAREs (SNAP receptors), 104, 105f, 106f, 108, 109
SNB (spinal nucleus of the bulbocavernosus), 638, 639f, 643
SOD1 (superoxide dismutase) mutations, 399–400
Sodium/calcium (Na^+/Ca^{2+}) exchangers, 89, 89f, 90f, 161
Sodium (Na^+) co-transporters, 138b
Sodium-glucose co-transporters (SGLTs), 363
Sodium (Na^+) ion channels, 46
 action potential generation and, 51–53, 51f
 anesthetics and, 307
 blockage of, 127b
 neurological diseases caused by altered, 82–83
 patch clamp measurements of, 71–73, 71f
 refractoriness of, 64
 structure of, 79f
 toxins that poison, 75b
 voltage-gated, 74, 74f

Sodium (Na^+) ions
 conductance, 59–62, 59f, 60f, 61f
 in early current, 57–58, 58f
 EPC to EPP changes and, 114–15, 115f
 membrane permeability to, 54–55
 translocation by Na^+ pumps, 88–89, 88f
Sodium (Na^+) pumps, 87–89, 87f, 88f
Soleus muscle, 384
Solitary chemosensory cells (SCCs), 364b
Soluble NSF-attachment proteins (SNAPs), 104, 105f, 106f, 107
Somatic motor nuclei, A–11, A–11f
Somatic motor system, of PNS, 18, 19f
Somatic stem cells, 519b
Somatosensory afferents (mechanoreceptors)
 anatomy and morphology of, 279–81, 279f
 animals with specialized, 296b, 296f
 endings formed by, 284–86, 284f, 286f
 mechanical forces on skin conveyed by, 278–79
 properties of, 283t
 for proprioception, 288–89, 288f
 rapidly adapting, 282, 283f
 receptive fields of, 282, 283f
 slowly adapting, 282, 283f
 spinal cord and subtypes of, 291, 291f
 taste and, 369–71, 370f
Somatosensory cortex, A–14
Somatosensory dysfunction, 299–300
Somatosensory map, 606t
Somatosensory system, 278–300
Somatosensory thalamus, A–18b
Somatostatin, 148f
Somatotopic maps, 294, 295f
Somites, 280, 534f
Songbird communication, 602–4, 606t
Sonic hedgehog (Shh)
 as chemoattractant or chemorepulsive signals, 567
 diseases associated with, 537–39
 distribution of, 536f
 inductive signaling pathways and, 534f
 neural development and, 535, 536f
 neuron migration and, 690
Sorge, W. A., 247b
Sound
 auditory perception and, 252
 distortion in, 247b
 ears generating, 249–50, 249f
 interaural level differences in localizing, 255–57, 255f, 257f
 interaural timing differences in localizing, 255–56, 255f, 256f
 localization, 606t
 neural codes for frequency of, 252–55, 253f, 254f
 properties of, 236–37, 237f, 238f

spectral cues in localizing, 255–56, 255f
 taste and, 371
 vision in localization of, 257–58, 258f
Sound waves, 236–37, 237f
Source-filter model of speech, 776
Sour taste, 356
 receptors detecting, 362f, 363
Sox2 transcription factor, 521
Sparks, David, 483
Sparse coding, of odorants, 346
Spasticity, 426–27, 427b
Spastic paraplegia, X-linked, 570
Spatial attention, 742–45, 742f, 743f, 744f
Spatial coding, visual compared to auditory, 258, 258f
Spatial memory, 723, 766, 766f
Spatial orientation, perception of, 276
Spearmint, 349, 351f
Specific Learning Disorder, 791
Spectral cues, 255–56, 255f
Spectrograms, 237, 237f, 238f
Speech
 cortical contributions to, 778–81
 elements of, 776–78, 777f
 language and, 775–95
 learning, 790–91, 790f
 N400 response and comprehension of, 783, 784f
 perception of, 237, 238f, 252, 254–55
 phonological processing of, 783, 784f
 premotor cortex and, 380
 production of, 418
 source-filter model of, 776
Sperry, Roger, 593–94, 787
Spike timing-dependent plasticity (STDP), 197, 198f, 200
Spike-triggered averaging, 410, 412f
Spinal accessory nerve, A–8t, A–9
Spinal accessory nucleus, A–10f
Spinal cord
 blood supply of, A–21–A–27, A–22f, A–24t
 brain-machine interfaces and injuries to, 416b, 416f
 central pattern generators in, 425
 cervical enlargement of, 382, 382f
 CNS and, 18, 19f
 corticospinal and corticobulbar tracts in, 407f
 damage, 659
 in discriminative pain pathways, 313f
 dissociated sensory loss and, 308
 external anatomy of, A–3f, A–4–A–5
 flexion-crossed extension reflex circuitry in, 394, 394f
 gray matter, 382, 382f, A–6f
 intermediolateral cell column of, 495
 internal anatomy of, A–5–A–7, A–5f, A–6f

interneurons, 10
lateral horn of, 495
local circuit neurons in, 382–83, 382f
mechanoreceptor subtypes and, 291, 291f
motor cortex pathways to, 425f
muscle stretch reflexes and, 389, 390f, 391
neuron populations in, 544
pathway formation along, 592, 593f
preganglionic neurons of, 495, 499f, 500
projections from brainstem to, 420f
projections from vestibular nuclei, 421
in touch pathways, 289–92, 290f, 291f
upper motor neuron control of, 403–28
ventral horn of, 381–82, 381f, 382f
white matter, 382, A–5, A–5f, A–7
Spinal nerves, A–4, A–5f
Spinal nucleus, 292
Spinal nucleus of the bulbocavernosus (SNB), 638, 639f, 643
Spinal shock, 426
Spinal trigeminal nucleus, 312, 313f, A–10f, A–12f
Spinocerebellar ataxia type 6 (SCA6), 82
Spinocerebellar pathways, 457
Spinocerebellum, 452, 452f, 453t
Spinocervical tract (SCT), 290f, 292
Spinothalamic tract, 311, 312f
Spiny stellate neurons, 227, 228f
Splanchnic nerves, 495, 500
Splice variants, 20
Split-brain patients, 786–88, 787f
Split-brain syndrome, 733–34, 786–88, 787f
Spongiform degeneration, 463
Spontaneous nystagmus, 271, 271f
Sports, casualties of, 676–77, 677f
Sprague effect, 750
Squid, giant neurons of, 47t, 48b, 49, 49f, 50b, 51f
SQUIDs (superconducting quantum interference devices), 31, 31f
SRY (sex-reversal gene on the Y chromosome), 631–32, 634, 650
SSRIs (selective serotonin reuptake inhibitors), 144, 812
Staggerer mutations, 466b, 467t
Stance phase locomotion, 395, 397f
Stapedius, 243
Stapes, 241f
STDP (spike timing-dependent plasticity), 197, 198f, 200
Stellate neurons
  cerebellar cortex, 459, 459f
  visual cortex, 227, 228f
Stem cells
  brain repair and, 690–91
  embryonic, 519b, 519f, 520–21, 520f, 535

foundational experiment on, 518, 518f, 520
induced pluripotential, 519b, 521–22, 521f
molecular basis of, 521–22, 521f
multipotent neural, 665
in nervous system development, 517–23, 518f
neural, 519b, 522–23, 522f, 685, 687, 688f, 689
neural development and, 534f, 535–36
neurogenic, 689
niches, 688–89, 689f
pluripotent, 517, 520
promise and peril of, 519b
somatic, 519b
specification of, 536f
symmetrical division of, 541, 542f
Stereocilia
  anatomy of, 245, 245f
  cochlea and, 242f
  deflection of, 246f
  movement of, 262
  sensitivity of, 246
  sound transduction and, 244f, 245
Stereognosis, 314
Stereopsis, 227
Steroid hormones. See also specific hormones
  cell-permeant signaling by, 157–58
  emotions and, 816
  neuronal effects of, 635f
  organizational effects of, 634
  production of, 634
Steroid-thyroid nuclear receptors, 636
STG (superior temporal gyrus), 250, 252, 252f, 255, A–13, A–14f
Stomach, 492t
Strabismus, 616–17, 617f
Stress, 606t
Stress hormones, emotions and, 816
Stretch reflexes, 389, 390f, 391
Striatal neuron inhibition, 442
Striate cortex. See Primary visual cortex
Striatum
  anatomy of, 431, 431f, A–19, A–20f
  functional subdivisions of, 434–35
  GABAergic neurons of, 439
Stria vascularis, 247
Striola, 263f, 264f, 265
Strittmatter, Warren, 768b
Stroke, 408–9, 655t
  focal, 661, 662f
  imaging of, A–27
  recovery after, 661–62, 662f, 663f
  treatment of, A–27
  types of, A–27
Stroop effect, 833, 833f
Stroop Test, 709b
Strychnine, 127b, 139, 140f
Strychnos nux-vomica, 127b
Stuttering, 654t
Styron, William, 812
Subarachnoid space, A–29, A–29f

Subgranular zone (SGZ), 687, 688f, 689
Subjective values, 828
Sublingual glands, 492t
Submandibular glands, 492t
Submucous plexus, 501, 501f
Substance abuse, 655t
Substance P, 147, 148f, 315
Substantia nigra, 435, A–7, A–12f
Substantia nigra pars compacta, 435, 828f
Substantia nigra pars reticulata, 431, 431f, 432f, 437–39, 438f
Subthalamic nucleus, of ventral thalamus, 441, 441f
Subthalamus, A–17b
Subventricular zone (SVZ), 687, 688f, 689, 689f, 691
Succinic semialdehyde dehydrogenase, 136
SUFU protein, 538–39
Sugar, mixture suppression and, 373
Sulci, A–4
Sulcus limitans, A–11, A–11f
Summation, synaptic potentials, 117–18, 118f
Superconducting quantum interference devices (SQUIDs), 31, 31f
Superior, definition of, A–2f
Superior cerebellar artery (SCA), A–25, A–26f
Superior cerebellar peduncle (brachium conjunctivum), 452f, 453, A–10f, A–12f
Superior colliculi, A–7, A–12f
Superior colliculus
  anatomy of, 222, 222f, 457f, 481, 485f, A–10f
  eye movement and, 457–58
  eye-movement related neurons in, 485f
  saccadic eye movement and, 481
  sensorimotor integration in, 482b, 482f, 483f
  upper motor neurons and, 379, 424–25, 453
  visual space of, 481f
Superior frontal gyrus, A–13, A–14f
Superior oblique muscles, 473
Superior olivary complex, 250, 251f
Superior parietal lobules
  anatomy of, A–14, A–14f
  attentional control and, 750f
  in locomotion control, 398f
Superior peduncles, A–7
Superior rectus muscles, 473
Superior sagittal sinus, A–21f, A–25
Superior salivatory nuclei, 500
Superior temporal gyrus (STG), 250, 252, 252f, 255, A–13, A–14f
Superoxide dismutase (SOD1) mutations, 399–400
Super-taster, 358
Suprachiasmatic nucleus (SCN), 222, 222f, 496b, 497f, 498f, 715–17, 716f
Supramodal attention, 741–42

Supraoptic nucleus, 497b, 497f, 498f
Sustentacular cells, 328f, 329
SVZ (subventricular zone), 687, 688f, 689, 689f, 691
Sweat gland innervation, 500
Sweet taste, 356
  as palatable, 372, 372f
  receptors detecting, 361–62, 362f
Swing phase locomotion, 395, 397f
Sylvian (lateral) fissure, A–4, A–4f
Sympathetic chain ganglia, 495
Sympathetic division, of visceral motor system
  anatomy and physiology of, 495, 496f, 498
  Horner's syndrome and, 504–5, 504f, 505f
  major functions of, 493t
  neurons of, 491, 494f
  organization of, 19–20, 490
Synapses
  age and counts of, 772–73
  calcium signals, 171b, 172f
  chemical, 93–96, 94f, 97f, 98
  classes of, 94–95
  discovery of, 3–4
  in early life in PNS, 583f
  electrical, 4, 93–96, 94f, 95f, 96f
  elimination of, 582–85, 584f
  excitatory, 171f
  formation of, 575, 577–80, 578f
  function of, 93
  glutamatergic, 172f
  pathway formation of, 592, 593f
  peripheral, regeneration of, 671–75, 673f, 674b
  silent, 193b, 193f, 194f
  tripartite, 119b
Synapsin, 104, 105f
Synaptic adhesion molecules (SynCAMs), 578f, 579
Synaptic cleft, 94, 98
Synaptic delay, 98
Synaptic depression, 180f, 181–82, 183, 184f
Synaptic facilitation, 180–81, 180f
Synaptic plasticity
  augmentation and, 180f, 181
  experience-driven, 617
  long-term, behavioral modification and, 182–85, 183f, 184f, 185f
  long-term, hippocampus and, 186–89, 187f, 188f, 189f
  LTD and, 176
  potentiation and, 181
  short-term, 179–82, 180f, 181f
  silent synapses and, 193b, 193f, 194f
  spike timing-dependent, 197, 198f, 200
  visual deprivation and, 615–16
Synaptic potentials
  description of, 11
  generation of, 38, 38f
  summation of, 117–18, 118f
Synaptic terminals, 599

Synaptic transmission
  chemical synapses and, 93–96, 94f,
    97f, 98
  electrical synapses and, 93–96, 94f,
    95f, 96f
  estrogen influencing, 646, 648f
  mechanisms of, 94f
  at neuromuscular junctions, 98, 99f
Synaptic vesicle cycle, 103, 103f
Synaptic vesicles
  ACh in, 102
  discovery of, 102
  electron microscopy of, 5f
  endocytosis mechanisms and,
    109, 110f
  exocytosis and, 98, 102f
  exocytosis mechanisms and, 104,
    106f, 109
  local recycling of, 103–4, 103f
  neurotransmitters and, 4, 94–95,
    94f, 98, 101–2, 102f
Synaptobrevin, 104, 105f, 106f
Synaptojanin, 109, 110f
Synaptotagmins, 104, 105f, 106f, 107,
    109, 181
SynCAMs (synaptic adhesion
    molecules), 578f, 579
α-Synuclein mutations, 444
Syntax, 776
Syntaxin, 104, 105f, 106f, 108
Synuclein proteins, A–31
System A transporter 2 (SAT2), 131

T

TAAR6 (trace amino acids) gene,
    169b
TAARs (trace amine-associated
    receptors), 334–36
Tachistoscopic presentation, 787f,
    788
Taeniopygia guttata (zebra finch),
    636–38, 637f, 686f, 687
Takeuchi, Akira, 114
Takeuchi, Noriko, 114
TANK-binding kinase 1 (TBK1)
    mutations, 400
Tartini, G., 247b
TAS1Rs, 361–63, 362f, 363f, 364b
TAS2Rs, 362, 362f, 363f, 364b
Taste
  auditory cues and, 371
  changing, 373–74, 374f
  coding of, 366–67, 366f
  conditioned taste aversion, 373–74
  COVID-19 and, 360, 361
  critical periods and, 606t
  extraoral taste receptors, 364b, 364f
  function of, 357–58
  hedonic value of, 372–73, 373f
  loss of, 360, 361
  meanings of, 356
  neural pathways in, 365–66
  odor integrated with, 369, 370f
  perception of, 356–58
  perturbation, 360, 361

plasticity, 373
  receptors, 360–65, 362f, 363f, 364b
  sensitivities, 358
  spatial coding of, 367, 368f
  temporal coding of, 367–68, 368f
  texture and, 369–71, 370f
  transduction, 360–65, 363f
  unitary view of coding, 368
  visual cues and, 371
Taste buds, 358–60, 359f, 360f
Taste papillae, 358, 359f
Taste system, organization of, 357f
TBIs (traumatic brain injuries),
    676–77, 677f
TBK1 (TANK-binding kinase 1)
    mutations, 400
T cells
  after brain injury, 683t
  maturation of, 682
TDF (testis-determining factor), 631
Tear gland innervation, 500
Tectorial membrane, 242f, 243, 244f
Tectospinal tract, 424
Tegmentum, 422b, A–9
Telencephalon, 528f, 529, A–2
Teleost fish, neurogenesis in, 685–87,
    686f
Temperature
  autonomic reflexes to, 315
  pathways to face, 312, 313f, 314
  sensation of, 314
Temporal association cortex, 705–7,
    706f
Temporal lobes
  anatomy of, A–4, A–4f
  declarative memory and, 762–66,
    770
  midsagittal view of, A–16f
Temporal–parietal junction (TPJ),
    735, 736f
Tensor tympani innervation, 242–43
Terminal fields, of axons, 4, 6
Testes
  development of, 588
  embryonic, 635, 635f
  hormone secretion by, 634, 643f
  "Testes-at-12," 650t, 651
Testicular feminization, 650–51, 650t
Testis-determining factor (TDF), 631
Testosterone
  lipophilic structure of, 635
  neuronal growth/differentiation
    and, 644f
  sexual development and, 634–35
  sexually dimorphic development
    and, 642–44
  structure of, 634f
Testosterone receptors, 635
Tetanus, 108, 180f, 182
Tetanus toxin, 108
Tetraethylammonium, 58, 58f
Tetrahydrocannabinol (THC), 150b,
    150f
Tetrodotoxin, 58, 58f, 75b
Texture, taste and, 369–71, 370f

TGF-β. See Transforming growth
    factor-β
Thalamic nuclei, 312, 458, 700
Thalamocortical pathways, 276f
Thalamocortical relations, A–17b,
    A–18f
Thalamocortical relays cells, 223
Thalamus
  anatomy of, 294f, A–16, A–16f,
    A–17b, A–17f
  cerebellum projections to, 455–58
  cortical input from, 699b
  deep brain stimulation targeting,
    445–46, 446f
  medial geniculate complex of, 250,
    251f, 252
  projections of, A–18b
  sleep interactions of cortex and,
    729, 730f, 731
  somatosensory pathways of, 294,
    294f
  subdivisions of, A–17b, A–17f
  super-direct pathway and, 441
  thalamocortical relations and,
    A–17b, A–18f
  VA/VL complex of, 453f, 457f
  vestibular pathways to, 276, 276f
THC (tetrahydrocannabinol), 150b,
    150f
Thermosensitive ion channels,
    76–77, 77f, 84–85, 85f
Thinking, decision making,
    planning and, 819–38
Thiopentone, 42
Third ventricle, 496b, 498f, 528f,
    A–30, A–30f, A–31f
Thoracic nerves, A–3f
Thoracic vertebrae, A–4, A–5f
Thorndike, Edward, 831
Threshold potentials, 39, 39f, 117,
    117f
Thr kinases, 165
Thrombotic stroke, A–27
Thyroid hormone receptors, 174
Thyroid hormones, cell-permeant
    signaling by, 158
Thyrotropin releasing hormone
    (TRH), 148f
Thyroxin, cell-permeant signaling
    by, 158
Tight junctions, A–28, A–28f
Tiling, axonal and dendritic, 573–75,
    576f
Timbre, 237
Tinnitus, 249
Tissue damage, inflammatory
    response to, 316f
TMC1 mutations, 248
TMC2 mutations, 248
TMEM132D gene, 170b
TMIE mutations, 248
TMS (transcranial magnetic
    stimulation), 26–27, 27f, 783, 784f,
    824
TNF-α. See Tumor necrosis factor α

Tobacco hawk moth (Manduca sexta),
    636, 637f, 638
Tobacco plant (Nicotinia tabacum),
    126b
Tolman, Edward, 770b
Tonotopic map, 606t
Tonotopic organization, of cochlea,
    250
Tonotopy, 244, 253
Topical spastic paraparesis (TSP), 68
Topographic logic, 367
Topographic maps
  formation of, 592–95, 594f, 595f
  neural systems and, 15
Toscanini, Arturo, 759
Touch
  pathways, 289–92, 290f, 291f
  somatosensory system and,
    278–300
Touch domes (Merkel cell-neurite
    complexes), 283t, 284f, 286
Tourette syndrome, 434b, 654t, 824
TPJ (temporal–parietal junction),
    735, 736f
Trace amine-associated receptors
    (TAARs), 334–36
Trace amino acids (TAAR6) gene,
    169b
Tracts, nervous system, 19
Tranquilizers, 42
Transcranial magnetic stimulation
    (TMS), 26–27, 27f, 783, 784f, 824
Transcriptional codes, 536
Transcription factors (transcriptional
    activator proteins)
  activation of, 192
  neural development and, 536f
  nuclear signaling and, 173
  patterning, 521, 530–31, 535–36
Transducin, 211
Transforming growth factor-β (TGF-β)
  after brain injury, 683f
  endogenous antagonists and, 533,
    534f
  neuron migration and, 690
Transgenic mice, 22, 23f, 24b, 25
Transgenic reporters, 16, 17f
Trans-Golgi network, 147
Transient receptor potential (TRP)
    channels, 306–7, 340f, 341
Transient receptor potential (TRP)
    gene families, 76–77, 77f
Transit amplifying cells, 541, 689–90
Translational movements, 261–62
7-Transmembrane G-protein-
    coupled receptors, 507
Transmissible spongiform
    encephalopathies (TSEs), 463
Transverse, definition of, A–2
Transverse pontine fibers
    (pontocerebellar fibers), 454
Transverse sinuses, A–21f, A–25
Tranylcypromine, 142
Traumatic brain injuries (TBIs),
    676–77, 677f

Traveling waves, cochlear, 243f, 244–45, 244f
TREM2 (triggering receptor expressed on myeloid cells 2), 683f
TRH (thyrotropin releasing hormone), 148f
Trichromatic vision, 214, 215f
Tricyclic antidepressants, 169b
Trigeminal brainstem complex, 292
Trigeminal fibers, 369–71
Trigeminal ganglion, 312
Trigeminal lemniscus, 290f, 292
Trigeminal motor nucleus, A–10f, A–12f
Trigeminal nerve, 292, 369, A–8t, A–9
Trigeminal nucleus, 293
Triggering receptor expressed on myeloid cells 2 (TREM2), 683f
Tripartite synapses, 119b
Triskelions, 109
Trisomy 21, 540
TrkA receptors, 175–76, 589, 590f, 669
TrkB receptors, 589, 590f, 625
TrkC receptors, 589, 590f
Trk (tyrosine kinase) receptors, 589–90, 590f
Trochlear nerve
    anatomy of, 474, A–9
    function of, A–8t
    injury to, 477
Trochlear nucleus, A–10f
Trophic interactions, 580–82, 581f, 585, 587–89, 587f
Trophic molecules, 564
Tropic molecules, 564
TRPA1 sensitivity, 307
TRP channels, 314
TRP (transient receptor potential) channels, 306–7, 340f, 341
TRPM8 channels, 314
TRPV1 channels, 84–85, 86f
TRPV1 (vanilloid) receptors, 306b, 307, 315
TRPV3 channels, 314
TRPV4 channels, 314
Trunk neural crest, 525f, 526
Trustworthiness, of facial expressions, 808, 809f
Tryptophan, 5-HT and, 144, 144f
TSCs (tuberosclerosis complex genes), 625
TSEs (transmissible spongiform encephalopathies), 463
Tsien, Roger, 163b
Tsimpli, Ianthi-Maria, 760b
TSP (topical spastic paraparesis), 68
Tuberosclerosis complex genes (TSCs), 625
Tubulin, 561–62, 563f
Tufted cells, external, 342f, 343
Tumor necrosis factor α (TNF-α)
    blockade of, 316
    after brain injury, 683f
    CNS response to injury and, 681, 681f

microglia production of, 317
    release of, 315
Tuning curves
    auditory nerve, 253f, 254, 254f
    visual cortex, 228, 229f
Tuning forks, 236, 237f, 243
Turner syndrome, 650, 650t
Twist genes, 547
Two-point discrimination measures, 282, 283f, 314
Tympanic membrane, 240, 241f, 242
Type 2 dopamine receptor (DRD2) gene, 169b
Tyrosine, 141f
Tyrosine hydroxylase, 141f, 177, 177f
Tyrosine kinase (Trk) receptors, 589–90, 590f

U
Ube3a, 625
Umami taste, 356, 358
    receptors detecting, 361–62, 362f
Unconditioned response, 757
Unconditioned stimulus, 757
Uncoordinated (Unc) genes, 564
Uncus, A–15, A–15f
Undershoot phase, of action potentials, 52b, 53
Unimodal attention, 741–42
Upper extremities, 492t
Upper motor neurons
    axons of, 403
    balance maintained by, 419–25
    brainstem control and, 379, 380, 403–28, 455
    corticospinal and corticobulbar tracts and, 405–9, 407f
    damage to, 425–28, 477–78
    descending motor control and, 404–5, 404f
    directional tuning of, 412, 413f
    eye movement and, 457–58
    firing rates of, 410
    frontal eye fields and, 481
    gaze oriented by, 419–25
    influence of, 410, 412f
    lesions, signs and symptoms of, 426t
    locomotion initiated by, 419–25
    in movement control, 380, 380f, 453
    muscle fields of, 410
    place codes, rate codes and, 485b
    posture governed by, 419–25
    sources of, 379, 382
    spinal cord control and, 403–28
    spinocerebellar pathways and, 457
Upper motor neuron syndrome, 425–28, 426t
Urbach-Wiethe disease, 807b
Ureter
    referred pain from, 309b
    visceral motor system functions and, 492t
Urethra, closure of, 510

Urinary bladder, referred pain from, 309b
Urination, regulation of, 510–12
Utricle, 262–65, 262f, 263f, 266f
Utricular macula, 264f, 265, 265f
Uveal tract, 206

V
V1Rs, 340–41, 340f
V2Rs, 340–41, 340f
VAChT (vesicular ACh transporter), 125, 125f
Vagal neural crest, 525f, 526
Vagina, contraction of, 512
Vagus nerve
    anatomy of, A–9
    dorsal motor nucleus of, 499f, 500, 510, A–10f, A–12f
    facial pain and, 312
    function of, A–8t
Valence, emotions and, 798
Valium (diazepam), 42f, 137, 170b
Valproic acid, 199
Value signals, 821
Vanilloid (TRPV1) receptors, 306b, 307, 315
Vascular endothelial growth factor (VEGF), 567, 630
Vas deferens, contraction of, 512
Vasoactive intestinal peptide (VIP), 148f
Vasoactive intestinal peptide (VIPR2) gene, 169b
Vasocorona, A–22f, A–26
Vasopressin, 148f, 717
Vasopressin receptors, 632b, 632f
VCR (vestibulocervical reflex), 273
VDAC1 gene, 170b
Vection, 269
Vecuronium, 43
VEGF (vascular endothelial growth factor), 567, 630
Ventral, definition of, A–1, A–2f
Ventral anterior nuclei, 438
Ventral columns, A–6
Ventral (anterior) corticospinal tract, 406–7
Ventral horns, A–5–A–6, A–5f, A–6t
Ventral lateral nuclei, 438
Ventral pallidum, 827
Ventral posterior lateral (VPL) nucleus
    anatomy of, 292, 293, 294f
    pain information from, 311, 312f
Ventral posterior medial (VPM) nucleus, 292, 294f, 312
Ventral posteromedial thalamic nucleus (VPMpc), 366
Ventral roots, A–4, A–5f
Ventral stream, 230–31, 231f
Ventral tegmental area (VTA), 372, 623
    dopamine and, 828b
    drug abuse and, 826–27, 826f, 827f
    innervation of, 633b, 633f

Ventral thalamus, subthalamic nucleus of, 441, 441f
Ventricles, in CSF, A–3
Ventricular system, A–30–A–32, A–31f
Ventrolateral columns, A–6
Ventrolateral prefrontal cortex (VLPFC), 824–25, 825f
Ventrolateral preoptic nucleus (VLPO), 729
Ventromedial nucleus, 497b, 497f, 498f
Ventromedial prefrontal cortex (VMPFC), 828. See also Orbitofrontal cortex
    activation of, 709
    decision making and, 810, 810f
Venus flytrap domains, 135, 135f, 139, 139f
Veratridine, 75b
Vergence eye movement, 475–76, 488
Vermis, 452–53, 452f
Vertebral arteries, A–21, A–21f
Vertebrates. See also specific types
    adult neurogenesis in non-mammalian, 685–87, 686f
    head development in, 527, 527f
    Hox genes in, 529
    sleep in, 718
Vesicular ACh transporter (VAChT), 125, 125f
Vesicular glutamate transporters (VGLUTs), 131, 131f
Vesicular inhibitory amino acid transporter (VIAAT), 136, 136f
Vesicular monoamine transporters (VMATs), 140, 142
Vestibular-cerebellar pathways, 274, 274f
Vestibular complex, 420, 455
Vestibular hair cells. See also Cochlear hair cells
    morphological polarization of, 262, 263f, 264f
    structure and function of, 262–63
Vestibular labyrinth, 261–62, 262f
Vestibular nerve ganglion, 268
Vestibular nerves
    anatomy of, 241f
    axonal responses of, 268, 268f
    cochlea and, 242f
    otolith organ response of, 266, 266f
Vestibular nuclei, 269, 270f, 273, 274f, A–10f
    anatomy of, 457f, A–12f
    projections from, 274f
    projections to, 269, 421
    sensory information to, 420
    upper motor neurons and, 379, 382
Vestibular system, 261–77
Vestibular system clinical evaluation, 271–72, 271f, 272f
Vestibule, 241f
Vestibulocerebellar syndrome, 468
Vestibulocerebellum
    anatomy of, 452f, 453, 453t

damage to, 468
projections from, 455–58
Vestibulocervical reflex (VCR), 273
Vestibulocochlear nerve. *See* Auditory nerve
Vestibulo-ocular eye movements, 269, 270*f*, 273
operational ranges of, 476, 476*f*
vestibular complex and, 455
Vestibulo-ocular reflex (VOR)
adaptive changes to, 274
clinical evaluation of, 271–72, 271*f*, 272*f*
corrective lenses and, 270*f*, 273
function of, 269, 270*f*
loss of, 269, 273
plasticity, 273
studies of, 464, 465*f*
Vestibulospinal reflex (VSR), 273
VGLUTs (vesicular glutamate transporters), 131, 131*f*
VIAAT (vesicular inhibitory amino acid transporter), 136, 136*f*
Viagra (sildenafil), 513
VIP (vasoactive intestinal peptide), 148*f*
VIPR2 (vasoactive intestinal peptide) gene, 169*b*
Visceral disorders, referred pain from, 309*b*
Visceral motor nuclei, A–11, A–11*f*
Visceral motor system. *See also* Parasympathetic division, of visceral motor system; Sympathetic division, of visceral motor system
central control of, 503–4, 503*f*
distinctive features of, 491
early studies of, 490–91
enteric division of, 20, 501, 501*f*
Horner's syndrome and, 504–5, 504*f*, 505*f*
hypothalamus and, 490
major functions of, 492*t*
neurons of, 491, 494*f*
neurotransmission in, 491, 507–8, 508*t*
organization of, 18–20, 19*f*
sensory components of, 502–3, 502*f*, 503*f*
Visceral pain, 310*b*, 310*f*, 502–3
Visible light, 205–6
Vision loss, 223*b*, 223*f*
Visual acuity, 207

Visual aftereffects, 732
Visual cortex
in locomotion control, 398*f*
memory and, 769–70, 770*f*
visually-guided reach-to-grasp and, 419*f*
Visual cues, taste and, 371
Visual deprivation, 611–21
Visual fields
active control of, 231–32
binocular, 208*f*, 209
deficits of, 223*b*, 223*f*, 224*f*
fixation points in, 233
foveation and, 471
hemispatial neglect and, 746, 748
image maintenance in, 233
neuronal, 216–17
of photoreceptors, 216–17
projection onto retinas, 207–9, 208*f*
scanning of, 486, 487*f*
spatial details in, 226, 226*f*
Visual grasp, 482*b*
Visual search, 745
Visual system, 205–34, 257–58, 258*f*, 604, 605*f*, 618–19. *See also* Eyes
Visual thalamus, A–18*b*
Visual word form area (VWFA), 792, 792*f*
Vitamin B$_6$, pyridoxal phosphate and, 136
Vitreous humor, 206
VLPFC (ventrolateral prefrontal cortex), 824–25, 825*f*
VLPO (ventrolateral preoptic nucleus), 729
VMATs (vesicular monoamine transporters), 140, 142
VMPFC. *See* Ventromedial prefrontal cortex
VNO (vomeronasal organ), 326–27, 326*f*, 328*f*, 331, 343, 344*f*
Vocal folds, 776, 777*f*
Volitional facial paresis, 803*b*, 803*f*
Voltage clamp method, 54–55, 55*b*, 56*f*
Voltage-gated ion channels. *See also specific channels*
cell adhesion signaling and, 625
functional states of, 74, 74*f*
hair cell, 247, 263
neurotransmitter release and, 100–101, 100*f*
structure of, 80, 81*f*
Voltage-sensitive fluorescent dyes, 12

Voltage sensors, 74, 74*f*
Vomeronasal organ (VNO), 326–27, 326*f*, 328*f*, 331, 343, 344*f*
Vomeronasal receptor neurons (VRNs), 331–33, 340–41
Vomeronasal receptors (VRs), 339–41, 340*f*
Vomeronasal system
accessory olfactory bulb in, 343–44, 344*f*
function of, 324
mechanisms of transduction in, 327
organization of, 326–27, 326*f*
transduction mechanisms in, 327, 339–41
VOR. *See* Vestibulo-ocular reflex
Vowel sounds, 776–78
VPL (ventral posterior lateral) nucleus, 292, 293, 294*f*
VPM (ventral posterior medial) nucleus, 292, 294*f*, 312
VPMpc (ventral posteromedial thalamic nucleus), 366
VRNs (vomeronasal receptor neurons), 331–33, 340–41
VRs (vomeronasal receptors), 339–41, 340*f*
VSR (vestibulospinal reflex), 273
VTA. *See* Ventral tegmental area
VWFA (visual word form area), 792, 792*f*

W

Waardenburg's syndrome, 540
Wada test, 785, 789*f*
Wakefulness, cellular mechanisms governing, 726, 727*f*, 729*t*
Wall, Patrick, 320
Wall eyes, 618
War, casualties of, 676–77
Wasp venom, 127*b*
Water, refractive index of, 206
Watts, James, 710*b*
Weaver mutations, 466*b*, 466*f*, 467*t*
Weber test, 239*b*, 243
Wernicke, Carl, 781
Wernicke's aphasia, 769, 779–81
Wernicke's area, 252*f*, 781, 785
Whisker barrels, 601, 602*f*, 623, 624*f*
White matter
description of, 19
growth of, 609
spinal cord, 382, A–5, A–5*f*, A–7

Whole-cell recording, 72*b*
Wide-dynamic-range neurons, 308
Wiesel, Torsten, 613, 614
Williams, H., 335*b*
Willis, Thomas, 129
Windup phenomenon, 317
Wingless mutation, 530*f*
Wisconsin Card Sorting Task, 708*b*, 709*f*, 823
Wnt family, 534*f*, 535, 567
Woolsey, Clinton, 411*b*
Word-processing tasks, 607, 608*f*
Working memory, 755, 756*f*
Wound healing, 319, 322
Writing deficits, 781
Wurtz, Robert, 439

X

Xanthines, 146
X chromosomes, 630–31, 631*f*, 650
X-linked disorders, 570
XO chromosomes, 650, 650*t*
X-rays, CT and, 27–28, 28*f*
XX genotype, 631
XXY phenotype (Klinefelter's syndrome), 631, 650, 650*t*
XY genotype, 631
XYY chromosomes, 650, 650*t*

Y

Yamanaka, Shinya, 518, 521
Yarbus, Alfred, 471–72, 741
Y chromosomes, 630–31, 631*f*, 650
Yin, Henry, 437*b*
Young, John Z., 48*b*

Z

ZEB genes, 547
Zebra finch (*Taeniopygia guttata*), 636–38, 637*f*, 686*f*, 687
Zebrafish (*Danio rerio*), 24*b*, 560*f*
Zic2 gene, 570*b*
Zicam nasal gel, 329–30, 329*f*
Zinc gluconate, 329*f*, 330
Zinc nasal spray, 329–30, 329*f*
Zinc sulfate, 330
Zofran (ondansetron), 145
Zoloft (sertraline), 812
Zolpidem (Ambien), 42, 42*f*, 137